SPACE SCIENCE SERIES

Tom Gehrels, General Editor

Planets, Stars and Nebulae, Studied with Photopolarimetry
Tom Gehrels, editor, 1974, 1133 pages

Jupiter
Tom Gehrels, editor, 1976, 1254 pages

Planetary Satellites
Joseph A. Burns, editor, 1977, 598 pages

Protostars and Planets
Tom Gehrels, editor, 1978, 756 pages

Asteroids
Tom Gehrels, editor, 1979, 1181 pages

Comets
Laurel L. Wilkening, editor, 1982, 766 pages

Satellites of Jupiter
David Morrison, editor, 1982, 972 pages

Venus
D.M. Hunten, L. Colin, T.M. Donahue and V.I. Moroz, editors, 1983, 1143 pages

Saturn
Tom Gehrels and Mildred S. Matthews, editors, 1984, 968 pages

Planetary Rings
Richard Greenberg and André Brahic, editors, 1984, 784 pages

Protostars and Planets II
David C. Black and Mildred S. Matthews, editors, 1985, 1293 pages

Satellites
Joseph A. Burns and Mildred S. Matthews, editors, 1986, 1021 pages

The Galaxy and the Solar System
Roman Smoluchowski, John N. Bahcall and Mildred S. Matthews, editors, 1986, 485 pages

Meteorites and the Early Solar System
John F. Kerridge and Mildred S. Matthews, editors, 1988, 1269 pages

Mercury
Faith Vilas, Clark R. Chapman and Mildred S. Matthews, editors, 1988, 794 pages

Origin and Evolution of Planetary and Satellite Atmospheres
S.K. Atreya, J.B. Pollack and M.S. Matthews, editors, 1989, 881 pages

Asteroids II
Richard P. Binzel, Tom Gehrels and Mildred S. Matthews, editors, 1989, 1258 pages

Uranus
Jay T. Bergstralh, Ellis D. Miner and Mildred S. Matthews, editors, 1991, 1076 pages

The Sun in Time
C.P. Sonett, M.S. Giampapa and M.S. Matthews, editors, 1991, 996 pages

Solar Interior and Atmosphere
A.N. Cox, W.C. Livingston and M.S. Matthews, editors, 1991, 1414 pages

Mars
H.H. Kieffer, B.M. Jakosky, C.W. Snyder, and M.S. Matthews, editors, 1992, 1536 pages

Protostars & Planets III
E.H. Levy, J.I. Lunine and M.S. Matthews, editors, 1992, in press

Resources of Near-Earth Space
J. Lewis and M.S. Matthews, editors, 1992, in press

This image of Mars was obtained with the Hubble Space Telescope's Wide Field/Planetary Camera on 1990 December 13, about two weeks after opposition. At the time Mars was about 85 million km from Earth and about 16.4 arcsec in angular diameter. This image is at 588 nm, and has an approximate surface resolution of 50 km. Syrtis Major Planitia dominates this face of the planet. Also visible at the top of the image is the north polar hood. (Image courtesy of P. B. James of The University of Toledo.)

MARS

Hugh H. Kieffer
Bruce M. Jakosky
Conway W. Snyder
Mildred S. Matthews

Editors

With 114 collaborating authors

THE UNIVERSITY OF ARIZONA PRESS
TUCSON & LONDON

About the cover:

A panorama obtained by the Viking 2 Lander. This mosaic was constructed from images taken at different times, so that the shadows of the Lander are discontinuous and the direction of rock shadows is inconsistent. This mosaic was acquired by camera 1, and covers the left and central portions of the region accessible to the Viking 2 sampler arm, which is shown (discontinuously) at the extreme right of the image. An accumulation of drift material can be seen just above and to the left of the sampler housing. The white object extending to above the horizon in this image is the meteorology boom; its shadow can be seen slightly to its right. The colored shadows of some rocks results from changing illumination direction between data acquisition in different bands. The curved horizon in the initial mosaic resulting from the tilt of Lander 2 has been removed from this image, revealing a very flat horizon. Presuming that the rocks in this region are approximately uniformly distributed, their variation in apparent density in the background indicates gently undulating terrain. These images were obtained early in the Lander 2 mission, and show two Lander trenches, one near the middle of the image to the left of the large rock informally named Centaur, and a smaller trench to the left of the rock "ICL" (see also Fig. 2 in the nomenclature appendix, which was acquired by Lander camera 2). Image processing by E. Guinness and M. Dale-Bannister at Washington University in St. Louis. Final image production at U.S.G.S., Flagstaff.

The University of Arizona Press

♾ This book is printed on acid-free, archival-quality paper.
Manufactured in the United States of America.

96 95 94 93 92 5 4 3 2 1

Library of Congress Cataloging-in-Publication Data

Mars / Hugh H. Kieffer . . . [et al.], editors : with 114 collaborating authors.
p. cm. — (Space science series)
Includes bibliographical references and index.
ISBN 0-8165-1257-4
1. Mars (Planet) I. Kieffer, Hugh H. II. Series.
QB641.M35 1992
522.4'3—dc20 92-10951
CIP

British Library Cataloguing in Publication data are available.

CONTENTS

PART III—BEDROCK GEOLOGY AND GEOLOGIC UNITS

PART IV—SURFACE PROPERTIES AND PROCESSES

PART V—CURRENT ATMOSPHERES

PART VI—BIOLOGY

PART VII—SATELLITES OF MARS

COLLABORATING AUTHORS

Arvidson, R. E., *626*
Baker, V. R., *493*
Balmino, G., *209*
Banerdt, W. B., *29, 249*
Banin, A., *594*
Barlow, N. G., *383*
Barnes, J. R., *835*
Barth, C. A., *1054*
Batson, R. M., *321, 1249*
Bauer, S. J., *1054*
Beish, J. D., *34*
Bell, J., *1257*
Biemann, K., *1221*
Bills, B. G., *209*
Bougher, S. W., *1054*
Brace, L. H., *1090*
Burns, J. A., *1283*
Carr, M. H., *120, 493, 1135*
Christensen, P. R., *686*
Clark, B. C., *594*
Clifford, S. M., *523*
Clow, G. D., *453*
Costard, F. M., *523*
Croft, S. K., *383*
Cruikshank, D., *1257*
Davies, M. E., *321*
Dollfus, A., *34*
Drake, M. J., *147*
Duxbury, T. C., *1249*
Edwards, K., *1249*
Esposito, P., *209*
Fanale, F. P., *1135*
Geissler, P. E., *453*
Golombek, M. P., *249*
Gooding, J. L., *626*
Greeley, R., *345, 730*
Gulick, V. C., *493*
Haberle, R. M., *835, 969*
Harmon, J. K., *652*
Herkenhoff, K., *767*
Holloway, J. R., *184*
Horowitz, N. H., *1221*
Howard, A., *767*
Hunten, D. M., *1054*
Iwasaki, K., *34*
Jakosky, B. M., *xi, 1, 969*
James, P. B., *34, 934*
Kahn, R. A., *1017*
Kieffer, H. H., *xi, 1, 934, 1180, 1305*
Klein, H. P., *1221*
Knittle, E., *184*
Kuzmin, R. O., *523*
Lancaster, N., *730*
Lee, S. W., *730, 1017*
Leovy, C. B., *835*
Lindal, G. F., *209*
Longhi, J., *184*
Lucchitta, B. K., *453*
Luhmann, J. G., *1090*
Lunine, J., *1257*
Mancinelli, R. L., *1234*
Marley, M. S., *493*
Martin, L. J., *34*
Martin, T. Z., *1017*
McEwen, A. S., *453*
McKay, C. P., *1234*
Moore, H. J., *686*
Moroz, V. I., *71*
Mouginis-Mark, P. J., *424*
Muhleman, D. O., *652*
Murray, B., *767*
Nagy, A. F., *1054*

PREFACE

For many years there has been a gap on the shelf of books about planetary objects. Between Venus and the asteroids reside two planets and three satellites; there is an abundance of books discussing the Earth and the Moon, but a conspicuous absence of a book summarizing what is known about Mars and its two satellites. We hope that this book will fill that gap for at least a few years.

Although the scientific literature about Mars is voluminous, and there have been several books oriented around images of Mars returned by spacecraft, no comprehensive treatment of the results of the age of spacecraft exploration has been available. (With our retrospectoscope, the editors now realize that this situation may have been due to others being more perceptive than we about the magnitude of the task.) A list of some prior books about Mars is included in chapter 1. A list of special journal issues about Mars can be found in chapter 3.

The objective of this book is to provide a summary of what is known about Mars, to present our understanding of the processes that govern its surface and atmosphere and to identify those intriguing areas where we know more than we understand, either where alternate hypotheses seem viable or where no solution seems straightforward. This book is meant to be a source book on all aspects of Mars; it is one step away from the most detailed information available, the full scientific literature. As such, it is at a level appropriate for graduate study.

This book is part of the Space Science Series of the University of Arizona Press and has been done in the style of that series. A large scientific organizing committee was constituted, and their advice sought at each step. A candidate list of chapters was constructed, and groups of authors were invited to write these chapters. To provide the detail and authority suitable for a research reference, scientists actively studying Mars were asked to author chapters; ultimately 114 authors were involved. An activity closely related to the generation of this book was the Fourth International Conference on Mars, held in Tucson in January 1989 and attended by 280 scientists from 14 countries. The prior conferences had been held in 1973, 1979 and 1981. Many of the chapters in this book were presented in part as review talks at this Fourth Conference. The chapter authors were encouraged to point out controversial issues; in some cases, co-authors of a chapter continue to hold opposing viewpoints and these are explicitly expressed (e.g., chapter 2).

The chapter manuscripts were first submitted between January 1989 and April 1990; reply to referees and revisions occurred over June 1989 to September 1991. We anticipate and hope that future observations and analysis will supersede these discussions, but also hope that this book will stand as a benchmark in time, documenting mankind's understanding of Mars at the junction between initial exploration and detailed survey and at a time when visits by humans could first be seriously considered.

This book was planned to consider information available prior to the renewal of spacecraft observations with the Soviet Phobos mission in 1989 and Mars Observer, scheduled for launch in 1992. Although the Phobos mission ended unexpectedly early, a complete and thorough consideration of its results could not be accomplished in time to be incorporated in this book. Some results from the Phobos mission are included, especially in chapters 3, 17, 31 and 37. The comparisons to Venus in this book largely predate analysis of the Magellan results.

We have tried to provide SI units for all quantities, even when non-SI units are most commonly used in the literature; e.g., surface pressure is given in both Pascal and the more common millibar (1 millibar = 100 Pa).

The text in places refers to "absolute ages" of various Martian events and surface units. These assigned ages are far from absolute, depending upon several model steps relating mass and velocity of impacting bodies to resulting crater size, and the abundance and orbital characteristics of such bodies in the past at Mars and at the Moon. Some discussion of the uncertainties is contained in chapter 12; errors of a factor of 3 in younger units are possible.

Viking Lander 1 has been formally designated the Thomas A. Mutch Memorial Station in honor of "Tim" Mutch, leader of the Viking Lander Imaging Team and NASA Associate Administrator for Space Sciences at the time of his death in a climbing accident in the Himalayas in 1980; some chapters refer to VL-1 as the Mutch Memorial Station, or MMS.

Some material is summarized in several places because it is related to several topics. For example, dust storms are summarized in chapters 2 (telescopic observations) and 22 (aeolian surface processes), but the major treatment is in chapter 29. A chapter on the history of spacecraft missions to Mars (chapter 3) takes advantage of the considerable experience of its authors (who have participated in nearly all such missions) to discuss in relatively greater detail Soviet missions, about which information is less readily available and whose results are not as widely known.

The chapter on atmospheric dynamics (chapter 26) is much longer than all others. This is the intentional result of combining what had been three chapters covering synoptic observations, Viking Lander meteorology, and numerical modeling and theory. This chapter and that on radar (chapter 20) have appendicies describing the symbols used therein.

The nature of Mars' surface varies considerably over the globe, indicating significantly different geologic histories. Also, Mars shows great topographic variation, with a wider range of elevations than known for any other object in the solar system, including consideration of the Earth's seafloor. For these and many other reasons, maps are an important aspect of the study of Mars and are included with the initial printing of this book.

Detailed global geologic and topographic maps of Mars have been published at 1:15,000,000 scale by the U.S. Geological Survey. The information contained in these maps could not be shown at a much smaller scale and could not be reproduced in the book without excessive segmentation. It was found to be most practicable to incorporate these maps at their original scale, exactly as originally published. The maps have been folded no more than necessary for distribution with the book.

The six maps are:

I-1802A Geology: western equatorial region
I-1802B Geology: eastern equatorial region
I-1802C Geology: polar regions

I-2160–1 Shaded relief: polar regions
I-2160–2 Shaded relief: eastern equatorial region
I-2160–3 Shaded relief: western equatorial region

On Mars, all longitudes are designated West; the Western hemisphere is taken to be longitudes 0° to 180°, and the eastern hemisphere between 180° and 360°.

Additional copies of these maps can be ordered from:

Branch of Distribution
U.S. Geological Survey
Box 25286, DFC, MS 306
Lakewood, Colorado 80225

Unfolded copies can be requested, but may no longer be available.

Some figures are utilized by several chapters; these are generally not repeated, but are placed with the chapter most closely related to the observations that are the basis of the figure: These include the following:

All color plates can be found in the Color Section.

All references cited are in a single bibliography. A definition or explanation of many technical terms or those peculiar to Mars is contained in a glossary at the back of the book. A description of the geographic type names used in Martian nomenclature and a brief history of the formal nomenclature system is contained in the Appendix. Readers not familiar with Martian geography will benefit by becoming acquainted with the nomenclature and the regional place-name map prior to study of the chapters dealing with geologic subjects.

The Fourth International Mars Conference (EOS, Trans. Amer. Geophys. Union, v. 70, no. 17, pp. 545 and 560, 1989) and the preparation of this book were supported by the National Aeronautics and Space Administration (NASA) through G. Briggs and W. Huntress, Directors of the Solar System Exploration Division; John Aaron, Director of the Exploration Division; W. Purdy and D. Evans, Project Managers of the Mars Observer mission; and J. Boyce, Manager of the Planetary Geology and Geophysics Program. This last Program also supported printing of the maps that accompany this book. Grants and contracts that supported individual authors are acknowledged in the Acknowledgments Section. The efforts of the staff of the University of Arizona Space Science Series contributed substantially to the success of the conference and to the production of this book; in particular, M. Magisos had a critical role in both endeavors. M. Guerrieri compiled the glossary and bibliography and helped coordinate much of the later stages of book production. We greatly appreciate the efforts of the approximately 100 referees.

H. Kieffer thanks A. Allison and V. Banning for their assistance in numerous ways, including forging through a complete change of word processing systems in "mid-book." B. Jakosky thanks A. Alfaro for administrative support and assistance through all phases of this effort. C. Snyder, who is retired from JPL, thanks the Mars Observer Project for providing access to electronic communications with the other authors.

Perhaps only after a gestation such as we have experienced can one appreciate why acknowledgement to close relations is seemingly ubiquitous in books of this kind; but we certainly have cause to add our thanks. We greatly appreciate the encouragement, support, and forbearance of our families.

H. Kieffer
B. Jakosky
C. Snyder
M. Matthews

PART I
Introduction

1. THE PLANET MARS: FROM ANTIQUITY TO THE PRESENT

HUGH H. KIEFFER
U.S. Geological Survey

BRUCE M. JAKOSKY
University of Colorado

CONWAY W. SNYDER
Jet Propulsion Laboratory (retired)

Mars has been a subject of wonder since earliest recorded time. Its red color sparked the imaginations of astrologers of early civilizations, and it was the symbol of fire and war in many mythologies. Telescopic observations, from those of Galileo in 1610 to the end of the nineteenth century, were summarized in two volumes by Flammarion (1892 and 1909). Modern observations began with Schiaparelli during the close opposition of 1877, the year when Hall discovered Mars' two satellites. The next 50 years saw the drawing of detailed maps by many astronomers, debates over the existence of features, and Lowell's advocacy of canals and intelligent life on Mars. Major compilations of knowledge of Mars were produced by Antoniadi (1930) and de Vaucouleurs (1954). Spectroscopic studies of the atmosphere led to the discovery of CO_2*, but until 1964 it was thought to be a minor component. The first reasonably accurate estimate of surface pressure, which was far below the then-accepted value, was made in the mid-sixties. Polar cap composition was debated until the surprising discovery by the Viking mission that the north and south perennial polar caps are composed of different materials. Spacecraft have yielded enormous amounts of data that indicate a rich and diverse geologic history as well as many unsolved puzzles. Following an overview of the remaining chapters, an annotated list of books about Mars is provided. Basic data about Mars are summarized in tabular form, and a guide to Martian seasons concludes the chapter.*

I. MARS AND ANCIENT CIVILIZATIONS

"The planet was known under the name of Harmakhis in ancient Egypt, where it was also called Har décher, that is to say the Red One, because of its color. The Chaldaeans gave it the name of Nergal—the Babylonian Mars, the great hero, the king of conflicts, the master of battles and the champion of the gods; and the planet was known on the banks of the Euphrates under the names of Allamou and Almou.

The Greeks called it Ares (Mars), name of the God of War, and apparently it was also known sometimes as Hercules. Moreover, it was regarded as fiery and impetuous. Proclus Diodochus called it "celestial fire," and added that it symbolized fire. The sign ♂ represents, so far as we know, the spear and the shield.

The Arabs, the Persians and the Turks named it Mirikh, which signifies a torch, iron, steel, and a long arrow thrown to a great distance. It was also known in Persia under the name of Bahram and Pahlavani Sipher, or the celestial warrior.

Among the Indians Mars was called Angaraka (from *angara,* burning coal), and they also called it Lohitanga, the red body, from *lohita,* red, and *anga,* body. Its astrological influence was regarded as being essentially malign."

Thus (as translated from the French) began the classic book, *The Planet Mars* (1930), by Eugenios M. Antoniadi, an astronomer at Meudon Observatory, France.

Mars was an object of great interest to ancient observers of the sky, but besides its obvious unique color, they knew only two more things about it—that it reappears about every two years and that it follows a strange, looping path across the sky. Only one important advance was made before the modern era, when Aristotle observed an occultation of Mars by the Moon and inferred that Mars was "higher up in the heavens" than the Moon, i.e., farther away.

II. THE FIRST PERIOD OF MARS OBSERVATION

The history of Mars observations was first treated in exhaustive detail in the two-volume tome of Nicolas Camille Flammarion. Volume 1, containing 608 pages, was published in 1892; Volume 2, with 595 pages, appeared in 1909. Volume 1 bears the appropriately ponderous title *La Planète Mars et ses Conditions d'Habitabilité. Synthèse Générale de Toutes les Observations. Climatologie, Météorologie, Aréographie, Continents, Mers et Rivages,*

Eaux et Neiges, Saisons, Variations Observées. Illustré de 580 Dessins Telescopique et 23 Cartes. It recounts in detail the observations of more than 200 people, not all of whom can be considered astronomers. [While checking references for this chapter after the text was completed, an English translation by Patrick Moore was discovered in the Lowell Observatory Library. This copy appears to be a complete translation of both volumes in typescript form, but probably never edited.]

On its initial page, the book credits the beginning of the historical study of Mars "par les astronomes de la Terre" to the telescopic observations of Francisco Fontana of Naples. Fontana published two sketches of Mars, neither showing actual surface detail. The second sketch, made on August 24, 1638, when Mars was at quadrature, was important because it depicts (albeit inaccurately) a gibbous phase. The unlighted portion actually never exceeds about one-seventh of the diameter, but Fontana's sketch shows it as nearly one-quarter. Apart from this sketch, he contributed nothing to our knowledge of the planet, as is emphasized by his statement that "except for the Sun, Mars is much the hottest of the stars."

Flammarion noted (p. 2) that Galileo Galilei had first observed Mars with his primitive telescope in 1609, and on December 30, 1610, he had written to his friend, Father P. Castelli, that "I dare not affirm that I was able to observe the phases of Mars; nevertheless, if I am not mistaken, I believe that I have seen that it is not perfectly round."

The Dutch astronomer Christiaan Huygens built telescopes of much improved quality, and on October 13, 1659, he produced what Percival Lowell later called "the first drawing of Mars worthy of the name ever made by man." It showed a large dark spot, which was almost certainly Syrtis Major. Observing it on successive rotations, he wrote (in Latin) in his journal on December 1 that "the rotation of Mars, like that of the Earth, seems to be in a period of twenty-four hours."

Three oppositions later, in 1666, Giovanni Domenico Cassini made about twenty sketches of the planet at the observatory in Bologna. He calculated the rotation period to be 24 hours 40 minutes and engaged in a controversy with Salvatore Serra of Rome, who insisted that it was 12 hours 20 minutes. Cassini's value was slightly corrected four decades later by his nephew, Giacomo Filippo Maraldi. The corrected value of 24 hours 39 minutes was not improved upon for another six decades until William Herschel, observing a surface feature on two dates 802 Earth days apart, obtained 24 hours 39 minutes 21.67 seconds. This value stood for another six decades.

Huygens was interested in the question of life on Mars, and his book on the subject, entitled *Cosmotheoros,* was published in 1698. In it he discussed what is required for a planet to be capable of supporting life, and he speculated on what intelligent extraterrestrials might be like. Interesting excerpts from the book were quoted by Ley and von Braun (1956).

Cassini, in 1666, was apparently the first to notice a polar cap. One of the crude sketches by Huygens in 1672 shows the south cap. Maraldi made observations during every opposition from 1672 to 1719, and he was lucky in that Mars came closer to Earth in 1719 than it would for another 284 years. Maraldi described both polar caps but stopped short of claiming them to be snow caps; he called them "taches blanches" (white spots) and discovered that the south cap is not centered on the rotational pole. He noted temporal changes both in the polar caps and in the equatorial dark areas. He also described a dark band around the edge of one cap and interpreted it as meltwater.

The surprisingly accurate incidental description of the Martian moons, Phobos and Deimos, by Jonathan Swift in *Gulliver's Travels* (1726), 150 years before their telescopic discovery, has often been noted:

> "They have likewise discovered two lesser Stars, or *Satellites,* which revolve about *Mars,* whereof the innermost is distant from the Center of the primary Planet exactly three of his Diameters, and the outermost five; the former revolves in the Space of ten Hours, and the latter in Twenty-one and a Half . . ."

Swift knew of Kepler's Third Law (published in 1619), as his Martian satellites explicitly obey it. However, because orbital radii were expressed in terms of the planetary diameter, to yield such short periods for the satellites, the density required for Mars would be 4.3 times that of Earth, regardless of the radius assumed for Mars! Although this density ratio could have been known to Swift (Newton's description of gravitation was published in 1687), that this ratio implied a density of 23.5 g cm^{-3}, greater than that of any material known then or now, would have had to wait until 1798 for Cavendish's experiment.

Frederick William Herschel, the organist turned astronomer, observed Mars during the oppositions from 1777 to 1783; the last was fairly close. He built the first telescopes that approached the quality of modern instruments, the largest having a 40-foot focal length and a 4-foot aperture. The Philosophical Transactions in 1774 published his paper entitled "On the remarkable appearances at the polar regions of the planet Mars, the inclination of its axis, the position of its poles, and its spheroidal figure; with a few hints relating to its real diameter and atmosphere." Herschel determined the inclination of the equator from the orbital plane to be about 30 degrees and concluded that Mars has seasons like Earth's. He guessed that the polar caps are thin layers of ice and snow, and he noted their seasonal changes. He searched unsuccessfully for satellites of Mars. Like several of his predecessors, he considered the dark areas to be oceans and the lighter regions land. Attributing the changes that he saw to "clouds and vapors," he concluded that Mars "has a considerable but modest atmosphere." He decided that the atmosphere is tenuous because he could see no effect on the near occultation of dim stars. He speculated that Martian "inhabitants probably enjoy a situation similar to our own."

It should be noted that in Herschel's time the existence of inhabitants on all planets (even on the Sun) was generally assumed by educated and uneducated people alike. Although, since Copernicus, Catholics had been forbidden to believe in extraterrestrial life, Protestants considered it unthinkable that God would have been so wasteful as to create planets without putting people on them.

Herschel was in regular correspondence with Johann Hieronymus Schröter of Bremen, who made careful and systematic observations of Mars from 1785 to 1802, using a Herschel telescope. Although little known today, Schröter should be ranked with Herschel as one of the half dozen preeminent observers of Mars. (Flammarion devoted about 15 pages of text to his work and displayed 65 of his sketches, compared with less than 10 pages and 35 sketches for Herschel.) Schröter improved upon several of Herschel's measurements, and his polar cap observations were superior. However, his great work *Areographische Fragmente* (447 pages, 230 drawings) lay untouched in his observatory until it was finally published in 1881; thus his contribution to Martian science was negligible.

Honoré Flaugergues, who made observations at Viviers in southeast France through eight oppositions, was the first to note (1809) the presence of "yellow clouds," which only much later were identified as dust.

Little more was learned of Mars in what Flammarion called the "First Period" of observations, which he considered to end in 1830. His summary of knowledge as of this date is given below, and some of the more interesting sentences are translated verbatim (in quotation marks). Modern values of the parameters are given in parentheses.

1. "The revolution of Mars has been fixed approximately since antiquity. Since Copernicus, we know that revolution takes place about the Sun. We know today that it is accomplished in 687 days. . . ." (686.96)
2. Distance from the Sun 1.5237 AU (1.52368); thus sunlight is 2.32 times less intense than on Earth. "But it is useful to remark that it is the constitution of the atmosphere that regulates the temperatures. The temperature of the surface of Mars could be equal and even superior to that of our world."
3. Diameter 0.5237 times Earth's; hence volume ratio 0.147. (0.150)
4. Mass ratio, Mars/Sun 1/2546320. "This is the mass obtained by Delambre from the perturbations of the Earth and adopted by Laplace in his Mécanique Céleste (1802). Today [1892] it is known with more precision from the movement of satellites and we know it to be 1/3093500." (1/3098710)
5. Density, from mass/volume, 0.711 times Earth's density. (0.715)

6. "The gravity at the surface [relative to Earth], derived from the mass and the radius of the planet, is 0.376." (0.3795 at the equator)
7. "The duration of the rotation was already, in 1830, known with a sufficiently great precision, and was evaluated as 24 hours 39 minutes." (24 hours 37 minutes 22.7 seconds)
8. "There are on Mars some spots, more or less dark. These spots are difficult to discern well. . . ." and they are variable. "We have seen pass under our eyes 191 views of the planet Mars, drawn by most dissimilar observers: these views constitute the primary base of our understanding of the world of Mars."
9. "There are also on Mars some white spots, marking its poles. These spots vary with the seasons, increasing in winter, diminishing in summer. They submit to the influence of the Sun like our polar ice. We can consider them to be ice or snow."
10. "These polar snows are not situated exactly at the extremities of the same diameter and do not mark absolutely the geographic poles. These poles are generally covered. But, at the epoch of minimum, they are reduced to a white point approximately circular that is removed a certain distance from the pole. . . ."
11. "The inclination of the axis of Mars does not differ much from that of the axis of the Earth, so that the seasons there are analogous, although nearly two times as long."
12. "There is on that planet a second order of seasons, caused by the eccentricity of the orbit, Mars is much closer to the Sun at perihelion than at aphelion, in the proportion of 1.3826 to 1.6658 or 10 to 12." (It is interesting to note that Antoniadi in 1930 was still quoting these figures although the correct values are 1.3815 and 1.6666.)
13. "The planet is surrounded by an atmosphere, in which are formed some snows about which there have been serious questions and in which float some white clouds and probably also some dark clouds."

III. THE NINETEENTH CENTURY

Beginning with the close encounter of Earth and Mars in 1830, there was a resurgence of interest in the planet, ushering in what Flammarion called the "Geographic Period" of investigation.

Preeminent in this period was Johann H. von Mädler of Berlin, whose observations, extending over six oppositions, were made in collaboration with the banker Wilhelm Beer, who owned the observatory. They made the first attempt to compile a complete map of the planet, and in 1840 they published an improved and expanded version. Few of their dark markings match anything known today, but their map had one lasting effect—it established a system of latitude and longitude and defined zero longitude by a meridian passing through a small, very dark spot. They made three determinations of the rotation period using baselines of 759, 1604, and 2234 days, the mean of which gives the value of 24 hours, 37 minutes, 22.6 seconds. (This value is within 0.1 second of the correct value, although each of their three values was more than 1 second off.) Noting that Herschel's value had been about 2 minutes greater, they reviewed his records and concluded that Mars had rotated one more time than Herschel had assumed; they corrected his value to 24 hours 37 minutes 26.27 seconds.

A generation later the pace of Mars investigations quickened with the increasing availability of better telescopes. Drawings worthy of note were made by Francois Arago (whose observations extended from 1811 to 1847), by the British amateur Warren de la Rue in the 1850s, by Joseph Norman Lockyer in 1862, by Frederick Kaiser in 1862 and 1864, by Flammarion from 1862 until the 1890s, and by his collaborator, E. M. Antoniadi, until the early 1900s. The papal astronomer Father Pierre Angelo Secchi in 1863 published the first known sketches in color and suggested that the colors of some features change. He reported seeing reddish brown, blue, yellow, and perhaps greenish hues.

Drawings were gradually incorporated into maps. William R. Dawes of Haddenham, England, made a fairly accurate map in 1864. In 1867, Richard Anthony Proctor combined the sketches of several observers into a map that was the first to carry names for the principal features. His nomenclature found little favor with other astronomers and was soon superseded; only his choice of the zero meridian (which he had taken from Mädler) has survived. Flammarion's 1876 map also carried place names that were quickly forgotten. However, he confirmed Secchi's suspicion about color changes, and in 1901 he collaborated with Antoniadi to produce a much improved map. Over the next quarter century, Antoniadi produced a series of excellent maps, the best available until the 1950s.

IV. THE MODERN ERA BEGINS

For Martian studies, 1877 was a banner year. Mars' close perihelion opposition enabled Giovanni Virginio Schiaparelli, director of the Observatory of Milan, to begin the modern era of Mars investigations. He was a meticulous observer and a competent illustrator, and his 1877 map was the first to approximate modern standards. He embellished it with a system of names for the features that he saw that was based on a romantic blend of biblical and mythical geography. He retained Herschel's designations of dark areas as "seas" and lighter areas as "land." His nomenclatural scheme quickly became accepted and has evolved into the one used today.

Schiaparelli's other major contribution was his popularization of the straight lines that he called "canali" (channels), perhaps because that was the name that Angelo Secchi had used in 1858. At least half a dozen earlier sketches of Mars had shown some of these lines, but Schiaparelli's first map showed forty, and he ultimately claimed to have seen about a hundred. No one else saw these lines in 1877, and indeed not until 1884, so the question of their existence immediately became a subject of great controversy that did not end until the 1960s, although Antoniadi really settled it in 1909.

Another landmark event of 1877 was the discovery of Phobos and Deimos by Asaph Hall at the U.S. Naval Observatory. He very nearly gave up too soon but was goaded by his wife into making one more attempt (Ley and von Braun 1956, p. 49–50). Also in this year came the observation by Nathaniel E. Green of white spots near the limb of Mars, which he concluded correctly to be morning and evening clouds high in the atmosphere. Also in 1877, the first attempt to photograph the planet was made by M. Gould, director of the observatory at Cordoba, Argentina. The first useful photographs of Mars were taken in 1890 and 1892 by William H. Pickering (no relation to the Jet Propulsion Laboratory director of the same name).

At the next opposition in 1879, Schiaparelli was surprised to find that one of his canali was now double—an example of the notorious phenomenon of "gemination." In 1881 and 1882 he revised his map, confirming all of his original canali and adding more, including 20 examples of gemination. He and the other astronomers who reported seeing the canali noted that they all ended in large, round, dark spots, which Schiaparelli called "lakes," and Pickering later added that the crossing points of canali were marked, by small, dark spots, which came to be called "oases." Schiaparelli never claimed that the canali were artificial (nor denied the idea either), and later he espoused the increasingly popular assumption that the dark regions were vegetation and not seas. This idea was engendered by three observations: Pickering's discovery of canali crossing the dark regions, that nobody ever saw a reflection of the Sun in any of the "seas," and that the seasonal changes in the shapes of the dark regions seemed suggestive of the growth and death of plants. The stage was now set for Percival Lowell and the burgeoning of interest in the planet, which has sometimes been called the "Mars mania."

Lowell made his first observations of Mars from his new observatory in Flagstaff, Arizona, on May 24, 1894. He continued observing until his death in 1916, making 917 sketches and mapping 437 "canals." That he insisted to the end of his days that they were artificial and proved the presence of technically advanced beings on the planet is too well known to require elaboration. Few astronomers accepted these ideas, two notable exceptions being Flammarion and Pickering. The two principal effects of Lowell's work were somewhat contradictory. He created enormous interest in Mars among the general public, but he gave planetary science such a bad reputation among American astronomers that it was almost completely neglected in this country during the first half of the 20th century. Although he established what is still a thriving and productive observatory where many useful planetary and ring observations were made during this period, Lowell may perhaps have done more harm than good to the cause of planetary science.

Other phenomena attributed to Mars by visual and photographic observers included "blue clearings" and the "wave of darkening." Blue clearings were occasional periods of a few days when a supposed blue or violet haze layer dissipated and allowed surface contrast to be seen at wavelengths below 450 nm. The wave of darkening was a seasonal lowering of surface albedo that allegedly progressed slowly outward from the springtime edge of the polar caps and reached across the equator (Michaux and Newburn 1972). Both of these phenomena were treated seriously until the Mariner 9 mission showed them to be illusory. The lack of blue contrast is now known to be intrinsic to the Martian surface, and irregular seasonal contrast changes are associated with changes in surface wind streaks following dust storms.

V. RESOLUTION OF OLD QUESTIONS AND CONTROVERSIES

In 1909, Flammarion published the second volume of his encyclopedia on Mars, containing 426 drawings and 16 maps from the period 1890 to 1901. From that time to this, the contributions from simply looking at or photographing the planet have been restricted largely to a record of clouds and seasonal surface changes. Major advances have been made only with increasingly sophisticated instrumentation. Ultimately, spacecraft observations have replaced uncertainty with facts in many areas, thus settling the heated controversies that marked the first half of the 20th century. We shall mention only four of these areas of controversy.

A. Canals

Many reputable observers claimed to have seen them at one time or another—notably Earl C. Slipher, who studied the planet at Lowell Observatory from 1904 to 1964. At the close opposition of 1939 he went to Africa to photograph Mars and reported (1940): "These African photographs record so many of the canals and oases in the position, form, and character as depicted in Lowell's maps of the planet that they should remove all doubts as to the reality of these markings." However, the pictures did not convince the skeptics. George Ellery Hale, who used the 60-inch reflector at Mt. Wilson, stated categorically that they could not be seen. Antoniadi produced excellent maps that did show many rather diffuse linear features, but he wrote (1930): "Nobody has ever seen a genuine canal on Mars, and the more or less rectilinear, single or double 'canals' of Schiaparelli do not exist as canals or as geometrical patterns."

Various other explanations for linear features were advanced. Pickering agreed with Lowell about the existence of "canals" but attributed them to natural cracks in the surface. The most bizarre explanation was advanced by Donald Lee Cyr in 1959. He claimed that the "oases" were actually deep meteor craters warm enough for copious vegetation and that the lines connecting them marked the "migratory fertility paths caused by the action of a mobile animal species." He did not attempt to describe these grazing fauna, but he gave them a name: "Marsitrons." Probably the best explanation, if indeed there is one, is that more or less linear alignments of indistinct dark spots tend to be fused by the human eye into a continuous line. This or a similar idea was espoused by Antoniadi, Nathanial Green, Vincenzio Cerulli, E. W. Maunder, Audouin Dollfus, and others.

B. Polar Caps

Early observers naturally assumed the seasonal polar caps to be water ice, by analogy with caps on Earth. Pickering wrote in 1899: "The polar caps are without the slightest doubt layers of crystallized water—probably more like white frost than solid ice or snow." At about that time, however, others, for example J. Joly, George Johnstone Stoney, and A. C. Ranyard, championed a composition of carbon dioxide because of the low temperature of the planet and the apparent absence of water. Even as late as 1950, two of the preeminent students of Mars, Gerard Kuiper and Audouin Dollfus, were sure that their observations proved that the caps were water ice. M. Trouvelot of the Harvard College Observatory wrote in 1882: "If the polar spots are composed of a white substance melting under the rays of the Sun, as seems altogether probable, its melting point must be above that of terrestrial snow." Apparently the possibility of sublimation was overlooked by most investigators until 1947, when de Vaucouleurs pointed out that this must be the process involved.

C. Surface Variation

In the early days, most of the subtle changes that were observed were explained as atmospheric phenomena. Emmanuel Liais, founder of the Rio de Janeiro observatory, appears to have been the first to suggest, in 1860, that the changes were due to vegetation. This hypothesis came to be generally accepted, and the controversies were centered on the nature of the plants. The certainty attached to this hypothesis is strikingly illustrated by Slipher (1962): "A third presence on Mars indicates a living world: vegetation. The evidence is in the blue-green areas and the changes in their appearance. Vegetation would present exactly the appearance shown, and nothing we know of but vegetation could. The seasonal change that sweeps over them is metabolic; that is, it shows both growth and decay and proclaims its organic constitution such as only vegetation could produce. . . . Since the theory of life on the planet was first enunciated some fifty years ago, every new fact discovered has been found to be accordant with it. Not a single thing has been discovered which it does not explain." Other suggested explanations for the dark regions included that of Svante Arrhenius in 1912 that they were hygroscopic salts that change with the humidity and that of D. B. McLaughlin that they were ash deposits from volcanoes.

D. Craters

One issue that might have stirred intense controversy seems to have been virtually ignored. It is reported that Edward Emerson Barnard at the Lick Observatory's 36-inch refractor saw craters in 1892 and that J. E. Mellish saw them with the 40-inch at Yerkes Observatory in 1915. Unfortunately no one seems to have paid attention, and the drawings of these observers have not survived to substantiate the discoveries. In 1950 E. J. Öpik published a paper entitled "Collision Probabilities with the Planets and the Distribution of Interplanetary Matter," in which he calculated the probability of meteorite impacts on Mars and concluded that there should be hundreds of thousands of craters exceeding in size the Meteor Crater in Arizona. Clyde Tombaugh (1968) states that he independently made a similar suggestion in 1950, and Fred Whipple supported the idea. Still, it got so little notice that nearly everyone was surprised by the findings of Mariner 4 in 1965.

VI. SPECTROSCOPY AND RADIOMETRY

Flammarion related (Vol. 1, p. 182) that "In the course of the year 1862, M. William Huggins, member of the Royal Astronomical Society of London, and M. A. Miller, professor of Chemistry at King's College of London, attempted for the first time to apply spectral analysis to the study of the planets Venus, Jupiter, Mars and Saturn. The results obtained have been published in the *Philosophical Transactions* in the year 1864. In the United States, Rutherfurd undertook at the same time the same research." L. M. Rutherfurd's paper appeared in the *American Journal of Science* in January 1863. Hermann Vogel in Germany attempted similar measurements at the same time. In 1867 Jules Janssen, who founded the Meudon Observatory, took his instruments to the top of Mt. Etna (3390 meters) to compare the spectra of Mars and the Moon. These investigators observed absorption bands produced by the Martian atmosphere and concluded that it contains considerable quantities of water vapor. Their instruments were crude by modern standards, however, and their results are no longer believed. It is interesting to note that Flammarion, when he was writing his book in 1892, was unaware of Janssen's work, and he later inserted an apologetic footnote.

It was not until the 1920s that the use of spectroscopes, polariscopes, and radiometers for investigating Mars began in earnest.

VII. MILESTONES IN THE MODERN STUDY OF MARS

Detailed discussions of the role of modern instrumentation in the progress of Martian research in the twentieth century are found in many places in this book, and the role of spacecraft missions is treated in chapter 3. Therefore, we shall end this historical introduction by simply mentioning some of the recent Earth-based investigations that have been most crucial to our current understanding of Mars.

A. The Atmosphere

The attempt to understand Mars' atmosphere—its abundance and composition—has engaged many investigators throughout this century, but the interdependence of its parameters has complicated their research. In the absence of data, it was customarily assumed that nitrogen and argon were the major components, and this assumption was difficult to refute. In 1952, Gerard Kuiper discovered two relatively strong bands of CO_2 near 1.6 μm on one of his pioneering spectrograms made in 1947 October 7. He found three more near 2.0 μm a few weeks later, and for two decades this gas was the only known atmospheric constituent—presumably a minor one. The Mariner 4 radio-occultation experiment in 1965 indicated that CO_2 is indeed the major constituent, and this was soon confirmed by the interferometric spectrometer (Kaplan et al. 1969) that also detected the presence of carbon monoxide.

Lowell was probably the first to suggest that Martian water vapor might be detected by using the Doppler shift to distinguish its spectral lines from those of terrestrial origin. The experiment was attempted at Lowell and Lick observatories with apparently positive results, but inadequate equipment and technique caused large uncertainties. Several attempts by Adams and coworkers at Mount Wilson between 1925 and 1943 fared little better (see de Vaucouleurs 1954). The first detection to be widely accepted was that of Spinrad et al. (1963), which measured 14 $\pm$ 7 precipitable micrometers of water. Many other measurements followed soon thereafter.

The search for the right value of the surface pressure of the atmosphere is a particularly interesting story. It began in 1926, when Donald Menzel at Lowell Observatory reported that his photometric observations indicated an upper limit of 66 millibars (mbar) but noted that, under different assumptions, the limit could be 26 mbar. In 1929, B. Lyot (1929) at the Meudon Observatory near Paris concluded from polarimetric data that the upper limit must be less than 24 mbar. Several attempts in the 1940s gave much higher values, and in his 1954 book Gerard de Vaucouleurs analyzed all the results and concluded that the right figure "cannot be far from 85 $\pm$ 4 mbar" (p. 125); this became the standard and generally accepted value as the spacecraft age dawned.

Unfortunately, analysis of the observational data in all these investigations was based on uncertain or erroneous assumptions. The most serious was the neglect or underestimate of the effect of light scattering by dust in comparison with Rayleigh scattering. A detailed critique of these measurements was given by Chamberlain and Hunten (1965).

The breakthrough came during the 1963 opposition of Mars when Kaplan et al. (1964), using the same spectrogram on which they had discovered water vapor, calculated the partial pressure of carbon dioxide to be 4 mbar. Still clinging to the assumption that it was a minor component, they concluded that the most probable value of the surface pressure was 25 ± 15 mbar. Shortly thereafter, the development of the Fourier spectrograph by Pierre and Jeanine Connes in 1966 made possible the detection of so many spectral lines, with such precision, that Louise Young was able to get the correct answer in 1971. Her value for the mean pressure over one hemisphere in April 1967 was 5.16 ± 0.64 mbar. Meanwhile, the radio occultation investigation on Mariner 4 confirmed that carbon dioxide is the major component of the atmosphere and that the pressure is indeed that low.

B. Polar Caps

From the time of Cassini (1666), it had usually been assumed that the seasonal polar caps were water ice. The observations of Kuiper in 1952 and Moroz in 1964 appeared to confirm this assumption, although there were a few dissenters (notably Alfred Russell Wallace). Once the question of the pressure was settled by Mariner 4, however, Leighton and Murray (1966) produced an atmospheric model that (1) indicated that the poles should get cold enough to condense carbon dioxide, and (2) predicted major seasonal variations in pressure as a result. Laboratory observations of infrared spectra of H_2O-CO_2 frosts by Kieffer (1970) indicated that very small amounts of water could mask the characteristics of the carbon dioxide spectrum and explained the reason for the results of Kuiper and Moroz.

C. Surface

In the first half of the 20th century, many studies by visual observation, photography, spectrometry, radiometry and polarimetry attempted to determine the chemical composition, mineralogy, texture, relief and temperature of Mars' surface. Most of the conclusions reached were imprecise or uncertain, and they were superseded if not contradicted by the spacecraft observations of more recent times. A major experimental milestone was the introduction in 1963 of radar observations pioneered by Goldstein and Gilmore in California at 12.5-cm wavelength and by Kotelnikov in the U.S.S.R. at 43 cm. The former covered a full rotation and produced the first brightness maps along latitude 13°N. Subsequent measurements were made by JPL, MIT, and Arecibo of reflectivity, roughness, and dielectric constant, and altitude profiles were obtained.

D. Interior

Even today our ideas about the interior of Mars come primarily from theoretical models. This discipline was pioneered by Harold Jeffries in 1937 and by K. E. Bullen in 1949. Harold Urey in his 1952 book on the planets discussed models of both interior structure and thermal history, and other theorists soon followed. All these early models were severely limited by the lack of accurate values of the radius and moment of inertia of Mars, which became available only with spacecraft data.

E. Literary Milestones

The outstanding summary of Martian science produced by Flammarion at the end of the 19th century has been noted. At the time of his death (1925) he was working on a third volume, but, unfortunately, his collaborators never finished it. Also in this century, three other works were particularly notable. The first was produced by Antoniadi (1930), who had worked with Flammarion at the turn of the century. De Vaucouleurs called it "the most complete and most representative summary of the analytical description of the surface of Mars, based essentially on visual observations and its pictorial reproduction." The second was by de Vaucouleurs himself (1954). He concentrated on the results of combining physics and astronomy and introduced the term "areophysics." Finally, just before Mariner 4, E. C. Slipher (1962) set forth the results of his half century of meticulous observations with detailed descriptions and beautiful photographs. This delightful work is still a joy to read, even though it is marred by his dogmatic defense of the hypotheses of artificial canals and growing plants.

F. Spacecraft Missions

The most spectacular milestones in the investigation of Mars were about a dozen successful spacecraft missions, most notably the Viking Orbiters and Landers. These are described in chapter 3. It is interesting that the early spacecraft imaging results were misleading, a coincidental result of the orbital dynamics of flyby missions and the global dichotomy of Mars. The observations of Mariner 4 (in 1965) and Mariner 6 and 7 (in 1969) were concentrated in the southern hemisphere, and they largely showed a heavily cratered surface, similar to the cratered highlands on the Moon, although there were hints of unusual terrain. The planned mission of Mariner 8, which failed on launch, was directed toward possible color and albedo changes in part based on the apparent phenomena of blue clearing and wave of darkening.

Global mapping by Mariner 9 (in 1971–72) revealed for the first time the great geologic diversity of the surface and the intriguing large scale of several Martian features, including volcanoes an order of magnitude larger than on Earth, a canyon system 3000 km long, evidence of catastrophic floods whose inferred flow rates exceeded those of the largest terrestrial rivers, and remnants of an integrated drainage system suggesting a vastly different climate in earlier epochs. (All of these features are discussed in later chapters.) Clearly, the Mars seen by Mariner 9 was tremendously different from that seen by earlier spacecraft.

The next major advance came with the Viking missions in 1976, when the Landers and Orbiters provided new insight into the Martian surface, atmosphere and history. The boustrophedonic scanning capability of the Orbiters allowed the remote sensing instruments to obtain uniform hemispheric or global coverage at many seasons. Many of the discussions in this book are based heavily on Viking data.

Of all the Viking results, the biology experiments had the greatest impact. Although they are still debated, the general consensus has been that the data are most consistent with the absence of biological material at the surface. Perhaps the strongest evidence against ongoing biological activity came from the gas chromatograph/mass spectrometer experiment on the Lander, which detected none of the organic material in the soil that would be expected as detritus from the presence of life.

G. Recent Milestones

Since the Viking missions, two developments have been especially significant. The first was the discovery of the "SNC" meteorites, which are thought to have come from Mars. They show evidence of recent volcanic activity, their oxygen isotope ratios are clearly not terrestrial, and they contain trapped gases that closely resemble the Martian atmosphere. Laboratory analyses are providing new insights into the chemistry and petrology of the Martian surface. The SNC's are discussed in chapters 4, 6 and 18.

The second development was the observation that the deuterium/hydrogen ratio in the Martian atmosphere is enhanced by a factor of 5 over that of Earth (Owen et al. 1987), suggesting the escape of the equivalent of a global layer of water 50 to few hundred meters thick. Although there is a consensus that a significant amount of water has been lost, the details are still being debated.

VIII. OUTSTANDING QUESTIONS AND PUZZLES

A large number of major questions still remain about the history of the planet; these are discussed in more detail in later chapters. Most of these questions have been formulated only since the beginning of spacecraft exploration, and many can be addressed only with additional data, both from spacecraft and from Earth-based observations. Some of the major questions requiring substantial additional data follow:

1. What were the conditions on early Mars in terms of thermal history, volcanic output, outgassing, etc.?
2. What caused the global dichotomy (southern old highlands and northern young lowlands) of the surface?
3. What were the volatile inventories of the material from which Mars accreted and of the newly formed Mars, and what is it today?
4. To what extent are current conditions representative or indicative of past conditions in terms of surface or interior processes, global climate, episodic climate variations, etc.?

5. To what extent does the material in the polar caps exchange with the rest of the planet on all different time scales.
6. Were past climatic conditions conducive to the formation of life, and is there fossil evidence that life did exist?

Finally, we list some of the true puzzles about Mars. These are issues that are not at all understood today even though we have a significant amount of relevant information.

a. If Mars had an early thick atmosphere, where is all the CO_2 now?
b. Why are the two perennial polar caps so different from each other, and to what extent do they vary from year to year?
c. Do great dust storms occur regularly or irregularly? Did they really begin to occur more often just as we began to explore Mars with spacecraft?
d. Has all the water that formed the channels and integrated drainage systems escaped to space, or is it sequestered in the regolith and polar caps?

IX. INTRODUCTION TO THE CHAPTERS

The 38 chapters and the appendix of this book each deal with a specific subject. Effort has been made to integrate the chapters, so that each major topic is generally covered in a single chapter, and the chapters are appropriately cross referenced. Although each chapter stands alone, the chapters form nine logical groups, as follows:

Introductory material (chapters 1–4); this group places the rest of the chapters in the context of our current and past understanding of Mars as a whole; these chapters consist of this introduction, the history of telescopic observations, the history of spacecraft observations, and a discussion of major outstanding questions about Mars.

Geophysics (chapters 5–10); chapters deal with origin and thermal history, composition and structure of the interior, observations and analyses of topography and gravity, state of stress at the surface, orbital and spin dynamics of the planet, and global geodesy.

Bedrock geology (chapters 11–16); chapters include discussions of the global stratigraphy and geologic history, the role of impact cratering, the role of volcanism, the canyon systems, channels and valley networks, and processes involving ice within the megaregolith.

Surface (chapters 17–23); chapters cover spectroscopic observations and analyses of the surface layer, the chemistry and mineralogy of the surface, chemical and physical weathering of the surface, radar observations of surface properties, physical properties of the near-surface layer, the role of aeolian processes, and geologic processes in the polar regions.

Atmosphere (chapters 24–30); chapters include an introduction to the atmosphere and climate system, composition of the atmosphere, structure, dynamics and meteorology of the atmosphere, the seasonal carbon dioxide cycle, the seasonal water cycle, the properties and processes pertaining to atmospheric dust, and the aeronomy and chemistry of the atmosphere.

Exosphere and magnetic field (chapter 31); linked closely to the evolution of the atmosphere, this chapter is a discussion of the intrinsic magnetic field and how this influences the interaction of the solar wind with the upper atmosphere.

Climate history (chapters 32–33); chapters deal with possible epochal climate changes on Mars, and quasi-periodic climate change.

Biology (chapters 34–35); chapters include a summary of the search for extant life on Mars and a discussion of the possible presence of life on Mars during earlier epochs.

Satellites (chapters 36–38); chapters include the geodesy of the satellites, their geologic processes, and their dynamical evolution.

Because in many chapters a basic understanding of the names and locations of surface features is assumed, an appendix to the book describes the formal system of the nomenclature used for Martian surface features, with a brief discussion of the informal names for rocks at the Viking Lander sites.

Chapter 2 is unique in that its authors present some of the flavor of the contributions made by the many amateurs whose observations have advanced our knowledge of the variations on Mars. The advent of spacecraft and new-technology, Earth-based observations requiring large facilities, such as radar, very large array passive microwave, or imaging spectroscopy, has not diminished the importance of extending the historical record of Mars' appearance. Mars and Earth are unique in having substantial surface weather systems (although little is known yet about possible temporal variations of weather on Venus and Titan). Even accounting for the low resolution and difficulty of interpretation associated with Earth-based observation, much more is known than is understood about Martian weather. Detailed observations provided by spacecraft are necessarily restricted in time, but when they are combined with a growing historical record from Earth's observers, the result will be the best basis against which theories of Martian weather and atmosphere-surface interactions should be tested. Some day, humans will step out onto the surface of Mars; by then the ability to forecast large dust storms will be more than an academic issue.

We also wish to point out subjects that are covered in more than one chapter or in a place that may not be obvious at first glance.

Three chapters serve as broad introductions to specific groups of chapters. In chapter 4, Pepin and Carr discuss the major outstanding questions in our understanding of Mars today, and they include an introduction to the role of the SNC meteorites in Martian science. The origin and implications of the SNC meteorites form an important topic that is brought into many chapters in this book. Their early introduction is appropriate, so a general discussion of their occurrence, clues to their Martian origin, and implications for the volatile inventory of the planet are contained in this chapter. The properties of the SNC's then play an important role in discussions of the thermal history of Mars (chapter 5 by Schubert et al.), the composition of the interior (chapter 6 by Longhi et al.), the composition and mineralogy of the surface (chapter 18 by Banin et al.), and the volatile inventory of the planet.

In chapter 11, Tanaka et al. describe the global stratigraphy and geologic history of Mars; thus this chapter serves as an introduction to the following chapters on specific geologic processes.

In chapter 24, Zurek introduces the subject of the Martian atmosphere, summarizes the major seasonal cycles, and discusses some of the outstanding questions of the behavior of the atmosphere and climate system.

The volatile inventory of Mars and mechanisms of loss of material from the atmosphere are of such broad interest and have such wide implications that discussion is spread over several chapters. Some discussions of volatile abundances based on the SNC meteorites are contained in chapters 4 and 6, as mentioned above. The noble gas abundances and isotopic ratios as measured in the current atmosphere, together with implications for atmospheric evolution, are discussed in the chapter on atmospheric composition (chapter 25 by Owen). The processes governing the loss of gases to space and a quantitative description of their implications are discussed along with the atmospheric chemistry and aeronomy (chapter 30 by Barth et al.). Discussions of the geologic constraints on volatile abundance, primarily water and carbon dioxide, are contained in the chapter on epochal climate change (chapter 32 by Fanale et al.). Loss of volatiles by "impact cratering" is discussed along with other impact processes (chapter 12 by Strom et al.), and loss via hydrodynamic escape is included in the chapter on major questions (chapter 4 by Pepin and Carr) and, to a lesser extent, in the chapter on interactions with the solar wind (chapter 31 by Luhmann et al.). The implications of the escape of volatiles to space for exchange with nonatmospheric reservoirs are described in the chapter on epochal climate change (chapter 32).

The history and current state of the interior of Mars are discussed from the points of view of the condensation from the early solar system and thermal evolution (chapter 5), inferred internal chemistry and density distributions (chapter 6), and the expression of internal statics and dynamics at the surface (chapters 7 by Esposito et al. and 8 by Banerdt et al.).

The physical properties of the Martian surface are discussed in two chapters. In chapter 20, Simpson et al. describe how radar measurements are used to determine surface properties such as roughness and radar reflectivity and the presence of scattering elements of discrete sizes. The implications of these results for geologic processes involved in the evolution of the near-surface layer, along with discussion of the results and implications of other *in situ* and remote-sensing observations, are discussed in chapter 21, by Christensen and Moore, on the near-surface layer.

Polar processes are also discussed in several chapters. In chapter 23, Thomas et al. describe the geologic properties of the polar regions and the processes inferred to be ongoing. Issues related directly to volatiles are described in chapters dealing with their seasonal cycles. The sublimation and condensation of water, which provide exchange between the polar caps and the atmosphere, are described in the chapter on the seasonal water cycle (chapter 28 by Jakosky and Haberle), as is the role of the polar caps in the seasonal water cycles of earlier epochs. Exchange of CO_2 between the polar caps and the atmosphere is described in the chapter on the seasonal carbon dioxide cycle (chapter 27 by James et al.). These seasonal processes respond to the changing insolation pattern due to variations in the orbital elements for Mars (chapter 9 by Ward); these longer term polar processes and their impact on the atmosphere are discussed in the chapter on quasi-periodic climate change by Kieffer and Zent (chapter 33).

X. BOOKS ABOUT MARS

The following list includes books that are primarily or entirely about Mars. It is intended to be representative, but it is unlikely to be complete. The technical level ranges from general interest (see, e.g., Richardson and Bonestell 1964; Caidin 1972; Moore 1977) to strictly technical (Antoniadi 1930; Michaux 1967; Kaplan 1988). Included are several books devoted to spacecraft imaging (see, e.g., Collins 1971; Hartmann and Raper 1974; Carr and Evans 1980; Viking Lander Imaging Team 1978; Viking Orbiter Imaging Team 1980). Many of these books are Special Publications produced by the National Aeronautics and Space Administration (NASA-SP series). The books by Lowell and his biography by Hoyt provide views of a man and his conviction that dominated the popular concept of Mars for half a century.

To understand the history of the investigation of Mars, there is no substitute for the two volumes of Flammarion for those who read French. The rest of us would do well to begin with the synthesis of telescopic observations by Antoniadi (1930), which is available in English translation. Following the tradition that these two scientists established, de Vaucouleurs (1954) wrote a scientific assessment of Mars which treated virtually all observations and measurements made to that time; although several conclusions have since proved erroneous, this work provides a complete view of the state of Martian studies at mid-century. The photographic compilation by Slipher (1962) provides an excellent comparison with the drawings of Antoniadi.

Two later summaries of knowledge about Mars: the works of Glasstone (1968) and Firsoff (1969, updated 1980) are primarily compendia.

Anonymous 1969. *Mariner-Mars 1969: A Preliminary Report* (Washington, D.C.:NASA SP-225) 145 pp.
A technical summary of the mission and its science instruments and summaries of the initial scientific results.
Anonymous 1974. *Mars Engineering Model* (Hampton, Virginia:NASA Viking Project Document M75-125-3) 331 pp.
A conservative reference model of Mars and its environment. Derived from the scientific literature, but with controversy removed, this was the basis for the engineering design of the Viking mission.
Anonymous 1976. *Viking 1: Early Results* (Washington, D.C.:NASA SP-408) 67 pp.
A moderately technical description of the landing site certification, entry and landing sequences, and initial results of Lander 1 investigations.
Anonymous 1984. *Viking: The Exploration of Mars* (Washington, D.C.:NASA Educational Publication 208) 55 pp.
Brief summary of the Viking Mission with many Orbiter and Lander photographs.
Antoniadi, E. M. 1930. *La Planète Mars, 1659–1929* (Paris:Librairie Scientifique Hermann et Cie) 239 pp.
A detailed synthesis of visual observations of the surface features and clouds on Mars. English translation, *The Planet Mars,* 1975, by Patrick Moore. (U.K., Shaldon Devon, Keith Reid Limited), 335 pp.
Averner, M. M., and MacElroy, R. D., eds. 1976. *On the Habitability of Mars* (Washington, D.C.:NASA SP-414) 105 pp. Subtitled "An Approach to Planetary Ecosynthesis."
Consideration of the viability of some terrestrial organisms on Mars, possible modification of the Martian environment to support terrestrial life forms, and the genetic engineering of life to tolerate Martian environments.
Baker, V. R. 1982. *The Channels of Mars* (Austin: University Texas Press) 198 pp.
Detailed discussion and hydraulic analysis with extensive maps of channel systems.
Batson, R. M., Bridges, P. M., and Inge, J. L. 1979. *Atlas of Mars: The 1:5M Map Series* (Washington, D.C.:NASA SP 438) 146 pp.
A complete set of the photomosaics, airbrush drawings, and contour maps that resulted from the Mariner 9 mission. In this bound atlas, the maps are published at a scale of 1:10M.
Biemann, Hans-Peter 1977. *The Vikings of '76* (Westford, Mass.:Murray Printing Company) 144 pp.
A photographic record of the Viking project personnel and their activity at the time of the Viking landings.
Blunck, Jürgen 1977. *Mars and Its Satellites* [2d ed., 1982] (Hicksville, N.Y.:Exposition Press) 222 pp.
Subtitled "A Detailed Commentary on the Nomenclature," both editions are dedicated entirely to Martian names and their origins.
Burgess, Eric 1978. *To the Red Planet* (New York City, N.Y.:Columbia University Press) 181 pp.
A general description of the events leading up to the Viking mission, the site-selection and landing process, and an overview of the mission results.
Caidin, Martin 1972. *Destination Mars* (Garden City, New York:Doubleday) 295 pp.
A general description of Mars and the exploration process through the Mariner 9 mission.
Carr, M. H. 1981. *The Surface of Mars* (New Haven, Conn.:Yale University Press) 232 pp.
A richly illustrated scientific description of the geology of Mars, with brief treatment of the atmosphere and the Viking search for life.
Carr, M. H., and Evans, Nancy 1980. *Images of Mars: The Viking Extended Mission* (Washington, D.C.:NASA SP-444) 32 pp.
A collection of some of the more spectacular images from the Viking Orbiter extended mission.
Collins, S. A. 1971. *The Mariner 6 and 7 Pictures of Mars.* (Washington, D.C.:NASA SP-263) 159 pp.
Every image acquired by Mariner 6 and 7.
Corliss, W. R. 1974. *The Viking Mission to Mars* (Washington, D.C.:NASA SP-334) 77 pp.
A description of the Viking mission, spacecraft, and landing scenario; written about two years before launch.

Duxbury, T. C., Calihan, J. D., and O'Campo, A. C. 1984. *Phobos: Close encounter imaging from the Viking Orbiters* (Washington, D.C.:NASA Reference Publication 1109) 51 pp.
A large-format presentation of the high-resolution imaging of Phobos related to the mapping coordinate system.
Ezell, E. C., and Ezell, L. N. 1984. *On Mars: Exploration of the Red Planet, 1958–1978.* (Washington, D.C.:NASA Special Publication 4212) 535 pp.
A detailed chronology of the Viking mission, including evolution of its concept through the early period of space exploration, and a narrative of the decision-making processes related to development of the spacecraft, selection of the landing site, and interpretation of the Lander results.
Firsoff, V. A. 1969. *The World of Mars* (Edinburgh: Oliver and Boyd) 128 pp.
A book aimed at the amateur scientist, with a summary of knowledge through about 1966 and a very brief treatment of the Mariner 4 images
Firsoff, V. A. 1980. *The New Face of Mars* (Hornchurch: Ian Henry) 159 pp.
Essentially an update of his 1969 book, incorporating the Viking mission results.
Flammarion, N. C. 1892 and 1909. *La planète Mars et ses Conditions d'Habitabilité* (Paris, Gauthier-Villars et Fils) 2 volumes of 608 and 595 pages, in French. [Lowell Observatory Library has a single copy of a draft translation by P. Moore]
Volume 1 summarizes or quotes in detail virtually everything that was written about Mars up to 1890. Volume 2 covers observations between 1890 and 1901.
French, Bevan 1977. *Mars: The Viking Discoveries* (Washington, D.C.:NASA) 36 pp.
A profusely illustrated brochure summarizing the Viking nominal mission.
Glasstone, Samuel 1968. *The Book of Mars* (Washington, D.C.:NASA SP-179) 315 pp.
Semitechnical compendium of what was known about Mars prior to the Mariner 9 mission, emphasizing the search for life on Mars.
Gornitz, V. 1979. *Geology of the Planet Mars.* Volume 48 of Benchmark Papers in Geology. (Stroudsbourg: Dowden, Hutchinson, and Ross, Inc.) 414 pp.
A collection of 26 "benchmark" journal articles about Mars, with emphasis on its surface.
Hartmann, W. K., and Raper, Odell 1974. *The New Mars: The Discoveries of Mariner 9* (Washington, D.C.:NASA SP-337) 179 pp.
A heavily illustrated discussion of the imaging results from Mariner 9, with little discussion of the other experiments.
Hoyt, W. G. 1976. *Lowell and Mars* (Tucson: University of Arizona Press) 376 pp.
A historian's examination of Lowell, his observatory, Lowell's "life on Mars" theme and its influence on the public for more than 50 years. Also includes an account of the discovery of Pluto.
Joëls, K. M. 1985. *The Mars One Crew Manual* (Ballentine Book) approx. 180 pp.
Hypothetical description of the first manned mission to Mars, but with attention to technical and scientific detail.
Kaplan, David 1988. *Environment of Mars* (Washington, D.C.:NASA Technical Memorandum 100470).
A summary from an engineer's standpoint of the Mars surface environment, based primarily on published analyses of Viking observations.
Kellogg, W. W., and Sagan, Carl 1961. *The Atmospheres of Mars and Venus* (Washington, D.C.:Publication 944, National Academy of Sciences-National Research Council) 151 pp.
An overview by a NASA Planetary Atmosphere Panel on the state of knowledge of the atmospheres of Mars and Venus prior to spacecraft exploration. Also includes several appendices on specific techniques and theory.
Leighton, R. B., Murray, B. C., Sharp, R. P., Allen, J. D., and Sloan, R. K. 1967. *Mariner Mars 1964 Project Report: Television Experiment; Part I, Investigators Report* (JPL Technical Report 32-884) 178 pp.
A detailed presentation and description of each of the 20 Mariner 4 images, including an airbrush drawing and interpretation of each frame. The imaging system is also described in detail.
Ley, Willy 1966. *Mariner IV to Mars* (New York and Toronto:A Signet Science Library Book, The New American Library) 157 pp.
An introduction to the history of Mars observations for the lay person, with a corresponding description of the Mariner 4 mission and its results.

Ley, W. and von Braun, W. 1956. *The Exploration of Mars* (London:Sidgwick and Jackson Ltd.) 184 pp.
Although aimed at the public, this is a serious summary of what was known about Mars, the history of study of Mars, and a possible mission by humans to Mars. Contains a lengthy bibliography of books about Mars and 21 paintings by Chesley Bonestell.
Liebes, Sidney, Jr. 1982. *Viking Lander Atlas of Mars* (NASA Scientific and Technical Information Office, NASA Contractor Report 3568) 289 pp.
A large-format, technical description of the Viking Lander images and the development of terrain profiles using the Lander stereo ability.
Lowell, Percival 1895. *Mars* (Houghton Mifflin Co.) (History of Astronomy Reprints, Limited edition, 1978) 228 pp.
A detailed description of the observations made at Lowell Observatory during the favorable opposition of 1894–1895. A congenial narrative liberally sprinkled with Lowell's interpretation of the canals and what they implied about life on Mars. Their artificiality was noted and Lowell began his speculation on the intelligence behind them.
Lowell, Percival 1906. *Mars and Its Canals* (New York:Macmillan) 393 pp.
A synthesis of the observations of Mars by Lowell and his predecessors, including a description of Lowell's observing program, the observations over several years, and the seasonal and atmospheric changes observed. A discussion of the interpretation of observations is largely confined to the later chapters. This is the basic exposition of the Lowellian view of Mars.
Lowell, Percival. 1910. *Mars as the Abode of Life* (New York:Macmillan) 288 pp.
A publication of eight lectures delivered by Lowell, building his case for life on Mars through extensive comparisons with life on Earth.
Mariner 9 Television Team 1974. *Mars as Viewed by Mariner 9* (Washington, D.C.:NASA SP-329) 225 pp.
A collection of annotated images representing all aspects of the Mariner 9 imaging discoveries. Comparison photographs of the Earth, Moon, and laboratory experiments are also included.
Meszaros, S. P. 1985. *Mars-Earth Geographical Comparisons: A Pictorial View* (Washington, D.C.:NASA Technical Memorandum no. 86166) 35 pp.
A brief pictorial comparison of Martian features at the same scale as similar terrestrial features or widely recognized political geographic outlines.
Michaux, C. M. 1967. *Handbook of the Physical Properties of the Planet Mars* (Washington, D.C.:U.S. Government Printing Office) 167 pp.
A technical synthesis of the Martian environment based on the scientific literature, with extensive references.
Michaux, C. M., and Newburn, R. L., Jr. 1972. *Mars Scientific Model* (Jet Propulsion Laboratory, Document 606-1).
This is a massive compilation of what was known about Mars in 1972, excluding the results of the Mariner 9 mission. No longer available, this large 3-ring binder edition includes colored maps with overlays of seasonal atmospheric activity and changes of surface markings. With an abundance of figures, graphs, tables, and charts, this was a veritable warehouse of information. Originally produced in 1968.
Moore, H. J., Hutton, R. E., Clow, G. D., and Spitzer, C. H. 1987. *Physical Properties of the Surface Materials at the Viking Landing Sites.* (U.S. Geological Survey Professional Paper 1389) 222 pp.
Exhaustive assessment of the nature of the Martian surface based upon its response to the landing process, soil sampler activity, and changes seen by Lander imaging.
Moore, Patrick 1977. *Guide to Mars* (New York:W. W. Norton and Company, Inc.) 214 pp.
Description of the history of exploration of Mars through the Viking mission, written at the level of the amateur astronomer.
Mutch, T. A., Arvidson, R. E., Head, J. W., Jones, K. L., and Saunders, R. S. 1976. *The Geology of Mars* (Princeton, New Jersey:Princeton University Press) 400 pp.
A profusely illustrated summary of our understanding of Mars with emphasis on geology based on observations through the Mariner 9 mission; at a level appropriate for advanced planetary geology students.
Pickering, W. H. 1921. *Mars* (Boston:Gorham Press) 173 pp.

The author supports the existence of canals with his own observations, but he disclaims the reported doubling of canals as being beyond theoretical resolution limits.

Reiber, D. B., ed. 1988. *The NASA Mars Conference* (July 21–23, 1986, at the National Academy of Sciences, Washington, D.C.) (San Diego, California:Univelt, Incorporated, for NASA and the American Astronautical Society, Technology Series, v. 71) 535 pp.
In approximately equal proportions, an overview of current understanding of Mars, discussion of planned or possible future unmanned missions, and an overview of the rationale and kinds of systems necessary for manned exploration of Mars.

Richardson, R. S. 1954. *Exploring Mars* (New York: McGraw-Hill) 261 pp.
An astronomer writes about Mars for the public. About half of the book concerns the rest of the solar system.

Richardson, R. S., and Bonestell, Chesley. 1964. *Mars* (New York: Harcourt, Brace and World, Inc.) 151 pp.
A semitechnical discussion of Mars and its possible exploration by spacecraft, written before the earliest Mars missions. Contains 12 color illustrations by the father of scientific space art.

Schiaparelli, G. V. 1877. *Osservazioni astronomiche e fisiche sull' asse di rotazione e sulla topografia del pianete Marte* (Atti della R. Academia dei Lincei, Memoria della el. di scienze fisiche. Mem. 1, ser. 3, v. 2) pp. 308–439.
The work that established much of modern Martian nomenclature. Additional reports followed in the next 11 years, but 1877 was the best opposition of this period.

Schroeter, J. H. 1881. *Areographische Fragmente, in Mathematisch-Physischer Hinsicht* [Aerographic contribution to the precise knowledge and interpretation of the planet Mars, with respect to mathematics and physics] (Leiden:E. J. Brill) 447 pp.
A record of Schroeter's work compiled from his notes by the director of the Leiden observatory nearly 100 years after Schroeter made his observations and drawings.

Slipher, E. C. 1962. *A Photographic History of Mars: 1905–1961* (Flagstaff, Arizona:Northland Press; also Cambridge, Mass.:Sky Publishing Corporation) 168 pp.
A careful presentation of photographs and drawings emphasizing the seasonal and secular changes of appearance of Mars, as well as the progression of specific atmospheric events including dust storms and "blue clearing."

Szego, K. 1991. *The Environmental Model of Mars* (New York: Pergamon Press) 160 pp.
A set of 22 papers from a 1990 conference concerned with design of future spacecraft and instruments for study of Mars.

Tyner, R. L., and Carroll, R. D. 1983. *A Catalog of Selected Viking Orbiter Images* (Washington,D.C.:NASA Reference Publication 1093) 400 pp.
A collection of Viking Orbiter photomosaics produced by the U.S. Geological Survey up to December 1978, with the geographical location of the individual frames.

University of Leeds 1982., Workshop Proceedings. *The Planet Mars* (Noordwijk, The Netherlands: ESA Scientific and Technical Publications Branch) 120 pp.
A set of 16 scientific papers addressing our knowledge of Mars in 1982 and the potential contributions of instruments being considered for a possible European spacecraft.

Vaucouleurs, G. de 1950. *The Planet Mars.* Translated by P. Moore (London: Faber and Faber Limited) 91 pp.
A semitechnical summary of the appearance of Mars and conditions there. It concludes with a synopsis of views in favor of and opposed to life on Mars. The author was skeptical.

Vaucouleurs, G. de 1954. *Physics of the Planet Mars: An Introduction to Areophysics* (London: Faber and Faber Limited) 365 pp.
An advanced and comprehensive summary of what was known about Mars at mid century. This is the last of the classic syntheses prior to the age of spacecraft exploration.

Viking Lander Imaging Team 1978. *The Martian Landscape* (Washington, D.C.:NASA SP-425) 160 pp.
A collection of 226 images representative of the Lander mission, many of which are explained. In large format with several images in color; includes a stereoscopic viewer. Provides an overview of development of the cameras and Lander operations.

Viking Orbiter Imaging Team 1980. *Viking Orbiter Views of Mars* (Washington:NASA SP-441) 182 pp.
A companion to Mariner 9 Television Team "Mars as viewed by Mariner 9." This is a series

of annotated individual frames, mosaics, and color composites representative of the Viking Orbiter Imaging experiment.
Virginia Polytechnic Institute 1965. *Conference on the Exploration of Mars and Venus* (Blacksburg, Virginia:Virginia Polytechnic Institute, and Ann Arbor, Michigan:University Microfilms) paged in 19 sections.
Proceedings of a technical conference on planetary exploration after the Mariner 4 mission. Includes a technical overview of future missions as perceived at that time.
Washburn, Mark 1977. *Mars at Last!* (New York:G. P. Putnam's Sons) 291 pp.
A summary of man's fascination with Mars and a humanistic description of spacecraft missions through Viking.
Wells, R. A. 1979. *Geophysics of Mars* (New York:Elsevier Scientific Publishing Company) 678 pp.
A technical discussion of our knowledge of Mars shortly after the Viking mission, emphasizing the author's interpretation of Mars' gravity field and tectonics.

XI. MARS' ORBIT AND SEASONS

Mars' orbit around the Sun can be expressed by the six classical Keplerian elements. These elements change slowly with time due to the influence of the other planets, primarily Jupiter. Seidleman and Standish (personal communication from the U.S. Naval Observatory, based on the work of Bretagnon 1982) have computed the mean value of these elements in the fixed equinox of J2000 and expressed them as slow functions of time. Here, T is measured in thousands of Julian years (365250 days) from J2000.00 (JD 2451545.0, which is 2000, January 1.5 Terrestrial Dynamical Time).

α semimajor axis of the orbit (astronomical units);

$$\alpha = 1.5236793419 + 3.\text{E-}10\, T$$

λ mean longitude of the planet;

$$\lambda = 355^\circ\!.432\,999\,58 + 689\,050\,774''\!.939\,88 T + 0''\!.942\,64\, T^2 - 0''\!.010\,43\, T^3$$

ε eccentricity of the orbit;

$$\varepsilon = 0.093\,400\,6477 + 0.000\,904\,8438\, T - 80\,641. \times 10^{-10}\, T^2 - 2\,519. \times 10^{-10}\, T^3 + 124. \times 10^{-10}\, T^4 - 10. \times 10^{-10}\, T^5$$

$\bar{\omega}$ longitude of perihelion, measured from the equinox along the ecliptic to the node, and then along the orbit from node to perihelion;

$$\bar{\omega} = 336^\circ\!.060\,233\,95 + 15\,980''\!.459\,08\, T - 62''\!.328\,00\, T^2 + 1''\!.864\,64\, T^3 - 0''\!.046\,03\, T^4 - 0''\!.001\,64\, T^5$$

i inclination of the orbit to the ecliptic;

$$i = 1^\circ\!.849\,726\,48 - 293''\!.317\,22\, T - 8''\!.118\,30\, T^2 - 0''\!.103\,26\, T^3 - 0''\!.001\,53\, T^4 + 0''\!.000\,48\, T^5$$

Ω longitude of the ascending node of the orbit on the ecliptic, measured from the equinox;

$$\Omega = 49^\circ\!.558\,093\,21 - 10\,620''\!.900\,88\, T - 230''\!.574\,16\, T^2 - 7''\!.069\,42\, T^3 - 0''\!.689\,20\, T^4 - 0''\!.058\,29\, T^5$$

The orientation of Mars' north pole adopted by the IAU (Davies et al. 1989) is:

$$\text{Right Ascension of North Pole} = 317^\circ\!.681 - 1^\circ\!.08\,T$$
$$\text{Declination of North Pole} = 52^\circ\!.886 - 0^\circ\!.61\,T$$

The location of the prime meridian, defined to pass through the center of the crater Airy-0, and specified as the angle W measured along the planet's equator eastward from the node of the planet's equator on the J2000 Earth equator to the 0° longitude, is:

$$W = 176^\circ\!.868 - 350^\circ\!.8919830\,d$$

where d is the interval in days from J2000.0.

The most accurate representation of Mars' orbit readily available is in the form of high-order Chebychev polynomials of the Cartesian position of Mars relative to the barycenter of the solar system. These have been derived in the process of generating a best estimate of the masses and orbits of all the planets by using spacecraft tracking, planetary radar, astrometric observations, and n-body mutual gravitational attraction. The J-2000 coordinate system is used (see the Supplement to the Astronomical Almanac 1984).

Software and tables allowing high-accuracy computation of the position of Mars are now available from two sources. The U.S. Naval Observatory Interactive Computer Ephemeris (ICE) is available in executable form for PC-compatible computers. The JPL Navigational Ancillary Information Facility (NAIF) software is available as source code that can be implemented on many computers.

Seasons on Mars are conventionally expressed in terms of "L_s," the areocentric longitude of the Sun. This is an angular measure of the apparent revolution of the Sun about Mars, measured from the intersection of Mars' equatorial plane with the plane of its orbit (the vernal equinox of Mars). Due to the eccentricity of Mars' orbit and the projection of its orbital true anomaly onto its equatorial plane, L_s is not linear with time, but its relation to time repeats each Martian year. In Fig. 1 is shown the progression of L_s and other parameters through a Martian year. Table I lists the calendar date and Julian date of $L_s = 0$ for several Mars years, derived by using ICE. By using this figure, L_s at any time can be estimated as $L_s = (\text{J.D.} - T_0)*(360.0/686.98) + \Delta L_s$, where T_0 is the closest preceding time from Table I and ΔL_s is from Fig. 1.

Detailed compilations of Mars' position, season, and appearance to terrestrial viewers are contained in the Astronomical Almanac for each year. Prior to 1981, this publication was issued jointly by the U.S. Naval Observatory (as the American Ephemeris and Nautical Almanac) and the Royal Greenwich Observatory (as the Astronomical Ephemeris).

The uncertainty of 2 milliseconds in the rotation period of Mars derived by radio tracking of the Viking Landers corresponds to an uncertainty of longitude of only 0.3 degree in a century. This is better accuracy than is

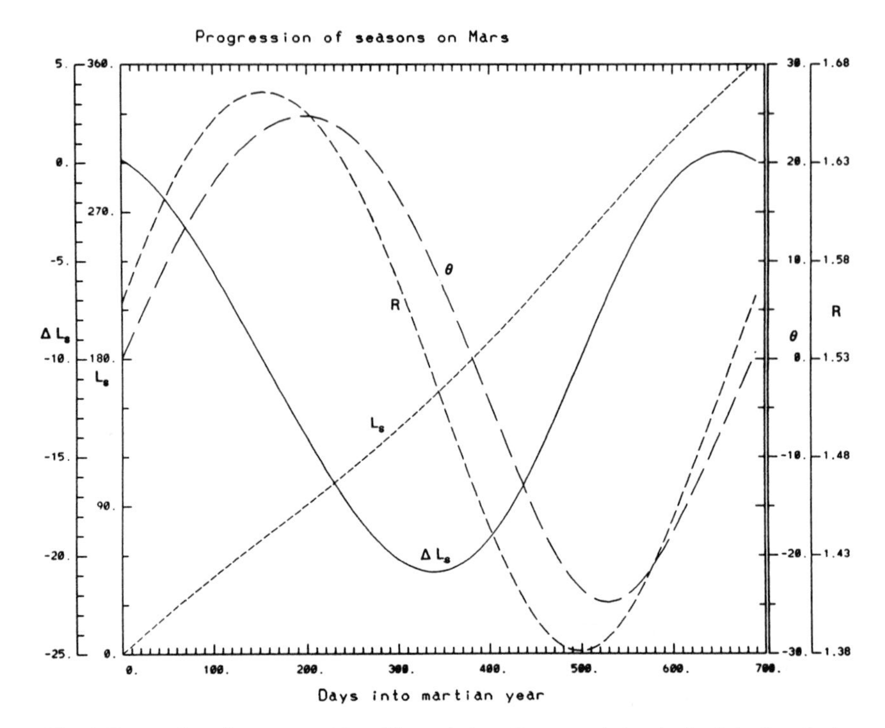

Fig. 1. Progression of seasons on Mars. The variation of aerocentric longitude of the Sun, L_s; its deviation from a linear relation, ΔL_s (see text); the sub-solar latitude on Mars, θ; and the heliocentric range, *R* (in astronomical units) are shown through a Martian year of 687 days. Calendar dates for day 0 are given in Table I.

possible from telescopic observations of Mars' rotation, even though their time base is over 300 years.

TABLE I

Date (Fractional Day in GMT) and Julian Day at which Mars' Seasonal Longitude (L_s) is 0 for 1870 to 2049[a]

Year	Month	Day	Julian Day	Year	Month	Day	Julian Day
1870	Aug	20.8	2404295.3	1960	Dec	1.4	2437269.9
1872	Jul	7.7	2404982.2	1962	Oct	19.4	2437956.9
1874	May	25.7	2405669.2	1964	Sep	5.3	2438643.8
1876	Apr	11.7	2406356.2	1966	Jul	24.3	2439330.8
1878	Feb	27.7	2407043.2	1968	Jun	10.2	2440017.7
1880	Jan	15.7	2407730.2	1970	Apr	28.2	2440704.7
1881	Dec	2.6	2408417.1	1972	Mar	15.2	2441391.7
1883	Oct	20.6	2409104.1	1974	Jan	31.2	2442078.7
1885	Sep	6.5	2409791.0	1975	Dec	19.2	2442765.7
1887	Jul	25.5	2410478.0	1977	Nov	5.1	2443452.6
1889	Jun	11.5	2411165.0	1979	Sep	23.1	2444139.6
1891	Apr	29.4	2411851.9	1981	Aug	10.1	2444826.6
1893	Mar	16.4	2412538.9	1983	Jun	28.0	2445513.5
1895	Feb	1.4	2413225.9	1985	May	15.0	2446200.5
1896	Dec	19.3	2413912.8	1987	Apr	2.0	2446887.5
1898	Nov	6.3	2414599.8	1989	Feb	16.9	2447574.4
1900	Sep	24.3	2415286.8	1991	Jan	4.9	2448261.4
1902	Aug	12.3	2415973.8	1992	Nov	21.9	2448948.4
1904	Jun	29.3	2416660.8	1994	Oct	9.9	2449635.4
1906	May	17.2	2417347.7	1996	Aug	26.9	2450322.4
1908	Apr	3.2	2418034.7	1998	Jul	14.8	2451009.3
1910	Feb	19.1	2418721.6	2000	May	31.8	2451696.3
1912	Jan	7.1	2419408.6	2002	Apr	18.7	2452383.2
1913	Nov	24.1	2420095.6	2004	Mar	5.7	2453070.2
1915	Oct	12.1	2420782.6	2006	Jan	21.7	2453757.2
1917	Aug	29.1	2421469.6	2007	Dec	9.7	2454444.2
1919	Jul	17.0	2422156.5	2009	Oct	26.6	2455131.1
1921	Jun	3.0	2422843.5	2011	Sep	13.6	2455818.1
1923	Apr	21.0	2423530.4	2013	Jul	31.6	2456505.1
1925	Mar	7.9	2424217.4	2015	Jun	18.5	2457192.0
1927	Jan	23.9	2424904.4	2017	May	5.5	2457879.0
1928	Dec	10.9	2425591.4	2019	Mar	23.5	2458566.0
1930	Oct	28.8	2426278.3	2021	Feb	7.5	2459253.0
1932	Sep	14.8	2426965.3	2022	Dec	26.4	2459939.9
1934	Aug	2.8	2427652.3	2024	Nov	12.4	2460626.9
1936	Jun	19.8	2428339.3	2026	Sep	30.3	2461313.8
1938	May	7.8	2429026.3	2028	Aug	17.3	2462000.8
1940	Mar	24.7	2429713.2	2030	Jul	5.3	2462687.8
1942	Feb	9.7	2430400.2	2032	May	22.3	2463374.8
1943	Dec	28.6	2431087.1	2034	Apr	9.2	2464061.7
1945	Nov	14.6	2431774.1	2036	Feb	25.2	2464748.7
1947	Oct	2.6	2432461.1	2038	Jan	12.2	2465435.7
1949	Aug	19.6	2433148.1	2039	Nov	30.1	2466122.6
1951	Jul	7.5	2433835.0	2041	Oct	17.1	2466809.6
1953	May	24.5	2434522.0	2043	Sep	4.1	2467496.6
1955	Apr	11.5	2435209.0	2045	Jul	22.1	2468183.6
1957	Feb	26.4	2435895.9	2047	Jun	9.0	2468870.5
1959	Jan	14.4	2436582.9	2049	Apr	26.0	2469557.5

[a]Based on Interactive Computer Ephemeris computations.

XII. BASIC PHYSICAL AND CHEMICAL DATA FOR MARS

We have collected here the values of many of the physical and chemical properties of Mars, nearly all of which are discussed in the following chapters. Some of these "constants" are known to vary in time, specifically the absolute composition and most other parameters of the atmosphere and the orbital and orientation elements.

Notes to specific values are in parentheses and appear at the end of the table; references are in brackets and appear as the last part of the table.

Orbital Characteristics

Heliocentric osculating elements referred to the mean ecliptic and equinox of J2000.0, for Julian Date 2448120.5, 1990 Aug 17.

Semimajor axis	1.52366 AU	[1]
Eccentricity	0.0934	[1]
Inclination	1°.8504	[1]
Longitude of ascending node	49°.59	[1]
Longitude of perihelion	335°.94	[1]
Mean daily motion	0°.52405/day	[1]
Mean longitude	0°.89	[1]
Mean orbital velocity around Sun (calculated from mean semimajor axis)	24.13 km s^{-1}	
L_s of perihelion	250°.87	

Orientation of Polar Axis

Right ascension	317°.61	[1]
Declination	52°.85	[1]
Obliquity relative to orbital plane	25°.19	[1]

Time Properties

Length of sidereal day	24 h, 37 m, 22.663 ± 0.002 s	[2]
	88642.663 ± 0.002 s	[2]
Mean orbital period (1)	686.98 Earth days	
	669.60 Mars solar days	
Length of mean solar day (2)	88775.2 s	

Notes:

(1) Calculated from mean semimajor axis and Kepler's third law.

(2) Calculated from length of sidereal day and mean orbital period.

Geophysical Parameters

Parameter	Value	Ref.
Mass		
GM	42828.3 ± 0.1 $km^3 s^{-2}$	[3]
mass	6.4185 × 10^{23} kg	[4]
Mean radius (radius of sphere of equal volume)	3389.92 ± 0.04 km	[5]
Radius and direction of principal axes of 2nd-degree triaxial ellipsoid: (given as radius, latitude, longitude)		[5]
3394.5 ± 0.3 km;	0°.7 ± 0.2, 18°.5 ± 0.8W	
3399.2 ± 0.3 km;	−2°.0 ± 0.2, 108°.4 ± 0.8W	
3376.1 ± 0.4 km;	87°.9 ± 0.2, 128°.8 ± 6.3W	
Volume (1)	1.6318 × 10^{22} m^3	[5]
Mean density	3.9335 ± 0.0004 g cm^{-3}	[4]
J_2	1960.454 × 10^{-6}	[1]
C/MR^2	0.345–0.365	[6]
Center of mass/center of figure offset	2.50 ± 0.07 km	[4]
cf. offset in direction of −62°.0 ± 3.7 lat, 87°.7 ± 3.0 (long W)		
Magnetic dipole moment	< 10^{-4} that of Earth <8 × 10^{11} T m^{-3}	[7]
Surface gravity at pole (2)	3.758 m s^{-2}	
Surface gravity at equator (2)	3.711 m s^{-2}	
Ratio of surface gravity at equator to Earth surface gravity at equator	0.3795	
Mean escape velocity (3)	5.027 km s^{-1}	
Total surface area (1)	1.4441 × 10^{14} m^2	[5]
Ratio of total surface area to that on Earth	0.2825	[8]
Ratio of total surface area to land surface area on Earth	0.976	[8]
Area of perennial polar caps		[24]
south polar region	88,000 km^2	
north polar region	837,000 km^2	
Area of polar layered terrain: (4)		[24]
south polar region	1,395,000 km^2	
north polar region	395,000 km^2	

Notes

(1) Calculated from radius of sphere of equal volume.

(2) Given by GM/R^2; equatorial value based on mean equatorial radius.

(3) Given by $\sqrt{2GM/R}$.

(4) Excluding area of perennial polar caps.

Atmospheric Composition (by Volume) [9]

Species	Abundance	Notes
CO_2	0.9532	(1)
N_2	0.027	
Ar	0.016	
O_2	0.0013	
CO	0.0007	
H_2O	0.0003	(2)
Ne	2.5 ppm	
Kr	0.3 ppm	
Xe	0.08 ppm	
O_3	0.04 to 0.2 ppm	(2)

Notes:

(1) Because CO_2 varies seasonally due to condensation at the polar caps, all abundances will vary seasonally by as much as about 30%; values are those measured at the Viking landing sites, and are relative to a total pressure of 7.5 mbar.

(2) Abundance varies with season and location. Annual global average of H_2O is $\sim 10 \times 10^{-3}$ kg m^{-2} or 0.00016 by volume [23]

Atmospheric Isotopic Ratios [9]

Ratio	Earth	Mars
$^{12}C/^{13}C$	89	90 ± 5
$^{14}N/^{15}N$	272	170 ± 15
$^{16}O/^{18}O$	489	490 ± 25 (measured in atmospheric CO_2)
$^{36}Ar/^{38}Ar$	5.3	5.5 ± 1.5
$^{40}Ar/^{36}Ar$	296	3000 ± 500
$^{129}Xe/^{132}Xe$	0.97	2.5^{+2}_{-1}
D/H	1.6×10^{-4}	$(9 \pm 4) \times 10^{-4}$

Representative Chemical Composition of Soil: [10]

Constituent (1)	Concentration (%)	
SiO_2	43.4	(2)
Fe_2O_3	18.2	(2)
Al_2O_3	7.2	(2)
SO_3	7.2	(2)
MgO	6.0	(2)

CaO	5.8	(2)
Na_2O	1.34	(3)
Cl	0.8	(2)
P_2O_5	0.68	(3)
TiO_2	0.6	(2)
MnO	0.45	(3)
Cr_2O	0.29	(3)
K_2O	0.10	(3)
CO_3	<2	(4)
H_2O	0–1	(5)

Notes:

(1) Based on elemental composition, expressed as oxides.
(2) Based on direct soil analyses from Viking X-ray Fluorescence Spectrometer.
(3) Based on SNC meteorite analyses.
(4) Based on terrestrial simulations of Viking Labelled Release experiment.
(5) Spatially and temporally variable.

Atmospheric Properties

Atmospheric pressure at surface (1)	5.6 mbar or 560 Pa	
Average columnar mass of atmosphere (2)	150 kg m^{-2}	
Average mass of atmosphere (3)	2.17×10^{16} kg	
Mass of seasonal polar cap deposits: (4)		[11]
north polar cap	3.5×10^{15} kg	
south polar cap	8.1×10^{15} kg	
Mean atmospheric scale height (at T = 210 K)	10.8 km	[12]
Dry adiabatic lapse rate	4.5 K km^{-1}	[12]
Mean atmospheric lapse rate (observed)	~ 2.5 K km^{-1}	[12]
Atmospheric visible optical depth (5)	0.1–10	[12]

Notes:

(1) Seasonally and spatially variable; this value is an estimate of the global annual average; see chapter 2.
(2) Calculated from 5.6-mbar surface pressure and global-average gravity.
(3) Calculated from mean global surface pressure of 5.6 mbar.
(4) Based on seasonal variations in atmospheric pressure.
(5) Temporally and spatially variable; values are approximate, integrated over entire visible or infrared spectrum, respectively.

Thermal Properties

Solar constant:		
at 1 AU (1)	1367.6 W m^{-2}	[13]
at mean Mars distance from Sun	588.98 W m^{-2}	
Average surface temperature (2)	210.1 K	
Range of surface temperature	~ 140–300 K	[14]
Surface bolometric albedo (3)		
mean	0.25	[15]
range	0.08–0.40	[15]
Surface thermal inertia: (4)		
mean	6.5×10^{-3} cal $cm^{-2}s^{-1/2}K^{-1}$ 272. J $m^{-2}s^{-1/2}K^{-1}$	[16]
range	$1 - 15 \times 10^{-3}$ cal $cm^{-2}s^{-1/2}K^{-1}$ 42.-628. J $m^{-2}s^{-1/2}K^{-1}$	[16]

Notes:

(1) Mean of measurements from SMM/ACRIM I experiment, 1980–1983.
(2) Equilibrium for a perfectly conducting sphere of albedo = 0.25, emissivity = 1.0 at Mars mean heliocentric distance.
(3) As measured at ~ 60-km spatial scale.
(4) As measured at ~ 100-km spatial scale.

Viking Landing Site Locations

	VL-1	VL-2	
Latitude (1)	22°.480 ± 0.003	47°.967 ± 0.003	[22]
Longitude (1)	47°.968 ± 0.1	225°.737 ± 0.1	[22]
Radius to surface	3389.32 ± 0.06	3381.86 ± 0.06 km	[22]
Elevation relative to geoid (2)	−1.54	−2.44 km	
Average annual surface pressure	~ 780	~ 870 Pa	

Notes:

(1) Areographic coordinates. Longitude depends upon the precise location of Airy-0. Inertial longitude known to +0°.002.
(2) Relative to fourth order and degree datum described in [17], using the radii at sites listed here. The approximate elevations shown on the U.S. Geological Survey (1991) topographic maps I-2160 are −1.9 km and −0.8 km, respectively. The landing sites were not used as part of the vertical control system for these maps.

Satellites

	Phobos	Deimos	Unit	
Radii of triaxial				
ellipsoid: a:	13.3 ± 0.3	7.6 ± 0.5	km	[18]
b:	11.1 ± 0.3	6.2 ± 0.5	km	[18]
c:	9.3 ± 0.3	5.4 ± 0.5	km	[18]
Volume	5680 ± 250	1052 ± 250	km^3	[19,18]
Mass:				
GM	$(7.22 \pm 0.05) \times 10^{-4}$	$(1.2 \pm 0.1) \times 10^{-4}$	$km^3\ s^{-2}$	[20,21]
M	1.08×10^{16}	1.8×10^{15}	kg	
Mean density	1.905 ± 0.053	1.7 ± 0.5	$g\ cm^{-3}$	
Orbital elements:				
Semimajor axis	9378.5	23458.8	km	[20]
Eccentricity	0.0152	0.0002		[20]
Inclination	1.03	1.83	deg	[20]
Orbital period	0.31891023	1.2624407	day	[1]
Rate of decrease of longitude of ascending node	158.8	6.614	$deg\ yr^{-1}$	[1]
Length of day	synchronous	synchronous		[1]

References cited in preceding tables:

[1] The Astronomical Almanac for 1990
[2] Mayo et al. 1977
[3] Null 1969
[4] Esposito et al. (Chapter 7)
[5] Bills and Ferrari 1978
[6] Bills 1989*a,b;* Kaula et al. 1989
[7] Luhmann et al. (Chapter 31)
[8] Stacey 1977
[9] Owen (Chapter 25)
[10] Banin et al. (Chapter 18)
[11] Hess et al. 1980
[12] Zurek et al. (Chapter 26)
[13] Willson 1984
[14] Kieffer et al. 1977
[15] Pleskot and Miner 1981
[16] Palluconi and Kieffer 1981
[17] Wu 1981
[18] Duxbury et al. (Chapter 36)
[19] Duxbury 1991
[20] Kolyka et al. 1991
[21] Williams et al. 1988
[22] Michael 1978
[23] Jakosky and Farmer 1982
[24] Tanaka et al. 1988

Acknowledgments. We appreciate a careful review of the historic overview by G. de Vaucouleurs.

2. TELESCOPIC OBSERVATIONS: VISUAL, PHOTOGRAPHIC, POLARIMETRIC

LEONARD J. MARTIN
Lowell Observatory

PHILIP B. JAMES
University of Toledo

AUDOUIN DOLLFUS
Observatoire de Paris

KYOSUKE IWASAKI
Kwasan Observatory

and

JEFFREY D. BEISH
Institute for Planetary Research Observations

The high points of telescopic observations of Mars have been roughly divided into three time periods: historical, missions support (recent), and present. The emphasis is on visual and photographic observations, with brief discussions of spectroscopic and polarization studies. (Photometric observations are not included, while radio and radar observations are also left to appropriate chapters.) Within the above sections, observations are discussed separately from results, although inevitably lines between subjects cannot be strictly drawn. In attempting to discuss several centuries of observations, there are many omissions, since space does not allow us to be either complete or evenly balanced. Major topics of Martian phenomena included are albedo features, polar caps, dust storms, and white clouds. These are features that have been, and continue to be successfully observed from Earth, although spacecraft data have been used to verify, interpret, and increase the significance of the telescopic observations. The interannual variability of the recessions of seasonal polar caps has been compared to dust storm activity, but this relationship remains uncertain. The seasonal dependence of dust storms and white clouds has become less well defined with an increasing data base, although some general trends still seem apparent. Some changes in albedo features that have been attributed to the seasons are questioned by some researchers and accepted by others. A discussion of canals is included because of their historical importance in the study of Mars. Only a very limited number of canals can be related to markings on the Viking images. The remainder are optical illusions created by observers pushing their perceived resolution beyond practical limits. Technical advancements in detection equipment (primarily CCD cameras) are making greater resolution possible, providing an increased data base for ongoing research and support for future space missions. The role of amateur observers is acknowledged, along with their increasing responsibilities.

We have not attempted to cover ground that has been well tilled by skilled historians. Abstracting interesting bits and pieces from older works has been well done by authors we refer to. Instead, this chapter attempts to fill the gaps of the less well-documented recent past and bring the reader to the present state of Earth-based observations. We also discuss the role that these telescopic observations have played in the space age, and how results from space missions have enabled us to understand better our Earth-based observations.

Many of the earliest observers were aware of changes in the dark albedo features and therefore attributed these features to clouds in the Martian atmosphere. It was not until later in the 18th century that the consensus swung toward interpreting the dark areas as oceans, seas and canals. Changes were interpreted as flooding and withdrawal of shorelines. Not until the end of the last century did the vegetation hypothesis take hold. Vegetation was favored over moving dust by many observers because it could account for the restoration of features that had temporarily disappeared. We should respect the efforts of those early pioneers and hope that those who follow us will do likewise.

I. HISTORICAL OBSERVATIONS

A. Huygens to Slipher

The two volumes by C. Flammarion (1892,1909) are very comprehensive collections of the observations, known data and theories on Mars from 1636 through 1901. These books include reproductions of the many drawings and maps produced through this period, beginning with those by F. Fontana in 1636 and C. Huygens in 1659, and including color reproductions of two of G. Schiaparelli's maps (Flammarion 1892) as well as a fold-out map, drawn by E. Antoniadi (Flammarion 1909). More condensed histories are found in volumes by Pickering (1921), Glasstone (1968), Hartmann and Raper (1974), Hoyt (1976*c*) and Mutch et al. (1976*c*). Probably the most quent descriptions of early observations of Mars can be found in a book by Moore (1977).

During the period following Flammarion, there have been only brief attempts at comprehensive and universal documentation of Mars observations. This may have been caused by the rift created by the canal controversy or by the increasing number of observations. But the more likely reason is that no one has stepped forward and assumed the task. Antoniadi's book (1930), which has been translated by P. Moore, gives us a thorough description of Mars, along with his interpretations and analyses of Mars and Mars research. However, it is based primarily upon his own observations. The same thing is true of Slipher's book (1962), although it covers more than fifty years of observations and is based more upon photographic than visual observations. Other books based on observations are by Lowell (1895,1906,1908), Pickering (1921), and de Vaucouleurs (1954). Antoniadi's book includes very detailed maps that have names for even the smallest features. Ebisawa (1960,1963) independently compiled similar maps, correcting, updating and adding features based upon various visual and photographic sources through the 1956 apparition. Other recent telescope-derived maps do not have as many named features (McKim 1986), and Ebisawa's names go far beyond the International Astronomical Union (IAU) list of adopted names, although many of these features can be identified on recent photographs and by experienced observers.

B. The International Mars Committee

In 1953, a cooperative organization, the International Mars Committee was formed to collect an optimum amount of data from the upcoming favorable apparitions of Mars. This group successfully promoted a worldwide observation effort that coordinated various programs, standardized procedures, and collected and exchanged data. A grant from the National Science Foundation provided partial funding, some of which was used to provide photographic plates to several observatories. The International Mars Committee also organized an International Photographic Patrol to assure adequate longitudinal coverage to keep Mars under near-continuous surveillance. In 1954 the seven participating observatories were Pic-du-Midi, France; Lamont-Hussey, Union of South Africa; Helwan, Egypt; Kodaikanal, India; Bosscha, Java; National University, Argentina; and Lowell Observatory, United States. Additional images were taken at Lick, Mt. Wilson and Union (South Africa) Observatories. By 1956 the number of cooperating observatories had grown to 13 and included Kwasan, Japan; Yale-Columbia Southern Station, Australia; Palomar; and Uttar Pradesh, India. The most productive program was conducted by E. Slipher, who observed from Lamont-Hussey Observatory in both 1954 and 1956, thanks to a grant from the National Geographic Society and support from Lowell Observatory. Slipher was also a prime organizer of the Committee and served as its chairman. Extensive reports were published after both the 1954 and 1956 apparitions (Slipher and Wilson 1955; Slipher 1964). These included listings of all the photographic observations and the individual reports from the many participants. What turned out to be a final Committee meeting was held on 1957 June 19, at Lowell Observatory. Plans for the 1958 apparition were discussed, but it would be another fourteen years until Mars was again at its closest; Slipher had already observed 25 oppositions.

C. The Planetary Research Center at Lowell Observatory

As a direct result of efforts of the International Mars Committee, the photographic collection at Lowell grew at an increasing rate. Mars images from many of the participating observatories were sent to Lowell, complementing the Lowell observations from Flagstaff and South Africa. Lowell Observatory had acquired the world's largest collection of planetary photography and was therefore selected as the Western Hemisphere site of the International Astronomical Union's (IAU) data center for planetary studies. A similar data center was established at the same time at Observatoire de Meudon in France. In 1965 the operation of the new IAU Planetary Research Center (PRC) began at Lowell Observatory. Under the direction of W. Baum, the PRC conducted an exchange program with the center at Meudon, further increasing the size of the collections both at Lowell and in Meudon. Hundreds of visitors have used the research facilities at the PRC, and many of them have made use of the photographic collection there. The PRC also was the headquarters for the International Planetary Patrol (see Sec.III.B). This and subsequent Mars monitoring programs have mushroomed the Mars collection at the PRC to well over one million images.

D. Association of Lunar and Planetary Observers

The Association of Lunar and Planetary Observers (ALPO) is an international group of students of the Sun, the Moon, bright and remote planets, minor planets, meteors and comets. The ALPO was founded in 1947 by W. Haas (New Mexico State University) and now enrolls approximately 800 members. Its goal is to stimulate, coordinate and generally promote the study of these bodies using methods and instruments available to planetary astronomers. The ALPO director appoints one or more astronomers, professionals and/or advanced amateurs as section "recorders" for each of the solar system objects. These recorders collect and study members' observations, correspond with observers, encourage beginners and contribute reports to *The Strolling Astronomer* (Journal of the ALPO) at appropriate intervals.

In 1969, planetary astronomer C. Capen assumed responsibilities as Mars Recorder for the ALPO and established a worldwide network of observers directed toward a 24-hr surveillance of Mars during each apparition. To this end, he created the International Mars Patrol (IMP) which contributes thousands of observations worldwide each apparition.

Under Capen's guidance the ALPO Mars Section strove to produce a systematic Mars observing program with standardized observing methods and quantitative results, with considerable emphasis on color-filter work, photography, micrometry and statistical analysis of data with modern computers. As a result, IMP astronomers have made several significant discoveries or rediscoveries concerning the "Red Planet." These include recording a number of secular changes in the albedo features of the planet, and Martian climatic variations over the past 25 yr. Since Capen's untimely death in 1986, his work has been carried forward by his assistants, J. Beish and D. Parker. The work of the ALPO Mars Section has grown to the extent of now having three assistant recorders.

E. Europe

Pic-du-Midi, in the French Pyrenees, has been the premier site for telescopic observations in Europe for many years. These observations began with testing by Lyot in 1941 (Lyot et al. 1943). In 1945, a 38-cm refractor was replaced by a 60-cm refractor, which is dedicated to planetary astronomy (Lyot 1953). In 1963, a 107-cm reflector was added. A systematic photographic survey of Mars covering 17 apparitions was conducted from 1945 to 1975. H. Camichel, J. Focas, A. Dollfus and co-workers collected more than 2000 selected plates of Mars images. These images were systematically composited and copied at a standard scale by the IAU Centre de Documentation Photographique sur les Planetes (Planetary Photographic Center) in Meudon. Established as a result of a 1961 Resolution of the IAU, this organization also collected and cataloged most of the high-quality plates from around the world, dating from the beginning of astronomical photography, including more than 6500 plates for Mars alone. This international center operated for 15 yr, making the data available to scientists from all parts of the world. It was managed by Focas, C. Boyer, and R. Servajean, under the direction of Dollfus.

European scientists from several countries took a major role in the research conducted at the Center at Meudon. Camichel (France) determined the coordinates of the Martian poles, the diameter of the planet, and the reference system for surface coordinates (Camichel 1954). De Mottoni y Palacios (1975) from Italy produced 32 Mercator projection maps of the surface albedo features of Mars, based entirely on photographs, for each apparition from 1907 to 1971. This map series remains a prime reference document in the field. Focas (Greece; 1961) made 7200 photometric measurements of the surface albedo features on 663 photographic images, in order to analyze seasonal and secular variations. He also studied the seasonal cloud evolution at the poles. Dollfus (France; 1965*a*, 1973) measured the polar caps to produce regression curves. The photographic and polarimetric analysis of clouds and dust storms was conducted by Dollfus, Bowell (Great Britain), and Ebisawa from Japan (Dollfus et al. 1984*a,b*).

As a supplement to the photographic work, very detailed drawings have been derived from visual observations using the Pic-du-Midi telescopes, primarily by Focas and Dollfus. Spots on the Martian surface as small as 50 km have been recorded (Dollfus 1953,1961*b;* Focas 1962). The aspect and recessions of polar caps; the configurations and changes of the albedo features; the positions of discrete clouds; the evolution, motion and classification of clouds; and the evolution and spreading of dust storms have all been visually recorded. Using the technique of double-image micrometry, the polar and equatorial radii were measured to an accuracy of 2×10^{-3} (Dollfus 1972). Actually, Dollfus' calculated equatorial value is within 1 km of the currently accepted value of 3397.2 km (U.S. Naval Observatory 1988).

There are several associations in Europe that have been encouraging and coordinating telescopic observations by amateurs for many years. The British Astronomical Association (BAA), founded in 1890, has a long history of involvement in Mars observations. The BAA Mars Section, presently headed by R. McKim, is the longest established cooperative amateur group of this type in the world. It has published numerous papers in the *Journal of the BAA*, based upon reports from observers from all over the world.

Société Astronomique de France (SAF) was founded in 1887 by Flammarion. Coordination of planetary observations has been among the long-standing goals of this association. Its Mars activity has benefitted from the leadership of G. Fournier, G. de Vaucouleurs, Dollfus, J. Dragesco, and, presently, D. Crussaire. Results are periodically published in the SAF journal *l'Astronomie* with specific notes in "Observations et travaux." A particular effort has been made to stimulate photographic work; in recent years, amateurs have had access to the 60-cm refractor at Pic-du-Midi.

In addition to these two major associations, there are several other groups in Europe involved in Mars observations. These include the Unione Astrotili Italiani, headed by M. Falorni, publishing in *Orione;* the Vereniging Voor Sterrenkunde, led by L. Aerts, in Belgium, publishing in *Ciel et Terre;* the Agrupacion Astronomica de Sabadell, led by M. Alamany, in Spain, publishing in *Astrum;* and the Société d'Astronomie Populaire, headed by J. Dijon, in France, publishing in *Pulsar.* The private observatory of G. Viscardy, in southern France, has produced a large collection of high-quality photographs of Mars over a 20-yr period.

F. Japan

Visual observations of Mars began at the Kwasan Observatory in Kyoto during the 1956 apparition (Miyamoto 1957). These observations continued through 1967, using the Cook 30-cm refractor (Miyamoto and Matsui 1960; Miyamoto 1963,1965,1968). In 1969, visual work at Kwasan began use of the Zeiss 45-cm refractor (Miyamoto 1970,1972,1974). In addition, photographic observations were carried out at Kwasan during the 1960–61 apparition, using the Tsugami 60-cm reflector (Miyamoto and Nakai 1961). After a lengthy hiatus, photographic observations were resumed at Kwasan during the 1977–78 apparition and continued through 1988 (Iwasaki et al. 1979,1982,1984,1988,1989*a,b*).

At the Hida Observatory in Gifu, photographic observations of Mars began during the 1969 apparition, using the Tsugami 60-cm reflector (Miyamoto 1970,1972). From the 1973 apparition and thereafter, the Zeiss 65-cm refractor has also been used (Hattori and Akabane 1974; Ebisawa 1975; Hattori et al. 1976; Akabane et al. 1980,1982). Ebisawa has also been visually observing Mars for many years at his private observatory in Tokyo.

An amateur group, the Oriental Astronomical Association (OAA), has been visually observing Mars from Japan since the apparition of 1935. Their results have been published in the OAA journal, *Heavens*.

G. Spectroscopic Observations

Spectroscopic observations of Mars began very shortly after photographic observations, probably in 1905, with Lowell and V. Slipher attempting to measure water vapor in the atmosphere (de Vaucouleurs 1954). As might be expected, the interpretation of their results was controversial. Other pioneering searches for atmospheric water took place at Lick Observatory and from the summit of Mt. Whitney. From 1925 to 1943 observations were pursued at Mt. Wilson with uncertain results (Hess 1951). Searching for evidence of life on Mars was another focus of historical spectroscopic observations. Although those efforts failed to find chlorophyll in the spectrum of Mars, Kuiper (1952) pointed out that it should not be expected from the types of vegetation most likely to survive on Mars, i.e., lichens and mosses. The presence of these forms of plant life was not incompatible with observations. Attempts to determine Mars' surface mineralogy also had early beginnings. Between 1909 and 1944, researchers in this field included Wilsing and Scheiner; Barabescheff, Semejkin, and Timoshenko; Scharonow; and Fessenkoff (see de Vaucouleurs 1954).

Through the years, there have been a number of renewed efforts to determine the properties of the surface and atmosphere with each technical advance in available instruments. Kuiper's (1952) observations in 1947 and 1948 provided the first detection of carbon dioxide in the atmosphere. Dollfus (1963,1964,1965*b*) and, independently, Kaplan et al. (1964) successfully identified water vapor in their spectra, which was confirmed by observations by Schorn et al. (1967). More complete descriptions of these early spectroscopic efforts can be found in the volumes by de Vaucouleurs (1954) and Michaux and Newburn (1972); see also chapters 17 and 25. Recent applications have been primarily concerned with more precise mineralogy of the surface.

H. Polarization Observations

Early polarization observations were carried out by Lyot (1929) during the apparitions of 1922, 1924 and 1926. These were essentially measurements of the whole disk. Polarimetry of Mars was resumed at Meudon Observatory in 1948 by Dollfus (1957,1961*a*). The visual fringe polarimeter adapted to the 60-cm Pic-du-Midi refractor provided measurements of Martian areas as small as 2 arcsec. Between 1948 and 1956 about 4000 polarization measurements of Mars were made by Dollfus and Focas. Additionally more than 1000 measurements localized over the Martian disk were made by Focas at Athens Observatory from 1954 to 1960. Observations at different phase angles and distances from the center of the disk made it possible to separate the polarization due to atmosphere, clouds, dust veils, polar caps and the surface itself and to produce interpretations based upon modeling, laboratory work and simulations (Dollfus 1958,1961*a*,1965*a*; Focas 1961).

Finally, a combined survey was developed among Pic-du-Midi, Athens and Meudon Observatories. Until 1965, this survey used 4 visual fringe polarimeters; in addition, since 1965, photoelectric polarimeters were employed at Meudon and at Pic-du-Midi. Accurate photoelectric observations were also made at Kiev Observatory by Morozhenko (1966). Altogether, more than 6000 measurements were made available for a very detailed analysis of the surface of Mars.

For the favorable opposition of 1971, Meudon Observatory coordinated a polarimetric and imaging survey with Pic-du-Midi involving 4 telescopes and 9 observers and covering 9 months. A similar program covering 9 months was conducted for the 1973–74 opposition with additional contributions by the Hida and Kwasan Observatories and by Ebisawa at Hida and in Tokyo. Local area measurements were made, maps of the degree of polarization over the Martian disk were recorded in 5 wavelengths, with measurements at the disk center in 6 wavelengths (Dollfus et al. 1984*a*,*b*; Ebisawa and Dollfus 1986). Additional good observations were also recorded at Kiev by Morozhenko (1973,1975), at Kharkov by Lupishko and Lupishko (1977) and at Cornell University by Veverka et al. (1973).

II. CONCEPTIONS AND MISCONCEPTIONS OF THE DATA

A. Nature of the Surface

An early telescopic evaluation of the Martian surface properties had been determined by polarimetry (Dollfus 1957,1958; Dollfus and Focas 1969; Dollfus et al. 1969). The highlights of those findings are described below.

The large ochre regions covering the major part of Mars show a degree of polarization related essentially to phase angles and wavelength. Very few substances have the type of polarization properties exhibited by the Martian regolith. The negative polarization implies numerous multiple diffusions between grains. Hence, the surface must be fine, separate grains. These grains must be opaque, in spite of their small diameters. The spectral variation is specific, and the pulverized substances reproducing these variations are iron oxide, especially goethite or, more specifically, limonite ($Fe_2O_3 \cdot nH_2O$), when finely divided into small grains.

The polarization from the dark areas is not markedly different from that of the bright regions. Its spectral variation is slightly less pronounced. The greenish or bluish colors attributed to these dark-hued areas are merely due to the contrast with the neighboring orange regions. One can obtain a surface with the proper polarization properties when pulverized limonite is mixed with a powder of very absorbing grains or partly covered by these grains.

B. Surface Albedo Features

The albedo features on Mars undergo nearly continuous change, although as far as we can determine none of the changes are exactly reproducible (Baum 1974). Each apparition shows features that differ somewhat from the previous apparition and all other apparitions. Nevertheless, the well-known perennial features do not move and tend to retain or at least return to their general configurations. New features may come and go, but usually if one digs deeply enough into the historical record it is found that many of these new features have appeared at least briefly during some previous apparition.

The appearance of Martian albedo features has been recorded for each apparition since before the turn of the century on composite drawings in the form of globes (Lowell) or maps. The map series by de Mottoni y Palacios (1975), using the Meudon photographic collection, is the most numerous, covering each apparition from 1907 through 1971. More recently, Inge et al. (1971) have rendered airbrush maps for apparitions 1967 through 1986, using the extensive Lowell photographic collection (see Fig. 1).

Many classical albedo features (see names map, Fig. 1, in Appendix and Map I-2160) can now be directly associated with topographic features determined from spacecraft (Coprates Canal is Coprates Chasma, Juventae Fons is Juventae Chasma, Hellas is Hellas Planitia). Most of the more prominent albedo features, however, show little apparent relationship to the topography as we now know it (Inge and Baum 1973); we cannot, however, rule out subtle elevation variation or minor differences in surface texture as possible influences on the distribution of dark and light areas.

Until spacecraft taught us otherwise, vegetation was accepted by many observers as the cause of change in albedos. The dark collar that is often seen at the perimeter of the receding polar caps led to additional seasonal associations seen by many as a "wave of darkening" (described below) and, in the extreme, Lowell's (1908) theory of intelligent life creating a planetwide irrigation system. More recent theories attribute albedo changes to windblown

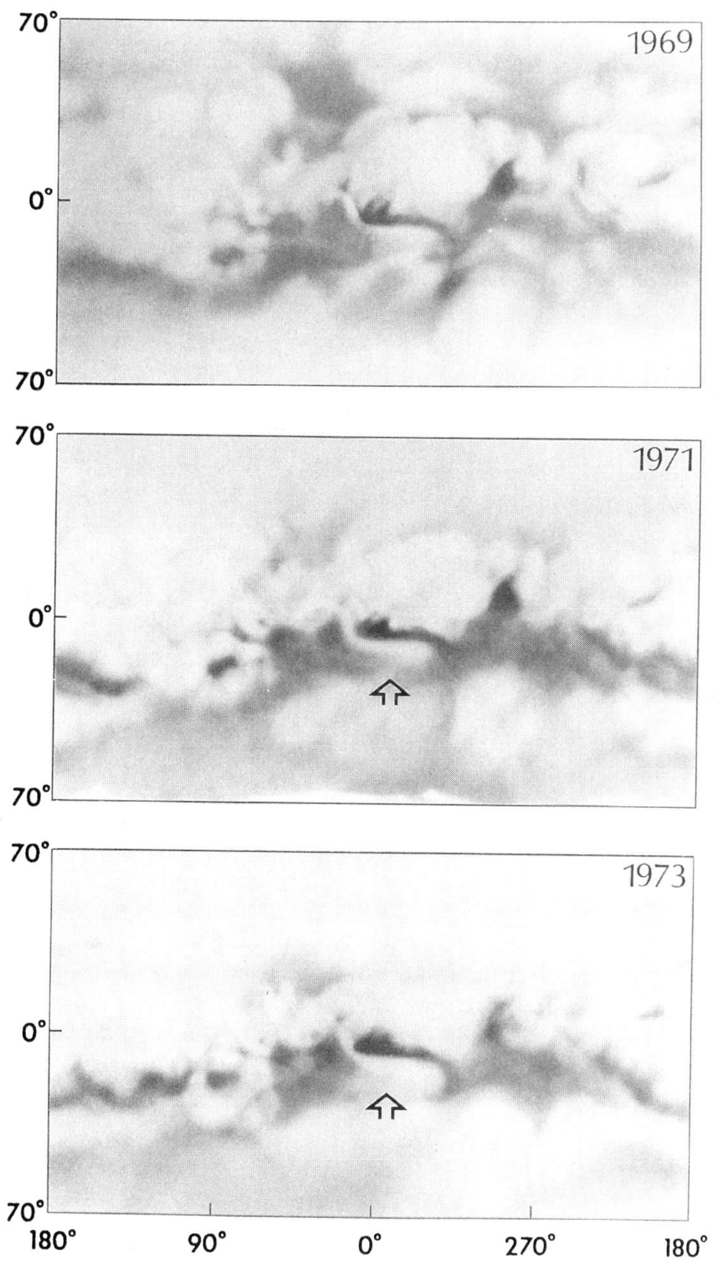

Fig. 1. Albedo maps are derived from photography taken during the first three Martian apparitions of the International Planetary Patrol program and rendered by airbrush onto a Mercator projection by J. Inge. This Lowell Map Series provides a composite documentation for each apparition from 1967 through 1986. Most of the changes seen between maps cannot be attributed to Martian seasons, although the amount of detail portrayed at higher latitudes for each map is determined by the direction and amount of latitudinal tilt that Mars displays towards the Earth. The arrows point out Pandorea Fretum, which is discussed in the text as an area subject to purported seasonal change. If that concept were true, this feature would have been darker in 1973 than it had been in 1971. Instead, just the opposite is true.

dust. In some cases, these models include seasonal wind patterns. Ebisawa (1963) discussed the possibility of windblown dust or sand causing secular changes.

Authors Dollfus and Martin express differing opinions on the classical "wave of darkening" theory (see quoted sentences below) in which Martian albedo features progressively darken, from the south polar region towards the equator during southern spring, and to some degree the cycle is repeated in the opposite direction in the northern hemisphere during spring. This phenomenon has been noted by numerous observers over the years, including Lowell (1906), Antoniadi (1930) and de Vaucouleurs (1954). Lowell and others interpreted this as evidence for seasonal vegetation changes, and many observers reported color changes as well as albedo darkenings. Focas (1961,1962) described these waves in detail after examining the data from eight apparitions, which provided an aggregate Martian year of combined Martian seasons. He used the photography in the IAU collection at Meudon to make photometric measurements of selected dark albedo features. Focas' data were statistically evaluated by Pollack et al. (1967), who concluded that "Although there are exceptions, a very significant correlation emerges between latitude and time of maximum darkening. . . . However, the actual evidence favoring the existence of a *wave* of darkening is not immediately compelling."

Dollfus states: "The long-term record of shape and contrast changes in the features on the surface of Mars by de Mottoni and Dollfus (1982) is based upon photographs taken throughout the world during 70 years and collected at the IAU Planetary Photographs Center of Meudon, France. They are summarized by 32 maps published by de Mottoni (1975). This analysis suggests that darkening has some relationship with the seasonal increase of solar radiation flux and surface heating. The classical tendency for a wave of darkening propagating from south pole toward equator at southern spring is apparently related with the increase with latitude of the seasonal amplitude of solar flux received. High solar heating may favor high-velocity winds which in turn can lift the soil dust and move the grains at the surface. Local, environmental, and topographic influences are noted; for instance, Hellas Planitia has a very specific behavior."

Syrtis Major is Mars' most prominent and therefore first known dark albedo feature. Although Antoniadi (1930) considered the changes seen there as seasonal, more recent observations do not support that theory. Slipher's book (1962) shows this feature changing through 24 apparitions. It is difficult to identify reliable seasonal trends in these changes (Fig. 2).

Slipher and others felt that, in general, low-latitude features darkened between southern spring and summer. The 1973 rotation sequence shown in Fig. 3 is an example of southern summer, when those features were not particularly dark. This sequence can be compared to 1971 (southern spring) and

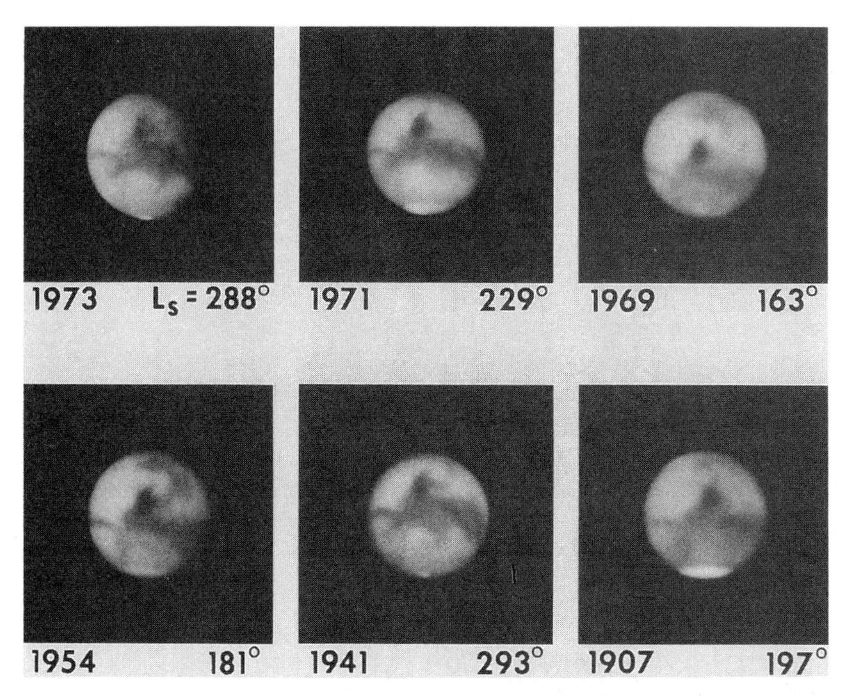

Fig. 2. A perennial albedo feature, Syrtis Major undergoes nearly continuous change. Some researchers have attributed these changes to season while others (including E. Slipher) considered them secular. We do not see a consistent long-term pattern in these changes. Note also the changes in Hellas, directly south of Syrtis as seen as a bright oval in 1941. Lowell Observatory photographs.

1986 (also southern spring) seen in Fig. 4. In 1986, all of the southern hemisphere was unusually dark. Solis Lacus appeared much as it had in 1926, although that apparition occurred during Mars' southern summer.

Probably the most convincing evidence for seasonal change in albedo features is Pandorae Fretum (15° S, 320°-360° W). In his book, Slipher (1962) showed this area in spring and summer through 4 seasonal cycles, although his 1941 example is weak. However, this change was not repeated during the following two seasonal cycles (1971–73 and 1986–88). In 1971 (spring), Pandorae was darker than in 1973 (summer; Fig. 1, arrows; see also Capen 1976).

Martin argues: "Seasonal influences in albedo changes are probably very minor and that the 'wave of darkening' is not real. Capen (1976) made 3960 photometric measurements on calibrated International Planetary Patrol film taken in 1971 and 1973. He concluded that the 'wave of darkening' does not exist, although he did find evidence for seasonal change. However, Capen's data show that the greater changes were in the light albedo areas and that the

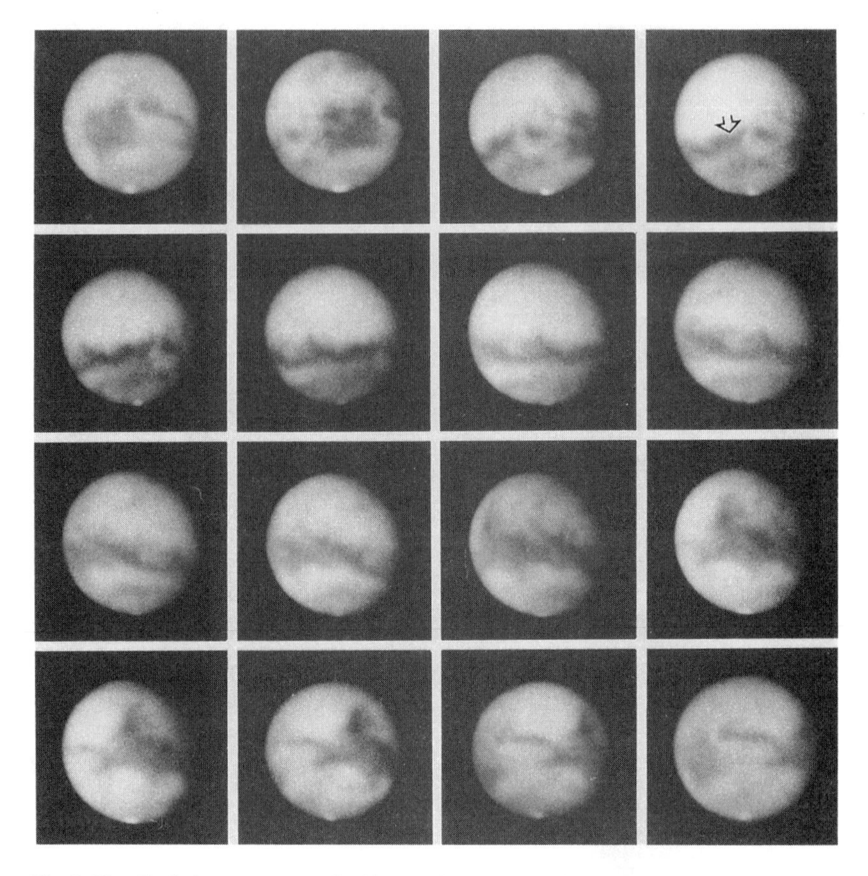

Fig. 3. The albedo features are seen in this "rotation" sequence from 1973 International Planetary Patrol red-filter photographs. The pictures are from different dates (September 22–October 17) and from different observatories. The season was in southern summer (L_s 287°–302°). They are arranged in rows of increasing central meridians beginning with 11° at the upper left and ending at the lower right with 345°. The most notable change was the appearance of the large darkening in the Claritas area, seen in the center of the upper right image (arrow). This feature re-appeared following the large 1973 dust storm. Pandorae Fretum (south of Sinus Sabaeus) failed to darken seasonally as would be expected according to Slipher (1962; see lower right image and Fig. 1).

only significant changes in the dark areas that follow the traditionally expected pattern were confined to the high southern latitudes. The low-latitude features (35° N to 35° S), which are those we know best, appear to be dominated by secular changes that generally last at least 2 or 3 Martian years. At higher latitudes the surface comes under direct influence of the polar caps and hoods, where the probability of seasonal change is more likely. It does seem generally true that, at least in the southern hemisphere, these high-latitude areas are lighter during their respective summers than during springs. In the

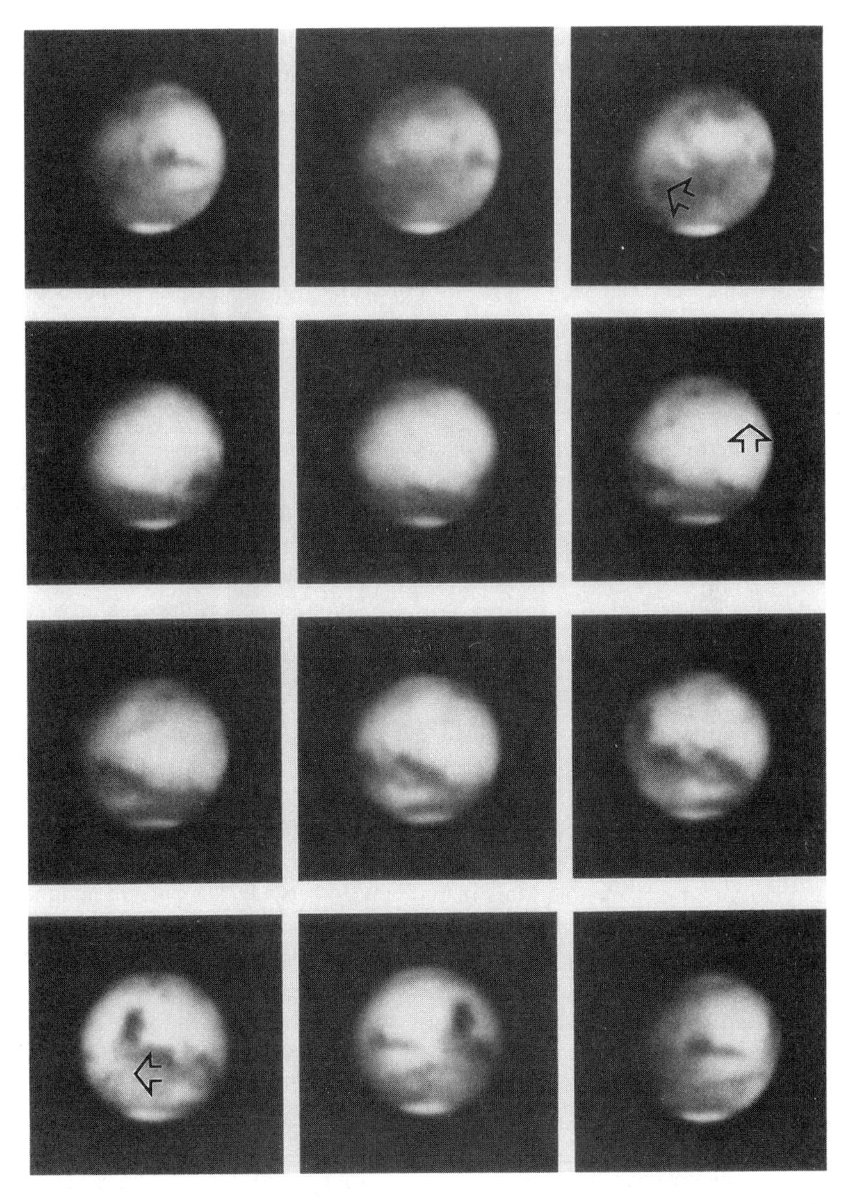

Fig. 4. This is a 1986 "artificial rotation." Although the season was in southern spring (L_s 202°–219°), most of the southern hemisphere was abnormally dark and not in accordance with the "wave of darkening" theory. Argyre was not bright, and Hellas (lower-left image) did not brighten until later in the apparition. Nix Olympica (the albedo feature) was bright in red light (see arrow on right image, second row) but not in blue. Measurements over several days showed that this brightening did not move, suggesting that this was *not* a cloud over Olympus Mons, but more likely a surface phenomenon. Lowell Observatory red-filter photographs from the Planetary Patrol Telescope at Mauna Kea Observatory.

northern hemisphere, there is much less area covered by dark albedo features, and these seasons are more difficult to observe. Possible seasonal influences in the north are therefore even less certain.

If seasonal patterns do exist, they are not reliable enough to make predictions as to how a particular feature might look during a coming apparition. A map of the previous apparition makes a better prognosticator than a map made from earlier apparitions that occurred during the same season."

Thompson (1973*b*) also made extensive photometric measurements of the International Planetary Patrol film and likewise came to the conclusion that the light areas on Mars underwent greater changes than the dark areas, although his study was independent of Capen's. Again, assuming that light areas are covered with movable dust, it is logical that this is where most of the activity is located.

Dust storms move dust, but the degree to which this affects albedo features is uncertain. Dollfus (1965*a*) and Lee (1986*b*) have demonstrated that changes occurred in the Solis Lacus area following the 1956 and 1977 storms, respectively; but Capen (1976) pointed out that the 1973 temporary darkening in the Claritas area survived the 1973 storm. This "new" feature persisted for several additional apparitions. Likewise, the prominent new dark feature that spanned 40° of latitude across Elysium Planitia during *Viking* was unscathed after both 1977 dust storms. However, many albedo changes apparently occur more frequently than large dust storms. If we assume that these changes are caused by moving dust, it must be concluded that most of this motion could often be on or near the surface (Baum 1972).

C. The Canals of Mars

Did anybody really get it right? Except for those by de Mottoni, Focas and Dollfus, every notable Mars map produced between Schiaparelli and the Mars Mariner missions included "canals" in one form or another. (Dollfus [1948] attributes the above exceptions to a breakthrough in resolution that occurred in the late forties with the availability of telescopes at Pic-du-Midi Observatory.) Lowell was not the first to see linear features on Mars, nor was his predecessor, Schiaparelli. This use of the term "canali" was not even originated by Schiaparelli (Hartmann and Raper 1974), although he adopted it to describe what he believed to be waterways (Pickering 1921). Although observers had already recorded a number of "canals" (in English this word does not necessarily infer artificial), it was Schiaparelli's observations during the favorable apparition of 1877, when he mapped far more than previously seen, that popularized these illusive features. During succeeding oppositions he continued recording the canals, as did many other observers. Lowell took the ball in 1894 and began mapping canals in ever-increasing numbers. A major difference between Schiaparelli's canali and Lowell's canals was that Lowell's crossed the dark areas (that he believed to be vegetation); whereas Schiaparelli's did not, since he still believed the dark areas to be bodies of water which were connected by the canals (Pickering 1921). (Pickering is given credit for the discovery that canals cross dark areas.) Schiaparelli, however, saw no evidence that the canals were artificial; and, after his retirement (due to poor eyesight) he even began to doubt their reality (Wells 1979). Most of the prominent Mars observers of the time drew at least some canals, including Flammarion and Antoniadi (on early drawings), who drew 36 canals on their map of the 1898–99 apparition (Flammarion 1909). But there were skeptics even prior to Lowell's declaration that the canals must be artificially constructed: E. Barnard at Lick Observatory failed to see canals in 1894 (Hartmann and Raper 1974), the same year that Lowell began his observations in Flagstaff.

Antoniadi's observations during the perihelic opposition of 1909, using the 83-cm refractor at Meudon, gave him the impression that, although Schiaparelli's canals had a basis for reality (Antoniadi 1930), they were either broader, irregular and diffuse; borders between tones; or partially connected, adjoining small dark patches. He discounted Lowell's numerous additional canals as simply unreal. His subsequent drawings and maps were greatly improved portrayals of the albedo features as we know them, although he continued to draw and label most of the Schiaparelli canals. However, he no longer drew them as sharp thin lines but rather as broader, fainter lines. Antoniadi's linear features and spots cannot always be identified on the Viking images, although some discrepancies can be attributed to the constant changes that take place on the Martian surface.

Probably in part because of the Lowell legacy, E. Slipher believed in the reality of the canals until his death in 1964. His career at Lowell Observatory overlapped Lowell's by nearly a dozen years and his drawings and map show canals nearly as boldly and numerous as Lowell's. Slipher felt that they could be identified on his photographs; and in order to convince others, he had a photographic print that included his best examples pasted into every copy of his Mars book. Mars is a difficult object to observe even under the best conditions, and it should be remembered that all of the researchers mentioned above were undoubtedly very acute observers.

D. Polar Caps

The Martian polar caps, the most conspicuous albedo features on the planet, were discovered and described by J. Cassini and C. Huygens. Herschel (1784) identified the waxing and waning of the Martian polar caps with a seasonal cycle caused by insolation redistribution during the Martian year and used observations of the seasonal cycle to locate the rotation axis of the planet. By analogy with the Earth, he assumed that the caps were composed of water-ice snow. This was the opinion of most Mars observers until the first space probes visited the planet and identified the contribution of carbon dioxide to this condensate.

The nature of the polar surface deposit has been sensed by polarimetry

(Dollfus 1957,1965*a*). The cap, observed near the edge of the planet, produces only weak polarization, although under this viewing angle, ice, hoarfrost and snow deposits exhibit a strong polarization produced by refraction inside the medium. But, in a vacuum chamber simulating the Martian environment and subjected to radiation flux comparable to the solar effect on Mars, an ice deposit begins to sublime and produces a surface pockmarked with pores. The polarization is then comparable to that of the Martian polar caps.

Many post-Herschel but pre-spacecraft observers charted the seasonal courses of the polar caps of Mars. A number of the visual observations are reported by Antoniadi (1930) and Slipher (1962). Fischbacher et al. (1969) compiled mean regression curves based upon the photographic material in the Lowell Observatory archive covering the oppositions from 1905 to 1965. Telescopic observations of subsequent north polar cap regressions have been reported in the literature by several observers: Dollfus (1973) described recessions in oppositions between 1946 and 1952; Miyamoto (1963) described the 1962–63 recession; and C. Capen and V. Capen (1970) reported observations between 1962 and 1968. The behavior of the north cap during the oppositions of the late 1970s and early 1980s was discussed by Iwasaki et al. (1979, 1982,1984) and by James et al. (1987*a*). The behavior of the south polar cap during the excellent oppositions of the mid-1950s was analyzed by Dollfus (1965*a*). The comprehensive coverage of the south cap recession provided by the International Planetary Patrol in 1971 and 1973 was analyzed by James and Lumme (1982); independent observations of the 1973 recession, including data earlier in the opposition, were published by Iwasaki et al. (1986). The recent 1986 and 1988 oppositions have been reported by Iwasaki et al. (1988,1989,1990) and by James et al. (1990).

Because the Martian poles are tipped sunward and therefore earthward during the respective spring and summer seasons but are tilted away from us during fall and winter, the recession phases are well documented, while the deposition phases are almost unknown. Since orbital perihelion is located near southern summer solstice, the resolution available for the south cap recession is much better than that for the north cap. An additional complication is the atmospheric obscuration which develops, particularly in the north, as the caps begin to form; these are called the polar hoods, and many observations of cap growth during fall and winter confuse the surface and atmospheric condensates.

Polarimetric analysis is a relevant approach to separate the two effects (Focas 1961). The polarization signature of the hood indicates that it is the same nature as the other white clouds, which are interpreted as ice crystals, although a slight increase of the inversion angle suggests a difference in crystal size (Dollfus 1957,1958,1961*a,b*). As early spring proceeds, the hood progressively clears away and the cloud polarization is more and more diluted by the almost unpolarized light from the surface cap. The polarimeter indicates that it is only by early summer that the white surface deposit is completely unveiled under clear sky.

The polar hoods are one of the least known dynamical phenomena on Mars. Some Mariner 9 (Briggs and Leovy 1974) but only a few Viking observations (Briggs et al. 1977) are relevant, so telescopic observations are the best information on the extent and duration of the hoods. The north polar hood is the better known, partly because it appears during favorable oppositions, but also it is the more persistent of the two. Present during most of the fall and winter seasons (L_s = 180° to 360°), the north polar hood is an extremely variable phenomenon (Martin and McKinney 1974), especially during periods of global dust storms (Martin 1975). This seems to be due to the effects of baroclinic traveling waves in the northern hemisphere (see chapter 26). The south polar hood is less-well understood; conflicts between observers who argue against such a phenomenon and those who report it (Fig. 5) seem to be resolved by the observation that the south hood forms only after

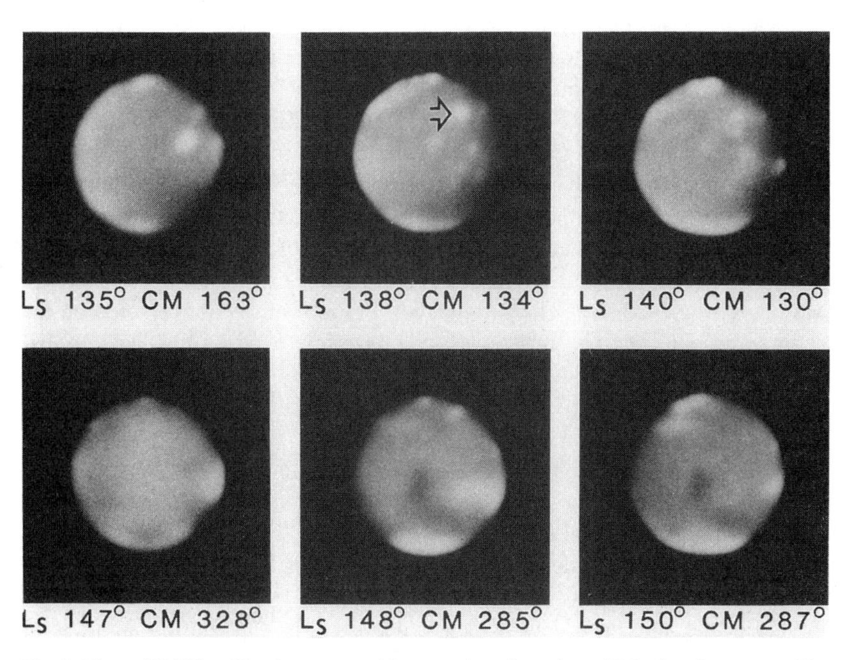

Fig. 5. These 1984 blue-filter images provide examples of a variety of Martian phenomena. The controversial south polar hood is seen at the bottom of each image. The receded north cap is at the top along with periphery clouds. The upper row shows the combination of clouds dubbed the "W" clouds. In the upper middle picture we can easily identify clouds over the three major Tharsis volcanoes, as well as over Olympus Mons and Alba Patera (the biggest and brightest cloud; see arrow). "Blue clearing" may be largely a contrast effect seen in all colors. A good example of blue clearing is seen on the lower-middle and right images based upon the high visibility of Syrtis Major, not normally seen in blue light. Limb clouds are seen on the three lower-row images. Lowell Observatory photographs.

mid-winter (L_s = 135°; Briggs et al. 1977; Ebisawa and Dollfus 1987; Ebisawa 1989).

Numerous irregularities are seen in both the north and south surface cap recessions (Veverka and Goguen 1973; Dobbins et al. 1988). Most of these can be seen in detail in the Viking mosaics of the south cap recession published in James et al. (1979). The most notable feature of this sort is the "Mountains of Mitchel" (Novissimal Thyle), which appear as a peninsula and then an outlier on the south polar cap at about 70° S, 330° W in mid-summer (Cutts et al. 1972); the time of appearance of this feature seems to vary from year to year. Finally, the core or residual cap, classically referred to as Hypernotius Mons, is discerned as a small but bright albedo feature. The south cap is surrounded by a dark band during its recession; once thought to be caused by melting ice, the feature is possibly produced by off-cap winds that remove the brighter surface dust component. The surface cap often seems to have a bright annulus near its outer edge; this possibly results from water released at the subliming cap edge and recondensed on the nearby surface cap components. The recession is not uniform in longitude, and the cap is off center with respect to the pole by as much as 6°5 at about the same time that the Mountains of Mitchel appear.

Neither polar cap sublimes completely during the respective summer season; residual caps remain during the summer in both hemispheres. Although some observers have reported the complete disappearance of the south cap in some years, there is generally no corroborating evidence from independent observers (Slipher 1962); the permanence of this deposit, now known to consist at least partially of CO_2 in some years, can have major effects on Martian climate change. The residual cap in the south is not centered on the geographic pole; the eccentric south cap recession leaves a residual cap displaced by about 3°5 from the pole. Although the north residual cap is more symmetric about the geographic pole, there are several large disconnected outliers associated with it (Dobbins et al. 1988).

The south seasonal polar cap is larger than its northern counterpart; a primary reason is the longer polar night during southern autumn and winter which occurs near orbital aphelion. The cap in late winter extends to between −50° and −55° latitude and reaches as far north as −45° in the low-elevation basins.

Both the amount of CO_2 which is deposited as caps during an arctic or antarctic winter and the rate of springtime sublimation at a given location depend upon factors such as atmospheric condensate and dust clouds, advection and surface deposition of dust which may differ from year to year. The reader is referred to chapter 27 on the seasonal CO_2 cycle for details. The repeatability of the seasonal migration of the polar cap boundaries is therefore one measure of the degree of interannual variability within the Martian climate cycle.

The recession of the south cap historically has been more regular than that of the north cap. However, some variations are apparent in the twentieth century collection of Lowell Observatory plates, and a comparison of spacecraft views from Mariner 9 and Viking established that variability occurs in the summer phases (James et al. 1979). The 1956 south-cap recession was advanced relative to the others (Dollfus 1965*a;* James et al. 1987*b*). Iwasaki et al. (1986) found that the 1973 regression was also faster than the mean during the southern spring, agreeing with the later 1973 data of James and Lumme (1982). Their analyses also showed that the springtime recessions of 1971 and 1977 were almost identical before summer solstice and near the mean of all recessions.

The synoptic coverage provided by terrestrial observations has reinforced the view that clouds and hazes are important phenomena in the regions surrounding and over the north polar cap (Fig. 5). There is a large amount of circumpolar obscuration until about $L_s = 20°$. Clouds seem to be less prevalent during the plateau in the recession but reappear with the renewed cap retreat, suggesting that they are associated with water vapor released at the edge of the cap (James et al. 1987*a*). The CO_2 cap virtually disappears by summer solstice, revealing the H_2O residual cap; clouds are often present also during this season (Iwasaki et al. 1984). In the south, condensate clouds are much less prevalent after the equinox, although dust clouds near the cap periphery seem to be common.

E. Yellow Clouds and Dust Storms

Martian dust storms have been known since the last century. Antoniadi (1930) reported that large obscurations occurred in 1877, 1892, 1909 and 1924. He proposed that these large storms were associated seasonally with perihelion. These reports were based upon visual observations which gave extensive descriptions of colors; but, significantly, Antoniadi never referred to the yellow clouds as "bright, " suggesting that he may not have seen them in their initial stage. Antoniadi also reported on several large yellow clouds that were observed during other seasons. E. Slipher of Lowell Observatory began his lifelong career of visual and photographic observations of Mars in 1905; in 1922 he photographed a "great dust storm" over a period of four days. This short-lived event was actually a relatively small cloud that showed some movement, but its expansion was limited. The only planet-encircling storm observed by Slipher in his career of nearly six decades was the 1956 event (see chapter 29 for a discussion of dust-storm terminology).

The 1956 dust storm was well documented by many observers from around the world due in part to the organizing efforts of the International Mars Committee. The initial "brilliant bar-like white cloud" was visually discovered in Japan both at the Kwasan Observatory in Kyoto and the National Science Museum Observatory in Tokyo on August 20 and also was photographed by Murayama and Ebisawa (Miyamoto 1957, 1966; de Vaucouleurs 1958; Ebisawa 1973; Ebisawa and Dollfus 1986). The earliest photographs of the 1956 storm were probably those from Lowell Observatory taken by S. Hess and P. Booth on August 19. On days that followed, the storm was photographed by a number of observers from around the world. These images followed the storm's expansion and progress for several months. Other observations that were communicated to the IMC are described in its report (Slipher 1964). At the IAU Planetary Photographic Center in Meudon, selected photographs from Mt. Wilson, Pic-du-Midi, Johannesburg, Lick, Bloemfontain, Rome, Volgograd, Milan, Tokyo and Genoa were analyzed by de Mottoni (1969), who compiled a series of maps describing the life of the storm from its first emergence until its disappearance. These data were analyzed by Dollfus, who, with Focas and Camichel, had observed the storm from Pic-du-Midi (Dollfus and Focas 1969).

Slipher (1962) photographed several other yellow clouds that showed some motion but none expanded much beyond their original dimensions. Few other dust clouds showing motion were photographed in the pre-spacecraft era. Most of the other reported storms were short lived, or in many cases obscurations were observed but bright clouds were not seen. Thus, until 1969, we had just two very different cases (1922 and 1956) of photographically documented storms that had bright clouds showing some motion and expansion.

In 1969 Capen (1971) observed and photographed a regional dust storm while using the McDonald Observatory 82-inch reflector. This storm began in the same location as the 1956 storm and lasted at least a week, expanding to the west as the other storms in that area (1956 and two in 1971). The season, however, was L_s 163°, a significant distance from perihelion (L_s 250°). This storm probably did not encircle the planet.

Based upon his knowledge of the historical record, Capen (1971) predicted a 1971 storm near the time of perihelion, warning of possible interference with observations from Mariner 9. Capen could hardly have been more correct. When Mariner 9 arrived at Mars, the planet was totally enveloped by the largest dust storm on record. It is the only known case of a completely planetwide storm and, because of the space mission, it also became the most publicized. Nearly two months earlier, the initial cloud had been identified on International Planetary Patrol images taken in South Africa on September 22 (Capen and Martin 1972) and on polarization measurements taken at Meudon (Dollfus et al. 1984*a*). The L_s was 260°, precisely midway between perihelion and southern summer soltice, convincing everyone that storms of great magnitude were regular, seasonal events (Hartmann and Raper 1974). Capen even got a bonus: In 1971 July, there was a precursor dust cloud that began in the same Hellespontus area at L_s 213° (Ebisawa 1973; Capen 1974). Like the 1969 storm, this one did not encircle the planet, although it lasted for about two weeks.

During the 1973 apparition still another Martian dust storm was observed (Hattori and Akabane 1974; Martin 1974*b*,1976; Miyamoto 1974; Ebisawa 1975; Dollfus et al. 1984*b*). This planet-encircling storm was not as large as the planetwide storm of 1971 and began farther west, in Solis Planum (Fig. 6) at L_s 300°, thus extending the dust storm "season" to 50° beyond perihelion.

Because it was observed by both groundbased telescopes and by Mariner 9 (although only marginally simultaneously), the larger 1971 event is the most studied dust storm to date (Larson and Minton 1971; Kirby and Robinson 1971; Capen and Martin 1972; Miyamoto 1972; Dollfus et al. 1984*a*). The early phases of this event were intensely photographed nearly continuously for several months by the worldwide network of the International Planetary Patrol. These images were extensively analyzed using Lowell Observatory's specially designed Planet Image Projector. Semi-hourly cloud positions were plotted onto daily maps to show the progress of the storm during its first 20 days (Martin 1974*a*). Similar studies were made of the 1971 July storm (Capen 1974) and the 1973 storm (Fig. 7).

The 1971 dust storm and its precursor event were also monitored in France during nine months of observations coordinated between Meudon and Pic-du-Midi Observatories, involving 9 observers and 6 telescopes, and the evolution of the Martian atmospheric dust loading was documented by exten-

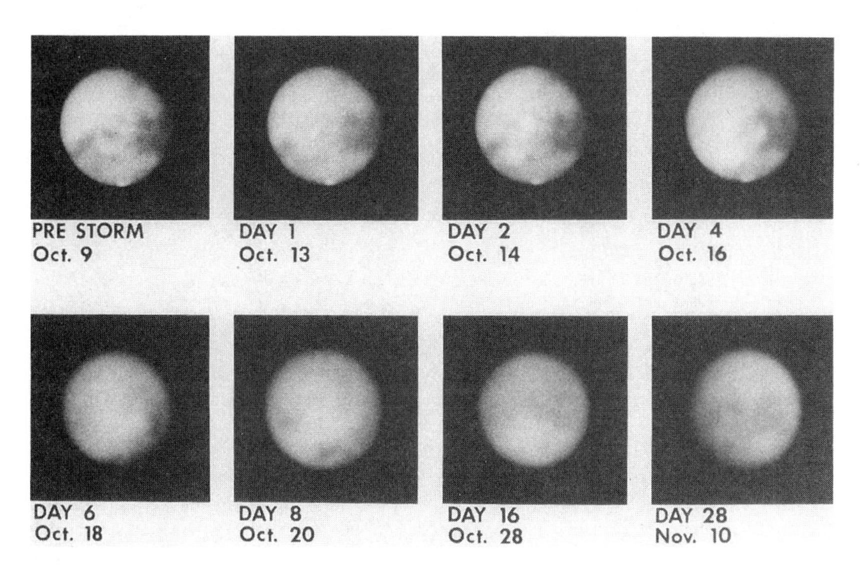

Fig. 6. International Planetary Patrol images in red light, showing the growth of the 1973 planet-encircling storm over a period of four weeks. These images were some of those used to compile the maps shown on Fig. 7. The same face of Mars is shown in all of these pictures, to demonstrate the progressive obscuration of the albedo features. This is the best-documented dust storm on record and the largest for so late in the season (L_s 300°).

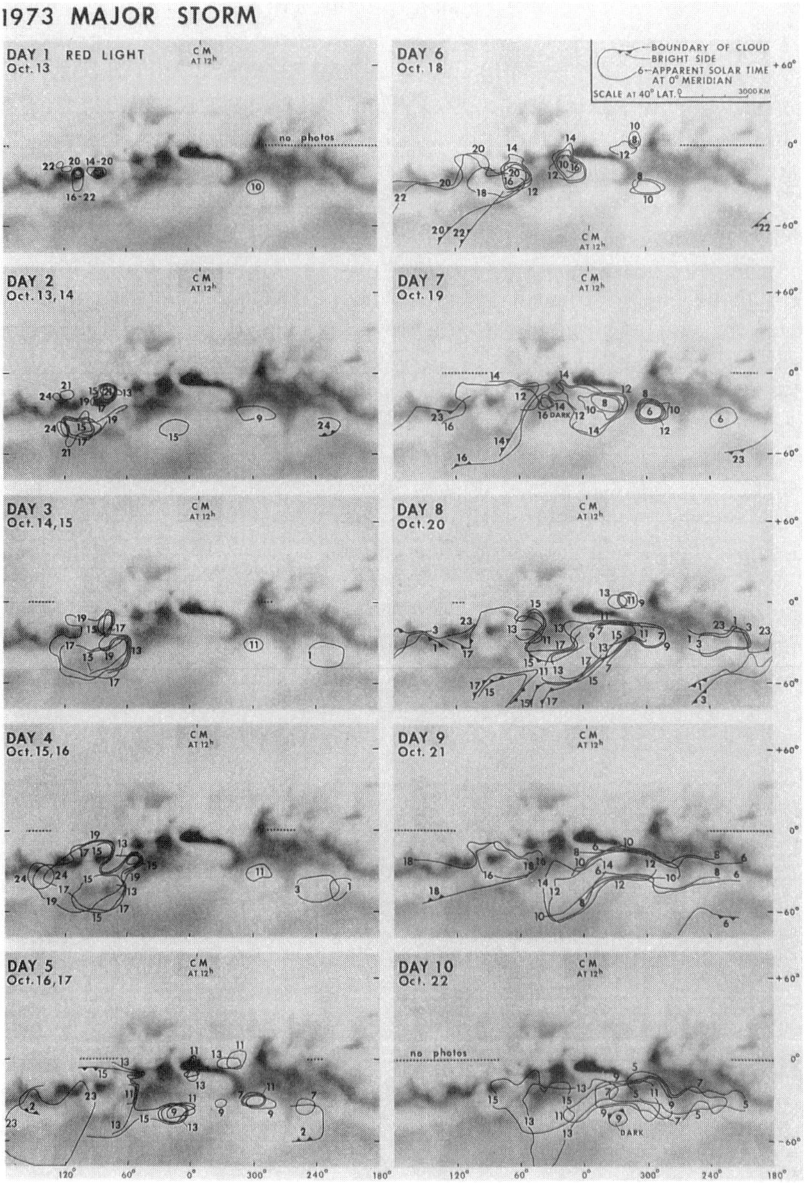

Fig. 7. The first 10 days of the 1973 dust storm were mapped to show the progress of the bright clouds every other hour. Additional maps show the second 10 days; and this 20-day series was repeated for blue-filter images in Martin (1976). Cloud positions were determined by using International Planetary Patrol photographs in the custom designed Planetary Image Projector at Lowell Observatory. Cloud outlines are numbered according to the nearest hour of Martian Apparent Solar Time (MAST); and each "day" is one Mars rotation, beginning with the first day of the storm. Thus each hour represents 15° of rotation, and 12 is the hour that the sub-solar point crossed the 0° meridian. The longitude where the sub-Earth point crossed at 12 hr MAST is shown at the top center of the maps. Note the number of different areas where individual dust clouds arose, in some cases creating new storm centers.

sive photography. These events were followed in real time visually by Dollfus, Ebisawa, D. Bonneau, and J. Murray.

Polarization has also been successfully used to observe dust storms. At the time of the dust storm of 1924 December, Lyot (1929) discovered that the polarization was noticeably reduced. During the 1956 dust storm, very opaque yellow clouds expanded and obscured the Martian features; the atmosphere remained dusty for some time. Observations at Pic-du-Midi showed a general decrease of the polarization, which correlated with the opacity of the clouds (Dollfus 1965*a*). The planetwide dust storm of 1971 and its precursor event were analyzed during 9 months of coordinated polarimetry and imaging observations at Meudon and Pic-du-Midi Observatories (Dollfus et al. 1984*a*). The decrease of polarization produced by the storm in 5 wavelengths is displayed in Fig. 8. Two consecutive dust storms in 1973–1974 were also extensively observed in a coordinated polarimetric and imaging survey that covered 8 months (Dollfus et al. 1984*b*). Based upon these observations, shapes for the polarization curves were derived for several wavelengths which are apparently consistent with small, solid, dark particles of a few μm diameter.

Large dust storms do not occur during a narrowly defined season, but rather over a period that includes at least one-third of a Martian year. This period is too long to observe well during a single apparition, but the evidence suggests that they do not occur during every Martian year. The main problem with assuming that dust storms are seasonal is that seasons thought to be lacking dust storms are poorly observed. In order to positively identify storms on average-quality photographs, Mars' apparent diameter should be at least 13 arcsec (Fig. 9). For aphelic apparitions this is a period of about 20° of L_s, or less than 1/7 of the known storm "season" around the time of perihelion. The odds of seeing a storm during that season are further reduced by the lack of interest by potential observers. When Mars will be more than 22 arcsec, many more people (amateurs and professionals alike) attempt observations than during apparitions when the maximum diameter reaches $<$ 14 arcsec. Of course, those that do observe not only do so for a shorter time, but also cannot get the resolution that they would during close apparitions. We cannot be sure just how seasonal these storms are until we have more data on the other seasons (Martin 1984). See chapters 26 and 29 for further discussions of dust storms, including spacecraft observations and results.

F. White Clouds and Hazes

More common than yellow clouds are the Martian "white" or "blue" clouds (Fig. 5). These are those bright clouds that show on ultraviolet-, blue-, and sometimes green-filter images, but are usually absent on red-filter images. They may show up as diffuse and widespread hazes or as well-delineated discrete clouds. These clouds have also been observed through most of the recorded history of telescopic observations. They are condensate

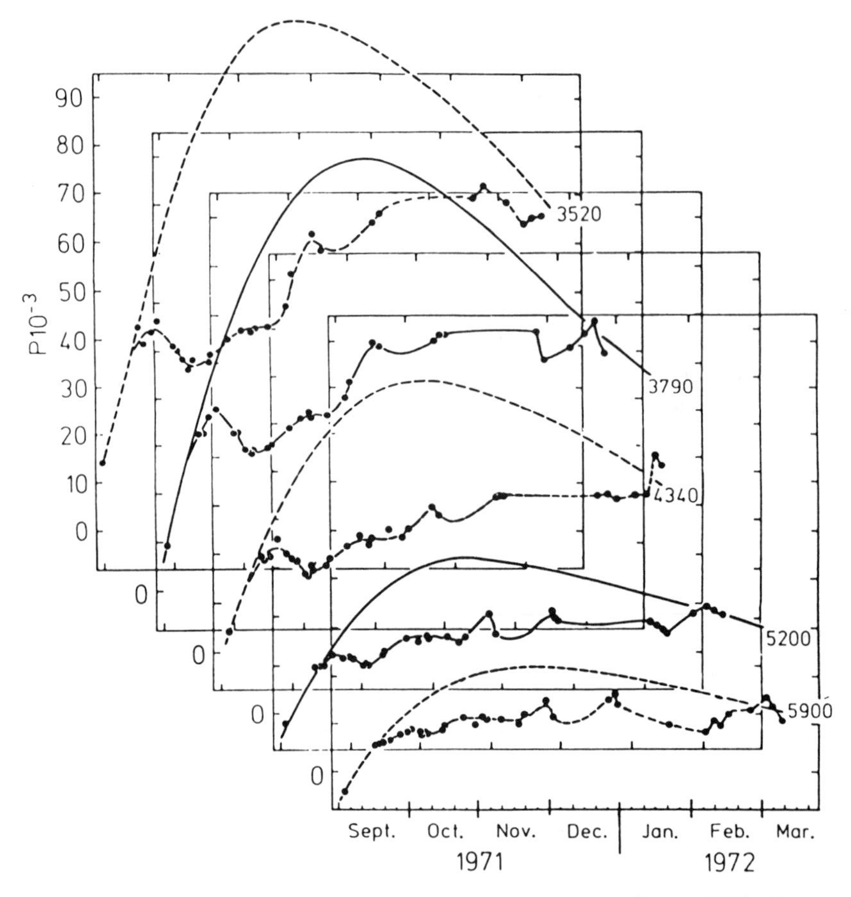

Fig. 8. Whole-disk polarization of Mars during the global dust storm of 1971–1972 in 5 wavelengths (curves labeled with wavelengths in Å). Plotted is the degree of polarization as a function of time. The reference (smooth) curves are the values which would be observed if the Martian atmosphere were free of airborne dust (Dollfus et al. 1984*a*).

clouds usually thought to be water-ice crystals, but some may be CO_2 (S. Smith and B. Smith 1972). In addition to condensates, many of these clouds may also contain dust and are often seen in association with dust clouds (Miyamoto 1957; Martin 1976). They also are often seen around the periphery of the polar caps, especially the north cap (James et al. 1987*a*). Discrete white clouds may appear singly or in groups. Most discrete clouds remain relatively stationary for days at a time, but those in dust storms display rapid movement. Mars displays limb and/or terminator hazes much of the time that are occasionally small but may be generally quite extensive, both north-south along the planet's edge, as well as extending out onto the bright disk (see Color Plate 1). At times they even form a band completely across the planet (Slipher 1962).

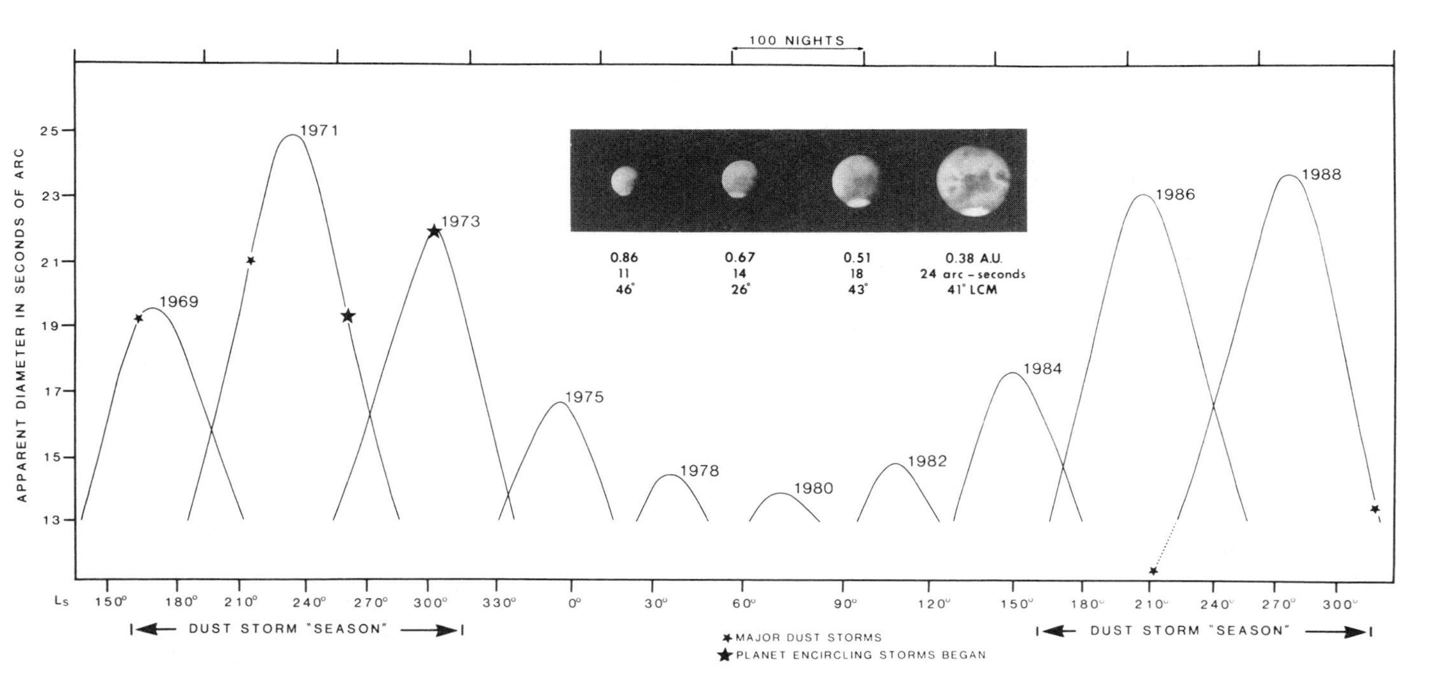

Fig. 9. The apparent diameter of Mars from Earth is plotted against time for the cycle of apparitions from 1969 to 1988. The linear scale across the top is the number of observing nights from Earth. The L_s scale (bottom) is a scale of Martian seasons, with 0° marking the vernal equinox for the northern hemisphere. This L_s scale overlaps differing parts of 10 adjoining Martian years as each apparition samples a different but consecutive orbit of Mars. The scale is compressed near perihelion (L_s 250°) because of Mars' eccentric orbit; thus more of that season is observable than the opposite season during an equal number of nights. This seasonal bias of observations repeats through every 15 to 17 yr apparition cycle. The 1971 Mars images demonstrate the visibility of features at various diameters but show much more detail than can be expected from photographs of average quality. "LCM" denotes the longitude of central meridian, indicating a similar face of Mars on each image.

The term "W" clouds was applied by E. Slipher to a group of clouds he observed while in Chile during 1907. The combination he saw that year formed a "W" pattern and the name stuck, although that same pattern has been seldom repeated (Martin and James 1985). Some combination of these clouds has been observed many times over the years, but the configuration usually is different during each apparition it is observed and may change from one day to the next. The same pattern may persist for several weeks, but the relative sizes and brightness of individual clouds may vary (Fig. 5). Some of them are over or displaced slightly from the great volcanic shields of Tharsis and/or Olympus Mons (Slipher's "W" did not include Olympus). Other clouds have associations with more subtle topographic features such as Alba Patera or smaller features: Dollfus (1965*a*) has observed clouds over the crater Schiaparelli; but some may not relate to topography at all. Slipher felt that the "W" clouds were strictly afternoon events, but there are some exceptions. When morning "W" clouds do occur, they tend to be weaker. These clouds are generally seasonal (Smith and Smith 1972), but the season is broad (L_s 80° to 240°) and some individual clouds or pairs of clouds appear outside that interval. Because of the difficulties mentioned above in observing some seasons, the "W" cloud season could be even longer but probably does not include the well-observed season after L_s 240°. Hunt et al. (1980) mistakenly attributed this seasonal void in "W" clouds to the presence of dust storms. However, in 1988, we observed neither over most of that period.

During the 1981–82 apparition, Akabane et al. (1987) measured the optical thickness of an Olympus Mons cloud. They determined that it had a peak value of about 0.5 near 14 hr Martian local time.

Polarization has been used as a sensitive method of monitoring various types of clouds and hazes because it can discriminate between ice crystals, dust grains and μm-size aerosols. For the white clouds, there is no trace of the strong positive polarization produced on liquid droplets by the rainbow effect. Thus, liquid aerosols are excluded. But icy clouds formed when blowing air above a flask containing liquid nitrogen reproduced the Martian cloud polarization. White veils on Mars behave like the icy crystals of the cirrus clouds (Dollfus 1957).

The polarimeter often detects morning hazes along the sunrise limb, even when too faint to be seen visually. They usually disappear a few hours after sunrise but may recur each morning over the same area for weeks. Hazes may appear also along the evening limb, where the air is cooling before sunset (Dollfus et al. 1984*a*). Statistics of seasonal variations are based upon 1216 polarization measurements covering 7 Martian apparitions (P. Lee, unpublished). In the northern hemisphere, there are more cloud occurrences in summer, at the time when the amount of atmospheric water vapor was shown to increase, according to Viking Mars Atmospheric Water Detector measurements. This relationship supports a water-ice nature for these clouds. Over the tropical regions, there are no strong seasonal variations. In the northern hemisphere, the seasonal effect may be hampered when dust storms occur.

G. Determination of Surface Pressure

Even by Lowell's time, it was recognized that Mars' atmosphere must be much thinner than the Earth's. Based upon Mars' gravity and apparent albedo, Lowell in 1908 calculated the surface pressure to be 87 mbar (Glasstone 1968). Results from subsequent improvements in observation methods and calculations by many investigators led Slipher to conclude that the value must be between 83 and 89 mbar, although several polarimetric studies had led to lower values. De Vaucouleurs (1954) had come to nearly the same conclusion, putting the value at 85 $\pm$ 4 mbar.

With the pending approach of the Mariner IV launch and more improvements in groundbased technology, the situation began to change rapidly. The story of the application of Fourier spectroscopy beginning in 1963 and subsequent developments is recounted by Michaux and Newburn (1972) starting with early and revolutionary determinations by Kaplan et al. (1964). After several years of observations and calculations by a number of investigators, the effort culminated when Young (1971) calculated the average surface pressure at 5.2 mbar. This was made possible by using observations from 1967 by Connes et al. (1969) who used a new spectrograph built at the Jet Propulsion Laboratory. By that time, the three Mariner flyby missions had taken place and made surface pressure measurements; but, because those were done by occultation, there were just two spots measured from each for a total of six locations on Mars. Consequently, Young's determination was still considered to be the best spatial average value at that time. Applying a seasonal correction derived from Viking Lander data gives a temporal average of about 5.6 mbar.

III. MISSIONS SUPPORT

A. Mars, Mariner and Viking Missions (1964–1980)

During the late 1950s, in anticipation of planetary exploration from space (the Mars Mariner and Viking missions, various U.S. government agencies began seeking data on Mars from a number of experienced observers and encouraged the expansion of observing programs. This included a series of Lunar and Planetary Colloquia and evolved into a number of contracts and grants supporting specific programs, including the publication of Slipher's book (1962; also supported by the National Geographic Society) and several new Air Force Mars maps drawn at Lowell Observatory (MEC-1, 1963; MEC-2, 1967). The NASA-funded IAU Planetary Research Center was established in 1965 at Lowell Observatory. Also, observing programs were supported at New Mexico State University; Catalina Observatory, Lunar and Planetary Laboratory, University of Arizona; McDonald Observatory, University of Texas; and Table Mountain Observatory, Jet Propulsion Laboratory. The latter resulted in a very comprehensive report on the 1964–65 apparition (Capen 1966). Other reports on contracted research include studies by de Vaucouleurs (1969) on a Martian coordinate system, by Fischbacher et al. (1969) on Martian polar cap boundaries, and by Martin and Baum (1969) on cloud motions. Following Slipher's book, NASA published two more on Mars (Glasstone 1968; Hartmann and Raper 1974). Both volumes include spacecraft data, the latter emphasizing results from Mariner 9. Jet Propulsion Laboratory published a *Mars Scientific Model* (Michaux and Newburn 1972), which contains a comprehensive blend of telescope and early spacecraft data.

A number of telescopic observers also became directly involved in the various Mars missions. They included W. Baum, B. Smith, R. Leighton, L. Martin, T. Owen, D. Thompson and S. Hess, among others.

B. The International Planetary Patrol Program (1969–1982)

The 1969 Mars apparition marked the beginning of an extensive worldwide observing program (Baum et al. 1970; Baum 1973) in which 7 observatories were involved in a coordinated effort to photograph designated planets on an hourly basis. Because the observing stations were distributed in longitude around the Earth, it was possible to monitor atmospheric and surface changes on Mars on a nearly continuous basis. The program was supported by NASA and managed by the Planetary Research Center at Lowell Observatory. It was initially directed by Baum, and later by R. Millis. The other observatories participating were Mauna Kea, Hawaii; Cerro Tololo Inter-American Observatory (CTIO), Chile; Republic Observatory, South Africa; Kavalur Station, Indian Institute of Astrophysics, India; Perth Observatory, Western Australia; and Mt. Stromlo Observatory, Australia. Kavalur and Perth did not begin participation until 1971, although New Mexico State University was involved during 1969. After 1973, the program was gradually reduced both in numbers of observatories and in length of scheduled observing runs. The San Vittore Observatory, an amateur group in Bologna, Italy, joined the network in 1977 in order to help fill the increasing gaps in the longitudinal coverage. By 1982, the only stations remaining were Lowell, Mauna Kea and Perth.

Uniformity in the observations was maintained by using identical camera systems, observing procedures, film and film processing. All telescopes were similar in size and used at nearly the same image scale; three identical telescopes were purchased for the project to augment those already available. Weather permitting, a set of sequences consisting of 14 images each in red, green, blue and ultraviolet light was taken at least once an hour. The film was then returned to Lowell Observatory for rigidly controlled processing. Tens of thousands of computer catalogued Mars sequences from this program are available for use at the Planetary Research Center.

C. Correlations with Viking

The Viking Orbiters were highly successful in performing the tasks for which they were primarily designed. However, the collection of albedo, atmospheric and polar cap data were not high-priority goals of the missions. Highly detailed information on the above phenomena was derived from the space missions, but data generally distributed in time and space were deficient. Even the "global" mosaics from near apapsis were unable to cover as much of the full disk as seen from Earth and do not show complete coverage in two colors. Usually two-color coverage was achieved by alternating image strips through red and violet filters. More important, this type of coverage was infrequent, unevenly distributed and, although the life of the spacecraft was unexpectedly long, spanned a limited time, ending nearly a decade ago. Especially in the case of albedo features, what was then, no longer is. Using Viking data, albedo maps have been produced from both the solar-band (0.3 to 3.5μm) Infrared Thermal Mapper (IRTM) data (Palluconi and Kieffer 1981) and the Viking Imaging System (VIS) data. The imaging albedo maps are highly detailed but incomplete, somewhat unbalanced between areas, and it is difficult to discern some major features (e.g., Solis Lacus). The IRTM maps come much closer to the albedos seen from Earth, but that instrument saw some areas differently than telescope cameras with red filters. This is also seen in the Solis Lacus area; however, that area shows little agreement between the VIS and IRTM maps. The resolution on the Viking IRTM maps may not be as good as seen on some of the recent Earth-based CCD images. It is important to be aware of these discrepancies but also to resolve why they exist in order to understand the nature of the Martian surface. The detailed renditions from VIS data and their correlation with topographic features provides mappers of Earth-based data some insights into what they are mapping although changes over time are great enough that Viking albedo maps are accurate only for that particular period.

"W" clouds were studied by Hunt et al. (1980) using Viking images. An Olympus Mons cloud has been identified on both Viking and Earth-based images taken during the same day (Martin 1983). Also, cloud surveys have been done with Viking data that include much smaller and fainter clouds than can be seen from the Earth (Kahn 1984). Statistical weakness arises because of the irregular nature of the sequencing of Viking imaging. The Viking images showed a wealth of cloud detail during the 1977 dust storms but was not programmed to image the initial clouds and much of their development and progress around the planet. Many of our ideas about these 1977 storms were derived from comparisons with studies of earlier storms seen on Earth-based photography (Martin and Baum 1978). Conversely, the Viking images provided detailed impressions of what the earlier storms may have looked like from up close.

Viking images have been used to measure one recession of the south polar cap (James et al. 1979) and two recessions of the north polar cap (James 1979,1982). Viking sequences established an accurate south cap recession curve for 1977, but no measurable images were obtained during the second southern recession in the mission. The Viking data for the north cap, however, have serious deficiencies which produce some scatter among a minimal number of data points. Nevertheless, the derived curves are sufficient for all three recessions to establish both the accuracy of telescopic measurements and the year-to-year variability of the recessions (Fig. 10). The evidence before the advent of spacecraft observations was equivocal. Antoniadi's observations suggested to him that there was a small effect on the size of the polar

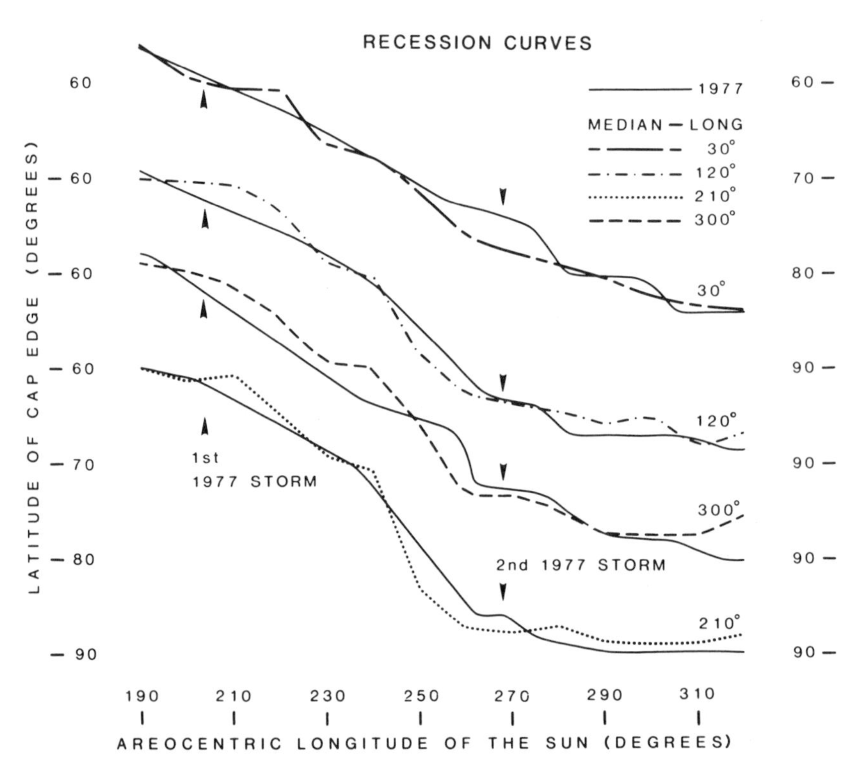

Fig. 10. The recession of the south polar cap, comparing the median values at four different Martian longitudes from the study of historical data by Fischbacher et al. (1969) and the 1977 measurements from Viking by James et al. (1979). Some of the largest deviations are attributed to gaps in the data points from Viking at some longitude and L_s combinations (James et al. 1987*b*).

caps due to solar activity; he found that the caps were larger near solar minima. Slipher (1962), on the other hand, considered the pre-spacecraft observations of the caps by many different observers and concluded that there was no compelling evidence for interannual variability.

The telescopic data of Iwasaki et al. (1979,1982) gave regression curves which are in excellent agreement with those representing the edge of the north cap from Viking images (James 1979,1982); that these visual measurements do correspond to the edge of the surface CO_2 deposits was established by comparisons with the Viking IRTM observations (Christensen and Zurek 1984). There are several outliers which are left behind by the north cap recession. A dark lane across the cap, Rima Tenuis, has been reported by some observers, most recently in 1980 after a 60-yr absence (Parker et al. 1983). There is, however, no hint of such a feature in the extensive Viking coverage of the cap.

It is difficult to separate the polar haze from the surface cap; red-filter pictures show surface albedo features, including a surface cap (if present), while blue-filter images reveal the hood. This led Iwasaki et al. (1979) to suggest on the basis of 1975–76 observations that at least the outer portions of the surface cap might not be present beneath the hood as it was not visible on red-filter images. Careful comparisons of visual images obtained with red and violet filters both telescopically and by Viking and IRTM infrared observations have confirmed the existence of both a surface cap, albeit patchy and/or thin near its edge (which is about 55° latitude), and clouds in late winter during the Viking mission (James 1982; Christensen and Zurek 1984).

IV. PRESENT OBSERVATIONS

A. Status

Photography continued to be the primary observing mode through the 1980s, although some visual work has continued, primarily among amateurs. Spectroscopy continues with improved techniques, and some charge-coupled device (CCD) work has been successfully pioneered (even by some amateurs). During the apparitions of 1984, 1986 and 1988, P. James of the University of Missouri at St. Louis and Martin of Lowell Observatory conducted a modest photographic monitoring program using International Planetary Patrol equipment and procedures. They were assisted by C. Gulley, N. Thomas, R. Sullivan, J. Melka and D. Parker. These observations were made at Lowell Observatory (1984 and 1988), at Mauna Kea Observatory (1984, 1986 and 1988; see Color Plate 1) and at CTIO in Chile (1986). Perth Observatory also participated during each apparition, and San Vittore Observatory contributed during 1984. The National Geographic Society supported this program in 1986 and 1988. (They had provided support for Lowell Observatory's Mars observing expeditions to South Africa in 1954 and 1956.) Also in 1988, de

Vaucouleurs and J. Dragesco, as guests at Lowell Observatory, made extensive photographic and visual observations using the 24-inch Clark refractor. In Japan, the photographic program described earlier continued through 1988 at the Kwasan and Hida Observatories and at Ebisawa's private observatory. During the apparitions of 1986 and 1988, Japanese observers also traveled to the Bosscha Observatory in Indonesia, where both the weather and latitude provided conditions superior to those in Japan (Iwasaki et al. 1988). Mars observations in France have continued using the telescopes at Meudon and at Pic-du-Midi, where a new 2-meter telescope is now in use. Reduced photographic programs also continue at Catalina Observatory, New Mexico State University Observatory and Table Mountain Observatory.

A number of CCD cameras were tried on Mars in 1988 at various observatories with very promising results. As this technology becomes more adapted to Mars observations and systems become more available, very useful telescopic observations can be expected for the future. In France, CCD technology with image processing has been made available at Pic-du-Midi by C. Buil and at Meudon Observatory by T. Faucounier and Dollfus. Exceptional images of Mars have been acquired at the 1-meter telescope at Pic-du-Midi by a group of observers including P. Laques, J. Lecacheux and P. Guérin (de Vaucouleurs 1989; see the cover of *Sky and Telescope,* December 1988). The resolution of these images exceeds that of the best Earth-based photographs, establishing this technique as a definite observational breakthrough. CCD imaging was also used at the 1-m telescope at Meudon. Other outstanding CCD images were obtained by S. Larson and G. Rosenbaum using the 61-inch telescope at Steward Observatory in Arizona (O'Meara 1988). At Lowell Observatory, 11 narrow-bandpass filters were used with a CCD system in an attempt to provide improved discrimination between wavelengths. For these observations, both the 24- and 42-inch reflectors were employed with Barlow lenses (Martin et al. 1989).

Spectroscopic observations using telescopes are also being pursued in the post-Viking era. High-resolution mineralogy studies are made possible by technological advances in instrumentation. Much of this work has been carried out at Mauna Kea Observatory in Hawaii under the direction of T. McCord (Bell et al. 1989*c*). Additional observations were made at the University of Arizona's Catalina Station by Singer et al. (1989) and at the Infrared Telescope Facility (IRTF) at Mauna Kea during the apparition of 1988 by Clark et al. (1989). Also, T. Martin of the Jet Propulsion Laboratory spectroscopically observed dust in the Martian atmosphere using the IRTF at Mauna Kea in 1988 (see chapters 17, 18 and 29 for discussions of contemporary observations).

Polarimetric analysis of the atmospheric turbidity, condensate clouds and dust storms are conducted each apparition as a coordinated survey between Tokyo and Meudon Observatories, essentially by Ebisawa (1984; Ebisawa and Dollfus 1986,1987). At Meudon, the new technology of CCD imaging polarimetry made operational by Dollfus produced direct images of the clouds, veils and hazes over the surface of the Martian disk.

B. Amateur Observations

Planetary observations by amateurs are a long-standing tradition. Until the 1970s, they consisted essentially of visual observations, a practice that developed some exceptional skills of talented observers. More recently, photographic technique has developed to a high level among amateurs, and the quality of the images now produced often competes with the best professional pictures (Dobbins et al. 1988). This achievement coincided with the time when professional astronomy had shifted priorities and reduced its planetary telescopic activity. These circumstances entrust amateurs with a heavier responsibility when conducting telescopic observations of Mars, although throughout history many of the most prominent planetary observers have come from the ranks of amateurs.

Currently, the Mars Section of the British Astronomical Association receives over 1000 observations each apparition from several dozen observers from all over the world. International cooperation is maintained with several other groups in Europe. At the end of each apparition, a report is prepared for the *Journal of the BAA* detailing the atmospheric activity, changes in the surface features, and the changes in the polar hoods and caps. For example, for the recent 1988 opposition, the BAA Mars group received reports and observations from 74 persons in Belgium, Brazil, France, India, Italy, Japan, Romania, Spain, United Kingdom and United States. There were 1164 visual drawings, 4 CCD images, and 339 high-quality photographs, a total of 1507 observations.

During the recent 1988 perihelic opposition, D. Crussaire headed a group of 10 experienced observers from Société Astronomique de France (SAF) who used the large 83-cm refractor at Meudon (the one used by Antoniadi). A total of 96 visual observations and more than 100 selected photographic images have been collected, plus several dozen CCD or video images, over a six-month survey. In addition, SAF received drawings by 23 other persons and more than 200 selected photographic images. Recession of the south polar cap, white cloud activity, dust storm occurrences, surface-marking variations, and periods of atmospheric clearing were analyzed. Foreign observers participating included R. McKim from United Kingdom and M. Falorni from Italy. These observational efforts were aided by Dollfus and Ebisawa.

Recent efforts of the Mars Recorders of ALPO have attracted older, experienced observers who had discontinued planetary observing during the Mariner and Viking explorations. Encouragement by professionals such as Capen and Martin from the Lowell Observatory, and James from University of Missouri at St. Louis has instilled renewed enthusiasm for groundbased Mars observations, stressing that the spacecraft data have provided better, more intelligent direction for this work. Several members have collaborated on a textbook for interested observers (Dobbins et al. 1988). The Mars Recorders continue to develop programs that better meet the needs of the professional community and are exploring new technologies, such as computerized image processing, for studying Mars during future apparitions.

The International Mars Patrol (IMP) by 1988 had grown to include over 250 observers in 38 countries and U.S. territories: Arabia, Argentina, Australia, Austria, Belgium, Bolivia, Brazil, Canada, Czechoslovakia, Colombia, Faroe Islands, France, Germany, Greece, India, Israel, Italy, Japan, Korea, Mexico, New Guinea, New Zealand, Norway, Nova Scotia, Okinawa, Philippines, Polynesia, Puerto Rico, Rwanda, St. Croix, Samoa, South Africa, Spain, Sweden, Taiwan, United Kingdom, United States and Venezuela. During the 1987–1989 apparition, the IMP has contributed over 5300 high-quality observations of Mars, including 1400 photographs, a number of CCD images and video tapes, and 200 micrometer measurements of the south polar cap.

For the future, it is anticipated that the systematic recording of changes in Martian surface albedo features will rely more and more on amateur observations with private telescopes and the development of new CCD detectors may initiate a new era for planetary surface surveys.

C. Increasing the Data Base

Since each apparition presents its own unique version of albedo features, a map of each new apparition provides a composite record of the observations. When a series of these maps is compiled using similar methods (Inge et al. 1971; de Mottoni y Palacios 1975), an invaluable data set is created for the analysis of albedo changes. A new map is therefore planned in the Lowell Observatory Series (Fig. 1) based upon photographic observations from 1988. This map will be compared to earlier maps for the same season as well as other seasons. An extended data base for changes over a number of years should yield more sophisticated insights as to the underlying mechanism for change (or is it simply arbitrarily shifting patterns controlled by local winds?). Because some areas tend to stay darker or return to configurations seen earlier, we can assume that the situation is complex; but, although we have tens of thousands of Orbiter images, these changes are not fully understood. Future maps based upon CCD imaging will provide more detailed renditions of these features.

Accurate spacecraft observations of a limited number of seasonal polar cap cycles have established the reality of interannual variability, but Earth-based observations continue to provide values for an extended data base. Micrometer observations of the north polar cap suggest a large amount of interannual variability in its spring recession (Parker et al. 1983). Observers agree that the regression curve exhibits a region of inflection or plateau in mid-spring, and analyses of photographs of Mars obtained from Japanese observatories have shown that there are year-to-year variations in the seasonal location of the plateau (Iwasaki et al. 1979,1982,1984). The 1986 and 1988 recessions of the south cap were found to have been slower than normal (Iwasaki et al. 1988,1989,1990; James et al. 1990).

In spite of some successful predictions, the occurrence of large dust storms remains very unpredictable. The expected planet-encircling storm looked for during the 1988 perihelic apparition never occurred. When a smaller storm was observed early in the apparition (Fig. 9), researchers believed it to be the usual precursor to a much larger storm nearer perihelion. This scenario had been developed by Capen after his study of the 1971 July storm. Later, Dollfus et al. (1984*b*) published their findings that an early storm had also taken place in 1973 prior to the planet-wide storm of that year. The precursor storm theory together with the recent history of perihelic storms (1971, 1973, 1977) seemed to indicate that 1988 would certainly be the year for another big one. Another storm did develop later in 1988 and, although it was of major proportions, it failed to last long enough to become planet encircling. This second storm was late in the apparition and at the end of the storm season at L_s 314°. It was first seen east of Solis Lacus on November 23 and photographed by Parker on November 28 and December 1 at his observatory in Coral Gables, Florida. The earlier storm, also photographed by Parker, was initially detected by G. Teichert in France on June 3 (Beish et al. 1989). These were the first major storms to be successfully photographed since 1973. Because these storms are difficult to predict, extensive monitoring is required to increase the data base.

V. CONCLUSION

In addition to the phenomena discussed above, there are a number of interesting features that show up on the Earth-based images that do not have obvious explanations in spite of Viking observations. These include the "violet haze" and its counterpart "violet clearing" which may be simply contrasts of surface features (Thompson 1973*a*), although some violet Viking images show veiling clouds which do not obscure the same features through red filters. A rarer phenomenon is the albedo reversal seen when features imaged as dark through red filters are lighter than their surroundings on blue-filter images (Thompson 1973*b;* Thomas and Veverka 1986). During large dust storms, new but short-lived dark features occasionally appear (Slipher 1962). Also, dark clouds have been seen in blue and violet (Slipher 1962), including during dust storms (Fig. 11). The bright *surface* feature Nix Olympica on Olympus Mons which only infrequently appears (Fig. 4), is not snow as the name implies; but what does cause it? And, although polar hoods are the most dynamic atmospheric features on the planet, only limited studies of them have been made. The polar surface caps are formed beneath the hoods, but the processes involved have not been observed. A number of observations

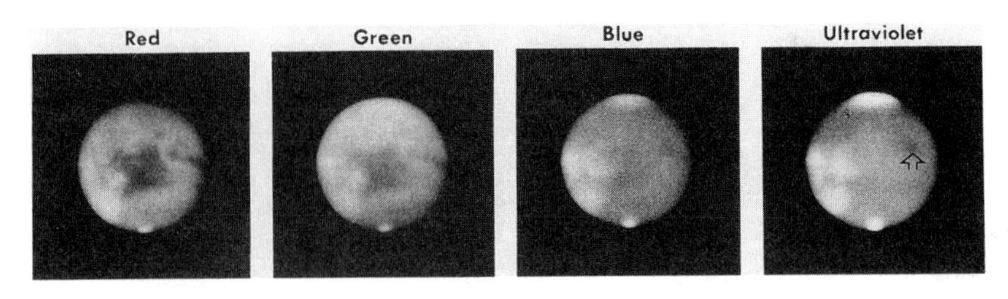

Fig. 11. Day 4 of the 1973 dust storm as seen at the same hour in four different colors on International Planetary Patrol photographs (Martin 1976). The central meridian was 72°. Note the three large dust clouds to the west are bright in all colors; whereas the obscuration over Meridiani Sinus (upper right) is dark in blue and ultraviolet (noted by arrow). It has been suggested that clouds composed purely of dust should be dark in blue and ultraviolet. In the following days, this cloud became bright in red light, expanded, moved to the southeast, and merged with the other storm clouds to the east (see Fig. 7, days 5–8).

of the polar caps suggest that they occasionally experience temporary expansions during the recession cycle. If these phenomena can be verified, they need explanation. Although the polar hoods and caps are not part of a massive irrigation system, they are major features of both the water and CO_2 cycles on Mars.

Telescopic observations continue to contribute to some areas of Mars research and also provide a historical record that can be useful in many other sectors. Spacecraft data have enabled us to more realistically interpret Earth-based observations, in many cases making them more useful than ever. These observations can often be used to supplement the spacecraft data, especially in cases where a longer time frame is needed or a wider angle of coverage is desirable. Ideas developed by researchers like E. Slipher, who concentrated on Mars for nearly 60 yr, seem archaic in many ways but also may offer valuable insights to unresolved problems. Taking advantage of their experience can be rewarding.

Acknowledgments. We are grateful to T. Akabane, W. Baum, S. Ebisawa, T. McCord, D. Parker, S. Saito and Y. Saito for helpful comments. We thank H. Horstman for a comprehensive editorial review. We would also like to express our gratitude to the many hundreds of observers who have contributed to our understanding of Mars. We have mentioned only a few of their names—and probably not always the most important. We apologize to the many others who have also forgone the comforts of a warm bed to pursue the secrets of the Red Planet.

3. SPACECRAFT EXPLORATION OF MARS

CONWAY W. SNYDER
Jet Propulsion Laboratory (retired)

and

VASSILI I. MOROZ
Space Research Institute, Moscow

In the past quarter century, spacecraft missions have replaced our earlier ideas about Mars, many of which were badly in error, with a great quantity of solid information, thus making this present volume possible. Both the USA and the USSR had hoped to begin this program of exploration in 1960, and between them they have made more than twenty attempts to send spacecraft to observe the planet. Most of the earliest Soviet attempts failed and were not announced, and some of the information about them is discussed here for the first time in English. Of the series of spacecraft that were announced, designated Mars 1 to Mars 7, none fulfilled all its scientific goals, but some good photographs and other important data were obtained. The earliest planned American missions never even got to the launch pad, and of the six spacecraft in the Mariner series, two failed, but Mariner 4 first revealed the cratered surface of Mars, and Mariner 9 discovered all the major geologic features, revolutionizing our understanding of the planet. The Viking mission, with its two Orbiters, two Landers, and its 6-yr duration, far surpassed in quantity and variety of data all other missions combined and provided most of the information on which this book is based. The very ambitious Phobos mission, which the Soviets had planned as the precursor of a new series, ended in two failures, but the second of the two spacecraft did acquire significant new information about Mars and Phobos. This chapter closes with an appendix listing special issues of journals containing collections of papers about Mars.

I. THE CLASSIC MARS

It is not easy to remember, now a quarter century after Mariner 4, how little we once knew about the planet Mars and how much of what we thought we knew was illusion. Most astronomers had much earlier ceased to credit Percival Lowell's canals and dying civilizations, but many other ideas that were considered to be firmly established were in fact quite wrong. A few quotations from the classic *Physics of the Planet Mars* by Gerard de Vaucouleurs (1954), which summarized the astronomical research up to the early 1950s, will emphasize the point. Observations, both visual and photographic, showed about three-quarters of the surface to consist of "bright regions," reddish or yellowish in color, that were assumed to be deserts. The numerous dark markings, often thought of as bluish or greenish (although actually red), were about half as bright and had shapes that were permanent in general but subject to changes in detail, some seasonal and some random. The variability was commonly attributed to some type of vegetation—definitely not chlorophyll plants but perhaps something similar to lichens. There were also the polar caps, which waxed and waned with the seasons in a very regular and logical manner and were naturally assumed to be water ice. Three types of clouds were also observed—yellow, blue and white. The first was correctly inferred to be dust, and the other two were assumed to be some condensate, probably water or possibly carbon dioxide.

No geographical features could be detected, and it seemed clear that "relief must be fairly moderate, since no permanent prominence has ever been observed at the terminator when they rise or set at the visible edge of the disc under oblique illumination, near the quadratures. If there are mountains on Mars they would not exceed 6 to 9 thousand feet at most (Lowell 1896). Almost the only suggestion of mountainous formations is afforded by a few small and brightish spots which have been observed from time to time in some fixed places (e.g., Nix Olympica discovered by Schiaparelli in 1879) and are suggestive of isolated peaks, high enough to induce condensation of frost or clouds" (de Vaucouleurs 1954,p.29).

Investigations with photometers, spectroscopes and polarimeters had led to other important conclusions. "The polar caps are without the slightest doubt layers of crystallized water—probably more like white frost than solid ice or snow—which condenses during the cold season" (ibid. p.28). As no water vapor could be detected in the atmosphere, it was concluded that this frost must be very thin. Dollfus (1951*a*) found the polarization curve of bright areas to match limonite (ferric oxide). Kuiper (1949) using spectrophotometry in the near infrared found a match with felsite, an igneous rock made up of a silicate of aluminum and potassium with quartz and other inclusions.

Over the previous five decades many investigations of the atmosphere had led to apparently well-founded conclusions, nearly all wrong, some of which were to have significant effects on the space program. Between 1926 and 1943 Adams and collaborators at the Mount Wilson Observatory searched repeatedly for oxygen and water vapor but found no trace of either (Adams and St. John 1926; Adams and Dunham 1934,1937; Adams 1941). In 1947 Kuiper (1949) first detected the infrared bands of carbon dioxide in the Martian atmosphere and concluded its pressure to be about 0.35 mbar, thus making it a minor constituent. De Vaucouleurs concluded, "These considerations point, by a process of elimination, to nitrogen and . . . argon as the main permanent constituents of the atmosphere of Mars, the latter being present only in small quantity. . . . In addition a small amount of carbon dioxide is definitely known to be present" (ibid., p.32). His proposed atmosphere (ibid., p.127) had in percentages nitrogen 98.5, oxygen < 0.1, argon 1.2?, and carbon dioxide 0.25.

Four Soviet photometric observations between 1940 and 1944 yielded values for the surface pressure of the atmosphere between 112 and 125 mbar. Three French observations with photometers and polarimeters between 1945 and 1951 yielded values between 80 and 95 mbar. De Vaucouleurs (ibid., pp.124–125) discussed these measurements and concluded with considerable assurance that the right figure "cannot be far from 85 ± 4 mbar." The reason that these "measurements" were so wide of the mark was that all these investigators, in interpreting the observed spectral features, assumed that carbon dioxide was a minor constituent and that the atmosphere was clear and scattered light in accordance with the Raleigh law. We now think that the atmosphere is probably never clear of dust.

II. EARLY SOVIET DISAPPOINTMENTS

Shortly after the space age dawned, with the launchings of Sputnik in October 1957 and of NASA in July 1958, the mission planners in both nations had ambitions to send two or more spacecraft towards Mars at every biennial opportunity. The USSR being about thirty months ahead of the USA in launch-vehicle development (according to a NASA estimate at about this time) was able to make their first attempt in 1960 (Table I).

The program of technical and scientific preparation for missions to Mars and Venus with Automatic Interplanetary Stations (AIS) began in the USSR immediately after the historic flight of the first Sputnik. The principal leaders were S. P. Korolev, responsible for spacecraft, and M. V. Keldish, who headed and coordinated science in space. These two were very enthusiastic about the possibilities of planetary exploration, but the general scientific community was not yet ready. Only a small group in Moscow University, led by A. I. Lebedisky, began developing instrumentation for the first planetary spacecraft. One idea that appeared promising at the time was to measure photometric profiles of Mars in a few spectral bands near 3.4 μm. Sinton (1957) had detected absorption bands on Mars in this region, which he attributed to the CH bond in organic molecules (Martian vegetation?). Consider-

TABLE I
Chronology of Mars Mission Attempts

	Name	Launched	Result
1960			
1	None	Oct 14	failed to reach Earth orbit
1962			
2	None	Oct 24	failed to leave Earth orbit
3	Mars 1	Nov 1	passed Mars 1963 June 19; telemetry failed
4	None	Nov 4	failed to leave Earth orbit
1964			
5	Mariner 3	Nov 5	shroud separation failure
6	Mariner 4	Nov 28	photographed Mars; measured atmosphere
7	Zond 2	Nov 30	contact lost after 5 months
1965			
8	Zond 3	Jul 18	photographed Moon; went to Mars's orbit
1969			
9	Mariner 6	Feb 24	flyby Jul 31; successful mission
10	Mariner 7	Mar 27	flyby Aug 5; successful mission
11	None	Mar 27	Proton booster failure
12	None	Apr 14	Proton booster failure
1971			
13	Mariner 8	May 8	Centaur booster failure
14	Kosmos 419	May 10	failed to leave Earth orbit
15	Mars 2	May 19	orbited Mars; DM[a] crashed
16	Mars 3	May 28	orbited Mars; DM[a] landed; transmitter failed
17	Mariner 9	May 30	orbited Mars Nov 13; long successful mission
1973			
18	Mars 4	Jul 21	failed to orbit Mars
19	Mars 5	Jul 25	orbited Mars; partially successful mission
20	Mars 6	Aug 5	flyby; DM[a] landed on Mars; very little data
21	Mars 7	Aug 9	flyby; DM[a] missed the planet
1975			
22	Viking 1	Aug 20	orbited 1976 Jun 19; landed Jul 20
23	Viking 2	Sep 5	orbited 1976 Aug 7; landed Sep 3
1988			
24	Phobos 1	July 7	lost telemetry August 30
25	Phobos 2	July 12	orbited Mars 1989 Jan 29; lost telemetry March 27

[a]DM = Descent Module, i.e., landing capsule.

able effort went into the creation of instruments for measuring these bands from Soviet spacecraft, but all the early Mars missions failed, and the idea was recognized later to be probably wrong. Rea et al. (1965) concluded that the bands were produced by deuterated water (HDO) in the Earth's atmosphere. No one now believes in the possibility of discovering life on Mars by remote sensing from fly-by or orbiting spacecraft. It was a tragic characteristic of planetary science in those early days (and often indeed even later) that scientists often labored long and hard for goals that were never achieved.

Several of the early attempts for a Mars mission by the USSR were failures, but there was not—and still is not (despite *glasnost*)—any published information on them. Our knowledge about them comes from personal communications from people who were involved and from surveillance by the US and the UK. Some of the information in this chapter has not previously appeared in the open literature in either language but is believed by the two authors to be accurate.

In 1962 the Administrator of NASA reported to Congress that there had been unsuccessful Soviet attempts to launch Mars spacecraft on 1960 Oct 10 and 14. However, Soviet scientists who were involved now say that only the latter was intended for Mars, and they have no knowledge of the former. The spacecraft was similar to Venera 1, launched in February of 1961, which weighed about 640 kg. Its 10-kg scientific payload included a magnetometer, ion traps for measuring solar plasma, cosmic-ray counters and micrometeoroid sensors. Contrary to American assumptions, it did not include a camera.

In the next launch period in 1962, Mars spacecraft were launched on Oct 24, Nov 1 and Nov 4. The first and third were stranded in Earth orbit, but the second became the first AIS to head for Mars, and it was thereupon named Mars 1. It weighed 894 kg, and its mission was described by Denisov and Alimov (1971) thus:

> "The Mars 1 interplanetary station was launched in our country on 1 November 1962; it was the world's first research vehicle sent to the planet Mars. Sixty-one radio contacts were made with this station and a great volume of scientific data was obtained, including data on the solar wind, magnetic fields and meteor streams. Radio communications were maintained to a distance of 106 million km, which for that time was a great achievement. Approach of the station to the planet Mars occurred on 19 June 1963. It passed the mysterious planet at a distance of 195,000 km from its surface, after which it went into a heliocentric orbit. Thus an interplanetary trail to Mars was blazed for the first time."

An irregularity in the orientation system that pointed the antenna toward Earth terminated communication with Mars 1 on 1963 Mar 21 (Trud 1963). Its primary objective was to photograph Mars in several colors. Scientific instrumentation included "magnetometers to detect the presence of a Martian magnetic field and magnetic fields in space, gas-discharge and scintillation counters to detect Martian radiation belts and study the spectrum of cosmic radiation, counters for the study of the nuclear component of primary cosmic radiation, . . . special instruments to record micrometeorite impacts, . . . a spectroreflectometer for detecting vegetation, and a spectrograph for study of the ozone absorption bands" (Priroda 1963). Of course, no data relative to Mars were acquired.

In 1964 two space vehicles were launched in the direction of Mars—Mariner 4 on Nov 28 and Zond 2 on Nov 30. The latter shared the fate of Mars 1. Although it was called a probe, apparently to de-emphasize the fact that it was intended to make measurements at Mars, it carried the same complement of instruments as on Mars 1 with the addition of a set of electrojet plasma engines. Pravda announced that its power output was only half the anticipated level, probably because of failure of one of the two solar panels, and communication was lost in early May at 150 $\times$ 10^6 km from Earth. It encountered the planet on 1965 Aug 6, missing the surface by only 1500 km but making no observations. Its only significant success was the testing of the electrojets under real spaceflight conditions (Johnson 1979).

A second identical spacecraft had been prepared, but it proved impossible to get it launched before the end of the Mars launch period. Consequently, on 1965 Jul 18, Zond 3 was sent on a trajectory to approach the orbit of Mars even though Mars was not in the vicinity at the time. The launch was timed so that Zond 3 could use its Mars camera to photograph the back side of the Moon. It succeeded in sending back pictures of it (Johnson 1979). The camera system recorded the images on film, which was then chemically processed, and the individual frames were scanned by a television camera and read out to telemetry.

III. EARLY AMERICAN DISAPPOINTMENTS

In its early days the American Mars program was almost as disappointing as the Soviets but in a different way and for a different reason. In December 1958, The Jet Propulsion Laboratory (JPL) of the California Institute of Technology was transferred from the sponsorship of the U.S. Army to the newly authorized National Aeronautics and Space Administration, and JPL recommended to its new chiefs a schedule of lunar and planetary missions, including 18 flights in five years, starting with a circumlunar probe and two Mars missions in 1960 and culminating in a manned circumlunar flight in 1964 and a manned flight around Mars in 1965. A scaled-down version of this ambitious plan was approved by NASA the following March and designated Project Vega. This project was canceled in December 1959 as it became clear that launch vehicles were not going to be available to carry it out (Koppes 1982).

In that same month, JPL was designated by NASA as its lead center for lunar and planetary missions. In the next five years, one after another, planetary projects were authorized by NASA and subsequently canceled—Mariner A (1960 Venus flyby), Mariner B (1961–63 Venus and Mars landers), Mariners E and F (1966 Mars flyby with small atmospheric probe), and an "Advanced Mariner" (1969 Mars orbiter and lander) (Ezell and Ezell 1984). All these spacecraft were to have been launched by the Atlas Centaur, but as the Centaur encountered a long series of development problems, the planetary program was reduced to a total of three missions based on the less capable Atlas Agena.

When Mariner A was canceled in August 1961, JPL hastily assembled a scaled-down substitute called Mariner R. The rocket carrying the first of these spacecraft (already named Mariner 1) experienced guidance problems and had to be destroyed, but Mariner 2 left Earth on 1962 Aug 27 and flew past Venus in December. Although less than one-fourth the weight of its contemporary, Mars 1, and much less complex, this spacecraft fulfilled all of its scientific and technical objectives and ushered in the era of planetary exploration.

The same scenario was repeated the following year as Mariner B, on which JPL had worked for nearly three years, was replaced by Atlas-Agena-launched Mariners C and D. The shroud enclosing the Mariner 3(C) spacecraft failed to deploy, resulting in a frantic three-week effort to understand and fix the problem before the end of the launch period. Mariner 4 headed for Mars on 1964 Nov 28. Weighing only 260 kg, it carried only one instrument specifically for planetary studies, the camera. The radio telemetry link provided a second planetary experiment by probing the atmosphere during Earth occultation.

During the 1963 Mars opposition, several Earth-based observations with improved instrumentation had shown that the conventional 85-mbar figure for the atmospheric pressure was much too high. The investigators included Moroz (1964), Kuiper (1964) and Kaplan et al. (1964). The latter made the first positive detection of water vapor and measured the partial pressure of carbon dioxide as 4 mbar, but still assuming that significant quantities of nitrogen and argon were present, concluded that the most probable value of the surface pressure was 25 ± 15 mbar.

IV. MARINER 4

Mariner 4 was launched on 1964 Nov 28, flew by Mars on 1965 Jul 14 at an altitude of 9850 km, and continued to operate in solar orbit and communicate with Earth until 1967 Dec 20, a total of 1118 days. The experiments carried are listed in Table II. Well before Mars encounter, the plasma probe had suffered some degradation and the Geiger counter had failed, but the other interplanetary experiments operated properly. The initial reports of

TABLE II
Scientific Experiments on Mariner 4

Experiment	Principal Investigator	Objective
TV subsystem	R. B. Leighton	photograph Martian surface
Helium magnetometer	E. J. Smith	planetary and interplanetary magnetic fields
Solar-plasma probe	H. L. Bridge	density and energy of positive-ion solar wind
Ionization chamber and geiger counter	H. V. Neher	corpuscular-radiation dose rate
Trapped-radiation detector	J. A. Van Allen	Mars radiation belts and charged particles in space
Cosmic-ray telescope	J. A. Simpson	high-energy charged particles
Cosmic-dust detector	W. M. Alexander	micrometeorite number and momentum
Radio-occultation experiment	A. J. Kliore	atmospheric pressure vs altitude

Mariner 4 scientific results were published together, in what is now a tradition for planetary missions. There have been many issues of journals that contain collections of scientific articles about specific missions at various stages. Those about missions to Mars, or containing other collections of articles about Mars, are listed in the appendix to this chapter.

The Mariner 4 camera focused its images onto a 5.5-mm square on the surface of a vidicon tube, whose output was recorded digitally on a 4-track tape recorder to be read out later. It commenced operation at a range of 16,900 km, taking a picture every 48 s until its tape recorder was full at 11,900 km range. The first of these is shown in Color Plate 2 and Fig 1. The last three of the 22 pictures were partly or wholly beyond the terminator; the first 19 made a discontinuous swath from latitude 40° S, longitude 96° W to 40° N, 187° W, centered near the Memnonia region. They covered about 1% of the surface. With each picture consisting of 200 lines of 200 pixels each and the minimum slant range being about 12,000 km, the minimum pixel spacing was slightly more than 1 km on the surface. With a telemetry rate of 8.33 bits s^{-1}, it took 4 days to transmit them to JPL.

The results were usually summarized in one word—moonlike. The public (and most astronomers except Tombaugh [1950]) had not expected Mars to be covered with impact craters, but these were just about the only features observed. Seventy craters were visible, ranging in diameter from 5 to 120 km. The magnetometer indicated that the spacecraft had been inside the solar-wind bow shock for about 2 hr, and the degraded plasma instrument

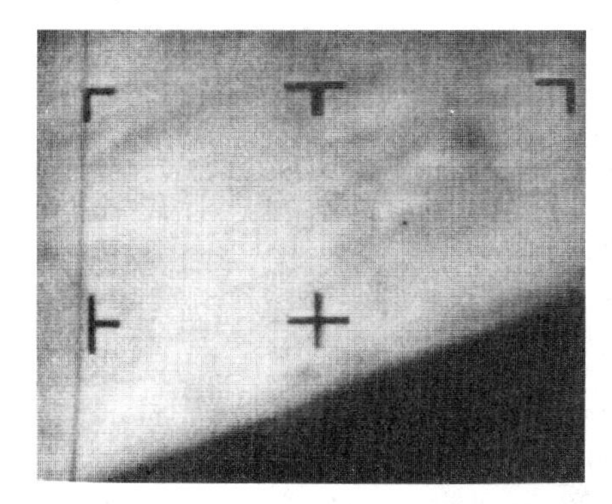

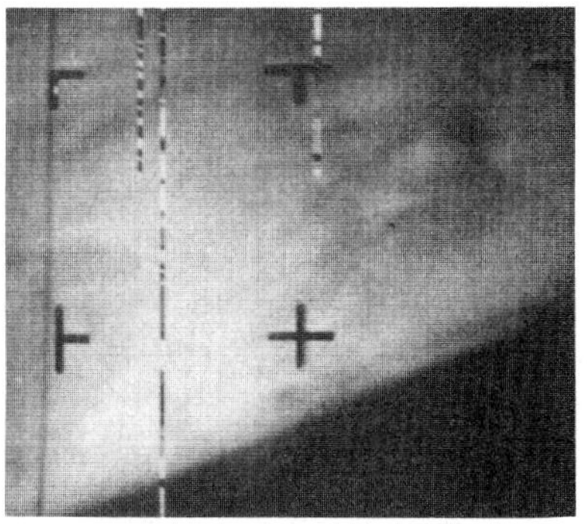

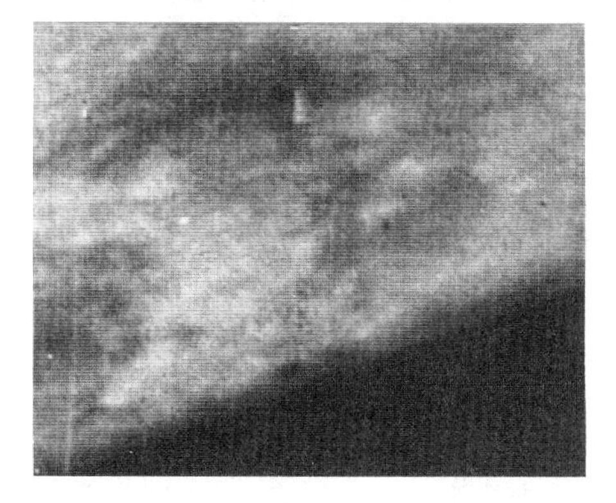

Fig. 1. Mariner 4 Picture Number 1 as it first emerged in raw form from the JPL photography laboratory. Taken from a slant range of 16,900 km, it is centered on 35° N latitude, 172° E longitude, showing the region of Trivium Charontis and Propontus II Phlegra on the limb of the planet. Subsequent processing to enhance the contrast and remove the fiducial marks considerably improved the quality. North is at the bottom. (Mariner 4 M-V01)

appeared to confirm this, but the data did not clearly demonstrate whether or not Mars had an intrinsic magnetic field.

The most significant scientific finding came from an experiment that had not been included in the mission as originally planned. Beginning in 1960, Von R. Eshleman and his colleagues at the Radioscience Laboratory, Stanford University, in numerous meeting papers and discussions, letters, proposals, and memoranda had been advocating the technique of "bistatic radar astronomy," which would utilize the propagation between two radio terminals, one on Earth and one on a spacecraft near a planet, to study several aspects of the target planet, including its atmosphere and ionosphere (Eshleman 1976). In early 1964 the Deep Space Network was in its infancy and just beginning to develop the sophisticated capabilities and precision that would later characterize it. Thus it was not clear whether an experiment that proposed to measure the propagation effects on the amplitude of the radio signal could yield significant data on the Martian atmosphere with the facilities that were available at the time. It was not proposed to use the precise measurements of the signal phase, which the DSN was already capable of doing, and for this and other reasons, Eshleman had not been able to get his proposal accepted by NASA.

As the launch period for Mariners 3 and 4 approached, D. L. Cain, an engineer in the Systems Analysis Section at JPL, had the job of generating the computer programs to correct the spacecraft navigation data for the effect of the Earth's atmosphere. Thus he was very familiar with the extreme accuracy of the newly developed radio Doppler system (the "microlock" phase-locked loop) that could detect shifts of a few centimeters in an orbit. It occurred to him that the effect of the Martian atmosphere on the radio signal as the spacecraft passed behind the planet might be easily detectable. He brought the idea to some of his associates, who immediately began to study it intensively and to urge the Project Manager to do the experiment.

It was no easy matter to persuade the Project Manager voluntarily to relinquish communication and control links to the spacecraft during the latter part of the critical Mars encounter phase, before the primary mission data (the TV pictures) had been played back and received (Schurmeier 1975). He eventually became convinced well after the launch of Mariner 4 but before the time when its trajectory could be changed to take it behind the planet. Thus it came about that one more experimenter team was organized for the Radio Science Experiment, including in its membership both JPL and Stanford people. Kliore, who had done most of the theoretical analysis for the investigation, was named the Principal Investigator.

The other surprising and perhaps most significant result of the mission came from the radio-occultation experiment. Mariner was hidden behind the planet for 54 min, and the variations in the frequency, phase and amplitude of the radio signal as it traversed the atmosphere in ingress and egress permitted calculations of the density, temperature and scale height as functions of altitude. From these data, the pressure at the surface was calculated to be between 4.1 and 7.0 mbar, depending upon the proportion of argon assumed to be mixed with the carbon dioxide (Kliore et al. 1965). The Martian atmosphere was even sparser than anyone had expected. This now incontrovertible fact invalidated all earlier designs for Mars landing probes and sent the designers back to their drawing boards. Achieving a Mars landing in the 1967 launch period was now seen clearly to be impossible.

In fact, in January 1967 the USA had no mission at all ready, being preoccupied with the Mariner 5 mission to Venus and preparations for Apollo. Neither did the USSR, also occupied with missions to Moon and Venus. Venera 4 in 1967 made the first *in situ* sounding of the atmosphere of Venus, and it was the real start of Soviet planetary space science.

V. SOVIET POLICY AND NOMENCLATURE

The policy of Soviet spacecraft designers has been to produce a small number of designs, use them for a variety of purposes, and gradually improve them. Thus at any given period, spacecraft that were very similar (or even identical) might be used for missions to Mars, Venus or interplanetary space and designated Mars, Venera or Zond. In 1960 and 1961, these craft weighed about 640 kg; in 1962, about 890 kg; from 1963 to 1965, 950 to 965 kg; from 1967 to 1970, 1100 to 1180 kg; from 1971, about 4650 kg; and beginning in 1988, 6200 kg (see Table III).

The policy for launch vehicle systems was similar. The early sputniks, beginning in 1957 were launched by a 1½-stage rocket. Two generations of upper stages were designed to be stacked on it for launching larger sputniks, early Luna spacecraft and Vostok manned satellites. An additional escape stage was added for Mars and Venera missions. No official Soviet name was applied to this launch system; its American designation was A-2-e. It was used for all Mars and Venus missions prior to 1969 and was still in frequent use into the 1980s, usually for Molniya communications satellites or military satellites (Sherman 1982). In 1965, the Soviets began testing a much more powerful rocket, which became known as the Proton, and by 1968 it had been fitted with an Earth-escape stage and was available for launching lunar and planetary missions (Johnson 1979).

Not unexpectedly, Russian nomenclature relating to space differs from the American. What in America is called a planetary spacecraft, the Soviets usually call an "automatic interplanetary station," (usually abbreviated AIS in English) or an "automatic Mars station" (AMS). An orbiter is usually called an "orbiting station," an "orbital compartment," or an "orbital module (OM)," and a lander is a "descent module (DM)" or a "descent apparatus."

TABLE III
The Growth of Spacecraft Weights

Spacecraft	Launched	Weight (kg)
(Unnamed)	1960	640
Mars 1	1962	895
Mariner 4	1964	260
Zond 2	1964	950
Zond 3	1965	950
Mariners 6, 7	1969	413
Mars 2, 3	1971	4650
Mariner 9	1971	980
Mars 4, 5, 6, 7	1973	4650
Viking 1, 2	1975	3527
Phobos 1, 2	1988	6200

VI. SUCCESSES AND FAILURES IN 1969

The first attempt to use the Proton for Mars missions was in 1969. Two spacecraft were planned to be inserted into Mars orbit, but no landers were prepared. The new Institute for Space Research (IKI) of the Academy of Science, which had been founded in 1967, was responsible for the scientific payload, including most of the instruments listed in Table V in Sec. VII. Although the Proton had been used in 1968 for three successful lunar missions (Zonds 4, 5 and 6), the Soviet Mars jinx was still at work, and both launches (March 27 and April 14) were unsuccessful and were never announced officially.

The USA fared better in 1969 and for the first time managed to get both of the pair of Mars spacecraft to the planet. Mariners 6 and 7 were launched on February 24 and March 27 and flew by Mars on July 31 and August 5, respectively. The Centaur, now being operational, was able to carry the 413-kg spacecraft, nearly twice as heavy as the earlier Mariners. This time all the instruments were intended for observing the planet; they are listed in Table IV.

The two vidicon cameras were essentially identical with that on Mariner 4, but with the availability of the data transmission rate nearly 2000 times higher than in 1964, some significant improvements were possible. The images on the vidicon photosurfaces were much larger (9.6 by 12.3 mm), and the number of pixels in each frame was 16 times greater (704 by 935). The cameras had different lens systems, focal lengths 50 and 500 mm, and there were two sessions of photography for each spacecraft, producing 200 times the picture data of Mariner 4. The far-encounter sequences, using only the long-focal-length, "narrow-angle" cameras, provided 117 photographs of the entire disk with pixel spacing between 11 and 43 km and 26 more pictures of

TABLE IV
Scientific Experiments on Mariners 6 and 7

Experiment	Principal Investigator	Objective
Television experiment	R. B. Leighton	photography both at long and short range
Ultraviolet spectrometer	C. A. Barth	composition of the upper atmosphere
Infrared spectrometer	G. C. Pimentel	composition of the lower atmosphere
Infrared radiometer	G. Neugebauer	surface temperature
S-band radio occultation	A. J. Kliore	atmospheric pressure and density
Celestial mechanics	J. D. Anderson	gravitational field

large parts of the disk with pixel spacings down to 4 km as the two spacecraft approached. The near-encounter sequences covered about 10% of the surface in 30 frames with pixel spacings between 1 and 2 km, within which were nested 29 frames with 10 times smaller pixels. The close-up pictures of Mariner 6 (Fig. 2) extended roughly from latitude 0°, longitude 60° W to 10° S, 320° W, covering the areas of Aurorae Sinus, Pyrrhae Regio and Deucalionis Regio. Mariner 7 took a short swath centered on the edge of the polar cap at 60° S, 0° and a long swath from 10° N, 20° W to 30° S, 105° W, including Thymiamata, Deucalionis Regio, Hellespontus, Hellas and Mare Hadriacum. This time the coverage was extensive enough to deliver a *coup de grace* to the canals.

A major result of the pictures was the identification of three distinct types of terrain—cratered, chaotic and featureless, the latter being observed only in the large circular basin named Hellas. The geographical features that were observed showed no correlation with the traditional bright and dark areas. A memorable moment in the mission occurred as R. Leighton, the TV science team leader, was describing the successive pictures on public television in real time as they appeared on his monitor in the Space Flight Operations Facility. He had been building suspense by predicting that the upcoming picture of Hellas would be spectacular and was astonished and disappointed to see that its interior looked blank. (Of course, we now know that the surface was concealed by dust.)

The far-encounter pictures (see Fig. 3) enabled the production of a much improved map and a truly topographically based control net as distinguished from a net based only on the light and dark features. Nevertheless, they were still not very clear, as is indicated by the erroneous conclusions that Nix Olympica, a feature well known from telescopic observations to be frequently cloud covered, was a large circular impact crater (Murray et al. 1971).

The early press conferences where the preliminary results were announced engendered considerable excitement and controversy. The Principal Investigator for the Infrared Spectrometer announced the detection of methane and ammonia near the edge of the polar cap and stated "The qualitative presence of methane and ammonia in the atmosphere gives no direct clue whatsoever concerning its origin. Nevertheless, one cannot restrain the speculation that it might be of biological origin." (Subsequent analysis showed that neither gas had been detected, and that the observed features were due to solid carbon dioxide.) He also claimed that the polar cap temperature was 203 K, indicating that it must be water ice. However, the Infrared Radiometer Principal Investigator contended that it was 150 K, almost exactly what would be expected for carbon-dioxide ice. The radiometer also showed that the dark areas of the surface were warmer than the bright areas, and measurements on the dark side of the terminator showed that the surface material had a very low thermal inertia.

The Ultraviolet Spectrometer saw no trace of nitrogen in the atmosphere,

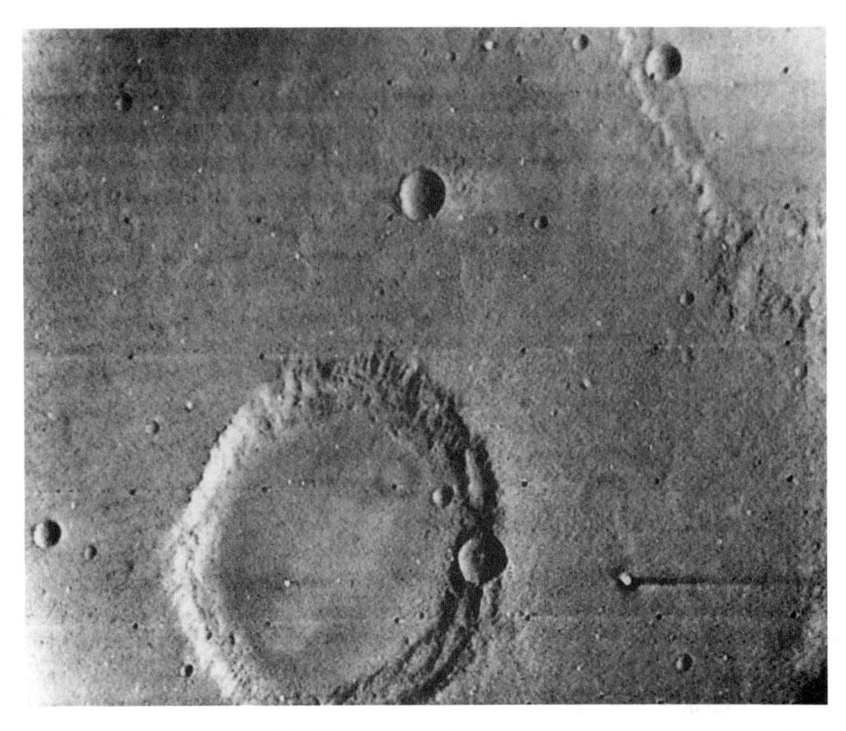

Fig. 2. Martian craters of widely different ages. The prominent crater near the top of the image has been modified by later impacts. The lower left corner of the crater shows the edge of a much larger, nearly obliterated ghost crater (15° S, 5° E). (Mariner 6 6N18)

a conclusion that prompted discussion about its implications for Martian biology. The ultraviolet instruments on both spacecraft detected carbon monoxide. The Radio Occultation Experiment confirmed the Mariner 4 findings on the atmospheric pressure, detected the presence of an ionosphere on the sunlit side but none on the dark side, and made measurements of the shape of the planet. The Celestial Mechanics Experiment provided a measurement of the planet's mass with unprecedented precision.

VII. THREE SPACECRAFT AT MARS IN 1971

The Mars launch period in 1971 began with two aborted missions. On May 8 the Centaur booster failed to put the Mariner 8 spacecraft into orbit, and two days later the Proton booster failed to eject from Earth orbit its payload, identified only as Kosmos 419. It has always been assumed in the US that this orbiter was a failed Mars probe, and a recent book (Markov 1989) has confirmed it. It was, in fact, intended to become the first artificial satellite of Mars. Unlike its companions, Mars 2 and 3, which were combined

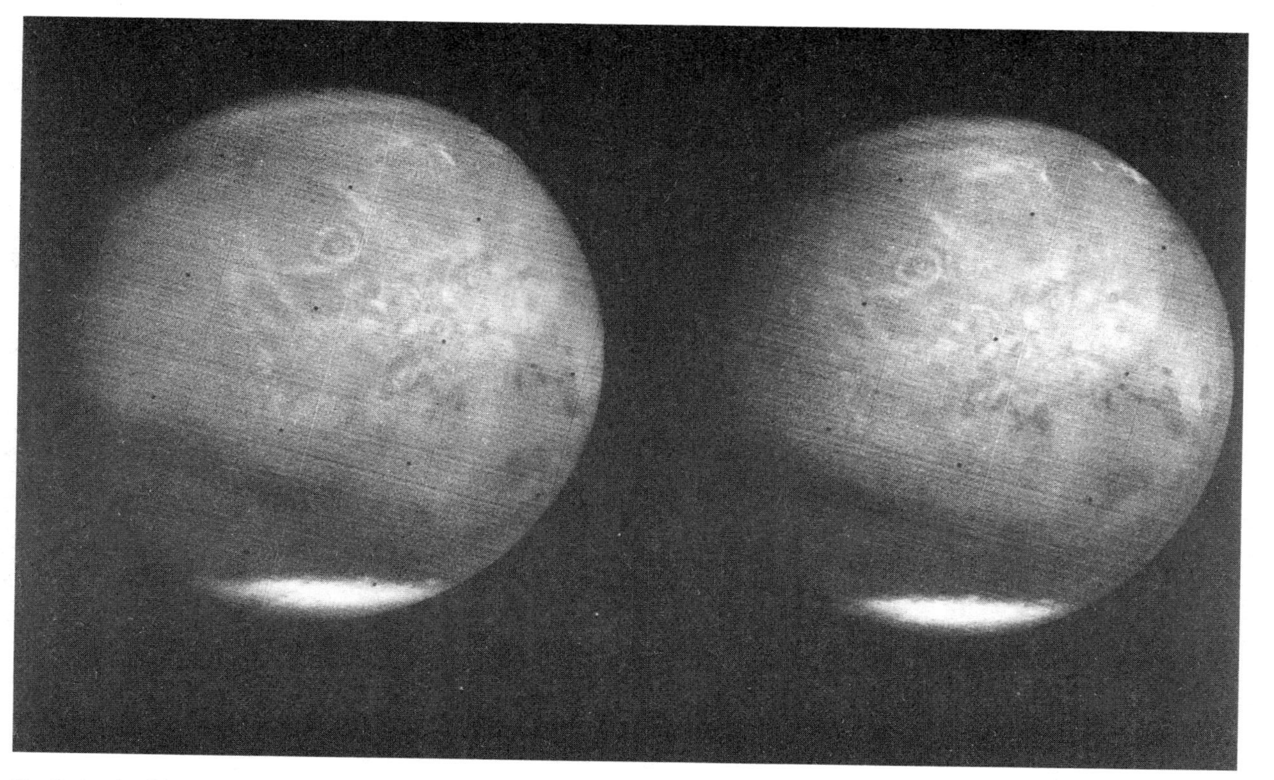

Fig. 3. A pair of far-encounter photographs from Mariner 7 taken at an interval of 47 minutes, during which time Mars rotated through an angle of 12°. They are presented here in a form suitable for viewing through a stereoscope. Note particularly the ring-shaped Nix Olympica, still not identifiable as a mountain. (Mariner 7 7F73 and 7F74)

orbiter/landers, Cosmos 419 consisted of an orbiter only. The resulting weight reduction would have allowed it to fly a faster trajectory ahead of the two Mariners, clinching another major first for the Soviet space program. The failure was attributed to "a most gross and unforgivable mistake" in a command sent to the on-board computer.

Mars 2 and Mars 3 were launched on May 19 and 28, respectively, and Mariner 9 (identical to Mariner 8) was launched on May 30. Mariner 9 went into orbit on November 14 and was followed a couple of weeks later by its two companions, setting the stage for the first international competition (and cooperation) in planetary exploration. Mariner weighed 998 kg, while each of the Soviet craft consisted of a carrier vehicle of 3440 kg with a 1210-kg descent module, enclosing a 350-kg landing capsule. A telephone "hot line" was set up between the Jet Propulsion Laboratory and the Institute for Space Research, over which results of the missions could be exchanged for mutual benefit.

Unfortunately, on September 22 a cloud appeared over the Noachis region and soon developed into the densest and most widespread dust storm ever recorded. As the picture taking sequences of the Mars spacecraft had been preprogramed, each proceeded to photograph the featureless planet until the sequence was completed, thus obtaining practically no information. Only one single picture was ever published, taken by Mars 3 at great distance.

The AIS Mars 2 went into Mars orbit on November 27, 4.5 hr after having released its descent module, which failed to function properly and crashed on the surface. Tass characterized the performance as a success, bragging that the DM "for the first time delivered to the planet Mars a capsule containing a pennant depicting the coat of arms of the Soviet Union" (Tass 1972).

Mars 3 went into orbit on December 2, having released its descent module 4.5 hr earlier. The DM operated flawlessly for a while, deploying its auxiliary parachute followed by the main parachute, then jettisoning its heat shield, detaching the parachute, and igniting the final braking rocket. The radar altimeter encountered some difficulties, but the DM landed (perhaps in a nonstandard fashion), the television camera was turned on, and the radio began transmitting to the orbital station a test image in preparation of photographing the surroundings. After 20 s, with only a portion of the sequence completed, the signal disappeared, so that no data from the other instruments were transmitted. The landing capsule was intended to make measurements of the temperature, pressure, wind velocity and atmospheric composition, measure the chemical and physical properties of the surface material, and look for organic compounds that might indicate current or former living organisms.

The orbital parameters of Mars 2 were: periapsis 1380 km, apapsis 24,900 km, period 18 hr. For Mars 3, they were 1530 km, 190,000 km, 12.7 days. This very unsatisfactory orbit resulted from a partial loss of fuel in the propulsion system and gave the spacecraft only seven opportunities for close-up observations, although both stations continued operating for about four months. Although the orbit of Mars 2 was very appropriate, its telemetry was of very low quality, and, except for the radio-occultation data, almost all the planetary data were lost.

The scientific payloads of Mars 2 and Mars 3 were identical (see Table V). The optical instruments in infrared and visual wavelengths on Mars 3 yielded good data. The diurnal variation of surface temperature indicated its low heat conductivity, the thermal inertia being characteristic of that for dry dust at a low pressure. Dark regions were 10° to 15° warmer than bright regions (confirming Mariner 9). At the north polar cap, temperatures were close to the carbon-dioxide condensation temperature, and condensation clouds were observed with particles of sub-μm dimensions. The altitudes measured from the carbon-dioxide bands were usually in satisfactory agreement with the results of Earth-based radar observations, and there was some

TABLE V
Scientific Instruments on Mars 2 and Mars 3

Experiment	Principal Investigators	Objective
Phototelevision units (PTU) (52 and 350 mm)	A. S. Selivanov	photograph Martian surface
Infrared radiometer (8-40 μm)	V. I. Moroz L. V. Ksanfomality	surface temperature
Radiotelescope and radiometer (3.4 cm)	N. N. Krupenio	surface temperature and dielectric constant
Photometer for 2-μm CO_2 bands	V. I. Moroz L. V. Ksanfomality	altitude profiles of surface
Photometer (0.35–0.7 μm)	V. I. Moroz L. V. Ksanfomality	albedo and color profiles
Photometer for water-vapor band (1.38μm)	V. I. Moroz A. E. Nadzhip	water-vapor content in atmosphere
Lyman-α sensor	V. G. Kurt	hydrogen in upper atmosphere
Magnetometer	Sh. Sh. Dolginov	magnetic field
Ion traps	K. I. Gringauz	solar wind and its interaction with planet
Narrow-angle electrostatic plasma sensor	O. A. Vaisberg	solar wind and its interaction with planet
Radio-occultation experiment	M. A. Kolosov	atmospheric density and ionosphere profiles
"Stereo"	(Y. L. Stainberg) (Mars 3 only)	solar bursts at meter wavelengths

correlation between altitudes and albedos, the dark regions tending to be higher. The altitude of the clouds during the dust storm was about 10 km, but the upper boundary was indefinite. During the dust storm, the water-vapor content of the atmosphere appeared to be very small (only a few precipitable μm) but in March it increased to 20 μm. The humidity was higher near the equator than in the northern circumpolar region. The infrared photometric profiles showed higher contrast at longer wavelengths during the storm, indicative of particles of approximately μm dimensions, and bright ultraviolet clouds were observed, evidently associated with still smaller particles. The total quantity of dust was estimated at $\sim 10^9$ tons (Moroz et al. 1973).

The ultraviolet photometers measured radiation scattered from the atmosphere, detecting the Lyman α line of hydrogen and the atomic oxygen triplet. Neutral hydrogen concentration was found to be much less than Mariner 6 and 7 detected, probably because it comes from water dissociation and escapes in a few days and there was less water in the atmosphere at this time (Dementyeva et al. 1972).

The measurements of the magnetic field and the plasma near the planet (Dolginov et al. 1973; Gringauz 1973) were interpreted to indicate a bow shock produced by an intrinsic planetary field with a magnetic moment ~ 4000 times smaller than Earth's. The measurements were much superior to those of Mariner 4, but still left some room for controversy (Ness and Bauer 1974).

The radio-occultation experiment on Mars 2 measured the ionosphere below approximately 330 km altitude; near 210 km a bend in the electron concentration profile separated the ionosphere into two regimes, the scale height below being 36 km and that above being 57 km. The neutral upper atmosphere had a temperature of 250 to 325 K and was composed mostly of carbon dioxide with about 2% atomic oxygen (Ivanov-Kolodny et al. 1973).

The radio-frequency polarimeter made measurements of the dielectric constant and temperature of subsurface layers (Krupenio et al. 1974).

Four final report papers collected all the results of processing and interpretation of Mars 3 data on surface temperature, thermal inertia and soil grain sizes (Moroz et al. 1975), on altitudes determined from the intensity of the 2-μm CO band (Ksanfomality et al. 1975), on water vapor (Moroz and Nadzhip 1975*c*), and on albedo variations and the optical properties of the dust-storm clouds (Moroz et al. 1975*b*).

VIII. MARINER 9 AND THE NEW MARS

Over the months before the launching of Mariner 8, the Television Science Team had hammered out detailed scenarios for the missions of the two Orbiters so that their observations would complement each other. The first spacecraft was to be placed into a high-inclination 12-hr orbit with periapsis synchronized over the Goldstone tracking station to conduct what was called the "fixed feature investigation." Its principal objective was contiguous coverage of about 70% of the surface to define the geology and topography. The second spacecraft, in a lower-inclination orbit with period 32.8 hr, was to carry out the "variable features investigation," a detailed study of variations of the surface. When they found themselves with only one spacecraft, the scientists had hurriedly to generate a new plan, and when they were confronted with a dust shrouded planet, yet another one (Fig. 4). It now appeared unlikely that many of their goals could be achieved. However, within a few weeks after orbit insertion, the dust had begun gradually to dissipate, and by 1972 March the atmosphere was quite clear (Fig. 5). As the orbiter continued to operate until its attitude-control gas was exhausted, Mariner 9 had 349 Earth days of operation and achieved all of its objectives. Its scientific payload is listed in Table VI.

On this and subsequent Mars missions, the quantity of data acquired was so vast that the findings can be discussed only in very sketchy outline in this chapter. Mariner 9 alone returned 7329 photographs and a total of 54×10^9 data bits, 27 times as many as the three preceding Mariner flybys combined. The startling revolution in our understanding of Mars that resulted from the photographs is well illustrated by a listing of some of the new things discovered (Figs 5, 6, and 7).

The great volcano at Nix Olympica, renamed Olympus Mons. . . . The immense Tharsis Bulge with its three more large volcanoes. . . . The canyon system, Valles Marineris, stretching about one-quarter of the way around the planet. . . . Fractures, faults, escarpments, graben. . . . The various configurations of craters. . . . The dichotomy of the surface, with about half consisting of ancient cratered terrain and half covered by younger features—plains, volcanic rocks and volcanoes. . . . The strange layered deposits at the poles. . . . The plethora of channels of five distinct types, many of them appearing to be ancient river beds. . . . The numerous evidences of aeolian erosion and deposition, including the dark and light tails downwind of topographic obstacles, which changed with the seasons. . . . Meteorological phenomena including local dust storms, weather fronts, the afternoon brightenings in the Tharsis region during summer, caused by ice clouds, morning ground fogs in the valleys, wave clouds over craters, and lee waves downwind of them. . . . The startling height of the top of the dust storm, 70 km. . . . The shapes and sizes of Phobos and Deimos and the contrast between their surface textures. . . . All these discoveries were made by the Television Experiment alone.

In addition, the photographic coverage provided a control net of 1645 points at 10-km accuracy covering about 80% of the surface and the data to produce maps of the entire surface at scales of 25 and 5 million and regional mosaics and maps at scales of 1 and 1/4 million.

Each of the other experiments also contributed significantly to the revelation of what Hartmann and Raper (1974) called "The New Mars." The

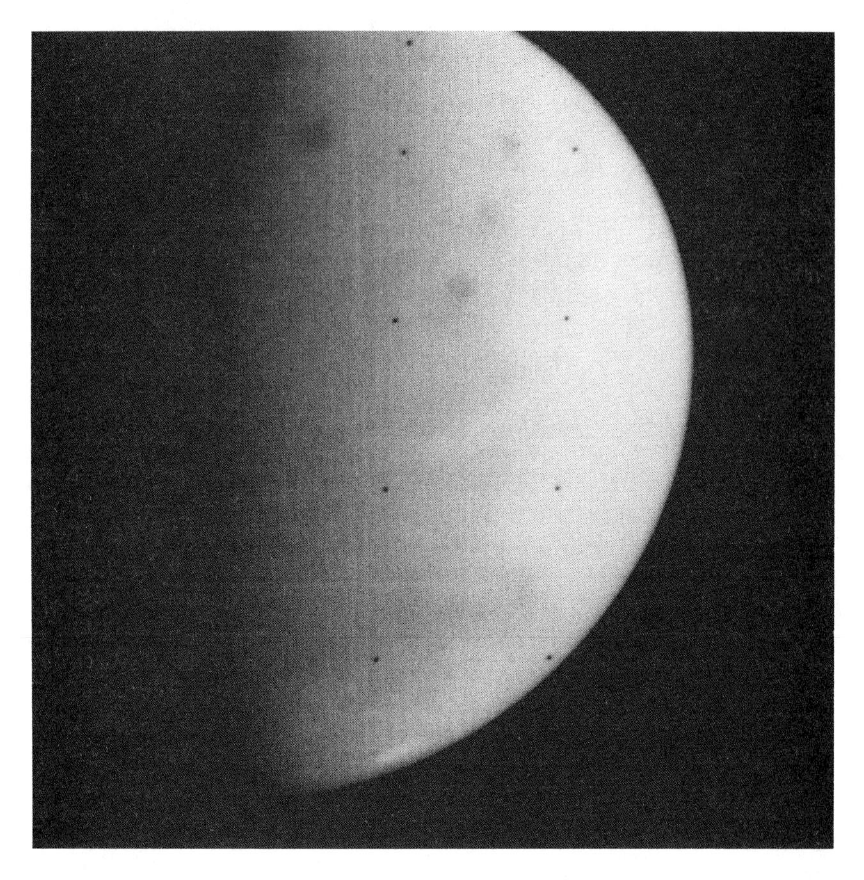

Fig. 4. Two days before orbit insertion, Mariner 9 saw a gibbous Mars completely blank except for three faint dark spots in a row and a larger spot to the left, which were soon identified as mountains projecting up into the dust cloud: Arsia, Pavonis, Ascraeus, and Olympus Montes.

Celestial Mechanics Experiment, by analyzing the metric information in the tracking data, determined a more accurate description of the gravity field of the planet, improved the ephemeris of Mars by an order of magnitude, and, by correlations with Earth-based radar data, obtained topographic information with an accuracy of 100 m.

The *S*-Band Occultation Experiment, with more than 300 occultations to analyze, observed the greatly reduced temperature gradient in the atmosphere produced by the dust storm. It measured the lower atmosphere temperature near both the poles to be close to the sublimation temperature of carbon dioxide in the winter and that near the north pole to be much warmer in the spring; it measured a large number of surface pressures, varying between 2.8 and 10.3 mbar, and it measured the peak electron density in the daytime ionosphere to be 1.5 to 1.7×10^5 cm^{-3} at altitude of 134 to 140 km.

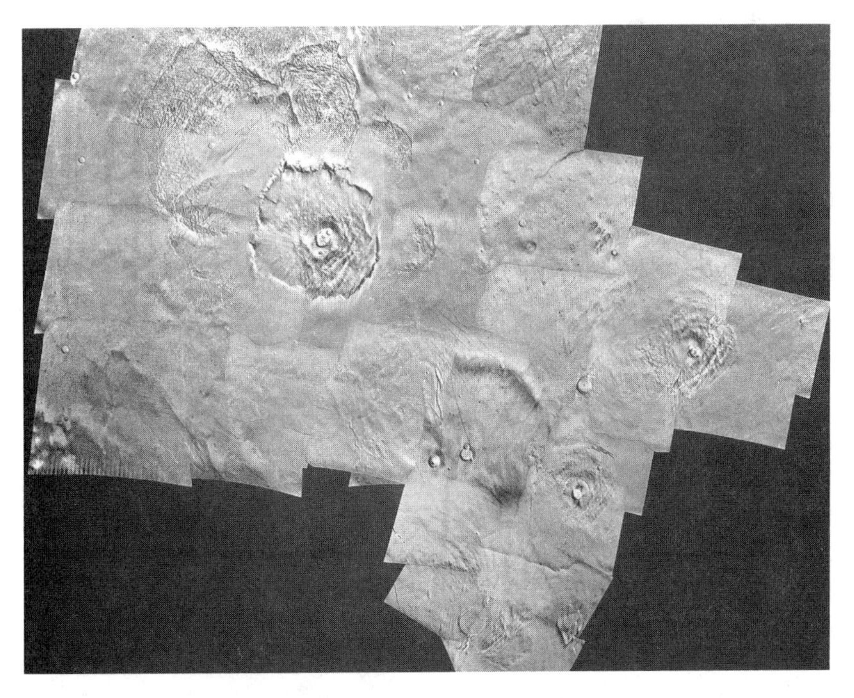

Fig. 5. The four large volcanoes of Tharsis complex, photographed after the dust had cleared.

TABLE VI
Scientific Experiments on Mariner 9

Experiment	Science Team Leader	Objective
Television	H. Masursky	photographic coverage of the entire planet surface
Ultraviolet spectroscopy	C. Barth	detect molecular species in atmosphere and surface relief
Infrared spectroscopy	R. Hanel	temperature, pressure and constituents of atmosphere, and types of surface material
Infrared radiometry	G. Neugebauer	brightness temperature of the surface
Celestial mechanics	J. Lorell	measure gravity field, improve ephemeris and obtain topographic data
S-band Occultation	A. Kliore	profiles of temperature and pressure in atmosphere and electrons in ionosphere

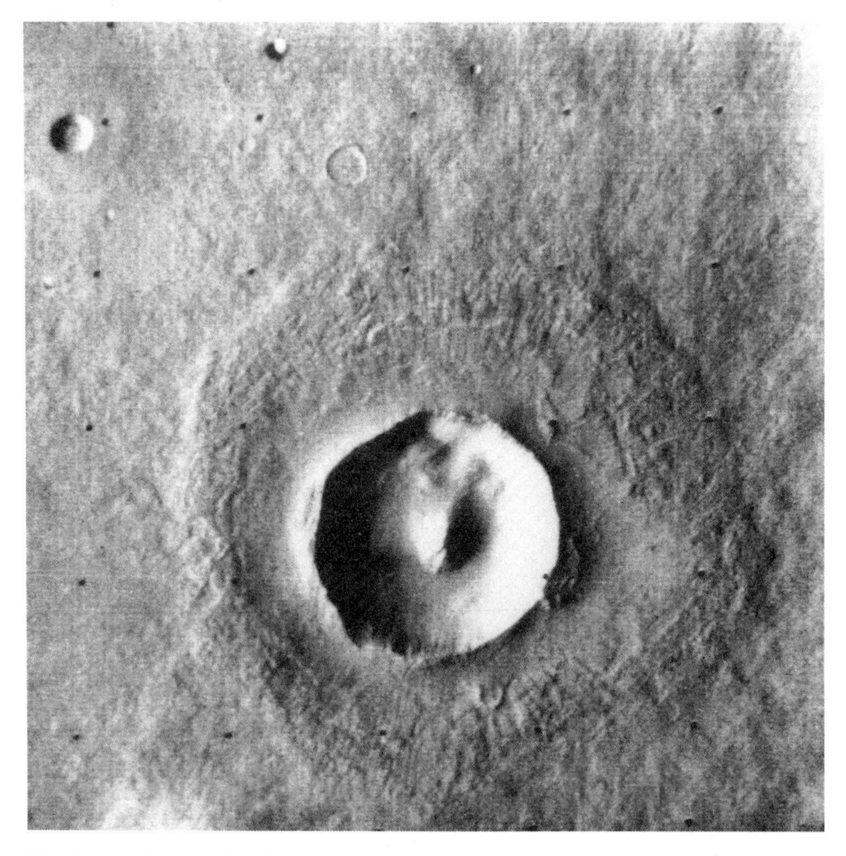

Fig. 6. A perfect example of a crater with a well-formed central peak and radially striated ejecta blanket, near 38° N, 260° W. This photograph demonstrates the remarkable detail provided by the Mariner 9 long-focal-length camera, which was the basis for the tremendous leap in understanding of the geology of the Martian surface. (Mariner 9 1433–210033)

The Infrared Radiometry Experiment measured the surface thermal inertia, and hence the effective mean particle size, over much of the planet, finding no correlation with visual albedos. It found no large hot spots (which might have indicated volcanic activity), and showed that the surfaces of Phobos and Deimos are very poor thermal conductors. It also provided an estimate of the height of Olympus Mons by estimating the slope of its sides based on their temperature.

The Infrared Spectrometry Experiment, with more than 20,000 spectra, showed features identified with gaseous carbon dioxide and water, entrained dust, and ice crystal clouds. Surface atmospheric pressures were determined at many sites, varying from < 1.5 to ~ 8 mbar. The atmospheric water-vapor

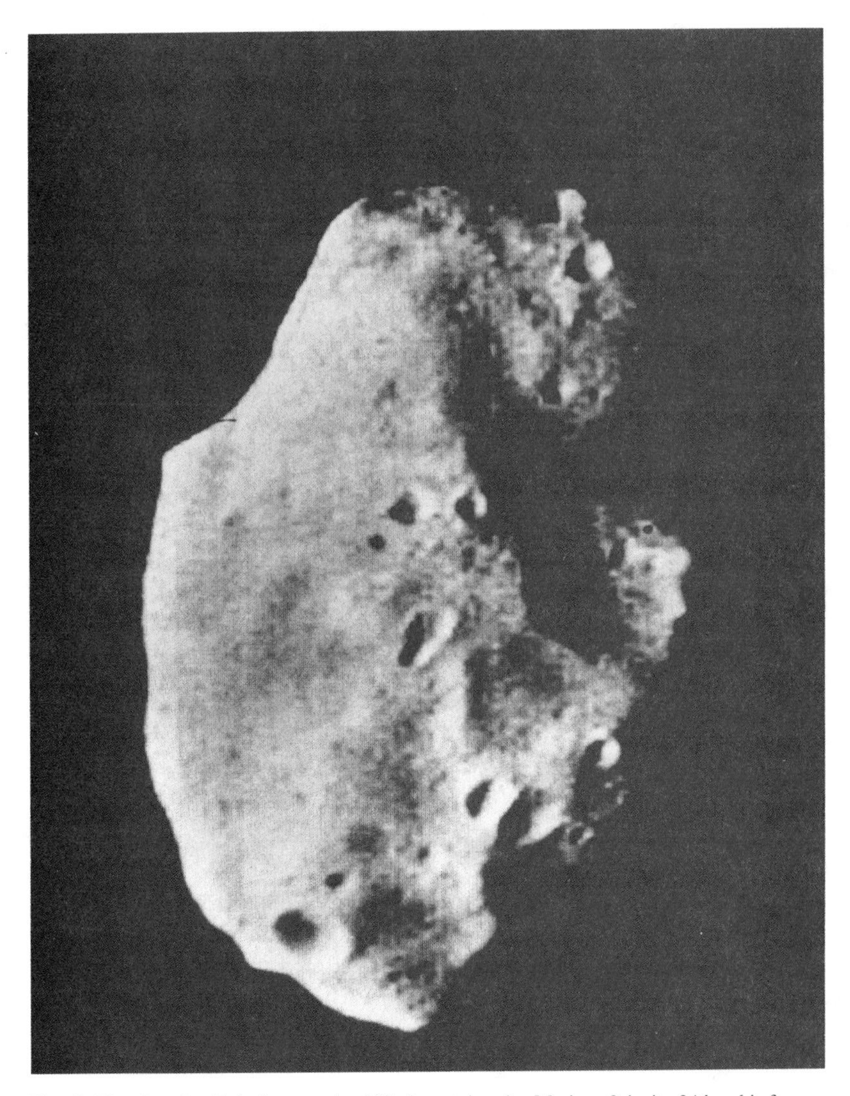

Fig. 7. The first detailed photograph of Phobus, taken by Mariner 9 in its 34th orbit from a distance of 5540 km.

content was found to be fairly constant at about 10 precipitable μm all over the planet during the dust storm (except near the north pole) and between 20 and 50 μm in the northern hemisphere later during its spring and summer seasons. Atmospheric temperatures were measured from the surface up to about the 0.1-mbar level; large diurnal variations in temperature were found at least up to 30 km; and migrating thermal waves in the atmosphere were discovered during the dust storm. It was found that the dust in the atmosphere was about 60% silicon dioxide and the surface consisted of fine-grained material with a composition essentially the same.

The Ultraviolet Spectrometer Experiment measured the local atmospheric pressure over a major portion of the planet and thus determined relative altitudes in many places, finding that the Valles Marineris were as much as 6 km deep in places and Olympus Mons rose 25 km above the surrounding plain. The detection and mapping of ozone was accomplished; none was found at low latitudes, and its appearance at high latitudes was seasonal. Yellow and white clouds and blue hazes were detected and studied, the latter as high as 70 km. Atomic hydrogen was measured in the exosphere and upper atmosphere, its density correlated with solar activity. An ultraviolet spectrum of Phobos was obtained (Mariner 1971).

IX. U.S. MISSES 1973

Both nations had grandiose plans for the 1973 opportunity. As early as 1964 a group at NASA's Langley Research Center (LRC) had been studying the problem of soft landing on the Martian surface, and in the middle of 1967 as the overly ambitious Voyager Project was being scuttled by the Congress and Langley's highly successful Lunar Orbiter mission was drawing to a close, the LRC organized a group to begin actively planning a Mars landing mission and attempting to persuade NASA to allow them to implement it. In January 1968, NASA assigned to JPL the management responsibility for the "Mariner Mars 1971" orbiters and to LRC that for the "Titan Mars 1973 Orbiter and Lander." Over the next eleven months the decisions as to just what the 1973 mission should include and who should manage what parts were hammered out, and the project, now called Viking, officially began in December 1968. LRC was to have overall direction of the project and responsibility for the Lander, to be built by an industrial contractor, and JPL was to have responsibility for the Orbiter and the mission operations. The primary scientific goal of the mission was to be the search for evidence of Martian life forms.

During the first year of the project, the Denver Division of Martin Marietta Aerospace was chosen to design and build the Lander, all the basic design decisions for both Lander and Orbiter were made, the scientific experiments to be carried and the science teams to design and implement them were chosen, and the project seemed well on its way. But on the last day of 1969, NASA learned that the Bureau of the Budget had uncovered a four-billion-dollar problem in the federal budget and was recommending a drastic cut in the Viking budget and the postponement of the mission until 1975. After a few days of soul searching, NASA decided not to change the mission plans except for stretching out the prelaunch activities by two years. So the 1973 opportunity came with no American mission ready.

X. SOVIET MISSIONS OF 1973–74

Orbiters Mars 4 and Mars 5 were launched on 1973 July 21 and 25 and lander carriers Mars 6 and Mars 7 on August 5 and 9, the Proton launch system performing faultlessly by now. All four arrived at Mars between 1974, February 10 and March 12. The scientific payloads of Mars 4 and Mars 5 are described in Table VII and those of Mars 6 and Mars 7 are described in Tables VIII and IX. Some of the instruments were new versions of those on Mars 2 and 3, and some were new to Soviet planetary missions.

On Mars 4, a loss of propellant prevented ignition of the braking motor, and the AIS flew by at 2200 km from the surface, but it did obtain some

TABLE VII
Scientific Instruments on Mars 4 and 5

Experiment	Principal Investigator	Objective
Phototelevision units (PTU) and survey "scanning device"	A. S. Selivanov	photograph Martian surface
Infrared radiometer (8–40 μm)	L. V. Ksanfomality	surface temperature
Radio telescope polarimeter (3.5 cm)	N. N. Krupenio	subsurface dielectric constant
Polarimeters (2) 0.32–0.70 μm)	L. V. Ksanfomality	surface texture
Photometer for 2 CO_2 bands	V. I. Moroz L. V. Ksanfomality	altitude profiles of surface
Photometer (0.35–0.7 μm)	L. V. Ksanfomality	albedo and color profiles
Photometer for water-vapor band (1.38 μm)	V. I. Moroz A. E. Nadzhip	water-vapor content in atmosphere
UV photometer (2600 and 2800 Å)	V. A. Krasnopolsky	ozone in the atmosphere
Spectrometer (0.3–0.8 μm)	V. A. Krasnopolsky	search for upper atmosphere emissions
Lyman-α sensor	V. G. Kurt J. Blamont	hydrogen in upper atmosphere
Magnetometer	Sh. Sh. Dolginov	magnetic field
Ion traps	K. I. Gringauz	solar wind and its interaction
Narrow-angle electrostatic plasma sensor	O. A. Vaisberg	solar wind and its interaction
Radio-occultation	M. A. Kolosov	atmospheric density profiles
Dual-frequency radio occultation experiment	N. A. Savich	ionospheric density profile
"Stereo-2"	Y. L. Steinberg	solar bursts at meter wavelength

TABLE VIII
Scientific Instruments on Mars 6 and 7 Flyby Modules

Experiment	Principal Investigator	Objective
Telephotometer	A. S. Selivanov	imaging of Martian surface
Lyman alpha sensor	V. G. Kurt	hydrogen in upper atmosphere
Magnetometer	Sh. Sh. Dolginov	magnetic field
Ion traps	K. K. Gringauz	solar wind and its interaction with Mars
Narrows-angle electrostatic plasma sensor	O. L. Vaisberg	solar wind and its interaction with Mars
Solar cosmic-ray sensors	N. F. Pissarenko	solar cosmic rays
Micrometeorite sensors	T. N. Nazarova	micrometeorite impacts
Solar radiometer	Y. L. Steinberg	solar long-wavelength radio emission
Radio occultation experiment	M. A. Kolosov N. A. Savich	atmospheric density and ionospheric profiles

TABLE IX
Scientific Instruments on Mars 6 and 7 Descent Modules

Experiment	Principal Investigator	Objective
Panoramic telephotometer	A. S. Selivanov	imaging of Martian surface around the module
Temperature and pressure sensors	V. S. Avduevsky M. Ya Marov	atmospheric temperature and pressure
Density and wind sensors	V. M. Linkin	atmospheric density and wind measurements at the surface
Accelerometer	Z. V. Cheremuchina	atmospheric density measurements during descent
Mass spectrometer	V. G. Istomin	atmospheric composition
Automatic laboratory for activation analysis	Yu. A. Surkov	soil composition
Mechanical sensors	A. L. Kemurdzhen	mechanical properties of soil

photographs (Fig. 8) and some radio-occultation data, which provided the first detection of the nighttime ionosphere. Mars 5 successfully entered an orbit with apsis altitudes 1755 and 32,555 km, inclination 35°.3, and period 24.88 hr. However, its lifetime was severely limited by the loss of pressurization in the housing of its transmitter after 22 orbits.

Mars 5 carried two types of photographic instruments, a film camera (Fig. 9) with a TV readout (called a phototelevision unit, PTU) like those on Mars 2 and 3 and a single-line scanning device for obtaining panoramic pictures, with a 30° angle of view and sensitivity in the visible and near infrared.

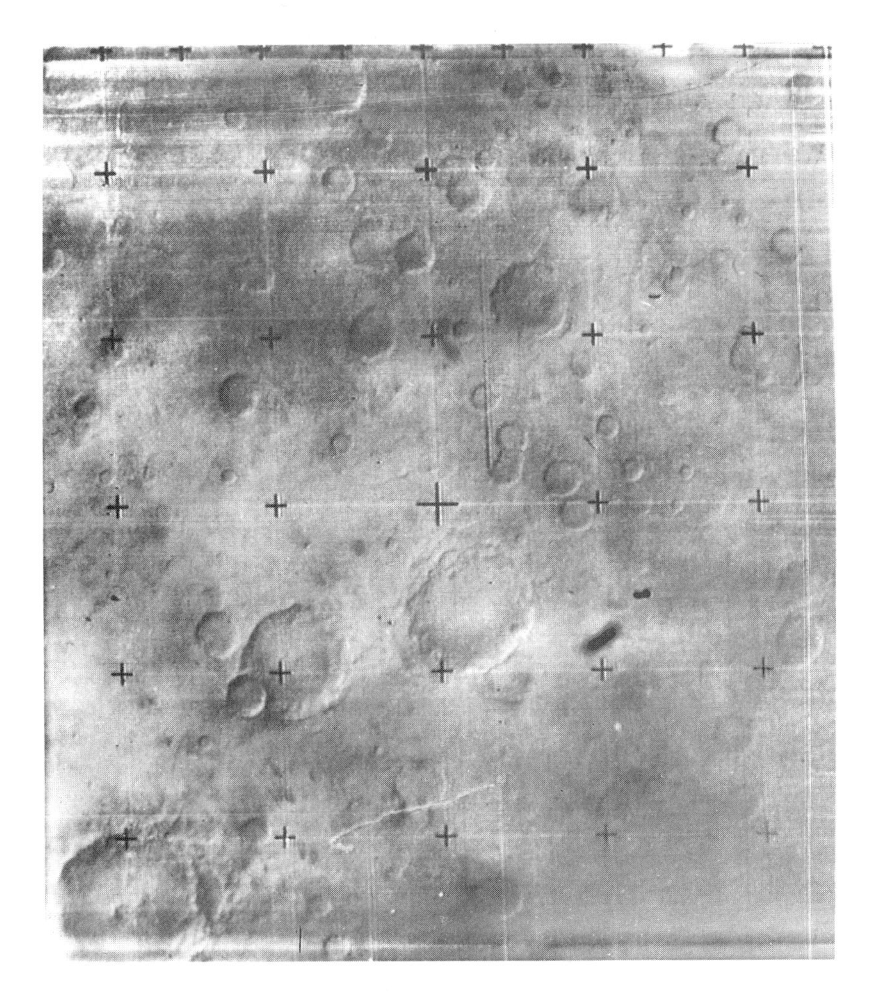

Fig. 8. A view from Mars 4 of cratered terrain at 35°.5 S, 14°.5 W, taken from a 1800 km range through a red filter. From the lower-left corner diagonally upward, the three craters are Lohse, Hartwig and Vogel.

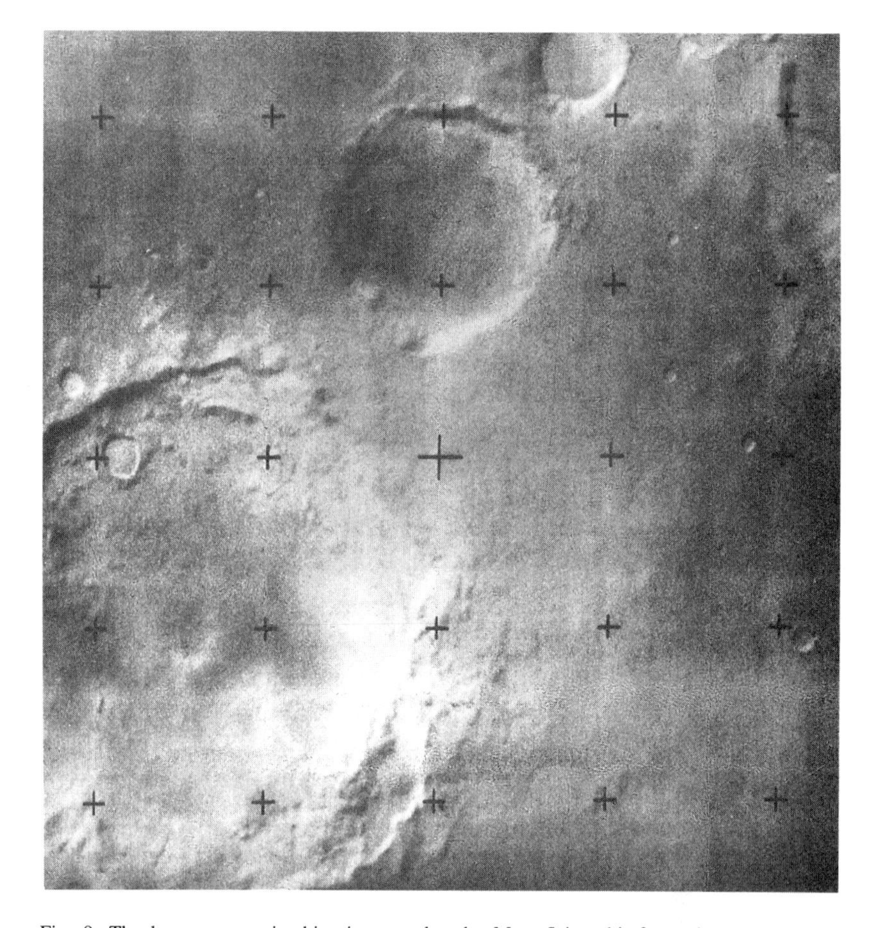

Fig. 9. The largest crater in this picture, taken by Mars 5 in orbit from about 1700 km, is Lampland at 36° S, 79° W.

About 60 photographs of the planet were taken in a 9-day period (including some in color), and all were in roughly the same area, extending from about latitude 5° N, longitude 330° W to 20° S, 130° W (Moroz 1975), very close to the track of Mariner 6. A portion of the photographed area near Argyre was analyzed by Florenskiy et al. (1975), and a map of the observed part of the planet was produced. The full collection of pictures from Mars 4 and 5 with accompanying papers on their geologic interpretation appeared later (Sidorenko 1980).

Near periapsis, several remote-sensing devices on Mars 5 measured different properties of the surface with spatial resolution of a few tens of km. The line-of-sight paths on the surface were 7 adjacent arcs south of Valles

Marineris. Along these lines, brightness in the visual and near infrared, polarization in visible and radio waves, and gamma radiation were measured. Also there were data on the intensity of carbon dioxide bands, giving relative altitudes.

The infrared radiometer recorded a maximum temperature of 272 K, with 230 K near the terminator and 200 K in the dark. Comparison with thermal models of the soil and corresponding values of thermal inertia implied soil grain sizes from 0.1 to 0.5 mm (Ksanfomality and Moroz 1975; Vdovin et al. 1980). The visual polarization data also relate to grain size, but their interpretation is much more difficult. Results from measurements in 9 wavelengths were published by Ksanfomality and Moroz (1975), and a careful step-by-step comparison with the pictures was made by Dollfus et al. (1983). The conclusion is that, for regions covered by aeolian deposits, typical grains are smaller than 0.04 mm and that fractured terrains processed by tectonic activity have a coarser soil, which includes larger grains and unmantled rocks.

Photometric profiles were found to correspond well in general to the Lambert law in wavelengths 5600 and 7000 Å but not in short wavelengths. Brightness near 3500 Å anticorrelated with heights obtained from the CO_2 photometric data (Ksanfomality and Krasovsky 1975) owing to atmospheric scattering, but in some places there were color anomalies probably belonging to the surface itself.

Analysis of data from the CO_2 photometer (Zassova et al. 1982) gave 6 altitude profiles along the path of Mars 5 between longitudes 20° and 120°, latitudes 20° and 40° S. These were in general agreement with Mariner 9 ultraviolet-spectrometer data and gave some additional details.

Polarization data at 3.5-cm wavelength yielded values of soil dielectric constant between 2.5 and 4 at depths of several tens of cm (Basharinov et al. 1976). Gamma-ray spectrometry in the range 1 to 8 MeV found that the content of U, Th and K is the same as in terrestrial mafic rock (Vinogradov et al. 1975).

Four instruments measured atmospheric composition: water vapor and ozone in the lower atmosphere, atomic hydrogen and other trace constituents in the upper.

An interference-polarization photometer of the same type as on Mars 2 and 3 measured the absorption by 1.38-μm bands of water vapor. Relatively large humidity (of the order of 100 precipitable μm) was found in the eastern part of the region studied, south of Tharsis (Moroz and Nadzip 1975*b*).

The ultraviolet filter-photometer on Mars 5 detected an ozone layer at about 40-km altitude in which the concentration was about 3 orders of magnitude less than on Earth. The altitude was a surprise, as a layer near or at the surface had been anticipated. Ozone was found in equatorial regions, whereas the Mariners had detected it only near the poles, where it is more abundant (Krasnopolsky et al. 1975). The density of aerosol particles (probably water-ice crystals) was also evaluated from the data of this experiment (Krasnopolsky 1979).

The Lyman-α photometer measured the temperature of the exosphere as 325 $\pm$ 30 K and that in the altitude range 200 to 87 km as 10 K lower (Bertaux et al. 1975).

The high-sensitivity spectrometer for detection of upper-atmosphere emissions in the range 0.3 to 0.8 μm found nothing, and only several upper limits were evaluated (Krasnopolsky and Krysko 1976).

The solar-plasma detectors on Mars 5 supplemented the data from Mars 2 and Mars 3 by discovering 3 plasma zones near the planet: (I) the undisturbed solar wind; (II) the thermalized plasma behind a bow shock; and (III) very small proton currents in the tail of the magnetosphere (Gringauz et al. 1976). These results coupled with those of the magnetometer supported the earlier conclusion that the solar wind is held off by an intrinsic planetary field. Its properties were concluded to be: magnetic moment 2.5×10^{22} gauss-cm^3 (2.5×10^{12} Tesla-m^3, about 0.03% that of Earth) and inclination 15° to 20° to the rotation axis (Dolginov et al. 1975). The height of the forward point of the bow shock above the Martian surface was about 350 km, and the Mach number of the solar wind was approximately 7 (Vaisberg et al. 1975).

Measurements of the atmosphere and the ionosphere by occultation had been made by Mars 2 as it orbited, using both single-frequency and coherent two-frequency radio systems. The experiment was repeated by Mars 4 and Mars 6 on their flybys and by Mars 5 on one orbit, making measurements on both the sunlit and dark sides. The nighttime ionosphere, which had not previously been detected, was found to have a peak electron concentration of 4.6×10^3 cm^{-3} at 110-km altitude (cf. Mariner 9 results, above). Also, the near-surface atmospheric pressure was found to be 6.7 mbar (Vasilsyev et al. 1975; Kolosov et al. 1975; Savich et al. 1976).

The descent module (DM) from Mars 7, released 4 hr before encounter, missed the planet by 1300 km. That from Mars 6 entered the atmosphere properly, transmitting data for 224 s "as it descended through the atmosphere." The radio signal stopped "in direct proximity to the surface," at which time the velocity of the DM was 61 m s^{-1} (Avduyevskiy et al. 1975).

The descent module of Mars 6, in contrast to Mars 2 and 3, transmitted data in real time to the flyby module, which relayed it immediately to Earth. The intention was to preclude the loss of data even in the case of destruction of the module by contact with the surface. It was a good decision, because actually no scientific data were received after landing. The DM did acquire the first *in situ* data about the profiles of the Mars atmosphere (Avduyevskiy et al. 1975). For 224 s during the transit of the atmosphere, data were received from pressure, temperature, radio-altimeter, accelerometric and Doppler velocity measurements, which enabled calculations of troposphere structure (with approximately linear temperature gradient between the surface boundary layer (230 K) and the base of the stratosphere (150 K) at 25 km), and atmospheric density from altitude 82 km to 12 km. The landing site at latitude 24°S, longitude 19°W was determined to be 3389 km from the center of Mars (Kerzhanovich 1977).

A big disappointment was the mass spectrometer. Its scientific data (being too voluminous for the real-time telemetry link) were recorded for playback after landing and thus were lost. An engineering parameter, the current to the vacuum pump on the spectrometer, was measured periodically, and its increase with time during the parachute descent was several times higher than had been expected. The hypothesis offered to explain this observation was that the atmosphere contained a significant portion of an inert gas (presumably argon) which could not be removed by the pump. The quantity was estimated to be 35 ± 10% (Istomin et al. 1975). Unfortunately, subsequent Viking measurements determined the argon abundance to be 1.6%.

The flyby modules (FBM) of Mars 6 and 7 flew by as intended, but they did not carry cameras or other instrumentation for planet observations. Thus the amount of information returned by the 4 missions was a minuscule fraction of that from Mariner 9 alone, and the opportunity to upstage the Viking Mission was lost. This disappointing performance marked the end of the Soviet effort at Mars exploration until the launch of the two Phobos spacecraft in July 1988. It was decided by those in charge of the planetary program that there was little probability of competing favorably with the Viking mission and that they should concentrate on Venus for the next decade. The result was the highly successful series of Venera 9-16 and Vega 1-2 spacecraft.

XI. THE VIKINGS: A QUADRUPLE MISSION

The two identical Viking spacecraft, each consisting of an Orbiter carrying a Lander, were launched on 1975 Aug 20 and Sep 9. Viking 1 was inserted into Mars orbit on 1976 Jun 19, and immediately began a photographic search for a landing site that appeared safe. It had been hoped to land on July 4, and a prime candidate site had been chosen in advance along with several contingency sites, based on the information from Mariner 9 photographs and Earth-based radar observations. The prime site (Fig. 10) was located at the place where the largest Mars channel debouches onto Chryse Planitia, it being considered to be the best area where water and near-surface ice might be found—the optimum place to look for complex organic molecules. By June 26, it had been decided to abandon the prime site because orbital imaging data showed it to be unexpectedly complex and also the first contingency site because Mariner 9 images indicated the presence of knobs and small craters. Photographic and radar observations continued, and by July 1 a site was found that looked very smooth in the pictures, but later radar observations showed it to be an area of anomalous radar scattering, interpreted as a surface that was very rough on the scale of the Lander or of very

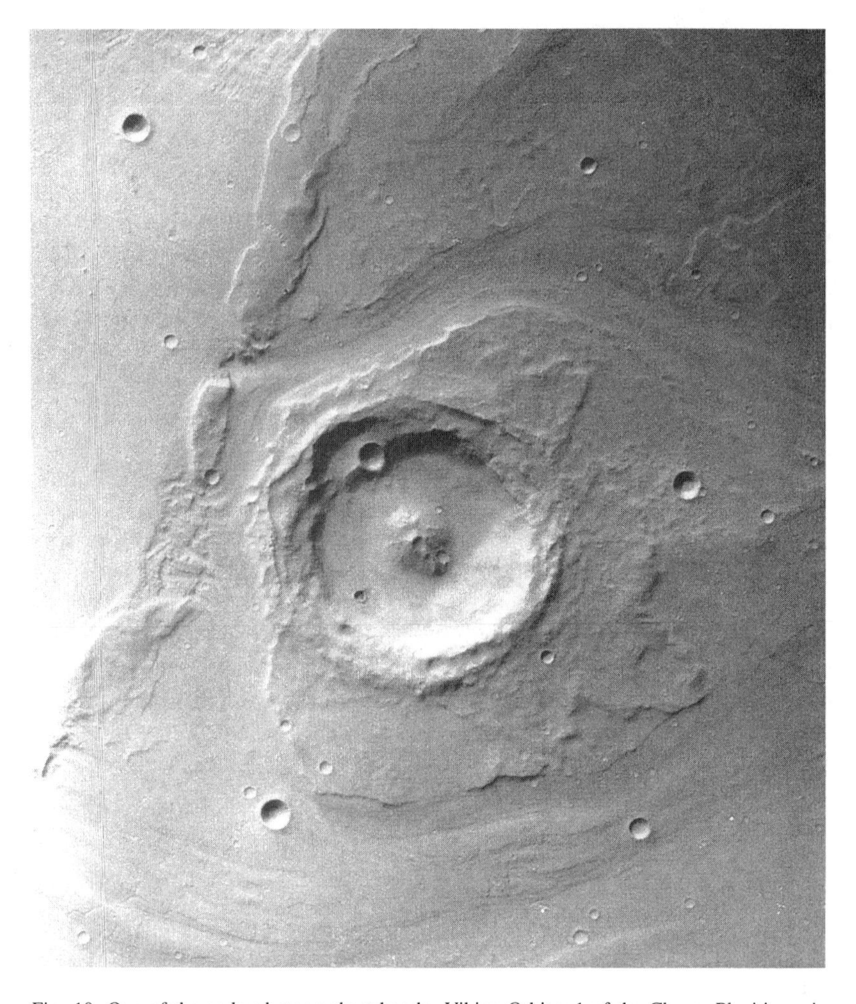

Fig. 10. One of the early photographs taken by Viking Orbiter 1 of the Chryse Planitia region pre-selected as the site for Lander 1. It shows the effect of a great flood on the impact crater Dromore at 20° N, 49° W. This was one of the pictures that quickly persuaded the Project Manager to search for a different site. (Viking 20A62)

low density or both. By July 9, about 600 orbiter images had been examined, a site had been found that appeared acceptable from both types of observation, and the orbit was trimmed to move the spacecraft over it. On July 14 the orbit was synchronized over the site, and the landing occurred on July 20—coincidentally the seventh anniversary of the first Apollo landing on the Moon (Masursky and Crabill 1976*a*). It was later christened the Mutch Memorial Station in honor of Tim Mutch, Team Leader of the Lander Imaging Team.

Viking 2 went into orbit on August 7, and because of the intense activity with Viking 1, its landing was deferred until September 3. In the interim a site-certification program similar to that for Viking 1 was carried out, the difference being that the photographic data were supplemented by IRTM and MAWD observations, as radar observations at the latitude chosen (40° to 50° N) were not possible (Masursky and Crabill 1976*b*).

The primary Viking mission was defined as 90 days, which was the design lifetime of the Landers, but as all four craft continued to perform well, the mission was repeatedly extended by NASA. The last successful communication session with Lander 1 occurred on 1982 Nov 13, after which it failed to respond to commands. Thereupon, NASA declared the mission officially terminated. Thus the total length of the mission at Mars was 6.40 Earth years or 3.40 Mars years. Transmissions ceased from Orbiter 2 on 1978 Jul 25, from Lander 2 on 1980 Apr 11, and from Orbiter 1 on 1980 Aug 7.

The scientific mission of Viking consisted of 4 investigations by each Orbiter and 9 by each Lander, as shown in Table X. The 13 science teams included a total of 78 members.

The Viking mission did not revolutionize our ideas about Mars as Mariner 9 had done, but, building on the foundation of its Mariner and Mars precursors, it far surpassed them in the variety, quality and quantity of its data because of four factors—the number of spacecraft, the number of experiments, the duration of the mission and the telemetry rate. Space limitations in this chapter preclude even a cursory account of its scientific results. Even the brief summary by Snyder (1979*b*) ran to 20,000 words. We shall merely attempt to enumerate what the 4 spacecraft did. This list will not be complete, nor will it convey any idea of the great importance of the many scientific discoveries. These are to be found in many of the chapters of this book.

Each Orbiter carried 2 cameras, which were similar to those on Mariners but much improved, and 6 selectable filters provided exceptional color capability. Each frame consisted of 1056 lines of 1182 pixels, and each tape recorder could record a picture in 8.96 s. By alternating cameras, pictures could be taken at 4.48-s intervals, providing good overlap of successive frames. As the orbit altitude was eventually reduced to 300 km, the pixel spacing on the highest-resolution pictures was 7.5 m, about 150 times better than those of Mariner 4. The Orbiters returned 52,603 pictures, which covered the entire planet at all Martian seasons (Figs. 11 and 12) as well as close-up views of both Martian moons. They provided the data to establish more than 9000 mapping control points with estimated accuracies of 6 km everywhere and better than 3 km in some regions.

The Mars Atmospheric Water Detectors (MAWD) produced global maps of the quantity of water vapor at all seasons, thus revealing the seasonal cycle of vapor transport and demonstrating that the residual northern polar cap in the summer was water ice.

The Infrared Thermal Mappers (IRTM) mapped the temperature, albedo

TABLE X
Viking Scientific Investigations

Investigations	Orbiter Instruments	Science Team Leaders
Orbiter imaging	two video cameras	M. H. Carr
Water-vapor mapping	infrared spectrometer	C. B. Farmer
Thermal mapping	infrared radiometer	H. H. Kieffer
Entry science		A. O. C. Nier
Ionospheric properties	retarding potential analyzer	
Atmospheric composition	mass spectrometer	
Atmospheric structure	pressure, temperature and acceleration sensors	
Investigations	**Lander Instruments**	**Science Team Leaders**
Lander imaging	2 facsimile cameras	T. A. Mutch
Biology	3 analyses for metabolism, growth, or photosynthesis	H. P. Klien
Molecular analysis	gas chromatograph mass spectrometer	K. Biemann
Inorganic analysis	X-ray fluorescence spectrometer	P. Toulmin
Meteorology	pressure, temperature, wind velocity sensors	S. L. Hess
Seismology	three-axis seismometer	D. L. Anderson
Magnetic properties	magnet on sampler observed by cameras	R. B. Hargraves
Physical properties	various engineering sensors	R. Shorthill
Radio science	Orbiter and Lander radio and radar systems	W. H. Michael
Celestial mechanics, atmospheric properties and test of general relativity		

and thermal inertia of the entire surface; they detected many local dust storms, and demonstrated (along with MAWD) that the northern polar cap in the summer was water ice; the residual south polar cap was covered with carbon dioxide frost.

Both Landers landed successfully, and (except for one of the two seismometers) all experiments operated essentially as planned for much longer than the design lifetime.

Each of the Landers was equipped with two scanning panoramic cameras capable of acquiring data from 40° above the horizon to 60° below, through nearly 360° of azimuth, with a best resolution of 0.°04 (Fig. 13). The Landers returned 4587 pictures of their surroundings, many in color, and recorded the variations in the opacity of the atmosphere. The capability of stereoscopic

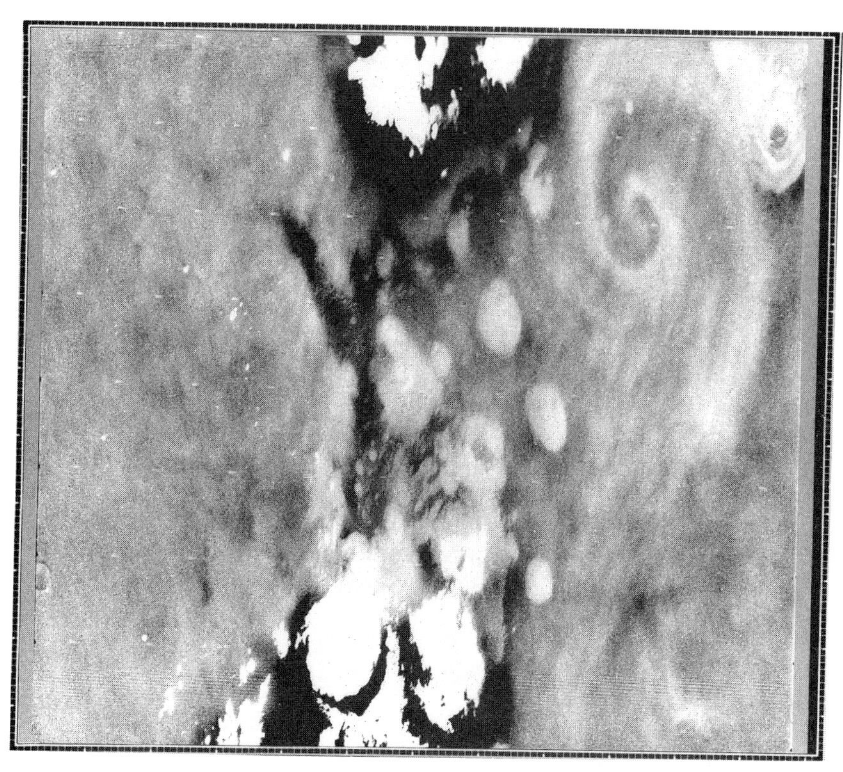

Fig. 11. Viking Orbiter 1 image showing a Martian storm system at high northern latitude (81° N, 160° W) early in summer (L_S = 105°). The prominent spiral structure is relatively rare on Mars; only four such storms were observed by Viking. The origin of these summertime spiral storms is still being debated. Gierasch et al. (1979) proposed a mechanism of radiative instability associated with the heating of dusty air, while Hunt and James (1979), noting the strong latitudinal gradiant of surface temperature just to the north, argued that this storm was due to baroclinic instability and so was similar to terrestrial extratropical cyclones. (Viking 01633)

photography made possible the generation of detailed maps of the landing areas. The passage of the shadow of Phobos over Lander 1 was recorded by both the Lander camera and the Orbiter camera (Fig. 14).

The Entry Science investigation twice determined the molecular and isotopic composition, thermal structure, and density of the atmosphere as functions of altitude below 200 km, and made the first *in situ* measurement of the ionosphere of another planet.

In the Biology investigation, 3 instruments on each Lander tested 4 different hypotheses about the characteristics of possible Martian microorganisms, found no unequivocal evidence for their existence, but discovered a surprising evolution of various gases from the soil when it was moistened.

In the Molecular Analysis investigation, the gas chromatograph-mass spectrometer tested for organic compounds in the soil but found none even

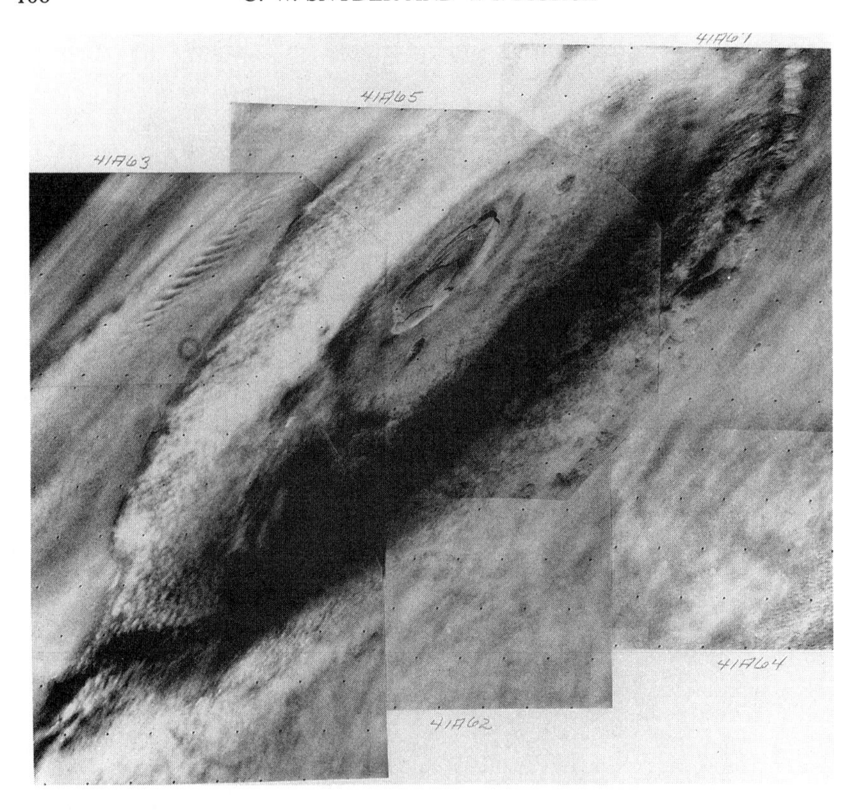

Fig. 12. This mosaic of Viking Orbiter photographs of Olympus Mons surrounded by clouds, after being hand colored by an artist, became so widely copied that it virtually became the symbol of the Viking missions. (Viking P-17444)

down to the ppb level. It also measured the composition of the atmosphere, including isotopic ratios for 5 of the components, and showed argon to be a minor constituent (1.6%) contrary to the interpretation of the Mars 6 mass-spectrometer data.

In the Inorganic Analysis investigation, the X-ray fluorescence spectrometers measured the concentrations in the surface material of 13 elements with atomic numbers from 12 to 40, finding Si and Fe to be the most abundant.

The meteorology sensors monitored the atmospheric pressure and temperature and the wind direction and velocity through more than 3 Mars years.

The seismometer detected vibrations of the Lander caused by the wind but detected no signals that could be positively identified as seismic.

The Magnetic Properties investigation concluded that the Martian regolith contains several percent of a highly magnetic material.

Fig. 13. The first photograph ever taken on the surface of Mars. The Viking 1 Lander camera began scanning the scene from left to right 25 seconds after landing and continued for 5 minutes. The picture was displayed to an amazed and excited audience in real time as it was received at the Space Flight Operations Facility at JPL. The dark band near the left end was caused by the dust cloud stirred up by the landing as it drifted downwind and momentarily came between the Lander and the Sun. (Viking P-17053)

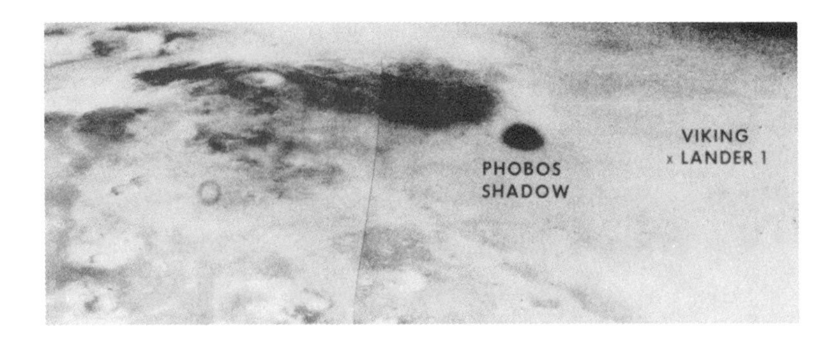

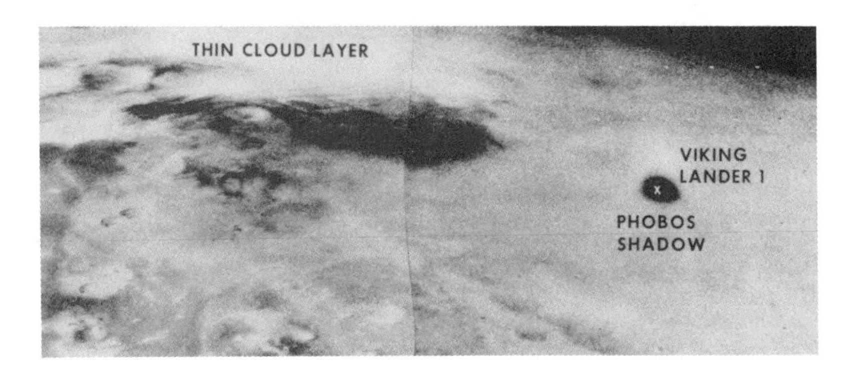

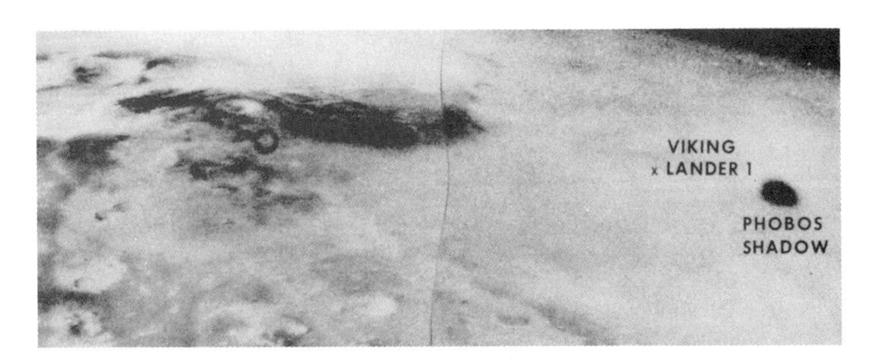

Fig. 14. These Viking Orbiter 1 pictures were taken during an experiment to locate the position of Viking Lander 1 on the Martian surface using the shadow of Phobos. As the passage of the shadow was detected also by the Lander, it became possible to determine its location to within 1 km. (Viking P-19552)

In the Physical Properties investigation, the surface sampler manipulated the surface material, digging trenches, pushing rocks, and forming piles, thus determining its texture and cohesiveness.

The Radio Science investigation, using the telemetery signals of both Landers and Orbiters, determined the locations of the two Landers, and made measurements of the dynamical, surface and internal properties of Mars (including spin axis orientation and motion, rotation rate, gravity field, figure and surface dielectric constant), the atmospheric and ionospheric properties of Mars (including altitude profiles of pressure, temperature, density and electron concentration), and solar system properties (including ephemerides of Mars and Earth, masses of Martian satellites, interplanetary medium, solar corona and tests of general relativity).

It is a sad commentary on space research in the 1980s that, more than two decades after the start of the Viking project, it still stands alone as preeminently the most successful of all the Mars missions. We fervently hope that the passage of another decade will not find this still be to true.

XII. THE PHOBOS MISSIONS

Almost thirteen years after the launch of the two Vikings, the next phase of Mars exploration began in 1988 with the launch of Soviet probes Phobos 1 (July 7) and Phobos 2 (July 12). The main target was not now Mars itself but one of its two tiny satellites, the name of which was given to the new missions. The selection of this goal was motivated partly by the increased interest in the planetary scientific community in studies of the small bodies in the solar system, which are believed to be the key to understanding the origin of the solar system. The Martian satellites are very appropriate targets for these studies; they are probably captured asteroids, and rendezvous with them is simpler than with any small body in solar orbit. Another factor was that Mars has been chosen as the main long-term goal for the Soviet planetary program after 1990, and the Phobos missions could be considered as the precursor of this new program.

Technical and scientific preparation of the missions took about eight years, and plans changed several times. Landing of the main spacecraft on Phobos, which could provide wide possibilities for many *in situ* experiments, was one of the preliminary versions considered; but it was concluded that such a landing posed serious problems of reliability. The scenario finally chosen was (1) to place the spacecraft in the elliptical, near-equatorial orbit around Mars; (2) to maneuver to an almost circular orbit very close to the orbit of Phobos; and (3) to approach to within 50 to 100 m for several tens of minutes. At the time of this closest approach, two small landing probes would separate and descend to the surface to perform analyses of its elemental composition, and two experiments on the main spacecraft would be activated—laser (LIMA-D) and ion (DION) guns that would evaporate small pieces of soil to give tiny samples of gaseous products for analysis by sensitive mass spectrometers.

On Phobos 2 was a small lander, called Hopper, which would have only a few days' lifetime but could change its position on the surface, moving as much as 20 m about ten times. Each AIS carried a lander, called a "long-term automatic station" (abbreviated DAS in Russian), which would have a nominal 1-yr lifetime but could not move.

The Phobos spacecraft were by far the heaviest yet used on planetary missions, with a total mass of 6200 kg (of which 3600 kg was the propulsion system) and a scientific payload of about 500 kg. From Mars 2 and 3 (1971) to Vega (1985) all Soviet planetary spacecraft had been of the same basic design. The Phobos spacecraft were the first of a new generation. This innovation incorporated many important improvements, but the limited familiarity with it held dangers for the project. Less than two months after launch (September 30) Phobos 1 received an erroneous command sequence, which switched off the orientation system, causing irrevocable loss of communication.

Phobos 2 reached Mars on 1989 January 29 and was inserted into an orbit with period 79.2 hr, periapsis altitude 865 km, and eccentricity 0.903. Observations of Mars and its plasma environment commenced soon thereafter. Two orbit changes in mid-February placed the orbiter in a nearly circular orbit somewhat outside the orbit of Phobos. Remote studies of Mars continued on this orbit, and the TV cameras started to obtain images of Phobos. Three additional corrections brought the spacecraft by March 21 into an orbit practically in synchrony with the moonlet between 200 and 600 km distance from it. On March 27 an apparent computer malfunction disabled the attitude control, and communication was lost just as the culmination of the mission was in sight.

Some but not all of the scientific instruments were duplicated on the two orbiters. The full list is given in Table XI; the instruments for which results have been reported are marked with an asterisk. Most of the experiments were designed on a broad international basis, involving scientific facilities and individual scientists not only of the USSR but also of Austria, Bulgaria, Czechoslovakia (CSSR), the European Space Agency (ESA), Finland, Federal Republic of Germany (FRG), France, German Democratic Republic (GDR), Hungary, Ireland, Poland, Sweden, Switzerland and USA. Coordination of all these participants was the responsibility of the International Scientific Council (ISC). This type of organization was introduced in the Vega missions to Comet Halley and successfully transplanted to the Phobos mission. It was a nontraditional procedure, especially for the USSR before *perestroika*. It was originated and energetically executed by R. Z. Sagdeev as chairman of ISC.

A more detailed history of the Phobos mission is contained in the article by Sagdeev and Zakharov (1989). The issue of *Nature* (1989 Oct 19) also contains fourteen papers discussing the preliminary results from the scientific experiments executed in the 52 active days of Phobos 2 in orbit. The list of nearly 200 authors of these papers strikingly illustrates the international nature of the project. A set of papers in Russian not exactly identical to those in *Nature* was published in Soviet Astronomical Journal Letters.

The French space agency CNES held an international symposium in October 1989, where somewhat more detailed papers were presented, although all the results were preliminary at that time. A brief review of the results is given below.

After the final orbit correction of March 21, the gravitational perturbation of the orbit was detectable, thus permitting a more accurate determination of the mass of Phobos, $1.08 \pm 0.01 \times 10^{16}$ kg. This led to a density of 1.95 ± 0.1 g cm^{-3}, a value that is less than typical for carbonaceous chondrites or black mafic material and suggests that the interior of Phobos may be more or less porous or contain ice (Avanesov et al. 1989).

The reflectance spectra of the Phobos surface were studied with space resolution of about 1 km by the KRFM (thermal-IR radiometer and UV/visible photometer) at 0.3 to 0.6 μm (Ksanfomality et al. 1989) and the ISM at 0.8 to 3.2 μm (Bibring et al. 1989). In neither range did the spectra match satisfactorily to carbonaceous chondrites, which had been supposed to be the most probable analogs for the material of the surface. Substantive local inhomogeneities in reflectance spectra were found by both instruments.

The KRFM also measured thermal radiation from Phobos in five bands between 5 and 60 μm. Maximum brightness temperatures of about 300 K were registered. Thermal inertia values from 0.8 to 20×10^{-3} cal cm^{-2} deg^{-1} s$^{-1/2}$ (33 to 837 J m^{-2} deg^{-1}s$^{-1/2}$) were evaluated, also with some local inhomogeneities (Ksanfomality et al. 1989).

Thirty-seven good-quality images of Phobos were obtained by the VSK (videospectrometric system) from distances of 200 to 1100 km. Some of these complement the data from Mariner 9 and Viking missions, providing higher-resolution (40 to 80 m per pixel) coverage of the area to the west of the crater Stickney (Avenesov et al. 1989) (Fig. 15).

Imaging of Mars at thermal infrared wavelengths (8 to 12 μm) was provided for the first time by the TERMOSKAN, a mechanically scanned radiometer with a nitrogen-cooled detector (Selivanov et al. 1989). It also obtained a pixel-by-pixel coincident image in the visible (0.5 to 0.95 μm). Most of the equatorial regions of the planet were covered with spatial resolution of $\sim$ 2 km. Thermal images show in general the same surface features as visible ones but with much better contrast, a spectacular demonstration of the importance of the constant atmospheric haze. Some examples of TERMOSKAN images were printed in *Planetary Reports* (Vol. 9, No. 4, p. 22, Jul/Aug 1989).

Multi-spectral images of Mars were provided also by the ISM (Bibring et al. 1989), a near-infrared mapping spectrometer of a type never before used on planetary missions. With a spectral resolving power of $\sim$ 50, it recorded

TABLE XI A
Scientific Experiments on Phobos Mission Orbiters[a]

Experiment	Principal Investigators	Objective
	Studies of Phobos Surface	
LIMA-D laser and mass spectrometer	G. Managadze (USSR) R. Pellinen (Finland) J. Kissel (FRG)	elemental composition of surface layers
DION ion gun and mass spectrometer	G. Managadze (USSR) M. Hammelen (France) W. Riedler (Austria)	elemental composition of surface layer
*RLK radar	N. A. Armand (USSR)	subsurface structure
	Studies of Phobos and Mars	
*GS-14 gamma-ray spectrometer	Yu. A. Surkov (USSR)	radioactive element content of surfaces
IPNN neutron spectrometer	Yu. A. Surkov (USSR)	search for water in surface layers (Phobos 1 only)
*VSK videospectrometric system (FREGAT) TV camera and spectrometer	G. A. Avanesov (USSR) D. Petkov (Bulgaria) V. Kempe (DDR)	surface mapping at 3 wavelengths
*ISM near-IR mapping spectrometer	J.-P. Bibring (France)	mineralogic and pressure/altitude
*KRFM Thermal-IR radiometer and UV/visible photometer	L. V. Ksanfomality (USSR)	temperature, thermal inertia, and photometery of surfaces
*THERMOSCAN mapping thermal-IR radiometer	A. S. Selivanov (USSR) M. K. Narayeva (USSR)	surface temperature mapping
	Studies of Mars Atmosphere	
*AUGUST solar occultation spectrometer (UV and near IR)	J. Blamont (France) V. A. Krasnopolsky (USSR)	altitude distribution of aerosols and small constituents

[a] Asterisks indicate results reported.

128 spectral pixels simultaneously for every spatial pixel. Spatial resolution was ~ 5 km from the first elliptical orbit and ~ 30 km from the circular one. The mapping covered the Tharsis volcanoes, Valles Marineris, and most of the major geologic formations with the exception of the polar caps. New altitude maps can be derived from these data using the 2-μm carbon dioxide band, and a detailed contour map of part of the caldera of Pavonis Mons was presented in the *Nature* (1989) article. The ISM data also contain mineralogic information, especially from the 3.1-μm band that appears in the reflection

TABLE XI B
Scientific Experiments on Phobos Mission Orbiters

Experiment	Principal Investigators	Objective
	Studies of Solar Wind, Plasma Environment and Magnetic Field of Mars[a]	
*ASPERA cosmic plasma scanning analyzer	R. Lundin (Sweden) A. Zakharov (USSR) W. Riedler (Austria)	ion mass composition, directional ion & electron distribution, structure & dynamics of magnetosphere
*SOVICOMS electrostatic analyzer with large geometric factor	H. Grunwald (FRG) K. Grungauz (USSR)	ion composition, energy distribution & spatial characteristics of plasma
*HARP electrostatic analyzer	P. Király (Hungary)	energy and angular distribution of ions and electrons
*TAUS electrostatic & magnetic analyzer	H. Rosenbauer (FRG) N. Schutte (USSR)	direction and velocity distribution of protons, α particles and heavy ions
*APV-F plasma wave analyzer	S. Klimov (USSR) R. Grard (ESA) J. Juchnevich (Poland)	plasma density and frequency spectrum of electrostatic and electromagnetic waves
*FMGG fluxgate magnetometer	D. Möhleman (GDR) V. Oraevsky (USSR)	magnetic field near Mars
*MAGMA fluxgate magnetometer	W. Riedler (Austria)	magnetic field near Mars

[a] Asterisks indicate results reported.

spectra of any type of hydrated rocks. Mineralogic local variations were observed at all spatial scales and were more pronounced than expected from groundbased observations.

The KRFM short-wavelength channel (0.32 to 0.55 μm), a prism spectrometer with multi-element detector array, obtained photometric limb-to-limb profiles of Mars in 8 wavelengths with about 25-km spatial resolution. These profiles show many features belonging mainly to aerosol formations (clouds, limb brightening, etc.), that have strong wavelength dependence. Preliminary analysis of a bright spot observed at Arsia Mons indicated that it was probably an icy cloud with a column density of order 10 μg cm^{-2}.

Coincident positionally with these profiles were the profiles obtained by

TABLE XI C
Scientific Experiments on Phobos Mission Orbiters

Studies of Solar and Cosmic Radiation		
Experiment	**Principal Investigators**	**Objective**
*RF-15 solar X- and gamma-ray analyzer	F. Farnik (CSSR) O. Lukin (USSR)	solar X- and γ-ray monitoring
*SURF UV photometer	T. V. Kazachevskaya (USSR)	solar EUV monitoring
*IPHIR high-precision photometer	C. Frohlich (Switzerland)	solar oscillation
*TEREK coronograph	I. I. Sobelman (USSR) B. Valnicek (CSSR)	solar corona in X rays and visible light (Phobos 1 only)
*LET cosmic-ray analyzer	R. G. Marsden (ESA) V. V. Afonin (USSR) A. Varga (Hungary)	high-energy solar cosmic rays
*SLED cosmic-ray analyzer	V. V. Afonin (USSR) A. Riechter (FRG)	low-energy solar cosmic rays
*VGS (APEX) gamma-ray burst monitor	C. Barat (France) I. Mitrofanov (USSR)	high-energy solar and galactic gamma-ray bursts (100 keV-10 MeV)
*LILAS gamma-ray burst monitor	C. Barat (France) R. Syunyaev (USSR)	low-energy solar and galactic gamma-ray bursts (3 keV-1 MeV)

*Asterisks indicate results reported.

the KRFM long-wavelength channel, a filter radiometer covering five bands between 5 and 60 μm at ~ 50 km resolution. One filter, centered on the 15-μm carbon dioxide band measured brightness temperatures in the atmosphere at altitudes from 10 to 30 km. Temperature profiles in this wavelength were ~ 30 K lower than others and much smoother.

Measurements of the weak gamma radiation of the Martian rocks were made by the gamma-ray spectrometers GS-14 (Surkov et al. 1989) and VGS (d'Uston et al. 1989) on the first 4 periapsis passages. Lines of U, Th and K are clearly visible on measured spectra, and probably Fe, Al, Si, Ca and Ti are present. The indicated abundance of basic rock-forming elements were in agreement with the results from the X-ray fluorescence spectrometers on Viking Landers, and the abundance of U, Th and K are nearly the same as measured by the gamma-ray spectrometer of Mars 5.

TABLE XI D
Scientific Experiments on Phobos Mission Landers

Experiment	Principal Investigators	Objective
	Long-Term Small Lander (DAS)	
TV camera	J. Blamont (France) V. V. Kerzhanovich (USSR)	microstructure of surface layer
ALPHAX spectrometer of α back-scattering and X-ray fluorescence	H. Wanke (FRG) L. M. Mukhin (USSR)	elemental composition of surface layer
LIBRATION sun-angle position sensor	J. Blamont (France) V. M. Linkin (USSR)	study of libration
Seismometer	V. M. Linkin (USSR)	internal structure of Phobos
Celestial mechanics experiment	P. Preston (USA) V. M. Linkin (USSR) J. Blamont (France)	orbital motion of Phobos
On-board microprocessor	L. Shabo (Hungary) V. Kerzhanovich (USSR)	control of other experiments and data compression
	Movable Small Lander (Hopper)	
X-ray fluorescence spectrometer	Yu A. Surkov	elemental composition of surface layer
Magnetometer	Sh. Sh. Dolginov	magnetic field & permeability
Gravimeter	L. V. Ksanfomality	free-fall acceleration
Temperature sensors	L. V. Ksanfomality	temperature of surface layers
Conductometer	V. V. Gromov	electrical conductivity of surface
Mechanical sensors	A. L. Kemurdzhan	mechanical properties of surface

The only experiment specifically atmospheric was done by the occultation spectrometer AUGUST, which operated in 2 spectral regions: ultraviolet-visible (Blamont et al. 1989) and infrared (1.9 and 3.7 μm) (Krasnopolsky et al. 1989). Observing the Sun as the orbiter went into and out of occultation, a complicated shape and day-to-day variability of the ozone profile was found in the ultraviolet, and pronounced altitude variation of the water-vapor mixing ratio was found in the infrared.

Phobos 2 made the most comprehensive survey to date of the interaction

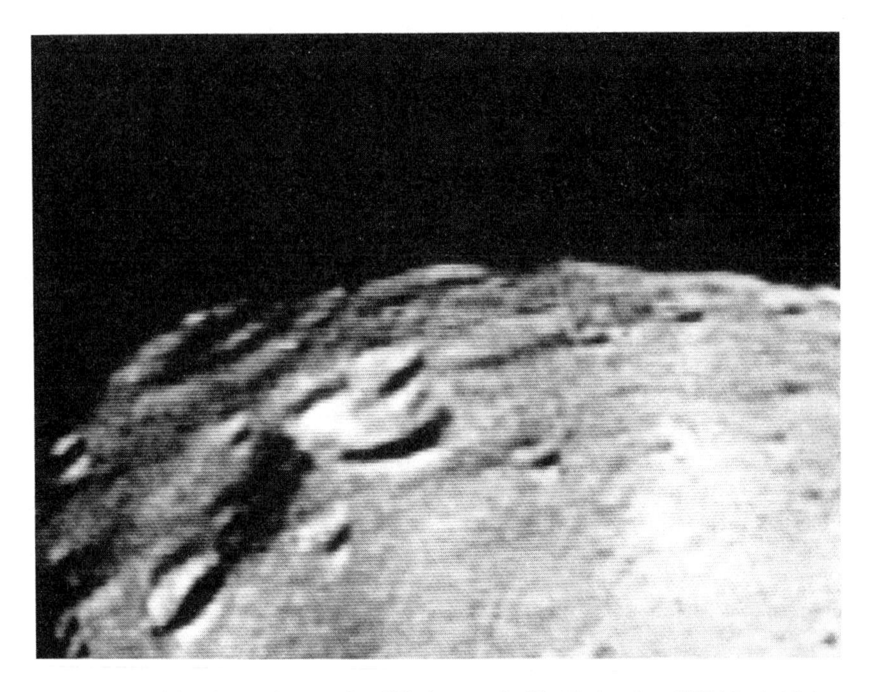

Fig. 15. One of the closest photographs of Phobos acquired by Phobos 2 on 1989 March 25, just two days before communication to the spacecraft was lost. This is a raw image with no computer enhancement. Taken from a range of 220 km, it has a resolution of 40 meters.

of the solar wind with the Martian magnetic field. The results were reported in 6 of the papers in *Nature* (1989) by Riedler et al.; Grard et al.; Lundin et al.; Rosenbauer et al.; Shutte et al., and Afonin et al.; they are discussed in chapter 31. Much new information was obtained about the plasma boundaries (bow shock, magnetopause and planetopause), the magnetospheric tail, and the plasma waves. However, the orbiter did not come close enough to Mars to settle the major question of whether or not its magnetic field is intrinsic.

A new process of atmospheric loss was found. It is an escape of tailward streaming planetary ions (Rosenbauer et al. (1989); Lundin et al. 1989). The magnitude of the outward flux of oxygen ions, of the order 10^{25} s^{-1}.

Although the failure of the very ambitious "culmination" of the mission with the close Phobos encounter was a staggering disappointment, the mission obviously produced significant results. A set of new methods was tested with the clear perspective to be applied in future planetary missions. It is hoped that this mission will indeed be the precursor of a new wave of studies by spacecraft, leading to the international manned flight to Mars.

APPENDIX: COLLECTIONS OF SCIENTIFIC PAPERS ON MARS MISSIONS

A majority of the published papers on the Mariner and Viking results during and shortly after the missions have appeared in special editions of various journals that were devoted entirely or primarily to these results.

Mariner 4

Science, Vol. 149, No. 3689, 10 September 1965. 6 papers on the initial results one month after Mars encounter.

Mariners 6 and 7

J. Geophys. Res., Vol. 76, No. 2, 10 January 1971.
Final report on scientific findings from Mariner 6 and 7 pictures of Mars. 12 papers.

Mariner 9

Icarus, Vol. 12, No. 1, January 1970. 7 papers on proposed investigations of Mariner 9 (prelaunch).

Icarus, Vol. 17, No. 2, October 1972. 11 papers on Mariner 9 data and 3 on data of Mars 2 and 3, presented at the Symposium on Planetary Atmospheres and Surfaces, held in Madrid, May 1972.

J. Geophys. Res., Vol. 78, No. 20, 10 July 1973 (entire issue). 35 papers on Mariner 9 results.

Icarus, Vol. 22, No. 3, July 1974 (entire issue). 11 papers on Mars, mostly from the International Colloquium on Mars, held in Pasadena, November 1973. Additional papers in Vol. 22, No. 1 and Vol. 23, No. 1.

J. Geophys. Res., Vol. 79, No. 26, 10 September 1974. 11 papers on Mars, all from the International Colloquium on Mars, held in Pasadena, November 1973. Additional papers in Vol. 79, No. 24.

Vikings 1 and 2

Icarus, Vol. 16, No. 1, February 1972 (entire issue). 17 papers on objectives and experiments of proposed Viking mission.

Science, Vol. 193, No. 4255, 27 August 1976. 13 papers on preliminary results up to sixth day after the first landing.

Science, Vol. 194, No. 4260, 1 October 1976. 14 papers on results of first month of Viking 1 at Mars, up to the landing of Lander 2.

Science, Vol. 194, No. 4271, 17 December 1976. 20 papers, mostly on Viking 2.

J. Geophys. Res., Vol. 82, No. 28, September 30, 1977. 53 Papers on the Viking Primary Mission.

Icarus, Vol. 34, No. 3, June 1978 (entire issue). 17 papers on Mars (mainly geology), mostly relating to Viking.

Science, Vol. 204, No. 4395, 25 May 1979. Summary of Lander imaging observations in the first year of the mission.

J. Geophys. Res. Vol. 84, No. B6, 10 June 1979. 17 papers on Viking results, mostly on Mars volatiles, meteorology and dust.

J. Geophys. Res. Vol. 84, No. B14, 30 December 1979 (entire issue). 50 papers on Viking results, generated by the Second International Colloquium on Mars, Pasadena, January 1979.

Icarus, Vol. 45, No. 1, January 1981. Special issue on the results of the Mars Data Analysis Program. 15 papers near the end of mission.

Icarus, Vol. 45, No. 2, February 1981. Special issue on the results of the Mars Data Analysis Program. 14 additional papers.

Mars 4, 5, 6 and 7

Kosmicheskiye Issledovaniya, Vol. 13, No. 1, January-February 1975, pages 3–128. Translations NASA-TT-F-16, 333 to 16,351. Microfiche N75–24611 to 24625.

Kosmicheskiye Issledovaniya, Vol. 13, No. 1, pp. 1–128 (January-February 1975 issue) was entirely devoted to 21 reports of preliminary results from Mars 4, Mars 5, Mars 6 and Mars 7. Fifteen of these are available in English in NASA Translations NASA-TT-F-16,333 to 16,351 and in microfiche N75–24611 to 24625. The initial paper by Moroz (TT-F-16,333; N 75–24615) summarizes the following twenty.

Space Sci. Rev. Vol. 19, 1976, pp. 763–843. Moroz, V. I. (1976). The Atmosphere of Mars.

Phobos Missions

Nature, Vol. 341, No. 6243, 19 October 1989. First results from Phobos 2 at Mars. 15 papers.

Planetary Space Sci., Vol. 39, No. 1–2, Jan–Feb 1991. 41 papers dealing primarily with Phobos 2 results on Mars and Phobos.

Collections of Papers on Mars in General

Icarus, Vol. 18, No. 1, January 1973 (entire issue). 17 papers on Mars, presented at the URSI-IAU-COSPAR Symposium on Planetary Atmospheres and Surfaces in Madrid, including 2 from Mars 2 and 3 from Mariner 9.

Icarus, Vol. 28, No. 2, June 1976. Special Pre-Viking Issue. 11 papers on Mars and 2 on the Mars 6 results on atmospheric argon.

Journal of Molecular Evolution, Vol. 14, No. 1–3, 1979. Special issue on the third anniversary of the Viking mission: The Viking Mission and the Question of Life on Mars. 21 papers.

J. Geophys. Res., 1979, Vol. 84, No. B14: General topics on Mars.

Icarus, Vol. 45, No. 1 & 2, January & February 1981, 30 papers from the Mars Data Analysis Program.

Icarus, Volume 50, No. 2 and 3, May and June 1982. 28 papers presented at the Workshop on Quasi-Periodic Climatic Changes on Earth and Mars, held at NASA's Ames Research Center, February 1981.

J. Geophys. Res., Vol. 87, No. B12, 30 November 1982 (entire issue). Proceedings of the Third International Colloquium on Mars: Dedicated to Thomas A. Mutch, 1931–1980, Pasadena, September 1981. 44 papers on all aspects of Mars.

Icarus, Vol. 66, No. 1, April 1986. 11 papers primarily on dust on Mars.

Icarus, Vol. 71, No. 2, August 1987 (entire issue). 10 papers on the evolution of Mars climate and atmosphere from the culminating symposium of the study project on the Evolution of Martian Atmosphere and Climate (MECA), Washington, DC, July 1986.

LPI Technical Report 8701. Abstracts and summaries of technical sessions of the MECA symposium.

J. Geophys. Res., Vol 95, No. B9, August 30, 1990. 53 papers on many aspects of Mars, many resulting from the Fourth International Conference on Mars; Tucson Jan 10–13, 1989.

4. MAJOR ISSUES AND OUTSTANDING QUESTIONS

ROBERT O. PEPIN
University of Minnesota

and

MICHAEL H. CARR
U.S. Geological Survey, Menlo Park

In this final introductory chapter, we introduce and summarize the physical, chemical, geologic and biological issues and questions, discussed in detail in the chapters that follow, lying at the core of current efforts to understand Mars. These central scientific inquiries emerge from the decades of telescopic observation and spacecraft exploration reviewed above, and from a broad range of comparative experimental and theoretical investigations of other bodies and of processes that have shaped the inner solar system as a whole. Thus, they focus not only on study of the present state and past history of this unique planet itself, but also on Mars as one member of the class of terrestrial planets, and as one product of evolution from the primordial accretion disk. The issue of origin and evolution of Mars' inventory of volatile elements is pivotal to all of these perspectives. It is therefore treated here in some depth, as are the SNC meteorites, both for what these putative Martian samples have to say about volatile distributions, and because of their profound impact on assessments of bulk chemical composition and the chronology of planetary differentiation and late-stage volcanism. We conclude with overviews of the major problems and puzzles arising from the suite of multidisciplinary investigations of this diverse and enigmatic planet that are described in the body of this book.

I. INTRODUCTION

Knowledge of Mars from telescopic observations and spacecraft measurements, and (probably) from the SNC meteorites, is now extensive enough to allow first-order characterizations of the present-day geophysical, chemical, geologic, atmospheric, and biological states of the planet and its interaction with the interplanetary environment. Discussions in the following chapters of this book deal with these characterizations and with the framing of focused questions about how the planet might have evolved to these contemporary states. The growing emphases on deciphering planetary history in the context of inner solar system origin and evolution as a whole, on delineating the nature of the primordial planet and the processes that drove it down its specific evolutionary track, and on assessing and augmenting the current data base in areas needed to address these issues, mark transitions in the scientific investigation of Mars from early reconnaissance phases of exploration to an era of detailed interpretation and informed modeling.

Nevertheless, it will be clear to readers of this book that while our present knowledge of Mars allows us to ask what now seem to be the right questions in intelligent ways, data in crucial areas of investigation are still too sparse to deliver many of the answers. It is of course true that this will always be the case; experience in attempting to understand our own planet demonstrates that each increment of progress generates further questions, demanding in turn more detailed and usually more inaccessible information for their resolution. But it is important to appreciate how early it really is in our study of Mars, with respect to access to the fundamental data needed for further progress on many of the major issues. The SNC meteorites aside, no sample of the planet is yet available for chemical and isotopic analysis in the laboratory; and even the SNC objects, as likely as their origin on Mars may be on geochemical grounds (see Sec. II.A below), are not *proven* Martian samples and in any event were ejected from unknown and geologically restricted sites on the planetary surface. *In situ* geophysical, geologic and meteorological information is limited to measurements by a single seismometer, images from two camera systems, and data from two meteorology stations on the Viking Landers, and to a variety of orbital magnetic field investigations (chapter 3) that still leave open the central question of whether an intrinsic planetary field exists, and if so, what its magnetic moment may be. There are no heat flow data.

However, this is not to say that this book is premature by the decade or more it will probably take to emplace a network of geophysical and meteorological stations across the surface of Mars, explore at least a few of its amazingly diverse geologic provinces with automated rovers, and return from them a representative suite of samples. As noted above, the data we do have in hand, in concert with increasingly sophisticated laboratory simulations, theoretical investigations, and modeling of processes that appear to be central to the evolution and present state of the planet, have not only told us much but also are sharpening the focus, in the detailed chapters that follow, on the many issues and questions that remain. This kind of perspective on Mars is particularly timely just now, at the beginning of what in intent and perhaps in actuality will be a multinational emphasis on Martian exploration. And in many of these areas of scientific inquiry, an additional and equally important perspective is emerging, i.e. that Mars does not exist in isolation, that many of its unknowns will have to be examined in the much broader context of the origin and evolution of all the terrestrial planets, the meteorites, and indeed the primordial accretion disk and the Sun itself.

What are the major scientific issues, and what new data are needed to address them? There is first the overall problem of the chemical and physical state of the present-day interior, surface and atmosphere, and the processes that drive each of these planetary domains and govern the interactions among them and with the interplanetary environment. Outstanding atmospheric questions are summarized in chapter 24. Here we note only a typical example of the impact of a single new datum on interpretations in nominally distinct areas of study: definitive confirmation of the presence or absence of an intrinsic planetary magnetic field would have profound implications for both the present and past structure and thermal regime of the deep interior and for interaction of the planet's ionosphere with the solar wind through time (chapters 5, 30 and 31). We may expect incremental progress in the chemical, geophysical and geologic characterization of the contemporary solid planet from orbital global multispectral, altimetric and gravity measurements and high-resolution visual imaging, but real breakthroughs await the advent of *in situ* network data, intensive automated study of geologically diverse sites on local scales, and sample return.

The histories and interrelation of the geophysical, geologic, and geochemical development of the interior and surface of the planet (chapters 5–23) and of atmospheric evolution, climate change and past biology (chapters 24–35) are two principal and coupled themes that run throughout this book. Each involves a series of outstanding questions. Perhaps the most basic issue involves the need to calibrate relative crater-count chronology by precise radiometric dating, and determine absolute ages for crucial surface signatures of evolutionary processing such as the heavily cratered uplands, volcanic eruptions, and putative sedimentary and evaporite deposits. The SNC meteorites address only the second of these, and that ambiguously in the context of the planet as a whole because their geologic provenance is unknown. However, their complex, intriguing radiometric chronologies—particularly that of the shergottites (Jagoutz and Wänke 1986; Chen and Wasserburg 1986; Jones 1986)—have led several workers to conclude that Mars has been thermally active within the last few 100 Myr. This important deduction of recent near-surface volcanism, together with the many other clues from SNC analyses relating to differentiation history and bulk chemistry (chapter 6), surface composition and weathering (chapters 18 and 19), and atmospheric composition and evolution (Sec. II below; chapters 25 and 32), clearly identify the matter of SNC origin as a central outstanding question (in the words of the author of chapter 25, ". . . do these wretched meteorites come from Mars or not?"). A definitive answer, and with it the validity of applying to Mars what by then will be decades of laboratory work on these objects, will come only when we have sampled the planet directly. As far as the SNC's are concerned, a sample of the current atmosphere by itself, allowing laboratory measurement of oxygen and noble gas isotopic distributions, would probably settle the issue. But addressing the fundamental questions of Martian chronology noted above will unquestionably require intelligent selection and return of solid materials from a wide variety of sites on the planet's surface.

The second of these interconnected evolutionary themes is at least equally rich with open questions. The overriding issue for atmospheric evolution, climate change and past biological activity is whether early Mars ever had a warm dense atmosphere, if so when, and what happened to it. Current approaches to the problem are both comparative and multidisciplinary, as they must be. The questions of initial volatile inventory and atmospheric escape processes involve data, theory, and modeling applied to and assessed against Earth, Venus and the volatile-rich meteorites as well as Mars, and here more detailed knowledge of isotopic distributions in the contemporary atmospheres of Venus and Mars (for Mars beyond that extractable with confidence, or at all, from the SNC's) will be essential for quantum jumps in our understanding in these particular directions. Another dimension of the ancient atmosphere-climate issue is the existence or absence of powerful regolith sinks for the carbon dioxide, nitrogen and water that would have comprised an early dense atmosphere-wet surface system but, if it were once extant, are now missing from observable reservoirs (chapter 32). While it might appear from assessments above that sample return is the *deus ex cosmos* for all too many of the questions we now ask of Mars (and, in fact, there is no doubt of its crucial importance in this area as well), this one can be addressed, at least in part, by remote orbital and Earth-based searches for spectral signatures of carbonates, hydrates and nitrites in surface and airborne dust (e.g., see Pollack et al. 1990*b*) and for near-surface regolith water. Assessment of concentrations of, say, carbonate and adsorbed CO_2 in the deeper megaregolith is another matter, however; this does seem to require returned materials, and moreover ones that will be quite challenging to acquire, by drilling or sampling appropriate crater ejecta, and to transport to Earth in pristine condition.

The question of past and present volatile abundances on and in the planet is central to many of the interpretations and models set out in this book. It crosscuts issues and problems relating to the origin, compositional history, and mass and dynamics of the atmosphere (chapters 6, 25 and 26); to the emergence of a coupled surface-atmosphere system and the evolution of the Martian climate and perhaps its biosphere as volatiles are released, sequestered, or lost to space (chapters 30, 32, 33 and 35); to contemporary surface morphology and its record of past geologic processing (chapters 11–16); to planetary rheology, differentiation and subsequent geophysical history (chapters 5, 7 and 8); and in general to the geochemical, mineralogic and petrologic development of the atmosphere-regolith-polar cap crustal system (chapters

18, 19), the mantle and the core. For these reasons, we will consider the volatiles question in some detail in Sec. II, and conclude in Sec. III with brief summaries of geologic, geophysical, geochemical and biological issues that are discussed more fully in later chapters.

II. INVENTORIES, ORIGIN AND EVOLUTION OF MARTIAN VOLATILES

A. Observable Inventories

In the overall context of deciphering Martian history, the most important volatile species are water, carbon dioxide, nitrogen and the noble gases. Assessments of current inventories, crustal storage reservoirs and capacities and modeling estimates for the present and past abundances of water, carbon dioxide and nitrogen are included in the chapters 32 and 33 discussion of climate change on Mars. The focus in this section is principally on the noble gases, with emphasis on clues and constraints provided by isotopic distributions.

There are two sources of current information on abundances and isotopic compositions of atmophilic species on Mars, today or in the relatively recent past: direct measurements by Earth-based spectroscopy and the Viking mass spectrometers (chapter 25), and indirect inferences from laboratory studies of trapped gases in the SNC meteorites—on the assumption that they are fragments of the Martian crust, ejected from the planet by impact(s) $\leq$200 Myr ago. The SNC measurements are far more precise and informative than Viking data for trace constituent noble gases, and so the status of the case for origin of these objects on Mars is a matter of some importance.

The SNC Meteorites. These meteorites comprise a group of 8 individual stones, collectively called the SNC group after Shergotty, Nakhla and Chassigny, the type specimens for 3 distinct subclasses. All 8 are clearly closely related to each other in petrological, chemical and isotopic characteristics, and as a group they differ in intriguing ways from other classes of stony meteorites. They are extraordinarily young, compared to the usual ~4.5 Gyr age of most such objects; their chronology is complicated, but they appear to have crystallized from melts within the past ~1.3 Gyr, and on this and other grounds they are most readily understood as products of differentiation within a large, thermally active planet. McSween and Stolper (1980) and Wasson and Wetherill (1979) concluded that Mars is their least improbable parent body, a view later supported by Wood and Ashwal's (1981) detailed inferential arguments. However, the basis for speculation of Martian origin for the SNC's was transformed from circumstantial to direct and quantitative experimental evidence by the discovery (Bogard 1982; Bogard and Johnson 1983; Bogard et al. 1984) that shock-generated glassy nodules embedded in the basaltic matrix of the EETA 79001 shergottite contain a noble gas component, not seen in other meteorites, that resembles in striking detail the compositional pattern for noble gases in the Martian atmosphere as measured by mass spectrometers on the Viking spacecraft. Subsequent laboratory work on EETA 79001 revealed a pronounced enrichment of ^{15}N, consistent with the isotopically heavy nitrogen that distinguishes the atmosphere of Mars from virtually all other volatile reservoirs in the solar system (Becker and Pepin 1984; Wiens et al. 1986; Nier and McElroy 1977). Noble gas measurements have been sharpened in precision and expanded to include other individuals in the SNC group (Ott and Begemann 1985; Swindle et al. 1986; Ott 1988), and the data base extended to include carbon dioxide (Wright et al. 1986) in addition to noble gases and nitrogen.

Viking analyses of Martian atmospheric constituents are set out in Table IA, and are compared there and in Fig. 1a with measurements of gases trapped in the glassy lithology of EETA 79001. Both relative and absolute abundances of noble gases and nitrogen in the two data sets are seen to be identical within their respective uncertainties, as are isotopic compositions (Bogard and Johnson 1983; Bogard et al. 1984; Becker and Pepin 1984; Wiens et al. 1986; Swindle et al. 1986) in the two cases ($^{40}Ar/^{36}Ar$ and $^{129}Xe/^{132}Xe$) measurable by Viking. Moreover, the hypothesis that these trapped EETA 79001 gases represent ambient Martian atmospheric species which were shock injected *without either elemental or isotopic fractionation* into the glassy phases of the meteorite is strongly supported by recent laboratory shock experiments (Bogard et al. 1986; Wiens and Pepin 1988; Wiens 1988). Fig. 1b compares absolute elemental abundances of gases emplaced into basalt with those in the gas phase in contact with the sample during shock. Note that except for a somewhat lower efficiency of emplacement in the laboratory experiment, Figs. 1a and b are virtually identical.

These striking agreements, together with the chemical similarity of shergottite material to the nonvolatile silicate fraction of Martian fines (chapter 18) and other, more indirect lines of evidence (Wood and Ashwal 1981; McSween 1985), constitute strong geochemical arguments for origin of EETA 79001, and by extension the SNC group, on Mars. The long-standing dynamical problem of ejecting relatively large SNC "parent" objects from the deep Martian gravitational potential well no longer appears intractable (Vickery and Melosh 1987), although questions still remain about the implications of their cosmic ray exposure age distribution, and the relative fluxes to Earth expected for impact ejecta from Mars vs Moon (Wetherill 1984).

The value of the SNC meteorites as probable Martian derivatives is obvious; they allow us to address questions about the planet that cannot be answered, or even adequately constrained, with only the data now in hand from Earth-based and spacecraft observations. Accordingly, many other physical, chemical, isotopic and chronologic studies of these meteorites have been carried out, and applied in a variety of ways to elucidate the origin,

TABLE I
Volatile Inventories

A. Absolute number densities, in molecules or atoms cm^{-3}, of species measured in the ground-level Martian atmosphere by Viking, and in lithology C (glass) of the SNC meteorite EETA 79001 (from Hunten et al. 1987).

	CO_2 [x10^{17}]	N_2 [x10^{15}]	^{20}Ne [x10^{11}]	^{36}Ar [x10^{12}]	^{40}Ar [x10^{15}]	^{84}Kr [x10^{10}]	^{130}Xe [x10^{8}]	^{132}Xe [x10^{9}]	$\frac{^{129}Xe}{^{132}Xe}$
Viking[a]	2.24	6.3	5.4	1.27	3.82	4.0	—	3.6	2.5
	±0.22	±1.3	[2.2–12.9]	±0.41	±0.75	[1.3–12.6]	—	[1.1–10.0]	[1.5–4.5]
EETA 79001[b]	4.0[c]	6.0[d]	5.3[ef]	1.44[d]	3.25[d]	5.08[fg]	3.87[fg]	2.50[fg]	2.39[f]
	±1.8	±2.0	±0.4	±0.30	±0.74	±0.29	±0.50	±0.32	±0.03

B. Martian inventories of atmospheric carbon dioxide, nitrogen, and nonradiogenic noble gases, in gram/gram-planet, calculated from Viking (CO_2, N_2) and EETA 79001 (noble gas) mixing ratios and footnote (*a*) data from Table IA.

CO_2 [x10^{-08}]	N_2 [x10^{-10}]	^{20}Ne [x10^{-14}]	^{36}Ar [x10^{-13}]	^{84}Kr [x10^{-14}]	^{130}Xe [x10^{-16}]
4.08 ± 0.73	7.3 ± 1.9	4.38 ± 0.74	2.16 ± 0.55	1.76 ± 0.28	2.08 ± 0.41

C. Isotopic compositions of carbon, nitrogen, and noble gases in terrestrial planet atmospheres (for carbon on Earth, in "juvenile" crustal/mantle reservoirs). Units are parts per thousand (‰) deviations from the standard AIR and PDB (Peedee belemnite formation) compositions for $\delta^{15}N_{AIR}$ and $\delta^{13}C_{PDB}$ respectively, and atom/atom for noble gas isotope ratios.

Argon, Neon, Nitrogen and Carbon

	$\frac{^{36}Ar}{^{38}Ar}$	$\frac{^{20}Ne}{^{22}Ne}$	$\frac{^{21}Ne}{^{22}Ne}$ [x10^{-2}]	$\frac{^{15}N^{14}N}{^{14}N_2}$ [x10^{-3}]	$\delta^{15}N_{AIR}$ [‰]	$\frac{^{13}C^{16}O_2}{^{12}C^{16}O_2}$ [x10^{-4}]	$\delta^{13}C_{PDB}$ [‰]
Venus Atmosphere[h]	5.56±0.62	11.8±0.7	?	7.3±1.5	~0±200	112±2	−3±18
Earth Atmosphere[i]	5.320±0.002	9.800±0.080	2.899±0.025	7.353±0.008	≡ 0	111.65±0.15	−6.4±1.3
Mars Atmosphere[j]	4.1±0.2	10.1±0.7	?	11.9±1.2	620±160	< 118	< 50

Krypton

	^{78}Kr	^{80}Kr	^{82}Kr	^{83}Kr	^{84}Kr	^{86}Kr
Earth Atmosphere[k]	0.6087 ± 0.0020	3.960 ± 0.002	20.217 ± 0.021	20.136 ± 0.021	≡ 100	30.524 ± 0.025
Mars Atmosphere[f,g]	0.637 ± 0.036	4.09[m]	20.54 ± 0.20	20.34 ± 0.18	≡ 100	30.06 ± 0.27

Xenon

	^{124}Xe	^{126}Xe	^{128}Xe	^{129}Xe	^{130}Xe	^{131}Xe	^{132}Xe	^{134}Xe	^{136}Xe
Contemporary	2.337	2.180	47.146	649.58	≡100	521.27	660.68	256.28	217.63
Earth Atmosphere[k]	±0.007	±0.011	±0.047	±0.58		±0.59	±0.53	±0.37	±0.22
Nonradiogenic	2.337	2.180	47.146	605.3	≡100	518.73	651.8	247.0	207.5
Earth Atmosphere[l]	±0.007	±0.011	±0.047	±2.9		±0.71	±1.3	±1.3	±1.3
Mars Atmosphere[f]	2.45	2.12	47.67	1640.0	≡100	514.7	646.0	258.7	229.4
	±0.24	±0.23	±1.03	±8.0		±3.7	±8.8	±2.4	±2.4

[a] Atmospheric mixing ratios from Owen et al. (1977) and Pollack and Black (1982). Atmospheric mass taken to be 2.7×10^{19} g, ground-level density 1.7×10^{-5} g cm^{-3}, and mean molecular weight 43.5 u.

[b] Assumed glass density = 3.3 g cm^{-3}.

[c] Wright et al. (1986).

[d] Calculated from measured two-component N-Ar mixing systematics in EETA 79001 (Wiens et al. 1986), assuming $\delta^{15}N = +620‰$ (Nier and McElroy 1977) in the trapped (atmospheric) component.

[e] Wiens et al. (1986).

[f] Swindle et al. (1986).

[g] Becker and Pepin (1984).

[h] von Zahn et al. (1983) for C and N; Donahue (1986) for Ne and Ar.

[i] Schwarcz et al. (1969) for C; Junk and Svec (1958) for N; Eberhardt et al. (1965) for Ne; Nier (1950) for Ar.

[j] Nier and McElroy (1977) for Viking C and N; Wiens et al. (1986) for EETA 79001 Ne and Ar.

[k] Basford et al. (1973).

[l] Pepin and Phinney (1978); Hunten et al. (1987).

[m] Assumed value; original ^{80}Kr contaminated by neutron capture in Br.

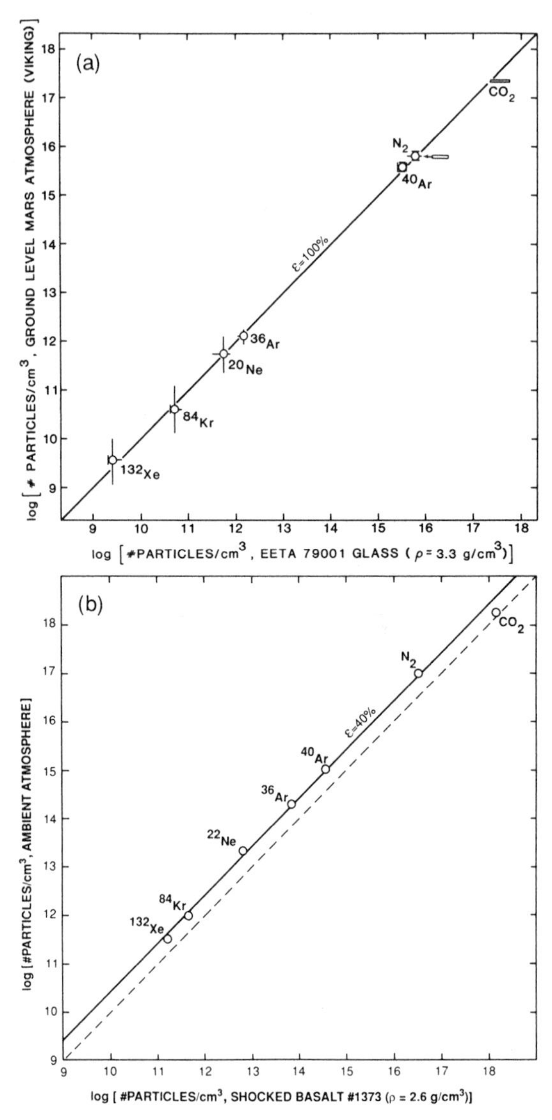

Fig. 1. (a) Martian atmospheric constituents plotted as the log of the number densities at ground level vs the log of the number of particles cm^{-3} in EETA 79001 glass (Pepin 1985; Wiens and Pepin 1988). The heavy line is the least-squares fit to the noble gas and N_2 data, and bisects the plot, giving an effective emplacement efficiency of 100%. N_2 in EETA 79001 is a two-component system; the atmospheric abundance is obtained by apportioning the total N_2 abundance (box) between an indigenous component with $\delta^{15}N \sim 0‰$ (Wiens et al. 1986; Becker and Pepin 1986) and a trapped Martian atmospheric component (plotted datum) with $\delta^{15}N$ assumed to be 620‰ (Nier and McElroy 1977). (b) Ambient atmospheric abundances vs noble gases, N_2 and CO_2 extracted from shocked basalt (Wiens and Pepin 1988), plotted with the same coordinates as (a). The heavy line represents an emplacement efficiency of 40% averaged over noble gases and N_2. The dashed line represents the correlation from (a).

evolution and current state of Mars (e.g., see chapters 5, 6, 7, 18, 32, 35; and Dreibus and Wänke 1985,1987,1989).

B. Elemental and Isotopic Distributions and Constraints

Best estimates of atmospheric CO_2, N_2 and nonradiogenic noble gas inventories on Mars, utilizing both Viking and SNC data sources, are given in Table IB and are compared in Fig. 2, normalized to silicon and relative to solar abundances, with chondritic and other terrestrial planet inventories (Pepin 1991). With the exception of carbon and nitrogen in the CI chondrites, depletions below solar abundances are very large, and are most extreme for Mars. Note, however, that atmospheric C and N on Mars are surely lower limits for initial planetary inventories, as neither possible sequestering of CO_2 in the megaregolith by adsorption or carbonate deposition nor N_2 loss by nonthermal escape to space (chapters 30 and 32) are included. Table IC lists isotope data on nonradiogenic planetary atmospheric species, for Mars from SNC measurements except for Viking determinations of $\delta^{13}C$ and $\delta^{15}N$. The Table IC isotopic distributions are plotted in Figs. 3 and 4 for the C-N and Ne-Ar pairs, together with C/N and Ne/Ar abundance ratios for meteorites, Earth and Venus (Pepin 1991), and in Figs. 5 and 6 for Kr and Xe.

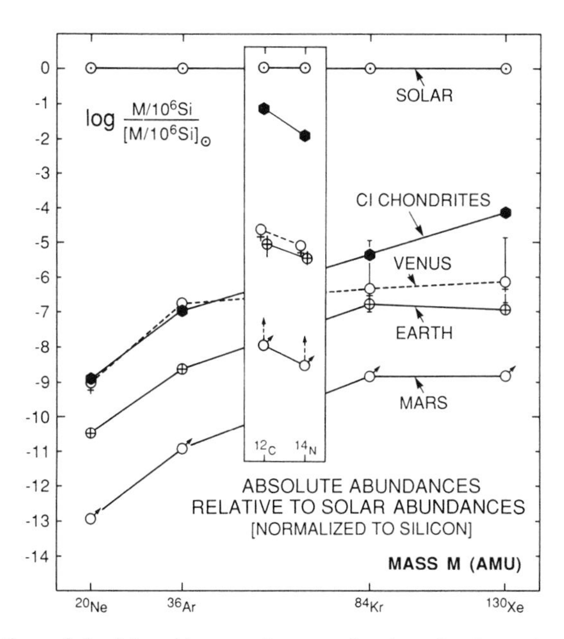

Fig. 2. Depletions of absolute noble gas, nitrogen and carbon abundances in terrestrial-planet atmospheres (atoms per 10^6 Si atoms-planet) and CI carbonaceous chondrites with respect to solar volatile/Si ratios. "Atmospheric" abundances for terrestrial C and N include estimates for crustal reservoirs (Walker 1977).

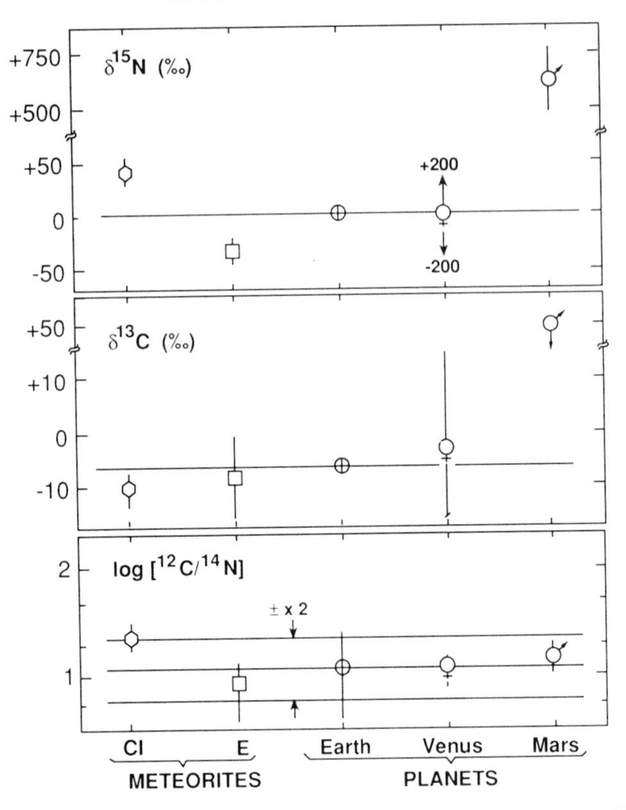

Fig. 3. Elemental and isotopic carbon and nitrogen compositions in carbonaceous (CI) and enstatite (E) chondritic meteorites and in terrestrial planet atmospheres (where the "atmospheric" reservoir for Earth is defined as in Fig. 2). The δ^{13}C value of $\sim +50$‰ for Mars is an upper limit (Nier and McElroy 1977). Units are per mil deviations from standard compositions for δ^{13}C and δ^{15}N, and g/g for the elemental C/N ratios.

Implications of these distributions for the origin and evolution of planetary noble gases have been discussed by Pepin (1989). The central and much-debated question (e.g., see Donahue and Pollack 1983) raised by the Ne-Ar comparisons in Fig. 4, for example, is whether the relatively tight grouping of the planetary Ne/Ar ratios points to a common accretional carrier of these gases, perhaps as bulk CI carbonaceous chondrite material, represented in Fig. 4 by the open CI symbol. The "Q"-component (filled symbol) ratio refers to gases sited on CI grain surfaces (Alaerts et al. 1979; Wieler et al. 1989; Pepin 1991), and is clearly lower, as is the E-chondrite ratio. However, isotopic considerations severely constrain this possibility: Venusian ^{20}Ne/^{22}Ne is $>2\sigma$ above the terrestrial value, and Martian ^{36}Ar/^{38}Ar is much too low. Rather than simple addition of a common "veneer" material, these data ap-

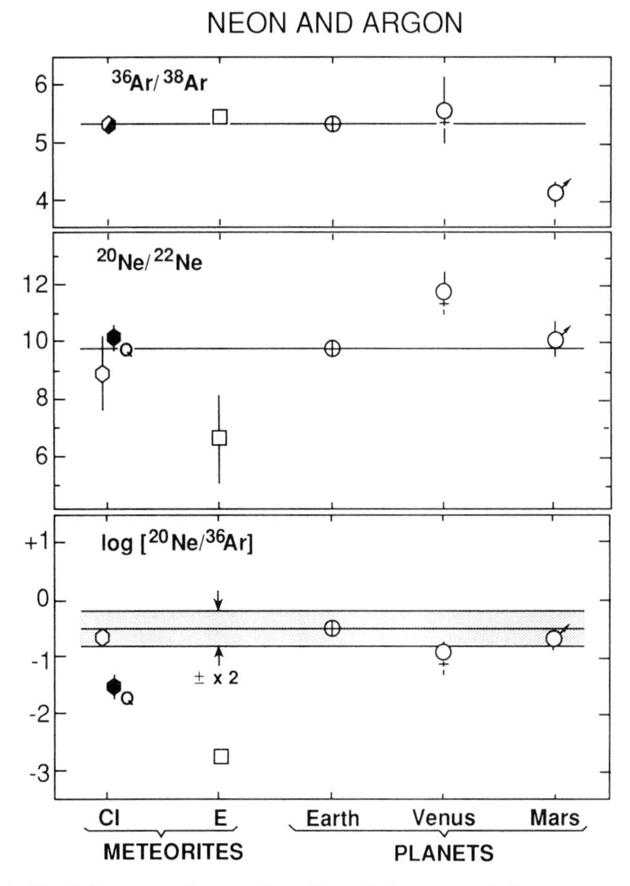

Fig. 4. As in Fig. 3, for neon and argon. Open CI symbols represent bulk gas contents. The filled symbols represent the "Q"-component ratio, which refers to gases sited on CI grain surfaces. Units are atom/atom for isotope ratios, and g/g for the elemental Ne/Ar ratios.

pear to call not only for mechanisms that fractionated isotopes, but also for differing degrees of such isotopic processing on the different planets or in their precursor materials. The same question arises with regard to the C-N distributions shown in Fig. 3. Again the elemental ratios could imply a common carrier, and again, to within factor ~2, it could be a CI-like precursor (however, it is important to note that the Mars atmospheric ratio must fall in the shaded band entirely by accident, as there is no reason to suppose that fractional depletions of original C and N inventories by losses of CO_2 to the megaregolith and N_2 to space would have been the same). Here the isotopic arguments are not yet as definitive as for Ne-Ar, since Martian $\delta^{13}C$ from Viking is an upper limit and $^{15}N/^{14}N$ has been enriched from an unknown primordial composition by nonthermal escape (Nier and McElroy 1977), and

Venusian C and N compositions are too imprecisely known to be diagnostic. One can only say that without isotopic fractionation somewhere in the evolutionary history of these elements, neither CI nor E chondrites are isotopically acceptable as a source of nitrogen, although an appropriate mixture of the two would serve.

Isotopic constraints on the composition of possible primordial reservoirs, and on the nature of evolutionary processing from primordial to present-day planetary compositions, are much firmer for the two heaviest noble gases. The first-order observation from Figs. 5 and 6 is that these meteoritic and planetary isotope distributions tend to be different. The difference is subtle, but still significant, for CI-Earth Kr and Mars-Earth Xe. The one exception may be the solar vs Martian Kr compositions shown in Fig. 5, which are not resolvable within present measurement uncertainties, indicated for solar Kr (Pepin 1991) by the shaded band. The generally good fits of the data to the simple mass-dependent relationships indicated by the various lines and curves in Figs. 5 and 6 point very strongly to the operation of mass-

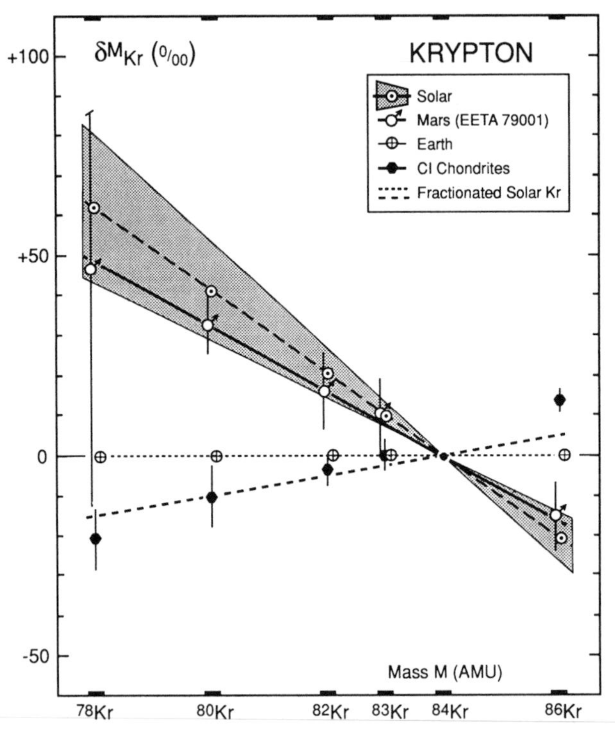

Fig. 5. Isotopic compositions of krypton in CI meteorites, planetary atmospheres and the solar wind, plotted as per mil deviations from terrestrial Kr isotope ratios. The shaded band represents present measurement uncertainties for solar Kr. References for data not given in Table IC are Eugster et al. (1967) for CI-Kr and Pepin (1991) for solar Kr.

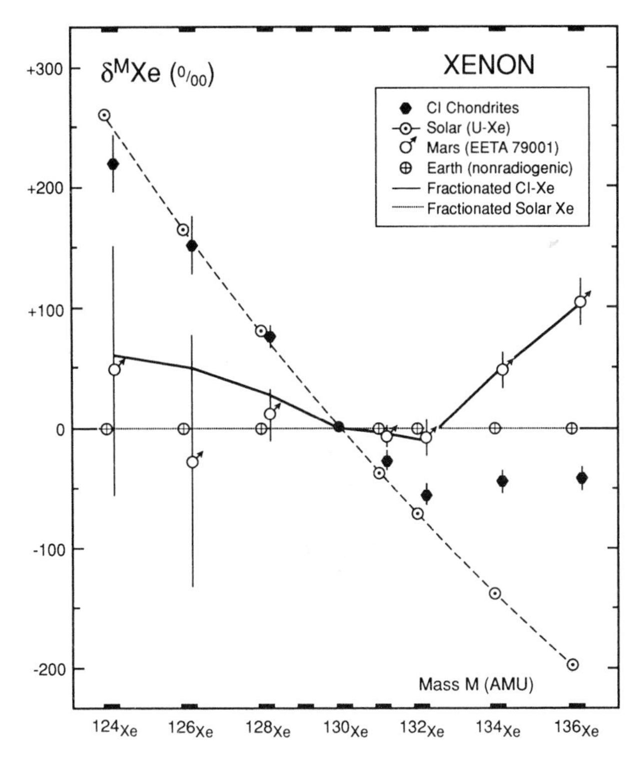

Fig. 6. Isotopic compositions of xenon in CI meteorites, planetary atmospheres and assumed for the primordial nebula (U-Xe), plotted as per mil deviations from the nonradiogenic terrestrial Xe isotope ratios listed in Table IC. References for data not given in Table IC are Pepin and Phinney (1978) and Pepin (1991) for CI-Xe and U-Xe.

fractionating processes; accretion of a common carrier with one fixed isotopic composition for Kr and Xe (e.g., a CI veneer) is equally clearly ruled out as a source for these gases on Earth and Mars (unfortunately there are no isotopic data as yet for these species on Venus). The most straightforward scenario for evolution of these planetary and meteoritic mass distributions from a common primordial parent would seem to be the following (Pepin 1989,1991): (i) For Kr in Fig. 5, generation of CI and terrestrial Kr by mass fractionation of initially solar-composition parental reservoirs, leaving the residual gases isotopically heavier. Kr on Mars apparently experienced a much smaller fractionation, or none at all, and accordingly is still compositionally close to solar. (ii) For Xe in Fig. 6, severe fractionation of solar into terrestrial Xe, and derivation of CI Xe by mixing unfractionated or slightly fractionated solar Xe with a nucleogenetic $^{131-136}$Xe component carried into the solar system by infalling dust grains. Xe on Mars differs from that on Earth in that it seems to have evolved from a nonsolar parent. Swindle et al. (1986) were the first to note that Martian Xe is most simply interpreted as fractionated CI Xe. This is illustrated in Fig. 6, where the solid line, representing a *calculated* composition of mass-fractionated CI Xe, is seen to be an excellent fit to the measured Martian isotope abundances (excluding ^{129}Xe, where a Mars-specific radiogenic contribution dominates). It is interesting that this genesis for Xe on Mars is fully compatible with Dreibus and Wänke's (1985,1987,1989) geochemical modeling (chapter 6), which calls for a large CI accretional component in the planet, and for a massive hydrodynamic blowoff of its primordial atmosphere that could have implemented the required Xe fractionation.

It is clear that the SNC meteorites, if one accepts the evidence for their provenance discussed above in Sec. II.A, have provided relatively precise data on the compositions of contemporary noble gases on Mars, and thus a set of constraints on their origin. These isotopic patterns are at first glance puzzling in that they seem to display contradictory signatures of origin. Xenon superficially resembles terrestrial Xe, but as noted above its detailed isotopic pattern indicates otherwise. The composition of krypton appears close to solar, but the uniquely low ^{36}Ar/^{38}Ar ratio of 4.1 is as far from the inferred solar ratio of $\sim$5.8 (Pepin 1991) as one sees in any nonspallogenic solar system reservoir, and ^{20}Ne/^{22}Ne is Earth-like (Fig. 4). In the following section we turn to an overview of modeling efforts to account for volatile abundances and these noble gas isotopic distributions in the atmospheres of the terrestrial planets.

C. Model Estimates of Past and Present Inventories

Noble gases and other species of high volatility were largely lost from the early planetary system when solid matter separated from the gas phase of the solar accretion disk and the disk's gaseous component later dissipated. Comets, meteorites, planetary atmospheres and of course the Sun are the principal surviving volatile reservoirs. "Solar" or "solar system" abundances of the elements (Cameron 1982; Anders and Ebihara 1982; Anders and Grevesse 1989), based on data from the Sun's photosphere, the solar wind, meteorites (for nonvolatile species), and extrasolar sources, represent our best estimates for the composition of the primordial solar nebula. Relative to these estimates, noble gases in primitive meteorites and the atmospheres of the terrestrial planets are grossly depleted, as shown in Fig. 2, by factors ranging from 4 to 9 orders of magnitude for xenon to 9 to 13 orders of magnitude for neon. There are no data as yet on noble gas abundances in comets. In addition to large and variable elemental depletions, we have seen that each of these contemporary volatile reservoirs displays its own set of isotopic signatures. Distributions of noble gas isotopes in the atmospheres of Venus, Earth and Mars are generally distinct from each other, from meteorites, and from inferred nebular (i.e., "solar") compositions.

Modeling the origin and evolution of terrestrial planet atmospheres, in ways that accommodate these observations in the context of realistic astrophysical environments and processes, is a classic problem in the planetary sciences. An excellent review of different classes of models, and their difficulties, has been given by Donahue and Pollack (1983). Attacks on the problem have tended in the past to focus on deriving the elemental abundance patterns of atmospheric noble gases from primordial meteoritic or nebular reservoirs. Comparatively little attention has been given to isotopic distributions, in particular those of the heavy species krypton and xenon shown in Figs. 5 and 6.

Following Donahue and Pollack (1983), one can divide theories developed up to that time into two general categories of planetary formation, in "gas-rich" or "gas-poor" environments. The first involves planetary growth in the presence of nebular gas, leading to gravitational capture of solar-composition primordial atmospheres, and includes Cameron's (1978,1983) giant gaseous protoplanet hypothesis and the massive, gravitationally captured early terrestrial atmosphere proposed by the Kyoto group (see Hayashi et al. 1985, and references therein). Neither of these has been examined rigorously and comprehensively from the point of view of elemental and isotopic evolution of atmospheric constituents during subsequent loss of the protoatmospheres. However, Sasaki and Nakazawa (1986,1988) have advanced the Kyoto model considerably in this respect by showing that present-day terrestrial Xe could be a fractionated residue of an early massive atmosphere dissipated by hydrodynamic escape.

"Gas-poor" scenarios, in which planetary accumulation occurs after loss of the nebula, can be further subdivided into "veneer" and "grain-accretion" theories (Donahue and Pollack 1983). The first of these model classes attributes atmophilic element inventories on one planet or another to late-stage accretion of cometary (Lewis 1974*a*; Owen 1986,1987; Hunten et al. 1988) or meteoritic (Anders and Owen 1977; see also chapter 32) volatile sources, the second primarily to sorption of nebular gases (Pollack and Black 1979,1982) or implantation of early solar wind gases (Wetherill 1981; McElroy and Prather 1981) in precursor materials from which the planets later accreted. Veneer theories are basically mixing models that aim to reproduce planetary volatile inventories by accretion of known (or occasionally hypothetical) meteorite types in specified proportions. Grain-accretion theories constructed to be broadly applicable to all three terrestrial planets (see, e.g., Pollack and Black 1982) also feature nonuniform mixing of host grains for particular kinds of meteoritic and solar wind gases, together with dust-grain carriers of variable amounts of sorbed nebular gases, to form planetary atmospheres. Bogard (1988), in a more general mixing model approach not tied to any specific view of atmospheric origin, has examined the ability of combinations of known (or suspected) solar system volatile components to replicate atmospheric relative abundance patterns of noble gases, carbon and nitrogen.

Many of the difficulties with these "gas-rich" and "gas-poor" theories are well summarized by Donahue and Pollack (1983). A point to be emphasized here is that the gas-poor veneer and grain-accretion models are intrinsically unable to account in a self-consistent way for the full range of *isotopic* variability seen in planetary atmospheres. The processes of sorption and mixing invoked in these models do not fractionate isotopes, contrary to the strong evidence in the patterns shown in Figs. 4, 5 and 6 that such fractionating mechanisms have been at work. For this reason, gas-rich scenarios that postulate the initial presence of primordial atmospheres, of whatever origin, on protoplanetary bodies or preplanetary planetesimals are fundamentally more attractive, because they offer the potential for isotopic fractionation in the process(es) that subsequently dissipated these atmospheres.

A new generation of evolutionary models is beginning to focus on two such processes. In the first of these, Donahue (1986) has explored a scenario in which mass fractionation results from Jeans escape of pure, solar-composition noble gas atmospheres from large planetesimals. Their atmospheres are assumed to derive from outgassing of nebular gases previously adsorbed on grain surfaces of planetesimal materials, and are subsequently fractionated to different degrees by losses that depend on the rates of planetesimal growth. These bodies later accumulate in various proportions to form the terrestrial planets and their atmospheres. Donahue's model accounts reasonably well for relative Ne:Ar:Kr elemental abundances and for Ne isotopic compositions in the atmospheres of Venus, Earth and Mars. Predicted $^{36}Ar/^{38}Ar$ ratios, however, are much lower than observed values, and variations in Kr and Xe elemental and isotopic compositions in different planetary reservoirs cannot be explained since Jeans escape of these heavy species from the parent planetesimals is essentially nil.

These problems are proving to be more tractable in the context of a related thermal loss mechanism, hydrodynamic escape, which is currently attracting much attention in atmospheric evolution circles (Zahnle and Kasting 1986; Zahnle et al. 1990*a*; Sasaki and Nakazawa 1986,1988; Pepin 1987*a,b*,1989,1991; Hunten et al. 1987,1988,1989). Here atmospheric loss is taken to occur from much larger bodies, i.e., partially or fully accreted planets. In this process hydrogen-rich primordial atmospheres are assumed to be heated at high altitudes by intense extreme-ultraviolet (EUV) radiation from the young Sun after the nebula has dissipated. Under these conditions fluxes of thermally escaping hydrogen can be large enough to exert upward drag forces on heavier atmospheric constituents sufficient to lift them out of the atmosphere, at rates that depend on their mole fractions and masses (Hunten et al. 1987). Lighter species are entrained and lost with the outflowing hydrogen more readily than heavier ones, leading to mass fractionation of the residual atmosphere. Hydrogen escape fluxes high enough to sweep out and fractionate atmospheric species as massive as Xe require energy inputs roughly 2 to 3 orders of magnitude greater than presently supplied to atmospheric exospheres by solar EUV radiation (Pepin 1991). Observations of similar or even larger enhancements in soft X-ray luminosities of solar-type pre-main-sequence stars (Walter et al. 1988; Feigelson et al. 1991) suggest that these levels of EUV activity may well have been reached on the early Sun. The hydrodynamic escape process is particularly attractive for its ability to generate, in an astrophysically realistic environment, large isotopic fractionations of the type displayed by terrestrial Xe with respect to solar Xe, and by Xe in the atmosphere of Mars relative to Xe in the CI meteorites (Fig. 6). One should note, however, that it may not be the only mechanism which could have implemented such fractionations. The possibility that Xe isotopes were gravitationally separated by diffusive equilibration of nebular Xe within porous regoliths on large preplanetary planetesimals was suggested a decade ago by Ozima and Nakazawa (1980), and is currently being re-examined (Zahnle et al. 1990*b*).

The potential power of the hydrodynamic loss mechanism to replicate observed mass fractionation patterns, given adequate supplies of hydrogen and energy, is well illustrated by Hunten et al.'s (1987) application of the process to several specific cases—among them the derivation of terrestrial Xe from solar Xe—and by Zahnle et al.'s (1990*a*) demonstration that isotopic evolution from solar compositions during escape can at least come close to accounting for the present-day compositions of Ne on Earth, and Ne and Ar on Mars. In an effort to assess the viability of hydrodynamic escape as a more general instrument of noble gas evolution, Pepin (1991) has applied it to the full range of elemental and isotopic mass distributions found in contemporary planetary atmospheres and volatile-rich meteorites, and examined the astrophysical and planetary conditions under which the process could reconcile the entire data base from Ne to Xe. This model involves a continuous, solar-driven episode of escape fractionation of primordial (accreted) and secondary (outgassed) atmospheres extending over $\sim$100 to 400 Myr on Venus, Earth and Mars. It appears promising, in that it is capable of generating compositions identical or close to those observed, provided that the EUV flux from the early Sun was intense enough to promote Xe loss (see above), and enough water was accreted (up to a few wt% in each planet) to supply the escaping hydrogen.

It is clear from this overview that two new elements have appeared in current efforts to model the origin and evolution of planetary noble gases and other volatiles—an emphasis on considering isotopic as well as elemental mass distributions, and a realization that at least part of the processing of atmospheric volatiles probably occurred very early, on the planets themselves and in an energetic and rapidly evolving astrophysical environment. One may expect that the evolution of atmospheres on the one hand, and that of the Sun and its accretion disk on the other, will be linked more and more firmly together as these models and their successors are developed.

III. GEOPHYSICAL, GEOLOGIC, AND BIOLOGICAL ISSUES

We have just seen from isotopic evidence that the young Mars may well have had an atmosphere that experienced significant chemical and isotopic fractionation during an early period of massive loss of hydrogen. This phase could have lasted for ~100 to 400 Myr. The state of the planet and its atmosphere at the end of this period is unclear as no recognized geologic record from this epoch is preserved; any such record would have been largely obliterated by the high impact rates. The oldest extant geologic record dates from the time that impact rates declined, which by analogy with the Moon is assumed to be around 3.8 Gyr ago when the planet had already acquired many of its distinguishing characteristics. The dichotomy between the dominantly southern highlands and the low-lying areas to the north had already been established. The Tharsis bulge appears to have already partly formed. Volcanism was widespread, and the valley networks were being eroded. During earlier eras, Mars had accreted, differentiated, and set off on its unique evolutionary path.

The SNC meteorites suggest that Mars accreted with a larger fraction of volatile components than the Earth, and chemical compatibility between the mantle and core further suggests that, in contrast to the Earth, Mars accreted homogeneously (Dreibus and Wänke 1985,1987,1989). Much of its initially large inventory of water probably reacted with Fe to produce FeO, part of which is now in the core, and hydrogen which outgassed to implement the hydrodynamic escape discussed above. Geologic evidence of extensive water action later in the planet's history indicates that not all of the water was lost by reaction with Fe. A major unanswered question is how one reconciles the apparent presence of abundant water on the surface with the apparent chemical equilibrium between the upper mantle and the core. One possibility is that Mars accreted a late veneer of water-rich material that was never folded into the planet because of the lack of plate tectonics.

Mars is presumed to have differentiated into a crust, mantle and core. From the mean density of the planet and its moment of inertia, estimates have been made of the size of the core and the mantle and core densities. The moment of inertia, however, is not well known. The most widely accepted value is 0.365, but it could be as low as 0.345, and this order of difference has significant implications for internal structure and composition (chapters 5 and 6). Models of planet formation, the moment of inertia, and the abundances of chalcophile and siderophile elements in SNC meteorites suggest that the Mars core is more sulphur-rich and less dense than the Earth core, and that Mars' mantle is more iron-rich than Earth's. But model compositions and core size are loosely constrained and will remain so until we have direct seismic measurements of core dimensions and more precise data on the moment of inertia. Related questions concern the magnetic field. If Mars has an intrinsic planetary field, its moment is very small (chapter 31). Nevertheless, as noted in Sec. I, we still need to determine if indeed it has an intrinsic field, and if so what its characteristics are. The planet could have had a larger field in the past, and its record may be preserved in magnetized rocks at the surface. SNC measurements (Cisowski 1986; Sugiura and Strangway 1988) suggest, very tentatively, that the ambient field strength at the time of crystallization of two of these objects was $\leq$1000 gamma ($\leq 1 \mu$T, or $\leq$2% of the Earth's present field).

Even less is known about the dimensions and composition of the crust. We assume, by analogy with the Earth and Moon, that a compositionally distinct crust exists on Mars. It appears almost certain, from morphologic (chapters 11 and 13) and chemical (chapter 18) evidence and from SNC meteorites, that basalts are present at the surface. However, they may be merely a thin veneer of younger volcanics overlaying a primitive crust. Whether this primitive crust is basaltic, anorthositic like the Moon's, granodioritic like the Earth's continents, or some other composition, is unknown. The most probable place for it to have been excavated by impact, and so exposed for direct observation and eventual sampling, is in the cratered highlands.

Crustal materials appear to be unevenly distributed. The cratered highlands, which cover almost two thirds of the planet including most of the southern hemisphere, mostly stand over 2 km above the Mars datum; in contrast, the low northern plains tend to lie at least 1 km below the datum (chapters 7 and 10). The cause of this global dichotomy remains one of the most important unanswered questions of Martian geology. One possibility is that the low-lying northern plains are a gigantic impact scar (Wilhelms and Squyres 1984). Another is that the southern highlands formed at the end of accretion or shortly thereafter as internal convection swept up most of the crustal material into one large protocontinental mass (Wise et al. 1979*b*). These two hypotheses do not exhaust the possibilities (chapters 5, 11 and 12). A better understanding of the nature of the dichotomy may be possible when Mars Observer accurately determines gravity and topographic changes across the boundary, and improves our knowledge of surface compositions.

Superimposed on the dichotomy is another anisotropy caused by the Tharsis and Elysium bulges. Formation of the Tharsis bulge, and to a lesser extent the Elysium bulge, had a pronounced influence on the evolution of the planet through major effects on the histories of rotation (chapter 9), volcanism and deformation (chapters 8, 11 and 13), heat dissipation from the interior (chapter 5), migration of groundwater, and flood eruptions (chapter 15). Yet the cause of the bulges, as well as the configurations of the crust and lithosphere beneath them and how they contrast with the rest of the planet, are almost completely unknown. Are they simply the result of intense volcanic activity and the consequent massive differentiation of the upper mantle? Or are the bulges, at least in part, the result of uplift, possibly induced by convection of the interior? Did their formation involve lateral redistribution of the materials of the planet, or merely vertical redistribution? When did the bulges form, and what effect did their formation have on the rotation of the planet? Is the Tharsis bulge in isostatic equilibrium? And to what extent is it supported by the lithosphere? All of these questions remain unanswered.

Formation of the Tharsis bulge is almost certainly connected with the internal dynamics of the planet and how it dissipates or has dissipated its internal heat. Thermal modeling (chapter 5) suggests that heat is transported from the interior toward the surface by cylinder-like plumes of hot mantle, and that this upward flow is balanced by extensive sheet-like downward flow of colder mantle. One uncertainty is the extent to which heat is lost advectively through volcanism as opposed to conduction. Since volcanic activity appears to have been concentrated in a few local areas such as Tharsis and Elysium throughout much of the planet's history, a major question is how the available heat has been internally focused on these regions. A related question concerns the zoning of the upper part of the planet. The upper regions of Earth are slowly mixed as internal heat is dissipated by subduction and seafloor spreading. Lack of plate tectonics (chapter 5 and 8) inhibits such mixing on Mars, and its crust may therefore have evolved with minimum chemical interaction with the rest of the planet. Does the Martian crust preserve chemical evidence of the late stages of accretion, evidence that has been largely destroyed on Earth by active interaction of the crust with the interior, and which is not preserved on the Moon because of its Earth-related origin?

Formation of Tharsis may have caused a change in the position of the rotational pole. The axis of rotation could also have been altered as a result of large impacts. Possible geologic evidence for polar wander on Mars is discussed in chapters 9 and 11. Although arguments for wander based on current evidence are not widely considered to be compelling, the question of the nature and timing of possible rotational changes, and what signatures these might have left in the geologic record, is very much open.

For much of Mars' history the surface continued to be reshaped by volcanism and tectonism (chapters 8, 11 and 13), and by the action of wind, water and ice (chapters 14–16, 21 and 22). There are many questions of geologic interest. How has the nature of volcanic activity varied with space and time? Is Mars presently volcanically and tectonically active? What is the present near-surface inventory of water and what are the major sinks? Is liquid water present anywhere close to the surface? What evidence is there in the geologic record of both secular and periodic changes in climate?

Perhaps the most intriguing questions about early Mars concern climatic conditions. We saw in Sec. II that during the first 400 Myr or so of the planet's history, Mars may have had a thick hydrogen-rich atmosphere of both primary and secondary origin. However, the nature of the atmosphere that survived until the time that the geologic record emerged ~600 to 700 Myr after the planet formed is very unclear. The main evidence that Mars had a thick atmosphere at this later time is the widespread occurrence of valley networks on terrains that date from this era (chapter 15), and indications that obliteration rates were then much higher than they were subsequently (chapter 12). The simplest explanation for the occurrence of valley networks in old terrains is that conditions for valley formation were commonly met early in Mars' history but only rarely at later times. However, we do not know what conditions were required. If the valleys are fluvial, i.e., eroded by streams of running water, then surface temperatures almost certainly had to be close to or above freezing. This is probably true irrespective of whether the water derived from the ground (sapping) or from the atmosphere (precipitation), because under subzero temperatures streams would have tended to accumulate ice and dam themselves, as they do here on Earth under such conditions. Greenhouse calculations show that surface CO_2 pressures of $\sim$1 to 3 bar would have been needed to enable surface temperatures near 0°C (chapter 32), and thus it has been postulated that ancient Mars had a substantial atmosphere and relatively warm temperatures. This hypothesis appears partly supported by estimates of erosion rates. Low rates (< 1 cm Myr^{-1}) at low latitudes are indicated by the preservation of small craters on surfaces that post-date the end of heavy bombardment ($\sim$3.5 Gyr ago if the lunar bombardment chronology is taken to apply to Mars). However, much higher obliteration rates, perhaps approaching 10 m Myr^{-1}, are inferred from the deficiency of craters < 20 km in the cratered highlands, and the infilling and lack of rims on many of the larger craters.

Both of these lines of evidence underlie the widely held supposition that early Mars had a thick atmosphere and moist surface conditions, and that most of the atmosphere was lost, partly by formation of carbonates, close to the end of heavy bombardment. However, this view rests heavily on the assumption of erosion by streams of water, and one must note that the mode of valley formation and hence the climatic conditions required are not that well understood. While typical valley networks loosely resemble terrestrial valley systems, there are numerous differences, and stream erosion may not be the only way such structures could have formed. Mass wasting, for example, perhaps abetted by ground ice, could have been important. Moreover the presence, albeit rarely, of young valley networks such as those on Alba and in Memnonia suggests that a simple secular change in climate cannot account for all the observations. The younger valleys may have formed in areas of high local heat flow, and the higher global heat flow and steep thermal gradients on early Mars may have contributed to widespread channel formation in ways not yet understood. In addition, although obliteration rates early in Mars' history seem to have been higher than at later times, they were still much lower than most terrestrial rates, and much of the obliteration may have been due to the high impact cratering rate and thus has little climatic relevance. It is difficult to understand how Mars could have maintained an atmosphere thick enough to stabilize liquid water at the surface for hundreds of Myr. CO_2 would be rapidly scavenged from the atmosphere if liquid water were present (chapter 32), and although CO_2 could have been reintroduced into the atmosphere by volcanism and impact, neither mechanism may have been efficient enough to counteract the efficient removal. Finally, if Mars did have a 1 to 3 bar CO_2 atmosphere ~3.5 Gyr ago, where is that CO_2 now? It has been widely postulated that the CO_2 is in carbonate deposits, but none have yet been identified.

In view of such uncertainties in arguments for a warm, wet early Mars, the nature of the ancient climate remains an unresolved and fundamental question. In fact, this is a crucial question for one of the most intriguing unknowns, the possible development of indigenous life on the planet (chapter 35). The present environment is emphatically hostile to life, and prospects for its presence today are poor. The Viking biology experiments failed to detect complex organic molecules or reduced carbon in the soil despite sensitivities in the parts-per-billion range (chapter 34). Moreover, while ground ice may be abundant (chapter 16 and 33), liquid water—universally accepted as essential for life—is unstable everywhere near the surface; the soil is oxidizing in character; the surface ultraviolet flux is high; and availability of nitrogen is limited.

But although the probability for extant Martian life is extremely small, the chance that it ever arose on the planet is more favorable. The oldest fossils on Earth date back to ~3.5 Gyr, by which time terrestrial life had achieved a fair degree of biological sophistication. When it initiated, at some earlier time, Earth's surface was characterized by 2 to 5 times the present global heat flow, high rates of volcanism, relatively high impact rates, oceans and continents, and was arguably surrounded by a N_2-CO_2 atmosphere comparable in thickness to its present atmosphere. The corresponding epoch on Mars *could* have been similar, perhaps with large lakes (which appear to have repeatedly formed) instead of oceans. Some indication of the habitats of the oldest terrestrial life can be derived from nucleotide sequencing on RNA. Phylogenetic trees have been developed that show the unity and relatedness of all life on Earth. Although interpretation of their evolutionary implications is controversial, there is an emerging consensus that the genomic ancestor of all life extant on Earth today was a thermophylic sulfur-metabolizing organism. Whether sulfur-rich hot springs were necessary for life to begin, or, having once emerged, to continue, is unclear. It seems likely, however, that both conditions would have been satisfied on early Mars: abundant evidence exists for extensive volcanism, presumably accompanied by hydrothermal activity as on Earth, and analysis of surface soils and geochemical models of the planet both suggest that sulfur should have been readily available.

As we do not yet know what conditions are required for the beginnings of life, even on Earth, one cannot judge whether they could have been met on ancient Mars. We can only observe that there might have been considerable environmental similarity early in the histories of the two planets, prior to the onset, from whatever cause, of what would have been a biologically catastrophic change in Martian climate from its past to current state. But even if life never did emerge on Mars, evidence may exist for pre-biotic organic evolution. Because of the warm wet conditions that prevailed for most of Earth's history, fossil organics have dehydrogenated and degraded, on geologically short time scales, to an insoluble macromolecular material called kerogen. Thus, chemical remains of early terrestrial life, and organic materials available at the surface prior to its emergence, have not survived. On Mars, however, early reversion of warm wet conditions, if they ever existed, to the present cold dry climate would have favored preservation of the ancient biological record, provided the organics were protected, perhaps in buried sediments, from the oxidizing conditions at the surface. So Mars could be a planet of enormous biological interest, either for the development of life itself, or for the formation of complex organic molecules that ultimately might have led to life.

PART II
Solid Body Geophysics

5. ORIGIN AND THERMAL EVOLUTION OF MARS

G. SCHUBERT
University of California, Los Angeles

S. C. SOLOMON
Massachusetts Institute of Technology

D. L. TURCOTTE
Cornell University

M. J. DRAKE
University of Arizona

and

N. H. SLEEP
Stanford University

The thermal evolution of Mars is governed by subsolidus mantle convection beneath a thick lithosphere. Models of the interior evolution are developed by parameterizing mantle convective heat transport in terms of mantle viscosity, the superadiabatic temperature rise across the mantle and mantle heat production. Geological, geophysical and geochemical observations of the composition and structure of the interior and of the timing of major events in Martian evolution, such as global differentiation, atmospheric outgassing and the formation of the hemispherical dichotomy and Tharsis, are used to constrain the model computations. Isotope systematics of SNC meteorites suggest core formation essentially contemporaneously with the completion of accretion. Ancient fluvial landforms and a high atmospheric D/H ratio imply substantial degassing and *atmosphere formation early in the history of Mars. Accretional considerations also favor initial melting of much of the outer portions of the planet. Initial conditions assumed for the thermal history calculations thus include full differentiation of a silicate mantle and metal-sulfide core and a mantle at a high, near-solidus temperature. The subsequent thermal evolution involves cooling of the mantle and core, differentiation of a crust and thickening of a rigid lithosphere. As a result of the removal of much of the initial interior heat by vigorous mantle convection, perhaps augmented by an upward concentration of heat-producing elements into the crust, the rate of decrease of mantle temperature and surface heat flow was much more rapid during the first 1 Gyr of Martian history than subsequent to that era. Crustal production rates were also much higher in the first 1 Gyr of Martian evolution than later in the planet's history. This temporal behavior is consistent with geologic evidence for a general decrease with time in the Martian volcanic flux and, following a brief initial period of massive crustal formation, early global contraction recorded in the widespread formation of wrinkle ridges on geologically ancient surface units. The thermal evolution models predict a present lithosphere several 100 km thick, in agreement with estimates of elastic lithosphere thickness inferred from the response to major surface loads. Numerical calculations of fully three-dimensional, spherical convection in a shell the size of the Martian mantle are carried out to explore plausible patterns of Martian mantle convection and to relate convective features, such as plumes, to surface features, such as Tharsis. The models have upwellings in the form of cylindrical-like plumes and downwellings in the form of interconnected sheets. There is no single dominant plume. Therefore, if Tharsis is associated with one or more mantle plumes, then the lithosphere beneath Tharsis must be thinned or cracked either to promote plume concentration in the region or to facilitate magma migration through the lithosphere. Thermal history models for Mars admit present core states similar to the Earth, but they also allow completely fluid cores and cores closer to complete freezing. The key parameter distinguishing among these possibilities is the weight fraction of sulfur in the core. If this fraction is $\sim$ 15% or more, then the present Martian core is completely fluid and there is no thermal convective dynamo. The unambiguous determination of the presence or absence of an intrinsic Martian magnetic field by future spacecraft missions will provide an essential constraint on interior structure and thermal evolution models. The eventual measurement of the natural magnetic remanence of rock samples on Mars with ages $>$ 3.5 Gyr will also provide an essential test of thermal history and core dynamo models.*

I. INTRODUCTION

The acceptance of Mars as the parent body of the SNC meteorites (McSween 1984; Bogard et al. 1984; Becker and Pepin 1984; chapter 4) has profoundly changed our view of the planet's evolution. Martian thermal history models of the late 1970s and early 1980s were largely dominated by the idea that the core of Mars formed subsequent to its accretion, after radioactive heating had raised the temperatures in the planet's interior sufficiently above the relatively cold initial temperatures to initiate melting and gravitational separation of Fe-FeS (Johnston et al. 1974; Solomon and Chaiken 1976; Johnston and Toksöz 1977; Toksöz and Johnston 1977; Toksöz et al. 1978;

Toksöz and Hsui 1978; Solomon 1978,1979; Arvidson et al. 1980; Davies and Arvidson 1981). Core formation in these models occurred as late as a few Gyr after Mars accreted, and the segregation of the core lasted for up to 1 Gyr. Late core formation was supported by the notion that Martian surface geology was dominated by extensional tectonics, requiring global heating and planetary expansion until late in the planet's evolution (Solomon and Chaiken 1976; Solomon 1978,1979). However, the U/Pb isotopic composition of SNC meteorites requires core formation at about 4.6 Gyr ago (Chen and Wasserburg 1986), either contemporaneous with accretion or within a few 100 Myr of the end of accretion. We must therefore abandon thermal evolution scenarios of Mars with cold initial temperatures and late core formation and instead adopt the view that accretional heating raised temperatures inside Mars sufficiently high that the core formed early, prior to the end of accretion or within a few 100 Myr thereof, and that Mars began its post-accretional evolution fully differentiated and hot. In this new view of Martian thermal history, early Mars was similar to the larger terrestrial planets Venus and Earth, whose cores formed early as a consequence of high accretional temperatures (see, e.g., Kaula 1979*b;* Wetherill 1985). The post-accretional evolution of Mars, like that of Venus and Earth, is one of secular cooling.

In the following, we discuss the case for a hot initial Mars and early core formation in greater detail. Numerous lines of evidence, in addition to the U-Pb isotopic composition of SNC meteorites, support an early hot, differentiated planet. The structure of the Martian interior is a major factor in determining the post-accretional thermal history of Mars, and we briefly discuss interior structural models inferred from geophysical and geochemical data (see also chapter 7). In accordance with the model of early core and crust formation, we assume that the major radial structure of Mars has been little changed since the end of accretion (except for the lithosphere, which has thickened with time).

The principal characteristics of Mars, i.e., the north-south crustal dichotomy, the Tharsis Rise, the center of mass-center of figure offset, global tectonic patterns and the possible absence of a magnetic field, must all be understood in terms of the thermal history, and we discuss how these characteristics might be accommodated in a model in which Mars steadily cools from a hot, differentiated start. We present quantitative models of Martian cooling history that parameterize heat transport by subsolidus mantle convection. These models allow us to estimate properties of Mars for which no direct measurements presently exist, such as the cooling rate of the planet, its present lithosphere thickness and surface heat flux, and the present extent of inner core solidification. Finally, we use the results of numerical calculations of fully three-dimensional, spherical convection to discuss convective patterns in the Martian mantle. These computations place constraints on possible convective models for the formation of the hemispheric crustal asymmetry and Tharsis.

II. INTERNAL STRUCTURE

Although the seismic experiment on Viking Lander 2 provided no direct information on Mars' internal structure (Anderson et al. 1977; Goins and Lazarewicz 1979; Toksöz 1979), it is likely that Mars is divided into a crust, mantle and core. However, the thicknesses, densities and compositions of these regions are uncertain (see chapter 7).

A. Core

Geophysical and geochemical data constrain the size and composition of the Martian core. The geophysical data include the mean density of Mars (3933 kg m^{-3}; Bills and Ferrari 1978) and its dimensionless axial moment of inertia (0.365, according to Reasenberg [1977] and Kaula [1979*a*]). The dimensionless axial moment of inertia C/MR^2, where C is the principal moment of inertia about the rotation axis, M is the mass of Mars (6.42×10^{23} kg) and R is the radius of Mars (3390 km) is obtained from the inferred value of J_{2H} (the second-degree zonal coefficient in the spherical harmonic representation of the hydrostatic part of the gravitational potential of Mars). To determine C/MR^2 it is necessary first to remove the relatively large nonhydrostatic contribution J_2' to the observed $J_2 (J_2 = J_2' + J_{2H})$. Estimation of J_2' is model dependent and, as a result, the axial moment of inertia of Mars is uncertain. According to Reasenberg (1977) and Kaula (1979*a*), J_2' is principally due to Tharsis (see also Binder and Davis 1973). Under the assumption that the nonhydrostatic contributions to the moments of inertia are symmetric about an equatorial axis through the center of Tharsis, it can be shown that $C/MR^2 = 0.365$ (Reasenberg 1977; Kaula 1979*a*). Bills (1989*a*) has recently suggested that the nonhydrostatic component of J_2 is more likely to be a maximally triaxial ellipsoid, i.e., with the intermediate moment of inertia exactly midway between the greatest and least moments. This assumption leads to a value of C/MR^2 equal to 0.345. Kaula et al. (1989) have argued that the larger value of C/MR^2 is more physically plausible: (1) because the rotation axis of Mars adjusts sufficiently rapidly compared to changes in nonhydrostatic density anomalies that it is the axis of maximum moment of inertia for the nonhydrostatic density field; (2) because the gravity and topography of Mars are dominated by Tharsis; and (3) because tectonic models of Tharsis require generation and support that are almost axisymmetric about a line from the center of Mars through Tharsis.

Goettel (1981) has explored the consequences of the mean density and moment of inertia I of Mars for the radius of the Martian core and the densities of the planet's core and mantle ($I = (C + B + A)/3$, where A and B are principal moments of inertia about two orthogonal axes in the equatorial plane; $I/MR^2 = C/MR^2 - 2J_2/3$; $J_2 = 1.96 \times 10^{-3}$). His results, based on $I/MR^2 = 0.365$, are summarized in Table I. (Spherically symmetric compositional models can only be constrained by I/MR^2 and other spherically aver-

TABLE I
Ranges in Properties of the Martian Core Allowed by the Mean Density and Dimensionless Mean Moment of Inertia[a]

Property	Range
Radius	1300–1900 km
Fractional mass	13–26%
Central pressure	45–37 GPa
Density	8900–5800 kg m^{-3}

[a] $I/MR^2 = 0.365$ assumed. Table after Goettel 1981.

aged observables.) Mars could possess either a small, iron-rich core of high density (a pure Fe core would have a density of 8090 kg m^{-3} and constitute 14.8% of the mass of Mars), or a large, low-density core with substantial S (or other light element). An FeS core would have a density of 5770 kg m^{-3} and comprise 26.3% of the mass of Mars. The core mass increases with its size despite the density decrease (Fig. 1). Goettel's (1981) results for core properties consistent with the mean density and the mean moment of inertia of Mars are in general agreement with the predictions of other models (Johnston and Toksöz 1977; Okal and Anderson 1978; Basaltic Volcanism Study Project 1981). On the basis of all these models, the probable radius of the Martian core lies in the range 1500 to 2000 km, and the fractional mass of the planet occupied by the core is likely to be between 15 and 30%.

Independent geochemical evidence on the mass of the Martian core is provided by the SNC meteorites. Treiman et al. (1987) have determined the

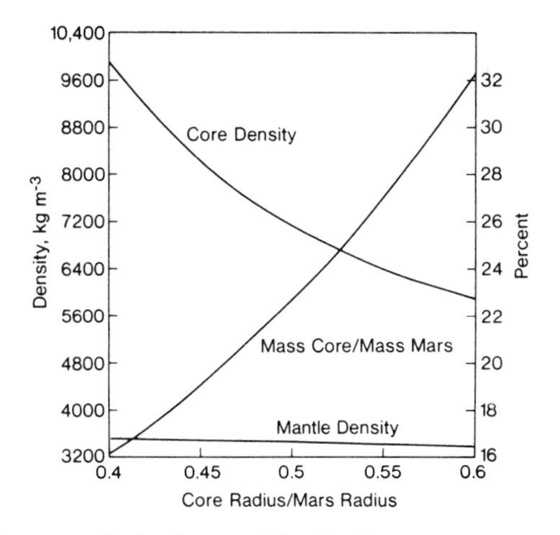

Fig. 1. Core density ρ_c, mantle density ρ_m and fractional core mass vs normalized core radius x for a two-layer model of Mars consistent with the mean density (3393 kg m^{-3}) and dimensionless moment of inertia (0.365). The curves are simultaneous solutions of the equations 3933 kg $m^{-3} = \rho_m + (\rho_c - \rho_m)x^3$ and (0.365)(3933)(2.5) kg $m^{-3} = \rho_m + (\rho_c - \rho_m)x^5$.

abundances of siderophile and chalcophile elements in the mantle of Mars from SNC meteorite abundances (using an element correlation method); they have modeled these abundances in terms of segregation of metal into the core. The abundances of siderophile and chalcophile elements may be substantially matched if Mars accreted homogeneously and a metallic core constituting 25 to 35 wt% of the planet formed in chemical equilibrium with the mantle. Best fits are achieved if the core material during differentiation consisted of ~ 50% solid Fe-Ni metal and 50% liquid Fe-Ni metal containing ~ 25 wt% S. The S content of such a core would be ~ 12.5 wt%. The ratio of Fe to Ni is poorly known, and the relative fractions of the solid and liquid components of both the early and modern core are unconstrained.

Laul et al. (1986) have also calculated the mass and composition of the core by assuming C I abundances of Fe and Ni and a 0.35 × C I abundance of S. From mass balance, and abundances of these elements inferred for the Martian mantle from the compositions of the SNC meteorites, they obtained a core comprising 21.7 wt% of the planet, with an S abundance of 14 wt%. Laul et al. (1986) also conclude that Mars accreted homogeneously and that the core is in equilibrium with the mantle. However, the low core mass that they compute does not yield as good a fit to the depletions of siderophile and chalcophile elements as the larger core masses of Treiman et al. (1987).

We will see in Sec.IV.A that the abundance of S in the core of Mars is a crucial parameter determining the extent to which the core solidifies as the planet cools through geologic time. Unfortunately, the S content of the Martian core must still be considered as uncertain. The agreement between the estimates of Laul et al. (1986) and Treiman et al. (1987) of the wt% S in the Martian core, while encouraging, may only reflect the use of the same limited geochemical data set by both groups.

B. Mantle

Consistent with the core properties summarized in Table I, the thickness of the Martian mantle ranges between ~ 1500 and 2100 km. The zero pressure density of the mantle, for $I/MR^2 = 0.365$, is between ~ 3400 and 3470 kg m^{-3} (Goettel 1981). If the olivine-spinel phase transition occurs in the Martian mantle, it would be found at depths greater than approximately 1000 km (Basaltic Volcanism Study Project 1981; see chapter 7). Still higher-pressure phase changes involving the formation of perovskite and magnesiowustite might occur in the deep lower mantle of Mars depending on the size of the Martian core (chapter 7).

C. Crust

Mars has a distinct low-density crust of variable thickness, as indicated by the partial to complete isostatic compensation of surface topography (Phillips et al. 1973; Phillips and Saunders 1975). The mean thickness of the crust, however, is poorly constrained. A minimum value for the average crustal thickness of 28 ± 4 km (depending on choice of crust-mantle density difference) was obtained by Bills and Ferrari (1978) by fitting a model crust of uniform density and variable thickness overlying a uniform mantle to topography and gravity expressed in spherical harmonics to degree and order 10; for this minimum mean thickness, the crust is of zero thickness beneath the Hellas basin. For a crustal thickness of 15 km at the Viking 2 Lander site, a value derived from arrival times of seismic phases of a possible Marsquake interpreted as waves reflected off the crust-mantle boundary (Anderson et al. 1977), Bills and Ferrari (1978) obtained a mean crustal thickness of 37 ± 3 km. The crustal thickness beneath the Hellas basin in this model is 9 ± 1 km while beneath Tharsis it is 69 ± 8 km. In all the models of Bills and Ferrari (1978), maximum crustal thickness occurs beneath Tharsis and minimum crustal thickness is found beneath the Hellas basin. Sjogren and Wimberly (1981) used topography and gravity data to infer that the Hellas basin is isostatistically compensated at a best-fitting compensation depth of 130 ± 30 km. From the models of Bills and Ferrari (1978), this value corresponds to a globally averaged crustal thickness of ~ 150 km.

The approximately hemispherical division of the Martian surface between the topographically lower and stratigraphically younger northern plains and the heavily cratered southern uplands, often termed the crustal dichotomy, is associated with thicker crust in the southern hemisphere (see, e.g., Janle 1983). The hemispheric crustal dichotomy contributes to the Martian center of mass-center of figure offset, as does the Tharsis bulge. The Martian center of figure is displaced from the center of mass by ~ 2.5 km toward a direction approximately midway between the centers of the southern highlands and the Tharsis bulge (see chapter 7).

III. CONSTRAINTS ON THERMAL EVOLUTION

Among the major constraints on thermal history models of Mars is the origin of the present internal structure of the planet. Viable thermal evolution models must account for the timing and formation of the core, the hemispheric crustal dichotomy, and the Tharsis Rise. Models should provide an explanation for the lack of magnetic field at present, or, if future observations should prove the existence of a small Martian magnetic field, the models should allow for the generation of that field. Thermal history models should be consistent with the evolution of surface stresses and strains as revealed by global tectonic patterns.

A. Core Formation

U-Pb data for several SNC meteorites intercept the concordia curve at about 4.5 Gyr as well as at a younger age variously thought to represent the crystallization age of the shergottites (Jones 1986) or the impact event that resulted in ejection of SNC material from the parent body (Chen and Wasserburg 1986). The 4.5 Gyr "age" indicated by U-Pb data, as well as whole-rock Rb-Sr model ages for SNC meteorites of about 4.6 Gyr (Shih et al. 1982), suggest early global differentiation, including formation of the core essentially contemporaneously with the completion of accretion. Differentiation of a core would heat Mars on average by as much as 300 K (Solomon 1979). A hot initial state for the planet is indicated by these results. Other indicators of a hot early Mars include: (1) the old age (≥4 Gyr) of the southern hemisphere highlands, suggesting early crustal differentiation (see below); (2) geologic (ancient, large flood features and valley networks; Carr 1987) and isotopic (high-atmospheric D/H ratio; Owen et al. 1988) evidence of early outgassing and an early atmosphere; and (3) tectonic evidence of global compression associated with planetary cooling over geologic time (see below). The interpretation of D/H in terms of early loss of water is not unique. Yung et al. (1988) and Jakosky (1990*a*) suggest loss of water over geologic time, while Zahnle et al. (1990*a*) argue for loss of water upon accretion.

Early core formation is made possible by the high accretional temperatures achieved through the burial of heat by large impacts (Kaula 1979*b*; Wetherill 1985). An example of accretional temperature profiles for Mars is shown in Fig. 2 (Coradini et al. 1983). In this example, there is a power-law distribution of impacting planetesimals by size, 30% of the impact kinetic energy is retained as heat, and a 100-Myr time scale of accretion is assumed. For this particular model, melting of a 360-km-thick shell occurs at the end of accretion. Models with higher temperatures and larger degrees of melting

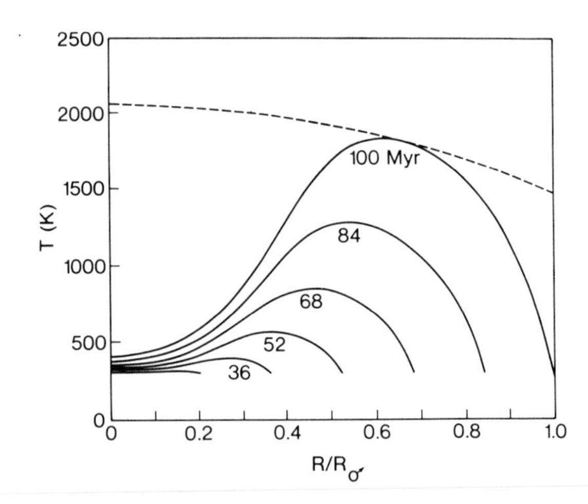

Fig. 2. Model accretional temperature profiles for Mars (figure after Coradini et al. 1983). The calculations are based on a 100 Myr accretion time scale and the assumption that 30% of the impact kinetic energy is retained as heat in the growing planet. The dashed curve is a model solidus temperature profile.

are possible, for example, by adoption of a larger value for the percent of impact kinetic energy retained as heat.

The importance of large impacts in the formation of the terrestrial planets and the validity of the 100-Myr accretional time scale are demonstrated by the planetary accumulation models of Wetherill (1985,1986). The spin and orbital properties of Mars and the other planets, especially their obliquities, are plausibly explained as consequences of impacts with large bodies during the process of planetary formation (Harris and Ward 1982; Hartmann and Vail 1986). A major problem in understanding the growth of Mars according to present models of planetary accretion is that the calculations typically result in objects that are several times more massive than Mars (Wetherill 1985,1986). The size of Mars may have been limited by disturbances to the planetesimal source population for Mars associated with outer-planet secular resonance regions, assuming of course that the formation of the outer planets preceded the growth of Mars (Wetherill 1986).

B. Magnetic Field

Mars may have a weak magnetic field, though this hypothesis is presently a matter of dispute (Russell et al. 1984; Dolginov 1987; see chapter 31). A fit of a dipole to the available Soviet magnetic field data gives a moment of 2×10^{12} Tesla-m^3 (2×10^{22} gauss-cm^3), or 3×10^{-4} times the Earth's moment, tilted $\sim$ 15° with respect to the rotation axis and oriented in the opposite sense to the Earth's dipole moment (Dolginov et al. 1973,1976). A re-evaluation by Russell (1978*b*) leads only to an upper bound on the moment of 2×10^{11} Tesla-m^3, an order of magnitude less than the value of Dolginov and co-workers. Recent Phobos observations of the Martian magnetotail give no indication of an intrinsic planetary magnetic field (Riedler et al. 1989*b;* Ong et al. 1989). Magnetic field measurements in the close vicinity of Mars are needed to resolve this issue.

Several SNC meteorites display natural remanent magnetization (NRM) consistent with a magnetizing field of less than $\sim$ 8 A/m (0.1 oe; Cisowski 1986). The timing and acquisition mechanism of the NRM are not well established, but it has been inferred that the magnetizing event was shock metamorphism during impact (Cisowski 1986). Whether an internal magnetic field on Mars is indicated at the time the SNC meteorites were ejected from the planet is unclear.

C. Crustal Dichotomy

The relative chronology of the various large-scale surface units on Mars is reasonably well established from stratigraphic and crater density relationships (see, e.g., Carr et al. 1973; Tanaka et al. 1988; also see chapters 11 and 12). The oldest units are the cratered terrain, an upland region as heavily cratered as the lunar highlands and occupying approximately the hemisphere south of a great circle inclined 35° to the equator (Mutch and Saunders 1976).

While the absolute chronology of the Martian surface is a matter of debate (see, e.g., Hartmann 1973*b*,1977; Soderblom et al. 1974; Neukum and Wise 1976; Neukum and Hiller 1981; see chapter 12), most workers are in agreement that the heavily cratered southern uplands probably record the terminal phase of heavy bombardment of the inner solar system, dated for the Moon at 3.9 to 4.0 Gyr ago. The bulk of crustal formation must have occurred prior to this time.

The approximately hemispherical dichotomy between the ancient southern highlands and the younger northern plains is generally held to be an ancient first-order feature of the Martian crust. The dichotomy has been ascribed variously to a very long-wavelength mantle convective planform (Lingenfelter and Schubert 1973; Wise et al. 1979*a*), to post-accretional core formation (Davies and Arvidson 1981), and to a giant impact (Wilhelms and Squyres 1984). In view of the evidence from SNC meteorites discussed above, that core separation occurred essentially contemporaneously with accretion, scenarios for the formation of the crustal dichotomy involving late core-mantle segregation (Wise et al. 1979*a;* Davies and Arvidson 1981) may be discounted. Whether the dichotomy was the result of endogenic or exogenic processes, however, remains unresolved despite considerable ongoing photogeological analysis of the dichotomy boundary region (see, e.g., McGill 1988; Wilhelms and Baldwin 1989*a*).

D. Volcanic Flux

Mars shows abundant evidence of surface volcanic activity spanning a wide range of relative geologic ages (see chapter 13). There are numerous volcanic constructs, including the large volcanoes of the Tharsis province, and extensive volcanic plains with apparent flow fronts and other features associated with volcanic units (Carr 1973,1974; Malin 1977; Carr et al. 1977). There are relatively old and relatively young examples of both plains and shields. While the stratigraphic sequence of major volcanic units has been reasonably well established from crater density and superposition relationships (Tanaka et al. 1988), estimates of the absolute ages of volcanic units depend upon knowledge of the scaling of cratering flux and impacting-object size vs crater size from the Moon to Mars, knowledge that is, at best, incomplete (see, e.g., Chapman 1974; Wetherill 1974; Hartmann et al. 1981). Published ages for the stratigraphically young surfaces of the Tharsis shields, for instance, range from 2.5 Gyr (Neukum and Wise 1976) to on the order of 0.1 Gyr (see, e.g., Hartmann 1977).

An important constraint on global thermal evolution is the volcanic flux through time. Estimates of the surface area of volcanic material at each major stratigraphic stage, including corrections for later burial, have been given by Greeley (1987) and Tanaka et al. (1988). Both find 2×10^8 km^2 of volcanic material, though the two analyses differ in detail, particularly in the relative strength of a "peak" in the flux curve at early Hesperian times (corresponding to the formation of the Martian ridged plains) ~ 3 to 3.5 Gyr ago (Tanaka 1986). In the synthesis of Tanaka et al. (1988), the early Hesperian "peak" is quite modest; a monotonic or nearly monotonic decrease of flux with time is implied. Greeley (1987) has suggested that the volume of volcanic material may be estimated by multiplying the area by an average thickness of ~ 1 km. A volume of 2×10^8 km^3 is equivalent to a global layer of volcanic material 1.5 km thick. Because the volume of igneous intrusions accompanying each eruption is generally larger than the volume of volcanic material (by a factor of 10 in continental regions on Earth [Crisp 1984] and perhaps even larger on Mars), a significant fraction of the Martian crust may have been added by igneous activity since the end of heavy bombardment.

E. Global Tectonic Deformation

Large-scale patterns of tectonic deformation on a planetary surface can be direct signatures of global thermal evolution (see chapter 8). For a planet with a globally continuous lithosphere, such as Mars, warming or cooling of the interior will give rise to net global expansion or contraction and thus, respectively, to extensional or compressional horizontal stress and strain near the planetary surface. The magnitude σ_t of thermal stress accumulated in any time interval is given by

$$\sigma_t = [E/(1 - \nu)]\Delta R/R \quad (1)$$

where E and ν are the Young's modulus and Poisson's ratio of near-surface material and $\Delta R/R$ is the fractional change in radius during that time interval (Solomon 1986). The fractional radius change is related to the radial distribution of temperature change $\Delta T(r)$ by

$$\Delta R/R = (1/R^3) \int_0^R r^2 \alpha(r) \Delta T(r) \mathrm{d}r \quad (2)$$

where α is the volumetric coefficient of thermal expansion. For $\alpha = 3 \times 10^{-5}\mathrm{K}^{-1}$, a change ΔT in average interior temperature of 100 K yields $\Delta R/R = 10^{-3}$ or $\Delta R = 3$ km. From Eq. (1), with $E = 80$ GPa and $\nu = 0.25$, such a radius change corresponds to $\sigma_t = 100$ MPa (1 kbar). Sufficiently large thermal stress and strain should be visible in globally distributed tectonic features whose timing and sense of deformation yield strong constraints on the history of internal temperatures (see, e.g., Solomon and Chaiken 1976). Large-scale tectonic features confined to a regional, rather than global, scale are also important indicators of thermal evolution, particularly of the characteristics of heat and strain imparted to the lithosphere by mantle dynamic processes.

The view of Martian tectonics that followed the Mariner 9 mission was that the planet experienced a prolonged period of lithospheric extension which gave rise to the extensive systems of graben mostly centered on the

Tharsis region (Hartmann 1973*c*; Carr 1974*b*), the Valles Marineris canyon system (Sharp 1973) and the pervasive volcanism (Carr 1973). Thermal history models consistent with this view of tectonic evolution involved net warming and global expansion over much of the Martian history. The net warming was attributed variously to late core formation (Solomon and Chaiken 1976), to radioactive heating of the mantle following low-temperature differentiation of a sulfur-rich core (Toksöz and Hsui 1978), or to degassing of the interior and a consequent stiffening of the mantle rheology (Tozer 1985). As noted above, a hot initial state is now indicated for Mars from lead isotope data and planetary accretion considerations, so scenarios involving an initially cool interior (see, e.g., Solomon and Chaiken 1976; Toksöz and Hsui 1978) are not viable. Further, thermal history calculations explicitly including interior degassing and the consequent effect on mantle viscosity (McGovern and Schubert 1989) do not show a secular warming of the mantle as suggested by Tozer (1985).

Much of the evidence for lithospheric extension on Mars is provided from tectonic features in and near the Tharsis area. Although the extensional fractures radiating from the center of Tharsis span a region more than 8000 km across, Tharsis may nonetheless be regarded as a regional feature rather than part of a response to global stress. Further, there are important compressional features located in the ridged plains of Tharsis and oriented approximately circumferential to the center of activity (Wise et al. 1979*b*). Considerable effort has gone into understanding the evolution of the Tharsis province from this regional perspective. The long-wavelength gravity and topography of the region are not consistent with complete isostatic compensation by a single mechanism, such as crustal thickness variations (Phillips and Saunders 1975). Complete local isostasy is possible, however, if a combination of Airy (crustal thickness variations) and Pratt (mantle density variations) mechanisms act in concert; but this is possible only if the crust is relatively thin (or is pervasively intruded by high-density plutonic material) beneath the Tharsis Rise and substantial density anomalies persist to at least 300 to 400 km depth (Sleep and Phillips 1979,1985; Finnerty et al. 1988). Alternatively, a portion of the high topography of Tharsis can be supported by membrane stresses in the elastic lithosphere (Banerdt et al. 1982; Willemann and Turcotte 1982).

These compensation models have been used to predict lithospheric stresses for comparison with the observed distribution of tectonic features. The isostatic model for Tharsis predicts stresses in approximate agreement with the distribution and orientation of extensional fractures in the central Tharsis region and of compressive wrinkle ridges, while the model involving lithospheric support of a topographic load predicts stresses consistent with the more distal extensional features in regions adjacent to the Tharsis Rise (Banerdt et al. 1982; Sleep and Phillips 1985). An evolution in the nature of the support of Tharsis topography has been suggested (Banerdt et al. 1982; Solomon and Head 1982), though the sequence depends upon the relative ages of distal and proximal tectonic features. If the distal features are older, then viscoelastic relaxation of stresses associated with an early episode of lithospheric loading may have led to an essentially isostatic state at present; if the distal features are younger, then a progression from local isostasy to lithospheric support as the Tharsis Rise was constructed may have been the natural consequence of global interior cooling and lithospheric thickening (Sleep and Phillips 1985). This distinction is complicated by the fact that superimposed global thermal stress is apparently required to account for the formation of many of the graben and wrinkle ridges, particularly in regions where both types of features are present.

Recent work on Martian tectonics may sharpen the constraints on global thermal stress. Chicarro et al. (1985) have utilized Viking images to map the global distribution of wrinkle ridges on Mars, including regions distant from Tharsis. They find that ridges occur commonly throughout ancient terrain. In volcanic plains, however, the distribution is highly uneven, with ridges strongly concentrated in the ridged plains units and in spotty occurrences in other regions. The lower Hesperian age (approximately 3 to 3.5 Gyr ago) for most major ridged plains units (Tanaka 1986) and the contrast in ridge density between cratered uplands and young volcanic plains (Chicarro et al. 1985) suggest that ridge formation may have been concentrated in a comparatively early state in Martian evolution (Watters and Maxwell 1986). Examining of crosscutting relations between ridges and graben also supports the view that most ridge formation in the Tharsis region was restricted to an early time period (Watters and Maxwell 1983).

The Martian tectonic history most consistent with all of these findings is one in which Tharsis evolved after the end of heavy bombardment from a primarily isostatic state to one with long-term lithospheric support. Superimposed on the stresses associated with the Tharsis Rise was a globally compressive stress produced by significant interior cooling in the interval 3 to 4 Gyr ago. Any additional cooling (or warming) in the last 3 Gyr has been sufficiently modest so that further changes in planetary volume have not led to widespread development of young compressive (or extensional) features.

F. Lithospheric Thickness

Estimates of lithospheric thickness on Mars provide important constraints on near-surface thermal gradients and thus on heat flux. The thickness of the thermal lithosphere may be inferred approximately from the heights of volcanic constructs, and the thickness of the elastic lithosphere may be inferred from the response to volcanic loads.

Volcanic constructs on Mars show a tendency to increase in height with time of formation, in that the oldest such features are a few kilometers high and the youngest shields are approximately 20 km high with respect to surrounding terrain (Carr 1974*b;* Blasius and Cutts 1976). This relationship has been ascribed to an increase in the hydrostatic head of the magma with time because of a progressive deepening of the source region (Vogt 1974; Carr 1976). Assuming a relative density contrast of 10% between magma and average overburden, and ignoring viscous head loss, these heights give depths to magma chambers varying from perhaps as little as a few tens of km to somewhat over 200 km over the history of Martian shield formation.

The thickness T_e of the elastic lithosphere of Mars has been estimated from the tectonic response to individual loads (Thurber and Toksöz 1978; Comer et al. 1985) and from the global response to the long-wavelength load of the Tharsis Rise (Banerdt et al. 1982; Willemann and Turcotte 1982). The spacing of graben circumferential to the major volcanoes Ascraeus Mons, Pavonis Mons, Arsia Mons, Alba Patera and Elysium Mons indicate values for T_e of 20 to 50 km—equivalently, values of flexural rigidity D of 10^{23} to 10^{24} N-m (10^{30} to 10^{31} dyn cm) at the times of graben formation (Comer et al. 1985). For the Isidis basin region, the elastic lithosphere thickness is inferred to have exceeded 120 km ($D > 10^{25}$ N-m) at the time of mascon loading and graben formation (Comer et al. 1985). The absence of circumferential graben around Olympus Mons requires the elastic lithosphere to have been at least 150 km thick ($D > 3 \times 10^{25}$ N-m) at the time of loading (Thurber and Toksöz 1978; Comer et al. 1985; Janle and Jannsen 1986). Models of the response of Mars to the long-wavelength topography of the Tharsis Rise provide a reasonable fit to the geoid and to the distribution of tectonic features in the Tharsis province if the elastic lithosphere of Mars is globally $\sim$ 100 to 400-km thick, corresponding to $D = 10^{25}$ to 7×10^{26} N-m (Banerdt et al. 1982; Willemann and Turcotte 1982).

The values for T_e derived for individual loads are not consistent with a simple progressive increase with time in the thickness of the elastic lithosphere of Mars. The largest estimates of T_e, for instance, are for perhaps the oldest (Isidis mascon) and youngest (Olympus Mons) features considered (Tanaka et al. 1988). Spatial variations in elastic lithosphere thickness must have been at least as important as temporal variations (Comer et al. 1985). In particular, there appears to have been a dichotomy in lithosphere thickness that spanned a significant interval of time, with comparatively thin elastic lithosphere (T_e = 20 to 50 km) beneath the central regions of major volcanic provinces and substantially thicker elastic lithosphere (T_e in excess of 100 km) beneath regions more distant from volcanic province centers and appropriate for the planet as a whole.

The values of T_e may be converted to estimates of the lithospheric thermal gradient and heat flow, given a representative strain rate and a flow law for ductile deformation of material in the lower lithosphere and estimates of lithospheric thermal conductivity. Under the assumption that the large values of elastic lithosphere thickness determined from the local response to the Isidis mascon and Olympus Mons and from the global response to the Tharsis Rise exceed the thickness of the Martian crust, the depth to the base of the mechanical lithosphere is determined by the ductile strength of the mantle. The minimum values of T_e for the Isidis mascon and Olympus Mons correspond, by this line of reasoning, to mean lithospheric thermal gradients of no greater than 5 to 6 K km^{-1} and heat flow values $<$ 17 to 24 mW m^{-2} (Solomon and Head 1990). For the Tharsis Montes and Alba Patera, the mechanical lithosphere thickness is likely governed by the strength of crustal material. The mean thermal gradients consistent with the values of T_e for these loads under this assumption fall in the range 10 to 14 K km^{-1} and heat flow values in the range 25 to 35 mW m^{-2} (Solomon and Head 1989). Essentially contemporaneous temperature differences of at least 300 K at 30 to 40 km depth are implied at a late stage in the development of the Tharsis province. Such temperature differences are too large to be solely the effect of large impacts that occurred some Gyr earlier (Bratt et al. 1985), but they are broadly similar to the temperature variations associated with lithospheric reheating beneath hot-spot volcanic centers on Earth (McNutt 1987). The temperature and heat-flow anomalies beneath the central regions of major volcanic provinces on Mars are presumably also related to mantle dynamic processes, such as convective upwelling plumes and magmatism. Lithospheric thinning beneath the central regions of major volcanic provinces by hot, upwelling mantle plumes can account for the different estimates of elastic lithosphere thickness in these regions as compared with the global average.

G. Mantle Heat Sources

Of particular importance to the thermal evolution of Mars are the concentrations of the radiogenic heat-producing elements K, Th and U in the Martian mantle. Estimates of the abundances of these elements have been made by Taylor (1986), Treiman et al. (1986) and Laul et al. (1986) using SNC meteorites. Taylor (1986) plots analyses of the meteorites on a K/U vs K diagram and shows that the SNC meteorite data are consistent with terrestrial data. The K/U ratio in Mars is not distinguishable from the Earth's, but this analysis does not yield information on the bulk K abundance in Mars.

Treiman et al. (1986) and Laul et al. (1986) have estimated abundances of K, Th and U using their correlation with other refractory lithophile elements. The database used by the two sets of authors is substantially the same. Correlations of K with La show that the mantle of Mars has a K/La ratio of $\sim$ 0.3 of the C I ratio. The corresponding Th/La and U/La ratios are $\sim$ 1.7 and 2 times the C I ratio, respectively (Treiman et al. 1986). Uncertainties in each estimate are factors of 2, 1.7 and 2, respectively. The abundances of K, Th and U, assuming that the Martian mantle has C I abundances of La, are summarized in Table II. If the Martian mantle has higher abundances of La, the abundance of K, Th and U must be scaled upward accordingly. The bulk Martian abundances must be reduced in proportion to the mass of the core, assuming that these elements are excluded from the core. The ratios of K/U

TABLE II
Radiogenic Element Abundance Estimates for the Mantles of Mars and Earth

	K (ppm)	Th (ppb)	U (ppb)	K/U	Th/U
Mars[a]	170	48	16	10^4	3
Mars[b]	315	56	16	2×10^4	3.5
Earth[c]	257	102.8	25.7	10^4	4

[a]Treiman et al. 1986.
[b]Laul et al. 1986.
[c]Turcotte and Schubert 1982.

and Th/U are $\sim 10^4$ and 3, respectively, the former being indistinguishable from the Earth's ratio while the latter is lower than the terrestrial value of ~ 4. There is uncertainty in each ratio of about a factor of 2 based on the scatter in the raw data and the lack of knowledge of the absolute values of refractory-element abundances such as La in Mars relative to C I chondrites. Given the uncertainties in these values, we conclude that radiogenic element abundances in the mantles of Mars and Earth are broadly similar.

IV. THERMAL HISTORY MODELS

On the basis of the constraints discussed in the previous section, we present quantitative simulations of Martian thermal history. In its initial state, Mars is hot and completely differentiated into a core and mantle. The mantle temperature is essentially at the solidus and the core is superliquidus. Radiogenic heat sources are assumed to be distributed uniformly through the mantle, although upward concentration of radioactive elements could accompany differentiation of an early crust. The subsequent evolution consists of a simple cooling, with monotonic decreases in temperature, heat flux and convective vigor and a monotonic increase in the viscosity of the mantle. We parameterize convective heat transport through the mantle by a simple Nusselt number-Rayleigh number relationship. The parameterization approach is well established as a way of investigating the thermal evolution of the planets (see, e.g., Schubert et al. 1979; Sharpe and Peltier 1979; Stevenson and Turner 1979; Cook and Turcotte 1981). We employ two different parameterization schemes. With the parameterization model of Stevenson et al. (1983) we focus on the cooling of the core, the extent of core solidification and the generation of a planetary magnetic field by a core dynamo. With the parameterization model of Turcotte et al. (1979), we emphasize crustal differentiation.

A. Coupled Core-Mantle Evolution

The planetary thermal history model of Stevenson et al. (1983) is employed in this section to study the consequences of core cooling for the thermal and physical state of the Martian core and the generation of a magnetic field by thermal or chemical compositional convection in the core. The model includes: (1) mantle radiogenic heat production; (2) a mantle viscosity directly proportional to exp(H/RT), where H is a constant activation enthalpy, T is a mantle temperature and R is the gas constant; (3) heat transfer by whole-mantle subsolidus convection parameterized by a Nusselt number-Rayleigh number relation; (4) coupled energy balance equations for the mantle and core; (5) possible inner core freezeout with exclusion of a light-alloying element (most likely sulfur) which then mixes uniformly through the outer core; and (6) realistic pressure- and composition-dependent freezing curves for the core. More detailed information about the model equations and parameter values can be found in Stevenson et al. (1983) and Schubert and Spohn (1990), who have recently extended the Mars model to include lithospheric growth and the magnetic dipole moment. The solutions we discuss below are actually new calculations, based on the Mars model of Schubert and Spohn (1990), of cases that are almost identical to ones presented in Stevenson et al. (1983). Parameter values are given in Stevenson et al. (1983) and Schubert and Spohn (1990).

Model cooling histories of Mars are typified by the results in Fig. 3, which shows mantle heat flow vs time for models with initial core sulfur concentrations x_S of 10 and 25 wt% (sulfur is assumed to be the light-alloying element in the core in these models). The models with x_S = 10% and 25%

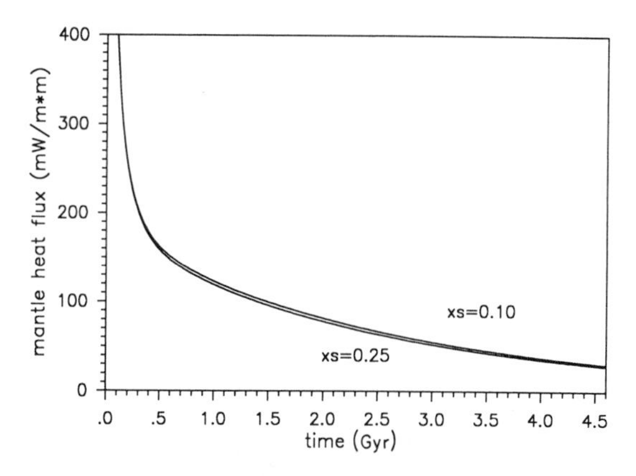

Fig. 3. Heat flow from the mantle vs time for two models of Martian thermal history. Details of the models are given in Stevenson et al. (1983). The models have initial core sulfur concentrations x_S of 10 and 25 wt%, respectively.

are essentially identical to models M1 and M2, respectively, of Stevenson et al. (1983). During the first few 100 Myr of evolution, when the planet is hot and its mantle is convecting particularly vigorously, there is a dramatic decrease in heat flow. Following this early period of rapid cooling is a phase of gradual, slow cooling lasting most of the geologic life of the planet. The present heat flow in these models is ~ 30 mW m^{-2}.

The decrease of mantle temperature with time in these models (Fig. 4) occurs, like the mantle heat flow, in an early period of short and dramatic temperature reduction followed by a decrease of only 200 to 300 K over the last 4 Gyr of the planet's history. The lithospheres in these models grow to thicknesses of ~ 100 km at the present (Fig. 5). Lithosphere thickness and mantle heat flow and temperature are largely independent of the sulfur concentration in the core. The models have thermal boundary layers at the base of the mantle that at present are ~ 100 km thick. Relatively small temperature increases across these bottom boundary layers raise the present core-mantle boundary temperatures of the models ~ 100 K above the mantle temperatures.

The present lithosphere thicknesses in the models of Fig. 5 are smaller, by perhaps a factor of 2 or more, than what would be expected based on our discussion of lithospheric thickness in the previous section. The lithosphere thickness calculated in the models is the thickness of the rheological lithosphere. It is assumed in the models of Fig. 5 that a temperature of 1073 K marks the base of this rheological lithosphere. At temperatures higher than 1073 K, the Martian mantle is taken to be readily deformable on a geologic time scale. Lithosphere thickness is strongly dependent on the concentration of radiogenic heat sources in the mantle. If the mantle heat-source density were half the value assumed in the models of Fig. 5, consistent with the

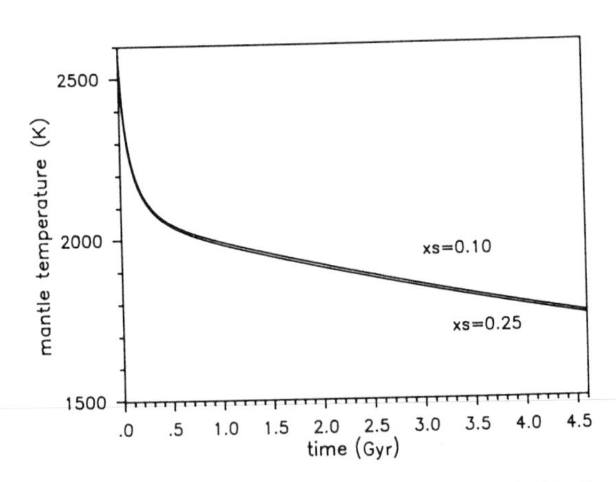

Fig. 4. The decrease of characteristic mantle temperature with time in the Martian thermal history models of Fig. 3.

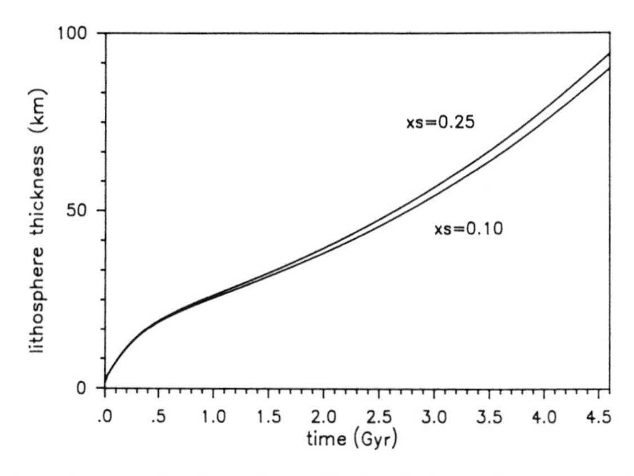

Fig. 5. Thickening of the Martian lithosphere with time in the cooling models of Fig. 3.

estimates of mantle heat-source densities from SNC meteorite data (refer to our discussion in the previous section), then the present lithosphere thickness would be about 200 km (Spohn 1990). Lithosphere thickness would also be increased by magmatic heat transfer (Spohn 1990) and by upward differentiation of radiogenic heat sources from the mantle into the crust. Thermal history models discussed later in this chapter (Sec. IV.B) demonstrate that depletion of mantle heat sources with time as a consequence of crustal differentiation results in a thicker lithosphere at present. The present lithosphere thicknesses calculated in the models of Schubert et al. (1979) are representative of maximum thicknesses as those models contained no mantle heat sources.

The decrease with time in the heat flow from the core is shown in Fig. 6

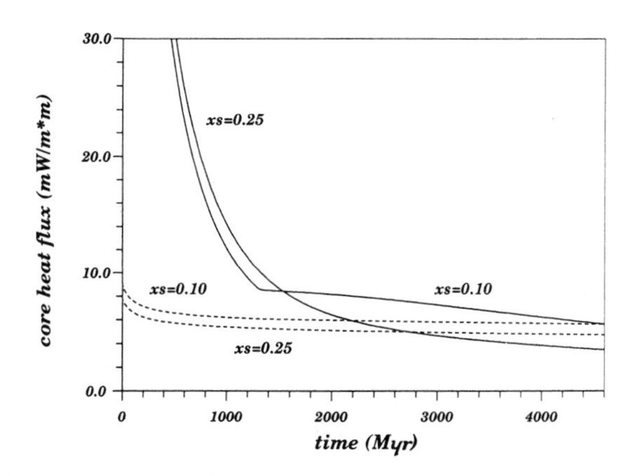

Fig. 6. Core heat flux as a function of time in the Martian thermal evolution models of Fig. 3. The dashed line marks the conductive heat flux along an adiabat in the liquid outer core. The sharp bend in the curve for the model with x_S = 0.10 denotes the onset of inner-core solidification.

for models with x_S = 10% and 25%. Inner-core growth in the x_S = 10% model is marked by the sudden change in the rate of core cooling at ~ 1.3 Gyr after core cooling begins. After 4.5 Gyr, the inner core in this model is about 920 km in radius, leaving an outer liquid core ~ 670-km thick. When inner-core growth in this model begins, the core heat flow is above the estimated value of heat flow conducted along the core adiabat (the upper dashed curve in Fig. 6). Convection in the outer core after ~ 1.3 Gyr is maintained by both thermal and chemical buoyancy, the latter arising from the release of gravitational potential energy upon concentration of the light-alloying element into the outer core. The slow decline in core heat flow after ~ 1.3 Gyr is mainly a consequence of latent heat and gravitational energy release upon inner-core growth. Core heat flow in the x_S = 10% model falls below the heat flow conducted along the adiabat just after 4.5 Gyr. Core convection beyond 4.5 Gyr would still occur but it would be driven entirely by compositional buoyancy. Thermal convective transport after 4.5 Gyr would actually be downward in the core, but the compositional buoyancy would be adequate to offset the slightly stable thermal state. In the x_S = 10% model, a magnetic field would be generated by thermal convection prior to inner core solidification at ~ 1.3 Gyr. Both thermal and chemical convection would support a dynamo for times between 1.3 and 4.5 Gyr. Subsequent to 4.5 Gyr, a magnetic field would be produced by compositionally driven convection associated with inner-core growth.

The sulfur-rich model (x_S = 25%) does not nucleate an inner core. Thermal convection ceases in this model after ~ 2.7 Gyr, when the core heat flux falls below the heat flux conducted along the core adiabat (lower dashed curve in Fig. 6). There is no dynamo action in the core or planetary magnetic field subsequent to 2.7 Gyr. Stevenson et al. (1983) have used the results of the model calculation with x_S = 25% to estimate the smallest initial sulfur fraction for which no inner-core freezeout would occur after 4.5 Gyr; they obtain a value of ~ 15 wt% S.

Schubert and Spohn (1990) have carried out a number of additional Martian thermal history calculations with a view toward more carefully delineating conditions for no inner-core freezeout. Their results are shown in Fig. 7, which gives the dependence of present inner-core radius (as a fraction of total core radius) on both present mantle viscosity and initial wt% sulfur x_S in the core. Present inner-core radius increases with decreasing x_S or decreasing present mantle viscosity because the former increases the core melting temperature and cooling is more rapid with the latter; both result in earlier inner-core freezeout and a longer period of inner-core growth. Fifteen wt% is a good estimate of the minimum core sulfur concentration required for no core freezing through geologic time, unless the present mantle viscosity is $\lesssim 10^{16}$ $m^2 s^{-1}$ (or ~ 3.5×10^{19} Pa s for a mantle density of 3500 kg m^{-3}).

Figure 8, after Schubert and Spohn (1990), shows inner-core radius vs time for three different initial S concentrations. Inner-core growth is very

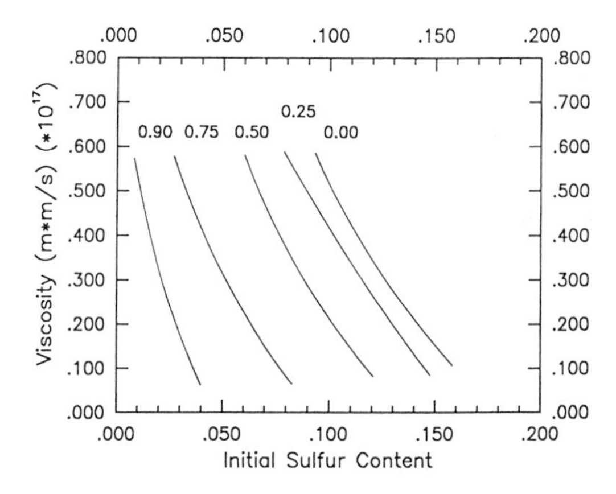

Fig. 7. Contours of fractional inner-core radius as a function of present mantle kinematic viscosity and initial weight fraction sulfur in the core (figure after Schubert and Spohn 1990).

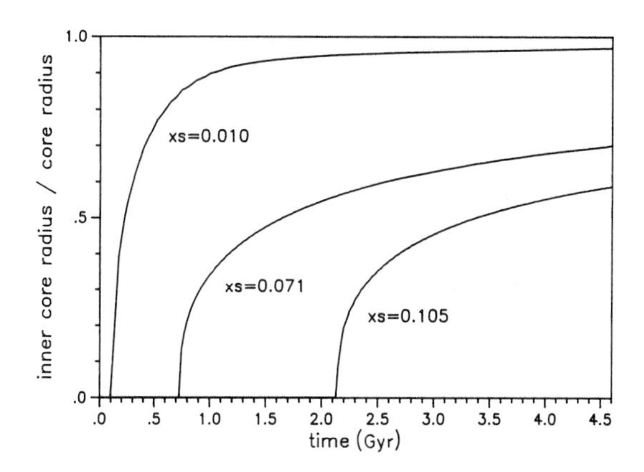

Fig. 8. Fractional inner-core radius vs time for model cooling histories of Mars with three different initial sulfur concentrations x_S in the core (figure after Schubert and Spohn 1990).

rapid once freezeout begins; the inner core is almost completely grown within 0.5 to 1 Gyr of initial freezeout. The depression of the melting temperature in the outer core as the sulfur concentrates there with progressive inner-core growth and the slowing of the cooling rate as the planet evolves both contribute to the reduction in inner-core growth rate with time. With increasing x_S, the time of initial core freezeout is delayed and the present inner-core radius decreases.

If the estimates of Laul et al. (1986) and Trieman et al. (1987) of 14 and 12.5 wt% S in the Martian core are taken to imply that x_S is $\lesssim$ 15 wt%, then based on Fig. 7, it is likely that Mars has a solid inner core. If Mars has no magnetic field, the explanation may then lie in the nearly complete solidification of the core (Young and Schubert 1974). On the other hand, if the Laul et al. (1986) and Treiman et al. (1987) predictions of wt% S in the Martian core are underestimates, then according to Fig. 7, Mars would not have a solid inner core at present, and the explanation for lack of a present Martian magnetic field might lie in the absence of a drive (a growing inner core) for chemical convection in a core that had cooled too far to convect thermally at present (Stevenson et al. 1983). Our knowledge of Mars is inadequate to distinguish unambiguously between these alternative possibilities, although complete freezing of the Martian core probably requires an unreasonably low content of S in the core. The thermal history models are also consistent with a Mars that presently has a growing solid inner core, a dynamo in its liquid outer core driven by chemical compositional convection, and a planetary magnetic field, albeit a small one. Figure 7 shows that if x_S is $\lesssim$ 15 wt%, but not too small, then the core of Mars would only be partially solidified at present and a Martian dynamo and magnetic field would be likely.

This is illustrated in Fig. 9, which shows the results of a model calculation of the Martian magnetic dipole moment μ_M, normalized with respect to the Earth's present magnetic dipole moment, as a function of time for the Mars model with $x_S = 10\%$ and 25%. The calculation of μ_M follows Stevenson et al. (1983) and Schubert and Spohn (1990). The magnetic dipole moment decreases rapidly during the first several 100 Myr of evolution concomitant with the early rapid cooling of the planet and the rapid decline in core heat flow. With $x_S = 10\%$, dynamo action occurs throughout the planet's evolution. Prior to 1.3 Gyr, the model Martian dynamo is driven by thermal convection in the core. There is a sudden increase in μ_M at ~ 1.25 Gyr coin-

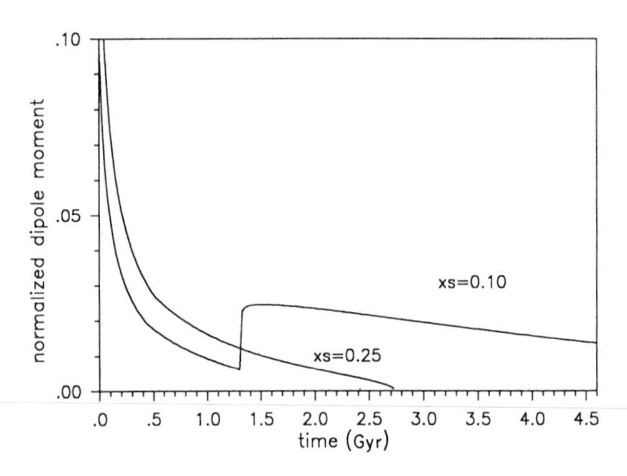

Fig. 9. Normalized magnetic dipole moment vs time for the models of Fig. 3. The dipole moment is normalized with respect to the value of the Earth's present magnetic dipole moment computed following the method presented in Schubert and Spohn (1990).

cident with inner-core formation. Subsequent to 1.3 Gyr, the dynamo is driven by both thermal and chemical compositional convection as the inner core solidifies and releases latent heat and gravitational potential energy. With $x_S = 25\%$, there is no inner-core solidification and the dynamo is driven solely by thermal convection until about $t = 2.7$ Gyr when dynamo action ceases because the core heat flux falls below the conductive heat flux along the core adiabat and thermal convection can no longer occur in the core. There is no present magnetic field in this model of Martian evolution because thermal convection is not possible in the present liquid core and there is no source of chemical convection in the core. Because of the uncertainty in the exact value of the conductive heat flux along the Martian core adiabat (Schubert and Spohn 1990), we cannot exclude the possibility that at present there is a weakly thermally convecting liquid core in Mars and a weak dynamo. If it should be determined that Mars has a small magnetic field, then such a model would provide a plausible explanation.

B. Crustal Differentiation

Magmatism in planetary interiors results in crustal formation and the removal of heat-producing radiogenic elements from the planetary mantle. The removal of heat sources reduces the vigor of mantle convection, allows the mantle to cool more rapidly, increases the lithosphere thickness, and reduces surface volcanism. In this section we provide quantitative estimates of the rate of crustal magmatism by extending the approach of Turcotte et al. (1979,1980) to the parameterization of convective cooling in the mantle of Mars. We refer the reader to Turcotte and Huang (1990) for the details of this approach and discuss here only the modifications necessary to simulate crustal production and the depletion of the mantle in radiogenic heat sources. A recent study of Martian thermal history with crustal differentiation has also been carried out by Spohn (1990).

In order to model the loss of heat-producing elements from the interior of Mars to its crust by magma transport, we take the rate of internal heat generation per unit mass H to be given by

$$\frac{\mathrm{d}H}{\mathrm{d}t} = -H\left(\lambda + \frac{\chi u}{R}\right) \tag{3}$$

where t is time, λ is the radioactive decay rate ($2.77 \times 10^{-10}\mathrm{yr}^{-1}$; Turcotte and Schubert 1982), u is a mean convective velocity in the mantle, R is the radius of Mars and χ is a crustal fractionation parameter. The parameter χ is the ratio of the characteristic turnover time for mantle convection to the characteristic time for crustal fractionation. Based on the present rate of formation of the oceanic crust on the Earth, we show below that $\chi = 0.01$ for the Earth. Of course, the crustal fractionation model developed here is not applicable to the Earth because crustal recycling through subduction is occurring.

If the crustal fractionation parameter is sufficiently large then the planet will be fully differentiated. If f is the fraction of crustal material available, then the maximum thickness of crust D_{co} that can be formed is

$$D_{co} = \frac{f\rho_m R}{3\rho_c}. \tag{4}$$

In deriving Eq. (4) we have neglected the volume of the core. With $f = 0.1$, mantle density $\rho_m = 3940$ kg m^{-3}, crustal density $\rho_c = 2900$ kg m^{-3} and $R = 3398$ km, we find that $D_{co} = 154$ km. The thickness of the evolving crust is assumed to be given by

$$D_c = D_{co}\left(1 - \frac{H}{H_e}\right) \tag{5}$$

where H_e is the rate of heat generation in the interior without any crustal extraction given by

$$H_e = H_o e^{-\lambda t} \tag{6}$$

and $H_o = 2.47 \times 10^{-11}$ W kg^{-1} (based on estimates for the Earth; Turcotte and Schubert 1982).

Removal of radioactive elements from the mantle through crustal differentiation affects the growth of the lithosphere by reducing the heat flux from the mantle that must be conducted across the lithosphere. We assume that the heat-producing elements in the crust are sufficiently near the surface that they do not influence the conductive gradient; the base of the lithosphere is defined to lie at a temperature T_r. With these assumptions, the thickness of the conductive lithosphere is given by

$$D_L = \left(T_r - T_s\right)\left[\frac{R}{3}\left(\frac{\rho_m H}{k} - \frac{1}{\kappa}\frac{dT}{dt}\right)\right]^{-1} \tag{7}$$

where k is the thermal conductivity, κ is the thermal diffusivity, T is the mean mantle temperature, and T_s is the surface temperature. The denominator is related to the heat flow through the lithosphere.

The volumetric rate $\dot{V}_v$ of addition of magma to the crust can be determined through its relation to the change in crustal thickness by

$$\dot{V}_v = 4\pi R^2 \frac{dD_c}{dt}. \tag{8}$$

The parameterized convection calculations can also be used to determine whether the radius of the planetary body is increasing or decreasing. The appropriate relation is

$$\frac{\delta R}{R} = -\frac{\alpha}{3}\left[T_o - T\left(1 - \frac{3}{2}\frac{D_L}{R}\right)\right] + \frac{\Delta\rho D_c}{3\rho_m D_{co}} \tag{9}$$

where T_o is the initial temperature of the planet, α is the volumetric coefficient of thermal expansion and $\Delta\rho$ is the decrease in density associated with crustal differentiation. The first term on the right of Eq. (9) follows from Eq. (2) and the second term represents the increase in volume associated with crustal formation.

We present results of planetary evolution calculations for the following parameter values: $\kappa = 10^{-6}$ m^2 s^{-1}, thermal conductivity $k = 4$ W m^{-1} K^{-1}, $T_r = 1000$ K, $T_s = 255$ K, $T_o = 2000$ K, $\alpha = 3 \times 10^{-5}$ K^{-1} and $\Delta\rho = 80$ kg m^{-3}. The mean mantle temperature, crustal thickness, rate of volcanism, lithosphere thickness, surface heat flow and fractional radius change are given in Figs. 10 through 15 as functions of time for various values of the crustal fractionation parameter χ in the range 0 to 10^{-2}.

The extraction of heat-producing elements into the crust can have important effects on the thermal evolution of Mars. The results for the mean mantle temperature given in Fig. 10 show a temperature reduction of about 250 K if the heat producing elements are removed. It is seen from Fig. 11

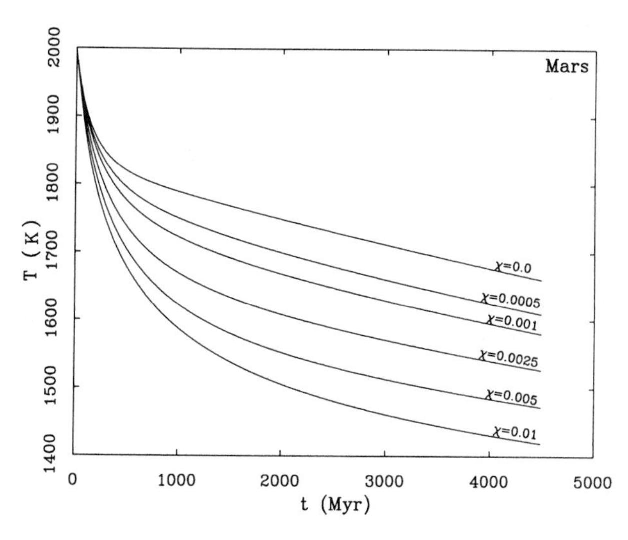

Fig. 10. Dependence of the mean mantle temperature T on time t for several values of the crustal fractionation parameter χ. See text for a discussion of the model used.

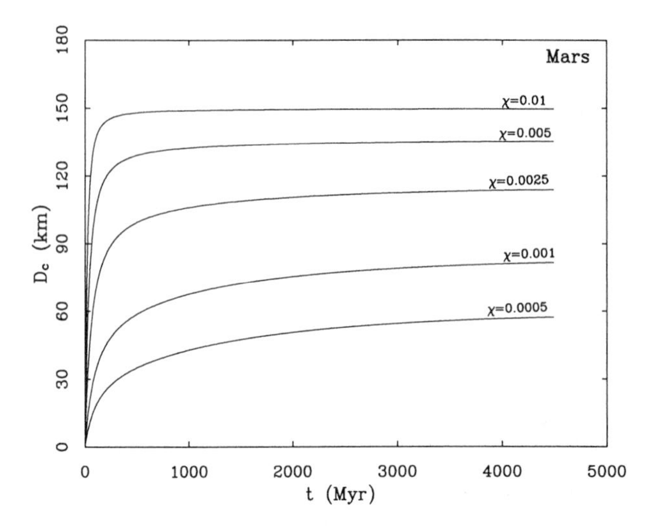

Fig. 11. Thickness of the crust D_c vs time t for a number of values of the crustal fractionation parameter χ. The crustal thickness for a fully differentiated Mars is D_{co} = 154 km.

that most of the crustal growth occurs early in the evolution of Mars, i.e., within several 100 Myr of the end of accretion. This is in accord with the evidence discussed earlier of an age $\geq$ 4 Gyr for the southern crustal highlands. The present mean crustal thickness depends on the choice of the crustal fractionation parameter χ. The volumetric flux of volcanics is given in Fig. 12. For a small crustal fractionation parameter (χ = 0.0005) the volcanic flux decreases rapidly early in the evolution, but it continues throughout the history of the planet. For larger values of the parameter χ, the interior is completely depleted and the volcanic flux drops to low levels.

With the extraction of the heat producing elements the lithosphere becomes considerably thicker (Fig. 13) and the mantle heat flux is reduced (Fig. 14). The heat flux given in Fig. 14 is the heat flux into the base of the Martian crust. The expansion and contraction of the planet is given in Fig. 15. With no crustal extraction (χ = 0), only thermal contraction occurs. As the rate of crustal extraction is increased (increasing χ), the initial expansion due to crustal formation increases and the rate of thermal contraction also increases.

We can estimate the value of the crustal fractionation parameter for the Earth. It is the ratio of the crustal fractionation time $V_e/\dot{V}_c$ to the characteristic mantle turnover time R/u ($\dot{V}_c$ is the volumetric fractionation rate of the mantle and V_e is the volume of the mantle). Taking the rate of formation of oceanic crust to be 2.8 km^2 yr^{-1} and the depth processed to be 60 km, we have $\dot{V}_c$ = 168 km^3 yr^{-1}. With V_e = 10^{12} km^3, we find that the characteristic time for creation of the oceanic crust is $V_e/\dot{V}_c$ = 6 Gyr. With R = 5800 km (approximately twice the mantle thickness) and u = 0.1 m yr^{-1}, we find that

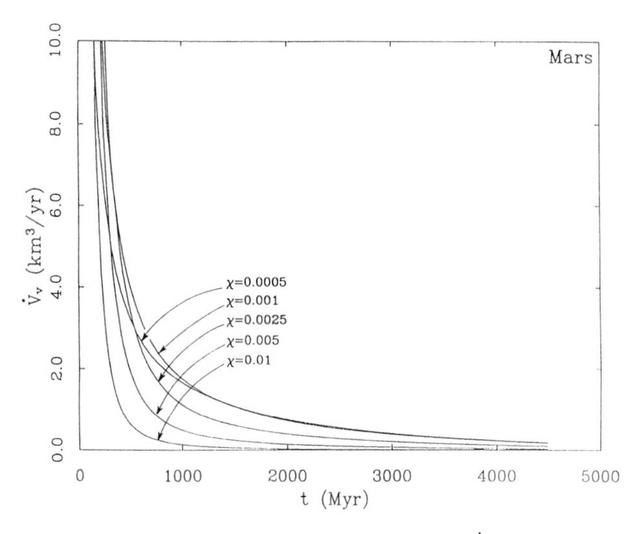

Fig. 12. Dependence of the volumetric flux of Martian volcanism $\dot{V}_v$ on time t for several values of the crustal fractionation parameter χ.

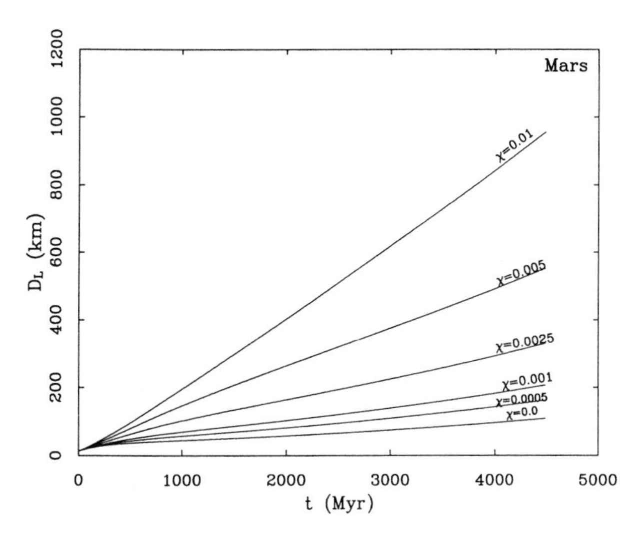

Fig. 13. Thickness of the Martian lithosphere D_L vs time t for several values of the crustal fractionation parameter χ.

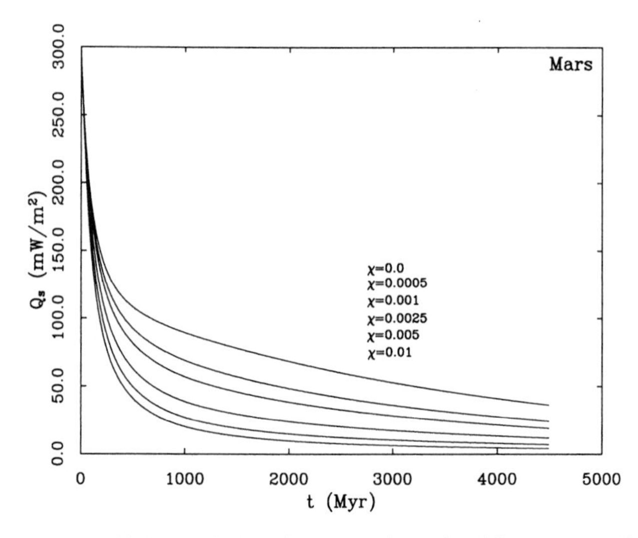

Fig. 14. Dependence of the mantle heat flux Q_S on time t for different values of the crustal fractionation parameter χ.

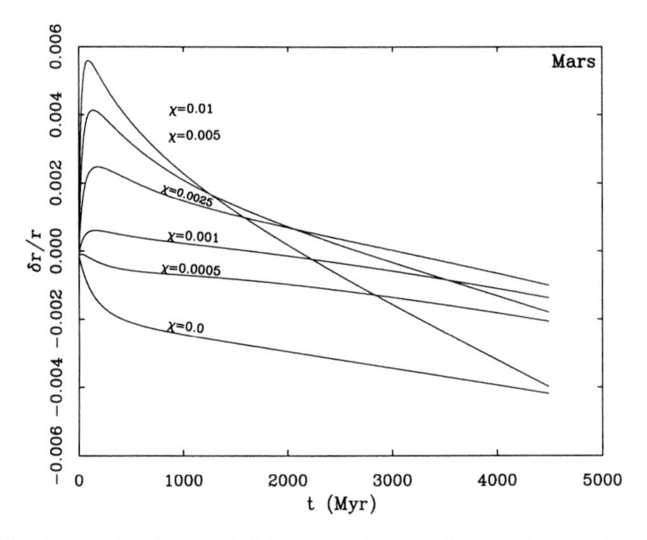

Fig. 15. Radial expansion (contraction) $\delta R/R$ as a function of time t for several values of the crustal fractionation parameter χ.

the characteristic mantle turnover time is R/u = 58 Myr. Thus we find that χ is about 0.01 for the Earth at the present time.

The crustal thickness on Mars has been estimated using gravity data and the assumption of Airy isostasy. For Hellas Planitia, as noted above, Sjogren and Wimberly (1981) estimated that the depth of compensation is 130 km. Using similar data and assumptions for the crater Antoniadi, Sjogren and Ritke (1982) found that the depth of compensation is 115 km. From gravity profiles across the highland-lowland escarpment, Janle (1983) found that the depth of compensation is also 115 km. Taking the crustal thickness to be D_c = 120 km, we find from Fig. 11 that the corresponding crustal fractionation factor is χ = 0.003, about a factor of 3 less than for the Earth. This estimate of χ is uncertain because of the poorly constrained value of the mean crustal thickness on Mars.

Accepting, for purposes of discussion, that the crustal fractionation factor for Mars is χ = 0.003, we can determine rates of crustal addition from Figs. 11 and 12. The average amount of crust added in the last 1 Gyr was 600 m, that added between 1 and 2 Gyr ago was 1.8 km, and that added between 2 and 3 Gyr ago was 2.9 km. The volumetric rate of crustal magmatism was much higher early in the evolution of Mars. The decline to the lower values of crustal production rate characteristic of most of Martian geologic history occurred with about a 100 Myr time scale.

The most recent volcanism on Mars during the Upper Amazonian period (Tanaka 1986) occurred in the Arcadia, Olympus Mons, Medusae Fossae and Tharsis Montes Formations, but the principal volcanics are flood basalts in the southern Elysium Planitia. These have an area of 10^5 km^2, but Tanaka (1986) suggests that the thickness is only a few 10 m. Taking a thickness of 50 m, this is only 0.03 m when averaged over the surface of Mars.

Greeley (1987) has estimated that 26×10^6 km^3 of volcanics erupted during the Middle and Upper Amazonian. This corresponds to a mean thickness of 200 m when averaged over the entire surface of Mars. From the estimate of cratering rates given by Hartmann et al. (1981), the Upper Amazonian extended from 0 to 0.7 Gyr ago and the Middle Amazonian from 0.7 to 2.3 Gyr ago. Thus, the volume of young volcanics on Mars is broadly consistent with the model results above as volcanism represents only a part of crustal addition. It should be emphasized, however, that there are considerable uncertainties in the absolute ages and in the crustal fractionation parameter.

Other predictions of our calculations using χ = 0.003 are a lithospheric thickness D_L = 400 km and a net global contraction corresponding to $\delta R/R = -0.001$. From Fig. 15 we see that a global expansion of δR = 10 km occurred in the first 200 Myr of the evolution of Mars. This expansion was caused by the density change associated with the generation of the early crust. For the remainder of the evolution of Mars, a nearly steady contraction occurred associated with the cooling of the interior. The total contraction subsequent to the period of early crustal differentiation was $\delta R = -13$ km. All models in Fig. 15 with substantial crustal fractionation show an early period of planetary expansion followed by a larger amount of planetary contraction.

As noted earlier, large-scale surface tectonic features of Mars include both extensional and compressional structures. The graben systems in and near the Tharsis region are likely to be the result of stresses generated by the Tharsis load. Compressive wrinkle ridges occur commonly throughout ancient terrains. These can be attributed to the early phase of rapid thermal contraction illustrated in Fig. 15.

Models of the response of Mars to the long-wavelength topographic load of the Tharsis Rise provide a reasonable fit to the observed gravity if the elastic lithosphere of Mars was globally in the range 100 to 400 km when the Tharsis construct formed (Banerdt et al. 1982; Willemann and Turcotte 1982). As the elastic lithosphere is generally ~1/2 the thickness of the thermal lithosphere, these values are consistent with those given in Fig. 13. All models in Fig. 13 with large lithosphere thicknesses show substantial depletion in mantle radioactivity due to crustal differentiation. For no crustal differentiation, the lithosphere thickness from Fig. 13 is ~ 100 km, in agreement with the result previously obtained in Fig. 5.

V. PATTERNS OF MANTLE CONVECTION

Numerical calculations of fully three-dimensional convection in a spherical shell were recently carried out by Schubert et al. (1990) to simulate possible convective planforms in the Martian mantle. These results have important implications for proposed convective origins for major geologic features on Mars, such as the crustal dichotomy. The reader is referred to that work for more detailed information about the calculations than can be provided here.

The spherical shell model of the Martian mantle consists of a Boussinesq fluid that is heated both internally and from below to account for secular cooling, radiogenic heating and heat flow from the core. The lower boundary of the shell is assumed to be isothermal and stress-free, as appropriate to the interface between the mantle and a liquid outer core. The upper boundary of the shell is rigid and isothermal, as appropriate to the base of a thick, immobile lithosphere. The ratio of the inner radius of the shell to its outer radius is 0.55, in accordance with possible core radii in Mars. We present results for two different modes of heating. In one case, 20% of the surface heat flow originates in the core, and in the other case the percentage of heating from below is 94%. The Rayleigh numbers of both cases are approximately 100 times the critical Rayleigh numbers that characterize the onset of convection in the constant-viscosity spherical shells. These Rayleigh numbers may be an order of magnitude or more smaller than the Rayleigh number of the Martian mantle. However, the Rayleigh number of the Martian mantle is unknown because of uncertainties in the thickness of the mantle and its material properties, viscosity in particular. The numerical approach is described in detail in Glatzmaier (1988) and Bercovici et al. (1989*a*). Table III lists the parameter values for the calculations discussed here.

The horizontal planforms of convection for both modes of heating are illustrated in Fig. 16 by contours of radial velocity on spherical surfaces midway through the shells. Meridional cross sections of entropy contours (equivalent to isotherms in these Boussinesq calculations) for both heating modes are shown in Fig. 17. The prominent form of upwelling in the Martian mantle is the cylindrical plume. The number of upwelling plumes is strongly influenced by the mode of heating; with only 20% heating from below, there are a dozen plumes, while 94% bottom heating produces only 6 plumes. There are fewer, stronger plumes as the proportion of bottom heating increases. Plumes carry the heat flow from the core and arise from instability of the lower thermal boundary layer at the core-mantle interface. In general, the fraction of mantle heating delivered from the core has probably decreased with time as the core cooled to temperatures not much greater than those of the lower mantle (see, e.g., Fig. 6). The isotherm cross sections of Fig. 17 show several plumes originating in the lower thermal boundary layer. Convective downwelling occurs in planar sheets that form an interconnected network surrounding the upwelling plumes. The downwellings also show cylindrical concentrations along the sheets and even distinct cylindrical downwellings.

The patterns of Fig. 16 have evolved through many overturns of the mantle and the solutions appear to be fundamentally time dependent. However, the basic nature of the convective planform, i.e., cylindrical upwelling plumes surrounded by planar downwelling sheets, does not change with time. Thus, we can expect major volcanic provinces on Mars, like Tharsis and Elysium, to reflect the cylindrical nature of upwelling mantle plumes, similar

TABLE III
Parameter Values for Three-Dimensional Spherical Convection Models of the Martian Mantle

Outer radius	3200 km
Inner radius	1762 km
Density	3450 kg m^{-3}
Core mass	1490×10^{20} kg
Kinematic viscosity	10^{18} m^2 s^{-1}
Thermal diffusivity	10^{-6} m^2 s^{-1}
Specific heat at constant pressure	1.2 kJ kg^{-1} K^{-1}
Thermal expansivity	2×10^{-5} K^{-1}
Temperature difference across the mantle	800 K
Internal heating rate (94% from below)	1.5×10^{-13} W kg^{-1}
Internal heating rate (20% from below)	5.3×10^{-12} W kg^{-1}

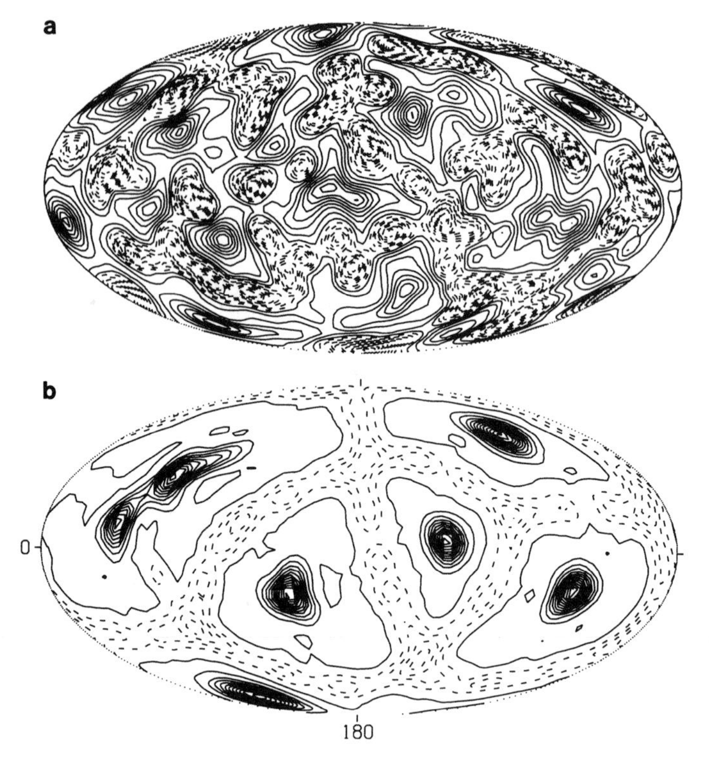

Fig. 16. Contours of radial velocity at mid-depth in the Martian mantle in numerical models of three-dimensional mantle convection with (a) 20% and (b) 94% heating from below. The projection is an equal-area projection extending 360° in longitude and over all latitudes. Model parameter values are listed in Table III (figure after Schubert et al. 1990). Solid contours indicate radially outward motion, dashed contours denote radially inward motion.

to hot spots on the Earth. There are no sheet-like upwelling features in the Martian mantle to produce a pattern similar to the linear global system of mid-ocean ridges on the Earth. Even the mid-ocean ridges on the Earth are not connected to deep sheet-like upwellings in the Earth's mantle (Bercovici et al. 1989*b*). The deep upwellings in models of convection in the Earth's mantle are also cylindrical plumes. The Earth's mid-ocean ridges are shallow, passive upwellings occurring in response to the tearing of lithospheric plates by the pull of descending slabs (Bercovici et al. 1989*b*). The non-Newtonian rheology of the Earth's lithosphere is essential for the occurrence of plate tectonics. Mars is a one-plate planet with a thick lithosphere (Solomon 1978; Schubert et al. 1979) beneath which mantle upwellings are in the form of cylindrical plumes.

The results of the spherical convection models have implications for proposed explanations of the crustal dichotomy and the concentrations of volcanism at Tharsis and Elysium. If the crustal dichotomy was caused by a

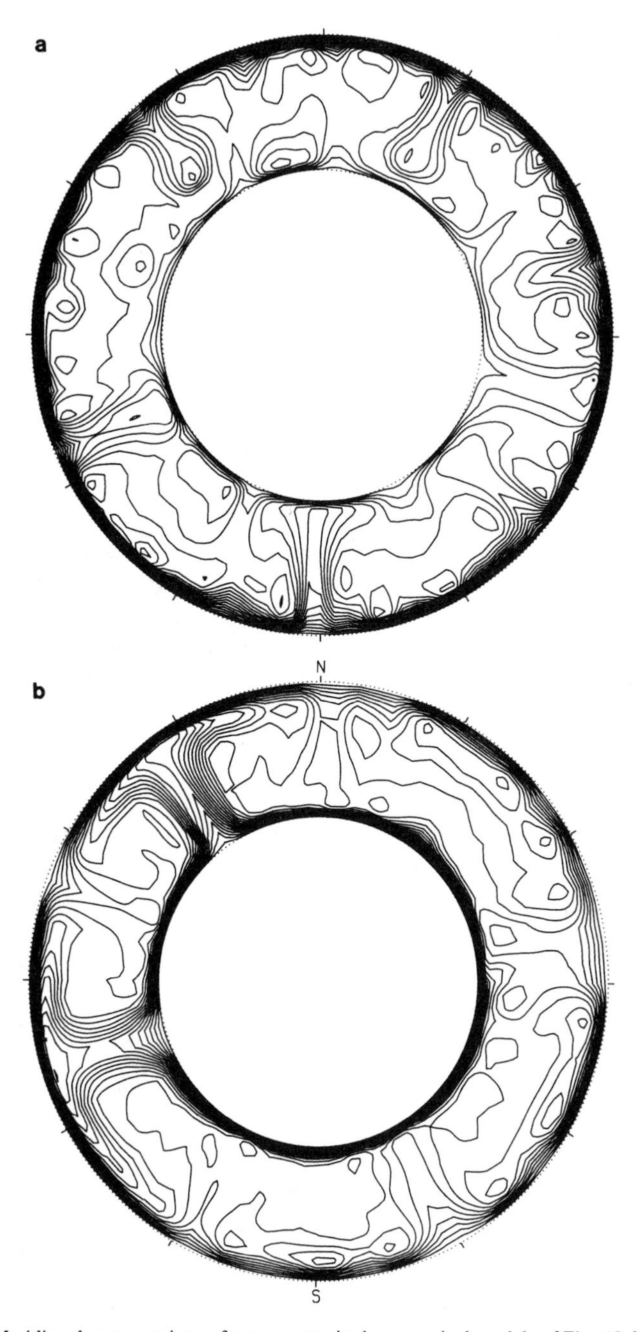

Fig. 17. Meridional cross sections of temperature in the numerical models of Fig. 13 for (a) 20% and (b) 94% heating from below.

convective system dominated by spherical harmonic degree $\ell = 1$ (Schubert and Lingenfelter 1973; Lingenfelter and Schubert 1973; Wise et al. 1979*a*) with upwelling under the northern hemisphere, then the convection must have been driven strongly from below (while our models produced 6 to 12 plumes, the number of plumes decreases with increasing percentage of bottom heating). Such strong heating concentrated deep within Mars can arise from the heating pulse accompanying core formation or from the flow of heat from a hot core. Indeed, the overturning accompanying core formation could in itself be an $\ell = 1$ mode (Stevenson 1980), obviating the need for a thermally driven motion.

Core size is another important factor in determining the number of convective plumes. With smaller cores there is a tendency toward fewer upwelling plumes (Zebib et al. 1983). This is confirmed by the results of Fig. 18, which shows mid-depth radial velocity contours and isotherms in meridional cross section in the mantle of a Mars model with a core radius of 0.2 times the radius of Mars (Schubert et al. 1990). The model has settled into a predominantly $\ell = 2$ (not $\ell = 1$, however) convection pattern. During the early stages of core formation, the effective core radius would have been smaller than the radius at present, favoring a thermally forced convection with perhaps just one dominant upwelling. Since the Martian core formed contemporaneously with accretion or within a few 100 Myr of the end of accretion, conditions favoring $\ell = 1$ convection, i.e., a small core and a deep heat source, occur very early in the evolution of Mars. If a convective mechanism is responsible for the crustal dichotomy, then the dichotomy must also be a very ancient feature.

It is not obvious from the models why there should be only two major volcanic centers (Tharsis and Elysium) on Mars. The models predict several to ~ 10 major mantle plumes. Perhaps the models are not realistic enough to predict the actual number of major hot spots on Mars. On the other hand, there may be many plumes in the Martian mantle, but the properties of the lithosphere may select only one or two of them for prominent surface expression. Plume activity could be focused beneath Tharsis if fracturing or thinning of the lithosphere in this region has facilitated magma and heat transport across the lithosphere. The temperature dependence of mantle viscosity will strongly influence the structure of plumes and their number, through the control that variable viscosity exerts on the nature and vigor of small-scale convective activity in the lower thermal boundary layer (Olson et al. 1987).

The numerical solutions discussed above can be used to infer that several km of dynamic topography could be associated with plumes in the Martian mantle (Schubert et al. 1990). Dynamic uplift is insufficient to account for the 10 km of topography in the Tharsis region. This large topographic excess must be largely the result of other processes, such as volcanic construction, magmatic thickening of the crust, or depletion of the underlying mantle (see, e.g., Sleep and Phillips 1979,1985; Solomon and Head 1982). Nevertheless,

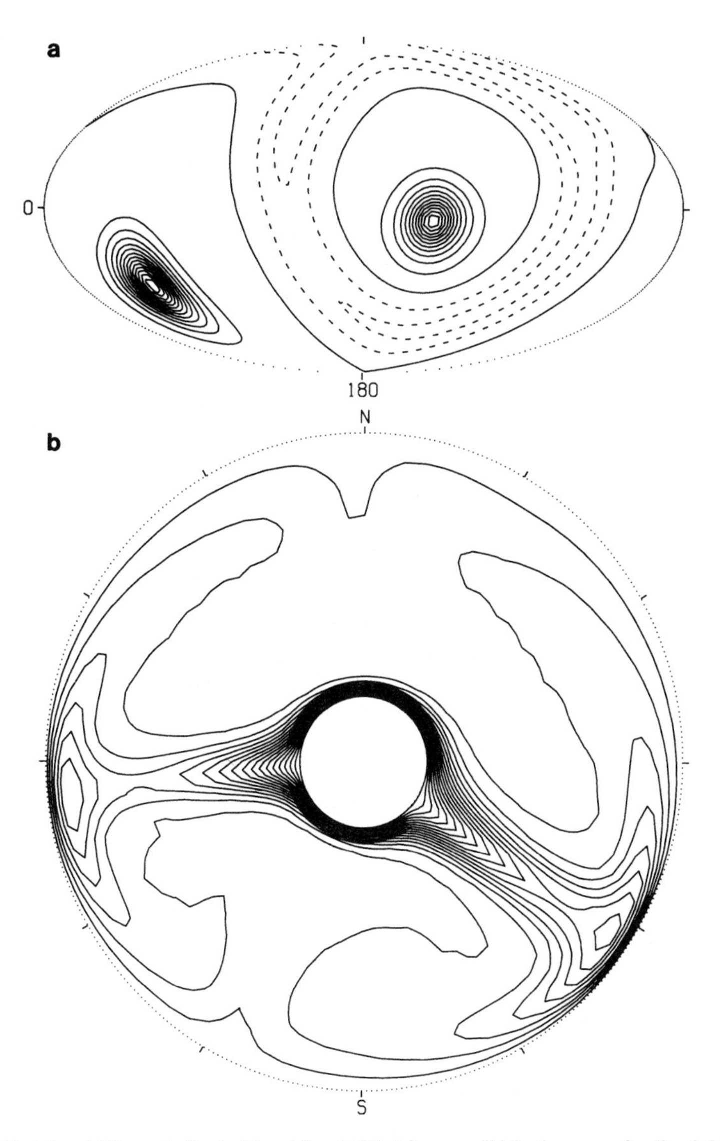

Fig. 18. (a) and (b) are similar to Figs. 16 and 17 but for a small Martian core of radius 0.2 times the radius of Mars. In the model 90% of the heating is from below.

it is likely that the Tharsis Rise and its volcanic constructs are a consequence of a strong mantle plume (or grouping of plumes) beneath the region.

VI. CONCLUSIONS

Several themes emerge from this overview of the thermal history of Mars. The first is the hot initial state of the planetary interior and the sharp contrasts that can be drawn between the first 1 Gyr of Martian history and the subsequent 3.5 Gyr. As a result of accretional heating and core formation essentially contemporaneous with planetary formation, the early history of Mars was characterized by high internal temperatures, a vigorously convecting mantle, and high surface fluxes of heat and magma. Outgassing contributed to an early atmosphere, and widespread magmatism may have helped trigger the release of subsurface water and large-scale floods. Parameterized convection models indicate, however, that on a time scale of only a few 100 Myr the mantle convective engine slowed, as primordial interior heat was lost and as radioactive heat production decayed or was concentrated into the shallow crust. Rapid interior cooling led to a globally thick lithosphere and was accompanied by global contraction, recorded in the pervasive formation of wrinkle ridges now preserved on ancient geologic units. The last 3.5 Gyr of Martian history was marked, in contrast, by slow cooling and by the concentration of volcanic and tectonic activity in ever more limited regions.

The second theme of Martian thermal history is the strong role of plumes expected for mantle convection. As long as a significant fraction of mantle heating comes from the core, three-dimensional convection calculations indicate that plumes dominate the upwelling portions of the flow. At least some of those plumes would be expected to have strong signatures in the surface topography and volcanic flux. The Tharsis and Elysium volcanic provinces are probably the consequences of plume-delivered heat and magma. The number and characteristics of plumes depend on the relative contributions of radioactivity and core cooling to the mantle heat budget, as well as the size of the core. While current models do not predict as few as 2 dominant plumes or plume families, development of Tharsis and Elysium very early in Martian history, when core cooling occurred at its highest rate, is favored.

A third theme of this overview is the dominating influence of core sulfur concentration on the thermal evolution of the core and the history of the Martian magnetic field. A core with more than $\sim$ 15 wt % S probably would not crystallize a solid inner core and probably would not be thermally convecting at present. This critical S concentration is tantalizingly close to estimates of the core sulfur content from elemental abundances of SNC meteorites. Therefore, a nearly completely fluid core that is nonconvecting or only weakly thermally or chemically convecting may provide an explanation for the lack of a present Martian magnetic field or the existence of a very weak one.

A final theme of this overview is the strong impact that new information from Mars would have on our understanding of the origin and thermal evolution of the planet. Detection of an internal magnetic field or evidence for a paleofield would strongly constrain the evolution of the Martian core. Seismic determination of the thickness of the crust would provide a crucial tie point to discussions of the planetary magmatic budget, and a seismic estimate of the radius and state of the core would enable a considerably more confident assessment of bulk composition. Measurement of surface heat flow would substantially constrain the present mantle heat production. Finally, the return of Martian igneous rocks to terrestrial laboratories would provide crucial knowledge of the detailed absolute time scale for the major events in the history of the planet.

Acknowledgments. We thank T. Spohn for recalculating the models of Figs. 3 through 6, partial results of which were originally presented in Stevenson et al. (1983). This work was supported by several grants from the National Aeronautics and Space Administration.

6. THE BULK COMPOSITION, MINERALOGY AND INTERNAL STRUCTURE OF MARS

JOHN LONGHI
Lamont-Doherty Geological Observatory

ELISE KNITTLE
University of California, Santa Cruz

JOHN R. HOLLOWAY
Arizona State University

and

HEINRICH WÄNKE
Max-Plank-Institut für Chimie

The only firm constraints on the bulk composition and internal structure of Mars are its size and mass. All previous estimates of Mars' composition and structure are based upon a cosmochemical model and an assumed value of the moment of inertia. However, the moment of inertia remains model dependent. The SNC meteorites (supposedly originating from Mars) are basaltic rocks that provide an independent means of estimating Mars' bulk composition and internal structure. Correlation of various pairs of elements is a means of estimating the ratios of key volatile, siderophile and chalcophile elements in the crust and mantle. By assuming approximately chondritic abundances for refractory elements (Al, Ca, Ti, Mg, Si, REE), Dreibus and Wänke (1985) have thus calculated absolute concentrations for a number of key elements. Their calculations show that the silicate portion of Mars is enriched in FeO, MnO, alkalis and halogens, but depleted in siderophile (Ni, Co) and chalcophile (Cu, Zn) elements relative to the Earth. These gross chemical characteristics of Mars are consistent with *accretion from more volatile-rich material than the Earth, more extensive oxidation of Fe-metal, and separation of a Fe-Ni-S core. Mass balance considerations suggest a core that is 21.7% of the mass of Mars with 14.5 wt% S. Recalculation of the silicate portion into likely mineral components yields an uncompressed mantle density of 3.52 Mg m^{-3}. The uppermost Martian mantle is likely to be dominated by olivine and orthopyroxene, as is the Earth's upper mantle, although the Martian mantle has a lower MgO/(MgO + FeO) ratio (0.74 vs 0.89). The olivine-peridotite layer extends to a depth of 900 to 1100 km where the transition to silicate spinel begins. This transition is at a greater depth than in the Earth because of the shallower pressure gradient in Mars. The extent of the perovskite-wüstite-stishovite layer is highly dependent on the composition and size of the core. In the present model, there is a 200-km thick perovskite layer above the core-mantle boundary at 1700 km. The state of the core is unknown, but the absence of a strong magnetic field is consistent with weak convection, as would be expected in a totally molten or totally solid core that was not releasing latent heat. Calculations of the high-pressure liquidus and solidus temperatures indicate that for the case of a molten core the minimum temperature at the core-mantle boundary is ~ 2000 K, whereas for the case of a solid core the maximum temperature is ~ 1800 K. Summation of the masses in the various layers of Mars yields a value of 0.353 for the dimensionless moment of inertia, which is intermediate between the generally accepted value of 0.365 and the value of 0.345 predicated on a nonaxisymmetric distribution of mass about the Tharsis plateau.*

I. INTRODUCTION

Our estimates of the nature and proportions of minerals in the Martian interior are based on three independent sets of observations: the Viking Lander soil analyses, the mean density and gravity field, and the compositions of the SNC (shergottites-nahklites-Chassigny) meteorites that we assume to be of Martian origin. These observations are the major parameters of models that produce a self-consistent picture of mass distribution, bulk composition and mineralogy.

Each set of observations is controversial in its own way: Do the soil compositions represent an average basalt composition or are they the result of extreme nonisochemical weathering of some now obscure protolith? How tightly do the mean density and gravity field constrain the mass distribution as expressed by the dimensionless moment of inertia? Do the SNC meteorites really come from Mars? The first question is discussed in chapters 18 and 19. The major arguments in favor of a Martian origin for the SNC's include a mineralogy distinct from lunar and other meteoritic basalts, relatively young crystallization ages ($\leq$ 1.3 Gyr; summarized by McSween 1985), and a rare-gas isotopic signature in impact melt similar to that observed in the Martian atmosphere (Bogard and Johnson 1983; Becker and Pepin 1984). Furthermore, there are important similarities between the Viking soil analysis at Sandy Flats and the magmas from which the SNC's crystallized. Although we cannot be certain that the concentration of FeO in the Sandy Flats soil ($>$ 20 wt% normalized; Toulmin et al. 1977) accurately reflects its basaltic precursor, it is nonetheless instructive to note that such high FeO concentrations are not common in terrestrial basalts, whereas FeO in excess of 18 wt% is the rule for lunar mare basalts (Basaltic Volcanism Study Project—BVSP 1981) and calculated FeO concentrations in the various parent magmas of the SNC meteorites range from 16 to 27 wt% (Longhi and Pan 1989). Thus at the present level of uncertainty, our major assumption, the Martian origin of the SNC meteorites, is consistent with the available data.

First we derive a bulk composition for Mars, then we convert this composition to a pressure-dependent mineralogy, and finally we compare the density distribution of our model with density distributions derived from the global gravity field.

II. BULK COMPOSITION

There are many published estimates of Martian bulk composition. Some of these are listed in Table I modified after BVSP (1981) along with a recent estimate of the Earth's mantle composition. Most of the estimates of the Martian bulk composition are based upon a cosmochemical model plus an allowance for adjustable parameters that bring the composition in line with the density distribution constraints imposed by the moment of inertia. The major adjustable parameters are FeO concentration in the mantle and the concentrations of light elements such as O and S in the core. As illustrated in Fig. 1 in chapter 5, there is a restricted range of mantle density, core density and core size for a given value of the moment of inertia such that a lower-density mantle correlates with a larger, less-dense core. However, estimates of the dimensionless moment of inertia have changed over the last twenty years, thus adding to the range of calculated Martian compositions. For example, the first two compositions in Table I were constructed to agree with the older estimates of the moment of inertia in the range 0.370 to 0.377 (Binder 1969; Binder and Davis 1973). These two compositions yielded denser, more iron-rich mantles and smaller cores than subsequent estimates based on the I/MR^2 value of 0.365 calculated by Reasenberg (1977) and Kaula (1979*a*). These first two compositions differ from one another in that the first (Ringwood 1981) is based upon a mixture of two meteorite components, an oxidized low-temperature component as represented by a C1 chondrite (Orgeil) and a high-temperature component devoid of volatile elements, whereas the second (Anderson 1972) is based upon a mixture of several classes of chondritic meteorites. Anderson's model produces a slightly less-dense mantle because higher SiO_2 produces a higher pyroxene/olivine ratio. However, the major difference between the two compositions is the presence of 18.7% oxygen as iron oxide in Ringwood's estimate (a core of iron oxide plus nickel sulfide). Experimental data discussed below suggest that incorporation of such amounts of oxygen in the Martian core is unlikely. The next two estimates listed in Table I have widely disparate concentrations of alkalis and H_2O in

TABLE I
Model Compositions of Mars

	1	2	3	4	5	6	7
Mantle + Crust							
SiO_2	36.8	40.0	43.9	41.6	39.4	44.4	45.1
TiO_2	0.2	0.1	0.16	0.3	0.6	0.1	0.2
Al_2O_3	3.1	3.1	3.2	6.4	3.1	3.0	4.0
Cr_2O_3	0.4	0.6	—	0.6	—	0.8	0.5
MgO	29.9	27.4	31.2	29.8	32.7	30.2	38.3
FeO	26.8	24.3	16.7	15.8	20.8[a]	17.9	7.8
MnO	0.1	0.2	—	0.15	—	0.5	0.1
CaO	2.4	2.5	3.0	5.2	2.7	2.4	3.5
Na_2O	0.2	0.8	1.4	0.1	0.5	0.5	0.3
H_2O	—	0.9	0.44	0.001	—	—[c]	—
K (ppm)	218	573	1199	59	1100	305	260
Mg/(Mg + Fe)	0.66	0.67	0.77	0.77	0.74	0.75	0.90
ρ(STP)	3.61	3.59	3.45	3.54	3.53	3.52	3.36
Core							
Fe	63.7	72.0	60.4	88.1[b]	—	77.8	—
Ni	8.2	9.3	5.8	8.0	—	8.0	—
S	9.3	18.6	33.8	3.5	—	14.2	—
O	18.7	—	—	—	—	—	—
ρ(STP)	5.82	6.51	5.53	7.51	7.40	6.8	7.16
Relative Masses							
Mantle + crust	81.8	88.1	74.3	81.0	85.0	78.3	67.6
Core	18.2	11.9	25.7	19.0	15.0	21.7	32.4
I/MR^2	0.373	0.375	0.368	0.363	0.368	0.353	0.331

[a] Including 0.8% Fe_2O_3.
[b] Also 0.4% Co.
[c] H_2O is uncertain; 0.004 wt% estimated by Dreibus and Wanke (1987) is lower limit.
1: 30% Orgueil C1 chrondrite + 70% high-termperature component (Ringwood 1981);
2: Mixture of chrondritic meteorites (Anderson 1972);
3: Modified equilibrium condensation (Weidenschilling 1976);
4: Four-component meteorite model (U, Fe, K, T1) (Morgan and Anders 1979);
5: Pyrolite + FeO (McGetchin and Smyth 1978);
6: SNC model (this study);
7: Terrestrial crust + mantle (Jagoutz et al. 1979).

the mantle and sulfur in the core. The first of these is an equilibrium condensation model modified by Weidenschilling (1976) to allow for the likelihood that planets accrete material from overlapping zones. In these models, the concentrations of volatile constituents such as alkalis, H_2O and sulfur increase with decreasing nebular temperature as determined by radial distance from the Sun. Mars is farther from the Sun than the Earth, so volatile concentrations higher than those of the Earth (column 7) are a consequence of this approach. Goettel (1981,1983) produced a similar estimate based on equilibrium condensation principles. He adjusted FeO to bring the Martian mantle density to 3.44 ± 0.06 Mg m^{-3}, a value that he considered most consistent with the 0.365 moment of inertia given the range of possible components in the core. The equilibrium condensation model is not generally favored by planetary scientists because it fails to account for the Earth's Moon and volatile-depleted meteorites, such as eucrites and diogenites, from the asteroid belt. The composition in column 4 (Morgan and Anders 1979) is based upon the Martian surface K/U ratio of 3000 measured by an orbiting γ-ray spectrometer (Surkov 1977). This value is similar to those of lunar rocks and eucrites and much lower than the terrestrial value of 10,000. Because K and U are both highly incompatible (i.e., both elements are strongly partitioned into basaltic liquid during melting), igneous processes do not change the K/U ratio significantly; however, because K is moderately volatile and U is refractory, the K/U ratio is a good measure of the volatile depletion in a planet relative to chondrites (K/U = 64,000). Accordingly, Morgan and Anders (1979) calculated volatile contents for Mars to be lower than those of the Earth and similar to those of the Moon. As discussed in chapter 4, the K/U ratio in the SNC meteorites is slightly higher than the terrestrial average, so if the SNC meteorites come from Mars, then either the orbital measurement of the K/U ratio on Mars is in error or some weathering process has altered the K/U ratio in the soil (leaching of alkalis is a possibility). To arrive at the composition in column 5, McGetchin and Smyth (1978) noted the high iron concentrations in the Viking Lander soil analyses and inferred an iron concentration in the Martian mantle higher than that of the Earth. They modeled the Martian mantle composition as a mixture of terrestrial pyrolite (Ringwood 1975) plus sufficient oxidized iron (FeO) to bring the mantle density to 3.55 Mg m^{-3}, a value determined by Johnston and Toksöz (1977) to be consistent with a moment of inertia of 0.365.

It is clear that estimates of the bulk composition have relied heavily on accepted values of the reduced moment of inertia. However, Bills (1989*a*) has pointed out that the currently accepted value of 0.365 is model dependent. Unlike the Earth, where the moment of inertia is derived from the response of the Earth to an applied torque, the Martian value is derived from harmonic coefficients of the gravity field that are based upon an assumption that the nonhydrostatic portion of the Martian mass distribution is symmetric about an equatorial axis through Tharsis. Any significant departure from this symmetric distribution leads to lower values of the moment, perhaps as low as 0.345 (Bills 1989*a*). Consequently, in this chapter we take a different approach from previous authors and estimate the bulk composition of Mars independent of the moment of inertia. We employ an assumption of overall chondritic proportions of refractory lithophile elements plus iron in Mars and then rely primarily on the compositions of the basaltic SNC meteorites to determine the concentrations of siderophile and volatile elements in the mantle. Mass balance then gives us the composition and size of the core.

Tests of this model are likely only after new orbiter gravity measurements tightly constrain the moment of inertia and new lander seismic measurements constrain the size of the core.

The basic approach is to estimate the absolute mantle abundances of a few key elements and then to make use of observed inter-element correlations to calculate the rest. Not all of the SNC meteorites are likely to retain equally strong chemical links to their source regions, however. Rocks such as the nakhlites and Chassigny were formed by the accumulation of crystals, whereas at least some of the shergottites have the compositions of basaltic liquids (see, e.g., McSween 1985). Therefore, in some cases we rely only on the elemental correlations among the shergottites; in other cases, particularly where two geochemically similar elements are involved, elemental correlations for all of the SNC's are useful.

In the following, we briefly discuss the basis of the compositional estimates for a few key elements, i.e., the oxyphile and siderophile elements, that have the greatest bearing on density. First, we make the very general assumption that Mg, Si, Al and all oxyphile refractory elements have C-1 abundance ratios (Anders and Ebihara 1982) in the Martian mantle. There is general agreement that for the Earth and Moon the deviation of these elements from C-1 values is less than 30%. Thus we restrict the problem to estimating the abundances of the remaining elements: oxidized iron, moderately siderophile, moderately volatile and volatile elements. Recent interpretations of spectroscopic measurements discussed by Anderson (1989) suggest that solar abundance ratios are different from those in C-1 chondrites (Fe/Si and Ca/Al are higher in the Sun). Although these interpretations have been disputed (Anders and Isnevesse 1989), Anderson's discussion begs the question of whether C-1 abundances are really the appropriate reference for constructing planetary compositions. Current accretion models (see, e.g., Wetherill 1976) predict that the terrestrial planets did not accrete directly from the solar nebula, but rather accreted from planetisimals of various sizes. This hypothesis allows for the possibility of chemical fractionations between the material from which the planets ultimately accreted and the portion of the nebula that collapsed to form the Sun. Our model is thus based upon the assumption that C-1 chondrites are the best representatives of material that accreted to form the terrestrial planets.

FeO and MnO. Under normal planetary conditions with olivine, orthopyroxene and clinopyroxene as the major FeO- and MnO-bearing phases, the liquid/solid partition coefficients (weight ratios) of FeO and MnO are very similar and only slightly above 1. FeO concentrations are similar in the five shergottite rocks (ALHA77005, EETA79001A, EETA79001B, Shergotty and Zagami) as are MnO concentrations with mean values of FeO = 18.9% and MnO = 0.48%. These average values are similar to estimates of FeO and MnO in the shergottite parent liquids made by Longhi and Pan (1989).

The C-1 MnO concentration normalized to Si is 0.46%. Hence, Mn is obviously not depleted in the Martian mantle. A similar situation holds for the eucrite parent body (Dreibus et al. 1977), whereas Mn is strongly depleted in the lunar and terrestrial mantles (Dreibus et al. 1977; Jagoutz et al. 1979). We therefore assume a C-1- and Si-normalized Martian mantle abundance of 1.00 for MnO. The FeO/MnO ratio of all five shergottite rocks is 39.5 ± 1.2. If we assume that the FeO/MnO ratio has approximately the same value in the Martian mantle, then dividing by the C-1 FeO/MnO ratio of 100.6 yields a chondrite-normalized abundance of 0.39 for FeO or an absolute concentration of 17.9 ± 0.6 wt% in the Martian mantle. The uncertainty of this estimate is undoubtedly larger than that generated by variation in the shergottite analyses; however, our estimate of FeO concentrations is well within the range of previous estimates (Table I) and is markedly higher than the FeO concentration in the terrestrial upper mantle (8%), derived by an analogous procedure. Since the MgO contents of the two mantles are assumed to be similar, we should accordingly expect substantial differences in the compositions of their ferromagnesian phases.

K, Rb and Cs. There is an excellent correlation of K and La concentrations in all SNC meteorites (Fig. 1). K and La are both incompatible ele-

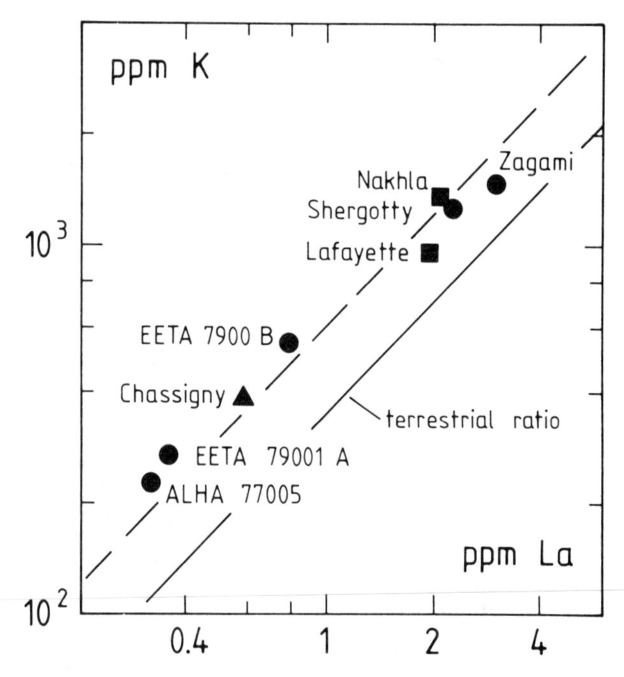

Fig. 1. Correlation of K (moderately volatile) with La (refractory) in the SNC meteorites (figure after Dreibus and Wänke 1985).

ments. Accordingly, their ratio will not change during most igneous processes. However, La is a highly refractory oxyphile element with a chondrite-normalized abundance of 1.00, whereas K is a moderately volatile element. Therefore, their ratio (655) provides not only a means to calculate the K content of the Martian mantle (0.31 × C-1; 0.16 wt% K_2O), but when compared to the K/La ratios of chondrites (2110) and the Earth (413), the Martian K/La ratio indicates that Mars is depleted in moderately volatile elements relative to chondrites, but enriched relative to the Earth. Good correlations of Rb and Cs with K yield chondrite-normalized abundances of 0.28 and 0.20, respectively for Rb and Cs.

Na and Ga. These moderately volatile elements correlate well with Al, a refractory element, in all the shergottites. Given a normalized abundance of 1.00 for Al, the average Na/Al ratio (0.235) and the Ga/Al ratio (0.00041) in shergottites yield C-1 normalized abundances of 0.38 and 0.37, respectively. It appears that despite large depletions relative to chondrites, Na and Ga have not fractionated from one another on Mars.

Co and Ni. As with lunar and terrestrial basalts, Co concentrations in shergottites are proportional to the sum MgO + FeO (Fig. 2a). For the Mar-

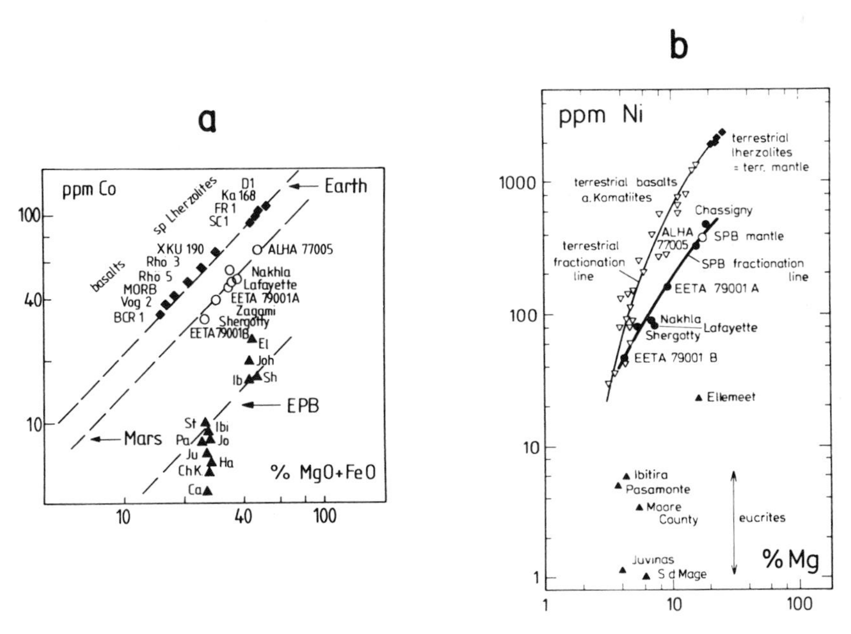

Fig. 2. (a) Cobalt vs MgO + FeO in eucrites (EPB = eucrite parent body), SNC meteorites, and in terrestrial basalts and upper mantle nodules (spinel lherzolites). (b) Ni/Mg fractionation lines for SNC meteorites and terrestrial basalts and upper mantle nodules (figure after Dreibus and Wänke 1985).

tian mantle value of MgO + FeO = 48.1% (Table I), the corresponding Co concentration is 68 ppm or 0.070 times C-1. Given a C-1 abundance for Mg, the Ni/Mg fractionation line in Fig. 2b yields Ni concentrations of 400 ppm or 0.019 times C-1 in the Martian mantle. By comparison, the Earth's mantle contains approximately 2108 ppm of Ni (Dreibus and Wänke 1987).

Cu and Zn. Rough estimates of concentrations of these compatible siderophile elements are taken directly from their observed concentrations in ALHA 77005, a cumulate shergottite with major and trace element abundances similar to our estimates of the Martian mantle itself. The observed concentrations are 5.5 ppm (0.027 times C-1) for Cu and 62 ppm (0.099 times C-1) for Zn.

The elemental abundances in the Martian mantle obtained by these various approaches are listed in Tables I and II along with estimates for other elements not discussed above. Dreibus and Wänke (1985,1987) provide details of the derivations of the abundances of these other elements. This bulk composition generally agrees with one derived by Treiman et al. (1986), who employed a similar approach, although there are some important discrepancies. In the Dreibus-Wänke model all the refractory lithophile elements (Ca, Al, Ti, Si, Mg, REE, U, Th) are assumed to have C-1 abundances in the Martian mantle. Treiman et al. (1986) observed that U/La, Th/La, and Al/Ti ratios in SNC's average approximately 2 times C-1, 2 times C-1, and 0.4 times C-1, respectively. They inferred that mantle abundances of U, Th, La, Ti and Al were likely to be similarly fractionated with respect to C-1 meteorites inasmuch as these ratios should not change much during melting processes. This inference has a significant impact on thermal modeling (chapter 4) because U and Th are important heat-producing elements. However, as difficult as it is to account for changes in U/La and Al/Ti during melting, it is more difficult to envision a series of nebular and/or accretion processes that would lead to planet-wide fractionations of the refractory lithophile elements. A resolution to this dilemma may lie in isotopic fractionations. There is clear evidence in the SNC meteorites of Nd-Sm fractionations 1.3 Gyr ago (Nakamura et al. 1982; Shih et al. 1982), whereas the U-Pb system shows no sign of disturbance at this time (Chen and Wassurburg 1986); consequently the 1.3 Gyr event may have involved a decoupling of U and REE and hence a fractionation of U/La. If the true crystallization ages of the shergottites are much younger than 1.3 Gyr, as suggested by Jagoutz and Wänke (1986) and Jones (1986), then the petrogenesis of the shergottites may have involved assimilation of the 1.3-Gyr old component. Given such a multistage history, it is possible to construct scenarios involving previous melting events in the presence of plagioclase or garnet that would fractionate Al/Ti as well. Consequently, we present the simpler Dreibus-Wänke compositional model here, although we note that the same multistage melting events may have also af-

TABLE II
Selected Minor and Trace Elements in the Mantles of Earth and Mars

		Earth (Jagoutz et al. 1979)	Mars (Wänke and Dreibus 1988)	Mars: Rel. to Si and C1
P	ppm		700	0.36
K		260	305	0.30
Rb		0.81	1.06	0.26
Cs		—	0.07	0.20
F		16.3	32	0.31
Cl		1.33	38	0.029
Br	ppb	11	145	0.029
I		—	32	0.029
Co	ppm	105	68	0.070
Ni		2110	400	0.019
Cu		28	5.5	0.027
Zn		50	62	0.099
Ga		3.0	6.6	0.37
Mo	ppb	—	118	0.066
In		—	14	0.090
La	ppm	0.63	0.48	1.00
Tl	ppb	—	3.6	0.013
W		16.4	105	0.6
Th		94	56	1.00
U		26	16	1.00

fected our estimates of Fe and K mantle abundances.

Listed in Tables I and II and illustrated in Fig. 3 is an estimate of the Earth's mantle composition for comparison with the Martian composition. There are important differences between the two planets. First, Mars appears to have a significantly higher concentration of oxidized iron. As a result, we should anticipate that at equivalent pressures the Martian mantle would be denser because of a lower Mg/Fe ratio and a higher olivine/pyroxene ratio. It may be that this distinction applies only for the upper terrestrial mantle inasmuch as Anderson (1989) has suggested that the Earth's lower mantle has approximately twice the FeO concentration as the upper mantle. Second, Mars also has significantly higher abundances of moderately volatile elements such as Ga, Na, P, K and Rb in its mantle (there are conflicting lines of evidence on H_2O abundance, discussed below). Figure 1 suggests that differences in the volatile elements transcend the uncertainties in the estimates of their absolute concentrations. Low $^{238}U/^{204}Pb$ ratios in shergottite minerals also indicate a more volatile-rich parent body than the Earth (Chen and Wasserburg 1986). A secondary effect on mantle density, caused by the higher alkali concentration, is an increase in the proportion of lower-density pyroxene relative to higher-density aluminous spinel and garnet: pyroxene incorporates alkalis coupled to Al, whereas aluminous spinel and garnet admit negligible alkalis, but are the major Al carriers; consequently, pyroxene

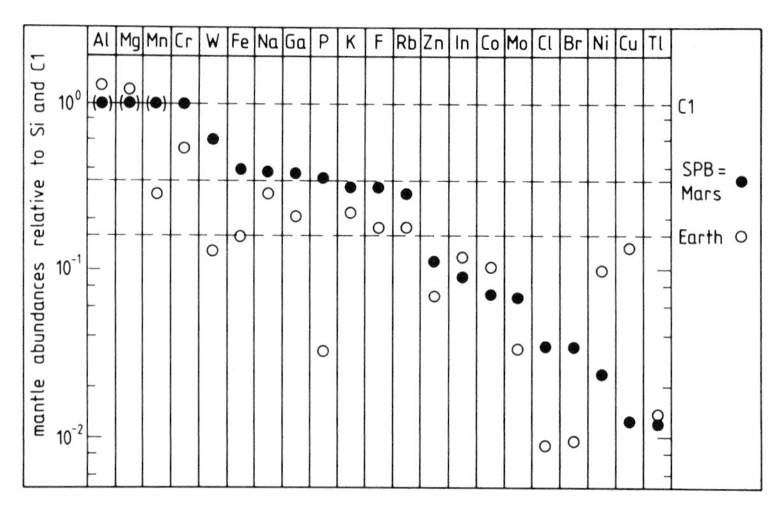

Fig. 3. Comparison of major and trace elements in the silicate portions of the Earth and Mars normalized to chondritic abundances (figure after Dreibus and Wänke 1985).

forms at the expense of spinel and garnet in order to accommodate the alkalis. Third, Mars has relatively lower abundances of certain siderophile elements (Co, Ni, Cu, In) that also have strong chalcophile affinities, yet higher abundances of siderophile elements (W, Cr, Mn) that have weak chalcophile affinities. The simplest explanation is that higher abundances of volatile components in the material from which Mars accreted led to more extensive oxidation of metal and thus higher concentrations of FeO and other siderophile elements in the mantle (this process may also have grossly depleted the main oxidant, H_2O, as discussed below). Separation of a FeS phase extracted the strongly chalcophile elements from the Martian mantle during core formation, but left relatively high levels of W, Cr and Mn. It is possible that formation of the Earth's core also involved separation of a sulfide phase, but if so, then it is necessary to invoke addition of the strongly chalcophile elements (Co, Ni, Cu, In) to the Earth's mantle following core formation to mask this earlier process.

To estimate the core composition, we first assume that Mars has an overall S abundance (0.35 times C-1) similar to those of other elements (Ga, Na, P, K and Rb) with similar volatility, and that nearly all of this sulfur is in the core. Second, we calculate the Fe and Ni concentrations of the core as the difference between the Fe and Ni in the mantle and the Fe and Ni concentrations of C-1 chondrites, which we assume to be those of Mars as a whole. This simple approach produces a core composition (14.2% S), density (6.8 Mg m^{-3}), and percent of the planet (21.7%) that differ only slightly from the more detailed modeling of Treiman et al. (1986) discussed in chapter 5 (core composition 12.5% S, core density 7.1 Mg m^{-3} and mass fraction 25 to 35%).

III. CRUSTAL THICKNESS AND COMPOSITION

There is only indirect evidence as to the thickness and composition of the Martian crust. Returns from a single possible Marsquake recorded by the Viking 2 Lander seismic experiment suggest, but do not unambiguously constrain, the crust in the vicinity of the lander site in Utopia Planitia to be 15-km thick (Anderson et al. 1977). Bills and Ferrari (1978) assumed that Bouger anomalies in the global gravity field were due to variations in crustal thickness and were able to infer mean global crustal thicknesses of 34 to 40 km, depending on the choice of crustal density (2.7 to 2.9 Mg m^{-3}). Their modeling revealed maximum thicknesses under the Tharsis plateau (61 to 77 km) and minimum thicknesses under the Hellas basin (8 to 10 km). If the SNC meteorites are typical products of Martian magmatism and if the crust is primarily basaltic, then Bills and Ferrari (1978) employed densities that were too low (even the least mafic sample, EETA79001B, has a computed density on the order of 3.0 Mg m^{-3}) and, consequently, their estimates of crustal thickness are a few kilometers too low. (Martian gravity anomalies apparently arise from density differences between mantle and crust: the lower the crustal density, the smaller the crustal thicknesses needed to produce the observed anomaly.) However, in addition to the possibility that there may have been lower-density basaltic magmas on Mars, there is also some circumstantial evidence that suggests that the crust may not be entirely basaltic. Longhi and Pan (1989) argued that the parent magmas of Shergotty and Zagami could be related to those of the Antarctic shergottites by assimilation of a component rich in potassium and other incompatible lithophile elements. Because there is no evidence of negative europium anomalies in the parent magma compositions, this component is apparently more like terrestrial continental crust than lunar KREEP, at least in terms of trace elements. Future studies of the SNC meteorites may define this component better, but determining whether Mars has a granitic component in its crust awaits new mapping and sampling missions.

IV. MANTLE MINERALOGY AND DENSITY

In this section we assume that the Martian mantle is chemically homogeneous. Given the observation of widespread volcanism and the evidence of long-term isotopic heterogeneities in the SNC source regions (Jones 1986), chemical homogeneity is almost certainly a vast oversimplification. However, inasmuch as there is no basis for specifying the magnitude or scale of heterogeneity, we must tenatively accept homogeneity as a necessary approximation.

It is convenient to subdivide the Martian mantle into two parts, an upper part extending to about 900 to 1100 km deep (with a corresponding pressure of 12 to 14 GPa), and a lower part extending from the base of the upper mantle to the core. The subdivision is made at the olivine/silicate-spinel phase transition (Akimoto et al. 1976; Akaogi et al. 1984). Although there is a considerable body of work on terrestrial mantle compositions, little has been done on the more iron-rich compositions indicated by the various estimates of the Martian mantle composition. However, the work on the more-magnesian terrestrial compositions does provide a useful guide and can be extrapolated to the Martian mantle in many cases.

Upper Mantle

Following the methods of Bertka and Holloway (1988), we have recast the mantle composition listed in Table I into various mineral components and then combined these components into model minerals. Although extrapolation to the pressures of the lowermost upper mantle is tenuous, the schematic phase diagrams of BVSP (1981) and McSween (1985) suggest that petrologically all of the Martian upper mantle is a lherzolite: olivine, orthopyroxene, clinopyroxene and minor aluminous phase. The compositions and wt% of the minerals present at 2.3 GPa are listed in Table III. The STP density is 3.52 Mg m^{-3}. The temperature-pressure ranges of the various aluminous phases are illustrated in Fig. 4. The calculated density is especially sensitive to uncertainties in composition. For example, variations in the FeO concentration of ± 1.0 wt% yield density variations of ± 0.06 Mg m^{-3}; increasing the Ca/Al ratio from the chondritic value of 0.72 to the new solar value of 0.98 (Breneman and Stone 1985), increases the clinopyroxene/orthopyroxene ratio, eliminates garnet from the mode at 2.3 GPa, and lowers the calculated density to 3.47 Mg m^{-3}.

Olivine is the dominant mineral in the Martian upper mantle, as it is on Earth. Orthopyroxene is the next most abundant mineral, also as it is on

TABLE III
Mantle Mode and Mineral Compositions for the Model Martian Mantle at 2.3GPa (23 kbar), 1400°C

Mineral	Model (Wt%)	Density (Mg m^{-3})	Composition (Atoms)
Olivine	49.5	3.41	$Mg_{1.50}Fe_{0.50}Si_1O_4$
Orthopyroxene	31.5	3.36	$Ca_{0.09}Mg_{1.35}Fe_{0.44}Al_{0.13}(Si_{1.86}Al_{0.14})O_6$
Clinopyroxene	18.8	3.35	$Na_{0.02}Ca_{0.32}Mg_{1.11}Fe_{0.41}Al_{0.13}(Si_{1.91}Al_{0.09})O_6$
Garnet	0.3	3.69	$Ca_{0.43}Mg_{1.81}Fe_{0.86}Al_{1.96}Si_{2.98}O_{12}$

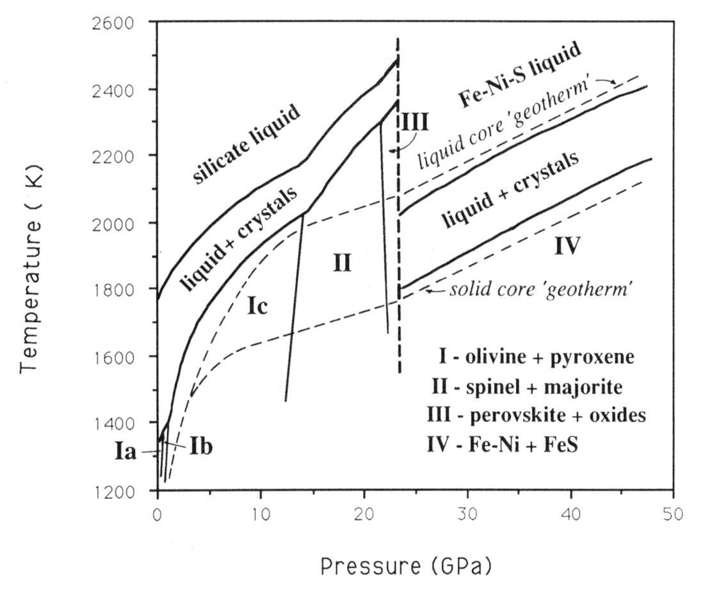

Fig. 4. Schematic phase equilibria relations for the Dreibus-Wänke Mars model and calculated temperature-depth relations. Low pressure silicate phase boundaries are based upon the 1 atm data of Leshin et al. (1988) and the 2.3 GPa data of Bertke and Holloway (1988); higher pressure boundaries are estimated from the work of Akimoto et al. (1976), Ito and Takahashi (1987), Ohtani et al. (1986) and Scarfe and Takahashi (1987). Subdivisions of region I refer to accessory aluminous phase: Ia—plagioclase; Ib—spinel [$(Mg,Fe)Al_2O_4$]; Ic—garnet. Vertical dashed line marks core-mantle boundary for the preferred compositional models. Phase boundaries for core materials are calculated from freezing point depression equations as discussed in text. Melting curves of Fe and FeS are shown in Fig. 5a.

Earth. Schematic phase diagrams (BVSP 1981) and thermodynamic calculations (Wood and Holloway 1982) showed that many earlier model Martian mantles lacked orthopyroxene at pressures > 2 GPa, so our model represents a significant departure in this respect. The aluminous phase (plagioclase, spinel, or garnet) is the least abundant of the major volatile-free minerals. Plagioclase is the low-pressure phase and is replaced by aluminous spinel with increasing pressure. The pressure of the plagioclase/spinel transition has not been measured for Martian compositions, but is probably in the 0.8 to 1.0 GPa range by analogy with terrestrial compositions. Spinel is stable through a small pressure interval and is replaced by garnet. Patera and Holloway (1982) have determined the position of the spinel/garnet transition experimentally. The amount of garnet in the model mode is $< 1\%$ at pressures ≤ 2.3 GPa; however, it probably increases in abundance with increasing pressure and extends into the lower mantle as majorite garnet (Ito and Takahashi 1987; Takahashi and Ito 1987).

It is very likely that the Martian upper mantle contains some hydrogen and carbon. The only reports of hydrous minerals are for trace amounts of amphibole in melt inclusions in Chassigny (Floran et al. 1978) and in Shergotty (Treiman 1985). There is also minor alteration of olivine to iddingsite in Nakhla (Reid and Bunch 1975), but otherwise there is little evidence of aqueous alteration. A corollary of these observations is that the flows and shallow intrusions in which the SNC magmas crystallized did not suffer significant hydrothermal alteration. Apparently, the SNC parent magmas contained no more H_2O than typical fresh terrestrial basalts ($\leq$ 1 wt%). In line with the higher end of this crude estimate, Johnson et al. (1990) recently estimated that 1 wt% H_2O was present in melt trapped in an olivine inclusion in Chassigny. On the other hand, Yang and Epstein (1985) have reported only 200 ppm H_2O in Shergotty, a value that is nearly an order of magnitude less than MORB levels. It appears unlikely that most of the indigenous H_2O was lost during shock heating, because there is no petrographic evidence of decrepitated minerals that once could have been large amphiboles or micas. It is quite possible, however, that most of whatever magmatic water that was present simply escaped as anhydrous minerals crystallized at low pressure. An interesting alternative is that the low H_2O content of Shergotty reflects a H_2O-poor interior. Dreibus and Wänke (1987) have contended that most of Mars' water was consumed in the primordial oxidation of Fe metal to FeO (H_2 released in this process has been lost along with most of Mars' early atmosphere). Thus despite higher concentrations of moderately volatile elements, Mars may actually have much less H_2O than the Earth. Their estimate of the bulk H_2O content, 36 ppm, is footnoted in Table I.

There is as yet no direct evidence for CO_2 in the parent magmas of the SNC meteorites. However, Longhi and Pan (1989) calculated a parent magma for Nakhla with unusually high CaO and unusually low Al_2O_3 which they attributed to high-pressure partial melting in the presence of CO_2. There are no experimental data on Martian mantle compositions containing H_2O and only preliminary results on carbonated Martian mantle (Odezynski and Holloway 1989). Nonetheless, since petrographic evidence from the SNC meteorites suggests that the near-surface, magmatic oxidation state of Mars is very similar to that of the Earth (Stolper and McSween 1979), we should expect that hornblende, phlogopite, diamond, graphite and probably carbonate are stable at least locally in the Martian upper mantle. The relatively minor amounts of ferric iron present are likely to be dissolved mainly in silicates.

Lower Mantle

The mineralogy and extent of the lower mantle depend upon the size of the core—the larger the core, the lower the pressure at the core-mantle boundary. Table IV lists the pressures at the core-mantle boundary for the range of model core compositions (discussed below). Extrapolation of the results of Akimoto et al. (1976) and Akaogi et al. (1984) suggests that the olivine $\rightarrow$ spinel transition begins at about 900 to 1100 km (12 to 14 GPa)

TABLE IV
Range of Models for the Martian Core

Mass % of Mars in the Core (%)	Composition (wt %)	Core Radius (km)	Depth to Core-Mantle Boundary (km)	Pressure of CMB (GPa)
30[a]	~ 66 Fe + Ni ~ 34 S	2020 (±70)	1370 (±70)	18.5 (±1.5)
22[b]	85.5 Fe + Ni 14.5 S	1695 (±70)	1695 (±70)	23.5 (±1.5)
15[c]	100 Fe + Ni	1400 (±70)	1990 (±70)	28.0 (±2.0)

[a] Wood, in *Basaltic Volcanism on the Terrestrial Planets* (1981). Note that Fe and Ni are grouped together in these simplified models because of their similar atomic weights.
[b] Wänke and Driebus (1988).
[c] McGetchin and Smyth (1978); Morgan and Anders (1979). Note that Morgan and Anders propose that the core of Mars contains a very small amount of sulfur (about 3.5 wt %) and that the core of Mars is about 19 mass percent of the Martian interior. However, their model may not satisfy the mass and moment of inertia constraints for Mars unless the mass percent of the core is reduced to ~ 16 % (cf. BVSP 1981).

in Mars (Fig. 4), so all of the models allow for several hundred kilometers of stable olivine-spinel. The phase diagram of Akimoto et al. (1976) shows that at equivalent temperatures, the more iron-rich Martian mantle would begin its transition at pressures several GPa lower than the terrestrial mantle. Along with the olivine → spinel transition is the conversion of pyroxene and garnet to an intermediate phase, majorite, that has garnet structure but composition closer to pyroxene. Because of the somewhat higher concentrations of alkalis in the Martian mantle, we might expect a small stabilization of pyroxene, which accomodates alkalis more easily than does the garnet structure, relative to the Earth's mantle. Overall, the density increase will probably be on the order of 10%, (to 3.72 Mg m^{-3}) as it is in the Earth's mantle; however, the olivine → spinel transition zone will be much less abrupt in Mars, in part because of the iron and alkali effects, but primarily because of the shallower pressure-depth gradient.

The transformations of olivine-spinel [$(Mg,Fe)_2SiO_4$] to perovskite [$(Mg,Fe)SiO_3$] plus magnesiowustite [(Mg,Fe)O] and of majorite to perovskite structure all occur in the range of 22 to 24 GPa (Liu 1979). Consequently, no perovskite would be encountered if the Martian core had the sulfur-rich composition discussed by BVSP (1981), because the core-mantle boundary occurs at ~ 20 GPa for this model. Both our preferred core composition (14.5 wt% S) and the Fe-Ni core composition models predict a significant layer of perovskite-rich lower mantle. The pressure at the core-mantle boundary in the Dreibus-Wänke model is marginally high enough to drive the transition to perovskite, as long as temperatures are above ~ 1500 K. The accompanying 10 to 12% increase in density (to the range of 4.09 to

4.17 Mg m^{-3}) should be readily resolvable at this depth by seismic wave velocity even for a layer < 200 km thick. More iron-rich core models predict an even more conspicuous perovskite layer 200 to 500 km thick.

Minerals in the perovskite layer will likely have very different Mg/Fe ratios than minerals in the upper mantle. Experiments on the transition to silicate-perovskite from olivine-spinel and pyroxene-majorite have demonstrated that the perovskite structure cannot accomodate more than 20 to 25 mole percent iron before the structure decomposes into its constituent oxides (Liu 1979; Williams et al. 1989). Additionally, iron partitions preferentially into the magnesiowustite (the distribution coefficient $K^{pv/mw}$ is 0.08 ± 0.03; Bell et al. 1979). Therefore, for a total iron content corresponding to a Mg/(Mg + Fe) ratio of 0.75, transformation of silicate-spinel to perovskite would produce highly magnesian perovskite [$(Mg_{0.94}Fe_{0.06})SiO_3$] and more ferroan magnesiowustite [$(Mg_{0.56}Fe_{0.44})O$]. Transformation of associated pyroxene-majorite to perovskite would produce similar phase compositions plus SiO_2-stishovite to satisfy mass balance.

V. CORE COMPOSITION, SIZE AND TEMPERATURE

Numerous workers have estimated the core composition of Mars: Anderson (1972), BVSP (1981), Morgan and Anders (1979), Ringwood (1981), Dreibus and Wänke (1985). Like the Earth, the Martian core is assumed to be predominantly iron and to contain a small percentage of nickel based on comparisons with iron meteorites. A range of compositional models is consistent with the gross geophysical and geochemical information on Mars. At one extreme, the core may be nearly pure iron-nickel with a radius of approximately 40% of the planet's radius (BVSP 1981; Morgan and Anders 1979). At the other extreme, the core could contain as much as 34 wt% sulfur (or some other light component) with a core radius as much as 60% of that of the planet (BVSP 1981). Table IV summarizes the likely range of simplified core compositions, core radii and pressures at the core mantle boundary for the various models. We can place meaningful bounds on Martian core temperature only if the physical state (solid, liquid or stratified mixture of the two) is known. However, lack of seismic data on the Martian interior makes a determination of the present state of the Martian core difficult. The only measurements relevant to this problem are those of the magnetic field: a strong magnetic field would presumably be generated by convection currents in a molten core. Spacecraft data, recast into models of solar wind interaction with Mars (Slavin and Holzer 1982; Dolginov 1978*a,b*), indicate a magnetic dipole moment 3 orders of magnitude weaker than that of the Earth. And, based on the quality of the data, some investigators question whether Mars has an internally generated field at present (Russell 1978; see also chapter 31). One model of the current Martian magnetic field suggests that, in its past, Mars had a field generated by a core dynamo mechanism that resulted in a magnetization of surface rocks; therefore it is remnant magnetism that interacts with the solar wind at the present time (Curtis and Ness 1988). The implication of this model is that the Martian core was (partially?) molten until approximately 1.3 Gyr ago at which time it solidified. The melting temperatures of the core constituents would thus provide not only an upper bounds for core temperatures, but also for temperatures in the lower mantle. However, we note that although convection of metallic liquid is probably necessary to generate a strong magnetic field in a terrestrial planet, the lack of a strong magnetic field may not necessarily dictate a solid core: in the absence of heat-producing elements, convection in a completely molten core would not be sufficiently vigorous to generate a magnetic field either (chapter 5). As it appears that the driving force for the geodynamo is precipitation of the inner solid core from the outer molten core (Gubbins et al. 1979), we are left with the possibility of either a completely molten or completely solid Martian core; as with the core radius, only seismic data will resolve the uncertainty. However, thermal modeling of the Martian interior (chapter 5) predicts a completely molten core for compositions with $\geq$ 15 wt% sulfur. Our independent estimate of sulfur concentration of 14.5 wt% is well within error of this limit, so it appears that our compositional estimate is most consistent with a liquid core. In order to accommodate both possibilities, we illustrate two possible temperature profiles in Fig. 4, one based on a solid core and the other based on a liquid one as inferred from the thermal modeling discussed in chapter 5.

Figure 5a shows a melting relations for Fe and FeS measured over the pressure range of the Martian interior (Williams et al. 1987; Williams and Jeanloz 1990). These curves represent lower bounds (liquid core) and upper bounds (solid core) on the temperatures of the 2 end-member core compositions: pure Fe(-Ni) metal and a mixture of 34 wt% sulfur and 66% iron (which corresponds approximately to FeS). Since FeS appears to melt congruently in the Fe-S system at high pressures, its liquidus is equivalent to the lower-temperature bounds of the most sulfur-rich models of the Martian core. If the Martian core is indeed liquid at this time, then these data imply that the Martian interior can be no cooler than 3000 $\pm$ 300 K for the end-member models. To estimate the temperature bounds of our preferred core composition with 14.5% sulfur, we have calculated its liquidus and solidus (also, the Fe-FeS eutectic) by the melting-point depression equations of Williams and Jeanloz (1990) and plotted these curves in Figs. 4 and 5b. In detail these curves are likely to overestimate temperature somewhat, inasmuch as we have ignored the effects of Ni solubility. Nevertheless, the maximum temperature that can be achieved for a completely solid Martian core with the Dreibus-Wänke composition is 2200 $\pm$ 250 K, whereas the minimum temperature for a liquid core is 2400 $\pm$ 250 K.

Most of the chemical models for the Martian core assume sulfur to be the most plausible light alloying component. This assumption is based upon sulfur's ability to dissolve in molten iron at the relatively low pressures and

temperatures likely to be encountered in an accreting planet (see, e.g., Usselman 1975). However, Ringwood and Clark (1971) and Ringwood (1981) have suggested that oxygen is an important constituent of the Martian core. The solubility of oxygen in iron metal is not known at core pressures; however, its solubility is insignificant at least to 10 GPa (Ohtani et al. 1986), so a substantial change in the chemical bonding properties of oxygen is required for oxygen to be dissolved in metal in the Martian core. Diamond-anvil and shock-wave experiments show that iron oxide becomes metallic only at pressures of 70 GPa and above (Knittle and Jeanloz 1986). These results suggest that comparable pressures are necessary for oxygen to have any significant solubility in iron metal or to be completely miscible in iron liquid. For comparison, the maximum central pressure in Mars is < 50 GPa. Therefore, if the core formed by gravitational sinking of iron-rich melt through the Martian interior, then oxygen would not be incorporated, at least initially, into the molten protocore. A second possible way to incorporate oxygen into the core is by reaction of silicate-perovskite with liquid iron at the core-mantle boundary. This reaction forms FeO and FeSi with release of oxygen which dissolves into the liquid iron (Knittle and Jeanloz 1986). At the terrestrial core-mantle boundary (136 GPa) this reaction is viable. However, at 23.5 GPa, the nominal pressure of the Martian core-mantle boundary, the solubility of oxygen in liquid iron is likely to be minimal and the reaction unlikely to be significant. Of course, if the core is solid, then the reaction is not possible. In summary, there is little experimental support for oxygen in the Martian core.

VI. TEMPERATURE GRADIENT AND MOMENT OF INERTIA

In order to calculate the density gradient, we must first calculate a temperature gradient. Our measurements of core-constituent melting relations provide us with temperature bounds on a range of Martian models, so we begin with the core. We do this calculation based on the solid-core model that provides us with upper bounds for temperature (2200 K) at depth. Using the Wiedermann-Franz law and electrical resistivity measurements (Matassov 1977; Keeler and Mitchell 1969), we estimate the thermal diffusivity of iron at high pressure to be 1 to 2×10^{-5} m^2 s^{-1}. This value plus trial assumptions that the core is adiabatic (with a temperature gradient on the order of 10K GPa^{-1}) and that its viscosity is comparable to that of the Earth's inner core (10^{10} Pa-sec; Jeanloz and Wenk 1988), if it is solid, and the Earth's outer core ($< 10^4$ Pa-sec; Stevenson 1981), if it is liquid, allow us to calculate Rayleigh numbers for the Martian core. For a solid core, the Rayleigh number is approximately 12 orders of magnitude greater than the critical value (2000) for the onset of convection, whereas in a liquid core the Rayleigh number exceeds the critical value by a factor of 10^{18}. This excess implies that the calculation is not sensitive to whether the core is solid or liquid or to the estimate of thermal diffusivity or to the assumption of an adiabatic gradient.

Most importantly, the Martian core has a near adiabatic temperature gradient maintained by convection no matter what its physical state. This gradient is given by:

$$\frac{dT}{dP} = \frac{\gamma T}{K_s} \tag{1}$$

where γ is the Gruneisen parameter, T the temperature at the center, and K_s the adiabatic bulk modulus at the center of Mars. Table V lists temperature gradients calculated for each of the core models as well as the input parameters. We employed third-order Eulerian finite-strain equations of state (Birch-Murnaghan) to calculate densities and adiabatic bulk moduli for a solid core. To obtain liquid densities, we used the Clapeyron equation, $dT/dP = \Delta V/\Delta S$, where dT/dP is the slope of the melting curve for Fe, FeS and Fe + 14.5% S. Assuming ΔS to be equivalent to the gas constant R allows us to calculate the volume change on melting (ΔV) for the various core models and correct the solid densities to obtain the liquid densities. As the densities of liquid and solid iron tend to approach each other at high pressure (Jeanloz 1979), this approach is far less uncertain than trying to estimate the bulk modulus and its pressure derivative for liquid Fe, FeS and Fe + 14.5% S at zero pressure and then calculating the equation of state to the appropriate pressure and temperature conditions of the Martian core. As shown in Table V, the differences in calculated density for solid and liquid of a given composition are small (0.2 to 0.3 Mg m^{-3}) compared to the differences produced by differences in composition. Accordingly, our estimates of the temperature increase from the core-mantle boundary to the center range from 350 to 510 K, depending on model composition. The upper bound on temperature at the core-mantle boundary for our preferred composition (14.5% S) is thus 1800 $\pm$ 350 K for a solid core and the lower bound for a liquid core is 2000 $\pm$ 350 K.

In the absence of any evidence of a heat source in the core, e.g., latent heat released from a solidifying inner core, there is no reason to expect a thermal boundary layer at the base of the Martian mantle. Consequently, we assume that the temperature at the base of the mantle is the same as at the top of the core. With $\gamma = 1.0 \pm 1.0$, $T = 2000 \pm 350$ K, and $K_s = 218 \pm 30$ GPa (silicate perovskite at 23 GPa), we have obtained a lower mantle gradient of 9 $\pm$ 4 K GPa^{-1} from Eq. (1) and have used this value to extrapolate the temperature from the core-mantle boundary to the uppermost mantle as illustrated in Fig. 4. The depth at which the temperature gradient ceases to be determined by convective flow and increases to the larger, conductive gradient expected for the lithosphere remains a primary uncertainty. Again, the lack of seismic data on the Martian interior precludes knowing with any certainty the depths at which the interior is partially molten. Given the absence of widespread, current volcanism on Mars, we assume a temperature profile

TABLE V
Thermodynamic Parameters of Candidate Martian Core Materials*

	γ-Fe (fcc-structure)	**FeS (high-pressure phase)**	**Fe + 14.5wt%S ($Fe_{0.78}S_{0.22}$)**
ρ_0(Mg m^{-3}):			
solid	8.0 (±0.1)[a]	5.6 (±0.1)[b]	7.0 (±0.1)
liquid	7.7 (±0.1)	5.4 (±0.1)	6.8 (±0.1)
K_{0S} (GPa)	183 (±10)[c]	140 (±18)[d]	177 (±25)
dK_{0S}/dP	4 (±1)[e]	4 (±1)	4 (±1)
$\langle\alpha\rangle_{300K}$ (K^{-1})	$3.2 (\pm 0.1) \times 10^{-5}$ [f]	$3.2 (\pm 0.1) \times 10^{-5}$	$3.2 (\pm 0.1) \times 10^{-5}$
$\langle\alpha\rangle_{2600 \pm 400K}$ (K^{-1})	$7.0 (\pm 0.5) \times 10^{-5}$ [f]	$7.0 (\pm 0.5) \times 10^{-5}$	$7.0 (\pm 0.5) \times 10^{-5}$
$K_{S\ 50\ GPa,\ 2600 \pm 400K}$	257 (±17)	220 (±26)	251 (±35)
γ	1.5(±1.0)[g]	1.5(±1.0)[d]	1.5(±1.0)
$T_{center\ of\ Mars}$ (K):			
solid core	<3000 (±300)	<3000 (±300)	<2200 (±250)
liquid core	>3000 (±300)	>3000 (±300)	>2400 (±250)
dT/dP (KGPa^{-1})	17.5 (±12)	20.5 (±15)	14.4 (±14)
ΔT (adiabatic temperature change across core)	350 (±230)	511 (±350)	solid: 396 (±300) liquid: 430 (±330)
ρ_{cmb} (Mg m^{-3})			
solid	9.5 (±0.2)	7.4 (±0.3)	8.3 (±0.3)
liquid	9.2 (±0.5)	7.1 (±0.5)	8.1 (±0.5)
g_{cmb} (m s^{-2})	3.0 (±0.3)	3.0 (±0.3)	3.0 (±0.3)
$\langle\alpha\rangle_{core}$ (K^{-1})	$1.0(\pm 1.0) \times 10^{-5}$ [h]	$1.0(\pm 1.0) \times 10^{-5}$	$1.0(\pm 1.0) \times 10^{-5}$
b (km)	1356 (±70)	2034 (±70)	1695 (±70)
μ (Pa-sec)			
solid	$\sim 10^{10}$ [i]	$\sim 10^{10}$	$\sim 10^{10}$
liquid	$< 10^{4}$ [k]	$< 10^{4}$	$< 10^{4}$
κ (m^{2}s^{-1})	$1\text{-}2 \times 10^{-5}$ [j]	$1\text{-}2 \times 10^{-5}$	$1\text{-}2 \times 10^{-5}$
Rayleigh number			
solid	2.4×10^{15}	9.5×10^{15}	5.1×10^{15}
liquid	2.4×10^{21}	9.5×10^{21}	5.1×10^{21}

* ρ_0 = uncompressed or 0 GPa density; K_S = adiabatic bulk modulus; K_{0S} = adiabatic bulk modulus at 0 GPa; α = thermal expansion; γ = Gruneisen parameter; g = acceleration due to gravity; b = core radius; μ = viscosity; κ = thermal diffusivity.

[a] Estimated to be between the values for α and ε iron. [b] Mao (1981). [c] Jephcoat et al. (1986). [d] Brown et al. (1984). [e] This value is shown to empirically correct for a wide variety of solids (cf. Jeanloz and Knittle 1986). [f] Andrews (1973). [g] Brown and McQueen (1986). [h] Estimated to be the same for all three compositions. [i] Jeanloz and Wenk (1988). [j] Matassov (1977), Keeler and Mitchell (1969). [k] Stevenson (1981).

analogous to that inferred beneath the terrestrial continents, i.e., one that passes well below the dry solidus.

With the compositions, mineralogies and temperature-depth relations discussed above, we have divided Mars into a series of shells (core, lower mantle, upper mantle, crust) and calculated a moment of inertia (I/MR^2). The value we obtained, 0.353 ± 0.003, is much smaller than the value of 0.365 proposed by Reasenberg (1977) and Kaula (1979*a*), yet is not as small as the number (0.345) suggested by Bills (1989*a*). This value of the moment is somewhat lower than might be expected in part because of the use of the high-density phase of FeS (ρ = 5.6 Mg m^{-3}) in our core model, rather than the low-density phase (ρ = 4.5 Mg m^{-3}) employed by some earlier workers, such as Okal and Anderson (1978). It is difficult to place meaningful uncertainties on our estimate of the moment of inertia. As liquid and solid core material do not have significantly different densities at their melting points, our moment of inertia calculation is not sensitive to the state of the core. The presence of a differentiated crust also has little effect upon the calculation. For example, the present calculation incorporates an extreme crustal model (100-km thick with density of 2.7 Mg m^{-3}) to satisfy mass balance constraints, yet the presence of this crust lowers the moment by only 0.002. If the basic chondritic model is correct, then within error of our calculation the FeO concentration in the mantle could be as low as 16.9 wt% and the core mass as high as 22.7% of the planet. These parameters would imply a smaller, less-dense mantle and a larger, more-dense core, which in turn would yield a lower moment (by a few thousandths). However, it seems unlikely that any chondritic model consistent with the SNC meteorites could account for the more extreme low values of the moment (~0.345) discussed by Bills (1989*a*).

VII. MAGMATIC SOURCE REGIONS

The estimate of bulk composition outlined above was designed to see through differentiation events and produce an average planetary composition and mineralogy. It is likely that there are significant local departures from the average composition, mineralogy and density in the upper mantle. The observation of volcanism on the surface requires differentiation in the interior. Predictably, the residues of melt extraction will be richer in MgO and poorer in Al_2O_3 than the primordial mantle. The residual ferromagnesian minerals will have higher Mg/Fe ratios and, therefore, will be less dense. Either the proportion of aluminous phase will be diminished or the aluminous phase will be exhausted completely. If volumetrically significant amounts of basalt have erupted over time, then it is likely that our computed upper mantle density is too high.

Isotopic evidence from the SNC meteorites shows that portions of the Martian interior have been extensively differentiated for long periods of time.

Nakhla was derived at 1.3 Gyr from an already depleted source (Nakamura et al. 1982). Although the precise crystallization ages of the shergottites are controversial because of shock effects, the more probable younger ages (< 200 Myr) also require depleted source regions (Jones 1986).

Petrologic evidence from the SNC meteorites is also consistent with a depleted interior. Figure 6 contrasts the compositions of the magmas from which the SNC meterorites crystallized, calculated by Longhi and Pan (1989) with the 2.3 GPa Martian minimum melt (BH) composition determined by Bertke and Holloway (1988). Projection from the olivine component (Ol) facilitates comparison of the non-olivine fractions of basaltic compositions. The low-pressure liquidus boundaries demonstrate that BH and the SNC parent magmas, which lie on opposite sides of a thermal divide, cannot be related by low-pressure differentiation. BH represents the initial melt of an undifferentiated mantle at high pressure, where a different set of liquidus equilibria (not shown) apply; it is in equilibrium with olivine, two pyroxenes and garnet. With increasing melting the garnet will be consumed and the liquid will migrate to the left. Melting in the range of 40 to 50% is necessary to bring the melt composition into the range of the SNC parent magmas. Such large degrees of melting require special circumstances to prevent the melt from separating from the surrounding crystals. It is rather more probable that melting occurred in stages, so that at the time the SNC magmas were gener-

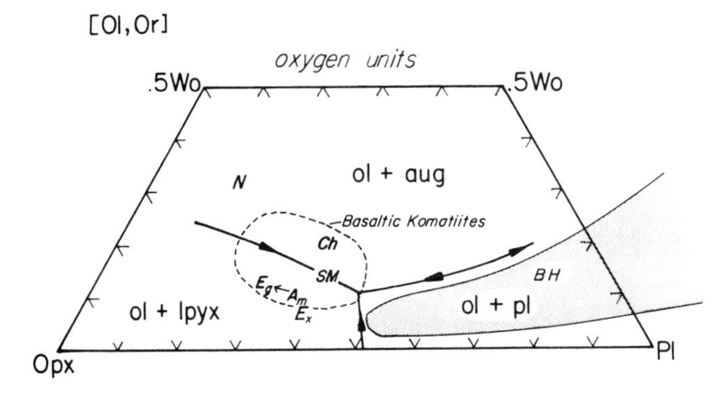

Fig. 6. Comparison of SNC parent magma compositions with expected melts from small degrees of partial melting of a lherzolitic (ol, opx, cpx, Al-phase) mantle. Diagram is a projection from olivine and orthoclase components onto a portion of the orthopyroxene-plagioclase-wollastonite plane. Heavy solid curves are low-pressure liquidus boundaries appropriate for terrestrial-MORB; arrows indicate direction of decreasing temperature. N (Nakhla), Ch (Chassigny), SM (Shergotty-Zagami), E_g (EETA79001A groundmass), E_x (EETA79001A xenocrysts), and A_m (ALHA79005) refer to parent magma compositions of SNC meteorites from Longhi and Pan (1989). BH is the minimum melt composition of a Dreibus-Wänke Martian mantle at 2.3 GPa and 1400°C from Bertke and Holloway (1988). Shaded region is the range of melt compositions expected from a lherzolitic terrestrial mantle from 0.5 to 4.0 GPa (Longhi and Pan 1989).

ated, their source region was already depleted in basalt component and probably did not contain an aluminous phase.

We have no way of knowing how extensive the SNC source regions are at present. We also do not know how much basalt has been erupted on Mars over time. Therefore we cannot estimate the scale or magnitude of departures from our average mantle composition or density.

VIII. SUMMARY

If the SNC meteorites come from Mars and if Mars has overall chondritic abundances of refractory lithophile and siderophile elements, then Mars has the following characteristics: a lower degree of depletion of moderately volatile elements than the Earth; a depleted mantle that is mineralogically similar to that of the Earth, but richer in iron and denser; a Fe-Ni core whose major light element is sulfur and whose radius is approximately half that of the planet; and a moment of inertia ~ 0.353. Because the Martian mantle is much shallower than the Earth's, the proportions of rock types are different: the lherzolite upper mantle is dominant; the olivine-spinel/majorite lower mantle is subordinate; and the perovskite layer is minor (no more than 200 to 500 km).

Acknowledgments. We thank R. Buck, G. Schubert, and Q. Williams for helpful discussions and comments. We also thank K. Burke and an anonymous reviewer for their suggestions. We gratefully acknowledge support from the National Aeronautics and Space Administration (JRH,JL), from the Petroleum Research Fund (EK), and from the Deutsche Forschungsgemeinshaft (HW).

7. GRAVITY AND TOPOGRAPHY

P.B.ESPOSITO, W.B. BANERDT, G.F. LINDAL, W.L. SJOGREN, M.A. SLADE
Jet Propulsion Laboratory

B.G. BILLS, D.E. SMITH
Goddard Space Flight Center

and

G. BALMINO
CNES/Groupe de Recherches de Geodesie Spatiale

The fundamental gravity field constants for Mars are summarized with a brief historical review of early determinations and current-day accurate estimates. These include the planetary gravitational constant, global figure, dynamical oblateness, mean density and rotational period. With the advent of orbiting spacecraft, such as Mariner 9 and Vikings 1 and 2, detailed and complex gravity-field models emerged. The evolution of the global model development is presented. The culmination of this work is the eighteenth degree and order model of Balmino et al. (1982) yielding a Mars geoid accurate to 15 m near the equator and 30 m in the polar regions. Applying the techniques similar to those used to discover lunar mascons, Sjogren developed a global line-of-sight acceleration map that emphasizes local gravity variation and their correlation with topographic features. Mars topography was first deduced from Earth-based radar observations. This review summarizes topographic results from data acquired from the 1967 opposition to the most recent 1988 opposition. Both global (spanning 360° in longitude) and selected, local topographic variations and features (e.g., south of Arsia Mons and craters Gusev and Williams) *are discussed. Radar observations continue with increasing accuracy both for new topographic results and Mars ephemerides. Spacecraft radio occultation measurements yield a global distribution of planetary radii; the error is within 1 km. A laser altimeter is included in the Mars Observer payload. A uniform and global distribution of measurements are planned with an overall accuracy of 20 to 30 m. By combining gravity and topography, basic inferences can be made about the interior of Mars. This includes a detailed discussion of the inertia tensor and the nonhydrostatic component of Mars. In particular, the dimensionless moment of inertia about the rotation axis Λ is 0.4 for a body of uniform density and 0.37621 if Mars were in hydrostatic equilibrium. Smaller values result under various assumptions of the nonhydrostatic nature of Mars. Finally, by comparing models of both gravity and topography, inferences are made about the degree and depth of compensation in the interior and stresses in the lithosphere.*

This chapter is divided into three parts with each section naturally leading to the next. In the first section, we review the current state of the art of the gravitational field of Mars both globally and locally. After a review of fundamental constants, the progress of the development of high-order gravity models is presented. All of this work relies on the analyses of Doppler data received primarily from orbiting spacecraft such as Mariner 9 and Vikings 1 and 2. With these models, a global geoid is deduced as well as power and error spectra. Local gravity-field modeling (by analysis of Doppler residuals) involves analysis of specific features responsible for gravity variations. The methodology is presented in detail and the correlation of gravity with topography is shown.

Because of the need to identify gravity variations with surface (or subsurface) features, the second section deals with Mars topography deduced from Earth-based radar and spacecraft measurements (especially occultations). Radar-deduced, global topography is shown (roughly within $\pm 25°$ of the equator) and the precision is within 150 m. In addition, local features are discussed and compared with spacecraft imaging results. The occultation measurements provide absolute radii estimates with an error of 1 km. We briefly summarize the laser altimeter on Mars Observer and expected altimetry results. These sections complement the results given in chapters 10 and 20.

Finally, the third section utilizes the results of the previous two in order to make basic inferences about the interior of Mars. It starts with equations for the gravitational potential and topographic heights from which figures of the geoid and topography are shown. An in-depth discussion follows, concerning the inertia tensor and the nonhydrostatic component of Mars' ellipticity. Thereafter, measures of the spectral variations of gravity and topography are defined in order to correlate the two. Inferences are made about the degree of compensation in the interior and stresses in the lithosphere. These results and hypotheses complement the review presented in chapter 8.

I. CHARACTERISTICS OF THE GRAVITATIONAL FIELD

A. Basic Results

Gravitational Constant and Oblateness. The earliest reliable estimates of the gravitational constant (product of the Newtonian gravitational constant and planetary mass, *GM*) and gravitational oblateness of Mars were obtained from observations of the orbital motions of its satellites Phobos and Deimos. Hall (1878) estimated $GM = 42900 \pm 70$ km^{3}s^{-2}, and Struve (1895) obtained $J_2 = 1951 \times 10^{-6}$. The best current *GM* estimate is based on analysis of Doppler data from the Mariner 4 flyby: $GM = 42828.3 \pm 0.1$ km^{3}s^{-2} (Null 1969). The currently accepted value of the gravitational oblateness is $J_2 = 1960.454 \times 10^{-6}$ (Balmino et al. 1982).

Global Figure. Obtaining accurate estimates of the size and shape of Mars has proven to be more difficult than for the other geodetic properties, and even now the global figure remains largely unconstrained. Classical estimates of the polar and equatorial radii are summarized by de Vaucouleurs (1964). The mean of 32 polar radius determinations spanning the period 1890 to 1958 yields 3378.0 ± 3.6 km. The mean of 68 measurements of the equatorial radius from the period 1879 to 1958 is equivalent to 3414.2 ± 3.6 km. It is interesting to note that these early polar radius estimates essentially agree with present estimates, whereas the equatorial radius was consistently overestimated. The best current estimates of Martian topography come from Earth-based radar (restricted to $\pm$ 25° latitude) and spacecraft data (principally occultations). Bills and Ferrari (1978) have used essentially all of the pre-Viking data to estimate spherical harmonic coefficients to degree and order 16. The mean radius thus obtained is $R = 3389.92 \pm 0.04$ km.

Mean Density. The degree zero harmonics of gravity and topography, together with a value for the Newtonian gravitational constant (Luther and Towler 1982) $G = (667.26 \pm 0.05) \times 10^{-13}$ m^3 kg^{-1} s^{-2} yield a mean density of 3933.5 ± 0.4 kg m^{-3}. The largest contributor to the error in this estimate is uncertainty in the Newtonian gravitational constant. This mean density is much closer to that of the Moon (3344.0 kg m^{-3}; Bills and Ferrari 1977) than that of the Earth (5514.8 kg m^{-3}; Rapp 1974) or Venus (5244.8 kg m^{-3}; Bills and Kobrick 1985). However, much of the density difference for the larger bodies is due to higher internal pressures. Estimates of the uncompressed densities of these bodies (Kovach and Anderson 1965) place Mars roughly midway between the Moon and the Earth.

Rotational Period. The rotation period of Mars has been known with impressive accuracy for a long time. By 1666, Cassini had determined a period of just over 24 hr. Bakhuyzen (1897) and Wislicenus (1886) used observations of the time of central meridian crossings of Syrtis Major, spanning the interval from 1659 to 1881, to derive a period of 88,642.655 ± 0.013 s. Radio tracking data from the Viking Landers give the best current estimate: 88,642.663 ± 0.002 s (Mayo et al. 1977). Exchange of mass between the atmosphere and the seasonal polar caps will produce annual and semi-annual variations in the rotation period with an amplitude of roughly 0.04 s (Reasenberg and King 1979).

Low-Degree Topography Harmonics. The first degree topography harmonics indicate that the center of figure of Mars is displaced from the center of mass by 2.50 ± 0.07 km towards (62.0 ± 3.7)° S latitude, (87.7 ± 3.0)° W longitude. This is slightly east and considerably south of the center of Tharsis. Note that the direction given here is exactly antipodal to the direction cited by Bills and Ferrari (1978). They erred in converting the cartesian coordinates to the (left-handed) spherical coordinates of Mars. The error was pointed out by Roth et al. (1981) but the erroneous earlier interpretation has not been successfully eradicated.

B. High-Degree, Global Gravity Field Models

Only after orbiting spacecraft were placed around Mars (Mariner 9, Vikings 1 and 2), did a detailed gravitational field structure and model emerge. The first detailed gravity field models were developed from the analysis of the Mariner 9 Doppler data (Lorell 1972; Lorell et al. 1973; Born 1974; Jordan and Lorell 1975; Sjogren et al. 1975; Reasenberg et al. 1975). Lower-order coefficients were evaluated first. These confirmed the large oblateness of Mars (which was in excellent agreement with that determined previously from Earth-based Phobos observations) and determined the equatorial ellipticity. Furthermore, knowledge of the orientation of the planet's rotation vector was improved.

As more Doppler data became available, models increased in complexity ranging primarily from sixth through ninth degree and order. The analysis methodology focused on three distinct procedures: (1) the standard, long-arc, direct determination of gravity coefficients (Lorell et al. 1973); (2) the analysis of classical orbit elements determined from a single orbit of data (near apoapsis) and extended to more than 200 consecutive and separate orbits (Born 1974); and (3) the sequential accumulation and analysis of short-arc, periapsis-only data (Gapcynski et al. 1977), coupled with modeling the gravity field by a network of point masses (Sjogren et al. 1975). In the last procedure, the point-mass distribution was transformed to standard gravity coefficients and compared with the results of the previous procedures. In all, the results were in general agreement. This can be demonstrated by converting the global gravity field model to an equivalent surface mass distribution (expressed as surface height contours) or to an equipotential surface. All models exhibited a large gravitational bulge in the Tharsis region. In addition, there was a correlation of some topographic and surface features with the gravity results (e.g., Hellas and the Acidalia Planitia).

After the two Viking spacecraft were established in orbit, gravitational field models were developed by combining radiometric (primarily Dopper but also range) data from Mariner 9 and the Viking 1 and 2 Orbiters. The first of these (Gapcynski et al. 1977) was based on a short-arc method and a limited data set (thirty-three, 4-hr data arcs centered on periapsis). The result was a sixth degree and order gravity field model and the associated contours of the equipotential and surface undulations. This new gravity field defined the equipotential surface to an accuracy of $\sim$ 75 m over a latitude range of 65° S to 65° N.

Recognition of the limitations in the previous model, along with the accumulation of additional Viking radiometric data, led to the development of a 12 $\times$ 12 gravity field model (Christensen and Balmino 1979). The methodology involved a multiple-arc technique (67 arcs) applied to data arcs ranging from 1 to 12 days in length. The derived field exhibited stronger correlations between gravity and topography than previous low-order models. In addition, the power and noise spectra of the gravity model was discussed, leading to the suggestion "that terms above degree seven may be suspect" (Christensen and Balmino 1979).

Balmino et al. (1982) included additional Viking Doppler data in this multiple-arc analysis procedure thus utilizing 366 data arcs. With the additional data, it was possible to expand the gravity field solution to an eighteenth degree and order model. However, it was noted that there remained a lack of low-altitude data below 60° S latitude. In order to interpret and correlate this model with surface features, a Mars geoid has been produced and is shown in Color Plate 3. A correlation with gross topographic features such as the Tharsis region, Hellas Planitia and the Isidis Planitia is evident (see also Sec. III). Geoid height errors have been computed and are uniformly small across the planet. These reach a maximum (30 m, 3σ), in the polar regions and are smaller (15 m, 3σ) near the equator. However, because of the global nature of these models, they smooth over local gravitational effects. Techniques used to delineate more clearly local gravity effects are described below in Sec. I.C.

Gravity Field Spectrum. Based upon the 18 $\times$ 18 gravity field, power and noise spectra have been produced and discussed (Balmino et al. 1982). The power in the Mars gravity field is an order of magnitude larger than that of the Earth's field. Considering the known irregular shape of Mars, that is probably to be expected. Inspection of the power spectrum shows that it can be represented by the relation

$$\left[\frac{1}{2l+1}\sum_{m=0}^{l}(\bar{C}_{lm}^{2}+\bar{S}_{lm}^{2})\right]^{1/2}=\frac{1.3\times10^{-4}}{l^{2}} \qquad (1)$$

where the coefficients are normalized and l,m refer to their degree and order (Balmino et al. 1982).

Future Earth-based Dopper measurements may be as accurate as 0.2 mm s^{-1} (1σ over a 60-s period). When applied to orbiters in low altitude, nearly circular orbits, the power and error spectra are expected to behave as shown in Fig. 1. We can see that at around degree 50 the errors in the coefficients are approaching 100%. Thus, to extract all the gravity information from the Doppler signal, it may be necessary to solve for a gravity field of degree and order fifty. This corresponds to a resolution of 7.2 deg on the globe or the wavelength of the smallest feature that can be resolved, which in turn corresponds to about 430 km or a half-wavelength of 215 km. If this result holds, the improvement in the knowledge of the Mars gravity field is likely to be at least an order of magnitude.

C. Regional and Local Gravity Variations

Background. Planetary spacecraft provide a unique viewing geometry which was not initially attainable with Earth-orbiting spacecraft. Spacecraft in planetary orbit can be tracked over 50 to 100% of each revolution with the radio communications link (with data acquisition limited only by occultations), whereas a single station track of a near-Earth satellite exists for only 10% of the orbit or less. Thus, for Earth orbiters many stations are required and only short arcs of data are obtained. For a planetary orbiter, a single tracking station yields a long continuous profile of the orbit and one can easily discern high-frequency signatures that are associated with the planetary gravity field. As a result, significant gravity information can be extracted from each profile or orbit of data in near real time. For the Earth, many small blocks of data from many tracking stations must be simultaneously reduced to produce a global gravity field estimate, taking many months of analysis and many hours of computer time.

This became especially evident in the analyses of the Lunar Orbiter Doppler data in the late 1960s. Each orbit of data was reduced independently and its Doppler residuals were fit with smooth polynomials. The polynomials were analytically differentiated to produce line-of-sight accelerations (the vector component of acceleration along the line from the spacecraft to the Earth tracking station). When these accelerations were plotted as a function of spacecraft position over the lunar surface, a very striking correlation between these gravity signatures and surface features became clearly evident. This led to the discovery of the lunar mascons (Muller and Sjogren 1968), and detailed analyses of many local features such as craters, mountains and individual mascons. Data from Apollo missions were especially interesting as spacecraft periapsis altitudes were very low (~10 to 15 km in many cases).

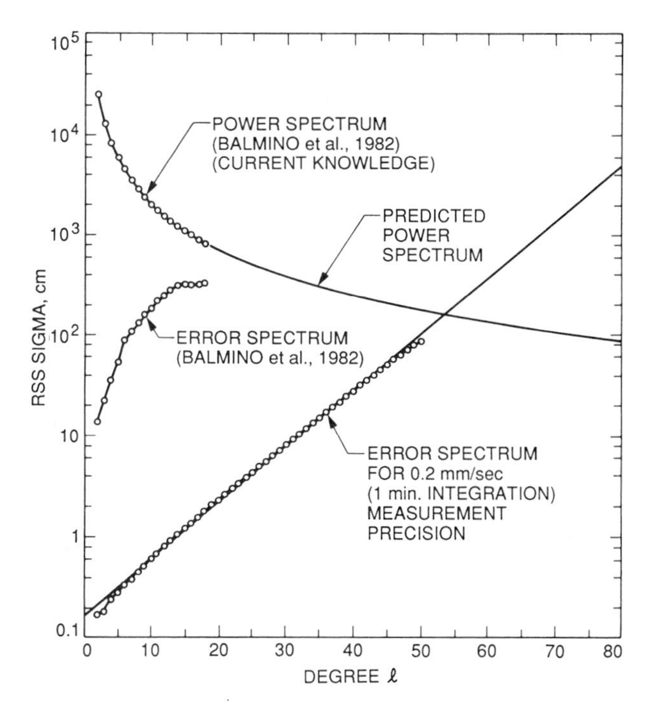

Fig. 1. Power and error spectra of the Mars gravity field. The current knowledge (Balmino et al. 1982) is contrasted with that expected from the Mars Observer mission. The horizontal axis is the degree l of the spherical harmonic; the vertical axis is the *RSS* of the coefficients of degree l multiplied by the radius of Mars.

Modeling Techniques. Following the success with the lunar data, similar analyses have been applied to the Viking low-altitude data (~300 km periapsis). Due to the huge Tharsis bulge, there is a very significant low degree and order spherical harmonic gravity field, so the Doppler residuals were extracted using a fourth degree and order spherical harmonic, reference gravity field model (Sjogren 1979). The 2-hr blocks of data within 1 hr of periapsis were independently reduced to produce line-of-sight acceleration profiles. These data blocks were selected from periods during which the spacecraft orbit plane was near the Earth-Mars line, so the line-of-sight accelerations represent a significant component of vertical gravity. These near-vertical accelerations were overlaid on a Mars topographic map for correlation with surface topography. The horizontal components of gravity are very difficult to interpret, for their contours are ambiguous and their amplitudes do not peak over a feature. Vertical gravity usually does peak over sizable topography.

Another approach that has been used for modeling the observed data is the representation of basins, craters and volcanoes by near-surface disks (Phillips et al. 1978). These features are essentially circular and have definite gravity anomalies associated with them, as is clearly evident from the line-of-sight acceleration mapping. A single orbit profile which passes directly over the feature of interest can be modeled with a disk at the surface to represent the visible topography and another disk at depth to account for moho relief or isostatic compensation. One can then estimate the mass, radius and depth of the subsurface disk and infer crustal structures that are consistent with the mass value obtained.

Present Results. On 1977 October 25, the periapsis altitude of Viking Orbiter 2 was lowered to 300 km and some 51 orbits of data were acquired that provided fair longitudinal coverage (Fig. 2). The line-of-sight accelerations from these orbits were plotted, contoured and overlaid on a Martian topographic map (Fig. 3) (Sjogren 1979). These contours display the correlation of the volcano highs (Olympus Mons, Alba Patera, Arsia Mons, Ascraeus Mons and Elysium Mons), the Isidis Planitia high and the Valles Marineris low.

The orbit which had the peak acceleration over a particular feature was then selected for modeling with disks, and one determined a more quantitative answer to the amount of isostatic compensation that may have occurred. As detailed topography for Mars is sorely lacking, only a few anomalies were studied and even these have significant uncertainty. That is, the global topography is available with 1-km resolution and comparable accuracy, but for geophysical interpretation 100 to 200 m resolution is required. Investigations of Olympus Mons, Isidis, Hellas Planitia and the crater Antoniadi have been reported in the literature (Sjogren 1979; Sjogren and Wimberly 1981; Sjogren and Ritke 1982). If the topographic values are reliable and a simple isostatic model is assumed, then Antoniadi and Hellas are compensated at a depth of 100 km or more. The analysis of Olympus Mons reveals hardly any compensation or else very deep compensation (>500 km), again under considerable topographic uncertainty. It is possible that Olympus Mons is a very young feature in an early stage of relaxation, implying a large degree of flexural support (see chapter 8).

More recently, Janle and Jannsen (1986) conducted detailed analyses of Olympus Mons, using better topography volume estimates determined by Wu et al. (1984). Their results indicate significant compensation as the volume estimate changed by 33%. They have modeled Airy, Pratt and elastic bending and have placed bounds on crustal and lithospheric thickness for the various models. Janle (1983) also investigated the highland/lowland escarpment and concluded that it was essentially in isostatic equilibrium. Janle and Ropers (1983) studied the Elysium dome and estimated a mean crustal thickness between 30 and 120 km. All of Janle's results are based on interior structure models that will produce accelerations on an orbiting spacecraft. After numerically integrating these accelerations on the Viking 2 orbit, a comparison was made with the real gravity data profiles. Malin (1986), using a disk

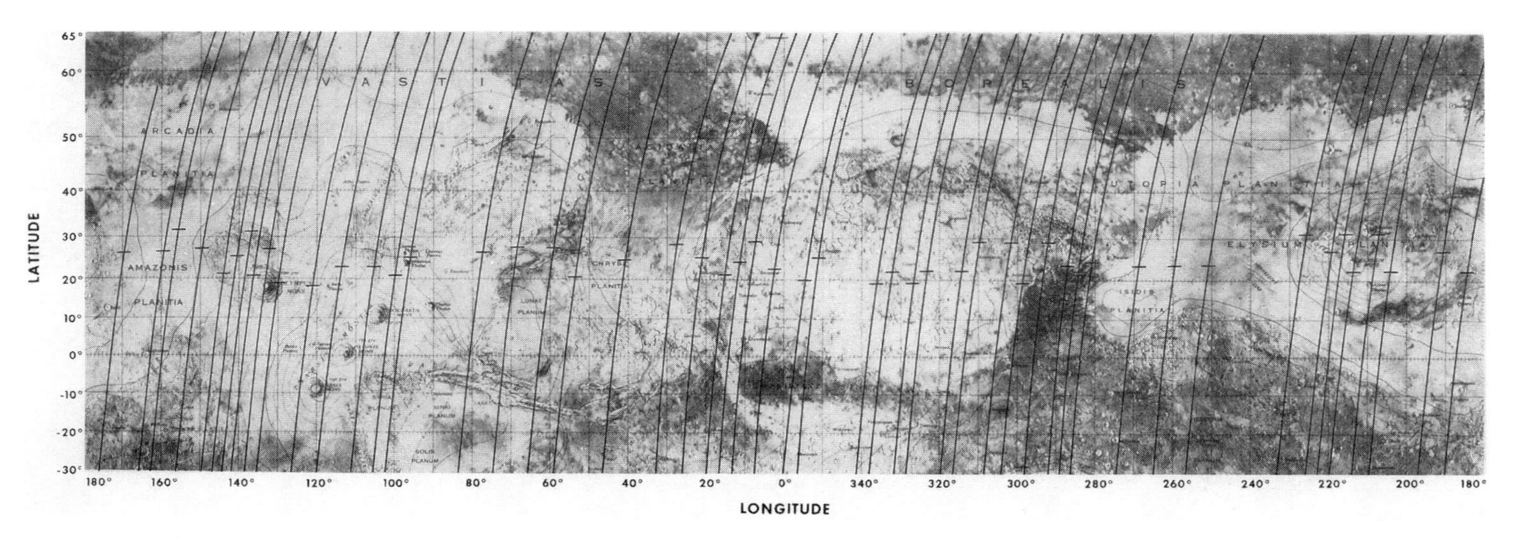

Fig. 2. Diagram of the data coverage (Sjogren 1979). Lines represent the surface tracks of the Viking Orbiter 2 spacecraft during 51 low-periapsis orbits of good Doppler data. The tick mark on each ground track indicates the periapsis location, which was at 31°N latitude at the beginning of the data set and moved linearly southward to 17°N at the end.

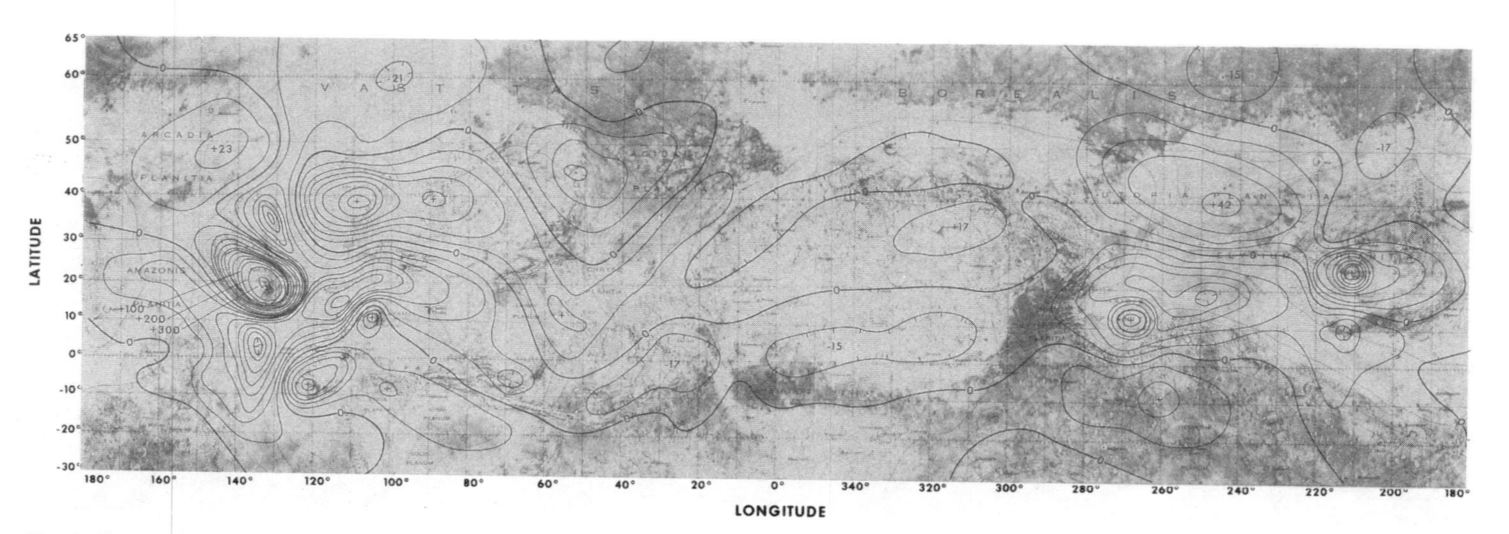

Fig. 3. Contoured line-of-sight (LOS) accelerations (Sjogren 1979). Contour intervals are 10 mgal except those notated at Olympus Mons (18°N, 133°W). These accelerations are referenced to a fourth degree and order background field. Accelerations are at spacecraft altitudes that were approximately 1000 km at 70°N, 300 km at 25°N and 1000 km at 30°S.

model of the north polar layered deposits, suggests a density of 1 g cm^{-3} for this terrain, having a composition of 5% silicate dust and 95% ice.

It is anticipated that data from Mars Observer will provide information over some 30 or more additional features and add more information to those presently observed. This will allow one to evaluate the compensation of feature types over the entire planet.

As an understanding of the gravity results is coupled to the topography, we next review Mars topography deduced from Earth radar and spacecraft observations.

II. MARS TOPOGRAPHY

A. Mars Topography from Earth-Based Radar

Introduction. The rapid rotation rate of Mars causes it to be a difficult radar target for topography determination from Earth. Direct unambiguous measurement of the topography of Mars by interferometric techniques, as has been done for the Moon (Zisk 1972) and Venus (Jurgens et al. 1980), is a formidable task. Mars is an "overspread target" in the jargon of the field; see chapter 20. The resources to perform such measurements would amount to a significant fraction of an interplanetary mission budget. However, radar time-delay measurements can be used to measure "profiles" of the topography along the locus of the sub-Earth point motion due to Mars' rotation and changing orbital geometry.

Roughly 42% of the surface of Mars is accessible by the latter technique (briefly called altimetry). In reality, observations have been most intensively performed during oppositions when Earth and Mars are closest to each other. Some oppositions are closer than others, with longer visibility for northern hemisphere observatories. These oppositions occur over a ~15 yr cycle: 1971–1973 and 1986–1988 are the years of these so-called perihelic oppositions. The observable region falls in a band of Martian latitudes between roughly ±25°, which samples the crustal dichotomy boundary, the Tharsis complex, Valles Marineris and other significant topographic features.

The topographic information in altimetry has some limitations that must be kept in mind during interpretation of radar topography. The range (i.e., the distance to the target) component of the radar echo provides an elevation measurement for a patch of the ground located somewhere within the resolution cell (sometimes referred to as the radar footprint). The average uncertainty in the determination of the echo delay to this patch is better than 100 m in, for example, 1973 Goldstone Mars data (Downs et al. 1975). This uncertainty governs the precision of the data. Global accuracy depends on the accuracy of the planetary ephemeris and on the choice of the reference figure. The measured patch of ground is located somewhere within a resolution cell. In Goldstone data, for example, the size of the resolution cell in the

1971, 1973, 1978, 1980, 1982 and 1988 data is 0°16 (~8 km) in longitude and about 1°30 (~80 km) in latitude. The ranging algorithm fits a template to the first response in the echo power to calculate the point closest to the Earth. The radar pulse is convolved with the reflection properties of the terrain within the radar footprint. This convolution means that topographic relief at size scales smaller than the resolution cell can be underestimated. Geometric effects also play a role (see Harmon et al. 1986, for more detail).

The complex interaction between the radar pulse and the planet's surface, however, also means that the depths of geologic features smaller than the length of the resolution cell or of features located off the nominal subradar track can be estimated. Independent photographic evidence must be utilized to insure an accurate location for the estimated radar topography. In those cases when topography away from the subradar track is estimated, the strongest return comes from the low-lying, apparently smooth floor, rather than from the elevated, rough rim. We stress that the radar-derived interior/exterior relief of concave/convex landforms has to be viewed as the minimum estimated relief.

History of Early Observations. The first paper utilizing Mars time-delay observations covered the 1967 opposition (Pettengill et al. 1969) as observed at 3.8 cm by Haystack; it emphasized the details of the new data acquisition techniques. This report necessarily presented only relative topography not reduced to any unified datum, as an accurate ephemeris for Mars was still being developed. The most significant results from this work were (1) the demonstration of the lack of correlation between radar/visual brightness and elevation, and (2) at the northern latitudes covered during the 1967 opposition, visually dark regions appeared to correlate with high radar reflectivity. These results were confirmed during the 1969 opposition by measurements from Haystack (Rogers et al. 1970) covering Martian latitudes 3° to 12°. Observations at 13 cm during the 1969 opposition were presented by Goldstein et al. (1970).

The usefulness of Earth-based radar altimetry for characterizing the geologic setting of features on Mars was first recognized by Pettengill et al. (1971), working with the Haystack instrument, and by Downs et al. (1971), working with the Goldstone facility. The Haystack observations were compared with Mariner 6 and 7 photographs. Numerous craters were identifiable, as well as a scarp at 41° W, 14°S.

The most extensive series of topographic measurements following these preliminary results were carried out by Downs and co-workers (Downs et al. 1971, 1973, 1975, 1978, 1982) during the perihelic opposition of 1971 and the intermediate opposition of 1973 (68 scans). (A radar scan refers to an arc of constant latitude on the planet, traversed by the subradar point during one ranging session.) The 1971 opposition also had observations at 3.8 cm from

Haystack (Pettengill et al. 1973). The program of observations from Goldstone was less extensive for the intermediate and aphelic oppositions of 1975 through 1982 (40 scans). The 1971–1973 series yielded some spectacular topographic profiles (see, e.g., Fig. 4 covering part of Arsia Mons). However, in the 1971–1973 data (Roth et al. 1989) the radar-derived topography does not readily distinguish between coarse geologic subdivisions, e.g., between units Hprg (ridged plains), Nplc (cratered plateau materials), and Nhc (hilly and cratered materials) of Scott and Carr (1978) (see: Color Plate 3 and Map I-1802-A). Exceptions might exist at smaller scales (cf. Schaber 1977). Systematic elevation differences are associated with the variable-albedo features within craters and perhaps with some topographic lows. On a regional scale, albedo and elevation appear to be unrelated, in agreement with earlier observations (Pettengill et al. 1969). We discuss below a few examples of radar-determined topography from the 1971–1973 data set.

Because of the wide variation in Mars' distance during various oppositions, the density of latitude coverage is quite variable. Only the 1971–1973 data set is uniform enough to portray global-scale topography accurately. Color Plate 4 shows an overview of the radar altimetry referenced to the 6 mbar surface. In order to produce these plots, we interpolated the altimetry to a regularly spaced grid. The interpolation is that of a two-dimensional thin plate spline. The resultant surface is approximately the minimum curvature surface. These radar scans covering the Martian equatorial regions are far from uniform in density of their coverage, particularly in the 1975–1982 observations that cover northern latitudes. Note in particular that the Tharsis complex has been covered in a very irregular fashion. While Arsia Mons is well represented (~longitude 125°), Olympus, Pavonis and Ascraeus Montes have almost no expression in the plots. In addition to density of coverage, one must keep in mind that the radar topography represents the minimum positive/negative relief. The regions of no-data coverage for the 1971–1973 observations are shown explicitly in Downs et al. (1975). As a rule of thumb, regions in which the colors change slowly and broadly have a low density of coverage. The accuracy associated with the topography covered in Plate 4 is better than 150 m (1σ).

While global trends are important, the 1971–1973 Goldstone data are extremely useful for deriving local and regional relief. For example, a topographic feature that would be difficult to identify from visual imagery alone is the sharp elevation increase seen in radar scans immediately west of crater Williams (164°.0, −18°.0) (Fig. 5). As in the case of the radar-identified Coprates rises (Roth et al. 1980), the sudden elevation change may be accompanied by a north-south-trending, low-albedo trough (Fig. 5a). The sharp dip at 182° longitude shown in Fig. 5b represents the echo from the floor of the south-north trending channel Ma'adim Vallis. Detection was possible because of the alignment of the channel trunk with the radar resolution cell.

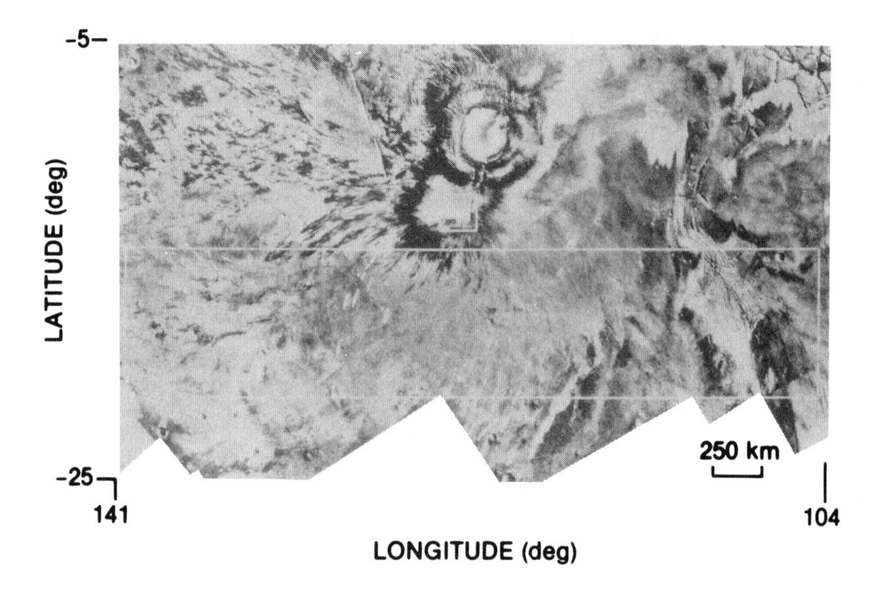

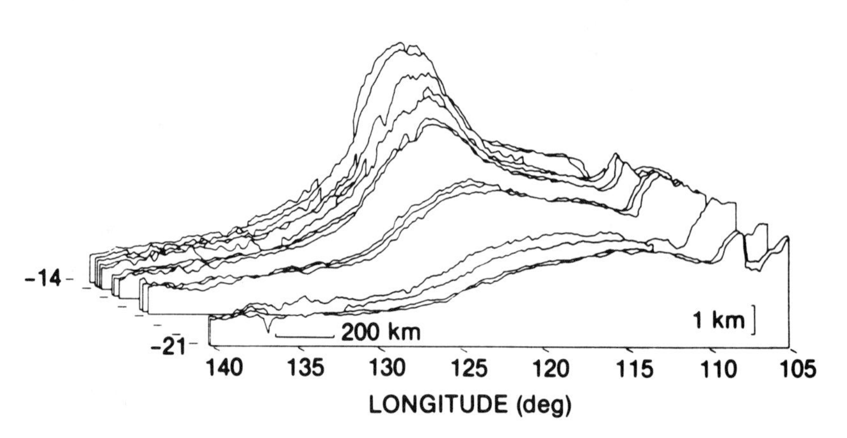

Fig. 4. Composite of topographic profiles south of Arsia Mons from 1971 and 1973 Goldstone observations at 13-cm wavelength (Roth et al. 1980). The central cone shows the margins of the hypothetical parasitic shield; the ridge to the east is part of the Claritas Fossae fracture zone.

Crater Gusev (184°5, −14°7) (Fig. 6) is the terminus of the Ma'adim Vallis channel system. Low-albedo material, visible in both the Viking Orbiter and Mariner 9 images (Fig. 6a), covers the southwest half of the crater floor. The crater was scanned six times. All six scans cut across the low-albedo region and, to various degrees, all six show the crater to have an inclined floor, with the dark segment of the floor always at a higher elevation

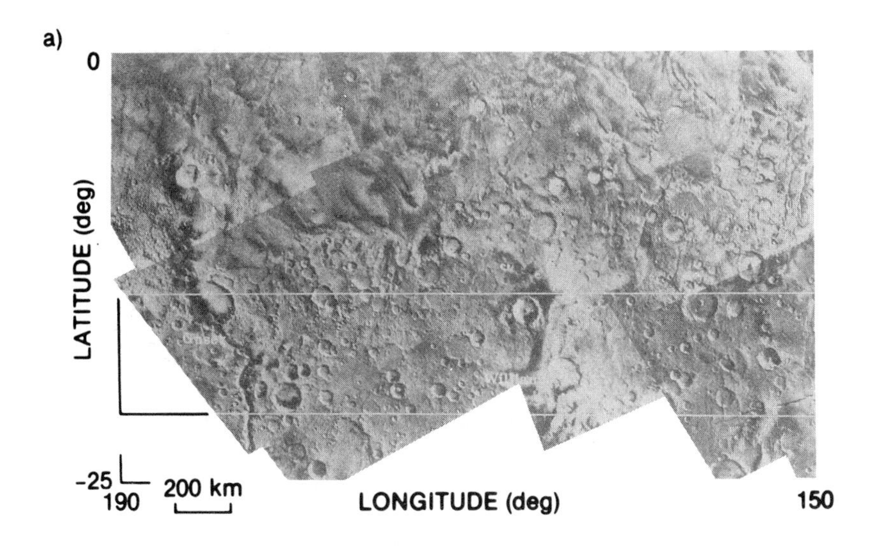

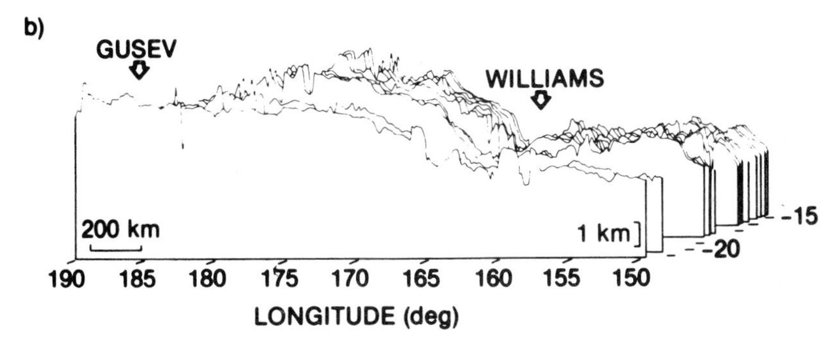

Fig. 5. (a) Photomosaic composite of segments of the Memnonia (MC-16) and Aeolis (MC-23) quadrangles. Goldstone radar coverage in 1971–1973 ends at the northern edge of the Memnonia-Mare Sirenum fault fields (Mutch and Morris 1979). Reconstruction of the topography enclosed in the box is shown below. Uncontrolled, partially completed Viking Orbiter Imaging Team photomosaic. (b) Radar composite topography in the Memnonia and Aeolis quadrangles. The figure provides a setting for craters Williams and Gusev. Vertical exaggeration is 100x. (14 radar scans between $-14\overset{\circ}{.}80$ and $-21\overset{\circ}{.}23$ latitude.)

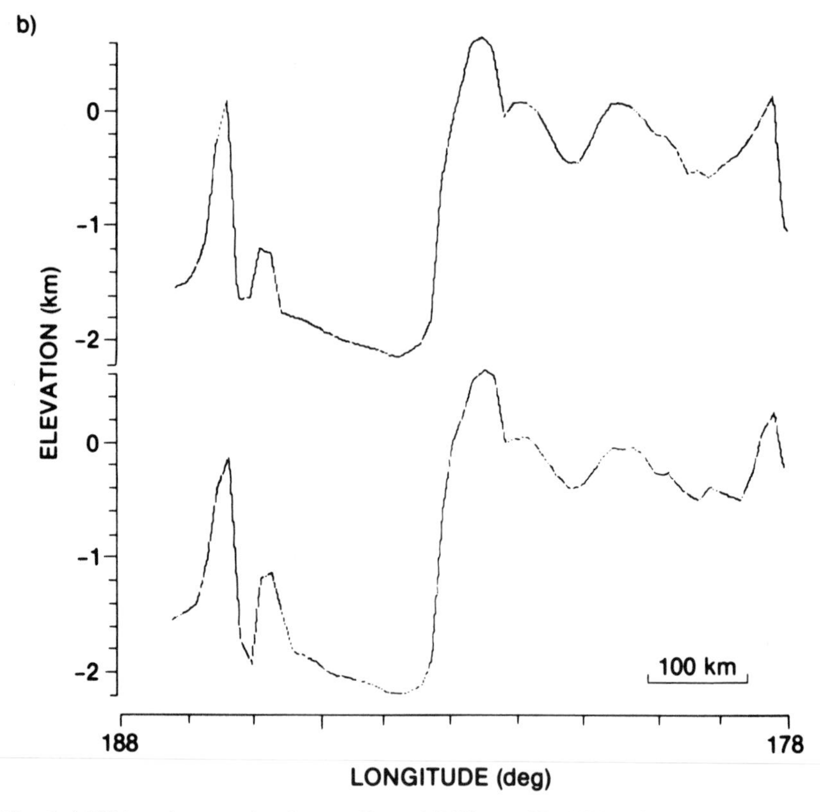

Fig. 6. (a)Viking photomosaic of crater Gusev (184°.5, −14°.7). Viking frame: 603A41. (b) Goldstone 1971 radar altimetry profiles of crater Gusev. Vertical exaggeration is 100x. Radar scans: −14°.80, 71 Sep 13, 0.442 AU; −14°.89, 71 Aug 10, 0.376 AU.

than the light segment. The two best scans (Fig. 6b) suggest the floor slope to be uniform and equal to <0.05 km/km (~3°0), with no observable discontinuity over the albedo boundary. The height of the west rim is ≲400 m above the level of the surrounding plains. The prominent topographic spike masquerading as the west rim is probably a composite profile of the rims of a swarm of superimposed craters located outside the Gusev perimeter. In contrast to crater Williams, the tilt of the Gusev floor appears to be associated with the tonal differences between materials that constitute the floor, rather than with the regional topographic trends. In analogy to Ladon basin, the floor of Gusev appears to be tilted toward the point where a major runoff channel (the Ma'adim Vallis) is entering the crater. If this is the lowest point of the crater floor, then the dark wedge superimposed on the light background material must have evolved since the time of cessation of the Ma'adim Vallis flows. An aeolian origin would imply entrapment of the material of the dark surface to a depth of about 400 m. Goldstone C-parameter data (Downs et al. 1975) indicate the dark material to be radar smooth and the light surface to be radar rough. The combined elevation uncertainty due to differences in radar roughness and to possible contributions from the neighboring Doppler cells is less than the measured elevation difference. Roth et al. (1989) conclude that the difference is probably real.

Intermediate-Quality Oppositions. The oppositions between 1973 and 1986 were much more distant oppositions. Data were taken at Goldstone in 1975–76, 1978 and 1980–82. The Goldstone radar data for these "intermediate" oppositions were summarized in Downs et al. (1982), and the radar-derived topography was compared there to the U.S. Geological Survey 1:25M Map (I-961).

Soviet radar observations at 39 cm during the 1980 opposition produced the first profile (Fig. 7) across Olympus Mons (Kotel'nikov et al. 1983). Their scan is compared against the Haystack 3.8 cm data, mentioned previously (Pettengill et al. 1969). The discrepancy over Olympus Mons is a good example of the peak elevations of resolution cells appearing rough at scales comparable to the radar wavelength, and scattering the reflected power into undetectability. The surface was evidently sufficiently smooth at the Soviet wavelength for the profile to be obtained, but too rough at 3.8 cm for the signal to noise available at Haystack during the 1967 opposition.

1988 Opposition: S- and X-band Ranging. The 1986 and 1988 oppositions were too far south in declination for Arecibo to make significant ranging observations. Goldstone was able to perform S-band (13 cm) and X-band (3.5 cm) ranging during the 1988 opposition only. The 1988 altimetry in the cratered highlands includes coverage of the Tyrrhena Patera volcano, the Claritas and Memnonia Fossae, the Nirgal Vallis watershed, and the Bakhuysen, Graff and Millochau impact craters/basins. While only a small amount

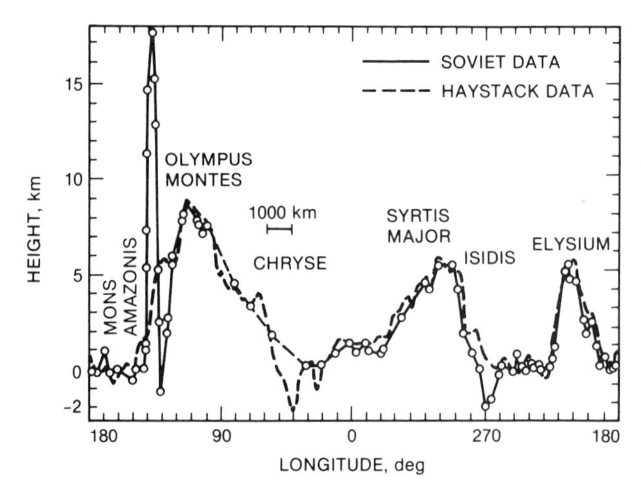

Fig. 7. Soviet 1980 radar profile at 20°.73 latitude compared with 1967 Haystack measurements at 21°.5.

of data processing has been done from this large data set, Fig. 8 shows two scans over and near Tyrrhena Patera taken on 1988 Aug. 30 and Oct. 2 placed on the same longitude scale as a geological sketch map adapted from Greeley and Crown (1990). These radar topography measurements confirm interpretations of Viking imagery indicating that Tyrrhena Patera is a low-relief and broad shield volcano constructed upon relatively flat plains material.

The 1988 Goldstone observations repeat many scans (with somewhat different longitude coverage) obtained during 1973. The so-called closure points thus generated for ephemeris improvement are valuable because uncertainty in the common topography can be removed, producing highly accurate differential ranges with a long time span between them.

B. Martian Topography Deduced From Spacecraft Data

Radio Occultation Measurements. A wealth of data on the atmosphere and topography of Mars have been obtained from radio propagation experiments utilizing the tracking links with Mariner and Viking spacecraft (Kliore et al. 1965; Fjeldbo and Eshleman 1968; Eshleman 1970; Fjeldbo et al. 1972; Kliore et al. 1973; Lindal et al. 1979). Measurements of the topography were conducted by recording the amplitude and phase of the carrier frequencies received from the spacecraft during occultations by the planet. The analysis of the measurements was carried out by using the limb diffraction effects observed during ingress and egress together with the spacecraft ephemerides to determine the radius at the locations where the tracking links grazed the Martian surface.

Figure 9 summarizes the topographic data acquired with the Viking Orbiters. It shows the radius and elevation of the Martian surface as a function

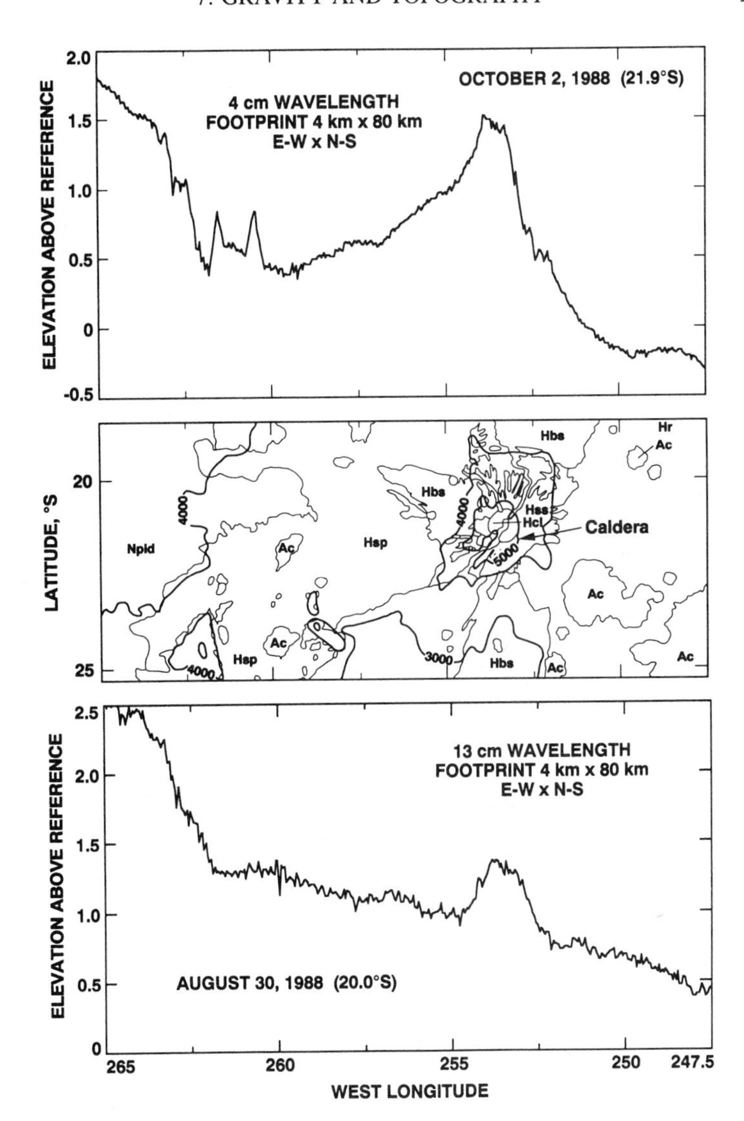

Fig. 8. 1988 Goldstone radar altimetry in the vicinity of Tyrrhena Patera. Errors in absolute height relative to radar datum are around 200 m. Two of the four scans that cover the basal(Hbs) or summit shield(Hss) are shown. Geologic sketch map with U.S.G.S. topographic contours superposed is from Greeley and Crown (1990). Only the 1988 Oct 2 scan passed directly through the caldera.

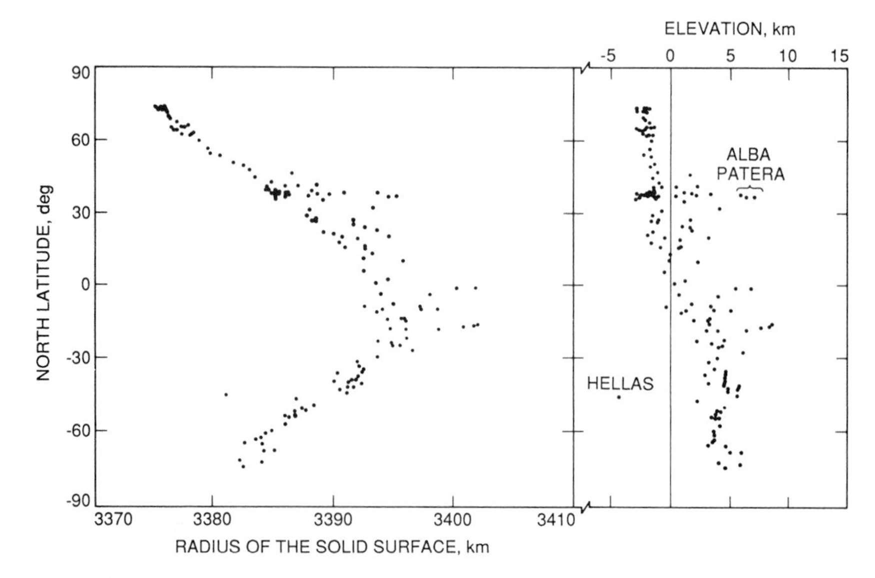

Fig. 9. Radius and elevation of the Martian surface as a function of areocentric latitude (Lindal et al. 1979). The data were obtained from occultation measurements with the Viking Orbiters at longitudes ranging from 0 through 360°.

of areocentric latitude. The radius was measured from the planetary center of mass, and the altitude was measured relative to a reference geodetic surface or geoid developed by Christensen (1975).

The main error source in the determination of the radius at each occultation point was the uncertainty in the position of the Viking orbiters in the plane of the sky orthogonal to the limb of Mars. In order to keep the errors in the radius measurements within 1 km, the pressure computed from the occultation data at the 5-km altitude level was monitored from revolution to revolution. As the pressure at an equipotential surface is expected to remain relatively constant from one orbital revolution to the next, large abrupt pressure deviations at the 5-km altitude level indicate significant ephemeris errors. This pressure correlation technique was therefore found to be useful in identifying and removing inaccurate radius values from the data base.

Topographic data obtained from occultation measurements with the Mariner 9 and Viking Orbiters show that the northern hemisphere of Mars consists of relatively smooth lowlands with elevations ranging from -1 to -3 km, interspersed with highlands. The southern hemisphere is comprised of rugged highlands with an average elevation of about 4 km. Numerical tabulations of the data giving the radius, elevation, areographic latitude and longitude, and link azimuth angle at each occultation point are available in previous publications (Kliore et al. 1972,1973; Lindal et al. 1979).

The occultation data may be used as elevation control points across the Martian surface. An example of this application is the U.S. Geological Survey's topographic map M 25M 3 RMC which was produced by using Mariner 9 data as control points (Wu 1978); see also chapter 10.

Additional Topographic Measurements. Topographic information is contained in the measurements made by the ultraviolet spectrometer (Hord et al. 1972) and the infrared interferometer spectrometer (Conrath et al. 1973) throughout the Mariner 9 mission. Changes in Mars topography are related to local pressure and temperature variations. However, the large instrument resolution (100 km) limits the precision of the location of the measurement (Jordan and Lorell 1975). Christensen (1975) gives a detailed account of analyzing various data (radio occultation, Earth-based radar, spacecraft spectral and optical) to derive a global topographic map of Mars. Topographic data sources used for Mars topographic mapping are covered in chapter 10.

Future Altimeter Measurements. Plans for the Mars Observer mission include a 1.06-μm laser altimeter. The system is expected to have an instrumental precision of $\sim$ 1 to 2 m over smooth flat Martian terrain and an overall accuracy of 20 to 30 m with respect to the center of mass of the planet, including the radial orbit error. The altimeter will illuminate a 100-m spot on the surface of the planet at a pulse rate of 10 s^{-1} making the spot locations approximately 330 m apart under the track of the spacecraft. It is anticipated that the altimeter will track continuously over most surfaces with slopes (or roughness) of 25° or less over the 100-m footprint, thus providing a capability to profile canyons, channels, craters, etc., even though the altimeter may not be able to follow the steepest of slopes. The altimeter is planned to be operated continuously throughout the life of the mission although it is not expected to be able to obtain return signals during heavy dust storms. If the instrument operates successfully for the planned one Martian year, it will map the planet's surface with an average (profile) separation at the equator of slightly $<$1.5 km.

With this topographic information, a significant improvement can be expected in the vertical control of surface features. Global control at the 2 or 3 tens of m on a grid size of a few km is a reasonable expectation for all of Mars except for the areas within approximately 150 km of the poles (Mars Observer will have an orbital inclination of 92.°8). When combined with the expected gravity information, more detailed studies of major geophysical problems, such as the hemispheric dichotomy, the formation of the Tharsis bulge, and thickness of the crust and lithosphere, will be possible.

III. MARTIAN GEOPHYSICAL GEODESY

A. Introduction

The objectives of planetary geodesy are to determine the gravity, topography and rotation of the planets. These measurements are put to two basic uses. In a somewhat pragmatic vein, the geodetic parameters are useful for navigational purposes: the motion of a spacecraft in orbit around a planet, relative to features of interest on the planetary surface, depends on the gravity, topography and rotation of the planet. On the other hand, these same geodetic parameters provide the only remotely accessible information on the internal distribution of mass and stress in the planet.

The techniques used and results obtained for these basic geodetic parameters were discussed earlier in this chapter. The primary purpose of this section is to discuss the basic geophysical inferences that can be drawn from the gravity, topography and rotation of Mars. In discussing long-wavelength phenomena, it is frequently convenient to describe the topography and gravitational potential in terms of their spherical harmonic coefficients.

The gravitational potential U and the topographic heights T with respect to some reference surface are expanded in normalized spherical harmonics as:

$$U(r,\phi,\lambda) = \frac{GM}{r} \sum_{l} \left(\frac{R}{r}\right)^{l} \sum_{m=o}^{l} (\bar{C}_{lm} \cos m\lambda + \bar{S}_{lm} \sin m\lambda)\bar{P}_{lm}(\sin\phi) \quad (2)$$

$$T(\phi,\lambda) = R\,[1 + \sum_{l>0} \sum_{m=0}^{l} (\bar{A}_{lm} \cos m\lambda + \bar{B}_{lm} \sin m\lambda)\bar{P}_{lm}(\sin\phi)] \quad (3)$$

where r,ϕ,λ = radial distance from center of planet, latitude and east longitude; $\bar{P}_{lm}(\sin\phi)$ = Legendre polynomials ($m=0$) and associated functions of first kind ($m>0$), fully normalized; $\bar{C}_{lm},\bar{S}_{lm}$ = normalized harmonic coefficients of U, of degree l and order m; $\bar{A}_{lm},\bar{B}_{lm}$ = normalized harmonic coefficients of T, of degree l and order m; and R,GM = reference mean radius of Mars and planetary gravitational constant. In addition, we define the following spectral components, for the potential:

$$S_l(U,U) = \sum_{m=o}^{l} \left(\bar{C}^2{}_{lm} + \bar{S}^2{}_{lm}\right) \quad (4)$$

and similarly for the topography:

$$S_l(T,T) = \sum_{m=o}^{l} \left(\bar{A}^2{}_{lm} + \bar{B}^2{}_{lm}\right). \quad (5)$$

The cross-spectrum components are also defined as

$$S_l(U,T) = \sum_{m=0}^{l} (\bar{C}_{lm}\,\bar{A}_{lm} + \bar{S}_{lm}\,\bar{B}_{lm}). \quad (6)$$

We, of course, have to limit the series to the common maximum degree l_{max} to which the coefficients are available in order to investigate correlative properties.

In the following, the gravity field model of Balmino et al. (1982) and the topographic model of Bills and Ferrari (1978) have been used, (resulting in a maximum degree $l_{max} = 16$). Figure 10 shows the nonhydrostatic part of the Mars geoid (that portion which is not due to rotation alone) evaluated from Eq. (2). The hydrostatic contributions to the measured $\bar{C}_{20}$ and $\bar{C}_{40}$ coefficients used here were estimated using a derivation by Sleep and Phillips (1985, Appendix B) assuming a dimensionless hydrostatic moment of 0.366 (see discussion below). Figure 11 shows Martian topography (less the rotational contribution) with respect to the geoid surface, calculated from Eq. (3). The geoid is the appropriate reference surface to use for geologic and geophysical applications, as departures from an equipotential surface provide the forces involved in dynamic processes such as water or lava flow and lithospheric flexure. Although the geoid is somewhat smoother than the topography (as is expected from potential theory), the correlation between gravity

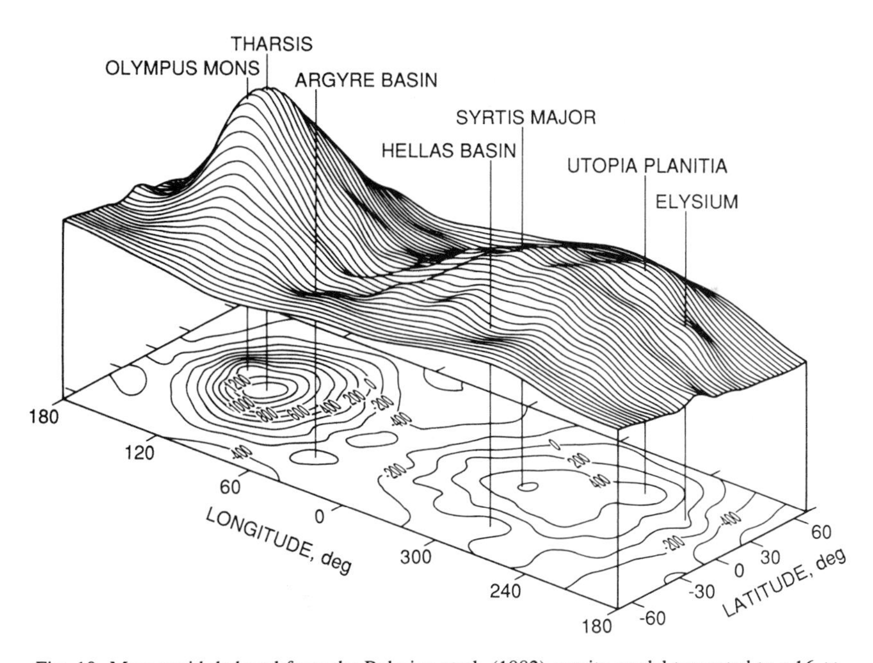

Fig. 10. Mars geoid deduced from the Balmino et al. (1982) gravity model truncated to a 16 × 16 field. A hydrostatic contribution, corresponding to $I/MR^2 = 0.366$, has been removed. The nonhydrostatic coefficients used here are the same as those given by Balmino et al. (1982) except $\bar{C}_{20} = -6.38 \times 10^{-5}$ and $\bar{C}_{40} = 0.36 \times 10^{-5}$. This Mercator projection covers 180°W to 180°E longitude, 65°S to 65°N latitude with contour intervals of 200 m.

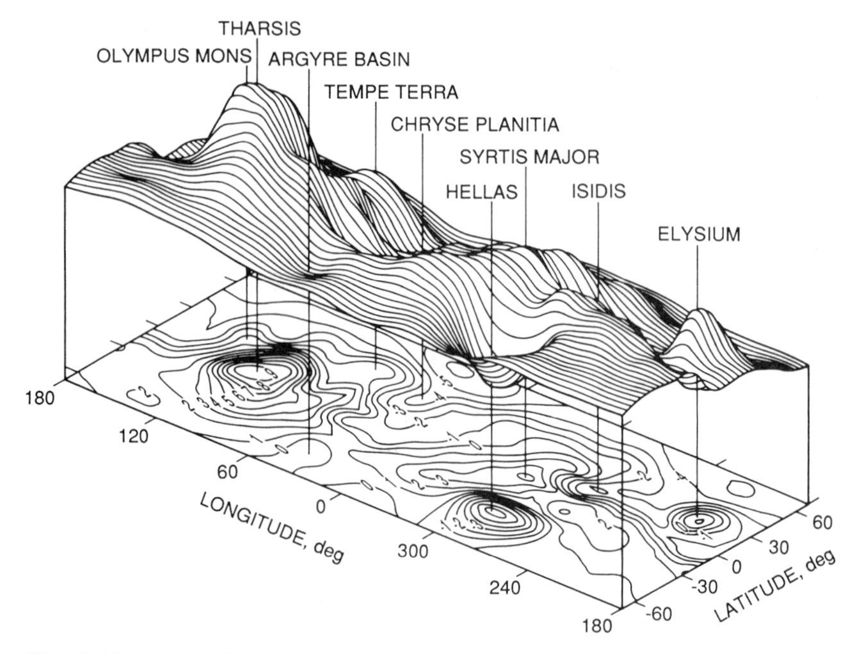

Fig. 11. Mars topography, as represented by the 16 × 16 field of Bills and Ferrari (1978) with rotational effects removed, relative to the geoid of Balmino et al. (1982). The centrifugal correction is performed by adding the term $R\Omega^2/(3\sqrt{5}\,g)$ to $\bar{A}_{20}$, where Ω is the rotation rate and g is the gravitational acceleration, resulting in $\bar{A}_{20} = -1142.8 \times 10^{-6}$. The contour interval is 1 km. The Mercator projection covers the same region as in Fig. 10.

and topography at long wavelengths can be clearly seen. This correlation is especially strong for the Tharsis Rise, that has by far the largest long-wavelength gravity anomaly. More subdued gravity signatures can be identified for the Hellas and Argyre basins, the Syrtis Major and Chryse planitiae and the Elysium Rise. More enigmatic features include Utopia Planitia, that has a pronounced circular gravity high associated with a broad, irregular topographic depression, and Tempe Terra and Isidis basin, that are major topographic features with indiscernible (at this resolution) gravity signals.

B. Moments and Products of Inertia

The moments and products of inertia have much in common with the gravitational harmonic coefficients (or Stokes coefficients) of degree two. Both are second degree radial moments of the internal density distribution $\rho(r,\phi,\lambda)$. The degree n moments are functionals of the form

$$F_{nm} = \int \rho(\xi,\mu,\lambda) f_{nm}(\mu,\lambda) \xi^n \, dV. \tag{7}$$

In this formula, (r,ϕ,λ) have been previously defined and $\xi = r/R$, $\mu = \sin(\phi)$ and f_{nm} is the weighting kernel.

Despite this intrinsic similarity, the inertial moments and Stokes coefficients differ in two important ways: the functional form of their kernels and the observations required to estimate their values. The gravitational harmonic coefficients are most easily determined from the perturbed motions of natural and artificial satellites. Thus, reasonably detailed models have been constructed for the gravitational fields of the Earth (Lerch et al. 1985), Moon (Bills and Ferrari 1980), Venus (Bills et al. 1987) and Mars (Balmino et al. 1982). Direct estimation of the inertial moments is more difficult, as they are determined from the rotational response of a planet to applied torques. Because of this difficulty, useful observations are available only for the Earth (Feissel 1980) and Moon (Ferrari et al. 1980).

Because of this limitation, the 5 Stokes coefficients of degree two are especially important to studies of Mars, because they determine the orientation of the principal inertial axes and the relative sizes of the principal moments of inertia. As the inertia tensor is symmetric, there are only 6 independent elements; the 3 diagonal terms (known as moments of inertia) and either the upper or lower triangular set of off-diagonal terms (known as products of inertia).

It is always possible to choose a set of coordinate axes in which the inertia tensor reduces to diagonal form. In this principal axis system, instead of the usual 5 Stokes coefficients of degree two, we have 3 Euler angles, which define the coordinate transformation, and 2 nonzero Stokes coefficients (Lorell 1972). If the principal moments of inertia are denoted $A \leq B \leq C$, the relationship with the Stokes coefficients takes the form (Soler 1984)

$$\begin{aligned} J_2 MR^2 &= C - (A+B)/2 \\ J_{22} MR^2 &= (B-A)/4 \end{aligned} \tag{8}$$

where $J_2 = -C_{20}$ and $J_{22} = (C^2_{22} + S^2_{22})^{1/2}$ (unnormalized).

The trace of the inertia tensor is invariant under coordinate rotations, and under a broad class of material deformations of the body itself (Rochester and Smylie 1974). The mean moment

$$I = \frac{1}{3} \sum_k I_{kk} \tag{9}$$

thus provides an important constraint on the radial density profile. Unfortunately, the gravity harmonics place no direct constraints on the value of the mean moment. An arbitrary internal density variation can be decomposed uniquely into two parts, a purely radial variation, and lateral deviations from the radial mean. The gravitational field yields little direct information about the spherically symmetric, radial density profile. In fact, the gravitational signature of such a density distribution is uniquely determined by its total mass (zero-order moment) and centroid location (first-order moment).

The observations required to determine unambiguously the mean moment of Mars are not presently available. However, because of the importance of this parameter, a number of schemes have been proposed to provide plausible estimates of its value. All of them rely on assumptions about the hydrostatic state of Mars. Below we review the pertinent facts.

For a body in hydrostatic equilibrium, isosurfaces of density and pressure coincide. If such a body is at rest, it will attain spherical symmetry. Any orthogonal set of axes will be principal axes, and all 3 moments of inertia will be equal. If, instead, the body is rotating at rate Ω, the symmetry changes from spherical to cylindrical. The rotation axis will be the unique principal axis of greatest inertia and any 2 orthogonal axes in the equatorial plane will be principal axes. In that case we have (Love 1911; Munk and MacDonald 1960; Lambeck 1980)

$$\begin{pmatrix} A \\ B \\ C \end{pmatrix} = I + \begin{pmatrix} \Delta A \\ \Delta B \\ \Delta C \end{pmatrix} \tag{10}$$

where axial symmetry requires that $\Delta A = \Delta B$, and invariance of the trace of the inertial tensor demands

$$\Delta A + \Delta B + \Delta C = 0. \tag{11}$$

The explicit form of the hydrostatic moment increment is

$$\Delta C = \frac{2}{9} k_s \, qMR^2. \tag{12}$$

Here k_s is a constant of proportionality known as a secular Love number (degree two Love number with zero rigidity) and

$$q = \Omega^2 R^3 / GM \tag{13}$$

is the ratio of centrifugal to gravitational acceleration on the (undeformed) equator. Using Eqs. (8) through (12) above, we see that the simple observational constraint is that

$$k_s = 3J_2/q. \tag{14}$$

For self-gravitating hydrostatic bodies that are not rotating too fast, i.e., $q << 1$ (Chandrasekhar 1969), there is a unique relationship between the secular Love number and the dimensionless moment of inertia about the rotation axis ($\Lambda = C/MR^2$). To lowest order the relationship is simply (Darwin 1899; de Sitter 1924)

$$3\Lambda/2 = 1 - 2Q/5 \tag{15}$$

where

$$Q^2 = \frac{4 - k_s}{1 + k_s}. \tag{16}$$

For a body of uniform density, the appropriate values are:

$$\begin{aligned} k_s &= 3/2 \\ \Lambda &= 2/5. \end{aligned} \tag{17}$$

Smaller values of these parameters would imply that the density increases with depth.

If we could safely assume that Mars is in hydrostatic equilibrium, it would be a simple exercise to use the observed values of J_2 and q to compute Λ. Using observed values of the appropriate parameters, we thus would find

$$\begin{aligned} k_s &= 1.2869 \pm 0.0001 \\ \Lambda &= 0.37621 \pm 0.00001. \end{aligned} \tag{18}$$

This is essentially the logic used and the result obtained by Lorell et al. (1973), Binder and Davis (1973), and Cook (1977). In the latter two works, the possibility of significant nonhydrostatic contributions to J_2 were considered, but discounted. This result was incorporated into a number of models of Martian internal structure and composition (Anderson 1972; Johnston et al. 1974; Cook 1977).

A major improvement was made by Reasenberg (1977) and Kaula (1979*a*) who recognized that it is possible to estimate the amount by which Mars is nonhydrostatic. Their basic assertion consists of two parts. First is the recognition that, for nonhydrostatic bodies, Eq. (10) should be amended to read

$$\begin{pmatrix} A \\ B \\ C \end{pmatrix} = I + \begin{pmatrix} \Delta A \\ \Delta B \\ \Delta C \end{pmatrix} + \begin{pmatrix} \delta A \\ \delta B \\ \delta C \end{pmatrix} \tag{19}$$

where δA, δB and δC are the nonhydrostatic contributions to the principal moments of inertia. In addition to the previously noted constraints on the hydrostatic contributions, we also require

$$\delta A + \delta B + \delta C = 0. \tag{20}$$

Whereas the hydrostatic configurations are oblate spheroids, the inertial ellipsoid of a nonhydrostatic body is in general triaxial. A convenient measure of triaxiality is provided by the parameter

$$f = (B - A)/(C - A). \tag{21}$$

The possible range of values is obviously $0 \leq f \leq 1$. For a prolate spheroid, $A = B < C$, so that $f = 0$. The opposite end member is an oblate spheroid, for which $A < B = C$ and $f = 1$. A nonhydrostatic ellipsoid could similarly be characterized by

$$\delta f = (\delta B - \delta A)/(\delta C - \delta A). \tag{22}$$

The hydrostatic value of J_2, corresponding to a given value of δf, can be estimated as

$$J_2^h = J_2 - 2\left(\frac{2 - \delta f}{\delta f}\right) J_{22}. \tag{23}$$

The second assertion, made implicitly by Reasenberg (1977) and explicitly by Kaula (1979*a*), is that the nonhydrostatic components are symmetric about an equatorial axis through Tharsis, so that

$$\delta B = \delta C. \tag{24}$$

This gives $\delta f = 1$, and ultimately yields a dimensionless moment value of $\Lambda = 0.3663$. As Kaula (1979*a*) was careful to point out, the errors associated with such an estimate have little to do with the accuracy of the measured quantities, but rather "depend entirely on the extent to which mass irregularities depart from axial symmetry about Tharsis."

Bills (1989*a*) has recently argued that the nonhydrostatic mass distribution of Mars is unlikely to be a prolate spheroid with an equatorial symmetry axis, and that a fully triaxial ellipsoid is a much more probable configuration. Goldreich and Toomre (1969) examined the probability density distribution of the moment difference ratio for bodies composed of random density inhomogeneities superimposed on a vastly more massive, homogeneous sphere. The most likely configuration is fully triaxial, with the intermediate principal moment B located roughly midway between A and C. It is only the nearly oblate ($f \sim 1$) or the almost prolate ($f \sim 0$) configurations that are comparatively rare. If the nonhydrostatic figure of Mars conforms to this pattern, we should expect a dimensionless moment of inertia close to $\Lambda = 0.3452$.

A more empirical basis for choosing an appropriate value of δf for Mars would be to examine the values for the Moon and Venus, where the moment differences are well known and clearly nonhydrostatic. For the Moon, the full inertia tensor is well known (Ferrari et al. 1980), and the rotation is slow enough ($q = 7.57 \times 10^{-6}$) that it makes no appreciable contribution. The resulting value for the moment difference ratio is $\delta f = 0.3610 \pm 0.0003$. If this value were applied to Mars, the corresponding dimensionless moment would be $\Lambda = 0.3276$. For Venus, only the Stokes coefficients are available (Bills et al. 1987), but the rotation is so slow $q = 6.11 \times 10^{-8}$ that the observed values are clearly entirely nonhydrostatic. The moment difference ratio is $\delta f = 0.490 \pm 0.034$. Again applying this value to Mars, we find $\Lambda = 0.3443$.

A final point of comparison is provided by the Earth, where the full inertia tensor is known and rotation is rapid ($q = 3.45 \times 10^{-3}$), but the internal structure is sufficiently well known that the hydrostatic component can be accurately removed (Nakiboglu 1982). The resulting nonhydrostatic moment difference ratio is $\delta f = 0.579$. Application of this value to Mars would suggest $\Lambda = 0.3512$.

Confirmation of a value in the vicinity of $\Lambda = 0.345$ would have quite profound implications for the internal structure and composition of Mars (Johnston and Toksoz 1977; McGetchin and Smyth 1978; Okal and Anderson 1978; Morgan and Anders 1979; Goettel 1981; see also chapter 6), as well as the obliquity history (Ward 1979; Ward et al. 1979; Borderies 1980). It should be pointed out, however, that the magnitude of the contribution of Tharsis plateau to the second degree gravity and topography components of Mars has no counterpart on any of the other planets thus far studied (Kaula et al. 1989). Therefore, these essentially statistical arguments may not be valid for this case (see, however, Bills 1989*b*).

The actual value of the moment of inertia of Mars will remain unknown until the axial precession rate is determined. That rate is simply $\alpha \cos(\theta)$, where θ is the obliquity and

$$\alpha = \frac{3}{2}\frac{n^2}{\Omega}\frac{J_2}{\Lambda} \tag{25}$$

where n is the orbital mean motion. Thus, the rate is roughly 8 arcsec yr^{-1}. Range measurements to at least three fixed points on the surface (to allow separation of orbital vs rotational effects) might be contemplated in the latter part of the decade. Ranging to the 2 Viking Landers (Mayo et al. 1977; Reasenberg and King 1979; Hellings et al. 1983; Williams 1984) have already provided significant and interesting results, but geodetic results comparable to those obtained via lunar laser ranging (Ferrari et al. 1980) will require a minimum of 3 reference points and a reasonably long observation campaign.

It has frequently been suggested (Murray and Malin 1973*b;* Mutch and Saunders 1976; Ward et al. 1979; Melosh 1980*b;* Schultz and Lutz-Garihan 1982) that the present equatorial location of Tharsis may not be merely coincidental. The basic idea is that a sufficiently large, nonhydrostatic, positive mass anomaly will cause Mars to re-orient itself relative to the spin axis, so that the anomaly will lie on the new equator. The argument is basically correct, and if the nonhydrostatic mass distribution were axially symmetric, no more could be concluded. Tharsis must indeed lie on the equator, but that constraint only determines the plane in which the rotation axis must lie. It appears to be less widely appreciated that a stronger statement can be made. The rotation axis of any "quasi-rigid body" (Goldreich and Toomre 1969) will coincide with the axis of least inertia of its nonhydrostatic mass distribution (see, however, the discussion of polar wander in chapter 8).

C. Higher-Degree Harmonics

If Mars were in hydrostatic equilibrium, the only nonzero harmonic coefficients of gravity and topography would be the even degree zonals (C_{20}, C_{40}, C_{60}, . . .) forming a rapidly decreasing series. Instead, the higher-degree harmonics are quite complex and decrease in amplitude rather more slowly. Because of the complexity, much of the following analysis will be essentially statistical in nature (see also Phillips and Lambeck 1980). The spatial variations in gravity and topography can be conveniently parameterized in terms of their separate degree variances given by Eqs. (4) and (5). These give a measure of the spectral variance in each wavelength. The cross variance is defined by Eq. (6) which gives a measure of the correlation between the gravity and the topography. An alternative representation of the variance spectrum that has been widely used in gravity analyses (Kaula 1968) is in terms of the degree variances $\sigma_l(U)$, where

$$\sigma_l(U) = \left[\frac{S_l(U,U)}{2l+1}\right]^{1/2} \tag{26}$$

and similarly for $\sigma_l(T)$ and $\sigma_l(U,T)$.

The correlation coefficients from Eq. (6) (with the zonal coefficients C_{20} and C_{40} related to rotational effects removed) are plotted in Fig. 12. It shows a high degree of correlation between gravity and topography at virtually all wavelengths, with a probability of $< 5\%$ that this correlation is a matter of chance (Lorell et al. 1972; Phillips and Lambeck 1980). This implies some sort of direct relationship between the density structure generating gravity anomalies and the processes responsible for topography. The correlation observed on Mars is in contrast to the situation on the Earth, for which there is

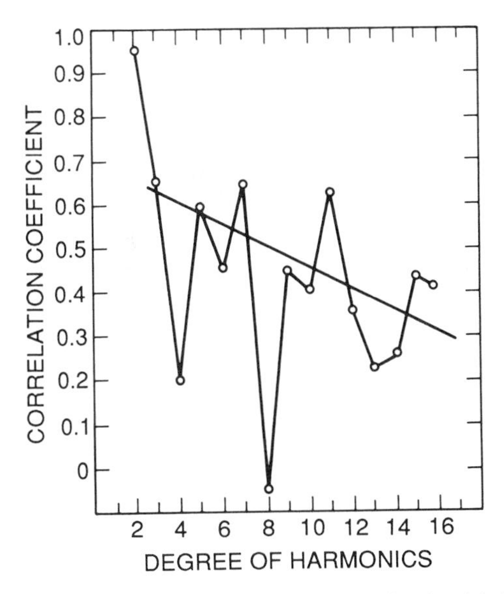

Fig. 12. Correlation coefficients between gravity and topography; the global correlation is 0.58 (figure from Balmino et al. 1982).

no significant low-degree correlation between gravity and topography and the gravity field is related to processes not directly associated with topography.

Observed gravitational anomalies may be ascribed to contributions from both surface topography and subsurface lateral density variations (including variations in the depth to a density interface such as the crust-mantle boundary). The direct gravitational effect of the topography is simply (MacRobert 1967)

$$\Delta G_{lm} = \frac{3\rho_c}{(2l + 1)\bar{\rho}} H_{lm} \tag{27}$$

where ρ_c is the density of the topography and $\bar{\rho}$ is the mean planet density. The symbol G_{lm} is shorthand notation for the set of gravitational coefficients C_{lm} and S_{lm} and likewise H_{lm} for the topographic coefficients A_{lm} and B_{lm}. If complete isostatic compensation occurs at a single interface at a mean depth D, the gravitational effect is reduced to

$$\Delta G_{lm} = \frac{3\rho_c}{(2l + 1)\bar{\rho}} H_{lm}(1 - \zeta^{l+1}) \tag{28}$$

where $\zeta = (R - D)/R$. It can be easily seen from Eq. (28) that the gravitational effect of fully compensated topography vanishes as the depth of the compensating zone approaches zero, and becomes identical to that of uncompensated topography as the compensation depth becomes large.

The Bouguer anomaly ΔG^B can be computed for long wavelengths by subtracting ΔG of Eq. (27) from the observed gravity coefficients. The correlation between the Bouguer gravity and topography, $\sigma_l(\Delta G^B, H)$ is shown in Fig. 13. The coefficients display a highly negative correlation that is generally indicative of a large amount of compensation. The variation of crustal thickness implied by the Bouguer anomaly can be estimated by assuming that all lateral density variations occur as undulations on the crust-mantle interface (Khan 1977; Bills and Ferrari 1978). Figure 14 shows such a theoretical crustal thickness map. It is characterized by a maximum thickness under Tharsis and a minimum thickness beneath the Hellas Basin.

Figure 15 provides a comparison in the spectral domain between the gravity and topography for the Moon, Mars, Venus and Earth. The lower curve in each panel represents the observed gravity, whereas the upper curve represents the gravity which would be produced by the topography if it were uncompensated, which is simply the variance spectrum of ΔG_{lm} in Eq. (27). There are a number of features to note. First, the observed gravity is consistently smaller than predicted from uncompensated topography. Thus, topography on all of the terrestrial planets appears to be at least partially compensated. Also, the discrepancy is greater on the large, high-density planets (Earth and Venus) than it is on the small, low-density bodies (Moon and Mars). This suggests that either the depths of compensation are greater for the smaller planets, or that some other mechanism of support (such as elastic support by the lithosphere) is more important for these bodies.

We also note that the spectral slopes for all the curves are similar (Bills 1978; Turcotte 1987). The variance spectra have the approximate forms

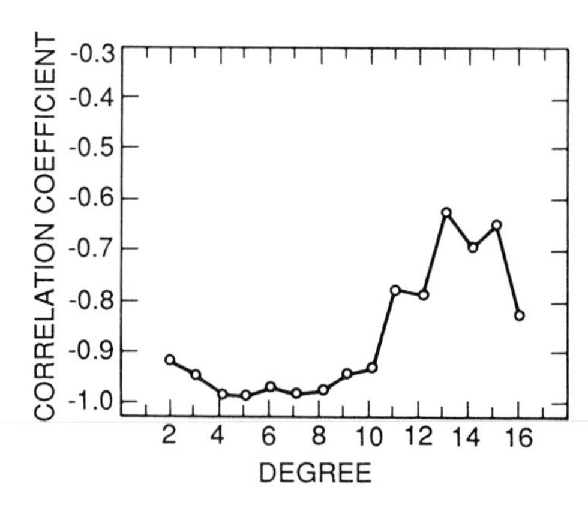

Fig. 13. Correlation coefficients between Bouguer gravity and topography (figure from Balmino et al. 1982).

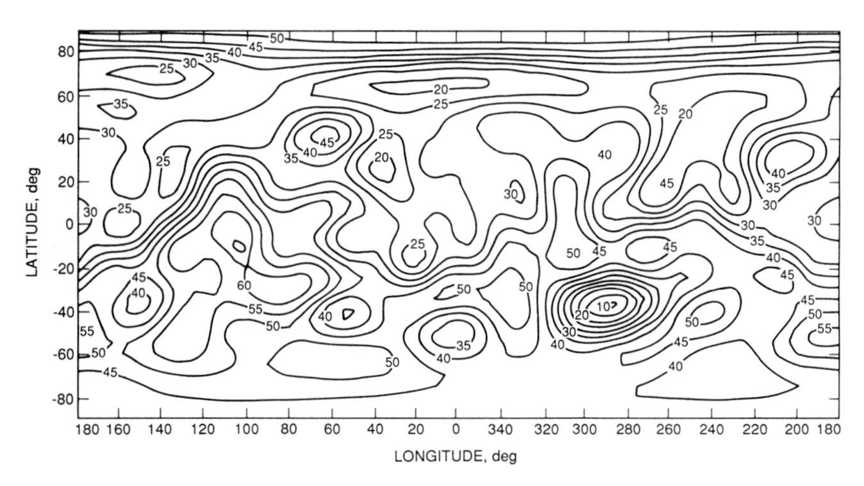

Fig. 14. Mars crustal thickness (km). This assumes gravity variations are entirely due to the surface topography of density 3.4 g cm^{-3} and an interface at a mean depth of 40 km with a density contrast of 0.5 g cm^{-3} (figure from Balmino et al. 1982).

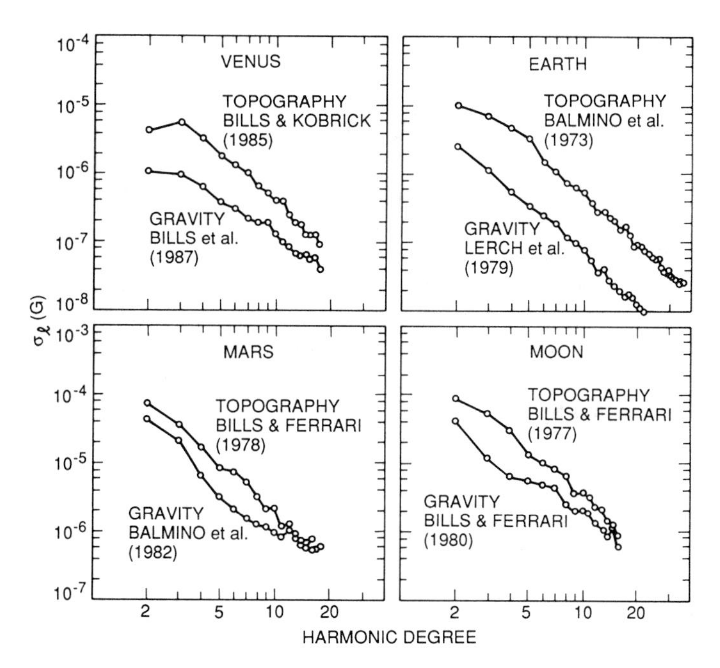

Fig. 15. Variance spectra of gravity and equivalent gravity from topography for terrestrial planets and the Moon.

$$S_l(U,U) \approx \frac{(2l+1)}{l^4} u^2 \tag{29}$$

$$S_l(T,T) \approx \frac{t^2}{l(l+1)} \tag{30}$$

where u and t are constants. These relations have long been recognized for the Earth, the first by Kaula (1968), the second by Vening-Meinesz (1951). These approximations can be used to estimate the stresses in the lithosphere of Mars relative to those of the Earth. Lambeck (1979*a*) showed that the shear stress variance τ for compensated topography is given by

$$S_l(\tau,\tau) \approx \frac{1}{4} (g\rho_c R)^2 S_l(T,T) \tag{31}$$

where g is the surface gravity. Using Eq. (31) to scale stresses between the planets and substituting into it Eq. (30), we can derive an expression for α, the ratio of the shear stress in the crust of Mars to that in the Earth:

$$\alpha = \frac{g_m R_m t_m \rho_m}{g_e R_e t_e \rho_e} \tag{32}$$

where the subscripts m and e refer to Mars and the Earth, respectively. Using the parameters in Table I and assuming the crustal densities are the same, we find that $\alpha = 0.35$. Thus, if the topography of Mars is compensated and all factors controlling isostasy are the same in the two planets, then the stresses in the Martian lithosphere are generally somewhat less than those in the Earth, and the topography of Mars can be considered to be isostatically supported (Lambeck 1979*a*).

The above result does not take into account the magnitude of the gravity spectrum, however. If the observed gravity is due entirely to density anomalies residing within the lithosphere, the stress variance is given by (Lambeck 1979*a*)

$$S_l(\tau,\tau) \approx \frac{1}{4} \left[\frac{g\bar{\rho}R(2\ l\ +\ 1)}{3(1\ -\ \zeta^{l+1})} \right]^2 S_l(U,U) \tag{33}$$

and α is given by

$$\alpha \approx \frac{g_m \rho_m D_e u_m}{g_e \rho_e D_m u_e} = \left(\frac{g_m}{g_e}\right)^2 \frac{R_e}{R_m} \frac{D_e}{D_m} \frac{u_m}{u_e} \tag{34}$$

where the compensation depth $D_i << R_i$ ($i = e,m$) has been assumed. If the compensation depths are the same, this gives $\alpha = 3.5$. For the Earth, density

TABLE I
Planetary Constants Used in Calculations

	Mars	Earth
Planet radius R (km)	3390	6371
Surface gravity g (m s^{-2})	3.72	9.81
Gravity spectral constant u	1.3×10^{-4}	1.0×10^{-5}
Topography spectral constant t	9.8×10^{-4}	5.6×10^{-4}

anomalies are generally assumed to reside below the lithosphere, as Eq. (33) gives excessively large stresses (Lambeck 1972). Thus, much greater compensation depths are required on Mars if the gravity is to be attributed to density anomalies within the lithosphere. Closer inspection of the details of the gravity spectra shows that the second and third degree terms, which are dominated by the contributions of the Tharsis Plateau, bias this result somewhat, and the higher-degree terms can be explained satisfactorily using depths of compensation of 100 to 200 km (Lambeck 1979*a*).

D. Implications for Internal Structure and Processes

Gravity and topography are virtually the only observations available which relate directly to the structure and state of Mars' interior, given the present lack of surface-based seismic and heat flow measurements. In the most general terms, lateral variations in the gravity field can be interpreted in terms of deviations from a state of mechanical equilibrium in a spherically stratified planet. These deviations imply interior processes acting in such a way as to maintain the lateral density variations that give rise to the gravity anomalies. Topography provides a direct measure of the surface contribution to the gravity field as well as a measure of the radial forces that must be present in the lithosphere. Thus, the gravity and topography are direct manifestations of processes active today in the outer portions of the planet, and comparison of these observations with geophysical models can help to elucidate these processes.

As discussed in the preceding section, one basic constraint on geophysical models is the spectral ratio of gravity to topography. A more systematic comparison between gravity and topography in the spectral domain can be made by assuming a linear relationship of the form

$$G_{lm} = F_l H_{lm} + E_{lm} \tag{35}$$

where F_l is a spectral admittance whose value will depend on the processes producing the gravity and topography (see, e.g., Dorman and Lewis 1970; Banks et al. 1977; Ricard et al. 1984; Richards and Hager 1984), and the remainder term E_{lm} is the residual anomaly. In mathematical terms, F_l is a measure of the correlated part of the spectra, and E_{lm} the uncorrelated part. Available estimates of the gravity and topography harmonics allow us to obtain least squares estimates of both the admittance

$$F_l = \frac{S_l(U,T)}{S_l(T,T)} \tag{36}$$

which places constraints on the global average compensation mechanism, and the residual gravity anomaly E_{lm}, which elucidates regional departures from the global average. The fact that E_{lm} is generally small compared to the first term in Eq. (35) suggests that relatively simple relationships exist between topography and the density variations that generate gravity anomalies.

For several reasons, the state of isostasy is generally used as the primary reference against which the admittance is compared. First, buoyancy is an extremely simple and effective mechanism for supporting long-wavelength loads on the surface. Second, topography will naturally tend toward a state of isostasy through stress relaxation from creep or fracture, as it is a state of minimum strain energy. Finally, isostasy has been shown to be an important mechanism on the Earth for the support of large-scale topography.

However, it is difficult to explain the relative amplitude of Martian gravity anomalies, as compared to the topography, in terms of a single, simple global compensation mechanism. Figure 16 compares the empirical admittance estimates with model predictions for local Airy (crustal thickness) compensation. The single-layer isostatic model of Fig. 16 satisfies the observed admittance for harmonic degree greater than or equal to four for a compensation depth between 100 and 200 km. But the second and third harmonics (which are associated almost wholly with the Tharsis bulge) require much deeper compensation. This presents two fundamental problems: (1) the different depths of compensation required for the low and high degrees violate a basic assumption of simple buoyant equilibrium for the topography; and (2) the implied depths of compensation of the low-degree harmonics are great enough that under any plausible thermal/rheological scenario, the material should not have sufficient strength to support the stresses required to maintain the density anomalies over any appreciable length of time. This complexity can also be observed in the spatial domain. Phillips and Saunders (1975) plotted point measurements of gravity and topography on a scatter diagram and found that Tharsis and its flanking lows had high gravity-to-topography ratios, whereas the rest of the planet had relatively low ratios. These observations led Phillips and Saunders (1975) and Phillips and Lambeck (1980) to suggest that most of the topography of Mars is isostatically compensated at a depth $\sim$ 100 km, whereas Tharsis (represented by the lowest harmonics) must be supported by some other mechanism.

A compensation depth of 100 to 200 km for the non-Tharsis regions of Mars is in agreement with compensation depths determined independently (at

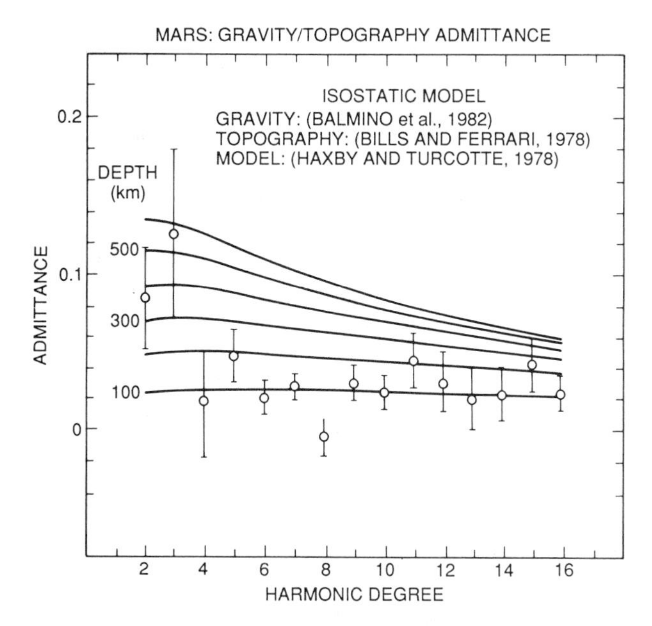

Fig. 16. Comparison of empirical admittance for Mars with the theoretical admittance due to the Airy isostatic model with depths of compensation from 100 to 600 km.

shorter wavelengths) by line-of-sight (LOS) modeling (described above), and could be indicative of a global mean crustal thickness of ~ 100 km. The mean crustal thickness has significant implications for the differentiation history of Mars and the strength of the lithosphere (see chapters 5, 6 and 8). Note that this thickness corresponds to a crustal volume more than 10 times greater than that of the Earth, normalized to planetary volume.

The support of Tharsis presents a more difficult problem. Stated plainly, there is a large excess of gravity at long wavelengths relative to that which can be explained by simple compensation models. There are a limited number of ways to accommodate this in a physically reasonable manner. One is isostatic compensation utilizing a density dipole, with a shallow positive anomaly (corresponding to a locally thinned crust) overlying a deeper negative anomaly (due to a lower-density region in the upper mantle; Sleep and Phillips 1979). By suitably choosing the density variations at these two (relatively shallow) levels, any apparent depth of compensation can be produced. Heuristically, this can be thought of as "burying" part of the topography at the crust-mantle boundary (thus decreasing the admittance) and providing support for both loads from a density deficit in the upper mantle. The petrological and tectonic implications of this type of structure have been investigated by a number of workers (Banerdt et al. 1982; Sleep and Phillips 1985; Finnerty et al. 1988; Phillips et al. 1990). This model appears to require massive removal of material from the base of the crust and static support by a thick (~200 to 400 km), immobile layer of magmatically depleted mantle.

The other approach to explaining the large free-air anomaly is to invoke an additional, nonbuoyant means of support, either statically by elastic flexure or dynamically by mantle convection. The admittance curve for a simple model for dynamic support (Bills et al. 1986) is compared with observations in Fig. 17. The model shown assumes that viscosity and perturbing density are uniform throughout the mantle, placing the density variations responsible for the gravity signature relatively deep in the interior. This model fits the admittance much better than the Airy model at the lowest harmonics. The significantly better agreement across the wavelength band as a whole can be attributed to the wavelength-dependent response of the lithosphere-to-mantle motions and to the fact that the density anomaly is confined to the deep interior. Static flexural models can also be constructed that fit the gravity observations (Banerdt et al. 1982; Willemann and Turcotte 1982; Sleep and Phillips 1985). In general, these models require a component of isostatic support in order to maintain a close fit to the topography/gravity boundary conditions.

These mechanisms both reduce the amount of compensating mass required in the lithosphere, resulting in a large net gravity signature as required by observations. However, they have extremely different implications for the

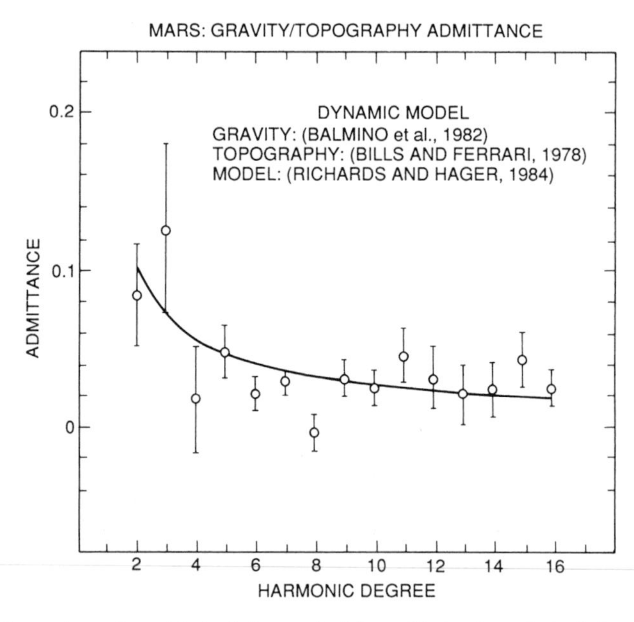

Fig. 17. Comparison of empirical admittance for Mars with the theoretical admittance due to a dynamic compensation model. The curve labeled topography represents the gravity which would be produced by the topography if it were uncompensated.

evolution of Tharsis. For the dynamic model, the topography is formed primarily by uplift of the surface. This implies that a single immense plume supports Tharsis, and that it has not moved significantly with respect to the lithosphere since the Noachian period. The viability of such a hypothesis has yet to be rigorously tested. For the flexural case, the highlands are formed by a combination of extrusive and intrusive volcanism, resulting in a crust that is about 30-km thicker than the planetary mean. This would make the formation of Tharsis a significant event in the differentiation of the Martian mantle. Also required is an elastic lithosphere at least 100 km thick that is capable of supporting stresses on the order of several hundred MPa, which places a constraint on the thermal structure of the interior.

Unfortunately, it is virtually impossible to discriminate among these three types of models on the basis of gravity and topography observations alone, because of the inherent nonuniqueness of potential field problems. This has led investigators to use other lines of evidence in order to further bound the problem. In particular, the use of surface stresses inferred from surface faulting, in combination with theoretical stress modeling, has proved useful in providing constraints on permissible models (Banerdt et al. 1982; Sleep and Phillips 1985). These models and their implications are discussed more fully in chapter 8.

IV. CONCLUSIONS

Doppler data received from planetary orbiters provide an excellent means of determining gravity fields both globally and locally. Such an analysis has been completed for the Moon, Venus, Earth and Mars. For Mars, the eighteenth degree and order model represents the current state of the art. Direct estimates include the planetary mass, center of mass offset, mean density (once the mean radius is known or established from spacecraft occultation measurements), planetary "geoid", power and error spectra and orientation of the rotation axis. The global gravity map shows strong correlations with major topographic features such as Tharsis, Hellas Planitia and Isidis Planitia.

Local gravity variations (or gravity anomalies) have been evaluated using the line-of-sight acceleration technique. Features such as Olympus Mons, Elysium Mons, Arsia Mons, Pavonis Mons, Ascraeus Mons, Alba Patera and Utopia Planitia are associated with gravity highs. Both Hellas Planitia and Valles Marineris correlate with gravity lows. Interestingly, Isidis Planitia appears as both a gravity peak (at its center) and low (near its edges).

Radar-derived topography has been used as control points in the equatorial zone for the new U.S. Geological Survey topographic maps. The radar altimetry has served as the basis for morphological studies, particularly of craters. Regional slopes over distances of the order of 100 to 1000 km, difficult to obtain by any other means, have been used to study the deformation of the surface of Mars over time. Future oppositions will enable higher-resolution studies, especially at the northern latitudes where coverage is now relatively sparse compared to the southern equatorial zone.

Correlative studies of gravity and topography indicate a complex relationship. Statistical studies on a global scale are in favor of several compensation mechanisms at work. The anticipated data from Mars Observer will significantly increase our knowledge of both gravity and topography and therefore of modeling the interior of Mars.

Acknowledgment. We acknowledge contributions to Sec. II by L. E. Roth (JPL), S. H. Zisk (University of Hawaii) and R. Mehlman (UCLA/IGPP). The research described in this chapter was carried out in part by the Jet Propulsion Laboratory, California Institute of Technology, under a contract with the National Aeronautics and Space Administration.

8. STRESS AND TECTONICS ON MARS

W. BRUCE BANERDT, MATTHEW P. GOLOMBEK
Jet Propulsion Laboratory

and

KENNETH L. TANAKA
U.S. Geological Survey

The tectonic record on Mars, in conjunction with theoretical stress modeling, has greatly contributed to our understanding of Mars and its evolution. In this chapter, we discuss deformation of the lithosphere and the various theoretical formulations used to model its behavior on local, regional and global scales. We provide an overview of the various classes of tectonic features found on Mars, and summarize the tectonic record on Mars, which provides the basic framework for interpreting theoretical thermomechanical models in terms of major tectonic events and provinces. For loads on the lithosphere with a horizontal extent which is small compared to planetary curvature, deformation can be modeled using flat plate or shallow shell theory. Investigation of local-scale loads can be used to derive information about the mechanical properties of the lithosphere, as stresses are almost solely due to bending. The thickness of the elastic lithosphere has been estimated using its response to local surface loads, and significant spatial variability is found, with values ranging from 20 to 300 km. These results can be analyzed in terms of a lithosphere with an elastic/plastic rheology derived from laboratory measurements on rocks to derive corresponding lithospheric thermal gradients at the time of load emplacement of 5 to 15 K km^{-1}. The study of larger-scale deformation requires mathematical formulations that take into account the spherical nature of the lithosphere. Regional and global analyses are most useful in studying the loading processes themselves, as deformation is less sensitive to the mechanical properties of the lithosphere at these wavelengths. Stresses are produced by a combination of *membrane and bending deformations, as well as by horizontal gradients in the load. Distinct stress patterns are produced for subsidence, uplift and isostasy. The thermal evolution of the planet may have significant implications for the state of stress in the lithosphere, as a 100 K change in average internal temperature can generate a horizontally isotropic stress of 150 MPa. Hypotheses regarding planetary despinning and polar wander have been tested, with little tectonic evidence found for major changes in the magnitude or orientation of the spin vector. Arguments involving the state of stress in the lithosphere support relatively high values for the moment of inertia. Relations between stress and faulting allow us to infer stress directions on a planet's surface. The wide distribution of extensional features, such as grabens and rifts, and compressional structures, such as wrinkle ridges, makes it possible to place firm constraints on theoretical stress models, allowing some insight into internal structure and processes. In contrast, only a few examples of strike-slip faults have been identified on Mars, and the role of this type of deformation in its tectonics is less well understood. The ubiquity of extensional features (found mostly around Tharsis) has generally been cited as evidence for net planetary heating and expansion. However, the recently recognized global distribution of compressional features allows for the possibility that the thermal history of Mars was dominated by cooling and contraction. Most tectonic activity since the end of early bombardment has been associated with the Tharsis region. The tectonic history of this region includes: (1) pre-Tharsis faulting in the Thaumasia region and at Acheron Fossae; (2) older main Tharsis (Syria)-centered radial faulting; (3) wrinkle ridges concentric to the Tharsis swell; (4) younger main Tharsis (Pavonis)-centered radial faulting; (5) faulting at Alba Patera, Tempe Terra and Tharsis Montes; and (6) radial and concentric grabens surrounding the Tharsis Montes. A comparison of the major radial graben systems with theoretical stress models clearly shows that more than one mechanism of lithospheric deformation is required to produce its enormous extent. However, detailed mapping shows that sets of radial fractures that extend from the central regions of Tharsis to its periphery appear to have formed coevally. It is not clear how to form such extensive radial fractures in a single event, when the stress models seem to require two distinct deformation events. This is a major problem in constructing self-consistent scenarios that link the observed tectonic features and the inferred stress systems with the evolution of the lithosphere and asthenosphere in the Tharsis region.*

I. INTRODUCTION

Mars occupies something of an intermediate position among the terrestrial planets. On one end of the scale are the small bodies such as Mercury and the Moon; on the other, Venus and the Earth. The members of the former group tend to have undistinguished tectonic histories, at least from the end of heavy bombardment to the present. Their observable tectonic records are characterized by global events, such as planetary contraction or despinning, punctuated by occasional local activity, generally in close association with impact basins. All major activity ceased roughly 3 to 4 Gyr ago. In addition, the surface manifestations of stress events are subdued by the strength of a thick lithosphere which resulted from relatively early cooling. As a consequence, stress investigations are of somewhat limited value in studying the structure and history of these objects. In stark contrast to this situation are the Earth and Venus, which have rich tectonic histories. The long-lived heat engines of these planets appear to have maintained to the present time both high levels of activity on all spatial scales and thin lithospheres that are easily deformed by that activity. This produces a confusing record, with older events overprinted and sometimes obliterated by younger ones, making interpretation extremely difficult (although geologists specializing on the Earth have a decided advantage in terms of the quality of observation possible).

Mars, on the other hand, appears to be unique in that it has experienced a tectonic history with enough variation to be interesting and informative, but which may have been simple enough to allow us to make some progress in producing a coherent story using our limited observations and relatively crude tools of interpretation and analysis. The global physiography is dominated by two features, the hemispheric dichotomy and the Tharsis rise. Whereas the dichotomy has little or no observable large-scale tectonic signature, fault and ridge systems associated with the formation and evolution of Tharsis overprint nearly half the planet. Thus, the tectonics of Tharsis is nearly synonymous with the global-scale tectonics of Mars. Another striking property of Mars is the high degree of correlation between topography and gravity, especially at the longest wavelengths (see chapter 7). This suggests that relatively straightforward connections exist between the density anomalies at depth which generate the potential field and the mechanisms responsible for maintaining the relief. Thus Mars provides an excellent laboratory for developing schemes for modeling tectonic processes on single-plate planets.

From a geophysical standpoint, studying the response of the lithosphere to regional-scale loads ($\gtrsim$ 1000 km on Mars) is primarily useful for providing information about the forces responsible for the deformation. This is because deformation of a lithosphere at these wavelengths is relatively insensitive to its mechanical properties, but strongly reflects the complexities of the loading mechanisms. Alternatively, analysis of deformation under loads of smaller lateral extent can be used as a probe of the rheology and density structure of the uppermost layers of a planet, as can the morphology of some tectonic features. The huge volcanic constructs of the Tharsis and Elysium regions, as well as a number of impact basins, are well suited for such investigations, and numerous well-developed extensional features and compressional ridge sets provide additional opportunities for determining lithospheric properties.

The correlation of theoretical stress orientations with observed tectonic features can be used to construct a thermo-tectonic history of a planet that reflects the evolution of its interior as well as processes restricted to the near-surface mechanical layers. Recently, a global geologic map series at 1:15 million scale based on high-quality Viking images has been completed (Scott and Tanaka 1986; Greeley and Guest 1987; Tanaka and Scott 1987). These maps are forming the basis for detailed global stratigraphic analyses (see chapter 11). Within this framework, detailed structural mapping on a regional scale provides specific information on the relative timing of the major events of Mars' tectonic history.

II. DEFORMATION OF THE LITHOSPHERE

The mechanical lithosphere can be defined generally as the outermost layer of a planet that can support stresses over a significant period of time. The precise model used to approximate its strength and rheological properties depends on the processes being investigated, as well as the level of sophistication desired. A useful definition from the standpoint of tectonic analyses is that of the elastic lithosphere, the layer that deforms in an essentially elastic fashion over the period of time relevant to the process or feature being studied. If predominantly elastic behavior is assumed, conventional engineering theory for flat plates and spherical shells can be adapted to investigate lithospheric deformation.

The model geometry (flat or spherical) required to study deformation processes on a planet depends on the dimensions of the load with respect to the radius of the planet and the thickness of the elastic lithosphere. Figure 1 shows the behavior of a nearly flat plate under a radially symmetric downward load. The radial stress σ_{rr} is induced almost wholly by the bending of the plate, with only a small contribution from stretching. At the upper surface σ_{rr} is compressive where the deflection is concave upward and tensile where concave downward; the lower surface experiences the opposite sense of

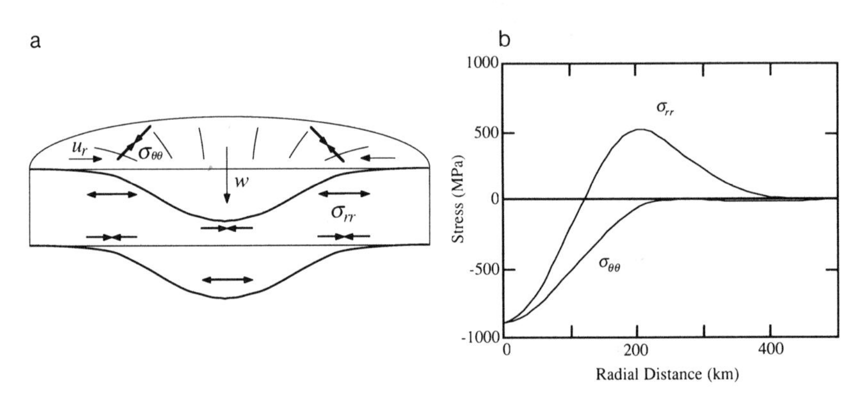

Fig. 1. (a) Schematic diagram of the behavior of a flat plate under an axisymmetric load. Dark arrows denote radial (σ_{rr}) and tangential ($\sigma_{\theta\theta}$) stresses (converging arrows represent compression, diverging arrows extension) and lighter arrows denote radial (u_r) and vertical (w) displacements. (b) Radial profiles of the stress from a local conical load on a shallow shell. The figure is adapted from Comer et al. (1985), who calculated the response of a 22-km-thick Martian elastic lithosphere ($E = 10^{11}$ nt m^{-2}, $\nu = 0.25$, $\Delta\rho = 3.5$ Mg m^{-3}) to a 200-km-radius load corresponding to Ascraeus Mons, the northernmost of the Tharsis Montes volcanoes. Tension is positive.

stress. The tangential stress (or "hoop" stress) $\sigma_{\theta\theta}$ is everywhere compressive; this is due to the fact that any given point is pulled toward the center by the increase in the plate curvature caused by the vertical deflection. This results in a decrease in the line length of a hoop-shaped volume element centered on the load, producing compression normal to the radial displacement. Changing the sense of the load (i.e., negative or upward vs positive or downward) will reverse the sense of σ_{rr}, but will leave the sense of $\sigma_{\theta\theta}$ unchanged.

If the lateral extent of the load is large enough that planetary curvature is important, quite different behavior results. The radial stress now has two components whose relative magnitudes depend on the wavelength of the load and the thickness of the lithosphere. The bending component is the same as the corresponding flat plate mode, and dominates for shorter wavelengths and thicker lithospheres. The membrane component is induced by in-plane forces in the lithosphere, and is uniform across the thickness of the shell. The bending and membrane contributions to the support of a load are roughly equal for $\lambda \approx 4\pi(R^3/t_e)^{1/2}$, where λ is the wavelength of the load (approximately twice its lateral extent), R is the undisturbed radius of the shell, and t_e is the thickness of the elastic lithosphere. For wider loads, membrane stresses dominate, whereas narrower loads are supported primarily by bending stresses (Turcotte et al. 1981). Membrane stresses in a spherical shell (illustrated in Fig. 2) depend on the vertical displacement as well as the change in the principal radii of curvature at a given point. The membrane component of σ_{rr} for a circularly symmetric downward load is everywhere compressional be-

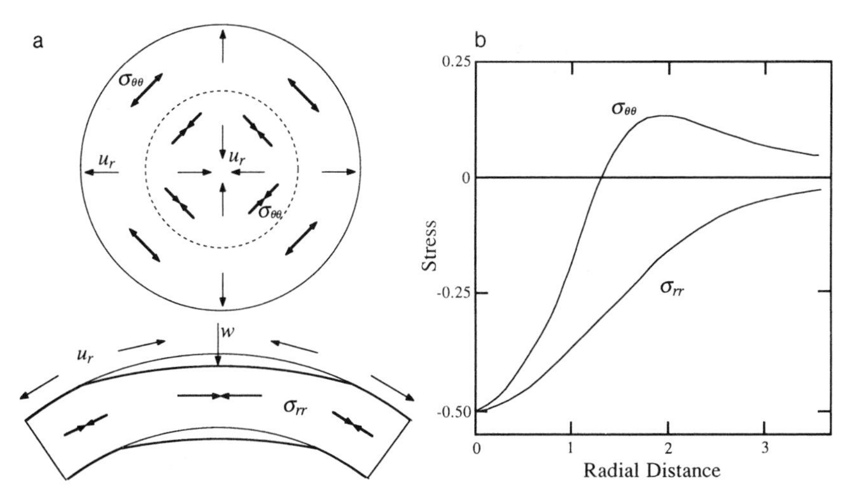

Fig. 2. (a) Schematic diagram of the membrane response of a spherical shell under a long-wavelength load. See Fig. 1 for an explanation of symbols. (b) Profiles of the radial and circumferential membrane stress components for a gaussian load given by $q = q_0 \exp[-(r/r_0)^2]$ on a spherical shell (figure adapted from Turcotte et al. 1981). Radial distance is normalized by r_0 and stress is normalized by q_0t_e/R. Tension is positive.

cause of the overall shortening of the lithosphere along lines radial to the load. The tangential (hoop) stress for a symmetric load has only a membrane component. It is compressional near the center of a downward load because of the lithospheric shortening caused by the downward displacement under the load. Beyond the edge of the load, the radius of an annular surface element is increased by outward displacement, resulting in circumferential tension. In contrast to the flat plate situation, a change in the sense of loading on a spherical shell will change the sense of all stress components.

Another effect which can be important for longer-wavelength loads is the stress generated by lateral variations in topography or density, sometimes referred to as isostatic stress. This results from the force acting down the topographic (or density) gradient in such a way as to tend to remove the variations (Artyushkov 1973,1974; Sleep and Phillips 1985). The "downslope" force induces a positive radial displacement within a positive circular load. The corresponding stress state (Fig. 3) is purely membrane. The radial stress is extensional within the load and becomes compressional outside, whereas the tangential stress is everywhere extensional. Note that this situation is, in a sense, complementary to the flexure case in that now $\sigma_{\theta\theta}$ is everywhere extensional, and σ_{rr} changes sign near the edge of the load, from tension to compression.

Loading by Local-Scale Features: Lithosphere Thickness

The bending response of a thin elastic plate or shallow spherical shell overlying an inviscid substrate is governed by the flexural parameter $\alpha = [D/(\Delta\rho g + Et_e/R^2)]^{1/4}$ (see, e.g., Brotchie 1971), where $D = Et_e^3/12(1 - \nu^2)$ is the flexural rigidity, E is Young's modulus, ν is Poisson's ratio, R is the radius of the planet ($= \infty$ for a flat plate), $\Delta\rho$ is the difference between the density of the substrate and the density of any material filling in the depression at the surface and g is gravitational acceleration ($= 3.72$ m s^{-2} for Mars). Loads with a wavelength much less than α will be supported by bending stresses, whereas longer-wavelength loads will be supported by buoyancy and/or membrane stresses, depending on the relative sizes of the two terms in the denominator of α (if $\Delta\rho g$ is much larger than Et_e/R^2, wide loads will be supported isostatically; if the latter term dominates, the membrane stress, or "arch support," mode will prevail). The densities and elastic parameters can generally be estimated reasonably well; the strongest dependence is on t_e. Thus, to a first approximation, the elastic lithosphere can be characterized by its thickness alone.

As can be seen in Fig. 1b, the radial stress exhibits a marked extensional peak. The position and magnitude of this peak is determined primarily by the dimensions of the load and t_e. For relatively small isolated features, such as volcanoes and filled impact basins, it is generally possible to make a reasonable estimate of the magnitude and extent of the load from topography and gravity data, as the subsurface density structure associated with them can be

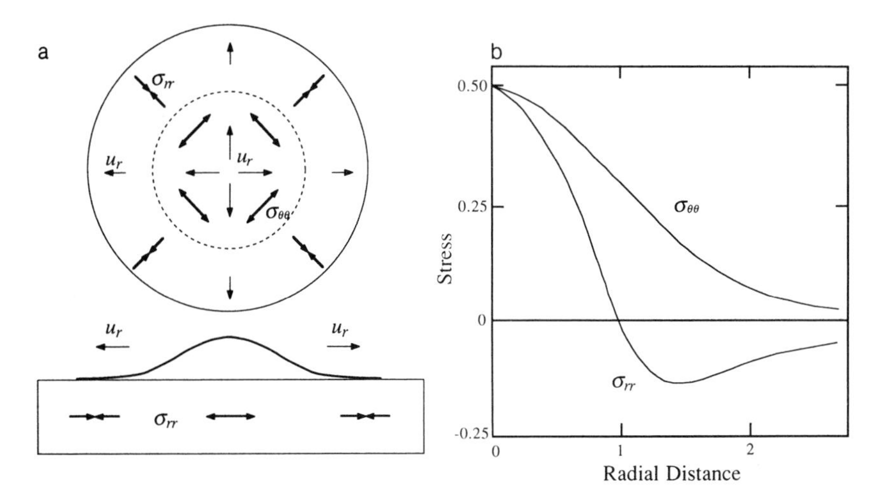

Fig. 3. (a) Schematic diagram of the response of a plate or shell to an isostatic, or gradient load. See Fig. 1 for an explanation of symbols. (b) Generalized profiles of the radial and circumferential membrane stress components for a regional isostatic load. Stress is normalized by $\rho g h_{max}$ and distance is normalized by load radius. Tension is positive.

assumed to be reasonably simple. The position of concentrically oriented fractures around such features can then be combined with elastic modeling to infer t_e. This, of course, presupposes uniform mechanical properties in the lithosphere and assumes that other deformation mechanisms are not operative. In the case of structures such as major volcanoes, estimates made using this method should be viewed with some caution, as mechanical inhomogeneities may be important and the massive movement of material from the interior may influence the displacement field.

A number of investigators have used elastic flexure theory to estimate t_e at various locations (see Table I). The lack of evidence for concentric extensional failure around Olympus Mons, the largest single load on Mars, led Thurber and Toksöz (1978) and Comer et al. (1985) to propose a lower limit on t_e of 150 km based on the apparent size of the volcanic load and lithospheric strength considerations. Thickness limits of 140 to 230 km were computed by Janle and Jannsen (1986) using a larger estimate of the volume of Olympus Mons by Wu et al. (1984). Interestingly, other smaller features on Mars do show evidence for flexural failure. Comer et al. (1985) used position and width of the zone of circumferential grabens about the Isidis basin and the large volcanoes of Tharsis and Elysium to constrain the position and width of the radial extensional stress peak. These constraints were formally inverted to produce bounds on t_e. Best-fitting thicknesses beneath Ascraeus Mons, Pavonis Mons, Arsia Mons, Alba Patera and Elysium Mons were found to be 20 to 50 km. The mascon in Isidis basin appeared to have formed on a lithosphere at least 120 km thick, and perhaps as thick as 300 km. Thus

TABLE I
Estimates of Effective Elastic Lithosphere Thickness and Lithospheric Thermal Gradient on Mars[a]

Feature	Age of Deformation[b]	t_e (km)[c]	dT/dz (K km^{-1})[d]
Arsia Mons	Upper Amazonian	18 (10–50)	14 (7–23)
Ascraeus Mons	Upper Amazonian	22 (8–50)	12 (7–27)
Pavonis Mons	Upper Amazonian	26 (10–50)	11 (7–21)
Alba Patera	Lower Amazonian	33 (19–85)	10 (8–14)
Elysium Mons	Lower Amazonian	54 (48–110)	7–13 (6–14)
Olympus Mons	Upper Amazonian	140–230	<5
Isidis Mascon	Upper Noachian	120–300	<6

[a] Table adapted from Solomon and Head (1990).
[b] Stratigraphic positions from Tanaka (1986).
[c] From Thurber and Toksöz (1978), Comer et al. (1985), and Janle and Jannsen (1986). Parentheses denote formal bounds on parameter values from Comer et al. (1985).
[d] From Solomon and Head (1990). Gradients are derived under the assumption that the strength of the mechanical lithosphere is determined by diabase for $t_e < 50$ km and by olivine for $t_e > 50$ km; for Elysium Mons, the ranges reflect the possibility that t_e may fall on either side of this boundary.

t_e near the center of Tharsis may have been less than 50 km, whereas globally it appears to be have been greater than 150 km. Note that these thickness estimates are relevant for the time at which the loads were emplaced, so that the age of the feature must be considered when using the thicknesses to infer thermal histories (see below).

The apparent elastic thickness can be related to the lithospheric structure (composition and temperature vs depth) via more realistic models of the lithosphere. Perhaps the most useful involves the concept of strength envelopes, in which it is assumed that the yield strength at a given depth can be represented by an empirical equation describing the weakest deformation process operating at that depth. Stresses outside the envelope result in deformation by the appropriate nonelastic yield process (such as faulting or solid-state creep), whereas stresses less than the yield strength produce elastic strain only.

Experience on the Earth indicates that the strength of rocks in the shallow lithosphere, where the temperature is relatively low, is generally limited by frictional sliding on pre-existing faults (Brace and Kohlstedt 1980). The coefficient of friction for this process is virtually independent of temperature, pressure (except for a small change at about 200 MPa), surficial properties of the fault and rock type (Byerlee 1978), but can be affected by the presence of a pressurized fluid, such as ground water. Frictional deformation is well described by Byerlee's law, a Coulomb criterion giving a yield stress that increases linearly with depth. Tectonic failure is manifested in shear faulting, with the type of faulting determined by the orientation of the maximum and minimum principal stresses (see below). In terms of the mechanical response of a heavily faulted layer, this results in an effectively plastic rheology for length scales much greater than the mean fault spacing.

The outermost portions of the Martian lithosphere are likely to be pervasively faulted by impact cratering and an active tectonic history; thus the assumption of Byerlee's law is justified for the general case. However, some near-surface rock units may be essentially intact due to late deposition or chemical cementation (Tanaka and Golombek 1989). In this case, the yield stress can be approximated by the Griffith failure criterion, which is derived from a simple model of unstable crack growth (see, e.g., Jaeger and Cook 1979, p. 277). This results in a similar, but somewhat stronger, yield curve and plastic rheology. Deformation in this case occurs by opening of tension cracks near the surface and shear failure at depth (see, e.g., Golombek and Banerdt 1990).

Both mechanisms described above result in a yield stress that increases with depth. At the higher temperatures characteristic of the lower portions of the lithosphere, however, ductile creep becomes the strength-limiting factor. In contrast to frictional failure, ductile flow laws are strongly dependent on temperature, rock type and (to a lesser extent) strain rate. The exponential dependence on temperature gives a rapid decrease in the ductile strength with depth. The maximum lithospheric strength occurs at the brittle-ductile transition, where the two deformation mechanisms are equally strong. In addition, ultramafic minerals characteristic of mantle rocks have a considerably higher creep resistance than those minerals that are common in crustal rocks. Thus, the lithospheric strength envelope may have two or more strength maxima separated by ductile layers if the lithosphere has a weak upper crust underlain by stronger (e.g., more mafic) crustal or mantle layers of different composition, as is the case for the Earth (Brace and Kohlstedt 1980). The existence, size and depth of these strength maxima will depend on the crustal thickness and thermal gradient.

An example of a strength envelope for Mars is shown in Fig. 4. It can be seen that if the crust is less than about 40 km thick (or, for a thicker crust, if the thermal gradient is significantly lower than the 9 K km^{-1} assumed here), there will be a single maximum and the strength of the lithosphere will be determined by the mantle properties. Conversely, if the crust is thicker (or the thermal gradient higher), a weak zone due to the lower creep strength of crustal rocks will develop in the lower crust. This will result in a second strength maximum, and deformation will depend more heavily on crustal parameters as the upper crust becomes increasingly decoupled from the mantle.

As the lithosphere is deformed at a given strain rate, stress will increase at a given depth until it equals the yield stress at that depth, after which it will deform at constant stress by motion on faults in the brittle region or by ductile flow in the high-temperature creep region (Goetze and Evans 1979).

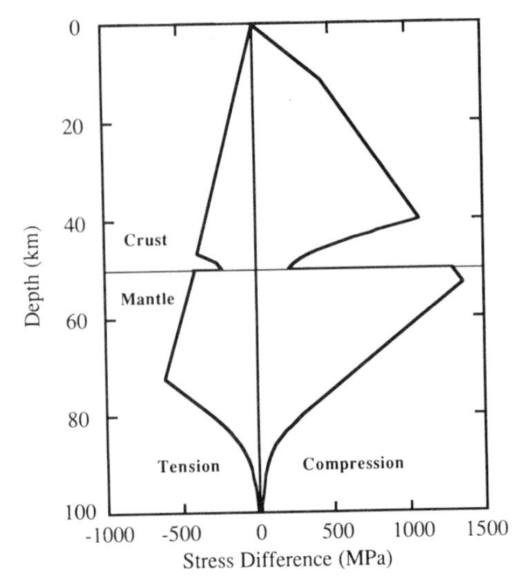

Fig. 4. Lithospheric strength envelope for Mars showing the maximum stress difference (horizontal minus vertical stress, or twice the maximum shear stress) that can be supported as a function of depth for both horizontal compression (positive stress differences) and extension (negative stress differences). The figure assumes a 50-km basaltic crust overlying an olivine mantle with a thermal gradient of 9 K km^{-1}, a surface temperature of 250 K, and a geologic strain rate of 10^{-15} s^{-1}. Note that for a crust thinner than 40 km, there will be a single strength maximum, and the elastic lithosphere will be less than 100 km thick. For crusts thicker than ~50 km, two strength maxima will occur, and the effective elastic thickness of the surface layer will only be ~50 to 60 km.

The lithosphere will behave elastically, however, for stresses less than the yield stress at a given depth and strain rate. Thus, the effective thickness of the elastic layer will decrease with increasing stress. Relations have been derived between the apparent elastic thickness and the compositional and thermal structure of the lithosphere for both membrane stress (see, e.g., McAdoo and Sandwell 1985; Banerdt and Golombek 1988) and plate bending conditions (see, e.g., McNutt 1984).

Solomon and Head (1990) used these concepts to convert the estimated elastic thicknesses into lithospheric thermal gradients. Their approach assumed that the smaller values of t_e (20 to 30 km) are less than the thickness of the Martian crust, given the lower bound on crustal thickness of 30 km derived from global gravity considerations (Bills and Ferrari 1978; see chapter 7). The larger values of t_e (> 100 km) were assumed to exceed the crustal thickness and computations were performed using material properties for the mantle. Their derived thermal gradients, along with the inferred ages of deformation are given in Table 1. Note that there is not a simple decrease in thermal gradient with time, implying that substantial spatial variations in subsurface thermal structure must have been superimposed on any progressive cooling of the lithosphere (Comer et al. 1985). In particular, higher values of heat flow appear to be associated with the central regions of major volcanic provinces, with lower values (which are more appropriate for the planet as a whole) being found in regions more distant from these areas.

Regional-Scale Loading on Mars: Tharsis Structure

Whereas local-scale loads provide information about the mechanical properties of the lithosphere, the regional response of a spherical planet is relatively insensitive to the details of lithospheric structure. This fact can be used to advantage in investigating the nature and history of large-scale forces acting upon the lithosphere by varying the model parameters associated with the load and holding the mechanical properties constant.

For the case of a volcanic construct, the net force can be considered simply as the weight of the observed volcanic pile and any material filling the flexural depression, less the counteracting buoyancy at the base of the crust. For features thousands of kilometers across, however, the situation is more problematic. The surface topography is obviously an unambiguous component of the load; but, as the horizontal scale is increased, the contributions of lateral density anomalies at depth (such as upper mantle density and crustal thickness variations) become an increasingly important factor, and dynamic forces due to solid-state convection in the mantle may contribute to the longest wavelengths of deformation. Indeed, it is often not even possible to determine by casual inspection whether the net force and deflection is directed upward or downward, as a topographic high can result either from piling of material on the surface (resulting in subsidence of the lithosphere), uplift of the surface from below by buoyancy or convective flow, or some combination of these processes. A detailed understanding of the stress patterns caused by various load models, coupled with careful interpretation of the geologic record, is required to unravel the tectonic history of a relatively active planet like Mars.

Static Models. Several different formulations of thin- and thick-shell theories have been used to investigate the long-wavelength loading of Mars. The differences between these approaches are summarized below. All assume laterally invariant mechanical properties (elastic constants and lithosphere thickness) over the entire globe and require spherical harmonic representations of the topography and gravity to construct the loading function, the mathematical representation of the combination of forces acting upon the lithosphere.

A thick-shell approach was taken by Banerdt et al. (1982) to study the mechanisms of support for Tharsis. The analysis involves the numerical integration of the full set of elastic equations in a spherical geometry after Alterman et al. (1959) and Arkani-Hamed (1973), modified to include a laterally varying density anomaly at depth (Kaula 1963) and an inviscid interior

(Longman 1963). Both topography and gravity are used as boundary conditions, with self-gravitation effects included. The laterally varying density anomaly at depth is included in the loading function, in addition to the topographic load, in order to match the observed gravity. Banerdt et al. (1982) placed this anomaly at the crust-mantle boundary in order to simulate a variable crustal thickness, but it could also be interpreted in terms of density changes in the lower crust or upper mantle. As that investigation was concerned with the problem of support for the Tharsis Rise (which contributes primarily to the 2nd and 3rd harmonics of the topography and gravity; see Phillips and Lambeck 1980), only harmonics to degree and order 4 were used. This thick-shell formulation gives an exact solution to the elastic problem, but the numerical code is unwieldy and has stability problems at high harmonic degrees. An analytical approach to the thick-shell problem (limited to axisymmetric loads) has recently been developed by Janes and Melosh (1990).

Thin-shell theory offers considerable computational advantages over the thick-shell approach, allowing more exhaustive investigations of parameter trade-offs. The thin-shell approximation requires that: (1) the shell is thin compared to its radius ($t_e/R \lesssim 1/10$); (2) deflections are small compared to the shell radius; and (3) the normal stress in the radial direction is negligible, i.e., deflections of the inner and outer surfaces of the shell are equal (Kraus 1967). Under these assumptions, the spherical elastic shell problem is reduced to a set of relatively easily solvable algebraic equations. For geologically reasonable situations, these conditions are met for Mars as long as the lithosphere is assumed to be $\lesssim$ 300-km thick. As discussed above, estimates for lithospheric thickness on Mars are within this limit.

A number of implementations of thin-shell theory have been published. Willemann and Turcotte (1982; see also Turcotte et al. 1981; Willemann and Turcotte 1981) applied the theory of Kraus (1967) to investigate the stresses required to support Tharsis. Their formulation includes bending and membrane stresses, but neglects horizontal tractions due to topographic gradients (isostatic stresses). The loading function consisted of an axisymmetric cosine approximation to the topography along with the buoyant response; gravity was not used as a formal boundary condition, but was computed from the final solution as a check on model validity.

Sleep and Phillips (1985) also used the theory of Kraus (1967) for global stress calculations. Their implementation utilizes the gravity boundary condition, but retains only the membrane stress components. The membrane stress approximation is justified because again they were interested primarily in the stresses due to Tharsis and included only the lowest harmonic terms up to degree 4. The inclusion of the gravity boundary condition allows for a more realistic loading function similar to that of Banerdt et al. (1982), with a laterally varying density structure at depth contributing to the net force on the lithosphere as well as to the potential. In addition, they added to the loading function a contribution from the tractions due to lateral gradients in the vertical load (the so-called slope stresses or isostatic stresses). This must be done explicitly in order to correct for errors in the calculated stresses at small deflections due to the third thin-shell assumption stated above.

Banerdt (1986) developed a planetary thin-shell formulation (originally applied to Venus) based on an alternative derivation of thin-shell theory by Vlasov (1964). It utilizes both topography and gravity as boundary conditions, includes lateral gradient loads, and incorporates a full thin-shell treatment including bending and membrane stresses. The main advantage of this formulation is in the generality and flexibility of the loading function. It allows lateral density anomalies at two depths (corresponding, for example, to a simultaneously varying crustal thickness and upper mantle density), and constraints can be placed on any of the variables (i.e., displacements and densities) either singly or in combination in order to simulate different mechanisms of formation. It is this formulation which has been used to generate the stress fields illustrated in Figs. 5–7 discussed below.

Despite their many differences, these studies all produced broadly similar results with regard to stresses generated by the Tharsis Rise (Banerdt and Golombek 1989). Mechanisms for the origin of Tharsis can be divided into three general classes: external loading, uplift and isostatic. The stress patterns resulting from each of these scenarios are illustrated in Figs. 5, 6 and 7. Whereas previous investigations used simple representations of the gravity and topography of Tharsis (such as single-wavelength cosine function [Willemann and Turcotte 1982] or a harmonic representation truncated at degree 4 [Banerdt et al. 1982; Sleep and Phillips 1985]), we have included in these calculations harmonic coefficients through degree and order 8, effectively doubling the "resolution" of the models. The general trends identified in the original studies (see below) are still apparent (as should be expected due to the fact that Tharsis is primarily described by the 2nd and 3rd degree harmonics), but the effects of local departures from an idealized axisymmetric dome can also be discriminated.

Downward displacement of the lithosphere beneath Tharsis due to external loading should produce generally radial compression within the Tharsis Rise and concentric extension farther out, as can be seen in Fig. 5. Upward displacement (doming) that is caused by support applied to the base of the lithosphere produces stresses with the same orientations but with the opposite sign, with inner extension and outer compression (Fig. 6). The isostatic case is illustrated in Fig. 7. For isostasy (corresponding to either complete relaxation of flexural stresses or zero net vertical displacement of the lithosphere), radial compression appears outside Tharsis, with concentric extension being developed on its flanks. The agreement between these predicted stress patterns and the observed tectonics is discussed below.

These different cases have widely divergent implications for subsurface structure as well as for tectonics. For the external loading (downward dis-

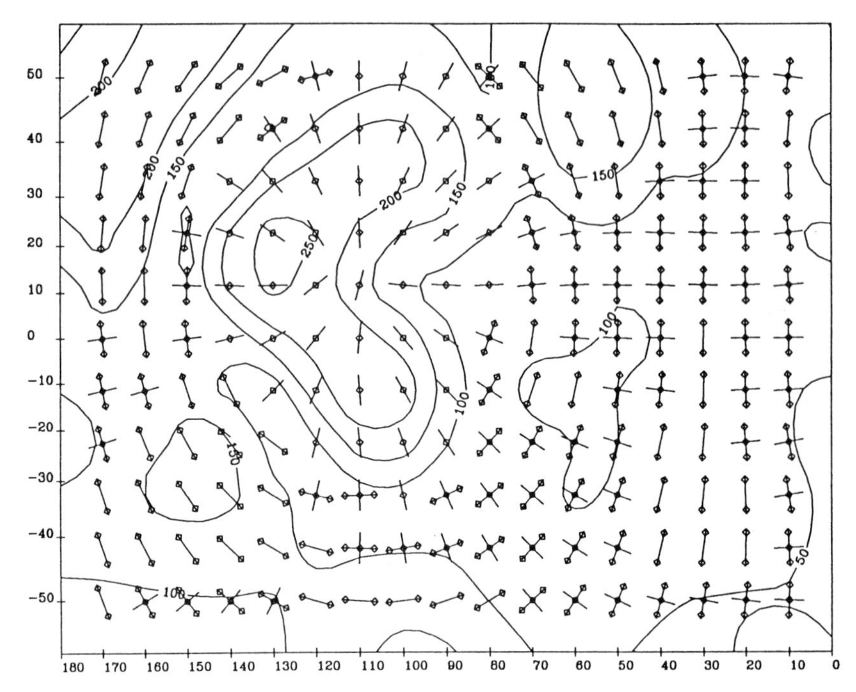

Fig. 5. External loading stresses for Tharsis, plotted on a mercator projection. Tharsis is centered at roughly 5° S, 105° W. Contours show the magnitude of the maximum stress difference at the surface in MPa. The stress orientation indicators with symbols at each end denote horizontal extension; those with symbols in the center denote compression. For this case, the topography of Tharsis is formed by adding material to the surface, causing a peak lithospheric deflection of 8 km. The local thickness of the crust is simultaneously adjusted in order to satisfy the observed gravity, reaching a maximum excess thickness of 28 km beneath Tharsis. Note the generally concentric extensional stresses outside the Tharsis load and the radially oriented compressional stresses inside. Stresses for this and the following figures were calculated using the thin-shell formulation of Banerdt (1986) assuming $t_e = 200$ km, mean crustal thickness $t_c = 100$ km, $E = 1.25 \times 10^{11}$ nt m^{-2}, $\nu = 0.25$ and $\Delta\rho = 0.5$ Mg m^{-3}. Topography and gravity harmonic coefficients through degree and order 8 (Balmino et al. 1982; Bills and Ferrari 1978) have been used as boundary conditions.

placement) case, the crust is significantly thicker beneath Tharsis due primarily to the great thickness of extrusives required. The thickness of this load ranges from ~ 15 km for a 200-km thick lithosphere to nearly 60 km for a 75-km thick lithosphere (Banerdt et al. 1982).

The isostatic case requires two density anomalies at different depths in order to satisfy the gravity boundary condition (Sleep and Phillips 1979). The shallower anomaly must be positive (corresponding, for example, to a locally thinner crust) and the deeper anomaly negative (perhaps due to higher tem-

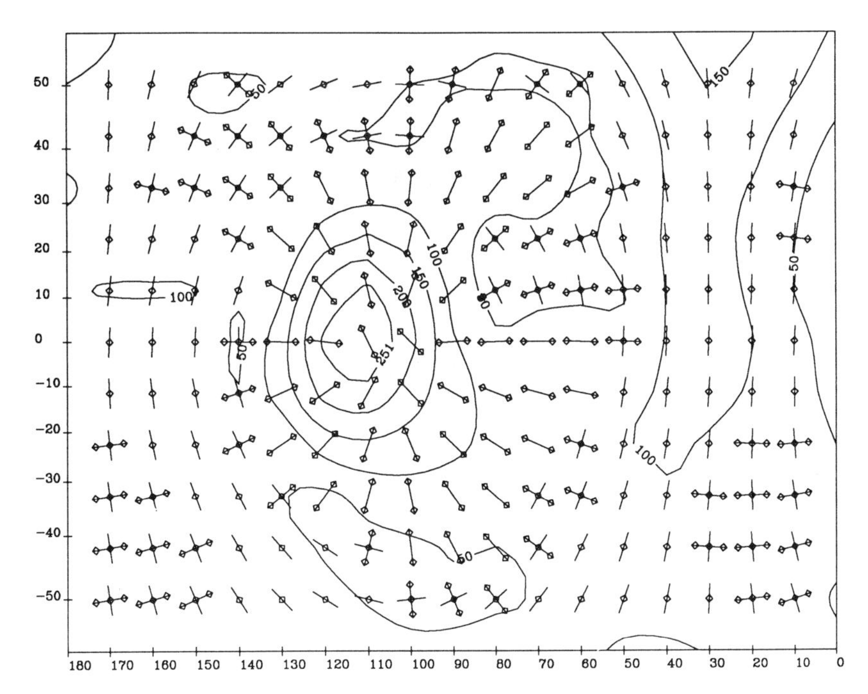

Fig. 6. Flexural uplift stresses for Tharsis. For this case, Tharsis is formed by the buoyant uplift of the lithosphere by a combination of locally thinning the crust and decreasing the density of the upper 350 km of the mantle, most of which is below the lithosphere. The dominant stress trends are radial extension within the rise and concentric compression in the periphery. See Fig. 5 for explanation of symbols and model assumptions.

perature or magmatically depleted material). The isostatic state is particularly significant, because it represents a minimum strain energy configuration for a given gravitational anomaly (Sleep and Phillips 1985); thus, the lithosphere will tend to approach this state over time if stresses are allowed to relax by faulting or creep. A thorough treatment of isostatic balance on a small planet can be found in Sleep and Phillips (1979).

Uplift can be accomplished while satisfying the gravity simply by overcompensating the topography starting from the isostatic model outlined above (although this results in unreasonable values for crustal thinning and the mantle density deficit). However, in that case, stresses result which are similar in direction to the isostatic stresses, but with larger magnitudes. This is because the gradient stresses induced by the large laterally varying density anomalies at depth dominate over the flexurally induced uplift stresses. No static model has yet been proposed that produces the type of stress configuration illustrated in Fig. 6 while satisfying the current gravity.

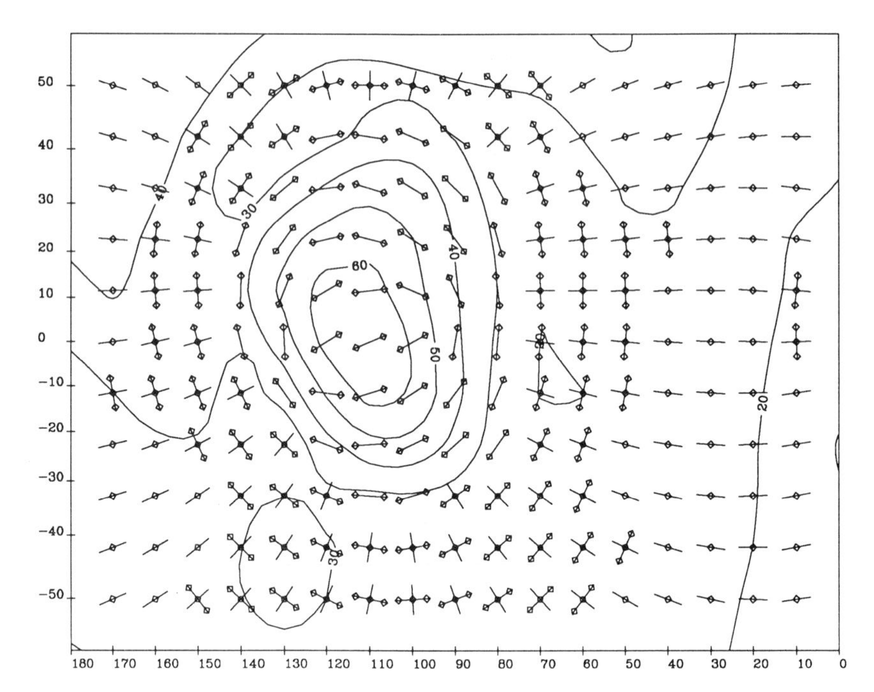

Fig. 7. Isostatic stresses for Tharsis. These are the stresses that would result from the relaxation of all stresses in the lithosphere except the minimum required to support the observed topography and gravity. The subsurface density structure consists of about 60 km of crustal thinning beneath Tharsis underlain by a negative mantle density anomaly of 0.18 Mg m^{-3}. The isostatic state is characterized by circumferential extension close to the load center and radial compression farther out. Note that the stress magnitudes are considerably lower for the isostatic case than for either of the flexural cases. See Fig. 5 for explanation of symbols and model assumptions.

Dynamic Models. The subject of dynamic support for long-wavelength topography and gravity anomalies is just beginning to be investigated quantitatively. Finite element calculations of thermal convection in a cylindrical geometry by Kiefer and Hager (1989) suggest that most of the current topography and gravity at long wavelengths could be maintained by convective support. Fully three-dimensional spherical simulations of Martian convection appear to support this conclusion (Schubert et al. 1990).

Whereas some work has been done on the support of long-wavelength features on Mars by mantle convection, there have not yet been any published calculations of the detailed stresses expected to result from the dynamic support of Tharsis. The stresses due to the viscous coupling of the lithosphere with convective motions in the mantle are the result of two components. The first, induced by the vertical component of the flow, depends only on the vertical displacement of the lithosphere, and is equivalent to the static shell stresses illustrated in Figs. 5, 6 and 7. The second is due to horizontal tractions at the base of the lithosphere. Around a region of mantle upwelling, these tractions will tend to produce either radially oriented extension or concentric compression, depending on the details of the flow field. The relative contribution of these components will depend on the thickness of the lithosphere and the amount of support derived from static buoyancy within the lithosphere itself.

The variables and boundary conditions for Banerdt's (1986) thin-shell formulation are posed in a form that is compatible with the upper-boundary-condition vector of a simplified model of mantle convection (Richards and Hager 1984), providing a capability for the calculation of stresses due to dynamical effects (see, e.g., Banerdt 1988). Preliminary calculations indicate that for moderate-to-thick lithospheres (> 25 km), the lateral displacements induced by vertical forces dominate over those due to viscous tractions and the stress orientations are similar to the static uplift case, with circumferential extensional stresses predicted within the rise and radial compression outside. For very thin lithospheres, the stresses due to tractions become important and stress orientations are rotated by 90°, becoming more nearly like the isostatic case, but with greater magnitudes.

Global-Scale Deformation

Analytically, global-scale deformation can be considered as the long-wavelength limit of the regional loading discussed above. However, the mechanisms underlying this class of deformation are distinct, in that they act on the planet as a whole. These processes are related to changes in the planet's radius or in the magnitude or orientation of its spin.

Planetary Expansion/Contraction. As the interior of a planet warms or cools, the planetary volume changes. Because the outer surface of the lithosphere remains at a roughly constant temperature (determined by the balance between surface insolation and radiation), this differential expansion results in an isotropic lithospheric stress induced by the net change in its surface area. Interior volume variations due to mineralogical phase changes will induce stress in the lithosphere in a similar fashion. The stress due to planetary expansion or contraction is given by $\sigma = \Delta R/R \cdot E/(1 - \mathrm{v})$, where ΔR is the net radius change; if this radius change is a result of thermal expansion, then

$$\sigma = \frac{(\Delta T_I - \Delta T_L)E\alpha_{\mathrm{v}}}{3(1 - \mathrm{v})} \tag{1}$$

where α_v is the volume coefficient of thermal expansion, ΔT_I is the volume-averaged interior temperature change, and ΔT_L is the average temperature change of the lithosphere (Turcotte 1983). This stress is intimately related to the thermal history and internal evolution of the planet via the coefficient of thermal expansion and the Clapeyron slope of any relevant phase changes (see, e.g., Solomon and Chaiken 1976; Solomon 1978; Turcotte 1983; see also chapter 5). For reasonable values of the material constants, an average interior temperature change of 100 K should result in a radius change of $\sim$ 3.5 km and an isotropic stress of about 150 MPa, assuming for simplicity that the average temperature of the lithosphere remains constant.

Because the thermally induced stresses are isotropic (for a uniform lithosphere), there is no preferred pattern of faulting, although locally an orthogonal grid might form, with the orientations determined by local structures and stress fields. Interior warming will cause extensional faulting to be generally favored over compression. Cooling will tend to favor compressional deformation. Note, however, that the lithosphere is considerably stronger in compression than in extension (Fig. 4), so that the effects of global cooling may not be as well developed in the tectonic record.

Despinning. When a planet loses spin angular momentum due to tidal dissipation, its oblateness (or equatorial bulge) also decreases. If the lithosphere was formed before the despinning episode, the change in shape will lead to stresses which may form a global system of fractures. The stresses induced in the lithosphere of a planet by this process were computed by Melosh (1977). He showed that despinning should result in a distinctive pattern of fractures, with east-west trending normal faults near the poles, north-south trending thrusts in the equatorial region, and conjugate strike-slips faults at midlatitudes. The size and character of these tectonic provinces may be modified by the superposition of global contraction or expansion stresses, but the azimuthal (east-west) stress will always be more compressive than the meridional (north-south) stress for a fault system formed by despinning.

Whereas there is some evidence for these patterns on Mercury (Melosh and Dzurisin 1978) and the Moon (Strom 1964), none has been identified on Mars. This is not surprising, as comparison of Mars' momentum density to that of the other planets suggests that Mars never had a significantly faster spin than at present (Kaula 1986, p. 223). Therefore, despinning would not be expected to play a major role in Martian tectonics.

Polar Wander. The reorientation of a planet's lithosphere with respect to the spin axis can also cause a diagnostic pattern of tectonic fractures due to the repositioning of the equatorial bulge. For example, as part of the lithosphere moves from a location at midlatitudes to a position nearer the equator, its principal radii of curvature will change, causing membrane stresses to develop. The stresses in a reoriented lithosphere were evaluated by Vening-

Meinesz (1947), and the resulting fracture patterns were determined by Melosh (1980*b*). Normal faults with north-south orientations will occur in two areas centered on the ancient poles. Thrust faults should form in the present polar regions, again with orientations roughly parallel to great circles connecting the former and present poles. Conjugate sets of strike-slip faults with northwest and northeast trends are predicted elsewhere. These orientations are all referenced to the present coordinate system. The maximum stress difference is given by $\Delta\sigma = 4\mu f(1 + \nu)/(5 + \nu) \cdot \sin\delta$, where μ is the shear modulus, f is the hydrostatic flattening, and δ is the reorientation angle (Melosh 1980*b*). Using values of f, ν, and μ appropriate for Mars yields $\Delta\sigma \cong 4\delta$, if δ and $\Delta\sigma$ are given in degrees and MPa, respectively. This maximum stress difference occurs in the strike-slip region flanking the ancient pole, and should be sufficient to cause observable faulting for δ of the order of 10° or more.

The question of reorientation is particularly germane to Mars because of the presence of the huge Tharsis free-air gravity anomaly centered near the equator. Any excess mass placed near a planet's surface will tend to reorient the planet such that the mass moves toward the equator. Either reorientation has shifted Tharsis from an initial position at higher latitudes, or else it was formed at its present equatorial location by coincidence. Melosh (1980*b*) estimated that Tharsis could have induced up to 25° of polar motion, based on the change in the moment of inertia tensor produced by removing the excess gravitational mass from the Tharsis region. A more recent calculation by Willemann (1984) includes the effects on polar stability of an elastic lithosphere and the partial compensation of the surface load. The best estimate for the maximum permissible reorientation angle using this model is only 3 to 9°.

Moment of Inertia. One of the major unsolved questions in Mars geophysics is the value of the mean moment of inertia of the planet (see the discussion in chapter 7). The value of this parameter has far-reaching implications for the internal composition and structure of Mars (see chapters 5 and 6). Simply stated, the problem is in separating that part of the observed gravitational oblateness due to the rotational deformation of the surface and internal radial density variations from that part which is supported by nonhydrostatic stresses, either elastically or by convective forces. Recently, Kaula et al. (1989) have suggested that consideration of the stress field may be useful in placing bounds on the magnitude of this nonhydrostatic component.

They cite two arguments. The first involves the minimization of the strain energy implicit in the gravity field. In the strictest application of this criterion, the nonhydrostatic oblateness would be set to zero, resulting in a dimensionless moment of inertia Λ ($\Lambda \equiv C/MR^2$, where C is the moment of inertia about the rotation axis and M is the mass of Mars) of about 0.370. Kaula et al. (1989) suggest that a more physically plausible approach is to minimize the stresses in the Tharsis region where the maximum stresses occur. This is roughly equivalent to adjusting the nonhydrostatic component such that the long-wavelength gravity best correlates with the topography of Tharsis, which is nearly circularly symmetric, and gives a value of $\Lambda = 0.365$ (see also Kaula 1979).

The second argument involves comparing the observed fracture patterns on the planet with predicted stresses derived from using a range of assumed moments of inertia as boundary conditions on theoretical stress calculations (described above). They found that whereas the larger dimensionless moments (say, $\Lambda \geq 0.360$) produced stresses that were in general agreement with observations, smaller moments produced stress fields with some characteristics that were inconsistent with significant portions of the global fracture pattern. Increasing the nonhydrostatic component of the gravitational oblateness (corresponding to lower mean moments of inertia) superimposes upon the stress fields described above (which were computed assuming $\Lambda = 0.365$) a stress pattern analogous to that due to despinning. Thus, east-west trending normal faulting becomes more favorable in the polar regions, north-south reverse faulting is promoted in the equatorial band, and a tendency toward strike-slip faulting is enhanced at mid-latitudes. As noted above, this type of fracture pattern has not been identified on Mars. Kaula et al. (1989), using the thin shell formulation of Sleep and Phillips (1985), calculated stress magnitudes in excess of 50 MPa in a number of tectonically featureless areas for $\Lambda \leq 0.355$.

Both these lines of evidence suggest a higher value of the mean moment of inertia, near $\Lambda = 0.360$ to 0.365, as do a number of other plausible means of estimating the moment (Reasenberg 1977; Kaula 1979; Kaula et al. 1989). However, it should be noted that the stress arguments, although persuasive, are probably not sufficient in themselves for constraining the value of the moment of inertia (Bills 1989*a,b*). For example, there are several other ways to minimize the strain energy, utilizing somewhat different assumptions with arguable plausibility, that can give different results. And although the alignment of a theoretical stress field with fracture patterns is compelling evidence for a given model, the absence of observed faulting must be considered a somewhat weaker constraint. This is because of our lack of detailed understanding of the processes which lead to widespread fracturing of the lithosphere, such as the effects of lateral inhomogeneities in temperature and lithosphere thickness and of regional tectonic activity.

III. SURFACE MANIFESTATION OF STRESS

Relations Between Stress and Faulting

The most obvious indicator of stress in planetary lithospheres, and the only one detectable from orbit, is the presence of tectonic features. Tectonic features form when stresses in all or part of the lithosphere exceed its yield strength, resulting in the deformation of surface materials. This deformation is commonly manifested as slip on faults that are recognizable in images of the planet's surface. In general, three types of faults are characteristic at the surface of a planet: normal, reverse and strike-slip faults. The type of fault formed in a given situation depends on which principal component of stress is vertical (Fig. 8). Normal faults form when the maximum principal com-

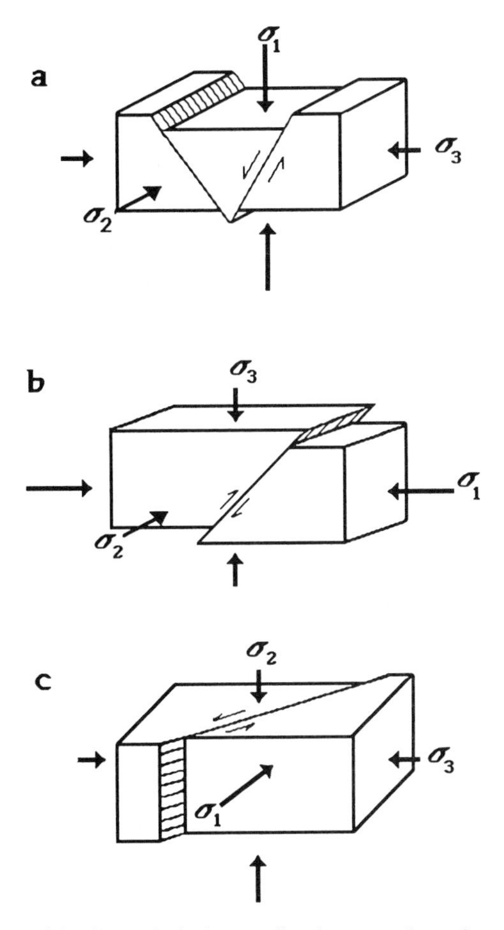

Fig. 8. Orientations of the three principal stress directions near the surface of a planet and the predicted types of shear failure. The maximum compressive stress is σ_1, the intermediate compressive stress is σ_2 and the minimum compressive stress is σ_3. (a) Normal faults and grabens are predicted when the maximum principal stress is vertical; the minimum principal stress is horizontal and perpendicular to the strike of the faults. Tension cracks also form under this stress configuration, except that the minimum compressive stress must be tensile. (b) Thrust faults are predicted when the minimum principal stress is vertical; the maximum compressive stress is horizontal and perpendicular to the strike of the fault. (c) Strike-slip faults are predicted when the intermediate principal stress is vertical.

pressive stress is vertical, causing the hanging wall of the fault to move down relative to the footwall. Tension cracks (narrow, deep vertical structures in which both sides separate without a floor) may also form under this stress system if the minimum stress is tensile. When the minimum compressive stress is vertical, reverse faults form, with the fault dipping beneath the raised block or the hanging wall of the fault moving up relative to the footwall. In both of these situations, the intermediate principal stress is parallel to the fault strike under conditions of plane strain. Strike-slip faults occur when the intermediate principal stress is vertical. In this case, movement along the fault is parallel to the strike of the fault, which is oblique to the maximum and minimum principal stress directions. Because the surface of a planet is stress-free, one of the three principal stresses at the surface is vertical and the other two are horizontal (Anderson 1951). In the absence of extreme relief, this will be approximately true at depth as well. As a result, each of the above types of faults at the surface of a planet uniquely defines the vertical principal stress and bounds the orientations of the horizontal principal stresses (Fig. 8). This allows the use of structural features observed on a planet's surface to be used to determine the orientations of the stresses responsible for their development at the time they formed. The geometry of the responsible stresses can be determined by mapping the orientations of these features on the surface, thereby providing information useful in constraining likely lithospheric deformation processes. The study of tectonic features therefore proceeds from understanding the geometry and kinematics of structures to placing constraints on the dynamics of relevant lithospheric deformation mechanisms.

Extensional Structures

Simple Grabens. The most common type of extensional tectonic feature on Mars is the simple graben. These grabens define the enormous radial fracture system around Tharsis (Wise et al. 1979*b*). Simple grabens are found in a variety of forms and settings within the Tharsis and Elysium regions of Mars. At their simplest, they are long (many hundreds of kilometers), narrow (a few km) troughs (Fig. 9a) with large spacings (tens of kilometers) between them (e.g., Memnonia, Sirenum and Icaria Fossae). In other places, the individual grabens are also generally narrow, although shorter in length (tens of kilometers), and spaced very close together (within a few km). Such terrains (e.g., Claritas and Ceraunius Fossae) have a highly disrupted appearance with ridge-groove topography; individual grabens are less well defined (Fig. 9b). In other places such as Alba Patera, grabens appear in swarms (Fig. 10). Individual structures are up to ten kilometers wide and have well-defined flat floors. The grabens typically intersect each other at small angles, with younger grabens utilizing older graben faults for short distances. In many places (e.g., Noctis Labyrinthus and Valles Marineris), graben structures are highly eroded by mass wasting or other processes, so that individual fault-bounded structures are either highly modified or difficult to define.

A simple graben is a special class of graben that is bounded by two inward dipping normal faults that have experienced equal displacements (on the order of tens to hundreds of meters) with a flat floor that is unbroken by subsidiary or antithetic faults (Fig. 8a). This simple geometry suggests that both faults are of equal importance and that both faults initiated at a common point at depth and propagated to the surface. Analysis of a variety of simple grabens indicates that the depth at which the faults initiate is typically controlled by a subsurface mechanical discontinuity where extensional stresses are concentrated, providing an explanation for the consistent graben widths and equal spacings typical between members of a set (McGill and Stromquist 1979; Golombek 1979; Golombek and McGill 1983). Dips of about 60° for the faults bounding grabens are indicated by observations of fault dip on the Moon (McGill 1971), mechanical scale-model studies, fault angle information from experimental work (see Golombek 1979, and discussion and references therein), failure criteria based on the frictional resistance to sliding on pre-existing faults applicable to the shallow crust of Mars (see Tanaka and Golombek 1989, and discussion and references therein), and direct measurements of unambiguous surface expressions of faults in the walls of troughs and valleys (Davis and Golombek 1990). For a 60° fault dip, faults bounding grabens in the western equatorial region of Mars intersect at depths of 0.5 to 5 km beneath the surface (Runyon and Golombek 1983; Tanaka and Davis 1988; Davis and Golombek 1990), with a distinct peak in frequency of intersections at about 1 km depth. This indicates failure of only the uppermost crust (*not* failure of the entire lithosphere), which effectively limits the maximum stress differences required for shear failure associated with simple grabens to tens of MPa (Fig. 4). The only type of fracture possible beneath the intersection of faults bounding grabens that does not violate their simple geometry is a tension crack, for which there is evidence in a small subset of Martian grabens (Tanaka and Golombek 1989).

Complex Grabens/Rifts. In many intensely faulted regions surrounding Tharsis, larger more complex grabens can be found. These complex grabens range in width from ~5 km to 100 km and typically have multiple border faults and deeper, multiply faulted floors (Figs. 10 and 11). The largest of this type of structure is located in Claritas Fossae (Fig. 11). It is 100 km wide, a few km deep, more than a thousand km long and resembles large continental rifts on the Earth. Many of the border faults are reactivated older faults and the floor is intensely fractured, with many tilted blocks. Other complex grabens share many of these characteristics except that they are narrower and shallower than this, yet still wider and deeper than simple grabens; obvious volcanism is absent. The inherited border faults probably propagate

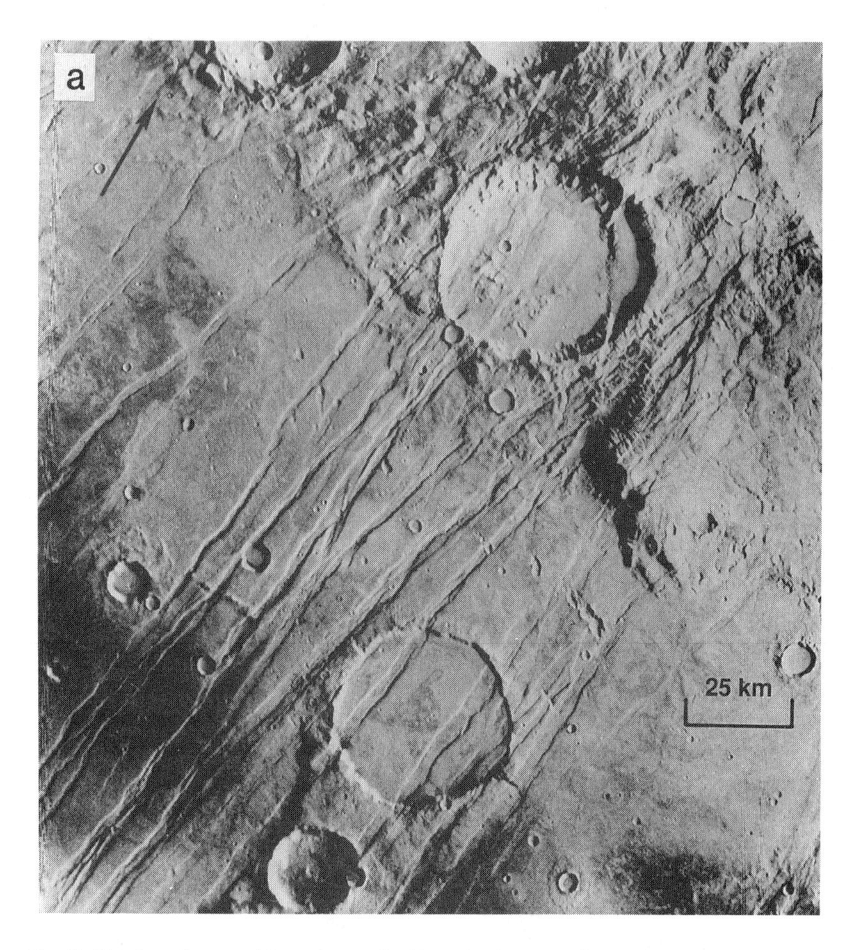

Fig. 9. Simple grabens on Mars. Arrows in this and subsequent figures denote north. (a) Fresh simple grabens in Thaumasia Fossae (97°W, 44°S). Grabens are less than 10 km wide (Viking Orbiter image 532A14). (b) Simple grabens in Ceraunius Fossae (107°W, 25°N) that have been heavily modified by subsequent erosion and deposition in the floors (Viking Orbiter image 516A29).

deeper into the crust as they become involved in deformation associated with larger, more complex structures. As a result, these large complex grabens probably mark the site of failure deeper into the lithosphere than simple grabens, with the depth of failure roughly equivalent to the width of the structure. It seems likely that the largest of these structures involves failure of the entire lithosphere, as is the case for terrestrial rifts.

Tension Cracks. Enlarged tension cracks and joints have also been identified in near-surface rocks and deep within the crust of Mars (Schumm

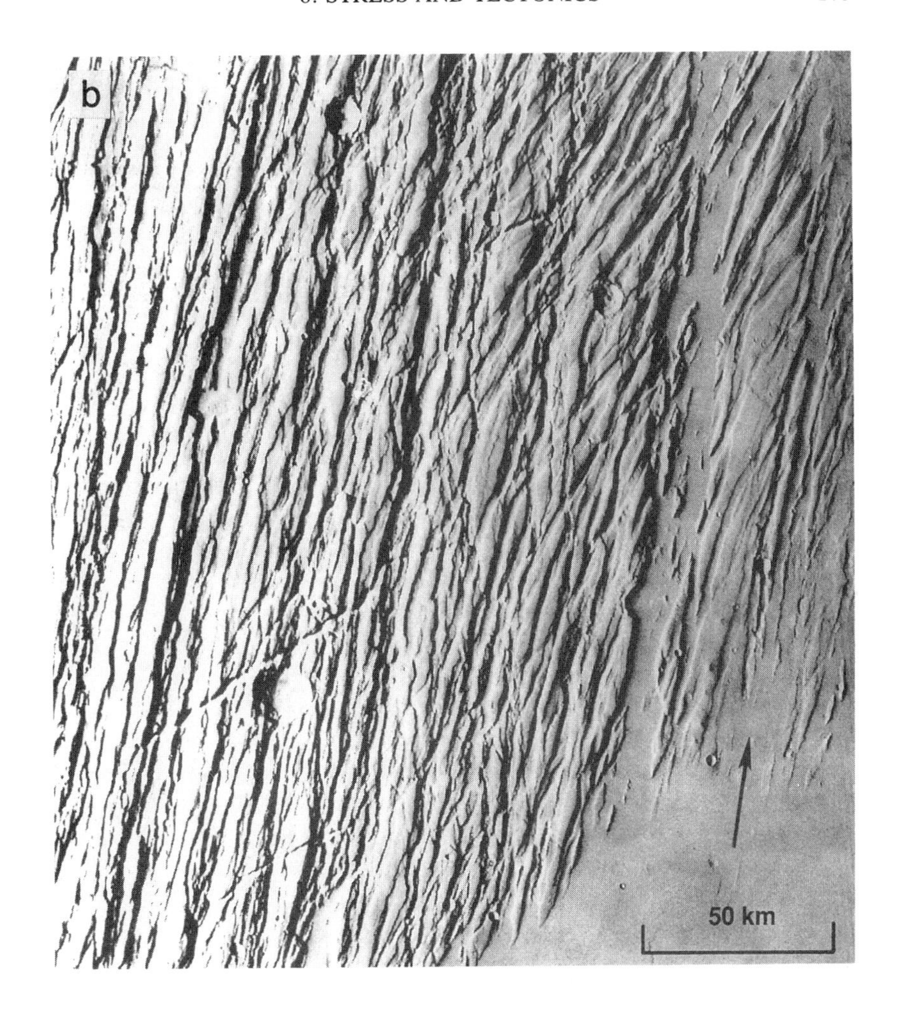

1974; Tanaka and Golombek 1989). Tension cracks are morphologically distinct from simple grabens in that they are typically narrow, deep structures without identifiable flat floors. Examples include grooves within near-surface rock units (Fig. 12a), some channel-like features, volcanic fissures (Fig. 12b) and subsurface tension cracks. Near Valles Marineris, the association of collapse pits and pit chains with simple grabens imply deep tension cracks beneath the grabens that accommodate some subsurface drainage of material (Figs. 13a and 13b). Pit chains within grabens in Tantalus Fossae (east side of Alba Patera) suggest that these grabens are also underlain by tension cracks (Fig. 10). Depending on the mechanical properties of the rocks and subsurface conditions, such as possible pore water pressure, these tension cracks

Fig. 10. Swarm of grabens southeast of Alba Patera (102°W, 40°N). Note some grabens, particularly those to the west, are wider, more complex structures (multiple interior faults), and note the two north-northeast striking grabens that have associated pits and pit chains (Mosaic MTM 40102).

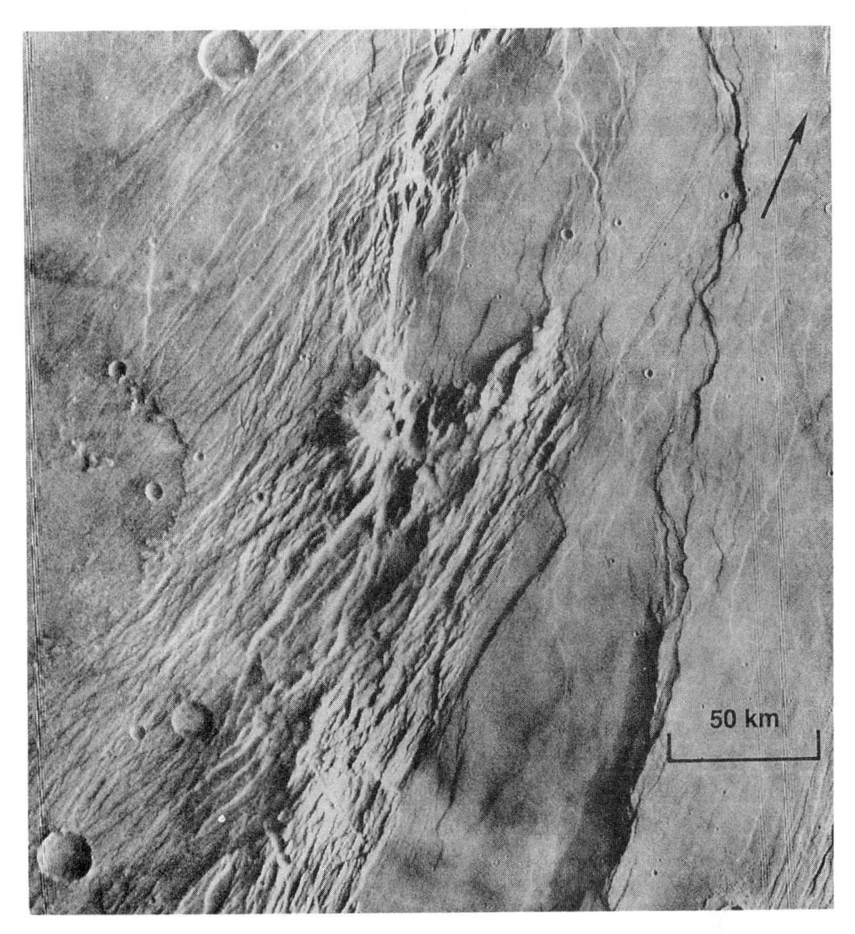

Fig. 11. Structure on Mars analogous to terrestrial rifts, located in Claritas Fossae (107°W, 21°S). The rift is ~100 km wide and a few kilometers deep. Note multiply faulted floor and sides (Viking Orbiter image 643A38).

could extend to substantial depths (tens of km), involving yield stresses in the hundreds of MPa.

Troughs. The identification of structural blocks defined by scarps and down-dropped surfaces suggests that many, if not most of the troughs making up Valles Marineris are block-faulted structures analogous to terrestrial rifts (Blasius et al. 1977; Frey 1979*a*; Schultz 1991; chapter 14). Although there has clearly been severe subsequent mass wasting and collapse, triangular facets along interior trough edges provide strong evidence for fault control of many of the troughs (Fig. 13a). There has been little detailed work on what these scarps imply for the subsurface structure of the troughs, but it seems

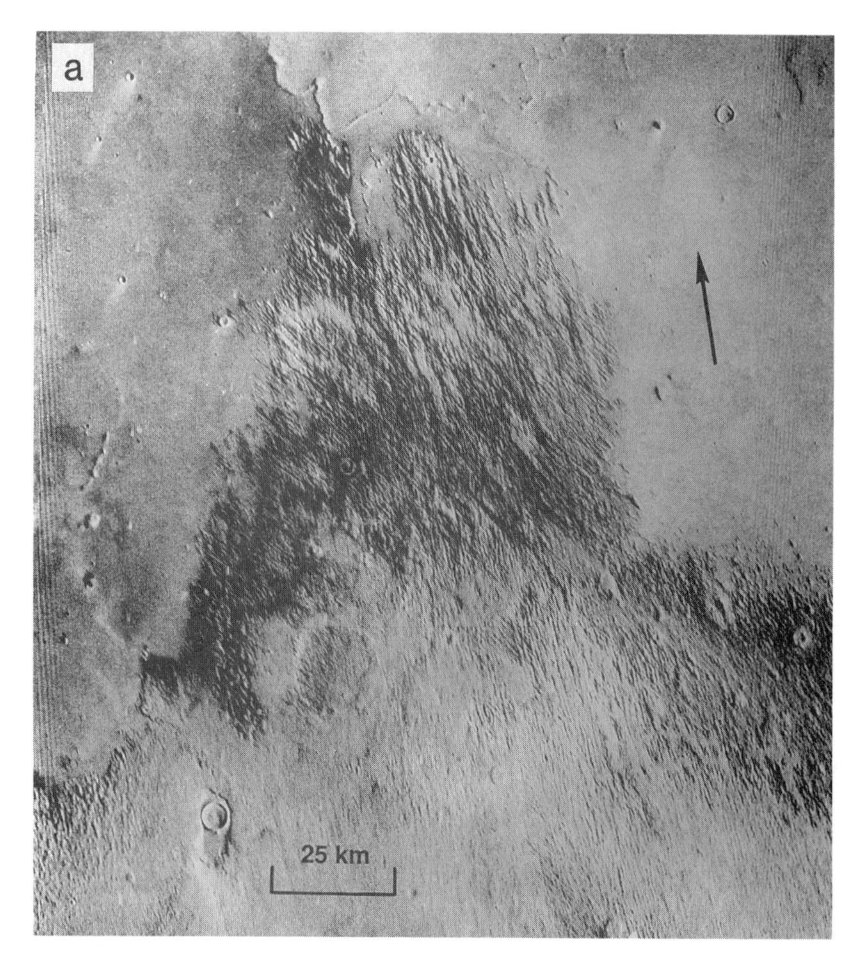

Fig. 12. Tension cracks on Mars. (a) Two sets of grooves in the weak Medusae Fossae Formation in southern Amazonis Planitia (159°W, 11°N). Grooves are interpreted as joints of a small dihedral angle (Muehlberger 1961) of about 40° that have been eroded out (Viking Orbiter image 886A11). (b) Volcanogenic fissures and pits northwest of Elysium Mons similar to dike-fed depressions on Earth (Viking Orbiter image 514A37).

likely that at least some of the troughs are modified fault-bounded valleys. If so, their widths (tens to hundreds of km wide) and sizes (individually, hundreds of km long; together, thousands of km long and many km deep; see Frey 1979*a*) suggest faulting of the entire lithosphere, again analogous to rifts on the Earth.

Polygonal Troughs. Another type of tectonic feature found on Mars is the enigmatic giant polygonal troughs of the northern plains. These troughs

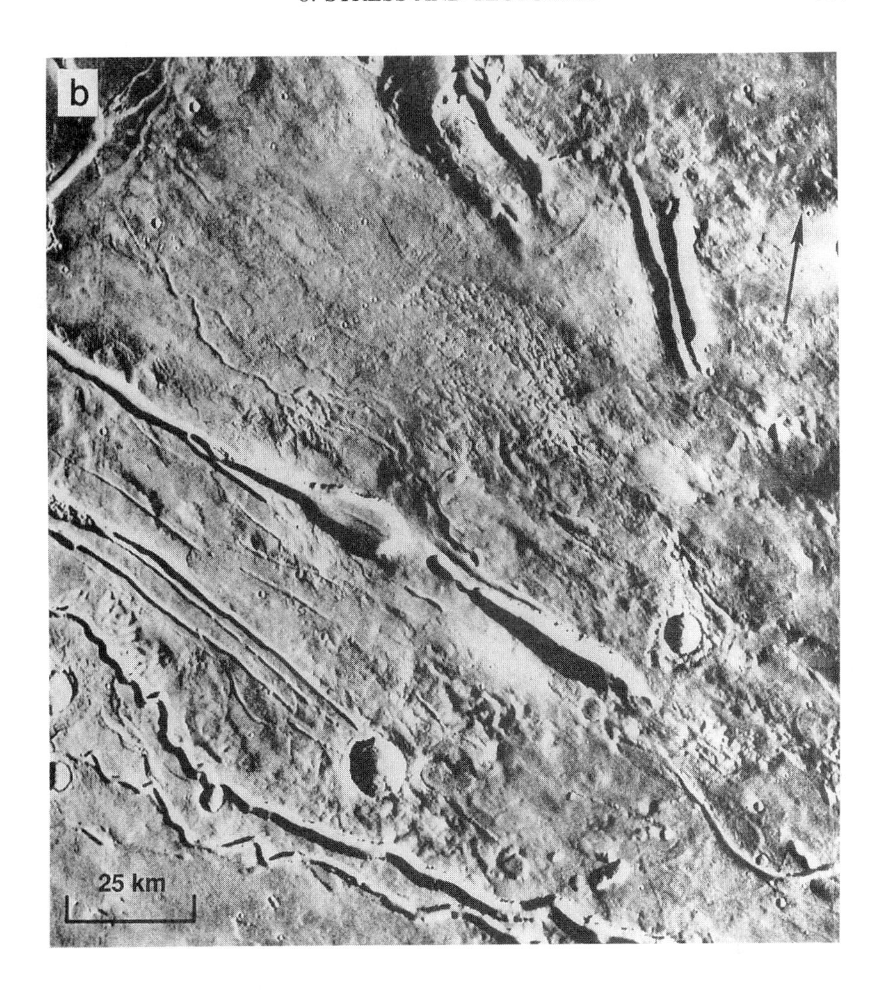

are 250 to 1000 m wide and are arranged in giant polygons about 5 km across. Initial ideas concerning their origin focused on the common Y-type or triple-junction intersections between troughs, suggestive of a tension-cracking mechanism similar to ice-wedge polygons, cooling volcanic crusts or dessication cracking of sediments. However, mechanical analyses show that the features are orders of magnitude too large to have formed in this manner (Pechmann 1980), implying that the troughs are small grabens (although different from any other grabens on Mars or the other terrestrial planets). McGill (1986) showed that the polygons were formed coevally with deposition of the polygonal terrain material, suggesting that they resulted from cooling of volcanics or shrinkage of sediments plus differential compaction over the rough surface of the underlying cratered highlands.

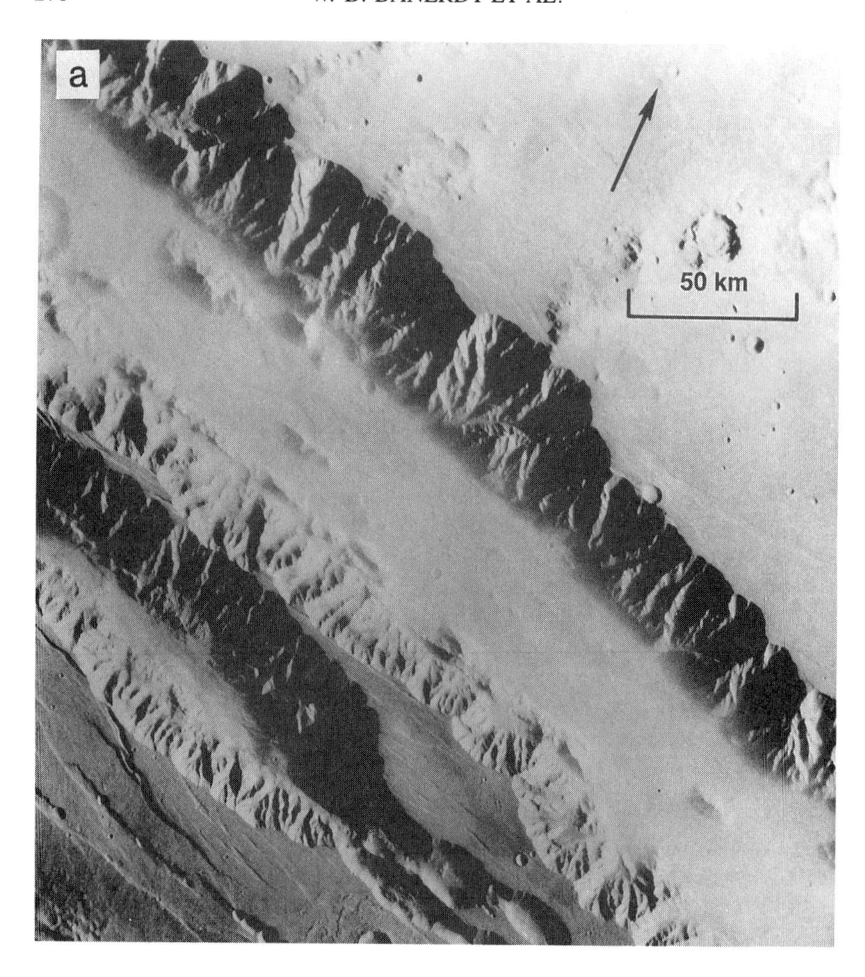

Fig. 13. Structures of Valles Marineris on Mars. (a) Southern edge of Coprates Chasma (61°W, 15°S). The coalescence of two pit chains into a trough (lower left of figure) implies a genetic relationship. The triangular facets at the base of the northern scarp of the main trough clearly indicate a bounding (normal) fault. The large crater in the trough floor indicates the surface making up the bottom of at least this part of the trough is the same surface as that making up the undeformed adjacent plains. These observations strongly argue for a rift-like structure for this part of Valles Marineris (Viking Orbiter image 610A06). (b) Part of Coprates Chasma immediately to the east (lower right) of Fig. 13a (the three pit chains match those in Fig. 13a). Note regular spacing of grabens and associated pit chains adjacent to Coprates Chasma. This regular spacing and association suggests that tension cracks underneath these grabens provided the space necessary for the subsurface drainage of material (Viking Orbiter image 610A07).

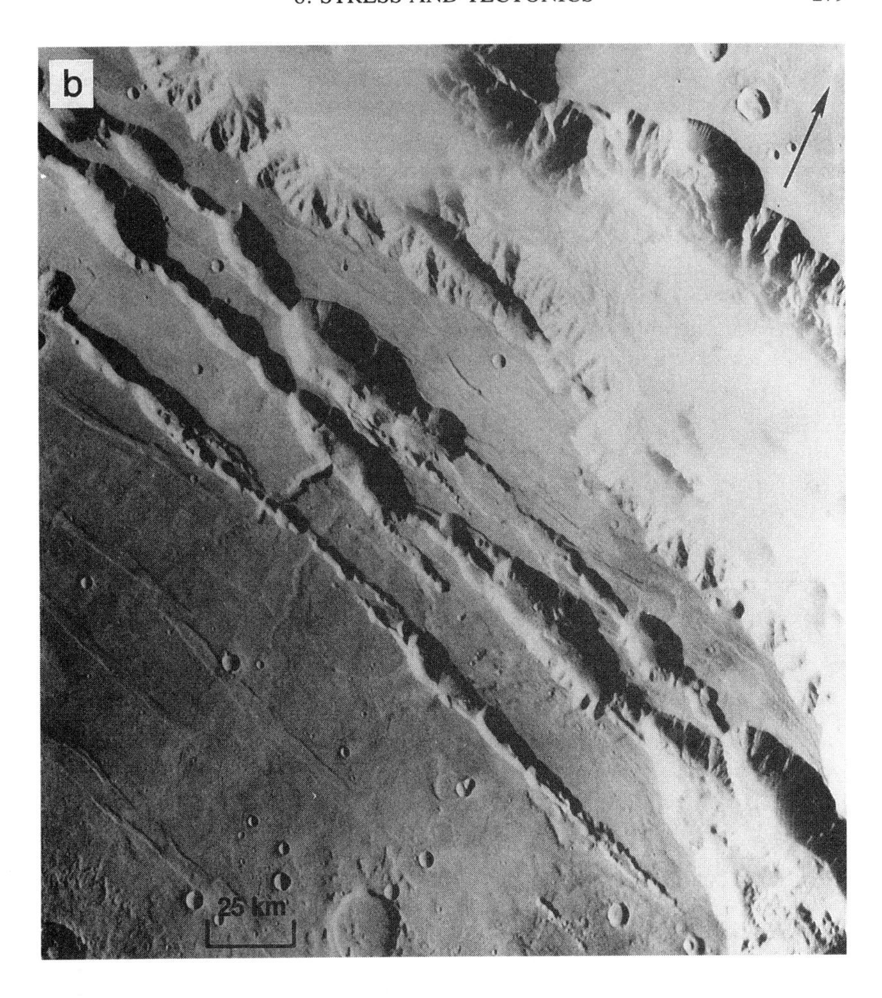

Summary. The most common extensional tectonic structure on Mars is the narrow simple graben, which provides direct evidence for failure of only the upper few km of the crust. Wider and more complex grabens are present (although fewer in number) that imply deeper failure and involvement of more of the lithosphere. The largest complex graben on Mars resembles terrestrial rifts, indicating the likely extensional failure of the entire lithosphere. Tension cracks are also present in surface materials on Mars, and collapse features associated with grabens suggest deep underlying tension cracks (tens of km deep) that could allow the observed subsurface drainage of material. Some troughs within Valles Marineris are probably highly modified fault-bounded valleys, whose size and loose analogy with rifts on Earth suggest extensional failure of the entire lithosphere. The polygonal troughs are

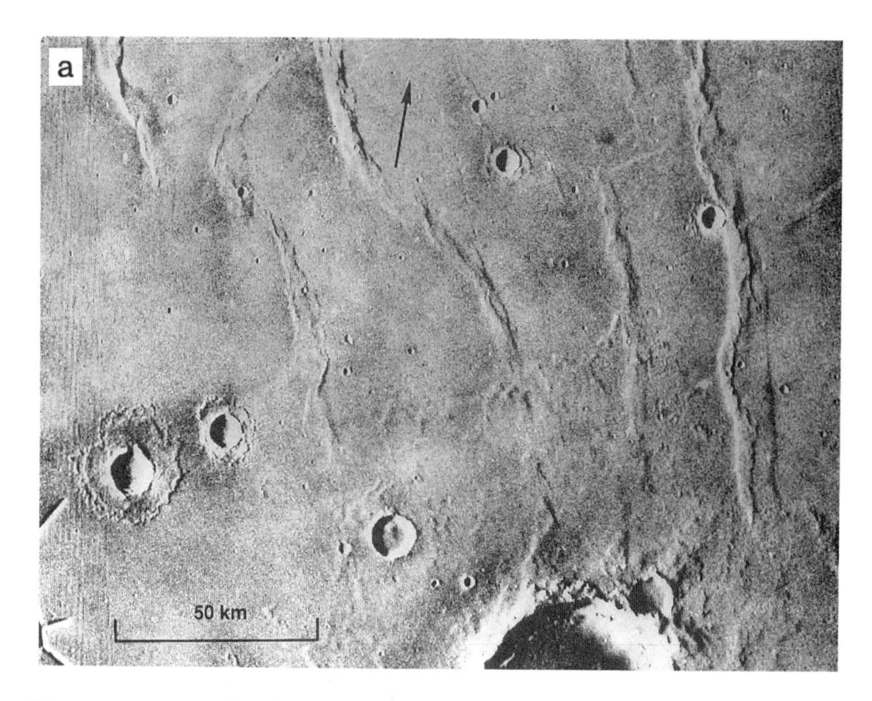

Fig. 14. Wrinkle ridges on Mars. Both examples show the basic physiographic elements that make up ridges: the broad rise, the superposed hill or ridge, and the crenulation. (a) Wrinkle ridges in northern Lunae Planum (69°W, 19°N) (Viking Orbiter image 519A23). (b) Close-up of wrinkle ridge located at 71°.5W, 21°N (Viking Orbiter image 664A16).

probably extensional in origin, although the actual mechanisms responsible for their formation are uncertain.

Compressional Structures

Wrinkle ridges are linear to arcuate asymmetric topographic highs that show a considerable degree of morphologic complexity (Fig. 14). They are found primarily on smooth plains units of Mars (Watters and Maxwell 1986; Scott and Dohm 1989), although similar features have been identified on most geologic units on Mars (Chicarro et al. 1985). Early work on wrinkle ridges defined a few basic physiographic elements from which they are typically composed (see, e.g., Strom 1972; Bryan 1973; Maxwell et al. 1975; Lucchitta 1977). These are the broad rise, a broad, gentle topographic rise (with widths of tens of kilometers and slopes of only a few degrees) that can typically only be seen in low Sun angle images; the superposed hill or ridge, a narrower hill (commonly less than ten kilometers wide) that comprises the ridge of wrinkle ridges; and the crenulation, a small wrinkle in the surface that makes up the wrinkle of wrinkle ridges. All or some of these morphologic elements may be present in a wrinkle ridge; broad rises, or arches, can be found by themselves. Subsequent classifications have broken out the broad

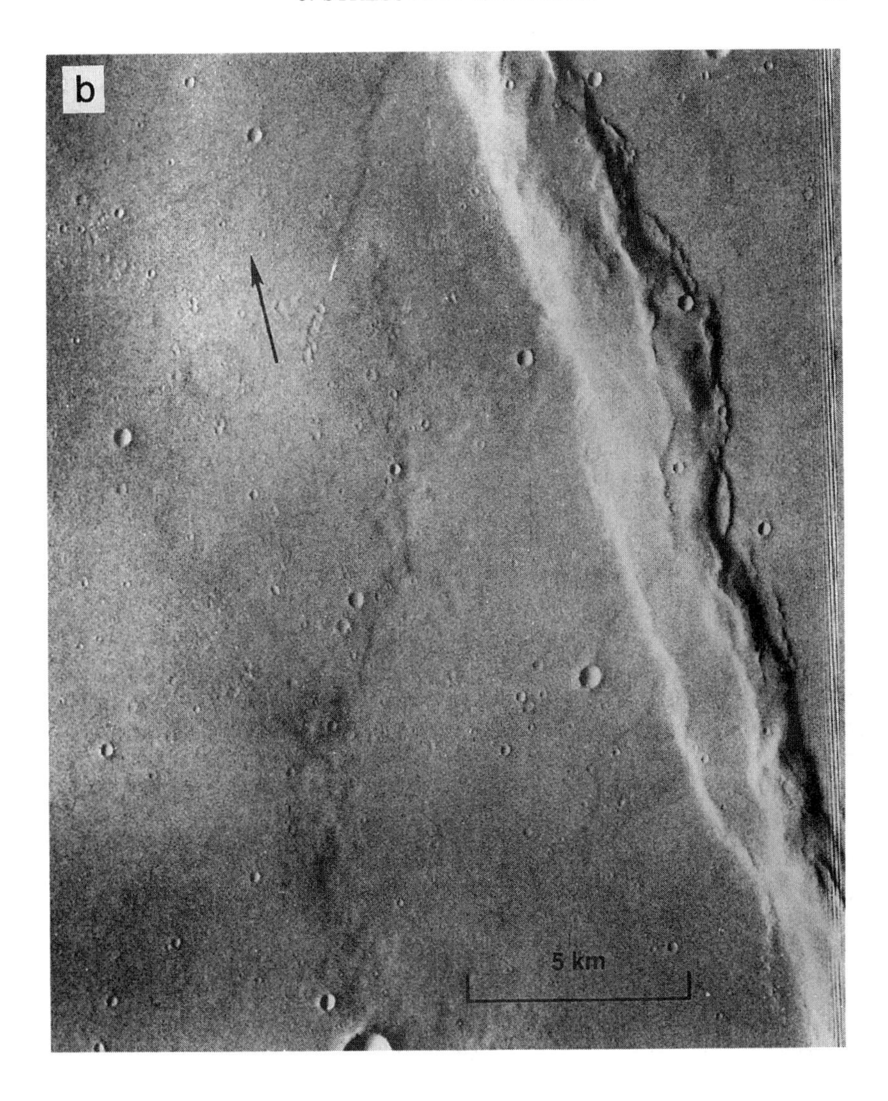

rise and different size ridges (arches and arches bounded by scarps of Lucchitta and Klockenbrink [1981]; arches and first-, second- and third-order ridges of Watters), as well as identifying certain geometric patterns among the elements (Aubele 1988).

Early hypotheses on the origin of wrinkle ridges stressed a volcanic origin in which the intrusion and extrusion of high-viscosity magma along linear-to-arcuate conduits built up to form the ridges (see, e.g., Quaide 1965; Strom 1972; Scott 1973), and this view still attracts some support (Scott 1989). More recent hypotheses have stressed the likely tectonic origin of wrinkle ridges by compressional faulting and folding (see, e.g., Ronca 1965;

Bryan 1973; Howard and Muehlberger 1973; Muehlberger 1974; Plescia and Golombek 1986; Watters 1991; Golombek et al. 1991). Combined volcanic/ tectonic hypotheses have also been suggested (Colton et al. 1972; Young et al. 1973). It should be noted that most of the work on the origin of wrinkle ridges has been done on lunar examples, where a variety of imaging and altimetric data sets have been used to constrain their formation (see, e.g., Sharpton and Head 1982,1988). Evidence for a volcanic origin include the association of some ridges with sinuous rilles (believed to be volcanic in origin), the sinuous appearance and patterns of surface features, the interpretation of the broad rise as a laccolith, and the possible flow-like modification of pre-existing craters (Quaide 1965; Strom 1972; Scott 1973). Evidence for a tectonic origin include the vertical offsets in mare surface across many ridges (Lucchitta 1976,1977; Golombek et al.), ridges extending into the highlands as fault scarps (Howard and Muehlberger 1973), ridges occurring along buried craters or along changes in mare basalt thickness where stresses would be concentrated (Maxwell et al. 1975), deformed and offset subsurface mare reflectors (Phillips and Maxwell 1978; Sharpton and Head 1982), and regional trends of wrinkle ridges (Fagin et al. 1978). Strong evidence for a tectonic origin of wrinkle ridges comes from observed offsets in pre-existing craters (Conel 1969; Sharpton and Head 1988) and the remarkable resemblance of wrinkle ridges to a variety of structures on the Earth that formed by compressional folding and faulting of surface materials (Plescia and Golombek 1986). Study of these Earth analogs has suggested two end-member kinematic models that loosely correspond to thick- and thin-skinned deformation mechanisms. In the thin-skinned model, folding and faulting is confined to the smooth plains materials and is decoupled from the rest of the lithosphere (Watters 1991). In the thick-skinned model, faulting extends through a significant portion of the lithosphere, with folding near the surface (Golombek et al. 1990; Zuber and Aist 1990). Regardless of which model is more applicable to the subsurface structure of Martian wrinkle ridges, the association of wrinkle ridges that extend over half of the planet and are concentric to Tharsis is additional evidence for a tectonic origin (Watters and Maxwell 1986).

Other compressional structures have also been suggested on Mars. Examples include a suggestion that the scarp surrounding Olympus Mons is a fault-propagation fold, formed from the weight of the shield and a suggestion that flank structures on the Martian shield volcanoes are also thrust faults (Thomas et al. 1990).

Strike-Slip Faults

Until quite recently, there has been little discussion of strike-slip faulting on Mars. At least part of the reason has been the lack of unambiguous identification of strike-slip offsets in pre-existing structures. (Note that the ubiquity of circular craters that show no evidence for lateral offset argues against strike-slip faulting being a common process on Mars; see, e.g., Golombek 1985.) Nevertheless in the past few years a number of strike-slip faults on Mars have been tentatively identified. Forsythe and Zimbelman (1988) have suggested that the Gordii Dorsum escarpment is an ancient strike-slip fault of lithospheric proportions. This escarpment is a northwest-trending structure in the equatorial transition zone southwest of Tharsis. An association of anticlinal folds, push-up structures and a variety of Reidel shears are cited as evidence for an underlying left-lateral strike-slip fault with 30 to 40 km of slip, suggesting an earlier period of possibly greater lithospheric mobility. Schultz (1989*a*) has hypothesized from the association of uplifted segments of wrinkle ridges and connecting faults that these structures are analogous to compressional steps and push-ups in terrestrial strike-slip faults. The geometric relations of the faults, their proposed slip sense, and their uplifts are consistent with the expected direction of horizontal maximum compressional stress radial to Tharsis in an area southeast of the eastern end of Valles Marineris (Fig. 15). Finally, Watters and Tuttle (1989) have tentatively identified some prospective strike-slip faults associated with wrinkle ridges on Mars based on a similar association of properly oriented strike-slip faults with the anticlinal ridges of the Columbia Plateau. More work on these and other structures will help elucidate the role of strike-slip faulting on Mars.

IV. THE TECTONIC RECORD ON MARS

In the previous sections we have shown that the lithosphere of Mars is composed of deformable media whose strength and elasticity vary as a function of composition, thickness, temperature, inhomogeneities, and character of applied stresses. The lithosphere behaves as a flat plate or spherical shell in which flexural and isostatic deformation occurs due to various loading phenomena. Resulting observed structural features include faults, tension fractures, folds, and broad depressions and rises. By studying the timing and distribution of such features and their relation to geologic history, we may decipher the tectonic history of the crust. There is no doubt that a variety of stress states have caused deformation of the Martian lithosphere. The causes of these stress states may have included global stresses provided by expansion or contraction due to planetary differentiation and by changes of spin axis orientation, regional or local stresses due to mantle thermal anomalies and lithospheric loading, and exogenic stresses produced by impacts. In this section, we review major stress events on Mars and their imprint on the planet's surface; a summary of these processes and the times at which they caused substantial deformation is outlined in Fig. 16. Figure 17 shows the distribution of faults, grabens and wrinkle ridges observed on Mars. This figure can be compared directly with the 1:15 M geologic map (USGS map I-1802) and Figs. 5–7. (See also Color Plates 5 and 6.)

Fig. 15. Suggested strike-slip faults on Mars in plains units south of the eastern end of Valles Marineris (54°W, 22°S). Left-lateral strike-slip faults have been proposed to link the bounded push-up structures (stippled dark) in the schematic drawing (figure adapted from Schultz 1989*a*) (Viking Orbiter image 610A44).

Global Processes

Global processes such as internal differentiation, accretion, bombardment and changes in orbital and rotational dynamics may result in lithospheric stresses sufficient to produce tectonic features. Differentiation and bombardment processes may focus stress and heat at particular areas of the crust, whereas dynamical changes would tend to produce global-scale tectonic patterns.

Planetary Differentiation and Thermal Evolution. Although the extent of core and mantle differentiation of Mars is poorly known (see chapters 5 and 6), the geologic history of the planet constructed from Mariner 9 and Viking images (see chapter 11) has the formation of the crust occurring prior

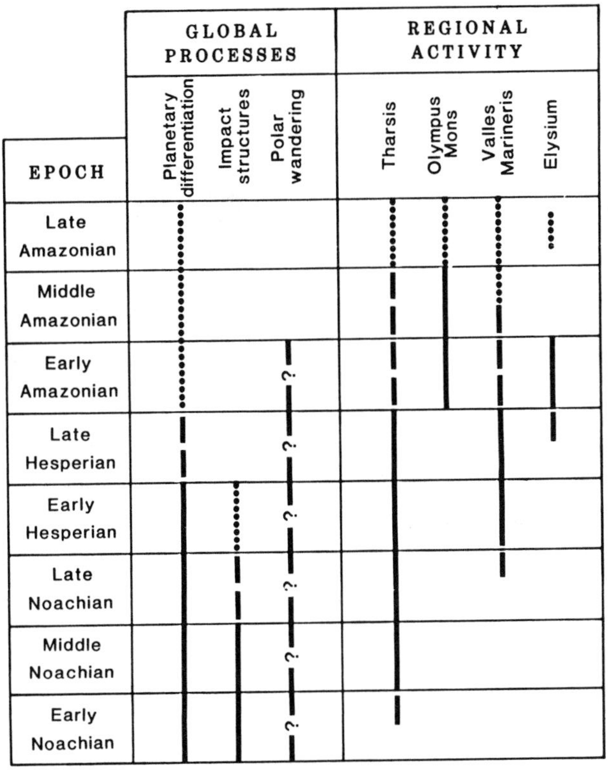

Fig. 16. Activity of global processes and deformation in tectonic regions through time on Mars. Solid lines show intense activity; dashed lines show moderate activity; dotted lines denote lingering activity; queried line indicates uncertain activity.

to the end of heavy bombardment, followed by extended volcanic and tectonic activity. Based on assumed initial thermal conditions and heat sources, Toksöz and Hsui (1978) proposed the following stages in Martian evolution:

1. 0 to 1 Gyr (following planet formation): core formation and crust differentiation;
2. 1 to 3 Gyr: heating, expansion, and mantle differentiation;
3. 3 to 4 Gyr: lithosphere thickens and partial melt zone deepens;
4. 4 Gyr to present: cooling.

This sequence is generally consistent with a tectonic history of the planet characterized by the extended activity of extensional tectonism. The ubiquity of extensional features, particularly in the Tharsis region (see below), has generally been cited as support for such a thermal history. However, the recent mapping of nearly ubiquitous, globally distributed ridges with a presumably compressional origin (Chicarro et al. 1985), along with the observation that global extension does not appear to be necessary for the formation of the

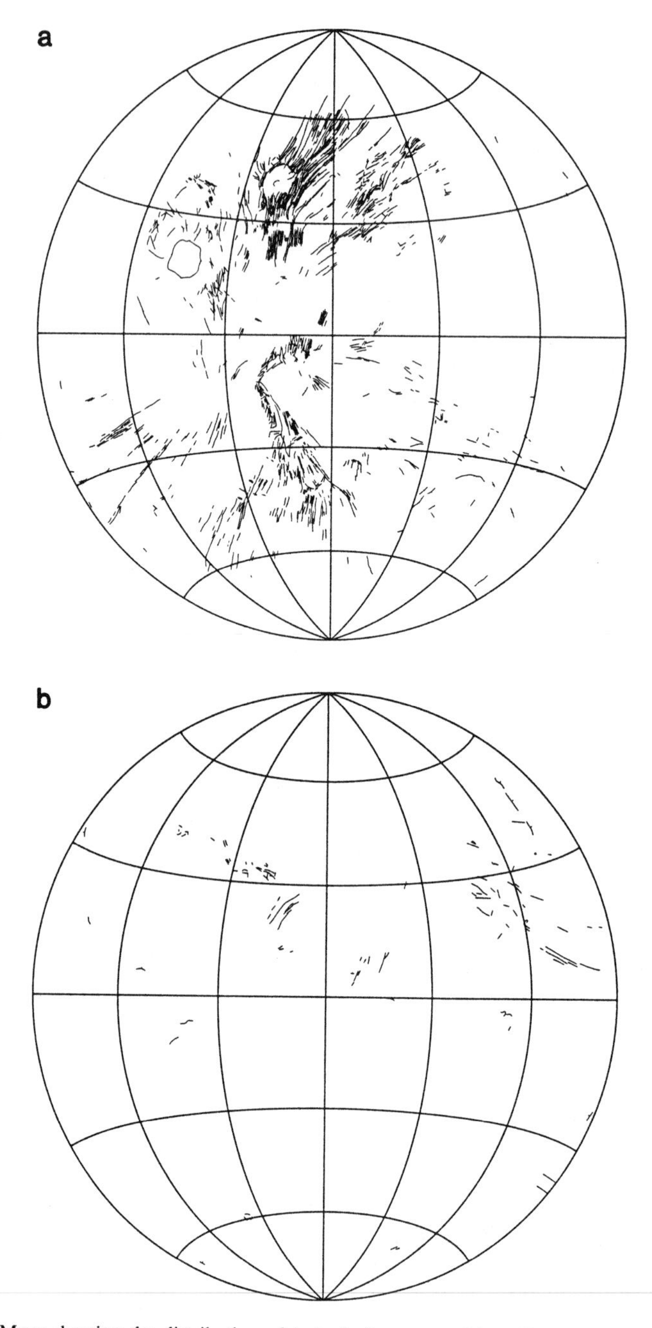

Fig. 17. Maps showing the distribution of tectonic features on Mars. Data are taken from the three 1:15 M-scale geologic maps of Mars by Scott and Tanaka (1986), Greeley and Guest (1987) and Tanaka and Scott (1987). Lambert equal area projection; the grid interval is 30°: (a) grabens (plain lines) and fault scarps (hachured lines) in the western hemisphere (longitudes 0° to 180°); (b) grabens and fault scarps in the eastern hemisphere (longitudes 180° to 360°); (c) wrinkle ridges in the western hemisphere; and (d) wrinkle ridges in the eastern hemisphere.

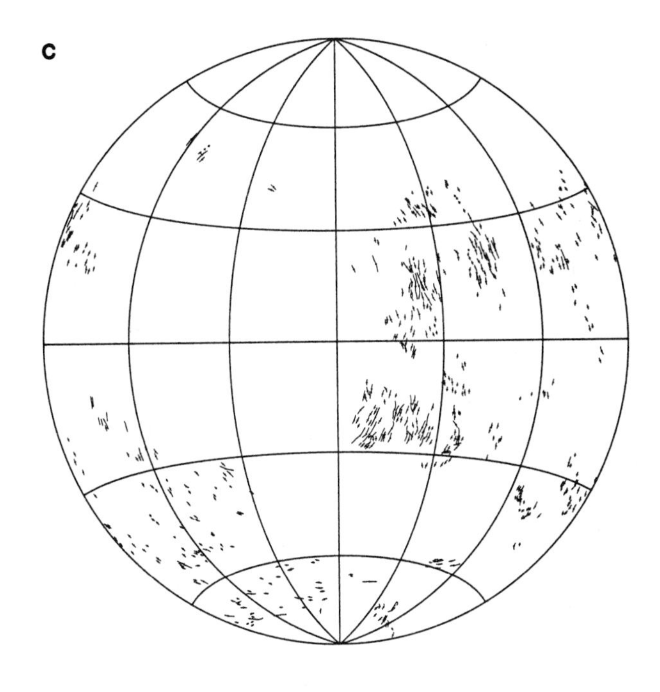
c

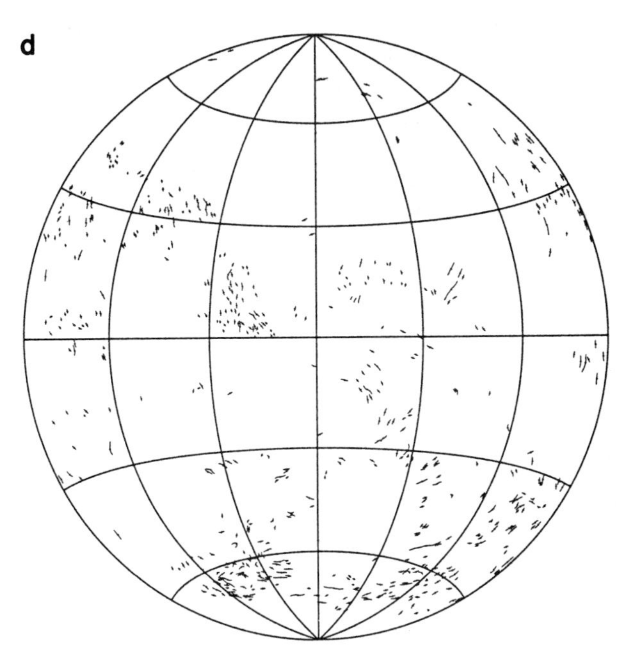
d

radial grabens around Tharsis (Wise et al. 1979*b*; Banerdt et al. 1982; Sleep and Phillips 1985), allows for the possibility that even the early stages of the thermal history of Mars were dominated by cooling (see chapter 5). Global contraction may result from cooling of the planet, and wrinkle ridges not accounted for by other tectonic forces may arise through this mechanism (Maxwell 1982).

The profound asymmetry in Martian topography expressed by the dichotomy between the northern lowlands and southern highlands may be related to early differentiation. The lowlands represent one third of the entire surface of the planet and lie, on average, ~3 km below the highlands. The highland-lowland boundary forms an irregular circle scalloped by local basins (e.g., Chryse and Isidis Planitia). The boundary zone in places appears to have retreated several 100 km or is overlain by volcanic rocks; only minor structural development occurred along it. Although largely resurfaced during Hesperian and Amazonian time, ancient impact craters occur in some parts of the northern plains, such as in northeastern Elysium Planitia.

Wise et al. (1979*a*) proposed that a single convection cell operated in the Martian interior during planetary differentiation, resulting in subcrustal erosion and foundering to produce the northern lowlands. Recent numerical calculations of convection in a spherical shell allow the possibility of a first-order convection system occurring in conjunction with core formation (Schubert et al. 1990). Crater counts of the oldest exposed plains surfaces that embay the highland boundary show an Early Hesperian age, similar to ages of fracturing in highland rocks (Wise et al. 1979*a;* Tanaka 1986; Wichman and Schultz 1985; Frey et al. 1988; Maxwell and McGill 1988). McGill (1989*a,b*) argued that the boundary has not migrated significantly and was formed by the end of the Noachian, coincident with a period of plains volcanism and sill intrusion on Mars (Wilhelms and Baldwin 1989*a*).

Impact Structures. Many large impact basins on Mars are readily observed or are inferred based on various landforms (see chapter 12). Large impacts produce evenly spaced ring faults along which the base of fault blocks commonly are rotated inward toward the basin center (see, e.g., Schultz et al. 1982). Such structures form zones of weakness along which renewed faulting, volcanism and accelerated erosion may take place. The impact bombardment of the stable lithosphere of Mars no doubt has produced extensive fracturing of the Martian basement. This fracturing probably extends to depths of 10 to 15 km (Clifford 1987*a;* MacKinnon and Tanaka 1989).

Impact basins are noted for their deflection of regional stress patterns and therefore exert control on the occurrence and orientations of structures like wrinkle ridges (Mouginis-Mark et al. 1980; Gifford 1981; Schultz et al. 1982; Chicarro et al. 1985). Wrinkle ridges in particular are deflected along basin rings and their formation appears inhibited just inside the rings; such relations are inferred for the proposed rings of Chryse basin (Chicarro et al. 1985).

An early mega-impact origin for the northern lowlands was proposed by Wilhelms and Squyres (1984). Their supporting evidence includes massifs along and ridges concentric with the proposed rim. This event would need to have been during the Early Noachian because some areas within the lowlands are intensely cratered (Tanaka 1986), and the proposed rim would largely be relaxed. Another proposal by Frey and Schultz (1988) likens the northern lowlands to the extensive mare on the nearside of the Moon. Based on statistical arguments of Martian basin populations, these authors suggest that the number of "missing" (unidentified) large impact scars on Mars is better accounted for if the lowlands were made up of multiple, overlapping basins rather than resulting from a single mega-impact. The origin of the northern lowlands is likely to remain a topic of debate until more definitive data are acquired (McGill and Squyres 1991).

Planetary Despinning. Despinning of Mars probably has not been an important process because of the modest levels of tidal force exerted on it. If it had been significant in the Martian past, midlatitude systems of conjugate strike-slip faults and equatorial thrust faults should have been produced as discussed above; such features have not been recognized (Shultz 1985).

Polar Wandering. It has been proposed that some Martian tectonic features such as Valles Marineris (McAdoo and Burns 1975) and certain fields of ridges and grabens (Schultz and Lutz 1988) were produced by polar wandering. Presumably the spin axis shifted because of widespread filling of basins by lavas and by the later formation of Tharsis (Melosh 1980*b*; Schultz and Lutz-Garihan 1981; Schultz and Lutz 1988). Melosh (1980*b*) and Grimm and Solomon (1986), however, calculated the stress patterns and magnitudes expected from polar wandering and found little support in the geologic record for their predictions. Nevertheless, Schultz and Lutz (1988) proposed that 120° of polar wandering has occurred in a series of short-lived shifts throughout much of Martian geologic history. They cite deposits similar to those found in the polar regions that may record paleopole sites (although other explanations for the origins of some of these deposits have been offered; see Scott and Tanaka 1982; Greeley and Guest 1987) and argue that the tectonic signature produced by polar wandering may have been obscured by the presence of other stress fields and features.

Tectonic Regions

The following tectonic regions developed primarily after the end of heavy bombardment on Mars; thus the record of deformation associated with them is fairly well preserved. These regions have provided the most fertile areas for tectonic mapping and analysis on Mars.

Tharsis. Dominating the western hemisphere of Mars, the Tharsis Rise (an elongate area centered on Syria Planum, ascending as much as 8 to 10 km above the datum; U.S.G.S 1989) is intensely fractured by long, narrow grabens that extend radially hundreds of km beyond the rise and is ringed by mostly concentric wrinkle ridges that formed over 2000 km from the center of the rise (Fig. 17). The Tharsis region records a long and extensive history of volcanism and tectonism (see chapter 11). Because of the complexity of the region and the varied scopes and approaches in the literature, we will first describe broad-scale studies and then look into local or more specific works.

What caused the formation of this unique planetary feature? Hartmann (1973*c*) and Carr (1974*b*) postulated a mantle plume developed in an area of convective upwelling to produce uplift and volcanism. This theory has recently found some support in the convection calculations of Schubert et al. (1990), who noted a tendency under Martian conditions for a convective system to take the form of a limited number of cells, or plumes. Wise et al. (1979*a,b*) proposed that material eroded from the base of the early crust in the northern plains was redeposited and intruded into the crust below the site of Tharsis, resulting in isostatic uplift. They further suggested that the descending mantle flow induced by core formation consolidated beneath Tharsis, concentrating the eroded subcrustal material and providing a localized heat source (Wise et al. 1979*a*). Alternatively, Schultz (1984) drew basin rings surrounding Syria Planum that aligned with the Tharsis Montes, suggesting that volcanism was concentrated along pre-existing basin structures. Plescia and Saunders (1980) have argued that it is uplifted and has only a thin veneer of volcanics, whereas Solomon and Head (1982) suggested that Tharsis formed and grew to its present size by an extended period of regional volcanism on a massive scale.

The broad tectonic history of the Tharsis region was first examined carefully by Wise et al. (1979*b*). Based on crater-count ages, they deduced the following fault sequence:

1. Earliest faulting followed development of cratered highlands and northern plains.
2. Following Lunae Planum plains-material deposition and accompanied by an initial rapid topographic rise, wrinkle ridges generally concentric to Tharsis formed, main radial faults to Tharsis developed, Claritas Fossae deformed, and Valles Marineris began opening.
3. Intense faulting in Alba Patera region occurred.

Scott and Tanaka (1980) indicated that most Tharsis faulting was associated with three major centers of uplift: Syria Planum, Alba Patera and Tharsis Montes (in chronologic but overlapping order). These also were among the most volcanically active areas within the Tharsis region. Based on an interpretation of cross-cutting relationships, most wrinkle ridges surrounding Tharsis apparently formed coeval or prior to local Tharsis radial faults and Valles Marineris (Watters and Maxwell 1983). Global-scale geologic mapping of Mars based on Viking images led to a more comprehensive synthesis of the tectonic history of the Tharsis region (Plescia and Saunders 1982; Scott and Tanaka 1986). The broad history now developed includes:

1. Pre-Tharsis faulting in the Thaumasia region and at Acheron Fossae;
2. Older main Tharsis (Syria-centered) radial faulting;
3. Wrinkle ridges concentric to the Tharsis swell;
4. Younger main Tharsis (Pavonis-centered) radial faulting;
5. Faulting at Alba Patera, Tempe Terra and Tharsis Montes;
6. Radial and concentric grabens surrounding the Tharsis Montes (Fig. 17).

The style and orientation of structures have been used to document paleostress histories in the Tharsis region. The most notable structures are the radially oriented grabens surrounding Tharsis that show dominant north and northeast trends. Carr (1974*b*) attributed the fracturing to one main uplift event in Syria Planum, and variations in fracture density to the irregular shape of the dome itself; subsequent work has shown, however, that uplift cannot account for the radial faults around Tharsis (see Fig. 6). An older, less pronounced fracture center in the Thaumasia region was proposed by Frey (1979*b*). Plescia and Saunders (1982) divided the development of main radial fracture systems into three episodes having centers in Syria Planum and near Pavonis Mons. Similarly, ridge systems in the western hemisphere of Mars are concentric about various centers, including Tharsis and Syria Planum (Wise et al. 1979*b*; Chicarro et al. 1985). Joint systems apparent in eroded rocks in southern Amazonis Planitia and in Kasei Valles also appear to reflect Tharsis-centered tectonism (Tanaka and Golombek 1989). A recent, more precise analysis of fault centers incorporating geometrical corrections confirmed these centers (Golombek 1989; Watters et al. 1990).

Complex faulting is observed in the vicinity of the Syria rise. Analysis of fault trends led Masson (1977) to conclude that domal uplift produced the intersecting network of grabens and troughs at Noctis Labyrinthus. Masursky et al. (1978) concluded that east-west extension formed grabens and normal faults at Claritas Fossae; then northeast compression south of Valles Marineris followed, forming wrinkle ridges. Masson (1980) generally supported this history. Recently, Tanaka and Davis (1988) defined six stages of deformation in the Syria Planum area, including: (1) east-west faulting; (2) radial faulting centered on Syria Planum; (3) faults oriented tangentially to the Syria rise center; (4) concentric grabens about Syria Planum; (5) troughs and wide grabens of Noctis Labyrinthus; and (6) faults in Noctis Fossae that may be due to Tharsis-centered activity, as well as normal faults in Claritas Fossae. Orientations of faults and grabens at Alba Patera (Rotto and Tanaka 1989) and Tempe Terra (Scott and Dohm 1989) also indicate both local and regional fault patterns. Ages of specific fracture systems are given by Tanaka (1986). Local, concentric faulting about the volcanic centers has been attributed to loading (Comer et al. 1985) or magma withdrawal (Tanaka and Davis 1988).

A comparison of the major radial graben systems of regional extent defined by Wise et al. (1979*a*) and Plescia and Saunders (1982) with the stress models produced by Banerdt et al. (1982) and Sleep and Phillips (1985) clearly shows that more than one mechanism of lithospheric deformation is required to produce the enormous extent of concentric extensional stresses required by the radial graben system. In particular, concentric extensional stresses on top of the Tharsis Rise are only produced by isostatic models, which show compressional stresses off of the rise (Fig. 7). Conversely, only flexural loading produces concentric extensional stresses off of the rise (Fig. 5). As a result, isostatic stresses are required to produce the observed radial grabens on the top of Tharsis and flexural loading stresses are required to produce the radial grabens on the flanks of the rise. The key nature of the relative timing of these stress states was originally recognized by Banerdt et al. (1982), and a number of attempts have been made at determining the sequence of these two systems based on the timing of graben formation around Tharsis (Golombek and Phillips 1983; Sleep and Phillips 1985; Banerdt and Golombek 1989). However, detailed mapping has revealed the puzzling result that sets of radial fractures that extend from the central regions of Tharsis to its periphery appear to have formed simultaneously, or at least closely interspersed in time. It is not clear how to form such extensive radial fractures in a single event, when the stress models seem to require two distinct events. Resolution of this apparent paradox is required before self-consistent scenarios can be constructed that attempt to link the stress systems with the evolution of the lithosphere and asthenosphere in the Tharsis region.

A closely related problem is the present state of Tharsis, which is, of course, one end-member of its evolution. Here the tectonic record is of limited help, as the most recent faulting is primarily local in character (Fig. 16). The strongest constraints come from gravity measurements which indicate a very large free-air anomaly (see chapter 7). Simple isostatic compensation is virtually ruled out, as the apparent depths of compensation for the longest wavelengths range from 550 to 1100 km (Phillips and Saunders 1975), well below depths at which static stresses can be maintained (Fig. 4). Direct support by mantle flow is generally considered unlikely due to the long time required for such a massive plume to be maintained in a single location (but see Kiefer and Hager 1989). Isostatic models with more reasonable compensation depths can be constructed by effectively "burying" part of the load (Sleep and Phillips 1979) by thinning the crust or requiring its density to be higher beneath Tharsis. However, in either case, this requires some mechanism to remove large amounts of original crustal material in order for it to be replaced by mantle or denser crust, because simple displacement of the lighter material will not allow the gravity constraint to be satisfied. Perhaps the most likely extant state is one in which Tharsis is partially supported by the elastic strength of the lithosphere, with the remainder of its support derived from the buoyancy of a crustal root at a depth of around 50 to 100 km. This implies an excess crustal thickness of around 25 to 30 km, and suggests a scenario in which a transient mode of support (e.g., a thermally or chemically induced density deficit in the upper mantle, or a convective plume) early in its history has been removed, leaving a superisostatic load on an increasingly cool, thick lithosphere.

Olympus Mons. Although generally considered part of the Tharsis region, Olympus Mons and its related deposits show very little Tharsis-centered deformation. Unlike the Tharsis Montes, Olympus Mons has no concentric fractures, suggesting that it is underlain by a thick lithosphere (Thurber and Toksöz 1978; Comer et al. 1985). Morris (1982), however, recognized from topographic data that aureole deposits of Olympus Mons are depressed near the shield, indicating that broad subsidence has occurred. Furthermore, Morris (1981*a*) postulated that the abrupt basal scarp of Olympus Mons (locally several kilometers high) was formed by thrust faulting at the base of the mountain, perhaps resulting in fault-propagation folding. Among various hypotheses for the origin of the Olympus Mons aureoles, those advocating gravity spreading describe the aureoles as tectonic thrust sheets (Francis and Wadge 1983; Tanaka 1985).

Valles Marineris. Some uncertainty exists regarding the relation between Tharsis and Valles Marineris, although it appears that stress systems associated with Tharsis were dominant at least in controlling the orientation of the canyon (Banerdt et al. 1982). The dominant east-northeast trend of the canyons and associated grabens of Valles Marineris are generally radial to Tharsis (Plescia and Saunders 1982), but this relation does not explain the great depth of the canyons. Localized crustal rifting (Sharp 1973*c*; Frey 1979*a*) and deep-seated tension fracturing (Tanaka and Golombek 1989) have been proposed to account for at least part of the development of the canyons. Some uplift centered on the central canyons of Valles Marineris is indicated by topographic data (Witbeck et al. 1991); however, the presence of this high has recently been questioned (Schultz 1991). Further details on the structure and history of Valles Marineris can be found in chapter 14.

Elysium. The Elysium Rise is a broad dome, roughly 2000 km across, that may have developed due to a relatively thin lithosphere where stress concentration and fracturing would more likely occur (perhaps similar to the

Tharsis Rise; Solomon and Head 1982). Regional northwest-trending grabens cut through young volcanic rocks on Elysium and local concentric faults formed on the flanks of Elysium Mons. Comer et al. (1985) and Hall et al. (1986) concluded that the concentric grabens were produced by volcanic loading of the lithosphere. Additionally, Hall et al. (1986) suggested that (1) regional-scale volcanic loading by the lava plains did not lead to any observed tectonic deformation; (2) the northwest-trending grabens may have resulted from flexural uplift of the lithosphere at the Elysium Rise; and (3) Tharsis flexural loading stresses contributed to the formation of Cerberus Rupes, a system of long, curvilinear fractures that trend southeast from the Elysium Rise and swerve east toward the Tharsis Rise. Thermal and mechanical arguments favor the hypothesis that flexural uplift preceded or was contemporaneous with the emplacement of most of the volcanic units (Hall et al. 1986). Gravity modeling suggests that Elysium is currently isostatically compensated at relatively shallow depths (Janle and Ropers 1983).

V. DISCUSSION

Given the present lack of surface-based geophysical and petrological measurements on Mars (excepting, perhaps, SNC meteorites), the principal observations that relate directly (albeit nonuniquely) to the structure and state of its interior are gravity and topography. In the most general terms, the gravity field can be interpreted in terms of deviations from a state of minimum potential energy relative to a spherically stratified planet. These deviations imply an interior state of stress acting in such a way as to oppose the maintenance of the lateral density variations that give rise to the gravity anomalies. Topography provides a direct measure of the surface contribution to the gravity field and the radial load acting on the lithosphere. Thus, the gravity and topography are direct manifestations of the tectonic driving forces present today in the outer portions of the planet.

There are two major problems in using these data sets to infer the interior structure and history of a planet. First, the interpretation of gravity anomalies in terms of interior density variations is intrinsically ambiguous. Second, we can only obtain the gravity and topography corresponding to the current state of the planet (although it may be possible to construct to a limited degree the paleotopography using slopes inferred from channel flow directions). Thus, it is necessary to invoke additional constraints on geophysical models in order to draw meaningful conclusions.

We have seen that the state of stress (magnitude and orientation) at the surface of the planet, as reflected in tectonic features, is particularly useful as such a constraint. Whereas interpretations of gravity and topography alone primarily address vertical forces and movements, faulting and other manifestations of lithospheric failure reflect horizontal strains, allowing additional bounds to be placed on deterministic models. Tectonic features also possess a "memory," wherein it is possible (at least in principle) to decipher the sequence of events that produced them. Thus, the stress field history inferred from the geologic record not only facilitates the determination of the present lithospheric structure and processes, but also give insight into their evolution through time.

Another line of evidence, which has been used to understand the development of major topographic features such as Tharsis, comes from igneous petrology. Finnerty et al. (1988) developed a quantitative petrologic model for the formation of the Tharsis plateau, based on laboratory observations of partitioning of melt products during the partial melting of geologic materials which are likely to predominate in Mars' mantle, for which the gravity and topography boundary conditions are satisfied. Partial melting of the mantle beneath Tharsis results in a local decrease in the mantle density due to the preferential partitioning of iron into the basaltic liquid. The basalt is erupted onto the surface or intruded at shallow levels in the crust, where it solidifies into low-pressure mineral assemblages. The net volume of the crust-mantle column is increased, resulting in a prominent topographic rise with no net increase in mass. This distribution of densities with depth is qualitatively consistent with the isostatic models of Sleep and Phillips (1979,1985). This investigation was extended by Phillips et al. (1989), who looked at the constraints imposed by petrologic and geologic observations in more detail. They derived expressions for total elevation and structural uplift in a magmatic system which may suffer lateral mass loss, and considered the difference between mass balance and isostatic (or force) balance. The structural uplift suggested by the circumferential grabens found in Claritas Fossae (Tanaka and Davis 1988) was interpreted to require permanent uplift of an ancient surface that was resisted flexurally by an elastic lithosphere. In that case, most of the magma generated in the mantle must end up as intrusive bodies in the crust and upper mantle, even though most of the support for the uplift is provided by the residuum in the source region. However, this model suffers from an apparent inability to satisfy gravity constraints in a straightforward manner.

As should be clear from the preceding sections, the application of stress modeling techniques to Mars has resulted in an improvement in our understanding of the physical state and history of its outermost layers. The thickness of the lithosphere has been estimated in several areas, indicating substantial spatial and temporal variability. It has been shown that the evolution of the Tharsis Rise was a complex affair requiring a combination of broad-scale mechanisms of support in order to explain the overall faulting patterns, with much local activity superimposed. Hypotheses regarding the evolution of its spin vector have been tested, with little evidence found for major changes in the past. However, many questions remain to be resolved, including lithospheric properties in areas far removed from the major tectonic regions, the actual sequence of tectonic events that formed Tharsis, the mechanisms responsible for Valles Marineris, and the details of local tectonic episodes.

Future Directions

Many avenues are open for further investigation of Mars via stress modeling. In general, further progress in understanding tectonic processes on Mars will require advances in both data quality and modeling sophistication.

Topography and Gravity. The uncertainty in absolute topography at large scale severely limits our ability to model adequately the loads on Mars' lithosphere, especially at regional scales. The current status of gravity measurements is better, but it is clear that a more accurate gravity field representation at higher resolution will have a positive impact on our ability to determine subsurface structure that will affect the stress field. The instrumentation for upcoming missions to Mars has been chosen with these factors in mind, and Mars Observer in particular should provide a vast improvement in the topography and gravity field.

Structural Mapping. One of the quantities that is needed to constrain stress models is the orientation and relative timing of stresses that are recorded on the surface of Mars. This is not "data" as such, but is a derived product obtained from the mapping and interpretation of structural features seen in orbital images. The global-scale stratigraphic framework now in place (see chapter 11) forms the basis for detailed mapping of areas which have undergone tectonic activity. These studies will ultimately provide the crucial information needed to construct the tectonic history of the planet.

Extensions to Theoretical Models. The dynamic support of topographic and gravitational anomalies should also be studied. Although it is unlikely that a long-lived feature such as Tharsis could be maintained dynamically over its entire history, convective processes in the mantle may well have contributed to its evolution for an appreciable period of time. Similarly, thermal stresses may have had an important, albeit temporally limited, impact.

Efforts to date have consisted basically of snapshots of various mechanical states with only minimal consideration as to whether or not any given stage will evolve plausibly into the next stage. The next obvious step in furthering our understanding of these problems is to develop methods to constrain the temporal sequence of events, both observationally and in a modeling sense. Thus, the ultimate key to deciphering the complex history of Mars may lie in integrated models of the tectonic, thermal and magmatic evolution of the planet.

Acknowledgments. We thank J. Plescia for his detailed suggestions for improving the manuscript, and R. Phillips for his review. Discussions during the MEVTV Workshops on Tharsis and on Tectonic Features on Mars helped to sharpen many of the ideas included in this chapter. The authors were funded by contracts from the Planetary Geology and Geophysics Program Office of the National Aeronautics and Space Administration with the Jet Propulsion Laboratory, California Institute of Technology (WBB and MPG) and the United States Geological Survey (KLT).

9. LONG-TERM ORBITAL AND SPIN DYNAMICS OF MARS

WILLIAM R. WARD
Jet Propulsion Laboratory

The long-term dynamical behavior of Mars is reviewed. The discussion covers three topics of interest: secular variations of the orbit, oscillations of the obliquity and polar wandering. Calculations of the large-scale obliquity oscillations of Mars are updated using the most recent orbit theory and contrasted with the Earth. The motion for Mars is characterized by $\sim 10^5$-yr oscillations driven by differential spin axis and orbit plane precession rates during which the obliquity may change by as much as $\sim 20°$. In addition, there are 10^6-yr modulations of this amplitude due to changes in the orbital inclination of Mars. The possible role of spin-orbit secular resonances to the spin axis histories of the Earth and Mars is also considered. Numerical integrations of the equations of motion indicate that Mars may have passed through resonance as little as 5 Myr ago and that obliquities approaching $\sim 45°$ could have been achieved during such an event. Future tidal evolution of the lunar orbit will eventually drive the Earth through such a resonance as well. The obliquity of the Earth could conceivably reach $\sim 60°$, with the annual insolation at the poles exceeding that of the equator.

I. INTRODUCTION

No object in the solar system is dynamically invariant. Orbital and spin motions slowly evolve as a consequence of gravitational interactions with the other bodies in the system. Adjustments in a planet's internal structure can also affect its dynamical state. These changes modify the insolation pattern and, if a body has an atmosphere, may provide an important driver of its climate system. Climatic change can alter the surface-atmosphere interaction and may leave a strong imprint on the surface geology. Thus, the history of a planet's motion becomes a key element in understanding its current condition.

Although all objects experience some dynamical variations, Mars is an exceptionally vigorous case. In particular, the combination of solar and planetary gravitational perturbations experienced by this planet cause surprisingly large excursions in its orbital eccentricity e and inclination I and in its obliquity θ, i.e., the angle between the spin axis and orbit normal. This is in marked contrast to the behavior of the Earth, for which obliquity variations are small.

Table I lists the current orbital and spin characteristics of the Earth and Mars together with some relevant physical parameters. A direct comparison of the motions of these objects helps identify specific traits responsible for their dissimilar behaviors. Of special importance is the orbital inclination, since this is ultimately the source of obliquity variations. Our discussion begins in Sec. II with a review of the secular variations of planetary eccentricities and inclinations. Obliquity oscillations are discussed in Sec. III while in Sec. IV we consider attendant changes in the insolation pattern. In Secs. V and VI, we speculate on the dynamical behavior of Mars and the Earth at other epochs and in Sec. VII, the case for polar wandering is presented.

TABLE I
Selected Physical and Dynamical Parameters of the Earth and Mars

Parameter	Earth	Mars
Rotation Period, D (hr)	23^{h}.9	24^{h}.6
Revolution period, P (yr)	1.00	1.88
Semimajor axis, a (AU)	1.000	1.524
Inclination to invariable plane, I	1°.58	1°.66
Eccentricity, e	0.017	0.093
Obliquity, θ	23°.5	25°.2
Second harmonic of gravity, J_2	0.001083	0.001960
Gyration constant, λ	0.3308	0.3663
Longitude of ascending node Ω_o on the invariable plane[a]	180°.0	249°.9
Azimuthal angle of spin axis Φ on the orbit plane[b]	162°.7	356°.3

[a] With respect to the descending node of the Earth's orbit.
[b] With respect to the ascending node on the invariable plane.

II. SECULAR VARIATIONS OF THE ORBIT

The simplest treatment of the secular problem is the Laplace-Lagrange solution (see, e.g., Brouwer and van Woerkom 1950). The secular portion of the disturbing function experienced by planet i due to planet j is expanded to second order in eccentricities and inclinations:

$$\mathcal{R}_{ij} = Gm_j\Big\{\mathcal{C}_{ij} + \mathcal{N}_{ij}[h_i^2 + h_j^2 + k_i^2 + k_j^2 - p_i^2 - p_j^2 - q_i^2 - q_j^2 + 2(p_ip_j + q_iq_j)] - 2\mathcal{P}_{ij}(h_ih_j + k_ik_j)\Big\} \tag{1a}$$

where m_j is the mass of the jth planet,

$$\begin{aligned} h_i &= e_i \sin \tilde{\omega}_i & k_i &= e_i \cos \tilde{\omega}_i \\ p_i &= \sin I_i \sin \Omega_i & q_i &= \sin I_i \cos \Omega_i . \end{aligned} \tag{1b}$$

$\tilde{\omega}_i$ and Ω_i are the longitudes of perihelion and the ascending node, and $a^*\mathcal{C}_{ij} = (1/2)b^{\circ}_{1/2}$, $a^*\mathcal{N}_{ij} = (1/8)\alpha b^1_{3/2}$, $a^*\mathcal{P}_{ij} = (1/8)b^2_{3/2}$, with $a^* = \max(a_i,a_j)$ and $\alpha = \min(a_i,a_j)/\max(a_i,a_j)$. The terms, b^m_s, are Laplace coefficients which are functions of α,

$$b^m_s(\alpha) = \frac{2}{\pi}\int_0^{\pi} (1 - 2\alpha\cos\theta + \alpha^2)^{-s} \cos m\theta \, d\theta. \tag{2}$$

The equations of motion are

$$\begin{aligned} \frac{dh_i}{dt} &= \frac{1}{n_ia_i^2}\frac{\partial}{\partial k_i}\sum_{j\neq i}\mathcal{R}_{ij} & \frac{dk_i}{dt} &= \frac{-1}{n_ia_i^2}\frac{\partial}{\partial h_i}\sum_{j\neq i}\mathcal{R}_{ij} \\ \frac{dp_i}{dt} &= \frac{1}{n_ia_i^2}\frac{\partial}{\partial q_i}\sum_{j\neq i}\mathcal{R}_{ij} & \frac{dq_i}{dt} &= \frac{-1}{n_ia_i^2}\frac{\partial}{\partial p_i}\sum_{j\neq i}\mathcal{R}_{ij}. \end{aligned} \tag{3}$$

Equations (3) define an eigenvalue problem with solutions of the form

$$\begin{aligned} h_i &= \Sigma\, M_{ij} \sin (s_jt + \varepsilon_j) & k_i &= \Sigma\, M_{ij} \cos (s_jt + \varepsilon_j) \\ p_i &= \Sigma\, N_{ij} \sin (s_j't + \delta_j) & q_i &= \Sigma\, N_{ij} \cos (s_j't + \delta_j) \end{aligned} \tag{4}$$

where s_j and s_j' are eigenfrequencies and (M_{ij}, ε_j), (N_{ij}, δ_j) are eigenvector amplitudes and phase constants that are evaluated from observed planetary positions. Eccentricities and inclinations are then obtained from $e_i = (h_i^2 + k_i^2)^{1/2}$ and $\sin I_i = (q_i^2 + p_i^2)^{1/2}$.

Table II lists the eigenfrequencies, amplitudes and phase constants for the inclinations I_i and longitudes of the ascending nodes Ω_i of the Earth and Mars as found by Brouwer and van Woerkom (1950; henceforth BVW). The resulting inclination variations are shown in Figs. 1a and 2a. Although $I_\oplus$ remains modest ($\lesssim 2\overset{\circ}{.}95$), the inclination of Mars periodically approaches $\sim 5\overset{\circ}{.}85$ on a time scale of ~ 2 Myr. Thus, while the current inclination of the Earth is rather typical, that of Mars is anomalously low. Not surprisingly, the strong $j = 3$ amplitude for Mars reveals that the principal perturber of its orbit plane is the Earth. However, the dominant $j = 4$ term may be primordial

TABLE II

BVW[a] Eigenfrequencies, Phase Constants and Amplitudes for the Orbital Inclinations and Longitudes of the Ascending Node of the Earth and Mars

j	s'_j (arcsec yr^{-1})	P_j (yr)	δ_j (deg)	$N_{\oplus j}$	$N_{\delta j}$
1	−5.202	249,000	272.06	0.00849	0.00180
2	−6.571	197,200	210.06	0.00810	0.00180
3	−18.744	69,100	147.39	0.02448	−0.03589
4	−17.633	73,500	188.92	0.00453	0.05025
5	−25.734	50,400	19.58	0.00281	0.00965
6	−2.903	446,500	207.48	−0.00173	−0.00126
7	−0.678	1,912,900	95.01	−0.00130	−0.00123

BVW[a] Eigenfrequencies, Phase Constants, and Amplitudes for the Orbital Eccentricity and Longitude of Perihelion of Mars

j	s_j (arcsec yr^{-1})	P_j (yr)	δ_j (deg)	$M_{\delta j}$
1	5.463	237,200	92.18	0.00064
2	7.345	176,400	196.88	0.00290
3	17.328	74,800	335.22	0.02972
4	18.002	72,000	317.95	0.07237
5	4.296	301,700	29.55	0.01861
6	27.774	46,700	125.12	0.01502
7	2.719	476,600	131.94	0.00059
8	0.633	1,954,800	69.02	0.00002
9	−19.182	67,600	−66.02	−0.00020
10	51.252	25,300	220.69	0.00064

[a]BVW = Brouwer and van Woerkom (1950).

in nature, possibly established during the planet's accretion or during the early evolution of the solar system. The Mars orbit is similarly characterized by large variations in its eccentricity (see, e.g., Murray et al. 1973; see also Figs. 5 and 6 below) which are also partially primordial in origin. Table II lists the eigenfrequencies, amplitudes and phase constants for the eccentricity and longitude of perihelion given by Brouwer and van Woerkom. [This part of their solution included the great inequality between Jupiter and Saturn (j = 9, 10) in addition to strictly secular terms.]

For the most part, the BVW Laplace-Lagrange solution of Table II is adequate to demonstrate the main features of the motions of the Earth and Mars. Because it is simple to apply, we will use it as the basis for much of our discussion. However, several improvements in long-term orbit theory have been made since this pioneering work. For example, a solution by Bretagnon (1974) complete to second order in the masses and degree 3 in eccentricities and inclinations was used by Berger (1976) to update calculations of the Earth's obliquity and by Ward (1979) for the obliquity of Mars (Fig. 5 below). For this chapter, the obliquity oscillations have been recalculated using the most recent orbit model devised by Laskar (1988). Laskar's theory is complete to second order in masses and degree 5 in eccentricities and inclinations. He also includes relativity and lunar effects. These results are displayed in Fig. 6 below. Although shifts in the calculated eigenfrequencies and amplitudes may have important consequences for specific details of the motion (in particular, for the possible passage through resonances; see Sec. V), the large-scale obliquity variations clearly persist between these theories.

A more substantive question is whether eigenfrequencies and amplitudes have suffered *real* changes through time, i.e., the long-term stability of the orbital motions. A possible source of change is through some physical alteration of the system's components. For example, Ward (1981) has shown that strong secular orbit-orbit resonances can scan through the inner solar system during dissipation of the primordial solar nebula. These resonances are associated with the giant planets (which contain most of the system's angular momentum) and can seriously disturb terrestrial planetary orbits. Alternatively, adjustments in the components of the inner solar system, (such as loss of mass from the asteroid belt or primordial Sun or even solar spin-down [Ward et al. 1976]) could also modify the secular solution. Finally, even the long-term stability of an unaltered planetary system is not assured. Indeed, recent numerical experiments by Laskar (1989) suggest that the inner solar system may exhibit chaotic behavior on time scales of order ~10 Myr. Similar conclusions were reached by Sussman and Wisdom (1988) regarding the orbit of Pluto. Thus, although it is likely that the Earth and Mars exhibited behaviors similar to those of Figs. 1 and 2 throughout most of their histories, specific solutions may only be valid over a small fraction of the age of the solar system.

III. OBLIQUITY OSCILLATIONS

The obliquity of Mars will not in general remain constant (Ward 1974) if the orbit plane is moving in inertial space. Figure 3 shows the projection of the orbit normal **n** of Mars onto the invariable plane of the solar system for the last 10 Myr according to the Brouwer and van Woerkom solution. In addition to the ~ 1.2 Myr inclination cycle, the orbit plane precesses with a period of $\sim 7 \times 10^4$ yr. If the spin axis **s** were itself fixed in space, the Mars obliquity would oscillate as shown in Fig. 1b. The long-term modulation is clearly controlled by the inclination while the shorter-term oscillation arises from orbit precession. A similar calculation for the Earth is shown in Fig. 2b for comparison.

Orbit evolution is not the whole story, however. The solar torque on the equatorial bulge of Mars generates a precession of the spin axis about the orbit normal in the same manner that combined lunar and solar torques cause the well-known precession of the equinoxes on the Earth. The equation of motion for the spin axis **s** is

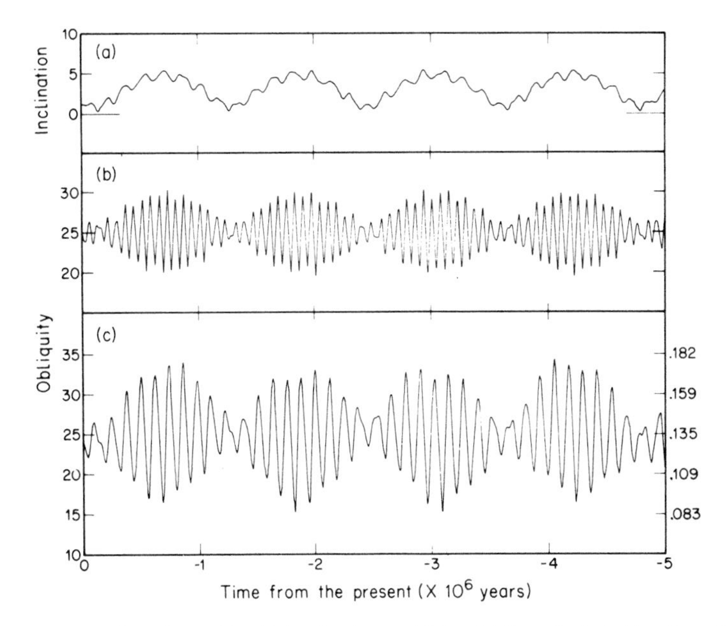

Fig. 1. (a) Inclination of the orbit of Mars for the past 5 Myr. (b) Hypothetical obliquity variations for a spin axis fixed in inertial space. (c) Actual variations in the obliquity of Mars taking into account the precession of the spin axis. Calculations are based on secular orbit theory of Brouwer and van Woerkom (1950).

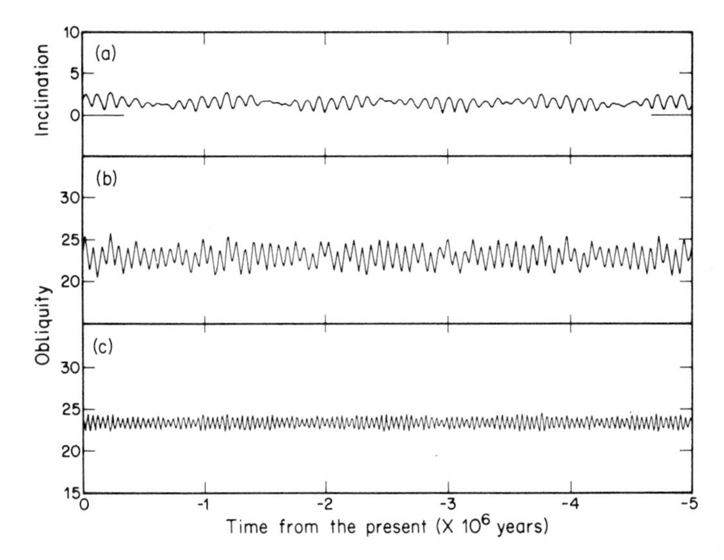

Fig. 2. (a) Past inclination of the Earth's orbit for 5 Myr. (b) Hypothetical obliquity variations for a spin axis fixed in inertial space. (c) Actual variations in the Earth's obliquity taking into account the precession of the equinoxes. Calculations are based on secular orbit theory of Brouwer and van Woerkom (1950).

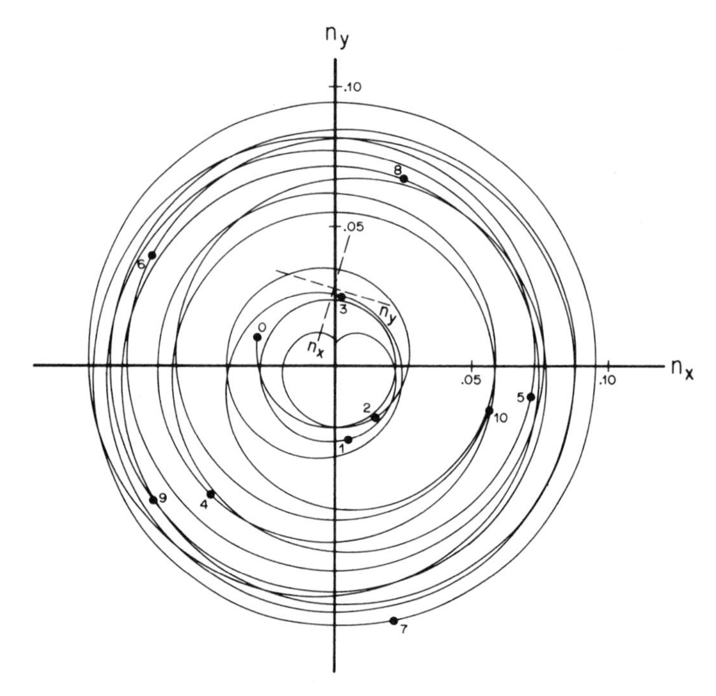

Fig. 3. The orbit normal of Mars resolved with respect to the invariable plane of the solar system. Ten Myr intervals into the past are shown.

$$\frac{\mathrm{d}\mathbf{s}}{\mathrm{d}t} = \alpha(\mathbf{n}\cdot\mathbf{s})\ (\mathbf{n}\mathrm{x}\mathbf{s}). \tag{5}$$

The precession period is $\sim 2\pi/\alpha\cos\theta$, where the precessional constant α is

$$\alpha = \frac{3\pi}{P}\left(\frac{D}{P}\right)\frac{J_2}{\lambda} \tag{6a}$$

P and D being the lengths of the Martian year and day, J_2 the coefficient of the second harmonic of the gravity field and λ the planetary gyration constant (Table I). For Mars, $\alpha = 8.25$ arcsec yr^{-1} and the precession period is $\sim$173,000 yr. The expression for α of the Earth

$$\alpha = \frac{3\pi}{P}\left(\frac{D}{P}\right)\frac{J_2}{\lambda}\left[1 + \frac{M_☾}{M_\odot}\left(\frac{r_\oplus}{r_☾}\right)^3\right] \tag{6b}$$

contains an additional factor to correct for the lunar torque, where $r_\oplus$, $r_☾$ are the semimajor axes of the Earth and lunar orbit, and $M_\odot$, $M_☾$ the solar and lunar masses, respectively. Using values in Table I yields a precession period of ~26,000 yr.

The overall behavior of the spin axis depends critically upon whether its precession rate is faster or slower than that of the orbit. Figure 4 illustrates two limiting cases of interest here. If the orbit-normal precession is much faster (case a), the spin axis will precess about the average position of **n**, i.e., the normal to the invariable plane **k**; if the spin axis precession is much faster (case b), its precessional cone will track the moving orbit normal, maintaining a nearly constant obliquity. If the two rates are nearly equal, the situation is called a secular spin-orbit resonance and obliquity oscillations can be quite severe.

The equations of motion (i.e., Eq. 5) for the direction of the spin vector as seen from the moving orbit frame are (Ward 1974),

$$\frac{d\theta}{dt} = -\sin I \cos\phi \frac{d\Omega}{dt} + \sin\phi \frac{dI}{dt} \tag{7}$$

$$\frac{d\phi}{dt} = -\alpha\cos\theta - (\cos I - \sin I \cot\theta \sin\phi)\frac{d\Omega}{dt} + \cot\theta\cos\phi\frac{dI}{dt} \tag{8}$$

where ϕ is the azimuthal angle measured with respect to the ascending node of the orbit on the invariable plane of the solar system. To first order accuracy in the inclination amplitudes,

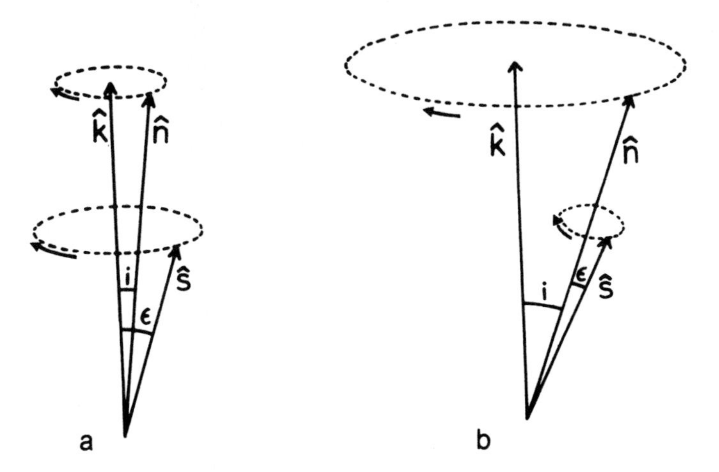

Fig. 4. Spin axis precession when the orbit plane also precesses. (a) If the motion of the orbit normal **n** is much faster than the motion of the spin axis **s**, then the latter precesses about the mean direction of the orbit normal, i.e., the normal to the invariable plane **k**. (b) If the motion of **s** is much faster than the motion of **n**, the spin axis can maintain a near-constant obliquity as it precesses about **n**.

$$\theta = \Theta - \Sigma N_j' \,[\sin(s_j' t + \alpha t \cos\Theta + \Delta_j) - \sin\Delta_j] \tag{9}$$

where

$$N_j' = s_j' N_j / (s_j' + \alpha \cos\Theta) \tag{10}$$

$\Theta = \theta\,(t = 0)$ and $\Delta_j = \delta_j - \Phi - \Omega_0$. Setting $\alpha \rightarrow 0$, recovers the behaviors shown in Figs. 1b and 2b to first-order accuracy. However, if $\alpha \rightarrow \infty$, oscillations are suppressed and $\theta \rightarrow \Theta$. In this case, the solid angle enclosed by the precessional cone of the spin axis is an adiabatic invariant (Ward et al. 1979). Finally, if $\alpha\cos\theta \sim -s_j'$, the oscillations can be greatly enhanced over the fixed spin axis case; i.e., $|N_j'| >> |N_j|$. Table III lists the BVW obliquity amplitudes and enhancement factors for the Earth and Mars. For the Earth, $|N_j'/N_j|$ is generally less than unity (with the exception of $j = 5$ which is a relatively weak term associated with Jupiter). For Mars, most enhancement factors are greater than unity (ignoring sign), including those for the strongest terms, $j = 3,4$. Figures 1c and 2c show the resulting obliquity variations experienced by each planet. Again, modulation by the inclination is apparent in Fig. 1c, while the short-term periodicity has been increased from $\sim 2\pi/s_{3,4}'$ to $\sim 2\pi/(s_{3,4}' + \alpha\cos\Theta) \approx 1.2 \times 10^5$ yr by the motion of the spin axis. The maximum obliquity variations for Mars are $\sim \pm 10\overset{\circ}{.}3$ centered on $\bar{\theta} = \Theta + \Sigma N_j' \sin\Delta_j = 25\overset{\circ}{.}2$. (Interestingly, the current obliquity is close to its long-term average both because the orbital inclination is low and because the spin axis is nearly aligned with the ascending node of the orbit on the invariable plane.)

If the Brouwer and van Woerkom solution is replaced by that of Bretagnon (Fig. 5), the maximum orbital inclination is increased from $\sim 5\overset{\circ}{.}8$ to $\sim 7\overset{\circ}{.}2$. This produces a corresponding increase in the calculated obliquity variation in the past 10 Myr: $\theta_{min} \sim 12°$ at $t \sim -2.06$ Myr, $\theta_{max} \sim 36°$ at $t \sim -4.65$ Myr, with a long-term average of $\bar{\theta} = 24\overset{\circ}{.}4$. Note also that in contrast to Fig. 1a, only every third node in the inclination appears to approach zero.

TABLE III
Obliquity Amplitudes and Enhancement Factors for the Earth and Mars

j	$N_{\oplus j}'$	$N_{\oplus j}'/N_{\oplus j}$	$N_{\male j}'$	$N_{\male j}'/N_{\male j}$
1	−0.00103	−0.1209	−0.00422	−2.342
2	−0.00128	−0.1578	−0.01393	−7.740
3	−0.01558	−0.6363	−0.05940	1.655
4	−0.00261	−0.5767	0.08678	1.727
5	−0.00322	−1.1455	0.01356	1.405
6	0.00011	−0.0641	0.00081	−0.643
7	0.00002	−0.0143	0.00012	−0.101

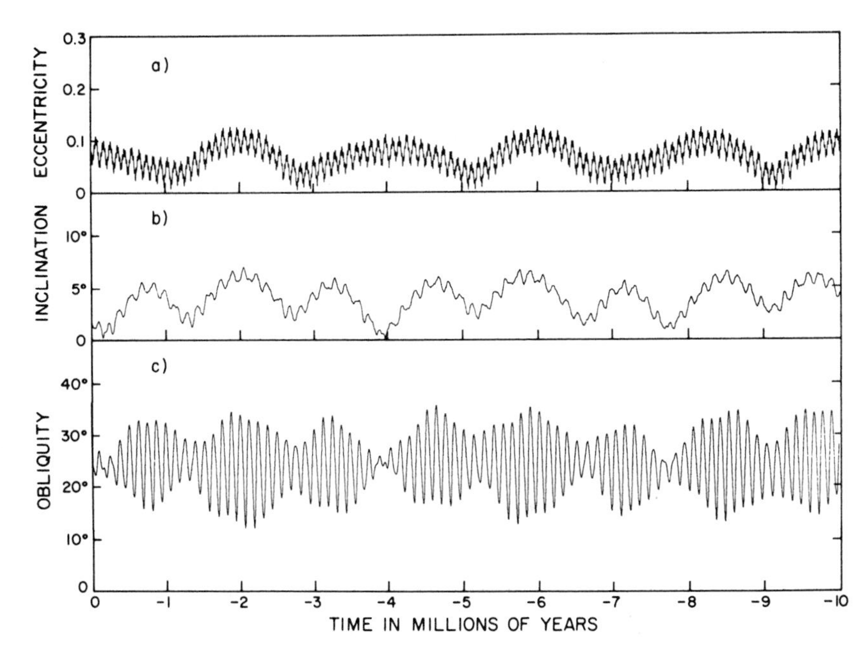

Fig. 5 (a) Eccentricity and (b) inclination of Mars for the past 10 Myr. (c) Variation in the obliquity taking into account precession of the spin axis. Calculations are based on the orbit theory of Bretagnon (1974) with $\alpha = 8.25$.

Figure 6 updates these calculations for the newer orbit theory of Laskar (1988). The maximum inclination is comparable to that given by Bretagnon's theory except that every other node appears to approach zero. The obliquity varies from $\theta_{min} \sim 13°$ at $t \sim -9$ Myr, to $\theta_{max} \sim 42°$ at $t \sim -5.5$ Myr. For the eccentricity, there is a long-term variation with a peak to peak amplitude of $\gtrsim 0.1$ and periodicity of order 2 Myr. A smaller, more rapid (i.e., $\delta e \sim 0.04$, $P \approx 10^5$ yr) oscillation is superimposed on this. The effect of the eccentricity on the spin-axis motion can be included by replacing α with $\alpha(1-e^2)^{-3/2}$ in Eq. (5)(Ward 1979). It was, in fact, this version that was integrated numerically to produce the curves in Figs. 5 and 6. Figure 6 also reveals a pronounced long-period drift in the "center" of the oscillation pattern that is associated with the close proximity of several Laskar frequencies to the frequency of spin axis precession. We consider this resonance state in more detail in Sec. V.

IV. INSOLATION PATTERN

We shall digress for a moment to illustrate briefly the importance of obliquity change to the climate through its modulation of the solar insolation

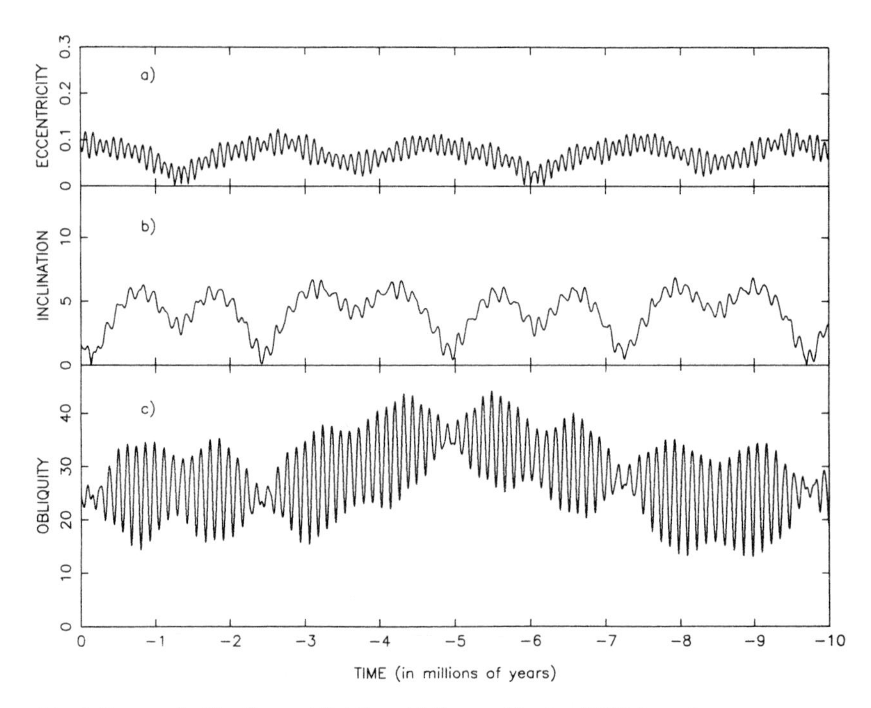

Fig. 6. Same as for Fig. 5 except that the orbit theory of Laskar (1988) is used.

pattern. (See chapter 33 for a more complete description of this issue.) Since the solar flux S^* varies with heliocentric distance r,

$$S^* = S\left(\frac{a}{r}\right)^2 = S\frac{(1 + e\cos f)^2}{(1 - e^2)^2} \tag{11}$$

where $S = 600\ \mathrm{Wm^{-2}\,s^{-1}}$ is the solar constant at $a = 1.52$ AU and f is the true anomaly of the Sun, there is a greater flux at perihelion $[S^*_{max} = S/(1-e)^2]$ than at aphelion $[S^*_{min} = S/(1+e)^2]$. This produces a strong seasonal difference at high eccentricity and is important for phenomena, such as dust storms, that may be triggered by threshold conditions. However, on a yearly basis, the excess flux at perihelion compensates for the flux drop at aphelion to first order in e, so that the yearly radiation budget is only a weak function of the eccentricity, i.e., $\langle S^* \rangle = S/(1-e^2)^{1/2}$ (Murray et al. 1973).

On the other hand, the latitudinal distribution of the yearly flux is a strong function of the obliquity. The local insolation is $I = S^*(\hat{\mathbf{r}}\cdot\hat{\mathbf{r}}_s)$ for $\hat{\mathbf{r}}\cdot\hat{\mathbf{r}}_s > 0$, otherwise $I = 0$, where $\hat{\mathbf{r}},\hat{\mathbf{r}}_s$ are unit vectors pointing from the center of the planet toward a surface element and the Sun, respectively. Averaging over both day and year yields (Ward 1974)

$$\langle I \rangle = \frac{\langle S^* \rangle}{2\pi^2} \int_0^{2\pi} \{1 - (\sin\delta \cos\theta - \cos\delta \sin\theta \sin\psi)^2\}^{1/2} d\psi \qquad (12)$$

where δ is the latitude. Figure 7 shows the yearly average insolation as a function of latitude for various values of the obliquity. The polar values are quite sensitive to θ; for $\delta = 90°$ Eq. (12) reduces to

$$\langle I \rangle_{\rm pole} = \frac{\langle S^* \rangle}{\pi} \sin\theta. \qquad (13)$$

This quantity differs by a factor of $\sin\theta_{max}/\sin\theta_{min} \sim 3$ between the maximum and minimum obliquities attained in Fig. 6. The obliquity becomes a critical quantity in the polar heat balance and its oscillation constitutes a strong driver of the Mars climate (Toon et al. 1980).

Curiously, we also learn from Eq. (12) that for high enough obliquities, the yearly insolation at the equator drops *below* that of the poles. Setting $\delta = 0°$ yields

$$\langle I \rangle_{\rm equator} = \frac{2\langle S^* \rangle}{\pi^2} E\left(\frac{\pi}{2} \backslash \theta\right) \qquad (14)$$

where $E(\phi \backslash \theta)$ is an elliptical integral of the second kind (see, e.g., Abramowitz and Stegun 1968). The critical angle, θ_c, for which $\langle I \rangle_{\rm pole} = \langle I \rangle_{\rm equator}$ is given by the condition (Ward 1974)

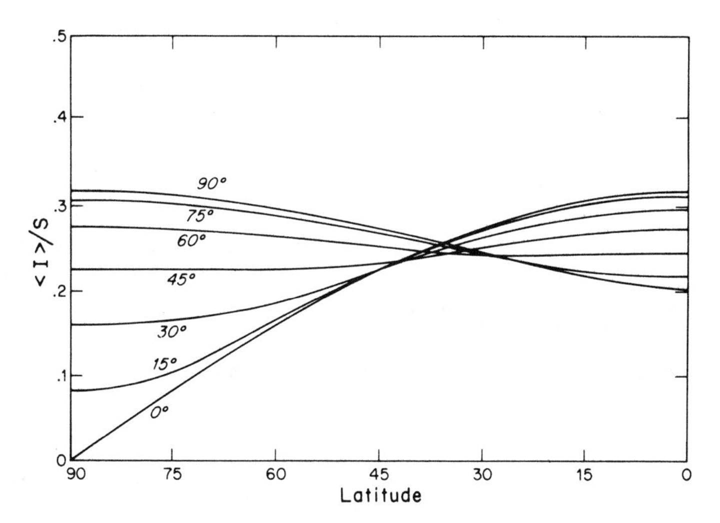

Fig. 7. Average annual insolation $\langle I \rangle$ as a function of latitude for various values of the obliquity. The insolation is normalized to the averaged solar flux $\langle S^* \rangle = S/(1-e^2)^{1/2}$.

$$\sin\theta_c = (2/\pi)E\left(\frac{\pi}{2}\backslash\theta_c\right) \approx 1 - \frac{1}{4}\sin^2\theta_c - \frac{1}{3}\left(\frac{9}{64}\right)\sin^4\theta_c - \cdots \quad (15)$$

and has a value $\theta_c \approx 54°$. (Current obliquities of 97°.9 for Uranus and its satellites and 118°.5 for the Pluto-Charon pair exceed this value.) We next inquire as to whether or not, at some other epoch, the obliquities of the Earth and Mars could substantially differ from their present range of values.

V. PAST OBLIQUITY OF MARS

We have seen that the response of the Mars spin axis to a jth orbital forcing term is a sensitive function of the ratio α/s_j'. It was remarked earlier that the long-term stability of the Laplace-Lagrange eigenfrequencies $\{s_j'\}$ is not guaranteed. This is also true of α. Ward et al. (1979) discuss a number of possible modifying mechanisms, the most important of which appear to be differentiation of the planet and the rise of the Tharsis construct. The former could decrease α by ~4.5% for a change of $\lambda = C/MR^2$ from ~0.4 to the current 0.366 over a time scale $\sim 10^{8-9}$ yr., while the latter implies an increase of ~6.5% over a similar, but probably later, interval.

Ward et al. (1979) speculated that the long-term average $\bar{\theta}$ may itself have suffered a secular change if the spin-axis frequency ever actually passed *through* the frequency of even a minor inclination term, i.e., a *secular spin-orbit resonance*. For Bretagnon's theory, the $s_2' = -6.77$ arcsec yr^{-1} and $s_{26}' = -7.76$ arcsec yr^{-1} terms were identified as principal candidates for such an event. The newer theory of Laskar changes this picture in two important ways: (1) the five terms in the Bretagnon solution lying between $\sim\alpha\cos\theta$ and 11 arcsec yr^{-1} are absent (or shifted) in Laskar's version; and (2) the remaining two Bretagnon terms lying just below $\alpha\cos\theta$ are each replaced by a cluster; the eight terms replacing s_2' being listed in Table IV. A comparison of the Laskar and Bretagnon spectra between 5 and 11 arcsec yr^{-1} is shown in Fig. 8.

The effect of an isolated resonance passage is represented schematically in Fig. 9. As $s_j' + \alpha\cos\Theta$ approaches zero for some j, the oscillations associated with N_j' become very slow (see Eq. 9). The other "high frequency" terms can be averaged over to define a short-term averaged obliquity $\langle\theta\rangle$,

$$\langle\theta\rangle = \bar{\theta} - N_j' \sin(s_j't + \alpha t\cos\Theta + \Delta_j). \quad (16)$$

This quantity slowly oscillates with amplitude $\delta\theta \sim N_j'$ about the long-term average $\bar{\theta}$. As the resonance is approached, Eq. (10) diverges and the linear form of the jth term breaks down. Nonlinear analysis (Ward et al. 1979; Ward 1982) reveals that the maximum oscillation amplitude can be approximated by

TABLE IV
Near Resonant Eigenfrequencies Phase Constants and Amplitudes for Mars Orbital Inclination From Laskar Theory

j	s'_j (arcsec yr^{-1})	P_j (yr)	δ_j (deg)	$N_{\delta j}$
25	−7.053108	183,748	144.957652	.00131766
35	−6.963110	186,123	311.798757	.00103666
41	−7.002513	185,076	118.034667	.00073136
44	−7.148319	181,301	327.622257	.00068832
47	−6.860594	188,904	298.460307	.00061120
49	−7.189558	180,261	301.393902	.00059400
61	−6.813234	190,218	330.778441	.00035216
75	−6.750999	191,971	285.604903	.00023228
		pseudo-pole		
current	−6.932192	186,954	214.81942	.00143651
reference	−7.021698	184,571	—	.00217334

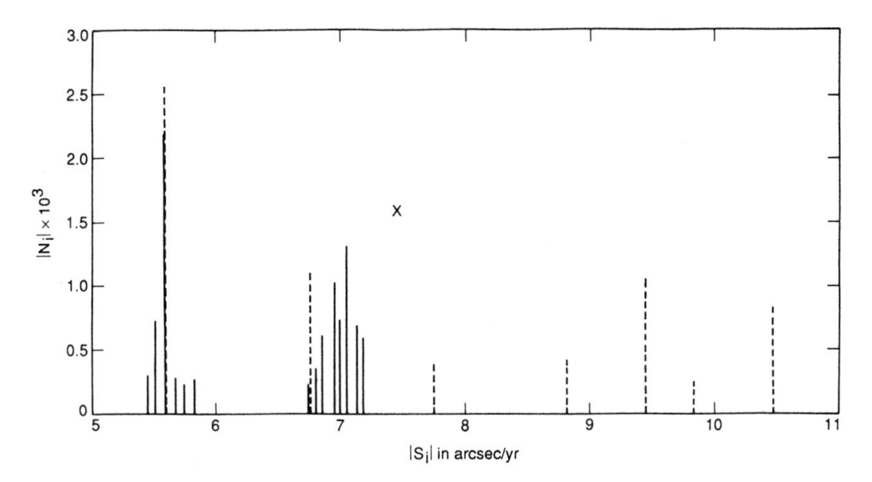

Fig. 8. Amplitudes and frequencies of the near inclination terms for the orbit of Mars from Bretagnon (1974; dashed lines) and Laskar (1988; solid lines). The current estimated value of $\alpha\cos\theta$ is indicated by x.

$$\delta\theta \approx (N_j/\tan\theta_t)^{1/2} \tag{17}$$

where θ_t represents a transition angle $\theta_t \sim \bar{\theta}_1 + \delta\theta + \bar{\theta}_2 - \delta\theta$, and $\bar{\theta}_{1(2)}$ denote long-term averages just above (below) resonance (i.e., Fig. 9). There is a discontinuity, $\Delta\bar{\theta} = \bar{\theta}_2 - \bar{\theta}_1 \approx 2\delta\theta$ in the long-term average across the resonance. A more careful treatment of the nonlinear problem reveals that oscillations of $\langle\theta\rangle$ are not quite centered on $\bar{\theta}_{1(2)}$ and the $\Delta\bar{\theta}$ is actually some-

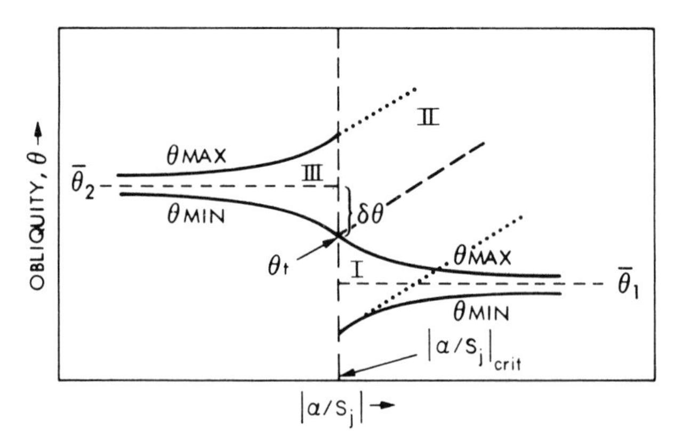

Fig. 9. Schematic diagram of resonance passage. The short-term averaged obliquity oscillates between θ_{max} and θ_{min} centered on the long-term mean, $\bar{\theta}$. At resonance, $|\alpha/s$, the motion undergoes a discontinuity and the long-term mean jumps by $\Delta\theta \equiv \bar{\theta}_2 - \bar{\theta}_1 \sim 2\delta\theta$, where $\delta\theta$ is the maximum oscillation amplitude.

what greater. Detailed analysis also reveals the existence of an alternative dynamical state above resonance. In this motion, the transition angle θ_t takes on the role of the long-term average and the short-term average oscillates about it with amplitude $\sim 2\delta\theta$. Furthermore, as α increases above exact resonance, the long-term average drifts upward, carrying the oscillation envelope with it (Ward et al. 1979). These classes of motion can be interpreted in the context of generalized Cassini states (Colombo 1966; Peale 1969; Ward 1975) for a uniformly precessing orbit. The spin axis, as viewed from a reference frame precessing with the orbit plane, traces out closed curves on the unit sphere. The curve that corresponds to exact resonance is a separatrix that partitions the unit sphere into three domains, each containing a stable Cassini state, which are points of stationary phase. The right-hand curves (class I motion above resonance) in Fig. 9 are associated with Cassini state ①, the left-hand curves (class III motion below resonance) with Cassini state ③. The alternative dynamical state (class II motion) above resonance is associated with Cassini state ②. The separatrix passes through a fourth, unstable Cassini point. The transition probabilities between these domains resulting from slow changes in α/s_j can be calculated analytically (Henrard and Murigande 1987). The probability of entering class II motion is generally small (i.e., ~8% for the $j = 2, 26$ terms of Bretagnon theory).

An indication of resonance isolation is furnished by a comparison of the spin-axis precession rates, $\alpha\cos(\theta_t \pm 2\delta\theta)$, on either side of a resonance with frequencies of adjacent terms $\{s'_k\}$. Isolation of the jth resonance requires $|s'_k - s'_j| >> 2\delta\theta\alpha\sin\theta_t$, or

$$\left|\frac{s_k' - s_j'}{s_j'}\right| >> 2\,(N_j \tan \theta_t)^{1/2} \tag{18}$$

where Eq. (17) has been employed. While the candidate Bretagnon terms ($|\Delta s'/s'| \sim 10^{-1}$) satisfied this criterion, the Laskar terms of Table IV ($|\Delta s'/s'| \sim 10^{-2}$) do not.

An alternative approach is to recombine the terms in Table IV into a single forcing function with a slowly varying amplitude and frequency, i.e.,

$$\mathcal{N}(t) = (\mathcal{P}^2 + \mathcal{Q}^2)^{1/2} \tag{19}$$

$$\omega(t) = (\dot{\mathcal{P}}\mathcal{Q} - \mathcal{P}\dot{\mathcal{Q}})/(\mathcal{P}^2 + \mathcal{Q}^2) \tag{20}$$

where $\mathcal{P} = \Sigma p_k$, $\mathcal{Q} = \Sigma q_k$ and the phase constant $\tan \delta^* = \mathcal{P}(0)/\mathcal{Q}(0)$. The summation is to be taken over the near resonant terms only. The present values of the amplitude and frequency are listed in Table IV, along with the *rms* amplitude, $\mathcal{N}_{rms} = \sqrt{\Sigma N_k^2}$, and a reference frequency, $\omega^* = \langle \dot{\mathcal{P}}\mathcal{Q} - \mathcal{P}\dot{\mathcal{Q}} \rangle / \langle \mathcal{N}^2 \rangle = \Sigma s_k' N_k^2 / \Sigma N_k^2$. The time variation of $\mathcal{N}(t)$ and $\omega(t)$ are displayed in Fig. 10.

A crude measure of the consequences of resonance passage can be obtained by considering a single forcing term of the form $\mathcal{N}_{rms}\exp\{i\omega^* t\}$ and applying the analysis for an isolated resonance. Setting $\bar{\theta} = 25\overset{\circ}{.}7$ for Laskar's model, Eq. (17) then yields $\delta\theta \sim 3\overset{\circ}{.}6$ and a transition angle of $28\overset{\circ}{.}6$, indicating a jump in the long-term obliquity to $\bar{\theta}_2 \sim 32\overset{\circ}{.}2$. The short-term average

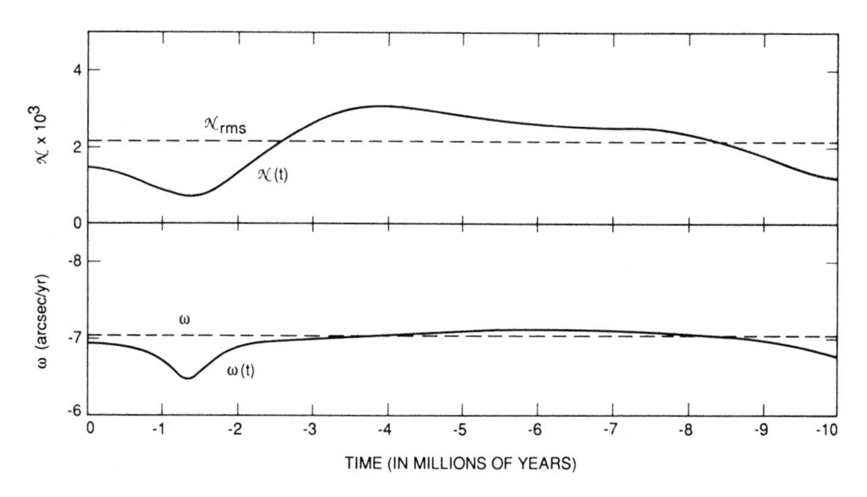

Fig. 10. Time variations of inclination $\mathcal{N}$ and frequency ω of the pseudo-pole defined by the near resonant terms of Table IV. Values of $\mathcal{N}_{rms}$ and the reference frequency ω^* are also shown.

obliquity would approach 35°8 on the other side of resonance. High-frequencies oscillations from nonresonant terms would then be superimposed on this value. If instead, the maximum frequency attained in Fig. 10 is used in this exercise ($\omega \sim -7.10$ arcsec yr^{-1}, $\mathcal{N} \sim 2.9 \times 10^{-3}$ at $t \sim -5$ Myr), the values are $\delta\theta \sim 4°3$ and $\bar{\theta}_2 \sim 34°3$, respectively.

The actual spin axis behavior is likely to be far more complex because time variations in $\omega(t)$ and $\mathcal{N}(t)$ can cause repeated resonance transitions (Ward and Rudy 1991). The adiabatic invariants used to derive Eq. (17) break down under these conditions and stochastic changes in $\bar{\theta}$ become possible. Although one cannot expect to reconstruct a precise history of the Mars obliquity under such circumstances, an appreciation for the role of resonance overlap can be acquired by employing a generalization of the "guiding center" approach outlined in Ward et al. (1979).

To a good approximation, the short-term averaged obliquity $\langle\theta\rangle$ can be calculated by numerically integrating an equation of motion for the guiding center $\mathbf{s}'$,

$$\frac{d\mathbf{s}'}{dt} = \alpha(\mathbf{s}'\cdot\underset{\sim}{\mathcal{N}})(\mathbf{s}'x\underset{\sim}{\mathcal{N}}) + \omega(t)(\mathbf{s}'x\mathbf{k}) \tag{21}$$

where $\underset{\sim}{\mathcal{N}}$ represents a pseudo-pole; $\mathcal{N}_x = 0$, $\mathcal{N}_y = -\mathcal{N}$, $\mathcal{N}_z = (1-\mathcal{N}^2)^{1/2}$. Equation (21) has been written for a reference frame rotating about the normal to the invariable plane at the instantaneous rate $\omega(t)$, so that the pseudo-pole always remains on the -y axis. The starting position of the guiding center is estimated by removing linear contributions of high-frequency terms to the current spin-axis location (Ward and Rudy 1991).

Figure 11 displays the time evolution of $\langle\theta\rangle$ for several values of α. There is an abrupt onset of resonance at about 8.33 arcsec yr^{-1}. Resonance passage is triggered by the increase in $\omega(t)$ shown in Fig. 10 and the various curves in Fig. 11 can be understood in terms of crossing a time varying separatrix (Ward and Rudy 1991). It is clear that small uncertainties or variations in α can result in substantially different behaviors. A similar result occurs in full integrations of Eq. (5) except for minor frequency shifts. Figure 12a shows the full obliquity oscillations for $\alpha = 8.31$ arcsec yr^{-1}.

Although taken at face value, the frequencies found by Laskar strengthen the argument that Mars could have passed through resonance sometime in its history—perhaps relatively recently—they also underscore the cautionary remark in Ward et al. (1979*b*) as to whether or not orbital frequencies are sufficiently well determined. Indeed, chaotic variations in the orbital elements suggested by Laskar (1989) also contribute to this uncertainty. Bills (1989*a*) has expressed additional concern over the estimated value of λ, which enters the calculation of α through Eq. (6). The gyration constant is related to the hydrostatic J_2° by Clairaut's equation, $\lambda =$

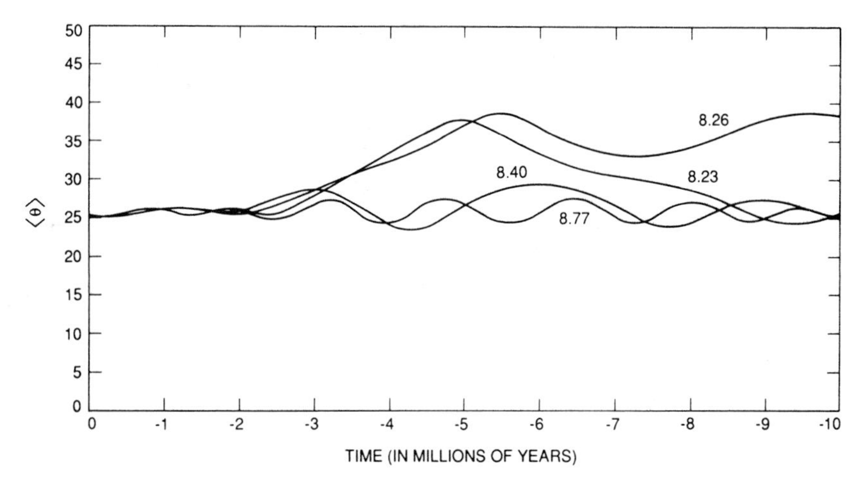

Fig. 11. Variation of short-term averaged obliquity $\langle\theta\rangle$ obtained by numerical integration of the spin-axis motion when perturbed by the pseudo-pole via Eq. (21). Curves are labeled by values of α.

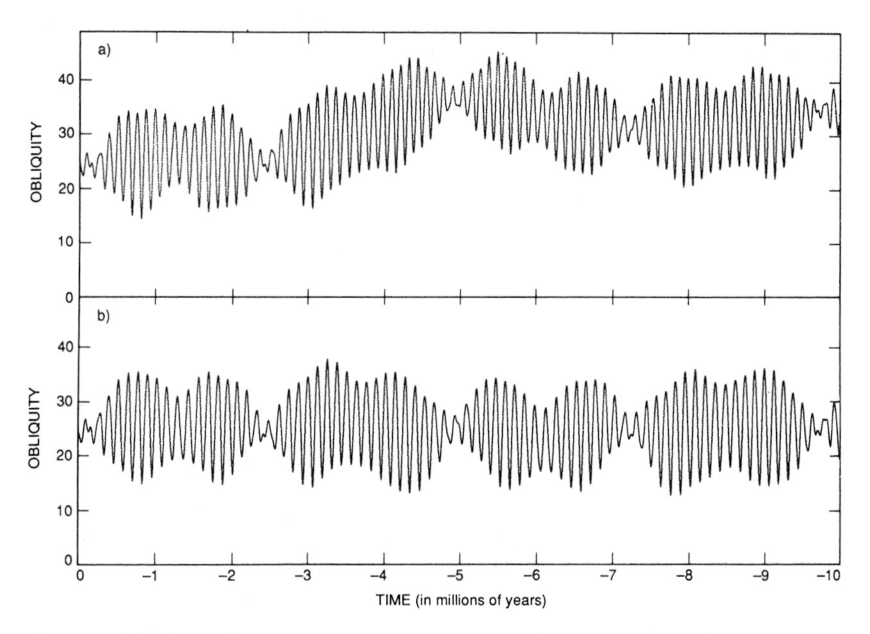

Fig. 12. Obliquity oscillations for (a) $\alpha = 8.31$ arcsec yr^{-1} (plot a) and $\alpha = 8.77$ arcsec yr^{-1} (plot b) using the orbit theory of Laskar.

$2/3 - (4/15)(1-4k_s)^{1/2}(1 + k_s)^{-1/2}$, where $k_s = 3J_2^\circ GM/R^3\psi^2$ is the secular Love number, R is the radius of Mars, and $\psi = 2\pi/D$ is its spin frequency. The value $J_{22} = (1/4)(B-A)/MR^2 = 6.3098 \times 10^{-5}$ obtained from Mars gravity data (Balmino et al. 1982) indicates the presence of a relatively large nonhydrostatic contribution, $\Delta J_2 = (3/2)(\delta C/MR^2)$. This must be subtracted from J_2 before λ can be found. An additional relationship among the nonhydrostatic moments, $\delta A + \delta B + \delta C = 1$, must be assumed to obtain a unique solution. Reasenberg (1977) and Kaula (1979*a*) assumed that the nonhydrostatic figure is nearly a prolate spheroid, dominated by the presence of the Tharsis construct; $\delta B \sim \delta C$, $\delta A \sim -2\delta C$. In this case, $\Delta J_2 \sim 2J_{22}$, leading to the value of λ listed in Table I. Bills (1989*a*) contends that a triaxial figure with $\delta B - \delta A \approx (1/2)(\delta C - \delta A)$ is more likely on statistical grounds, in which case, $\delta B \sim 0$, $\delta A \sim -\delta C$ and $\Delta J_2 \sim 6J_{22}$. The resulting lower value of J_2° decreases λ to 0.345 and raises α to 8.77 arcsec yr^{-1}. A curve for this value has been included in Fig. 11. In this case, Mars is located fairly far from resonance and contributions from Table IV terms are relatively modest (Fig. 12b). However, this claim has been greeted with some skepticism by Kaula et al. (1989) who have since reviewed a number of lines of evidence (*viz.*, cosmochemical data, gravity correlation, topography correlation, fracture patterns, etc.) that support almost axial symmetry of Tharsis tectonics, consistent with the earlier estimates. Unfortunately, direct measurement of the spin-axis precession rate is not currently feasible.

VI. OBLIQUITY OF THE EARTH

Why are the behaviors of the Earth and Mars so different? Again, this can be partly traced to the relatively large primordial component of the inclination of Mars. For instance, the combined strengths of $N_{\delta 3} = 0.03589$ and $N_{\delta 4} = 0.05025$ in the BVW solution of Table II is some 3.5 times the strength of $N_{\oplus 3} = 0.02448$, the principal Earth term. However, enhancement factors (Table III) are also important; $|s'_{3,4}/(s'_{3,4} + \alpha\cos\theta)| \sim 1.7$ for Mars, $|s'_3/(s'_3 + \alpha\cos\theta)| \sim 0.6$ for the Earth, introducing another factor of ~ 2.8 in Mars' favor. Indeed, we have seen in the last section that at exact resonance, the enhancement factor can be quite large, $\sim \delta\theta/N_j \sim (N_j \tan\theta_r)^{-1/2}$. In fact, if the Earth resided in exact resonance with its $j = 3$ orbital term, its enhancement would be $N'_3/N_3 \approx 9$ and obliquity oscillations would rival those of Mars in spite of its smaller inclination.

Interestingly, the higher value of α for the Earth is due to the presence of the Moon. Without the assistance of the lunar torque, the precessional period of the Earth would be $\sim 81{,}000$ yr and $N'_3/N_3 \sim 6.8$. Thus, obliquity oscillations of the Earth would be quite Mars-like were it not for the Moon's stabilizing influence. Of course, the lunar orbit has not always been at its present ~ 60 Earth radii. Earth-Moon tides have expanded the lunar orbit and lengthened the day as well. Tracing the system backwards in time reveals that α was larger in the past and θ more stable. On the other hand, continued tidal evolution of the system *will* eventually drive the Earth into exact secular spin-orbit resonance (Ward 1982). Using the Laskar orbit theory, the Earth first encounters the $s'_{12} = -26.33042$ arcsec yr^{-1} ($N_{12} = 0.0026714$) resonance at $\sim$ 66.5 radii with a transition angle of $\sim$ 30° and an oscillation amplitude of $\delta\theta \sim 3.^{\circ}9$. The long-term average jumps from $\bar{\theta}_1 \sim 26°$ (having already been increased from 23° by direct tidal action) to $\bar{\theta}_2 \sim 34°$ across the resonance. As the lunar orbit expands further, oscillations of the short-term average weaken but the long-term average continues to drift up due to tidal friction. There is a cluster of some 28 terms packed closely between -16.67225 and -19.93956 arcsec yr^{-1} that replace the two local terms of the BVW solution. These terms will be encountered at $\approx$ 68 Earth radii. Using their *rms* amplitude, $\mathcal{N}_{rms} = 0.01922034$, and the reference frequency $\omega^* = -18.62998$ arcsec yr^{-1} as representative of their perturbations, yields a transition angle $\theta_t \sim 43°$ and $\delta\theta \sim 8.^{\circ}2$, implying a post-passage long-term average of $\bar{\theta}_2 \sim 51°$. Short-term oscillations about this mean may produce obliquities near 60°. This exceeds the critical value $\theta_c \sim 54°$ for which the equator receives a smaller yearly insolation than the poles. Again, since there are many overlapping resonances, the overall motion will be quite complex. This should have monumental consequences for the Earth's climate since even the current modest obliquity oscillations are thought responsible for important climatic shifts (see, e.g., Hays et al. 1976; Evans and Freeland 1977).

The time scale for these events is controlled by the tidal rates. Scaled to the current rate of decrease in the lunar mean motion [-23.8 arcsec (century)$^{-2}$; (Williams et al. 1978)], the $j = 12$ term and the principal resonance cluster are some $\sim$ 1.5 Gyr and 2 Gyr away, respectively. Curiously, had the same scaling applied throughout all of Earth-Moon history, these resonances would have been encountered long ago. Modification in the tidal dissipation rate (i.e., a larger Q in the past) seems a likely candidate for the apparent "slowing" of the system's early evolution.

VII. POLAR WANDER

We now turn to a distinctly different, but possibly complimentary motion of Mars: polar wandering. This is a torque-free response of the spin motion to adjustments in the inertial tensor. Since there is no change in the angular-momentum vector, its angle to the orbit normal is unaffected. However, there is generally change in both the magnitudes and directions of the principal axes, $A < B < C$, with respect to the body of the planet.

Goldreich and Toomre (1969) showed that an object executing principal axis rotation will continue to do so, even if its principal axis slowly migrates large distances through the body. For slow migration, adiabatic invariance prevents the excitation of any nonprincipal axis rotation, i.e., wobble, and the spin axis and principal axis remain closely aligned. Here, "slow" means that the polar-wander rate must be much less than the wobble frequency,

$$\Omega_{pw} << \Omega_{wobble} \sim \dot{\psi}\,[(C - A)(C - B)]^{1/2}C^{-1}. \tag{22}$$

Because Mars spins rapidly and has large hydrostatic-moment differences, the wobble period is quite short and condition (Eq. 22) easily satisfied, i.e., $\approx 2\pi/\Omega_{wobble} \sim D(\lambda/J_2) \sim O(10^2)$ days, which is comparable to the period of the Earth's Chandler wobble. [This is in contrast to a slow rotator like Venus where $D \sim 243$ days and $J_2 \approx 10^{-6}$, yielding a long wobble time of order $O(10^5)$ yr (Yoder and Ward 1979).] In addition, energy losses associated with the migration of the second-degree rotational distortion provide for an efficient damping ($\sim$ 14 yr) of the Chandler wobble on Earth and should work similarly for Mars.

Changes in the inertial tensor can come from both internal and external causes. Although for a rapidly spinning planet, rotational flattening dominates the moment differences, relaxation and re-alignment of the hydrostatic figure about the instantaneous axis may leave the nonhydrostatic portion as the actual steering component (Gold 1955). For the Earth, nonhydrostatic contributions are predominately due to density inhomogeneities associated with mantle convection. The nonhydrostatic principal axes become essentially uncorrelated with their previous orientation over a time scale comparable to the lifetime of a typical convection element. As a result, the pole can in principle drift much faster than a single element, provided hydrostatic adjustment is not too sluggish. This latter condition is much more restrictive than Eq. (22). If the mantle is modeled as a homogeneous fluid with constant kinematic viscosity ν, this requires a "viscous polar-wandering rate" at least equal to the polar-wandering rate (Gold 1955; Goldreich and Toomre 1969):

$$\Omega_{pw} \lesssim \Omega_{\nu} \approx I_{1,3}/[\tau_M(C - A)] \tag{23}$$

where $I_{1,3}$ is the largest product of inertia, $C - A$ is the hydrostatic moment difference and $\tau_M \approx 2(\nu/gR)$ is the viscous relaxation time for a second-harmonic distortion, with g and R denoting surface gravity and planetary radius, respectively.

On the Earth, nonhydrostatic-moment differences greatly exceed contributions from surface topography. On Mars, however, partially compensated relief is an important source of the nonhydrostatic inertial tensor, the most prominent example being Tharsis. Melosh (1980*b*) has argued that the rise of the Tharsis construct could have re-oriented Mars by $\lesssim 25°$, although Willeman (1984) argues for $\lesssim 18°$ on grounds that the rotational flattening of an elastic lithosphere may be only partially relaxed.

The first suggestion of polar wander on Mars based on observational evidence from surface geology was that of Murray and Malin (1973), who speculated that the quasi-circular polar deposits may record past [i.e., $O(10^8)$ yr] pole positions. McAdoo and Burns (1975) offered a complementary hypothesis that the Coprates trough complex represented an additional ramification of a shifting pole. They modeled the formation of the trough as a failure due to brittle fracture brought on by changes in the curvature of the lithosphere. The orientation of the fracture trajectory curves as the pole migrates, producing the anti-symmetric shape of Coprates. However, a problem with this picture is the absence of a similar antipodal feature, since first-order treatment predicts stress symmetry about the wander path.

The most detailed attempt to make a case for surface manifestations of polar wander is that presented by Schultz and Lutz (1988). These authors contend that thick unconformable deposits in the equatorial provinces of Mesogaea and Arabia bear strong resemblance to present-day polar deposits and proposed that they are, in fact, relicts of previous pole locations. Figure 13 shows their proposed wander path connecting five suspected past positions of the poles. The relative ages of the circumpolar deposits are inferred from crater count statistics. (They also cite as an independent line of evidence, sites of hypothesized grazing impacts, interpreted as a record of now defunct equatorial satellites that tidally decayed in the early Martian history [Schultz and Lutz-Garihan 1982]. Inferred pole positions from these data are also shown, numbered in an increasing age sequence.)

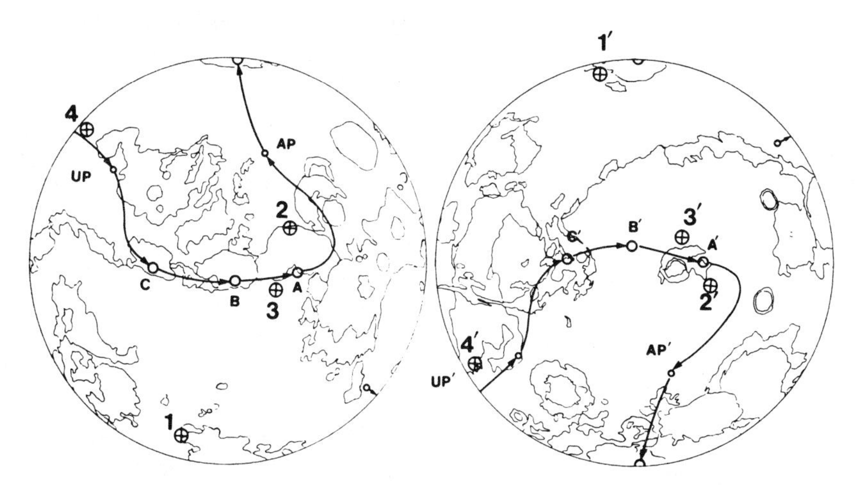

Fig. 13. Proposed polar-wandering path of Schultz and Lutz (1988). The sequence in preservation of the equatorial layered deposits (C through AP) and in the number of pedestal craters suggests a systematic change in the time of deposition interpreted here as a change in the geographic position of the pole. The locations of pole points from the record of grazing impacts are also shown from the oldest (4,4′) to the youngest (1,1′). The oldest position near Utopia Planitia (UP) is very uncertain.

In general, the proposed scenario of Schultz and Lutz ". . . keeps the north polar deposits largely in the lower-lying and relatively uncratered regions of Elysium, whereas the south polar deposits are distributed across the heavily cratered uplands." The global occurrence of thick unconformable deposits similar to the Mesogaea region are not found to be uniformly distributed, but are claimed to exist as age-correlated antipodal features in widely different geologic settings. While this constitutes their main evidence for the polar wander, global distribution of ice-related features and various tectonic signatures are also claimed to be reasonably consistent with the sketched polar sequence. However, Grimm and Soloman (1986) have disputed this last claim by computing the predicted lithospheric stress field due to spin-axis reorientation and comparing these results with surface features in the vicinity of the most recent suspected pole positions. They find that the minimally expected tectonic signature (i.e., the formation of normal faults or graben in a broad region around the former pole positions) is *not* present. They conclude that either no substantial polar wander occurred, or that the last episode of significant wander occurred prior to the emplacement of most of the preserved surface units. Since Tharsis is a major contributor to the nonhydrostatic figure, it must have either always been located near the equator or began to dominate the nonhydrostatic figure before the end of the heavy bombardment.

VIII. SUMMARY

Mars has experienced a complex dynamical history. Planetary perturbations generate relatively large excursions in the orbital eccentricity and inclination. These, in turn, result in large-scale oscillations in Mars' obliquity and furnish an important driver of the climate system. In the past, additional variations in the obliquity may have been induced by resonant amplification of key terms in the time-varying orbital inclination. Calculations employing the recent orbit theory of Laskar suggest that resonance passage may have occurred as little as 5 Myr ago. Numerical integrations of the equations of motion yield obliquities approaching 45° in such an event. In addition, Mars could have exhibited some degree of polar wandering due to evolution of its interior. It is likely that these phenomena have left a strong imprint on the surface geology of the planet.

Acknowledgments. This research was performed at the Jet Propulsion Laboratory under contract with NASA. The computational assistance of D. Rudy is gratefully acknowledged. The author also thanks the Department of Physics, Harvey Mudd College for their hospitality during a portion of this project.

10. GEODESY AND CARTOGRAPHY

MERTON E. DAVIES
RAND

RAYMOND M. BATSON and SHERMAN S. C. WU
U.S. Geological Survey

The first maps of Mars were published more than a century ago; they were compiled from drawings made at the telescope. Definitions of the coordinate system parameters and early work on control networks go back to the telescopic period. The modern exploration began with the Mariner 4, 6 and 7 flyby missions, followed by the Mariner 9 and Viking missions that mapped the entire surface of Mars. The primary modern changes to the coordinate system have dealt with improved measurements of the rotational period, the direction of the spin axis, and the size and shape of Mars. A reference spheroid has been adopted for cartographic products following the tradition established for the Earth. Although there were early publications of control networks based on telescopic and flyby space missions, it was not until the Mariner 9 and Viking data became available that planet-wide control was achieved. Over the years, the density of points in the control network has increased and a corresponding improvement in the accuracy of the coordinates of the points realized. Planimetric mapping based on Mariner 9 pictures began with a 1:25,000,000-scale (1:25M-scale) sheet and thirty 1:5M-scale sheets that covered the entire surface of Mars. The quality of the Viking Orbiter pictures was improved greatly over Mariner 9 and led to the publication of 140 controlled photomosaic sheets at a scale of 1:2M. The sheet layout was simply a subdivision of the 1:5M-scale series. Large-scale maps of areas of special interest were prepared using high-resolution Viking Orbiter images. Vertical control was tied to the horizontal positions of the horizontal control and the radio-occultation radii from both Mariner 9 and Viking. Earth-based radar measurements were also incorporated. Topographic maps of Mars have been produced at a variety of scales.

The early maps were based on Mariner 9 data including, in addition to imaging, S-band radio occultation, ultraviolet spectrometer, infrared interferometer spectrometer and Earth-based radar measurements. When the Viking data became available the vertical control network was devised by analytical photogrammetry; 4502 control points were measured on 1157 Viking pictures with a pixel size at the surface of about 1 km. Maps at scales of 1:2M, 1:1M, 1:0.5M and larger have been produced. The entire series of 140 maps at a scale of 1:2M that cover the entire surface of Mars have been published. All contour lines on the topographic maps are measured relative to the Mars 6.1 mbar topographic datum. Digital mapping is becoming increasingly important and new applications of this form of data set continue to evolve. Two digital data bases have been compiled for Mars, the digital image model (DIM) and the digital terrain model (DTM). The formats of both models are the same; however, the DIM is an array of brightnesses and the DTM is an array of elevations. Conventional maps can be derived from these digital models for hard-copy reproduction.

Mars cartography began in the 19th century when astronomers produced maps from drawings of markings on Mars observed as the planet rotated about its axis. Measurements of its rotation period and the direction of its rotation axis in space were required to establish the coordinate system for maps. Nomenclature became an integral part of those maps as identifiable features were observed year after year. Early maps were published by Beer and Maedler (1840), Kaiser (1864), Flammarion (1876) and Green (1877); these four early maps may be found reprinted in Lowell (1906). However, it was G. Schiaparelli who published a series of maps covering the observational period 1877 to 1888. He introduced the classical nomenclature that is still in use and mapped linear markings he called "canali." These "canali" were reportedly observed by other astronomers of the period; others, like Barnard, disavowed their existence.

In 1894 while observing from Flagstaff, Arizona, P. Lowell immediately saw the "canali," which he interpreted as canals built by an intelligent population. Lowell wrote three popular books on Mars and compiled many maps and globes based on his observations.

E. M. Antoniadi observed Mars for many years beginning in 1909. He failed to see the canals reported by Schiaparelli and Lowell and portrayed Mars the way he saw it. He was an excellent artist and in 1930 published a book, *La Planete Mars,* that contained a summary of his observations and many beautiful maps; he adopted and extended the nomenclature of Schiaparelli (Glasstone 1968; Hartmann and Raper 1974). (Further discussion of the history of telescopic observations is contained in chapters 1 and 2.)

G. de Vaucouleurs identified 26 control points on telescopic drawings of Mars made between 1659 and 1971. This project was carried out over a ten-year period and the results were published in de Vaucouleurs (1980).

Mars' exploration by spacecraft began with the Mariner 4 flyby mission, followed by two more flyby missions, Mariners 6 and 7. However, it was the Mariner 9 orbiter mission that returned more than 7300 images and mapped the entire surface of Mars, revealing the nature and characteristics of the entire planet. This mission was followed by the Viking Orbiters and Landers that greatly extended the exploration; the Orbiters took more than 55,000 pictures and obtained full coverage at medium and high resolution. These missions provided the basic data sets for the modern cartographic program.

I. THE COORDINATE SYSTEM

In 1970, the question of definition of the coordinate systems for the planets and satellites was brought before the International Astronomical Union (IAU) General Assembly. Results from the Mariner 6 and 7 missions were in hand, maps of Mars based on the resulting data were in preparation, and Mariners 8 and 9 were about to be launched. It was clear that the exploration of the solar system was under way, with Mariner 10 scheduled to go to Mercury and Pioneers 10 and 11 and Voyagers 1 and 2 headed for the outer planets. Mapping the planets and satellites would be part of the exploration.

After considerable discussion and debate, two guiding principles were adopted:

1. The rotational pole of a planet or satellite that lies on the north side of the solar system invariable plane shall be called north, and northern latitudes shall be designated as positive.
2. The planetographic longitude of the central meridian, as observed from a direction fixed with respect to an inertial coordinate system, shall increase with time. The range of longitudes shall extend from 0° to 360°.

These conventions are consistent with those selected by Schiaparelli for his maps of Mars and were followed by most of the astronomers who prepared maps of Mars. The only serious objection to these principles concerned the use of west longitudes on Mars (which has direct rotation) as this results in a left-hand coordinate system.

The Geodesy/Cartography Group of the Mariner 9 Television Team selected the small crater named Airy-O to define the 0° meridian on Mars; the reference spheroid for defining areographic coordinates had an equatorial radius of 3393.4 km and a polar axis of 3375.8 km (de Vaucouleurs et al. 1973). The rotational period and the direction of the spin axis were to be updated as improved measurements became available.

It was observed that two coordinate systems are commonly used on Mars—areocentric and areographic. Longitude in both systems is identical. Both use west longitudes; they differ only in the definition of latitude. Most computations are carried out with areocentric latitudes, although maps use areographic latitudes. The origin of both systems is at the center of mass of

Mars with the z axis coincident with the axis of rotation in the northern direction.

Areocentric coordinates of a point—latitude ϕ, longitude λ and radius R, are converted to Cartesian coordinates (X, Y, Z) by

$$\begin{aligned} X &= R \cos \phi \sin \lambda \\ Y &= R \cos \phi \cos \lambda \\ Z &= R \sin \phi . \end{aligned} \tag{1}$$

If a point lies on the reference spheroid, the areographic latitude ϕ' of the point is the angle that a line perpendicular to the tangent to the reference spheroid at the point makes with the plane of the equator. In this case, the areocentric latitude ϕ, and the areographic latitude ϕ' of the point are related by

$$\tan \phi = \left(\frac{3375.8}{3393.4}\right)^2 \tan \phi' . \tag{2}$$

When the point does not lie on the reference spheroid, the line that defines areographic latitude ϕ' must be normal to the tangent to the spheroid at the point that it crosses the spheroid. In this case, it is easiest to solve for ϕ' by successive approximation. Frequently, the deviation from the reference spheroid is sufficiently small so that it can be ignored and the equation above will give sufficiently accurate values for the areographic latitude. It should be noted that the elevations represented on the topographic maps are not measured from the reference spheroid but from a specially defined topographic datum described in Sec. VI.

Analysis of the radio tracking data from the Viking Landers led to very accurate measurements of Mars' rotational period (24.623 hr), the direction of the spin axis ($\alpha = 317\overset{\circ}{.}341$, $\delta = 52\overset{\circ}{.}711$), and the Lander coordinates relative to the inertial frame (Mayo et al. 1977; Michael 1979; Borderies et al. 1980). The location of Viking 1 Lander site was identified on 2 high-resolution frames, thus tying the control network into the coordinate system (Morris and Jones 1980). For this reason the coordinates of control points in this region are very accurate, with errors < 300 m.

In 1976, a Working Group on Cartographic Coordinates and Rotational Elements of the Planets and Satellites was established at the IAU General Assembly; the Group publishes reports every 3 yr (Davies et al. 1980,1983,1986,1989). The reports give the best current solutions for the right ascension and declination of the north poles of the planets, including Mars, and the location of the prime meridian defined by the angle W. The angle W is measured from the node defined by the intersection of the J2000 Earth equator and the Mars equator, along the Mars equator to the Mars prime meridian (see Fig. 1).

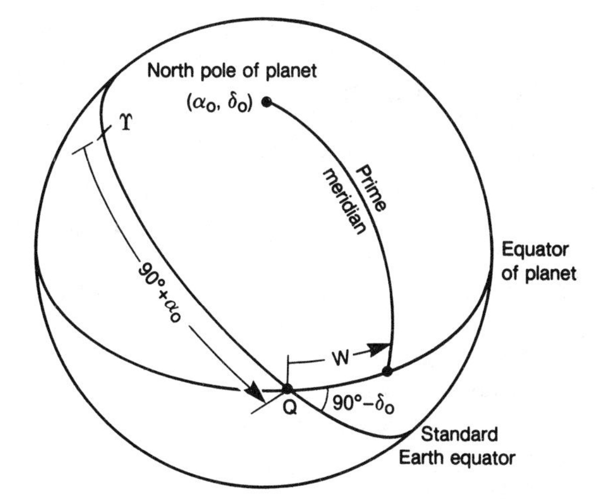

Fig. 1. The coordinate system of Mars is defined by the right ascension α_o and declination δ_o of its north pole and its prime meridian is defined by the angle W measured along Mars' equator from the ascending node of Mars' equator on the standard J2000 equator. The equations for Mars are $\alpha_o = 317\overset{\circ}{.}681 - 0\overset{\circ}{.}108T$, $\delta_o = 52\overset{\circ}{.}886 - 0\overset{\circ}{.}061T$, and $W = 176\overset{\circ}{.}868 + 350\overset{\circ}{.}8919830d$, where T is measured in centuries and d is measured in days from JD 2451545.0TDB.

II. MARS NOMENCLATURE

The nomenclature of features on Mars started in the telescopic period with names for albedo marking and was greatly expanded to include topographic features when the surface was mapped in detail by spacecraft instruments. Nomenclature is an integral part of the cartographic program.

Schiaparelli was a student of Greek mythology and named areas on his maps of Mars after regions of the ancient world. This scheme of nomenclature was adopted and extended by later astronomers such as Lowell and Antoniadi.

At the 1970 IAU General Assembly, a Working Group on Martian Nomenclature was appointed to define province boundaries and to apply principles of topographic nomenclature to areas observed in Mariner 6, 7 and 9 pictures (de Vaucouleurs et al. 1975). At the 1973 IAU General Assembly, it was apparent that many of the planets of the solar system would be explored in the next decade or so. Thus, the IAU Working Group for Planetary System Nomenclature was established. Details of the formal nomenclature system are discussed in the Appendix.

III. THE HORIZONTAL CONTROL NETWORK

G. de Vaucouleurs carried out an extensive program to establish a telescopic control network based on measurements of points identified on drawings and photographs of Mars. The results of his work were reported in de Vaucouleurs (1962,1980) and were the source of positional information for the 1971 Mariner 9 planning chart (scale 1:25M).

In 1969, Mariners 6 and 7 flew by Mars, taking pictures as they approached the planet to observe its rotation and taking high-resolution pictures as they passed close to the planet. Pixel coordinates of selected points on the pictures were measured, and latitudes and longitudes computed (Davies and Berg 1971; Davies 1972), resulting in the first control network for Mars based on spacecraft data. A planet-wide 1:25M-scale Mars chart was produced by the U.S. Army Topographic Command using the control network and the Mariner 6 and 7 pictures. Cross (1971*b*) published a 1:10M chart based on the near-encounter pictures and control.

The modern geodetic and cartographic program began with the Mariner 9 data set. The pictures were of sufficient quality and coverage—almost the entire surface of Mars—so that the topographic character of the planet was revealed. Mariner 9 took many thousand pictures from orbit, many in sequences of mosaics as the planet rotated. From this a new understanding of Mars emerged.

To establish geodetic control, it is necessary to find common features (frequently craters) in the overlapping area of adjacent pictures. The location of these features or points on the pictures is measured, and the measurements entered into an analytical triangulation program. The program, designed to compute the latitude and longitude of the measured points, consists of a least-squares best fit to the data. This set of control points can then be used to position the features on a map. Many points and pictures were measured to obtain a planet-wide control network. Because of the large amount of data and the limitations of the available computer, the first network was broken into 5 regional networks that were then tied together (Davies and Arthur 1973). All subsequent published planet-wide control networks were computed in single analytical triangulations by introducing more efficient algorithms for solving linear equations and using more powerful computers.

The analytical triangulation attempts to improve the values of the latitude and longitude of each point as well as the 3 orientation angles of the camera when the picture was taken. Thus, each point contributes 2 unknowns and each picture 3, thereby determining the number of simultaneous linear equations that must be solved. The size of the control network has continued to expand (see Table I).

Viking pictures have been used to create high-resolution photogrammetric strips encircling Mars at the equator and at 60° N and S latitudes. The strips have been tied together by meridian strips about every 90° of longitude. The strips were developed by selecting overlapping pictures from the 1:2M-scale series of maps with resolution of 150 m to 250 m per pixel. Control points were selected in the overlapping areas so that the pictures were rigidly tied together photogrammetrically.

TABLE I
The Growth of the Mars Horizontal Control Network

Computation Date	Points	Pictures		Reference
		Mariner 9	Viking	
1972	1205	598		Davies and Arthur (1973)
1974	2061	762		Davies (1974)
1977	3037	928		Davies (1978*b*)
1978	4138	1009	204	Davies et al. (1978)
1982	6853	1054	757	Davies and Katayama (1983)
1989	9236	1054	1396	

In 1972, the accuracy of the coordinates of the control network was estimated to be in the range of 10 to 20 km depending upon the location (Davies and Arthur 1973). With the introduction of the Viking data, the estimated accuracy of the control points increased to 6 km everywhere and in some regions to 3 km (Davies and Katayama 1983). In general, the errors are thought to be < 3 km along the photogrammetric strips.

IV. PLANIMETRIC MAPPING

The selection of scales at which Mars was to be mapped, the projections used, and the division of the planet into quadrangles were originally based on the resolution of Mariner 9 pictures (Batson 1973; Batson et al. 1979). A quadrangle scheme similar to that previously used for the Moon (Kopal and Carder 1974), consisting of conformal projections divided by parallels of latitude, was selected for user convenience and familiarity. In this scheme, the polar stereographic projection was selected for the polar regions, Lambert conformal conic for intermediate latitudes, and the Mercator projection for the equatorial regions (see Snyder [1982,1987] for a detailed discussion of the mathematics of the projections used).

In planning the Mariner 9 cartography, it was stipulated that the scale of the quadrangles be sufficiently large that any feature appearing on the Mariner 9 pictures that could be classified geologically could also be annotated legibly at map scale. Coverage of 70% of Mars with image resolution of $\sim$ 1 km per pixel was expected at the time the mapping plan was designed (Masursky et al. 1970). A scale of 1:5M was selected based on the assumptions that 5 to 8 picture elements are required to classify a feature geologically and that such features should have minimum dimensions of 1 mm on a map. It was further stipulated that sheets should have approximate dimensions of 0.5 m square, and that map scales should be identical at projection boundaries. These criteria resulted in the design of a regional mapping scheme for Mars consisting of 30 segments of quadrangles (Fig. 2). The selection of the equator as the

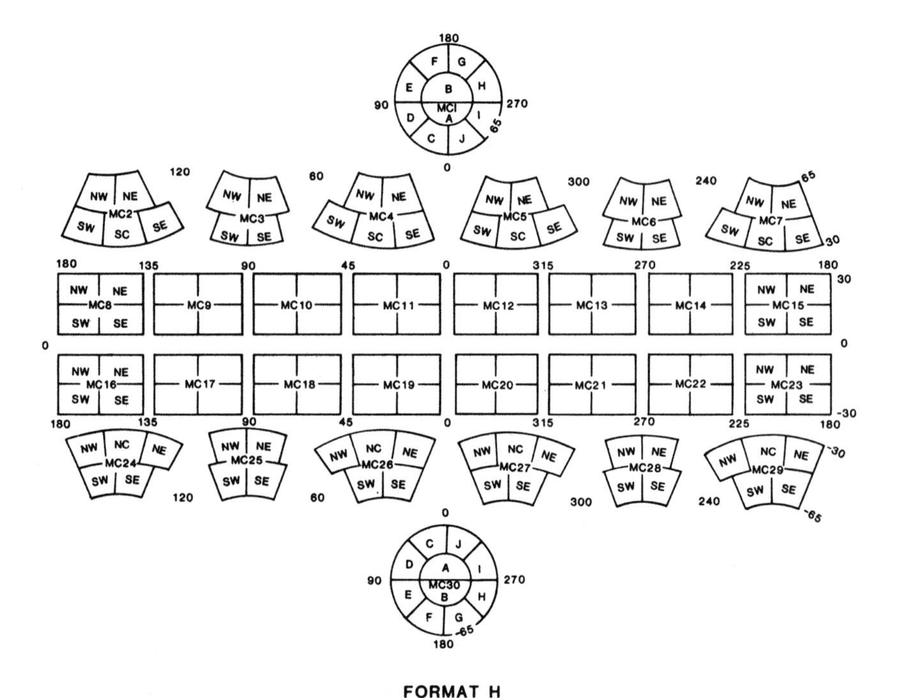

Fig. 2. The 1:5M-scale series of maps is published in 30 sheets with designations running from MC-1 through MC-30. Of these sheets, 16 use Mercator projection, 12 use Lambert conformal projection, and 2 use polar stereographic projection. The 1:2M-scale series of maps is published in 140 sheets that are subdivisions of the 1:5M-scale layout.

latitude of true scale for the series resulted in a series of maps with scales > 1:5M everywhere but at the equator.

Cartographic studies with Mariner 6 and 7 images of Mars demonstrated that photomosaics alone are inadequate for planetary mapping. Any given image displays unique information about any given landform, but only one image of that landform can appear in a mosaic. A second kind of map is therefore made with an airbrush, incorporating all possible detail from all sources, showing idealized relief shading and, where feasible, surface coloration or albedo patterns (Inge and Bridges 1976; Batson 1978). Some of the latter are visible from the Earth, and have fascinated and sometimes misled astronomers and laymen for nearly 400 yr. Airbrush mapping requires such unique skills that the number of available practitioners of this art is very small. The technique is therefore used only in synoptic and special-area mapping where photomosaics are clearly inadequate.

Although Mariner 9 did not cover all of Mars (particularly the northern hemisphere) with high enough resolution to justify the 1:5M scale, the number, size and scale of the sheets have proved to be optimum for reconnaissance mapping. This scale and variants of the Mars quadrangle layout have therefore been adopted for most of the planets and satellites explored to date. Planet-wide maps of Mars produced at 1:15M and 1:25M scale have proved to be very popular.

Viking Orbiter spacecraft returned more than 55,000 television pictures of Mars, with resolutions ranging from 7 m per pixel to more than 1 km per pixel. This data set contains, but is not limited to, a "medium resolution" (4 to 6 pixels per km) global survey for comprehensive planimetric and geologic mapping, a multispectral survey of at least 80% of the planet in 3 bands (0.45, 0.53 and 0.59 μm) with spatial resolutions of ~ 1 km per pixel, and a photogrammetric survey of at least 80% of the planet containing convergent stereoscopic image pairs with resolutions better than 1 km per pixel. Nearly half of the remaining images have resolutions between 7 and 100 m per pixel, providing the basis for detailed geologic studies and mapping that continue to produce exciting new discoveries on Mars, years after the images were returned to Earth.

The medium-resolution data set is a collection of approximately 5000 pictures taken specifically for planimetric and geologic mapping. The pictures were taken when the Sun was about 20° above the horizon and when the spacecraft altitude was such that the images have resolutions of 150 to 250 m per pixel. This data set was used to compile a planet-wide series of 140 controlled photomosaics at 1:2M scale and to revise the 1:5M shaded relief maps. The 140 quadrangle layout is simply a subdivision of the 1:5M layout, with longitude boundaries revised in parts of the band of Lambert conformal conics to produce roughly equal sheet sizes (Fig. 2). The same set of map projections and standard parallels were used. A series of controlled photomosaics was compiled manually in the format. The true scale was stipulated to lie on the standard parallels of the conics (35°.83 and 59°.17). To preserve scale joints at projection boundaries, the Mercators must have true scales at 27°.476 and the stereographics at 75°.008.

Geodetic controls on both the original and the revised 1:5M shaded relief maps are based on photogrammetric triangulations with Mariner 9 data (Davies 1973; Davies and Arthur 1973), whereas the 1:2M controlled photomosaics are tied to a more recent net (Davies et al. 1978; Davies and Katayama 1983). There are slight positional differences between the two types of maps.

Large-scale maps for topical studies of areas of special scientific interest use high-resolution Viking Orbiter images. Nearly 25,000 Viking Orbiter frames have resolutions better than 100 m per pixel and display some of the most fascinating surface structure found on Mars. A large-scale mapping scheme containing 1964 quadrangles (5° latitude by 5° longitude in equatorial and intermediate latitudes) designed at 1:5M scale exploits the information in these images. Larger-scale mapping is possible in special cases, but no mapping scheme has been specified. The current plan is to compile selected blocks of quadrangles at 1:5M (or larger) only in areas of exceptional scientific interest, including candidate landing sites for future Mars missions. Geologic mapping on these maps will examine particular problems in deciphering the geology of Mars.

Stereoscopic image pairs and other kinds of data have made possible the quantitative mapping of surface relief. The density of contour lines is sufficient to delineate landforms on most areas on Earth, but data available for mapping Mars support only generalized contouring for delineating regional or continental trends. The contour lines are most useful for depiction of structural information by overprinting on photomosaics or on airbrush maps. Wu and Doyle (1990) have described this compilation of topographic maps of the planets.

V. THE VERTICAL CONTROL NETWORK

To provide controls for the systematic mapping of Mars topography, 1157 high-altitude Viking Orbiter pictures were combined to establish a planet-wide control network of Mars (Wu and Schafer 1982,1984; Wu et al. 1984). Analytical photogrammetric methods used the known positions and orientations of camera stations of the orbiting spacecraft. The adjustment was performed with the U.S. Geological Survey's General Integrated Analytical Triangulation (GIANT) program (Elassal and Malhotra 1987); 4502 control points cover the entire Martian surface. The block adjustment also assured internal consistency of the camera positions and orientations that enable the establishment of stereomodels for photogrammetric compilation on analytical stereoplotters, despite the fact that the Viking Orbiter pictures have extremely narrow fields of view. The vertical control net was tied to the horizontal positions of Davies' primary control net (Davies et al. 1978), with elevations measured from the Mars 6.1 mbar topographic datum (Wu 1978,1981). This pressure level, which corresponds to the triple point of water, was chosen before detailed knowledge of the absolute position of Mars surface was known. Current topographic mapping places the median elevation of the Mars surface at 1.88 km above this datum.

The Mars vertical control computations were carried out in 2 steps. The first analytical triangulation contained 3172 control points on 715 Viking pictures. Most of these pictures were taken at high altitudes where the pixel size corresponds to about 1 km on the surface. The pictures cover an equatorial band around Mars between 30° N and S latitudes and wide bands of pictures along 120° and 300° longitude covering the north and south poles. The second step was to fill the coverage gaps with 442 Viking pictures and 1330 additional control points using 4 small single-block analytical triangulations (Wu et al. 1984).

Input for the GIANT analytical triangulation program included the Image-, Frame- and Ground-files. The Image-file contains all of the photo coordinate measurements made on a comparator. In addition to the decalibration, all measurements were further corrected by a third-degree polynomial, which also uses calibrated reseau coordinates. In general, through decalibration processing, pictures produced by the digital film-writing device have a residual of 9 μm (about one half of a pixel). The Frame-file includes 3 rotational and 3 positional components for each camera station derived from the Viking orbital parameters. The Ground-file includes geodetic coordinates of those points that are used for primary ground controls and those that have estimated coordinates. Different weights are assigned to the controls depending on their uncertainties or reliabilities. Primary ground controls include: occultation measurements from the *S*-band radio experiment on both Mariner 9 and Viking missions (Kliore et al. 1970,1972,1973; Lindal 1978; Lindal et al. 1979); elevations derived from Earth-based radar observations of Mars (Downs et al. 1971,1973,1975; Goldstein et al. 1970); and horizontal coordinates from the primary control networks (Davies et al. 1978).

The *S*-band occultation measurements determine the radius of Mars at the instant of spacecraft orbital occultations at both entry and exit by recording the time immediately before and after an occultation. There are 256 and 155 usable points, respectively, from the Mariner 9 and Viking missions scattered over the entire Martian surface. The uncertainties of the occultation measurements range from 0.25 to 2.1 km (Kliore et al. 1973; Christensen 1975; Lindal et al. 1979). The Earth-based radar calculates altitudes of the Martian surface from signal time delay. The precision of the radar height measurements ranges from 75 to 200 m (Pettengill et al. 1971,1973; Downs et al. 1971,1973; Goldstein et al. 1970). About 1000 Goldstone radar points of Mars have been used for the control network adjustment. About 70% of the pass points of the vertical control net have the same locations as those produced from Davies' primary control net (Davies et al. 1978), so that the vertical control net can be tied to Davies' horizontal positions.

The adjustment of the vertical control network has residuals of 4 km and 800 m, respectively, for the horizontal and vertical coordinates. The vertical control network enables the accurate compilation of the Mars 1:2M and other various Mars topographic maps at larger scales. The data from the vertical control network are also valuable for: (1) improving the horizontal locations of occultation points on both the Mariner 9 and Viking missions; (2) calibrating locations of Earth-based radar profiles; (3) improving elevations of Davies' horizontal control network; and (4) adjusting the internal consistency of camera positions and orientations. In fact, the adjusted internal consistency of camera stations is the only means by which the extremely narrow field-of-view Viking pictures can be used to establish stereomodels on the analytical stereoplotter for photogrammetric compilation. Stereomodels are established by pre-computing model parameters using adjusted camera positions and orientations (Wu and Schafer 1982).

VI. TOPOGRAPHIC MAPPING

Mars topographic maps provide a quantitative representation of the landforms and relief of the Martian surface and lead to a better understandanding of geologic and tectonic histories of Mars. They are also used to interpret the nature of gravitational fields and the distribution of gravitational anomalies and the properties of Mars' crust, and are important for planning future Mars missions.

Topographic mapping of Mars, however, differs in many ways from the mapping of Earth. It involves solving unprecedented problems: the absence of oceans to provide a zero-elevation reference surface, the lack of precise ground controls or methods of data acquisition, and so forth. These unconventional factors require the development of new methodologies and special equipment.

As Mars has no seas and hence no sea level, the most appropriate definition of its topographic datum (the zero-elevation surface) for altitudes is a gravity surface (Wu 1981; Wu and Schafer 1984). Using Mariner 9 radiotracking data, topographic datum of Mars is defined by its gravity field described in terms of fourth-degree and fourth-order spherical harmonics (Jordan and Lorell 1975; Christensen 1975) with a 6.1-mbar pressure surface with a mean radius of 3382.9 km. This datum has been used for compiling all Mars topographic maps of various scales. The datum can be approximated by a triaxial ellipsoid with semimajor axes of $A = 3394.6$ km and $B = 3393.3$ km, and a semiminor axis of $C = 3376.3$ km. The semimajor axis A intersects the Martian surface at longitude 105°. Radii of the datum as a function of latitude and longitude can be found in Wu (1981).

On a topographic map, areographic latitudes are computed with respect to the reference spheroid described earlier. However, the elevations are measured relative to the topographic datum.

Topographic data sources used for Mars topographic mapping are derived from remotely sensed data, imaging or nonimaging, obtained from various devices over a broad wavelength spectrum. Using nonimaging data, for example, a global topographic map (1:25M) and 30 quadrangles of 1:5M-scale contour maps of Mars were compiled between 1971 and 1975 by the synthesis of measurements obtained by various scientific experimental devices (Wu 1975,1978). They include the S-band radio occultation, the ultraviolet spectrometer (UVS), the infrared radiometer (IRR), and the infrared interferometer spectrometer (IRIS) on board the Mariner 9 spacecraft, and the Earth-based radar oppositions of Mars. The UVS experiment was to measure the surface pressure and composition of the Martian atmosphere at the surface of Mars (Barth et al. 1972). The variations of local surface pressure can be interpreted to measure the Martian topography (Barth and Hord 1971). There are almost 7500 measurements of elevation provided by the UVS experiment covering the Martian surface from 60° S latitude to 45° N latitude for Mars topographic mapping (Wu 1978). The IRIS experiment, on the other hand, infers the vertical temperature structure, which also provides topographic information through the absorption of certain bands of CO_2. The temperature map compiled from the IRR experiment correlates with topographic variations (Cunningham and Schurmeir 1969). There are about 4600 elevation points provided by the IRIS and IRR for the compilation of the Mars 1:25M-scale contour map (Wu 1978).

From radar observations, altitudes on the Martian surface are calculated from signal delay. In other words, variations in radar travel time to and return from Mars are associated with the topographic relief on the Martian surface. The relative precision of height measurements, which is simply a direct translation from precision of time measurements, ranges from 75 m to 200 m; as knowledge of the planetary ephemeris has improved, the accuracy of radar topography has reached this same magnitude. More than 15,000 radar data points on Mars topography observed from Goldstone and the Haystack observatories were used, together with various sensor data from Mariner 9 for the compilation of the Mars 1:25M-scale global topographic map (Wu 1975,1978). Radar points play an important role for the topographic control network.

Television imaging systems were major instruments (along with other experiments) on the Mars exploration spacecraft, Mariners 4, 6, 7 and 9 (Masursky et al. 1970); but they were not designed with topographic mapping as the primary objective. It was the two Viking missions that, with about 55,000 pictures sent back to Earth, made it possible to map systematically Mars in great detail (Wu and Schafer 1982). By using adjusted camera positions and orientations from the control network, stereomodel orientation parameters can be established for the analytical stereoplotters. This development makes it possible to use the Viking Orbiter pictures at various ranges of resolution to map stereoscopically Mars topography at various scales—1:2M, 1:1M, 1:0.5M, and larger. Figure 3, for example, is a typical 1:2M-scale topographic map of Mars. Digital elevation data from this map, combined with digital image data of the same coverage through digital image processing techniques, result in a perspective view of the mapped area (Fig. 4). A total of 140 quadrangles of the 1:2M-scale topographic maps cover the entire Martian surface. With topographic data extracted from the 1:2M-scale maps, the Mars global topographic map has been revised at a scale of 1:15M (U.S.G.S. 1989). Topographic maps of Mars are also compiled at high resolution and at large scales for areas of special geologic interest and for future mission planning, particularly for those involving spacecraft landings on the Martian surface. Such maps include Olympus Mons, Tithonium Chasma, Candor Chasma, Kasei Valles, etc. (with scales from 1:0.5M to 1:1M) with a contour interval of 200 m. Such high-resolution large-scale topographic maps have been found useful for other purposes. For example, from the topographic map (1:1M), Olympus Mons is found to have a volume of 2.594×10^6 km^3 for

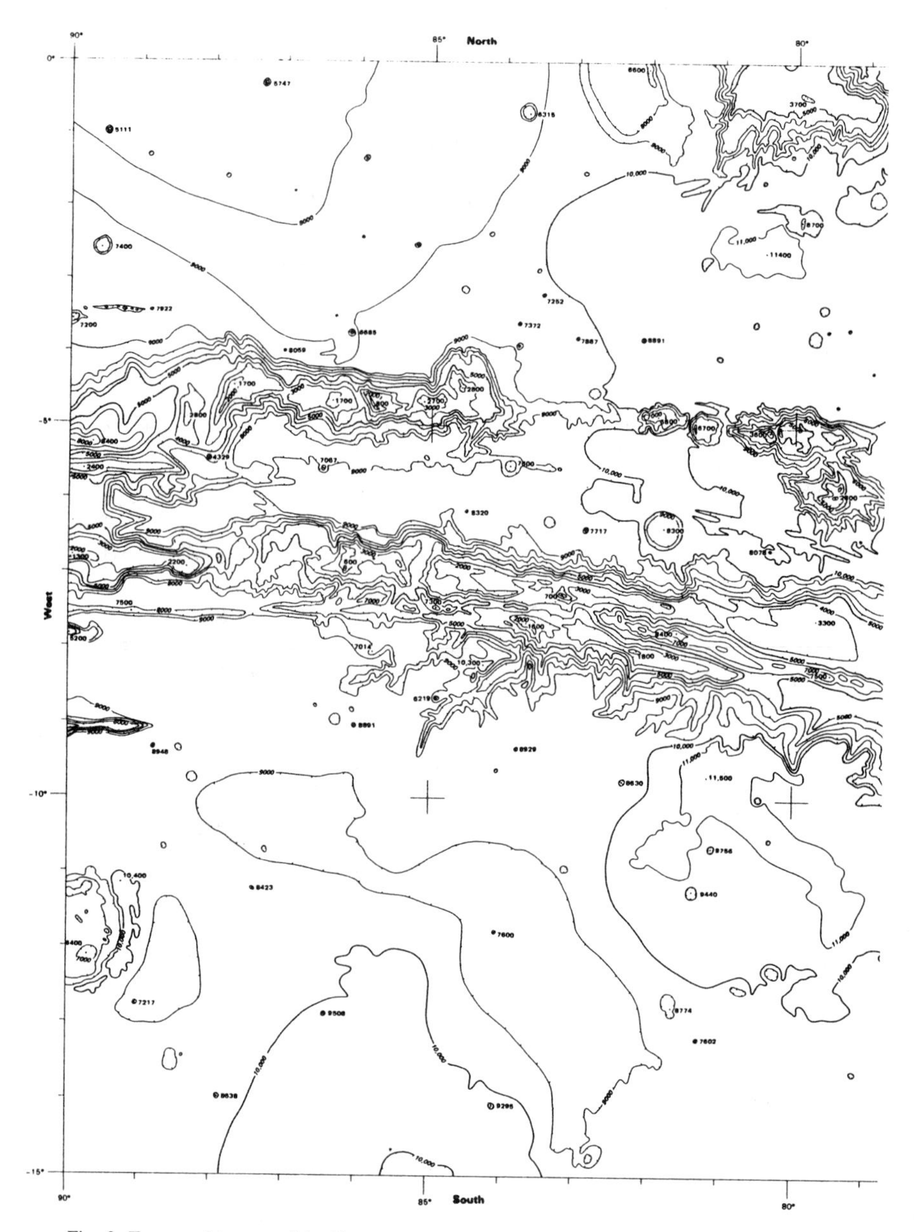

Fig. 3. Topographic map of the Coprates (MC-18) NW quadrangle of Mars. The map is a typical format of Mars 1:2M-scale topographic map series in the equatorial belt. Each map covers an area of 22.°5 longitude by 15.°0 latitude. Contour interval is 1 km. Original map was compiled at 1:2M. This is perhaps the widest place (600 km) of the Mars canyonland. The depth of the canyon can be 8 to 9 km and the slope of the canyon wall ranges from 20° to 30°.

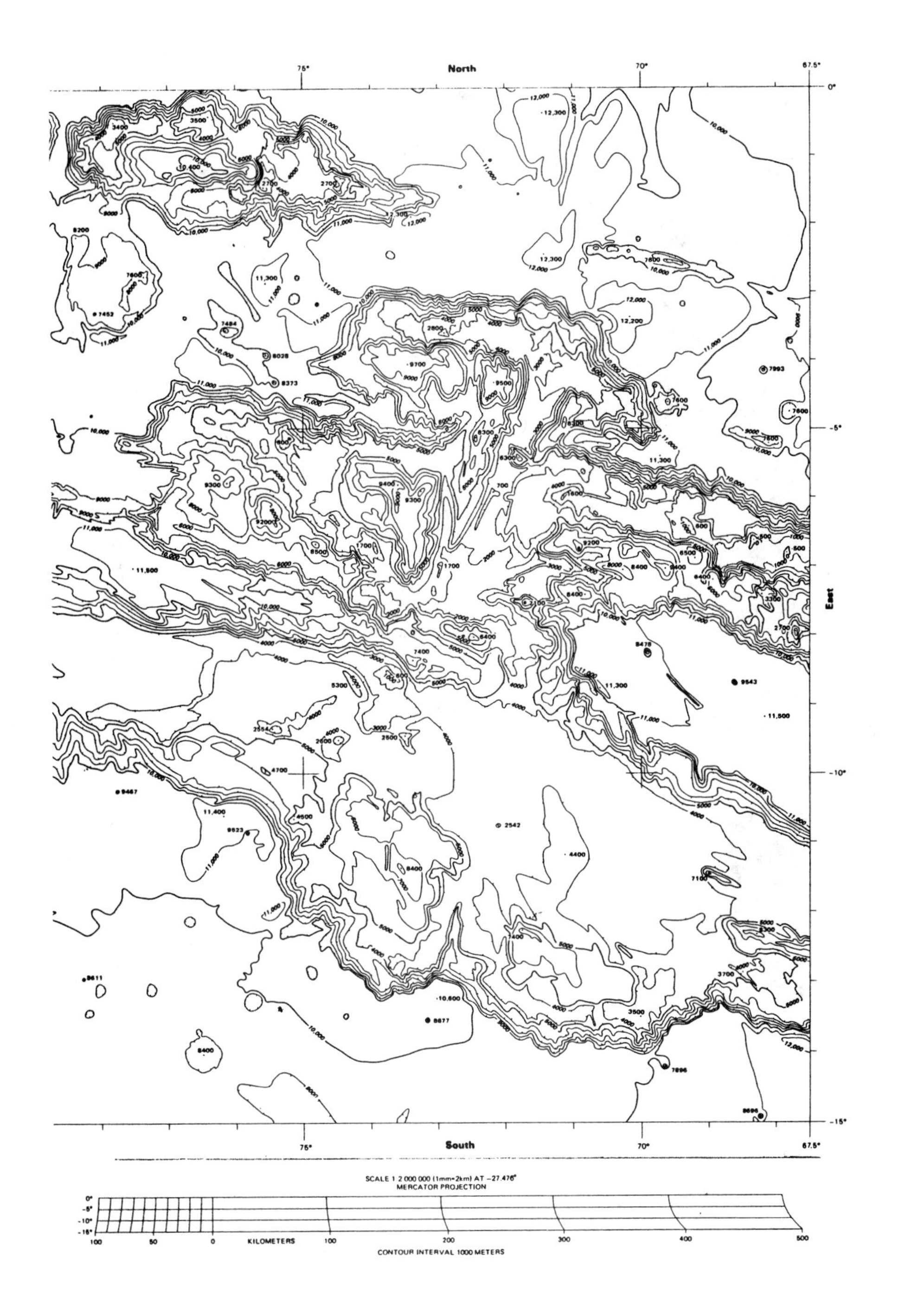
North
South
East
75°
70°
67.5°
0°
-5°
-10°
-15°
SCALE 1:2 000 000 (1mm=2km) AT -27.476°
MERCATOR PROJECTION
KILOMETERS
100
50
0
100
200
300
400
500
CONTOUR INTERVAL 1000 METERS

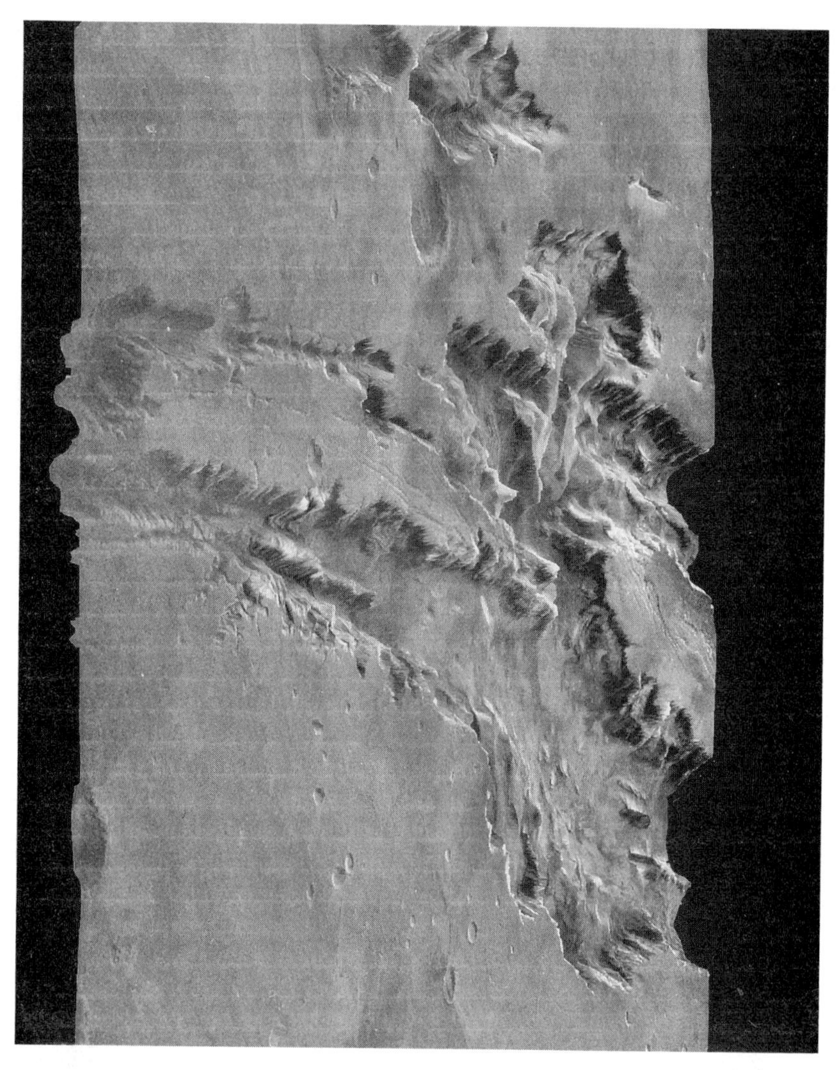

Fig. 4. A perspective view of a small portion of the Mars canyonland shown. Picture was generated by digital image-processing techniques from Viking Orbiter pictures 663A44, 663A42 and 613A61. Relief depiction is controlled by digital elevation data derived from the 1:2M-scale topographic map of MC-18NW shown in Fig. 3. View is toward due east at an angle of 25° from the horizon with a 5 times vertical exaggeration.

the part of the volcano above the 5-km elevation, and a total volume above the 2-km contour of 3.862×10^6 km^3. By using the same map, Olympus is found to have slopes of 2° to 3° and 8° to 24°, respectively, between elevations 2 to 7 km and 7 to 16 km. The upper slopes, from 13 km to the outside rim of the caldera at 24 km, measure 2°5 to 6°5 (Wu et al. 1985).

Because the Mars missions were not specifically planned for making three-dimensional photogrammetric measurements, stereomodels in many cases can be constructed only with pictures that were taken from camera stations in different orbits. As a consequence, the pictures of the stereomodels have different flight heights, a different appearance of the same surface area caused by shadowing effects from different Sun angles, and unusual model geometry. These attributes sometimes result in poor precision of topographic maps.

VII. MAP SHEET IDENTIFICATION

Each 1:5M sheet is identified in four ways:

1. By the name of a conspicuous feature within its boundaries. These names were selected prior to the Mariner 9 mission by the Geodesy/Cartography Group of the Mariner 9 television team. The features are albedo markings observed through Earth-based telescopes. Following IAU recommendations, this name also identifies provinces to which a crater lettering scheme is tied. For example, within the Argyre quadrangle, all double or triple letters that designate craters are prefixed by Argyre (Arg).
2. By an arbitrary numbering system from 1 through 30 preceded by the letters MC for Mars Chart. The MC numbering system is concise and convenient for quadrangles in a systematic mapping scheme, but has little mnemonic value, especially as the number of maps in a series increases.
3. By a serial number assigned by the U.S. Geological Survey as maps are published. This U.S. Geological Survey-assigned number is not restricted to planetary maps. The set of maps with "I-" numbers contains geologic maps of the Earth, Moon and planets, along with topographic maps of the planets, in sequential order of publication. This number must be used to order a map from the U.S. Geological Survey map distribution center.
4. By a five-part alphanumeric map designator code, that includes:
   a. A prefix identifying the planet. The system is applied to all mapped bodies in the solar system except the Earth and the Moon. Mars maps are identified by the letter M. Satellites have two-letter designations, in which the first (capitalized) is that of the primary and the second (lower case) is that of the satellite. A map of Phobos thus has the prefix Mp.
   b. The scale of the map. For example, a 1:5,000,000 map is designated 5M and a 1:500,000 map is designated 500k.
   c. The location of the area mapped, designated by the center coordinates

of the quadrangle rounded to the nearest degree. On planet-wide maps this designation is replaced by a simple numeric code.

d. The kind of map. A code consisting of three letters or less, e.g., R for shaded relief.
e. The year of publication.

According to this system, a 1:5M shaded relief map of Mars with boundaries of 0° and 30° S and 45° and 90° W and published in 1975 would be designated M 5M -15/68R 1975.

VIII. DIGITAL MAPPING

Most planetary images are converted to black-and-white photographs for making base maps (Batson 1982) and for systematic geologic analysis. The fact is, however, that digital spacecraft images are more than photographs; each is a complex array of data with potential application far beyond the display capability of film. Not only is it becoming essential to pursue major research tasks with basic digital data, but the analytical value of data acquired during future missions by advanced systems such as mapping spectrometers will demand a fully digital system and the ready availability of maps in digital format prepared by image processing methods.

With these concepts in mind, a system for continuing the mapping of Mars by digital methods has been implemented (Batson 1987; Edwards 1987). The system will create a standardized cartographic data base for distribution in digital form, and conventional maps can be derived from it for hard-copy reproduction. The process includes radiometric correction of Viking Orbiter images to remove instrument signatures and other artifacts; geometric processing to correct camera distortions, to transform images to a common projection, and to position them accurately in their correct locations on Mars; and photometric modeling to normalize image contrast and base tones as a function of solar incidence angle and viewing angles, so that each image has similar contrast whether it was taken near the terminator or the subsolar point, regardless of the color filter. Finally, all the processed images are placed into a digital mosaic array.

These digital data bases are referred to as digital image models (DIM's). They are compatible with digital terrain models (DTM's) that are being derived for Mars from photogrammetric compilations (Wu and Howington 1986). DTM formats are the same as those for the DIM except that the former are arrays of elevations and the latter are arrays of brightnesses.

The simplest form of a digital model (DM) is one in which each image element's value is stored in a bin labeled in terms of latitude and longitude. For computer work, the concept of map projection is irrelevant; it is only necessary that each bin be readily accessible. In compiling and describing the DM, however, it is convenient to discuss a digital array in terms of map projections. The simplest projection is one in which each image line, or row of bins, is a parallel of latitude and each column of samples, or bins, is a meridian. This "simple cylindrical" or "square" projection is appealing in its simplicity, even though the higher latitudes are oversampled (e.g., the pole of a planet, in reality, is a point, but is represented digitally by an image line with as many samples as that for the equator, all with the same value). Several planetary consortia, consisting of geologic, geochemical and geophysical data bases in this format, have employed the format for several years for the Moon, Mars, Venus and the Galilean satellites (Johnson et al. 1983; Soderblom et al. 1978; Pettengill et al. 1980; Kieffer et al. 1981). The total storage required for this kind of array is only about 60% more ($\pi/2$) than if each element represented the same size area on the planet, and is therefore not prohibitive. However, this projection does present an operational problem, in that a simple cylindrical projection of a single spacecraft image containing the north or the south pole has an unmanageable number of pixels in each image line. The sinusoidal equal-area projection was therefore selected for compiling the planetary DM. The conversions between simple cylindrical and sinusoidal equal-area geometry are so computationally trivial that the two formats are nearly twins.

In the sinusoidal equal-area projection, each parallel of latitude is an image line. The length of each line is compressed by the cosine of its latitude (Fig. 5). Although the sinusoidal is not a particularly suitable projection on which to publish maps, "quick-look" images of the nonpolar parts of the data base can be made without extensive resampling simply by sliding image lines to make a straight line of the meridian in the center of any area of interest. The polar regions are best transformed to a polar projection, even for quick-look application.

Instead of defining pixel dimensions in m or km on the surface of a target, the DM are encoded so that the number of lines in a global DM is an integer. It is therefore more convenient to specify DM resolution in terms of planetocentric degrees than in linear units. The size of pixels in DM is specified as some negative power of 2 (e.g., 1/4, 1/8, 1/16 . . . 1/256, etc.) degrees per pixel. Resolutions intermediate to these values are not used. Data bases can be registered in scale simply by successively doubling or halving the pixel sizes by interpolation or averaging.

The DIM's of Mars are being compiled at three digital scales. A *low-resolution* DIM has been compiled at 1/64°/pixel (about 1 km per pixel) to serve as a geometric control base for all other Mars digital mapping. This concept was developed to eliminate the difficulties associated with identification and location of established control points in images of differing resolution and lighting angles. The low-resolution DIM was made with images and camera orientation matrices derived for controlling topographic mapping. The DIM was compiled solely as a control base, and not for general distribution. It contains gores and no cosmetic processing was performed. A

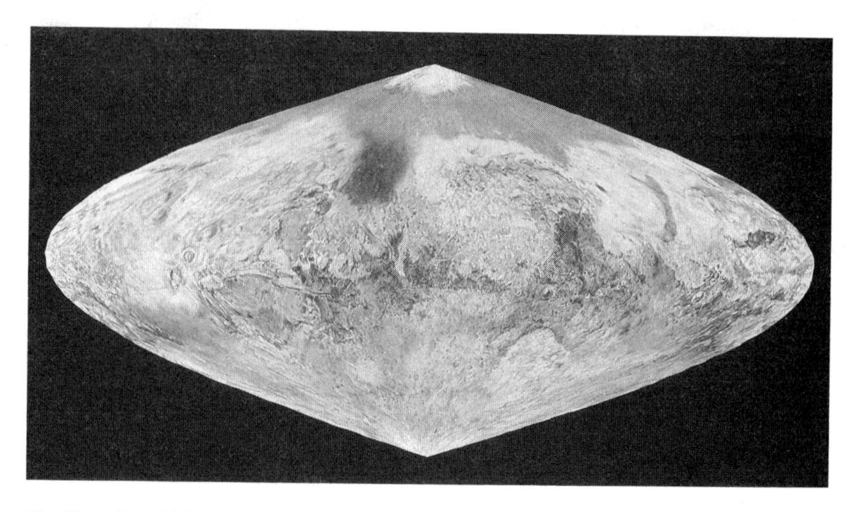

Fig. 5. A sinusoidal equal-area projection map of Mars.

medium-resolution DIM is being compiled at 1/256°/pixel (231 m per pixel) to support the Mars Observer and other future Mars missions. Roughly 5000 Viking Orbiter images are included in this product; most of the same images were contained in the published 1:2M controlled photomosaic series. Whereas the 1:2M mosaic series was controlled by discrete points in a control net, the medium-resolution DIM is tied to the low-resolution DIM. The medium-resolution DIM will replace the low-resolution DIM as the planimetric datum for Mars. It is important to note that the published standard error of the lateral topographic control on which the low-resolution DIM is based is about 5 km; the absolute accuracy of any Mars DIM cannot, therefore, be better than 5 km.

A DIM consisting of multispectral Viking Orbiter images of Mars is nearly complete at 1/64°/pixel (personal communication, L. Soderblom). Selected high-resolution (1/512° to 1/1024°/pixel) DIM's also being made will support special science studies, including evaluation of potential landing sites.

The DIM's described above are being used to make conventional image maps of 100 or more controlled photomosaics with color image and contour line overlays, and 50 or more selected 1:500,000-scale controlled photomosaics will be base maps for special-area geologic mapping.

Digital maps are in high demand by Mars researchers. A global-scale digital terrain model (DTM) has been derived from the Mars 1:15M-scale topographic maps. Figures 6a and b are the digital presentation at global scale of Mars in the form of a shaded relief map produced by computer digital processing techniques. Elevation values are interpolated into a grid raster at

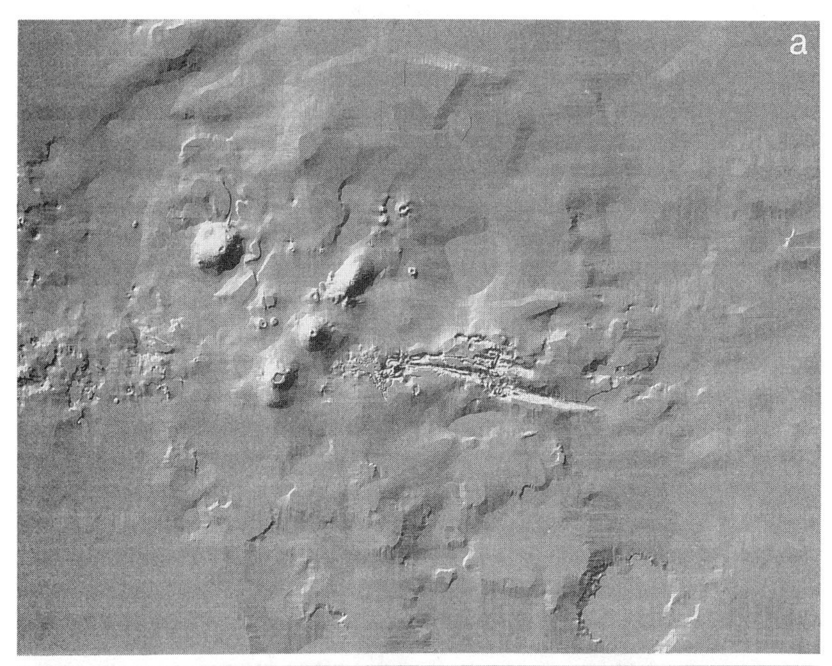

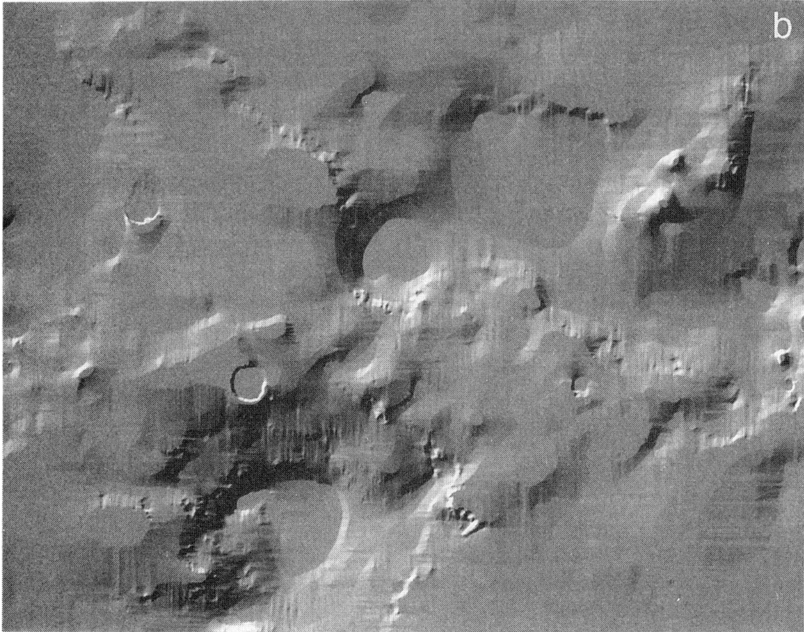

Fig. 6. Shaded relief maps of Mars produced from digital elevation data of the 1:15M-scale global topographic map. Images are in Mercator projection. Lighting direction is from northwest. Vertical exaggeration is $\sim$ 5 times: (a) western hemisphere (long 0° to 180°); (b) eastern hemisphere (long 180° to 360°).

PART III

Bedrock Geology and Geologic Units

II. GLOBAL STRATIGRAPHY

KENNETH L. TANAKA, DAVID H. SCOTT
U.S. Geological Survey

and

RONALD GREELEY
Arizona State University

Recent major advances have been made in the definition and documentation of Martian stratigraphy and geology. Mariner 9 provided the images for the first global geologic mapping program, resulting in the recognition of the major geologic processes that have operated on the planet and in definition of the three major chronostratigraphic divisions: the Noachian, Hesperian and Amazonian Systems. These systems are based on rock sequences and represent the major periods of geologic activity. Viking Orbiter images permitted the recognition of additional geologic units and the formal naming of many formations. Each system was then subdivided into series that correspond to time units (epochs). Epochs are assigned absolute ages based on the densities of superposed craters and crater-flux models. The southern Martian highlands consist of rocks of the Noachian and Hesperian Systems. Rock units include the materials of impact craters and several dozen multi-ringed impact basins. An uncertain, perhaps major, proportion of Noachian highland rocks are of volcanic origin. During Late Noachian and Hesperian time, the highlands were locally incised by channels, marked by ridges and low volcanic paterae, infilled by smooth intercrater plains materials, or eroded by the wind. The northern lowlands may have been formed by one or more huge impacts or by tectonism. They are covered mainly by vast areas of Hesperian and Amazonian plains materials of volcanic and sedimentary origins, but in the northernmost areas the materials have been greatly modified by secondary processes. The lowlands are bordered by a zone of degraded highland material. In Late Hesperian and Amazonian time, immense flow fields, shields and domes developed in the Tharsis and Elysium regions. Associated with the tectonic development of the immense canyon system of Valles Marineris were huge outflow channels that originated in the canyons and chaotic terrain in the highlands and extended to the lowland plains.The long-term history of polar deposits is uncertain; the south polar surface appears to be older than the north polar surface. Mars has had a lively geologic history, as shown by impact craters of Noachian and younger ages and by intermediate-age volcanism, tectonic deformation and outflow-channel formation. Although such processes have waned during the past 2 Gyr, aeolian activity continues; faulting, volcanism and flooding may also occur in the future. This rich geologic history has bequeathed to us many unsolved geologic problems whose solutions will require new data and approaches.

I INTRODUCTION

A. Importance of Martian Stratigraphy

Stratigraphy forms the framework of historical geology. It deals with the overall relations of rock units, their areal distribution, the processes by which they were formed and modified, and the geologic history they record. The development of the Martian stratigraphic record is based on the identification and delineation of geologic (lithostratigraphic or rock-stratigraphic) units (including groups, formations and members) and the relative ages of the units. Geologic units are identified by their topographic, morphologic and spectral properties as displayed on spacecraft images and then delineated on the basis of these properties. Relative ages are established by the traditional principles of superposition and intersection and by the concentration of impact craters superposed on the geologic units.

Thus, Martian stratigraphy provides the basis for assessing the history of local regions as well as the geologic evolution of the entire planet. Stratigraphy also provides the context for describing specific processes and events, such as early and post-accretional impact rates, episodes of volcanism and tectonism, climate changes and release of volatiles, and the resurfacing of large areas of Mars. Martian stratigraphy will make important contributions to future exploration and missions, particularly in the selection of sample-return sites and the interpretation of the geologic significance of radiometrically dated samples.

B. Historical Overview

Telescopic observations made over the years suggested that Mars is an Earth-like planet, with permanent surface markings, polar ice caps that wax and wane with the seasons, surface color changes, atmospheric hazes and clouds, and even possible canals (chapter 2). Our first close-up views of Mars, and indeed of any planetary surface other than the Earth and Moon, were provided by the Mariner 4 spacecraft during its flyby encounter in 1965. However, the 19 images that it transmitted to Earth covered only about 1%

of the Martian surface, mostly in the highly cratered southern highlands of the western hemisphere (chapter 3). The Moon-like appearance of Mars provided by this small pictorial sample was surprising at the time, but in retrospect it is understandable (Hartmann and Raper 1974,p.22).

In 1969, the Mariner 6 and 7 spacecraft flew by Mars and returned 199 images of the planet that extended coverage to about 10% of the surface. Their much higher resolution (300 m/pixel) was provided by a narrow-angle vidicon camera boresighted with a wide-angle camera. These images revealed new features, including light and dark markings, chaotic terrain and smooth plains in Hellas Planitia (Hartmann and Raper 1974,p.24), but they show no evidence of canals. The imaging tracks of both spacecraft were also largely within the southern highlands, extending from the equatorial zone across the southern polar cap. As a result, the great dichotomy in terrain types and crater frequencies that distinguishes the northern and southern hemispheres of Mars was not recognized. However, the returned data enabled production of prototype regional geologic maps of a part of Mars (U.S. Geological Survey 1972).

In 1971–1972, the Mariner 9 Orbiter provided the first global coverage (over 7000 images) of Mars. Many new, spectacular terrain types could be distinguished and identified. The vast northern plains (Fig.1), Nix Olympica (now known as Olympus Mons), Valles Marineris, large channel systems, and a myriad of other tectonic, volcanic, aeolian and degradational terrains and features were revealed (McCauley et al. 1972). Meaningful geologic and stratigraphic analyses and compilation of preliminary geologic maps at local and global scales became possible (see, e.g., Carr et al. 1973; Soderblom et al. 1974; Mutch et al. 1976,ch.9). Production of a geologic map series of 30 quadrangles at 1:5M scale for the entire Martian surface was begun. Relative ages were related to impact-crater degradation states (see, e.g., Masursky et al. 1978; Wise 1979; Moore 1980) and, for three quadrangles (Milton 1974*b*; Masursky et al. 1978; Wise 1979), absolute ages were related to impact-crater densities based on crater-flux models (Soderblom et al. 1974; Neukum and Wise 1976; chapter 12). The geologic map series was used as a basis for the first formal global geologic map of Mars at 1:25M scale (Scott and Carr 1978). In this map, the three time-stratigraphic systems were established based on the relative positions of major rock units and the densities of craters 4 to 10 km in diameter (Condit 1978).

With the advent of high-resolution (10 to 100 m/pixel) Viking Orbiter images, more detailed stratigraphic investigations were pursued (see, e.g., Masursky et al. 1977; Plescia and Saunders 1979; Hartmann et al. 1981; Neukum and Hiller 1981; Scott and Tanaka 1981). The new 1:2M-scale photomosaic series was used for thematic maps, including a series of 16 maps portraying lava flows and geologic units in the Tharsis region (Scott et al. 1981*a*). More recently, a set of three revised geologic maps at 1:15M scale

Fig. 1. Shaded-relief map of northern lowlands of Mars, showing approximate location of highland/lowland boundary centered at 50° N, 190°. Lambert Equal-Area projection; 30° grid spacing. Circle shows approximate outline of proposed mega-impact basin.

was completed that covers the entire Martian surface (Scott and Tanaka 1986; Greeley and Guest 1987; Tanaka and Scott 1987; Color Plates 5 and 6; Maps I-1082-A, -B, -C). Because these maps were based on high-resolution Viking Orbiter images, they show nearly four times as many map units as the previous global map, and many relative-age assignments could thus be modified. Tanaka (1986) used the new maps to subdivide the three time-stratigraphic systems into eight series and to assign relative ages to all major geologic units and features on Mars. Barlow (1988*c*) mapped all craters on Mars > 8 km in diameter and delineated seven crater-density units to determine ages of surfaces relative to the heavy bombardment (see chapter 12). Overall, great improvements in the deciphering and description of Martian stratigraphy have resulted from photogeologic studies of Viking Orbiter images.

II. FORMAL MARTIAN STRATIGRAPHIC CLASSIFICATIONS

Although Martian stratigraphy is largely based on photogeologic interpretation, the strict guidelines provided by the North American Commission on Stratigraphic Nomenclature (1983) and its predecessors for the definition of formal stratigraphic units were followed to the extent possible. For the recent global map set, names were reviewed by the U.S. Geological Survey's Geologic Names Committee. The Martian stratigraphic nomenclature is broadly similar to that previously adopted for the Moon (see Wilhelms 1987,ch.7). *Geologic (rock or lithostratigraphic) units* are mappable and can be defined by their topographic, morphologic and spectral characteristics (which are used to interpret lithic properties) and by their stratigraphic position. *Time-rock (chronostratigraphic) units* are based on reference rocks (referents) that represent all rock units formed during the same interval of time. A given rock unit may fall within more than one time-rock unit. *Time (chronologic) units* are divisions of time represented by the time span of an established time-stratigraphic unit. The following sections describe how these units have been applied on Mars; for a more detailed overview of the rationale, methods and conventions of planetary geologic mapping, see Wilhelms (1990).

A. Geologic (Rock or Lithostratigraphic) Units

The 1:15M-scale geologic maps of Mars contain a mixture of informal and formal geologic units; these units supersede those used in previous global mapping (Scott and Carr 1978). The formal names are limited to 18 formations; most are distinctive volcanic or plains deposits. These formations are generally well defined in stratigraphic position and occurrence and commonly consist of two or more members. Each member is assigned a type area if widespread. Additional formal names may be adopted on future regional- and local-scale maps where appropriate. Informal names are used to describe other rock units that are less well defined in age, origin, or extent; these units include a variety of surficial materials. Informal group names are also used to describe geographic or geomorphic associations (e.g., lowland terrain materials, plateau sequence and northern plains assemblage). Color plate 7 shows all the map units used but with a generalized format and names. Many major or typical geologic units are illustrated in Figs. 2, 3 and 4.

Relative ages of craters (and their ejecta) on the 1:15M-scale maps are shown in two ways: the crater material either superposes or is partly buried by surrounding materials; crater-morphologic ages are avoided. Although crater morphology commonly is used on Mars maps to distinguish relative age, this method has problems at global scale. Dating schemes differ considerably, even on maps of the same scale (Tanaka 1986), and crater-degradation rates may have varied locally. Crater materials, as well as other materials

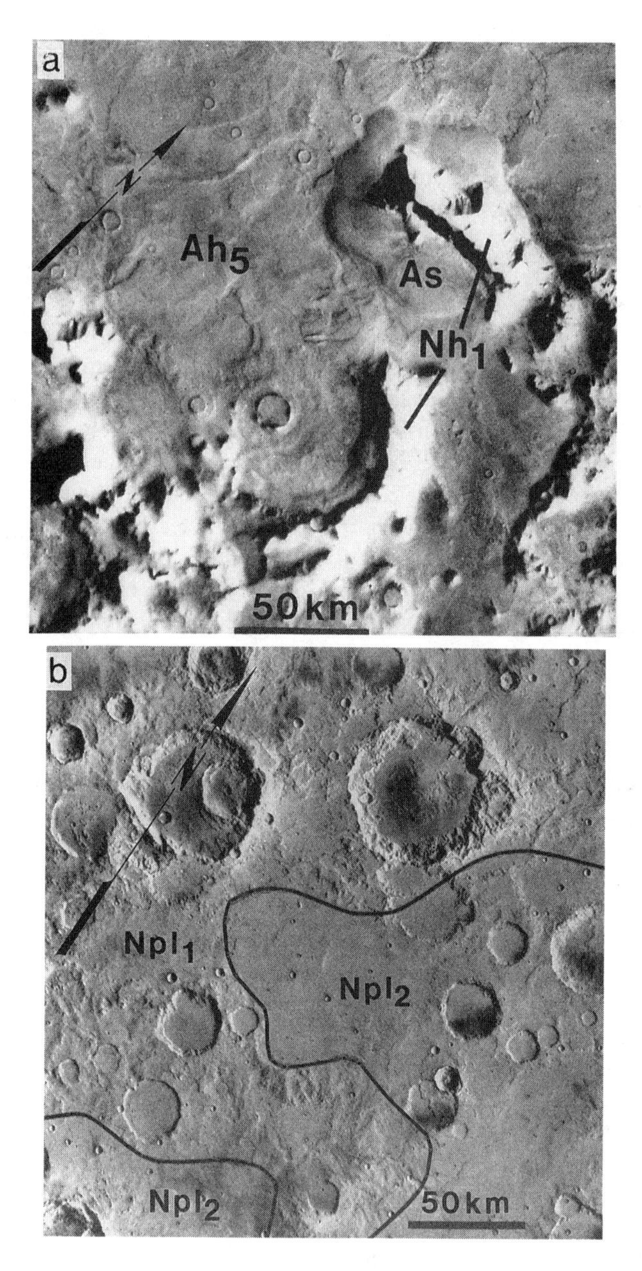

Fig. 2.(a) View of Noachian geologic units. Lower Noachian rim material of Hellas basin (unit Nh_1) embayed by basin-fill material (channeled plains rim unit, unit Ah_5). Young aprons of slide material (unit As) surround larger massifs. (Viking image 97A61.) (b) View of Noachian geologic units. Middle Noachian cratered unit of plateau sequence (unit Npl_1) in Terra Sirenum embayed by Upper Noachian subdued cratered unit (unit Npl_2). (Viking image 597A75.)

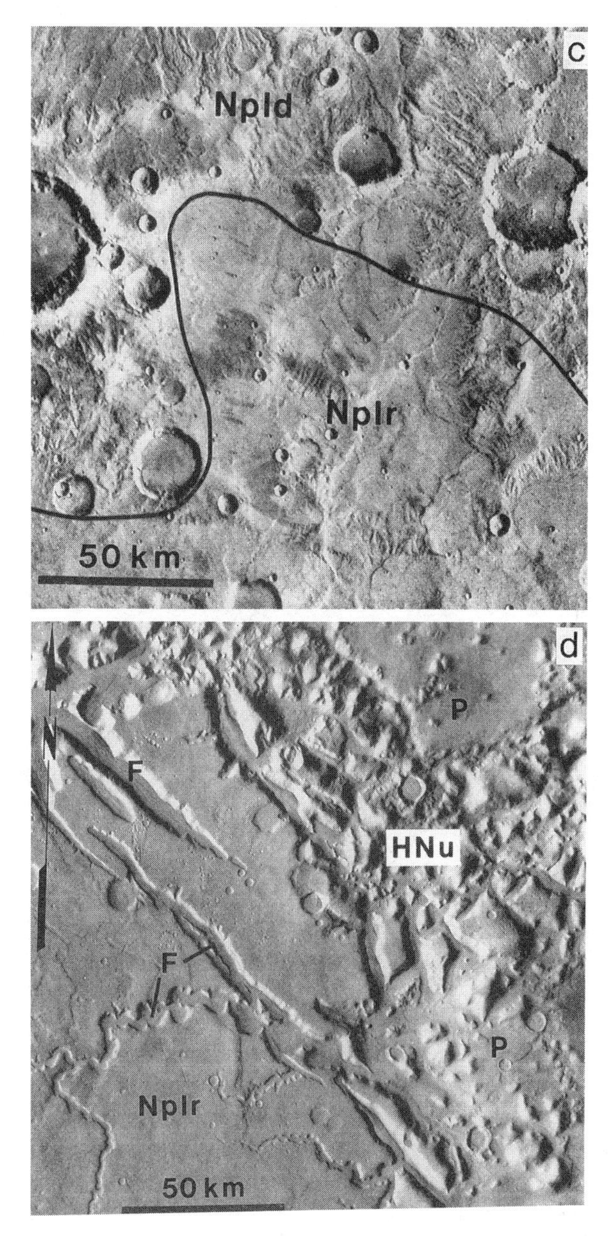

Fig. 2.(c) View of Noachian geologic units. Middle Noachian dissected unit (unit Npld) in Noachis Terra embayed by Middle to Upper Noachian ridged unit (unit Nplr). (Viking image 84A17.) (d) View of Noachian geologic units. Middle to Upper Noachian ridged unit (unit Nplr) along highland/lowland boundary cut by fretted channels and troughs (F). Broken-up mesas and knobs of plateau material, mapped as undivided material (unit HNu), make up part of Protonilus Mensae. Mesas and knobs are embayed by Hesperian and Amazonian plains material (P). (Viking image 569A20.)

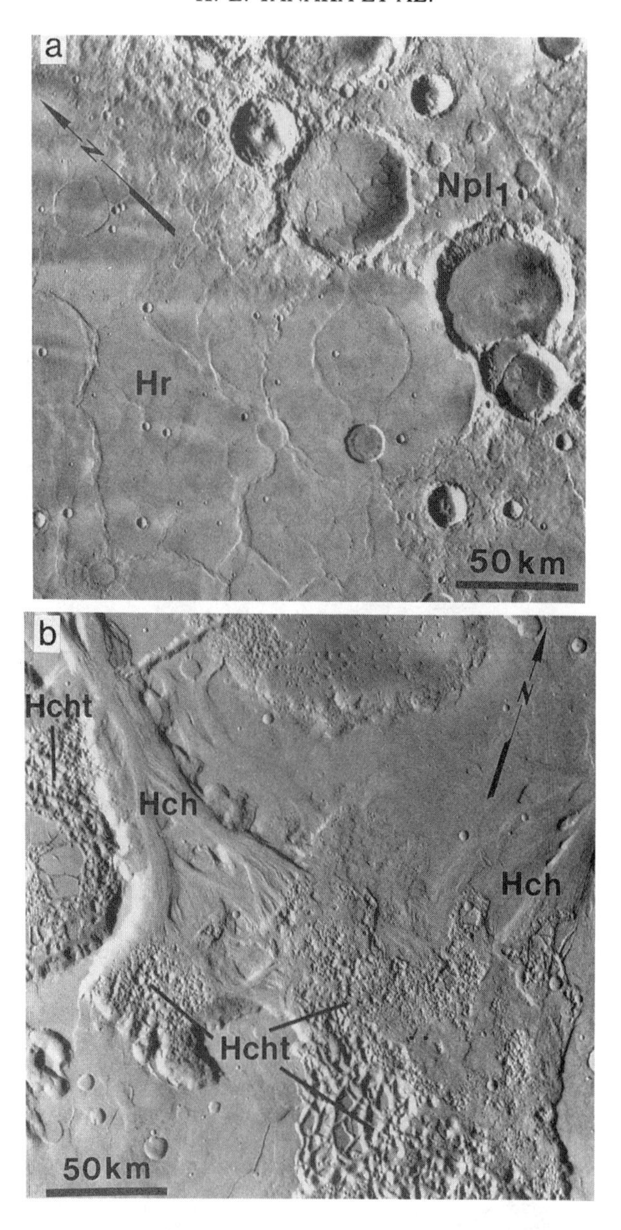

Fig. 3.(a) View of Hesperian geologic units. Lower Hesperian ridged plains material (unit Hr) in Hesperia Planum embaying older Noachian cratered terrain (unit Npl_1). (Viking image 365S65.) (b) View of Hesperian geologic units. Upper Hesperian chaotic material (unit Hcht) near heads of Tiu and Ares Valles outflow channels (filled by unit Hch). Note streamlined features on channel floor. (Viking image 366S68.)

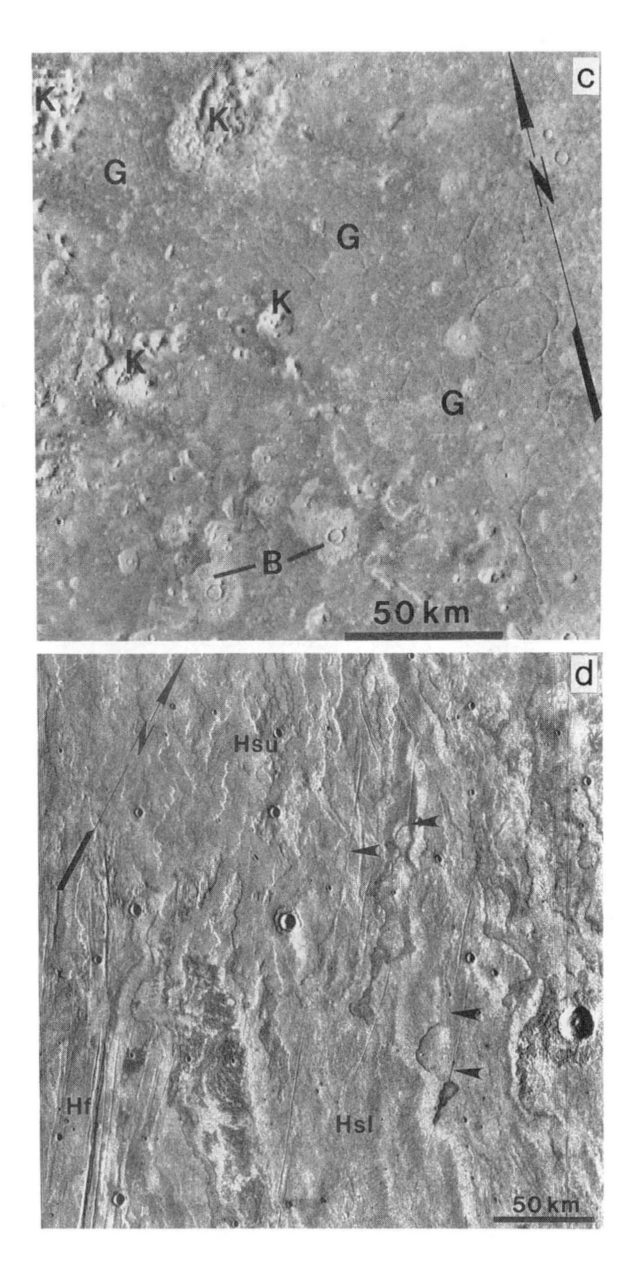

Fig. 3.(c) View of upper Hesperian plains materials making up Vastitas Borealis Formation in Acidalia Planitia. Note older rocks degraded into knobs (K) and embayed by dark, grooved plains (G). Impact-crater blankets (B) are light colored; hence plains appear mottled. (Viking image 670B18) (d) View of upper Hesperian lava flows of Syria Planum Formation. Younger flows (unit Hsu) bury older, fractured flows (units Hf and Hsl); some flows appear to emanate from fissures (arrows). (Viking image 643A64.)

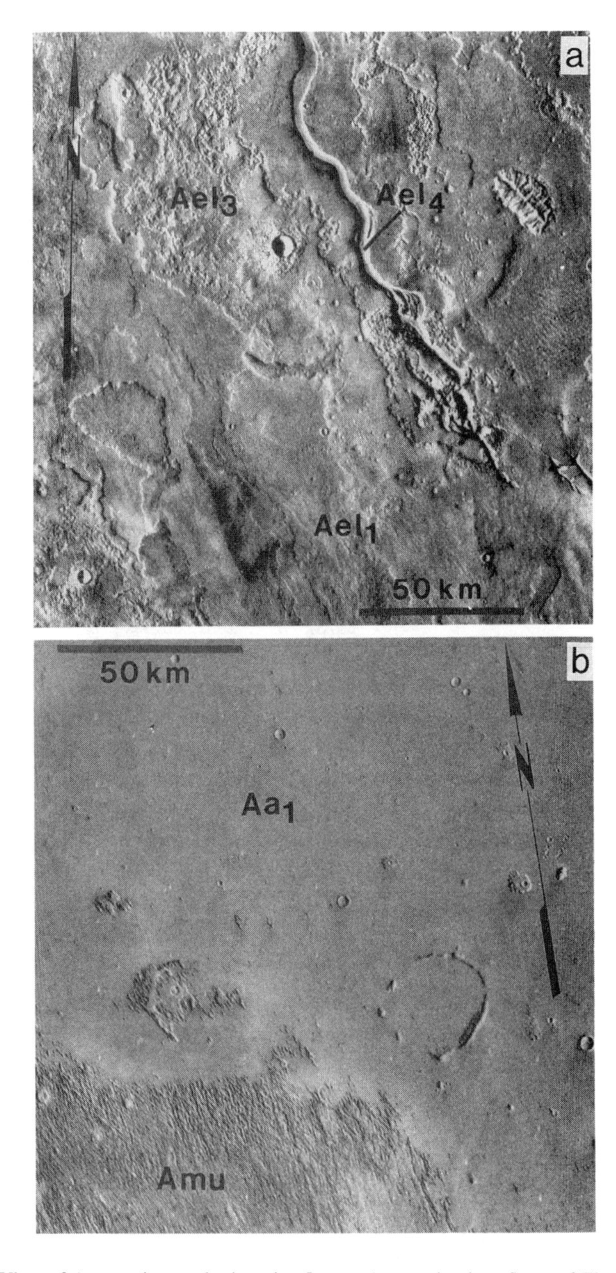

Fig. 4.(a) View of Amazonian geologic units. Lower Amazonian lava flows of Elysium Formation (unit Ael_1) northwest of Elysium Mons. Hrad Vallis (filled by unit Ael_4) originates near base of flows, possibly from northwest-trending fractures of set exposed in region. Rough, degraded flows (unit Ael_3) may be lahars. (Viking image 541A10.) (b) View of Amazonian geologic units. Lower Amazonian smooth plains material of Arcadia Formation (unit Aa_1) in Amazonis Planitia superposed by Upper Amazonian deeply eroded and grooved material of Medusae Fossae formation (unit Amu). Linear grooves trending north and northwest may be eroded complementary joint sets (Viking image 886A13.)

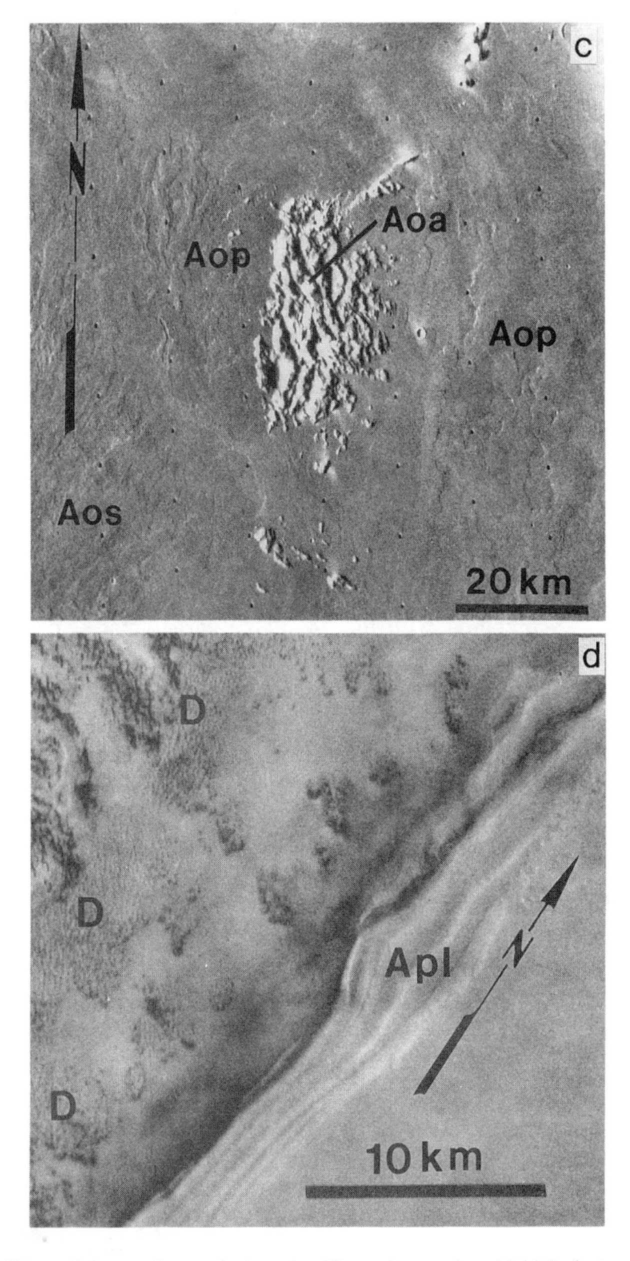

Fig. 4.(c) View of Amazonian geologic units. Upper Amazonian shield (unit Aos; northeast-trending flows) and plains (unit Aop; north-trending flows) units of Olympus Mons Formation burying rugged aureole material (unit Aoa). Note scarcity of impact craters. (Viking image 47B29). (d) View of Amazonian geologic units. Polar layered material (unit Apl) and dune fields (D) in north polar region. (Viking image 541B29.)

whose ages or makeup are not well known, such as mountain, channel-bar, dome and undifferentiated volcanic materials, are mapped as features without defined stratigraphic position rather than as stratigraphic units.

B. Time-Rock (Chronostratigraphic) Units

Formal definition of Martian time-stratigraphic units began with the 1:25M-scale geologic map of Mars by Scott and Carr (1978). The Noachian System is represented by materials of the ancient cratered and rugged terrain in the Noachis quadrangle (see Fig. 2b,c); the base of the system is not exposed. Ridged plains material (as on Hesperia Planum; Fig. 3a) forms the base of the Hesperian System and overlies Noachian rocks. The Amazonian System consists largely of smooth plains materials; the base of the Arcadia Formation exposed in the Amazonis quadrangle defines the base of the system (Fig. 4b). On most geologic maps of Mars published by the U.S. Geologic Survey after 1978, the system containing a rock unit is indicated by the first letter of the unit's symbol. (Two letters are given if a unit straddles a system boundary.) The set of 1:15M-scale maps adheres to this nomenclature and shows the correlation of geologic units and systems with cumulative densities of craters of 2-, 5- and 16-km diameters. This map set also shows 90 geologic units, compared with the 24 of the earlier map (Scott and Carr 1978). Many of the units in the map set have relative ages not accommodated in the earlier scheme. For example, most materials in the northern plains, as well as circum-Chryse outflow-channel materials, postdate ridged plains material but predate smooth plains in Amazonis Planitia. Thus, the plains materials as a group (named the Vastitas Borealis Formation) are the referent for the Upper Hesperian Series. On the basis of such refinements, Tanaka (1986) divided the three Martian systems into eight series (Table I) based on referents shown on the 1:15M-scale map set.

TABLE I
Martian Stratigraphic Series and Reference Materials[a]

Series	Reference Unit(s)[b]	Type Locality
Upper Amazonian	Achu	Southern Elysium Planitia
Middle Amazonian	Aa_2, Aa_3	Amazonis Planitia
Lower Amazonian	Aa_1	Acidalia Planitia
Upper Hesperian	Hvk, Hvg, Hvr, Hvm	Vastitas Borealis
Lower Hesperian	Hr	Hesperia Planum
Upper Noachian	Npl_2	East of Argyre Planitia
Middle Noachian	Npl_1	Noachis quadrangle
Lower Noachian	Nb	Charitum and Nereidum Montes

[a] Table modified from Tanaka (1986).
[b] See Color Plate 7.

C. Time (Chronologic) Units

The Martian systems and series correspond to time spans known as periods and epochs (Table II). Although direct measurements of absolute age are unavailable for Martian rocks, model ages based on estimated relative impact-crater flux between the Moon and Mars have been calculated. Ages attributed to a given density among existing models may vary widely (see chapter 12). Commonly used models include those of Neukum and Wise (1976) and Hartmann et al. (1981), in conjunction with the classification of Tanaka (1986). These models have been applied to the Martian stratigraphic column to indicate possible absolute ages of epoch boundaries; see Table II (Tanaka 1986; Tanaka et al. 1988).

III. GEOLOGY OF MARS

A. Introduction

This overview of the stratigraphy of Mars is arranged according to major physiographic provinces and assemblages of geologic materials as shown on the set of 1:15M-scale maps (Scott and Tanaka 1986; Greeley and Guest 1987; Tanaka and Scott 1987). Table III shows the major divisions of geologic units on Mars, their ranges in stratigraphic position, and their areal extents. We offer lithologic interpretations for many geologic units and recount the geologic histories of major features. In most cases, we only summarize interpretive models regarding possible geologic processes; detailed explanations are available in other chapters of this book.

TABLE II
Martian Epochs, Crater-Density Ranges and Absolute-Age Ranges[a]

	Crater-Density Range N(D) = no.>D/10^6 km^2			Absolute-Age Range (Gyr)	
Epoch	**N(2)**	**N(5)**	**N(16)**	**HT**	**NW**
Late Amazonian	<40			9.25–0.00	0.70–0.00
Middle Amazonian	40-150	<25		0.70–0.25	2.50–0.70
Early Amazonian	150-400	25-67		1.80–0.70	3.55–2.50
Late Hesperian	400-750	67-125		3.10–1.80	3.70–3.55
Early Hesperian	750-1200	125-200	<25	3.50–3.10	3.80–3.70
Late Noachian		200-400	25-100	3.85–3.50	4.30–3.80
Middle Noachian		>400	100-200	3.92–3.85	4.50–4.30
Early Noachian			>200	4.60–3.92	4.60–4.50

[a]Data from Tanaka (1986) and Tanaka et al. (1988) based on Hartmann-Tanaka (HT) and Neukum-Wise (NW) ages which represent the two relatively different time-scale models.

TABLE III
Major Geologic Units on Mars and Their Stratigraphic Positions and Areal Extents[a]

Geologic Unit	Stratigraphic Position[b]	Area (10^3 km^2)
SURFICIAL MATERIALS	UH-UA	5993
Aeolian and dune materials	UA	832
Mantle material	UA	1802
Slide material	MA-UA	644
Polar deposits	UH?-UA	2715
CHANNEL-SYSTEM MATERIALS	UH-UA	4326
Channel materials	UH-UA	1563
Floodplain materials	UH-UA	1073
Chaotic material	UH	829
Younger channel and floodplain materials, undivided	UA	808
LOWLAND TERRAIN MATERIALS	UH-UA	30912
Northern Plains Assemblage	UH-UA	25450
Arcadia Formation	LA-UA	6623
Medusae Fossae Formation	MA-UA	2487
Vastitas Borealis Formation	UH	12,896
Smooth plains material	LA-UA	2970
Etched plains material	UH-MA	474
Eastern Volcanic Assemblage	UH-LA	5454
Elysium Formation	LA	4050
Albor Tholus Formation	UH-LA	19
Hecates Tholus Formation	UH	28
Syrtis Major Formation	UH	1357
HIGHLAND TERRAIN MATERIALS	LN-UA	81570
Western volcanic assemblage	UH-UA	14102
Olympus Mons Formation	LA-UA	1746
Tharsis Montes Formation	UH-UA	6897
Alba Patera Formation	LH-LA	3167
Ceraunius Fossae Formation	UH-LA	462
Syria Planum Formation	UH	1830
Hellas assemblage	LN-MA	4255
Plateau and high-plains assemblage	LN-MA	63,213
Valles Marineris interior deposits	UH-MA	324
Highland paterae	UH	745
Tempe Terra Formation	LH-UH	417
Dorsa Argentea Formation	LH-UH	1255
Plateau sequence	LN-UH	53473
smooth unit	LH	3505

TABLE III Continued

Geologic Unit	Stratigraphic Position[b]	Area (10^3 km^2)
mottled smooth plains unit	LH-UH	268
subdued cratered unit	UN	8270
cratered unit	MN	18866
dissected unit	MN	12961
etched unit	MN	2557
ridged unit	MN-UN	4462
hilly unit	LN	2584
Highly deformed terrain materials	LN-UH	3075
Undivided material	LN-LH	3351
Highland volcanoes	MN-UH	323
Mountain material	LN-MN	250
MATERIALS THROUGHOUT MAP AREA	LN-UA	21,197
Knobby plains material	LA-MA	2386
Ridged plains material	LH	13,302
Impact-crater materials	LN-UA	5509
TOTAL	LN-UA	143998

[a]Table modified from Tanaka et al. (1988).
[b]N = Noachian; H; = Hesperian; A = Amazonian; L = Lower; M = Middle; U = Upper.

The major geologic division of Mars' surface is known as the crustal dichotomy, which separates the physiographically and geologically distinct southern highlands and northern lowlands (Fig. 1). This highland-lowland boundary is a large, irregular circle centered at latitude 50° N, longitude 190°; most of it is marked by a gentle, irregular scarp and clusters of low knobby hills. In places the boundary scarp is covered by younger lava flows or is deeply indented by large basins that appear to be evidence of earlier impacts.

The southern highlands include assemblages of materials covering more than 60% of the surface of Mars. Most of the highland terrain consists of ancient, rough, densely cratered rocks (largely impact breccias) formed early in the planet's history when impact rates were high. Extensive lava flows have covered large areas within the highlands. These flows, together with aeolian and fluvial processes, have mantled or subdued parts of the older and rougher terrain. Although highland rocks range in age from Early Noachian to Late Amazonian, most are grouped in the plateau sequence, largely formed or emplaced during the Noachian Period.

The northern lowland plains lie, for the most part, below the 0-km elevation datum of Mars. They extend to Planum Boreum, which covers the north polar region. The plains are mostly covered by lava flows and sediments of Late Hesperian to Late Amazonian age, and thus they are smoother and less cratered than the southern highlands. However, in some places such as Phlegra Montes and Acidalia Planitia, remnants of eroded and hilly terrain resembling highland material are surrounded and embayed by plains units. Fluvial deposits occur in basins along the highland-lowland boundary; these deposits and morphologic evidence of shorelines in places (Fig. 5a) suggest that large paleolakes or even oceans may have accumulated from the runoff of outflow channels originating in the highlands (Parker et al. 1989).

The origin of the highland-lowland dichotomy is unknown; it predates Lower Hesperian lava flows that extend across the boundary scarp. Several hypotheses have been advanced to account for the dichotomy; they involve subcrustal erosion due to mantle convection (Wise et al. 1979*a*; McGill and Dimitriou 1990), a large impact (Wilhelms and Squyres 1984), and several impact events (Frey and Schultz 1988). Despite the profound dichotomy between the hemispheres, the polar regions are similar in that both have vast accumulations of aeolian materials and perennial ice caps.

B. Highland Rocks

1. Impact Basins. Large impact basins are among the oldest recognizable features on the surface of Mars. Even before Mariner 9 revealed its complex and diverse geology, telescopic observations from Earth showed circular white zones during winter in the southern hemisphere. With spacecraft imaging, these were seen to be the conspicuous impact basins, Hellas and Argyre, the floors of which are frost covered in winter. Although Mariner 9 images (McCauley et al. 1972) permitted the detection of several other impact basins (such as Isidis), they seemed to be less abundant on Mars than the Moon and Mercury, particularly multi-ringed basins. On Mars, their identification is difficult because the ancient terrains have been heavily eroded by wind and water as well as covered by sediments and volcanic deposits. However, the images of the Viking Orbiter cameras have enabled more than 50 large, mostly multi-ringed basins to be tentatively identified (Table IV).

The Hellas basin has a rim diameter of $\sim$ 2300 km and is among the largest impact scars in the solar system. It is the dominant surface feature of the Martian southern highlands. The Hellas assemblage (Greeley and Guest 1987) consists of units in and around the basin (e.g., Fig. 2a). Its lowermost unit, the Noachian *basin-rim unit,* consists of massifs as large as 180 km across that form an incomplete ring defining the basin. Rim material is most abundant in the northwestern part of the ring. The northeastern part of Hellas has been breached by fluvial channels whose deposits partly fill the basin. The southwestern part of the rim is covered with volcanic materials associated with Amphitrites and Peneus Paterae, and southeastern massifs of the basin rim are partly degraded and surrounded by aprons of *slide material*

TABLE IV
Impact Basins on Mars[a]

Designation	Center (lat., long.)		Ring Diameter(s) (km)
Acidalia	60° N	30°	1000, 1500, 1950
Al Qahira	20° S	190°	141?, 353, 715, 1034
Al Qahira A	13° S	184°	335?, 530, 731, 994
Amazonis	6° N	168°	800
Antoniadi	22° N	299°	200?, 400
Aram Chaos	3° N	22°	140, 250, 440, 550
Arcadia A	37° N	167°	380, 600
Arcadia B	32° N	167°	880, 1463, 1925
Argyre	50° S	42°	540, 1140, 1850
Borealis	50° N	190°	7700
Cassini	24° N	328°	321, 547, 930
Cassini A	14° N	324°	354, 653, 928, 1204
Chryse	22° N	47°	891, 1534, 2596, 3600, 4600
Daedalia A	26° S	125°	1800
Daedalia B	15° S	127°	1475, 2540, 3960
Deuteronilus A	44° N	342°	44, 80, 220, 280
Deuteronilus B	43° N	338°	55, 201
Elysium	33° N	201°	800?, 1540, 2000, 3600, 4970
Gale	5° S	222°	85?, 150
Galle	51° S	31°	100, 220
Hellas	43° S	291°	1350?, 2295, 4200
Herschel	15° S	230°	160?, 290
Holden	25° S	32°	260, 580
Huygens	14° S	304°	250, 460
Isidis	13° N	273°	1100, 1900
Kaiser	46° S	340°	100?, 200
Kepler	47° S	218°	110, 210
Ladon	18° S	29°	270, 470, 580, 975, 1300, 1700
Liu Hsin	55° S	171°	60, 135
Lowell	52° S	81°	90, 190
Lyot	50° N	330°	100, 200
Mangala	0° N	147°	300, 570
Memnonia	22° S	166°	1593, 2065
Molesworth	28° S	210°	80?, 180
near Columbus	25° S	164°	70, 145
Nilosyrtis Mensae	33° N	282°	380
North Tharsis	11° N	98°	1455, 2330, 3650, 4500
overlapped by Newcomb	22° S	4°	380, 800
overlapped by Schiaparelli	5° S	347°	140, 560
overlapped by South Crater	73° S	213°	300, 680
Phillips	67° S	44°	90?, 175
Ptolemaeus	46° S	157°	65?, 150
Schiaparelli	3° S	344°	230, 470
Scopulus	5° N	278°	1900, 2700
Sirenum basin	44° S	166°	500, 710, 1000, 1548

TABLE IV Continued

Designation	Center (lat., long.)	Ring Diameter(s) (km)
southeast of Hellas	58° S 273°	225, 500
southeast of Ma'adim Vallis	30° S 180°	180, 340, 1000
South Hesperia	32° S 255°	900, 1255
south of Hephaestus Fossae	10° N 233°	500, 1000
south of Lyot	42° N 322°	60, 145, 260, 400, 480, 570
south of Renaudot	38° N 297°	600
South polar	83° S 267°	850
Utopia	48° N 240°	3300, 4715
west of Le Verrier	37° S 356°	430
West Tempe	56° N 78°	830?

[a]Data from Wood and Head (1976), Schultz et al. (1982), Frey and Schultz (1988), McGill (1989*a*), and Schultz and Frey (1990).

(Fig. 2a). The basin-rim unit is also overlain by the *ridged plains floor unit,* presumed to be flood lavas contemporaneous with the extensive Hesperian-age ridged plains found elsewhere on Mars. The Hellas basin also contains a Hesperian *dissected floor unit* that is considered to consist of modified aeolian, fluvial and lava-flow materials.

Partial filling, mantling and modification of the Hellas basin continued through the middle of the Amazonian Period to produce the younger units of the Hellas assemblage. Most of these units occur near the basin rim and are considered to be mantling material, some of which is eroded. The *lineated floor unit* and the *channeled plains rim unit* (Fig. 2a) have similar crater-count ages and may be only slightly modified parts of a mantling unit. The floor unit is characterized by lineaments within smooth plains material and may have formed by tectonic modification of mantling material. The channeled plains unit has a subdued appearance and low remnant mesas and narrow channels. The other younger Hellas units form plains distinguished by low-relief hills that may reflect differential erosion of mantling material. Included are a *reticulate floor unit* characterized by its pattern of ridges, a *rugged floor unit* whose undulatory surface has km-scale relief, and a *knobby plains floor unit* having low knobs a few km across.

Unlike large impact basins on the Moon and Mercury, the Hellas basin lacks prominent ring features such as fractures and multiple rims. However, Peterson (1978) sketched possible ring fractures outside the rim and suggested that the locations of several highland volcanic paterae were controlled by the fracture system, as discussed below. Most of the fracture structures associated with Hellas, as well as with Argyre and Isidis, are sufficiently cratered (Wichman and Schultz 1989) to indicate that they formed during Early to Middle Noachian time.

The Argyre basin is centered at about 51° S, 43° and has a diameter in excess of 1500 km. Its rim is composed of uplifted fractured Noachian plateau materials and ejecta (the *hilly unit of the plateau sequence*), and it is somewhat more rugged than the Hellas basin. The floor of the basin is mantled by several units, including an etched, cratered unit characterized by deep erosion; the unit is interpreted to be fill that has been sculptured by wind, water and periglacial processes. The floor also contains an Hesperian-age ridged unit considered to be volcanic or sedimentary and a younger, smooth unit that may consist of interbedded lavas and windblown sediments. No large volcanic constructs or outer ring-fracture systems appear associated with the Argyre basin.

The Isidis basin straddles the cratered highlands-lowlands plains boundary at 16° N, 272°. Rugged *mountainous material* and a *hilly unit* inferred to be the basin rim are on the southern third of the structure. Together with ring fractures that form arcs on the northwest side, they define a basin about 1100 km in diameter. Additional discontinuous fractures define an outer ring 1900 km in diameter. As discussed by Schaber (1982) and Comer et al. (1985), extensive lava flows on Syrtis Major Planum appear to have been erupted from complex calderas centered over ring fractures of the Isidis basin.

In addition to the three impact basins described above, many other large features in various states of preservation appear to be of impact origin, although the origin of some is controversial (Table IV). Among the features more recently proposed to be impact structures are the Daedalia, Utopia and Borealis basins. Craddock et al. (1990) assessed the distribution and orientation of grooved terrain and highland massifs in the cratered highlands surrounding Daedalia Planum in the southwestern part of the Tharsis province; they found that the focus of the grooves is a zone centered at 26° S, 125°. On the basis of these and other considerations, they proposed that an impact basin some 1800 km in diameter underlies much of Daedalia Planum.

McGill (1989*a*) mapped the distribution of knobs, partly buried craters, ring fractures and mesas in Utopia Planitia and determined that a major circular depression about 3300 km in diameter has persisted there during most of Mars' geologic history. The center of this proposed impact basin is roughly coincident with a large mascon. The highland/lowland boundary southwest of the basin center may be part of an outer ring.

Perhaps the most intriguing of the recently proposed Martian basins is the "Borealis basin." Wilhelms and Squyres (1984) suggested that the northern lowlands can be explained as an enormous impact scar to which they gave this name. (Other proposed origins have been mentioned above.) If this feature is indeed an impact structure, it is the largest observed thus far in solar system exploration.

2. Plateau and High-Plains Assemblage. The majority of Martian highland rocks constitute the plateau and high-plains assemblage. Some materials in this assemblage also form hills and small knobs in scattered patches, both large and small, in lowland areas along and adjacent to the highland/lowland boundary. Rock units in the assemblage range in age from Early Noachian to Late Amazonian; the younger units, however, consist largely of surficial deposits of small areal extent. In the western equatorial region, the assemblage is partly overlain by other highland terrain materials consisting of lava flows of the western volcanic assemblage that have been erupted from the large volcanic centers of Tharsis Montes, Alba Patera and Olympus Mons.

The oldest (Early Noachian) identifiable material not directly associated with known ancient impact basins is the *basement complex* (Scott and King 1984). The unit has been recognized only in the western hemisphere of Mars, where it underlies and is transitional with highly deformed terrain materials, particularly *older fractured material.* Basement rocks have high relief and are intensely faulted and cratered; they probably consist of impact breccias with interbedded lava flows. Relief on the basement material may be due to erosion of an old, regionally high surface as well as to structural uplift along normal faults resulting from tectonism and possibly impact deformation. The *hilly unit,* which makes up basin rims and other ancient, rugged mountains, has somewhat arbitrarily been grouped with a series of units forming the plateau sequence, the most widespread rocks in the highlands. Seven other rock units in this sequence range in age from Middle Noachian to Late Hesperian; these units display morphologic variations that are due to various geologic processes acting over extended periods. The *cratered unit* is characterized by a multitude of partly buried and superposed craters $>$ 10 km in diameter (Figs. 2b and 3a). Intercrater areas are rough but without the high relief of the hilly unit. Where highly dissected by small channels or etched by the wind into irregular grooves and hollows, the modified materials are classified as the *dissected unit* (Fig. 2c) and the *etched unit,* respectively. Patterns of channels in the dissected unit resemble those of terrestrial streams; the channels are thought to have been carved by runoff from groundwater seepage (Pieri 1976) or rainfall (Masursky et al. 1977) (see chapter 15). Large areas are plains marked by subdued and partly buried old crater rims (*subdued cratered unit;* Fig. 2b) or characterized by a flat and relatively featureless surface (*smooth unit*) or a mottled appearance (*mottled smooth plains unit*). These units locally embay other deposits of the plateau sequence and are interpreted to consist of interbedded lava flows and aeolian deposits that bury or partly bury underlying rocks. The *ridged unit* (Middle Noachian) of the plateau sequence (Fig. 2c,d) and *ridged plains material* (Lower Hesperian; Fig. 3a) are broadly similar in morphology, and the ridges that distinguish them generally follow similar regional trends. However, surfaces of the ridged plains material are less cratered, ridges are commonly smaller and of the lunar mare type (wrinkle ridges), and the unit locally contains lava-flow lobes (among the oldest recognized on Mars). Noachian ridges may be due to normal or thrust faulting (Scott and Tanaka 1986); wrinkle ridges are generally associated with volcanic rocks and probably result from folding or thrust faulting (see chapter 8).

The plateau and high-plains assemblage also contains two other important units. *Undivided material* of Noachian-Hesperian age forms hills and small knobs adjacent to the highland-lowland boundary scarp (Fig. 2d). The *Tempe Terra Formation* consists of intermediate-age (Middle Hesperian) lava flows extruded from small volcanoes and fissures on the Tempe Terra plateau.

C. Lowland Rocks

1. Degraded, Ancient Rocks Along the Highland/Lowland Boundary Zone. Ancient cratered terrain in the northern plains of Mars is largely buried by Hesperian and Amazonian plains materials. Where exposed, the ancient cratered materials are typically degraded into knobs (shown on the 1:15M-scale maps as *undivided material*). Older structures are largely modified beyond recognition, with the exception of crater rims $\gtrsim$ 5 km in diameter. In the eastern hemisphere, the highland/lowland boundary is broad (as much as 300 km across) and consists of plains materials that embay scattered mesas and knobs of ancient terrain.

Degradation of ancient materials in the northern lowlands and boundary zone was intense during the Late Noachian (and perhaps earlier), forming fretted channels and the knobby terrains (Fig. 2d; Squyres 1978; Lucchitta 1984*a*; Frey et al. 1988; Maxwell and McGill 1988). Degradation of lowland and boundary materials since the Noachian has been largely limited to mass wasting of slopes. Hesperian and Amazonian eroded plains material, possibly made up of eroded highland debris, is designated the *etched plains material.* The *knobby plains unit* consists mainly of young Amazonian plains deposits surrounding knobby inliers of older material. The cessation of intense erosion along the highland/lowland boundary at the end of the Noachian Period was accompanied by a similar decrease in erosion of highland surfaces, as reflected by waning widespread valley-network formation and crater degradation (see, e.g., Carr and Clow 1981); some continued relaxation of crater forms and other evidence indicate that minor gradation has persisted (Squyres and Carr 1986). These changes may have been due to cooling of the climate or reduced endogenic heat flow, which would have reduced the intensity of fluvial and aeolian activity (Tanaka 1986).

2. Northern Plains Assemblage. Plains-forming materials cover the greater part of the northern lowlands of Mars. On Mariner 9 images the plains appear flat to gently rolling with few topographic irregularities, and they were generally interpreted to be lava flows with varied amounts of aeolian cover (Scott 1979). Viking images have revealed much more detail and have allowed the subdivision and age reclassification of many plain units. Their interpretation, in many cases, remains speculative; volcanic, alluvial and aeolian origins all appear likely. Most of the plains consist of the northern plains assemblage that is composed of three formations (Scott and Tanaka 1986), but large areas are covered by smooth and etched plains materials mainly of aeolian origin.

The oldest and most extensive rocks in the northern plains assemblage make up the Late Hesperian *Vastitas Borealis Formation*. The formation consists of four members distinguished by albedo and textural differences (Fig. 3c) that probably represent secondary morphologic characteristics produced by interactions of erosion, compaction and ground ice (Carr and Schaber 1977; Pechmann 1980; McGill 1986; see chapter 16). The surface of the formation is windswept and degraded in places, and many small craters (<5-km diameter) have been severely degraded or even obliterated. The members intergrade with one another. Large areas of the formation have many conical hills whose crests are darker than their flanks; some hills have summit craters and may be volcanoes and cinder cones. Where the hills are closely spaced or coalesce into clusters, they are mapped as the *knobby member*. Other members of the Vastitas Borealis Formation are distinguished by concentric, whorl-like patterns of narrow ridges resembling fingerprints (*ridged member*); grooves and troughs forming curvilinear and polygonal patterns as much as 20 km across (*grooved member*); and contrasting high- and low-albedo patches due to impact craters whose ejecta deposits are brighter than intercrater areas (*mottled member*). All members of the formation, in whole or in part, were previously mapped from Mariner 9 images as mottled terrain. The unit was assigned a Noachian age, because it was thought possibly to correlate with old materials along the boundary scarp in highland terrain. An unknown portion of the formation likely was deposited as alluvial sediments from outflow channels that debouch into Chryse Planitia and perhaps into Utopia Planitia (Lucchitta et al. 1986); in other areas, a volcanic origin is consistent with sparse flow features and possible cinder cones (Frey et al. 1979).

The *Arcadia Formation* occurs as low-lying plains in Arcadia, Amazonis and Acidalia Planitiae (Fig. 4b). It comprises five members whose age range defines and spans the Amazonian Period. All of the members are exposed within Arcadia Planitia, and they are separated from the more rugged, plateau-forming highlands to the south by the highland/lowland boundary scarp. Stratigraphic relations between the Arcadia and Vastitas Borealis Formations are unclear except in Acidalia Planitia, where the lowermost member (*member 1*) of the Arcadia overlies the grooved and mottled members of the Vastitas Borealis; member 1 also embays the highland margin in places and partly buries the large outflow channels where they emptied into Chryse Planitia. Members are distinguished by their morphology, albedo and crater density. The common boundaries of the older members are poorly defined and, in places, they are mapped arbitrarily on the basis of variations in crater density or slight differences in texture and albedo of the bounded units. Landforms that are commonly visible on high-resolution images resemble lobate fronts, pressure ridges, small volcanoes, distributary channels cutting fan deposits (possible pyroclastic deposits), and collapsed lava tubes or lava channels. The younger members (*members 4 and 5*) have dark, fresh-appearing flows with few superposed craters. On Mariner 9 maps (Morris and Dwornik 1978; Morris and Howard 1981), the materials now included in the Arcadia Formation were interpreted to consist of thick sequences of lava flows, a conclusion supported in most areas by the Viking mapping.

A series of voluminous, areally extensive rock units occurs along an irregular east-west zone following the highland-lowland boundary in the Amazonis, Memnonia and Aeolis quadrangles. The units have been postulated to be ash-flow tuffs (Scott and Tanaka 1982), pyroclastic or aeolian materials (Greeley and Guest 1987), or paleopolar deposits (Schultz and Lutz 1988). These rocks make up the *Medusae Fossae Formation,* which consists of lower, middle and upper members. Their surfaces are level, flat to gently undulatory and, unlike lava flows, do not show pressure ridges or lobate fronts; they have been etched and serrated by the wind, particularly along their edges (Fig. 4b). As seen on high-resolution Viking images, the members are massive deposits without visible bedding, although differential erosion (possibly of resistant and friable zones) gives the appearance of layering in places. Crater counts indicate an age range of Early to Late Amazonian. In places some of the members exhibit complementary joint sets (Fig. 4b). The *lower member* has smooth to rough, highly eroded surfaces darker than those of other members and may include some dark lava flows. It is most widespread in the eastern equatorial region. The *middle* and *upper members* have smoother, rolling surfaces and occur mostly in the western equatorial region. All of the Medusae Fossae Formation is less hilly and cratered than materials of the highlands, but it has more relief and is lighter in color than the contemporaneous lava flows of the Arcadia Formation.

In the Utopia and Elysium Planitiae regions of the eastern hemisphere, *smooth plains material* of Amazonian age covers large areas in the lowlands. Aeolian deposits that mantle underlying terrain (perhaps alluvial or volcanic materials) are thought to make up a large part of the smooth plains material.

D. Volcanic and Tectonic Regions

1. Highland Paterae. The term patera was adopted for a low-relief, circular to elliptical feature first imaged by Mariner 9 that has channel-like structures radiating from central depressions. Although most paterae were considered to be a type of Martian volcano, subsequent data obtained by Viking show that some are clearly of nonvolcanic origins (e.g., Orcus Patera is an elongate impact crater), while others are simply modified or partly buried shield volcanoes (such as Apollinaris Patera and paterae in the Tharsis region). Consequently, the term highland paterae was applied to those features that constitute a distinctive type of Martian volcano (Plescia and Saunders 1979). As modified slightly by Greeley and Spudis (1981), this category includes Tyrrhena, Hadriaca, Amphitrites, Peneus and Tempe Paterae.

Highland paterae formed during the Hesperian (Greeley and Guest 1987) and appear to represent the earliest "central-vent" volcanism on Mars. All earlier volcanic units that have been identified are flood lavas whose vents are presumed to have been fissures. In addition, the paterae may represent the earliest explosive volcanism postulated for Mars.

Four of the five identified highland paterae (all but Tempe Patera) occur near the Hellas basin. First mapped by Potter (1976), Peterson (1977) and King (1978) from Mariner 9 images, the paterae were thought to have been associated in some way with the Hellas impact. Peterson (1978) suggested that ring fractures formed as post-impact crustal adjustments and that magma subsequently flowed along the fractures to the surface. Study of lunar volcanism shows similar associations with some multi-ringed basins; the hypothesis may be valid for Mars as well.

Of the highland paterae near the Hellas basin, Tyrrhena is the best imaged. Centered at 23° S, 255°, the volcanic materials of Tyrrhena Patera are northeast of Hellas. The volcano is marked by an irregular summit caldera more than 40-km long by 12-km wide. An oval, discontinuous ring graben about 80-km long cuts through the caldera on the southwest part of the summit area. Tyrrhena Patera consists of at least two principal units, a series of basal shield-forming flows and a younger unit comprising the summit area. Both of these units are composed of material that has been easily eroded to form broad channels, scalloped margins and erosional remnant mesas. Some of the channels may be large volcano-tectonic features that have been modified by fluvial processes. One such channel originates in the summit caldera and can be traced westward more than 200 km to where it merges with the surrounding volcanic ridged plains. The floors of the caldera and the channel are covered with lava flows.

Hadriaca Patera is on the northeast rim of the Hellas basin. It has a central caldera about 70 km in diameter from which flows as long as 400 km can be traced. Hadriaca flows embay the surrounding heavily cratered terrain and flow downslope toward the basin floor. The morphometry and morphology of Tyrrhena and Hadriaca Paterae suggest that they are largely ash shields, perhaps produced by phreato-magmatic eruptions when rising magma explosively interacted with groundwater (Pike 1978; Greeley and Spudis 1981; Greeley and Crown 1990; see chapter 13).

Amphitrites and Peneus Paterae are southwest of the Hellas basin. Both have erupted flows that flooded the basin rim and the surrounding older terrain. In addition, both have ring fractures surrounding large central calderas. Although Tanaka and Scott (1987) interpreted the ridged plains of Malea Planum (southwest of Hellas) to have originated from Amphitrites and Peneus Paterae, there is no clear evidence to suggest such an origin.

Tempe Patera, first identified by Plescia and Saunders (1979), is centered at 44° N, 62°, at the northeast end of the Tharsis province. It is superposed on older cratered terrain of Noachian age and is in the middle of an area over 200 km across that is heavily eroded by radial channels. Smooth deposits on the patera floor may be late-stage volcanic material, or fluvial or aeolian fills.

2. *Tharsis Region.* The volcanoes and fracture systems of the Tharsis region were among the most spectacular and intriguing discoveries of Mariner 9 (McCauley et al. 1972; Carr 1973,1974*b*). Broadly defined, the Tharsis region of Mars extends from the northern lowland plains bordering Mareotis Fossae southward to Solis Planum and from Amazonis and Arcadia Planitiae eastward to Lunae Planum (see map I-2160-2). In this region, the boundary between highlands and lowlands is covered by lava flows of intermediate and young age (Hesperian-Amazonian) and the scarp, if it exists, is not visible. The Tharsis region encompasses a somewhat elongate, irregular dome of enormous size ($>6.5 \times 10^6$ km^2) that rises $\sim$ 10 km above the 0-elevation datum. Volcanic materials associated with Tharsis that make up the western volcanic assemblage cover some 14×10^6 km^2; older, degraded and fractured terrains exposed in the Tharsis region may also consist largely of volcanic rocks.

Theories on the origin of this dome are speculative (see chapters 5 and 8 for more thorough discussion); proposed models include isostatic uplift followed by flexural loading (Banerdt et al. 1982), thick volcanic accumulations extruded through a locally thin lithosphere (Solomon and Head 1982), and crustal thickening by intrusion (Willemann and Turcotte 1982). Volcanic and fault-history studies (see, e.g., Wise et al. 1979*a*; Scott and Tanaka 1980,1981; Plescia and Saunders 1982; Tanaka and Davis 1988; Scott and Dohm 1990*a,b;* Tanaka 1990) have provided constraints to test and refine such models. The history of fault and ridge formation, as well as their distribution and orientation, defines the local and regional stress-field histories.

Present mapping (Scott and Tanaka 1986) suggests that development of the Tharsis dome involved a complex history of episodic tectonism, closely associated with volcanism, on local and regional scales. The most intense deformation around Tharsis occurred during the Noachian and Hesperian Periods, resulting in fault systems that are radial and concentric to centers at Tharsis Montes, Thaumasia Fossae, Syria Planum, Alba Patera, Acheron Fossae, Valles Marineris and locally on the Tempe Terra plateau. These fault systems cut *older* (Noachian) and *younger* (Hesperian) *fractured materials* that are mostly made up of volcanic rocks erupted during early stages of

Tharsis activity. Systems of ridges concentric to the Tharsis rise were formed largely during the Late Noachian and Early Hesperian Epochs and occur mainly more than 2000 km from the center of the rise. Amazonian faulting was largely limited to the active volcanic centers of Alba Patera, Tharsis Montes, Olympus Mons and Elysium Mons (whose regional fault pattern is attributed to Tharsis stresses; Hall et al. 1986).

Three immense volcanic shields (Arsia, Pavonis and Ascraeus Montes) form the Tharsis Montes, a linear chain of volcanoes extending northeastward across the Tharsis rise. These three shields have gentle slopes of a few degrees (the upper slopes are commonly steeper than the lower slopes), wide calderas and flank vents. The shields appear to be formed largely of basaltic flows and are similar in morphology to terrestrial shields (Pike 1978; Plescia and Saunders 1979). The Martian shields crest 10 to 18 km above the Tharsis rise and attain elevations of 18 to 26 km. Along the Tharsis axial trend, volcanoes stretch from Arsia Mons to near Tempe Patera, some 4000 km. Six units of lava flows that were erupted from the Tharsis Montes and surrounding vents make up the *Tharsis Montes Formation* of Late Hesperian to Late Amazonian age; these flows cover nearly 7×10^6 km^2. The characteristic lobate sheet flows of the formation (Schaber et al. 1978) extend nearly 1500 km from the shields, embaying older highland rocks and flooding part of Kasei Valles and part of the northern plains at the mouth of Mangala Valles. On the west flank of each shield are young lobate deposits (arbitrarily mapped as slide material) whose origin is uncertain; the deposit on Arsia Mons is $\sim$ 500 km across. Near the Tharsis Montes are several smaller shields and domical constructs, including Biblis, Ulysses, and Uranius Paterae and Uranius, Ceraunius, and Tharsis Tholi. These volcanoes commonly exceed 100 km in diameter and attain heights of 2 to 6 km above the surrounding plains. The morphology of the Tholi and Ulysses Patera suggest that they may be composite volcanoes (Plescia and Saunders 1979).

North of the Tharsis Montes (and largely within the northern plains), extensive flows were erupted during the Hesperian and Amazonian Periods to form the *Alba Patera, Ceraunius Fossae* and *Olympus Mons Formations*. The Alba Patera flows built a broad, circular plateau some 500 km across and several km high; the flows extend about 1000 km from the center of this plateau. Early channeled deposits north of the patera may be pyroclastic material (Mouginis-Mark et al. 1988), whereas crested (tube-fed) and sheet flows make up much of the patera and appear to be lava flows that reflect diverse eruptive conditions (Schneeberger and Pieri 1991; chapter 13). The plateau is surrounded by extensive circumferential grabens and contains calderas, wrinkle ridges and linear ridges interpreted as spatter ridges (Cattermole 1986). The Ceraunius Fossae flows, on the other hand, are mainly sheet flows that were erupted from local fissures south of Alba Patera.

Olympus Mons is a shield volcano nearly 600 km in diameter and over

26-km high, the tallest mountain known in the solar system. Flows of the *shield member* and surrounding *plains member* that were erupted from fissures east of the volcano are among the youngest flows on Mars (Fig. 4c). The volcano is marked by a basal scarp and surrounded by unusual overlapping, lobate aureole deposits (*aureole members* of the Olympus Mons Formation, Fig. 4c) partly overlain by lava flows. The aureole deposits are as much as 2-km thick and extend as far as 700 km from the shield. The origins of the scarp and aureole deposits are uncertain, and a variety of suggestions have been offered involving volcanic, tectonic and mass-wasting processes (see, e.g., Hodges and Moore 1979; Lopes et al. 1980; Morris 1982; Francis and Wadge 1983; Tanaka 1985).

South of the Tharsis Montes, lava flows emanating from fissures at the crest of Syria Planum during the Late Hesperian Epoch formed the *Syria Planum Formation* (Fig. 3d). These flows appear to have flooded Solis and Sinai Plana, covering ridged plains material to the east and embaying intensely fractured materials to the south, and they were coeval with periods of Syria Planum-centered faulting (Tanaka and Davis 1988).

3. Valles Marineris. Here, we offer a brief synopsis of the geologic history of Valles Marineris largely based on recent geologic mapping at 1:2M scale (Witbeck et al. 1991); more detailed descriptive and interpretive information is found in chapter 14. Overall, the canyon system of Valles Marineris is about equal in length (4000 km) to the Tharsis volcanic chain. However, the longest nearly linear valley of the system extends from the west end of Ius Chasma to the east end of Coprates Chasma, a distance of about 2300 km. Somewhat smaller, parallel valleys coalesce in places to form broad chasmata whose maximum width is $\sim$ 600 km. The canyon system and associated outflow channels and chaotic terrain developed mainly during the Hesperian Period. All of the canyons occur along grabens having parallel strike; in places, the valley faults were associated with volcanic activity (Lucchitta 1987*b*). In comparison, East African rift valleys are narrower and have extensive associated volcanoes. A gravity profile over the center of Valles Marineris shows a negative free-air anomaly that extends more than 1000 km (chapter 7). This anomaly is probably due to a lack of compensation of the rift valleys (Tanaka et al. 1986); a comparable zone of negative anomalies occurs across East African rift valleys.

The canyons are incised in a thick stack of plateau-sequence rocks, capped in their western part by ridged plains material and the Syria Planum Formation. Rifting, magma withdrawal and tension fracturing have been proposed as possible processes involved in the initiation and development of the canyons (Sharp 1973*c*; Frey 1979*a;* Tanaka and Golombek 1989). Significant canyon-scarp retreat appears to have followed emplacement of the ridged plains cap rock, but it is not known if all of these units were breached by the developing canyon system. Possibly the ridged plains material of Lunae and Solis Plana was in part lava flows extruded from fissures along faults that later formed Valles Marineris. Compared with Noctis Labyrinthus, the large canyons of Valles Marineris are more highly developed and probably older, possibly Early Hesperian.

During the Late Hesperian, the canyons were filled with *layered material,* which occurs in light and dark horizontal beds. Possibly they accumulated as waterlaid sediments in large lakes within the canyons (McCauley 1978; Nedell et al. 1987) or as ash-fall deposits (Peterson 1981). Following aeolian and perhaps fluvial erosion of much of the layered material, only local high-standing mesas and hills of the material were preserved. The canyons continued to expand by faulting and mass wasting through Amazonian time (Blasius et al. 1977); also deposited were *floor material* made up of colluvium, *slide materials* emplaced by enormous landslides and debris flows (Lucchitta 1979,1987*c*), and possible dark volcanic materials (Lucchitta 1987*b*).

4. Elysium Region. The eastern volcanic assemblage, defined by Greeley and Guest (1987), consists of four formations in the Elysium region and at Syrtis Major. First described by McCauley et al. (1972) from their discovery in Mariner 9 images, three principal volcanoes make up the Elysium region. Hecates Tholus is the oldest and northernmost; Albor Tholus is intermediate in age and is southernmost; and Elysium Mons is the youngest and largest. As the Elysium region developed, complex interactions of volcanism, tectonism and ground ice resulted in a variety of landforms (Mouginis-Mark et al. 1984).

The *Hecates Tholus Formation* was produced during the Hesperian Period by eruptions that ultimately formed a shield volcano about 180 km across and 6 km high. Its summit is marked by a caldera complex about 11 km across, and mantling of the volcano west of the summit may be the result of airfall products of an explosive eruption (Mouginis-Mark et al. 1982*b*). Linear depressions inferred to be channels and collapsed lava tubes (Greeley 1973) radiate from the summit area and extend down the flanks. The lower flanks are deeply incised by channels that indicate erosion by pyroclastic flows (Reimers and Komar 1979) or water runoff (Gulick and Baker 1988*d*). The areal extent of the lava flows from Hecates Tholus is not known, because the distal parts of the volcano are buried by younger flows from Elysium Mons.

Data from crater counts suggest that in the waning stages of volcanic activity at Hecates Tholus, eruptions shifted some 850 km to the south to form the *Albor Tholus Formation.* Albor Tholus is the smallest of the Elysium volcanoes, and because it is poorly covered by Mariner 9 and Viking images, few of its details are known. It is a dome-shaped volcano about 30 km across and has a summit caldera some 7 km in diameter. As is the case at Hecates

Tholus, the extent of flows from Albor cannot be determined because flows from Elysium completely surround the small dome.

Elysium Mons is the major volcano of the Elysium volcanic province. The Amazonian *Elysium Formation* comprises four distinct members that are associated with development of the shield (Greeley and Guest 1987). The oldest and most extensive, *member 1,* consists of hundreds of flows that radiate from the central construct. Individual flows and flow units encircle Hecates and Albor Tholi and can be traced as far as 1300 km to the east, where they form ridged plains and embay older, cratered terrain. *Member 2* makes up the main Elysium Mons construct, which is nearly 500 km across and stands some 9 km high. Member 2 is composed of flows that are generally younger than the surrounding plains lavas described above. A caldera ~ 14 km across marks the summit and was apparently the source of many tube- and channel-fed lava flows. The general morphology of Elysium led Malin (1977) to suggest that it is similar to Emi Koussi of the Tibesti highlands in Africa, a volcano consisting of lava flows and ash deposits of intermediate composition. A series of ring grabens surrounds the main construct of Elysium Mons. In places, the grabens are buried by plains lavas that embay some of its flank flows. Although the sources for these younger flows are not identified, they may have been fissures that formed in association with loading from the growth of Elysium Mons and subsequent fracturing of the crust (Comer et al. 1985).

Member 3 of the Elysium Formation includes a zone 800-km wide by 1800-km long that extends northwest from Elysium Mons. This enigmatic unit forms plains of rugged, hummocky relief; local lobate flows can be seen on high-resolution (~20 m/pixel) images (Fig. 4a). The unit is considered to be composed of lava flows and volcanoclastic materials (including lahars; see Christiansen 1989) that have been modified by aeolian, fluvial and periglacial processes. *Member 4* of the Elysium Formation occurs primarily within the plains formed by member 3 and consists of channel materials (Fig. 4a). Several fracture systems, the Elysium Fossae, occur in the area northwest of Elysium Mons and appear to be sources for some of these channel materials. MacKinnon and Tanaka (1989) suggested that some of the channel materials could be the result of catastrophic floods associated with the Elysium Fossae.

The *Syrtis Major Formation* is also part of the eastern volcanic assemblage. Centered at latitude 10° N, longitude 290°, Syrtis Major was described by Schaber (1982) as a low-relief shield volcano whose flows make up a plateau more than 1000 km across. The flows are of Hesperian age and appear to have originated from a series of complex calderas. Arcuate segments of grabens in the summit area define a depression 280 km in diameter interpreted to have formed by foundering of a magma chamber (Schaber 1982). The caldera complex lies on extrapolations of the ring grabens associated with the Isidis impact basin, described above; thus Syrtis Major volcanism may have been triggered by or at least associated with post-impact adjustments of the

Martian crust. Radar data (see chapter 20) across Syrtis Major (Schaber 1982) show that flows spilled into the Isidis Planitia depression to the east and buried or embayed older, cratered terrain to the west. Older flows are ridged and younger ones are lobate.

E. Channel Systems

A variety of channel-system units, mostly associated with outflow-channel development, occur (1) from Valles Marineris to Chryse Planitia; (2) along Mangala Valles; and (3) in southern Elysium Planitia. Channel-system units cover more than 4×10^6 km^2. Their channels were active during the Hesperian and Amazonian Periods, postdating development of valley networks that dissected older highland rocks. Other large channels (e.g., Al-Qahira and Ma'adim Valles) locally cut highland rocks or were associated with Elysium volcanism (see above). Interpretations of the processes leading to the origin and sculpturing of the outflow channels are given in chapter 15; an origin involving water or water and ice seems most likely (Baker and Milton 1974; Sharp and Malin 1975; Nummedal and Prior 1981; Lucchitta 1982*a*; Mars Channel Working Group 1983; MacKinnon and Tanaka 1989).

Most of the channels originating near Valles Marineris emanate from areas of *chaotic material* (Fig. 3b); some of this material occupies the floors of chasmata as much as several km deep. Chaotic material is generally made up of km-size knobs and irregular mesas apparently derived from the breakup of plateau rocks. *Older channel and floodplain materials* were deposited within and along the outflow-channel courses, some of which exceed 1000 km in length. Erosional islands or depositional streamlined bars are commonly found within the channels (Fig. 3b). Carr (1979) postulated that floods originating from outbreaks of an aquifer system produced the outflow channels. In Chryse and Acidalia Planitiae, where the channels grade into lowland plains, relict crescentic depressions are seen that are nearly identical in both shape and size with meander patterns of terrestrial streams. Here, as elsewhere in the northern plains (Scott 1983), the location and stratigraphic position of these depressions support the concept of widespread flooding in the lowlands during Late Hesperian time (Lucchitta et al. 1986). Although the evidence for large paleolakes is thus far sparse, some possible paleoshore lines have been identified in the Deuteronilus Mensae region (Fig. 5; Parker et al. 1989).

At Mangala Valles and southern Elysium Planitia, outflow channels appear to originate from fractures (Carr 1981; MacKinnon and Tanaka 1989). Several major episodes of flooding during Late Hesperian to Early Amazonian time are apparent at Mangala Valles (Masursky et al. 1986*a*). The vast (nearly 1×10^6 km^2 in area) *younger channel and floodplain materials* in southern Elysium Planitia are virtually uncratered by km-sized impacts and form the referent for the Upper Amazonian Series (Table I).

Fig. 5.(a) Photomosaic of plains near western Deuteronilus Mensae. Deposits having a striped or thumbprint texture fill part of canyon. Benches and breaks in slope on canyon wall are numbered 1 to 3. Bench 3 ends at A. Benches and thumbprint texture may be shoreline features bordering series of paleolake deposits. Subdued crater at lower left ~ 10 km across; north at top. (Viking images 458B65-70.) (b) Aerial photomosaic of part of Lake Bonneville region of Utah. Arcuate beach ridges and bars show changing lake levels. Scene width ~ 50 km; north at bottom (figure courtesy of T. J. Parker, Jet Propulsion Laboratory).

F. Polar Regions

1. Physiography. Vastitas Borealis, the north polar plains, are mostly near or below the Martian elevation datum (Map I-2030-C). They surround the subcircular polar plateau of layered deposits and ice known as Planum Boreum, which is more than 1000 km across. Estimates of the height of this plateau based on low-resolution stereoimaging range from 1 to 2 km (U.S. Geological Survey 1989) to 4 to 6 km (Dzurisin and Blasius 1975). The plateau is dissected by long, arcuate troughs that form a counterclockwise spiral pattern. Along Planum Boreum, between longitude 120° and 240°, a broad belt of linear dunes forms Olympia Planitia. To the southeast, Scandia Colles, which appear to be remnants of highland materials, make up a vast field of knobs.

In contrast, the south polar region is typified by cratered highland terrain rising several km above the Martian datum. This terrain includes the southern rims of Argyre and Hellas impact basins, as well as Promethei Rupes, which partly encircles another basin near the south pole. Within the highlands are many areas of relatively smooth intercrater plains (such as Malea Planum). Like the region around the north pole, the south polar region is capped by an oblong plateau of layered deposits and ice known as Planum Australe, which is ~ 1600-km long and 2 to 5 km high. Pitted terrains mostly south of latitude 70° S are known as Angusti and Sysiphi Cavi.

2. South Polar Intercrater Plains and Pitted Terrain. Following heavy bombardment, in Late Noachian and Hesperian time, smooth and ridged intercrater plains were emplaced over about one-fourth of the south polar region. On Malea Planum, ridged plains material apparently consists of lava flows, originally of low viscosity, from three paterae 100 to 300 km across of the *Amphitrites Formation.* The Malea ridges have a typical wrinkle-ridge morphology and various orientations. However, the Hesperian *Dorsa Argentea Formation* consists of flows and smooth plains material marked by long, sinuous, bifurcating ridges of uniform width that follow apparent flow directions. Lava-flow, mud-flow and glacial origins for the Dorsa Argentea materials have been proposed (see Tanaka and Scott 1987).

The Dorsa Argentea Formation surrounds the km-deep pits of Angusti and Sisyphi Cavi. If the formation was once rich in ice, the cavi could have formed by sublimation and aeolian erosion (Sharp 1973*b;* Plaut et al. 1988). However, if the formation consists of lava flows and the cavi formed in ice-rich material below them, the cavi may result from melting due to igneous activity (Tanaka and Scott 1987).

3. North Polar Dunes and Mantle. Unlike the south polar highlands, the north polar plains surrounding the polar plateau are largely covered by Late Amazonian aeolian deposits. Viking images reveal vast, mature dune fields (Fig. 4d; Breed et al. 1979; Tsoar et al. 1979; chapter 22). In particular, the *linear dune material* of Olympia Planitia forms an enormous erg, or sand sea, of about 2×10^5 km^2; *crescentic dune material* covers more than 4×10^5 km^2. Such huge deposits of dune materials indicate that layered deposits or sediments in the northern plains yield sand-sized particles for aeolian redistribution. Furthermore, 1.8×10^6 km^2 of northernmost Vastitas Borealis is buried by *mantle material* that obscures low-lying features beneath it (Squyres 1979*b*). The northern lowlands may serve as a trap for saltating particles; remote-sensing data indicate very little dust or duricrust exposed in the northern plains (see chapters 17 and 18).

4. Polar Layered and Ice Deposits. Each of the polar regions contains a thick sequence of *polar layered deposits,* consisting of alternating light and dark layers capped by H_2O and CO_2 ice (Fig. 4d). *Polar ice deposits* of the northern cap (and probably of the southern cap) are made up of dirty water ice seasonally covered by carbon dioxide frost (Kieffer et al. 1976*b*). The north polar ice covers more than 8×10^5 km^2; the south polar ice, nearly 0.9×10^5 km^2. Existing layered deposits appear to have accumulated as dusty ice whose dust-to-ice ratio may relate to periodic climatic oscillations. These oscillations are thought to vary according to cycles of rotational obliquity and orbital eccentricity having periodicities of 51,000 to 2,000,000 yr (chapters 9 and 33). The lack of intense degradation of south polar layered materials over the past few hundred Myr (as indicated by crater counts) suggests that either climatic oscillations of several Myr are responsible for layering or that accumulation of layered material ceased by Late Amazonian time (Plaut et al. 1988). The time of initial accumulation of the layered deposits is largely uncertain. For example, layered materials of the Amazonian Medusae Fossae Formation have been interpreted to be paleopolar deposits, which would suggest that dramatic excursions of the planet's axial pole position have occurred (Schultz and Lutz 1988), whereas south polar pitted terrain may indicate ancient polar deposition consistent with the present orientation of Mars. (See chapter 28 for discussion of seasonal water cycles that control polar ice deposition and removal; see chapter 23 for more information on polar materials and processes.)

IV. RESURFACING HISTORY OF MARS

No planetary body in the solar system has had such a varied, long-lived, and yet well-documented resurfacing history as Mars. Some bodies, such as Earth, Venus and Io, have undergone such extensive and geologically recent resurfacing that their ancient histories are largely obliterated. Other bodies, such as the Moon and Mercury, became virtually dormant following early outpourings of lava and impact bombardment. Outer-planet satellites also have had generally short periods of activity, confined to their early or recent history. On the other hand, Mars displays a rich geologic record that spans the time from heavy bombardment to the present day and indicates a great variety of geologic processes.

The stratigraphy and geologic interpretations described above were used to compile the following overviews of the major processes that have resurfaced Mars (see Table V and Fig. 6; Tanaka et al. 1988). Our descriptions are in general agreement with earlier assessments by Arvidson et al. (1980), Greeley and Spudis (1981) and Greeley (1987).

Resurfacing as used here signifies obliteration of the previous surface to the extent that the new surface is the dominant one. Thus burial, erosion and surficial reworking of materials are involved. Depth of resurfacing has not been measured, but at the resolutions used for 1:15M-scale mapping (generally ~ 200 m), it appears that depths of more than 100 m are reached. More recent resurfacing may obscure the older resurfacing record. Nevertheless, Tanaka et al. (1988) attempted to reconstruct the resurfacing record on the basis of exposed materials and what likely underlies them and was modified by processes acting in the past.

A. Volcanic Resurfacing

Volcanic rocks cover about 60% of Mars (Table V). These rocks include lava flows, pyroclastic deposits and probably portions of old cratered highlands and northern plains where diagnostic landforms are lacking. If we assume that most heavily cratered terrains are volcanic, which is consistent with theoretical thermal histories of Mars (see chapter 5), areas of volcanic resurfacing apparently have steadily decreased from ~ 1 km^2 yr^{-1} to ~ 10^{-2} km^2 yr^{-1}, as volcanism evolved from pervasive to local activity. Most recent Mar-

TABLE V
Extent of Resurfaced Areas (in 10^3 km^2) Exposed on Mars According to Age and Process[a]

Epoch	Volcanic	Aeolian	Fluvial	Periglacial	Impact[b]	Total	(%)
Late Amazonian	3280	4878	1245	322	574	10299	7
Middle Amazonian	7272	2076	326	1021	574	11269	8
Early Amazonian	12676	1569	354	394	574	15567	11
Late Hesperian	11503	983	7235	6501	574	26796	19
Early Hesperian	19267	1306	1099	209	783	22664	16
Late Noachian	7702	2432	5353	209	1130	16826	12
Middle Noachian	20019	320	3241	209	10766	34555	24
Early Noachian	2074	—	—	209	3741	6024	4
Total	83793	13564	18853	9074	18716	144000	
(%)	58	9	13	6	13		

[a]Data from Tanaka et al. (1988).
[b]Only craters >150 km in diameter were included.

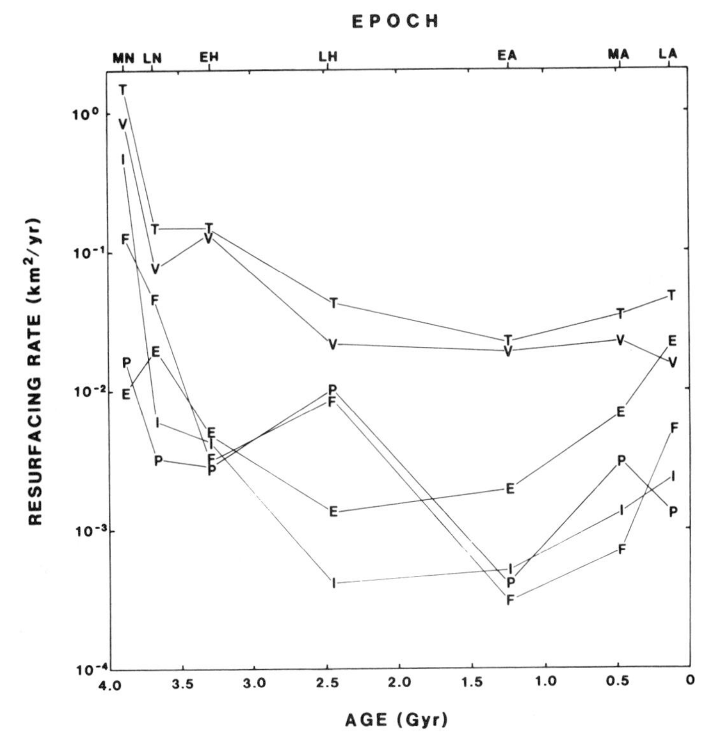

Fig. 6. Semi-log graph showing average rates of resurfacing for total (T), volcanic (V), fluvial (F), aeolian (eolian = E), periglacial (P), and impact (I) processes for each Martian epoch (except Early Noachian). (Note corrections in rates from Tanaka et al. [1988,Fig. 3].)

tian flows are much larger than terrestrial flows: areas of many Olympus Mons flows are 100 km^2 and larger. Therefore, given the average resurfacing rate and size of flows, we estimate that a lava flow has been extruded about once every 10^4 yr for the past few hundred Myr.

B. Fluvial Resurfacing

Fluvial action has had a disproportionately large influence on the Martian surface, particularly because rainfall probably has not been widespread, at least since the period of heavy bombardment (for a thorough discussion, see chapters 15 and 32). However, large quantities of water may have been emplaced in the Martian crust during outgassing. The most widespread fluvial activity occurred during the Noachian Period, when valley networks were incised into highland terrain, perhaps when surface temperatures were relatively high. Based on populations of rimless craters in the Amenthes and Tyrrhena region, Craddock and Maxwell (1990) estimated that this erosion stripped as much as 1 km of material from some highland areas. Outflow channels and flood plains developed primarily during the Late Hesperian, apparently as a result of catastrophic discharges of water in intimate association with intense volcano-tectonic activity at Valles Marineris, Tharsis Montes and Elysium Mons. At least one very young outflow channel and flood plain formed in southern Elysium Planitia contemporaneous with local volcano-tectonic activity.

C. Aeolian Resurfacing

Aeolian resurfacing is not well documented for Mars because of cyclic climate changes and the ephemeral nature of aeolian deposits. Permanent or semi-permanent aeolian deposits may be relatively rare. Also, very little aeolian erosion of ancient landforms is evident, indicating that Martian aeolian materials may be weak, capable of only minor abrasion (Greeley et al. 1982). Evidence for ancient aeolian activity is limited to some Noachian and Hesperian etched and pitted terrains that appear to result from aeolian deposition and erosion of poorly indurated materials (Grant and Schultz 1988; Plaut et al. 1988). Late Amazonian aeolian resurfacing at the poles, on the other hand, is well documented by dune, mantle and layered deposits, and resurfacing during this epoch amounts to $\sim 10^{-2}$ km^2 yr^{-1}. However, if the existing north polar aeolian deposits are less than a few Myr old, aeolian resurfacing may be effectively a hundred-fold greater.

D. Periglacial Modification

Periglacial activity over much of its history resulted from the prevalence of groundwater and ice on Mars, together with freezing temperatures and topographic relief. As groundwater was discharged from highland areas and settled, along with sediments, in lowland regions over geologic time, periglacial modifications have occurred along and in channels and in lowland basins. Ground-ice sapping may have contributed to the breakdown of possibly ice-rich, old cratered terrain in the northern lowlands and in some areas of the south polar highlands, forming fretted, knobby and pitted terrains. Because the plains of Vastitas Borealis formed at about the same time as outflow-channel activity, it is likely that many of the northern plains materials were deposited in or with water. Their grooved, mottled, knobby and ridged morphologies are attributed to periglacial modifications (Tanaka et al. 1988).

E. Impact Resurfacing

Most Martian impact craters and basins and their clastic deposits were formed during the Noachian Period, contributing to the formation of a megaregolith 1 km or so thick (see chapter 16). This megaregolith has contributed to other resurfacing processes by supplying fine, clastic material readily eroded and transported by aeolian, fluvial and mass-wasting processes. The resurfacing rates given in Table V and Fig. 6 are based on craters $>$ 150 km in diameter only, and thus they should be understood as relative; see chapter 12 for a thorough review of impact rates and their interpretation.

V. CONCLUSIONS

Mariner 9 must be regarded as the first important benchmark in understanding the complex geology of Mars. It revealed the products of extensive volcanism and the fluvial channels, aeolian deposits and tectonic patterns indicative of the geologic evolution of Mars. Its images provided the basis for establishing a broad stratigraphic framework. Viking added significantly to our understanding of these features and permitted refinement of complex stratigraphic relations.

However, we must remember that the study of Mars is still in its infancy. Much more can be done even with existing information and that anticipated in the near future. Some of the key geologic problems in which advances can be made include the following:

1. Crustal stratigraphy and structure. Are there regional or planet-wide vertical zonations in Martian basement rocks? What is the volume and distribution history of volatiles and carbonate material in the shallow crust? Are shallow igneous intrusions such as sills and dikes prevalent? What is the stratigraphy below the widespread, young northern plains materials?
2. The highland-lowland boundary. Does the crustal dichotomy have an exogenic impact or endogenic tectonic origin?
3. The Tharsis Rise. Is it primarily due to uplift, or is it largely a construct? What was its early history? What were the relative roles of local vs regional volcanic and tectonic activity?
4. Valles Marineris. What is the relation of this feature to the Tharsis province? Where is all of the material formerly contained within the canyon system? What is the nature of the deposits that fill parts of the canyon system? What is the role of volcanism in the evolution of the Valles Marineris?
5. Channels and valley networks. What were the sources of water? What were the relative roles of water and ice in forming these features? Was free-standing surface water involved? What is the precise age range of outflow-channel and valley-network formation? What were the primary erosional processes? Did they include other agents of formation such as extremely fluid lava flows?
6. Martian oceans, lakes and ponds. Did they exist? If so, when and where did they occur? What was their relation to channels and valley networks?
7. Medusae Fossae Formation. Is it primarily volcanic pyroclastic deposits, paleopolar deposits, mantles of windblown sand and dust, or something else?

Significant advances in answering these and many other questions may still result through the use of existing data sets, particularly through application of new techniques and studies such as image processing, derivation of topography from photoclinometry, laboratory experiments and theoretical modeling. However, most major advances in understanding the geology of Mars must await new data. The Mars Observer mission will provide new insights into the chemical compositions of rock-stratigraphic units as well as into thermal properties of the surface. However, as shown by the impact of the Apollo program on our understanding of lunar geology, many important geologic questions can be answered only by chemical and mineralogical data from returned samples of rocks and *in situ* geophysical measurements.

Acknowledgments. The authors wish to thank M. Carr, H. Moore and S. Saunders for thoughtful reviews and comments. T. Parker generously supplied the photomosaics making up Fig. 5. We are indebted to A. Acosta for producing the Color Plates of the geologic map of Mars.

12. THE MARTIAN IMPACT CRATERING RECORD

ROBERT G. STROM, STEVEN K. CROFT
University of Arizona

and

NADINE G. BARLOW
Lunar and Planetary Institute

The Martian impact cratering record is more diverse than that of any other planet or satellite in the solar system. Martian crater morphology is unique and has probably been strongly influenced by subsurface volatiles, but significant effects by the atmosphere on ejecta blanket emplacement cannot yet be ruled out. Large impact basins have influenced subsequent geologic processes, e.g., the concentration of valley networks on ancient basin ejecta blankets. More speculatively, the Martian crustal dichotomy may be the result of a gigantic impact that occurred at the end of the accretion phase of Mars, and the numerous elliptical craters may be due to a circum-Martian swarm of planetesimals that accumulated from ejected material and later impacted the surface when their orbits decayed. Mars has two production crater populations: one represents the period of late heavy bombardment, possibly from accretional remnants left over from the formation of the terrestrial planets, whereas the other has accumulated on younger surfaces since the end of heavy bombardment from asteroids and comets, with asteroids dominating over comets. Based on these two populations and their crater densities, the Martian surface units are assigned ages relative to the period of late heavy bombardment. Absolute ages have been derived for surface features by a number of authors, but there are several uncertainties, including assumptions about crater production rates and origins, which can lead to rather large errors. Absolute ages may range from 4.2 Gyr for the ancient cratered terrain to as young as 0.3 Gyr for Olympus Mons (Neukum-Hiller Model 2). The polar deposits may be even younger.

I. INTRODUCTION

No planet or satellite has a more diverse impact cratering record than Mars. In terms of crater morphology, modification and crater density, there are greater variations than observed on any other solar system body. Furthermore, differences in crater populations based on variations of the crater size/frequency distributions are better displayed because of the wide range in surface ages and more reliable crater statistics resulting from the greater counting areas. These factors greatly aid in deciphering the times at which various geologic events occurred, the types of geologic processes that affected the surface, the geologic conditions (both surface and subsurface) at the time of impact, and the origin of the objects responsible for the cratering record on Mars and elsewhere in the solar system.

In general, the surface of Mars can be divided into two approximately hemispherical regions based on the abundance of craters (Fig. 1). The southern region, bordered by a small circle inclined ~30° to the equator, contains a high crater density with a large range in crater diameters that records the period of late heavy bombardment in the solar system. The northern region is dominated by younger plains with a low crater abundance and a crater size/frequency distribution that differs significantly from that of the heavily cra-

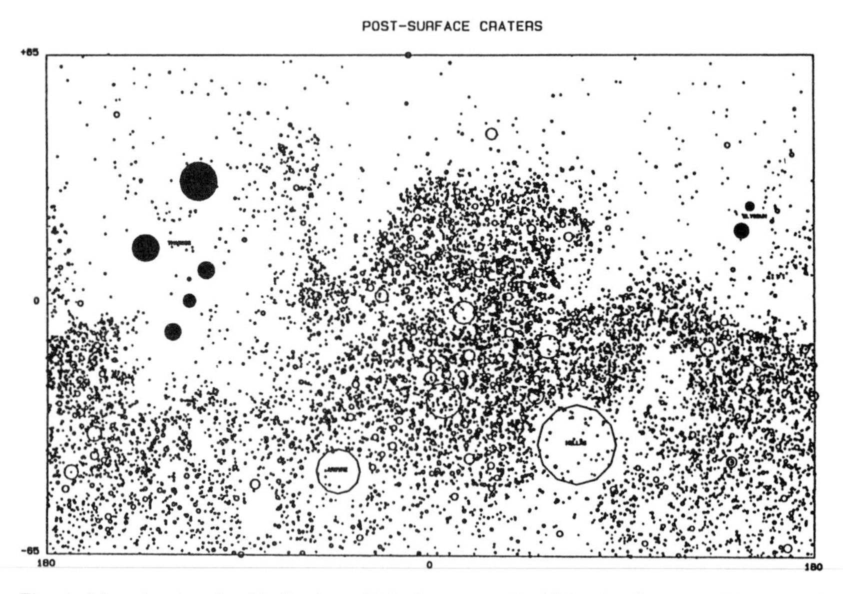

Fig. 1. Map showing the distribution of Martian craters ≧ 15 km in diameter within ± 65° latitude. All craters shown here superpose the surface on which they appear (no buried craters are included). The large volcanoes in the Tharsis and Elysium regions and the Hellas and Argyre impact basins are included for reference.

tered terrain. Within these two broad regions are smaller areas with a variety of crater densities. This two-fold division of Mars' physiography has been termed the "Martian crustal dichotomy" and may itself be the result of a giant impact to be discussed later.

In this chapter the crater size/frequency distribution is displayed as an "*R* plot" (Crater Analysis Techniques Working Group 1979). The "*R*" plot displays information on the differential size distribution, and is the ratio of the observed distribution to the function $dN \propto \bar{D}^3 dD$. The value of *R* is

$$R = \frac{\bar{D}^3 N}{A(b_2 - b_1)} \tag{1}$$

where $\bar{D}$ is the geometric mean of the size bin $\sqrt{b_1 b_2}$, *A* is the area counted, b_1 is the lower limit of the size bin, and b_2 is the upper limit of the size bin. $\sqrt{2}$- size bins with a minimum diameter of 8 km are used in all plots shown here, although other size bins and minimum diameters are possible. The confidence interval, $\pm 1\sigma$, is log $[R \pm (R/\sqrt{N})]$. Because most large crater populations have slope or population indices within the range of ± 1 of the function D^{-3}, they plot as nonsloping or moderately sloping lines on these log/log plots. On an *R* plot, a horizontal line has a differential -3 slope index; one sloping down to the left at an angle of 45° has a differential -2 slope index and one sloping down to the right at 45° has a differential -4 slope index. For a given single-sloped cumulative slope index, the equivalent differential slope index is decreased by 1, e.g., a cumulative -2 is equivalent to a differential -3 slope index, a cumulative -1 is a differential -2, etc. The vertical position of the curve on an *R* plot is a measure of crater density (number of craters per unit area): the higher the curve, the greater the crater density and, therefore, the greater the age of the surface it represents. The *R* plot has the great advantage of better showing the structure of the crater size/frequency distribution, as well as differences in crater density and therefore relative age.

II. CRATER MORPHOLOGY AND MORPHOMETRY

The variety and complexity of Martian crater forms is greater than found on any other terrestrial planet. As a result, difficulties and disagreements are encountered even in the fundamental exercise of morphologic classification. As shown below, inconsistencies exist between various observational studies. The physical models invoked to explain the observed morphologies are, not unexpectedly, equally diverse. It is useful to consider the observational problems separately from the interpretive ones. Therefore, primarily descriptive material on Martian craters is presented in this section. Analyses and interpretations of the observed morphologies in terms of impact physics are presented in Sec. III.

The morphologies of fresh craters on Mars include features both similar and dissimilar to fresh crater morphologies found on the other terrestrial planets. The basic sequence of changing morphologies with increasing crater diameter, simple craters to complex craters to multi-ring basins, best described for the Earth's Moon (see, e.g., Schultz 1976; Croft 1981*b;* Wilhelms 1987, and references therein), applies to Martian craters as well (see, e.g., Hartmann 1973*b;* Wilhelms 1973; Mutch et al. 1976*c*; Wood et al. 1978). Differences, described in detail below, are primarily second-order variations, presumably due to unique features of the Martian environment.

Rim and Interior Morphology

Small impact craters with pristine morphology display the bowl shaped appearance (Fig. 2a) typical of simple craters on the Moon and Mercury. Similarly, craters larger than ~5 km in diameter (Pike 1980*a*) display the interior features characteristic of complex craters, including central peaks and wall terraces (Fig. 2b). Central peaks have been found in Martian craters as small as 1.5 km in diameter (Mouginis-Mark 1979), reach 50% occurrence in craters around 5 to 7 km in diameter (Wood et al. 1978; Pike 1980*a*), and reach virtually 100% occurrence in craters greater than 20 km in diameter. Hale and Head (1981) found that the relative size of Martian central peaks is larger than for lunar or Mercurian central peaks in similar-sized craters. They suggest that the presence of subsurface volatiles or other terrain characteristics may be responsible for this relationship. Wood et al. (1978) have noted a terrain dependence in the onset diameter for central peak craters: central peak craters ≤ 10 km in diameter are concentrated in the heavily cratered units whereas craters superposed on plains units begin showing central peaks in the 10 to 20 km diameter range. Based on Mariner 9 data, Cordell et al. (1974) found higher concentrations of central peak craters in the south polar region compared to equatorial regions; however, this result is questioned based on Viking data analyses (Bradley and Barlow 1988).

The pattern of wall failure in Martian complex craters is similar to the pattern in lunar craters. Slumping and wall scalloping develop first in craters larger than ~5 km, grading into terraces as the dominant form of wall failure in craters larger than ~15 km (Wood et al. 1978). Wall failure tends to occur at smaller diameters in craters on the (presumably basalt-covered) ridged plains than in craters on the cratered highlands, indicating some form of lithologic control; yet terrace dimensions in Martian craters (Posin 1989) behave the same as lunar terrace dimensions (Croft 1985*a*), increasing linearly with crater diameter but showing no correlation with terrain, latitude or ejecta morphology.

Insofar as they can be discerned, interior features of the largest Martian impact structures, the impact basins, also parallel their counterparts on the Moon and the other terrestrial planets. Martian basins have been listed and described in a number of papers, including those by Wilhelms (1973), Hart-

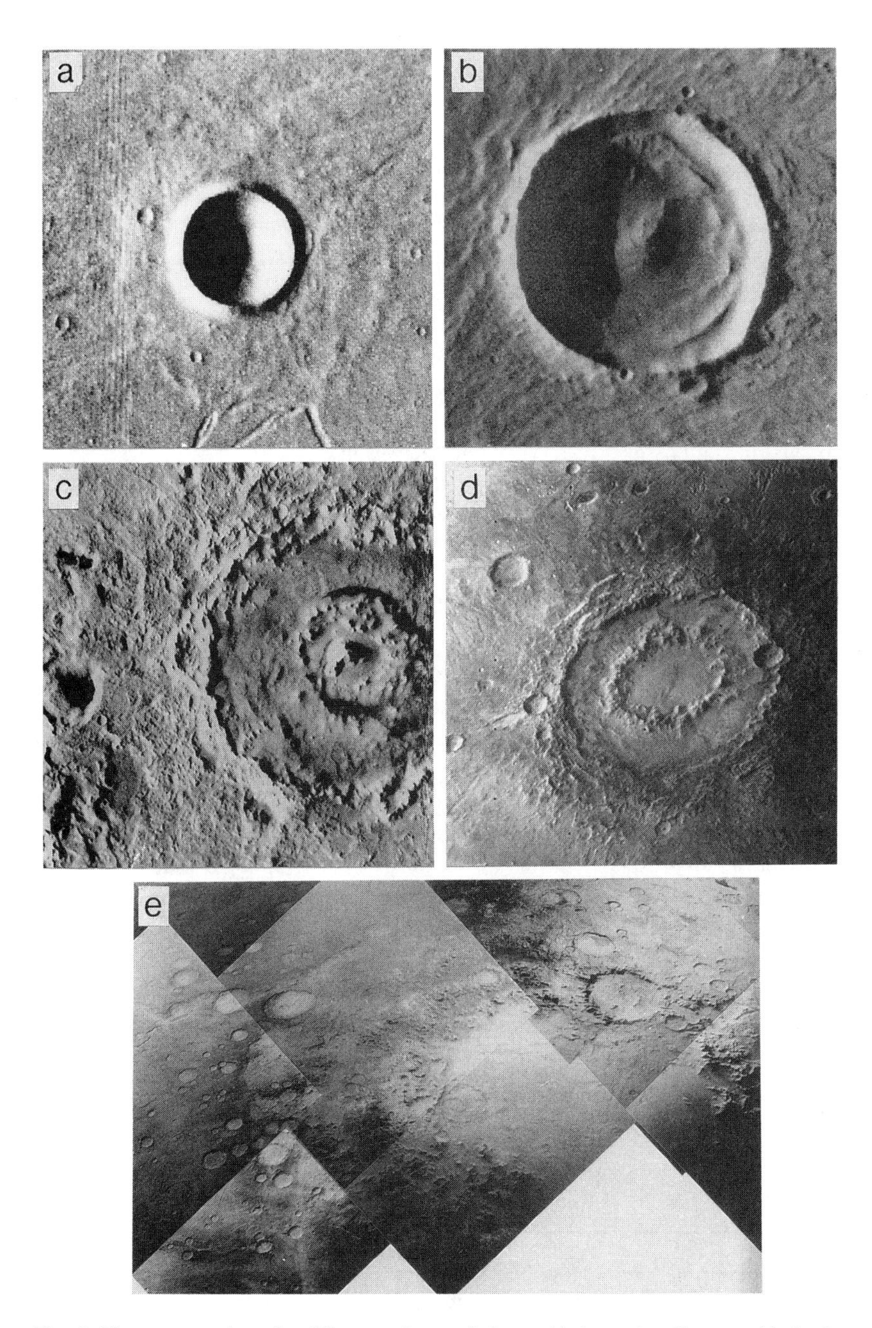

Fig. 2. These craters show the differences in morphology with increasing diameter: (a) simple crater (3.6 km in diameter, 28°.1 N, 66°.6 W, Viking Frame # 665A25); (b) complex crater (Lunae Palus Qt, 9.6 km, # 664A57); (c) central peak basin (Liu Hsin, 140 km, # 526A42); (d) peak ring basin (Lowell, 196 km, # 34A37); (e) multi-ring basin (Argyre, 740 km diameter, JPL mosaic 211–53040).

mann (1973*b*), Malin (1976*a*), Wood and Head (1976), Croft (1979), Wood (1980), Pike and Spudis (1987) and, most recently, by Schultz and Frey (1990). Examples of the morphological classes of basins are shown in Fig. 2: central peak basins (the protobasins of Pike and Spudis) in Fig. 2c, peak ring basins (the two-ring basins of Pike and Spudis) in Fig. 2d, and multi-ring basins in Fig. 2e. The transition from complex craters to basins begins at 50 to 70 km diameter on Mars (Wood 1980; Pike and Spudis 1987). Table I lists the onset diameters for basin morphology on the Moon, Mars and Mercury. Unlike the lunar case, central peak and peak ring basin morphologies are mixed over a large diameter interval. At present, the reason(s) for the overlap are poorly understood, but may be related to unusual crustal properties of Mars or the velocities or compositions of the impactors. Alternatively, weathering may have uncovered (or buried) structures that ordinarily remain buried (or visible) in basins on Mercury and the Moon.

Reconstructions of the pristine state of large basins on Mars are much more difficult than on Mercury or the Moon because of active Martian weathering systems that have altered the basins by both erosion and deposition (see, e.g., Mouginis-Mark et al. 1981). Until about 1980, only three multi-ring basins were generally recognized on Mars: Argyre, Isidis and Hellas. Subsequently, detailed large-scale mapping using Viking mosaics revealed a number of large to very large Martian basins that had been nearly completely obscured by weathering and subsequent tectonic and volcanic modification (Schultz and Glicken 1979; Schultz et al. 1982; Schultz 1984; Pike and Spudis 1987; Schultz and Frey 1990). These degraded basins were mapped with as many as six individual rings. The poor preservation state of these degraded basins makes it difficult to correlate their ring structures with the ring structures in better preserved basins on the basis of morphology. Indeed, Schultz and Frey (1990) think that ring morphology changes substantially with increasing basin diameter on Mars, which would make correlations between rings in basins of different diameters somewhat moot. Alternatively, Pike and Spudis (1987) suggest that the ring structures of both fresh and degraded basins on Mars can be correlated with ring structures in basins on the other terrestrial planets primarily on the basis of numerical relationships. Unfortu-

TABLE I
Onset Diameter (km) of Basin Morphologies[a]

	Mercury	Moon	Mars
Central peak basins	70	130	60
Peak ring basins	150	300	50
Multi-ring basins	300	400	300

[a]Data from Pike and Spudis (1987).

nately, all of these studies are limited by the ambiguities engendered by the poor preservation states of these structures.

Somewhat more unusual interior features in Martian complex craters are the so-called central pits. Two main types of pit morphology are found on Mars (Wood et al. 1978): (1) summit pits (Fig. 3a), where the pit is situated atop a central peak, and floor (or central) pits, which can be further subdivided (Awwiller and Croft 1986) into symmetric (Fig. 3b) and asymmetric types (Fig. 3c). At least two craters appear to have both a central pit and a peak ring: Wahoo (23°4N,33°7W; Fig. 3d) and Bakhuysen (23°1S,344°2W), indicating that pits and peak rings are not structurally related. Wood et al. (1978) suggested a progression from central peak to central pit morphology with increasing crater diameter on heavily cratered terrain

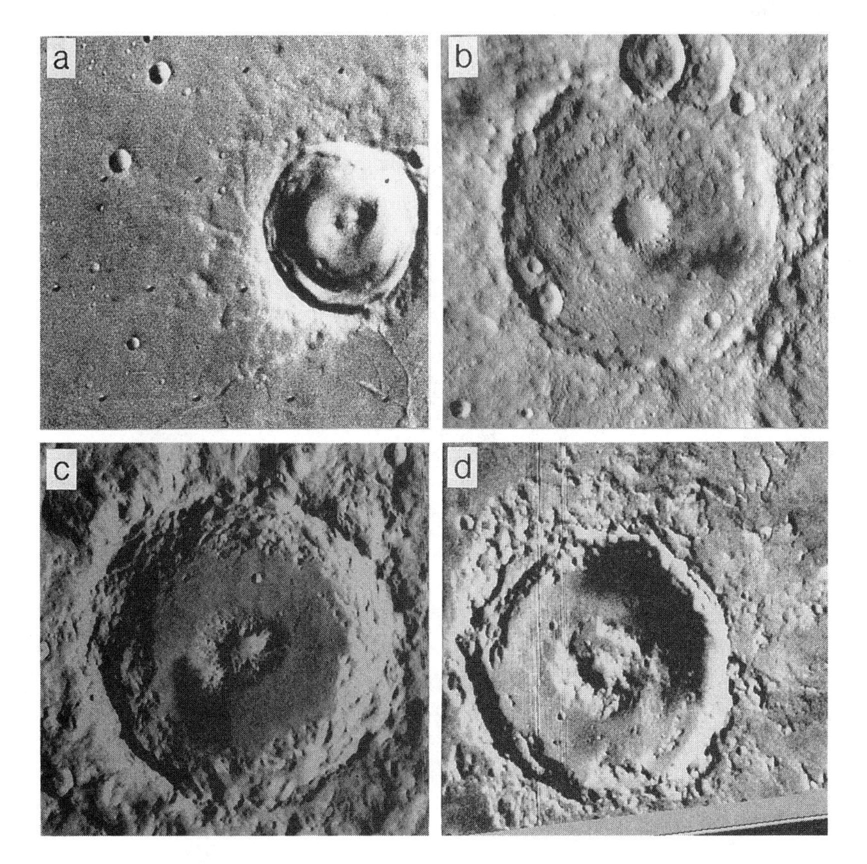

Fig. 3. Examples of pit crater classifications: (a) summit pit (9.5 km in diameter, 19°9 N, 51°8 W, Viking Frame # 20A34); (b) symmetric floor pit (Syrtis Major Mu, 64 km, # 341S10); (c) asymmetric floor pit (Syrtis Major Jv, 62 km, # 534A46); (d) mixed pit/peak morphology (Wahoo, 60 km, # 827A24).

units, the units where pit craters are concentrated. Hodges et al. (1980) and Hale (1982) corroborated these results based on larger data sets. The most recent analyses of Martian pit craters were carried out using data derived from 1:2 M photomosaics of Mars by Awwiller and Croft (1986) and Bradley and Barlow (1988). They found that the three morphological classes are gradational with each other, supporting the suggestion that all central pits are the result of a common physical process. The relative abundance of the three morphologic types shows no variation with either latitude or terrain type, again suggesting a common origin. However, the relative abundance of pit craters appears to be related to age: on average, pits are roughly twice as abundant in craters on terrains that formed before and during heavy bombardment than on surfaces (planitia and young volcanic shields) that formed after the end of heavy bombardment. Even then, the areal distribution of pit craters is not uniform. Fig. 4 shows the ratio of pit craters to all craters with identifiable internal structures averaged in $10^\circ \times 10^\circ$ bins over most of Mars' surface. The global distribution is quite irregular except for a general drop in pit abundance toward the poles (which may be an observational effect). Awwiller and Croft suggested one possible correlation: areas of high concentrations of pit craters tend to cluster in the middle to outer bands of the five largest ($>$ 2000 km in diameter) basins on Mars: Hellas, Chryse, Isidis and two unnamed basins proposed by Schultz (1984).

Pit craters are not unique to Mars. They are ubiquitous on the icy satellites Ganymede and Callisto and a few are found on the Moon and Mercury (cf. Croft 1981*a;* Schultz 1988). The lunar pit craters are features that long fueled the impact-volcanic controversy concerning the origin of lunar craters (cf. Warner 1962; Allen 1975; Pike 1980*b*), a controversy finally settled in favor of impacts. Many of the lunar so-called pit craters are simply small late impacts on originally unpitted central peaks, and prior to receiving high-resolution images of Martian pit craters, opinion had swung to assigning all lunar pit craters to this origin (Pike 1980*b*). However, close comparison of the Martian and the lunar (and Mercurian) pits strongly indicates that at least some of the lunar pits are morphologically indistinguishable from the Martian summit pits (Croft 1981*a*), suggesting a similar origin. Pit populations, however, are not the same on Mars, the Moon, Ganymede and Callisto. First, few, if any, lunar analogs to Martian floor pits are found, although the entire suite of pit morphologies is found on Ganymede and Callisto. Second, the fraction of all craters with identifiable central structures that contain pits is about 31% on Mars, only about 2 to 3% on the Moon (based on data in Wood [1968], Allen [1975] and Croft [1981*a*]), but approaching 100% for larger craters on Ganymede and Callisto.

Ejecta Deposits

Lunar and Mercurian craters are surrounded by ballistically emplaced ejecta and secondary craters. In contrast to having fairly similar interior crater

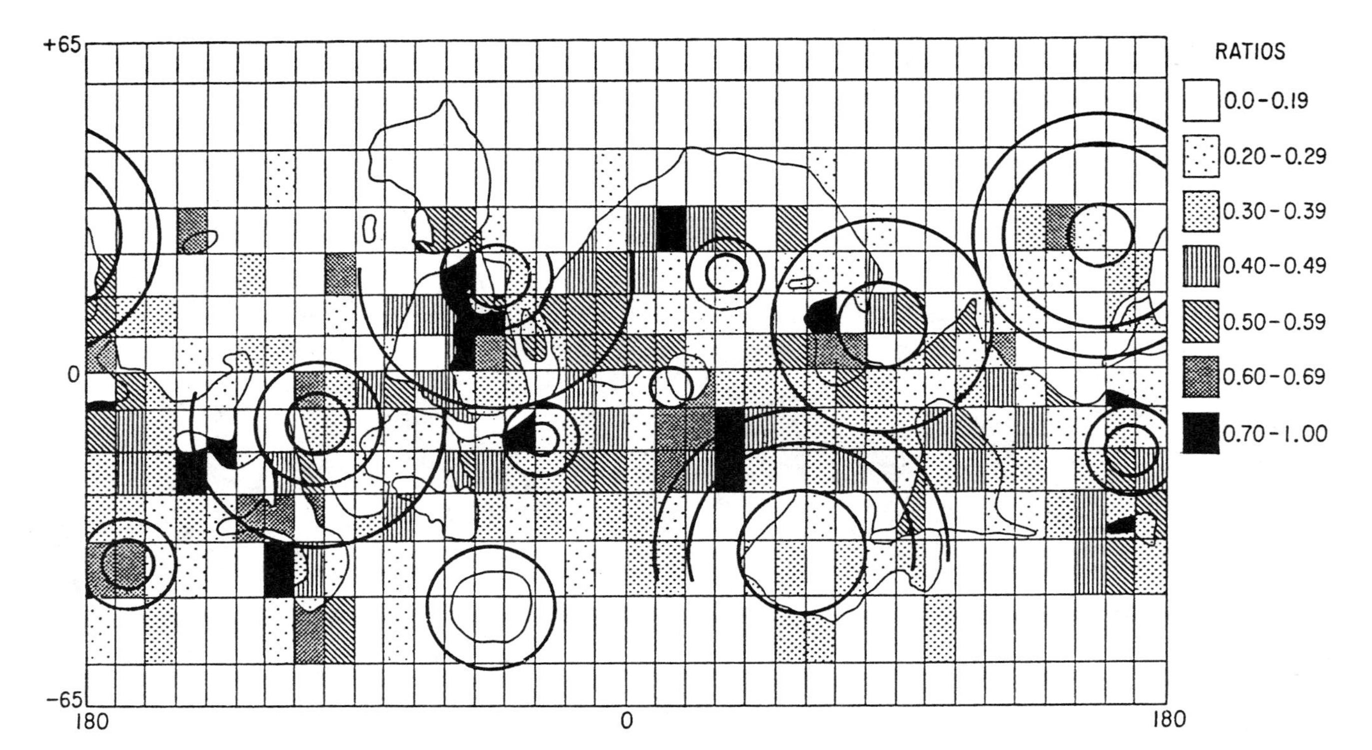

Fig. 4. Map showing the areal distribution of pit craters. Patterns in 10° × 10° boxes show the fraction of all craters with identifiable central structures that are pit craters. Circles show approximate locations of the largest basins on Mars. The highest concentrations of pit craters appear to correlate roughly with the largest basins (figure after Awwiller and Croft 1986).

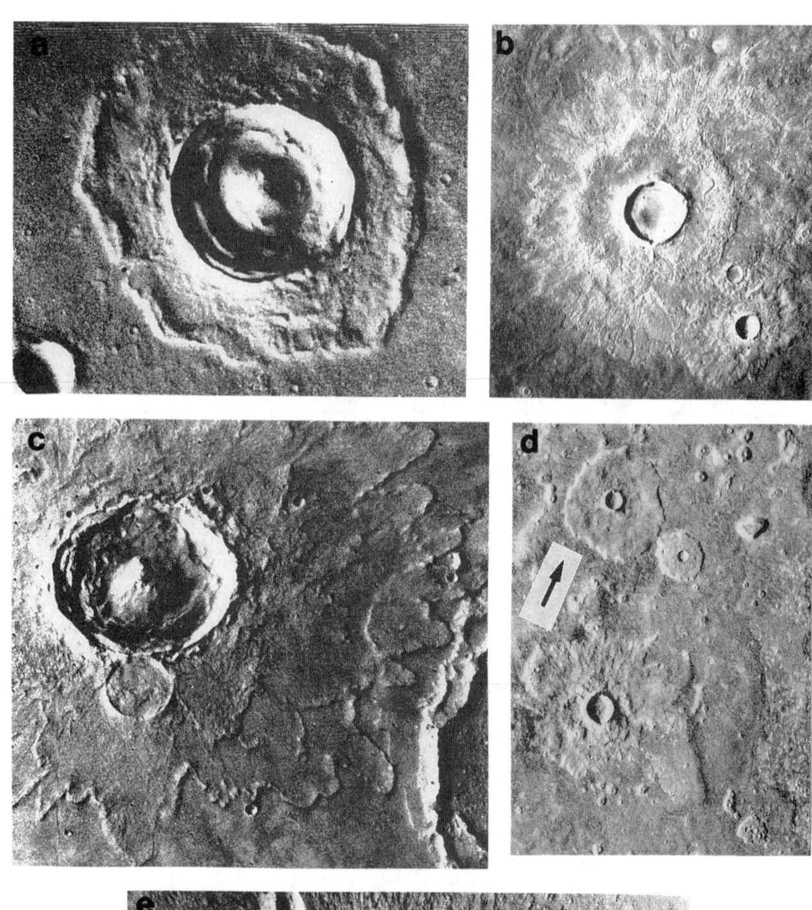
a
b
c
d

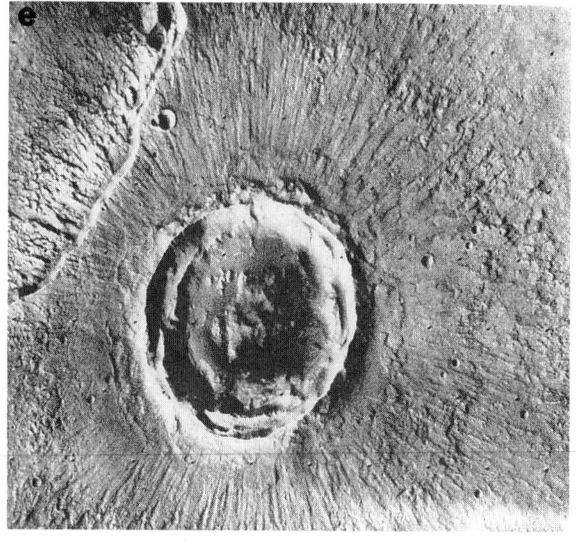
e

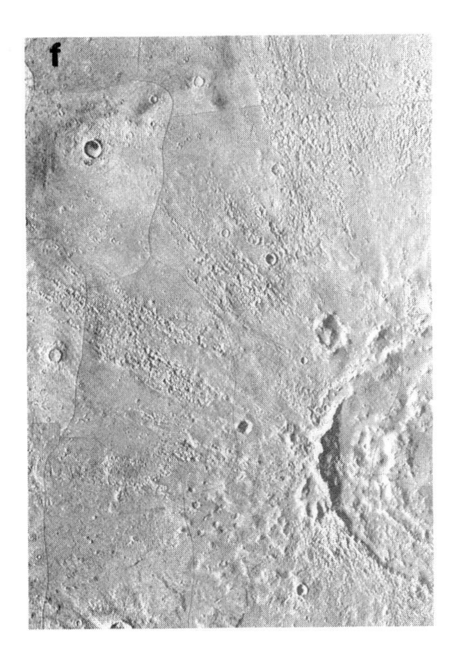

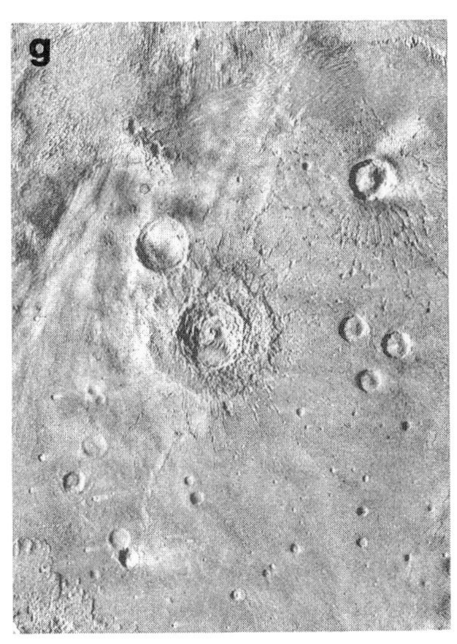

Fig. 5. Examples of fluidized ejecta morphologies associated with Martian impact craters: (a) single lobe rampart (crater diameter = 7 km; 21°.8 N, 31°.5 W; # 06A93); (b) double-lobe rampart (Casius Jh, crater diameter = 16 km, # 10B81); (c) multiple-lobe rampart (Oxia Palus Ut; crater diameter = 18 km; # 022A54); (d) pancake ejecta (arrowed crater = 3 km diameter, 48°.8 N, 349°.0 W, # 060A53); (e) radial fluid ejecta (Lunae Palus Du; crater diameter = 20 km); (f) ballistic ejecta with strings of secondaries around Lyot (crater diameter = 227 km, 1:2 M photomosaic MC05NW); (g) diverse ejecta blankets showing both ballistic and lobate ejecta (Memnonia Yg, crater diameter = 37 km, # 635A82).

structures on all three planets, the majority of impact crater ejecta morphologies on Mars are quite unusual, being lobate in form, some with ejecta terminating in a flat-topped lip ("pancake") whereas others end with a distal ridge (Fig. 5a,c,d). Some craters have single lobes (Fig. 5a), while others show 2 or more overlapping sets of lobes (Fig. 5b,c). Other craters, including most with diameters $\lesssim$ 4 km, have a lunar-like ballistic ejecta morphology with radial lineations and a thin, irregular boundary (Fig. 5e). Finally, some ejecta blankets are diverse (Fig. 5g), showing both lobate and ballistic elements. The variety of ejecta morphologies has resulted in a diversity of names in different studies, such as splosh, flower, rampart, pedestal and pancake crater. The plethora of morphologic names for Martian craters in different studies indicates the inevitable subjectivity of morphologic classification. The subjectivity inherent in classification of individual craters is intensified on Mars because of the abundance of transitional cases and the variable preservation state of even fresh crater ejecta. Thus, even though studies of the dependence of ejecta morphology on latitude, diameter, terrain and elevation abound, they are often contradictory in their conclusions. The subjectivity of classification also makes it difficult to choose the "best" value for a particular parameter from a set of mutually inconsistent ones.

In spite of these problems, it is generally agreed that ejecta morphology shows characteristic changes with increasing crater diameter (although estimates of transition diameters vary). Mouginis-Mark (1979) found that single-lobe craters are generally smaller than 20 km in diameter whereas multiple-lobe, radial and diverse morphologies are associated with craters $\geq$ 10 km and often $\geq$ 50 km in diameter. Double-lobe craters are $>$ 30 km and the pancake ejecta class occurs for crater diameters $\leq$ 8 km in this study. Barlow (1988*b*) found similar diameter/morphology dependencies: all pancake craters are $<$ 15 km in diameter, single-lobe and double-lobe craters dominate at diameters $\leq$ 30 km, multiple-lobe frequency exceeds single-lobe frequency between 25 and 40-km diameter, and ballistic and diverse ejecta patterns dominate at diameters $\geq$ 45 km. The largest craters included in these ejecta morphology studies are about 100 km in diameter, because of the poor preservation states of the ejecta of larger craters (cf. Mouginis-Mark et al. 1981). Thus, a comparison between the ejecta blankets of large basins on the Moon and Mars is difficult.

There is less agreement in the extent and reality of latitudinal variations in ejecta morphology. Johansen (1979) suggested a definite latitudinal dependence for ejecta morphologies, with more fluid-appearing craters concentrated at high latitudes and radial ejecta morphologies dominating in the equatorial regions. But, based on an analysis of 1558 craters, Mouginis-Mark (1979) found no latitude correlation for 5 of his 6 ejecta classes, the sole exception being pancake craters (craters with convex termini) which increased in frequency at latitudes poleward of $\pm$ 40°. A study by Mouginis-Mark and Cloutis (1983) found no increase in ejecta area with latitude for craters superposed on ridged plains. Kargel (1986), however, found an increase in ejecta sinuosity with increasing latitude. Barlow (1988*b*), using data on 3819 craters across the entire surface of Mars, found a slight latitudinal dependence for single-lobe and multiple-lobe craters and that double-lobe craters appear concentrated in the 40 to 50° latitude range. Horner and Barlow (1988) found that the transition from ballistic to rampart and back to ballistic ejecta with increasing crater diameter shows a distinct latitudinal variation at the small diameter end within heavily cratered terrain, with the onset diameter for rampart craters decreasing as one moves poleward. However, the transition diameter from lobate to ballistic ejecta at larger diameters apparently remains approximately constant throughout the southern highlands.

Crater Morphometry

Dimension data for Martian craters are given by Cintala and Mouginis-Mark (1980) and by Pike (1980*a*). Simplified depth/diameter curves are given

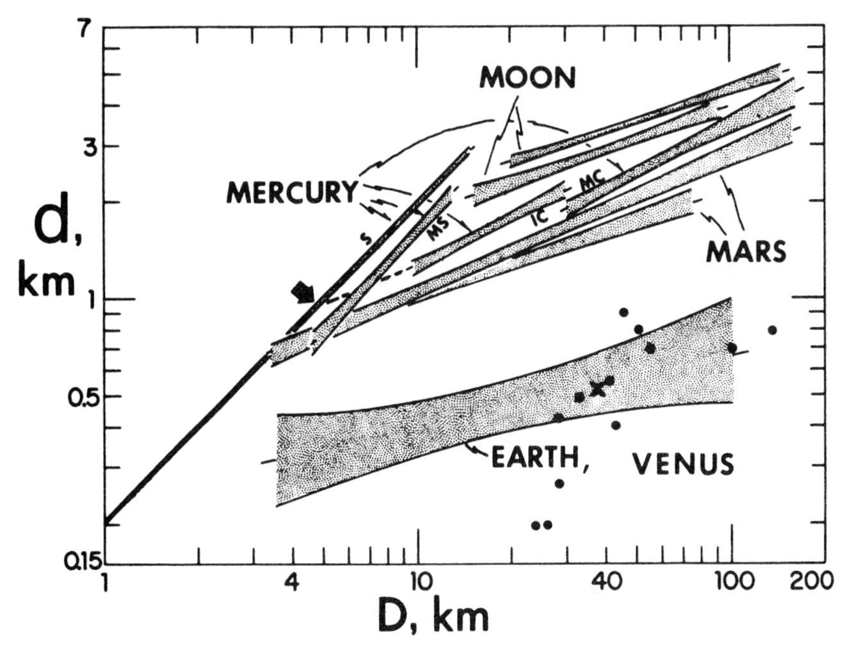

Fig. 6. Depth (d)/diameter (D) relationship for craters on Mars, Mercury, Moon, Earth and Venus (dots). The stippled areas are the 95% confidence intervals. The curves labeled S, MS, IC and MC are for simple, mature-simple, immature-complex, and mature-complex craters, respectively. The dots are 12 fresh impact craters on Venus (figure from Pike 1988).

for all terrestrial planets in Fig. 6. The Martian craters follow the same general trends found for fresh crater populations on all the other terrestrial planets: simple craters are geometrically similar with a constant depth/diameter ratio of ~0.20, while complex craters become shallower with increasing diameter. The logarithmic slope for the depth/diameter ratio for Martian complex craters is ~0.4, somewhat higher than the 0.3 for lunar and terrestrial complex craters. The simple/complex morphology transition on each planet corresponds roughly with the intersection of the steep line defining the depth/diameter relation for simple craters (the *s* line in Fig. 6, roughly the same for all terrestrial planets) with the less steep lines defining the depth/diameter relations for complex craters on the various planets. The simple/complex transition occurs for craters ~5-km diameter on Mars compared to 18 km on the Moon and 10 km on Mercury (Pike 1988). Figure 7 shows the simple/complex transition diameters for the terrestrial planets as a function of surface gravity. The transition diameter is seen to be roughly inversely dependent on surface gravity, with some second-order variations due to other causes. As the simple/complex transitions are almost certainly a result of material failure during the modification stage of the cratering process (see below), the generally shallow depths of fresh Martian craters relative to the Moon and Mer-

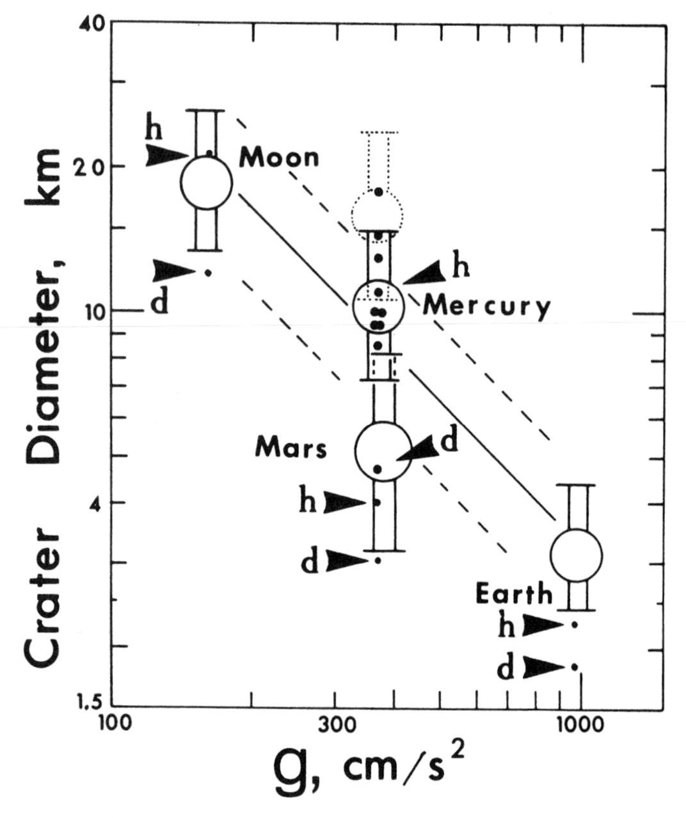

Fig. 7. Plot of the simple-to-complex transition crater diameter vs surface gravity for the Moon, Mercury, Mars and Earth. The dotted symbol is old Mercury data. Mars' anomalous position may reflect a weak layered or volatile-laden crust. The arrowed dots are intersection diameters of the rim height/diameter ratio h and the depth/diameter ratio d, regressions. The vertical bars are 1σ uncertainty limits (figure from Pike 1988). Data for the Earth are for crystalline rocks; the displacement of Mars below the trend formed by the other three bodies is roughly consistent with the difference between craters in crystalline and sedimentary rocks on the Earth.

cury (Leighton et al. 1969; Pike and Arthur 1979) are almost certainly due to impact processes and not to any long-term erosion or relaxation. Basin transition diameters, given in Table I, also show a crude inverse gravity dependence.

III. IMPACT PHYSICS: MARTIAN CRATER MORPHOLOGY AND VOLATILES

The overall similarities between craters on Mars and the other terrestrial planets imply that the same basic cratering process operates on all of them. The key question is, what are the unique elements of the Martian environment that produced the "Martian" variations on basic crater morphology described in the previous section, such as the central pits and the variety of ejecta blankets? To understand some of the suggestions which have been made, we give a brief summary of the physics of impact cratering.

Impact Processes

Understanding of the process of impact cratering has grown enormously over the last two decades. Detailed descriptions of the various aspects of cratering are readily available for those desiring more information (see, e.g., Roddy et al. 1977; Croft 1980; Grieve et al. 1981; Croft 1981*b*; Chapman and McKinnon 1986; Melosh 1989). Impact crater formation may be divided into three stages:

1. Short high-pressure phase;
2. Longer cratering-flow phase;
3. Modification phase.

The high-pressure phase includes the initial contact of the impactor with the ground and the transfer (coupling) of the impactor's kinetic energy into the planetary surface. Very high pressures prevail (1 to 50 Mbar, depending on the impact velocity) near the point of impact, decaying rapidly with distance. If the impact velocity is high enough, melting and vaporization of the projectile and surface materials occur. The extent of phase changes in the surface depends on the composition (e.g., ice is easier to vaporize than rock). The outgoing shock wave imparts a radially outward velocity field to the surface materials that is partially deflected by rarefactions from the free surface into a characteristic curvilinear material flow that produces the roughly parabolic-shaped transient cavity and the overturned flap of the ejecta blanket (phase 2). The transient cavity is then modified (phase 3) by slumping and rebound to form the final crater. If the transient cavity is small enough, only minor adjustments occur and a bowl-shaped simple crater results. If the transient cavity is large enough, the crater floor rebounds to form a central peak and terraces form as the rim fails, resulting in a complex crater. For still larger transient craters, the floor rebound and rim collapse results in a series of rings, producing a multi-ring basin.

The dependence of crater diameter on the conditions of impact, defined by the impacting body (radius a, mass m, density d, impact velocity U) and the planetary surface (surface gravity g, density δ and strength Y), is generally estimated by some form of scaling which employs both theory and small-scale measurements to extrapolate to the large scales of observed craters and basins. Again, several recent papers should be consulted for details (see, e.g., Holsapple 1987; Holsapple and Schmidt 1987; Schmidt and Housen 1987; Croft 1985*b*; O'Keefe and Ahrens 1981, and references therein). Scaling relations can be divided into three diameter regimes. The first regime is that of strength scaling, where the cohesive strength of the coherent pre-impact surface dominates. This regime is important on Mars only at crater diameters smaller than a few tens of meters, well below the sizes of most observed craters. The second regime is that of gravity scaling, where the crater is large enough that surface gravity results in the retention of a significant amount of crushed material inside the crater. Gravity scaling governs the size of the transient cavity in all large craters. Scaling theory suggests that the diameter D_t of the transient cavity is given by

$$D_t = Aa^{1-\alpha}\left(\frac{d}{\delta}\right)^{\beta}\left(\frac{U^2}{g}\right)^{\alpha} \tag{2}$$

where A, α and β are material constants. Laboratory experiments indicate that α is near 0.2 (Schmidt and Housen 1987), thus the transient cavity diameter increases with increasing impactor diameter, increasing impact velocity, and decreasing surface gravity. Dependence of D_t on material properties other than density is obtained through α and β, which vary slightly with target porosity, and the constant A. As noted above, simple craters are only slightly modified from the transient cavity, so gravity scaling applies directly to simple craters. The third regime is that of "modification" scaling (Croft 1985*b*), which describes the progressive enlargement of complex craters and basins relative to their transient cavities with increasing diameter

$$D_t = D_Q^{0.15}\, D_r^{0.85} \tag{3}$$

where D_Q is the simple-complex transition diameter and D_r is the final rim diameter. The enlargement can exceed a factor of 2 at basin diameters. Environmental dependencies in modification scaling are obtained through D_Q, which as seen in Fig. 7 depends both on gravity and other, presumably material (see below), parameters.

Applications to Mars

Based on our knowledge of impact mechanics, some obvious differences in the impact environment between Mars and the other terrestrial planets include differences in surface gravity, impact velocity, the presence of an atmosphere on Mars, and the probable abundance of water or ice in the Martian subsurface. The latter two factors have been frequently invoked to explain the unusual features of Martian crater morphology.

Central pits were suggested to be the result of explosive decompression of subsurface volatiles on Mars during the impact process (Wood et al. 1978; Schultz 1987). Croft (1981*a*) suggested that pits resulted from drainage of shock-melted water into the fractured zone beneath the final crater. Alternatively, Hodges (1978) drew a parallel to Gosses Bluff on Earth and suggested pits to be the result of impact excavation in a layered medium, where a soft layer underlies a harder one. Aside from the possibility that the central hole in Gosses Bluff may be an erosional artifact, if it is a pit, it too may be the result of water in the sedimentary rocks. Yet another possible origin for pits is that they are due to low-velocity impactors (Schultz 1988). Assuming that water is abundant in the crust of Mars (see chapter 15), the morphological and statistical variations for pits in craters on Mars, the Moon and Ganymede are consistent with an association of subsurface volatiles and pits. However, the same observations suggest that water is not necessary for the initiation of central pits, but that its presence greatly enhances pit formation (Croft 1981*a*).

Lobate ejecta morphologies in particular were suggested to have formed as a result of interactions with volatiles in the environment. Carr et al. (1977*b*) suggested the lobes were due to surface flow phenomena associated with entrainment of gas or impact-melted ice in the ejected material. This idea is supported by impact experiments into viscous-liquid materials that produced ejecta deposits that duplicated many morphological elements of lobate craters (Gault and Greeley 1978; Wohletz and Sheridan 1983).

Alternatively, the lobes were suggested to result from ejecta interaction with the atmosphere (Schultz and Gault 1979). This model is also supported by experimental impacts in a Martian-like atmosphere that produced rampart-like morphologies (Schultz and Gault 1984; Schultz 1989). Subsequently, numerous suggestions have associated various characteristics of rampart craters with volatiles and/or atmospheric effects. For example, variations in morphology with latitude were suggested to reflect changes in content and/or physical state of subsurface volatiles (Johansen 1979; Horner and Barlow 1988). The radial structure of some Martian craters was suggested to result from excavation into volatile poor material (Kuzmin 1980*b*; Horner and Barlow 1988) or decreased interaction of the ejecta with the atmosphere (Schultz and Gault 1979). Horner and Barlow (1988) interpreted the latitudinal variation in the diameter of ballistic to rampart and rampart to ballistic ejecta morphology changes as a change in depth to the upper boundary of a buried volatile-rich layer with latitude but a relatively constant lower depth. Early studies suggested that the pancake or pedestal morphology resulted from aeolian erosion of a lobe ejecta blanket (McCauley 1973; Arvidson et al. 1976); more recent studies suggest that this morphology is a primary structure caused by impact interaction with the atmosphere, groundwater or ice, and/or grain size differences (Head and Roth 1976; Mutch and Woronow 1980). Kargel (1986) suggested that an observed increase in the sinuosity of lobate ejecta boundaries towards the poles was due to desiccation of near-surface volatiles in equatorial regions. Cintala and Mouginis-Mark (1980) suggested that the relatively steep logarithmic slope of the depth/diameter relation for Martian complex craters might be due to the effects of subsurface volatiles on cratering efficiency.

Terrain properties related to volatiles or atmospheric coupling properties were also suggested to play a role in the formation of at least some ejecta morphologies. Mouginis-Mark (1979) found that multiple-lobe craters were preferentially located on ridged plains and that radial patterns primarily form around craters on young lavas and rolling plains. Pancake craters were found to be concentrated on terrains with probable high volatile contents such as channel floors, fractured terrain, and smooth plains. Barlow (1988*b*) found less morphology-terrain correlation: double-lobe craters were found to preferentially form on plains units in both northern and southern high latitudes and pancake craters were commonly found on channels and ridged plains. Mouginis-Mark (1979) detected a decrease in ejecta extent with increasing altitude and attributed this to either decreasing volatile content at high elevations or to a decrease in atmospheric pressure with height, decreasing the drag effects necessary for ejecta flow. Horner and Greeley (1987) found that elevation affected the distribution of some ejecta types in ridged plains regions but not in heavily cratered regions, contrary to results of earlier studies in the same regions by Mouginis-Mark and Cloutis (1983), who found no correlation between altitude and ejecta in the ridged plains. Schultz (1989) suggested that the various morphologies of the near, middle and distal regions of ejecta deposits around Martian craters may be affected by a whole range of properties: impactor (velocity, size, composition, angle), target (particle/size distribution, water, composition), and environment (gravity and atmospheric pressure).

Finally, the simple/complex morphology transition diameter on Mars may be affected by volatiles. Pike (1980*a*,1988) noted that the Martian transition diameter fell significantly below the inverse gravity curve defined by the Moon, Mercury and crystalline rocks on the Earth (Fig. 7). The Martian transition is, however, roughly consistent with an inverse gravity line defined by craters in sedimentary rocks on the Earth. Terrestrial sedimentary rocks are generally physically weaker than crystalline rocks and can be more porous. Thus terrestrial sedimentary rocks generally have a higher volatile content than crystalline rocks. If the volatile content and/or "weakness" of the sedimentary rocks relative to the crystalline rocks accounts for the lower terrestrial transition diameter in sedimentary rocks, then the relatively low transition diameter on Mars may be indicative of crustal volatiles or noncrystalline rocks there.

Unfortunately, in spite of all of the suggestions and indications of the effects of volatiles and atmosphere on crater formation, no firm conclusions can be drawn at this time. This is due in part to the inconsistencies between the data sets noted above: Do latitudinal variations exist, and which are real? There are other inconsistencies: Are lobate ejecta due to volatiles or atmospheric interactions? Horner and Greeley (1982) suggested that pedestal craters might be present on Ganymede, where volatiles are certainly present and an atmosphere is not. Unfortunately, the relevant structures are just at the limit of resolution and thus the extent of morphological correlation with Martian craters is unknown. Ejecta blankets of impact craters on Earth, where an atmosphere is definitely present, are not well enough preserved to show atmospheric effects for craters of the relevant sizes.

Are pits due to volatiles? Simplified theoretical calculations (Croft 1983) suggest that observed pit sizes are consistent with calculated dimensions of water-ice melt zones, but detailed calculations relevant to the dynamics of pit formation have not been done. Further, pit occurrence does not correlate in any obvious way with lobate ejecta structures (Hale 1982; Awwiller and Croft 1986), which appears inconsistent if both are due to crustal volatiles. Considering that ejecta materials originate from shallower levels than the material of the central uplifts in which pits occur, it is possible that separate reservoirs of volatiles exist at different depths with different area distributions which could account for this disparity. If central pits are indeed indicators of subsurface water on Mars, then many important planetological inferences can be drawn. For instance, Awwiller and Croft (1986) suggested that the lower abundance of pit craters in later geologic units indicates a long-term drying of the upper Martian crust, and that the concentration of pit craters in the ring structure of large basins was possibly due to water collected in the pores of the ejecta down-faulted between the rings. On the other hand, the apparent occurrence of summit pits on the Moon indicates, as noted above, that at least the rudiments of the processes that formed pits on Mars were operable on the Moon in the absence of volatiles. Given the uncertainties, it is possible that either central pits or lobate ejecta blankets or both are due entirely to factors other than volatiles, which would certainly account for the lack of correlation between the two features.

Although a definitive conclusion on the origin of Martian crater morphology cannot be made, a provisional statement may be made on the basis of three fairly well-founded observations:

1. Ejecta morphology correlates with crater diameter;
2. For smaller craters, the transition diameter from ballistic to lobate ejecta decreases toward the south pole in the highlands;
3. For larger craters, the transition diameter from lobate back to ballistic ejecta is constant and independent of latitude.

Observation (2) appears inconsistent with an atmospheric effect as being the primary cause of lobate ejecta blankets. Observations (2) and (3) taken together can be interpreted to mean that the depth to a volatile-rich layer decreases with increasing latitude, while the depth to the bottom of this layer remains constant. Until more data are forthcoming and the inconsistencies between existing data sets are resolved (radar images of Venus by Magellan, images of Mars from Mars Observer, and images of Ganymede from Galileo will all be relevant), we suggest that the unique Martian crater morphology and geologic evidence for subsurface water or ice probably indicates that the Martian crater morphology is largely governed by subsurface volatiles, and that, if atmospheric effects have occurred, they were subordinate. In any case, caution should be exercised at this time in drawing conclusions about the presence or distribution of Martian groundwater from impact crater morphologies.

IV. IMPACT STRUCTURES AND MARTIAN GEOLOGY

Erosion and Crater Degradation

Martian impact craters display a variety of morphologies and degradational states. The preservational state of these craters ranges from pristine to almost completely obliterated. In contrast to the lunar and Mercurian cases where crater obliteration is primarily due to degradation by other impacts (and volcanic burial), degradation processes on Mars are primarily "traditional" fluvial, aeolian and possibly glacial erosion as well as volcanic modification. The degree of crater degradation has been used to estimate the temporal and areal extent of erosional episodes throughout Martian history (Hartmann 1973*b*; Soderblom et al. 1974; Chapman 1974; Jones 1974; Barlow 1988*a*). The low rims and shallow depths of many craters in the generally older southern highlands has led to the theory that obliteration rates were higher in the past than at present. These early high obliteration rates have been linked to the higher impact flux (Chapman et al. 1969; Murray et al. 1971; Soderblom et al. 1974), increased volcanic activity (Greeley and Spudis 1981), and/or a thicker atmosphere and its associated increase in erosional activity (Sagan et al. 1973*a;* Pollack et al. 1987). At diameters $\lesssim$ 30 km, the Martian highlands crater size/frequency distribution shows a marked paucity of craters relative to the lunar highlands indicating a significant degree of crater loss due to erosion and/or deposition (see Sec. V on crater statistics). Based on statistical and morphological studies of moderate-sized craters, Chapman and Jones (1977) conclude that the erosional episode was a "pulse" that occurred near the end of the Noachian and contemporaneous with valley network formation in the southern highlands. This is consistent with atmospheric models (Pollack et al. 1987) which suggest that Mars had a warm and wet climate with from 1 to 5 bar CO_2 surface pressure early in its history. Furthermore, a recent model of an ocean in the northern plains and an ice sheet in the southern hemisphere (Baker et al. 1991; Kargel and Strom 1992) may suggest such a climate, linked with the erosional pulse. Squyres and Carr (1986) suggest that creep deformation associated with deeply buried ice may result in quasi-viscous relaxation of topography at latitudes poleward of 30°, and the low rims and shallow floors of impact craters may reflect the operation of this terrain-softening process. Controversy exists over the relative importance of terrain softening vs aeolian activity in explaining concentric crater fill (Zimbelman et al. 1989). Nevertheless, studies of Martian erosional history based on statistical analyses of crater degradational classes may need to be re-evaluated in light of the terrain-softening model.

Structural Imprints

Impact craters generate a number of structural features in the region in and around an impact that affect the region's subsequent geologic development: (1) highly porous and easily erodable ejecta blankets that may, by extrapolation from lunar basins (see Cordell 1978; Housen et al. 1983), reach thicknesses of tens of km in troughs between basin rings for the largest basins; (2) topographic depressions capable of trapping both aeolian deposits and surface water flows (see, e.g., Zimbelman et al. 1989); (3) brecciated and fractured zones usually 2 to 3 times the size and depth of the visible crater (cf. terrestrial impact papers in Roddy et al. [1977]) which act as zones of tectonic weakness; and (4) topographically high rims which can direct the flow of subsequent aeolian deposits (see, e.g., Ward et al. 1985). In Mars' active geologic environment, craters can be and are buried to varying degrees, serving as subsurface inhomogeneities around which later surface structures form. For example, several amphitheater-like bays in the walls of the canyons of Valles Marineris are rimmed with lobate ejecta blankets (e.g., north wall of Tithonium Chasma, north and south walls of Gangis Chasma), apparently the remains of impact craters whose fracture zones promoted erosional collapse. Similarly, several craters in the Valles Marineris highlands show extensive subsurface collapse (e.g., Coprates Lx and Ku), indicating localization of subsurface erosion (cf. Croft 1989). Major flow channels have their origin in impact craters (e.g., Coprates Az and Ds). Numerous craters show erosional breakup of smooth (aeolian, lacustrine or volcanic?) deposits on their floors (Schultz and Glicken 1979). Larger features may be affected: the concentric structure of Noctis Labyrinthus may be an impact imprint, and extensive valley networks appear localized in ancient basin ejecta blankets (Schultz and Glicken 1979; Schultz et al. 1982; Schultz and Frey 1990). Naturally, large impact basins leave the largest, deepest and most long-lasting imprints: Schultz et al. (1982) suggest basin-related origins for several large geologic discontinuities and fracture systems (including Valles Marineris); several major volcanic centers are located in the ring structures of large basins (Schultz 1984; Greeley and Guest 1987); the inner depressions of Argyre, Isidis, Hellas and Chryse basins are the debris catch basins of the largest channel systems, and Utopia Planitia may simply be a flow debris-filled basin some 3000 km in diameter (McGill 1988*a*, 1989*a*).

Crustal Dichotomy

Impact cratering may also have affected the geology of Mars at the very largest scales. Mars is divided into two major topographic regions consisting of the northern lowlands mainly occupied by young plains with a low crater density, and the southern uplands with a generally high crater density. The mean elevation difference between these two regions is $\sim$ 3 km. This dichotomy, of course, excludes the Tharsis uplift and associated volcanics which are later events. The dichotomy is most likely related to a major difference in crustal thickness of ~ 21 km between the two regions (McGill and Dimitriou 1990). Explanations offered for this difference in thickness are: (1) initial crustal inhomogeneity (Mutch et al. 1976*c*); (2) subcrustal erosion by a global-scale convection cell (Wise et al. 1979*a*); (3) multiple large basin-forming impacts (Frey and Schultz 1988); and (4) one giant impact (Wilhelms and Squyres 1984). There is no good reason for assuming that Mars initially formed with a crustal inhomogeneity as there is no evidence for this on other terrestrial bodies such as the Moon or Mercury. An endogenic origin is a possibility which is discussed by Wise et al. (1979*a*) and McGill and Dimitriou (1990). Although multiple large impacts could conceivably account for the crustal dichotomy, all of the largest impacts on Mars would have had to occur on less than a third of the planet. This is unlikely from orbital dynamic considerations of impacting objects and is inconsistent with the cratering record on the Moon where the largest basins are more or less randomly distributed across the surface.

The giant impact hypothesis of Wilhelms and Squyres (1984) has been criticized on the basis of a lack of an associated ejecta blanket and radial or concentric structures in the adjacent highlands, a younger age of the proposed rim than that of the basin itself (Frey and Schultz 1988), and peak volcanic activity in the northern lowlands well after the proposed impact event (McGill and Dimitriou 1990). The latter objection assumes that the main period of volcanism is related to the formation of the dichotomy. The proposed Borealis basin is centered at 50° N latitude and 190° W longitude and encompasses about 80% of the northern lowlands. Its diameter is ~ 7700 km measured along the surface curvature. The location and diameter are based on the location of very widely spaced massifs that are interpreted to be remnants of the basin rim. This basin is proposed to have formed about 4.2 Gyr ago or earlier (during the period of late heavy bombardment) by analogy with Nectarian or pre-Nectarian basins on the Moon.

An extension of this hypothesis, also involving a large impact, would alleviate the objections concerning the Borealis basin. In this case, an even more energetic impact occurs near the geometric center of the present northern lowlands at the end of the final accretion phase of Mars. Three-dimensional simulations of terrestrial planet formation by Wetherill (1986, 1988) have shown that the natural orbital and collisional evolution of planetesimals in the inner solar system not only result in the number and approximate size and orbits of the terrestrial planets, but also in large objects that eventually collide with these planets. The simulations start with 500 objects ranging in mass from 5.7×10^{24} to 1.1×10^{26} g which accumulate into the terrestrial planets and produce by-products ranging in mass up to 3 times the mass of Mars which collide with the planets at velocities of ~ 9 km s^{-1}. After about 64 Myr, 90% of the initial mass resides in the planets and the rest has formed over 20 objects with masses $\gtrsim 5 \times 10^{24}$ g, and with semimajor axes between 1 and 2 AU (Mars semimajor axis = 1.5 AU) and eccentricities of $\sim$ 0.2 to 0.6. Such objects have been proposed to account for the origin of the Moon by a Mars-size object colliding with Earth (Hartmann et al. eds. 1986), the anomalously high mean density of Mercury by a collision which removes much of the silicate mantle (Cameron et al. 1988; Wetherill 1988), and possibly the slow retrograde rotation of Venus. If large collisions occurred on these planets, then it is likely that a similar collision occurred on Mars. Such a collision may have been responsible for the Martian crustal dichotomy. In this case, the object would have collided with Mars at the end of accretion and redistributed the excavated crustal material over most of the southern hemisphere. Excavated material that went into low orbit would have reaccreted uniformly over Mars. This event would have occurred well before the period of late heavy bombardment that eventually resulted in the heavily cratered surfaces observed today in the southern hemisphere. Such heavily cratered surfaces in the northern hemisphere would eventually be buried by later volcanism preferentially concentrated in the region of thinner crust caused by the giant collision. Schultz and Lutz-Garihan (1982) interpret the anomalously large number of elliptical craters on Mars as the result of a circum-Martian swarm of planetesimals that collided with low impact velocities and angles on Mars (see below). If this interpretation is correct, then possibly these objects were formed from the accumulation of material ejected into high Martian orbit by the giant collision and which later impacted with Mars when their orbits decayed. Most of these craters were formed during the period of late heavy bombardment early in Martian history (see below).

Although rigorous computer simulations similar to those for the collisional origin of the Moon need to be performed, it seems likely that such a collision occurred on Mars. This explanation for the Martian crustal dichotomy ameliorates the problems raised by the Wilhelms and Squyres hypothesis by having an even larger impact occur earlier in solar system history. The concentration of volcanism in the northern hemisphere is the natural consequence of a thinner crust that allowed easier access to the surface throughout most of Martian history.

Elliptical Craters

Many Martian impact craters display an asymmetric ejecta pattern and elliptical crater form indicative of impact angles within 15° of horizontal (Gault and Wedekind 1978). An example is shown in Fig. 8. Over 170 elliptical craters > 3 km in diameter exist on Mars compared to only 4 or 5 on the Moon. The most detailed study of the distribution and degree of degradation of this class of craters was done by Schultz and Lutz-Garihan (1982). They found that many elliptical craters were oriented along great circles and that the youngest examples of such grazing impacts generally impacted in east-west directions whereas older grazers displayed a predominantly north-south orientation. They proposed that the large number of grazing impacts

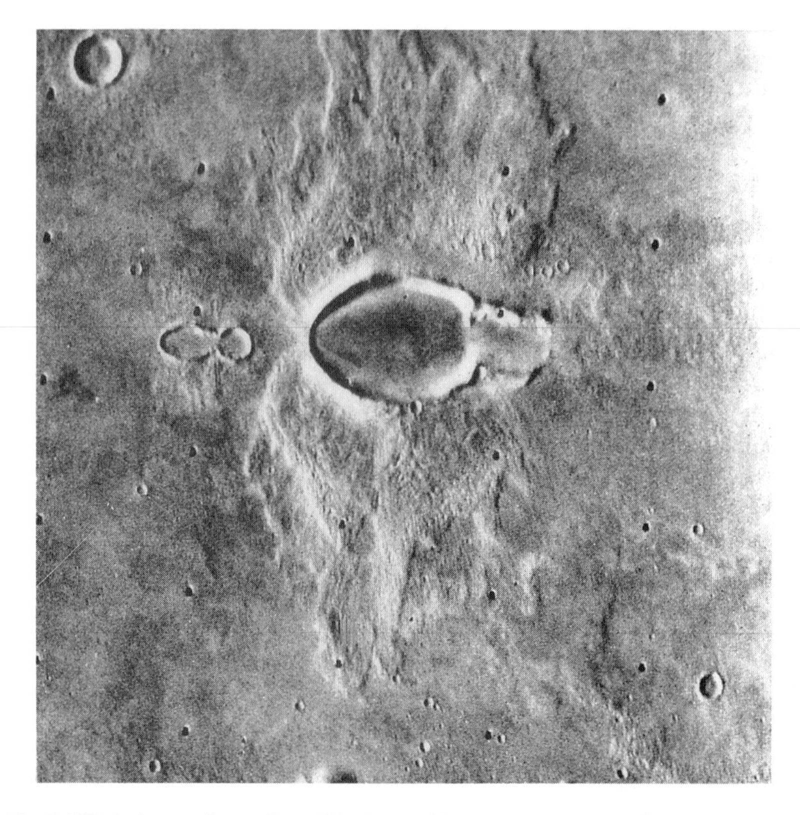

Fig. 8. Elliptical crater (long axis = 10 km) caused by an impact angle $\lesssim 10°$ from the horizontal (40°.9 N, 137°.5 W, # 039B13).

originated as satellites whose orbits tidally decayed with time, similar to the orbital decay which Phobos is currently undergoing, and that the spin axis of Mars changed with time. Over 93% of the satellite swarm impacted prior to the formation of the ridged plains regions, which were emplaced during the period of late heavy bombardment (see Sec. VI on relative chronology). Elliptical craters have been the center of much attention in recent years because one or more of them have been suggested as the originating site of the SNC meteorites (Wood and Ashwal 1981; Nyquist 1983).

V. CRATER STATISTICS

The Martian cratering record exhibits two distinct crater size/frequency distributions of different ages. The heavily cratered highlands has a complex multi-sloped curve which is similar, but not identical, to those of the heavily cratered terrains on the Moon and Mercury (Fig. 9). Below a diameter of

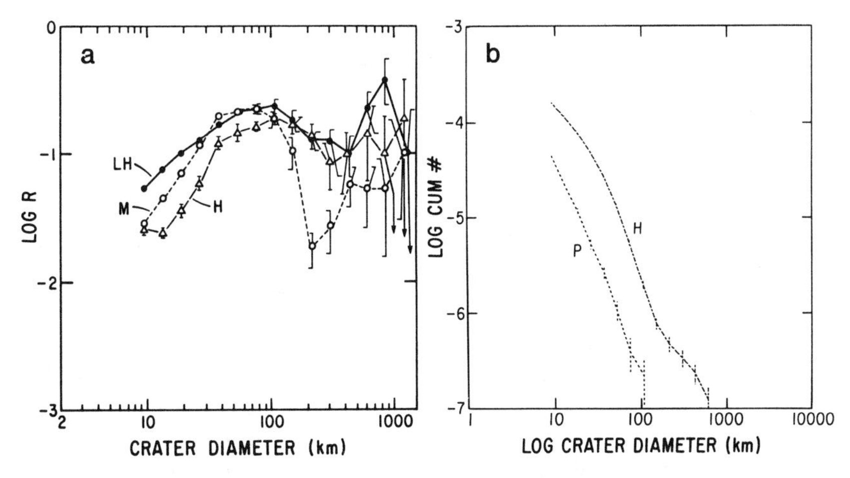

Fig. 9. (a) R plot of the crater size/frequency distribution for the lunar highlands (LH), the Martian heavily cratered terrain M, and Mercury's cratered terrain H. The paucity of craters $\lesssim$ 30 km diameter on Mars and Mercury relative to the Moon is due to erosion and deposition on Mars and intercrater plains emplacement on Mercury (see text for explanation). (b) Cumulative crater plot of the Martian highlands (upper curve) and the Martian northern plains (lower curve).These cumulative curves correspond, respectively, to curve M in Fig. 9a and curve YMP in Fig. 10. Although the difference in crater density is well illustrated, the difference in the shapes of the curves is not readily apparent in this type of plot.

$\sim$ 40 km, the slope is roughly a differential -2, but bends over to roughly a differential -5 between 100 and 200 km diameter. At large diameters the slope may change to ~ -2, but the statistics are very poor and consequently the uncertainties large.

Although the overall shape and crater density of the Martian highlands size/frequency distribution is similar to that of the lunar highlands, there are three differences. At diameters $\lesssim$ 30 km, the curve is less negative indicating that the paucity of craters increases with decreasing diameter relative to the lunar highlands. This paucity is almost certainly the result of obliteration of a significant number of smaller craters by a variety of processes, including emplacement of intercrater plains and atmospheric, fluvial and glacial erosion and deposition. The Mercurian highlands crater curve shows a similar paucity of craters relative to the lunar curve (Fig. 9). This paucity also has been attributed to the emplacement of intercrater plains, the most widespread terrain unit on Mercury (Strom 1977). The second difference occurs at large crater diameters. There appears to be a paucity of Martian craters larger than $\sim$ 100 km, particularly between 150 to 300km diameter, relative to the Moon and Mercury (Fig. 9). The reason for this paucity is not clear. The statistical significance of craters between 100 and 300 km diameters is relatively good and includes the very subdued basins identified by Pike and Spudis (1987). The paucity could be an observational loss due to the difficulty of identifying degraded craters larger than ~ 150 km diameter (see Sec. VIII on impact structures and Martian geology), or the impacting population may have been deficient in large objects. The final difference between the lunar and Martian highlands curves is that the Martian curve appears to be laterally displaced to smaller diameters by a factor of ~ 1.4. This displacement is probably due to differences in impact velocities between the Moon and Mars for objects in heliocentric orbits because such objects, on average, impact at higher velocities on the Moon than on Mars, and therefore produce larger craters on the Moon (see scaling equation in Sec. VII). The significance of this displacement is discussed in the next section.

The issue of crater saturation or equilibrium has been extensively discussed and debated in the literature (Woronow 1977,1978; Hartmann 1984; Chapman and McKinnon 1986; Strom and Neukum 1988), and is not discussed in detail here. One of the central issues concerning saturation is whether the heavily cratered terrains of the terrestrial planets essentially display a production crater population or whether the production function has been changed by attaining saturation. Woronow (1977,1978) and Chapman and McKinnon (1986) have shown through Monte Carlo computer simulations that changes in the production size/frequency distribution at saturation are highly dependent on the slope of the initial production population. For example, initial production populations with a slope less negative than a differential −3, e.g., −2 or −1, essentially retain their production size/frequency distribution at saturation. An initial multi-sloped production size/frequency distribution such as that observed in the lunar, Mercurian, and Martian highlands also retains its production function at or near saturation. Initial single-slope production populations with slopes more negative than a differential −3 decrease their slopes by ~ 1 at saturation, e.g., a −4 slope becomes a −3 slope at saturation. The crater density at which saturation occurs is still a subject of debate. Woronow's (1977,1978) Monte Carlo-Markov chain computer simulations suggest that saturation occurs at about twice the crater density observed for the lunar highlands. Chapman and McKinnon (1986), using slightly different parameters, find that saturation occurs at a lower crater density, but still slightly above the observed lunar highlands density. However, saturation crater densities are only quasi-constant because they oscillate about some mean value in response to the largest cratering events. The most heavily cratered surface in the solar system is Rhea and it has a crater density about 1.5 times greater than the lunar highlands (Strom 1987). Therefore, the truth may lie somewhere between the two models. In any event, most investigators who have studied the problem quantitatively, believe that the heavily cratered uplands of the Moon basically record the shape of the crater production population even though it may be near saturation (Woronow 1977,1978; Chapman and McKinnon 1986; Strom and Neukum 1988).

The Martian cratering record provides compelling evidence that the heavily cratered surfaces in the inner solar system (particularly on the Moon) record essentially the production size/frequency distribution of the objects responsible for the period of late heavy bombardment in the inner solar system. Unlike the Moon and Mercury, Mars has numerous surfaces of varying ages and superposed crater densities. Furthermore, the surfaces are extensive enough to record a relatively large number of superposed impacts providing much better statistics than are possible for younger surfaces on the Moon and Mercury. Surfaces of intermediate age (curve OMP in Fig. 10) have superposed crater densities that are considerably less than those of the heavily cratered lunar, Martian and Mercurian highlands. These surfaces exhibit a crater production population whose size/frequency distribution is statistically indistinguishable from that of the lunar highlands over the same diameter range (Fig. 10). This strongly confirms that the lunar highlands size distribution is basically a production population. As previously mentioned, however, the Martian and Mercurian highland curves at diameters less than ~ 30

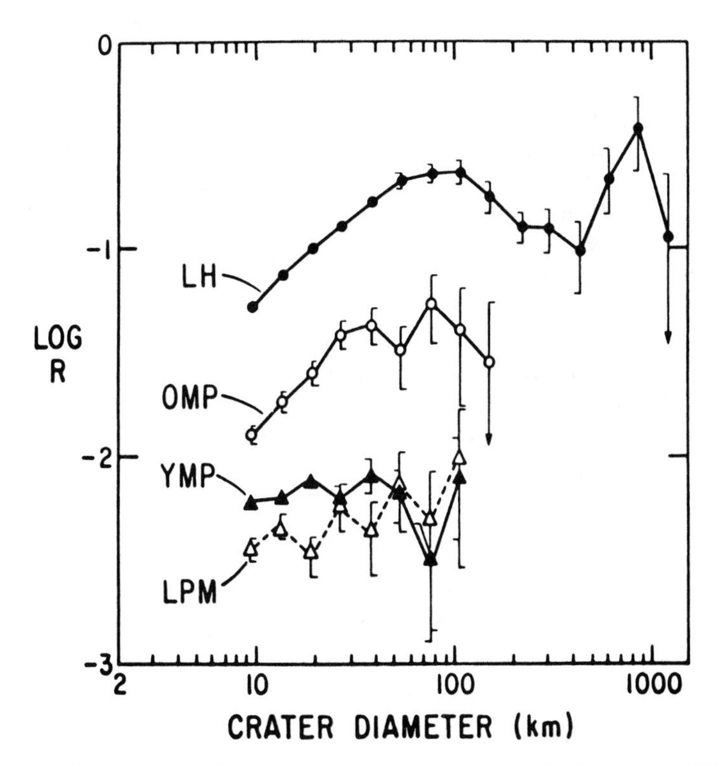

Fig. 10. R plot of the crater size/frequency distribution for the older Martian plains (OMP), the younger Martian plains (YMP) and the lunar post-mare craters (LPM). The lunar highlands crater curve (LH) is shown for reference. The older Martian plains production population is different from the younger production populations (YMP and LPM), but the same as the lunar highlands (see text for discussion).

km have been altered by intercrater plains emplacement, and atmospheric, fluvial and glacial erosion and deposition on Mars.

The younger surfaces on Mars (curve YMP in Fig. 10) also have a superposed crater production population, but these surfaces exhibit a size/frequency distribution that is very different from that of the highlands or surfaces of intermediate age. On these younger surfaces, the crater size distribution has a differential -3 slope compared to a -2 slope for surfaces of intermediate age. Statistical tests (chi squared and Wold-Wolfowitz Runs tests) indicate that the two crater populations are different at the 95 to 99% confidence level (Barlow 1988*c*). All surfaces with a crater density $\log R < -2$ have a crater size distribution of slope -3, while those with crater densities of $\log R > -2$ have size distributions of slope about -2 in the 8 to 70 km diameter range. On the Moon, the post-mare crater population also has a size/frequency distribution similar to that on the younger Martian surfaces (Fig. 10). The youngest surfaces on Mercury (smooth plains in and around the Caloris basin) have a crater density similar to the intermediate age surfaces on Mars and also show a similar size/frequency distribution with a slope of about -2 (Strom 1987; Strom and Neukum 1988).

These data strongly suggest that two populations of impacting objects are predominantly responsible for the cratering record in the inner solar system. Surfaces that show a crater size/frequency distribution with a -2 slope were formed during the period of late heavy bombardment, while surfaces having a size/frequency distribution with a -3 slope were formed after the period of late heavy bombardment. This provides a means of identifying surfaces relative to the period of late heavy bombardment. As Mars exhibits such a large variety of surface ages (crater densities), this method is useful for dating geologic units with respect to the end of late heavy bombardment (see Sec. VI below). If the period of late heavy bombardment ended at the same time on Mars as it did on the Moon (3.8 Gyr ago), then surfaces showing the signature of late heavy bombardment (-2 slope) are older than 3.8 Gyr, while other surfaces are younger than 3.8 Gyr. Unfortunately, the duration and end of the period of late heavy bombardment on Mars depends on assumptions about the origin of the objects responsible for it. In one scenario for accretional remnants, Wetherill (1977) suggests that the period of late heavy bombardment could have ended 1 Gyr later on Mars than on the Moon. This time dependence on the origin of impacting objects is discussed in Sec. VII.

VI. RELATIVE CHRONOLOGY OF MARTIAN SURFACE UNITS

Crater statistical techniques, together with stratigraphy, are the major tools by which relative chronologies of most planetary surfaces are derived. The shape of the crater size/frequency distribution curve provides important information about the population of impacting objects responsible for the cratering record at a particular period of time (Fig. 11). Relative chronologies for Mars indicate that Martian terrain units formed during two major cratering periods: late heavy bombardment and post late heavy bombardment (Hartmann 1973; Soderblom et al. 1974; Mutch et al. 1976*c*; Mutch and Saunders 1976; Scott and Carr 1978; Condit 1978; Arvidson et al. 1980; Tanaka 1986; Barlow 1988*c;* see chapter 11). The analyses use different data sets (Mariner 9 data for all studies except those by Arvidson et al [1980]; Tanaka [1986] and Barlow [1988*c*] who use Viking data) and different crater diameter ranges ($\geq$ 64 km for Hartmann [1973*b*]; 4 to 10 km for Soderblom et al. [1974] and Condit [1978]; 1,2,5,16 km for Tanaka [1986], and $\geq$ 8 km for Barlow [1988*c*]). Although these studies differ in their details, the relative age chronologies are fairly consistent.

The Martian cratering record (Table II) begins in the Noachian-aged southern highlands which display multi-sloped size/frequency distribution curves and crater densities $N(8) \geq 400$ (400 or more craters/10^6 km^2 > 8-km diameter). The oldest units include the hilly and cratered uplands and rim material of several large basins (i.e., Isidis, Hellas and possibly Argyre [Tan-

TABLE II
Relative Ages of Terrain Units Listed in Order of Decreasing Age

Relative Age	Unit
Late heavy bombardment	uplands intercrater plains mountains fractured uplands fretted terrain exhumed cratered uplands dissected plateau knobby terrain Hellas volcanics
End of late heavy bombardment	basin floor deposits ridged plains
Post heavy bombardment	cratered/fractured plains plains chaotic terrain mottled plains outflow channels equatorial layered deposits volcanic plains volcanic shields aureole canyon floor polar caps polar layered deposits

aka 1986; Barlow 1988*c*]). These old units are overlain in many areas by slightly younger intercrater plains. Superposition relations and crater density studies indicate that the valley networks are contemporaneous with much of the intercrater plains (Baker and Partridge 1986; see chapter 15). The existence of valley networks coupled with the degraded appearance of the highlands crater population led to the theory that high obliteration rates, perhaps associated with a thicker atmosphere (Sagan et al. 1973*a;* Pollack et al. 1987), occurred near the end of the period of late heavy bombardment (Hartmann 1973*b;* Mutch and Saunders 1976; Chapman and Jones 1977; Tanaka 1986). Crater data from the highland patera and many of the smaller volcanic constructs in the Tharsis and Elysium regions indicate that volcanic activity during late heavy bombardment was widespread and not confined to plains formation (Neukum and Wise 1976; Greeley and Spudis 1981; Neukum and Hiller 1981; Barlow 1988*c*). Knobby terrain displays a range of crater densities, all of late heavy bombardment age. The appearance and age of this unit has led several investigators to suggest that knobby terrain represents highland material embayed but not completely buried by later plains (Scott 1978; McGill 1986). The knobby terrain provides age constraints for the process(es) creating the crustal dichotomy (Mutch and Saunders 1976; McGill and Dimitriou 1990; Barlow 1988*c*). The youngest large basin on Mars is the Argyre basin: Tanaka (1986) places this impact in the Early Noachian Epoch, prior to intercrater plains development whereas Barlow (1988*c*) finds that Argyre formed subsequent to the major phase of intercrater plains formation.

Impact rates declined throughout the Noachian and Early Hesperian, but the highly structured crater size/frequency distribution curves indicative of the late heavy bombardment population of impactors is imprinted on the ridged plains formed in the Early Hesperian (Barlow 1988*c*). The high obliteration rates that are enhanced toward the end of the Noachian period (Chapman and Jones 1977) decline rapidly during the Early Hesperian (Tanaka 1986).

The crater size/frequency distribution curves change from a -2 to a -3 differential slope at about a log R value of -2, suggesting that the late heavy bombardment population of impactors no longer dominated the cratering record (Fig. 11). Units of Late Hesperian and Amazonian age display distribution curves indicative of the post-late-heavy-bombardment population of impactors.

The Late Hesperian Epoch is dominated by formation of northern plains units, especially in the eastern hemisphere of the planet. Fracturing and volcanism were prevalent in the Tharsis and Elysium provinces, particularly around Alba Patera. This period also marked the major occurrence of chaotic terrain and outflow channel formation, although multiple stages of flooding are suggested from stratigraphic analyses (Baker 1982; chapter 15).

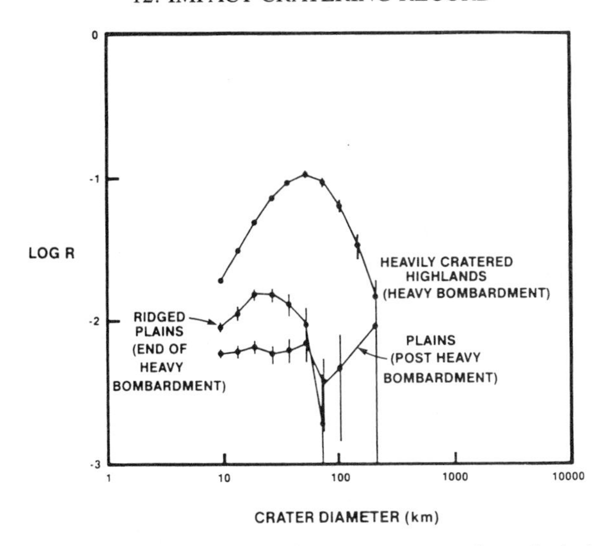

Fig. 11. *R* plot showing representative Martian crater curves used as a basis for determining relative ages of various surface geologic units (see also Table II and Color Plate 8).

Northern plains deposits, shield volcanism in the Tharsis region, mass wasting in Valles Marineris, and low levels of aeolian erosion typify the Amazonian Epoch (Scott and Carr 1978; Tanaka 1986). Crater densities are low (generally $N(8) \leq 50$ craters/10^6 km^2) and the relative crater size/frequency distribution curves are horizontal within the error bars (differential -3 slope) (Barlow 1988*c*). Extremely low crater densities indicate that the youngest geologic units on the planet are volcanic flows on and around Olympus Mons, landslides in Valles Marineris and around the Olympus Mons basal scarp, and the polar deposits (Lucchitta 1979; Tanaka 1986; Barlow 1988*c*). The polar-layered deposits display very low crater densities which may indicate either recent deposition or the ease with which craters in these deposits are eroded (Mutch et al. 1976*c*; Tanaka 1986; Plaut et al. 1988; chapter 23).

The relative chronology for Mars derived from analysis of the shapes and crater densities of its various geological units is listed in Table III and shown in Plate 8 in the Color Section. The wide range in age of Martian terrain units indicated by this chronology will provide better constraints on questions such as the occurrence of fluctuations in the cratering rate once a radiometric absolute chronology is obtained.

VII. ORIGINS OF IMPACTING OBJECTS

The origin of the objects responsible for the cratering record observed on solar system bodies is largely an unsolved problem. This is particularly true for the objects responsible for the period of late heavy bombardment.

TABLE III
Crater Densities for Martian Stratigraphic Boundaries[a]

Series	N (1)	N (2)	N (5)	N (16)	N (4–10)
Upper Amazonian	< 160	< 40			
Middle Amazonian	160–600	40–150	< 25		< 33
Lower Amazonian	600–1600	150–400	25–67		33–88
Upper Hesperian	1600–3000	400–750	65–125		88–165
Lower Hesperian	3000–4800	750–1200	125–200	< 25	165–260
Upper Noachian			200–400	25–100	> 260
Middle Noachian			> 400	100–200	
Lower Noachian				> 200	

[a] Table from Tanaka (1986). N = number of craters > (x) km diameter/10^6km^2.

There are only two classes of objects that presently cross the orbits of the planets and collide with them and their satellites: comets and asteroids. However, early in solar system history there may have been other objects, e.g., accretional remnants, that contributed significantly (perhaps overwhelmingly) to the cratering record. Assignments of absolute ages to surfaces based on crater densities are highly dependent on one's assumptions about the origin of the objects that created them. The cratering rate must be known to derive absolute ages and this in turn depends on the origin, i.e., the orbital elements, of the impacting objects. The cratering records on various solar system bodies provide some constraints on the origin of the objects responsible for them. In this section the constraints on the origin of impacting objects imposed by the cratering record are discussed, with major emphasis on the inner solar system.

As discussed in Sec. V, two crater production populations of different ages are recognized in the inner solar system (Strom and Whitaker 1976; Strom 1987; Barlow 1988*c*; also see Figs. 10 and 11). The oldest population occurs in the heavily cratered terrains of the Moon, Mercury and Mars and represents the period of late heavy bombardment. As this population has the same size/frequency distribution spanning the entire range of inner planet distances, it must represent the same family of objects, probably in heliocentric orbits. Strom and Neukum (1988) found that the crater curves (where they are probably unaffected by erosion and the statistics are good) are laterally displaced with respect to each other so that higher impact velocities are required at planets with smaller heliocentric distances, i.e., larger craters on Mercury and smaller craters on Mars compared to a given size crater on the Moon (see Strom and Neukum 1988, Fig. 18). This is consistent with impacting objects being in heliocentric orbits. Furthermore, the impact velocity ratios between Mercury and the Moon, and Mars and the Moon require small semimajor axes that confine them to the inner solar system (Strom and Neukum 1988). At least on the Moon, this population of objects dominated the cratering record for the first 700 Myr of solar system history and then became extinct or evolved into the younger population $\sim$ 3.8 Gyr ago.

The other crater population is superposed on younger surfaces on the Moon and Mars and has been accumulating there since the end of late heavy bombardment. On Mercury, the relatively young plains within and surrounding the Caloris basin record the same crater population as the heavily cratered terrains, but at the same crater density as the older Martian plains which also show this population (Strom and Neukum 1988). If the young crater population is present on Mercury, its numbers are too low to be recognized. The two crater populations are distinguished on the basis of their significantly different size/frequency distributions, which implies that the impactors also had different size distributions. Either the populations had different origins or the older population evolved through time into the younger one.

The Galilean satellites Ganymede and Callisto have a crater population that is completely different from that in the inner solar system. It is characterized by a severe paucity of craters $\gtrsim$ 60-km diameter and an overabundance of craters $\lesssim$ 40-km diameter relative to the heavily cratered regions in the inner solar system (Fig. 12). Studies by Woronow and Strom (1981), Chapman and McKinnon (1986) and Strom (1987) indicate that this crater population is also a production population. The reader is referred to these sources for details of the analyses. Other different crater populations may be present on the satellites of Saturn and Uranus (Strom 1987), but the discussion here is confined to the Jovian satellites and the terrestrial planets where the cratering record is better defined.

The cratering record at these locations in the solar system imposes two important constraints on any proposed origin of impacting objects: (1) the different size distributions strongly imply that the objects responsible for them had three different origins; and (2) because the terrestrial planet crater populations are not seen at Jupiter, the objects responsible for them were probably confined to the inner solar system. The latter constraint is consistent with orbital restrictions imposed by the lateral shifts of the inner planet late heavy bombardment crater curves mentioned earlier. In one form or another, the solar system cratering record has been attributed to comets, asteroids, accretional remnants or some combination of these objects. Accretional remnants are considered to be objects left over from the accretional formation of the planets. In a sense, asteroids can be considered to be accretional remnants as they are collisionally evolved accreted planetesimals that failed to form a planet because of the influence of Jupiter's strong gravitational field.

The origin of the objects responsible for the period of late heavy bombardment in the inner solar system is uncertain. At least on the Moon, the cratering rate of these objects rapidly declined during the first 700 Myr of solar system history and then they became extinct about 3.8 Gyr ago. Accretional remnants left over from the formation of the terrestrial planets is an

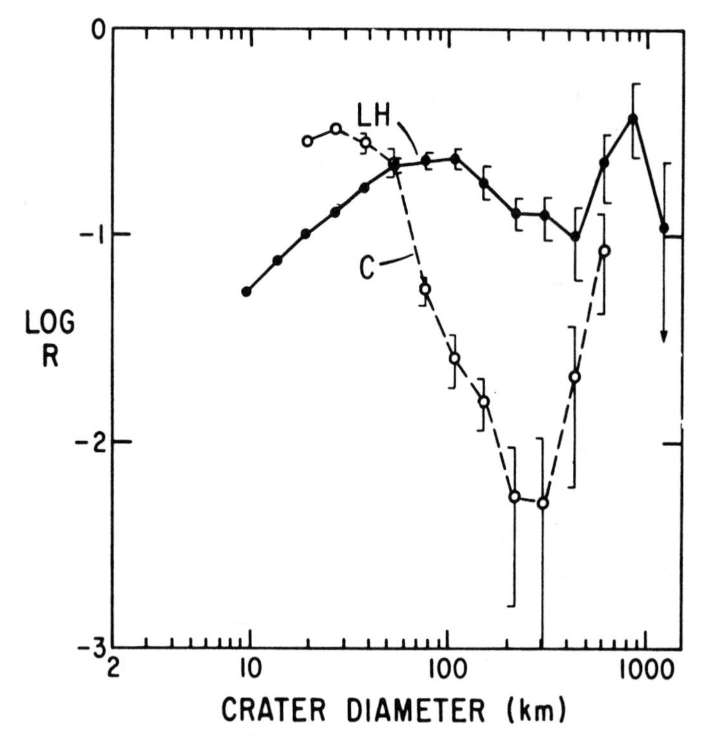

Fig. 12. *R* plot of the size/frequency distribution for the crater production population on Callisto (C) compared to the lunar highlands (LH) (see text for explanation).

attractive origin for these objects because it meets the constraint that they were confined to the inner solar system. A major obstacle to this origin has been the very short sweep-up time for accretional remnants: about 10 Myr rather than the 700 Myr derived from lunar data. Wetherill (1977), however, has shown that the sweep-up time can be extended to about the required time by perturbations of the objects into the ν_6 secular resonance. This results in an extended period of late heavy bombardment on Mars of $\sim$ 1 Gyr compared to the other terrestrial planets.

An intense bombardment by comets early in solar system history has also been suggested to have been responsible for the period of late heavy bombardment (Shoemaker and Wolfe 1982). It is unlikely that comets could have been largely responsible for the heavily cratered surfaces in the inner solar system because the crater populations at Jupiter and in the inner solar system are radically different. If comets were responsible then they should have left a comparable signature at Jupiter and the terrestrial planets. Although comets can become smaller with time by losing mass in the inner solar system, this cannot account for the different size distributions between Jupiter and the terrestrial planets. For comets to account for this difference, their nuclei would have to be on average ~ 4 times larger in the inner solar system where they rapidly lose mass than at Jupiter where their mass is conserved (Strom 1987). If the crater population at Jupiter and elsewhere in the outer solar system is due to comets, then they either hit the terrestrial planets before the period of late heavy bombardment in the inner solar system, or they were so thoroughly mixed with the other population that they are not recognized as a separate population today. In either case, the heavily cratered terrains in the inner solar system are unlikely to have been primarily produced by comets.

The origin of the objects responsible for the younger crater population on the terrestrial planets is much more certain. Because this population has accumulated on younger surfaces after the period of late heavy bombardment, it is almost surely due to a mixture of comets and asteroids. At Mars the present cratering rate estimated from current Mars-crossing asteroids is ~ 7 times that of comets (Basaltic Volcanism Study Project 1981). If this proportion of asteroid to comet impacts was more or less constant since the end of late heavy bombardment, then the young crater population at Mars should be dominated by craters formed from asteroidal impacts.

Davis et al. (1985) performed numerical simulations of the collisional evolution of hypothetical initial asteroid populations and derived the size distribution of each. They found that the initial population that gave the best agreement with the current asteroid size distribution was a runaway-growth planetesimal accretion size distribution characterized by a few large bodies beginning to run away in size from the much smaller bodies that contain most of the mass. Although their simulations produced asteroid size distributions down to ~ 1 km diameter, they could compare them only with the observed asteroid diameters down to ~ 20 km where the data are fairly reliable. The young Martian crater population provides an opportunity to estimate the size distribution of asteroids down to ~ 1 km diameter by using impact velocities derived from the orbital elements of current Mars-crossing asteroids in an appropriate crater scaling law.

Figure 13a is the "asteroid" size distribution derived from the young crater population by a Monte Carlo simulation which randomly selects Mars-crossing asteroid impact velocities (214 as of May 1988) for use in a modified Holsapple-Schmidt crater scaling law (Holsapple and Schmidt 1982; Holsapple 1987). The modified scaling law for the gravity regime is

$$d = \left[\frac{D_e g^{1/6}}{k\{c\mathrm{v}[1 - 0.095(1 - \sin A)]\}^{1/3}}\right]^{1.2} \qquad (4)$$

where d = projectile diameter, D_e = excavation diameter, g = surface gravity, k = coupling factor (~ 4.8, experimentally determined), c = ratio of projectile to target density, v = impact velocity and A = impact angle from the horizontal, which can vary from 1 to 90° (most impacts occur at 45°). The

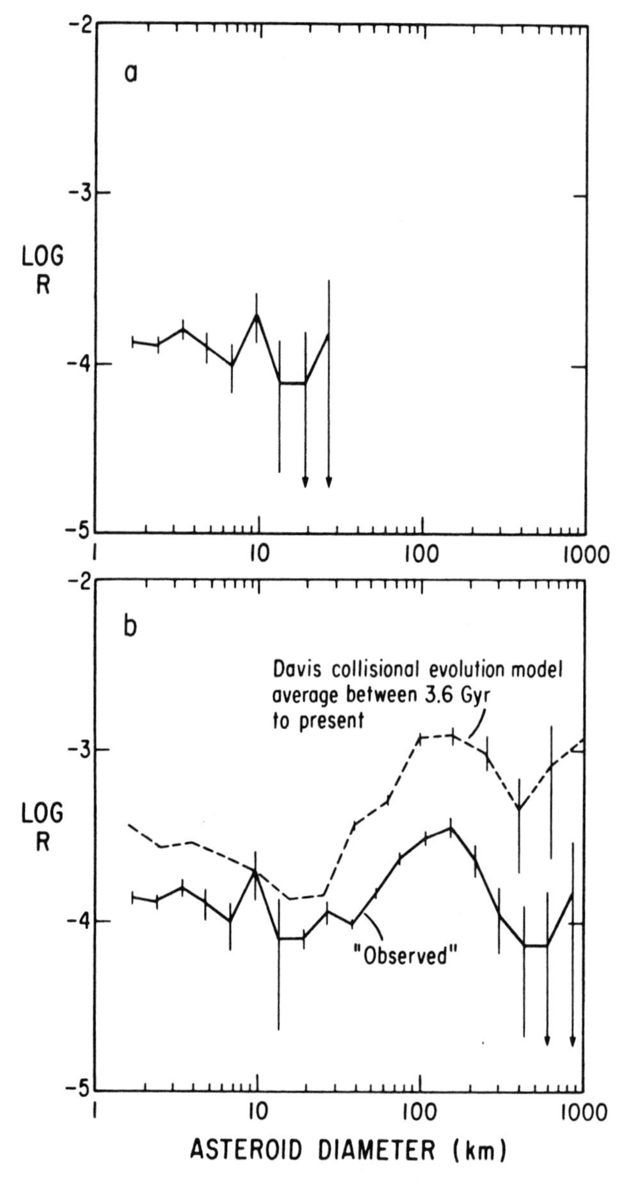

Fig. 13 (a). *R* plot of the approximate size/frequency distribution of asteroids derived from the young Martian plains crater population (curve YMP in Fig. 10) using a modified Hosapple-Schmidt crater scaling law (Eq. 4; see text for explanation). (b) *R* plot of the asteroid size/frequency distribution shown in Fig. 13a combined with the size distribution of asteroids $\gtrsim 20$ km diameter from Earth-based and IRAS data ("observed" curve). The dotted upper curve is the size distribution of asteroids derived by Davis et al. (1985) from a collisional evolution model of runaway-growth. The similarity between the two curves is remarkably good (see text).

excavation crater diameter D_e for complex craters is from Eq. 3, assuming the excavation diameter is equal to the transient cavity diameter. For simple craters, $D_e = D_r$. The impact angle and asteroid density are also randomly selected from a table of impact angle probabilities, where the probability of impact at angles less than a given angle A is $\sin^2 A$, and a table of densities weighted for low densities based on meteorite data. Figure 13b is the "observed" size/frequency distribution for asteroids: diameters > 20 km are taken directly from IRAS and Earth-based data; and diameters < 20 km are derived from the young Martian crater population. Also shown in Fig. 13b is the size distribution of asteroids derived by Davis et al. for the runaway-growth model averaged between 3.6 Gyr ago and the present (the approximate time since the end of late heavy bombardment). The agreement between the model curve and the observed and Martian-derived curve is good. This suggests that the Davis et al. model of collisional evolution is correct and that the young Martian crater population is primarily asteroidal in origin. The fact that the old highland crater size distribution is similar to the asteroid size distribution at asteroid diameters $\gtrsim 20$ km suggests that the objects responsible for the period of late heavy bombardment may have had a relative size distribution similar to the current asteroid population, but an order of magnitude smaller in absolute size.

VIII. MARTIAN ABSOLUTE CHRONOLOGIES

Absolute chronologies can be obtained by extrapolation to the lunar cratering chronology provided the following conditions are met: (1) the impacting populations at the Moon and the planet are the same; and (2) the origins of the impacting objects are known, thus allowing the cratering rate through time to be modeled. The method for deriving absolute ages from cratering records involves determining the relationship between the differential distribution (number of craters per unit area per diameter at time *t*) and the differential cratering rate (number of craters per unit area per diameter per time at time *t*). For the inner planets, the cratering rate is usually extrapolated from that derived from dated surfaces on the Moon. This method is discussed in detail by Strom and Neukum (1988), and the reader is referred to this reference for the detailed methodology. Several studies have derived absolute chronologies for Martian terrain units using various assumptions for the above parameters (Hartmann 1973*b;* Soderblom et al. 1974; Neukum and Wise 1976; Neukum and Hiller 1981). Viking data are used in the two chronologies discussed by Neukum and Hiller (1981). Slight modifications to original chronologies are found in Hartmann (1978) and Neukum et al. (1978). All models assume that the Moon and Mars have been exposed to identical impact populations and that the temporal evolution of the cratering rate is identical between the two bodies. The primary source of disagreement between the models results from the estimate of the impact cratering rate.

Wetherill (1974) discusses some of the problems associated with estimating the Martian cratering rate. Hartmann (1973*b*), using the estimate of a Martian asteroid flux 25 times the lunar rate and making adjustments for impact velocity differences, determines that the crater-forming rate on Mars is 6.2 times the lunar rate for the same size crater. Hartmann's 1978 paper lowers the Martian crater production rate to twice the lunar rate based on better statistics of Mars-crossing asteroids (Shoemaker and Helin 1977). Soderblom et al. (1974) consider that the cratering rate on Mars and the Moon have been the same over the past 4 Myr because crater densities on lunar and Martian old plains are similar.

Neukum and his colleagues have derived a number of absolute chronologies based on the comparison of Martian and lunar standard production curves (Neukum and Wise 1976; Neukum et al. 1978; Neukum and Hiller 1981). All models consist of a standard crater production curve derived from homogeneously cratered areas that have not experienced recent endogenic obliterative events. An inflection in the lunar and Martian cumulative crater size/frequency distribution curves near $D = 2$ km is interpreted as part of the production population rather than being a contribution from secondary craters (Soderblom et al. 1974). The curves are then shifted so as to bring the Martian and lunar distributions into concurrence. Differences in impact velocity and target characteristics cause the distribution curve to shift left or right (diameter shift) whereas age or cross-sectional effects cause vertical (abundance) shifts. Neukum and Wise (1976) considered that the impact mass fluxes at the Moon and Mars were identical based on comparisons of Martian crater frequencies with those on the Moon and Phobos and corrections for impact velocity differences on final crater size. Neukum and Hiller (1981) used observational results of the numbers of Mars-crossing asteroids to obtain a model where the Martian impact mass flux has been a factor of 2 higher than the lunar rate in the past 3 Gyr but was essentially identical earlier. The Martian production size/frequency distribution is flatter than the lunar distribution resulting in 1-km crater frequency values for Mars to be 4.5 (Neukum and Wise 1976) to 3.0 (Neukum et al. 1978) times lower than the corresponding lunar values for surfaces of identical age. This allows determination of a Martian model cratering chronology from the lunar crater density vs absolute age relationship.

A comparison of the absolute ages determined for several Martian geologic units or features by the above models is shown in Table IV. The lack of agreement between the various models emphasizes the difficulties inherent in deriving Martian absolute chronologies from crater counts. These difficulties result from (1) the uncertainty in the ratio of Martian to lunar impact mass flux; (2) the question of whether fluctuations in the impact cratering rate have occurred; and (3) the uncertainty in the time dependence of the cratering rate throughout the inner solar system.

The ratio of the Martian to lunar crater production rates is among the

TABLE IV
Comparison of Absolute Ages (Gyr) for Selected Features

	Hartmann (1973)	Soderblom et al. (1974)	Neukum and Wise (1976)	Neukum and Hiller I (1981)	Neukum and Hiller II (1981)
Highlands	3.0–4.0	≥3.2	4.4	3.8–4.2	4.0–4.2
Ridged plains	[a]	2.1–1.8	3.9	3.9	2.7–4.0
Elysium vol.	[a]	1.5–0.8	3.7	3.2–3.9	1.7–4.0
Alba Patera	[a]	[a]	3.7	0.5–3.8	0.3–3.7
Olympus Mons	0.1	≤1.7	2.5	0.3–3.3	0.3–2.0

[a]Feature not dated.

most controversial of these uncertainties. Estimates range from 25 times the lunar rate (Hartmann 1973*b*) to identical Martian and lunar rates (Soderblom et al. 1974; Neukum and Wise 1976). The estimated crater production rates depend on the population of impacting objects: the production rate at Mars is 4 times higher than the lunar rate if 80% of the impactors are Mars-crossing asteroids, but only 0.9 times the lunar rate in the unlikely event that the impacting population is composed entirely of periodic comets (Basaltic Volcanism Study Project 1981, p.1080). The range in crater production rates for Mars is estimated to be between 1 and 4 times the lunar rate, with twice the lunar rate being the most likely value. Statistics of Mars-crossing asteroids (Shoemaker and Helin 1977) and the collisional lifetimes of these asteroids (Wetherill 1975) have resulted in most estimates converging on a Martian cratering rate about twice that for the Moon during the post-late-heavy-bombardment period. However, this cratering rate probably was very different during the period of late heavy bombardment, as discussed in Neukum and Hiller (1981).

Martian absolute chronologies could be dramatically affected if significant fluctuations in the cratering rate occur through time. Studies of biological mass extinctions on Earth have been interpreted to mean that increased impact rates occur approximately every 30 Myr (Raup and Sepkoski 1984; Rampino and Stothers 1984). These fluctuations in the cratering rate have been attributed to comet showers caused by the Sun's motion through the galactic plane (Raup and Sepkoski 1984) or by the perturbations from an unseen stellar companion (Whitmire and Jackson 1984; Davis et al. 1984). Some investigations find support for these periodicities in the sparse terrestrial cratering record (Alvarez and Mueller 1984), but the uncertainties in dating these features and the paucity of preserved terrestrial impact structures raise serious doubts about these results (Grieve et al. 1987). Attempts to resolve the question of cratering rate variations by using lunar crater data have also met with controversy because of the lack of large expanses of young terrains and the age uncertainty of most lunar features (Neukum et al. 1975; Neukum and Konig 1976; Konig et al. 1977; Guinness and Arvidson 1977; Young 1977). If cratering rate fluctuations exist but are unaccounted for in the lunar crater chronology, features dated solely by crater density analyses may not be as old or as young as they appear. Extrapolations of the lunar cratering chronology to Mars propagates these uncertainties, resulting in possible large errors in the age of Martian geologic features.

All the Martian absolute chronologies discussed assume the time dependence of the impact flux is identical throughout the inner solar system. This assumption of identical temporal variations throughout the inner solar system implies that the period of late heavy bombardment ceased around 3.8 Gyr ago on Mercury, Venus and Mars as well as the Moon. The orbital evolution of the impacting objects depends on the origins of these bodies, which, for late heavy bombardment impactors in particular, are very uncertain. Although most dynamical models indicate that the end of late heavy bombardment was approximately contemporaneous throughout the inner solar system, at least one scenario indicates that the end of late heavy bombardment may have been extended by ~ 1 Gyr at Mars by perturbations of the impactors in the ν_6 secular resonance (Wetherill 1977). Such a change in the timing of heavy bombardment termination would decrease the ages of the post late heavy bombardment terrain units, dramatically affecting the thermal history models for Mars derived from the absolute chronologies.

Absolute chronologies allow detailed analysis of thermal and geologic evolutionary models. However, their reliance on the extrapolation of the lunar cratering chronology and the associated uncertainties in the cratering rate and temporal dependence require that these chronologies be used with caution. A reliable absolute chronology will be available only when radiometrically datable samples of known location are returned from the Martian surface. Until then, and keeping in mind the large uncertainties discussed above, probably the most reliable absolute age estimates are those derived from Model 2 of Neukum and Hiller (1981). As estimates of crater production rates from asteroids and comets improve by new discoveries of Mars crossers, the younger surfaces should become more reliably dated.

IX. CONCLUSIONS

Although there are uncertainties in the interpretation of the Martian cratering record, some tentative conclusions seem warranted:

1. The major differences in impact crater morphology and morphometry between Mars and the Moon and Mercury are probably largely the result of subsurface volatiles on Mars. In general, the depth to these volatiles may decrease with increasing latitude in the southern hemisphere, but the base of this layer may be at a more or less constant depth.

2. The Martian crustal dichotomy could have been the result of a very large impact near the end of the accretion of Mars. Monte Carlo computer simulations suggest that such an impact was not only possible, but likely.
3. The Martian highland cratering record shows a marked paucity of craters $\lesssim$ 30 km in diameter relative to the lunar highlands. This paucity of craters was probably primarily the result of the obliteration of craters by an early period of intense erosion and deposition by aeolian, fluvial and glacial processes.
4. There are two crater populations of different ages on Mars distinguished by their different size/frequency distributions. The older population represents the period of late heavy bombardment that may have been due to remnants left over from the accretion of the terrestrial planets. The younger crater population has accumulated on Mars since the end of heavy bombardment and probably represents primarily impacts by asteroids.
5. Martian surface units span a large range in ages, from heavily cratered terrains that date from the period of late heavy bombardment $\sim$ 4.2 to 4.0 Gyr ago, to very lightly cratered volcanic units that may be as young as 0.3 Gyr. The polar-layered terrains are almost crater free suggesting they are younger still, and may be undergoing significant continuing modification.

13. THE PHYSICAL VOLCANOLOGY OF MARS

PETER J. MOUGINIS-MARK, LIONEL WILSON
University of Hawaii

and

MARIA T. ZUBER
NASA/Goddard Space Flight Center

Several types of constructional volcanic landforms (central volcanoes, tholi, the highland paterae, Alba Patera and small domes), as well as plains units of probable volcanic origin, exist in both the northern and southern hemispheres of Mars. Volcanism in the central highlands appears to have been explosive in character, while most of the constructional activity in the northern plains was effusive. In addition, highlands volcanism appears to be relatively old compared to that in the northern hemisphere. Alba Patera appears to be unique in preserving both styles of volcanism and being transitional in age. Volcanism has had a significant influence on both the Martian cryosphere and atmosphere; numerous examples of volcano/volatile interactions exist within Elysium Planitia and to the south of Hadriaca Patera, and volcanism is believed to have contributed large volumes of both water vapor and other gases to the evolving Martian atmosphere. Theoretical models can be used to predict the style of activity and the extent of the deposits produced by eruptions of various plausible magma/volatile combinations on Mars; if the magmas were rich in volatiles there would have been a greater likelihood of explosive activity taking place on Mars than on the Earth. There is evidence for the existence of large magma chambers and very high effusion rate eruptions on Mars, based on the volume of observed calderas and lava flows, and the occurrence of sinuous rilles. Tectonic deformation associated with volcanic constructs is primarily a consequence of loading and magma transport, while deformation in the volcanic plains reflects stresses associated with Tharsis and major impact basins. De-

spite the wealth of information that has been derived about Martian physical volcanology from Viking Orbiter data and related numerical modeling, several key questions remain unanswered. The lack of knowledge about the range of magma chemistries, the frequency and duration of individual eruptions, the longevity of activity at individual volcanoes, the reason for the lack of recent constructional volcanism in the southern highlands, and a lack of detailed knowledge of the meter- to decameter-scale topography and regional slopes of Martian volcanic landforms precludes a truly rigorous analysis of volcanic landforms on Mars. Some of these gaps in our current knowledge may be filled by continuing studies of high-resolution images and new data from the Mars Observer mission, but additional information from carefully selected sample-return sites will be needed to substantially improve our understanding of the physical volcanology of Mars.

In this review, the physical volcanology of Mars is considered to include the diversity of volcanic landforms, the implied styles of eruption associated with the construction of these landforms, the inferred internal structure of the volcanoes, and the influence that the eruptions have had on the Martian environment (both local and global in scale). The regional tectonics of volcanic landforms are considered in chapter 8, but the important deformational features locally associated with the volcanic constructs and plains are considered here because they provide insights into the internal structure of Martian volcanoes and the volumes and mechanical properties of the plains. Similarly, the regional stratigraphy of Mars, which contains abundant information on the rates of volcanic resurfacing through time, is described in chapter 11; the rate of production of volcanic materials is in turn related to the thermal history of the planet (see chapter 5). However, several of the specific geologic units that can be identified from this global mapping help to define the likely diversity of volcanic activity on Mars, as well as the rates at which magma and juvenile volatiles were released. These aspects of global volcanic stratigraphy are also considered here. Global spectroscopic data (chapter 17) together with measurements from the Viking Landers (Arvidson et al. 1989*a*) and geochemical models (McGetchin and Smyth 1978) support the concept originally proposed on morphological grounds that Mars has been dominated by mafic and ultramafic volcanism. The volcanic plains of the southern highlands are not given comprehensive treatment here, as few details are known of their eruptive processes and rates; Greeley and Spudis (1981) provide an excellent and still current review of Martian plains volcanism.

I. MORPHOLOGIC TYPES OF MARTIAN VOLCANIC FEATURES

Much of our knowledge of the physical volcanology of Mars comes from the 15 years of morphologic analysis of Mariner 9 and Viking Orbiter images (cf. Carr 1973,1981). The morphologies of Martian volcanoes in general have strong analogies to those of volcanic landforms on Earth, and Martian features are described using terrestrial terminology (see, e.g., Greeley and Spudis [1981], Cas and Wright [1987, pp. 27–30], and the glossary in this volume for definitions). In a particularly good early Viking-based review, Greeley and Spudis (1981) subdivided the main morphological types of volcanic features on Mars as listed in the following subsections.

A. Central Volcanoes

1. Shields. The Tharsis Montes (e.g., Arsia Mons) and Olympus Mons are the best known of this category. Early in the 1970s the Mariner 9 mission revealed that members of this class of volcano have many general similarities to basaltic shields in Hawaii (Carr 1973), possessing nested summit calderas and having numerous lobate lava flows on their flanks. Such Martian volcanoes differ from their terrestrial equivalents by virtue of their great size: Olympus Mons rises more than 25 km above the surrounding plains (Fig. 1) and has a basal diameter in excess of 600 km (Wu et al. 1984). Based on the overall topography of the volcano Elysium Mons, which apparently has shallow sloping outer flanks and a steeper summit region than Olympus Mons, Malin (1977) and Pike (1978) suggested that some Martian montes may be composite volcanoes rather than shields analogous to those found in Hawaii.

2. Domes (Tholi). These are smaller volcanoes than the montes. The tholi are generally dome shaped, and have relatively steep peripheral flanks (in some cases, flanks exceed 8° in slope). Typically the lower flanks of the tholi have been buried by flows of the surrounding plains, so that the true areal extent and vertical dimension of the volcano are not known. An analysis of the slopes and the sizes of the summit calderas of tholi in the Tharsis region suggest that these volcanoes may have been partially buried by up to about 4 km of lava (Whitford-Stark 1982). In some instances, such as Hecates Tholus (Fig. 2), these volcanoes appear to have experienced explosive activity (Mouginis-Mark et al. 1982*b*).

3. Highland Paterae. These are low relief shields with irregular summit craters and numerous radial channels on their flanks. West (1974) first attributed their topographic profiles to explosive volcanism, suggesting that they were made of ash flow deposits rather than lava flows. Subsequently, Greeley and Spudis (1981) and Greeley and Crown (1990) have carried out more detailed mapping of Tyrrhena Patera and, with the exception of a few lava flows close to the summit, agree that the flanks of this volcano are most likely composed of unconsolidated materials, probably ash deposits (Fig. 3). Earth-based radar topography of Tyrrhena Patera (Fig. 4) shows that the northern flanks have very shallow slopes ($<0\overset{\circ}{.}25$) over a distance in excess

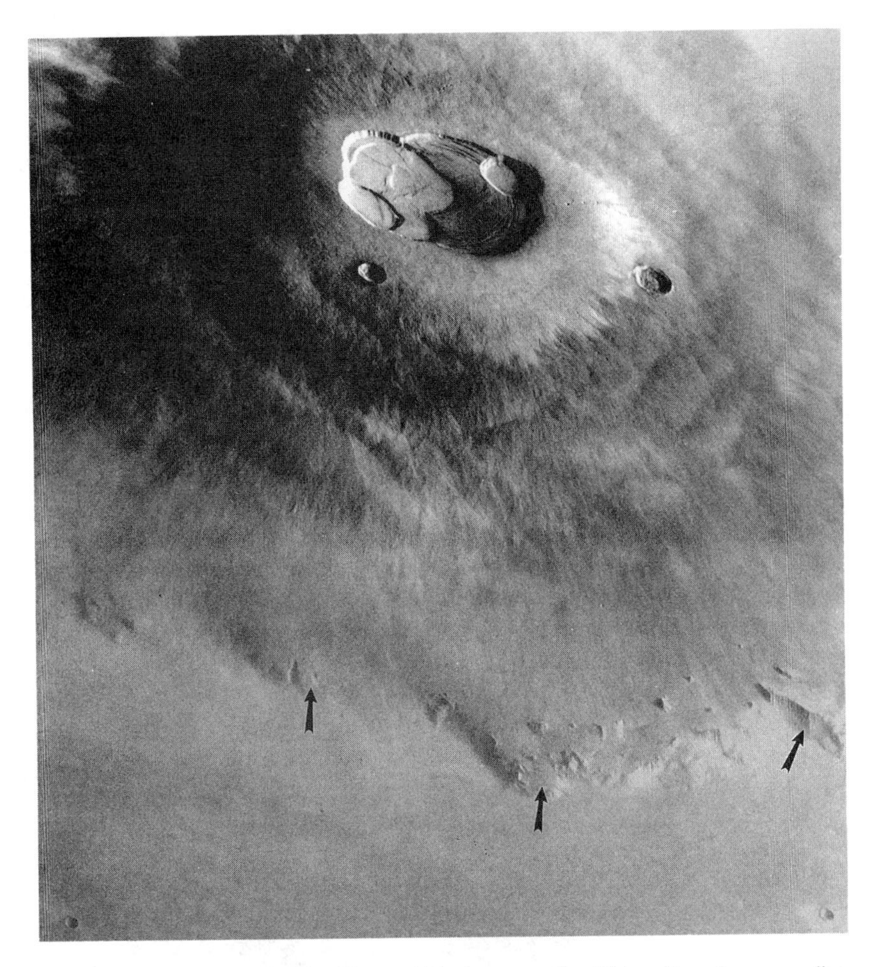

Fig. 1. Oblique view of Olympus Mons, which rises more than 25 km above the surrounding plains. In this view (from approximately south of the volcano), the 8-km high Olympus Mons escarpment (arrow) and the 60 × 90 km diameter nested summit caldera are clearly visible. (Viking Orbiter frame 641A52.)

of 100 km, and that the summit of the volcano probably rises < 1 km above the surrounding plains.

4. Alba Patera. This volcanic landform appears to be unique on Mars (Carr 1973; Greeley and Spudis 1981). Although many of the flanks (Carr et al. 1977*a*) and the summit area (Cattermole 1987) consist of lava flows (Fig. 5), numerous digitate channels are found on the northern flanks that, together with thermal inertia data, suggest a pyroclastic origin for these materials (Fig.

Fig. 2. Photomosaic of the central portion of Hecates Tholus volcano. The summit caldera is indicated by the arrow. Note that, at an image resolution of ~50 m/pixel, there is an asymmetry in the distribution of presumed impact craters and channel networks, with the area to the west (left) of the summit caldera devoid of such features. This distribution is interpreted to be the result of a recent airfall deposit that now mantles this part of the volcano (Mouginis-Mark et al. 1982*b*). Width of mosaic is equivalent to ~170 km. (Viking Orbiter frames 651A15–23.)

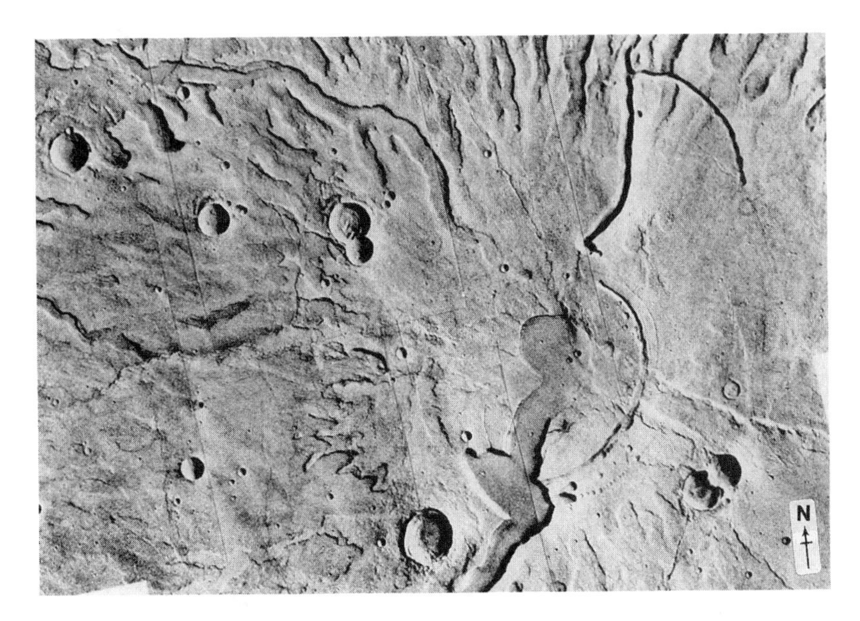

Fig. 3. Tyrrhena Patera is one of several highland volcanoes that appears to have formed predominantly by ash eruptions. Greeley and Spudis (1981) and Greeley and Crown (1990) interpret the summit area to be covered by lava flows, but prominent flow lobes and diagnostic surface morphologies are absent here, so that the summit may comprise welded-ash deposits. Prominent on the flanks of the volcano is a set of deeply eroded channels that emanate from the summit. Image width is equivalent to ~130 km. (Part of JPL Photomosaic 211–5730.)

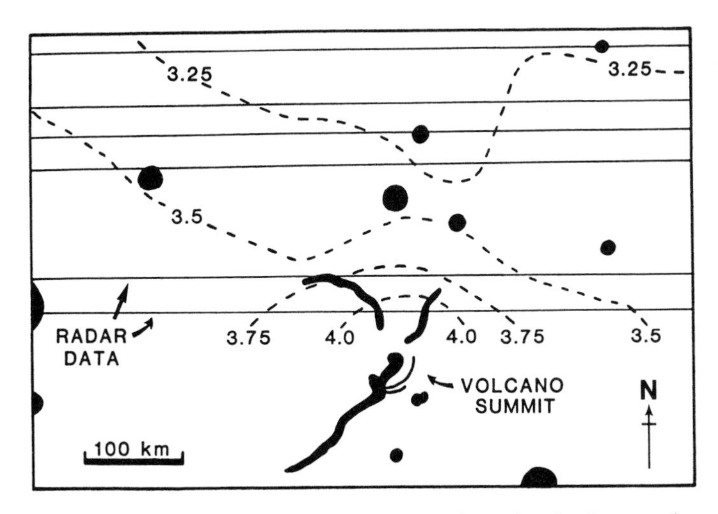

Fig. 4. Earth-based radar topography of Tyrrhena Patera shows that the slopes on the northern flanks (top of figure) of the volcano are very low. Although the volcano has a basal diameter of ~600 km, the summit rises < 1 km above the surrounding plains. Note that the contours are interpolated between groundtracks, and assume an azimuthally symmetric volcanic cone. Contours are in km, solid horizontal lines mark positions of radar groundtracks. Larger adjacent impact craters (solid circles) are also shown. Data collected by Downs et al. (1975).

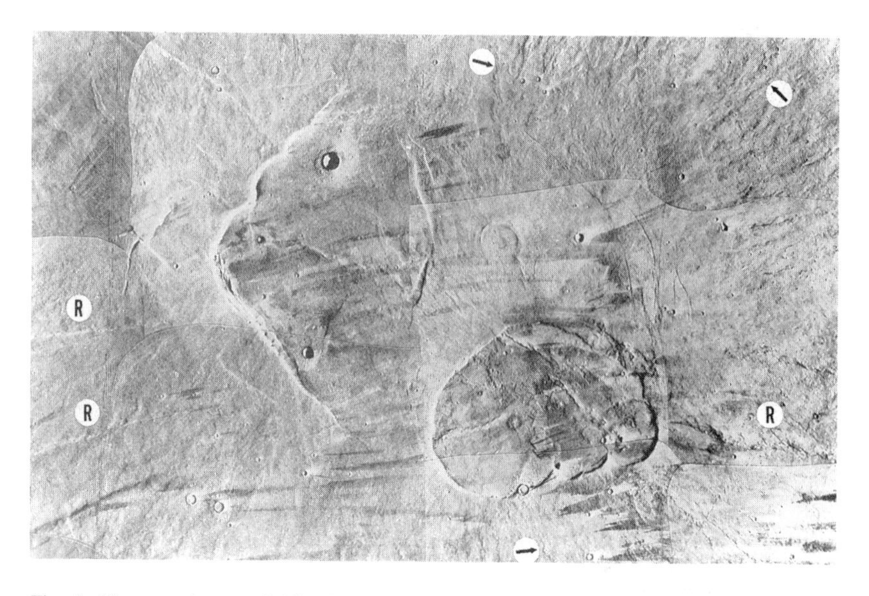

Fig. 5. The summit area of Alba Patera, centered at 40°.6 N, 110°.1 W. Numerous small lava flows (arrow) and wrinkle ridges (R) can be found on the near-summit flanks, supporting the idea of late-stage effusive volcanism associated with tectonic deformation. Image width is equivalent to 220 km. North is to the top. (Part of Viking Orbiter Photomosaics MTM-40107 and -40112).

6; Mouginis-Mark et al. 1988). Together with the low aspect ratio of the flanks (Alba Patera has a basal diameter > 1600 km, but rises only ~3 to 4 km above the surrounding plain), it appears likely that the unique nature of Alba Patera is due to its polygenic style of activity, with early explosive eruptions followed by subsequent effusive activity (Mouginis-Mark et al. 1988).

5. Numerous Small Domes. Many have been identified across Mars. Most enigmatic are the thousands of subkilometer-sized hills that exist within the northern plains (Frey et al. 1979; Frey and Jarosewich 1982). In addition to these lowlands plains hills, Plescia (1981) and Tanaka and Davis (1988) have identified several breached cones in Tempe Terra and Syria Planum that are morphologically similar to cinder cones produced by pyroclastic activity on Earth. Some of these appear to have lava flows emanating from them, strongly suggesting a volcanic origin for the cones.

B. Volcanic Plains

1. Simple Flows. These materials form areally extensive plains that generally contain wrinkle ridges similar in morphology to terrestrial and lunar mare ridges (Lucchitta and Klockenbrink 1981; Plescia and Golombek 1986; Sharpton and Head 1988; Watters 1991). There has been much debate about

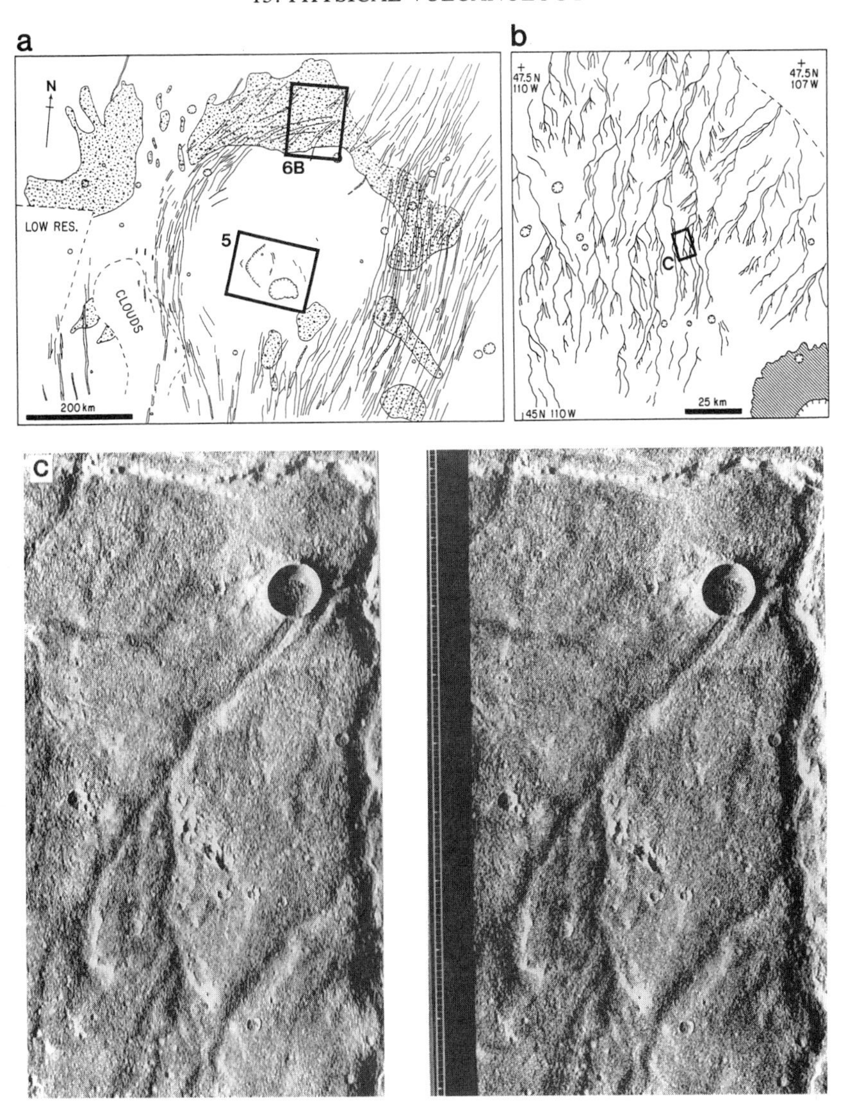

Fig. 6. (a) Map of Alba Patera, showing the area from 32°.5 N to 47°.5 N and 100°.0 W to 122°.5 W, and the grabens that almost surround the summit. Outlined are areas shown in Figs. 5 and 6B. Also shown are the parts of the volcano where the digitate channels are found (stippled area), large meteorite craters (barbed circles), and areas where cloud cover or poor resolution precluded positive identification of certain surface features (dashed lines). (Mapped from U.S. Geological Survey Viking Orbiter photomosaics MTM-35102, -35107, -35112, -35117, -40102, -40107, -40112, -40117, -45102, -45107, -45112, -45117 and JPL Mosaic 211–5071.) (b) Map of channel networks on the northeast flank of Alba Patera. Outlined area *C* denotes the location of the stereo pair shown in Fig. 6c (figure from Mouginis-Mark et al. 1988). (c) High-resolution stereo pair showing a digitate channel network northeast of the summit caldera of Alba Patera (see Fig. 6B for location). Direction of flow is towards top of image. Note the lack of evidence of lava flows, which argues against these channels being of volcanic origin. Image resolution is ~8 m/pixel. (Viking Orbiter frames 445B07 (right) and 445B08.)

the origin of the ridged plains. A volcanic origin is preferred, based primarily on morphologic similarity to the lunar maria (Fig. 7). In a few places, sinuous rilles can be found in intimate association with the ridged plains, such as in southern Lunae Planum and Syrtis Major (Schaber 1982), which also suggest that these plains are volcanic. The term "simple flows" implies single cooling units (Walker 1972), although it is not clear whether these Martian plains consist of high-volume, single-cooling units (possibly flood lavas) of great extent (Greeley and Spudis 1981).

Recent depolarized Earth-based radar data for much of the equatorial belt of Mars (Thompson and Moore 1989*a*) reveal that the ridged plains have

Fig. 7. Ridged plains in Hesperia Planum closely resemble the lunar maria, suggesting a volcanic origin for these Martian plains (Scott and Carr 1978). North is to the top left; image width is equivalent to 107 km. (Viking Orbiter image 418S39; centered at 28.°56 S, 240.°94 W.)

a much lower radar echo strength than the lava plains surrounding the shields. Such data may indicate that the ridged plains are comparable to terrestrial pahoehoe flows, while the lava plains may comprise topographically rougher aa flows (Gaddis et al. 1989). If this interpretation is correct, then radar cross sections may provide a useful indirect method for estimating the mass eruption rates of different volcanic materials due to the correlation between flow roughness and eruption conditions (Rowland and Walker 1990). However, these radar data are ambiguous, because the ridged plains have radar cross sections comparable to most of the other (probably sedimentary) plains units on Mars (Table I).

Thicknesses of the ridged plains have been measured to constrain eruptive volumes. Thickness estimates for Lunae Planum, determined using either

TABLE I
Radar Depolarized Echo Strengths for Selected Martian Volcanoes, Lava Flow Fields and Plains Units.[a]

Unit Name	Echo Strength
Volcanic Constructs with Lava Flows on Flanks	
Ascraeus Mons	0.175
Arsia Mons	0.165
Olympus Mons	0.100
Alba Patera	0.090
Elysium Mons	0.070
Volcanic Constructs with Channelized Flanks	
Hecates Tholus	0.070
Tyrrhena Patera	0.010
Hadriaca Patera	0.010
Lava Flows on Volcano Flanks	
Olympus Mons Base	0.160
Elysium flows	0.090
Alba Northwest	0.080
Alba Southwest	0.080
Distal Alba flows	0.030
Ridged Plains	
Syria Planum	0.010
Syrtis Major	0.010
Hesperia Planum	0.010
Plains Materials of Questionable Origin	
Arcadia Planitia	0.030
Acidalia Planitia	0.025
Isidis Planitia	0.010

[a]Note that volcanic constructs with lava flows (e.g. Olympus Mons) have echo strengths more than an order of magnitude larger than the highland patera, and that most simple flow units have much lower returns than complex flow materials (table from Thompson and Moore 1989*a*).

[b]Echo strength is variable A, defined by Thompson and Moore, and is given in equivalent total radar cross section.

the exposed rim heights of partially buried impact craters or the smallest diameter crater that survived a resurfacing event, range from about 250 m to 1.5 km (De Hon 1981,1982,1988; Saunders and Gregory 1980; Saunders et al. 1981; Frey et al. 1988). Independent estimates of the thicknesses of these units have also been made where they are exposed on the walls of the Valles Marineris and Kasei Valles. Values range from <1 km to approximately 2 km (Nedell et al. 1987; Robinson and Tanaka 1988; Watters 1991; chapter 14).

2. Complex Flows. These materials have multiple overlapping flow lobes, and are usually found around the periphery of the shield volcanoes (Fig. 8). Greeley and Spudis (1981) suggested that this type of plains material resulted from the eruption of more sporadic, lower effusion-rate eruptions compared to the simple flows. Detailed geologic maps of flows of this type within the Tharsis region have been prepared by Scott and Tanaka (1986), and the identification of many flows originating several hundreds of km from the large volcanic constructs (Schaber et al. 1978; Mouginis-Mark et al.

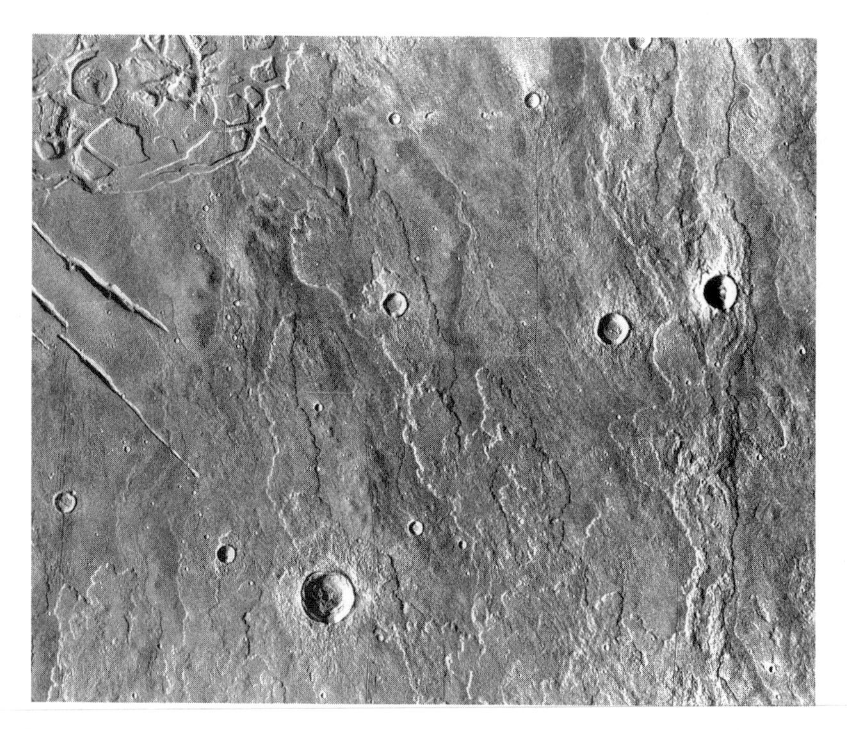

Fig. 8. A sequence of lava flows to the west of Hecates Tholus in Elysium Planitia. The direction of flow is towards the top of this image. Image resolution is 52 m/pixel. The width of the image is equivalent to 115 km. North is towards the top of the image. (Mosaic of Viking Orbiter images 651A07–12; centered at 33°N, 215°W.)

1982*a*) suggests that vents, fissures and feeder dikes are quite numerous within much of Tharsis. The complex flows can extend for >1000 km from the summit calderas of the shields, but have smaller areal extent than the simple flow plains. High-resolution (10 to 20 m/pixel) images of these flows show numerous festoon ridges and lava channels (Theilig and Greeley 1986), and Earth-based radar-scattering data indicate that these flows are very rough at a radar wavelength of 12 cm (Schaber 1980).

3. Undifferentiated Flows. There are several instances where plains materials are located close to a volcanic construct, but the materials are of uncertain origin. Greeley and Spudis (1981) discuss the plains units around the margins of Tharsis and Elysium, and note that although there is no unequivocal evidence for a volcanic origin for these plains, there is also no compelling evidence for an origin by other processes. The northern plains of Mars constitute most of this class, and in several places, such as Ismenius Lacus (Lucchitta 1978*c*), wrinkle ridges and occasional flow lobes are seen.

4. Plains of Questionable Origin. Particularly within the southern highlands, there are many eroded-plains materials (approximately 42% of the total surface area of the cratered hemisphere of Mars; Greeley and Spudis 1978) that may have a volcanic origin but are too heavily eroded to permit the confident identification of flow fronts and other diagnostic features. The heavily fractured plains around Tharsis is a second example of questionable materials that may be of volcanic origin.

II. STRUCTURE OF MARTIAN VOLCANOES

Considerable information regarding the internal structure of terrestrial volcanoes has been gleaned by analysis of seismic data (Ryan 1988), by the detailed mapping of vent distributions (Porter 1972; Nakamura 1977; Munro and Mouginis-Mark 1990), from analysis of exposed dike swarms (Walker 1987; Knight and Walker 1988), from laboratory models of volcanic edifices (Fiske and Jackson 1972) and from numerical modeling of eruptive episodes (Wilson and Head 1988*a;* Head and Wilson 1989). For Martian volcanoes, the dominant observation is that the giant montes have relatively simple morphological structures compared with their terrestrial counterparts, apparently being mainly controlled by deep-seated regional tectonic forces. This is in contrast to terrestrial volcanic centers (e.g., Hawaii, Galapagos Islands, Reunion Island), where multiple volcanoes may be active in close proximity to each other, with the resultant formation of rift zones due to the interference of local stress fields (Fiske and Jackson 1972), and the production of flanks liable to collapse due to oversteepening (Duffield et al. 1982). Maps of the distribution of vents on Arsia, Pavonis and Ascraeus Montes (Carr et al. 1977*a;* Crumpler and Aubele 1978) have nevertheless helped reveal incipient rift zones trending roughly perpendicular to the major NE-SW trend of the Tharsis ridge volcanoes.

The summit areas of the Martian shields also reveal information on the internal structure of the volcanoes (Fig. 9). The large (>60-km diameter, 2- to 3-km deep) nested summit calderas of Olympus Mons and Ascraeus Mons show that multiple collapse events were associated with each volcano, presumably due to large-volume flank eruptions partially evacuating each magma chamber (Mouginis-Mark 1981). The topography of the summit area of Olympus Mons (Fig. 9b) reveals that the rim of the youngest collapse caldera is also topographically higher (by >2 km) than the other parts of the caldera rim. This correlation between rim elevation and age of collapse, and the absence of young lava flows infilling older adjacent segments of the caldera, suggest that dike intrusions have played a major role in volcano growth in a similar manner to certain terrestrial shields (Walker 1987,1988; Mouginis-Mark and Mathews 1987). Only small-volume lava flows can be identified at the summits of Olympus Mons and Ascraeus Mons (Mouginis-Mark 1981; Zimbelman 1985), but one of the more enigmatic aspects of Martian volcanism is the implied size of the parent magma chambers. Due to the very large volume of many of the lava flows associated with the Tharsis volcanoes (>100 km^3; Wood 1984*a;* Cattermole 1987), magma chambers of Martian volcanoes are believed to be up to 3 orders of magnitude larger in volume than their terrestrial counterparts. The spacing of volcanic centers in Tharsis suggested to Whitford-Stark (1982) that the size of each volcano directly correlates to the distance from its nearest neighbors, and Whitford-Stark proposed that in the case of the groups of smaller Tharsis volcanoes (Biblis/Ulysses Patera, Ceraunius Tholus/Uranius Tholus/Uranius Patera), these volcanoes shared common magma sources that inhibited the large-scale growth of any of the individual volcanoes. The relatively large diameters of the summit calderas of Biblis, Ulysses and Uranius Paterae nevertheless suggest that the size of the magma chamber within these volcanoes was probably as large as those within other Martian volcanoes such as Pavonis Mons (Wood 1984*b*).

To the west of Elysium Mons there is a complex area of hummocky terrain which includes lava flows, vents, fissures, collapse pits and sinuous rilles (Mouginis-Mark et al. 1984). This area appears to be a likely candidate for a failed volcanic construct where, perhaps because of the high regional slope or the high mass eruption rate of the magmas, the volcano was unable to produce the more common shield topography.

III. ERUPTIONS THROUGH TIME

A. Age of Volcanoes

Several different cratering curves have been developed that place the ages of Martian volcanic landforms in relative chronologies (see Tanaka et

Fig. 9. (a) The summit caldera of Olympus Mons has experienced at least six episodes of caldera collapse. The size of the caldera complex is ~90 × 60 km. North is towards the top of the image. (Viking Orbiter image 890A68.) (b) Inferred chronology of collapse events for Olympus Mons caldera (Mouginis-Mark 1981). 1 is the oldest, 6 youngest. (c) Topography of Olympus Mons summit area, derived from stereographic data by Wu et al. (1984). Elevations are in km above 6.1 mbar pressure surface of Mars; contour interval is 200 m. (d) Distribution of wrinkle ridges on the floor of Olympus Mons caldera. (e) Distribution of grabens on the floor of Olympus Mons caldera.

al. 1988; chapter 12). Tanaka et al. (1988) developed a resurfacing chronology for Mars, and inferred that volcanic surfaces ($\sim$84 $\times$ 10^6 km^2) cover more than half of Mars. Plescia and Saunders (1979) and Neukum and Hiller (1981) have also carried out crater counts for the flanks of the Martian volcanoes, and both identified Olympus Mons as one of the youngest features and the highland paterae as being generally old. Of particular interest is the range of intermediate ages of parts of the flanks of Alba Patera, indicating that this volcano had a protracted history that spanned the period of early southern-hemisphere volcanism and late-stage shield building (Neukum and Hiller 1981). Extremes in the age of volcanic events include the speculative identification of highland volcanoes in the western hemisphere of Mars (Scott 1982), and the possible very recent activity on Hecates Tholus (Mouginis-Mark et al. 1982*b*) and in the floor of Valles Marineris (Lucchitta 1987*a,b*).

B. Mechanism of Volcanic Resurfacing

The origin of the major young volcanic centers on Mars remains enigmatic. Other chapters address in detail the geophysical (chapter 5) and petrologic (chapter 6) constraints that can be placed on the early volcanic activity on Mars. Here, we note two key observations: the location of the most recent volcanism on Mars in the northern lowlands, and the absence of lava shield-building activity in the southern highlands. In addition, any mechanism proposed to account for the extensive resurfacing in the north should also explain the approximately 3-km elevation difference between the northern and southern hemispheres. If the elevation change is isostatic, as suggested by gravity models (Phillips 1988), then it is appealing to invoke an endogenic mechanism such as subcrustal erosion, occurring perhaps in response to vigorous mantle convection (see, e.g., Wise et al. 1979; Turcotte 1988). Such a mechanism would "thin" the crust above a broad thermal anomaly and would be consistent with a concentration of volcanism in the lowlands. Alternatively, it has been proposed that a giant impact (Wilhelms and Squyres 1984) or multiple impacts (Frey and Schultz 1988) could have created thermal and mechanical weakening in the deep lithosphere that could have promoted continued lowlands volcanism.

IV. INFLUENCE OF VOLCANISM ON THE MARTIAN ENVIRONMENT

A. Input to the Atmosphere

Volcanoes on the Earth typically inject large quantities of volatiles (water, carbon dioxide, sulfur dioxide and lesser amounts of other compounds including halogens) into the atmosphere, and it is likely that Martian volcanoes had a similar influence. Slow release of sulfur compounds and carbon dioxide also commonly takes place from shallow magma reservoirs through overlying fracture systems during repose periods between eruptive episodes.

Settle (1979) considered the formation and deposition of juvenile volcanic sulfate aerosols on Mars, and concluded that volcanic SO_2 released at the surface would result in sulfate aerosol formation, and that the upper atmosphere circulation would cause these particles to be distributed on a global or hemispheric scale. Such a process could account for the high concentrations of sulfur (~3 to 4 wt%) within the surficial soils studied at the two Viking Lander sites. The release of water from effusive activity on Mars as a function of geologic time was addressed by Greeley (1987), who estimated that for the preserved geologic record a total amount of water equivalent to a layer ~50-m thick could have been degassed from the plains and constructional volcanic materials preserved on the Martian surface. Based on the exposed surface area of these materials, Greeley (1987) estimated that the greatest volume of juvenile water (equivalent to a layer 16.3-m thick distributed planet-wide) was probably released during Early Hesperian times, with significant additional amounts during Late Hesperian (~11.1 m) and Early Amazonian (~7.7 m) (Table II).

The effect of volcanic outgassing of water vapor and sulfur dioxide on the early Martian atmosphere has also been considered by Postawko and Kuhn (1986). Although it is likely that sulfur dioxide has been released into the Martian atmosphere at various times from volcanic eruptions (Settle 1979), the actual amount is uncertain and, due to the low atmospheric density, the lifetime of sulfur dioxide in the atmosphere may have been as short as ~10 yr. Because it is a net absorber of solar radiation, the effect of this sulfur dioxide in the atmosphere would most likely have been to act as a trigger initially to increase the amount of water vapor in the atmosphere by greenhouse warming (Postawko and Kuhn 1986).

TABLE II

Estimated Amounts of Water Released by Volcanism as a Function of Geologic Age[a]

Age	Volcanic Materials (10^6 km^3)	Water Layer[b] (m)
Late Amazonian	5.3	1.2
Middle Amazonian	20.78	4.8
Early Amazonian	33.87	7.7
Late Hesperian	48.81	11.1
Early Hesperian	71.62	16.3
Late Noachian	21.03	4.8
Middle Noachian	0.28	0.1
Totals	201.69	46.0

[a]Table after Greeley 1987.

[b]Values are given for the thickness of a water layer that would completely cover the planet. A layer of water 1-m thick is equivalent to a water volume of 0.144×10^6 km^3.

Specific amounts of released volatiles from individual volcanoes have also been estimated by Wilson and Mouginis-Mark (1987) for Alba Patera. The morphology of the volcano flanks suggests that, early in the eruptive history of Alba Patera, explosive volcanism characterized this volcano (Mouginis-Mark et al. 1988). The explosive eruptions were probably driven by volatiles that were both juvenile and terrigenous in origin. If the main explosive activity was due to juvenile volatiles, then all of the eruptions of Alba Patera would have injected a volume of water into the Martian atmosphere that was equivalent to a layer ~5 mm to 5 cm deep distributed over the surface of the entire planet (Wilson and Mouginis-Mark 1987). This volume is small compared to the 46 m calculated by Greeley (1987) for all water released by volcanism on Mars, but indicates the contribution of individual volcanoes to the global volatile inventory. Wind-directed fume from the summit of Alba Patera has been proposed as a possible mechanism for the asymmetric distribution of the channel networks on the northern flanks of the volcano (Mouginis-Mark et al. 1988).

B. Volcano and Near-Surface Volatile Interactions

The profusion of large volcanic plains and shields, together with the morphologic (Carr 1986) and thermal evidence (Kieffer et al. 1976*b*) for water and ice on Mars, strongly suggest that magma and water or ice would have interacted during the evolution of the Martian landscape. Allen (1979) and Hodges and Moore (1979) each suggested that certain features on Mars resemble landforms found in Iceland that are due to the interaction of basaltic magma and ice, and the basal materials of Olympus Mons have variously been described as subglacial in origin (Hodges and Moore 1979), or the result of early pyroclastic volcanism (King and Riehle 1974), perhaps akin to the early explosive volcanism associated with Alba Patera (Mouginis-Mark et al. 1988). Greeley and Spudis (1981) and Greeley and Crown (1990) have also proposed that phreatomagmatic eruptions were responsible for much of the construction of Tyrrhena Patera.

More direct evidence for the interaction between magma and volatiles within the Martian regolith comes from the large numbers of probable fluvial channels that exist to the west of Elysium Mons (Mouginis-Mark 1985) and to the south of Hadriaca Patera (Squyres et al. 1987). In the case of the Elysium channels (Fig. 10), melt-water release appears to have been responsible for the generation of the large volumes of sediment that now partially infill the northern plains in Utopia Planitia (Lucchitta et al. 1986). Intrusive activity probably played a key role in the release of water adjacent to Hadriaca Patera (Squyres et al. 1987), and there are additional small (10 to 50 km long, <2 km wide) channels immediately to the east of Olympus Mons that could also be the result of water released by intrusive activity (Mouginis-Mark 1990).

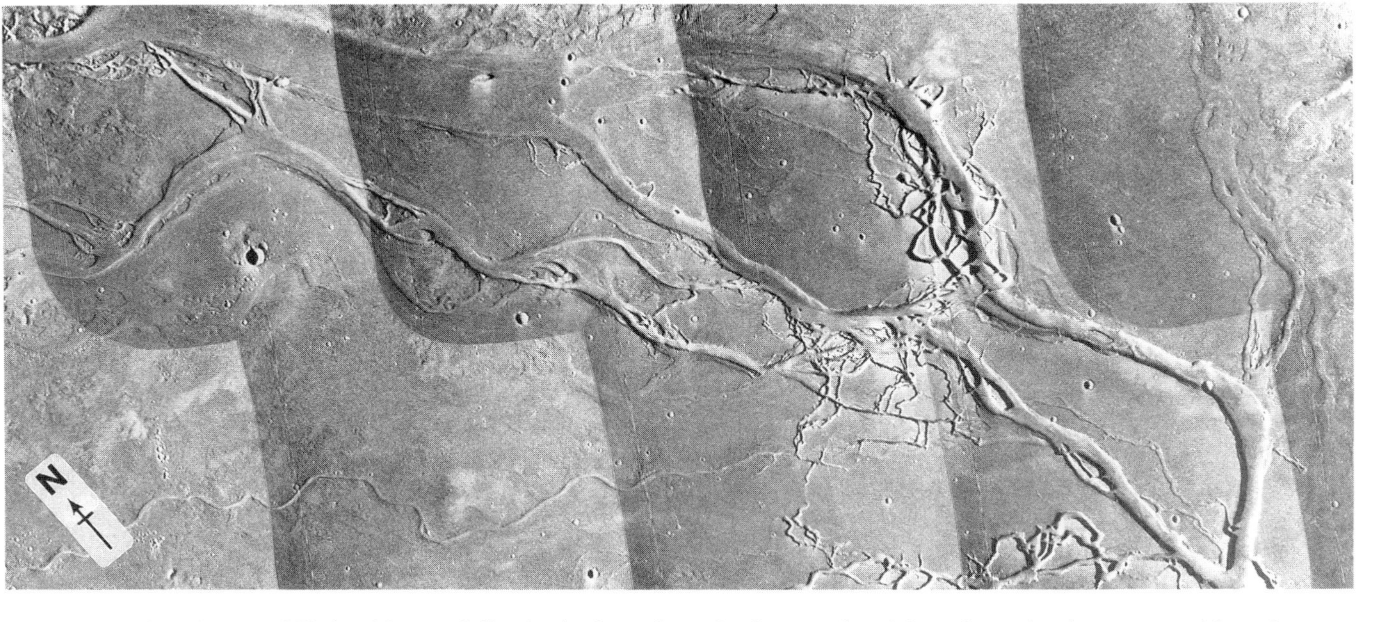

Fig. 10. Channels to the west of Elysium Mons are believed to be the products of melt water released due to interactions between ground ice and magma. Multiple channel levels and streamlined islands indicate that the fluid flow probably was a repetitive phenomenon. Image width is equivalent to 150 km. (Mosaic of Viking Orbiter frames 649A11–188; centered at 28°N, 225°W.)

V. THEORETICAL MODELING OF ERUPTION PROCESSES

A. General Considerations

The range of styles of volcanic activity on any planet are a complex function of the chemistry, and hence rheology, of the magmas produced at depth by partial melting, and of the environmental conditions acting on magmas as they approach the surface—mainly the acceleration due to gravity and the atmospheric pressure (Wilson and Head 1983; Wilson 1984). The relatively low gravity on Mars compared with the Earth's gravity reduces the buoyancy forces acting on partial-melt bodies, and so reduces the ascent speed of a melt body of a given size. This reduction in the ascent speed requires the formation of systematically larger diapiric bodies ascending through the deep mantle if such bodies are to avoid excessive cooling. The consequence should be the accumulation of systematically larger magma reservoirs in the shallow lithosphere. The low gravity should also lead to the formation of systematically wider dikes connecting the subsurface reservoirs to the surface (Wilson and Head 1988*b;* Wilson and Parfitt 1989), encouraging the more rapid effusion of magma of a given rheology.

Another consequence of the lower Martian gravity is that any specific lithostatic pressure is reached at a greater depth on Mars than on Earth. Since magma volatile solubility is mainly pressure dependent, gas-bubble nucleation and gas exsolution will begin at greater depths than on the Earth, encouraging efficient volatile release, especially of low solubility species like CO_2 (Table III). The low atmospheric pressure will lead to thorough degassing of Martian magmas as they approach the surface and lead to enhanced gas loss from lavas and pyroclastics immediately after their eruption. Any

TABLE III
Calculated Gas Nucleation Depths and Magna Fragmentation Depths on Mars[a]

Total Volatile Content (wt. %)	Nucleation Depth: H_2O in Basalt (m)	Nucleation Depth: H_2O in Rhyolite (m)	Nucleation Depth: CO_2 (km)	Fragmentation Depth: H_2O in Basalt (m)	Fragmentation Depth: H_2O in Rhyolite (m)	Fragmentation Depth: CO_2 (km)
5	21,000	13,000	196	1,800	1,460	901
3	10,000	4,800	118	1,000	771	713
1	2,100	530	39	318	181	173
0.3	380	48	12	83	29	51
0.1	80	5	4	24	4	17
0.03	14	0.4	1.2	6	0.3	5

[a] Values are given as a function of total volatile content of magma for three magma/volatile combinations. The solubility of CO_2 is essentially the same in basalts and rhyolites (table from Wilson et al. 1982).

magma containing >5 to 10 ppm by weight of a relatively volatile substance (such as H_2O) is likely to be completely disrupted in a fire fountain, and will form pyroclastic particles entrained in their own released gas (Wilson and Head 1981*a*). As a result, given similar volatile contents, a larger fraction of all volcanic eruptions is more likely to be explosive on Mars than on the Earth, although this does not preclude the formation of extensive lava flows since coalescence of hot pyroclasts around vents can produce low-viscosity lavas (Head and Wilson 1989).

B. Explosive Activity

In view of the morphologic (Carr et al. 1977*a;* Carr 1981), spectroscopic (chapter 17), and Viking Lander (Arvidson et al. 1989*a*) evidence suggesting that Martian magmas were commonly mafic, it is important to explore the sizes expected for the common products of explosive activity in such magmas. Strombolian activity (defined as the intermittent explosive discharge of material due to coalescence of gas bubbles in magmas rising at relatively slow speeds in narrow vents [Wilson and Head 1981*a*]), will lead to a range of gas pressures and pyroclast speeds on Mars similar to the range for the Earth (Wilson et al. 1982). The dispersal of these clasts in the Martian atmosphere and gravity field should produce cinder cones with widths up to 200 m. The more nearly continuous accumulation of pyroclasts from Hawaiian-style fire fountains may produce ash and scoria cones with diameters in the range 200 m to 10 km, and some of the low cone-shaped to shield-shaped features found in Tempe Terrae (Plescia 1981) and Syria Planum (Tanaka and Davis 1988) may represent the products of such activity.

On Earth, the majority of large-scale explosive eruptions driven by magmatic volatiles, especially of the plinian and ignimbrite-forming types, involve andesitic or silicic magmas (Walker 1973*a*). Francis and Wood (1982) have reviewed the morphologic and geochemical evidence for silicic volcanism on Mars, and concluded that there is no compelling evidence for such activity. However, the factors mentioned above that lead to more vigorous gas release and higher eruption rates in all magmas on Mars compared with those on Earth imply that it may have been common on Mars for basaltic magmas to produce plinian eruptions (the 1886 eruption of Tarawera, New Zealand, is a terrestrial example of such activity; Walker et al. 1984), or to generate ignimbrites. Additionally, the whole spectrum of smaller-scale explosive activity styles encountered on Earth in mafic magmas (Hawaiian, Strombolian, vulcanian, plinian) is also expected on Mars (Wilson and Head 1981*b*,1983; Wilson et al. 1982).

Strong morphologic evidence for ash-fall deposits from plinian activity has been identified on Hecates Tholus (Mouginis-Mark et al. 1982*b*). The rise height of a plinian eruption cloud is expected theoretically to be a function of the heat-injection rate at its base (which is in turn proportional to the mass-eruption rate of magma) and the pressure and temperature distribution in the planetary atmosphere (Morton et al. 1956). This relation has been verified using observational data for terrestrial plinian eruption clouds (Settle 1978; Wilson et al. 1978). When the mean atmospheric pressure and temperature profiles of the Earth and Mars are inserted into the theoretical relation it is found (Wilson et al. 1982) that plumes fed by eruptions with a given discharge rate of magma from the vent are expected to rise to heights about 5 times greater on Mars than on Earth (Fig. 11). Models of the dispersal of pyroclastic particles falling from the plumes (Mouginis-Mark et al. 1988) show that mm-sized particles will commonly be carried 100 km downwind from the vent, and that 50 μm diameter particles may travel up to 1000 km.

Reimers and Komar (1979) recognized that several Martian volcanoes (such as Ceraunius Tholus, Uranius Tholus, Uranius Patera and Hecates Tholus) have channel systems on their flanks, which they interpreted to be due to massive density currents, possibly pyroclastic flows, scouring the flanks of these volcanoes. Although many channels of this type are now believed to be of fluvial origin (Mouginis-Mark et al. 1982*b*,1988; Gulick and Baker 1987*a*), other examples of channels on the partly eroded flanks of Alba Patera suggest that flank deposits of this type may have been formed by pyroclastic volcanism early in the volcano's history (Mouginis-Mark et al. 1988).

More speculative debate concerns possible ignimbrites in Amazonis Planitia. These deposits remain controversial because their characterization as ignimbrites relies on their high albedo, easily eroded nature, large volume ($\sim 3.85 \times 10^6 km^3$), and their morphologic resemblance to sequences of welded and nonwelded terrestrial ash flows (Scott and Tanaka 1982). Other interpretations have been suggested (Schultz 1985), but the key problems in describing them as ignimbrites lie in the implied silicic magma chemistry (Francis and Wood 1982), the poorly defined source areas, and their very large surface areas and, hence, implied great travel distances. Run-out distances in excess of 2500 km would be needed for some of the older flows identified by Scott and Tanaka (1982) if they are indeed single ignimbrite units, whereas theoretical considerations (Wilson et al. 1982; Mouginis-Mark et al. 1988) suggest that maximum runout distances of 400 km for ignimbrite flows are more likely on Mars.

Finally, a variety of volcanic-explosion styles on Earth involve the interaction between magma and near-surface volatiles such as liquid water and ice. Both water and ice may be involved in such activity on Mars. When intruding magma vaporizes volatile materials trapped within an edifice, or lava advances over volatile-rich ground, the peak pressures reached are controlled by the strengths of the surrounding country rock layers or the cooling upper surface of the magma. Since these strengths are likely to be similar in all silicate rocks, it is expected that the initial velocities of explosion products will be essentially the same on Mars as on the Earth (Wilson 1980). The maximum travel distances of the products will be influenced by gravity and atmospheric drag, however, and discrete explosions on Mars should produce

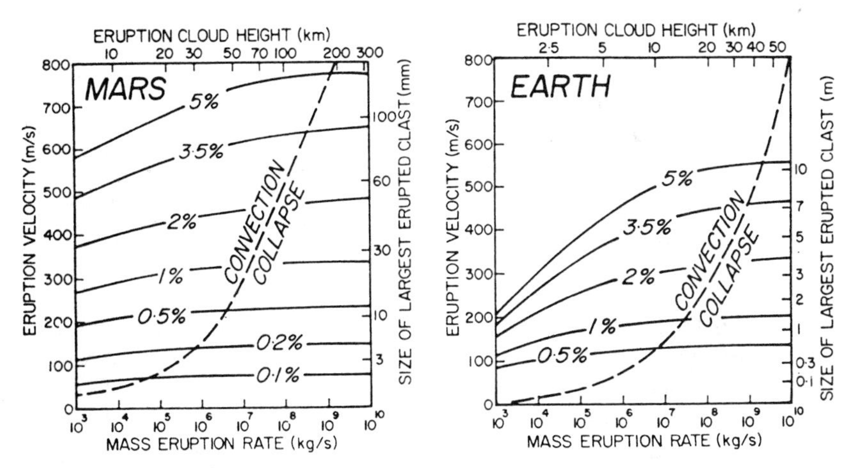

Fig. 11. Shown separately for plinian eruptions on Mars and Earth are the velocity of gas and small pyroclasts in the vent (left side) as a function of mass-eruption rate (lower edge) and exsolved magma content (curves labeled in wt%). The upper edge shows the eruption-cloud heights corresponding to the mass-eruption rate at the lower edge. The right side shows the size of the largest clast having density 1000 kg m^{-3} which can be carried out of the vent at the corresponding gas speed on the left side. Dashed line marks the boundary between eruption conditions leading to a stable convecting eruption cloud (to left of line) and a collapsed fountain over the vent feeding pyroclastic flows. Note that Martian plumes are less dense than their terrestrial equivalents at all heights in the atmosphere because the lower pressure leads to a lower density of the gas component of the plume, and this in turn leads to a smaller particle load as ever smaller clasts are shed by the rising plume. Martian plumes must consist almost entirely of gas by the time that they reach elevations of the order of 100 km (figure from Wilson et al. 1982).

ballistic debris distributed over ranges up to 10 km. The fact that fallout of small clasts from any associated eruption clouds will also occur over a greater area (since the clouds rise higher on Mars) leads to the expectation that the landforms produced by such events will be wider and have less relief than on the Earth, making them harder to recognize in spacecraft images.

C. Lava-Producing Activity

Recognition of the existence of very long lava flows on Mars (see, e.g., Carr et al. 1977*a;* Pieri et al. 1985) has been an important spur in attempts to relate lava-flow length, width, planimetric area and thickness to lava rheology and effusion rate (Hulme 1976; Moore et al. 1978*b;* Zimbelman 1985; Baloga and Pieri 1986; Pieri and Baloga 1986). These relationships are not well understood, even empirically, for the Earth (cf. Walker 1973*b;* Malin 1980; Pinkerton and Wilson 1988) and there is a strong incentive to develop a completely general model of lava-flow motion for application to all of the terrestrial planets (Wilson and Head 1983). Existing models suggest that the lengths of the longest flows are dictated mainly by magma eruption rates and that, for a wide range of rheologies, it may not be safe to use the inferred rheology to make a further inference about magma chemistry. The rate of magma effusion feeding flows on Mars was commonly very high by terrestrial standards ($>10^6$ m^3s^{-1}; Baloga and Pieri 1985; Cattermole 1987), being comparable to that of terrestrial flood basalt or lunar mare eruptions. Most theoretical models can be made to fit ultra-basic to andesitic compositions (Zimbelman 1985), provided that the volatile contents of Martian magmas were similar to those of terrestrial magmas with the same major element composition.

Though many Martian flows were probably fed from long fissure-type vents (see, e.g., Cattermole 1986), local concentration of eruptive activity into shorter lengths of the fissures would have provided the conditions for more localized fire-fountains to form. In such eruptions, the combinations of particle-size distribution, clast-ejection velocity and total particle flux would lead to optically dense conditions, pyroclasts being so closely crowded that little radiative cooling to the surroundings could have occurred (Wilson and Head 1981*a;* Head and Wilson 1989). These pyroclasts would have coalesced on landing near the vents into lava ponds at near-magmatic temperatures which in turn would have overflowed to feed lava flows. Analysis of the lava motion in such ponds and flows (Wilson and Head 1981*a;* Wilson et al. 1982) shows that turbulent conditions would prevail at the higher end of the inferred range of effusion rates for Newtonian magmas (Fig. 12). The consequent

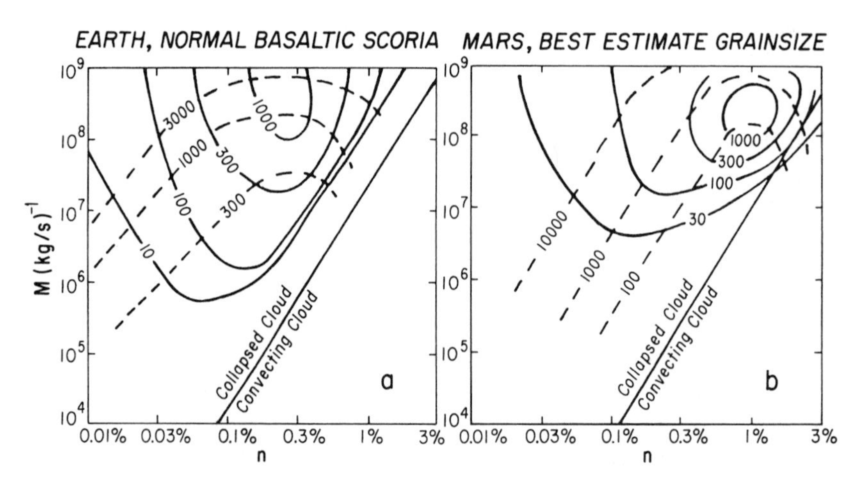

Fig. 12. Eruption conditions in terrestrial basaltic scoria (left) and Martian (right) fire-fountain eruptions that result in thermal erosion of bedrock to form sinuous rilles. The solid curves show, as a function of mass eruption rate M, through the vent and exsolved magma content n, the radius in meters of the zone within which magma clots would land hot enough to coalesce into a lava flow. The dashed lines are contours of equal Reynolds number for the motion of the lava draining from the resulting lava pond. If the Reynolds number exceeds about 1000, turbulent motion and rille formation will occur (figure from Wilson et al. 1982).

efficient transfer of heat to the underlying surface (Hulme 1973) would maximize the chance of thermal erosion taking place (Hulme and Fielder 1977) to create sinuous-rille channels and their characteristic circular or elongate source craters (Head and Wilson 1981). A total of 81 Martian sinuous rilles have been identified planet-wide (Wilson and Mouginis-Mark 1984), of which 37 are located in Elysium Planitia and several other examples are concentrated in Syrtis Major (Schaber 1982). The Elysium Planitia rilles (Fig. 13) range in length from 20 to 200 km, and for the larger (>120 km in length) examples, the inferred mass eruption rates are in the range 1.2 to 23.0 × 10^8 kg s^{-1} (Wilson and Mouginis-Mark 1984).

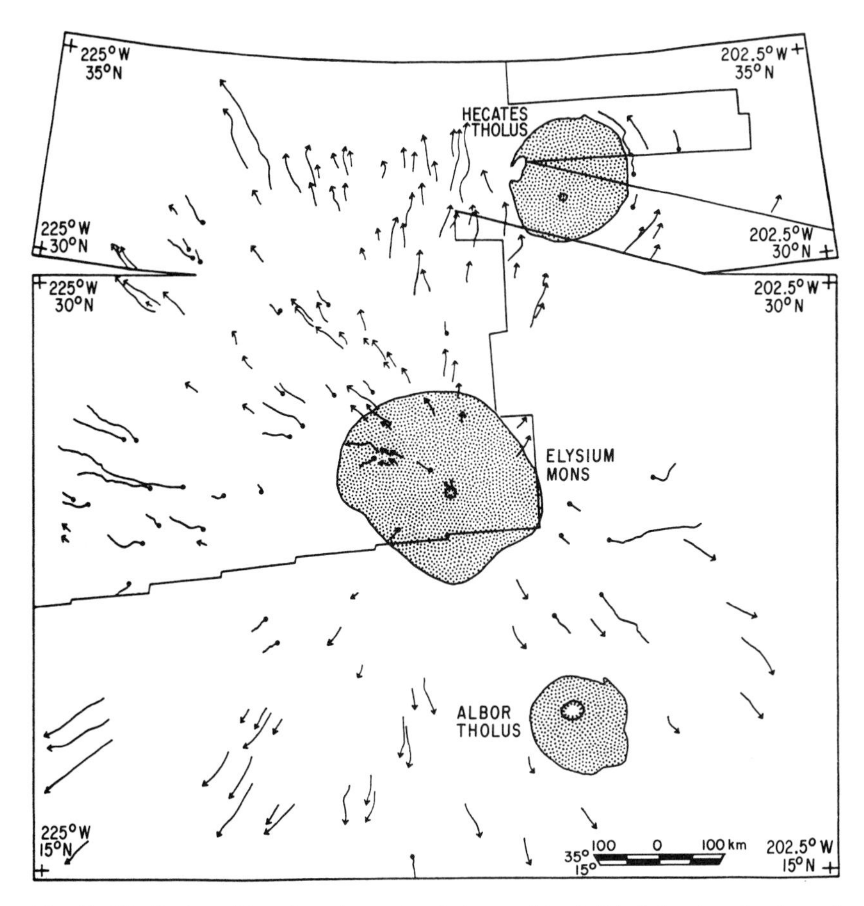

Fig. 13. Distribution of lava flows (arrows) and sinuous rilles (solid circles show locations of source craters) within Elysium Planitia. Note that rilles occur on the flanks of Elysium Mons and the adjacent plains. Such a distribution probably indicates more than one near-surface magma chamber in order to accommodate the high-mass eruption rates believed to be associated with these eruptions (figure from Wilson and Mouginis-Mark 1984).

VI. TECTONICS ASSOCIATED WITH MARTIAN VOLCANISM

A. Volcanic Constructs

Concentric graben that surround major volcanic constructs are interpreted to have formed due to loading of the Martian lithosphere (Comer et al. 1985). Knowledge of the mass of the load and the radial distance of the graben from the volcano center provides an indication of the thickness of the elastic lithosphere at the time of graben formation. Comer et al. (1985) and Hall et al. (1986) have analyzed the major Martian volcanic loads and found marked spatial heterogeneities in elastic thickness of the elastic lithosphere of Mars. These lithosphere thickness variations have implications for Martian internal structure and thermal evolution which are discussed in chapter 8.

Two of the most enigmatic features on Mars are the Olympus Mons aureole and basal escarpment (Fig. 14). The aureole consists of lobes of ridged materials distributed primarily to the north and northwest of the volcano that extend for distances of up to 1000 km. This feature has been interpreted as the erosional remnant of a pre-existing volcanic structure (Carr et al. 1973), and as a series of subglacial lava flows (Hodges and Moore 1979) or pyroclastic flows (Morris 1982). Other workers have proposed the aureole to have formed due to gravity sliding of material away from the basal escarpment (Harris 1977; Lopes et al. 1980,1982; Francis and Wadge 1983; Head and Carras 1985; Tanaka 1985). The nature of the escarpment, which has a height of up to six kilometers above the surroundings, is also a matter of debate. Interpretations include both erosional (King and Riehle 1974; Head et al. 1976; Lopes et al. 1980; Tanaka 1981; Morris 1982) and tectonic (Harris 1977; Morris 1981*b;* Borgia et al. 1990) mechanisms.

The summit caldera complex of Olympus Mons represents one of the clearest examples of the relationship between tectonic and volcanic processes (Fig. 9). The central portion of the caldera, which is topographically low (Wu et al. 1984), is marked by numerous ridges, the largest of which are morphologically similar to wrinkle ridges found in the ridged plains and lunar maria. Inside the caldera rim, at a topographically high level, are circumferential graben (Mouginis-Mark et al. 1990; Watters and Chadwick 1990). The relationship of the summit topography to the tectonic features, in combination with photogeologic evidence for basalt-like resurfacing of the caldera floor (Greeley and Spudis 1981), is believed to indicate that a large lava lake cooled and subsided as magma in the underlying chamber was withdrawn by flank eruptions (Mouginis-Mark 1981). The state of stress implied by certain tectonic features within the caldera, which presumably formed due to floor subsidence associated with magma withdrawal, has been used to estimate the depth to the top of the magma chamber (Zuber and Mouginis-Mark 1990). The best-fit depth range (<16 km) constrains the magma chamber to have been within the volcanic edifice at the time of the subsidence episode. This

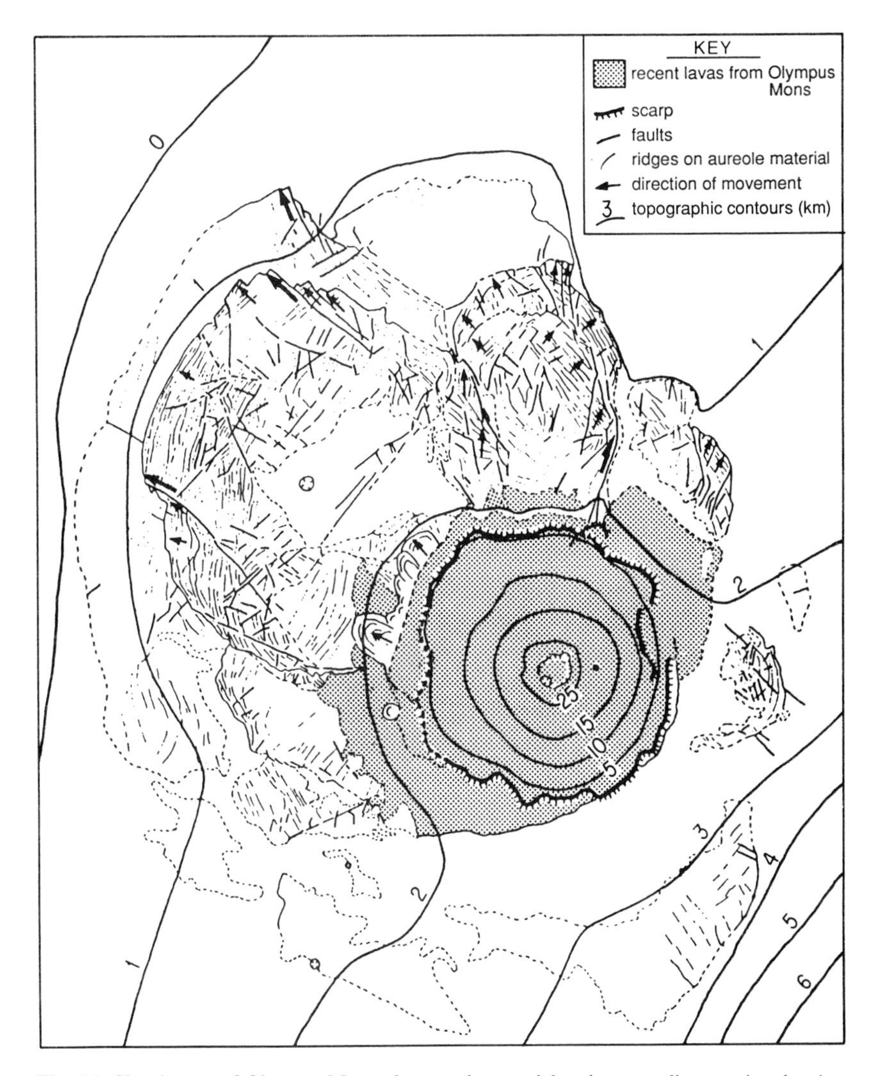

Fig. 14. Sketch map of Olympus Mons, the aureole material and surrounding terrain, showing the pattern of faults and ridges from which the local directions of movement were inferred (Lopes et al. 1982).

depth is much shallower than the predicted source depth of Martian magmas (Bertka and Holloway 1989). However, it is consistent with a result suggested from gravity scaling of the depths of magma chambers of terrestrial shields, assuming that chamber depth is controlled by magma buoyancy (Wilson and Head 1990; Ryan 1987).

The caldera complex at the summit of Ascraeus Mons consists of eight separate craters in various stages of preservation. The floor of the largest (and youngest) crater contains numerous ridges that may be tectonic in nature, while the caldera rim exhibits mare-type wrinkle ridges and circumferential graben that may be related to detumescence and slumping of the caldera walls (Mouginis-Mark 1981). The summit caldera of Arsia Mons is infilled by lava flows and there are no graben or ridges on the caldera floor; however, circumferential graben occur away from the rim (Crumpler and Aubele 1978). The calderas of Hadriaca Patera, Amphitrites Patera and Syrtis Major show similar structures. The two-level summit caldera of Pavonis Mons exhibits radial ridges within the caldera and concentric graben outside the rim (Crumpler and Aubele 1978) that may reflect subsidence of the summit. Details of the stress regimes and the possible implications for near-summit volcanic processes have not been quantitatively addressed for any of these structures.

On the flanks of Olympus, Ascraeus, Arsia and Pavonis Montes are a series of terraces that are distributed concentrically about the summit calderas. Morphological analysis and models suggest that these features may be thrust faults caused by self-compression of volcanic edifice (Thomas et al. 1990). Subsequent models have shown that flexure of the lithosphere beneath a volcano due to the rapid emplacement of the volcanic construct would modify the magnitudes and principal directions of edifice stresses with time (McGovern and Solomon 1990). These models predict stress states consistent with summit eruptions shortly after load emplacement and flank eruptions at later times. Yet to be quantitatively addressed are the effects of volcano growth and associated lithospheric flexure for more realistic loading time scales, in which the state of stress of the volcano and surroundings changes during the shield-building process.

B. Volcanic Plains

Tectonic processes associated with probable volcanic flows are manifest in the ridged plains (Fig. 7). Wrinkle ridges in the plains can be explained by "global-scale" compressional stresses related to the development of Tharsis (Phillips and Lambeck 1980; Banerdt et al. 1982; Sleep and Phillips 1985; Watters and Maxwell 1986) and regional scale stresses associated with impact basins (Chicarro et al. 1985). While ridges are found in other geologic units, they are most common and best developed in the plains (Chicarro et al. 1985), which suggests that the mechanical properties of this unit facilitated ridge nucleation.

In many areas, the ridges form a subparallel or concentric pattern with a regular spacing (Wise et al. 1979; Saunders and Gregory 1980; Saunders et al. 1981; Maxwell 1982; Watters 1991) that provides a constraint on relationships between the competence and thickness of the volcanic materials and the structure of the underlying lithosphere (Saunders and Gregory 1980; Saunders et al. 1980, 1981; Watters 1991; Zuber and Aist 1990). For current estimates of ridged plains thickness, strength contrasts between the plains and underlying megaregolith from 1 to 3 orders of magnitude are allowable, which does not meaningfully limit the range of possible rheologies of these units. Improved observational estimates of flow thicknesses would further constrain both the mechanical properties of the plains at the time of ridge formation and the volume of plains-forming volcanism.

VII. FUTURE STUDIES

As the Mars community awaits new data from the Mars Observer mission, it is useful to consider what issues in Martian volcanology have remained unresolved with Viking and Earth-based data. A selection of appropriate questions that we hope will occupy the next stage in our attempts to understand volcanic processes on Mars include the following.

1. What was the range of magma chemistries that produced the volcanic landforms that are now exposed at the surface? In particular, the discovery of silicic materials on Mars would not only have significant impact on our understanding of magmatic evolution on Mars, but would also greatly increase the expected diversity of explosive volcanic deposits that should exist in Mars (Francis and Wood 1982). Conversely, if the leading candidate deposits that could be silicic (such as the Amazonis deposits described by Scott and Tanaka [1982]) are found to be basaltic, then further considerations of the physics of explosive basaltic activity must be pursued.
2. What are the absolute ages of volcanic rocks associated with the major eruptive episodes on Mars? Radiometric ages would establish the amount, frequency and consequence of the release of juvenile volatiles into the evolving Martian atmosphere. A better understanding of the frequency of volcanic activity on Mars, and the probable time intervals between successive eruptions of individual volcanoes (yr or Myr?) would provide climate models with important constraints on the evolution of the Martian atmosphere through geologic time.
3. What are the implications of volcano growth with time for the local-stress history of the lithosphere in the vicinity of the major volcanic constructs, and how did this changing load affect the internal structure of each volcano? Loading of the volcanic pile may well have altered the internal plumbing system, promoting a redistribution of dikes, vents and fissures, as well as a change in the geometry and location of the magma chamber.
4. What was the origin of the Tharsis and Elysium Provinces? Understanding the origin of these young volcanic centers may provide an explanation as to why constructional volcanic activity evidently terminated at an earlier time in the Martian southern hemisphere than in the north, and why only pyroclastic volcanoes appear to have formed in this region, as opposed to the probable polygenic activity in the northern plains.
5. What information can the distribution of tectonic features within summit calderas provide about caldera formation and the evolution of Martian magma chambers?

It is likely that the current debates concerning the origin of the Olympus Mons escarpment, and the probable volcanic origin of the ridged-plains materials, will also continue. Several of the above specific questions will probably only be answered with the return to Earth of carefully selected fresh Martian samples. We nevertheless eagerly await the thermal infrared, high-resolution camera, laser altimetry, gravity and gamma-ray data from Mars Observer, and also feel that the careful use of Earth-based radar and spectral data, along with the continued analysis of the Viking image data set, will provide additional insights into the physical volcanology of Mars.

Acknowledgments. This research was supported by grants from the National Aeronautics and Space Administration's Planetary Geology and Geophysics Program.

14. THE CANYON SYSTEM ON MARS

B. K. LUCCHITTA, A. S. McEWEN
U.S. Geological Survey, Flagstaff

G. D. CLOW
U.S. Geological Survey, Menlo Park

P. E. GEISSLER, R. B. SINGER
University of Arizona

R. A. SCHULTZ
University of Nevada, Reno

and

S. W. SQUYRES
Cornell University

The equatorial canyon system of Mars extends approximately east-west for about 4000 km, from longitude 40° to 110°. It includes the valley network of Noctis Labyrinthus in the west, the linear troughs and wide depressions of the Valles Marineris chasmata in the center, and wide troughs, transitional to outflow channels, in the east. Individual chasmata are as much as 100-km wide but coalesce to form a depression as much as 600-km wide in the central part of the canyon system. Their 8- to 10-km depths offer excellent three-dimensional views into the upper Martian crust. The Valles Marineris troughs opened in Early Hesperian time, probably due to tectonism. Their opening was apparently preceded by the formation of wrinkle ridges trending perpendicular to the troughs, and it coincided with the formation of shallow grabens trending par-

allel to the troughs. Early scarps were soon eroded and developed spur-and-gully morphology. After Early Hesperian time, interior layered deposits first filled the troughs and were then partially removed by erosional or displaced by structural processes. Landslides fell into the newly created voids. Later, irregular floor materials were emplaced that may be as young as Late Amazonian. Deflation and aeolian deposition continue to the present. The rock forming the trough walls is probably composed of impact breccia; slope-stability analyses suggest that it contained only minor amounts of water or ice. Interior layered deposits may be of lacustrine or volcanic origin; the later irregular floor deposits, associated locally with dark, apparently mafic materials, are more likely volcanic. Landslides may have been emplaced as dry-rock avalanches, yet they contained some water or ice that gave rise to alluvial materials. The Valles Marineris are tectonic grabens located on the flanks of the Tharsis rise. Their opening was influenced by a tensional stress regime that surrounded Tharsis and was caused by isostatic adjustments or subsidence from loading. However, why the Valles Marineris were faulted to form deep troughs is not known. The cause(s) may have been (1) collapse of near-surface rock due to withdrawal of material or opening of tension fractures at depth; (2) development of keystone grabens at the crest of a bulge; or (3) the drifting and subsequent failure of plates. Furthermore, reduced crustal strength because of subsurface aquifers may have localized strain. Erosional widening has been relatively minor. Because of the diverse geology of the Valles Marineris, future instrumental or manned missions to this region should yield important information on compositions, processes and climate history.

I. INTRODUCTION

The Valles Marineris system (Fig. 1) includes a series of troughs just south of the Martian equator that trend N 75° W for a distance of 4000 km, from longitude 40° to 110°. The system includes the Noctis Labyrinthus network of troughs in the west (Fig. 2a), the Valles Marineris in the center (Figs. 2b-d), and troughs merging with chaotic terrains and outflow channels in the east (Fig. 2e). Individual troughs are generally on the order of 50 to 100 km wide, but in the central part they merge to form a depression 600 km wide. Depths below the plateau surface of as much as 8 km are common, and locally depths of 10 km are reached (U.S. Geological Survey 1986). The troughs are interconnected and open toward the east, except for Hebes Chasma, which is entirely enclosed, and Echus and Juventae Chasmata, which open toward the north. All lie on a regional high extending eastward from the Tharsis rise (Christensen 1975; U.S. Geological Survey 1986) and are paralleled by many shallow grabens that are part of a system radial to Tharsis (Carr 1974*b*; Blasius et al. 1977; Wise et al. 1979*b*; Plescia and Saunders 1982).

Although the troughs are commonly referred to as canyons, this designation is not entirely appropriate, because it implies fluvial processes. Sharp (1973*c*) preferred the use of troughs, and we follow his recommendation even though we retain the word "canyon" in the chapter title for the sake of continuity with previous usage. The International Astronomical Union uses the

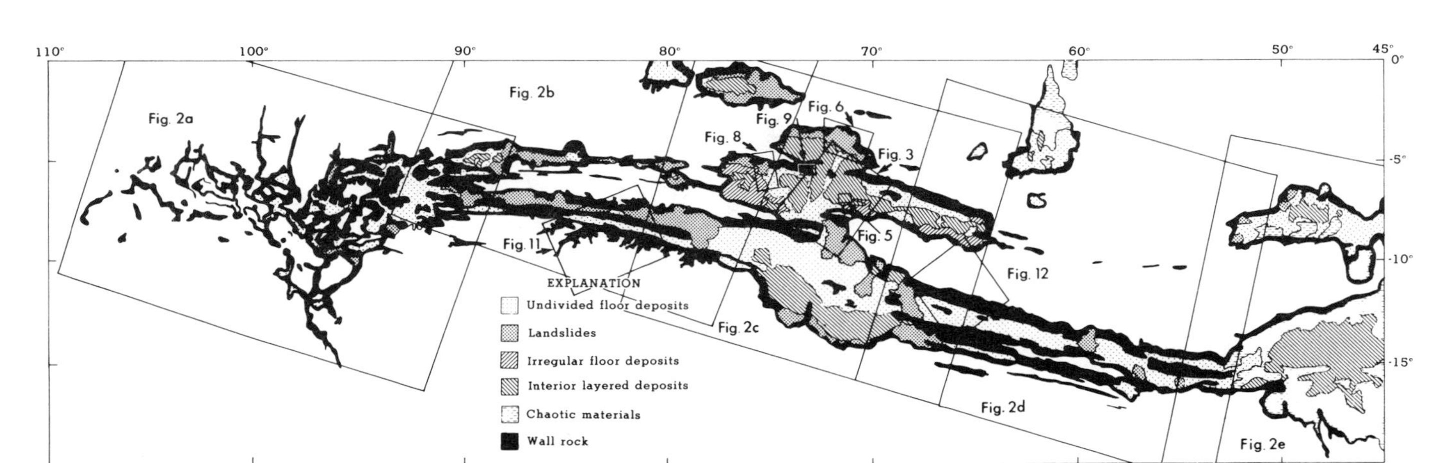

Fig. 1. Geologic map of Valles Marineris canyon system, showing boundaries of included figures. (Some figures encompass larger areas than shown here.) Plateau surfaces are unpatterned. Geology modified from Witbeck et al. (1991).

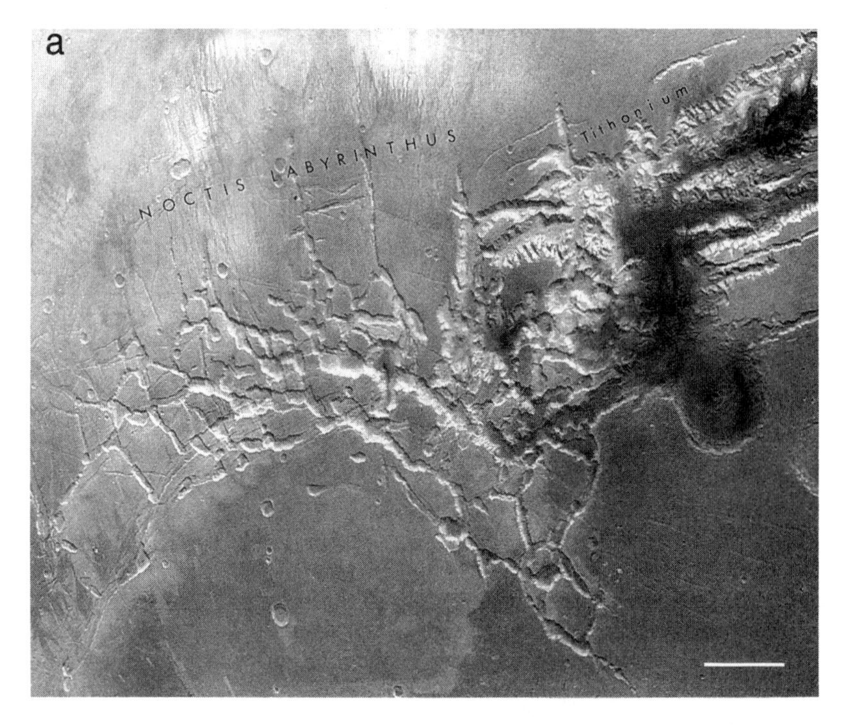

Fig. 2. The Valles Marineris system of troughs. Excerpts of mosaic of apoapsis red-filter images with 800 m/pixel resolution (McEwen and Soderblom 1989). In all frames (a-e), scale bars (white lines) are 100 km long; north is toward upper left. (a) Noctis Labyrinthus and transition to western troughs. Labyrinth is composed of a rectangular pattern of troughs that merge with pits and pit chains. Unconnected depressions of western pits widen and merge eastward, where they connect with continuous troughs of Valles Marineris.

word *chasma* (plural *chasmata*) to mean an elongate, steep-sided depression. The troughs occupy the area of the classical dark markings of Tithonius Lacus, Coprates and the westernmost part of Aurorae Sinus, and some of the individual troughs are named after these markings. However, the main trough system, the Valles Marineris (Mariner valleys), was named in honor of the achievements of the Mariner 9 mission (Blasius et al. 1977).

This chapter deals only with the equatorial system of troughs; we do not discuss other shallow graben systems, such as those radial to Tharsis (chapters 8 and 11), troughs peripheral to ancient basins (Wichman and Schultz 1989), or erosional canyons of the fretted terrain (Sharp 1973*a*).

The equatorial system of troughs was first seen on Mariner 9 images and was one of the most spectacular discoveries of this mission. In papers based on Mariner images, several workers (McCauley et al. 1972; Sharp 1973*b*; Lucchitta 1978*a*, McCauley 1978) emphasized the role of ice and water in

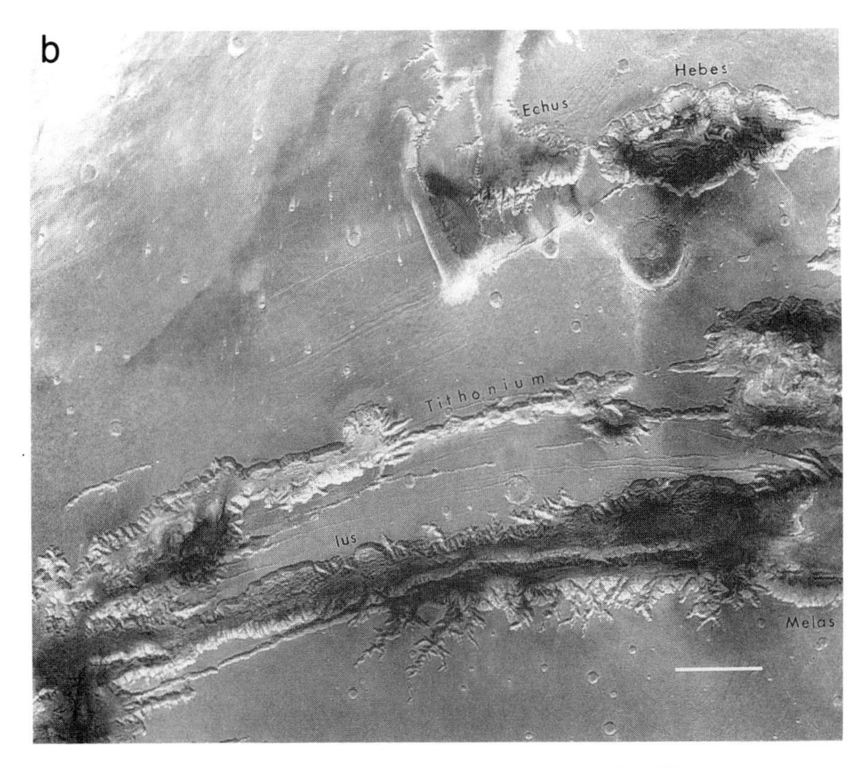

Fig. 2.(b) Western and northwestern Valles Marineris troughs. Tithonium Chasma narrows eastward and becomes a thin, scalloped depression, whereas Ius Chasma widens toward Melas Chasma. Chain craters parallel troughs on intervening plateau. Network of tributary canyons is conspicuous on south side of Ius Chasma (see also Fig. 11). Echus Chasma is source of Kasei Valles and is partly filled with deposit postdating Kasei. Hebes Chasma is entirely enclosed.

the formation and erosional widening of the canyons, even though tectonic processes were also considered for their origin (Courtillot et al. 1975; Masson 1977). A major paper summarizing new insights on the Valles Marineris was published soon after the Viking Primary Mission (Blasius et al. 1977); this paper made a strong case for the main origin of the troughs as prolonged tectonism caused by crustal extension and subsidence, possibly lasting into recent times. Another summary of the equatorial troughs is given by Carr (1981, pp.124–134).

Since these early works, research has focused on the detailed morphologic and stratigraphic relations of the Valles Marineris. In addition, because the troughs offer a three-dimensional view into the upper Martian crust, they reveal potentially important compositional and structural information.

In this chapter we give a general description of the individual troughs,

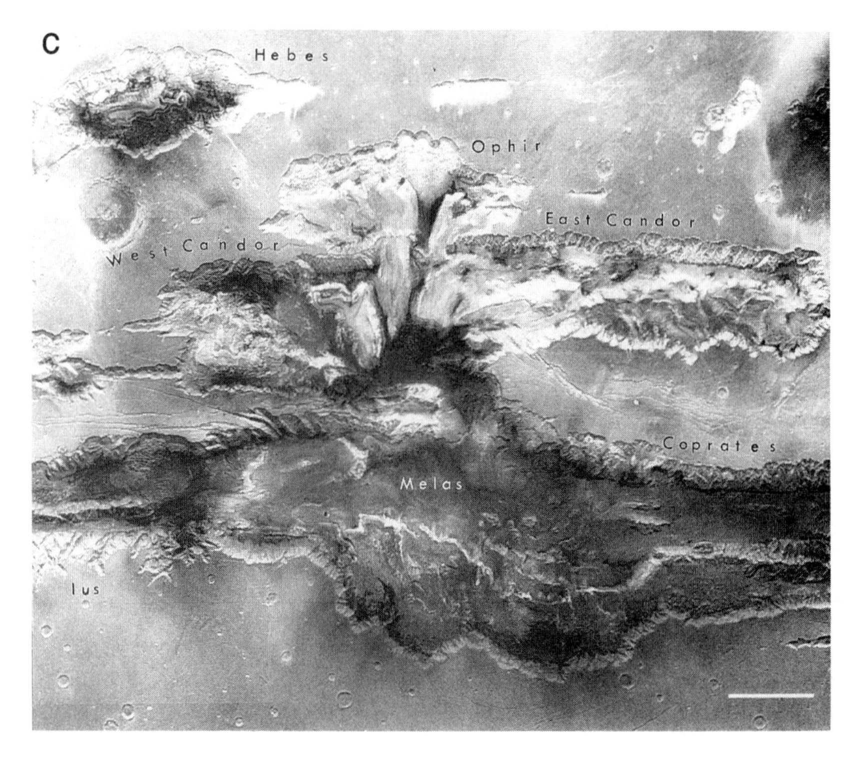

Fig. 2.(c) Central and north-central Valles Marineris troughs. Ophir, Candor and Melas Chasmata are separated only by narrow, eroded ridges of plateau and wall rock that project into troughs; ridges are breached in their centers. Ophir and Candor Chasmata have blunt ends nearly orthogonal to their trends. Melas Chasma has curved re-entrant on south side, paralleled by curved grabens on plateau. Interior layered deposits (Fig. 1) form conspicuous tablelands in Hebes, Ophir and Candor Chasmata. Dark materials litter chasma floors.

followed by discussions of their stratigraphy, geomorphology and structure. Then we address possible origins and the overall sequence of events and conclude with a summary of our current state of knowledge.

II. PHYSIOGRAPHY

The trough system can be divided into segments that have distinctive characteristics. Noctis Labyrinthus (Fig. 2a) forming the westernmost segment lies on the northern and northeastern flanks of a regional high centered on Syria Planum. The labyrinth consists of a network of intersecting valleys that divide the cratered uplands into a mosaic of blocks. Scarps bounding the individual valleys range in plan from very straight to strongly cuspate. Long, deep valleys merge with pits and pit chains, forming a rectangular pattern. Pits occur mostly at trough intersections. Valleys of the western labyrinth are

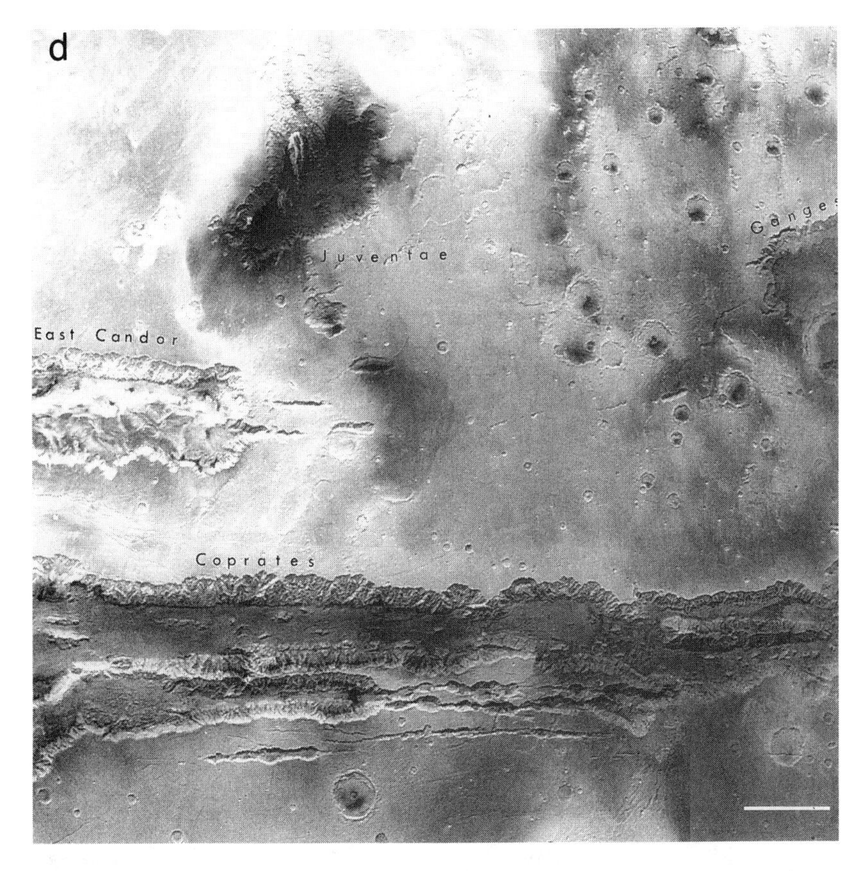

Fig. 2.(d) Eastern and northeastern Valles Marineris troughs. Floor of Juventae Chasma is partially covered by interior layered deposits and chaotic terrain (Fig. 1). Northern trough of Coprates Chasma is bounded on north side by steep, straight fault scarp. Southern trough of Coprates merges eastward with two scalloped depressions; additional linear depressions parallel Coprates on its south side.

unconnected, but eastward they widen, merge and ultimately connect with the continuous troughs of the Valles Marineris. Widths of the valleys range from 3 to 20 km, depths from 1 to 2 km. The valleys follow dominantly north-south, north-northeast, east-northeast, and east-southeast structural trends (Masson 1977,1980).

The west-central troughs of the Valles Marineris trend nearly east-west and are composed of a broad trough that splits eastward into two segments, Tithonium and Ius Chasmata (Fig. 2b). Both troughs are choked with landslide debris in their wider sections. Interior deposits occur in only one place, in the widest portion of Tithonium Chasma near its west end (Fig. 1). Quasi-dendritic and orthogonal tributary canyons extend over 100 km into the up-

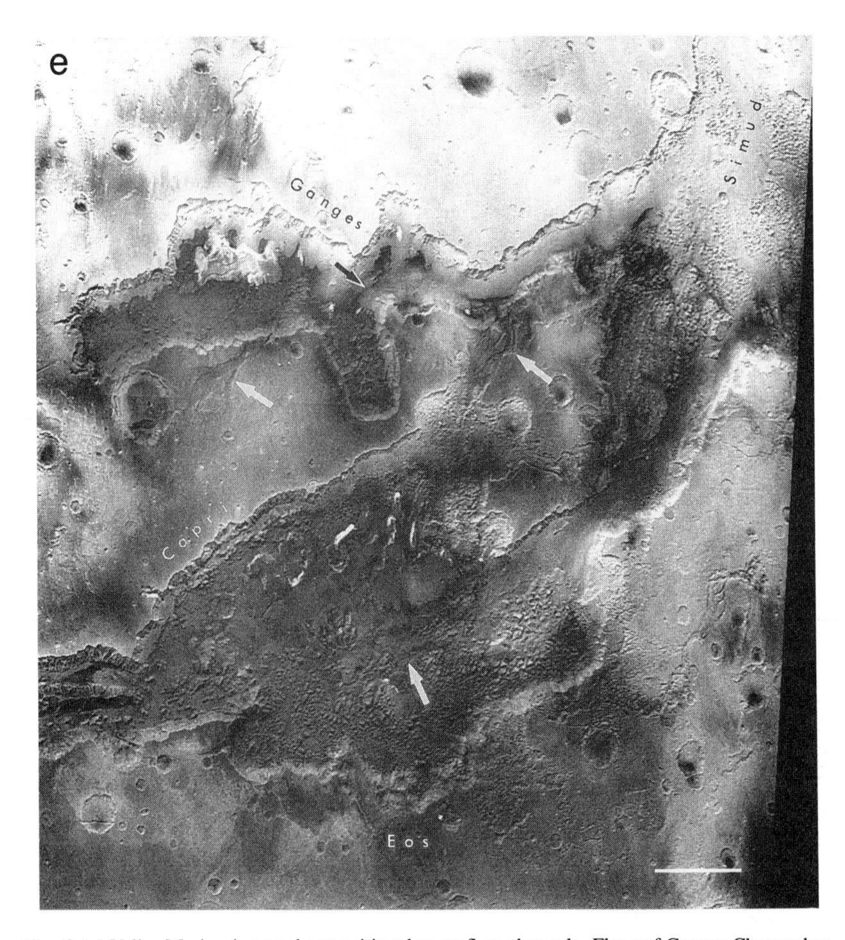

Fig. 2.(e) Valles Marineris troughs transitional to outflow channels. Floor of Ganges Chasma has mostly chaotic terrain and some interior layered deposits (Fig. 1). Interior beds of Capri Chasma, also surrounded by chaotic terrain, appear to be composed of down-faulted plateau rock mantled by layered deposits. Channels (arrows) appear on plateau and inside troughs, which merge eastward to become Simud outflow channel.

land plateaus to the south of Ius Chasma. Shallow grabens and chain craters (catenae), located mostly north of Ius Chasma, parallel the troughs in this region.

The central troughs follow a more southeasterly trend and are interconnected (Fig. 2c). Locally, narrow troughs and chain craters emerge from the blunt ends of Ophir and Candor Chasmata. The bulk of the interior layered deposits occurs in the three central troughs (Fig. 1). Irregular floor deposits are thick in western Candor Chasma, and landslide deposits are abundant along the north walls of Ophir and western Candor Chasmata.

The south boundaries of the north-central troughs Echus and Juventae Chasmata and the elongation of Hebes Chasma are approximately collinear and parallel to the main troughs, suggesting a common structural control (Figs. 2b-d). Echus and Juventae Chasmata, 4 and 3 km deep, respectively, are open to the north and give rise to the northward-sloping outflow channels Kasei and Maja Valles. The floor of Echus Chasma is smooth, but that of Juventae is rugged and diverse, consisting of irregular, chaotically arranged hills, interior layered deposits, and a thick mantle of dark material. The latter gives rise to a swath of thin, wind-drifted deposits on the adjacent plateau surface to the south (Thomas 1982). Hebes, 8 km deep and entirely enclosed, has an irregular outline largely because of deep landslide re-entrants on its north and south walls. A central tableland rises 6 km above its floor and locally reaches the elevation of the surrounding rim.

Coprates Chasma consists of two parallel troughs in the west, separated by a ridge of wall rock (Fig. 2d). The floor of Coprates is relatively smooth; interior layered deposits and landslides are scarce except in the westernmost part. Even though the surrounding plateau drops eastward more than 4 km in this region, the trough floor remains relatively level, so that the Coprates trough walls decrease in height from ~ 8 km in the west to only ~ 4 km in the east.

East of longitude 53° the character of the troughs changes dramatically: the wall height is reduced to 3 to 5 km, the floors are littered with small tablelands and chaotically arranged hills and, at about longitude 44°, the troughs merge with smooth-floored depressions bearing streamlined forms and ultimately become the Simud outflow channel (Fig. 2e). Capri Chasma contains a low-lying tableland that is pockmarked with ancient craters up to 25 km in diameter; this mesa is probably a downdropped plateau segment capped by layered deposits (Fig. 1). Ganges Chasma has an embayment on its south side whose floor also consists of downdropped fractured plateau material, and central Ganges Chasma contains chaotic hills and a mesa of well-developed layered deposits (Fig. 1). Several outflow channels on the surrounding plateau apparently dropped into Ganges Chasma, and streamlined forms appear on its floor as far west as longitude 48°.

III. STRATIGRAPHY

The sequence of emplacement of different units inside the Valles Marineris (Fig. 1) has been established with reasonable confidence by superposition and cross-cutting relations, but the compositions of units remain conjectural. Major stratigraphic units, in order of decreasing age, include wall rock (shallow crust of the surrounding plateaus), interior layered deposits, landslide deposits, irregular floor deposits, fractured floor material and surficial deposits.

A. Wall Rock

Wall rock is exposed in the walls bounding the troughs, in ridges projecting into the troughs as erosional remnants of plateaus, in isolated mountain masses on the floors of the troughs, and probably in conical hills of chaotic terrain in the eastern troughs.

On the uppermost walls, the plateau edge is generally lined by a 1-km-high steep slope. Horizontal layering in this slope locally shows as bands of differing albedo or as aligned protrusions or indentations indicating layers of differing competence (Lucchitta 1978*a*). Four layers can be recognized in many places, and as many as 8 can be seen locally. Many of the layers are extensive, and some can be traced for hundreds of km. The layers may be the exposed edge of the material that resurfaced the plateaus of Lunae, Syria and Sinai Plana, and they are probably composed of mafic volcanic rock, most likely basalt (Scott and Carr 1978; Greeley and Spudis 1981; Tanaka 1986; Scott and Tanaka 1986; see also chapter 11). However, the resurfacing layers are inferred to be only ~ 0.4 to 0.6 km thick (DeHon 1985; Frey et al. 1989; Davis and Golombek 1990), and probably only half of the steep slope is underlain by the resurfacing material. The discontinuity at the base of the 1-km-thick section of steep slope may mark the base of the cryosphere (Fanale 1976; Rossbacher and Judson 1981; see chapter 16), where differences in mechanical or chemical processes resulted in differences in weathering of the rock (Soderblom and Wenner 1978; Battistini 1985; MacKinnon and Tanaka 1990; Davis and Golombek 1990). The lower wall rock commonly does not show layering, except for a dark horizontal layer in the south wall of Coprates Chasma.

The wall rocks are conspicuously different from the interior layered deposits; wall rocks generally show layers only near the top, display no evidence of wind erosion, and have spur-and-gully morphology (Fig. 3). Another characteristic is that wall rock typically gives rise to extensive landslides. The crest lines of most wall-rock ridges extending from the plateaus into the troughs are marked by zones of dense faults, which apparently support the ridges. The zones appear more competent than the wall rock; perhaps they are intruded by dikes or the faults are lithified (Lucchitta 1987*c*).

The composition of the wall rocks is unknown, but its spatial uniformity suggests compositional uniformity and perhaps massive bedding. The most likely rock is impact breccia of the megaregolith, an inference based on the wall rock's stratigraphic position in the upper part of the Martian crust (Carr 1979,1986). The megaregolith is estimated by some workers to be ~ 2 to 3 km thick on Mars (Woronow 1988); underneath the megaregolith probably lies a fractured basement of significant porosity and high permeability (MacKinnon and Tanaka 1990). Other estimates for the thickness of the megaregolith are as much as 10 km (Carr 1979,1986). The morphologic uniformity of the Valles Marineris walls suggests that the megaregolith is thick,

Fig. 3. Central Candor Chasma. Wall rock (w) typically displays bifurcating and irregular spurs and gullies. Interior layered deposits (il), by contrast, have light-colored, smooth surfaces and distinctive parallel flutes on some slopes. Interior layered deposits butt against and overtop erosional remnants of wall rock (curved arrow). (Mosaic of Viking Orbiter images 912A–914A; 220 km wide; north toward upper left.)

or that the transition from megaregolith to fractured basement is gradual and results in rocks of similar competence. Water or ice may be trapped in pores in the breccia (Carr 1979).

Thermal-inertia measurements with a spatial resolution of as little as 5 km were made in a few traverses across the Valles Marineris (Skinner and Zimbelman 1986; Zimbelman et al. 1987) with the Viking Infrared Thermal Mapper (Zimbelman and Kieffer 1979). The measurements show that the wall rock has intermediate thermal inertias (190 to 290 J m^{-2} $s^{-1/2}$ K^{-1}) corresponding to fine sand (if uniform grain sizes are assumed, a strong simplification for real surfaces; Kieffer et al. 1977); thus the walls may be mantled by relatively fine-grained materials.

Broadband spectral data (Fig. 4) provided by Viking images (bandpasses 0.45, 0.53 and 0.59 μm; resolution 600 to 800 m/pixel) show that wall rock is relatively red and similar to landslide deposits (Geissler et al. 1989), which are clearly derived from the walls. Wall rock is also nearly as red as the surrounding plateaus; both may be mantled by reddish dust, or the dust on the plateaus may be derived from wall rock.

Spectral properties of the dark layer in the walls of Coprates Chasma have been studied by Geissler et al. (1989,1990). The layer is several hundred m thick and crops out about two-thirds of the way up the slope over a distance of ~ 400 km (McEwen and Soderblom 1989; Geissler et al. 1990). Comparisons between (1) three-color reflectances of this layer from Viking Orbiter images (corrected for photometric and atmospheric effects), and (2) laboratory spectra of terrestrial rocks, convolved to the same spectral resolution (see chapter 17), show that the material at the core of the Coprates layer may be relatively pristine mafic glass. This interpretation is consistent with the observed friability of the layer and its incompetence as inferred from apparent erosional characteristics seen on images. The layer may be intruded. If it was extruded, surface deposits must be several km thick in this region.

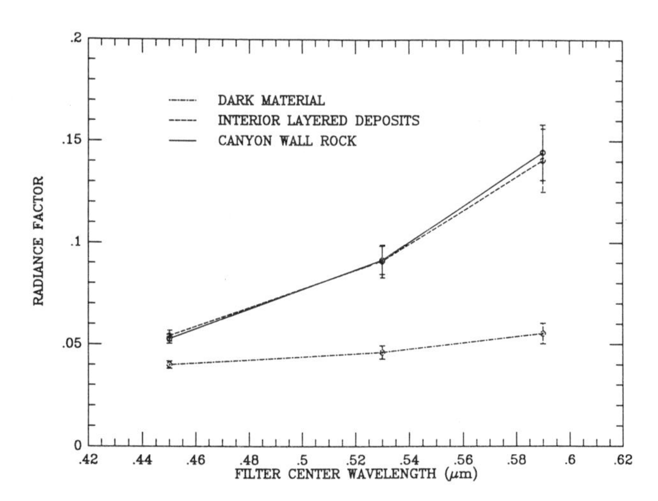

Fig. 4. Broadband spectral reflectances of some Valles Marineris units as measured by Viking Orbiter 1 on orbit 586. Spectra corrected for atmospheric and photometric effects. Error bars represent range of intrinsic variability in diverse exposures of these units at the one standard deviation level. Note low slope on dark material, indicating near-colorless substance. This spectrum is compatible with spectra of terrestrial mafic materials.

B. Interior Layered Deposits

Layered deposits occur in most of the central and northern troughs and in Capri and Eos Chasmata (Figs. 1 and 2). They are thickest in Hebes and Ophir Chasmata and in parts of Candor Chasma, and they thin southward toward Melas Chasma. They have a total volume of 1.3×10^5 km^3 in the central troughs (Nedell et al. 1987). Although grouped together here for purposes of discussion, the layers differ considerably in color, albedo, thickness and competence, and they may well be of diverse origins.

In Ophir, eastern Candor and Melas Chasmata, the layered beds (Fig. 1) occupy southern benches; in western Candor and Hebes Chasmata, the beds form free-standing tablelands (Fig. 2c); in Juventae, Ganges and western Tithonium Chasmata, they form smaller tablelands that stand isolated from the walls; and in Eos Chasma, a wide central plateau appears to be capped by interior deposits (Nedell et al. 1987; Witbeck et al. 1991).The benches and mesas rise to elevations locally less than 500 m below the surrounding plateau rims, and they have scarps as high as 6 km.

Early workers noticed that some interior deposits are evenly layered and that alternating light and dark beds have great lateral continuity (McCauley et al. 1972; Sharp 1973*c*; McCauley 1978). Whereas such "rhythmic" layering is visible in Juventae and Ganges Chasmata, the regularity is not nearly as obvious in places where high-resolution pictures show the detailed structure. In eastern Candor Chasma, the lower beds are composed of sequences 300 m thick of alternating light and dark deposits (Nedell et al. 1987), but the upper layers, averaging 70 m in thickness, are composed of alternating layers of varied thickness and resistance to erosion (Fig. 5). (Thickness values were derived stereo-photogrammetrically.) The beds exposed in the slopes of Candor Mensa (average thickness of 100 to 200 m) and in Hebes Chasma are also varied: they appear to be locally massive or thin, they are lighter or darker than their surroundings, and they seem to differ in competence.

Dark layers are conspicuously interbedded in places. Near the top of the tableland inside Hebes Chasma, two dark layers, both at least 100 km long, are less resistant to erosion than adjacent beds. On Candor Mensa, dark layers that are $\sim$ 30 km long shed talus and form resistant ledges that cap mesas or form benches on slopes. Layers in light materials are more difficult to trace, perhaps because they are less resistant to erosion or because they blend in with Mars' general color and albedo.

Spectral data from color images were obtained on bright talus slopes where exposed units are extensive enough to yield information at the 600- to 800-m resolution of the color images (Geissler et al. 1989). Three-point spectra from the Viking Orbiter images show that the interior layered deposits are relatively red and among the brightest materials in the troughs, and that, in spite of morphologic differences due to erosion, the spectral signature of these

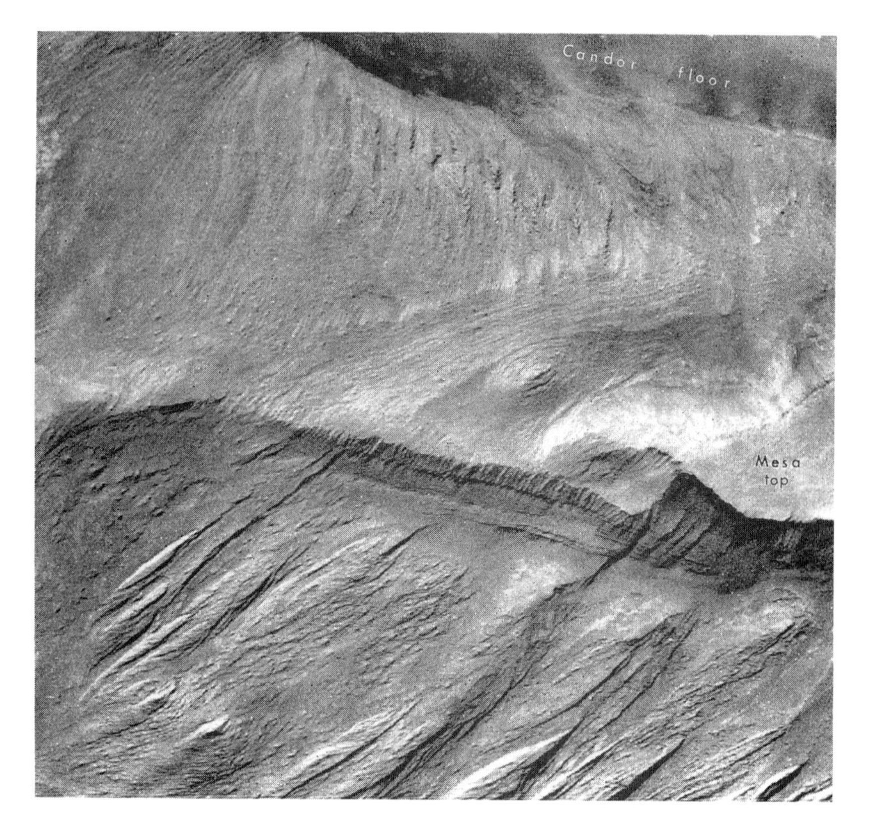

Fig. 5. Interior layered deposits in eastern Candor Chasma. Lower layers are light colored and massively bedded and display distinctive fluting. Upper layers are of varied thickness and competence. Resistant cap rock tops this eroded mesa. (Viking Orbiter image 815A58; ~ 25 km wide; north toward top.)

units is within the range of those of the canyon wall rock (Fig. 4). These observations suggest either that both units are mantled by similar reddish deposits, probably atmospheric fallout, or that at least some of the layered deposits are composed of reworked wall material. McEwen and Soderblom (1989) found bright layers in interior layered deposits in Ganges and Capri Chasmata. Low red-to-violet ratios (~ 1.5) show that this bright material is more neutral in color than most other bright units on Mars.

Thermal inertia measurements of layered deposits (Skinner and Zimbelman 1986; Zimbelman et al. 1987) show values ranging from 100 to 190 J $m^{-2} s^{-1/2} K^{-1}$ for level top surfaces to 190 to 240 J $m^{-2} s^{-1/2} K^{-1}$ for slopes. Accordingly, the top surfaces appear to be fine-grained or mantled by fine-grained atmospheric dust fallout; the slope values are similar to those of wall rock, supporting the color observations that suggest similar compositions.

Eroded surfaces of the deposits typically show a distinctive fluted topography that is strikingly different from the morphology of the canyon walls (Figs. 3 and 5). Wind erosion apparently created these parallel, yardang-like flutes, perhaps by transforming mass-wasted gullies into parallel ridges and troughs. Most of the flutes are on slopes and trend roughly downhill, but some occur on level surfaces. The fluting is more common on light-colored units, which suggests that the light materials are highly susceptible to wind erosion.

The presence of angular unconformities within layered deposits is controversial. Nedell et al. (1987) asserted that they could not recognize any, even though they measured dips of as much as $\sim 4°$ on some layers. By contrast, Peterson (1981) and Lucchitta (1985*b*) thought that they detected several. In addition, the layered deposits butt against spurs and gullies on wall rock (Fig. 3), a relation showing that at least the upper layered deposits were emplaced after the chasma walls had eroded substantially and that thus a significant unconformity exists between the wall rock and these deposits. In Juventae and Ganges Chasmata, layered deposits overlie conical hills that are similar to chaotic terrain farther west (Witbeck et al. 1991). The formation of chaotic terrain has generally been attributed to loss of material in the source areas of catastrophic floods (McCauley et al. 1972; Sharp 1973*a*; Masursky et al. 1977; Carr 1979; chapter 15). The superposition of layered deposits on chaotic material in Juventae and Ganges Chasmata therefore implies that these layered deposits were emplaced later than the outflow channels that emerge from these troughs, and that perhaps the interior layered deposits are younger than the outflow channels. In the central troughs, the interior layered deposits clearly predate the emplacement of landslides that spilled around mesas and benches carved from these deposits.

The origin of the interior layered deposits is a significant outstanding question. At least 5 hypotheses have been proposed: that the deposits are (1) erosional remnants of the same material that makes up the canyon walls; (2) aeolian deposits; (3) mass-wasted material; (4) lacustrine material; or (5) volcanic material.

The hypothesis that the deposits are erosional remnants of the canyon-wall material (Malin 1976*b*) can be eliminated because of superposition relations, the lack of similar layering in the canyon walls, and the different erosional styles of the two materials (Peterson 1981; Lucchitta 1987*c*; Nedell et al. 1987). The hypothesis of a purely aeolian origin is flawed, because some of the deposits are much too regularly bedded and too diverse in albedo, competence and erosional characteristics to be formed by this mechanism alone. Also, because the layered deposits are found only within the canyons and not on the surrounding plateaus, some trapping mechanism has to be invoked. Trapping of material on ice-covered lakes (Nedell et al. 1987), of saltating sand-sized particles in lows, or perhaps of material in denser, more stagnant air pools would permit the selective deposition of aeolian materials.

Thus, even though a purely aeolian origin is unlikely, an aeolian contribution may well be present, especially in the massive-bedded, fluted layers (Peterson 1981).

An origin by mass wasting from wall rock alone (Sharp 1973*c*) is also unlikely, because (1) at least the upper layered deposits embay already-eroded wall rock; (2) the thickness and volume of the deposits are too large for them to be derived only from eroded walls (unless the troughs are entirely of erosional origin); and (3) the deposits are too diverse in albedo and competence to be derived from wall rock alone, which appears to be fairly uniform in composition. However, some material obtained from the walls by mass wasting or flushed into a lake by subterranean drainage (piping) may have contributed to the interior layers.

The lacustrine hypothesis has found favor with several investigators (Lucchitta 1982*b*; Nedell et al. 1987) ever since McCauley (1978) suggested that only deposition within a low-energy, liquid-water environment could readily explain the horizontal bedding and lateral continuity of individual layers and their differences in reflectance and competence. Also, there is ample evidence that, at the time of canyon formation, substantial subsurface aquifers may have provided large quantities of water (Sharp 1973*a*; Carr 1979). When deep, perhaps tectonic, depressions first formed, water could have partially filled them. The climate was probably sufficiently cold that the water was quickly capped by a thick cover of ice (Carr 1983).

Although deposition in standing water might account for the characteristics of the layers, the source and composition of the materials is an open question. No large channels that might have carried enough sediments to form the thick layered deposits are seen to drain into the troughs. However, a few small valleys are observed on the surrounding plateaus, and several channels drain into Ganges Chasma (Witbeck et al., 1991). Thick accumulations of aeolian sediments, causing foundering of an ice cover or penetrating through it by Rayleigh-Taylor instability, might also have contributed to the layers. Another possibility is that some of the layers are carbonates. McKay and Nedell (1988) suggested that perennially ice-covered lakes on early Mars were in many ways an ideal environment for precipitation of atmospheric CO_2 as carbonates, and McEwen and Soderblom (1989) suggested that the bright, colorless layers in Ganges and Ius Chasmata could be carbonates.

Also not to be ruled out is a volcanic origin (Peterson 1981; Lucchitta 1990; Witbeck et al. 1991), which would be compatible with the tectonic setting of the troughs; volcanism is associated with most tectonic features on Earth that come close to the magnitude of the Valles Marineris. Compatible with a volcanic origin are the diversity of layers, the low albedo and high competence of some layers, and the fluting on light-colored deposits that is similar to fluting on tuffs in terrestrial desert regions. The dark, resistant ledges could be mafic flows; the light-colored, massive units could be tuffs. An ash-flow origin appears more likely than an origin by airfall, because the layers are restricted to the inside of the troughs. The main evidence against a volcanic origin is the great regularity and continuity of the deposits and the lack of obvious vent edifices or calderas (Nedell et al. 1987). Lucchitta (1990), however, argued that possible vents are present, that some irregular depressions may be calderas, that probable necks and dikes can be seen, and that other source vents may be buried by thick volcanic deposits. Nedell et al. admit the possibility of subaqueous volcanic eruptions, because volcanic vents more likely would be buried in an underwater environment, where more effective redistribution of material would favor the great lateral continuity and the horizontality of the layers.

The interior layered deposits, now preserved in benches and tablelands, give the impression of having been eroded so that they now stand as isolated remnants. They are surrounded by depressions informally referred to as moats (Nedell et al. 1987). The mechanism by which these moats formed is poorly understood. Several origins have been proposed, but all are controversial. Erosion involving mass wasting, sublimation, aeolian transport and removal by water have been suggested, as has a structural origin. If the moats formed by mass wasting alone, the material must have been removed later by deflation or evaporation. Deflation would have required complete breakdown of the mass-wasted material into fine grain sizes, or a lag would have formed (Nedell et al. 1987). Evaporation alone is unlikely, because this hypothesis would require so much ice or water in the walls that they would collapse in a matter of years (see Sec. IV.B). Rapidly released lake water (McCauley et al. 1972) could have eroded only deposits in the interconnected chasmata (see Sec. VI.A). A combination of processes involving evaporation of ice and aeolian and fluvial transport may provide a better explanation than any one of these processes alone.

Locally, the moats may be of structural origin (Lucchitta and Bertolini 1989). The best evidence comes from eastern Candor Chasma, which is divided lengthwise into a northern trough underlain by wall rock and a southern bench underlain by layered deposits; the two rock types are closely juxtaposed. The relation suggests a fault contact running the length of eastern Candor Chasma. Perhaps a southern trough, now occupied by the layered deposits in the bench, formed first, followed by a subsidiary trough in the north that formed later than the layered beds. The free-standing mesas are much more problematical, however, because they would have to be structural horsts of large vertical relief.

C. Landslide Deposits

Conspicuous are the many landslide deposits that fell from chasma walls into the moats (Lucchitta 1978*b*,1979; see Fig. 6). The deposits are especially abundant in Ius, Ophir and Hebes Chasmata, but they also occur in other troughs, mostly on their north walls (Fig. 1). The landslide scars form conspicuous re-entrants in the walls. Many scars are curved, but some are

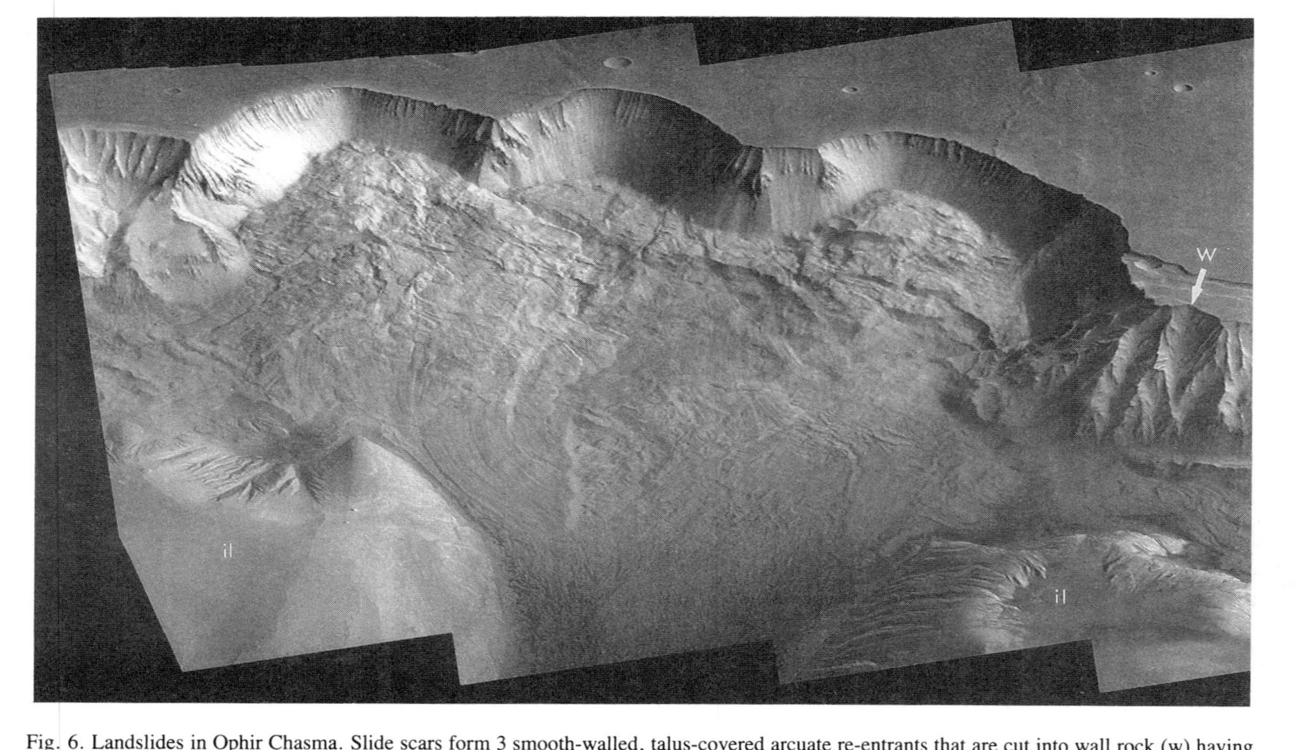

Fig. 6. Landslides in Ophir Chasma. Slide scars form 3 smooth-walled, talus-covered arcuate re-entrants that are cut into wall rock (w) having spur-and-gully morphology. Landslides spilled into pre-existing trough (center of mosaic) entrenched in interior layered deposits (il). Absence of impact craters on landslide deposits suggests that they are young. Oblique view with twice vertical exaggeration of mosaic. (Viking Orbiter images 913A09–12; mosaic ~ 150 km wide; north toward top.)

straight, especially where they are long and where they either parallel grabens on the plateau surface or lie along the extension of faults. The scars have heights of as much as 5 km and are commonly covered by smooth-looking talus. The landslide deposits bury the lower parts of the walls to depths that can reach several km. Estimated volumes of some individual landslides exceed 1000 km^3 (Lucchitta 1979; McEwen 1989). The greatest distance traveled, measured from the head of the slide to its toe, is ~ 100 km. Many small landslides, several km long, emerge from walls without obvious scars; their deposits consist of smooth aprons, aprons with pronounced rim ridges, or hummocky material throughout.

The landslides typically have slump blocks at their heads, which generally do not extend far beyond the landslide re-entrant. Below the slump blocks, most unconfined landslides have a vast apron of longitudinally ridged or smooth material (Lucchitta 1978*b*), which is so thin that the subjacent topography can be recognized in places. The ridges generally form fan-shaped patterns and range in width from the limit of resolution to ~ 500 m. Mutually overlapping lobes, apparently originating from the same slump mass at the head, indicate that several pulses occurred during major sliding events. Thermal inertia measurements indicate that landslide deposits are coarser and more blocky than their surroundings (Zimbelman et al. 1987). Also, they show little evidence of deflation.

The longitudinal ridge pattern on Martian landslides differs from that on terrestrial slides, which is generally transverse. However, exceptions are the Alaskan landslides that moved over glacier ice, particularly the Sherman landslide (Shreve 1966; Marangunic and Bull 1968). Longitudinal patterns in debris arise from spreading and differential shear between rapidly moving debris streams (Shreve 1966). Evidently, on Mars and in the Alaskan slides that moved over ice, the pattern persisted after the debris came to rest, because the high-speed flow was not impeded significantly by friction with the ground or compressed by obstacles in its path (Lucchitta 1979).

It is not clear whether landslides occurred early in the history of the Valles Marineris. Some ancient wall re-entrants, displaying spur-and-gully morphology and embayed by interior layered deposits, could be ancient landslide scars, but no deposits are now recognizable. The combined landslide deposits of Ius and Tithonium Chasmata have 570 ± 130 craters ≥ 1 km per 10^6 km^2 (Lucchitta 1979), and thus they are approximately equivalent in age to plains-lava eruptions on the Tharsis plateau to the west (Schaber et al. 1978; Scott and Tanaka 1981,1986; see chapters 11 and 12). Three conspicuous landslides on the north slope of Ophir Chasma (Fig. 6), however, have so few craters superposed that they may be considerably younger.

Lucchitta (1987*c*) suggested that the landslides were gigantic wet-debris flows, which implies that the mass was saturated with water or was a water-clay slurry, thus eliminating or greatly reducing grain-to-grain contacts. It has also been suggested that the landslides were subaqueous, collapsing into lakes (Shaller et al. 1989). However, because the landslides are geologically young (Lucchitta 1979) and date from a time when liquid water on the surface of Mars could not be maintained, release into a short-lived water body with a thick protective ice cover would be required. A significant problem for this hypothesis may be the lack of morphologic evidence for lakes in the Valles Marineris at the time of landslide formation.

In contrast, McEwen (1989) has concluded that the landslides were probably dry-rock avalanches in which the mass was unsaturated and particle-to-particle friction was not significantly reduced by water or ice. McEwen's conclusion was reached on the basis of yield-strength estimates and the size-mobility trend of the landslides. From the heights of flow fronts, McEwen estimated yield strengths of 10^{-2} to 10^{-1} MPa for the Valles Marineris landslide deposits, which are comparable to those of terrestrial dry-rock avalanches (typically 10^{-2} MPa) and much higher than that of typical wet-debris flows (10^{-4} to 10^{-3} MPa).

McEwen measured H/L (height of drop/length of runout) for Martian landslides ranging in volume from 0.1 to 10,000 km^3, and he found a trend of decreasing H/L with increasing volume that is parallel to but that lies above a similar trend for terrestrial dry-rock avalanches (Scheidegger 1973; Hsü 1975) (Fig. 7). The offset between the Martian and terrestrial trends may be explained by the effects of reduced gravity on flows with high yield strengths. The low values of H/L of the large Martian landslides are also similar to terrestrial wet debris flows supported by a water-clay matrix (Johnson 1984). However, because such debris flows would be expected to plot below the trend for terrestrial dry-rock avalanches, McEwen's observation does not support the hypothesis of wet debris-flow origin.

Nevertheless, some evidence suggests that the landslide deposits in the Valles Marineris contained water. At least one landslide apparently gave rise to a channel whose fluid flowed for 250 km, negotiated several bends, and apparently contained blocks of ice (Lucchitta 1987*c*). Perhaps the Martian slides were similar to the terrestrial slides on Mt. Huascaran (Plafker and Erickson 1978) or Mount St. Helens (Voight et al. 1981), which incorporated water and ice that formerly capped the mountain. These slides started as dry avalanches but lower down became secondary mudflows.

The abundance of water or ice in the Martian landslides has important implications for volatile abundances on Mars. Within 30° of the equator, mean surface temperatures are well above the atmospheric frost point, and ice is unstable (Farmer and Doms 1979; chapter 28). If the landslides were dry-rock avalanches, then the region of Mars within about 30° from the equator may be largely desiccated today. This result would be consistent with models for a water-poor Mars and the suggestion that Mars remains rich in ice today only at high latitudes (Squyres and Carr 1986). On the other hand, if the avalanches were mud flows or gave rise to channels, ice may be present now in the equatorial area at depth (Smoluchowski 1968; Kuzmin 1980*b*;

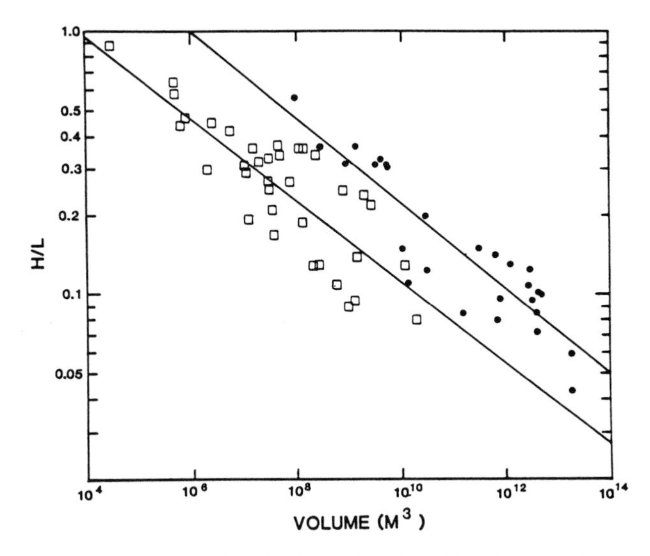

Fig. 7. Plot of *H*/*L* (height of drop/length of runout) vs landslide volume. Circles are data points from Valles Marineris (McEwen 1989); squares are data points for terrestrial dry-rock avalanches of nonvolcanic origin (Scheidegger 1973; Hsü 1975). Lines are linear least-square fits to each data set.

Kuzmin et al. 1988*a*). The latter hypothesis implies that the equatorial crust has remained water rich into recent geologic time (Lucchitta 1987*c*), in spite of having previously released large quantities of water to the outflow channels (Carr 1979).

D. Irregular Floor Deposits

After the troughs had nearly reached their present configuration, a new series of deposits was emplaced. These deposits, overlying eroded wall rock, interior layered deposits and landslide debris, attest to the presence of a second major unconformity in the troughs (Lucchitta 1990).

This young sequence of floor deposits consists of three main units: (a) deposits having rough surface textures (on a scale of tens of m), lobate fronts and mottled albedo patterns; (b) light-colored, thick, smooth-looking deposits; and (c) dark material that forms small lobes or patches with diffuse or wispy edges, which characteristically occurs along faults (Lucchitta 1987*b*).

The mottled, rough surface unit is as much as 3-km thick in western Candor Chasma, where it embays tributary canyons and buries landslide deposits. The unit has a conspicuous front that forms light-colored lobes and cuts across beds on the chasma floor that appear to have been previously tilted and eroded (Fig. 8). A dearth of superposed impact craters and the superposition on landslides suggest that the unit is young.

The light-colored, smooth-looking material occurs in the southeastern

Fig. 8. Irregular floor deposits in western Candor Chasma. Light-colored, fluted, lobate deposits (arrows) are superposed on tilted beds (tb) and landslide deposit (ls; upper center of image). (Viking Orbiter images 917A12 and 14; images ~ 70 km wide; north toward top.)

part of central Candor Chasma, where it overlies dissected mesas of older interior layered deposits. It is associated with a peculiar nested set of rimmed craters consisting of a circular crater ~ 7 km in diameter inside an oblong depression. The inner crater is deep, but both craters lack the rugged ejecta blankets that might be expected around deep, young impact craters; the craters could be of volcanic origin (Lucchitta 1990).

The dark lobe- or patch-forming material, where not composed of aeolian drift (see Sec. III.F, below) may be erupted from young volcanic vents, particularly where the patches are aligned along chasma faults (Lucchitta 1987*b*). Such vents may occur along a line on the north wall of Coprates Chasma, where wispy, dark material is associated with two apparent fault scarps. This dark material also occurs on top of and apparently penetrated through young landslide deposits in the chasma (for landslide age see Sec. III.C), suggesting that that dark material is also young. Linear exposures of dark material are also common elsewhere along the base of the trough walls, where trough boundary faults are inferred (Lucchitta 1990). Such dark exposures could be dark material concentrated by the wind against the steep slopes of the walls. However, the outcrop relations of dark material along the base of a wall-rock ridge separating western Ophir from western Candor Chasma (Fig. 9) suggest that this dark, linear exposure is competent material and thus that other dark exposures along faults may also be bedrock and not aeolian aggregates.

Viking Orbiter three-color spectral measurements on dark patches in Juventae Chasma and on the dark layer in the walls of Coprates Chasma (Fig. 4) indicate that these materials are relatively colorless and compatible with the signature of mafic materials (Geissler et al. 1990). The reflectance of these materials falls below 5% after calibrated images are corrected for the effects of atmospheric scattering and attenuation, and for differences in illumination and viewing geometry. Few terrestrial materials are this dark; mafic glass is one likely possibility. Thermal inertia measurements of dark material in Juventae Chasma give values of 570 to $\geq$860 J m^{-2} $s^{-1/2}$ K^{-1}, possibly indicating relatively large grain sizes and rock abundances (Christensen 1986). These values are the highest thermal inertia values yet found on Mars in high-resolution data (Zimbelman et al. 1987).

The origin of the irregular floor deposits is unknown, as is the origin of the older, layered interior deposits. The floor deposits are not readily explained as being of lacustrine origin, because they were emplaced relatively late in Martian history when the existence of water in the Valles Marineris was unlikely (Carr 1983). Further evidence of the relative youth of the floor deposits and of their postdating the presence of surface water is their embayment of tributary canyons. A debris-flow origin is also unlikely, because the material has no obvious source areas and buries other debris such as landslides. The lobate fronts and embayments suggest a flow emplacement. A volcanic origin (Lucchitta 1990) is suggested by the apparent flow lobes, by

Fig. 9. Dark material at base of wall-rock ridge. Dark material (arrows) is competent unit holding up lower spurs on wall-rock ridge (w) that separates western Ophir Chasma from western Candor Chasma. Dark material forms mound (m) plugging entrance to valley (v) separating wall rock on left from irregular layered deposits (il) on right. (Viking Orbiter image 814A46; ~ 30 km wide; north toward top.)

the spectral signature of the dark material that suggests mafic composition, by the association of the dark material with structures, and by fine fluting on the light-colored deposits similar to fluting on terrestrial tuffs in arid regions. Also, the light material is associated with peculiar craters that might be volcanic (Lucchitta 1990).

E. Fractured Floor Material

Deposits in central and eastern Candor Chasma have fairly level surfaces and a texture of irregular plaques and hummocks traversed by many cracks. The deposits occur in the lowest areas of the troughs, where fluids would have ponded and embayed the surrounding terrain. They appear to be alluvium derived from draining trough walls, landslides, or other trough units. The cracks may have developed when the water-charged deposits froze while settling over uneven ground (Lucchitta and Ferguson 1983; McGill 1986), or when they were desiccated after the ice sublimated.

F. Surficial Deposits

Wind and atmospheric fallout have contributed to the surficial deposits in the troughs. Aeolian drifts on the trough floors are generally dark. They have migrated along the long axes of the troughs and are concentrated in low areas (Thomas 1982; Lucchitta 1990). Barchan or parabolic dunes are conspicuous where the deposits are thick. The concentration of dark material in dunes or low areas suggests that it moved by saltation and is therefore sand sized (Cutts and Smith 1973; Greeley et al. 1980). Elsewhere, as noted above, dark material covers the trough floors with a thin mantle. The dunes and the dark mantle appear to be derived from outcrops of dark, possibly mafic materials inside the troughs (Lucchitta 1987*b*, 1990; Geissler et al. 1990).

Atmospheric fallout from dust storms locally and temporarily obscures albedo markings (Sagan et al. 1973*b*) in the Valles Marineris (McEwen 1985). It may form a thin layer on many of the other deposits. In some places on the trough floors, accumulated atmospheric fallout may be thick enough to bury underlying units completely.

IV. GEOMORPHOLOGY

In this section, we discuss chasma walls and their development and stability, pits and pit chains, tributary canyons, and the transition from troughs to channels.

A. Chasma Walls

The chasma walls have spur-and-gully morphology or smooth talus slopes (Sharp 1973*c;* Blasius et al. 1977; Lucchitta 1978*a*). The spur-and-gully morphology is well displayed in Coprates and Ius Chasmata and on interior ridges paralleling the chasma walls. The upper parts of the walls commonly have a steep slope that is dissected into vertical, subparallel ribs. Lower on the walls, sharp-crested, downward-branching spurs descend to the chasma floors. The intervening gullies are shallow to deeply incised (Sharp 1973*c*). Spurs on the walls of Ius Chasma have gradients of 15° to 20°; intervening gullies, as much as 30° (U.S. Geological Survey 1980). Lower gradients are common in some gullies that are incised obliquely to the main trend of the troughs. Spur-and-gully morphology on Martian chasma walls is similar in many respects to that on high scarps in terrestrial alpine and desert environments (Rapp 1960; Sharp 1973*c*). Walls of this type, however, are also similar to walls developed on submarine canyons, an observation consistent with the interpretation that perhaps a lake once filled the Valles Marineris (McCauley 1978).

Smooth talus slopes occur dominantly on landslide scars and on fault scarps on lower trough walls. Most talus slopes are also rimmed by steep slopes. The talus slopes typically have angles near 30° (U.S. Geological Survey 1980), close to their angle of repose.

Spurs and gullies formed early in the history of the Valles Marineris (Lucchitta 1987*c*). Whether they are forming now is conjectural, because they have not been observed on young wall surfaces such as landslide scars and young fault scarps. The scarcity of transitional forms between spurs and gullies and smooth talus slopes may indicate that the spur-and-gully morphology is an old erosional form, perhaps related to a different climate (Lucchitta 1984*b*), or that it results from a different local environment, possibly a subaqueous one. Or, faulting of the troughs may have recurred recently, forming many new, steep scarps that may not yet have developed spurs and gullies. Another idea was developed by Patton (1981,1984,1985), who explained wall-slope development in the Valles Marineris entirely by dry mass-wasting processes governed by limitations in transport: on active fault scarps, whether ancient or recent, spurs and gullies developed because the scarps remained oversteepened at the base, permitting transport of material on and away from them; on inactive scarps, talus slopes developed because no material was transported away. Talus slopes, Patton suggested, are now seen at the heads of landslides; inside pits, pit chains and tributary canyons; and on the walls of narrow troughs. However, Patton's model implies that overall spur-and-gully development is an active, ongoing, and therefore more recent process than the development of talus slopes. His inference is not borne out by the observations that landslide scars having talus slopes developed at the expense of spurs and gullies, and that talus-covered fault scarps appear to have developed more recently than scarps having spur-and-gully morphology.

B. Wall Stability

The Valles Marineris walls now stand at great heights because the absence of recent pluvial and fluvial activities on Mars prevented rapid infilling of the depressions. Clow et al. (1988*b*) analyzed the mechanical stability of some walls and of failure surfaces on large landslides occurring on these walls, and they suggested possible constraints on the physical properties of wall rock in those areas. Slopes in Noctis Labyrinthus and in Ius and Tithonium Chasmata, which are covered by the most detailed topographic maps (U.S. Geological Survey 1980), show that many canyon walls are near their limit of mechanical stability, having slope angles as great as 35° and heights of 5 km or more.

A mathematical inversion of the least stable topographic profiles (Clow et al. 1988*b*), utilizing the slope-stability model of Baker and Garber (1978), indicates that the cohesion of the wall materials must be between 0.05 and

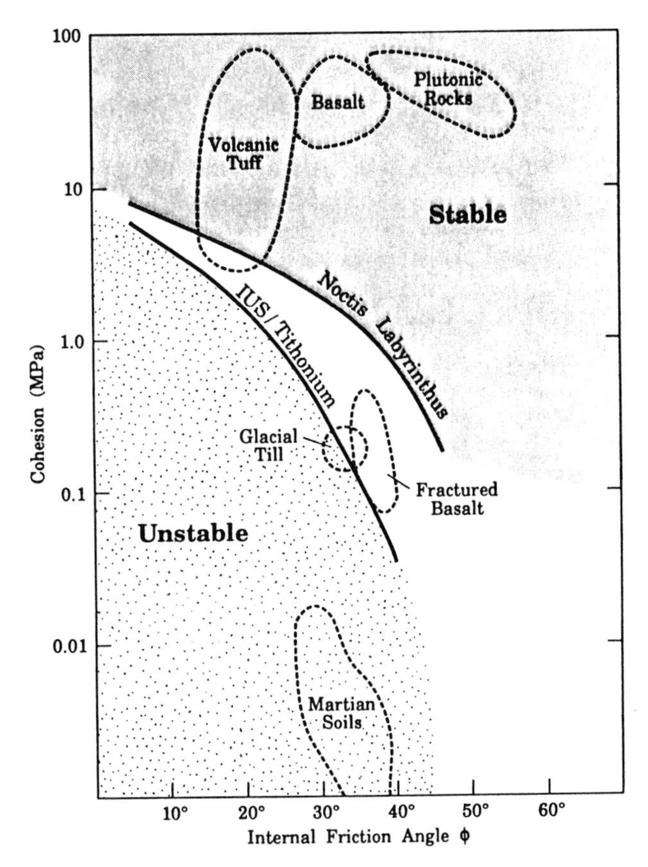

Fig. 10. Mechanical properties (cohesion C and internal friction angle ϕ) that would provide marginal long-term stability for observed canyon walls in Noctis Labyrinthus and in Ius and Tithonium Chasmata. Combinations of C and ϕ below solid lines would be mechanically unstable and thus could not support existing canyon walls. Also shown are ranges of mechanical properties measured for several terrestrial geologic materials in their dry state (Kulhawy 1975; Hoek and Bray 1977; Marachi et al. 1972). Glacial tills and weak volcanic tuffs may have mechanical properties similar to impact breccias, which are thought to underlie chasma walls. Fractured basalt may represent strength of basement rock; carbonates have approximately same strength as competent basalt. Mechanical properties of soils at Viking Lander sites are shown for comparison (Moore et al. 1987).

5.00 MPa to produce internal friction angles between 15° and 40° (Fig. 10). Carbonates, basalts, plutonic rocks and other competent geologic materials would be in many sections. However, the strength properties of possible mechanical analogs of impact breccia and highly fractured basement rock would also provide the required long-term stability of the trough walls, at least for Ius and Tithonium Chasmata; glacial till, a material having a mixed grain-size distribution including clays and silts (Fig. 10), may provide a mechanical analog for an impact breccia, whereas intensely fractured basalt may simulate highly fractured basement rock.

The failures in Valles Marineris are deep seated. For the major failure in Ius Chasma at latitude 7°S, longitude 85°, the internal friction angle must be $< 20°$. The only common terrestrial material with a friction angle this low is volcanic tuff. But impact ejecta from volatile-rich target rocks may produce rocks similar to tuffs in terms of particle sizes and degassing history (Newsom et al. 1986), which suggests that the walls may be underlain by such material.

The materials discussed above cannot maintain their high porosities under lithostatic loads in the dry state; when lithostatic stresses exceed the tensile strength of the material involved, all pore space should be eliminated at depths approaching 10 km (Clifford and Hillel 1983; Warren and Rasmussen 1987; MacKinnon and Tanaka 1989). However, if liquid water or ice were to fill the interstitial pores, the effective pressures tending to close the pores would be greatly reduced, allowing these materials to maintain their porosity to much greater depths. Yet, the long-term strength of intensely fractured basement rock or coarse-grained impact breccia should be unaffected by the presence of ice in the joints and pores, because coarse-grained materials with ice-filled pores develop normal frictional resistance under long-term stresses; the interstitial ice flows under pressure, allowing the interlocking of grains and blocks.

By contrast, when the pores of fine-grained materials such as clays and silts are filled with ice, these materials fail under long-term loads (strain rates $< 10^{-6}\ s^{-1}$) by dislocation creep rather than by brittle fracture. Also, if the ice content of a porous material is so high that the ice prevents the development of grain-to-grain frictional resistance, the material will rapidly fail under typical crustal loads. Spencer and Croft (1986) estimated that the mean lifetime of the canyon walls would be only ~ 50 yr if the ice content were as high as 50% by volume with shear stresses of 40 to 50 MPa.

The collapse of canyon walls in many places by translational and rotational failure indicates that mechanical stability has indeed been exceeded. The most likely causes of the instability are (1) oversteepening of the canyon walls by tectonic or erosional processes; (2) a reduction of cohesion through weathering or hydrothermal alteration; or (3) a reduction of shear strength through the effects of hydraulic pore pressures in the presence of an active groundwater system. Oversteepening of the walls by faulting has certainly occurred (see Sec. V).

The interior layered deposits are as steep and high in places as the outer walls, but no large slope failures or landslides are apparent (Lucchitta 1987*c;* Nedell et al. 1987). Their absence is a significant constraint on the composition and origin of these deposits, suggesting that, if the deposits are fine grained as proposed by several workers (see Sec. III.B), cementing by carbonates or other minerals is needed to explain the slope stability. An alternative is that the layered deposits consist primarily of competent volcanic flows.

C. Pits and Pit Chains

Pits, coalesced pits and pit chains occur on the plateaus dominantly in the Noctis Labyrinthus area, north of Ius Chasma, and south of Corprates Chasma in linear arrays that roughly parallel the Valles Marineris. The pits are generally rimless, smooth-walled depressions having talus-covered slopes with slope angles around 30°, but angles can be as steep as 40°. They are 1- to 10 km wide and 1 to 2 km deep (Tanaka and Golombek 1989; Davis and Golombek 1990). Some pits are nested. Pit floors, if present, are narrow and flat. The pits are circular or elliptical to irregular in outline, and locally they are elongate in the direction of structures that they truncate. Pit chains are continuous or interrupted, have an en echelon configuration in places (Schultz 1989*a,b*), and may reach hundreds of km in length. Pit chains extend from the blunt ends of several chasmata, but, with the exception of the small, irregular trough Hydrae Chasma (Tanaka and Golombek 1989), they do not continue from the plateaus onto adjacent trough floors (Blasius et al. 1977). Depressions that do occur on the trough floors are more irregular than those on the plateaus and are not aligned, so it is unlikely that they are the equivalent of pit chains.

The rectangular pattern of Noctis Labyrinthus and the alignment of chains on or parallel to shallow grabens suggest structural control (Blasius et al. 1977; Masson 1977). Sharp (1973*c*) favored an origin of pits by collapse due to magma withdrawal, but he also considered withdrawal of ice. Lucchitta (1978*a*) and Battistini (1985) preferred the idea that ice withdrew along fault planes and caused collapse, and Croft (1989*a,b*) and Spencer and Fanale (1990) thought that carbonates were dissolved along the planes. Davis and Golombek (1990) suggested that the common depths near 1 km may have been predetermined by the base of the permafrost (the interface between ice-laden regolith above and dry, water-laden, or cemented regolith below; see chapter 16). Bousquet, in Masson (1985), thought that perhaps explosive steam venting and subsequent collapse may have been responsible for the pits, whereas Tanaka and Golombek (1989) envisioned tension cracks in the subsurface into which surficial materials collapsed. Schultz (1989*a*) also proposed that the pits originated in cracks near the surface, perhaps accompanied by dike intrusion at depth. Diagnostic criteria for the age of the pits are scarce, but Schultz (1989*b*), on the basis of superposition relations, thought that locally they predate major trough development.

D. Tributary Canyons

Tributary canyons dissect the chasma walls in places. The majority occur in dense networks on the south wall of Ius Chasma (Fig. 11), but elsewhere tributary canyons are mostly solitary valleys. Single valleys and the main trunk tributaries generally trend perpendicular to the chasma walls (Davis and Golombek 1990), but networks, commonly rectilinear in plan view, trend

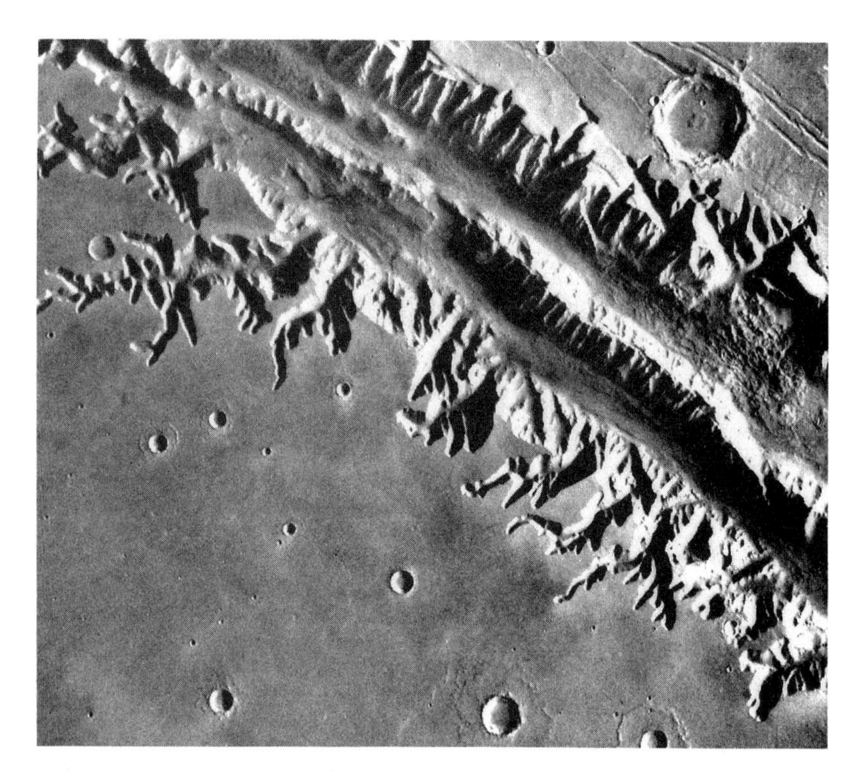

Fig. 11. Tributary canyons on south wall of Ius Chasma. Canyons have blunt heads and even gradients; most join trough floors concordantly. Canyons apparently developed along preexisting joint pattern, and their origin is commonly attributed to sapping processes. (Viking Orbiter image 645A59; ~ 300 km wide; north toward upper right.)

obliquely to the walls, a configuration that has been attributed to fault and joint control (Sharp 1973*c;* Sharp and Malin 1975; Lucchitta 1978*a;* Kochel and Capar 1982).

The morphologic characteristics of tributary canyons have been described by Blasius et al. (1977), Lucchitta (1978*a*), Kochel et al. (1985) and Davis and Golombek (1990). The canyons are 30 to over 100 km long and up to 10 km wide. Branches tend to be short, canyon segments tend to be straight but joined at high-angle bends, and interchannel, undissected areas tend to be extensive (Kochel et al. 1985). Valley heads typically are blunt, cirque-like bowls ("theaters"), and few drainage networks are seen on the plateau surfaces above them. Either the valleys are of nearly equal width throughout, or they widen and deepen from their heads to their mouths. They have V-shaped cross profiles with smooth, talus-covered walls sloping at an average angle of 28° (U.S. Geological Survey 1980; Davis and Golombek 1990). Davis and Golombek also determined that the average depth of the theater-shaped valley heads is ~ 1 km and the average gradient of the valleys on the north wall of Ius Chasma is 5°.6, almost twice the 3°.2 valley gradient of the south wall, where some valleys have gradients as low as 2°.

Tributary canyons probably formed by sapping processes acting preferentially along structures (Sharp 1973*c;* Kochel and Piper 1986). In a detailed study of the morphometry of Valles Marineris tributaries and terrestrial analogs and in model studies, Kochel et al. (1985) and Kochel and Piper (1986) concluded that the tributary canyons have many of the morphometric indicators of sapping valleys on Earth. By analogy with canyon asymmetries on the Colorado Plateau of Utah, the north-south asymmetry in distribution of tributary canyons to the Valles Marineris may be due to a regional dip to the north of aquifer layers (Sharp and Malin 1975; Laity and Malin 1985). Davis and Golombek (1990), however, suggested that the difference between canyon distribution on the south and north sides of Ius Chasma is more likely due to different tectonic styles. These workers and Soderblom and Wenner (1978) suggested that the consistent depth of ~ 1 km of valley heads and other features may indicate that the depths of the sapped aquifers coincide with the depth of permafrost in this area (Fanale 1976; Rossbacher and Judson 1981; see chapter 16).

Most of the tributary canyons have even gradients and merge with the main trough at the level of its floor. Most side branches also merge with tributary canyons accordantly, but some intersect the trunk canyon as hanging valleys (Lucchitta 1978*a*). Locally, the canyon floors have hummocky, lobate, or longitudinally streaked deposits; a small channel empties from one canyon mouth. All these observations suggest that the tributary canyons were eroded by fluids or perhaps by ice-lubricated debris such as rock glaciers (Lucchitta 1978*a*). These observations also imply that ice or water was present in the chasma walls at a depth of ~ 1 km at the time of tributary-canyon formation.

The age of the tributary canyons is questionable. Their V-shaped profiles suggest choking with talus, because terrestrial sapping valleys have U-shaped cross profiles (Sharp 1973*c;* Laity and Malin 1985; Kochel and Piper 1986); therefore, the canyons would be relict and ancient (Patton 1984). Also, tributary networks are poorly integrated, suggesting that they are immature and that their development may have been arrested by the drying up of aquifers after an early climate change (Kochel et al. 1985). Even though some tributaries are truncated by landslide scars, suggesting a pre-landslide age, the uniform gradients and accordance with the main troughs suggest processes that may have been active more recently. Witbeck et al. (1991) placed the development of tributary canyons in the Early Amazonian, predating the landslides.

E. Transition from Troughs to Channels

Even though troughs and outflow channels are connected (Sharp 1973*c;* see chapter 15), the two forms appear to be genetically distinct. Whereas the troughs are dominantly of structural origin, bounded by fault scarps (see Secs. V and VI), the outflow channels are largely erosional features (Blasius et al. 1977). In the eastern troughs, the transition from troughs to channels occurs in the eastern parts of Ganges, Capri and Eos Chasmata. Blasius et al. (1977) considered the western parts of these chasmata to be largely controlled by structures, but chaotic terrain (indicative of channel source regions), and streamlined forms appear on their floors. These observations imply that once these chasmata may have served as source areas for outflow channels, but that their major configuration is controlled by structures. That some troughs were linked to the origin of the channels is evident not only in the eastern troughs, but also in the north in Juventae and Echus Chasmata, which gave rise to Maja and Kasei Valles (Carr 1979). Juventae Chasma, like the eastern chasmata, has chaotic terrain on its floor, but Echus' smooth floor suggests that it is filled by a deposit.

Whether the remaining troughs served as source areas for the outflow channels is less certain. The troughs are most likely of tectonic origin (see below), and the role of water may have been dominant only in trough modifications. Hebes Chasma cannot have served as a source for water because it is entirely enclosed. Because the trough floors do not have consistent gradients toward the outflow channels, flow might have been impeded. Furthermore, some of the layered interior deposits, which may be lacustrine, are younger than the chaotic terrain, which formed at the heads of outflow channels (as discussed in Sec. III.B). Clearly, it is questionable whether lakes within the central troughs served as sources for the major Chryse channels.

V. STRUCTURE

The Valles Marineris troughs lie along a structural trend that is nearly radial to the center of the Tharsis Rise (McCauley et al. 1972; Hartmann 1973*c;* Sharp 1973*c;* Carr 1974*b;* Blasius et al. 1977; Masson 1977; see chapter 8). This trend is expressed in many shallow grabens on the surrounding plateaus, in the main boundary faults of the troughs, and in some subsidiary faults inside the troughs. The trend of wrinkle ridges on the plateaus is also influenced by Tharsis stress patterns and is generally perpendicular to that of the shallow grabens (Watters and Maxwell 1983,1986; Chicarro et al. 1985; Zuber and Aist 1990). The wrinkle-ridge trends parallel the blunt ends of troughs and many smaller faults and lineations inside them. However, many structures within the troughs and on the surrounding plateaus diverge from these general trends (Schultz 1989*a*). An analysis of the adjacent grabens and wrinkle ridges suggests that wrinkle ridges developed first, grabens second (Watters and Maxwell 1983,1986), and the Valles Marineris either at the same time as the grabens, or last (Schultz 1989*a,b*).

Within the troughs, linear boundary faults at the base of the walls extend for hundreds of km and locally cut lower spurs to form triangular facets (Blasius et al. 1977; Spencer 1984). Such scarps are conspicuous on the north slopes of Coprates (Fig. 12) and eastern Candor Chasmata. More common are straight wall segments whose trends line up across wide landslide reentrants; the basal break in slope along the straight segments is inferred to mark the trace of the trough-boundary faults. Examples occur in Ius, Ophir and Hebes Chasmata (Fig. 2b and c). Schultz (1989*a*) noted that the boundary faults on the south side of Coprates Chasma have an en echelon pattern. Irregular spur-and-gully scarps, such as those on the south wall of eastern Candor Chasma, appear to reflect erosional backwasting from the fault scarps. By contrast, curved wall segments may reflect structures rather than erosional backwasting, the structures having been influenced by buried craters or basin rings. An example is the curved south wall of Melas Chasma, where the chasma wall is paralleled on the north by a curved, north-facing scarp on interior layered deposits and on the south by curved grabens on the adjacent plateau (Lucchitta and Bertolini 1989; Peulvast and Costard 1989; see Fig. 2c).

Faults paralleling the boundary faults also occur inside the Valles Marineris (Schultz and Frey 1988; Schultz 1989*a*), locally resulting in subsidiary troughs parallel to the major ones (Lucchitta and Bertolini 1989). Dense fault zones occur locally on the crests of interior ridges. Other faults have diverse

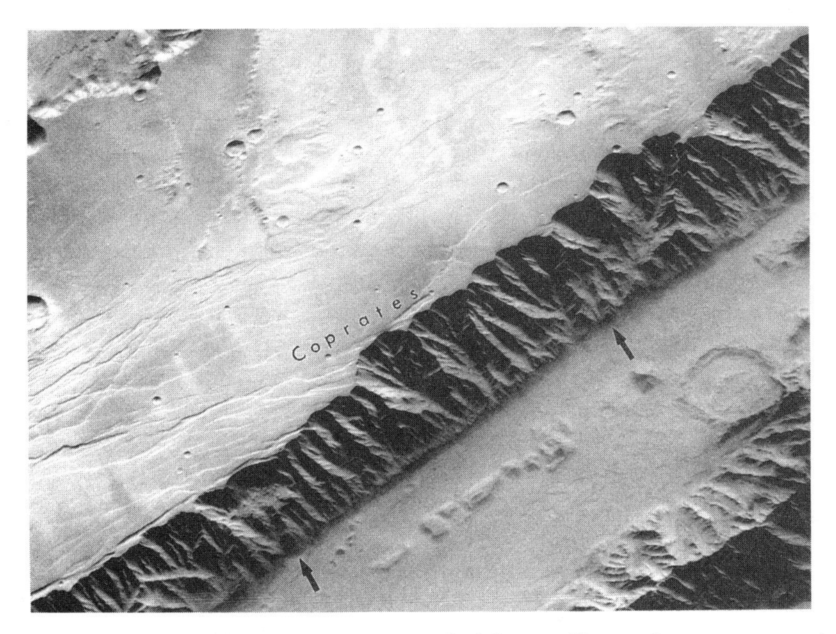

Fig. 12. Fault scarp (arrows) at base of north wall of Coprates Chasma. Scarp cuts spurs and gullies of wall and forms triangular facets. (Viking Orbiter image 608A54; ~ 250 km wide; north toward upper left.)

trends and displace trough floors and interior deposits; an example is the fault that offsets mesa surfaces in central Candor Chasma (Lucchitta 1990). Faults cutting landslide deposits attest to continued tectonic activity into relatively recent time (Blasius et al. 1977; Lucchitta 1979).

The attitude of the faults is not certain. Carr (1981) noted that some faults appear to be nearly vertical. Davis and Golombek (1990) found 60° dips from photoclinometric measurements, suggesting opposing fault scarps associated perhaps with simple grabens and a discontinuity at depth (Golombek 1979,1985). If the Valles Marineris faults are near vertical, some form of collapse may have occurred (Tanaka and Golombek 1989). Near-vertical faults, however, may also form by tectonism of vertical blocks, as envisioned for thick crusts by Milanovsky and Nikishin (1984).

Tilted beds on the surrounding plateau were reported by Blasius et al. (1977), but such beds also occur inside the Valles Marineris (Lucchitta 1987*a*). Tilting is best recognized on interior layered deposits such as those in western Ophir and eastern Candor Chasmata, where a few eroded monoclinal folds appear to be hogbacks. These folds may drape over steeply dipping faults in the subsurface. Tilted beds may also occur on some trough walls where crudely developed hogbacks appear to be present, but the absence of distinct stratification in wall rock makes identification tentative.

Structural relief can be determined by identifying some wall or plateau rock on the floors of the troughs. A recession in the south side of Ganges Chasma is underlain by downdropped plateau material, and the floors of Ganges, Capri and Eos Chasmata display small tablelands and chaotic hills probably composed of wall rock. Furthermore, a tableland in Eos Chasma is pockmarked by large craters, suggesting underlying plateau material. Accordingly, the eastern troughs appear to be structurally shallow, with relief on the order of 2 to 3 km. In the central troughs, apparent plateau rock, cut by shallow grabens, occurs on the trough floor between Candor and Melas Chasmata; other wall rock is exposed in the northern part of eastern Candor Chasma. The presence of ancient craters and a relatively high crater density on the floor of western Coprates Chasma (Schultz 1990) also suggest that plateau rock is exposed on the floor. Accordingly, structural relief of some of the central troughs is on the order of 7 to 8 km (U.S. Geological Survey 1986). Elsewhere in the central troughs, interior layered deposits, irregular floor and landslide deposits, and other alluvial or windblown materials obscure the floors, and the plateau rock appears to be faulted to depths > 8 km. The 10-km depth that may exist in the Valles Marineris, however, would not be unusual for deep terrestrial grabens.

The Valles Marineris void is apparently not compensated, as shown by a negative line-of-sight gravity acceleration (Phillips and Lambeck 1980; Tanaka et al. 1986; see chapter 7). A minor topographic rising towards the trough edge observed in several places could be caused by incipient isostatic adjustments to compensate for the void.

VI. ORIGIN

The origin of the Valles Marineris remains controversial. The major hypotheses involve origin by erosion, collapse and tectonism.

A. Erosional Origin

Removal of material by thermokarst processes or by water or wind has been implicated in creating the troughs. A thermokarst origin was advocated by McCauley et al. (1972), Sharp (1973*c*) and McCauley (1978), who envisioned the melting of ground ice along shallow, parallel grabens that widened and deepened to form individual troughs harboring lakes. McCauley (1978) proposed that headward erosion by processes such as landsliding, artesian sapping, and possibly erosion by running water enlarged each of the chasma lakes with consequent capture by other lakes. Eventually, capture resulted in breakout of enormous volumes of water through the outflow channels. However, canyons that are entirely enclosed compelled these authors also to consider other origins as well. Another problem with thermokarst is that a substantial amount of ground ice had to be present in the equatorial region of Mars to create the void. It is difficult to form that much segregated ice (Sharp 1973*a*), and it is difficult to remove the solid slag, which, if not removed by running water, had to be removed by wind. Also, slope-stability considerations (see above) prohibit the presence of massive ice in the walls. However, thermokarst was probably involved in the formation of pits and chains, tributary canyons and some of the troughs of Noctis Labyrinthus (Sharp 1973*c;* Masson 1977; Tanaka and Davis 1988; see Sec. IV).

Spencer and Croft (1986), Croft (1989*a,b*) and Spencer and Fanale (1990) have proposed that the material that was dissolved and removed from the troughs was carbonates that underlay the trough walls. They argued that rock would better support the high, steep chasma walls than would ice-charged breccia. However, because it would be difficult to remove all the trough materials by carbonate solution alone, Croft (1989*a,b*) considered that the main troughs are tectonic and that solution processes only acted along structural planes of weakness to permit erosional widening and to form pit chains and tributary canyons. Problems are that carbonate spectral features on Mars are weak, at best (Blaney and McCord 1990*b*), and that the formation of substantial carbonate deposits requires an active early hydrosphere to form the bodies of water necessary to precipitate the carbonates (chapter 32); the existence of ocean-sized bodies of water on Mars at any time is questionable.

Erosion by water from a draining lake that simultaneously formed the outflow channels (McCauley 1978; Lucchitta and Ferguson 1983) may have helped to enlarge some troughs or erode some interior layered deposits, but it did not form the chasmata. Water erosion may have occurred in Ophir, Candor and Melas Chasmata, which connect with the outflow channels, but it did not take place in Hebes Chasma, which is entirely enclosed. Also, chaotic hills and streamlined forms that are commonly associated with water erosion on Mars occur only in the troughs transitional to channels.

Origin of the troughs by deflation alone is also untenable, because all trough materials would have had to be fine grained or reduced to fine-grain sizes that could be picked up easily by the wind; this proposition is unlikely, especially when we consider that probable volcanic cap rock would also have had to be removed by this process. Furthermore, trough walls show no evidence of wind erosion.

B. Collapse Origin

A collapse origin is predicated on the idea that an apparent morphologic progression exists between individual pits to large coalesced pits that form chains to troughs, suggesting that the pits enlarged to become the major troughs (Sharp 1973*c;* Tanaka and Golombek 1989; Davis and Golombek 1990). One proposed origin to explain this progression is collapse due to withdrawal of magma in the subsurface. Transfer of magma into the region of the Tharsis Rise has been proposed by McCauley et al. (1972), Sharp (1973*c*) and Schonfeld (1979), but the hypothesis has never been analyzed in detail. Another hypothesis is the collapse of materials into voids that are created at depth by the opening of tension fractures (Tanaka and Golombek 1989). However, these authors admitted that additional processes of erosion or rifting were needed to create the voids of the major troughs.

A collapse origin for the major troughs was questioned by several authors. Masson (1985) thought that troughs having straight walls are distinct from pits or coalesced pits having scalloped walls. Furthermore, studies of the trends, morphology and morphometry of pit chains and their transition to the major troughs (Schultz 1989*a;* Lucchitta et al. 1990) showed that pit chains and troughs are different and their developments may be unrelated, and that trough-boundary faults imply orderly collapse of large tectonic blocks rather than chaotic collapse of material into subsurface voids.

C. Tectonic Origin

Blasius et al. (1977) advanced the theory that the Valles Marineris are tectonic grabens. Their position on the flanks of the Tharsis Rise and radial to its center (see chapter 8) led many investigators to conclude that they are related to a tensional stress regime surrounding the rise; the same regime also formed the shallow grabens that parallel the troughs (see Sec. V). The decrease in structural relief toward the east end of Coprates Chasma is also consistent with waning stresses toward the outer periphery of the Tharsis Rise. Many early stress models postulated that structural uplift of Tharsis caused the tensional regime (Wise et al. 1979*b;* Plescia and Saunders 1982), but subsidence due to the superimposed volcanic load was also considered (Solomon and Head 1982). More recent models propose that isostatic adjustments of Tharsis would cause circumferential tensional stresses in the western part of Valles Marineris, whereas external loading (flexing an elastic lithosphere downward) would cause such stresses near the eastern chasmata (Banerdt et al. 1982; Willeman and Turcotte 1982; Sleep and Phillips 1985; Phillips et al. 1990; see chapter 8). Thus, stress models require two distinct events, and it is not clear how the Valles Marineris could have formed in a single event. Perhaps different parts opened at different times. A possible younger age has been postulated by Tanaka and Davis (1988) for the western troughs, as well as for the Noctis Labyrinthus valleys that may have opened due to local uplift (Masson 1980) in Late Hesperian to Early Amazonian time. Different ages have also been proposed by Lucchitta and Bertolini (1989) for the formation of subsidiary troughs parallel to but within the major troughs. Local structural inhomogeneities probably also played a role by perturbing the regional stress systems and causing reactivation of older structures at different times.

Another unresolved problem is why only the Valles Marineris—not other fractures radial to Tharsis—opened to form deep troughs. The great depth suggests failure involving the entire lithosphere. An apparent limit to this depth near 10 km (Lucchitta et al. 1990) suggests a limiting stress in the region. Perhaps the stress in the trough area was relieved by a few major, deep faults, rather than by distributed shear. Perhaps the strain was localized, because the crust contained major subsurface aquifers that reduced its strength (Carr 1981). Perhaps the troughs opened into deep grabens because they lie along the crest of an elongated bulge (Christensen 1975; Blasius et al. 1977; U.S. Geological Survey 1986; Map I-2030A). Wise et al. (1979*b*) suggested that the troughs represent a "key-stone" collapse of the crest of this bulge. (However, the magnitude and significance of this bulge has recently been questioned [R. Schultz 1989*b*; R. Schultz and Frey 1989].) Perhaps the ancestral eastern troughs were concentric to the Chryse basin (P. Schultz et al. 1982; Wichman and P. Schultz 1989).

Because of the size of the grabens and their location along a possible arch, the origin of the Valles Marineris has also been ascribed to aborted planetary rifting (Hartmann 1973*c;* Blasius et al. 1977; Masson 1977; Frey 1979*a;* Schonfeld 1979). Arches are characteristic of terrestrial rifts (Baker et al. 1972; Frey 1979*a*). However, in detail the Valles Marineris differ from terrestrial plate-tectonic rifts. Most terrestrial rifts have many parallel faults in en echelon pattern that taper out along strike; beds are tilted (Frey 1979*a*). By contrast, the Valles Marineris faults are more widely spaced, trough ends are blunt and bounded by cross faults, fault planes appear to be very steep, and tilted beds are relatively few. Perhaps the difference is due to a thick crust on Mars (Milanovsky and Nikishin 1984; see chapters 5 and 6); Frey (1979*a*) estimated the Martian crust to be 50% thicker than the African rift-zone crust on Earth. Also, most terrestrial rifts are associated with voluminous volcanism, whereas the occurrence of volcanism inside the Valles Marineris (Peterson 1981; Lucchitta 1987*b*,1990; Nedell et al. 1987; Witbeck et al. 1991) is still conjectural.

VII. HISTORICAL SUMMARY

The Valles Marineris troughs were formed as tectonic grabens responding to stresses associated with the Tharsis Rise and possibly to additional stresses from the formation of a local bulge or incipient but aborted planetary rifting. The Valles Marineris opening was probably preceded by the formation of wrinkle ridges perpendicular to the trend of the troughs, and it coincided with the formation of parallel shallow grabens, both wrinkle ridges and grabens having formed in response to Tharsis stresses (Banerdt et al. 1982; Watters and Maxwell 1983,1986; see chapter 8). The troughs disrupted the Lunae and Syria Planum plateaus of Early Hesperian age (Scott and Tanaka 1986; Tanaka 1986). These plateaus were probably capped by lavas, and the troughs must have opened after the lavas were emplaced. Accordingly, the Valles Marineris are younger than Early Hesperian (Fig. 13). On the other hand, the Noctis Labyrinthus troughs, disrupting Late Hesperian or Early Amazonian lavas (Tanaka and Davis 1988), may be younger. After the opening of the troughs, the walls were eroded into spurs and gullies, locally widening the troughs to 3 times the width of the original structural grabens (Schultz 1989*b*). At the same time or later, interior layered deposits were emplaced, embaying the walls and locally overtopping erosional remnants of wall rock. The layered deposits may have been emplaced in lakes as sediments or volcanic rocks, with contributions of mass-wasted or aeolian materials. Deposition of interior layered deposits may have extended through Late Hesperian and perhaps even into Early Amazonian time. This interval also saw the formation of the Simud and Tiu outflow channels farther east and the Maja and Kasei Valles farther north (Masursky et al. 1977; Scott and Tanaka 1986). This general coincidence in time supports a genetic link among the Valles Marineris, the layered deposits and these channels, even though the detailed relations remain problematic (see Sec. III.B). After the troughs were filled nearly to their rims by layered deposits, unknown processes removed parts of these deposits, leaving many as isolated tablelands surrounded by newly formed depressions, the so-called moats. The removal process may have been erosional, but structural disruption is a likely alternative. In the latter case, interior deposits may have filled only older, partial troughs, and the moats now occur where later faulting widened the original depressions.

Formation of the moats heightened the relief of the trough walls, and landslides fell into the voids. These landslides are the only material inside the Valles Marineris dated by crater density; they were emplaced in the last third of Valles Marineris history, in Middle to Late Amazonian time. According to the relative- to absolute-age conversion scale of Neukum and Hiller (1981), the landslides occurred $\sim$ 1 Gyr ago, but some may be recent.

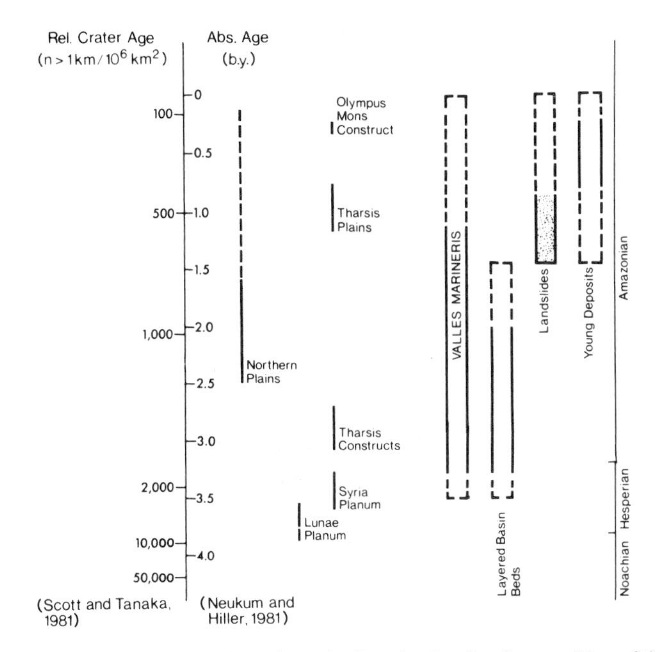

Fig. 13. Ages of Valles Marineris deposits and selected units elsewhere on Mars. Stippled area in landslide column denotes crater-dated time span. Dashed lines indicate uncertainty in age assignment. Young deposits include irregular floor deposits and possible colluvial, alluvial and aeolian materials.

After some of the landslides formed, a new unit, several km thick in places, was deposited on the Valles Marineris floors. Some dark patches associated with this unit may be mafic volcanic material; the origin of the remainder is unknown, but it may also be volcanic. Low areas of the chasmata were filled with possible colluvium from mass-wasted material and alluvium from fluids drained from late-stage landslides. The latest events were fallout of light-colored atmospheric dust and transport of dark-colored saltating material that locally formed dark dune fields on the trough floors. Tectonism apparently occurred throughout the history of the Valles Marineris. If some of the interior deposits are of volcanic origin, volcanism may have been similarly long lasting.

VIII. CONCLUSIONS

Major questions remain. Even though we understand the first-order sequence of events that formed the Valles Marineris and the morphology of the interior deposits, we still do not fully understand the mechanism responsible for the formation of the troughs. A critical problem is to determine how the Tharsis Rise stress patterns affected the evolution of the troughs. Another problem is the lack of information on the detailed elevations and dimensions of the arch or topographic bulge associated with the Valles Marineris. Furthermore, we know little of the composition of the materials, about the processes that formed the interior deposits, and the extent to which water and ice were involved in the original formation of the troughs and in their later modification. The role of ice in recent morphologic modifications is important, because it bears on the amount of ground ice present on Mars and on the question of whether the equatorial area is now desiccated. We do not know if lakes existed within the Valles Marineris or whether such lakes, if present, gave rise to the outflow channels. The recognition and investigation of paleolakes would enable study of climate change and would significantly enhance the prospects of our discovery of ancient life on Mars.

The role of volcanism is another problem. Deep rifts on Earth have usually been accompanied by volcanism. If no volcanic rocks exist inside the Valles Marineris, this anomaly needs explanation. On the other hand, if the interior deposits are largely volcanic, they probably incorporate ashflow tuffs. If the tuffs are silicic, this discovery would alter our perception of volcanism on Mars. However, they may be basaltic pyroclastic rocks: the low atmospheric pressure on Mars favors pyroclastic eruptions and potentially leads to more widespread deposits than on Earth (Wilson and Head 1983; chapter 13).

The Valles Marineris offer a potential surface-exploration site of unusual interest. Although the lack of extensive surfaces that can be dated by crater statistics is a detriment, other factors outweigh this disadvantage. The troughs offer an opportunity to sample (1) the walls, composed of very old rocks from deep within the Martian crust; (2) the interior layered deposits, formed of intermediate-age rocks that may be lake deposits; and (3) the irregular floor deposits that may be very young, possibly volcanic. Perhaps some of the layers are carbonates, giving us the opportunity to study past climates and possibly to discover fossil life. The role of volatiles in the wall rock, interior deposits, and landslides could also be evaluated. The Valles Marineris would offer a scientifically rewarding and visually spectacular backdrop for a manned landing. The Earth-like complexities of the local geology would be most suitable for investigation by humans.

Acknowledgment. We are grateful for thorough reviews by J. R. Zimbelman and T. R. Watters that significantly improved the manuscript. The research was funded by the Planetary Geology and Geophysics Program of the National Aeronautics and Space Administration.

15. CHANNELS AND VALLEY NETWORKS

VICTOR R. BAKER
University of Arizona

MICHAEL H. CARR
U.S. Geological Survey

VIRGINIA C. GULICK, CAMERON R. WILLIAMS
University of Arizona

and

MARK S. MARLEY
NASA/Ames Research Center

Channels, valleys, and related features of aqueous origin on Mars are of profound importance in comparative planetology. Martian outflow channels formed by large-scale fluid outflow from subsurface sources. Elements of cataclysmic flooding, debris and ice flowage, as modified by volcanism and wind action, seem best to explain observed morphologies. Martian valley networks show their greatest similarity to terrestrial networks formed by sapping as water emanated from seeps and springs. Exogenetic sources of recharge (precipitation) or endogenetic cycling (hydrothermal systems) would be required to achieve the indicated extensive valley development during the heavy bombardment phase of Martian history. Recent studies of valley network development associated with volcanism suggest that endogenetic cycling is likely and that it may not be necessary to invoke extensive epochs of warm, wet atmospheric conditions to explain valley network formation on Mars.

I. INTRODUCTION

Mars is the only planet besides Earth known to manifest the dynamic workings of a hydrological cycle. The channels and valleys of Mars are a bold testament to the past operation of that cycle and its profound implications for environmental change on that planet. However, to understand environmental change on Mars, one must have an exact understanding of the genesis of channels and valleys. The importance of this problem traces back to James Hutton, the acknowledged founder of geology, who ascribed the origin of valleys on Earth to the prolonged denudational action of rivers (Hutton 1788,1795).

To ascertain the origin of Martian channels and valleys, genesis is first inferred from detailed study of landforms imaged by remote sensing devices on spacecraft. Hypotheses are generated by the reasoning process of abduction (Engelhardt and Zimmerman 1988), whereby local details of morphology are used to infer probable mechanical processes, geologic controls, and ultimately the exogenetic and endogenetic environments of formation. This is a nontrivial exercise involving a type of reasoning (Gilbert 1886) that lies at the heart of geology. Application of the methodology in planetary studies is discussed by Mutch (1979) and Baker (1984,1985).

Although most generally recognized from imagery of the 1972 Mariner 9 mission (Masursky 1973; McCauley et al. 1972; Milton 1973), the fluvial forms on Mars were also inferred from studies of Mariner 6 and 7 images showing signs of crater dissection by valleys (Schultz and Ingerson 1973). Because of the incompatibility of the implied liquid water with the modern Martian environment, it was quickly hypothesized that the forms must be relict from an ancient warm, dense atmosphere (Sagan et al. 1973*a;* Sharp and Malin 1975; Mutch et al. 1976*c*). Indeed, the valley networks in particular have become a principal element of evidence that the Martian atmosphere evolved from an early volatile-rich state to its present condition (Walker 1978; Pollack 1979; Cess et al. 1980; Squyres 1984; Kahn 1985; Carr 1987; Pollack et al. 1987; see chapter 32). Because of the importance of this inference, the present review will examine the evidence in some detail.

II. OUTFLOW CHANNELS

Martian outflow channels are large-scale complexes of fluid-eroded troughs. The flows which formed these channels appear to have emanated from discrete collapse zones known as chaotic terrain (Fig. 1). The channels are immense by terrestrial standards, as much as 100 km wide and 2000 km in length. Gradients of channel floors range from nearly zero to 2.5 m km^{-1}. Many reaches of outflow channels appear to include closed depressions. Morphological attributes of the channels are extensively discussed by Baker (1982), whose conclusions are briefly reviewed and updated here.

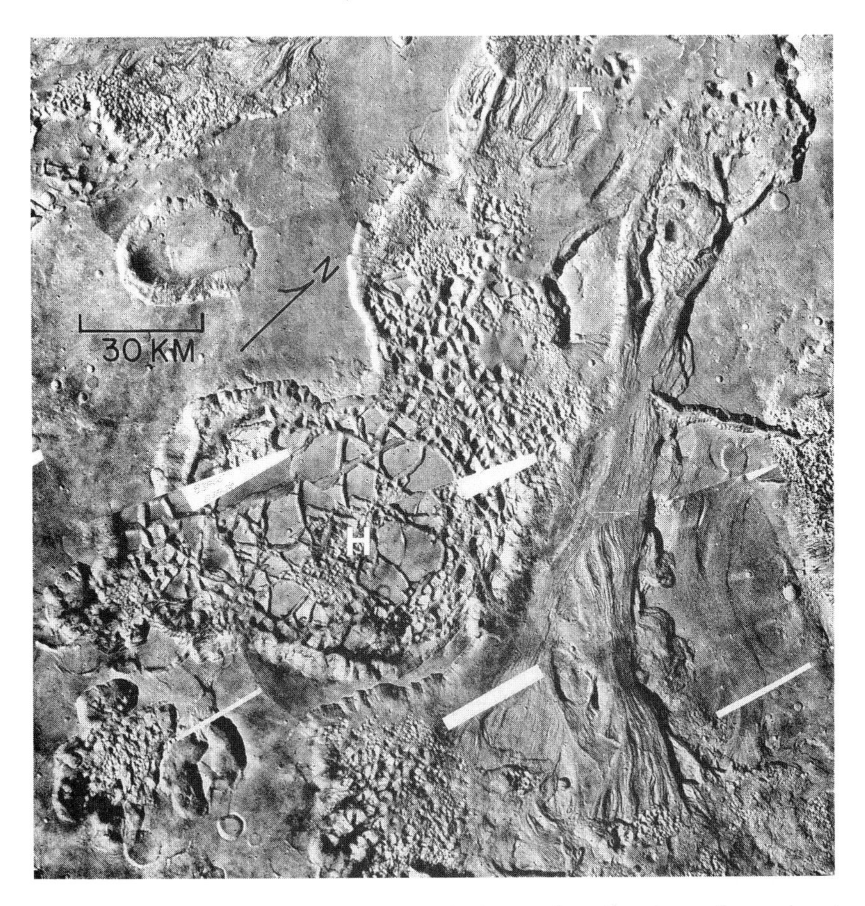

Fig. 1. Chaotic terrain at Hydaspis Chaos (H). The large outflow channel extending northward from the chaos zone is Tiu Vallis (T). (The images are a portion of JPL Viking Orbiter Mosaic 211–5556.)

Macroscale Features

The major elements of outflow-channel patterns are the large-scale flow features (Table I). A regional anastomosing pattern is noted for many of the outflow channels including Ares, Simud, Tiu (Baker and Milton 1974), Kasei and Maja Valles (Baker and Kochel 1979). Other examples of anastomosing or braiding of outflow channels are described by Carr (1974*a*,1979), Mars Channel Working Group (1983), Komar (1980), Sharp and Malin (1975) and Hartmann (1974).

The issuance of channels from discrete source regions is a feature common to almost all outflow channels. These discrete source regions can take the form of fractures or elongate depressions (Schultz et al. 1979; Wise 1984;

TABLE I

Macro- and Mesoscale Features of Outflow-Channel Morphology and their Consistency (x) or Inconsistency (−) with Hypothesized Genetic Mechanisms[a]

MACROSCALE Channel Pattern	MESOSCALE Bedforms
1. Anastomosing pattern of channels	1. Longitudinal grooves
2. Discrete source areas, such as chaotic terrain	2. Inner channels
3. Residual uplands separating channels	3. Recessional headcuts (cataracts)
4. Many uplands partly smoothed and streamlined by fluid flows, especially on their upstream ends	4. Scabland
5. Pronounced flow expansions and constrictions	5. Scour marks near flow obstacles
6. Distinct upper elevational limit to eroded terrain	6. Pendant forms (small-scale streamlined hills, which may be either residual or depositional)
7. Transected divides and hanging valleys	7. Expansion bar complexes, or fan deltas
8. Erosion of diverse rock types thousands of kilometers from probable fluid source areas	
9. Low sinuosity	
10. High width-depth ratio	
11. Differential erosion of terrain controlled by structure and lithology	
12. Indistinct channel terminus, including the lack of obvious large-scale fans and deltas	

Morphological Features	Wind	Mud Flow	Glacier	Lava	Catastrophic Flood
Anastomosis	?	X	X	X	X
Streamlined uplands	X	X	X	?	X
Longitudinal grooves	X	X	X	?	X
Scour marks	X	?	?	?	X
Scabland	?	—	?	—	X
Inner channels	?	X	?	X	X
Lack of solidified fluid at channel mouth	X	—	X	—	X
Localized source region	?	X	X	X	X
Flow for thousands of kilometers	X	—	X	X	X
Bar-like bedforms	?	?	?	?	X
Pronounced upper limit to fluid erosion	—	X	X	X	X
Consistent downhill fluid flow	?	X	X	X	X
Sinuous channels	?	X	X	X	X
High width-depth ratio	X	X	X	—	X
Headcuts	—	?	X	—	X

[a]Table modified from Baker (1982).

Christiansen 1985; Carr 1974*a*, 1979) or they can be areas of jumbled blocks on the floors of large vaguely circular depressions termed chaotic terrain (Sharp 1973*a*). Chaotic terrain consists of slump and collapse blocks in steep-walled arcuate depressions at channel heads. It is generally assumed that removal of fluid from below the surface causes loss of support, and the material collapses under its own weight, leaving irregular blocks of varying sizes on the depression floor.

Residual uplands separating channels and partially to fully streamlined upland remnants are common in outflow channels (Fig. 2). These residual uplands take two forms: one roughly similar to a single lemniscate loop, usually wider toward the upstream end and tapering downstream narrowing to a point (Baker and Kochel 1978*b*). The other common form is rhombic or diamond shaped. Other macroscale features that have been described include expansions and constrictions to flow (Baker and Milton 1974; Baker and Kochel 1978*a*, 1979; Baker 1982), upper elevational limits on eroded terrain,

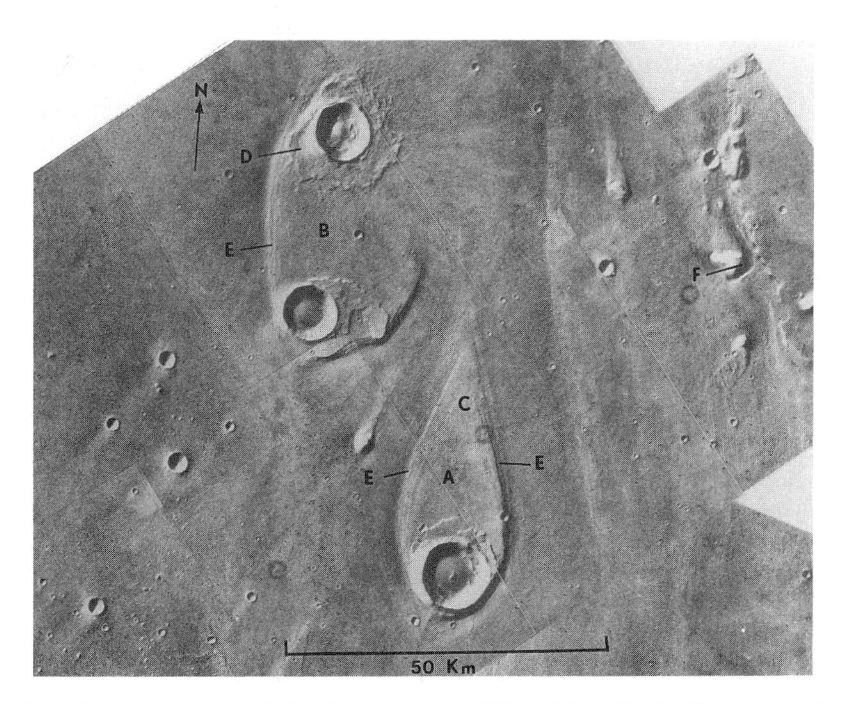

Fig. 2. Streamlined uplands (A,B) at the mouth of Ares Vallis in Chryse Planitia. Note the tapered downstream tail (C) and the erosion of ejecta blankets from craters at upstream (southern) ends of the uplands. An uneroded crater can be compared at (D). The erosive fluid flowed north in massive quantities, scouring bedrock as shown by the terrace-like benches on the upland flanks (E). A prominent crescent-shaped scour hole occurs at the upstream end of the small island at (F). The scarp surrounding the lemniscate upland (A) is about 600-m high, and that surrounding island (B) is about 400-m high. (JPL Viking Orbiter Mosaic 211–4985.)

transected divides, hanging valleys, low sinuosities, and high width-to-depth ratios (Baker and Milton 1974; Baker 1982; Baker and Kochel 1979; Mars Channel Working Group 1983).

Mesoscale Features

A variety of bedforms are located on the floors of the outflow channels (Table I). Inner channels with either residual or recessional headcuts, scour marks around presumed flow obstacles, and longitudinal grooves (Baker and Milton 1974; Baker 1979,1982; Baker and Kochel 1978*a*,1979; Carr 1979, 1986; Carr and Clow 1981; Komar 1983,1984; Lucchitta 1982*a;* Lucchitta et al. 1981; Mars Channel Working Group 1983; Cutts and Blasius 1981; Thompson 1979) which parallel the presumed flow direction are particularly common on the floors of the larger outflow channels (Baker 1978*a*,1982; Baker and Kochel 1978*a*,1979). These grooves converge in areas of channel constriction, and diverge around flow obstacles and in areas of channel expansion (Fig. 3).

Small streamlined features or pendant forms (Baker 1982) are present in many channels. From their morphology, it is impossible to determine if they are residuals similar to the larger streamlined hills, or if they are some type of depositional bar (Baker 1982). Other bedforms, such as expansion-bar complexes and fan deltas (Baker 1982; Baker and Kochel 1978*a*,1979) are probably depositional in nature. The expansion bars appear to have been generated in areas where the flow diverged and the velocity decreased.

On the floor of some channels, irregular bright patches occur that are similar to the scabland topography described by Bretz et al. (1956). Scabland results from erosional plucking wherein pieces of surface material are stripped away due to high-velocity turbulent fluid flow (Baker 1982).

Alternative Hypotheses for Channel Genesis

The morphology and scale of the outflow channels clearly indicates genesis by fluid flows of immense magnitude. Several fluids have been suggested, including liquid hydrocarbons (Yung and Pinto 1978), lavas (Carr 1974*a;* Schonfeld 1976), glaciers or ice streams (Lucchitta 1982*a;* Lucchitta et al. 1981; Lucchitta and Ferguson 1983), winds (Cutts and Blasius 1981), debris flows or mud flows (Nummedal 1978; Nummedal and Prior 1981; Thompson 1979), or cataclysmic water floods (Baker and Milton 1974; Baker 1978*a*,*b*,1979,1982; Carr 1979; Mars Channel Working Group 1983; Komar 1979,1980). Each of these mechanisms can produce some of the observed morphological features associated with the channels (see Baker 1982, Table I), but the hypothesis which most parsimoniously accounts for the entire suite of features is cataclysmic flooding.

Lava erosion was originally proposed as a mechanism of outflow channel genesis because of channel occurrences in volcanically produced terrains and apparent near contemporaneity of channelization and lava emplacement in

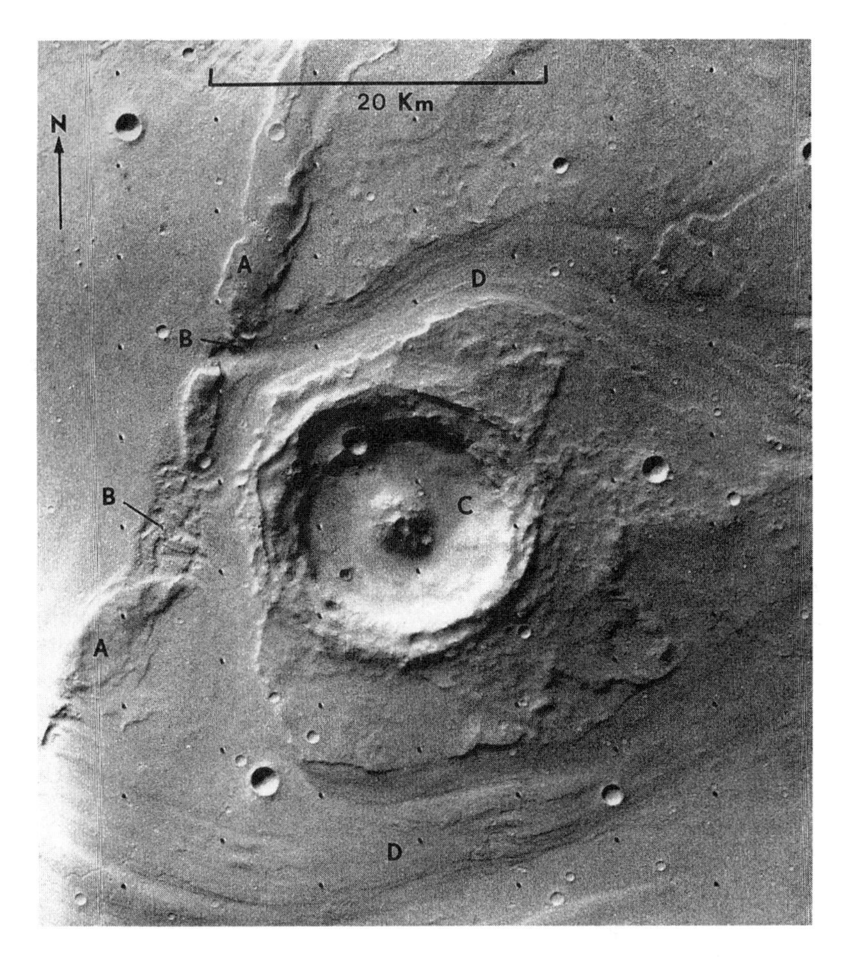

Fig. 3. A portion of Maja Vallis in Chryse Planitia. Fluid flow followed the topographic gradient sloping to the east (right on the photo). Fluid ponded upstream (west) of the mare-like ridge (A) and spilled through gaps (B) as it overflowed low points in the ridge. It then spilled around the crater (C), scouring channels and grooves (D) into the ejecta blanket. (Viking Orbiter frame 20A62.)

some regions. Typical terrestrial lava flows do not produce observed channel bedforms, but low-viscosity flows (Carr 1974*a;* Schonfeld 1976) could move turbulently and erode into underlying and surrounding material by heating it and entraining it in the flow. In theory, low-viscosity lava flows might mimic fluvially produced features. The most obvious advantages of this model include the abundance of volcanic features on Mars, the stability of lava in the current Martian environment, and local channel occurrences in volcanic terrains and even on the flanks of volcanic constructs. Disadvantages of this model include: (1) the lack of large lava deposits at channel terminations; (2)

the fact that lava channels usually have distributaries at their distal reaches which are lacking in outflow channels; and (3) regional anastomosis of the channels is not seen in terrestrial lava channels, since flowing lava, once channelized, tends not to move laterally.

Another proposal is that winds carrying saltating grains may have produced the outflow channels (Cutts and Blasius 1981). Advantages of this aeolian hypothesis include documented aeolian activity in the current environment and the lack of sediment source/sink problems (Nummedal et al. 1983; Baker 1982; Cutts and Blasius 1981; Murty et al. 1984). The source for saltating grains could presumably have been ubiquitous, although at the two Viking Lander sites there was a marked absence of grains in the proper range for saltation (Nummedal et al. 1983). The sinks for the wind-blown sediments could be the great polar ergs, or the material could have been widely dispersed by the transporting winds. Streamlined hills could be analogous to terrestrial yardangs, and longitudinal grooving similar to that within the channels is observed on Earth in desert regions like the Sahara. However, there are many problems with this model. Winds on Earth do not cut channels, and it is difficult to envision how Martian winds could be localized enough to cut a channel and yet leave the surrounding terrain unmodified (Nummedal et al. 1983; Murty et al. 1984). Also, winds do not have to follow topographic gradients, yet the outflow channels all indicate flow directions down topographic gradient. (Baker 1978*a,b,*1982; Lucchitta and Ferguson 1983). Finally, features like trim lines and indicated fluid spillage into craters document that the eroding agent had a free upper surface (Baker and Kochel 1978*a,*1979), which is difficult to reconcile with an aeolian origin (Murty et al. 1984; Baker 1982; Nummedal et al. 1983). Although wind is highly unlikely as an originator of outflow channels, secondary modifications to already existing channel features by aeolian processes are much more tenable (Lucchitta 1982*a;* Baker 1982).

A fluid which does have a free upper boundary and may be locally stable under current Martian conditions is ice. Lucchitta and Anderson (1980) proposed that the outflow channels were produced via glacier or ice-stream erosion, and subsequent papers extend this model (Lucchitta 1982*a;* Lucchitta and Ferguson 1983; Lucchitta et al. 1981). These authors demonstrated that glaciers may be stable and move on the surface of Mars, but flow under current conditions might be extremely slow. Additional effects, like the addition of brine (Lucchitta et al. 1981), frictional heating due to travel down a steep topographic gradient (Lucchitta 1982*a*) or subglacial water lubrication (Lucchitta 1982*a;* Lucchitta et al. 1981) are required for the increased-glacial-mobility hypothesis. A change in the Martian climate could also change the velocity at which glaciers could travel down the channels. The advantages of the glacial/ice-stream model are the surface stability of ice for present Martian conditions, the similar scales of terrestrial glacial features and Martian channel features, and similarities of channel bedforms and glacial features. The latter include similarities among streamlined hills and drumlins or Antarctic subglacial forms, longitudinal grooving similar to that seen in areas of Canada, anastomosing patterns, scouring, and U-shaped channel cross sections (Lucchitta 1982*a;* Lucchitta et al. 1981). The difficulties with an Earth-like glacial model include the slow movement of hypothesized glaciers on Mars given current conditions, a lack of cirques at the heads of the channels, the lack of glacial deposits at channel ends, and the difficulty of getting precipitation to produce the glacier on the surface. Lucchitta et al. (1981) and Lucchitta (1982*a*) suggested that the glaciers could have been grown from subsurface water feeding up onto the surface, and that this might also produce the chaotic terrain at the channel heads. Although not strictly glaciers by definition, such seepage-fed ice masses might, in theory, move downgradient and be intimately associated with impounded and released water flows (floods).

Nummedal (1978) and Nummedal and Prior (1981) proposed that mud flows or debris flows could have cut the Martian outflow channels, suggesting that the high-fluid content in Martian surface rocks may lead to induced liquefaction (loss of cohesion) via shock or strain-induced pore-pressure increase followed by rapid flow, as observed terrestrially in Scandinavian quick clays. The chaotic terrain is seen as the equivalent to quick-clay collapse depressions (Nummedal 1978; Nummedal and Prior 1981). Grooving is observed in areas where quick clays flow (Thompson 1979). An interesting related model was proposed (Komar 1979,1980) suggesting an analogy between outflow channels and features observed in submarine turbidity currents and debris flows due to the similar effects of Mars' lesser surface gravity and buoyancy in terrestrial submarine settings. Erosion by turbidity currents on Earth produces features of similar scale to outflow channels as well as many similar bedforms. Debris flows begin their movement as laminar flows, but increase in turbulence as they travel downchannel and lose transported material, producing features consistent with regional anastomosis and partial to fully streamlined residuals downchannel from the sources. In debris flows and mud flows, longitudinal roller vortices can become stable and could produce longitudinal grooves consistent with those observed in channels. The problems with this mechanism are the small size of source regions compared to the amount of mud or debris required to have eroded hundreds to thousands of kilometers, and the apparent lack of vast debris deposits (Baker 1982). Moreover, the high turbulence suggested by scabland-stripped zones, scouring around flow obstacles, and recessional inner channel headcuts are difficult to reconcile with flows saturated or supersaturated by debris and mud (Baker 1979,1982).

The Cataclysmic Flood Model

Baker and Milton (1974) pointed out the many similarities between the Martian outflow channels and the Channeled Scabland in Washington State formed by breakout flooding from prehistoric Lake Missoula. By analogy they proposed that the outflow channels were the products of catastrophic floods of immense proportions. Numerous other workers have since analyzed and expanded upon this hypothesis. Analytic studies have been made of erosion and hydraulics for channelized flows, chaotic terrain, longitudinal grooving, scouring and streamlining of residual terrain. Highly turbulent catastrophic flows are extremely effective erosion agents (Baker 1982). The entire suite of outflow-channel landforms is also observed in the Channeled Scabland, and the scale of the features, although not equal, is certainly closer than that of usual terrestrial fluvial features. Catastrophic floods are characterized by high-velocity, high-density, low-viscosity water flows with large discharges (Baker 1973,1982; Baker and Komar 1987). Usually, catastrophic floods produce channels that are not very sinuous, with high width-to-depth ratios. Features that are produced by catastrophic floods include anastomosis, streamlined remnants, longitudinal grooving, inner channels with recessional headcuts, scouring around flow obstacles, scabland-plucked erosional scars, and many other features similar to those observed in the Martian outflow channels. Komar (1983,1984) argues that some streamlining within the outflow channels could only be produced by water flow.

The advantage of the catastrophic flood model is that all of the Martian channel forms are also seen in analogous terrestrial catastrophic flood regions (Baker and Milton 1974; Baker 1982; Komar 1979). The major difficulties associated with this model include:

1. The general instability of water on Mars' surface under current climatic conditions;
2. The lack of obvious fluvial depositional areas at channel termini;
3. The source areas seem to be too small to account for the amount of water required to fill and erode the outflow channels.

In relation to difficulty (3), Baker (1982) proposed that the chaotic terrain zones probably represent only the final stages of progressive channel growth by headward growth as more and more terrain collapses. Each subsequent collapse releases new water which erodes the downchannel chaotic terrain from an earlier phase of the flood, so that evidence for the early sources of the channel water are obliterated by subsequent outflows down the same channel. Associated ice and debris must have also contributed to the genesis of the evolving channel system. Low-sinuosity channels headed by chaotic-collapse terrain have been modeled by Manker and Johnson (1982) who constructed scaled-down physical models of Martian near-surface materials infused with ground ice. When tilted and heated from below, the models yielded collapse-headed channels of low sinuosity similar to observed chaotic terrain.

Carr (1979) proposed a mechanism for outflow in which confined ground water is released when an aquifer is breached by some rupturing or fracturing event. Masursky et al. (1986*b*) proposed that the water stored as ground ice in Martian regolith might be melted by localized heating due to intrusive magma emplacement and may escape catastrophically creating the channels. Because release processes occurred in the Martian subsurface, these and other mechanisms (Clark 1978; Milton 1974*a;* Peale et al. 1975) are very difficult to confirm with present information.

The problem of maintaining surface-water flows on Mars under present conditions does not seem particularly serious for short-duration floods. Even prolonged flows could be maintained as ice-covered rivers (Lingenfelter et al. 1968; Wallace and Sagan 1979) or seepage flows (Carr 1983). Freezing-point depressants could be present (Ingersoll 1970; Brass 1980). Moreover, the low atmospheric pressure would actually facilitate the effective flood erosional process of cavitation (Baker 1979,1982; Baker and Cotta 1987). Cavitation occurs when dynamic-pressure variations in a flow produce vapor bubbles whose subsequent collapse shatters channel-bed materials which are then removed by the flow (Baker 1979). Scabland etching is probably produced by cavitation and the plucking action of kolks and is controlled by inherent bedrock structure (Baker 1979).

Komar (1979) examined Martian outflow-channel hydraulics by comparison to similar scaled terrestrial features, concluding that outflow channels and the Channeled Scablands were hydraulically similar to terrestrial turbidity currents in terms of scale, velocity, discharge, stress and sediment transport capacity. Blocks over one meter in diameter may be carried in suspension in these flows, demonstrating their enormous erosion potentials. Large amounts of sediment can be transported as wash load in the Martian floods (Komar 1980) allowing very rapid erosion. The paucity of depositional features near outflow-channel mouths and the great extent of channels compared to source-region sizes may both be as a result of extensive wash-load transport (Komar 1980).

Longitudinal grooving seen in Martian outflow channels was analyzed by Thompson (1979) and Baker (1979) who conclude that it was most likely produced by longitudinal roller vortices which are generated in high-velocity flows (Baker 1979) or flows with vertically stratified viscosities (Thompson 1979). Scouring around flow obstacles was studied by Komar (1985) in flume experiments. Scour marks produced around lemniscate-shaped flow obstacles in the flume compare favorably to scour marks observed in outflow channels on Mars. Komar (1985) indicates that flow strengths easily exceed those required to entrain channel sediments.

Cataclysmic flooding also seems consistent with the quantitative details of observed streamlined forms. Flow obstacles are modified into forms which are the least resistant to the flow by a progressive process of erosion and deposition into lemniscate (airfoil) shapes. Minimization of total drag around an obstacle is achieved by a compromise between drag due to an obstacle's form and the friction over its surface (Baker 1979). Komar (1983,1984) examined streamlining of various features such as glacial drumlins, aeolian yardangs and fluvial tear-drop islands. In general, drumlins are more elliptical than riverine lemniscates, whereas yardangs may have much greater length-to-width ratios (although these range widely) compared to those produced fluvially. Comparisons between Channeled Scabland and lemniscate forms in Martian outflow channels show good agreement, suggesting that streamlined residuals within Martian outflow channels were produced by a high-velocity, very turbulent water flow.

Because of the difficulties in explaining all the complexities of outflow channels with a single, Earth-based analogy model, it may be that the unique Martian environment modified flow processes. Conditions of enhanced wash-load sediment transport (Komar 1980) plus ice formation, with ice flowage (Lucchitta 1982*a*), may have acted in complex combination. These events could even have produced interlayered ice-rich debris flows in troughs subsequently invaded by lava and modified by wind deflation.

Fate of Flood Water and Sediment

The large channels around Chryse and northwest of Elysium debouch onto the low-lying northern plains, and all traces of the channels are lost between latitudes 45°N to 65°N. The plains in these areas display a variety of distinctive features that have been attributed to the presence of ground ice (Carr and Schaber 1977; Rossbacher and Judson 1981; Lucchitta 1981). The most pervasive and striking characteristics are a widespread polygonal pattern of fractures and a mottled appearance caused by bright crater ejecta superimposed on a dark surface. McGill (1985) and Lucchitta et. al.(1986) suggest that the distinctive features of these areas are caused by the presence of sedimentary deposits from the large floods. Parker et al. (1989) describe the regional extent of northern plains features that might reflect the accumulated ponding of outflow sediment and water discharges.

It seems clear that the floodwaters that cut the outflow channels must have pooled in low areas at the ends of the channels. If, when the floods occurred, climatic conditions were the same as at present, then the pooled water would have immediately frozen over to form an ice-covered lake. Sediment would have settled out and freezing would have continued until an ice deposit was left over frozen sediments. The amount of water that would have sublimed into the atmosphere during the flood and while the terminal lake was freezing was probably trivial compared to the size of the flood (Carr 1983). Recent modeling (Carr 1990) indicates that the ultimate fate of the water from these high-latitude ice deposits would depend largely on whether the deposit became covered with debris. If the ice were continually swept free of debris then the ice would slowly sublime into the atmosphere, and the water would be trapped at the poles. On the other hand, if the ice became covered with a few meters of material, then at these high latitudes the ice would be permanently stable and the ice deposit would remain for the life of the planet (Farmer and Doms 1979; Zent et al. 1986). The recurrence of dust storms, the likelihood of a sediment-lag accumulation on the ice, and photogeologic evidence of the accumulation of debris at high latitudes (Soderblom et al. 1974), all suggest that the ice deposits would have quickly become covered with debris and stabilized. A similar reasoning applies to the deposits at the ends of the large channels in Hellas. In contrast, the terminal ice deposits from those channels, such as Mangala Vallis, that end at low latitudes would have been permanently unstable. The ice from these deposits would have continued to sublime into the atmosphere until all the water had been lost.

III. MARTIAN VALLEYS

Martian valleys are distinguished from channels by the absence of bedforms which are direct indicators of fluid flow (Mars Channel Working Group 1983). Although the valleys may contain channels, only in rare instances can the latter be detected with the resolution of the existing imagery. Most (perhaps 98%) of the valleys are located in the old cratered-terrain regions. Younger valleys are located on the south wall of Ius Chasma and on the flanks of some Martian volcanoes. Valleys are concentrated in the 65°S to 65°N latitude belt, and decrease in number toward the high latitudes. Because most of the valley networks are located only in the oldest terrains, the valley networks themselves are thought to be old (Pieri 1976; Carr and Clow 1981), with formation ceasing just after the end of heavy bombardment, approximately 3.8 to 3.9 Gyr. Alternative explanations for the almost exclusive formation of valleys in the older terrains are that these regions were more easily eroded by valleys and that water was preferentially located in the ancient terrains (Carr 1984). A notable group of young Martian valleys is located on Alba Patera (Gulick and Baker 1989, 1990). These valleys formed well after the period of heavy bombardment and thus have important paleoclimatic implications for Mars. Many origins which are similar to those proposed for the outflow channels have also been considered for the formation of the Martian valleys, including wind, water, lava and volcanic density currents. As in the case of the outflow channels, erosion by running water is considered to be the primary mechanism responsible for the formation of valley networks (Mars Channel Working Group 1983).

General Morphology

The Martian valleys have widths ranging from < 1 km to nearly 10 km and lengths ranging from < 5 km to nearly 1000 km (Mars Channel Working Group 1983). Unlike the channels, valleys exhibit a wide variety of drainage patterns ranging from well-integrated to mono-filament networks. Cross valley profiles range from V-shaped to U-shaped to valleys with steep, nearly vertical walls and broad, flat floors. Although there is a wide variety of valley types, most individual valleys have some characteristics in common. Upper reaches frequently appear degraded and have higher drainage densities than the lower reaches (Baker and Partridge 1986). In general, drainage densities for the Martian valley networks are much lower than those for terrestrial networks (Pieri 1980*a*). Notable exceptions are the fluvial valleys located on the Martian volcanoes Ceraunius Tholus, Hecates Tholus and Alba Patera. These valleys are morphologically similar and have similar drainage densities to those formed on the Hawaiian volcanoes (Gulick and Baker 1986). Another characteristic common to most of the valley networks is that tributaries commonly have blunt, theater-headed terminations. Networks have numerous stubby first-order tributaries. Drainage patterns often are structurally controlled (Schultz et al. 1982; Gulick 1986), with valleys commonly following fractures and other linear features. Interfluves remain largely undissected, reflecting the inefficiency of valley networks at filling space within their own drainage basin (Pieri 1980*a,b;* Baker 1982). Unlike most terrestrial river valleys, tributaries often appear to be as deep as the main trunk. Most Martian valleys have steep sidewalls with a relatively constant downvalley width. Tributaries join main trunk valleys at low mean junction angles when compared to terrestrial drainages (Pieri 1980*a*).

Martian valleys have been classified in many different ways: by planimetric pattern (Pieri 1980*a*), by size and general location (Baker 1982) and by combined planimetric patterns/cross valley profiles (Brakenridge et al. 1985). For the sake of discussion, we will use a modified version of the classification scheme presented in Baker (1982).

Valley Types

Longitudinal Valley Systems. Longitudinal or elongate valley systems (Fig. 4) are by far the largest of the valley networks. Examples of this type are Nirgal, Nanedi, Bahram, Ma'adim and Al Qahira Valles. These valleys are typically several hundreds of kilometers in length and several tens of kilometers in width. Upper reaches usually have short theater-headed tributaries with the exception of Ma'adim Vallis which has long branching tributaries. Lower reaches are usually sinuous and have broad, flat floors. Valley walls in the lower reaches have a scalloped appearance which is attributed to subsequent modification by landsliding, undercutting and other mass-wasting processes (Baker 1982). There is a general lack of dissection of adjacent uplands by these valley systems which are usually located in the old cratered uplands region. It is thought that these valleys may have been initiated as small valleys and then became enlarged by wall retreat as lower courses became deeply incised (Baker 1982). The undissected nature of the interfluves and the short accordant tributaries with width-to-depth ratios equalling that of the main trunk are thought to provide strong arguments against a formation by direct rainfall and surface runoff relationships (Pieri 1980*b;* Baker 1982).

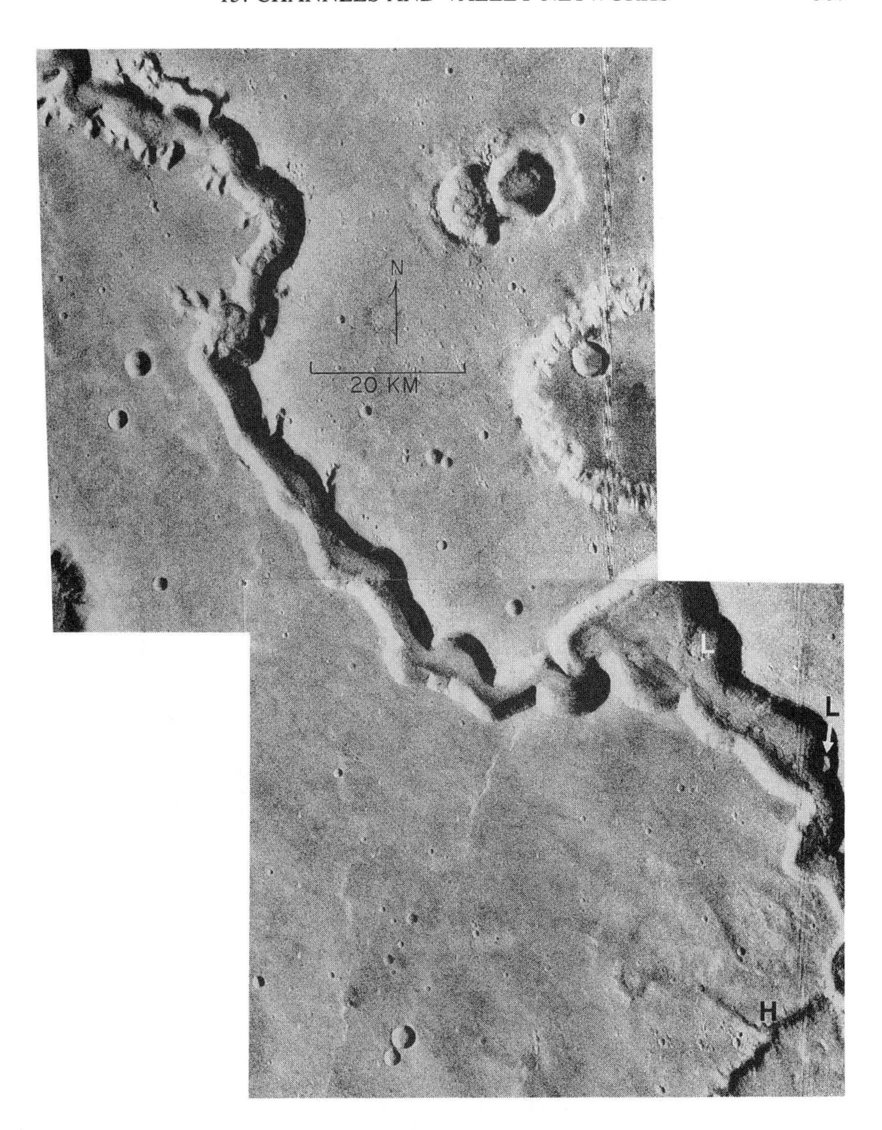

Fig. 4. Downstream portion of Nirgal Vallis, an example of a large longitudinal valley system. Note the short, stubby tributaries upstream (northwest) and the irregular widening downstream. Landslide deposits (L) and a hanging valley (H) also are present. (Viking Orbiter frames 466A61 and 466A64.)

Small Valley Systems. Small valley systems are ubiquitous in the heavily cratered terrain. However, an unusually large number of small valley types are concentrated in the Margaritifer Sinus region (Grant 1987) (Fig. 5). Many of these valleys are located in the hilly and cratered terrain, which formed during the period of heavy bombardment (Carr 1984), and drain into the cratered plateau region. The cratered plateau consists of smooth intercrater

Fig. 5. Small valleys in the heavily crater terrain. Mono-filament valleys radiate out from the crater rim. Integrated systems are located adjacent to crater in the intercrater terrain. (The images are a portion of JPL Viking Orbiter Mosaic 211–5207.)

plains which postdate the hilly and cratered terrain. The intercrater plains have buried or partly buried some craters. Linear valleys, also referred to as endogenic depressions (Baker 1982), are concentrated along fractures and drain into linear troughs. Small parallel valleys are concentrated along crater rims suggesting outward drainage of fluids. These valleys are common around large craters and clearly postdate the period of heavy bombardment (Baker 1982). Small valleys in the heavily cratered terrain, in general, exhibit a wide variety of morphologic patterns (Pieri 1980*a*). Valleys forming rectangular drainage patterns are structurally controlled, probably by underlying fracture systems. These valleys generally have low tributary junction angles and very low drainage densities. Monofilament parallel valleys are concentrated along crater rims and have little or no tributary development. Digitate to transitional dendritic valleys, which are more similar to terrestrial river valleys, have the highest drainage densities in the heavily cratered terrain.

Small valley systems also dissect some Martian volcanoes (Fig. 6). Both lava (Carr 1974*a*) and volcanic density-flow (Reimers and Komar 1979) origins have been proposed for the volcano valleys. Recent studies (Mouginis-Mark et al. 1982*b*,1988; Wilson and Mouginis-Mark 1987; Gulick and Baker 1990, 1991) show that some valleys are probably fluvial. The morphology of most valleys present on the volcanoes Hecates Tholus, Ceraunius Tholus and Alba Patera is compatible with a fluvial origin (Gulick and Baker 1989, 1990). These valleys formed in regions of subdued lava-flow morphology where the surface appears to be composed of fine-grained sediments, probably ash. They are inset into the surrounding land surface, widen slightly in

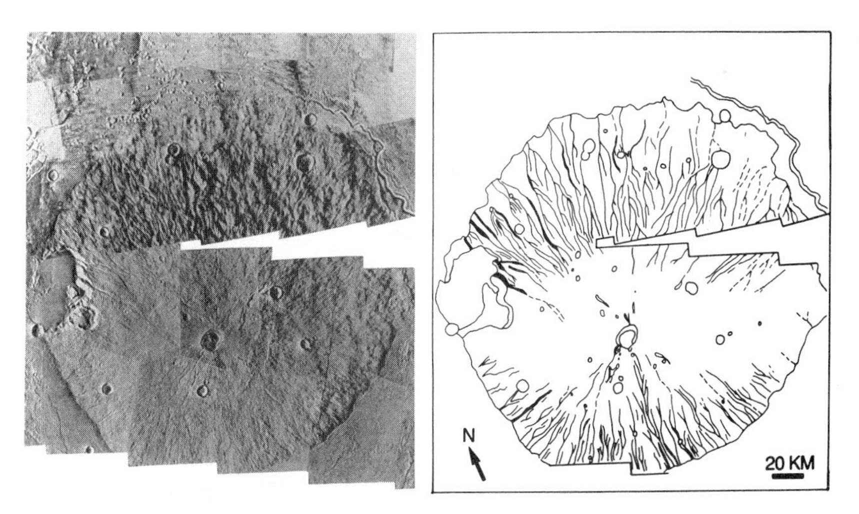

Fig. 6. Small fluvial valleys dissecting the volcano Hecates Tholus. Heavy black line denotes areas of valley enlargement, probably by sapping. The figure is taken from Gulick and Baker (1989*d*). (The images are a portion of JPL Viking Orbiter Mosaics 211–5601 and 211–5787.)

the downstream direction, form integrated tributary networks, and have drainage-density values which are comparable to those of fluvial valleys on the Hawaiian volcanoes (Gulick and Baker 1989). Valleys are also present on the volcanoes Hadriaca Patera, Apollinaris Patera and Tyrrhena Patera. The morphology of valleys on these particular volcanoes suggests that a combination of several genetic processes (water, volcanic density flows and lava) may have been important in their formation (Gulick and Baker 1990). However, extensive subsequent modification of these valleys by one or more of these processes has obscured the original morphology.

Other small valleys, referred to as slope systems by Baker (1982), formed along the walls of Valles Marineris in the Ius Chasma region. These valleys display primitive, angular, dendritic-drainage patterns. The apparent structural control of these patterns in addition to the theater-headed tributary terminations provide a strong argument for a sapping origin (Kochel et al. 1985).

Local Genetic Processes

Although it is widely accepted that the Martian valleys formed by fluvial processes (Mars Channel Working Group 1983), the way in which fluvial processes form valleys has a profound impact on the resulting morphology. Runoff valleys form from the transport of flowing water across the surface, while sapping valleys form from the outflow of ground water into the surface environment. Both processes exhibit unique morphologic characteristics. Below is a brief summary of how valleys are formed by surface runoff and ground-water sapping.

Surface Runoff. Fluvial valleys and channels, resulting from surface-runoff processes, form in locations where water runoff concentrates and flows with enough force to erode the land surface. In order to have water available at the surface for channel cutting, the rate of runoff over the surface must exceed the rate of infiltration. Once these conditions are met, water from dispersed source areas migrates across the surface and collects in topographically low areas. This action concentrates the flow of water moving downslope, thereby increasing the ability of water to erode the surface and establish a channel. Once established, channels provide a locus for the convergence of surface flow leading to further incision. This process is particularly well described by Knighton (1984). As drainage of the land surface becomes more efficient, channels integrate and eventually form a complex system of networks.

Fluvial runoff valleys and channels are erosional features which are inset into the land surface. Tributaries are present except during the early stage of development where channels form simple, trough-shaped patterns (Macdonald et al. 1983). In plan view, these drainages have tapered tributary heads, a continuous valley form and an increasing valley width in the downstream direction.

Ground-water Sapping. Sapping valleys are formed by the undermining of rock and sediment by ground-water outflow. In the model of sapping valley network formation proposed by Dunne (1980), subsurface water emerging from valley sides and headwalls disrupts ground-water flow patterns. Ground water converges and concentrates along weakness planes removing material which supports the overlying surface. This removal of support eventually results in the collapse of overburden into the valley, causing valley sides to widen and the headwall to retreat. The rates of headward erosion in these valleys are faster than the rates of valley widening because the headwall is the region of greatest flow convergence. The process is self-enhancing in that the greater the amount of erosion that takes place, the more flow convergence is produced and the higher the rate of retreat of the headwall. This process eventually halts when the flow of ground water becomes insufficient to maintain further sapping.

Sapping valleys have theater-headed tributaries (Baker 1982) and anomalously large valley width-to-depth ratios when compared to runoff valleys. Because ground-water flow tends to exploit planes of weakness in the terrain, sapping valleys will often form along fractures and joints in the bedrock, resulting in a rectilinear or structurally controlled pattern. Sapping processes will often modify the morphology of runoff valleys which have downcut into the ground-water system. This modification results in valley enlargement in which the V-shaped, cross-valley profiles typical of runoff-dominated systems on volcanic landscapes are eroded into broad U-shaped or flat-floored valleys with steep walls. The overall effect of sapping modification to an existing drainage system is to change the network pattern to one that is simpler and less integrated. Baker (1990) reviews the origin of valleys by sapping.

Numerous comparative planetology studies (Laity and Malin 1985; Kochel and Piper 1986; Gulick and Baker 1987*b*,1989,1990) have recognized the importance of both surface runoff and ground-water sapping processes in the valley development and have distinguished the morphologic characteristics resulting from each of these processes. In addition to these field-based studies, a fairly extensive series of experimental flume studies (see, e.g., Kochel and Piper 1986; Kochel et al. 1985,1988) of drainage networks formed by sapping processes have been conducted in soft and weakly consolidated, layered sediments. These flume studies have been able to replicate much of the sapping morphologic characteristics and controls in drainage networks, including, theater-headed tributaries, joint control of ground-water flow, subsequent orientation of drainages and the importance of lithology (i.e., presence of permeable and impermeable strata) in controlling ground-water sapping processes. The results of these experimental studies are being used to develop numerical simulation models of the sapping process (Howard 1988; Howard and McLane 1988).

Regional Genetic Conditions

While the morphology of the large, longitudinal valleys is generally consistent with a sapping origin, the genesis of the small valleys is less clear. Valleys in the heavily cratered terrain exhibit a complex morphology (Baker and Partridge 1986). These valley systems have relatively fresh-appearing, low-density network segments in their lower reaches and higher-density network segments in their upper reaches. Baker and Partridge (1986) concluded that the dense upper networks formed during the period of heavy bombardment and the pristine lower reaches formed after the emplacement of the intercrater plains. The pristine valley segments subsequently evolved by headward growth into the regions occupied by the degraded valleys. Baker and Partridge suggested that the relatively higher-network density of the degraded valleys may imply a greater influence of surface runoff processes in their genesis.

A similar complex morphologic pattern is exhibited by the fluvial valleys developed on older Martian volcanoes (Gulick and Baker 1990). Fluvial valleys present on volcanoes Ceraunius and Hecates Tholi, which formed during the period of heavy bombardment (Barlow 1988*c;* Neukum and Hiller 1981), exhibit a parallel drainage pattern of integrated valley systems. Some valleys on Ceraunius appear pristine and less integrated while others appear degraded and better integrated. The pristine valleys appear re-activated. Valleys on Hecates also exhibit a parallel drainage pattern, but the valley systems are better integrated than those on Ceraunius. Unlike the valleys on Ceraunius, however, only the lower reaches of some valleys appear pristine and enlarged. Fluvial valleys developed on Alba Patera, a volcano which formed well after the period of heavy bombardment (Barlow 1988*c;* Neukum and Hiller 1981), exhibit only the better-integrated upper-reach morphology (Fig. 7). Indeed, these valleys are the most integrated and most terrestrial-like fluvial valleys on the surface of Mars. The Alba valleys have tapered tributary heads that blend in gradually with the surrounding landscape and V-shaped cross-valley profiles. The Alba valleys appear morphologically similar to those formed by rainfall-runoff processes on the Hawaiian volcanoes. In addition, this same complex morphologic pattern of better-integrated upper reaches and less-integrated lower reaches is also apparent on the Hawaiian volcanoes. Recent morphologic studies of the Hawaiian valleys (Gulick 1987; Gulick and Baker 1987*b;* Kochel and Piper 1986) concluded that Hawaiian valleys formed initially by surface runoff and were subsequently enlarged by ground-water sapping. Based on comparison morphologic studies with the Hawaiian valleys, valleys on the Martian volcanoes probably evolved in a similar manner (Gulick and Baker 1990).

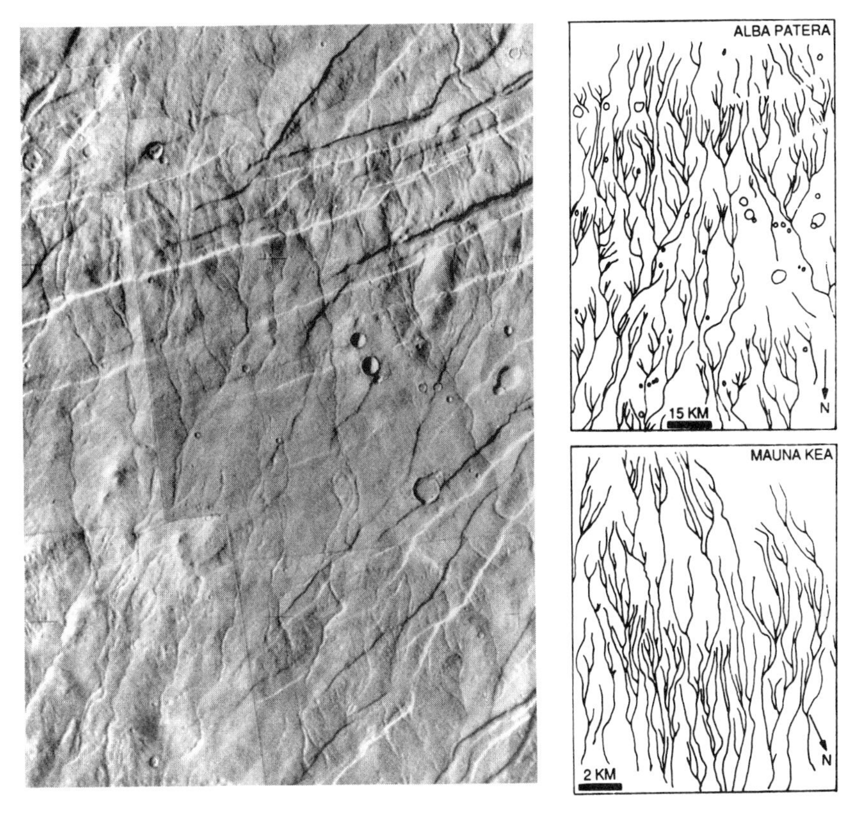

Fig. 7. Fluvial valleys on the northern flank of Alba Patera. Valleys exhibit a well-integrated drainage pattern which is morphologically similar to fluvial valleys formed by surface runoff on the Hawaiian volcanoes (e.g., Mauna Kea volcano, Hawaii). The Alba valley systems are the best developed and the most terrestrial-like fluvial networks on the surface of Mars. The figure is taken from Gulick and Baker (1990). The images are from Viking Orbiter Mosaics MTM 45107.)

IV. RELATED LANDFORMS

Erosional processes modified all varieties of Martian landforms, including channels and valleys. Mass wasting, aeolian, thermokarst and periglacial processes have subsequently acted to modify the walls and floors of channels and valleys originally carved by fluvial processes. The most extreme modification processes may have formed the fretted channels and fretted terrain (Fig. 8). The alteration of the original landforms has been so complete, however, that it is not known if the fretted channels formed by modifying pre-existing valleys or not. In this section, we first examine the fretted channels and terrain and then consider the other secondary processes which have modified the channels and valleys.

Fig. 8. Fretted terrain in the Nilosyrtis region near 34°N, 282°W. The isolated upland massifs are surrounded by valleys that appear to be filled with debris derived from adjacent slopes. Subparallel ridges and grooves imply flowage of the debris. (JPL Viking Orbiter Mosaic 211–5207.)

Fretted Terrain and Channels

Fretted terrain is defined as a complex of smooth, flat lowland areas separated by abrupt escarpments from relatively heavily cratered uplands (Sharp 1973*a*). The scarps are typically 1 to 2 km high and may indicate the depth of a geologic unit, perhaps the megaregolith or the transition to ice-rich permafrost (Sharp 1973*a;* Davis and Golombek 1989; chapter 16). The terrain is best developed along the cratered upland and northern plains boundary (Mutch et al. 1976*c*). Fretted channels are steep-walled valleys with wide flat floors and indented walls.

Both the fretted terrain and channels exhibit complex planimetric patterns with isolated butte and mesa outliers (Baker 1982). The channels exhibit structural control but no fluid flow bedforms (Baker 1982). The fretted channels are generally located in the region of the fretted terrain but occasionally extend hundreds of kilometers back from the main scarps into the older heavily cratered terrain (Baker 1982). The geomorphology of the fretted terrain is discussed in detail by Kochel and Peake (1984).

A common aspect of both the floors of the fretted channels and the lowlands of the fretted terrain is the presence of debris flows (Carr and Schaber 1977; Squyres 1978). In the fretted terrain, debris flows emanate from the scarps and flow onto the surrounding lowlands, often forming aprons around isolated mesas. The lobate debris flows characteristic of the fretted terrain are found in two planet-wide 25° latitude swaths centered at 40°N and 45°S

(Squyres 1979*a*). When unconfined, these flows may reach lengths of 10 to 20 km. There may be a correlation between scarp face orientation and apron width (Eppler and Malin 1982; Kochel and Peake 1984). The floors of some of the fretted channels are also covered by lineated deposits which appear to be debris flows. The debris aprons generally have a convex profile (Squyres 1978), although some channel and outlier escarpments exhibit a moat-like depression or swale on the apron at the base of the escarpment (Weiss and Fagan 1982). In a crater-counting study, Squyres (1978) found that both lineated and unlineated debris are younger than the uplands, escarpments and surrounding lowland. This may imply that the debris-forming process may still be operating today. It is unknown whether mass wasting and debris flowage are secondary-modification processes acting on features formed by another mechanism, or whether they are the primary mechanism that formed the fretted terrain.

The best evidence that flow of debris indeed took place is the presence of numerous striae in the debris deposits. In the fretted terrain the striae are oriented at right angles to scarp faces, and they diverge and converge around obstacles. These characteristics are generally accepted as implying flow from the scarp faces. In the fretted channels, where the striae typically run parallel to the channel length (Fig. 8), the interpretation of the striae is more problematic. Squyres (1978) interpreted the striae orientation as implying flow from the channel walls meeting in the center, the resulting compression-producing striae parallel with the walls. Squyres states that little or no downvalley flow has taken place. Lucchitta (1983*a*) argued against this interpretation, noting that side-canyon debris bends downvalley, that striae split at obstacles, and that striae are always oriented downvalley, even at the base of mesas where right-angle-oriented flow would be expected. The main problem with Lucchitta's interpretation of considerable downvalley flow is the lack of depositional features at the mouths of most channels.

Fretted channels typically extend into plateaus and have stubby tributaries. They are best developed along the boundaries of the northern terrain and other older elevated terrains. There is typically a gradation from channels cutting plateau to fretted terrains to plains with plateau outliers (Carr 1981). As distance from the cratered uplands and northern plains boundary scarp increases, the number and size of the outliers becomes smaller while the areal volume of the plains material increases (Kochel and Peake 1984). The similarity of the channels and terrain led Baker (1982) to conclude that both were formed by the same erosional process. Under such a scenario the channels would erode back into escarpments along zones of least resistance. As the channels widened and dissected the highlands, the landscape would eventually take the form of the fretted terrain. Kochel and Peake (1984) also concluded that progressive degradation and retreat of the boundary scarp was responsible for the fretted terrain. While an appealing concept, no thorough study of this fretted-terrain degradation process has yet been made.

Beyond the questions of downvalley transport and fretted-terrain evolution is the simple question of what caused the scarp-forming and debris-transport process in the first place. Mass-wasting processes probably undermined escarpments and provided materials which flowed down onto the surrounding lowlands. Sharp (1973*a*) proposed dry sapping as the principle mass-wasting process. Dry sapping proceeds as exposed ice is sublimated from scarp faces. As ice is lost, support is removed from overlying material which then collapses. Squyres (1978) proposed more conventional mass-wasting processes in which material was supplied from eroding angle of repose scarps. Both processes require an initial cliff or valley to provide the face from which the ice sublimates or mass wasting occurs. Wet sapping is another possibility that cannot be ruled out (Sharp and Malin 1975; Baker 1982). Resolution of this question will probably require much higher-resolution imagery and topography, or actual field reconnaissance.

Once the debris material has been wasted off the surrounding scarps, it must then be transported away from the scarp and perhaps down the valley. This transport removes the talus and allows mass wasting from the scarp to continue (Carr 1984). Most authors agree that ice facilitates the flow, but there is less agreement on the source of the ice. Squyres (1978) favors seasonal frost deposits being covered over by subsequent mass wasting. This process would result in interstitial ice similar to that in terrestrial rock glaciers (for discussions of terrestrial rock glaciers, see the collection edited by Giardino et al. [1987]). Lucchitta and Persky (1982) argue instead that the ice contained in the ground over which the debris flows is sufficient to lubricate the transport. The problem with this mechanism is that, in similar situations on Earth, a basal shear of approximately 1 bar is required to initiate the flow. On Mars this requires flows 2 to 5 km thick (Lucchitta 1983*a*), which is thicker than observed. However, the terrestrial criteria may be inapplicable because of the long time scale needed to maintain Martian flows (Lucchitta 1983*a*). Another possible mechanism for removing debris from scarp bases is weathering followed by aeolian deflation (Squyres 1978). Finally, the most provocative mechanism suggested for escarpment formation and debris removal is wave action at the shores of an ancient Martian ocean (Parker et al. 1989).

The fretted channels and terrain may represent important examples of an ongoing Martian geologic process. Lucchitta (1983*a*) notes that, in those latitudes where fretted terrain is located, the current climate allows for the retention of ice in the ground. Thus, ground ice is present to be exposed at scarps for dry sapping and to lubricate movement of ice-rich materials. At high latitudes with colder temperatures, it may be too cold for rock glaciers to be mobilized. Since the observed fretted terrain corresponds with those regions where present conditions allow ice both to be retained in the ground and to lubricate the flow of material, fretting may still be occurring.

Since the channels seem to be modified at low temperatures by a process which is probably associated with ice, the fretting process may be an archetypical Martian mechanism. As such, the process and the associated features deserve more study. Specifically, the mass-transport process and the question of whether downvalley transport is operating should be resolved. If these processes become better understood, they may help provide a record of the recent Martian climate. High-resolution imagery of the escarpments, tributary junctions, valley mouths and topographic information is needed in order to answer such questions.

Other Secondary Processes

Once a valley or channel has been formed, it represents a system of steep walls and valley floors which other secondary geologic processes can exploit. The results of secondary processes can be found in and along most channels and valleys. Wind, for example, can act as an erosive agent by transporting saltating particles. While inadequate to actually carve channels or valleys, aeolian processes can subsequently modify the landforms. Valleys can also provide sheltered areas for deposition by wind and the formation of dunes. In the mid to low latitudes, dunes are present mainly in the protected areas of canyons and valleys (Carr 1984). Another secondary process, mass wasting, removes material from scarps and slopes under the influence of gravity. Most slopes surrounding channels and valleys have probably been modified by these processes, as they have been on Earth. Sapping and hillslope-retreat processes may have enlarged the width of the outflow channels (Baker and Kochel 1979). Such processes are important, for if they are not taken into account when considering channel volume, anomalously large quantities of water may be derived as necessary for channel formation (Baker 1982). Mass wasting is also important in the formation or modification of the fretted landforms.

Processes which depend on the melting of ground ice can also modify channels and valleys. The very large landslides found around the margins of some of the outflow channels were probably lubricated by ground ice-derived water (Mars Channel Working Group 1983). Thermokarst topography forms when melting of ground ice produces local collapse depressions which should not be confused with channels or valleys. Chapter 16 discusses these processes in more detail.

V. IMPLICATIONS FOR PALEOENVIRONMENTAL CHANGE

Global Martian Water Budget

The channels provide a crude way of estimating the total amount of water outgassed from the planet (Carr 1986,1987). The calculation utilizes the volumes of material removed to form the various canyons, channels and chaotic terrain around the Chryse basin. Ignoring the volumes of Ius and Coprates canyons, in which faulting is most evident, then there is a negative regolith volume of roughly 5×10^5 km^3 in the remaining canyon, channels and chaotic terrain. If this volume was all removed by water, and the water carried its maximum sediment load, or 40% by volume (Komar 1980), then 7.5×10^6 km^3 of water would be required, which is the equivalent of 50-m spread over the entire planet. Clearly there is considerable uncertainty in this number since it is unknown precisely how much of the negative volume of the canyons is due to erosion and how much is due to tectonic forces. Moreover, it is unlikely that all the water carried the maximum sediment load.

Further assumptions are required to estimate how much water has outgassed from the planet. The planet-wide distribution of valley networks (Pieri 1976; Carr and Clow 1981) suggests that, early in the planet's history, water was distributed globally and not concentrated in the local areas such as around Chryse Planitia. The concentration of floods around Chryse probably represents later ground-water accumulation in the region as a result of slow migration of ground water from the surrounding higher terrains (Carr 1979). The drainage basin around Chryse Planitia can be roughly outlined from drainage patterns and topography. It constitutes roughly one sixth to one eighth of the planet's surface. Thus, assuming that water was initially distributed evenly over the planet, and that the Chryse drainage basin contained at least 50 m of water, the indicated global inventory is at least 300 to 400 m. This is a very rough estimate. It would be too high if the same water passed several times through the circum-Chryse channels. However, most of the circum-Chryse outflow channels formed after the period of intense valley formation for which warmer climatic conditions have been proposed. Climatic conditions during formation of the outflow channels are likely to have been similar to modern conditions, which cause a thick, planet-wide permafrost. As described above, the water that eroded the channels probably formed permanent ice deposits in the low-lying northern plains, and was not recycled through some global aquifer system. The estimate of 300 to 400 m may also be too low because it ignores the ground water within the Chryse drainage basin that failed to reach the surface. For comparison, the Earth is estimated to have outgassed 3 km of water (Turekian and Clark 1975).

From the volume of volcanic rocks that have accumulated on the surface, Greeley (1987) estimated that the planet has outgassed approximately 50 m of water in the last 3.8 Gyr. This number is entirely consistent with the 500 m estimated (Carr 1987) for the total outgassed water since most of the outgassing is believed to have occurred very early in the history of the planet, before the geologic record was retained. Most geochemical estimates (summarized in Pepin 1987*a*) are lower than both the Greeley and Carr estimates but can be reconciled if Mars lost part of its early atmosphere by impact erosion (Cameron 1983; Melosh and Vickery 1989) or hydrodynamic escape

(Hunten et al. 1989). The volatile inventory and evolution is reviewed in chapters 4, 6 and 32.

Paleoclimatic Implications

The valley networks have been widely cited as evidence of a former thicker atmosphere and surface temperature substantially warmer than those at present (see, e.g., Sagan et al. 1973*a;* Masursky 1973; Pollack 1979; Toon et al. 1980; Pollack et al. 1987). However, the climatic conditions required to form channels and valleys are unclear. Outflow channels could probably form under present climatic conditions. They involve floods of such magnitude that the amount of freezing under present climatic conditions would be trivial (Lingenfelter et al. 1968; Carr 1979). Cold climatic conditions with temperature well below freezing may even be required for the formation of those floods caused by eruption of ground water. A thick permafrost may have been needed to contain the ground water, thereby allowing artesian pressures to build and eventually cause massive outflows (Carr 1979).

The valley networks are much smaller than the flood outflow channels and are presumed to have formed from correspondingly smaller discharges. Moreover, the valleys divide into smaller valleys upstream. If temperatures were well below freezing, it is reasonable to assume that, irrespective of the source of the water, small streams in the distal parts of the networks would freeze, thereby cutting off flow into the larger channels downstream and arresting further development of the valleys. The presumed warmer conditions are believed to have occurred mainly very early in the planet's history. The valley networks are almost entirely restricted to the oldest terrains on Mars, in contrast to the outflow channels which cut into materials with a wide range of ages (Carr and Clow 1981; Baker and Partridge 1986). The simplest explanation of the almost complete restriction of valleys to the oldest terrains is that the valleys themselves are old, and that conditions required for valley formation were commonly met early in the planet's history but only rarely throughout the planet's subsequent history. These two inferences, that warm conditions are required for valley formation and that the valleys are mostly old, have led to a perception that early Mars was warm and wet, and that climatic conditions then changed such that conditions similar to those prevailing today were maintained for much of the later planetary history. We caution that these conclusions are by no means proven.

The climatic conditions required for valley formation are not clear. The assumption of warm wet conditions is based on the premise that the valleys formed by slow erosion of running water. The arguments for water as the erosive agent are based on analogy with terrestrial valleys and the abundant corroborative evidence, in addition to the valleys, for the presence of water at the Martian surface. Agents other than water or ice, such as wind, carbon-dioxide ice and lava are very unlikely for a number of reasons (for summaries, see Baker 1982; Carr 1981). However, as described above, many of the networks have a distinctly different appearance from terrestrial river valleys. Because of these differences and the difficulty of maintaining liquid water at the Martian surface, and, given the abundant evidence for water and ice, the possibility should be left open that many, perhaps most valleys formed not by Earth-like fluvial processes, but by some other mechanism, such as mass wasting, that involves water or ice.

If the valleys were formed by running water then the water could come from two possible sources, the ground or the atmosphere. Many valley networks have characteristics suggestive of ground-water sapping (Pieri 1980*a;* Baker 1982, Higgins 1982; Kochel et al. 1985). One requirement for ground-water sapping is that liquid water is stable sufficiently close to the surface that it can seep onto the surface. This implies that temperatures are above the freezing point at depths comparable to the scale of local topographic relief. This can be achieved in two ways: (1) the surface temperature may be above freezing; or (2) if surface temperatures are below freezing, the temperature gradient is large enough to allow liquid water close to the surface. Steep thermal gradients are likely on Mars. Heat of accretion and core formation would have resulted in heat flows more than a factor of 10 larger than at present during the first few hundred Myr of the planet's history (see chapter 5). This, combined with low conductivities expected of the brecciated mega-regolith could have resulted in liquid water at shallow depths early in the planet's history irrespective of the surface temperatures.

If surface temperatures were below freezing and water reached the surface, then it may have been possible for streams to survive and cut the valley networks. Modeling of the freezing of small streams (Wallace and Sagan 1979; Carr 1983) suggests that if a stream 1-m deep or more can be started then the water could flow for a few 100 km, depending on slope and other factors, before it froze solid. However, such calculations are highly idealized and do not take into account the vagaries of natural systems. Streams from terrestrial springs under arctic conditions are typically arrested within a relatively short distance of the spring as a result of formation of icings (Carey 1973; Childers et al. 1977; Sloan et al. 1976). Thus, while theoretical calculations suggest streams could survive and cut channels under present conditions, analogy with terrestrial streams indicates the contrary. Moreover, if temperatures were permanently below freezing, then the water released onto the surface could not readily re-enter the ground-water system and recharge. Any recharging would have to be from juvenile sources.

If the valleys formed from precipitation, then surface temperatures most likely had to be above freezing. While it has been suggested that ice could be precipitated at low latitudes during periods of high obliquity (Jakosky and Carr 1985) and that the ice could melt and generate runoff despite low temperatures (Clow 1987), the amount of liquid water generated under these conditions is very small, and it is doubtful that it would be sufficient to cut the valleys.

Endogenetic Hydrologic Cycling

Local water budgets must be balanced to achieve the development of valley networks. For water to flow at the surface, it must be available at source areas. For spring sapping, a hydraulic gradient is ultimately achieved for terrestrial examples as solar energy generates an atmospheric hydrological cycle that transports evaporated or transpired water as vapor back to river headwaters, where it precipitates and infiltrates to recharge aquifer systems. However, it may also be possible to supply energy geothermally to recycle water in the subsurface.

The ancient valleys of the heavily cratered terrains are associated with very high cratering rates and with widespread volcanism. The former association led Brakenridge et al. (1985) to hypothesize that some of the valleys might have originated through the interaction of ground ice and hot springs located along the semi-permeable fringes of slowly cooling impact melts. They concluded that, with widespread operation of impact-related hydrothermal systems, the possible existence of which was proposed by Newsom (1980) and Schultz et al. (1982), it is not necessary to infer major atmospheric change to explain valley formation. Gulick (1992) estimated lifetimes of hydrothermal systems associated with impact crater and volcano formation. She concluded that hydrothermal systems could remain active for long enough periods to form the valley networks that are associated with these features.

Hydrothermal systems are an inevitable consequence of volcano or crater formation, tectonism, sill intrusion, or other plutonic intrusions into permeable, liquid water or ice-rich subsurface environments. Such geologic events generate local thermal anomalies which induce density perturbations in the ground water as heat is dissipated out into the surrounding country rock. Water near the anomaly migrates upward in response to buoyancy forces. This action results in the flow of ground water towards the thermal anomaly. Hydrothermal systems can focus, transport and recirculate large amounts of ground water to dynamic surface water environments for periods in excess of 10^5 yr (Gulick 1992). As pointed out by Gulick and Baker (1987*a*) water flow transported to the surface, via seeps and springs, may initiate valleys if there is a low permeability, erodible surface (e.g., ash) and sufficient runoff. However, if the surface is permeable (e.g., basalt flows), water will infiltrate and recharge near surface (perched) aquifers. These aquifers may eventually intersect with the surface farther downslope and initiate sapping at these locations. However, some of the upward migrating water may recharge high-level aquifers without ever flowing on the surface.

Squyres (1989*b*) proposes that the higher heat flow and steeper thermal gradient that existed early in Mars' history would have resulted in the melting of ground ice to a minimum depth of 350 m (see also chapter 32). However, in order to form sapping valleys, water would have to intersect with the surface environment and exploit a pre-existing (runoff) valley, joint, fault or fracture system as indicated by several terrestrial sapping process studies. The remaining problem, then, is in getting the subsurface water up into the surface environment, so that valleys can be initiated and eroded down to depths of the melted-ground ice reservoirs. The generation of hydrothermal systems, as discussed above, may provide a way of transporting and circulating this water through the surface environment.

The complex sequences of ancient cratered plateaus on Mars apparently include interstratified impact breccia, reworked aeolian sediments, lava flows or sills, and ice (Tanaka 1986; Wilhelms and Baldwin 1989*b*). The coincidence of very heavy cratering, extensive volcanism and regional valley formation suggests an association with endogenetic energy sources driving a dynamic hydrological system. As with the outflow channels, the subsurface character of this hypothesized system precludes its direct study.

VI. CONCLUSIONS

The spectacular relict channels and valley networks of Mars represent ancient hydrological conditions greatly different from those seen to be active on the planet today. The outflow channels are relatively young, late Hesperian or Amazonian in age (Tanaka 1986). They formed by immense outbursts of fluid from subsurface sources. Complexity in outflow-channel morphology was generated by varying amounts of sediment and ice in the aqueous-fluid flow systems. The overall cataclysmic-flood morphology thus may be locally transitional to morphologies generated by ice and debris flowage. Moreover, secondary processes, including wind, lava flows and mass movement, extensively modified some channel systems.

Although local areas of valley networks, such as on Alba Patera, formed coevally with outflow channel activity, regionally extensive networks dominate in the heavily cratered terrains. These networks are Noachian and early Hesperian in age. The morphology of many valleys suggests genesis by ground-water sapping; for some valleys, surface runoff may have been more important. The morphologic evidence is consistent with but does not require atmospheric or climatic change for its explanation. Endogenic cycling of water, as in volcanic or impact-related hydrothermal systems, provides an alternative explanation.

Acknowledgments. We thank R. C. Kochel and P. D. Komar for their insightful reviews of the manuscript. Partial support for the preparation of this chapter was provided by the National Aeronautics and Space Administration, Planetary Geology and Geophysics Program.

16. ICE IN THE MARTIAN REGOLITH

S. W. SQUYRES
Cornell University

S. M. CLIFFORD
Lunar and Planetary Institute

R. O. KUZMIN
V.I. Vernadsky Institute

J. R. ZIMBELMAN
Smithsonian Institution

and

F. M. COSTARD
Laboratoire de Géographie Physique

Geologic evidence indicates that the Martian surface has been substantially modified by the action of liquid water, and that much of that water still resides beneath the surface as ground ice. The pore volume of the Martian regolith is substantial, and a large amount of this volume can be expected to be at temperatures cold enough for ice to be present. Calculations of the thermodynamic stability of ground ice on Mars suggest that it can exist very close to the surface at high latitudes, but can persist only at substantial depths near the equator. Impact craters with distinctive lobate ejecta deposits are common on Mars. These rampart craters apparently owe their morphology to fluidization of subsurface materials, perhaps by the melting of ground ice, during impact events. If this interpretation is correct, then the size-frequency distribution of rampart craters is broadly consistent with the depth distribution of ice inferred from stability calculations. A variety of observed Martian landforms can be attributed to creep of the Martian regolith abetted by deformation of ground ice. Global mapping of creep features also supports the idea that ice is present in near-surface materials at latitudes higher than ± 30°, and suggests that ice is largely absent from such materials at lower latitudes. Other morphologic features on Mars that may result from the present or former existence of ground ice include chaotic terrain, thermokarst and patterned ground.

One of the most significant realizations to come from the exploration of Mars has been that a very important role has been played in the planet's history by water and ice. Many of Mars' major geologic features, particularly the spectacular channels, appear to have formed by release of water from beneath the ground. Large amounts of water were involved in forming these and other features, and there is no obvious way that so much water could have been lost from the planet since they developed. It is commonly concluded, therefore, that a significant quantity of water resides on Mars, locked up beneath the surface as ground ice. Ground ice is the largest prospective reservoir for Martian H_2O, and if we can determine the total amount of H_2O present as ground ice, we will have learned much about the planet's total water inventory. As we discuss below, the presence of ground ice appears to have had significant effects on the morphology of surface features. And, if we are ever to send humans to Mars, ground ice may present one of the best hopes for finding a water resource on the planet. In this chapter, we outline the physical mechanisms that control the distribution of ice in the Martian regolith, and discuss the geologic features that provide evidence concerning this distribution.

I. THE MARTIAN MEGAREGOLITH

Impact processes have played a major role in the structural evolution of the Martian crust (Soderblom et al. 1974; Gurnis 1981). Field studies of terrestrial impact craters (Shoemaker 1963; Short 1970; Dence et al. 1977) and theoretical models of the cratering process (Melosh 1980*a*; O'Keefe and Ahrens 1981) have shown that impacts modify a planetary surface by the production and dispersal of large quantities of ejecta, and through the intense fracturing of the surrounding and underlying basement. Fanale (1976) has estimated that over the course of Martian geologic history, the volume of ejecta produced by impacts was sufficient to have created a global blanket of debris up to 2 km thick. As noted by Carr (1979), it is likely that this ejecta layer is discontinuously interbedded with volcanic flows, weathering products, and sedimentary deposits, all overlying a heavily fractured basement (Fig. 1). This description of the near-surface structure of the Martian crust is very similar to that proposed for the Moon (Hartmann 1973*a*,1980), where an early period of intense bombardment resulted in the production of a

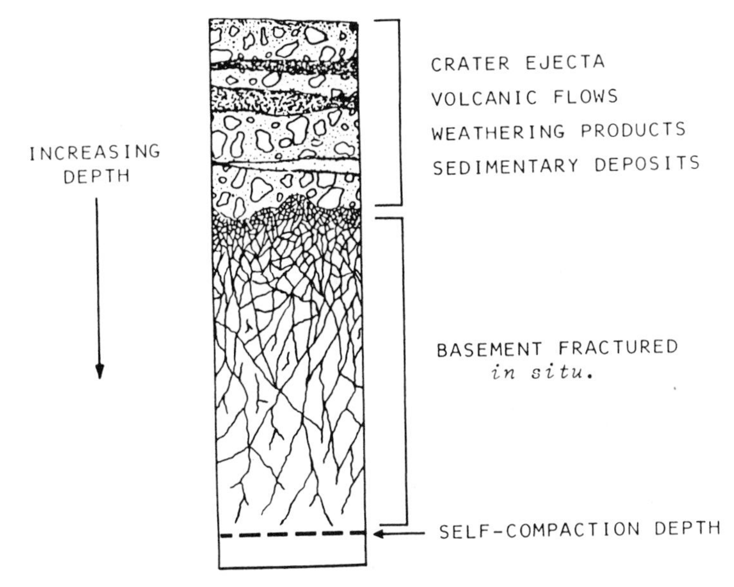

Fig. 1. An idealized stratigraphic column of the Martian crust (figure from Clifford 1981). The first km or so may consist of interbedded layers of different materials in a sequence which depends upon local geologic history.

blocky, porous megaregolith that extends to considerable depth—a view supported by the seismic propagation characteristics of the lunar crust (Toksöz 1979).

A. Porosity versus Depth

For the near surface of the Moon, seismic *P*-wave velocities increase with depth until they reach a local maximum at about 20 km (Toksöz 1979). This behavior is consistent with a reduction in crustal porosity with increasing depth, and hence lithostatic pressure. The transition between fractured and coherent lunar basement is believed to coincide with the beginning of a constant velocity zone (> 20 km), where lithostatic pressure is thought to be sufficient (> 1 kbar $= 10^8$ Pa) to close virtually all fracture and intergranular pore space (Toksöz 1979; Binder and Lange 1980).

Binder and Lange (1980) suggest that the seismic properties of the lunar crust are best explained by an exponential decline in porosity with depth. According to this model, the porosity at a depth z is given by

$$\Phi(z) = \Phi(0)\exp(-z/K) \tag{1}$$

where $\Phi(0)$ is the porosity at the lunar surface and K is a decay constant. Binder and Lange found that a porosity decay constant of 6.5 km best fit the lunar data. Assuming comparable crustal densities, the appropriate gravitationally scaled value for Mars is 2.8 km, i.e., $K_{\text{mars}} = K_{\text{moon}}(g_{\text{moon}}/g_{\text{mars}})$ (Clifford 1981).

Two potential porosity profiles of the Martian crust are illustrated in Fig. 2. The first is based on a surface porosity of 20%, the same value assumed for the Moon by Binder and Lange (1980), and one that agrees reasonably well with the measured porosity of lunar breccias (Warren and Rasmussen 1987). This model yields a self-compaction depth (the depth at which crustal porosity becomes $< 1\%$, a value at which pore spaces typically become discontinuous) of approximately 8.5 km, and a total pore volume of roughly 8×10^7 km^3, a volume sufficient to store a global layer of water approximately 550 m deep (Clifford 1981,1984).

In the second profile, a surface porosity of 50% is assumed, a value consistent with estimates of the bulk porosity of Martian soil as analyzed by the Viking Landers (Clark et al. 1976). A surface porosity this large conceivably could be appropriate if the regolith has undergone a significant degree of weathering (see, e.g., Malin 1974; Huguenin 1976; Gooding 1978) or if it contains large segregated bodies of ground ice. The self-compaction depth predicted by this model is roughly 11 km, while its total pore volume is approximately 2×10^8 km^3, a volume equivalent to a global ocean some 1.4 km deep (Clifford 1984). It is doubtful, however, that weathering or large-scale ice segregation will affect more than the upper few km of the crust; therefore, below this depth, the porosity profile will most likely resemble the gravitationally scaled lunar curve.

The presence of groundwater could affect estimates of the self-compaction depth and total pore volume in several ways. As discussed by Hubert and Rubey (1959) and Byerlee and Brace (1968), the hydrostatic pressure of water within a pore can partially offset the lithostatic pressure acting to close it. For a "wet" Mars, crustal porosity may persist to depths roughly a third greater than those indicated by the "dry" models in Fig. 2. Ground-

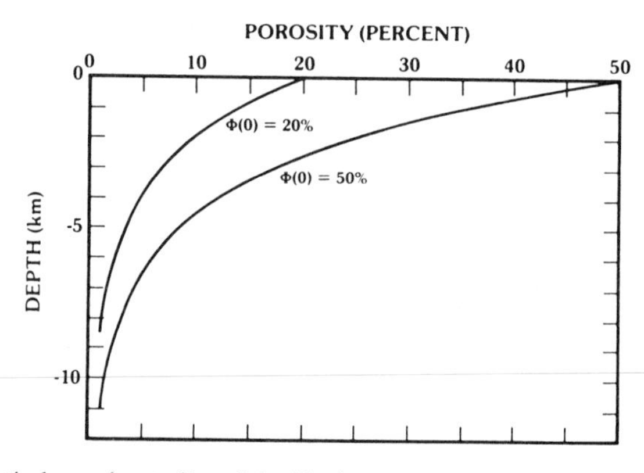

Fig. 2. Theoretical porosity profiles of the Martian crust based on a lunar analog (figure from Clifford 1981,1984).

water can also reduce crustal porosity by solution, compaction, and cementation (Maxwell 1964; Rittenhouse 1971). Such processes are of greatest significance under conditions of high temperature and pressure (Maxwell 1964; Pettijohn 1975). Currently these conditions are likely to exist only at great depth on Mars, where their impact on the overall porosity of the megaregolith is expected to be small. However, conditions were almost certainly different in the past. Models of the thermal history of Mars suggest that 4 Gyr ago the planet's internal heat flow and geothermal gradient were ~ 5 times larger than they are today (see chapter 5). In addition, an early greenhouse may have permitted liquid water to exist at or near the surface, where it may have reacted with atmospheric CO_2 to form carbonates within the crust (Kahn 1985; Postawko and Kuhn 1986; Pollack et al. 1987; see also chapter 32). The extent to which such processes have affected the porosity and depth of the megaregolith cannot be assessed.

B. Thermal Structure

The zone of permafrost on Mars is, like permafrost on Earth, defined as that portion of the crust where the temperature remains continuously below the freezing point of water. Because the term permafrost is so often misused, however, we instead adopt for this region the unambiguous term cryolithosphere, which is widely used in the Soviet literature (see, e.g., Kuzmin 1977). If we assume a solute-free freezing point of 273 K, then temperatures are below freezing at or immediately below the surface at every latitude, defining the cryolithosphere's upper bound. (While temperatures in the uppermost diurnal skin depth of Mars can exceed 273 K, this "active layer" of the Martian cryolithosphere is expected to be thoroughly desiccated.) The depth to the lower bound of the cryolithosphere is considerably less certain, because neither the magnitude of the Martian geothermal heat flux nor the thermal conductivity of the crust are known.

The depth z to the base of the cryolithosphere can be calculated from the steady-state one-dimensional heat conduction equation

$$z = k\frac{T_{mp} - T_{ms}}{Q_g} \tag{2}$$

where k is the thermal conductivity of the regolith, T_{ms} is the mean annual surface temperature, T_{mp} is the melting point temperature and Q_g is the value of the geothermal heat flux (Fanale 1976). At present, only the current latitudinal range of mean annual surface temperatures is known to any accuracy (~ 160 to 220 ± 5 K), while the remaining variables have associated uncertainties of 25 to 100%. Plausible ranges for each of these variables are discussed below.

Four principal factors influence the thermal conductivity of terrestrial permafrost: bulk density, degree of pore saturation, particle size and temperature (Clifford and Fanale 1985). An increase in bulk density and/or pore saturation increases the thermal conductivity of permafrost because the conductivities of rock and ice are significantly higher than that of the air they displace. The effects of particle size and temperature are more complicated. Experiments have shown that thin films of adsorbed water remain unfrozen on silicate surfaces down to very low temperatures (Anderson et al. 1967; Anderson and Tice 1973), particularly in the presence of salts (Banin and Anderson 1974). Because the thermal conductivity of unfrozen water (~ 0.54 W m^{-1} K^{-1}; Penner 1970) is lower than that of ice, its presence will decrease the effective thermal conductivity of silicate-ice mixtures. As the quantity of H_2O adsorbed per unit surface area is roughly constant for all mineral soils, the conductivity of a frozen soil is related directly to its content of high specific surface-area clay; the effects on thermal conductivity of unfrozen water are generally seen only when a significant clay fraction is present. As the temperature of a frozen soil declines, so too does its content of adsorbed water. This results in an increase in the soil's effective conductivity, an increase that is compounded by the temperature dependence of the thermal conductivity of ice, which rises from 2.25 W m^{-1} K^{-1} at 273 K, to 4.42 W m^{-1} K^{-1} at 160 K (Ratcliffe 1962).

Remote thermal measurements at the two Viking landing sites indicate soil thermal conductivities in the range of 0.075 to 0.11 W m^{-1} K^{-1} (Kieffer 1976). However, in addition to the fine materials present, Viking Lander and Orbiter images also show that volcanics and other massive rocks make a significant contribution to the volume and mechanical properties of the outer portion of the Martian crust (see, e.g., Greeley and Spudis 1981). In light of this observation, it is important to note that terrestrial basalts typically have conductivities in the range of 1.5 to 3.5 W m^{-2} K^{-1} (Clifford and Fanale 1985). Given the range of possible variations in regolith composition and lithology, a thermal conductivity of 2 W m^{-1} K^{-1} (± 1.0 W m^{-1} K^{-1}) appears to be a reasonable guess for the top few km of the Martian crust (Clifford and Fanale 1985).

Another factor that influences the depth to the base of the cryolithosphere is the melting temperature of the ice. The melting point can be depressed below 273 K by both pressure and solute effects. The effect of pressure is minor (~ 7.43×10^{-8} K Pa^{-1}; Hobbs 1974), while the effect of dissolved salts can be quite large. The existence of various salts in the regolith is supported by the discovery of a duricrust layer at both Viking Lander sites and by the elemental composition of this duricrust as determined by the inorganic chemical analysis experiments on board each spacecraft (Toulmin et al. 1977; Clark 1978; Clark and Van Hart 1981). Among the most commonly cited candidate salts are NaCl, $MgCl_2$ and $CaCl_2$, which have associated freezing points (at their eutectic) of 252 K, 238 K and 218 K, respectively (Clark and Van Hart 1981). Multicomponent salt solutions can have eutectic temperatures as low as 210 K (Brass 1980). Although the presence of $CaCl_2$ and $MgCl_2$ cannot be ruled out, serious questions have been raised concerning their chemical and thermodynamic stability under ambient Martian conditions, particularly in the presence of abundant sulfates (Clark and Van Hart 1981). Given these arguments, brines based on NaCl are probably the most likely candidates to be found on Mars.

Based on the assumption that Mars has the same K, U and Th concentrations as chondrites, Fanale (1976) has estimated a current geothermal heat flux on the order of 3×10^{-2} W m^{-2}. More detailed thermal modeling by Toksöz and Hsui (1978) and Davies and Arvidson (1981) has yielded the slightly higher estimates of 3.5×10^{-2} W m^{-2} and 4×10^{-2} W m^{-2}, respectively. As noted previously, the geothermal heat flow of Mars was probably much greater in the past—particularly during the planet's first half Gyr of geologic history, the period in which it was radiating away most of its accretional heat.

Substituting the best current estimates of Martian heat flow, crustal thermal conductivity, and melting temperature into Eq. (2), the depth to the base of the cryolithosphere is found to vary from 1 to 3 km at the equator to approximately 3 to 8 km at the poles (Fig. 3)—the poleward increase in depth reflecting the corresponding latitudinal decline in mean annual surface temperature (Fanale 1976; Rossbacher and Judson 1981; Crescenti 1984; Clifford 1984,1987*b*). By integrating the crustal porosity profile given by Eq. (1) down to the melting isotherm depths calculated from Eq. (2), the pore volume of the Martian cryolithosphere can be estimated. For the likely physical and thermal properties of the megaregolith, this amounts to a volume of 10^7 to 10^8 km^3. (Note that this volume is distinct from the *total* megaregolith pore volume calculated above.) The quantity of H_2O required to saturate this pore volume is equivalent to a global layer of water approximately 70 to 700 m deep (Clifford 1984,1987*b*).

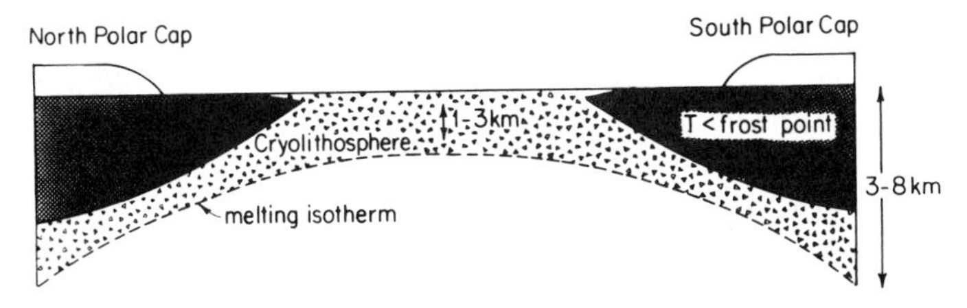

Fig. 3. A pole-to-pole cross section of the Martian crust illustrating the theoretical latitudinal variation in depth of the 273 K isotherm. Ground ice can exist in equilibrium with the atmosphere only at those latitudes and depths where crustal temperatures are below the frost point of atmospheric water vapor (~ 198 K; Farmer and Doms 1979). Outside these locations, ground ice can only survive if it is diffusively isolated from the atmosphere by a regolith of low gaseous permeability (see, e.g., Smoluchowski 1968; Clifford and Hillel 1983; Fanale et al. 1986) (Figure adapted from Fanale 1976 and Rossbacher and Judson 1981.)

C. Ground Ice Stability

Although mean annual surface temperatures are below freezing everywhere on Mars, observations made by the Viking Orbiter Mars Atmospheric Water Detectors (MAWD) indicate a global frost point temperature of roughly 200 K. As a result, the existence of ground ice in equilibrium with the water vapor content of the atmosphere is restricted to the colder latitudes poleward of about $\pm$ 40° (Farmer and Doms 1979).

Despite the current instability, there is a large body of morphologic evidence that suggests that ground ice has existed in the equatorial regolith throughout much of Martian geologic history (Carr and Schaber 1977; Mouginis-Mark 1979; Allen 1979; Rossbacher and Judson 1981; Squyres 1984; Carr 1986; Lucchitta 1987*c*). This evidence presents a problem, for it is difficult to account for both the initial origin and continued survival of ground ice in a region where the present mean annual temperature exceeds the frost point by more than 20 K. However, the problem may be resolved if equatorial ground ice is a relic of a former climate that has been preserved by the diffusion-limiting properties of a fine-grained regolith (Smoluchowski 1968; Kuzmin 1978; Clifford and Hillel 1983; Fanale et al. 1986).

The stability of equatorial ground ice is governed by the rate at which H_2O molecules can diffuse through the regolith and into the atmosphere. This process is complicated by the fact that the mean free path of an H_2O molecule in the Martian atmosphere is $\sim$ 10 μm. When the ratio of the pore radius r to the mean free path λ of the diffusing molecules is large ($r/\lambda > 10$), bulk molecular diffusion is the dominant mode of transport; the movement of molecules through the pore system occurs in response to repeated collisions with other molecules present in the pores. On the other hand, for very small pores ($r/\lambda < 0.1$), collisions between the diffusing molecules and the pore walls greatly outnumber those that occur with other molecules, leading to the process known as Knudsen diffusion. Because the frequency of pore wall collisions increases with decreasing pore size, small pores can substantially reduce the efficiency of the transport process. For pores of intermediate size ($0.1 < r/\lambda < 10$), the contributions of both processes must be taken into account. The effect is illustrated in Fig. 4, where the effective diffusion coefficient of H_2O is shown as a function of pore size.

The survival of equatorial ground ice was considered in detail by Clifford and Hillel (1983) and Fanale et al. (1986). They found that near the equator the regolith has probably been desiccated to a depth of several hundred m over the past 3.5 Gyr, assuming that our current knowledge of the quasi-periodic changes in Martian obliquity and orbital elements is accurate (Ward 1979; Toon et al. 1980; chapter 9). However, because the sublimation rate of H_2O is sensitively dependent on temperature, the amount of ice lost at higher latitudes is expected to be much less, falling to perhaps a few tens of m at 35° latitude (Fig. 5; see also chapter 33). Greater depths of desiccation might

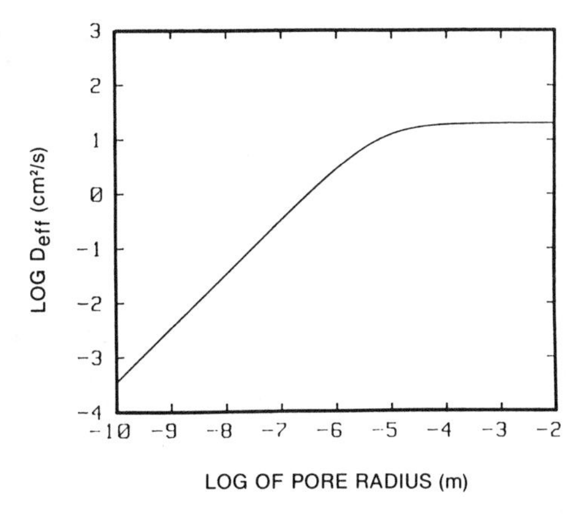

Fig. 4. A graph of the effective diffusion coefficient of H_2O as a function of pore size. The sloping part of the curve represents Knudsen diffusion, while the horizontal branch on the right side of the graph depicts the region of bulk molecular diffusion (figure from Clifford and Hillel 1983, Fig 3).

be possible if the effective pore size of the regolith is much greater than 10 μm, or if the regolith has a specific surface area ≥ 10^3 m^2 g^{-1}, which would give rise to a diffusive surface flux greater than any likely pore gas flux (Clifford and Hillel 1983). The general picture expected, then, is that ground ice will be found fairly close to the surface at high latitudes, but only hundreds of meters beneath the surface near the equator.

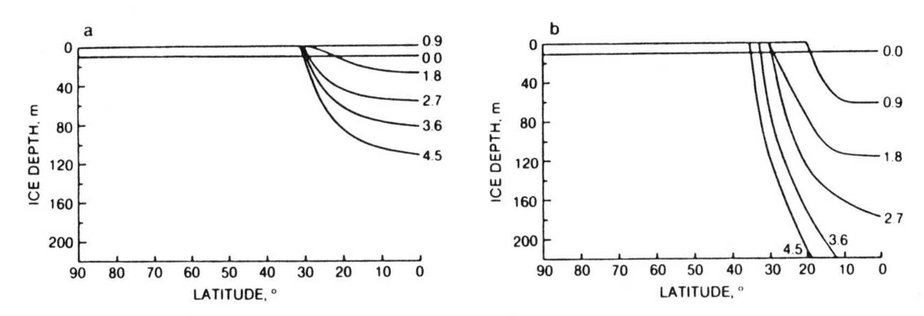

Fig. 5. The depth of ground ice on Mars as a function of latitude and time (Gyr). The regolith models adopted for these calculations differed only in assumed pore size: (a) 1 μm; (b) 10 μm (figure from Fanale et al. 1986, Fig. 5).

D. Groundwater

As discussed by Carr (1979,1986), Baker (1982) and Squyres (1984), the Martian outflow channels provide persuasive evidence that a large reservoir of groundwater was stored in the Martian crust throughout its first Gyr of geologic history (chapter 15). Much of this water may now be stored as ice in the cryolithosphere. However, once the pore volume of the cryolithosphere has been saturated with ice, any additional subsurface H_2O will inevitably be stored as groundwater.

Under the influence of gravity, groundwater will drain to saturate the lowest porous regions of the crust. Therefore, if we assume that Mars is uniform with a porosity profile everywhere as described by Eq. (1), we can readily calculate the saturated thickness of an aquifer that results from any given volume of water. For example, a quantity of water equivalent to a global layer 10-m deep is sufficient to saturate the lowermost 0.85 km of the porosity profiles seen in Fig. 2; while a quantity of water equivalent to a 100-m layer would create a global aquifer nearly 4.3-km deep (Clifford 1984,1987*b*). However, Mars is not uniform; therefore, the actual self-compaction depth of the megaregolith is expected to have a high degree of variability, restricting the areal coverage of any groundwater system to something $< 100\%$. This restriction in turn means that, for a given quantity of water, the resulting aquifer thickness will locally be greater than that calculated on the basis of global coverage.

If the present inventory of H_2O on Mars exceeds the quantity required to saturate the pore volume of the cryolithosphere, then a subpermafrost groundwater system of substantial proportions could result. This condition may be satisfied by a planetary inventory of H_2O as small as 100 m; however, in no event does it appear that it would require more than 1000 m (Clifford 1984,1987*b*).

E. Permeability

On Earth, the effective permeability of crustal rocks is critically dependent on the scale of the sample under consideration. For example, laboratory measurements of samples several cm in size generally indicate only the minimum permeability of a rock mass (Brace 1980). However, at the scale of interest for most hydrogeologic studies (generally 10^2 to 10^3 m), virtually all crustal rocks have undergone a considerable degree of fracturing, dramatically increasing their effective permeabilities (Streltsova 1976; Legrand 1979; Seeburger and Zoback 1982; Brace 1980,1984).

Despite the relatively low porosity of fractured rock, its permeability can often be quite high (Legrand 1979; Brace 1980; Fetter 1980). This is because the pore geometry of a fracture is inherently more efficient at conducting a fluid, per unit porosity, than the geometry of an intergranular pore network. A rock with a fracture porosity of only 0.011% can have the same permeability as a silt with a porosity of 50% (Snow 1968). The pervasive nature of crustal fractures means that few, if any, geologic materials can be considered impermeable (de Marsily et al. 1977; Fetter 1980; Brace 1980,1984). Brace (1980,1984) has summarized the results of *in situ* borehole permeability measurements and permeabilities inferred from large-scale geologic phenomena (such as earthquakes triggered by fluid injection from nearby wells) and concludes that on a size scale of $\sim$ 1 km, the average permeability of the top 10 km of the Earth's crust is roughly 10 millidarcies (where 1 darcy is defined as the permeability that will permit a specific discharge of 1 cm s^{-1} for a fluid with a viscosity of 1 centipoise under a hydraulic pressure gradient of 1 bar cm^{-1}; 1 darcy $= 9.87 \times 10^{-13}$ m^2).

Estimates of the permeability of the Martian crust have generally been much higher. Based on the assumption that each of the Martian outflow channels was carved by a single catastrophic outbreak of groundwater, Carr (1979) calculated the required volume of discharge and necessary permeability of the crust. His calculations suggest that permeabilities as high as 10^2 to 10^3 darcies would be necessary over distances of $\sim$ 10^3 km and to depths of 1 to 2 km. He concluded that permeabilities this high were only reasonable if the Martian crust was intensely fractured and/or possessed numerous unobstructed lava tubes, characteristics similar to those of certain Hawaiian basalts (Davis 1969). However, Carr (1979) also noted that the required discharge and crustal permeability could be reduced 1 to 2 orders of magnitude if the observed relief of the channels was created by shallower flows acting over a longer period of time, or by multiple episodes of groundwater outbreak and erosion.

F. Rheology

As we discuss below, there is morphologic evidence that the Martian regolith in some areas has undergone viscous creep induced by deformation of interstitial ice. Three factors have been found to influence strongly the strain rate of frozen ground undergoing steady-state creep: stress, temperature and regolith structure. The stress dependence of the rheology of ice depends on the deformation mechanism operating at the atomic scale. Two fundamental deformation mechanisms are possible: diffusion of atoms and movement of dislocations in the crystal lattice. For diffusion creep, the strain rate is roughly proportional to the applied stress; i.e., the material has a Newtonian rheology. For dislocation creep, the strain rate is proportional to stress to some power n, where n is typically $\sim$ 3. Whether diffusion or dislocation creep dominates in ice depends on stress, on temperature and on grain size (Shoji and Higashi 1978). Both mechanisms could operate in the Martian regolith during the deformation of a given topographic feature, perhaps at different times and at different depths.

The temperature dependence of ice and frozen soil rheology is relatively well understood. The rate of steady-state creep is generally proportional to $\exp(-Q/kT)$, where T is temperature, k is Boltzmann's constant and Q is the activation energy of the deformation mechanism. This temperature dependence should have a significant influence on regolith rheology as a function of latitude, with mobility substantially inhibited at latitudes higher than 50 to 60° (Lucchitta 1984*a;* Squyres and Carr 1986; Squyres 1989*c*).

The most important determinant of the rheology of an ice-silicate mixture can be the structure of the mix; that is, the relative proportions of ice and silicate and the way in which the grains interact. Depending on structure, the material can be as mobile as pure ice, or as rigid as pure rock. At small silicate concentrations in an ice-silicate mixture, grains of rock act as particles in suspension in the ice. They do not deform appreciably in response to stress, and a comparatively higher strain rate in the ice is required to provide a given bulk deformation rate for the mix as a whole. When the volume fraction of silicates exceeds about 0.5 to 0.6, however, the physical manner in which the grains and ice come into contact (the regolith "structure") becomes crucially important in controlling the rheology of the mix. If the silicate material is thoroughly comminuted and disaggregated, forming a true soil, then significant mobility still is possible at very high silicate fractions (see, e.g., Andersland and Akili 1967; Ersoy and Togrol 1978; Haynes 1978). However, other possible structures can have substantial ice content, but have rheologic properties controlled by the rheology of the rocky component. Fractured bedrock, in which large fractures have been produced but complete disaggregation, comminution and re-orientation of blocks have not taken place, can have such characteristics. As the transition is made from comminuted soil near the surface to solid bedrock below, structural effects will cause mobility of the Martian regolith to decrease with depth, dropping it to zero at some depth where the rheology becomes controlled by rock rather than ice.

II. POTENTIAL MORPHOLOGIC INDICATORS OF GROUND ICE

Many physical properties of the Martian regolith can be expected to have an impact on the present distribution of ice beneath the surface. Based on expected regolith properties and estimates of the total Martian H_2O inventory, one may infer a likely distribution of ground ice like the one shown in Fig. 5. However, the distribution may also be inferred directly from observation of morphologic features on the planet's surface. In the sections that follow, we discuss several Martian geomorphic features that may provide evidence for the former or present existence of ground ice. We also describe the geographic distributions of the major features, and compare them to the predictions of thermodynamic models.

A. Rampart Craters

1. Morphology. Most large craters on Mars for which distinct ejecta blankets can be observed have ejecta morphologies that are very different from those observed on, for example, the Moon. Large Martian craters typically have an ejecta sheet with a pronounced low ridge or escarpment at its outer edge. The ejecta sheet commonly has a distinctly lobate outer margin, giving the appearance of a flow formed by the rapid outward spread of a highly mobile fluid. Larger craters may have several overlapping sets of flow lobes. The flows can bury a variety of topographic features in the underlying terrain, and appear massive enough to contain the majority of the total ejecta from the crater. Craters possessing such ejecta are referred to as rampart craters (chapter 12). The morphology of rampart craters strongly suggests that ejecta were emplaced primarily as radially directed surface flow, rather than ballistically. The most convincing evidence for fluid emplacement is the geometry of the flow lobes, and particularly the observation that they can be diverted around relatively small pre-existing topographic obstacles, rather than draped over them (Carr et al. 1977*b*).

The most likely explanation for the morphology of rampart craters is that their ejecta were emplaced as a mud flow. The water entrained in the ejecta apparently was derived from beneath the Martian surface. In fact, a number of the basic morphologic characteristics of the craters can be duplicated by high-velocity experimental impacts into mud (Gault and Greeley 1978). Significant questions still concern the role of the atmosphere in the development of fluidized ejecta deposits, and it has been suggested that certain aspects of rampart crater morphology are due to atmospheric effects rather than to the presence of subsurface H_2O (Schultz and Gault 1979; Schultz 1986). It also is not clear whether the subsurface H_2O that may have been involved was in the liquid or solid state at the time of the impact (See the discussion of lobate ejecta craters in chapter 12.) For the purposes of our discussion here, however, we will accept what we believe to be the most probable explanation for rampart craters, namely that they result from melting of subsurface ice during the impact process.

2. Distribution. Rampart craters with a wide range in diameters constitute a considerable fraction of all the craters on Mars. As has been found by many investigators (Carr et al. 1977*b;* Allen 1978; Mouginis-Mark 1979; Kuzmin 1980*a*, 1983; Horner and Barlow 1988; Costard 1989), rampart craters are present in almost every major geologic unit on Mars and at all latitudes from the equator to the edges of the polar caps. They have formed at altitudes ranging from nearly the lowest on Mars to over 8 km above the planetary datum. Despite their ubiquitous nature, however, they show important regional and global trends in their morphology and morphometry that may be used to make a number of inferences about the subsurface distribution of H_2O on Mars.

From the physics of the impact process (Croft et al. 1979; Croft 1984; Kieffer and Ahrens 1980; Chapman and McKinnon 1986; Melosh 1989) it is known that vaporization and melting of ground ice at the moment of impact may occur in the part of the subsurface that is shocked to roughly 10 to 100 GPa. The volume of such a zone is directly related to the crater's size; it becomes larger with an increase in the crater's diameter (Ivanov 1986). If the origin of the fluidized ejecta is connected with the vaporization and the melting of ground ice, then fluidized ejecta may be produced in cases where crater excavation depths are greater than the depth to the top of an ice-rich layer in the regolith.

A fundamental observation concerning rampart craters is that, in a given area, a certain critical crater size exists (Boyce 1980; Kuzmin 1980*a*); craters smaller than this size lack fluidized ejecta, while craters larger have it. This crater size, called the onset diameter, is in some sense an indicator of the depth to which an impact event must excavate to reach significant quantities of ice. Mapping of onset diameters as a function of geographic location on Mars therefore may allow a good semi-quantitative test of the ground ice distributions predicted by the thermodynamic models discussed above.

The most complete statistical studies of the geographic distribution of rampart craters have been performed recently by Costard (1988,1989) and Kuzmin et al. (1988*b*,1989). These studies involved characterizing the morphology, morphometry and location of over 10,000 Martian rampart craters. Figure 6 shows the geographic distribution of onset diameters. The most important finding is that there is a pronounced dependence of onset diameter on latitude. Near the equator, onset diameters are typically 4 to 7 km. At latitudes of 50 to 60°, however, they decrease to just 1 to 2 km. Apparently, the depth to ground ice (i.e., the thickness of the desiccated upper layer of the regolith) is substantially greater near the equator than it is at high latitudes. This result is in agreement with predictions of thermodynamic models: loss of ground ice has proceeded to the greatest depths at the lowest latitudes.

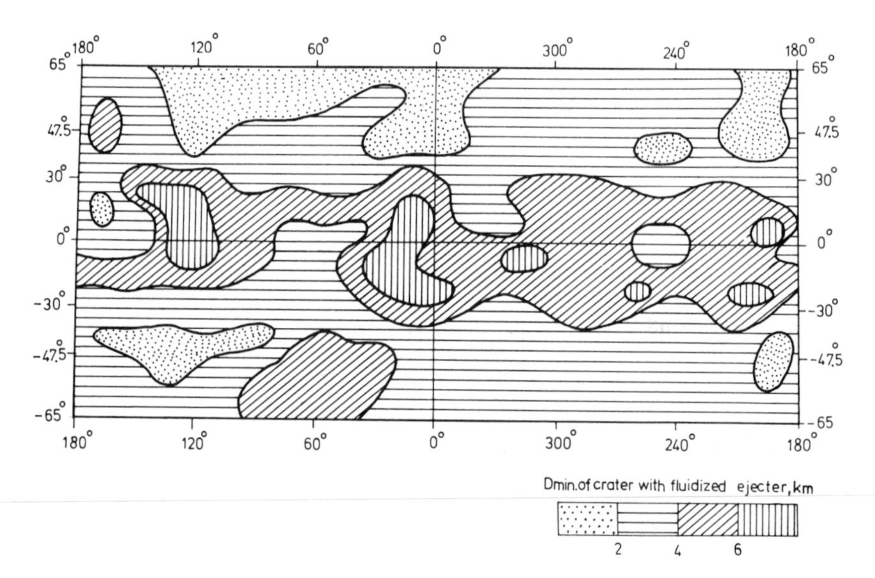

Fig. 6. The geographic distribution of rampart crater onset diameters on Mars. Onset diameters are greatest near the equator, indicating that the depth to ground ice is greatest in this region.

Interpreting the geographic distribution of onset diameters in terms of an absolute depth to the top of an ice-rich layer is a somewhat more difficult problem. The issue is complicated by several factors. First, it is not immediately clear how deeply the impact must excavate into the ice-rich layer for an observable fluidized ejecta deposit to develop. The answer probably depends on the concentration of ice in the lower layer; deeper penetration of this layer will be required for lower ice content. A second problem is that thermodynamic considerations suggest that the thickness of the upper desiccated zone should increase gradually over time. Since the craters in a given area may have formed over some significant time interval, they may sample a range of ice depths, blurring the quantitative conclusions reached from onset diameters. Nevertheless, attempting to infer ice depth from onset diameter is instructive, and this approach has been taken by Kuzmin et al. (1988*b*, 1989). For this work, a relationship between observed crater diameter and the depth of the crater's transient cavity was used, derived from the physics of the impact process (Stoffler et al. 1975; Croft 1980,1984; Ivanov 1988*a*). It was assumed that any penetration of the ice-rich layer by the transient cavity would be sufficient for fluidized ejecta formation. This assumption should yield an upper limit on the depth to the ice-rich layer at the time of impact. A map of the depth to ground ice on Mars derived in this manner is shown in Fig. 7. Under the assumptions made, the thickness of the desiccated upper

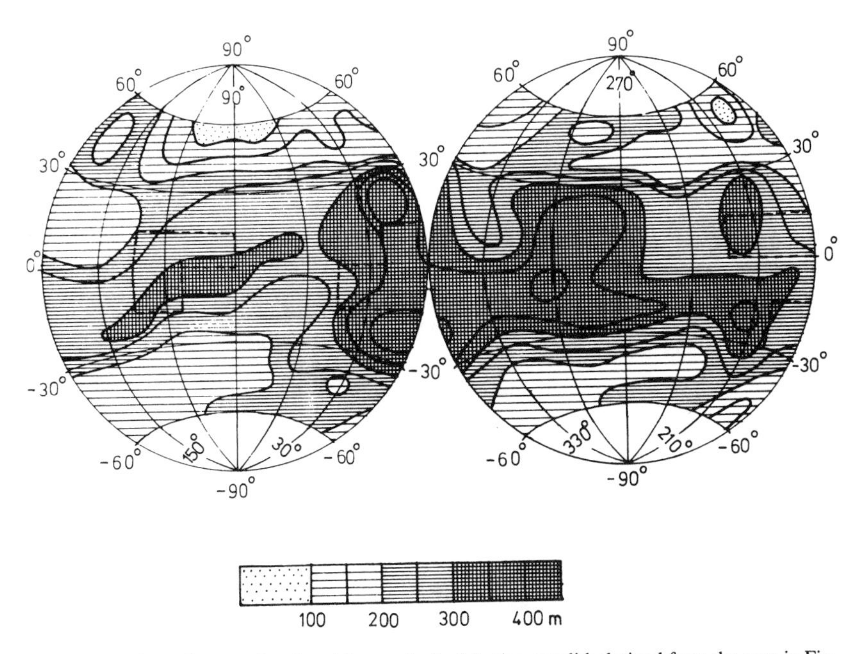

Fig. 7. Depths to the top of an ice-rich zone in the Martian regolith derived from the map in Fig. 6 and assumptions about the nature of the rampart crater formation process.

zone of the regolith is typically 300 to 400 m near the equator. At 30° latitude the roof of the ice-rich zone rises to 200 to 250 m, and at 50° latitude to 100 m. As a rule, the maximum roof depths in equatorial regions are localized in the more elevated areas of the planet (e.g., Syria Planum, Noctis Labyrinthus, Elysium Planum) or in the oldest and highest areas of the cratered terrain within the southern highlands. Roof depths are slightly greater in the southern hemisphere than in the northern hemisphere.

It is difficult to determine quantitatively the ice content of the subsurface materials. An assessment of the relative ice content may be possible, however. The basic technique that has been applied to this problem is measurement of the ratio of the fluidized ejecta blanket diameter to the crater diameter (Kuzmin et al. 1988*b;* Costard 1988). The assumption made is that relatively more extensive fluidized ejecta deposits indicate a higher concentration of ice in the ice-rich subsurface layer. Figure 8 shows the variation of this ratio as a function of latitude for two different ranges of crater diameter. These data suggest that the concentration of ice in the ice-rich subsurface materials increases with increasing latitude. They also suggest that the regolith of the northern hemisphere is somewhat more ice-rich than that of the southern hemisphere. The reason for this asymmetry may be that the northern plains of Mars were the primary areas of accumulation for the great Martian outflow channels. As the water from these floods pooled and soaked into the ground, it could have substantially enriched the ice content of the underlying regolith.

In summary, study of rampart craters and mapping of their morphologic

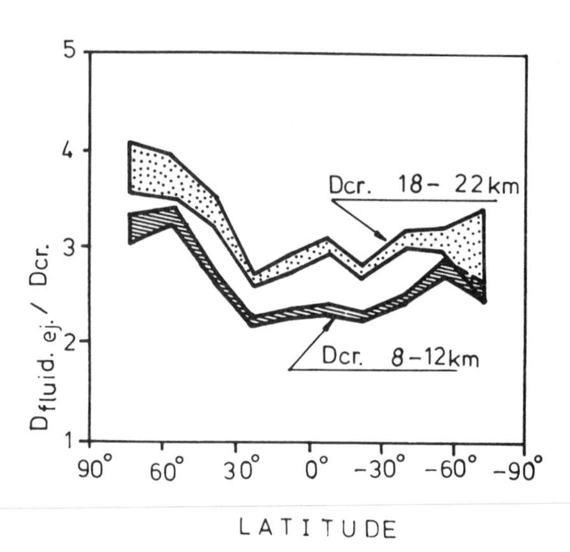

Fig. 8. Ratio of fluidized ejecta sheet radius to crater radius, as a function of latitude, for two size ranges of Martian craters. The ratio increases with latitude, suggesting that subsurface regions may be relatively richer in ice at higher latitudes.

and morphometric parameters shows that the main feature of the Martian cryolithosphere is a latitudinal zonality, with prominent desiccation of shallow layers at the equator. This inferred distribution of ice is in excellent qualitative agreement with the predictions of thermodynamic models. The ice concentration appears greater and the spatial distribution somewhat more complex in the northern hemisphere than the southern one, perhaps reflecting the processes that have been responsible for emplacement of the ground ice in the north.

B. Debris Flows and Terrain Softening

Another important way in which subsurface ice can influence the morphology of a planetary surface is by promoting slow creep and flow in the solid state. By geologic standards, ice is an extremely mobile material. Ice-rich permafrost on Earth can undergo substantial deformation as the ice that cements the soil deforms and flows in the solid state (Thompson and Sayles 1972; Phukan 1983; Sadovsky and Bondarenko 1983; Weerdenburg and Morgenstern 1983). Rock glaciers, which are found in many arctic and alpine environments, are rock-ice masses that deform and flow due to solid-state creep in the ice that cements the rocky debris; again, no thawing is involved (Ives 1940; Wahrhaftig and Cox 1959; Thompson 1962; White 1976). We discuss here two classes of Martian landforms that result or may result from viscous creep of near-surface materials. The first class is formed of materials that have been loosened from their original locations, transported down an escarpment by mass-wasting, accumulated at the base of the escarpment, and undergone subsequent creeping flow. These debris flows include *lobate debris aprons, lineated valley fill,* and *concentric crater fill* (Squyres 1978,1979*a*). They may be directly analogous to terrestrial rock glaciers. The second class, called *terrain softening,* is really a style of landform degradation produced by viscous deformation of material *in situ.* It is more directly analogous to the solid-state deformation observed in terrestrial permafrost, or to the relaxation of topography that is observed on some icy satellites in the outer solar system.

1. Debris Flows. Lobate debris aprons (Fig. 9) are thick accumulations of debris at the bases of escarpments. Their most noteworthy morphologic characteristic is their pronounced convex topography. The surface of the debris apron slopes gently away from the source escarpment, and then steepens to form a distinct flow terminus. This morphology is a clear indication that deformation and flow have taken place throughout a substantial thickness of the deposits. Lobate debris aprons can exhibit distinctive surface lineations (Squyres 1978), both parallel and transverse to the flow, that have counterparts in terrestrial rock glaciers (Wahrhaftig and Cox 1959).

Lobate debris aprons are very common in some areas, particularly in the fretted terrain separating the northern lowlands from the southern highlands between 280 and 350° longitude. In locations where flows commonly run up

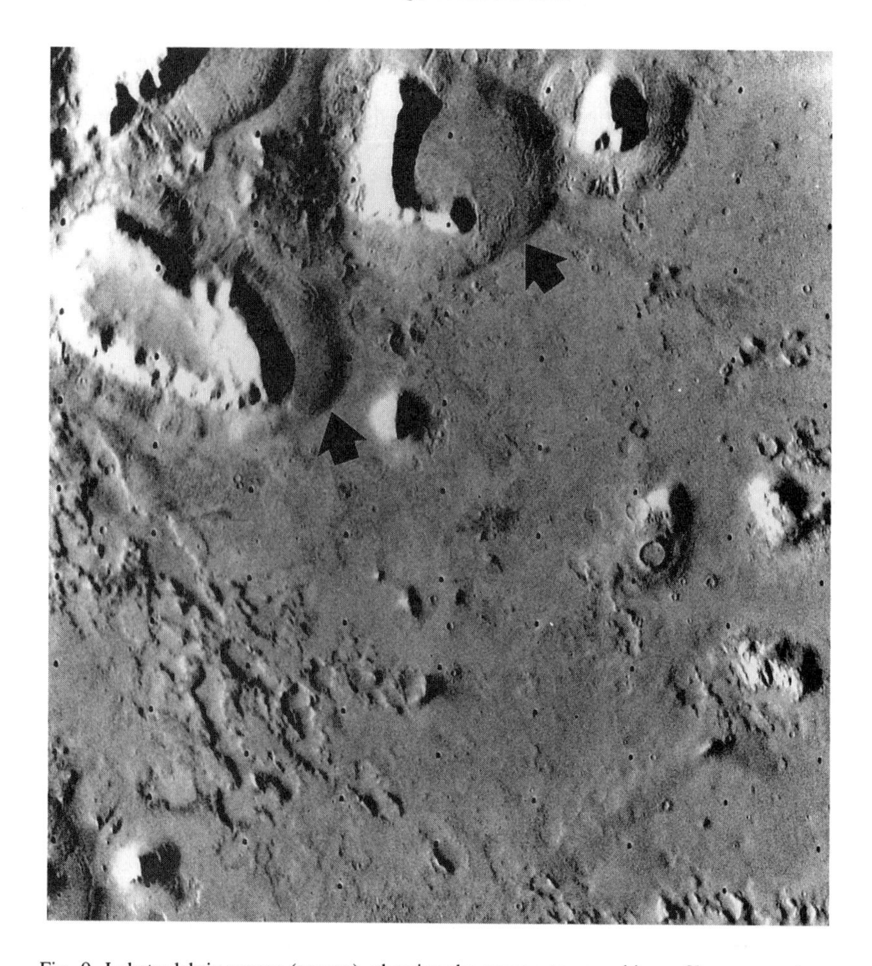

Fig. 9. Lobate debris aprons (arrows), showing the convex topographic profiles commonly exhibited by these features. (Viking Orbiter image 58B51; latitude 47°, longitude 325°. Image width is ~ 90 km.)

against opposing escarpments or against one another, complex patterns of flow and compressional lineations can form. An extreme case is shown in Fig. 10. Here the debris has been confined entirely in narrow valleys. The pervasive obstruction of flow outward from the valley walls has produced many compressional lineations parallel to those walls. These valley-floor deposits have been termed lineated valley fill (Squyres 1978). They are composed of the same material that forms the lobate debris aprons, and are the other end-member of a continuum of landforms produced by this material in the fretted terrain.

Another class of landforms that has been interpreted to be formed from flow of ice-cemented erosional debris is shown in Fig. 11. Material is ob-

Fig. 10. Lineated valley fill, apparently composed of the same kind of material that makes up lobate debris aprons. (Viking Orbiter image 84A73; latitude 34°, longitude 290°.)

served on the floors of craters with a distinct, crudely concentric pattern of ridges and troughs. Such deposits have been called concentric crater fill (Squyres 1979*a*), and the topography has been interpreted as resulting from compressional stresses generated by inward flow of ice-rich material from the crater walls.

The interpretation of most concentric crater fill as having resulted from ice-aided flow is less clear cut than is the case for lobate debris aprons and lineated valley fill. Arguments in support of this view include the morphologic similarity between concentric crater fill and the other types of flows, the observation of landforms transitional between lobate debris aprons and concentric crater fill (Squyres 1989*c*), and the fact that the geographic distributions of all three types of features are similar and consistent with theoretical predictions of ground-ice stability (see below). An alternative view of con-

Fig. 11. Impact craters in Utopia Planitia containing concentric crater fill. (Viking Orbiter image 10B70; latitude 42°, longitude 272°.)

centric crater fill, however, is that much of the observed morphology is aeolian in origin. High-resolution (< 10 m/pixel) images of materials both within and surrounding craters in the Utopia region (45°N, 271°W) show an eroded, multiple-layer deposit emplaced on the cratered topography, most likely by aeolian processes (Zimbelman et al. 1989). Such layering would not be preserved well if the material moved by slow deformation throughout its thickness. It is likely that both ice-related and aeolian processes have been active throughout the Martian midlatitudes, so that care must be exercised in attributing concentric crater fill morphology to ground ice in areas where high-resolution images are lacking.

2. Terrain Softening. Terrain softening is a distinctive style of landform degradation observed in some regions on Mars. It affects landforms of all types, but one type of landform on which it is particularly evident is impact craters. An example is given in Fig. 12. The first image (a) of the pair shows craters in normal, i.e., unsoftened, terrain. The second image (b) presents a similar scene in softened terrain. The differences are striking. In the unsoftened terrain, the topography is crisp and rugged. Crater rims are sharp. Crater walls are blocky and angular, with slope profiles that are dominantly straight or concave upward. Between large craters are low but angular ridges and hills. In the softened terrain, however, crater rims are broad and gently rounded. Crater walls preserve their blocky nature to some degree, but the individual blocks are softly rounded rather than angular. Slope profiles of crater walls are dominantly convex upward. The terrain between large craters has a smooth rolling character and lacks the angular ridges and hills characteristic of unsoftened terrain. Terrain softening is by no means limited to impact craters, and affects all manner of topographic features (Squyres 1989*c*). In all cases, the effects are the same: terrain softening produces broad rather than sharp slope inflections, and convex rather than concave slope segments.

Before attributing terrain softening to ice-induced creep, however, alternative causes for the observations must be considered. Processes that might contribute include atmospheric obscuration and burial by aeolian debris. Suspended aerosols can degrade the detail visible in Viking images at high spatial frequencies, and periods of high opacity should be avoided in statistical studies (Kahn et al. 1986). However, examination of images of unsoftened terrain acquired during times of even very high atmospheric opacity still clearly show the crisp nature of the topography (Squyres 1989*c*). Aeolian mantling may contribute significantly to a general muting of topography, and clearly has taken place in some areas on the planet. Soderblom et al. (1973*b*) concluded on the basis of Mariner 9 images that the middle-to-high latitudes of Mars are largely blanketed with aeolian debris, and layered deposits observed within and around some impact craters appear to confirm that widespread, perhaps cyclic mantling has taken place in some midlatitude regions. High-resolution Viking images show that both creep and mantling processes have operated on Mars, with creep being most important at the middle latitudes and mantling most important at higher latitudes. Discriminators between the effects of mantling and creep include (a) the observation that aeolian debris tends not to accumulate readily on sharp, exposed ridge crests, so that such features commonly remain sharp even in mantled terrains; and (b) the tendency of mantling to most effectively remove the smallest topographic features, while creep preserves small features as the stresses they generate are insufficient to lead to their relaxation. Because of the importance of aeolian blankets on Mars, these discriminators must be applied with care when inferring the distribution of subsurface ice on Mars from terrain softening.

Color Plate 9 presents a map of the distribution of viscous creep features on Mars (Squyres and Carr 1986), compiled with consideration of the key

Fig. 12. Images showing unsoftened (a) and softened (b) terrain. (Viking Orbiter image 423S10; latitude − 32°, longitude 227° with image width ~ 160 km. Viking Orbiter image 195S20; latitude 33°, longitude 313° with image width ~ 60 km.)

morphologic discriminators for creep deformation. The features observed are grouped into three classes: (1) lobate debris aprons and lineated valley fill; (2) concentric crater fill; and (3) creep-related terrain softening. The distribution shows a pronounced variation with latitude. Virtually no creep features are found within 30° of the equator. In the northern hemisphere, lobate debris aprons and lineated valley fill are most common in Tempe Fossae, Mareotis Fossae, the Phlegra Montes, in a small region of old cratered terrain north of Olympus Mons, and particularly in regions of fretted terrain. Concentric crater fill is found in most of these areas, and also is common on Utopia Planitia. In the southern hemisphere, lobate debris aprons/lineated valley fill and con-

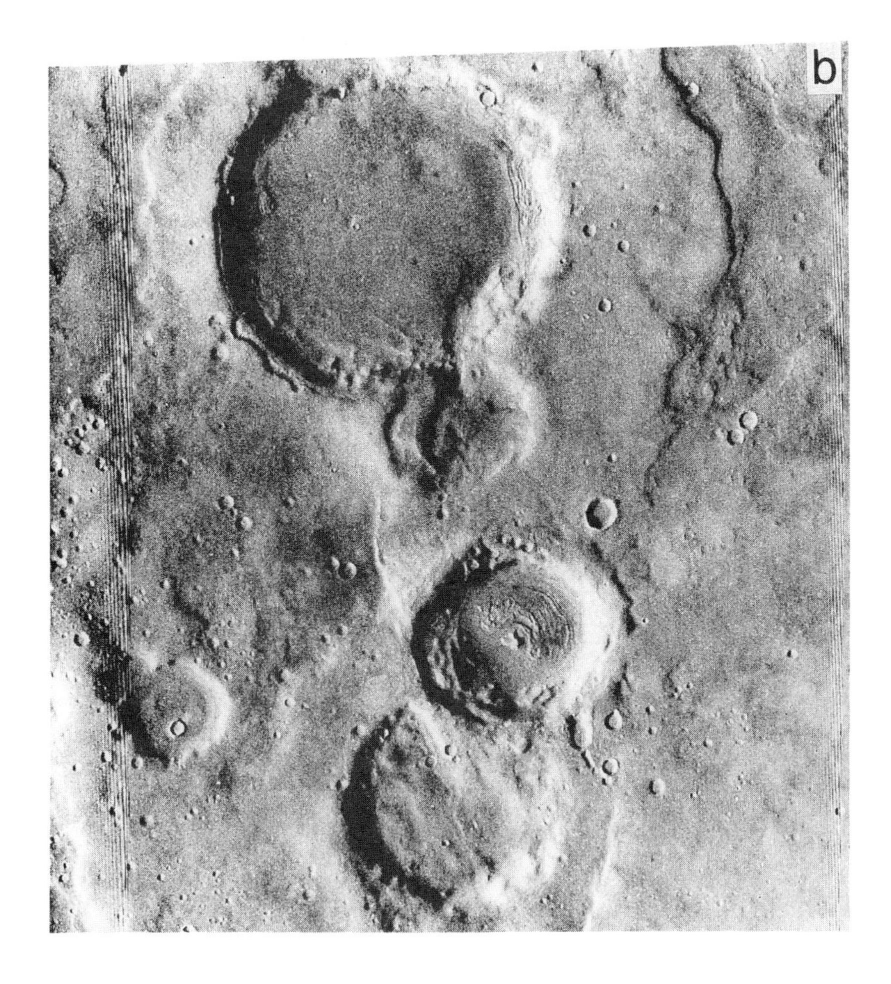

centric crater fill are most common in regions adjacent to the Hellas and Argyre basins. Creep-related terrain softening in the northern hemisphere is present in all areas where the ancient cratered highlands extend north of 30° latitude. Cratered highlands dominate the southern hemisphere, and most regions observed south of $-30°$ latitude exhibit some degree of creep-related terrain softening. It is ubiquitous in the midlatitudes, and becomes less pronounced toward the south pole. The distribution of creep features in Plate 9 again is broadly consistent with thermodynamic predictions of ground-ice stability. Where the calculations indicate that ground ice should be stable close to the Martian surface, creep features are observed in abundance. Where they indicate that ice is unstable near the surface, creep features are essentially absent.

There is also a change in the style and prevalence of creep-related terrain softening over the latitudes where it is observed. At middle latitudes (30 to ~ 55°), it is common and has the morphologic characteristics already described. At higher latitudes, however, it becomes much less widespread and pronounced, and other creep features (lobate debris aprons, etc.) disappear altogether. Crater rims are sharper than at the middle latitudes, and intercrater areas are rougher. The decrease in the observed frequency and extent of creep-related terrain softening at high latitudes is most likely due to the colder temperatures and higher viscosities found there.

Images of softened terrain convey the impression that the creep that apparently produces it is a relatively near-surface phenomenon. This impression can be investigated with a simple finite-element model of crater relaxation in a viscous medium overlying a rigid substrate (Squyres 1989*c*). Results are shown in Fig. 13. At the top of the figure is the profile of a fresh 20-km crater before deformation. The lower part shows the crater after roughly equivalent amounts of relaxation for 4 thicknesses of the deforming layer: 20, 5, 2.5 and 2 km. With relaxation in a deep viscous layer, the longest topographic wavelengths are removed most rapidly, producing a pronounced upwarping of the crater floor and leaving the crater rims sharp. As the depth to the rigid substrate becomes smaller, however, deep rim-to-bowl flow is suppressed, and the crater floor remains flat. The rim is lowered, and is broadened significantly. The profile of the crater walls is altered markedly, from concave to distinctly convex. This is the style of deformation observed in creep-related terrain softening. These results indicate that the thickness of the deforming

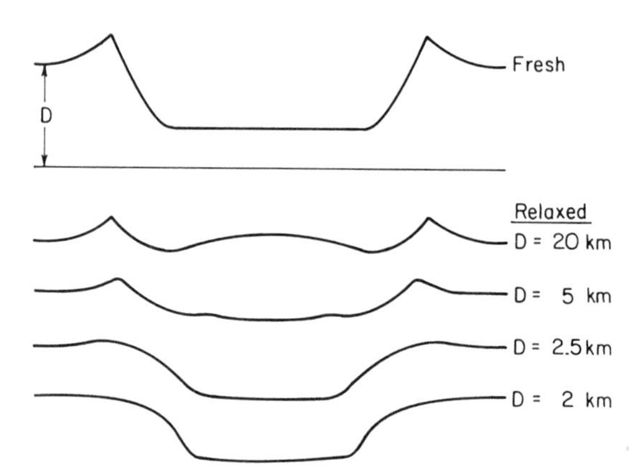

Fig. 13. Results of finite-element calculations of relaxation of a crater in a viscous material overlying a rigid substrate. Upper profile shows the topography of the initial crater, with a vertical exaggeration of 2.5 times. Lower 4 profiles show topography at the same vertical exaggeration after comparable amounts of relaxation for 4 different thicknesses of the viscous layer. In order to produce topography like that observed in softened terrain, the viscous layer thickness must be a small fraction of the crater diameter.

layer must be a very small fraction of the diameter of the relaxed crater to produce the observed morphology by viscous relaxation.

C. Other Possible Morphologic Indicators of Ground Ice

Several other morphologic features on Mars have been proposed as potential indicators of the presence of H_2O within the near-surface materials. Many such features involve the former presence of zones within the cryolithosphere (taliks) where temperatures were above the freezing point, and liquid water could have been present. These features are thoroughly reviewed elsewhere (see e.g., Carr and Schaber 1977; Rossbacher and Judson 1981; Lucchitta 1985*a*) and only some of the more widely cited landforms are discussed here. Along with a description of each feature, we include a brief discussion of the primary arguments for and against a direct association with subsurface volatiles.

1. Chaotic Terrain. This landform consists of lowland areas separated from cratered uplands by abrupt escarpments and displaying a rough floor topography with a haphazard jumble of large angular blocks and arc-shaped slump blocks (Sharp 1973*a*). The interpreted formation process involves collapse induced by removal of subsurface water or ice (Carr and Schaber 1977). Evidence in support of an origin related to ground ice includes several examples of large outflow channels that extend downslope from areas of chaotic terrain, indicating a probable causal relationship between the eroding fluid and the chaotic terrain. Sharp (1973*a*) reviews several possible hypotheses for the formation of chaotic terrain, including withdrawal of either subsurface ice or magma and the flow of groundwater, and concludes that all three mechanisms could have been involved. Carr (1979) supports the dominant role of groundwater by suggesting that release of water from a confined aquifer could explain the common association between chaotic terrain and outflow channels.

2. Thermokarst. This is the process whereby melting of ground ice causes local collapse and formation of rimless depressions (alases) which generally are roughly circular in outline and flat floored (Carr and Schaber 1977). Features interpreted to be possible alases have been observed in Chryse Planitia and near Maja Vallis (Theilig and Greeley 1979) and Ares Vallis (Costard 1987). An example is shown in Fig. 14. Along the irregular outline of the cliff which may have resulted from coalescing alases is a band of higher albedo that may relate to a change in the tableland materials in proximity to the cliff (Carr and Schaber 1977). Discrete ice lenses within the Martian surface material could enhance the efficiency of thermokarst development (Lucchitta 1985*a;* Costard and Dollfus 1987). The presence of isolated, buried ice deposits at equatorial latitudes is consistent with the interpretation of entrained ice in some fluidized landslides within Valles Marineris

Fig. 14. Possible thermokarst terrain in the southern part of Chryse Planitia, as identified by Carr and Schaber (1977). Small depressions in the plateau area appear to enlarge and merge as the plateau is destroyed, quite possibly through the removal of ground ice with subsequent slope failure. An albedo feature that follows the outline of the cliff may indicate some change in the plateau materials caused by proximity to the cliff. (Viking Orbiter image 8A70, latitude 23°, longitude 36°. Image width is ~ 50 km.)

(Lucchitta and Ferguson 1983; Lucchitta 1987*c*), although these features are not associated with a thermokarst source. Devolatilization from ice lenses exposed by channel formation may contribute to the cuspate banks observed along some Martian channels, analogous to thermocirques present in Siberia (Czudek and Demek 1970).

Arguments in favor of the features cited above being thermokarst are numerous well-documented cases of terrestrial features similar in size and morphology that occur in regions possessing 80 to 90% (by volume) of ground ice (Washburn 1973; Soloviev 1973). The main argument against a ground-ice origin is the possibility of alternative methods for eroding the tablelands without the required presence of a subsurface volatile, such as dry mass wasting in a vertical section possessing variable competency.

3. Patterned Ground. This category consists of a variety of features including circles, polygons, nets and stripes, all of which are associated on Earth with ice wedging in periglacial regions (Washburn 1973). Polygons are the dominant form of patterned ground identified on Mars. The polygons on Mars found in close proximity to polar ice deposits (Fig. 15a) typically are at least an order of magnitude larger than comparable ice-related features on Earth (Rossbacher and Judson 1981). The troughs defining them have upturned edges similar to cracks associated with active terrestrial ice-wedge polygons (Lucchitta 1981).

Mechanical considerations make it unlikely that the giant polygons, with diameters of up to 20 km, are the result of thermal cracking (Pechmann 1980). Strong arguments have been presented that the giant polygons in the northern plains of Mars may be due to fracturing of a layer of sediment deposited over ancient heavily cratered terrain, a scenario not requiring the presence of ice wedges (McGill 1986). This scenario is consistent with other features in the northern lowlands that are interpreted to be sediments derived from the southern highlands, transported through the numerous outflow channels that debouch on the northern lowland plains (Lucchitta et al. 1986).

Polygons in the size range consistent with terrestrial periglacial polygons also have been identified in high-resolution images of Mars (Fig. 15b; Evans and Rossbacher 1980; Lucchitta 1983). These small features cover a much more restricted region than the more prominent large polygons. They have good potential for being related to periglacial processes, but the origin of the giant polygons is ambiguous.

4. Pseudocraters. These landforms are closely spaced cinder cones, generally $<$ 100 m across, derived from phreatic eruptions where lavas have been emplaced over materials containing either abundant groundwater or ice (Thorarinsson 1953). In the Cydonia Mensae region of Mars, numerous small, aligned domes are present, some of which have summit depressions (Frey et al. 1979; Allen 1979). The mounds are near the limit of resolution in most images, so that a phreatic eruption origin cannot be directly verified. Pseudocraters may be useful in providing supporting evidence for the presence of subsurface volatiles, but their uncertain origin makes them an ambiguous indicator by themselves.

5. Pingos. These landforms are domelike mounds with ice-cored interiors that result from the injection of groundwater, perhaps aided by freezing of the surrounding ground (Flint 1971, p. 280). Degraded pingos are characterized by summit craters (Rossbacher and Judson 1981), and the collapse depressions on pingos usually have an irregular shape (Lucchitta 1985).

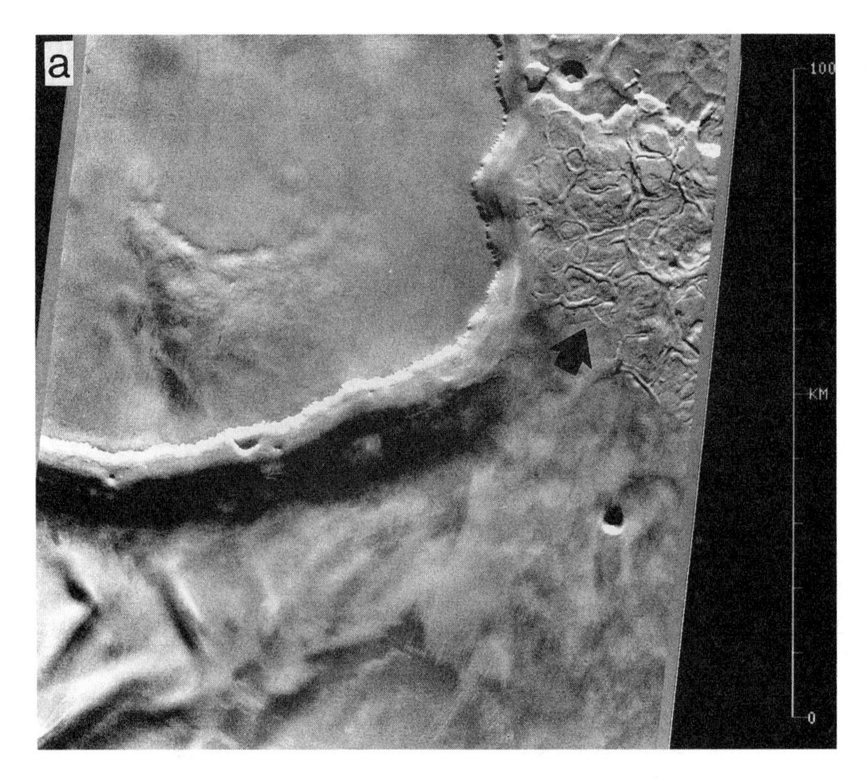

Fig. 15. (a) Polygonally fractured patterned ground near the margin of the north polar ice cap on Mars. Note that the troughs making up the polygons have raised edges, similar to active ice-wedge polygons on Earth (Lucchitta 1981). Individual polygons range from 2 to 10 km in diameter. (Viking Orbiter image 56OB42, latitude 81°, longitude 63°. Image width is ~ 115 km.) (b) Polygonally fractured patterned ground in the Deuteronilus Mensae region of Mars. The polygons range from 50 to 300 m in diameter, much closer in scale to polygonal ground on Earth than the polygons in (a). (Viking orbiter image 458B67; latitude 47°, longitude 346°.)

These characteristics are similar to those of the possible pseudocrater mounds on Mars discussed above, but once again the Viking image resolution is insufficient to provide conclusive evidence for or against an ice-related origin (Rossbacher and Judson 1981; Lucchitta 1981, 1985). Support for the possibility of pingos on Mars comes from the association of a few mounds with alases (Theilig and Greeley 1979), but most Martian mounds with summit craters are considerably larger than the largest pingos on Earth (Coradini and Flamini 1979). Thermodynamic constraints (Coradini and Flamini 1979) make it unlikely that significant quantities of near-surface water would be available for injection into Martian pingos, and indeed suggest that near-surface freeze thaw should not play a significant geomorphic role under the present Martian climate.

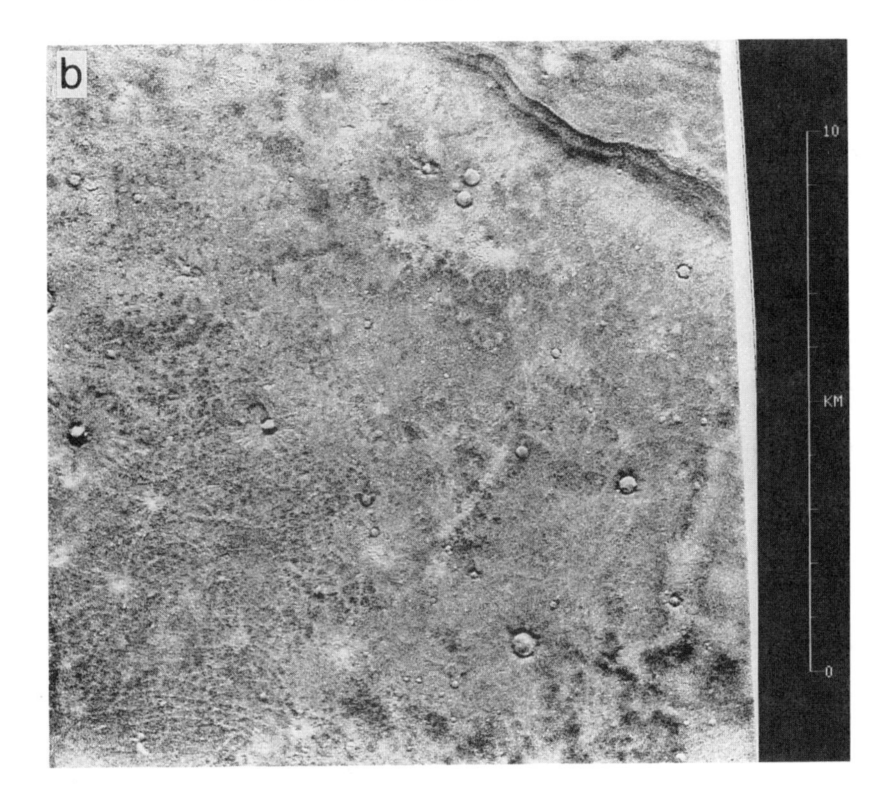

6. Table Mountains. A common landform in Iceland, these form when volcanic eruptions occur beneath a thick cover of ice, producing steep-sided mountains topped by gently sloped volcanic flows (Jones 1970). Features similar to table mountains have been identified in the northern plains of Mars (Hodges and Moore 1979; Allen 1979) but, with the rather extreme exception of Olympus Mons (Hodges and Moore 1979), definite evidence of volcanic materials on top of the Martian mesas is lacking. While a large amount of H_2O may exist on Mars, it is not likely that a great deal of this water was ever on the surface at one time or frozen into km-thick sheets of ice. Without clear evidence of a volcanic center on the Martian mesas, then, these features cannot be considered reliable indicators of the presence of ice.

7. Curvilinear Ridges. Ridges in concentric arcs with lengths of tens of km and spacing of 2 to 5 km are common at certain locations on Mars, particularly in the northern lowland plains. The curvilinear pattern gives a "fingerprint" appearance to the plains (e.g., Fig. 16). The origin of these features is problematic at best (Rossbacher and Judson 1981). There is no direct evidence of involvement of volatiles in the generation of the curvilinear features, but they have been related to erosion of midlatitude mantling mate-

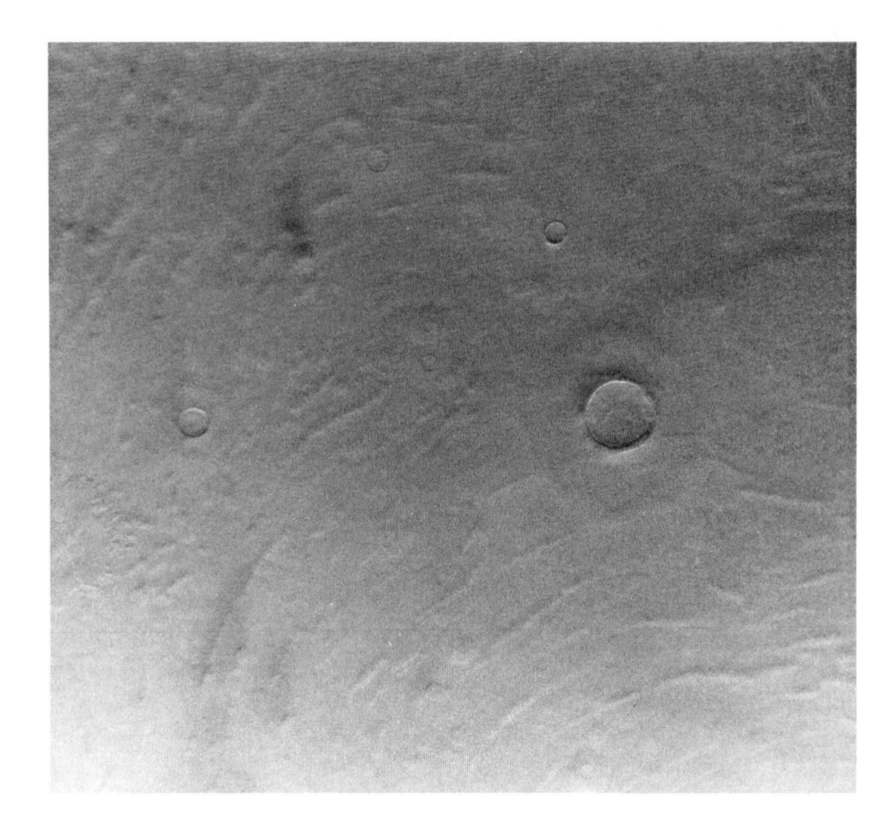

Fig. 16. Curvilinear ridges on the northern plains, arranged in patterns similar to what is observed in some periglacial patterned ground on Earth. (Viking Orbiter image 57B56, latitude 50°, longitude 287°; scale across frame is about 90 km.)

rial (Carr and Schaber 1977), glacial moraines (Lucchitta 1981), and an ancient volatile-rich debris layer in Isidis Planitia (Grizzaffi and Schultz 1989).

Finally, a whole range of other surface features have been related to possible localized reservoirs of subsurface H_2O. Examples include fluidized landslide deposits with associated fluvial channels present within Valles Marineris (Lucchitta 1987*c*), debris aprons next to the northwest flanks of the large shield volcanoes of Tharsis interpreted to result from glacier-like creep of localized ice deposits (Lucchitta 1981), crater chains near Valles Marineris resulting from collapse over a zone of preferential melting of ground ice (Lucchitta 1981), and ridges with bounding depressions, both near the mouths of Martian outflow channels (Lucchitta et al. 1986) and on smooth plains within an impact basin (Grizzaffi and Schultz 1989).

III. DISCUSSION AND SUMMARY

We have discussed above a number of possible morphologic indicators of ground ice on Mars, concentrating on rampart craters and terrain softening. An important observation is that detailed mapping of these two classes of features appears at first glance to lead to different conclusions. Rampart craters are seen at all latitudes, while terrain softening is only observed poleward of ± 30°. In order to form a consistent picture of the distribution of ground ice on Mars, we need to resolve this apparent inconsistency.

As discussed above, the Martian regolith is primarily a product of intensive impact bombardment coupled with volcanic and weathering activity. Near the surface, materials in heavily cratered terrains are dominated by disaggregated and pulverized impact ejecta. This grades downward into bedrock that has been extensively fractured in place and, finally, into solid bedrock below. Because regolith structure can exert a dominant influence on the rheology of ice-rich ground, one would expect such a structure to lead to an abrupt decrease in regolith mobility at some depth, perhaps not more than several hundred m below the surface. If this is the case, then the differences between the distributions of rampart craters and terrain softening can be resolved in a straightforward manner.

Figure 17 shows a hypothetical and schematic pole-to-pole cross section through the Martian regolith that appears to be broadly consistent with all the current data and theoretical considerations that bear on the issue of ground-ice distribution. The shaded region indicates the locations where ground ice is present. Ice extends very close to the surface at high latitudes. The top of the ground ice becomes deeper with decreasing latitude, dropping off fairly sharply in the vicinity of ± 30°. Ground ice is present at equatorial latitudes, but only at substantial depths. This ice distribution is in agreement with both theoretical considerations and observations of rampart craters.

The dashed line in Fig. 16 shows the depth at which the effective transition takes place from pulverized, thoroughly disaggregated debris near the surface to fractured bedrock below. The transition of course will be gradual

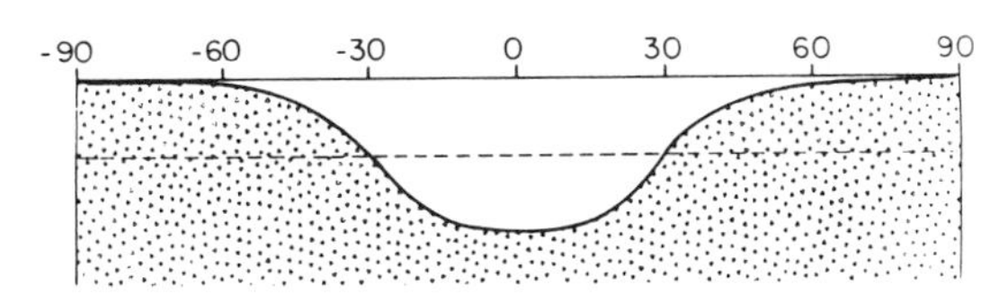

Fig. 17. Schematic representation of the distribution of ground ice in the Martian subsurface, as deduced from both terrain softening and rampart craters. Ice in the equatorial region has been removed to a substantial depth by sublimation and diffusion. Dashed line indicates the depth above which the crustal material has been sufficiently disaggregated by impacts that it is able to flow and produce terrain softening. Current understanding is inadequate to allow accurate labeling of the depth scale on the vertical axis.

rather than abrupt, but the effect on the mobility of materials will be qualitatively the same. Both materials are porous and can contain significant amounts of ice, but only the upper one has a structure that will allow creep deformation to occur. At latitudes higher than ± 30°, ice is present in near-surface, thoroughly disaggregated materials, and terrain softening can take place. At lower latitudes, ice only persists in deep, heavily fractured bedrock, still allowing rampart craters to form for large impacts, but preventing terrain softening. This picture is consistent with the observed distribution of ice-related features, with our knowledge of the rheology of ice-silicate mixtures, and with the expected vertical structure of the upper few km of the Martian surface. The vertical scale on Fig. 16 is left unlabeled intentionally. We do not know the depth to which terrain softening extends with any real accuracy; nor can we determine the depth to the top of the ice-rich layer with precision. It will be left to future investigations to fully characterize its regional variability and its quantitative details.

Despite the large uncertainties that still exist, a general picture has emerged of the subsurface distribution of ice on Mars that appears to be generally consistent with our understanding of the major physical processes involved. This general picture is certainly consistent with the view that H_2O has played a major role in the geologic evolution of the planet's surface. It may also have significant geophysical implications: the strong latitudinal dependences observed, coupled with the long time scales for migration of ground ice on Mars, may place severe constraints on hypotheses of significant polar wander on the planet. However, there is great potential for future investigation of ground ice on Mars. Techniques that could be applied, in order of increasing capability to probe to depth, include gamma-ray spectroscopy, direct sampling using penetrators or rover-mounted drills, electromagnetic sounding and seismology. All of these could contribute substantially to our understanding of the quantity of Martian ground ice, of its geographic distribution and of the physical processes that have controlled its evolution.

PART IV

Surface Properties and Processes

17. THE COMPOSITION AND MINERALOGY OF THE MARTIAN SURFACE FROM SPECTROSCOPIC OBSERVATIONS: 0.3 μm to 50 μm

LAURENCE A. SODERBLOM
United States Geological Survey

Reflection spectra of Mars show a 3 μm absorption due to H_2O in some form distributed ubiquitously in surface materials in an abundance of roughly 1%. The spectrum of the retreating south polar cap displays numerous CO_2 ice features (1.2 to 2.4 μm), whereas that of the retreating north cap shows a 1.5 μm H_2O ice feature. As small amounts of H_2O ice can spectrally mask CO_2, the H_2O may only be a surficial lag left as the annual CO_2 deposit sublimes. Spacecraft images show soils and rocks to fall into three groups: bright red units, probably fine-grained dust; intermediate-albedo, red materials, perhaps older, more indurated aeolian debris; and dark units occuring as dune masses or coarse rocky debris. Overall, the bright- and dark-region spectra can be matched by basalts with varied degrees of oxidation and grain size. Spectra of both types of regions display a deep absorption in the ultraviolet to blue and rise rapidly through the visible, giving Mars its red-ocher color. Although this is generally attributed to ferric iron, other strong absorptions would be expected for crystalline iron oxides. Such absorptions are extremely weak, suggesting hematite in minute amounts if it is coarsely crystalline, or abundant if it is in nanocrystalline form. Mariner 9 infrared spectra (5 to 50 μm) show absorptions from atmospheric dust near 10 and 20 μm. These data suggest the dust to have a low abundance of crystalline iron oxides, an SiO_2 content of roughly 60% (which is consistent with intermediate igneous rocks), and possibly abundant clay materials. Crystalline clays, like the oxides, display a variety of other spectral features in the visible and near-infrared (2.2 to 2.4 μm) and mid-infrared (near 20 μm) that are not observed or only weakly evident. A weak absorption feature at 2.35 μm was interpreted as due to an Mg-OH bond in poorly crystalline clay. It is also possible that the surface materials are largely amorphous, containing little crystalline material. If so, palagonites are *currently the best analogs to explain both the visible and near-infrared reflectance of bright regions and the 10 and 20 μm transmission of the airborne dust. Dark surface materials are varied in color, ranging from extremely red to more neutral shades; the latter probably indicate less oxidation. An absorption in dark-region spectra near 1 μm is commonly interpreted as due to Fe^{2+} in Ca-rich clinopyroxene. Variations in this band suggest varied Ca and Fe content in the clinopyroxenes; orthopyroxenes also may be present. Olivine is evidently not the dominant mafic component. Carbonate abundance is limited to a few percent by the absence of strong absorptions at 4 μm in surface reflection spectra and at 7 μm in infrared transmission spectra of the dust. Thermal emission spectra collected recently from the Kuiper Airborne Observatory show a weak absorption near 6.7 μm, consistent with carbonate in an abundance of ≤3%. These spectra also show absorptions near 8.7 and 9.8 μm suggestive of sulfate or bisulfate with an abundance of 10 to 15%. Reflectance spectra acquired recently with the NASA Infrared Telescope Facility (IRTF) also show an absorption near 4.5 μm consistent with sulfate or bisulfate. High-resolution IRTF spectra of the 2–2.5 μm region show a series of features that may be due to scapolite, a mineral closely related to plagioclase that commonly occurs with metamorphosed limestones on Earth. Because the features are extremely weak and some coincide with those of atmospheric CO, conclusive identification is difficult.*

A broad collection of spectroscopic and multispectral imaging observations of Mars (0.3 to 50 μm) has been acquired from both Earth and spacecraft over the last four decades. The subjects of this chapter are the composition and global distribution of materials exposed at the surface of Mars that can be determined from such observations. These materials include permanent units, such as soils and bedrock generated by various impact, volcanic, weathering, erosional and depositional processes, as well as mobile windblown dust, polar ices and frost that give dramatic seasonal variability to Mars' global appearance.

Atmospheric components (including dust and ice particles) are considered here only where they interfere with or complement the interpretation of the nature and composition of surface materials.

The chapter is organized to give a historical account of the development and refinement of observations and interpretations beginning in the early 1960s. These early studies generated our basic ideas about the composition and physical nature of Martian surface materials: a low H_2O concentration in some form distributed ubiquitously through the soils; bright materials consisting of largely amorphous, iron-rich alteration products such as palagonites; dark materials, most consistent with oxidized basalts containing pyroxenes; polar volatiles consisting of CO_2 and H_2O ices; and atmospheric dust with an SiO_2 content typical of intermediate igneous rocks and carbonates and crystalline iron oxides limited to minor constituents. Information from the Viking mission was generally consistent with these early ideas. Recent, more highly refined spectroscopic observation, analysis and laboratory modeling are then discussed, including evidence for nanocrystalline iron oxides, clays, carbonates, sulfates and scapolite.

I. WATER IN THE MARTIAN SOILS

Earth-based observations of the reflection spectrum of the Martian surface are hampered by many absorption bands in both the terrestrial and Martian atmospheres. The terrestrial absorptions are dominated by H_2O with contributions of O_2, O_3 and CO_2 (Fig. 1). The primary atmospheric absorber in the Martian atmosphere is CO_2; the dominant absorptions are indicated in Fig. 2c.

The earliest attempt to construct an overall spectral geometric albedo through the visible and near infrared (0.4 to 4.0 μm) for the red planet was made by Moroz (1964). He combined observations by Woolley (1953) for the 0.4–0.6 μm region with those of Esipov and Moroz (1963) for the 0.8–1.0 μm region and with his own observations for specific wavelengths through telluric windows (1.30, 1.61 and 2.13 μm), coupled with a crude estimate of the reflectance between 3 and 4 μm obtained by correcting for thermal emission by assuming a surface temperature of 290 K (Fig. 2a). From these observations, Moroz suggested that the depression in the spectrum between 2 and 4 μm is due to a strong 3 μm absorption of H_2O, which he termed "water of crystallization"; the modern interpretation is still that this feature is largely due to H_2O in some form.

Subsequently, Sinton (1967) telescopically observed the Martian globe with a birefringent spectrometer and began to resolve the 3 to 4 μm region.

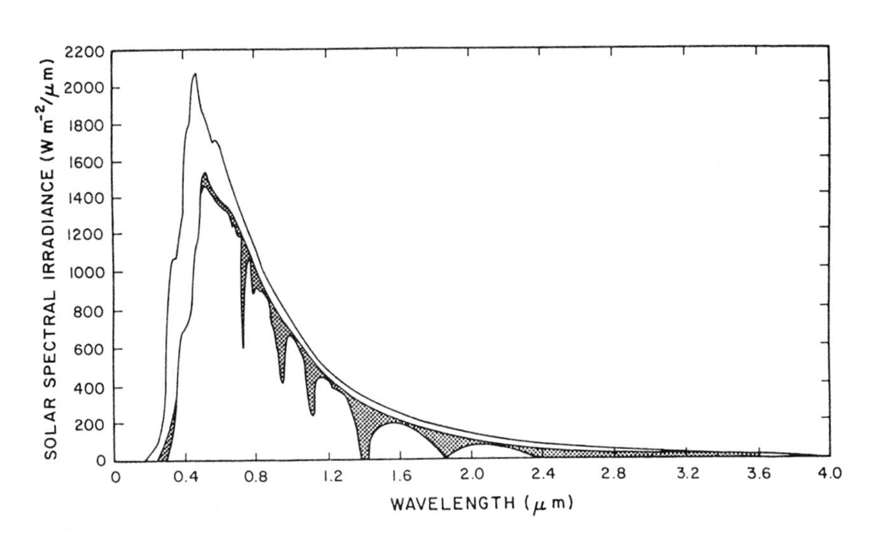

Fig. 1. Absorption features in the Earth's atmosphere due to a variety of gases that interfere with telescopic spectral observations of Mars. The upper curve represents the incident solar irradiance at the top of the Earth's atmosphere; the lower curve represents the Sun's irradiance at the surface of the Earth; shaded areas represent absorption by various gases including O_3, O_2, H_2O and CO_2 (figure from Liou 1980).

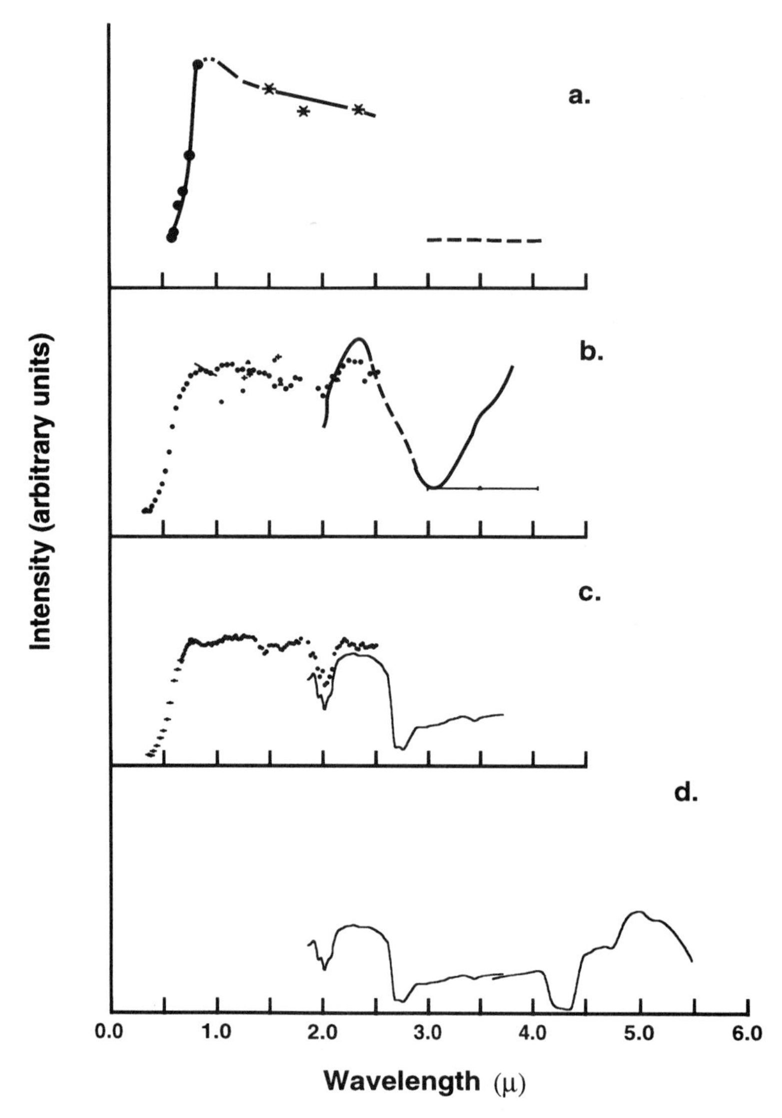

Fig. 2. A historical collection of the visible and near-infrared spectra. (a) Earliest attempt, by Moroz (1964), to estimate Mars' spectral geometric albedo of 0.4–4.0 μm (see text). (b) Composite of published spectra produced by McCord et al. (1971) from data taken from Binder and Cruikshank (1963), Esipov and Moroz (1963), Moroz (1964), Tull (1966), Sinton (1967), and McCord and Westphal (1971). (c) Composite of typical bright-region spectra in which discrete points are from Singer et al. (1979) and the solid line is from Pimentel et al. (1974). (d) Composite of Mariner 1969 IRS spectra for the 2–5 μm region, and data from Pimentel et al (1974) for the 1.8–3.7 μm region and data from McKay and Nedell (1988) for the 3.5–5.5 μm region.

Like Moroz, he attributed a broad absorption in this region to "water of crystallization" after Coblentz (1906), who found such a feature at 3 μm in several minerals containing H_2O. Sinton's spectrum (Fig. 2b) was combined by McCord et al. (1971) with their shorter-wavelength spectrum. The dashed line from 2.4 to 2.9 μm indicates a poorly observed region in Sinton's spectrum due to strong H_2O and CO_2 absorptions in the Earth's atmosphere. Houck et al. (1973) more completely observed the shape of the absorption from aircraft at altitudes where the telluric H_2O and CO_2 only minimally interfere. They suggested that the band is consistent with a content of about 1% by mass of bound H_2O in the Martian soils.

Pimentel et al. (1974) analyzed data from the Mariner 6 and 7 Infrared Spectrometers (IRS), which clearly resolved the 2–4 μm region as a composite of relatively sharp CO_2 gas bands in the 2.6–2.7 μm region and a broad band extending from this region out to about 3.3 μm (Figs. 2c and d). They suggested that this feature is attributable to water of hydration and/or ice. McCord et al. (1978), on the basis of a tentatively identified absorption at 1.5 μm, also suggested that the ubiquitous H_2O in the Martian soils might be at least partially in the form of H_2O ice. This point is debatable; Esposito and Jones (1988) suggested that more detailed modeling of the contribution of Martian CO_2 and scattering can simulate the 1.5 μm absorption without recourse to H_2O ice.

More recent spectra obtained from Mars' orbit by the French imaging spectrometer experiment (ISM), which was carried by the Soviet Phobos 2 spacecraft, have been used to map the variation of the 3-μm H_2O absorption feature in several mid-latitude and equatorial regions with a spatial resolution of about 25 km. These studies suggest that the abundance of H_2O in both volcanic and canyon terrains varies by as much as 20% (Bibring et al. 1990*a*, *b;* Erard et al. 1990). Although the exact nature and the variability of H_2O in the Martian soils remain debatable, it is clear that H_2O in some form—absorbed, in hydration, or possibly as ice—is ubiquitous in varied amounts in the range of a few tenths of a percent to a few percent.

II. POLAR VOLATILES

Determination of the composition of the polar ice caps by spectroscopy alone is difficult at best. As soon as the polar hoods dissipate, the caps begin to recede (Fig. 3). The composition of the cap surfaces undergo continual change during this period owing to the different volatilities of H_2O and CO_2. By the time the recession is complete, the residual or permanent caps are difficult or even impossible to observe from Earth.

Telescopic spectral observations do, however, show the compositions of the receding caps to be different. Using a high-spectral-resolution Fourier spectrometer, Larson and Fink (1972) identified 11 sharp features in the 1.2–2.4 μm region of the south polar cap spectrum that they attributed to solid

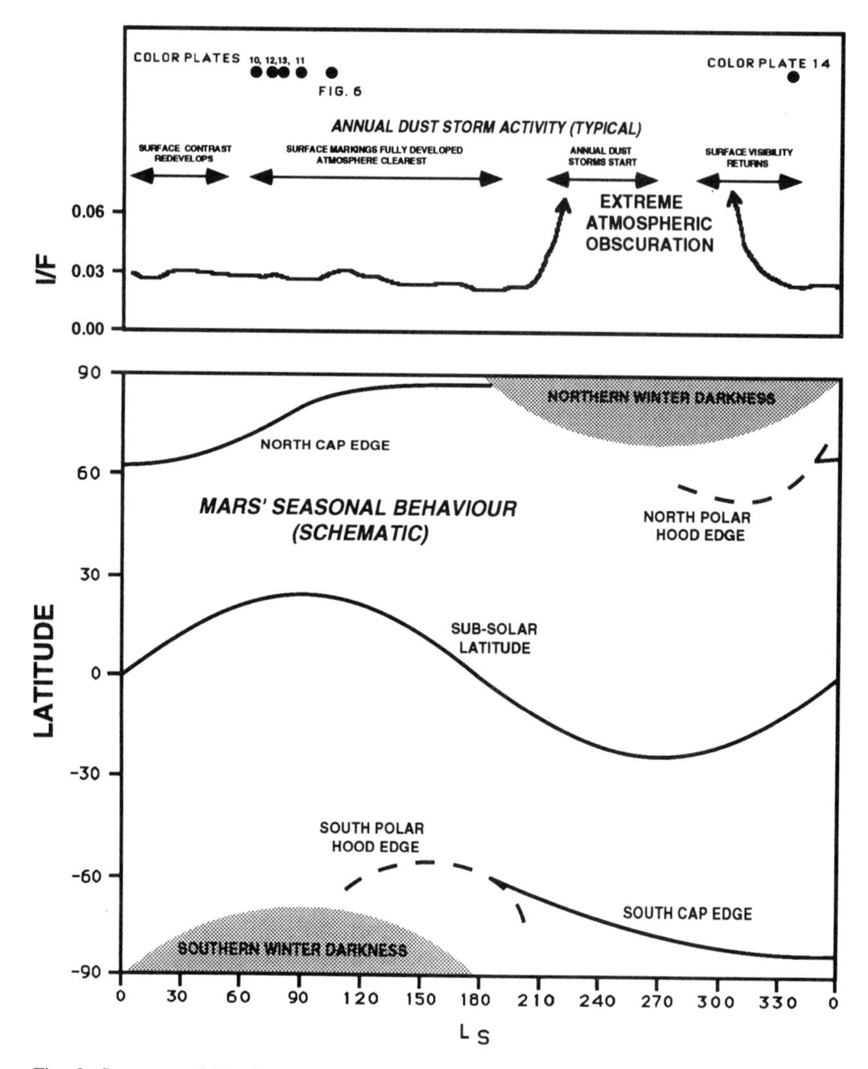

Fig. 3. Summary of Martian seasonal phenomena as they relate to imaging and spectroscopic observations of the surface. Shown are the seasons (L_S) in which the Viking Orbiter images in Color Plates 10–14 and Fig. 6 were acquired. I/F is the spectral radiance of the atmosphere seen from orbit in the Viking Orbiter red-filter bandpass (0.59 μm).

CO_2. These spectra show no noticeable H_2O features. This was consistent with the identification of solid CO_2 by Herr and Pimentel (1969) from the recognition of two CO_2 ice features in the Mariner 7 IRS data.

The earliest Earth-based infrared observations of the reflectance of the Martian north polar cap in the 1.4 to 1.8 μm region were made by Kuiper (1952) and Moroz (1964). In this region there is strong absorption near 1.4

μm due to a combination of H_2O and CO_2 in Earth's atmosphere and of CO_2 in Mars' atmosphere. Both workers noted a shift toward longer wavelengths of this absorption in spectra of the north polar cap as compared with those of equatorial regions (cf. Fig. 4). Recognizing that a strong 1.5 μm absorption feature is characteristic of H_2O ice or snow, both concluded that these observations are consistent with a dominantly H_2O composition.

On the basis of higher-quality north-cap spectra from McCord et al. (1978), Clark and McCord (1982) concluded that the composition of the cap surface is consistent with a laboratory simulation of 60% H_2O ice and 40% gray material (Fig. 5). Although no CO_2 ice features were detected in these north polar observations, their absence does not require that the north cap be composed dominantly of H_2O. Kieffer (1968, 1970*a,b*) showed that low concentrations of H_2O frost mixed with CO_2 frost would mask the CO_2 spectral

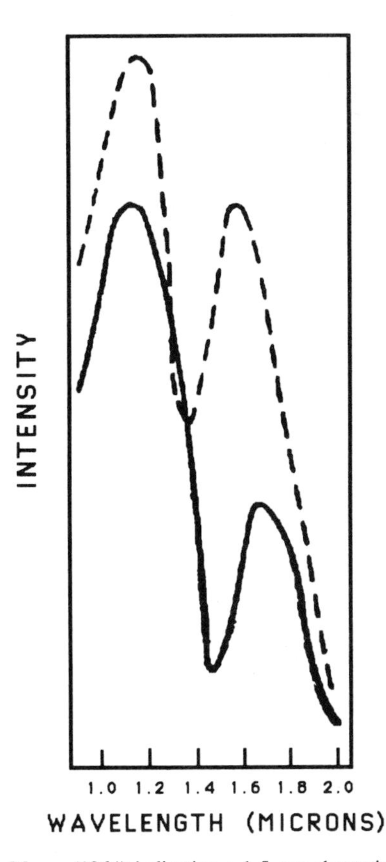

Fig. 4. Observations by Moroz (1964) indicating a 1.5 μm absorption feature in the reflection spectrum of the north polar ice cap due to H_2O ice. The dashed curve is for the equatorial regions showing prominent absorption by telluric H_2O at 1.4 μm. The solid curve, for the north cap, shows a major shift in the position of the absorption inferred to be due to H_2O ice.

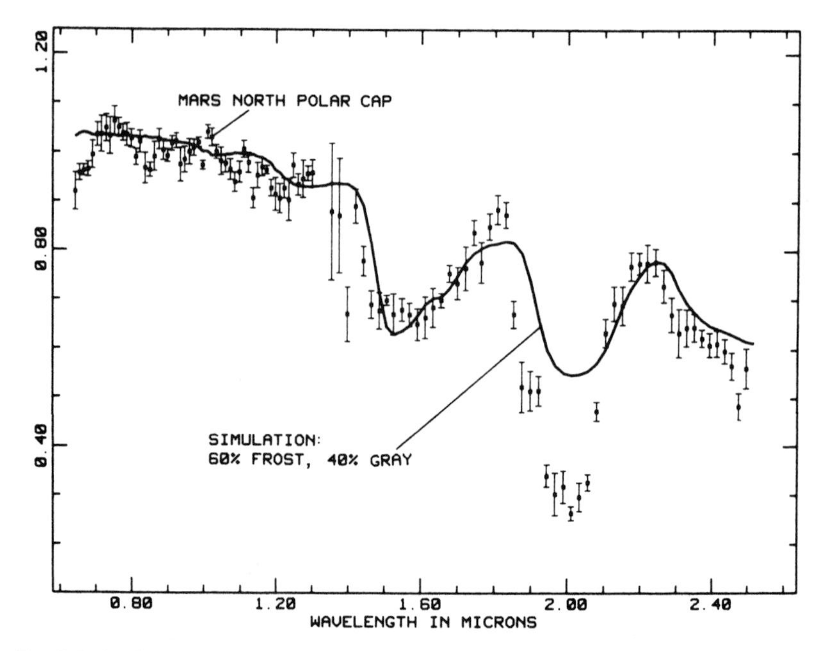

Fig. 5. Reflection spectrum of the retreating north polar cap (L_S = 50) derived from observations by McCord et al. (1982) along with a laboratory spectrum for a mixture of 60% H_2O frost mixed with 40% gray material. From Clark and McCord (1982*b*).

features, resulting in a spectrum characteristic of effectively pure H_2O ice; he noted that Moroz' and Kuiper's conclusion that the north cap is H_2O ice is not supportable. Also, all three of these sets of observations were made while the north cap was about midway through its annual recession (L_S = 50° to 65°). Jakosky (1983*c*) suggested that the H_2O might be present as a lag on the surface as the annual CO_2 deposit sublimes.

III. MULTICOLOR IMAGING: EVIDENCE FOR REGIONAL-SCALE COMPOSITIONAL HETEROGENEITY

Seen on a global scale, Mars displays a rich contrast in color and albedo and an appearance that continually changes through the Martian year. Sandwiched between the active polar regions where the bright polar hoods and ice caps seasonally form and retreat (cf. Fig. 3), the rest of the planet's surface is made up of the bright red-ocher regions and the darker (and, on average, less red) regions that to many early astronomers appeared to have a slight greenish tinge (cf. Slipher 1962). The bright regions have geometric albedos in the red (near 0.6 μm) as high as 0.4; the dark regions, as low as 0.1. In the blue (near 0.4 μm), the contrast between the bright and dark regions nearly vanishes; each has a geometric albedo typically near 0.05.

The contrast between bright and dark regions seasonally fades and re-emerges with nearly annual regularity, an effect which some early astronomers attributed to seasonally changing vegetation. We now know the contrast to be caused by intense global dust storms that typically begin just prior to the southern summer solstice ($L_S \approx 210°$ to $270°$), at a time that roughly coincides with perihelion of Mars' eccentric orbit. Following the dust storms, the surface displays little contrast due to coatings of bright dust and the dusty atmosphere (Christensen 1986*a*). The contrast gradually increases and, by the middle of northern spring, the variations in color and albedo have fully redeveloped. Slipher (1962), comparing modern photographs with maps drawn in the 18th and 19th centuries, noted that these annually emerging global patterns have remained the same for at least 200 yr (cf. Fig. 6).

In Viking Orbiter color views (Color Plates 10–13), the classical bright and dark regions appear as three discrete units. In addition to the typical bright red-ocher and darker, grayer units, a third unit that is intermediate in albedo can be easily distinguished in the equatorial and mid-latitude regions in those views. This third type of unit has been pointed out in several studies (Soderblom et al. 1978; Kieffer et al. 1981; McCord et al. 1982*b*; Arvidson et al. 1982, 1989*b*). Arvidson et al. (1989*b*) argued that the color and albedo of the intermediate material suggest that it is a mixture of the bright red and the more neutral dark materials. The three types of units can best be seen in Color Plate 11. Centered in that view is Arabia, a classical large, bright region, roughly circular and about 2000 km in diameter. West of Arabia in the Oxia Palus region is the largest occurrence of the intermediate albedo unit. This unit is bounded on the south by Sinus Meridiani and extends northward and then eastward around the margin of Arabia. Within this intermediate-albedo region, many large craters exhibit bright materials around their rims, and some craters have tails of dark material that extend from their floors across their rims to the south.

Kieffer et al. (1981) noted that the intermediate-albedo materials are also thermally anomalous. In general, there is a strong correlation between thermal inertia and albedo, displayed as a homologous distribution between two end-members: bright units are low and dark units are high in thermal inertia (see also chapter 21; Arvidson et al. 1982). This is attributed to bright regions being composed of much finer material than that of the dark regions. The intermediate-albedo materials fall outside the general distribution, displaying thermal inertias similar to those of the dark regions, which suggests that they, too, are coarser or indurated materials.

With respect to discussions that follow, neither in the historical observations of the classical markings nor in more recent Earth-based spectroscopic observations are the intermediate-albedo units discriminated as a group. This arises from several factors: (1) the bright and dark units dominate in terms of area; (2) many intermediate units occur in boundary zones between bright and dark units and are thus difficult to distinguish with telescopic

Fig. 6. Color variations within Martian dark regions. (a) An airbrush drawing taken from Inge et al. (1971) shows the typical pattern of historically observed bright and dark patterns that emerge each Martian spring—coverage of parts (b) and (c) marked by rectangle. (b) Part of a global mosaic acquired by Viking Orbiter 2 on approach through the red filter (0.59 μm) showing one of the conspicuous concentrations of dark regions in the south equatorial belt of the eastern hemisphere (figure from Soderblom t al. 1978). (c) The ratio of the red-to-violet images (0.59 μm/0.45 μm) for the same area shows major variations in the color of this cluster of dark regions (red/violet ranging from about 2 to 3).

resolution; and (3) boundaries between bright and intermediate units are commonly gradational, the two occurring as mixed patches. The detailed relations of the intermediate units to brighter and darker units await examination of future observations.

Dark areas are concentrated in two belts (Fig. 6a). The larger, south equatorial belt is situated between the equator and about 40°S and extends completely around the planet. The smaller, north belt is at about 50°N latitude and consists of two large, dark regions centered at roughly 30° and 270° longitude (also see Color Plates 10 and 12). Some of the dark material clearly occurs as sand dunes and sand sheets, particularly in a northern circumpolar dark band and trapped in the floors of many craters and depressions in the dark belts. In other places the dark materials are apparently exposed bedrock or coarse debris locally derived therefrom.

The dark materials themselves display significant heterogeneity in color. During the final stages of the approach to Mars, prior to injection into Mars orbit, the cameras aboard Viking 2 acquired global multicolor coverage of the planet as it rotated. The season ($L_S \approx 105°$) was ideal to study the fully developed contrast within the dark regions (cf. Fig. 3). The observations were made with a phase angle of about 105°; images were acquired in three bandpasses (centers: 0.45 μm, 0.53 μm and 0.59 μm) with resolutions between 10 and 20 km/line-pair. From these data Soderblom et al. (1978) produced a global digital multicolor map of the equatorial region from about 30°N to 40°S. About 30% of this coverage was obscured by clouds, particularly near the terminator portion of individual frames. Figures 6b and c, taken from that multicolor map, show a section of the equatorial region dominated by conspicuous dark regions, including Sinus Sabaeus, Syrtis Major, Iapygia and Mare Tyrrhenum. Comparison of the geometric albedo image in the red 0.59 μm band (Fig. 6b) with the red/violet color ratio image (0.59 μm/0.45 μm) (Fig. 6c) reveals substantial variation in the color of the dark units, particularly in the eastern part of the section shown. Recent Earth-based narrowband telescopic imaging by Pinet and Chevrel (1990) shows substantial variation also in spectral reflectance among dark areas in the 0.7–1.1 μm region.

In the multicolor images, the dark materials generally fall into two discrete groups: one is among the reddest on Mars (red/violet ratio ≈ 3); the other is among the least red (red/violet ratio ≈ 2). The relatively young volcanic shields found in the Tharsis region are among the very red dark materials. In one particular area of the dark south equatorial belt situated in highland terrain, the boundary between the two types of dark unit clearly correlates with differences in geomorphology and geologic age. Here the very red dark unit is correlated with ancient highland plateaus and ridges that are riddled with dendritic channels and saturated with normal faults. By contrast, the least red or the gray dark units of the south equatorial belt coincide with younger volcanic plains, which resemble the lunar maria with many wrinkle ridges. This suggests that the color variations, at least in those regions, are directly related to differences in bedrock mineralogy rather than simply to aeolian deposits, veneers and coatings. The very red dark materials, occurring in the ancient highlands and younger volcanics, are richer in Fe^{3+} than the less red, mare-like volcanic plains. McCord et al. (1982*b*) correlated the Viking approach multicolor map with telescopic spectra, which they compared with spectral properties of analog materials measured in the laboratory materials. They suggested that the bright materials are consistent with palagonites or iron-rich silica gels (discussed in Sec. VII), that the less red dark volcanic plains could be basalts with thin oxidized coatings, and that the very red dark units are consistent with nonhydroxylated, more hematitic material.

In summary, regional- and global-scale multicolor images in visible wavelengths reveal that bedrock and soils on the Martian surface exhibit substantial variation in composition and/or physical properties. The classical bright and dark regions are seen to fall into three discrete albedo classes: (1) the brightest red-ocher materials that are probably the fine, windblown material involved in annual dust storms; (2) intermediate-albedo red materials that are the least common among the three classes and may be duricrusts of aeolian debris, perhaps like those observed at the Viking landing sites; and (3) dark materials, some as red as the two brighter classes, which could be more hematitic or heavily oxidized basalts, and others more neutral in color that may be less oxidized basaltic material. It is important to keep this heterogeneity in mind for the discussions of Earth-based spectra that follow.

IV. OXIDIZED BASALTS AND CRYSTALLINE IRON OXIDES

The reflectance spectra of both bright and dark regions display a deep absorption in the blue to ultraviolet (0.3–0.6 μm), rising out of the near-ultraviolet into the visible (cf. Fig. 7), giving the planet its rich red-ocher color. This characteristic has traditionally been attributed to ferric iron arising from the wings of a pair of intense $Fe^{3+} - O^{2-}$ charge-transfer absorptions centered in the near-ultraviolet around 0.34 μm (cf. Huguenin et al. 1977). In what mineralogical forms(s) the ferric ion resides has been a major issue, whether as crystalline iron oxide or clay or as amorphous material such as palagonite and silica glass.

Crystalline iron oxides were early suggestions as primary candidates for the surface materials (cf. Wildt 1934; Sharonov 1961). Moroz (1964) suggested that the visible reflectance is consistent with that of limonitic samples of brown iron ore. Following up on Moroz's suggestion, Younkin (1966) and Sinton (1967) telescopically observed Mars in the spectral region where limonite exhibits a strong 0.88 μm absorption feature and observed no such features; this provided an upper limit of about 2% for the abundance of limonite.

In analyzing an integral disk reflectance spectrum of Mars collected by Tull (1966) in the 0.5–1.2 μm region, Adams (1968) suggested alternatively

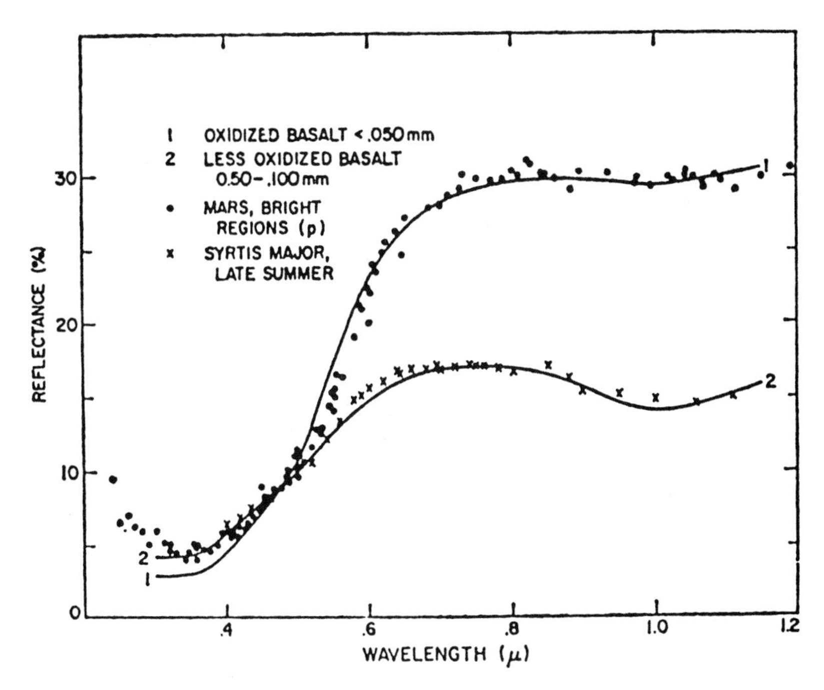

Fig. 7. Comparison of the reflectivity of Martian bright areas (dots) with laboratory simulations of basalts (figure from Adams and McCord 1969).

that the best match for the Martian spectrum is that of an oxidized basalt. Adams and McCord (1969) showed that the difference between bright- and dark-region spectra could be modeled with basalts whose surfaces were oxidized to different degrees (Fig. 7). They concluded that the reflectance spectra of the bright regions are consistent with their being more heavily oxidized and consisting of finer (<50 μm) particles.

Singer and McCord (1979), noting that Mariner 9 and Viking images clearly show localized deposits of bright materials in the dark regions, estimated that the groundbased reflection spectra for these regions are contaminated typically by about 20% bright-region material. They approximated the pure dark-region spectrum by simple subtraction of a contribution of a bright-material spectrum, assuming linear mixing (patchy coverage of bright material). The result (Fig. 8) shows the overall dark-region spectrum to have negative slope in the infrared. Singer (1980) subsequently explored the types of coatings on basalts that could produce both the deep ferric absorption in the 0.3–0.6 μm region and the decreasing albedo out into the mid-infrared. He showed that a thin Fe^{3+}-rich coating over a dark substrate such as basalt could account for (1) the deep blue to ultraviolet absorption, (2) the rapid rise in reflectance to 0.75 μm where the coating becomes bright, optically thick, and scattering, and (3) the fall in albedo longward of 1 to 1.5 μm where the

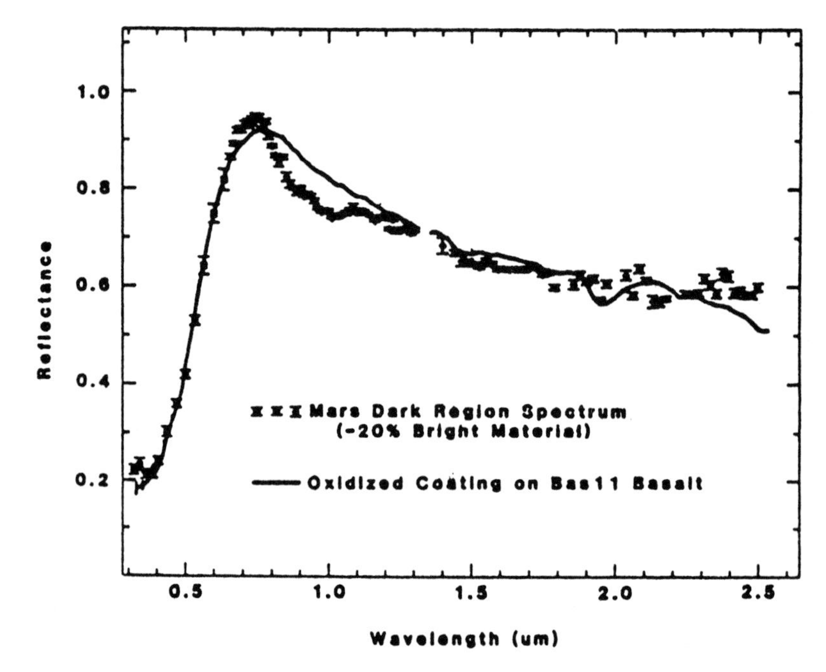

Fig. 8. Comparison of a theoretically pure dark-region spectrum (obtained by subtracting an assumed bright-region material contribution of 20%) with a Mauna Kea basalt having an extremely thin palagonitic coating (figure from Singer 1980).

coating rapidly becomes transparent, optically thin and nonscattering, allowing the dark substrate to dominate (Fig. 8).

Spectra of two iron oxides, hematite and goethite, which are typical of crystalline iron minerals that might be expected to occur on the Martian surface, are shown in Fig. 9. These minerals exhibit a variety of Fe^{3+} electron transition absorptions near 0.9 μm and at 0.4 μm and 0.7 μm along the steep slope of the $Fe^{3+} - O^{2-}$ charge-transfer band (Sherman et al. 1982; Singer 1982, 1985; Morris et al. 1985). Various workers have searched for these spectral features. McCord and Adams (1969) first suggested detection of a very weak absorption in bright-region spectra near 0.85 μm. Adams and McCord (1969) noted that this feature is attributable to a ferric iron oxide but that it certainly is not unique to limonite. McCord et al. (1977) and Singer et al. (1979) acquired higher-quality spectra in 1973, 1976 and 1978 of bright and dark areas and dust clouds. Composite bright- and dark-region spectra taken from these observations are shown in Fig. 10. A weak absorption near 0.9 μm and perhaps a subtle break in slope near 0.6 μm can be discerned in both spectra, although more clearly in the bright-region spectrum. From these data, Huguenin et al. (1977) concluded that the bright soils and dust could

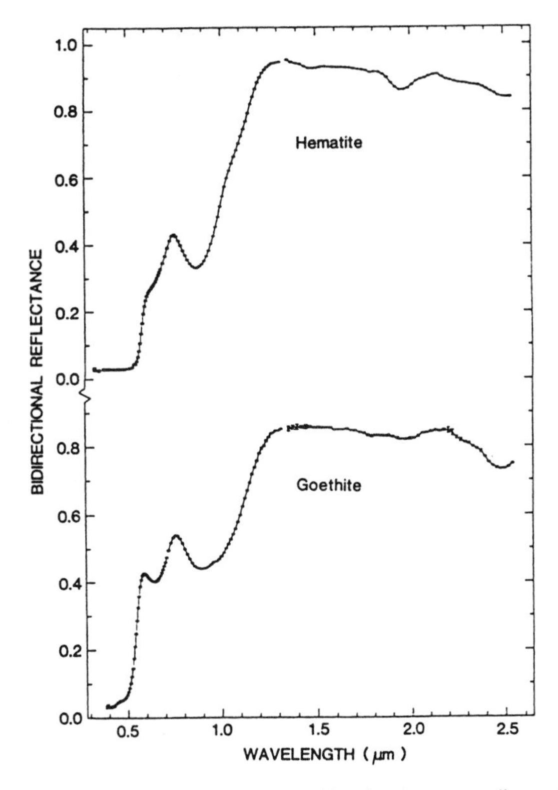

Fig. 9. Reflection spectra of two crystalline iron oxides showing strong diagnostic Fe^{3+} absorptions in the 0.5–1.0 μm spectral region (figure from Singer 1985).

contain at most a few percent of crystalline iron oxides. From spectra of laboratory samples, Singer (1982) showed that as little as 1% hematite or goethite would produce a band in this region of greater strength than is seen in the Martian spectra. Recent laboratory work by Morris et al. (1989) and Morris and Lauer (1990) suggests that the best interpretation of the subtle features near 0.6 and 0.9 μm is that they are, in fact, due to hematite. This work shows that the Martian data are consistent with either a percent or so of coarsely crystalline hematite or abundant very finely crystalline hematite, which they termed nanocrystalline (< 10 nanometers). Recent high-quality telescopic observations by Bell et al. (1989*e,* 1990*b,c*) with a spectral resolution of about 1.5% for the 0.3–1.0 μm region clearly show subtle features (Fig. 11). These observations are consistent with the suggestion of Morris and colleagues that the features are indicative of substantial, abundant finely crystalline hematite, enough to explain the 18% Fe_2O_3 measured by the Viking Landers in the Martian soil.

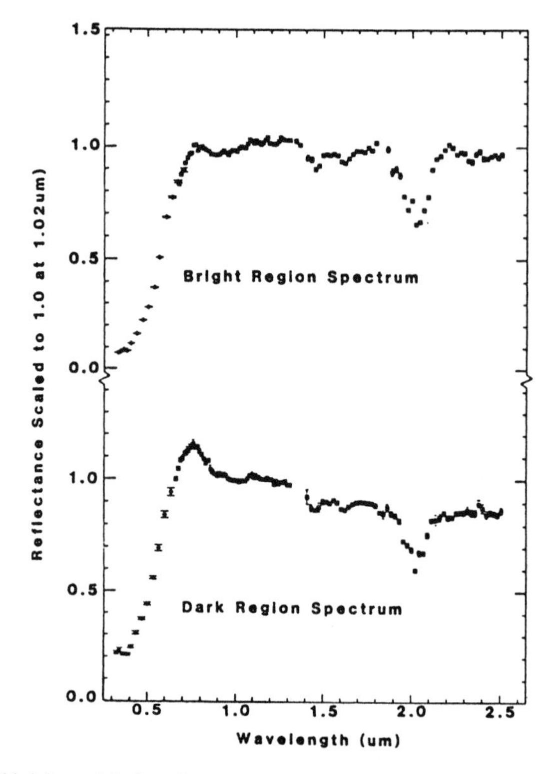

Fig. 10. Typical bright- and dark-region spectra. Although these spectra have been scaled to unity at 1.02 μm, the actual ratio of bright to dark albedo at 1 μm is typically about 4:1. The bright-region spectrum is an average of bright areas observed in 1973 and 1978. The dark spectrum is a composite of spectra of two adjacent dark regions observed in 1973 and 1978 (figure from Singer et al. 1979).

Mariner 9 carried an infrared interferometric spectrometer (IRIS) that operated in the 200 to 2000 cm^{-1} (5 to 50 μm) region with a spectral resolution of about 2.4 cm^{-1}. This instrument collected a substantial quantity of spectra during the 1971 Martian dust storm. In these data, thermal emission from the surface is partially absorbed by atmospheric dust (and gases) and can be used to analyze the composition of the dust by comparison with transmission spectra of various materials. Early analyses by Hanel et al. (1972*a*, *b*) led to the conclusion that the dust had a high silica content (as much as 60% SiO_2) on the basis of two broad absorptions between 400 and 600 cm^{-1} and between 900 and 1300 cm^{-1} (Fig. 12). These features arise from bending and stretching modes of SiO_4 tetrahedra, respectively (Farmer 1974; Gadsden 1975). Hunt et al. (1973) and Toon et al. (1977) noted that crystalline iron oxides typically have multiple deep absorptions between 200 and 500 cm^{-1} (cf. Fig. 13), which are not evident in the Mariner 9 spectra; thus materials such as limonite, hematite and goethite could be only minor components.

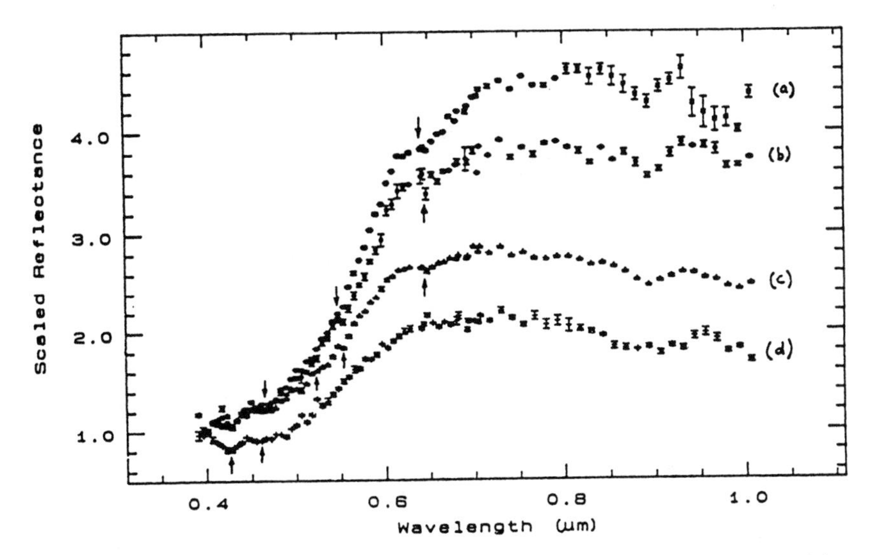

Fig. 11. Martian reflection spectra showing clear indications of subtle absorption features giving rise to slope changes in the 0.4–0.8 μm region. These are attributable either to very low concentrations of coarsely crystalline iron oxides or substantial concentrations of nanocrystalline iron oxides (figure from Bell et al. 1989*e*).

In summary, spectral evidence in both the visible and far-infrared regions precludes substantial quantities of coarsely crystalline iron oxides (e.g., hematite, goethite, limonite) in the Martian soils and dust. Whether substantial nanocrystalline hematite, proposed as a possibility by Morris and coworkers, should have been detected in the Mariner 9 IRIS spectra of the airborne dust remains an open question.

V. PYROXENE AND OLIVINE

Adams (1968) recognized an absorption feature near 1 μm in the integral disk spectrum obtained by Tull (1966) and suggested that it is due to Fe^{2+} in olivine or iron- and calcium-bearing clinopyroxene or both. McCord and Adams (1969) and McCord and Westphal (1971) confirmed the existence of a broad, shallow feature near 1 μm and demonstrated that it occurs dominantly in spectra of dark regions (Fig. 14).

Building on the work of Burns (1965), Adams (1974) developed a spectral-analysis scheme to derive the composition of pyroxenes from absorption bands that occur near 1 μm and 2 μm. A summary figure from that work is reproduced in Fig. 15. Adams noted that the positions of these two bands are strong functions of the concentrations of Fe^{2+} and Ca^{2+}. Unfortunately, the strong CO_2 absorption by the Martian atmosphere at 2.0 μm (cf. Fig. 2) makes the 2 μm pyroxene feature difficult to observe. The position of the

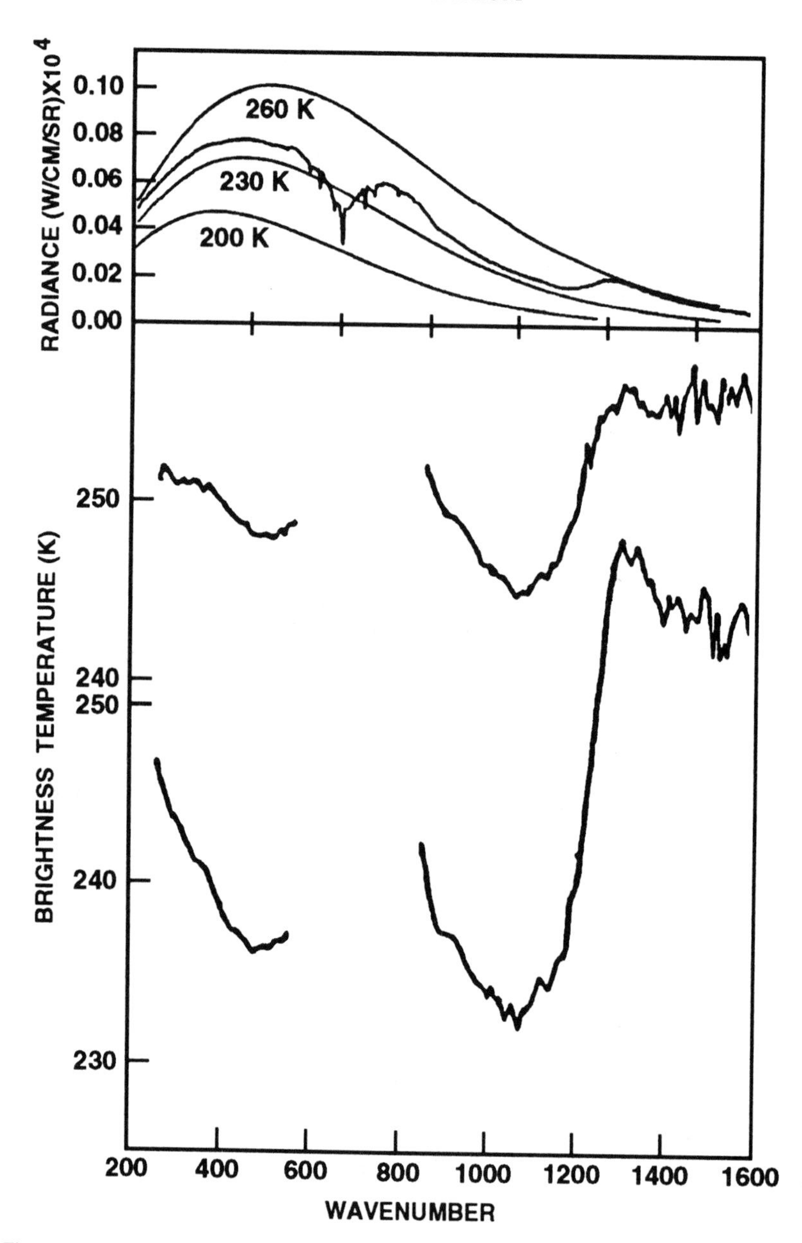

Fig. 12. Mariner 9 IRIS spectra acquired during the 1971 Martian dust storm. (*Top*) Example of an average equatorial emission spectrum in radiance taken from orbit 8 compared with three representative blackbody spectra (figure from Hanel et al. 1972*a*). (*Bottom*) Emission spectra from orbits 8 (upper) and 56 (lower) expressed in brightness temperature (figure from Toon et al. 1977).

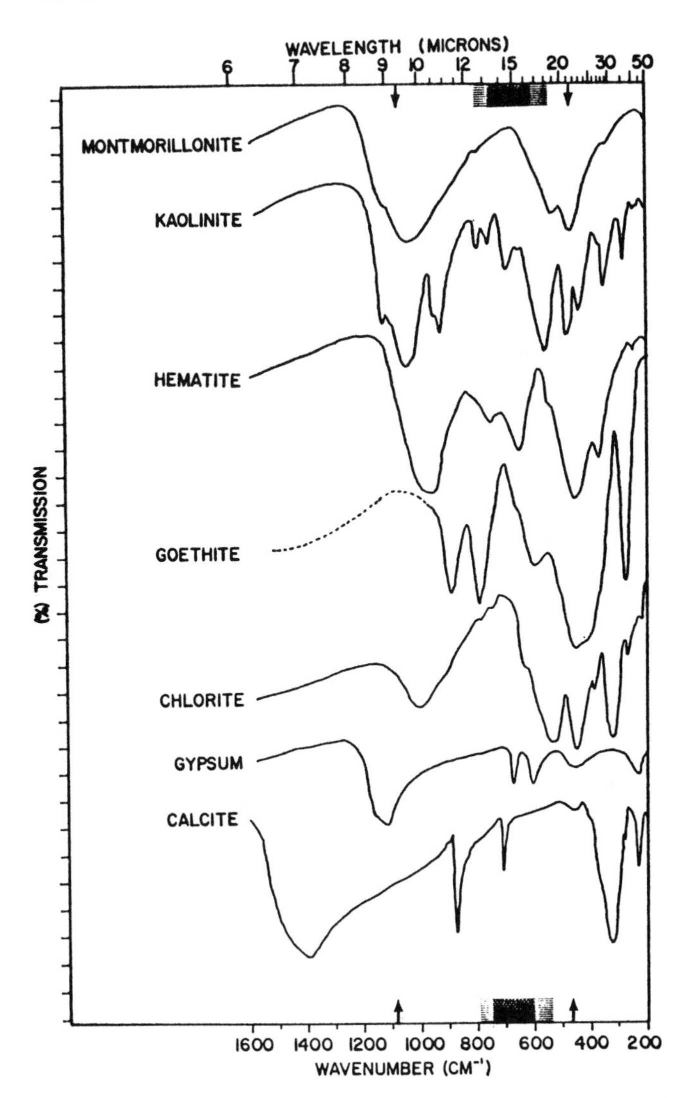

Fig. 13. Infrared transmission spectra of common major secondary minerals (weathering and alteration products) that might be reasonable analogs of Martian materials. Each tic is 10% (figure from Hunt et al. 1973).

1 μm absorption feature, however, is very sensitive to Ca^{2+} abundance in the case of clinopyroxene and to Fe^{2+} abundance in the case of orthopyroxene. These variations in band position with composition can also be seen in Fig. 16. The spectra of Fig. 16 exhibit many features due to Fe^{3+} and H_2O as well.

Huguenin et al. (1977, 1978) examined a collection of spectra covering

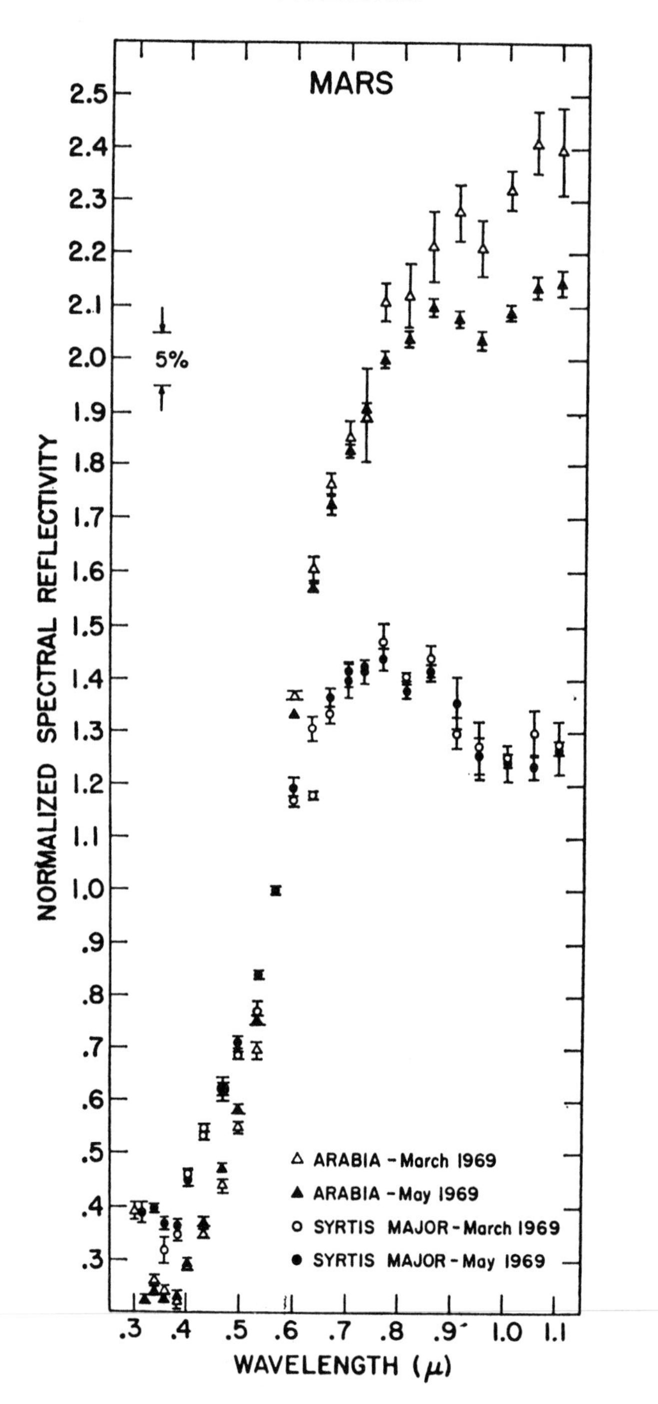

Fig. 14. Early spectral reflectance observations showing a broad absorption feature near 1 μm that occurs dominantly in dark-region spectra (figure from McCord and Westphal 1971).

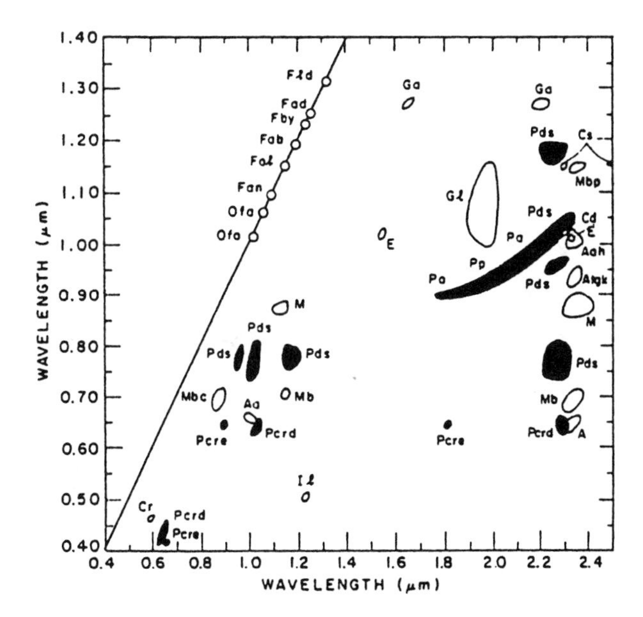

Fig. 15. Positions of diagnostic absorption bands in reflection spectra of common minerals. Minerals falling on the line have a single band. Of interest in interpretation of the Martian spectra are the pyroxenes (shaded) (P, pyroxene; Pa, augite; Pcrd, chrome diopside; Pcre, chrome enstatite; Pds, diopside-salite; Po, orthopyroxene; Pp, pigeonite) and olivines (Ofo, fosterite and Ofa, fayalite) (figure from Adams 1974).

the 0.3–1.1 μm region acquired for a variety of dark areas by McCord and coworkers in 1969 and 1973. These spectra show substantial variation in the detailed nature and shape of the absorptions in the 0.8–1.1 μm region from area to area, indicating major differences in dark-region mineralogy. Huguenin and coworkers proposed that variations in the concentrations of different types of pyroxene, of olivine, and of bright dust containing iron oxides could explain the dark-region variability. These authors concluded that the variation is consistent with a range in composition from mafic basalts to ultramafic rocks rich in olivine.

Because the spectra used in these early studies extended to only 1.1 μm (for example, in Fig. 14), the unequivocal identification of substantial olivine was not possible. Figure 16 shows that the absorption feature in olivine centered near 1.05 μm (dominant in the fayalite end-member) is a very broad feature extending to well beyond 1.5 μm. When spectra extending out to 2.5 μm (cf. Fig. 10) were acquired by McCord and coworkers in 1976 and 1978 (McCord et al. 1978; Singer et al. 1979), this broad absorption was not observed. Singer (1982) re-examined all available dark-region spectra and concluded that olivine is not conclusively apparent in any of the data. The evidence indicated that olivine is not the dominant mafic mineral, at least in the large regions that had been observed telescopically.

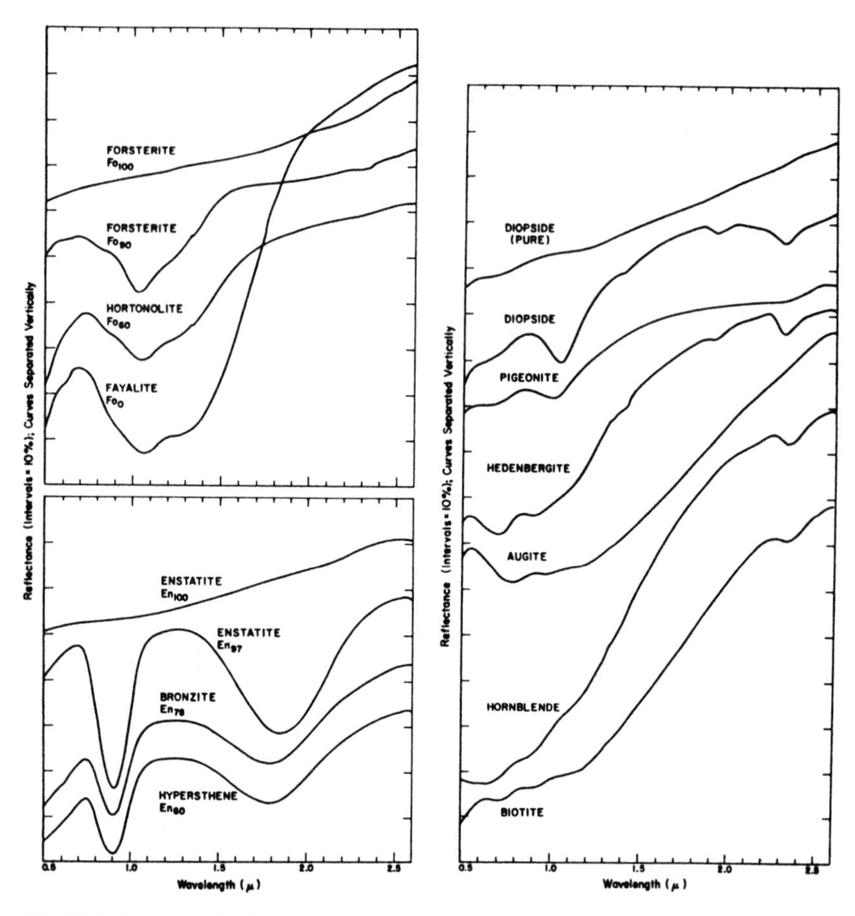

Fig. 16. Laboratory reflection spectra of representative iron-bearing, rock-forming minerals: olivines (upper left), orthopyroxenes (lower left), and clinopyroxenes (right) (figure from Adams 1968).

Singer (1980) examined in detail features in the 0.7–1.3 μm region of the best available dark-area spectrum (for Iapygia) from the 1978 observations. He was able to resolve the position of two weak bands at 0.88 μm and 0.99 μm (Fig. 17). On the basis of work by Adams (1974) (cf. Fig. 15), Singer attributed the 0.88 μm band to ferric oxide in the dark-material coating and concluded that the 0.99 μm band is due to an augite clinopyroxene containing 30 to 45% mol Ca^{2+}. Again, conclusive evidence for olivine was not apparent.

As mentioned earlier, the absorption band in pyroxene spectra near 2 μm is difficult to observe for the Martian surface due to the CO_2 triplet centered in the same region. With new spectra for Hesperia, Iapygia and Syrtis Major acquired in April 1980, Singer and Roush (1985) identified what they

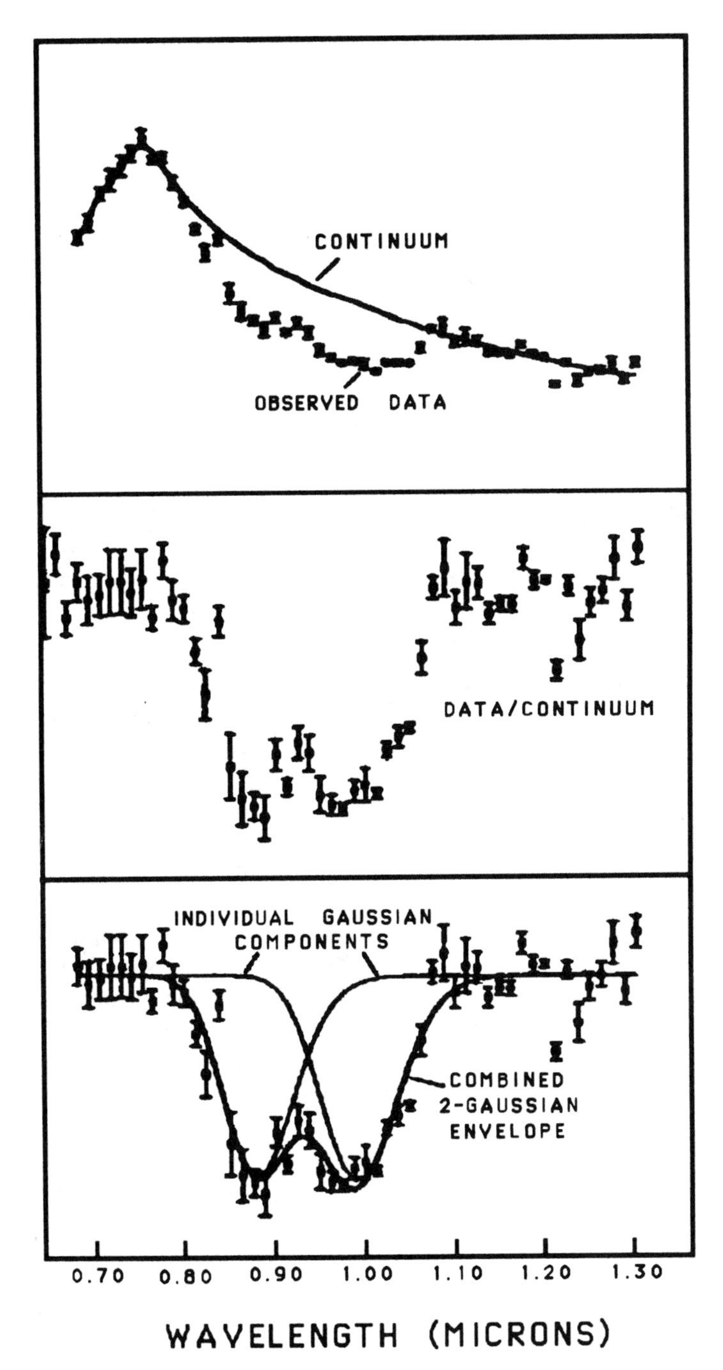

Fig. 17. Double absorption structure in the 0.8–1.1 μm region of the reflection spectrum for the dark-region Iapygia attributed to a ferric-oxide coating and an augite clinopyroxene, respectively (see text) (figure from Singer 1980).

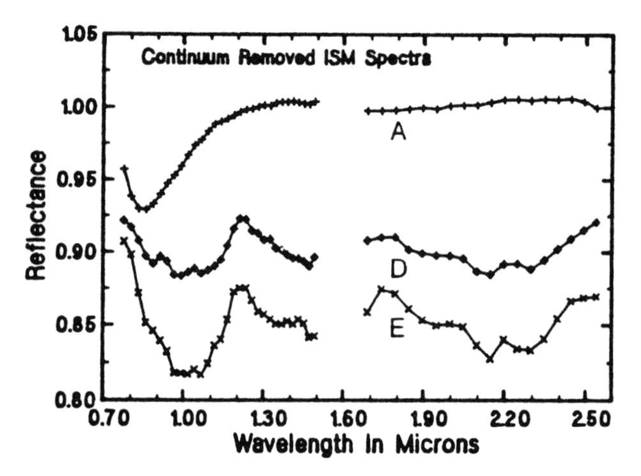

Fig. 18. Normalized reflectance derived from Phobos 2 ISM spectra for a bright region, Isidis (A), and two dark regions, Syrtis East (D) and Syrtis West (E). After a model of atmospheric absorption was removed from the spectra, they were normalized to telescopic observation by forcing bright regions in the ISM and telescopic data to match. Finally, a simple straight-line continuum was removed (figure from Mustard et al. 1990).

believed to be a broad band superposed on the atmospheric CO_2 and estimated it to be centered near 2.1 μm. From this evidence and from the position of the shorter-wavelength band (0.95 to 0.97 μm) in these spectra, they suggested augite, low-Ca orthopyroxene, or pigeonite as the candidate pyroxenes. More recently, in the analysis of spectra collected by the ISM experiment aboard Phobos 2, Erard et al. (1990) and Mustard et al. (1990) identified similar broad absorption bands centered between 2.1 μm and 2.3 μm, respectively (cf. Fig. 18). These authors used a range of techniques to suppress the CO_2 feature—normalization to bright-region ISM spectra, modeling the atmospheric band, and ratios with groundbased telescopic spectra. These results are all consistent with a variety of pyroxenes but dominantly high-Ca^{2+} clinopyroxene in the Martian dark regions.

The most recent spectroscopic measurements of the 1 μm region confirm suggestions from multispectral imaging (Soderblom et al. 1978; McCord et al. 1982) that the composition of dark regions is variable. From new spectra acquired during the 1988 opposition, Singer et al. (1990*a,b*) noted that the position of the 1 μm pyroxene band ranges from 0.93 μm to 0.97 μm across the dark areas; in other spectra, no Fe^{2+} pyroxene absorption was observed at all. Pinet and Chevrel (1990) acquired narrowband ($\Delta\lambda/\lambda \sim 1\%$) telescopic images of Mars at 0.73 μm, 0.91 μm, 0.98 μm and 1.02 μm. Although identification of specific minerals (e.g., clinopyroxene, orthopyroxene, or olivine) is impossible with such images alone, they show substantial differences in the dark-region absorptions in the 1 μm region down to the limits of resolution, suggesting major variation in both composition and abundance for the mafic minerals in the dark areas. The Phobos 2 ISM spectra also show substantial variation in the 1 μm region within dark areas down to the spatial resolution (~25km/pixel) of the spectral maps. Bibring et al. (1990*a,b*) noted that, although in some dark-region spectra the absorption is centered near 1.05 μm, the presence of olivine as the dominant mafic mineral is not consistent with those data (compare Figs. 16 and 18).

In summary, reflection spectroscopy for Martian dark areas clearly shows an absorption near 1 μm and, with less certainty due to interference by CO_2 absorption, another band near 2 μm. The position of the 1 μm feature varies among dark regions from 0.93 to 1.05 μm and may even be absent from some dark areas. The consensus is that these features are consistent with pyroxenes of varied abundance and Ca^{2+} and Fe^{2+} content. Although clinopyroxenes such as augite and pigeonite provide the best candidates, orthopyroxenes also may well be present. Although no convincing evidence for olivine has yet been found, some olivine would be expected on other grounds. Abundant pyroxene could spectrally mask olivine in low-to-moderate abundance.

VI. CLAYS

The presence of bound H_2O in some form in the Martian soil, as evidenced by the 3 μm feature, together with the suggestions for abundance alteration and weathering products, make clays likely candidates for spectroscopic searches. In fact, Toulmin et al. (1977) suggested abundant clay minerals, and in particular nontronite, an iron-rich smectite clay, as the best candidates that are consistent with the result of the X-ray fluorescence experiment aboard the Viking Landers.

Hunt et al. (1973) suggested that the broad silicate absorption band centered near 1100 cm^{-1} (9 μm) observed in the Mariner 9 IRIS data could be explained by airborne dust being composed largely of the clay mineral montmorillonite (compare Figs. 12 and 13). Toon et al. (1977) were able to better resolve the longer-wavelength dust-absorption feature centered between 400 and 500 cm^{-1} (Fig. 12, bottom), showing it, like the shorter-wavelength feature, to be broad and devoid of sharp features. Clays typically show a doublet absorption feature in this spectral region, as can be seen clearly in a transmission spectrum for montmorillonite from Salisbury et al. (1987) (Fig. 19). Toon et al. (1977) and Roush et al. (1989*a*) concluded that, although mixtures of materials with clay minerals provide a reasonable overall fit to the IRIS spectra, the absence of a spectral structure near 20 μm is in conflict with the identification of clay as the major ingredient.

Clays also have various diagnostic weak absorption features that should be observed in the near-infrared reflection spectrum of Mars (Fig. 20). These are generally due to a combination of metal-ion-OH lattice modes and O-H or H-O-H vibrational modes (cf. Singer 1985). The stronger features near 1.4

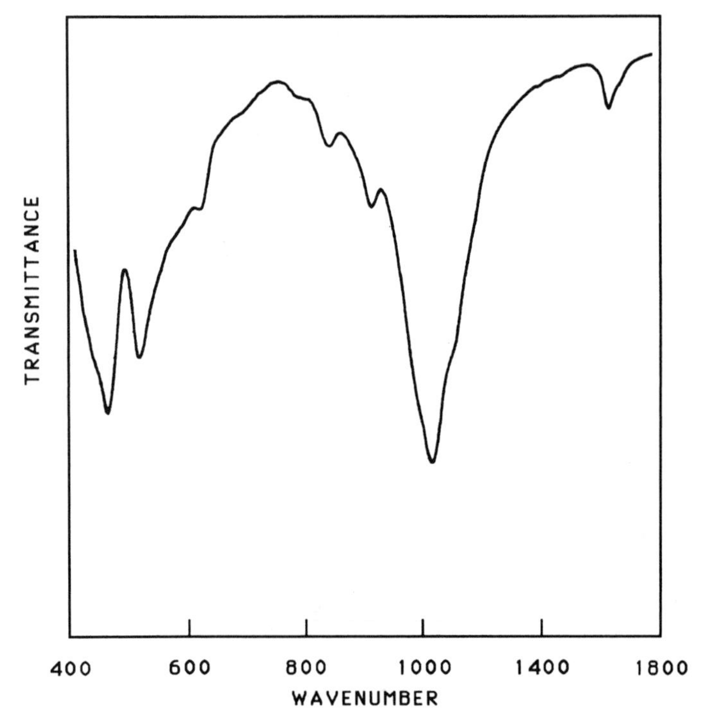

Fig. 19. Transmission spectra of montmorillonite from Apache Co., Arizona (figure adapted from Salisbury et al. 1987).

μm and 1.9 μm in Fig. 20, due to O-H stretching and H-O-H bending modes, are difficult to observe for Mars owing to interference by absorption bands in the atmospheres of the Earth and Mars. The metal-OH features in the 2.0–2.5 μm region are important, as this region is one of the few that are free of dominating CO_2 and telluric H_2O absorptions. Aluminous clays such as kaolinite and montmorillonite exhibit Al-OH vibrations at 2.2 μm; iron-rich clays such as nontronite exhibit Fe-OH absorption near 2.3 μm, and trioctohedral minerals with Mg-OH structure exhibit absorption near 2.3 to 2.35 μm (Singer 1985).

The 1978 observations of McCord et al. (1982*b*) suggested very weak spectral features near 2.3 μm, but those data were marginal for confident identification. From an analysis of newer, higher-quality spectra acquired in 1980, Singer et al. (1984) detected no feature near 2.2 μm, which precludes aluminous clays such as montmorillonite. They did, however, confidently identify a very weak absorption near 2.35 μm (Fig. 21), noting that the feature was several times weaker than what might be expected for an abundant crystalline clay. The high-albedo regions (Utopia, Elysium and Arabia) show

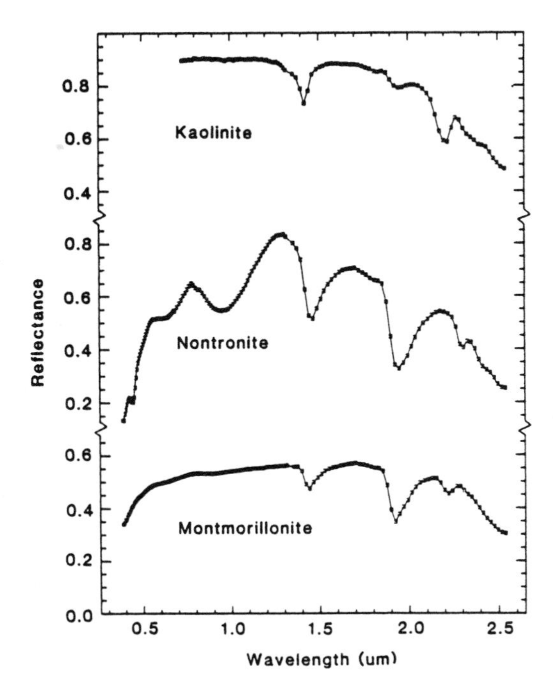

Fig. 20. Representative spectra of clay minerals. Weak features in the 2.2–2.4 μm region are diagnostic of different metal-OH bonds in these materials (figure from Singer 1982).

the strongest absorption as compared with the dark-region Iapygia, as might be expected if the bright regions are more heavily altered. Singer (1985) suggested an explanation for this feature: low abundance of a desiccated magnesian clay. The iron-rich clay, nontronite, also shows an absorption at 2.3 μm (Fig. 20) and would also be a candidate minor constituent.

To sum up, evidence from both the near-infrared reflectance spectra and mid-infrared transmission spectra would allow low concentrations of clay minerals in the bright regions and dust. Crystalline clays typically exhibit greater spectral structure than is seen in the Martian dust spectrum, particularly near 2.0 μm. Abundant crystalline aluminous clays (e.g., montmorillonite and kaolinite) would exhibit a feature near 2.2 μm that is not observed in the Martian data, and these clays are thus precluded. A subtle absorption in the reflection spectrum near 2.35 μm has been suggested as due to a poorly crystalline magnesian (or iron-rich) clay, although higher-resolution spectra suggest an alternate candidate (scapolite), discussed in Sec. X. The observations are inconsistent with crystalline clays being the dominant component of the Martian surface; spectroscopic evidence for clay minerals remains inconclusive.

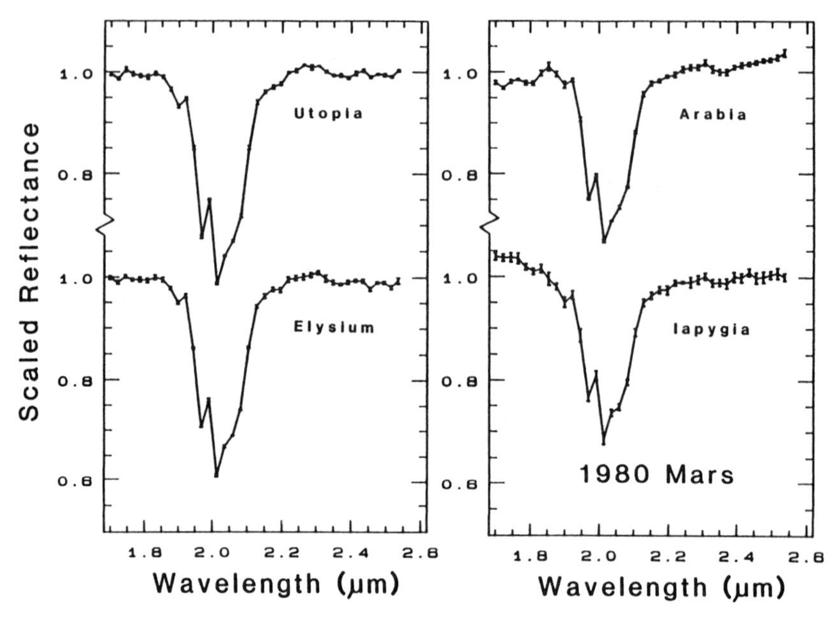

Fig. 21. Reflection spectra for three bright regions and one dark region (Iapygia) showing a weak absorption feature at 2.35 μm attributed to a possible Mg-OH vibration in a weakly crystalline clay (figure from Singer et al. 1984).

VII. PALAGONITES

Both the visible and near-infrared reflectance spectra of the surface and the mid-infrared transmission spectra of the dust have posed a general problem for interpretation. In fact, most crystalline minerals exhibit a variety of absorption features in both spectral regions (cf. Figs. 9, 13, 16, 19 and 20) that should have been detected if such materials were present in abundance. An alternative suggestion is that the surface materials are largely amorphous or poorly crystalline, as are palagonites or altered volcanic glasses (Toulmin et al. 1977; Gooding and Keil 1978; Soderblom and Wenner 1978).

Palagonite, an amorphous ferric-iron silica gel, can be formed by volcanic eruption of basalt into H_2O-rich environments or by subsequent alteration of basaltic ash. In Fig. 22, reflection spectra of three Antarctic palagonites are compared with reflection spectra for bright and dark areas. These palagonites were formed when normally dark basalts were erupted into ice; the absorption bands at 1.4 μm and 1.9 μm are due to substantial H_2O in the samples. Much like the Martian soils, these iron-rich materials display a rapid rise in reflectance from 0.3 to 0.7 μm (along the wing of the deep charge-transfer band), but they do not show the variety of absorption features between 0.5 and 1.0 μm that are characteristic of crystalline iron oxides (Fig. 9).

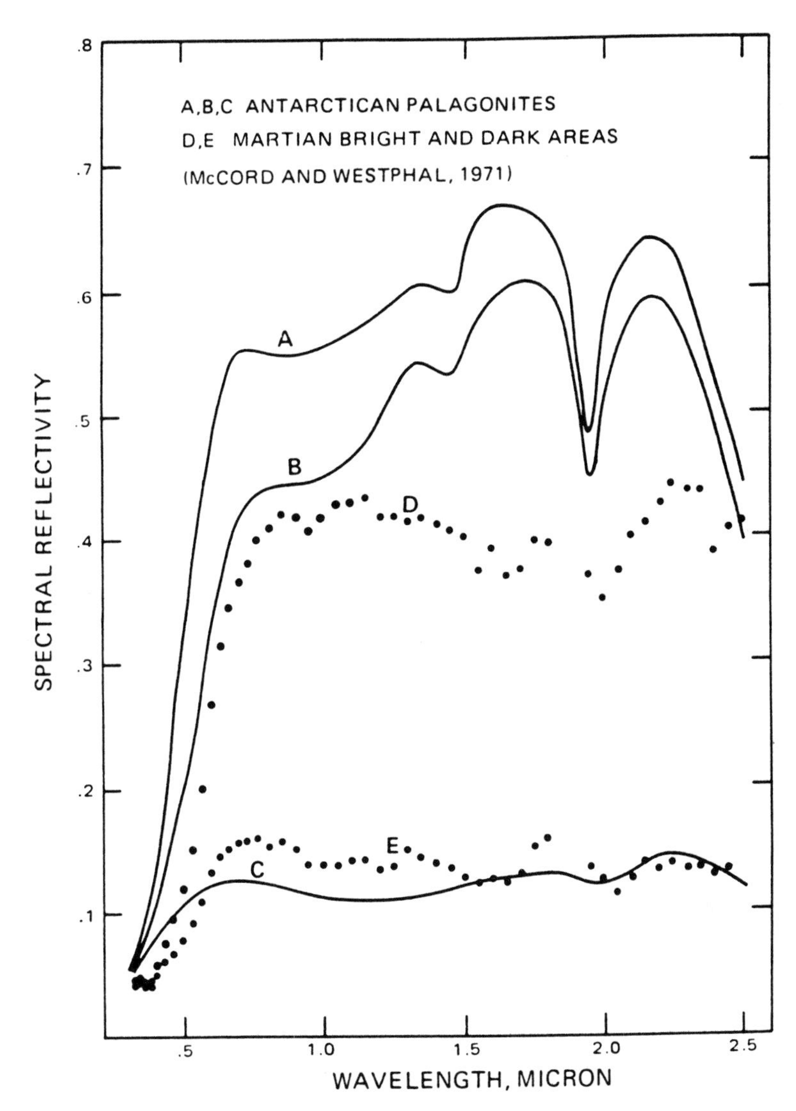

Fig. 22. Comparison of spectral reflectance for three Antarctic palagonites with Martian reflection spectra. The samples were strongly (A,B) and weakly (C) palagonitic. The bright-area (D) and dark-area (E) spectra are from McCord and Westphal (1971); comparison from Soderblom and Wenner (1978).

Synthetic iron-silica gels (Evans and Adams 1980) and a range of natural palagonitic and poorly crystalline weathered soils have been studied as potential analogs (Evans and Evans 1979; Allen et al. 1981; Singer 1982). The four Hawaiian palagonites shown in Fig. 23 are the result of weathering of mafic volcanic glass in semi-arid environments. These amorphous altered materials, again like the Martian reflectance spectra, show only very weak evidence for the 2.2–2.4 μm features characteristic of clays.

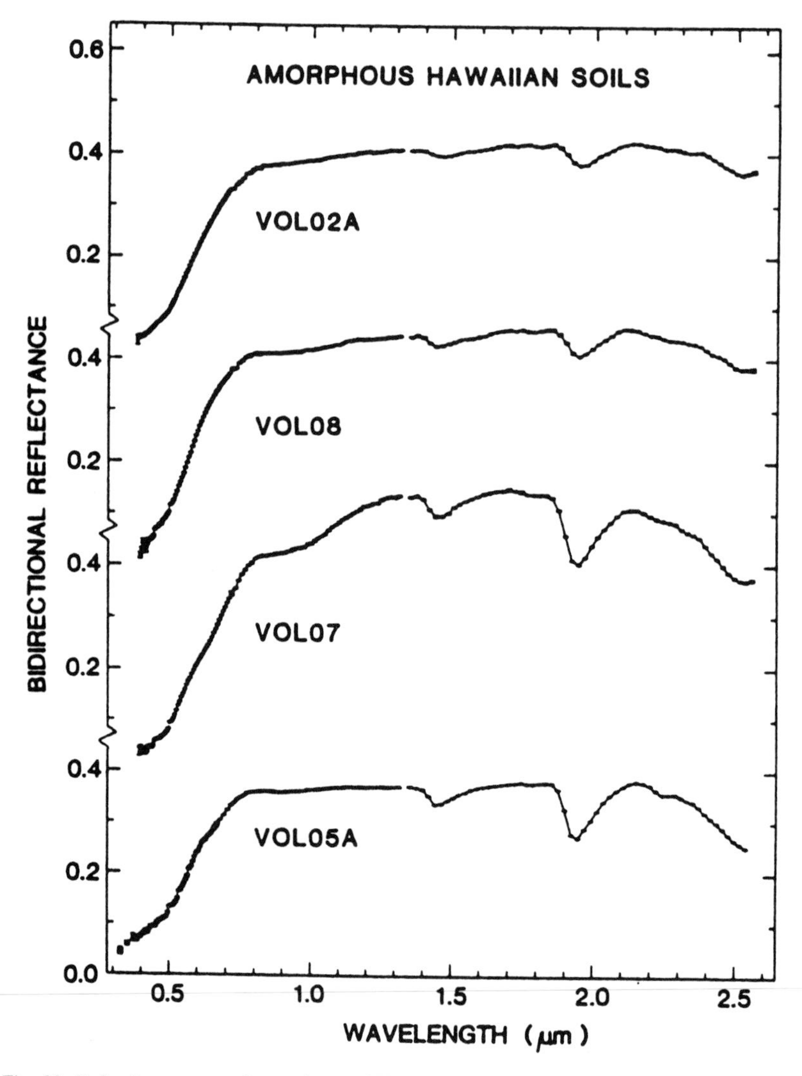

Fig. 23. Reflection spectra of several natural Hawaiian palagonitic soils generated by weathering of mafic volcanic glasses (figure from Singer 1982).

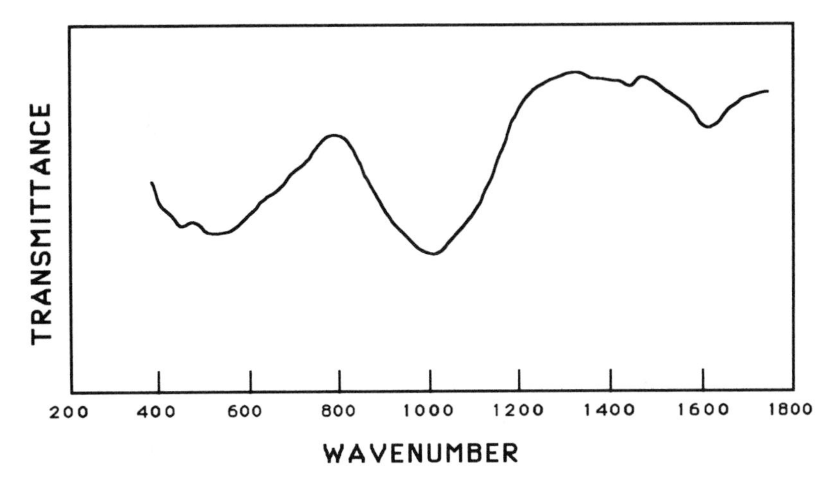

Fig. 24. Infrared transmission spectrum of a Hawaiian palagonite (VOL15) that was produced from weathering of basaltic glass (figure from Roush 1989).

Roush (1989) compared the transmission spectra of iron-rich smectite clays and palagonitic weathered volcanic glass with the Mariner 9 spectra. He found that palagonites are generally better matches to the Martian dust spectra than are the smectite clays, exhibiting much less spectral structure in the 400 to 600 cm^{-1} region (cf. Fig. 24) than clays. Thus, in all three spectral regions—the visible, near-infrared and far-infrared—palagonites constitute an excellent spectral analog for the bulk composition of Martian bright soils and dust.

VIII. CARBONATES

Typical carbonates exhibit a variety of absorption features in spectral reflectance from 2 to 5 μm, including a combination band at 4.0 μm just short of the strong 4.2 μm CO_2 atmospheric band. Recent analysis of the Mariner 6 and 7 IRS data covering this spectral region by McKay and Nedell (1988) (cf. Fig. 2) gave no indication of carbonates. While noting that the IRS coverage did not include the most likely geologic settings for carbonates, such as the layered deposits in the canyon systems, they were able to place an upper limit for carbonate abundance of about 10%. Blaney and McCord (1990*a*) examined the 3.2 to 4.2 μm region with new, high-quality spectra collected with the NASA Infrared Telescope Facility (IRTF) and also found no evidence of the strong carbonate fundamental near 4.0 μm. They concluded that carbonate abundance is limited to <3 to 5% in the regions observed.

Carbonates also exhibit strong features in the thermal infrared, in particular near 7 μm (cf. Fig. 13). Analyses of the Mariner 9 IRIS transmission

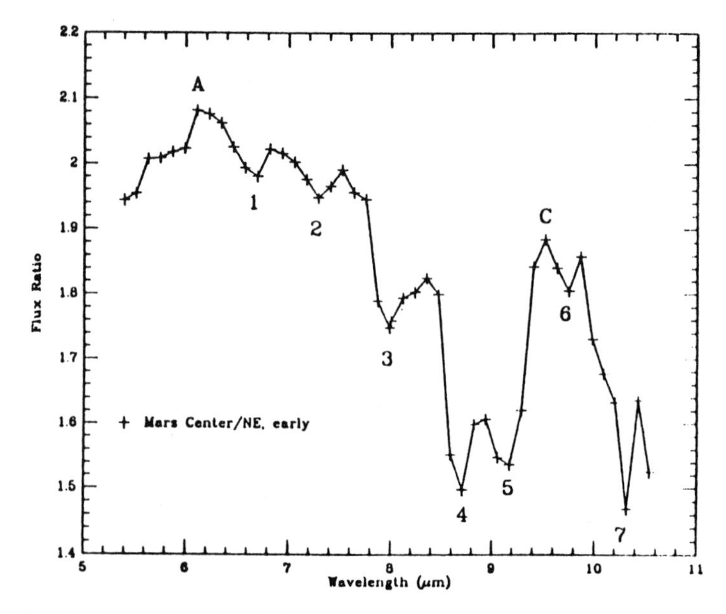

Fig. 25. Ratio of two thermal-emission spectra of Mars. Numerator is a spectrum with higher thermal emission and about the same (low) air mass as the denominator. The result is enhancement of thermal-emission features due to silicates in the surface (A and C), CO_2 absorptions in the atmosphere (2, 3, 5 and 7), and absorption by atmospheric dust (1, 4 and 6). Feature 1 is attributable to carbonate or bicarbonate; feature 4 (and possibly 6) to sulfate or bisulfate (figure from Roush et al. 1989*b*).

spectra of the Martian dust in the far infrared (cf. Fig. 12) were used to place upper limits on the carbonate content of 10% by Hunt et al. (1973) and a few percent by Toon et al. (1977).

More recent observations of Mars in the 5.4–10.5 μm spectral region were acquired in 1988 by Roush et al. (1989*b*) from the Kuiper Airborne Observatory (KAO). These spectra show a variety of thermal-emission features from materials on the surface as well as absorption features from atmospheric gases and dust suspended therein. By ratioing spectra for regions that have large thermal contrast but similar air masses, these features are enhanced (Fig. 25). Roush et al. (1989*b*) interpreted one of these features located near 6.7 μm (feature 1 in Fig. 25) to be due to a carbonate more complex than calcite and having distorted crystallographic structure. Pollack et al. (1990*b*) derived an abundance of 1 to 3% for carbonate in the airborne dust. Bibring et al. (1990*b*) detected a feature near 2.51 μm in ISM spectra in the region from Ascraeus up to Lunae Planum that they tentatively attributed to carbonate. If so, they may have detected one of the locations richer in this material.

Recent work by Calvin and King (1991) has shown that the absorption features in *hydrous* carbonates are very weak compared with those of anhydrous carbonate that have been conventionally studied. Their work suggests that hydrous carbonate could be present in significant abundance and remain undetectable.

IX. SULFATES

The surface chemistry experiments of the Viking Lander detected several percent sulfur, thought to be in the form of sulfate in soils in an abundance of 5 to 10% (Toulmin et al. 1977). Sulfates exhibit absorptions in both the thermal infrared and reflected mid-infrared that should be detectable. For example, gypsum ($CaSO_4 \cdot 2H_2O$) shows a strong absorption near 1100 cm^{-1} (9 μm) (Fig. 13), which is characteristic of sulfates. Roush et al. (1989*b*) and Pollack et al. (1990*b*) identified two features at 8.7 μm and 9.8 μm (features 4 and 6 in Fig. 25), which they attributed to sulfate or bisulfate (the first confidently and the second tentatively). Pollack and coworkers concluded that these data are consistent with sulfates in the airborne dust in an abundance of 10 to 15%.

Sulfates also exhibit an absorption in the mid-infrared near 4.5 μm which might be detected on the long-wavelength shoulder of the strong atmospheric 4.2 μm CO_2 band. Blaney and McCord (1990*b*) collected spectra in the 4.4–5.1 μm region from the IRTF for a variety of Martian regions during the 1988 opposition. Some of these spectra, one of which is shown in Fig. 26, exhibit an absorption at 4.5 μm which Blaney and McCord attributed to sulfate. They noted further that there is large regional variability in the strength of this feature, suggesting substantial variation in sulfate content.

X. SCAPOLITE

Recent high-spectral resolution observations of the 2–2.5 μm spectral range were acquired by Clark et al. (1989, 1990) using the IRTF, during the 1988 Mars opposition. Superposed on the broad 2.35 μm feature that was originally attributed to a poorly crystalline magnesian clay by Singer et al. (1984) is a series of narrow spectral features that Clark et al. (1989, 1990) suggested are due to the presence of scapolite (Fig. 27). Scapolite is a common metamorphic mineral produced from limestones and is closely related to plagioclase. The scapolite series is complex, ranging from sodium-rich (marialite) to calcium-rich (meionite) minerals.

Encrenaz and Lellouch (1990) suggested that these narrow spectral features might well be due to atmospheric CO. Clark et al. (1990) agreed that CO absorptions are responsible for some of the structure, but because features vary with location on the planet and because a few specific bands are not consistent with CO alone, they maintained that at least some of the information is surface related. It is clear that, if real, the features are weak, diluted in the amorphous materials. The confident detection and compositional as-

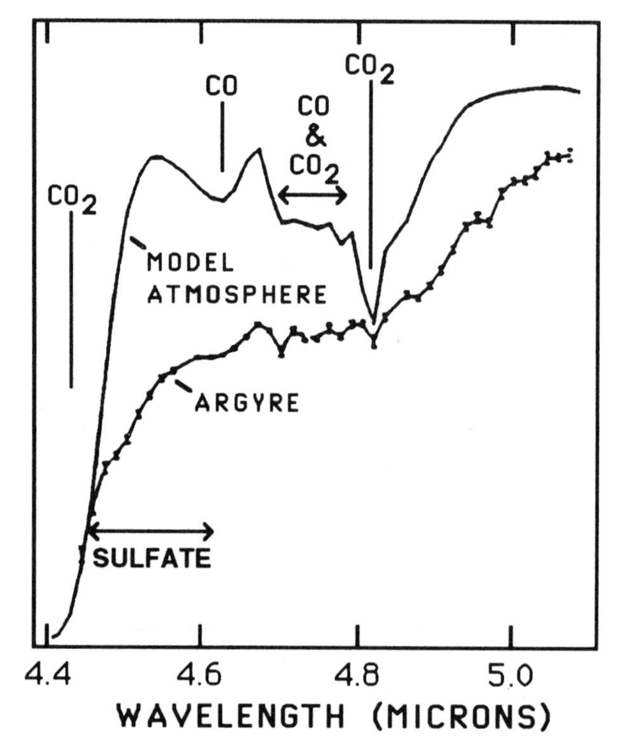

Fig. 26. Reflection spectrum of the Agyre region showing an absorption feature near 4.5 μm attributed to sulfate or bisulfate. Solid line is a synthetic spectrum to model absorptions by atmospheric CO_2 and CO in this region (figure from Blaney and McCord (1990*a*).

signment of these weak spectral features offer exciting prospects for further observation.

XI. SUMMARY

Spectroscopic observations (0.3 to 50 μm) coupled with multispectral imaging yield the following insights into the composition, mineralogy, physical nature and distribution of solid materials exposed at the surface and suspended in the atmosphere of Mars:

1. Spectroscopic observations dating back to the early 1960s show a strong absorption in the 3 μm region due to H_2O in some form distributed ubiquitously through the Martian soils in an abundance of a few tenths of a percent to a few percent.
2. The spectrum of the retreating south polar cap displays many absorption features (1.2 to 2.4 μm) due to CO_2 ice. The retreating north cap's spec-

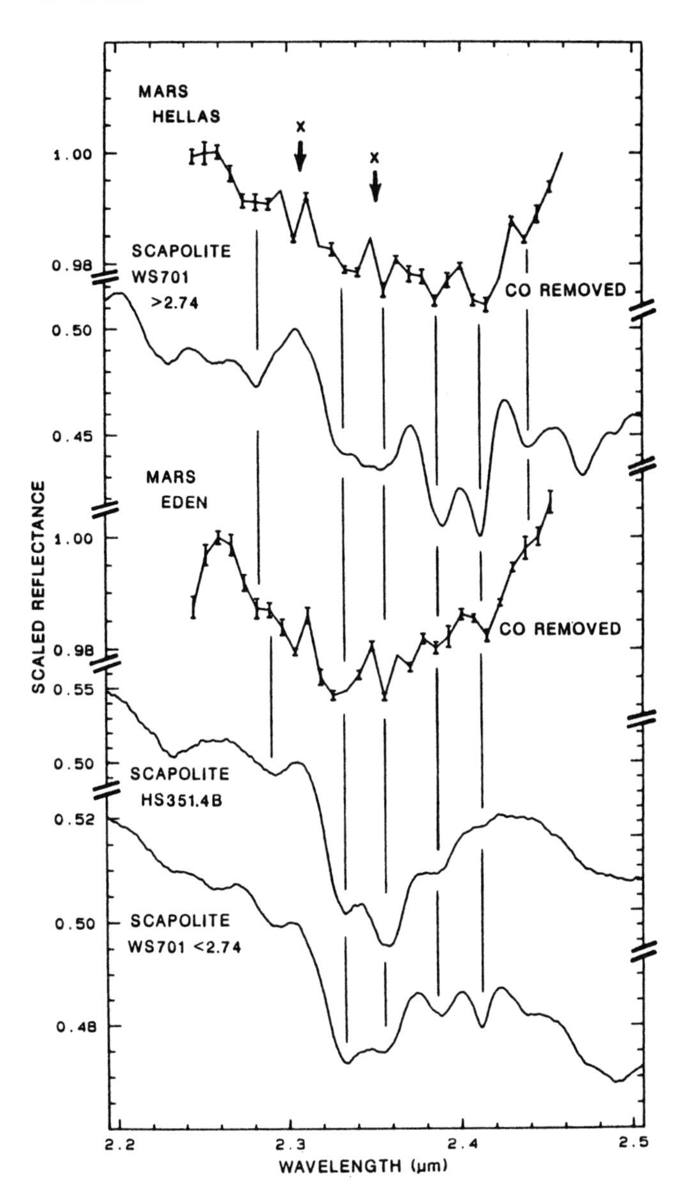

Fig. 27. Comparison of IRTF spectra of two Martian regions with three laboratory samples of the mineral scapolite. The features marked "x" correspond to areas where uncertainties were introduced in removal of CO features (figure from Clark et al. 1990).

trum shows an absorption feature at 1.5 μm due to H_2O ice. This may represent a lag that develops as the annual CO_2 deposit sublimes.

3. Viking Orbiter multicolor images show three discrete albedo classes: bright red units, probably fine-grained materials that are moved in the annual global dust storms; intermediate-albedo, red materials that may be older, indurated aeolian deposits, perhaps akin to the duricrusts found at the Viking landing sites; and dark units occurring as dune masses and sand sheets and as exposures of bedrock or of coarse debris locally derived from the bedrock. The dark units range in color from extremely red (as red as the brighter materials) to more neutral, probably representing various degrees of iron oxidation.
4. Early laboratory studies showed that spectra of the bright and dark materials could be simulated by basalts with different degrees of oxidation, the more heavily oxidized and finer material being exposed in the bright areas.
5. Reflection spectra for dark regions show an absorption near 1 μm and a more tentative one near 2 μm. Evidence suggests that the position of the 1 μm feature varies among dark regions from 0.93 to 1.05 μm and may even be absent for some dark areas. The consensus is that these features are consistent with clinopyroxenes of varied abundance and Ca^{2+} and Fe^{2+} content. Although clinopyroxenes such as augite and pigeonite are the candidates most often suggested, orthopyroxenes may well be present also. No convincing evidence for olivine has yet been found.
6. Although, in all probability, ferric iron controls Mars' rich red-ocher color, coarsely crystalline iron oxides cannot constitute more than a percent or so of the surface material or airborne dust. If they did, significant absorption features would be easily detected, both in the visible reflectance of the surface and in the infrared transmission of the airborne dust. Recent laboratory studies show, however, that subtle features detected in the surface reflection spectrum (0.3–0.7 μm) could be due to abundant crystalline hematite (as much as the 18% Fe_2O_3 estimated from the Viking Lander data) if this material occurs in very fine crystalline or "nanocrystalline" form (<10 nanometers).
7. Although the mid-infrared transmission spectra of Martian airborne dust were early suggested as due to clays (e.g., montmorillonite or iron-smectite clay), crystalline clays typically exhibit greater spectral structure than is seen in the Martian spectrum, particularly near 20 μm. Abundant crystalline aluminous clays (e.g., montmorillonite and kaolinite) exhibit a feature near 2.2 μm that is not observed in the Mars reflection spectra, and thus they are precluded. A subtle absorption in the spectrum near 2.35 μm has been suggested as due to a poorly crystalline magnesian (or iron-rich) clay, although higher resolution spectra suggest an alternative candidate (scapolite). Spectroscopic evidence for clay minerals remains inconclusive.

8. An alternative explanation to crystalline oxides and clays or silicates in general is that the bright soils are highly amorphous, consisting of iron-rich palagonites or silica glasses. Laboratory measurements of palagonites show them as the best analogs to explain the broad, smooth absorptions in the blue to ultraviolet in the surface reflectance spectra as well at 10 μm and 20 μm in the dust-transmission spectra.
9. The abundance of carbonates on a regional scale is limited to less than about 5%, as determined by the absence of strong absorptions at 4 μm in the Mariner 6 and 7 IRS spectra of the surface and at 7 μm in Mariner 9 IRIS transmission spectra of the dust. A dust-absorption feature near 6.7 μm in recent KAO spectra suggests 1 to 3% of a carbonate more complex than calcite. Phobos ISM spectra suggest a weak feature near 2.51 μm tentatively assigned to carbonate. Thus carbonates appear to exist at the percent level on regional scales and may well be more abundant locally.
10. Sulfate or bisulfate has been identified in recent KAO transmission spectra of airborne dust from an absorption feature near 8.7 μm and more tentatively from a feature near 9.8 μm. This observation is consistent with an abundance of 10 to 15%, roughly consistent with that expected from modal analysis of elemental sulfur from Viking Lander data. An absorption feature in the surface reflection spectrum, detected in IRTF observations near 4.5 μm on the long-wavelength edge of the 4.2 μm atmospheric CO_2 band, has also been attributed to sulfate or bisulfate. This feature apparently varies among areas, indicating substantial regional variation in sulfate abundance.
11. Recent IRTF spectra of higher spectral resolution resolved a series of narrow spectral features in the weak absorption band at 2.35 μm that earlier had been attributed to poorly crystalline magnesian clay. These new, higher-resolution features have been attributed to the mineral scapolite. The assignment is debated, as some but not all of these features coincide with atmospheric CO bands.

Acknowledgments. Helpful reviews and criticism of this chapter were provided by D. Blaney, C. Pieters, B. Jakosky, D. MacKinnon, and A. McEwen.

18. SURFACE CHEMISTRY AND MINERALOGY

A. BANIN
The Hebrew University

B. C. CLARK
Martin-Marietta Astronautics

and

H. WÄNKE
Max Planck Institut für Chemie

A single "geological unit" consisting of fine, apparently weathered soil material covers large portions of the surface of Mars. The chemical-elemental composition of the surface materials has been directly measured by the Viking Landers. Positive detection of Si, Al, Fe, Mg, Ca, Ti, S, Cl and Br was achieved. Compared to most basalts, Fe is high and the MgO/Al_2O_3 *ratio is uncommonly close to one. Potassium concentration is relatively low. SNC meteorite analyses supply additional elemental-concentration data, broadening considerably our chemical database on the surface materials. A compositional model for Mars soil, giving selected average elemental concentrations of major and trace elements, is suggested. It has been constructed by combining the Viking Lander data, the SNC meteorite analyses, and other related analyses. The mineralogy of the surface materials on Mars has not yet been directly measured. By use of various indirect approaches, including chemical correspondence to the surface analyses, spectral analogies, simulations of Viking Lander experiments and various modeling efforts, the mineralogical composition has been somewhat constrained. It is suggested that the fine surface materials on Mars are a multicomponent mixture of weathered and nonweathered minerals. Smectite clays, silicate mineraloids similar to palagonite, and scapolite have been suggested* *as possible major candidate components among the weathered minerals. Iron is present as amorphous iron oxyhydroxides mixed with small amounts of crystallized iron oxides and oxyhydroxides, having extremely small particle sizes (nanophases). As accessory minerals, it is likely that the soil contains various sulfate minerals and chloride salts. If present, carbonates are likely to be at very low concentrations, although siderite ($FeCO_3$) may be present at somewhat higher concentrations. Organic matter is lacking.*

I. INTRODUCTION

Large portions of the surface of Mars are covered by fine-particle, nonconsolidated, weathered material in a homogeneous, presumably thin, blanket. Rocks and boulders are exposed on the surface and are surrounded and covered by fine, loose soil material as well as cemented, crusty material.

The fine-soil materials are strongly involved in present-day, global, surface-atmospheric interactions affecting the Martian climate via dust storms, volatile redistribution and albedo changes. The volcanic history of Mars and the evolving weathering conditions on its surface will have left their fingerprints on the soil. Furthermore, the soil may hold clues to the processes that shaped the early history of Mars including abiotic chemical evolution and the possibilities of primitive biotic evolution. In future exploration of Mars, these widespread soil materials may be an important resource for essential raw materials, as well as shielding and building materials.

A general definition of soil on Earth is *the top weathered layer of the terrestrial lithosphere that is exposed to atmospheric and biotic effects*. It constitutes a mineral matrix containing a mixture of weathered and nonweathered lithospheric rock grains. The weathering agents are both physical, such as thermal expansion-contraction and freeze-thaw cycles, and chemical, mostly hydrolytic processes but also redox, chelation and complexation reactions. By definition, the particle size of soils is smaller than 2 mm. Since only relatively weak interparticle forces bind the soil particles, they are prone to be easily moved by wind or water shear. The solid soil particles form a nonconsolidated porous body with a typical porosity of about 50% of the total volume. In this pore system the soil fluids, water and air, reside.

Soil on Mars can be analogously defined as *the top nonconsolidated layer of weathered and partly weathered rocks of the Martian lithosphere that is or was exposed to atmospheric effects*. It is likely that most of the weathered soil material had been formed during the early part of Mars' history at a period when Mars is believed to have been much warmer and wetter (Carr 1986; Pollack et al. 1987; see chapter 32). At that stage the conditions on the surface of Mars are believed to have been much like those on the primitive Earth so that abiotic and even biotic evolution could have taken place also (McKay 1986; see chapter 35). Thus, the fine Martian surface materials have several attributes which render them a soil rather than a planetary regolith. (In contrast, the lunar regolith contains primary-rock particles which were never weathered hydrolytically, and have never been exposed to atmospheric effects.)

The Martian soil covers large portions of the planetary surface as one soil unit (Singer et al. 1979; McCord et al. 1982*a*). The soil material appears to have been thoroughly mixed and homogenized on the global scale by repeated dust storms over extremely long periods of time, perhaps hundreds of Myr (but see chapters 21 and 22 for discussion of aeolian sorting). One evidence for homogeneity is the similarity of spectral reflectance fingerprints of large areas on Mars designated as the bright regions that appear to represent the more oxidized and weathered rock materials (McCord et al. 1982*a*; Singer et al. 1979). Even more striking is the almost identical physical-mechanical properties (Moore et al. 1977,1979,1982), and bulk-elemental chemical composition (Toulmin et al. 1977; Clark et al. 1982), of the fine soil studied at the two Viking landing sites.

In this chapter, we review the accumulated knowledge on the chemistry and mineralogy of the Martian surface materials. We review and analyze pertinent information obtained by direct analyses of the soil on Mars by the Viking Landers, by remote sensing of Mars from flyby and orbiting spacecraft, by telescopic observations from Earth, and through the detailed analyses of the SNC meteorites presumed to be Martian rocks. On the basis of the current knowledge, we suggest a chemical model and a mineralogical model for the surface materials of Mars intended to be a working model for future exploratory missions to Mars and in related studies of its surface.

II. ELEMENTAL-CHEMICAL COMPOSITION

A. Direct Elemental Analyses

Considerably after the instrument selection for the Viking Lander science payload, it was realized that a serious shortcoming was the lack of suitable instrumentation to determine the chemical composition of the soil that the life detection experiments would be analyzing. A 2-yr delay of the mission allowed rectification of this deficiency. After a competition (Clark and Baird 1973,1974), a special miniaturized energy-dispersive *X*-ray fluorescence spectrometer (XRFS) design was selected. It was considered to be beyond the scope of the mission objectives to provide for deployment of this instrument onto the surface. Rather, fines were to be delivered in the same manner as to the other two analytical instruments on board (Biology experiments and GC/MS). Nonetheless, in the hope that pebble-sized rocks might be present on the Martian surface, the Viking XRFS sample inlet was designed to accommodate the possibility of receiving fragments that could pass a screen with 1.3-cm wire spacings (the screen was necessary to minimize the possibility of plugging the 2.5-cm diameter inlet tube, that would prevent further samples from being received). Although the mission plan called for only three samples to be delivered to this instrument, the sample dump cavity was enlarged to the limit imposed by surrounding instruments. At the Viking Lander 1 site (Chryse Planetia), 13 samples were acquired and analyzed. Mission operations and the lack of compositional variability in the repeated analyses limited the number of samples obtained at the Utopia Planetia site to eight.

The XRFS instrument design provided analysis of samples that were dumped into an inlet funnel with the Viking sampler scoop. After free-falling down a tube, the material was captured in a square cross-section cavity that contained two ultrathin polymer windows. At least 25 cm^3 of material was required to fill the cavity. *X*-ray fluorescence analysis was accomplished using radioisotope excitation sources (^{55}Fe and ^{109}Cd), one per window, and taking the energy spectrum of fluorescent *X*-rays as detected in four gas-filled proportional counters, three with thin beryllium windows and one with a grid-supported, 4-μm aluminum window to enhance transmission to Mg and Al fluorescent *X*-rays. The sensors and analysis section was optimized for geologic samples and the engineering constraints imposed by the mission, which precluded using cooled high-resolution silicon detectors. A flip-in calibration target was included with samples of Ca and Al to allow stability and resolution measurements, and to aid in the deconvolution of the overlapping Mg/Al and K/Ca peaks. The simplified scanning pulse-height analyzer increased the necessary sample analysis time. Digital encoding of the spectra was performed inside the XRFS, precluding data errors in transmission. All flight instruments were characterized on the same suite of 25 rock standards prior to installation on the spacecraft.

The elemental composition of 6 "protected" samples of fines, so-called because they were taken from beneath rocks after they were pushed away, or from holes dug by the sampling arm, is given in Table I. Results of all 21 samples analyzed can be found in Clark et al. (1982). The results are expressed as oxides, in the oxidation states expected for Mars, even though only the elemental concentrations were measured. The justification for this reporting methodology is in keeping with long-standing tradition whereby rock compositions are reported as simple oxides because of wet chemical procedures that produce oxide products. The sum of oxides is considerably less than 100%. Several reasons for this shortfall can be given. Light elements ($Z < 12$) could not be detected and therefore adsorbed water, and some hydroxide, carbonate and nitrate minerals could be present without being tallied. Certain elements such as Na, P and Mn could not be detected because of interferences from major Mg, Cl and Fe peaks. Finally, grain sizes and mineral heterogeneities have been demonstrated in pre-flight experiments to cause lower-than-expected peak heights for major elements due to weathering rinds or coating of large grains by smaller ones. Because no processing for increasing sample homogeneity on Mars was allowed prior to the XRF analyses, these effects could have also been present.

TABLE I
Average Compositions of Mars Soil Classified by Sample Types

Sample Type	Protected Fines		Surface Fines		Clods
Samples	Chryse	Utopia	Chryse	Utopia	Chryse
Averaged	C-6, -11	U-2, -4, -6, -7	C-1, -7, -8	U-1, -3, -5	C-2, -5, -13
Constituent			(% by Weight)		
SiO_2	44	43	43	43	42
Al_2O_3	7.3	(7)[a]	7.3	(7)[a]	7
Fe_2O_3	17.5	17.3	18.5	17.8	17.6
MgO	6	(6)[a]	6	(6)[a]	7
CaO	5.7	5.7	5.9	5.7	5.5
K_2O	<0.15	<0.15	<0.15	<0.15	<0.15
TiO_2	0.62	0.54	0.66	0.56	0.59
SO_3	6.7	7.9	6.6	8.1	9.2
Cl	0.8	0.4	0.7	0.5	0.8

[a]Not measured because of instrument difficulties; assumed to be the same as in Chryse samples.

From Table I, it is obvious that the chemical compositions are quite similar at the two landing sites, in spite of the remoteness of their locations from one another, consistent with the fact that the spectral signatures of these two regions of the planet are similar (Guinness 1981). However, it must be kept in mind that many of these samples were taken from some depth and cannot merely represent the contemporary aeolian superficial deposit from recent windstorms. The possibility that much of Mars is covered with a material of relatively uniform composition cannot be excluded, and indeed is likely.

That these fines are not simply ground-up basalt with a minor oxidation of iron minerals, as was widely expected prior to the landing, is quite evident from the bar chart in Fig. 1. The low Al, high Fe, moderate Mg, very high S, unexpected Cl, and lack of clearly detectable concentrations of one or more of the trace elements Rb, Sr, Y or Zr, yield a compositional profile that is quite unique and unusual. Computer searches into a library of over 500 terrestrial, lunar and meteorite rock and soil compositions gave no satisfactory matches within the measurement accuracy of the instrument. To date, no terrestrial natural sample with the same element composition profile as Martian fines has been identified. The very high Fe coupled with the low levels of expected trace elements and low K/Ca ratio pointed to a mafic or ultramafic source material. With S some 2 orders of magnitude higher than for typical igneous rock and for the lithospheres of Earth and Moon, it was supposed either that the fines were enriched in salt leachate (Toulmin et al. 1977) or that the lithosphere of Mars could be higher in S content (Clark and Baird 1979*a,b*). A more complete discussion of the mineralogical phases in the Martian soil is given in Sec. III.

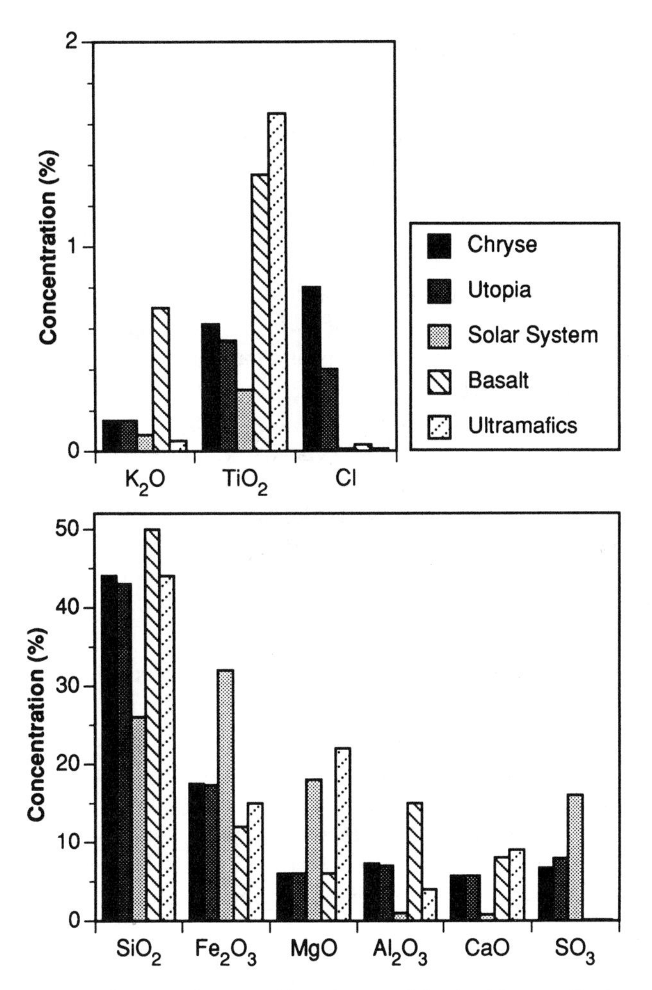

Fig. 1. Comparison of Mars soil compositions (protected fines) with normalized average solar system (Anders and Grevesse 1989) and compositionally averaged terrestrial rocks (Poldervaart 1955).

Attempts to collect lithic fragments for analysis, which presumably would not contain such high S and Cl concentrations, met with frustration. The methodology was to scoop a large quantity of material with the sampler, invert the sampler head, and then operate the jaw chopping motion to facilitate removal of fines through the built-in 2-mm screen. Laboratory tests had demonstrated this to be an effective means for separating out pebbles using samples from mixed test beds. This procedure was not successful at the Utopia landing site; no material remained in spite of several attempts. At Chryse, this approach was successful for all four attempts. The compositions measured were, however, distinctly not that of purely rock materials. As seen in Table I, the S and Cl contents in these captured "clods" were considerably higher than that for fines. Crusty material is observed at many places at both Viking Lander sites, in terms of planar surfaces with crack features, as well as clods produced by digging operations with the sampler (see chapter 21). It is therefore widely believed that the attempts at obtaining rock samples succeeded rather in the collection of crusted material, presumably formed by cementation due to its higher salt content. All efforts at collecting a handful of pebbles being unsuccessful has been interpreted by some to indicate a surprising lack of small stones at the two sites investigated, particularly in view of the observed size distributions of what appear to be abundant, often-vesiculated, igneous rocks.

Samples C-1 and C-6, taken at the surface and at 23-cm depth into what appears to be the edge of the drift deposits at Chryse, are identical for all practical purposes (Table II) (see chapter 21 for a more detailed description of the surface units at the Viking Lander sites). This includes the S and Cl content, indicating no vertical concentration gradient in salts within the measured depth range, if these elements are associated with salt compounds (see Sec. III.D). A surprising detection of bromine occurred in a few cases. The most striking example is shown in Fig. 2.

Laboratory simulation tests of the Martian results using artificial mixtures of various materials show that the Fe, S and Cl concentrations are particularly susceptible to error if the mineral phases in which these elements occur are present as large grains (tens to hundreds of μm in diameter). In each case, the already unexpectedly large values for these elements would be increased by 30% or more of the reported values. This could be one reason that the oxide representation of the Martian soil totals to only about 90%. Alternatively, the soils could be high in carbonates or nitrates, although spec-

TABLE II
Comparison of Composition of Surface and Deep Fines

Sample Type Sample[a]	Surface C-1	Deep Fines C-6
Constituent	**(% by Weight)**	
SiO_2	43	44
Al_2O_3	7.5	7.3
Fe_2O_3	17.6	17.3
MgO	6	6
CaO	6	6.0
TiO_2	0.65	0.61
SO_3	7	6.7
Cl	0.7	0.8

[a] Samples taken at the Chryse Planitia landing site: Sandy Flats, a drift deposit area. C-1 from the topmost layer of soils and C-6 from Deep Hole at 23-cm depth.

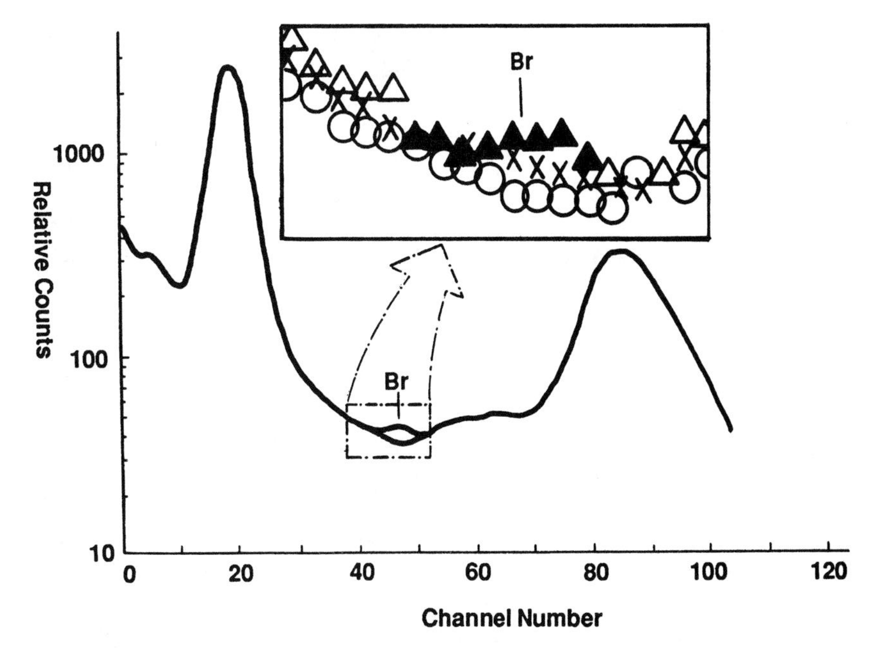

Fig. 2. Detection of bromine in sample C-5 (duricrust) compared to sample C-8 (fines from Rocky Flats). Triangles are measures from Chryse-5; the filled triangles delineate Br region. The crosses and circles indicate two other samples from Viking 1.

troscopic observations and laboratory simulations seem to preclude levels this high (see Sec. III.C).

Attempts to correlate the average composition of Martian fines with terrestrial materials have failed. When the composition is corrected for the possible introduction of exogenous salts, in the form of Mg and/or Ca sulfates and chlorides, the resultant also does not correlate with known materials. This resultant also produces normative quartz (Toulmin et al. 1977), so that it is difficult to derive a mineralogical interpretation that is consistent with evidence for the mafic character of the Martian surface and lithosphere (see review by Arvidson et al. 1989*a*).

Another approach has yielded more interesting results. It was suggested (Clark and Baird 1979*b*; Settle 1979) that the S and Cl content result from the addition of these elements as volcanic gases, either with or without H cation. Correcting the Viking XRFS data in this manner produces a much different igneous composition, that closely matches the Shergotty meteorite, as first pointed out by McSween and Stolper (1980), and confirmed later in more detail (Baird and Clark 1981).

B. SNC Meteorite Analyses

If one assumes that the SNC meteorites are indeed rocks from the Martian surface (see chapters 4 and 6), they contain valuable information about this planet. Wood and Ashwal (1981) were the first to correlate systematically data on SNC meteorites with the knowledge about Mars obtained by space missions. They also used the SNC meteorites to obtain new insight into the chemistry and evolution of Mars.

1. Petrography and Mineralogy of the SNC Meteorites. The SNC achondrites group (shergottites, nakhlites and chassigny), includes eight meteorites (Table III), that are very similar to terrestrial basaltic and ultramafic rocks.

Shergottites comprise Shergotty, Zagami, EETA 79001 and ALHA 77005. Shergotty and Zagami resemble terrestrial diabases in terms of texture and mineralogy (Duke 1968; Stolper and McSween 1979; Smith and Hervig 1979). Both stones consist mainly of the pyroxenes pigeonite and augite, and of maskelynite, a shocked plagioclase glass (see Table IV). The two pyroxenes are zoned; they have homogeneous Mg-rich cores ($En_{60}Fs_{28}Wo_{12}$ and $En_{48}Fs_{20}Wo_{32}$ for pigeonite and augite, respectively, where En = enstatite, Fs = ferrosilite, Wo = wollastonite), and Fe-rich rims (average composition for pigeonite $En_{21}Fs_{61}Wo_{18}$ and for augite $En_{25}Fs_{47}Wo_{28}$). The maskelynite is also zoned, varying from $An_{57}Ab_{42}Or_1$ cores to $An_{43}Ab_{53}Or_4$ rims (An = anorthite, Ab = albite, Or = orthoclase). Titanomagnetite, ilmenite, pyrrhotite, fayalite, trydimite, whitlockite, chlorapatite and baddeleyite are the accessory phases. The co-existence of titanomagnetite and ilmenite show equilibration under conditions close to the quartz-fayalite-magnetite buffer which indicates that Shergotty and Zagami crystallized under relatively oxidized conditions (Stolper and McSween 1979). Treiman (1985) reported melt inclusions in pyroxenes consisting of fine-grained aluminous spinel and kaersutite. EETA 79001, another member of the shergottites, is the only known layered meteorite, comprising three lithologies: two igeneous lithologies A

TABLE III
The SNC Meteorites

Group	Meteorite	Fall	Location	Mass in kg
Shergottites				
	Shergotty	1865, August 25	India	5
	Zagami	1962, October 3	Nigeria	23
	ALHA 77005	found, 1977	Antarctica	0.48
	EETA 79001	found, 1979	Antarctica	7.9
Nakhlites				
	Nakhla	1911, June 28	Egypt	40
	Lafayette	found, 1931	USA	0.60
	Governador Valadares	found, 1958	Brazil	0.16
Chassignites				
	Chassigny	1815, October 3	France	4

TABLE IV
Modal Compositions of SNC Meteorites (% by vol.)[a]

	Shergotty[1]	Zagami[1]	EETA 79001A[2]	EETA 79001B[2]	ALHA 77005[3,4]	Nakhla[5]	Lafayette[6]	Governador Valadares[7]	Chassigny[8]
Pigeonite	36.3	36.5	59.3	39.5	26				
Augite	33.5	36.5	6.1	20.0	11	78.6	major	major	3.8
Ortho-pyroxene			5.4		+				4.0
Olivine			8.9		52	15.5	minor	minor	88.5
Plagioclase	23.3*	21.7*	17.1*	29.1*	10*	3.7	+	+	2.6
K-Feldspar						1.1	+	+	+
Titano-magnetite	2.3	2.1	2.0	3.5		1.9	+	+	
Chromite			1.0		1.0				1.2
Sulfides	+	+	+	+	+	+	+	+	+
Phosphates	+	+	0.2	0.4	+	+	+	+	+
Mesostasis	4.0	2.1	0.1	0.7	+				

[a](*) = maskelynite; (+) = trace. References: 1 = Stolper and McSween 1979; 2 = McSween and Jarosewich 1983; 3 = Ma et al. 1981; 4 = McSween et al. 1979; 5 = Bunch and Reid 1975; 6 = Boctor et al. 1976; 7 = Burragato et al. 1975; 8 = Nehru et al. 1983.

and B, which differ from each other in chemical composition but are both similar to basaltic rocks (McSween and Jarosewich 1983), and lithology C consisting of shock-melted glass. The last member of the shergottites, the ALHA 77005, is similar to the terrestrial plagioclase harzburgites (McSween et al. 1979). It differs from Shergotty and Zagami in having abundant cumulate olivine and chromite which in many cases are enclosed poikilitically by pyroxene. The chemical composition of the olivine is uniform with an average of Fo_{74} (Fo = forsterite).

The three nakhlites (Nakhla, Lafayette and Governador Valadares) show many petrographic and mineralogical similarities to each other (Bunch and Reid 1975; Berkeley et al. 1980). They resemble terrestrial clinopyroxenites. The total absence of maskelynite indicates that this group did not experience significant shock metamorphism. The major phase in the nakhlites is cumulus augite ($En_{38}Fs_{23}Wo_{39}$). However, minor coarse-grained cumulus olivine (Fa_{65-67}) is also present (Fa = fayalite).

The Chassigny meteorite is a dunitic rock (Table IV) with suggestive cumulate texture (Floran et al. 1978). It consists mainly of homogeneous fayalitic olivine (Fa_{32}) that has inclusions of melt (silicate-glass). Other phases are augite and orthopyroxene with poikilitic texture.

2. Chemical Composition of the SNC Meteorites and of the Martian Crust as their Parent Body. The SNC meteorites are thought to represent surface rocks of Mars, ejected into space by large impacts. Their composition (Table V) is in agreement with the conclusion of the rather mafic character of the Martian crust based on the composition of the Martian soil as discussed above (Sec. II.A; Baird et al. 1976; Toulmin et al. 1977; Clark et al. 1982). Five of the eight SNC meteorites, namely Chassigny, the three nakhlites, and the shergottite ALHA 77005 have mafic to ultramafic composition. On Earth, a rock like ALHA 77005 with 27% MgO would be classified as komatiite (Baird and Clark 1981). Treiman (1986) has shown that obviously ultrabasic volcanism was more important on the SNC parent body than on the Moon, the eucrite parent body, or on present-day Earth. The closest terrestrial analogues to the nakhlites are pyroxene cumulate layers in different picrite and komatiite flows. From the cumulus and intercumulus phases of Nakhla, Treiman (1986) estimated the composition of the intercumulus magma, suggesting an ultrabasic composition with a Fe/Mg ratio higher than that of terrestrial ultrabasic magmas. Estimates of the composition of the Martian mantle inferred from SNC meteorites composition by Dreibus and Wänke (1984) also yielded a high Fe/Mg ratio (see chapter 6). The intercumulus magma of Nakhla was found to be considerably enriched in refractory incompatible elements. This enrichment is in contradiction to the ultrabasic major element composition. Mixing or assimilation of trace-element rich evolved magmas,

TABLE V
Chemical Composition of the SNC Meteorites[a]

		Sher- gotty	Zagami	EETA 79001 A	EETA 79001 B	ALHA 77005	Nakhla	Lafayette	Gov. Valadares	Chassigny
MgO	%	9.28	11.0[a]	16.31	7.38	27.69	11.82	12.9[a]	10.9[a]	31.6
Al_2O_3		7.06	5.67[a]	5.37	9.93	2.59	1.64	1.55[a]	1.74[a]	0.69
SiO_2		51.36	50.8[a]	48.58	49.03	43.08	49.33	46.9[a]	49.5[a]	38.16
CaO		10.0	10.8[a]	7.05	10.99	3.35	14.3	13.4[a]	15.8[a]	0.60
TiO_2		0.87	0.77[a]	0.64	1.12	0.44	0.35	0.33[a]	0.35[a]	0.10
FeO		19.41	18.00[a]	18.32	17.74	19.95	21.70	22.7[a]	19.74[a]	27.1
Na_2O		1.29	0.99[a]	0.82	1.66	0.44	0.57	0.36[a]	0.82[a]	0.13
P_2O_5		0.80	—	0.54	1.31	0.36	0.103	—	—	0.058
S		0.133	—	0.16	0.19	0.060	0.025	—	—	0.012
K_2O		0.189	0.14[a]	0.033	0.065	0.027	0.166	0.09[a]	0.43[a]	0.041
Cr_2O_3		0.203	0.30[a]	0.589	0.183	0.963	0.250	0.18[a]	0.21[a]	0.63
MnO		0.525	0.50[a]	0.469	0.452	0.46	0.550	0.79[a]	0.67[a]	0.526
Σ		100.1	99.0	98.9	100.1	99.4	100.8	99.2	100.2	99.6
Li	ppm	5.6	2.0	4.54	2.21	1.31	3.91	—	—	1.3
C		620	—	36	98	82	696	—	—	847
F		41.6	41	39	30.9	21.9	57	—	—	15
Cl		108	137	26	48	14	1145	—	—	34
Sc		58.9	53[c]	36.1	50.5	21.1	55	53	—	5.4
V		263	312[c]	200[c]	199[c]	158[c]	192	—	—	—
Co		39.0	36[c]	47.3	31.1	69.5	54	49	—	126
Ni		83	50[c]	158	46	335	90	85	—	480
Cu		26.0	—	—	—	5.5	6.7	<100	—	2.6
Zn		83.0	62[c]	81	120	71	220	91	—	74
Ga		14.7	14[c]	12.6	24.4	7.5	2.86	3.0	—	0.7
As		0.025	0.046[c]	0.005	0.012	0.022	0.015	<0.15	—	0.008
Se		0.41	0.32[c]	<0.8	—	<0.4	—	<0.3	—	—
Br		0.89	0.87	0.189	0.287	0.085	4.08	0.56	—	0.097
Rb		6.84	5.7[c]	1.04	1.78	0.63	—	<4	—	—
Sr		51	46[c]	57	67	<100	—	75	—	—
Zr		67.9[e]	88.3[e]	29.4[e]	64.8[e]	19.5[e]	9.46[e]	9.44[e]	—	2.74[e]

TABLE V Continued

		Sher-gotty	Zagami	EETA 79001 A	EETA 79001 B	ALHA 77005	Nakhla	Lafayette	Gov. Valadares	Chassigny
Nb		4.6[e]	5.5[e]	0.68[e]	1.7[e]	0.72[e]	1.57[e]	1.46[e]	—	0.36[e]
Mo		0.37	—	—	—	0.20	—	—	—	—
In	ppb	24[b]	22[c]	46[c]	68[c]	11[c]	24.4[d]	20.3[d]	—	—
Sb	ppm	0.005	12[c]	0.01	0.03	0.06	—	—	—	—
I		0.036	<0.005	<0.1	0.96	1.72	<0.01	—	—	<0.01
Cs		0.405	0.38[c]	0.075	0.131	0.04	0.43	<0.3	—	—
Ba		35.5	25[c]	<10	14.0	—	34	27	—	—
La		2.29	1.60[f]	0.37	0.80	0.32	2.14	1.96	—	0.59
Ce		5.54	3.75[f]	1.4	3.1	1.09	5.6	4.76	—	—
Nd		4.5	2.89[f]	1.4	2.9	1.15	2.85	2.8	—	0.7
Sm		1.37	1.17[f]	0.75	1.56	0.42	0.78	0.83	—	0.16
Eu		0.56	0.48[f]	0.35	0.73	0.20	0.23	0.24	—	0.052
Tb		0.44	—	0.30	0.64	0.17	0.13	0.14	—	0.04
Dy		2.94	2.66[f]	2.11	4.58	0.96	—	1.0	—	0.27
Ho		0.56	—	0.50	0.99	0.22	0.17	0.2	—	0.058
Tm		0.30	—	0.21	0.37	0.08	—	—	—	—
Yb		1.69	1.38[f]	1.12	2.14	0.52	0.40	0.33	—	0.12
Lu		0.25	0.20[f]	0.15	0.30	0.073	0.062	0.050	—	0.018
Hf		1.97	1.7[c]	0.93	1.93	0.55	0.29	0.28	—	<0.1
Ta		0.25	0.2[c]	0.03	0.09	0.026	0.09	0.080	—	<0.02
W	ppb	480	420	83	155	84	176	400	—	46
Ir	ppm	0.4	<3	<2	<3	3.5	<2	<2	—	2.4
Au		8	6	2.8	1.1	0.3	2.9	<1	—	1.0
Tl		11[c]	12[c]	6.9[c]	7.9[c]	1.7[c]	3.1[d]	7.2[d]	—	—
Bi		1.6[c]	5.1[c]	0.67[c]	0.76[c]	<0.7[c]	0.5[d]	0.5[d]	—	—
Th		390	—	<100	<200	<100	—	150	—	<200
U		116	154[c]	<60	<100	<50	—	49	—	<100

[a] All data from the Mainz laboratory (Burghele et al. 1983; Dreibus et al. 1982) except (a) Wood and Ashwal 1981; (b) Laul et al. 1986; (c) Smith et al. 1984; (d) Laul et al. 1972; (e) Palme and K. P. Jochum, unpublished data; (f) Shih et al. 1982.

or their crystallization products, with a hot ultramafic magma may be a possible explanation.

Moderately Volatile Elements. The rather low abundance of highly incompatible trace elements in all SNC meteorites suggests a low degree of fractionation of the Martian crust. Although the K/La ratio of the SNC meteorites is 1.5 times higher than the terrestrial ratio, the concentration of K in the SNC meteorites is low. It ranges from 224 to 1430 ppm with a calculated mean of 787 ppm (Table V), and is in striking contrast to the mean concentration of 1.7% K (Wedepohl 1981) in the terrestrial mantle. As pointed out by Dreibus and Wänke (1987), this difference may explain the difference of the release factors of radiogenic ^{40}Ar of Earth and Mars (Anders and Owen 1977). The higher abundance of moderately volatile elements in the Martian mantle (see chapter 6) as compared to Earth is shown by the higher K/La and K/U ratios observed in SNC meteorites and also by the high content of Rb-derived radiogenic Sr and low radiogenic lead on Mars as compared to the analog terrestrial isotope systems.

Phosphorus. A major difference between terrestrial and Martian rocks as inferred from the SNC meteorite composition (Table V) is the high phosphorus content of the latter, especially if compared with the concentration of other elements with similar degree of geochemical incompatibility. The high abundance of phosphorus in the SNC meteorites is in all likelihood the consequence of high P abundance in the Martian mantle as compared to the terrestrial mantle (see chapter 6).

Volatile Elements. The chemical composition of the SNC meteorites further suggests that compared to Earth, Mars is richer in volatile elements such as the halogens (Dreibus and Wänke 1985). This observation is in striking contrast to the obviously low abundances of the most volatile elements, i.e., the primordial rare gases, and of water. The amount of ^{36}Ar per gram planet mass in the Martian atmosphere is 130 times smaller than the corresponding terrestrial figure. This difference might be explained only partly by a smaller release factor. Clark and Baird (1979*a*) have speculated that the Martian soil may contain some of the missing volatile elements, but that their concentration is below the detection limits of the Viking elemental analyzers.

Water. The SNC meteorites can be used in various ways to estimate the amount of water on Mars. The most straightforward estimate is based on the measured concentration of water in SNC meteorites. Yang and Epstein (1985) measured 180 ppm H_2O in the Shergotty meteorite. This meteorite is enriched in La by about a factor of 5 relative to its mantle source. Assuming a similar enrichment for H_2O, i.e., assuming that water is as incompatible as La, water concentration in the mantle is 36 ppm (Dreibus and Wänke 1989). This is exactly the value obtained earlier by Dreibus and Wänke (1987), by comparing the solubility of H_2O and HCl in basaltic melts and using the observed Cl/La and Br/La ratio in SNC meteorites and assuming that Cl and H_2O were added to the planet in C1-abundance ratio. A mantle concentration of 36 ppm

H_2O would, under the assumption of 100% release, yield a water layer of 130 m depth covering the whole planet. Other approaches were used to estimate the water content of Mars, yielding a wide range of values (see, e.g., Anders and Owen 1977; Carr 1986; see chapter 4).

Hydrated Minerals. There is some controversy about the occurrence of hydrated minerals in SNC meteorites. Bunch and Reid (1975) observed a hydrated alteration phase in Nakhla and Lafayette similar to terrestrial iddingsite. As mentioned above, halite and sulfate were found in the interior as well as in the fusion crust of the Nakhla meteorite by Wentworth and Gooding (1988*a,b*), who favored therefore a terrestrial origin of these minerals. Huge amounts of Cl and Br were measured by Dreibus and Wänke (1987) who also suspected terrestrial contamination as the Cl/Br and Cl/I ratios were close to that in ocean water. In terrestrial ocean water, iodine is depleted relative to Cl and Br, as well as relative to its chondritic abundance, due to its precipitation and concentration in ocean sediments. Hence, if the halogens measured in Nakhla are of Martian origin, a depletion process of iodine relative to Cl and Br would have to be postulated for Mars. Recently, Ott (1988) found excess of ^{80}Kr in Nakhla which was ascribed to neutron capture by ^{79}Br. The amount of excess ^{80}Kr produced during the cosmic-ray exposure of Nakhla seems to require a Br concentration in the range of that measured by Dreibus and Wänke (1987). Thus, the ^{80}Kr excess may indicate a Martian origin of the unusually high halogen concentration found in the Nakhla meteorite. Because chlorine and bromine, as well as sulfur, have been measured in considerable concentrations in the Martian soil by the Viking Landers (see Sec. II.A, Fig. 2), excess halogens and sulfate from water soluble salts in SNC meteorites are not too unexpected.

The question of what is Martian and what is terrestrial contamination is still not resolved. The most reliable observations of Martian secondary minerals in SNC meteorites are those by Gooding and co-workers (Gooding and Muenow 1986; Gooding et al. 1988) who found a sulfate component plus sulfur- and chlorine-rich aluminosilicates, and calcium-magnesium carbonate in lithology C (glass inclusion) of EETA 79001, but not in the other major parts of this meteorite (lithology A and B). Detailed carbon and oxygen analyses of the carbonates in EETA 79001 (Wright et al. 1988; Clayton and Mayeda 1988) showed them to be different from terrestrial carbonates, and similar to those found in Nakhla (Carr et al. 1985) supporting their Mars origin. Trapped noble gases, nitrogen, and CO_2 in elemental and isotope ratios identical to that of the Martian atmosphere have been previously found in the glass inclusions of EETA 79001 (Bogard and Johnson 1983; Becker and Pepin 1984; Carr et al. 1985). Bulk elemental composition of the two aluminosilicates in lithology C can be arithmetically mixed to produce a composition that agrees with that of Martian surface fines at the Viking landing sites including the sulfur/chlorine ratio (Gooding and Muenow 1986).

Oxygen Isotopes. In the oxygen isotope diagram of Clayton et al.

(1976), the SNC meteorites form a distinct group and plot above the terrestrial fractionation line relative to which they have an ^{16}O deficiency of about 0.6‰. In light of the above discussion, this value is to be expected for Mars as well. It is interesting to note that the samples from the eucrite parent body plot below the terrestrial oxygen isotope fractionation line (Fig. 3). Among the three objects, eucrite parent body, Earth and SNC parent body (or Mars), the eucrite parent body has the lowest and the SNC parent body the highest oxygen fugacity. Hence, at least the sequence seems to be correct.

3. Time of Crystallization and Ejection of SNC Meteorites. Up to about 1982, it was thought that all SNC meteorites had crystallization ages close to 1.3 Gyr (see Wood and Ashwal 1981 for references). This was in line with the time scale based on crater statistics (Hartmann 1973*b*; Soderblom 1977; Neukum and Hiller 1981; chapter 12), according to which volcanism on Mars generally extended into the last third of Mars history and in the Olympus Mons area possibly even into the most recent history. However, Nyquist et al. (1979) reported, based on the Rb-Sr system, a crystallization age of Shergotty of 650 Myr, while Jagoutz and Wänke (1986) reported, based on the Sm-Nd system, a crystallization age of 360 ± 16 Myr.

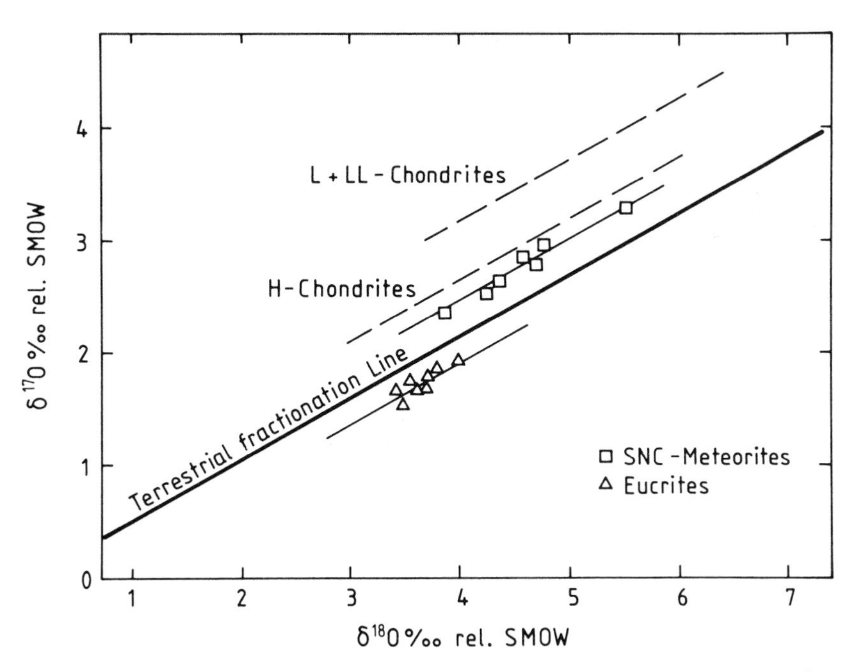

Fig. 3. $^{18}O/^{16}O$ vs $^{17}O/^{16}O$ plot for achondrites. The SNC meteorites clearly form a separate line from the eucrites and are separated from the terrestrial fractionation line, indicating separated reservoirs for the two groups (figure courtesy of Clayton and Mayeda 1983).

Observed disturbances of the Rb-Sr system at 180 Myr were previously ascribed to the shock event which ejected Shergotty, Zagami and ALHA 77005 from their parent planet (Shih et al. 1982). Jagoutz and Wänke (1986) interpreted this event in Shergotty, which they dated by Rb-Sr to 167 Myr and which only affected the isotope system of plagioclase and mesostasis, as an endogenic thermal process. In contrast, Jones (1986) interpreted the 180 Myr event observed in the Rb-Sr system of all four shergottites to be their true crystallization age. Chen and Wasserburg (1986), using the U-Th-Pb system, were the first to suggest that the shock event which transferred feldspar to maskelynite might be associated with the final breakup of this meteorite at 2.5 Myr as dated by cosmic-ray exposure age.

Jagoutz (1989) found a Rb/Sr crystallization age for augite and pigeonite in shergottite ALHA 77005 of 154 $\pm$ 6 Myr; this is almost identical to the 167 Myr event observed in Shergotty. He interpreted this as the crystallization age of the intercumulus phase. However, in the same meteorite, Jagoutz (1989) found a Rb/Sr age of 15 $\pm$ 15 Myr for the plagioclase recrystallized from the shock melt. Within the uncertainty, this shock age is identical to the cosmic-ray exposure age of 2.5 Myr of this meteorite (Nishiizumi et al. 1986). Thus, it is possible that the large differences in the reported crystallization ages of SNC meteorites result from the fact that different isotope systems in different mineral phases date different events.

Using rare gas data for SNC meteorites, Ott (1988) argued that the introduction of the (Martian) atmosphere component by shock must have occurred rather recently and cannot be ascribed to a 180 Myr event. This contradicts the model originally proposed by Nyquist et al. (1979) according to which the SNC meteorites were ejected from the parent body in a single major impact event 180 Myr ago in fragments large enough to be shielded from cosmic-ray exposure since that time. The new evidence suggests that it is more likely that the SNC meteorites were ejected from Mars in three considerably smaller impact events at times corresponding to the three groups of cosmic-ray exposure ages, i.e., 0.5 Myr ago ejection of EETA 79001, 2.6 Myr ago ejection of Shergotty, Zagami and ALHA 77005, and 11-Myr ago ejection of the nakhlites and Chassigny (Bogard et al. 1984). In respect to these three groups of exposure ages reflecting three ejection events, it is worthwhile to remember that the three nakhlites of the 11-Myr group are almost identical in their chemistry and mineralogy. This is also the case for the pair Shergotty and Zagami.

For Nakhla, a crystallization age of 1.3 Gyr has been measured by Nakamura et al. (1982). However, it seems that all the observations on the crystallization ages of shergottites arrive at values considerably below the value of 1.3 Gyr which was suggested initially. They range from about 160 Myr to several hundred Myr. As the shergottites seem to have been derived from two different impact events, most likely from two areas with considerable distance from each other, the question arises as to whether their young crystallization ages can be brought in accordance with the time scale of formation of the Martian surface as inferred from crater statistics. In other words, is it likely that there exist two areas on Mars where lava flows have occurred a few hundred Myr ago and which are large enough so that half of the SNC meteorites found (i.e., the shergottites), can be expected to be derived from these areas with finite probability? The occurrence of late, thin lava flows that do not influence the crater statistics might be one possible explanation. In this respect, the scenario to explain the different crystallization ages observed in one and the same SNC meteorite proposed by Jagoutz (1989) should be mentioned. In this scenario for the formation of Shergotty, ALHA 77005 and possibly other SNC meteorites, an older regolith is thought to have been covered by late, possibly more acidic, flows which cooled rapidly at the surface of Mars. Recent reports (Plescia 1990) have given photogeologic evidence for the occurrence of such flows on Mars.

C. Proposed Elemental Composition Model for Mars Soil

The rather compelling evidence for Mars being the parent body of the SNC meteorites provides a basis for supplementing the directly measured compositional average data for the Martian soil (Table I) with selected averages for various elements in the SNC meteorites (Table V) to obtain a representative chemical model for the Martian soil.

In the Viking XRFS analysis of the soil, the elements analyzed accounted for about 50% of the soil weight; if all detected elements, except Cl, were assumed to be in their common oxide forms, the total would account for about 90% of the soil weight (Sec. II.A). The remainder was attributed (Clark et al. 1977, 1982) to two groups of compounds. The first are compounds of elements that could have been detected but could not be determined unambiguously and accurately by the Viking XRFS. These include compounds of the elements P, Mn, Cr and Na. (Potassium was tentatively reported, but was also below the detection limit of the instrument [Clark et al. 1977].) The total amount of these elements as oxides is estimated to be about 2 to 3%. The second group consists of compounds of elements that could not be detected by the Viking XRFS (elements lighter than Na, atomic number <11). These may include water, carbonates and nitrates and may account for the remainder of the soil weight, although particle size-distribution effects may have biased the results for the major elements as well.

In Table VI, we present a composition model for the Mars soil, giving selected averages for the major components based on Viking XRFS analyses. As observed by McSween and Stolper (1980), the major-element composition of the Shergotty meteorite closely matches the Viking soil analyses. Hence, Shergotty data (Table V) were used to constrain those elements (K, P, Mn, Na and Cr) that were at or below the detection limits of the Viking XRFS. Estimates of the light-element compounds are based on the Viking GC-MS results (Biemann et al. 1977) for water, and simulations of the Labeled Re-

TABLE VI
Representative Chemical Composition of Mars Soil[a]

Constituent	Selected Average Concentration (%)	Source
SiO	43.0	direct soil analysis by Viking XRFS
Al_2O_3	7.2	"
Fe_2O_3	18.0	"
MgO	6.0	"
CaO	5.8	"
TiO_2	0.6	"
K_2O	0.2	analyses of Shergotty meteorite
P_2O_5	0.8	"
MnO	0.5	"
Na_2O	1.3	"
Cr_2O_3	0.2	"
SO_3	7.2	direct soil analysis by Viking XRFS
Cl	0.6	"
Sum of above	91.4	
CO_3	<2	estimated from Labeled Release simulation
NO_3	?	
H_2O	0-1	varying content; direct soil analysis by Viking GC-MS.

[a]Selected average concentrations of elements, generally given as oxides (see text).

lease Biology experiment on Viking, for carbonates (Banin et al. 1981). Using the same assumptions, it may also be possible to employ data given in Table V to obtain estimates of the concentration of the various trace elements in the Mars soil.

III. SOIL MINERALS AND OTHER COMPONENTS

Among the prime, unresolved and often-debated questions related to the modern study of Mars is the question of the structure and mineral composition of its fine, dusty surface materials, the soils of Mars. Many mineral candidates have been suggested as components of the soil, and our perception of its mineralogy has gone through a rather tortuous path since the first flyby of spacecraft in the late 1960s.

Because no direct analysis of the mineralogy of the soil on Mars has been possible thus far, various indirect approaches have been used to constrain its mineralogical composition. These were based on chemical correspondence with the elemental analyses of the soil, the spectral remote sensing observations of Mars, simulations of Viking Biology and other Viking Lander experiments, and various thermodynamic modeling efforts. A compilation of mineral candidates proposed on the basis of these studies is presented in Table VII. The abundance and variety of candidates is the result of, and evidence for, the lack of a true and conclusive definition of the soil's mineralogy.

A. Silicates

Minerals of the silicate group constitute the major components of the soil as shown by its chemical analysis (Table I). Among the numerous candidates suggested, two materials appear to be most accepted currently, although by no means agreed upon; these are smectite clays and palagonites. Scapolites have been recently suggested (Clark et al. 1990) as the most abundant silicate mineral on Mars, but evidence is still in development.

Smectite clays such as montmorillonite and the iron-rich nontronite have been initially suggested as best satisfying the spectral (Hunt et al. 1973; Toon et al. 1977) and chemical (Clark et al. 1976, 1977; Toulmin et al. 1977) properties of dust and soils on Mars. Further strong support was obtained from simulations of the Viking Biology experiments (Banin and Rishpon, 1978, 1979) and the modeling of the weathering processes on Mars (Gooding and Keil, 1978; Zolotov et al. 1983; Zolensky et al. 1987). Smectites do not, however, show the magnetic susceptibility measured on Mars (Hargraves et al. 1977; Moskowitz and Hargraves 1982). Palagonite, a poorly defined weathering product of basalts, is another candidate that satisfies spectral and compositional constraints to some extent (Allen et al. 1981; Singer 1982; Adams et al. 1986; Guinness et al. 1987; Morris et al. 1990). It fails, however, in the simulation of the reactivity shown in the Viking Biology experiments (Banin and Margulies 1983; Banin et al. 1988*b*).

Part of the controversy is real, and part is merely semantic. Whereas clays are well-defined silicate mineral entities, palagonite is very loosely defined. Some clarification of the nature of palagonite is thus pertinent in the context of this review and may be helpful in providing better definition of the open questions regarding Mars soil mineralogy.

Palagonitization. Since the term palagonite was introduced about 150 yr ago by Von Waltershausen (1845) to describe a brown material associated with altered tuffs and glassy lavas, it has been assigned to various materials, occurring in a wide range of terrestrial environments: submarine, subaereal, and near-surface hydrothermal in both cold and warm climates (see, e.g., Honnorez 1980). The use of the term is proliferating particularly in recent years, with the advent of the Deep Sea Drilling Project and other projects. In a number of cases the term has been used implicitly or explicitly to identify a mineral, while in other cases it was considered more like a rock (see, e.g., Bonatti 1965; Hay and Iijima 1968*a,b*; Honnorez 1978, 1980; Furnes 1984; Allen et al. 1981; Staudigel and Hart 1983; Guinness et al. 1987). A wide range of minerals have been identified in palagonitized basalts: nontronite, montmorillonite and other smectites; various zeolites; amorphous and cryp-

TABLE VII
Some Proposed Secondary Mineral Components for Mars Soil

METHOD AND REFERENCE	MINERAL COMPONENTS
MODELING	
XRFS Modeling	
Baird et al. (1976); Toulmin et al. (1977); Clark et al. (1976; 1982).	Mostly (80%) phyllosicates; some sulfates, carbonates; nontronite (47%) montmorillonite (17%) saponite (15%) kieserite (13%) calcite (7%) rutile (1%)
Clark & Van Hart (1981).	Salts of (Mg, Na)-sulfate; NaCl
Baird & Clark (1981).	Igneous minerals; some unaltered and others altered by acidic volcanic gases
Thermodynamic Modeling: Rock Atmosphere Equilibrium	
Gooding (1978); Gooding & Keil (1978).	Mostly oxides, carbonates and sulfates; residual nonweathered igneous minerals; no smectites; possibly kaolinite
Zolotov et al. (1983).	Combination of oxides and phyllosilicates; some sulfates; no carbonates; quartz (21%) montmorillonite (24%) talc (28%) hematite (10%) anhydrite (13%) rutile (0.6%) kieserite (0.2%)
Zolensky et al. (1987).	Smectites (nontronite, beidellite); carbonates (calcite, magnesite, siderite)
Kahn (1985).	Ca-Mg carbonates (5% or more)
Geological Scenarios	
Soderblom & Wenner (1987).	Subsurface hydrothermal alteration-palagonite
Allen (1979,1980,1981).	Subglacial hydrothermal alteration-palagonite
Boslough (1986,1987).	Shock-modified nontronite
Newsom et al. (1986).	Impact activated suevite, hydrothermally altered to clays; Fe-oxides
Burns (1986,1987); Burns & Fisher (1989).	Gossaniferous ferrihydrite-jarosite-opal, clay silicates
SPECTROSCOPIC PROPERTIES	
VIS-IR	
Landers	
Viking Imagery (VIS-NIR)	
Bragg (1977).	Nontronite
Evans & Adams (1979;1980).	Amorphous Fe oxides
Guinness et al. (1987).	Mafic rocks coated with palagonite
Orbiters	
Mariner 9-IRIS (IR)	
Hunt et al. (1972*b*).	Montmorillonite
Hanel et al. (1977).	Mixture of silicate minerals
Toon et al. (1977).	SiO_2-rich mineral; no carbonates
Viking Orbiter (VIS)	
Evans & Adams (1980).	Amorphous Fe oxides
Earth-based	
Telescope-Reflectance (VIS-NIR)	
Singer (1982).	Palagonite
Singer (1985).	Amorphous Fe oxides; magnesian clay
Clark et al. (1989).	Scapolite

TABLE VII Continued

METHOD AND REFERENCE	MINERAL COMPONENTS
Simulations (NIR)	
Morris et al. (1983).	Fe-doped aluminum oxyhydroxide
Banin et al. (1985;1988).	Smectites: adsorbed and short-range-ordered Fe oxyhydroxides
Morris et al. (1985,1987,1988*b*).	Cryptocrystalline hematite
UV	
Orbiters	
Mariner 9-UV Spectrometer	
Egan et al. (1975); Pang & Ajello (1977).	Anatase (TiO_2)
Earth-based	
Telescope-UV absorbance	
Abadi & Wickramasinghe (1977).	Complex interstellar organic molecules
Simulations	
Banin et al. (1985).	Smectites; adsorbed and short-range-ordered Fe oxyhydroxides
SIMULATION OF VIKING EXPERIMENT	
Viking Biology	
Labeled Release (LR)	
Banin et al. (1978,1979,1981, 1983,1985,1988*a*,*b*; Banin 1989*a*).	Smectites (Fe-montmorillonite, Fe-nontronite); Fe-oxyhydroxides; no carbonates of Ca, Mg (less than 0.5%); possibly siderite ($FeCO_3$)
Pyrolytic Release (PR)	
Hubbard (1979).	Smectites (Fe forms); no Ca nor Mg carbonates
Gas Exchange (GEX)	
Oyama et al. (1978).	Polymeric carbon-sub-oxide (C_3O_2)
Blackburn et al. (1979).	Manganese oxide (MnO_2)
All 3 experiments	
Huguenin (1982).	Mafic silicates
Magnetic properties	
Hargraves et al. (1977); Moskowitz & Hargraves (1982).	1 to 7% maghemite or magnetite
Morris (1988).	Palagonite
SNC METEORITES MINERALOGY	
Nakhla	
Bunch & Reid (1975).	Iddingsite (terrestrial origin?)
Wentworth & Gooding (1988*a*,*b*;1989).	Silicate rust; calcium carbonate (calcite), halite; calcium sulfate (terrestrial origin?)
Shergottite—EETA 79001	
Gooding & Muenow (1986).	Sulfate salts and sulfur- and chlorine-rich alumosilicates
Gooding et al. (1988).	Calcium carbonate and calcium sulfate (pre-terrestrial origin)

tocrystalline phyllosilicates; various iron oxyhydroxides; serpentine, hydrotalcite, plagioclase, olivine, calcite and opal. Apparently these represent the products of a continuous, time/temperature-dependent but sometimes overlapping series of interlocking processes, from initial hydration and selective dissolution of the parent sideromelane (basaltic glass), to the completed precipitation of secondary minerals. The mineral phases identified with palagonite by the individual observer thus merely reflect specific stages of a time and space continuum.

Formation of palagonite from sideromelane takes place by a process which is not yet fully clear. Terms used to describe it have included hydration (Peacock and Fuller 1928), diffusion-hydration (Moore 1966), microsolution-precipitation (Hay and Iijima 1968*a,b*) and isomolar exchange of matrix ions with water and K_2O (Staudigel and Hart 1983). Large gains in H_2O, relative gains in Fe and Ti, and losses in Na and Ca are most common. Potassium, Rb, Cs are gained in submarine and lost in subaereal palagonite. Losses in other components vary, according to the particular stage in the palagonitization process (Furnes 1980). The Fe^{3+}/Fe^{2+} ratio increases very significantly. The replacement processes are not isovolumetric.

Three phenomenological stages have been proposed in the process on the basis of presence or absence of the source glass and/or the product authigenic materials (Honnorez 1978). In the initial stage, the source glass is still abundant, but some of it is altered to residual glass (palagonite); authigenic minerals are deposited only in the intergranular spaces. In the more advanced-mature stage, all the fresh glass is altered and authigenic minerals form inside altered glass granules and replace them. In the final stage, the whole rock is converted to an aggregate of authigenic minerals that have lost almost all the grain boundaries of its source rock.

Scanning electron microscope studies show palagonites to be mostly noncrystalline (see, e.g., Furnes 1980; Furnes and El-Anbaawy 1980). On the sub-μm scale, a structure of small spherical bodies (0.2 to 0.8 μm) dispersed in a groundmass is revealed. This micromorphology is attributed to preservation of the structure of the sideromelane and to result from liquid-liquid immiscibility in the molten lava. Embedded in the noncrystalline mass, many well-defined crystals of authigenic minerals are seen, mostly of clay minerals and zeolites. Microprobe elemental analyses with resolutions of 1 to 10 μm show that the palagonite phases are enriched in iron and depleted in Si and Al (Furnes and El-Anbaawy 1980).

A general observation is that during the advancement of palagonitization, the network-forming ions Si and Al are eluted while Fe and Ti passively accumulate in the palagonite phase. If maturation of the palagonite is measured by Ti enrichment, it is found (Staudigel and Hart 1983) that iron is first passively enriched almost in parallel to Ti, while Si and Al are leached out. It is also found that Ca and Mn are leached out very effectively at the early stages of the interaction with water and of Ti enrichment, while Mg is less efficiently leached out. Extended leaching or changes in redox potential may result in increased leaching of Fe. It is commonly observed that the leached-out elements are precipitated as smectites, carbonates and zeolites that eventually replace the magmatic phases (Honnorez 1980; Furnes 1980; Mehegan et al. 1982; Crovisier et al. 1983).

The palagonitization process may be quite fast. In laboratory studies on artificial basaltic glass weathering in artificial seawater at 60°C and 1 bar, iron oxyhydroxides precipitates were observed after 4 days, hydrotalcite ($Mg_6Al_2CO_3(OH)_{16}4H_2O$) crystals after 5 days and poorly crystalline smectites after 120 days (at 50°C) (Crovisier et al. 1987). The rate of dissolution of the glass measured by silica appearance in solution was 0.1 g $m^{-2}day^{-1}$, corresponding to a rate of recession of the glass/palagonite interface of 30 to 40 nm day^{-1}. Under natural conditions, palagonitization of basalt tuff in the Surtsey Volcano, Iceland, was virtually completed within 10 to 12 yr after intrusion of dikes that caused the temperature to rise to 120–150°C. Nontronite formed at 120°C, and other authigenic minerals including phillipsite and anhydrite crystallized at temperatures in the range of 25 to 150°C in this intensively active hydrothermal system (Jakobsson 1972, 1978; Jakobsson and Moore 1986).

In most cases palagonites appear to be buffered at high *p*H's (>7 to 8). Detailed studies of the kinetics of weathering of crystalline basalt and basalt glass at low temperatures under laboratory conditions, and field observations on the interactions of water with similar basaltic formations in Iceland (Gislason and Eugster 1987*a,b*), have shown: (a) that when sealed from the atmosphere the *p*H of the system increases rapidly from about 5–6 to 9–10 where it tends to level off; (b) when open to the atmosphere the *p*H levels off at a lower range of 7 to 8. This results from the consumption of protons in the dissolution of the "basic" basalt and the buffering of the system at various *p*H's depending on availability of CO_2. The relatively rapid dissolution rate of basalt glass supports relatively high dissolved silica concentrations. In the absence of CO_2, the solution is buffered at high *p*H (9 to 10) by silicic acid dissociation. However, upon exposure to the atmosphere, silica is more efficiently leached on the one hand, and enough CO_2 is dissolved on the other hand, to buffer the system at a lower *p*H range of 7 to 8, by the CO_2-H_2O-CaO-MgO system. The basic *p*H (>7 to 8) of palagonites may explain their limited reactivity observed in the simulation of the Viking Labeled Release experiments (Fig. 4; Banin and Margulies 1983; Banin et al. 1988*b*). Thus, terrestrial palagonites as such cannot be considered as completely satisfactory analogs to the Martian soil. As long as traces of palagonite or the basaltic glass are present in the palagonitized rock or soil, they are capable of buffering it at a high *p*H, not commensurate with the purported more acidic *p*H range of the Martian soil.

Although it is difficult to determine at present whether the soil on Mars is in the initial or the advanced stages of palagonitization (or, even, whether

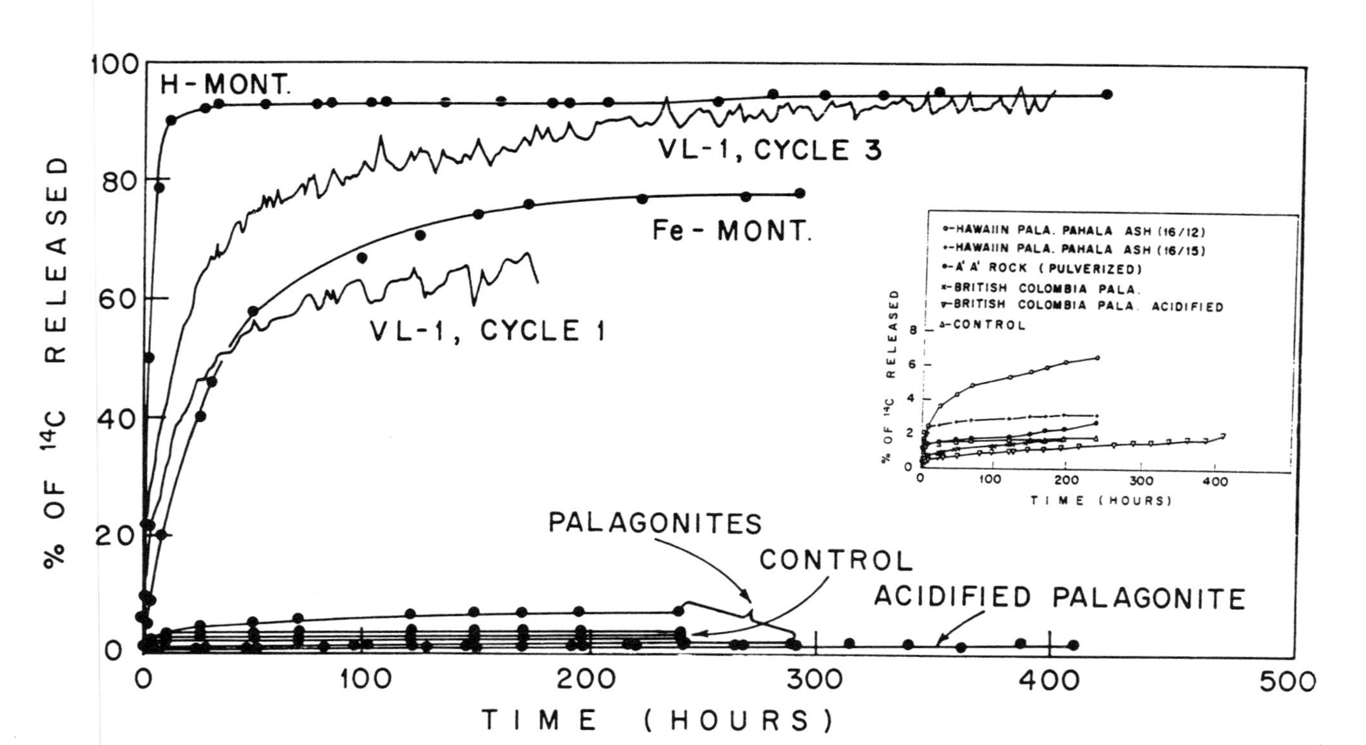

Fig. 4. Comparison of the reactivity of iron-enriched smectites (mont.) and various palagonites as Mars soil analogs, to that measured in the Viking Labeled Release Biology experiment on Mars, Viking Lander 1 (figure courtesy of Banin et al. 1988*b*).

it is at all a product of the palagonitization process), several lines of evidence suggest that the fine soil material has been weathered and matured beyond the initial stages of the process. The particle size of the dusty material is in the μm-size range (Pollack et al. 1977,1979*b*), ruling out the presence of large glass particles and phenocrysts, and the particles may be disk shaped (Murphy et al. 1990), suggesting the presence of phyllosilicates. The reflectance spectra in the visible range of Martian soil and dust do not show the spectral features of the characteristic silicate minerals of metamorphosed rocks (McCord et al. 1982*a*; Singer 1982) but do show the absorptions in the visible range of amorphous or short-range-ordered iron oxides and oxyhydroxides (see chapter 17). The unique chemical composition and the environmental conditions on the Martian surface led Toulmin et al. (1977) to suggest that the fine-soil material is of palagonitic origin but highly enriched in smectite, i.e., at an advanced stage of maturation. The lack of pronounced carbonate spectral features (Singer 1985) in the infrared and near infrared could indicate also that advanced weathering took place dissolving calcite and converting calcium to other mineral or salt forms (sulfates?).

One explanation of these observations is that the weathering processes that produced the Mars soil (see chapter 19) may have involved more intensive leaching than that typical for palagonites on Earth. These could have led on Mars to enrichment in clay minerals, through continued evolution of the weathered minerals under mild acidic leaching, beyond the stage known on Earth as "palagonite." Such continued leaching could be promoted by periodic inputs of volcanic volatiles such as water vapor and compounds of sulfur and chlorine, as evidenced by the relatively high content of these elements in the Martian soil.

There are a number of other environments in which relatively fast alteration and weathering of volcanic glass can take place (chapter 19). Clark (1978) suggested smectites could form in transient hydrothermal zones created by hypervelocity meteoroid impact into water- or ice-containing regolith. Newsom et al. (1986) and Boslough (1986,1987) have proposed that much of the weathered material on Mars is produced from what is loosely defined as "shock-activated" glass (also termed in some cases "maskelynite"). In a crater on Earth (the Ries Crater in West Germany), it was estimated that of the 4 to 5-km^3 volume of rocks melted by the impact, 0.2 to 0.6 km^3 remained as glass, constituting a fraction of 4–16% of the melted rocks; this fraction agrees well with the fraction of altered clay minerals found in suevite deposits surrounding the crater (Newsom et al. 1986). The altered clay was pure, iron-rich montmorillonite, with only a small fraction of illite, indicating low-temperature alteration regime of the impact deposit. This alteration scenario appears as a plausible alternative to the one envisioning palagonite formation in subhydral or subglacial volcanic eruptions. The differences in mineralogy and chemical reactivity of the products formed by these two weathering scenarios are not yet fully known.

The possibility that the silicate rust or iddingsite found in the interior of the Nakhla meteorite (Bunch and Reid 1975; Wentworth and Gooding 1988*a,b*,1989) is of preterrestrial origin has not been completely ruled out as Nakhla is a meteorite fall (Table III) with very limited terrestrial weathering. The rust, as it were, is a silicate alteration product produced perhaps by the reaction of olivine with fluids enriched in halides. The very limited data are consistent with di-, trioctahedral phyllosilicates with excess Fe and Si (Wentworth and Gooding 1989).

The argument between clays (smectite) and palagonite as Mars soil analog minerals can be partly reconciled by defining more clearly and specifically the two terms, and, primarily, by obtaining more detailed information on the soil.

B. Iron Minerals

Iron is a major chemical component in the Martian soil, the second most abundant after Si (Table I). Its minerals strongly affect the spectral characteristics of the soil in the visible range, lending the soil its characteristic reddish coloration (see chapter 17). Reflectance spectra of the bright regions of Mars, believed to contain the more weathered components of the soil, show an intense though relatively featureless absorption edge from 0.75 μm to the near ultraviolet (Singer et al. 1979; McCord et al. 1982*a*; Sherman et al. 1982). Several mineral candidates have been shown to satisfy these spectral constraints (Table VII; Fig. 5) including amorphous iron oxides (Evans and Adams 1979,1980), palagonite (Singer 1982,1985; Adams et al. 1986; Guinness et al. 1987), aluminum oxyhydroxide doped with iron (Morris et al. 1983), jarosite (Burns 1986), cryptocrystalline hematite deposited in silica matrix (Morris et al. 1985; Morris 1988; Morris and Lauer 1988) and iron-enriched smectites (Banin et al. 1985,1988*a*). The major conclusion from the abundance of spectral analogs is that unambiguous identification of the iron mineral(s) in Mars soil, on the basis of reflectance only, is not possible at present. The spectral evidence does imply, however, that iron in the weathered component of Mars soil is mostly present in the oxidized form (Fe^{3+}) either in poorly crystallized clusters of iron oxides or oxyhydroxides, or as crystalline minerals but in extremely small particle-size range (nanocrystals or nanophase iron oxides and oxyhydroxides). Oxidative weathering of basaltic rocks by acidic groundwater has been proposed as a mechanism for the production of the nanophase iron-oxides on the Mars surface (Burns 1987; Burns and Fisher 1990). It is suggested that strongly acidic and sulfate-rich underground water mobilize iron from basaltic rocks. Above the water table the dissolved iron is precipitated, initially as jarosite, and eventually as the more oxidized goethite.

Smectite clays chemically enriched with ferrous iron, which is then oxidized and polymerized *in situ* (Gerstl and Banin 1980), show strong spectral

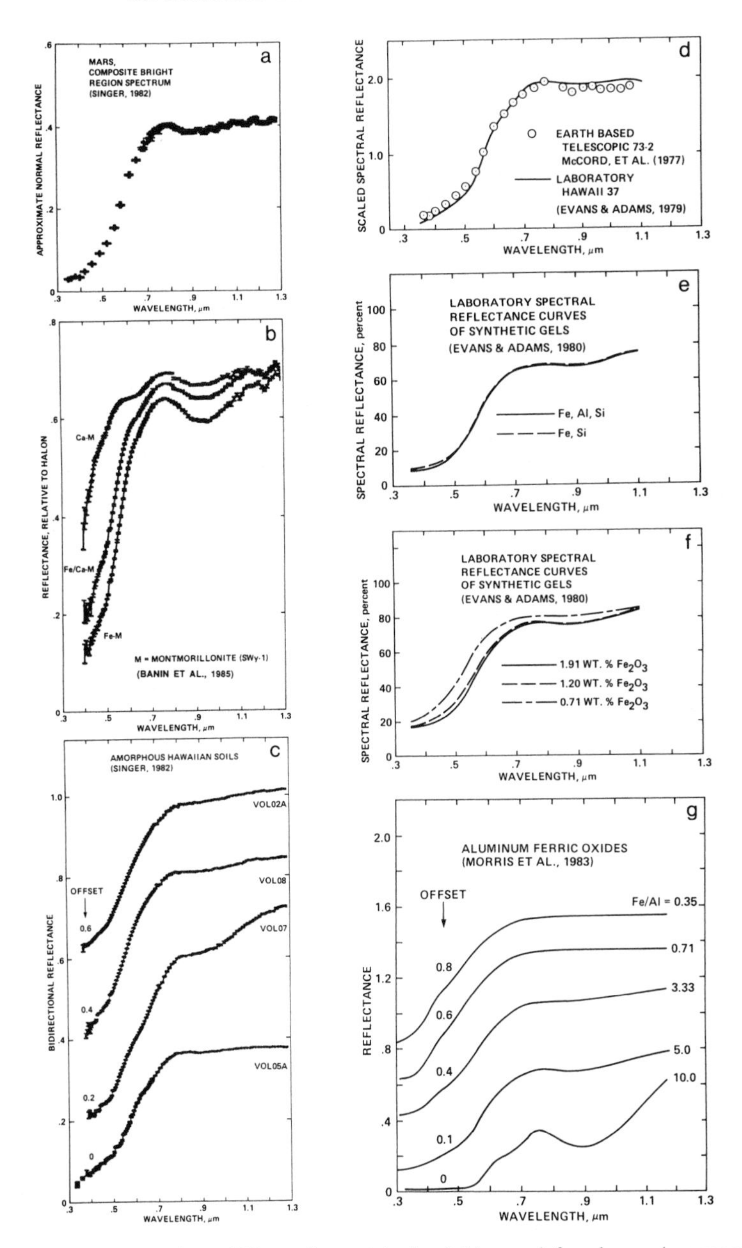

Fig. 5. Spectral simulations of Mars reflectance in the visible near-infrared range (sources of data as indicated in the figures).

analogy with Mars (Banin et al. 1988*a,b*). Compositionally, these preparations are full chemical analogs of the Mars soil with respect to iron oxide content. *X*-ray diffractometry, chemical extractions and electron micrographs show that iron in these preparations is in poorly crystalline or cryptocrystalline forms and appears in particles of size range 1 to 10 nm (A. Banin and T. Ben-Shlomo, unpublished data). In these Mars soil analogs, the crystallization of iron oxide minerals and the growth of their crystals is apparently prevented: (a) by the presence of Si and Al in the equilibrium solution, known to affect Fe oxide crystallization in soil environments (Schwertmann 1988); and/or (b) by the presence of an adsorbing surface of the smectite that immobilizes colloidal iron oxide and oxyhydroxide particles and reduces iron concentration in solution, thus slowing down or preventing the precipitation and crystallization of pure iron minerals. Similar conditions might prevail in the Martian soil, explaining its spectral properties and chemical reactivity. The causes for the somewhat puzzling observation that no well-crystallized secondary iron minerals have been clearly observed spectroscopically in Mars soil and dust (although recent telescopic observations [Bell et al. 1990*b*] suggest the presence of at least traces of crystalline oxides), will probably be understood only after direct micro-analytical mineralogical analyses are performed on the soil.

C. Minerals of Salt-Forming Elements and Carbonates

The presence of high concentrations of both S and Cl in the Martian soil analyzed by the Viking Landers (Table I) leads to consideration of the hypothesis that salt compounds are present, a situation common in desert environments. Higher concentration in the clod samples than in loose fines is further evidence because salt cementation is a common mode of formation of crusty material. The probable form of S is as sulfate because of the evidence against sulfides (Toulmin et al. 1977), although higher oxidation states should not be ruled out. The chlorine is most likely present as chloride, although once again the highly oxidizing environment of Mars could produce other salt forms. Arguments based upon cation concentrations, solubility of salt compounds and salt/salt and salt/igneous mineral reactions has implicated magnesium sulfate and sodium chloride as the major salt components in Martian soil (Clark and Van Hart 1981). A model of Martian soil as a two-component mixture in which the only variations are the relative concentration of a salt component and a silicate component can explain most element concentration variations observed among all the samples taken (B. C. Clark, unpublished data). Nitrates or carbonates may also be present in the Martian soil, but if the source of sulfates and chlorides are volcanogenic acidic gases, then both compounds will be readily attacked even in the cold, dry Martian ambient conditions (Clark et al. 1979). This process releases CO_2 and NO_x back to the atmosphere.

No clear spectral evidence for carbonates on the surface of Mars has been found (Singer 1985; McKay and Nedell 1988), although recent telescopic evidence has detected the presence of 6.7 μm absorption band, that was tentatively assigned to carbonate or bicarbonate anions in crystalline states in the atmospheric dust (Pollack et al. 1990*b*). However, considerations of Mars climate evolution has led to suggestions that carbonate minerals should have been deposited on Mars (see, e.g., Kahn 1985; Warren 1987) and might, therefore, be present in the soil and dust on Mars. Laboratory studies have shown that synthesis of carbonate is possible under Mars-like atmospheric and thermal conditions (Booth and Kieffer 1978). Simulations of the Viking Labeled Release (LR) Biology experiments using mixtures of the active ingredient (iron-enriched smectite clays) with various carbonate minerals (Banin et al. 1981; Banin 1989*b*) have shown that even low amounts (0.5 to 1.0%) of added calcite reduce considerably the decomposition activity of the smectite and diminish the release of $^{14}CO_2$, making the mixture an unlikely candidate for Mars soil. A much higher content of siderite ($FeCO_3$) can be tolerated, although the stability would depend also on the redox potential of the soil. Thus, it is possible that the carbonate component, if present in Mars soil, may be siderite rather than calcite, magnesite or dolomite.

Direct evidence for aqueous low-temperature alteration on Mars comes from recent analyses of the petrography and mineralogy of the SNC meteorites. The discovery of traces of typical salt-minerals such as sulfur- and chlorine-rich aluminosilicate mineraloid and Ca sulfate in the lithology C of EETA 79001 (Gooding and Muenow 1986), and halite, sulfate and carbonate in Nakhla (Wentworth and Gooding 1988*a,b,*1989), and lithology C of EETA 79001 (Gooding et al. 1988), constitutes the first direct evidence for the low-temperature alteration on Mars if indeed the SNC meteorites are of Mars origin. Carbon and oxygen isotope ratios in the carbonates support the extraterrestrial origin (Wright et al. 1988; Clayton and Mayeda 1988). The proposed scenario (Gooding et al. 1988) is that during impact, surface rocks that contained disseminated crystals of secondary minerals have only partially melted and some of the more refractory minerals (e.g., carbonates, sulfates) remained trapped in the maskelinized matrix.

The carbonate contents are low and were estimated to be between 30 to 100 ppm C as carbonate in EETA 79001 (Wright et al. 1988). The carbonate crystals appear in small, 10 to 20-μm size, rounded and partially decrepitated grains disseminated in the quench-textured pyroxene and the glass phases (Gooding et al. 1988). Structural evidence strongly supports the suggestion that the carbonate was deposited before the event that produced the glass, i.e., on the surface of the parent body of the meteorites. Sulfate phases, mostly Ca, are intimately associated with the carbonates but their concentration is usually much higher (Gooding et al. 1988); whether *all* of the sulfate minerals are of extraterrestrial origin or some were formed through weathering on Earth is still not clear but it appears that at least some of the sulfate grains pre-date the shock event.

D. Soil Formation Scenario and Mineralogical Model

The following is one geological-pedological scenario for the formation of the Martian soil. Secondary silicate minerals, including smectite clays, are the product of alteration of volcanic materials and basic igneous rocks in various environments. On Earth, their formation involves the hydrolytic action of water and usually takes place in marine or lacustrine environments and under hydrothermal conditions. Ample evidence for the presence of liquid water on Mars and for its past activities has accumulated (see, e.g., Carr 1986; chapters 15 and 16). Secondary silicate minerals may thus have been formed in the past at periods and locations on Mars where enough water was present for sufficient time to leach and weather volcanic rocks.

Even on a frozen planet, a number of niches can develop in which locally and temporarily restricted weathering of basaltic rocks can take place at an enhanced rate. Such niches are, for example, volcanic eruptions through and onto ice (Toulmin et al. 1977), catastrophic flooding events (Baker 1982) enhanced weathering of shock-activated materials produced by meteorite impacts (Newsom et al. 1986; Boslough 1987) and weathering of the Martian lithosphere by fossil water at the liquid-ice interface (Soderblom and Wenner 1978). Under any of such conditions, enhanced weathering can result in basaltic minerals and glass alteration and formation of clays and amorphous iron oxides (see chapter 19). Thus, it is possible that the formation of the secondary silicate minerals now found in the Mars soil mostly took place early in the history of Mars at the same time and under the same climatic, tectonic and planet-forming conditions at which the intensive fluvial and volcanic activities that shaped the planet's surface, occurred. Under such weathering conditions on Earth, iron tends to oxidize and form oxide and oxyhydroxide compounds which are *X*-ray amorphous or with short-range crystalline ordering only (Schwertmann 1988; A. Banin and T. Ben-Shlomo, unpublished data). The interference with iron crystallization and crystal growth is attributed to the presence of Si and Al in the solution from which the authigenic minerals are formed, and perhaps due to limited solubility of iron in the high *p*H prevailing in the micro-environment of weathering. Under these conditions, carbonates and zeolites were precipitated, also as observed in weathering environments of volcanic glass on Earth. At that stage in Mars history, as on Earth, organic matter may have been present and accumulated to significant concentrations on the surface. This organic matter was imported by carbonaceous meteorites and comets impacting the planet, and possibly also synthesized under the highly reactive conditions in the atmosphere and on the surface. Chemical interactions between the organic and the silicate-mineral materials may have proceeded on a route of abiotic evolution paralleling that on Earth which culminated in the origin of life. Mars, however, cooled considerably by about 0.5 to 1.0 Gyr after accretion, and these processes were halted (McKay 1986; chapter 35). Rock weathering slowed down, but some alteration continued. The less acid-stable minerals (carbonates and zeolites) were dissolved by inputs of volcanic acidic volatiles of sulfur and chlorine, and salts of these elements were deposited in the soil. Consequently, the more acid-stable clays and amorphous iron oxides and oxyhydroxides were enriched in the top-soil layer. As Mars desiccated by freezing out of most of its water, the fine-soil particles were readily carried by winds, mixed in the atmosphere in planet-wide dust storms, and thoroughly homogenized. The organic matter in the soil decomposed through the effects of oxidation, desiccation, ultraviolet radiation and catalytic reactions on the clay surfaces (Oro and Holzer 1979). Since meteoritic importation slowed down and then practically ceased (except for small flux of micrometeorites), and synthesis of new organic matter also probably slowed down as temperatures dropped, no organic matter remained in the fine-soil material. Mixing of the loose-soil material continued by wind activity, and the dust settled and formed the thin blanket of fine, silicate- and iron-rich soil material covering much of the planet's surface. This is evidenced by the identical spectral characteristics, physical properties, chemical composition and chemical reactivity of the soils in the two distant Viking landing sites.

With the limited decisive information at hand, we may speculatively suggest that the fine soil on Mars is likely to be a multicomponent mixture of weathered and nonweathered minerals. Smectite clays are suggested as important components among the weathered minerals, assumed to be adsorbed and coated with amorphous iron oxides and oxyhydroxides and mixed with small amounts of crystallized iron minerals (ferrihydrite, goethite and hematite) as separate phases, but having extremely small crystal sizes. Some form of sparingly soluble phosphates such as apatite or carbonic apatite, may be present. As accessory minerals, it is likely that the soil contains various sulfate and chloride salt-forming minerals. If present, carbonates are likely to be at very low concentrations, although siderite ($FeCO_3$) may be present at somewhat higher concentrations, depending on the redox potential of the soil. The chemical reactivity of this multicomponent system is complex to predict. It may be dominated by the ability of the smectite clays to adsorb volatiles due to their extensive specific surface area (up to 800 $m^2\ g^{-1}$) and their catalytic reactivity in chemical and photochemical reactions. Furthermore, it is likely that the presence of the nanocrystalline iron oxide and oxyhydroxide particles, with their large specific surface area, will increase the adsorption capacity of the system and enhance its reactivity.

Acknowledgments. Partial support for this review (A.B.) was obtained from the Exobiology and Solar System Exploration Programs of NASA through MERC-ARC. The help of S. Salomon and H. Prager in preparing the manuscript is greatly appreciated.

19. PHYSICAL AND CHEMICAL WEATHERING

JAMES L. GOODING
NASA Johnson Space Center

RAYMOND E. ARVIDSON
Washington University

and

MIKHAIL YU. ZOLOTOV
V. I. Vernadsky Institute

Although Viking Lander data suggest oxidized, palagonite-like weathering rinds on some rocks and bulk surface sediments possibly composed of clay-salt mixtures, studies of Martian weathering have remained largely theoretical with few observational constraints. Physical weathering may have included rock splitting through growth of ice, salt or secondary silicate crystals in voids. Chemical weathering probably involved reactions of minerals with water, oxygen and carbon dioxide although predicted products vary sensitively with the abundance and physical form postulated for the water. Dominance of crystalline clay minerals (rather than clay-like mineraloids such as palagonite) among silicate weathering products would be a critical indicator of extensive weathering in liquid water. Based on kinetics data for hydration of rock glass on Earth, the rate of weathering-rind formation on glass-bearing Martian volcanic rocks is tentatively estimated to have been on the order of 0.1 to 4.5 cm Gyr^{-1}; lower rates would be expected for crystalline rocks. Despite hypotheses concerning rapid chemical oxidation of iron by ultraviolet irradiation of minerals, its significance on Mars has been cast into doubt by experimental studies producing conflicting results.

I. GEOLOGIC CONTEXT AND OVERVIEW

Weathering is the collection of processes which, through interactions between surface and atmosphere, leads to the decomposition or alteration of rocks, minerals or mineraloids and the possible formation of new phases (Gooding 1986*a*). On Mars, it probably includes both physical (mechanical) and chemical processes (Table I) that disintegrate rocks to provide the progenitors of soils as well as debris that can be transported and deposited elsewhere as sediment. In a broad sense, weathering can be extended to include transient influences such as hydrothermal systems that might be spawned by volcano-ice interactions or by impact cratering in volatile-rich target rocks.

We can contrast *weathering* with *erosion* by the mass-movement of material that is essential to definition of the latter process. In general, weathering can decompose a rock without removing component grains from the immediate vicinity. In addition, weathering operates at the scale of mineral grains (micrometers to millimeters) whereas erosion operates at the scale of outcrops (meters to kilometers). In reality, of course, weathering and erosion can be closely associated; *in situ* weathering can make a rock much more susceptible to mass loss through subsequent erosion. Wind abrasion occupies a special position as a process that possesses attributes of both weathering and erosion.

TABLE I
Possible Attributes of Weathering on Mars

PHYSICAL WEATHERING PROCESSES	
Frost riving	
Secondary mineral(oid) riving	
Wind abrasion (dust or ice projectiles)	
CHEMICAL WEATHERING PROCESSES	
Oxidation:	uptake of oxygen to form oxides or more highly oxidized silicates
Hydration:	uptake of water to form minerals with structural OH^- or H_2O
Carbonation:	uptake of CO_2 to form carbonate minerals
Solution:	dissolving of mineral in water
STYLES THAT VARY WITH ENVIRONMENTAL CONDITIONS	
Surface weathering (controlled by mineral surfaces exposed to atmosphere)	
Gas-solid (present climate, +/− UV photocatalysis)	
Liquid-solid (ancient climates with liquid water)	
Pedogenic weathering (within soil profile below free surface)	
Liquid-like films of unfrozen water	
Pore-gas pressures higher than atmospheric pressure	
Brines	
Hydrothermal alteration (groundwater circulation driven by transient heat source)	
Meteorite impact crater	
Volcanic eruption	

Weathering on Mars has undoubtedly been a major influence on production of the fine-grained, windblown dust that blankets much of the surface (chapter 22) and on the mineralogy and chemistry of surface and subsurface soils (chapter 18). Because weathering is climate dependent, by definition, products of weathering should serve as valuable indicators of climate changes on Mars (chapter 32). In addition, products of chemical weathering might have become major sinks for H_2O, CO_2, as well as for other volatile compounds that are thought to be missing from the current atmosphere/hydrosphere cycles on Mars. Transformation of CO_2 into carbonate rocks, for example, would have necessarily proceeded through chemical weathering.

In the absence of samples of weathering products that can be directly analyzed, knowledge of Martian weathering remains largely conjectural. In this chapter, we review weathering processes that might be important on Mars and summarize the limited observations, including Viking results and laboratory simulations, that bear on the issue. No concrete answers are offered because the definitive analyses, involving Martian samples and environmental measurements, remain to be done.

Weathering products on Mars are likely to be a mixture of relics from previous climatic epochs and more recent products of contemporary processes. If ancient Mars possessed more Earth-like surface inventories of H_2O, CO_2, and other volatile compounds, the dominant volume of weathering products discoverable on Mars might be "fossil" clays, oxides, carbonates and salts that have little relationship to modern weathering processes (O'Connor 1968*b*).

Figure 1 shows a schematic model for relative magnitudes of weathering, erosion and deposition over geologic time for Mars. It implicitly includes exogenic processes, such as impact cratering, and endogenic processes, such as the interaction of magma with ice and ground water. The model assumes that rates of weathering were higher in early geologic time, when impact rates were higher, volcanism was more prevalent, and the efficacy of chemical and mechanical weathering by a higher-density atmosphere was probably greater, as inferred from stratigraphy (chapter 11). Evidence for an early torrential bombardment by impactors is clear from the Noachian System of rocks which includes surfaces that have more than an order of magnitude higher areal crater abundance than the most heavily cratered plains units. Thus, cratered terrain might have accumulated a megaregolith of mechanically and chemically altered debris. The greatest areal exposure of (presumably volcanic) plains on Mars corresponds to rocks of the Hesperian System that date from declining stages of heavy bombardment to perhaps 1 Gyr in age. Thus, volcanism on Mars peaked during the first half of geologic time and any hydrothermal alteration caused by interaction of magma with ice or ground water (e.g., formation of palagonites) would also have peaked during the first half of geologic time (Soderblom and Wenner 1978). Settle (1979) also suggested that sulfuric acid aerosols produced by volcanic eruptions could have accel-

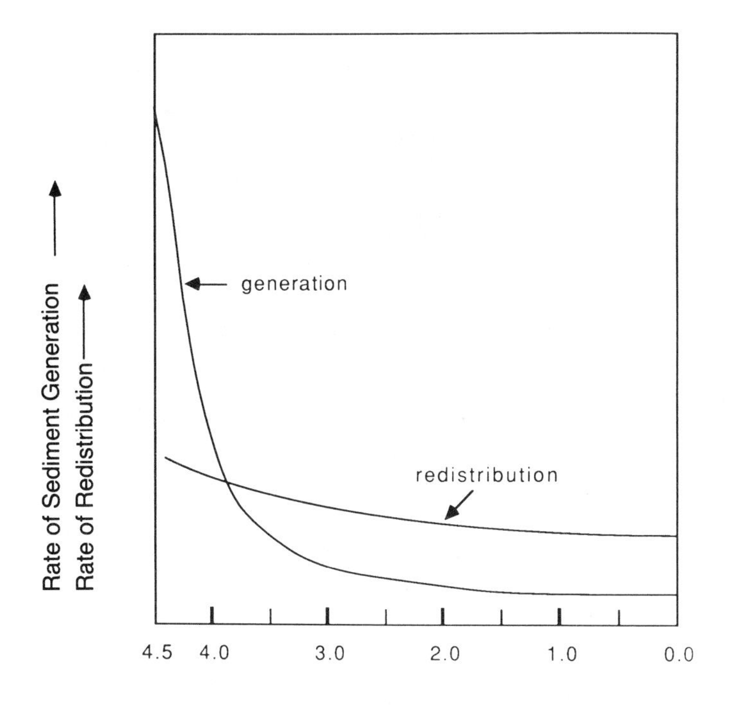

Fig. 1. Hypothetical rates of generation and recycling of sediment vs time on Mars.

erated weathering by attacking surface rocks. Finally, it is postulated in the model that the atmosphere was denser and that subaerial weathering rates were also higher in early time. This assumption is based on the observation that valley networks are mainly found on Noachian System rocks. Regardless of whether the valley networks were formed by rainfall and runoff or by ground-water sapping, their extensive development suggests conditions of higher atmospheric pressure that sustained liquid water at the surface. Figure 1 also shows rates of aeolian redistribution of sediments decreasing with time, but not as dramatically as the rate of weathering, based on substantial evidence for ongoing differential erosion on Mars. The model suggests that the greatest volume of weathering products were produced early in geologic time and that active redistribution of those materials has continued until the present.

The remainder of this chapter reviews possible weathering processes on Mars, realizing that modern rates and mechanisms may differ significantly from those that prevailed during the first half of geologic time. Possible variations are listed in Table I. It should be emphasized that, for a given climate, various weathering environments can be postulated. Actual *surface weather-*

ing should be the most sensitive to climate variations because it is typified by exposure of bare rock to the prevailing atmosphere, including solar insolation. Whether or not the atmosphere supports condensation of liquid water will profoundly affect the course of surface weathering. Even in climates not supportive of condensation, however, it is possible that *pedogenic weathering* might involve water phases that would not be found at the free surface, as a consequence of the physical chemistry of wet geologic substrates (Sec. II.A) and pore-gas pressures that might exceed those in the free atmosphere. Accordingly, pedogenic weathering products might differ significantly from surface weathering products. Finally, *hydrothermal alteration* might result from geologic incidents (volcanic eruptions or meteorite impacts) having no relationship to the prevailing climate. By definition, hydrothermal alteration would be controlled by liquid water or subcritical aqueous fluids and its products would differ drastically from those of either surface or pedogenic weathering. Again, none of the possible styles of Martian weathering will be understood until future experiments are accomplished. Nonetheless, a useful way to anticipate key differences is to focus on the role of water in making each weathering style unique.

II. SIGNIFICANCE OF WATER AND WIND

Our collective knowledge of weathering on terrestrial-type planets is restricted to that on Earth, a water-rich planet where water-based processes are totally dominant. The rates and styles of both physical and chemical weathering vary with the availability of liquid water, which controls solution, transportation and precipitation processes. Water currently exists on Mars mostly as ice; atmospheric water vapor is the potentially most reactive form observed. Still, it can be expected that even small amounts of liquid water would profoundly affect Martian weathering. Similarly, exposure of fresh mineral surfaces is necessary to sustain weathering over long time periods. Blowing dust is the only agent known to be currently active on Mars that might fulfill the required process of surface renewal by abrading weathering rinds.

A. Phase Behavior of Water

It is generally acknowledged that bulk liquid water should not be a stable phase at the surface under current Martian climate conditions. For pure water, the equilibrium triple point occurs at higher temperature and pressure values than are found at nearly all locations on Mars. Accordingly, any liquid water exposed at the Martian surface should either evaporate or freeze. As pointed out by Farmer (1976), however, liquid water can occur as a transient phase during melting of ice covered by dust that would retard evaporation. Indeed, a thermal model for Mars (Kieffer et al. 1977) predicts that, for the respective northern and southern summers (L_s = 90 to 180° and 270 to 360°), various portions of the Martian surface between latitudes of 30°N and 60°S could reach maximum temperatures >273 K. However, models for distribution of ground ice (chapter 16) disfavor ice within a few meters of the surface at temperate latitudes (45°N to 45°S). Consequently, the localities on Mars where melting of ice might occur by solar heating should be rare. Although melting at the base of the northern polar cap has been hypothesized (Clifford 1987*b*), it is not clear that such melt water would be accessible to the surface to participate in weathering. Therefore, we are left without a major source of liquid water to drive weathering on present-day Mars according to the style displayed on Earth.

Based on terrestrial field and laboratory experience, however, we can expect that water on Mars might also exist on mineral surfaces or in pore spaces as liquid-like films of a few molecular thicknesses. The appropriate phases can be referred to as unfrozen capillary-pore water, unfrozen interfacial water and undercooled crystal-plane water.

Unfrozen capillary-pore water is a well-established phenomenon in terrestrial soils (Anderson 1968) and, in its most rudimentary analytical formulation, can be described as

$$T_{in} = T_e[1 + (\sigma \cos \alpha)/(\Delta H)r] \tag{1}$$

where T_{in} is the actual temperature of ice nucleation, T_e is the equilibrium freezing temperature (ordinarily 273 K), σ is the water/rock surface tension, α is the water/rock contact angle, ΔH is the enthalpy of freezing (algebraic sign neglected), and r is the radius of the pore. It follows that the largest depression of the freezing temperature is favored by surfaces that are not wet by water (i.e., $\alpha>90°$) and by small capillary radii. Still, the magnitude of the effect is usually $<$ 1 to 2 K.

All other factors being constant, the survivability of unfrozen interfacial water varies with character of the substrate because of variations in specific surface area that are related to particle size and porosity (McGaw and Tice 1976). For a given temperature and total soil water content, there will be some fraction of the water inventory that is unfrozen. The curves in Fig. 2, which describe the thicknesses of liquid-like water films, follow the McGaw-Tice equation,

$$d_{uw} = k_1(T) + k_2(T)A_s/k_3 \tag{2}$$

where d_{uw} is the film thickness, A_s is the specific surface area of the mineral substrate, and $k_1 - k_3$ are empirical constants, the first two of which vary with temperature T. The McGaw-Tice equation was derived from freezing data for a variety of terrestrial soils of different water contents and different specific surface areas. Thermodynamically liquid-like properties are only observed if the interfacial films exceed about 15 molecular thicknesses (Pruppacher and Klett 1978, p 119), a condition that develops only at temperatures $>$ 263 K

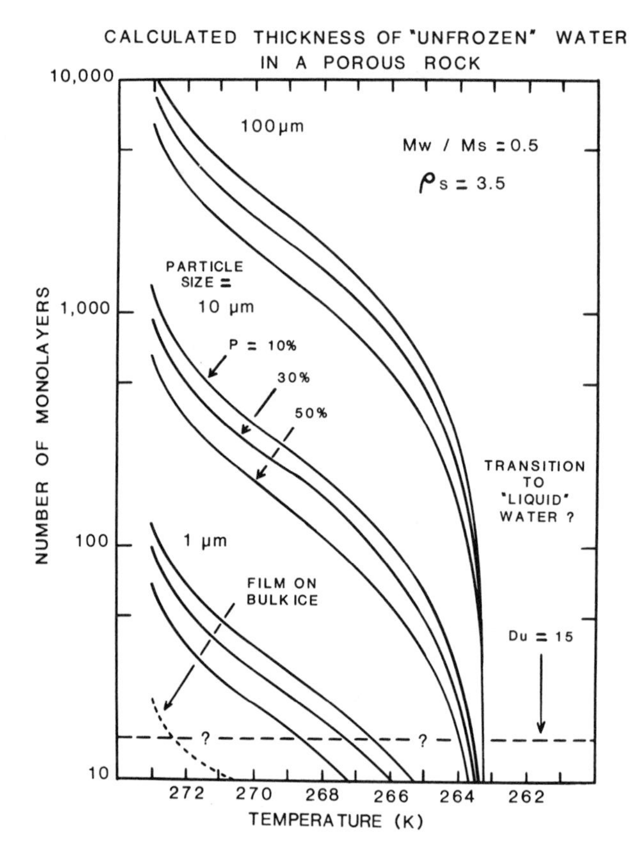

Fig. 2. Thickness of unfrozen water layers as a function of rock/sediment particle size, porosity and temperature (figure after Gooding 1986*c*).

(Fig. 2). At Martian temperatures, even very hygroscopic minerals should not be able to sustain unfrozen water layers of more than 1 to 2 molecular thicknesses; solid water ice would remain the most stable phase (Anderson et al. 1978).

Undercooled crystal-plane water can develop as a consequence of inhibited nucleation of ice on mineral surfaces. Experimental studies of ice-nucleation temperature, T_{in}, have shown that

$$T_{in} \propto (1 + |\delta|)^{-1} \tag{3}$$

where the ice/mineral crystal disregistry $|\delta|$ is the absolute fractional value of the structural mismatch between the most favorable growth plane for a water-ice crystal and the available crystal plane of a mineral substrate. Although the thickness of crystal-plane water is not readily predictable, freezing depressions on the order of 10 K or more have been demonstrated (see, e.g., Gooding 1986*b*).

B. Rock Splitting by Ice and Secondary Minerals

An important mechanism of physical weathering on Earth involves destructive growth of ice, clay or salt crystals in fractures within or between mineral grains in rocks. The physical phenomenon is the volumetric expansion entailed by the organization of molecules or ions into a solid at the expense of a homogeneous solution. In addition to overall increases in specific volume, differential growth of various crystal planes can produce large and uneven stresses on surfaces they encounter. The net result is a force acting to enlarge the space in which the late-formed crystal grew. During such frost or secondary-mineral *riving* (Gooding 1986*a*), repeated cycles of crystallization progressively widen the fractures, leading to eventual disintegration of the rock. In contrast with ice riving, secondary-mineral riving would not require repetitious freeze-thaw cycles—only cyclical wet-dry periods. In principle, precipitation of secondary minerals could occur exclusively at temperatures below the freezing point of pure water as long as the parental solutions were replenished from some source.

Figure 3 summarizes the crystallization pressures expected for precipitation of both ice and candidate minerals that might be relevant to Martian conditions. Clearly, water ice is the most powerful splitting agent although it is closely rivaled by highly supersaturated systems of a few common salt minerals. Riving powers of clay minerals are more difficult to estimate but appear to be lower than those of either ice or common sulfate and chloride salts. Riving by water ice would be limited by the availability of liquid water during freeze-thaw cycles; therefore, it may be of little importance on present-day Mars. Riving by secondary minerals, though, might have operated through much of Martian history, providing that interfacial films of water were sufficiently thick to foster mineral precipitates.

Malin (1974) first suggested that salt riving might occur on Mars. Indeed, the high sulfur and chlorine concentrations in Martian surface fines (chapter 18) would be consistent with sulfate and chloride minerals of the types that might be expected to foster salt riving. Still, sufficient liquid water would be required to form the supersaturated salt solutions that are essential to the salt-riving process.

C. Abrasion by Windblown Dust

Significant differential erosion can be observed in high-resolution Viking Orbiter images of the type shown in Fig. 4. For example, the Mutch Memorial Station (Viking Lander 1) site has a crater size-frequency distribution that is consistent with crater production (and little crater erosion) down to the resolution limit of tens of meters (Arvidson et al. 1979). This implies a rock

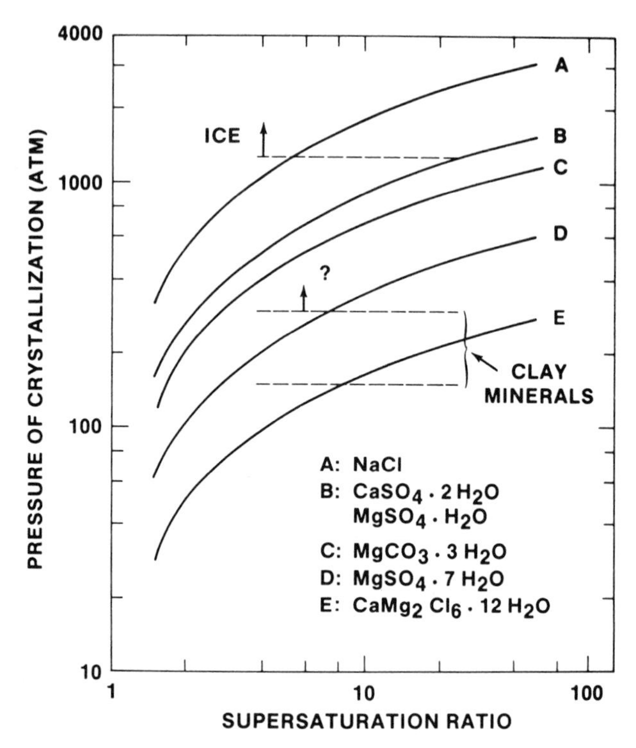

Fig. 3. Crystallization pressures of ice and secondary minerals. Results for salt minerals were computed by the method of Winkler and Singer (1972) whereas the estimated range for clay minerals (zone between dashed lines) represents a measurement by Madsen and Muller-Vonmoos (1985).

breakdown and removal rate of much less than 1 μm yr^{-1}, since the relatively fresh-looking surface (Fig. 4) has a crater population with a model age of 1 to 3 Gyr. On the other hand, regions within Acidalia Planitia and other locations at higher latitudes have clearly been stripped of fine-grained and friable material by winds. The rate of stripping must be orders of magnitude higher than the rate of rock breakdown. For example, Arvidson et al. (1979) estimated a rate of removal of older sediments of approximately 1 μm yr^{-1} in Utopia Planitia, based on the depth of erosion and the areal abundance of impact craters. The bright streaks and dark splotches evident in the mosaic depicted in Fig. 5 are aeolian deposits and the overall pattern is correlated with local (e.g., craters, hills) and regional topography (e.g., Acidalia is a basin swept clean of dust). The patterns suggest cover by an aeolian sedimentary veneer that is largely decoupled from bedrock (Presley and Arvidson

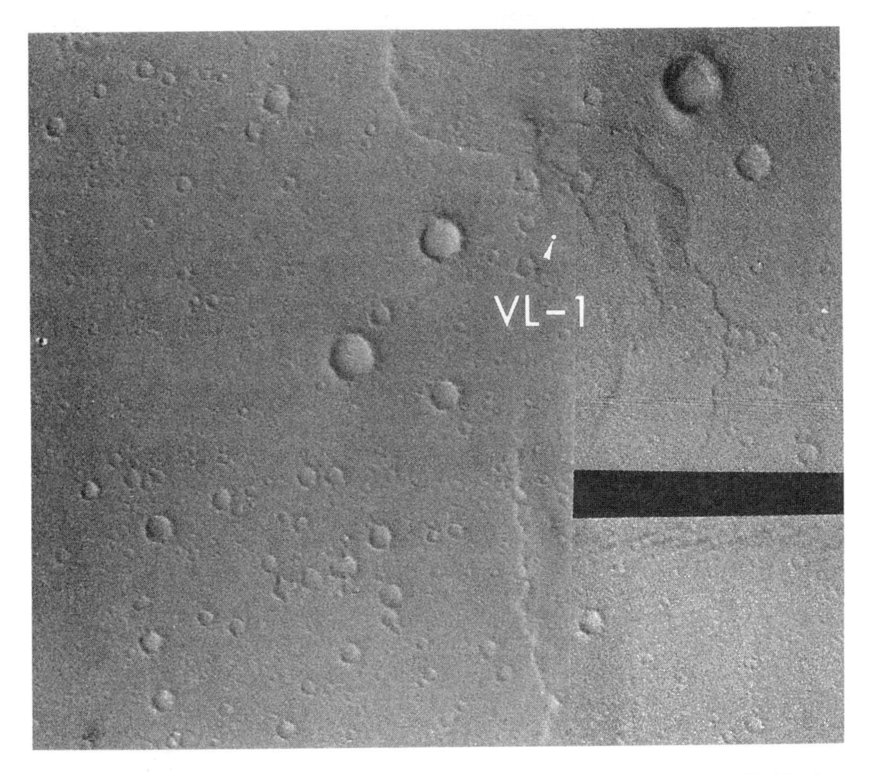

Fig. 4. High-resolution Viking Orbiter image (Frame 452,1310,11,12) of Chryse Planitia, including Mutch Memorial Station (Viking Lander 1), and showing a relatively fresh, uneroded surface. North is toward top and Sun is from southwest. Width of frame is about 10 km.

1988). Most of the sediments were probably generated early in geologic time and have been recycled as part of the climate shifts associated with the orbital oscillations on the scale of 10^4 to 10^6 yr.

The Viking Orbiter perspective demonstrates that differential erosion by wind and recycling of sediment dominates the sedimentary cycle on Mars today. At the Lander scale, there is evidence for a range of preservation states of loose blocks on the surface. For example, Fig. 6 is a Mutch Memorial Station stereo image pair that shows a rock that has apparently broken in place; small fragments litter the surrounding surface (see also Fig. 7). In addition, evident in the image pair are rocks that have been sculpted by winds into ventifacts. Pristine-looking rocks also can be seen. Since most of the rocks have probably been added to the scene as ejecta from surrounding impact craters, the variety of preservation states is probably related to age of block emplacement in addition to intrinsic rock properties. Again, this suggests that the rock breakdown rate has been very low over the age of plains rocks exposed at the Station. Thus, for 1 Gyr or so, rocks have been added

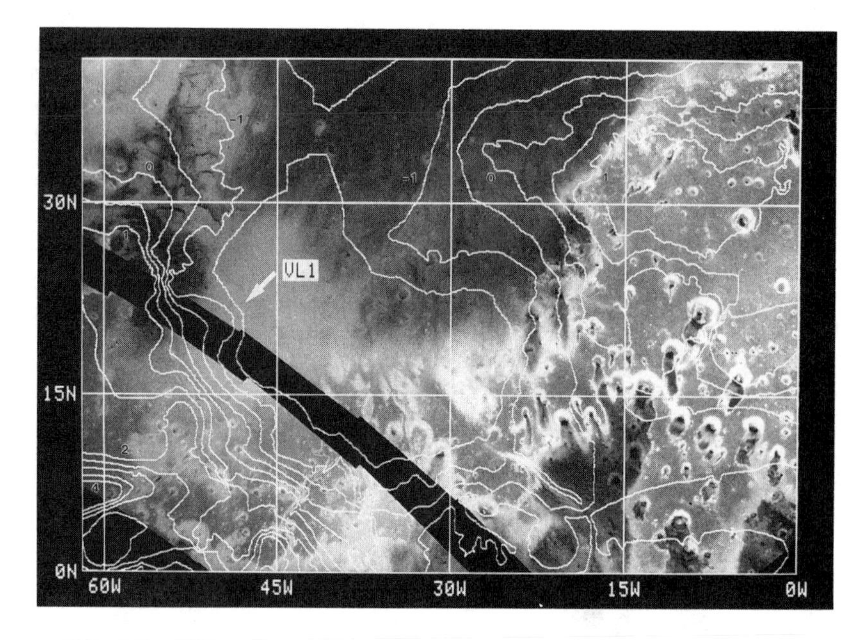

Fig. 5. Mosaic of Viking Orbiter images acquired using a red filter during the northern spring season (L_s = 37°) over Lunae Planum, Acidalia Planitia, Chryse Planitia and Oxia Palus. Topographic contours are in km. The location of the Mutch Memorial Station (Viking Lander 1) is shown. The reflectance patterns observed in the data can be explained by aeolian mixtures of two end-members: a bright dust component and a darker, coarser-grained (sand size) component. Intermediate areas represent a mixture of these two materials, sometimes accompanied by rougher microtopography, and are interpreted as cemented mixtures of dust and sand (i.e., similar to the duricrust observed at the Mutch Memorial Station). (Figure from Arvidson et al. 1989*b*.)

and slowly weathered, while aeolian sediments have accumulated, become cemented, and been removed, probably over 1 Myr time scales.

III. MODELS FOR CHEMICAL WEATHERING

Given some knowledge about the composition of the atmosphere and surface materials on Mars, it is possible to explore chemical weathering processes on Mars using conventional thermochemical computations. Only four basic types of chemical reactions are anticipated (Table I) although they must be tested for many different minerals and under various conditions of temperature, pressure and atmosphere or solution chemistry. In adopting such an approach, however, it is essential to distinguish clearly mechanisms from rates in judging the relative importance of candidate processes. For example, at the very low temperatures that occur on Mars, a given chemical reaction could be thermodynamically "spontaneous" but ponderously slow to the point

Fig. 6. Possible evidence for physical rock weathering at the Mutch Memorial Station (Viking Lander 1 site). Stereo image pair created from morning mosaic. Distance to horizon is ~ 3 km. Boulder in foreground (arrow) may have been broken by *in situ* mechanical weathering.

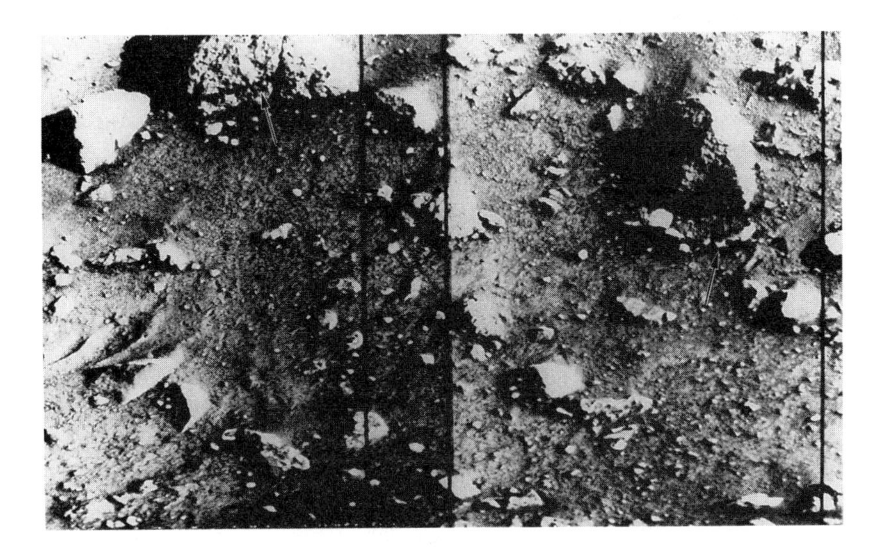

Fig. 7. Possible evidence for physical rock weathering at Mutch Memorial Station (Viking Lander 1 site). Near-field image of degraded rocks (arrows) that include fractures (upper left) and possible debris aprons (middle right). The field of view is ~ 1-m wide. (Viking Lander Image 12A081/012.)

of being unmeasurable in ordinary experiments. Still, over geologic time, the same reaction could conceivably predominate over all other weathering reactions.

Two different approaches have been taken to computing the thermodynamically stable chemical weathering products that might be expected on Mars: mineral-reaction diagrams and free-energy minimization of multi-element systems. The first approach considers decomposition/formation of individual minerals through reaction with H_2O, CO_2 and O_2 and is strongly influenced by geologic reasoning regarding candidate phases. The second approach follows the traditional chemical-engineering treatment of reaction processes and implicitly assumes that no barriers exist to complete reaction among elements. Not surprisingly, these two competing approaches can produce somewhat different predictions but, for Martian weathering, both approaches have produced mutually supportive results.

Most work to date has implicitly addressed surface weathering as the variety for which environmental conditions are best known. Hydrothermal alteration (if it occurred on Mars) might not differ greatly from similar alteration on Earth and computational conditions for pedogenic weathering remain undefined because of our collective ignorance about the Martian subsurface.

A. Mineral-Reaction Computations

Mineral-reaction diagrams appropriate for Mars were computed by O'Connor (1968*a*), Gooding (1978), and Gooding and Keil (1978). Although most work emphasized reactions involving water vapor, sufficient work was done to contrast important differences that would result from participation by liquid water.

In the absence of liquid water, the equilibrium weathering products should be dominated by anhydrous oxides and carbonates whereas, with liquid water as a reactant, hydrated minerals become significant, including layer-structured "clay" minerals (Table II). The contrast is illustrated for reaction of a pyroxene in Fig. 8 where the gas-solid case predicts no clay-mineral products but the liquid-solid case includes a Mg-phyllosilicate as a major alteration product. Therefore, products of chemical weathering expected on Mars depend critically on the availability of liquid water during weathering reactions.

A second major consideration, especially for production of clay-mineral or other aluminosilicate phases, is the availability of silicate glass as a reactant during weathering. As illustrated in Fig. 9 for gas-solid reactions, albite (Na-plagioclase) should be stable with respect to low-temperature chemical alteration by gaseous H_2O and CO_2. In contrast, glass of albite composition is predicted to react under Martian conditions to form Na-beidellite (a montmorillonite clay) and accessory products. Given the reactant/product trends exhibited by feldspars and their glassy equivalents, Gooding and Keil (1978) argued that basaltic glass might be the most likely source of clay minerals on

TABLE II
Chemical Weathering Products on Mars Predicted by Mineral-Reaction Computations[a]

Mineral Reactant	Gas-Solid Reaction[b] at 240 K	Liquid-Solid Reaction[c] at 273 K
Olivine	$MgCO_3 + Fe_2O_3 + SiO_2$	Mg-phyllosilicates + $MgCO_3$ + FeO (OH) + SiO_2
Pyroxene	(Mg,Ca) $CO_3 + Fe_2O_3 + SiO_2 + Al_2O_3$	Mg-phyllosilicates + $CaCO_3 + SiO_2 + Al_2O_3$ + FeO(OH)
Ca-plagioclase	Ca-beidellite (smectite) + $CaCO_3 + Al_2O_3$	Ca-beidellite (smectite) + $CaCO_3 + Al_2O_3$
Alkali feldspar	stable	(Na,K)-beidellite (smectite) + $(Na,K)_2CO_3 + SiO_2$
Feldspar glass	(Ca,Na,K)-beidellites (smectites) + kaolinite + $(Ca,Na,K)CO_3 + SiO_2 + Al_2O_3$	not computed
Magnetite	Fe_2O_3	FeO(OH)
Troilite, Pyrrhotite	$FeSO_4 + FeSO_4 \cdot H_2O \pm S$	FeO(OH) + S
Apatite	stable	$Ca_3(PO_4)_2 + CaHPO_4 + CaCO_3$

[a] Table after Gooding 1978; Gooding and Keil 1978.
[b] Assuming H_2O as vapor.
[c] Assuming liquid H_2O.

present-day Mars if gas-solid reactions were dominant. Indeed, glass would be an essential ingredient in weathering if palagonitization has ever been an important process on Mars (see, e.g., Soderblom and Wenner 1978; Allen et al. 1981). By definition, the mineraloid palagonite is a chemical alteration product of basalt glass and does not form by weathering of holocrystalline basalts.

It should be emphasized that we do not purport that mineral-reaction diagrams supplant observation with computational ideality. Given a set of total reactions involving a mineral reactant and its possible decomposition products, mineral-reaction diagrams summarize the trend toward equilibrium. Such diagrams do not, however, override the fact that many geochemical reactions do not reach equilibrium and that metastable intermediate reaction products are commonly observed as "final" products. Instead, mineral-reaction diagrams provide the equilibrium benchmarks that are essential to recognizing and diagnosing observed assemblages of metastable or disequilibrium weathering products. In addition, mineral-reaction diagrams are subject to revision based on new or improved values of thermochemical data for reactants or products.

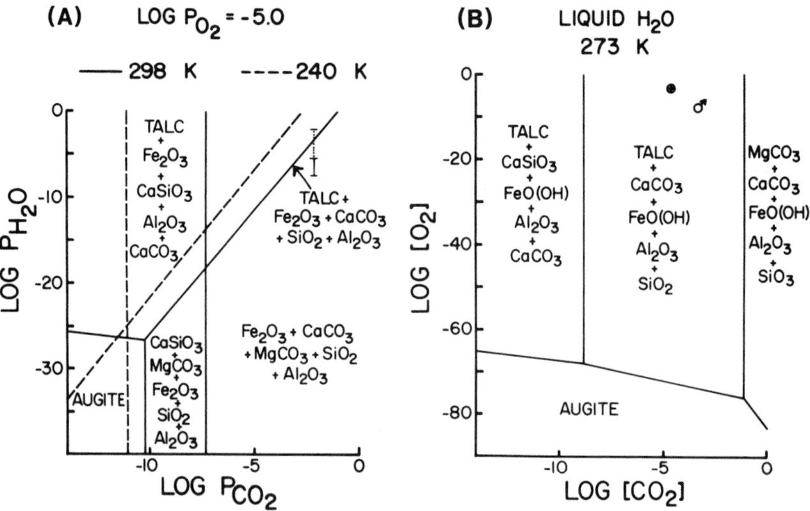

Fig. 8. Mineral-reaction diagrams comparing weathering of pyroxene by gas-solid and liquid-solid reactions (figure after Gooding 1978). Solid boundary lines define stability fields expected at 298 K whereas dashed lines define stability fields expected at 240 K. (A) Stability fields of augite and its decomposition products in equilibrium with gaseous H_2O and CO_2 and the partial pressure of O_2 in the Martian atmosphere and at two different temperatures. Vertical bracketed lines show range of H_2O partial pressures expected in Martian atmosphere. (B) Stability fields of augite and its decomposition products in equilibrium with liquid water containing dissolved O_2 and CO_2 at 273 K. Reference points are shown for gas-saturated aqueous solutions that would be in equilibrium with atmospheres of Earth and Mars, respectively.

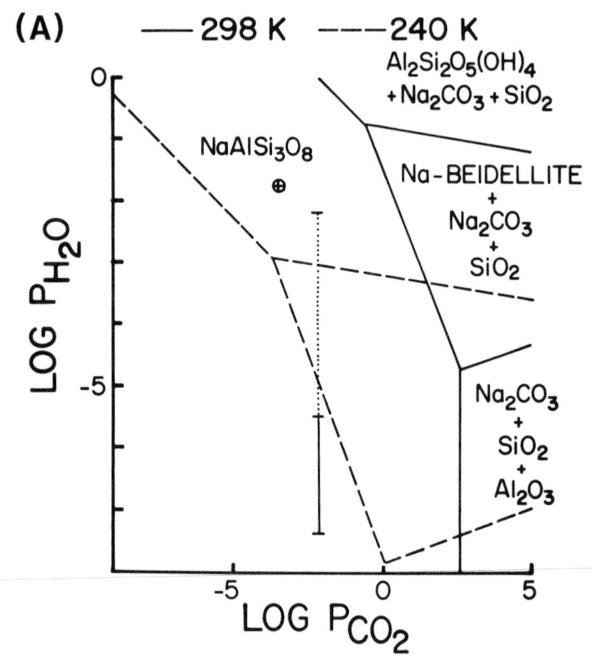

Fig. 9. Mineral-reaction diagrams comparing gas-solid weathering of Na-feldspar (albite) and its glassy equivalent (figure after Gooding 1978; Gooding and Keil 1978). (A) Stability fields of albite and its decomposition products, following the same conventions as in Fig. 8a. (B) and (C) Stability fields of glass of albite composition, at two different temperatures, with other factors being the same as in (A).

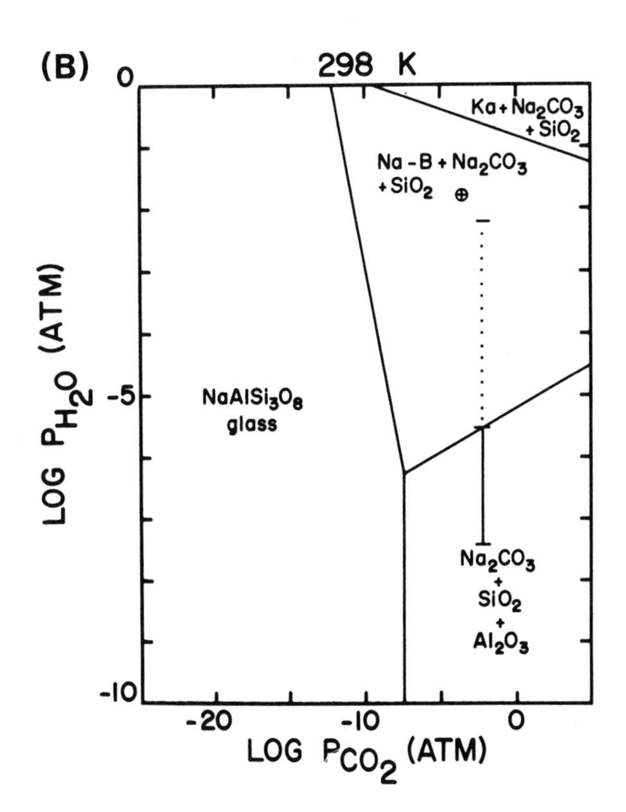
(B)
298 K
Ka + Na_2CO_3 + SiO_2
Na - B + Na_2CO_3 + SiO_2
$NaAlSi_3O_8$ glass
Na_2CO_3 + SiO_2 + Al_2O_3
0
-5
-10
-20
-10
0
LOG P_{H_2O} (ATM)
LOG P_{CO_2} (ATM)

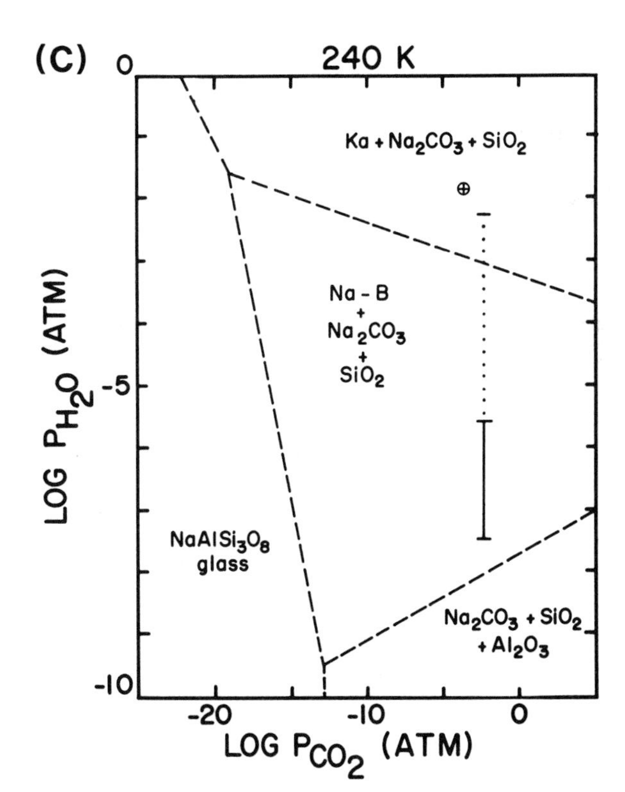
(C)
240 K
Ka + Na_2CO_3 + SiO_2
Na - B + Na_2CO_3 + SiO_2
$NaAlSi_3O_8$ glass
Na_2CO_3 + SiO_2 + Al_2O_3
0
-5
-10
-20
-10
0
LOG P_{H_2O} (ATM)
LOG P_{CO_2} (ATM)

For major rock-forming minerals expected to occur on Mars, the equilibrium products of weathering by gas-solid and liquid-solid reactions are summarized in Table II. Abundant clay minerals are not favored products of the weathering of crystalline igneous rocks on Mars unless liquid water can be invoked as a reactant, although availability of igneous or impact-produced rock glasses should make clay formation significantly more favorable. Ferric oxides are favored in nearly all reactions even for the very low concentrations of free oxygen in the present-day Martian atmosphere; however, anhydrous forms (e.g., hematite) rather than hydrous forms (e.g., goethite) are expected to dominate unless liquid water participates as a reactant. Sulfates are expected to form by weathering of igneous sulfide minerals in both gas-solid and liquid-solid reactions.

The expected high susceptibility of sulfides to weathering inspired Burns (1987,1988) to advocate weathering of primary iron sulfides to various sulfate assemblages as the dominant weathering process on Mars. Burns used Eh-pH diagrams, a special variety of mineral-reaction diagram that assumes liquid water conditions, and predicted a wide variety of ferric sulfate weathering products based on analogy with terrestrial gossans (supergene ore deposits). However, the relative stabilities of the candidate minerals were not evaluated with respect to Martian environmental conditions where liquid water is absent.

B. Free-Energy Minimization Computations

Free-energy minimization computations for Martian weathering were first explored by Zolotov et al. (1983), Sidorov and Zolotov (1986) and later, by Zolensky et al. (1988). In this approach, a large number of chemical elements or compounds, which satisfy an initial postulated model composition for the total system, are allowed to establish simultaneous reaction relationships, until the total Gibbs free energy of the system reaches a minimum for a specified temperature and pressure. At that point, the dominant phases in the system are presumed to represent the most favorable final assemblage for the postulated conditions. Although free-energy minimization is, in some ways, more elegant than the reaction-diagram approach, its application to low-temperature geochemical systems implicitly assumes material transportability that is usually achieved only in homogeneous gas-phase or liquid-mediated reactions.

Sidorov and Zolotov (1986) examined three different model compositions (basalt, komatiite and Viking Lander surface sediments) and reactions involving 17 different chemical elements, including 5 different model atmospheres. All computations were made for Martian atmospheric pressure and a temperature of 240 K. Basic assumptions included chemical equilibrium and complete reaction without kinetic constraints. The model that was most relevant to current knowledge of Mars, which was based on the Viking sediment bulk-elemental composition and O_2, CO_2 and H_2O as gaseous reactants, produced results that resembled those derived by the reaction-diagram approach of Gooding (1978), including hematite as the most stable form of iron. Compared with Gooding's results, however, Sidorov and Zolotov found montmorillonite and talc to be more important as reaction products (Table III). At least for montmorillonite, differences between Sidorov-Zolotov and Gooding results is attributable to the availability of aluminum and magnesium. In multicomponent free-energy minimization, all mineral-forming elements were available simultaneously (without kinetic restrictions) whereas mineral-reaction diagrams constrained elemental compositions according to the progenitor mineral that was being weathered. In both cases, though, the two approaches predicted Mg,Al-rich phyllosilicates (beidellite/montmorillonite and talc) rather than Fe-rich phyllosilicates (e.g., nontronite) as the equilibrium weathering products.

Sidorov and Zolotov (1986) also considered reactions between minerals and gaseous SO_2 on the presumption that, on early Mars, volcanic gases might have contributed to surface weathering, as suggested by Settle (1979). Although SO_2 might be expected, on thermodynamic grounds, to attack carbonate and halide minerals, Sidorov and Zolotov used available kinetics data to conclude that such reactions should be negligible at the low temperatures on present-day Mars. Effectiveness of the subject reactions in the past would imply higher ambient temperatures during those epochs.

Zolensky et al. (1988) applied modified free-energy minimization calculations to the case of a shergottite meteorite in contact with a liquid water reservoir containing dissolved CO_2 and O_2 at temperatures as low as 273 K. Shock-metamorphosed plagioclase glass (maskelynite) was treated as crystal-

TABLE III
Chemical Weathering Products on Mars Predicted by Free-Energy Minimization Computations for Gas-Solid Equilibria at 240 K[a]

Mineral	Vol %
Quartz	20.8
Montmorillonite	23.9
Talc	27.9
Hematite	10.5
Rutile	0.6
Pyrolusite	0.2
Anhydrite	13.4
Kieserite	0.2
Halite	0.3
Sylvite	0.5
Bischofite	1.1
Phosphates	0.4

[a] Bulk chemical composition taken as that of surface fines analyzed by Viking Landers, reacting with atmospheric CO_2, O_2, and H_2O (after Sidorov and Zolotov 1986).

line plagioclase in the starting solids and equilibrium was computed incrementally as a function of degree of reaction of the solids. Zolensky et al. found that hematite, gibbsite, and kaolinite should be the earliest reaction products but that, after a few percent of the rock had reacted, nontronite (an Fe-rich smectite clay) should predominate among the weathering products, with calcite as an accessory. They noted with interest, as well, that if less than 0.5% of the rock reacts, the weathering products would be dominated by clay minerals without carbonates.

The results of Zolensky et al. (1988) accentuate the differences to be expected between hydrothermal alteration and low-temperature surface weathering of the type examined by Sidorov and Zolotov (1986). During liquid-solid reactions at elevated (above freezing) temperatures, true clay minerals become significant alteration products whereas in cold gas-solid reactions they remain only marginally important. In that regard, the computational results agree with observational comparisons of surficially weathered rocks with those altered by hydrothermal systems associated with impact craters on Earth. The most abundant and well-crystallized clay minerals are found in the altered impact melt rocks and breccias (Allen et al. 1982) whereas surficially weathered rocks are dominated by poorly crystallized mineraloids (Allen et al. 1981).

The difference between clay-like mineraloids and genuine phyllosilicate clays as weathering products has several important implications. Not only would crystalline clays imply sustained geochemical systems dominated by liquid water but they would provide reactive surfaces of major consequence. Burt (1989) suggested that coupled substitutions of $Fe^{3+}O$ for $Fe^{2+}OH$ in hydrated ferroan clays could act alternately as a sink or source for hydrogen on Mars. In the Burt model, humidification of dehydrogenated clays could lead to spontaneous release of oxygen. Furthermore, catalytic properties of phyllosilicate grains might profoundly affect the geochemical cycle of any organic carbon on Mars. Therefore, understanding whether or how chemical weathering produced clay minerals on Mars remains a cornerstone of Martian weathering studies.

The free-energy minimization studies of Martian weathering have generally supported earlier conclusions derived by the reaction-diagram approach. In addition, they illuminated the variety and complexity of weathering products that can be expected when liquid water is a reactant. Overall, the computational models have remained complementary with few significant discrepancies. All agree that the abundance and phase identity of water is the most important determinant of the products to be expected from chemical weathering on Mars.

C. Special Effects Related to Ice

As pointed out by Gooding (1986*c*, 1989), chemical weathering under freezing conditions could possess special attributes, based on effects related to the physical chemistry of water in contact with minerals at or below the equilibrium freezing temperature (see Sec. II.A). Because water is the most important reactant in weathering reactions, the different abilities of various minerals to nucleate water ice and, therefore, render it less reactive, should influence the relative susceptibilities of those minerals to aqueous attack. If all other factors were constant, the most effective nucleators of ice should have greater immunity to chemical attack from unfrozen water.

Figure 10 compares, for various minerals, the thermodynamic impulse for reaction (with liquid water) with the ability to arrest water by freezing into ice. For a given value of thermodynamic drive toward reaction (y-axis), increasing degree of crystallographic mismatch between the mineral and

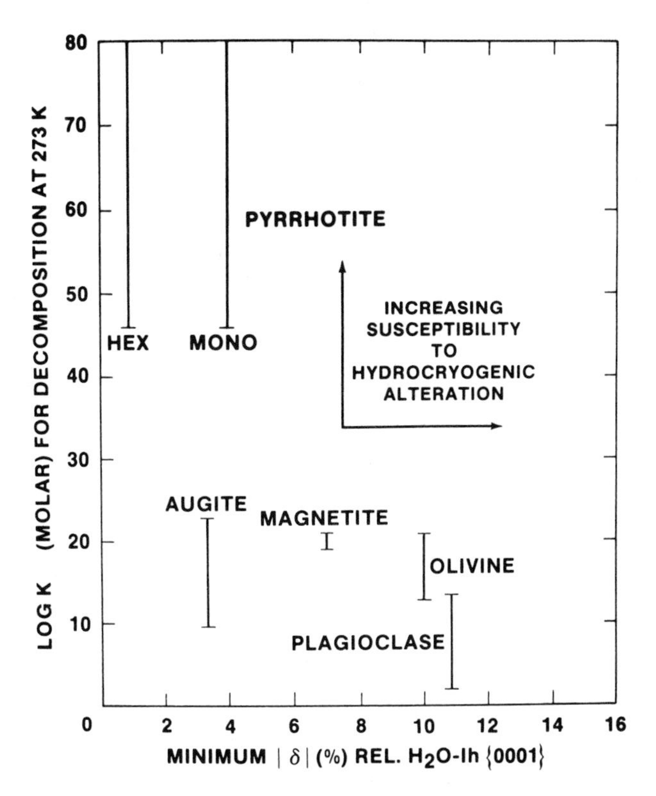

Fig. 10. Relative susceptibilities of primary igneous minerals to chemical weathering under freezing conditions, as functions of thermodynamic impulse for reaction (y-axis) and efficiency of ice nucleation (x-axis). At temperatures above freezing (> 273 K), where water is liquid, the y-axis position determines the relative disposition of each mineral toward weathering. At subfreezing temperatures (< 273 K), where heterogeneous nucleation of ice depends on mineral substrates, propensity for weathering is also influenced by the effectiveness with which a mineral can immobilize water by promoting ice nucleation. Prospects for survival of undercooled liquid water, and therefore liquid-based weathering reactions, increases with increasing value of the δ parameter.

common water ice (x-axis) implies increasing susceptibility to chemical reaction with unfrozen water. Figure 10 predicts that igneous sulfides (e.g., pyrrhotite) are more susceptible to reaction with water than are igneous silicates. Among the silicates, however, olivine and plagioclase should be more susceptible than pyroxenes. Indeed, olivine and plagioclase could have similar susceptibilities despite their differences in iron content. Glass, of course, would have indeterminately high mineral/ice mismatch and should be the silicate phase most susceptible to reaction with water. Applicability of the model in Fig. 10 to Mars remains to be tested. However, it has shown promise in preliminary attempts to explain the weathering behavior of stony meteorites found in Antarctica (Gooding 1986*c*).

Huguenin (1982) suggested a more exotic role for water frost during Martian weathering as the key reactant in a four-step sequence of reactions leading to formation of hydrogen peroxide, H_2O_2. Although the frost-weathering scheme was proposed to explain results of the Viking biology experiments (see Sec. IV), it was also described as a process that would chemically reduce silicate materials through addition of H^+ ions. In order to explain the obvious abundance of oxidized ferric (Fe^{3+}) iron on Mars, Huguenin invoked a photocatalytic oxidation process (see Sec. IV) that would operate independently but in parallel with frost weathering.

D. Reaction Rates

Although computation of possible reaction pathways is relatively straightforward, prediction of their probable rates on Mars is exceedingly difficult. Reaction rates are ultimately rooted in experimental studies but little work has been done on the complex mineral systems under low-temperature conditions appropriate for Mars. By their nature, most of the candidate reactions are heterogeneous (i.e., mixtures of solid and gaseous reactants and products) and are expected to proceed by multiple steps with ample opportunity for significant rate variations between steps. Net rates may be too low to measure accurately except over baselines of many years.

In the absence of kinetics data, Gooding (1978) reasoned that for mineral-gas reactions relative rates of competing reactions should be controlled by the effective reactivities of the respective gases. For surface-controlled reactions, the dominant effect might be relative adsorptivities of the gases which, on Mars, should be $H_2O > CO_2 > O_2$. For reactions controlled by diffusion of gas into the solid, reactivities should occur in the order, $H_2O \geq O_2 > CO_2$. One qualitative conclusion might be that hydration would occur faster than either oxidation or carbonation.

Shock-metamorphic enhancement of dissolution rates might further distinguish Martian weathering from weathering on Earth, given the greater density of impact craters on Mars. Boslough and Cygan (1988) showed experimentally that silicate minerals subjected to shock pressures expected in meteorite impacts dissolve in aqueous solutions at rates up to 20 times faster than their unshocked equivalents. Although glasses produced by shock melting are obvious candidates for rapid weathering, the materials studied by Boslough and Cygan were essentially unmelted products that might have been activated toward enhanced weathering through crystal dislocations.

Perhaps, an upper limit to the order-of-magnitude for chemical weathering rates on Mars can be estimated by considering the most likely process that would operate on the least stable solid: hydration of basalt glass. Hydration of rock glass occurs according to the rate law

$$x = (kt)^{1/2} \qquad k = A\, e^{-E/RT} \tag{4}$$

where x is the thickness of the hydration layer that forms after time t and k is a constant that varies with activation energy E, gas constant R, and temperature T according to the Arrhenius equation. For Hawaiian submarine basalts, $k = 480$–$2000\ \mu m^2\ Kyr^{-1}$ but can be as low as $81\ \mu m^2\ Kyr^{-1}$ for subaqueous basalts in fresh water (Moore 1966). Obsidian (rhyolite glass) hydrates much more slowly with typical values of $k = 2$ to $16\ \mu m^2\ Kyr^{-1}$ under Earth surface conditions (Michels et al. 1983). Using the slow end of the basalt hydration range ($k = 81$) leads to a prediction of hydration at the rate of 0.9 cm Gyr^{-1}. For a very conservative value of $k = 1$, the predicted rate would be 0.1 cm Gyr^{-1}. If the aeolian abrasion rate on Mars is as high as 1 $\mu m\ yr^{-1}$ (see Sec. II.C), the extrapolated rate of 10^5 cm Gyr^{-1} implies that the rate of hydration would be trivial relative to the probable rate of abrasive removal of hydration rinds. Even the fast end of the basalt hydration range ($k = 2000$) would translate to a rate of hydration of only 4.5 cm Gyr^{-1}.

Although these estimates imply that basalt glass should not survive over geologic times scales on Mars, they further imply that chemical weathering may have been subservient to mechanical weathering over much of Martian history. This conclusion, of course, requires observational verification. The estimated abrasion rate probably applies only to loose sediments and may be far higher than the rate applicable to solid rocks (Sec. II.C). If, in fact, the rate of rock abrasion is much less than the rate of chemical weathering, then Martian rocks may have accumulated weathering rinds as quasi-permanent features. Indeed, further growth of the weathering rinds might have become self-limiting, as would be the case with diffusion-controlled processes. Dynamical equilibrium between chemical and physical weathering, for a chemical weathering rate of 1 cm Gyr^{-1}, would imply an aeolian abrasion rate of $10^{-5}\ \mu m\ yr^{-1}$.

IV. OBSERVATIONS AND EXPERIMENTAL TESTS

A. Viking Lander Imaging

Lander images were acquired in 6 broad wavelength intervals covering 0.4 to 1.0 μm, with a spatial resolution of several centimeters. Data were also acquired at three times better spatial resolution using a broadband filter. The combination of multispectral coverage at a variety of incidence, emission, and phase angles, coupled with high-spatial-resolution data, allow analyses that place first-order constraints on the types of materials exposed at the Lander sites. The radiance factors and scattering properties of exposed sediments are consistent with the presence of ferric iron-rich, dust-size particles of poor crystallinity (Guinness et al. 1987; Arvidson et al. 1989*a,b*). A variety of data imply that some of the sediments are cemented, including a correlation between mechanical properties and chlorine and sulfur content (see, e.g., Arvidson et al. 1989*a*). The best spectral match is to Hawaiian palagonites (Guinness et al. 1987) for which the spectrally dominant component may be nanophase hematite (Morris et al. 1989; Morris and Lauer 1990). Furthermore, the Lander image data demonstrate that most rocks are coated with the ferric iron-rich material, although some rocks have radiance-factor signatures similar to thinly coated basalts and andesites. Rocks have probably been added to each Lander site by impact events and have slowly weathered over time as aeolian sediments more rapidly accumulate and erode. Nonetheless, the indications of palagonite-like materials and Fe^{3+}-bearing phases clearly imply that both hydration and oxidation have been important aspects of chemical weathering on Mars. If the palagonite-like materials are, in fact, analogous to terrestrial palagonites, then basaltic glass has played an important role as a reactant during Martian weathering.

As shown in Figs. 6 and 7, some rocks on the Mars surface exhibit physical evidence for mechanical weathering. No direct evidence exists to link the degradation to frost riving or secondary-mineral riving, although the high salt contents in the sediments (inferred from S and Cl concentrations) would at least be permissive of schemes for riving by sulfate and chloride minerals.

B. Viking Lander Chemistry

Although the Viking Landers did not directly address weathering environments at the Martian surface, significant information that is relevant to chemical weathering on Mars can be inferred from results of the X-ray fluorescence spectrometer (XRFS), gas chromatograph/mass spectrometer (GCMS), and biology experiments (see review by Arvidson et al. 1989*a*). No minerals were identified by Viking but model interpretations of bulk-elemental compositions measured for surface fines by XRFS favored a mixture dominated by smectite clay minerals mixed with sulfur- and chlorine-bearing salts (chapter 18). The GCMS results indicated at least 1 wt% water in the bulk fines and absence of sulfur gases after pyrolysis at 773 K showed that the sulfur resided in an oxidized form such as sulfate. Therefore, the circumstantial evidence for clay minerals and evaporite salts implied evolution of the Martian fines through aqueous geochemical processes that included oxidation and hydration.

At both Viking landing sites, all samples that were analyzed by the biology experiments displayed strong chemical reactivity (see review by Arvidson et al. 1989*a;* also, chapter 34). The samples released O_2 upon contact with water, oxidized organic compounds upon innoculation with nutrient media, and fixed carbon from artificially introduced CO_2. Given negative GCMS results for organic compounds in the same samples, the reactivity observed in the biology experiments was almost certainly nonbiological in origin. Traces (parts per million or billion) of inorganic peroxides or superoxides became favored candidates for the unidentified reactants although later laboratory simulations demonstrated similar reactivities for iron-exchanged smectite clays and maghemite (γ-Fe_2O_3). In the context of chemical weathering, however, the biology results also supported interpretation of Martian surface fines as highly oxidized materials. Indeed, attention focused on reactions involving peroxides or superoxides as candidate processes of chemical weathering on Mars.

C. Laboratory Simulations

Post-Viking laboratory simulations of the Lander biology results emphasized the search for metastable inorganic oxidants that might be produced by gas-mineral reactions under the influence of ultraviolet light. That approach was largely motivated by results of the pre-Viking laboratory study by Huguenin (1973*a,b*) who reported remarkably high, ultraviolet-stimulated oxidation rates for magnetite in a simulated Martian atmosphere. Similar high rates of oxidation were later reported for ultraviolet-irradiated basalt glass samples (Huguenin 1974). Unfortunately, Huguenin's results were not reproducible by other research groups and led to a debate over the significance of ultraviolet radiation in Martian chemical weathering.

Morris and Lauer (1980) attempted specifically to reproduce Huguenin's results for magnetite but without success. Huguenin reported that particulate magnetite samples became visibly reddened after only a few minutes of ultraviolet irradiation and that 16% by volume of a sample could be oxidized after 0.5 hr of ultraviolet exposure. In contrast, Morris and Lauer (1980) tested nine different magnetite samples but found no measurable oxidation after hundreds of hours of irradiation. They pointed out, however, that oxidation can occur if an unfiltered infrared component from the ultraviolet light source is allowed to excessively raise the sample temperature by radiant heating. The clear implication of their work was that the magnetite oxidation reported by Huguenin was an artifact of radiant heating and had no basis in ultraviolet-photostimulated chemistry. Using the same approach and an independent set of negative results, Morris and Lauer (1981) also challenged Huguenin's (1976) claim that any hydrated ferric oxides (e.g., goethite, α-FeOOH) on Mars would become dehydrated by ultraviolet-photostimulated reactions.

Other research groups also reported negative evidence for ultraviolet-stimulated oxidation. Booth and Kieffer (1978) reported evidence for incipient formation of carbonates by reaction of powdered basalt with gaseous CO_2 under simulated Martian conditions. They concluded that photochemistry of adsorbed H_2O, rather than direct ultraviolet irradiation, was the dominant influence on carbonate formation. However, Booth and Kieffer specifically noted absence of sample reddening or oxidation of the type reported by Huguenin (1974). Clark et al. (1979) exposed basalt glass, magnetite, nontronite (a smectite clay) and a variety of salt minerals to hundreds of hours of ultraviolet irradiation in a Mars-like atmosphere but found no detectable yields of either solid or gaseous reaction products. In separate experiments, the only detectable reactions involved decomposition of carbonate minerals (to sulfates) in the presence of gaseous SO_2. Only Blackburn et al. (1979) reported success in oxidizing minerals under simulated Martian conditions; however, their experiments were restricted to powdered pyrolusite (β-MnO_2), rather than Fe-bearing minerals. Still, they concluded that ultraviolet-stimulated oxidation was not a direct process but operated indirectly through the photochemistry of H_2O on mineral surfaces.

The net result of the several different studies of ultraviolet-stimulated weathering under Martian conditions was that its significance for Mars is much less than once believed in the immediate light of Viking results. The incredibly rapid oxidation process originally proposed by Huguenin appears to have been an artifact of experimental procedure. Instead of oxidizing minerals directly, the importance of ultraviolet irradiation is more likely to be as an influence on the photochemistry of H_2O and its decomposition products. In that regard, the laboratory simulations served to reiterate the point made repeatedly in earlier sections: the most important ingredient in weathering on Mars should be water.

V. WEATHERING STUDIES AS PART OF FUTURE EXPLORATION

The greatest contribution to understanding Martian weathering will come from direct analyses of samples that have been weathered to different degrees and under different climatic conditions. Weathering products must be properly identified before their paragenetic relationships can be determined and their testimony obtained with regard to Martian geologic history. Ideally, intact profiles (surface to interior) of rocks of similar composition, but with the widest possible range of ages, should be mineralogically analyzed to determine how surface weathering processes might have changed through Martian history. Comparable analyses would be desirable for soil profiles so that pedogenic weathering could be compared with surface weathering. If ground ice samples could also be obtained, oxygen and hydrogen stable-isotopic analyses of the ice and of hydrated weathering products might permit derivation of paleotemperatures of weathering products of different ages. In addition, gas analyses of weathering products that act as natural sorbents (e.g., clay minerals and zeolites) might reveal changes in the composition of the Martian atmosphere through time (Gooding et al. 1989). Finally, analysis of cosmogenic nuclides and particle tracks induced in rock surfaces by cosmic-ray bombardment could reveal information about variations in atmospheric density, including changes in aeolian abrasion rate (Arvidson et al. 1981*b*). Therefore, studies of weathering products should be expected to provide much relevant information regarding climate change on Mars.

Specific sample-analysis goals should include determination of whether aluminosilicate weathering products are dominated by poorly crystalline mineraloids, such as palagonite, or by well-crystallized clay minerals, such as nontronite. Each of those two possibilities would imply very different histories of chemical weathering on Mars. Identification and age-dating of any residual glass in regolith samples should help constrain rates of chemical weathering. Of course, determining relative proportions of aluminosilicate, oxide and salt-mineral weathering products should indicate the effectiveness of specific chemical-weathering processes, such as carbonate formation, relative to other processes. Detailed studies of intact weathering rinds from a variety of rocks should help reveal pathways of chemical weathering.

Field experiments might include long-duration exposure of test rocks and minerals to the ambient surface environment to determine contemporary rates of physical and chemical weathering. Such experiments are commonly performed by geomorphologists on Earth. Because weathering rates are expected to be very low, however, such experiments would be useful on Mars only if very long-lived surface stations were available or if test samples could be deployed in an early mission and retrieved by a much later mission for return to Earth. Other *in situ* measurements should include trace-component analyses of the near-surface atmosphere and subsurface probes to determine the abundance and physical state of free water. The gas analyses would identify volatile compounds that might be active agents of chemical weathering. The soil-water analyses would indicate whether liquid-based weathering reactions might be supportable in the contemporary pedogenic environment.

To set sample studies in global perspective, the entire Martian surface should be mineralogically mapped by a visible-infrared imaging spectrometer using high spectral and spatial resolution. High spatial resolution is needed to distinguish deposits of weathered sediments, for example, from exposures of unweathered bedrock. High spectral resolution is needed to maximize prospects for unambiguous mineral identifications. With available Viking Orbiter data, only red dust and (unidentified) dark material can be distinguished; intermediate colors are attributable to mixtures of those two components and surface roughness variations. With high-resolution multispectral data, prospects would be good for distinguishing subtypes of both weathered cover and bedrock.

20. RADAR DETERMINATION OF MARS SURFACE PROPERTIES

RICHARD A. SIMPSON
Stanford University

JOHN K. HARMON
Arecibo Observatory

STANLEY H. ZISK
University of Hawaii

T. W. THOMPSON
Jet Propulsion Laboratory

and

DUANE O. MUHLEMAN
California Institute of Technology

*Radar studies of Mars have provided measurements of surface texture on scales of centimeters to hundreds of meters and measurements of surface material properties. Texture (*rms *surface tilts and estimates of small-scale roughness) may be inferred from dispersion and/or polarization of the radar echo; material properties (reflectivity or dielectric constant) are derived from echo strength. Mars is a diverse target; depending on location,* rms *surface tilts have been found to vary over the range 0°25 to 10° while reflectivity covers at least 3 to 13%. Plains units are the most variable, having both the smoothest and roughest surfaces; cratered terrain can be considered predictable and "average" by comparison. Recent data identify scattering by small structures (perhaps rocks on or near the surface) as playing a more important role than previously rec-*

ognized. Scattering by the residual ice cap near Mars' south pole is particularly unusual. The present state of radar surface studies is summarized as an introduction to the broader discussion of the surface layer in chapter 21.

I. INTRODUCTION

Goldstein and Gillmore (1963) obtained the first Mars radar echoes during the close Earth-Mars approach of 1963. At approximately 26-month intervals since, additional data have been acquired, increasing in volume, quality and complexity as planetary radar systems have become more advanced and powerful. The data have been used by the original experimenters for a variety of purposes from estimating the astronomical unit to helping select sites for Viking Landers. In approximate order of increasing difficulty, radar measurements of Mars yield:

1. Earth-Mars distance and altimetry of Mars' surface;
2. Roughness of the surface near the subradar point (suprawavelength scales);
3. Reflectivity and/or radar cross section of the target area;
4. Estimates of small-scale (subwavelength-sized) surface structure.

In addition, a number of later studies have been based in whole or in part on these data, including studies of Martian volcanic processes and their histories, surface physics and the chemistry and history of Martian volatiles. The measurements listed above are obtained, respectively, from:

a. Round-trip echo flight time;
b. Shape of the echo (especially its dispersion);
c. Strength of the echo;
d. Polarization of the echo and variations in (b) and (c).

Following a discussion of the scattering process (Sec. II) and radar as a remote-sensing probe (Sec. III), the remainder of this chapter will be a synopsis of scattering models (Sec. IV) and a review of the Mars data, particularly as they relate to inference of surface properties. The emphasis here will be on derivation of surface material density and surface texture from radar measurements. Altimetry is mentioned in passing, but the reader interested in a comprehensive discussion of those data should see chapter 7.

Space does not permit an extensive discussion of current research topics, nor even a very thorough coverage of uncertainties in our methodologies. For example, a full review of radar signal penetration *into* planetary surfaces and methods one might employ to distinguish those (volume scattering) echoes from scattering at the surface proper could be a small chapter by itself. An evaluation of realistic (as opposed to idealized) models for inferring density of surface materials would also be relevant. The models used in practice almost invariably assume sharp boundaries between hypothetical material layers, whereas gradations are more realistic; the difference can have an important bearing on data interpretation. The focus here will be on generally accepted models and approaches to interpretation of Mars radar data with complications kept to a minimum.

The reader interested in additional detail on scattering concepts, techniques and interpretation may find the work of Evans and Hagfors (1968), Ruck et al. (1970) and Elachi (1987) useful. Relatively recent reviews of planetary radar astronomy, including discussion of Mars data, have been written by Pettengill (1978) and Ostro (1987). Key data sets include those described by Downs et al. (1975), which has become the *de facto* standard for many Mars studies, by Simpson et al. (1984), which is largely oblique scattering data, and by Harmon and Ostro (1985), which includes polarization results. Where there is allusion to narrow topics in the sections which follow, the reader is encouraged to consult the additional references cited.

II. THE SCATTERING PROCESS

We imagine planetary surfaces to be spherical, but modulated by craters, mountains, grabens and smaller structure such as magma flows, dunes, blocks and pebbles. The different scales of surface texture influence radio wave scattering in different ways. We identify three broad classes of texture here. Topographic variations on scales *much* larger than the radio wavelength λ (e.g., km) will be called regional *slope;* regional slope and associated topography will not be discussed extensively here (see chapter 7). Moderate-scale variations (topographic variations on the scale of the radio wavelength and somewhat larger—e.g., tens of cm to perhaps a hundred m) will be called *tilt.* Variations on wavelength and smaller scales (e.g., cm and less) will be called *roughness.* Analysis of the scattered signal yields various properties of the target, including measures of surface texture over these three broad scale ranges.

A. Types of Scatter

If the surface is gently undulating on wavelength (λ) scales, we expect mirror-like (specular) reflections (Fig. 1a), where the local angles of incidence and reflection are equal. A distant observer will see scattered energy over only a limited range of viewing angles (Fig. 1b) and with a polarization determined by that of the incident wave and the average scattering geometry. This "quasi-specular" scatter is well understood theoretically and is an important component of radar echoes from the terrestrial planets.

If the surface is highly irregular, the scattering will be distributed broadly in angle (Fig. 1c). The composite pattern will show little or no preference in scattering direction (Fig. 1d). An echo, albeit relatively weak, will be observed at virtually any viewing angle. Its polarization generally indicates a complex interaction. This "diffuse" scatter is understood only poorly in terms of empirical models and limited theory. It is an important component

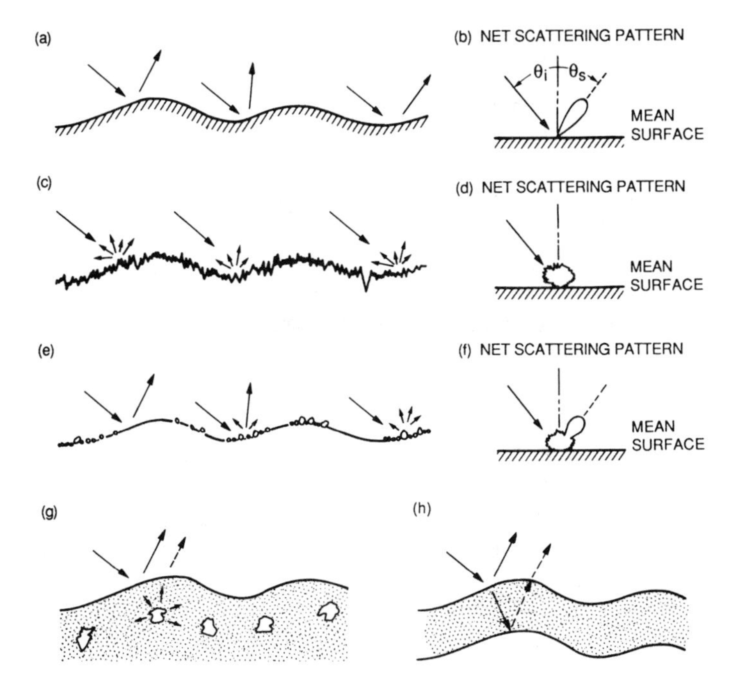

Fig. 1. Representative surfaces for (a) quasi-specular scatter, (c) diffuse scatter and (e) a combination of scattering. Companions (b, d and f, respectively) show macroscopic behavior, the scattering pattern that would be seen by a distant observer. Angles θ_i and θ_s are measured with respect to the normal to the mean surface and denote angles of incidence and scattering, respectively. Panels (a)-(f) deal with scattering which occurs at the "top" surface. If the material is homogeneous or if it is lossy, little from below the surface will be detected. In some materials penetration and subsurface scattering may occur: in (g) by subsurface inhomogeneities which add to the top-surface diffuse component; and in (h) by layers that can add to the quasi-specular component.

of Mars echoes, particularly under nonspecular viewing geometries. There is general agreement that surface and subsurface rocks and wavelength-scale roughness are primary sources of diffuse scatter but less apparent sources, such as cracks in an otherwise smooth plane, may also contribute.

Intermediate to these two cases is a surface which is gently undulating in its large-scale behavior but which has finer structure superimposed (Fig. 1e). Each block or other small irregularity scatters a small fraction of the incident wave diffusely, but when the density of these perturbations is low, the overall behavior appears to be quasi-specular, with a low level of diffuse background scatter which is relatively independent of angle. As the density of irregularities increases, the diffuse background rises and the quasi-specular lobe becomes weaker and may broaden slightly. The composite scattering pattern (Fig. 1f) shows both the quasi-specular lobe and the more nearly isotropic background.

B. Subsurface Effects

There are other factors which we note here, then discuss more thoroughly in Sec. IV. If the wave penetrates the surface and interacts with subsurface structure, there are additional contributions to the echo. In many instances, we expect this subsurface structure to be irregular, in which case the additional contribution will add to the diffuse echo (Fig. 1g). If, on the other hand, layering is present, an enhancement of the quasi-specular echo would be expected (Fig. 1h). If the wave is scattered twice (or more) or if the scattering is by nonsymmetric objects, its polarization is affected, often strongly.

The radar wavelength λ plays a role in scattering from subsurface structure. Natural soils and rocks are characterized by a complex dielectric constant. The real and dominant part determines how well the signal is transmitted into the surface; the imaginary part, sometimes expressed as the loss tangent, determines effectiveness with which the signal is attenuated in the material. The loss tangent and, to some extent, the "real" dielectric constant are strongly affected by liquid-water content, as compared with other ordinary geologic materials. Typical loss tangents for dry terrestrial materials allow probing to depths of 10λ or so; longer wavelengths thus probe more deeply. Highly dessicated materials may permit greater penetration.

C. Scales

The radar wavelength is also important as compared with the scale of surface structure (or the dimensions of a discrete object). When the characteristic dimension a is small compared to the wavelength λ, scattering efficiency varies as a^6/λ^4; small objects and structures very quickly become "invisible," unless the population increases proportionately. When the dimension is large *and the scatterer acts coherently* (e.g., a distant, perfect mirror), scattering varies as a^4/λ^2. Curvature generally destroys coherence, so most large, natural objects appear to scatter as one or more smaller objects; surfaces with significant tilt have many "glints," for example, the strength of each glint in turn depending on the ratio of its curvature radius to wavelength. The sum usually scatters as a^2. The wavelength thus imposes a filter on effective surface textures; structure which is small compared with λ is not important, while large structure breaks up into smaller quasi-specular elements with dimensions of a few λ (depending on characteristic curvature). Changing the wavelength readjusts the filter and permits probing of the surface over different scales. This phenomenon is well known from lunar radar studies, which now span the wavelength range 8.6 mm to 25 m. Lunar diffuse scattering has been observed to become relatively much more important as the radar wavelength decreases; at 8.6 mm in backscatter, the lunar surface approaches the uniformly bright appearance of the optically observed full Moon with a relatively minor quasi-specular peak in the center of the disk.

D. Polarization Effects

Radar echoes which have interacted with a surface only once generally have a predictable polarization; signals propagating along similar paths to and from the target are modified in similar ways, and echo polarization can be determined from knowledge of the incident polarization and the scattering geometry. When multiple scattering occurs, many paths between transmitter and receiver are possible. Each scattering event changes polarization in a different way; the polarization of the sum of signals scattered two or more times is thus randomized. The degree to which an echo has deterministic vs random polarization is thus another clue to the physical state of the planet's surface.

E. Experimental Configurations

Figures 1b and 1f show that the average scattering angle θ_s will be equal to the incidence angle θ_i for gently undulating surfaces. If the transmitter and receiver are separated (a bistatic configuration), the quasi-specular component will be measured if the receiver is generally in the direction θ_s, and the diffuse component will be measured otherwise. When the transmitter and receiver are colocated (a monostatic configuration; $\theta = \theta_i = -\theta_s$ in Fig. 1), the radar measures backscatter. For the geometry in Fig. 1f, a monostatic radar would sense primarily diffuse scattering unless $\theta = \theta_i \sim 0°$. The range over which quasi-specular scattering can be observed (in either configuration) depends on the distribution of surface tilts, sometimes expressed as the root-mean-square (*rms*) surface tilt θ_r. Typically, the greater the *rms* surface tilt, the broader the quasi-specular lobe. Away from the main lobe, diffuse scattering dominates since we expect relatively few quasi-specular surface elements to have high tilts.

III. RADAR AS A REMOTE-SENSING PROBE

Radar is unique among remote sensing instruments in the degree of experimental control it provides. Observing parameters such as wavelength, amplitude, timing and polarization of illumination can be selected by the experimenter. Radar is also unique in sensing the dynamical electric field of the scattered wave rather than its time-averaged intensity, so that phase as well as amplitude of the echo can be measured. To the extent that the radar echo signal is coherent as compared to the naturally occurring radiation which is also present, a high degree of determinism can be built into the analysis. It is this determinism, plus the fact that the radar waves respond to the surface at a scale that is comparable to the radar wavelength, that distinguishes radar from other remote-sensing techniques.

TABLE I
Mars Radar Facilities and General Characteristics

Facility (Location)	Operation	Antenna Diameter (m)	Wavelength (cm)
Goldstone (Barstow, CA)	1963–present	64, 70	3.6, 12.6
Canberra (Australia)[a]	1977–78	64	12.6
Arecibo (Puerto Rico)	1965–present	305	12.6, 70
Haystack (Tyngsboro, MA)	1967–1973	37	3.8
Crimea (USSR)	1963, 1971, 1980	70	39
VLA (Socorro, NM)[a]	1988	26 at 25	3.6

[a] Bistatic only.

A. Facilities

Monostatic Mars radar observations have been reported by three U.S. facilities and one in the Soviet Union (Table I). Bistatic radar observations have been conducted over a large range of scattering geometries using Viking Orbiters and Earth stations of NASA's Deep Space Network (Goldstone and Canberra). A novel, Earth-based, bistatic configuration has been used by Muhleman et al. (1991), who transmitted from Goldstone, California, and received at the Very Large Array (VLA) near Socorro, NM. Table II lists selected papers reporting Mars radar observations.

B. Types of Measurements

A simple radar emits a burst of energy which, after being backscattered by a target, is received at some later time $\Delta t = 2d/c$, where d is target distance and c is the speed of light (Fig. 2). During the target interaction, the signal frequency will be Doppler shifted up or down by an amount $f_d = -2\mathrm{v}/\lambda$ where v is the radial component of target velocity (positive away from the radar). Time and frequency are thus the two principal domains for analysis of radar echoes. Targets with the size and rotation of Mars cannot be represented by single delay and Doppler values; each scattering element contributes to the echo at its own time delay and frequency offset. A signal incident on Mars will thus, upon scattering, be dispersed in time and frequency; additionally, there will, in general, be a conversion and randomization of its polarization. The objective in planetary radar studies is to measure the absolute power in the echo along with some estimate of the dispersion(s) and the polarization and to combine these with knowledge of the transmitted signal in order to infer the target properties responsible for the changes.

Radar Cross Section. The most basic property of the target itself is the efficiency with which it scatters energy, quantified by its radar cross section σ. Technically, σ is the cross-sectional area of a perfectly isotropic scatterer which, if positioned at the target location, would give the observed echo

TABLE II
Mars Radar Observations

Opposition Year	Wavelength (cm)	Subradar Latitude	Selected Source Publications
1963	12.6	13°N	Goldstein and Gillmore (1963)
1965	12.6	22°N	Goldstein (1965) Sagan et al. (1967)
	70.	22°N	Dyce (1965) Dyce et al. (1967)
1967	12.6	21°N	Carpenter (1967)
	3.8	19-24°N	Pettengill et al. (1969)
1969	3.8	3-12°N	Rogers et al. (1970)
	12.6	3-12°N	Goldstein et al. (1970)
1971	3.8	14-22°S	Pettengill et al. (1971, 1973)
	12.6	13-18°S	Downs et al. (1971, 1973)
1973	12.6	14-22°S	Downs et al. (1975)
	12.6, 70.	16-22°S	Simpson et al. (1977)
1975–6	12.6, 3.5	6°S-20°N	Downs et al. (1978)
	12.6	12°S-24°N	Simpson et al. (1978a,b)
1978	12.6	24°N	Simpson and Tyler (1980)
	13.1, 3.6	[a]	Simpson and Tyler (1981)
1978, 80	12.6, 13.1	[a]	Simpson et al. (1979, 1982)
1980	12.6	22°N	Harmon et al. (1982)
	39.	21°N	Kotel'nikov et al. (1983)
1982	12.6	24°N	Harmon and Ostro (1985)
1986	12.6	7°S	Thompson and Moore (1989*a*)
1988	3.5	[a]	Muhleman et al. (1991)

[a] Not applicable.

strength at the receiver. For backscatter from large objects, this is the same as the projected area of a perfectly conducting sphere. The value of σ integrated over an entire planet of radius R_p is often expressed as a dimensionless quantity, normalized to the projected disk area πR_p^2. An important incremental quantity, the specific cross section per unit surface area $\sigma_0(\theta)$, is a function of the angle of incidence of the radar wave to the unit surface. This function of θ is often called the scattering function or scattering law, and its amplitude and shape contain information about the dielectric constant and morphology (tilts and roughness) of the surface.

Delay Measurements. An observer who transmits a short pulse toward a planet will receive an echo which is dispersed in time by an amount equal to the two-way range depth $\tau_0 = 2R_p/c$ of the planet (Fig. 3). From an average of successive echoes, one can obtain a profile $\sigma(\tau)$ of radar cross section as a function of time delay. Absolute time delay Δt of the leading edge gives the

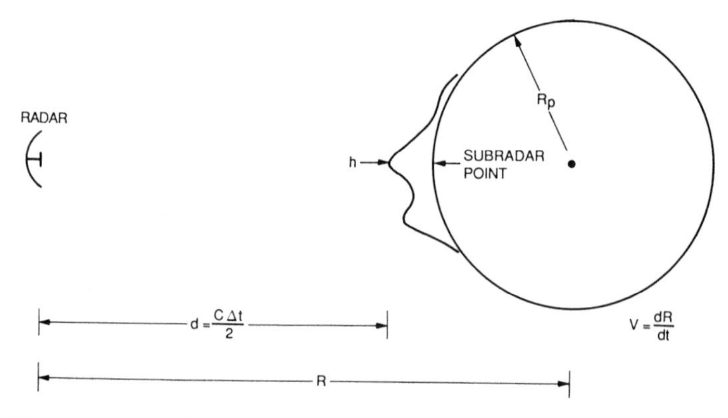

Fig. 2. Range to a target d and the time required for a radar echo Δt are proportional; c is the speed of light. The center-of-mass position and velocity (R and v, respectively) can be determined (e.g., as planetary ephemerides). Then the local topography at the subradar point is $h = R - d - R_\rho$. Duration (time dispersion) τ_0 of the echo from a planet is proportional to its radius (see also Fig. 3.)

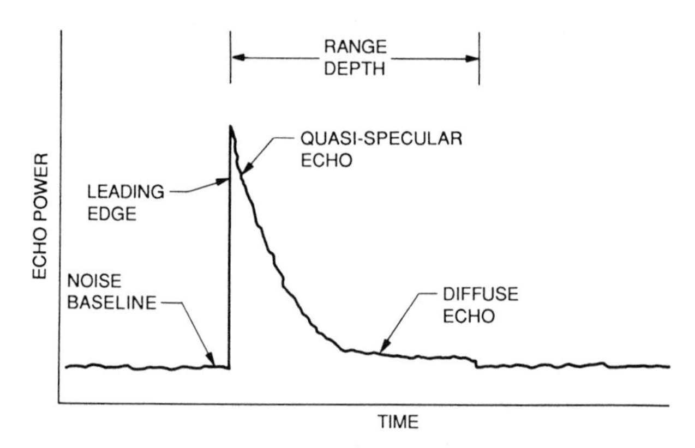

Fig. 3. Time profile $\sigma(\tau)$ of a typical planetary echo. The "leading edge" results from initial contact with the planet's closest point(s) (see Fig. 2); mirror-like reflections at and near the subradar point are strong because the mean planetary surface is perpendicular to the incoming wave. Progressively weaker returns come from more distant (and more highly tilted, but scarcer) scattering elements all the way to the limb. The "range depth", $\tau_0 = 2R_p/c$, is the length of the total echo. The earliest echoes come from the area at the center of the visible disk, while the latest echoes arise from areas at the limb. Quasi-specular scattering dominates near the leading edge, while diffuse mechanisms are most important at large angles (see text). Ability to remove the noise baseline is critical to accurate determination of diffuse echo strength and, thereby, the total echo power and target radar cross section. The dielectric constant (and, from that, the density) of the surface material can be estimated from strength of the quasi-specular echo under certain conditions.

distance d to the closest point on the planet's surface. Ordinarily this point will be very close to the subradar point, the intersection of the planet's surface with a line between the radar and the planet's center of mass (Fig. 2). If the radar to center-of-mass distance is known from ephemerides and is designated R, the local topography at the subradar point is $h = R - c\Delta t/2 - R_p$ (see, e.g., Fig. 4).

For scattering studies a simple delay dispersion measurement has the advantage that the delay profile $\sigma(\tau)$ and the scattering function $\sigma_0(\theta)$ (for an assumed homogeneous surface) are related by

$$\sigma(\tau) = \frac{2\pi R_P^2}{\tau_0}\sigma_0(\theta) \tag{1}$$

with $\theta = \arccos[1 - (\tau/\tau_0)]$.

Doppler Measurements. Alternatively, the observer may transmit a monochromatic wave and measure the Doppler spreading imposed by planet rotation. The shape of the echo Doppler spectrum provides information on wavelength-scale tilts and small-scale roughness. Loci of constant Doppler frequency correspond to straight lines on the projected disk which are parallel to the apparent spin axis (Fig. 5). Thus, for a homogeneous planet, the Doppler spectrum and scattering function are related through the integral transform

$$\sigma(f) = \frac{4R_P^2}{f_{LL}} \int_{\theta_L}^{\pi/2} \sigma_0(\theta) \left[\sin^2\theta - \left(\frac{2f}{f_{LL}}\right)^2\right]^{-1/2} \sin\theta \, d\theta \tag{2}$$

where the spectrum $\sigma(f)$ is the cross section per unit frequency, f_{LL} is the planet's limb-to-limb Doppler bandwidth, and $\theta_L = \arcsin(2f/f_{LL})$. In principle, $\sigma_0(\theta)$ can be obtained from $\sigma(f)$ with an inverse transform; in practice, the usual approach is to make model fits to $\sigma(f)$ using a functional form for $\sigma_0(\theta)$ and integrating the equation above. Despite this extra complication, Doppler measurements have been preferred over delay measurements for most radar scattering studies because (1) requirements for the radar hardware are simpler, and (2) higher average power can be obtained from many radar transmitters when operated in the continuous-wave (monochromatic) mode.

Delay-Doppler Measurements. If one transmits a sequence of pulses and maintains coherence from pulse to pulse, the echo can (in principle) be resolved in delay and Doppler frequency simultaneously and a two-dimensional map of planet radar properties can be constructed, subject only to a north-south ambiguity in delay-Doppler coordinates. For a rapidly rotating planet, however, the large Doppler bandwidth requires a high-pulse repetition rate for unaliased Doppler analysis. This is incompatible (except at

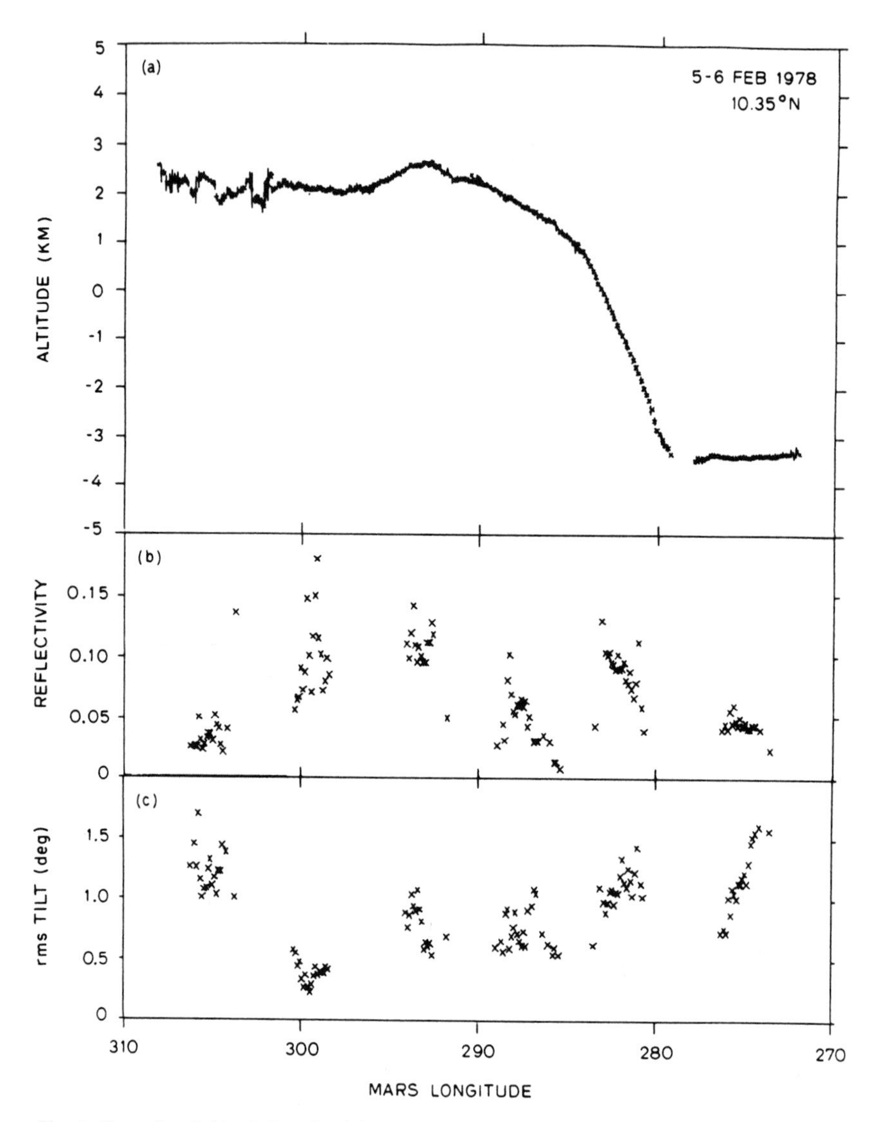

Fig. 4. Example of altitude h, reflectivity ρ_0, and *rms* surface tilt θ_r derived from Arecibo quasi-specular echo data obtained at $\lambda = 12.6$ cm wavelength. The subradar point traversed longitudes shown (abscissa) at latitude 10°35 N during this set of observations. Elevations are referenced to a sphere of radius $R_p = 3393.4$ km, but center-of-mass adjustments have not been made (expect ± 1 km absolute errors, ± 150 m relative errors along the ground track shown). Reflectivity $\rho_0 = 0.10$ corresponds to dielectric constant $\epsilon = 3.7$, a possible mixture of soil and rock. The *rms* surface tilt $\theta_r = 0°25$ near 300°W is the lowest obtained from planetary radar studies to date. Data gaps in the lower two panels result from cycling between transmit and receive in the radar system. Altitudes in several adjacent frequency strips (see Fig. 5) can be obtained simultaneously, allowing gaps to be filled in the upper panel from a single set of measurements.

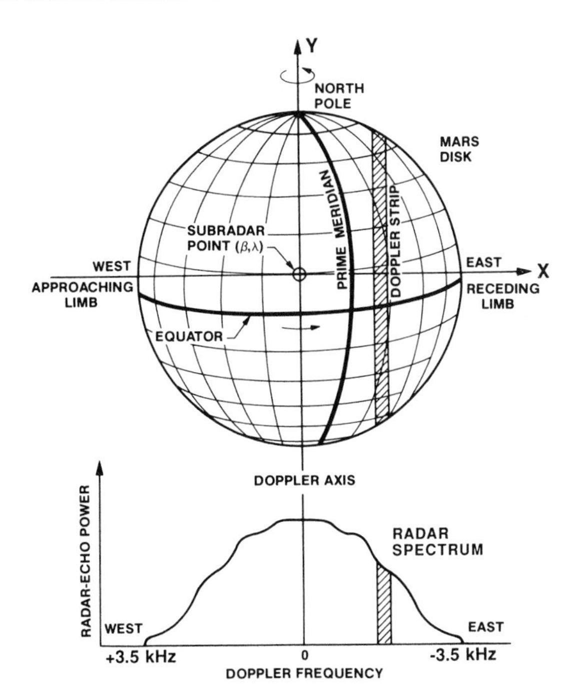

Fig. 5. Martian disk as viewed from the Earth showing mapping from surface points into the Doppler spectrum. Shaded strip on the planet is responsible for the shaded frequency range in the spectrum. Zero Doppler shift corresponds to Mars' rotation axis; frequency limits shown here are appropriate for $\lambda = 12.6$ cm. The area under the spectrum represents the planet's total radar cross section. Estimation and removal of the spectrum-noise baseline (not shown here; see Fig. 3) is essential for accurate cross-section measurement.

impractically long operating wavelengths) with the requirement that there be only one radar pulse on the planet at a time, to prevent ambiguity in range measurement. This is referred to as an overspread condition ($f_{LL}\tau_0 > 1$). One solution to both the ambiguity and overspreading problems for Mars currently being explored is the use of large antenna arrays for receiving that are capable of resolving the planet.

Delay-Doppler mapping has proved very useful on Mars. The quasi-specular echo component arises from scattering elements near the subradar point. This component may itself not be overspread in which case range-Doppler analysis may be applied to it. Surface properties of the quasi-specular surface (only) and the associated regional topography at the subradar point may be inferred.

Polarization. Echo polarization offers an additional source of information for scattering studies. Radar observations of Mars use circular rather than linear polarization (see Sec. IV) in order to avoid the confusing effects of Faraday rotation in the Earth's ionosphere and to ensure uniform polarization incident at all points on the target surface. Often reception is made only in the circular polarization which is opposite to the transmitted sense. This opposite circular (OC, or polarized) component is the polarization sense which results from a purely specular reflection. In recent years, an increasing number of observations have been made in which the same-sense circular (SC, or depolarized) component has also been measured. The Mars SC echo contains power associated with polarization distortion due to irregularities in the shape or morphology of the scatterer, multiple scattering or subsurface transmission effects. Most of the dual polarization analysis to date has involved separate estimation of the OC and SC Doppler spectra $\sigma_{OC}(f)$ and $\sigma_{SC}(f)$. Although much of the depolarized power associated with $\sigma_{SC}(f)$ may be a product of polarization randomization, the true random or unpolarized component can only be determined from computation of the cross spectrum of the coherently obtained OC and SC complex voltages. The above applies when a single circularly polarized signal is transmitted; if, in addition, one varies the polarization of the transmitted wave, then a complete description of the scattered field is possible (Hagfors 1967).

Mapping with Large Antenna Arrays. As was noted above, one solution to the overspreading problem for Mars is the use of large antenna arrays capable of resolving the planet. The 27 antennas of the National Radio Astronomical Observatory's Very Large Array (VLA) in New Mexico were instrumented in 1988 with very sensitive, dual-polarization receivers at $\lambda = 3.6$ cm. While signals are transmitted from another site, such as the 70-m NASA antenna at Goldstone (Table I), a narrow gaussian-like beam may be synthesized at the VLA by correlating outputs of all antenna pairs. The detailed shape of the beam depends on the declination of the planet, with the best beams being created when Mars has northern declination. The beam size can be roughly characterized by the diffraction formula $\phi = \lambda/B$ where B is the maximum antenna spacing. This gives a $0''.2$ beam for the largest VLA configuration, yielding a beam spot at the Mars sub-Earth point of about 60 km at opposition. For each independently formed beam, the measurements described above (radar cross section, polarization, etc.) may be obtained from the output signal with additional calculation.

C. Observing Constraints and Planet Coverage

Because of limited system sensitivities and the R^{-4} dependence of radar echo strength, most Earth-based Mars radar observations are made around the time of opposition, which occurs (on average) every 780 days. Mars' eccentric orbit produces a large (factor of 1.8) difference in Earth-Mars distance between most- and least-favorable oppositions. This translates to a 10:1 signal-strength difference from best to worst oppositions. The cycle time between favorable oppositions is 15 to 17 years (7–8 oppositions). Over this cycle, Mars' subradar-point latitude (at opposition) varies between 25°N and 25°S latitude; latitudes are near the southern limit of this range during the most favorable oppositions and near the northern limit during the least favorable.

The $\pm 25°$ equatorial band defines that portion of Mars that is accessible to those Earth-based Mars radar studies (altimetry and quasi-specular scattering) which use the strong echo from the subradar region. Because the subradar latitude varies little during any given opposition, coverage of the full latitude range can require the better part of a decade. On the other hand, Mars' relatively rapid rotation permits the observer to obtain full longitudinal coverage in a single opposition; approximately one third of Mars can be observed in a single night, and the 37-min difference in rotational periods of the Earth and Mars yields a 9°.3 shift in longitude every day.

In a forward-scattering bistatic configuration, location of the specular point is a function of relative positions of the spacecraft, Mars and Earth so that polar regions can also be probed. Echo strength varies as $R^{-2}r_s^{-2}$, where r_s is the Mars-to-spacecraft distance, making these results less sensitive to a single range. Forward scattering is oblique; angle of incidence/reflection varies with location of the specular point, increasing toward the limb as viewed from either antenna. At near-grazing angles, shadowing and diffraction become factors, but neither of these subjects has been studied extensively.

IV. SCATTERING MODELS

A. Specular and Quasi-Specular Scattering

A plane, conducting, specular surface will behave as a mirror; any wave incident will be reflected with complete fidelity. The reflected field strength will be undiminished by the interaction—only its direction will be changed, while its polarization will be unchanged if linear and reversed if circular. Perfect conductors are useful modeling tools; they obviate the need to consider subsurface structure that will interfere with the reflection, but they are not realistic.

Smooth Dielectric Interfaces (Fresnel Reflectivity). Most planetary surfaces are lossy dielectrics. If smooth and homogeneous, they can reflect specularly, albeit with lowered intensity; the remainder of the wave propagates into the surface where it is eventually absorbed. The ratio of reflected to incident field strength (the Fresnel reflection coefficient) depends on the dielectric constant, the angle of incidence and the polarization of the incident wave.

We denote the polarization state of the incident field by the vector $(E_H, E_V)_i$, where the subscripts H and V refer to the horizontal and vertical components of the electric field (these correspond to the components "perpendicular" and "parallel," respectively, used in certain other disciplines). For a purely specular interaction, the reflected field is given by

$$\begin{bmatrix} E_H \\ E_V \end{bmatrix}_r = \begin{bmatrix} R_H & 0 \\ 0 & R_V \end{bmatrix} \begin{bmatrix} E_H \\ E_V \end{bmatrix}_i \tag{3}$$

where R_H and R_V are the Fresnel reflection coefficients

$$R_H = \frac{\cos\theta_i - \sqrt{\varepsilon - \sin^2\theta_i}}{\cos\theta_i + \sqrt{\varepsilon - \sin^2\theta_i}} \tag{4}$$

$$R_V = \frac{\varepsilon\cos\theta_i - \sqrt{\varepsilon - \sin^2\theta_i}}{\varepsilon\cos\theta_i + \sqrt{\varepsilon - \sin^2\theta_i}} \tag{5}$$

ε is the material dielectric constant, and $\theta_i = \theta_s$ is the incidence/reflection angle on the surface (Fig. 6). Strictly speaking, ε should be considered a complex quantity; the imaginary part, which is responsible for loss or absorption, is small for most materials encountered in planetary studies, so the error introduced by equating ε to its real part in these expressions is negligible. Note that for backscatter these expressions must be evaluated for $\theta_i = \theta_s \equiv 0°$.

If the incident and reflected fields are, instead, given in terms of right (E_R) and left (E_L) circularly polarized components, then we have

$$\begin{bmatrix} E_R \\ E_L \end{bmatrix}_r = \begin{bmatrix} R_{sc} & R_{oc} \\ R_{oc} & R_{sc} \end{bmatrix} \begin{bmatrix} E_R \\ E_L \end{bmatrix}_i \tag{6}$$

where $R_{sc} = (R_V + R_H)/2$ and $R_{oc} = (R_V - R_H)/2$. Thus, specular scattering processes do not distinguish between right- and left-circular polarization; for a given incident wave, the results can be described in terms of scattering in the same (SC) and in the opposite (OC) senses of circular.

For the special case of normal incidence (e.g., backscatter at the subradar point), we have

$$\begin{aligned} R_{oc} &= R_V = -R_H = \frac{\sqrt{\varepsilon} - 1}{\sqrt{\varepsilon} + 1} \\ R_{sc} &= 0. \end{aligned} \tag{7}$$

That is, all of the reflection appears in the opposite circular sense. Another special case occurs at the Brewster angle

$$\theta_B = \arctan(\sqrt{\varepsilon}) \tag{8}$$

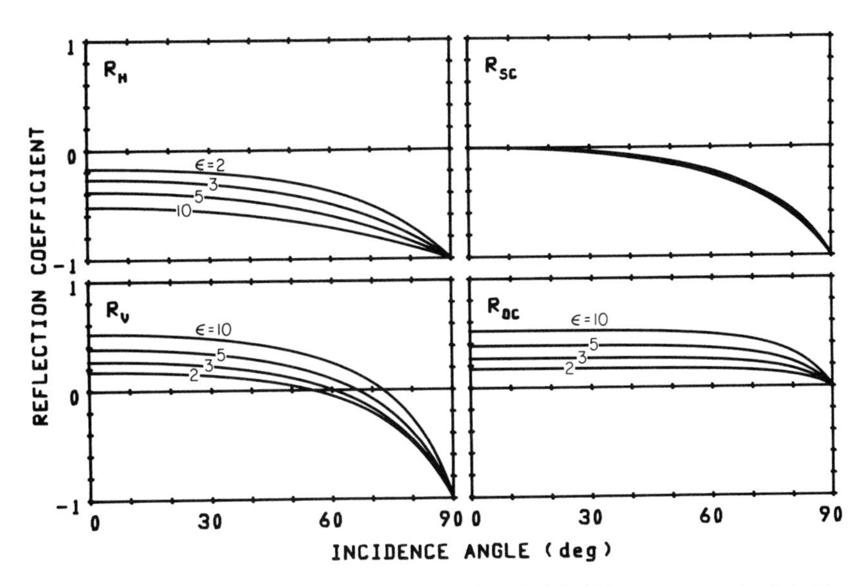

Fig. 6. Reflection coefficients for horizontal (R_H) and vertical (R_V) linear components of electric field and for opposite-sense (R_{OC}) and same-sense (R_{SC}) components if field is defined in circular terms. Reflectivity ρ_0 is the square of any of these reflection coefficients. The curves are plotted for $\epsilon = 2, 3, 5$, and 10, with ϵ increasing in the direction away from the horizontal center line in each panel at incidence angle $\theta_i = 0°$.

where

$$R_{sc} = -R_{oc} = R_H/2 \qquad (9)$$
$$R_V = 0.$$

In many cases, it is sufficient to know the *intensity* of the reflected wave. For normal incidence this is given by the reflectivity ρ_0, which is simply related to the dielectric constant by

$$\rho_0 = \left[\frac{1-\sqrt{\varepsilon}}{1+\sqrt{\varepsilon}}\right]^2 \qquad (10)$$

which can be solved to obtain

$$\varepsilon = \left[\frac{1+\sqrt{\rho_0}}{1-\sqrt{\rho_0}}\right]^2 . \qquad (11)$$

For a smooth, lossy, dielectric sphere with radius $a >> \lambda$ the radar cross section is given by $\sigma = \rho_0 \pi a^2$ or, if σ is expressed as a normalized cross section, by $\sigma = \rho_0$. Thus, the radar cross section for specular scatter can be used as an estimate of the reflectivity ρ_0 and, through Eq. (11) above, the dielectric constant ε.

A solid-rock surface will have $\varepsilon \sim 5$ to 9, which corresponds to $\rho_0 \sim 0.15$ to 0.25; typical soils are in the range $\varepsilon \sim 2$ to 3, which translates to $\rho_0 \sim 0.03$ to 0.07 (Campbell and Ulrichs 1969). In general, lower material density means lower reflectivity for a given composition. A useful relationship is the "Rayleigh mixing formula" (Campbell and Ulrichs 1969; Evans 1969) which relates the dielectric constant ε_p of a rock powder at density d_p to its solid-rock values ε_s and d_s by

$$\frac{(\varepsilon_p - 1)}{d_p\,(\varepsilon_p + 2)} = \frac{(\varepsilon_s - 1)}{d_s\,(\varepsilon_s + 2)}\,. \tag{12}$$

Other relationships have been proposed on both theoretical and empirical grounds; many of these have been discussed by Garvin et al. (1985).

Subsurface Transmission and Reflection. That portion of the incident wave which is not reflected will be transmitted into the surface where it will usually be absorbed. The penetration depth, the distance over which the *power* decreases by a factor e^{-1} is

$$L_P = \frac{\lambda}{2\pi\sqrt{\varepsilon}\,\tan\delta} \tag{13}$$

where $\tan\delta$, the loss tangent, is the ratio of imaginary to real parts of ε (ratio assumed $<<1$). Loss tangents compiled by Campbell and Ulrichs (1969) for terrestrial samples give typical penetration lengths $L_p \sim 2$ to 3λ for solid rock and $L_p \sim 10\lambda$ for soils (materials in which L_p is at least several times λ are sometimes called lossy dielectrics). Lunar samples, which are highly dessicated and may be more representative of Mars surface material, show penetration depths about ten times longer (Gold et al. 1970; see also chapter 21 for additional discussion). If the surface material is homogeneous over at least a few times L_p, then the wave transmitted into the surface will be dissipated as heat. If, however, there is a second interface or other inhomogeneity, then a portion of this subsurface wave can be scattered back to the observer.

Quasi-Specular Roughness Models. Most real surfaces are not smooth. To the extent that *elements* of the surface are smooth, however, on dimensions larger than a few wavelengths, the Fresnel concepts above can be applied to the quasi-specular component of the echo. Quasi-specular scattering itself has several forms. All assume that the surface can be approximated by locally flat elements on which either *geometrical* or *physical optics* may be applied. In the former only geometry matters. Incoming waves may be considered as parallel rays; each ray is reflected specularly at the surface. Modeling is considerably simplified in geometrical optics; only the distribution of surface tilt is required to specify the scattering pattern completely.

In physical optics, electric and magnetic boundary conditions must be satisfied on the surface. With certain simplifying assumptions (such as that of a gaussian height distribution and a well-behaved surface autocorrelation function), the physical-optics solution reduces to geometrical optics; that is, the scattering pattern is completely determined by the surface tilt distribution.

One special case of physical optics deserves separate attention because of its successful application to planetary radar data analysis. Hagfors (1964) proposed a scattering law developed with physical optics but using an exponential surface autocorrelation function. Strictly speaking, an exponential is not well behaved because its first derivative is discontinuous at the origin (Barrick 1970). Hagfors (1968) and others since, have argued that the autocorrelation function's behavior at very small arguments is determined by the finest-scale surface structure. Structure with dimensions below a few tenths of a wavelength is sensed only very weakly by the probing radio wave (Sec. II.C). Thus the *effective* autocorrelation function is actually well behaved at the origin and there is no inconsistency.

The form of Hagfors' filtering function and the role (or lack thereof) of the small ($a<\lambda$) structure have been the subject of considerable debate over the years, with little satisfactory resolution. It is worth noting, however, that Chan and Fung (1978) have obtained a similar result to Hagfors' using numerical techniques. In the end, the function proposed by Hagfors (1964)

$$\sigma_0(\theta) = \frac{\rho_0 C}{2}\,[\cos^4\theta + C\sin^2\theta]^{-3/2} \tag{14}$$

has become a widely adopted analytical model because of its excellent agreement with data. The quantity $\theta_r = C^{-1/2}$ has, with some justification (Lincoln Laboratories 1967), been interpreted as the *rms* surface tilt (in radians) responsible for at least the shape of the quasi-specular echo component.

A planet with no diffuse scattering will have a normalized radar cross section $\sigma = g\rho_0$, where g is a backscatter gain given approximately by $g = 1 + \theta_r^2$ (Hagfors 1964). Hence, for a gently undulating surface, $g \approx 1$ and the measured normalized cross section will be a good approximation to the reflectivity ρ_0.

B. Diffuse Scattering and Depolarization

The quasi-specular model seems to provide a good description for an important part of the typical Mars radar echo; areas near the subradar point

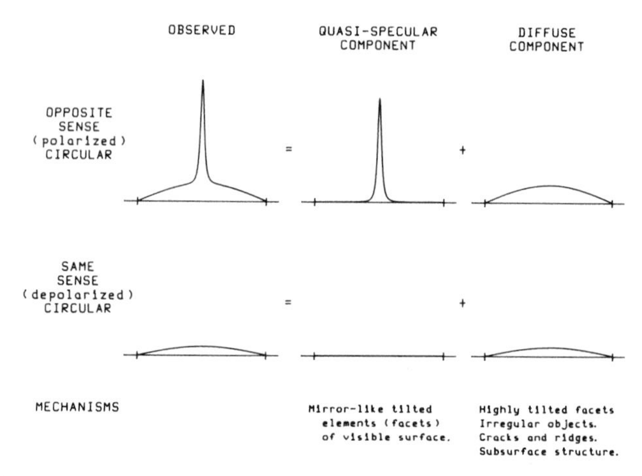

Fig. 7. Backscatter echo from a circularly polarized transmission has both opposite (OC) and same (SC) sense components. The OC echo includes contributions from quasi-specular and diffuse processes. The SC echo will have quasi-specular contributions only if there is multiple scatter from large surface elements; these are not expected and no current model includes them. Thus the SC echo can be attributed entirely to diffuse mechanisms.

yield high OC (polarized) echoes in both time and Doppler observations. When subjected to a scattering-law fit, these echoes give θ_r of a few degrees or less. A typical Mars OC spectrum, however, also has a substantial fraction of echo power in the form of broad, low wings which extend all the way to the limb frequencies (Fig. 7). Also, Mars' SC (depolarized) Doppler spectrum is found to contain a nonnegligible amount of power and has a shape which more or less mimics that of the wings of the OC spectrum. This broadband, partially depolarized component of the echo is commonly referred to as the diffuse component.

Diffuse Roughness Models. The visual impression given by many of the lunar and Mars Landers (Surveyor, Apollo, Viking) is that of an undulating surface strewn with rocks; details differ, of course, particularly with respect to rock populations (see chapter 21). Several studies have been made in which diffuse cross sections for the Moon and Mars are compared with "predicted" cross sections based on rock abundances derived from Lander photos or thermal-inertia data. Most of these have assumed that only single scattering is important and have used rock backscatter-efficiency (radar cross section normalized by πa^2) estimates derived directly or indirectly from Mie theory for scatterers of radius a. A useful rule of thumb derived from Mie calculations is that a distribution of spherical rocks with radii in the range $\lambda/3 < a < 3\lambda$ will have an effective backscatter efficiency of roughly unity; in other words, an estimate of the fractional surface coverage by wavelength-scale rocks provides a rough estimate of the normalized diffuse radar cross section expected from that surface (Thompson et al. 1970; Harmon and Ostro 1985).

This suggests that a relatively modest rock coverage can give rise to a substantial diffuse echo.

While the simple, rock-based Mie approach provides a useful starting point for modeling diffuse scatter, complications and limitations soon arise. The most important of these are: (1) the possibility that there is a substantial population of buried rocks not seen by the Landers or that the Lander sites are not representative; (2) the possibility that the assumed backscatter efficiencies are biased by the assumption of rock sphericity or by failure to account for rock-rock and/or rock-regolith interactions; (3) the failure of spherical-rock models to predict single scattering depolarization; and (4) the inability of such simple models to reproduce the angular scattering law of the diffuse echo. Points (1) and (2) have been discussed at length by Thompson et al. (1970) and Harmon and Ostro (1985).

Depolarization. An important distinguishing characteristic of Mars' diffuse component is that it is partially depolarized ($\sigma_{sc} \neq 0$). Two important depolarization mechanisms are single scattering from irregular wavelength-scale objects and multiple scattering. For single scattering, the most extreme model is one in which the structure is described by an ensemble of randomly oriented dipoles. In this case, the polarization of the wave will be randomized such that the echo power is equally distributed between the two circulars ($\mu_c \equiv \sigma_{sc}/\sigma_{oc} = 1$) for a circularly polarized incident wave (Hagfors 1967). The depolarization will be somewhat weaker for the more realistic case of scattering by irregular (nonspherical) rocks. Experiments show that irregular, wavelength-size rocks give $\mu_c \sim 0.3$ to 0.5 upon single scattering (Cuzzi and Pollack 1978), a result which agrees with average μ_c values for diffuse echoes from the Moon and terrestrial planets (Mercury, Venus, Mars). On the other hand, a few locations on Venus and Mars show essentially complete depolarization ($\mu_c \cong 1$). Perhaps the most likely model for such regions is one that has multiple scattering by surfaces or regoliths which are rock saturated or otherwise extremely chaotic. Elachi et al. (1980) proposed multiple scattering to explain the strong depolarization measured from airborne radar studies of extremely rough aa-type terrestrial lava flows. Multiple scattering can also occur if one has a sufficiently dense population of rocks embedded in regolith. This mechanism has been studied by Pollack and Whitehill (1972) to model possible multiple scattering of radar waves in the ejecta of fresh lunar craters.

Diffuse Scattering Laws. Model fitting to the shape of the diffuse echo spectrum is sometimes done to aid in partitioning the radar cross section between quasi-specular and diffuse contributions or to draw tentative conclusions about the scattering mechanism itself. The model which is almost universally used is the so-called cosine law, which has the form

$$\sigma_0(\theta) = A \cos^n\theta \tag{15}$$

where n is a shape parameter and A is an amplitude factor which is related to the total (disk-integrated) diffuse cross section σ^D by $\sigma^D = 2A/(n+1)$. Mapping of this function into the frequency domain (Doppler spreading of the echo) often shows reasonable agreement with the shape of the SC spectrum or with the shape of the diffuse wings of the OC spectrum with fitted n values in the range $1<n<2$. Unlike quasi-specular scattering laws such as the Hagfors' law, the cosine law has no *a priori* physical basis. It has found favor simply because it describes two real classes of rough surfaces often encountered at visible wavelengths. The case $n=1$ describes a surface such as the full Moon which shows a uniformly bright disk in backscatter. The case $n=2$ gives a limb-darkened disk. The $n=1$ law may be typical of a surface that is rough on large ($>>\lambda$) scales (where shadowing is important), while the $n=2$ law may be associated with diffractive scatter from wavelength-scale structure (Evans 1962,1969). Unfortunately, this otherwise useful distinction has not been thoroughly tested either theoretically or experimentally.

V. MARS RADAR OBSERVATIONS

Early experiments yielded estimates of radar cross section and regional topography from OC echos. An average radar cross section was obtained ($\sigma \sim 0.07$) but there were repeatable variations ($0.03 < \sigma < 0.13$) which correlated with subradar point longitude (Dyce et al. 1967). Analysis of echo shape also suggested considerable variation in *rms* surface tilts (see, e.g., Goldstein 1965). Topography was also found to vary; within the 19° to 24°N band alone, it ranged over 15 km in elevation (Pettengill et al. 1969). Mars had thus been established firmly by 1970 as a diverse radar target. Subsequent experiments have refined and expanded on those early conclusions.

A typical experiment consists of transmitting a right circularly polarized wave toward Mars and acquiring the echo with receivers connected to antenna feeds sensing both the same circular (SC) and opposite circular (OC) polarizations. The quasi-specular contribution to the echo is strong, polarized (OC) and concentrated in time and frequency to ranges corresponding to the surface region around the subradar point. From the quasi-specular echo one can determine regional topography and estimate wavelength-scale and larger surface tilts and the dielectric constant of the surface material. The diffuse echo appears in both polarizations and arises from scattering elements that are widely distributed across Mars' projected disk; from the diffuse echo, we may qualitatively estimate small-scale roughness. We illustrate the present understanding of Mars' surface properties determined from radar data by focusing on a small number of relatively recent experiments.

A. Observations of Doppler Spectra

SC and OC spectra from Mars are shown in Figs. 8 and 9. It is apparent that these echoes are neither as well behaved nor as uniform as the model

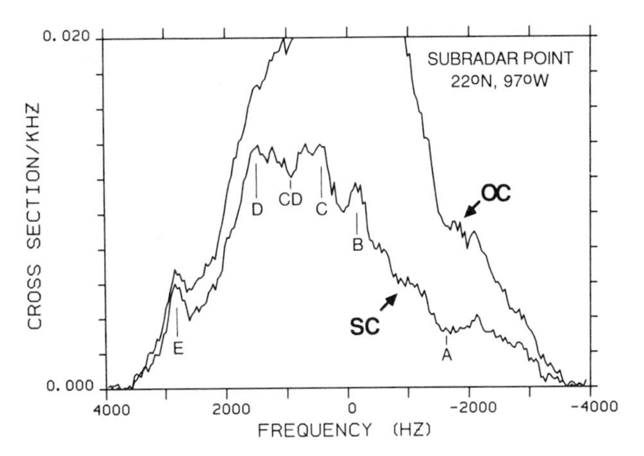

Fig. 8. Polarized (OC) and depolarized (SC) echoes from Arecibo Observatory at $\lambda = 12.6$ cm, illustrating structure in the SC echo within a generally enhanced Tharsis response. The spectral features, "A," "B," etc. were identified by Harmon et al. (1982); estimated source locations were derived by tracking their movement through spectra as Mars rotated (Harmon and Ostro, 1985).

spectra (Fig. 7). The broad maximum in the SC spectrum of Fig. 8 may be associated generally with Tharsis; the lettered "features" represent local areas of especially strong SC scatter. The OC spectrum in Fig. 9, in addition to the considerable quasi-specular spike in the center, has both a substantial diffuse component and a feature F. The latter is unique in being the only off-center OC feature for which no corresponding SC feature has been identified in the Harmon et al. (1982) data set.

Most spectra do not show the wealth of structure that these have, and it is possible to fit scattering functions for shape and amplitude. Harmon et al. (1982) fitted a composite scattering law

$$\sigma_{oc}(\theta) = \frac{\rho_0 C}{2} [\cos^4\theta + C\sin^2\theta]^{-3/2} + A_{oc}\cos^{n_{oc}}\theta \tag{16}$$

to OC spectra to obtain estimates of *rms* surface tilt ($\theta_r = C^{-1/2}$), reflectivity ρ_0, and the exponent n_{oc} and coefficient A_{oc} of the diffuse term. The relative magnitudes of the two terms in this expression can be used to infer the approximate areal coverage of the two types of scatterers; the fitted ρ_o represents a lower limit on the reflectivity in the absence of diffuse scatterers. From the fit results, Harmon et al. (1982) could then estimate the quasi-specular σ^Q and diffuse $\sigma_{oc}{}^D$ contributions to the total σ_{oc}

$$\sigma_{oc} = \sigma^Q + \sigma_{oc}{}^D . \tag{17}$$

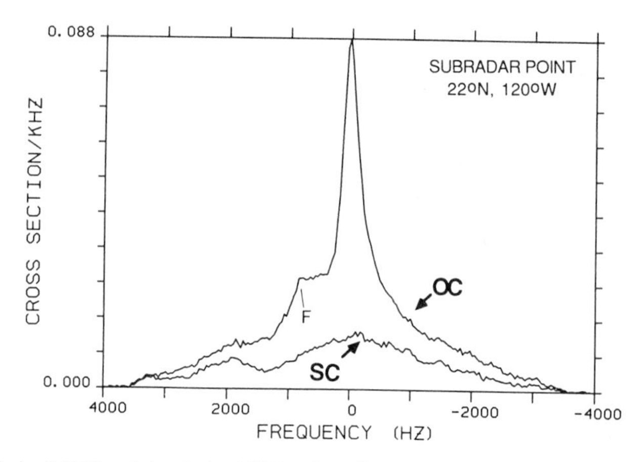

Fig. 9. Polarized (OC) and depolarized (SC) echoes from Arecibo Observatory at $\lambda = 12.6$ cm. Feature *F* is unique in being the only off-center purely OC feature identified in the data sets of Harmon et al. (1982) and Harmon and Ostro (1985).

They also fitted

$$\sigma_{sc}(\theta) = A_{sc}\cos^{n_{sc}}\theta \tag{18}$$

to SC spectra. The total SC cross section is diffuse so one can obtain an estimate of the total diffuse cross section from

$$\sigma^D = \sigma_{oc}{}^D + \sigma_{sc} . \tag{19}$$

These values are listed in Table III along with additional values reported by Harmon and Ostro (1985) for subsequent observations.

The striking point about the numbers in Table III is the large role played by diffuse scattering σ^D in determining the total radar cross section. The lowest ratio, $\sigma^D/(\sigma^D + \sigma^Q)$, in this list, admittedly not a random sample, is 41% for the (23°1N, 39°8 W) data. In the most extreme case (22°N, 124°7 W) 85% of the echo is diffuse. If we accept the 24°N values as being representative of the planet as a whole, the average echo is 45% diffuse. Even within the OC component, 38% of the echo is diffuse. These figures mean that a significant fraction of the scattered power from Mars' surface comes from wavelength and smaller (cm) roughness. One must conclude that a large fraction of the surface has irregular, wavelength-scale structure (either intrinsically or because it is littered with debris having this effective size), or there is a significant contribution from subsurface scattering, or both. From the Harmon et al. (1982) results, it is clear that within Tharsis this small-scale structure is particularly common.

TABLE III
Results of Fits to Composite Scattering Law

Latitude (°N)	Longitude (°W)	θ_r	ρ_o	n_{oc}	n_{sc}	σ^Q	$\sigma_{oc}{}^D$	σ^D	σ^D/σ^Q
23.1	39.8	5.7	0.13	1.5	0.7	0.10	0.049	0.070	0.69
25.2	55.7	4.0	0.11	2.2	1.0	0.094	0.066	0.092	0.98
22	78.8	5.3	0.064	2.1	1.5	0.051	0.057	0.089	1.7
22	85.3	4.9	0.060	2.6	1.8	0.049	0.068	0.106	2.2
22	88.3	4.5	0.037	3.0	1.9	0.031	0.047	0.074	2.4
22	91.3	7.2	0.076	2.4	2.3	0.056	0.074	0.122	2.2
22	93.9	7.5	0.036	2.4	2.4	0.027	0.045	0.072	2.7
22	97.4	3.8	0.048	3.5	3.0	0.041	0.081	0.128	3.1
22	100.1	4.0	0.037	3.0	3.2	0.031	0.071	0.113	3.6
22	101.2	3.7	0.028	3.1	3.4	0.025	0.047	0.076	3.1
22	103.4	2.2	0.031	4.2	3.8	0.029	0.065	0.098	3.4
22	106.3	3.9	0.047	2.8	4.0	0.040	0.074	0.119	3.0
22	107.5	3.8	0.041	3.0	4.3	0.035	0.065	0.105	3.0
22	112.5	1.8	0.030	4.4	4.7	0.028	0.071	0.107	3.8
22	113.9	1.9	0.038	4.1	4.6	0.035	0.088	0.135	3.8
22	120.2	2.6	0.035	3.9	4.0	0.032	0.080	0.123	3.9
22	124.7	2.0	0.021	5.0	3.3	0.020	0.071	0.107	5.5
24.1	244.9	1.9	0.10	3.1	1.1	0.095	0.059	0.080	0.84
24.1	250.7	1.9	0.12	3.3	1.0	0.11	0.067	0.087	0.78
24.1	256.4	2.9	0.10	2.3	1.0	0.087	0.050	0.068	0.78
24.1	261.8	2.5	0.087	3.0	0.9	0.079	0.050	0.066	0.83
23.7	330.2	3.2	0.061	2.4	1.1	0.054	0.039	0.049	0.90

[a] Adapted from Harmon et al. (1982) and Harmon and Ostro (1985). Spectra acquired in dual circular polarizations at Arecibo Observatory (λ = 12.6 cm). Coordinates are for subradar point at center of observing interval. Other columns give model fits of *rms* surface tilt in degrees θ_r, reflectivity ρ_o, exponents for power-law diffuse component in OC and SC polarizations (n_{oc} and n_{sc}, respectively), normalized quasi-specular radar cross section σ^Q, normalized diffuse OC cross section $\sigma_{oc}{}^D$, normalized total diffuse cross section σ^D, and the ratio of total diffuse to total quasi-specular cross sections.

The degree to which the diffusely scattering material actually reduces the quasi-specular scattering is difficult to estimate, since several assumptions are required before a quantitative estimate is possible. It is probably fair, based on simple models such as those used by Evans and Pettengill (1963) for lunar data, to assume that quasi-specular estimates of *rms* tilt and reflectivity are not unduly affected when diffuse scatter is 50% of the total but that considerable caution should be exercised when the percentage climbs as high as 85%. Pettengill et al. (1988) have dealt with a related question using Pioneer Venus data by using high-angle (diffuse) measurements to adjust quasi-specular values.

Harmon et al. (1982) mapped the diffuse features identified in Figs. 8 and 9 onto the surface by tracking their motion through a series of spectra. If one assumes that the material responsible for features scatters isotropically and without much dependence on angle, the spectral signature may be used to estimate surface coordinates. There is good reason to believe that feature *D* (Fig. 8) arises from Arsia Mons. Feature *F* (Fig. 9) appears to be within or just south of Acheron Fossae (Harmon and Ostro 1985) and thus north of Olympus Mons. This feature is distinctive because it appears in OC but not in SC, implying presence of high-angle tilted facets scattering specularly rather than irregular, wavelength-scale structure.

Thompson and Moore (1989*a*) collected all the depolarized spectra published to date, compiling them into a global set of spectra that give a complete description of the drift of features as Mars rotates. They have also compiled a map, based on geologic, infrared and radar considerations (including some unpublished, recent estimates of total radar cross section from Goldstone experiments), showing flows and other surface features which might be expected to depolarize. They then constructed synthetic spectra, after assigning scattering "efficiencies" to the various areas, which match the data quite well. Although their solution is far from unique, they have succeeded in demonstrating that diffuse-scattering mechanisms (as we understand them and as they may be related to photogeologic units) can produce the results observed. Of particular note here are the high cross sections assigned to Arsia Mons (observed) and the other giant volcanoes (which do not stand out as much on an individual basis).

It is worth noting that estimates of radar cross section comparable to those shown in Table III were reported from $\lambda = 3.8$ cm data by Pettengill et al. (1969) at rather similar latitudes and longitudes. It is particularly relevant that they reported whole-disk cross sections as well as cross sections from spectra peaks and from ranging measurements (both of which are much more responsive to quasi-specular processes). They concluded that ". . . an increase in the peak narrowband or short-pulse cross section is *not* obtained at the expense of contributions from outlying regions of the (echo or delay) spectrum and vice versa. The primary source of the variations in cross section is thus located within an area extending no more than a few degrees of surface arc from the subradar point." It was thus apparent in 1969 that the quasi-specular component, $\sigma^Q \approx 0.05$ and highly variable, rode on a rather stable base of diffuse power $\sigma^D \approx 0.05$ to 0.10. The data of Harmon et al. (1982) and Harmon and Ostro (1985) pin down the numbers and show that the base also varies but do not contradict the basic conclusion of an important "pedestal" of diffuse cross-sectional energy.

B. Range-Doppler Observation of Subradar-Point Characteristics

A large data set was collected, reduced and distributed from the Goldstone Mars observations of 1971 and 1973 (Downs et al. 1975). These, collectively, represent a circling of the planet more than seven full times at latitudes 13° to 22°S. The data contain, for each measured point on Mars, values for topography, regional slope, *rms* surface tilt and reflectivity/dielectric constant derived from observations in the OC polarization. Additional radar data have been collected, processed and distributed for the Mars-opposition years after 1973, but they have been reduced only to values of topography and *rms* tilt; the reflectivity calculations, which require careful integration with calibration data, have not yet been carried out. The 1971–73 data were obtained in the range-Doppler mode from a 5° radius "front cap" (the circular region surrounding the subradar point) at $\lambda = 12.6$ cm. Because of overspreading, echo contributions in time and frequency outside the 5° limit are aliased (spectrally folded) into the 5° window, adding to the noise and diffuse pedestal already there and upon which the quasi-specular signal is superimposed. In normal processing the pedestal is interpreted to be a combination of system and sky thermal noise and is discarded (see Fig. 3 caption). Depending on the surface texture in the subradar region, the recent results of Harmon and co-workers suggest that much of the discarded signal may actually consist of the aliased diffuse echo from other regions of the planet. For smooth surfaces, this aliased signal would not mimic the time-frequency dependence of the quasi-specular echo (as per the conclusion of Pettengill et al. [1969] above) and would have only a minor effect on the reliability of the results. For rough surfaces, the possibility of errors in the calculated reflectivities becomes much more significant.

The Downs et al. (1973) estimate of average reflectivity $\rho_0 = 0.064 \pm 0.012$ applies only to the quasi-specular echo and only to that part captured by the 5° range-Doppler processing window. For smooth surfaces ($\theta_r \leq 3°$) their values of ρ_0 are probably reasonable. For rougher quasi-specular surfaces, a significant part of the echo falls outside the processing window and is either lost or folded back as part of the noise pedestal. The average ρ_0 quoted by Downs et al. (1973) and updated to $\rho_0 \approx 0.082$ for the "smoother regions" by Downs et al. (1975) must be considered lower bounds when inferring properties of the surface material, such as its density. It should be noted that Downs et al. (1973) cautioned readers on exactly this point.

Because the 5° processing window clips the quasi-specular echo width for rougher surfaces, the estimates of θ_r from Downs et al. (1973,1975) must also be considered lower bounds. The magnitude of the error here is considerably less than for ρ_0 and one should only be concerned when estimates of θ_r exceed 3°. Values of θ_r shown in Table III are not dramatically different from those reported by Downs and co-workers, so it appears that the *shape* of the quasi-specular echo is not affected greatly by the presence of diffuse scattering.

Simpson et al. (1977) examined the Downs et al. (1973,1975) data as well as data acquired at Arecibo Observatory and Haystack Observatory during the same time periods. The three data sets were obtained at 12.6, 70 and 3.8 cm wavelengths, respectively, but there appeared to be very little (if any) detectable wavelength dependence in the scattering, in contrast to results over similar wavelength ranges in lunar radar data (see Evans and Hagfors 1968). In particular, the very weak echoes reported near and west of Tharsis by Downs et al. (1973) were apparent in each data set. This observation plus the high percentage of diffuse scattering and low thermal inertia suggest an inhomogenous, low-density surface material.

Downs et al. (1975) searched 18 areas on Mars for SC (depolarized) echo. They used the same range-Doppler mode, however, and reported only one depolarized signal, an SC echo $3.3 \pm 2.1\%$ times the strength of OC in a 2° region centered on (17°S, 72°W). Given our present understanding from the Arecibo observations of Harmon and co-workers, it seems likely that Downs et al. (1975) should have detected SC at each location. Disguised as noise, the multiply aliased echo probably was discarded each time.

C. Viking Bistatic Observations

Simpson et al. (1979) reported a Brewster angle measurement in the Hellas basin using Viking Orbiter 2 at $\lambda = 13.1$ cm in a bistatic mode; this configuration bypasses some of the difficulties encountered in conventional range-Doppler analysis. The value, $\theta_B = 60^\circ\!.6$, was determined by finding the conditions under which equal circular powers were received when a right-circular wave was transmitted (see Sec. IV). This Brewster angle corresponds to $\epsilon = 3.1$. Since the measurement depends on the *ratio* of the two powers rather than the absolute value of either, it is not subject to the same errors in estimating the diffuse component. It applies, however, only to the surface elements contributing to the quasi-specular echo.

Where bistatic ground tracks crossed tracks from Goldstone and Arecibo observations near the equator, good agreement was found between values of θ_r obtained in the two configurations. Near the spot where Downs et al. (1975) reported their single detection of *de*polarized echo, Simpson et al. (1984) reported observation of *un*polarized echo power. In oblique forward-scattering geometries, both OC and SC are predicted from quasi-specular scatter, so depolarization is not easily obtained; cross correlation of the OC and SC signals permits measurement of the coherence in the echo and, hence, its random (unpolarized) part.

D. VLA Bistatic Observations

Muhleman et al. (1991) reported image data using VLA beam resolution to discriminate among regions on Mars' surface. Data were taken in both OC and SC polarizations for circularly polarized signals transmitted from Goldstone. OC images were dominated by the quasi-specular, nearly phase coherent peak that always appears in the region around the subradar point. Since this specular spike was only slightly resolved by any pair of antennas in the array, it could be used for both phase-closure and amplitude-closure calibrations for the phase and amplitude of each antenna for each record. The SC images, on the other hand, yield much better maps, revealing the geologic structure of the surface and subsurface of Mars (Fig. 10).

All SC images show a background of radar brightness to the limbs which is approximately proportional to the cosine of the radar incidence angle. Except for a small feature at (41°.0 ± 0°.6 S, 93° ± 1°W) and another at the residual south polar ice cap, all of the radar anomalous structures are confined to a rectangular region bounded by 10°S, 30°N, 90°W and 190°W.

The radar features in the rectangular region can be divided into two classes: (1) those associated with distinct geologic features, such as the cald-

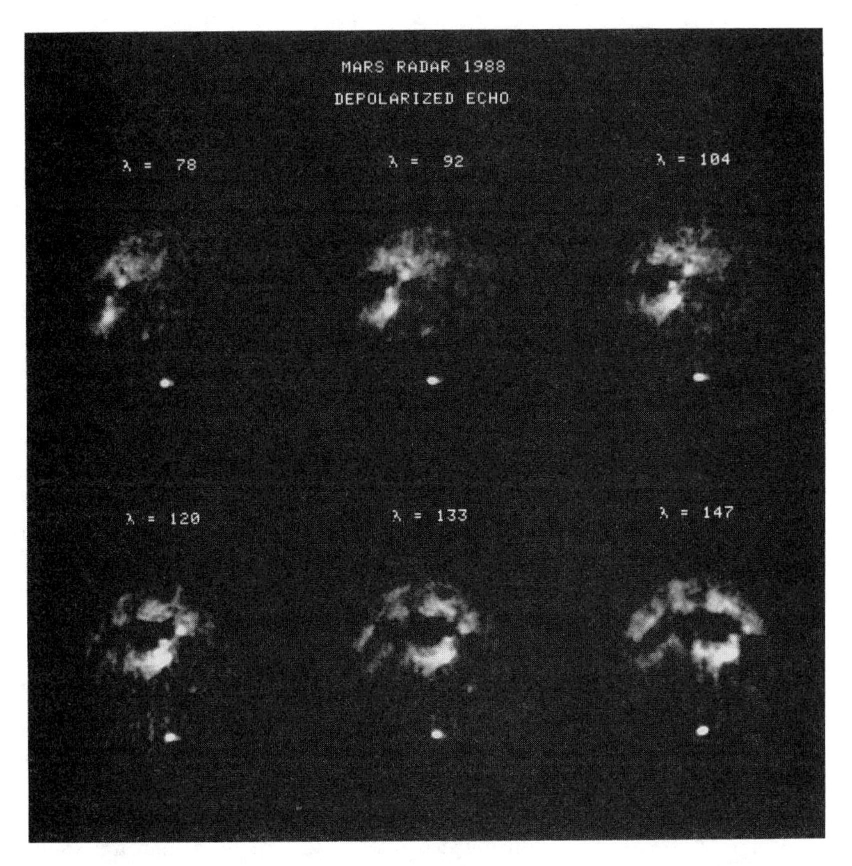

Fig. 10. SC (depolarized) image data from the VLA at approximately 1-hr intervals. Images were acquired on 1988 October 22; each represents about 10 min of integration time. Sub-Earth latitude was 23°.7 S; sub-Earth longitude (denoted here by λ) ranged from 78°W to 147°W. Average resolution is about 0″.4, or about 150 km on Mars' surface at the sub-Earth point. The bright spot at the bottom of each image is from the residual south-polar ice cap. The remaining bright regions are almost exclusively within an area bounded by 10°S, 30°N, 90°W and 190°W (see text).

eras of Arsia and Ascraeus Mons, and (2) diffuse features which cannot be so related, such as the bright region at (22° ± 1°S, 123° ± 2°W). The region centered at (5°N, 123°W) is anomalous in a different way. No SC echo was detected from this region throughout the six hour experiment; that makes it a factor of at least 30 weaker than the bright SC features noted above and significantly weaker than the general background echo flux seen on the disk. The strongest SC radar return was from the south pole. Careful tracking of this feature showed that it was centered on the residual ice cap (86°.5 ± 1°.4 S, 53°.8 ± 16°.3 W). Backscatter radar cross section per unit area was found to be the same as for Jupiter's satellite Ganymede.

VI. SURFACE PROPERTIES

Conventional Earth-based (monostatic) radar studies of Mars provide access to about 40% of the surface of Mars. Bistatic configurations, especially those with imaging capability such as from the VLA, permit study of additional areas although at generally lower resolution. It is not possible in a review of this scope to translate the data that have been acquired (even from the monostatic experiments) into meaningful, specific surface-properties estimates valid for the entire planet. Mars is simply too diverse. Radar cross-section variations of at least $0.03 < \sigma < 0.13$ were noted in the previous section, *rms* tilts derived from quasi-specular echoes range over $0°.25 < \theta_r < 10°$, and the scattering mechanism itself varies from almost entirely specular (smooth plains) to inexplicable (south-polar residual ice cap). We have instead chosen a limited number of areas for which we believe representative characterizations can be made. The reader is referred to chapter 21 for a more extensive integration of these results with conclusions drawn from other studies and for a more thorough discussion of the implications for surface geology.

Syria and Sinai Plana. These high plains, bordered on the west by Tharsis and on the north by Valles Marineris, are among the smoothest surfaces on Mars. Typical values θ_r = 0°.5 to 2°.0 were found at both $\lambda = 13$ cm and $\lambda = 70$ cm. Tilts at the longer wavelength were marginally smaller over 75° to 110° W at 15°.8 S, suggesting a weak wavelength dependence in the scattering behavior. At both wavelengths θ_r decreases over this region east to west. There appears to be no unusual diffuse scattering, though the single identification of depolarization in range-Doppler studies is found here.

Of considerable interest has been the detection of an apparent seasonal variation in both the *rms* surface tilt and reflectivity at $\lambda = 12.6$ cm by Zisk and Mouginis-Mark (1980). By comparing θ_r and ρ_0 values by Downs et al. (1975) from essentially the same ground tracks at different seasons L_S and during different oppositions, Zisk and Mouginis-Mark found variations as large as 20% in ρ_0. They postulated development of a layer of moisture (or liquid) at about a meter's depth from summer insolation; this could dramatically alter the scattering behavior of a regolith. Zent and Fanale (1986) later suggested that certain brines could provide the mechanism, but Zent et al. (1988) subsequently concluded that the thermal phasing was not correct for the differences in radar properties reported. Wind-blown dust has also been suggested as the cause of the reflectivity variation; however, the dust was being deposited at the time when reflectivity was increasing, which is opposite to the desired effect.

Tharsis and Olympus Mons. Tharsis as a region is a weak quasi-specular target. There are isolated areas south of Arsia Mons that show good specular return (Downs et al. 1973), while the area north of Ascraeus Mons appears to be smooth (Harmon et al. 1982), but most of the construct is rough. Simpson et al. (1978*a*), for example, found $\theta_r \sim 10°$ west of (1°3S, 112°W), among the highest θ_r values reported from any analysis of quasi-specular scatter. The extreme θ_r complicates determination of reflectivity because the quasi-specular signal is widely dispersed and therefore weak from any single surface area. Those measurements which have been made successfully, however, show that the reflectivity is low; values of $\rho_0 \sim 0.03$ to 0.04 are common in Table III and from the Goldstone data set (Downs et al. 1973,1975). The combination of large θ_r and small ρ_0 has complicated altimetry measurements considerably; despite repeated attempts to determine the heights of the major volcanos, only flanks (no summits) have been measured (see Kotel'nikov et al. 1983).

As has been reiterated here, recent measurements of diffuse and depolarized scatter imply a high degree of roughness at wavelength scales. The diffuse radar cross section from Tharsis longitudes $\sigma^D \sim 0.12$ is among the highest on Mars, while the ratio of diffuse to total radar cross section reaches 85%. There are apparent concentrations of this texture around Arsia Mons. Given the fact that this is a region of major volcanic activity, it is fair to assume that the upper few meters of surface is rough flow material with a high proportion of blocks and other irregularities. Infrared data, however, imply low thermal inertias. Jakosky and Muhleman (1981) have proposed density variations to explain partial correlation between thermal and radar data. Christensen (1986*a*) proposes up to a two-meter-thick dust blanket over Tharsis with a very rough, rocky surface underneath. Calvin et al. (1988) suggest buried rocks. Schaber (1980) shows that radar roughness and aeolian mantling hypotheses are consistent with high-resolution (10 to 15 m/pixel) Viking Orbiter image interpretation. Low values of ρ_0 are consistent with a dust mantle in the Tharsis region.

The region north of Olympus Mons, extending perhaps into Acheron Fossae, is unusual in producing a strong high-angle return in the OC polarization only. Since this is the sense expected from planar facets, rather than from blocks or irregular structure, one must conclude that the surface in this region looks more like a highly fractured eggshell than a boulder-strewn plain.

West of Tharsis lies the broad, lower Amazonis region. It also is a poor quasi-specular scatterer, with relatively large θ_r and small ρ_0, and has low thermal inertia. Arecibo data (Harmon et al. 1982) point toward it as a scattering center for SC echoes. The direct observations of Muhleman et al. (1991) are in general agreement, so far as the scattering center is concerned. This area may be like Tharsis; buried rocks with a low-density mantle. Farther south, Muhleman et al. (1991) show no echo in either polarization at incidence angles $\theta > 20°$. The quasi-specular echo is also weak, as well as broad (Simpson et al. 1978*a*). This region may have very low density to depths of several times the $\lambda = 3.5$ cm and $\lambda = 12.6$ cm radar wavelengths. Further, the material would have to be free of subsurface volume scatterers (e.g. rocks)—to sufficient depth—otherwise significant backscatter would be observed. Being at lower elevations than Tharsis, this could be flow material that originated elsewhere. Scott and Tanaka (1982) have mapped part of this region as being covered by ash flows. Schaber (1980) has found evidence toward the west for aeolian modification.

Chryse. Chryse Planitia, landing area for Viking 1, has been one of the most intensively studied regions on Mars. Radar data, in fact, were an important input to the site selection process (Masursky and Crabill 1976*a*; Tyler et al. 1976). The quasi-specular echo is broad in Chryse implying meter-scale *rms* surface tilts $\theta_r \approx 4°$ to $6°$, decreasing to the west. Surface-based stereo estimates of local surface texture are consistent with the low end of this range at VL-1 (S. Liebes, personal communication, 1980). Harmon and Ostro (1985) (see Table III here) find quasi-specular reflectivity $\rho_0 = 0.13$ in Chryse, the highest in their sample; this translates to $\epsilon \approx 4.5$ which may most easily be interpreted as indicating the presence of at least some rock at or near the surface. Harmon and Ostro (1985) inferred that surface coverage at the landing site by rocks with radii $\lambda/3 < a < 3\lambda$ was about 9%, higher than for lunar landing sites and consistent with the fact that Mars has a much higher σ^D than the Moon.

Syrtis Major. Downs et al. (1978) found generally smooth surfaces in Syrtis Major but the $\theta_r = 1.^\circ1$ at $\lambda = 12.6$ cm and $\theta_r = 1.^\circ6$ at $\lambda = 3.5$ cm indicated a wavelength dependence in the scattering, a phenomenon that has been surprisingly rare on Mars. Comparing the Syrtis Major data with that from the Moon, where wavelength variation is common, Simpson et al. (1982) concluded that cratering cannot be controlling the meter-scale and larger tilts, at least in Syrtis Major, and proposed aeolian control as an alternative. Simpson et al. (1982) also reported a dielectric constant $\epsilon = 4$ from bistatic quasi-specular data (see also Fig. 4b for Arecibo estimates of ρ_0) and the lowest quasi-specular *rms* tilt yet found, $\theta_r = 0\overset{\circ}{.}25$, on the western edge of Syrtis Major (Fig. 4c).

Cratered Terrain. The cratered terrain is an "average" landform on Mars so far as its radar properties are concerned. Whereas on the Moon cratered areas are among the roughest units in radar scatter, on Mars they represent an intermediate (and surprisingly uniform) terrain type with typical quasi-specular *rms* tilt $\theta_r \approx 3°$, less than half the lunar value for comparable experimental conditions (Simpson and Tyler 1982). By contrast, plains on Mars reach extremes with $\theta_r = 0\overset{\circ}{.}25$ in Syrtis Major and values 5° to 10° west of Tharsis. Cratered terrain on Mars also has intermediate values of reflectivity and generally low diffuse radar cross section (Table III). Individual craters themselves represent exceptions; Downs et al. (1978) detected no echo from the interior of Schiaparelli at either 3.6 or 12.6 cm wavelength. This suggests that the floor which spans more than 300 km at the $2\overset{\circ}{.}9$ S radar ground track is unusually rough. For the most part, however, the cratered terrain appears to be among the most predictable and well behaved (to radar) on Mars.

Polar Regions. Several Viking Orbiter 2 bistatic ground tracks crossed the north polar region (Simpson and Tyler 1981; data were acquired during early northern hemisphere spring: $1° < L_s < 62°$). The ice itself appears to have a very uniform response with values $\theta_r \approx 2°$. Estimates of dielectric constant were not well determined but were consistent with what one would expect of loosely packed snow or ice (of unknown composition). Toward the edges of the polar cap scattering becomes more variable. Near Chasma Boreale a region with $\theta_r \leq 1°$ was identified in $\lambda = 12.6$ cm data. This was so smooth that a bistatic echo from Viking's 0.1 watt transmitter at $\lambda = 3.6$ cm was also detected.

Goldstone/VLA Earth-based bistatic imaging shows an entirely different scattering behavior from the residual south polar cap (Muhleman et al., 1991). Backscattered energy is stronger than that expected from a perfectly reflecting spherical target at that location and comparable to the scattering observed from Jupiter's icy moon Ganymede (Campbell et al. 1978; Goldstein and Green 1980). No mechanism is known to produce such echo strengths, but several candidates are being studied. The phenomenon appears to be uniquely associated with ice.

VII. CONCLUSIONS

Radar has provided and will continue to provide unique data about the upper surface of Mars. Because experimenter control over key parameters makes radar very much an active remote-sensing instrument and because coherent high-precision data acquisition and processing is possible, a considerable amount of determinism can be built into radar data analysis and interpretation. Modeling remains a difficult step, but models are being developed to understand a suite of scattering mechanisms.

Mars itself represents an attractive target in this context. Its diverse surface displays a wide range of quasi-specular and diffuse scattering behavior; some regions may, in fact, be time variable. Where photometric studies at visible and infrared wavelengths provide information on Mars' surface at μm scales and high-quality orbiter images show surface morphology with pixel sizes ~10 m and larger, radar scattering responds to cm-m scale structure, an important range for both engineering of lander/rover vehicles and geologic study.

In specific areas, radar data show a diverse surface. Data from Syrtis Major, in conjunction with image analysis, have shown that aeolian processes control Mars surface texture on scales of cm to m. Radar data in conjunction with thermal-inertia results near Tharsis suggest mantling by dust layers over rough rocky surfaces; further west, the same combination of data suggests homogeneous low-density materials to considerable depth. The bright, depolarized return from Mars' south-polar residual cap resembles behavior seen on the icy Galilean satellites of Jupiter more than any terrestrial analog. As remote probing of Mars' surface continues, we expect radar studies to contribute in significant and fundamental ways. The next chapter integrates the radar results with inferences from other data types to give a more comprehensive picture of Mars' surface properties than has been possible here.

APPENDIX
LIST OF SYMBOLS AND ABBREVIATIONS

a	characteristic dimension of discrete target or surface structure
A	coefficient of diffuse scattering cosine law
B	antenna spacing (in array)
c	velocity of light
C	Hagfors roughness parameter
d	one-way distance to nearest point on target
d_p	density of rock powder (soil)
d_s	density of solid rock
$\mathrm{E_H}$	electric field strength, horizontal component
$\mathrm{E_L}$	electric field strength, left circular component
$\mathrm{E_R}$	electric field strength, right circular component
$\mathrm{E_V}$	electric field strength, vertical component
f	frequency
f_d	Doppler shift in frequency
f_{LL}	limb-to-limb Doppler bandwidth of planet
g	backscatter gain

h	local planet vertical relief (topography)
L_p	absorption length
L_s	areocentric solar seasonal longitude
n	exponent (shape factor) of diffuse scattering cosine law
OC	opposite-sense circular polarization (from that transmitted)
r_s	spacecraft distance from planet center
R	distance between Earth-based radar and planet center of mass
R_H	Fresnel reflection coefficient (horizontally polarized incident wave)
R_{oc}	reflection coefficient for opposite-sense circular polarization
R_p	planet radius
R_{sc}	reflection coefficient for same-sense circular polarization
R_V	Fresnel reflection coefficient (vertically polarized incident wave)
SC	same-sense circular polarization (as that transmitted)
Δt	two-way delay time (range) of target echo
v	target (radial) velocity
VLA	Very Large Array
β	latitude
$\tan\delta$	material loss tangent
ϵ	dielectric constant
ϵ_p	dielectric constant of powdered rock (soil)
ϵ_s	dielectric constant of solid rock
μ_c	ratio σ_{sc}/σ_{oc}
ρ	mean target albedo (average over hemisphere)
ρ_0	reflectivity
λ	radar wavelength
λ	longitude (Figs. 5 and 10, only)
σ	target radar cross section, or normalized cross section
$\sigma(f)$	target Doppler spectrum (radar cross section per unit frequency)
$\sigma(\theta)$	scattering function or scattering law
$\sigma_0(\theta)$	specific radar cross section (per unit area)
$\sigma(\tau)$	target delay profile (radar cross section per unit time delay)
σ^D	diffuse component of radar cross section
σ^Q	quasi-specular component of radar cross section
θ	incidence/scattering angle in backscatter geometry
θ_B	Brewster angle
θ_i	incidence angle (with respect to mean surface normal)
θ_r	*rms* surface tilt
θ_s	scattering angle (with respect to mean surface normal)
τ	time delay
τ_0	two-way time dispersion (range depth) of target
ϕ	antenna/array beam width

21. THE MARTIAN SURFACE LAYER

PHILIP R. CHRISTENSEN
Arizona State University

and

HENRY J. MOORE
U.S. Geological Survey

The surface materials on Mars provide important constraints on the nature, timing and duration of surface process and can be used to study the history of erosion, transport and deposition of sediments and the relationships between surface sediments and bedrock. Perhaps most importantly, they may contain a record of changes in environmental conditions. These materials are readily accessible to a variety of remote-sensing observations, as well as direct, in situ *observations from landed spacecraft. Remote observations include: (1) diurnal thermal measurements, used to determine average particle size, rock abundance and the presence of crusts; (2) radar observations, used to estimate the surface slope distributions, wavelength-scale roughness and density; (3) radio emission observations, used to estimate subsurface density; (4) broadband albedo measurements, used to study the time variation of surface brightness and dust deposition and removal; and (5) color observations, used to infer composition, mixing and the presence of crusts. Thermal data reveal the presence of large low- and high-inertia regions in the northern hemisphere, with much of the south covered by material of moderate inertia. There is a strong anticorrelation between inertia and albedo, a correlation between inertia and rock abundance and, over much of the planet, a correlation of radar-derived density with inertia. The correlation between density and inertia might be due to the presence of subsurface crusts which would simultaneously increase both of these properties. Viking Orbiter color data indicate the presence of three major surface materials: low-inertia, bright-red material that is presumably dust; high-*

inertia, dark-grey material interpreted to be lithic material mixed with palagonite-like dust; and moderate-inertia, dark-red material that is rough at subpixel scales and interpreted to be indurated. Observations from the Viking landing sites show rocks, fines of varying cohesion and crusts. These sites have indications of aeolian erosion and deposition in the recent past. Large rocks have been exhumed from beneath fines, suggesting one or more cycles of deposition and erosion. Taken together, the remote and in-situ *data suggest that much of the surface can be characterized by 4 basic units. Unit 1 is covered by fine, bright dust, with few rocks exposed at the surface. Radar observations indicate that much of this unit is very rough, suggesting a rough surface that is mostly buried beneath several meters of fine dust. This unit may be a recent deposit, whose location may be linked to periodic climate changes. Unit 2 is also active, with the motion of particles keeping the surface free of dust, resulting in a dark, coarse-grained surface with abundant rocks. Unit 3 has intermediate inertia, albedo and color, and is interpreted to be a rough, indurated surface. Unit 4 is a relatively minor unit, but it contains both Viking landing sites. It is characterized by relatively high inertia and high albedo, suggesting that a thin layer of dust may have accumulated. The landing sites appear to be representative of the types of processes that have occurred globally, but are not completely representative of the major units. They have not accumulated significant amounts of dust, nor are they experiencing active transport and erosion by sand-sized particles. Crusts are present, but not to the degree that may be present elsewhere. Surface characteristics, including dust deposition, erosion, aeolian transport and sorting, and crust formation, indicate that the surface of Mars has been recently active. The available data suggest that cyclic changes in sedimentary processes may occur over several time scales associated with periodic climate changes. Large dust deposits presently occur in the north, with maximum winds and dust storm activity in the south. Under different environmental conditions, these deposits may be eroded and transported elsewhere. Much of the present surface appears young and may have been continually reworked. The continued erosion and redeposition of this loose material, rather than erosion of fresh surfaces, may provide the material for the high rates of aeolian activity. As a consequences, much of the fine material on the surface may have been globally homogenized and essentially decoupled from the underlying bedrock.*

I. INTRODUCTION

The upper few meters of the surface materials of Mars represent a minute fraction of the total volume of the planet, yet they play a significant role in its geologic history. These materials provide important clues toward the understanding of a wide range of questions related to the processes and history of Mars. In particular, these materials are keys to understanding: (1) the processes that are currently or recently active; (2) the rates and degree of modification of the surface by current processes; (3) the history of erosion, transport and deposition of surface sediments; (4) the relations between surface sediments and bedrock; (5) how surface processes relate to environmental conditions; and (6) changes in environmental conditions through time.

At present our understanding of Martian properties and processes comes from the surface observations at two locations over a period of six Earth-years and from remote-sensing observations from orbit and Earth. These observations have been acquired at wavelengths from centimeters (radar and radio) through submicron (imaging and visible/near infrared), which is a range over which many physical phenomena may be observed. With these observations it has been possible to determine a great deal about processes currently or recently active on the Martian surface and to begin to understand how the surface processes and environmental conditions may have varied in the past.

The intent of this chapter is to discuss the global characteristics of the Martian surface layer. However, much of the thermal, albedo, color and radar data used exist only for the region between approximately 60° S and 60° N. These limitations are due primarily to difficulties in obtaining thermal and reflectance data in the polar regions and in obtaining Earth-based radar observations from areas outside the sub-Earth region. For these reasons, little discussion is provided on the polar regions of Mars; the interested reader is referred to chapter 23 for a discussion of these areas.

II. LANDER OBSERVATIONS

The Viking mission placed two spacecraft on the surface. Lander 1 (Lander 1 has been renamed the Mutch Memorial Station in honor of Timothy Mutch and will be referred to throughout as MMS) is located at 22°.5 N, 48°.0 W and Lander 2 (VL2) at 48°.0 N, 225°.7 W (Michael 1978). MMS observed Mars from 1976 July 20 to 1982 November 20, while VL2 observed from 1976 September 3 to 1980 March 6. During this period these spacecraft provided a detailed view of the surface (see, e.g., Mutch et al. 1977; Binder et al. 1977; Moore et al. 1977; Sharp and Malin 1984; Moore et al. 1987) as well as a time history of surface changes (Sagan et al. 1977; Jones et al. 1979; Guinness et al. 1982; Arvidson et al. 1983), and a record of dust storms, dust opacity and changing atmospheric conditions (Hess et al. 1977; Pollack et al. 1979*b;* Tillman et al. 1979). A description of the Viking Lander missions, Landers, surface samples, and other aspects are given in Moore et al. (1987), Soffen (1977) and Snyder (1979*a*). The landing sites are briefly described below.

A. Regional Character

MMS landed on a cratered plain in Chryse Planitia. About 400 km to the southwest of the site, Maja Valles debouches into Chryse Planitia, and streamlined islands are as close at 150 km to the southwest (Carr 1979). The MMS site has been mapped as Hesperian-age flood plain material (Scott and Tanaka 1986; unit Hchp). However, the cratered plain resembles the lava-flow surface of lunar maria when viewed at the resolution of the Viking Orbiter images (50 to 100 m/pixel). An arcuate wrinkle ridge, ~1.5 km to the north and east of the site, is similar to ridges on lunar basalts and is suggestive of a lava surface. Thus, the bedrock at the MMS site may be lava flows like those of the Arcadia Formation (Arvidson et al. 1989*a*), which is mapped 480 km to the northeast of the site (Scott and Tanaka 1986; unit Aa1). Impact craters within 1.5 km of the MMS site have diameters up to 390 m across and ejecta from them must have reached the site (Moore et al. 1987). The number of superposed craters in the general area of the site represent a duration measured in Gyr (Moore et al. 1987; Tanaka 1986).

The MMS has fines, some of which are in dune-like drifts, that are superposed on a rocky substrate (Mutch et al. 1976*a,b*; Binder et al. 1977; Sagan et al. 1977) (see Figs. 1A and 2). Close inspection of the drifts shows that they have been deflated by wind erosion, exposing internal cross-laminations. Surfaces of the substrate between the rocks are covered with a veneer of fines or littered with small mm- to cm-size clods. In the distance, craters with rocky rims rise above the local surroundings. Some of these craters can be identified in Orbiter images (Morris and Jones 1980).

VL2 landed on a cratered plain in Utopia Planitia some 180 km west of the ~100-km diameter crater Mie (Mutch et al. 1977). The cratered plain has broad bulges and swales ~3 to 4 km across. Broad lobes radial to Mie occur to the southeast of the site. Also present are pedestal craters which have variously been interpreted to be the result of volcanic or aeolian processes. The site has been mapped as the Hesperian Knobby member of the Vastitas Borealis Formation (Greeley and Guest 1987; unit Hvk); a wide range of origins for the Knobby member are possible (volcanic flows, aeolian mantles, etc.). The probable locations of the Lander cross the radial facies of the ejecta of Mie so that a considerable thickness of ejecta from Mie, admixed with local debris, should be present (Mutch et al. 1977; Moore et al. 1987).

The surface viewed by VL2 is a rock-strewn plain with a flat monotonous horizon (Fig. 1B) except for a rocky ridge to the south (Mutch et al. 1977; Moore et al. 1987). Large rocks (>0.14 m) are more abundant at the VL2 site than they are at the MMS site. Surfaces between the rocks are covered with a veneer of fines or littered with small mm- to cm-size clods. Locally, smooth surfaces of crusts transected by fractures and polygonal mosaics of crust have been exposed by wind erosion. Drifts are both scarce and small; they occur in patches between clusters of rocks and as small windtails. Interconnected troughs about 1 m across are unique to VL2 and may represent the surface expressions of ice-wedge polygons or desiccated units of near-surface materials (Arvidson et al. 1989*a*). One of the troughs contains drifts and ripples that have been interpreted to be composed of granule-size particles that moved by traction induced by saltating sand-sized particles (Sharp and Malin 1984) (see Fig. 3D).

Regional erosion rates at both sites have been estimated based on crater statistics and the degree of surface modification (Arvidson et al. 1979). At

Fig. 1. Panoramic views from Viking Landers. (A) View looking toward the southeast from MMS. Note the cross-laminated drifts superposed on rocky substrate (1); sample trench in soil-like drift material (2); and rocks (3). Soil-like blocky material is present between rocks (4). In the far field see blocky rim of impact crater (5); bright dunes (6); and outcrop of rock (7). (A high-resolution view of drift is shown in Fig. 2A and outcrop in Fig. 2C.) Spacecraft parts appear in the foreground. (Camera 1 high-resolution mosaic of morning images; see Levinthal and Jones 1980.) (B) View looking toward the northwest from VL 2. Note the platy crusts displaced by engine exhausts (1) and rocks (2). Soil-like crusty-to-cloddy material along with small clods (3) is present between rocks. VL 2 rocks are typically larger than MMS rocks. Bright ridge on horizon (4) may include ejecta from Mie. Large rock and small drift (5) and ripple-like forms (6) are present in the far field. (A high-resolution view of large rock and drift is shown in Fig. 3C and ripple-like forms in Fig. 3D.) Spacecraft parts appear in foreground. (Camera 1 high-resolution mosaic of afternoon images; see Levinthal and Jones 1980.)

MMS only several meters of erosion has occurred on surfaces estimated to be 3.6 Gyr old; at VL2 as much as several hundred meters of erosion has occurred during a similar timespan. These rates of <1 μm yr^{-1} are consistent with the current rate of accumulation of dust and the longer-term accumulation of drift materials. Drifts are currently oriented parallel to the present dust-storm-period winds, suggesting that they formed during the past 15,000 yr when the storm winds have had their current orientation (Arvidson et al. 1979).

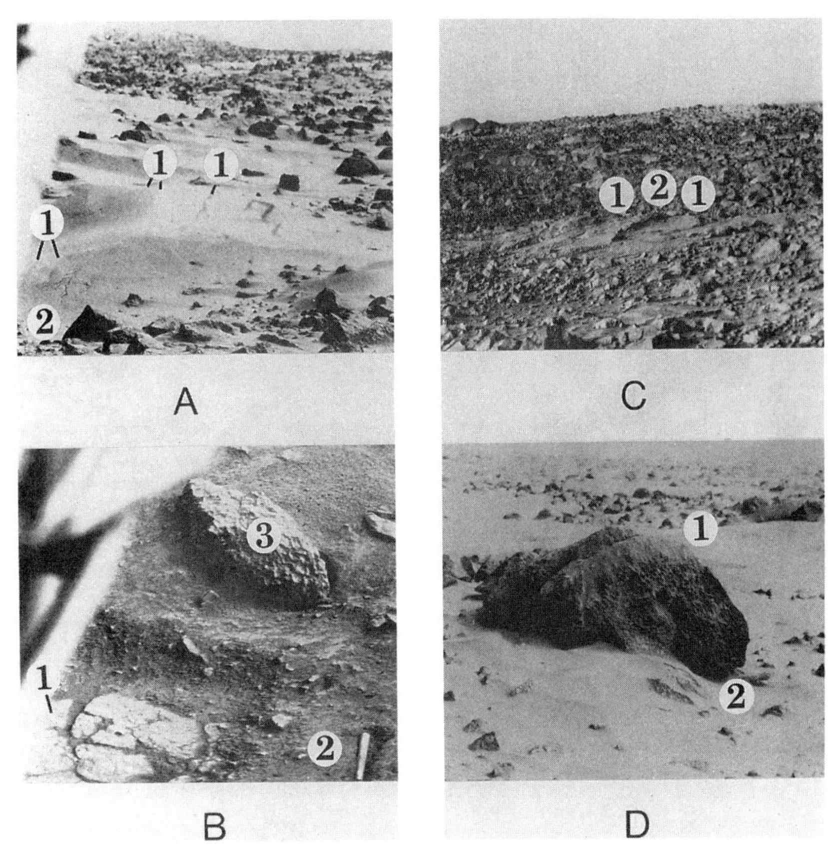

Fig. 2. Collage of selected high-resolution images of MMS. (A) Layering in drift about 10 m east of MMS. Layers (1) have been interpreted as cross bedding. Note the open fractures (2) in drift. Pointed rock near (2) is about 0.25-m wide. Part of meteorology boom is at left. (Frame 11C162/239.) (B) Blocky material (1) (also called duricrust) is exposed by engine exhaust erosion. Note the planar surfaces of and fractures in blocky material. Rod-shaped object (2) is sampler latch pin lying within impact crater that it produced; crater is 2.4 cm wide. Rock (3) is 0.23 m long. Base of sampler housing and mounting bracket are at left. (Frame 12C158/232.) (C) Probable outcrop of rock ~10 m south of MMS. Note the continuity of planar surfaces (1) and dike-like ridge that transects outcrop (2). Outcrop is ~4 m long; width of frame at outcrop is 3.5 m. (Frame 11A158/027.) (D) Large rock, called Big Joe, is northeast of MMS. Note the deposit of fine-grained material on top of the rock (1) and the miniature landslide on drift near the base of the rock (2). Big Joe is 8 to 10 m from MMS and 1 m high. (Frame 11C162/239.)

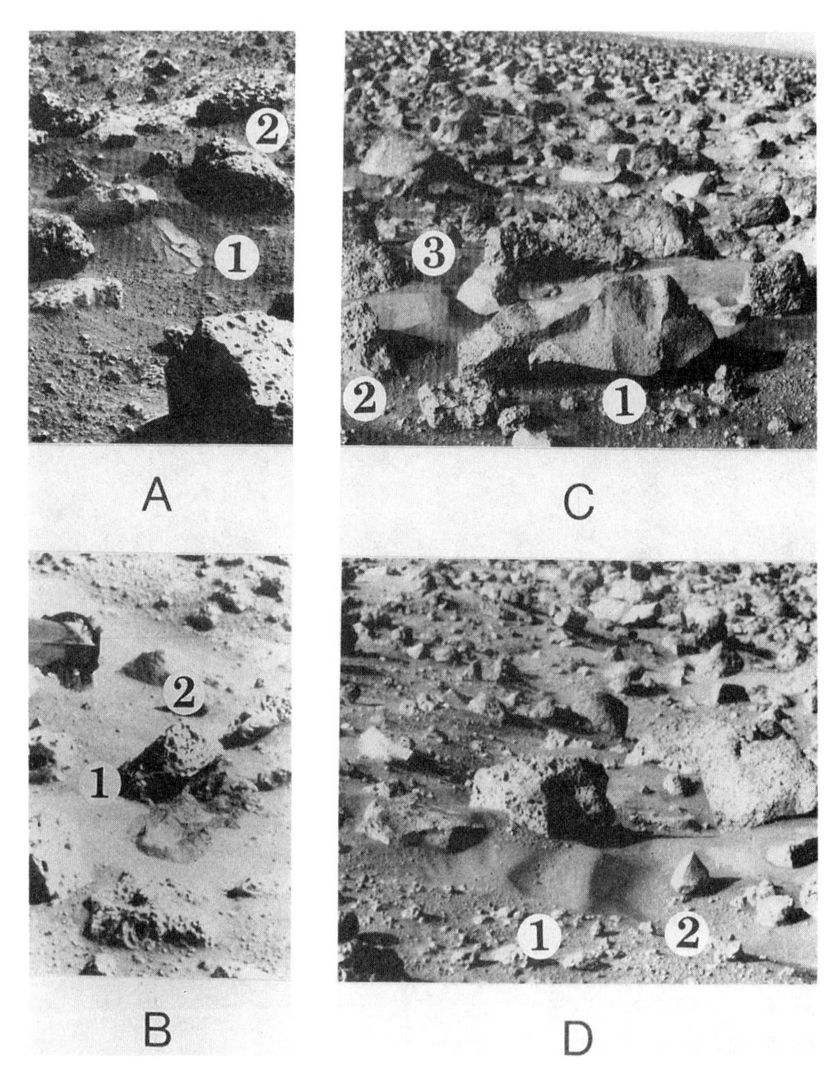

Fig. 3. Collage of selected high-resolution images of VL 2. (A) Exposure of bright crust (1). Note the open fractures transecting the crust and surrounding materials that overlie the crust. The sampler excavated a trench here (see Fig. 4C). Rock (2) is ~0.25 m long. (Frame 22A007001.) (B) "Water-line" ledge of crust (1) adhering to rock (2) is displaced by sampler; rock (2), called Badger, is ~0.25 m wide. (Frame 22B030/034.) (C) Dense (1) and pitted (2) rocks and drift (3). Pits have been interpreted to be vesicles in volcanic rocks, but pits on some rocks may be flutes produced by aeolian erosion. Drift is bright and appears to be fine grained. Rock (1) is ~1 m long. (Frame 22A011/001.) (D) Ripple-like forms (1) in trough and possible ventifact (2). Note the small craters on ripple-like forms produced by impact of debris from engine exhausts. Ripple-like forms have been interpreted to be composed of granules, but they are too distant for camera to resolve the granules. Possible ventifact is ~0.2 m wide. (Frame 22A011/001.)

B. Surface Materials

There are 4 major types of surface materials that were sampled at the landing sites: (1) rock (MMS, VL2); (2) drift (MMS); (3) crusty-to-cloddy (VL2); and (4) blocky (MMS) (Moore et al. 1982,1987). The chemical compositions of the soil-like materials at the two landing sites are remarkably similar (see chapter 18) despite the fact that they are separated by some 6500 km. This similarity may be due to a homogenizing global process, such as dust storms (Toulmin et al. 1977), or to volcanic processes that may have introduced the large amounts of sulfur and chlorine in the soil-like materials (Settle 1979). Descriptions of these materials, their inferred physical and mechanical properties, and the methods used to infer these properties have been given in detail (Moore et al. 1982,1987; Moore and Jakosky 1989). The major materials are briefly described below with emphasis on their properties that affect surface processes. Their estimated mechanical and physical properties are given in Table I.

1. Rock. Both landing sites have relatively large concentrations of rocks at the surface, but their provenances, compositions and physical properties can only be inferred. At MMS, most of the rocks are probably ejecta from local impact craters, but flood detritus from Maja Valles is also possible. The impact craters may have excavated rocks from local lava flows or even indurated sediments. Other rocks may have come from some distant crater that excavated rocks not represented locally because the crater immediately to the east of the Lander appears be a large secondary impact crater. Many of the rocks at the VL2 site may be ejecta from the crater Mie, but many rocks may have been derived locally as a result of secondary cratering by Mie ejecta and local primary impacts.

No demonstrable rocks or rock fragments were sampled and analyzed for inorganic elements by the X-ray fluorescence spectrometers (XRFS) of the Landers (Clark et al. 1982; Moore et al. 1987). Small rock fragments (<0.01 to 0.02 m) are inferred to be scarce to absent and the vast majority of cm-sized objects seen at both sites are interpreted to be clods. Colors of rock surfaces may lead to incorrect conclusions about rock compositions because some rock surfaces resemble local soil-like materials while others resemble a Hawaiian palagonite (Guinness et al. 1987; Arvidson et al. 1989*a*). However, several dark objects deposited on the MMS XRFS delivery port had colorations compatible with mafic igneous rocks (Dale-Bannister et al. 1988).

Physical properties of the rocks must be inferred from terrestrial experience (Table I). Some of the rocks appear to be dense and fine grained, some may be breccias, and others appear to be vesicular (Sharp and Malin 1984; Garvin et al. 1981) (see Figs. 2 and 3). The surface sampler was unable to chip, scratch, or spall any rock surfaces, indicating that they do not have

TABLE I
Estimated Mechanical Properties and Remote Sensing Signatures of the Surface Materials in the Sample Fields and at the Viking Landing Sites[e]

	Grain Size (μm)	Bulk Density (kg m^{-3})	Cohesion (kPa)	Angle of Internal Friction (degrees)	Fraction of Area Covered	Thermal Inertia (10^{-3} cal cm^{-2} s$^{-1/2}$K^{-1})	Dielectric Constant
LANDER 1							
Drift	0.01–10.0	1150 ± 150	1.6 ± 1.2	18.0 ± 2.4	0.14	3	2.4
Material			0–3.7			2.1–2.6	
Blocky	0.1–1500	1600 ± 400	5.1 ± 2.7	30.8 ± 2.4	0.78	9.3 ± 0.5[a]	3.3
Material			2.2–10.6				2.4–4.5
Rocks	35×10^3	2600	1000–	40–60	0.08	40	8
	240×10^3		10,000				3.3
Sample Field		1600			1		2.6–4.6
		1300–2000				9.0 ± 0.5	3.3 ± 0.7[c]
Remote Sensing		1300–1900[c]					3.0[d]
		1500[d]					4.0–4.6[d]
		2000[d]					
LANDER 2							
Crusty-to-Cloddy	0.1–10.0	1400 ± 200	1.1 ± 0.8	34.5 ± 4.7	0.86	6.3 ± 1.5[b]	2.8
Material			0–3.2				2.4–3.8
Rocks	35×10^3	2600	1000–	40–60	0.14	40	8
	450×10^3		10,000				
Sample Field		1600	—	—	1	3.2	
		1400–1700				2.8–3.6	
Remote Sensing			—	—	0.20 ± 0.10	8.0 ± 1.5	2.8–12.5
						8.3–8.8	

[a] Thermal inertia is 8.2 ± 1.4 if fraction of area covered by rock is taken as 0.15 ± 0.05 (see text).
[b] Thermal inertia is 5.6 ± 1.4 if fraction of area covered by rock is taken as 0.20 ± 0.10 (see text).
[c] Derived from Viking Lander-Orbiter radio link.
[d] Derived from Earth-base radar.
[e] Table from Moore and Jakosky (1989).

weak rinds where exposed to the atmosphere or buried (Moore et al. 1977, 1978*a*,1987).

Rocks cover such a large fraction of the surfaces at both sites that they tend to reduce wind stresses at the surface (Marshall 1971; Moore et al. 1979). About 8% of the surface immediately in front of MMS is covered by rocks with diameters between 0.035 and 0.24 m, but there are larger rocks in the distance. About 14% of the surface immediately in front of VL2 is covered by rocks with diameters between 0.035 and 0.45 m, but if rocks in the farfield are included, about 19% of the surface is rock (Moore and Jakosky 1989).

The degree of erosion of rocks remains uncertain. Many rocks at both sites appear to be uneroded and unworn, with sharp, angular edges (Sharp and Malin 1984). Other rocks appear to have vesicles and pits, either modified or produced entirely by wind erosion (McCauley et al. 1979; Garvin et al. 1981; Sharp and Malin 1984). Even the presence of some degree of rock erosion is consistent with the modest, regional erosion rates discussed above. Possible explanations for the low erosion rates include: (1) a limited supply of sand; (2) insufficient wind velocities to initiate particle motion; or (3) protection of the rocks by burial.

The rocks at the MMS site appear to be partly to completely exhumed from previous overlying deposits (Fig. 1). Some rocks protrude through drifts, some are capped by thick layers of sediments, and some have wind scours on one side and windtails of deflated drift material on the other side. Although many of the rocks at the VL2 site appear to rest on the surface, many of them are partly buried and it seems probable that there are completely buried rocks (Fig. 3). Thus, many VL2 rocks have also been exhumed, but none are capped by thick layers of fine sediments. There are several large exposures of rock at the MMS site that have been interpreted to be outcrops, some of which appear to consist of layered rocks (Binder et al. 1977).

2. Drift Material (MMS). Drift material, which is weak and porous (Table I), has three properties that are important for surface processes: (1) it is fine grained, (2) it is cohesive, and (3) it is relatively unfractured. These properties make drift material resistant to aeolian erosion. The specific surface of drift material (17 m^2g^{-1}) implies a grain size of 0.14 μm (Ballou et al. 1978). However, the range in grain size may be from 0.1 to 10 μm because the physical sizes of grains can be 10 to 100 times larger than those inferred from specific surfaces (Fanale and Cannon 1971). The cohesion of drift material, while variable, is quite small ($\sim$1.6 kPa) (Table I), and it is possible that cohesions are smaller near the surface than they are at depths of 4 or 5 cm. Even a small cohesion is sufficient to inhibit aeolian erosion. Although the drift material is cross-laminated, it is relatively unfractured and unjointed (Moore et al. 1987). Contributions of the properties above to the stability of drift material are illustrated by its response to the engine exhaust gases during landing. Despite the fact that the shear stresses of the exhaust gases (Romine et al. 1973) at the surface exceed those expected for Martian conditions, drift material was not extensively eroded by the engine exhaust gases (Moore et al. 1979,1987; Hutton et al. 1980).

3. Crusty-to-Cloddy Material (VL2). Although crusty-to-cloddy material was extensively eroded by the engine exhaust gases during landing, it also has properties that should make it relatively resistant to most Martian winds. Erosion by the engine exhaust gases was facilitated by the large stresses of the exhaust gases and the presence of the closely spaced fractures and joints that produce polygonal mosaics of in-place crusts and clods a few cm across and the veneer of clods at the surface (Fig. 4). Like drift material, crusty-to-cloddy material has a grain size near 0.1 to 10 μm (Oyama and Berdahl 1977; Moore et al. 1987; Moore and Jakosky 1989) and its cohesion (~1.1 kPa) is comparable to that of drift material (Table I). Because crusts are locally layered and adhere to the edges of rocks the cohesion must be partly due to cementation (Moore et al. 1978*a*,1987). Cementation may have been produced by some sulfur and chlorine compounds (Clark et al. 1982). Some aspects of the surface in front of VL2, such as exposed fractured crusts, mosaics of in-place prismatic units of the material, and concentrations of mm- to cm-size clods are similar to terrestrial soils that have formed stable surfaces to wind erosion (Chepil 1950; Chepil and Woodruff 1963). These occurrences suggest that crusty-to-cloddy materials are not easily eroded by Martian winds.

4. Blocky Material (MMS). The uppermost surface of blocky material was extensively eroded by the engine exhaust gases during landing and the resulting erosion crater was flat-floored (Fig. 2B). Again, this erosion was facilitated by the large stresses of the engine exhaust gases and the fine-scale heterogeneous character of the material. The material exposed on the floor of the erosion crater has been called duricrust, but it is the same blocky material excavated in sample trenches 0.05 m or so deep. Because samples of blocky material were never analyzed by the Biology Gas Exchange experiment (Oyama and Berdahl 1977; Moore et al. 1987), its grain size is unknown. Smooth surfaces produced by the sampler tamping and rubbing blocky material argue for significant amounts of silt-size or smaller grains, but unusually large motor currents measured while grinding it argue for the presence of significant amounts of very hard or indurated mm-size grains (Moore et al. 1987). Blocky material has the largest cohesion (~5.1 kPa) of the three soil-like materials (Table I), and the large amounts of sulfur and chlorine in the coarse fraction of blocky material (Clark et al. 1982) suggest that cementation is the major contributor to the cohesion of blocky material. Because of its large cohesion, blocky material is not easily eroded by Martian winds.

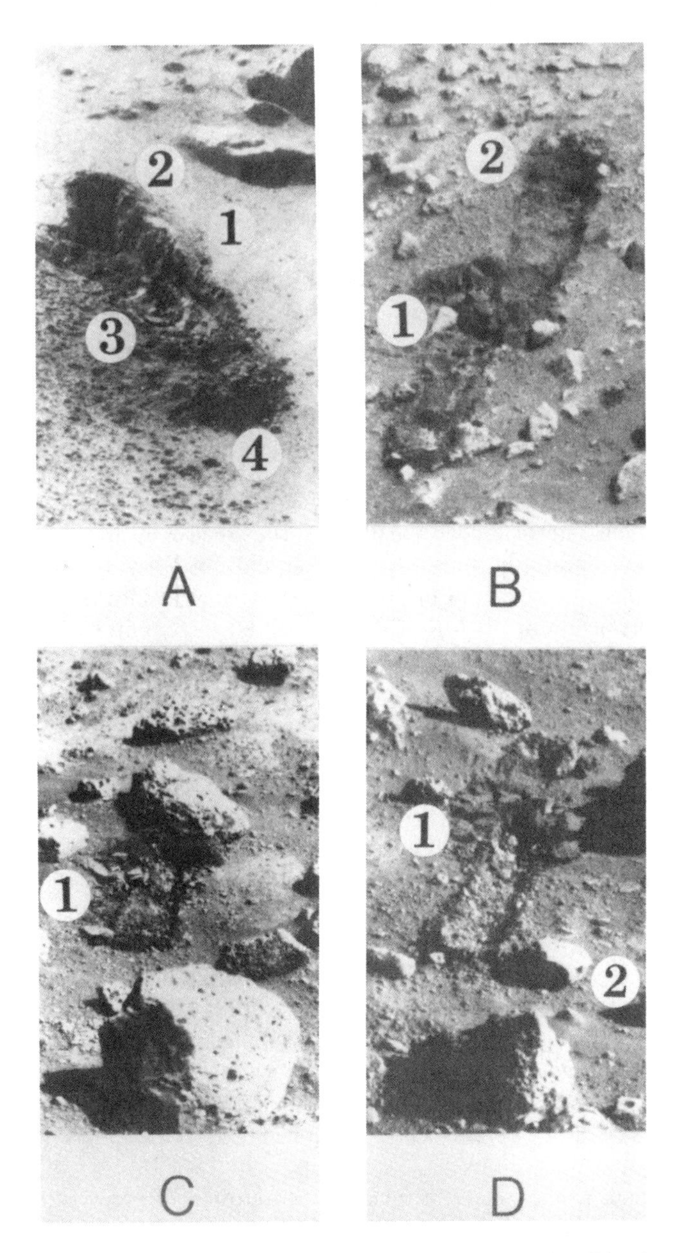

Fig. 4. Trenches excavated in soil-like materials by the samplers of the Viking Landers. (A) Trench in drift material at MMS site is ~0.1 m wide and 0.6-m long. Note the domed surface (1), steep walls (2), bright, smooth surfaces tamped by sampler (3) and lumpy tailings (4). (Frame 11A065/008.) (B) Trenches in blocky material at MMS site are ~0.07 to 0.09 m wide. Note the disrupted area with blocky clods up to 0.04 m across (1) and blocky clods at tip of other trench (2). (Frame 12B138/053.) (C) Trenches in crusty material at the VL 2 site are ~0.12 m wide. Note the bright pieces of crust that are 0.04 m wide and 0.01 m thick (1). (Frame 22B047/037.) (D) Trench in cloddy material at VL 2 site is ~0.07 m wide. Note the prismatic clods of disrupted surface (1) and mosaic of in-place units of cloddy material (2). (Frame 22C053/058.)

C. Active Surface Processes

Deflation of drifts, scoured moats on the upwind sides of rocks, deflation windtails of the lee sides of rocks, residues of surface clods, etched crusts, and what may be ripples attest to aeolian activity at both sites in the recent past. Yet, there are numerous indications that the surfaces at the sites are relatively stable in the current wind regime. During the lifetime of VL2 and most of MMS, there was no evidence for significant modifications of natural undisturbed surfaces by the wind. At the MMS site, materials that were disturbed by the Lander and its sampler, such as small piles of the soil-like materials and trenches made by the sampler (Moore et al. 1979,1987), were stable and unmodified for about 1742 sols (Moore 1985; Arvidson et al. 1983). In the late winter of the third Martian year ($\sim$ sol 1742), strong winds of a local dust storm eroded small piles constructed by MMS and some trenches, but undisturbed natural materials did not show similar erosion. Winds of this local dust storm probably achieved velocities of $\sim$ 40 to 50 m s^{-1} at the height of the MMS Meteorology boom (1.6 m) (Moore 1985), whereas winds of previous storms were $\sim$ 25 to 30 m s^{-1} (Ryan and Henry 1979; Ryan et al. 1981). At VL2, no significant erosion of disturbed or natural materials was observed during the 1212 sols that VL2 observed the surface (Moore et al. 1979,1987; Arvidson et al. 1983).

Reasons for the long-term stability of the materials to wind erosion include the presence of rocks that shield the surface, concentrations of nonerodible elements such as clods and soil-like crusts, and the general resistance of very fine-grained cohesive materials such as drift material (Moore et al. 1979). The importance of nonerodible elements in stabilizing the surface is illustrated by the behavior of piles of loose drift and blocky material. Loose drift material was blown away, whereas piles of blocky material were reduced to mounds of small clods ($\leq$0.7 cm) (Moore 1985). It has been suggested that the present stability at MMS may be disrupted periodically during periods of higher atmospheric pressure. The subsurface crust layer exposed by the engine exhausts may represent an erosional discontinuity that is periodically exhumed during periods when strong winds scour the surface (Arvidson et al. 1983).

Although neither site experienced wind erosion of pre-existing undisturbed surface materials, both sites underwent the deposition and removal of layers of bright, red dust 10 to 100 μm thick (Arvidson et al. 1983). At MMS, this dust was deposited from the global dust storms during the fall and winter of the first Martian year, producing an increase in surface brightness and a decrease in contrast. An additional layer was deposited during the fall of the second Martian year (Guinness et al. 1982). These thin layers of dust are periodically removed and distributed by mild winds (Arvidson et al. 1983), suggesting very low particle cohesion.

Although substantial wind erosion did not occur, there was evidence for surface activity in the form of two natural miniature landslides on the slopes of drifts well beyond the reach of the MMS sampler (Jones et al. 1979; Guinness et al. 1982). These features may represent an important step in the dust cycle on Mars because they produce surface configurations and disrupted materials that are no longer in equilibrium with respect to the Martian winds (Moore 1986), as evidenced by the erosion of a slide by the dust storm of sol 1742. These slides may occur when loose dust accumulated on the protected, downwind slopes of rocks reaches a critical thickness, resulting in failure (Moore 1986). They could also be triggered by positive pore pressures generated by desorption of CO_2 (Huguenin et al. 1986), unusual wind stresses, Mars-quakes, or dissolution of a cementing agent (Moore 1986).

In summary, rocks, fines of varying cohesion, and crusts are present at both Lander sites. Both sites show signs of aeolian erosion and transport in the recent past, with evidence of exhumation of rocks from overlying fines. Surface rocks at both sites show only moderate degrees of pitting and erosion. Large-scale features also suggest relatively little erosion at these sites over long periods. At present, neither of the sites appears to experience net dust accumulation, erosion, or significant transport of sand-sized particles.

D. Remote-Sensing Signatures of Surface Materials

Moore and Jakosky (1989) have estimated the dielectric constants and thermal inertias of the materials at the landing sites (Table I) and compared them with dielectric constants inferred from normal reflectivities of quasi-specular echoes from Earth-based radar observations and thermal inertias inferred from Viking Orbiter thermal observations. These estimates are based on the physical and mechanical properties of the materials, experimental data on the electrical properties (Campbell and Ulrichs 1969; Hoekstra and Delaney 1974) and thermal properties of rocks, powders and soils (Wechsler and Glaser 1965; Wechsler et al. 1972; Fountain and West 1970), and a mixing model (Kieffer et al. 1977). The comparison of the dielectric constants and thermal inertias of the materials at the landing sites with those obtained remotely by Earth-based radar and Viking Orbiter thermal observations suggest that the materials at the landing sites are good analogs for materials elsewhere on Mars, as discussed in Sec. IV.C.

III. REMOTE-SENSING OBSERVATIONS

Remote-sensing observations from orbit and Earth provide a means for examining the surface properties over the entire planet, but are necessarily more difficult to interpret than observations from the surface. The principal data available are: (1) diurnal thermal measurements, which are used to infer average particle size, rock abundance and the presence of subsurface crusts;

(2) radar observations, which are used to estimate the surface and near-surface slope probability distributions, wavelength-scale roughness and dielectric constant; (3) radio emission observations, which are used to estimate dielectric constant and subsurface density; (4) broadband albedo measurements, which are used to study the time variation of bright and dark surfaces, possibly related to variable dust deposition and removal; and (5) color observations, which are used to infer composition, mixing of materials and the presence of crusts.

The basic methods and interpretations of individual data sets are summarized below and the reader is referred to the cited literature and to chapters 2, 17 and 20 for more detailed discussions. Section IV discusses the constraints on surface properties and processes provided by analysis of the combined suite of *in situ* and remote observations.

A. Thermal Observations

1. Thermal Inertia. The amplitude of the diurnal temperature variation of the surface is determined by the thermal inertia of the upper few centimeters of material. For a smooth, homogeneous surface, thermal inertia is given by $(K\rho C_p)^{1/2}$ (given here in units of 10^{-3} cal cm^{-2} s$^{-1/2}$ K^{-1}), where K is the thermal conductivity, ρ is the density, and C_p is the specific heat. For typical geologic materials in the Martian environment, density varies by a factor of 4, the specific heat varies by only 10 to 20% due to differences in composition, whereas the conductivity can vary by up to 3 orders of magnitude (Wechsler and Glaser 1965; Neugebauer et al. 1971; Kieffer et al. 1973). The variation in conductivity can be produced by changes in particle size and porosity of cohesionless materials (Kieffer et al. 1977; Jakosky 1986), or the degree of bonding (cementation) of the surface materials. Conductivity, density and therefore thermal inertia are lowest for small particles with small cohesion, such as loose dust, and highest for solid rock.

Laboratory measurements provide the link between thermal inertia and particle size (Wechsler and Glaser 1965; Kieffer et al. 1973). It is important to note, however, that thermal inertia only provides information on the average particle size of the surface. Thus, unimodal, medium sand, a mixture of fine sand and pebbles, or crusted fines could all have identical, intermediate thermal inertias. Low-inertia materials are less ambiguous, with only very fine, unbonded materials having inertias of ~2 to 3. High-inertia surfaces, however, can be produced by a range of surface properties, including the abundance of rocks, the bonding of surface materials and the presence of multiple components (Kieffer et al. 1977; Kieffer 1976; Jakosky 1979; Palluconi and Kieffer 1981; Christensen 1986*b*). When combined with other data, such as radar observations, diurnal, multiwavelength temperature measurements, and high-resolution images, a better estimate of the characteristics of the materials, such as particle size distribution and degree of bonding, can be made. Interpretations based on these combined observations are discussed in Sec. IV.

Thermal observations of Mars have been determined from a number of spacecraft missions including Mariner 6 and 7 (Neugebauer et al. 1971), Mariner 9 (Kieffer et al. 1973), Mars 3 (Moroz and Ksanfomality 1972; Moroz et al. 1976), Mars 5 (Ksanfomality and Moroz 1975) and the Viking Infrared Thermal Mapper (IRTM) (Kieffer et al. 1977). Of these, only the IRTM instrument provided sufficient diurnal and spatial coverage to produce global maps of thermophysical properties.

Thermal inertias have been determined for Mars using IRTM measurements of the thermal emission from the Martian surface as a function of time of day (Kieffer et al. 1977; Palluconi and Kieffer 1981). This instrument had four thermal channels to determine surface temperatures and the wavelength dependence of thermal emission, one channel to determine atmospheric temperatures, and one channel to measure solar reflectance (Chase et al. 1978). During the Viking mission much of the planet was viewed at multiple times of day, season and viewing geometry at a surface resolution of ~30 km (Kieffer et al. 1977).

Temperature measurements from the IRTM, together with direct measurements of the albedo of the surface, have been used to derive thermal inertia with the aid of thermal models (Kieffer et al. 1977). The best approach is to fit diurnal temperature measurements to model-derived temperature curves (Palluconi and Kieffer 1981). This technique provides a best-fit determination of thermal inertia and reveals any deviations in diurnal temperature behavior produced by nonuniform surface and atmospheric properties, such as subsurface layering, surface roughness, temperature-dependent thermal properties and diurnally varying atmospheric radiation (Jakosky 1979; Ditteon 1982; Haberle and Jakosky 1991). To provide diurnal coverage, however, data from multiple observations must be binned to a coarser spatial resolution.

A second approach is to determine the thermal inertia by fitting the thermal model to a single, pre-dawn temperature measurement. These observations allow the intrinsic resolution of a single observation to be retained, permitting a global map with a resolution of 25 to 40 km to be made. This approach must assume a value for the albedo, however, producing uncertainties of up to 1 inertia unit. Zimbelman and Kieffer (1979) initially presented the results from this technique, using a compilation of individual IRTM observations acquired by VO2 between L_S 9° and 40° (L_S is the aerocentric longitude of the Sun). Subsequently, Christensen and Malin (1988) have presented similar data covering L_S 7°.6 to 102°.4 in image format, revealing the presence of small-scale (~30 km) thermal variability in all terrains (Fig. 5).

Several global studies of thermal inertia using both of these approaches have been performed and provide a means of studying the spatial variations

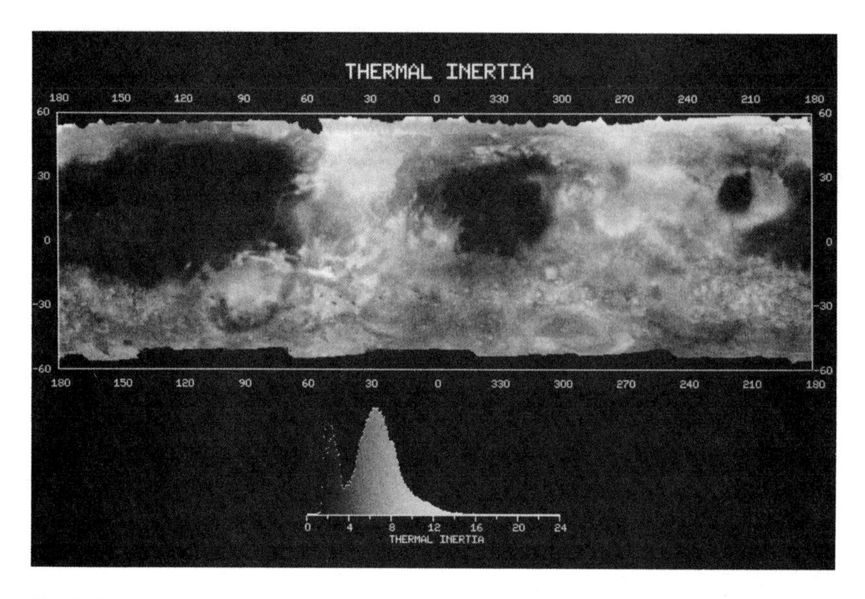

Fig. 5. Thermal inertia. Image constructed from moderate-resolution (~30 km) night-time IRTM observations. Values are expressed in units of 10^{-3} cal cm^{-2} s$^{-1/2}$ K^{-1}. Note the presence of three major low-inertia regions located primarily in the northern hemisphere and the large degree of spatial variation. Also see Color Plate 18.

across the planet (Kieffer et al. 1977; Palluconi and Kieffer 1981; Christensen and Malin 1988) (see Fig. 5). At a scale of 2° × 2° in latitude and longitude, the surface inertias vary from 1.5 to 14, with a bimodal distribution peaked at 2.5 and 6.5 (Palluconi and Kieffer 1981). For homogeneous, flat-lying surfaces, these values correspond to cohesionless particles with sizes ranging from 20 μm to 35 mm (Kieffer et al. 1973).

At 0°.5 × 0°.5 resolution, the inertia covers a wider range from 0.5 to 25, corresponding to particles from ~10 μm to several centimeters (Christensen and Malin 1988). The highest-resolution data obtained by the IRTM (3–5 km) revealed local inertia maxima of up to 26, corresponding to average grain sizes of several centimeters or larger (Christensen 1983; Zimbelman and Leshin 1987). These data indicate that the Martian surface is variable at all spatial scales.

The global observations show the presence of three large regions of low thermal inertia located in the Tharsis (20° S to 50° N, 60° to 190° W); Arabia (5° S to 30° N, 300° to 369° W); and Elysium (10° to 30° N, 210° to 225° W) regions. These areas have inertias between 2 and 3, indicating surfaces covered by dust particles ~40 μm or smaller (Palluconi and Kieffer 1981). These low-inertia regions have very uniform thermophysical properties over their entire extent, with less variability in inertia at 5 to 30 km scales than is observed elsewhere (Zimbelman and Greeley 1982). The boundaries of the low-inertia regions are relatively sharp, typically 75 to 150 km in width, and do not correspond well to observed morphologic or geologic boundaries (Zimbelman and Kieffer 1979). From their thermal inertia, these regions have been proposed to be accumulations of fine dust (Kieffer et al. 1977; Zimbelman and Kieffer 1979; Palluconi and Kieffer 1981; Christensen 1986*a*).

The major regions of high inertia ($\geq$ 8) occur in the northern low plain of Acidalia Planitia (20° to 50° W, 20° to 60° N); Utopia Planitia (240° to 280° W, 30° to 50° N); Arcadia (150° to 190° W, 40° to 55° N); Isidis and Elysium Planitia (240° to 280° W, 5° to 35° N) (see Fig. 5). These regions are generally topographic lows, suggesting that they may be traps for coarse particles or lag deposits of rocks, coarse particles, and crusts or clods remaining after the removal of fines by the wind.

The southern cratered highlands have moderate inertias that range from 6 to 8 (Fig. 5). These uplands lack both of the extremes in inertias found in the north, although there are localized areas of high inertias associated with crater floors and other topographic features that have been interpreted to be local accumulations of wind-blown sediment (Christensen 1983; Zimbelman and Greeley 1982).

One of the major findings of the analyses of thermal observations is the general lack of correlation between inertia and the regional surface morphology observed from Orbiter images (Kieffer et al. 1977; Palluconi and Kieffer 1981; Zimbelman and Kieffer 1979; Zimbelman and Leshin 1987). Boundaries between the major inertia units crosscut virtually all types of unit characterizations, including geology, morphology, age and crater distribution. Thus, the surface materials, at least to a depth of several thermal-skin depths (~5–30 cm), are different than the materials of the underlying "bedrock."

Analysis of the moderate resolution (~30 km) global data set indicates that the thermal inertia variations in Lunae Planum and Oxia Palus occur at discrete boundaries, in many cases at the spatial resolution of the observations (Christensen and Malin 1988). Typically, inertias increase one or two inertia units eastward from the low-inertia region in Arabia and westward from the extensive low-inertia materials west of Tharsis into the region of Chryse Basin/Acidalia Planitia. The pattern of inertia values, and the discrete changes in values, suggest that a nonuniform process has produced the present surface in these regions, rather than continuous mixing of materials of varying grain size. A possibility, which is discussed in Sec. IV, is the presence of crusts of varying degree of induration.

Additional studies using the IRTM thermal-inertia data have focused on regional variations (Christensen and Kieffer 1979; Zimbelman and Kieffer 1979; Zimbelman and Greeley 1982) and the properties of specific types of material, including wind streaks (Peterfreund 1981; Zimbelman 1986), intracrater materials (Christensen 1983), and polar materials (Paige and Kieffer 1986). Regional studies generally confirm the results of global analyses, with little correlation between inertia and geology and morphology. Where correlations of inertia and morphology do occur, it is probable that the inertia variations have been produced after the formation of the major surface morphology. For example, features such as Kasei, Aris and other major channel systems serve to focus winds and locally affect the transport and deposition of the mobile surface fraction, resulting in accumulations of dark, high-inertia material (Christensen and Kieffer 1979; Zimbelman and Leshin 1987). Thus, it seems likely that aeolian processes play an important role in creating thermal inertia contrasts and in enhancing the pre-existing contrasts by sorting and redistributing the mobile surface material.

Further evidence for aeolian control comes from studies of streak and intracrater materials. Approximately one-fourth of all craters larger than 25 km have dark "splotches" on their floors (Thomas 1984; Arvidson 1974). These materials have high inertias relative to their surroundings; grain sizes are inferred to range from 0.1 mm to 1 cm and average 0.9 mm (Christensen 1983). The downwind location (Sagan et al. 1973*b;* Arvidson 1974), low albedo, color (Thomas 1984) and inferred grain size and rock abundance (Christensen 1983) of splotches all suggest local enhancement of particle size due to sorting (see Sec. IV.D).

Martian wind streaks have been divided into three general types: (1) bright streaks formed by deposition of dust on lee side of topographic obstacles; (2) dark streaks formed by erosion of bright material; and (3) dark streaks formed by deposition of dark material (Thomas et al. 1981; Thomas and Veverka 1979*b;* Greeley et al. 1974; Peterfreund 1981). Limited, high-resolution IRTM data indicate that some bright streaks have relatively low inertias, suggesting the presence of dust (Peterfreund 1981). In some cases, such as the Pettit crater streaks, the bright dust layer must be relatively thin (<several cm) because there is no discernable difference in inertia from the darker surroundings (Zimbelman 1986).

2. Rock Abundance. Diurnal temperature measurements alone provide information only on the average thermal inertia and particle size of the surface material, and do not give any indication of the distribution of particle sizes. This ambiguity can be resolved to a certain extent by incorporating additional, multiwavelength thermal observations to resolve the surface materials into fine and rock components. This modeling is based on the fact that the temperatures of high-thermal-inertia rocks and low-thermal-inertia fines differ by up to 60 K at night. The energy emitted from a surface of materials at different kinetic temperatures is not blackbody in nature, and the observed thermal spectrum can be inverted to determine the fraction of the surface covered by each temperature component.

In order to retrieve the areal coverage of rocks and fines, it is necessary to model the temperatures of the components (Kieffer et al. 1977; Kieffer 1976; Christensen 1986*b*). To keep the number of free parameters small, yet provide useful surface information, models used to date assume only two components: rocks and fines. The thermal inertia of the rock component was chosen to be 30, close to that of three-dimensional rocks approximately 10 to 15 cm in diameter (Kieffer 1977). This size is significant because rocks this size and larger have essentially the same temperature, whereas for rocks smaller than 10 cm the temperatures depend strongly upon size. This model is therefore only weakly sensitive to the exact size of the rocks, and the parameter thus determined is the total fraction of the surface covered by rocks 10 cm in diameter and larger.

This approach has been used to determine the global distribution of surface rocks on Mars from Viking IRTM observations taken at 7 and 20 μm (Christensen 1986*b*) (see Fig. 6). These data were binned at 1° × 1° in latitude and longitude (60 km at the equator) using data collected on multiple orbits. The relative uncertainty in the 7 and 20 μm temperatures was ~0.2 K which, together with uncertainties in the variations in emitted energy due to nonunit surface emissivities and atmospheric dust, produces uncertainties in the rock abundance determinations. The estimated accuracy of the surface rock cover determined from the present model is approximately ±20% of the derived value.

Results from this modeling indicate that the modal surface rock cover is 6%, with abundances ranging from several percent to ~35%. The rock abundances of the regions surrounding the Viking 1 and 2 landing sites, determined from this model, are ~10% and 20%, respectively, in good agreement with the rock abundances observed from the Landers, as discussed previously.

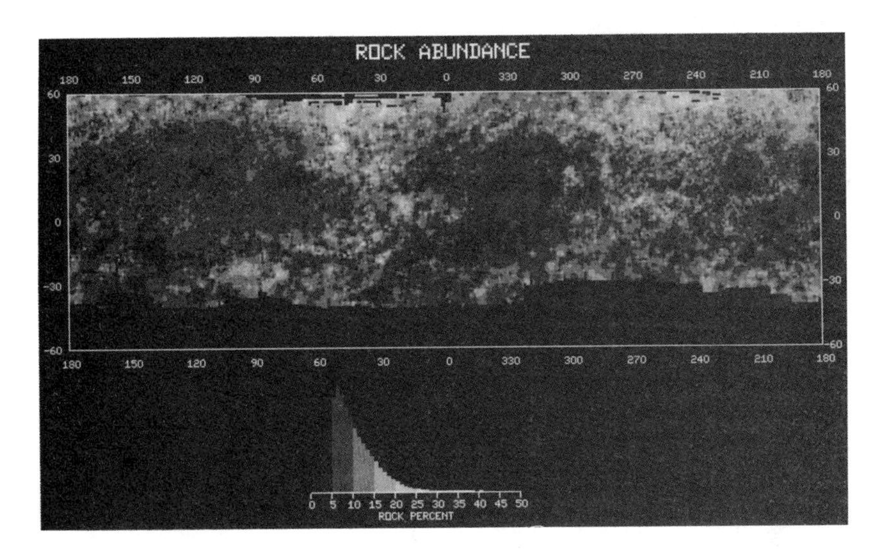

Fig. 6. The abundance of surface rocks. Rock abundance is given as the fraction of the surface covered by material of inertia 30. Values were determined by modeling pre-dawn 7 and 20 μm IRTM temperatures (figure from Christensen 1986*b*).

Thus, in retrospect, it can be seen that both sites have above average rock abundances, and the VL2 site is one of the rockiest regions on the planet.

Using this model, the thermal inertia observations can be reinterpreted in terms of a rock component and a fine component. The diurnal temperatures used to derive thermal inertia provide only an average particle size of the surface, which could increase by addition of rocks, keeping the particle size of the fine component fixed, or vice versa. Once the abundance of rocks has been estimated from the multiwavelength data, the inertia of the remaining fine component can be determined based on the constraint that the inertia of the fines, together with the abundance of rocks, must result in a bulk inertia equal to that derived from diurnal temperature measurements (Kieffer et al. 1977).

The resulting fine-component inertias are generally 1 unit lower than the bulk inertia (Christensen 1986*b*), but have essentially the same spatial pattern and variability as the bulk thermal inertias. These results indicate that global variations in the properties of the fine materials, rather than variations in the abundance of rocks, are the major contributors to the observed thermal inertia variations.

3. Layered Surfaces. Subsurface layering results in a diurnal variation in surface temperatures that deviates from that of an ideal, homogeneous surface (Kieffer et al. 1977; Jakosky 1979; Ditteon 1982). Thus, the diurnal temperature measurements from the IRTM provide a means to investigate the presence and distribution of layered surfaces (Ditteon 1982; Jakosky and Christensen 1986*b*). The IRTM data indicate that there are systematic temperature deviations from a homogeneous model, especially in the afternoon. During this period the observed temperatures are lower than predicted, resulting in a phenomena that has been termed "anomalous afternoon cooling" (Kieffer et al. 1973,1977; Jakosky 1979; Ditteon 1982). The most pronounced afternoon cooling occurs within the low-inertia areas.

Detailed modeling of the anomalous afternoon temperatures indicates that a layered subsurface could explain the observed temperatures to some degree. The subsurface layers must have inertias of approximately 30 or larger and be buried at depths of 2 cm or less in order to produce the measured effect (Ditteon 1982). One of the potential problems with this model is the fact that the inertia and depth of burial were required to vary together in order to produce the proper result. In addition, while subsurface layering can account for a substantial fraction of the long-period temperature deviations, there are variations between the data and a homogeneous model that occur over short ($\sim$2 hr) time scales that even a layered model cannot reproduce (Jakosky and Christensen 1986*b*). Finally, the crust model requires relatively high-density material near the surface, but, as will be discussed in the following section, the radar observations suggest that the material in low-inertia regions is relatively unbonded. Therefore, it appears that some additional process is necessary to fully account for the observed afternoon cooling. Potential processes include temperature-dependent thermal properties, incident-angle-dependent albedo, slopes and roughness, sensible heat transfer between the surface and atmosphere, shadowing by rocks and the radiative effects of dust and water-ice clouds (Ditteon 1982; Jakosky and Christensen 1986*b;* Haberle and Jakosky 1991).

In summary, diurnal temperature data can be used to study variations in properties within the upper few skin depths (5 to 30 cm). The direct observation of crusts at the Lander sites suggests that layered materials may be common on Mars, but at present their detection using thermal data is debated. Other techniques, including the radar and radio emission measurements discussed in Secs. III.B and C, offer additional means for probing the Martian subsurface.

B. Radar Observations

Radar observations of Mars from Earth and from Mars orbit provide 4 kinds of information about the surface and near surface: (1) root-mean-square (*rms*) surface tilts (slopes) on a scale comparable to or larger than the radar wavelength (meters); (2) surface reflectivity and dielectric constant; (3) fine-scale roughness on a scale of the radar wavelength and smaller (centimeters); and (4) elevations and relief on a kilometer scale. The details of radar observations and theory are discussed fully in chapter 20; here we will only summarize the results as they relate to derived surface properties.

1. Root-Mean-Square Slope. Quantitative estimates of surface slopes can be obtained from measurements of the dispersion of radar echoes in time and/or frequency. The *rms* slope is a statistical measure of the probability distribution of surface slopes derived from the quasi-specular component of the echo. In practice, *rms* slopes are obtained by modeling a slope probability distribution and adjusting this distribution to fit the angular spread of the observed quasi-specular echo (Hagfors 1964; chapter 20). These distributions may be exponential, gaussian, or multimodal depending on the actual distribution of slopes producing them. Quasi-specular echoes arise from surface facets on the scale of the radar wavelength and larger that are properly oriented for mirror-like reflection. This is the component of surface texture that is gently undulating. Small tilts are most common, with the majority of the reflecting facets only angled a few degrees from the mean surface slope.

In general, Mars appears to have relatively low *rms* slopes as measured at radar wavelengths (chapter 20). Typical values range from 2° to 4°, indicating that Mars is smoother on average than the Moon. Mars has more extreme values than the Moon, however, particularly in areas where there has been extensive volcanic activity. Syria and Sinai Planum southwest of Tharsis are relatively smooth, with *rms* slopes from 0°.5 to 2°.0 (Downs et al. 1975). Western Syrtis Major also appears very smooth, with *rms* slopes of ~0°.25

(Simpson et al. 1982). In general, the cratered highlands have average (~3°) *rms* slopes, whereas the volcanic plains range in value from the lowest in Syrtis Major to highest values observed to date (5°–10°) in the Tharsis region (Simpson et al 1982; chapter 20).

2. *Radar Cross Section and the Dielectric Constant.* The reflectivity, or equivalently, the radar cross section, may be used to estimate the dielectric constant of the surface materials (Dyce et al. 1967; Pettengill et al. 1969,1973; Downs et al. 1973,1975; Simpson et al. 1977,1978*a,b;* chapter 20). Reflectivity is the ratio of the strength of the radar echo to the incident signal. For mirror-like reflections the reflectivity of the quasi-specular echo can be directly related to the dielectric constant (ε) of that fraction of the surface producing the echo. However, if only part of the surface reflects in a specular manner, then a correction factor must be applied.

Variations in ε are primarily due to changes in the bulk density of the surface, with lower-density material generally having a lower ε and lower reflectivity (Campbell and Ulrichs 1969; Jakosky and Muhleman 1981). Under unusual conditions, however, it might be possible to have compositional variations in ε due to exposures of high-ε materials, such as metals or ices. Solid-rock materials have values of ε ranging from 5 to 9, corresponding to reflectivities from 0.15 to 0.25. The dielectric constant of a powder can be related to that of a solid rock of the same composition through an expression such as the Rayleigh mixing formula (Campbell and Ulrichs 1969), with typical values of ε ranging from 2 to 3, corresponding to reflectivities from 0.03 to 0.07.

Downs et al. (1973) reported an "average" reflectivity for Mars of 6.4% based on an extensive series of measurements in the south equatorial region (14° to 22° S). They noted considerable variation, however, in particular the extremely low reflectivities in Tharsis and the region to the west. A reflectivity of 6.4% corresponds to a dielectric constant of 2.8 and a density of approximately 1.5 g cm^{-3}, which is consistent with powdered rock values. Higher values ($\varepsilon = 4$) have been reported in Syrtis Major (Simpson et al. 1982) and elsewhere, suggesting the presence of solid rock on the surface.

In practice, a measured reflectivity value comes from both surface reflection and scattering from some depth within the regolith. The depth of penetration increases with increasing wavelength and decreasing absorption coefficient. The fact that there may be more than one interface contributing to the echo signal introduces some uncertainty into the interpretation. For example, radar signals at 12.6 cm wavelength may be reflected in part by a bedrock layer a meter or more below the surface; such a layer might not be apparent from either image or infrared data.

3. *Small-Scale Roughness.* That part of the radar echo that cannot be attributed to mirror-like reflection from large facets is called the "diffuse"

component. It arises from scattering by irregular structure in the surface, perturbations on the facets and from multiple scattering, typically at scales of 0.3 to 3 times the radar wavelength (see chapter 20). This scattering results in a depolarized echo in addition to the quasi-specular, polarized return. At present no quantitative model has been developed that fully incorporates multiple scattering at and below the surface, but several semi-quantitative approaches to this analysis have been attempted.

Dual-polarization radar observations at 12.6 cm have detected the diffuse reflection, with a concentration of data in the Tharsis region at 20° to 25° N. These observations indicate that the Tharsis region has very large concentrations of surface to near-surface roughness elements (Harmon et al. 1982; Harmon and Ostro 1985).

Calvin et al. (1988) have modeled the diffuse reflectance using the derived surface rock abundance with a model that assumed isotropic scattering from surface rocks only. The scattering cross section was assumed equal to the geometric cross section, which was modeled either as three-dimensional objects whose scattering efficiency varied with their projected area, or as flat-lying plates with no increase in scattering cross section with viewing geometry (Calvin et al. 1988). Reasonable values of total reflectance were obtained, but neither the variations of depolarized echo cross section with longitude, nor the spectra of depolarized radar reflection with phase angle, were well reproduced.

In a second analysis, Thompson and Moore (1989*b*) devised a model for 12.5-cm depolarized radar echoes backscattered from Mars that: (1) reproduces the variations of the total cross sections with longitude observed by the Goldstone radar in 1986 along 7° S; (2) yields larger magnitudes of total radar cross section along 22° N than those along 7° S, in agreement with the 1980–82 Arecibo observations (Harmon et al. 1982; Harmon and Ostro 1985); and (3) produces depolarized echo spectra that broadly match those observed by the Arecibo radar (Harmon et al. 1982; Harmon and Ostro 1985). In their model the surface of Mars was divided into 61 radar map units with varying depolarized echo strengths. These radar map units were defined, first, by generalized geologic map units and, then, by thermal inertia map units. Depolarized echo strengths were assigned to the map units and then adjusted until the model cross sections and spectra agreed as well as possible with the observed cross sections and spectra. Although these results are not unique they do provide insight into possible variations in surface properties.

Modeled depolarized cross sections of the radar units vary from 0.007 to 0.175. Cross sections for the cratered uplands and other old terrains are 0.010, but those for the northern plains are higher and range from 0.015 to 0.040. Young volcanoes have cross sections that range from 0.070 to 0.175, and those of the surrounding lava plains range from 0.050 to 0.160. The concentrations of roughness elements such a slags and blocks on lava flows or rocks range from 3% to 76% areal cover. Large concentrations of roughness elements (22% to 70%) are inferred for the lava plains of the Tharsis, Amazonis and Elysium regions, where low thermal inertias suggest that the roughness elements are buried in places by loose dust. Inferred concentrations of roughness elements in northern low plains, such as Chryse, Arcadia and Acidalia Planitiae and Vastitas Borealis range from 6.5% to 15%. The high thermal inertias of the northern plains suggest that the roughness elements are chiefly at the surface. Inferred concentration of roughness elements for the cratered uplands and some older terrains are 2.7% to 3.8%.

Muhleman et al. (1989) have obtained a series of 3.6-cm wavelength images of Mars using bistatic, depolarized echoes generated at the Goldstone facility and received at the Very Large Array (VLA) (see chapter 20). This method can resolve a 60-km spot at the sub-Earth point and can be used to determine the radar cross section and diffuse return at this scale. This technique provides a powerful tool for studying the surface properties. In these data, the Tharsis region has a strong return, indicating a high density of small-scale scatterers. The area to the west, however, gives essentially no radar reflection. These images also reveal a bright spot on the residual south polar cap, suggesting peculiar scattering by ice (Muhleman et al. 1989).

C. Radio Emission Observations

Radio emission observations of Mars have been made from Earth at wavelengths of 2 and 6 cm (Rudy et al. 1987), 2.8 cm (Cuzzi and Muhleman 1972; Doherty et al. 1979; Jakosky and Muhleman 1980), 8 mm (Kuzmin and Losovskii 1984) and 3.5 mm (Epstein et al. 1983). These data consist of brightness-temperature measurements as a function of the sub-Earth latitude and longitude of Mars for the entire Martian disk. As different portions of the planet are visible, the whole-disk temperatures are observed to vary.

Radio-brightness temperature depends on surface properties in a variety of ways. First, for homogeneous surfaces of a given albedo the brightness temperature varies with inertia and radio wavelength because the subsurface temperatures vary with inertia and the depth of sampling varies with wavelength. Thus, a high-inertia surface will have higher average temperatures at depth and will appear warmer at longer radio wavelengths than a low-inertia surface (Jakosky and Muhleman 1980). Second, the brightness temperature depends on radio emissivity or dielectric constant, which in turn depends primarily on the bulk density of the near-surface material (Campbell and Ulrichs 1969). Finally, the presence of buried rocks can scatter radio emissions, altering both the emissivity and the subsurface weighting function (Jakosky and Muhleman 1980; Keihm 1982; Epstein et al. 1983).

Analysis of the data involves modeling the thermal emission from the surface and subsurface, using measured surface albedo and inertia, and deriving best-fit values for the subsurface properties (Cuzzi and Muhleman 1972; Jakosky and Muhleman 1980). Results from an analysis and comparison of the whole-disk measurements indicate that a lunar-like surface with a loss tangent of 0.0075 is consistent with the data (Cuzzi and Muhleman 1972). The globally averaged value of dielectric constant was found to be 2.5, along with a loss tangent of 0.003 to 0.015, an albedo of 0.25, a thermal inertia of 6 and a thermal emissivity of 0.90 (Cuzzi and Muhleman 1972).

Observations from the VLA have been used to provide a spatially resolved view of radio emission from Mars at wavelengths of 2 and 6 cm (Rudy et al. 1987). These observations included simultaneous measurements of the polarized and unpolarized energy, and a separate estimate of the whole-disk dielectric constant can be made for each. This analysis gave whole-disk dielectric constants of 2.34 ± 0.05 and 2.70 ± 0.10 at 2 and 6 cm, respectively, which correspond to subsurface densities at these wavelengths of 1.25 ± 0.11 and 1.45 ± 0.10 g cm^{-3}.

A latitudinal variation in density derived from the VLA observations indicates that density varies between ~1 and 1.8 (± 0.5) at both wavelengths. The density values derived from 2- and 6-cm observations are relatively consistent in the northern low- to mid-latitudes, varying from ~1.4 at the equator to ~1.8 at 40° N. The radio absorption length for these wavelengths was assumed to be 14, in very close agreement with values derived for the Moon (Muhleman 1972), indicating that the densities measured are average values over depths of roughly 30 to 90 cm at 2 and 6 cm wavelengths, respectively. These observations indicate that the surface to a depth of several wavelengths is not dominated by rocks on the scale of several wavelengths. The higher densities determined at 6 cm suggest that the density of the subsurface increases with depth.

These observations suggest that the Martian surface is dominated by fine, dry, particulate material with a density of 1.0 to 1.6 g cm^{-3}. In addition, there is a better correlation between thermal inertia derived from the IRTM data and the 2-cm observations, than with the 6-cm data, indicating that the properties of the subsurface are slightly different from those at the surface, and suggesting the possibility of subsurface layering.

D. Albedo Observations

Albedo observations have been used to study the reflectance properties of the surface using broadband Viking IRTM measurements from 0.3 to 3.0 μm (Pleskot and Miner 1981) (see Fig. 7) and Viking camera multiple-filter observations (Thorpe 1981). These measurements provide information on the brightness and phase function, which in turn can be used to study the compositional variability, particle size, packing, porosity and macroscale roughness of the uppermost micrometers of the surface layer.

IRTM observations have also been used to determine the spatial and temporal variations of the albedo of the Martian surface and atmosphere throughout the Viking mission (Pleskot and Miner 1981; Christensen 1988). The classical bright and dark regions are readily apparent (Fig. 7), varying in albedo from 0.10 to 0.36. The major changes in albedo with time were as-

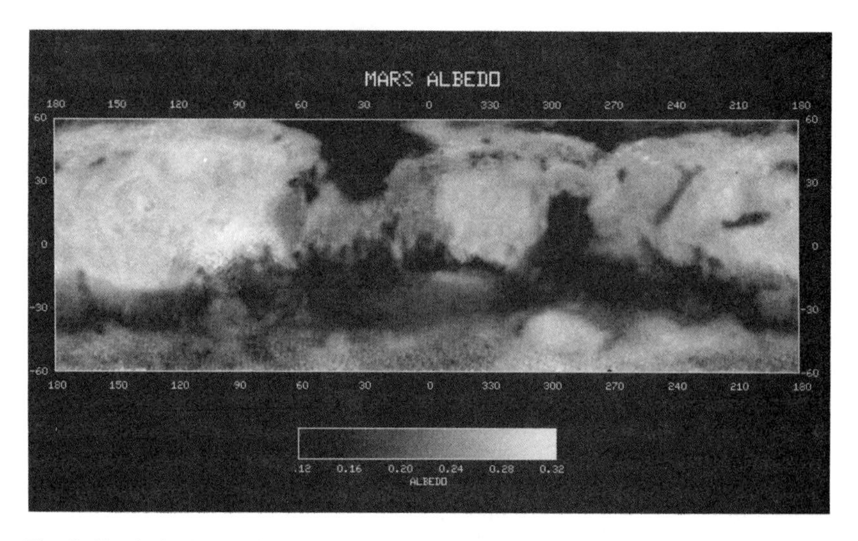

Fig. 7. Nominal, clear period albedo. Data were acquired by the IRTM instrument and were compiled and described by Pleskot and Miner (1981).

sociated with the two global dust storms of 1977, beginning near L_S 205° and 274°. For optically thick atmospheric dust, the observed albedo was 0.37 to 0.40. Over all regions of the planet the atmosphere cleared substantially between the two storms, with many southern hemisphere regions returning to their pre-storm albedos. In general, the northern hemisphere atmosphere had a higher dust content throughout the storm phase, and retained dust longer during the decay phases, than the southern hemisphere.

Observations of changes in absolute surface reflectance throughout a Martian year provide information on the deposition and removal of bright dust from different regions (Christensen 1988). Southern-hemisphere dark regions were not measurably brighter after the atmosphere had cleared by L_S 330°, suggesting little net deposition of dust (Fig. 8a). In contrast, the northern-hemisphere dark regions of Syrtis Major and Acidalia Planitia were measurably brighter following the storm (Fig. 8b), indicating the deposition of ~7 to 45 μm of dust per year (Christensen 1988). These surfaces subsequently darkened over the following months, returning to pre-storm albedo values prior to the next dust storm season. The material removed from these surfaces may account for the general dust haze observed throughout the year. Based on the ubiquitous presence of dust in the northern hemisphere during the decay phase of storms, it is assumed that the bright dust is also deposited by fallout on bright, low-inertia surfaces, where it may remain.

E. Color Observations

Color observations have been used to infer compositional units on Mars (Soderblom et al. 1978; McCord et al. 1982*b*) as is discussed in detail in

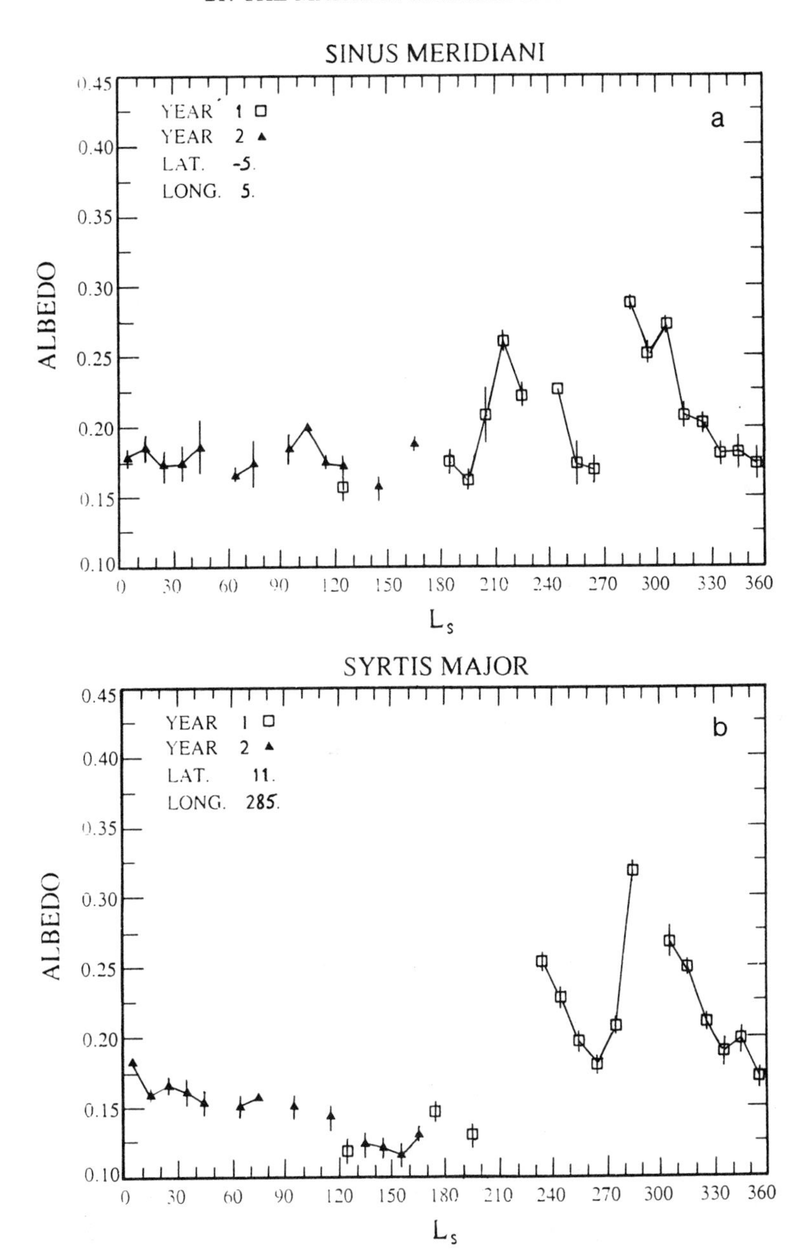

Fig. 8. Albedo variations with season. (a) Seasonal variation of phase-corrected albedo of Sinus Meridiani. Note the return to pre-storm values by L_S 330° (figure from Christensen 1988). (b) Seasonal variation of phase-corrected albedo of Syrtis Major. Note that the albedo remains above pre-storm value until L_S 150° (figure from Christensen 1988).

chapter 17. The discussion here will focus on the physical properties of the surface determined from color observations. Of particular utility are studies of the mixing of surface materials using mixing trends identified in multi-spectral images (Arvidson et al. 1982,1989*b;* Presley and Arvidson 1988; Adams et al. 1986; see Color Plate 15).

Among the most significant complicating factors in color analyses are the presence of sky radiance and the effects of surface scattering. Arvidson et al. (1989*b*) have addressed these problems using *in situ* Lander observations to separate the atmospheric effect from surface reflectance in Viking Orbiter color images, and the Hapke photometric function to correct to constant viewing and illumination geometries. From these results, it was found that the multiplicative and additive terms of sky brightness canceled one another for conditions pertaining to the data that were utilized. Once these corrections were performed, the surficial dust is indistinguishable from Earth-based spectra of Martian bright areas.

This study showed that the region around the MMS site consists of two major types of material, red and gray, with the red unit subdivided into bright and dark subcomponents (Arvidson et al. 1989*b*). There is no evidence to suggest that any of these units are directly related to the underlying bedrock. The bright red material is similar to the dust found at the MMS site and is interpreted to be wind-blown dust. The gray unit appears similar to mafic rock materials mixed with small amounts of bright-red dust. The dark-red unit has a radiance factor suggesting that it is a mixture of the bright-red and dark-gray materials (Arvidson et al. 1989*b*). Based on the photometric behavior with phase angle, this unit appears to be relatively rough at subpixel scales, and was interpreted to be an indurated deposit (Arvidson et al. 1989*b;* Kieffer et al. 1981). This study demonstrated that roughness plays an important role in color variations, and that many color units represent physical, rather than compositional, differences.

F. Imaging Observations

High-resolution orbiter images are important to the interpretation of remote-sensing observations. The images of Mars portray landforms similar to those shown in terrestrial and lunar photographs. Geologic experience can be utilized to infer kinds of surface materials present, and estimates of thickness and relative times of formation of surficial deposits can be made using geologic techniques.

The major difficulty in relating existing Mariner 9 and Viking Orbiter images to surficial properties and processes is the difference in the nature of processes producing surface features at different scales. Often these processes differ greatly in energy and in timing. Meter- to decameter-scale landforms may have been produced by ancient processes, such as channel formation and impact cratering. Conversely, the millimeter- to meter-scale properties sensed by nonimaging techniques are produced by less energetic, but more recently active, processes such as aeolian erosion and deposition. Because of these differences, we will not deal with images and their interpretations at length, but an example of a comparative analysis between morphologic and remote-sensing observations can be found in Schaber (1980).

IV. DISCUSSION

A. Major Correlations in Global Data Sets

Important correlations between inertia, albedo, color, radar properties and elevation have been identified. These correlations have been discussed by several authors (Kieffer et al. 1977; Palluconi and Kieffer 1981; Arvidson et al. 1982,1989*b;* Jakosky and Christensen 1986*b*); only a brief review is presented here.

1. Thermal Inertia and Albedo. Inertia and albedo are inversely correlated at all spatial scales and over most of the surface (Kieffer et al. 1977; Palluconi and Kieffer 1981; Zimbelman and Leshin 1987) (see Fig. 9). This correlation is most evident for materials with the highest and lowest inertias. Comparison of data collected in 2° × 2° latitude-longitude bins reveals that materials with inertias < 4 have measured albedos > 0.24, with a mode at 0.28, materials of inertia 6.5 have a modal albedo of 0.21, and surfaces with

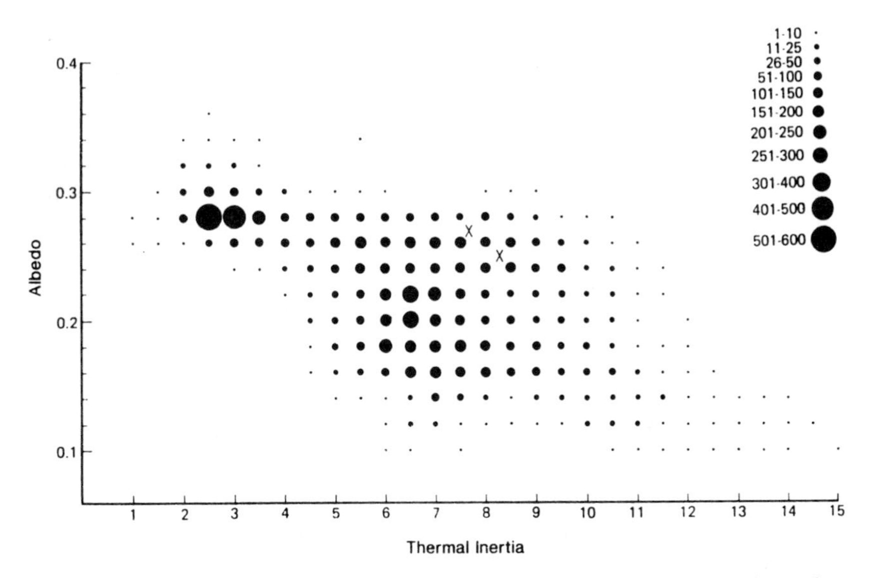

Fig. 9. Correlation of thermal inertia with albedo. The size of each circle represents the number of 2° × 2° latitude-longitude bins which have a given combination of albedo and inertia. Crosses indicate the values at the two Lander sites (figure from Jakosky and Christensen 1986*a*).

inertias of 14 to 15 have albedos from 0.10 to 0.14 (Palluconi and Kieffer 1981; Jakosky and Christensen 1986*b*) (see Fig. 9).

The most likely explanation for this correlation is that the low-inertia material is composed of bright dust, with the inference that the high-inertia materials are dark, dust-free lithic fragments (Kieffer et al. 1977). The intermediate albedo and inertia materials exhibit less correlation, although the general trend is still present (Fig. 9). Materials with intermediate inertias and albedos may possess a wide range of properties in both particle size and bonding; materials with similar inertias may have substantially different compositions, degree of dust cover, degree of induration and origin (see Sec. IV.D).

2. *Thermal Inertia and Color.* Global, factor-analysis studies using three-filter Viking Orbiter images, thermal inertia and albedo as a five-component system have identified clusters associated with major surface units (Arvidson et al. 1982; Presley and Arvidson 1988). These studies showed two major correlations. The first was related to color properties alone, in which changes in surface brightness occur simultaneously, but not uniformly, at red, green and blue wavelengths (Arvidson et al. 1982). The second trend confirmed results from the IRTM data analysis of a strong correlation between inertia and reflectance (Kieffer et al. 1977,1981; Palluconi and Kieffer 1981). Once these trends were reduced using principal component techniques, eight statistically significant clusters were identified. There appeared to be mixing between clusters, indicating that the boundaries were gradational (Arvidson et al. 1982). The three major components of these clusters consist of: (1) bright-red materials of low inertia; (2) dark, gray materials of high inertia; and (3) materials that are intermediate in all variables. Mixing between these materials produces the eight clusters and results in the gradational nature of the boundaries between units. This mixing suggests that further subdivision of the surface materials into a larger number of units may not be significant (Arvidson et al. 1982).

3. *Thermal Inertia and Rock Abundance.* Although the variation of rock abundance does not account for the variation in thermal inertia, there is a weak correlation between these parameters, with high-inertia surfaces having more rocks. This correlation is most apparent in the comparison of low-inertia surfaces (<3) that have a mean rock abundance of approximately 4%, with high-inertia surfaces (>8) that have rock abundances of 12% to 20% (Christensen 1986*b*). These results are consistent with the presence of a mantle of fine-grained material of low inertia that covers some fraction of the near-surface rocks. Rocks in low-inertia areas may be partially buried, and high-inertia surfaces have more rocks, coarse particles and perhaps crusts. However, in limited cases, such as in Syrtis Major, high-inertia surfaces have

few rocks, possibly due to mantling by extensive deposits of sand (Simpson et al. 1982; Christensen 1986*a*).

4. Thermal Inertia and Radar Reflectivity. Thermal inertia and dielectric constant, determined from radar reflectivity at wavelengths of 3.8, 12.5 and 70 cm, are correlated for approximately 75% of the longitudes observed at a latitude of +22° N (Jakosky and Muhleman 1981) (see Fig. 10). Regions that do not correlate are within the dark region of Chryse Planitia (Jakosky and Muhleman 1981). Study of a larger radar data set acquired at 12.5 cm (Downs et al. 1973,1975) also shows a correlation between inertia and radar reflectivity (Jakosky and Christensen 1986*a*). A similar direct correlation between inertia and dielectric constant appears probable for the Goldstone 1986 observations along ~7° S (Thompson and Moore 1989*b*).

Jakosky and Christensen (1986*a,b*) discussed possible explanations for these correlations. Their preferred model is one in which grain-to-grain bonding of surface materials results in an increase in both the density and the inertia. In the simplest case, increases in particle size of unbonded materials are not accompanied by increases in density, so particle size alone will not account for the observed correlations. Although conductivity and radar reflectivity increase with bulk density or grain packing, the radar data indicate densities of 0.9 to 1.7 g cm^{-3} for the low- and high-inertia surfaces, respectively, so that grain packing alone is not sufficient to produce the observed increase in inertia. High rock abundances (~50%) are required to produce the correlation (Jakosky and Muhleman 1981), which does not agree with values derived from thermal data.

Thermal conductivity does increase with increased conductivity at grain-to-grain contacts and through the filling of pores with cement of a higher thermal conductivity than the gas. The radar dielectric constant and density also increase due to the filling of pores by cement or fine material. Thus, varying degrees of bonding within the upper meter can explain the observed correlation between density and inertia. Both of these explanations appear qualitatively reasonable, but the necessary laboratory measurements of thermal conductivity with grain-to-grain bonding to address this problem in a more quantitative manner do not currently exist.

5. Thermal Inertia and Elevation. A weak anticorrelation is observed between inertia and elevation, with inertia decreasing with increasing elevation (Kieffer et al. 1977; Jakosky 1979; Palluconi and Kieffer 1981). An important component of this relationship may result from the variation of gas pressure with elevation. For gas pressures between 1 and 10 mbar, the thermal conductivity varies by approximately the square root of gas pressure, resulting in a fourth-root variation in inertia (Kieffer et al. 1973). This mechanism can produce a small variation in inertia of approximately 0.16 per 1 km of

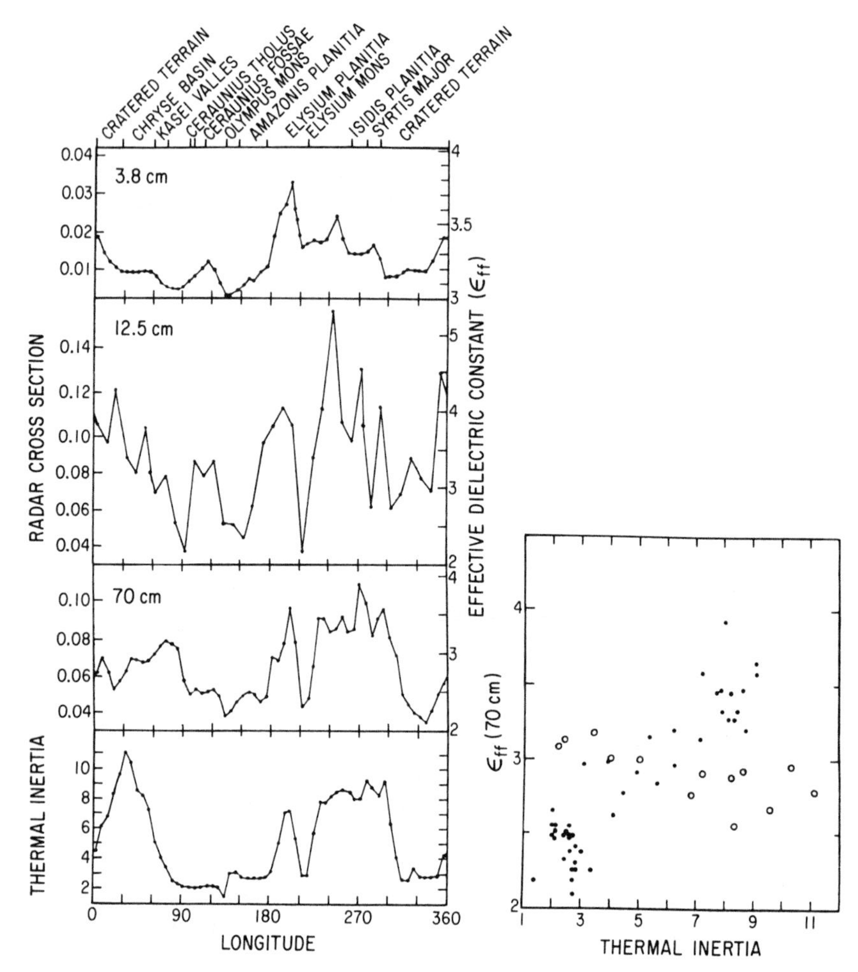

Fig. 10. Comparison of thermal inertia and radar cross section. (a) Longitudinal variation in inertia and cross section (and inferred dielectric constant) along latitude 20° N. Note the strong correlation over much of the planet at 70-cm radar and the lesser correlation at shorter wavelengths (figure from Jakosky and Muhleman 1981). (b) Correlation of 70-cm radar dielectric constant and thermal inertia. Each point represents an average over 6° of longitude at 22° N latitude. Open circles indicate points located between 10° and 90° longitude; closed circles are points from the remaining longitudes (figure from Jakosky and Muhleman 1981).

elevation (Jakosky 1979). This effect alone can explain much of the observed extremes in inertia, especially the variation in inertia over the Tharsis volcanoes (Zimbelman 1984).

Within basins such as Chryse, Acidalia Planitia, Hellas and Huygens, the observed increases in inertia with elevation could not be produced by this effect, indicating real variations in surface properties. The high inertias in these basins may be due to the presence of coarser-grained materials or cementation, although cementation may be less important because the low albedos, high rock abundances, and poor correlation of dielectric constant with inertia suggest the presence of rocks and coarse, mobile particles. This result provides evidence that the low-elevation regions on Mars are not regions of accumulation of fine, dust-storm fallout. It appears that the higher atmospheric pressures result in the removal of fines from the surface, whereas fine material appears to collect at higher elevations, where it is more difficult to set in motion due to the decreased atmospheric pressure (Zimbelman and Kieffer 1979; Christensen 1986*a*).

B. Distribution and Origin of Global Surface Units

Each of the data sets discussed above provides a different measure of the surface properties and, perhaps more importantly, each probes to a different depth into the surface. In particular, the albedo and color data are relevant to the upper few tens of microns; the thermal inertia data provide information on the upper few thermal skin depths (2 to 10 cm); the rock abundance data provide direct information on the abundance of rocks at the surface and indirect information for the thickness of surface mantles of fine material; and the radar and radio observations provide information of the upper few radio skin depths (0.1 to 10 m). Together these observations can be used to identify four major surface units and provide insight into their origin. The distribution of the units based on inertia and albedo properties alone is shown in Fig. 11; regions that did not fit into any of the specified classifications or for which there were no data are shown in black.

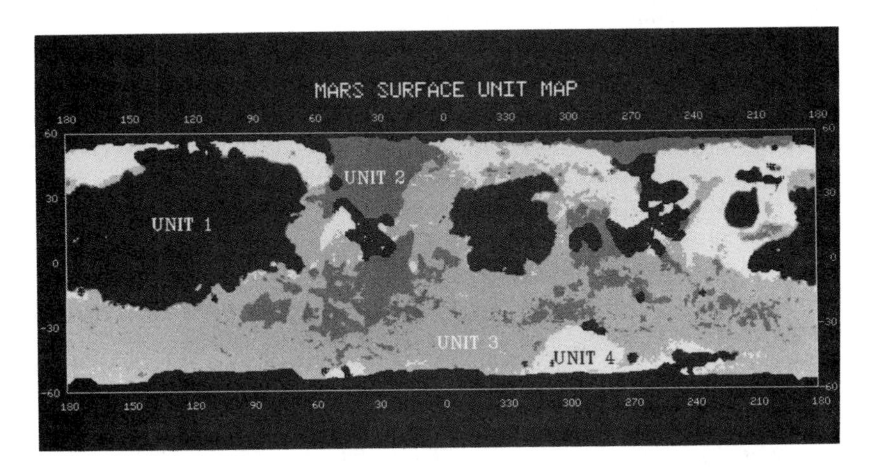

Fig. 11. Mars surface material unit map. Units are based on values of thermal inertia I and albedo A. Values of inertia and albedo used to separate units are: Unit 1, $I = 1$–3.5, and $A = 0.26$–0.40; Unit 2, $I = 7.7$–20 and $A = 0.1$–0.2; Unit 3, $I = 3.5$–7.7 and $A = 0.15$–0.25; Unit 4, $I = 5$–9 and $A = 0.25$–0.3. The origin and nature of these units are discussed in detail in the text.

1. Unit 1: Active Dust Deposits. This unit is defined as having low inertia (1 to 3.5), high albedo (0.26 to 0.40), low rock abundance, low density, and bright-red color (Kieffer et al. 1977; Palluconi and Kieffer 1981; Arvidson et al., 1982,1989*b;* Christensen 1986*a;* Jakosky and Christensen 1986*b*) (see Fig. 11). The effective particle size is less than ~40 μm, indicating a predominance of loose dust to a depth of at least 2 cm. Of particular importance in determining the age and origin of this unit is the presence of some surface rocks, typically <5% areal cover, indicating incomplete burial of all rocks by a relatively thin mantle of fine material (Christensen 1986*a*).

Radar observations provide further insight into the properties of Unit 1. Of particular importance are the low reflectivities and low inferred dielectric constants, which imply the prevalence of material with low bulk density, consistent with loose dust. The very strong depolarized radar echoes from the Tharsis region, however, indicate that a major portion of Unit 1 is rough on the scale of the radar wavelength (12 cm) (Harmon et al. 1982). These observations can be reconciled with the thermal results by postulating a layer of loose dust overlying a buried, rough, lava-flow surface. Estimates of the thickness of this layer have been made from the electrical properties of dry, powdered rock and the attenuation of the returned radar signal (Christensen 1986*a*). Assuming attenuation properties similar to the Moon and a dielectric constant of 3, the thickness of the dust layer is approximately 1 to 2 m. For a dielectric constant of 2, which may be more appropriate for Tharsis, this depth is ~6 m. These thicknesses are consistent with the presence of some rocks exposed within Unit 1, indicating incomplete burial and a relatively thin mantle. Dust would conform to meter-scale topography on the flows, filling depressions and leaving ridges exposed. This depth and degree of mantling are also consistent with the interpretation of imaging observations, which indicate the presence of mantles of dust a few to ~6-m deep on what appear to be rough, lava-flow surfaces in Tharsis (Schaber 1980; Zimbelman 1984). Thus, Unit 1 appears to consist of a dust mantle 2 cm to 2–6 m in thickness overlying possibly a wide range of other types of surfaces and covering a significant fraction of the northern hemisphere.

Unit 1 is interpreted to be areas of current, active dust accumulation (Zimbelman and Kieffer 1979; Palluconi and Kieffer 1981; Christensen 1986*a;* Arvidson et al. 1989*b*). The primary distinctions between these surfaces and others at the present are differences in wind velocity and surface shear stress (Christensen 1986*b;* Pollack et al. 1981). Global circulation models suggest strong topographic control of atmospheric motion, producing low-level convergence and low wind velocities over most of the present low-inertia regions (Webster 1977; Pollack et al. 1981).

Whatever the mechanism for initiating these deposits, the accumulation of fine materials would act as a positive-feedback mechanism, with the burial of sand and rocks and the in-place bonding of fine particles making it increasingly difficult to raise dust (Zimbelman and Kieffer 1979; Christensen

1986*b*). Wind-tunnel studies have shown that 40-μm particles require wind velocities twice as high as sand-sized (120 μm) particles in order to be moved (Greeley et al. 1980; chapter 22). It is likely, therefore, that burial of sand-sized particles is a major factor in allowing further accumulation of fines and the development of the low-inertia deposits to their current depth and extent (Zimbelman and Kieffer 1979; Christensen 1986*a*).

The age of this unit has been estimated, based on estimates of current and previous dust-deposition rates (Christensen 1986*a*). During the 1977 global storms, approximately 10 to 40 μm of dust was deposited globally (Pollack et al. 1979*b;* Christensen 1986*a*). Over geologic time, cyclic variations in orbital parameters (Ward 1974,1979) certainly produce variations in solar insolation, atmospheric pressure and wind velocities (see chapter 33 for a review), although the effect of these variations on dust-storm activity remains uncertain. Periods of high pressure may produce nearly continuous storms, with perhaps a factor of 10 higher dust sedimentation rate than at present, whereas during low-pressure periods the winds may be incapable of raising dust at all (Pollack and Toon 1982). Therefore, over geologic time the dust sedimentation rate may have varied from 0 to ~250 μm per year, but the timing and duration of these limits remains uncertain. These rates, together with an estimated thickness of several meters for the surfaces in Unit 1, imply that this unit must be relatively young, on the order of 10^5 to 10^6 yr (Christensen 1986*a*).

2. *Unit 2: Active, Coarse Material.* This unit is characterized by high inertia (7.7 to 20) and low albedo (0.1 to 0.2), with rock abundance and *rms* slope typically high, but variable. Major occurrences are in the northern hemisphere and in the southern equatorial zone, generally within dark regions (Fig. 11). Despite variations in some properties, surfaces within this unit do have consistent characteristics and appear to have been formed by similar processes. The low albedos suggest dust-free surfaces. In the present environment, there is significant dust deposition on most, if not all, northern hemisphere surfaces (Christensen 1988), and the accumulation of dust can only be prevented by active removal. Dust deposition and subsequent removal is confirmed by Lander (Arvidson et al. 1983) and global albedo (Christensen 1988) observations.

Thermal data indicate coarse materials on these dark surfaces, with average particle sizes in the range from 0.1 mm to 1 cm. Wind tunnel measurements show that 0.12 mm particles, which are within this range, are the ones that are most easily set in motion in the present Martian environment (Greeley et al. 1980; chapter 22). These observations suggest a simple explanation for the lack of dust on these surfaces; sand-sized particles are set in motion by the wind, which in turn erodes dust from the surface. Once in motion the dust is easily carried in suspension over long distances. This process can continue until the dust is deposited on a surface from which it cannot be eroded, where it remains and accumulates. Alternatively, the dust could be set in motion directly by the wind, although this would require wind velocities a factor of approximately 2 higher (Greeley et al. 1980). This process appears to have occurred at the Viking Lander sites, where dust was removed with little or no disturbance of other surface materials (Arvidson et al. 1983).

The presence of numerous rocks on most of Unit 2 is consistent with the presence of coarse, active fines. Locally, however, there are dark, moderate-to-high-inertia surfaces that have few rocks on the surface and are regions of very low *rms* slopes determined from radar observations. Of particular interest is a region in Syrtis Major from 0° to 15° N, 290° to 300° W that is very dark with sand-sized particles, yet has few rocks (<7%) and extremely low (0°.5 to 2°.0) *rms* slopes (Simpson et al. 1982). This smooth surface was suggested to be due to resurfacing by low-viscosity lavas or to relatively smooth sand surfaces (Simpson et al. 1982). Dark dunes have been observed between 0° to 10° N, 290° to 295° W (Peterfreund 1981) and may be pervasive over the entire low-albedo surface of Syrtis Major (Lee 1986*a*). These observations support the presence of a mantle of dark sand deposits, which could account for the low albedo, the high inertia, the low rock abundance and the low *rms* slope. Thus, regions of low slopes and few rocks may differ from other surfaces in Unit 2 only in the presence of a higher abundance of sand that can bury rocks and mantle the surface to sufficient depth to produce a relatively flat-lying upper boundary (Christensen 1986*b*).

In summary, the surfaces in Unit 2 appear to be active, with particle motion preventing net accumulation of dust. Dark, coarse grains and rock material are exposed at the surface. The rock abundance is generally high but locally reduced due to mantling by abundant sand-sized material. Within this unit, aeolian resurfacing, rather than the underlying geology, may be the most important factor for controlling the abundance of surface rocks and roughness.

3. Unit 3: Indurated (?) Surfaces. The combination of thermal and radar observations, together with Viking Orbiter color observations, provide evidence for the presence of a third type of surface. This surface is average with regard to its thermophysical properties, with moderate inertias (3.5 to 7.7) and albedos (0.15 to 0.25) (Fig. 11). However, it does not appear to be simply a transition region between the low-inertia Unit 1 and the high-inertia Unit 2. Instead it is unique in that it is characterized by a correlation between thermal conductivity, as derived from thermal inertia, and density, as inferred from radar reflectivity determined at wavelengths of 3.8, 12.5 and 70 cm (Jakosky and Muhleman 1981).

As discussed previously, the correlation of radar and thermal data support the presence of a surface or subsurface crust of bonded material. This correlation does not occur within Unit 2, consistent with the argument that that unit is characterized by unbonded, active, coarse materials. Low values of inertia and density are found in Unit 1, again consistent with loose, unbonded materials at the surface, although there is the possibility of partial bonding at depth (Ditteon 1982).

The color data also support the possibility of crusts based on mixing trends between materials of different colors (Arvidson et al. 1982,1989*b;* Presley and Arvidson 1988), as discussed in Sec. III.E. Regions of intermediate inertia and color in Unit 3 appear to be mixtures of bright-red and dark gray material, but require a relatively rough surface at subpixel scales. Thermal inertia data suggest that these materials should be easily moved by winds, yet no evidence for transport was found and these surfaces appear to be substrates across which other materials migrate. This apparent lack of mobility was suggested to be due to induration, increasing the inertia and decreasing the mobility of the surface material (Arvidson et al. 1989*b*).

Finally, the occurrence of discrete increases in inertia within this unit may be related to the presence of surfaces of differing degrees of crust formation (see Sec. III.A). If these changes are due to crust formation, they provide additional evidence for the importance of induration in this unit.

In summary, Unit 3 consists of surfaces of moderate inertia, albedo and color that may be characterized by the presence of surface or subsurface crusts. In this regard, it differs from Units 1 and 2, which are composed primarily of unbonded material that is actively deposited or transported by aeolian activity. Unit 3 occurs around the margin of Unit 1 in the northern hemisphere and may also form a major portion of the southern hemisphere.

4. Unit 4: Mixed Coarse and Dust Materials. Unit 4 is characterized by moderate-to-high inertia (5 to 9) but is unique in having a high albedo (0.25 to 0.30) (Fig. 11). This unit is uncommon, but includes both of the Viking landing sites and appears to be different in several aspects from much of the rest of the planet (Jakosky and Christensen 1986*a*). In particular, in the region of the MMS site there is not a correlation between inertia and radar reflectivity as seen elsewhere. This observation suggests that crusts may be thinner or discontinuous within this unit than in Unit 3. In addition, the high albedos indicate the presence of a layer of bright dust, which does not have sufficient thickness ($\lesssim$1 cm) to affect the thermal properties. This observation is also consistent with observations at MMS, where a thin layer of dust was observed to be deposited and subsequently removed.

C. Comparison of Global and Local Properties

The Viking Lander sites currently provide the only direct tie between the global remote sensing measurements and direct, *in situ* observations of the surface of Mars. It appears that the Lander sites are representative of the processes that occur globally but are located on a relatively uncommon surface type (Unit 4). They have not accumulated dust to the same degree as Unit 1; they do not appear as active as Unit 2; and the MMS site is not characterized by a strong correlation between radar-inferred density and inertia, suggesting that the MMS site at least is not dominated by near-surface crusts as suggested for Unit 3 (Jakosky and Christensen 1986*a*). Materials within the sites are, however, good analogs for materials elsewhere on Mars (Moore and Jakosky 1989). In particular, rocks, fines and crusts are present, but in different proportions to the major units. Drift material has a low thermal inertia, low dielectric constant and fine grain size, and should be similar in its mechanical properties to the loose dust of Unit 1. The Lander sites may be similar to regions with substantial aeolian activity and exposed rocks and sand, but are currently less active, are forming crusts, and have accumulated small amounts of surface dust (Jakosky and Christensen 1986*a*).

D. Active Surface Processes

Some interesting observations and conjectures regarding the nature of the processes currently operating on Mars may be drawn from the properties of the global surface units. One of the major conclusions that comes from the available suite of observations is that material in the upper meter of the surface over much of the planet is active at the present time.

1. Dust Deposition. Active dust deposition appears to be occurring over a substantial fraction of the northern hemisphere, with net accumulation in the current low-inertia, bright regions. This process is very large in scale, and may be dominated by atmospheric circulation, rather than surface properties. Evidence for its occurrence was discussed in Sec. IV.C.

2. Erosion. To support the current deposition, active erosion of surface materials must be occurring in some locations. It is unclear whether the source regions for the global dust storms are local to the dust-storm-initiation sites, or are globally distributed. In addition, the apparent nature of the dust deposits suggests that erosion of this material has occurred at some point in the past (Christensen 1986*a;* Arvidson et al. 1989*b*). The properties of these dust deposits make them difficult to erode, however, with 20 to 40 μm particles requiring wind velocities a factor of 2 higher than the most readily moved particles (Greeley et al. 1980). Several possible mechanisms have been proposed, including incomplete burial of sand and exposed rocks, the formation of coarse particles by aggregation of dust, the occurrence of atmospheric vortices (dust devils) of various scales, and climatic changes in the general circulation (see the review by Christensen 1986*a*).

Of these possible processes, dust devils may be important because they form due to convection in an atmosphere with a superadiabatic lapse rate (Ryan 1972), so they preferentially occur over hot surfaces. They would be most pronounced over low-inertia surfaces during day time, and may be an effective mechanism for eroding fine material during periods of maximum summer insolation. Dust devils were observed from the Lander sites (Ryan and Lucich 1983) and from Orbiter images (Thomas and Gierasch 1985). Of the 118 vortices measured by the Landers, 4 had inferred maximum wind velocities >30 m s^{-1}, which represented a factor of 2 to 3 enhancement of the background wind velocity and may have been strong enough to raise dust (Ryan and Lucich 1983; Greeley et al. 1981).

Local enhancements in the global circulation, including slope winds and local topography, may also be important for eroding the surface and producing local and global dust storms (Lee et al. 1982; chapter 22). Changes in the obliquity and eccentricity may produce changes in atmospheric circulation, with regions of current erosion becoming sites of deposition. A detailed study of these changes must await further development and application of circulation models incorporating the full effects of topography, atmospheric dust and surface properties.

An important question concerns the degree of rock erosion at the two Lander sites. Some rocks have sharp, angular facets, suggesting that abrasion has been relatively limited (Sharp and Malin 1984). Material does appear to move through these sites, however, as evidenced by the presence of wind scours, windtails, drifts and granular ripples, and disturbed surface materials were eroded by strong winds at MMS. It is possible that the rocks have been protected by burial by fines for a substantial period of time, consistent with the presence of drift material covering large rocks at the MMS site.

3. Crust Formation. Chemical analyses suggest that the crusts are similar in composition to unbonded fines except that the amounts of S and Cl are twice as large in the crusts, suggesting that sulfate salts act as a bonding agent (Clark et al. 1982; chapter 18). The formation of crusted materials is generally thought to be enhanced by the presence of liquid water (Fuller and Hargraves 1978). In Antarctica, it appears that a monolayer of unfrozen, adsorbed water is capable of transporting dissolved ions through soils even at temperatures below freezing (Ugolini and Anderson 1973; Cary and Mayland 1972). This process may be relevant to Mars, where liquid water is not currently stable at the surface (Ingersoll 1970; Farmer 1976) but a monolayer of adsorbed water does appear to exist in the upper layer, even in the equatorial regions (Fanale and Cannon 1974; Houck et al. 1973; Jakosky 1983*a*). The ion diffusion rates of Ugolini and Anderson (1973) imply that salts can migrate several meters in the upper surface layer in a time period of 10^5 yr (Jakosky and Christensen 1986*b*). Diffusion of salts may be facilitated by the diffusion of water into and out of the regolith on a time scale of 10^5 to 10^6 yr (Fanale and Jakosky, 1982*a*; Fanale et al. 1982*a*; Lindner and Jakosky 1985). Thus, it may be possible for crusts to form near the surface of Mars on these time scales (Jakosky and Christensen 1986*b*). These times are similar to those estimated for the age of the current dust deposits. This similarity in rates is consistent with the hypothesis that dust deposition, bonding, erosion and subsequent deposition is a cyclic process occurring in the northern (and southern ?) hemisphere in the recent past (Christensen 1986*a;* Arvidson et al. 1989*b*).

4. Aeolian Sorting. Aeolian activity appears to be the dominant, active process on Mars, based upon global remote sensing observations, and on *in situ* views from the Viking Landers. At a local scale, aeolian processes are primarily responsible for sediment transport through saltation, traction and suspension (Christensen 1983; Zimbelman and Kieffer 1979; Thomas 1984; chapter 22). A possible mechanism for the accumulation of high-inertia, dark material may be aeolian redistribution of mobile material within topographic traps. Marginally mobile material may be trapped within craters and other topographic lows due to the asymmetry in slope profiles. There must be some size particle that can be marginally transported up the gentle exterior slope of craters and other obstacles, either by saltation or by traction, but which cannot be transported back up the steeper, interior slope (Christensen 1983). Thus, local enhancements of particle size within topographic depressions will occur regardless of the exact nature of the wind velocity because of the differential nature of transport of material into, versus out of, these areas.

This process may serve to trap locally sediments being transported across the Martian surface, and thus may provide a means of sampling the composition of the surrounding bedrock. Once these coarse sediments are collected, they often form dunes indicating saltation, perhaps active at the present time. This activity is likely responsible for the dark, apparently dust-free surfaces of these dune and splotch forms (Thomas 1984; Christensen 1983).

5. Cyclic Deposition and Erosion. It appears that Mars is currently undergoing active deposition and erosion. There are indications that this activity extends back in time, with at least one episode of burial and exhumation evident at the MMS site and indications of older, indurated surfaces suggested from the pattern of thermal inertia observed in Lunae Planum and Arabia (Christensen and Malin 1988; Arvidson et al. 1989*b*). Alternating periods of deposition and erosion on a much larger scale are indicated by the occurrence of craters and other landforms that appear to be undergoing exhumation from beneath overlying mantles of sediments (McCauley 1973; Arvidson 1974), and by the polar laminated terrains (chapter 23).

Current dust accumulations appear to have a relatively young age (Christensen 1986*a;* Arvidson et al. 1989*b*) but it is unlikely that deposition in these regions was initiated for the first time in Martian history within the past 1 Myr. A more plausible explanation for the young age of these surfaces is the cyclic or episodic deposition and removal of surficial materials, with the deposits of Unit 1 representing the most recent phase of activity.

The most likely cause for cyclic variations are the cyclic changes in climatic conditions, presumably produced by changes in the orbital parameters (Murray et al. 1973; Ward 1974,1979; chapters 9 and 33). Polar-layered terrains are generally accepted to be evidence for these processes and it is reasonable that such changes would affect sedimentary processes in the equatorial region as well. The current distribution of dust deposits is consistent with such a climate-controlled model. At present, the strongest winds occur in the southern hemisphere during the season when atmospheric circulation models predict the presence of strong cross-equatorial Hadley circulation capable of transporting material from south to north (Haberle et al. 1982; Haberle 1986*b;* chapter 26). Because of the close connection between dust transport and atmospheric circulation at present, it is possible that under reversed conditions of north-to-south circulation during epochs when the maximum solar heating occurs in the north, the current deposits in the north would be eroded and transported back to the south.

The time scale for significant changes in atmospheric conditions is controlled by the obliquity and eccentricity variations, which combine to give north-south oscillations in the maximum solar insolation with a period of 5.3×10^4 yr (Ward 1979; Pollack and Toon 1982). The approximate agreement between this time scale and the estimated age of the Unit 1 deposits may be coincidental, but could indicate the existence of astronomically driven climatic controls on the deposition and erosion in the equatorial region.

Changes in sedimentary conditions and reworking of surface materials are consistent with the presence of buried boulders at MMS that have subsequently been partially exhumed. It may also explain the discrepancy between the apparent strength of aeolian activity but the lack of a significant degree of surface erosion; much of the erosive power of the wind may be expended in the continuous erosion and redeposition of a relatively thin layer of dust, preventing substantial new erosion of fresh, bedrock surfaces. Continual redistribution of material in the equatorial zone dating much further back in Martian history is recorded by the lack of mantling of small craters within the regions poleward of 40° (Soderblom et al. 1974). Based on analysis of Mariner 9 images (which may contain a bias because periapsis was located at ~25° S), the zone with the least mantling lies near 25° S. This area corresponds to the current zone of highest winds and has been interpreted to represent the most recent phase of a continual reworking of sediments within the zone from 25° S to 25° N (Soderblom et al. 1974).

If changes in environmental factors play an important role in controlling surface processes, then the upper layer may be an important paleoclimatic indicator, providing a record of surface-atmosphere interactions over the past few Myr. Older surfaces, either buried or exposed outside the current extent of the accumulating dust deposits, may provide a similar record extending further back in time.

An important, outstanding question concerns the origin and nature of the substantial mantles produced early in Mars' history and the relationship of the processes that produced them to those operating today. It has been proposed that there are large regions of Mars that have undergone episodes of substantial deposition and erosion of sediments that may be up to several kilometers thick (Soderblom et al. 1973*a;* Pieri 1976; Arvidson et al. 1984). At present it is difficult to determine the relationship between the small-scale deposition recorded in the upper few meters and the much thicker mantles apparently produced in an earlier stage in Martian history. Unfortunately the processes that produced these earlier mantles are poorly understood, and it is likely that they were significantly different from the processes operating today. There may, however, be some connection, with the earlier mantles consisting of wind-deposited materials produced when atmospheric conditions permitted more rapid erosion and transport, or when impact-related processes produced an environment different from that observed today.

V. CONCLUSIONS

The surface layer of Mars holds the key to our understanding of a wide range of surface processes that have shaped the surface over time and may be active today. Our knowledge of the surface properties comes from two sources, global remote sensing and *in situ* observations from the two Viking Landers. The remote sensing observations include thermal, radar, radio emission, albedo and color data. These data have shown that there exists a wide range of surface materials, but that much of the surface can be characterized by 4 basic surface units. Unit 1 is covered by fine, bright dust that appears to be actively accumulating, with few rocks exposed at the surface. Radar observations of this unit indicate a very rough surface that is most likely buried beneath several meters of fine dust. This unit is interpreted to be a recent deposit, whose existence and extent may be linked to periodic climate changes. Unit 2 is also active, with the motion of particles preventing dust accumulation and resulting in a dark, coarse surface with numerous rocks. Unit 3 has intermediate inertia, albedo and color, and is interpreted to be a rough, indurated surface over which other materials can be transported. Finally, Unit 4, which contains both Viking landing sites, is characterized by relatively high inertia, but also has a high albedo, suggesting that dust may have accumulated, but has not reached sufficient thickness to affect the thermal properties.

The Viking sites have rocks, fines of varying cohesion and crusts. Aeolian processes are evident at both sites, with indications of erosion and deposition in the recent past. Large rocks have been exhumed from beneath fines, suggesting that one or more cycles of deposition and erosion have occurred. Large-scale features, however, indicate that little net erosion has occurred, suggesting that the major activity has been reworking of material, rather than erosion of bedrock. The landing sites are representative of the types of processes that have occurred globally, but are not exactly representative of any of the major units. They have not accumulated significant amounts of dust, nor are they experiencing active transport and erosion by sand-sized particles. Crusts are present, but not to the degree that may be present elsewhere. The materials present, in particular rocks, fines and crusts, are, however, good analogs for materials inferred elsewhere from the remote-sensing data.

Numerous, recently active processes are inferred from the surface characteristics, including dust deposition, erosion, aeolian transport and sorting, and crust formation. In addition, the global data, together with the Lander observations, suggest that cyclic changes in sedimentary processes may occur associated with periodic climate changes. At present, regional dust deposits occur in the north and the maximum wind velocity and dust storm activity occurs in the south, with net south-to-north transport. At a different stage in orbital evolution the maximum wind activity should occur in the north and these deposits may be eroded and transported elsewhere.

Perhaps the most significant conclusion that can be drawn from our current understanding of the upper layer is that much of the present surface is young and has been continually reworked. The continued erosion and redeposition of this loose material, rather than erosion of fresh surfaces, may provide the material for the high rates of aeolian activity. As a consequence, much of the fine material on the surface may have been globally homogenized and may be essentially decoupled from the underlying bedrock.

Acknowledgment. We wish to thank R. Arvidson, B. Jakosky, F. Palluconi, R. Simpson and C. Snyder for their thoughtful reviews of this manuscript which contributed significantly to its improvement. Discussions with numerous other individuals, including H. Kieffer, S. Lee and R. Haberle, have contributed to the ideas expressed here. This work was supported by the National Aeronautics and Space Administration.

22. MARTIAN AEOLIAN PROCESSES, SEDIMENTS, AND FEATURES

RONALD GREELEY, NICHOLAS LANCASTER
Arizona State University

STEVEN LEE
University of Colorado

and

PETER THOMAS
Cornell University

In the present environment of Mars, the most active, and perhaps the only, significant surface-modifying process is the wind. Moreover, on the assumption that Mars has had an atmosphere throughout most of its 4.6 Gyr history, aeolian processes have probably always played a significant role in the evolution of its surface. Vast dune fields, various albedo patterns that change with time, wind-eroded hills (yardangs), and drifts of fine-grained material observed at the Viking landing sites are all attributed to aeolian processes. Large parts of the cratered uplands and smooth terrains in both polar regions are considered to be composed of deposits of windblown particles. These deposits may be important reservoirs for volatiles and may influence climate change via changes in regional albedo. Because aeolian processes and products (e.g., sediments) are important today and in the past, they must be taken into account in models of surface evolution and for planning future missions to Mars.

I. INTRODUCTION

Even before exploration of Mars via spacecraft, there was speculation that the Red Planet was swept by dust storms. Perhaps the first person to consider aeolian processes on Mars in the geologic context was McLaughlin (1954) who suggested that the changing albedo patterns observed telescopically from Earth resulted from windblown dust. Unfortunately, McLaughlin's work was not generally recognized and a decade passed before there were serious considerations of Martian dust storms (Ryan 1964) and predictions for wind speeds needed to set particles into motion in the thin Martian atmosphere (Sagan and Pollack 1969).

Confirmation of aeolian processes on Mars came with the Mariner 9 mission in the early 1970s. Not only did the spacecraft arrive during one of the largest dust storms observed on Mars, but as the dust cleared from the atmosphere, surface features of indisputable aeolian origin were revealed, including dune fields and wind-eroded hills (McCauley 1973). The most abundant aeolian feature discovered via Mariner 9 were the so-called variable features (Sagan et al. 1972,1973). These are albedo patterns that appear, disappear, and change their size, shape, and orientation with time. Many of these features are associated with local topographic obstructions and appear to be formed by the erosion and/or deposition of windblown material. Variable features of both the "bright wind streak" and "dark wind streak" varieties serve as local wind vanes, indicating the prevailing wind direction at the time of their formation.

Although the Viking mission (1976–1981) provided much more detailed information on Martian aeolian features and processes (Fig. 1; see, e.g., Veverka et al. 1977), many fundamental questions remain unanswered. For example, sand dunes occur in nearly every region of Mars. Are the dunes composed of "sand" in the terrestrial sense? What is the origin of the grains that form the dunes? Are the dunes currently active? These and other questions remain unresolved, as discussed later in this chapter.

In the absence of liquid water, aeolian processes (and, to some extent, downslope movement of debris) currently dominate the modification of the surface of Mars. Moreover, it is likely that aeolian processes have operated throughout Martian history. Perhaps sand and dust storms were even more important in the past when atmospheric density may have been greater. As shown in Fig. 2, the minimum wind speed needed to set particles into motion is partly a function of atmospheric density. Several lines of evidence show that the climate of Mars in the past may have involved a denser atmosphere (reviewed by Clifford et al. 1988; see also chapters 32 and 33). Consequently, aeolian processes may have been even more effective, perhaps leading to the deposition of some mantling plains of the Martian upland which appear to be of aeolian origin (Scott and Carr 1978).

In this chapter, we review the aeolian regime on Mars. First, we consider the sources and characteristics of the particles that are involved in aeolian processes and the winds that are required to set grains into motion. We then review dust storms and assess previous observations and the mechanisms of dust-storm generation. Various aeolian features, including dunes and albedo features, as well as windblown mantle deposits, are then discussed. We con-

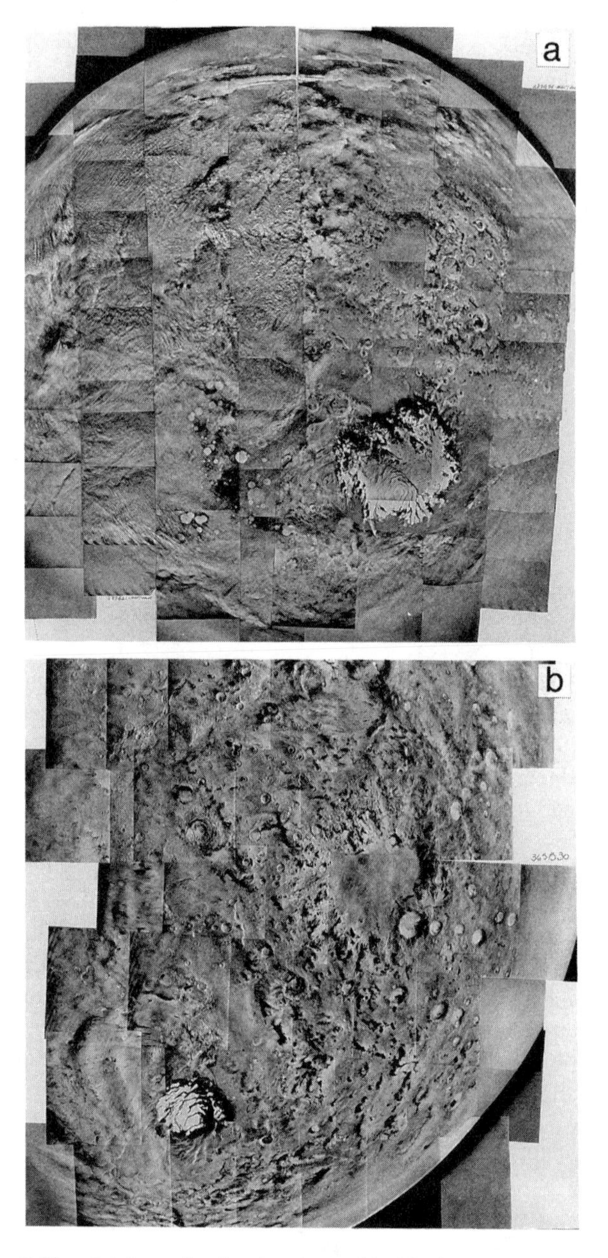

Fig. 1. Viking Orbiter global mosaics showing views of the Martian southern hemisphere during and after the great dust storms of 1977. (a) This mosaic was created from images obtained during the 1977b storm (Viking Orbiter revolution 287B; L_s ~280°, alternating red and violet filters); note the general obscuration of surface features by dust clouds. (b) This mosaic illustrates the clearing of the atmosphere following the 1977b storm (Viking Orbiter revolution 365B; L_s ~330°, red filter).

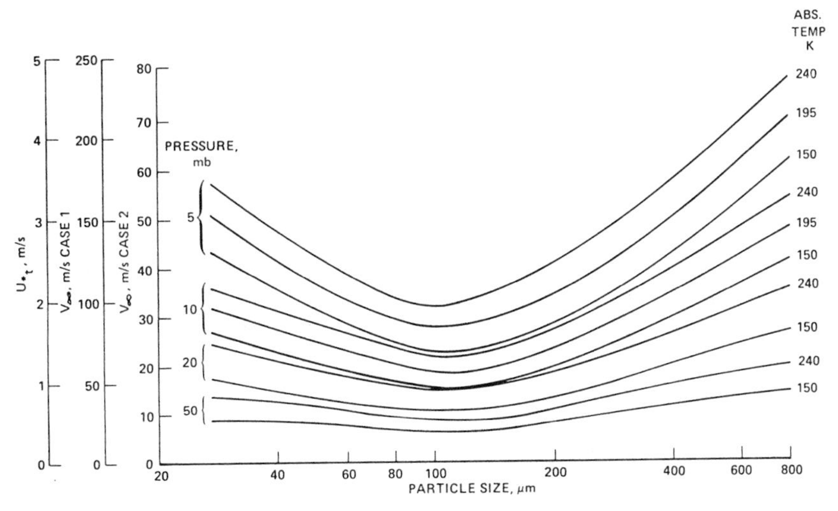

Fig. 2. Particle threshold curves as a function of particle size for Martian surface pressures (5 to 50 mbar) and for representative Martian temperatures. Case 1 scale is free-stream velocity (above boundary layer) for winds blowing over a flat surface of erodible grains; case 2 surface containing cobbles and small boulders, such as at the Viking landing sites (from Greeley et al. 1980).

clude with a consideration of the needs for future work on aeolian processes on Mars and consider some of the key measurements that could be made via future missions to Mars.

II. PARTICLES AND THEIR ENTRAINMENT BY THE WIND

There are two requirements for aeolian processes to occur: there must be a supply of loose particles (Table I) and there must be winds of sufficient strength to move the particles. In this section we review the potential supply of windblown particles, winds on Mars, and the requirements for particle entrainment by the wind.

A. Windblown Particles

Particles moved by the wind on Mars range in size from less than a micron, for suspended dust, to perhaps as large as a centimeter in diameter. The principal mode of transport and size ranges for particles are shown in Fig. 3. On Mars the particle size most easily moved is about 100 μm in diameter, or fine sand. This material is set into saltation in trajectories averaging about 1 m long and 10 to 20 cm high (White 1979). Saltating sand grains may strike larger grains (<1 cm) and push them along the surface in creep. Saltation may also dislodge small, dust-size (<20 μm) grains and set them into suspension.

TABLE I
Characteristics of Windblown Particles on Earth and (Probably) Mars

Term	Size (μm)	Transport: Dominant Mode	Transport: Height[a]	Transport: Approximate Range, km[a]	Deposit/Bedform
Granules	2000–4000	creep	ground level	10^{-3} to 10^{-2}	megaripples
Sand					
Fine	60–200	suspension/ saltation	above ground		
Dune	200–400	saltation	near ground		
Coarse	400–2000	creep/ saltation	at/near ground	10^{-3} to 10^{0}	sheets/ripples/ dunes
Dust					
Clay	<4	suspension	upper troposphere	?	
Aerosolic dust	1–10	suspension	troposphere middle-upper	10^{4} to 10^{6}	mantles
Loess	10–50	suspension	lower troposphere	10^{-1} to 10^{2}	
Silt	4–60	suspension	—	—	
Aggregates					
Dune parna	400	saltation	at/near ground	10^{-3} to 10^{0}	dunes
Sheet parna	100	suspension	lower troposphere	10^{1} to 10^{3}	mantles
Volcanic	250–500	suspension	lower troposphere	10^{0} to 10^{3}	mantles

[a] On Earth.

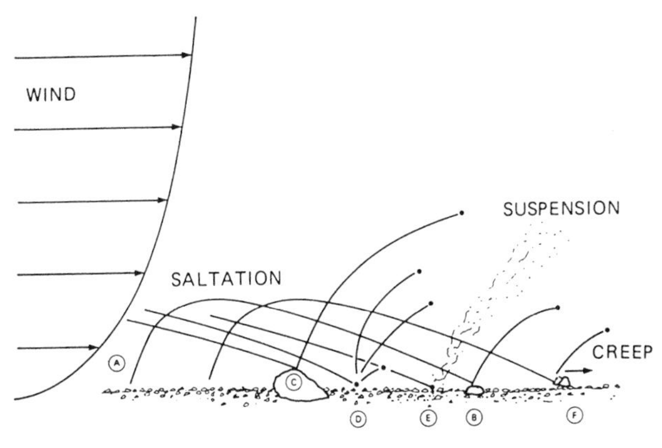

Fig. 3. Diagram showing the three principal modes of aeolian transport of grains: surface shear stress exerted by the wind causes grain (A) to lift off the surface, carries it downwind as it falls to the surface where it bounces (B) back into flight; this motion is termed *saltation;* grain at (C) hits a large rock—possibly causes some erosion—and elastically rebounds to a relatively high saltation trajectory; grain at (D) strikes the surface and "triggers" other grains into saltation; grain at (E) strikes the surface containing very fine particles (too fine to be moved by the wind alone in this case; see threshold curve, Fig. 2) and sprays them into the wind where they are carried by turbulence in *suspension;* grain at (F) strikes larger grain and pushes it downwind a short distance in a mode of transport termed *impact creep,* or *traction* (figure from Greeley and Iversen 1985).

There is a wide variety of processes that can generate particles in the sizes appropriate for entrainment by the wind. These processes include fragmentation by impact cratering and volcanism, comminution in landslides, physical and chemical weathering, and tectonic processes (such as crushing along fault planes), all of which are appropriate for Mars, and erosion by running water, which may have been possible on Mars in the past. Most of the information on the nature of small particles on Mars is derived from Viking Lander observations. Images at both landing sites show a surface littered with rocks as large as 2 m (Mutch et al. 1976*a*). Between the rocks are "drifts" of material presumably deposited and later sculpted by the wind. Drifts appear to be composed of fine grains $\lesssim$ 100 μm in diameter (Sharp and Malin 1984). Close to Lander 2, fine material was blown away by the retrorockets to expose a subsurface cemented layer termed *duricrust* by Binder et al. (1977). Duricrust is considered to be composed of fine-grained particles cemented together by soluble salts such as magnesium sulfate. Duricrust is easily broken into clods, or pellet-like aggregates of a wide range of sizes, including material of "sand" size that would be more easily moved by the wind than larger and smaller particles (Fig. 2). Pollack et al. (1979*b*) assessed the atmospheric opacity during both clear periods and during dust storms and estimated that the size of dust in the atmosphere is on the order of 1 μm. This particle size is extremely difficult to set into motion by wind shear alone (Fig. 2), but on Earth, is commonly set into motion by the impact of saltating sand.

The composition of the fine material at the landing sites was determined via the X-ray fluorescence experiment (Clark et al. 1982) and inferred from remote-sensing observations (Singer and McCord 1979; Singer et al. 1979; reviewed in chapters 17 and 18). Results show it to consist of iron-rich clays, including nontronite. The clays are probably derived from the chemical weathering of mafic rocks, such as basalts. The composition of Martian rocks and their weathering products raises an important question regarding the observed sand dunes on Mars, and the supply of sand grains in general. On Earth, most sand is composed of quartz, a mineral highly resistant to physical and chemical weathering, and derived ultimately from granites. Granite is a silica-rich rock generated by extensive crustal recycling. Thus far in Martian exploration, there is little or no evidence for Earth-like crustal recycling nor for chemically evolved rocks such as granite. Consequently, it has been suggested that quartz sand may be rare or even absent on Mars (Smalley and Krinsley 1979). Could the sand inferred to exist on Mars be composed of some material other than quartz? Basalt, inferred to be a common rock on Mars, does form sand-size grains on Earth under some circumstances. The best known are the black sand beaches of Hawaii which result from lavas pouring into the sea and their fragmentation by quenching and by wave action. Less well known are the basaltic sand dunes of the Columbia River Plateau, a region of predominantly basaltic volcanism. The low albedo of the Martian north polar sand sea would support a basaltic composition for the Martian dune sands. On Earth, however, basaltic sands quickly break down chemically and physically to clay minerals. Although it is difficult to extrapolate terrestrial weathering rates to Mars, if the sands on Mars are basaltic, then the dunes probably represent relatively recent formation, an interpretation borne out by the lack of superposed impact craters on Martian dunes.

B. Particle Threshold

The physics of windblown particles are reviewed by Greeley and Iversen (1985). When wind blows across the surface, kinetic energy is transferred to the surface through shear stress. If sufficient energy (i.e., sufficiently strong wind) is involved, then loose material such as sand can be entrained. For laminar flow over a surface, the shear stress is related to the vertical velocity gradient of the wind

$$\tau = \mu \frac{\partial U}{\partial z} \tag{1}$$

where τ is the surface shear stress, μ the viscosity coefficient, U the fluid velocity and z the distance above the surface. If a surface is smooth, laminar flow develops in a very thin surface layer, even for flows in which most of the boundary layer is turbulent. The velocity within the laminar sublayer varies with distance above the surface, so that

$$U = \tau z/\mu = u_*^2 z/\nu, \; u_* z/\nu \leq 10. \tag{2}$$

The surface friction speed u_* is the square root of the ratio of shear stress to fluid density ($u_* = (\tau/\rho)^{1/2}$). The kinematic viscosity ν is the ratio of absolute viscosity μ to fluid density, ρ, ($\nu = \mu/\rho$). The laminar sublayer is estimated to be < 9 mm on Mars for very smooth, homogeneous surfaces composed entirely of fine grains (White 1981). Thus, very fine particles, such as dust, would be completely immersed in the laminar sublayer and would be extremely difficult to entrain by the wind unless disturbed, as by saltation impact of larger, more easily moved grains.

Above the laminar sublayer, the shear stress is sustained by turbulent exchange, and the coefficient of proportionality in Eq. (1) is a function of the flow rather than of the fluid. Above the sublayer, the wind speed becomes a logarithmic function of height, written:

$$\frac{U}{u_*} = \frac{1}{0.4} \ln \frac{9 u_* z}{\nu}. \tag{3}$$

This equation is valid for values of surface friction Reynolds number $Ru_* = uu_* D_p/\nu \leq 5$, when the surface is said to be aerodynamically smooth and to consist only of a smooth, quiescent sand surface of mean particle diameter D_p. If $R_* \geq 70$, the surface roughness (pebbles, rocks, etc.) become larger than the thickness of the laminar sublayer, the laminar sublayer ceases to exist and the surface is said to be aerodynamically rough. The velocity profile then becomes independent of viscosity, such that

$$\frac{U}{u_*} = \frac{1}{0.4} \ln \frac{z}{z_o} \tag{4}$$

where z_o is the equivalent roughness height. For a quiescent sand surface, z_o is about 1/30 the sand particle diameter (Bagnold 1941). For most rough surfaces, whether they consist of pebbles, rocks or lava flows, the equivalent roughness height under neutral atmospheric conditions must be determined by measurement of the wind speed at two or more heights above the surface.

The critical point at which particles begin to move was termed by Bagnold (1941) as the fluid threshold (U_{*t})

$$U_{*t} = A \left(\frac{\rho_p - \rho_a}{\rho_a} g D_p \right)^{-2} \tag{5}$$

in which A is an empirical coefficient, ρ_p is grain density, ρ_a is atmospheric density, g is gravitational acceleration and D_p is grain diameter. Figure 2 shows the threshold friction velocity required to move different size particles with a density of 2.73 g cm^{-3}; curves are shown for a range of atmospheric surface pressures and temperatures appropriate to Mars. The increase in friction velocity to entrain particles smaller than about 100 μm is due to aerodynamic effects (e.g., immersion in laminar sublayer) and interparticle forces. Interparticle forces may be due to moisture, electrostatic forces and other forces of cohesion which are known to be relatively more important for small particles and relatively independent of particle density (Iversen et al. 1976). Particles of any solid, if small enough, cohere on contact, even when thoroughly dry; this is particularly true in a vacuum. Electrostatic forces appear to be especially important for grain cohesion on both Earth and Mars (Greeley and Leach 1978).

C. Alternative Methods of Dust Raising

Estimates for particle size in Martian dust storms yield values of a few microns and smaller in diameter (Pollack et al., 1979*b*). Estimates of threshold wind speeds for particles of this size show that markedly higher winds are required for particles progressively smaller than the optimum size of ~ 100 μm (Fig. 2). However, once threshold is attained, very low winds can keep dust aloft. Because threshold scales inversely with atmospheric density, the winds required to mobilize dust on Mars are exceedingly high; for example, minimum winds required to move 10-μm grains exceed the speed of sound and are far higher than those measured (or predicted) on Mars. Thus, mechanisms other than wind shear alone have been proposed for raising dust on Mars. These include: (a) dust fountaining by desorbed CO_2 (Johnson et al. 1975) and H_2O (Huguenin et al. 1979; Greeley and Leach 1979); (b) dust devils (Neubauer 1966; Sagan and Pollack 1969; Thomas and Gierasch 1985) or perhaps "tornados" (Grant and Schultz 1987); (c) presence of triggering particles (Peterfreund 1981); (d) clumping of fine grains to produce particles of larger, more easily moved sizes (aggregates, chunks of duricrust, etc., as reviewed by Moore and Jakosky [1989]); and (e) winds during unstable atmospheric conditions.

Johnson et al. (1975) conducted experiments with 1 to 10 μm silica particles in a CO_2 atmosphere which was cooled until CO_2 was absorbed onto the particles; the system was then heated so that desorption occurred. The surface initially formed a crust which acted as a barrier, causing buildup of subsurface gas pressure until the crust ruptured; the gas jets ejected dust several cm above the surface. They suggested that similar processes on Mars could inject very fine particles into the atmosphere where they could be carried aloft by relatively gentle winds.

A similar mechanism could occur involving water (Huguenin et al. 1979; Greeley and Leach 1979). Exploratory experiments show the effects of water vaporization on the movement of dust. In bell-jar tests, as the atmospheric pressure was reduced below 10 mbar (temperature ~ 24°C) absorbed water vaporized and ejected particles by one of two processes: (1) vent holes and fissures developed in the surface, followed by a fountain-like spray of particles as high as 20 cm above the surface; or (2) violent eruptions occurred in a boiling fashion; the smaller the grain size, the more violent the eruption. Some activity increased with depth of particle bed, with 20-cm high fountains occurring for beds 10-cm deep; some activity was also observed in beds as shallow as 1 mm. The amount of absorbed water affected the activity, with ejection occurring with water contents as low as 0.75% by weight. Similar effects were observed in tests conducted in the wind tunnel. However, in some experiments the particle bed remained stable until a low-velocity wind passed over the surface at which time injection of dust was triggered. Although there is no clear explanation for this effect, the triggering could be related to lift, produced by slight differences in pressure resulting from the wind. In some cases, desorption of water did not eject dust, but caused the surface to crack. Erosion began as the wind picked up crust-like sections of the particles; the wind speed of ~25 m s^{-1} was substantially lower than threshold for undisturbed particles. Evidently, the fissures sufficiently roughened the surface to lower the threshold speed.

Dust devils are local cyclonic winds that result from atmospheric instabilities. They have been suggested as a dust-raising mechanism on Mars, an idea enhanced by the discovery of possible dust devils observed on Viking Orbiter images (Ryan and Lucich 1983; Thomas and Gierasch 1985). Field studies of dust devils on Earth (Sinclair 1966; Carrol and Ryan 1970) show that a wide range of particle sizes can be raised even when unidirectional winds are very gentle. Laboratory experiments (Greeley et al. 1981) suggest that the entrainment mechanism for raising particles via dust devils is markedly different than for a uniform wind boundary-layer case. In dust devils, the primary factor may be the difference in pressure between the top and bottom of the particle layer, and the size and density of the particles seem to have little importance in threshold; consequently, even the very small grains are easily raised.

Bagnold (1941) defined two types of threshold. *Fluid threshold* is defined as the wind speed at which continuous particle motion first occurs as described above. *Impact threshold* occurs if particles are introduced from upstream; continuous movement of particles from the initially quiescent surface begins at a lower wind speed. Impact threshold is ~80% of the fluid threshold. Several authors have suggested that dust on Mars could be entrained by saltation impact of larger (e.g., sand), more easily moved grains (Peterfreund 1981; Christensen 1983). In order to assess this mechanism, wind tunnel experiments conducted at Martian atmospheric pressures were run involving dust particles (< 20 μm) that were impacted by saltating sand grains (120-μm diameter). Although some dust was entrained upon impact by the sand, the effect was not as pronounced as expected. Many of the grains bounced off the surface of the dust bed or became buried without dislodging any dust. Moreover, some of the dust that was dislodged was redeposited within a meter from its source.

Any process that could increase the effective diameter of dust particles could lower the threshold necessary for entrainment. Aggregation of small particles by electrostatic bonding occurs in dust storms (Kamra 1972) and volcanic eruptions (Sorem 1982), and has been proposed for Mars (Greeley 1979). Moore et al. (1987) have shown that clods of small particles occur at the Viking landing sites. Experiments show that threshold velocities for aggregate clods of basaltic silt are ~25% lower than the same nonaggregated silt grains. Even though the aggregates are very fragile and do not survive saltation impact, the soils are sufficiently rough that clumps are more easily picked up by the wind. Disrupted crusts developed on deposits of fine grains may also produce clumps that are more easily entrained by low winds. Gillette et al. (1982) note significantly lower threshold values for silt and clay on Earth where crusts on the deposits have been disturbed. For application to Mars, duricrust and cloddy soils have been described; disturbance by events such as the slumping in drift deposits at "Big Joe" could provide local surface roughness which would lower the effective threshold wind speed for Martian dust.

Atmospheric conditions can have a marked influence on threshold velocity, particularly for dust-size grains. The values given in Fig. 2 are for neutral atmospheric conditions. However, for unstable atmospheres, as occur during surface heating, vertical turbulent motions are enhanced by buoyancy. To assess this potential effect on Mars, preliminary experiments (White et al. 1989) at low atmospheric density (7 mbar, "Earth" air) were run in which the surface of a bed of dust (particles <10 μm) was heated to produce a ΔT of 25K above ambient. Threshold measurements were then made and it was found that friction velocities were reduced by 25 to 40% below those for neutral atmospheric conditions.

In summary, static threshold for very small Martian dust grains appears to be difficult to achieve. The observations of abundant and frequent dust storms suggest that one or more of the mechanisms described above, or other, unknown mechanisms, must be occurring on Mars.

III. MARTIAN DUST STORMS

Dust storms probably have a greater short-term influence on the appearance and evolution of the Martian surface at present than any other phenomenon, as evidenced by numerous variable albedo features that are attributed to aeolian processes. Long-term atmospheric dust transport may be responsible for extensive sedimentary deposits, particularly in the polar regions where condensation of volatiles may lead to rapid deposition of suspended dust. Investigation of Martian aeolian processes and features requires an understanding of the location, genesis, timing and frequency of dust storms.

A. Earth-Based Observations

Yellow clouds obscuring part or all of the Martian surface have been reported by telescopic observers as long ago as 1800 (cf. Slipher 1962; Briggs et al. 1979; Zurek 1982; Martin 1984; see chapter 2). Even though high-quality photographs of Mars date from early in this century, Earth-based views at opposition (when viewing is optimized) and at the same season on Mars are repeatable only every 15 yr. Thus, the historical record of dust storms is variable in spatial and temporal coverage. It is evident, however, that while localized dust clouds appear during all seasons, global or great dust storms are most common when Mars is near perihelion. This period of maximum insolation occurs currently in late southern spring near $L_s \sim 250°$ (Capen 1974; Peterfreund and Kieffer 1979; Martin 1984).

The great dust storms typically originate as localized dust clouds in any of several areas in the southern temperate zone, such as Hellespontus, Isidis Planitia/Syrtis Major and Claritas Fossae/Solis Planum (Fig. 4; Zurek 1982; Martin 1984). These storms expand slowly over a few days, after which they may decline or rapidly intensify into globe-encircling dust hazes (Fig. 1) that may obscure the entire planet (Martin 1974*a*,1976,1984). The atmosphere clears in a few weeks or months, and large-scale albedo features again become discernable. Only a handful of storms have become truly global in extent, including those of 1922, 1956, 1969, two in 1971, 1973 and two observed via Viking in 1977 (Martin 1984).

B. Spacecraft Observations

Mariner 9 arrived at Mars during the slow decay of the second global dust storm of 1971. During the early weeks of the mission, only the summits

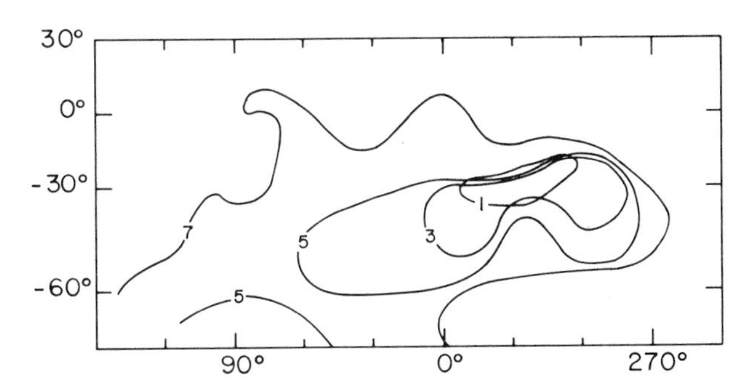

Fig. 4. Variations in the positions and size of dust clouds during the first 7 days (1971 September 22–29) of growth of the great dust storm of 1971. The numbered outlines delineate the extent of the clouds on particular days (figure simplified from Martin 1974*a*).

of the Tharsis volcanoes were visible above the dust pall. Images of the planet's limb revealed dust loading to heights of ~60 km (Leovy et al. 1972). The atmosphere slowly cleared as viewed from orbit, and surface albedo markings became increasingly visible over the course of several weeks (Sagan et al. 1973*b*)

The Viking Orbiters were operating in Mars orbit well before the major 1977 dust storms, providing observations of both the growth and decay phases of storm activity. The first storm (designated 1977*a*) began at L_s ~205° and grew to global proportions within a few days; it had decayed significantly by the time the second great storm (1977*b*) arose at L_s ~275° (Briggs et al. 1979; Pollack et al. 1979*b*). Numerous local storms were seen before and after this period, particularly at high southern latitudes and near the retreating southern polar cap (Peterfreund and Kieffer 1979; Peterfreund 1985).

The Viking Landers provided *in situ* observations of the growth and decay of the major 1977 storms, and provided a time base of observations during nonstorm conditions. As seen from the Landers, the dust storms were detected by increases in atmospheric opacity over a period of a few days, followed by an asymptotic decline (Fig. 5). "Background" opacities of ≤0.5 were typical for most of the year, but increased to as high as ~9.0 during dust storms (Pollack et al. 1979*b;* Thorpe 1981; Zurek 1981). During the storms, wind velocities at the landing sites remained below ~30 m s^{-1} (Ryan and Henry 1979) at the height of the sensor (~1.6 m) and no surface mate-

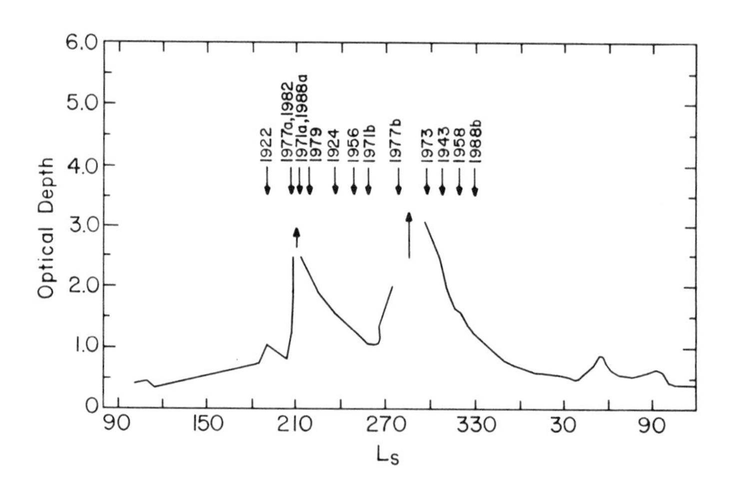

Fig. 5. Optical depth of the Martian atmosphere for the first Mars year of Viking observations (simplified from Zurek 1982). Optical depth values were determined by imaging the solar disk with the Viking 1 imaging system. Upper bounds for the first (1977a) and second (1977b) great dust storms were ~3.2 and ~9.0, respectively (Pollack et al. 1979*b*). The L_s values of the onset of historical great dust storms (denoted by the year of occurrence) are also marked.

rials within view of the Landers were seen to move. Following the storms, a gradual brightening and decrease in contrast of the surface was detected via the Landers due to deposition from the atmosphere (Guinness et al. 1982). After the first Martian year of observations, no global storms were detected from orbit, although increased opacity detected from Lander 1 during the third year may have been associated with dust raising by local storms (Arvidson et al. 1983).

C. Dust Storm Frequency

Global dust storms apparently occur when seasonal heating of the surface and atmosphere induces convection of sufficient strength to raise dust from the surface; currently, such conditions occur near southern hemisphere spring. Thermal effects of the suspended dust probably increase convection, lifting dust to greater altitudes and diffusing it planet-wide (Leovy et al. 1973*b;* Gierasch 1974; Zurek 1982). The frequency of great dust storms, however, is much less certain (Zurek and Haberle 1988*a*). Zurek (1982) noted that Earth- and spacecraft-based data indicate that planetary-scale storms could have occurred nearly every Martian year. Leovy et al. (1985) noted that even during the Viking years, two very different weather patterns were seen in years with and without global storms. Thus, the occurrence of storms may be determined by randomly occurring variations in the atmosphere/dust/polar cap system during earlier seasons. Haberle (1986*b*) suggests that the occurrence of a major dust storm one year may suppress the generation of such a storm the following year.

A major question is how representative the dust storms observed at present are of those occurring over the long term. The timing of great dust storms is related to the time of Mars' perihelion, which follows the ~51,000 yr Martian obliquity cycle. Therefore, the hemisphere in which maximum insolation occurs will reverse every ~25,000 years (see chapter 9). It has been suggested that the regions which currently appear to contain windblown dust deposits (Tharsis, Arabia and Elysium, all in the northern hemisphere) will serve as sources for dust storms when maximum insolation has shifted to that hemisphere (Christensen 1986*a,b*). However, many of the parameters controlling dust storms (such as surface properties and slope) in the source regions are not well constrained or understood. Confident extrapolation from the present (apparently seasonal) dust storm cycle (Fig. 5) to much longer climatic cycles requires more data than are available (Zurek 1982).

Variations in dust storm activity have been observed. The 1971 global dust storm was the most extensive in history (Martin 1974*a*, 1984). A global storm occurred during the first Mars year of Viking, but little dust storm activity was apparent during the subsequent two Martian years. It is not known how close this range in activity is to the maxima and minima over longer time scales, but present theory restricts the expected intensity and duration of global storms to near the maximum event observed in 1971 (Gierasch 1974; Zurek 1982). The aeolian features resulting from current dust storms are probably not unique to this period in Martian history, and thus may give clues to the features resulting from aeolian activity expected over the long term.

D. Dust-Storm Fallout, Surface Albedo Changes and Geologic Significance

Spacecraft observations allow the size, vertical distribution and time scale of suspension of dust particles in the atmosphere to be estimated. Throughout the year, particles with effective radii of ~2.5 μm predominate (Toon et al. 1977; Pollack et al. 1979*b*), although this value is somewhat model dependent and there is probably a range of particle sizes. The dust appears to be well mixed through several scale heights, and the size distribution of 1 to 10 μm particles remains essentially constant with the decay of the major storms (Conrath 1975; Toon et al. 1977). Based on the 1977 dust storms, Pollack et al. (1979*b*) estimated an average global sedimentation rate of $\sim 2 \times 10^{-3}$ g cm^{-2} yr^{-1}. Much of this material may be deposited in the north polar region due to enhanced sedimentation of dust acting as condensation nuclei for CO_2 and water; thus, deposition rates at the poles (constituting only ~2% of the planet's surface) could be very rapid on geologic time scales. Assuming widespread sources and significant trapping of dust in polar deposits, Pollack et al. (1979*b*) derived global erosion rates of ~7 m Myr^{-1}. Such high rates are inconsistent with the preservation of small craters in many areas (Arvidson et al. 1979); hence, either a few source regions for dust storms must be undergoing higher erosion rates, or significant recycling of dust, even from the polar regions, must be taking place, as discussed below.

Some of the colors of Martian albedo features can be interpreted as arising from variable amounts of dust covering the surface (Soderblom et al. 1978; Singer and McCord 1979; McCord et al. 1982*a,b*). There is little difference between the color of dust storms and the brightest regions of the surface (McCord and Westphal 1971; McCord et al. 1982*b*). The general appearance, thermal properties and color of two such regions (Arabia and Tharsis) are consistent with the presence of surface mantling by dust (McCord et al. 1982*b;* Christensen 1983). Viking Orbiter color measurements of dust clouds (Briggs et al. 1979) and parts of Arabia (Thorpe 1982) are consistent with the interpretation of the similarity between dust on the surface and that contained in dust clouds.

Experiments by Wells et al. (1984) show that small amounts of dust deposited from the atmosphere are sufficient to alter surface albedo features (Fig. 6). Deposition of only 10^{-4} g cm^{-2} of bright dust on a dark area should increase the albedo by several tens of percent of its initial value. Such deposition amounts to an average accumulation of about 0.5 to 10 μm (depending upon particle size, particle density and porosity), and is much less than the annual sedimentation rate (2×10^{-3} g cm^{-2} of Pollack et al. (1979*b*). A

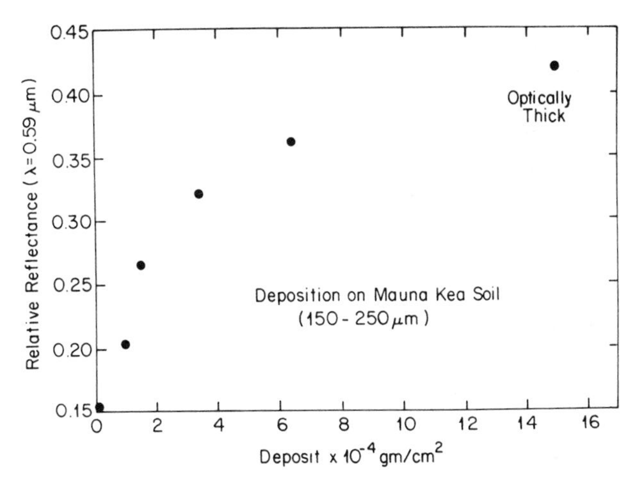

Fig. 6. Albedo changes caused by deposition of dust on bright and dark substrates. The dust is Mauna Kea volcanic soil, airfall sorted into particles <5μm diameter. The black substrate has approximately the reflectance of some of the darker regions of Mars. Note that significant contrast changes occur with much less than 10^{-3} g cm^{-2} deposited (figure after Wells et al. 1984).

photometric and color study by Thomas et al. (1984) indicates that most individual wind streaks exhibit contrasts relative to their surroundings of <20%, with contrasts of only a few percent resulting in readily visible features. Thus, the dust fallout expected to occur over a single Martian year is more than sufficient to change the albedo patterns observed on the planet.

Study of regional albedo features may provide clues to the timing, location and rate of sediment transport involved in the dust-storm cycle. Variations in albedo patterns are indicative of sediment transport through a region (Lee et al. 1982; Lee 1984,1987); where regional slopes are present, as in Tharsis, daily slope winds may dominate the global circulation pattern and govern the direction of sediment transport (Lee et al. 1982; Magalhães and Gierasch 1982). Other pertinent observations using thermal inertia data indicate the degree of surface mantling by dust (Kieffer et al. 1977; Palluconi and Kieffer 1981; Christensen 1982,1986*a*,*b*,1988; Jakosky 1986). The visual and thermal data are therefore diagnostic of whether net erosion or deposition of dust-storm fallout is taking place currently and whether such processes have been active in a region over the long term. Combined with experiments on albedo changes due to dust fallout (Thomas et al. 1984; Wells et al. 1984), such data allow constraints to be placed on the regional sediment transport rates involved in the observed variations of surface albedo features. Regional differences in sedimentary patterns are readily detected. For example, the Solis Lacus albedo feature associated with the Solis Planum region becomes dark during dust-storm season and brightens during the rest of the year (indicating that surface dust is mobilized primarily by local dust storms), whereas the Syrtis Major albedo feature associated with Syrtis Major Planitia becomes steadily darker following dust storms (consistent with a sandy surface from which active saltation will eject dust into suspension throughout the year). The Solis Planum region, therefore, is acting as a pulsatory dust source only during dust storms, whereas the mobile surface and regional atmospheric circulation pattern in Syrtis Major Planitia provides a relatively continuous source of dust for enhanced deposition in neighboring Arabia (Lee 1987). These detailed regional observations are consistent with global studies (Christensen 1988) that indicate that dust is removed from dark regions in the northern hemisphere subsequent to global dust storms. Cyclic deposition from the atmospheric dust load (amounting to, at most, a few tens of microns of dust-storm fallout annually) and subsequent redistribution of thin dust deposits by near-surface winds can explain the observed patterns and variability of large-scale albedo features. Transport of dust into or out of regions over longer time periods will yield geologically significant amounts of sediment deposition or erosion; such activity may well determine the geomorphic evolution and character of much of the Martian surface.

IV. AEOLIAN SURFACE FEATURES

A wide variety of surface features attributed to aeolian processes have been identified on Mars. Most common are various wind streaks, or variable features. These include forms that are attributed to both erosion and deposition. Sand dunes, observed in all regions of Mars, and drifts seen at both landing sites are clearly deposits of windblown particles. Less certain aeolian deposits are large areas of the planet inferred to be mantled by windblown sand and dust. Erosional features in the form of yardangs and possible ventifacts are also present.

A. Wind Streaks

Wind streaks are albedo markings interpreted to be formed by aeolian action on surface materials. Most are elongate and allow an interpretation of effective wind directions. Many streaks are time variable and thus provide information on seasonal or long-term changes in surface wind directions and strengths (Sagan et al. 1973*b*). Observations of wind streaks from Mariners 6, 7 and 9, and Viking Orbiter images allow mapping of their global occurrence and temporal variability.

Classification of wind streaks is based on three main characteristics: their upwind sources (sediment deposit or topographic obstacle); their albedo contrast (darker or brighter than surroundings); and special morphologic or compositional features (cf. Thomas et al. 1981). The interpretation of their genesis is based on variability, source feature, photometry (including colors), comparison to terrestrial features, and results from wind tunnel experiments and atmospheric modeling (Greeley et al. 1974). There are four principal types of wind streaks: bright depositional streaks (considered to be formed by deposition of particles), dark erosional streaks (considered to result from erosional processes), dark depositional streaks, and frost streaks.

1. Bright depositional streaks are the most common and are formed downwind from crater rims, mesas, knobs and other positive topographic features (Fig. 7). Most show only modest temporal variations, although those on the flanks of the Tharsis and Elysium volcanoes show substantial seasonal changes (Lee et al. 1982). These streaks probably form by deposition of dust from suspension during times of high atmospheric dust content and high static stability of the atmosphere (Veverka et al. 1981). Photometric and color analysis of the streaks show that they are consistent with a mantle of an optically thin layer of bright red dust, or with patches of optically thick dust over a few percent of the surface (Fig. 8). The amounts of dust indicated ($<10^{-3}$ g cm^{-2}) are easily accommodated by the annual dust loading of the order of 10^{-3} g cm^{-2} (Figs. 6 and 8; Wells et al. 1984; Thomas et al. 1984). Some bright features may be thick (>few mm) because they correlate with thermal inertia anomalies (Zimbelman and Greeley 1982; Zimbelman 1986).

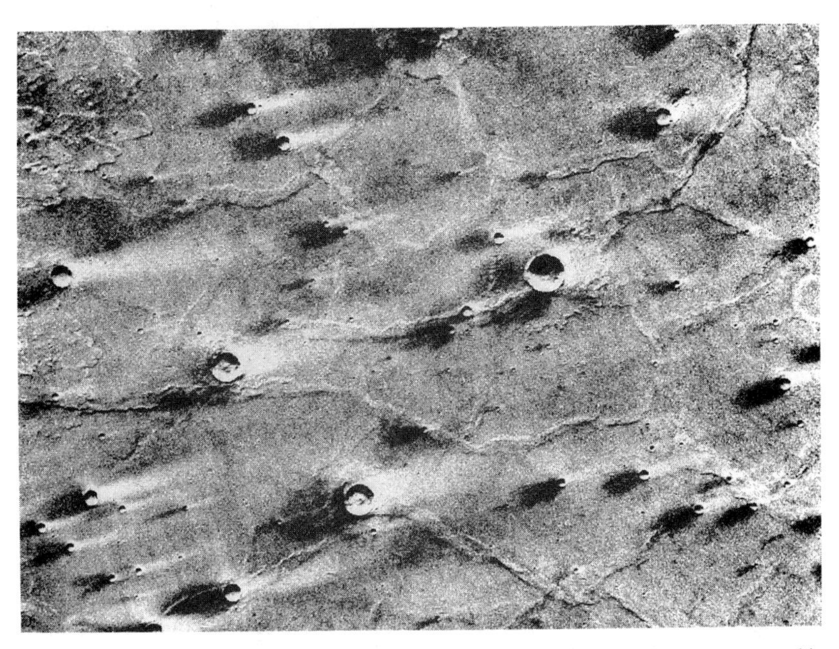

Fig. 7. Bright and dark streaks from craters in Hesperia Planum. Because the same topographic obstacle can cause both erosion and deposition at different times, meteorology is inferred to be the variable affecting erosion or deposition (Viking Orbiter image 553S54, red filter, located about 28°S, 245°W, L_s = 23°; image width = 200 km).

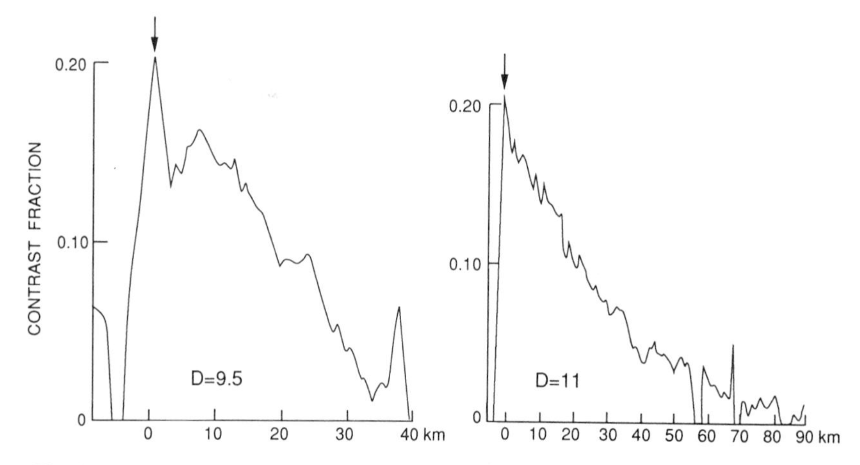

Fig. 8. Amount of ground covered by optically thick patches necessary to make contrast changes in bright wind streaks. The plots show the calculated fraction of ground covered by dust in transverses downwind (parallel to the wind) from the source craters. Arrows show crater rim positions; D = crater diameter (figure after Thomas et al. 1984).

These features might indicate some binding of deposited dust particles (e.g., duricrust formation) or deposits of dust in areas swept clear of some fraction of intermediate-sized particles.

2. *Dark erosional streaks* also form downwind from topographic obstacles, but are more restricted in location than bright streaks. They show great seasonal variability; most form in late southern summer or fall and are obliterated by later dust-storm activity. Dark streaks apparently form when the topographic obstacle generates local turbulence and increases wind stress on the surface (Greeley et al. 1974). Both dark and light streaks can form from the same obstacle (Fig. 7) which demonstrates meteorological control of erosion or deposition of windblown material. Photometry of the dark streaks indicates that they are consistent with removal of dust from an area covered by optically thin dust or with a few percent of optically thick dust (Fig. 9).

3. *Dark depositional streaks* emerge from sediment deposits ("splotches") that are usually inside large craters. Some of these "splotches" are dune fields, and thermal inertia data indicate that they are composed of coarse-grained sediment (sand) rather than dust (Christensen 1983). The streaks from these deposits range up to 200 km in length and are somewhat time variable (Thomas and Veverka 1979*b*). Their colors indicate some mixture of the darker materials (sand?) with a minimal dust cover (Thomas 1984; Presley and Arvidson 1988). These streaks are interpreted to be caused by sand blown from the source areas which either covers the surface or causes dust to be thoroughly cleaned from it. They provide additional wind-direction

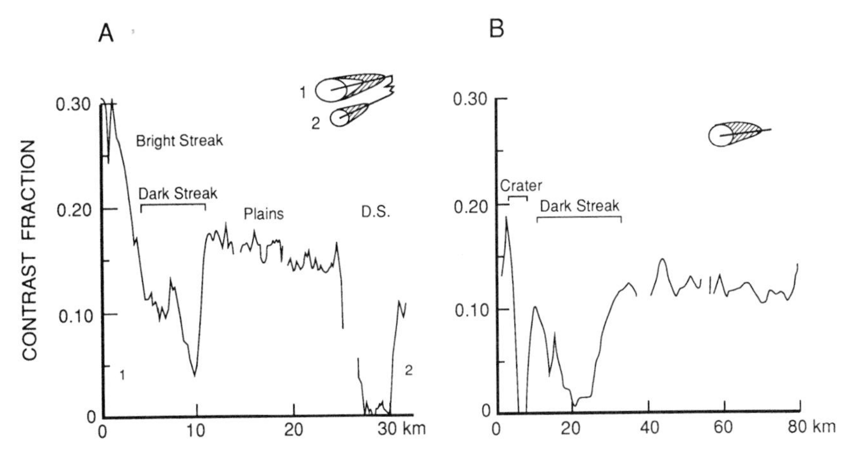

Fig. 9. Assessment of dust cover in dark streaks; the darkest area is assumed to be free of dust. The albedo of the streaks can be explained by patches of optically thick dust that are fewest in the dark streaks, intermediate in areas on the plains and greatest in the bright streak areas (figure after Thomas et al. 1984).

information and they suggest that some of the dune fields and splotches are areas of active saltation. Thus, they may be losing material at present.

4. Frost streaks. Frost from the annual polar caps is also formed into wind streaks behind crater rims. Frost streaks provide data on surface winds during formation of the caps. Unlike the dust streaks, frost streaks can accumulate significant mass (80 g cm^{-2}), as estimated by the length of time the streaks survive in the spring beyond the surrounding frost (Thomas et al. 1979*a*).

B. Dunes

Dunes are among the most distinctive aeolian features on Mars. They were first recognized on Mariner 9 images of the southern hemisphere by McCauley et al. (1972) and Cutts and Smith (1973), primarily by comparison with terrestrial sand dunes. Large dune fields in the north polar region were identified on Viking orbiter images by Cutts et al. (1976). Most Martian dunes are similar in form to barchan and transverse dunes on Earth. This morphologic similarity remains the best evidence to indicate that Martian dunes are composed of sand-size material, although the source and composition of the sand remain controversial.

Dunes are found in three main settings on Mars. The largest concentration occurs in a broad belt that partly surrounds the north polar ice cap (Fig. 10). Four major sand seas cover a total area of 6.79×10^5 km^2 between 75 and 80° N and 40 to 280° W (Tsoar et al. 1979; Dial 1984; Lancaster and Greeley 1989) and have an estimated sediment thickness that ranges from

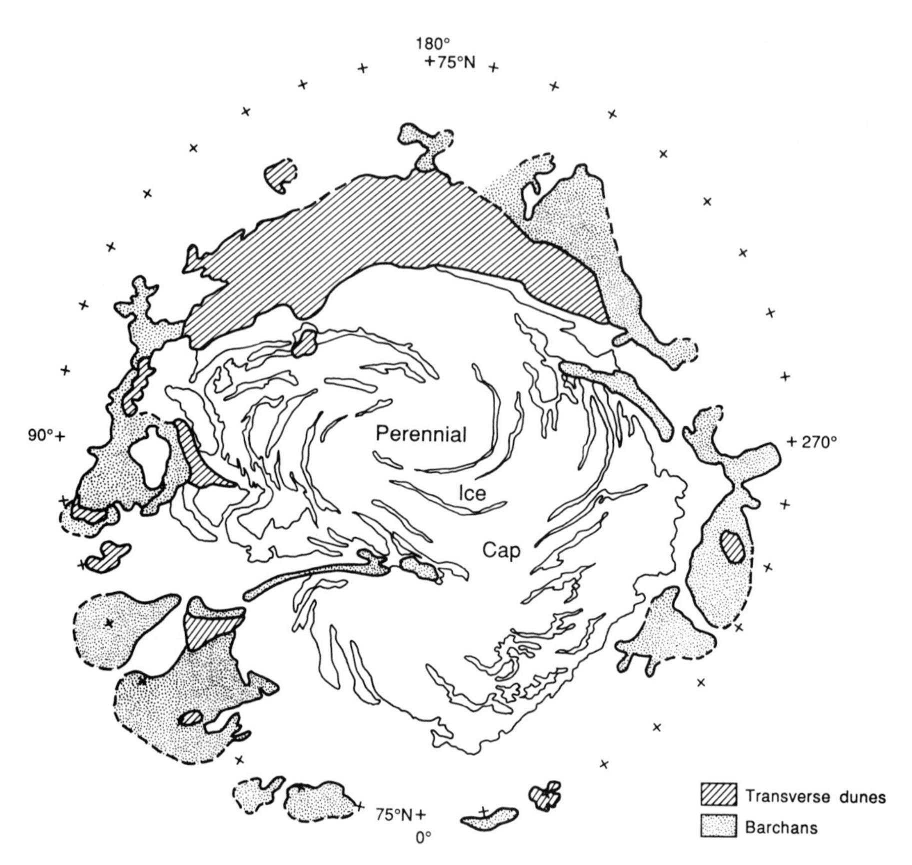

Fig. 10. Distribution of dunes and sand seas in the north polar region (figure after Tsoar et al. 1979). Note major sand sea between 110 and 260° W, and lack of dunes in sector between 310 and 30°.

0.10 to 0.50 m on their margins to 6.1 m (mean thickness = 1.81 m). The total volume of sediment contained in the north polar sand seas is estimated at 1200 km^3, of which 78% lies between 110 and 260° W.

The north polar dunes have been described by Breed et al. (1979), Tsoar et al. (1979) and Lancaster and Greeley (1990). Small simple barchan dunes (Fig. 11a) and large mega-barchans occur on the margins of the sand seas and in areas of thin sand cover. Approximately 50% of the area of the north polar sand seas is composed of various transverse dunes (Fig. 11b), especially in the area between 76 to 83° W and 110 to 260° W, which is the largest continuous area of dunes on Mars, covering some 4.7×10^5 km^3. Many of the transverse dunes appear to be modified by secondary winds from directions that are perpendicular or opposed to the primary dune-forming wind, giving rise to oblique, braided or reversing patterns (Tsoar et al. 1979). Evidence for possible seasonal reversal of lee-face orientations and dune-forming wind directions suggest that these dunes are active today. Additional evidence for contemporary dune activity is provided by the perfect form of many dunes, and by albedo streaks from barchan dune horns that indicate sediment is eroding from the dunes (Tsoar et al. 1979). However, the dunes may be inactive in the winter when they are covered by CO_2 frost (Ward and Doyle 1983). The crest-to-crest spacing of dunes varies widely, from <200 m in areas of closely packed transverse ridges to >1000 m in areas of scattered barchans. Mean dune spacing is 512 m, with most dunes being spaced 300 to 800 m apart. Comparisons with terrestrial dunes of a similar spacing indicate that the north polar dunes are 10 to 25 m high, with a mean height of 22 m (Lancaster and Greeley 1990).

The second group of Martian dunes occurs in the high latitudes of the southern hemisphere (Thomas 1981; Lancaster and Greeley 1987). A large cluster of intracrater dune fields with areas ranging from 40 to 3600 km^2 lies in Noachis Terra between latitudes -40 to $-48°$ and longitudes 330 to 340°. Largest accumulations are in the craters Kaiser, Proctor and Rabe. These dunes were among the first to be recognized (Cutts and Smith 1973; Breed 1977) and typically consist of straight-to-wavy crescentic ridges with a large dune wall or rampart forming the margin of the dune field, suggesting a sediment accumulation of as much as 100 m (Thomas 1982). The other dunes consist of isolated patches of transverse dunes and barchans on the interbasin plains of the southern hemisphere, and in troughs and channels in equatorial regions (Ward et al. 1985).

There are two varieties of dunes in southern hemisphere intracrater dune fields (Lancaster and Greeley 1987). The most common type is massed straight to slightly wavy transverse dunes (Fig. 12a) with a spacing between 700 to 1200 m, similar to those described by Breed (1977) and Breed et al. (1979), and rather larger than dunes of similar form in the north polar sand seas. The second type of dunes are clusters of large, widely spaced straight or curved ridges, which often intersect to create rectilinear patterns (Fig. 12b). Dunes are typically spaced 1600 to 4000 m apart with a mean spacing of 2400 m. On the margins of some of these dune fields, there are small (500 to 900 m spacing) transverse dunes together with widely spaced barchans. Pyramidally shaped dunes are evident in some dune fields and many dunes appear to have multiple slip faces which face northwest or south.

The occurrence of transverse, but not linear dunes on Mars, indicates that most dune fields occur in areas that experience near-unidirectional sand-moving winds although seasonal reversals of wind directions appear to be important. The existence of larger, more complex dune forms in the southern hemisphere intracrater dune fields suggests a more complex wind regime, with local topography modifying the regional strong southeasterly spring and southwesterly winter winds suggested by Thomas (1981).

Dunes in the north polar regions indicate two dominant formative winds: easterly north of 80° N, and westerly south of 80° N (Ward and Doyle 1983).

Fig. 11 (a)Barchans in the north polar sand sea. Dunes range in size from less than 100 m across to a maximum of ~750 m in this area. Formative winds from northwest (Viking Orbiter image 525B15; lat. 73°47, long. 57°52). (b) Massed, straight-crested transverse dunes in the north polar sand sea. Dune crest-to-crest spacing is approximately 480 m (Viking Orbiter image 514B61; lat. 80°91, long. 144°74).

Tsoar et al. (1979) modeled winds as summer off-pole easterly and on-pole westerly, winter on-pole westerly, and spring on-pole westerly; whereas Ward and Doyle (1983) inferred a pattern of strong off-pole west winds in fall, moderate west winds in winter, weak to strong off-pole northeast winds in spring, and weak west winds in summer.

C. Other Aeolian Bedforms

Viking Lander images indicate that the Martian surface is strongly affected by small-scale wind erosion and deposition. The drifts of fine grained, homogeneous deposits described by Mutch et al. (1976*a*) are probably aeolian in origin (Sharp and Malin 1984). Possible granule ripples and abundant "tails" (Fig. 13) in the lee of boulders suggest sand in transport, but little accumulation of windblown sand (Sharp and Malin 1984).

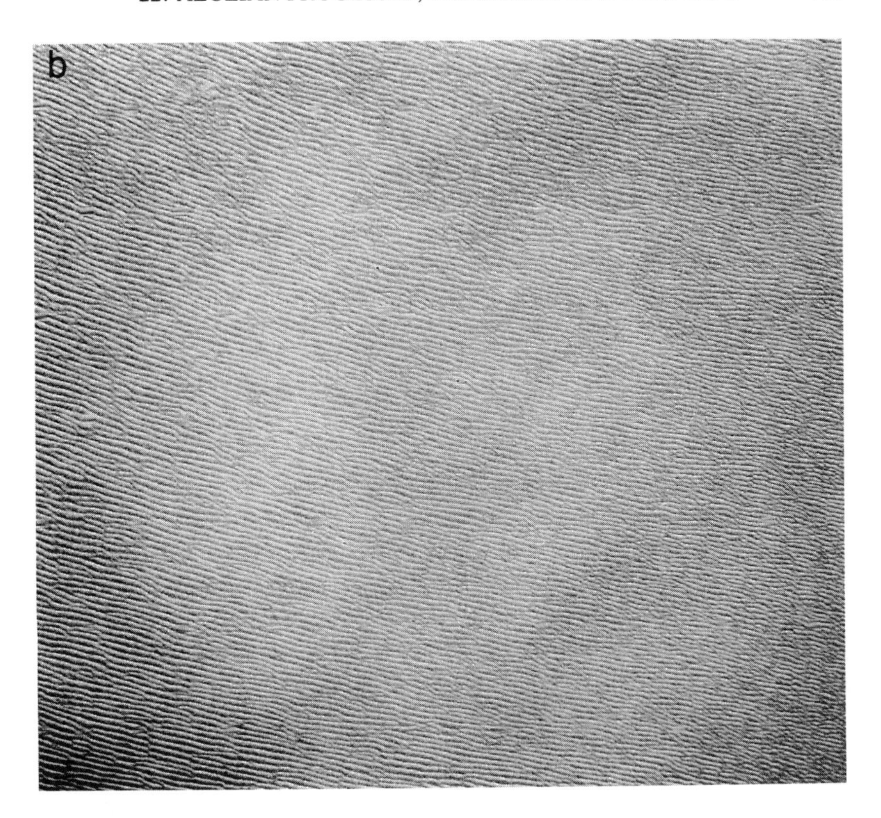

Sand surfaces that are without recognizable dunes probably occur in two main forms on Mars. Infrared Thermal Mapper (IRTM) data suggest that many of the dark intracrater "splotches" are composed of grains 0.2 to 2 mm (Christensen 1983; Jakosky 1986). The absence of dunes in most cases suggests that these surfaces form sand sheets or low zibars (dunes without slip faces), as a result of the trapping of coarse, barely mobile material within craters. Breed et al. (1987*a*) described gently rolling topography consisting of giant ripple-like bedforms with a spacing of 200 m in the Tharsis province 500 km northwest of Olympus Mons. The bedforms are oriented transverse to the direction of strong slope winds and are developed on areas mapped as mantled plains (Zimbelman and Kieffer 1979; Scott et al. 1981*b*). Similarities to bedforms developed in coarse sand sheets in the eastern Sahara suggest that these forms may be partly composed of coarse material.

D. Erosional Features

A variety of wind erosion features have been recognized on Mars. These include yardangs, pits and grooves in the equatorial regions (McCauley 1973; Ward 1979; Ward et al. 1985), and wind-eroded pits in the southern hemisphere polar regions (Sharp 1973*b*). Small-scale wind erosion features (ven-

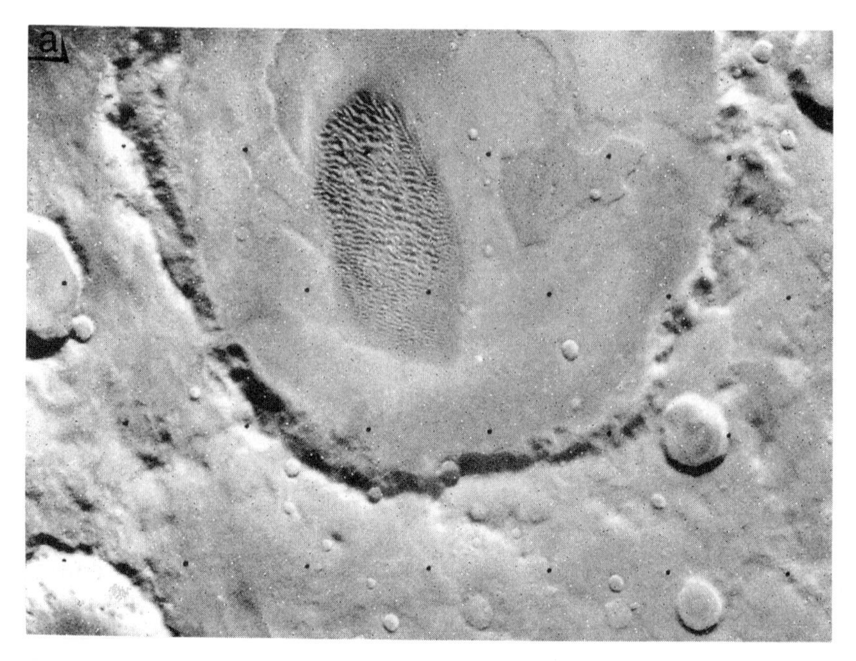

Fig. 12 (a) Intracrater dunes in southern hemisphere high latitudes. Major dune field is in crater Proctor; maximum crest spacing of dunes is ~4 km. Note other smaller dune fields in adjacent craters (arrowed) (Viking Orbiter image 094A47; lat. −45°.88, long. 327°.83; area shown is ~150 km by 220 km). (b) Detail of southern hemisphere intracrater dunes in crater Kaiser. Note smaller transverse dunes and barchans on margins of dune field (arrowed), and large symmetrical ridges (reversing dunes?) (Viking Orbiter image 575B60; lat −47°.30, long. 340°.49; area shown is ~50 km by 50 km).

tifacts) have been proposed at the Viking Lander sites (El-Baz et al. 1979; McCauley et al. 1979; Sharp and Malin 1984).

Yardangs (streamlined ridges separated by parallel troughs) have been described by McCauley (1973) and Ward (1979) from the equatorial regions of Mars. Yardangs form on Earth by wind abrasion, deflation and erosion (Whitney 1983; Breed et al. 1989) of fine-grained sedimentary rocks in areas of strong, unidirectional winds. On Mars, the best-developed yardangs occur in the Amazonis region (Fig. 14) where they are as long as 50 km, with troughs 1-km wide and 200-m deep (Ward 1979; Ward et al. 1985). These yardangs have a southwest-northeast orientation, which parallels regional wind streaks, and are carved in geologically young layered plains units that have been suggested to be composed of volcanic ash (see, e.g., Scott and Tanaka 1982). Other important areas of yardangs, parallel grooves and deflation pits are in the Olympus Mons aureole and Aeolis, Ares Valles and Iapygia regions (Ward 1979; Ward et al. 1985). In both the Amazonis and Aeolis regions, multiple sets of yardangs on different trends have been recognized,

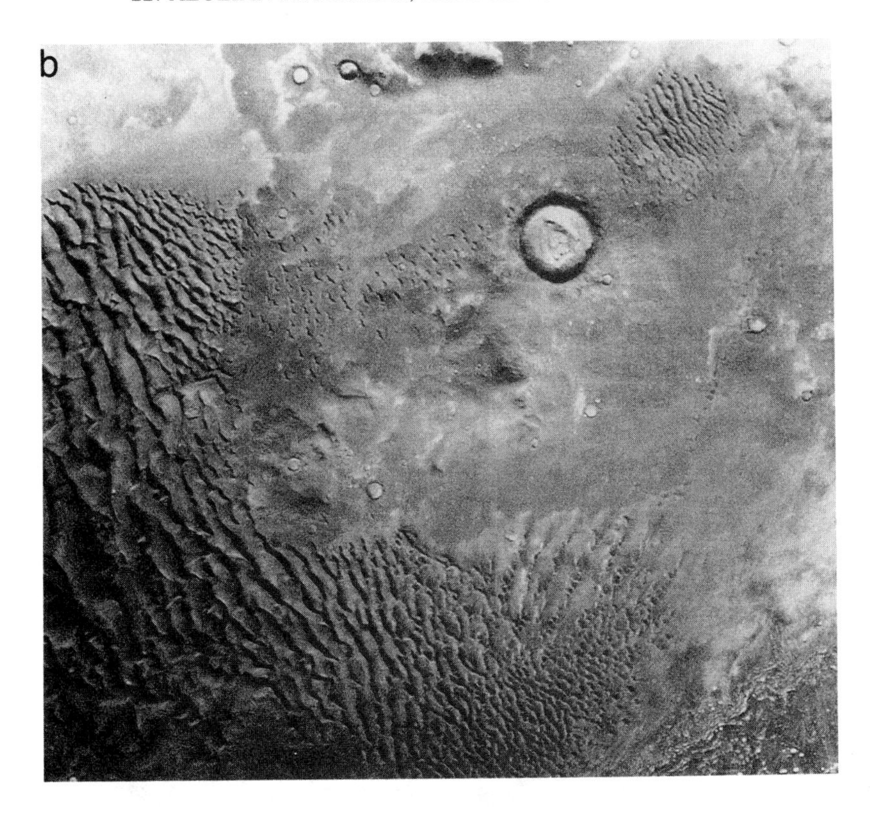

suggesting changes in the regional wind regime (Ward 1979; Zimbelman and Wells 1987).

Many rocks at the Viking Lander sites are pitted and fluted. Some have interpreted these features as primary vesicles that have been modified by wind abrasion, but others attribute them entirely to wind etching of pits in homogeneous rocks, as is common on ventifacts on the Earth (El-Baz et al. 1979; McCauley et al. 1979; Sharp and Malin 1984). Wind faceted rocks are also seen (Sharp and Malin 1984). These features, in combination with wind scoured drifts and basal scouring of material from around rocks, indicate that wind erosion is effective at these locations. Garvin et al. (1981) and Sharp and Malin (1984) consider wind action to be intense, whereas Mutch et al. (1977), Arvidson et al. (1981*a*) and Greeley et al. (1982) suggest that wind erosion is limited by low wind velocities and lack of abundant holocrystalline sand-size particles. Greeley et al. (1982,1985) estimated rates of wind erosion from laboratory experiments as 2.1×10^{-2} cm yr^{-1}, or an order of magnitude higher than those indicated by the observed surface features, and suggested that abrasion rates were limited by a shortage of effective material for abrasion.

Fig. 13. The Viking Lander 1 site, showing drifts of fine-grained material. Note perched nature of rocks, suggesting undercutting by the wind, and "sand tails" in their lee (Viking Lander image 77/03/25/152623).

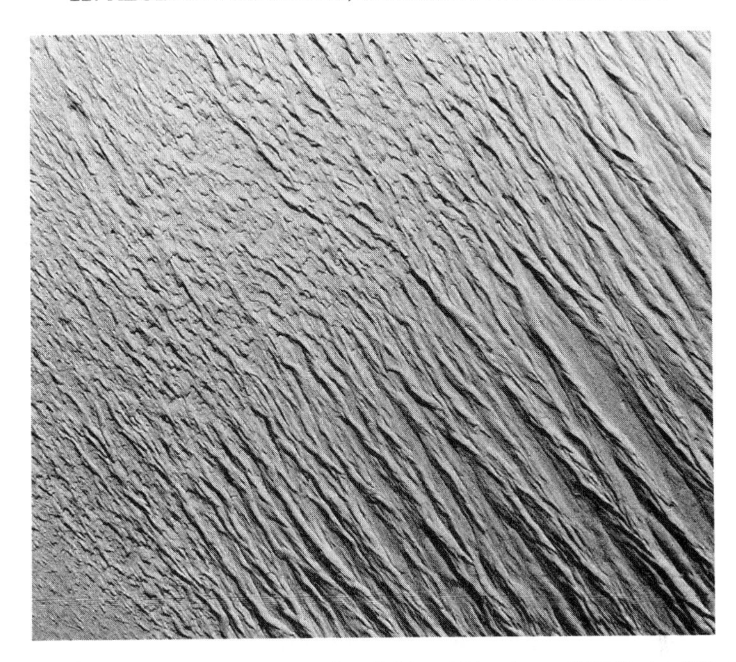

Fig. 14. Yardangs in the Biblis Patera region (Viking Orbiter image 732A58; lat 4°30, long. 137°80; area shown is ~21 km by 24 km).

V. MARTIAN AEOLIAN REGIME

Wind is currently redistributing large quantities of material on Mars, and probably has been active throughout the evolution of the surface. In this section, the possible source regions and deposition sites of windblown material are discussed. Although the emphasis is on present conditions, consideration is also given to potential previous aeolian regimes.

A. Regional Aeolian Deposits

Several regional and numerous local deposits of windblown sediment can be identified. In addition to the various dune fields and sand seas described in Sec. IV, vast regions of Mars appear to be mantled with aeolian deposits. The two polar deposits, dealt with more fully in chapter 23, are approximately 1300 km in diameter and are offset slightly from the poles. Maximum thicknesses may be ~5 km (Malin 1986). Although the composition of the deposits is unknown, Malin (1986) estimated the density of the northern polar deposits to be about 1.0 g cm^{-3}, which would imply that they are largely ice (the result has substantial uncertainty and permits substantial mineral and/or void space). The colors of the layered deposits suggest that they include material similar to the dust found on much of Mars. The deposits also contain some darker, less red material that can be correlated with dune-forming materials (Herkenhoff and Murray 1990*a;* Thomas and Weitz 1989). The layering and unconformities within the deposits suggest cyclical variations such as different ratios of dust to ice caused by astronomically driven changes in dust-storm seasons and intensities (Cutts 1973*a*; Cutts et al. 1979; Toon et al. 1980; Howard et al. 1982*a*).

Christensen (1986*a*) has used Viking Orbiter thermal inertia data and Earth-based radar data to estimate the thickness of dust in the Tharsis area as 0.1 to 2 m; effective particle sizes are 2 to 40 μm. Arabia Terra is the site of dust deposits nearly 1000 km across. In much of this region, there is morphologic evidence of mantling that may be much thicker than the 2-m limit from radar data, and may represent long-term deposition (Arvidson et al. 1982). In addition, there is evidence that removal of dust from Syrtis Major Planitia provides enhanced deposition in neighboring Arabia (Lee 1987). Consequently, the Arabia deposits probably represent both modern dust deposition and deposits from earlier aeolian regimes.

There are other deposits identified by their morphology that are variously ascribed to polar or volcanic processes, such as materials near 0° N, 180° W (Schultz and Lutz 1988; Scott 1979). Regardless of origin, these deposits have at least been modified by aeolian processes. The style of their erosion suggests that they consist of fine-grained material easily transported by the wind and therefore easily carried by wind into streamlined hills, or yardangs (Fig. 14). In addition, many regions of Mars show large mantled units attributed to aeolian deposits (Plaut et al. 1988).

B. Relation of Aeolian Features to Wind Patterns

This section compares the present distribution of aeolian deposits and other features (Ward et al. 1975) with surface wind patterns derived from both wind streaks and general circulation models. Wind streaks show some global patterns and significant topographic influences. The bright streaks, which occur at mid to low latitudes, probably form during late southern summer and fall. They show a consistent north to south wind flow (Fig. 15) with a gradual rotation from an easterly to westerly component. This pattern is interrupted by the Tharsis and Elysium volcanoes. Only a few bright dust streaks occur at high latitudes, and their season of formation is not known (Veverka et al. 1981; Magalhães 1987).

Dark erosional streaks are concentrated at low southern latitudes and show winds mostly from east to west, apparently redistributing particles after the dust storms (Christensen 1988). They also show significant downslope winds in Tharsis and on the volcanoes (Lee et al. 1982). Dark depositional streaks at mid to low latitudes follow the same directions as the bright dust streaks, and probably indicate movement of material in southern summer. However, the dark depositional streaks at mid to high southern latitude show winds from southeast to northwest. Late southern spring wind flow (partly driven by spring sublimation of the polar cap) is consistent with these direc-

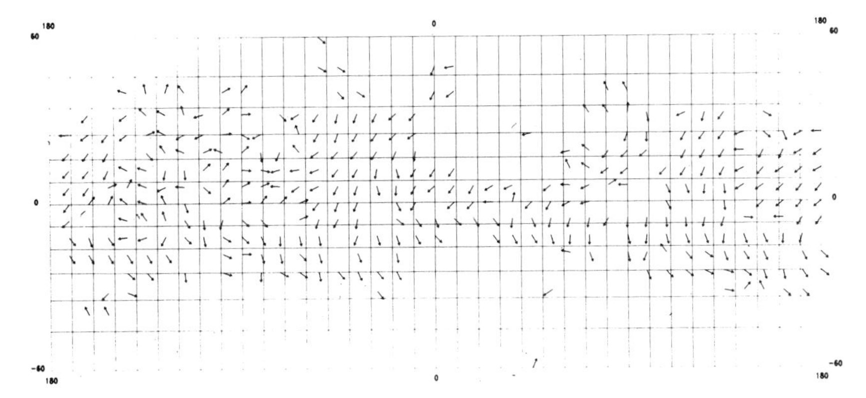

Fig. 15. Crater-related bright streaks (compiled by Thomas et al. 1984). Streak azimuths are binned within each GCM (7°.5 × 9°.0) cell, arrows indicate inferred downwind direction.

tions (French and Gierasch 1979). Frost streaks and some other erosional streaks show additional wind patterns that are discussed in French and Gierasch (1979), Howard (1980), and Thomas (1981).

Present dust deposits, apart from polar ones, appear to correlate better with topography and surface properties than with the global wind patterns. Christensen (1986*a*, 1988) used IRTM albedo data to assess dust activity during the Viking missions. Dust deposited on the bright, low thermal inertia regions tends to remain in place after the storms. In contrast, northern hemisphere dark regions such as Syrtis Major and Acidalia Planitia appear to lose the dust which had been deposited during the storm. The lack of significant dust deposition in the southern dark areas indicates a general south-to-north

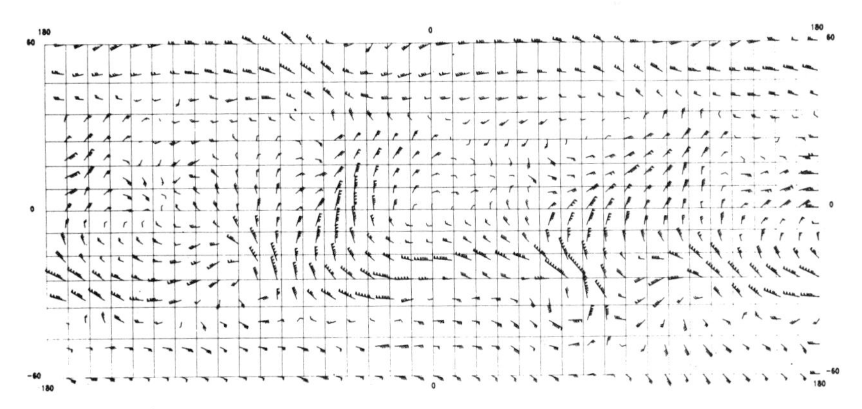

Fig. 16. GCM surface stress vectors averaged over a 10 sol period during dust-storm season (L_s = 281°.02 to 286°.73). Barb points indicate downwind direction, flags indicate magnitude: triangle = 50 units, long slash = 10, short slash = 5. Stress units are $N/m^2 \times 10^{-4}$. Saltation threshold for 25 μm particles at 5 mbar is approximately 550 $N/m^2 \times 10^{-4}$.

dust transport, consistent with the upper-level Hadley circulation expected in southern summer.

General circulation models (GCM) for the atmosphere have been developed for Mars. The GCM of Pollack et al. (1990*a*) allows the magnitude and direction of the wind surface shear stress (τ) to be determined and takes into account such factors as atmospheric density and temperature, global topography, Martian season and atmospheric dust loading. Maps based on the GCM show wind-shear stress vectors for areas 7°5 latitude by 9° longitude for 6 Martian seasons. Preliminary comparisons of GCM runs and various aeolian features, such as wind streaks, show several correlations (Greeley et al. 1989). A good correlation exists between bright wind streaks and surface winds inferred from GCM results for the southern hemisphere summer, L_s 280°, particularly in the Chryse Planitia, Margaritifer Sinus and southern Elysium areas (Fig. 16). In general, both the GCM and the bright-streak patterns correspond to an expected Hadley cell circulation near the surface. A poor correlation, however, exists between GCM results and bright streak orientations in the Tharsis area (Fig. 16), and is attributed to the local topography of the shield volcanoes and the Tharsis bulge for which the GCM does not adequately account. Consequently, the bright streaks are probably better indicators of near-surface winds than the GCM near the volcanoes or in other areas where local topography has a strong influence on wind patterns; these results are in accord with the observations of Lee et al. (1982) and models of winds over regional slopes (Magalhães and Gierasch 1982). In contrast to the good bright-streak-GCM correlation, most dark streaks do not match the GCM patterns. This suggests that most bright streaks tend to be representative of global wind patterns, whereas dark streaks appear to be more influenced by local conditions such as topography.

GCM results have also been compared to the "rock abundance" distribution for Mars as derived from Viking IRTM data by Christensen (1986*b*). Results (Fig. 17) show a correlation between high surface shear stress and high "rock" abundance ("rocks" include fragments a few cm and larger, bedrock surfaces, and surfaces of indurated sediments such as duricrust). This correlation is consistent with a surface that would be swept free of most fine material such as dust and loose sand. We note, however, that in rocky areas not subjected to high wind shear, dust probably is trapped among the rocks, at least until such time that the rocks become buried.

The time scale of dune formation and the considerable influence of topography makes the relation of dunes to wind patterns indicated by wind streaks or by GCM calculations complex. The low-latitude dunes occur in deep depressions and their orientations are difficult to relate to global wind flow, although streaks from these dune fields suggest winds similar to those in southern summer. The wind patterns summarized in Fig. 15 suggest that the polar dune concentrations are partly controlled by seasonally reversing winds. This apparent control does not mean that the dunes are completely

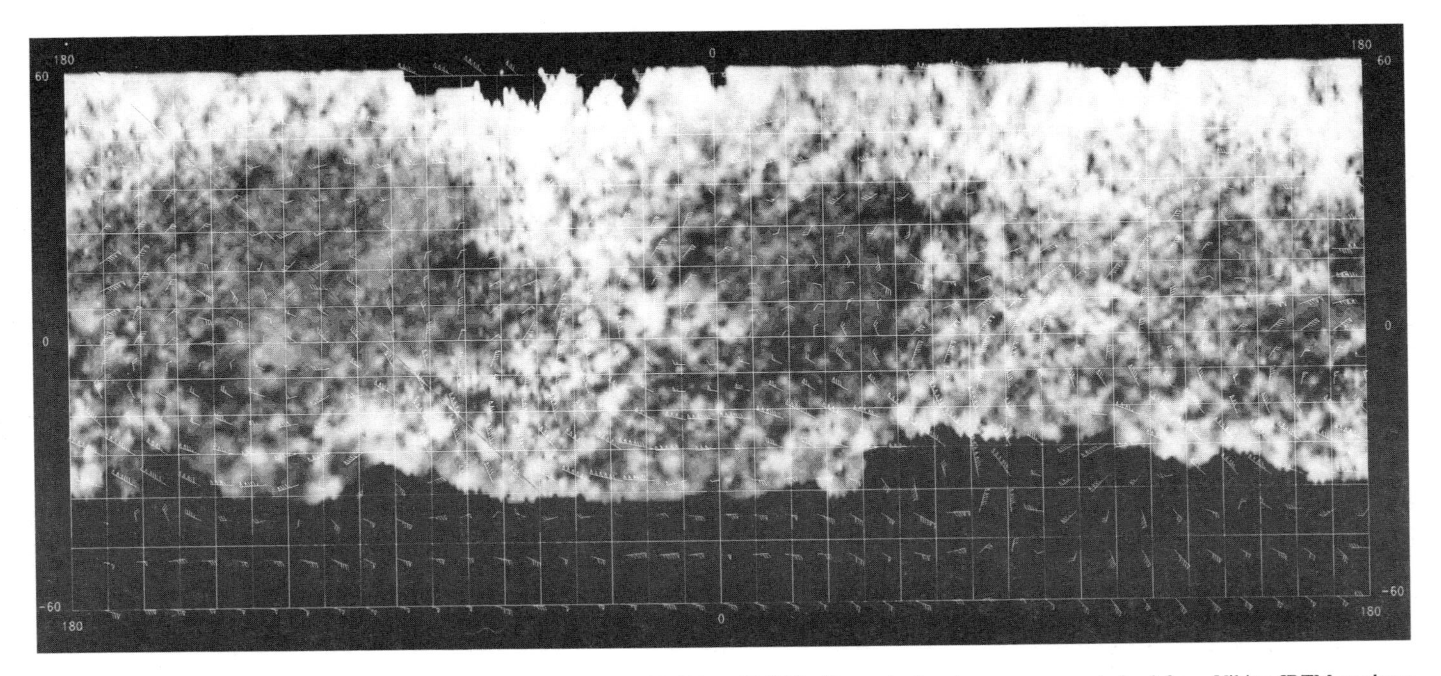

Fig. 17. Mars global rock abundance and GCM surface stress (L_s = 281°.02 to 286°.73). The rock abundance map was derived from Viking IRTM predawn temperature measurements (Christensen 1986*b;* see chapter 21) assuming the surface cover consists solely of cobbles 10 cm on a side and dust. Dark areas indicate low (0 to 10%) rock abundance; light areas indicate high rock abundance (30 to 35%).

confined to these zones, only that their motion through these latitude bands is so slow that any geologic "snapshot" should have more dunes in these belts than in surrounding areas of more uniform wind directions. The morphology of the dunes suggests complex wind patterns in both polar regions (Tsoar et al. 1979; Lancaster and Greeley 1990), but also that winds blow away from the poles at both north and south polar latitudes. In the north, wind directions shown by dune orientations in the southern part of the sand seas indicate a component of wind from the south. In the south polar area, a poleward component appears to be almost entirely absent, perhaps because most dunes there are confined within craters (or in other topographic traps). The regional winds indicated there are mostly from the southeast, in accordance with the wind-streak data at high southern latitudes (Thomas et al. 1979*a*).

C. Time Scales for Formation of Aeolian Deposits

The time involved in the formation of aeolian deposits on Mars can be estimated for different scales of features. For wind streaks and thin dust deposits, the dust loading of the atmosphere and the photometry of the surface allows estimates to be made of their formation times. In some cases, image comparisons provide direct information on times of streak formation. For thick dust deposits, atmospheric loading and assumptions of net deposition/erosion rates must be extrapolated over longer (geologic) times. For polar layered deposits, the time of layer formation can be predicted from likely astronomical influences on climate. The time required for dune formation or movement is very difficult to estimate because of the poor knowledge of surface wind velocities.

The amount of dust raised in the atmosphere was discussed in Sec. III. Seasonal and spatial patterns of dust deposition are discussed by Christensen (1988) and Pollack et al. (1979*b*). The average dust deposition rate is $\sim 4 \times 10^{-3}$ g cm^{-2} per Mars year but this varies seasonally and spatially. In the regional dust-deposit model (Christensen 1988), dust is largely deposited between 0 to 60° N, or an average of 4.6×10^{-3} g cm^{-2} yr^{-1}. This is much larger than the amount needed to mask the background albedo (Wells et al. 1984); consequently, much dust probably falls in patches that are already high-albedo zones. This deposition rate could form the present equatorial and temperate aeolian deposits in 10^5 to 10^6 yr.

Calculation of the accumulation rates of the polar deposits is subject to similar assumptions, except that correlations of the dust layers with particular astronomical time scales may also be attempted (Cutts et al. 1979; Pollack et al. 1979*b*). If it is assumed that the layers of a few tens of meters thickness correlate with time scales of 10^5 yr, then the deposition would be about 0.3 mm yr^{-1}. If all of the global dust loading were at the poles, then polar dust deposition would be about 0.2 mm yr^{-1}. Unfortunately, it is not known if the 30-m layers are typical (Blasius et al. 1982), nor is it known if the polar material is mostly dust or mostly ice. Moreover, the present dust budget of the polar regions is not known (James 1988; Kieffer 1990). Thus, although a consistent model of dust movement at a particular order of magnitude scale can be derived, it is insufficiently documented to allow definitive conclusions (see chapter 23 for more details). A deposition rate of 0.3 mm yr^{-1} (300 km Gyr^{-1}) obviously cannot apply for very much of Martian history without some mechanism of removal (Pollack et al. 1979*b*; Clifford 1988).

The observed global dust loading also is difficult to apply as an average "bedrock" erosion rate for much of Mars' history. The average rate of about 10^{-3} cm yr^{-1} would be 10 km Gyr^{-1}. Because most of Mars appears more than 2 Gyr old and shows little erosion, application of this erosion rate to the smaller young surfaces would give very large (and unobserved) erosion. The clear implication is that much of the dust is simply recycled material and not new erosion of bedrock, a conclusion supported by the work of Arvidson et al. (1979) and Christensen (1988).

In one sense, calculating dune sediment transport is easier than calculations for dust; if the density of the particles is known, then the physics of their transport in saltation could provide transport rates. Application of saltation transport-rate equations to terrestrial meteorological data has had some success in describing dune movement (Wilson 1971). Unfortunately, few wind data are available for Mars, and there is a controversy as to sand density. However, some limiting cases can be useful. The latitudinal distribution of dunes on Mars implies a time scale of formation at least equal to that needed to move small sand dunes over several degrees of latitude. Thomas (1982) estimated the time scales of formation of latitudinal banding of dunes with current winds and particles of density 2.5 g cm^{-3} and found that several Myr are required, well in excess of the period of orbitally driven climate variations. The size and distribution of dunes in the north polar sand seas appear, by analogy with sand seas on Earth, to have required at least 1 Myr to form (Breed et al. 1979). The orientations of intracrater dune fields in the southern hemisphere are approximately aligned with present winds, and might be predicted to change from the northwest sides of craters to the northeast over the 51,000-yr cycle of season of perihelion (Thomas 1981). However, unless the winds were very much stronger than at present, moving some of these large dune masses 20 km requires times much longer than the assumed climate cycle. There may be asymmetries in the effective transport vectors that persist through the climate cycle. If the sand were fluffy aggregates (Greeley 1979; Saunders et al. 1985), this problem could be alleviated, but continuing investigation of dune thermal inertias suggests that they are solid particles (Paige and Kieffer 1987; P. Christensen, personal communication), as strongly implied by the requirement that they survive repeated saltation impacts over the time of dune migration (Breed et al. 1979).

VI. CONCLUSIONS

Here we present a model for the aeolian part of the sedimentary cycle, derived in part from Greeley (1986), as shown in Fig. 18. The proposed aeolian cycle consists of three main processes: (a) entrainment; (b) transportation; and (c) dust settling, or deposition; two associated processes are also included: (d) particle removal and (e) the addition of new particles to the cycle. Perhaps the least understood aspect of Martian aeolian cycles is the mechanism(s) for setting particles into suspension. Because of the low-density atmosphere, extremely high winds are required to raise loose dust of the size inferred for Martian dust ($\sim$ a few μm). Although some dust may be raised by these high winds, most investigators invoke other threshold mechanisms. Dust devils may be one such mechanism; laboratory simulations show that vortical wind shear is very efficient in raising particles (Greeley et al. 1981). With the discovery of dust devils on Mars (Thomas and Gierasch 1985), this possible mode of dust raising is enhanced. Outgassing of volatiles (CO_2 or H_2O) from the regolith in response to changes in atmospheric temperature/pressure has also been suggested as a means of injecting dust into the atmosphere (Greeley and Leach 1979; Huguenin et al. 1979). Laboratory experiments show that this process can involve rapid venting of gasses which carry dust particles tens of cm above the surface, depending upon the rapidity of outgassing. Although slow outgassing does not cause dust injection, it can lead to surface fracturing and fissuring, which may be important in the "ag-

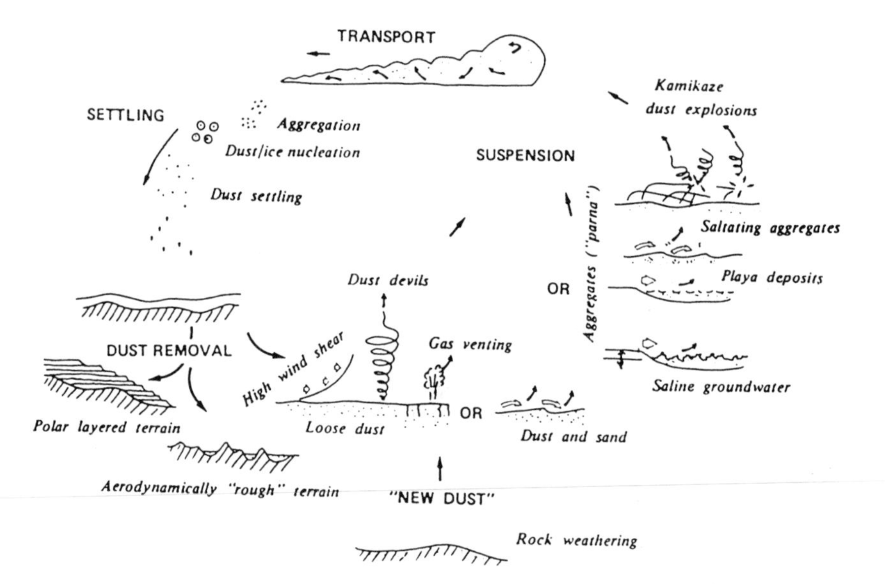

Fig. 18. The Martian aeolian cycle (figure from Greeley 1986).

gregate" mode of dust raising discussed below. However, it is doubtful that adequate CO_2 or H_2O are present in near surface materials for this to be an effective mechanism for dust raising by itself.

Because sand-size grains are more easily moved than dust, it has been suggested that saltating sands could act as triggers to set dust into suspension (Peterfreund 1981; Christensen 1983). Laboratory experiments show this to occur, although in some circumstances, the saltating sands merely indent the surface without dislodging the dust (Greeley 1985). Another mechanism of dust raising involves *aggregates* (Greeley 1979); dust-size grains collected into sand- and larger-size clumps. Four sources of aggregates may occur on Mars: (a) deposits settled from the atmosphere as part of the dust cycle; (b) ancient lake bed deposits, similar to clay pellets derived from playas on Earth; (c) aggregates derived from frothy/crusty surfaces generated by fluctuations of saline groundwater; and (d) crusty/cloddy weathering surfaces that may be disrupted, as by outgassing, discussed above. Aggregate strength is variable, depending on mode of bonding (electrostatic, cementation, etc.). Because the wind speeds on Mars required to set particles into motion are high, McCauley (1973) and Sagan et al. (1977) proposed that holocrystalline grains would be pulverized into fine grains by the "kamikaze" effect as they collide with each other and as they impact bedrock surfaces. If aggregates were involved, the self-destruction would be even more effective because of their low strength. Wind tunnel experiments with aggregates show that as they lift above the surface, they rupture into dust clouds (Greeley 1985). Thus on Mars, disaggregation may occur via "kamikaze dust explosions."

The movement of dust in both local clouds and as global transport is documented through Earth-based and spacecraft observations. Pollack et al. (1979*b*,1981) have developed models to predict the patterns of transport and provide the key to this part of the dust cycle. With a decrease in driving energy for dust clouds, in time particles settle from the atmosphere. However, because of possible feedback mechanisms (Pollack et al. 1981), one of the problems in the dust cycle is how to shut down dust storms. One mechanism involves the accretion of ice on dust grains (Pollack 1979). Aggregation of grains may also occur via electrostatic bonding, a mechanism demonstrated in laboratory experiments and observed in volcanic dust clouds. Lee (1985) has shown, however, that the settling rate of larger grains (either ice/dust or dust/dust) is a function of the grain density and diameter, both of which change during accretion.

With settling and deposition, the cycle may be completed and the stage set for dust raising to initiate the next cycle. However, in addition to the dust threshold, transport and deposition components of the cycle, dust may be removed from the cycle, and new dust may be added. On Earth, the largest sinks for dust are the oceans. Although these sinks are lacking on Mars, there are other areas where dust may be removed from the cycle. These include the polar regions (layered deposits) and surfaces that are aerodynamically rough such as impact ejecta fields, terrain that has been severely fractured by tectonic and other processes, and some lava flows. Once deposited on these surfaces, the dust would infiltrate cracks and crevices in rocks as accretion mantles (Breed et al. 1987*b*). Such trapped dust would be difficult to remove until the roughness elements were effectively buried. The formation of duricrust may also remove dust from the active aeolian regime. New dust may be added to the cycle by a variety of mechanisms on Mars, including chemical and physical weathering (clays are the normal end product of weathering on Earth), impact cratering, volcanism and tectonism.

In planning for future missions to Mars, aeolian processes must be taken into account. Surface modifications by the wind and windblown deposits can influence remote-sensing observations, affect sampling strategies, and have detrimental effects on manned and unmanned spacecraft on the surface. In many desert regions on Earth, rocks are commonly coated with a veneer of dark material, termed desert varnish. Although the origin of desert varnish is controversial, one commonly accepted model involves windblown dust, which adheres to the surface and reacts with dew to form a veneer. Compositional measurements obtained via remote sensing or through shallow surface sampling thus may represent only the varnish, which is commonly very different from the host rock. Although desert dew or its equivalent is unlikely on Mars today, a type of Martian varnish may form through other processes. In addition, fine material is mobilized by the wind, transported and redeposited in other areas. In the process, windblown materials become sorted by grain size and density. The surface which remains is also altered; large or high-density particles may remain as a lag deposit, contributing to the development of desert pavement surfaces. Fine windblown materials may accumulate to a thickness of several meters by accretion beneath a surface layer of apparently continuous rock, as on lava flows in deserts on Earth. On the other hand, very fine material is carried into the atmosphere via dust clouds and has the potential for becoming thoroughly mixed. Estimated to be less than a few microns in diameter, Martian dust-cloud material is globally transported. Thus, dust derived from diverse geologic sources may become homogenized in the dust cloud and, as such, may reflect a type of selected planetary compositional average. Erosion of the surface by the wind and deflation of weathered materials in some areas may enhance sampling efforts by future missions. Low-albedo surfaces, including those found in the lee zones of some craters, may represent eroded and deflated bedrock surfaces and could be high-priority sites for obtaining relatively fresh samples. In designing future lander spacecraft for Mars, consideration must be given to the infiltration of fine dust into spacecraft components, abrasion of optics by windblown sand, and the possible contamination of life support systems by dust in human exploration programs.

23. POLAR DEPOSITS OF MARS

P. THOMAS, S. SQUYRES
Cornell University

K. HERKENHOFF
Jet Propulsion Laboratory

A. HOWARD
University of Virginia

and

B. MURRAY
California Institute of Technology

Layered deposits unique to the poles of Mars indicate long-term winter deposition of ice and sediment modulated by periodic climate change. The layered deposits contain at least two kinds of nonvolatile materials: one that is similar to the bright red dust raised in Martian dust storms, and one similar to dark sand dune materials observed at low and mid latitudes. The ice fraction of the deposits remains unknown. Because the present polar dust and ice budgets are poorly constrained, correlation of layer thicknesses with specific astronomical cycles remains very uncertain. Layered deposits in both polar regions appear to have been slightly more widespread in the past. Other sediments near the poles, and the estimated ages of the layered deposits, suggest that the polar sedimentary regime has evolved over the geologic history of Mars. Acquisition of good data on the vertical sequence of layer characteristics and better crater retention ages should allow revealing comparisons with astronomically driven cycles and secular atmospheric changes.

I. INTRODUCTION

Seasonal polar frosts on Mars have been observed for more than a century and were important in early efforts to interpret the Martian climate. In the last 25 years, images returned from spacecraft have revealed distinctive layered deposits at both Martian poles. The presence of layered sediments in these climatically unique and sensitive areas suggests that they contain records of climate change on Mars. Despite much circumstantial evidence linking the deposits to climate-controlled processes, inferring a climate record from them has been far from straightforward.

This chapter examines the geologic and climatic context of the polar deposits. We first examine the major geologic units, their topography, and possible compositions. Then, the present CO_2, H_2O and dust budgets of the polar region are reviewed. Finally, we compare the expected depositional and erosional mechanisms and rates with the observed geologic and topographic characteristics of the polar deposits.

II. MAJOR GEOLOGIC UNITS

The Martian polar regions exhibit a complex stratigraphy, but the simplest classification is: (1) layered deposits and associated ices and frosts; and (2) a variety of "mantled" terrains peripheral to the regions of layered deposits. The northern and southern layered deposits are much more similar to one another than are the northern and southern mantling units. The layered deposits lie atop the basal plains materials, which are ancient and heavily cratered in the south, and younger and smoother in the north. Residual ices overlie much of the layered deposits at each pole. The entire polar region is in turn overlain by the seasonal frosts. The approximate boundaries of layered terrains in the north and south polar regions are shown in Figs. 1 through 4; photomosaics of areas poleward of 77° are shown in Fig. 5. Our treatment groups geologic units for an overview; the more detailed classification of Tanaka and Scott (1987) is shown in Figs. 3 and 4 (see maps I-1802C and I-2160–1).

A. Layered Deposits

The layered deposits were first recognized in Mariner 9 images of Mars (Murray et al. 1972; Soderblom et al. 1973*b*; Cutts 1973*a*) and are the most distinctive polar sedimentary units. They are found at both poles, and extend equatorward to ~ 80° in the north and 70° latitude in the south. They consist of sequences of essentially horizontal layers (nearly 20 layers are visible in some areas), each pair, called band pairs, is about ~ 10 to 50 m thick (Blasius et al. 1982). The thicknesses of layers appear to be more regular than would be expected from a random process, but the small number of layers in any single trough wall and uncertainties in topographic profiles have frustrated

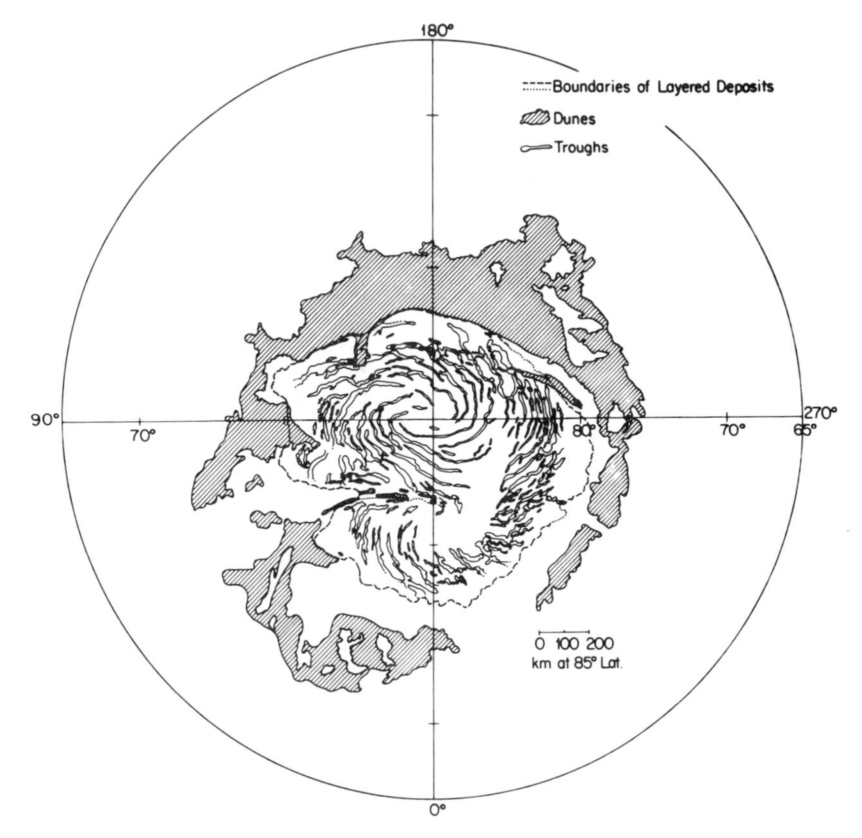

Fig. 1. Simplified map of main polar deposits—north polar region. The area of dunes is somewhat interpretive as the fraction of ground covered by dunes varies. A slightly different mapping of the dune cover is given in Tsoar et al. (1979) and Lancaster and Greeley (1990). Note that dunes encroach on the layered deposits in several places.

efforts to test statistically for regularity in layer thickness. Blasius et al. (1982) suggest that there may exist layers in a spectrum of thicknesses below the thinnest observable layers. However, bowl-shaped erosional depressions occur locally in trough bottoms, thereby exposing elliptical outcrop patterns. The gradient approaches zero in these locations, but individual layer exposures become wider rather than breaking into sublayers as would be expected if image resolution had limited layer discernment. Therefore, if there are layers thinner than the observed 10 to 50 m range, they must be several times thinner. Individual layers commonly are laterally continuous for tens of kilometers, but the complex topography and cover of residual ice prevents physical correlation of any layers beyond what amounts to local scales in relation to the entire deposits which are ~ 1200 km across.

The very low slope angles (2 to 5°) of the exposures of the layered

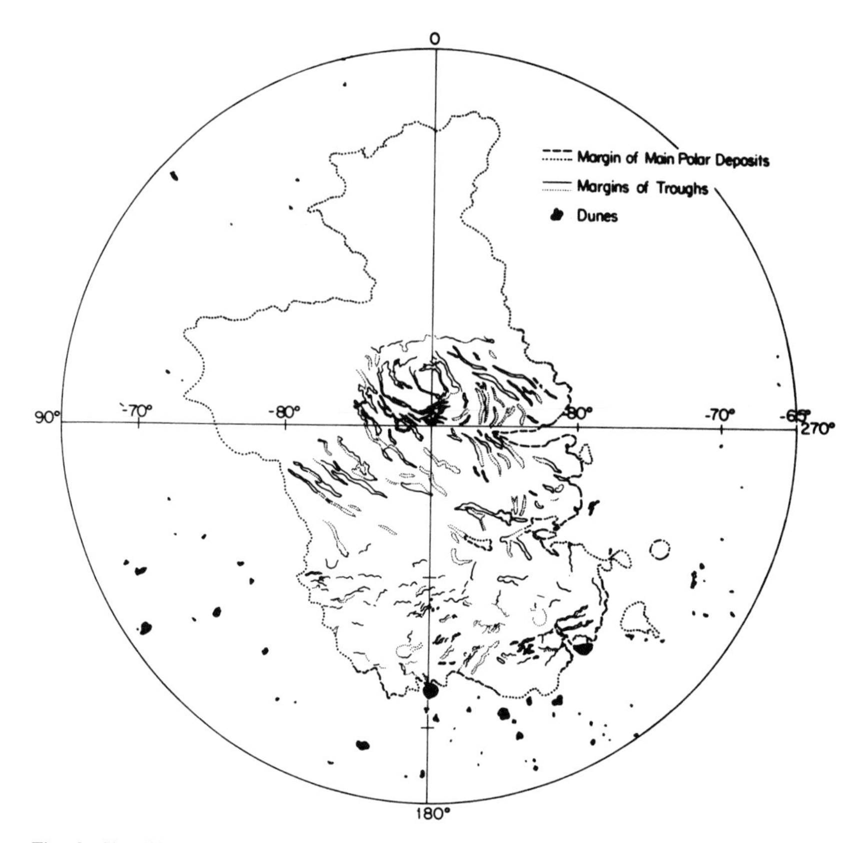

Fig. 2. Simplified map of main polar deposits—south polar region. The part of the deposits conspicuously lacking troughs (chiefly 320°W to 90°W) is pitted terrain. Note that the intracrater dune fields are largely near the layered deposits rather than near the section of pitted terrain.

deposits combined with the sharpness of the layers suggests that variations in the layer thicknesses are less than a few meters over horizontal distances of a few kilometers. The striped appearance of the layers appears to reflect both differences in topography and albedo (Blasius et al. 1982; Howard et al. 1982*b*, Herkenhoff and Murray 1990*b*).

Most exposures of the layers are in the slopes of troughs within the deposits (Figs. 6–9). While the usual exposures are conformable layers cut by nearly planar surfaces, unconformable layering is exposed in some places (Fig. 8). Additionally, poleward-facing parts of troughs in the northern deposits include poorly defined and gradational layers (Fig. 7; Howard et al. 1982*a*). Layers are also exposed near the edge of the northern deposits in irregular, erosional topography called striped layers (Howard et al. 1982*a*) and in areas of yardang development in the southern layered deposits.

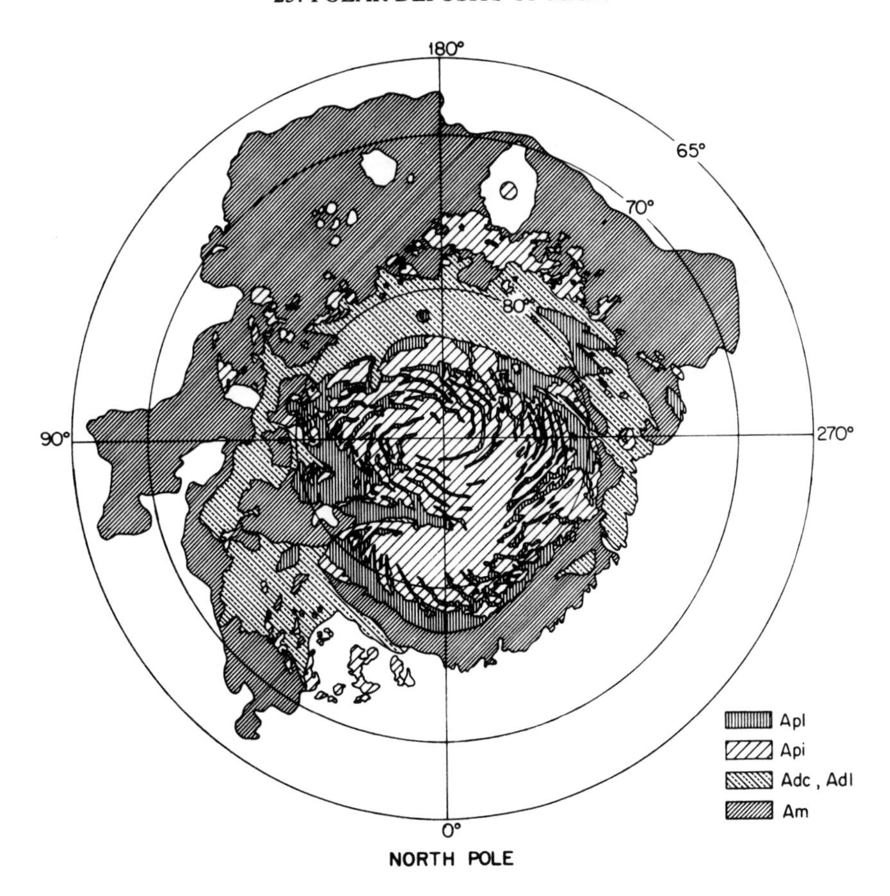

Fig. 3. North Polar region after Tanaka and Scott (1987). Api is residual ice; Apl is layered deposits. Adc and Adl are crescentic and linear dunes. Am is mantle material: smooth to hummocky deposit mapped where thick enough (up to a few hundred meters) to obscure features characteristic of Vastis Borealis Formation. Blank areas are plains units that do not have specific polar associations (cf. Fig. 1).

The total thicknesses of the deposits are difficult to determine because of the uncertainties in both the topography of the deposits' surfaces and in the topography of underlying units. Dzurisin and Blasius (1975) estimated the northern deposits to be 4 to 6 km thick and the southern ones to be 1 to 2 km. Malin (1986) re-analyzed some of the data of Dzurisin and Blasius and concluded that the northern deposits reached maximum thickness of 5 km along a ridge at ~ 87°N and averaged 2 km thick. The most recent contour maps of the polar regions (USGS 1989; map I-2160-3) that have listed uncertainties of 1.5 km suggest that the northern deposits may average ~ 2 km thick. Estimating thicknesses of the southern deposits from surface contours

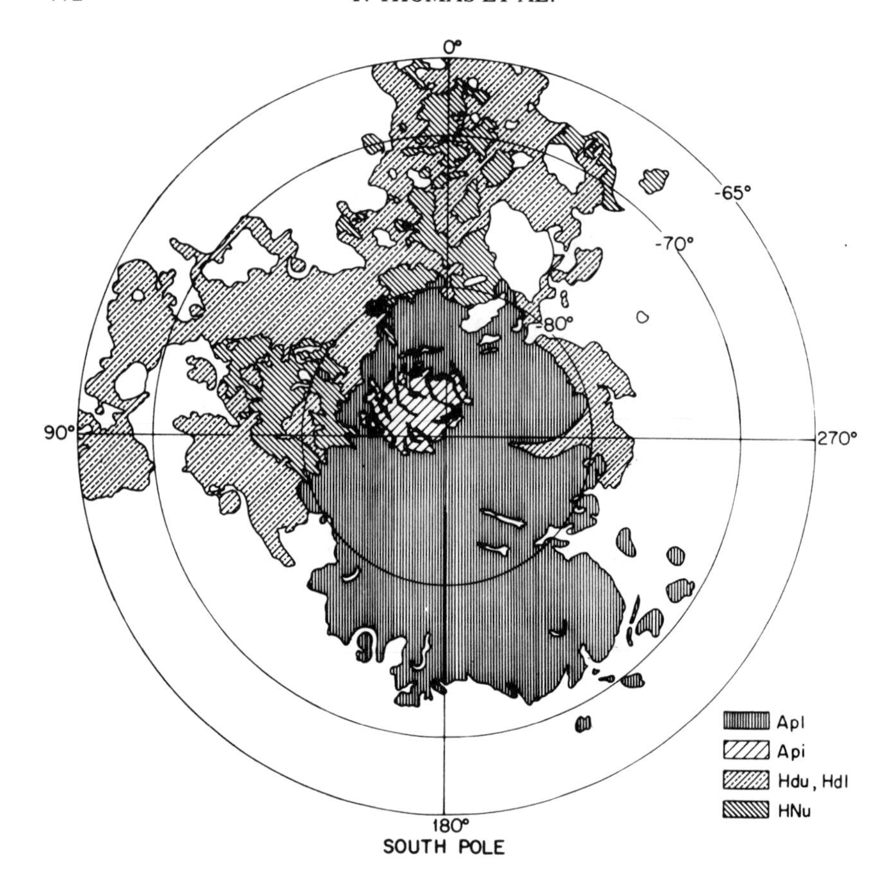

Fig. 4. South polar region after Tanaka and Scott (1987). Api is residual ice; Apl is layered deposits. Hdu and Hdl are upper and lower members of the Dorsa Argentea Formation with smooth-to-pitted surfaces, some areas with sinuous ridges, interpreted as either volcanic or aeolian. HNu is undivided material that has degradation caused by removal of ground ice, mass wasting, aeolian removal. Blank areas are plains units that do not have specific polar associations (cf. Fig. 2).

is even more uncertain because of the underlying large impact basin, but a 2-km average thickness would also be reasonable for these deposits.

B. Polar Ices

The layered deposits are locally overlain by summer residual ice and are seasonally covered by CO_2 frosts. Details of the study of the ices and their environments are covered in chapters 27 and 28; here we summarize the observations of the ices critical for evaluating their significance for polar sediments.

The north and south polar residual ices differ in their extent relative to

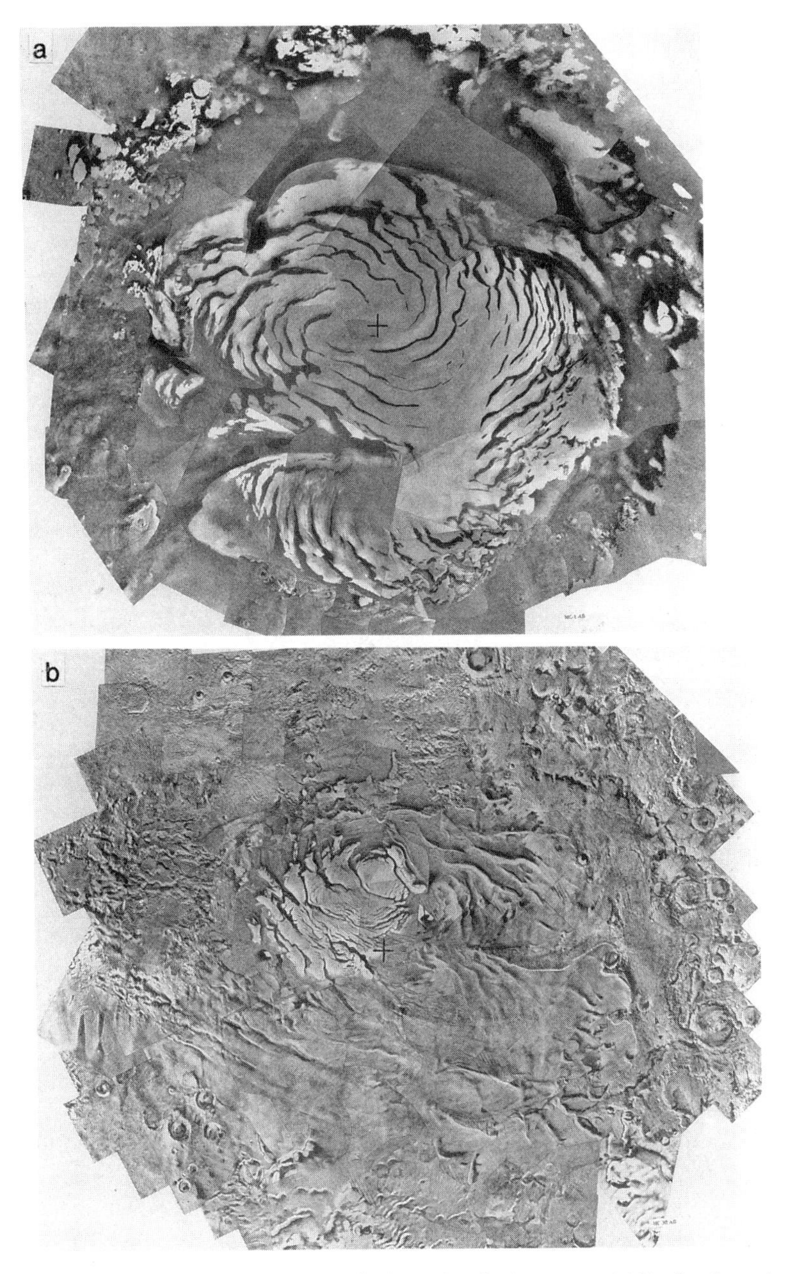

Fig. 5. Mosaics of Viking Orbiter images of polar regions in the summer. (a) North polar region; (b) south polar region; see also Color Plate 14.

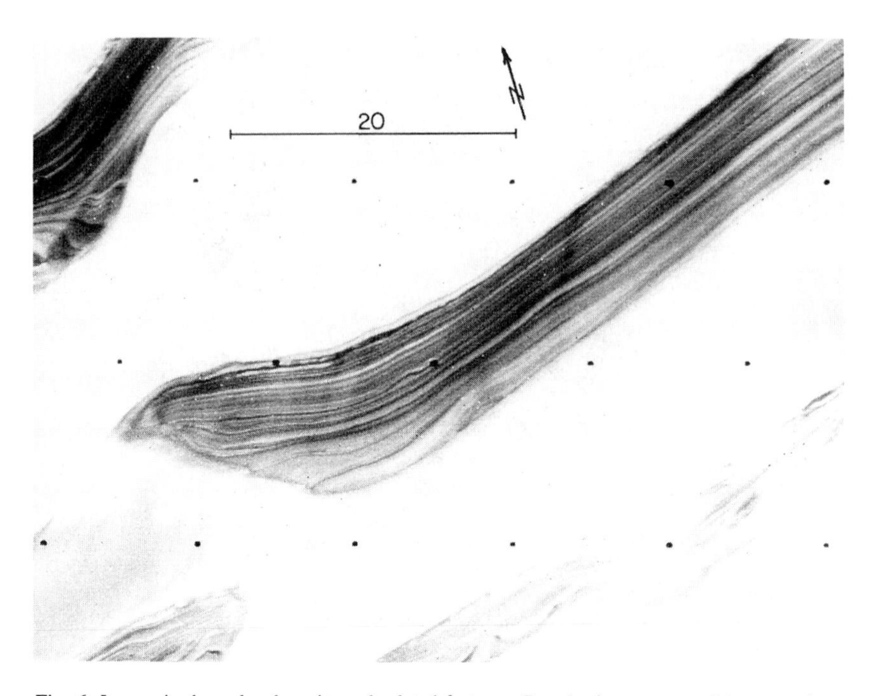

Fig. 6. Layers in the polar deposits and related features. Regular layers exposed in a trough at 82°N, 265°W. Wind redistribution of frost reduces the contrast between layers in some parts of the trough. (Image 65B67, clear filter, L_s = 138°). Bar indicates distance of 20 km.

the layered deposits, and during at least some years they differ in composition. In the north polar region, the residual ice reaches almost to the perimeter of the layered deposits, while in the south it covers a substantially smaller area (Fig. 5). The perennial ice in the south is not centered on the pole, but is offset by two or three degrees. At both poles, the local coverage by residual ice appears to be controlled at least in part by the slope angle; slopes receiving greater insolation are more likely to be free of residual ice (Howard et al. 1982*b*).

Viking Orbiter Infrared Thermal Mapper (IRTM) data show the northern residual ice to be H_2O (Kieffer et al. 1976*b*), in accord with large, near-saturation amounts of atmospheric water vapor detected near the north pole during the summer (Farmer et al. 1976; chapter 28). The albedo of the ice is ~ 43%, indicating an admixture of a small (but poorly known) amount of dust or other darker debris. In the south polar region the situation is more complex (Kieffer 1979; Paige et al. 1990): in at least two years the residual ice was CO_2 and temperatures stayed well below the sublimation temperature of water ice.

The thickness of the northern residual ice deposits is not determined.

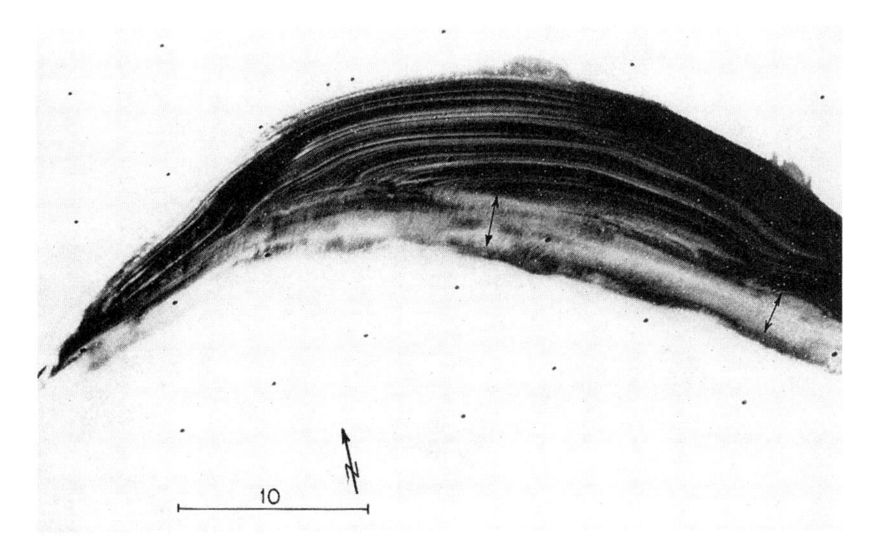

Fig. 7. Layers in the polar deposits and related features. Example of layered and banded terrain at the north pole. Upper boundary of scarp of layered terrain is near the top of the image. The unconformable boundary with banded terrain at the trough bottom occurs near the center of the image. Extent of banded terrain on north-facing trough slope is shown by arrows. (Detail from image 77B57, located at about 83°.8N, 306°W.) Bar indicates distance of 10 km.

Because it is not known if they may grade into the underlying layered deposits, it may not be meaningful to speak of a discrete thickness of residual ice. Thermal inertia data (Paige and Kieffer 1986) show the northern residual cap area to be consistent with dense ice; no information on dust or rock fractions can be obtained from the thermal data. Southern residual CO_2 ice may be distinct from the bulk of the layered materials, but unless the processes that built up the layers never included occasional residual CO_2, then it need not be distinguished from those layers. It is not known if the southern residual CO_2 ice is underlain by H_2O ice analogous to the northern residual ice.

The seasonal CO_2 caps extend well beyond the area of layered deposits and involve deposition of several tens of g cm^{-2} per winter. The seasonal advance and retreat of the caps is largely repeated from year to year (chapter 27), though interannual variations do occur. One particularly interesting aspect of seasonal cap retreat is that outliers of cap material may persist in some locations for substantial periods after the main body of the cap has dissipated. These outliers are commonly associated with impact craters and other topographic features, and are probably locally thicker deposits of frost caused by aeolian redistribution (James et al. 1979) and by shading from small-scale topography (Herkenhoff and Murray 1990*a;* Svitek and Murray 1990). Seasonal frost outliers in the north are less extensive than in the south, probably due to smoother topography.

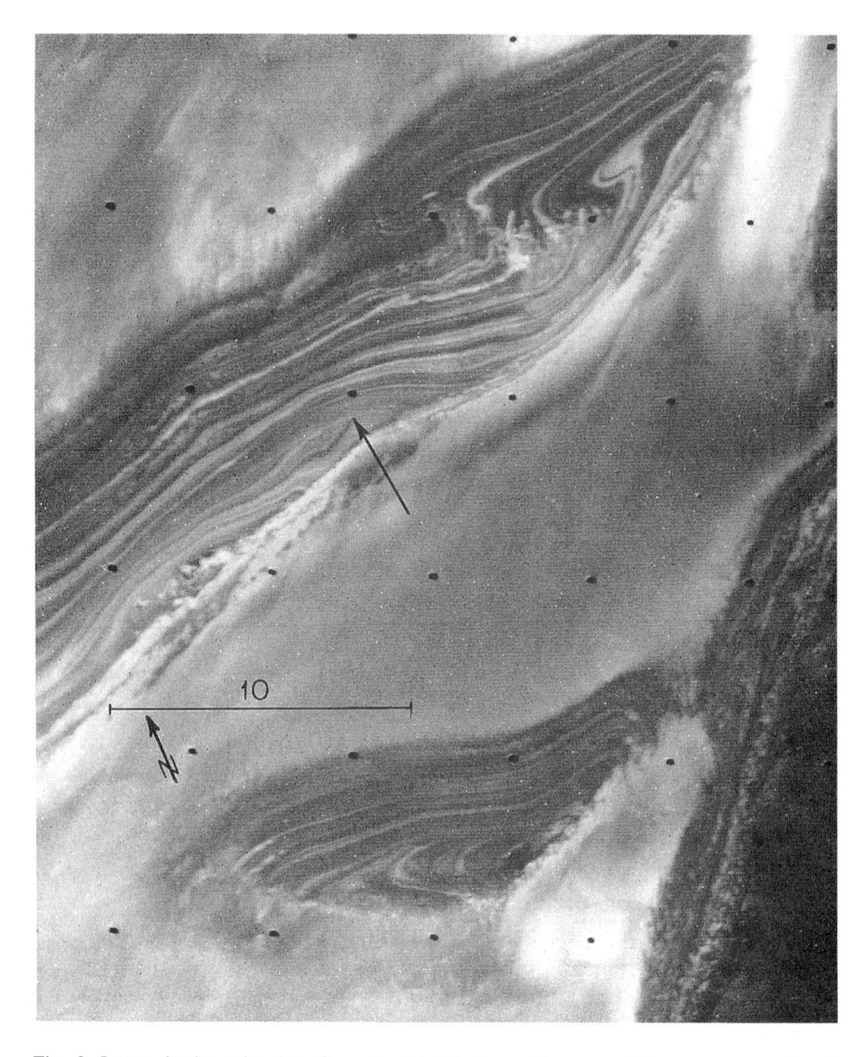

Fig. 8. Layers in the polar deposits and related features. Exposure of layers near margins of deposits at 82°N, 84°W. Arrow indicates unconformity. (Image 59B77, clear filter, L_s = 135°.)

C. Peripheral Deposits

Three other depositional units lie near or on the major units described above: a northern debris mantle, dune fields in both polar areas, and a variety of plains units in the south polar region. With the exception of the distinctive dune morphologies, these units have been difficult to classify, and as a result, interpretations and maps have varied widely. The maps in Figs. 1 and 2 in-

Fig. 9. Layers in the polar deposits and related features. Complexity of some exposures of layers in north polar deposits, centered at 81°N, 336°W. Note the complex albedo patterns on the residual frost areas, the variety of shapes of defrosted areas, and "closed contour" layer exposures. (Image 56B86, clear filter, L_s = 133°.)

clude our generalized interpretations; Figs. 3 and 4 present the polar units from Tanaka and Scott 1987.

Scattered debris mantles surround the north polar layered deposits down to the middle latitudes (Fig. 3; see Soderblom et al. 1973*b*; Squyres and Carr 1986; Kahn et al. 1986; Tanaka and Scott 1987). Mantling is indicated by the partial burial of impact craters; inferred depths of the mantle are < 200 m (Squyres 1979*b*). The debris mantle is quite smooth, and appears uniformly draped over most materials immediately equatorward of the layered deposits. The contacts between layered deposits and debris mantle are complex. In some exposures, it appears that the debris mantle unconformably overlies the layered deposits, while in others the units grade smoothly into one another.

Deposits surrounding the south polar layered materials and unconformably overlying the old cratered terrains are much more voluminous than those in the north (Figs. 1–4). Inferred origins of these units vary from volcanic to aeolian; relationships to the layered deposits consequently are not well constrained. Materials at the periphery of the southern layered deposits have been broadly classified into plains units and pitted terrain by Plaut et al. (1988). The plains units are essentially coincident with the Dorsa Argentea formation of Tanaka and Scott (Hdl and Hdu of Fig. 4) and are distinguished by crater densities much lower than those on the cratered terrain and by surfaces that are smooth with some coverage by braided ridges.

The pitted terrain of Plaut et al. (1988) was termed pitted plains by Murray et al. (1973) and pitted and etched terrain by Sharp (1973*b*). The unit was not separately mapped by Tanaka and Scott, but is included in their Hdl, Hdu, and HNu units (Fig. 4). It occurs primarily between longitudes 340° west to 100°, and between latitudes −70° to −85°, and makes up much of the south polar deposits in Fig. 2 that are not cut by troughs (i.e., upper left of Fig. 2). This deposit is characterized by a smooth surface broken by irregular, steep-sided depressions (Fig. 10).

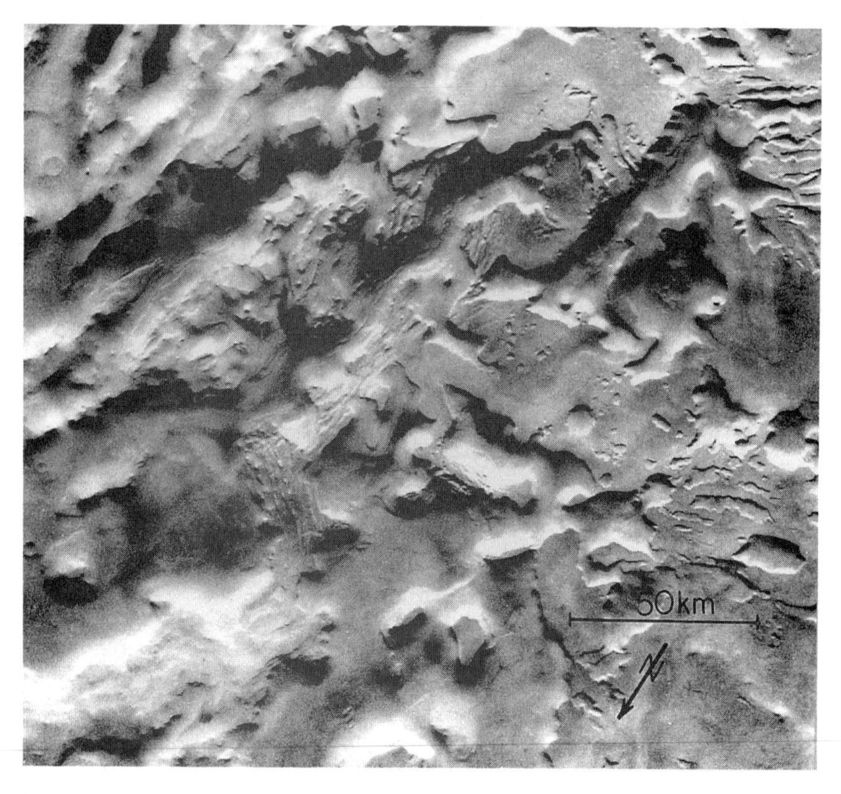

Fig. 10. Example of pitted terrain centered at 78°S, 69°W. Note the variety of structures that influence pit shapes, including some impact craters. (Image 390B88.)

Plaut et al. find that the pitted deposits are superposed on the plains units, and have substantially smaller crater densities. They find that the layered deposits in turn are superposed on pitted terrain and are thus clearly younger. Several workers (Scott and Carr 1978; Condit and Soderblom 1978; Plaut et al. 1988) interpret the pitted terrain as remnants of a debris deposit emplaced unconformably over the plains. The thickness of the deposits can be estimated from pit depths which are commonly 500 to 1000 m (Plaut et al. 1988).

Dunes are prominent in both polar regions (Figs. 1–4). The northern polar deposits are nearly ringed by dunes overlying the smooth plains; they can be subdivided into four major ergs (Lancaster and Greeley 1990). In the south, dunes near the polar deposits are topographically trapped in craters, by wrinkle ridges, or in some other types of closed depressions (Figs. 1–4). These dunes cover much less area than do the northern ergs, but many appear to be extremely thick accumulations. Measurements of areas and probable depths (Thomas 1982) show that the south polar dunes are likely to be of the same order of magnitude of volume as the north polar erg. The volume estimates of Lancaster and Greeley (1990) for the northern erg even raise the possibility that the south polar dune material might be the more voluminous.

In both the north and south there is evidence the dunes have some sources within the layered deposits. In the north, the four major groupings of dunes have identifiable source areas within the layered deposits (Lancaster and Greeley 1990; Thomas and Weitz 1989). These sources are strongly correlated with the steep erosional scarps examined by Howard et al. (1982*a*). In the south, dunes appear to have sources within some closed depressions in the layered deposits, but cannot be associated with specific erosional scarps as in the north.

III. COMPOSITION OF THE POLAR DEPOSITS

The composition of the layered deposits is poorly constrained. Malin (1986) used gravity and topographic data to estimate the density of the northern deposits to be 1 g cm^{-3} ($\pm$ about 0.5 g cm^{-3}). The uncertainty of 50% allows for a wide range of combinations of dust, sand, ice and void space.

Color data suggest components of bright red dust in both polar layered deposits (Thomas and Weitz 1989; Herkenhoff and Murray 1990*a*). A sample of the Viking Orbiter color data is given in Fig. 11 for northern polar layered deposits, residual frost (L_s = 137°) and dunes at ~ 85°N. Photometric corrections allow comparisons with other Martian units such as low latitude dark dunes and the ubiquitous "dust." The layered deposits appear to be compatible with typical "red" dust, either mixed with some darker component or having a variety of local phase coefficients that reduce the reflectances by a factor of as much as 2. We suspect that much of the range of reflectances and

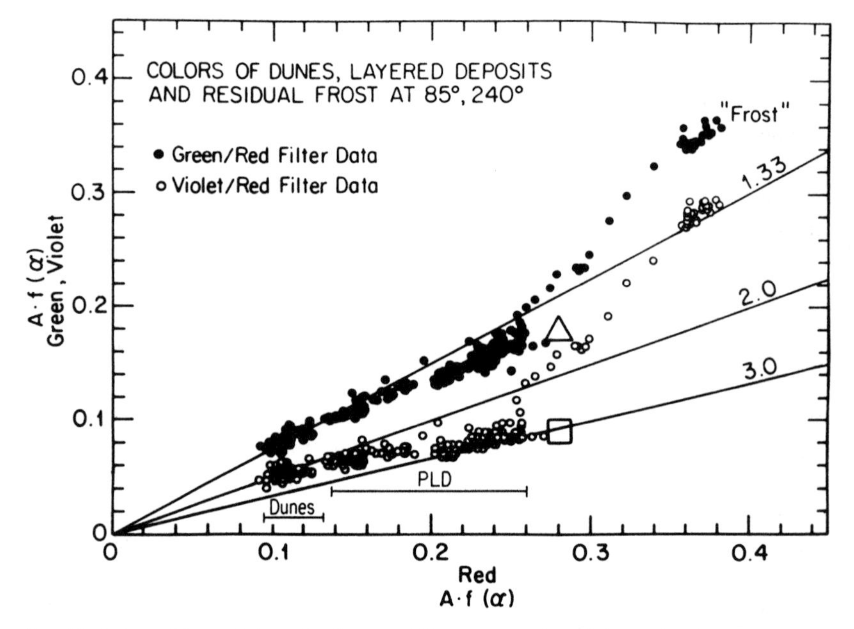

Fig. 11. Colors of layered deposits, dunes and frosts at the edge of the north polar residual cap. Data have approximate atmospheric and surface photometric function removed for comparison with other Martian data. Solid lines show the reflectance ratios to the red data. Triangle and square are typical green and violet data, respectively, for dust-covered ground. The layered deposits approach the dust in color and albedo, but include much darker and less red materials. The dunes are the same colors and albedo as low latitude dunes, and some of the range of layered-deposit data is probably due to mixing with the dark dune materials; see Thomas and Weitz (1989) for details of the photometric corrections.

change in color (brighter parts are much redder than darker parts) is due to mixing with darker materials similar to those in many Martian dune fields below the resolution limit (50 m in this instance).

Although the surface of the layered deposits has a similar albedo and color to "dust", the layers are not necessarily pure dust. We discuss the problem of dust/ice ratios below in Sec. VI, but here note that the work of Kieffer (1990) and others shows that the albedo of a dust/ice mixture can be essentially that of the dust even for very small fractions (10^{-2} to 10^{-3}) of dust if the grain size of the ice is significantly larger than that of the dust.

The surface material may also be a lag deposit from both sublimation and aeolian sorting, and therefore a translation of a surface color component into a dust component of the cap cannot be made. The possibility of darker sand or dust mixed with the bright dust component also complicates the interpretation of the ice fraction.

The dunes in both polar areas have colors and albedos similar to dark dunes at low latitudes (Thomas and Weitz 1989; Herkenhoff and Murray 1990*a*). Lambert albedo in the red (0.59 μm) is ~ 0.07, and Viking red/violet ratios are ~ 2. These colors and albedos are approximately the same for all major ergs on Mars and are dramatically different from the widespread bright dust on Mars.

Mineralogically the dust is likely to include clays and iron oxides (see chapter 18). The dunes might be minimally weathered solid basaltic particles. The sand-sized dune materials, might, however, be made up of agglomerations of finer materials such as magnetite and maghemite (see Herkenhoff and Murray 1990*b*).

The composition of the polar dunes may have considerable sedimentologic implications. If they are low-density agglomerations of fine particles that could have been carried to the polar region in suspension, then the sedimentary history could be similar to that of the brighter dust. If the dunes are composed of dense basaltic sand, then they would have to have been transported to the poles at the surface during part of the formation of the layered deposits. This wind regime would probably be very different from the present one. Surface wind markings in the area of layered deposits currently show only outward flow (Thomas 1981,1982). Herkenhoff (1990) prefers a hypothesis in which dark, magnetic dust grains preferentially form sand-sized particles upon erosion of the layered deposits. Saunders et al. (1985,1986) showed that such filamentary sublimation residue particles can be formed by sublimation of dust/ice mixtures and will survive saltation.

IV. TOPOGRAPHY OF THE POLAR DEPOSITS

Regional topography of the north polar deposits is known crudely from stereogrammetry (Dzurisin and Blasius 1975; USGS 1989). Most of the area underlain by layered deposits appears to be characterized by nearly featureless smooth terrain. Malin (1986) interpreted the deposits to control the topography much more than do any underlying regional features. The contours presented in Malin (1986) show the highest area to be a ridge 5 km above the reference surface near 87°N between 270° and 360°W, and that the steepest regional slopes, ~ 1°, occur near the interior of the deposits and not near the margins. The topography given in USGS (1989; see map I-2160–1) is slightly different but suggests about the same general heights and shape. Perhaps the most important feature of the USGS topography is that the highest part of the north polar deposits is only offset 2° from the pole, and the highest part of the south polar deposits is centered on the pole.

Local topography superimposed on these broad domes of material can be grouped in six categories: (1) broad curvilinear reentrants, or chasmae; (2) troughs, which form the large, spiraling patterns of dark bands in both polar regions; (3) gentle undulations on the smooth upper part of the layered deposits; (4) steep scarps within the layered deposits; (5) marginal convex slopes; and (6) pits in the southern polar deposits.

Both polar deposits have large reentrants which are tens of km wide, as much as 2 km deep, and over 500 km long (Figs. 1–5). These are not simple valleys cut into the deposits; the troughs in the adjacent layered deposits nearly parallel the chasmae (Figs. 1–5). This relation shows some development of troughs after chasmae formation began. Both polar chasmae expose the underlying plains units in their outer sections; both have their upper sections floored by remnants of layered deposits or other mantling debris. The Chasma Boreale heads at a steep scarp (Howard et al. 1982*a*), and has some other scarps, ~ 200 to 400 m high, 100 km from the head of the Chasma. A very pronounced pattern of yardangs roughly parallel the trend of Chasma Australe.

The most prominent features of the polar terrains in both hemispheres are crudely spiraling bands (Figs. 1–5). They are formed by gentle troughs whose equator-facing slopes expose layered terrain and are mostly defrosted in summer. These bands roughly parallel one another, although junctions are common in some areas. They average ~ 10 km wide and are typically separated by ~ 50 km. These troughs also extend into the portions of the south polar cap terrains that do not have residual ice. The pole-facing slopes are somewhat gentler than the equator-facing slopes of the troughs (all are $< 5°$; Dzurisin and Blasius 1975) but rise to roughly the same elevations. Pole-facing slopes tend to retain residual ice in areas where the smooth terrain also retains frost.

The smooth terrain at the north pole commonly exhibits low-relief, wave-like undulations spaced 2 to 10 km apart (Cutts et al. 1979; Howard et al. 1982*a*). The crests of the undulations are curvilinear and generally parallel to the troughs, suggesting a genetic relationship.

Steep (20 to 30°) arcuate scarps, up to a few 100 m high and up to 50 km in length, occur in two linear zones in the north polar deposits (Howard et al. 1982*a*; Thomas and Weitz 1989). They may mark areas of rapid erosion that expose materials that are subsequently formed into dunes.

Slopes at the margins of the deposits often are convex, and high resolution imaging shows that the layered deposits have very sharp bases where they rest on underlying plains. In the south polar region, closed depressions in the layered materials that expose underlying plains usually have convex slopes and sharp lower contacts with the underlying units.

In the pitted terrain (see Sec. II.C above) the depressions usually have sharp crests at the upper surface with concave slopes leading to flat floors (Fig. 10). They have depths up to 1 km (Plaut et al. 1988). They have horizontal dimensions from resolution limits (300 m) to tens of kilometers and occur singly and in clusters. Associated topography includes ridges that resemble exposed dikes (Sharp 1973*b*).

Other landforms in the polar deposits include a variety of yardangs, crescentic forms on the smooth terrain that have been interpreted as ice dunes (Cutts et al. 1976), and faint ridges that parallel parts of the margins of the northern layered deposits (Fig. 12; Howard et al. 1982*a*).

V. PRESENT ENVIRONMENT OF MARS POLAR REGIONS

The present environment of the poles is discussed in detail in chapters 27 and 28 (see also Kieffer 1990; Pollack et al. 1990*a*). Here we summarize the properties and interpretations of the CO_2, H_2O and dust seasonal cycles relevant to the geologic and/or paleoclimatic interpretation of the polar deposits.

A. CO_2 Cap

Mars' seasonal frost caps led early observers to postulate an Earth-like atmosphere on the planet, even with liquid water present. Spacecraft data, however, showed it to be most unearth-like in that the caps are carbon dioxide frost and that their solid-vapor equilibrium is controlled by the annual balance at the poles between absorbed solar energy and reradiated thermal energy (Leighton and Murray 1966).

The amounts of CO_2 frost can be estimated from atmospheric pressure measurements and thermal data (Kieffer et al. 1976*b*; Hess et al. 1979; Paige and Ingersoll 1985). Paige and Ingersoll calculate that within areas of the residual caps during the winters observed by Viking Orbiter, there was 75 ± 12 g cm^{-2} of CO_2 frost deposited in the north and 110 ± 7 g cm^{-2} deposited in the south. The difference in deposition arises chiefly from the longer southern winter. The cycling of $\sim 25\%$ of the atmosphere through the seasonal polar caps is expected to have a profound effect on global winds (see chapter 26).

The seasonal CO_2 sublimation is very asymmetric between poles, apparently due to different cap albedos: the northern cap is darker and thus sublimates more efficiently than does the southern one (Paige and Ingersoll 1985). Dust concentration may affect cap albedo, but as pointed out by numerous authors (Paige and Ingersoll 1985; Kieffer 1990) a full explanation of any difference in dust content of polar ices may not be straightforward.

The most spectacular polar asymmetry in sublimation of the CO_2 frost is its persistence through the summer in the southern residual cap for at least some seasons. Sublimating CO_2 remained on the surface of the southern cap throughout the 1976–1977 summer and also during 1971–1972 (Paige et al. 1990). However, in 1969 a high global water vapor abundance developed in the atmosphere during the southern summer (Barker et al. 1970) which has been interpreted by Jakosky and Barker (1984) as indicating that during that summer, CO_2 was depleted by sublimation and an underlying residual water

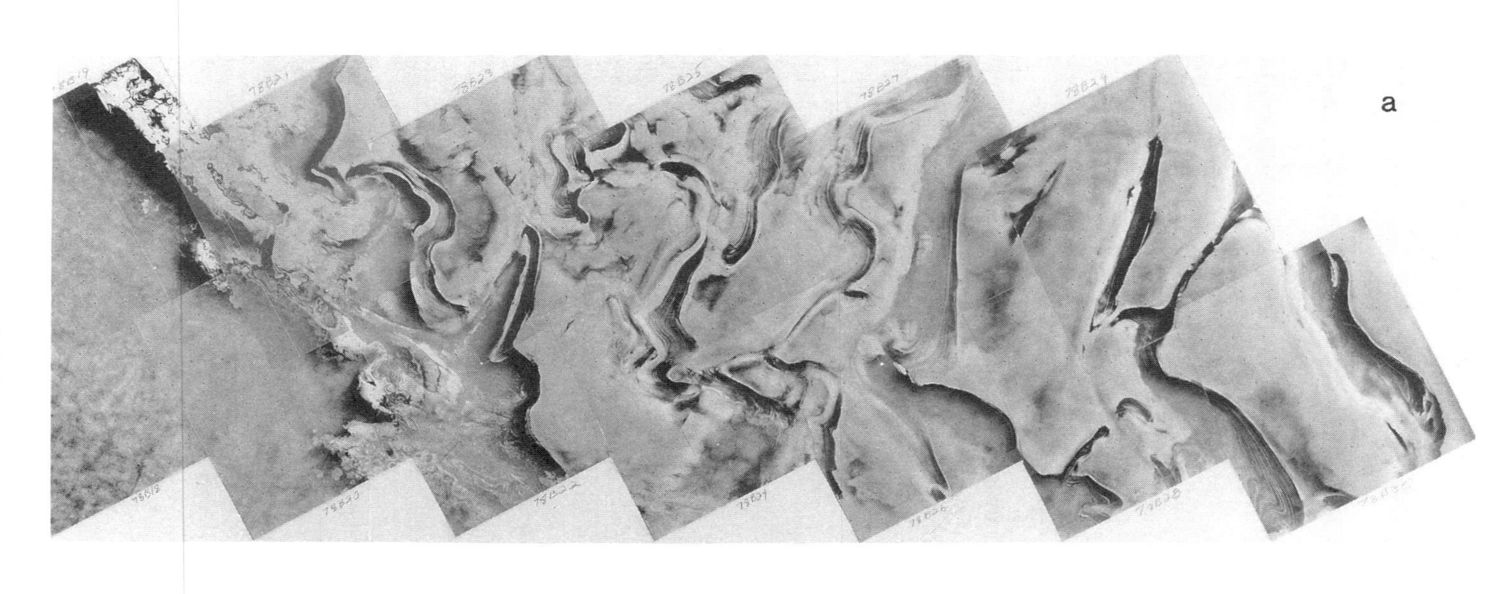

Fig. 12. Mosaic of summer images of part of the north polar cap at about 80°N, 350°W. (a) Orthographic mosaic of images 78B19–30; (b) stratigraphic map; (c) physiographic map. Figure after Howard et al. 1982*a*, Fig. 5.

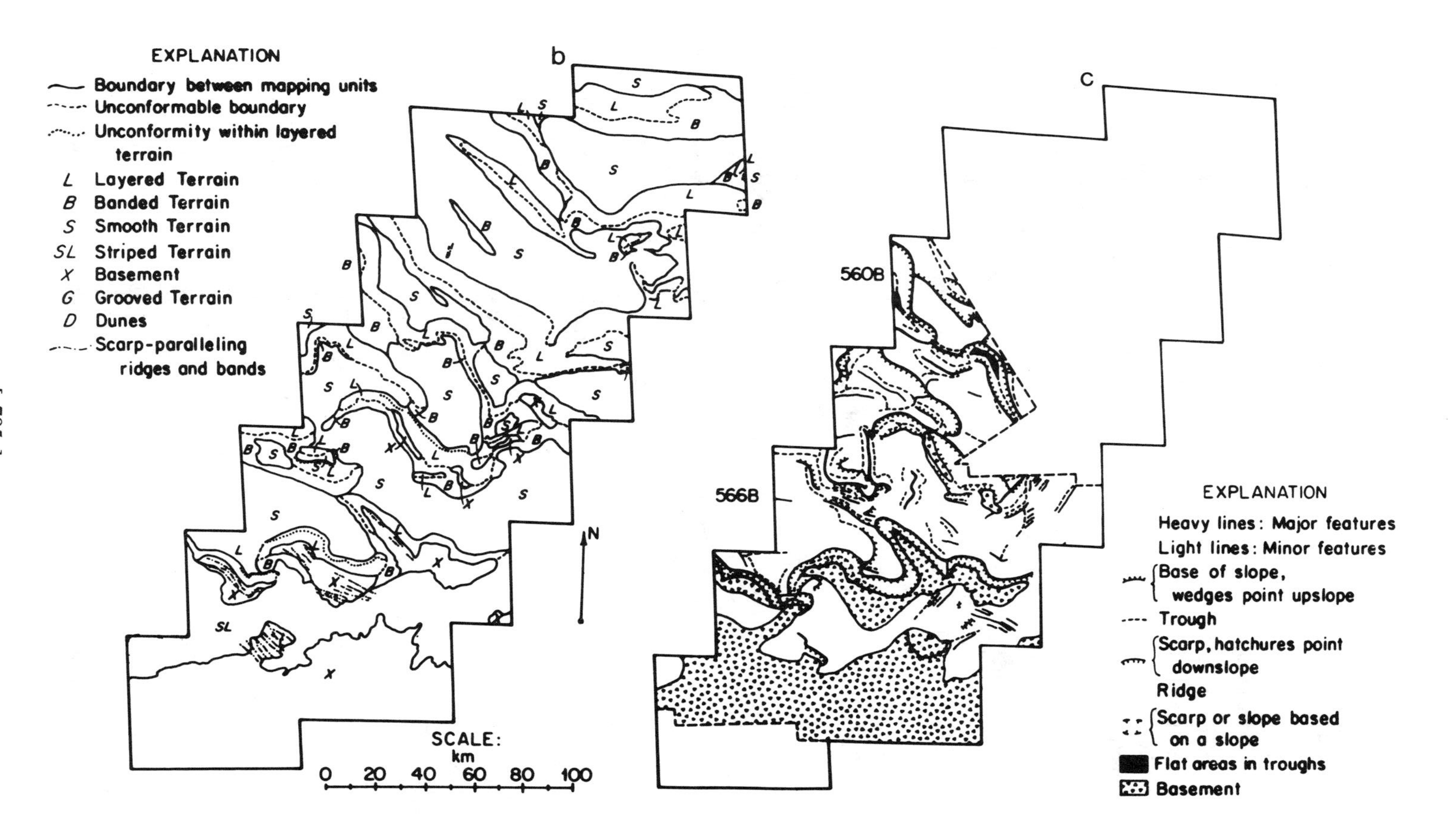

b
EXPLANATION
Boundary between mapping units
Unconformable boundary
Unconformity within layered terrain
L Layered Terrain
B Banded Terrain
S Smooth Terrain
SL Striped Terrain
X Basement
G Grooved Terrain
D Dunes
Scarp-paralleling ridges and bands
N
SCALE: km
0 20 40 60 80 100
c
560B
566B
EXPLANATION
Heavy lines: Major features
Light lines: Minor features
Base of slope, wedges point upslope
Trough
Scarp, hatchures point downslope
Ridge
Scarp or slope based on a slope
Flat areas in troughs
Basement

ice cap was able to warm to its radiometric equilibrium temperature. The feedback mechanisms that may allow shifting from CO_2 to water-ice residual frosts are dealt with by Jakosky and Haberle (1990*a*) and in chapter 28.

B. H_2O Budget

The polar water budget is critical for understanding of the layered deposits because they may be composed largely of water ice. Relating the MAWD observations of water vapor to surface processes, and obtaining a generalized net water budget is, however, a very challenging task (Jakosky 1983*b*; chapter 28). Haberle and Jakosky (1990) calculate that the northern residual cap area can sublimate 1 to 8 $\times$ 10^{-2} g cm^{-2} per summer, and that the observed northern water-vapor abundance may require additional sources of water in the regolith. They further conclude that atmospheric transport is most likely to be inadequate to resupply this amount of ice to the cap and thus there may be a current net loss of water ice from the north. Uncertainties in the various transport mechanisms and in the role of the regolith require caution in declaring a net deficit, however.

Much of the observed water vapor may originate in areas other than the residual cap such as older ice in the darker, warmer, exposed layered deposits (Kieffer 1990). This is important because although the residual cap area may be able to supply much of the observed atmospheric water vapor, the atmosphere may not be able to resupply the water to the cap in the winter. Kieffer points out that using the global average water-vapor abundance, and the amount of CO_2 frost deposited in the north polar cap, should amount to only $\sim$ 3.7 $\times$ 10^{-3} g cm^{-2} yr^{-1} water ice deposited, a value well below that calculated from the estimated sublimation.

Other complications of the water budget may include local redistribution and chemical bonding. Howard et al. (1982*b*) and Svitek and Murray (1990) noted evidence for local redistribution and cold trapping of water vapor in the polar regions. The layered deposits might allow some chemical weathering in the present Martian environment. Although the H_2O in the layered deposits is generally well below freezing temperatures, thin films of water can surround fine-grained sediment more than 20°C below the freezing point of water (see, e.g., Tsytovich 1975). Temperatures on low-albedo equator-facing trough walls may reach such levels. Temperatures may reach melting at depths greater than $\sim$ 2 km (Clifford 1987*b*). Because the dust and water of the polar deposits probably have been extensively recycled during Martian history, any polar weathering products might be globally important.

C. Dust Budget

The dust budget at the poles is no less complicated than the water budget. Pollack et al. (1979) used the observed average annual dust loading for the first Viking Mars year to estimate an average of 2 $\times$ 10^{-2} g cm^{-2} per Mars year deposition if half the global dust is concentrated above 60°N in winter frost deposition. There is little evidence that this efficient transport of sediment actually occurs. Kieffer (1990) estimates dust deposition potential from the atmospheric dust loading and the amount of CO_2 actually deposited, and obtains 1.8×10^{-3} g cm^{-2} yr^{-1} (in the north residual cap area), an order of magnitude less than the estimate of Pollack et al. Additionally, it has been very difficult to obtain transport of dust to the poles in atmospheric circulation models (Haberle et al. 1982; Barnes 1990*b*; chapter 26). As Kieffer (1990) points out, however, it is difficult to see how the dust could be segregated from the CO_2 that does precipitate, so some must fall on the polar area. The amount of dust falling on the south polar cap may be much less because the average dust loading in southern winter is lower, and the higher albedo of the southern cap supports this inference (Paige and Ingersoll 1985).

The interannual variability of Martian dust storms has been examined by Zurek (1982) and Haberle (1986*b*). There are some years without great dust storms and there is a strong indication that during the last two decades (those with spacecraft observations) there may have been a greater than average frequency of global dust storms (see chapter 29). Thus the "present" conditions used as the basis for many calculations may be very unusual.

D. Present Aeolian Environment at the Poles of Mars

The polar regions exhibit an active wind regime that moves material along the surface in addition to the aeolian action involved in phase changes and precipitation. The condensation and sublimation of about one-fourth of the Martian atmosphere generates significant winds by mass transfer alone (Leovy and Mintz 1969; French and Gierasch 1979; Pollack et al. 1990*a*; chapter 24). Thermal contrasts caused by latent heat effects of ice, and to lesser degree albedo differences between areas of frost cover and plains or dunes, can generate very strong regional winds. The topography of the polar areas may be of sufficient length to generate slope winds capable of significant sediment transport and perhaps erosion (Magalhães and Gierasch 1982).

Winds near the poles are indicated by a variety of frost and dust streaks, and by the sand dunes. Most inferred winds have a large component away from the pole; spring winds generally have components to the west; fall winds to the east. The dunes are inferred to be active from their lack of dust cover, and from observations of dark streamers from some dunes. Breed et al. (1979) interpreted some polar dunes to be cut by erosional grooves. These possibly degraded dunes would imply great variation in present dune sediment budgets. Cloud patterns show diverse winds, including cyclonic storms near the north polar deposits (Gierasch et al. 1979; Hunt and James 1979). Local dust storms commonly arise in the spring near the edge of the southern cap (Briggs et al. 1979).

The frost streaks that form in the seasonal caps suggest that the wind might redistribute substantial amounts of frost, and could cause nonuniform net deposition on the cap of both ice and dust. The frost streaks associated with the northern residual cap suggest spring or summer winds are effective at moving residual cap materials. This redistribution almost certainly includes some dust because some of these streaks remain visible longer than one year (Thomas 1981). The dunes show that in the present geologic era material is being removed from the area of layered deposits (Thomas and Weitz 1989). Yardangs, especially in the south, show that some winds are strong enough, and carry enough (sand-sized?) material, to erode the deposits by abrasion.

VI. ORIGINS OF THE POLAR DEPOSITS

A. Formation of the Deposits

A major question regarding deposit origin is: what is the relationship to the present dust and ice environment; i.e., are the residual caps merely the tops of the sediment piles or are they something fundamentally different from the layered deposits?

The annual evolution of ice/dust deposits involves the condensation, precipitation, sublimation, and annealing of ice crystals, and possible migration of dust particles within the ice. One of the key elements is the change in crystal size of the water-ice particles, dealt with in detail by Kieffer (1990). There is little change expected while the seasonal CO_2 cap buffers temperatures at $\sim$ 145 K; after loss of the CO_2 the temperature can rise above 200 K and initiate sublimation and crystal growth in the near-surface region. The modeling done by Kieffer suggests that in either net sublimation or net accumulation after one Martian year water-ice crystals will be a minimum of several tens of microns in diameter. The practical importance of this prediction in understanding the residual ice and layered deposits is that if the ice grains become much larger than the included dust grains (a few microns), the albedo of the deposit will approach that of the dust. Albedo of the residual frosts are 40 to 50%, the layers are $\sim$ 20%, and pure frost would be $>$ 90% (Kieffer 1990; Herkenhoff and Murray 1990*a;* Thomas and Weitz 1989). If Kieffer's model is adopted, the intermediate albedo of the residual cap can be explained only by either (1) very clean, old ice; (2) young and very fine-grained ice with a moderate dust component; or (3) an intermediate balance of particle sizes, porosity and ice grain sizes. The first alternative is not consistent with present atmospheric dust/water ratios (by orders of magnitude); the second implies accumulation of water ice on the residual cap. Accumulation scenarios would require the water to come from defrosted layered deposits, and probably would imply a shrinking of the residual cap area (see also chapter 28). The third alternative is not favored by Kieffer as it might require too delicate a balance of nonlinear effects to maintain a nearly uniform albedo over the residual cap.

The difference between the residual cap and the layered deposits then might be largely a matter of age, or current net ice budget. There may be no fundamental difference between the residual ices and the layered deposits except that the layered materials exposed in troughs have been buried under tens to hundreds of meters of material and then exposed by erosion and/or sublimation.

Another factor in the annual development of residual cap material is that the water ice and dust are most likely well distributed through the CO_2 deposit (at least partly supported by MAWD observations; see chapter 28) and if they are to remain part of a local deposit they must survive the summer sublimation wind that is sufficient to suspend dust particles of a few microns radius (Kieffer 1990).

B. Origins of the Layers

The fine layering observed in the Martian polar deposits suggests some kind of periodic modulation of deposition and erosion at the poles. There may be alternating aspects of deposition and erosion, or there may simply be cyclic fluctuations in the dust/ice ratio of the deposited material. Because our present understanding of the deposition process suggests that layer formation requires stability of H_2O ice at the surface in the polar regions and injection of dust into the atmosphere at lower latitudes, the appearance of the layers may indicate cyclic fluctuations in ice stability at the poles, in atmospheric dust loading, or both.

Mars undergoes a number of variations in its orbital motion and axial orientation that may result in substantial cyclic variations in climate. The variations consist of oscillations in obliquity, oscillations in orbital eccentricity, and precession of the equinoxes (see chapter 9). Precession and eccentricity control the seasonal distribution of insolation, while obliquity affects annual mean insolation as a function of latitude (chapter 33). Eccentricity varies in cycles of $\sim 10^5$ and 1 to 2×10^6 yr, and the precessional cycle has a period of 5.1×10^4 yr. The obliquity oscillates with a period of 1.2×10^5 yr, modulated by a longer cycle of 1.3×10^6 yr (Ward 1979; chapter 9). The combined effect of the several different periods involved is a myriad of potential climatic cycles (Cutts and Lewis 1982).

The obliquity cycle appears to be particularly important. Obliquity excursions can be quite large, from a minimum of 15° to a maximum of 35°. At high obliquity, the annual mean insolation at high latitudes is substantially greater than at low obliquity. These variations in insolation may drive large fluctuations in atmospheric CO_2 pressure. For example, Toon et al. (1980) calculated that at low obliquity the mean atmospheric pressure would be < 1 mbar, while at maximum obliquity desorption of CO_2 from the regolith at high latitudes could drive it to 20 mbar or more.

Large variations in high-latitude insolation and mean atmospheric pressure could affect polar deposition in several ways. When the atmospheric pressure is at its lowest, it may be insufficient to initiate dust storms, reducing atmospheric dust loading. With higher pressure, dust loading and poleward transport of dust may be significantly greater than they are at present. The surface stability of H_2O ice in the polar regions may be affected by the amount of polar insolation, and higher pressures may significantly enhance rates of atmospheric water vapor transport (chapter 28). Especially at times like the present, when the obliquity is at an intermediate value, variations in eccentricity and equinoctial precession may also be important in controlling ice stability, water-vapor transport, and the timing and severity of dust storms. The details of the interplay between the orbital and axial element fluctuations and the formation of polar layers are still far from understood, but a connection seems secure. It is noteworthy that the oscillations experienced by Mars are similar to, but substantially larger than, oscillations suggested to have driven the Earth's ice ages (Toon et al. 1980).

C. Origins of The Troughs in Polar Layered Deposits

The polar troughed topography (Figs. 5–9 and 12) is unlike any extant polar or glacial landscape of Earth, and suggests a unique Martian polar environment and history. Most recent models of layer development envision an interplay between ablational erosion of the equatorward-facing scarps and deposition on the inter-trough flats and poleward-facing trough walls. The models differ on the relative importance of depositional and ablational processes, and the amount of poleward migration of the troughs.

Study of high-resolution Viking Orbiter images of the north polar deposits suggests that erosion can occur on equatorward-facing slopes, while poleward facing slopes retain residual frost that may aid deposition of dust (Squyres 1979*b*; Howard et al. 1982*a*). These processes may result in poleward migration of the topographic troughs in the layered deposits (Squyres 1979*b*) as supported by stratigraphic relationships exposed in trough junctions (Howard et al. 1982*a*). Figure 13 shows schematic cross sections of inferred layering and trough development in different parts of the northern deposits. These interpretations of layer geometry require erosion of pole-facing slopes to create some of the unconformities, and the migration of troughs toward the pole.

The troughs, however, have complex patterns, and scenarios for their origins must allow for many troughs that trend essentially north-south (Fig. 1: 10 to 30°W, 82°N; Fig. 2: 60°W and 240°W), and for the crossing of many troughs at nearly right angles. Additionally, there are strong indications that structures in the deposits influence locations of the troughs (Thomas and Weitz 1989). Troughs in the north polar area often terminate along linear trends (Figs. 1 and 5a), and some southern troughs align with linear rows of pits in the pitted terrain (Fig. 5b) and even with ridges on surrounding plains. The role of faults, folds, or other structures in the deposits in controlling trough formation or migration is not known, however, and none are revealed in layer exposures. Wind flow is locally nearly perpendicular to trough walls and may influence the formation and evolution of the troughs (Howard 1980).

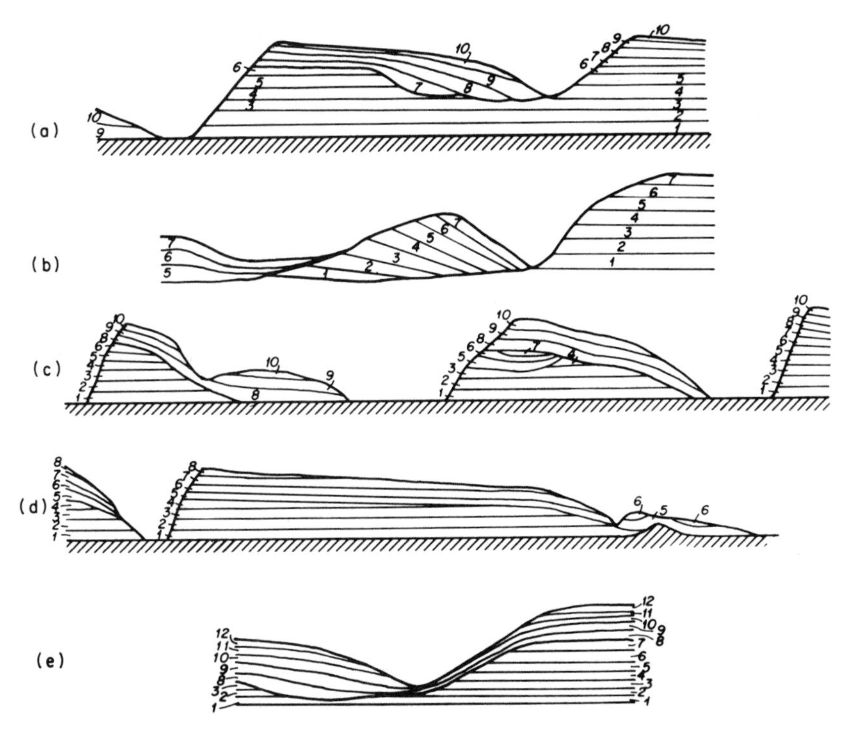

Fig. 13. Interpretations of sections across several kinds of troughs in the northern layered deposits. Horizontal scale is ~ 20 km for all sections; vertical dimensions are ~ 300 to 500 m. The pole is toward the right. The troughs are interpreted to be eroding on the right (equator-facing slopes) and receiving net deposition on the pole-facing slopes (modified from figures in Howard et al. 1982*a*).

Exposure of layers in troughs by ablation or erosion is not simple to model. Ice and mineral mixtures tend to make complex albedo patterns with considerable local feedback effects. Thick lag accumulations can be effective insulating blankets (see, e.g., Bozhinskiy et al. 1986; Drake 1981; Nakawo and Young 1981,1982; Warren 1984). Solid-state greenhouse effects may also be important on translucent frost/ice deposits (Matson and Brown 1989; Clow 1987). Ablational processes may alter the physical properties of layer sediments, including chemical weathering, grain cementation, and development of frothy residues (Saunders and Blewett 1987).

The origin of the broad reentrants, Chasmae Boreale and Australe (Fig. 1–4), found on each polar cap is uncertain. Howard (1980) favored a wind erosional origin due to concentration of katabatic winds, citing the prevalence of yardang-like forms, steep erosional scarps, and dunes along the axis of the north polar chasma. Similar yardangs also occur in the south polar chasma. Clifford (1987*b*) has suggested that basal melting of the polar caps may produce episodic flooding that scours the polar cap deposits.

Most studies have concluded that glacial flow has not played a role in creation of the polar topography. Clark and Mullen (1976) suggested CO_2 ice flow occurred in the polar layered deposits, but Viking observations ruled out CO_2 as a significant component of the layered deposits. More recently Budd et al. (1986) suggested that wave-like instabilities in flow of water ice has created the low-relief undulations of the north polar cap, and that selective ablational enlargement of some of these undulations has created the troughs. Howard et al. (1982*a*) noted that the scarp-paralleling ridges near parts of the northern layered deposits are morphologically similar to morainal deposits. However, no evidence of faults or filled crevasses is found in layered deposit exposures (Howard et al. 1982*a*). Also arguing against substantial flow is the near horizontality of exposed layers, nearly constant layer thickness with depth in observed exposures, and local exposure of basement rocks in troughs near the polar cap margin.

VII. AGES, TIME SCALES AND CYCLES OF THE POLAR DEPOSITS

A. Age and Formation Rates of the Polar Deposits

The polar deposits appear to be among the youngest surfaces on the planet. Cutts et al. (1976) found no craters on the north polar layered terrain; image resolution restricted the search for craters $>$ 300 m in diameter. Converting these data to limits on ages depends on the highly uncertain crater production models as well as on estimates of crater obliteration. Keeping in mind these uncertainties, the apparent lack of craters in the north suggests surface ages of order 10 Myr or less.

Plaut et al. (1988) derived a steady-state model of the crater population for the southern layered deposits that suggests upper limits of sedimentation of 8 km Gyr^{-1} (8×10^{-4} cm yr^{-1}). From the thickness of the deposits and the time since formation of the underlying pitted terrain, the average accumulation rate of the layered deposits is calculated at $\sim$ 1 km Gyr^{-1} (10^{-4} cm yr^{-1}). For the northern deposits a similar minimum rate can be estimated from the likely maximum age of deposits of $\sim$ 3 Gyr and thicknesses of order 3 km.

The effective accumulation rate of the south polar layered deposits cited by Plaut et al. (1988) is based upon their finding of 15 craters on these deposits. However, some of these craters occur within troughs exposing layered deposits, and might therefore be exhumed.

B. Deposition Rates and Time Scales of Periodic Layering

For comparison to these inferred long-term deposition rates, possible current deposition rates can be estimated from the summary of environmental factors given in Sec. V above. Sublimation rates can give estimates of the maximum water ice deposition possible. Jakosky and Haberle (chapter 28) estimate current sublimation of 0.03 g cm^{-2} yr^{-1} (Earth years) in the north polar region. This value may change by 2 orders of magnitude depending on the obliquity. This maximum current rate, $\sim 3 \times 10^{-2}$ cm yr^{-1} (assuming density is 1 g cm^{-3}) is very large compared to the "geological" rate of 10^{-4} cm yr^{-1} given above. The approach of Kieffer (1990) that uses the average atmospheric water loading gives a water-ice deposition rate of $\sim 2 \times 10^{-3}$ cm yr^{-1}. This also is not a net annual deposition rate, which must be lower.

The current north polar dust deposition rate estimated by Kieffer (1990) is $\sim 1 \times 10^{-3}$ g yr^{-1}; that from Pollack et al. (1979) is 20 times greater, 2×10^{-2} g cm^{-2} yr^{-1}. The huge difference comes from the assumption by Pollack et al. that half the global dust is deposited north of 60°N, an effective concentration of $\sim$ 10 times the average dust loading of 2×10^{-3} g cm^{-2} yr^{-1}, and from Kieffer's use of a lower average dust loading of 0.5×10^{-3} g cm^{-2} per Mars year in combination with a concentration factor of 4.5 (4.5 atmospheric columns to condense the observed CO_2 mass at the pole). Much effort can be expended on choosing between these models (basically, Kieffer's concentration factor and Pollack's observed loading would seem the most defensible approach and gives 9×10^{-3} g cm^{-2} yr^{-1} at the north pole), yet the large disagreement on the appropriate dust sedimentation rate is useful in reminding us that we know little about what actually is deposited seasonally within areas of the residual caps.

Thicknesses of band pairs in the north range from 14 to 46 m, with an average value of 30 m (Blasius et al. 1982). While there is a suspicion that thinner layers might be revealed at higher image resolutions, these values are taken as the "typical" ones to explain. Modeling of possible stratigraphic sequences by Cutts and Lewis (1982) showed that layer thicknesses might be expected to be quite variable; consequently, use of average layer thicknesses may be more misleading than just the measurement errors would imply.

The ice deposition rate obtained from current sublimation, 3×10^{-2} cm yr^{-1}, would form a 30-m band in only $\sim 10^5$ yr. The combined ice/dust deposition from Kieffer's estimates would be $\sim 3 \times 10^{-3}$ cm yr^{-1}, or 10^6 yr for a 30 m layer. Deposition rates from Pollack et al.'s parameters would give a 30-m band in under 10^5 yr. The water ice, and possibly the dust component, are subject to reduction by sublimation or aeolian transport, and thus the net rates for geologic calculations are likely to be substantially less than these (widely varying) estimates of maximum deposition. On the basis of inferred present ice/dust deposition, we remain unable to select a period within a factor of 10 corresponding to the 30-m layers.

The net deposition rate of 10^{-4} cm yr^{-1} inferred for the south polar layered deposits from crater densities by Plaut et al. (1988) suggests that layers have been formed on time scales of at least 10^6 yr. This time scale for layers would imply interaction of the obliquity and eccentricity cycles to produce the long period climate oscillations rather than simple obliquity or eccentricity forcing.

A complication of applying "present" rates to layer thicknesses derives from the nature of visible layering: something has changed over time to produce the layers. Unless one knows to which part of a layer sequence a rate applies, extrapolations with the rate are of limited significance. The present rate might be applicable to a subset of the visible layers, to thin layers not visible, or to unconformities in the deposits. Image resolution at present is inadequate to define what the physical changes shown by the layers are; such information might help in assigning present processes to specific parts of the deposits.

C. Unconformities and Layer Formation

The unconformities in the layered deposits suggest major interruptions in, or long-period modulation of, the basic cycle of layer formation. Howard et al. (1982*a*) concluded that episodes of widespread deposition of 5 to 10 major layers over the polar cap alternate with episodes of pronounced erosion of the troughs. Such major changes in the polar sedimentary regime, if cyclic, would have to occur with periods of many million years if the geologic deposition rates are assumed to apply, and on periods as low as 10^6 yr if the greatest estimated present deposition rate applies over long periods.

The unconformities need to be considered when asking if the residual ices are really distinguishable from the layered deposits. Unconformities appear to be an integral part of the history of the deposits. Present conditions, which include different residual ices at each pole, cannot be eliminated as being the conditions that create unconformities in the deposits (this could apply to one pole but not the other). Thus, even if the present residual ices are not representative of deposition of most of the layered materials, the conditions of their formation could still occur periodically as an integral part of overall deposit history. If this condition derived from periods of stable, intermediate obliquity (the recent situation), then one could speculate that the major unconformities form about every 3.9 Myr on the basis of the plots of obliquity history given by Ward (1979).

The overall erosional state of the deposits has been treated by Cutts et al. (1976), Thomas (1982), and Lancaster and Greeley (1990). The presence of significant outliers and large areas of striped terrain (Howard et al. 1982*a*) suggest that the present deposits are not at their maximum extent. A time scale for this erosion cannot be calculated reliably at present, but presumably would be one of the longest cycles or a secular trend.

VIII. SUMMARY

The polar deposits remain mysterious after many years of close scrutiny. Here we summarize knowledge of them by short lists of what is reasonably safe to conclude and what is definitely lacking in our present knowledge.

Fortunately many of the items in this second list will be directly addressed with Mars Observer data.

A. What is Known About the Polar Deposits

1. Thick sequences of layered deposits are present at each pole, capped by an incomplete cover of high-albedo residual frost. The residual frost is H_2O in the north, and is CO_2 in at least some summers in the south.
2. Polar layered deposits include exposed sequences of up to 20 layers, each a few tens of meters thick. Layers are mostly nearly horizontal and many extend for tens of kilometers.
3. There are some unconformities between sets of layers.
4. The layers probably include bright dust similar to that in Martian dust storms, and a darker component that is concentrated in dunes. Water ice is also inferred to be a component.
5. Both polar layered deposits appear to be fundamentally the same kind of sedimentary units, although they exhibit some morphologic differences.
6. Peripheral deposits are very different in the south from those in the north.

B. What is Not Known About Polar Deposits

1. The composition of the layered deposits is not known; the ratio of ice to dust or dust/sand is not known to several orders of magnitude.
2. The physical characteristics that cause the layers are not known (i.e., different dust/ice ratios, other compositional or textural variations).
3. It is not known to what extent the layered deposits are compositionally distinct from the residual frosts.
4. The vertical sequence of thicknesses of layers is not known.
5. The present water ice and dust budgets at either pole are not known in either sign or magnitude. Reasonable estimates of the magnitude of dust and ice annual net budgets vary by nearly 2 orders of magnitude.
6. Correlation of layer thicknesses with expected climate cycles cannot be made.

PART V
Current Atmospheres

24. COMPARATIVE ASPECTS OF THE CLIMATE OF MARS: AN INTRODUCTION TO THE CURRENT ATMOSPHERE

RICHARD W. ZUREK
Jet Propulsion Laboratory

Investigation of the atmosphere and climate of Mars continues to be framed largely by comparison with Earth. This is appropriate, given that there are fundamental similarities in rotation, seasonality and the thermal driving of the two atmosphere-surface systems, and yet these similarities sometimes obscure unique aspects of Martian phenomena. The historical perspective of the Martian climate is reviewed in this chapter, beginning with the early view of inexorable climate change on an older, but otherwise very Earth-like planet and continuing through the period when Earth-based spectroscopy and the Mariner 4, 6 and 7 flyby missions portrayed a Moon-like body, heavily cratered and almost airless by comparison with Earth. The arrival of Mariner 9 during a global dust storm and extensive observations by the Viking Orbiters, Viking Landers and Mars spacecraft have emphasized again the dynamic nature of the atmosphere and climate of Mars. This augmented database, together with application of sophisticated radiative, dynamical and photochemical models, has led to renewed comparison of the atmospheres of Earth and Mars. The most general features of these two atmospheres are discussed and compared in this chapter, with detailed discussion to be found in the chapters that follow on the atmosphere of Mars and its short-term interaction with the surface.

The study of the climate of Mars has long been framed by both emphasizing its fundamentally Earth-like nature and seeking to understand what its differences tell us about Earth. Lorenz (1978) once noted that while the study of Earth's climate and its physical processes would be a worthy pursuit even if there were no other planets in the solar system, the fact that there *are* other planets with their individual climates should accelerate the learning process.

Mars, with its accessible surface and climatic similarity to Earth, has already served in this role. Mars and Earth are sufficiently similar that much of the theory and methodology developed for the study of radiative, dynamic, photochemical and geologic processes on Earth should be applicable to Mars. And yet Mars is unique enough that such application can say something profound about the theories and their assumptions.

At the beginning of this century, there was hope that Mars could provide a glimpse into the future of Earth's climate; the hope today is that the current study of Mars can tell us something about present and past climates of both Earth and Mars. The historical perspective is reviewed here, followed by a brief introduction to the chapters on the atmosphere which follow in this book section. It is in these chapters that the current knowledge and understanding of the present atmosphere and its interactions with the surface are discussed in detail. The remainder of this chapter compares certain aspects of what we presently know about the current climates of Earth and Mars.

I. HISTORICAL PERSPECTIVE

Even the early telescopic observations of Mars were adequate to show that Mars had an atmosphere. The observations of yellowish clouds and veils (reported first by H. Flaugergues, early in the 19th century) and of the white clouds (since the observations of P. A. Secchi in 1858; see Maggini 1939) provided direct evidence of the presence of a significant atmosphere. Indirect evidence was provided by the observed seasonal changes of the planet's surface: the waxing and waning of the polar caps and the seasonal "wave of darkening." Mars had a surface area comparable to the land area of the Earth, it had seasonal changes and, as first noted by C. Huygens (entered in his diary, in 1659), it had a similar length of day. Thus, Mars was regarded as a very Earth-like planet (early telescopic observations are summarized in chapter 1; modern telescopic observations are treated in chapter 2).

A. Lowell's View: Mars as an Older Earth

At the beginning of this century, Mars and other planets were popularly regarded as being inhabited. Even some of those who doubted the existence of the famed Martian "canals" felt that, of all the planets in the solar system, Mars was the most likely planet, other than Earth of course, to have life. In this manner, Mars—or rather our view of Mars—had a profound effect early in this century on the debate about life, about the evolution and ultimately the demise of the planets themselves, and about what all this meant for our own planet Earth.

These early ideas were most eloquently expressed by P. Lowell. Not all the ideas originated with him, but he synthesized them into a unified perspective that even today encompasses our approach to the study of planets. And it was he who first coined the word "planetology," by which he meant the scientific study of planetary evolution, the "life-histories of planetary bodies" (Lowell 1908).

Lowell combined elements of the "nebular theory" of planet formation together with the Darwinian theory of evolution. Planets, he believed, condensed out of the nebular material raised from the close passage of another star with the Sun. Planetary bodies then passed through the following sequence, cooling with time and gradually losing their atmospheres to space (Lowell 1908): (1) Sun Stage (hot enough to emit light); (2) Molten Stage (hot, but lightless; e.g., Neptune, Uranus, Saturn, Jupiter); (3) Solidifying Stage (solid surface formed, great basins); (4) Terraqueous Stage (oceans formed; e.g., Earth); (5) Terrestrial Stage (oceans disappeared; e.g., Mars); (6) Dead Stage (air departed; e.g., Mercury).

Lowell argued that Mars, being a smaller world than Earth, would have cooled first and would thus be further along the evolutionary path of planets: Mars was well into the Terrestrial Stage and had already lost its oceans. Lowell insisted that Earth was losing its oceans, too, and he cited as evidence of this desertification the petrified forests of northern Arizona and north Africa and the apparent inundation of vast regions of North America and Europe in the geologic past. Thus, the canals that Lowell saw on Mars were not just evidence of life—he, and many others, took the existence of such life almost for granted—but of an intelligent race, struggling heroically against the inevitable desiccation of their planetary climate.

Even in his own time and apart from the controversy about the existence of Martian canals and intelligent life, Lowell was attacked for being altogether too selective in his citation of the geologic and climatological records (Hoyt 1976). However, many of the themes he brought so eloquently to the attention of the general public still frame much of the current scientific discussion about Mars, i.e., the following:

1. Theories of planetary evolution have to account for *all* of the planetary bodies, including the major satellites;
2. Planetary climates will change. In particular, water vapor escapes to space (or, Lowell argued, possibly into the cooling interior) and does so more rapidly on smaller bodies, with the implication that Mars had a wetter past;
3. Mars serves as a test of hypotheses about long-term climate change on Earth;
4. Mars was a planet on which life had evolved.

With regard to the third point above, Lowell (1908) noted that Croll (1875) had hypothesized that glacial epochs occurred when variations in the eccentricity of the Earth's orbit exceeded a critical value; the ice ages themselves would alternate from the northern to the southern hemisphere on the shorter precessional time scale, with the glaciated hemisphere being the one for which winter occurred near aphelion. If this theory were correct, argued Lowell, how does one explain the Mars situation, because Mars today has a greater orbital eccentricity than Earth ever achieved and yet the (southern) hemisphere with the short, hot summer and long, cold winter has the smallest permanent ice cap.

Lowell ignored Croll's conjecture that the changes in radiation due to orbital variations were not sufficient in themselves to produce an ice age, but that they probably served as a trigger for changes in oceanic heat transports (Imbrie and Imbrie 1986). However, Mars may still prove a definitive test of the Milankovitch theory of climate change (see chapter 33).

B. The Beginnings of Martian Dynamical Climatology

Dynamical climatology is that branch of meteorology which aims to understand quantitatively the overall thermal structure and circulation of the atmosphere and the means by which its main features (e.g., the global pattern of zonally averaged surface winds) are generated and maintained. Together, these features are known as the general circulation (Lorenz 1967). In practice, study of the dynamical climatology of Mars begins with that of Earth and the assertion that Mars is *dynamically* similar to Earth, as both are rapidly rotating planets with shallow, relatively clear atmospheres heated by seasonally varying insolation.

The first global map of the atmospheric "climatology" of northern winter and southern summer on Mars was drawn by Hess (1950), while at Lowell Observatory. Hess used the distributions of surface temperature derived by Coblentz and Lampland (1927), together with terrestrial experience and with limited wind data: some 18 wind directions obtained from tracking clouds on the terminator and limb of Mars during the oppositions of 1894, 1896 and 1924. (Hess' [1950] paper was the first dealing with a planet other than Earth to be published in the *Journal of Meteorology,* now the *Journal of the Atmospheric Sciences.*)

In a remarkably prescient appendix to a later book, Mintz (1961) used terrestrial meteorological theory to argue that the Mars atmospheric general circulation should have the following characteristics:

1. At solstice, the daily averaged temperature would decrease monotonically from the summer to the winter pole, as it does in the Earth's stratosphere, but not in Earth's troposphere with its greater opacity and powerful latent heat release from condensing atmospheric water vapor.
2. The winter hemisphere would have the Rossby regime, with wave-like disturbances imbedded in westerly (i.e., eastward) winds prevailing in mid-latitudes; the largest-amplitude waves would typically have three maxima and minima around a latitude circle, but this could change with season, as the atmospheric static stability changed.
3. As one consequence of the "reversed" temperature gradient, with the summer polar regions warmer than low latitudes, Mintz predicted that the summer hemisphere would be dominated by a Hadley circulation, with easterlies near the surface even in mid-latitudes.

4. The large diurnal variation of surface temperature could drive a significant global-scale daily oscillation of air temperature, wind and pressure; i.e., a thermal atmospheric tide.

(Mintz's [1961] paper is difficult to read for the nonspecialist, but Hess [1968] provides the context and a nice summary of the major points.)

C. The Martian Opposition of 1956

In 1956, during the closest approach of Mars to Earth in this century, a major yellow cloud, or dust storm, was observed. The size of this storm, which encircled the planet and spread into both hemispheres, was judged to be quite remarkable even by experienced observers (Slipher 1962; Kuiper 1957). Its full importance for both dynamical meteorology and for spacecraft operations at Mars was not noted at that time.

However, the obvious ability of winds to move dust on Mars and the lack of vivid color contrasts motivated Kuiper (1957) to argue that the dark areas on Mars might be lava fields and that the "seasonal wave of darkening" could be just the emplacement and removal by seasonally changing winds of a thin veneer of dust. This struck at the heart of the argument that only vegetation could explain the re-emergence of the dark areas after dust storms and their darkening each spring (e.g., see Rea 1964).

D. Mariners 4, 6 and 7: Mars the Moon-like Planet

The concept of Mars as an older, but very Earth-like planet, began to unravel as Earth-based spectroscopic measures of CO_2 in the Mars atmosphere suggested (Kaplan et al. 1964) that the CO_2 partial pressure and probably the total atmospheric surface pressure was considerably less than the 80 to 100 mbar value upon which previous techniques, based on the difficult photometry and polarimetry of the Mars atmosphere-surface system, had converged (de Vaucouleurs 1954). However, the same Earth-based spectra also confirmed that there were significant amounts of water vapor in the Martian atmosphere (Spinrad et al. 1963). Later spectra revealed that the water vapor amounts varied significantly with season and with place on the planet (Schorn et al. 1967).

The flyby of Mariner 4 past Mars in 1965 radically changed our perception of that planet. Its swath of photographs revealed a heavily cratered terrain. Analysis of the radio-occultation data (Kliore et al. 1965) indicated that the surface pressure of Mars at the two occultation points might be only 4 or 5 mbar. Earth-based spectroscopic data suggested that these were globally representative and that the air was probably all CO_2 (Young 1971; chapter 25). Such low pressures were just below the triple point of water, so that on Mars today, water is not stable as a liquid on its surface.

Cold temperatures and a thin CO_2 atmosphere would make Mars different from Earth in another way, too. Using a one-dimensional surface energy balance equation, Leighton and Murray (1966) computed the surface temperatures for Mars as a function of latitude and season. They found that temperatures in the winter polar regions should fall far below the 145 K value at which the partial pressure of CO_2 gas above CO_2 ice on the surface would equal 4 mbar. They thus concluded that the seasonal caps of Mars, and quite likely the permanent polar caps as well, were composed mainly of CO_2, and not water ice; furthermore, the surface pressure of the mainly CO_2 atmosphere would vary significantly over the Martian year and would have two maxima and two minima each year as CO_2 was exchanged between the poles (see chapters 26 and 27).

Early in this century, Wallace (1907) had performed a similar calculation, showing that Mars should be cold and raising the possibility of CO_2 ice; his primary aim was to refute Lowell's claims of an inhabited and hospitable Mars. It was Leighton and Murray's calculation, however, that showed just how substantial the surface pressure variation could be for a CO_2 Martian atmosphere. Furthermore, major variations could occur, not only on the seasonal time scale (chapter 27), but also on the much longer time scales in which the orbital variations of Mars altered the amount and latitudinal distribution of solar radiation reaching its surface (chapter 33).

The flybys of Mariners 6 and 7 past Mars in 1969 extended visual coverage of the planet from Mariner 4's 1% to more than 10% of the planet, at spatial resolutions $\leq$3 km. The near-encounter observations cut across the equatorial regions into the same southern hemisphere observed by Mariner 4. As before, the terrain was heavily cratered, except for some intriguing chaotic terrain observed near the equator and for the apparently featureless floor of the giant Hellas basin.

The general impression left by these early Mariner flybys of Mars was of a very old, almost dead planet. Not only were there no traces of the canals, there were no rivers, few clouds, perhaps no water (even as ice) on the surface at all. The atmosphere was thin and cold. Modeling studies (see, e.g., Gierasch and Goody 1968) showed just how strongly radiation prescribed the temperature fields (radiative time constants of only 2 to 3 days) and how difficult it was for atmospheric heat transport to affect the high latitudes (see, e.g., Leovy and Mintz 1969). Studies also indicated that to raise dust from the surface in such a thin atmosphere would require strong winds (Sagan and Pollack 1969) or intense whirlwinds (Ryan 1964).

E. Mariner 9, Mars and Viking Spacecraft: A Dynamic Planet Again

Mariner 9 transformed the image of Mars once again. Having arrived in the midst of the most global dust storm yet seen on Mars, the spacecraft provided dramatic images of how extensive, long-lived and opaque dust storms on Mars could be. Observations of infrared radiances showed that the atmospheric thermal structure of Mars, and presumably its circulation, was altered by solar heating of the airborne dust to a degree that is quite remarkable when compared to Earth (Hanel et al. 1972*b*). Once the airborne dust had cleared sufficiently, the Mariner 9 and later the Mars 5 imaging systems revealed great volcanic mountains, enormous valleys and basins and mysterious layered terrains, particularly in the polar regions. Perhaps most intriguing were the sinuous channels, valley networks and streamlined landforms, all pointing to a past when water ran on the surface of the planet: Schiaparelli's *canali* reborn (Baker 1982; chapter 15).

The topographic variations on Mars, as revealed first by Earth-based radar (Pettengill et al. 1969) and then globally by the Mars, Mariner 9 and Viking spacecraft, are remarkable in many ways. The elevation differences that influence the atmosphere on Mars at planetary scale are larger than on Earth, particularly since the oceans on Earth fill approximately half the elevation difference of the solid surface. The immense volcanic shields of the Tharsis plateau clearly influence the local meteorology; e.g., orographic heating during the day draws air up the slopes until the air reaches its saturation level and clouds form. Such influences explain the historically observed tendencies for cloud formation in particular regions and seasons. Such orographic control of cloud patterns and of the winds responsible for at least some surface albedo changes had been anticipated (see, e.g., Sagan et al. 1971*a*).

F. New Insight into Old Mysteries

Observations by spacecraft, in orbit around Mars and on its surface, have provided new insight into the classical questions about Mars. No evidence of Martian life was found; Lowell's canals were nowhere to be seen. Atmospheric water vapor (chapter 28) and carbon dioxide (chapter 27) do vary seasonally. The Viking Orbiters observed the global variation of water vapor for all of one and parts of the next annual cycle. Surface pressure was measured at Viking Lander 1 for more than 3 Mars years. The permanent north polar cap has been identified as water ice, the south permanent cap more tentatively as CO_2. As judged from the Viking data, there is remarkably little variation in the water and CO_2 cycles from year to year, but some Earth-based data indicate more substantial interannual differences, including the possible disappearance of the south polar cap (chapters 27 and 28) and an enhanced release of water vapor in the southern hemisphere (chapter 28).

Airborne dust appeared to be always present above the two Viking Lander sites, with vertical column visual dust opacities always exceeding a few tenths of an optical depth. The presence of such airborne dust would help to explain why surface pressures were over-estimated by Earth-based observers using photometric techniques and assuming scattering only by atmospheric gases. The effect of this background dust on the differential spectral absorption of solar radiation reflected from the planet/atmosphere system could also account for the strong diurnal variation of water vapor reported by Earth-based observers (Davies 1979*c*). The "blue clearings" and "opposition effects" (chapter 2) appear to be due to the phase angle effect of scattering by airborne dust (see, e.g., Thorpe 1978).

Finally, the seasonal "wave of darkening," which seemed to point so surely to the presence of vegetation, is better explained by the seasonal movement of a fine veneer of relatively bright dust (chapters 22 and 29). Lowell (1908) and others attributed this wave to the release of water in the polar regions and its equatorward transport; the response of vegetation to this water supply then accounted for the time lag between polar cap melting and the observed darkening. Ironically, there is a seasonal wave of water vapor, in that the maximum amount of water outside the tropics occurs at progressively lower latitudes during (northern) summer. Whether or not this is due to atmospheric transport of the water vapor is still debated (chapter 28).

Unlike the seasonal cycles of water and CO_2 observed during the Viking mission, the amount and distribution of airborne dust varied tremendously. Two planet-encircling dust storms occurred during the first (Mars) year and another during the fourth year of Viking observations; the dust loading at other times and places was also quite variable (chapter 29). Interestingly enough, the historical data suggest—despite their inherent bias due to the changing distance between Earth and Mars—that the long-lived, planet-encircling dust storms have occurred mostly during the period of spacecraft missions to Mars, rather than earlier in this century (Zurek and Martin 1992).

Both visual and infrared observations from spacecraft have revealed details of dust storms, those classical (chapter 2) yellow clouds of Mars (chapter 29). The various apparitions of the famous "W" cloud and more general blue-white clouds in the vicinity of Olympus Mons and Tharsis have been identified spectrally as water-ice clouds, as have the cap-edge clouds that follow the northern seasonal polar cap retreat. The relative contributions of water ice and of CO_2 ice to the polar hoods is still debated; models indicate that CO_2 ice clouds should form in the polar night, particularly if dust is present (chapters 26 and 27).

With regard to atmospheric dynamics, many of Mintz's (1961) theoretical predictions have been borne out, first by numerical experiments beginning in the 1960s when the state-of-the-art UCLA terrestrial general-circulation model was first adapted to Mars (Leovy and Mintz 1969) and then by actual observation of the atmospheric temperature field by orbiting spacecraft and the acquisition of pressure, temperature, wind and opacity data by the two Viking Landers (chapter 26). The meteorological time series acquired by the Viking Landers on the surface of Mars is unprecedented for planets other than Earth in terms of its longevity and its temporal resolution. As seen in Fig. 1, the surface pressure measurements spanned more than three Mars years (at Viking Lander 1) and captured variations on time scales ranging from less than a Martian day (or sol) to interannual.

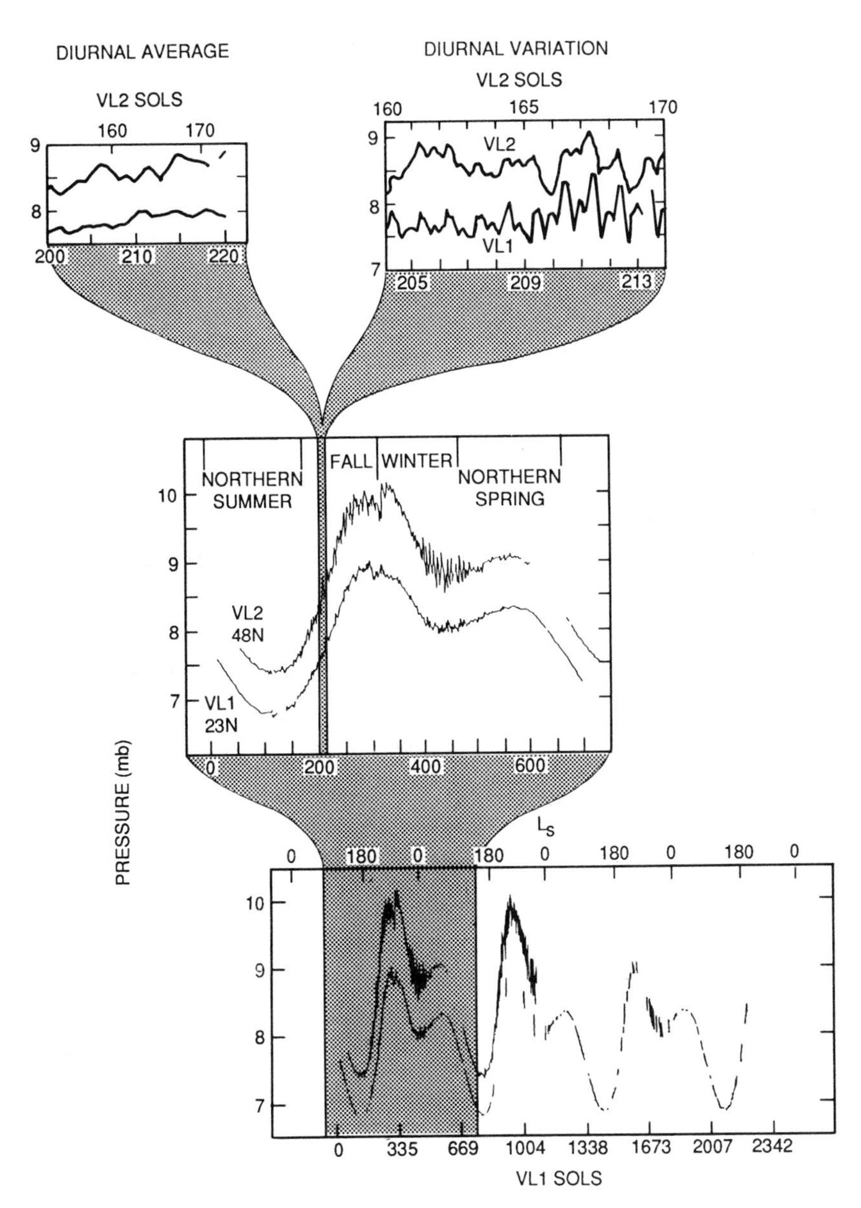

Fig. 1. The different time scales of variation of the surface pressure field measured at the two Viking Lander (VL) sites and for parts of four Mars years (1976–1982). (Bottom) The full record of daily averaged surface pressures, showing interannual and seasonal variations. The offset between curves is due to the elevation difference (~ 1.1 km) between the two sites. The time scales are in sols (Mars solar days) after arrival of VL 1 or VL 2 (which arrived 45 sols after VL 1) and in areocentric longitude L_s, an angular measure of the Mars year with 0°, 90°, 180° and 270° marking the beginning of northern spring, summer, fall and winter. (Middle) An expanded view of the first annual curve. (Top, left) Day-to-day variations. The larger variations at VL 2 represent storm systems passing by the more northern site. (Top, right) Hourly variations showing the enhanced diurnal and semidiurnal (once and twice per sol) components of the surface pressures measured following the arrival of a planet-encircling dust storm (figures adapted from Tillman 1988). See Color Plate 16.

II. THE PRESENT STATE OF KNOWLEDGE

The rest of this book section on the atmosphere of Mars deals with our present understanding of that atmosphere, of its structure and processes, and of its interactions with the surface. Chapter 25 describes the composition of the Mars atmosphere and how that composition imposes constraints on theories of its evolution. Chapter 26 deals with the dynamical meteorology of Mars, including the nature and theory of the large-scale circulation and temperature fields, of boundary layer processes, and of the thermodynamic impact of airborne dust. The next three chapters deal with the seasonal cycles of carbon dioxide (chapter 27), of water vapor and ice (chapter 28) and of airborne dust (chapter 29). In each case the contributions of surface sources and sinks and of atmospheric transport to the annual variations of these fields are discussed, together with implications for their long-term variations. Chapter 30 then describes the structure, circulation and photochemistry of the high-altitude atmosphere.

III. COMPARING THE ATMOSPHERES OF MARS AND EARTH

The final segment of this introduction addresses briefly some of the comparative aspects of the atmospheres of Mars and Earth.

A. The Standard Atmospheres

Figure 2 shows a "standard" atmosphere for Earth and for Mars, which is meant to be representative of globally averaged conditions. For Earth temperature decreases with height near the surface at a rate determined by a combination of dry and wet adiabatic processes. That is, on Earth the vertical temperature gradient that would be produced by dry convection from a strongly heated surface is modified by the latent heat release of saturated air parcels. The region of decreasing temperature nearest the surface on Earth is called the troposphere and its top, which varies from 18 to 8 km with latitude, is called the tropopause. Above the tropopause on Earth, temperatures cease to decrease and then begin to increase with height because of solar heating due to the ozone layer. This region of isothermal or increasing temperature is the stratosphere. Such a region has a high static stability, in that air parcels displaced upward (downward) from their initial heights will cool (warm) due to adiabatic expansion (compression) at a rate that makes them heavy (light) compared to their surroundings and so they will sink (rise) back to their original levels.

On Mars, the thin, cold atmosphere can hold little water vapor even when saturated. The total amount of atmospheric water on Mars is equivalent to 1 to 2 km^3 of ice (chapter 28). The terrestrial atmosphere holds the equivalent of 13,000 km^3 of ice, but its stratosphere holds an amount comparable to Mars ($\approx$ 1 km^3 of ice). However, the relative humidity of the stratosphere

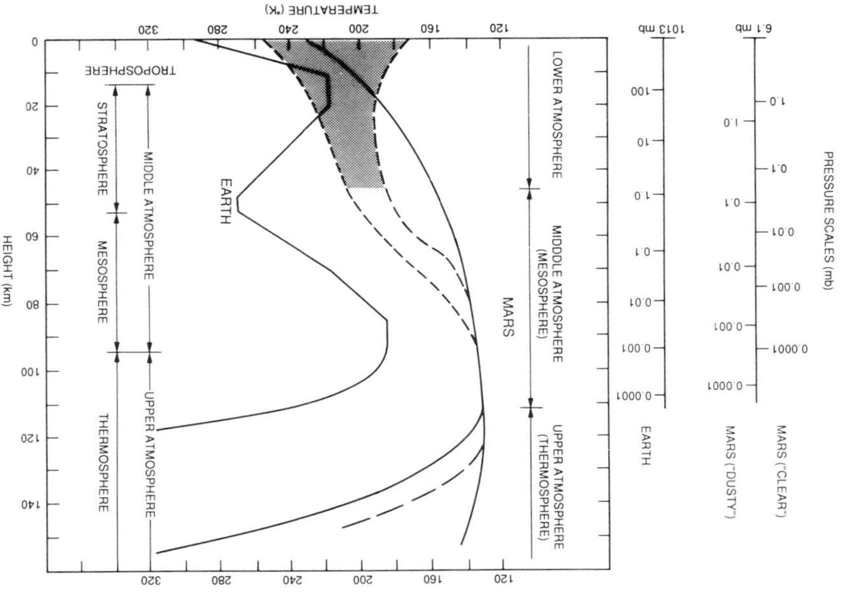

Fig. 2. The U.S. standard atmosphere (1976) for Earth and representative temperature profiles for Mars during relatively clear conditions (solid line, based on the Viking entry profiles [Seiff and Kirk 1977]) and during dusty conditions (range of values indicated by dashed lines; based on Mariner 9 IRIS and Viking radio-occultation data covering the shaded region [Hanel et al. 1972*b*; Lindal et al. 1979]). The high temperature branch for the Mars upper atmosphere is based on Mariner 6 and 7 UVS data and reflects effects of the 11-yr solar cycle; the upper-level dashed line indicates the possible displacement of pressure surfaces due to dust heating in the lower atmosphere (see Stewart and Hanson 1982). The height scales are essentially geometric height referenced to sea level for Earth and to the 6.1 mbar surface for Mars. Representative variations of pressure for the various temperature profiles are indicated at the left. The nomenclature is discussed in the text.

is much lower than that of the less massive (by a factor of ~ 5) Martian atmosphere. Thus, the terrestrial stratosphere is dry, while the thin, cold Martian atmosphere is close to saturation, despite the paucity of water vapor.

With so little water on Mars, the condensation of *all* the water vapor in an air column results in a latent heating that is negligible when compared to radiative effects. Thus, one would expect the temperature variation in the lower Mars atmosphere to be close to the dry adiabatic value of about -4.5 K km^{-1}. However, like ozone in the terrestrial atmosphere, airborne dust in the Martian atmosphere absorbs incoming solar radiation and thermal emission from the surface. This can reduce or even reverse the decrease with height of temperature in the lower Martian atmosphere, producing a region of high static stability.

Temperatures and pressures in the terrestrial stratosphere are comparable to those in the Martian lower atmosphere. Thermodynamically, the terrestrial stratosphere and the dusty lower Martian atmosphere are similar in that the local diabatic heating is dominated by a radiatively active atmospheric trace constituent. However, the total amount and spatial distribution of ozone on Earth is far less variable than is that of dust on Mars, so that the region of greater static stability is permanent on Earth, but ephemeral on Mars.

Ozone (O_3) is photochemically produced on Mars by the photolysis of CO_2 and, as on Earth, reaction of atomic oxygen O with O_2 (chapter 30). The destruction of O_3, and therefore its equilibrium value, is more difficult to determine. There is less O_3, more than 300 times less when integrated globally, in the atmosphere of Mars than of Earth, because there is less molecular oxygen on Mars; this reflects the terrestrial biological sources of O_2. Furthermore, the Martian lower atmosphere is wet by comparison to Earth's stratosphere. In fact, the catalytic destruction of ozone on Mars by reactive hydrogen-containing molecules produced by the photolysis of water vapor limits ozone primarily to the winter high latitudes where cold temperatures limit atmospheric water vapor abundances. In spring, the high latitudes warm and water vapor can return to the atmosphere, catalytically destroying the "mountain" of polar ozone (chapter 30).

On Earth, the springtime emergence of the polar regions into light can also lead to the catalytic destruction of ozone. In this case, the catalytic agent is believed to be reactive chlorine-containing molecules, produced by photolysis in the stratosphere of man-made compounds and sequestered by the special conditions of the cold polar regions from other trace gases that would convert the chlorine to a less reactive form. In the absence of strong mixing of polar air and air from lower latitudes, the catalytic destruction is sufficiently rapid to produce the large springtime deficit of ozone known as the "ozone hole."

It has become conventional to refer to the stratosphere and mesosphere of the Earth together as the middle atmosphere. In the Earth's mesosphere (the region above the stratopause), globally averaged temperatures are controlled largely by the radiative emission and absorption of CO_2. The rate of temperature decrease with height in the terrestrial mesosphere is affected in part by non local thermodynamic equilibrium (non-LTE) of the CO_2 vibrational states; that is, at low enough pressures collisions between molecules are too infrequent to thermalize absorbed photons before they are re-radiated, decreasing local heating. A similar situation should occur in the atmosphere of Mars.

A major remaining puzzle is why the atmosphere of Mars has so little CO and O. These gases are produced by photolysis of CO_2 and can be recombined only in the lower atmosphere, below 30 km. The vertical mixing of the global atmosphere required to preserve the observed equilibrium of gases, given present photochemical theory, appears too large even when the effects of wave breaking are taken into account (chapters 26 and 30).

On Earth, the mesopause separates the mesosphere from the thermosphere where temperature increases with height, as heat is carried by molecular conduction down from the region where the solar extreme ultraviolet (EUV) radiation is absorbed. The mesopause is also near the homopause, defined as that height ($\sim$ 105 km on Earth; $\sim$ 125 km on Mars [Nier and McElroy 1977]) above which the atmospheric gases begin to separate diffusively, rather than to form an homogeneous mixture. Diffusive separation occurs because molecular kinematic viscosity, as well as thermal diffusivity, is inversely proportional to density and thus increases exponentially with height. Temperatures in the Mars thermosphere appear remarkably variable (Nier and McElroy 1977); sometimes, as during the Viking Lander 2 entry, the Mars thermosphere is very cold, with little or no increase of temperature (chapter 30). Above the thermospheres are the exospheres, where molecules or atoms may escape to space.

On Earth, then, the atmospheric regions are distinguished by the vertical variation of temperature due largely to the presence of ozone. On Mars, the comparable absorber is airborne dust, but its distribution and column amounts can change enormously and unpredictably. The height of any dust haze on Mars will vary seasonally, regionally and episodically. Thus, the terrestrial nomenclature fails, in that the thermal structure of a relatively dust-free lower Martian atmosphere resembles the terrestrial troposphere, while the dusty Martian atmosphere resembles the terrestrial stratosphere (but bounded by a solid surface) with its more nearly isothermal temperature profiles.

Given this tremendous variability, the region below the Martian thermosphere in Fig. 2 has been divided arbitrarily into two layers: (1) *the lower atmosphere* defined as the region from the ground to $\sim$ 45 km above the reference geoid. This region is influenced strongly in its lower reaches by radiative or sensible heat exchange with the ground; it is also the region most strongly heated by airborne dust. Most of the currently available temperature observations are of this region, sometimes called the Martian troposphere; (2) *the middle atmosphere* defined as the region extending from the lower atmosphere to the base of the thermosphere (~ 45 to 110 km). This region, sometimes called the Martian mesosphere, is most likely to be affected by the "breaking" of vertically propagating waves (chapter 26).

B. Atmospheric Circulation

The mechanisms by which momentum, heat, trace gases and aerosols are transported about the planet are still not fully characterized for Mars. Zonally symmetric circulations appear to be more important components of the Martian circulation than is the case on Earth. In some sense, the Hadley circulation on Earth, a zonal-mean overturning with rising air at low latitudes and sinking air in the subtropics, exists only in the climatological or averaged sense. Dynamical theory and the observation of atmospheric temperatures and of surface wind streaks suggest that on Mars there is a more physically coherent, zonally symmetric Hadley-type overturning of air (chapter 26). This reflects both the strong radiative heating in the often dusty Martian atmosphere and, unlike Earth, the seasonal condensation and sublimation on Mars of a significant fraction of the total atmospheric mass. On Mars, atmospheric pressure gradients quickly arise in response to this mass loss or gain to drive motions (the "condensation/sublimation wind") that transport CO_2 as needed between the two polar regions.

On Earth, the pattern of surface winds at solstice has easterlies at low latitudes and westerlies in the mid-latitudes of both hemispheres. A Martian regime at solstice of summer easterlies and winter westerly winds resembles the terrestrial stratosphere. Hess (1968) noted that the wind patterns on Earth produce a nearly negligible net frictional torque on the solid planet, but that the torque on Mars should be greater if, as expected, the surface easterlies are more extensive. Momentum of the solid planet will also be changed by the seasonal redistribution of mass. All this leads to quite small, but measurable, variations of the length of the Martian day (chapter 26). On Earth, the location of certain climate regimes, like the subtropical high-pressure zones, are remarkably unchanged from year to year. On Mars, they are much more variable; for example, the high-pressure zones associated with Hadley-type circulations on Mars can be moved from subtropical to high mid-latitudes during a planet-encircling dust storm.

Even though Mars does not have the land-sea differences that drive planetary-scale components of Earth's atmospheric circulation, there are continental-scale differences in surface thermal-heat capacity and albedo that can drive regional, or longitudinal (Walker-type), circulations on Mars. In effect, Mars has "thermal" continents: vast regions of high surface albedo and low thermal inertia, which are most strikingly apparent as the coldest regions in global pre-dawn surface temperatures (see Color Plate 18). At low latitudes, these thermal continents encompass the highest terrains and may generate both stationary and traveling waves through the thermal and mechanical effects of the orography (chapter 26).

Traveling waves can also be generated through atmospheric instabilities in the presence of latitudinal variations in the zonal flow and in zonal-mean temperatures. The barotropic-instability mechanism involves transfers of kinetic energy from the zonal-mean flow to the wave; baroclinic instabilities convert the potential (gravitational) energy of the zonal-mean state to wave potential, and then wave kinetic, energy. One class of baroclinic waves, namely the mid-latitude winter storms, come in a remarkably regular procession on Mars; the terrestrial storm systems are more irregular.

The middle atmosphere on Mars remains virtually unexplored (chapters 26 and 30). As the breaking at high altitude of vertically propagating gravity waves and thermal tides are known to alter the circulation of the Earth's mesosphere, it is likely that they will also modify the middle atmosphere on Mars. Martian orography should generate a multitude of gravity waves, and there may be interesting consequences of the dichotomy of the Martian topography, with the higher and more cratered terrain in low latitudes and in the southern hemisphere. On Earth, of course, the northern hemisphere, with its greater land extent, has the more extensive spatial variations in terrain.

C. Seasonal Cycles of Carbon Dioxide, Water and Dust

The major feature of the observed CO_2 cycle on Mars is the seasonal condensation and sublimation in the polar regions of as much as 25% of the total atmospheric mass (chapter 27). As noted earlier, this cycle is driven almost entirely by the radiative budget at high latitudes. This enormous seasonal change in atmospheric mass on Mars has no counterpart on Earth, and it may produce in the Mars atmosphere significant meridional transports of heat, of momentum and of atmospheric trace constituents like water (see chapters 26–28).

Figures 3 and 4 contrast our knowledge of the seasonal cycles of water on Earth and on Mars. On Earth, only the subtropical regions provide, on an annual basis, sources for atmospheric water; elsewhere precipitation exceeds evaporation (Fig. 3). The implied atmospheric transport needed to move water vapor from the source latitudes to those where it is lost is consistent with the computed transport, although the observational uncertainties for both are quite large.

The mechanisms for carrying out this transport are the Hadley circulation and wave transport. The Hadley circulation moves water vapor into the equatorial regions in its lower branch. Adiabatic cooling of the moist air in the rising branch removes vapor through precipitation, while the release of latent heat helps drive the upward motion. In the subtropics the sinking air is warmed by adiabatic compression, reducing the relative, but not the specific, humidity. This produces a cloud-free region in which relatively more sunlight reaches the sea surface, producing more evaporation. The transport to higher latitudes is accomplished by the correlation of longitudinal variations in the poleward (equatorward) winds with regions of greater (lesser) water vapor;

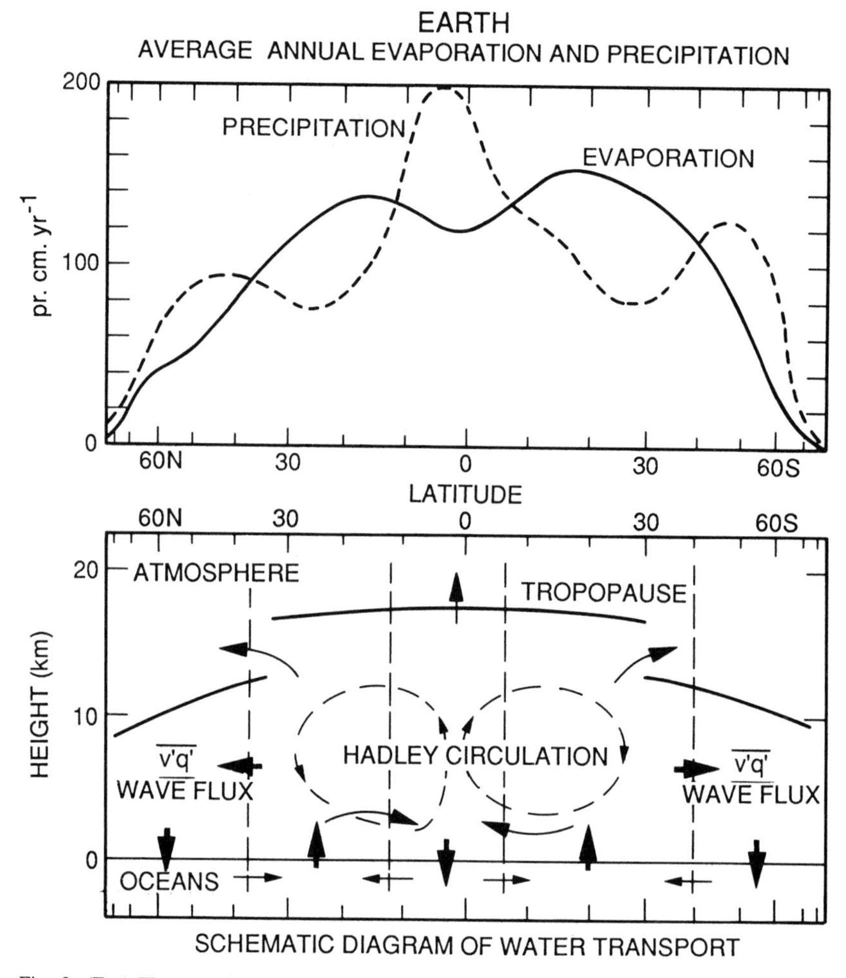

Fig. 3. (Top) The annual average evaporation and precipitation per unit area, as a function of latitude for Earth (as shown by Lorenz 1967). The excess of evaporation over precipitation in the subtropics means that water vapor is being exported to other regions. (Bottom) A schematic diagram of water transport on Earth; note that the surface sources are replenished by the flow of oceanic water and ice from the regions of surface accumulation.

this is the so-called "eddy flux," as it depends upon the character of eddies (i.e., longitudinal variations) in the meteorological fields. Note that on Earth the surface sources are automatically replenished on relatively short time scales by the flow of water (and ice) from continents to oceans and within oceans to the subtropical regions.

Now, what of the water cycle on Mars (Fig. 4)? The residual north polar water-ice cap is obviously a source during late northern spring and early summer when it is illuminated and free of its seasonal CO_2 frost cover. How-

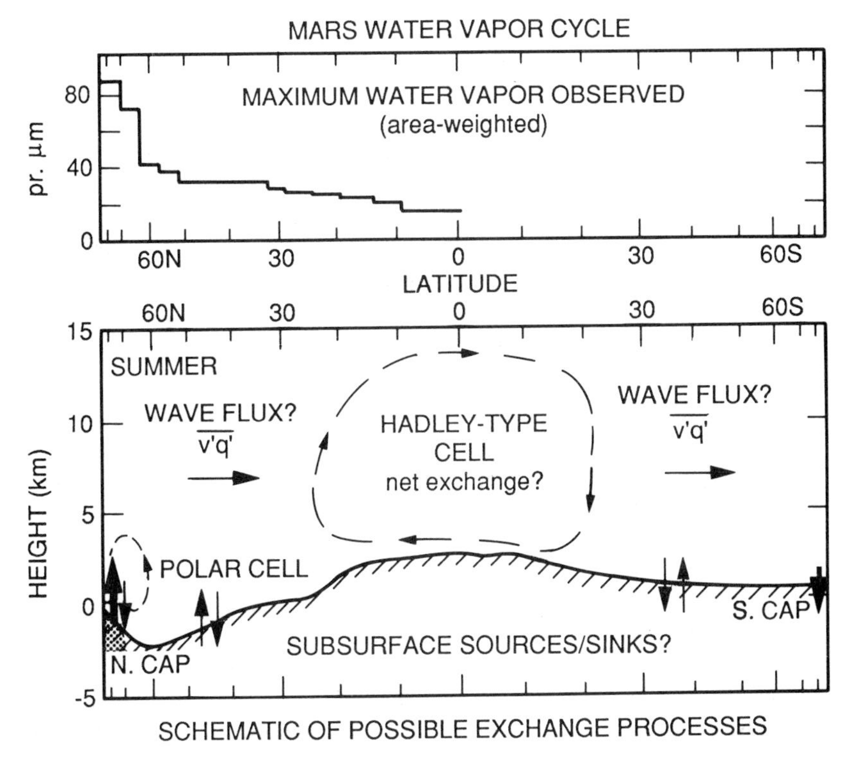

Fig. 4. (Top) Maximum observed column amounts of water vapor, as derived from Viking MAWD radiances averaged over equal-area latitude zones. (Observed maxima for the southern hemisphere are not shown due to observational effects of the two planet-encircling dust storms in 1977.) Note that no graph comparable to the upper panel in Fig 3 exists for Mars. (Bottom) Schematic diagram of possible exchange mechanisms for water on Mars.

ever, the local atmospheric transport is not known, so the issue of net transfer between the permanent ice caps and the atmosphere on time scales of a year or more is unresolved (chapter 28). The one known annual sink is the permanent CO_2-ice south polar cap, which acts as a cold trap for water vapor. However, the area of this cap is small and water vapor or ice must be transported to it before the water can be incorporated into the residual cap. Furthermore, the CO_2 cap itself may disappear completely in some years (chapters 2 and 28), so that it may not be a permanent sink of water vapor.

The unfrosted ground (i.e., the regolith) can serve as both a source and as a sink for water vapor, as water vapor is desorbed when the ground warms seasonally during spring and is adsorbed as ground temperatures cool in late summer and in fall. The efficiency of sorption of water vapor depends on the surface concentration of water vapor (which is extremely difficult to measure remotely) and its general variation in the lower atmosphere. Sublimation of water ice depends upon atmospheric mixing and winds near the surface. Condensation depends on relative humidity and thus on atmospheric temperatures. Atmospheric transport depends upon the horizontal and vertical distributions of water vapor (chapter 28).

Thus, the water cycle on Mars is closed locally only in the absence of net atmospheric transport. In the presence of transport, parts of the surface will serve as sinks and other parts as sources. Unlike Earth with its surface flow of water (and ice), Mars has no mechanism to replenish surface sources on time scales of a few to a few hundred years and so to come rapidly to a steady state. Furthermore, water vapor has no means, other than the formation of clouds and ice fields, to alter the circulation that is transporting it.

For Mars, it is the dust cycle which involves a radiatively active trace constituent (chapters 26 and 29). To the extent that the dust cycle is dominated by the planetary-scale dust storms, observations suggest that the southern subtropics are probably losing dust to the rest of the planet. However, the role of less intense regional transports during the long periods without great dust storms is unknown (chapter 29).

D. Coupling between Cycles

The cycles are intertwined in several respects. Mixing dust into the seasonal polar caps or the presence of ice clouds at critical times can alter the radiation balance in the polar regions, affecting the CO_2 flux and potentially the water vapor flux. By acting as condensation nuclei, airborne dust can aid the atmospheric condensation of water and CO_2; the growth and fall of such condensates provide an efficient scavenging of the dust. And, CO_2 surface ice acts as a cold trap for atmospheric water (chapters 26–28).

The occurrence of regional and planet-encircling dust storms may depend upon the "condensation wind" or upon a precursor level of dustiness produced by local dust storms near the polar-cap edge (chapter 26). Because the ability to lift dust from the surface depends in part on the total surface pressure (chapter 22), the ability of dust storms to occur at all depends on the long-term variation of CO_2. Kahn (1985) has argued that it is not a coincidence that the present surface pressure should be close to the triple-point pressure of water; this is the expected limiting state because the depletion of atmospheric CO_2 by carbonate formation should continue until liquid water can no longer form on the surface, even in disequilibrium (chapter 32).

The transport of water, either as vapor or condensates, depends upon the "condensation winds" and other components of the atmospheric circulation, including those that will respond to variations in the distribution and amount of airborne dust. Water vapor as well as dust may be exchanged across the equator by Hadley-type circulations, depending upon the vertical mixing of the dust and vapor. Following planet-encircling dust storms, Walker-type circulations may redistribute dust longitudinally to produce suspected long-term dust deposits (Christensen 1986*a*). Given the chaotic nature of the great or major dust storms (meaning those that are regional or planet encircling in extent; see chapter 29), the coupling between the seasonal cycles of volatiles and dust means that it will be difficult to extrapolate the net effects of the presently observed seasonal cycles to much longer time scales (chapter 29).

IV. WHAT NEXT?

Throughout the period of spacecraft observation of Mars, the scale and detail of atmospheric modeling has closely followed that of terrestrial meteorology and climate studies. Driven first by Earth-based observations and then augmented by the Mariner and Viking spacecraft data, such modeling is now again outdistancing constraints that the existing data can provide and so one looks forward to a new era of comprehensive, systematic observation of the Earth's dynamical neighbor. Such an era can commence, but hopefully will not end, with the Mars Observer and Mars 94 spacecraft observing Mars in the mid-1990s. The study of the atmosphere and climate of Mars can help us solve some of the riddles of climate change on both Earth and Mars, but new data are required if Mars is to fulfill the ancient hope of showing us a clear view into its past and into our future.

Acknowledgment. I wish to thank R. Haberle, B. Jakosky and R. Kahn for their comments and encouragement. This work was supported by NASA's Planetary Atmospheres Program and was performed at the Jet Propulsion Laboratory, California Institute of Technology, under contract with the National Aeronautics and Space Administration.

25. THE COMPOSITION AND EARLY HISTORY OF THE ATMOSPHERE OF MARS

TOBIAS OWEN
University of Hawaii

Carbon dioxide was the only known constituent of the Martian atmosphere from its discovery in 1947 until 1963, when water vapor was identified in the planet's spectrum. High-resolution groundbased spectroscopy and spacecraft observations in the next decade added CO, O_2, O_3, and showed that the atmospheric surface pressure on Mars is more than 2 orders of magnitude lower than it is on Earth. The Viking investigations in 1976 included the detection of N_2, NO, Ne, Ar, Kr and Xe and measurements of isotopic ratios in all species except hydrogen. Omitting uncertainties, $^{15}N/^{14}N$ is 1.6 times, $^{40}Ar/^{36}Ar$ 10 times, $^{129}Xe/^{132}Xe$ 2.5 times terrestrial, and the other ratios are indistinguishable from terrestrial values within their uncertainties. Subsequent Earth-based studies led to the discovery of HDO, a determination that D/H is 5 times terrestrial and provided new measurements of oxygen isotope ratios that suggest lighter oxygen on Mars than on Earth. Studies of SNC meteorites offer another possible source of information on Martian volatiles, but the history of these rocks and their relationship to the planet are still not completely clear. There are several independent arguments based on the atmospheric data suggesting that Mars once had a much thicker atmosphere than it does today. However, these arguments require additional support in the form of new observations, including direct investigations of carefully selected Martian surface materials, before they can be considered conclusive.

I. EARLY OBSERVATIONS AND INTERPRETATIONS

The presence of an atmosphere on Mars was noted by early telescopic observers who could watch the seasonal changes of the polar caps, follow the onset and dissipation of global dust storms, and discern both hazes at the limb and white clouds over various regions on the disk (see de Vaucouleurs [1954] for a review). Repeated attempts to determine the composition of that atmosphere by means of groundbased spectroscopy met with failure, however. For example, no evidence of water vapor, oxygen or carbon dioxide was found in high-resolution photographic spectra recorded at Mt. Wilson during the 1930s, although such spectra led to the discovery of CO_2 in the atmosphere of Venus (Dunham 1952).

The first detection of an identifiable constituent in the Martian atmosphere was the discovery of carbon dioxide by Kuiper in 1947 (Kuiper 1952). He observed that the CO_2 bands near 1.6 μm and 2 μm were stronger in spectra of Mars than in comparable spectra of the Moon, indicating an addition to the telluric absorption caused by Martian CO_2. Using lunar spectra obtained at different air masses, Kuiper deduced a carbon dioxide abundance in the Martian atmosphere of 4.4 meter amagats (m am). This was approximately twice the column abundance in the terrestrial atmosphere.

Grandjean and Goody (1955) pointed out that the amount of Martian CO_2 implied by Kuiper's observations depended critically on the atmospheric pressure assumed for Mars. Kuiper's method of calibration implicitly assumed a Martian surface pressure close to one-half the terrestrial surface pressure at his telescope. Unfortunately, the atmospheric pressure on Mars was not well known. Indirect methods such as studies of the polarization of reflected sunlight from Mars (Dollfus 1951*a*) indicated values that were optimistically summarized by de Vaucouleurs (1954) as 85 ± 4 mbar, far below the roughly 450 mbar corresponding to Kuiper's observations.

In their analysis, Grandjean and Goody (1955) adopted a value of P_s = 100 mbar for Mars, and derived a CO_2 abundance of 30 m am. The point here is that the absorption in the strong CO_2 bands observed by Kuiper depends on the square root of the product of the column abundance and the ambient pressure. The higher the pressure, the less CO_2 is required to obtain a given absorption. Conversely, as the pressure decreases, the absorption lines become saturated, so much more gas is required to produce a given amount of absorption. Without independent knowledge of either the pressure or the amount of gas, it is not possible to interpret unambiguously a strong absorption band, unless the resolving power is sufficient to reveal the shapes of individual lines.

Eight years later, Spinrad, Münch, and Kaplan (1963) succeeded in finding water vapor in the Martian atmosphere. Using an improved photographic emulsion hypersensitized to record very-near-infrared radiation, they detected several Doppler-shifted absorption lines in the region of the 8200 Å band on a spectrogram obtained at Mount Wilson. In a subsequent paper, they reported that a weak CO_2 band was visible in this same spectrogram at 8689 Å, thereby permitting a nearly pressure-independent evaluation of the CO_2 abundance (Kaplan, Münch, and Spinrad 1964). The result was 55 ± 20 m am. Using this number to interpret the pressure-dependent absorptions observed by Kuiper, Kaplan et al. derived a Martian surface pressure of 25 ± 15 mbar. The corresponding column abundance of water vapor was equivalent to 14 ± 7 μm of precipitable water. This watershed paper marked the beginning of a new era in Martian studies.

Analysis of new spectra of the strong CO_2 bands obtained by Kuiper (1964) confirmed this surprising result (Owen and Kuiper 1964). Subsequent studies of the methods that had been used to obtain the much higher surface pressure that was in vogue up to that time demonstrated that indeed these indirect techniques were subject to strong systematic errors (Chamberlain and Hunten 1965; Cann et al. 1965).

In the wake of these discoveries, several independent observations of the near-infrared spectrum of Mars were carried out during the next opposition of the planet in 1965. The results indicated that the atmospheric CO_2 abundance was somewhat larger and the total atmospheric pressure correspondingly lower than the initial values deduced by Kaplan et al. (Spinrad et al. 1966; Owen 1966). These conclusions were greatly strengthened by a determination of the total pressure from the effect of the Martian atmosphere on radio transmission from the Mariner 4 spacecraft when it passed behind the planet in 1965. This radio-occultation experiment indicated a Martian surface pressure in the range of 4.1 to 6.0 mbar (Kliore et al. 1965). Since 80 m am of CO_2 would produce a surface pressure of 6 mbar on Mars, it appeared likely that at least 90% of the atmosphere consisted of this gas. Belton et al. (1968) confirmed these findings through observations at the 1967 opposition.

II. ABUNDANCES OF MINOR CONSTITUENTS

Carbon monoxide was first definitively detected spectroscopically by Kaplan et al. (1969) who were able to identify several lines in the (2,0) and (3,0) bands. They deduced a CO abundance of 5.6 cm am, a surface pressure of 5.3 mbar, and a CO mixing ratio of 8×10^{-4}. Young and Young (1977) re-analyzed these observations and the subsequent work of several other investigators, obtaining a final result for both bands of 5.6 ± 0.5 cm am. In the same year, Kakar et al. (1977) reported the first microwave detection of CO, observing the $J = 0 \rightarrow 1$ rotational transition. This result, as well as that of Good and Schloerb (1981), who suggested possible temporal variations, were re-analyzed by Clancy et al. (1983) along with their own observations of the $J = 1 \rightarrow 2$ transition. They found evidence for a possible increase in Martian CO content by as much as a factor 2, but noted that the uncertainties in the observations allowed them to be consistent with no increase at all.

Ozone was discovered by Barth and Hord (1971) from observations with an ultraviolet spectrometer on the Mariner 7 spacecraft. The amount was found to be highly variable over the Martian disk, being strongly correlated with low surface temperature. This correlation was interpreted as an indication that the presence of water vapor inhibited the formation of ozone. Abundances $< 3 \times 10^{-4}$ cm am were recorded over most of the planet, while values ranging from 3×10^{-4} to 60×10^{-4} cm am were found over the cold polar caps (Barth et al. 1973; Barth 1974; also see chapter 30).

After unsuccessful attempts by many different observers, molecular oxygen was finally detected by Barker (1972) and by Carleton and Traub (1972), using Doppler-shifted lines in the *A*-band at 7600 Å. The derived column abundance was 9.5 ± 0.6 cm am and 10.4 ± 1.0 cm am, respectively. The mixing ratio was estimated as 1.3×10^{-3} by both sets of authors. Eleven years later, Trauger and Lunine (1983) found a column abundance of 8.5 ± 0.1 cm am. Spectroscopic observations of carbon dioxide and water vapor continued throughout this period. Seasonal and local variations in the abundances of both of these constituents were detected and recorded (see, e.g., Schorn et al. 1967; Owen and Mason 1969; Barker et al. 1970; Young 1971; Schorn 1971; Barker 1976; also see Sagan et al. [1971*b*] and Woszczyk and Iwaniszewska [1974] for additional references). Infrared results from Mariner 9 were summarized by Conrath et al. (1973).

Spectroscopic searches for a number of other atmospheric constituents met with no success. The suggestion by Kiess et al. (1960) that the low ultraviolet reflectivity of Mars was caused by the presence of oxides of nitrogen in the atmosphere was considered to be unlikely on chemical grounds (Sagan et al. 1965) and failed to find confirmation in accurate spectrophotometry of the planet (Owen et al. 1975). Lists of upper limits on various other constituents have been published by Beer et al. (1971), Horn et al. (1972), Owen and Sagan (1972), Barth (1974) and Maguire (1977). An edited version of Maguire's list is reproduced here as Table I.

With reference to the possibility of an indigenous Martion biota, the low upper limits on methane (0.02 ppm), hydrogen sulfide (0.1 ppm), formaldehyde (0.6 ppm), and other gases that would be far from chemical equilibrium in the Martian atmospere are especially noteworthy. *In situ* measurements by the Viking instruments (see below) were not capable of improving on these limits. The terrestrial abundance of 1.8 ppm CH_4 has long been recognized as biogenic in origin. Our entire atmospheric inventory of methane would be lost through oxidation in just a few years, if it were not continually replaced. No comparable case of disequilibrium has been found on Mars (see chapter 30).

III. THE VIKING MEASUREMENTS

Besides nitrogen, argon was the gas most often considered to be a possibly significant constituent of the Martian atmosphere. As ^{40}Ar is a product of the decay of radioactive potassium and as potassium must be present in Martian rocks, it seemed reasonable to assume that this heavy, chemically inert gas would accumulate in the Martian atmosphere during geologic time. Brown (1952) had speculated that it might even be the major constituent of

TABLE I
Upper Limits to Possible Minor Constituents[a]

Gas	Band (cm^{-1})	Upper Limit (cm-am)	Upper Limit (ppm)[b]
C_2H_2	ν_5 729	2×10^{-5}	0.002
C_2H_4	ν_{12} 1443	4×10^{-3}	0.5
C_2H_6	ν_9 821	3×10^{-3}	0.4
CH_4	ν_4 1306	2×10^{-4}	0.02
N_2O	ν_1 1285	1×10^{-3}	0.1
NO_2	ν_3 1621	1×10^{-4}	0.01
NH_3	ν_2 968	4×10^{-5}	0.005
PH_3	ν_4 1122	1×10^{-3}	0.1
SO_2	ν_3 1362	7×10^{-4}	0.1
OCS	ν_1 859	1×10^{-4}	0.01
HCl	ν_0 2890	1×10^{-3}	0.1
HCOH	ν_5 2843	5×10^{-3}	0.6
H_2S	Rydberg (UV)	1×10^{-3}	0.1

[a]Table adapted from Maguire (1977), with HCl and HCOH limits from Beer et al. (1971) and H_2S limit from Owen and Sagan (1972).
[b]The abundance limits in parts per million are based on a CO_2 column abundance of 84 m-amagat (m-am).

the atmosphere, but the spectroscopic and radio-occultation observations appeared to rule out this possibility. Owen (1974) pointed out that a high concentration of ^{40}Ar would provide evidence that the total outgassed atmosphere of Mars might have been much denser than the atmosphere we observe today. An indirect measurement from the Soviet space probe Mars 6 in 1974, based on the behavior of a mass spectrometer ion pump seemed to indicate that as much as 35% of the atmosphere might be a noble gas (Isotomin and Grechnev 1976). Argon was certainly the most likely candidate, and the uncertainties in the spectroscopic measurements were such that this amount could just be accommodated (Moroz 1976). Because the molecular weight of radiogenic argon is so close to that of carbon dioxide, this mixture would have been consistent with the radio-occultation observations as well.

This was the situation when the Viking Spacecraft reached Mars in the summer of 1976. The first instrument to sample the Martian atmosphere was the mass spectrometer on the aeroshell. Analysis of these observations revealed that both nitrogen and ^{40}Ar were present, but there was far less argon than predicted from the Mars 6 experiment (Nier et al. 1976*a*). Molecular oxygen, carbon monoxide and atomic oxygen were confirmed in the upper atmosphere. The mass spectrometer on the first Lander verified the presence of nitrogen and argon and detected ^{36}Ar and ^{38}Ar (Owen and Biemann 1976). Subsequent analysis and data from the second spacecraft added nitric oxide to the list of species in the upper atmosphere, discovered an enrichment of $^{15}N/^{14}N$ compared with Earth, and added krypton, xenon and neon to the list of bulk atmospheric constituents (Nier et al. 1976*b;* Biemann et al. 1976*a*; Owen et al. 1976,1977; McElroy et al. 1976*b*; see chapter 30 for further discussion of upper atmosphere results).

Water vapor in the Martian atmosphere was monitored from orbit (Farmer et al. 1977). These data have been extensively analyzed (Farmer and Doms 1979; Jakosky and Farmer 1982) and compared with groundbased observations (Jakosky and Barker 1984). Discussion of water vapor and its variations is the subject of chapter 28.

The total atmospheric pressure at the first landing site was 7.65 mbar at the time of landing, but it was found to be gradually decreasing with time (Hess et al. 1976*a*). The same situation existed at the second landing site (Hess et al. 1976*b*,1977). Such a decrease had been predicted by Leighton and Murray (1966), who pointed out that it would be a necessary consequence of the seasonal exchange of carbon dioxide between the two polar caps. The pressure cycle was tracked through a maximum at the northern autumnal equinox of 9 mbar at the first landing site and 10 mbar at the second (Leovy 1982; see chapter 27 and Color Plate 16).

This variation is troublesome for a proper evaluation of mixing ratios of noncondensible minor constituents; because the total atmospheric pressure varies seasonally with carbon dioxide deposition and sublimation, the concentrations (but not column abundances) of the other gases will vary with time. Ozone and water vapor present additional problems, as discussed above. Viking results are based on a mean atmospheric pressure of 7.5 mbar which corresponds to conditions during the first few Martian days after landing at each of the two sites. Table II is a summary of the atmospheric composition as we understand it from the Viking measurements. Additional *in situ* analyses of the Martian atmosphere will be required to bring about substantial improvements. While waiting for these, the growing consensus that the SNC meteorites come from Mars allows us to use the relative abundances derived from gases trapped in them to refine the listings in Table II (see chapter 4). High-precision monitoring of O_2, CO, H_2O and CO_2 from Earth may also reveal coupled variations in the abundances of these species.

IV. ISOTOPIC RATIOS

Groundbased and spacecraft studies of carbon dioxide and carbon monoxide bands established that $^{12}C/^{13}C$ and $^{16}O/^{18}O$ were the same on Mars and Earth to within $\pm$ 15% (Kaplan et al. 1969; Young 1971; Maguire 1977). This result was confirmed and the uncertainty reduced to $\pm$ 5% by the Viking entry mass-spectrometer measurements of CO_2 at m/e = 44 to 46 (Nier and McElroy 1977). An apparently conflicting measurement for isotopes of oxygen in Martian water vapor has been reported by Bjoraker et al. (1989). Using

TABLE II
The Composition of the Martian Lower Atmosphere

Gas	Abundance	Ref.[b]
CO_2	95.32%	a
N_2	2.7	a
^{40}Ar	1.6	a
O_2	0.13	a
CO	0.07	a
H_2O	0.03[a]	b
$^{36+38}Ar$	5.3 ppm	a
Ne	2.5 ppm	a
Kr	0.3 ppm	a
Xe	0.08 ppm	a
O_3	0.04–0.2 ppm[a]	c

Isotope Ratios

Ratio	Earth	Mars	Ref.[b]
$^2H/^1H$	1.56×10^{-4}	$9 \pm 4 \times 10^{-4}$	d
		$7.8 \pm 0.3 \times 10^{-4}$	e
$^{12}C/^{13}C$	89	90 ± 5	f
$^{14}N/^{15}N$	272	170 ± 15	f
$^{16}O/^{18}O$	489	490 ± 25	f
		545 ± 20	e
$^{16}O/^{17}O$	2520	2655 ± 25	e
$^{36}Ar/^{38}Ar$	5.3	5.5 ± 1.5	g
$^{40}Ar/^{36}Ar$	296	3000 ± 500	a
$^{129}Xe/^{132}Xe$	0.97	2.5^{+2}_{-1}	a

[a] Variable with season and location (see text).
[b] References: (a) Owen et al. (1977); (b) Farmer and Doms (1979); (c) Barth (1974); (d) Owen et al. (1988); (e) Bjoraker et al. (1989); (f) Nier and McElroy (1977); (g) Biemann and Owen (1976).

high-resolution infrared spectra of Mars obtained with the Kuiper Airborne Observatory, they found Martian values of $^{17}O/^{16}O = 0.95 \pm 0.01$ times terrestrial and $^{18}O/^{16}O = 0.090 \pm 0.03$ times terrestrial. (The Viking mass spectrometers did not determine $^{17}O/^{16}O$ because of interference with ^{13}C at $m/e = 45$.) This surprising result needs to be confirmed by additional measurements in both CO_2 and H_2O, to see if there is indeed a systematic difference in the oxygen isotope ratios in these two species. Both gases can be investigated by means of Earth-based observations.

It would also be of great interest to know the value(s) of these ratios in the gases trapped in SNC meteorites. The closest analog to the Martian atmosphere as described by the Viking measurements that has been found so far is the mixture of gases residing in glassy inclusions in the Shergottite EETA 79001 (Becker and Pepin 1984; Wiens et al. 1986). Unfortunately, the oxygen isotopes in these gases have not yet been measured. Existing analyses only deal with the oxygen in carbonate, where an enrichment of $\delta^{18}O$ = + 21‰ has been reported in EETA 79001 (Wright et al. 1988; Clayton and Mayeda 1988). [$\delta^{18}O = 1000\ ({}^{18}O/{}^{16}O_{sample} - {}^{18}O/{}^{16}O)_{standard})/{}^{18}O/{}^{16}O_{standard})$.] If this carbonate was in fact produced by weathering on the Martian surface, Clayton and Mayeda (1988) suggest that both H_2O and CO_2 must have been present, with an oxygen atom ratio of $H_2O/CO_2 \approx 4$. However, the substantial amount of nitrogen that was liberated during combustion of the carbonate-rich fraction of EETA 79001 showed no evidence of the isotopic enrichment found in the Martian atmosphere (see below) and in gas trapped in the glassy inclusions in this meteorite. Evidently this nitrogen was not introduced by the same weathering process that produced the carbonate or the carbonate was not formed on Mars. A determination of $\delta^{17}O$ in the carbonate should provide a definitive test of the latter possibility, but has not yet been carried out (Clayton 1990, personal communication).

Brinkman (1971) originally suggested nitrogen could be lost from Mars by atmospheric escape. McElroy (1972) predicted that the nonthermal escape of nitrogen and oxygen from Mars should lead to the selective enrichment of the heavier isotopes of these two elements. This prediction was confirmed for nitrogen by the Viking entry mass spectrometers which found ${}^{15}N/{}^{14}N$ enhanced by factor 1.6 ± 0.2 over the telluric value (Nier et al. 1976*b*; Nier and McElroy 1977). Given the current nitrogen escape rate, McElroy et al. (1977) calculated that the present enrichment of ${}^{15}N$ on Mars implied a starting partial pressure of 1.3 to 30 mbar of nitrogen. The high value included chemical fixing of nitrogen in the soil as well as the nonthermal escape from the upper atmosphere. These numbers can be compared with the present partial pressure of N_2 on Mars, which is 0.13 mbar.

This problem was subsequently addressed by Fox and Dalgarno (1983), who made a comprehensive analysis of all processes leading to nitrogen escape. They concluded that if the escape rate has been constant over the last 4.5 Gyr, the initial N_2 partial pressure would have been at least 2 mbar and an enhancement of ${}^{15}N/{}^{14}N$ by a factor 2.5 over the telluric ratio would have occurred, much higher than the observed value of 1.6. Fox (1989) has revised this argument further on the basis of new laboratory studies of the dissociative recombination of nitrogen. She finds that the isotopic enhancement is lower (a factor 2.1), and that the observed value can be reached within the inherent uncertainties of the population of excited states in the nitrogen atoms in response to the time-integrated solar flux. The initial column abundance corresponds to a partial pressure of 1.5 mbar, close to the original estimate of McElroy (1972).

Fox and Dalgarno (1983) pointed out that the much higher EUV (extreme ultraviolet) flux expected from the young Sun would have increased the nitrogen escape rate but probably reduced the fractionation of the isotopes. In contrast, an early, dense CO_2 atmosphere would diminish the escape rate by reducing the concentration of N_2 at the exobase. These two effects can be played off against each other in a time-dependent model; however, this has not yet been done with the new data presented by Fox (1989).

It is useful to note the implications of this discussion for the overall abundances of outgassed volatiles on Mars. On Venus, where it is generally assumed that all the gases (except water vapor) released by the planet are presently in the atmosphere, we find a ratio of C/N = 14. On Earth, this ratio is 24, if we assess all the volatiles outgassed over geologic time. To reach C/N = 19 ± 5 on Mars with an initial partial pressure of 30 mbar of N_2, we require 1800 ± 400 mbar of CO_2. Thus the assumption of nitrogen fixation in the soil in addition to nonthermal escape carries with it the implication of an early, dense atmosphere, if the initial Martian volatile inventory resembled that of Earth and Venus. On the other hand, if the initial partial pressure of N_2 was only 1.3 mbar, the case in which nitrogen depletion occurred only by escape, the corresponding amount of CO_2 would be 78 mbar.

Clark et al. (1979,1981) and Zolotov and Sidorov (1986) have suggested that nitrates may not be stable in the Martian environment. However, their arguments depend on the availability of SO_2 and/or HCl, preferably during periods of relatively high water activity on the planet. As Table I demonstrates, this is not a problem today. One can certainly imagine conditions in which nitrate deposits would be sufficiently shielded to endure episodic, destabilizing events associated, e.g., with volcanic activity. The search for such deposits has high priority in any list of experiments designed to unravel the past history of Martian volatiles.

In the case of oxygen, McElroy and Yung (1976) demonstrated that the absence of an enrichment of ^{18}O in parallel with that found for ^{15}N must indicate the presence of a reservoir of oxygen on Mars that could exchange with the atmosphere. Given the Viking constraint on $^{18}O/^{16}O$, McElroy et al. (1977) concluded that *at least* 4.5×10^{25} atoms cm^{-2} of oxygen would be required in such a reservoir. Both carbon dioxide and water were potential candidates, with water considered to be more likely. The abundance of water corresponding to this oxygen reservoir would provide a layer of liquid over the surface of Mars that was at least 13 m deep. The apparent *depletion* of the heavy isotopes of oxygen in H_2O as reported by Bjoraker et al. (1989) may drive the size of the required oxygen reservoir even higher. Alternatively, it could indicate a systematic difference between the oxygen in water and that in carbon dioxide, requiring the former to be a relatively recent arrival on the time scale of isotopic exchange between the two species.

The noble gas isotopes were evaluated by the mass spectrometer (GCMS) on the Lander. The nonradiogenic isotopes of argon, krypton and xenon appeared to exhibit terrestrial values within the uncertainties in the measurements. In the text of the review of all the lower-atmosphere Viking results by Owen et al. (1977), the $^{36}Ar/^{38}Ar$ determination was mistakenly given an uncertainty of ± 10%, appropriate only to the non-noble gases. In fact, the range of uncertainty on this ratio extends from $4 \leq {}^{36}Ar/{}^{38}Ar \leq 7$, as originally reported by Biemann et al. (1976*a*). The terrestrial value of 5.3 falls near the center of this range of uncertainty, but the SNC value of 4.1 ± 0.2 (Wiens et al. 1986) can just be accommodated by the lower bound.

The two anomalies discovered in the noble gas data were the unusually high ratios of ${}^{40}Ar/{}^{36}Ar$ and ${}^{129}Xe/{}^{132}Xe$ compared with the terrestrial atmosphere. These two radiogenic isotopes are the decay products of ${}^{40}K$ and ${}^{129}I$, respectively. The high values could be explained in terms of different mixtures of materials from which the Earth and Mars were composed, but other evidence argues against that interpretation. For example, Surkov (personal communication) has just reported K/U values for Mars measured by the PHOBOS spacecraft that are in agreement with the terrestrial ratio.

The likelihood of large-scale impacts on Mars as discussed in the next section offers an alternative solution that is more attractive at the present time. In this scenario, the mass of the planet's atmosphere is repeatedly reduced by impacts. The escape of gases from such processes occurs in a nonfractionating way (Melosh and Vickery 1989). As this reduction in the total inventory of outgassed volatiles takes place early in the planet's history, subsequent degassing of radiogenic species can enrich the latter with respect to the background. This accounts especially well for the ~ 10 times enrichment of ${}^{40}Ar/{}^{36}Ar$ over the terrestrial value (Turcotte and Schubert 1988).

It is important to remember that the *total* outgassed ${}^{40}Ar$ on Mars is a factor 16 less (per gram of rock) than the terrestrial value. Thus, even though outgassing continued on Mars after the early impacts, it was evidently not as vigorous as it has been on the Earth. This is entirely consistent with the generally diminished tectonic activity on Mars compared with the Earth, which is the real reason for the failure of Brown's (1952) suggestion that argon could be the major component of the Martian atmosphere. It is interesting to note in this context that the total ${}^{40}Ar$ per gram of rock outgassed by Venus is just 4 times greater than that of Mars (Hoffman et al. 1980). Venus again shows less tectonic activity than Earth, despite the similar sizes of these two planets.

The Viking mass spectrometers were unable to provide a measurement of the abundance of deuterium on Mars. With improvements in observing sites and instrumentation, however, this experiment can now be carried out from Earth. Owen et al. (1988) reported the first detection of deuterium on Mars by recording the fundamental ν_1 band of HDO observed at 3.7 μm from Mauna Kea. Lines of Martian H_2O were recorded at the same time in order to derive $D/H = 1/2(HDO/H_2O) = (9 \pm 4) \times 10^{-4}$. This represents an enrichment of a factor 6 ± 3 over the terrestrial value of 1.5×10^{-4} corresponding to standard mean ocean water (SMOW). Bjoraker et al. (1989) have subsequently improved the accuracy of this result, finding $D/H = (7.8 \pm 0.3) \times 10^{-4}$ from airborne observations, corresponding to an enrichment of 5.2 ± 0.2 times terrestrial D/H.

V. THE EARLY ATMOSPHERE: WAS THERE AN ANCIENT EDEN?

It would be wonderful if one could use these isotopic ratios to determine the evolutionary history of the Martian atmosphere. One could then hope to discover the characteristics of the planet's early climate, and employ this knowledge to interpret the erosional history of the surface. It seems obvious that a massive change in climate has occurred, but we simply do not have enough data yet to create a model that uniquely interprets the atmospheric and geologic evidence for this change.

If the SNC meteorites really do come from Mars, one might hope to gain some insight into the evolution of the D/H enrichment presently observed in the Martian atmosphere by determining the value of D/H in these meteorites. Unfortunately, it seems unlikely that this approach will succeed. Kerridge (1988) has reported measurements in HDO produced by combustion of samples from the SNC meteorites Shergotty and Lafayette (a nakhlite) at two different temperatures. The results indicated enrichments of 1.9 and 1.5 times the terrestrial value of D/H, respectively. Kerridge concluded that these values actually corresponded to lower limits because of the background inherent to the analysis and the possibility of isotopic exchange with terrestrial hydrogen. For this and other reasons, he concluded that the data ". . . are consistent with the presently observed Martian D/H ratios, but are not informative about the evolutionary route by which that ratio was achieved."

We must therefore resort to theory to try to determine what combination of processes led to this enrichment. Owen et al. (1988) assumed that the similarity of D/H in the ice in Halley's comet, in the water of hydration in meteorites and (apparently) in condensed matter in the outer solar system to the terrestrial value suggests that Mars also started with this same value (see discussion in Sec. VI). Thermal escape of hydrogen will lead to the enrichment of deuterium, but to produce the observed enrichment of 5 times terrestrial, over 99% of the water available to the Martian surface must have been destroyed with subsequent hydrogen escape.

This poses a problem in reconciling the known escape rate of hydrogen at the present time with the amount of water required in the early history of Mars to produce observed erosional features. The present hydrogen escape rate corresponds to the loss of the equivalent of a layer of water 2.5 m deep over the entire planet during its 4.5 Gyr history (McElroy et al. 1977). Owen et al. (1988) point out that the deuterium enrichment then suggests that this was essentially all of the water that was ever available at the Martian surface. Yung et al. (1988) have carried this analysis further with a detailed study of deuterium fractionation on Mars by the escape of hydrogen. They concluded that $\sim$3.4 m of H_2O has escaped from the planet, leaving only 0.2 m behind. In contrast to these low estimates, Carr's (1986) interpretation of erosional and ice-sculpted features suggested that the equivalent of a planet-wide layer of water between 500 and 1000 m deep must have been present. Looking just at the volcanic features on the planet, Greeley (1987) concluded that they alone produced an amount of water equivalent to a layer on the order of 46 m thick. Jakosky (1990) has also queried the low value of the current H_2O reservoir deduced by Yung et al. (1988), as he finds it highly probable that much more water than this is exchanging between the polar deposits during the present era.

To reconcile the large amounts of water required by the geologic evidence with the observed deuterium enrichment and the present escape rate, Owen et al. (1988) proposed that escape must have been more rapid in the past. They offered three possible mechanisms: oxidation of the crust, enhanced solar EUV, and a general warming of the atmosphere, concluding that only the latter had the necessary potential.

A relatively modest example of this latter effect has been studied by Jakosky (1990) who pointed out that cyclical changes in the planet's obliquity will lead to higher abundances of atmospheric water vapor on average than those we find today, thus enabling higher escape rates. He concluded that at least 60 m of water has escaped over geologic time. To get beyond this, we need a dense, warm early atmosphere that would contain much more water vapor, therefore allowing the possibility for much more substantial hydrogen escape. The difficulty in this proposal is to show that the tropopause in such an atmosphere does not constitute a cold trap that would prevent the H_2O from reaching high enough altitudes for efficient photodissociation. Alternatively, some other noncondensible species could carry the hydrogen up to these levels, if such a species were present. For example, oxidation of the crust would liberate H_2 at the surface of Mars. However, Owen et al. (1988) have calculated that the complete oxidation of a 1 km basalt regolith on Mars would only consume the equivalent of a 27-m layer of H_2O. Yet another possibility is massive hydrodynamic escape of hydrogen during the planet's early history, but this approach to D/H enrichment has not yet been conclusively demonstrated (Zahnle et al. 1990*a*).

The deuterium enrichment carries one other message that relates to the early atmosphere, as Owen et al. (1988) emphasized. At the present time, one oxygen atom escapes from Mars with each two atoms of hydrogen (McElroy 1972; Liu and Donahue 1976; Donahue and Hunten 1976). The coupled escape of these two gases means that water itself is leaving the planet. Yet as we saw above, unlike hydrogen and nitrogen, the heavy isotopes of oxygen either exhibit no enrichment or are actually depleted compared with terrestrial abundances. Given the enrichment of deuterium, we must conclude that oxygen would also be fractionated if water were its sole source (except for the relatively small amount sequestered through oxidation). It therefore seems likely that CO_2 rather than H_2O constitutes the normalizing reservoir. Following the work of McElroy et al. (1977), the amount required corresponds to an atmosphere with $P_s \geq 600$ mbar. This lower limit is approximately one-half the value required to provide an atmospheric greenhouse capable of allowing liquid water to exist at the planet's surface (Cess et al. 1980; Pollack et al. 1987).

We thus have a general consistency; the deuterium enrichment requires a large fraction of the water available at the Martian surface to have escaped. To reconcile this requirement with the geologic need for a large volume of water to cut the observed erosional features, one possible solution seems to be a denser, warmer ancient Martian atmosphere, that would provide a *potential* increase for hydrogen escape by over 2 orders of magnitude (Krasnopolsky 1986). In such an atmosphere, the nonthermal escape of oxygen would no longer be coupled to hydrogen loss, hence oxygen isotope fractionation would not occur.

This period would lead to some enrichment of D/H, but less than the present value. Formation of carbonates and nitrates would gradually reduce the mass of the atmosphere, decrease the greenhouse effect and leave less water in the exchangeable reservoir as chemical reactions proceeded and temperatures declined. As the flux of escaping hydrogen atoms decreased and nonthermal escape of oxygen became important, CO_2 in the thinning atmosphere would provide the buffer for the oxygen isotopes, maintaining their normality. Given the virtual certainty of a high EUV flux from the young Sun, the observed enrichment of ^{15}N could probably be accommodated, provided a substantial amount of the original nitrogen is now deposited in surface materials. This dense, warm, wet atmosphere would obviously be conducive to the presence of open bodies of water on the Martian surface, and would also explain the enhanced erosion of small topographic features in ancient times (Chapman and Jones 1977).

While all of this is certainly encouraging, we should recall Emerson's (1841) apothegm: "A foolish consistency is the hobgoblin of little minds. . . ." We still lack a rigorous, time-dependent model for the early Martian atmosphere and surface demonstrating that all of these requirements can be satisfied. Present data on the oxygen isotopes in H_2O and CO_2 require some additional process(es). Furthermore, the history of water activity on Mars has not yet been rigorously constrained (see, e.g., Carr 1981; Mouginis-Mark 1990; chapter 15). Carr (1990) has argued that episodic flooding, volcanism and cometary impacts are likely to reset the value of D/H in atmospheric water vapor, which will then be different from the value of D/H in the bulk water on the planet. He concluded that a measurement of atmospheric D/H therefore tells us nothing about the planet's early history. While some resetting must certainly occur, the degree to which the atmospheric isotope ratio differs from that in the main water reservoirs will depend on how well these reservoirs actually exchange water with the atmosphere and how recent and massive the resetting event was. The age of the polar layered terrains becomes a key issue, as these probably constitute the largest sink for water injected into the atmosphere. Carr (1990) suggested that these terrains are older and more stable than previous workers have thought. Jakosky (1991) has criticized Carr's conclusion on the grounds that geologic evidence pushes the episodes of flooding and volcanism into the distant past, so the cratering ages of the polar deposits require more water to have been produced during the last 10^8 yr than such evidence warrants. Instead, he favors exchange of water between the two poles, as mentioned above, as the means of providing the relatively young ages of these terrains.

Jakosky's (1990*b*,1991) own interpretation of the stable isotope data currently available from the Viking measurements, remote spectroscopy and the SNC meteorites suggests that polar water ice is the buffer for isotope fractionaton by nonthermal escape of oxygen. He favors exchange of oxygen between atmospheric CO_2 and H_2O to explain the oxygen isotope fractionation reported by Bjoraker et al. (1989). One aspect of his model that is difficult to understand is the requirement for an excess of CO_2 over H_2O in the original reservoir. Given elemental and molecular abundances in the solar nebula, the meteorites, the comets and the Earth, this seems unlikely. We review the evidence we have for the source(s) of Martian volatiles and the model(s) of their delivery and loss in the next section.

VI. ORIGIN

Anders and Owen (1977) presented a model invoking a meteoritic source for Martian and terrestrial volatiles in the wake of the Viking discoveries. Perhaps the most dramatic new information and ideas that have occurred since then are the detailed data concerning composition of the atmosphere of Venus resulting from Pioneer Venus and Venera probes (Donahue and Pollack 1983), the suggestion that SNC meteorites originated on Mars, and the evocation of large-scale impacts to explain, for example, the origin of the Earth's Moon. As Cameron (1983) has emphasized, such impacts are bound to affect the atmosphere of the target planet. Here, a personal appraisal of the significance of the first and third of these items for the Anders-Owen model is offered. The second is discussed in detail in chapter 4, which also contains a review of alternative models for atmospheric origin.

A basic tenet of our original work is that the volatile inventories of the three inner planets with atmospheres were established by meteoritic bombardment in the form of a late-accreting veneer. This assumption was buttressed by the similarity of the noble-gas abundance patterns in the atmospheres of Earth and Mars and in the so-called planetary pattern in meteorites. At the time of our paper (Anders and Owen 1977), the available information about volatile abundances on Venus was entirely consistent with this approach.

The new Venus data indicate that the atmosphere of that planet differs from the predictions of our model in two important respects: the abundances of neon and argon per gram of planet are excessively high, and the pattern of noble-gas abundances is distinctly non-"planetary." That is, while [Ne]/[Ar] is similar on Venus and Earth, [Ar]/[Kr] more nearly resembles the solar pattern (Hoffman et al. 1980; Donahue and Pollack 1983). Despite these differences, the amounts of carbon and nitrogen presently in the atmosphere of Venus are close to those outgassed by the Earth over its geologic history. One question we obviously need to answer is which of these comparisons is/are telling us something significant and which is/are simply coincidence(s)?

The importance of large-scale impacts for the atmospheric evolution of the terrestrial planets has been studied in some detail for Mars by Melosh and Vickery (1989). These authors found that the vapor plume created by a sufficiently large projectile could eject the entire air mass above the plane tangent to the point of impact. No fractionation of constituents occurs during such impact erosion. Using a plausible distribution function for the masses and frequency of early impactors, they found that the early atmosphere of Mars must have been at least 100 times more massive than the 7.5 mbar we see today. Mars is particularly vulnerable to impact erosion of its atmosphere because of its small radius and low gravitational acceleration. Nevertheless, Melosh and Vickery (1989) argue that impact velocities during accretion were sufficiently low that such erosion did not occur. In other words, one can bring in volatiles by bombardment during accretion, but lose them through major impacts subsequent to the formation of the planet.

Anders and Owen (1977) stressed that even a reconstruction of the Martian atmosphere based on the present abundances of the noble gases leads to a total volatile inventory for the planet that is surprisingly meager, close to the lower limit of 750 mbar set by Melosh and Vickery (1989). A comparison with the volatile inventories of Earth and Venus suggests that Mars must have lost a large fraction of its original endowment. Impact erosion seems to provide a satisfactory solution to this problem, despite our original aversion to it. An early history of intense bombardment can also explain the relative excess of ^{129}Xe and ^{40}Ar (compared with other Xe and Ar isotopes) by episodically reducing the abundances of primordial species before these radiogenic isotopes were released into the atmosphere. Ott and Begemann (1985; see also Ott 1988) have argued that a Martian origin for all of the SNC meteorites requires a higher abundance of radiogenic nuclides in the Martian atmosphere than in Martian rocks. This is contrary to the terrestrial situation, which is the expected one for any closed system. Impact erosion provides a neat solution to this paradox, *viz.*, Mars was not a closed system during the outgassing of most of its atmosphere.

This new perspective is still consistent with the basic idea that the volatiles on the inner planets were mainly contributed by a late-accreting veneer. The major uncertainty is the nature of the material. The C3V source invoked by Anders and Owen (1977) is clearly too restrictive, but it is probably unwise to substitute some other particular class of meteorites in its place (cf. chapter 4). Clearly a variety of materials was raining down on the young inner planets and it may never be possible to deduce the exact proportions of the various components of the rain from the meteorites studied in our laboratories.

Comets must also have played a significant role in bringing in volatiles (e.g., Oró 1961; Lewis 1974*b*; Sill and Wilkening 1978; Grinspoon and Lewis 1988; Chyba et al. 1990). For example, both Chyba (1987, 1990) and Ip and Fernandez (1988), using different approaches, found that icy planetesimals could have brought in all the water now present in Earth's oceans. It is hard to see how this flux could have avoided Mars. For scaling purposes, it is worth noting that the impact of one Halley-size comet nucleus every 100 million years would nicely compensate for the loss of a 2.5 m thick global layer of H_2O as deduced from the current escape rate of hydrogen. The relatively recent impact of an even smaller comet might provide a solution to the D/H paradox: an enrichment of deuterium in the water delivered by this impact could occur with a much less stringent requirement on the size of the reservoir required to prevent the depletion of ^{16}O by escape, since the total amount of water involved is now very small. Alternatively, one could imagine the comet itself contributing water slightly enriched in D/H compared with the terrestrial value. Accurate determinations of both the hydrogen and oxygen isotope ratios in several comets would obviously help to constrain this problem. Owen et al. (1991) have suggested that the impact of a 16-km diameter comet nucleus (slightly larger than P/Halley) could supply all of the krypton and non-radiogenic argon in the Martian atmosphere. Thus it may ultimately be possible to set some limits on the total cometary contributions to Martian volatiles since the last major impact event. At the present time, however, we are constrained by the fact that such calculations are only based on laboratory and theoretical studies of gas trapping in ice. We still don't know what noble gases (if any) are contained in pristine comet nuclei.

To proceed further with these ideas, we need more information. Additional Earth-based measurements of the oxygen isotopes in both water vapor and carbon dioxide in the Martian atmosphere will be extremely helpful. If Mars once had a dense, CO_2-rich atmosphere, there should be extensive deposits of carbonates and a large amount of nitrogen sequestered on or near the surface today. These deposits must be found and evaluated to test this model. An assay of the total amount of water remaining on the planet is another key measurement. It is critical to determine oxygen and hydrogen isotope ratios in ice deposits and in rocks: e.g., hydrated silicates and carbonates, to try to establish a time history for the D/H buildup and to identify the oxygen reservoir. Trapped gases in breccias would also help with this analysis, adding the nitrogen isotopes (also measurable in nitrate deposits) to the picture. Accurate ages for the fluvial features would give us a chronology for the early history of water activity. Detailed chemical analyses of ancient, protected sediments might provide additional clues about the composition and mass of the early Martian atmosphere. More work can be done on the SNC meteorites, such as the determination of oxygen isotope ratios in trapped

CO_2. Conversely, an accurate measurement of $^{36}Ar/^{38}Ar$ in the Martian atmosphere would offer a definitive test of the hypothesis that these meteorites originated on Mars. Finally, if we are to understand the role of comets in bringing volatiles to the inner planets, we must know much more about the composition (e.g., noble gas abundances, major element isotope ratios) and size distribution of these elusive objects. The Comet Rendezvous/Asteroid Flyby mission should provide some of these answers, but we may have to wait for the return of a sample from a comet nucleus to get the complete story.

Acknowledgments. I thank R. Clayton, J. Fox, B. Jakosky and M. Mumma for helpful comments.

26. DYNAMICS OF THE ATMOSPHERE OF MARS

RICHARD W. ZUREK
Jet Propulsion Laboratory

JEFFREY R. BARNES
Oregon State University

ROBERT M. HABERLE, JAMES B. POLLACK,
NASA Ames Research Center

JAMES E. TILLMAN and CONWAY B. LEOVY
University of Washington

Observations by the Mariner, Mars and Viking spacecraft have provided detailed insights into the thermal structure and dynamics of the atmosphere of Mars. While these observations have confirmed many of the expectations drawn from terrestrial experience about the atmosphere's general circulation and wave activity, there are many aspects which are due to unique attributes of Mars. These attributes include the role of condensation and sublimation of a sizeable fraction of the atmospheric mass (mainly CO_2), the radiative effects of variable dust loading, and seasonal asymmetries in insolation due to the current eccentricity of the orbit of Mars and the seasonal phasing of its perihelion. The zonally symmetric circulation appears to be a more coherent and prominent component of the general circulation on Mars than it is on Earth, although its exact nature is still debated. Traveling atmospheric waves are known to be prominent, and remarkably regular, in northern mid-latitudes during fall and winter. Variability in the amounts and distribution of dust suspended in the atmosphere produces significant diurnal, seasonal and interannual variability in atmospheric thermal structure and circulation. The effects of suspended dust are greatest during episodic planet-encircling dust storms. Regions of the at-

mosphere on Mars which remain relatively unobserved include the planetary boundary layer, the middle atmosphere (≈ 45–90 km) and the southern mid-latitudes during winter. Elsewhere for Mars, the data now available are too sparse to define the general circulation diagnostically, but can still provide powerful insights when viewed through the constraints of angular momentum, energy and mass balances of a shallow atmosphere on a rapidly rotating planet. Further advances in the understanding of the dynamics of the Martian atmosphere will require future acquisition of global, long-term data records describing the weather and climate of Mars.

I. INTRODUCTION

A central goal in the comparative study of the dynamics of planetary atmospheres is to understand the response of atmospheric circulations to the factors which shape and drive them, especially rotation and heating. Mars is particularly interesting in this respect because of its similarities to Earth (Tables I and II). Earth and Mars have very similar rotation rates and axial tilts. Both have relatively transparent atmospheres through which solar radiation penetrates to the surface much of the time. Carbon dioxide is an important radiating gas in the thermal infrared for both planets and, although Mars is colder than Earth, its temperature regime is more like that of Earth than of any other planet. The temperature and pressure ranges of the Martian lower atmosphere are comparable to those of Earth's middle stratosphere.

There are important differences as well. Temperature variations of Earth's lower atmosphere are moderated by the high thermal inertia of the oceans; Mars has no oceans and its surface is covered by material having low thermal inertia. As a result, its atmosphere experiences much larger seasonal and diurnal temperature variations than Earth's troposphere. Seasonal variations in the lower atmosphere of Mars are similar to those of Earth's stratosphere, in that the temperature pattern at solstice is characterized by a general decrease from the summer to the winter polar regions. This reflects the comparable obliquities of the two planets and similar degrees of radiative control. Mars differs from Earth in that orbital eccentricity and its seasonal phasing are important factors for Mars' present climate, but of relatively minor importance for Earth today.

Surface topography influences circulation on both planets. The large-scale orography on Mars has greater vertical relief, due again in part to the absence of oceans, and the thermal effect of the orography on Mars may be significantly greater than on Earth, because comparable surface heating is driving a thinner atmosphere on Mars. While Mars lacks the spatial contrast in heating provided by the land-sea distribution on Earth, it does exhibit "thermal continents," meaning vast regions of its surface in which high albedo is correlated with low thermal inertia and, to a lesser extent, with high terrain.

In Earth's troposphere, evaporation and condensation of water are powerful means of redistributing heat. On Mars, as in the Earth's stratosphere, there is too little atmospheric water for its latent heat to be significant thermodynamically. However, on Mars the condensation and sublimation of carbon dioxide constrain the temperatures of the polar regions and drive a seasonal mass flow from subliming to condensing polar cap. Dust blown into the atmosphere by strong winds at the surface of Mars intercepts solar radiation, and provides a powerful internal heat source somewhat analogous to the latent heat of condensed water in Earth's troposphere.

A second goal in the study of Martian atmospheric dynamics is to understand how the general circulation and the associated boundary layer processes interact with the surface (see, e.g., Haberle 1986*a*). The surface of Mars has undergone, and apparently continues to undergo, extensive modification as a result of aeolian erosion, redistribution of dust, and weathering (chapters 21–23, 29). The atmosphere also transports water and carbon dioxide between various surface reservoirs (chapters 27 and 28). If the present is to be the key to the past or future, the mechanisms controlling the present seasonal injection, transport and removal of dust, water vapor and carbon dioxide, together with their interannual variations, must be understood quantitatively and in detail.

In this chapter, the current state of understanding of the Martian atmospheric circulation and boundary layer is described. This understanding is of necessity based on models as well as observations, as the data in hand are far too sparse to define the general circulation alone. A great deal has been learned about various components of the general circulation through the judicious use of the available data with a variety of models and by taking full advantage of the many dynamical similarities with Earth.

In Sec. II, the meteorological data in hand and the first-order constraints that can be derived from them are reviewed. In Sec. III, pertinent aspects of atmospheric radiation on Mars are briefly presented, the main features of which are the short radiative time constants on Mars, as compared to Earth, and the important role of suspended dust in providing a potent thermal drive for the atmosphere.

Sections IV and V present a discussion of our current understanding of the Martian atmospheric circulation and its various components: the zonal-mean (i.e., longitudinally averaged) zonal and meridional flows, stationary and traveling planetary waves, atmospheric thermal tides, topographic wind systems, free modes, and gravity waves. The models used to study these phenomena span the range from two- and limited three-dimensional radiative-dynamic process models to the most comprehensive of the existing general circulation models (GCMs).

In Sec. VI, our current, and quite circumscribed, understanding of the planetary boundary layer (PBL) on Mars is reviewed. The Viking Landers have provided our first detailed observations of this region; elsewhere on Mars only qualitative constraints have been possible. Particular attention is given in this section to the Viking Orbiter and Lander observations relating to the winds required for raising dust from the surface; the lifting mechanisms themselves are discussed in chapter 22. In Sec. VII, our current knowledge and theory of the origin and decay of great dust storms on Mars, meaning those dust storms which attain regional scale or encircle the planet, are reviewed. Much remains to be learned about these episodic events, which can so strongly impact the present climate of Mars and whose occurrence may have long-term effects as well (chapter 29). Finally, a brief summary of the chapter is given in Sec. VIII.

II. METEOROLOGICAL FIELDS: OBSERVED AND DEDUCED

In this section, data on temperature, wind and surface pressure are summarized. On Earth routine observations of the meteorological fields of atmospheric and surface temperatures, pressure, horizontal winds, relative humidity, of clouds and of key trace species like ozone form the basis of a detailed climatology. A complete climatology for Mars would include the distribution of atmospheric dust, in addition to the conventional meteorological fields.

Presently, no three-dimensional meteorological field on Mars has been observed globally for a full Mars year. A time series of nearly synoptic global views of the Martian atmosphere simply does not exist. Fields observed with the most complete coverage in time and space include Viking Orbiter observations of surface temperatures, column abundances of atmospheric water vapor, column dust opacity in the infrared, and a weighted average of air temperatures at altitudes of 20 to 35 km. Typically, one month was required to acquire complete longitudinal coverage for these fields.

Temperatures retrieved from infrared radiances observed remotely from the Mariner 9 Orbiter form the meteorological data set with the most complete horizontal and vertical coverage; radio occultation data has yielded temperature profiles with the best vertical resolution, but only at a limited number of sites. The Viking Lander meteorological data records, including that of visual opacity, have the best time resolution and the Viking Lander 1 pressure and temperature records have the longest span of any nearly continuous data set, but just at that site.

Atmospheric winds have been measured near the surface at the Viking Lander sites, derived from the Viking entry data, estimated from the form and displacement of cloud features, and inferred from aeolian features seen on the surface. The global circulation pattern and its regional variation are essentially unobserved. Fortunately, observations of surface pressure and of the spatial variation of temperature provide strong constraints on the vertical variation of wind. The following sections review the current meteorological data bases for Mars.

A. Surface Pressure

Prior to the Viking mission, the principal goals of those making surface pressure measurements (whether from Earth-based observatories or from spacecraft) were to determine, first the bulk mass of the Martian atmosphere (chapter 25) and then later, the large-scale orography of Mars (Conrath et al. 1973; Hord et al. 1974; chapter 7). The only *in situ* meteorological measurements of surface pressure on Mars were made by instruments (Chamberlain et al. 1976) onboard the two Viking Landers in low-lying regional basins on opposite sides of the Northern Hemisphere: Viking Lander 1 (the Mutch Memorial Station) at 22°.5 N, 48°.0 W and Viking Lander 2 at 48°.0 N, 225°.7 W (areographic latitudes, after Michael 1978).

The Viking Lander surface pressure measurement records (Fig. 1) with their precision (limited by digitization to $\approx$ 0.08 mbar, or 8 Pa), their frequent sampling, and their unanticipated longevity have made it possible to

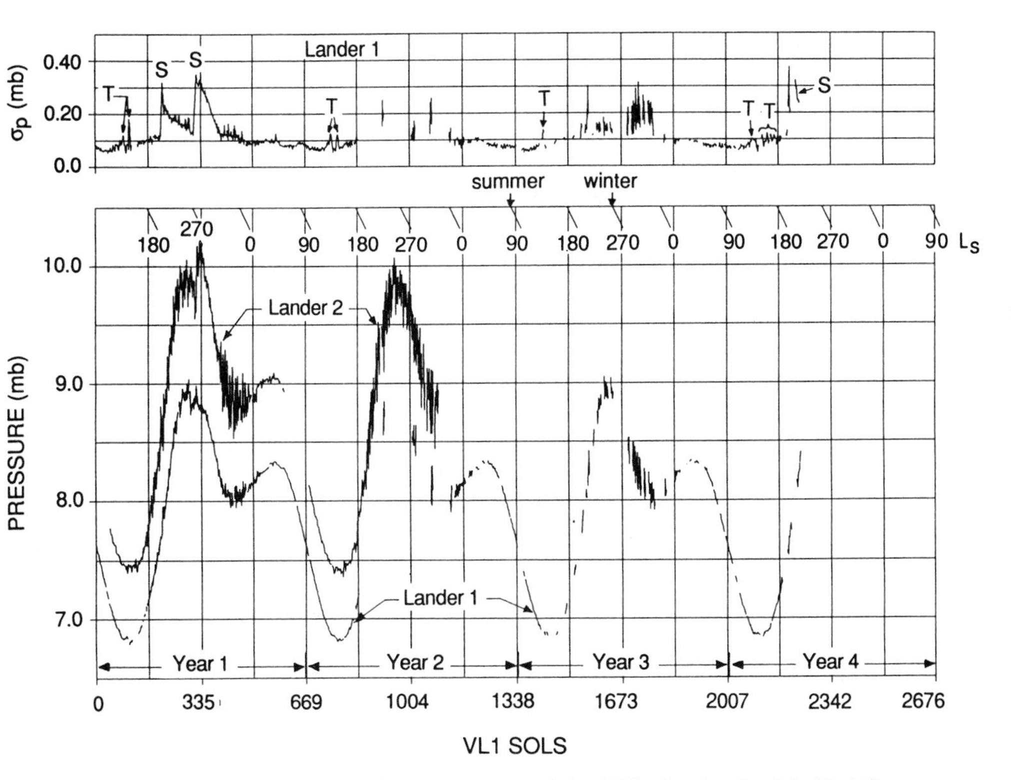

Fig. 1. (Bottom) Sol-averaged surface pressures recorded at Viking Landers 1 and 2. (Top) The standard deviation of surface pressure about the sol mean at VL1. Both for 3.3 Mars years. Note the signatures (S) of the two 1977 and the 1982 great dust storms and the presence of several transient events (T) (figure from Tillman 1988). See also Color Plate 16.

look at meteorological phenomena on time scales ranging from less than a day to interannual. Because surface pressure measurements represent vertically integrated samples of the atmospheric column above the site, they contain information primarily about those components of the circulation which have large vertical scales, and these are typically characterized by large horizontal scales, as well.

Viking Lander 2 operations ended in early 1980 after one and a half Mars years when the Viking Orbiters, having exhausted their attitude control fuel, were no longer able to relay data to Earth; Viking Lander 2 had lost its direct link in October, 1977. Viking Lander 1 made measurements for more than 3 Mars years, from July 1976 until the unfortunate loss of its direct communications link in late 1982. Color Plate 16 overlays the daily averaged surface pressure at both Lander sites and the daily standard deviation at Viking Lander 1 for each Mars year in which data was taken (Tillman 1985). The timelines in Fig. 1 and Plate 16 are given in sols (Mars solar days from the day of landing of Lander 1 or of Lander 2, 45 sols later than Lander 1). The aerocentric longitude of the Sun L_s is also used, with $L_s = 0°, 90°, 180°$ and $270°$ marking the beginning of northern spring, summer, fall and winter, respectively. (Note that, since Mars has a significantly elliptical orbit, L_s is not quite linear in time.)

There are several striking features in the figures. First, the offset between the pressure curves of the two Landers (Fig. 1) is due to the elevation difference ($\approx$ 1.2 km) between them. Second, the highly repeatable large-amplitude seasonal oscillations are due to the condensation of carbon dioxide to form seasonal polar caps during fall and winter and their subsequent sublimation during spring and summer (Fig. 1, Color Plate 16; Hess et al. 1980). Differences between the two seasonal maxima or minima reflect effects of the orbital eccentricity and phasing on the seasonal energy balance of the two polar regions (chapter 27).

Third, the sizeable high-frequency variability of daily mean pressure during fall, winter and spring is due to traveling planetary waves, similar to the mid-latitude storm systems on Earth. These waves were particularly evident at the more northern latitude of Lander 2, but also were prominent at Lander 1 during the third winter of observations (Fig. 1). Fourth, the seasonal variation of the (daily) standard deviation about the sol-mean at Viking Lander 1 (upper panels, Fig. 1; Color Plate 16) is due principally to variations in the amplitudes of thermally driven, once- and twice-per-sol, global oscillations known as the diurnal and semidiurnal atmospheric tides.

Variations in the tidal amplitudes are due largely to variations in the direct solar heating of an atmosphere whose dust content varies seasonally and from year to year. The largest changes in atmospheric dust loading are due to the episodic occurrence of planet-encircling dust storms. Two of these occurred during the first Mars year of observations and another arose during the last days of Lander 1 operations. These events, during which vast dust hazes spread rapidly over the planet and then gradually cleared, are denoted in subsequent sections as the 1977a, 1977b and 1982 great dust storms.

The sudden onset (at $L_s \approx 205°$, 279° and 205°, respectively) and slow clearing of these dust storms are reflected prominently in the daily pressure variance at Lander 1 (Fig. 1), the pressure harmonic amplitudes (Fig. 2) and in the pressure curves themselves. An example of the change of meteorological regimes which can occur (at the surface) during a planet-encircling dust storm is shown in Fig. 3, where the onset of the 1977b great dust storm at the mid-latitude Lander 2 site is marked by transition from a regime of baroclinic waves having several-sol periods to one dominated by semidiurnal and diurnal variations. Note also the rise in pressure (see also Plate 16; $L_s \approx 280°$-320°) and the shift in wind direction during the transition.

B. Temperature

Early Earth-based infrared radiance measurements were interpreted as blackbody temperatures of the surface (Coblentz and Lampland 1927; Sinton and Strong 1960), and showed a large diurnal temperature range consistent not only with a finely powdered surface, but a powdered surface with very low atmospheric pore pressure (Leovy 1966*b*). This thermal behavior was consistent with high-resolution spectra (Kaplan et al. 1964; Connes and Connes 1966) which indicated low atmospheric pressures. Spacecraft data have now detailed the distributions of surface temperature and associated surface properties (chapter 21; Color Plate 18), and have also yielded detailed information on atmospheric temperatures. The most extensive atmospheric temperature determinations have come from the Mariner 9 Infrared Interferometer Spectrometer (IRIS; cf: Hanel et al. 1972*b*), the Viking Orbiter Infrared Thermal Mappers (IRTM; cf Kieffer et al. 1977; T. Martin and Kieffer 1979; T. Martin 1981), and the Mars 2, 3 and 5, Mariner 9 and Viking Orbiter radio occultation experiments (Kerzhanovich 1977; Kliore et al. 1972,1973; Lindal et al. 1979).

These data sets were characterized by very different vertical resolutions and spatial coverages. The IRIS measurements in the 15-μm carbon dioxide band provided temperature profiles up to ≈45 km altitude (≈ 0.1 mbar level) with approximately one scale height vertical resolution. The IRTM, also measuring in the 15-μm band, produced horizontal distributions of vertically integrated temperatures, as defined by a weighting function peaking near the 0.5 mbar level (≈ 28 km) and broadly spanning the 20 to 35 km altitude range (see, e.g., T. Martin 1981). Radio occultation retrievals yield relatively high vertical resolution (≈ 1 km) over the lowest two or three scale heights of the atmosphere, but are few in number and quite restricted in time and space.

Figure 4 shows the distribution of IRIS temperature profiles with latitude and time of year (given by L_s). Together, the large number of profiles cover most latitudes and longitudes, but not systematically. Furthermore, there are

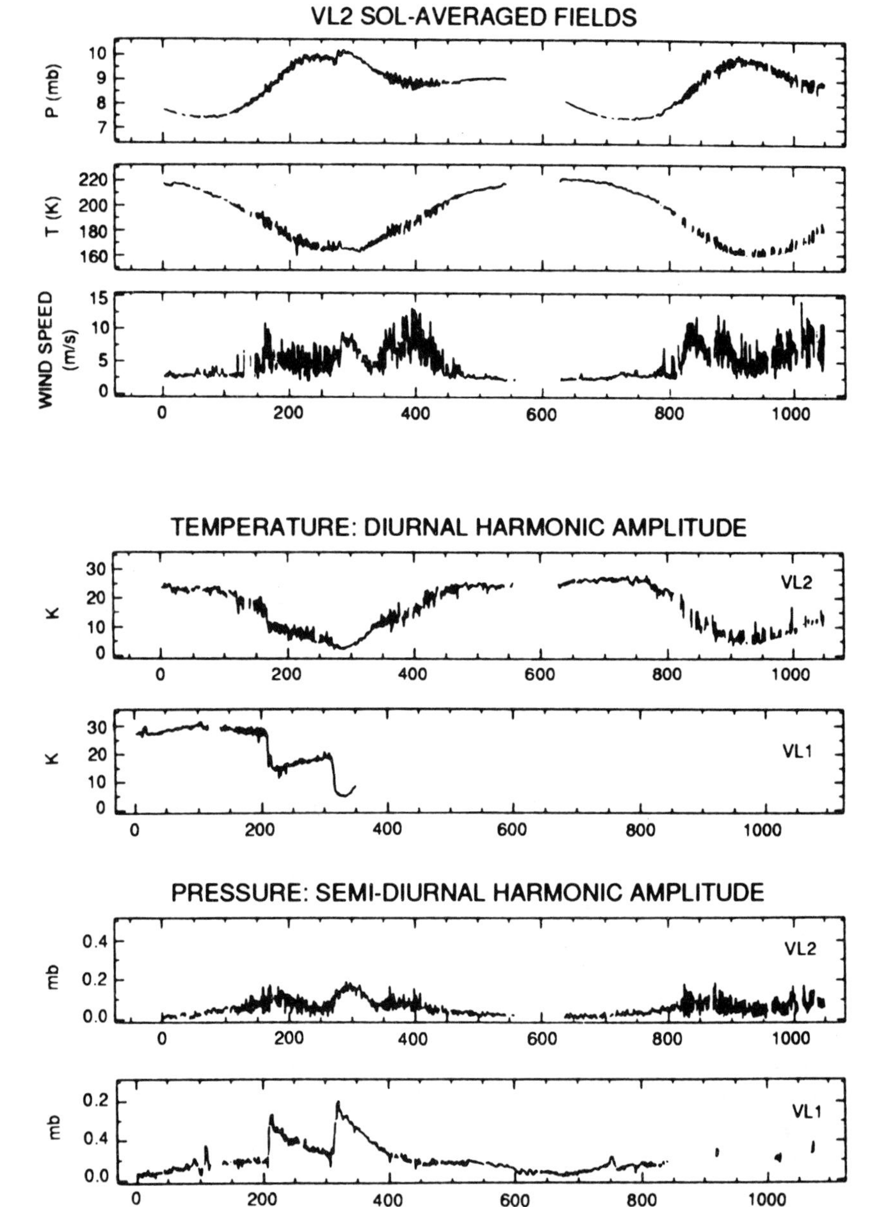

Fig. 2. (Top) Sol-averaged pressure, temperature and wind speed at Viking Lander 2. (Bottom) Diurnal harmonic of temperature and semidiurnal harmonic of pressure at Landers 1 and 2 for the first 1050 sols of observations. Note the saw-toothed signatures of the 1977 dust storms in the semidiurnal harmonic in surface pressure at Lander 1 (figure from Tillman and Johnson).

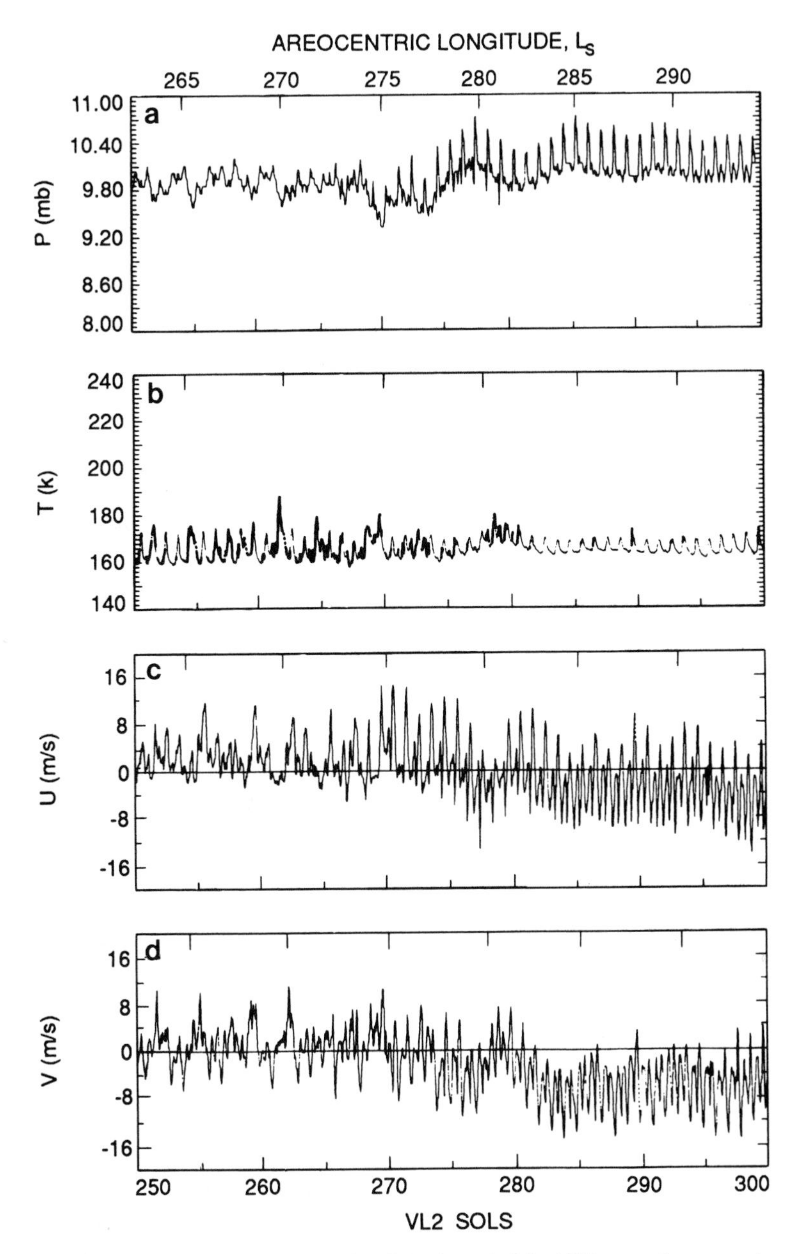

Fig. 3. Viking Lander 2 meteorological data from the end of the 1977a great dust storm through the onset of the 1977b storm: (a) pressure, (b) temperature, (c) eastward wind, (d) northward wind. Dotted line structure indicates missing data replaced by cubic splining and data affected by Lander interference (figure from Tillman and Johnson).

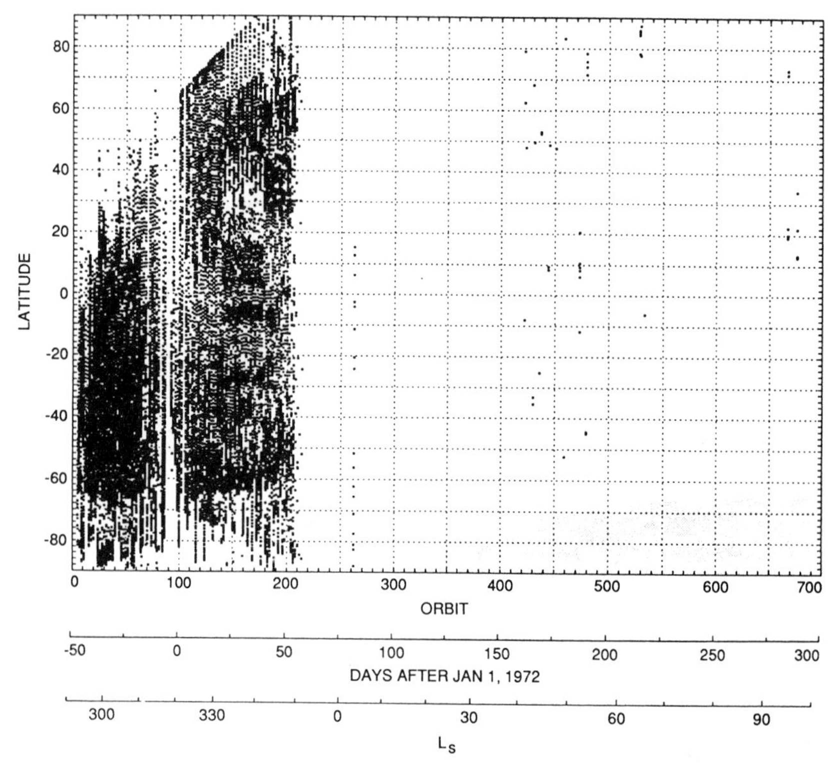

Fig. 4. Latitude-L_s coverage of Mariner 9 IRIS fields of view (filled more than 50% by Mars). Note the sparse coverage after L_s = 350° (figure from Paige et al. 1990a).

only a few seasonal dates at which meaningful meridional cross sections covering most latitudes in both hemispheres can be constructed. None of these are zonal averages, because the meridional cross sections are either aliased by the mixing of latitude and longitude information, or they represent only a single longitude.

Meridional profiles of temperatures are of particular interest because of the information they can potentially provide on zonal winds. Figure 5 shows two cross sections of temperature obtained from Mariner 9 IRIS retrievals. Perhaps the best spatial coverage achieved by IRIS was provided by a daily set of latitude strings across the northern hemisphere during the period L_s = 330 to 350° (late northern winter). A retrieved temperature cross section from this period is shown in Fig. 6. Data from this series were also used by Conrath (1981) to study planetary-scale waves.

Figure 7 shows the latitudinal and seasonal distribution of atmospheric temperatures derived from the IRTM 15-μm channel by averaging data blocked in seasonal and latitudinal bins (data from T. Martin, personal communication; see T. Martin 1981). These data show the seasonal increase in

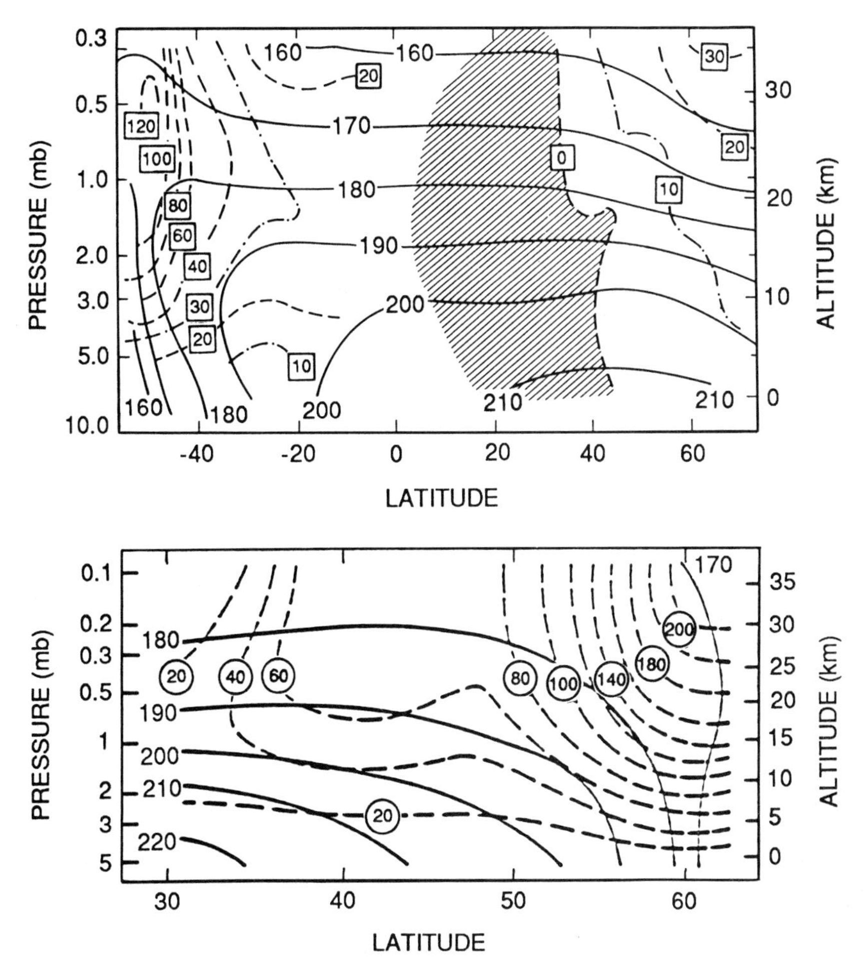

Fig. 5. Atmospheric temperatures (K, solid lines) retrieved from Mariner 9 IRIS data and geostrophic eastward winds (m s^{-1}, dashed lines) computed assuming no wind at the surface. (Top) Latitude-pressure cross section derived for northern midspring, L_s = 43°–54° (orbits 422–473). Shaded area indicates ill-defined region of westward winds. (Bottom) Cross section for late northern winter, L_s = 347° (orbit 192). (Figure from Leovy [1982], generated from data provided by B. Conrath and Mariner 9 IRIS team.)

atmospheric temperature over the thick layer sampled by IRTM as Mars approaches perihelion. This warming is enhanced by the large amount of suspended dust present at this season. The temperature maxima near L_s = 220° and L_s = 270° occur during major dust storms.

Polar warmings were observed shortly after the onsets of the 1977a,b planet-encircling dust storms (T. Martin and Kieffer 1979; T. Martin 1981; Jakosky and T. Martin 1987). During the 1977b storm, IRTM 15-μm brightness temperatures increased during a 30-day period (L_s = 274° – 292°) by at

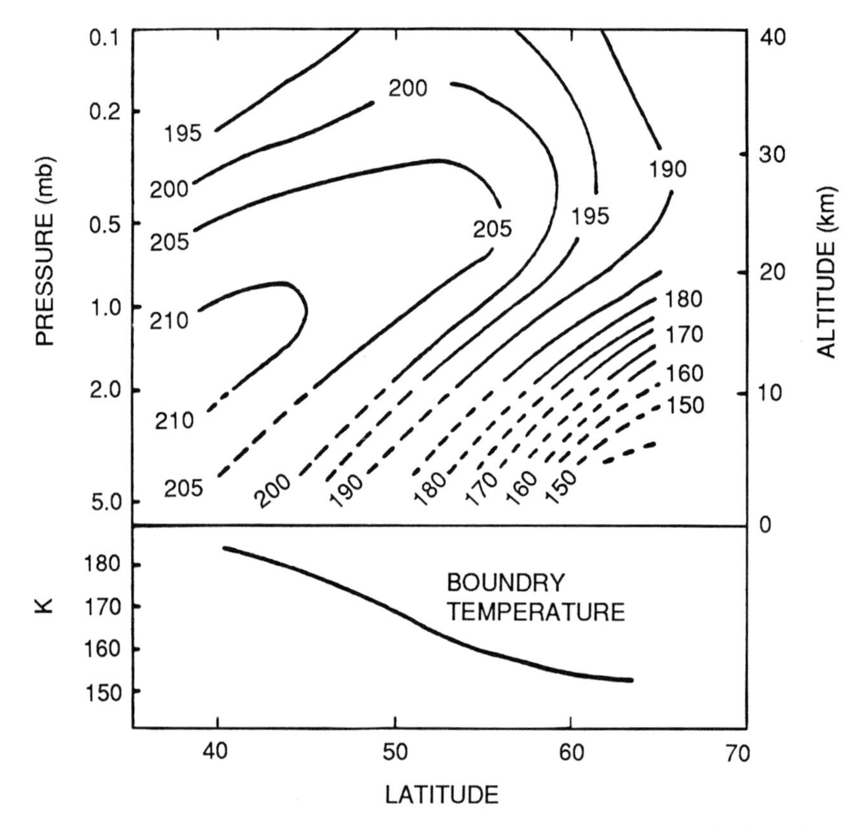

Fig. 6. Height-latitude cross section retrieved from Mariner 9 IRIS data (orbit 102, mid-northern winter) by R. Hanel et al. Dashed lines are uncertain due to possible influence of clouds (figure from Briggs and Leovy 1974).

least 30 K poleward of 45°N, and as much as 80 K at 65°N (Fig. 8). The warming event observed during the first storm of 1977 was not nearly as large (T. Martin 1981). It is clear that, at least for the second storm, some dynamical process is involved, since the warming occurs well into the polar night with temperatures far above their radiative condensation-equilibrium values.

Figure 9 shows several smoothed temperature profiles derived from radio occultation data (original data from Lindal et al. [1979]). Even in these smoothed profiles, there are sharp, shallow temperature inversions that would be difficult to retrieve from Orbiter radiance data. The profiles in Fig. 9 show seasonal changes in the static stability of the lower atmosphere (bottom four profiles), low-level inversions over the north polar seasonal carbon dioxide cap (two profiles at upper left), and very deep and strong temperature inversions during the clearing phase of the 1977b great dust storm (two profiles at

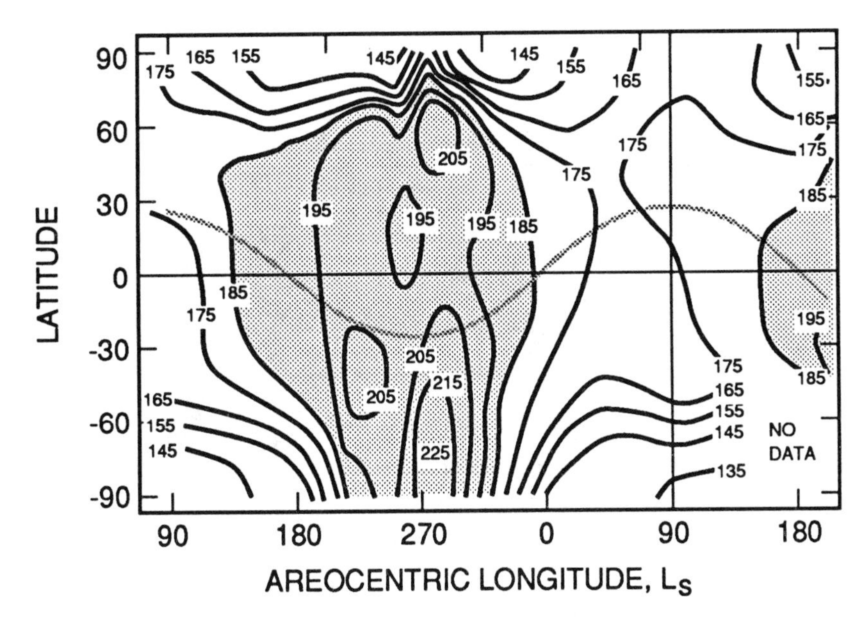

Fig. 7. Latitude-L_S cross section of zonally binned temperatures from the Viking IRTM 15-μm channel, sampling a layer ≈ 20–35 km high in the Martian atmosphere. Dashed line indicates subsolar latitude. The trace of the sub-solar latitude is indicated. (Data from T. Martin, personal communication, as published in Leovy [1985].)

upper right). The deep inversions at mid-latitudes during the decay of the 1977b storm may be due both to local solar heating of the suspended dust and to large-scale subsidence at upper levels.

Temperatures above 40 km have been inferred from Mariner 9 and the Mars 2 and 3 ultraviolet spectrometer measurements of airglow (Stewart et al. 1972; Dementyeva et al. 1972), Mars 5 and Viking entry measurements (Kerzhanovich 1977; Seiff and Kirk 1977; Nier and McElroy 1977), and Mariner 9 and Viking occultation measurements of ionospheric scale height (Kliore et al. 1972; Lindal 1979). Figure 10 shows the vertical temperature distribution derived from Viking entry measurements. Viking arrived near a minimum in the solar cycle, when thermospheric temperatures were relatively low (chapter 30). The large amplitude temperature oscillations in the 20 to 100 km region shown in the figure are probably due to the adiabatic compressional heating and expansional cooling associated with vertically propagating atmospheric thermal tides.

The temperatures of a layer 60 to 80 km high in the atmosphere of Mars have been studied using laser heterodyne techniques (Deming et al. 1986). These results indicate that temperatures at those altitudes are colder at low latitudes than at higher latitudes. This is reminiscent of the temperature dis-

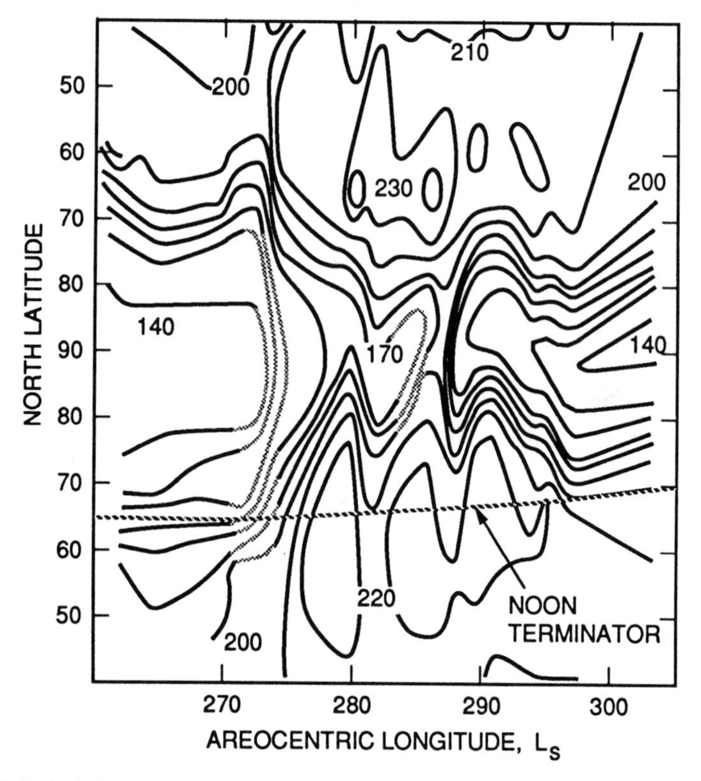

Fig. 8. Latitude-L_s cross section showing polar warming following the onset of the 1977b great dust storm. Upper and lower parts of the diagram are displaced roughly 180° in longitude; i.e., across the pole (figure from Jakosky and T. Martin 1987).

tribution found near 80 km in the Earth's mesosphere, where the breaking of gravity waves propagating from below drives zonal-mean meridional and vertical motions that warm the winter polar atmosphere by compressional (adiabatic) heating due to downward motion there.

The interannual variation of the vertical profile of globally averaged temperatures has been monitored from Earth using microwave techniques (Clancy et al. 1990). These results suggest that the global temperature profile can on occasion be significantly colder than indicated by the Viking entry data. Colder temperatures could occur if the widespread dust haze that was present throughout the Viking mission dissipated at times.

Atmospheric temperatures have been measured at the surface at the two Viking Lander sites. In conjunction with the Lander pressure and wind data, these provide a wealth of information about time-dependent meteorological phenomena. Of particular interest in these measurements are the seasonal

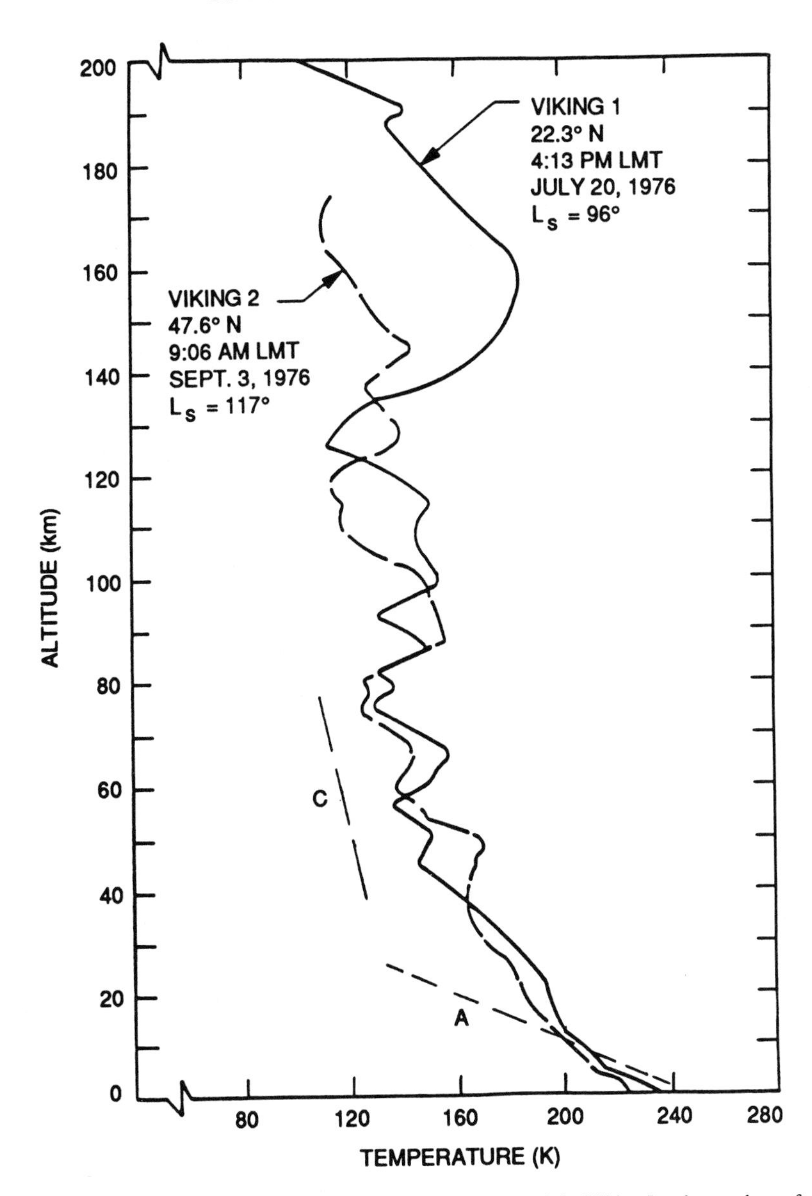

Fig. 9. Temperatures derived from data taken during descent of the Viking Landers to the surface of Mars (early northern summer). Dashed straight lines indicate approximate temperature variations corresponding to adiabatic lapse rates (A) and CO_2 condensation temperatures (C). Note the cold thermosphere and the often anti-correlated wave structure in the two profiles (figure from Seiff and Kirk 1977).

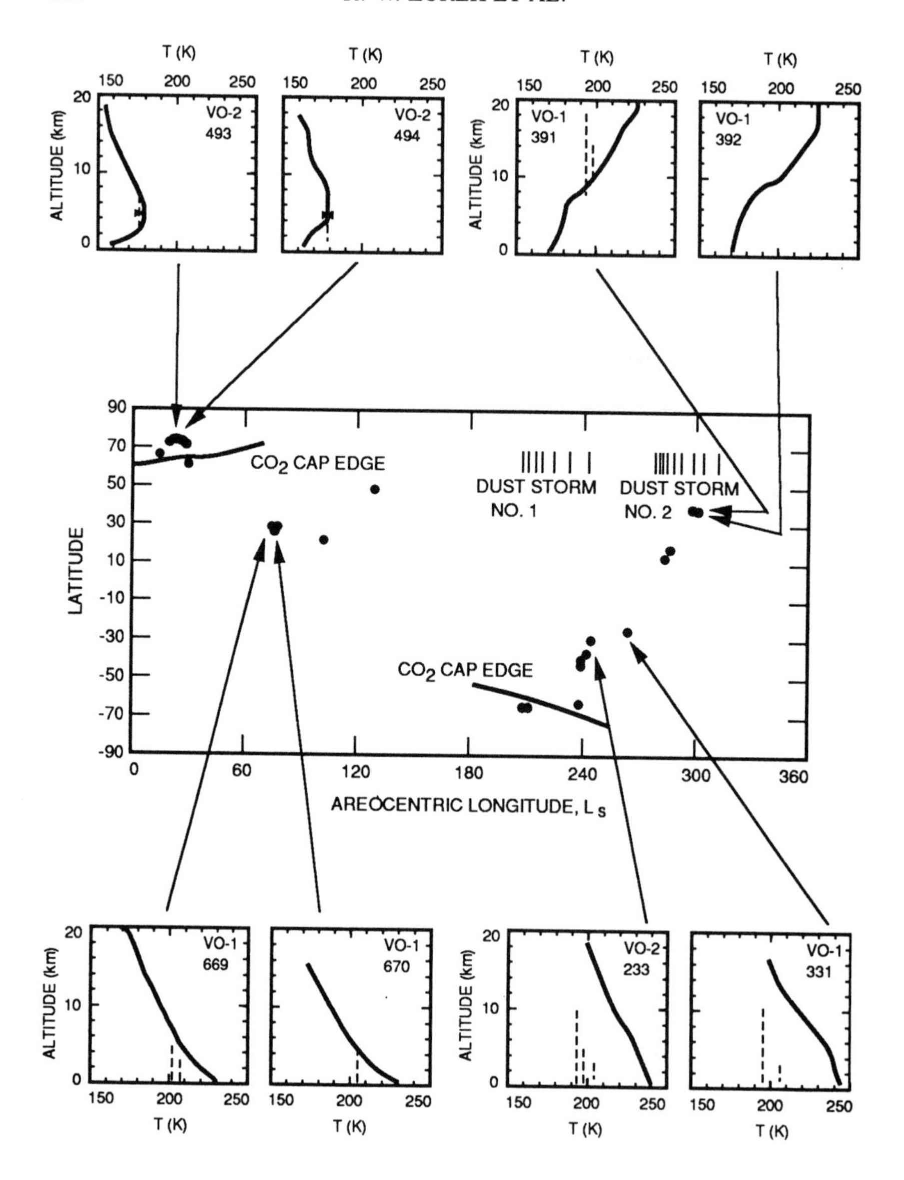

Fig. 10. Smoothed versions of temperature profiles retrieved from Viking Orbiter radio occultation data. Central graph shows latitude-L_s coverage of the analyzed data. Selected profiles are shown. Vertical dashed lines indicate layers where the atmosphere can hold the column abundance of water vapor observed at that time and place by the Viking Orbiter MAWD. Radio-occultation data are from Lindal et al. (1979); figure adapted from Davies (1979*b*). Local times are similar for each pair of occultation profiles: upper left, 3.0 hr; upper right, 19.8 hr; lower left, 16.0 hr; lower right 20.9 hr.

variations of daily mean temperature, the evident dependence of mean temperature and especially daily temperature range on the overhead opacity (Fig. 11), and the short-period variations in daily average temperature. These temperature variations are highly correlated with the high-frequency components of the daily averaged pressure field discussed earlier (e.g., Fig. 3) and are due to the same traveling planetary waves (Tillman 1977; Ryan and Henry 1979; Barnes 1980,1981; Sharman and Ryan 1980; Murphy et al. 1990*b*).

C. Wind

Direct estimates of wind are extremely sparse for Mars. In addition to Viking entry and Lander-site wind measurements, winds have been deduced from cloud motions and morphology (see, e.g., Kahn 1983), from observed spatial gradients of temperature (see, e.g., Leovy 1982), and from the orientation of a variety of aeolian surface features (see, e.g., Thomas et al. 1981). These are all limited by the lack of systematic global coverage. In addition, winds derived from temperature gradients suffer from the lack of

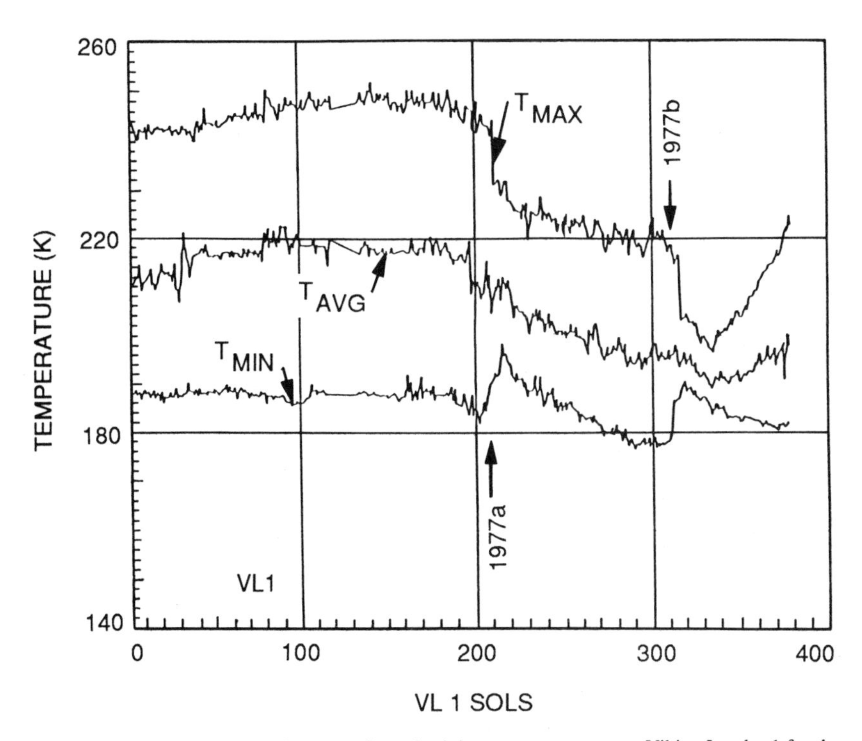

Fig. 11. Maximum, mean (sol-averaged), and minimum temperatures at Viking Lander 1 for the period from landing until after the 1977b great dust storm; times of dust arrival above Lander 1 are indicated for the two (1977a,b) storms. Note the reduced diurnal temperature range during the dust storms (figure from Ryan and Henry 1979).

reference-level winds (see below), while winds derived from cloud observations are limited by the unreliable occurrence of observable targets and the difficulty of assigning altitudes. The lack of a clear understanding of how and when surface aeolian features are formed and modified limits their usefulness as indicators of winds (chapter 22).

1. Direct Wind Measurements. The two Viking entries and the Viking Lander meteorological instruments provided by far the most extensive set of direct wind determinations. Entry wind profiles were obtained by Seiff and Kirk (1977) and Seiff (personal communication, 1989). The Viking Landers measured winds at 1.6 m above the surface (see, e.g., Hess et al. 1976*c*,1977), and analysis of these data have provided considerable insight into large-scale processes (Barnes 1980,1981; Ryan et al. 1978; Tillman et al. 1979; Leovy and Zurek 1979; Ryan and Sharman 1981*a*; Murphy et al. 1990*b*). For instance, the traveling planetary waves that are responsible for the short-period fluctuations in daily averaged surface pressure and temperature discussed earlier are also associated with relatively strong and highly variable winds. This is shown in Fig. 2, which displays the full record of daily averaged fields from Lander 2, where the traveling waves are more prominent.

Representativeness is always an issue for measured surface winds, as they are influenced by boundary layer turbulence, surface roughness, and both regional and local topography, all in a way that is difficult to assess fully. Furthermore, with data only from a single site at a given latitude, it is not possible to distinguish unambiguously between traveling and stationary waves. Even so, winds observed at the Viking 1.6-m height can be extrapolated upward to provide estimates of the geostrophic wind at the top of the boundary layer (see Sec. VI). As geostrophic winds are related to pressure *gradients* (see Sec. II.C.3), observations of pressure and estimates of geostrophic wind allow the inference of the spatial scales (e.g., zonal wavelengths) and propagation (phase) speeds of traveling waves. Barnes (1980,1981) and Murphy et al. (1990*b*) used this approach to determine wavelengths and phase speeds for traveling planetary waves apparently present at each Viking Lander site. The results of these analyses are discussed in Sec. V.A.

2. Winds Derived from Cloud Observations. Early telescopic observers were occasionally able to observe the drift of cloud features (see, e.g., Slipher 1962). Groundbased observers have continued to monitor Mars at favorable oppositions and have obtained some cloud drift measurements (chapter 2). Most cloud drift measurements have been obtained from features classified by observers as "yellow clouds," presumably wind-blown dust (Gifford 1964; L. Martin and Baum 1969). Even in the early data, there was a suggestion of a southern summer cross-equatorial circulation, switching from northeasterly to northwesterly flow in the lowland regions east of Syrtis Major.

Wind direction can be derived from some of the cloud structures seen in Mariner 9 and Viking Orbiter images. Lee wave clouds reveal an unambiguous wind direction, if the obstacle producing the wave can be identified in the images. On Mars, the obstacle is usually a crater rim. An extensive compilation of seasonally dependent wind directions obtained from cloud morphology in Viking and Mariner 9 Orbiter images has been published by Kahn (1983); these wind directions are summarized in Fig. 12. The major features include winter mid-latitude westerlies, particularly in the northern hemisphere; low-latitude wind systems controlled by topography; a seasonally reversing cross-equatorial circulation; and some evidence for outflow winds from the receding south polar cap during spring.

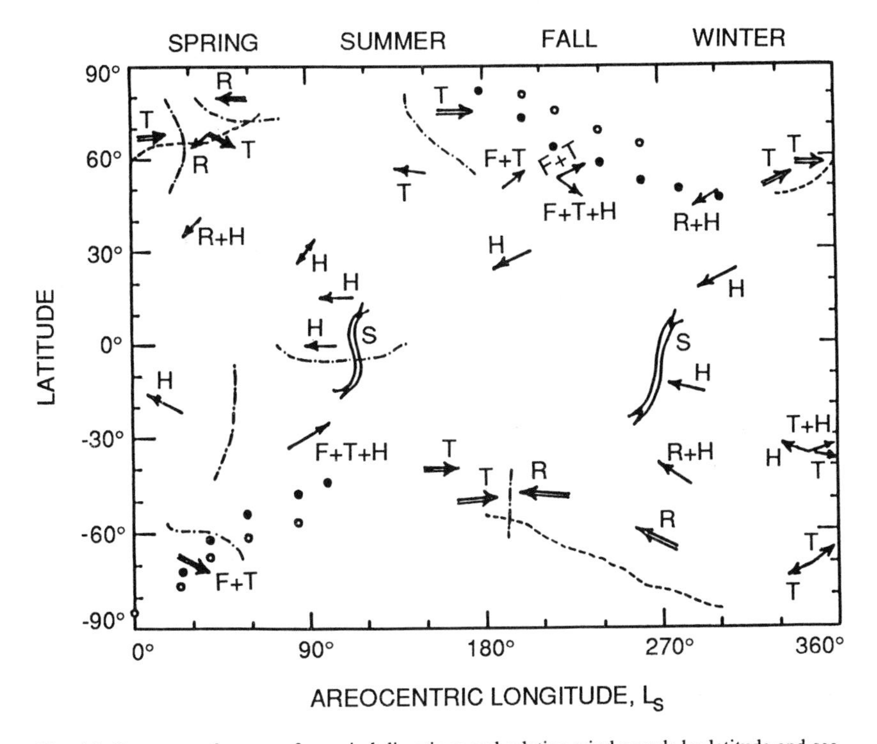

Fig. 12. Summary of near-surface wind directions and relative wind speeds by latitude and season; derived from observations compiled by Kahn (1983). Possible mechanisms for zonal-mean wind generation (see Sec. IV.C) are also indicated: *R/F*, cap recession/formation flow; *T*, large-scale geostrophic wind; *H*, (viscous) Hadley flow; *S*, slope winds. Arrow lengths are arbitrary; double arrows imply relatively strong winds. Dot-dash lines separate regions with different derived wind directions. Dashed lines indicate polar cap edges during recession; open and filled circles indicate two theoretical predictions of the cap edge during formation (figure from Kahn 1984).

3. Winds Derived From Temperatures. Fortunately, temperature, wind and surface pressure are closely coupled in the following sense. Large-scale atmospheric motions, including almost all motions down to scales of a few kilometers, can be assumed to satisfy the hydrostatic equation:

$$-\frac{1}{\rho}\frac{\partial p}{\partial z} = g \equiv \frac{\partial \phi}{\partial z} \tag{1}$$

with pressure p, density ρ, gravity g, gravitational potential (i.e., "geopotential") ϕ and geometric height z. Integrating Eq. (1) from the surface to infinity quickly shows that the surface pressure p_s is proportional to the mass in the overlying vertical column, $\langle\rho\rangle$, i.e., $\langle\rho\rangle = p_s/g$. Thus, surface pressure changes as mass is exchanged with the surface (this is important for Mars, but not for Earth) or as mass is advected horizontally into or out of the local atmospheric column.

The hydrostatic law (Eq. 1) permits a transformation from geometric height to the following pressure coordinate:

$$z^* \equiv H_r \ln \frac{p_r}{p} \tag{2}$$

where p_r is a constant reference pressure and H_r a constant atmospheric scale height, both chosen so that $z^* \approx z$. (Typically for Mars, $p_r = 6.1$ mbar and $H_r = 10$ km.) Using Eq. (2), the equations relating horizontal winds to pressure forces can be written:

$$\frac{\partial u}{\partial t} + \omega \frac{\partial u}{\partial z^*} - \zeta v = -\frac{\partial\left[\phi + \frac{1}{2}\left(u^2 + v^2\right)\right]}{\partial x} + F_x \tag{3}$$

$$\frac{\partial v}{\partial t} + \omega \frac{\partial v}{\partial z^*} + \zeta u = -\frac{\partial\left[\phi + \frac{1}{2}\left(u^2 + v^2\right)\right]}{\partial y} + F_y \tag{4}$$

with time t, vertical velocity $\omega \equiv dz^*/dt$, eastward and northward wind components u, v. In spherical coordinates $dy = a\, d\theta$ and $dx = a \cos\theta\, d\lambda$, with latitude θ, east longitude λ and planetary radius a. (These "horizontal" derivatives are taken along constant z^* and thus constant pressure surfaces.) The quantity ζ is the absolute vorticity

$$\zeta \equiv f + \left\{\frac{\partial v}{\partial x} - \frac{\partial u}{\partial y} + u\frac{\tan\theta}{a}\right\} \tag{5}$$

which includes planetary vorticity, defined by the Coriolis parameter $f \equiv 2\Omega \sin \theta$ with planetary rotation rate Ω, and atmospheric relative vorticity, given by the terms in braces.

In Eqs. (3) and (4) pressure forces per unit mass appear as gradients of ϕ. Atmospheric density has been removed as an explicit variable by using the coordinate defined in Eq. (2); density changes now appear as vertical displacements of the pressure surfaces. The geometric height of those pressure surfaces can be expressed in terms of ϕ which, from Eqs. (1) and (2) and the ideal gas law $p = \rho RT$, can be computed from temperature T as a function of pressure:

$$\phi(t,x,y,z^*) = \phi_{bc} + \int_{z^*_{bc}}^{z^*} R\,T \frac{dz^*}{H_r} . \tag{6}$$

Here R is the molecular gas constant (Table I). In a hydrostatic atmosphere, the temperature distribution completely determines ϕ and thus the pressure distribution, save for a spatially two-dimensional boundary field $\phi_{bc} \equiv \phi(t, x, y, z^* = z^*_{bc})$ at some altitude z^*_{bc}. On Earth, it is convenient and appropriate to take this boundary altitude as mean sea level. On Mars, the 6.1 mbar reference surface can serve a similar purpose, but whatever surface is chosen, the distributions of pressure and gravitational potential (i.e., both z^*_{bc} and ϕ_{bc}) must be known there.

TABLE I
Planetary Constants for Mars and Earth

	PARAMETER	MARS	EARTH	units
a	Planetary radius (equatorial)	3394	6378	km
g	Gravity (surface)	3.72	9.81	m s^{-2}
m	Mean molecular weight	43.4	29.0	g mole^{-1}
R	Molecular gas constant	192	287	J K^{-1}kg^{-1}
c_p	Specific heat at constant p	860	1000	J K^{-1}kg^{-1}
$\kappa = R/c_p$	R/specific heat at constant p	0.223	0.287	
γ	Ratio of specific heats, c_p/c_v	1.3	1.4	
R/g		0.0515	0.0293	km K^{-1}
Δ	Solar day (sol)	88775	86400	s
Ω	Planetary rotation rate	0.7088	0.7294	$\times 10^{-4}$s^{-1}
Year	Sols (Earth days)	669 (687)	365 (365)	
	Orbital eccentricity	0.093	0.017	
	Orbital inclination	25°	23°.5	
	Sun-planet distance	1.38–1.67	0.98–1.02	AU
p_s	Surface pressure[a]	6-8	1013	mbar(hPa)
p_s/g	Atmospheric mass[a]	160-220	10,300	kg m^{-2}.

[a] Surface pressure and atmospheric mass vary seasonally on Mars by 25 to 30%.

In Eqs. (3) and (4), F_x and F_y are "frictional" forces (per unit mass). These are important in the thermosphere where they correspond to molecular viscosity, in the planetary boundary layer where they correspond to turbulent viscosity generated by thermal convection and by wind shear, and in certain regions of the intervening atmosphere (thought to be mainly above 40 km for Mars) where they correspond to the statistical effects of momentum mixing produced by the turbulent eddies generated from "breaking" of vertically propagating gravity waves and atmospheric tides.

Terrestrial experience indicates that often the time rate-of-change terms $\partial u/\partial t$, $\partial v/\partial t$ are small in Eqs. (3) and (4); the major exceptions are the short-period gravity waves and atmospheric tides and during transient events like the onset of major dust storms. Also, vertical advection tends to be small compared to horizontal advection when motions are nearly nondivergent or the atmospheric static stability is high. If so, pressure forces are balanced by Coriolis torques and horizontal momentum advection; the horizontal wind components defined by this nonlinear balance are sometimes called *balance winds*. For long-period and large-scale motions, all advection can often be neglected and the resulting nearly linear balance between Coriolis torques and pressure forces reduces to:

$$u \approx u_g = -f^{*-1}\frac{\partial\phi}{\partial y}; \quad v \approx v_g = f^{*-1}\frac{\partial\phi}{\partial x}. \tag{7}$$

Equation (7) defines the *geostrophic* winds if $f* = f$ and a planetary-scale approximation to the *gradient* winds if the planetary curvature or cyclostrophic term is included: $f* = f + u(\tan\theta/a)$. Computing winds from Eq. (7) is straightforward, and terrestrial experience indicates that they are useful approximations on rapidly rotating planets like Earth and Mars to the true wind for long-period, large-scale motions away from the equator.

Given observations of temperature as a function of pressure on Mars, the shear, or variation with height, of the geostrophic or balance winds can be computed. If the boundary field ϕ_{bc} in Eq. (6), or equivalently the wind at any given (geopotential) height, is known, the horizontal winds themselves can be computed. When ϕ_{bc} is not known, it is typically assumed that $u = v = 0$ at $z = 0$. Evidence that surface winds are small comes from the Viking Lander data (Hess et al. 1977; Ryan and Henry 1979) and from constraints imposed by the observation of resonant gravity waves (Briggs and Leovy 1974). However, modeling studies indicate that this assumption is not valid at all latitudes and seasons (Pollack et al. 1981,1990*a;* Haberle et al. 1982). Furthermore, the wind at the surface is important in its own right, even when it is relatively small compared to the interior field.

Two examples of zonal winds computed geostrophically, assuming negligible surface winds, are shown in Fig. 5. (Because of the limited longitudinal coverage of the Mariner 9 IRIS measurements, some caution is warranted in describing the IRIS cross sections in Fig. 5 as zonally averaged fields.) In later northern winter, winds appear to increase rapidly with height near the edge of the seasonal (CO_2) polar cap, while winds derived from the northern springtime temperatures show generally weak zonal flows throughout the tropics and most of the northern hemisphere. At this season, jet-like features appear only at high southern latitudes, where significant horizontal temperature gradients exist. These seasonal changes have occurred in < 100 sols, with estimated winds in northern mid-latitudes having decreased from more than 200 m s^{-1} in late winter to $\approx$20 m s^{-1} in mid-spring.

When lee waves are observed in conjunction with temperature distributions, the lower boundary wind can be estimated, so that the entire vertical profile of wind can be derived. This is because the occurrence of resonant lee waves, readily observable as a train of uniformly spaced wave clouds in the lee of an obstacle, is controlled by the vertical distribution of the wave propagation (or Scorer) parameter:

$$\xi^2 = \frac{N^2}{(u-c)^2} - j^2 . \tag{8}$$

Here u is the wind velocity normal to the wave crests, c is the wave phase speed (zero for stationary lee waves), j is the observable downwind wavenumber and N is the buoyancy frequency (Table II). Equation (8) assumes that the gravity wave velocity field varies with height z as $e^{i\xi z}$, with $i \equiv \sqrt{(-1)}$ and vertical scale $1/\xi$. For resonant gravity waves to occur, ξ^2 in Eq. (8) must change from positive near the surface to negative aloft; typically, this requires strong wind shear with low values of u near the surface (Briggs and Leovy 1974; Pirraglia 1976). Given temperature information to constrain the vertical distribution of N and the geostrophic wind shear, the near-surface wind can be estimated. An example of a wind profile determined in this way from Mariner 9 data is shown in Fig. 13.

The fact that wind fields can be derived even approximately from knowledge of temperature and of surface pressure is of great importance, since observations of Martian temperatures have extensive spatial coverage, while direct information on winds is rare. Even for future Mars missions, it is the temperature field which is most likely to be measured globally and seasonally by remote sensing. As yet, there are no plans to fly instruments that could measure Martian winds remotely from orbit, although such instruments have been being built to observe the Earth's atmosphere (see, e.g., Reber 1990).

D. Atmospheric Dust

Dust suspended in the atmosphere is a major driver of atmospheric motions at all scales. Its effects on the thermal structure have been amply documented by spacecraft observations (e.g., Figs. 9, 11 and 14). The observed distribution and optical properties of suspended dust are reviewed in chapter

TABLE II
Planetary Parameters for Mars and Earth

	PARAMETER	MARS	EARTH	units
T_e	Planetary equilibrium temp.	210	256	K
$H = RT_e/g$	Mean scale height	10.8	7.5	km
Γ_a	Adiabatic lapse rate[a] $\Gamma_a = g/c_p$	4.5	9.8	K km^{-1}
$\Gamma = -dT/dz$	Mean lapse rate (lower atmos.)	~2.5	6.5	K km^{-1}
N	Brunt-Väisälä frequency	~0.60	1.12	10^{-2} s^{-1}
S	Stability $= N^2H = R(\Gamma_a - \Gamma)$	~0.38	0.95	m s^{-2}
$a\Omega$	Diurnal phase speed (equatorial)	241	465	m s^{-1}
$\sqrt{gH}$	External gravity wave speed	201	271	m s^{-1}
$\sqrt{\gamma gH}$	Speed of sound	229	321	m s^{-1}
NH	Internal gravity wave speed	65	84	m s^{-1}
$4\Omega^2a^2/g$	Lamb's parameter	62.23	88.47	km
$L_r = NH/\Omega$	Rossby deformation radius	920	1150	km
$\sqrt{F_e} = \sqrt{gH}/\Omega a$	External Froude number	0.83	0.58	
$F_i = NH/\Omega a$	Internal Froude number	0.27	0.18	
τ	Atmos. visible optical depth[b]	0.1–10	0.2–100	
ε	Atmos. thermal IR emissivity[b]	0.15–0.8	0.4–1.0	
$c_pT_ep_s/g$	Atmos. heat storage	3.4	260	10^7 J m^{-2}
$\sigma_BT_e^4$	Potential atmos. col. cooling rate	110	240	W m^{-2}
$\varepsilon\sigma_BT_e^4$	with $\varepsilon = 0.32$ (Mars), 0.7 (Earth)	35	170	W m^{-2}
t_c	$(c_pT_ep_s/g)/(\varepsilon\,\sigma_BT_e^4)$	~11	~180	sols
t_r	Rad. damping times (lower atmos.)	~2	>20	sols
$F_r^{-1} = \Omega t_c$		70	1100	
$R_T = F_eF_r$	Thermal Rossby number	0.01	0.0003	

[a] Assumes no condensation.
[b] τ and ε variations on Mars are due to dust storms; on Earth to water clouds.

29, which discusses the Martian dust cycle in detail. Key points with regard to the effect of suspended dust on the atmosphere are: (1) dust is an effective absorber of the incoming solar radiation and both absorbs and emits thermal radiation; (2) the amount and possibly the optical properties of suspended dust are highly variable in space and time; and (3) dust hazes can persist for very long periods. During the Viking mission, visible optical depths of the background dust haze above the Viking Lander sites always exceeded a few tenths (Colburn et al. 1989; chapter 29), and Orbiter imaging suggests that this may have been true over much of the planet (see, e.g., Thorpe 1977).

E. Condensate Clouds and Hazes

Condensate clouds and hazes known as polar hoods have been observed to cover both polar regions during winter. White clouds have also been observed in the tropics during the northern summer, particularly over the Tharsis

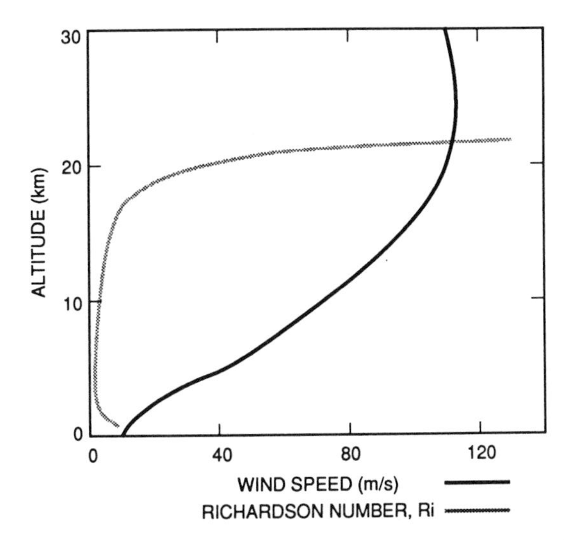

Fig. 13. Vertical profile of the zonal wind U and the Richardson number $Ri = N^2(\partial U/\partial z)^{-2}$, both calculated using Mariner 9 IRIS temperatures for 45°–50°N, as shown in Fig. 6 (mid-northern winter). Surface wind of $U_s \approx 10$ m s^{-1} was inferred from the character of an observed lee wave (Sec. II.C.3). Regions of small Ri may indicate presence of shear-induced turbulence (figure from Briggs and Leovy 1974).

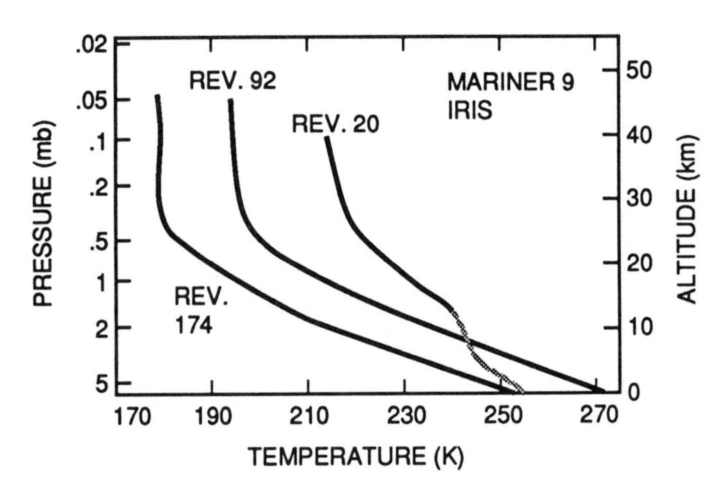

Fig. 14. Vertical profiles of temperature retrieved from Mariner 9 IRIS data taken in early to midsummer near 30°S during the clearing phase of the 1971 global dust storm. Dust opacities were lower for later observations, indicated by larger orbit (rev) numbers (Hanel et al. 1972*b*).

and Elysium uplands, and during the afternoon (chapter 2; Leovy et al. 1971,1973*a*; Smith and Smith 1972; Briggs et al. 1977). Analysis of IRIS and IRTM observations of clouds over the Tharsis plateau region indicate that they are composed of micron-sized water-ice particles (Curran et al. 1973; Kieffer et al. 1977; Christensen and Zurek 1984).

Cloud motions and morphology observed from spacecraft have also been used to infer characteristics of the large-scale circulation and atmospheric stability (Leovy et al. 1972,1973*a*; Briggs and Leovy 1974; Briggs et al. 1977; French et al. 1981; Kahn 1983,1984). Using Mariner 9 and Viking imaging data, French et al. (1981) systematically cataloged cloud features as various types, including fogs, plumes, streaky clouds, cloud streets, wave clouds and lee waves; they then examined the spatial and seasonal variability of these types. Kahn (1984) rebuilt this catalog, using all Mariner 9 and Viking Orbiter images and adding three categories of thin, moderate and thick haze; mosaics were used to define the context of features in individual images. Kahn then used this data base to constrain winds, static stability and, of course, humidity.

Water-ice clouds occur predominantly in two regional and seasonal groups, in the northern subpolar region during winter and over tropical uplands during northern summer. They also occur in the southern subpolar region during southern winter, but are much less common than in the north. Wave clouds, particularly lee wave clouds, are common in the subpolar regions and during morning hours in the tropical regions. Streaky and patchy cloud forms of uncertain origin appear in both regions. Based on observed temperatures, these clouds are inferred to be water ice, and they comprise at least the lower latitude portions of the polar hoods. Carbon dioxide clouds probably occur in the more poleward portions of the polar hood regions, as well as at high levels in other latitudes and seasons (Leovy et al. 1972; Pollack et al. 1990*a*). Limb hazes have been seen (Leovy et al. 1971; Anderson and Leovy 1978; Jaquin et al. 1986), and these may be composed of either water ice or carbon dioxide ice.

The water cycle is discussed in chapter 28. For atmospheric dynamics, the key point is that water vapor is present in amounts too small (Farmer et al. 1977; Jakosky and Farmer 1982) to affect atmospheric thermodynamics through latent heat release. This is true even though the relative humidity of the thin, cold Martian atmosphere is relatively high. The dashed vertical lines in Fig. 9 indicate layers warm enough to hold the entire amount of column water vapor observed at those times and places (Davies 1979*b*). Generally, the southern hemisphere is relatively dry, while northern hemisphere values are closer to saturation. Dynamically, water vapor may be most useful as a tracer of atmospheric motions (see, e.g., Haberle and Jakosky 1990).

Clouds appearing less frequently, but still of meteorological interest, include early morning clouds resembling atmospheric solitary waves in the vicinity of the Tharsis volcanoes (Pickersgill and Hunt 1981; Kahn and Gierasch 1982). Unique images of mesoscale spiral cloud structures, resembling the cloud forms in terrestrial polar cyclones, have been observed near the periphery of the summer residual north polar cap (Gierasch et al. 1979; Hunt and James 1979; see chapter 3, Fig. 11).

III. ATMOSPHERIC DIABATIC HEATING

On Mars, radiation is the primary source of diabatic heating and cooling of the atmosphere, except in the polar night where the latent heating of CO_2 is important. When present, suspended dust is the dominant absorber and scatterer of sunlight in the atmosphere. In the absence of dust, carbon dioxide is the main atmospheric absorber of sunlight. In the thermal infrared, dust and carbon dioxide are again the dominant sources of opacity. The Martian atmosphere is too thin for molecular scattering to be of great importance; it has too little ozone or water vapor for either to influence solar or thermal radiation very strongly. While the radiative effects of condensate clouds are not well constrained, observations of cloud extent and theoretical calculations of cloud radiative properties indicate that condensate clouds are likely to be most important in the winter polar regions.

A. Radiative Heating and Cooling in the Clear Atmosphere

A few percent of the incident sunlight is absorbed by the near-infrared bands of gaseous carbon dioxide, while the 15-μm vibrational fundamental dominates the transfer of thermal radiation. Atmospheric transmission has been evaluated for Mars using classical band models (see, e.g., Gierasch and Goody 1968), as well as more detailed models based on line-by-line calculations (see, e.g., Pollack et al. 1990*a;* Crisp 1990). Doppler broadening of the individual rotational lines becomes increasingly important with height (particularly above 20 km). In the middle atmosphere (45 to 110 km; see chapter 24), collisions become too infrequent to maintain local thermodynamic equilibrium (LTE) for the major CO_2 vibrational bands. However, Bougher and Dickinson (1988) showed that the 15-μm fundamental is sufficiently opaque that photons that are emitted before being thermalized are locally reabsorbed, and so LTE for this band holds up to $\approx$90 km. Furthermore, near-infrared heating above 50 km consists mainly of thermalization of 15-μm vibrational quanta, so that both heating and cooling at these altitudes are coupled to the CO_2 15-μm source term.

One means of testing radiative transfer models for the Martian atmosphere is to calculate radiative equilibrium temperatures by balancing solar and thermal radiative heating rates as a function of altitude. Typically, this is done with a one-dimensional radiative-convective model, which allows for thermal (free) convection by performing a convective adjustment (see e.g., Gierasch and Goody 1967,1968) to remove superadiabatic regions, while preserving the total energy of the adjusted region. As expected, such adjustments are most commonly needed to simulate the lowest regions of the atmosphere during the warmer times of day, when significant amounts of heat are being convectively exchanged between the atmosphere and the ground. However, the radiative time constants for the thin CO_2 Martian atmosphere are quite short by terrestrial standards ($\approx$2 days; Goody and Belton 1967), and this means that radiative exchange between the atmosphere and surface will be important and that there will be large diurnal changes in the atmosphere driven by the daily heating and cooling of the surface (Gierasch and Goody 1967,1968).

B. Radiative Heating and Cooling in a Dusty Atmosphere

The effects of solar radiation absorbed by suspended dust were first included in a one-dimensional model by Gierasch and Goody (1972). For very dusty conditions, they obtained the nearly isothermal vertical profiles and much enhanced diurnal range of temperatures (away from the surface) observed by the Mariner 9 instruments (Fig. 14). Moriyama (1974,1975) noted that suspended dust would affect both solar heating and thermal cooling of the atmosphere. His one-dimensional simulation included both absorbing and scattering properties of the suspended dust, as computed using the optical constants of Earth analogs (Moriyama 1976). Among other things, Moriyama found that strong daytime inversions could occur given paths with large optical depth and that the diurnal range of surface temperatures should be significantly constricted due to reduced solar heating of the ground and greater downward thermal infrared flux at night. Both effects were observed when planet-wide dust storms occurred during the Viking mission (Fig. 9, upper right-hand corner; Fig. 11).

Pollack et al. (1979*b*) used the optical properties of dust derived from Mariner 9 and Viking Lander data in a one-dimensional radiative-convective model. Results from that model are compared in Fig. 15 to temperatures derived from data taken during descent of Viking Lander 1 to the surface (Seiff and Kirk 1977). The (dashed) theoretical curve at 1600 follows the observed values in the bottom 5 km of the atmosphere, which corresponds roughly with the planetary boundary layer; at higher altitudes the curve deviates from the observed values by as much as 20 K. A good match at these heights can be obtained if one allows for adiabatic cooling due to the rising motions in the ascending branch of the Martian Hadley circulation that may exist over the northern subtropics in early summer. This comparison demonstrates both the limitations of a one-dimensional model and the sensitivity of atmospheric temperatures to large-scale motions. Simulations with the Mars GCM (see Sec. IV.C) have shown the potential impact of global dust hazes of various opacities (Pollack et al. 1990*a*). Obviously, suspended dust provides a potent thermodynamic drive for the atmosphere of Mars.

Radiative heating rate calculations have not yet included rigorously the joint action of dust and carbon dioxide on the transfer of radiation through

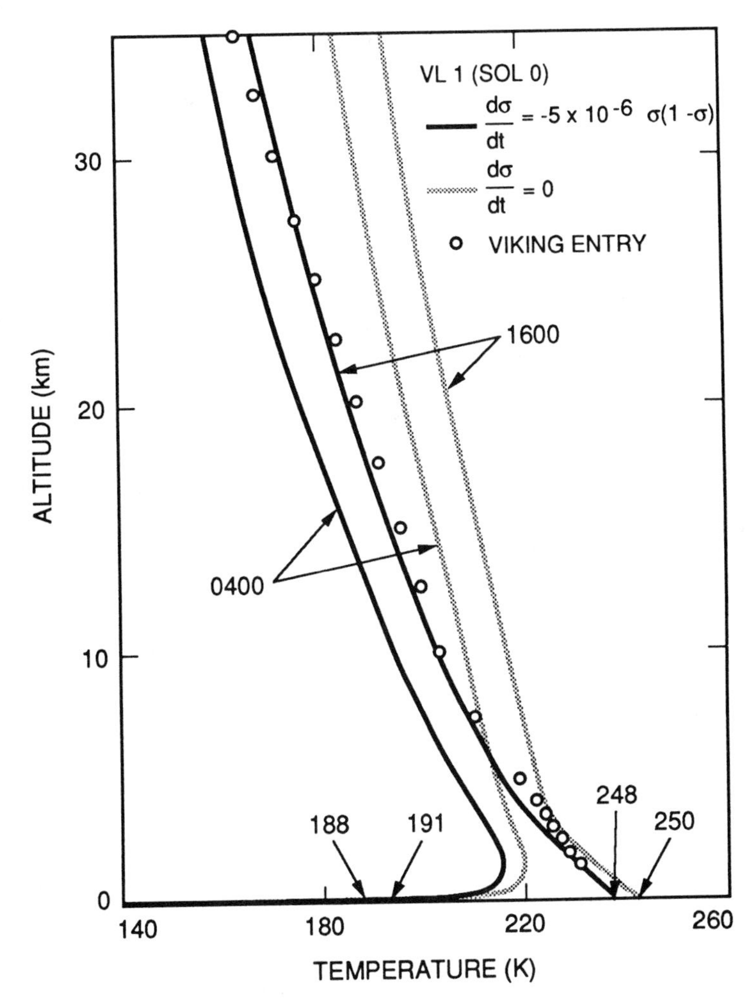

Fig. 15. Temperature profiles computed for two local times by a one-dimensional radiative-convective model for early summer at the subtropical Lander 1 site without (light lines) and with (dark lines) inclusion of a simple profile of adiabatic cooling due to regional upwelling ($d\sigma/dt < 0$, where σ = air pressure normalized by surface air pressure). Profiles were computed assuming a background dust opacity of 0.5 in the visible. Ground temperatures are indicated on the abscissa. Below 10 km the circles indicate temperatures derived from Viking Lander 1 entry data, taken at ≈ 16:00 local time; at higher altitudes the circles represent a smoothly varying "mean" profile, which ignores the vertical wave structure shown in Fig. 10 (figure from Pollack et al. 1979*b*).

the dusty Martian atmosphere. Rather, the influences of atmospheric gases and dust have been evaluated separately and then combined in reasonable, but approximate, ways (see, e.g., Moriyama 1976; Haberle et al. 1982; Pollack et al. 1979*b*, 1990*a*). The major uncertainties in the calculation of the radiative effects of dust suspended in the atmosphere, however, are due to the lack of information about the distribution and the optical properties of the dust itself. The amount of suspended dust is known to vary tremendously with both space and time (chapter 29). It is likely that the size distribution of the suspended dust, and perhaps its composition as well, will vary also (see, e.g., Murphy et al. 1990*a*). If so, the radiative properties of the suspended dust must change with time and space as well (chapter 29).

C. Other Atmospheric Heating Effects

Unlike Earth, ozone on Mars is not generally present in concentrations great enough to contribute significantly to atmospheric radiation, except perhaps at the edge of the polar night (Kuhn et al. 1979; Lindner 1990). Latitudinally, ozone is anti-correlated with water vapor, as photolysis of the latter produces hydrogen-containing species that are catalysts for ozone destruction. Thus, ozone is a maximum near the polar night where photolysis can lead to the generation of ozone, but the atmosphere is sufficiently cold that little water vapor is present to lead to its catalytic destruction (see chapter 30).

Water vapor is also present in amounts too small to affect significantly atmospheric heating and cooling either through its direct radiative effects at infrared wavelengths or through latent heating. To contrast the likely latent heating by condensation of water vapor to the energy absorbed directly by dust in the Martian atmosphere, let α be the fraction of the incoming solar radiation I_0 absorbed by suspended dust; α increases as the dust visible optical depth τ increases. For the optical parameters derived by Pollack et al. (1979*b*), a delta-Eddington solar heating code yields estimates of $\alpha \approx 7\%$ for $\tau \approx 0.5$, typical of the background dust haze, and $\alpha \approx 24\%$ for $\tau \approx 1.5$, typical of dustier periods (Zurek 1978). By contrast, in a clear CO_2 atmosphere only a few percent of the insolation is absorbed by the atmosphere; the rest goes to the surface.

Now, suppose an amount M_w of water vapor (in precipitable microns [pr μm]) is condensed in the atmosphere. The time δ required for atmospheric radiative heating to supply a comparable amount of heat can be defined by:

$$\delta[\mathrm{min}] \equiv \frac{M_w\, L_w}{\alpha \dfrac{I_0}{\pi}} \left(10^{-3}\ \mathrm{kg\ m^{-2}}\right) \approx 0.25\, \frac{M_w[\mathrm{pr}\ \mu\mathrm{m}]}{\alpha} \tag{9}$$

with the latent heat of sublimation L_w and with the insolation averaged over a day approximated by $\alpha\, I_0\, \pi^{-1}$, where I_0 is the incident solar flux at Mars.

Typical values are $\alpha \approx 0.1$, appropriate to a relatively clear atmosphere, $I_0 \approx 600$ W m^{-2} and $M_w \approx 2$ pr μm, equivalent to several percent of a typical atmospheric column abundance at low latitudes or to enough water ice to make an optically opaque cloud (see, e.g., Curran et al. 1973). These values yield $\delta \approx 5$ min, equivalent to $< 1\%$ of the sol-integrated radiative heating of even a relatively clear atmosphere.

The formation of ice clouds can affect radiative heating of the atmosphere and surface, and this provides a mechanism for atmospheric water to affect the circulation. Because the extent of ice clouds, except for the polar hoods, is quite limited in space and time, it has been generally assumed that their effect on large-scale motions is not significant. Optically thick ice clouds are most extensive and persistent in the polar night above the condensing seasonal polar cap. These polar hood clouds, whatever their mixture of water and carbon dioxide ice, affect the polar radiative balance by increasing the downward infrared radiative flux to the surface and thus reducing the rate of condensation of CO_2 in winter (see, e.g., Pollack et al. 1990*a;* Sec. IV.C). By scavenging suspended dust from the atmosphere, ice-cloud formation can affect atmospheric cooling in the polar night, while the scavenged dust may alter the surface visible albedo and thus the rate of CO_2 sublimation when sunlight returns to these regions at the end of winter (Pollack et al. 1979*b*,1990*a;* Paige and Ingersoll 1985; chapter 27).

IV. THE GENERAL CIRCULATION

The term "general circulation" denotes the larger-scale aspects of the atmospheric circulation, thermal structure and pressure distribution, particularly when averaged longitudinally and/or when averaged over periods long enough to remove variations associated with daily fluctuations and with individual weather systems, but still short compared to seasonal variations.

As on Earth (see, e.g., Lorenz 1967), the main features of the general circulation include zonally averaged zonal winds and meridional circulations, quasi-stationary planetary waves, and seasonally varying regions of significant activity due to traveling waves. The effects of smaller scale components of the flow, like internal gravity waves, must also be considered because of their great potential for altering the large-scale flows through momentum transport. Because of the fast response of the Martian atmosphere to radiative heating, the effects of thermally driven short-period, but large-scale atmospheric tides must also be considered. There is also one major component on Mars not present on Earth: the condensation flow associated with the seasonal condensation and sublimation in the polar regions of Mars of a large fraction of the total atmospheric mass.

For Mars, as for Earth, the general circulation is fundamentally driven by differential solar heating. Because of the short radiative response time of the Martian atmosphere, the low thermal inertia of its surface, the large eccentricity (0.093) and obliquity of its rotational axis (25°), and the episodic occurrence of great dust storms, the diabatic forcing and the circulation it drives vary significantly on diurnal, monthly, seasonal and interannual time scales. These changes are substantially greater than those in the ocean-buffered troposphere of the Earth. They are also larger than those found in the terrestrial stratosphere, where radiative heating is dominated by an ozone layer whose distribution is far less variable with time than is that of suspended dust on Mars.

This section summarizes our present understanding of the general circulation of the Martian atmosphere, with emphasis on the region above the boundary layer. (For a historical overview, see Leovy [1979,1985].) At all scales, much of our current understanding has come from modeling studies of various types. The sparseness of the current meteorological data for Mars does not allow the general circulation to be defined diagnostically, to the extent that it has been defined on Earth. However, the available data provide powerful insights into many aspects of the general circulation when viewed through the constraints of angular momentum, mass and energy balance of a shallow atmosphere on a rapidly rotating planet. These balance constraints are expressed mathematically to varying degrees of completeness in models ranging from two- or three-dimensional process models, up to the most comprehensive of the existing general circulation models (GCMs).

The GCM is an especially powerful, although unwieldy, tool when it combines fully three-dimensional, nonlinear mathematical representations of atmospheric dynamical and thermodynamical processes with consistent treatments of radiative processes and of surface influences. Ideally, a GCM applies to the entire globe and is based on the meteorological primitive equations, so-called because only the hydrostatic assumption, and not higher-order approximations (e.g., geostrophic), has been made.

Following convention, the discussion of the general circulation is divided here into one of zonal-mean (i.e., zonally averaged or zonally symmetric) flow components and one of departures from the zonal-mean flow, denoted as eddies or waves. The large-scale waves themselves are discussed in Sec. V. When wave activity is prominent, the waves and the zonal-mean flow are strongly coupled through the nonlinearity of atmospheric advection of momentum and heat.

In this section the zonal-mean circulation is discussed in some detail, making liberal use of the body of theory developed to understand such circulations in the atmosphere of Earth. Theory (including numerical simulation) suggests that the zonally symmetric component of the Mars atmospheric circulation may be a more vigorous component of the general circulation than is its terrestrial counterpart. The zonal-mean effects of the longitudinally varying components of the circulation are then discussed. This section then closes with a discussion of the zonal-mean budgets of momentum, mass and heat, as simulated numerically by a recent version of the Mars GCM developed at NASA Ames Research Center.

A. The Zonal-Mean Circulation

The zonal-mean state of the Martian atmosphere consists of the variation with height (or pressure), latitude and time of the zonal-mean zonal wind $\bar{u}$, meridional wind $\bar{v}$, vertical wind $\bar{\omega}$, and temperature. Overbars indicate longitudinal averages. These fields are not independent of one another, because of conservation of momentum, mass and heat. In the z^* coordinate system (Eq. 2), these conservation laws are given by the zonally averaged versions of Eqs. (3), (4), the zonally averaged mass continuity equation

$$\frac{\partial \bar{v}}{\partial y} - \bar{v}\,\frac{\tan\theta}{a} + \frac{\partial \bar{\omega}}{\partial z^*} - \frac{\bar{\omega}}{H_r} = 0 \tag{10}$$

and the zonally averaged thermodynamic energy equation

$$R\frac{\partial \bar{T}}{\partial t} + \bar{v}\,R\frac{\partial \bar{T}}{\partial y} + S\bar{\omega} = \kappa\,\bar{Q} - E_T \tag{11}$$

with $S \equiv R\,(\kappa\bar{T}/H_r + d\bar{T}/dz^*)$ the zonal-mean static stability, $\kappa \equiv R/c_p$ the ratio of gas constant to specific heat at constant pressure, $\bar{Q}$ the zonal-mean diabatic heating rate per unit mass (due to net radiation, sensible heat transfer and latent heating by CO_2 condensation) and E_T the combined divergences of zonal-mean vertical and horizontal heat fluxes due to waves of all scales, from planetary waves to turbulent eddies. The zonal average of the zonal momentum Eq. (3) can be written in terms of the total (planetary surface rotational plus atmospheric) angular momentum per unit mass

$$M \equiv \Omega\,a^2\cos^2\theta + \bar{u}\,a\cos\theta \tag{12}$$

and for steady-state conditions, as follows (in log-pressure z^* coordinates):

$$\nabla\cdot(\mathbf{v}\,M) = -E_M\,. \tag{13}$$

Here $\mathbf{v} \equiv (\bar{v},\bar{\omega})$ is the vector of zonal-mean meridional and vertical velocities, while E_M represents the divergence of zonal-mean fluxes of angular momentum per unit mass, due to waves and to viscous effects. The constraint of thermal wind balance is more approximate and is derived using Eq. (6) in Eq. (7), the geostrophic or gradient wind approximation to Eq. (4). Zonal averaging then yields:

$$f_* \, \bar{u}(z^*) = f_* \, \bar{u}(z^*_{bc}) - \int_{z^*_{bc}}^{z^*} R \frac{\partial \bar{T}}{\partial y} \frac{dz'}{H_r} \; . \tag{14}$$

While this thermal wind approach can be used to estimate zonal-mean zonal winds, it cannot be used to determine the mean meridional circulation **v** because the zonally averaged geostrophic wind has no meridional or vertical component.

In the study of the zonally symmetric circulation of an hydrostatic atmospheric on a rapidly rotating planet, the central question is then: By what mechanisms are the twin constraints of angular momentum conservation (Eq. 13) and thermal wind balance (Eq. 14) satisfied? In an atmosphere whose major gases are not subject to condensation and sublimation there are two mechanisms for driving the mean meridional circulation: (1) thermal forcing by zonal-mean heating (i.e., $\bar{Q}$ and E_T); and (2) mechanical forcing by zonal-mean components of the eddy fluxes of momentum (i.e., E_M).

For Earth's atmosphere, there is often a high degree of compensation between the large-scale eddy contribution to heating (which depends upon the zonal-mean correlation of the wave temperature and meridional or vertical velocity fields) and the contribution of advection by the mean meridional circulation. In addition, mechanical forcing tends to drive poleward (equatorward) motions in regions of wave momentum flux divergence (flux convergence); these zonal-mean motions then compensate the wave-induced torques per unit mass through advection of the planet's rotational angular momentum.

For this and other reasons, it proves to be convenient to transform the components of the Eulerian mean meridional circulation **v** by subtracting the thermodynamic effects of the large-scale eddy heat transport. The resulting circulation (v^*, ω^*), called the residual-mean meridional circulation, can be computed by replacing $\mathbf{v} \equiv (\bar{v},\bar{\omega})$ by (v^*,ω^*) in Eq. (11) with E_T set to zero and in Eq. (13) with E_M transformed into a combination of both heat and momentum flux divergences due to waves called the Eliassen-Palm flux divergence (see, e.g., Boyd 1976; Edmon et al. 1980). This residual-mean circulation has two practical advantages (see the discussion in Andrews et al. [1987]).

First, it can be derived directly from energy balance and mass continuity (Eqs. 10–11) using observed temperatures, if the diabatic heating can be accurately computed and if, as suspected, the vertical component of the wave flux divergence E_T is negligible. When the Richardson number is large (i.e., the static stability is large compared to the square of the vertical shear of the wind), the situation is even simpler as Eq. (11) can then be approximated by

$$\omega^* \cong \kappa \, \bar{Q}/S. \tag{15}$$

(In this case the circulation v^*, ω^* is sometimes called the diabatic circulation.) In stably stratified atmospheres such as that of Mars, then, thermal forcing tends to drive upward motions in heated regions and downward motions in cooled regions so that the thermal forcing is compensated by expansional cooling or compressional heating. Second, for a wide range of conditions, the residual-mean circulation is a good approximation to the Lagrangian mean meridional circulation, which fully describes the transport of fluid parcels in the meridional plane (Andrews and McIntyre 1978).

In Earth's troposphere, the low-latitude zonal-mean meridional circulation is primarily thermally driven and is referred to as the Hadley circulation. A much weaker mean meridional circulation, called the Ferrel circulation, is sometimes found at high latitudes and is mechanically driven (see, e.g., Lorenz 1967). In Earth's upper stratosphere and mesosphere, the dominant meridional circulation is a seasonally reversing pole-to-pole flow driven by a complex interplay between thermal and mechanical forcing (see, e.g., Andrews et al. 1987).

On Mars, thermal driving of the lower atmosphere by radiation and convection is generally much stronger than by the large-scale eddy flux of heat, so that the distinction between the Eulerian and residual-mean circulations is comparatively small. Given future observations of the global fields of suspended dust concentration and $\bar{T}$, the diabatic forcing $\bar{Q}$ can be computed and the residual-mean circulation determined directly from energy balance and mass continuity (Eqs. 10 and 11). This approach has been used to estimate the zonal-mean circulation of the middle atmosphere of the Earth (see, e.g., Hartmann 1976; Hitchman and Leovy 1986; Gille et al. 1987).

Aside from the two processes described above, the seasonal condensation and sublimation of CO_2 in the polar regions of Mars provides a unique mechanism for forcing mean meridional circulations. The injection and removal of mass drives a significant pole-to-pole circulation which reverses seasonally. As discussed below, this mass flow appears to be an important component of the general circulation of the Martian atmosphere.

On Earth, the main features of the zonal-mean circulation are provided by longitudinally averaging the observed horizontal wind and temperature fields and then computing zonal-mean vertical winds from mass continuity (see, e.g., Oort and Rasmusson 1971; Oort 1983). For Mars, directly measured winds are limited to two sites and two entry profiles, while derived winds suffer from incomplete observational coverage (Sec. II.C).

The existence of the zonal-mean meridional circulation on Mars is inferred from the large excess of winter mid-latitude temperatures above their radiative equilibrium values (Figs. 5 and 16), from the few wind directions determined from cloud drifts and morphology (Kahn 1983; Fig. 12), from the zonal winds derived from latitudinal temperature gradients (see, e.g., Leovy 1982; Fig. 5), and from the observed distribution of so-called "Type I(b)" bright streaks, believed to reflect the directions of near-surface winds during the decay of the planet-wide dust storms in 1977 (see chapters 22 and 29).

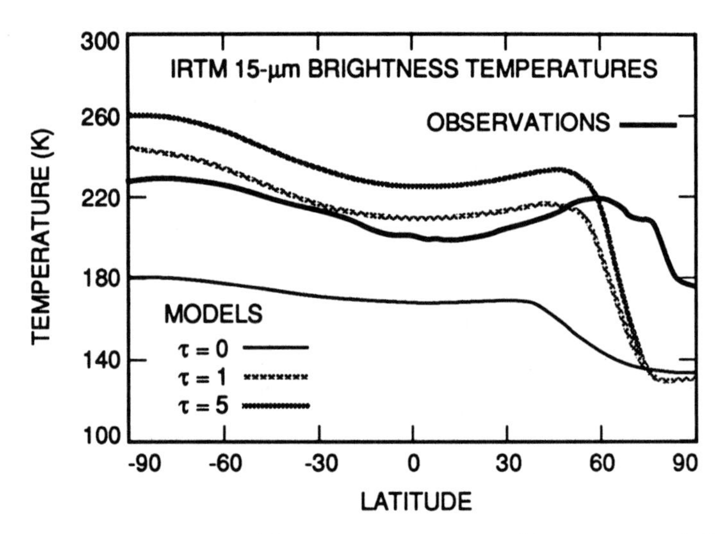

Fig. 16. Synthetic IRTM 15-μm brightness temperatures generated from temperature fields computed with a two-dimensional zonally symmetric model assuming different visible optical depths τ of a global dust haze and compared with averaged Viking IRTM 15-μm observations (heavy solid curve). Note the local maxima near 60°N (figure from Haberle et al. [1982] using IRTM data provided by T. Z. Martin).

The streak orientation clearly indicates that a well-organized and nearly zonally symmetric global-scale wind field exists during the great dust storms (Fig. 17); the "hook-like" pattern in the subtropics is consistent with the expectation that the mean-meridional circulation is dominated by a single cross-equatorial cell during this southern spring and summer season, with northeasterlies in the northern hemisphere and northwesterlies in the southern hemisphere.

The intense polar warmings during great dust storms discussed in Sec. II.B (T. Martin and Kieffer 1979; T. Martin 1981; Jakosky and T. Martin 1987) must be due to intense subsidence in a zonal-mean meridional circulation. Haberle et al. (1982) suggested that the polar warmings could be due to a greatly expanded Hadley cell in a very dusty atmosphere, if the descending branch moved north of Lander 2 during the storm. This interpretation would be consistent with the observed pressure increase and shift of winds from westerly to northeasterly at the Lander 2 site (48°N) (Tillman et al. 1979; Ryan and Henry 1979) and the decrease of surface pressure and enhancement of northerly winds at Lander 1 (23°N) (Murphy et al. 1990*b*). The bright streak data (Fig. 17) suggest that northeasterlies extend at least to 40°N during many great dust storms. Beyond this latitude, however, the frequency of streaks is sharply reduced and their zonal symmetry is less pronounced; thus, it is difficult to judge to what degree the behavior of the winds and pressure at Lander 2 during these storms is representative of the zonal mean.

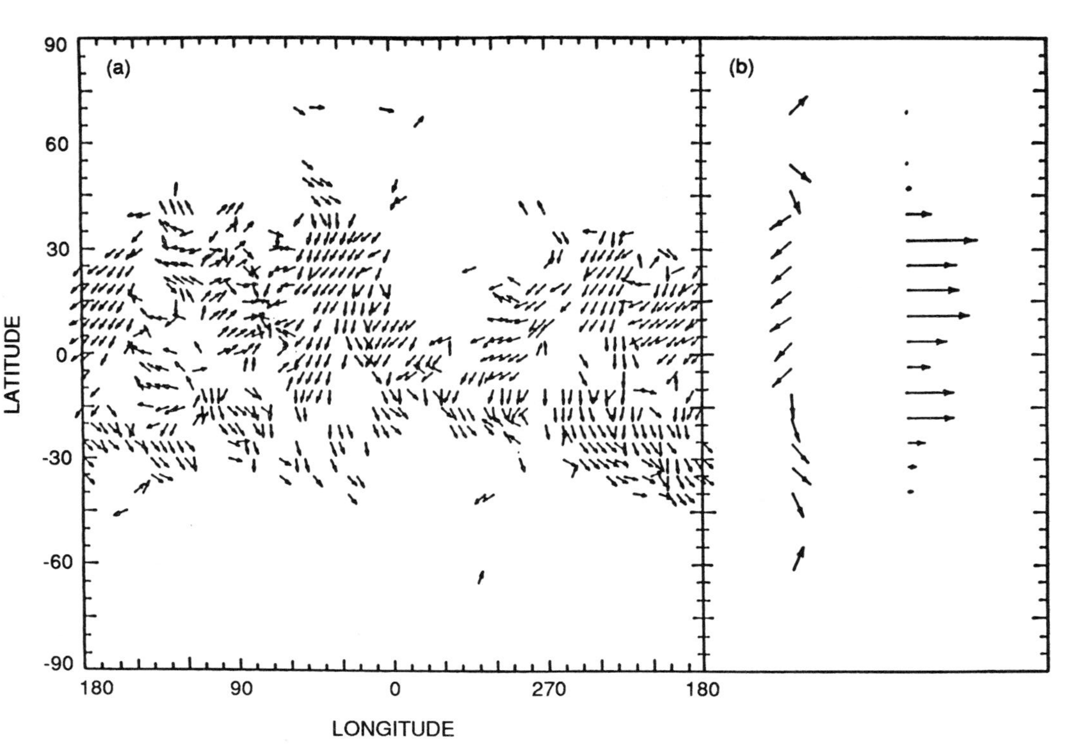

Fig. 17. Latitude-longitude distribution of Type I(b) "bright" surface streak orientations, derived from Viking Orbiter medium resolution mosaics. (a) Average azimuths for 5° × 5° bins. (b) Zonally averaged azimuths in 2° latitude bins (data shown for every fourth bin). Horizontal arrows indicate relative number of streaks in each bin, with the longest arrow representing 700 streaks (figure from J. Magalhaes [1987], using data provided by P. Thomas).

B. The Nature of Zonally Symmetric Circulations

The differences between the various concepts and models of zonally symmetric circulations are due largely to the different assumptions about how the total angular momentum balance (Eq. 13) is maintained. The right-hand side of Eq. (13) is often parameterized as a down-gradient vertical diffusion:

$$E_M = -\frac{\partial}{\partial z^*}\left(\nu \frac{\partial M}{\partial z^*}\right) \tag{16}$$

with viscosity coefficient ν. In the thermosphere E_M is dominated by molecular viscosity and Eq. (16) is valid with ν the molecular kinematic viscosity coefficient. Elsewhere, the representation in Eq. (16) is more problematic, and a fundamental question is: What are the processes that generate E_M, and how does the circulation in the meridional plane respond to E_M sources? Before examining these questions, some insights can be gained by considering the highly idealized case in which $E_M \approx 0$. Because of the representation in Eq. (16), the angular momentum balance in this case is termed "nearly inviscid."

1. Properties of Nearly Inviscid Circulations. The theory of nearly inviscid Hadley circulations is relevant for Mars because the zonally symmetric circulation in its lower atmosphere may be dominated by the zonally symmetric diabatic heating $\bar{Q}$. Consider an inviscid atmosphere in which temperatures are given by $\bar{T} = T_e$ everywhere, where T_e is the radiative equilibrium temperature. Because insolation is greatest at low latitudes, T_e is relatively warm there. However, $\bar{Q} = 0$ everywhere, so $\mathbf{v} \equiv 0$ and $\bar{u}$ is given by Eq. (14) with $\bar{u}(z^*_{bc}) = 0$. In the absence of viscosity (i.e., $E_M = 0$), this is a self-consistent solution at middle latitudes. More precisely, it is a self-consistent solution there in the limit of low viscosity, because angular momentum loss due to vanishingly small viscosity acting on the zonal wind shear can be balanced by advection of (planetary) angular momentum by a vanishingly small mean meridional circulation.

Near the equator, however, Held and Hou (1980) have pointed out that radiative equilibrium cannot be a solution, because relatively warm equatorial temperatures would imply, by the thermal wind equation, an increase with height of atmospheric angular momentum per unit mass near the equator. On the equator, both vertical diffusion and advection would act to destroy this increase, and so no steady-state balance would be possible there for *any* finite value of viscosity, however small. Thus, Held and Hou have argued that the radiative equilibrium solution at high latitudes must give way near the equator to a circulation in which $\bar{T} \neq T_e$ and $\mathbf{v} \neq 0$, even in the limit of vanishing viscosity.

The characteristics of this equatorial circulation are those of the nearly inviscid Hadley circulation. To investigate its properties in an idealized framework, suppose that the circulation occupies a depth H^* and that, because of surface friction, the zonal flow is negligible at the lower boundary near the latitude θ_0 of the rising branch. The angular momentum there is given by $M \approx M_0 \equiv a^2 \, \Omega \cos^2\theta_0$. Because the flow away from the surface is assumed to be inviscid, zonal angular momentum must be conserved in the upper branch of the circulation, where the meridional flow is opposite in direction to the meridional flow near the surface. Thus, $M \approx M_0$ at all latitudes in the upper branch and so, by Eq. (12):

$$\begin{aligned} \bar{u}(H^*,\theta) &= a\Omega \, (\cos^2\theta_0 - \cos^2\theta)/\cos\theta \\ &\equiv a\Omega \, (\sin^2\theta - \sin^2\theta_0) \, / \, \cos\theta. \end{aligned} \tag{17}$$

Note that Eq. (17) is in accord with Hide's (1969) theorem, which states that for a zonally symmetric circulation (forced by down-gradient fluxes of momentum) there can be no maxima or minima in angular momentum within the atmosphere and, in particular, no westerlies above the equator. Now, $\bar{u}(H^*,\theta)$ is symmetric with respect to the equator and, by the thermal wind equation, so is the vertically integrated temperature field $\langle \bar{T} \rangle$, again assuming that $\bar{u}(H^*,\theta) >> \bar{u}(0,\theta)$. However, this symmetry can occur only at those latitudes covered by the circulation cell in both hemispheres (Schneider 1983; Lindzen and Hou 1988). Asymmetries arise if the circulation cell extends farther into one hemisphere than the other.

At some poleward latitude, a nearly inviscid circulation will give way (almost discontinuously in $\bar{u}$, but not $\bar{T}$) to the radiative equilibrium solutions. That there must be some transition at high latitude is evident from Eq. (17) which shows that $\bar{u}(H^*,\theta) \rightarrow \infty$ as $\theta \rightarrow 90°$. The limit of this circulation depends on $\bar{Q}$, as does the intensity of the circulation. Held and Hou (1980) have approximated the energy balance (Eq. 15) as follows:

$$\bar{\omega} S \cong \kappa \, \bar{Q} \approx \frac{(T_e - \bar{T})}{\tau_r} \tag{18}$$

where τ_r is the radiative damping time (Table II). Mass continuity in the circulation cell then requires (for static stability parameter S and radiative time τ_r independent of latitude):

$$\int_{\theta+}^{\theta^*} \left[T_e - \bar{T} \right] \cos\theta \, d\theta = 0 \tag{19}$$

where θ^+ and θ^* are the latitudinal limits of the Hadley regime. By continuity, $\bar{T} = T_e$ at the limits of the cell, so $\Delta\langle \bar{T} \rangle = \Delta\langle T_e \rangle$ where "$\Delta\langle \, \rangle$" indicates the vertically integrated contrast across the cell. The combination of constant angular momentum (Eq. 17) and thermal wind balance (Eq. 14) yields:

$$F_e \frac{\Delta \langle T_e \rangle}{T_r} \approx -\frac{(\sin^2\theta^* - \sin^2\theta_o)^2}{2\cos^2\theta^*} \tag{20}$$

with $F_e \equiv gH^*/(\Omega a)^2$ (see Table II). The extent of the cell is determined by the height of the heated region and the meridional variation of the radiative heating $\bar{Q}$. As noted by Haberle et al. (1982) and evident in Eq. (20), it is very hard to extend the limit θ^* of the Hadley cell into high latitudes, given realistic gradients in heating and in the depth of the circulation. Although a nearly inviscid Hadley circulation on Mars can push into mid-latitudes during a planet-encircling dust storm, it is unlikely to move much beyond 60° because of the twin constraints of constant angular momentum and thermal wind balance.

Schneider (1983) applied the concepts of the vertically integrated, nearly inviscid Hadley circulation to Mars. He explored solstice situations, where the latitude θ_1 of maximum solar heating was offset from the equator. Schneider assumed that $\theta_1 = \theta_0$, the latitude of the rising branch. His model results indicated that localized heating offset from the equator could produce both local and global Hadley cells. Furthermore, the transition from local to global scale would be almost discontinuous, and this he identified with the onset of a global dust storm (see Sec. VII). Schneider also found that motions in the Hadley cell would be stronger at solstice than at equinox for the same heating. Lindzen and Hou (1988) showed that (for Earth) motions were amplified significantly in the cross-equatorial Hadley cell even when the maximum heating moved as little as 1 to 2° in latitude off the equator. In their simulations two cells were always generated, but when the maximum heating was not located at the equator ($\theta_1 \neq 0°$), the latitude θ_0 of the rising branch moved even farther poleward into the summer hemisphere in response to the twin requirements of energy conservation and continuity of the temperature fields. (When the heating is symmetric about the equator, the second requirement is automatically satisfied and $\theta_1 = \theta_0 = 0°$.) The cross-equatorial cell that developed encompassed the region of maximum heating and thus of maximum vertical velocity; this drove a stronger meridional flow into the winter, as opposed to deeper into the summer, hemisphere.

2. *Application of the Nearly Inviscid Theory to Mars.* Some of the features of the highly idealized models of Held and Hou, Schneider, and Lindzen and Hou are in qualitative agreement with observations. Figure 16 compares a zonal-mean Viking IRTM 15-μm brightness temperature profile obtained during the intense phase of a great dust storm with synthetic (brightness) temperature profiles calculated by the weakly viscid zonally symmetric circulation model of Haberle et al. (1982). The model simulations indicate that the radiative heating of a dusty atmosphere can expand and intensify a nearly inviscid circulation, leading to high-latitude subsidence and adiabatic

warming which is in qualitative agreement with the Viking IRTM and Mariner 9 IRIS measurements (Figs. 6 and 16).

Quantitatively, however, there are large differences between the nearly inviscid theory and the observations: (1) observations indicate that the zonal jet (i.e., the maximum in the zonal wind speed), as implied here by the thermal wind relation, is farther poleward than the simulations predict; and (2) the maximum wind speed inferred geostrophically from the Mariner 9 IRIS measurements, taken during the clearing of a global dust storm, is less than it would be if angular momentum were conserved in an inviscid circulation. More will be said about this in Sec. IV.B. For now, we note that the action of eddies of various scales must be invoked to explain these discrepancies.

3. Role of Friction. One possibility is that waves induce mixing which acts as an eddy viscosity to dissipate the zonal flow. Held and Hou (1980) compared the behavior of "viscous" flows simulated in a numerical model with the nearly inviscid solution. They used a vertical viscosity parameterization of the form given by Eq. (16). Results for equatorially symmetric forcing are shown in Fig. 18. The curves marked "theory" are approximate analytical solutions for the inviscid limit; Held and Hou were unable to obtain steady-state numerical solutions for vanishing viscosity. Schneider (1984) argued that this was a physical effect and that the existence of steady-state solutions required some arbitrarily small, though finite, viscosity ν. Lindzen and Hou (1988) noted that the effect of such minimal viscosity would be to mix angular momentum across the finite width of the rising branch and to produce, for decreasing ν, an asymptotic value of the angular momentum which was a representative average of the latitudes underlying the rising air (e.g., $\theta \approx 15°$ in Fig. 18). Because the behavior of the solutions did not depend on the value of ν, as long as it was small, the circulation was still "nearly inviscid."

It is clear, however, that for ν sufficiently large, the amplitude of the circulation and of associated transports is determined by the strength of the assumed friction (Fig. 18). The effects of vertical viscosity and of the even simpler dissipative parameterization, Rayleigh friction, can be incorporated into the linearized steady-state version of the zonal equation of motion by

$$f\bar{v} = -\frac{\partial}{\partial z^*}\left[\nu \frac{\partial \bar{u}}{\partial z^*}\right] \tag{21}$$

$$f\bar{v} = \eta\, \bar{u} \tag{22}$$

where η is a Rayleigh friction coefficient (Pirraglia and Conrath 1974; Pirraglia 1975; French and Gierasch 1979). In these equations, friction balances

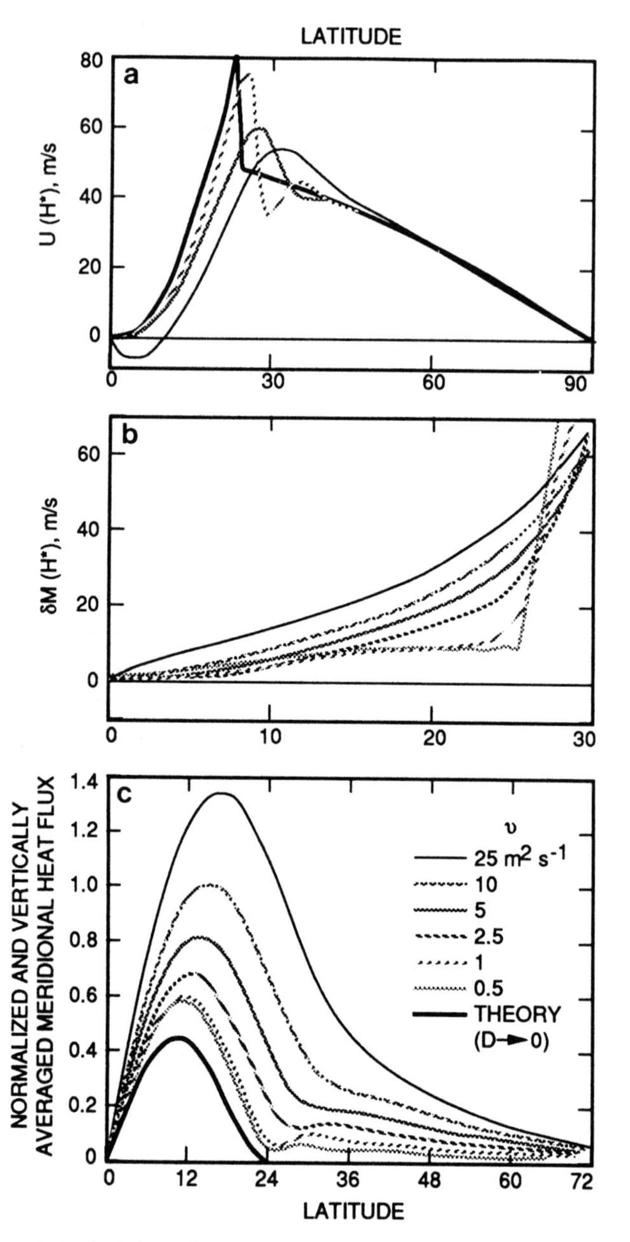

Fig. 18. Numerical calculations of Hadley circulations for different assumed values of the eddy viscosity υ. (a) Zonal winds at $z = H^*$, in the poleward branch. Solid curve gives prediction for nearly inviscid circulation ($\upsilon \rightarrow 0$). (b) The momentum excess ($\delta M \equiv \Omega a \sin^2\theta - u \cos\theta$) at $z = H^*$. The straight line at $\delta M = 0$ represents constant angular momentum. (c) Meridional heat fluxes computed for various υ and for inviscid theory (solid line). (Figure adapted from Held and Hou 1980.)

the Coriolis force; i.e., friction balances planetary, not total, angular momentum. The character of solutions for the dynamical system given by Eqs. (10), (14), (18) and (21) or (22) takes the form of a low-latitude boundary layer, although for large enough ν, the latitudinal "boundary" layer may be very thick. In either Eq. (21) or (22), friction determines the meridional circulation $\bar{v}$ and, through mass conservation, $\bar{\omega}$. By contrast, angular momentum balance of a nearly inviscid circulation does not constrain the meridional plane velocity $\mathbf{v}$, which is determined by the diabatic heating $\bar{Q}$ through the energy balance constraint (Held and Hou 1980).

Eddy forcing of the mean flow may or may not resemble vertical eddy viscosity or Rayleigh friction. Forcing of the zonal-mean flow by large-scale waves has been studied for three classes of waves: atmospheric thermal tides, transient planetary waves, and internal gravity waves. Much of this work has focused on how these waves might play a role in the polar warming observed on Mars during the 1977b storm.

4. Role of Planetary-Scale Waves. Atmospheric tides (Sec. V.B) are expected to break as they propagate vertically (Lindzen 1981). At altitudes >35 km over low latitudes and >45 km over mid-latitudes, the tides will generate temperature variations and wind shears that at certain times of day are convectively or mechanically unstable and so will "break," producing local turbulence and strong vertical mixing (Zurek 1976). Simple scaling arguments suggest an effective viscosity $\nu \approx 0.5 \times 10^4 m^2 s^{-1}$ in the region of breaking, and this value has often been cited in work using one-dimensional photochemical models (see, e.g., Rodrigo et al. 1990). However, at the lower altitudes where breaking first occurs, the implied breaking region is relatively narrowly confined in longitude and latitude, its cross section expanding with height. Meanwhile, half a planet away, the tide will reinforce the background static stability, thereby inhibiting breaking (Zurek 1986). Thus, the global mean effective eddy viscosity should be smaller than is suggested by simple scaling arguments. The atmospheric tide and other vertically propagating planetary waves can also modify the breaking of smaller-scale gravity waves by locally altering the temperatures and winds in regions through which the gravity waves pass (Walterscheid 1981).

Hamilton (1982) demonstrated that even nonbreaking atmospheric tides generated thermally during a global dust storm could produce zonal-mean flux divergences as large as 100 m s^{-1} day^{-1}. Zurek (1986) refined these estimates, using more realistic heating and temperature fields, but the flux divergences were still large (30 m s^{-1} day^{-1}) even below the heights at which the tides may break ($\approx$ 40 km). Furthermore, the pattern of the computed forcing was consistent with an expanded Hadley cell, in that the tides produced a broad belt of large easterly zonal-mean forcing per unit mass in the middle atmosphere (30 km) at low latitudes. However, in a two-dimensional numerical simulation Zurek and Haberle (1988) found that the zonal-mean tidal flux divergences extended the Hadley circulation in a dusty atmosphere from 55° to only 65°N; this was not far enough to account for the observed warming in the interior of the polar region. Their results did indicate that at low latitudes, tidal forcing could modify the zonal-mean flow and thus its advection of suspended dust into the northern hemisphere.

One promising mechanism for explaining the polar warming is that proposed by Barnes and Hollingsworth (1987) which involves forcing by planetary waves. The mechanism is fundamentally similar to that used to explain rapidly evolving (i.e., "sudden") warmings in the Earth's stratosphere (see, e.g., Holton 1980). On Mars during a great dust storm, the Hadley circulation intensifies and expands poleward. This expansion may modify the forcing of planetary-scale waves, as well as enhance their ability to propagate vertically. At upper levels vertically propagating planetary waves could produce large, transient fluxes of heat and momentum into high latitudes. If strong enough, these transient fluxes could induce a zonal-mean circulation which would adiabatically reverse the latitudinal temperature gradient. Maximum temperatures would then occur at the pole, with an accompanying reversal of zonal winds, as occurs during sudden stratospheric warmings in Earth's northern hemisphere. A better analog for Mars may be Earth's southern polar latitudes, where reversal of the zonal winds does not occur because of the presence of a very strong westerly jet stream.

The fluxes of heat and angular momentum generated by waves that are steady and conservative (i.e., waves that are adiabatic, inviscid, and have essentially fixed amplitude), will not accelerate the zonal-mean zonal flow nor change the temperature of the zonal mean state (Andrews and McIntyre 1978; Boyd 1976). Instead, the waves induce a steady-state component of the zonal mean circulation that exactly cancels (because it is driven by) the zonal mean wave flux divergences. The implication of this noninteraction theorem is that the occurrence of events like polar warmings require: (1) transience in the generation of the waves or in their transmission through the atmosphere; (2) longitudinally varying thermal forcing; or (3) dissipation (e.g., radiative damping).

Barnes and Hollingsworth (1987) specifically examined the mechanism of wave transience, assuming the transience to be produced by acceleration of surface zonal flow over topography. Their model was able to produce polar warmings by forcing zonal wavenumber 1 at the surface provided the vertical velocity prescribed there (and meant to represent flow over planetary-scale slopes) exceeded the rather large threshold value of 2 cm s^{-1}. They were unable to generate polar warmings by forcing zonal wavenumber 2 because disturbances of this (or higher) zonal wavenumber were trapped at relatively low altitudes.

For nonconservative tracers like dust and water vapor, there is also a tendency for advection by the zonal-mean flow to balance the tracer flux divergences due to waves; the more conservative the tracer and the waves, the more exact the cancellation. This means, for example, that tracer concentrations (e.g., dust in the polar atmosphere) are most likely to be changed under the same conditions, during the same periods, and by the same wave mechanisms by which temperature and wind fields are changed. Based on numerical simulations, Barnes (1990*b*) concluded that significant amounts of suspended dust (and possibly atmospheric water) could be transported into polar latitudes during a polar warming (such as the one observed during the 1977b great dust storm) if the warming is produced by planetary wave transience.

5. The Role of Internal Gravity Waves. The zonal flow can also be forced by the breaking of vertically propagating internal gravity waves. These buoyancy waves can exist wherever the atmosphere is stably stratified, so that the vertical displacement of air parcels produces density changes that cause gravity to act as a restoring force. Breaking of these waves is analogous to the breaking of surface water waves; the vertical profile of the wave steepens drastically and the wave crests overturn. As for atmospheric tides, breaking results from the generation of unstable temperature gradients associated with the exponential growth of the gravity-wave amplitudes as they propagate upward into regions of exponentially decreasing density (Lindzen 1981; Walterscheid and Schubert 1990). Wherever the waves break, they deposit their wave momentum flux into the background flow, driving it toward the horizontal phase speed of the wave (see, e.g., Killworth and McIntyre 1985). Waves may also break as their vertical wavelengths shorten due to vertical propagation toward critical levels, regions in which the wave horizontal phase speed approaches the speed of the background wind.

Barnes (1990*a*) has studied the possible effects of gravity-wave breaking on the zonal mean flow using a simplified model of the evolving zonal flow together with a gravity-wave parameterization based on the work of Lindzen (1981) and Holton (1982) for Earth's atmosphere. Zonally propagating gravity waves were assumed to be excited by flow over topographic obstacles. The breaking heights for these waves depend mostly on their initial amplitudes and the structure of the mean flow through which they propagate.

Topographically excited gravity waves have small phase speeds with respect to the ground, and they tend to drive the mean flow towards these speeds, thus acting as a drag on strong zonal flows. This drag is nonlinearly dependent on the flow-shifted phase speed at the breaking levels (Holton 1982). Dust storms may produce local turbulence beneath an overlying, stably stratified layer and so provide another source of gravity waves having small phase speeds.

Barnes (1990*a*) showed that gravity waves of intermediate scale (horizontal wavelengths $\approx$ 100 to 1000 km) may break between 40 and 80 km at the higher latitudes of the winter hemisphere, above the region where observations detected the polar warming during the 1977b storm. For topographically forced gravity waves to break at much lower heights, so as to produce such a warming, either very large amplitudes or very long horizontal wavelengths were required. Barnes concluded that nontopographic forcing, perhaps resulting from dust heating, was a more plausible source for very large amplitude gravity waves.

To quantify the effects of gravity-wave breaking, the breaking levels and excitation mechanisms of the waves need to be known. Even then, relatively sophisticated models of the wave generation, propagation, breaking and associated mixing may be required (see, e.g., Rind et al. 1988) to determine their effect on the large-scale circulation. Spacecraft measurements could help quantify the effects of wave breaking by providing evidence from cloud images of the regions where waves are being generated, and possibly also through the identification of very long internal gravity waves in temperature cross sections. In addition, atmospheric sounding can provide temperatures for estimating winds geostrophically. If dust distributions are obtained as well, so that the diabatic heating can be estimated, then the residual-mean circulation in the meridional plane can be estimated, as discussed above. Evaluation of the angular momentum advection by that circulation, together with evaluation of the momentum transport effects of the observable planetary-scale waves, permits an estimate of the gravity-wave drag as a residual in the angular momentum balance; such a technique has been applied to Earth's stratosphere (Smith and Lyjak 1985).

6. Condensation Flows. Previous discussions focused on the "overturning" component of the mean-meridional circulation; i.e., that part of the zonal-mean circulation having no vertically integrated component. A unique aspect of Martian meteorology is the existence of a nonzero vertically averaged meridional motion, due to the condensation and sublimation at high latitudes of a significant fraction of the CO_2 atmosphere. Approximately 25% of the Martian atmospheric mass is cycled into and out of the seasonal polar caps (Hess et al. 1979; Tillman 1988). This mass change is substantial enough to detectably affect the planet's rotation rate (Philip 1979; Reasenberg and King 1979; Chao and Rubincam 1990). The redistribution of mass globally can produce vertically and zonally integrated meridional mass flows as high as 0.5 m s^{-1} (Leovy et al. 1973*b*; James and North 1982; see chapter 27 for a detailed discussion of the seasonal CO_2 cycle).

The strength of the condensation flow and its effect on the circulation depend on season. At the edge of the CO_2 cap, a thermally driven "seabreeze" type circulation may produce equatorward and easterly surface winds. Coriolis forces operating on the mass outflow (inflow) generated by sublimation (condensation) of CO_2 during spring (fall and winter) could enhance these winds by as much as 50%, producing surface easterlies (westerlies) of 20 m s^{-1} or more (Haberle et al. 1979). Sublimation-enhanced surface winds associated with the retreating south polar cap may help generate the many local dust storms seen just equatorward of the edge of the seasonal cap

(Peterfreund and Kieffer 1979), as well as various aeolian surface features (French and Gierasch 1979).

C. Results of Zonally Symmetric Models

Haberle et al. (1982) computed the zonal-mean circulation driven by zonally symmetric radiative, convective and latent (CO_2) heating. Their two-dimensional model contained heating algorithms appropriate for a dusty CO_2 atmosphere and was based on the full primitive equations, but ignored thermal or mechanical forcing of the zonal-mean flow by zonally asymmetric waves. While vertical and horizontal diffusion terms were included to control computational instability, these were kept small: $\nu \approx 2 \times 10^2$ m^2 s^{-1} for vertical diffusion; $\nu_H \approx 5 \times 10^4$ m^2 s^{-2} for horizontal diffusion of momentum and heat. Thus, it is appropriate to compare their results with the nearly inviscid theory.

As noted earlier, Haberle et al. (1982) experimented with progressively increasing opacities for a uniformly mixed and time-independent atmospheric dust load at northern winter solstice. Their results showed a significant strengthening and expansion of the cross-equatorial Hadley circulation, consistent with the expectations of inviscid theory and, qualitatively, with the Viking IRTM data (Fig. 16). Quantitatively, however, the simulated circulation did not extend far enough poleward. Thus, model temperatures near the pole remained near the CO_2 frost point, and winds at the Lander 2 latitude remained westerly, rather than shifting as observed to northeasterly.

Haberle et al. (1982) noted that the failure to penetrate farther poleward in their model simulations is consistent with the nearly inviscid theory. Further poleward displacement would have produced zonal winds there which, through total angular momentum conservation (Eq. 17), would have been much too strong to be consistent with observed temperature gradients, unless the nearly inviscid circulation were very deep (i.e., $H^* \approx 40$ to 60 km). However, both the Viking IRTM and the Mariner 9 data (Figs. 16 and 6) indicate a reversal in the meridional temperature gradient above 20 km equatorward of 55°N. (Reversal of the cross-equatorial temperature gradient near 20 km is also suggested by comparing the two right-hand Viking radio-occultation temperature profiles in Fig. 9.) According to the thermal wind relation (Eq. 14), zonal winds must decrease above 20 km equatorward of 55°N, and this limits the depth of any inviscid circulation to $H^* \approx 20$ km, because the level of the zonal wind maximum corresponds to the jet level H^* in the nearly inviscid theory, with the temperature variation $\Delta<\bar{T}>$ situated below that level. Haberle et al. concluded, therefore, that nonaxisymmetric processes were important to the polar warming phenomenon.

Magalhaes (1987) used a viscous model of the Martian Hadley circulation to study the circulation during different seasons. His results for a global dust storm period showed that a simple thermally forced, but viscous, Hadley circulation could produce a polar warming, provided the eddy forcing, represented as vertical eddy diffusion, was large: $\nu \approx 10^3$ to 10^4 m^2 s^{-1}. However, this simple viscous model was unable to reproduce the warm polar temperatures and the relatively broad temperature maximum centered at 60°N near 25 km of altitude simultaneously.

Because this reversal of the latitudinal gradient of temperature near 60°N is more readily reproduced by weakly viscous models (see e.g., Haberle et al. 1982; Schneider 1983; Lindzen and Hou 1988), Magalhaes speculated that the warm pole and the temperature maximum 60°N might be reproduced with a model in which the eddy viscosity for vertical diffusion varies with height, increasing from $\nu \leq 10$ m^2 s^{-1} in the lower atmosphere to $\nu \approx 10^4$ m^2 s^{-1} above 30 km. Such a distribution of ν would be consistent with the estimates that gravity waves and atmospheric tides will break mostly above 35 km (Barnes 1990*a*; Zurek 1976,1986) and with the observation of discrete limb hazes at other times (Kahn 1990), but not necessarily with the vertical mixing occurring during the dust storms (Conrath 1975) nor with the global values apparently required to stabilize the CO_2 atmosphere of Mars against photodissociation (McElroy and Donahue 1972; chapter 30).

Another constraint on the zonally symmetric circulation is provided by observed near-surface wind directions. Using the framework of the French and Gierasch (1979) viscous model, Kahn (1983) assessed the relative contributions of the CO_2 mass flow and of viscously driven Hadley circulations, the latter constrained by the latitude and seasonally varying temperature gradients constructed from Viking IRTM data, Mariner 9 IRIS analyses and one-dimensional radiative-convective modeling. By matching the modeled wind direction to that derived from the cloud data base, Kahn inferred which of the modeled mechanisms dominated the apparent zonal-mean flow. His summary was shown in Fig. 12.

The validity of using near-surface winds to constrain the zonal-mean winds depends on how well the observed wind directions represent the zonal-mean fields. It also depends on the nature of the zonally symmetric circulations. Neither Magalhaes nor Haberle et al. were able to reproduce the steady northeasterly surface winds at higher latitudes during the 1977b dust storm. In the model of Haberle et al. (1982) northeasterlies extended to 40°N, consistent with the Type 1b bright streak data, but were much weaker than implied in the Lander 2 data. Magalhaes's calculations, on the other hand, showed them to extend to 30°N at most.

There are different reasons for the inability of these two models to reproduce the northeasterlies at Lander 2. The sign of the near-surface zonal-wind component of nearly inviscid models is determined by the Coriolis force due to the meridional flow; the weakly viscous Hadley circulation computed by Haberle et al. (1982), with its near-surface northerly wind, simply did not extend far enough north. In a viscously controlled Hadley circulation, the direction of the surface zonal wind depends on the relative magnitude of momentum transport by viscosity into the surface layer and its removal there by surface drag. In Magalhaes' (1987) model, northerly flow did occur at high midlatitudes, but the turbulent flux of momentum into the surface layer dominated the surface drag; thus, westerly, rather than the observed easterly, surface winds were required in order to provide a corresponding momentum sink.

Kahn (1983) and Magalhaes (1987) suggested that sublimation of the seasonal north polar CO_2 cap during the major polar warming that occurred during the 1977b dust storm could produce northeasterly winds at Lander 2 at that time. Magalhaes estimated that a southward flux of CO_2 from the north polar cap as large as 10^8 kg s^{-1} was required, *if* the thermally driven component of the mean flow produced no zonal-mean surface wind component. Otherwise, larger southward sublimation flows were required.

The minimum estimated rate is consistent with rough estimates of CO_2 sublimation during the polar warming (T. Martin and Kieffer 1979); larger rates are probably *not* consistent with Viking Lander measurements of surface pressure (Hess et al. 1977). Kahn (1983,1989) noted that reducing the wintertime deposition of CO_2, and thus the rate of sublimation in the following spring, could help explain an apparent hemispheric seasonal asymmetry in that zonal-mean westerlies at high latitudes appear to persist into mid spring in the north, but not in the south. Recent general circulation model calculations with variable dust loadings lend support to the idea that higher dust opacities during winter will lead to less condensation of CO_2 in the seasonal cap (Pollack et al. 1990*a*).

D. General Circulation Model Simulations

General circulation models (GCMs) developed over the last four decades have successfully simulated many aspects of the circulation and thermal structure of the atmosphere of Earth. In principle, these models should be directly applicable to the study of Mars. In fact, the use of GCMs to study Mars has paralleled their development for terrestrial studies almost from the beginning of the era of large-scale numerical modeling.

Meteorological fields generated by a GCM provide one perspective of what existing observations tell us about the physics of the Martian atmosphere and of what terrestrial experience suggests should be true. Due to the sparseness of the existing data set, the degree to which this perspective is correct will be truly tested only when future missions yield systematic observations of the Martian atmosphere and surface. However, the climatological aspects of the model-generated heat and momentum balances (i.e., zonally and time-averaged features for particular seasons and atmospheric dust opacities) are likely to be robust, and it is these model "balances" which are emphasized here.

Table III abstracts the elements of several GCMs used to study Mars since the pioneering work of Leovy and Mintz (1969). Mass and Sagan (1976) and Moriyama and Iwashima (1980) used quasi-geostrophic and linear

TABLE III
General Circulation Modeling of the Atmosphere of Mars

Model Study	Leovy & Mintz 1969	Mass & Sagan 1976	Moriyama & Iwashima 1980	Pollack et al. 1976	Pollack et al. 1981	Pollack et al. 1990
Domain	global	35°S–90°S	southern hemisphere	global	global	global
Resolution						
Vertical Layers	2	2	3	3	3	13
Lat × Long	7° × 9°	polar stereographic proj: 33 × 27 pts.	spectral 3 modes × 6 waves	5° × 6°	5° × 6°	7.5° × 9°
Dynamics	PE[a]	quasi-geostrophic	linear balance model	PE[a]	PE[a]	PE[a]
Radiation CO_2	IR: Prabhakara-Hogan (1965) VIS: Houghton	IR: Tgrnd/Trad prescribed seasonally VIS: BGW[b] (1973)	BGW[b] (1973)	equiv. widths	equiv. widths	fit to line-by-line
Dust[c]	no	no	yes four-stream IR: quartz	no	no	yes doubling/adding Pollack et al. (1979*b*)

TABLE III *(continued)*

Condensation Flow[d]	yes	no	no	yes	yes	yes
Orography[e]	no	yes	yes	M9	M9	MC
Variable Surface Thermal Inertia[e]	no	—	no	no	no	MC
Seasons	solstices, equinoxes	S. summer, spring, winter	equinox	N. summer/ S. winter	N. summer/ S. winter	full seasonal range
Primary Goal	atmospheric heat transport to poles	effect of orography (mechanical) on dust-raising	comparative effects of orography and dust heating	winds at Viking Lander sites	effect of orography	effects of dust and orography

[a]PE => full primitive equation model.
[b]BGW => Blumsack et al. (1973).
[c]Models with airborne dust have time-invariant opacity; dust is usually uniformly mixed throughout the atmosphere.
[d]Allows for variable atmospheric mass due to condensation/sublimation of CO_2.
[e]M9: early version from Mariner 9 data; MC: Mars Consortium Data Base (Kieffer et al. 1981).

balance approximations, respectively, to the dynamical equations. The series of Mars GCMs derived from the terrestrial GCMs developed at UCLA by Mintz and Arakawa have all been based on the full system of primitive equations (Leovy and Mintz 1969; Pollack et al. 1976*b*,1981,1990*a*).

1. Current Version of the Mars General Circulation Model. The model fields discussed below were generated by a recent version of the still evolving Mars GCM (henceforth, MGCM) developed at NASA Ames Research Center (Pollack et al. 1990*a*). Quasi-steady-state fields were generated by averaging the last 20 to 30 days of 50-day runs. In a given simulation, the mass mixing ratio of dust was constant with time and space. The model simulated a full diurnal cycle each day and included the seasonal variation of insolation. With 13 vertical levels, the model domain extends from the surface to an altitude $\approx$ 47 km, where a rigid lid separates this region from a radiatively active, but dynamically inactive, top layer.

Such vertical resolution and range is required on several counts: (1) the presence of dust in the atmosphere results in a deeper circulation; (2) high vertical resolution (1–2 km) near the surface is needed to represent the parameterized physics of the planetary boundary layer; and (3) good vertical resolution ($\approx$ 5 km) throughout the domain is needed to resolve baroclinic eddies properly and to approximate vertically propagating waves like atmospheric tides. Such vertical resolution has been achieved at the expense of a somewhat coarser horizontal resolution than in earlier models: $7.^\circ5 \times 9^\circ$ in latitude and longitude.

The MGCM includes radiative effects of dust at both solar and thermal wavelengths. As in the older versions of the GCM, heat and momentum exchange between the surface and atmosphere are parameterized in terms of heat and drag coefficients that are computed from the static stability and wind shear close to the surface. The bulk parameterization scheme is based on one used in terrestrial GCMs, derived from theory and observations of the Earth's boundary layer (Deardorff 1972). When the lapse rate becomes superadiabatic, velocities and temperatures are convectively adjusted in a momentum and energy conserving fashion (see, e.g., Pollack et al. 1981). Finally, a Rayleigh friction drag (i.e., proportional to the wind speed) is introduced into the upper several layers of the model. This crudely simulates the effects of wave breaking in these layers, minimizes wave reflection from the rigid model top, and helps maintain the numerical stability of the model.

2. Momentum Balance. Three-dimensional dynamical models can compute explicitly the contribution to the zonal-mean total angular momentum balance by the waves they resolve; unresolved, or "sub-grid," eddies are parameterized, often using forms like Eq. (16). Differences between three-dimensional models with the same physics are then due to their inherent spatial resolutions and to their treatment of sub-grid diffusion.

In the Leovy and Mintz (1969) solstice simulation, the net zonal-mean poleward flux of angular momentum at midlatitudes in the winter hemisphere was due almost entirely to the condensation flow. Transports due to other major components of the circulation were of equal magnitude, but tended to cancel: transports due to the zonally averaged circulation (minus the condensation flow) were poleward, while zonal-mean transports due to waves were equatorward, opposite to the direction normally found for wave transport in the Earth's atmosphere. Thus, the condensation flow appeared to be the major process for maintaining the mid-latitude westerlies in the winter hemisphere. Calculations from more recent models (e.g., the MGCM of Pollack et al. [1990*a*]) show less compensation between the eddies and zonal-mean flow at midlatitudes, although this depends on season. At higher latitudes, all three components of the circulation transport momentum poleward, although the condensation flow dominates.

The short response time of the Martian atmosphere-surface system should constrain the latitude of the rising branch of the Hadley circulation to be near the subsolar latitude, and this is evident in the MGCM simulations (Pollack et al. 1990*a*), as illustrated in Fig. 19. At the solstices, the low latitude circulation is dominated by a single cross-equatorial Hadley cell that extends between $\pm 40°$ low in the atmosphere. Near the equinoxes, two cells develop that are roughly symmetric about the equator but, as suggested by the inviscid theory, they have peak intensities that are much less than the solstice case. Also evident in Fig. 19 are high-latitude indirect (Ferrel) cells which are most apparent in the winter hemisphere of the solstice simulation. As on Earth, the Martian Ferrel cells are thermally indirect in that they are forced mechanically by the convergence of eddy-momentum fluxes in the equatorward branch and thermally by the convergence of eddy-heat fluxes in the rising branch.

The change in intensity of the zonal-mean circulation as Mars progresses around the Sun is due in part to the eccentricity of its orbit. For the current epoch, perihelion occurs at $L_S = 251°$, and the subsolar insolation at southern summer solstice is 43% larger than at northern summer solstice. However, the intensity of the simulated cross-equatorial Hadley circulation during southern summer is about a factor of 2 greater than during northern summer (Fig. 19). This suggests the influence of other factors, such as orography or the effects of the relatively vigorous wave activity at northern midlatitudes during winter.

3. Heat Balance. The general features (Fig. 20) of the zonal-mean temperatures computed by the MGCM for the northern spring case are in good agreement with the observed features shown in Fig. 5, except for the meridional temperature gradient above 20 km. The stronger easterlies predicted by the MGCM for the tropics are consistent with the model's temperature minimum at those latitudes. Both the GCM and the data show strong

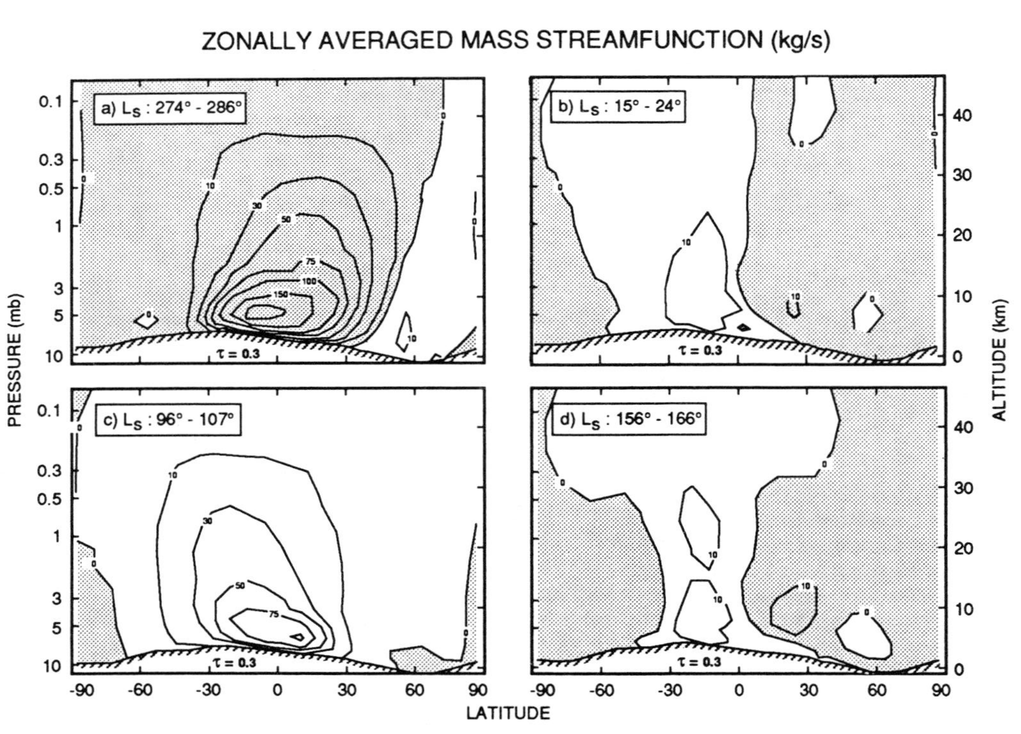

Fig. 19. Mass-weighted stream functions computed by the Mars-GCM (Pollack et al. 1990*a*) for early northern winter (a: $L_s \approx 280°$), early northern spring (b: $L_s \approx 20°$), early northern summer (c: $L_s \approx 103°$) and late northern summer (d: $L_s \approx 161°$). A background dust opacity of $\tau = 0.3$ was assumed. Flow in the meridional plane is clockwise around minima (negative values are shaded) and anti-clockwise around maxima; winds are strongest where contours are closest.

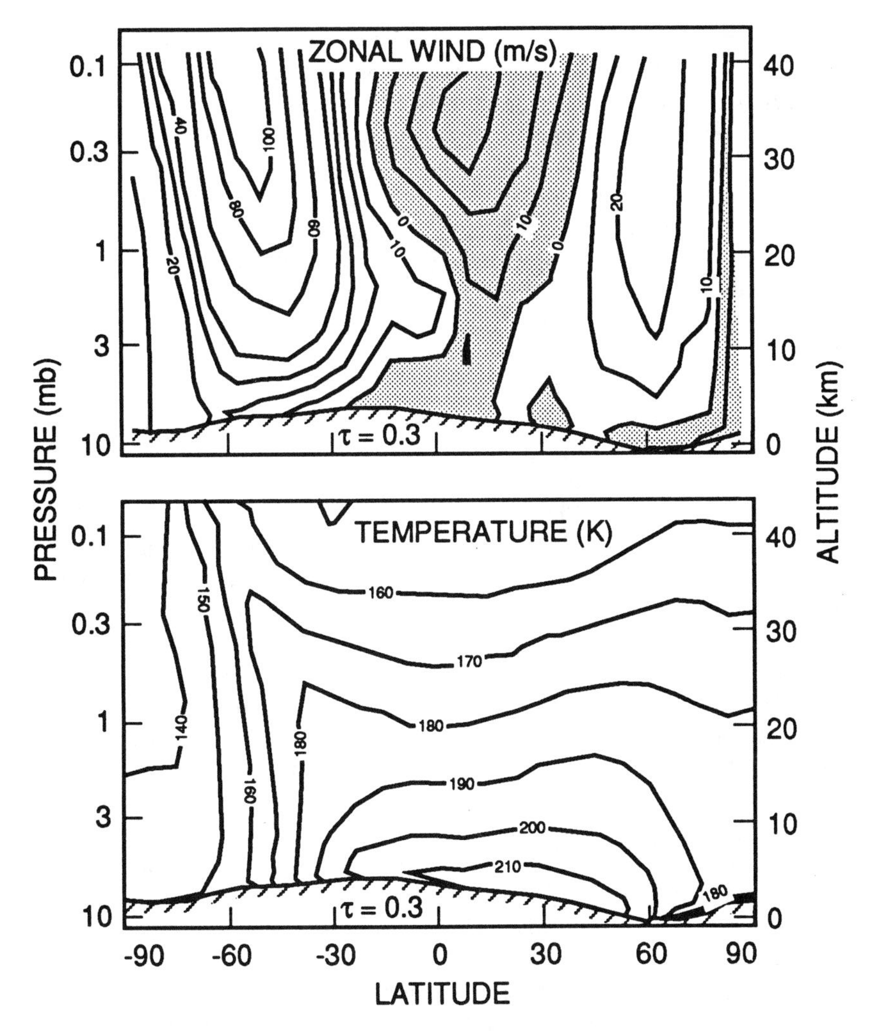

Fig. 20. Height-latitude cross sections of: (top) zonally averaged eastward wind (m s^{-1}; shaded region shows westward winds) and (bottom) zonal-mean temperature (K), as computed by the Mars-GCM (Pollack et al. 1990*a*) for mid-northern spring ($L_s \approx 44°$) and for a background dust opacity $\tau = 0.3$. Note the strong jet and a local maximum in temperature over the southern midlatitudes (figure from Pollack et al. 1990*a*).

horizontal gradients in temperature at high southern latitudes, with a relative temperature maximum aloft. As these temperatures are well above their radiative equilibrium values, dynamical processes are clearly implied.

Three major processes contribute to the heat balance of the zonally integrated atmospheric column and ground at different latitudes on Mars: (1) net radiation, (2) advection of heat by atmospheric motions, and (3) the release or uptake of latent heat by the condensation or sublimation of carbon dioxide. Heat advection in the thin Martian atmosphere is too weak to prevent temperatures from dropping to the frost point temperature of carbon dioxide at the ground in the winter polar regions (Leovy 1966*a*; Leovy and Mintz 1969). Latent heating due to the condensation of carbon dioxide is the primary heating source in these regions. Elsewhere on the planet, the balance is between net radiation and atmosphere heat advection. Much less important components of the heat budget include the frictional dissipation of winds at the surface and the resultant loss of internal energy.

Figures 21a-f show the zonally averaged heat balances for the combined atmospheric column and surface as computed by the MGCM for 6 seasonal dates spanning a full Martian year (Pollack et al. 1990*a*). Global dust hazes were specified for each of the seasonal dates, based on optical depths measured at the Viking Lander sites during 1976–78. In these simulations, the importance and vigor of atmospheric advection of heat depends primarily on the dust optical depth. When the optical depth is large, as it is for the simulations at $L_S = 214°$, 279° and 342°, atmospheric advection is larger at low latitudes and becomes important in the heat balance of the outer portion of the winter polar cap, although no "polar warming" occurs in the model. The rate of CO_2 condensation is decreased significantly in the north polar region, as compared to the rate for lower optical depths.

The atmospheric heat transport can be divided into zonal-mean and eddy contributions. The zonal-mean component can be further separated into that part due to the mass flow generated by condensation/sublimation and the remaining zonal-mean flow, known as the overturning circulation. The eddy contribution can be separated into parts due to traveling and to stationary waves. The eddy contributions to the zonal-mean fields depend upon the zonally averaged correlations between longitudinally varying temperatures and velocities (e.g., $\overline{v'T'}$, where v', T' denote spatial deviations of velocity and temperature at a given latitude from their zonal means).

As seen in Fig. 22, the relative importance of the various components of the circulation for advecting heat varies considerably with season (contrast north and south), with latitude and with dust opacity. At lower latitudes, the overturning circulation always carries a significant fraction of the heat. The condensation flow can contribute a comparable or greater amount during some seasons. At midlatitudes, transient eddies are sometimes quite important in the winter hemisphere. A surprising hemispherical asymmetry in the current MGCM simulations is that the eddies generated in southern midlatitudes during winter are considerably weaker than eddies generated in northern midlatitudes during winter, although the weakness of the former may be due to errors in the GCM topography (see Sec. V). Finally, with a few exceptions (Fig. 22), stationary eddies appear to play a secondary role in transporting heat.

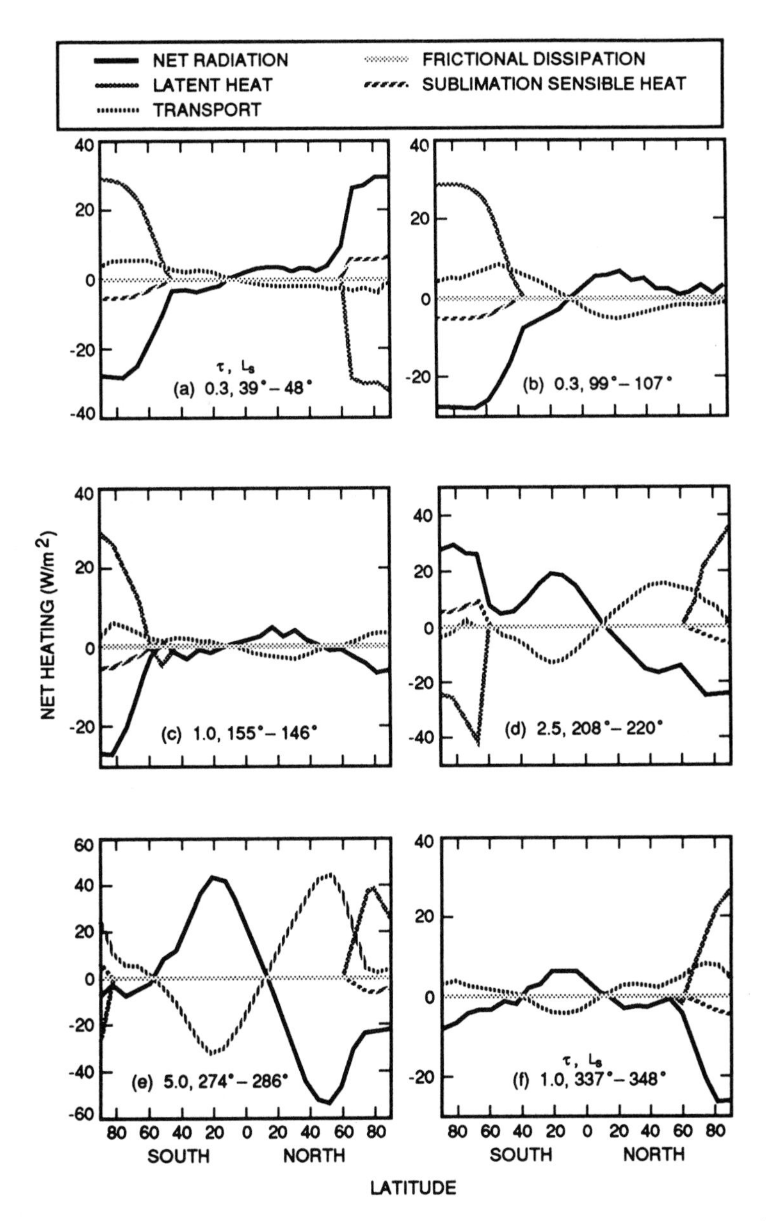

Fig. 21. Contributions to the energy budget for the combined atmospheric column and surface. Zonally averaged heating rate terms were computed by the Mars-GCM for six seasonal dates (L_s) spanning the first year of Viking observations and have been averaged over the last 20 days of each Mars-GCM simulation. In the simulation, the opacity (τ) of a global dust haze was taken to be that at Viking Lander 1 for the corresponding dates. τ, L_s: (a) 0.3, 44°; (b) 0.3, 103°; (c) 1.0, 160°; (d) 2.5, 214°; (e) 5.0, 279°; (f) 1.0, 342°. Note changes in ordinate scales (figure from Pollack et al. 1990*a*).

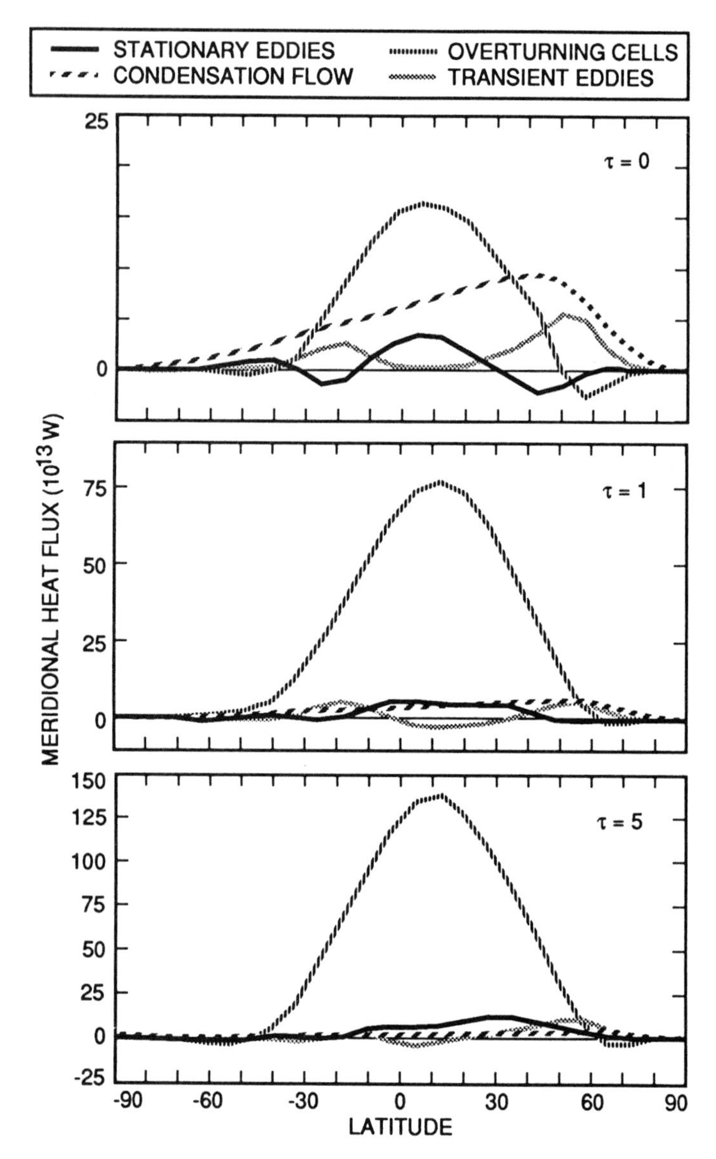

Fig. 22. Contributions of various atmospheric components (see text for details) to the atmospheric transport of energy, as computed by the Mars-GCM (Pollack et al. 1990*a*) for southern summer solstice ($L_s \approx 280°$) and for different optical depths τ of a global dust haze.

4. Atmospheric Mass (CO_2) Budget. Carbon dioxide condenses both in the atmosphere and on the ground in the winter polar region. The rate of atmospheric condensation is governed by the rate of radiative cooling of the polar atmosphere and therefore the emissivity of the polar atmosphere in the thermal infrared. At low winter polar temperatures, addition of dust particles to the winter polar atmosphere should produce a large increase in the polar cooling rate and hence the rate of condensation of CO_2 in the winter polar atmosphere. Model calculations indicate that increasing the dust optical depth from 0 to 5 in the winter polar atmosphere produces a transition from a situation in which almost all the carbon dioxide condenses on the ground to one in which it nearly all condenses in the atmosphere. The rate of CO_2 condensation in the polar atmosphere varies considerably in space and time (Pollack et al. 1990*a*) due to variations in heat advection by planetary waves, which are prominent during northern winter.

The spatial, transient and seasonal behavior of atmospheric carbon dioxide condensation simulated with the MGCM (Pollack et al. 1990*a*) is consistent with regions of anomalously low 20-μm brightness temperature measured by the Viking IRTM (Kieffer et al. 1977; Paige et al. 1991) and with the properties of the polar hoods (Briggs and Leovy 1974). CO_2 ice clouds may play an important role in the polar heat budget by reducing the thermal radiation emitted to space and hence the total (atmosphere plus surface) condensation rate and by scavenging dust particles from the atmosphere to the seasonal polar deposits through the nucleation and sedimentation of ice particles. Dust particles in polar surface ice may, in turn, control the albedos of the polar caps and hence their stability (Paige and Ingersoll 1985). A major observational uncertainty is just how much dust is advected into the winter polar regions from lower latitudes (chapter 29).

V. LARGE-SCALE WAVES IN THE MARTIAN ATMOSPHERE

Observations, both remotely sensed and *in situ*, have revealed the presence of various large-scale wave motions in the atmosphere of Mars. Numerical modeling has indicated that other types of such motions are almost certainly present as well, awaiting detection in future data. In this section, our current understanding of the characteristics and nature of large-scale waves in the Martian atmosphere is reviewed. Such waves are important in several contexts: (1) their intrinsic nature, especially in the framework of comparison with similar waves in the terrestrial atmosphere; (2) their role in forcing the zonal-mean flow, as discussed above; and (3) their role in the transport of suspended dust, water and other atmospheric constituents.

Large-scale waves are generated by atmospheric instabilities and zonally asymmetric "external" forcings. In the context of zonal-mean and zonally varying components, instabilities are generally of two types. Barotropic instability involves the transfer of kinetic energy from the zonal-mean flow to the wave; this exchange depends most sensitively on horizontal, not vertical, variations in the wind field. Baroclinic instability involves transformation of (gravitational) potential energy into wave kinetic energy and is highly dependent on horizontal variations in temperature and thus on flows which have vertical variations in winds. Baroclinic energy conversions are most efficient near the surface where air motions are forced to move across temperature gradients (Charney and Stern 1962). External forcings include diurnally varying solar heating, deflection of winds by large-amplitude Martian topography, and differential heating due to continental-scale variations in surface elevation, albedo and thermal inertia.

A. Transient Baroclinic Waves

1. Early Theory and Observations. Hess (1950) first speculated that traveling cyclones and anticyclones, similar to those in middle latitudes on Earth, were present in the Martian atmosphere. Mintz (1961) and Leovy (1969*b*) employed simple two-level/two-layer baroclinic models and utilized results from radiative equilibrium calculations to constrain thermal winds and static stability; they estimated the zonal wavelengths and growth rates of waves that would grow most rapidly and would thus be most likely to be seen. Both studies showed that these properties would change significantly with season, although the potential role of dustiness in producing such changes was not foreseen. For mid-winter conditions the most unstable scales were found to be relatively long with $m \leq 2$, where the zonal wavenumber m indicates the number of maxima (or minima) around a latitude circle. For fall and spring these scales were expected to shorten to $m \approx 3$ to 4 due to decreased static stability.

For both Earth and Mars, the most unstable zonal scale in simple baroclinic models is proportional to the Rossby radius of deformation, which is itself proportional to the Brunt-Väisälä frequency N, a measure of the atmosphere's static stability. The relatively low zonal wavenumbers of maximum instability on Mars, as opposed to Earth, are due to the smaller planetary radius, as the deformation radius is quite similar for the two planets (Table II). Mintz speculated that baroclinic waves might be completely stabilized in winter by very high static stabilities (due in part to the waves themselves), and that unstable waves might exist in an easterly wind regime in the summer hemisphere, as suggested by simple models of baroclinic instability. As more realistic models indicated that baroclinically unstable waves in the summertime easterlies would be confined very close to the surface, Leovy argued that summer easterly waves would not be present.

Leovy (1969*b*) also showed that strong radiative damping ($\tau_r \approx 1$ day) decreased growth rates of the most unstable (i.e., rapidly growing) waves only slightly. Blumsack and Gierasch (1972) found that Martian topography which sloped in an opposite sense to the isentropic surfaces, as it does at 50°

to 65°N during winter, would decrease significantly the wavelengths and growth rates of the most unstable disturbances.

Leovy and Mintz (1969) generated fully nonlinear, primitive equation simulations of baroclinic waves. In their model, transient baroclinic eddies occurred in midlatitudes in the winter hemisphere for the solstice experiment and in both hemispheres for the equinox simulation. The dominant zonal wavenumbers of the transient eddies at equinox were $m = 3$ to 4, but $m = 2$ to 3 for solstice conditions. Observational evidence for such waves was provided by Mariner 9 visual images showing front-like cloud features in northern winter midlatitudes (Leovy et al. 1972; Briggs and Leovy 1974). Propagation speeds of ≈ 5 to 15 m s^{-1} were estimated for the disturbances. In several instances, local dust storms appeared to be associated with the frontal systems. Mariner 9 IRIS temperature measurements and images of small-scale gravity lee waves implied strong westerly winds (Fig. 13; see Sec. II.C.3), indicating a background atmosphere highly susceptible to baroclinic instability (Briggs and Leovy 1974).

2. The Viking Meteorology Data. The Viking Lander *in situ* measurements of pressure, wind and temperature revealed the presence of highly coherent variations with periods of several sols or longer during the northern fall, winter and spring seasons. At the midlatitude Lander 2 site (48°N) these had the basic characteristics of transient baroclinic waves, as discussed below. The several-sol oscillations can be clearly seen in Fig. 23, where short-period variations ($<$ 1 sol) have been removed by filtering. At times, the regularity of these oscillations is quite striking.

Figure 24 shows power spectra of the pressure, wind and temperature data from Lander 2 for the northern winter-spring period of the first Mars year of Viking observations and for northern fall-winter of the second year. During the first period, prominent spectral peaks are present at periods of $\approx$ 6 to 8 sols and $\approx$ 3 sols, with lesser peaks at several shorter periods. Similar periodicities have been found in independent analyses of pressure for the initial Viking year (Tillman 1977; Tillman et al. 1979; Sharman and Ryan 1980; Niver and Hess 1982), and in all the meteorological variables during the first two Viking years (Barnes 1980,1981). While the longer-period variations dominated following the 1977b dust storm, shorter-period ($\approx$ 2 to 4 sols) disturbances were typically more prominent during the second year, although the longer-period waves were still present (Figs. 23 and 24). Barnes (1981) speculated that this reflects a generally clearer Mars atmosphere during the second year, with smaller static stabilities leading to stronger baroclinic instability at shorter wavelengths.

Relationships between the pressure, wind and temperature variations observed at Lander 2 allowed the inference of key properties of the wave disturbances (Ryan et al. 1978; Tillman et al. 1979; Barnes 1980,1981; Murphy

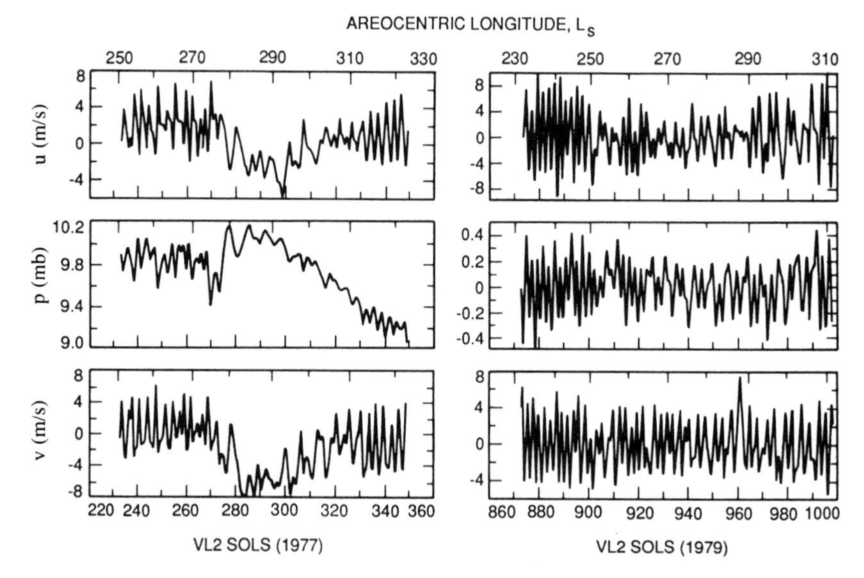

Fig. 23. Low-pass filtered pressure and wind data at Viking Lander 2 for: (left panel) northern fall-winter (L_s = 242°–330°) and (right panel) approximately one Mars year later (L_s = 223°–312°). The time series in the right panel have also been detrended. Units are: *p* [mbar], u,v [m s^{-1}]. Note the effect of the 1977b great dust storm during L_S275°–320° of the first year (figure from Barnes 1980,1981).

et al. 1990*b*). Examining data from the first-year fall and winter, Ryan et al. noted that the correlation of falling/rising pressures with southerly/northerly winds was indicative of eastward propagation and that the zonal winds implied the disturbances were centered somewhat north of the Lander 2 site. Utilizing the geostrophic relationship between pressure and meridional wind (discussed in Sec. II.C.1), they estimated zonal wavenumbers and wave phase speeds for the succession of wave disturbances of $\approx$ 4 to 6 and $\approx$ 5 to 15 m s^{-1}. These values are comparable to the theoretically expected values for baroclinic waves, although the zonal wavenumbers are slightly high. Based on a detailed analysis of the passage of one particular weather system during the first-year fall, Tillman et al. (1979) found evidence for the presence of a frontal structure.

Detailed cross-spectral analyses by Barnes of first and second year data from Lander 2 (Barnes 1980,1981) quantified the phase relationships and the (very high) coherences between the meteorological fields. Utilizing previous analyses (Leovy 1981) of tidal wind oscillations to relate the observed winds to the geostrophic flow, Barnes calculated values of the wavenumbers and phase speeds for the different dominant disturbance "modes".

The results indicated that the waves at Lander 2 propagate zonally, with (eastward) phase speeds $\approx$ 10 to 20 m s^{-1}. Waves of long period (6 to 8 sols) had relatively low zonal wavenumber ($m \leq 2$) during fall and winter of both

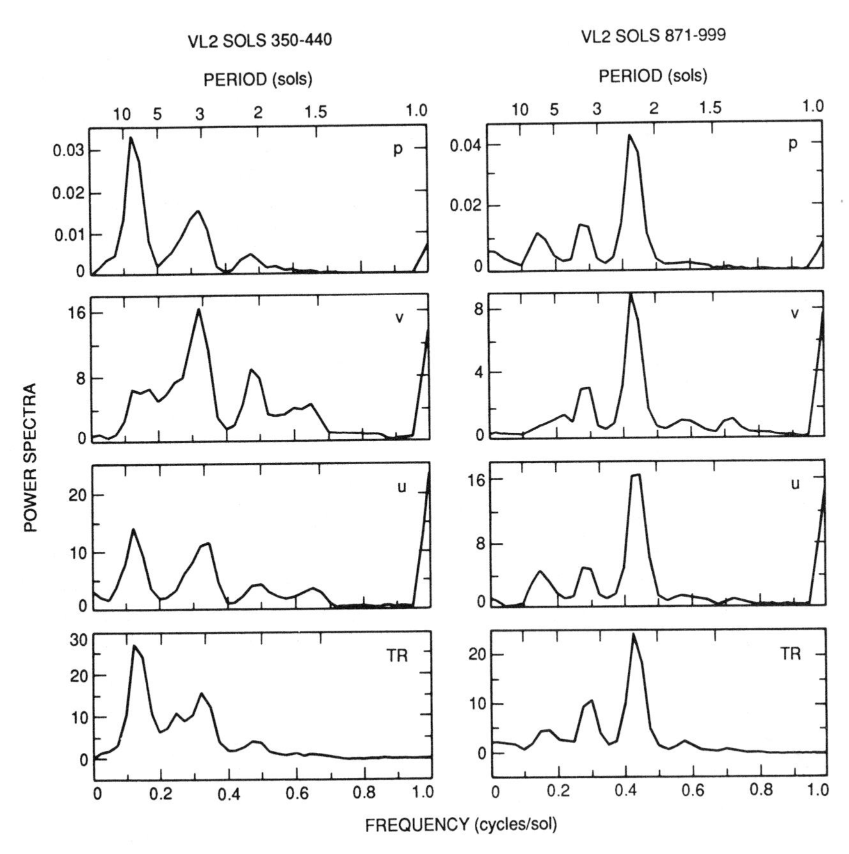

Fig. 24. Power spectra of pressure p, zonal and meridional winds u,v and of reference temperature TR computed for Viking Lander 2 data and the periods: (left) L_s = 324°–12°, Lander sols 350–440, in late northern winter-early spring; (right) L_s = 223°–312°, Lander 2 sols 871–999, late northern fall-early winter of the second year of Viking observations (see right-hand panel in Fig. 23). Note prominence of the short-period waves in the second year (figure from Barnes 1980, 1981).

Mars years of Viking data, except during the midwinter period following the 1977b great dust storm when $m \approx 3$. The shorter-period (2 to 4 sols) disturbances were generally associated with higher zonal wavenumbers ($m \approx 3$ to 5). Lander 1 data (at 23°N) revealed disturbances with similar periods, but substantially reduced amplitudes, as compared to those at Lander 2.

The Viking meteorology data, obtained over portions of four annual cycles, revealed significant seasonal and interannual variability in the transient baroclinic waves (Ryan et al. 1978; Barnes 1980,1981; Leovy et al. 1985). Observations following the winter solstice dust storm of the first Viking year (1977b) showed very anomalous high pressure and a substantial reduction in the wave amplitudes (Fig. 23). This period, which lasted through midwinter, has been attributed, at least in part, to a northward shift of the baroclinic zone, due to an expansion of the Hadley circulation driven in a dusty Martian atmosphere. Recent GCM simulations have reproduced this result (Pollack et al. 1990*a*).

As noted above, waves during the second year were predominantly of shorter period, higher wavenumber, and larger amplitude than those in the first year. Disturbances present in the third year Lander 1 data were of very large amplitude, with some appearing to be associated with the strongest observed winds at the site (see Sec. VI.D). Leovy et al. (1985) characterized the interannual variations in terms of two types of northern winters: one with planetary-scale dust storms and reduced baroclinic wave activity, and a second without great dust storms but with very vigorous waves which could raise considerable dust locally in northern subtropical and middle latitudes.

3. More Recent Modeling Studies. Modeling studies have gone significantly beyond the early work in predicting the properties of baroclinic waves in the Martian atmosphere. Gadian (1978) performed linear instability calculations with a β-plane model incorporating density stratification and radiative damping for a dust-free atmosphere; taking his basic-state parameters from Mariner 9 IRIS observations, he obtained maximum growth rates at zonal wavenumbers $m \approx 3$ to 6, with phase speeds ≈ 10 to 15 m s^{-1}. Using several parameterizations, Gadian also made estimates of poleward heat fluxes associated with such waves.

Barnes (1984) modeled linear baroclinic instability in both β-plane and spherical geometries. (β-plane models approximate the Coriolis parameter $f = 2\ \Omega \sin \theta$ as the sum of a constant f_0 and a linear term: $f \approx f_0 + \beta y$, $y = a\theta$.) Barnes' calculations allowed for more realistic vertical structure of the waves and incorporated radiative damping and zonally symmetric topography, as well; basic states were constrained from Mariner 9 and Viking data. Boundary layer frictional effects were incorporated in the form of Ekman pumping, in which the vertical velocity at the lower boundary is proportional to the geostrophic vorticity (e.g., see Holton 1979).

Growth rates and phase speeds were similar to those from earlier studies, with somewhat stronger instability at relatively short scales. A significant finding was a relative insensitivity of the most unstable scale to modest changes in static stability. However, growth rates for a highly stable dust-storm basic state were found to be quite low. As in earlier work, the topography in northern midlatitudes, sloping downward toward to the pole, was found to be strongly stabilizing. Also noteworthy was the absence of significant wave amplitude in low latitudes, as observed at Lander 1.

Using a simplified β-plane model, Barnes (1986) simulated the interaction between a single baroclinic wave and a zonal-mean background flow. In these simulations, strong radiative damping promoted a relatively "regular" wave-mean flow evolution and equilibration. The computed wave amplitudes were comparable near the surface to those observed by Viking and were relatively large near 20 km, consistent with the vertical thermal structure of a wave detected in Mariner 9 IRIS data (Conrath 1981).

Fully nonlinear GCM simulations of the Martian atmosphere, including effects of realistic dissipation and topography, have generated baroclinic waves. An early simulation for southern winter conditions (Pollack et al. 1981) showed that transient baroclinic eddies analogous to those inferred from the Viking data were present in winter middle and high latitudes. The heat and momentum transports, and energetics, of the GCM waves were quite similar to those of terrestrial baroclinic eddies. More recent MGCM simulations have incorporated much enhanced vertical resolution, the spatially varying topography, albedo and thermal inertia of the Mars Consortium, and the radiative effects of global dust hazes (Pollack et al. 1990*a*; Table III). In these results baroclinic eddies are very prominent during northern fall, winter and spring, for a wide range of dustiness. The transient eddy heat fluxes shown in Fig. 22 are due primarily to baroclinic eddies. However, the simulated transient eddies are quite weak during the corresponding seasons in the south. This may be due to the current MGCM topography at high southern latitudes, which is uncertain. The MGCM topography slopes steeply downward toward the pole, and, as discussed above, this is a highly stabilizing configuration.

B. Thermal Tides and Free Modes

Global oscillations of the wind, pressure and temperature fields of a planetary atmosphere are known as atmospheric tides, by analogy with the problem of describing the response of a global ocean to gravitational forcing. Depending upon the nature of their excitation, atmospheric tides are characterized as free or forced modes.

Forced modes are those components which are driven by nonrandom thermal or gravitational forcings; for these tides their frequency and spatial variation are closely tied to that of the forcing function. The forced tides of most interest are those responding to the largest components in the time-space domain of thermal or gravitational forcing. Because the Martian satellites have small masses and because the Sun's gravitational forcing is small compared to its thermal forcing, it is the thermally driven atmospheric tides which are of greatest interest for Mars.

Free modes are those global atmospheric oscillations which are driven by more or less random forcing. Their frequencies and spatial structures are determined, not by those of the forcing, but by the resonant properties of the atmosphere. In an inviscid, thermally undamped atmosphere, resonance would produce an infinite response in the meteorological fields for arbitrary forcing at the resonant frequency. In real atmospheres, free modes can be defined as those which would have the largest amplitudes for equal forcing at all frequencies. On rapidly rotating planets like Earth and Mars, the large-scale resonant properties are determined primarily by the atmospheric thermal structure and secondarily by the distribution of atmospheric winds. For this reason, observation of variations at the free-mode frequencies provide a sensitive indicator of the background thermal state of a planetary atmosphere.

On Earth, free modes, particularly at short periods ($\approx$ 1 day), tend to have small amplitudes compared to forced modes. However, because of the regularity of their oscillations, they can be detected in very long time series of surface pressure oscillations recorded at tropical and subtropical stations (records spanning more than 75 yr exist on Earth) where their amplitudes tend to be largest (Hamilton and Garcia 1986). On Mars, the thermal structure at low latitudes and thus the free-mode (i.e., resonant) frequencies vary to a much greater degree than on Earth (Zurek 1988*a*). Thus, free modes that are weak in amplitude compared to other transients cannot be isolated as readily in the Viking Lander pressure data, despite its multi-year length.

The largest amplitude atmospheric tides, of course, are those that have the optimum combination of strong forcing and efficient (i.e., resonant) response. Mars is particularly interesting in this regard, because its most strongly driven atmospheric tides have frequencies close to the resonant frequencies (Zurek 1976,1988*a*; Hamilton and Garcia 1986)

1. Thermal Tides. Atmospheric tides are a larger component of the total longitudinal and temporal variability of the Martian atmosphere than of Earth's, primarily because of the stronger daily varying solar heating (per unit mass). In a relatively clear atmosphere, the large daily variation produced in the ground temperature generates a thermotidal drive by the convective heat flux during the day and by radiative exchange between the atmosphere and ground during both day and night. In a dusty atmosphere, the incoming solar radiation is directly absorbed to generate a thermotidal drive which is significant even for the background dust hazes and which can be enormous during a great dust storm. The temperature and velocity variations associated with vertically propagating tidal components will grow with height, as air density decreases, and will contribute significantly to the time and space variation in the middle Martian atmosphere, even when the tides are relatively small near the ground.

Even though it used only a two-level model and no atmospheric dust, the original Mars GCM experiment (Leovy and Mintz 1969) generated a significant atmospheric tide. The straightforward application of the classical theory (Chapman and Lindzen 1970) which had been so successful in explaining quantitatively the observed atmospheric tides on Earth was limited by two considerations: the poorly known (and possibly adiabatic) temperature structure of the lower Martian atmosphere and the effects of the sizeable large-scale orography on Mars (Lindzen 1970).

Orography can modify the daily wave of solar heating so as to generate significant tidal components which do not move at the same rate, or even in the same direction, as the Sun's apparent motion. One of these, a diurnal Kelvin mode, is close to atmospheric resonance (Zurek 1976). (Kelvin modes are eastward propagating gravity waves which have maximum amplitudes at low latitudes, but distinctively small meridional velocities everywhere.) Its efficient response makes it the most likely of the orographically induced tides to be observed, and Conrath (1976) found evidence for such a wave in his analysis of the Mariner 9 IRIS temperature data.

Spacecraft observations of Mars have established the existence and quantified the size of the tidal components in three ways. The Viking Landers during entry detected vertical variations of temperature that were consistent with the rapid traverse of a vertically propagating atmospheric tide (Fig. 10). Once on the ground, the Viking Lander meteorology packages recorded a daily variation of surface pressure and of near-surface wind and temperature (Fig. 2; Hess et al. 1977; Tillman et al. 1979; Ryan and Henry 1979; Leovy and Zurek 1979; Leovy 1981; Tillman 1988). Finally, atmospheric temperatures derived from the Mariner 9 IRIS data in 1971 (Conrath et al. 1973) and the Viking Orbiter IRTM 15-μm channel during 1977–78 (T. Martin and Kieffer 1979) showed a significant variation with local time, particularly during dusty periods (Pirraglia and Conrath 1974; T. Martin 1981).

During two planet-encircling dust storms in 1977, the correlation of changes in the surface pressure field with changes in atmospheric dust loading was generally consistent with the theoretical predictions using classical theory of atmospheric tides driven by the daily solar heating of a dusty atmosphere (see, e.g., Zurek and Leovy 1981). As on Earth, the semidiurnal surface pressure oscillation at low latitudes was larger than its diurnal counterpart, even though the diurnal component of the heating was larger. On both planets, the semidiurnal tide has a long vertical wavelength and responds efficiently to elevated and vertically extended heating sources, while the diurnal tide has a short vertical wavelength and suffers from destructive interference (Lindzen 1966; Zurek 1980). On Mars, such a heating source occurs only during the periods of greatest dust opacity and is temporary; on Earth it is due to heating by stratospheric ozone.

Models of the correlation between visible optical depth and the surface pressure variation are good enough (Zurek 1981; Zurek and Leovy 1981) to be used with Lander 1 pressure data to estimate the peak opacity of great dust storms when the Lander Sun-diode measurements of opacity must be extrapolated or are absent altogether (see, e.g., Zurek 1982; Leovy et al. 1985; see chapter 29).

The effect of the elevated heating is most pronounced for surface pressure and wind. Away from the surface, the diurnal component of the meteorological fields is dominant in the middle latitudes, while the diurnal and semidiurnal fields are more comparable at lower latitudes. In a clear Martian atmosphere (when the thermotidal forcing is concentrated near the ground), the diurnal tide should be generally dominant.

Unlike the case for surface pressure, the daily variations in near-surface winds reflect local influences, as well as the inviscid velocities at the top of the boundary layer. However, even these can be modeled for the dust storm period with a relatively simple model of an oscillating Ekman boundary layer, driven by the inviscid winds in balance with the planetary-scale surface pressure gradients (Leovy 1981; see Sec. VI.D).

2. Free Modes. Tillman (1985,1988) identified several short-term enhancements in the multi-year Viking Lander surface pressure records of apparently large-scale waves having nearly diurnal or semidiurnal periods. One set of these transient features occurred during mid-northern summer ($L_S \approx 150°$); another set appeared during the onsets of the 1977b and 1982 great dust storms.

The midsummer transients detected at the Viking Landers lasted only a few sols. There are hints of these transients in each Mars year (*T* in Fig. 1, upper panel) at the time of the seasonal pressure minimum, when the south polar seasonal cap nears its maximum extent. In the first two years of data, the transients appeared as well-defined pairs of events, separated by some 20 sols. Using a maximum entropy method of time-series analysis, Tillman (1988) showed that these transients could be resolved into time-spectral components whose frequencies were close to, but distinct from, those of the diurnal and semidiurnal harmonics. Based on the free mode frequencies computed by Hamilton and Garcia (1986) for an isothermal atmosphere, Tillman (1988) proposed that these were nearly diurnal and semidiurnal Kelvin modes.

The short-lived prominence of the midsummer transients and their tendency to occur in pairs suggest that the atmospheric thermal structure moves relatively quickly through, and then back through, a state favorable to Kelvin mode resonance. The lack of representative temperature profiles for this period precludes a definitive test of the resonant (i.e., free-mode) hypothesis for the midsummer transients. Classical tidal theory (Zurek 1988*a*) indicates that the temperature trends shown by the Viking IRTM 15-μm brightness temperatures would move the atmosphere away from a resonant state during midsummer. The same theory, however, indicates that changes in the static stability below 15 km, undetectable by the Viking IRTM, might play a role.

The most dramatic transient event was the virtual disappearance of the diurnal surface pressure oscillation and a rapid change in its phase at Lander 1, but not at the more northern Lander 2, just prior to the 1977b storm (Leovy 1981; Tillman 1988). Zurek and Leovy (1981) suggested that this was due to the destructive interference at Lander 1 of thermally driven, but orographically modified tides. Tillman (1988) noted that the continuous phase shift could also be interpreted as a shift in the frequency of the dominant tidal mode from exactly diurnal to 1.1 cycles per sol, suggesting the explosive growth of a free mode (see Sec. VII).

Free-mode periods longer than a day have yet to be discovered in the Viking Lander time series, and yet terrestrial experience suggests (see, e.g., Salby 1984) that systematic observations of middle atmospheric temperatures will reveal such modes. The most likely mode is a 4-day Rossby wave (Zurek 1988*a*).

C. Other Transient Waves

Despite the current lack of observational evidence, there may well be large-scale transient waves in the Martian atmosphere in addition to baroclinic eddies, thermal tides and free modes. While Mariner 9 IRIS and Viking IRTM data hint at transient planetary-scale waves (T. Martin and Kieffer 1979; Conrath 1981), the asynoptic nature of these data does not permit a distinction even between traveling and stationary waves, so that their true nature remains obscure.

Modeling studies indicate that large-scale transient waves can be generated by atmospheric instabilities away from the surface. Barotropic instability may occur in winter near the very strong circumpolar jet (Barnes 1984; Michelangeli et al. 1987; Fig. 5). Michelangeli et al. calculated that the most unstable barotropic modes would have zonal wavenumbers 1 and 2 and periods of 1 to 3 days. As the speed of the zonal jet increases, the most unstable modes may move from the poleward to the equatorward side of the zonal wind maximum, suggesting change with season and dustiness.

Zonally symmetric disturbances associated with inertial instability have been found in numerical simulations (Haberle et al. 1982), and might be expected to exist in the Martian atmosphere in view of the strength of the zonal-mean circulation (especially under dusty conditions). Other types of large-scale transient waves that might be present (and detectable with future instrumentation) include inertia-gravity waves (with wavelengths of order 1000 km or more) and equatorially trapped waves (Kelvin and Rossby-gravity waves). These latter waves may propagate vertically and so exhibit relatively large amplitudes in the middle atmosphere.

D. Forced Quasi-Stationary Waves

Quasi-stationary waves of planetary scale should be present in the Martian atmosphere due to the mechanical lifting of air over orography and to the thermal forcing associated with topography. On Earth, such thermal forcing is due in large part to land-sea differences in diabatic heating of the atmosphere. On Mars, the laterally variable thermal forcing can be produced by elevation differences, in that heat transferred from the ground has a larger effect on the thinner atmosphere above high ground. Variations of ground temperature due to surface variations in albedo and thermal inertia may also contribute to variations in atmospheric heating, especially as there are "thermal continents"; i.e., vast regions of high ground (e.g., Tharsis and Arabia) that have relatively high surface albedo and relatively low thermal inertia (chapter 21).

Quasi-stationary waves are prominent in simulations generated both by GCMs and by more simplified, mechanistic models. The nature of the quasi-stationary waves should differ between two domains: summer tropical latitudes and winter extratropics. The discussion here is divided accordingly. Presently, there is little evidence for such waves, but this is almost certainly due to the limitations of the current data.

1. The Tropics and Summer Latitudes. Using a two-level linear model, Webster (1977) examined the steady circulation generated by mechanical lifting over Martian orography and by a simple parameterization of diurnally averaged orographic thermal forcing. Although complicated by effects of horizontal advection, the steady solutions obtained by Webster exhibit rising motion (and adiabatic cooling) in the regions of strongest heating (Tharsis) and sinking in lowland regions of cooling. At lower levels the upland (lowland) regions are warm (cool) and have low (high) pressure, with the pressure pattern reversed at upper levels. Horizontal flow converges into the upland areas at lower levels, with divergent, return flow aloft. Such a strong east-west overturning flow is similar to the Walker circulation, prominent over the equatorial Pacific in the Earth's atmosphere. A noteworthy feature of Webster's circulation was the fairly zonal character of the horizontal winds, due to the predominantly (steady) Kelvin mode response to stationary forcing in the presence of a basic state easterly flow.

There is little direct evidence for (or against) such a circulation. Earth-based historical data suggest that clouds are less prevalent over low-lying regions (such as Chryse) than over higher regions (Tharsis) in low latitudes, but the coverage is incomplete. Relatively opaque water-ice clouds do have an infrared signature which can be detected in the Viking IRTM data (Christensen and Zurek 1984). The pattern of occurrence of these signatures is consistent with a large-scale circulation pattern characterized by rising motions over uplands and sinking over lowlands, although this interpretation of the IRTM data cannot be definitive in the absence of supporting information on atmospheric temperatures and surface emissivities (Zurek and Christensen 1990).

Simulations with the Mars GCM have revealed a qualitatively similar standing wave field in lower latitudes (Pollack et al. 1981,1990*a*). More recent simulations indicate that continental-scale variations in the surface thermal inertia can play a role in forcing stationary waves. However, the GCM simulations also indicate that the response to the effects of variable thermal inertia and of elevated terrain need not linearly superimpose, even though high terrain tends to have low thermal inertia (but not necessarily vice versa; see Jakosky 1979). Unlike Webster's (1977) simulations, the MGCM appears to produce strong standing eddies in at least part of the summer hemisphere.

2. The Winter Extratropics. Mariner 9 IRIS data (Conrath 1981) suggest the presence of quasi-stationary waves in the Martian winter extratropics. These data, from late winter in the northern hemisphere, show a large-scale wave structure with maximum amplitude at upper levels (~0.5 to 1 mbar) and high latitudes (~65°N). Unfortunately, the asynoptic nature of the observations precludes a unique identification of the wavenumber and frequency. A stationary wave with zonal wavenumber $m = 2$ is one possibility, as is a quite long-period (~30 days) $m = 1$ disturbance. However, $m = 1,2$ and 3 disturbances traveling at phase speeds of ~5 to 10 m s^{-1} eastward also fit the data; such values are characteristic of the transient baroclinic eddies, as discussed above.

Viking IRTM data (from the 15-μm channel) show the presence of very large amplitude wave structure at high northern latitudes during the polar warming following the 1977b storm (T. Martin and Kieffer 1979; Jakosky and T. Martin 1987). These data are sparse and asynoptic, making it impossible to determine wavenumber and frequency, although it is clear that there is significant longitudinal variability.

Webster (1977) found large-amplitude planetary-scale disturbances in his simulations of the midlatitudes of both northern and southern winter hemispheres, with both mechanical and thermal orographic forcing playing a role. These simulated stationary waves were quasi-barotropic (having little variation of phase with height), with lows and highs tending to align with the topographic depressions and ridges, respectively. Earlier, Mass and Sagan (1976) obtained somewhat similar, mechanically forced quasi-stationary disturbances in a nonlinear model for southern winter conditions.

Incorporating the mechanical influence of southern hemisphere topography, Moriyama and Iwashima (1980) found strong enhancements of the quasi-stationary disturbances under dusty, equinox conditions. More recently, Hollingsworth and Barnes (1989) carried out linear studies with a high-resolution, primitive-equation model, simulating mechanically and thermally forced stationary waves in both winter hemispheres. For realistic basic states with latitudinal and vertical shear, they found that both zonal wavenumbers $m = 1$ and 2 can propagate vertically and so attain large amplitudes at upper levels (~20 to 40 km and above) in middle and high latitudes; this was consistent with Gadian's (1978) results for a mechanically forced $m = 2$ wave.

The GCM simulations of Pollack et al. (1981,1990*a*) have exhibited large-amplitude stationary waves in the winter extratropics of both hemispheres. Earlier experiments (Pollack et al. 1981) with a three-level version of the model, with topography and spatially varying albedo, showed stationary waves (for southern winter) qualitatively similar to those obtained by Webster, whose simulated waves were strongly dissipated (thermal damping times ~ 2 days). Recent MGCM experiments for various seasons have included spatially varying thermal inertia, and the vertical extent and resolution of the model have been greatly increased. Stationary waves are prominent in the MGCM simulations in middle and high winter latitudes and exhibit considerable vertical propagation, with large amplitudes at upper levels (~20 to 50 km).

VI. THE ATMOSPHERE NEAR THE GROUND

The accurate definition of meteorological and associated parameters at the surface-atmosphere interface is critically important for Mars, as these fields provide a fundamental set of scientific observations, as well as a lower boundary and "ground truth" for orbital observations and inferred parameters. Many outstanding questions about the climate of Mars, past and present, have to do with the exchange across the air-surface interface of volatile carbon dioxide and water and with the removal, reworking and emplacement of dust on the surface (chapters 21, 22, 23, 27, 28 and 29). A knowledge of how such materials are exchanged between the surface and the interior atmosphere, through the planetary boundary layer (PBL), is fundamental to answering those questions.

Furthermore, PBL fluxes of heat, mass and momentum are potent drivers for the interior circulation of the atmosphere. Thermodynamically, the PBL is that region where turbulence generated by thermal convection induced by intense heating of the surface or by mechanical instability of strong wind shears near the ground produces a temperature offset at the air-ground interface and rapid variation with height of temperature and wind. Finally, the PBL is the working environment at the surface, and a knowledge of PBL characteristics on Mars is needed for mission design and efficient operation during future exploration activities, whether robotic or manned, on the surface of Mars.

Generally, PBL fields cannot be fully resolved by remote sensing from orbit, as large variations in atmospheric fields take place near the ground in a short vertical distance. Of key interest are the vertical fluxes of heat, momentum and of trace constituents like dust and water vapor. The turbulent nature and fine structure of the PBL means that such fluxes must be inferred from surface measurements by indirect techniques or derived by fast sampling at one or more heights. The latter requires meteorological sensors of moderate complexity. The direct measurement of fluxes also requires quite careful considerations of instrument configurations, siting and deployment to avoid the effects of contamination by the platform itself. In practice, surface and PBL observations must be blended into models to derive the required PBL properties. However, validation of this approach ultimately requires rather complete field measurements from at least a few carefully chosen sites.

Most modeling work for Mars (including the GCMs) has parameterized the PBL using relationships extrapolated from terrestrial experience. Recently, PBL theoretical models with the requisite very fine vertical resolution and high-order closure schemes for nonlinear moments have been constructed for the study of Mars. All PBL modeling for Mars has been limited by the paucity of data specific to the PBL. Observations from orbit have defined certain key properties of the underlying surface (e.g., albedo and thermal inertia) and of the overlying interior atmosphere. Measurements of the PBL itself are largely restricted to a few high-vertical-resolution temperature profiles from spacecraft entry and radio occultation experiments and to data taken by the two landed Viking meteorology packages. Inferences regarding PBL fields have also been drawn from Orbiter imaging of cloud morphology and occurrence.

In the sections below, the likely vertical variation of PBL fields such as (potential) temperature and wind is presented, based on the application of terrestrial similarity theory and on preliminary model simulations of PBL fields. This is followed by a discussion of the observed PBL wind systems (VI.B), of constraints on water vapor near the surface (VI.C) and of the movement of dust in the PBL (VI.D). The section closes with a few remarks about the prospects for improving our understanding of this inadequately observed region of Mars (VI.E).

A. Structure of the Planetary Boundary Layer

Existing data do suggest that the planetary boundary layer on Mars *is* similar to that of Earth. Figure 25 uses the vertical variations of temperature T and of potential temperature

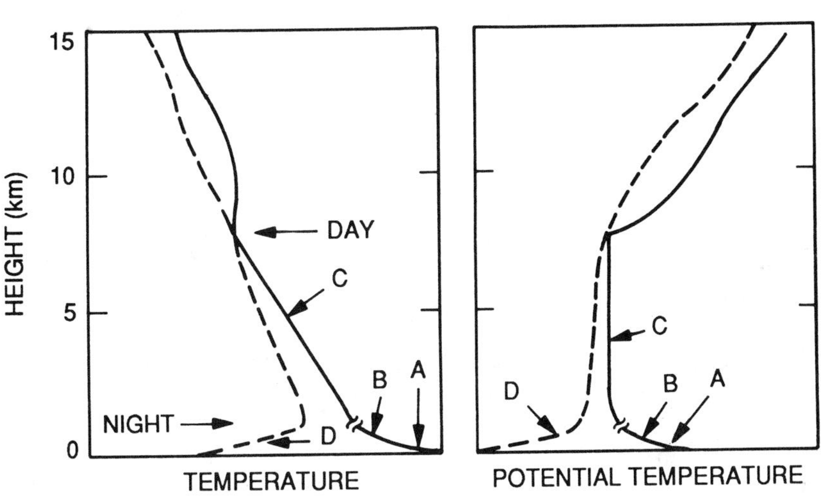

Fig. 25. Schematic description of the structure of the planetary boundary layer (PBL) on Mars (Sec. VI.A). The variations with height of temperature and of potential temperature are sketched for a relatively clear atmosphere. Solid lines represent mid-afternoon conditions, when the PBL is likely to attain its maximum depth; dashed lines indicate nighttime conditions. Horizontal arrows indicate the PBL top for the day and night profiles. Regions A: molecular conduction layer; B: unstable surface layer; C: convective part of the PBL; D: (nighttime) radiatively produced surface inversion. The disconnection indicates that the heights of A and B are greatly exaggerated (figure adapted from Leovy 1982).

$$\Theta \equiv T\left(\frac{p_s}{p}\right)^{\kappa} \tag{23}$$

to show schematically the structure of the atmosphere at and near the surface. In Eq. (23), p_s = surface pressure. Potential temperature is changed only by diabatic processes, because its pressure factor automatically takes into account temperature changes due to adiabatic expansion and contraction in a compressible atmosphere. Respectively, stable, neutral and unstable regions are characterized by increasing, constant and decreasing Θ with height z, or equivalently, lapse rates $-dT/dz$ which are less than, equal to or greater than the adiabatic value ≈ 4.5 K km^{-1}.

The lowest layer of the PBL (*A* in Fig. 25) is the molecular conduction layer in contact with the ground. Here, exchanges are controlled by the atmospheric molecular diffusivity and thermal conductivity. On Mars, this layer is ≈ 1 cm deep, about 100 times larger than on Earth (Sutton et al. 1978).

The next layer contains the greatest variation with height of temperature and wind speed. During the day, this is the unstable surface layer portion of the PBL (*B* in Fig. 25). This surface layer probably extends a few, to a few hundred, meters above the surface. The effects of Coriolis forces are generally smaller here than in the overlying PBL (Atreya 1988). After sunset, radiational cooling of the surface may lead to a surface inversion (*D* in Fig. 25). This can be up to a few hundred meters thick, except in flat terrain where it may be much thinner under calm conditions.

The convective region (*C* in Fig. 25) above the surface layer extends to the top of the well-mixed (constant Θ) zone. This top is also the top of the PBL, above which potential temperature increases with height and turbulent mixing is greatly reduced. Thermal convection typical of a clear, summer afternoon at low latitudes causes an unstable, turbulent PBL to grow several kilometers thick; the exact depth depends on the history of solar insolation earlier in the day and during previous sols, as well as the details of the wind and temperatures in the overlying, larger scale meteorological systems (Gierasch and Goody 1968; Tillman 1977; Pollack et al. 1979*b*).

At night on Earth (and probably on Mars), winds between the level of the nighttime surface inversion and the height corresponding to the top of the daytime PBL are often larger than the highest PBL winds at other times of day, because the surface-based inversion decouples the upper region from surface friction. The resulting wind, or "nocturnal jet," is thus strongly influenced by large-scale pressure gradients (Andre et al. 1978). Within the very stable surface inversion, turbulence can be produced mechanically by drawing energy from the shear of the nocturnal jet or from topographically produced drainage winds (Hess et al. 1976*c*).

1. The Surface-Layer Portion of the Planetary Boundary Layer. On Earth, wind and temperature profiles in both the molecular and surface layers are found to be logarithmic functions of height under neutral conditions, and modified by stability functions during stable and unstable conditions (Arya 1988):

$$\frac{U}{u_*} = \frac{1}{k}\left[\ln\frac{z}{z_o} - \Psi_m\left(\frac{z}{L}\right)\right] \tag{24}$$

$$\frac{\Theta - \Theta_o}{\Theta_*} = \frac{1}{k}\left[\ln\frac{z}{z_o} - \Psi_h\left(\frac{z}{L}\right)\right] \tag{25}$$

with geometric height z, velocity U, von Karman's constant $k = 0.4$, a characteristic surface roughness height z_o, a friction velocity u_*, $\Theta_o \equiv \Theta(z_o)$, $\Theta_* \equiv u_*^2\, \Theta_o/(k\, L\, g)$, and gravity g. The Monin-Obukhov "length" L characterizes the height of the surface layer; i.e., the height of transition from B to C in Fig. 25. In Eqs. (24) and (25), Ψ_m and Ψ_h are empirically determined functions of z/L; they have different forms for stable and unstable conditions and vanish for neutral conditions. Physically, u_*^2 represents the surface shear stress, or friction, per unit mass.

Using these relationships and given some knowledge of z_o and atmospheric static stability, profiles of wind and temperature can be generated from z_o to the top of the surface layer using the mean wind speed or the temperature measured at only a single height, as for the Viking Landers. Sutton et al. (1978) integrated a similar set of surface layer equations using a diurnally varying surface temperature estimated from Orbiter measurements (Kieffer 1976), together with Lander temperature and wind measurements, to estimate the stability, heat and momentum fluxes during the first 45 sols of Lander observations.

The Monin-Obukhov length L is inversely proportional to the vertical heat flux and is proportional to u_*^3. During midday at Lander 1, $-L$ dropped to 10 m due to strong insolation and low wind speeds (Fig. 26e; by convention, $L < 0$ for unstable conditions and $L > 0$ for stable conditions). Friction velocities u_* are shown in Fig. 26 for two different assumptions of surface roughness. The peak u_* values correspond to the peak wind speeds of each Lander's wind hodograph (Fig. 27). The higher u_* values at Lander 1 are due to the larger downslope winds of the Chryse Planitia site.

Heat fluxes were also correspondingly larger at Lander 1, although heat fluxes are more sensitive to the assumed z_0 values than u_* or L (Fig. 26c,d). Following Leovy (1969*a*), Sutton et al. (1978) matched the molecular layer to the surface layer to obtain better estimates of these parameters by explicitly considering the temperature drop across the molecular, as well as the surface, layer. This decreased the effects of different z_o on the heat flux estimates, now giving peak values of approximately 15 and 20 W m^{-2} for Lander 1 (Fig.

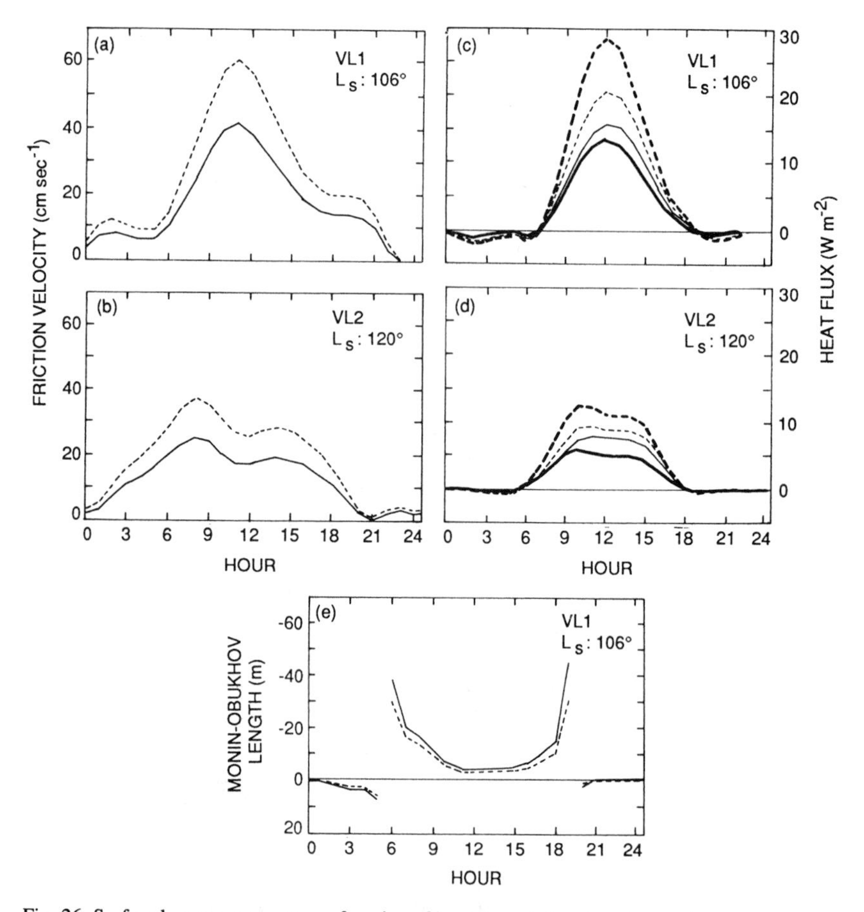

Fig. 26. Surface layer parameters as a function of local time of day (from local midnight), derived from Viking Lander data taken during early summer and averaged for several sols: (a,b) friction velocities u^* at Lander 1 (sols 16–30; $L_s \approx 106°$) and Lander 2 (sols 0–15; $L_s \approx 120°$), assuming surface roughness height $z_0 = 0.1$ cm (solid curve) and 1.0 cm (dashed). (c,d) heat fluxes derived for the same periods and z_0, with (thin lines) and without (heavy lines) inclusion of the molecular conduction sublayer. (e) Monin-Obukhov length for Lander 1 (sols 16–20), with molecular heat conduction taken into account (figure from Sutton et al. 1978).

26c) and 7 and 9 W m^{-2} for Lander 2 (Fig. 26d), for z_o values of 0.1 and 1.0 cm, respectively. Nighttime heat fluxes are much lower than the daytime ones at both sites; u_* is also lower then, particularly at Lander 2.

Tillman (1972) developed a technique for estimating heat fluxes exclusively from temperature statistics, and a crude application gives values of 10 W m^{-2} for Lander 1. Potentially, this technique could determine static stability and u_* solely from temperature statistics during unstable conditions, when the fluxes are the largest.

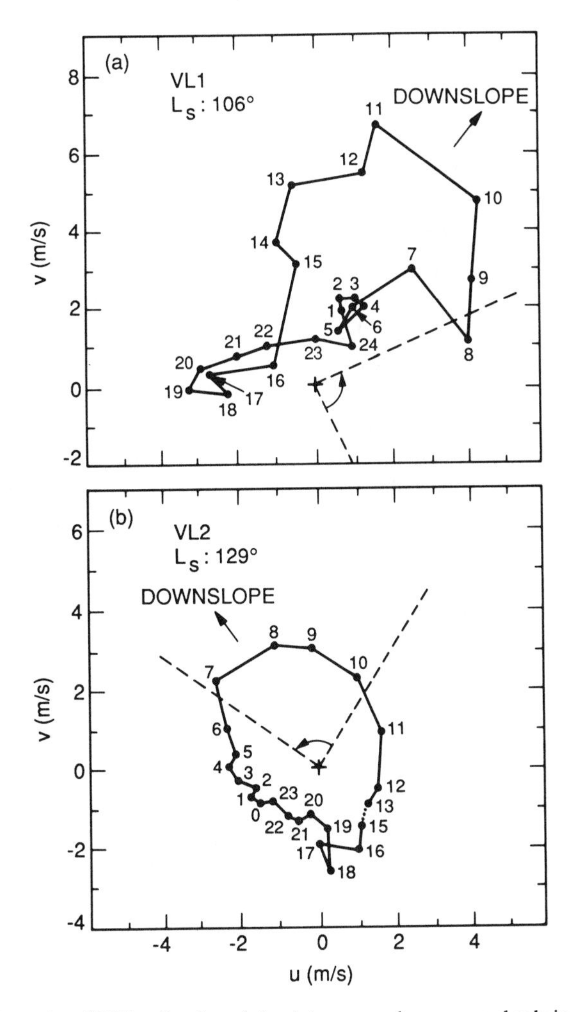

Fig. 27. Hodographs of Viking Lander winds, data averaged over several sols in hourly bins. A line drawn from the cross to the labeled points indicates the vector wind components (u, eastward; v, northward) measured at that hour (local time) and at ≈ 1.6-m height. Dashed lines indicate sector of Lander interference (wind blowing across the Lander to the sensor). Bold arrows point downslope (regional slope ≈ 0.015 at Lander 1, 0.002 at Lander 2). (a) First 44 sols at Lander 1. (b) First 50 sols at Lander 2 (figure from Hess et al. 1977).

2. Modeling the Martian Surface Layer. The potential for comparing data and model results is well illustrated by the terrestrial work of Andre et al. (1978). Using a high-order turbulence model, they compare the predicted evolution of temperature, wind, humidity and other parameters over a 24-hr period to terrestrial field observations of the PBL. For Mars, no equivalent comparison can be made due to the lack of appropriate data.

Results of the PBL model of Haberle (unpublished), based on the second-order closure scheme of Mellor and Yamada (1982, and references therein), are shown in Fig. 28 to illustrate the possible vertical variation of winds and potential temperature for the Lander 1 site at the most stable and unstable times of day during early northern summer. A westerly geostrophic wind of 10 m s^{-1}, based on a Mars GCM run for this season, has been imposed at the top of the modeled PBL and the speed in Fig. 28 has been forced to zero at z_o for purposes of plotting. Note the strong nighttime inversion and the nocturnal jet, whose 12 m s^{-1} speed exceeds the geostrophic wind imposed at the top of the PBL. Also shown in Fig. 28 are the predictions of the similarity theory (Eqs. 24 and 25), illustrating both their utility for the surface layer and their sensitivity to key parameters.

Diurnal variations computed by Haberle's model (not shown here) indicate that the surface layer typically becomes neutral by 8 a.m., unstable by 9 a.m. and increasingly unstable from 10 a.m. to 1 p.m. The PBL depth grows from $<$ 100 m at 7 a.m to $\approx$ 5 km by 3 p.m. This latter depth is similar to the maximum convective layer depth derived for summer in analytical and model comparisons of daytime flows over sloping terrains for Earth and Mars by Ye et al. (1990). From 6 p.m. through 6 a.m., the model simulation shows the formation of a shallow surface inversion, and the temperature at 1.5 m, close to the height of the Viking Lander measurements, decreases from 241 K to 190 K (cf. Fig. 11).

3. Measurements of the Vertical Variation of the PBL. Spacecraft entry and radio-occultation experiments provide the only direct measurements on the vertical variation of temperature and winds throughout the PBL. Imaging from orbiting spacecraft of convective clouds provides qualitative indications of the depth of the daytime PBL in some cases (Kahn 1984). Tillman (1977) suggests that surface-based convection extends to 2.5 km or greater based on cumulus cloud shadows; this is consistent with the model results discussed earlier.

During their entry in northern summer, Landers 1 and 2 measured temperatures using their footpad sensors from 3.5 km down to 1.5 km above the surface, where the terminal descent rocket engines were fired (Seiff and Kirk 1977). As expected, the near-surface atmosphere at these altitudes was relatively stable in the morning and nearly neutral in the late afternoon. Landing in Chryse near 4 p.m. local time, Lander 1 measured a lapse rate $-dT/dz$ which was slightly stable (3.7 K km^{-1} vs the adiabatic value of 4.5 K km^{-1}). Lander 2 measured a much more stable lapse rate of $-dT/dz \approx 0.97$ K km^{-1} around 10 a.m. Earlier, during northern spring, Viking radio-occultation measurements revealed a nearly neutral boundary layer in Chryse around 4 p.m.; there was also a suggestion of an unstable, superadiabatic variation of temperature within 1 km of the surface (Lindal et al. 1979).

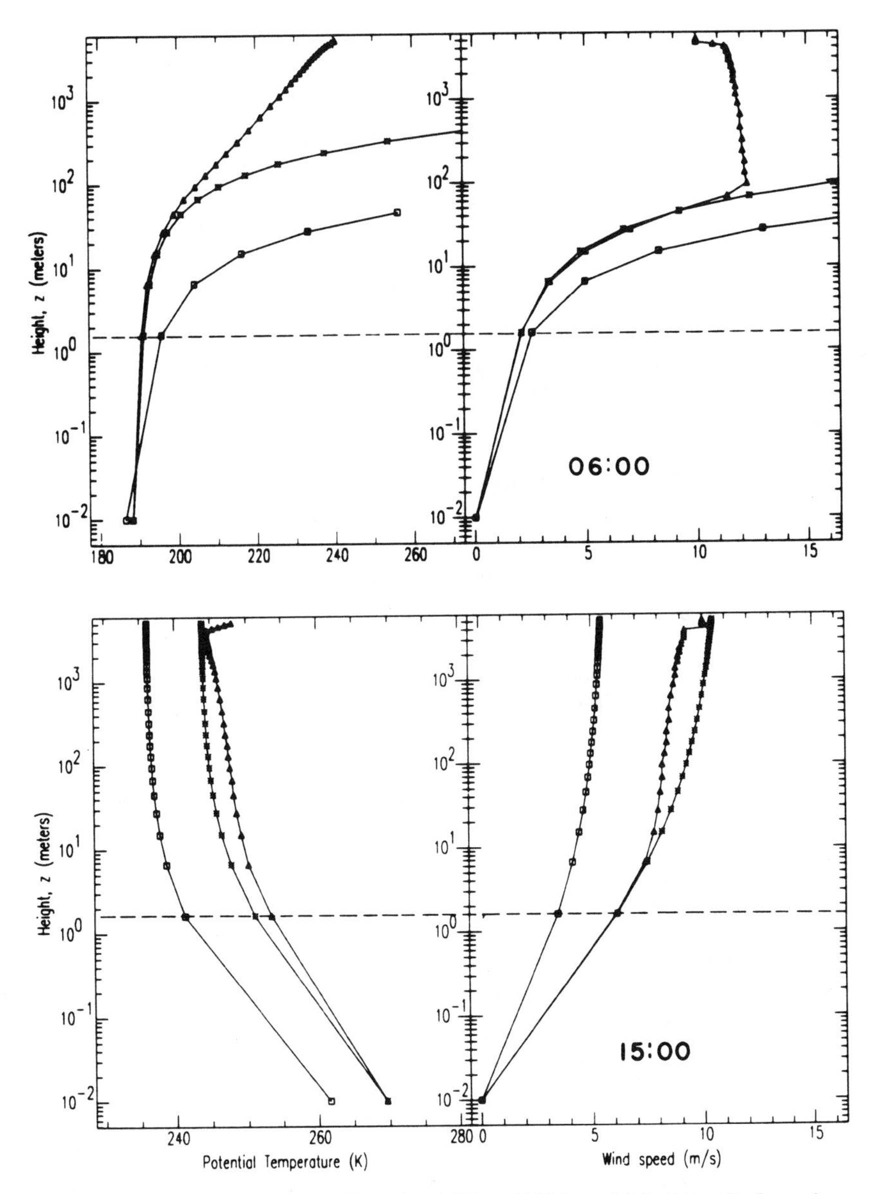

Fig. 28. Fields generated by a one-dimensional (32-level) PBL model for Mars. Surface values are plotted at z_0; dashed line approximates the 1.6-m height of the Viking Lander meteorological sensors. The conditions assumed are L_s = 110° (early summer), dust opacity τ = 0.4, z_0 = 1 cm and a top-level (geostrophic) wind U_g = 10 m s^{-1}, V_g = 0. (Top) Potential temperature and wind speed vs height from z_0 to 5 km at 06:00 local time. (Δ) indicates Haberle PBL model results; (■) indicates similarity profile with constants from Haberle model; (□) indicates similarity profile with constants and surface thermal model from Sutton et al. (1978). (Bottom) Same, except for 15:00 local time. Similarity estimates are from Tillman and Johnson (unpublished).

B. Winds in the Martian Planetary Boundary Layer

1. Summertime Climatology at Viking Landers 1 and 2. The local meteorology, as governed by regional surface properties and topography and as modulated by large-scale atmospheric tides, is best examined during summer when the transient wave activity is smallest; the Viking mission timeline was ideal in this regard. Lander 1 began operating on the Martian surface shortly after summer solstice (L_s = 97°). As on Earth during summer in the subtropics, the conditions were very repeatable from sol to sol (Hess et al. 1977; Leovy et al. 1985), except for the monotonic decrease of pressure due to CO_2 condensation in the south polar regions (Plate 16). There are essentially no transient large-scale flows (e.g., baroclinic weather systems) at either of the Lander sites during this season.

Strong diurnal variations of the wind were observed during early summer at both Lander sites (Fig. 27). At Lander 1, however, the diurnal variation is dominated by a strong upslope-downslope component (Hess et al. 1976*c*) and is consistent with the pattern expected over a strongly heated slope (Hess et al. 1977). These types of flow have been treated analytically (Blumsack et al. 1973; Hess et al. 1977). At Lander 2, the diurnal variation clearly prevails during summer. At both sites, wind speeds (at the 1.6-m height) were generally light at night (2 m s^{-1}) and stronger during the day (6 to 8 m s^{-1}); the highest speeds during this period are 5.3 and 7.4 m s^{-1} at Landers 1 and 2, respectively.

2. Slope Winds. The large-scale sloping terrain present over much of Mars can significantly affect the steady and diurnally varying wind and temperature fields, especially near the surface (Blumsack 1971*a,b;* Blumsack et al. 1973; Ye et al. 1990). During the day, buoyancy forces generated by strong surface heating have an upslope component, while at night they are oppositely directed. This simple pattern is modified by the Coriolis force and diurnal eddy mixing within the boundary layer. Blumsack et al. modeled these processes and found that significant boundary layer wind oscillations ($\approx$ 20 m s^{-1}) could occur over typical Martian slopes ($\approx$ 0.005).

Magalhaes and Gierasch (1982) have presented a model of Martian slope winds to explain the occurrence of dark streaks that form in the Tharsis and Elysium regions (Lee et al. 1982), and in the southern hemisphere as well (Thomas et al. 1981). Their model differs from Blumsack et al. (1973) in that it is valid only for the nighttime downslope case when the flow achieves a momentum balance between the slope-directed buoyancy force and friction, and a thermal balance between adiabatic warming and thermal diffusion. Such balances are readily achieved on relatively steep slopes (10%), or on moderately steep slopes ($\approx$2%) with a long fetch. In these areas, maximum downslope winds of 20 to 30 m s^{-1} are predicted to develop within the lowest 100 m just before dawn. Magalhaes and Gierasch conclude that these winds could provide a mechanism for the production of the dark streaks which have few visible topographic points of origin (see chapters 22 and 29).

3. Modeling PBL Winds for the 1977 Dust Storms. During the 1977 dust storms the semidiurnal component of the daily variation of surface pressure and of wind observed at the Viking Landers was amplified significantly (Leovy and Zurek 1979; Leovy 1981; Tillman 1988). Leovy found that the planetary-scale semidiurnal tidal wind could be simply related to the (boundary layer) winds measured at the height of 1.6 m. Using an Ekman boundary layer model (see, e.g., Holton 1979) with turbulent mixing parameterized by a constant eddy viscosity and driven at the top by an inviscid wind computed from the observed semidiurnal pressure variation, he reproduced the observed wind at 1.6 m given the choice of a single free parameter, namely the ratio of a (constant) PBL depth to a surface layer length scale, given by u_*/σ where $\sigma = 2$ per sol for the semidiurnal tide. If this parameter is assumed to be independent of the frequency σ of the driving atmospheric tide, it yields an estimate of the ratio of the zero frequency component of the Lander winds (at 1.6 m) to the geostrophic (not tidal) wind at the *top* of the PBL. Values of this ratio are in the range 0.3 to 0.6, with higher values tending to occur during the dustier periods when semidiurnal winds were strongest.

C. Water Vapor in the Planetary Boundary Layer

No direct measurements of water vapor were made by the Viking Landers. Humidity is an important but difficult *in situ* measurement on Mars. Ryan and Sharman (1981*a*) argued that inflections observed in the nighttime temperature decrease were in many instances best explained by radiative effects due to the formation of water-ice fogs. Assuming this to be the case, Ryan et al. (1982) found a seasonal variation of water-vapor mass mixing ratios during the first year of Viking measurements on the order of 0.6×10^{-5} to 10^{-4} at Lander 1 and 10^{-8} to 10^{-4} at Lander 2. Assuming uniform mixing with height (and an atmospheric scale height of 10 km), these mass mixing ratios imply column abundances ranging up to 20 precipitable μm (i.e., the depth of a water layer equivalent in mass per unit area to the column mass of water vapor: 1 pr μm $\leftrightarrow 10^{-3}$ kg m^{-2}), as compared to column abundances (during midday) of 30 to 35 pr μm derived from the Viking Orbiter MAWD data (Jakosky and Farmer 1982).

Although this indirect technique is subject to many sources of error, Ryan et al. (1982) presented circumstantial evidence for its reliability between sols 119 and 129 ($L_s \approx 172°$) at Lander 2 by noting the correlation between periods of higher inferred moisture and of winds from the south. More direct evidence was provided by optical depths derived from Viking Lander imaging of Phobos at night which indicated increased opacity between 2 and 4 a.m. and of the Sun which indicated additional morning, as opposed to afternoon, opacity. Pollack et al. (1977) attributed the enhanced morning opacity to ground fogs of water ice, consistent with theoretical predictions (Flasar and Goody 1976).

Latent heating of a layer of air due to the condensation there of water vapor can produce only a limited temperature change:

$$\Delta T < \chi_m \frac{L_w}{c_p} \approx 3300\ \chi_m \leq 0.3\ \mathrm{K} \tag{26}$$

with L_w the latent heat of sublimation, c_p the specific heat at constant pressure and χ_m the mass ratio of condensing water vapor to heated air. In Eq. (26), $\chi_m \leq 10^{-4}$ was used, as indicated by Ryan et al. (1982). Even so, the ΔT limit is an order of magnitude too low to explain the inflection points in the nighttime Viking Lander temperature data ($\Delta T \approx 2$ K). Using a one-dimensional numerical model, Ryan and Sharman (1981*a*) demonstrated that the radiative effects of a water-ice fog could produce temperature inflections in the nighttime PBL, though their relatively simple model was unable to produce such inflections at altitudes as low as the 1.6-m height of the Viking Lander observations.

The observed cloud distribution indicates that saturation of water vapor occurs most frequently at northern midlatitudes during spring and summer (Kahn 1984). These are the times and places when the Viking Orbiter MAWD experiments measured the greatest vertical column abundances of water vapor (chapter 28). However, as seen visually by the Mariner 9 and Viking Orbiters, localized fogs were most prevalent in the southern hemisphere, particularly during summer (Kahn 1984).

At the Viking Landers, the overhead opacities measured in the morning were generally larger by a few tenths than those measured in the afternoon. These differences were ascribed on the basis of one-dimensional radiative-convective model simulations to the condensation of water-ice aerosols at night and their evaporation in midday. These model simulations by Colburn et al. (1989) indicated that the observed seasonal enhancements of the diurnal opacity difference could be explained in terms of seasonal variations of water column abundances and insolation and, in a less straightforward way, of the effects of atmospheric dust. Their simulations indicated that nighttime condensation occurred most frequently above 25 km and not at the ground, as proposed by Ryan and Sharman (1981*a*). More sophisticated modeling of the planetary boundary layer may clarify these issues, but it seems likely that more detailed observations of humidity and opacity near the ground, perhaps by surface meteorological packages, are required.

D. Dust Movement in the Planetary Boundary Layer

1. The Saltation Layer. The saltation layer is the layer next to the ground where sand-sized particles are temporarily raised and blown about once surface winds exceed a certain threshold. The impact at the surface of these larger particles can lead to the raising and suspension of fine-grained dust particles. The majority of the large-particle motion should take place within 20 cm of the surface, for sand particles having diameters around 100 μm, and the particle mass flux may peak at the 5-cm height. The velocities of these particles generally reach only 30% of the free stream velocity (see chapter 22).

Calculations of the threshold wind speeds needed for saltation to occur, as well as many aspects of erosion on the Martian surface, are extensively treated and compared with that on other planets by Greeley and Iversen (1985) and are discussed with regard to Mars in chapter 22. The threshold u_* for the most easily moved (~100-μm sized) particles is ~1.5 m s^{-1}, while friction velocities needed to raise directly both larger and smaller particles are significantly greater; for example, $u_* > 4$ m s^{-1} is required for the 1 to 10 μm sized dust particles observed suspended in the atmosphere during major dust storms (chapter 22). Using Eq. (24), u_* ~1.5 m s^{-1} is equivalent to winds of ≈20 and 30 m s^{-1} at the 1.6-m height of the Viking Lander wind sensors, assuming surface roughness values of $z_0 = 1$ and 0.1 cm, respectively, as suggested by Sutton et al. (1978). These winds are comparable to the strongest gusts seen at the Viking Landers (Ryan et al. 1981; Arvidson et al. 1983).

A local dust storm was observed at Lander 1 during the clearing of the 1977b great dust storm on a day when winds measured at 1.6 m gusted to peak values of 25 to 30 m s^{-1}. (Hourly averages peaked at 15 m s^{-1}.) Viking Orbiter imaging (James and Evans 1981) and Lander data (Ryan et al. 1981) indicated this local dust storm was due to the passage of a baroclinic wave. Figure 29 shows the variation of near-surface meteorological fields during the time of this local dust storm. (It was not possible to test correlation of the wind gusts with the visible structure of the storm.) Apparently, no significant alteration of the surface occurred during this storm (Arvidson et al. 1983). Wind gusts reached peaks of 26 m s^{-1} at Lander 1 (but only 14 m s^{-1} at Lander 2) during the arrival of both the 1977a,b great dust storms (Ryan and Henry 1979). This suggests that dust was raised locally at those times, at least at Lander 1.

Arvidson et al. (1983) noted that the most significant disturbance of surface material at Lander 1 apparently occurred during the winter of the third year of Lander 1 observations, when bright dust deposits left by earlier dust storms were eroded and when piles of soil, artificially created using the Lander sample arms, were modified and in some cases removed. They suggested that wind speeds slightly above the threshold wind velocities (i.e., winds ≥ 30 m s^{-1}) could, after initiating saltation, account for the observed erosion of the piles and apparent disruption of clods 4 to 5 mm in size. Winds this strong were not observed in prior years when the Lander 1 wind sensor was still operating normally. However, the pressure fluctuations measured during this third winter were the largest at Lander 1 in the multi-year record (Plate

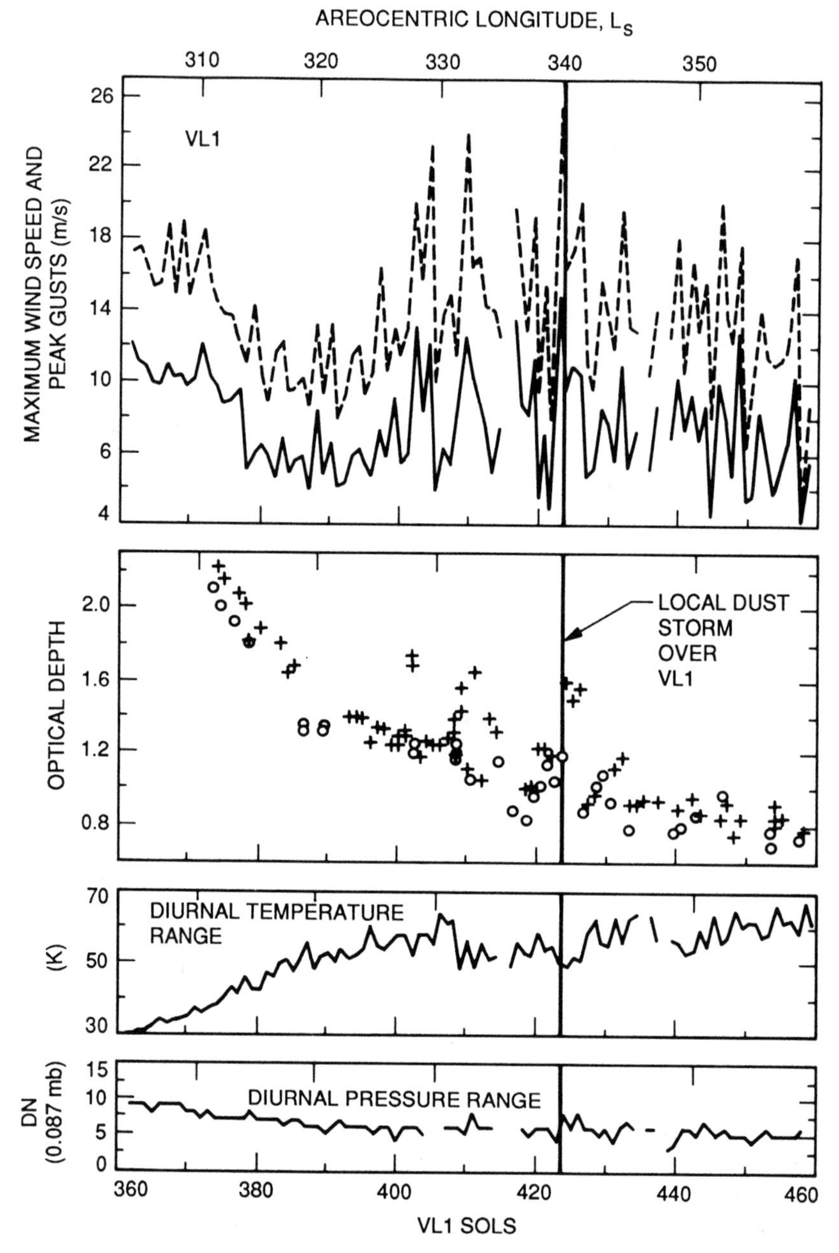

Fig. 29. Maximum wind speed (≈ hourly averaged, solid curve), peak gust (from point-by-point data, dashed curve), optical depth, daily temperature and pressure ranges observed at Viking Lander 1, all for each sol during a period spanning the observed passage (vertical line) of a local dust storm over the site. Optical depths are from Pollack et al. (1979*b*); + morning values, o afternoon values (figure from Ryan et al. 1981).

16). Their nearly 2-sol wave period is reminiscent of baroclinic waves observed at Lander 2 and of smaller amplitude waves by factors of 2 to 3, seen at Lander 1 in previous years (Barnes 1981). This suggests that winds associated with baroclinic waves were indeed responsible for the significant movement of surface material observed at Lander 1.

Taken together, the observations at the Viking Landers indicate that gusting winds of 25 to 30 m s^{-1} at a height of 1.6 m are required to raise dust into the atmosphere. As discussed above, these winds are consistent with values established by theory and by laboratory experimentation for the threshold friction velocities needed to raise small dust particles ($\leq$ 10 μm) or to cause erosion of unconsolidated surface material by saltation of the most easily moved ($\approx$ 100 μm) particles. Such winds were achieved only relatively rarely at the Viking Lander sites and, except during the planetary-scale dust storms, were associated with the apparent passage of baroclinic waves.

2. *Dust Devils.* Dust devils provide one means of raising dust even when background winds are light (Ryan 1964; Greeley and Iversen 1985). On Earth, these vortical winds, rendered visible by dust and sand picked up from the surface, typically have dust columns $<$ 30 m in diameter, extending $\lesssim$ 700 m in altitude, and lasting only a few minutes (see, e.g., Sinclair 1969; Ryan and Carroll 1970). There is evidence that very large dust devils ($>$ 100 m in diameter, $>$ 1 km in height) may occur more frequently on Mars.

Thomas and Gierasch (1985) detected columns of dust and cone- or funnel-shaped clouds in a few high-resolution ($\approx$ 70 m per pixel) Viking orbiter images of Mars. None of the 99 events so documented were near the Viking Lander sites; all occurred in the latitude range 33° to 43°N, nearly all in Arcadia Planitia. The Martian dust devils typically extended 1 to 3 km in height (the tallest reached 6.8 km). Their widths varied with height. All the cloud bases appeared to be narrow, with many at the limit of resolution, while some dust clouds were nearly a kilometer across near their tops. Smaller dust devils, which terrestrial experience suggests should be more frequent, would not have been detected in these Viking Orbiter images. All of those detected were observed between 2 and 3 p.m. local time and in summer (L_s $= 122° - 124°$, 139°) and over relatively smooth areas. This suggests that the vortices developed from convective plumes, although such transformations are not well understood. The greater size may reflect a deeper planetary boundary layer during the afternoon on Mars and a more superadiabatic surface layer (Ryan 1972).

Ryan and Lucich (1983) found evidence in the Viking Lander meteorology data for the passage of $\approx$120 (vertically oriented) vortices over the Lander meteorology sensors. Seven of these vortices had peak gusting velocities u_{max} $>$ 25 m s^{-1}; three were estimated to have produced peak gusts nearby (at a height of 1.6 m) of 36 to 44 m s^{-1}. As discussed above, these velocities appear sufficient to raise dust by saltation, suggesting that the vortices were indeed dust devils. While such speeds could initiate saltation of large ($\approx$ 100-μm sized) particles, the low core pressure of a dust devil may raise particles of smaller size as well (see chapter 22).

The cores of the dust devils inferred from the Viking Lander data were estimated to be in the range $\approx$ 200 to 600 m, consistent with the dust devils imaged in Arcadia Planitia (Thomas and Gierasch 1985). Smaller vortices inferred from the Lander data were associated with weaker winds and thus not classified as "dust devils." As expected, vortical winds were most frequent during early afternoon in spring and summer (occurring on $\approx$ 50% of the sols examined), and they seemed equally likely to occur at both Lander sites. However, all but one of the suspected dust devils (i.e., vortices with $u_{max} > 25$ m s^{-1}) occurred at Lander 1 during the dusty period following the onset of the 1977b great dust storm; the exception occurred at Lander 2 during late winter.

The data are too limited to estimate the true frequency and spatial range of occurrence of dust devils, but they may contribute significantly to the background dust haze detected above the Viking Landers during northern summer (see chapter 29).

E. Future Prospects for Planetary Boundary Layer Measurements

As noted several times, there is very little direct data on the structure and variation of the PBL on Mars. Needed are measurements of the vertical profiles of winds (speed and direction), temperature, the direct or indirect measurement of surface fluxes, and the determination of surface roughness, thermal and radiative properties, all at a number of sites. Much could be learned from surface stations measuring pressure, opacity and temperature, particularly if techniques for estimating many key PBL parameters from temperature statistics alone can be refined and validated (see, e.g., Tillman 1972). The addition of wind and humidity instruments to a network of stations would be ideal. While conventional wind and humidity instruments have been difficult to accommodate given practical limitations on the payload, deployment and operation of small surface stations, new technical approaches appear promising (e.g., ion-discharge anemometers; J. Tillman, unpublished). Given an adequate number of stations (> 20), a network of even relatively simple meteorological stations would provide some data needed to characterize the PBL at diverse sites, as well as to study regional and global meteorology.

To fully characterize the PBL requires vertical profiling from at least a few key sites. Future missions offer some exciting possibilities. Spacecraft may, for instance, release balloons into the Mars atmosphere which are tethered at a site or which descend to the surface after sunset and ascend after sunrise. Such balloons would provide unique platforms for sampling the variation with height of the PBL. Radio-occultation experiments could continue to provide key measurements with vertical resolutions of the PBL as fine as

1 km at certain sites. Advanced meteorological sounding stations, capable of detailed and prolonged vertical profiling of the lower atmosphere, including the PBL, could be deployed as part of more ambitious missions to the surface of Mars. Such stations would provide for Mars the equivalent of the PBL field experiments that have proven so useful for validating models and analysis techniques on Earth, while providing "ground truth" for remote sensing observations from orbiting spacecraft.

VII. GREAT DUST STORMS ON MARS

Dust in the atmosphere of Mars is the major absorber of solar radiation whenever its opacity exceeds even a few tenths (Sec. III). The changes that can be produced in the structure and circulation of the atmosphere of Mars are quite remarkable by comparison with Earth. These changes are largest during the episodic planet-encircling dust storms, when dust suspended in the atmosphere obscures much of the planet and extends up to 45 km or more (see, e.g., Leovy et al. 1972; Anderson and Leovy 1978; Briggs et al. 1979; chapter 2).

The Martian dust cycle, including both seasonal and long-term aspects, is discussed in detail in chapter 29. Here the discussion is focused on the great dust storms (i.e., the regional and planet-encircling events; see chapter 29), which can alter significantly the atmospheric circulation and thermal structure. Golitsyn (1973) and Gierasch (1974) provide reviews of the early work on the evolution of great dust storms; Zurek (1982) provides an update, based on data from the Viking mission.

A. Radiative-Dynamic Feedbacks

Dust heating affects temperatures (Sec. III), which alters the pressure fields and thus winds. By modifying the atmosphere's static stability, suspended dust indirectly affects the transfer of sensible heat from the surface and potentially the strength of the near-surface winds. Suspended dust also modifies the internal mixing of the atmosphere by providing a direct thermal drive for vertically propagating waves which may eventually "break" (Sec. IV.A). Alterations in the atmospheric static stability may amplify lee waves generated by mountain ridges or crater rims (Pirraglia 1976). Variations in atmospheric static stability and circulation will affect the ability of waves generated near the surface to propagate vertically, thus altering the supply of energy and momentum to the upper atmosphere (Barnes 1990*a*).

All theories for the onsets of the great dust storms invoke, at some stage, the positive feedback provided by the raising and heating of suspended dust: strong winds raise dust from the surface into the atmosphere, radiative heating of the suspended dust amplifies the diabatic drive for winds, and the winds increase in speed, raising more dust. Because suspended dust affects the thermal structure of the atmosphere, there also exists the potential for a negative feedback: dust raised into the atmosphere increases the static stability near the surface, possibly preventing the mixing of momentum in the upper boundary layer to the surface, thus decreasing surface winds and shutting off the supply of dust. This negative feedback is highly speculative, given the lack of near-surface atmospheric data in the dust-raising regions and the uncertainties in modeling the planetary boundary layer on Mars (Sec. VI).

B. Onset of Planet-Encircling Dust Storms

Spacecraft have yet to provide a comprehensive global view of the onset of a great dust storm. Earth-based observations have provided the best "synoptic" views (L. Martin 1974*a*,*b*,1976; chapter 2), but with relatively coarse spatial resolution. Even during the most global of dust storms there are observed to be large-scale variations in the dust concentrations during both the onset and decay phases (see, e.g., L. Martin 1974*a*,*b*,1976; T. Martin 1986, 1992; Thorpe 1979,1981). This spatial variability may reflect a variable distribution of source regions on the surface, as well as variable surface winds and properties of the planetary boundary layer.

Planet-encircling dust storms appear to result from expansion of one or more centers of activity; no dust storm covering vast regions has yet been observed to appear spontaneously. Present data have not revealed a significant difference in the morphology of those local dust storms that grow or coalesce into major dust storms from those that do not. Viking images of the beginning of the 1977a storm (Fig. 30) show an extensive, highly structured dust cloud wrapping around the high ground of Claritas Fossae; this cloud is reminiscent of Viking visual images of local dust storms that did not expand to this scale. To the right in Fig. 30, the dust activity is more mottled and less linear in form. Such structures could reflect the pattern of dust raising, but they may also be due to the convective action induced by the heating of the suspended dust itself.

Any model of dust storm onset must address the following observed attributes of the planet-encircling dust storms (also see chapter 29).

1. Expansion. Most local dust storms do not grow to large scale. Those that do expand to planetary-scale, appear to do so in stages: from longitudinally confined sectors of a latitudinal zone, to most or all of the zone and then to other latitudes. The expansion is primarily in an east-west direction at first, though modified by topography (Fig. 30). There may be significant diurnal variation in this expansion, with daily contraction and regeneration (L. Martin 1974*a*,*b*,1976; Gierasch 1974).

There is tremendous variability in the maximum areal extents attained by these storms. As might be expected, the longest-lived storms tend to cover the most area. Opacity and surface pressure measurements from Viking Lander 1 during the 1977a,b dust storms, suggest that the longest-lived storms attain the greatest opacities, as well.

Fig. 30. Viking Orbiter 2 visual imaging mosaic (red filter) of initial stages of onset of the 1977a great dust storm (L_s = 207°; orbit 178). The organized cloud at left extends around the southwest and southern edges of the Tharsis plateau (Claritas and Thaumasia Fossae). Vallis Marineris is visible in upper right corner (figure published in Briggs et al. 1979).

2. Location of Origin. Great dust storms originate mainly in the southern subtropics and in particular longitudinal sectors, namely the regions encompassing Solis, Syria and Sinai Planae; and the Hellas-Hellespontus-Noachis sector. These are also preferred regions for local dust storm development, but local storms do originate elsewhere. For instance, Viking Orbiter visual imaging indicated that the 1977b storm began in southern midlatitudes (Thorpe 1981). Interestingly, the thermal inertias derived from the Viking IRTM observations of regions like Solis Planum are characteristic of surfaces having large particles, rather than μm-sized dust particles; this suggests that saltation may play a role in raising dust from the surface (Peterfreund and Kieffer 1979; Peterfreund 1981; Sec. VI.D; chapters 21 and 22).

3. Time of Occurrence. Planet-encircling dust storms have been observed to occur during southern spring and summer, all within 90° in L_s of perihelion. No planet-encircling storm has been observed during northern spring and summer, a period when opacities were lowest at the northern hemisphere Viking Lander sites (Colburn et al. 1989). The most global of the observed dust storms occurred immediately after perihelion in 1971. One or two regional or planet-encircling dust storms may occur in any given Mars year, but there are many years when no such storm has occurred. Notwithstanding the observational biases in the historical record, due principally to the cyclical variation of distance between Earth and Mars, the Earth-based data do hint that planet-encircling dust storms have been more frequent in the last 40 yr than earlier in this century (Zurek and L. Martin 1989, 1992; chapter 29).

C. Dust Storm Decay

As seen at the Viking Landers, the column dust opacity begins to decline almost immediately after the onset of the planet-encircling dust storms (Pollack et al. 1979*b;* Colburn et al. 1989). Since neither of the Landers may be in a region of sustained dust raising, the rapid increase and subsequent decrease of dust opacity during the 1977a,b storms is consistent with the passage of a dust front (that is, the overhead appearance of the edge of the dust haze), followed by a clearing of that dust haze due to particle loss at the surface coupled with insufficient resupply of dust to the northern hemisphere. The question then becomes why do the dust source regions become inactive.

Pollack et al. (1979*b*) noted that there appeared to be a distinct change in the rate of clearing of atmospheric dust during the two 1977 planet-encircling dust storms, as the vertical column opacity of the suspended dust declined to values $\tau \approx 1.2$. The early stages of decay had exponential time constants of 75 and 51 days at Lander 1 for the 1977a and 1977b dust storms, respectively. Time scales for the latter stages of decay were 171 and 119 days, respectively. Conrath (1975) inferred an exponential time constant of 60 days for the latter stage of decay of the 1971 global storm in the zone 20° to 30° S and at both the 0.3 and 2 mbar level. Anderson and Leovy (1978) noted that this 1971 dust storm cleared first at high latitudes and more slowly at low latitudes.

D. Mechanisms for Dust Storm Onset

Several theories have been proposed for the onset of a regional or planet-encircling dust storm. All of these mechanisms utilize the positive feedback between dust and winds discussed in Sec. VII.A.

1. The Dusty Hurricane. Gierasch and Goody (1973) proposed that solar heating of suspended dust plays a role analogous to latent heating by condensing water vapor in a hurricane on Earth. In theory, the storm begins with a weak vortex in the presence of relatively light surface winds. As air convergences to form a dusty core, it is heated and rises, pulling in more air at low levels. Conserving angular momentum, near-surface air speeds up, picking up more dust, as it spirals into the increasingly dusty center. At higher levels, dusty air flows out in a reversed spiral structure. Given the importance of local solar heating, the dusty hurricane should have a strong diurnal variation.

The strongest argument against this theory is that local, regional and planet-encircling dust storms do not exhibit the expected spiral structure (Fig. 30, Plate 17). In fact, only a very few clouds have exhibited any spiral structure. These few were seen at high midlatitudes during northern summer (Gierasch et al. 1979; Hunt and James 1979), a time and place where surface winds are thought to be light (Hess et al. 1976*c;* Pollack et al. 1976*b*,1981).

2. Superposition of Planetary-Scale Circulations. Leovy et al. (1973*b*) proposed that superposition of the seasonally varying Hadley, atmospheric tidal and planetary-scale topographic winds would produce speeds great enough to raise dust from the surface. The success of this superposition depends upon a seasonal maximum in atmospheric heating during southern spring and summer to intensify the solar-driven Hadley and atmospheric tidal circulations. The enhancement would be partly due to the increased insolation near perihelion, but may also require greater atmospheric opacity, probably due to local dust storms. Increasing opacity was observed at the Viking Landers in the northern hemisphere in early southern spring, prior to the 1977a dust storm. Local dust storms had occurred in southern midlatitudes and near the edges of the seasonal south polar cap where winds are expected to be strong (Peterfreund and Kieffer 1979; Peterfreund 1981).

The atmospheric tidal circulation has its strongest winds in the subtropics of the summer hemispheres, and its strength is sensitive to seasonal changes in atmospheric dust opacity. Within the subtropics, the diabatic and mechanical effects of the large-scale orography, or potentially of longitudinally varying dust hazes, may generate eastward traveling diurnal Kelvin modes (Zurek 1976). Interference between this mode and the normal westward traveling diurnal tide would amplify response at some longitudes and reduce it at others (Zurek and Leovy 1981). Destructive interference at Lander 1 is suggested by the virtual disappearance of the diurnal surface pressure oscillation there just prior to the onset of the 1977b storm (Leovy 1981; Tillman 1988). If Lander 1 is near the longitude where eastward and westward traveling diurnal tides cancel, these same tides would reinforce one another to generate maximum winds at the top of the PBL near 135°W and 320°W, close to longitudes where planet-encircling dust storms have originated. Even in relatively clear periods, near-surface tidal winds there are computed to exceed 20 m s^{-1} (Zurek 1976). Diurnal variation is inherent in the mechanism.

As dust is spread throughout the subtropics, the Hadley circulation becomes much stronger and advects the dust towards the equator and into the northern hemisphere. How far north it goes depends on the magnitude of the diabatic heating and thus on the amount of dust raised first in the southern hemisphere (Haberle et al. 1982; Schneider 1983).

3. Transitional Hadley Circulation. Schneider (1983) proposed that zonally symmetric models could produce dust-storm behavior even without the aid of tides or topographic winds. Using a steady-state, nearly inviscid model, he explored situations in which the maximum solar heating was offset from the equator at latitude θ_0 (see Sec. IV.A.2). His model results indicated that there was a critical value Q_c of the total diabatic forcing $\bar{Q}$, in which the Hadley cell remained in the summer hemisphere when $\bar{Q} < Q_c$, extended up to the equator when $\bar{Q} \approx Q_c$, and extended discontinuously into the high latitudes of the winter hemisphere for any incremental heating ($\bar{Q} > Q_c$). Schneider identified this discontinuous extension to a global domain with the transition from a local to a global dust storm and argued that it occurred because the "front" separating the Hadley regime from the radiative equilibrium region was statically unstable, with cold, dense air above warm, less dense air in the zone between the equator and $-\theta_0$.

Since $\bar{Q}$ could be interpreted in Schneider's idealized model as the product of the area and height of the region of net diabiatic heating, global expansion could occur when the local storm grew big enough or tall enough. Larger values of Q_c were required as the initial zone of heating was moved poleward, so that an incipient dust storm in low latitudes was most effective in expanding to global scale for a given heating. However, the modeled surface friction velocities tended to increase strongly when the initial storm was moved farther from the equator, so that the "optimum" location of the initial storm was in the subtropics.

The possible roles of background or initial dust hazes in amplifying the Mars Hadley circulation introduce two factors by which interannual variability can occur in dust storm evolution: the stochastic nature of local dust storm occurrence and the vagaries of producing a hemispheric or global background haze. The Hadley circulation itself may produce such variation in the following way (Haberle 1986*b*): when a global dust storm occurs, dust is transported into the northern subtropics, where it may be raised locally in the following year (by baroclinic waves, for instance) to produce a northern background dust haze. The zonal-mean circulation produced by this northern haze might weaken the development of the solstitial cross-equatorial Hadley circulation, thus preventing a planet-encircling storm from occurring or perhaps limiting it to the southern hemisphere. Enough dust may then be lost from the northern subtropics, or sequestered in regions where it cannot be resuspended (Christensen 1986*a*), that the countervailing circulation will be too weak in subsequent years to prevent a planet-encircling dust storm.

4. Radiative-Dynamic Instability. Radiative-dynamic instability models (Houben 1981; Ghan and Covey 1990) assume that the spatial distribution of suspended dust is not generated by some independent means, but is controlled by the transport associated with the wave that is being generated through the solar heating of the dust haze. In these models, the dust storm begins as an initial perturbation of a basic-state dust distribution that varies with height or longitude. The diabatic heating produced by this perturbed dust distribution thermally drives an atmospheric wave whose longitudinal structure and phase is such that advection by the wave, linearized about the basic state, amplifies the initial perturbation in the dust distribution. This linear instability represents the onset of the dust storm.

Decay of the dust storm presumably occurs as the unstable wave grows large and the advection of dust becomes fully nonlinear, disrupting the dust pattern which led to its growth. Thus far, these models have been highly simplified, treating storms as linearized perturbations on prescribed basic states. It is not known whether radiative-dynamic instability will occur in models with more degrees or freedom. A fully integrated dust transport equation is required for Mars GCMs, or any other three-dimensional model, to test this mechanism in detail. Such models are currently in development.

5. Free-Mode Triggering. Tillman (1988) proposed a variation on the Leovy et al. (1973*b*) mechanism, by hypothesizing that the atmosphere passed through a resonant state, generating a free mode, rather than a thermally forced tide, which enabled the large-scale circulation to initiate a great dust storm. The best evidence for this mechanism was the observation during the onset of 1977b storm that much of the daily variance appeared to be concentrated in modes whose periods were almost, but not quite, diurnal and semidiurnal.

Zurek (1988*a*) raised two problems with this free-mode resonance serving as the trigger. First, the free modes most likely to be resonant are the short-period Kelvin modes, whose surface winds are strong throughout low latitudes, not just in the subtropics. Second, his tidal calculations suggested that resonance would be more effective in the colder, less dusty atmosphere of northern, rather than southern, spring and summer. If so, the absence of an enhanced background dust haze prior to a great dust storm might favor free-mode triggering, while its presence would favor the thermally forced tides.

The onset of a planet-encircling dust storm may require the optimal combination of efficient response (i.e., near-resonance) and amplified thermotidal forcing. Zurek (1976) found this to be the case when assessing the contribution of a thermally forced diurnal Kelvin mode to dust-storm onset, as discussed earlier.

E. Mechanisms for Dust-Storm Decay

If a planet-encircling storm decays simply because the dust supply in its source region is exhausted, dust must be resupplied into that region on timescales of a few years, as several storms appear to originate in the same regions. The resupply of localized regions must be by local weathering, which is likely to be too slow (chapters 21 and 22), by the movement of saltating particles (e.g., sand) into locales within the region where dust is still present, or by atmospheric transport. Atmospheric resupply would require a relatively long period, as dust transport into the region is likely to be weaker than the vigorous export which occurs during the planet-encircling storms themselves. While much of the dust raised in a given region may, of course, fall back to the surface there, a significant fraction must be exported elsewhere. It is conceivable that the regions from which dust is initially exported actually gain dust during planet-encircling storms from other regions which were initially inactive. If so, decay of the dust storms must be due to radiative-dynamic mechanisms.

In the dusty hurricane model dust-raising ceases once the high-level outflow of dusty air destroys the differential heating around the dusty core. This would seem to occur almost as soon as the initial local dust storm expanded significantly. Leovy et al. (1973*b*) proposed no mechanism to turn off the raising of dust. A negative feedback has been suggested (Pollack et al. 1979*b*) in which increased static stability near the surface suppresses surface winds, perhaps by decoupling them from winds at higher levels. Initial simulations of this mechanism are not encouraging (Haberle, unpublished).

The 1971 storm cleared first at high latitudes. This has been ascribed to scavenging of the dust by condensation of water vapor and CO_2 in the polar regions (Pollack et al. 1979*b*) or to the vertical mixing induced at low latitudes by atmospheric tides (Zurek 1976). Murphy et al. (1990*a*) found that other dynamical processes would also affect clearing rates; two key processes were lofting of the dust by upward motion in the dust storm source regions and the cross-equatorial resupply of suspended dust to northern sites.

Whatever the means of dust storm decay, Pollack et al. (1979*b*) suggested that there was a threshold dust opacity of $\tau \approx 1.2$, below which dust raising would begin again. As evidence, they cited the slower decay rates characterizing the later stages of the two 1977 dust storms and noted that several local storms were seen during the last stages of the 1977a storm, prior to the onset of the second planet-encircling storm.

If, as Tillman (1988) proposed, a resonant (free) mode is the trigger for dust raising, then dust-storm decay would be the result of a quick passage through resonance due to warming of the increasingly dusty atmosphere moving to a less favorable resonant state (Zurek 1988*a*). If the strength of the tidal winds was due in part to direct thermal forcing of a diurnal Kelvin wave, as well as efficient response (Zurek and Leovy 1981), then dust-storm decay will commence as dust is spread more uniformly with longitude.

In radiative-dynamic instability models for dust-storm onset, the winds associated with the growing wave become strong and destroy the special suspended dust distribution which forced the wave. Winds decrease, and the original source regions become inactive.

F. Interannual Variability

A difficulty for any of the dynamical models proposed for the origin and evolution of planet-encircling dust storms is accounting for the fact that such storms have occurred repeatedly, but not regularly. There is also great variation in their ultimate spatial extent. No reliable seasonal trigger is apparent, as there are many years without a planet-encircling dust storm (L. Martin 1984; Zurek and L. Martin 1989, 1992; chapter 29).

Perhaps great dust storms are examples (yet another class) of a chaotic dynamical system in which highly nonlinear feedbacks produce aperiodic, but recurring, responses (Zurek and Haberle 1989). (A terrestrial analogue may be the El Niño-Southern Oscillation phenomenon [see, e.g., Vallis 1986]). Leovy et al. (1985) suggested that relatively small random fluctuations in the atmosphere dust-polar cap system could push the atmosphere into one of two climatic regimes, one with and one without planetary-scale dust storms. Whatever the cause of the aperiodicity, a means must be found of weighting the effects of a year with Martian regional or planet-encircling dust storms relative to those without such storms. Otherwise, it may not be possible to ascertain the net effect over decades and longer time spans, particularly the very long periods of variation in the orbital elements (chapters 29 and 33).

Unlike Earth with its oceans, Mars would appear to lack a relatively long-term heat reservoir. In one sense, however, the ability to change atmospheric temperatures is "stored" wherever dust on the surface can be mobilized. The location of such areas will change, in response to changes in the availability of movable dust and perhaps of coarse particles likely to initiate saltation, in the strength of atmospheric winds, and in the patterns of their optimal combination.

VIII. SUMMARY

On Mars, variations of atmospheric heating due to suspended dust produce significant seasonal and interannual variations of the general circulation. As a result, observations over time of the general circulation and of the dustiness of the Mars atmosphere can define quantitatively the relationships between thermal forcing and atmospheric response. In effect, Mars provides not one, but rather a wide range, of realizations of the general circulation of a planetary atmosphere, and it does so on the time scale of a few years.

The detailed insights provided by the Mariner 9, Mars and Viking spacecraft into the thermal structure and dynamics of the atmosphere of Mars have confirmed many of the expectations about its general circulation and its wave activity that were drawn from terrestrial experience, while taking into account the unique attributes of Mars, namely the role of condensation and sublimation of CO_2, the radiative effects of variable dust loading, and the eccentricity and phasing of its orbit.

With regard to the Martian general circulation, the outstanding issues continue to be the strength, extent and nature of Hadley-type circulations; the relative contributions of the condensation flow, overturning circulations, stationary and traveling waves to the zonal-mean state; and the character of the flow in the virtually unobserved middle atmosphere region at 45 to 110 km altitude.

With regard to atmospheric waves, the outstanding issues to be resolved are the strength of (baroclinic) wave activity at high southern latitudes during winter, the existence and strength of topographically induced planetary waves and Walker-type (i.e., longitudinal) circulations, the level of gravity-wave activity and its effect on the momentum budget at various altitudes, and of course, the true nature of large transient events like dust storms and polar warmings.

The mechanisms by which dust storms on Mars evolve to large scale, often with almost explosive growth, are still being debated. There is still no comprehensive theory which explains all observed features of these events. A major problem is how to explain the marked interannual variability of regional and planet-encircling dust storms.

A better understanding of the mechanisms for seasonal and interannual variability in the atmospheric circulation is required to illuminate the roles of atmospheric transport in seasonal and longer-term cycles of water vapor, of carbon dioxide and of dust itself. Quantitative portrayals of these constituent budgets and of the thermodynamical balances for heat and momentum will also require more information about the planetary boundary layer and the physical interactions between the atmosphere and surface.

Atmospheric and climate models currently being applied to Mars are comparable in their sophistication to standard models used to understand the weather and climate of Earth. The Mars-GCM developed at NASA Ames, for example, is even now being extended to accommodate gravity-wave parameterizations and to allow for the interactive transport of dust. This will allow investigation of a wide range of radiative-dynamic feedbacks believed to be important to the structure and circulation of the atmosphere. Unfortunately, it also will open more pathways for the simulations to diverge from reality, so that acquisition of global meteorological data for the validation of model simulations becomes even more imperative.

Future MGCM developments are needed in the areas of more sophisticated boundary-layer parameterizations, of extension of the model tops to higher altitudes, and of simulations with higher spatial resolutions. Such improvements require faster computers and more clever computing algorithms. With regard to modeling the diabatic forcing of the atmosphere, the limiting uncertainty is now the lack of detailed knowledge of the time-dependent, spatially varying amounts and optical properties of dust suspended in the atmosphere.

Further advances in our understanding of the dynamics of the Martian atmosphere depend critically on future acquisition of global, long-term data records describing the weather and climate of Mars. Future Mars missions can provide the required systematic and global and nearly synoptic mapping of the atmosphere and surface by deploying appropriate instrumentation on orbiting spacecraft and on the surface itself. These data, together with still more sophisticated models, will yield a much more complete picture of the Martian general circulation and boundary layer and will fulfill the promise of a truly comparative study of the atmospheres and climates of Earth and Mars.

Acknowledgments. We wish to thank B. Conrath, P. Gierasch, A. Ingersoll, B. Jakosky and R. Kahn for their many helpful comments and for their perseverance in reviewing the manuscript. Special support was provided by Neal Johnson and the staff of the U. W. Viking Computer Facility. This work was supported by the Planetary Atmospheres Program of the National Aeronautics and Space Administration. The work was performed in part at Oregon State University, the University of Washington, NASA Ames Research Center and at the Jet Propulsion Laboratory, California Institute of Technology, under contract with NASA.

APPENDIX
LIST OF SYMBOLS

p atmospheric pressure
p_s atmospheric surface pressure
p_r (constant) reference pressure (e.g., 6.1 mb)
ρ atmospheric density
g gravity
ϕ gravitational potential (geopotential)

z	geometric height
z^*	pressure-based vertical coordinate $\equiv H_r \ln(p_r/p)$
R	gas constant for the Mars atmosphere
T	atmospheric temperature
Q	net diabatic heating per unit mass
Θ	potential temperature
H	atmospheric scale height $= RT/g$
H_r	reference H, $\equiv$ 10 km
H^*	depth of an idealized Hadley circulation
c_p	specific heat at constant pressure
κ	$= R / c_p$
$< >$	$=>$ vertically integrated quantity
$\overline{(\ )}$	overbar $=>$ zonal average
$()'$	prime $=>$ deviation from zonal average; e.g., T'
∂	$=>$partial derivative; i.e., $\partial/\partial t$
u	eastward (zonal) velocity component
v	northward (meridional) velocity component
ω	upward velocity $\equiv dz^*/dt$
t	time
z_{bc}	reference pressure level in z^* coordinates
x	west-to-east cordinate: $dx/d\lambda = a \cos \theta$
y	south-to-north coordinate: $dy = a\, d\theta$
a	planetary radius
λ	east longitude
θ	latitude
f	Coriolis parameter $f = 2\, \Omega \sin \theta$
Ω	planetary rotation frequency $= 2\pi$/day
β	Beta parameter, i.e., $f \approx$ constant $f_o + \beta y$
M	angular momentum per unit mass $\equiv a^2\Omega\cos^2\theta + u\, a \cos \theta$
ζ	absolute vorticity $= f + (\partial v/\partial x) - (\partial u/\partial y) + (u \tan \theta)/a$
f^*	$= f$ assuming geostrophic balance $= f + (u \tan \theta)/a$ assuming cyclostrophic balance
u_g	geostrophic or (planetary-scale) gradient zonal wind $= -(\partial\phi/\partial y)/f^*$
v_g	geostrophic or (planetary-scale) gradient meridional wind $= (\partial\phi/\partial x)/f^*$
v^*	residual-mean meridional velocity
ω^*	residual-mean vertical velocity
S	static stability $= R\, (\kappa T/H_r + dT/dz^*)$
N^2	Brunt-Väisälä (buoyancy) frequency $\approx S/H$
c	phase speed of a wave
ξ^{-1}	vertical e-folding distance or vertical wavelength (divided by $2\,\pi$) for vertically propagating wave
j	downwind horizontal wavenumber

m	zonal wavenumber (number of maxima around a latitude circle)
δ	time required for atmospheric radiative heating to be comparable to latent heat release of water sublimation
I_0	incoming solar radiation at the top of the atmosphere
α	fraction of I_0 absorbed by suspended dust
τ	dust optical depth at visible wavelengths
τ_r	radiative damping time
L_w	latent heat of sublimation
M_w	column water amount
χ_m	mass ratio of condensing water vapor to CO_2 on Mars
F_x, F_y	frictional forces per unit mass
E_T	divergence of zonal-mean heat fluxes
E_M	divergence of zonal-mean fluxes of angular momentum per unit mass
ν	viscosity coefficient
ν_H	horizontal eddy viscosity or mixing coefficient
η	Rayleigh friction coefficient
T_e	radiative equilibrium temperature
θ_0	latitude of the rising branch of the Hadley circulation
θ^*	poleward limit of the Hadley cell; low-latitude limit of Hadley cell
θ_1	latitude of maximum solar heating
$\Delta<>$	vertically averaged contrast across the Hadley cell
L	Monin-Obukhov length, a characteristic surface layer height
u^*	surface friction velocity
z_0	surface roughness height
k	von Karman's constant ≈ 0.4

27. THE SEASONAL CYCLE OF CARBON DIOXIDE ON MARS

PHILIP B. JAMES
University of Toledo

HUGH H. KIEFFER
U.S. Geological Survey

and

DAVID A. PAIGE
University of California, Los Angeles

Carbon dioxide, the major ingredient of the Martian atmosphere, condenses in the polar regions of the planet during their respective polar nights. The latent heat released is the major energy source during these periods. Condensed CO_2 sublimes during the spring and summer seasons in response to solar radiation. As much as 30% of the atmosphere takes part in this seasonal CO_2 cycle. The extent of the solid CO_2 deposits is controlled by their equilibrium with the predominantly CO_2 atmosphere which, in turn, depends upon physical properties such as their albedo and infrared emissivity. If the albedo of the condensed CO_2 is sufficiently large, condensation may equal or exceed sublimation during an annual cycle; permanent (or residual) CO_2 caps which survive the entire seasonal cycle are therefore possible. The existence of such deposits has great significance for the Martian climate as they would determine the mean annual atmospheric pressure through their equilibrium with the atmosphere and therefore control the response to changes in the insolation at the poles produced, for example, by obliquity changes. One of the major discoveries of Viking was a small, residual CO_2 cap in the southern hemisphere. The results of Viking investigations relevant to the CO_2 cycle are presented, and the extensive modeling efforts directed towards understanding this cycle and its couplings to the sea-

sonal cycles of water and dust are reviewed. The many questions that still remain are summarized together with possible experiments, observations and calculations which may help to resolve these puzzles.

I. INTRODUCTION

Carbon dioxide is the major ingredient in the Martian atmosphere. CO_2 was spectroscopically identified in this planet's atmosphere in 1948 by Kuiper (1952). Because accepted values for the total atmospheric pressure at that time were about 80 mbar, it was thought that CO_2 represented only a trace ingredient in a largely N_2 atmosphere. An accurate measurement of CO_2 partial pressure as well as a major reduction in the measured total pressure were made by Kaplan et al. (1964). Mariner 4 established that the total atmospheric pressure on the planet is only a few mbar, and Mariners 6 and 7 showed that CO_2 is the major ingredient. Viking measurements revealed that CO_2 constituted 95.3% of the atmosphere at the surface although, as will be seen, this value is seasonally variable (Owen et al. 1977).

The seasonal polar caps of Mars, known since the 17th century, were thought by most scientists to be composed of water-ice frost by analogy with the Earth and by correlation with the seasonal albedo changes of the Martian surface, which were thought to be related to water. The early spectroscopic studies led to estimates of the partial pressure of CO_2 which were too low to permit the condensation of dry ice within the anticipated range of Martian temperatures, so this belief persisted until the dawn of the space age despite some dissonant voices which suggested that the caps were composed of CO_2 or even more exotic substances. Following Mariner 4, Leighton and Murray (1966) used a simple thermal model to show that, for reasonable assumptions about the properties of the surface and atmosphere, CO_2 would accumulate in the Martian polar regions during the periods near winter solstices when insolation vanished or was very small. Mariner 7 subsequently established that the temperature of the seasonal south polar cap is consistent with CO_2 as predicted by Leighton and Murray (Neugebauer et al. 1971).

The seasonal carbon dioxide cycle on Mars is, to first approximation, the exchange of CO_2 between the atmosphere and seasonal polar caps in response to the changes in insolation distribution which result as Mars makes its annual orbit of the Sun. The energy balance of a volume element extending from the surface to the top of the atmosphere relates the net energy fluxes due to insolation, conduction from subsurface material, advection and thermal radiation. The diurnally averaged insolation is a function of latitude, position of Mars in its orbit (conventionally measured by the areocentric solar longitude L_S), and the planet's orbital parameters; seasonal changes are especially large near the poles, where times near winter solstices correspond to zero insolation. The energy lost at the top of the atmosphere due to thermal radiation cannot then be maintained on Mars by the advection or conduction terms, and the temperature drops until it reaches the condensation temperature of

CO_2 which, at the typical surface pressure, is about 148 K. CO_2 condenses at a rate necessary to supply the energy deficit through release of latent heat. When the insolation increases after equinox, the surface temperature remains at the CO_2 condensation temperature as the absorbed insolation is converted back into latent heat of sublimation. The condensation of the principal atmospheric constituent, a process not significantly limited by diffusion gradients, effectively sets a lower limit for temperatures on the planet's surface. The corresponding condition on earth would be the condensation of liquid nitrogen on exposed surfaces at 77 K, a situation reached only in the laboratory.

If an adequate amount of CO_2 is available to the atmosphere-frost system, a permanent polar cap with a surface temperature that yields a precise balance between annual averages of absorbed insolation and thermal emission will be attained on a rapid time scale (hundreds of years) and will subsequently control the average surface pressure of CO_2 (see chapter 33). For, if such a permanent cap of solid CO_2 remains throughout the seasonal cycle at one or both poles, then a perturbation of the atmosphere-cap system which causes a net accumulation of CO_2 in the polar caps during a Martian year will, through the resulting atmospheric-pressure decrease, increase the net sublimation from the permanent cap(s). Conversely, a net annual sublimation of solid CO_2 deposits will increase the pressure and retard sublimation. Such a reservoir would therefore contribute to the seasonal cycle only to the extent that condensation and sublimation were out of balance and would drive the system towards an average pressure at which they were in equilibrium. Permanent CO_2 caps are potentially important because of this ability to buffer the atmospheric pressure, making the average pressure a function of the insolation distribution and physical parameters such as the albedo.

Another reservoir in which CO_2 may be stored during portions of the seasonal cycle are clouds. There is some evidence that clouds may participate in the condensation of CO_2 during the polar winter. There is no evidence that CO_2 clouds outside the polar region contain significant amounts of CO_2 at any season. The potentially large amount of CO_2 adsorbed in the regolith of the planet (Fanale and Cannon 1974,1979) may also contribute to the seasonal CO_2 cycle, though its primary importance is on longer time scales. The regolith refers to the surface layers of the planet which have been fragmented by meteoric impacts. Because of their large effective surface areas and presumed permeabilities, regoliths are relatively easy to modify chemically by interactions with volatiles, etc. Depending on the regolith composition and depth, this reservoir may contain most of the carbon outgassed over the history of the planet and may be very important over time scales longer than one Martian year. For a discussion of this and other processes which will contribute to the CO_2 inventory on time scales long compared to one seasonal cycle, the reader is referred to the chapters on climate history (chapters 32 and 33).

This chapter will summarize our knowledge of the CO_2 seasonal cycle after more than a decade of study of the Viking mission results. The behavior of CO_2 on Mars is interdependent with those of dust and H_2O; there are many possible couplings and feedbacks that relate these three seasonal cycles, and exploration of these effects is quite incomplete. One should read this chapter in conjunction with those describing the water cycle (chapter 28) and dust cycle (chapter 29) in order to get a complete picture of our current knowledge of the present climate regime on Mars. Understanding the current Martian climate, which is essentially synonymous with these three cycles, is prerequisite to confident extrapolation to past climates of the planet.

II. OBSERVATIONS OF SEASONAL POLAR PHENOMENA

A. History

The best information on the nature of the Martian polar caps has come from the Mariner 7 and 9 and Viking space missions. The caps have been observed telescopically from Earth for over 300 yr, but quantitative records of the seasonal cycle date mostly from this century. The observations from Earth-based telescopes provide a long-term record which can be used to supplement the spacecraft observations for cases where variations having periods more than a year are of interest. They are limited by the relatively poor resolution and by the fifteen year cycle of the Martian season which pertains at oppositions. In addition, useful observations of the surface caps are not possible during polar night when the caps are tilted away from Earth and are unilluminated. A discussion of Earth-based observations of the polar caps is contained in chapter 2.

The Mariner 7 flyby of Mars occurred during the early spring season in the south polar region ($L_S = 201°$). Observations of temperatures of the cap by the Infrared Radiometer (IRR) were consistent with a CO_2 composition (Neugebauer et al. 1969,1971); and infrared spectroscopy identified CO_2 ice on the surface (Herr and Pimentel 1969). Mariner 7 also supplied the first opportunity to see the seasonal cap at good resolution, and images revealed the irregularity of its boundary as well as the variations in albedo and frost cover in the cap's interior. Since this was a single flyby mission, there was no opportunity to observe any seasonal cap changes. Mariner 6 did not observe the polar regions; see chapter 3 for more details of these missions.

Mariner 9 observed both Martian polar caps with an imaging system and IRR as well as with the Infrared Interferometer Spectrometer (IRIS). The south polar cap was observed by the Mariner 9 instruments during most of the summer season and during the earliest days of autumn. This included only the last phases of the seasonal-cap recession and the residual-cap phase. The spring recession of the north polar cap was imaged from $L_S = 40°$ to shortly after summer solstice (Soderblom et al. 1973*b*). The wide-angle camera provided several good synoptic views during this season. The IRIS observations have recently been interpreted to imply that CO_2 ice was present in the south polar cap during the entire southern summer; this will be discussed more extensively below.

The Viking orbiters made extensive observations of both polar regions during the more than two Martian years of observations. These included TV images (VIS), thermal infrared measurements (IRTM), and measurements of water vapor abundance (MAWD). VIS observations included: synoptic observations of a complete seasonal cycle at the south pole from Viking Orbiter 2, which had a high inclination orbit; high-resolution imaging of the residual north polar cap in two consecutive summer seasons by Viking Orbiter 2; high-resolution images of the seasonal north cap during one spring season from Viking Orbiter 2; and oblique views of both seasonal caps during two years by Viking Orbiter 1. The infrared thermal experiment acquired a large amount of data on the temperature distribution in the polar regions during all seasons as well as broadband albedo measurements during the illuminated hours.

B. Seasonal South Polar Cap

Viking observations of the south polar cap during midwinter (L_S = 130°) confirm terrestrial observations that the surface frost extends to −40° latitude, that the boundary is irregular, and that frost covers the relatively low-elevation Hellas and Argyre basins (Briggs et al. 1977). The observed extensions of the seasonal cap into the basins is consistent with the expectation that the surface pressure is elevated in the low basins and that, as a result, the temperature at which CO_2 condenses and the radiated energy are higher. The rate of CO_2 condensation needed to sustain the radiation is therefore greater, enhancing the formation of surface cap in the basins. For the Hellas basin, which is about 4 km below reference level, the surface will radiate up to 2.5 W m^{-2} in excess of surrounding regions during winter, thereby enhancing the rate of nighttime condensation and the probability of frost retention and accumulation. The effect should be smaller for Argyre, which is only ~1 km below reference. There are many bright frost streaks in the polar region which, if interpreted to be due to enhanced deposition in the lee of craters and other obstacles, are consistent with the wind patterns expected during cap condensation (Thomas et al. 1979*a*).

The 1977 spring-summer recession of the south polar cap was monitored in detail by Viking Orbiter 2 (James et al. 1979; Kieffer 1979). The polar cap remained fairly symmetric about the geographic pole until midspring, as seen, for example, in Fig. 1. There are numerous high- and low-albedo areas near the edge and within the cap, many of which can be readily identified with features noted by terrestrial observers (Veverka and Goguen 1973). The cap seems to have an annular structure, also consistent with terrestrial obser-

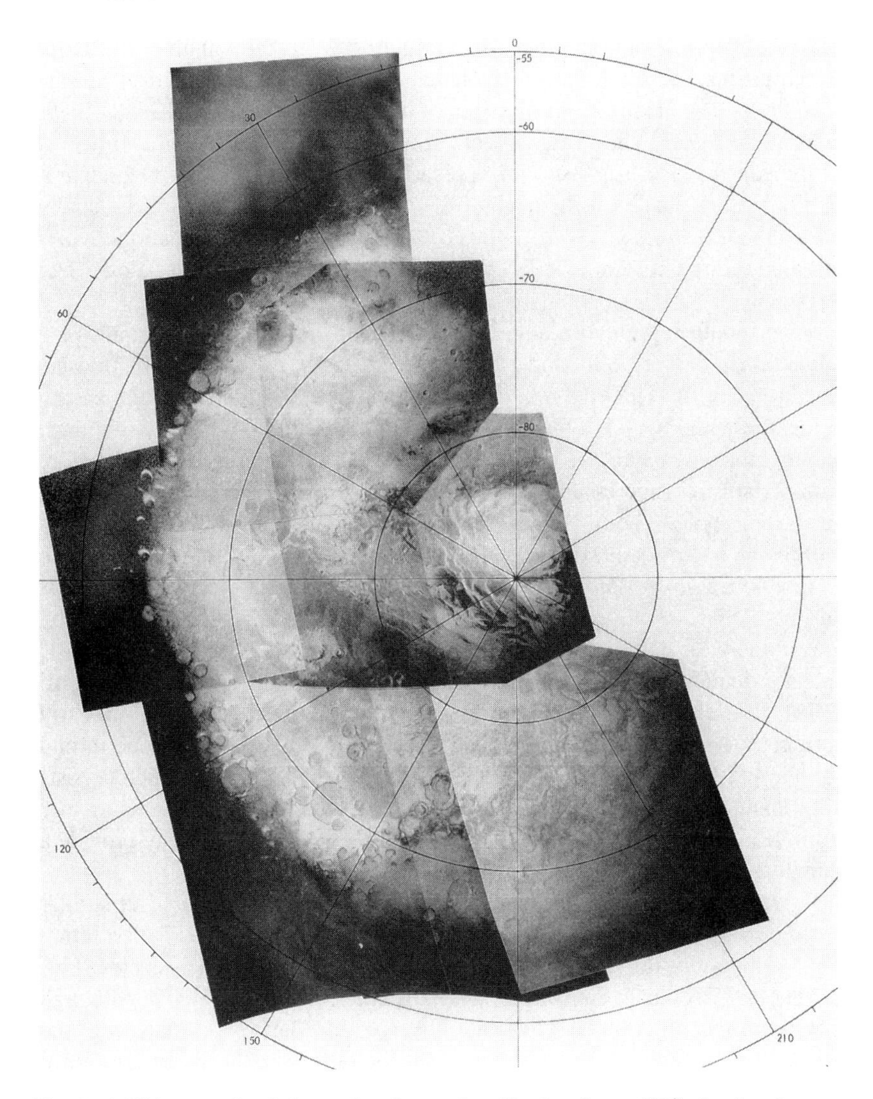

Fig. 1. A Viking mosaic of the south polar cap in midspring (L_S = 221°) showing the very inhomogeneous appearance of the seasonal cap. Of special interest is the central region, corresponding to the residual cap, that is very bright despite its proximity to the terminator (James et al. 1979). The mosaics in Figs. 1–3 were formed from images which were individually corrected and stretched; thus apparent albedo discontinuities at picture boundaries are artifacts of the assembly process and do not represent real effects. See also Color Plate 14 for a photometrically corrected image of the residual south polar cap.

vations, with a bright periphery surrounding a darker annulus; the core, which ultimately remains as the residual cap, is much brighter than its surroundings. The albedo measurements of the IRTM instrument (Kieffer 1979, his Fig. 4) clearly shows this effect. This annular structure seems to persist as the cap recedes, suggesting that resurfacing, possibly due to recondensation of water released at the edge of the cap or to removal of surface dust by off-cap sublimation winds, is involved. Another possible explanation involving a coupling between albedo and insolation will be discussed below (Sec. VII).

The south cap remains circular and centered close to the geographic pole until shortly before perihelion passage ($L_S = 251°$), as in the Viking mosaic shown in Fig. 2. The region between 160 and 300° longitude is quite patchy compared with the rest of the cap, and frost in these regions rapidly sublimes, leading to a very asymmetric cap. This asymmetry grows until, shortly before summer solstice, the center of the cap is displaced by 6°.5 latitude from the geographic pole. The displacement decreases as the remaining seasonal cap recedes to its residual configuration; the residual cap is displaced from the geographic pole 3°.5 along the 30° meridian. The major feature of interest during this late-spring period is generally called the Mountains of Mitchel by astronomers, although this nomenclature has not been approved by the IAU and the feature does not actually appear to be mountainous (Cutts et al. 1972). It appears as an unusually bright feature in the cap's interior in early spring, as a projection from the cap between 315 and 330° longitude in mid spring, and finally, as seen in Fig. 3, as a disconnected outlier which persists until about summer solstice. The exact timing of these events seems to vary from year to year, and the behavior of the cap regression then may provide a sensitive test for interannual variability (James et al. 1990).

The summer phases of the south polar cap were monitored at similar resolution by Mariner 9 in 1972 and by Viking Orbiter 2 in 1977, permitting a unique opportunity for detailed comparisons of two Martian cycles (Fig. 4). The 1977 remnant cap was larger at all times than that in 1972. Although the spring recession of 1971 was not observed by Mariner 9, analysis of the excellent telescopic data set for that year acquired by the Planetary Patrol showed that until $L_S = 260°$, the regression curves for the two years, shown in Fig. 5a, were indistinguishable (James and Lumme 1982); this suggests that the seasonal caps consisted of similar distributions of CO_2 in the two years. The summer differences should then be attributed either to factors that affected the energy balance over the cap differently in late spring or early summer in the two years, such as different dust optical-depth histories, or to differences intrinsic to the residual cap, such as a net accumulation of volatiles between the two years.

Observations of the condensation phase of the polar cap are complicated by the near coincidence of the terminator with the edge of the cap. Numerous mapping and monitoring sequences were acquired in the southern hemisphere

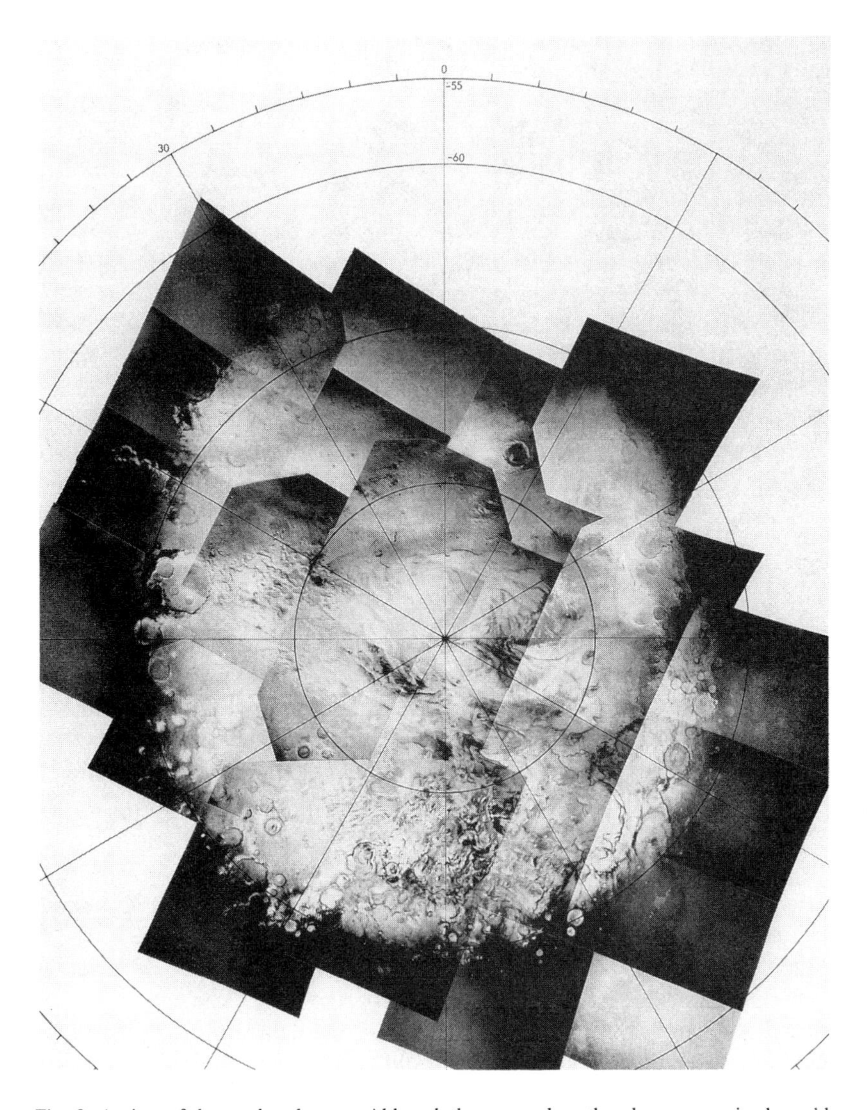

Fig. 2. A view of the south polar cap. Although the seasonal south polar cap remained roughly circular and centered on the geographic pole at L_S = 237°, large regions between 150° and 300° were already free of surface frost, forecasting the large asymmetry of the subsequent recession. The period from L_S = 230 to 250° seems to be the most sensitive to interannual variability (James et al. 1979).

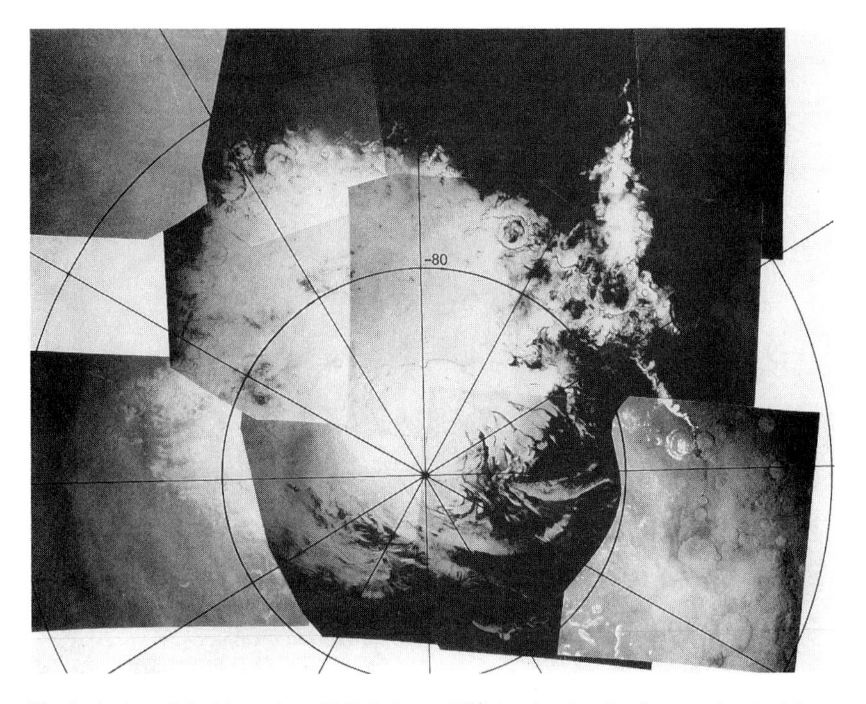

Fig. 3. A view of the Mountains of Mitchel near 320° showing that they became detached from the main cap shortly before summer solstice (L_S = 263°). This view of the cap also shows the extreme longitudinal asymmetry that characterized the south cap near summer solstice, when the center was displaced by about 6° from the geographic pole. The brightness of the residual cap relative to adjacent portions of the seasonal cap is also evident (James et al. 1979).

during autumn by Viking Orbiter 2 (James 1983), and there are also observations during the period by IRTM which, of course, are not restricted to illuminated areas. Surface frost could be identified because of enhanced albedo in the visual images. In general, the location of the cap edge using this method agreed with that determined from IRTM data (Kieffer et al. 1977), and both were consistent with what would be expected from energy-balance considerations. The bright frost streaks observed during spring recession could be recognized in the earliest condensation phases, and the bottoms of craters containing dark (dune) deposits during summer seem to be preferred sites for new deposition. This suggests that the physical properties of the surface may control the very large irregularities in frost cover seen within the polar caps. The dark streaks are consistent with the retrograde, off-cap flow anticipated when these areas are defrosted (Thomas et al. 1979*a*). The emission temperatures observed by IRTM were found to be substantially lower than those expected for CO_2 condensation in various regions of both caps (Kieffer et al. 1977); possible explanations will be discussed below (Sec. VIII).

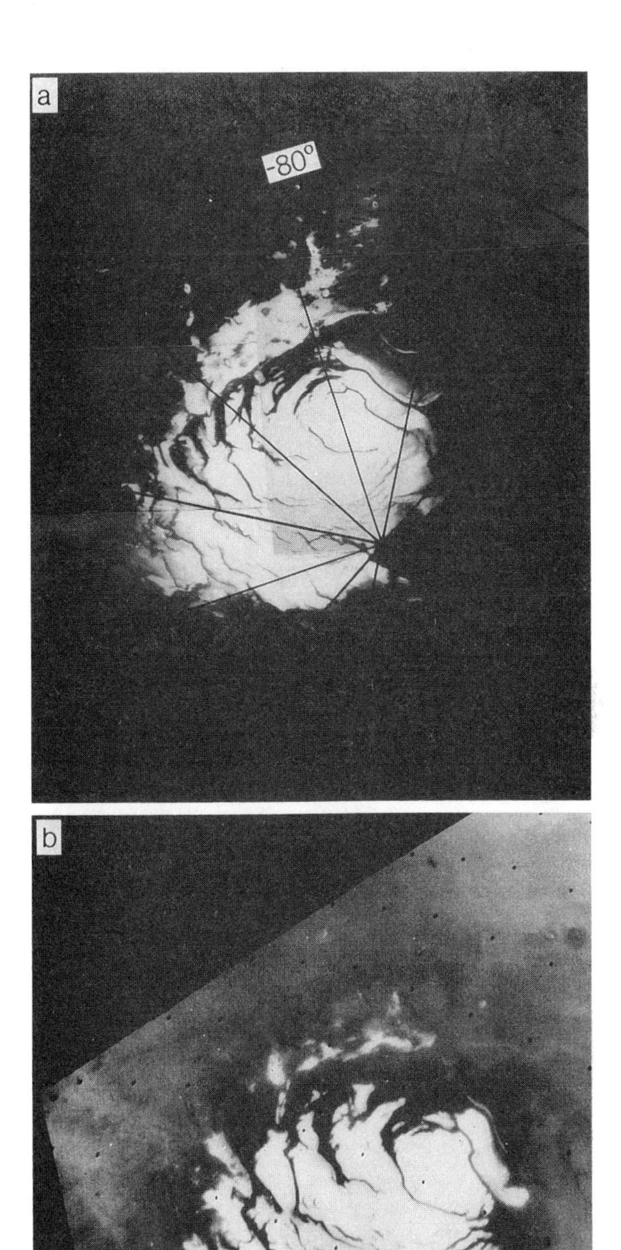

Fig. 4. Views of the south residual cap acquired by Viking in 1977 (a) and Mariner 9 in 1972 (b) at identical seasonal dates (L_S = 306°) clearly showing that the cap was geographically more extensive during 1977. The identity within statistics of the spring regression curves for 1971 and 1977 suggests that this could represent interannual growth of the cap during this period. Unfortunately, no subsequent images were acquired during the Viking mission (James et al. 1979).

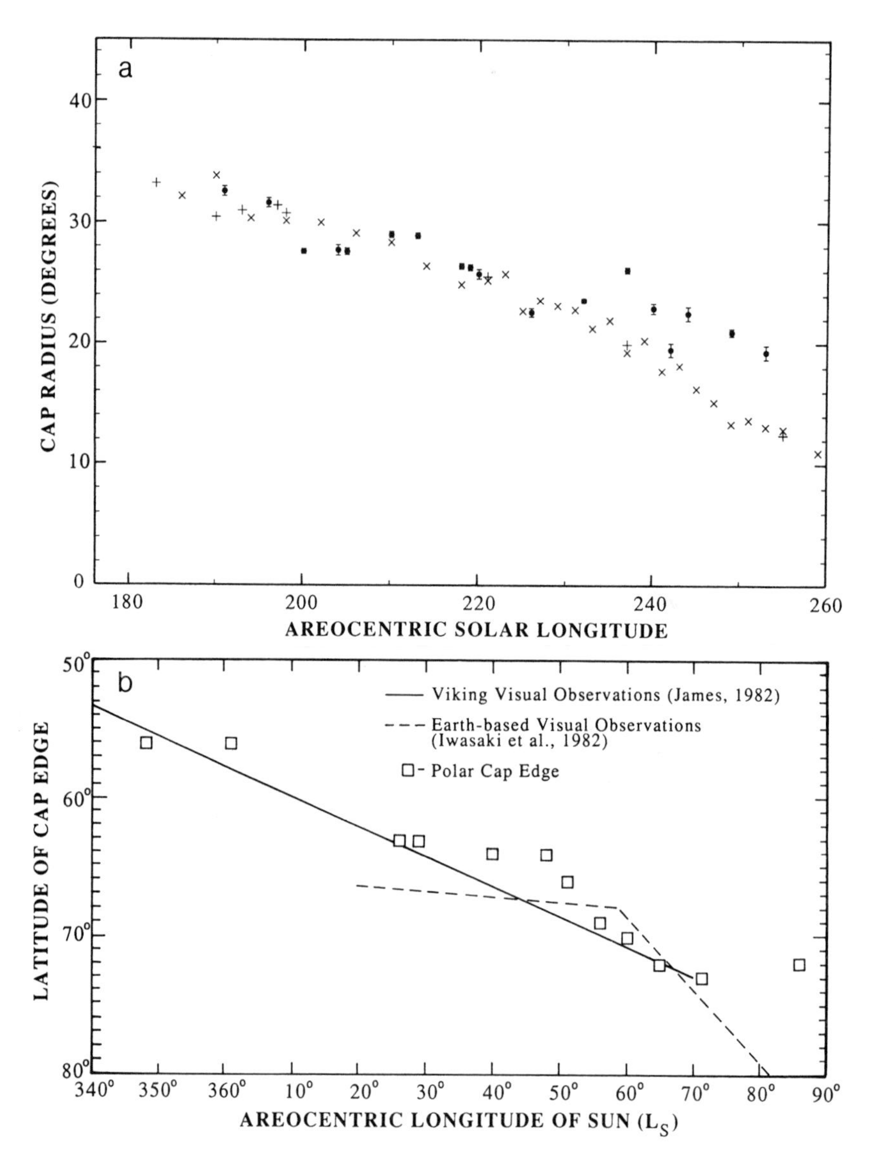

Fig. 5. A plot displaying regression curves for the south (a) and north (b) polar caps during their respective spring seasons. The former data are from the 1971 Planetary Patrol (×'s) and 1977 Viking(+'s) observations; the data points with error bars, from 1986, suggest statistically significant differences between 1986 and 1971/1977 recessions after $L_s = 240°$ (James et al. 1990). The north cap data include IRTM (boxes), Orbiter imaging (linear regression), and telescopic (dotted line) observations from 1980 (Christensen and Zurek 1984). Note the apparent plateau between $L_s = 35°$ and $L_s = 50°$.

C. Seasonal North Polar Cap

The seasonal behavior of the north polar cap is qualitatively different from that of the south cap. Before L_S = 20°, the cap is obscured by clouds which make observations difficult; the appearance of some albedo features beneath the clouds in red light has led some terrestrial observers to suggest that the surface cap is only deposited at the vernal equinox (Iwasaki et al. 1979). However, other images acquired from the Earth with red and violet filters support the results of analyses of Viking images (James 1979,1982) and IRTM data (Christensen and Zurek 1984) that show a surface cap extending to roughly 55° during the late winter season accompanied by (for Mars) relatively optically thick clouds. A possible resolution of this discrepancy is the thin, patchy nature of the frost deposits in the cap between 55 and 65°. Both spacecraft and telescopic observations of the north polar cap recession have indicated that the recession of the cap edge pauses between L_S = 20° and L_S = 45° at ~66° latitude (Fig. 5b); the exact dates vary from year to year (see chapter 2 for discussion and references). This "plateau" in the regression curve was confirmed by the IRTM observations of the cap edge. The dynamical significance of this plateau is uncertain. It coincides with a hiatus in the expected off-cap sublimation winds (Kahn 1983); water ice clouds seem to be absent during the "standstill" but reappear when the recession recommences (James et al. 1987).

D. Seasonal Pressure Variations

Because of the seasonal mass flux of CO_2 between the polar caps, there is a seasonal variation in the total atmospheric pressure on the planet. The amounts of carbon dioxide which condense on the two caps are very different. Because of the large eccentricity of the Martian orbit, the southern condensation season is appreciably longer than that in the north; and, as a first approximation, the amount of CO_2 condensed should be proportional to the length of time without insolation. Dust and condensate particles in polar hoods affect the radiative properties of the polar atmosphere and, therefore, the amount of CO_2 deposited; both observations and modeling suggest that this is an important effect, especially in the north.

The observed pressure variations at both Viking Lander sites display a large seasonally varying component which, if corrected for elevation differences, agree closely and reflect the condensation and sublimation of CO_2 at the poles (Color Plate 16) (Tillman 1988). The responses at the two landing locations, which were separated by 26 ° of latitude and by 180° of longitude, are essentially simultaneous; this contrasts with the slow diffusion of a trace substance such as water vapor. The interpretation of the pressure curve is complicated by the fact that the sublimation of the south cap occurs simultaneously with the condensation of the north cap, etc. However, there are seasons when there is only one active cap, such as that when the pressure minimum corresponding to the south cap condensation occurs; and these points, where there is no ambiguity, make good benchmarks for determination of interannual variability of the cycle.

The reproducibility of the phase and amplitude of the pressure curves between several Martian years with different meteorological histories which is apparent in Color Plate 16 is one of the most difficult boundary conditions for the interannual variations on the planet. Some interannual variability in the residual caps has been established as discussed above, and comparisons of the telescopic records of the recessions of both polar caps show considerable variability between spring seasons (see chapter 2). There is a great deal of variability in the pattern of global dust storm occurrence from year to year (see chapter 29); couplings between the dust cycle and CO_2 cycle, e.g., through the effects of aerosols on the insolation reaching the surface, on the emissivity of the atmosphere, and on the albedos of surface volatiles, have been popular explanations for the differences between recessions. There was a great deal of variability in the sequence of global dust storms during the years in which the Viking Landers monitored pressure, but only small, subtle differences are seen when the pressure curves from various years are compared.

E. Polar Hood Clouds

Terrestrial astronomers recognized for many years that the brightenings which were observed near the fall and winter poles, which are tipped away from the Earth, were due to atmospheric condensates rather than surface deposits. These large clouded regions have been generally referred to as polar hoods. The north polar hood is usually seen in astronomical photographs, but this is not the case in the south (see chapter 2 for a discussion of telescopic observations). Spacecraft observations of the hoods are quite limited. There seems to be a qualitative difference between the two hemispheres in the respective fall seasons: an extensive condensate hood was observed in the northern hemisphere, while the south was generally clear except for cloud streaks near the terminator. Clouds were observed near midwinter in the south, but they had largely dispersed by the equinox. In the north, clouds persisted into early spring. Interhemispheric asymmetries in the seasonal CO_2 cycle (James 1990) and the seasonal differences in aerosol concentrations (Pollack et al. 1990*a*) have been suggested to explain the asymmetry between the north and south polar hoods. Although the portions of the hoods which are equatorward of the Martian arctic and antarctic circles are composed primarily of water ice (Briggs and Leovy 1974), the authors noted that CO_2 condensation at some levels in the atmosphere may be possible. The general question of CO_2 clouds is discussed below.

III. CARBON DIOXIDE CLOUDS

A direct observation of CO_2 clouds was made by the Mariner 6 and 7 infrared spectrometers; three bright limb crossings yielded spectra with a spike at 4.26 μm which was attributed to reflection in the strong ν_3 band from solid CO_2 at ~ 25 km altitude (Herr and Pimentel 1970). However, the range of occurrence of CO_2 clouds on Mars is still largely uncertain. Recent theoretical studies of these clouds and the processes that can lead to their formation as well as data that are pertinent to these questions are discussed below.

The first theoretical study of Martian CO_2 clouds by Gierasch and Goody (1968) illustrated many of the important aspects of their behavior. CO_2 should begin to condense in the atmosphere when atmospheric temperatures drop to the local CO_2 frost condensation temperature. The cloud formed will remain at a temperature fixed by the CO_2 partial pressure; cloud temperatures near zero elevation will be the same as that of surface CO_2 frost and will decrease by ~ 1 K per kilometer altitude. Gierasch and Goody's calculations led them to conclude that the atmospheric temperatures on Mars are generally too high to permit atmospheric CO_2 condensation except, perhaps, at high latitudes during polar night. This reinforced the conclusion that most white clouds observed on Mars are composed of water ice.

The first general circulation model (GCM) calculations for Mars (Leovy and Mintz 1969) simulated orbital conditions at northern spring equinox and at northern winter solstice. Their calculated atmospheric temperatures were not low enough for atmospheric CO_2 condensation. Twelve years later, Pollack et al. (1981) published the results of a more sophisticated GCM model which included the effects of topography but did not include atmospheric dust. For the orbital conditions of southern winter solstice, the model predicted that atmospheric temperatures in the middle (of three) layer reached the CO_2 frost point south of $-50°$ latitude, but the condensation and precipitation processes were not simulated.

The conditions that are likely to give rise to atmospheric CO_2 condensation were further studied by Paige (1985) using a one-dimensional polar radiative model. Since diurnal insolation variations are absent during the polar night, the polar atmosphere is generally close to being in a state of thermal equilibrium. Atmospheric temperatures are determined by the competing effects of heating due to dynamical motions and cooling due to thermal emission. Paige's model showed that the presence of dust or water-ice clouds significantly enhances the probability that atmospheric temperatures will fall to the CO_2 frost point, since the net radiative effect of either dust or water ice crystals is to increase the infrared emissivity of the atmosphere and, therefore, to decrease the atmospheric temperature. More recently, Pollack et al. (1990*a*) have presented a new set of GCM calculations which include the radiative effects of atmospheric dust, though water ice is not included. The dust optical depth is assumed to be independent of location or time for each numerical experiment. Their results, discussed in more detail below, suggest that a substantial fraction of the total CO_2 condensation will occur in the atmosphere if the dust optical depth is > 1.

Obtaining definitive observational evidence for the presence of CO_2 clouds on Mars has been a difficult problem. To date, it has only been possible to make a fairly strong circumstantial case based upon the thermal structure of the Martian atmosphere combined with possible visual and spectral signatures. Briggs and Leovy (1974) examined Mariner 9 photographs of the north polar hood obtained during 21 consecutive days during northern winter and concluded that convective clouds observed at the edge of the north seasonal polar cap could be composed of CO_2 ice. Mariner 9 IRIS observations showing that temperatures in the lower atmosphere were close to CO_2 condensation temperatures during this season (Hanel et al. 1972*b*; Conrath et al. 1973) were also consistent with this hypothesis.

The Viking Infrared Thermal Mapper (IRTM) instruments observed both polar regions throughout their winter seasons. Brightness temperatures in the 20-μm channel showed considerable spatial and temporal structure, with minimum winter brightness temperatures < 134 K at the south pole (Kieffer et al. 1976*a*) and 128 K at the north pole (Palluconi 1977). Candidates for processes that might explain these observations include low surface emissivity (Kieffer et al. 1976*a;* Ditteon and Kieffer 1979; Warren et al. 1990), depressed vapor-solid equilibrium temperatures due to depressed CO_2 partial pressures at the surface (Kieffer et al. 1976*a*; 1977), and the presence of clouds (Kieffer et al. 1976*a*, Hunt 1980; Paige 1985). Hess (1979) argued that the molecular weight inversion produced in the second option was not dynamically stable. Paige (1985) has argued against the low surface emissivity and has shown that the spatial and temporal occurrence of the low brightness temperatures observed by IRTM are consistent with the notion that they are due to CO_2 clouds.

Ultimately, simultaneous observations of infrared emission and vertical temperature profiles will be needed to determine whether or not CO_2 clouds play an important role in the polar energy balance. Attempts to date, using Viking IRTM data (Paige 1985; Jakosky and Martin 1987) have not been complete because the IRTM 15-μm channel observations provided only the temperature in the middle atmosphere. The 15-μm channel observations during the north polar fall and winter seasons analyzed by Martin and Kieffer (1979), Paige (1985) and Jakosky and Martin (1987) indicate that temperatures at this level exceed the CO_2 frost point. A recent analysis of Mariner 9 IRIS observations has revealed low brightness temperatures in the north polar region during the 1971–72 period (Paige 1989). The low temperatures are only observed when lower atmospheric temperatures are near the local frost point, which is further evidence that they may be due to CO_2 clouds.

IV. RESIDUAL CAPS

The concept of a permanent polar CO_2 deposit which would buffer or regulate the atmospheric pressure in response to changes in orbital parameters, especially inclination, was of great interest at the start of the Viking mission. The obliquity of Mars, currently 25°.2, varies from 10°.8 to 38°.0 in a cycle dominated by periods of 1.2×10^5 and 1.3×10^6 yr (Ward 1979; chapter 9). The average annual insolation at the poles is proportional to the sine of the obliquity. If CO_2 remains at the pole throughout a year with no net accumulation, the annual balance of major energy sources and sinks then requires the temperature of the deposits to be roughly proportional to the 4th root of the sine of the obliquity

$$\frac{(1-A)Q_0 \sin\alpha}{\pi\sqrt{1-\varepsilon^2}} = \langle e\sigma T^4 \rangle \tag{1}$$

where A and e are the radiometric albedo and emissivity of the ice deposits. Q_0 is the solar constant, α and ε are the obliquity and eccentricity of the orbit, and T is the temperature of the ice. Because the CO_2 pressure is related to the sublimation temperature of the deposits through the Clausius-Clapeyron equation, the average pressure is therefore a sensitive function of the obliquity. If the albedo of the buffering deposits is adjusted to the value necessary to agree with the current pressure, one finds that the pressure will vary between ~ 0.4 and 35 mbar at the high and low obliquity limits, respectively (Ward et al. 1974; Pollack and Toon 1982), assuming that there is enough CO_2 in the permanent deposits to supply 35 mbar. This variation would be of great climatological significance and has been used to explain the evidence for short-term variations seen in the polar layered terrain (see chapter 33).

Terrestrial observers have established that residual caps are present at both poles in most years (see chapter 2). The north residual cap is considerably larger than its southern counterpart; it is roughly 1000 km in diameter, compared to ~ 350 km for the south cap, and is therefore an order of magnitude larger in surface area. The caps are neither circular nor centered on their respective pole; Borealis Chasma splits the north cap into two almost disconnected segments, and the south residual cap is centered almost 3°.5 from the geographic pole. The traditional explanation of the offset has been that the cap's location minimizes insolation, but if the cap albedo increases with insolation (see below) the opposite may be true. Topography near the south pole constructed by attributing the low winter brightness temperatures observed by IRTM to altitude (Paige 1985) suggests the latter (high-albedo areas tilted away from the pole), but his calculated contours do not agree with the topography derived photogrammetrically (Dzurisin and Blasius 1975; USGS 1989). Both caps are terraced and have a swirl texture which may migrate poleward by preferential removal of frost from equatorial-facing slopes (Howard et al. 1982*a*); there is at present no accepted explanation for the swirl pattern. Both residual caps appear to be elevated by 1 to 2 km above their surroundings, which are considerably higher in the south polar region than in the north. However, there are large uncertainties in the topography of the polar regions which are a serious obstacle to understanding the physics of the CO_2 caps. For more details of the topography and geology of the polar regions see chapters 7 and 23, respectively.

The compositions of the two residual polar caps has been the subject of a lively debate; though the Viking mission settled the issue for the time period of its observations, the physics needed to explain the observations and possible interannual differences remain in doubt. Murray and Malin (1973*a*) examined evidence available after Mariner 9 and concluded that the observed residual caps consisted of water ice but that the probability for some solid CO_2 deposits buried under a portion of the visible north cap was substantial. Ingersoll (1974) pointed out that a buried reservoir in vapor communication with the atmosphere would be unstable because the average temperature at the surface would exceed that of the buried deposit, leading to conduction into it and eventual sublimation. Briggs (1974) concluded from energetic considerations that the residual caps were probably H_2O, although neglected factors, such as wind drifting, could conspire to maintain CO_2 at either cap. The possibility that the cap could consist of carbon dioxide clathrate ($CO_2{\cdot}6H_2O$) was suggested by Dobrovolskis and Ingersoll (1975); the buffering capability of such a cap would be much less than that of a pure CO_2 residual cap.

Viking Orbiter 2 arrived at Mars shortly after summer solstice in the northern hemisphere. Observations of the Mars Atmospheric Water Detector (Farmer et al. 1976) and the Infrared Thermal Mapper (Kieffer et al. 1976) indicated that the residual north polar cap was composed entirely of water ice and that the amounts of water vapor in the atmosphere were consistent with equilibrium with an ice surface of the temperature indicated. The observed summer cap temperatures lead to an annual average temperature that is too high to permit even clathrate deposits.

A permanent cap at one pole will lose CO_2 to a cap at the other pole which is identical except for lower elevation due to the greater radiation and CO_2 condensation there. Thus, it was generally anticipated that the higher south residual cap would also be water ice. Analysis of the visual differences between the 1972 and 1977 caps suggested but did not prove conclusively that the residual cap retained CO_2 throughout that summer (James et al. 1979). This was confirmed by the IRTM (Kieffer 1979) and MAWD (Davies and Wainio 1981) observations of the south cap. Temperatures near the CO_2 frost point were observed throughout the summer, although interpretation was complicated by the presence of atmospheric dust, by uncertain optical properties of solid CO_2, and by lanes of dark, warmer material within the cap. The relative lack of water vapor over the south polar cap during summer was also inconsistent with even high-albedo deposits of water ice which were not stabilized by at least proximate contact with CO_2 deposits. Subsequent analysis of the Mariner 9 IRIS data by Paige et al. (1990) has shown that CO_2 was also present in the south residual cap during 1972.

The existence of a residual CO_2 cap at the south pole violates our intuition concerning the stability of solid CO_2 during the perihelion summer because of the relatively large insolation then (but see the discussion concerning insolation-dependent albedo in Sec. VII). There is evidence (Jakosky and Barker 1984) that suggests that there was much more water vapor over the south polar region in 1969 than in subsequent years; this suggests that the CO_2 ice could have been absent in that year. The questions related to stability will be discussed in more detail in Sec. V on the heat balance of the caps. Another peculiarity of the residual south polar cap is its displacement from the geographic pole by 3°.5 along the 30° meridian; although this location was thought to be near the summit of a relative elevation of several km (Dzurisin and Blasius 1975), a condition which would not at first glance be conducive to CO_2 stability, the residual cap has no obvious correlation in longitude with the latest topographic mapping (U.S. Geological Survey 1989). Finally, the visual appearance of the cap does not suggest that the frost deposits are particularly thick (Fig. 6). Coupled with the small area occupied by the cap, this raises numerous questions concerning its long-term stability and its role in controlling the atmospheric pressure.

V. THERMAL BALANCE OF THE RESIDUAL CAPS

The partitioning of available CO_2 between the polar caps, the atmosphere and other possible reservoirs is strongly influenced by the annual heat balance of the Martian residual polar caps. As noted above, during the year of Viking observations CO_2 frost was stable at the south residual cap throughout the year but was not stable at the north residual cap. Since both poles receive the same total incident sunlight over the course of a year, the present asymmetry in the behavior of CO_2 frost at the Martian residual caps must be due to a north-south asymmetry in the annual heat balance at these two locations. Understanding the causes of the present asymmetric budgets of the caps should provide important clues to the processes that influence the Martian CO_2 cycle over seasonal and climatic time scales.

The basic principle that underlies heat-balance studies is that the total net flux of energy into a region at its boundaries must be equal to the rate at which energy is stored inside the region. For a given region, the magnitudes of these energy fluxes depend on a variety of internal and external factors and can vary enormously with time of day and season. Polar regions are particularly favorable locations to conduct heat-balance studies because they do not experience diurnal variations in solar heating rates; therefore, variations in their heat balance parameters are purely a function of season.

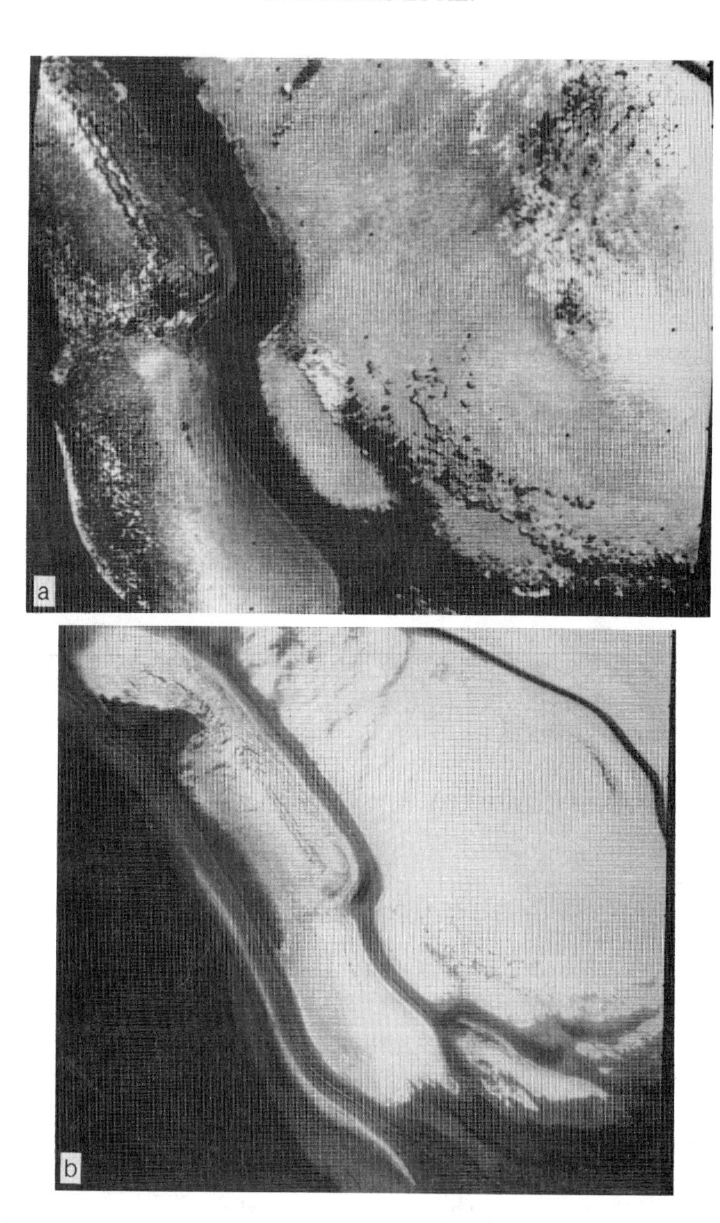

Fig. 6. The central portions of the residual south cap seen in a Mariner 9 B frame (a) acquired at L_S = 328°. The large holes in the frost cover seen in the cap were smaller and less significant in a similar Viking view (b). Although one cannot eliminate the possibility that topography is responsible for the "Swiss cheese" appearance in 1972, the pictures suggest that the frost was not very thick in 1972 and hint at additional deposition during the interval between 1972 and 1977 (James et al. 1979).

Figure 7 is a cartoon of the major energy fluxes into and out of a volume located at one of the Martian poles. In this case, the region is defined as a column extending from the top of the atmosphere to the base of the seasonal CO_2 frost deposits. In roughly decreasing order of importance, the energy fluxes at the boundaries of such a region could be due to:

a. incident solar radiation at the top of the atmosphere;
b. reflected solar radiation and emitted infrared radiation at the top of the atmosphere;
c. conductive heating at the base of the seasonal frost deposits due to seasonal heat storage and release within underlying permanent polar cap deposits;
d. horizontal heat transport (including the possibility of latent heat transport in the form of clouds) due to atmospheric motions.

The net energy flux through the boundaries of the region must be equal to the change in energy storage; this can be in the form of latent heat storage in condensing or sublimating CO_2 on the surface or in the atmosphere, or in the form of total potential energy storage within the atmospheric column (a term which is small and has not been shown in Fig. 7). During most seasons, the

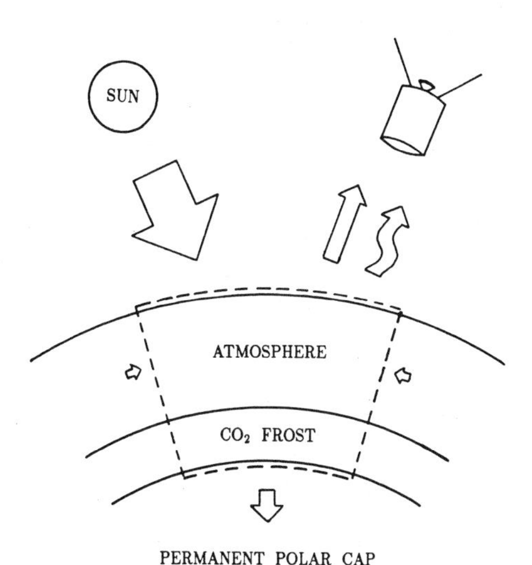

Fig. 7. Diagrammatic representation of the major energy fluxes into and out of a region located at one of the Martian poles. The sizes of the arrows are approximate indicators of the magnitudes of the various terms. They include: 1: solar radiation incident at the top of the atmosphere; 2: reflected solar radiation at the top of the atmosphere; 3: emitted infrared radiation at the top of the atmosphere; 4: downward heat conduction into permanent polar cap deposits (or soil); 5: convergence of horizontal energy transport in the atmosphere.

instantaneous polar heat balance is a balance between the net radiative flux at the top of the atmosphere and latent heat storage in the seasonal CO_2 frost. Subsurface heat conduction can play an important role during some seasons, but only if the CO_2 cap completely sublimes so that the subsurface, particularly a water ice cap with large heat capacity, is exposed. Because of the small mass of the Martian atmosphere, horizontal heat transport and total potential energy storage are minor terms.

Many of the important components of the polar heat balance can be determined from spacecraft observations. In the 1976–1979 period, the Viking Infrared Thermal Mapper instruments obtained a set of solar reflectance and infrared emission measurements of the Martian north and south polar regions that span an entire Martian year. These observations were used to determine annual radiation budgets and infer annual CO_2 frost budgets for the core regions of the north and south residual caps (Paige 1985; Paige and Ingersoll 1985).

Results of the Paige and Ingersoll polar-heat budget are summarized in Figs. 8 and 9 for the south and north, respectively. Figures 8a and 9a show ten-day averaged radiation budgets at the top of the atmosphere, including the flux of incident solar radiation, the solar radiation absorbed by the surface and atmosphere, and the outward flux of infrared radiation. Figures 8b and 9b show derived CO_2 latent-heat storage rates and conductive heat fluxes from the permanent caps. The latent-heat storage rates are proportional to the rate at which CO_2 frost is condensing (negative) or subliming within these regions. Figures 8b and 9b also show calculated latent-heat storage rates from a simple energy-balance model like that used by Leighton and Murray with parameters tuned to match the observed behavior at the south residual cap. The model assumes that both regions have identical, constant radiative properties and that CO_2 frost is stable at both residual caps throughout the year. Fitting the derived annual heat budgets at the south residual cap required a frost albedo of 0.78, a frost temperature of 142 K, a frost infrared emissivity of 0.85, and a constant horizontal atmospheric heating rate of 2 W m^{-2}, also assumed for the derived annual heat budgets.

The heat-budget results for the south residual cap were interpreted as follows. CO_2 frost condensed at a relatively constant rate throughout the fall and winter seasons and then sublimed during the spring and summer. To within the accuracy imposed by the measurements of the solar reflectance and infrared emission and by estimates of minor terms in the energy-balance equation, inferred total annual frost condensation was balanced by total annual sublimation, confirming the stability of CO_2 frost at the south residual cap throughout at least one Martian year. In other words, the radiative properties of the south residual cap during the sunlit seasons balanced with those during polar night, pointing strongly to the stability of CO_2 frost during this particular year and indicating that horizontal transport of latent heat was unimportant.

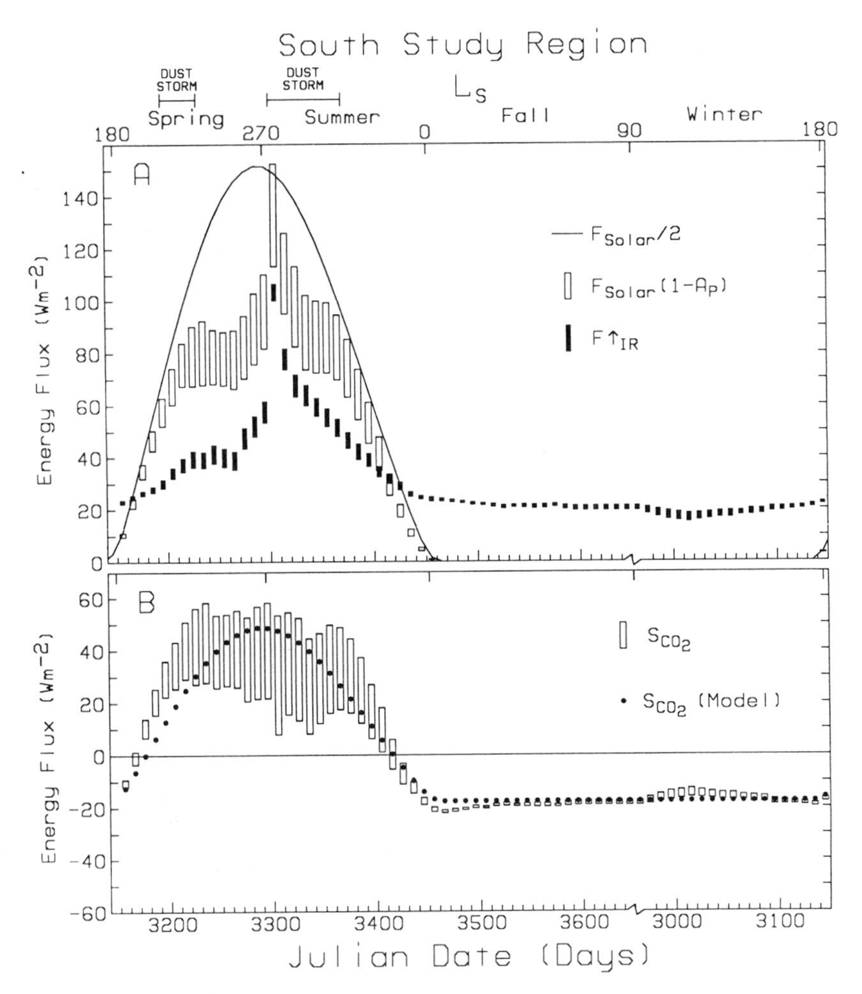

Fig. 8. (a) Ten-day average annual radiation budget at the top of the atmosphere for the south residual cap derived from observations of Viking Orbiters (Paige and Ingersoll 1985). F_{solar} is the flux of incident solar radiation at the top of the atmosphere; $F_{solar}(1-A_p)$ is the rate of solar energy absorption below the top of the atmosphere. F_{ir} is the flux of emitted infrared radiation at the top of the atmosphere. (b) CO_2 latent-heat storage rates at the south residual cap from Viking observations and from a simple Leighton/Murray model described in the text.

The CO_2 sublimation rates observed at the south residual cap during spring and summer were much smaller than those predicted by pre-Viking models. These low sublimation rates could be due to high surface frost albedo and/or the radiative effects of dust in the polar atmosphere. The radiative effects of dust were investigated by Davies (1979*a*) who showed that dust increases sublimation rates for high-albedo frosts and decreases sublimation rates for low-albedo surface deposits. Toon et al. (1980) argued that the ef-

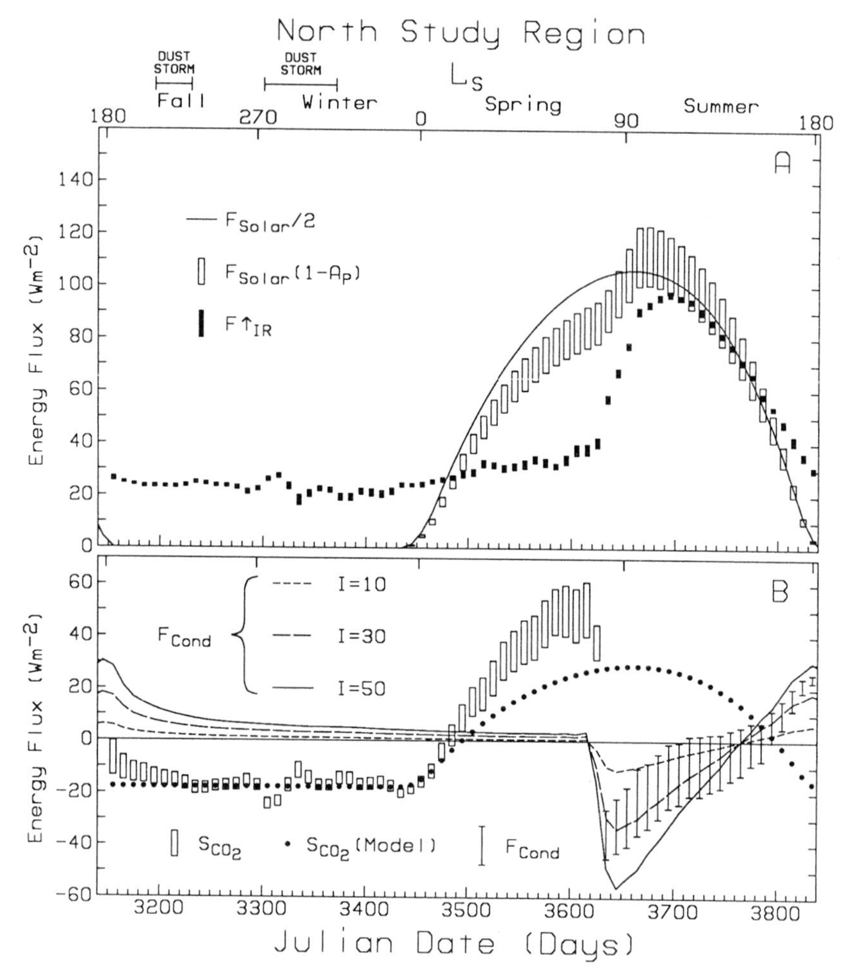

Fig. 9. (a) Ten-day average annual radiation budget at the top of the atmosphere for the north residual cap derived from observations of Viking Orbiters (Paige and Ingersoll 1985). F_{solar} is the flux of incident solar radiation at the top of the atmosphere; F_{solar} $(1-A_p)$ is the rate of solar energy absorption below the top of the atmosphere. F_{ir} is the flux of emitted infrared radiation at the top of the atmosphere. (b) CO_2 latent-heat storage rates at the north residual cap from Viking observations and from a simple Leighton/Murray model described in the text. Also shown are model values for F_{cond}, the downward conducted heat flux into the north polar cap, for three values of thermal inertia. Measured values for F_{cond} during summer, when CO_2 frost is absent, are also shown.

fects of atmospheric dust on surface frost sublimation rates should not be nearly as important as that of the albedo of the surface frosts. Paige and Ingersoll (1985) and Paige (1985) used a one-dimensional atmospheric model to show that frost albedo was indeed the determining factor and showed that the observed radiation balance at the south residual cap during spring could only be explained if the surface albedos of the seasonal frosts were between 0.7 and 0.95, with 0.82 ± 0.06 being most probable.

The heat-budget results for the north residual cap were interpreted as follows. The observed rates of CO_2 frost accumulation during the fall and winter seasons in the north were roughly comparable to those in the south despite the fact that accumulation rates in the north were reduced by the underlying, warmer permanent water-ice cap. If the permanent cap had not been exposed during the summer season, the total accumulation of CO_2 at the end of winter would have been 10 to 15% greater. Figure 9b shows that CO_2 sublimation rates during the spring season were significantly higher than those predicted by the simple model, implying that the radiative properties of the north polar region were different from those in the south during this season. Paige and Ingersoll (1985) used a one-dimensional radiative model to show that the observed heat balance during northern spring could be explained only if the surface albedos of the seasonal frost deposits were between 0.38 and 0.75, with an albedo of 0.64 ± 0.06 being the most probable; they concluded that the present asymmetric behavior of seasonal frost in the two residual caps was due primarily to a significant difference in spring season frost albedos.

Whether the residual cap behavior observed by Viking was representative of the behavior in other years is still an open question. Jakosky and Barker (1984) have interpreted 1969 Earth-based observations of elevated global water abundance in the southern hemisphere during late summer as evidence that seasonal and permanent CO_2 deposits there may have completely sublimed, exposing a permanent water-ice cap as in the north. Comparisons of 1972 and 1977 images of the residual cap (James et al. 1979) show that the residual cap grew during this period, so one must consider the possibility that the observed summer CO_2 deposits are ephemeral rather than permanent. Jakosky and Haberle (1990*b*) have suggested that both configurations are possible and that perturbations in the energy balance due, for example, to albedo changes caused by dust or water contamination can cause the residual cap to "jump" between compositional states; this scenario could explain periods with high humidity in the south and alleviate the necessity of having a very precise annual balance. It is easier to identify mechanisms that will destroy a CO_2 cap than those that will restore it when it is exposed; the sensible heat stored in the remaining water ice will significantly delay seasonal frost condensation during succeeding years and make it difficult to reform a permanent CO_2 cap.

A fundamental understanding of the processes responsible for the present asymmetric properties of frost deposits at the north and south residual caps has not yet been achieved. As discussed in Sec. VII, the reflectivities of the frost deposits are influenced by many factors including grain size and the presence of contaminants such as dust and water ice. The 20% difference between the reflectivities of the seasonal frost deposits in the north and south could be due to any of these factors. Determination of which processes are responsible for the difference and why the north is different from the south is prerequisite to understanding the current Martian climate.

VI. PHYSICAL PROPERTIES OF CO_2

Carbon dioxide is a linear, symmetric molecule; the solid is cubic (Wychoff 1963,p.368). The complex index of refraction for solid CO_2 has been compiled by Warren (1986). The spectral reflectivity of CO_2 frosts from 0.8 to 3.2 μm has been measured in the laboratory for a range of grain sizes (Kieffer 1970). The major bands in the solar reflectance range (0.3 to 3.5 μm) are a triplet at 2.0 μm (medium) and a doublet at 2.8 μm (strong). The $\nu 2$ (bending) fundamental at 15 μm is the major band in the thermal-emission region. The $\nu 3$ (asymmetric stretch) fundamental is at 4.3 μm; since this is in the crossover region between reflection and emission it has little effect on the Martian radiative balance.

Because strong intermolecular bonding does not occur in the solid, the vibrational frequencies in the solid and gas are almost identical and the spectral lines of the solid are shifted little from those of the gas (Osberg and Hornig 1952). As a result, thermal emission from CO_2 ice grains is largely blocked by the Martian atmosphere. Also, spectroscopic detection of solid CO_2 is based primarily on relatively weak "forbidden" transitions and bands which appear on the sides of the vibration bands involving combination with the solid lattice vibrations at 68 and 114 cm^{-1} (Kieffer 1970). Figure 10 shows the dependence of the spectral reflectance of CO_2 frost on grain size calculated using the complex index of refraction compiled by Warren (1986).

The relatively few measurements of the thermodynamic properties of CO_2 at conditions appropriate for Mars have been published in a rich variety of units. The values given here are all in SI units, with the exception of pressure for which the tradition of using millibars (100 Pascals) appears to continue in the Martian literature. The specific heat of solid CO_2 over 73 to 198 K is $(349 + 4.8\,T)$ J kg^{-1} K^{-1} (Washburn 1948). The triple point for CO_2 is 216 K, 5.18×10^5 Pa; the critical point is 304 K, 7.4×10^6 Pa (Meyer and Van Dusen 1933).

The most important characteristic of CO_2 for Mars is the vapor pressure curve, because this relation controls the sublimation temperature and, therefore, the amount of condensed CO_2. Across the temperature range relevant for Mars, the sublimation curve closely follows the Clausius-Clapeyron relation:

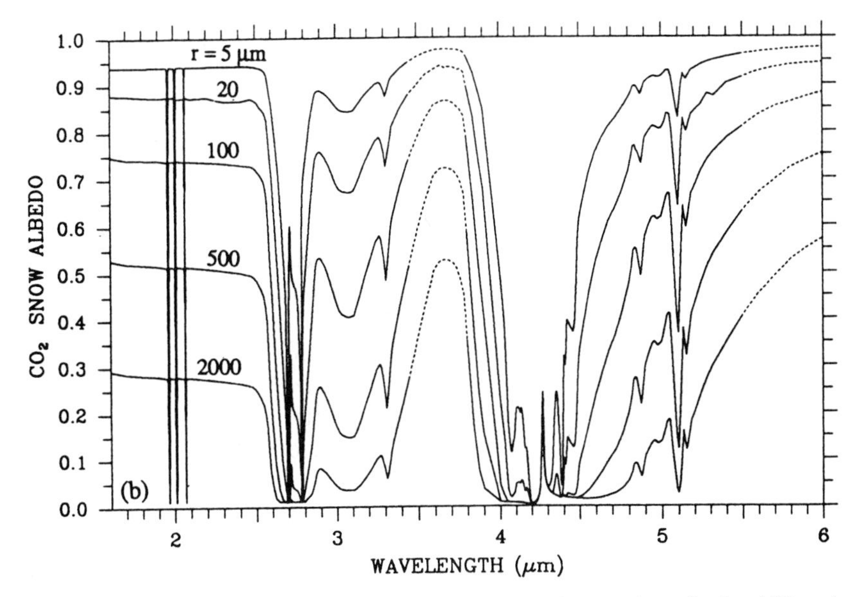

Fig. 10. The calculated albedo for CO_2 snow for several grain sizes are shown for the visible and infrared regions. These values were computed using Mie theory for nearly monodispersed size distribution and the properties of CO_2 presented by Warren (1986). (Figure from Warren 1990.)

$$\ln p = a - \frac{b}{T} \tag{2}$$

where p is the pressure and T is in Kelvins. The values for a mentioned here are appropriate for p measured in mbar; b is proportional to the molar heat of vaporization. The vapor pressure relations in many reference books are based on the work of Meyer and Van Dusen (1933). They measured the vapor pressure of solid CO_2 between 194.3 and 216.5 K and analyzed the results of prior measurements ranging from 90 to 194.7 K. They fit all of these data with a 4-term equation which is the basis for their vapor pressure tables. Mullins et al. (1963, now difficult to obtain) analyzed all available measurements of vapor pressure and other thermodynamic properties to derive a consistent thermodynamic model of solid CO_2, tabulating the computed vapor pressure and latent heat of sublimation at 1 K increments. Their results for vapor pressure differ from the Meyer and Van Dusen equation by < 0.15 K between 105 and 195 K (Mullins et al. 1963, Fig. 5). We have fit their values at 5 K increments with continuous functions. From 120 to 160 K (0.04 to 31.4 mbar), the Clausius-Clapeyron form with $a = 23.3494$ and $b = 3182.48$ has a maximum deviation in $\ln p$ of 0.011, corresponding to a temperature difference of < 0.1 K and well within the scatter of measurements. From 100 to 170 K, $\ln p = 25.2194 - 3311.57\, T^{-1} - 6.71 \times 10^{-3}\, T$ has a maximum deviation of 0.0007 in ln p; agreement is limited by the 4 significant figures in the printed table.

The Mullins et al. tabulated data for the latent heat of sublimation from 120 to 160 K can be expressed as $6.5231 \times 10^{+5} - 371.3\, T$ J kg^{-1} K^{-1} with a maximum deviation of 374 or 0.06 %. From 100 to 170 K, the form $6.21225 \times 10^{+5} + 78.153\, T - 1.61084\, T^{+2}$ has a maximum deviation of 59; again the fit is limited by the precision of the printed table and the residuals are trivial relative to the uncertainty of the measurements.

Brown and Ziegler (1980) expressed the logarithm of the vapor pressure of several substances as power series in $1/T$ with sufficient terms to "represent the vapor pressure of the substances within the uncertainty of the physical data used in the original tabular calculations." Their expressions for solid CO_2, covering 40 to 194.7 K, are based on the tables of Mullins et al. (1963). Six terms are used for vapor pressure; they also express the latent heat of sublimation as a 5-term power series in T.

Several different relations for CO_2 vapor pressure have been used in the planetary literature. The International Critical Tables list $a = 23.102$ and $b = 3148.0$ (Washburn 1948); these values were used by Ward et al. (1974) and by James and North (1982). Fanale et al. (1982*a*) used $a = 23.23$, $b = 3167.8$ with no source citation.

Miller and Smythe (1970) used the Meyer and Van Dusen compilation between 123 and 173 K to obtain

$$\ln p = 26.1228 - \frac{3385.26}{T} - 9.4461 \times 10^{-3}\, T \tag{3}$$

and this relation was used by Dobrovolskis and Ingersoll (1975). Haberle et al. (1982) used a nonphysical relation equivalent to $\ln p = T/6.48 - 21.022$, apparently to linearize a dynamical model.

The carbon dioxide clathrate or hydrate $CO_2 \cdot 6H_2O$ is stable at somewhat higher temperatures than pure CO_2. Miller and Smythe (1970) measured its dissociation pressure between 151 and 192 K; they give

$$\ln p = 24.313 - \frac{3081.8}{T} - 1.516 \times 10^{-2}\, T \tag{4}$$

for the clathrate. Their data also fit the Clausius-Clapeyron form, which is easy to invert, with $a = 18.827$, $b = 2638.7$ and a maximum error of 2%. The clathrate would be in equilibrium with the CO_2 atmosphere at a temperature 5 K higher than pure CO_2 ice. The average temperature of the north polar residual cap is 158.7, ~ 5 K above the stability region for clathrate (Paige 1985). Clathrates could be stable in the south residual cap, but the amount of CO_2 contained in a permanent clathrate deposit is much less than could be contained in a pure CO_2 deposit of the same size; therefore the potential buffering effect of the cap would be substantially reduced.

Dust is probably a common contaminant of Martian frosts and is of fundamental importance in its ability to modify the albedo of polar deposits. The difference in the molecular properties of CO_2 and H_2O is sufficiently great that the behavior of dirty frost on Mars is substantially different than that which is inferred from intuition based on the properties of terrestrial frosts. CO_2 frosts on Mars are of uniform and constant temperature because of the buffering relation with the atmospheric pressure. Thus, all processes related to the CO_2 caps are isothermal. On Earth, frost-ice surface temperatures undergo substantial diurnal and annual variations; the vapor pressure and sublimation rate of ice are strongly nonlinear with temperature, and the highest temperatures tend to dominate processes. H_2O is a polar molecule; melting rather than sublimation is the major phase change, and small dust grains are trapped by surface tension in the melting layer. The CO_2 molecule is not polar, and there is no liquid (except under extremely high pressures); apart from geometric entrapment and possible electrostatic binding, dust grains will not remain attached to CO_2 molecules during sublimation.

VII. CO_2 FROST ALBEDO

The albedo of the solid CO_2 deposits in the Martian polar caps is a crucial parameter in the energy balance and stability considerations. The albedo which appears in such calculations is technically the hemispherical albedo defined by Hapke (1981), although the closely related Bond albedo is often substituted; the two albedos are equal for Lambertian surfaces. As noted above, Paige and Ingersoll (1985) and Paige (1985) estimated the polar albedos directly from the energy balance. Because of the very limited phase angles for the caps as seen from Earth, extraction of a reliable albedo for the caps using telescopic data is very difficult and model dependent. Dollfus (1965*a*) found a midspring south polar cap reflectivity in 1956 which would, if the phase function were Lambertian, imply a Bond albedo of 0.67 at 0.58 μm. Lumme and James (1984) used the 1971 photographs acquired by the International Planetary Patrol to estimate the south cap albedo in early spring; a Bond albedo of between 0.64 (360 nm) and 0.79 (620 nm) was estimated assuming that the phase function is similar to that of Europa. Viking images included only a limited range of phase angles; using an essentially Lambertian phase function, James et al. (1979) found that the Bond albedo of a bright region within the seasonal south cap was 0.55 to 0.6 using the Viking red filter and 0.36 to 0.38 using violet. The albedo of the residual south cap was higher than that of the seasonal frost deposits at all wavelengths; lower (τ = 0) limits in violet and red, respectively, were 0.52 and 0.65. The albedo of the residual frost deposits was less dependent on wavelength than that of the seasonal frosts, which are fairly reddish; even the residual cap was not perfectly white, however. The albedo of the polar caps is not constant in time.

Interpretation of the latent heat storage rates for the residual cap areas discussed above (Figs. 9b and 10b) indicates that the albedo in the vicinity of both residual caps increases through the summer and appears to be nearly a linear function of insolation during this time, as seen in Fig. 11 (Paige 1985). Because atmospheric dust would generally be expected to increase the sublimation rate of a high-albedo cap by increasing the thermal emission of the atmosphere, the conclusion that the surface albedo is an increasing function of insolation is relatively independent of the particular treatment of the radiative effects of dust used in this calculation. This is the opposite of the relation for terrestrial frosts and is opposite to the relation expected for semi-infinite isotropic scattering media. Consideration of the microphysical properties of dust and CO_2 frost suggests possible causal mechanisms which can result in such an albedo as a function of time but which would not be expected to be effective for a water-ice cap. Radiation incident on dust grains imbedded in a CO_2 frost will rapidly heat the grains since the thermal time-constant for conduction for a 10-μm-radius grain is < 1 ms. The grain cannot remain in contact with solid CO_2 once its temperature exceeds the saturation temperature; a thin gas layer will be sublimed and support and insulate the grain from the bulk solid. The grain will then be mechanically disconnected from the ice and will "bore" into the ice in whatever direction gravity and gas flow take it. The smallest dust particles will float away on the sublimation gas flow; for a typical cap albedo and summer insolation, the net upward gas velocity is $\sim$ 0.002 m s^{-1}, sufficient to support particles < 5 μm in radius. Larger grains will bore their way deeper into the deposit at a rate which decreases as the insolation is attenuated; as they sink to a greater depth, their effect on the albedo is diminished.

The imaginary part of the index of refraction of CO_2 is much smaller for visible wavelengths than for thermal emission (Warren 1986). Therefore, solar radiation will be attenuated less rapidly than thermal emission. Because the frost must be isothermal, the net energy balance will favor sublimation at the depth where the insolation is absorbed, but will cause condensation nearer the surface where the infrared radiation originates. This can result in the growth of pristine frost deposits on the surface of a bulk deposit which is actually experiencing net sublimation. Since the layer of the atmosphere which is actually in contact with the ice deposits will be pure CO_2 (apart from floating dust which will be warmer than the condensation temperature), this hoar-frost formation will be clean. Thus, a top surface layer of clean, high-albedo frost, can be maintained.

Laboratory measurements of the emissivity of thick layers of solid CO_2 suggest that this quantity would be between 0.62 and 0.89 in the 20 μm IRTM band (Ditteon and Kieffer 1979; Warren et al. 1990). On the other

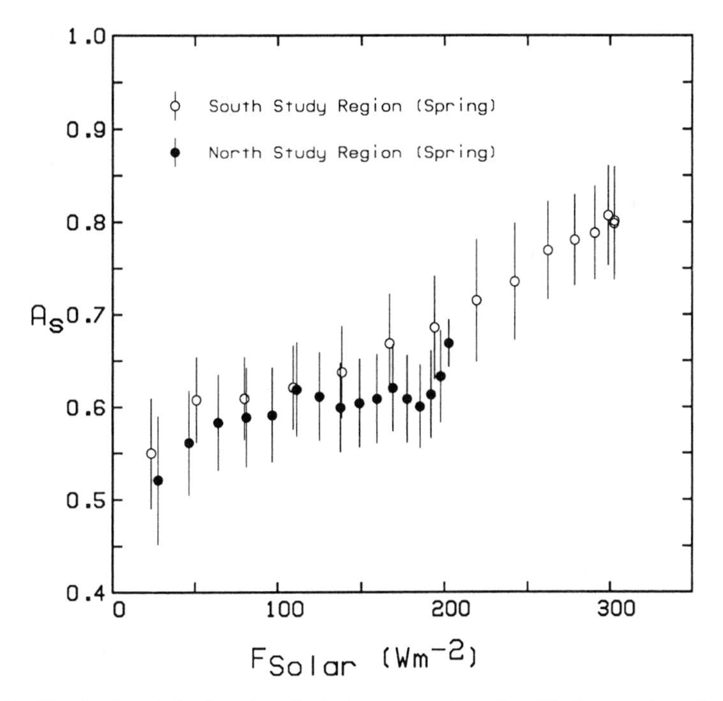

Fig. 11. The albedos of the frost deposits in the north and south residual cap regions shown as functions of the incident solar flux (Paige 1985). The linear increase of the albedo suggests a feedback effect which has important consequences for long-term climate variations due to obliquity oscillations.

hand, analysis of Mariner 9 IRIS observations gives no indication of nonunit emissivity. The actual emissivity of the deposits depends on the physical structure of the deposits (e.g., grain size) as well as on the molecular absorptions. It seems reasonable that, if the growth of deposits can proceed in alternative ways, the form which maximizes the emissivity will be favored since such deposits will grow at the maximum rate.

VIII. MODELS FOR CO_2 CYCLE

As was noted in the introduction, the precipitation and subsequent sublimation of CO_2 on Mars is a consequence of the energy balance in the polar regions. In addition to the CO_2 latent-heat term, potentially important contributions to the net energy flux into a volume element such as that shown in Fig. 7 include: incident and reflected insolation at the top of the column; infrared thermal radiation exiting the atmosphere; convergence or divergence of horizontal advection; latent heat transported by CO_2 clouds; energy exchanged with the subsurface layers due to conduction; and frictional dissipation of winds at the surface. Relating these processes to the amount of CO_2 which ultimately changes state requires that the effects of the atmosphere on the exchange of energy between the surface and the top of the element be correctly modeled. In addition, there are a number of variable or poorly known parameters which may affect the energy balance such as the optical depth of aerosols, the albedos and emissivities of surface components, and the properties and behaviors of potential condensate clouds. There have been a large number of models which examined the CO_2 cycle either as a primary goal or as an inevitable consequence of asking other meteorological questions. These models differ in the care with which particular portions of the total energy balance are treated, but they all fall into one of two major classes: zonally symmetric models, in which the longitudinal degree of freedom is suppressed, and full three-dimensional models based ultimately on the primitive meteorological equations. There is a great variation in the attention paid to the vertical degrees of freedom, particularly in the first group which includes models that virtually neglect the atmosphere (one-dimensional). In this chapter, we concentrate on models that focus primarily on the seasonal CO_2 cycle; applications of these models in other studies, such as those dealing with long-term climate changes, are discussed in chapters 26 and 33.

Leighton and Murray (1966) proposed a thermal model for the Martian climate that is the prototype for the first group of modeling efforts. Ignoring energy advection by the atmosphere, they showed that surface temperatures for a "moon-like" Mars, with no condensible gases, would fall well below 145 K in the polar regions of the planet. At the CO_2 partial pressure of $\cong$ 4 mbar observed by Mariner 4, 145 K represents the temperature at which CO_2 changes phase from gas to solid, releasing latent heat (6.0×10^5 J kg^{-1}) which stabilizes the temperature at the sublimation point. Leighton and Murray predicted a permanent CO_2 north polar cap, but this turned out to be an artifact of a simplified description for Mars's orbital variations which erroneously led to unequal average annual insolation at the two poles. Cross (1971*a*) repeated the previous calculation and showed that, as far as the CO_2 cycle is concerned, the heat conduction with the surface can be neglected compared to other terms when CO_2 is present throughout the year. However, heat conduction and storage can play an important role in interannual variability of the residual cap (Jakosky and Haberle 1990*b*).

This energy balance model was embellished over the next few years in response to the acquisition of new data from Mariner 9 and from Viking. Briggs (1974) included heat exchange between the atmosphere and the surface in order to match the observed seasonal migrations of the cap edge. He concluded that the observed seasonal polar-cap behavior required a greater downgoing radiance due to the atmosphere than expected for clear conditions; enhancements in this energy component of about 50 and 200% were needed for the south and north caps, respectively. These were attributed to the effects of the polar hood clouds, discussed above (Sec. II.E). Davies et al. (1977) used a model of the Leighton and Murray variety which incorporated some of the earliest results of the Viking mission. These included the lower-than-expected polar temperatures observed by IRTM and the water-vapor measurements in the north polar region. Dzurisin and Ingersoll (1975) studied the effects of an adsorbed reservoir of CO_2 as an alternative buffer to the solid CO_2 deposits previously considered. Finally, the IRTM thermal model (Kieffer et al. 1977) included an especially careful treatment of the planet's surface for use in interpreting the thermal measurements in terms of surface properties.

Later Viking results included measurements of the pressure at the Viking Landers for an entire seasonal cycle and identification of the residual south cap as carbon dioxide. Hess et al. (1979) compared the earliest Lander pressure measurements to predictions of existing models. James and North (1982) developed a model primarily constrained by the full seasonal pressure cycle observed at the Viking landing sites. The radiative effects of dust storms were modeled using a "look-up table" calculated by Davies (1979*a*) from a Monte Carlo photon propagation model. The horizontal heat flux, modeled as a macrodiffusion process, was small ($\cong 1$ W m^{-2}) except near the edge of the seasonal cap. A good fit to the seasonal pressure variation was obtained using the 1977 optical-depth history. The suppression of CO_2 condensation relative to that expected for a surface blackbody is represented in this model by a fitted surface-frost emissivity of roughly 0.7.

One of the most interesting manifestations of the seasonal CO_2 cycle is the meridional wind produced by the mass flux of CO_2 into the polar regions. This aspect of the seasonal climate is rather accurately predicted by the simple models because it essentially depends only on the latent heat contributions to the energy balance; however, these models suppress details of the vertical and azimuthal distributions of the winds. If the mass flux is uniformly distributed vertically and azimuthally, the largest wind speeds are $\cong 0.5$ m s^{-1} at the periphery of the subliming south polar cap in midspring. Concentration of the off-cap winds into the boundary layer and/or into certain longitudes by topography can augment their effects substantially, making them a potentially important factor in the development of the many small dust storms seen near the edge of the retreating cap in spring. It has also been suggested that these winds could be important in the transport of trace ingredients (James 1985).

A more sophisticated model which includes atmospheric circulation via assumed vertical profiles was proposed by Coffin (1967), who derived a seasonal pressure variation qualitatively similar to that measured much later. A zonally symmetric model that includes CO_2 condensation and sublimation has been used by Haberle et al. (1982) to study the dispersal of dust by the meridional circulation and to understand the feedback of the dust on the Hadley circulation. The model predicts an expansion of the Hadley circulation during dust storms which is effective in bringing dust to northern midlatitudes and which can explain the observed jump in pressure at Viking Lander 2 at the start of the second 1977 dust storm. Because of the improved treatment of the vertical structure of the atmosphere, this model can distinguish between atmospheric and surface condensation of CO_2. The model is especially useful in assessing the effects of dust on the CO_2 cycle, and it suggests that CO_2 condensation is reduced during periods of high optical depth. However, various processes within the polar regions do not seem to be adequately modeled, suggesting that waves may make an important contribution to energy and material transport at high latitudes: more sophisticated models may therefore be necessary to model the condensation process adequately.

Models which solve the full three-dimensional set of primitive equations on a grid have the advantage that they most accurately portray the physics of the problem. For example, inherently three-dimensional processes such as baroclinic waves are naturally included. The first such attempt to study the general circulation of Mars using the discoveries of the early spaceflights was due to Leovy and Mintz (1969) who applied a two-level model developed at UCLA by Mintz and Arakawa to the Martian circulation. They were especially interested in whether heat transport due to the meridional circulation could negate the CO_2 condensation in the polar regions which had been predicted by Leighton and Murray. They discovered that it could not, and in the process identified the seasonal CO_2 mass flux and its effects on the seasonal circulation pattern.

The most recent model is also descended from the Mintz-Arakawa model developed at UCLA. This model has much improved resolution in both the vertical and horizontal coordinates. It includes a more realistic treatment of the radiative properties of dust and of CO_2 than has been used previously but does not include water-ice clouds, which potentially can affect CO_2 condensation. The first major application of the model is to the CO_2 cycle (Pollack et al. 1990*a*). This is the first model to permit a detailed investigation of CO_2 condensation in the atmosphere, including the possible occurrence of CO_2 clouds. This model predicts that the fraction of the total CO_2 condensation that takes place in the atmosphere increases with optical depth because of the increased emissivity of the atmosphere produced by the dust. The polar CO_2 clouds in this model have the short time-scale variability which seems to characterize Viking IRTM observations of low polar brightness temperatures; this model therefore supports scattering by CO_2 clouds as the source of the low brightness-temperature anomalies. This also suggests that the portions of polar hood north of the arctic and antarctic circles are CO_2, although, since water is not included in the model, water ice is not ruled out as an associated component. One would expect large seasonal asymmetries between the two polar regions to result from the effects of much different optical depths in the two winter seasons; these asymmetries may be reflected in the observed hood asymmetries as well as in the different seasonal cap behaviors in the two polar regions. The model has not yet provided an answer to the question of how processes conspire to keep the pressure variations nearly identical in years with very different dust histories.

IX. CONCLUSIONS

Each space mission which has studied the polar regions has expanded our knowledge of the seasonal cycle of CO_2 on Mars. However, each new discovery has also raised as many new questions as it has answered. With regard to the seasonal CO_2 cap, the general forms of the cap evolution and of the seasonal pressure variation are in agreement with simple models based on energy balance. However, many details of the processes are not yet entirely understood:

1. Winter condensation is suppressed either because the solid CO_2 deposits are inefficient radiators or because of the scattering effects of CO_2 clouds. Some evidence favors each process. Better understanding of the emissivity of solid CO_2 is needed; this problem is amenable to terrestrial laboratory and theoretical analyses.
2. The significant spatial inhomogeneity in the seasonal polar caps is not completely understood. Is it due to only partial surface coverage in some areas, to differing frost thickness, to differing surface properties of the frost due, for example, to insolation-dependent albedo, or to some other mechanism? Are such inhomogeneities in the physical distribution and/or properties of the frost produced by topography, properties of the planet's surface, winds (which may be influenced by topography), clouds or some other factor? Spacecraft data with much better spatial and temporal resolution in the thermal infrared and much better topographic knowledge are needed to address these questions.
3. The qualitative difference between the spring regression curves and between the wind systems for the two seasonal caps is puzzling. One possibility is that the greater amount of aerosols in the atmosphere during northern winter results in a greater proportion of atmospheric condensation there than in the south. But why then does the plateau in the curve seem to occur in years with vastly different dust-storm history?
4. Expanding upon the preceding, the seasonal pressure curves obtained by the Viking Landers during four Martian years are remarkably similar. These years had very different storm histories, suggesting either that the CO_2 cycle is not sensitive to atmospheric dust and the meteorological variations accompanying global storms or that some subtle cancellation between different mechanisms suppresses variations.

The problems regarding the residual caps are even more intriguing:

1. The very different natures of the two residual caps can be understood in terms of the generally higher albedo of the southern deposits, particularly

of the residual south cap. The albedos individually present puzzles, however. North cap deposits are more reflective than one would expect for contaminated water ice (Kieffer 1990), and the south cap seems almost unaffected by dust contamination. Elimination of low-albedo dust contaminants in CO_2 ice due to a variety of mechanisms is a plausible hypothesis that can explain the relative stability of CO_2 in the south, where insolation is currently larger, but the idea has not been tested in the laboratory.

2. The offset of the south residual cap from the pole is a puzzle. Although the residual cap configuration is approached as a continuous limit by the seasonal cap recession, the unique topography of the environs of the residual cap suggests some topographical influence on atmospheric dynamics during the polar night which is not yet understood. Better topographical information seems essential to resolve the question of how the offset is related to topography.
3. The stability of residual CO_2 deposits in the south over periods of more than a few annual cycles is uncertain. A related question is whether the cap does actually control the atmospheric pressure at this time; does the permanent cap act as a buffer; is the current pressure simply determined by the amount of CO_2 available; or is the pressure controlled by some other process, such as regolith adsorption or carbonate formation (Kahn 1985)? There are not many data relevant to these points at present, and theoretical analyses have clarified the issues but have not resolved these questions.

Additional laboratory and theoretical analyses, spacecraft measurements with higher spatial and temporal resolution, and more systematic monitoring of the seasonal cycle, particularly of the residual south cap, over several years may be necessary to answer all of these questions. The polar caps have a history of confounding attempts at interpretation, so more surprises and puzzles should be expected before a final understanding of the seasonal CO_2 cycle on Mars is obtained.

Acknowledgments. The authors express appreciation to S. Clifford, B. Jakosky and C. Leovy for especially useful reviews of the original version of this manuscript. We also gratefully acknowledge the sustained support of our research efforts by the National Aeronautics and Space Administration.

28. THE SEASONAL BEHAVIOR OF WATER ON MARS

BRUCE M. JAKOSKY
University of Colorado

and

ROBERT M. HABERLE
NASA/Ames Research Center

Seasonal and spatial variations in the column abundance of water vapor in the Martian atmosphere are due to the combined effects of exchange of water with non-atmospheric reservoirs and transport of water within the atmosphere. The non-atmospheric reservoirs include the seasonal polar caps, the residual polar caps, adsorbed water within the regolith, and possible surface or near-surface ground ice. Atmospheric transport occurs via the global circulation, which consists primarily of the cross-equatorial Hadley cell, winter-hemisphere baroclinic waves, global-scale waves, and vertical mixing via small-scale processes. Transport is primarily as vapor, although some transport as condensate occurs. The mechanism which drives most of these processes is the solar heating; latitudinal variations in insolation drive the atmospheric circulation, while seasonal changes drive the exchange with the non-atmospheric reservoirs. Observations of the seasonal changes in atmospheric water content, combined with quantitative numerical models of the exchange and transport processes, have led to our present understanding of the importance of each process. These observations and models are summarized in this chapter, along with the inferences concerning the behavior of the seasonal water cycle. These results also allow us to infer the response to seasonal forcing at other epochs, and to discuss the long-term behavior of water on Mars.

I. INTRODUCTION

The behavior of water in the Martian atmosphere and in the near-surface part of the regolith is tied closely to the general nature of the Martian climate. Water responds to other climatic effects, such as the seasonal CO_2 and dust cycles, due to the strong temperature dependence of condensation and adsorption phenomena and to the role of CO_2 and dust in affecting atmospheric and surface temperatures as well as in driving atmospheric circulation. Although water is primarily a tracer of atmospheric motions, it can also influence the climate system, due to the potential influence of condensates and of water itself on the absorption of sunlight and the distribution of solar energy and on the removal of CO_2 and dust from the atmosphere via condensation. As one of the three major seasonal cycles involved in the current climate (with the carbon-dioxide and atmospheric-dust cycles being the other two), the water cycle is important in understanding the processes that can affect climate. This includes how these processes will respond to the seasonal variations in solar forcing, how these processes might vary from one year to the next in the absence of any interannual variation in external forcing, and how they respond to longer-term variations in the solar forcing due to variations in Mars' orbital elements.

Water is especially important as an indicator of global climate change on a long time scale. This is suggested by the presence of layers within the polar ice deposits that presumably consist of alternate layers of dust and ice or of dust and ice in varying proportions (see, e.g., Soderblom et al. 1973; Murray et al. 1972; chapter 23). These layers are generally thought to form over the 10^5- and 10^6-yr time scale of variations in the orbital obliquity (Ward 1974,1979; chapters 9 and 33) as a response to the incremental transport of water and dust to or removal from the polar regions occurring during each year (see, e.g., Cutts 1973; Toon et al. 1980; Pollack and Toon 1982; Jakosky 1985). On the seasonal time scale, the behavior of the polar regions plays an important role in controlling the global climate. For water, this results from the role of the polar caps as a summertime source and wintertime sink for water, with a dynamic equilibrium determining the amount of water vapor in the atmosphere at any given time (Jakosky 1983*b*,1985; Haberle and Jakosky 1990); although exchange with the regolith may be important in the seasonal variation of the atmospheric water content, the global behavior appears still to be controlled by the forcing at the poles (Jakosky 1983*a*,*b*,1985).

Interactions between the seasonal cycle of water and those of carbon dioxide and airborne dust may be important at the current epoch. For example, condensation of water vapor onto airborne dust grains may cause the dust to settle onto the surface in the polar regions more rapidly than it otherwise might (Pollack et al. 1977,1979*b*); or the complete disappearance of the south-polar residual covering of CO_2 frost in some years may allow an underlying water-ice cap to sublime into the atmosphere (Jakosky and Barker 1984;

Jakosky and Haberle 1990*b*); or the seasonal net meridional wind due to the condensation and sublimation of CO_2 at the polar caps may significantly influence the ability of the atmosphere to transport water or dust into the polar regions (James 1985). By defining the behavior of the water cycle, and of the processes that may play important roles in the cycle, we can better understand the interplay between the various seasonal cycles; this is of special importance not only for understanding the current climate, but also for understanding the behavior at other epochs, when each process and each cycle will behave differently.

Finally, we mention briefly the importance of water to the possible present or past existence of life at the Martian surface. Water is essential for the existence of life on Earth, and it is thought to be equally important for life on Mars (see chapters 34 and 35). Even though liquid water is not stable currently at the Martian surface, it may be present as a transient phase long enough to be accessible to any Martian organisms (Farmer 1976). The seasonal exchange of water between the regolith and atmosphere may play an important role in providing a source for regolith water.

Our goal in this chapter is to summarize what we know about the seasonal water cycle on Mars. A relatively recent review by Jakosky (1985) summarizes in more detail than is possible here some observations of the seasonal cycle and the physical mechanisms for exchange of water with non-atmospheric reservoirs. The next section contains a brief history of the observations of water vapor in the Martian atmosphere, along with an outline of the seasonal water cycle. This brief summary is not meant to be definitive, but rather serves to provide a basis for the more detailed discussions of the individual processes in the remainder of the chapter. The following sections contain discussion of the possible exchange of atmospheric water with non-atmospheric reservoirs of water; the role of transport of water through the atmosphere either as vapor or as condensate via the atmospheric circulation; and inferences about the behavior of the seasonal water cycle at other epochs, when the solar forcing would have been different from that at the present due to Mars having had different orbital elements.

II. HISTORY OF OBSERVATIONS AND SUMMARY OF THE SEASONAL WATER CYCLE

The first true detection of water vapor in the Martian atmosphere was made by Spinrad et al. (1963). By observing a water-vapor absorption line in the near infrared, they determined that the globally averaged column abundance was $\sim$ 10 precipitable micrometers (10 pr μm, or 10^{-3} g cm^{-2}); this is the equivalent thickness of liquid water if all of the atmospheric water were to be condensed onto the surface. For comparison, the column water abundance in the Earth's atmosphere is several precipitable centimeters, or almost 10^4 times greater.

Telescopic observations through the end of the decade showed that there was a distinct seasonal cycle of atmospheric water vapor (see, e.g., Barker et al. 1970). The column water abundance varied by about a factor of 2, reaching seasonal maxima just after each of the two solstices each year. These were primarily global measurements, however, with little information on the distribution of water with latitude; because of the viewing geometry imposed by the terrestrial and Martian orbits, observations were preferentially of each summer hemisphere. The similarity of the northern summer and southern summer observations led Barker et al. (1970) to infer that both residual polar caps were composed of water ice and that water was able to sublime into the atmosphere when the seasonal covering of carbon dioxide frost disappeared near the summer solstice. As discussed later, these observations differ from those in later years.

Earth-based telescopic observations have continued to the present, providing information on 8 different Martian years. Barker (1976) summarized the 1972–1974 Earth-based observations; these showed the entire seasonal cycle for a single year and provided some information on the distribution of water with latitude. Jakosky and Barker (1984) compared all of the then-existing Earth-based data with Viking spacecraft data and discussed variations in the seasonal cycle. Most recently, continued observations by Rizk et al. (1991) continue to extend the coverage of different Mars years.

More detailed information during some of those years has been obtained from orbiting spacecraft. The Mariner 9 spacecraft returned data for 6 months during 1971–1972; the thermal infrared interferometer spectrometer provided water-vapor data for $\sim$ 1/4 of a Martian year, although derived water abundances have a large uncertainty due to the unknown vertical distribution of water within the atmosphere (Hanel et al. 1972*b*; Conrath et al. 1973). The Soviet Mars 3 and Mars 5 Orbiters also provided some measurements of atmospheric water, but the data were limited in temporal and spatial extent (Moroz and Nadzhip 1975*b*).

The Viking orbiters provided a wealth of information on both spatial and seasonal variations in atmospheric water for $\sim$ 1¼ Martian years during the period 1976–1979 (Farmer et al. 1977; Farmer and Doms 1979; Jakosky and Farmer 1982), and less detailed information for the rest of the second Martian year (Zurek 1988). The Viking Mars Atmospheric Water Detection (MAWD) experiments consisted of near-infrared reflectance spectrometers operating in water-vapor absorption bands and continuum regions near 1.4 μm. The water vapor abundance along the observational path is determined from the relative absorption, and is translated to a vertical column abundance using the known observational geometry. Scattering by dust or other aerosols in the atmosphere can bias the measurement because the observed sunlight will not have sampled the entire atmospheric column; except during the global dust storms and dusty time periods, atmospheric dust probably reduces the observed column by $< 10\%$. Also, uncertainties in the assumed effective pressure and temperature of line formation will lead to uncertainties in the column abundance; errors can be as large as 30%, but are probably most often < 15%. The experiment, sources of error and data-reduction techniques are described in detail by Farmer and LaPorte (1972), Farmer et al. (1977), Davies (1981), Jakosky and Farmer (1982) and Jakosky et al. (1988).

In addition, the two Viking Landers provided *in situ* measurements relevant to the presence of water as near-surface atmospheric vapor and condensate (Ryan and Sharman 1981*a*; Ryan et al. 1982; Pollack et al. 1979*b*), surface condensate (Jones et al. 1979; Wall 1981; Hart and Jakosky 1986; Svitek and Murray 1990), and subsurface bound water (Anderson and Tice 1979).

The overall behavior of atmospheric water vapor as a function of latitude and season for the time period observed by Viking is shown in Fig. 1. (Season is given by L_s, the areocentric longitude of the Sun, with $L_s = 0°$ corresponding to the vernal equinox, $L_s = 90°$ to the following solstice, etc.) The total amount of water vapor in the atmosphere varies seasonally between ~ 1 and 2×10^{15} g, equivalent to a total volume of ~ 1 to 2 km^3 H_2O. The largest column water abundances occur over the north-polar residual cap at ~ L_s = 120°, and appear to result from the sublimation of water into the atmosphere from a residual polar cap composed of water ice (Farmer et al. 1976,1977;

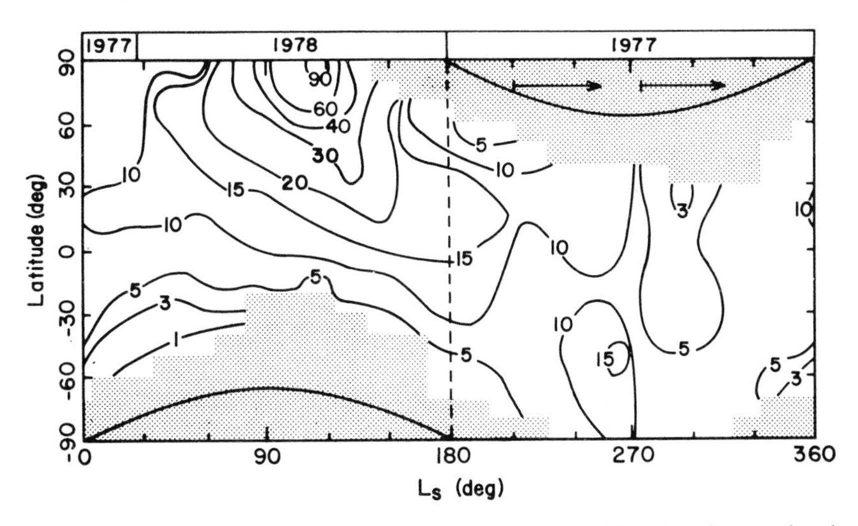

Fig. 1. Contours of column water-vapor abundance as a function of latitude and season, based on Viking MAWD observations, in units of precipitable micrometers (pr μm). Data span one Mars year covering the Earth years 1977–1978, and are spliced at L_s = 180° (vertical dashed line). Shaded regions indicate no observations, and the smooth curves are the latitude poleward of which the Sun will not rise above the horizon. Horizontal arrows mark the times of the two 1977 dust storms as observed by the Viking Landers and Orbiters. Earth years are shown along the top of the figure. This is a re-display of the data inversion of Jakosky and Farmer (1982).

Jakosky and Farmer 1982). The increase in atmospheric water vapor actually begins prior to the exposure of the residual cap at $L_s = 80°$, suggesting that water also is supplied to the atmosphere either from the edge of the retreating seasonal polar cap or from the regolith. Unique distinction between these two sources is difficult as both would be forced by the changing surface temperatures in response to the retreat of the seasonal cap and the disappearance of the CO_2 frost with its colder temperatures.

During the period $L_s = 90$ to $150°$, the peak column abundance at each latitude all the way down to the equator occurs at a progressively later time relative to the peak at the pole. The orientation of the contours of water abundance (Fig. 1) gives the impression that water is subliming into the atmosphere near the north pole and is being transported through the atmosphere toward the equator. The behavior is seen as well in Fig. 2 for three different latitude bands: 0–30°, 30–60° and 60–90°. Clearly, the increase at about $L_s = 80°$ is apparent in the 60–90° band and probably represents the rapid sublimation of water from the residual polar cap after the disappearance of the seasonal CO_2 frost. The increase at $L_s = 110°$ in the 30–60° latitude band has been used to argue in favor of transport from the polar regions (Jakosky and Farmer 1982; Jakosky 1985). Calculations of the transport of water from the polar region suggest, however, that the circulation is too sluggish to account for this increase, and that much of the water must be coming from the regolith (Haberle and Jakosky 1990). The gradual buildup of water at these latitudes prior to $L_s = 110°$ is, again, probably due to the release of water from the retreating seasonal polar cap and from the exposed regolith.

The behavior of water during the southern summer season as viewed from Viking differs markedly from that during the northern summer. The peak vertical-column abundance in the south is ~ 15 pr μm, and occurs well away from the pole during southern summer, at $\sim L_s = 270°$ (Fig. 1). This peak is a factor of 6 smaller than the northern-hemisphere summer peak, and is more like the global and annual average water vapor abundance. If the south residual polar cap were an exposed water-ice deposit, as the north residual cap is, it would be expected to heat up to sufficiently high temperatures that large amounts of water would sublime into the atmosphere (Farmer et al. 1977; Davies et al. 1977; Farmer and Doms 1979; Jakosky and Farmer 1982); atmospheric column abundances would be comparable to or greater than those in the north, in the range 100 to 400 pr μm (Davies et al. 1977). Combined with the thermal infrared observations, this suggests that the south-polar residual cap never loses its covering of CO_2 frost during the winter (Kieffer 1979; James et al. 1979). Because water is less volatile than carbon dioxide, no significant sublimation of water occurs.

Figure 3 shows the globally integrated water vapor abundance as a function of Martian season during northern summer, along with the integrated abundance within the northern and southern hemispheres separately. One interesting point is the evidence for cross-equatorial transport of water vapor

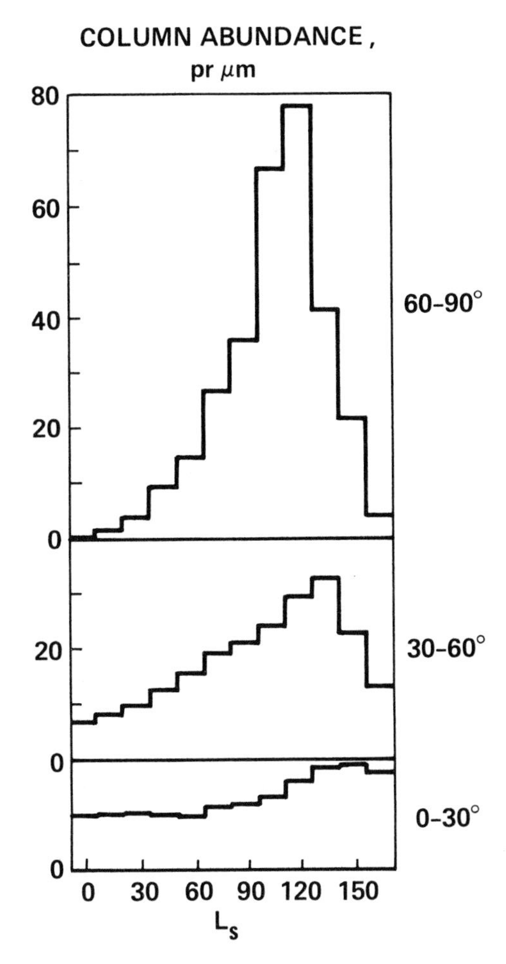

Fig. 2. Zonally averaged column water-vapor abundance during the northern hemisphere spring season for the latitude bands 0–30°, 30–60° and 60–90° (figure from Haberle and Jakosky 1990).

during the northern summer season. Just after the beginning of northern summer, the water-vapor abundance in the southern hemisphere begins increasing. At this season there should be no southern hemisphere sources for atmospheric water, so that this increase must be the result of transport from the north. The corresponding decrease in northern hemisphere water vapor further supports this interpretation. However, the rate of fall in the north is greater than the rate of rise in the south, which indicates that additional sinks for atmospheric water must exist; much of the northern hemisphere water vapor may be cycling back into the regolith during the latter part of the summer. The disappearance of the atmospheric water occurs more rapidly than is

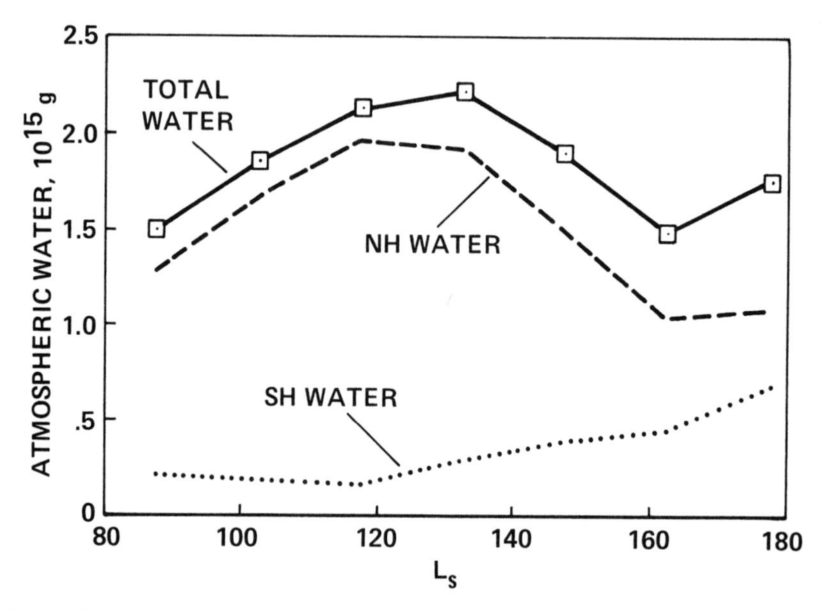

Fig. 3. Total atmospheric water-vapor content during the northern hemisphere spring season for the global atmosphere (solid line and boxes), and for the northern and southern hemispheres separately (dashed and dotted lines, respectively) (figure from Haberle and Jakosky 1990).

expected for diffusion into the regolith, suggesting that the water might be forced by an additional mechanism such as high-altitude condensation which would concentrate the water lower in the atmosphere (Jakosky 1983*a,b*). Although the data are not conclusive, the behavior of atmospheric hazes supports this suggestion, with the altitude of the base of the hazes decreasing steadily during the fall season (Kahn 1990).

Because the south polar cap does not lose its covering of CO_2 frost during the summer, at least during the Viking year, it acts as an efficient cold trap for removing any water from the atmosphere which comes into contact with the cap. The north cap acts as a summertime source for water, so that there will be a net loss of water from the north cap over the course of a year and a net gain onto the south cap (Jakosky and Farmer 1982; Jakosky 1983*b*). The magnitude of this loss is extremely uncertain. Transport northward during the dust storms or at other seasons might balance or exceed this net southward transport, although numerical modeling of the atmospheric circulation suggests that the cross-equatorial Hadley cell and winter hemisphere wave activity are both insufficient to transport enough water into the north polar region during winter to balance the summertime loss (Barnes and Hollingsworth 1987; Barnes 1990*b;* Houben et al. 1988). However, the models are not comprehensive and have not yet attempted full annual simulations. It remains, therefore, an open question as to the magnitude, and, to some extent, the direction of any net transport over the course of the Martian year observed by Viking.

Further, the year observed by Viking may not be typical of the behavior at the current epoch. Observations of atmospheric water vapor made from Earth since the 1960s suggest significant year-to-year variability which may be tied to variability of the seasonal cycle or to the possible disappearance of the residual CO_2 deposit in some years. Again, we are left uncertain as to the typical or average behavior at the current epoch, and as to the average net annual transport of water between the two polar caps.

Figure 4 shows a schematic summary of the seasonally accessible reservoirs for water and the exchange and transport of water between them. Water is supplied to the atmosphere during the northern hemisphere spring and summer seasons from the north residual polar cap and from adsorbed water in the northern hemisphere regolith. Some of this water is transported across the equator into the southern hemisphere atmosphere and may be lost by condensation onto the south polar cap. Half a year later, during the northern fall and winter seasons, water is removed from the northern hemisphere atmosphere into the regolith and the north polar cap. There may be some transport from the southern hemisphere atmosphere across the equator into the northern hemisphere, but the magnitude of such transport is uncertain. At the same time, water is lost (at least in the years observed by Viking) from

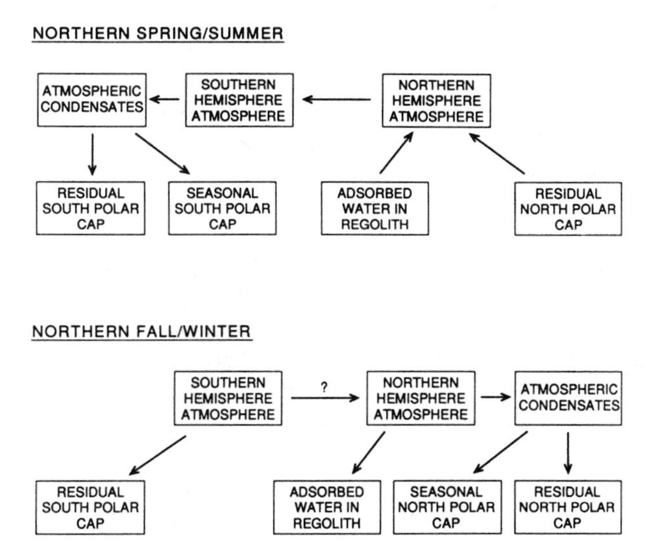

Fig. 4. Schematic summary of the seasonal water cycle. Boxes represent places where water can reside, and include both atmospheric and non-atmospheric reservoirs. Arrows represent the exchange of water between reservoirs at that season. See text for more detailed discussion.

the southern hemisphere atmosphere to the south-polar residual cap; again, the magnitude of this loss is uncertain.

The long-term evolution of the polar ice deposits, as is presumably reflected in the polar layered terrain, depends on the incremental annual exchange of water between the polar caps, the atmosphere and the regolith. Although the processes of summertime sublimation and wintertime condensation clearly are involved in the formation of the layered deposits (see, e.g., Toon et al. 1980), the exact role is not at present understood. An understanding of how each of the processes that involve water acts at the present epoch is required in order to extrapolate to other epochs when the solar forcing is different. Much of the remainder of this chapter deals with the issues of the present-day processes and the extrapolation to other time periods.

III. EXCHANGE WITH NON-ATMOSPHERIC RESERVOIRS

The seasonal variations in the globally integrated amount of atmospheric water in the Martian atmosphere require supply of water to the atmosphere from a non-atmospheric source region and subsequent removal from the atmosphere. The likely candidates for allowing significant exchange include the seasonal polar caps (which consist primarily of CO_2 frost but which also contain water frost), the residual polar caps and the regolith. Although the regolith may contain liquid water, either as a transient pure phase (see, e.g., Farmer 1976) or brine (Clark 1978; Brass 1980; Clark and Van Hart 1981; Zent and Fanale 1986), there is little evidence to support the existence of such a phase; we will generally confine our discussion to water which is bound either physically or chemically to the regolith grains. The discussion that follows is broken into separate discussions of exchange of water with the polar caps and with the regolith. The possible presence of liquid water is described in the latter.

A. The Seasonal and Residual Polar Caps

Ever since the discussions of the polar cap seasonal cycle by Leighton and Murray (1966) and Leovy (1966*a*), it has been generally recognized that CO_2 frost forms the bulk of the seasonal caps. Although no model to date has been capable of matching the exact recession of the caps, they do closely mimic the general behavior, suggesting that the basic physics has probably been incorporated (see, e.g., Cross 1971*a;* Briggs 1974; Kieffer et al. 1977; Hess et al. 1979; James and North 1982; Paige and Ingersoll 1985).

Carbon dioxide is more volatile than water, so that the surface acts as a very efficient cold trap for water when CO_2 frost is present. At 150 K, the approximate temperature at which CO_2 will condense at the ambient Martian pressure of 6 mbar, water vapor in equilibrium with frost will have a partial pressure of $\sim 10^{-7}$ mbar; this is roughly equivalent to a column water-vapor abundance of 10^{-3} pr μm, and is $\sim$ 4 orders of magnitude less than the typical atmospheric water abundance (Ingersoll 1974; Farmer 1976). Thus, when any ambient atmosphere is mixed to near the surface of the seasonal polar cap and cooled to cap temperatures, the water will condense onto the cap. As the cap retreats when the CO_2 frost sublimes during spring, surface temperatures rise well above the H_2O frost point, and the water ice that is present will be released back into the atmosphere. Models of this condensation and sublimation process, based on an early atmospheric circulation model, demonstrated that this could be an important process in the seasonal water cycle, although the observations suggested that it is not the dominant process (Leovy 1973; see Sec. IV.B).

Evidence for the presence of water ice within the seasonal cap is obtained from near-infrared spectral observations of the seasonal cap and from *in situ* observations made from the Viking Landers. Earth-based reflectance measurements of the north-polar seasonal cap during its springtime retreat phase clearly show the 1.4- and 1.9-μm absorptions due to water frost (see Fig. 5). The cap probably contains > 20 pr μm of water mixed in with the CO_2; this is a lower limit, as there may be additional water frost deeper within the CO_2 frost than is sensed by reflected sunlight (Jakosky 1983*c*). It is not clear whether this amount of water being released from the retreating seasonal cap dominates over release from the regolith in controlling the seasonal water cycle.

The south-polar seasonal cap is more problematic. Limited Earth-based spectral observations in mid-spring show no evidence for water ice mixed in with the CO_2 frost, with an upper limit of perhaps 15 pr μm (Larson and Fink 1972). Spacecraft observations at about the same season but during a different year, made from the Mariner 6 and 7 flybys, however, show clear evidence of water frost at the cap surface (Pimentel et al. 1974). Again, the amounts of water involved are uncertain, and the relative importance of this process in the seasonal water cycle is also unknown. The apparent release of water from the retreating seasonal cap has been observed in the Viking MAWD data, although the total amount of water involved is uncertain (Davies and Wainio 1981).

Finally, a transient surface brightening was observed during the northern hemisphere winter from the Viking 2 Lander, located at $+48°$ latitude and within the region expected to be covered by the north-polar seasonal frost deposits. This brightening lasted for about a sixth of the Martian year. The color of the deposits was most consistent with surface frost, rather than dust deposits (Wall 1981). Also, the temporal behavior suggested surface frost due to the sudden appearance of the surface patches, their gradual disappearance, and their apparent greater stability in shadows (Jones et al. 1979; Wall 1981; Svitek and Murray 1990). The frost formed initially when atmospheric temperatures were near the CO_2 frost point, and remained on the surface after temperatures rose, leading Jones et al. (1979) to conclude that it was initially a mixture of CO_2 and H_2O frost and that the CO_2 sublimed rapidly and the

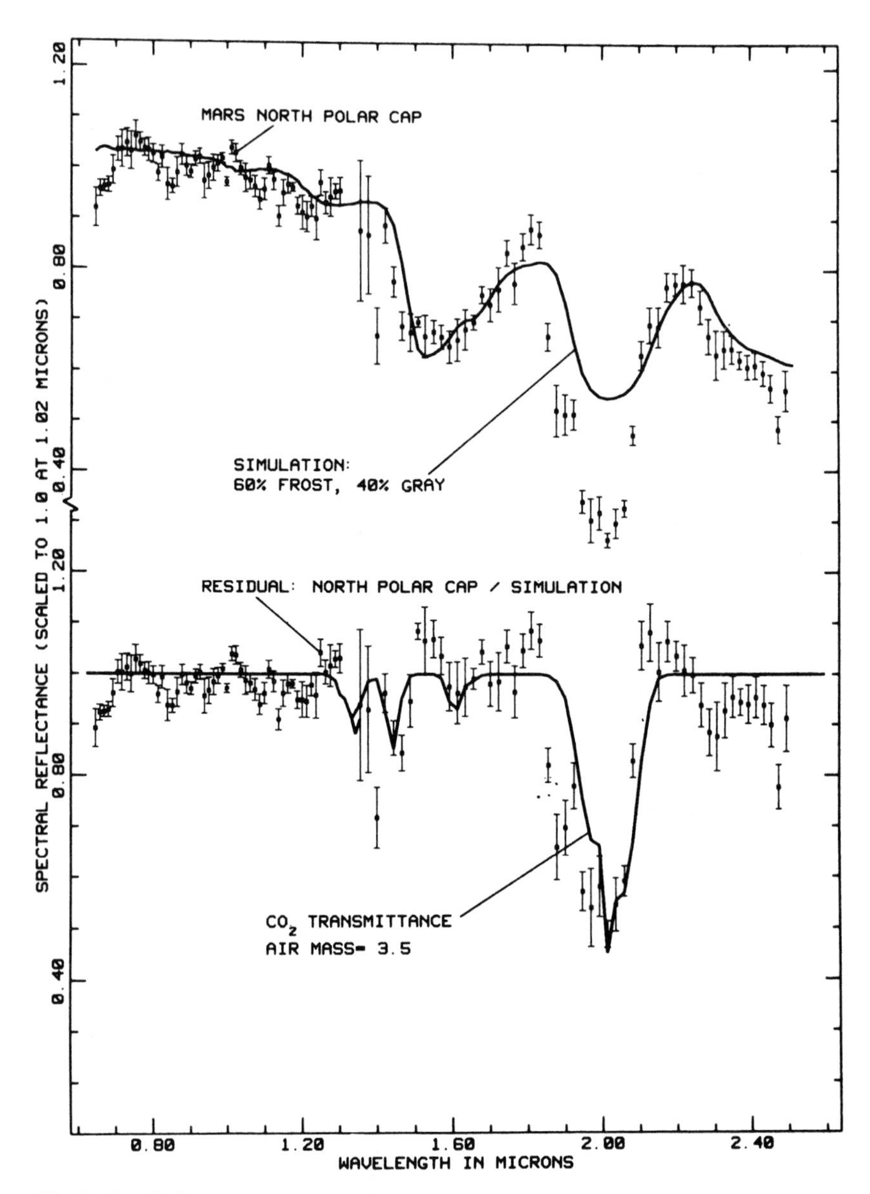

Fig. 5. (Top) Reflectance spectrum on the retreating north polar seasonal cap, made at $L_s = 50°$. The solid line is a model spectrum based on the reflectance spectrum of water ice. (Bottom) Same spectrum with the model spectrum removed; the residual spectrum shows features due to CO_2 atmospheric transmission. Deep water-ice features are seen at 1.4 and 1.9 μm, suggesting that a significant amount of water ice is mixed in with the CO_2 frost on the retreating cap (figure from Clark and McCord 1982).

H_2O more gradually. Analysis of the sublimation rates for both CO_2 and H_2O frosts during this entire time period showed that CO_2 frost would not be stable on the surface at any time throughout the day, but that H_2O frost would; as temperatures rose toward spring, water-ice deposits ceased to be stable, consistent with both the disappearance of the frost and the appearance of water vapor in the atmosphere (Hart and Jakosky 1986; Svitek and Murray 1990). The average surface coverage by frost was estimated to be equivalent to $\sim$ 10 pr μm of water, again yielding an equivocal answer as to the importance of the seasonal cap in the water cycle (Hart and Jakosky 1986).

The residual polar caps are important in the Martian water cycle due to their ability to act as either sources or sinks of atmospheric water. The north residual cap as seen from Viking consisted of a dirty water ice, with the CO_2 frost disappearing at $\sim L_s = 80°$ (Kieffer et al. 1976*b;* Paige and Ingersoll 1985; Kieffer 1990). After the CO_2 frost disappears and temperatures are no longer buffered to near 150 K, the temperature of the ice deposits rises to near 205 K, and that of the intermixed dark lanes and surrounding debris deposits rises to near 240 K (Kieffer et al. 1976*b;* Paige and Ingersoll 1985). At temperatures near 205 K, water sublimation into the atmosphere occurs at a rate of $\sim$ 10 μm day^{-1} (see, e.g., Toon et al 1980; Ingersoll 1970).

Using surface temperatures measured from the Viking Orbiter IRTM experiment, Haberle and Jakosky (1990) calculated the total sublimation of water into the atmosphere during the summer season. They concluded that the residual cap can sublime the observed 7×10^{14} g of water that appears in the northern hemisphere atmosphere after $L_s = 80°$ if near-surface winds are higher than expected (as high as $\sim$ 10 m s^{-1}); the cap infrared emissivity is lower than expected ($\sim$ 0.9); or the exposed soil in the dark lanes or circumpolar debris deposits contains ice that is accessible to the atmosphere. These requirements seem unlikely, so there are probably other sources for the summertime northern hemisphere water in addition to the cap. This conclusion is further supported by the results of two-dimensional transport simulations which show that most of the water sublimed from the cap remains at high latitudes; the polar circulation simply lacks the intensity and scale to move water very far from the cap (see Sec. IV). A weak circulation places additional constraints on cap sublimation rates as abundant clouds and significant surface ice deposits would form just equatorward of the cap edge, according to the model, if the cap is allowed to sublime as much as 7×10^{14} g H_2O; such predictions are inconsistent with available data, although some summertime cloudiness is observed. With the largest uncertainty being the ability of the atmospheric circulation to transport water away from the polar region, Haberle and Jakosky (1990) estimate that the polar cap loses between $\sim$ 0.1 and 0.8 mm of ice (0.01 to 0.08 g cm^{-2}) to sublimation over the course of a summer season.

The south residual polar cap is more puzzling. It never lost its covering of CO_2 frost throughout the year observed by Viking (Kieffer 1979; James et al. 1979), and, in fact, seemed to be in energy equilibrium averaged over a Martian year, neither losing nor gaining any significant net amount of CO_2 frost (Paige and Ingersoll 1985). Certainly, the residual CO_2 frost serves as a cold trap for water; it is not possible from available data, however, to determine the rate at which ice is incorporated or how much water ice may actually be present. There is some evidence that the CO_2 ice covering may disappear in some years to reveal the underlying water-ice cap. Figure 6 shows a comparison between the Viking and the Earth-based measurements, where the former have been spatially integrated to approximate the resolution of the latter. The 1969 observations show a factor of $\sim$ 6 times more water during the southern hemisphere summer season than was seen by Viking at the comparable season (Jakosky and Barker 1984). Jakosky and Farmer (1982) and Jakosky and Barker (1984) suggested that the CO_2 frost covering the south residual cap may have disappeared that year, exposing an underlying water-ice cap which could then sublime water into the atmosphere. Modeling of the seasonal behavior of the south-polar cap by Jakosky and Haberle (1990*b*) demonstrated that it is possible for the cap to have a CO_2 frost covering all year in one year and for the covering to disappear the next year, or vice versa, and that the cap can be stable if it is in either state.

B. The Regolith

Certain observations of atmospheric water suggest that there is seasonal exchange between atmospheric water and water in the regolith. These include the fact that the initial springtime increase in column water-vapor abundance occurs at latitudes significantly equatorward of the edge of the retreating seasonal polar cap (Jakosky and Farmer 1982; Jakosky 1983*a*); and the apparent inability of sublimation of water from the residual north-polar cap to supply all of the observed increase in atmospheric water during northern summer, after the seasonal polar cap has already disappeared (see the previous section). The most likely reservoir in each case is water molecules which have been adsorbed onto individual regolith grains in the near-surface layer. Further, such exchange would be expected if the regolith is at all permeable to water vapor diffusing through the pore spaces between grains. Below, we discuss the physics of adsorption and thermally and nonthermally driven exchange of water. We also discuss the ability of water to exist in the regolith as a pure liquid, a brine, or as solid ice, and its ability to exchange with the atmosphere.

Adsorption of water molecules onto regolith grains involves hydrogen bonds, with binding energies on the order of 10 kcal mole^{-1} (see, e.g., Brunauer et al. 1938,1967; deBoer 1968; see Jakosky [1985] for a more detailed discussion as applicable to Mars). Gas molecules in the pore spaces within the regolith will strike the regolith surface at random and bond or stick to the surface. Molecules that are already bound to the surface will vibrate thermally, with a Maxwellian distribution of energies, and some of them will have enough energy to break free of their bond and become free gas molecules in the pore space. A dynamical steady state will establish itself rapidly, where an equal number of molecules will strike the surface and stick as will become a free gas molecule in the pore space; the equilibrium number of molecules per unit area of grain will be defined uniquely for a given regolith material by the temperature and the water vapor partial pressure within the pore space. Changes in the surface and subsurface temperatures or changes in the partial pressure of water vapor in the atmosphere, along with diffusion of water molecules between the atmosphere and the pore spaces between grains, are capable of driving an exchange of water between the atmosphere and adsorbed water in the regolith.

Figure 7 shows adsorption isotherms for water on both basalt grains and montmorillonite grains, based on the laboratory measurements of Fanale and Cannon (1974) and Mooney et al. (1952*a*), respectively. In each case, choosing a partial pressure of water vapor and a temperature allows the steady-state amount of adsorbed water to be read off of the ordinate. Values range up to $\sim$ 10 mg H_2O g^{-1} for basalt; on clay minerals such as montmorillonite, which have a different molecular structure, as much as several 100 mg H_2O g^{-1} can adsorb (Fanale and Cannon 1974; Mooney et al. 1952*a*; Anderson et al. 1967,1978).

A simple one-dimensional model constructed by Jakosky (1983*a*) demonstrated the possible importance of exchange with the regolith. The model used seasonally varying surface and subsurface temperatures to drive adsorption and desorption within the regolith. The changing difference between the water-vapor partial pressure in the atmosphere and in the regolith causes diffusion of water between the surface and atmosphere, resulting in a changing atmospheric column water-vapor abundance. Figure 8 shows the seasonal variation of the atmospheric water predicted for a variety of plausible regolith vapor-diffusion constants. Clearly, regolith exchange may be an important process in the seasonal water cycle.

The role of processes other than simple vertical exchange, including the possible release of water from the polar cap and transport within the atmosphere, can be approximately included by using the observed seasonally changing atmospheric water-vapor abundance as a boundary condition in the models of exchange with the regolith. Figure 9 shows the resulting water-vapor exchange, assuming the atmospheric water to be distributed uniformly with altitude. Because the water vapor that diffused from the regolith does not exactly match the observed atmospheric water-vapor variations, transport of water, along with possible seasonal variations in the vertical distribution of the atmospheric water vapor, must play an important role in the seasonal cycle (Jakosky 1983*b*; Haberle and Jakosky 1990).

The presence of ice or liquid water within the near-surface regolith is

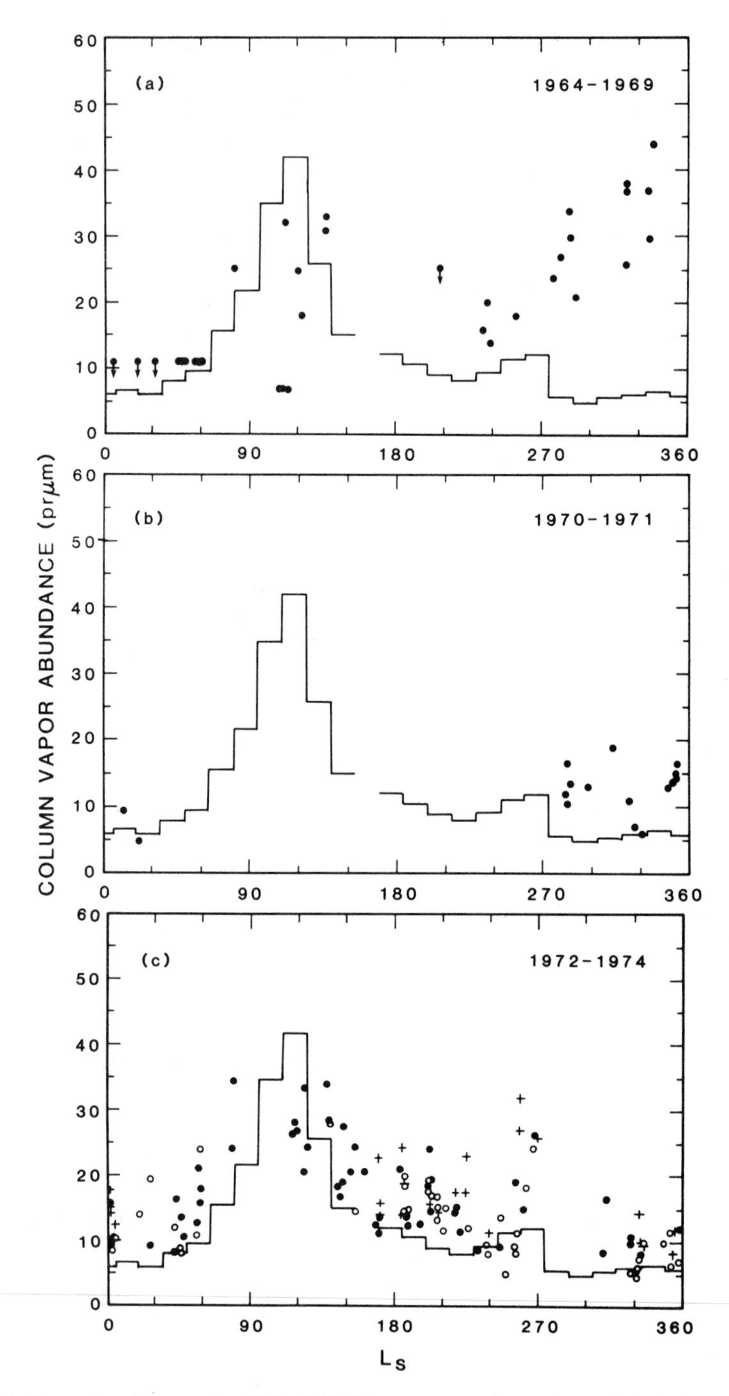

Fig. 6. Comparison between the Viking MAWD water-vapor observations and Earth-based telescopic observations made between 1964 and 1983. The solid histogram shows the Viking data, spatially integrated to approximate what would be seen from Earth, and the dots are the Earth-based data (figure from Jakosky and Barker 1984).

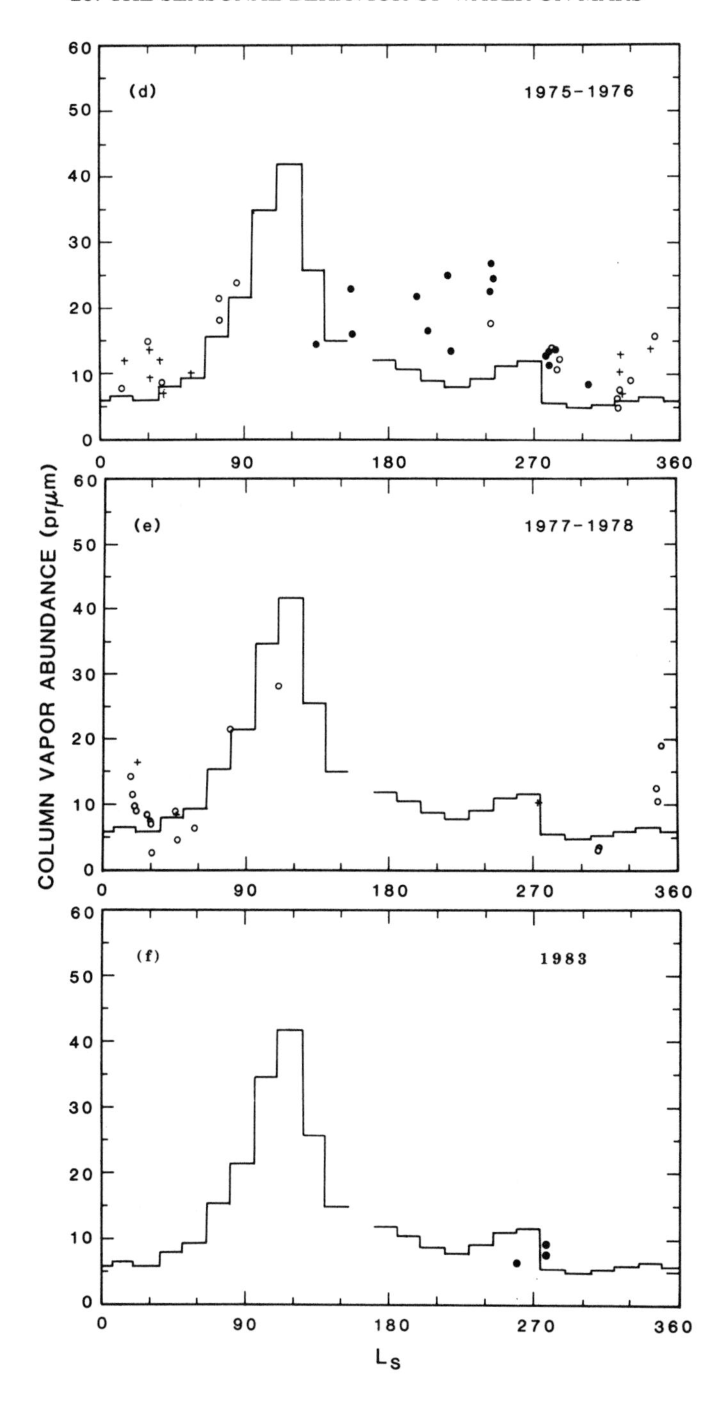
(d)
1975-1976
(e)
1977-1978
(f)
1983
COLUMN VAPOR ABUNDANCE (prμm)
0
10
20
30
40
50
60
0
90
180
270
360
L_S

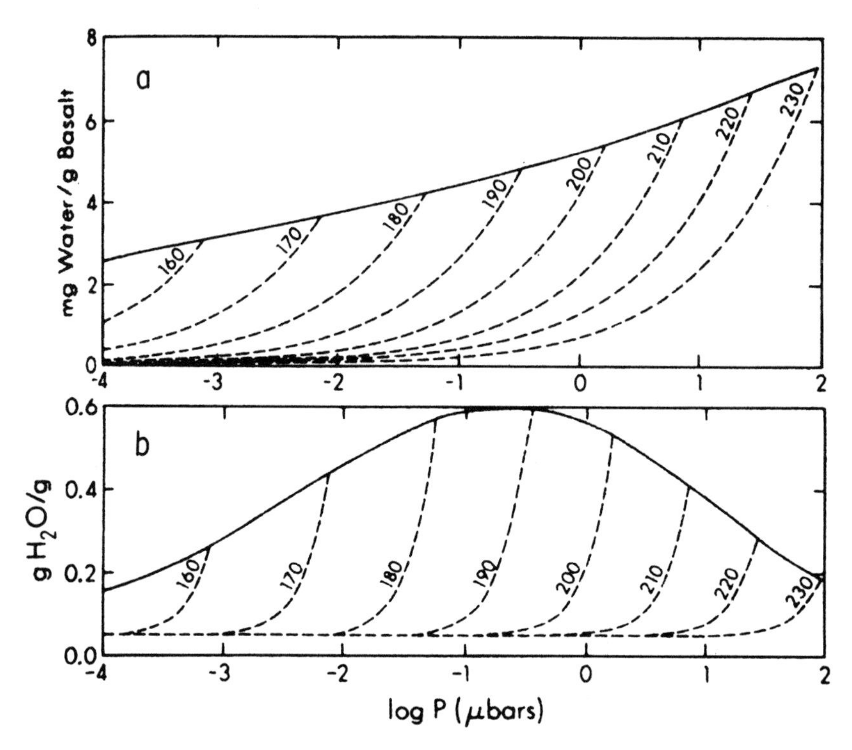

Fig. 7. Adsorption isotherms for water adsorbed on basalt grains (a) and montmorillonite grains (b). For a given water-vapor partial pressure (along the abscissa) and regolith temperature (dotted lines), the amount of adsorbed water in equilibrium can be read from the ordinate. The solid line represents the peak water which can adsorb; additional water would be present as ice (figure from Zent et al. 1986).

somewhat more problematic. Ice will be stable at a given location if the temperature there is lower than the condensation temperature corresponding to the atmospheric water-vapor partial pressure. For an average atmospheric vertical-column water-vapor abundance of ~ 10 pr μm, assumed to be distributed uniformly with altitude, the condensation temperature at the surface is ~ 200 K (Leighton and Murray 1966; Farmer and Doms 1979). Figure 10 shows the stable locations of ground ice predicted from nominal calculations of the surface and subsurface temperatures; properly including the seasonal variation of column water abundance and the marked depletion of water vapor during wintertime near the surface of the seasonal polar cap does not dramatically change the locations where ice would be stable all year or for part of the year.

More detailed calculations have been done by Zent et al. (1986), including the thermally driven interchange of water between ice, vapor and adsorbed water within the regolith. The seasonally varying subsurface temperatures drive an exchange between subsurface ice and adsorbed water. At the

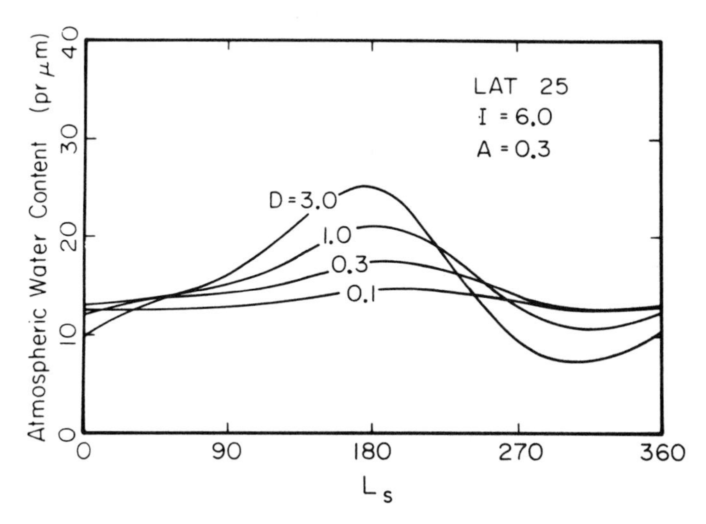

Fig. 8. Seasonal variation of atmospheric water-vapor abundance based on a simple one-dimensional model of exchange between the atmosphere and regolith. Models are for a latitude of 25°, surface thermal inertia of 0.006 cal $cm^{-2}s^{-1/2}K^{-1}$, albedo of 0.3, and regolith effective diffusion constant varying between 0.1 and 3.0 cm^2s^{-1} (figure from Jakosky 1983*a*).

low wintertime temperatures, the adsorptive capacity of the regolith is low and adsorbed water will be driven into ice (see Fig. 7, showing the adsorption isotherms). At higher summertime temperatures, the adsorptive capacity increases, and the water molecules in the ice will shift back to adsorption sites. Figure 11 shows an example of the ice content within the regolith as a function of season at a depth of 20 cm for different latitudes. Interestingly, ice will also be stable at lower latitudes in a montmorillonite regolith during the summer, when the condensation point is also reached at a lower amount of adsorbed water, this phenomenon does not occur in basalt due to the different shapes of the isotherms, as shown in Fig. 7 (Zent et al. 1986). Of course, ice in low-latitude regions will only be stable as long as the water vapor in the pore spaces can be confined within the regolith; diffusion to the atmosphere will otherwise rapidly deplete the near-surface regolith of ice (Smoluchowski 1968; Clifford and Hillel 1983).

Seasonal exchange of water between the atmosphere and any high-latitude subsurface ice deposit is unlikely to be important in the seasonal cycle. Exchange of adsorbed water molecules in the regolith occurs for water within only the uppermost 5 to 10 cm of the regolith (Jakosky 1983*a*). The depth at which a water-ice layer will exist is generally between 10 cm and 1 m (Farmer and Doms 1979; Zent et al. 1986), so that water molecules will not be able to diffuse deeply enough into the regolith on a yearly basis to provide significant exchange. On the other hand, one-way diffusion of water, from an unstable ice layer within the near-equatorial regolith into the atmosphere is possible (Smoluchowski 1968; Clifford and Hillel 1983; Toon et al.

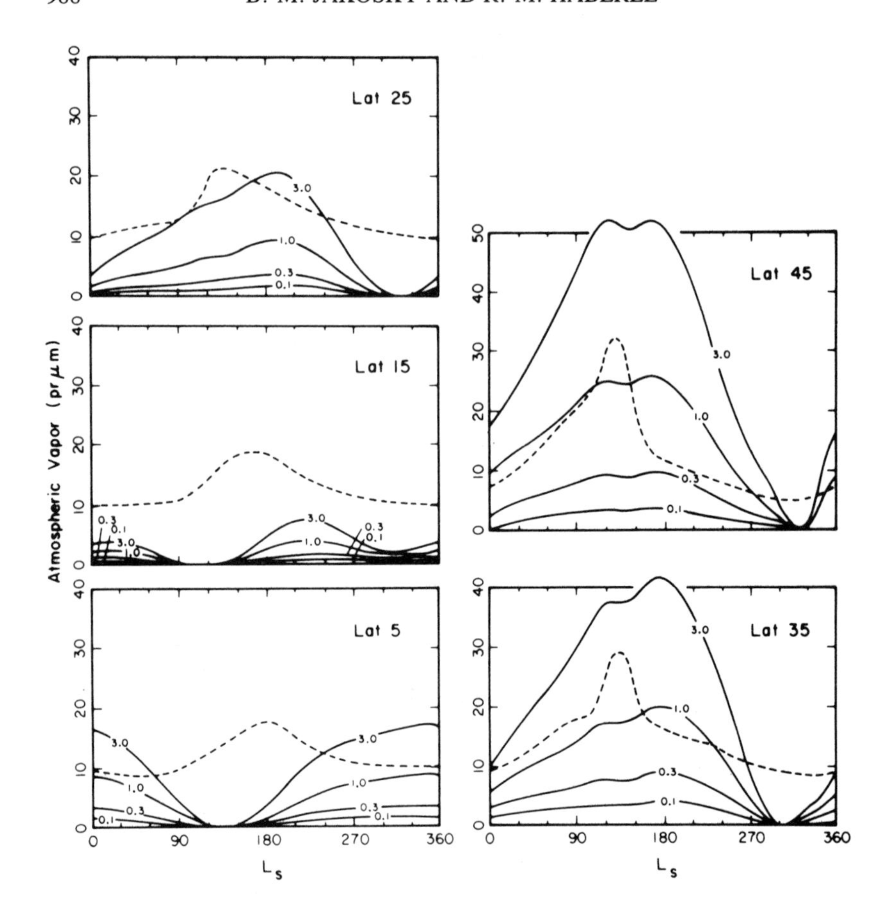

1980), providing a source of water to the atmosphere throughout the year. Such ice may be a relic from periods of high obliquity when near-surface equatorial ice may have been stable (Jakosky and Carr 1985; Jakosky 1985; see Sec. V below); there are no observations or models that either suggest or require such a source, however.

Pure liquid water may exist within the regolith at temperatures above 273 K. At that high a temperature, however, sublimation or diffusion into the atmosphere will be extremely rapid (see, e.g., Ingersoll 1970; Jakosky 1985), so that liquid water will not be stable. Dissolved salts will lower the freezing point of the liquid, but even the most extreme case would not result in an eutectic point (at which the last liquid would freeze) lower than ~ 220 K (Brass 1980; Clark and Van Hart 1981); again, the vapor pressure at this temperature is high enough that a brine would not be stable with respect to sublimation into the atmosphere (see, e.g., Clark 1978; Zent and Fanale 1986).

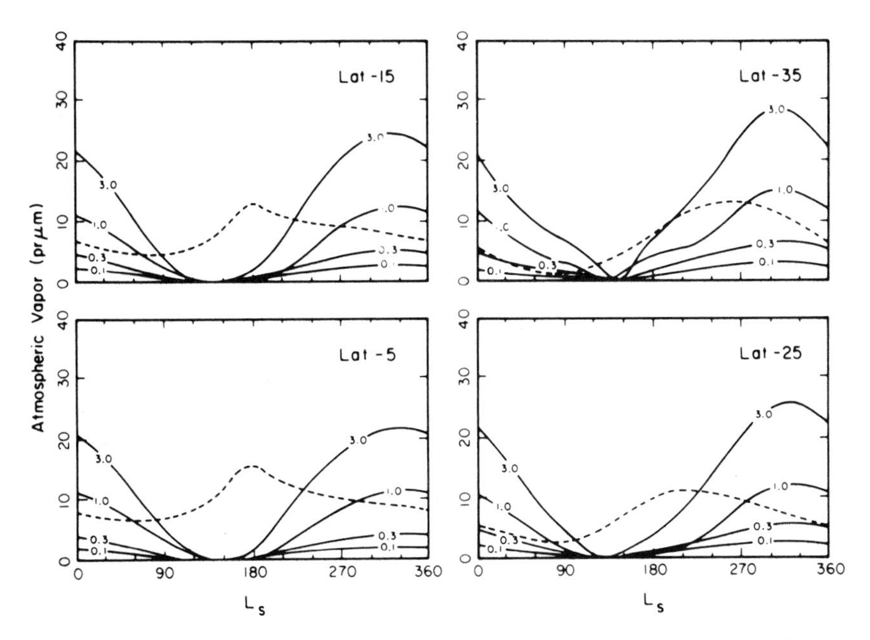

Fig. 9. Seasonal exchange of water in pr μm with the regolith at different latitudes, using the observed column water-vapor abundance as the atmospheric boundary condition and assuming a constant uniform vertical distribution of water within the atmosphere. Dashed line represents a smoothed version of the Viking MAWD observations of atmospheric abundance used as the boundary condition. Solid lines represent the amount of water exchanged with the regolith for several values of the effective diffusion constant (in $cm^2 s^{-1}$) (figure from Jakosky 1983*a*).

Bound water within the regolith has been detected using spectroscopic techniques from Earth (Sinton 1967; Houck et al. 1973) and from spacecraft (Herr and Pimentel 1969; Pimentel et al. 1974; see chapter 17). *In situ* analysis from Viking Landers also suggests the presence of bound water (Anderson and Tice 1979). These techniques generally do not distinguish between chemically and physically bound water. Chemically bound water is bonded more strongly than adsorbed water, and is incorporated from free water within the regolith. The presence of bound water, therefore, combined with the models of regolith physics, supports the idea that there is exchange of water between the regolith and the atmosphere on a seasonal basis. The largest uncertainty in assessing its importance in the seasonal water cycle is the ability of water vapor to diffuse through the regolith. Where the regolith is relatively nondiffusing, such as in the vicinity of a duricrust like that observed at the Viking Lander sites (see chapter 21), exchange will be small (but probably not zero); where the regolith is generally composed of loose, unconsolidated material, as much of it is thought to be, then significant exchange will occur and will be important in the seasonal cycle.

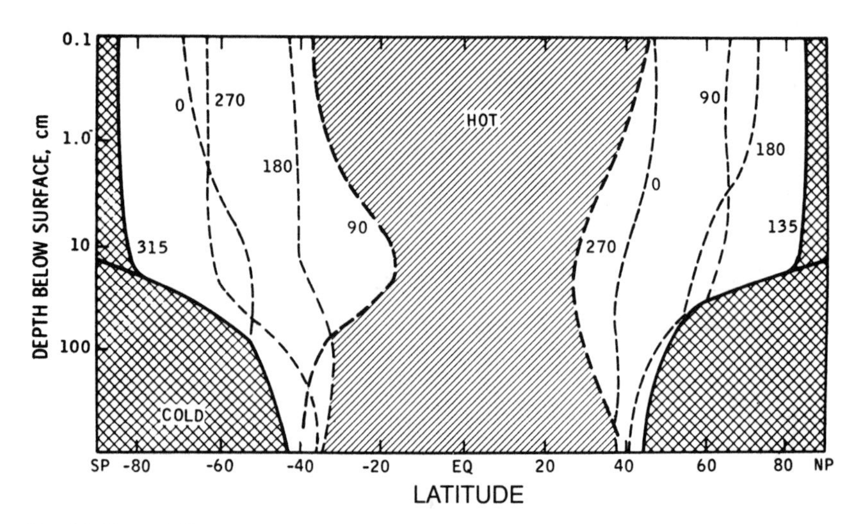

Fig. 10. The stability of water ice within the near-surface regolith. The cross-hatched region is cold enough all year for water ice to be stable with respect to sublimation into the atmosphere; because of the observations of the existence of residual polar caps, the cold region has been continued to the surface at both poles despite the fact that the thermal model does not predict water ice. Shaded region is too hot all year for ice to be stable. Remaining regions will have ice stable at some seasons, but not throughout the year; dashed lines show the equatorward limit of stability at different seasons (L_s) (figure from Farmer and Doms 1979).

IV. THE ROLE OF ATMOSPHERIC TRANSPORT

Large-scale motions within the atmosphere are driven by seasonal, latitudinal and diurnal variations in insolation. These motions move parcels of atmosphere over large distances and mix them vertically; clearly, they are capable of redistributing water vapor on a global scale. This redistribution of water is discussed in this section, with emphasis on the implications for the seasonal and annual behavior of the water. Below, we summarize the atmospheric circulation, discuss the potential for transport by the various components of the circulation, describe the distribution of atmospheric condensates and their effects on the seasonal water cycle, and discuss in detail the relationship between the atmospheric circulation and the net annual transport of water vapor.

A. Summary of the Atmospheric Circulation

The role of atmospheric transport in the seasonal water cycle is to provide a conduit through which water is exchanged between its various surface and/or subsurface reservoirs. This transport is accomplished by motions which occur on a variety of temporal and spatial scales and which are described in detail in chapter 26. Here, we present a brief summary.

On planetary scales, the Martian circulation is conveniently viewed in

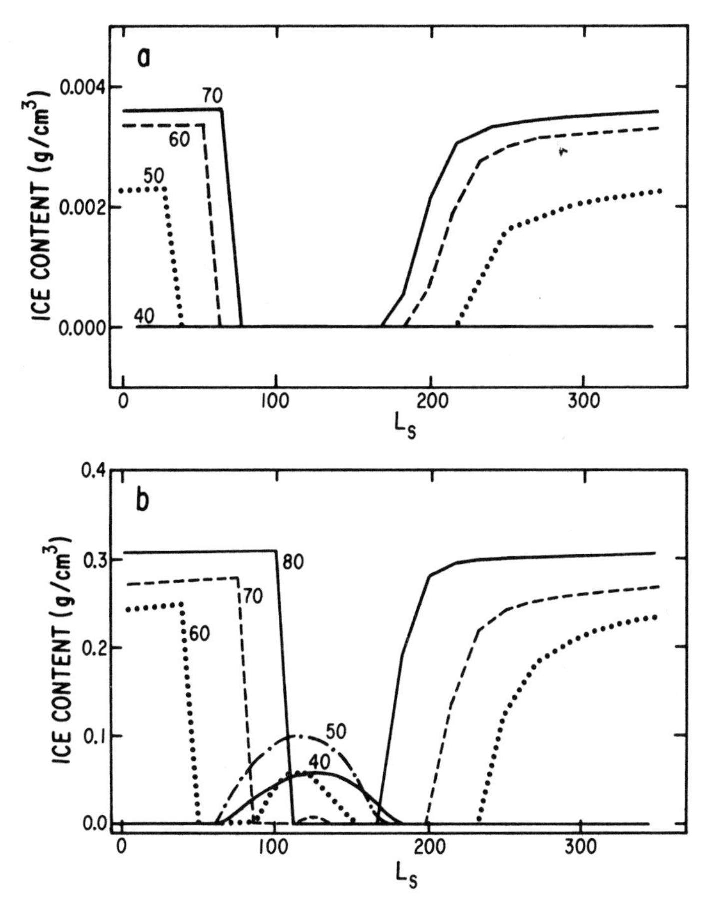

Fig. 11. Seasonal variation of water-ice content within the regolith at several latitudes and at a depth of 20 cm, for a basalt regolith (a) and a montmorillonite regolith (b). Variation is due to seasonal exchange between ice and adsorbed water, assuming no significant exchange between the subsurface and atmosphere (figure from Zent et al. 1986).

terms of a time- and zonal-averaged component and departures from it (Fig. 12). The former is dominated by overturning and condensation flows that are driven by the differential heating and the CO_2 cycle, respectively. The overturning circulation consists primarily of a Hadley cell which is similar in structure to that found on Earth, with rising motions near the subsolar latitude. On Mars, however, it varies significantly with season and dust loading (Leovy and Mintz 1969; Haberle et al. 1982; Pollack et al. 1990*a*). This is particularly important for the water cycle because inter-hemispheric exchange is facilitated on Mars by the development of large cross-equatorial Hadley cells during the solstice seasons.

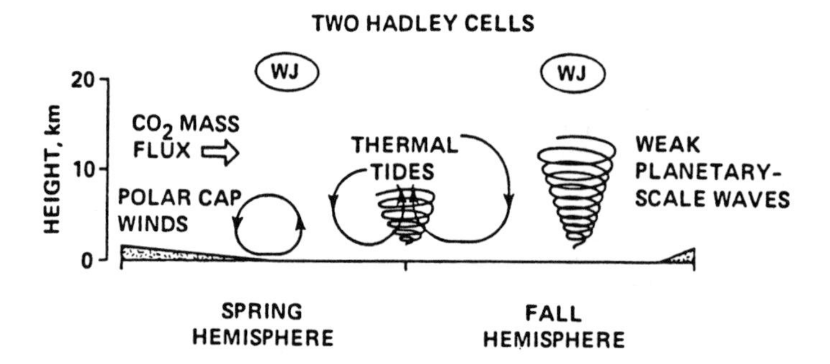

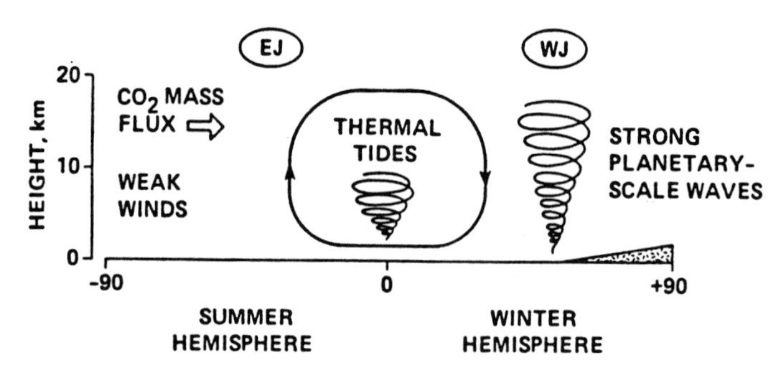

Fig. 12. Cartoon showing the main features of the atmospheric circulation at the equinox (*top*) and at the solstice (*bottom*). EJ and WJ represent the locations of the easterly and westerly jets, respectively. Slanting lines at the surface represent the seasonal polar caps. Arrows indicate zonally symmetric overturning, and spirals indicate mixing by atmospheric waves. For more discussion, see text or chapter 26.

Departures from the zonal-mean circulation include high-latitude eastward-propagating traveling waves, which are believed to be manifestations of baroclinic instability (Ryan et al. 1978; Barnes 1980,1981; Sharman and Ryan 1980); low-latitude westward-propagating thermal tides, which are driven by the daily solar heating cycle (Zurek 1976; Leovy and Zurek 1979); and quasi-stationary planetary waves, which can be found at both high and low latitudes, and which arise from the mechanical and thermal effects of large-scale orography (Webster 1977; Pollack et al. 1981; see chapter 26).

Smaller-scale motions may also be important for the water cycle. Intense "sea-breeze"-type circulations develop near the edge of the retreating polar caps during spring (Burk 1976; Haberle et al. 1979; French and Gierasch 1979) which may help redistribute water locally (Christensen and Zurek 1984). Other potentially important small-scale motions are localized circulations that arise from sloping topography or regional differences in surface properties (albedo, thermal inertia), local dust storms and dust devils, and turbulence within the boundary layer.

B. Transport of Water by the Various Circulation Mechanisms

The current data base is insufficient to diagnose transport processes within the atmosphere. Although a good first-order definition of the seasonal and latitudinal behavior of column water vapor does exist, equivalent knowledge of the simultaneous wind field and the vertical distribution of water does not. Consequently, models of the transport processes have been constructed. Some treat transport in a highly parameterized form (Davies 1981; Jakosky 1983*a,b;* James 1985,1990), while others address specific components of the circulation (Leovy 1973; Haberle and Jakosky 1990; Barnes 1990*b*). In this section we discuss the fundamental mechanisms by which various components of the circulation can bring about transport and how they have been modeled.

1. Zonal Mean Circulations. As on Earth, the low-latitude circulation on Mars is dominated on seasonal time scales by Hadley circulations. Hadley circulations can transport water vapor vertically or meridionally depending on the distribution of water vapor within the atmosphere. On Earth, the Hadley circulation transports water vapor toward the equator, which balances the excess there of precipitation over evaporation. This is accomplished by a circulation whose rising branch is near the equator and within which the time and zonally averaged water vapor mixing ratio decreases with height. As a consequence, the equatorward-moving low-level flow carries more moisture than does the poleward-moving upper-level flow, and a net equatorward transport results.

The ability of a Hadley cell to transport water meridionally, therefore, depends on the vertical distribution of water vapor and the intensity of the cell. On Earth, the vertical distribution of water vapor within the Hadley cell is maintained by saturation processes associated with precipitation. On Mars, water vapor is more likely to be more uniformly mixed because precipitation, if it occurs at all, is not likely to be significant. Furthermore, the tropics appear to be far from saturation throughout the bulk of the atmospheric column even though clouds do form (see Sec. IV.C).

However, some transport by the Martian Hadley circulation is indicated in the MAWD data. As described in Sec. II, the total amount of water vapor in the southern hemisphere begins to increase just after the beginning of northern summer. Since at this season there should be no southern hemisphere sources for atmospheric water, this increase must be the result of transport from the north. The wind system most likely to bring about this transport is the Hadley circulation.

To demonstrate this, we use the model of Haberle and Jakosky (1990), based on the circulation model of Haberle et al. (1982), to simulate the time evolution of the Hadley circulation for the entire northern summer season. The stream function, indicating the mass flow, corresponding to the middle and end of the season is shown in Fig. 13. After a modest spin-up period (20 days) the model is initialized with the observed latitudinal distribution of water vapor, assuming it to be uniformly mixed in the vertical. Because of this assumption, there is no net meridional transport initially. Instead, winds conspire to move water vertically, producing vertical gradients in the mixing ratio. This is precisely the vertical distribution needed for the Hadley circulation to transport water into the southern hemisphere. Interestingly, the transport occurs primarily in the upper branch of the Hadley cell. As illustrated in Fig. 14, the model begins to transport water into the southern hemisphere at about the same time as indicated in the observations, and at about the same rate. The good agreement between model and observations may be fortuitous, however, because other relevant physical processes (convective mixing and longitudinally asymmetric motions, for example) are not included. Nevertheless, it demonstrates the plausibility that cross-equatorial transport by the Hadley circulation can occur at this season.

A unique feature of Martian meteorology is the wind systems associated with the seasonal polar caps. Almost 25% of the Martian atmosphere is cycled into and out of the polar regions annually, and the resulting net mass flow can significantly influence the magnitude and direction of high-latitude surface winds (Haberle et al. 1979; French and Gierasch 1979; Kahn 1983; James 1985; chapters 26 and 27). In addition, the large temperature contrast between the cap and bare ground during spring can give rise to "sea-breeze"-like circulations in the vicinity of the cap edge (Burk 1976; Haberle et al. 1979), as well as vigorous baroclinic wave activity (Barnes 1981). Because the caps are important seasonal reservoirs for water, the circulations associated with them will influence the distribution and abundance of water stored in them.

During fall and winter, water is transported into the polar regions by the net mass flow toward the condensing cap and by mid-latitude storm systems (eddies, see below). Transport of water into the north polar region may be further enhanced by global dust storms (Farmer and Doms 1979; Davies 1981; Barnes 1990*b*). Although the amount transported is uncertain, some preliminary calculations with three-dimensional dynamical models indicate that it may not be sufficient to make up the amount lost during summer (Houben et al. 1988; Barnes 1990*b*). In any case, any water stored in the seasonal caps during fall and winter will be released back into the atmosphere during spring as the cap retreats. The formation of hazes that followed the

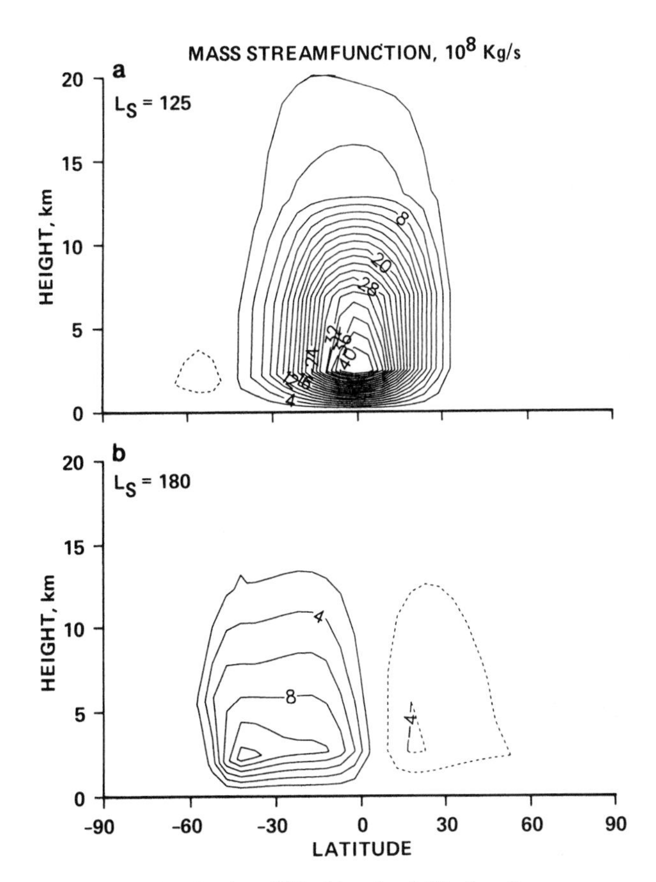

Fig. 13. Mass stream function for the middle (a) and end (b) of northern summer as simulated by the model of Haberle and Jakosky (1990). At any given latitude and height, the value of the mass stream function represents the net northward transport of mass between the surface and that point. Negative values represent southward transport. At the equator during early summer, for example, the stream function is positive at all altitudes indicating that the vertically integrated mass flow is northward at all altitudes. However, because the winds blow northward in the lower levels and southward in the upper levels, the stream function increases then decreases with height. The units of the stream function are kg s^{-1} and thus represent the total transport across a latitude circle. Thus, the sense of the circulation is counterclockwise around positive values, and clockwise around negative values. The contour interval is 2×10^8 kg s^{-1}. Notice the transition from a single intense Hadley cell to two comparatively weak Hadley cells.

retreating north polar cap during the Viking mission (Christensen and Zurek 1984) and which were seen in Earth-based observations (Iwasaki et al. 1979) are consistent with this interpretation. The Viking hazes were optically thick in visible light and contain between 1 and 2 pr μm of water. Whether the water comes from the cap itself or from freshly exposed regolith is difficult to determine. However, it is likely that the circulation associated with the

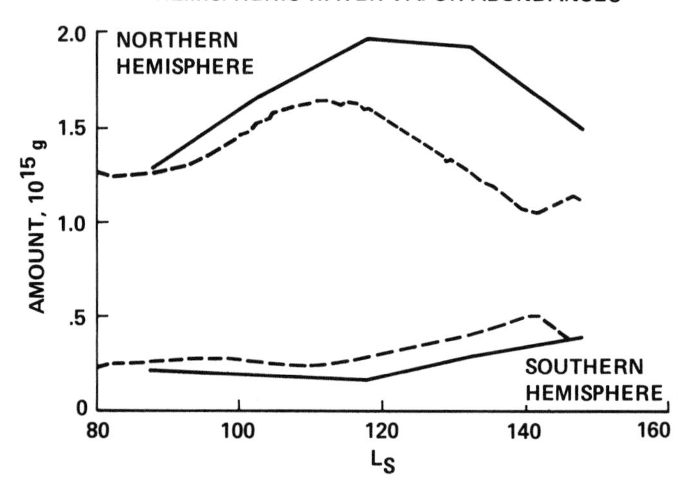

Fig. 14. Observed (solid lines; from Fig. 3) and simulated (dashed lines) hemispheric water-vapor abundance during northern summer. Simulated results are based on the model of Haberle and Jakosky (1990).

retreating cap, which has equatorward-directed winds in the lowest several km and poleward-directed winds at higher levels (Haberle et al. 1979), transports some of this water back toward the cap where cooler temperatures lead to saturation and haze formation (Christensen and Zurek 1984). This same process may be important in transporting water to the residual south polar cap during the springtime retreat of the south seasonal cap.

2. Eddy Motions. Outside the tropics, the large-scale circulation is dominated by wave motions due to traveling baroclinic disturbances (Ryan et al. 1978; Barnes 1980,1981; Sharman and Ryan 1980). In these regions, transport may no longer be controlled by the mean-meridional circulation. Waves can bring about transport if a correlation between the wave and moisture fields exists. For baroclinic waves, such a correlation exists because equatorward-moving air is generally colder, and hence drier, than poleward-moving air. Averaged over time and longitude, these waves produce a net poleward transport of moisture.

Leovy (1973) used this property of baroclinic waves to estimate the magnitude of transport of water into the polar caps. He assumed that the departure of the water-vapor mixing ratio from its zonal average was closely related to the temperature departure through the saturation constraint. In doing so, he was able to relate the moisture flux to the eddy heat flux, which was taken from the early general circulation calculations of Leovy and Mintz (1969). In his model, most of the water condensed near the edge of the fully developed polar cap. This result was insensitive to the assumption of transport in the vapor phase only, and was instead the result of the large decrease in atmospheric temperatures near the cap edge. As a consequence, the model predicted that large amounts of water would be released into the atmosphere during spring when the polar caps were retreating. This result was inconsistent with the observations of Barker et al. (1970), which showed a more gradual rise in water abundance and peak abundances occurring during midsummer. Leovy (1973) concluded that exchange between atmospheric water vapor and a subtropical permafrost was more likely to explain the observations. Of course, this neglects release of water from interior regions of the retreating seasonal cap as well as the summertime supply of water from the residual polar cap.

Recent calculations with the Mars General Circulation Model by Pollack et al. (1990*a*) show that eddy heat transport near the edge of the winter polar cap may not dominate as much as assumed by Leovy (1973). They find that the mean meridional circulation and condensation flow make significant contributions to the total poleward heat flux in this zone. This is particularly true for southern hemisphere winter, during which eddy activity is computed to be much weaker than in the corresponding season in the north; the difference may in part be due to differences in topography, however, which are very uncertain (see chapter 26).

Barnes (1990*b*) has constructed a model to examine the role of forced stationary planetary waves in the transport and deposition of dust and water at the pole. His analysis focuses on the transport that might occur during a polar atmospheric warming event. Such events appear to be associated with the onset of major global dust storms. The observational basis for the phenomenon is the increase in 15-μm brightness temperatures (corresponding roughly to atmospheric temperatures at 25-km altitude) over the north polar region during the second global dust storm of 1977 (Martin and Kieffer 1979; Jakosky and Martin 1987). Some dynamical process must have caused the warming as this region of the planet was in darkness at the time; the exact cause remains unknown (see chapter 26). However, as Barnes (1990*b*) points out, the heating of the polar atmosphere is necessarily the result of net sinking motion, which implies through mass continuity that a poleward transport of air exists at some level. Thus, on general dynamical considerations some meridional transport of water (and other tracers) must occur during these events.

To model this transport, Barnes (1990*b*) coupled the wind field produced by a dynamical model, whose wave forcing was chosen to reproduce successfully the observed warming (Barnes and Hollingsworth 1987), to a transport scheme similar to that developed by Garcia and Hartmann (1980) to study ozone transport in the Earth's stratosphere. The results indicated that substantial quantities of water could be transported into the polar regions during a warming event only if removal at subpolar latitudes by condensation and precipitation could be neglected. Given the overall warmth of the atmosphere during global dust storms, this assumption seems plausible.

It must be remembered that the question of the amount of water transported by winter hemisphere wave activity depends not only on the correlation of the water vapor (or condensate) field with the wind field and wave activity, but also on the amount of water present in the atmosphere. Most of the northern hemisphere water vapor has already been removed from the atmosphere by the time that the winter hemisphere wave activity occurs (Jakosky and Farmer 1982), and may not be available for transport. Implications for the net annual transport of water and the long-term stability and evolution of the polar cap are discussed in Sec. IV.D.

3. Boundary Layer. The boundary layer plays a particularly critical role in the water cycle as water must pass through it when exchanging between the surface and free atmosphere. Transport within the boundary layer is predominantly vertical and occurs via turbulent eddies which have a variety of length scales. During the day, turbulence is driven principally by free convection, and the boundary layer can grow to great heights (5–15 km) depending on season, latitude, dust load, large-scale dynamics and surface properties (Gierasch and Goody 1968; Sutton et al. 1978; Pollack et al. 1979*b;* Haberle and Hertenstein 1987). At night, the lower atmosphere cools and stabilizes, and turbulence is confined to a relatively shallow layer where mechanical shear stresses dominate.

Transport of water within the boundary layer has been modeled by Flasar and Goody (1976), treating exchange between the atmosphere and subsurface for the first time. Their goal was to explain Barker's (1974,1976) observations of an apparent diurnal variation in column water vapor. Although it has been subsequently demonstrated that Barker's observations could be the result of scattering by suspended dust particles (Davies 1979*c*), it is nevertheless instructive to consider the physical processes that Flasar and Goody were modeling in order to gain some insight into the exchange process.

The model of Flasar and Goody (1976) was one-dimensional (vertical) and contained algorithms to compute the movement of water through the soil and atmosphere. In the regolith, water could exist in three states (vapor, condensate, or sorbed), with transport occurring via diffusion in the vapor state only. In the atmosphere, both vapor and ice were considered. The entire system was driven by daily temperature changes due to radiative-convective processes in the atmosphere and conduction in the soil.

An important nonlinearity in the model was the dependence of the atmospheric exchange coefficients on local stability. The static stability was low during the day, and the exchange coefficient was large. As a consequence, rapid vertical mixing throughout the lowest scale height was predicted. However, above 2 km, the predicted temperatures were well below the frost points that would correspond to a well-mixed atmosphere. As extensive upper-level ice clouds are not observed, Flasar and Goody imposed a zero-flux boundary condition at the 2-km level in their model. This illustrates a potentially important role played by dust in raising atmospheric temperatures: the radiative properties of dust were not considered by Flasar and Goody (1976) but have since been shown to play an important role in determining the atmospheric temperature structure (Gierasch and Goody 1972; Pollack et al. 1979*b;* Haberle et al. 1982).

Except near the surface, where molecular conduction can be important, water vapor in the boundary layer is transported through the action of turbulent eddies that are driven by mechanical and/or buoyancy forces depending on temperature structure and wind shear. Evaporation of water from the surface is regulated by both turbulent and molecular processes. The theory of evaporation on Earth based on these processes is fairly well developed (see, e.g., Brutsaert 1982) but has not yet been applied to Mars. Instead, nearly all studies of sublimation on Mars (Ingersoll 1970,1974; Toon et al. 1980; Jakosky and Carr 1985; Hart and Jakosky 1986; Kahn 1985; Haberle and Jakosky 1990) have utilized two parameterizations of uncertain validity. The first was designed to represent the turbulence-driven sublimation mentioned above. However, because it was based on an attempt to describe the moisture flux over the Earth's oceans (Jacobs 1942), its applicability to Mars is questionable. For example, the transfer coefficient in Jacobs' relationship is based on stability conditions appropriate for northern oceans. Sublimation rates are strongly affected by the atmospheric stability which can be quite different on Mars than over the Earth's oceans. Furthermore, the interaction between the atmosphere and ocean is much different than that for a solid surface. Over the oceans, the principal source of moisture is sea spray, not water diffusing into the atmosphere from a stationary surface. Finally, the character of the ocean surface depends on wind speed which introduces a nonlinearity not appropriate for solid surfaces.

The second parameterization was also designed to describe a turbulence-driven flux, but instead of thermal or mechanical origin the turbulence arises from vertical gradients in the molecular weight due to the buildup of a saturated layer of water vapor near the surface. Such a process would be more important for Mars than Earth because of the low densities of the background gas. Indeed, at 200 K the associated sublimation rates are comparable to those due to purely wind-driven turbulence (Ingersoll 1970). However, for the self-buoyancy mechanism to work, calm winds and stable stratification are required because, otherwise, turbulent eddies would be generated that would tend to mix out the near-surface saturated layer. Thus, it is not clear to what extent this process actually occurs on Mars. At present, therefore, our knowledge of sublimation on Mars is fairly limited.

C. The Vertical Distribution of Atmospheric Water and the Role of Clouds

Water-ice clouds and hazes are frequently present in the Martian atmosphere; their occurrence and variability (both spatial and temporal) depend on the abundance and vertical distribution of water within the atmosphere. The Mariner 9 and Viking Orbiter imaging observations provide the most detailed information; these have been reviewed by Leovy et al. (1972,1973*a*), Briggs and Leovy (1974), Briggs et al. (1977,1979), French et al. (1981) and Kahn (1984). Telescopic observations also provide important information (Pettit and Richardson 1955; Slipher 1962; Smith and Smith 1972; chapter 2). In addition, clouds of CO_2 ice are also believed to exist. In this section we review the observational and theoretical understanding of water-ice clouds and how they might influence the seasonal water cycle. Although it is clearly relevant to the distribution of clouds and hazes, we will not specifically discuss the relative humidity of the atmosphere; see Jakosky (1985) and Davies (1979*b*) for details.

1. Vertical Distribution of Water Vapor. As discussed earlier, the vertical distribution of water vapor plays an important role in determining the transport through the atmosphere and the exchange of water between the atmosphere and regolith. The key processes that control the vertical distribution of water vapor on Mars are likely to be large-scale transport, turbulent mixing, and the microphysics of nucleation, growth and sedimentation of ice crystals. Photochemical reactions can also play a role, but are not significant in the lower atmosphere (Kong and McElroy 1977).

There are several lines of evidence that suggest that, away from the vicinity of the polar caps, water vapor is generally uniformly mixed in the lower atmosphere (to within a factor of 2, which still allows for significant transport via the Hadley cell): (1) Davies (1979*c*) compared the angular variation of the observed column water-vapor abundance from the MAWD experiment with calculations from a model of light scattering within a dusty atmosphere, and concluded that the water vapor was distributed with approximately the same scale height as the dust, which itself is nearly uniformly mixed (Ajello et al. 1973; Pollack et al. 1977; Kahn et al. 1981); (2) Farmer and Doms (1979) found that the effective temperature of line formation in the MAWD reflectance spectrum is similar to the mean value expected in the lowest several scale heights of the atmosphere rather than, say, to the temperature near the surface; (3) estimates of the altitude to saturation based on Mariner 9 and Viking IRTM temperature profiles and MAWD water abundances suggest that saturation would occur at relatively high levels (10–40 km) in a uniformly mixed atmosphere (Jakosky 1985), consistent with the altitude of hazes observed at the limb (Jaquin et al. 1986; Kahn 1990); this result is consistent with uniform mixing within the lowermost 1 to 3 atmospheric scale heights;

(4) there is a good correlation between surface topography and column vapor abundance that is consistent with a well-mixed distribution (Jakosky and Farmer 1982); (5) precipitation appears to be a weak mechanism of redistributing water vapor vertically, so that total water should be uniformly mixed (Rossow 1978).

Most recently, Hart (1989) and Hart and Jakosky (1989) have used the Viking MAWD data to determine the effective pressure of line formation for water vapor within the atmosphere at several latitudes and seasons. Essentially, the effective pressure is the average pressure in the atmosphere weighted by the number density of water-vapor molecules. As such, it provides information on where in the atmosphere the water is located; for instance, an effective pressure equal to the surface pressure implies water vapor concentrated near the surface, while an effective pressure of half the surface pressure would be consistent with water vapor uniformly mixed in the atmosphere. Because the effective pressure is only a single value, it does not allow one to determine the vertical profile of water vapor in detail or of small departures from a nominal vertical profile; it does allow one to distinguish, however, between water vapor that is concentrated in the atmosphere near to the surface, that is uniformly mixed, or that is concentrated high above the surface. Figure 15 shows the derived effective pressure for two latitude bands (0–40° N and 0–40° S) for the seasonal range L_s = 0–140°. In the northern hemisphere, the water begins the spring season nearly uniformly mixed (to within ~ 30%); as spring and summer progress, the water appears to become concentrated nearer to the surface. The water vapor in the southern hemisphere is more nearly uniform during this entire time period. The mechanism for the changing vertical distribution in the northern hemisphere is likely to be either supply of water from the regolith into the lower atmosphere, effectively concentrating the water nearer to the surface, or condensation at higher altitudes which essentially limits the water vapor to the lower atmosphere (Hart 1989; Hart and Jakosky 1989). Unfortunately, the data do not allow an adequate determination of the detailed spatial and seasonal variations in the vertical distribution of water vapor.

Clearly, the existence of sources and sinks for water at the surface, and the global redistribution of water by either the upper or lower branch of the Hadley cell, for example, may result in significant departures from uniform mixing.

There are several implications of a nearly uniform mixed vertical distribution of water vapor on Mars, for regions where such is the case. First, the time scale for local vertical mixing must be short compared to the time scale for exchange with sources and sinks. Second, a nearly uniform distribution within the tropics and subtropics implies that transport by zonal mean circulations, the Hadley circulation in particular, is much less efficient on Mars than it is on Earth where water is strongly concentrated near the surface.

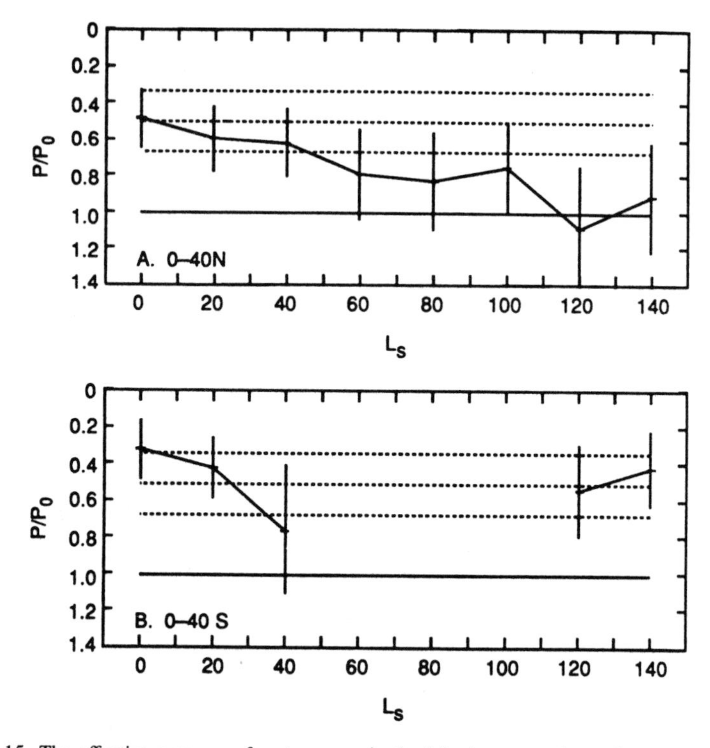

Fig. 15. The effective pressure of water vapor in the Martian atmosphere, from Hart (1989). Values were derived using data placed into 40° bins, with the results normalized to the average surface pressure within each bin and then averaged zonally. The error bars represent the uncertainties due to instrument noise as well as uncertainties in atmospheric temperature (which are the larger effect). Also shown for comparison are dotted lines at pressures of ⅓, ½ and ⅔ of the surface pressure. The inversions ignored scattering by atmospheric aerosols, which will have a small effect here (see Hart [1989] for details).

However, this does not mean that the Hadley circulation does not transport climatically significant amounts of water. To account for the implied cross-equatorial transport discussed in Secs. II and IV.B, for example, only a modest 30% departure from uniform mixing is required. Such a distribution is still consistent with the above observations, which can accommodate up to a factor of 2 departure from uniform mixing.

2. Cloud and Haze Distribution. Cloud opacities and the vertical distribution of both condensate and vapor will depend on the relative time scales for vertical mixing of water and growth and sedimentation of ice particles. Models which combine some elements of these processes have been constructed (Hess 1976; Krasnopolsky 1979; Kulikov and Rykhletskii 1984; Jaquin et al. 1986; Kahn 1990). They are steady-state models that assume that water vapor is uniformly mixed between the surface and the altitude z_s where water first saturates, and that above z_s a balance exists between the upward flux of total water (vapor plus condensate) by diffusion and the downward flux of condensed water by gravitational settling. This approach is appropriate because the atmospheric temperature generally drops more rapidly with altitude than does the water-vapor saturation temperature (Jakosky 1985; Davies 1979*b*). Following Hess (1976), this balance can be expressed as

$$K\frac{\partial^2 q_s}{\partial z^2} + K\frac{\partial^2 w}{\partial z^2} - \mathrm{v}\frac{\partial w}{\partial z} = 0 \tag{1}$$

where K is the eddy diffusion coefficient, q_s is the saturation mixing ratio of water vapor, w is the condensate mixing ratio, and v is the fall speed of the condensed particles. The distribution of condensate given by Eq. (1) depends on the value of the eddy diffusion coefficient, the total water in the column, the temperature structure of the atmosphere, and the fall speed of the particles, which depends on height (gas density) and particle radius r (which can also depend on height).

For a given amount of water and assumed temperature profile, solutions to Eq. (1) depend on the ratio K/v, which is the haze thickness Δh that results when the diffusion time ($t_d = (\Delta h)^2/K$) is comparable to the fall time ($t_f = \Delta h/\mathrm{v}$). When these time scales are not comparable, Δh is determined by turbulence if $t_d << t_f$, or by ability of the particles to nucleate, grow and settle if $t_d \lesssim t_f$. The ultimate size to which the particles grow is critical in the latter regime because of the dependence of fall speed on particle size. To date, only one observation has been used to infer particle size (Curran et al. 1973). In the $t_d << t_f$ regime, the particle profiles depend on the value of the diffusion coefficient, which is also poorly constrained at typical haze levels. Thus, solutions to Eq. (1) depend on both K and r, neither of which is adequately known.

Some loose constraints can be placed on these parameters by comparing observed haze opacities with those predicted from solutions to Eq. (1). The latter can be approximated by assuming that the calculated column cloud mass Δm consists of a population of particles of radius r, density ρ_p, and a scattering efficiency of 2, which is appropriate in the geometric limit. As shown by Hess (1976), the result

$$\Delta\tau = \frac{3}{2r\rho_p}\Delta m \tag{2}$$

demonstrates that the haze opacity is directly proportional to cloud mass and inversely proportional to particle size. As cloud mass, hence $\Delta\tau$, increases with mixing in this simple model, an upper limit for K can be obtained. To match Smith et al.'s (1970) observed haze opacity of 0.05, Hess (1976) concluded that the particles were < 10 μm, and that the corresponding eddy diffusion coefficient was $< 10^5$ cm^{2}s^{-1}. This conclusion, however, is partly the result of the adiabatic temperature profile he assumed; as a consequence, most of the water in his model was concentrated in the lowest scale height. Kulikov and Rykhletskii (1984), on the other hand, used a more realistic temperature profile based on Mariner 9 and Viking observations. Because the temperature decrease with altitude in their profile was less than what Hess (1976) assumed, they found that saturation occurred at a much higher level (25 km). However, they reached the same basic conclusion regarding the diffusion coefficient—that it does not exceed 3×10^5 cm^{2}s^{-1} at haze altitudes. This results because they assumed a particle radius of 1.5 μm, which is more in line with that inferred by Curran et al. (1973). This illustrates the difficulty of estimating K and r based on opacity measurements alone. The issue is further complicated if the water is nonuniformly mixed in the lower atmosphere (Krasnopolsky 1979).

Recently, Kahn (1990) has used both the opacity and vertical thickness of the haze to better constrain both K and r. In principle, knowledge of the haze thickness as well as opacity would allow complete determination of these parameters as the particle concentration ambiguity would be removed. This assumes, of course, that the temperature and moisture content are known. Kahn (1990) used Viking IRTM and MAWD data to constrain these fields, and Jaquin et al.'s (1986) Viking Orbiter limb haze observations to estimate haze thickness and opacity. Three different hazes were examined using the same basic approach as in Hess (1976) and Kulikov and Rykhletskii (1984). Kahn (1990) found that the eddy diffusion coefficient was similar to the earlier studies (10^5 cm^{2}s^{-1}) for each of the three cases he examined, but the particle size varied from several 0.1 μm for the high-altitude haze (at ~50 km), to 6 μm for the low-altitude haze (~20 km). The particles are larger in the low-altitude haze because the sedimentation velocity is smaller due to the higher atmospheric density and larger particles are required in order to balance the upward vapor flux.

The principle conclusion from these simple models is that, in the absence of sources or sinks for water, the notion of a well-mixed lower atmosphere, capped by an optically thin ice haze whose base varies between 20 and 50 km and whose particle sizes are on the order of μm, is consistent with available observations. The implied low values for the eddy-mixing coefficient may be inconsistent, however, with the much higher values (10^8 cm^{2}s^{-1}) suggested by Zurek (1976) for breaking thermal tides and by McElroy and Donahue (1972) for maintaining the stability of a CO_2 atmosphere against photodissociation. In the case of tides, the large values occur at very high levels (>40 km), and are most prevalent in the equatorial regions. The photochemical models require large mixing to lower altitudes as well (see chapter 30); indeed, an effective mixing agent between the boundary layer and the levels where tides and/or gravity waves can break has not been identified. However, the long mixing times implied by the low derived diffusion coefficients call into question the steady-state assumption upon which these models are based. Indeed, recent calculations by Michelangeli et al. (1989) have shown that the location and thickness of high-altitude hazes can be independent of the eddy diffusion coefficient when the steady-state assumption is relaxed and a more detailed treatment of cloud microphysical processes is considered. Thus, the physics of Martian hazes may not be adequately represented by the simpler models, and any conclusions about the inferred mixing times may be premature.

Colburn et al. (1989) used the difference between optical depth as measured at the Viking Lander sites in the morning and in the afternoon in combination with models of atmospheric structure to infer the daily formation and dissipation of a high-altitude haze component. The magnitude of the difference between the AM and PM optical depths was typically 0.1 to 0.2, which corresponds to < 1 pr μm of water vapor (Pollack et al. 1977; Colburn et al. 1989). At this season, the MAWD observations suggest water abundances at the Lander sites in the vicinity of 30 pr μm. The diurnally varying component of these hazes, therefore, do not constitute a significant reservoir for water.

In contrast to the high-altitude hazes discussed above, the presence of morning hazes suggests formation at lower levels where the diurnal temperature amplitude is more pronounced. From nighttime observations of Phobos, Pollack et al. (1977) inferred the nightly formation of a low-lying fog. Fog production is also implied by the Lander temperature measurements (Ryan and Sharman 1981*a*; Ryan et al. 1982), and morning fogs have been observed in some low-lying areas from orbit (see, e.g., Kahn 1984). Such a fog is consistent with theoretical predictions (Flasar and Goody 1976; Hess 1976).

Discrete clouds also form in the Martian atmosphere. They are generally wave-like or cumuliform, depending on the local meteorological environment (Briggs and Leovy 1974; French et al. 1981; Kahn 1984). Both cloud types can be optically thick. However, they tend to be short-lived, and are not as widespread and ubiquitous as the hazes. Consequently, discrete clouds are more likely to impact the local (rather than global) water budget.

It is appropriate at this point to discuss the possibility of precipitation. Precipitation from clouds, meaning the vertical sedimentation of water in the form of condensate to the ground, would occur on Mars if the time scale for evaporation greatly exceeded the time scale for sedimentation. Rossow (1978) has estimated these time scales for a low-lying ice cloud with a water content of ~ 3 pr μm, typical of the hazes discussed above. As indicated in Fig. 16, the time scale for evaporation depends on the saturation state of the underlying atmosphere (note that the figure uses τ for our t). For a dry atmosphere, the particles must grow to several 100 μm in size before they can survive the fall to the surface. However, there is not enough time to build

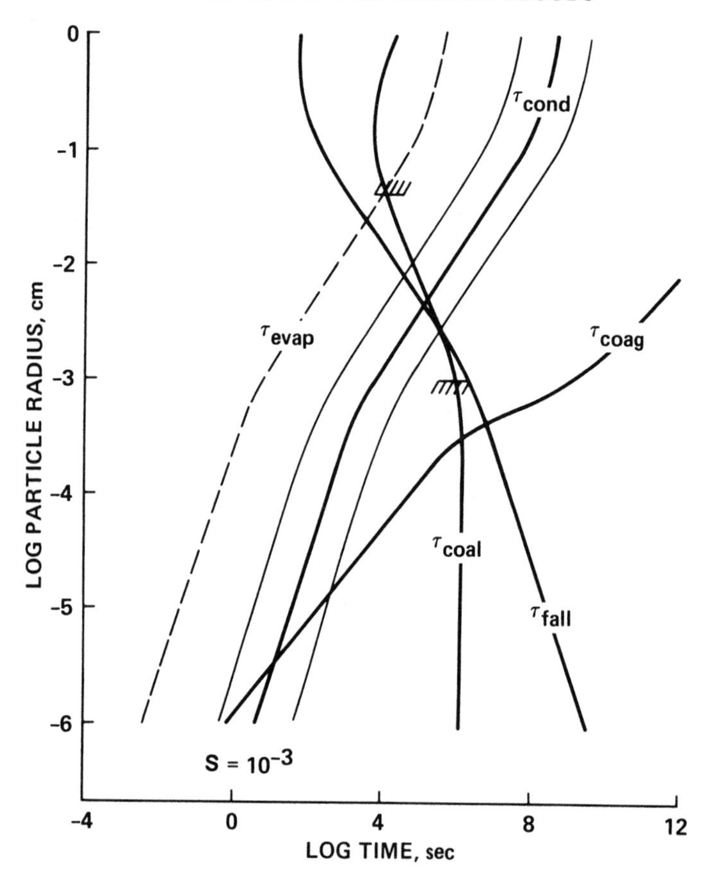

Fig. 16. Microphysical time scales for a near-surface water-ice cloud on Mars as a function of particle size. Time scales associated with growing particles are condensation (τ_{cond}), coagulation (τ_{coag}) and coalescence (τ_{coal}); evaporation (τ_{evap}) and gravitational settling (τ_{fall}) represent loss processes. The three condensation curves represent growth times associated with supersaturations (S) one order of magnitude larger and smaller than the value indicated. The evaporation curve is a lower limit as it assumes a dry atmosphere (figure from Rossow 1978). In the text t is used in place of τ.

particles this big because the fastest growth process (coalescence) is still slower than sedimentation. In this particular example, precipitation is possible only for relatively large supersaturations and even then it is marginal. Of course, conditions vary widely on Mars such that precipitation at some locations and seasons is possible (Kahn 1990). Observations of surface frosts (Jones et al. 1979) and recent microphysical studies (Michelangeli et al. 1989) support this possibility. However, for the planet as a whole, precipitation reaching the surface is much less common on Mars than it is on Earth.

Because precipitation to the surface appears to be unlikely on Mars, the transport of water in the condensed phase may be important. In Leovy's (1973) model of moisture transport by baroclinic eddies (see Sec. IV.B.2), the total amount of water that condensed in the inner regions of the polar cap (poleward of 70°) during winter increased by a factor of 10 when condensed-phase transport was included. Even though the bulk of the moisture remained concentrated in the cap periphery for this case, which Leovy was trying to avoid, it still demonstrates the potential impact of condensed-phase transport on the global water cycle. This potential has been further demonstrated by James (1990), who used a one-dimensional (latitude) advective-diffusive model to simulate the Viking MAWD data. James (1990) found better agreement with the data in simulations that included cloud transport; without it, unrealistically large amounts of surface-ice deposits accumulated at northern mid-latitudes. Furthermore, with cloud transport, the seasonal and spatial distribution of the simulated clouds were qualitatively similar to the observed polar hoods, which suggests that these clouds, which are haze-like in character, may be important to the water cycle.

D. Net Annual Transport of Water

When the MAWD data are averaged over a year and around latitude circles, there is a clear monotonic decline in water vapor from north to south (Fig. 17). This trend remains even when topographical effects are removed (Jakosky and Farmer 1982). Assuming that this distribution is typical of the current epoch, a question which naturally arises is, what are the processes responsible for maintaining this gradient? The answer has implications for the net annual transport of water. In this section, we review this aspect of the current Martian water cycle.

It is possible that such a gradient can exist without net transport. Davies (1981), for example, has constructed a model which maintains the hemispheric asymmetry shown in Fig. 17 and which also reproduces some of the observed seasonal and latitudinal variations of column water vapor. The model is one-dimensional (latitude) and atmospheric transport is parameterized in terms of diffusive processes with a seasonally dependent exchange coefficient based on atmospheric dust behavior. It is primarily this time dependence of the exchange coefficient which yields the hemispheric asymmetry. The justification for this dependence is the occurrence of global dust storms during northern fall and winter (see, e.g., Briggs et al. 1979) which have been shown to increase dramatically the intensity of the general circulation (Haberle et al. 1982; Pollack et al. 1990*a*). Thus, in Davies' (1981) model, a permanent water ice cap forms at the north pole and not at the south pole because the circulation is strongest during northern winter. Furthermore, there is no net annual transport; water lost by the cap during summer is returned in winter.

The conclusions reached by Davies (1981) are clearly dependent on

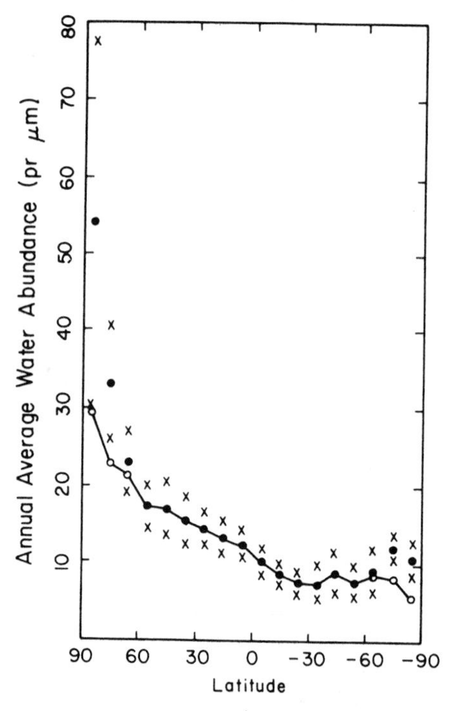

Fig. 17. Zonally and seasonally averaged atmospheric column water-vapor abundance as a function of latitude. The open circles represent the best estimates in the polar regions, where year-round observations are not available and the "*x*" values represent the standard deviation of the zonal values (figure from Jakosky and Farmer 1982).

transport. Because this is treated as diffusive in his model, it is difficult to judge the reality of his simulations. To reproduce the observed asymmetry, however, Davies requires a fifty-fold increase in atmospheric mixing rate during global dust storms. This corresponds to meridional wind speeds in excess of 100 m s^{-1} over the entire planet. An increase in the intensity of the circulation during global dust storms is expected, but not to the level required. In addition, global dust storms do not necessarily occur every Mars year as was originally thought (Martin 1984), and should not be relied on to close the seasonal water cycle (chapter 29).

James (1985) has proposed an alternate mechanism for maintaining the observed asymmetry. Because of Mars' large orbital eccentricity, the CO_2 cycle is not symmetric with respect to season. At the current epoch, southern winters are longer than northern winters, and the peak insolation is less. Consequently, the condensation winds accompanying the growth and retreat of the polar caps differ between the hemispheres. The difference is most pronounced during spring, when the outflow winds in the south can exceed those in the north by a factor of 2. James (1985) suggested that asymmetric advection of water vapor by the condensation winds acting on the 10^5- to 10^6-yr time scales associated with Mars' orbital variations could effectively act as a pump that would maintain the observed north-south asymmetry against the normal tendency of atmospheric motions to diffuse away gradients. He constructed a simple one-dimensional model in which column water vapor was transported both by diffusion and by advection via the condensation wind, with the latter being calculated from the model of James and North (1982). The transport model was run to steady state starting from symmetric distributions of vapor and polar surface ice. In all experiments, the southern hemisphere ended up drier than the northern hemisphere, confirming James' suggestion that the meridional condensation winds are potentially important.

A strong argument in favor of some net annual transport can be made, however, because of the composition of the residual south polar cap. Unlike its northern counterpart, the residual south polar cap retains a covering of CO_2 frost year round (Kieffer 1979). The year-round presence of CO_2 frost at the south pole would act as a cold trap for water vapor such that any water vapor brought into contact with the residual cap would then be permanently removed from the atmosphere, as described in Sec. III. Thus, if CO_2 persists at the south pole year after year, then a net annual transport must result. Jakosky (1983*a,b*) has suggested that the observed hemispheric asymmetry in atmospheric water vapor results from the asymmetry in polar-cap behavior and implies a net annual transport of water vapor from north to south. Water could be lost to a nonpolar subsurface reservoir, although there is no evidence for the existence of such a reservoir (see, e.g., Zent et al. 1986); otherwise water vapor lost by the north cap during summer would be ultimately gained by the south cap during the course of a Martian year.

Recently, Haberle and Jakosky (1990) have examined the role of the residual north polar cap in the current water cycle (see Sec. IV.B). In contrast to previous models, their model treats transport as an advective process in the meridional plane using the residual circulation approach. This approach is superior to the one-dimensional diffusion used in previous models, but is less accurate than a full-scale general circulation calculation. They found that the high-latitude summertime circulation was too sluggish to move water out of the polar environment during the summer season. As a consequence, water subliming from the cap remained at high latitudes. Eventually the polar atmosphere became saturated, which not only limited further increases because of removal by precipitation, but also reduced the sublimation flux because of its inverse dependence on relative humidity. An example of their model's predicted increase in northern hemisphere water compared with observations is given in Fig. 18. Although most of the water was found to remain in the polar regions, not all of it was recirculated back to the cap. At the end of summer, Haberle and Jakosky (1990) found that between 0.2 and 0.8 mm of water was removed from the polar cap and deposited in the adjacent terrain (65 to 75° latitude band). Thus, the net loss can still be significant.

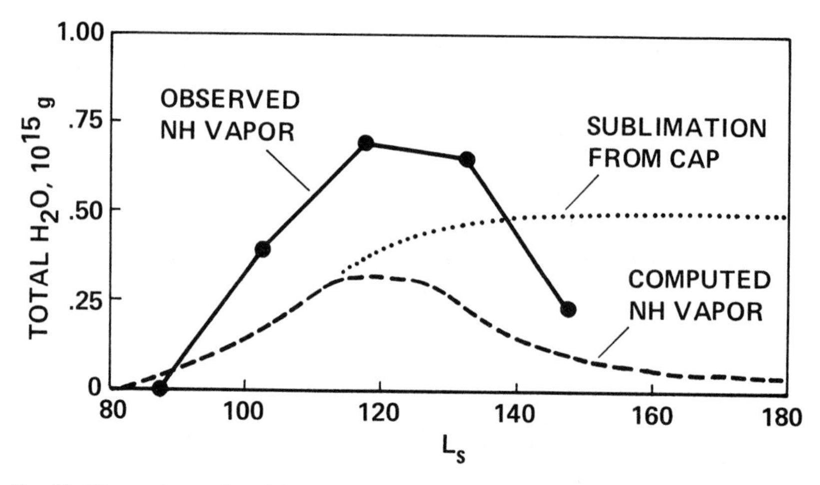

Fig. 18. Observations and model predictions of the integrated northern hemisphere water-vapor abundance during the spring and summer seasons. Dots and solid line are the observations. Dotted line is the total amount of water sublimated from the north polar cap, and the dashed line is that water which remains in the atmosphere as vapor (figure from Haberle and Jakosky 1990).

Barnes (1990*b*) and Houben et al. (1988) used models of the eddy component of the wintertime circulation to calculate the amount of water which could be returned to the north polar cap during the winter season (see Sec. IV.B). Again, this approach is based on the advection which results from the actual circulation, rather than treating transport as a diffusive process. They conclude that as much as 10^{14} g H_2O could be transported to the region poleward of 60° N latitude.

The net transport of water integrated over a Martian year represents the sum of transport due to the various components of the circulation as convolved with the supply and removal of water from seasonal nonatmospheric sources and sinks of water. The diffusive models of Davies (1981) and James (1985) both yield no net transport for the same reason. Each model has a time-dependent transport term, with a more rapid transport occurring during southern hemisphere summer. Thus, water is transported to the north more effectively, and requires a smaller gradient in atmospheric water abundance, compared to the southward transport occurring at other seasons. Both models iterate until a steady state is reached, and thereby result in no net transport despite the net gradient in atmospheric water vapor. The diffusive model of Jakosky (1983*a,b*), on the other hand, predicts a net transport of water between the poles. This conclusion is a direct result of: (1) the assumption that the eddy mixing coefficient is constant with both time and location; and (2) the presence of a gradient in atmospheric water-vapor column abundance which results from a source of water at the north pole and a sink at the south pole.

The models of Haberle and Jakosky (1990), Barnes (1990*b*) and Houben et al. (1988), in contrast, all calculate the transport resulting from the circulation patterns at a given season; the first calculates the loss of water from the cap during northern summer, and the second two calculate the return to the cap resulting from wave activity during the winter season. In some sense, these models must be considered preliminary, as not all components of the circulation are understood and not all dynamical phenomena are readily explained (see chapter 26). The results of these models suggest, however, that only a small fraction of the water lost from the cap during summer will be returned during winter. On this basis, the north polar cap is probably undergoing a net annual loss of water from the surface. The ability of the springtime circulation, near the edge of the retreating seasonal polar cap, to return water to the poles is uncertain; observations of water-ice hazes at the edge of the cap suggest that some water is recycled this way (Christensen and Zurek 1984). While the time scale for global transport is longer than a Martian year, it is likely that any water lost from the north cap in the current climate will ultimately make its way to the south polar cap, where it condenses despite the small size of the cap.

Two additional caveats must be applied to this conclusion. First, if the south polar cap loses its CO_2 covering in some years (Sec. III.A), then net transport in those years might be toward the north. The average behavior at the current epoch is therefore uncertain, in sign as well as magnitude. Second, Kieffer (1990) points out that the north residual water-ice cap has a relatively high albedo, indicating that only a small amount of dust is intermixed. If the cap is losing water each year, then there must be a mechanism of also removing dust in order to keep the albedo this high; Kieffer (1990) suggests that the water-ice portion of the residual cap may actually be gaining water each year, although no source for this additional water has been observed.

If the behavior seen by Viking is typical, however, and if most of the water sublimed from the north cap during summer is ultimately condensed onto the south cap, then we can estimate the long-term effect. Haberle and Jakosky (1990) calculate that between $\sim$ 0.2 and 0.8 mm H_2O sublimes from the north cap during summer, with the best estimate being $\sim$ 0.6 mm; the uncertainty reflects the uncertainties in both surface sublimation and atmospheric transport. Over 5×10^4 yr, corresponding to half of an obliquity oscillation cycle, as much as $\sim$ 15 m of ice may be lost from the north cap. This thickness is consistent with those of individual layers in the polar layered terrain, suggesting that sublimation and transport of water may be important in the evolution of the polar deposits. This transport rate is much larger than that obtained for the polar caps over the last 100 Myr using cratering rates and crater abundances (Plaut et al. 1988). The results can be made consistent, however, if only a small fraction of the polar ice is continually recycled back and forth between the caps.

V. THE SEASONAL WATER CYCLE AT OTHER EPOCHS

The Martian axial obliquity oscillates with periods of ~ 10^5 and 10^6 yr (see, e.g., Ward 1979; chapter 9). At low obliquities, the Sun never rises more than ~ 15° above the horizon at the poles, while at high obliquities it gets to as high as ~ 35° (as compared to the current value of ~ 25°). Prior to the formation of Tharsis, the obliquity may have been as low as 9° or as high as 45° (Ward et al. 1979). As a result, the summertime sublimation of water from the polar caps varies dramatically, as will the global atmospheric water content. These seasonal processes at other obliquity values are described in this section. Long-term exchange with the regolith as well as the implications for formation and evolution of the polar layered terrain are discussed in chapters 23 and 33.

At an obliquity higher than the current value, both summer polar caps are not likely to retain any CO_2 frost on the surface (see, e.g., Toon et al. 1980). No longer being buffered to the CO_2 condensation point, temperatures will rise significantly. In the following discussion, we assume that both caps have a significant reservoir of water ice, and that their thermophysical properties are not unlike those of the water-ice deposit currently seen at the north pole (Kieffer et al. 1976*b;* Paige and Ingersoll 1985). Mid-summer temperatures can rise to values as high as 230 to 250 K at 35° obliquity (Toon et al. 1980; Jakosky and Carr 1985); the range depends on the specific value of the orbital eccentricity and the argument of perihelion, with the pole having summer near perihelion having the higher temperatures. At these temperatures, as much as 10 to 20 g cm^{-2} of water can sublime into the atmosphere over the course of the summer season, as compared to the current summer value in the north of 0.02 to 0.08 g cm^{-2}. At 45° obliquity, the largest value likely to have obtained at any time in Martian history, temperatures as high as 260 K may have resulted in the sublimation of up to 50 g cm^{-2} of water ice in a summer season, equivalent to a layer 0.5 m thick (Toon et al. 1980; Jakosky and Carr 1985). In the approximately 10^4 yr spent at the highest obliquities in the cycle, then, as much as 5 km of ice may have sublimed into the atmosphere; this probably exceeds or is equivalent to the entire thickness of the polar deposits (see chapter 23).

There are several effects which may limit the sublimation, however. If sublimation removes water while leaving behind a residuum of dust, the dust will both decrease the albedo of the surface and act as a diffusive barrier to inhibit further sublimation. A 1-m-thick layer of dust will be an efficient barrier, and will, in fact, have a larger impact on the sublimation rate than would the resulting albedo decrease (Toon et al. 1980). If global dust storms occur more often, due to the larger atmospheric pressure at higher obliquities (Fanale et al. 1982*a*), there may be additional dust effects in the polar regions. Clearly, these processes may be important in the origin and evolution of the polar layered terrain (Toon et al. 1980; Pollack and Toon 1982; chapter 33).

Alternatively, if there is not a significant amount of dust in the polar deposits, or if the dust is efficiently removed along with the water ice, tremendous quantities of water may sublime into the atmosphere. As the polar caps ultimately govern the water content of the entire global atmosphere, there may be a significant transport of water away from the polar regions through the atmosphere (Jakosky and Carr 1985). The water content of the nonpolar atmosphere may exceed the saturation holding capacity, and water-ice deposits may become more stable in the equatorial and mid-latitudes than at the pole. Nonpolar ice deposits as thick as tens of meters may result, with this water possibly populating the near-surface regolith and contributing to the formation of water- and ice-related geomorphological features (Jakosky and Carr 1985).

These calculations of sublimation at high obliquity are extremely uncertain, however. As described in Sec. IV.D, the sublimation at the current epoch is uncertain by a factor of several due to uncertainties in the surface properties and in the ability of the atmosphere to transport the water away from the cap. Even the magnitude of any net loss at the current epoch is uncertain. At high obliquity, these uncertainties must be even larger.

At low obliquities, the poles do not warm up as much as they do currently and will retain a year-round covering of CO_2 frost. The caps then act as an efficient cold trap for water, removing most of it from the atmosphere (Ingersoll 1974; Toon et al. 1980; Fanale and Jakosky 1982). An atmosphere in equilibrium with a polar cap at a temperature of 150 K will hold 4 orders of magnitude less water than will the current atmosphere (Toon et al. 1980; Lindner and Jakosky 1985). It is unlikely, however, that both polar caps would retain a year-round covering of CO_2 frost (Henderson and Jakosky 1990). As the permanent presence of CO_2 frost depends on a balance between winter condensation and summer sublimation, as described by Leighton and Murray (1966), it is extremely unlikely that both caps would have just the right combinations of albedo, emissivity and radiation temperature to perfectly balance each other; one cap would be favored at the expense of the other, and one cap or the other would expose a summertime water-ice deposit. Depending on which cap was exposed in the summer (the one with summer at perihelion or the one at aphelion), temperatures would rise to between $\sim$ 185 and 205 K for an obliquity of 15°; the corresponding equilibrium atmospheric water content would be between $\sim$ 3 and 50 pr μm, no more than about an order of magnitude less than the current value (Henderson and Jakosky 1990). At the extreme low obliquity of about 10°, temperatures of 170 to 185 K give rise to a water content of $>$ 0.15 pr μm. Thus, although transport between the polar caps and exchange with the regolith will be lower than current values at low obliquities, it will not be negligible.

VI. SUMMARY OF THE SEASONAL CYCLE

The seasonal water cycle on Mars at the current epoch results from seasonally changing surface and subsurface temperatures, including the condensation and sublimation of the seasonal polar caps. The visible manifestation of the cycle is primarily the atmospheric water vapor abundance. This shows dramatic variations as a function of both location on the planet and season; the factor of 2 variation in the total atmospheric abundance indicates exchange of water between the atmosphere and non-atmospheric sources and sinks of water. Further, transport of water within the atmosphere appears to play a significant role in the seasonal water cycle.

The dominant processes responsible for the varying atmospheric water abundance are probably: sublimation of ice from the retreating seasonal polar caps and from the residual polar cap(s) during the spring and summer season; condensation onto the polar cap during the fall and winter season (and throughout the summer on the south residual cap, which retains its CO_2 frost covering); desorption of water from the regolith grains into the pore spaces in response to seasonally varying subsurface temperatures, followed by diffusion of water vapor into the atmosphere; adsorption of water molecules from the pore spaces onto the regolith grains, followed by diffusion of water from the atmosphere into the pore spaces. These processes are illustrated schematically in Fig. 4. Additionally, there may be a freeze-thaw cycle within the regolith, either of pure liquid water or of a brine which has a lowered freezing point, and there may be one-way diffusion of water into the atmosphere from subsurface ice or brine which had been deposited at an earlier epoch.

Although each of these processes appears to play a role in the seasonal water cycle, the residual polar caps ultimately control the behavior of the atmospheric water. This results from their ability to pump water into the atmosphere during the summer and to remove water during the winter. This seasonal supply of water sets up the gradient of water between the poles, and the regolith and seasonal polar caps respond to this gradient by, in turn, taking up or releasing water. The regolith is capable of reaching an equilibrium with the atmospheric water abundance in only a few years, with a seasonal forcing superimposed on the general equilibrium. Similarly, the transport of water within the atmosphere occurs on a time scale of a few years. The polar caps, however, do not generally respond in a significant way to either transport or the regolith; their behavior is essentially independent of the water cycle, being governed primarily by the energy balance within the CO_2 cycle.

In this sense, the water cycle at the current epoch or at other epochs is determined by the behavior of the polar caps and, therefore, by the seasonal CO_2 cycle. Subtle differences in the energy balance at the poles can dramatically change the seasonal behavior of atmospheric water (Jakosky and Haberle 1990*b*).

There may be direct interactions between the two cycles as well. Condensation of water and CO_2 together may be responsible for depositing the former in the polar regions (Pollack et al. 1979*b*), although the importance of such a process is uncertain. Also, the condensation wind associated with the CO_2 cycle may be important in the net transport of water vapor (James 1985).

Similarly, there are clear interactions between the atmospheric water and airborne dust, and between the water and dust cycles. Dust affects the atmospheric temperatures, dramatically influencing the atmospheric holding capacity and the degree of saturation of the water within the atmosphere. Transport of dust and water together may be important, with condensation for water and sedimentation for dust being the only ways to separate the two species; neither process appears to act with great efficiency.

Finally, in summarizing the seasonal water cycle, we have to turn to what is not known. Based on the discussion in this chapter, and on the analysis done since the Viking observations were obtained, it is tempting to conclude that, although we do not understand their relative importances, we do know what processes contribute to the seasonal water cycle. Thus, sublimation of water from the north polar residual cap certainly occurs, but the best model of the process still leaves a factor of several uncertainty in the quantity of water sublimed. Similarly, the regolith is certainly capable of adsorbing and desorbing water, thereby modulating to some degree the amount of water within the atmosphere; the efficacy of this process depends, however, on the composition of the regolith and on the ability of water molecules to diffuse through the uppermost centimeters of regolith, and both of these are very uncertain. The next generation of spacecraft will certainly address these issues. Below, we summarize the measurements and analyses which will play an important role in understanding the seasonal water cycle:

1. Monitoring of the seasonal cycle of atmospheric water over a period of many Martian years will allow us to get a more representative idea of the nature of the cycle at the current epoch, and will allow us to determine the possible year-to-year variations of the seasonal cycle. This monitoring can be done both from spacecraft and from Earth-based observations.
2. Global synoptic measurements of the atmospheric water abundance will provide much more information than is currently available, and will feed into our understanding of transport of water within the atmosphere and exchange with non-atmospheric reservoirs.
3. Global measurements of the vertical distribution of atmospheric water vapor will provide constraints on seasonal exchange with the regolith and polar caps, and on boundary-layer processes relevant to exchange, as well as on the amount of transport that results from the various components of the atmospheric circulation.
4. Determination of the near-surface abundance of water within the regolith,

and of its spatial variation, will immediately determine whether and where exchange with adsorbed water occurs (Drake et al. 1988).

5. Measurement of the amount of water contained within the seasonal polar cap, and of variations with latitude and season, will constrain the efficacy of exchange with the seasonal cap; by inference, then, this will constrain the role of exchange with the regolith (Jakosky 1985).
6. The presence or absence of near-surface liquid water can be inferred via microwave techniques; although such deposits are unlikely, their existence would have important implications for the long-term evolution of near-surface and atmospheric water.

Additional observations of other species will be important in understanding the behavior of water, due primarily to interactions between the seasonal cycles. We wish to specifically mention the importance of observations of atmospheric condensates (both water ice and CO_2 ice), the seasonal CO_2 cycle and possible year-to-year variations, the atmospheric dust cycle and possible year-to-year variations, and atmospheric winds and their importance to transport.

Acknowledgments. Discussions with our colleagues have been especially enjoyable and productive, and we wish to specifically thank R. Zurek, H. Kieffer, D. Paige, F. Fanale, A. Zent, R. Kahn, B. Toon, D. Michelangeli and J. Pollack. We also thank R. Zurek, A. Zent and R. Kahn for careful reviews of our manuscript. This research and analysis has been supported by NASA at the University of Colorado and at Ames Research Center through the Planetary Atmospheres Program. Administrative and manuscript support at CU from A. Alfaro, B. Hotard and L. Laubisch is very much appreciated.

29. THE MARTIAN DUST CYCLE

R. A. KAHN, T. Z. MARTIN, R. W. ZUREK
Jet Propulsion Laboratory

and

S. W. LEE
University of Colorado

Martian dust plays a key role in determining the current climate of Mars, and is suspected of having had a major influence on the evolution of the surface and the history of climatic conditions on the planet. From spacecraft and ground-based observations, some understanding has been acquired of the styles and frequency of dust motion in the atmosphere and at the surface, of atmospheric dust loading, dust particle properties, and their variability. Limited constraints have also been derived for the volumes and ages of some potential dust reservoirs on the surface. These data are adequate to frame questions about the seasonal and long-term cycles of dust on Mars. Additional data are needed to place quantitative constraints on components of these cycles, on their possible linkage to one another, and to draw conclusions from the dust budget about the climate history of Mars.

I. INTRODUCTION

It is twenty years since Mariner 9 first orbited a dust-shrouded Mars. Observations from Earth, from orbit about Mars and from the surface of the planet itself have shown that airborne dust is a highly variable and thermodynamically significant component of the Martian atmosphere (see chapter 26). A major finding of the Viking mission was the discovery of a persistent background dust opacity ranging from a few tenths to more than 1.0 at visible wavelengths (Pollack et al. 1979; Colburn et al. 1989), at least over the low-lying plains and great basins of Mars sampled by the Viking Landers. Periods of significant clearing of this background dust load are deduced from hemispheric-scale atmospheric temperature profiles derived from Earth-based microwave observations, found to be significantly colder than those typical of Mariner 9 or Viking observations (Clancy et al. 1989). The polar atmosphere, too, often appears to be relatively clear of airborne dust; this may be due in part to scavenging of the airborne dust particles by the condensation of CO_2 and perhaps water in the atmosphere during the polar night (Pollack et al. 1979,1990*a*).

Local dust clouds, the classical "yellow" clouds observed on Mars by Earth-based astronomers, are opaque clouds extending over areas as large as a few 10^6 km^2. They appear every Martian year and probably in every season. In some years, but not in every year, one or more local dust clouds expands to obscure much of a zonal corridor in the southern subtropics; this dusty corridor may then expand meridionally to cover most of one or both hemispheres. These are the planet-encircling dust storms of Mars. Dust raised during these storms is spread over much of the planet and eventually falls to the surface, causing changes in the local surface albedo that last for periods ranging from weeks to many years. The amount of dust injected into the polar regions, where it could become incorporated into the polar ice or layered terrains, is uncertain. Over millions of years, dust raised and redistributed by planet-wide storms may accumulate in long-term deposits at both high and low latitudes.

In this chapter, we summarize the observations of the various components of the dust cycle, focus on the microphysical properties of the dust, and discuss our current understanding of the dust cycle as it affects the surface of Mars on seasonal and geologic time scales.

II. EARTH-BASED OBSERVATIONS

Mars has been observed for hundreds of years from the Earth, and for portions of the last three decades by a series of spacecraft. Even so, temporal coverage of the Martian atmosphere and surface has been sporadic at best, due in large part to the fact that Mars has the largest synodic period of any planet. Moreover, because of the significant eccentricity of its orbit, the apparent size of Mars as seen from Earth even at opposition varies substantially in a 15 to 17 yr cycle, with the best viewing occurring when Mars is both at opposition and at perihelion. Since perihelion occurs just prior to southern summer solstice, at L_s 251° during the present epoch, there is an unavoidable bias to the Earth-based observations that must be kept in mind.

In recent years, more systematic coverage and documentation of changes in the atmosphere of Mars and on its surface was provided by the International Planetary Patrol (Baum 1973), a cooperative effort involving several observatories, motivated in part to support the ongoing series of spacecraft missions to Mars. Paradoxically, none of the spacecraft that have flown by or orbited Mars has systematically observed the development of a large-scale dust storm. Also, the longest-lived spacecraft observation of Mars, the meteorological measurements made by Viking Lander 1, lasted just over 3 Mars yr. Thus, we are still dependent on Earth-based observing to monitor Mars and to provide us with a synoptic view of changes there.

The long history of Earth-based visual, photographic and photometric observations of the Martian atmosphere and surface is discussed in detail in chapter 2. Briefly, yellowish dust clouds and veils, as opposed to "white" or "bluish" water-ice clouds, were inferred by H. Flaugergues as a result of his observations between 1796 and 1809 (cited by Capen and L. Martin 1971). An experimental program of photographing Mars began at Lowell Observatory as early as 1901. Yellow clouds were extensively photographed during the oppositions of 1920 and 1922. As noted by Slipher (1962), such yellow clouds were infrequent; typically they covered only a small fraction of the visible disk of Mars. A remarkable change in the Martian climate may have occurred in 1956, when a dust storm was observed, for perhaps the first time, to encircle the planet. Since then, a number of planet-encircling dust storms have occurred, with the 1971 storm being the most global in extent.

Quantitative dust opacity determinations have not been made from Earth-based measurements in a systematic way. It is difficult to establish opacity from photographic records, which have generally poor photometric accuracy and, for much of the time, poor spatial resolution. As noted in chapter 2, a commonly used criterion for the value of Mars photographic observations is that the planet exceed 13 arcsec in diameter, a condition met only $\sim$ 10% of the time. (1 arcsec seeing corresponds to an upper bound of about 500 km resolution on the planet.) Both well-calibrated data and modeling of the scattering by dust are required to extract opacity values from the somewhat greater brightness of dust palls relative to underlying bright surface regions. In practice this has not yet been done, although the large number of quality CCD images obtained in the last two oppositions (1986 and 1988) makes this approach promising (Hartmann 1989).

A new infrared approach to detecting atmospheric dust hazes from Earth (T. Martin 1989) utilizes the same 9-μm silicate band that was observed by Mariners 7, 9 and Viking (see Sec. III.A and D). The band falls in an atmospheric window for the Earth that is wide enough to measure both the band center and continuum regions on either side, near 8 and 12 μm. The depth of the band can be measured by making calibrated observations of Mars in conjunction with late-type stars that lack silicate shells. The atmospheric temperature cannot be measured in this way, as the Martian 15-μm band is masked behind its telluric counterpart. If the magnitude of the thermal contrast between dust haze and the surface is assumed based on past spacecraft observations or is taken from models, the optical depths of the detected hazes can be estimated from the depth of the observed 9-μm absorption. One advantage of a thermal infrared approach to monitoring opacity is that meaningful data can be obtained even when Mars is available only in the daytime sky.

Observations of the Martian atmosphere using groundbased microwave techniques provide estimates of the globally averaged atmospheric temperature. These produce an indirect constraint on the amount of atmospheric dust, because a dustier atmosphere is generally warmer and more isothermal. Observations by Clancy et al. (1989) indicate that there are times when the atmosphere is clearer than when it was observed during the Viking mission.

III. SPACECRAFT OBSERVATIONS OF DUST

A. Mariners 6, 7 and 9

The two flyby missions, Mariner 6 and 7, encountered Mars near L_s 200° in 1969. Images from the TV cameras showed a relatively clear atmosphere; the south polar cap edge and many dark features were prominent (Leovy et al. 1971). None of the specific clouds observed was attributed to dust, although the lack of features in Hellas suggested that airborne dust was present.

The Mariner 7 Infrared Spectrometer (IRS) obtained data in the thermal infrared out to 14 μm, including the silicate band near 9–11 μm (Horn et al. 1972). Midday spectra acquired in equatorial regions clearly show the silicate absorption feature induced by atmospheric dust overlying a warmer surface. Airborne dust is approximately at the atmospheric temperature; during the day, that temperature is cooler than the surface, resulting in an absorption band. The strength of the absorption is proportional to atmospheric opacity for any fixed, monotonic vertical temperature profile. Absorption at the band center near 9.5 μm amounts to about 25% for the Mariner 7 spectra. There may be some contributions to the silicate band from varying surface emissivity.

The Mariner 9 Orbiter carried an infrared spectrometer as well, the Infrared Interferometer Spectrometer (IRIS), that measured the silicate band region with greater precision and coverage than the IRS on Mariner 7. Variations in the depth of the silicate band were evident as the global storm of 1971 dissipated (Hanel et al. 1972*b*). At revolution 92 (near L_s 310°), when the thermal effects of dust on the atmosphere had diminished to about half their peak (Conrath 1975), the depth of the silicate band was about 40%. Systematic calculation of infrared dust opacities can be carried out from both the Mariner 7 and 9 spectral data, but this has not been done to date. Vertical temperature profiles, which are needed for accurate derivation of opacity information, can be derived from measurements in the 15-μm band by both instruments.

The visible appearance of the planet as Mariner 9 approached and orbited Mars offered dramatic evidence for atmospheric dust; the pall produced by the most severe dust storm ever seen made it impossible to discern surface detail except in the south polar region and near the tops of the high Tharsis volcanoes. Among the many Martian surface features observed first by Mariner 9 were "wind streaks"—thousands of streamlined bright or dark markings scattered over the surface and extending from impact craters, hills and intracrater deposits (Sagan et al. 1972,1973*b*). Their elongate and often teardrop shapes, their variability with time and the similarity of their orientations within regions, led to their identification as markers of near-surface aeolian activity. The time variability of streaks provides one of the few sources of information on temporal and seasonal changes in effective directions and strength of near-surface winds, allowing patterns of dust transport within a region to be inferred (see chapter 22).

B. Viking Lander

The two Viking Landers carried cameras that were capable of looking at the Sun as well as at the local environment. By observing the Sun, it was possible to determine the line-of-sight extinction opacity of the atmosphere on a regular basis (Pollack et al. 1977,1979; Colburn et al. 1989). These measurements constitute a prime data set for the study of opacity during the Viking Mission (Fig. 1). In 1977 a planet-encircling storm began near L_s 205°. Dust loading over the Lander sites had decreased to near pre-storm values when a more extensive storm began near L_s 274°. The dust raised by the second storm took somewhat longer to settle. Valuable information about the decay rate of opacity and the opacity levels preceding storms has been gained from Lander site opacities. Additional information about the properties of dust particles was obtained from sky brightness measurements (see below).

The data acquired by the Lander meteorology experiment provide an indirect but useful measure of opacity. During major dust storms, the constricted diurnal range of near-surface air temperatures reflects the decrease in surface insolation and increase in down-welling thermal radiation due to overhead opacity (Ryan and Henry 1979). Meanwhile, the enhanced semi-diurnal variation of pressure reflects the global expansion of the atmosphere produced by daily heating of a widespread (and opaque) dust haze. Zurek (1981) computed the opacity required for the solar heating of a global dust haze to reproduce the tidal surface pressure amplitudes observed at Viking Lander 1 during the two 1977 dust storms and found that it was comparable to the visual opacity measured locally using the Lander cameras (Fig. 1).

C. Viking Orbiter Cameras

The Viking Orbiter (VO) cameras provided images of both global and local dust storms. However, the coverage was serendipitous, and individual regions were rarely observed more than once a day, or on successive days. It is possible to assess the atmospheric opacity in a quantitative manner from

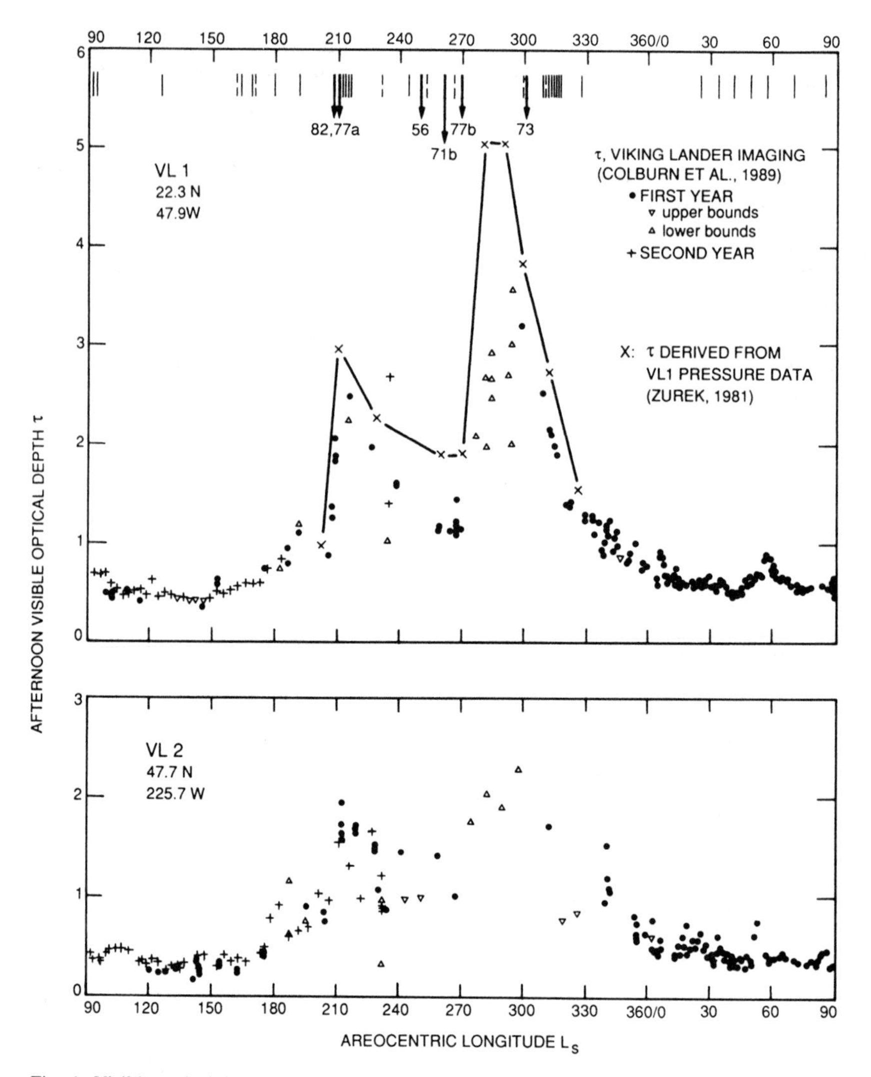

Fig. 1. Visible optical depths derived from the Sun-diode measurements made at the two Viking Landers (Colburn et al 1989). Line indicates the opacity of a global haze inferred from Viking Lander 1 surface-pressure data (Zurek 1981). Initial appearance or detection of all observed regional dust clouds, hazes or obscurations (vertical lines) and of planet-encircling dust storms (arrows) are indicated at the top of the upper panel. Dust events are listed in Table III (figure adapted from Zurek and L. Martin 1992).

images by observing shadows produced by surface relief features. The brightness and contrast of shaded regions depends upon scattering by dust in the optical path. Thorpe (1979,1981) applied this approach to images covering the southern hemisphere during the major storms of 1977, and in the northern hemisphere as well over a Martian year. During the period of the major dust storms, Thorpe's derived visual opacities are smaller by a factor of 2 than those obtained by Pollack et al. (1979). (See Sec. IV below.)

The Viking images provide particularly dramatic snapshots of the growth phase of the first 1977 storm (Briggs et al. 1979; Color Plate 17). The storm expansion was quantified by Thorpe (1979,1981) (Fig. 2). The zonal spreading shown compares favorably with the characteristics of typical storm growth, as documented by Earth-based observers (see, e.g., L. Martin 1974*a*). Local dust-storm morphology, frequency and spatial distribution have also been documented to some extent in the Viking Orbiter images (Briggs et al. 1979; Peterfreund and Kieffer 1979; Peterfreund 1985). During clearer periods, Viking observations of the atmosphere at the limb of Mars provide quantitative constraints on the seasonal variation of the vertical distribution of airborne dust (Jaquin et al. 1986).

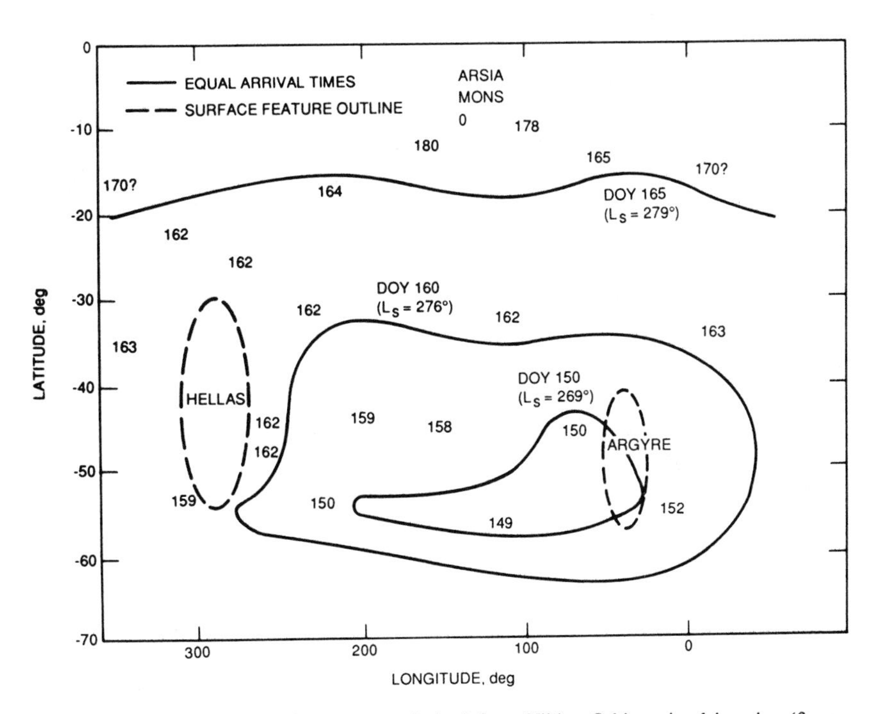

Fig. 2. Expansion of the 1977b storm, as derived from Viking Orbiter visual imaging (figure from Thorpe 1979). Arrival dates of maximum opacity change are shown in day of year, 1977; areocentric longitude L_s is given in parentheses.

Several columnar dust-devil type clouds have been noted in images (Thomas and Gierasch 1985), providing additional data on the scale of observed Martian dust cloud phenomena, and offering clues about the near-surface dynamical environment.

D. Viking Orbiter Infrared Thermal Mapper

The Infrared Thermal Mapper (IRTM) instruments on the two Viking Orbiters produced extensive maps of the thermal radiation from the surface and atmosphere, as well as determination of the broadband (0.3 to 3.0 μm) albedo of the planet. There are several ways that atmospheric opacity can be derived from these measurements. The surface albedo of the planet was determined by selecting measurements made when the atmosphere was relatively clear and by removing empirically the atmospheric contribution to the visual band brightness. The apparent brightness of dark regions is sensitive to the amount of overlying dust, and one can track the varying atmospheric opacity in this way once the underlying surface albedo is known (Pleskot and Miner 1981; Christensen 1988).

Several of the IRTM spectral bands were placed in the vicinity of the silicate vibrational feature that extends from about 9 to 11 μm. T. Martin (1986) applied a technique in which the thermal infrared opacity was derived from a comparison of 8–10 μm band brightness temperatures with those in the 6–8 μm band, less affected by dust. An atmospheric temperature profile was constructed from knowledge of surface temperature, an estimate of the surface temperature discontinuity based on experience from the Viking Lander measurements, and simultaneous measurement in the IRTM 15-μm band of temperature near the 0.6-mbar level. Opacities derived in this way are about 50% as large as those measured in the visible region by the Viking Lander. The difference here is assumed to arise from the variation of the dust optical properties with wavelength.

The IRTM spectral approach was applied to zonal mean IRTM data covering more than a Mars year, giving a picture of zonal mean opacity variation with latitude and time. Opacity was apparently greater in southern and equatorial regions than at the Viking Lander 1 site during the second major storm of 1977, whereas the lesser storm 1977*a* produced a more uniform latitudinal distribution (Fig. 3). The analysis also shows the atmosphere to be clearest in the northern hemisphere between $L_s \approx 20°$ and 150°; Kahn (1984) found similar results by studying his catalog of clouds in Viking Orbiter images, although his technique was unable to discriminate between dust and ice clouds. A histogram of zonal mean opacity made from the global coverage for 1 Mars yr is shown in Fig. 4. The large asymmetry to the data shows that the most probable condition on Mars, averaged over time and location, is a relatively clear atmosphere. However, the effects of major storms are large, and, by creating a high-opacity tail to the distribution, they cause the mean opacity, 0.51 at 9 μm, to be much greater than the modal value of 0.056. In

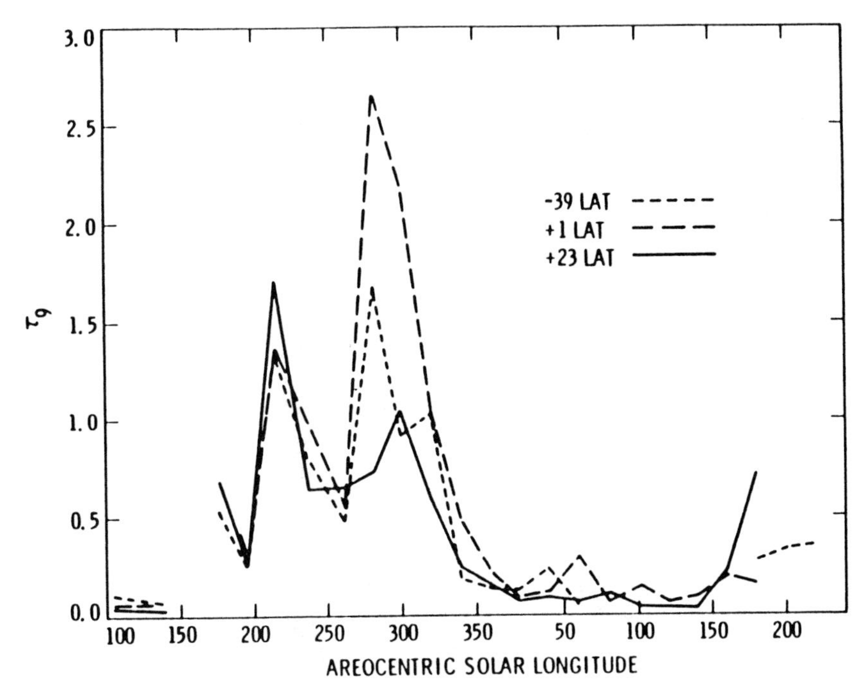

Fig. 3. Variation of zonal-mean 9-μm band opacity with season L_s for three 10°-wide latitude bands, derived from Viking Orbiter IRTM data. The typical size of an L_s bin is 20° (figure from T. Martin 1986).

years when no major storms occur, it is likely that the mean opacity is much closer to the low modal value shown in Fig. 4. T. Martin (1986, 1992) also created a series of opacity maps of the planet using IRTM data. These maps reveal regional storm activity in Hellas prior to the 1977a event; a lack of involvement of south polar latitudes in the 1977b storm; and evidence for dust-raising in high northern latitudes during northern spring and summer.

The Viking Orbiter IRTM also provided valuable observations of regional dust transport and surface properties. Thermal inertia data derived from the IRTM have been used to infer the degree of surface mantling by dust deposits (Kieffer et al. 1977; Christensen 1982,1986*a,b*,1988; Jakosky 1986; chapter 21); low thermal inertia values are indicative of surfaces covered with fine particles or highly vesicular material, while high values imply surfaces consisting of bonded fine particles (Jakosky and Christensen 1986*a,b*) or covered with coarse particles or a mixture of fines, blocks and exposed bedrock. Mapping of the regional albedo and variations therein, measured by IRTM, combined with experimental work in albedo changes due to dust fallout (Thomas et al. 1984; Wells et al. 1984), allow constraints to be placed on the sediment transport, such as whether net removal or deposition is occurring, and how much dust is involved (see below). See Color Plate 18.

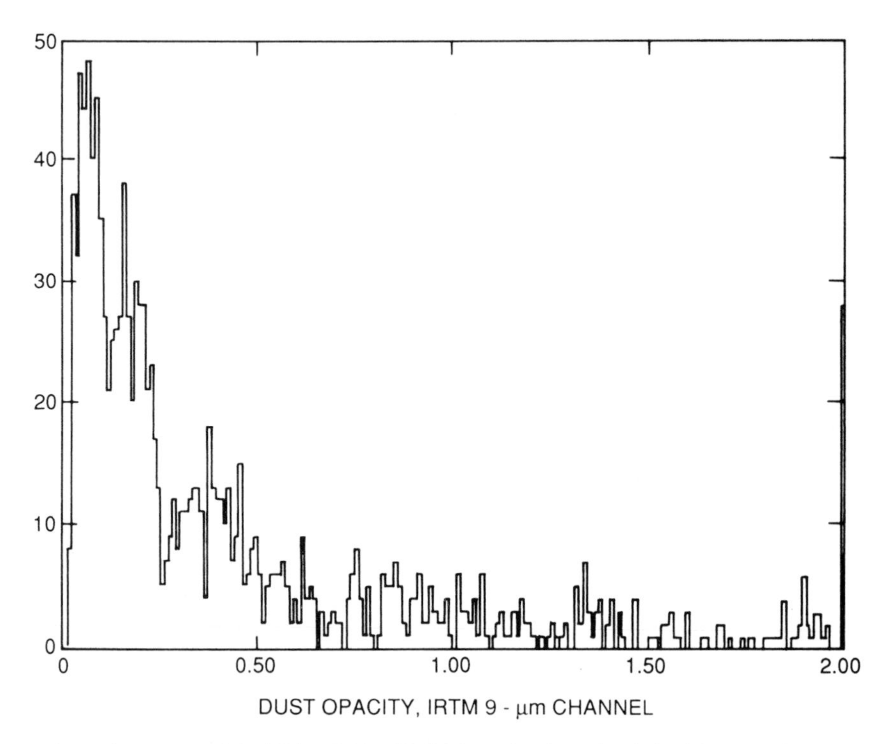

Fig. 4. Histogram of the Viking Orbiter IRTM 9-μm band opacity derived for more than one Mars year from global measurements. The mean value is 0.51 and the mode 0.056 (figure from T. Martin 1986). Bins of the data set are 2° latitude by about 20° in L_s.

IV. MICROPHYSICAL PROPERTIES OF AIRBORNE DUST ON MARS

One striking point about the properties of Martian dust particles is that they seem to vary vertically, horizontally and temporally on many scales. The problem of measuring these properties is compounded by the presence of diurnally varying ground fogs and higher-level water-ice hazes (Pollack et al. 1977; Colburn et al. 1989), as well as extensive and possibly more persistent high-level ice hazes (Leovy et al. 1972; Anderson and Leovy 1978; Jaquin et al. 1986; Kahn 1990). In addition, remote sensing measurements made in different parts of the spectrum are sensitive to different particle size ranges and sample different depths in the atmosphere, even for similar viewing geometries and constant climatic conditions. A review of the subject of properties of Martian dust particles was given by Zurek (1982).

Particle composition, size, and shape determine the optical properties of the dust. Observations affected by the optical properties of dust particles have been exploited in an attempt to constrain particle physical properties on Mars.

The web of interrelated phenomena and measurement techniques is illustrated in Table I. Particle composition determines the imaginary and real parts of the index of refraction as a function of wavelength, $n_i(\lambda)$ and $n_r(\lambda)$. Given $n_i(\lambda)$ and $n_r(\lambda)$, together with particle size $N(r)$ and shape distributions, as functions of particle radius r, the extinction efficiency Q_{ext} and the particle scattering phase function, usually characterized by a single scattering albedo ω_0 and a scattering asymmetry parameter g, can then be computed in principle for all wavelengths. The relationships linking particle physical and optical properties are well known today only for spherical particles (Mie theory), some ellipses, and a few other shapes; otherwise semi-empirical relationships are used (see, e.g., Pollack and Cuzzi 1979). While $Q_{ext}(\lambda)$ gives the variation of optical depth τ with wavelength, the total particle number density must be known to compute τ itself for a representative volume.

A. Dust Particle Composition

Sky color measurements across the visible and near infrared are particularly sensitive to $n_i(\lambda)$, and, together with mineralogical context (see chapter 21), they have been used to show that the characteristic "red" of Martian dust can be accounted for if the particles contain about 1% by volume of an iron oxide such as magnetite (Pollack et al. 1977,1979). This determination was made as part of a comprehensive, self-consistent solution for a suite of particle properties using measurements by the Viking Lander 1 cameras in early northern summer, at 22° latitude.

Unlike the visible optical properties, which appear to be dominated by trace elements, the thermal infrared part of the spectrum is more sensitive to the silicate content of the particles; based in part on surface compositional considerations, silicate is likely to be the main particle constituent. Toon et

TABLE I
Properties and Effects of Atmospheric Dust

Particle Physical Properties $n_i(\lambda)$, $n_r(\lambda)$, $N(r)$, shape (r)	**Particle Optical Properties Measured by Remote Sensing** $\omega_0(\lambda)$, $g(\lambda)$, $\tau(\lambda)$
Measurements	**Measurements**
Chemical analysis of particle composition	Extinction at surface of visible light from the Sun or moon(s) at varying angles
Direct sampling of particle population: sizes, structures, shapes	Visibility of surface features from space
	Variation with phase angle of surface feature visibility (opposition effect)
	Measurement of skylight contribution to surface illumination
	Measurement of emergent infrared radiation and balance with extinction of solar input
	Measurement of sky brightness

al. (1977) compared Mariner 9 IRIS spectra between 5 and 50 μm with mineralogical samples, and found a mixture of basalt and clay minerals with at least 60% SiO_2 to be consistent with the data. They obtained a best fit to the IRIS spectra shortward of the 15-μm band using a clay mineral sample, montmorillonite 219b. The IRIS spectra used sampled a wide area in the southern hemisphere, and were selected from data acquired during the 1971–1972 global dust storm to minimize the surface contribution to the observed radiance. Kahn (1980) used Viking Orbiter IRTM observations in four channels, taken in both the surroundings and the interior of a southern hemisphere low-latitude local dust storm, to constrain infrared dust-particle properties in the vicinity of the storm. Although the data only allow determination of ratios of opacity in different broad wavelength bands, these differed from the ratios of particle extinction found for particles in the global storm by Toon et al. (1977). Another attempt to obtain some compositional information was made by Hunt (1979), who used Viking Orbiter IRTM data to try to sort out the contributions of water ice and dust. He favored a composition for the dust of 25% montmorillonite 219b and 75% of a standard basalt sample in order to account for infrared radiances near 20 μm.

B. Dust Particle Size and Shape

The direct measurement of particle size distribution *N(r)* and particle shape distribution requires *in situ* sampling of atmospheric particles, which has never been done for Mars. Toon et al. (1977) and Pollack et al. (1979*b*) fitted *N(r)* as part of their self-consistent schemes, using the modified gamma function which is popular for modeling terrestrial cloud particle size distributions. They found that it is possible to match the observations of airborne dust during the 1971 global dust storm in the infrared and measurements of the ubiquitous dust above the Viking Landers during 1977 in the visible (Pollack et al.)—two radically different climatic conditions—with the same parameters in *N(r),* their particle size distributions. Toon et al. and Conrath (1975) also detected no variation in *N(r)* as the global dust storm of 1971–1972 cleared, a deduction which implies that simple gravitational settling of particles cannot be the only mechanism removing dust from the atmosphere at that time.

Several size-independent dust removal processes have been proposed. Pollack et al. (1979*b*,1990*a*) suggested that particles in the polar regions are removed after they become coated with water and CO_2 ice. Lee (1985) pointed out that the condensation of small amounts of water ice onto the particles can reduce the particle settling rate. This occurs because although the condensation of ice increases the mass of the particles, it also lowers the mean particle density and increases the particle cross section. If more than a few precipitable μm of water or CO_2 condense on the particles, the increased mass compensates for the other effects, and the coated particles fall faster. Effects of particle shape upon the particle fall rate have been considered: disk-shaped particles fall more slowly than spheres, but the models that have been used are still unable to retain the larger particles in the atmosphere, as suggested by the Mariner 9 IRIS observations (Murphy et al. 1990*a*). Recently, the model simulations of Murphy et al. (1990*a*) found that the particle size could be kept fairly constant, near the subsolar regions in the southern hemisphere that were viewed by IRIS, if the upward vertical motion generated by solar heating of the airborne dust was large enough; the velocities required were comparable to those generated in a zonally symmetric model of a global dust storm (Haberle et al. 1982).

Given typical particle size distributions, the IRIS measurements are most sensitive to particles in the size range 1 to a few tens of μm. A mean particle diameter near 2 μm was deduced from the changing thermal structure of the Mars atmosphere together with implied particle settling rate, during the clearing phase of the 1971–1972 global dust storm, as measured by the Mariner 9 IRIS instrument (Conrath 1975). Conrath assumed a monodisperse particle size distribution, and his derived value for mean particle radius involves a combination of effective particle settling cross section and infrared optical cross section. The scheme of Pollack et al. (1979*b*) gave an optical cross-section mean weighted particle radius (relative to the solar spectrum) in northern summer of about 2.5 μm, in agreement with Toon et al. (1977). The mode value for particle radius using this distribution is about 0.4 μm and is highly sensitive to the longer-wavelength data used (see Pollack et al. 1979). Work with the Mariner 9 Ultraviolet Spectrometer under global dust-storm conditions provided an estimate of the mean particle radius, weighted by cross section in that spectral range, of ~ 1.5 μm (Pang et al. 1976; Pang and Ajello 1977). The ultraviolet analysis, which also produced a deduction that TiO_2 is a constituent of the particles, has been called into question on the grounds that multiple-scattering contributions, and the possibility of non-spherical particles, were not considered (Pollack et al. 1977; Chylek and Grams 1978; Zurek 1982).

C. Dust Optical Constants

The mean single-scattering properties themselves have been inferred from Orbiter and Lander measurements, using physical phenomena ranging from the variation in sky brightness with scattering angle to the opposition effect. Some of these determinations are listed in Table II. Note the differences in spectral range and season of the observations. No systematic monitoring of the variation in particle properties, except the atmospheric optical depth, has yet been made for Mars. There are wide differences among determinations of ω_0, varying from about 0.85 to 0.95 in red or broadband visible light. Some of this variation could reflect seasonal and spatial variations in dust particle properties, as well as varying contributions from ice haze to the observed radiance. The techniques so far applied are inadequate to separate the effects of geographic location, elevation, time and method.

TABLE II
Determinations of Mars Atmospheric Particle Properties in the Visible[a]

Method	ω_0	g	Season	Reference
Mariner 9 images (clear)	~0.7	—	dust storm	Leovy et al. 1972
VO opposition effect	0.50–0.85	≤0.6	dust storm	Thorpe 1978
VL sky brightness (clear)	0.86	0.79	early n. summer	Pollack et al. 1979
VO limb images (red)	~0.94	~0.6	late dust storm	Jaquin et al. 1986
VO visibility (clear)	~0.95	~0.8	mid n. spring	Kahn et al. 1986
VO IRTM-EPF (albedo)	~0.90	0.55	all year; mid-low lat.	Clancy and Lee 1990, 1991
VO IRTM-EPF (albedo)	1.0	0.55	n. spring, high lat.	Clancy and Lee 1990

[a] Spectral ranges vary over the visible, and are given in parentheses; ω_0 for Thorpe: violet is 0.5 and red is 0.85 μm; IRTM-EPF = Infrared Thermal Mapper-Emission Phase Function sequences; the IRTM albedo channel ranges from 0.3 to 3.0 μm; VO = Viking Orbiter; VL = Viking Lander.

Other observations add to the complexity of the picture. In the limb images from the Viking Orbiter, there are indications that high-level ice hazes are very common, and contribute to the mean particle properties derived from spacecraft observations (Anderson and Leovy 1978; Jaquin et al. 1986; Kahn 1990). Variations in g during global dust storms were reported by Thorpe (1979,1981), who concluded that in red light the parameter increased from near zero to 0.6 as the storm intensified, and decreased as the storm decayed, implying that more forward-scattering particles (possibly larger ones) were present at the peak of the storm. The analysis by Thorpe is limited by the difficulty of properly accounting for multiple scattering, important when the dust opacity is high; this could be particularly important as the dust opacity increases (see Zurek 1982). Zurek (1981) used a coupled radiative-dynamical tidal model which gave $g = 0.5$ at the storm onset, a lower value than that deduced at less dusty times. Although this suggests smaller particles were present at dust storm onset than later in the storm's evolution, it is more likely that there was a change in the vertical distribution of the dust (Zurek 1982). Recently, Clancy and Lee (1990, 1991) began to analyze Viking Orbiter IRTM sequences which sampled a range of emission angles for fixed points on the surface. They used high- and low-emission-angle data from the very broadband (visible to near infrared) albedo channel to isolate atmospheric particle emission from the surface contribution, and derived larger particle albedos during dustier times than during clearer seasons at mid and low latitudes (see Table II). They also found more ice-like particle albedos in high northern latitudes in spring.

The best determined atmospheric dust property is the column optical depth above the two Viking Landers, which was measured at 0.67-μm wavelength by a special Sun diode on the Viking Lander cameras (Colburn et al. 1989; Pollack et al. 1977,1979). Pollack et al. (1979) determined a value for Q_e of 2.74, averaged over the solar spectrum. Estimates of the IRTM 9-μm band optical depth (T. Martin and Kieffer 1979; T. Martin 1986) suggest that the visible-to-infrared ratio of opacity, and thus of Q_e, is systematically ≈ 2; this is lower than expected for typical clay or basalt particles with the expected size distribution (the matter is discussed in detail by Zurek 1982). Uncertainties in the size distribution and in particle composition could account for the discrepancy (T. Martin 1986).

Thorpe's (1979,1981) opacity estimates derived from Viking Orbiter imaging of surface shadowing agree with the Viking Lander measurements before the two 1977 dust storms, but are lower systematically by a factor of 2 during very dusty periods. While Thorpe's estimates are more consistent with the expected ratio of visible and infrared optical depths than the IRTM analysis, this is probably fortuitous, given the difficulty of accounting properly for multiple scattering when analyzing the surface contrasts (Zurek 1982). Kahn et al. (1981) found that observations of the twilight with the Viking Lander cameras required the atmospheric particles to be more blue (less red) than the midday determinations made from the same platform by Pollack et al. (1979). Both diurnal changes in atmospheric ice particles and vertical variations of particulates combined with differences in the line-of-sight weighting could contribute to this difference.

In summary, the available remote-sensing information about atmospheric dust-particle composition is consistent with geologically reasonable silicate particles with some added iron oxide. No data have yet been collected that allow critical determination of the dust-particle composition and water-ice fraction. Airborne dust particles typically have a mean size in the range of a few μm. Particle shape, size distribution, composition and the mix of ice and dust seem to vary, although there are too few measurements to confirm any systematic behavior in these properties. A simple device that could monitor opacity and some aspects of the particle size distribution from vantage points on the surface could provide diagnostic information valuable for understanding other climatic effects, as well as offering a better-defined set of parameters for modeling Mars' atmospheric radiation.

V. MARTIAN SURFACE ALBEDO FEATURES

There is direct evidence for dust redistribution on the surface of Mars, but only tiny amounts of dust are required to move to explain the observed changes. Terrestrial observers have long noted the seasonal and interannual variability of classical Martian albedo features (Slipher 1962; de Mottoni y Palacios 1975; chapter 2). Spacecraft observations of such features have shown their variability to be related to aeolian transport of bright dust into and out of regions primarily in association with major dust storms (Sagan et al. 1973*b*; Thomas and Veverka 1979*b*; Lee et al. 1982; chapter 22). Long-term atmospheric dust transport may be responsible for the formation of extensive regional sediment deposits, which in turn may provide a source of wind-blown material for subsequent redistribution to other areas.

Both Earth- and spacecraft-based observations have established that some of the range in colors of Martian albedo features can be interpreted as arising from variable amounts of dust covering the surface (Soderblom et al. 1978; Singer and McCord 1979; McCord et al. 1982*a,b*). Color measurements of atmospheric dust in dust storms and the brightest regions on the surface are very similar (McCord and Westphal 1971; McCord et al. 1982*b*). In particular, similarity in the color properties of dust clouds (Briggs et al. 1979) and parts of Arabia (Thorpe 1982) support the interpretation that dust residing at these locations and that contained in dust clouds are derived from the same source. The gross surface morphology, thermal properties, and color of two such regions in particular (Arabia and Tharsis) are consistent with the presence of surface mantling by dust deposits (chapter 21).

Experimental studies by Wells et al. (1984) indicate that even very small quantities of dust deposited from the atmospheric dust load are sufficient to alter dramatically surface albedo features. Deposition of only 10^{-4} g cm^{-2} of bright dust on an average dark area should increase the surface albedo by several tens of percent. This amounts to an average dust accumulation of only a few tenths of a μm, which is comparable to the column dust opacity of the background dust haze above the Viking Lander sites and is much less than the sedimentation typical of the two 1977 major dust storms (Pollack et al. 1979). Most individual Martian albedo features (wind streaks) exhibit albedo contrasts relative to their surroundings of $< 20\%$, with contrasts of only a few percent resulting in readily visible features (Thomas et al. 1984). Thus, the dust fallout expected to occur over a single Martian year is more than sufficient to explain the variability observed in the planet's albedo patterns. Detailed analysis of the patterns and variability of regional dust transport, coupled with studies of regional surface properties, may be diagnostic of whether net erosion or deposition of dust-storm fallout is taking place today and whether such processes have been active in a region over the long term (Lee 1987; Christensen 1988).

Wind streaks have been classified on the basis of fundamental differences in appearance and formation mechanisms (Thomas et al. 1981; Veverka et al. 1981). The prevalent types are bright depositional and dark erosional (see Fig. 5). As discussed in chapter 22, bright depositional streaks (of higher albedo than their surroundings) form downwind of crater rims and other pos-

Fig. 5. Examples of Martian wind streaks: (Top) dark erosional streaks in Memnonia (Viking Orbiter image 450S33, clear filter, image center latitude −12°, longitude 150°); (Bottom) bright depositional streaks in Amazonis (Viking Orbiter image 545A54, red filter, image center latitude 20°, longitude 185°).

itive topographic features; except in areas of significant slope (such as the Tharsis volcanoes), they show little temporal variability. Bright streaks apparently form by deposition of dust from suspension during times of high dust loading and high static stability of the atmosphere (Veverka et al. 1981); such conditions occur during the clearing stages of great dust storms. Dark erosional streaks (of lower albedo than their surroundings) apparently form by removal of dust downwind of obstacles during times of low static stability (following the decay of great dust storms). Thus, wind streaks, areas of enhanced dust cover (bright streaks) or reduced dust cover (dark streaks), are indicative of local wind directions at the time of streak formation, and can be used to infer local sediment transport directions (Thomas 1982; Lee 1984).

VI. FREQUENCY AND LOCATION OF DUST STORMS

The earliest observers of Mars found it useful to distinguish between minor and major storms, the former being localized clouds that remained local. After the major storms of 1956 and 1971 much attention was focused on "global" dust storms (Gierasch 1974). Zurek (1982) replaced the adjective "global" by "great" to include storms which, though they covered vast regions, were confined largely to one hemisphere. A more detailed nomenclature was developed by L. Martin (1984), and revised by L. Martin and Zurek (1992). They categorized the dust storm activity on Mars as observed from Earth both by size (local, regional or planet-encircling) and by morphology (obscuration, dust cloud or storm; Tables III and IV). Observations to date suggest that the size categories are successive, in that even the largest storms appear to begin as one or more local storms. "Planet-encircling" distinguishes those "major" or "great" (i.e., nonlocal) dust storms that expanded to cover a latitudinal corridor. Unfortunately, it has not been possible to observe Mars systematically enough to remove the bias of incomplete coverage from these categorizations. To the extent that estimates can be made, the longest-lived

TABLE III
List of Nonlocal Martian Dust Storms[a]

Year	Dates	L_s[b]	Classification[c]	Comments
1877	Sept–Dec	265	R2O?	
1894	Oct 10–31	296	R2O?	south cap disappeared
1907	Jul 29–Aug 2	213	R2Sb	
1909	Jul	232	R2O?	
1909	Aug–Sept	252	R2O?	
1911	Sept–Oct	298	R2O?	
1911	Oct 11–20	313	R1Cb	"white" per Antoniadi
1911	Nov 3–Dec 23	326	R2Sh	
1922	May	161	R2O?	
1922	July 10–13	192	R3Sb	

TABLE III, continued

Year	Dates	L_s[b]	Classification[c]	Comments
1924–25	Dec 5–Jan	310	R3O?	polarimetry; encircling?
1926	Oct 25–26	311	R2Sb	may have been different clouds
1941	Nov 12–17	312	R3Sb	
1943	Sept 28–Oct 6	311	R1Sb	
1954	Jun 10–14	179	R3Ch	polarimetry
1956	Aug 19–Nov	249	E3Sb	first planet-encircling storm?
1958	Oct 13–18	309	R3Sb	
1961	Jan 19–23	24	R3Sb	
1963	Feb 24–28	59	R3Ch	polarimetry
1969	May 28–Jun 4	163	R3Sb	probably relates to storm below
1969	Jun 4–14	168	R3Sb	initial date from IPP images
1971	Jul 10–22	213	R3Sb	
1971–72	Sept 22–Jan	260	E3Sb	only truly "global" storm
1973	Jul–Sept	244	R3Sb	polarimetry
1973	Oct 13–Dec	300	E3Sb	several initial clouds
1977	Feb 15–Apr	204	E3Sb	Viking Lander (VL) and Orbiter data
1977	May 27–Oct	268	E3Sb	Viking Lander and Orbiter (VO) data
1978	Jan 30–May	40	R3Ch	VO; Earth-based polarimetry; not by VL
1978	May 14–Aug	86	R3Ch	VO; Earth-based polarimetry; not by VL
1979	Jan–Mar	212	R3Sh	Viking Lander meteorology data
1981	Nov 20–25	48	R2Sb	Mars < 6″; nothing from Viking
1982	May 5–16	122	R3Sb	detected visually and by polarimetry
1982	Oct	208	E3S?[d]	Viking Lander pressure data and imaging
1983–84	Nov 27–Jan	70	R3Sb	
1984	Jan 29–Feb 15	93	R2Sb	
1984	Mar 3–22	112	R3Cb	
1984	Jun 24–30	169	R3Cb	
1988	May 5–7	189	R3Sb	Hellas, Libya
1988	Jun 3–30	214	R3Sb	also detected by polarimetry
1988	Nov 23–Dec	314	R3Sb	also detected by polarimetry

[a] Table adapted from L. Martin and Zurek (1992).
[b] L_s is for earliest date shown when given; otherwise, the 15th of that month is used.
[c] Classification scheme: XYZz, where
(i) $X = R$ or E and $Z = O$, C or S, as described in Table IV. Also,
$z = h$ = haze, meaning ill-defined edges
$z = b$ = bright clouds
$z = ?$ = existence doubted or nature ambiguous
(ii) $Y = 1$ = phenomenon not described as yellow; may not be dust
2 = visual observation only
3 = some Earth-based photography or polarimetry, or spacecraft data
[d] Storm was not observed to move or to encircle planet, but the twice-daily pressure signature was as large or larger than that meaured during earlier planet-encircling storms, which were observed by the Viking Orbiters.

TABLE IV
Atmospheric Dust Phenomena Classified by Size and Morphology[a]

Classification by Morphology	
O: Obscurations	loss of contrast; may be due to changes in surface albedo contrasts, as well as to the presence of atmospheric hazes
C: Clouds	atmospheric phenomena, either well-defined clouds or hazes; distinguished from storms by absence of observed motion or expansion
S: Storms	moving and/or expanding clouds and hazes (observed motion clearly distinguishes atmospheric from surface phenomena, particularly for Earth-based observations)
Classification by Size	
L: Local	localized in area, with the long axis of the affected area < 2000 km
R: Regional	regional in area, with the long axis of the affected area > 2000 km, but did not encircle the planet
E: Planet-Encircling	area affected encircled the planet, usually in the east-west direction. Storm covers much of one or both hemispheres

[a]Table adapted from L. Martin and Zurek 1992.

of the dust storms seem to attain the largest opacities for the associated widespread dust haze; these storms may also be the largest in areal extent. Opacities vary spatially and are expected to be greatest in the source regions where dust is raised into the atmosphere.

A. Local Dust Storms

Yellow clouds and storms have appeared in all seasons and in both hemispheres (Table V and Fig. 6), although they have been observed more frequently during southern spring and summer, when Mars is closest to the Sun. Since this is also the season when Mars can be most easily viewed from Earth, it is not known whether this represents a true seasonal variation, at least for local dust storms that cover a few 10^6 km^2 (Table IV). Latitudinally, local dust storms tend to occur in three regions: broad zones equatorward of the seasonal polar cap edge in each hemisphere and in the southern subtropics (Fig. 6). Longitudinally (not shown), local storms are most common in Hellas, Noachis-Hellespontus, Argyre and the Solis, Sinai and Syria Plani regions, all in the southern hemisphere. In the northern hemisphere local dust clouds and storms tend to occur most frequently in Chryse-Acidalia, Isidis-Syrtis Major and Cerberus (Peterfreund 1985).

TABLE V
Frequency of Occurrence of Yellow Clouds[a]

Mars Year Season (L_s)		Mars Year 1963–65	1966–68	1968–69	1970–72	1972–73	1974–76	1977–78	1979–81	1981–83	1983–85	AVG
N. Spring	(0°–90°)	0	0	*	*	0	9	5	2	3	36	6
N. Summer	(90°–180°)	0	0	11	42	*	*	0	3	13	37	13
N. Fall	(180°–270°)	*	0	0	23	37	5	*	*	0	0	10
N. Winter	(270°–360°)	*	0	*	100[c]	43	9	0	0	*	0	25

[a] Table based on °L_s in which yellow clouds occurred, normalized by °L_s of all observations reported by Association of Lunar and Planetary Observers (A.L.P.O.) planetary astronomers.[b] Summarized from the tables of Beish et al. (1987). The asterisks indicate no observations were reported. Mars was generally near conjunction (i.e., on the far side of the Sun from Earth) during these periods.

[b] A.L.P.O. is an organization of professional and amateur astronomers; the contribution of the amateurs has been critical to extending observations of Mars throughout this period.

[c] This is the period of the 1971b global dust storm observed by Mariner 9.

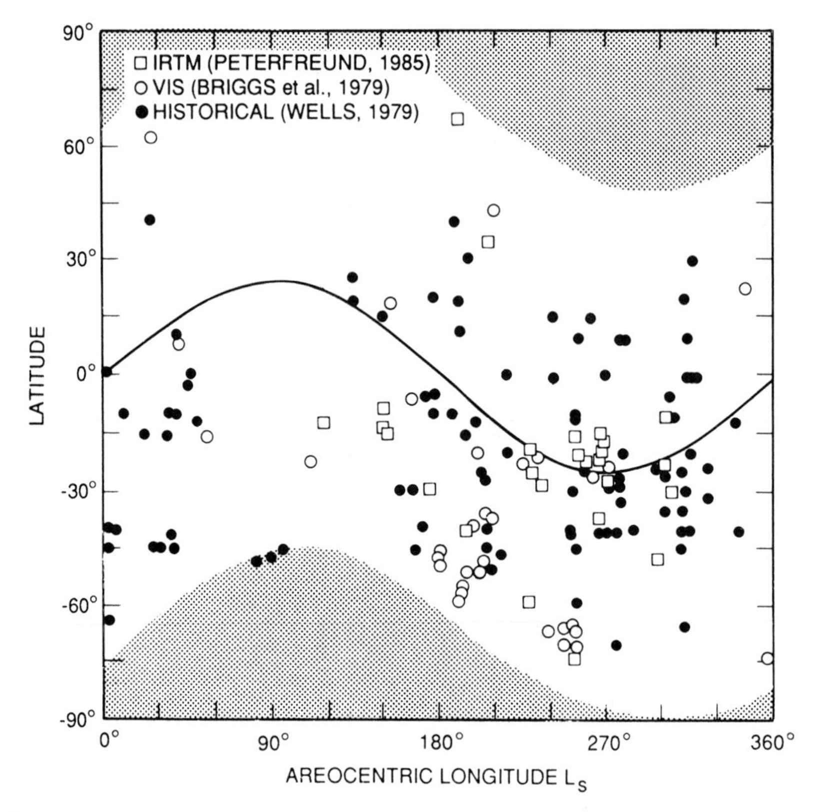

Fig. 6. Latitude and time-of-year occurrence of local dust storms detected by the Viking Orbiters, from either infrared or visible imaging observations, and those observed from Earth, as compiled by Wells (1979). Solid line represents the sub-solar latitude (figure from Peterfreund 1985).

B. Regional and Planet-Encircling Dust Storms

While observations suggest that larger dust storms occur in a more restricted range of season and latitude than do local storms, this may simply reflect the effects of spotty observational coverage and the smaller number of large storms. The largest of the observed regional dust storms and all of the planet-encircling dust storms have occurred within $\pm 90°$ in L_s of perihelion, i.e., almost exclusively in southern spring and summer (Fig. 7). Furthermore, the most common sites of origin for these storms have been the Noachis-Hellespontus and Solis Planum-Argyre regions.

Viking Lander imaging and meteorological measurements provide the most complete seasonal coverage of atmospheric dust opacity. During the first year of Viking observations at Mars, both the Lander sun-diode measure-

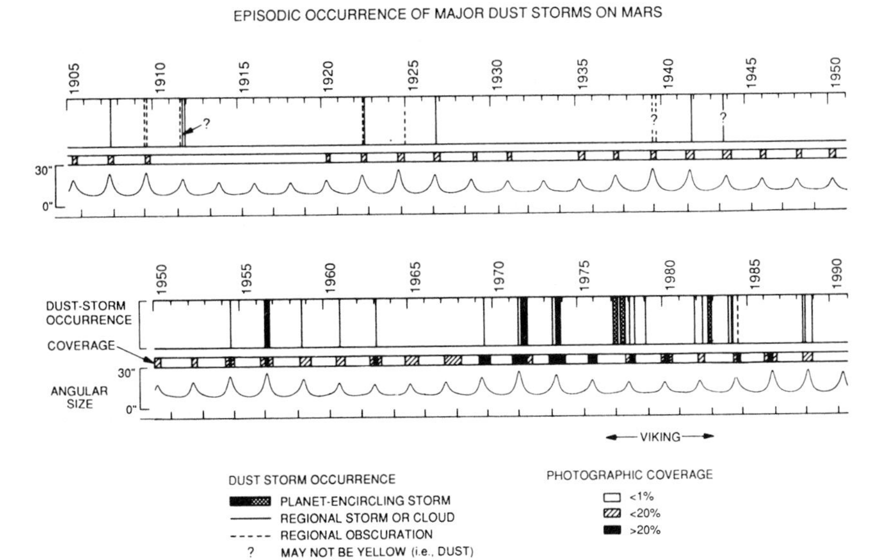

Fig. 7. Timeline of the detection of regional and planet-encircling obscurations, clouds and storms. These events are listed in Table III. Earth dates are indicated at the top, and perihelion (and thus Mars years) at the bottom of the dust-storm timeline. The second timeline indicates periods of photographic coverage of Mars, defined in terms of the percentage of L_s degrees that photographs were taken. Coverages of <1%, of 1 to 20% and >20% are indicated. The third timeline indicates the apparent size of Mars, as seen from Earth, on a scale of 0 to 30 seconds of arc (figure from Zurek and L. Martin 1992).

ments of solar intensity and the meteorological measurements showed that two great dust storms occurred in that one year (Fig. 1). Preliminary data from the Viking Landers during the next Mars year indicated significant opacity (Colburn et al. 1989) and a large diurnal range in surface pressure at nearly the same seasonal date (Ryan and Sharman 1981*b*), which suggested that a major dust storm had occurred during the second year. (Unfortunately, the Viking Lander sun-diode opacity measurements ceased shortly thereafter.) Given the difficulty of observing Mars from Earth and the appearance of planet-encircling storms during the most favorable of the oppositions of Mars and Earth, Zurek (1982) suggested that a regional, although not necessarily planet-encircling, dust storm occurred each Mars year.

L. Martin (1984) noted that there was little evidence in the Earth-based record that planet-encircling storms occurred yearly. Analysis of the four-Mars-year-long pressure record from Viking Lander 1 suggests that planet-encircling dust storms occurred during the first and fourth Mars years, that nothing larger than a regional storm occurred during the second year, and that nothing larger than local-scale storms occurred during the third (Tillman

1985; Leovy et al. 1985). Earth-based observations since 1982 reveal only localized or regional dust storms. In Fig. 7, the possible absence of planet-encircling storms in the first half of this century (L. Martin 1984; Zurek and L. Martin 1992) may be due to the observational bias described earlier (see also chapter 2). However, the detection of regional and smaller dust storms during the more favorable oppositions, and the more recent Earth-based data, imply that the period between 1969 and 1982 was exceptional with regard to the number of planet-encircling dust storms that occurred.

Based on the evidence for interannual variability of great dust storms (Fig. 7) and their characteristics (Table III), three major points can be made, keeping in mind the biases due to incomplete observational coverage:

1. One or more dust storms of regional or larger size can occur in any given Mars year. When they do occur, there is significant variation in their duration, areal extent and general opacity.
2. There are many Mars years in which no planet-encircling dust storm has occurred; less frequently, there may be years in which not even regional storms occur.
3. The relatively rare planet-encircling storms may have occurred more frequently during the last two decades (the period of spacecraft exploration of Mars) than earlier in the century; these storms all occurred within 90° L_s of perihelion.

The observed interannual variability of the great dust storms has several implications: (a) any enhanced meridional transport that may be associated with the planet-encircling dust storms cannot "close" on an annual basis the seasonal cycle of water, as proposed by Davies (1981; see chapter 28); (b) there is no reliable seasonal "trigger" for dust storm onset, although there may be seasonal precursors that could signal the occurrence of a large storm (see chapter 26); (c) to assess the long-term impacts of dust storm activity, we need a means of scaling from the short-term variations of the dust cycle to much longer periods, such as those associated with the astronomical forcing (see chapters 9 and 33).

VII. MECHANISMS FOR RAISING AND TRANSPORTING DUST

The thermodynamics of local dust storm generation and the mechanisms by which local dust storms become planet-encircling storms remain obscure. The ways dust may be moved from the surface into the atmosphere, and the meteorological phenomena associated with dust raising, are discussed in chapters 22 and 26. The presentation here emphasizes those aspects critical to the dust cycle.

It has been argued that low-latitude local storms are initiated either by free convection, where direct heating of the surface may cause dust-devil type events (Gierasch and Goody 1973), or that regional winds superimposed upon the general circulation produce the initial shear stress that mobilizes surface dust (Leovy et al. 1973*b*). Free convection is favored by low near-surface horizontal winds and low static stability; several types of observations indicate the presence of dust devils or convective vortices at times and places where these boundary-layer conditions are expected to occur (Thomas and Gierasch 1985; Ryan and Lucich 1983). On theoretical grounds, free convection near the surface is likely to be common on Mars, as it is in terrestrial deserts, when the surface is most strongly heated, as in summer.

Thomas and Gierasch (1985) detected 99 columnar cloud forms, identified by their shadows and transitory nature, in sporadic high-resolution (60 to 80 meters/pixel) Viking Orbiter imaging coverage of the northern mid latitudes in northern summer. Dust columns up to 6.8-km high and as much as 1 km in width were observed between 14 and 15 hr local time; they occurred on smooth plains 13 to 20° from the subsolar latitude, where they may be quite common. By correlating the temporal behavior of wind speed, wind direction and temperature in Viking Lander meteorology data, Ryan and Lucich (1983) deduced that convective vortices were common at the two Lander sites. A few of these vortices had associated wind gusts thought to be strong enough to raise dust; all but one of these were found at Viking Lander 1 (see also chapter 26).

Dust clouds associated with vortical motions can redistribute dust locally. While it has been suggested that local, and even regional, dust storms can be organized as vortices (Gierasch and Goody 1973), no dust cloud larger than a few km across has exhibited the expected spiral structure. The one local dust storm that was observed to pass over Viking Lander 1 appeared to be associated with a baroclinic storm system (a wave) passing over the site (James and Evans 1981; Ryan and Sharman 1981*b*; see also chapter 26). The local winds observed during this event and during the period when artificially created piles of soil at the Viking Lander 1 site disappeared (Arvidson et al. 1983) suggest that winds greater than 30 m s^{-1} (at 1.6-m height) are required to generate the shear stress needed to lift dust into the atmosphere. Peak velocities at the Viking Landers are associated with short-lived gusts (chapter 26), which may explain why current versions of the coarse-resolution Mars general circulation model are unable to produce sufficient shear stress to lift dust off the surface in low latitudes (Greeley et al. 1988).

The hypothesis that local storms are generated by regional winds is favored by the observed strong lineation in existing images of local storms (Color Plate 17), and the likely occurrence of slope winds in regions preferred for local storm formation (Peterfreund and Kieffer 1979). Regional winds may arise from a variety of effects, including slopes and strong horizontal surface temperature variation (as at polar-cap edges and areas of large albedo or thermal inertia contrast). The local effect of global winds (e.g., the seasonal mass flow, Hadley-type circulations and atmospheric tides) may be important additive factors for initiating dust motion, but may be even more critical to the expansion of local storms to larger scales. The role of solar heating of airborne dust is an essential part of all models for generating dust storms. Hypotheses involving both positive and negative feedbacks between regional to global scale winds and atmospheric dust loading have been proposed (see chapter 26).

The cessation of storm activity may be initiated by the negative feedback between high-opacity conditions and the shear stress generated at the surface by winds aloft. The static stability of the atmosphere is increased as dust in the atmosphere absorbs sunlight and lowers the temperature lapse rate. Transfer of momentum from high-level winds to the surface is inhibited under such conditions. This and other negative feedback effects on surface shear stress are thought to occur when the atmospheric opacity at visible wavelengths is somewhat greater than 1 (Pollack et al. 1977,1979). Local-storm generation appears to be inhibited when the dust opacity is high, resulting in a two-stage decrease in atmospheric optical depth: the early, more rapid decrease occurs during fallout when local storm activity may be reduced, and a more gradual decrease that occurs once the regional opacity nears unity, allowing resumption of local storm activity (Pollack et al. 1979).

The hypotheses about local and global storm behavior have yet to be adequately tested by observations. Global coverage of the planet at spatial scales of a few km and at daily frequency, both prior to and during the growth phases of a global storm, as well as during the decay period, will be needed to observe the operation of these mechanisms. Quantitative understanding will require measurements of the vertical and horizontal structure of temperature, wind vector and dust abundance as well.

Atmospheric circulation patterns derived from wind streaks show global patterns and a possible correlation with the global circulation (see chapter 22). Bright depositional streaks, primarily found at low to mid latitudes, indicate a consistent north to south direction of flow; dark erosional streaks, concentrated at low southern latitudes, indicate winds mostly from east to west (Fig. 8). There is excellent correlation between the Mars General Circulation Model (GCM) wind directions for southern summer and the orientations of bright streaks (Kahn 1983; Greeley et al. 1988), except in areas of significant regional slopes such as Tharsis (Lee et al. 1982; Magalhães and Gierasch 1982). The GCM results for this season, however, are not well matched by the dark streak patterns, which may mean that dark streaks form more in response to local conditions than to the global circulation, or that the current Mars GCM runs have not adequately reproduced the cross-equatorial Hadley cell (Kahn 1983).

Formal treatment of particle settling in most models follows the method of Prandtl (1952), which accounts for viscous and dynamical drag on spher-

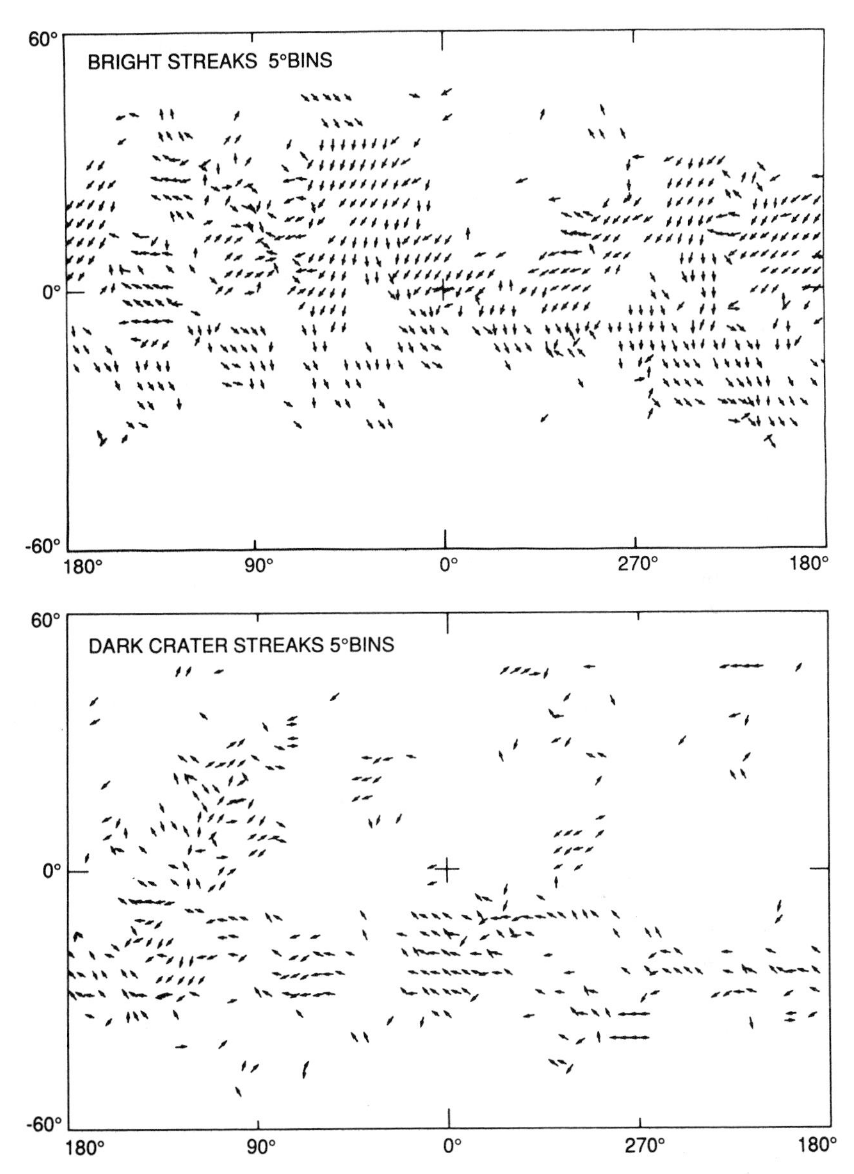

Fig. 8. Global orientations of (top) bright depositional streaks, and (bottom) dark erosional streaks. The maps indicate the streak orientations averaged over 5° latitude/longitude bins, and were derived from the Viking Orbiter medium resolution photomosaics (see Thomas et al. 1984).

ical particles falling in a gravitational field. For the thin Martian atmosphere, a correction that increases this fall speed, scaled to the ratio of the mean free path of the atmospheric gas to the particle radius, is typically applied (see, e.g., Ryan 1964; Rossow 1978). Changes in the particle fall rate due to nonspherical shape have been treated only in cases where a combination of optical and dynamical constraints were available (see e.g., Murphy et al. 1990*a*). Typical fall times for μm-sized particles are shown in Fig. 10. Available data are inadequate to provide unique measures of particle size and shape distributions, as mentioned earlier. Horizontal transport of particles is modeled according to the formalism of Bagnold (1941), and is discussed in detail in chapter 22.

VIII. REGIONAL SEDIMENT SOURCES AND SINKS

Temporal differences in regional sedimentation patterns are readily detected. For example, dramatic seasonal variability is characteristic of the Syrtis Major albedo feature (see Fig. 9). Lying on the low-albedo slopes of a vast volcanic plain, the darkest portion of the region (albedo ~0.1) is closely associated with a mass of dunes located near the crest of the shield. High thermal inertia values ($\sim 8 \times 10^{-3}$ cal cm^{-2} s$^{-1/2}$ K^{-1}) are also consistent with a sandy surface (particles coarser than ~100 μm; Kieffer et al. 1977; Jakosky 1986; see chapter 21). Syrtis Major increases in albedo immediately following global dust storms, then darkens steadily through the rest of the year until reaching its pre-storm albedo (Lee 1987; Christensen 1988). The observed trend of bright and dark streaks implies that they respond to winds generally directed upslope and to the west. The dust-transport cycle consistent with these observations is: (1) Enhanced deposition from global dust storms increases the regional albedo. (2) Relatively mobile surface material coupled with effective regional winds (possibly reinforced by the global circulation) results in removal of dust from the surface (possibly ejected by a saltation-triggering mechanism; see chapter 22). Net transport is to the west during the remainder of the year, yielding a decreasing regional albedo. Such a transport pattern may provide a source for significantly enhanced deposition in the neighboring low thermal inertia region of Arabia (Christensen 1982). (3) Occurrence of a subsequent great dust storm that begins the process again. According to this scenario, Syrtis Major acts as a dust source region through most of the year, but need not be a net source of dust when averaged over several years.

A somewhat different pattern of sedimentation is observed in the Solis Planum albedo feature. It is highly variable in extent and contrast with its surroundings, generally being most distinct (lowest albedo and highest contrast) during southern spring and summer and less distinct (higher albedo and lower contrast) during southern fall and winter. The regional thermal inertia values (~ 8 to 10×10^{-3} cal cm^{-2} s$^{-1/2}$ K^{-1}) may indicate a surface covered

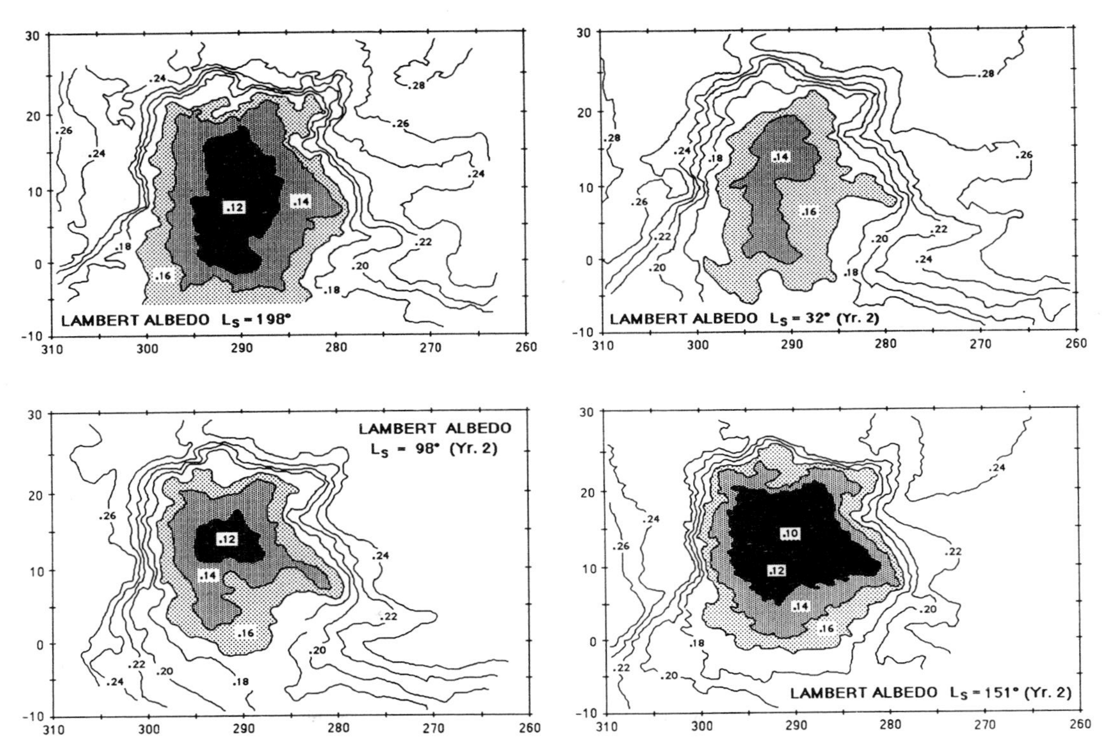

Fig. 9. Albedo maps of Syrtis Major as a function of time. Values of Lambert albedo, derived from Viking IRTM visual brightness measurements, were contoured for several time periods spanning the 1977a and 1977b global dust storms. No correction for the effects of atmospheric opacity was included in this analysis (figure from Lee 1987). Note the general brightening following the dust storms and the subsequent regional darkening.

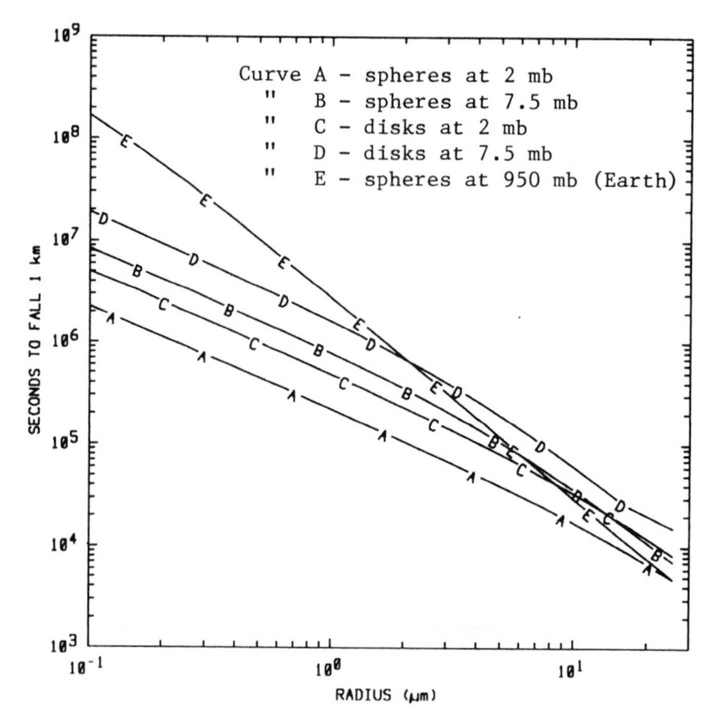

Fig. 10. Fall times of spherical and equivalent volume disk-shaped (thickness-to-diameter ratio of 0.1) dust particles at two vertical levels in the Martian atmosphere. Also shown are the fall times for spherical particles in Earth's atmosphere (figure from Murphy et al. [1990]; used with permission).

by sand-sized particles (Kieffer et al. 1977; Peterfreund 1985; Jakosky 1986; see chapter 21). Lee (1987) proposed a seasonal dust-transport cycle to explain these observations: (1) During late southern spring and summer, bright dust is eroded from these surfaces, where dust is relatively easy to mobilize, and transported from the region by local dust storms (Fig. 6). Removal of dust over a wide area results in the dark, distinct, Solis Lacus feature. (2) Following cessation of dust-storm activity, sedimentation from the atmospheric dust load occurs over the entire region, decreasing the contrast of the albedo feature with its surroundings. (3) The cycle may be renewed by local dust-storm activity the following year. The retention of some albedo features throughout the year, plus the constancy of the regional thermal inertia, requires that variations in the albedo feature not involve erosion or deposition of substantial amounts of material; cycling of at most a few tens of μm of dust is indicated. Thus, Solis Planum serves as a source of dust only during the limited period (dust-storm season) when regional winds are capable of inducing dust ejection from the surface. Contributions to the global dust load may occur only during major dust storms; differences in time of occurrence, severity and longevity of dust-storm activity may lead to the observed year-to-year variability in the Solis Lacus albedo feature.

IX. THE CURRENT SEASONAL CYCLE OF DUST ON MARS

An overview of the processes thought to be involved in the seasonal cycle of dust on Mars is given in Fig. 11. Dust can be raised from the surface into the atmosphere, as evidenced by the change of surface albedo features, by the occurrence of local, regional and global dust storms, by the measurements of sky color and of the direct solar beam as imaged from the Martian surface, and by the observations of limb hazes. The present data do not discriminate critically among the specific mechanisms that have been proposed for lifting dust. Airborne dust may be transported over short or long distances, depending upon regional meteorological conditions. Certainly, larger particles will tend to be removed more quickly by gravitational sedimenta-

SEASONAL MERIDIONAL DUST TRANSPORT ON MARS

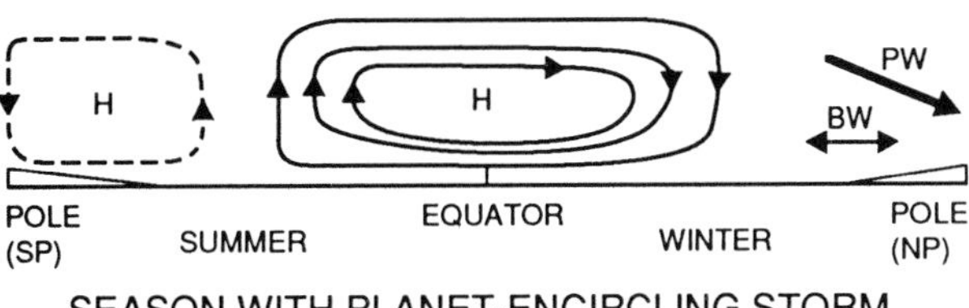

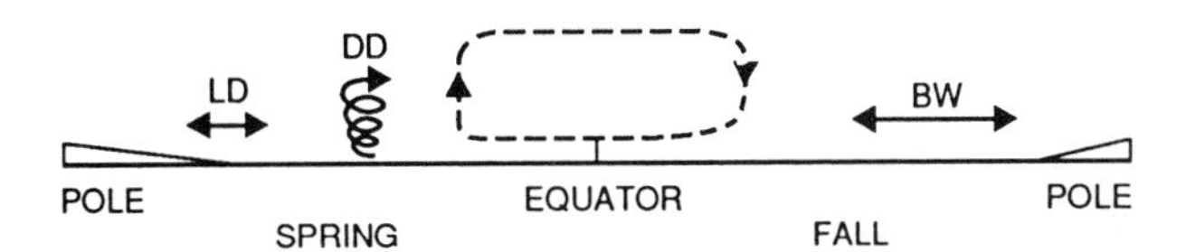

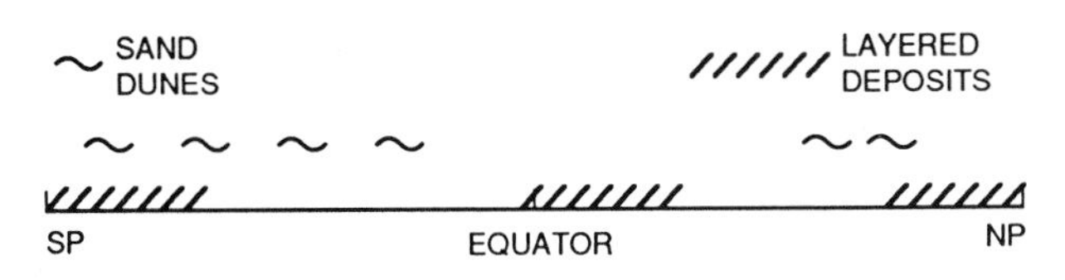

Fig. 11. Schematic diagram showing components of the meridional transport of dust during a year with a major dust storm (top) and without (middle). The cross-equatorial Hadley-type circulation *H*, the effects of local dust storms *LD*, dust devils *DD*, baroclinic waves *BW* and planetary waves *PW* are indicated. The bottom panel indicates schematically the meridional distribution of sand dunes and layered deposits.

tion, but spatially and temporally varying vertical mixing and large-scale upwelling can oppose their fall. Because horizontal wind speed tends to increase with height, the higher the dust is raised, the farther it is likely to travel horizontally.

The spatial distributions of dust storms and of changes in surface albedo suggest that dust can be removed from the surface almost anywhere on the planet, including the high slopes (and low surface pressure) of the Tharsis and Elysium volcanoes (Lee et al. 1982). There may be long periods when little dust moves at all at some locations (e.g., Viking Lander 1 site), while there appear to be other regions where dust is raised more frequently. This is consistent with our terrestrial expectation that dust raising should occur most frequently where there are strong near-surface winds or convection, or a favorable combination of winds and of movable particles. The apparent preference for both large and small dust storms to occur in the southern subtropics and of local dust storms to form near the edges of the retreating seasonal polar caps suggests that these regions may be net sources of dust to the atmosphere.

It is important to remember that the amount of dust needed to account for the observed atmospheric haze and the changes in surface albedo is relatively small—equivalent to a surface layer of a few μm even for the largest storms. The data in hand are not adequate to tell us diagnostically whether there has been net accumulation or loss of dust at any given place on Mars in recent times. The main observational evidence that there is a net transfer between regions on the planet today is provided by the observed evolution and seasonality of the largest dust storms. When a large dust storm occurs, dust spreads nonuniformly over large distances. The sense of this transport for planet-encircling storms is that dust will be moved about within the southern hemisphere or from the south to the north over several weeks. As the largest dust storms have been observed only during southern spring and summer, there is no equivalent dust storm near the northern summer solstice to reverse this cross-equatorial transport. (Thus, the labeling of the summer and winter poles in the dust-storm transport panel of Fig. 11.) However, it is possible that other dust transport mechanisms, affecting smaller quantities of dust at any one time but operating more frequently, could reverse over several years the net dust transports of the largest storms.

If some regions are perpetual sources or sinks, losing or gaining the equivalent of a few μm a year, meters of material can be removed or accumulated on the time scales over which the orbital elements of Mars change ($\approx$1 Myr). Two major net sinks of dust have been proposed on theoretical grounds. One sink involves transporting the airborne dust into the north polar region, where it may become incorporated into the residual polar cap (Pollack et al. 1979). While the Hadley-type circulation appears to be incapable of doing this directly (Haberle et al. 1982; Zurek and Haberle 1988), there is evidence that baroclinic waves are transporting dust poleward (Tillman et al.

1979). Furthermore, the polar warmings that occurred during each of the two 1977 dust storms (Jakosky and T. Martin 1987) may be caused by an amplified planetary wave that should efficiently transport dust as well as heat into the polar regions (Barnes 1990*b*). However, there is little indication that airborne dust is present in the north polar region at the altitudes sampled by the Viking IRTM (Jakosky and T. Martin 1987), and the polar cap albedo seems remarkably high for dirty water ice (Kieffer 1990). The south polar region is assumed to be a less likely sink for dust in the current epoch because great dust storms are not known to occur in the season when the south polar cap is condensing, and because surface features suggest that net erosion is currently taking place there (see chapter 23).

The mechanism proposed for the second sink is based on the assumption that the observed darkening of regions after the dissipation of the largest storms is accomplished by the removal of dust raised locally and moved into neighboring regions. Christensen (1986*a*) argues that in the case of Syrtis Major, dust is moved into Arabia and Tharsis, *where it remains,* in part because their surfaces are not as rough or conducive to saltation as in Syrtis Major, so that deposited dust is not as readily re-mobilized. According to this scenario, dust is transported by global dust storms into the northern subtropics, where it is redistributed longitudinally into the bright, low-thermal-inertia regions. Whether there is actual net accumulation of dust in these regions at the present time cannot be determined from existing observations.

Dust can be transported meridionally even in years without a nonlocal dust storm (Fig. 11). Baroclinic wave activity, which is capable of effecting meridional dust transport (Tillman et al. 1979), was more prominent at the Viking Lander 1 site during fall and winter of the years without planet-encircling dust storms (Leovy et al. 1985). Dust raised at the polar-cap edges in spring by local dust storms, near subsolar latitudes in summer by dust devils, and elsewhere by local dust storms, can be moved about by the general circulation patterns. At present, local dust-storm activity seems to be greater in the southern hemisphere near the cap edge and in the subtropics. Although this may be an artifact of the incomplete observational coverage, it is possible that part of the background dust haze seen above the Viking Landers may indeed come from the southern hemisphere.

In summary, the current seasonal cycle of dust on Mars consists of the redistribution about the planet of the equivalent of a dust veneer a few μm thick. As indicated by the observations of local dust storms, dust can be raised over much of the planet, particularly in the southern low latitudes and near the edges of the seasonal polar caps in both hemispheres. Dust spreads over a large part of the planet in just several weeks during planet-encircling dust storms, but changes in surface albedo patterns indicate that dust transport also occurs in seasons and years without such large storms. Deposition of dust cycled through the atmosphere, and subsequent local redistribution of the thin deposits by near-surface winds, can explain the observed patterns and variability of large-scale albedo features, including the disappearance and re-emergence of the classical dark albedo features. This can be accomplished without net accumulation anywhere on the planet. On theoretical grounds, the two most likely long-term sinks of dust during the current epoch are the bright, low-thermal-inertia regions in the northern low latitudes, and the northern polar residual cap. In recent years, the occurrence of several planet-encircling storms suggests that a net transfer of dust has occurred from the southern hemisphere to the north, consistent with the proposed dust sinks. Whether this trend is reversed or augmented by less dramatic, but perhaps more persistent, transport mechanisms is not known.

X. LONG-TERM ASPECTS OF THE DUST CYCLE

There is no evidence from existing data that the current cycling of Martian dust involves *net* deposition on time scales ranging over several years. However, net transport of dust into or out of regions could yield geologically significant accumulation or loss of surface material; such activity may well determine the geomorphologic evolution and character of much of the Martian surface. The presence of polar-layered terrains distributed fairly symmetrically about the geographic poles, as well as low-latitude layered terrains and other debris deposits, suggests that net deposition of airborne material is very likely either on longer time scales or under earlier conditions.

A schematic representation of sources, sinks, reservoirs and processes that may be involved in the long-term dust budget for Mars is given in Fig. 12. The areal extent of many of the potential dust surface reservoirs has been obtained from spacecraft data (see chapters 21, 22 and 23); to calculate the volumes, the vertical extent is required as well, but this is notoriously difficult to estimate with remote-sensing observations. For this reason, attention has focused on the polar layered terrains, where fairly strong constraints on the volume and the age of these units have been obtained (this subject is reviewed in chapter 23). Discussion earlier (see Secs. VI and IX) has highlighted the size and variability of the atmospheric dust reservoir.

Thinking about the interplay among dust reservoirs has been dominated by the dramatic climatic effects of global dust storms in the current epoch. Perhaps there is a relationship between the location and timing of major dust-storm events and the presence of regional sediment deposits. As noted previously, the occurrence of great dust storms is apparently related to the timing of Mars' perihelion. The season of perihelion follows the ~51,000-yr Martian obliquity cycle; the hemisphere receiving maximum insolation (currently the southern hemisphere) will reverse every ~25,000 yr (see chapter 9). It has been suggested that when maximum insolation has shifted to the northern hemisphere, the regions that currently appear to contain significant wind-blown sediment deposits (Tharsis, Arabia and Elysium—all in the northern hemisphere) will serve as source regions for dust storms (Christensen

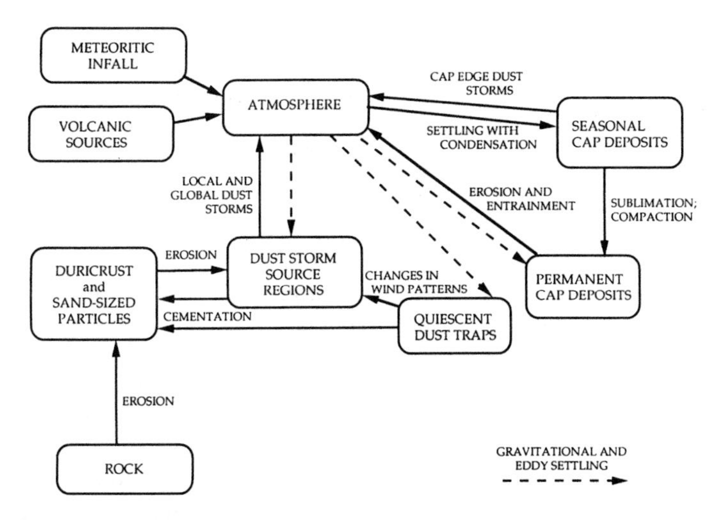

Fig. 12. Schematic diagram outlining the components of the Martian dust cycle.

1986*a,b*; chapter 33), and the south pole, rather than the north pole, may serve as a sink for dust raised into the atmosphere.

Viking observations of the 1977 dust storms led Pollack et al. (1979) to estimate that the observed atmospheric dust loading leads to a globally averaged sedimentation rate of $\sim 2 \times 10^{-3}$ g cm^{-2} yr^{-1} (≈ 7 μm yr^{-1}) for that year. Assuming that dust grains act as condensation nuclei for water and CO_2, the sedimentation rate could be significantly enhanced in the polar regions, yielding polar deposition rates that are very rapid on geologic time scales, comparable to the rate required to deposit one of the observed layers in the 25,000 yr allowed by the primary oscillation in the planetary obliquity (see chapter 23). Assuming that the source regions cover about 10% of the planetary surface, that the 1977 sedimentation rate was representative of all years, and that there is significant preferential deposition of dust at the poles, Pollack et al. (1979) derived erosion rates of ~ 70 m Myr^{-1} at the sources. Christensen (1986*a*) noted that the rough surface seen in dual-polarization radar measurements and the thermal inertia values derived from the Viking IRTM indicate that the bright, low-thermal-inertia areas of Arabia and around Tharsis and Elysium were consistent with a dust mantle 0.1 to 2 m thick. Assuming that these accumulated over 10^5 to 10^6 yr yields annual average dust deposition rates of 0.1 to 20 μm yr^{-1}.

Such a high average rate of mid- to low-latitude dust accumulation implies behavior different from that observed currently at the Viking Lander sites (Arvidson et al. 1979). The suggested high polar net deposition rate, if it occurs on the south pole, is inconsistent with the preservation of small craters that are found on the south polar layered terrain (Plaut et al. 1988). Currently observed regional albedo variations described above may be accomplished without net deposition of dust, and the occurrence of very large dust storms themselves appears to be intermittent. Taken together, these observations suggest that either the dust cycle has operated differently in the past, or that there are elements missing in our view of current dust behavior. Both are likely.

XI. SUMMARY

Dust is of great significance to the current climate of Mars, and there are strong indications that dust has had a long-term effect on both the climate and the surficial geology. When suspended, dust alters the atmospheric thermal structure and circulation and affects our ability to observe the planet remotely. The composition, size and shape of dust particles raised into the atmosphere and redeposited onto the surface appear to vary spatially and with time; certainly, the opacity of the atmosphere is variable in all these dimensions. Many of the parameters controlling the location, severity and frequency of dust-raising activity are not well constrained or understood.

The present seasonal cycle of dust appears to consist of seasonally modulated dust-raising, local redistribution, and regional exchange in years without major dust storms. During the planet-encircling storms, there must be net exchange between regions and, during the present epoch, possibly a net annual transport of dust from the southern to the northern hemisphere. Whether this transport is compensated by any exchange occurring during years without major storms remains to be determined. Local and regional redistribution of dust following the major dust storms accounts for the pattern of surface wind streaks and the re-emergence of the classical dark features of Mars; dust may also be lost into possible long-term sinks in the northern low latitudes and in the north polar region during the current epoch. All observed changes in surface albedo are consistent with the seasonal and interannual redistribution of very small amounts of dust, equivalent to a layer a few μm in depth; no quantitative evidence exists requiring any net regional redistribution of dust on the planet on seasonal or decadal time scales. Further progress in understanding the seasonal cycle of dust on Mars will require more systematic and detailed observation. Reliable extrapolation from the present seasonal dust-storm cycle to longer climatic cycles requires more observational data than are now available.

The most powerful constraints on the long-term cycle of dust on Mars are integral constraints on net deposition and loss provided by the volumes and ages of dust reservoirs. Much has been learned about climate history of Earth from the record of sedimentation processes; the observations of polar lamina and layering in terrains elsewhere on Mars imply that similar information may be available there. In the process of further constraining the dust cycle of Mars, we will also shed some light on the history of variations in the climate of the planet. More work on these questions is possible, even with existing data, and particularly once high-resolution images of the surface are available. Detailed mapping of sedimentary deposits may further constrain the volumes and ages of potential surface reservoirs of dust, and together with any additional information, the rates at which dust is formed, carried around the planet, and removed from the dust cycle.

Acknowledgments. The work of R. K., T. Z. M. and R.W. Z. is supported by the Planetary Atmospheres Program of the National Aeronautics and Space Administration, under contract with the Jet Propulsion Laboratory/ California Institute of Technology. The work of S. W. L. is supported by the Planetary Geology Program of the National Aeronautics and Space Administration.

30. AERONOMY OF THE CURRENT MARTIAN ATMOSPHERE

C. A. BARTH, A. I. F. STEWART
University of Colorado

S. W. BOUGHER, D. M. HUNTEN
University of Arizona

S. J. BAUER
University of Graz

and

A. F. NAGY
University of Michigan

The Mars atmosphere consists of a troposphere, an isothermal mesosphere, and a thermosphere where the daytime temperature at the top of the thermosphere may vary depending on the time in the solar cycle. Photochemistry occurs throughout the entire Mars atmosphere, from the top of the thermosphere to the bottom of the troposphere. The photodissociation of the major constituent carbon dioxide produces carbon monoxide and atomic oxygen. The recombination of CO and O takes place in the troposphere through catalytic reactions involving the odd-hydrogen species, H, OH and HO_2. Observational tests that confirm the general validity of the odd-hydrogen model are: (1) the observed abundances of molecular oxygen and carbon monoxide; (2) the observed seasonal variation of ozone over the polar caps of Mars; and (3) the observed density of atomic hydrogen in the thermosphere. Odd nitrogen is produced in the upper atmosphere from the reaction of $N(^2D)$ and CO_2. Viking mass spectrometers have measured neutral nitric oxide molecules in the Mars atmosphere. Photochemical processes control the behavior of the ionosphere of Mars. The prin-

cipal ion is O_2^+ *with smaller amounts of* CO_2^+ *and* O^+ *at the ionospheric peak near 135 km. The density of the* O^+ *ion becomes comparable to the density of the* O_2^+ *ion near the top of the atmosphere. The most intense emissions in the ultraviolet airglow of Mars are the carbon monoxide Cameron bands. Strong ultraviolet emissions from* CO_2^+ *are the result of photoionization of carbon dioxide. The atomic oxygen airglow at 1304 Å is used to determine the density of atomic oxygen and the 1216 Å Lyman* α *line is used to calculate the density of atomic hydrogen and, when coupled with the temperature measurement, the escape flux of atomic hydrogen. The most intense airglow is the infrared atmospheric band of* O_2 *at 1.27* μm *that results from the photodissociation of ozone. The escape mechanism for atomic hydrogen is thermal, or Jeans, escape while atomic oxygen escape is caused by a nonthermal process, namely, the dissociative recombination of* O_2^+*. The ratio of deuterium to hydrogen is enriched by a factor of 6. This observation may be used to study the past history of water on Mars. Three-dimensional models of the Mars thermospheric circulation show that planetary rotation has a significant influence on the wind, composition and temperature structure. There is upward flow on the day side, downward flow on the night side. Observations of the thermospheric temperatures near solar minimum from Viking showed a mean dayside temperature of 195 K and near solar maximum from Mariners 6 and 7, a value of 310 K. The challenge for the future is to make observations to study the diurnal, seasonal and solar cycle changes that occur in the Mars atmosphere.*

The Mars atmosphere is composed principally of carbon dioxide with small amounts of molecular nitrogen (2.7%) and argon (1.6%). The atmospheric pressure at the surface is low, $\sim$ 6 mbar, and it varies with season and with the topography. The surface temperature is also low, a mean temperature of 220 K, and it varies with time of day, season and location on the planet from a minimum near 147 K in the polar night to a maximum around 300 K near the subsolar point at perihelion.

Mars aeronomy includes those physical and chemical processes that occur in the Martian atmosphere from the surface up to the base of the exosphere (200 to 250 km). Photodissociation of carbon dioxide, the major constituent, occurs throughout the entire atmosphere all the way down to the surface. Water vapor is seasonally injected into and removed from the atmosphere. Chemical reactions between the various odd-hydrogen, odd-oxygen and odd-nitrogen species take place throughout the atmosphere and with the surface as well. Atomic hydrogen, atomic oxygen and atomic nitrogen all escape from the top of the atmosphere. Photoionization and ionic reactions occur in the atmosphere above 100 km in the ionosphere. Chemical and ionic reactions emit radiation (the airglow) which may be observed from planetary spacecraft or Earth-orbiting or Earth-based telescopes. The diurnal and seasonal variations in the photodissociation, photoionization and heating processes in the upper atmosphere lead to a thermospheric circulation.

Observations and *in situ* measurements relevant to the aeronomy of Mars have been made by a number of planetary missions: Mariners 6 and 7 in 1969, Mariner 9 in 1971, Mars 4 and 5 in 1973, Vikings 1 and 2 in 1976, and Phobos in 1989 (see chapter 3). These spacecraft observations and some

Earth-based telescopic observations form the basis of our knowledge of aeronomy on Mars.

This chapter first discusses the thermal structure of the atmosphere which varies diurnally, seasonally and episodically. Initially, the thermal structure is described in terms of its one-dimensional structure. Later, in this chapter, three-dimensional models of the Mars upper atmosphere (thermosphere) are discussed. An extensive section on photochemistry follows the thermal structure section. Photochemistry alters the composition of the atmosphere and leads to the creation of an ionosphere which is discussed next. Photochemistry also leads to the excitation of an airglow which alters the energy balance of the atmosphere and serves as a tool for remote sensing. There is a section on atmospheric escape which describes present-day processes that are operating on Mars and speculates about the past evolution of the atmosphere. Finally, there are suggestions as to the future course of research on Mars aeronomy.

I. THERMAL STRUCTURE

According to its thermal structure, the Mars atmosphere may be divided into a troposphere (lapse rate of 2 to 3 K km^{-1}, compared to the adiabatic lapse rate of 4.5 K km^{-1}), an approximately isothermal mesosphere and a thermosphere. The troposphere may be from 20 to 50 km deep, and temperatures at the surface may vary from 300 K near the subsolar point at perihelion to about 147 K (the 6 mbar frost point of CO_2) over the seasonal polar caps. The thermal structure of the atmosphere is discussed more extensively in chapter 26. Large dust storms can produce approximately isothermal layers that extend from the surface to as high as 50 km, greatly expanding the entire atmosphere. The role of dust in the atmosphere is described in chapter 29. The lower boundary of the thermosphere occurs near or above 100 km.

There is considerable evidence for planetary and orographic waves in the troposphere and for vertically propagating gravity waves throughout the atmosphere, at least at some seasons. There is also evidence that tropospheric and mesospheric temperatures depend noticeably on heliocentric distance (because of Mars' significant eccentricity, insolation varies by $\pm 20\%$ around its orbit). Finally, there is evidence that thermospheric temperatures also depend on heliocentric distance, but their dependence on short-term solar extreme ultraviolet (EUV) variations is less clear. Figure 1 shows vertical temperature profiles for a variety of conditions. Included are two profiles from entry-science experiments of Viking Landers 1 and 2, for which the solar longitude $L_s \simeq 106$, and the heliocentric distance $r \simeq 1.63$ AU (Seiff and Kirk 1977). Also shown is a Mariner 9 radio-occultation profile (Kliore et al. 1972) for a dust-filled lower atmosphere (Orbiter revolution 9) and a theoretical thermospheric profile that is consistent with early Mariner 9 airglow measurements (revolution 34; Stewart et al. 1972). Mariner 9 measurements were made with $L_s \simeq 310$, $r \simeq 1.44$ AU.

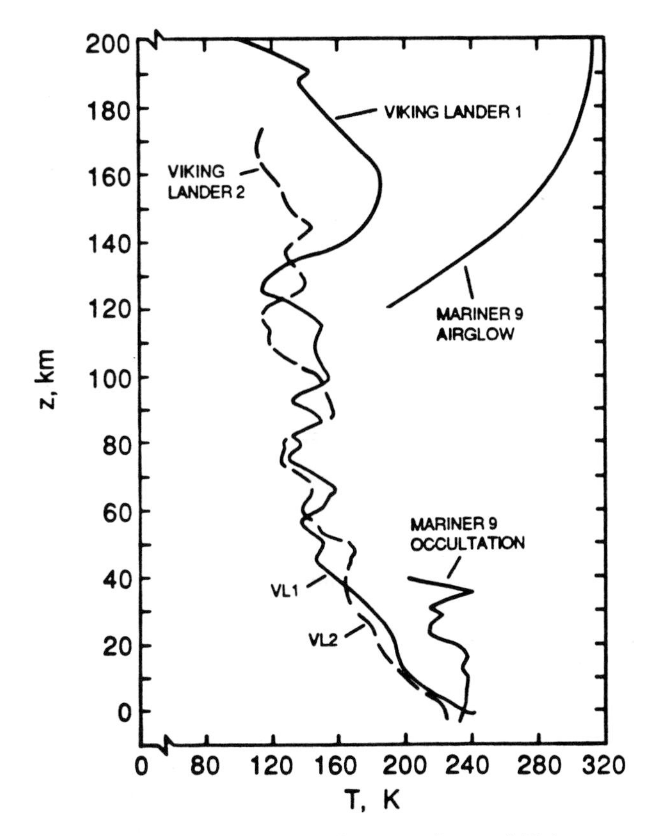

Fig. 1. Temperature profiles. The Viking entry science experiments yielded temperature/pressure profiles from the exobase to the surface, under clear-atmosphere, solar minimum conditions. Mariner 9 radio occultations provided profiles in a dusty lower atmosphere, and ultraviolet airglow measurements yielded thermospheric profiles at moderate solar activity.

Theoretical radiative-convective and general circulation models reproduce the general features of the thermal structure in the troposphere and mesosphere quite successfully, and they predict seasonal, local time and latitudinal variations. In the thermosphere, the observed temperatures were unexpectedly low, and the reasons for these low temperatures are still not fully understood. To model the observations, it has been necessary to assume that the efficiency of conversion of solar EUV radiation to heat is very low (15 to 20%, compared to 30 to 40% on Earth) and that non-LTE cooling in the 15-μm band of CO_2 is efficiently stimulated by collisions with ambient atomic oxygen and by excitation transfer from metastable atomic oxygen (Stewart 1972; Bittner and Fricke 1987; Bougher and Dickinson 1988). Support for low heating efficiency comes from the large intensity of the ultraviolet airglow.

The Mars thermosphere is the region of the upper atmosphere above approximately 100 km where the temperature increases with increasing altitude. Temporal variations in thermospheric temperatures are not well understood. Missions at or near solar minima (Mariner 4, Viking) have yielded temperatures (more strictly, plasma scale heights) that are about a factor of 2 smaller than those obtained by missions at or near solar maxima (Mariners 6, 7 and 9), and this long-term effect can be successfully modeled as arising from variations in insolation. Day-to-day variations within a given mission (Mariner 9, Viking), however, are not correlated with solar activity, as would be expected by analogy with the Earth. This has led to the suggestion that day-to-day temperatures are strongly affected by gravity-wave activity such as that seen in the Viking entry profiles. The circulation of the Mars thermosphere is described in Sec. VI.B.

II. PHOTOCHEMISTRY

Photochemistry occurs throughout the entire Mars atmosphere. Solar ultraviolet radiation penetrates all the way to the surface photodissociating the major constituent of the atmosphere, carbon dioxide, which produces carbon monoxide and atomic oxygen. The photochemistry is complicated by the episodic injection of water vapor into the atmosphere. The presence of molecular nitrogen introduces odd-nitrogen photochemistry.

A. CO_2 Photolysis and Transport

Photolysis of CO_2 has a threshold at 2275 Å, where there is a weak (spin-forbidden) absorption yielding ground-state $CO(X^1\Sigma^+)$ and $O(^3P)$. A stronger, permitted absorption yielding metastable $O(^1D)$ begins at 1671 Å, and the O may be in the (^1S) state at wavelengths below 1288 Å. Triplet metastable CO appears at 1081 Å; the lowest-lying triplet state is $(a^3\Pi)$, the upper level of the important Cameron bands. The threshold for photoionization is at 899 Å, and this channel quickly dominates dissociation. The dissociation channels predominantly yield an ionized fragment below 650 Å.

Although the continuum that produces ground-state CO and O is weak, it dominates throughout most of the middle and lower atmosphere because of the larger solar fluxes longward of 1671 Å, and it is responsible for about 90% of all "odd-oxygen" production in the entire atmosphere. The remaining 10% is approximately evenly shared by the CO + $O(^1D)$ continuum, which peaks in the upper mesosphere, and the lowest-energy continuum of O_2, which is only important in the first scale height above the surface. The planet-averaged column production rate of O and CO is $\sim 2.5 \times 10^{12}$ cm^{-2} s^{-1}, of which about 1% occurs in the thermosphere above 100 km.

Figure 2 shows a typical calculation of the vertical distribution of the photolysis processes described above. Table I gives values of the dissociation coefficients above the Martian atmosphere (for r = 1.52 AU) and for solar minimum (Levine 1985). At solar maximum the solar EUV may be a factor of $\sim$ 3 more intense (Torr et al. 1979; Torr and Torr 1985) and the shorter-wavelength coefficients will reflect this. The table also contains the approxi-

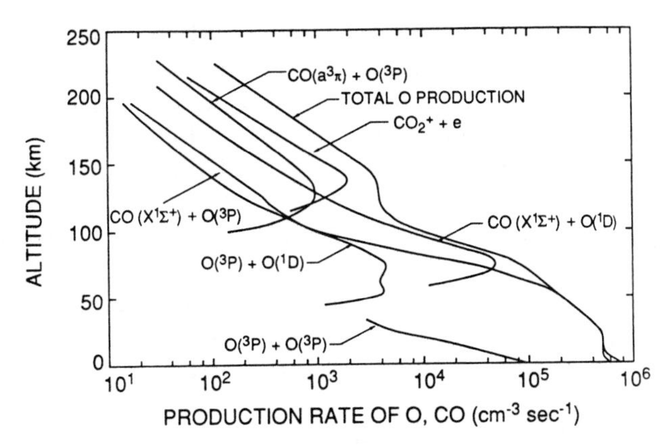

Fig.2. Production of CO and O by photolysis: planetary-averaged altitude profiles for the major photodissociation processes. The recombination of CO_2^+ leads to varying amounts of singlet and triplet CO and triplet and singlet O depending on whether the ion recombines dissociatively or reacts with O to form O_2^+ (figure from McElroy and McConnell 1971).

mate planet-averaged column photolysis rates above the surface and above 100 km.

B. The CO_2-H_2O-O_3 System

Essentially all the CO and O produced in the thermosphere is swept down to lower altitudes. Although much of the O is converted to O_2 in the middle atmosphere, further recombination is limited to the bottom few tens of km. As discussed below, this recombination is catalyzed by odd hydrogen so efficiently that the mixing ratios of O_2 and CO are maintained at a very low level, around one part per thousand (chapter 25). Ozone is much rarer

TABLE I
Production of CO and O by Photolysis[a]

Process		Threshold Å	J_∞ s^{-1}	Column Production: Above Surface $cm^{-2}s^{-1}$	Above 100 km $cm^{-2}s^{-1}$	Peak Production km
$CO_2 + h\nu$	$\rightarrow CO(X^1\Sigma^+) + O(^3P)$	2275	7.3(−9)	2.3(12)	1.0(9)	0—40
	$\rightarrow CO(X^1\Sigma^+) + O(^1D)$	1671	4.0(−7)	1.3(11)	1.0(10)	60—80
	$\rightarrow CO(a^3\Pi) + O(^3P)$	1082	1.2(−7)	3.0(9)	3.0(9)	120—150
	$\rightarrow CO_2^+ + e$	899	3.5(−7)	7.0(9)	7.0(9)	120—150
$O_2 + h\nu$	$\rightarrow O(^3P) + O(^3P)$	2424	2.6(−8)	8.0(10)	1.5(7)	0—10
	$\rightarrow O(^3P) + O(^1D)$	1759	1.8(−6)	1.5(10)	1.0(9)	50—80
Total				2.5(12)	2.2(10)	

[a]The dissociation coefficients J are given for the top of the atmosphere at r = 1.52 AU under solar minimum conditions. At solar maximum, the coefficients for processes with thresholds below Lyman α may be up to a factor of 3 larger. Numbers within parentheses in the three central columns are powers of 10.

yet, being undetectable except in polar regions, which during the fall-winter season are exceptionally deficient in odd hydrogen. If it were not for the odd-hydrogen catalytic chemistry, the atmosphere would be entirely different, with CO and O_2 mixing ratios in the range of 1 to 10%.

Odd hydrogen consists of the radicals H, OH and HO_2 which are conveniently treated as a unit because they are involved in a rapid chemical cycle that sets their relative abundances under any set of daytime conditions. The actual oxidation of CO and organic molecules is carried out by the OH. Production and loss of odd H occur on much longer time scales and can therefore be treated separately. The system is illustrated in Fig. 3; earlier versions of this diagram (Hunten 1974,1979*b*) show H_2O instead of H_2 in the topmost box. Both sources do in fact exist, but the one involving H_2 is more important. The principal source of odd hydrogen is photolysis of water vapor, the subject of the next section. Oxidants in the same family rapidly destroy any organic molecules volatile enough to reside in the atmosphere, and these oxidants also permeate the soil.

1. Water Vapor. As discussed in chapter 28, water vapor follows a seasonal cycle that is basically similar to that of the Earth, although there is no analog of the humid equatorial regions. At higher latitudes, the very low winter temperatures freeze out the water, which, however, becomes much

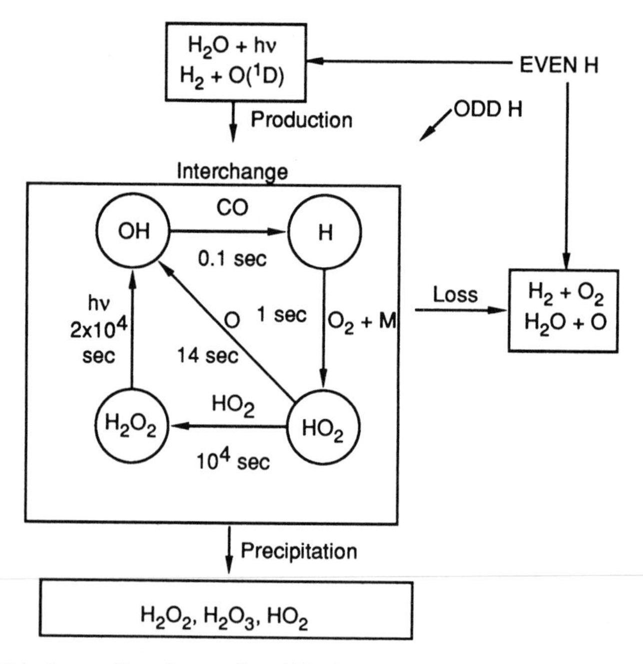

Fig.3. Odd hydrogen: flow diagram for odd hydrogen near the surface of Mars. Beside each arrow is the reaction time in seconds (figure from Chamberlain and Hunten 1987 adapted from Hunten 1979*b*).

more abundant in midsummer. An analysis of Viking Orbiter data indicates that the water vapor over the Viking 1 landing site is uniformly mixed with the atmosphere (Davies 1979*c*). For locations elsewhere on the planet and at different times, the best that can be done is to take a temperature profile, either measured or calculated, and assume that the water vapor number density at each height does not exceed the saturation vapor pressure.

Estimates along this line were made by Hunten and McElroy (1970), using temperature profiles calculated by Gierasch and Goody (1968). The results are shown in Fig. 4 for dawn and sunset at equatorial and midwinter at 45° latitude. The dashed line, with the label "observed," corresponds to an abundance of 15-μm precipitable water vapor, arbitrarily taken to be a uniform layer 5-km deep. The observations in question were obtained from Earth, but were well confirmed by the Viking MAWD (Mars Atmospheric Water Detector) experiments (Farmer et al. 1977). An independent evaluation, by Kong and McElroy (1977*b*), is very similar, with an abundance of 12.5 μm. Their adopted distribution was uniform up to 4 km, dropping off at greater altitudes according to the vapor-pressure curve. In a second paper, Kong and McElroy (1977*a*) used a latitude-dependent distribution, illustrated in Fig. 5.

The dissociation rate per molecule, and the corresponding production rate of odd hydrogen, depend strongly on altitude, because the relevant radiation is absorbed by CO_2. According to Hunten and McElroy (1970), its surface value with the Sun at the zenith is $1.3 \times 10^{-10}\ s^{-1}$, reduced by a factor of 4.3×10^4 from its value high in the atmosphere. With the Sun at a 60° zenith angle, it is less by another factor of 2.8. Because of this absorption,

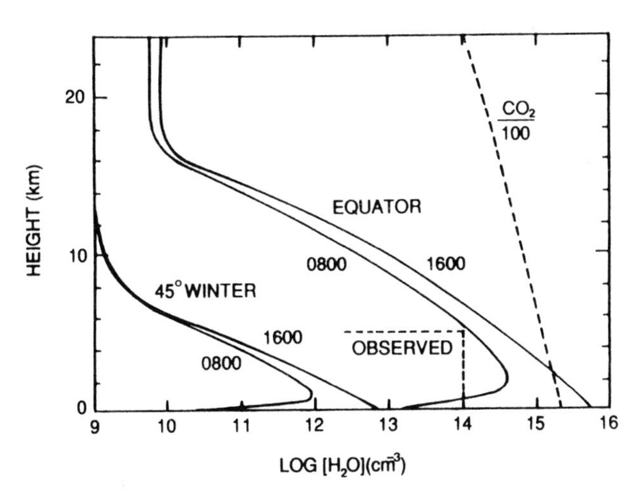

Fig.4. Vapor densities of H_2O: saturation vapor densities of H_2O calculated for several of the temperature profiles given by Gierasch and Goody (1968). Also shown is a typical observed amount, 15 μm precipitable or $5 \times 10^{19}\ cm^{-2}$, arbitrarily distributed in a uniform layer 5 km deep (figure from Hunten and McElroy 1970).

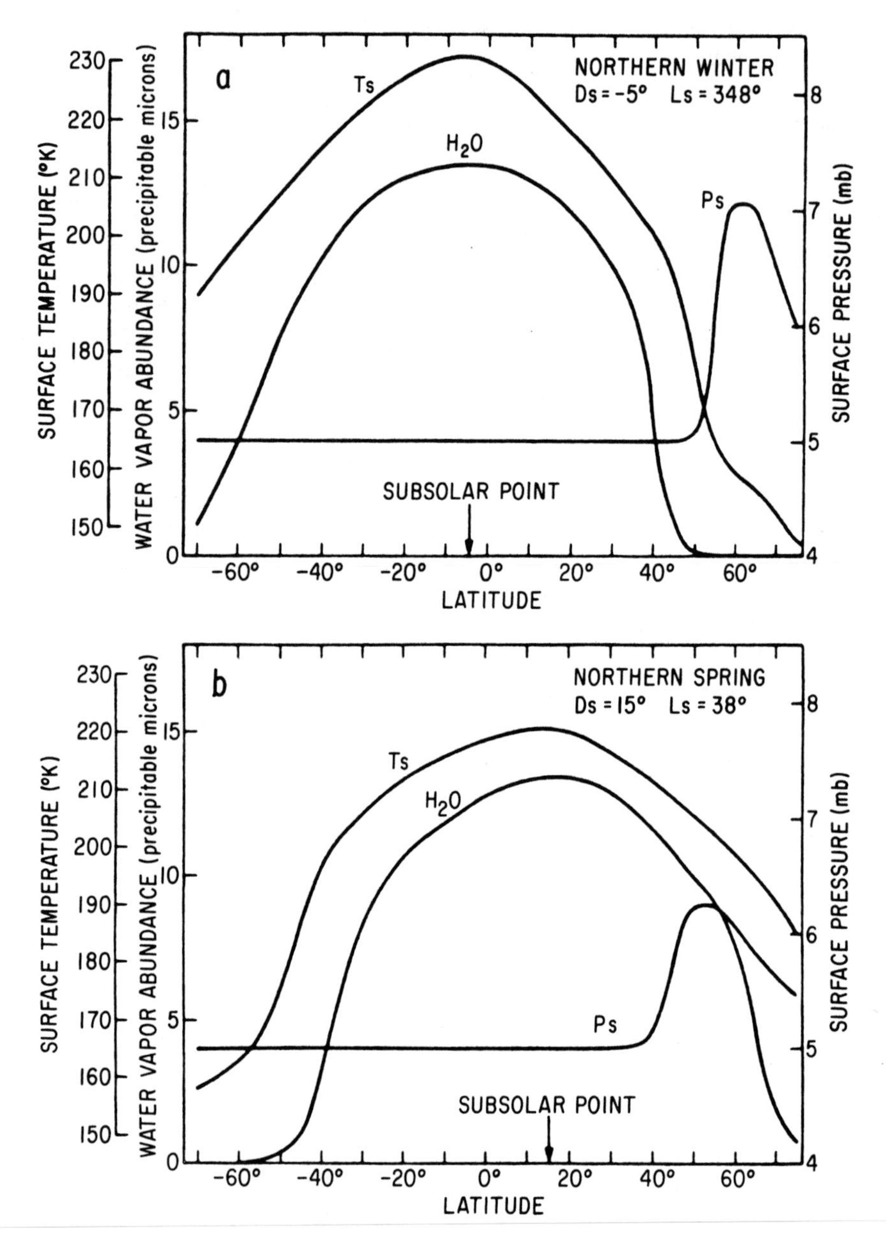

Fig.5.(a) Water vapor distribution during northern winter: surface temperatures (T_s), surface pressures (P_s), and integrated water vapor abundances for Mars' northern winter season with $D_S = -5°$ and $L_s = 348°$. Note that P_s does not reflect current knowledge of global topography (see chapter 10). (b) Water vapor distribution during northern spring with $D_S = 15°$ and $L_s = 38°$ (figure from Kong and McElroy 1977*a*).

the production of odd hydrogen depends critically on the vertical distribution of water vapor, it is not enough to know the geographical distribution of total column abundance.

2. Molecular Hydrogen. The second major source of odd hydrogen is the reaction of $O(^1D)$ with H_2, important because it still functions when all the water vapor is frozen out. $O(^1D)$ is a metastable excited state of the oxygen atom, produced in every photolysis of ozone. It is quenched on essentially every collision, but if the collision partner is H_2 or H_2O, the result is a reaction yielding H + OH or two OH radicals. H_2 is a product of the odd-hydrogen chemistry (right-hand box in Fig. 3), but its abundance has never been measured. According to Kong and McElroy (1977*b*), the mixing ratio is 8.3 ppm (parts per million by volume), and this figure is consistent with the H densities measured in the upper atmosphere, as discussed below.

3. Odd Hydrogen. Contours of production and loss rates are shown on a latitude-height grid for two seasons by Kong and McElroy (1977*a*); see Fig. 6. At low latitudes, the production rate (cm^{-3}) is a little over 20,000 at the surface, 5000 at 8 km, and 200 at 15 km; the height integral is about 10^{10} cm^{-2} s^{-1}. The rates become much smaller poleward of 50° in fall and winter. The losses extend over a much greater height range, with a broad peak somewhat above 20 km.

The large box in Fig. 3 shows the rapid cycling among the three forms of odd hydrogen. Beside each arrow is shown the reaction partner and an estimate of the reaction time (sec) in daytime near the surface. The slowest reaction is $HO_2 + O \rightarrow OH + O_2$; its 14-s time is the time-limiting step for the whole cycle. Hydrogen peroxide H_2O_2 is shown in the odd-hydrogen box because it photolyzes to a pair of OH radicals in a few hours of daylight. It is produced by reaction of a pair of HO_2 radicals.

The net effect of the fast cycle is $CO + O \rightarrow CO_2$, with no effect on the amount of odd hydrogen; however, it is limited by the supply of O atoms. The outer cycle is equivalent to $2CO + O_2 \rightarrow 2CO_2$, with O atoms supplied by the photolysis of H_2O_2. Unfortunately, the amount of this molecule cannot be confidently calculated, for several reasons: (1) the time constants for its production and photolysis are around half the duration of daylight; (2) at night the photolysis is shut off, and much of the odd hydrogen is converted to the peroxide; (3) the peroxide condenses roughly as easily as water vapor, and some of it presumably deposits on the ground, especially at night or in winter. All these effects were modeled by Kong and McElroy (1977*a*), but such modeling necessarily involves many simplifying assumptions. Hunten (1974) discusses, in a qualitative way, some of the interesting effects that could arise, especially when a winter's accumulation begins to be released.

Loss of odd hydrogen occurs in two ways: conversion back to even hydrogen or deposition on the surface. Generation of even hydrogen requires

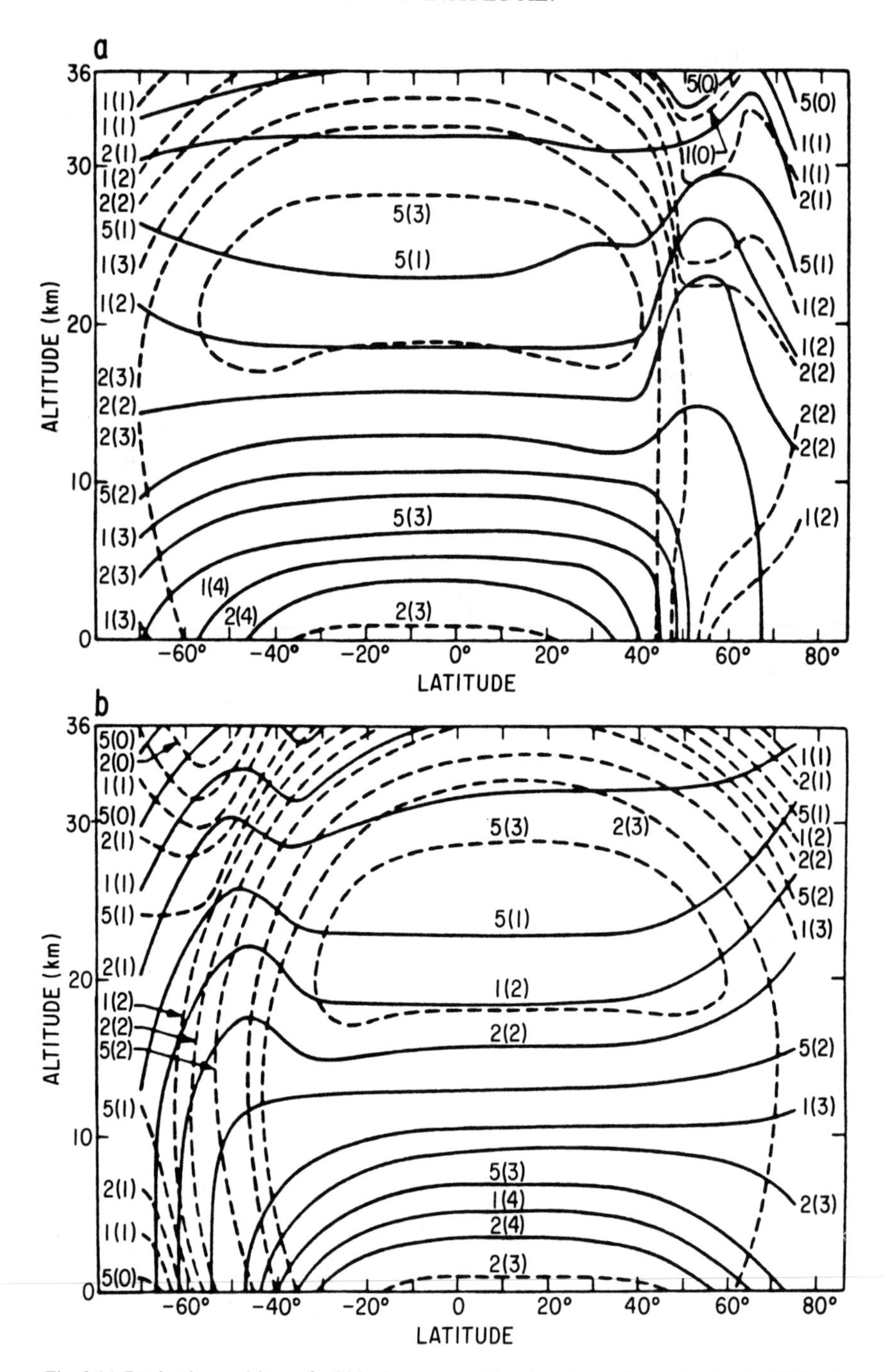

Fig. 6. (a) Production and loss of odd hydrogen: meridional contours for production (solid lines) and loss (dashed lines) of odd hydrogen ($cm^{-3}\ s^{-1}$) (a) during northern winter ($D_S = -5°$) (b) during northern spring ($D_S = 15°$) (figure from Kong and McElroy 1977*a*).

the reaction of two scarce odd hydrogens, either $H + HO_2$ or $OH + HO_2$. One path for the first reaction generates a pair of OH radicals and can be ignored except in laboratory measurements. The other paths produce H_2 or H_2O. Kong and McElroy stress the uncertainties in the measured rate coefficients and try various reasonable choices. As discussed below, fits to observations of Mars can be obtained under a variety of assumptions.

4. Surface Deposition. The bottom box of Fig. 3 schematically shows the precipitation (and condensation) of oxidants on the surface. The behavior of H_2O is well studied, but other molecules and radicals may also condense at the very low temperatures of night and winter. A considerable body of work on "frozen free radicals" is summarized by Hunten (1974); in general, such experiments subject a flow of a gas to a discharge and then condense some of the products on a cold finger located downstream. When the gas contains water vapor and the deposit is slowly warmed, a variety of products is released, and exothermic reactions are observed in some temperature ranges. It can be imagined that both H_2O_2 and HO_2 are among the major molecules condensed in a matrix of ordinary ice and that subsequent reactions produce various polyoxides such as the H_2O_3, indicated in the figure. The analogy to Mars is obvious, as is the possible relevance to the unexpected results of several Viking biology experiments (Hunten 1979*b*).

Some of the models by Kong and McElroy (1977*b*) were set up in such a way that the flux of peroxide could be estimated as 5×10^{16} molecules cm^{-2} per Martian year, or around 10^9 cm^{-2} s^{-1}. As they point out, this is a significant fraction of the total production of odd hydrogen, 6.5×10^9 cm^{-2} s^{-1}, especially since production of a peroxide molecule uses up two odd-hydrogen atoms. According to Klein (1977), the oxygen released at the Viking 1 site was 770 nanomoles per cubic centimeter of soil, and it was less by a factor of 4 at Viking 2. If the layer sampled was 1-cm deep, the burden was between 1 and 4.6×10^{17} molecules per cm^2, which could be supplied by the atmosphere in 2 to 10 Martian years. Such a lifetime does not seem unreasonable in the cold Martian soil. If the oxidants permeate to much greater depths, say a few hundred meters, times approaching 1 Myr would be suitable. Additional oxidants may be produced by photo-oxidation on surfaces of soil particles (Huguenin 1974); such materials may, however, be much less mobile.

There has been occasional speculation about "oases" that might be less hostile to organic molecules, and to living organisms, than the two Viking sites. In fact, moist areas are sites of greater than average production of oxidants and are probably unusually hostile. It is interesting to consider why such highly oxidizing environments do not occur on Earth; one major factor is the widespread occurrence of liquid water, which promotes rapid destruction of oxidants.

5. Recombination on Aerosols. Atreya and Blamont (1990) have revived the suggestion that significant recombination of CO and O_2 may be catalyzed by aerosol particles. Their specific suggestion is that the particles are water ice and might behave as they do in the formation of the Antarctic ozone hole on Earth. Their suggestion was prompted by the observation from the Phobos spacecraft that the 2.35 μm absorption due to CO is weaker relative to a nearby CO_2 band above Olympus Mons (25 km) than in nearby lower areas (Rosenqvist et al. 1990).

6. Observational Tests. Thorough tests of aeronomical models are difficult even on Earth, where data are much more abundant than for Mars and where it is possible to deploy many generations of experiments, each benefiting from previous ones. Nevertheless, a few tests are possible. The odd-hydrogen model was designed to explain the observed amounts of CO and O_2, which it does very well and without any forced assumptions. This must be considered at least a partial test (Parkinson and Hunten 1972; McElroy and Donahue 1972).

The odd-hydrogen model also accounts for the observed abundance of ozone and for the seasonal and latitudinal variation of ozone. Figure 7 shows the seasonal variation of ozone in the vicinity of the polar caps of Mars as measured from Mariner 9 (Barth et al. 1973; Lane et al. 1973; Barth 1985). Ozone has its maximum abundance in the winter when the atmosphere above the polar cap is dry because of the very low temperature, near 147 K. During the spring the ozone abundance decreases until the beginning of summer when it disappears because of the rising water vapor content of the atmosphere. After the carbon dioxide has sublimed away, the temperature of the residual cap rises to 190 to 200 K, and water vapor begins subliming to produce a column abundance as high as 90 μm precipitable (Jakosky and Farmer

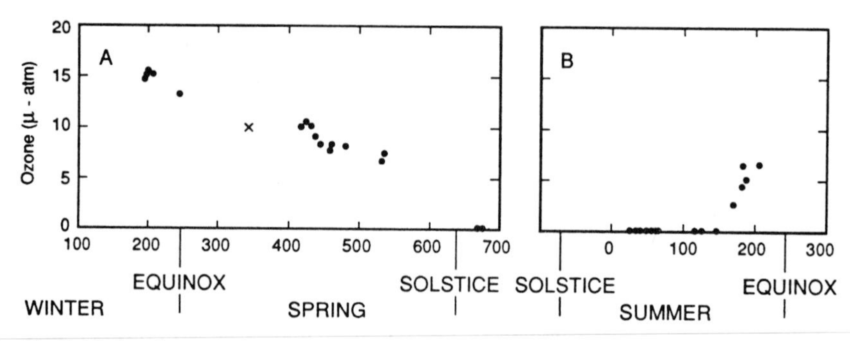

Fig.7. Mariner 9 ozone observations. (A) Amount of ozone observed over the north polar cap during the winter, spring and beginning of summer. The cross refers to the Mariner 7 measurement which was made over the south polar cap in 1969. (B) Amount of ozone observed over the south polar cap during the summer. (figure from Barth et al. 1973).

1982; chapter 28). In the late summer and fall, the temperature falls, the water vapor content decreases, and ozone reappears above the polar cap.

Another test is the predicted abundance of thermospheric atomic H, which can be checked against observations of Lyman α from Mariners 6, 7 and 9 (Hunten and McElroy 1970; Barth et al. 1972*b*). Water vapor and odd hydrogen are confined to low altitudes because of the low temperatures. The principal carrier of hydrogen to the upper atmosphere is H_2. Reactions with the ions O_2^+ and CO_2^+ convert most of the molecules to atoms, which then escape, as discussed at the end of this chapter. A major uncertainty is the mean escape velocity, which when multiplied by the density gives the escape flux and which is very sensitive to the exospheric temperature. This uncertainty, within perhaps a factor of 2 either way, combines with the chemical uncertainties discussed above; agreement of observed and calculated densities is entirely satisfactory.

C. Odd Nitrogen

Atomic nitrogen is produced in the upper atmosphere by the dissociation of molecular nitrogen; the most important processes are predissociation by solar photons and photoelectron impact dissociation. If the atoms are in the metastable ^2D state, they may produce nitric oxide by reacting with the dominant constituent, carbon dioxide, or they may be quenched by carbon monoxide or by atomic oxygen. Ground-state ^4S atoms can destroy nitric oxide, reacting with them to reform molecular nitrogen. Both atomic nitrogen and nitric oxide may react with the dominant ion, O_2^+, to produce NO^+, and the dissociative recombination of this ion produces predominantly $N(^2D)$; these reactions act to recycle some $N(^4S)$ into $N(^2D)$ and then into nitric oxide. The important odd-nitrogen reactions in the ionosphere are shown in Fig. 8. The N_2 at the top of the diagram is dissociated by electrons (*e*) and ultraviolet photons ($h\nu$) to produce the $N(^2D)$ and $N(^4S)$ atoms. The flow is downward to reform molecular nitrogen. At lower altitudes, where no dissociation of N_2 takes place, nitric oxide itself is still subject to predissociation, with a net loss of two molecules because the product, atomic nitrogen, will attack another nitric oxide molecule.

On Venus, Earth and Mars, odd nitrogen plays an important role in ionospheric chemistry (NO^+ is the terminal ion in all three cases). The Earth is rather a special case, because N_2 is the dominant constituent and because of the high abundance of O_2, with which not only $N(^2D)$ but also $N(^4S)$ can react to form NO; thus, there is an important channel by which $N(^4S)$ can be recycled to NO. Venus is similar in composition to Mars, except that O and CO are about an order of magnitude more abundant. On Venus, the quenching of $N(^2D)$ by O and CO is sufficient to suppress the formation of NO and make $N(^4S)$ the dominant form of odd nitrogen; most of the NO formed is consumed by reaction with $N(^4S)$, and the abundance of the surplus $N(^4S)$ is

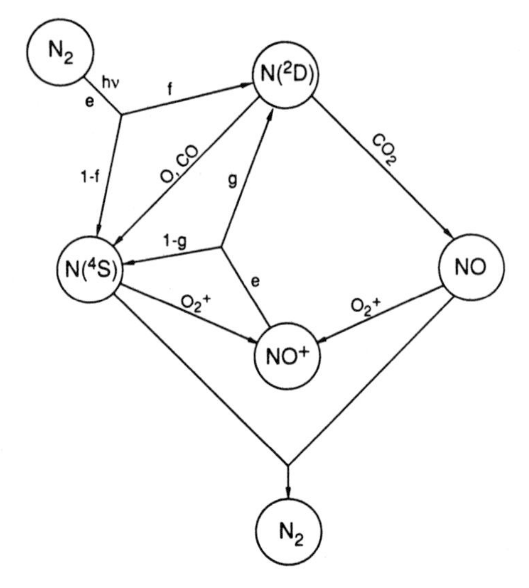

Fig.8. Chemistry of odd nitrogen. The reactions of N(^{4}S) and NO with O_2^+ to and from NO^+ provides a pathway to convert N(^{4}S) to NO and vice versa. Lacking this pathway, the branching ratio *f* would solely determine the dominant form of odd nitrogen. In either case, the dominant form keeps the less abundant form close to photochemical equilibrium, while its own density is controlled by transport.

controlled not by photochemistry (as on Earth) but by transport. A large fraction of the N is carried to the night side by the thermospheric general circulation, where it eventually recombines radiatively with atomic oxygen as the circulation descends to higher pressures. This picture is supported by measurements of abundant N on the day side, and of chemiluminescence of the recombination of the atoms on the night side (Stewart and Barth 1979). Calculations coupling the Venus odd-nitrogen chemistry and global thermospheric dynamics confirm the role of N-atom transport in producing the observed NO δ-band nightglow (Bougher et al. 1990).

On Mars the only relevant observation is the measurement of NO in the thermosphere by the Viking entry science mass spectrometer (Nier and McElroy 1977). The measurements indicate a mixing ratio of approximately 10^{-4} at altitudes above 120 km. Calculations of the formation of nitric oxide in the Mars upper atmosphere have been performed by McElroy et al. (1976*a*). However, they ignore the quenching of N(^{2}D) by O and CO and assume only downward eddy mixing for transport. A re-interpretation with current chemistry and circulation models is needed, similar to the work of Bougher et al. (1990) on Venus.

A calculation of the abundance of odd nitrogen in the lower atmosphere has been undertaken by Yung et al. (1977), with a view to examining the atmospheric sources of fixed nitrogen. They found that NO should be the dominant form, with abundances on the order of 10 ppb, depending on the reactivity of the surface towards HNO_2 and HNO_3.

In the Earth's upper atmosphere, the abundance of nitric oxide varies by as much as a factor of 10 (Barth et al. 1988). These large changes are caused by changes in the intensity of the solar EUV and X-radiation. The highly variable solar X rays produce photoelectrons that control the nitric oxide density through the electron impact dissociation of molecular nitrogen. As these same processes operate in the atmosphere of Mars, it is expected that the nitric oxide density in the Martian atmosphere varies in response to changes in solar activity.

Nitric oxide in the Martian atmosphere has not yet been measured by ultraviolet spectrometers using the technique of fluorescent scattering that has been used extensively in mapping nitric oxide in the Earth's upper atmosphere.

III. THE IONOSPHERE OF MARS

Photochemical processes control the behavior of the main ionospheric layer of Mars, just as is the case for Venus (cf. Nagy et al. 1983) and in a manner similar to the terrestrial F_1 layer (cf. Bauer 1983). Radio-occultation observations from US and USSR Mars missions have provided the bulk of the electron density data for the Mars ionosphere. A typical value for the peak plasma density N_m is $\sim 2 \times 10^5$ cm^{-3}, and the height of the maximum h_m is about 135 km. The first, and up to now the only, *in situ* measurements of the ionosphere of Mars were carried out by the Retarding Potential Analyzer (RPA) instrument, carried aboard the Viking Landers (Hanson et al. 1977). One of the ion density profiles obtained by these instruments is shown in Fig. 9. These RPA measurements clearly established that the principal ion in the Martian ionosphere does not correspond to the main ionizable neutral constituent CO_2 but is O_2^+, in a manner totally analogous to the case of the Venus ionosphere. The predominance of O_2^+ is the result of the ion molecule reactions (1) and (2) below, which transform very rapidly the originally produced CO_2^+ and O^+ ions to O_2^+:

$$CO_2^+ + O \rightarrow O_2^+ + CO \tag{1}$$

$$O^+ + CO_2 \rightarrow O_2^+ + CO. \tag{2}$$

This result and its explanation were deduced earlier from the analysis of the Mariner 6 and 7 ultraviolet spectrometer experiment (Stewart 1972). In addition to the principal ion O_2^+, the Viking RPA results also established the presence of CO_2^+ and O^+, O^+ becoming comparable in concentration to that of O_2^+ at altitudes above $\sim$ 280 km (see Fig. 9). The chemical reaction

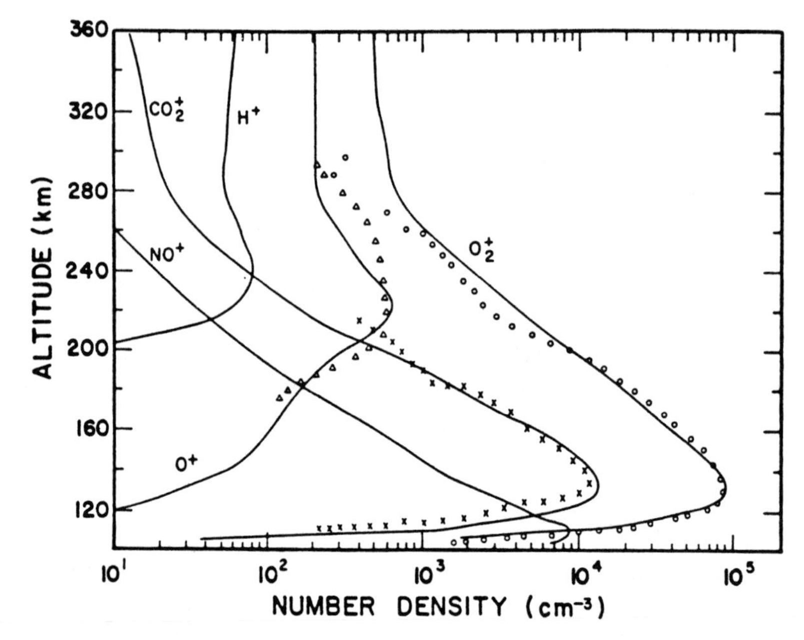

Fig.9. Ion density profiles: ion density profiles in which the solid lines are calculated profiles. The circles are O_2^+ data, the triangles are O^+ data and the crosses are CO_2^+ data; the values were by Chen et al. (1978) from the Viking 1 RPA experiment (Hanson et al. 1977) (figure from Chen et al. 1978).

scheme suggested to be responsible for both the observed and the expected ion species is shown in Fig. 10. The photochemistry of Mars is believed to be very similar to that of Venus; a recent summary of the more relevant reaction rates is shown in Table II (Kim et al. 1989). Note that the dissociative recombination rate of O_2^+ is electron-temperature dependent and therefore the altitude variation of the electron density is given by the following relation (Cravens et al. 1981):

$$n_e = 520\sqrt{P}T_e^{0.275} \tag{3}$$

where P is the total ionization rate and T_e is the electron temperature. This means that even in the photochemically controlled region, the neutral gas temperature can only be deduced from the electron density profiles corresponding to the very limited altitude range where $T_e = T_n$.

Analysis of the appropriate time constants or more sophisticated models indicate that transport processes take over from photochemical ones somewhere between 170 and 200 km in the dayside ionosphere of Mars. Comprehensive models, which describe the behavior of the ionosphere in both the photochemical and transport controlled regions, have been developed. Chen

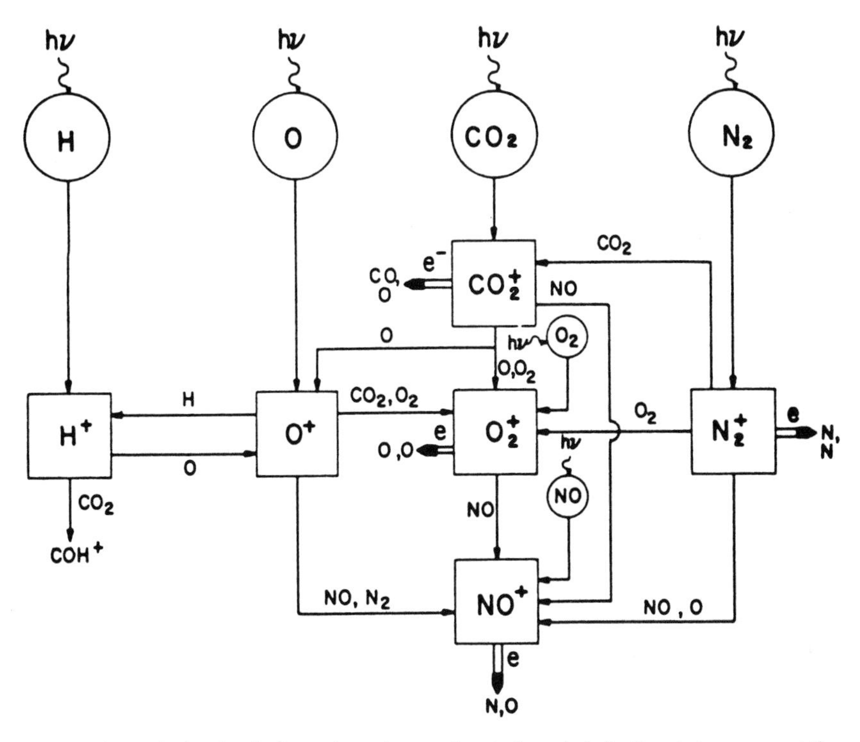

Fig.10. Ionospheric chemical reaction scheme: chemical symbols in the circles represent the number density of the neutral atmospheric constituents. The symbols in the boxes represent the equilibrium density of the ions. The arrows show the direction of the production and loss processes with the chemical symbol next to the arrow indicating the specie reacting with the ion (figure from Chen et al. 1978).

et al. (1978) solved the coupled continuity, momentum and energy equations to study the photochemistry and energetics of the Martian ionosphere. Their model was successful in matching the observed ion densities as indicated in Fig. 11. However, given the uncertainty in the atomic oxygen densities, they selected an oxygen profile to get a best fit for the ion densities. Furthermore, the Chen et al. (1978) calculations were carried out using mixed upper-boundary flow conditions. The model results are relatively independent of these upper-boundary values below ~ 250 km, but at higher altitudes the densities depend strongly on transport for which no data are available at this time.

The daytime ion temperatures measured by the RPAs (Hanson et al. 1977) depart from the neutral gas temperature ($T_n < 200$ K) at altitudes above the ionospheric peak (~135 km), reaching a value of ~ 3000 K near 300 km (see Fig. 12). Chen et al. (1978) found that EUV heating alone predicted ion temperatures considerably smaller than the values measured with the RPAs (Hanson et al. 1977). They had to assume a topside heat source

TABLE II
Ion-Neutral Chemical Reaction Rates

	Reaction			Rate Constant (cm^3s^{-1})
R1	$CO_2^+ + O$	→	$O_2^+ + CO$	1.64×10^{-10}
R2	$CO_2^+ + O$	→	$O^+ + CO_2$	1.0×10^{-10}
R3	$CO_2^+ + e$	→	$CO + O$	$1.4 \times 10^{-4}/T_e$
R4	$CO_2^+ + O_2$	→	$O_2^+ + CO_2$	6.4×10^{-11}
R5	$CO_2^+ + N$	→	$NO^+ + CO$	1.0×10^{-11}
R6	$CO_2^+ + N$	→	$CO^+ + NO$	1.0×10^{-11}
R7	$CO_2^+ + NO$	→	$NO^+ + CO_2$	1.2×10^{-10}
R8	$O_2^+ + e$	→	$O + O$	$1.6 \times 10^{-7}(300/T_e)^{0.55}$
R9	$O_2^+ + NO$	→	$NO^+ + O_2$	4.5×10^{-10}
R10	$O_2^+ + N$	→	$NO^+ + O$	1.2×10^{-10}
R11	$O^+ + CO_2$	→	$O_2^+ + CO$	9.4×10^{-10}
R12	$O^+ + N_2$	→	$NO^+ + N$	$1.2 \times 10^{-12}(300/T_n)$
R13	$O^+ + H$	→	$H^+ + O$	$2.5 \times 10^{-11}\ T_n^{0.5}$
R14	$O^+ + O_2$	→	$O_2^+ + O$	$2.0 \times 10^{-11}(300/T_n)^{0.55}$
R15	$O^+ + CO_2$	→	$CO^+ + O_2$	9.4×10^{-10}
R16	$H^+ + O$	→	$O^+ + H$	$2.2 \times 10^{-11}\ (T_i)^{0.5}$
R17	$H^+ + CO_2$	→	$CHO^+ + O$	3.0×10^{-9}
R18	$H^+ + NO$	→	$NO^+ + H$	1.9×10^{-9}
R19	$NO^+ + e$	→	$N + O$	$2.3 \times 10^{-7}(300/T_e)^{0.5}$
R20	$CO^+ + CO_2$	→	$CO_2^+ + CO$	1.0×10^{-9}
R21	$CO^+ + O$	→	$CO + O^+$	1.4×10^{-10}
R22	$CO^+ + NO$	→	$CO + NO^+$	3.3×10^{-10}
R23	$CO^+ + e$	→	$C + O$	$2.0 \times 10^{-7}(300/T_e)^{0.48}$

flowing directly into the ions, possibly associated with some ionosphere-solar wind interaction process, to match the observed ion temperature values. Chen et al. (1978) used standard thermal conductivity values in their calculation, but Johnson (1978) drew attention to the fact that if the thermal conductivity is reduced, because of a nearly horizontal magnetic field, the calculated temperatures would be close to the observed ones, even with a much reduced topside heat flow. Rohrbaugh et al. (1979) also carried out detailed energy calculations, including contributions from exothermic chemical reactions, and found good agreement between the calculated and observed ion temperatures (Fig. 12). The effect of Joule heating (Chen et al. 1978) is similar to that of exothermic reactions around the 200 km region; therefore, at this time, with insufficient data to constrain the models, either or both processes may be important in the Martian ionosphere. Careful analysis of the RPA electron data allowed Hanson and Mantas (1988) to deduce the altitude variation of the "thermal electron" temperature plus two higher-energy electron populations as shown in Fig. 12. The first of these higher-energy components is approximately Maxwellian, with a temperature of $T_{e2} \simeq 20{,}000$ K, attributable to fresh photoelectrons; the other high-energy population was found to

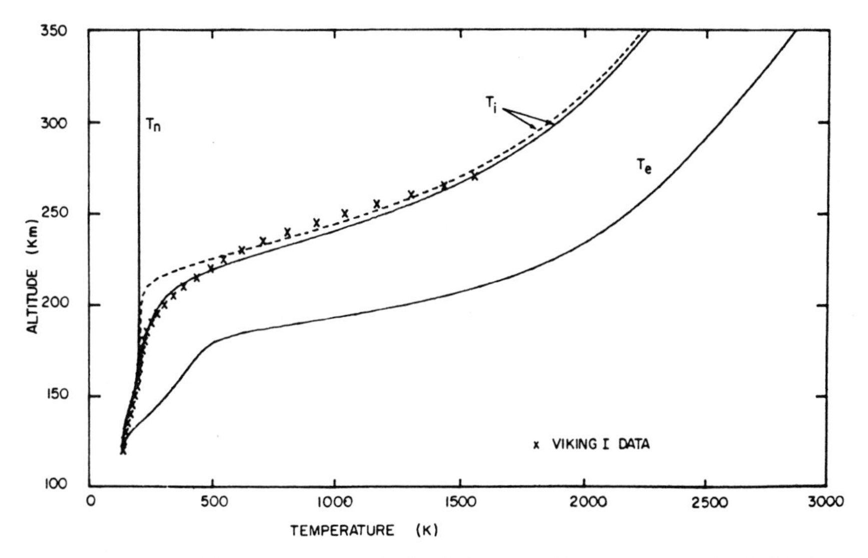

Fig. 11. Electron and ion temperatures: calculated electron and ion temperatures for the Martian ionosphere assuming an inclined magnetic field (12° dip angle) and downward ion and electric heat flows of 6×10^6 and 5×10^8 eV cm^{-2} s^{-1}, respectively, at the upper boundary. The dashed curve results from the same calculations, but with the heating from energetic O_2^+ neglected (from Rohrbaugh et al. 1979). The crosses indicate the ion temperatures measured by the Viking 1 Retarding Potential Analyzer (Hanson et al. 1977).

be present above 200 km and is possibly of solar wind origin. The measured thermal electron temperature profile agrees quite well with the calculations of Chen et al. (1978) and Rohrbaugh et al. (1979).

The only information on the nightside ionosphere of Mars is that obtained by radio-occultation measurements with Mars 4 and 5 (Savich et al. 1976) and from radio-occultation experiments from the Viking orbiters in 1977 (Zhang et al. 1990*a*). Two profiles were obtained from Mars 4 and 5, indicating a peak electron density of $\sim 5 \times 10^3$ cm^{-3}, with the height of the maxima at $\sim$ 110 and 130 km, respectively. There have been no published attempts to model the nightside ionosphere of Mars. The rotation period of Mars is relatively small, close to that of the Earth; therefore, the observed small densities do not seem to be especially difficult to account for, although they may still imply a contribution from a direct nighttime ionization source, similar to the Venus conditions.

The ionosphere of Mars, in the region around the electron density peak is, to first order, a Chapman-type layer (cf. Bauer 1983). A Chapman layer is the electron density distribution that results from photochemical equilibrium with the electron recombination rate proportional to the square of the electron density (Bauer 1973). However, a more detailed analysis of the data shows important departures from the simple Chapman theory. The departures

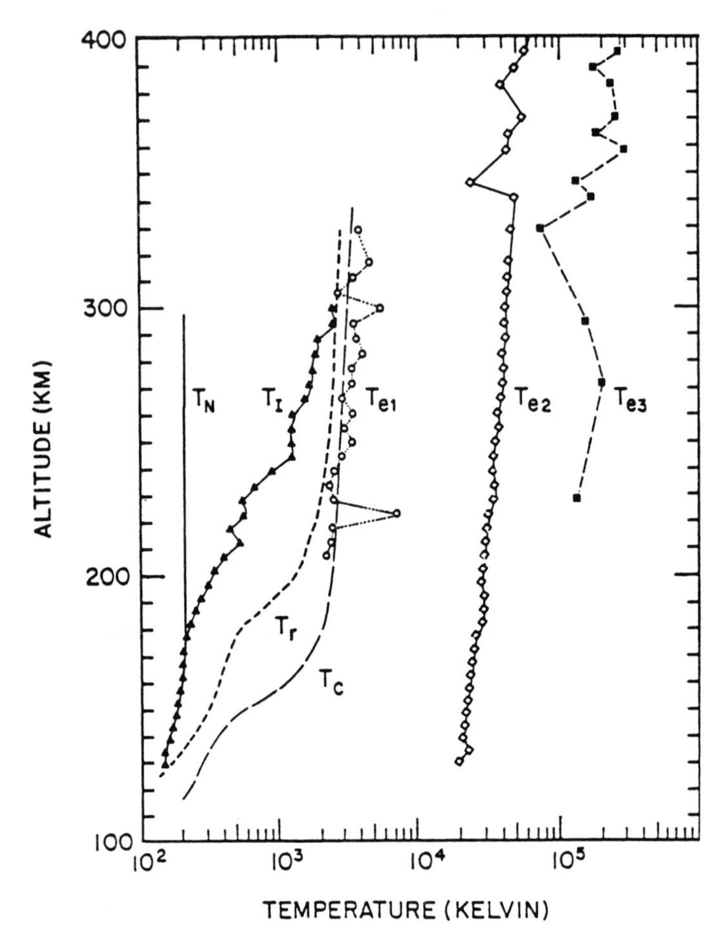

Fig. 12. Ion temperatures vs altitude: plots of the various plasma temperatures vs altitude. The values of T_i, the ion temperature and T_{e1}, T_{e2} and T_{e3}, the three energetically distinct electron temperatures, all of which were measured by the Viking RPA experiment, are plotted. T_n is a neutral temperature profile that is consistent with the Viking 1 measurements and T_r and T_c are model electron temperature profiles calculated by Rohrbaugh et al. (1979) and Chen et al. (1978) (figure from Hanson and Mantas 1988).

exhibit themselves in two ways (Bauer and Hantsch 1989). First, the solar zenith angle χ dependence of the peak electron density was found to be $N_m \simeq \cos^k \chi$, with $k > 0.5$ (compared to $k = 0.5$ for a simple Chapman layer); this is due to the fact that the principal ion O_2^+ does not correspond to the main, ionizable neutral-gas constituent. The solar cycle dependence of the peak electron density on Mars is $d(\ln N_0)/d(nF_{10.7}) \simeq 0.36$, which is different than that expected for a simple Chapman layer (0.5) but is very similar to that found for Venus (Bauer 1983; Kliore and Mullen 1989). A reassessment of the dayside radio occultation data from Mariner 9 and Viking 1 and 2 (Zhang et al. 1990*b*) shows how the altitude of the peak electron density varies with solar zenith angle. This study shows that when the incident solar wind dynamic pressure exceeds the peak ionospheric thermal pressure, the ionosphere peaks at a higher altitude at large solar zenith angle (see chapter 31).

Knowledge of the ionospheric charged-particle densities and temperatures is of particular importance to the, as yet still open, question concerning the existence and magnitude of a Martian intrinsic magnetic field (see chapter 31). Hanson and Mantas (1988) calculated (using their RPA results) the dayside ionospheric plasma pressures. They obtained a value of $\sim 5 \times 10^{-10}$ dyne cm^{-2} for 350 km, the highest altitude for which data exist. This ionospheric pressure is insufficient to balance the shocked solar wind pressure near 1000 km, thus implying the presence of a magnetic field of $\sim$40 nT within the topside ionosphere. Such a field can be either an intrinsic planetary field and/or an induced field. Shinagawa and Cravens (1989) used their magnetohydrodynamic ionospheric model and, by comparing calculated and measured ionospheric densities, they concluded that the necessary magnetic field of about 40 nT is most likely an induced field. Luhmann et al. (1987) came to similar conclusions using the Venus observations of the Pioneer Venus Orbiter and comparing that data base with what is known about Mars (for further discussion of this subject see chapter 31).

IV. AIRGLOW

The most intense emission in the ultraviolet airglow of Mars is the CO $a^3\Pi$–$X^1\Sigma^+$ band system of carbon monoxide. This band system appears in the spectrum in Fig. 13 between 1900 and 2700 Å (Barth et al. 1971). These Cameron bands are produced by several sources: (1) the photodissociation of molecular carbon dioxide by solar ultraviolet radiation with wavelengths shortward of 1080 Å; (2) the electron impact dissociation of carbon dioxide by photoelectrons with energy > 11.5 eV; and (3) the dissociative recombination of ionized carbon dioxide, one of the constituents of the ionosphere. The intensity variation of the Cameron bands as a function of altitude has been used to determine the structure of the neutral atmosphere above 120 km and from the scale height, the temperature of the thermosphere (Stewart 1972; Stewart et al. 1972). Two band systems of ionized carbon dioxide appear in the ultraviolet airglow of Mars: the CO_2^+ ($B^2\Sigma_u^+ - X^2\Pi_g$) at 2890 Å and the CO_2^+ ($A^2\Sigma_u$–$X^2\Pi_g$) between 3000 and 4000 Å (See Fig. 13). Both emissions are produced by the photoionization of carbon dioxide, the B–X bands from solar radiation with wavelengths shortward of 690 Å and the A–X bands from wavelengths shortward of 720 Å. The measurement of altitude variation of the airglow from these CO_2^+ bands, particularly the B–X bands, is a method of measuring the rate of ionization of the Martian ionosphere as a function of altitude. The CO_2^+(A–X) bands are also produced by fluores-

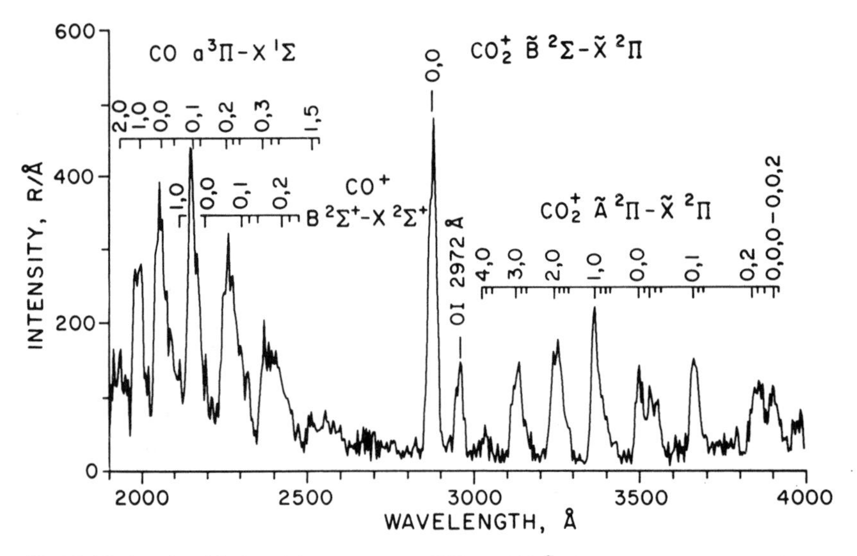

Fig.13. Mariner 6 and 7 ultraviolet spectrum at 1900 to 4000 Å of the upper atmosphere of Mars at 20 Å resolution. Spectrum was obtained by observing the atmosphere tangentially at an altitude between 160 and 180 km. The spectrum was compiled from the sum of four individual observations (figure from Barth et al. 1971).

cent scattering of solar radiation. Above 150 km, the measurement of the A–X bands is a method of measuring the density of CO_2^+ ions in the ionosphere (Fox and Dalgarno 1979). The 2972 Å line of atomic oxygen appears in the Mars airglow (see Fig. 13), and the 5577 Å line must be present as well. The excited $O(^1S)$ atoms are produced by the photodissociation of carbon dioxide by solar ultraviolet radiation of wavelengths shortward of 1200 Å. The intense Lyman α radiation at 1216 Å in the solar spectrum excites the atomic oxygen airglow in the Martian atmosphere at altitudes as low as 80 km (Fox and Dalgarno 1979).

The airglow spectrum between 1400 and 1800 Å consists mainly of the carbon monoxide fourth positive bands, $A^1\Pi–X^1\Sigma^+$. This airglow emission is produced by the following processes: (1) the photodissociation of carbon dioxide by solar ultraviolet radiation shortward of 920 Å; (2) the electron impact dissociation of carbon dioxide by photoelectrons with energy > 13.5 eV; and (3) the dissociative recombination of ionized carbon dioxide.

The atomic oxygen line at 1304 Å is present in the Martian airglow. This airglow line is produced by two processes: (1) the resonant scattering of solar ultraviolet radiation by atomic oxygen; and (2) the electron impact excitation of atomic oxygen by photoelectrons with energy > 9.5 eV. A radiative transfer analysis of the 1304 Å airglow has been used to determine the density of atomic oxygen in the Martian upper atmosphere (Strickland et al. 1972).

The strongest atomic line in the Martian ultraviolet airglow is the 1216

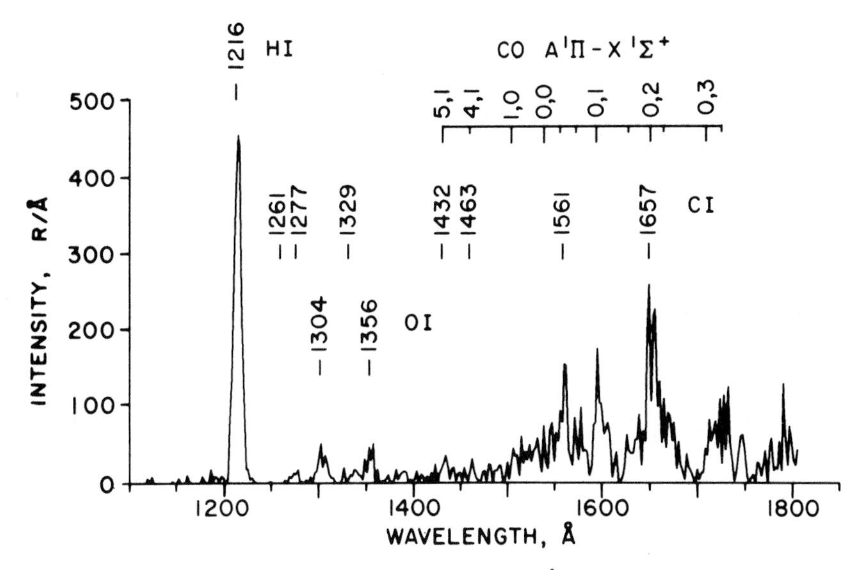

Fig. 14. Mariner 6 and 7 ultraviolet spectrum at 1100 to 1800 Å of the upper atmosphere of Mars at 10 Å resolution. Spectrum was obtained by observing the atmosphere tangentially at an altitude between 140 and 160 km. This spectrum was compiled from the sum of four individual observations (figure from Barth et al. 1971).

Å Lyman α line of atomic hydrogen. This airglow emission is produced by the resonant scattering of the Lyman α line in the solar spectrum. The measurement of resonantly scattered Lyman α radiation has been used by a number of spacecraft missions to determine the density and escape flux of atomic hydrogen.

Variations do occur in the intensity of the several airglow emissions. The carbon monoxide Cameron bands and the ionized carbon dioxide B–X bands respond to changes in the solar EUV radiation (Stewart et al. 1972). The solar EUV radiation directly affects these emissions through the photodissociation and photoionization processes and indirectly through the flux of photoelectrons. The atomic oxygen 1304 Å lines vary in response to variations in the solar 1304 Å lines and to changes in the photoelectron flux. During the time of the Mariner 9 observations of Mars in late 1971 and early 1972, the Lyman α emission from the Martian atmosphere did vary by ±10%. Measurements of the intensity of the solar Lyman α flux show variations of this magnitude and so, it is expected that variations in the intensity of the Lyman α airglow will occur. Because of the changing solar radiation, it is not known whether significant variations in the density of atomic hydrogen occur and whether changes in the escape flux take place.

As on Earth and Venus, the most intense airglow of all is the infrared atmospheric band of O_2 at 1.27 μm, with intensities varying from 3 to 26 MR

(Mega Rayleigh). It has been detected from Earth both by Fabry-Perot and Fourier spectroscopy (Noxon et al. 1976). As on Earth, the principal excitation is by photolysis of ozone, and its intensity can be used as a sensitive detector of this molecule (see Thomas et al. 1984). On Venus, by contrast, the principal source is recombination of O atoms, and it is a useful tracer of thermospheric motions. The principal uncertainty for Mars is the quenching by CO_2 of the metastable upper state, which has a radiative lifetime of an hour. Thus, the amount of ozone deduced for a given intensity depends on its altitude: more is required if the ozone is low than if it is high (Noxon et al. 1976; Traub et al. 1979). The latter authors obtained spatially resolved observations, showing clearly that ozone is present even at low latitudes, but greatly enhanced in the winter polar region where catalytic destruction is at a minimum. This set of observations agrees with the more extensive observations obtained by Mariner 9.

V. ATMOSPHERIC ESCAPE

As discussed in chapter 25, the atmosphere of Mars exhibits a number of striking isotopic effects that are probably the signatures of fractionation during escape: they are found in hydrogen, nitrogen and various noble gases. On the other hand, there is no measurable effect for carbon or oxygen, undoubtedly because the gases in the atmosphere are only a small part of the much larger exchangeable reservoir which greatly dilutes the effect of isotopically selective escape fluxes (McElroy and Yung 1976; chapter 32).

Because of the modest gravity well of Mars, it is affected by mechanisms that are much less important on Earth and Venus. Among these is nonthermal escape, reviewed by Hunten (1982), Hunten et al. (1989) and Chamberlain and Hunten (1987). Thermal (or Jeans) escape is dominant for hydrogen and helium (Hunten and Donahue 1976). These processes are summarized in Table III, along with two others likely to have been important very early in the evolution of Mars. Fractionation in blowoff (or hydrodynamic escape) is a potentially powerful means of affecting elements and isotopes in ways resembling those observed (Hunten et al. 1987; Pepin 1989). Impact erosion (Melosh and Vickery 1989) is another process that can be particularly effective on a small planet like Mars; it seems likely to be totally unselective in terms of composition. A general discussion is given by Hunten et al. (1989).

Table III indicates the dominant escape process for *oxygen,* dissociative recombination of O_2^+. The fast atoms are unlikely to penetrate to the top of the atmosphere unless they are produced in the exosphere, defined as the region where collisions can be neglected for many purposes. As only one of the atoms is directed upwards, the escape flux can be estimated by counting the number of dissociation events occurring in the exosphere. Because the

TABLE III
Atmospheric Loss Processes for Mars[a]

Process		Important for:
Thermal		H, He
Nonthermal		N, O, [C]
e.g.,	$O_2^+ + e \rightarrow O^* + O^*$	
	$N_2^+ + e \rightarrow N^* + N^*$	
Blowoff Fractionation		H, Ne, Ar, Kr, Xe
Impact Erosion		All gases

[a]The first two act throughout the life of the planet, while the others are expected to be important at an early stage only. An asterisk indicates kinetic energy.

two products have equal momentum, heavier isotopes have slightly lower velocities than lighter ones, but the consequent fractionation is very small. The major factor is the diffusive separation that enriches a large region in lighter atoms, molecules and isotopes. These processes were discussed by McElroy and Yung (1976), and Hunten (1982) has given a simplified treatment. For conditions typical of Mars, a lighter isotope is enriched in the escaping component by ~18% per mass unit; somewhat surprisingly, this factor depends on the mass difference, not on the ratio. As already mentioned, there has been no detectable effect within ~5%, on the isotopic composition of the remaining oxygen.

The total escape flux is calculated to be very close to half that of H atoms. That this is no accident was pointed out by McElroy (1972) and studied in more detail by Liu and Donahue (1976) and McElroy and Yung (1976). The loss rate of O is determined by the structure and chemistry of the ionosphere. The odd-hydrogen chemistry discussed in Sec. II.B regulates the amount of H_2 so that it carries just the right amount of hydrogen to the top of the atmosphere. Thus, the escape does not change the state of oxidation.

Nitrogen is particularly interesting because it does show a large enrichment of the $^{15}N/^{14}N$ ratio, a factor of 1.62 relative to Earth. Several nonthermal processes are important, including dissociative recombination, shown in Table III (Fox and Dalgarno 1980,1983; Fox 1989). Otherwise the physics is similar to that for O atoms. Discussions by Nier et al. (1976*b*) and McElroy et al. (1977) show that the original partial pressure of N_2 could have been between 7 and 50 mbar; the present amount is 0.08 mbar. It is also required that a large fraction of the nitrogen must have been degassed very early. Other studies of the escape flux of atomic nitrogen lead to lower estimates of the original partial pressure of molecular nitrogen, 0.4 to 1.4 mbar (Wallis 1978) and 0.5 to 0.9 mbar (Wallis 1989).

The escape flux of *hydrogen* is controlled by its ability to diffuse through the lower thermosphere, which depends primarily on the mixing ratio (Hunten 1973; Hunten and Donahue 1976; McElroy and Yung 1976). H_2, the dominant carrier in the middle atmosphere, is mostly converted to H atoms in the ionosphere by reaction with O_2^+ and CO_2^+; some 10% may escape as molecules. However, the nature of the diffusion limit is such that the total flux, expressed in atoms $cm^{-2} s^{-1}$, is insensitive to the proportions. If the rate has not varied appreciably, the total amount lost over geologic time corresponds to an ice layer 2 to 3 m deep. An early hydrodynamic escape episode would have required far more.

The sixfold enrichment in the D/H ratio has been interpreted by its discoverers (Owen et al. 1988) as implying an initial endowment of water at least 100 times greater than the present amount, which implies a greatly enhanced loss during the first 1 Gyr or so. They emphasize the great uncertainty and model dependence in their estimates. The only fractionation mechanism they consider is diffusive separation, which enriches the lighter isotope at the exobase, and they adopt the approximate treatment of Hunten (1982), mentioned above. A detailed treatment has been given by Yung et al. (1988), who show that additional fractionation processes are likely to be important, in particular that H_2 escapes much more readily than HD. The consequence is that the required initial quantity is considerably lower, equivalent to 3.6 m of liquid water. The study is limited to times when the atmosphere was similar to the contemporary one; in particular, it includes the constraint that the total loss of hydrogen atoms is twice that of oxygen atoms. A substantial crustal sink of oxygen could change the assessment.

Jakosky (1990*a*) has pointed out that this small initial quantity of water is inconsistent with the current theory of the polar volatiles. Estimates of the depth of the north polar cap are 3 km, all of which could be water ice. This amount is equivalent to a global layer of 15 m. Jakosky (1990*a*) also points out that during past epochs of high obliquity the polar cap sublimation would be 500 times the present rate. If the polar cap water exchanges with the atmosphere, then about 60 m of water may have escaped during the history of Mars.

Helium has not been detected on Mars; an upper limit to its mixing ratio is 100 ppm (Nier and McElroy 1977). On Earth, the major sink is thermal escape of ions in the polar wind (Axford 1968), and a similar process may be significant on Mars. However, such a process is limited by the rate at which ions can be produced. Thermal escape of neutral atoms is also rather slow, and may be less important.

Noble gases have been discussed in a preliminary way by Hunten et al. (1987) and in detail by Pepin (1989) (see chapter 4). (Helium is excluded because it is still escaping.) A sufficiently rapid outflow of H_2 from a primitive atmosphere can drag other gases with it. The drag force is almost independent of the species, but is opposed by a gravitational force proportional to mass. The result is a strong, but linear, selectivity, along with the ability to process a large quantity of gas. There is a cross-over mass, above which there is no escape; for a given planet, it is proportional to the hydrogen flux. Hunten et al. (1987) find that the elemental abundance data for Mars can be fitted, but only with models requiring an enormous hydrogen inventory, between 24 and 200 km of equivalent liquid water. The accretion model of Dreibus and Wänke (1985) suggests that about 90 km was available, enough to accommodate the smaller of these without difficulty.

Pepin (1991) has greatly extended the treatment, and has fitted isotopic patterns as well as the elemental ones. The hydrogen requirements (expressed as the equivalent depth of liquid water) range from 62 to 178 km. The early solar ultraviolet flux is assumed to decay exponentially with a time constant of 90 Myr. It is assumed that water molecules are reduced to hydrogen by reactions in the crust or are photodissociated in the atmosphere. The escape flux is a few times 10^{14} cm^{-2} s^{-1} and declines with the same law as the ultraviolet flux. The duration of the episode is between 100 and 270 Myr for the various scenarios.

Impact erosion is another process that benefits from Mars' modest gravity well. Impacts below a certain kinetic energy eject very little, if any, gas (Walker 1986; Hunten et al. 1989). Those of medium size can strip away the line-of-sight atmosphere, a fraction $H/2a$ of the total, where H is the scale height and a the planetary radius; for the current atmosphere the fraction is 0.0015, but it would be 22 times greater for a hydrogen atmosphere at the same temperature. Melosh and Vickery (1989) have treated the likely effects of a late heavy bombardment and found that it is likely to strip away the entire atmosphere. They make the interesting point that the decrease of mass can have a linear, rather than exponential, behavior because the mass of the smallest effective projectile decreases along with the mass of the atmosphere. It has been suggested (Cameron 1983) that a single huge impact could result in total loss, but there is no quantitative treatment justifying this.

It seems unlikely that such rapid ejections would have any mass selectivity. Their time scale is around $10H/\mathrm{v}$ = 100 s, where v is the escape velocity and the factor of 10 allows for the distance to the horizon. Times for diffusion, which is the most likely mechanism for any selectivity, are orders of magnitude greater. The bombardment is spread over 1 Gyr or so, much longer than any episode of hydrodynamic escape. It is not at all clear how both mechanisms could have operated, because each one is supposed to remove nearly the entire atmosphere.

Another aspect of impact is the gas and volatiles carried in by the impacting objects. Lange and Ahrens (1981) have studied the case of water and found that the atmosphere of an Earth-like body tends toward a steady state, with amounts comparable to the present ocean. The amount retained by Mars would presumably be much smaller. It is not at all clear that the noble-gas patterns could be explained.

VI. THERMOSPHERIC CIRCULATION

A. Mars One-Dimensional Model Precursors

Modeling efforts have just begun to address the state of the Mars upper atmosphere under various solar, seasonal and orbital conditions. One-dimensional models of Bittner and Fricke (1987) and Bougher and Dickinson (1988) have recently been constructed to examine the global mean heat balances necessary to reproduce the solar cycle variation in observed exospheric temperatures. Dayside mean temperatures are similarly calculated assuming minimal influence from global winds. Bittner and Fricke (1987) suggest that a long-term solar cycle temperature variation of roughly 200 to 400 K exists due to solar forcing, but is modulated by a stochastic component arising from variable eddy mixing and seasonal forcing (± 50 to 60 K). The Bougher and Dickinson (1988) one-dimensional model shows a similar long-term exospheric temperature variation extending over 200 to 350 K, for which a large rate of atomic oxygen collisional deactivation of CO_2 is used to promote 15-μm cooling in non-LTE situations. This one-dimensional model shows that the primary heat balance above 130 km occurs between EUV heating and molecular thermal conduction (much like Earth), with little influence from 15-μm cooling. A reasonable match of the observed and calculated dayside Mars exospheric temperature variation can thus be achieved (see Fig. 15). It is apparent that Mars dayside exospheric temperatures vary strongly as a function of solar activity (200 to 350 K), while those of Venus do not (240 to 300 K) (Bougher and Dickinson 1988).

These one-dimensional model calculations suggest that, to first order, solar EUV heating should drive the global winds of the Mars thermosphere. Effects due to auroral-joule heating are not expected to be important, as Mars apparently lacks a significant intrinsic magnetic field. Also, mechanical heating/cooling, due to tides and gravity waves propagating upward from below, may be important as a secondary influence upon the thermospheric structure. One-dimensional models, however, cannot provide the self-consistently calculated transport and adiabatic heating/cooling terms which drive departures from radiative equilibrium.

B. Mars Thermospheric Circulation Model

Further study of the Mars upper atmosphere requires the examination of three-dimensional circulation effects on dayside and nightside temperatures and densities. An initial study, examining solar heating alone, was done by Bougher et al. (1988*b*) using the NCAR thermospheric general circulation model (TGCM) (Dickinson et al. 1984). Global mean calculations (Bougher and Dickinson 1988) were used to develop a three-dimensional coupled chemical-dynamical model based upon experience gained from previous Venus thermospheric studies (Bougher et al. 1988*a*). This initial version of the

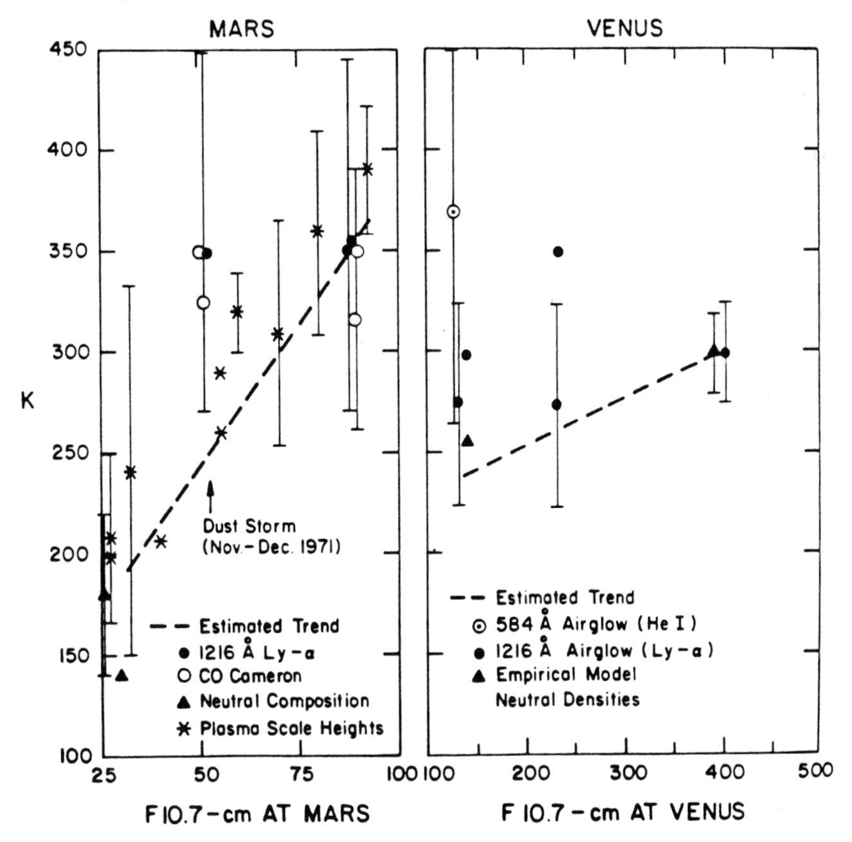

Fig. 15. Model exospheric temperatures: one-dimensional model estimates of dayside exospheric temperatures vs solar activity for Venus and Mars. Observed or inferred temperatures are plotted as a function of the $F_{10.7}$- cm flux received at each planet. Notice the large solar cycle variation for Mars and the rather small one for Venus. Dashed lines indicate the solar cycle trend calculated for Mars (Bougher and Dickinson 1988) and Venus (Dickinson and Bougher 1986). $F_{10.7}$ fluxes given are appropriate to specific observations, not monthly means (figure taken from Bougher and Dickinson 1988).

Mars TGCM predicted the state of the Mars upper atmosphere for mean orbital, seasonal and solar conditions (Bougher et al. 1988*b*), appropriate to the time of arrival of the Soviet Phobos 2 spacecraft at Mars in early 1989.

The NCAR TGCM code is based on the primitive equations in pressure coordinates of lower-atmosphere dynamic meteorology. The physical processes incorporated into the model, however, are those appropriate to thermospheric dynamics; i.e., fast molecular vertical diffusion of heat, momentum and constituents at thermospheric heights (see Dickinson et al. 1984). The Mars TGCM calculates global distributions of CO_2, CO and O consistent with the model day-night temperature and density contrasts and corresponding large-scale winds. Coriolis terms are included corresponding to the Mars rotational period of 88,800 s (24 hr 40 min). The same CO_2 chemistry and global CO_2 15-μm cooling parameterization previously used for Venus (Bougher et al. 1988*a*) are adopted for Mars. Results from the solar-driven Mars TGCM reveal that the Martian thermosphere has aspects found in both the Earth and Venus thermospheres. Planetary rotation has a significant influence on the Mars wind, composition and temperature structure, much like the Earth. In Fig. 16A, a pressure-level slice of temperature contours is superimposed on vectors representing the magnitude and direction of the calculated horizontal winds. The $z = 6$ surface corresponds to an average altitude of 210 km over the globe. The fields are presented on a latitude-local-time grid and illustrate a strong diurnal variation in temperature, with a maximum value (272 K) at $LT = 15$ hr, and a minimum value (161 K) at $LT = 05$ hr. A day-night temperature contrast of ~110 K is calculated. Such a basic diurnal behavior (offset from noon and midnight) is also seen for the Earth (see, e.g., Dickinson et al. 1981,1984). Winds with a maximum divergence near $LT = 16$ hr increase to roughly 230 m s^{-1} across the poles and the morning and evening terminators, then slow down to multiple convergence points near the morning terminator ($LT = 05$ to 07 hr) and on the night side ($LT = 20$ to 22 hr). Very strong gradients in horizontal and vertical winds are seen at these locations as a result of this three-dimensional flow convergence. No nightside cryosphere of the type observed for Venus (i.e., temperatures decreasing above the nightside mesopause) is suggested by this Mars TGCM calculation which neglects external momentum drag terms. The Martian rotation appears to preclude any effective isolation of the dayside and nightside thermospheres.

Vertical wind magnitudes are similarly plotted in Fig. 16B; upward flow is indicated primarily on the day side (<4 m s^{-1}); downward flow is generally seen on the night side (<9.6 m s^{-1}). These vertical winds give rise to strong adiabatic heating (nightside) and cooling (dayside) which is pronounced due to Mars' small radius. The descending vertical winds reach their maximum at the points of convergence of the horizontal winds ($LT = 06$ to 07 hr and $LT = 20$ to 22 hr) and provide a secondary warming effect which results in a "semi-diurnal" global temperature component in addition to the primary diurnal solar-driven temperature structure. Nightside heating sets up a pressure gradient that drives winds back toward the day side. Dayside upwelling winds, strongest at the equator, act to reduce net afternoon heating so that temperatures peak at $LT = 15$ hr near ±30° latitude, rather than at the equator where the solar heating is greatest. Significant departures from radiative equilibrium appear throughout Mars' thermosphere due to dayside divergence (upwelling) and nightside convergence (subsidence) of the winds.

Equatorial (2°.5 N) slices are given on a local-time vs log-pressure grid in Fig. 17 for temperature and density fields. The offset of maximum dayside

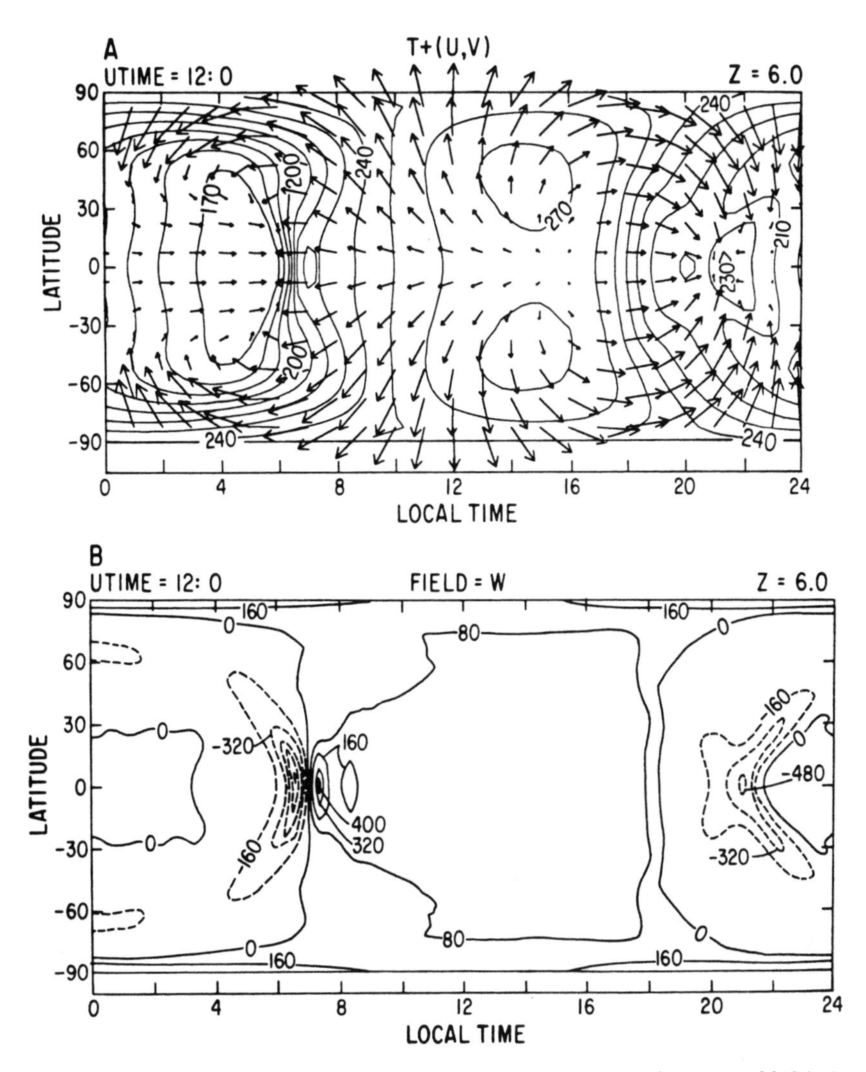

Fig. 16. Model pressure levels: Mars TGCM pressure level slices at $z = 6$ (average of 210 km) for: (A) T (temperature) + (u, v) (horizontal winds); temperature contours from 170 to 270 K with 10 K intervals; maximum wind vector of 230 m s^{-1}; and (B) w (vertical velocity); contours from -960 to 400 cm s^{-1} with 160 cm s^{-1} intervals.

temperatures from the subsolar point is not only a result of upwelling flow but is also due to the diurnal variation of solar heating and consequent thermal lags resulting from Mars' planetary rotation. The strongest gradients in horizontal temperatures appear at LT = 06 to 07 hr, where Fig. 16 indicates strong dynamical heating from convergent and descending winds. The maximum dayside exospheric temperature (269 K at the equator) is only 21 K

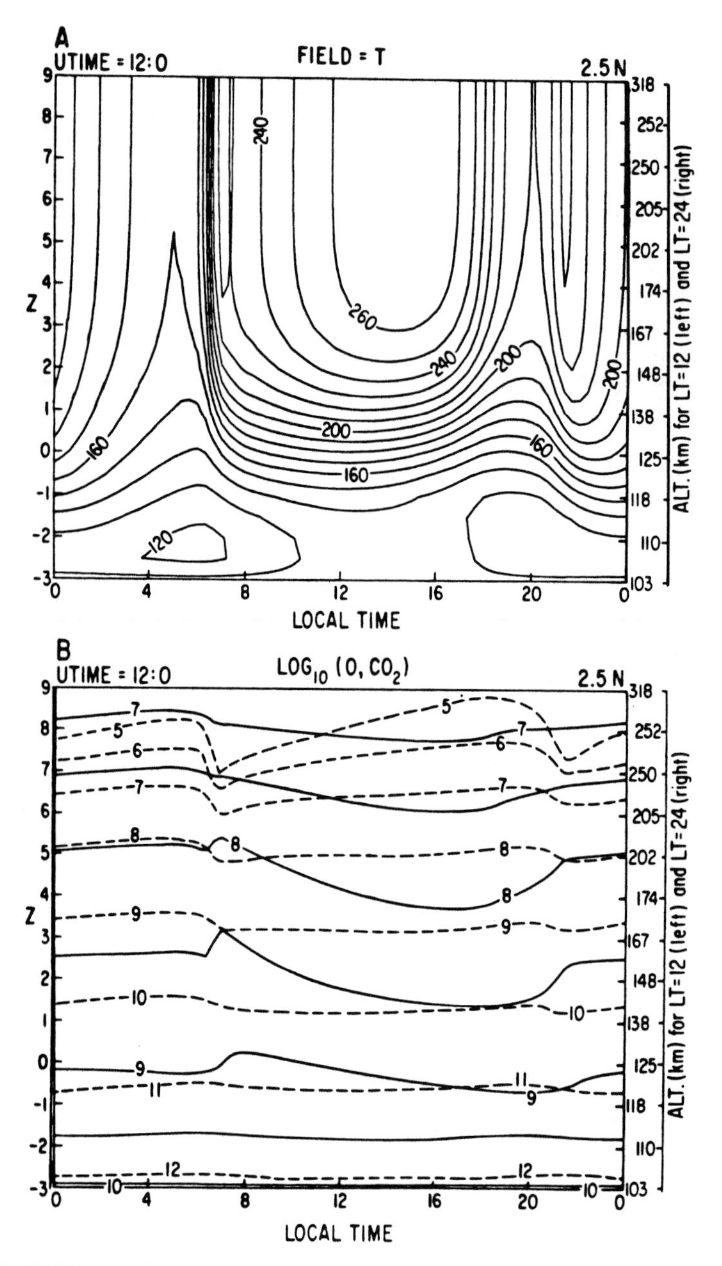

Fig. 17. Model equatorial calculations: Mars TGCM equatorial slices for: (A) T (temperature); contours from 120 to 260 K with 10 K intervals; and (B) $\log_{10}$ number density of atomic O (solid lines) and CO_2 (dashed lines). Atomic O contours from 7.0 to 10.0 with 0.5 intervals; CO_2 contours from 5.0 to 12.0 with 1.0 intervals. The vertical grid is log-pressure; corresponding local noon (left) and midnight (right): altitude scales are shown at the right margin.

warmer than the previously calculated global mean value of 248 K. This indicates that dynamics play a very important role in Mars' dayside thermal balance.

The calculated densities respond to the wind and temperature structure much as do those of Venus, with atomic oxygen being subject to global redistribution. Mars TGCM calculated densities are presented in Fig. 17B on the same latitude-pressure grid. Dayside CO_2 densities generally follow the temperature structure, showing maximum values from LT = 16 to 20 hr. Overall, CO_2 tracks the inverse of the O + CO densities, as it is calculated passively as the remainder of the total number density. Atomic oxygen, however, is more effectively transported by winds than CO_2 because of its lower atomic weight and correspondingly larger scale height. Calculated O densities show a maximum depletion in the region of peak divergence of the winds at LT = 16 to 20 hr, and a general nightside bulge consistent with the overall convergence of the winds (LT = 22 to 06 hr). The nightside O densities along $z = 6$ are as much as a factor of 2 greater than dayside values. Densities generally drop from the morning to the evening terminators (consistent with the large-scale flow), while temperatures rise by ~90 K due to solar heating. The diurnal O/CO_2 ratio along the $z = 0$ pressure surface varies from a minimum of 1.21% (LT = 19 hr) to a maximum of 2.23% (LT = 08 hr), giving an average F_1-peak value of 1.75%. Atomic oxygen surpasses CO_2 as the dominant species above ~210 km. A similar Mars TGCM calculation for Viking conditions (Bougher et al. 1989) yields an average F_1-peak value of ~1%. Hanson et al. (1977) estimated 1.25% from Retarding Potential Analyzer results.

This calculated day-night variation of atomic oxygen can be compared to Mariner 9 1304 Å airglow observations. Strickland et al. (1972) suggested that the rise of observed airglow intensity across the day side from the morning to evening terminator could be due to increasing O densities, increasing temperatures, or both. Mars TGCM calculations shown that divergent dayside winds deplete the afternoon sector of atomic oxygen, and redistribute it to the night side. Therefore, increasing temperatures across the day side seem to be required to give the 1304 Å airglow distribution observed.

The Mars TGCM CO_2 and O densities illustrated are maintained by the addition of limited eddy diffusion using coefficients of 2000 $m^2 s^{-1}$ or less. This suggests that the strong mixing of Mars' thermosphere that controls density profiles below the homopause (~125 km) is a product of both large-scale winds and small-scale processes, much like Venus (see, e.g., Bougher et al. 1988*a*). The calculated O/CO_2 mixing ratio at the F_1-ionospheric peak (1 to 2%) depends on the combined influence of large-scale global winds and reduced global mean eddy diffusion and is consistent with the range of proxy estimates of atomic oxygen derived from various measurements made over the solar cycle. The concept of a sharp homopause, defined as the boundary between molecular- and eddy-diffusion-controlled regions, needs to be modified on Venus and Mars to account for significant effects of large-scale winds (see, e.g., Bougher et al. 1988*a*).

Finally, a comparison of Mars TGCM derived exospheric temperatures over all local times is given in Fig. 18 for solar maximum (Mariner 6 and 7), solar minimum (Viking) and solar medium (Phobos) cases (Bougher et al. 1989). Dayside values are reduced somewhat from near radiative equilibrium model temperatures given in Fig. 18, due to global wind effects. Nevertheless, dayside mean Viking values of 195 K and Mariner 6 and 7 values of 310 K at LT = 15 hr are still obtained by the CO_2 Mars TGCM. A reduction of Mars TGCM calculated global winds by about a factor of 2 would raise the dayside temperatures on the solar medium case by ~25 K (see e.g., Bougher et al. 1988*b*). Mean nightside temperatures are predicted to vary from 145 to 185 K over the solar cycle.

This Mars TGCM benchmark for solar heating provides a basis for future Mars studies incorporating detailed tidal parameterizations for examining the influence of the lower atmosphere on the thermosphere. The Dickinson and Ridley (1977) Venus dynamical model, prior to Pioneer Venus observations, predicted strong nightside convergence and descending winds giving significant adiabatic heating. Later models suggested that momentum sink terms were needed to match the Pioneer Venus day-night temperature and density contrasts (see, e.g., Bougher et al. 1988*a*), thereby reducing the strong nightside heating effect and atomic oxygen bulge. Mars day-night ther-

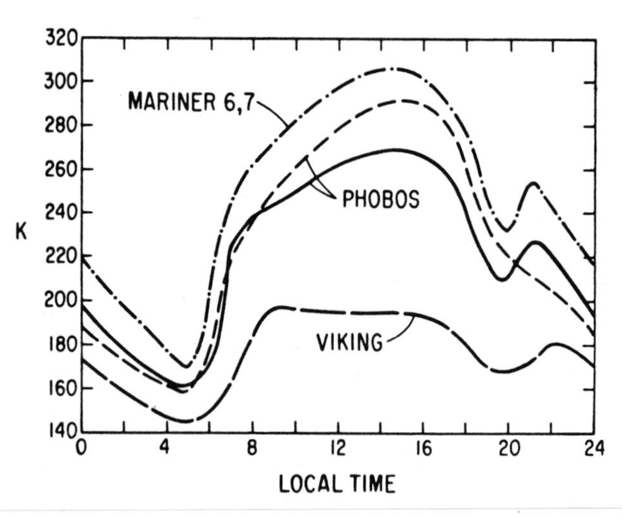

Fig. 18. Model exospheric temperatures: Mars TGCM calculated exospheric temperature variations over local time for three observations periods (a) Mariner 6 and 7 (solar maximum), (b) Phobos (solar medium) and (c) Viking (solar minimum). Two Phobos curves are presented: (1) solid—for no-wave drag, and (2) dashed—for Venus level wave drag (Bougher et al. 1988*a*) which reduces global winds by a factor of 2.

mospheric temperature and density contrasts may also be subject to similar forcings resulting from upward propagating gravity waves or tides (Zurek 1976). Significant effects are expected, as Mars tidal heating, driven by dust loading of the lower atmosphere, is known to be important. Thermospheric responses for various seasonal, orbital and solar conditions also need to be examined.

VII. THE FUTURE

The theory of the photochemistry, ion chemistry and airglow of Mars is now well developed and tested for the specific times and locations of spacecraft observations. The thermal structure is understood, although some of the parameters have been determined empirically. There is now the beginning of a thermospheric general circulation model which, however, has been tested only at a few points. The challenge for the future is to develop theories and models to describe the diurnal and seasonal changes that occur over the entire Martian globe. A critical test of such models would be observations made from orbiting spacecraft over a time period that is a substantial part of the solar cycle. The solar energy output varies with various periods: the daily rotation of Mars; the seasons of Mars; the 27-day rotation period of the Sun; the 11-yr cycle of solar activity; and episodic events associated with solar flares. Such tests of aeronomic models with such changing conditions will be severe, but the reward will be the ability to understand the past and to predict the future with greater confidence. This better understanding is most important for those processes that lead to the escape of atmospheric constituents and to the evolution of the atmosphere.

31. THE INTRINSIC MAGNETIC FIELD AND SOLAR-WIND INTERACTION OF MARS

J. G. LUHMANN, C. T. RUSSELL
University of California, Los Angeles

L. H. BRACE
University of Michigan

and
O. L. VAISBERG
Space Research Institute, Moscow

Plasma and field measurements on various spacecraft have demonstrated that Mars has an intrinsic dipole magnetic field that is no more than $\sim 10^{-4}$ *times as strong as that of the Earth. This difference is attributed to the weakness or absence of dynamo activity in the Martian core. An important consequence of the weak intrinsic field is a Venus-like interaction between the solar wind and the atmosphere. It has been found that this interaction leads to losses of Martian atmosphere constituents that are important for the atmosphere's evolution.*

INTRODUCTION

A. Intrinsic Planetary Magnetic Fields

It has been known for over 500 yr that the Earth possesses an intrinsic magnetic field, approximately dipolar, like that of a bar magnet (Gilbert 1600). This field is generated by a dynamo produced by convection in the electrically conducting, rotating, molten core (Gubbins 1974). In the last century astronomers have discovered that the Sun as well as stars in general also generate their own magnetic fields. One of the first discoveries of radio astronomy was the existence of emissions from Jupiter (Burke and Franklin 1955) which enabled astronomers to discover and remotely measure the magnetic field of Jupiter. Since the advent of interplanetary space flight, we have had the opportunity to measure directly the magnetic fields at Jupiter and many of the other planets. Some planets, like Jupiter, Saturn and Uranus, have intrinsic magnetic moments much larger than that of the Earth. Others, like Mercury, Venus and Mars, possess much weaker magnetic moments. The reasons for these differences, which have to do with the properties of the interiors of these planets and the nature of their dynamos, are subjects of continuing study.

Of particular interest are the differences between the fields of the terrestrial planets Earth, Venus and Mars. The history and current state of the interiors of our two neighboring planets appear to be considerably different from those of Earth. Venus is distinguished by the fact that it is a slower rotator, (243-day period) while it is similar in size to Earth (6050-km radius vs Earth's 6370 km). Mars is distinguished by its smaller size (1 $R_M \approx$ 3390-km radius) while it rotates at about the same rate as Earth ($\sim$24.6-hr period). The question of how these differences translate into different internal properties has been under investigation, but the uncertainty in the results of these studies is compounded by the lack of seismic data from Venus and Mars. In any event, it is clear that the weak magnetic field of Mars would present a puzzle if size and rotation rate were the only dominant factors controlling planetary dynamos because Mars is not unlike Earth in those respects.

B. Solar-Wind Interactions

When a planetary magnetic field is weak, like that of Mars, the solar wind strongly distorts its configuration. For this reason, it is difficult to infer the strength of the intrinsic magnetic field from observations without understanding the solar-wind interaction.

The expanding magnetized plasma atmosphere of the Sun, the solar wind, produces a quite different environment when it interacts directly with a planetary atmosphere than when it interacts with a planetary magnetic field. Nevertheless, there are a few elements common to both. These common elements, which are illustrated in Fig. 1, include a bow shock standing in the solar wind upstream of the "obstacle" that slows and deflects the flow, and a region between the shock and the obstacle, known as the magnetosheath, in which the solar-wind plasma and magnetic field are compressed. If the interplanetary magnetic field is oriented at a substantial angle to the upstream flow, the field appears to drape around the obstacle. The differences between the atmospheric and magnetic field obstacle, which are more numerous, are schematically illustrated in Fig. 2.

Figure 2 shows the basic features of the interactions of several kinds of planetary obstacles with the solar wind. In the case of a dipole field, the

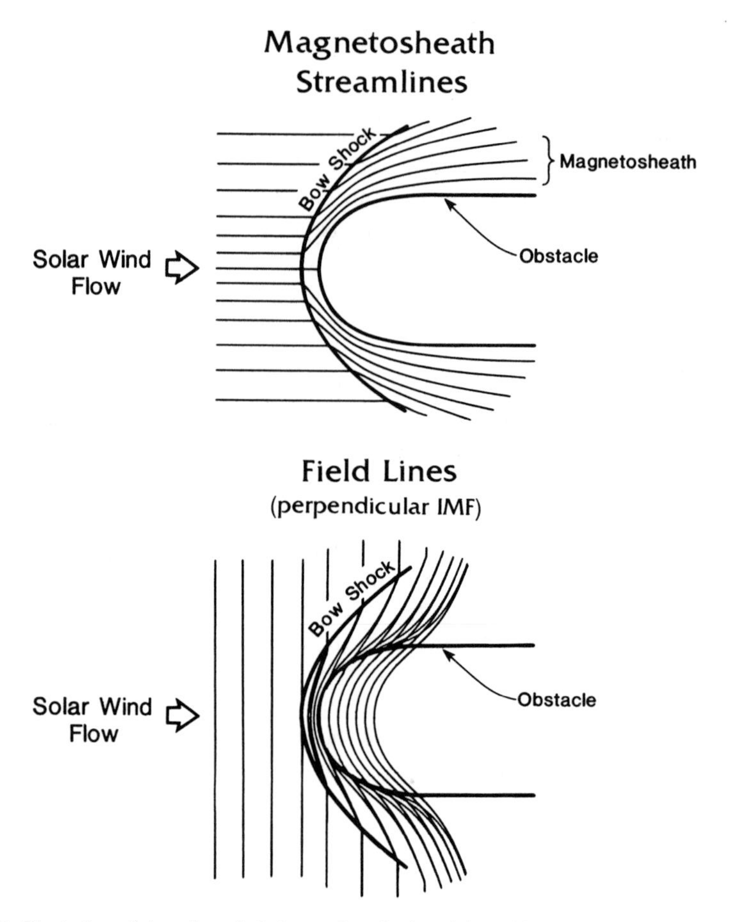

Fig. 1. Illustration of the solar-wind plasma flow (top) and draped interplanetary magnetic field (bottom) in a planetary magnetosheath.

obstacle is defined by the boundary between the shocked solar wind and the region dominated by the dipole field where the component of the solar-wind pressure normal to the surface equals the magnetic pressure of the internal field (Fig. 2a). This boundary is called a magnetopause. In the case where the atmosphere of a weakly magnetized planet is responsible for the analogous boundary, an ionopause forms where the pressure of the ionospheric plasma balances the incident solar-wind pressure (Fig. 2b). The latter is transformed to magnetic pressure above the ionopause where it forms an effective magnetic barrier between the solar-wind plasma and the relatively cold ionospheric plasma (although ionospheric ions are still produced within the barrier). If the solar-wind pressure exceeds the ionospheric pressure, the location of the ionopause can be modified by a magnetic field that is induced

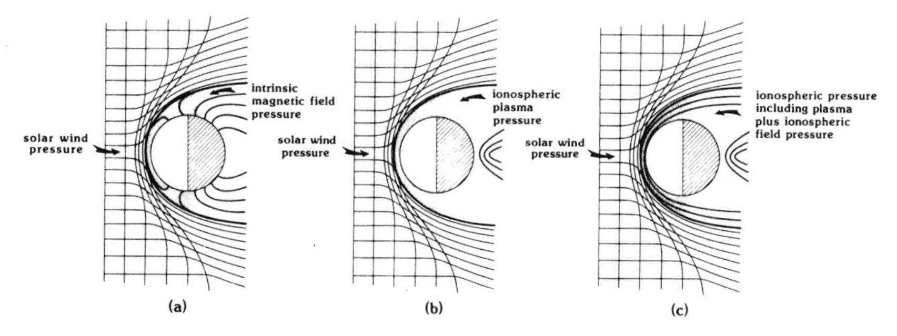

Fig. 2. Schematic sketch showing three types of planetary obstacles to the solar wind: (a) a magnetospheric obstacle where the incident solar-wind dynamic pressure is balanced by the pressure of the internal dipole field; (b) an atmospheric or ionospheric obstacle where the thermal pressure of the ionospheric plasma balances the solar-wind pressure; and (c) an ionospheric obstacle whose internal pressure is supplemented by an ionospheric magnetic field of interplanetary origin.

in the ionosphere by the solar-wind interaction (Fig. 2c). Since the scale length of gradients of pressure for planetary magnetic fields is the order of a planetary radius, while the scale length of a planetary atmosphere or ionosphere is only 10s to 100s of km, we expect the obstacle associated with a weakly magnetized planet to be much closer to the planet and much less compressible than the obstacle created by a magnetized planet. If an intrinsic magnetic field of marginal strength is present, the boundary is likely to be complicated by characteristics of both magnetospheric and atmospheric obstacles.

Different types of "tails" form in the wakes of magnetospheric and atmospheric obstacles. The intrinsic field obstacle has a tail composed of stretched-out dipole field lines attached to the magnetic polar regions of the planet, while the atmospheric obstacle has a comet-like tail composed of interplanetary or solar-wind field lines which appear to have become hung up in the ionosphere as they slip around it. Moreover, the flowing solar wind has an associated electric field that removes the ionospheric ions produced in the upper atmosphere above the ionopause, creating both an upper boundary to the ionosphere and a wake of planetary ions. Intrinsic planetary fields often shield a planetary atmosphere against this type of solar-wind "scavenging" (although it can still occur if the magnetopause is low enough). Below, we will see how observations related to these and other features of the solar-wind interaction have been used to argue for and against appreciable Mars magnetism.

Our position to assess the situation at Mars has been improved considerably by the Pioneer Venus Orbiter. As a result of this long-lived mission with its low-altitude phase, we have an unprecedented picture of the atmospheric obstacle-solar wind interaction with which to compare our more intimate knowledge of the terrestrial magnetic field obstacle-solar wind interaction. Analogies with Venus and Earth have formed the basis of many of the studies of the far more limited database for Mars. It will become clear that, at Mars, our understanding is, in the end, limited by our relatively inadequate *in situ* observations.

C. History of Measurements

Although there has recently been renewed activity in the space-based exploration of Mars with the launch of the two Phobos spacecraft (see the special Phobos issue of *Nature,* October, 1989), many earlier reviews of the results of Mars missions are still quite complete in terms of their scientific conclusions (see e.g., Vaisberg and Bogdanov 1974; Bogdanov and Vaisberg 1975; Russell 1979,1981; Intriligator and Smith 1979; Breus and Gringauz 1980; Slavin and Holzer 1982). Table I lists those missions during which *in situ* measurements of magnetic fields or other experiments with some direct relevance to the subjects of the magnetic field, ionosphere and solar-wind interaction have been carried out, including the latest Phobos 2 mission (Phobos 1 failed prior to arrival at Mars). Primarily because of radio-occultation experiments, which require minimal onboard experiment-specific instrumentation, this table constitutes a comprehensive list of almost all of the missions to Mars.

To get an idea of how thoroughly the space around Mars has been probed, it is useful to review the orbits of the spacecraft which made the *in situ* measurements. The spacecraft trajectory is typically displayed in a coordinate system where x is along the Mars-Sun direction and the orthogonal coordinate is the perpendicular distance of the spacecraft from that axis. Figure 3 contains a collection of such plots from various sources. From Fig. 3 and the periapsis altitudes in Table I, it is seen that, with rare exceptions, *in situ* data are not available from altitudes below $\sim$800 km and from the wake region. The exceptions are the Viking mission Landers, which made direct ionospheric measurements during their entry although those measurements did not include the magnetic field, and Phobos 2, which was the first spacecraft to probe the central wake. Remote sensing measurements of the ionosphere by radio-occultation methods are possible whenever the line of sight between the spacecraft and the Earth intersects the ionosphere. However, this line-of-sight requirement does not permit coverage of the subsolar ionosphere (solar zenith angles $\leq$45°) or antisolar ionosphere because of the Mars-Earth-Sun geometry.

D. Organization of this Chapter

Central to the present chapter is the issue of the Martian intrinsic magnetic field. Although another chapter (chapter 5) deals with the details of the interior of Mars and its associated dynamo activity, a brief overview of the work that has been done on the subject, independent of solar-wind interaction

TABLE I
Summary of Mars Measurements Relevant to the Magnetic Field and Solar-Wind Interaction

Spacecraft	Year of Encounter	Closest Approach	Relevant Instruments/ Experiments	Features Detected
Mariner 4 (USA)	1965	~13200 km (flyby)	magnetometer; energetic-particle detectors; radio-occulation experiment	bow shock; magnetosheath; ionosphere density profiles
Mariner 6, 7 (USA)	1970	flyby	radio-occultation experiment	ionosphere density profiles
Mars 2, 3 (USSR)	1971	~1100 km (orbiters)	magnetometer; ion/electron traps; plasma analyzer; radio-occultation experiment	bow shock; magnetosheath; planetary ion mantle; intrinsic field (?); ionosphere density profiles
Mariner 9 (USA)	1972	flyby	radio-occultation experiment	ionosphere density profiles
Mars 5 (USSR)	1974	~1800 km (orbiter)	magnetometer; ion/electron traps; plasma analyzer; radio-occultation experiment	bow shock; magnetosheath; planetary ion wake; magnetotail (?); ionosphere density profiles
Viking 1, 2 (USA)	1976	Landers	retarding-potential analyzer; radio-occultation experiment	ionosphere density, temperature, composition profiles
Phobos 2 (USSR)	1989	~850 km (orbiter)	magnetometer; plasma composition analyzers; Langmuir probe; energetic particle detector; plasma wave detector	bow shock; magnetotail; planetary ion mantle and wake

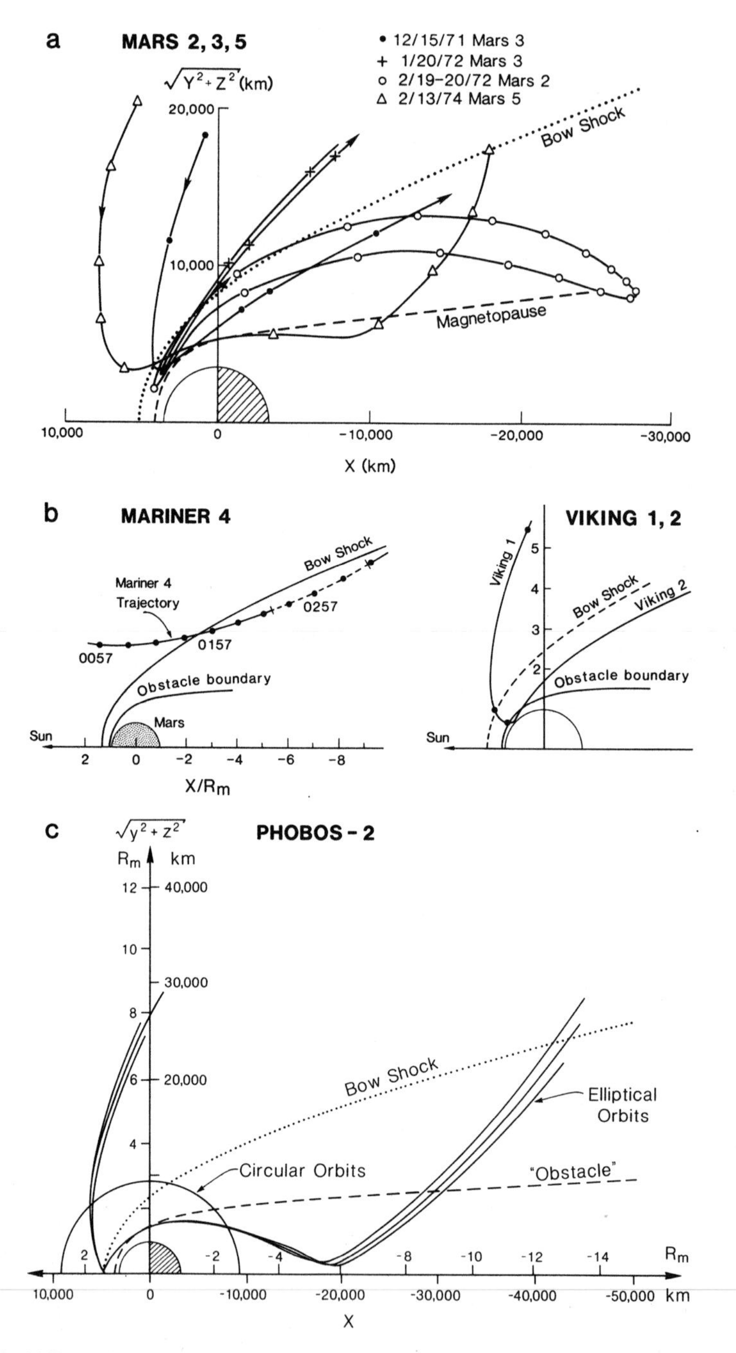

Fig. 3. (a) Examples of orbits of the Mars series of spacecraft in a cylindrical coordinate system where x points toward the Sun and $(y^2 + z^2)^{1/2}$ is the distance from the x axis. (b) Trajectories of the Mariner 4 flyby and Viking 1 and 2 Landers. (c) Trajectory of the Phobos 2 spacecraft during three of its elliptical orbits and in its circular orbit phase. Data from over 50 circular orbits were obtained.

observations and interpretations, is given here for the sake of perspective (Sec. II). This work deals principally with predictions based on dynamo theories of field generation in a rotating fluid sphere undergoing turbulent convection in its interior, and with theories concerning the properties of the cores of planets. A review of the inferences of Martian intrinsic magnetism from both *in situ* probes and meteorites follows.

The largest section of this chapter (Sec. III), like the largest proportion of published material, is devoted to the solar-wind interaction with Mars. The major features of the solar-wind interaction—the bow shock, magnetosheath, obstacle and tail, and their implications for the intrinsic magnetic field—are considered. Subsequent sections, which are devoted to the comparisons of Mars with Venus and Earth observations (Sec. III.G), emphasize the importance of comparative analyses in our ability to interpret what has been seen at Mars. Lastly, the related issues of the solar-wind interaction with the Martian satellites Phobos and Deimos are discussed with an eye toward their possible perturbations of the Mars environment (Sec. IV).

The discussion is selective in coverage as the aim of this chapter is to provide an overview of the current state of understanding of the subject and the basis for that understanding, rather than to provide an annotated bibliography. However, an effort has been made in the citations to maintain an accurate historical perspective.

II. MAGNETIC FIELD OF MARS

A. Dynamo Theory

For many years, in the absence of an adequate theory of planetary dynamos, it was conjectured that there was a "magnetic Bode's law," also called the Shuster or the Blackett hypothesis, which established that the magnetic moments of the planets (and stars) were proportional to their angular momenta (see Russell 1987). As shown in Fig. 4a, the magnetic moments of Jupiter and the Earth both fall on a line of proportionality on a plot of magnetic moment vs angular momentum if the angular momentum of the Earth-Moon system is used for the Earth. The fact that it is not obvious whether to use the angular momentum of the Earth or the Earth-Moon system is a clue to the lack of a physical basis for this hypothesis. The other planets with intrinsic fields lie significantly below (Saturn and Uranus) or significantly above (Mercury) the line. The two weakly magnetized planets (Mars and Venus) lie far below the line.

The concept of the dynamo generation of magnetic fields was first developed for the Earth (see review by Gubbins 1974). While no completely satisfactory dynamo theory currently exists, our present understanding suggests that the existence of a dynamo-generated planetary field requires rotation, an electrically conducting fluid core, and a significant source of energy in that core. It is thought that turbulent or nonaxisymmetric convection in the

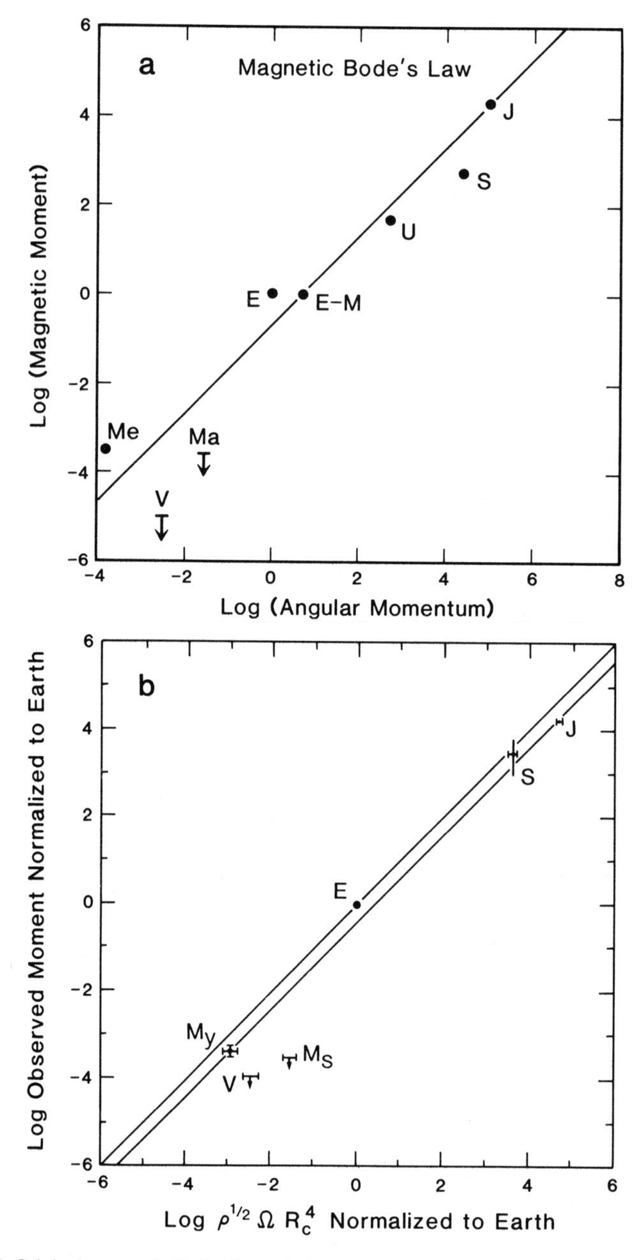

Fig. 4. (a) Original magnetic Bodes law relationship between planetary magnetic moment and angular momentum. (b) Dynamo theory-based modification derived by Busse (1976).

conducting core in the presence of rotation is essential to maintain the dynamo process. Busse (1976) applied one model of the dynamo process to create the physically based scaling law shown in Fig. 4b. This scaling law predicts that the planetary magnetic moment will be proportional to the rotation rate times the fourth power of the core radius. It is somewhat better at predicting magnetic moments than the original magnetic Bode's law. However, as before, Venus and Mars have magnetic moments far less than would be expected for active dynamos.

Planetary cores are quite inaccessible to empirical investigation by means other than modeling based on magnetic, seismic and gravitational measurements. Magnetic probing involves either theoretical interpretation of the measured intrinsic field in terms of dynamo models or electromagnetic "sounding" wherein the response of the near-planet field to interplanetary magnetic-field variability is measured (see Russell 1987). Seismic probing involves the inversion of seismographic data from instruments distributed on the planet's surface. As neither of these data sets are available for Mars, the properties of its core are estimated from other information. For example, the mean density of Mars, obtained from its radius and gravitational field, is only 3.9 g cm^{-3}, compared to 5.5 g cm^{-3} for Earth and 4.9 g cm^{-3} for Venus. If one presumes a mantle similar to that of the Earth, this implies a core radius of 1500 to 2000 km (Johnston and Toksoz 1977). Suggestions regarding the phase state and dynamical properties of the core range from a frozen core (Young and Schubert 1974) to a liquid core with minimal turbulent convection (Toksoz and Hsui 1978); however, these conclusions rely on the knowledge that the magnetic field is weak rather than on independent information about the core.

Because so little is known about the core of Mars, one is free to adjust its assumed properties constrained only by the size, mass, rotation rate and moment of inertia. Given these quantities, and the above core size estimate, the magnetic scaling law tells us that Mars should have a magnetic dipole moment of $\sim 10^{-2}$ times that of Earth if it behaves in a manner consistent with many of the other planets of the solar system. If Mars does not follow the trend of the other planets, it must have some distinctive internal properties that preclude the normal planetary dynamo action. As noted earlier, successful planetary dynamo operation requires an appropriately vigorous convection in the core and a long-lived internal energy or heat source to sustain that convection. In planets like the Earth, the most efficient source of energy for convective motion in the core is the release of gravitational and chemical energy during the course of formation of a solid inner core (Stevenson 1974; Stevenson et al. 1983). The conclusion is that Mars, which as we shall see has a "deficient" magnetic moment of $<10^{-4}$ that of Earth, is a planet with an entirely fluid or a completely solid core, or whose core solidification has ceased. Of course, it is possible that Mars had an active dynamo long ago, and if so, at least parts of the surface of Mars should possess remanent magnetization. However, it will be shown that remanent magnetization seems insufficient to play a major role in the solar-wind interaction with Mars.

B. Spacecraft Observations: Direct Observations

The first reports of the possible direct detection of a Martian magnetic field by a spacecraft experiment came from investigators analyzing data from the Soviet Mars spacecraft (see chapter 3). Dolginov and co-workers concluded that they observed the intrinsic field of Mars near the point of closest approach of the Mars 3 spacecraft (Dolginov 1976) and near the wake of the planet on the Mars 5 spacecraft (Dolginov 1978*a*). However, while supporting a detection of field lines directly connected to the planet, subsequent investigations made drastically different inferences as to the orientation and magnitude of the intrinsic field of Mars. For example, one analysis concluded that there was a feature similar to the Earth's magnetospheric cusp in the Mars 3 data, which implies a near-equatorial magnetic pole (Smirnov et al. 1978), while others resulted in the conclusion that the magnetic axis was nearly coincident with the rotation axis (see, e.g., Dolginov 1976). Meanwhile, the basic matter of the identification of the measured fields as planetary in origin was called into question by Wallis (1975) and Russell (1978*a,b*). The interpretations of these early observations continue to be controversial.

In particular, we know that no matter what the nature of the Mars obstacle to the solar wind is, it must have a magnetosheath as depicted in Fig. 1. As Wallis (1975) and Russell (1978*a,b*) have emphasized, allowance for the existence of this magnetosheath must be made in any accurate interpretation of the observed magnetic field. The magnetosheath field can also exhibit considerable structure and boundary-like features because the interplanetary magnetic-field orientation and strength are often variable, so particular care must be taken in determining the location of its inner boundary. One must accurately model the pile up of interplanetary magnetic field that occurs in a planetary magnetosheath to determine which features of the magnetic field observed at Mars *cannot* be explained by the solar-wind interaction. With one exception (see Russell et al. 1984), this was not done with the Mars spacecraft data. The latest Phobos 2 spacecraft magnetic observations, which provided unambiguous magnetotail data, indicate that at least above $\sim$2.7 R_M the structure consists primarily of draped interplanetary fields as in the "induced" magnetotail of Venus (Riedler et al. 1989*a;* Yeroshenko et al. 1990). Thus, direct observations of a planetary field remain elusive.

C. Meteoritic Evidence

Three families of meteorites appear to have originated in the Martian crust. These are the Shergottites, Nakhlites and Chassigny, the so-called SNC meteorites. The evidence for their Martian origin has been reviewed by McSween (1985; see also chapter 4). The SNC meteorites are similar in mineralogy and chemistry to terrestrial rocks but have distinctive chemical and isotopic compositions. Trapped gases in Shergottite shock melts have compositions similar to those measured by the Viking spacecraft on Mars (Bogard and Johnson 1983) These meteorites were all ejected from their parent body relatively recently. They appear to have crystallized at shallow levels in the crust of their parent body ~1.3 Gyr ago or less from magmas produced by partial melting of previously fractionated source regions. All Shergottites appear to have experienced a shock ~180 Myr ago, which may be the time they were ejected into space following the impact of a large projectile. However, cosmic-ray exposure ages are significantly less than this, ranging from 0.5 to 2.4 Myr (Bogard et al. 1984). Since calculations indicate that 2/3 of the material reaching the Earth from a Martian ejection takes longer than 10 Myr to arrive (Wetherill 1984), these ages suggest that the meteorites spent much of their time in space as part of larger objects, perhaps >6 m in diameter.

Iron-bearing rocks like these meteorites acquire a natural thermal remanent magnetization when they cool through their blocking temperature in a magnetic field. They also can be magnetized when shocked or when sitting for long periods in a strong magnetic field such as that of the Earth. The latter magnetization should be "soft" and easily removed. Cooling in a background magnetic field should give a very "hard" remanent magnetization. A hard component of magnetization is found in some of the Shergottites. For the Shergottite EETA 79001, Collinson (1986) estimates the magnetizing paleofield intensity to be from 10^3 to 10^4 nT. For the Shergotty meteorite (the first Shergottite identified) itself, Cisowski (1986) estimates a paleofield strength of 250 to 1000 nT. While conceivably this hard remanence could have been imposed at the time of the shock event that liberated the meteorite from Mars, it seems more likely that the remanence was acquired at the time of crystallization 1.3 Gyr ago. Shock magnetization would require a significant magnetizing field much greater than the present-day limits on the global Martian magnetic field to be present only 180 Myr ago.

Of the estimates of paleointensity, the most certain appears to be that of Cisowski (1986), who used a modification of a particularly reliable method called the Thellier-Thellier technique. Even this technique does not prove that Mars had an active dynamo when the Shergotty meteorite crystallized, because a field as high as 250 nT could conceivably be provided by remanent magnetic fields acquired by the surrounding crust billions of years earlier. However, the most plausible, albeit by no means proven, explanation for the Shergotty paleointensity is the presence of an active dynamo 1.3 Gyr ago, when Mars was more geologically active, which generated a magnetic moment greater than 10^{13} T-m^3 (Earth has a moment of 8×10^{15} T-m^3). We note that the implied surface field strength of 250 to 1000 nT is more similar to that of Mercury than present-day surface fields of the Earth, Jupiter, Saturn or Uranus.

III. SOLAR-WIND INTERACTION

A. Solar Wind at the Orbit of Mars

The properties of the solar-wind plasma at the orbit of Mars at 1.5 AU heliocentric distance can be obtained by extrapolation of its well-determined parameters at 1 AU. The average radial velocity of $\sim$400 km s^{-1} is expected to remain constant with radius. The density, which follows a $1/r^2$ law for the assumed spherical expansion, should be reduced to 1–2 ions cm^{-3} at Mars, while the proton temperature cools to $\sim 4 \times 10^4$ K. According to the Archimedean Spiral model of the interplanetary field, the magnetic field of $\sim$1 to 5 nT should lie at an angle of $\sim$ 56° from the Mars-Sun line, approximately in the ecliptic plane. These extrapolations have been more or less confirmed by plasma instruments on the Mars spacecraft (Gringauz et al. 1974,1976). If Mars was unprotected from the solar-wind, its cross section to the flow would intercept $\sim 5 \times 10^{25}$ solar-wind protons (of energies $\sim$ keV) per second. As noted earlier, a critical quantity in defining the planetary obstacle is the solar-wind pressure. The incident momentum flux or dynamic pressure at Mars is nominally $\sim 9 \times 10^{-9}$ dyne cm^{-2} at the subsolar point, but this, like the other quantities, has a range of variability as suggested by Fig. 5, which was derived from measurements at Venus (Phillips et al. 1984) and scaled for Mars. The peak thermal pressure of the Martian ionosphere from the Viking Lander observations of Hanson and Mantas (1988) is shown on this figure for comparison. The solar-wind pressure usually exceeds the ionospheric thermal pressure at Mars, at least for the solar activity minimum conditions that the Viking Lander observations represented.

Of course, as at the Earth, the solar wind at Mars will be sporadically perturbed by the passage of interplanetary shock waves and other solar-wind disturbances. Mars will also be occasionally subjected to an influx of energetic ($\sim$10 keV to 10 MeV) ions and electrons associated with these shock waves or with flares on the Sun. These events are noteworthy because the weak Martian magnetic field cannot deflect the high-energy particles away from the planet as does the field at Earth; some will penetrate the thin Martian atmosphere and deposit the bulk of their energy in the planet's surface. The absorption of energetic particles was observed on Phobos 2 as a "shadow" or disappearance in their flux in the hemisphere opposite that where the unperturbed upstream interplanetary field would intersect the planet (Afonin et al. 1989).

B. Bow Shock

1. Observations. Probably the best-observed feature of the Mars-solar wind interaction, and the most-used intrinsic field strength diagnostic, is the bow shock: a standing magnetosonic wave in the solar-wind plasma that forms to force the solar wind to be slowed and diverted around an obstacle to

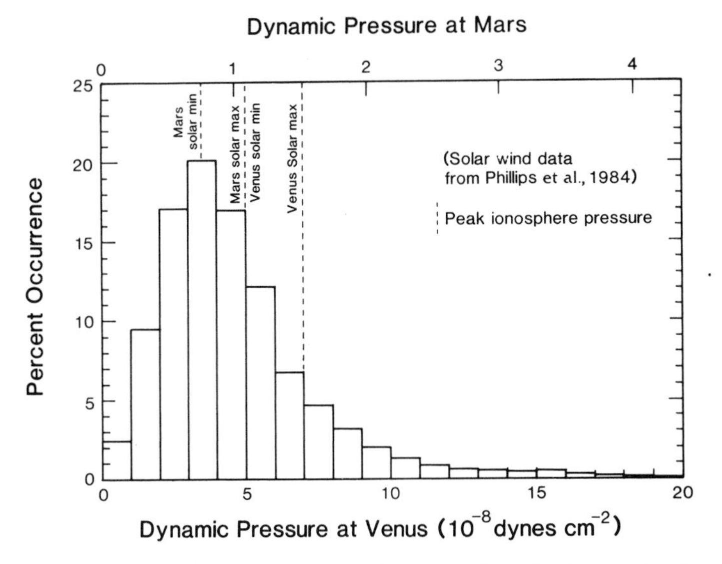

Fig. 5. Statistics of solar-wind dynamic pressure observed at Venus (bottom scale) and extrapolated to Mars (top scale). The dashed lines indicate how often the solar-wind pressure exceeds the peak ionospheric (thermal) pressure at both planets. The "Mars solar min(imum)" value was obtained from the Viking Lander measurements (Hanson and Mantas 1988). The peak thermal pressure at Venus at solar minimum should be comparable to that of Mars at solar maximum.

its flow. The bow shock was easily observed in the solar-wind plasma upstream of the planet by both particle and field experiments on the various spacecraft visiting Mars; see, e.g., the reviews by Slavin and Holzer (1982) and Russell (1985). In fact, the bow-shock crossings observed in the data from the magnetic-field experiment on Mariner 4 (Smith et al. 1965), reproduced here in Fig. 6, were the first unambiguous indication of the small size of the Mars obstacle to the solar wind and hence the weakness of the Martian intrinsic field. The magnetometers detected the bow shock as a distinctive jump (of up to a factor of ~4) in the magnetic-field magnitude (Smith et al. 1965; Dolginov 1976,1978*a*). Particle experiments capable of detecting plasma at solar-wind energies saw corresponding sudden increases in density and temperature, and decreases in flow velocity (Vaisberg 1976*a,b;* Gringauz 1976). Other plasma experiments not designed for detecting protons of solar-wind energy (~keV) detected shock crossings as changes in their backgrounds (see, e.g., Cragin et al. 1982), which are sensitive to the properties of the plasma surrounding the spacecraft.

The various bow-shock observations at Mars collectively provide an indication of the shape of its surface of revolution as shown in Fig. 7. A standard precedure in bow-shock studies is to fit the shock data with a conic section curve of elliptical form:

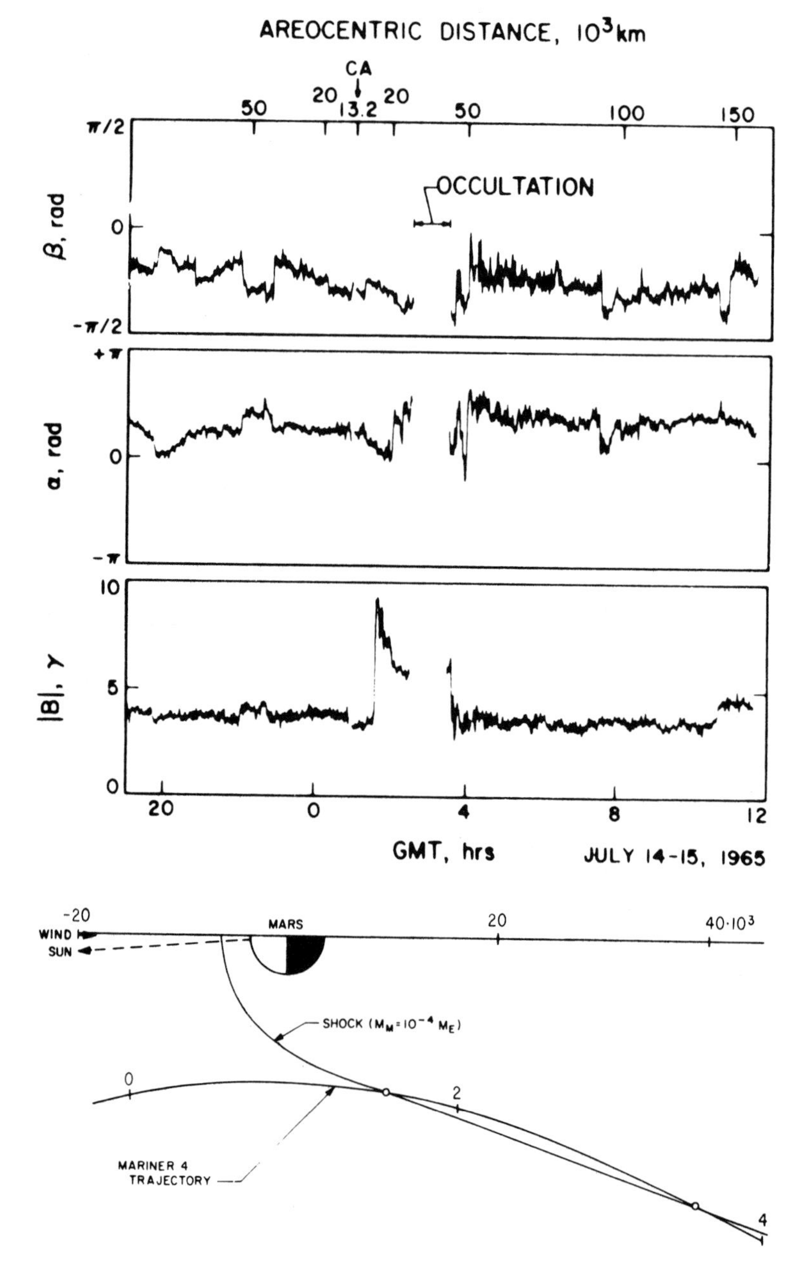

Fig. 6. First spacecraft observation of the perturbation of the solar wind by Mars. These magnetometer data from Smith et al. (1965) showed that the Martian obstacle to the solar wind must be on the order of the planet size. The panels show the polar and azimuthal angles of the field direction and the field magnitude. The spacecraft trajectory is labeled with GMT hr.

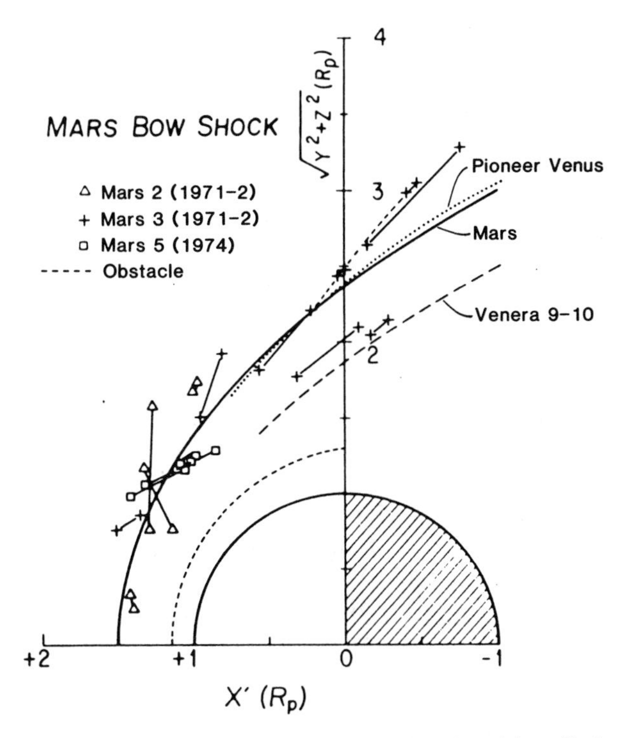

Fig. 7. Compilation of observed Martian bow-shock crossings adapted from Slavin and Holzer (1982) showing the best-fit conic section, the obstacle shape inferred from gas-dynamic models, and the relative positions of the Venus bow shock observed at solar maximum (from Pioneer Venus Orbiter) and minimum (from Venera).

$$R = \frac{K}{1 + \varepsilon cos\theta} \tag{1}$$

where R is the radial distance from the center of Mars, K is the semi-latus rectum or radius of the shock at the terminator (day-night boundary) plane, ε is the eccentricity, and θ is the solar zenith angle (the angle between the Mars-Sun line and the radius vector to the point of interest). Many investigations have used curves of this form to fit the bow-shock data (see, e.g., Slavin and Holzer 1981,1982; Slavin et al. 1983; Russell 1985); however, as the authors of these studies noted, the inferred shape is subject to orbital sampling bias. In the most recent study by Slavin and Holzer (1982), care was taken to eliminate those shock crossings farthest from the Sun, which the investigators felt were the most strongly biased. Their best-fit curve, shown superposed on the data in Fig. 7, has an eccentricity of 0.94 and a semi-latus rectum of 1.94 planetary radii. These values were derived with the additional assumption that the origin could be moved from the center of Mars. Thus, they refer to a coordinate system with its origin 0.5 planetary radii in front of the center of Mars so that $K = 1.94$ R_M is the distance from the x axis at $x = 0.5$ R_M (instead of the terminator plane $x = 0$). As seen in Fig. 7, their subsolar shock position is at $\sim x = 1.5$ R_M and the terminator shock is at ~ 2.4 R_M. Using essentially the same data, Vaisberg et al. (1976) derived a shock that was closer to the subsolar point by ~ 0.1 R_M. The most recent observations from Phobos 2 show that at encounter the subsolar distance of the bow shock from the center of Mars was also ~ 1.5 R_M, but the terminator distance was ~ 2.7 R_M (Schwingenschuh et al. 1990). The terminator-radius difference could be related to the effect of the level of solar activity (which we discuss below).

2. Gas-Dynamic Models. Another standard procedure in studies of the bow-shock shape is to compare the empirically determined shape with models of shocks formed by hypersonic hydrodynamic or gas-dynamic flow around cylindrically symmetric obstacles. The pioneers in this area were Spreiter and his co-workers, who first used the hypersonic-flow analogy for the bow shock of the Earth; later it was applied to the Mariner 4 observations (Spreiter and Rizzi 1972; also see Dryer and Heckman 1967) for the purpose of deducing the obstacle size at Mars and hence inferring its intrinsic field strength.

The reason why the gas-dynamic flow analogy is applicable to magnetosonic bow shocks is that the effect of the magnetic field on the shock (which is what distinguishes it from a gas-dynamic shock) is negligible when the upstream Alfvén Mach number (the square root of the ratio of plasma kinetic-energy density to the magnetic-energy density) is ≥ 8. Under these circumstances, one can simply assume that the magnetic field is convected and compressed with the flow. At Mars, the magnetosonic Mach number (the square root of the ratio of plasma kinetic-energy density to the thermal plus magnetic-energy density) of ~ 7 justifies the use of this approximation and hence the use of the gas-dynamic model to infer the obstacle size and shape.

From their early analyses of the Mariner 4 shock observations, Spreiter and Rizzi concluded that the size of the Mars obstacle to the solar wind was consistent with what would be expected if the planetary ionosphere alone deflected the flow (also see Spreiter et al. 1970). As mentioned in the introduction, the key to determining the obstacle is the consideration of pressure balance of the solar-wind pressure by either magnetic pressure ($B^2/8\pi$, where B is the field in nT and the pressure is in dyne cm^{-2}), plasma thermal pressure (nkT, where n is density, k is Boltzmann's constant and T is the sum of the ion and electron temperatures, all in cgs units), or a combination of both. The solar-wind pressure can be evaluated in the upstream region where it is practically all dynamic pressure (ρV^2 where ρ is mass density and V is the flow speed) with a correction for the angle of incidence (as only the pressure normal to the boundary surface matters in the obstacle determination). The authors calculated an obstacle shape for an ionosphere with a constant scale height (e.g., with a pressure profile that decreases exponentially with altitude) and then calculated the shock shape that it would produce. Thus, in order to conclude that the observed shock was consistent with an ionospheric obstacle, they had to make some assumptions about the ionosphere. Guidance was provided by the radio-occultation experiment on Mariner 4, which gave information about the ionospheric electron densities and plasma temperatures (from the scale height).

In the same way, more definitive assessments of the obstacle shape were made later, as the statistics from other spacecraft measurements better defined the shock shape. In the most recent study by Slavin and Holzer (1982), mentioned above, gas-dynamic models from a library of calculations carried out by Spreiter and Stahara (1980) were compared to the collected shock data. Their best fit, for a Mach number 7.2 and normalized scale-height obstacle shape factor H/r_0 of 0.03, is shown in Fig. 7. As the authors noted, the obstacle for this case, also shown in Fig. 7, has a subsolar altitude of ~400 km. At the time that this result was obtained, the Viking Landers had measured the ionospheric electron density and ion temperature along two trajectories through the dayside ionosphere. In addition to the observation that a negligible ionospheric plasma density remained at 400 km, the measured ion temperature and an inferred electron temperature of twice that value gave a pressure at that altitude of <50% of the estimated incident solar-wind pressure. Thus, in contrast to the earlier analysis by Spreiter and Rizzi, this latest study concluded that the obstacle size inferred from the observed bow-shock shape and gas-dynamic models could *not* be produced by the ionosphere alone. Slavin and Holzer further deduced that the magnetic dipole moment that would produce a field sufficient to balance the average solar-wind dynamic pressure at Mars at 400 km subsolar altitude was 1.4×10^{22} G cm^{-3} or 1.4×10^{12} T-m^{-3}. This conclusion was roughly consistent with the results derived in the host of intervening studies based on similar arguments, the history of which is illustrated in Fig. 8. (The three high values at the outset were deduced from the observed absence of trapped energetic particles on Mariner 4 instead of from the shock position.) Similar results obtained from fitting the Phobos 2 bow shock have now further corroborated these earlier estimates of the subsolar obstacle altitude and inferred intrinsic field strength (Riedler et al. 1989*a*; Schwingenschuh et al. 1990).

C. Magnetosheath: Observations and Gas-Dynamic Models

In the previous discussion of the interpretation of the Mars spacecraft magnetic-field data in terms of an intrinsic field, we emphasized the importance of accurately modeling the draped magnetosheath magnetic field. The same gas dynamic model used in studies of the bow shock data is appropriate for this purpose, but only one investigation (Russell et al. 1984) applied the model accordingly. The result of that effort is illustrated in Fig. 9, which shows the authors' favorable comparison between the model and the Mars 5

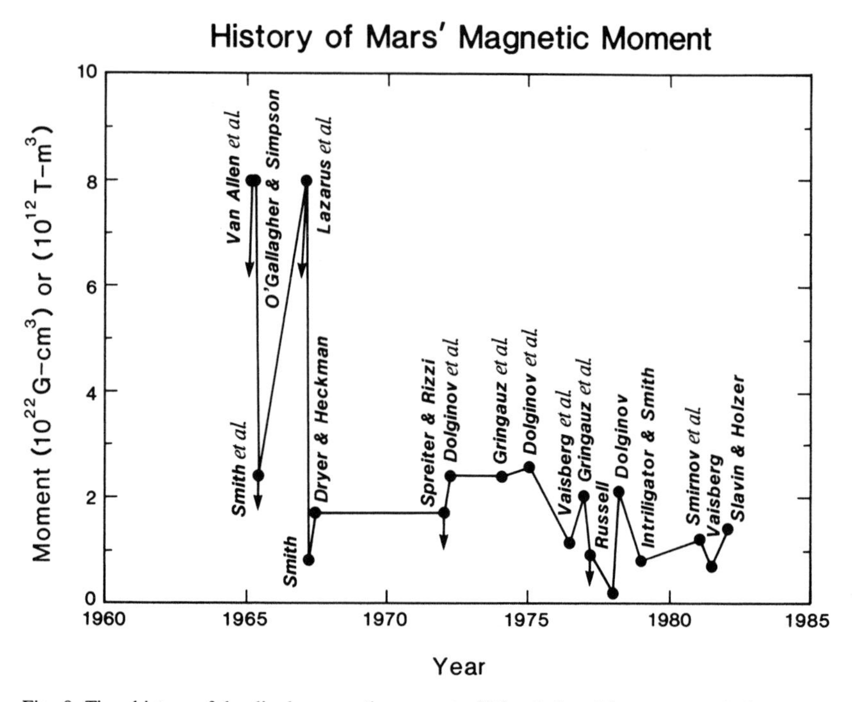

Fig. 8. Time history of the dipole magnetic moment of Mars inferred from spacecraft data.

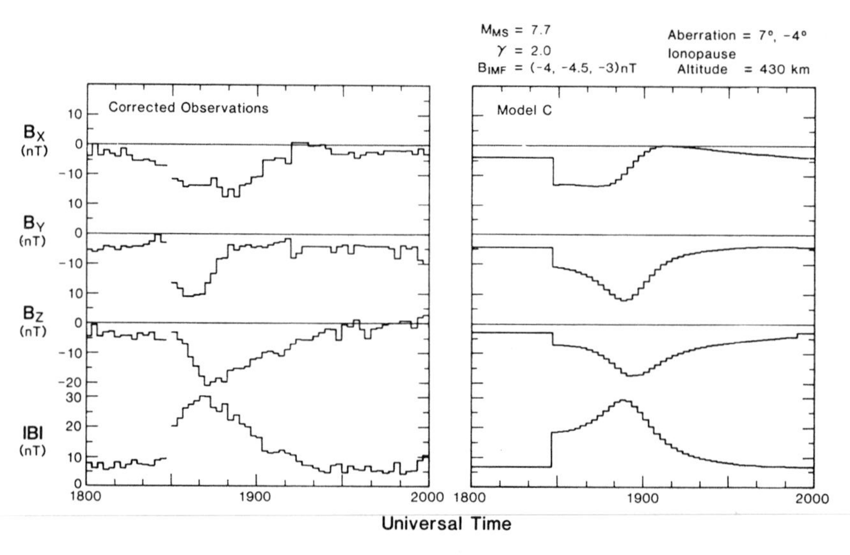

Fig. 9. Comparison of the magnetic-field vectors measured during a passage of the Mars 5 spacecraft through the Mars magnetosheath with a simulation of the interval using a gas-dynamic model (figure from Russell et al. 1984).

data. However, this comparison has been criticized by the Mars 5 principal investigator because it involved a rotation of the coordinate system of the magnetic field observations inconsistent with his own assessment (Dolginov 1986; see also Russell 1986). Nevertheless, we know both from observations outside the magnetospheric obstacle of Earth (Crooker et al. 1985), and from observations above the ionospheric obstacle of Venus (Luhmann et al. 1986*a*) that the gas-dynamic model should provide a good description of the magnetic field draping around Mars under steady solar-wind and interplanetary-field conditions regardless of the nature of the obstacle. Practically the only exception, besides that of disturbed interplanetary conditions, should be when the interplanetary field is aligned with the solar-wind velocity. Under such circumstances, the magnetosheath magnetic field will appear turbulent because of the turbulent nature of the subsolar bow shock when the upstream field is normal to its surface (a characteristic of this type of shock which is referred to as a "quasiparallel" shock) (see Luhmann et al. 1983). However, this condition is likely to be rarer at Mars than at Venus or Earth, because the oblique angle of the magnetic field increases with distance from the Sun.

An analogous comparison of the plasma data from Mars-5 with a gas-dynamic model is shown in Fig. 10. Although this is the only example available where such a comparison with plasma data was made, it illustrates the extent to which the magnetosheath-model variations in velocity and temper-

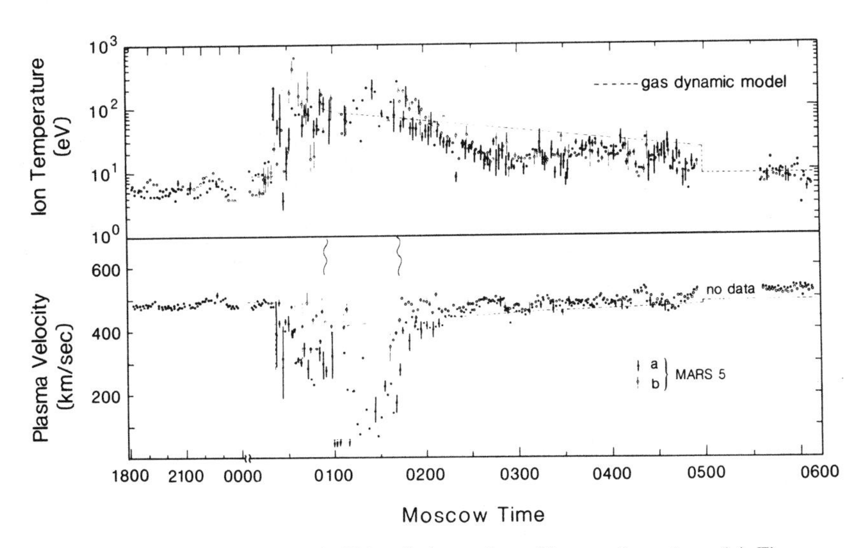

Fig. 10. Comparison of an interval of Mars 5 plasma data with a gas-dynamic model. The wavy lines mark the interval of disagreement that is considered evidence of entry into the Martian "magnetosphere." The closed symbols (a) are from the sensor that detected both light and heavy ions, while the open symbols (b) are from the sensor that detected light ions only. (The data are from Vaisberg et al. 1976*a*.)

ature along the spacecraft trajectory agree with the observations. As in the case of the magnetic field, such comparisons are particularly valuable for evaluating those plasma measurements that reportedly show evidence of passage into the obstacle. For example, the interval of disagreement in Fig. 10 marked by the wavy lines was considered evidence of entry into the Martian "magnetosphere" (Vaisberg et al. 1976*a*).

In spite of their success, gas-dynamic models of planetary magnetosheaths are expected ultimately to fail to describe the observations on several scores. One problem area is in the innermost layer of the magnetosheath where the convected, draped interplanetary magnetic field piles up to an extent that the field strength exerts control over the flow of the compressed solar-wind plasma. In this magnetic "barrier" or magnetic "cushion" region, one expects to find a depleted plasma density which is not predicted by the gas-dynamic frozen-field model (Zwan and Wolf 1976). Furthermore, in both the barrier and in an extended region of the magnetosheath, heavy planetary ions like O^+ will be produced by photoionization of that part of the neutral exosphere that extends above the obstacle boundary (Nagy and Cravens 1988). Impact ionization of these particles by solar-wind particles and charge exchange between solar-wind protons and the neutral atoms also add to the ion population (Russell et al. 1983). These processes constitute a source of mass in the magnetosheath plasma flow. The effects of mass loading on the shape of the bow shock have been shown to be significant at Venus (Alexander and Russell 1985), where the shock is found to be at larger distances from the planet than expected for an obstacle in the shape of Venus' ionopause, and its position depends on the ionizing solar EUV flux. Gas-dynamic models which include a source of mass have been developed (cf. Breus 1986; Berlotserkovskii et al. 1987), but these do not take into account some important physical details such as the manner in which the heavy ions are incorporated into the flow. A possible conceptual explanation of the solar-cycle effect could be that the barrier region, not the ionopause, forms the effective obstacle, and that the barrier size depends on the amount of mass addition. Thus it is notable that observations at Mars with the plasma analyzers on both the Mars 2 and Mars 3 spacecraft indicated the presence of a region of slowly moving plasma and increased magnetic field strength at approximately the location of the obstacle that would produce the observed bow-shock shape (Vaisberg 1976; Vaisberg et al. 1990). The existence of this barrier region provides an explanation for the interval of disagreement with the gas-dynamic model identified in Fig. 10. The detection of an inner boundary of the solar plasma at ~2000 km altitude near the terminator (Vaisberg and Smirnov 1986), and of heavy ions (possibly O^+) in a tail boundary layer illustrated in Fig. 11 by detectors on Mars 5 (Vaisberg et al. 1976*a,b*) was also reported. Although the heavy-ion interpretation of the early plasma data was contested (Bezrukikh et al. 1978), these observations taken together with the dayside observations are consistent with the idea that a planetary ion-laden mantle,

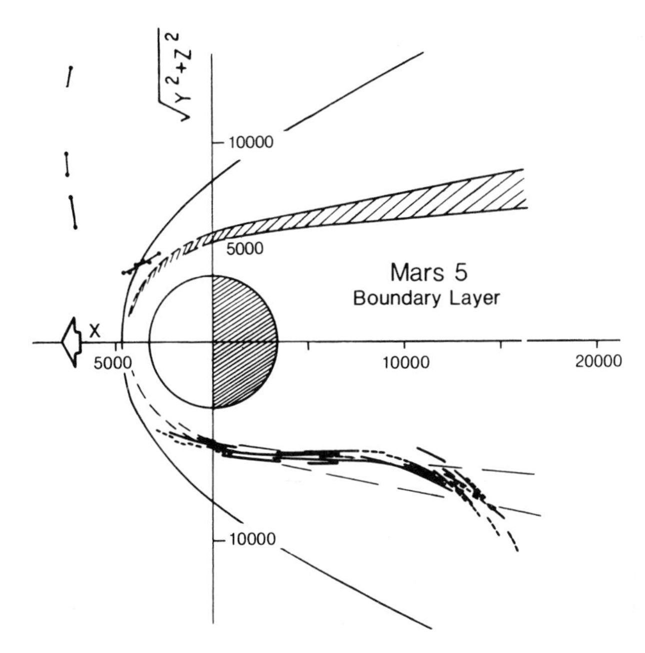

Fig. 11. Sketch of the location of the magnetosheath/magnetotail boundary layer suggested by plasma measurements on Mars 5 (figure from Vaisberg et al. 1976*a,b*).

threaded by stagnating piled up interplanetary field, determines the obstacle shape at Mars. Moreover, this picture is now reinforced by Phobos 2 results, which unambiguously show a heavy O^+ ion layer in the inner magnetosheath around which the solar-wind plasma appears to be deflected (Lundin et al. 1989; Rosenbauer et al. 1989; Riedler et al. 1989*a*). The outer boundary of this layer has been called the planetopause.

One aspect of mass loading that the modified gas-dynamic models do not include is related to the details of how at least some of the planetary ions are "picked up" by the solar wind. A convection electric field is associated with the moving magnetized solar-wind plasma. This field is distorted but persists in the magnetosheath where the solar-wind flow is diverted around the obstacle. If there is no turbulence in the magnetosheath to scatter the ions, they are smoothly accelerated by the electric field to energies that oscillate between zero and the kinetic energy associated with up to twice the velocity of the background flow. The important consideration here is the cycloidal geometry of their trajectories as they are accelerated. Planetary ions like O^+ are so heavy that their gyroradii can be comparable to or larger than the diameter of Mars. Thus, their trajectories can take them either into the lower atmosphere where they are absorbed or outward and tailward where they escape from the planet (Cloutier et al. 1974; Wallis 1982). The approximate appearance of the resulting "pickup ion exosphere" and associated "ion tail" can be examined by calculating the motions of test particles in the gas-dynamic magnetosheath magnetic field and convection electric field model. Figure 12 illustrates the behavior of exospheric O^+ ions near Mars that are expected to form an asymmetrical ion tail as a result of this finite gyroradius effect. Luhmann and Schwingenschuh (1990) and Luhmann (1990) discuss the energy distributions of these ions, some of which are accelerated almost as if they were in the unperturbed solar wind (with a nominal maximum energy of ~60 keV for an interplanetary field perpendicular to the velocity), while others which are diverted into the wake attain much lower energies. The Phobos 2 observations show that there are copious fluxes of planetary ions in the magnetosheath and wake of Mars (Lundin et al. 1990*a,b;* Rosenbauer et al. 1990), although it is not yet known whether their properties are consistent with the model in Fig. 12. In any case, the consequences of such a population on the flow and field may need to be considered when applying unperturbed gas-dynamic models to observations around Mars.

In a similar vein, kinetic treatments of at least the solar-wind protons may be necessary for studying the solar-wind interaction itself since the scale of the proton gyroradius can be comparable to the thickness of the subsolar Martian magnetosheath. In addition to a thick "foot" upstream of the subsolar shock (Moses et al. 1988), solar-proton absorption in the lower atmosphere analogous to the absorption of the pick-up ions can occur. The details of the effects of such finite proton gyroradius scaling on the bow shock and magnetosheath can be deduced only through the application of models such as global hybrid models of the solar-wind interaction which are currently under development (Brecht 1989). In the meantime, the favorable comparisons between the gross features of the gas-dynamic models and the observations mentioned above indicate that for many purposes, they can be used to describe the bulk plasma and field behavior around Mars.

D. Aeronomical Consequences of the Solar-Wind Interaction

This book includes a separate chapter on the aeronomy of Mars (chapter 30), but, as was the case for the planetary dynamo, it is appropriate to review here those aspects of Martian aeronomy that pertain to the intrinsic magnetic field and solar-wind interaction. The subjects of interest in the present context are the observed aeronomical properties of Mars, in particular of its ionosphere, that tell us about the local magnetic field, and the effects, both long-term and short-term, of the solar-wind interaction on the atmosphere.

1. Ionosphere Observations: Viking Landers. The only *in situ* measurements of the Martian ionosphere are from the retarding potential analyzers on the two Viking Landers (Hanson et al. 1977). These instruments provided altitude profiles of the electron and ion density, ion composition, ion and electron temperatures, and ion velocities at two dayside locations during conditions of low solar activity. The collected data on densities and tempera-

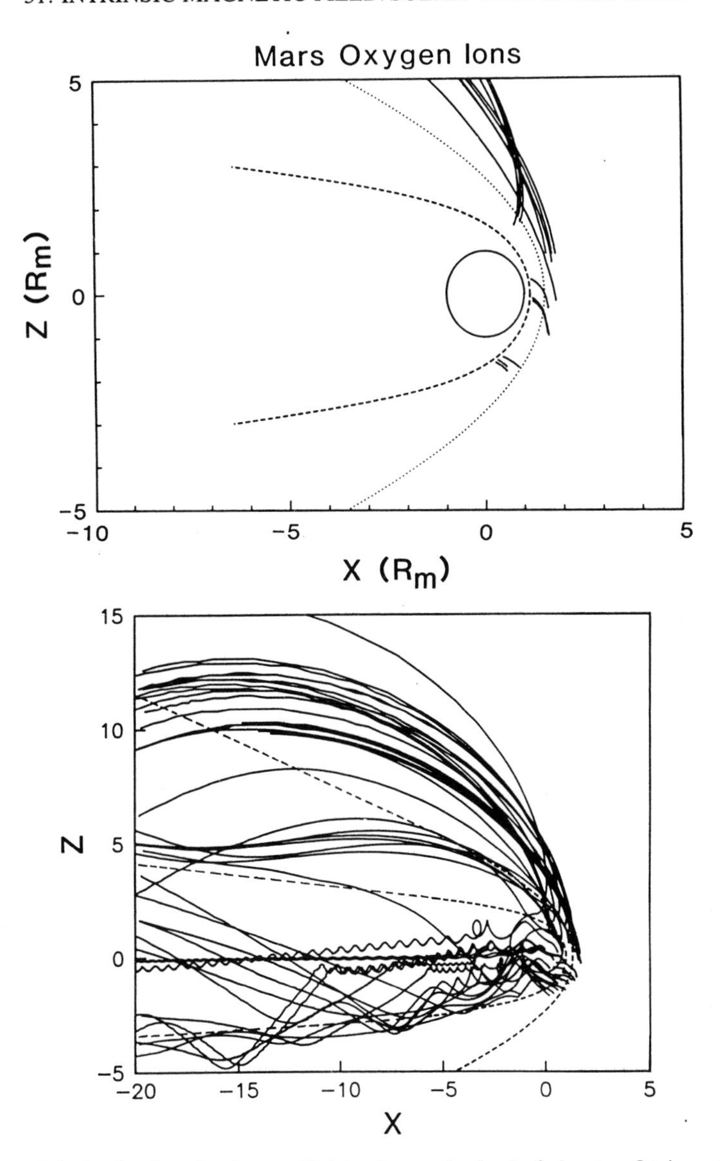

Fig. 12. Calculated trajectories (noon-midnight plane projections) of planetary O^+ ions picked up by the solar wind flowing past Mars. The upper frame shows representative trajectories outside of the obstacle surface, which either intersect the obstacle or reach large distances above the obstacle surface as the ion undergoes cycloidal motion. These were calculated using the magnetosheath-field model of Spreiter and Stahara (1980). The lower frame shows some trajectories from a model which includes a magnetic-field model for the wake inside of the obstacle (figure from Luhmann 1990).

tures are displayed in Figs. 9, 11 and 12 of chapter 30. As mentioned earlier in the discussion of the bow-shock shape, these data were used to compute the thermal-plasma pressure for comparison with the nominal solar wind pressure at Mars (Intriligator and Smith 1979; Slavin and Holzer 1982). The conclusion, reached even before the Viking investigators published their results on the electron temperatures (Hanson and Mantas 1988), was that during the Viking entry observations, the solar-wind pressure exceeded the thermal pressure of the Martian ionospheric plasma at the inferred obstacle altitude. The final thermal pressure profile deduced by Hanson and Mantas (1988) is shown in Fig. 13 where the average solar-wind pressure (see Fig. 5) is also noted. As the bow-shock studies indicated that the subsolar obstacle should be located near 400-km altitude, the weak ionosphere presented a problem that these early authors sought to solve by invoking the internal pressure of an intrinsic magnetic field of Mars. The absence of clear evidence of an ionopause in the Viking Lander density profiles strengthened their conviction that the ionosphere of Mars was magnetically protected from direct solar-wind influence. The ion velocity data were not conclusive because they showed a variability with altitude that was not consistent with any of the proposed models for magnetospheric or plasmaspheric convection (Bauer and Hartle 1973; Rassbach et al. 1974).

Several other analyses of the Viking Lander data yielded evidence for a magnetic field in the Martian ionosphere. Chen et al. (1978), Johnson (1978) and Rohrbach et al. (1979) used ionospheric models based on the Lander observations of the neutral atmosphere to argue that the ion temperatures

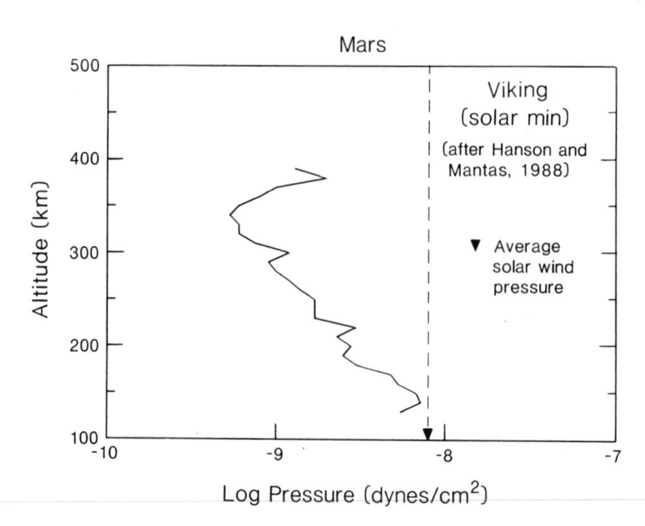

Fig. 13. Altitude profile of the thermal plasma pressure in the Martian dayside ionosphere as measured on the Viking Landers (figure adapted from Hanson and Mantas 1988). The dashed line denotes the typical solar-wind dynamic pressure at Mars. The Viking Landers arrived at Mars during a period of minimum solar activity.

measured on the probes were too large to be produced by solar radiation incident on a field-free atmosphere. By introducing near-horizontal fields of 10s of nT to inhibit vertical heat transport, they were able to approach more nearly the observed ion temperature gradients. Hanson and Mantas (1988), in their recent analysis of the electron temperature data, reached the same conclusion. However, all authors agreed that it is not possible to determine whether the required field was intrinsic or somehow induced by the solar-wind interaction. It is also not possible to rule out a contribution to the observed gradients by either local or topside heating of some kind (e.g., see Taylor et al. 1979).

The matter of induced ionospheric magnetic fields will come up again when we discuss the analogy with Venus (Sec. III.F). Models of induced magnetic fields in the ionospheres of the weakly magnetized planets, including Mars, had been formulated even before the Viking mission data were obtained (Cloutier and Daniell 1973). These fields were envisioned to be related to currents driven by solar-wind electric fields imposed across the ionosphere because the atmosphere absorbed a fraction of the incident solar wind (also see Cloutier and Daniell 1979). After the Pioneer Venus mission showed that induced fields, whatever their nature, were indeed present at Venus (Russell et al. 1979; Elphic et al. 1980), Slavin and Holzer (1982) "revisited" the inferred obstacle-size problem, asking the question of whether a Venus-like induced ionospheric field pressure could combine with the thermal pressure of the Martian ionosphere to produce the inferred obstacle at 400 km altitude. Their answer was negative, but, as we shall see later, that conclusion was premature based on later analyses of the Venus results (Luhmann et al. 1987). Thus the interpretation of the Viking measurements of the Martian ionosphere and their implications for an intrinsic magnetic field remain unresolved, largely because a magnetometer was not included on the Landers. Nevertheless, from the Landers we obtained two key pieces of information of interest to the present chapter. These concerned the magnitude of the ionospheric thermal pressure at Mars, and the inference of a large-inclination magnetic field of some 10s of nT in the ionosphere from the measurements of the plasma temperature gradients.

2. *Ionosphere Observations: Radio Occultation Experiments.* Radio-occultation techniques provide us with the most widespread coverage of the Martian ionosphere, spanning the solar zenith-angle range from ~42° to 138°. Virtually every mission to Mars (see Table I) has returned radio-occultation altitude profiles of electron density. This history allows us to examine such questions as the solar-cycle variability of the ionosphere; however, from the viewpoint of studying the solar-wind interaction, analyses of these data have been minimal. Because no ionopause-like boundary features were clearly distinguishable in the Mars radio-occultation profiles (Kliore et al. 1972; Lindal et al. 1979), efforts have generally concentrated instead on

the behavior of the electron density peak and the near-peak topside scale heights that tell us something about the neutral atmosphere (see, e.g., Fjeldbo and Eshleman 1968; Fjeldbo et al. 1970; Hogan et al. 1972; Kliore et al. 1972,1973; Lindal et al. 1979).

Except for an early attempt by Cloutier et al. (1969) to attribute the observed small topside ionosphere scale heights to solar-wind effects, which turned out to be unnecessary (see Hogan et al. 1972), Slavin and Holzer (1982) presented one of the first analyses of radio-occultation results aimed at addressing the solar-wind interaction problem. Using the electron density profiles available in the literature, they mapped the altitude of the "top" of the Martian ionosphere as a function of solar zenith angle (a study impossible with the available *in situ* data). Their result, reproduced here in Fig. 14, illustrated that the top of the ionosphere, so defined, was several 100s of km lower than the Mars obstacle altitude inferred from the observed bow-shock position and gas-dynamic models. These data were interpreted as further support for their intrinsic magnetic field argument. With such results by themselves, this conclusion was natural, but it must also be considered in light of the Venus observations, where neither the solar-minimum or solar-maximum shock position can be produced by an obstacle shaped like the ionopause. In any case, the archive of radio-occultation data may still contain some results of value for solar-wind interaction studies (e.g., see Zhang et al. 1990*a,b*).

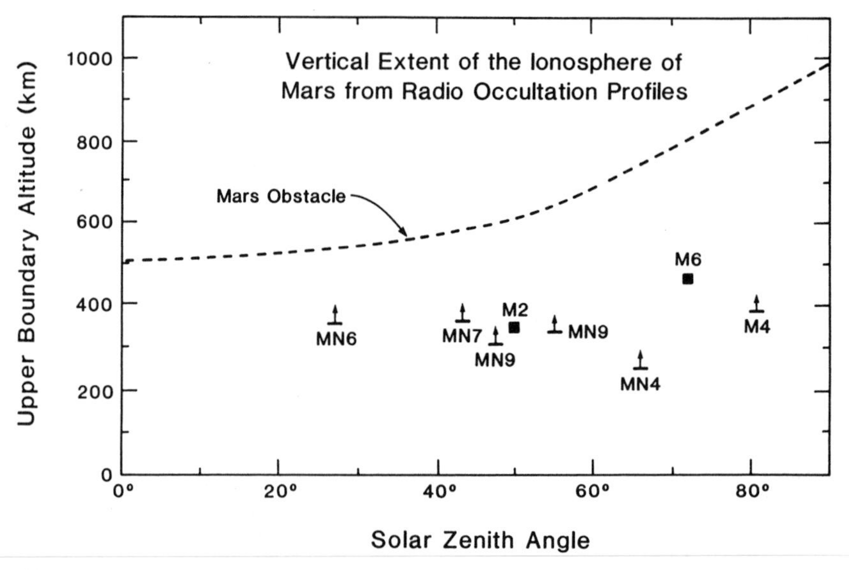

Fig. 14. Locations of the top of the Martian ionosphere from radio-occultation profiles compared to the obstacle height inferred from the bow-shock observations (figure from Slavin and Holzer 1982).

3. Implications for Neutral Atmosphere Evolution. Whether or not Mars has an intrinsic magnetic field of significance to the solar-wind interaction, Mars still presents little more than a planet-sized obstacle to the solar-wind flow as inferred from the observed bow-shock shape. Nagy and Cravens (1988) and Ip (1988) recently constructed models of the Martian neutral exospheres of oxygen and hydrogen that show these exospheres extend well above the 400-km inferred subsolar obstacle altitude. This means that, unlike the situation at Earth and like that at Venus, the particles in the upper neutral atmosphere at Mars can be ionized and "picked up" by the solar wind and magnetosheath convection electric field as described earlier. The picked-up ions illustrated in Fig. 12 constitute a source of mass for the solar wind, but a sink for the neutral atmosphere.

One can estimate the amount of solar-wind scavenging that occurs on time scales of the age of the solar system by integrating the total number of photoions produced per unit time from exospheric neutrals above the obstacle boundary and multiplying by 4.5 Gyr. Of course this presumes that all of the ions produced are indeed carried away from the planet (for a discussion of a possible limit see Michel [1971]), and that the exosphere, solar-ionizing radiation, and solar wind have been constant over this long time period. This loss mechanism applies to all constituents of the upper atmosphere, but the constituents of water are of special interest because the weakly magnetized planets seem to have a shortage of it. For a model exosphere of the type proposed by Nagy and Cravens (1988), a photoionization rate of 4×10^{-7} s^{-1}, and an obstacle located at ~400 km altitude over the dayside hemisphere, roughly 10^{23} oxygen ions s^{-1} and 10^{24} hydrogen ions s^{-1} are removed from Mars by the solar wind. Over 4.5 Gyr, the lost oxygen is enough to have contributed to under a meter of water over the planet's surface. The facts that other ionization mechanisms such as impact by solar-wind electrons and charge exchange between protons and neutrals are thought to increase the ion production rate by ~ a few times, and that the exosphere density is uncertain by at least a factor of ~5 (see Ip 1988), make this a conservative estimate. Moreover, if the exosphere was once hotter, and therefore more extensive, or if the ionizing solar photon flux was higher, these numbers could have once been considerably larger, making this mechanism much more significant in determining the present composition of the Martian atmosphere (e.g., see McElroy et al. 1977). On the other hand, a colder exosphere or lower ionizing solar-photon flux in the past, a historically weaker solar-wind dynamic pressure, or a substantial intrinsic magnetic field from some long-ago dynamo activity in the Martian core, would limit its effectiveness. There are also other solar wind scavenging mechanisms that one could propose (see, e.g., Brace et al. 1987; Perez-de-Tejada 1987), which may be necessary if the Mars/Phobos 2 spacecraft investigators' estimates of oxygen ion-loss rates of $\sim 10^{25}$ s^{-1} (Vaisberg 1976; Lundin et al. 1989; Rosenbauer et al. 1989) are accurate. With our lack of knowledge of the past Martian exosphere and intrinsic field, and of the history of the solar wind, we are left to speculate as to whether the weakness of the magnetic field of Mars is intimately connected with the current state of its atmosphere (e.g., see McElroy et al. 1977; Cordell 1980). Nevertheless, the Phobos 2 observations tell us something about the present ion-scavenging loss rates.

Some alternative evaporative escape processes that remove *neutral* constituents of the exosphere have little to do with the solar wind (cf. McElroy 1972; Hunten and Donahue 1976; Wallis 1978), but Watson et al. (1980) proposed that sputtering of the atmosphere by solar-wind particles reaching the Martian exobase (at ~200 km) could also have provided a significant loss of atmospheric neutrals over time. These authors suggested that this mechanism might be particularly important for explaining some of the anomalous isotope and rare-gas abundances observed in spectroscopic studies of the Martian atmosphere. However, all of the aforementioned studies indicate that, at least at present, the bulk of the solar-wind plasma is effectively slowed and deflected around Mars. On the other hand, this does not rule out the possible importance of sputtering by the accelerated pick-up ions that impact the lower atmosphere (Luhmann and Kozyra 1991). Measurements at Mars pertaining to the loss of neutrals such as oxygen and its causes remain to be done.

E. Wake and Tail

1. Models. A number of models have been proposed for the origin and configuration of the Martian magnetotail. As none of these have been developed to the point where they are numerical or analytical, they are generally presented in the form of cartoons showing the source of the field lines that extend into the wake of the planet. Figure 15 displays three that have appeared in the literature. A schematic of Venus' "induced" magnetotail, which presents another alternative for Mars, is shown in the upper left for contrast. As one can see, virtually all of the models proposed for Mars in the past incorporated a planetary dipole field.

2. Observations. The orbit diagrams in Fig. 3 best communicate the marginal quality of the presently available observations of the Martian wake and tail prior to the Phobos 2 mission. Nevertheless, the Mars 5 spacecraft missions produced the result in Fig. 16, which is a map of a "tail boundary." The identification of this boundary is based on observations of a transition in plasma properties suggesting a passage from a hot, rapidly flowing magnetosheath plasma to a cooler, more slowly flowing medium (Vaisberg and Smirnov 1986). As mentioned earlier, a layer of heavy ions has also been reported in the vicinity. A similarly shaped boundary was identified by Dolginov et al. (1976) on the basis of magnetic measurements. While these observations may indeed be of the Mars tail boundary or of the innermost edge of the magnetosheath, it is quite likely that they suffered from an orbital bias

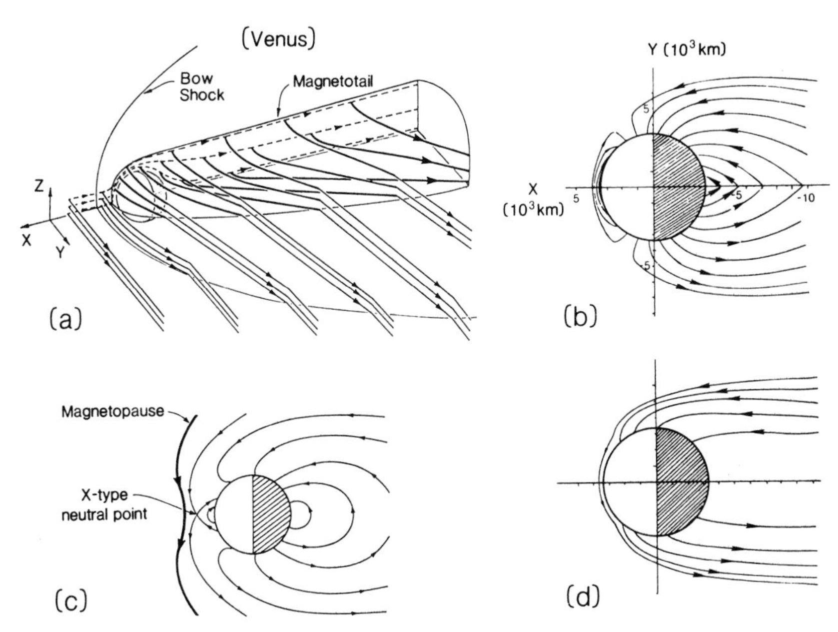

Fig. 15. The variety of possible configurations of the Martian magnetotail envisioned by various authors: (a) an "induced" magnetotail like that of Venus (from Saunders and Russell 1986; see also Vaisberg and Zeleny 1984); (b) a traditional Earth-like magnetotail; (c) a magnetotail wherein dayside magnetic merging plays a role (figure from Rassbach et al. 1974); and (d) a magnetotail that encompasses all of the flux leaving the day side of the planet so that a "bald" spot or band results.

problem like the bow-shock observations. On the other hand, except at the terminator, they are fairly consistent with the tail locations of the planetopause identified by the Phobos 2 investigators (Riedler et al. 1989; Rosenbauer et al. 1989*a*).

The plasma properties and magnetic field observed by the Phobos 2 spacecraft inside of the inferred obstacle boundary are especially valuable for investigating the question of an intrinsic magnetic field. In particular, the observations by Riedler et al. (1989*a*) of magnetotail lobe polarities that change with the orientation of the interplanetary magnetic field (also see Yeroghenko et al. 1990) suggest that any intrinsic field is too weak to have much effect at ~2.8 R_M downstream in the plane of the orbit of the natural satellite Phobos. These magnetic field observations, together with the absence of solar-wind protons, are very reminiscent of what has been observed behind Venus (see Saunders and Russell 1986; Mihalov and Barnes 1982). As comparisons with Venus are a key factor in the analysis of these observations, this is an appropriate point at which to begin the discussion of the Venus analogies.

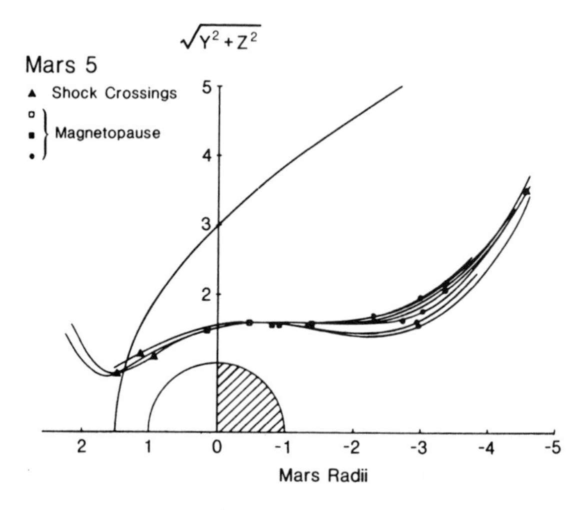

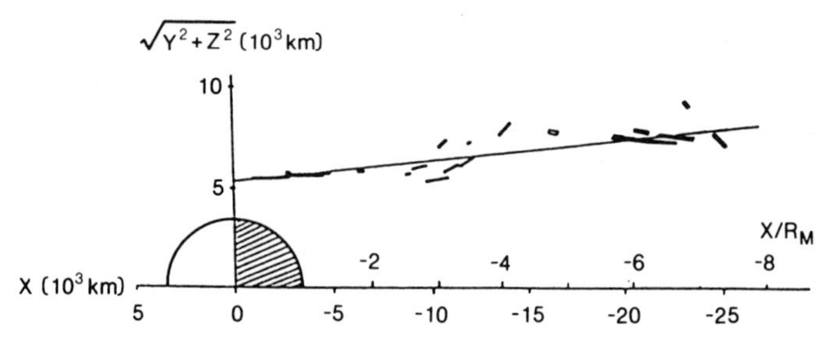

Fig. 16. The tail boundary at Mars inferred from Mars 2 and Mars 5 plasma measurements that show a change in flow and temperature characteristics at these locations (figure adapted from Vaisberg and Smirnov 1986).

F. Mars-Venus Comparisons

Although one should not jump to the conclusion that Mars is more like Venus than any other planet, there are compelling reasons for drawing analogies. For one, both are weakly magnetized obstacles to the solar wind that are little more than planet sized, as mentioned earlier. Both are terrestrial planets with some similarities in their surfaces and atmospheres, and they both reside in the inner solar system where the solar-radiation flux and the solar-wind density are substantial. Thus it is no surprise that comparative analyses of the Mars and Venus environments have proven illuminating. An assessment of our state of knowledge for Venus can be found in a review paper by Luhmann (1986). In this section, we return to the basic features of the solar-wind interaction considered earlier, but this time focus on the contrasts and similarities between our two neighboring planets.

1. Bow Shock. In their analysis of the bow-shock shape of Mars, Slavin and Holzer (1982) made the comparison with the bow shock of Venus after scaling to make the planetary radii equal. This comparison is included in Fig. 7, where it is seen that the Mars near-terminator bow shock determined from the Mars spacecraft data lies at about the same relative distance from the planet as the near-terminator shock at Venus at solar maximum, but well outside of the bow shock at Venus at solar minimum. As most of the Mars data used by these authors was from the approach to solar minimum, and as Venus at solar minimum is more Mars-like in its weakness of ionosphere pressure compared to solar-wind pressure, the result points to a difference in the two solar wind interactions. The recent observation by Phobos 2 of a near-solar-maximum terminator shock position of $\sim 2.7\ R_M$ (Schwingenschuh et al. 1990), well above the Venus solar maximum position, confirms this. The inferred Martian obstacle must be significantly more flared at the terminator than the obstacle at Venus under similar solar activity conditions. However, (at least for high solar activity at Venus) the bow shocks of both planets share the characteristic that their inferred dayside obstacle from gas-dynamic models is larger than the observed ionopause.

In another study, Slavin and Holzer (1981) presented data showing the dependence of the Mars bow shock radius in the terminator plane on incident solar-wind dynamic pressure, which they compared with the expected behavior for a purely dipole field obstacle in the solar wind. The size of a magnetic obstacle, and hence the bow shock, is expected to respond much more sensitively to solar-wind pressure than the size of a relatively incompressible ionospheric obstacle like Venus. Their display, which is shown in Fig. 17a, does not make a good case for dipole-field obstacle behavior. Rather, it suggests that although the bow shock position is quite variable, this variability is not clearly related to solar-wind dynamic pressure. An analogous plot for Venus, reproduced in Fig. 17b (cf. Luhmann et al. 1987), implied a like insensitivity of Venus' obstacle to the solar-wind dynamic pressure, thus pointing to a similarity between the planets that is in contrast to purely magnetospheric behavior.

Other aspects of the bow shock comparisons, such as a search for a Venus-like solar-cycle dependence of the Martian bow shock position (Alexander and Russell 1985), can now be considered because there are data for a wide range of solar activity. Figure 18 shows where the various missions to Mars have occurred with respect to the sunspot cycle. A comparison between the terminator bow-shock location observed during the Mars missions with that observed by Phobos 2 should give an indication of the effective Martian obstacle's variation with solar EUV radiation. As mentioned before, Schwingenschuh et al. (1990) report an average terminator bow-shock location for the near-solar-maximum conditions of Phobos 2 observations of $\sim 2.7\ R_M$. The earlier results shown in Fig. 7 give a distance of $\sim 2.4\ R_M$. A change of ~ 0.3 planetary radii was what was found at Venus where the terminator shock position moved from ~ 2.1 to ~ 2.4 planetary radii (Alexander and Russell 1985). So we are left with mixed results. On the one hand, the Mars

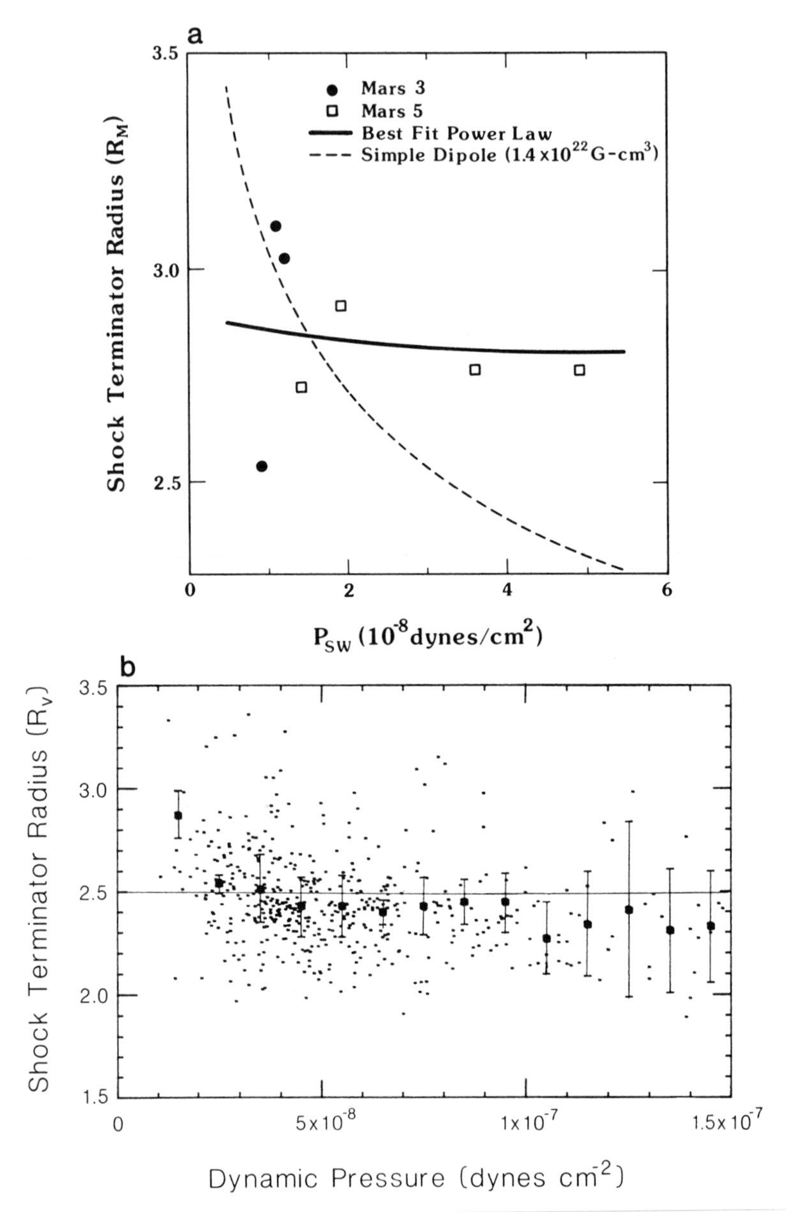

Fig. 17. (a) Radial distance of the observed Martian bow shock projected to the terminator plane vs solar-wind dynamic pressure. The dashed line shows the expected response for a weak dipole-field obstacle. The solid line describes the best fit of a low-order polynomial to the data (figure from Slavin et al. 1983). (b) Corresponding plot from Pioneer Venus Orbiter bow-shock observations that shows a similar insensitivity of the Venus obstacle to solar-wind pressure.

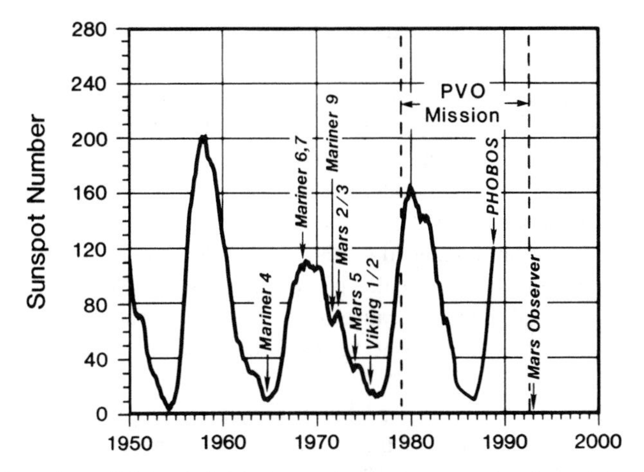

Fig. 18. A chronology of Mars missions referenced to the sunspot cycle. The period covered by the Pioneer Venus Orbiter (PVO) mission is indicated for comparison.

bow-shock insensitivity to the solar wind and its sensitivity to solar EUV argues that Mars is a Venus-like obstacle with negligible intrinsic magnetic field. On the other hand, the bow-shock position comparison raises the question of why the Martian bow shock and hence obstacle boundary are relatively farther from the planet than they are at Venus. Nevertheless, we have gained insight as to the differences that must be explained. One possibility is that the Martian atmosphere has a sufficiently different dependence on altitude and solar zenith angle that the effective obstacle is in fact a different shape (e.g., see Zhang et al. 1990*b*). An alternative explanation is that there are plasma physical processes at work in the solar-wind interaction with Mars (e.g., the finite gyroradius effects mentioned earlier, or charge exchange between solar wind protons and exospheric hydrogen atoms) that modify the magnetosheath in ways that are not as important at Venus; and of course there is the possibility that magnetic fields of either internal or induced origin are determining the different obstacle shape for Mars. The weight of evidence from further observations or analyses is needed to select from among these or other options.

2. Magnetosheath. The Russell et al. (1984) comparison of the Martian magnetosheath magnetic field with the gas-dynamic model discussed earlier had also been carried out for many cases of Pioneer Venus observations of the Venus magnetosheath with similarly good agreement (see Luhmann et al. 1986*b*). Vaisberg et al. (1976*a,b*) presented the plasma measurements along the Mars 5 spacecraft orbit in a form that demonstrated an analogous favorable comparison of some model and observed plasma properties (see Fig. 10), as was done for Venus by Spreiter and Stahara (1980) and Mihalov and Barnes (1982). Thus it seems safe to say that, for both planets, the general features of the magnetosheaths are understood. An understanding of the details, including the formation and properties of a magnetic barrier at the inner boundary, and the effects of the picked-up planetary ions, awaits more sophisticated observational analyses (e.g., of the Phobos 2 data) combined with sophisticated global models. For example, it is known that both planets have extensive oxygen exospheres (Nagy and Cravens 1988). Phillips et al. (1987) found evidence that the magnetosheath magnetic-field strength and draping at Venus are enhanced on the side of the magnetosheath where most of the associated asymmetrically picked-up ions should be found. This evidence was detectable only because of the extensive spatial and temporal coverage provided by the Pioneer Venus Orbiter database, which still has no equal at Mars.

Planetary pick-up ions, probably O^+, have also been detected in the Venus magnetosheath around the terminator (Mihalov and Barnes 1981; Phillips et al. 1987; Intriligator 1989) and downstream (Mihalov and Barnes 1982). Together with the aforementioned heavy-ion observations at Mars (Vaisberg and Smirnov 1986; Lundin et al. 1989; Rosenbauer et al. 1989; Lundin et al. 1990*a,b*) and the observation that the bow shocks of both Mars and Venus require a larger obstacle than the ionosphere to explain their shapes, these observations suggest that a magnetic field and heavy ion mantle in the inner magnetosheath may be the effective obstacle to the solar wind at both planets. (See Wallis [1982] for a discussion of pick-up ion effects at Venus, and Wallis and Ip [1982] for additional discussion of the comparison with Mars.) The mixture of solar wind and ionospheric electrons observed in the mantle regions of both planets (Spenner et al. 1980; Shutte et al. 1989; Nagy et al. 1990) indicate that at least solar-wind electrons find the boundary of this obstacle permeable.

Earlier, we alluded to the observation that the draped magnetosheath field becomes disturbed by convected fluctuations when the subsolar bow shock is quasiparallel. This effect was easily seen at Venus where the magnetic field throughout the magnetosheath became highly variable on such occasions (see Luhmann et al. 1983). Quasiparallel subsolar bow shocks may also be responsible for some of the observed field fluctuations in the Martian magnetosheath (see e.g., Dolginov 1976; Riedler et al. 1989*a*). However, since the proton gyroradius scale is much more important at Mars, there may also be additional sources of plasma instability that contribute to the local waves and turbulence.

3. Ionosphere. Although the ionospheres of Venus and Mars have been individually observed and modeled, their common features have been compared in only a few reviews without any particular emphasis on analogies (see e.g., Schunk and Nagy 1980; Mahajan and Kar 1988). Yet, both ionospheres exhibit the distinguishing features of O_2^+ compositional domination at altitudes below ~200 km (see Fig. 9 of chapter 30), H^+ and O^+ domination above 200 km, and ion and electron temperatures that increase with altitude far more rapidly than those that would result from heating by solar radiation alone (see Fig. 11 of chapter 30).

As discussed earlier, after the Viking Lander results became available, several investigators (Chen et al. 1978; Rohrbaugh et al. 1979; Johnson 1978) attempted to explain the observed temperature profiles using a combination of uniform, near-horizontal magnetic fields and topside and local heating. Cravens et al. (1979,1980) performed nearly the same analyses for the temperatures in the Venus ionosphere as measured on the Pioneer Venus Orbiter. As a whole, these studies demonstrated that with some additional heating and magnetic field, one can approximate the observations, but there is considerable ambiguity as to the appropriate combination of heating and magnetic field. Nevertheless, the necessity of similar modifications to the ionospheric models for both planets make the Venus *in situ* observations of the ionospheric magnetic field particularly relevant.

Specific comparisons of the Venus and Mars ionospheric properties have been discussed by Luhmann et al. (1987). These authors took into account the fact that, during the low-altitude phase of the Pioneer Venus Orbiter mission (when solar maximum conditions prevailed), Venus should be most like Mars on the few occasions when the solar-wind dynamic pressure at Venus exceeded the maximum thermal pressure in the ionosphere (see Fig. 5). The Venus ionosphere properties under these conditions are illustrated, together with the corresponding Viking Lander data from Hanson et al. (1977), in Fig. 19. Both ionospheres exhibit comparably steep ion-temperature gradients (see Fig. 19a), and the lack of a well-defined ionopause cutoff in the density profile (see Fig. 19b), but the most important aspect of this comparison is the contemporaneous behavior of the magnetic field at Venus. The induced horizontal ionospheric magnetic field was found to increase in magnitude, up to $\sim$150 nT, as necessary to produce pressure balance with the incident solar wind. Luhmann et al. (1987) suggested that the weak Martian ionosphere could stand off the solar wind in the same way. In fact, only a few 10s of nT of ionospheric field would perform this function on a typical day on Mars. The magnetic field needed to explain the Viking temperature measurements would be a natural part of such a scenario. Mars-like conditions may in fact be the norm during solar minimum at Venus when the ionospheric thermal pressure is expected to be low compared to the solar-wind pressure (Knudsen et al. 1987; Luhmann et al. 1990). Although there are no *in situ* ionospheric data for Venus at solar minimum, the comparison in Fig. 20 of radio-occultation profiles obtained from Mars and from Venus at solar minimum, showing a like absence of a distinct ionopause on their topsides (also see Luhmann et al. 1990; Zhang et al. 1990*b*), supports this idea. Shinagawa and Cravens (1989) have now adapted a magnetohydrodynamic model of the magnetization of the ionosphere of Venus to Mars with this analogy in mind. Of course, on occasions when the solar-wind dynamic pressure is extraordinarily

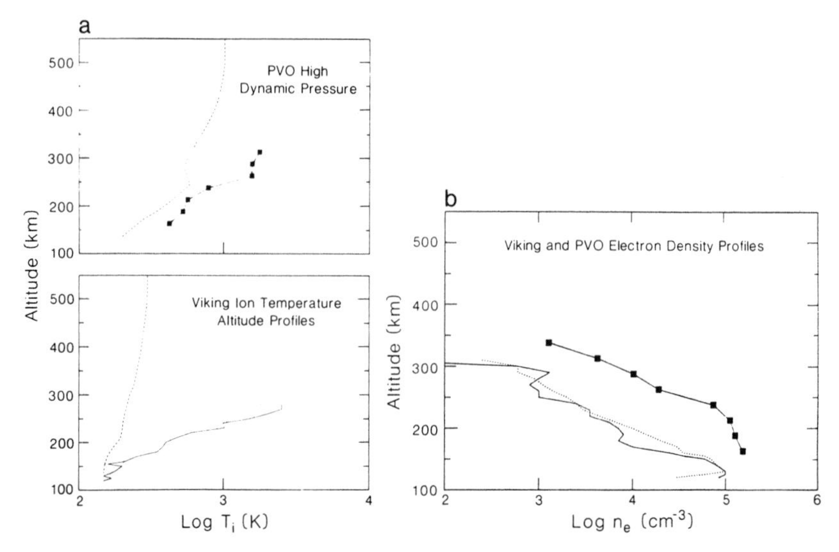

Fig. 19. (a) (bottom) Comparison of ion temperature profiles observed in the Martian ionosphere by the Viking Landers (Hanson et al. 1977) and (top) at Venus by Pioneer Venus Orbiter (PVO) when the solar-wind pressure was high compared to the ionospheric thermal pressure as at Mars. The dashed lines show the profiles expected for solar radiation heating alone. (b) Comparison of electron-density profiles from the same data sets (figure from Luhmann et al. 1987). In both (a) and (b) the black squares represent medians of PVO data. The data from the Viking Landers are shown by solid and dotted lines.

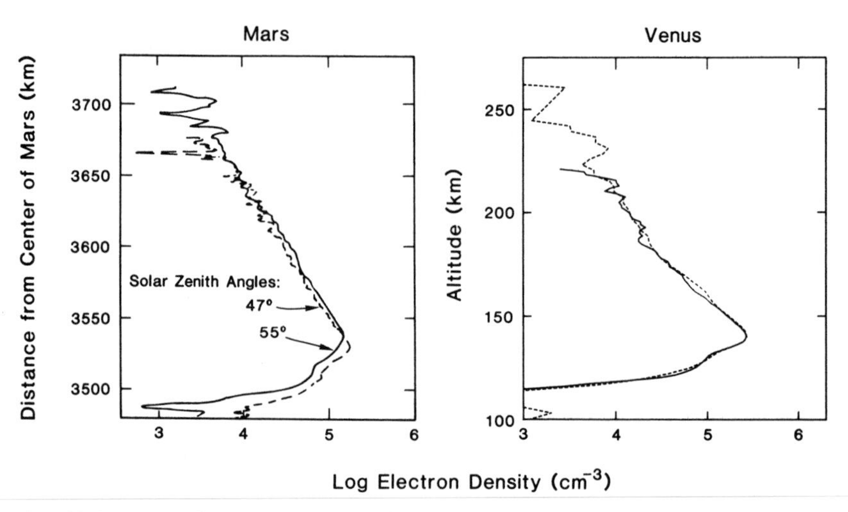

Fig. 20. Examples of radio-occultation profiles obtained at Mars and for the day side of Venus during solar minimum. The Mars data are from Mariner 9 and the Venus data are from Pioneer Venus Orbiter (figure adapted from Kliore et al. 1972).

low, or the solar EUV flux is extraordinarily high (e.g., at solar maximum), Martian ionospheric thermal pressure alone might be sufficient to stand off the solar wind in a Venus-like manner. However, only extensive *in situ* or radio-occultation measurements of the ionosphere with supporting solar-wind measurements would allow us to witness and prove such occurrences at Mars.

Another feature of the ionosphere of Venus that corresponds to conditions at Mars is the occasional depletion of nightside ionization. Radio-occultation observations at Mars show that the nightside ionosphere is generally rarefied although variable (e.g., see Breus and Gringauz 1980). Venus is known to have a substantial nightside upper ionosphere at solar maximum when dayside ionization is transported freely to the night side by pressure gradient forces (Knudsen et al. 1980). However, when the solar-wind dynamic pressure is high, or Mars-like conditions prevail, the nightward ion flow is choked off because the terminator ionopause altitude is low. On these occasions, the nightside ionosphere is confined to the region below ~200 km where the source is thought to be particle precipitation. At these times, Venus is said to have a disappearing ionosphere on the night side because the *in situ* data, from altitudes above ~150 km, show only weak, highly variable densities (see Cravens et al. 1982). According to the Venus *in situ* and radio-occultation experiment results, this condition also seems to prevail during solar minimum, when the dayside ionosphere is always weak compared to the solar-wind pressure (Knudsen et al. 1987; Knudsen 1988). For the same reason, the nightside ionosphere of Mars may always be "disappearing," or absent, except when the solar-wind pressure is extraordinarily low or the solar EUV flux is extraordinarily high (see, e.g., Zhang et al. 1990*a*). Of course, this interpretation assumes that the particle precipitation source at Mars is weak. While the more rapid rotation of Mars should have a small effect on the nightside ion densities, recombination is so rapid that it is expected to be minor.

It is worth noting that prior to the Pioneer Venus Orbiter *in situ* mission, when most of the Mars data were being interpreted, we did not know about the behavior of ionospheric magnetic fields at weakly magnetized planets; we did not know about "disappearing" ionospheres; and we did not know in general how ionospheric obstacles to the solar wind respond to excessive solar-wind pressures. The observations at Venus have allowed us to assess the Mars observations from an informed perspective (e.g., see Zhang et al. 1990*a,b*).

4. Wake and Tail. From the terrestrial example, we know that the wake of a magnetized planet is a region of low-density plasma dominated by magnetic fields with a large sunward or antisunward component. These fields have a fixed polarity consistent with the permanent magnetic field of the planet. A hot plasma sheet separates the regions of opposite magnetic polarity, which are referred to as tail lobes. In contrast, we know from the Venusian example that an induced magnetotail has a magnetic field with a similar structure consisting of two lobes of sunward or antisunward fields, but in this case the polarity is controlled by the orientation of the transverse component of the interplanetary magnetic field (e.g., see Fig. 15). Moreover, while the tail of Venus has an associated solar-wind plasma "void" like the tail of a magnetosphere (Milahov and Barnes 1982), it may also have a boundary layer and possibly a plasma sheet of cold, heavy, presumably ionospheric ions. Vaisberg and Smirnov (1986) discussed the *in situ* measurements of the plasma in the vicinity of the Martian wake from the viewpoint of comparisons with Venus. On the one hand, these authors stress the similarities in the plasma behavior at Mars and Venus, with particular reference to the boundary layer of possibly heavy, low-energy ions observed at the transition between the solar wind and the tail cavity at both planets. On the other hand, they point out that the boundary shape from the plasma measurements, shown in Fig. 16, is quite different from that at Venus, with the terminator radius of the Martian tail of $\sim$1.5 R_M significantly greater than the Venus tail terminator radius of only $\sim$1.2 planetary radii. This difference is supported by the Phobos 2 observations (Riedler et al. 1989; Rosenbauer et al. 1989).

The question of the magnetic field behavior inside the tail boundary at Mars compared to that at Venus was difficult to examine prior to the Phobos 2 mission because we did not have adequate deep-tail data for Mars. At Venus, such a database provided unambiguous evidence that the polarity of the field is controlled by the orientation of the interplanetary magnetic field (Saunders and Russell 1986). Vaisberg and Smirnov (1986) considered that what was observed around the tail boundary from the Mars spacecraft was sometimes influenced by the direction of the interplanetary field, as reported for Venus, but sometimes behaved independently. These authors suggested that a compromise was in order to explain the tail observations. They proposed that the tail field configuration may be some hybrid combination of intrinsic field emanating from the poles of a weak dipole and Venus-like draped interplanetary field supplying the field in the neighborhood of the magnetic equator. We are now more constrained by observations, which show that the magnetotail at the orbit of the natural satellite Phobos (at $\sim$2.8 R_M) and beyond is "induced" like that of Venus (Riedler et al. 1989; Yeroshenko et al. 1990). However, because of the lack of measurements in the polar regions of the magnetotail, one still cannot rule out the hybrid-tail possibility for Mars.

G. Mars-Earth Comparisons

1. Bow Shock. The shape of the Martian obstacle as inferred from the bow-shock shape is more flared than the corresponding obstacle at Venus, and is in fact more consistent with the shape of a magnetopause like that of the Earth. However, the contrast between the sizes of the two bow shocks, as illustrated by Fig. 21, is striking. Slavin and Holzer (1981) included Venus in this figure for the purpose of emphasizing the comparison of the Martian bow shock with those of known atmospheric and magnetospheric obstacles. Because the bow shocks of Mars and Venus are both much closer to the planet than that of Earth, and much smaller in absolute scale, finite solar-wind proton gyroradius effects and planetary ion pick-up effects may produce distinctive shock structure not found at the terrestrial bow shock (e.g., see Moses et al. 1988), although such features have not yet been identified at Venus.

2. Magnetosheath. As was discussed earlier, one of the consequences of a quasiparallel subsolar bow shock at Venus is that the magnetosheath is filled with convected turbulent magnetic fields. A similar effect of the quasi-

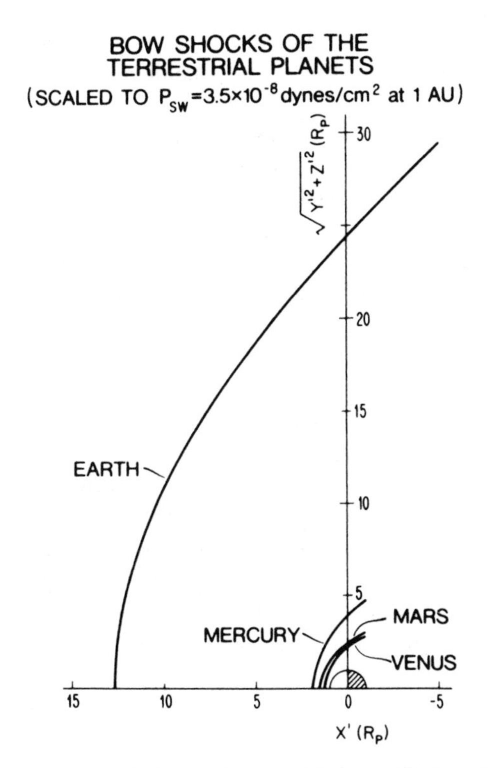

Fig. 21. Comparison of bow-shock sizes at the terrestrial planets. The bow shock positions were scaled to a common solar-wind dynamic pressure of 3.5×10^{-8} dyne cm^{-2} (figure from Slavin and Holzer 1981).

parallel shock on magnetosheath fields is observed at Earth (see Luhmann et al. 1986*b*), but the scale of the turbulent fluctuations relative to the subsolar magnetosheath thickness and the obstacle is much smaller. In addition, the distance for attenuation of the turbulent fields between the bow shock and the obstacle is much greater at Earth. Thus Mars is in this sense less like Earth than like Venus. However, it is worth repeating here that the large-scale magnetosheath fields for all of these obstacles appear to be at least roughly consistent with the gas-dynamic models of field draping.

3. Ionosphere. If Mars had a significant intrinsic dipole field, the ionosphere of Mars should be similar to that of Earth in at least one respect. Both planets have short rotation periods (~24 hr) that would, as at Earth, produce a particular diurnal asymmetry in the ionosphere. The electron density at middle latitudes at the Earth does not decay rapidly after sunset because of the large reservoir of mainly hydrogen ions that are stored in the overlying plasmasphere. This dipolar magnetic field reservoir is refilled during the daytime by upward ion flow and partially depleted at night by downward ion flow that helps to maintain the nightside ionosphere, especially near dusk, against the loss of ions by recombination at low altitudes (see Schunk and Nagy 1980).

An intrinsic magnetic field at Mars could either inhibit or promote the formation of a nightside ionosphere, depending upon its strength. An intrinsic field that is at a large angle to the nominal plasma velocity should inhibit the horizontal transport of ions from the day side, thus preventing the formation of a dense nightside upper ionosphere sensitive to incident solar-wind dynamic pressure like that of Venus at solar maximum (see Knudsen et al. 1980). On the other hand, if the magnetic field strength was large enough, an Earth-like plasmasphere should form, creating a reservoir large enough to maintain an ionosphere well after sunset. Thus the detection of dawn/dusk asymmetries in the nightside ionosphere could provide clues to the existence of a substantial intrinsic magnetic field. The available radio-occultation data of the nightside ionosphere have not yet been used to investigate these possibilities. However, they do tell us that the nightside ionospheric density is somewhat lower at Mars than at Venus (see Lindal et al. 1979; Knudsen et al. 1987). Although this suggests the possibility of a dipole field that is too weak to form a significant plasmasphere but strong enough to inhibit nightward flow, we know that at Mars ionosphere conditions are more like Venus at solar minimum where the nightward flow is effectively shut-off by the low terminator ionopause (Knudsen 1988). The weakness of the Martian nightside ionosphere is thus, by itself, inconclusive.

4. Wake and Tail. Some experimenters (notably Gringauz et al. 1976) considered that the plasma properties observed by the Mars spacecraft at the edge of the Martian tail resembled the plasma in the terrestrial plasma sheet, which fills the closed high-latitude dipole field lines that stretch out into the terrestrial magnetotail. Others (Vaisberg 1976; Vaisberg and Smirnov 1986) argued strongly for a Venus-like boundary layer interpretation. But, as discussed previously, the key to establishing the presence of an intrinsic magnetotail with its fixed polarity field anchored to the planet lies in central wake magnetic-field measurements.

Dolginov (1976) believed that Mars spacecraft magnetic measurements near the edge of the Martian tail showed evidence for behavior that was independent of the interplanetary field orientation. However, his analyses were based on the comparison of x or radial (sunward/antisunward) components of the field upstream near the tail boundary. Because an induced tail consists of the draped *transverse* or perpendicular (to the flow) component of the interplanetary field (see Fig. 15), an argument based on the radial component is not really valid (Vaisberg and Smirnov 1986). The interpretation of Dolginov also did not allow for rotations of the perpendicular component of the interplanetary field which can create a layered structure of sequentially draped fields. After Phobos 2, we know that at least the magnetic properties of the Martian tail near the orbit of Phobos are more like those of Venus than those of Earth (Yeroshenko et al. 1990). Nevertheless, Vaisberg and Smirnov (1986) found grounds for arguing in favor of at least some Earth-like intrinsic magnetotail contribution to the Martian tail because it could explain the greater cross section of the Mars obstacle at the terminator relative to that at Venus. Likewise, some of the Phobos 2 plasma investigators consider that there are Earth analogies to the ion "beams" seen in the Martian tail (see Lundin et al. 1989).

IV. SOLAR-WIND INTERACTION WITH PHOBOS AND DEIMOS

The natural satellites of Mars, Phobos and Deimos, are discussed at length in chapters 36, 37 and 38. They are much like the other minor bodies in the solar system in that they appear to be small, irregular, atmosphereless fragments of some larger body. These bodies orbit Mars near its equatorial plane at ~2.8 planetary radii and ~7 planetary radii, respectively. In these orbits, Phobos and Deimos regularly traverse the upstream solar wind, move through the Martian magnetosheath, and cross the wake and tail region. Their interaction with, and influence on, the plasma environment of Mars is probably not important for our basic understanding of the Martian magnetic field and solar-wind interaction, but it nonetheless is of intrinsic interest in our effort to become knowledgeable about the Mars system as a whole.

The interaction of a flowing plasma with Phobos and Deimos should be similar to that of asteroids in the solar wind, a subject on which little work has been done. The Moon is the closest prototype for which there exists both observations and theories. In particular, Spreiter et al. (1969) formulated a gas-dynamic model for the solar-wind interaction with a spherical nonconducting body, with no atmosphere, that absorbed the incident solar-wind plasma. The characteristics of that interaction were the absence of a shock wave in front of the body, and a wake void in the solar-wind plasma with a boundary in which the interplanetary magnetic field was perturbed after appearing to pass freely through the body. What limits the applicability of this picture to Phobos and Deimos are the sizes of these bodies compared to the Moon. All are much larger than the $\sim$10 m Debye length of the solar-wind plasma, a condition under which the plasma can be treated as a fluid as in gas-dynamic theory. However, the proton gyroradius of $\sim$600 km in the solar wind, which is much larger than Phobos or Deimos with their diameters of $\sim$22 km and $\sim$13 km, respectively, is such that the particle nature of the solar-wind plasma must be considered in a description of their interactions. As far as the environment of Mars is concerned, one can speculate that these bodies are so small compared to the gyroradii of solar-wind ions that their ion wakes are minimal and hence their perturbing effect on the plasma around Mars is negligible.

In spite of this assessment, it should be mentioned that at least one author (Bogdanov 1981) believed that the Mars 3 spacecraft detected evidence of an unexpectedly strong perturbation of the solar wind by Deimos. His explanation for the strength of the interaction was that Deimos was outgassing and that the emitted particles were producing the perturbation. There has been no other evidence for strong outgassing activity on Deimos or on Phobos, although the Phobos 2 investigators detected electric and magnetic-field perturbations when the spacecraft crossed the orbit of Phobos at a considerable distance from the satellite (Riedler et al. 1989*a*; Grard et al. 1989; Dubinin et al. 1990). It is certainly possible that these bodies are a source of atoms, ions and possibly dust in the Martian environment. Atoms and ions could result from the sputtering of their surfaces by energetic solar particles and Martian exospheric pick-up ions, while dust could conceivably be produced by micrometeoroid impact (see, e.g., Ip and Banaszkiewicz 1990). The ions should be carried away by the solar wind in the same manner as the exospheric ions, while the atoms produce a neutral torus around Mars (see Ip 1988). The smaller (submicron) dust particles would undergo quite complicated motions because they are subject to both gravitational forces and electromagnetic forces as they become charged in the Martian plasma environment (Horanyi et al. 1990), while larger particles would remain in the orbit of the satellite, forming a "ring." The latter could, in principle, produce the field perturbations observed on Phobos 2, but these ideas remain to be tested in light of the negative dust ring search using Viking data (Duxbury and Ocampo 1988).

V. CONCLUDING DISCUSSION

A. Summary of our Current Understanding

Estimates of the magnetic field of Mars from its solar-wind interaction have experienced a fairly consistent history as indicated by Fig. 8. However, at the time of writing this review we know with certainty only the following facts relating to the Martian field and solar-wind interaction:

1. The bow shock and magnetosheath of Mars indicate the presence of an obstacle to the solar wind that is somewhat larger than the size of the planet and its observed ionosphere, and also relatively larger than the Venus obstacle under comparable conditions.
2. The intrinsic magnetic field of Mars must be no greater than 1.5×10^{12} T m^{-3} or $\sim 10^{-4}$ times as strong as that of Earth (whose dipole moment is $\sim 8 \times 10^{15}$ T m^{-3}) to produce an obstacle of such small size.
3. At least for solar minimum conditions, like those prevailing at the time of the Viking Landers, the ionospheric plasma (thermal) pressure is insufficient to balance the incident solar-wind pressure by itself.
4. The ion and electron temperatures in the Martian ionosphere indicate the presence of local horizontal magnetic fields and heat sources in excess of solar radiation alone.
5. The plasma and field properties observed around Mars resemble those observed around Venus. In particular, planetary O^+ ions scavenged by the solar wind have been detected at both planets.
6. The magnetic field in the equatorial wake of Mars, at radial distances at and beyond the orbit of Phobos (~ 2.8 R_M) appears to be of predominantly interplanetary origin like that in the induced magnetotail of Venus.

The rest is speculation and inference. Using well-probed Venus as a model, one can find analogies that lead one to conclude that the two planets are much alike with their mutually weak intrinsic fields and atmospheric obstacles to the solar wind. From the observations in hand, it is easy to say that Mars probably has a Venus-like ionospheric magnetic field to help it stand off the solar wind, but we cannot know that this is true until we further probe the regions of space inside the dayside Martian obstacle boundary. Similarly, we do not know whether there is a field of planetary origin in the low-altitude wake. At this point, one cannot rule out the possibility that there is an intrinsic Martian dipole field of marginal importance that is responsible for the differences between Mars and Venus that we observe—in other words, that Mars is a "hybrid" obstacle.

B. Future Prospects

At the time of this writing, we are fortunate to be on the threshold of a new era of Mars exploration. The Soviet Phobos 2 spacecraft arrived at Mars in January of 1989. It carried instruments capable of measuring the plasma and field environments of both Mars and Phobos. Even though the mission came to a premature end, the magnetic field experiment has given us an unprecedented database from which to extract information on the sensitivity of the deep tail field polarities to interplanetary conditions. As shown by Fig. 18, Phobos 2 provided the first measurements of the Martian space environment near the maximum phase of the solar cycle. Phobos 2 also carried sophisticated experiments capable of measuring picked-up planetary ions over a broad energy range, thus leading to empirical estimates of the importance of solar-wind scavenging for atmosphere evolution and giving information about the local energetic particle environment. The U. S. Mars Observer spacecraft, scheduled for launch in 1992, will obtain a large magnetic field database, although the circular ~350 km polar orbit is not well suited for measurements free from the interference of the complex near-planet plasma environment. (At Venus, tail "rays" of ionospheric plasma and complicated magnetic-field structures associated with the solar-wind interaction were observed in the low altitude wake [see Brace et al. 1987].) Nevertheless, Mars Observer will be the lowest altitude probe to carry a magnetic-field experiment with the possible exception of the Soviet Mars 94 spacecraft. The latter is planned to provide aeronomy and particles and fields observations along an elliptical orbit that will probe Mars space to altitudes lower than the Mars Observer orbit.

Thus, plans are underway for a next salvo of Mars spacecraft to further define the upper atmosphere and space environment of Mars. The U.S. Mars Aeronomy Orbiter (described in NASA Tech Memo 89202, JPL, October, 1986) has yet to gain approval, but it now has an additional important application as a precursor mission to human landings in the possible future space exploration initiative. Together, these missions will replicate, and improve upon, the mission to Venus undertaken by the Pioneer Venus Orbiter. If these plans proceed as anticipated, many of the outstanding questions in this chapter can be answered within the decade.

32. MARS: EPOCHAL CLIMATE CHANGE AND VOLATILE HISTORY

FRASER P. FANALE, SUSAN E. POSTAWKO
University of Hawaii

JAMES B. POLLACK
NASA Ames Research Center

MICHAEL H. CARR
U.S. Geological Survey, Menlo Park

and

ROBERT O. PEPIN
University of Minnesota

The morphology and distribution of valley networks on Mars clearly indicates a difference in erosional style ~3.8 Gyr ago vs mid-to-late Martian history. Liquid water was certainly involved in network formation, but sapping processes rather than rainfall seem indicated, although the mechanism of recharge of the groundwater might possibly require precipitation. If enough CO_2 (1 to 5 bar) is in the atmosphere the surface temperature could be raised to near the freezing point ~3.8 Gyr despite a weak early Sun—at least at the equator and for the most favorable part of the orbital and axial cycle. Several lines of evidence suggest that Mars may have degassed enough CO_2 prior to ~4 Gyr ago to produce such a greenhouse effect. However, some estimates imply degassed CO_2 of < 1 bar. In addition, it is possible that hydrodynamic sweeping associated with H escape, together with impact erosion of volatiles, could possibly have resulted in an inventory of total degassed CO_2 of < 1 bar at the time of *network formation. Furthermore, removal of CO_2 by carbonates must be countervailed by carbonate recycling. Hence, the possibility of valley formation in the absence of a massive CO_2 atmosphere must also be addressed. In this context, it has been argued that higher internal regolith temperatures, associated with a much higher heat flow ~3.8 Gyr than at present, could have caused the 273 K isotherm to be close enough to the surface to account for the change in erosional style with time and might help to widen the latitudinal dispersion of the networks. The effectiveness of both these mechanisms is entirely dependent on a high early heat flow. In the case of the atmospheric greenhouse, this is because the atmospheric P_{CO_2} would have been dependent on the recycling time for regolith carbonate as well as the instantaneous supply of juvenile CO_2, and this recycling time can be quantitatively related to the heat flow. Thus, heat flow can affect internal temperatures and the depth to the 273° isotherm both by altering the thermal gradient and by altering the surface boundary condition via the atmospheric greenhouse.*

The possibility that long-term climate changes have occurred on Mars is important for understanding the geologic evolution of Mars and its surface environment, as well as for evaluating the possibility that some form of life could have originated on Mars. Such long-term climate change is suggested by an apparent difference in the erosional style exhibited by the ancient cratered terrain as opposed to terrain of later origin. In this chapter, we address the following questions: (1) How clearly does the morphology point to long-term climate change? (2) What mechanisms could have brought about such long-term change in erosional style? (3) In particular, could a higher early heat flow and/or an atmospheric greenhouse effect caused by high early CO_2 pressures have played a role? (4) Was there enough CO_2 degassed from Mars and retained until 3.8 Gyr ago to create such a greenhouse effect? And finally, (5) If both high early internal heat flow and an atmospheric greenhouse effect could have played a significant role, under what set of circumstances or properties of the Mars regolith-atmosphere system would each have dominated?

I. EVIDENCE FOR LONG-TERM CLIMATE CHANGE

Valley networks (as opposed to the large outflow channels) provide the main reason for suspecting that near surface conditions very early in Mars' history were very different from those that prevailed for most of the planet's subsequent history. The valley networks, common throughout the ancient cratered highlands, roughly resemble terrestrial river valley systems. They clearly are the result of some form of surface drainage. Most of the valleys have tributaries and increase in size downstream. They generally have flat floors and steep walls in their lower reaches, but upstream, particularly where incised into steep slopes, they tend to have V-shaped cross sections. They form networks that are generally much more open than is typical of terrestrial drainage systems. The valleys are mostly no more than several hundred kilometers long, draining into local depressions. Large drainage basins in which one main trunk stream dominates are rare (for a more complete treatment of valley network features on Mars, see chapter 15).

The mode of formation of the valleys is poorly understood. Liquid water is the most probable erosive agent, although ice could also have been involved. Agents other than water, such as lava and wind, are unlikely for a wide variety of reasons (for summaries see Carr 1981; Baker 1982). Liquid water on the surface can be derived from only two sources: either from the atmosphere or from the ground. Amphitheater-like terminations of tributaries, low drainage densities, parallelism of valleys, lack of competition for undissected interfluves, and junction angles between tributaries all suggest that groundwater sapping rather than surface drainage is the dominant process leading to formation of the valleys (Pieri 1979, 1980*b*).

The evidence for sapping is only suggestive, and does not constitute proof. Moreover, an origin by sapping requires some method of recharging the groundwater system so that flow can be sustained. One possibility is that a global groundwater circulation system was maintained by basal melting of the polar ice caps to offset losses of groundwater by sapping at low latitudes (Clifford 1987*b*). Another possibility is that losses by sapping were slowly replenished by release of juvenile water from the planet's interior. However, it is not clear that either of these mechanisms could supply enough water to erode the valley networks that are observed. Precipitation may have been required.

The networks are almost entirely confined to the heavily cratered terrain. By analogy with the Moon, this terrain is believed to have survived from early in the planet's history, probably from prior to 3.8 Gyr ago when the planet was subjected to heavy meteoritic bombardment. Almost all surfaces that are heavily cratered are extensively dissected by valley networks (Pieri 1976; Carr and Clow 1981; chapter 11). The network density falls off at high latitudes, but it is not clear whether fewer valleys formed at these latitudes, or whether the valleys there were more easily destroyed. Higher rates of destruction at high latitudes could result from the enhanced mobility of surface materials (Squyres and Carr 1986) or from burial (Soderblom et al. 1973*a*). In contrast to the almost universal presence of valleys in cratered terrain, surfaces that formed after the cratering rate declined are almost completely devoid of valley networks. The lower Hesperian plains of Lunae Planum, for example, have no valley networks, yet these plains are among the oldest that postdate the decline in cratering rates. There are a few exceptions to the general absence of valley networks on post-heavy-bombardment surfaces. Parts of the flanks of Alba Patera are, for example, dissected (Mouginis-Mark et al. 1988), and relatively young valleys have been identified in the Mangala region (Masursky et al. 1987). Nevertheless, the contrast between the pre- and post heavy bombardment surfaces is striking.

The almost complete restriction of the valley networks to the oldest terrains can be explained in at least two ways: The first possibility is that the valley networks are mostly ancient and that the conditions required for valley formation were commonly met very early in Mars' history but only rarely at later times. The second possibility is that the networks have a wide spread of ages but they formed more readily in the cratered terrain because it was more erodible, because water could access the surface more easily, or for some other reason. While the second possibility may be true in part, intersection relations throughout the cratered highlands indicate that the dominant reason that most valley networks are restricted to the oldest terrains is that most valleys themselves are old. Conditions during and shortly after heavy bombardment, therefore, appear to have greatly favored formation of valley networks as compared with later times.

The supposition that conditions on early Mars were different from subsequent conditions is supported by estimates of erosion rates. Baker and Partridge (1986) noted a wide range in preservation of valley networks. Although most of the valley networks are very ancient, and must have a similar age, some are perfectly preserved while others are highly degraded. This suggests that substantial erosion occurred during the relatively short period during which the surviving valley networks formed; whereas almost no erosion occurred following the intense valley-forming epoch, which comprises most of the planet's history. Schultz and Britt (1986) suggested that the change in channel morphology could correlate with the formation of the Argyre basin, and proposed that formation of the basin could have affected global conditions. Similar conclusions are reached from the survival of craters. A general deficiency of craters less than 30 km in diameter as compared with larger-size craters in the cratered highlands, coupled with estimates of present-day erosion rates, has also suggested to several authors that erosion rates very early in the planet's history were substantially higher than present erosion rates (Arvidson et al. 1980; Chapman and Jones 1977; Plaut et al. 1988). Thus, the early period of enhanced valley formation appears to also have been a period of enhanced obliteration of surface features.

As indicated above, groundwater sapping appears to be, at least in part, responsible for valley formation. The climatic conditions and ground temperatures required for sapping are unclear. It has generally been assumed that if the networks were formed as a result of erosion by liquid water then surface temperatures could not be lower than about 273 K; otherwise the streams would freeze. However, in Arctic regions on Earth streams can be maintained by groundwater seepage throughout winter when temperatures are subzero. Rates of freezing are such that the streams can flow several kilometers before freezing arrests flow (Childers et al. 1977; Carey 1973). Theoretical calculations suggest that if a stream 1-m deep could be initiated on Mars, then, even under present climatic conditions, flow could continue under a protective ice cover for tens of km depending on slope and other factors (Carr 1983). These calculations are, however, idealized. They do not take into account turbulence of the stream which would tend to inhibit formation of an ice cover. Nor do they take into account formation of ice dams and other constrictions that would tend to divert flow out of the main stream and accelerate freezing. Even in the ideal case the distance that flow could be sustained is relatively short compared to the length of the observed valleys, unless stream depths are at least a few m deep at the source, which implies very large seepage rates compared with most terrestrial springs. There is, additionally, the possibility that the water could be briny with a freezing point below 273 K (Brass 1980; Zent and Fanale 1986).

Although large streams may be able to flow considerable distances under a protective ice cover, such streams would be very difficult to initiate under present climatic conditions. With average surface temperatures close to 215 K near the equator, the base of the permafrost is likely to be a few hundred m deep to about a kilometer deep for plausible values of the heat flow (around 30 mW m^{-2}), thermal conductivity (around 500 mW $m^{-1}K^{-1}$) and salinity of the groundwater. Water may therefore be prevented from reaching the surface by an impenetrable seal. Water may have, on occasion, erupted catastrophically from below the permafrost seal to form the large outflow channels (Carr 1979) but slow sustained seepage from such depths is unlikely. Groundwater sapping probably requires a thinner permafrost zone than that which prevailed during the formation of the large outflow channels.

If precipitation is required for valley formation, either to form the valleys directly or to replenish the groundwater system, then surface temperatures in excess of 273 K are almost certainly required. Jakosky and Carr (1985) suggested that ice could accumulate at low latitudes at high obliquities during the present epoch. Clow (1987) showed that water could accumulate, temporarily, during the day under such an ice pack even under present climatic conditions, provided other conditions were satisfied. Of particular importance is the dust content, and hence the optical properties, of the ice. However, although liquid water might be produced in this way, a narrow set of conditions are required and the amounts produced are small—so small that liquid produced would re-freeze overnight. It appears unlikely that sufficient water could be produced in this way to form semi-permanent streams that could cut the channels observed. In order to form streams by precipitation, warm conditions appear to be required.

In summary, the following observations suggest that climatic conditions on early Mars were very different from those that prevailed for the rest of the planet's history: (1) the difficulty of forming valley networks by any agent other than running water; (2) the almost complete restriction of valley networks to the ancient cratered highlands; (3) the requirement that shallow regolith temperatures be close to 273 K for valleys to form irrespective of origin by sapping or precipitation; and (4) evidence of enhanced obliteration of surface features early in the planet's history.

Two factors could have contributed toward making early conditions more favorable to formation of valley networks. First, the higher heat flows expected for early Mars would cause groundwater to be closer to the surface than at present. Second, surface temperatures may have been higher because of a much thicker CO_2 atmosphere.

II. MECHANISMS FOR LONG-TERM CLIMATE CHANGE

A. Heat Flow on Early Mars

Higher heat flows are expected early in Mars' history primarily because of dissipation of the original heat of accretion, although a higher rate of production of radiogenic heat is also a factor. Schubert et al. (1979) estimated, as a function of time, the surface heat flow on Mars that results solely from cooling of a hot mantle. They calculated that this component of the heat flow should have been about 70 mW m^{-2} around 3.8 Gyr ago when most of the channels formed. If we add to this the radiogenic component, the total heat flow is likely to have been $\geq$100 to 300 mW m^{-2}. (The radiogenic component is difficult to estimate accurately because of the unknown bulk composition of the planet and the unknown distribution of radiogenic elements between crust and mantle.)

The thermal conductivity can be estimated by analogy with the Moon. On the lunar highlands, an incompetent megaregolith with an estimated porosity of 5 to 20% extends down to a depth of about 2 to 3 km. We should expect a similar megaregolith on the cratered highlands on Mars, but the upper low-porosity layer should extend down to only $\sim$ 1 km because of the higher gravity. Mitzutani and Osako (1974) determined the conductivity of an anorthositic breccia with 15% porosity to be 194 mW $m^{-1}K^{-1}$ under vacuum conditions. We do not know the composition of the Martian megaregolith. It could range from anorthositic as on the Moon to one much richer in Fe/Mg silicates. If composed entirely of Fe/Mg silicates the conductivity would be increased by a factor of $\sim$ 2.2. Presence of CO_2 in the pores could further increase the conductivity by a factor of 1.3 to 1.5 (Horai and Winkler 1980). A reasonable range for the conductivity of the upper porous layer of the megaregolith is 250 to 1000 mW $m^{-1}K^{-1}$ (these values are in the low-to-mid range of conductivities proposed by Clifford and Fanale (1985). These values, coupled with a nominal heat flow of 100 mW m^{-2}, would give a thermal gradient of 0.1 to 0.4 K m^{-1}.

Based on Viking images which suggest the depth of incision of the channels, Squyres (1989*a*) has estimated that a change in the depth to the 273 K isotherm from < 350 m to significantly > 350 m could account for the morphological differences observed between very early and later Martian terrains. He has also shown that, assuming reasonable values for ancient Martian highlands regolith conductivity, such a change in depth to the 273 K isotherm might be accounted for by the heat flow differences between $\sim$3.8 Gyr ago and much later times, which are implied in thermal models like those of Schubert et al. (1990) (see also chapter 5).

In addition, as shall be shown in Sec. IV, a high early heat flow may have played a second role in bringing about morphological changes by acting to recycle CO_2 in carbonates back into the atmosphere, which would sustain an atmospheric greenhouse effect.

B. Atmospheric Greenhouse on Early Mars

There are several tantalizing pieces of information about conditions on Mars during its first billion or so years of history that suggest that a thicker atmosphere may have existed then, and that the climate may have been warmer. As discussed previously, valley networks, which are largely restricted to a period 3.5 to 4.0 Gyr ago, the earliest discernible geologic times on Mars, provide evidence of liquid water occurring at the surface in many times and places during this epoch. While some valley networks appear to have been formed by sapping processes and therefore do not require temperatures above the freezing point of water at the surface, even they seem to require temperatures near 273 K close to the surface. Furthermore, some valley networks show branching patterns tied to surface elevations that do not appear to be due to sapping, and could have been caused by precipitation runoff.

In addition, the erosional environment was markedly different at these early times. In particular, erosion and depositional rates seem to have been much higher during the epoch of valley network formation than in subsequent epochs (Chapman and Jones 1977; Arvidson et al. 1980; Plaut et al. 1988). A denser atmosphere may be required to account for the higher rates during this early epoch.

As discussed in more detail below, enhancements of heavy isotopes of nitrogen ($^{15}N/^{14}N$) and hydrogen (D/H) in the Martian atmospheric gases may require the occurrence of a denser atmosphere in Mars' early history, perhaps one on the order of a bar or so of CO_2 (Fox and Dalgarno 1983; Owen et al. 1988). Finally, there seems to be evidence for the occurrence of carbonates among the Martian surface material from both spectroscopic data (Pollack et al. 1990*b*; Clark et al. 1990) and from mineralogical investigations of SNC meteorites (Gooding et al. 1988). These carbonates probably require an aqueous environment for their formation.

Taken together, the above data and their interpretation indicate that a much denser atmosphere may have existed on early Mars, perhaps one capable, at certain times and places, of elevating surface temperature to values in excess of 273 K. Below, we will examine the ways in which such wet, warm environments might have arisen, and their plausibility. However, we wish to emphasize that the present data do not demand that such environments did in fact occur.

1. Greenhouse Models. Optically active gases in planetary atmospheres produce a greenhouse warming of the surface. The nature of this warming can best be appreciated by considering the globally averaged heat budget at the surface. In the absence of an atmosphere, this budget basically consists of the heating of the surface by absorbed solar radiation balancing the cooling due to the thermal radiation emitted by the surface. As the latter varies approximately as the fourth power of the surface temperature, this balance determines the globally averaged surface temperature. The surface heat balance is altered in several ways when an atmosphere is present. Gases, particles and clouds generally decrease the amount of sunlight reaching the surface, thereby tending to cool the surface. However, if some of the gases are optically active in the thermal infrared portion of the spectrum, this cooling is usually more than compensated by thermal radiation incident on the surface from the atmosphere above it.

The magnitudes of the greenhouse heating of the surfaces of the terrestrial planets differ by sizable factors. Surface temperatures on Mercury, Mars, Earth and Venus are elevated by approximately 0, 7, 35 and 500 K, respectively, above the values they would have in the absence of an atmosphere and with the same fraction of sunlight absorbed by the atmosphere/surface system (see, e.g., Pollack 1979). These differences correlate with the mass of the atmospheres in the sense that the more massive atmospheres have the larger greenhouse effects. This correlation is due to two factors: optically active gases absorb over a larger range of wavelengths at higher total atmospheric pressures, due to greater line widths at higher pressures; also, the absolute abundance of optically active gases in a rough sense is larger in the more massive atmospheres and therefore some of the weaker transitions of the optically active molecules become opaque.

The very modest greenhouse warming of present-day Mars is due to a combination of its low atmospheric pressure (about 7 mbar) and its distance from the Sun. Thus, even though the major constituent of the atmosphere, carbon dioxide, is optically active in the thermal infrared, only its 15-μm vibrational fundamental has a strong interaction with thermal radiation. Its shorter-wavelength bands (e.g., the 9.4 and 10.4 μm hot bands) are either too weak, or they occur at wavelengths where there is little thermal energy (e.g., the 4.3 μm fundamental). Furthermore, the symmetry of this molecule prevents its having a pure rotational spectrum, which is important at wavelengths longward of 20 μm. These lines are activated only as pressure-induced transitions, which are important in Venus' high pressure atmosphere. At Mars' distance from the Sun and in the presence of a very modest greenhouse warming of the surface, the amount of water vapor in the atmosphere, as limited by its vapor pressure curve, is too small to have an important effect in the thermal infrared.

At first glance, it might appear that the surface of Mars would have been somewhat colder in the past due to a predicted lower solar luminosity then. A basic feature of nearly every solar model is an increase in solar luminosity with time. As pointed out by Newman and Rood (1977), the complexity of stellar evolution models frequently raises questions as to their accuracy. However, the issue of increasing luminosity is actually dependent only on the assumption that the energy source of the Sun is the fusion of hydrogen into helium. Unless we wish to throw out all that we believe we understand about how stars work, we must contend with the unavoidable conclusion that just after the Sun reached the main sequence, $\sim$ 4.5 Gyr, its luminosity was 25 to 30% smaller than its current value (Newman and Rood 1977; Gough 1981). This situation would also have affected the early Earth, and leads to the so-called "early, faint young Sun paradox." If the Earth's atmosphere had the same composition in the past as it does today (except for a water abundance controlled by its vapor pressure), the Earth should have been an ice-covered planet during its first 2.5 Gyr of history, contrary to the geologic record (Sagan and Mullen 1972). The solution to this paradox for the Earth appears to be an altered atmospheric composition, in which there was a higher abundance of optically active gases. For reasons that are discussed below, the most likely gas species to meet this requirement is carbon dioxide (Owen et al. 1979). Enhancement of the present carbon dioxide partial pressure by several orders of magnitude is required.

If the surface temperatures on early Mars were substantially elevated over their current values, despite a lower solar luminosity, then there are several candidate greenhouse gases that could have played key roles. These include ammonia, methane, sulfur dioxide, carbon dioxide and water vapor. The amount of greenhouse warming these various gases engender, either individually or in concert, depends on the fraction of the thermal infrared over which they have significant opacity. At Mars' current globally and annually averaged temperature of about 220 K, the relevant spectral region ranges from $\sim$ 7.5 to 100 μm. When permitted vibrational and rotational transitions occur in this spectral domain, comparatively small pressures of the absorbing gas species (a mbar or so) are required to blanket effectively the spectral region near these transitions—but only near these transitions. Thus, it is also useful for the total atmospheric pressure to be high, even if the dominant gas species is a poor absorber, because the major gas can then pressure broaden significantly the individual rotational lines that make up the permitted transitions and hence decrease the transparency of the atmosphere at wavelengths situated between rotational lines. When pressure-induced absorption occurs within the thermal infrared spectral region of interest, generally large total pressures (a bar or more) and significant partial pressures of the absorbing gas species are needed to make the atmosphere opaque across a vibrational-rotational band. The individual rotational lines are so strongly broadened in a pressure-induced transition that it appears as continuum absorption. Naturally, several optically active gases can produce a significantly larger greenhouse effect than any one of them, especially if their bands occur in non-overlapping portions of the thermal infrared. Carbon dioxide, water vapor, and sulfur dioxide, the three key greenhouse gases in Venus' atmosphere, represent one such complementary set of greenhouse gases.

Ammonia is an especially powerful greenhouse agent, due to its vibrational fundamental near 10 μm, its weaker 16-μm band, and its rotational transitions longward of 20 μm. Methane is far less effective in producing greenhouse warmings on Mars because its only permitted transition in the thermal infrared is centered near 7.7 μm, which lies on the Wien tail of the blackbody function for Mars. However, when the methane partial pressure approaches a bar, its pressure-induced opacity longward of 20 μm can abet the greenhouse warming in a significant way (see, e.g., Pollack 1979). Similarly, large amounts of hydrogen are required for its pressure-induced rotational lines at 17 and 28 μm and its translational component at longer wavelengths to produce a significant blockage of thermal infrared radiation. Sulfur dioxide is a potentially powerful greenhouse agent, as it has permitted rotational transitions at wavelengths longward of about 40 μm, and permitted vibrational transitions near 7, 9 and 19 μm. As mentioned earlier, carbon dioxide has a strong vibrational band near 15 μm and pressure-induced transitions longward of 20 μm and near 7 μm. Finally, water vapor has permitted rotational transitions longward of about 20 μm at the temperatures of interest, and a weak continuum in the 8 to 12 μm region.

Over the last decade, a variety of models have been constructed to assess the abundances of various greenhouse gases needed to elevate the surface temperature of Mars significantly above its current mean value. They can be grouped into three basic categories:

1. Global energy-balance models;
2. One-dimensional radiative-convective models;
3. Latitudinal energy-balance models.

The first of these models involves balancing the solar energy absorbed by the entire planet with the energy it radiates to space. As the energy radiated to space is a strong function of the planet's temperature, one can use this balance to determine the globally averaged surface temperature, given some assumption about the relationship between the temperatures within the atmosphere and the surface temperature (see, e.g., Pollack 1979). In a one-dimensional radiative-convective model, solar and thermal heating rates are balanced throughout the atmosphere, but the lapse rate is set equal to the adiabatic value in places where the radiative equilibrium value exceeds the adiabatic value (see, e.g., Pollack et al. 1987). Finally, a latitudinal energy-balance model is similar to a global energy-balance model in that energy is balanced for the entire atmospheric column, but now this balance is performed separately at all latitudes and therefore allowance is made for the contribution to this balance from heat advected by the atmosphere (or more precisely, the convergence of atmospheric heat transport) (see, e.g., Postawko and Kuhn 1986).

One of the first paleoclimate greenhouse calculations for the Earth was performed by Sagan and Mullen (1972). They pointed out the early faint young Sun paradox, and proposed that ammonia mixing ratios of only a few ppm could effectively counteract a reduced solar luminosity in the past. They used a global energy-balance model, with the absorption properties of various gases in the thermal infrared parameterized in terms of wavelength-integrated emissivities. Sagan (1977) pointed out that the calculations of Sagan and Mullen (1972) might also be relevant for Mars. In particular, small amounts of ammonia (several tens of ppm) in a hydrogen-rich atmosphere could have elevated the surface temperature on early Mars above the freezing point of water.

Using a global energy-balance model, Pollack (1979) performed the first nongray greenhouse calculations for Mars, in which the warming engendered by both highly reducing and fully oxidized atmospheres were considered. Dividing the thermal spectrum into a series of wavelength intervals, and evaluating the gaseous opacities in each interval, allowed him to describe much better the combined effect of several gas species than was possible with an emissivity representation. However, no allowance was made for possible changes in the planetary albedo, and crude band models, based on laboratory measurements and theoretical transmission calculations, were used to represent the gaseous opacity. Pollack (1979) found that atmospheres containing methane as the major gas species and ammonia as the minor gas species required total pressures ranging from $\sim$ 0.1 to 1 bar to raise the surface temperature to the melting point of water for the current solar insolation, with the lower values of the required total pressure corresponding to atmospheres with higher ammonia mixing ratios. The required total pressure increased by nearly an order of magnitude when a solar luminosity appropriate for 4.5 Gyr ago was considered.

Pollack (1979) also considered the greenhouse warming of Mars produced by atmospheres containing carbon dioxide and water vapor, with the abundance of water vapor being limited by its saturation vapor pressure. He found that carbon dioxide pressures of about 2 bar and 10 bar were required to raise the globally averaged surface temperature of Mars to the melting point of water ice for the current solar luminosity and the initial solar luminosity, respectively.

Cess et al. (1980) also explored the ability of massive carbon dioxide atmospheres on early Mars to engender a strong greenhouse warming. Their calculations represented improvements over that of Pollack (1979) in that they used more accurate band models, they employed a one-dimensional radiative-convective model, and they considered the negative feedback effect of enhanced Rayleigh scattering and hence an enhanced planetary albedo for the more massive atmospheres. They found that surface temperatures of 273 K could be realized with CO_2 pressures that were several times smaller than those of Pollack (1979). Unfortunately, an inadvertent error in the computer code used for these calculations vitiates these numerical results (V. Ramanathan, personal communication).

The first latitudinal energy-balance calculations for greenhouse warming on Mars were performed by Hoffert et al. (1981). They parameterized atmospheric heat transport in terms of an eddy diffusion coefficient, based on models of baroclinic eddies. However, baroclinic eddies, at least on present-day Mars, are present only at mid-latitudes of the winter hemisphere and even there they are not the only important component of the circulation (see, e.g., Pollack et al. 1990*a*). They found that globally averaged temperatures well in excess of 273 K could be realized with a 1 bar CO_2 atmosphere, which also contained saturation-pressure-limited amounts of water vapor. Even though this result refers to the present solar luminosity, it seems that the magnitude of the greenhouse warming is stronger than suggested by most other models. The reason for this apparent difference is not clear, although it may be related to the neglect of convection by Hoffert et al. (1981): radiative lapse rates can exceed the adiabatic value in the troposphere and lead to elevated surface temperatures for a fixed effective temperature.

Postawko and Kuhn (1986) used a somewhat more sophisticated latitudinal energy balance model to evaluate the greenhouse warming produced by carbon dioxide atmospheres alone, and CO_2 atmospheres that contained water vapor. They parameterized the atmospheric heat transport in terms of a latitude-independent eddy diffusion coefficient, and an altitude-integrated, longitudinally averaged meridional velocity, based on estimates of winds on current-day Mars. Accurately estimating either of these two wind components is a difficult exercise. A narrowband model was used of gas opacities in the thermal infrared. In addition to calculating latitudinal temperatures that would be realized for various CO_2/H_2O atmospheres, they also calculated temperatures across the planet for a CO_2/H_2O atmosphere which contained minor amounts of sulfur dioxide.

Figure 1 illustrates the potential importance of atmospheric heat transport in diminishing surface temperatures at low latitudes for a putative 3-bar atmosphere on Mars at times when the solar luminosity was 100% (heavy lines) and 75% (light lines) of its current value (Postawko and Kuhn 1986). Neither water vapor nor sulfur dioxide are present in the model atmospheres of this figure. For each solar luminosity, three curves are shown, corresponding to cases with no eddy transport, and with eddy coefficients of 10^8 and 10^9 $cm^2\ s^{-1}$, respectively. Clearly, if the eddy diffusivities are this high, temperatures at low latitudes are greatly reduced from their values when transport can be neglected. More sophisticated dynamical calculations are needed to explore this possibility.

Figure 2 shows the global greenhouse warming found by Postawko and

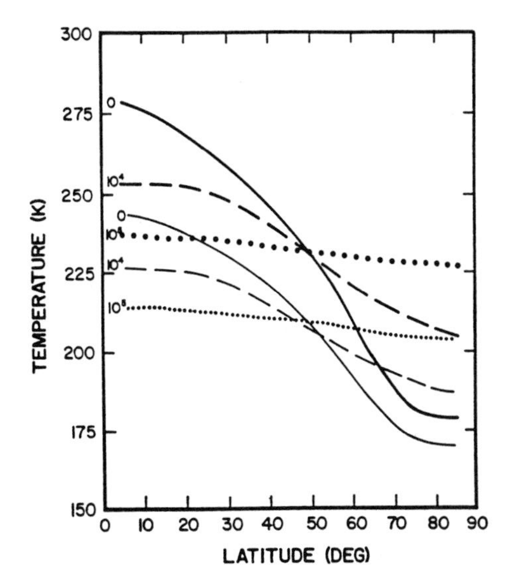

Fig. 1. Surface temperatures for a 3-bar CO_2 (only) atmosphere (no water vapor). Heavy lines correspond to calculations for present-day solar flux, and lighter lines are for a solar luminosity 75% of present. Temperatures are calculated for cases of no atmospheric transport (solid lines), and eddy diffusivities of 10^8 (dashed lines) and 10^9 (dotted lines) $cm^2\ s^{-1}$. Global albedo is 18% (after Postawko and Kuhn 1986, Fig. 4).

Kuhn (1986) due to the various greenhouse gases. When the atmospheric abundance of sulfur dioxide reaches about 1000 ppm, it can significantly augment the greenhouse warming due to CO_2 and H_2O. These authors conclude that horizontal heat transport within a 1 to 3 bar paleoatmosphere on Mars would be sufficiently vigorous to prevent surface temperatures from reaching 273 K at any latitude at times when the solar luminosity was only 75% of its current value. However, it should be noted that these calculations underestimated the greenhouse effect of massive carbon dioxide atmospheres by not including the pressure induced transitions of CO_2 longwards of 20 μm. Although dynamical heat transport was only crudely parameterized, the calculations point out the need for more sophisticated models to take into account the importance of heat transport in postulated massive atmospheres on early Mars.

Finally, Pollack et al. (1987) carried out the most recent and detailed treatment of gaseous opacities for putative massive CO_2 atmospheres on early Mars with a one-dimensional radiative-convective model. Consideration was given to the greenhouse warming produced jointly by CO_2 and water vapor, with the water vapor abundance being limited by its saturation pressure. Correlated k distributions were used to represent the opacity of these two species within a number of spectral intervals that spanned the near and thermal in-

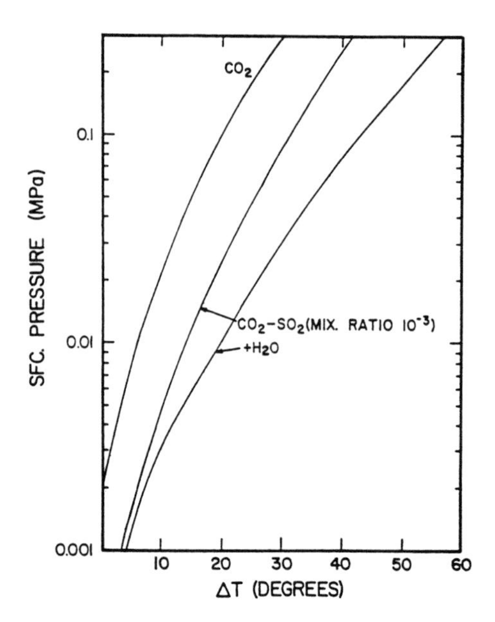

Fig. 2. Temperature increases due to atmospheric greenhouse heating for a pure CO_2 atmosphere, a predominantly CO_2 atmosphere with small amounts of SO_2, and a predominantly CO_2 atmosphere with both SO_2 and H_2O, with water vapor abundance being limited by its saturation pressure (after Postawko and Kuhn 1986, Fig. 6) (note: 0.1 MPa = 1 bar).

frared, with these distributions being derived from line-by-line calculations based on the data given in the AFGL HITRAN data base (Rothman et al. 1987). In addition, allowance was made for the pressure-induced transitions of CO_2 longwards of 20 μm and near 7 μm, for the 8 to 12 μm water-vapor continuum, and for weak lines of both gases in the visible and near infrared. Allowance was made for the effects of gas absorption and scattering on solar fluxes, but no attempt was made to include the radiative effects of water clouds, and dust and atmospheric heat transport was not explicitly included.

Figure 3 illustrates the variation of surface temperature as a function of the surface pressure of CO_2 for several alternative choices of incident solar fluxes and surface albedos, according to the calculations of Pollack et al. (1987). For an incident solar flux equal to the present globally and annually averaged value *S*, a CO_2 pressure at the surface of 2 bar would be required to raise the surface temperature to the freezing point of water. With pressures above this value, liquid water would commonly occur, while for smaller values it would not. There are a number of reasons that the incident solar flux for the cases of interest here may be different from the current value. First, as mentioned earlier, the solar luminosity was lower in the past than it is today. For greenhouse warming with 0.7 *S*, appropriate for Mars when it had just finished accreting, i.e., 4.5 Gyr ago (note that a value of 0.7 is within

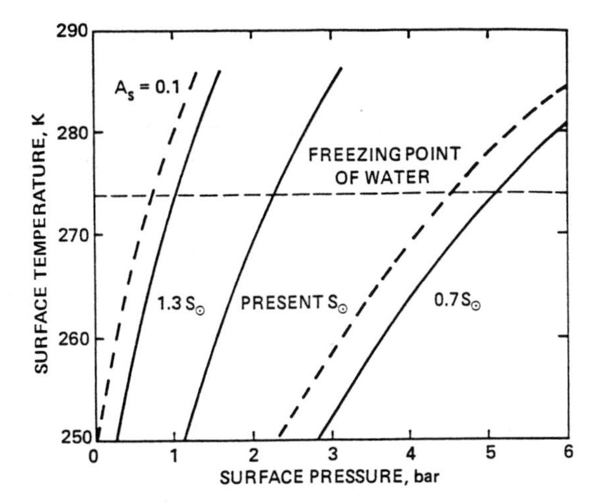

Fig. 3. Surface temperature on early Mars as a function of surface pressure of CO_2, with saturated water vapor. Solid curves refer to results for a surface albedo of 0.215 and alternative choices of incident solar flux. Present *S* corresponds to the current annually and globally averaged insolation. Dashed curves show results for a surface albedo of 0.1.

the range of uncertainty of solar evolution models) ~ 5 bar of CO_2 are needed to raise the surface temperature to the freezing point of water. However, larger insolation values are possible at later times, and for certain latitudes, orbital positions, and seasons. For example, while the globally and annually averaged insolation was 0.75 to 0.8 3.5 Gyr ago, the globally averaged insolation is 22% higher at perihelion for the maximum eccentricity achieved by Mars (0.14) than the annually averaged value, the equatorial regions receive about 40% more insolation than the globally averaged value, and the summer solstice pole could have received up to a factor of 3 times the annually and globally averaged insolation at times of high obliquity (the factor of 3 applies to an inclination of about 45°, corresponding to a possible pre-Tharsis maximum; see Toon et al. 1980). See chapters 9 and 33 for more detailed discussions of orbital variation and insolation distribution.

A value for the insolation of 1.3 *S* corresponds to conditions 4.5 Gyr ago at the equator and at perihelion for the maximum eccentricity. According to Fig. 3, about 1 bar of CO_2 is needed to raise the surface temperature to the freezing point of water under these insolation conditions. However, in practice, a somewhat larger pressure would be needed when account is taken of atmospheric heat advection and the finite heat capacity and hence response time constant of the atmosphere (see Pollack et al. 1987). Finally, the surface on early Mars may have been less oxidized than it is today, and hence a somewhat lower surface albedo may be appropriate for these early Mars situations. The dashed lines show the effects on the greenhouse warming of lowering the surface albedo from 0.215 to 0.1 for insolations of 0.7 and 1.3

S. In the latter case, the CO_2 pressure needed to achieve 273 K is lowered from about 1 to 0.75 bar.

So far, we have focused on greenhouse models of early Mars with emphasis on temperature conditions at low latitudes, where high temperatures can usually be more easily realized. However, there are occasions when temperatures above 273 K might be achieved with more modest CO_2 atmospheres. Toon et al. (1980) used a one-dimensional radiative-convective model to assess the potency of CO_2 greenhouses at both polar and equatorial latitudes for the present solar luminosity. As the axial obliquity occasionally reaches 35° for current astronomical perturbations, and might even have been as high as 45° before the Tharsis bulge formed, the summertime insolation near the poles at times of high obliquity can be a factor of several times larger than the globally and annually averaged insolation. Thus, summertime temperatures can reach and exceed 273 K near the poles at these times with CO_2 pressures that are in the range of about several tens to hundreds of mbar, provided that the ice-cap albedos are not too close to unity, and that atmospheric heat transport can be neglected to a first approximation (Toon et al. 1980). Toon et al. (1980) also showed that CO_2 partial pressures of several hundreds of mbar sufficed to raise the temperature near the equator to 273 K for the current solar luminosity, provided that atmospheric heat transport can be neglected. It should be emphasized, however, that the relevance of these particular results to early Mars is limited by the fact that the lower solar luminosity was not taken into account.

As the surface temperature approached 273 K, the abundance of water vapor in the Martian atmosphere became much larger than its present abundance. Consequently, optically thick water clouds would have formed over part of the globe and would have strongly interacted with solar and thermal radiation, just as is the case for the Earth's present climate. Also, latent heat released or taken up in phase changes of water would have made an important contribution to the overall energy budget. In most of the greenhouse calculations discussed above, the radiative effects of water clouds were ignored. In some cases, the latent heat effects were included in the sense that a wet adiabat was used in the troposphere. For the present Earth, clouds act as a negative feedback on the surface temperature in that the increase in the planetary albedo they produce by scattering some sunlight back to space is more important than the impact of their infrared opacity on the thermal radiation emitted to space. However, the sign of this effect depends sensitively on the height distribution of the clouds. Toon et al. (1980) and Postawko and Kuhn (1986) showed that high water clouds in paleoatmospheres of Mars could have enhanced the greenhouse warming of the surface. Once improved parameterizations of water clouds have been developed for Earth, including physically based estimates of their height distribution, it will be valuable to revisit their impact on Martian paleoclimates.

Finally, we have assumed in all the discussion above that permanent carbon dioxide ice caps do not form in the polar regions and strongly limit the amount of CO_2 in the atmosphere. The present climatic state and possible recent past climatic states on Mars serve as a starting point for assessing the realism of this important assumption. At present, the perennial northern polar cap is made of water ice, while the perennial southern polar cap has at least a top layer of CO_2 during some years. Even if the southern cap is made entirely of CO_2 and were present during all years, its mass would probably be small compared to that of the atmospheric carbon dioxide. At times of higher obliquity, a perennial CO_2 cap very likely does not occur in either hemisphere, while at times of lower obliquity such a cap or caps are probably present and exert a strong control on the atmospheric pressure (Toon et al. 1980; Fanale et al. 1982*a*; chapter 33).

Let us assume for the moment that the albedos of perennial CO_2 ice caps on ancient Mars were the same as today's values, and scale the above climatic states to early Mars. At very early times when the solar luminosity had a value of about 75% of its current value, the insolation incident on the summer poles at an obliquity of about 35° would have been about the same as that incident on the summer poles at present with an obliquity of 25°. If the maximum axial obliquity on early Mars was 45°, as might have occurred when the Tharsis bulge was not present, then perennial CO_2 ice caps may have been dissipated at times of high obliquity and a reduced solar luminosity. If the maximum obliquity was 35°, its current upper limit, then permanent CO_2 caps may have been a feature of earliest times on Mars and have been dissipated only after the solar luminosity increased somewhat from its initial main-sequence value. Once such caps had been dissipated, CO_2 ice caps may not have been able to reform at times of lower obliquity: if the atmospheric pressure of the liberated CO_2 was on the order of a bar, atmospheric heat transport from the equator to the poles would have strongly inhibited the formation of perennial CO_2 caps (Toon et al. 1980).

In summary, current greenhouse models indicate that CO_2 surface pressures of between about 0.75 and 5 bar are needed to raise the surface temperature on early Mars to the freezing point of water. Only slightly lower pressures characterize greenhouse warmings that are 10 to 20 K cooler. Finally, much more modest amounts of CO_2 are capable of substantially elevating the surface temperatures in the polar regions near summer solstice at times of very high obliquity, and several hundreds of mbar may suffice for high temperatures to be achieved near the equator with the present solar luminosity.

Carbon dioxide acting in concert with vapor-pressure-limited amounts of water vapor is not the only combination of greenhouse gases capable of causing a substantial warming of the surface during past epochs on Mars. Sulfur dioxide abundances, on the order of a 1000 ppm or greater, can significantly augment the greenhouse warming of atmospheres containing on the order of a bar of CO_2 and vapor-limited amounts of water vapor. Small amounts of ammonia (tens to hundreds of ppm) and large amounts of methane and/or hydrogen (on the order of a bar) can also generate substantial greenhouse warmings on early Mars.

Several caveats should be borne in mind about the above estimates of the amounts of various gases required to produce a substantial greenhouse effect. First, although gas opacity has been treated quite accurately in the more recent calculations, the radiative effects of water clouds and dust have either been ignored so far, or the highly idealized case of a high-altitude water cloud has been considered (see, e.g., Toon et al. 1980; Postawko and Kuhn 1986). Indeed, the magnitude and even the sign of the feedback effect of water clouds on altered climates of the Earth remains a major stumbling block to predicting accurately past and future climates for our planet. Second, the parameterizations of meridional heat transport used in latitudinal energy-balance models are highly idealized and therefore highly suspect. Third, the possible occurrence of massive perennial CO_2 ice caps on early Mars deserves careful scrutiny.

2. Greenhouse Gas Loss Processes. Greenhouse calculations that have been performed for possible early atmospheres on Mars indicate that several gas species could have caused a substantial elevation of the surface temperature, perhaps to values approaching or exceeding the melting point of water. To help evaluate these various possibilities, we need to consider the longevity of the proposed greenhouse gases, as there are a variety of processes by which their abundances may have been severely limited, especially on geologic time scales of 10^8 to 10^9 yr. These processes include photochemical reactions, driven by solar ultraviolet radiation, that can convert them into other gas or particulate species; various mechanisms, including hydrodynamical escape and atmospheric impact erosion, that may cause their loss to space; and chemical weathering processes that can convert them into components of surface rocks.

We first consider the stability of various greenhouse gases to photochemical reactions that can convert them irreversibly into other compounds, thereby potentially leading to an elimination of these gases in the absence of a mechanism for resupplying them. Ammonia is readily dissociated by solar ultraviolet at wavelengths shortward of about 0.23 μm. The dissociation rate of NH_3 is about 5×10^{12} molecules cm^{-2} s^{-1} for the current solar ultraviolet flux at Mars' distance from the Sun (Yung and Pinto 1978; Atreya et al. 1978). Only half, at most, of the dissociated NH_3 would have been recycled back into ammonia, with the rest being irreversibly converted into molecular nitrogen and hydrogen. The latter would have readily escaped to space. During very early times (the first several 10^8 yr), the solar ultraviolet output was much higher, by as much as a factor of several hundred (Zahnle and Walker 1982), and hence the NH_3 net destruction rate on early Mars would have been much greater than several times 10^{12} molecules cm^{-2} s^{-1}. This means that 1 mbar of NH_3 on early Mars would be irreversibly lost in < 100 yr. There is no plausible mechanism compensating for such an enormous loss rate.

Photolytic destruction may also greatly limit the ability of methane and sulfur dioxide to act as important greenhouse gases in the early Martian atmosphere. Photochemical reactions driven by the current solar ultraviolet flux can irreversibly convert one bar of methane on Mars to higher order hydrocarbons on a time scale of about 100 Myr (Yung and Pinto 1978). Much shorter time scales would have applied to early times when the solar ultraviolet output was substantially higher. Furthermore, oxygen generated from the photodissociation of water vapor, followed by the escape of H to space, would have oxidized the hydrocarbons, eventually converting them into carbon dioxide on a time scale of 10^8 to 10^9 yr (Yung and Pinto 1978).

Sulfur gases injected into the Earth's atmosphere by large volcanic explosions are irreversibly converted to sulfuric acid particles on a time scale of months to a year (Toon and Farlow 1981). The OH radical, created by the photodissociation of water, plays a key role in this conversion. Similar rates could apply on a hot early Mars.

Carbon dioxide is photolytically stable in the present Martian atmosphere and most past ones (except possibly ones of very low surface pressures) in the sense that CO_2, rather than its dissociation products CO and O_2, is the major constituent of the lower atmosphere. This occurs because of OH catalytically induced recombination of the dissociation products. The OH is derived from water or hydrogen peroxide photolysis. However, there are still major uncertainties connected with understanding in detail the reasons that CO_2 is stable in the present Martian atmosphere. It seems that large eddy diffusion coefficients, large water abundances, or the presence of water-ice particles, are required to ensure this stability (see, e.g., Atreya and Blamont 1990). Until this problem is better understood, it is difficult to be certain that CO_2 was stable in putative past atmospheres.

Gases can also be lost to space by a variety of processes that include loss by hydrodynamical escape, impact erosion, and exothermic chemical reactions (chapter 30). Hydrodynamic escape refers to a large outflow of gases from a planetary atmosphere in which the flux is sufficiently high that it is essentially a continuum flow. This is in comparison to a process such as Jeans escape, which occurs at elevations where the mean free path is comparable to the scale height for molecules whose thermal energy exceeds gravitational potential energy. In Jeans escape only the most energetic molecules are able to escape, and the time scale for Jeans escape increases exponentially with increasing mass of the gaseous species. Thus, for the terrestrial planets, loss of atmospheric species by Jeans escape is limited to the lightest gases.

Hydrodynamic escape, however, occurs at altitudes where the mean free path is less than the atmospheric scale height at the transonic point (see, e.g., Watson et al. 1981; Hunten et al. 1989). In hydrodynamic escape, the upward flux of light gases such as H and H_2 can carry along heavier species. The loss of heavier gases due to hydrodynamic escape is dependent on the flux of light gas, and declines linearly with increasing mass of the gaseous species. Such a vigorous escape of atmospheric gases may have occurred during the earliest history of the solar system when hydrogen may have been the dominant gas species in the atmospheres of the terrestrial planets and the solar ultraviolet radiation was tens to hundreds of times its present value. It is the solar ultraviolet flux that powers the escape by heating the gases in the uppermost part of the atmosphere sufficiently to enable them to escape out of the planet's gravitational well.

Scaling estimates of hydrodynamical escape for Earth (Watson et al. 1981) to Mars, we find that the current solar ultraviolet output is capable of powering the escape of a hydrogen-dominated Martian atmosphere at a rate equal to about 1.5×10^{12} H atoms cm^{-2} s^{-1}, or about the amount of hydrogen contained in the equivalent of 5 terrestrial oceans of water in 1 Gyr. Even more dramatic loss rates could have occurred at early times when the solar ultraviolet output was much higher. These calculations make it clear that hydrogen was probably never a major constituent of the Martian atmosphere (except possibly during its accretion when reactions between water and metallic iron may have generated copious quantities of hydrogen; Dreibus and Wänke 1987). Heavier gases, such as carbon dioxide, would have suffered much smaller losses by hydrodynamical escape (again, except possibly during accretion when hydrogen may have been the dominant gas) as the ultraviolet-energy-limited escape flux scales inversely with the molecular weight of the dominant atmospheric gas species, and as CO_2 and its photochemical derivative CO are potent radiators in the thermal infrared (and so the absorbed ultraviolet energy may be radiated back to space before molecular escape can occur).

Another potentially potent process by which large amounts of gases can be lost to space is impact erosion of atmospheres (see, e.g., Walker 1986; Melosh and Vickery 1989). Large impacting objects (say larger than a few km) may transfer enough of their kinetic energy to the atmosphere above the impact point to allow some or all of the atmosphere above the tangent plane to escape to space. When the impact velocity is larger than about twice the planet's escape velocity, the dominant mechanism by which this escape occurs may be the generation of a hot vapor plume during the impact event, which subsequently expands outwards and transfers some of its momentum to the overlying atmosphere (Melosh and Vickery 1989). When the impact velocity is smaller than twice the escape velocity, the amount of ejected atmosphere is generally much smaller and may result from the exchange of momentum between the impacting object and the atmosphere prior to impact (Walker 1986). Impact erosion is proceeding at a very slow rate for the current bombardment flux, but may have been much larger during the first 0.7

Gyr or so of Mars' history when the impacting flux may have been several to many orders of magnitude larger than its current value.

The amount of atmosphere that is removed by impact erosion depends sensitively on the distribution of impact velocities, for reasons explained above, on the time histories of atmospheric outgassing and impact bombardment, and on the size distribution of the impactors. In the case of low impact velocities, the amount of atmosphere ejected is limited to less than a few times the atmospheric mass traversed by the projectile (Walker 1986; Melosh and Vickery 1989). If the projectile has a great enough impact velocity so that a vapor plume forms, then significantly more atmosphere can be lost. Melosh and Vickery (1989) calculate that the minimum impact velocity for a vapor plume to form and exceed escape velocity on Mars is ~ 11.1 km s^{-1} for icy projectiles and ~ 14.3 km s^{-1} for silicate projectiles. In addition, if the impactor is massive enough, the entire mass of atmosphere above the tangent plane may be ejected. For present-day Mars, Melosh and Vickery (1989) estimate that the smallest projectile capable of removing the entire airmass above the tangent plane is on the order of 4×10^{13} kg. In general, larger atmospheres require a larger value for the minimum size of the impactor that is capable of eroding them, as the escaping gas must receive enough momentum from the impactor to reach the escape velocity. As much as a bar or so of CO_2 could have been lost to space by impact bombardment over the lifetime of Mars if the CO_2 was emplaced into the atmosphere during or shortly after the planet accreted and much of it remained there during the first Gyr or so (Melosh and Vickery 1989). Almost all of this erosion occurs during the first 700 Myr of Martian history, with a strong weighting towards the early portion of this period, when the impacting flux may have been largest (cf. Fig. 4).

Finally, exothermic chemical reactions can lead to a selective loss of atmospheric constituents, although the amount of gas removed in this way is quite modest compared to the amounts that can be lost by hydrodynamical escape and impact erosion. Ions produced by photochemical reactions initiated by solar ultraviolet radiation can recombine with electrons and dissociate to produce "hot" atoms. These atoms can have high enough velocities to escape from the modest gravitational well of Mars, and do so when the reaction occurs above the exobase (see, e.g., McElroy 1972). For example, the coupled H and O chemistry of the Martian atmosphere results in the joint escape of hydrogen and oxygen atoms from Mars by this process at a present rate of about 0.7 m of water ice per Gyr (McElroy 1972).

The Viking mass spectrometer found that the $^{15}N/^{14}N$ ratio in the Martian atmosphere is 60% larger than the terrestrial value (Nier et al. 1976*b*). The enhancement of heavy nitrogen on Mars has commonly been attributed to the selective escape of ^{14}N from the Martian atmosphere by exothermic chemical reaction (diffusive separation of the isotopes is the key factor in causing the

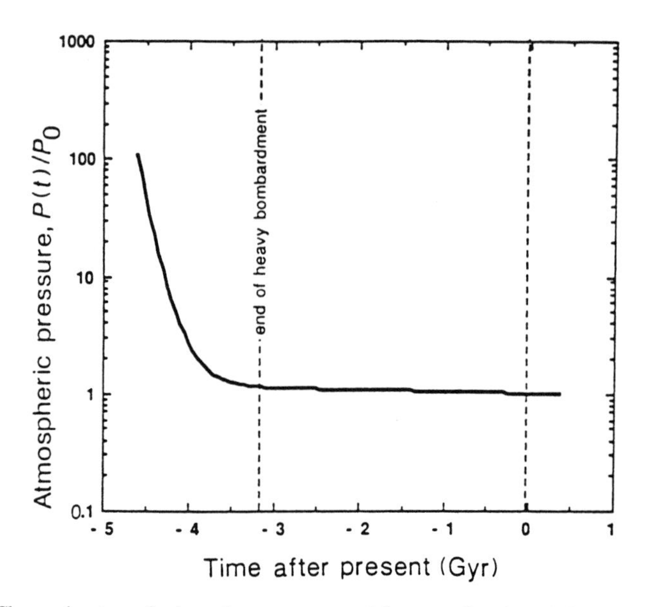

Fig. 4. Change in atmospheric surface pressure on Mars as a function of time, relative to its present-day value. The figure shows rapid decline in pressure during the period of late heavy bombardment (figure after Melosh and Vickery 1989).

fractionation) (see, e.g., McElroy et al. 1977; Fox and Dalgarno 1983). At its current escape rate, a few mbar of nitrogen have been lost from Mars over the age of the solar system. Fox and Dalgarno (1983) found that the observed degree of fractionation could be reproduced only by assuming that nitrogen alone was outgassed late in the planet's history, an unlikely possibility, or that a dense, early atmosphere was present and only gradually decreased over the planet's first several Gyr (e.g., a carbon dioxide dominated atmosphere that had an initial pressure of 1 bar and an exponential time constant of 0.7 to 1 Gyr). This latter possibility provides some support for the presence of a substantial atmosphere during Mars' early history, but this conclusion requires further scrutiny for its uniqueness.

Owen et al. (1988) found that the D/H ratio of Martian water vapor is about 6 times larger than that of terrestrial water. Because current escape processes may not be able to generate as much fractionation as is observed, Owen et al. (1988) suggested that the fractionation may have occurred at times of a warmer climate, when the water vapor abundance in the atmosphere was much higher and its dissociation rate was much higher as well. Whether the escape rate of H would have been higher is not clear (it may have been limited by a coupling of the O and H escape rates) and, even if the escape rate *was* higher, it is not clear that the D/H ratio would have tended to a larger value than occurs with present processes. More study of this important constraint is needed (see chapter 25).

Next, we consider the loss of atmospheric gases by the chemical weathering of surface rocks. Here, we focus particularly on the loss of atmospheric CO_2 by this means at times when it may have had contact with liquid water (for a more thorough treatment of chemical weathering on Mars, see chapter 19).

Atmospheric CO_2 on present-day Earth is being continually depleted by its chemical interaction with continental rocks that ultimately leads to the formation of carbonate rocks on the sea floor. This chemical weathering is occurring at a rate that is sufficient to eliminate almost all the atmospheric CO_2 on a time scale of 10^4 yr in the absence of CO_2 sources, such as volcanic outgassing (Holland 1978). The time constant for elimination of all the CO_2 from the combined atmosphere-ocean system is $\approx 4 \times 10^5$ yr. As shown schematically in Fig. 5, CO_2 dissolves in rain and in surface water to form a mild acid, carbonic acid. This acid interacts with silicate rocks to place cations such as Ca and Mg in solution, with CO_2 now being primarily in the form of HCO_3 anions. The dissolved cations and anions are carried by running water to the oceans. There, their concentration increases until carbonate rocks, such as calcite and dolomite, can be formed. On Earth, much of this formation occurs through the intermediary of ocean shell-forming biota, with some of the shells surviving dissolution in the lower ocean and creating a carbonate rock reservoir on the ocean bottom. Note, however, that in a steady state situation, the net production rate of carbonate rocks and hence the net loss rate of atmospheric CO_2 is entirely determined by the rate of continental weathering. In the absence of ocean biota, the cation and anion concentra-

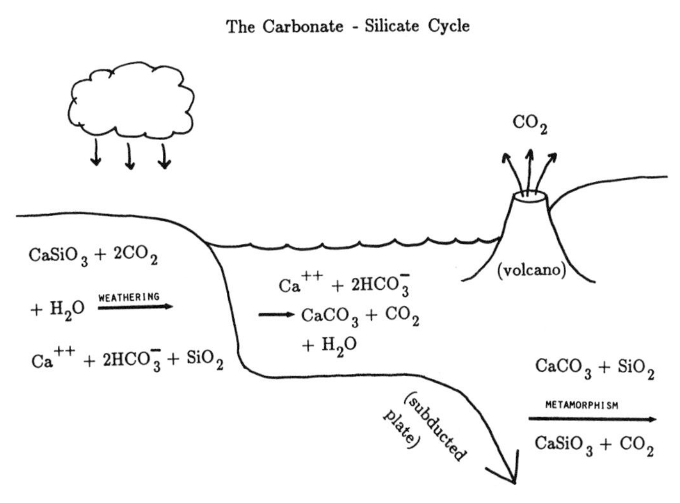

Fig. 5. A schematic representation of chemical weathering reactions of terrestrial continental silicate rocks by CO_2 dissolved in water, the subduction of the resultant carbonate rocks, their thermal decomposition at depth, and the outgassing of the released CO_2.

tions would increase until carbonates were formed at the same rate on the ocean bottom as rivers were bringing in new supplies of the cations and anions.

According to the above discussion, liquid water plays a vital role in the loss of atmospheric CO_2 through chemical weathering of silicate rocks. Thus, the occurrence of enough atmospheric CO_2 on early Mars to elevate the surface temperature to above 273 K could be self-limiting in the sense that the loss rate of CO_2 would have dramatically increased due to chemical weathering. Fanale et al. (1982*a*) were the first to estimate the chemical loss time constant on a wet warm Mars by simply scaling the terrestrial loss rate by the mass of CO_2 above a unit area of the surface. They obtained a weathering time scale of $\approx$10 Myr for a 1 to 2 bar CO_2 atmosphere. More sophisticated calculations, as described below, arrive at similar time scales. Thus, in the absence of resupply mechanisms, a wet, warm climate on early Mars would either be very short lived or at best would hover close to the freezing point of water.

In actuality, the chemical weathering rate of CO_2 depends on a number of factors. These include a dependence on the silicate rock mineralogy (the fractional abundance of cations that can be leached and form carbonates); pressure and temperature at which chemical reactions occur; the fractions of the surface covered by exposed rocks (the cation source) and liquid bodies of water (the source of running water); the rainfall rate (chiefly a function of temperature since precipitation and evaporation are in balance); and the fraction of time that liquid water exists on the surface (Pollack et al. 1987).

Figure 6 shows the chemical weathering time scales on a hypothetical early, wet, warm Mars as a function of the CO_2 pressure at the surface for

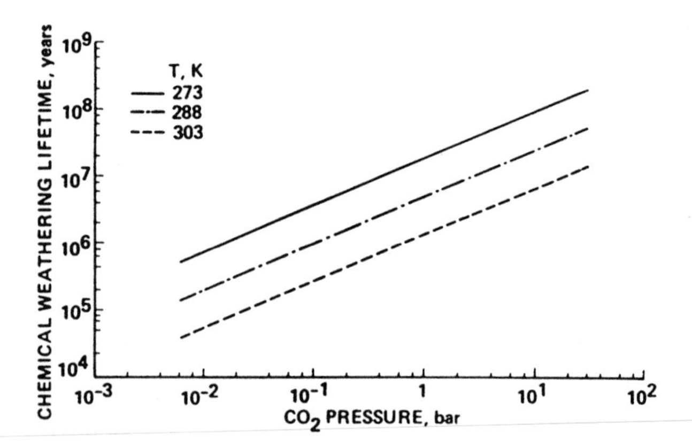

Fig. 6. Chemical weathering time scale for the elimination of a dense CO_2 atmosphere on early Mars as a function of CO_2 surface pressure. Liquid water is assumed to cover 0.05 of the surface, and to be present at all times. Results are shown for alternative choices of surface temperature.

several alternative choices of surface temperature (Pollack et al. 1987). These time scales were derived by taking account of the above factors and scaling from the chemical weathering rate on Earth. Nominal parameter choices for these calculations included a value of 0.05 for the fraction of area covered by liquid water, and a value of 1 for the fraction of time liquid water was present. According to this figure, a 1-bar CO_2 atmosphere has a chemical weathering lifetime of several tens of Myr for a surface temperature at the freezing point of water. Shorter lifetimes characterize cases with higher surface temperature, due chiefly to the dependence of the precipitation rate on temperature, and longer time scales apply to more massive atmospheres.

The above calculations indicate that the chemical weathering time scales for CO_2 depend most sensitively on parameters that control the amount of running water. In the absence of vigorous sources and the effective operation of other loss processes, one can expect the surface temperature to hover around the freezing point of water in the sense that the fractions of area and time when liquid water was present will adjust themselves to sufficiently small values for chemical loss to be in balance with resupply mechanisms. Naturally, if the resupply rate becomes essentially zero, the CO_2 pressure will fall to a sufficiently small value to guarantee that liquid water does not occur at all (see, e.g., Kahn 1985).

The discussion so far has focused on the weathering rates that could have occurred on early Mars at times that dense CO_2 atmospheres produced large greenhouse warmings of the surface. Kahn (1985) addressed the evolution of a CO_2 atmosphere subsequent to any early epoch during which liquid water may have been stable on the surface of the planet. Kahn proposed a mechanism which (1) removes CO_2 from the atmosphere of Mars at times when the CO_2 pressure is too low to produce adequate greenhouse warming to stabilize liquid water at equilibrium at the surface, but (2) ceases operating when the CO_2 pressure reaches a value around 7 mbar, the current average surface pressure on Mars.

Kahn (1985) argued that if ice at 273 K is heated enough so that the rate at which water is released from the surface is at least as great as the rate at which it diffuses away through the overburden gas, transitory (disequilibrium) liquid water will form before the molecules completely evaporate. The ability to form transitory liquid should cease when the overburden pressure of CO_2 approaches a value near the triple point pressure of water (6 mbar), because then the partial pressure of water will exceed the overburden pressure, and water molecules can leave the surface by hydrodynamic flow rather than by diffusion. The lower limit under actual Martian conditions depends on many factors, such as the physical and chemical state of the soil.

According to this scenario, even in the absence of stable liquid water, carbonates would have continued to form wherever local conditions allowed transitory liquid water to form, removing atmospheric CO_2 and creating a monotonic decrease in overburden pressure. The process would have continued until the overburden pressure of CO_2 became so low that it was no longer possible to form liquid water even in disequilibrium. Except for possible outgassing events, the atmospheric pressure would then remain at the limiting value. In support of the theory, Kahn (1985) pointed out that the occurance of liquid water in disequilibrium on the Martian surface is marginal, as would be expected if the process has gone to completion. He demonstrated this by comparing the evaporative and radiative cooling rates of water ice at 273 K with the solar heating rate. For current Martian conditions, the maximum surface heating rate would require about 20 mbar overburden pressure to form transitory liquid from pure water ice exposed at the surface.

Booth and Kieffer (1978) have argued that, even in the absence of liquid water, carbonate formation may be an important sink for atmospheric CO_2. They created conditions under which the only stable water available was in the form of vapor, and noted that carbonate formation was still able to take place. It must be pointed out, however, that the Booth and Kieffer conclusions were made only on the basis that *equilibrium* conditions for liquid water were not met. They did not test for, and cannot rule out, the presence of disequilibrium water in their experiments. Kahn (1985) noted that conditions for transitory water were met in the Booth and Kieffer experiments. Stephens and Stevenson (1990) have questioned the reaction rate measured by Booth and Kieffer (1978), noting that the ability of the CO_2 to diffuse through an incipient carbonate rind on a weathering grain may be a rate-limiting process.

Several uncertainties must be resolved before the above intriguing hypothesis can be accepted, the main issue being whether or not the necessary water, heat and cations occur simultaneously at the surface. In general, the warmest locations on the planet are the least likely to retain ice, and it is not clear that surface patches of water ice formed at low latitudes on Mars during the last several Gyr. Certainly there was no evidence of water frost on the surface at the Viking Lander 1 site at 22° north latitude at any seasonal date. Such frost did occur at the Viking Lander 2 site at 48° north latitude (Hart and Jakosky 1986), but only during the winter season when the ground temperature was always far below 273 K. Temperatures in excess of 273 K can occur at low latitudes at places having a sufficiently low value of thermal inertia, although it is likely that patches of water ice would have higher thermal inertias than dust covered ground due to their having more contiguous thermal contact between elements of the surface. The increased surface albedo of ice patches would also lower the temperatures actually realized, although Clow (1987) has shown that liquid water can form within dirty snow under energy balance conditions which would not support a free surface of liquid water.

On the other hand, several tens of microns of water vapor appear in the atmosphere at low northern latitudes in early to mid spring, as part of the seasonal cycle. IRTM measurements (Martin 1981) indicate that maximum daytime temperatures may exceed 273 K in these regions. If any ice is present, the temperature in the regolith is not likely to be uniform because the icy patches will be buffered at 273 K by the latent heat of sublimation. Because the surrounding regolith is warmer than 273 K, heat should diffuse toward the icy patch, increasing the heating rate. The size of the effect will depend on the temperature difference and the geometry of the ice in the soil. Nevertheless, even if small amounts of liquid water did occur for short times, it is not clear that a large enough number of particles would be chemically weathered in an efficient manner and significant quantities of carbonate rock formed.

If chemical weathering of surface silicates has occurred on Mars, at times of an early wet environment and/or at later times in a more episodic and transitory fashion, sedimentary rocks would have formed, including carbonates. There is now some limited evidence that carbonate rocks are present on Mars. In particular, a spectroscopic feature in the thermal emission spectra of Mars near 6.7 μm, and features in the reflection spectra near 2.3 μm may be due to carbonate anions (Pollack et al. 1990*b*; Clark et al. 1990). Additionally, there is good evidence that small amounts of native carbonates are present in SNC meteorites, which are believed to have originated on Mars (Gooding et al. 1988). These studies suggest that carbonates constitute somewhere between a few tenths of a percent to a few percent of the surface materials. If future studies support these identifications, then carbonates may constitute an important, perhaps the most important, reservoir of CO_2 on Mars at present, as advocated on theoretical grounds by Pollack and Yung (1980), Toon et al. (1980), and most thoroughly by Kahn (1985) on the basis of mass-budget arguments.

In summary, weathering time scales for atmospheric CO_2 indicate that a 1-bar atmosphere would be depleted on a time scale of several tens of Myr for a mean surface temperature of 273 K. Shorter time scales apply to higher surface temperatures, and longer ones to higher CO_2 pressures (cf. Fig. 6).

3. Resupply of CO_2. Here, we examine mechanisms that might prolong the lifetime of a massive CO_2 atmosphere on early Mars. In particular, we consider the demands on both juvenile and recycling sources of fresh CO_2. The latter source refers to processes that liberate the CO_2 contained in sedimentary carbonates and emplace the liberated CO_2 back into the atmosphere.

Potential sources of juvenile CO_2 include C-bearing material contained in the planetesimals from which the planet accreted that were not released upon impacting the growing planet, and C-bearing material contained in impacting objects during the late heavy bombardment phase that were impact released. In the former case, interior melting during the planet's early history mobilizes the C-containing volatiles, ultimately releasing them as gases into the atmosphere. In the absence of a metallic iron buffer, CO_2 would have probably been the chief C-containing gas species (see, e.g., Pollack and Yung 1980). For a juvenile source to maintain a CO_2 atmosphere having a pressure P over a time scale T_h, it must provide CO_2 at a rate inversely proportional to the weathering time scale T_w. The total equivalent pressure of CO_2 that must be released to maintain the atmosphere is given by PT_h/T_w. For example, to maintain a 1-bar atmosphere for 1 Gyr, with a weathering time scale of 20 Myr, the juvenile source must supply about 50 bar of CO_2. Similar amounts are necessary for other values of surface pressure, because the weathering time scales roughly as P for a given surface temperature. In the example given, a ground temperature of 273 K was assumed.

The inferred total amount of juvenile CO_2 needed to sustain a wet, warm climate over 1 Gyr is significantly larger than implied by a simple mass/area scaling of the Earth's inventory, and interior volatiles may have been much less efficiently released from the interior of Mars than from the Earth (Jakosky 1989). Nonetheless, it is quite possible that Mars was more heavily endowed with volatiles than the Earth (Dreibus and Wänke 1989). Even though much water may have been lost in homogeneous accretion scenarios, through iron reduction, such losses would not affect C-bearing volatiles.

Towards the end of the late heavy bombardment period, 3.9 Gyr ago, the impact rate was about 100 times higher than the current value (Wilhelms 1987). Based on the known crater production rate on the Moon during the Nectarian period, if one assumes that craters of a given diameter were produced at the same rate on Mars as on the Moon, that the bombarding objects consisted solely of C1 type carbonaceous material, and that all the meteoritic C was oxidized to CO_2, then about 0.3 bar of CO_2 per 10 Myr would be supplied by the late heavy bombarding objects at 3.9 Gyr ago (Carr 1989). This supply rate comes within a factor of 2 in compensating for the loss of 1 bar of CO_2 at a ground temperature of 273 K by weathering reactions. However, it seems unlikely that the average composition of the impacting objects had a C content as large as that of C1 meteorites.

On Earth, loss of atmospheric CO_2 by chemical weathering is balanced on geologic time scales by the recycling of the CO_2 content of carbonate rocks back into the atmosphere (Walker 1977; Holland 1978) (cf. Fig. 5). Carbonate rocks that were formed or emplaced on the ocean bottom are carried by plate tectonics to subduction zones, where they are carried to great enough depths to thermally decompose (carbonate rocks can also be plated on continental margins during subduction). The released CO_2 is then vented back into the atmosphere at volcanic ducts. Thus, there is a long-term balance between the rate at which CO_2 is removed from the atmosphere by chemical weathering and the rate it is resupplied by carbonate subduction and decomposition. The former rate depends on the atmospheric pressure explicitly and implicitly (through the temperature dependence of the weathering rate), and the latter rate depends on the vigor of plate tectonics. Note that as long as the latter occurs at some minimal rate (dictated by nonsubaerial loss processes), an ice-covered Earth is obviated since then the subaerial loss rate would go to zero and CO_2 would build up until higher temperatures were realized (Walker et al. 1978).

Global scale volcanism and heavy bombardment represent two additive mechanisms for recycling CO_2 on early Mars (Pollack et al. 1987; Carr 1989). In both cases, the recycling is chiefly accomplished by burying carbonate rocks to great enough depths (about 10 km) for them to thermally decompose. Temperatures of about 1000 K are needed to decompose carbonates, although the decomposition temperature varies by a few hundred K among the more common sedimentary carbonates (cf. Fig. 7). The depth at which the decomposition temperature is reached depends on both the thermal conductivity of the subsurface material and the interior heat flow. The thermal conductivity may be substantially reduced from its zero-porosity value in the porous and brecciated top layers of the surface and hence the depth to decomposition may be substantially reduced from that obtained by simply using conductivities appropriate for well-consolidated materials (cf. Fig. 7). Currently, the interior heat flux on Mars is probably $\sim$ 30 mW m^{-2} (Davies and Arvidson 1981). On early Mars, it may have been as high as about 100 to 200 mW m^{-2} due to a combination of accretional heat, radioactive heat and heat derived from core formation (Davies and Arvidson 1981; Schubert et al. 1990).

Pollack et al. (1987) estimated the recycling time scale T_r for burial of carbonate rocks by global-scale volcanism. They partitioned the heat reaching the surface between a thermal conduction term, which governs the rate at which temperature increases with depth and thus the depth to which carbonate rocks must be buried in order to thermally decompose, and a term due to the latent heat content of volcanic lava, which governs the rate of burial. As shown in Fig. 8, for a fixed total heat flux the volcanic recycling time is insensitive to variations in the partitioning coefficient, α (which specifically represents the fraction of the surface heat flux contributed by thermal conduction). The reason for this is that, while an increase in the volcanic term (that is, a smaller α) implies a greater burial rate, it also means a smaller conduction term and thus a greater depth to the decomposition point of the carbonate rock.

In a steady state between loss by weathering and resupply by recycling, the ratio R of the amount of CO_2 in the atmosphere to the amount in carbonate rocks is equal to the ratio of the corresponding time scales, i.e., $R = T_w/T_r$. For example, if P = 1 bar and T_w = 20 Myr, then the steady-state abundance of carbonate rocks would equal 7.5 and 1.8 bar, respectively, for interior heat fluxes of 100 and 200 mW m^{-2}. The total amount of CO_2 involved in this system would equal 8.5 and 2.8 bar, respectively. Thus, much smaller amounts of outgassed CO_2 are required to sustain a massive CO_2 atmosphere with this recycling mechanism on early Mars than with juvenile sources.

Carr (1989) has argued that the volcanic burial source was far less potent

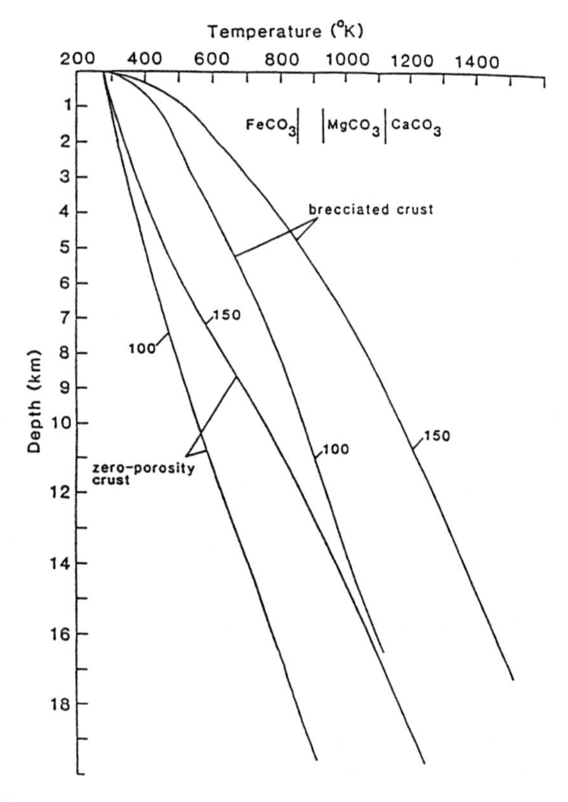

Fig. 7. Subsurface temperature profiles on early Mars for alternative choices of interior heat flux (numbers next to curve in units of mW m^{-2}) and subsurface porosity. The latter affects the thermal conductivity profile. The vertical lines near the top of the figure show the temperature at which various carbonates thermally decompose.

and less important compared to an impact ejecta burial mechanism near the end of the period of heavy bombardment on Mars. This is based on several arguments. First, some crater rims dating to this period have survived up until the present. This limits the rate of extrusive volcanism (although it does not limit the rate of intrusive volcanism). Second, there appears to be a near coincidence between the end of the heavy bombardment period and the end of the formation of valley networks.

Using a total amount of outgassed CO_2 of 10 bar and assuming that carbonates needed to be buried to a depth of 15 km for them to decompose, Carr (1989) found that between 0.15 and 0.3 bar of CO_2 would be thermally decomposed and hence resupplied every 10 Myr. These numbers apply to an interior heat flux of about 100 mW m^{-2}. The spread in values reflect uncertainties in the modeling of the burial process. Significantly higher resupply rates would apply to higher values of the heat flux. For the nominal values of the resupply rate found by Carr (1989), the ground temperature would have

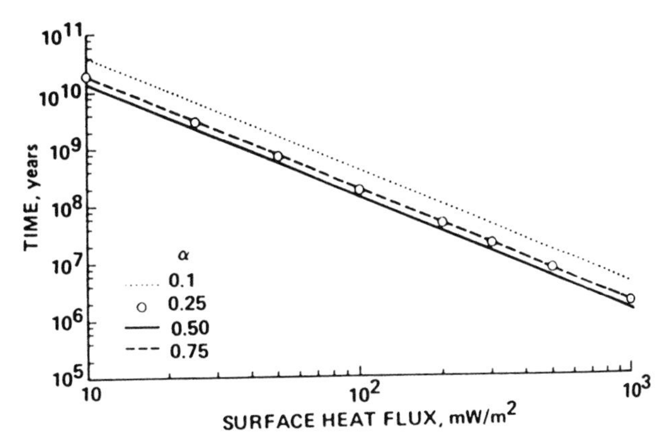

Fig. 8. Time scale for global scale volcanism on early Mars to recycle the CO_2 back into the atmosphere through the burial of carbonate rocks and their thermal decomposition at depth. Alternative curves show results for alternative choices of α, the fraction of the heat flux carried by thermal conduction, as opposed to the heat content of magmas. The horizontal axis is the total interior heat flux.

hovered near 273 K, but would have lain below this value most of the time so that the weathering rate could have been sufficiently reduced to balance the resupply rate. A slightly wetter and warmer climate would have occurred when the volcanic burial rate is included and especially when a higher interior heat flux is used.

III. ESTIMATES OF WATER AND CARBON DIOXIDE ON MARS

Loss processes aside, was Mars endowed with enough surface volatiles to produce a warm, wet climate? Abundances of relatively heavy and inert atmospheric species per gram of planetary mass are very much lower on Mars than on Earth and Venus, e.g., by factors of about 160 and 12,000, respectively for ^{36}Ar. The Martian volatile system is open to loss of molecular constituents such as nitrogen by isotope-dependent nonthermal escape to space (see previous section). Moreover, carbon dioxide and water are very likely sequestered in the cold megaregolith, in amounts which are unknown but probably large compared to contemporary atmosphere-polar cap abundances. Therefore we cannot assume that present-day atmospheric inventories of these species represent the abundances—and, at least for nitrogen, the isotopic composition—originally generated on the planet by early accretional and evolutionary processes.

A. Degassed and Interior Inventories of Water and Carbon Dioxide

1. Reservoirs and Storage Capacities. The principal *observable* reservoirs for these species on Mars today are the perennial north polar cap and the atmosphere. Water in the north cap system, if distributed globally, would cover the surface to a depth of ~10 to 40 m, depending on estimates of cap volume, dust content, and porosity. Most of the observable carbon dioxide resides in the present-day ~7 mbar CO_2 atmosphere, with (probably) minor amounts frozen out in the residual south polar cap.

By such measures alone, Mars is indeed volatile poor. But these are clearly minimum contemporary inventories. To assess present and past near-surface abundances of these species on Mars, one must consider (1) the existence and capacities of crustal reservoirs for storage of currently unobservable water or CO_2; (2) the extent to which these reservoirs might now be populated; and (3) the loss of primordial volatiles to space over the history of the planet.

For various estimates of melting-point temperatures and porosities, Clifford (1984) estimates that ~70 to 700 m of water planet-wide could be stored in a regolith permafrost cryosphere extending to depths up to ~3 km, and that the total pore volume of the crust to a self-compaction depth of ~10 km, if saturated, would provide storage capacity for ~1000 ± 400 m of globally distributed water. Occurrence of subsurface ground ice in excess of local porosities (Carr 1986) would of course increase these estimates of maximum megaregolith-carrying capacities (see chapter 16).

Carbon dioxide could be stored in the regolith by adsorption or fixation as carbonates. Adsorption of CO_2 on cold regolithic grain surfaces has been considered in most detail by Fanale and his co-workers (see, e.g., Fanale and Cannon 1971, 1979; Fanale et al. 1982*a*; Zent et al. 1987). Laboratory measurements of maximum adsorptive capacities from Fanale et al. (1982*a*) and earlier work by this group lead to estimates ranging from an upper limit near 1 bar of degassed CO_2 (adsorbed in a 1 km deep planet-wide nontronite clay regolith, all at near-polar thermal conditions and thus unlikely) to equivalent pressures of hundreds of mbar or less for more plausible models (Fanale et al. 1982*a*) of regolith composition and latitudinal temperatures vs depth profiles. Aqueous precipitation of CO_2 as carbonate rock would sequester ~45 mbar of equivalent CO_2 pressure per m of globally deposited calcite. It seems likely that this putative reservoir would by now have been largely comminuted to fine regolith particles if it were laid down prior to ~3.8 Gyr ago, during the epoch of heavy bombardment. Both Viking soil analyses (Clark and Van Hart 1982) and recent Earth-based spectroscopic measurements (Pollack et al. 1990*b*) estimate that carbonates comprise ~1 to 3% by volume of airborne Martian dust. If this mixing ratio is assumed to characterize the subsurface megaregolith as well, the abundance of CO_2 sequestered as carbonate could be on the order of 1 bar for each km of regolith depth.

2. *Model estimates of inventories.* A variety of modeling scenarios have been developed to estimate present and past abundances of surficial and interior water and carbon dioxide. All adopt one or more of four different

approaches to the problem: (1) assumption of Earth-like ratios for outgassed $H_2O/CO_2/N_2$ on Mars; (2) construction of geochemical models to assess the present bulk and crustal composition of the planet, including volatile constituents, and derivation of release factors for volatile degassing from observed atmospheric abundances of the radiogenic and nonradiogenic argon isotopes; (3) geochemical modeling, based most recently on SNC meteorite chemistry (chapter 4), of Mars' primordial (accretional) composition, and estimation of the abundances of accreted volatiles that might have survived loss of a primordial atmosphere by hydrodynamic blowoff or impact ejection to space; and (4) physical modeling of these two atmospheric escape processes to infer past volatile inventories, which essentially attempts to solve an inversion problem by backward modeling from isotopic and elemental abundance distributions in the contemporary atmosphere. Examples of these various models are briefly discussed below, and their estimates for water and CO_2 abundances plotted in Figs. 9 and 10 together with inventories in observable reservoirs and assessments of storage capacities.

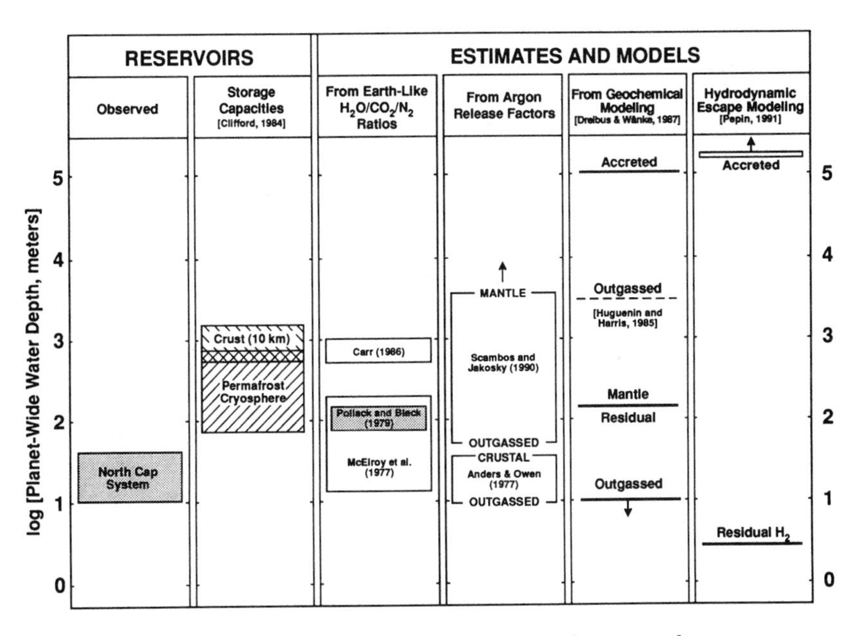

Fig. 9. Water storage capacities and abundances on Mars, inferred from observation of visible reservoirs and calculated from a representative set of models for the origin and evolution of Martian volatiles. Modeling estimates are for water outgassed to the surface unless otherwise designated; those labeled "crustal" and "mantle" refer, respectively, to near-surface but undegassed water and to amounts sequestered in the deeper interior; and "accreted" abundances represent estimates of primordial planetary inventories. All abundances are plotted as a corresponding depth, in m, of a globally distributed surface water layer.

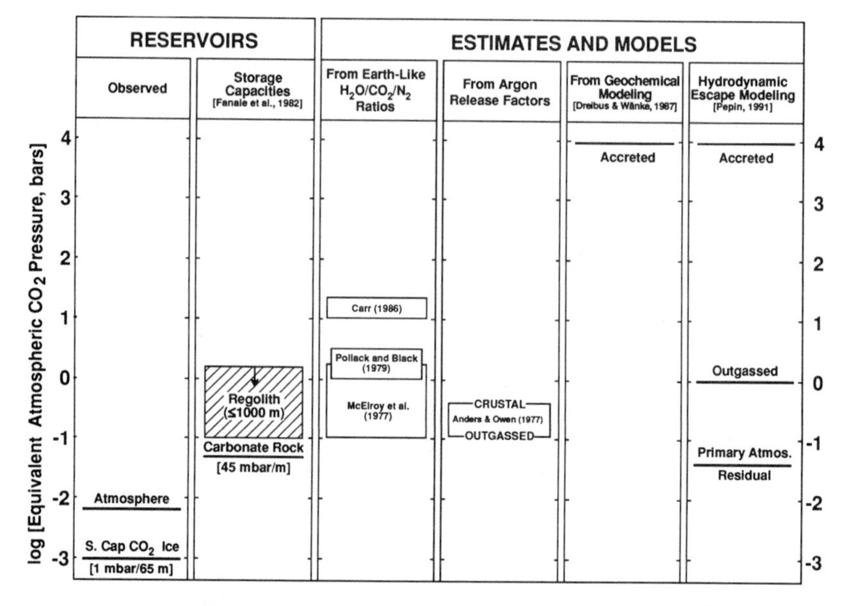

Fig. 10. As in Fig. 9, but for carbon dioxide on Mars. Possible contributions of south polar cap CO_2 ice to the observed atmospheric reservoir, and of carbonate rock to storage capacity, are shown as abundances per m of ice or rock thickness. All other abundances are plotted as a surface pressure, in bars, exerted by the corresponding amount of atmospheric carbon dioxide.

Calculating inventories by assuming Earth-like $H_2O/CO_2/N_2$ ratios requires the present (or past) abundance of one of these species on Mars to be estimated in some way. McElroy et al.'s (1977) modeling of present-day ^{15}N enrichment by nonthermal nitrogen escape to space over geologic time implies an initial atmospheric N_2 inventory ~15 to 500 times larger than the current concentration. This provides the basis for their assessments (and for those of Pollack and Black [1979], who adopt a somewhat narrower range of initial N_2 abundances) of the amounts of water and CO_2 originally outgassed on the planet. Carr's (1986) higher values are based on geologic and morphologic arguments. His estimates of the volumes of water required to cut the valley networks and outflow channels, and of ground ice underlying several types of surface terrain, point to early outgassing of at least 500 to 1000 m of water—of which he considers that ~400 to 500 m planet-wide is probably still extant in polar terrains and as subsurface ice deposits—accompanied by corresponding (Earth-scaled) amounts of CO_2 and N_2. Note that the inventories shown in Figs. 9 and 10 are calculated by the above authors using different estimates for terrestrial ratios in the atmosphere, oceans and sediments: $H_2O/CO_2/N_2$ by weight ~360/80/1 (McElroy et al. 1977) and 120/65/1 (Pollack and Black 1979; Carr 1986).

Anders and Owen (1977) infer a bulk composition for Mars by utilizing a chondrite-based multicomponent mixing model comprising five element groups of differing geochemical behavior, with educated guesses for the Martian abundances of certain "index" elements that establish the relative mixing ratios of the various groups. Their principal source of H, C, N and noble gases, on both Earth and Mars, is a late-accreting veneer of carbonaceous chondrite (C3V) material that has partitioned similarly into the crusts of the two planets. Therefore, as in the models above, crustal $H_2O/CO_2/N_2$ ratios on Mars and Earth are taken to be the same. Absolute abundances on Mars, however, are considered to be much lower; they depend directly on the assumed Martian concentration of the index element (Tl) for the most volatile group of elements, and here the authors argue for a 35-fold depletion compared to Earth. With this assumption, they derive a set of predicted global and crustal volatile element abundances for Mars, and, from these and observed atmospheric inventories, a set of relative release factors for degassing of atmophilic species from crust to atmosphere. These release factors suggest that about 25% of the crustal (and ~20% of the global) endowment of ^{36}Ar has been outgassed. This estimate for argon is taken to apply to release of crustal water and CO_2 as well, yielding ~10 m of globally distributed degassed water and ~100 mbar of CO_2 (presumably now mostly sequestered in the megaregolith), with ~3 to 4 times these amounts still retained in the crust. The Anders-Owen model, by its choice of a very low thallium abundance on Mars, thus implies an extremely volatile-poor planet. A somewhat similar approach (in that it again requires an estimate of the abundance of a particular element, potassium, and is based on derivation of release factors for argon) has been taken by Scambos and Jakosky (1990), but with considerably different results. These authors assume that the relative magnitude of outgassing with time on Mars was congruent with Greeley's (1987) estimate for the volume of erupted volcanics vs time. With this assumption, and a range of ^{40}Ar release factors calculated from an SNC-compatible range of bulk K estimates and the known atmospheric concentration of ^{40}Ar, they develop an expression, corrected for radiogenic growth of ^{40}Ar from potassium in the volatile source reservoir (here taken to be the mantle), for the fractional degassing of nonradiogenic volatile species from this reservoir over Greeley's volcanic history of the planet. Integrated release factors for nonradiogenic volatiles determined in this way are up to a factor of ~2 larger than those for ^{40}Ar. Applied to water, for which Scambos and Jakosky argue a minimum degassed inventory of 50 to 60 m, their calculated range of release factors (0.017 to 0.080) yields a contemporary *mantle* abundance, globally distributed, of at least 600 to 3500 m, substantially larger than the interior water content deduced from the Anders-Owen model. No estimates are made for carbon dioxide, but one might assume, for illustration, that carbon is present in the interior source as CO_2 with, say, the McElroy et al. (1977) terrestrial H_2O/CO_2 ratio of 360/80. This is close to the ratio chosen by Anders and

Owen (1977) for the crusts of Earth and Mars. If so, and if the two species did not fractionate with respect to each other on release, then degassed CO_2 amounted to ~400 mbar at least, and ~5 to 30 bar remains in the mantle reservoir. A stipulation common to both the Anders-Owen and Scambos-Jakosky estimates for interior concentrations of nonradiogenic volatiles is that present-day surface abundances reflect actual outgassed inventories, undepleted by subsequent escape to space, and unaugmented by surviving remnants of a co-accreted primordial atmosphere, or, in the latter model, by more substantive early outgassing than that suggested by Greeley's (1987) chronology of volcanic activity. In this respect these scenarios differ from those invoking partial loss of both primordial and rapidly degassed atmophilic volatiles from Mars by impact erosion (Melosh and Vickery 1989) or hydrodynamic escape (see, e.g., Zahnle et al. 1990*a*; Pepin 1991).

A very different perspective on primordial volatile abundances emerges from Dreibus and Wänke's (1985,1987,1989) geochemical modeling. They consider that both Earth and Mars accreted from just two principal components, one (A) which was reduced and volatile poor, with nonvolatile elements present in the abundance ratios found in contemporary C1 carbonaceous chondrites, and the other (B) which was oxidized and volatile rich with all elements in C1 proportions. In this model, the bulk chemical composition of Mars, derived from SNC meteorite compositions, is well reproduced by homogeneous accretion of the two components, with an A:B mixing ratio of ~60:40. The 40% contribution of C1 material results in enormous initial volatile endowments in the bulk planet: on the order of 100 km of globally distributed water, for example, and carbon and nitrogen equivalent to ~9 kbar of CO_2 and ~100 bar of N_2. However, all highly volatile (atmophilic) species are subsequently expelled almost completely from the planet. Oxidation of component A's reduced iron and other metallic phases to FeO and other oxides by component B's water generates immense quantities of H_2 that exhaust to the surface, extracting other gases with it, to form a massive primordial atmosphere that escapes to space by hydrodynamic blowoff and/or erosion by continuing accretional impacts. Dreibus and Wänke do not discuss the fate of accreted carbon, but they consider that virtually the only water surviving on Mars would be that dissolved in magma in contact with a temporary, water-rich atmosphere during accretion. They estimate retention of only about 36 ppm H_2O by equilibrium solubility in the present mantle, corresponding to a global water layer ~130 m deep. If we assume a release factor ≤0.08 from Scambos and Jakosky (1990), ≤10 m of this would have been outgassed to the surface. Huguenin and Harris (1985,1986) take a different view on water in the Dreibus-Wänke scenario. They argue, from Lewis' (1973) equilibrium condensation model, that iron at Mars' radial distance would have been almost completely oxidized in the nebula prior to accretion. Therefore no significant amount of metallic iron was available in the planet's interior to react with C1 water, and most of the accreted water is still there, in hydrated silicates (if so, Mars contains ~2.4 wt% H_2O—using Kerridge's [1985] average of 6700 ppm H in the C1's—if the A:B mixing ratio was 60:40). Huguenin and Harris calculate that ~2800 m of this, planet-wide, would have been degassed if the release factor was 0.026 (near the lower end of Scambos and Jakosky's [1990] estimated range).

There are many other volatile models, discussed in chapter 4, which deal primarily with the nonradiogenic noble gases. Much of the current activity in this area is centered on developing physical models of processes that could have acted, in planetary or preplanetary environments, to generate the isotopic patterns displayed by noble gases in meteorites and in the contemporary atmospheres of the terrestrial planets. Several of these compositional signatures, including those derived for Mars from the SNC meteorites, indicate that the mechanisms responsible for evolution of present-day noble gas distributions from primordial compositions were effective in fractionating isotopes as well as elements. This observation has directed attention to intrinsically mass-fractionating processes such as the hydrodynamic loss of an early, H_2-rich planetary atmosphere postulated by Dreibus and Wänke (1987). Two mathematical models of noble gas fractionation in hydrodynamic escape have been applied to Mars; one (Zahnle et al. 1990*a*) focused on Ne and Ar isotopes, and the other (Pepin 1991) on isotopic evolution of Ne, Ar, Kr and Xe from primordial volatile components accreted in the Dreibus-Wänke geochemical model. The total amount of H_2 which must escape hydrodynamically from the planet to generate the required noble gas fractionations in Pepin's scenario sets a lower bound on the abundance of initially accreted water; as seen in Fig. 9 it is comparable to the water supplied by Dreibus and Wänke's 40% C1 material. Although little H_2 remains at the end of hydrodynamic loss, we note below that this does not necessarily imply a very dry post-escape Mars. Regarding CO_2, Pepin (1991) concludes that virtually all of the large initial CO_2 inventory in the Dreibus-Wänke model would have been lost in hydrodynamic blowoff of the primary atmosphere. However, ~1 bar of CO_2 (estimated from noble gas isotope systematics, with an uncertainty of at least a factor of 2) was later outgassed from a small reservoir of C1 volatiles retained in the planet's interior.

During early epochs of high bombardment flux on Mars, impact-induced atmospheric erosion (Cameron 1983; Watkins and Lewis 1986; Walker 1986; Ahrens and O'Keefe 1987; Melosh and Vickery 1989) could have led to significant loss of atmophilic volatiles in the scenarios discussed above, as noted by Carr (1986), Dreibus and Wänke (1987), Scambos and Jakosky (1990) and Pepin (1991). Melosh and Vickery (1989) investigate the physical dependence of this ejection mechanism on impactor velocity and mass, and on the pressure and scale height of the impacted atmosphere, and derive from these considerations a simple analytic expression for depletion of atmospheric mass with time on a planetary body under given conditions of bombarding projectile flux and size distribution. They utilize this model to extrapolate from present to past atmospheric masses on Mars, and conclude that the current atmosphere could be an eroded remnant of a ~1 bar primordial CO_2 atmosphere existing on the planet ~4.6 Gyr ago, if Mars experienced the same intensity, chronology and size distribution of heavy bombardment flux as that recorded on the Moon, and provided there were no other sources or sinks of CO_2. This ~1 bar atmosphere was present only in very ancient times (it would have eroded to ~100 mbar by ~4.3 Gyr ago, and to only slightly (~30%) above Mars' present-day pressure by the end of heavy bombardment at ~3.5 Gyr ago) and so was a truly primordial product of volatile co-accretion or early degassing from the interior.

3. Conclusions. So, what limits can reasonably be set on water and carbon dioxide abundances, present and past, on Mars? There are no really firm estimates except perhaps those for directly observable present-day inventories. Let us consider those shown in Figs. 9 and 10, noting the sensitivities of the various models to their assumptions but without arguing the validity of one vs another.

Clifford's (1984) derivation of a ~600 to 1400 m maximum water storage capacity in megaregolith pore volumes seems likely to stand, absent a challenge to his estimate of ~20 to 50% surface porosity. Storage as ground ice in excess of porosity is not included, but even to double Clifford's median value would require amounts of excess subsurface ice equivalent to a >1 km thick layer if distributed globally, confined to the ~1 to 3 km of regolith above the probable melting isotherm (Clifford 1984) and thus constituting an extremely large fraction of the upper regolith by volume. So an overall maximum capacity of ≤2 km appears plausible. Of the model-derived estimates for outgassed H_2O in Fig. 9, only Huguenin and Harris's (1985) extreme variant of the Dreibus-Wänke scenario predicts more, and even this not significantly more because uncertain degassing factors are involved.

Storage capacities do not help in setting *lower* limits for outgassed water; modeling estimates down to the few tens of m sequestered in the north cap system are defensible. Several of those in Fig. 9 do approach the minimum north cap abundance near 10 m, and it is interesting to ask how sensitive these inventories are to uncertainties in model parameters. Anders and Owen (1977), for example, argue for a low abundance of Tl and thus of the entire Tl-group of highly volatile elements, primarily on the basis of the low present-day abundance of atmospheric ^{36}Ar. However, the Anders-Owen model was developed before it was recognized that the SNC meteorites are probably samples of Mars (chapter 4). Dreibus and Wänke's (1987) estimate for Martian Tl from SNC data is some 25 times higher, and if this value were chosen the Anders-Owen crustal water and ^{36}Ar inventories would increase by the same factor, for water to ~1000 m. Still only ~10 m of this would have been outgassed, as release factors for ^{36}Ar (and water) in their approach would now be ~25 times lower, if degassed argon was not later lost. But, in light of current thoughts about volatile loss mechanisms, one must consider the possibility of preferential argon depletion by hydrodynamic escape or impact erosion, particularly if outgassed water was condensed (Melosh and Vickery 1989) and argon was not.

The low abundance of residual mantle water in Dreibus and Wänke's (1987) model is due entirely to their assumption that all accreted water, except that dissolved in magma, was reduced to H_2 and lost from Mars. But only a slight reaction disequilibrium between H_2O and metallic Fe could increase this inventory dramatically. Each unreduced 1% of the 6% C1 water content would have contributed ~0.024 wt% H_2O to the planet, amounting to ~1000 m globally, either carried to the surface quantitatively by early H_2 degassing or later outgassed in amounts ranging from ~20 to 80 m according to Scambos and Jakosky's (1990) release factors. So degassing from an unreacted mantle reservoir comprising just ~10% of the accreted water would supply amounts resembling the Carr (1986) inventory. (The Huguenin-Harris proposition that essentially *none* of the accreted water was reduced clearly represents the upper limit to this argument.) We can augment hydrodynamic escape models with unreacted water in the same way. In Pepin's (1991) scenario, for example, reduction of ~4 wt% H_2O/g-planet is needed to supply escaping H_2, but here again accretion of a ~10% excess over this amount would leave the planet water rich. Without considering other possible sources such as cometary water, the hydrogen content of a 40% C1 accretional component would then have to be ~12,000 ppm (equivalent to ~11% H_2O), somewhat high for a contemporary C1, but comparable to abundances in a few other carbonaceous meteorites (Kerridge 1985).

Of the remaining estimates for water in Fig. 9, those of McElroy et al. (1977) and Pollack and Black (1979) assume Earth-like $H_2O/CO_2/N_2$ ratios for outgassed species, and so are sensitive not only to the possibility of nonterrestrial primordial ratios (e.g., $H_2O/CO_2/N_2 \approx 40/90/1$ in Dreibus and Wänke's C1 component) but also, because of differing geochemical affinities and condensabilities, to fractionation in degassing and decoupling in escape processes even if the source reservoir was compositionally terrestrial. Scambos and Jakosky's (1990) outgassed inventory of ~50 to 60 m is taken to be a minimum, and therefore their range of derived mantle abundances is likewise a lower limit. More outgassing implies more in the interior. For example, scaled to Carr's (1986) higher estimate of degassed water, the Scambos-Jakosky mantle water content rises to ~6 to 60 km, or ~0.14 to 1.4 wt% H_2O/g-planet. Perhaps the content is this large, or alternatively the Carr inventory, if correct, may derive in part from surficial accreted water or from early outgassing epochs not reflected in Greeley's (1987) history of volcanic activity.

Turning to carbon dioxide in Fig. 10, it seems unlikely from Fanale et al.'s (1982*a*) models that the CO_2 adsorptive capacity of the regolith exceeds a few hundred mbar at most. Storage above this level appears to require carbonate deposits, either in-place or more likely comminuted to fine particles and dispersed through the regolith by impact. Confirmation or disproval of Pollack et al.'s (1990*b*) observational estimate of ~1 to 3% carbonate grains in airborne Martian dust should be relatively straightforward, but the question of whether and how far this mixing ratio might extend to depth in the megaregolith will clearly be much more difficult to answer. If it does, say to ~5 km, ~2 to 7 bar of CO_2 is stored as regolithic carbonate. Carr's (1986) somewhat higher estimate of ~10 to 20 bar is pegged to water via the terrestrial CO_2/H_2O ratio. Some of the intrinsic uncertainties in this assumption are noted above. Recall also that McElroy et al.'s (1977) estimate of terrestrial CO_2/H_2O is smaller than that used by both Carr and Pollack and Black (1979) by a factor of ~2.5; Walker's (1977) choice for Earth's crustal reservoirs ($H_2O/CO_2/N_2 \simeq 310/40/1$) is even lower. While there is no consensus on relative terrestrial abundances, note that with these smaller ratios the Carr and Pollack-Black CO_2 inventories decline; Carr's to somewhere between ~2 and 8 bar, and Pollack and Black's to ≤1 bar. At the lower end of estimates in Fig. 10, the ~400 mbar of crustal CO_2 from Anders and Owen (1977) would rise to ~10 bar if the SNC-derived Tl estimate for Mars were incorporated into their model. Whether their ~100 mbar estimate for outgassed CO_2 would also increase depends on how release factors are interpreted and calculated, as discussed above for water. Finally, two estimates for early atmospheric CO_2 pressure derived from escape modeling are both near 1 bar. However, neither of these is at all firm, as they involve either arguments close to or within the error bars of present noble gas isotopic data (Pepin 1991) or an oversimplification of the probable crust-atmosphere exchange cycle for Martian carbon (Melosh and Vickery 1989).

At the moment, from these evaluations, our best estimates for the globally distributed abundances of water and carbon dioxide deposited (and retained) in surface reservoirs on Mars by accretional or outgassing processes lie roughly in the range of a few hundred m to ≤2 km of H_2O, and from several hundred mbar to perhaps 5 to 10 bar of CO_2. Lower surface inventories are possible but appear less likely; higher ones seem improbable. As for interior abundances, several types of modeling approaches suggest that the Martian mantle was rich in water and carbon dioxide during accretion and early differentiation, an important consideration in modeling the overall geochemical, petrologic and geophysical evolution of the planet. Moreover, it is possible that the mantle still retains relatively large concentrations of these and other volatiles.

IV. RELATIVE ROLES OF SURFACE GREENHOUSE AND INTERNAL HEAT FLOW

The preceding discussion makes it clear that current estimates of the quantity of CO_2 degassed prior to the epoch of valley formation cover a wide range, from a few hundred mbar to ~10 bar. In addition, we do not know the degree to which the effects of impact erosion and hydrodynamic escape might have further reduced this inventory prior to valley network formation. Thus, to explore properly the parametric space and likely early conditions, we must consider not only those cases where the total CO_2 in the atmosphere is assumed to be sufficient to raise surface temperatures to the freezing point, but also those where it is not. We give one example of each. These results are quantitative in nature, but may be only of qualitative significance owing to the number of parameters (e.g., regolith conductivity) that must be estimated. In both cases, we consider the roles of both surface insolation and internal heat flow, although in the "high total CO_2" cases the results simplify to produce models resembling those of Pollack et al. (1987), and the primary role of internal heat flow may be to widen the latitudinal band in which valley formation can occur. In the cases where we assume the lowest values of total degassed and retained CO_2, the high early heat flow must play the dominant role in valley formation and the CO_2 greenhouse effect is relatively impotent.

Although the above two mechanisms (internal heat flow and surface greenhouse) have often been regarded as alternatives to explain long-term morphological change, a relationship exists between them, which may be stated as follows. The CO_2 mean residence time in the Martian atmosphere, although not well known, is almost certainly much shorter than the total time span over which early climate differences are thought to have been sustained (Sec. II.B). Therefore recycling of previously degassed CO_2 quickly becomes more important than ongoing supply of juvenile CO_2. If so, then the atmospheric CO_2 pressure, and therefore the surface temperature, may be approximated (mathematically) as a function of both the total degassed CO_2 in the atmosphere plus buried material and the ratio of the atmospheric to regolith mean residence times. However, the latter ratio has also been quantitatively expressed as a function of heat flow (Sec. II.B.2). Hence, it follows that the surface temperature may also be expressed (given assumptions as to regolith conductivity) as a function of heat flow and the total amount of "available" CO_2. Assumption of uniform regolith conductivity implies that temperature is a linear function of depth (seasonal variations are ignored), and this temperature profile and the abundance of CO_2, along with the CO_2 atmospheric residence time, set the atmospheric pressure. The resulting greenhouse affects surface temperature.

Figure 11 illustrates that for a given amount of total available CO_2, regolith conductivity and atmospheric mean residence time of CO_2, the relative roles of internal heat flow and surface insolation increase are inextricably interlocked. For fixed values of the above parameters, only certain sets of values for internal thermal gradient, depth to the 273 K isotherm and surface greenhouse effect are allowed simultaneously. The atmospheric greenhouse effect affects the depth to the 273 K isotherm only by changing the surface boundary condition, but the internal heat flow affects it both by increasing

the internal gradient directly and by shifting the surface boundary condition as the result of lowering the recycling time for the available CO_2. Thus the values given for the surface temperatures and depth to the 273 K isotherm are *simultaneously* dictated by the value of heat flow Q chosen.

In Fig. 11a, we examine the case where the total CO_2 is 3.5 bar. These calculations were made for the most favorable conditions as described by Pollack et al. (1987). They point out that at maximum orbital eccentricity of Mars, the insolation at perihelion would be 35% larger than its orbitally averaged value, and that equatorial regions receive 40% more insolation than the planet as a whole. Thus, in spite of the early faint Sun, the insolation at the Martian equator at perihelion at maximum eccentricity would have been 1.3 times the *present* globally averaged insolation. While a value of 3.5 bar for the total CO_2 may be somewhat optimistic, it is within range of estimates for CO_2 as discussed in the previous section. In this situation, because we have assumed so much total CO_2, an internal heat flow of about 150 mW m^{-2} (a mid-range value as predicted by Schubert et al. 1979) is more than sufficient to recycle enough CO_2 to keep surface temperatures above freezing. In this case, therefore, the atmospheric greenhouse plays a dominant role, at least near the equator and for favorable orbital parameters. However, if one accepts Squyres' (1989*a*) contention that having a 273 K isotherm at $<$ 350 m depth is sufficient to permit widespread sapping, it is clear that the internal thermal gradient also plays a major role in widening the latitudes and time bands in which sapping would be prevelant.

In Fig. 11b we examine the less frequently considered case where total CO_2 is only one bar. This is important because it illustrates the locked interplay between surface temperature increments and internal gradient increments. It is also important because, as was seen in Sec. III of this chapter, in many plausible versions of the early Mars volatile inventory, it is entirely possible that the total available CO_2 at 3.8 Gyr might have been only 1 bar, or even less. We examine a case for a cool Sun, favorable orbital parameters and an equatorial site. In this case, we find that the atmospheric greenhouse effect plays almost no role. This is because the low total CO_2 abundance requires fast recycling in order to keep any significant abundance in the atmosphere. Thus for plausible values of earlier heat flow, the surface temperature changes by only a few degrees. On the other hand, the early heat flow, although insufficient to recycle the CO_2 fast enough, produces an enormous direct effect on the depth to the 273 K isotherm despite the near constancy of surface temperature; at Q = 30 mW m^{-2}, the depth is over a km, but at Q = 100 mW m^{-2} (less than the Schubert et al. nominal value for 3.8 Gyr ago), it is $<$ 300 m and near the critical depth suggested by Squyres (1989).

Calculations are sensitive to the values chosen for the (unknown) regolith conductivity, the CO_2 atmospheric mean residence time, and especially the total CO_2 inventory assumed. Thus it is important to explore the sensitivity of the conclusions to the values chosen. Although our exploration of the parametric space has been limited, the qualitative characteristics of the system are apparent.

V. SUMMARY

1. Long-term climate change on Mars is indicated most directly by the presence, age and distribution of the valley networks. Valley networks (as opposed to the large outflow channels) are confined almost exclusively to the ancient, heavily cratered terrain and clearly indicate some profound difference in the early Martian thermal regime either on or near the surface vs that which prevailed throughout most of Mars' later history. They were almost certainly formed by running water, but it seems more likely that they were formed by groundwater sapping than by rainfall. Nonetheless, the sapping hypothesis leaves unanswered the question of how the ground was repeatedly recharged with water. If this recharging requires extensive lateral transport of water, then precipitation may be indicated. Obliteration rates were also apparently higher in early Mars history.
2. It has been demonstrated that it is physically plausible that a higher early intensity of surface insolation caused by a CO_2 greenhouse effect could have overcompensated for an early weak Sun and raised temperatures to the freezing point near the equator under favorable conditions of obliquity and eccentricity. This could account for the morphological changes. This scenario requires at least from 1 to 5 bar of atmospheric CO_2, and has been invoked for the early Earth as well. Indirect estimates of the total degassed CO_2 inventory of Mars suggest that there may well have been from 1 to several bars of CO_2 available in the atmosphere-regolith system even after the effects of atmospheric impact erosion and hydrodynamic escape. However, both the total degassed CO_2 inventory and the amount lost by hydrodynamic escape and impact erosion are highly uncertain, and it is also possible that the available CO_2 inventory was < 1 bar.
3. It is also generally believed that the atmospheric mean residence time with respect to carbonate formation in the early history of Mars may have been short compared to the period of time over which network formation extended. If so, this means that the rate of recycling of previously degassed CO_2 was more important than the instantaneous rate of release of juvenile CO_2. Both are determined by the heat flow; and the expected much higher heat flow at $\sim$3.8 Gyr relative to later Mars history, and its subsequent decline, also makes the atmospheric greenhouse effect a good candidate mechanism for changing the erosional style with time.
4. That these erosional differences could have been caused, or at least abetted, by significant differences in the depth to the 273 K isotherm is also plausible. However, the absolute differences between a regolith thermal regime which would allow widespread sapping and one which would preclude it can only be estimated. Adding to the credibility of this hypothesis

are thermal models of early Mars which suggest that the heat flow at the time of network formation probably was 6 times or more than at present.

5. We can evaluate the relative roles of each of the two mechanisms (atmospheric greenhouse and higher heat flow) if we accept both Schubert et al.'s estimate of the time dependence of heat flow and Squyres' morphologically derived estimate of the critical depth (~350 m) below which a water table would be ineffective in producing networks and above which it would be effective. Assuming these estimates to be approximately correct, we find that if the total available CO_2 has always been ~3 bar or more (including carbonates), then the atmospheric greenhouse effect can easily account for the change in erosional style and the primary role of the heat flow is to raise groundwater temperatures and increase the latitudinal spread of valley occurrence. This result is in keeping with the earlier results of Pollack et al. (1987). On the other hand, if the *total* CO_2 (including carbonates) were <1 bar, the atmospheric greenhouse effect does not raise the surface temperature by more than a few degrees; but for plausible values of regolith conductivity, the change in internal gradient accompanying higher early heat flow can still easily decrease the depth of the 273° isotherm by a large factor, from well over a km to < 350 m, possibly enabling network formation at 3.8 Gyr. In any event, the high regolith temperatures would probably widen the latitudinal distribution of the networks. In either scenario, temperature must be near 273 K near the surface, and if precipitation was not involved there remains the problem of recharging the (groundwater) reservoir.

Acknowledgments. We acknowledge J. Kasting, H. Kieffer, B. Jakosky and R. Kahn for many useful discussions and suggestions for improvement of the manuscript. This research was supported by grants from the Planetary Geology Program, NASA.

33. QUASI-PERIODIC CLIMATE CHANGE ON MARS

HUGH H. KIEFFER
U.S. Geological Survey

and

AARON P. ZENT
NASA Ames Research Center

Evidence that the Martian climate undergoes quasi-periodic variations includes the polar layered terrain, differences between the residual polar caps, and the current net southward flow of H_2O. The driving functions for these variations are oscillations in the elements of the Martian orbit coupled with precession of the Martian spin axis. These "astronomic variations" control the distribution of insolation, which in turn influences the partition of volatiles between atmospheric and surface reservoirs. There are probably complex feedbacks involving the total surface pressure, the amount of dust and clouds in the atmosphere, the global circulation patterns, and the albedo of the polar caps. When the obliquity is at or below its current value, the atmospheric pressure is controlled by the radiative balance of the polar caps; the most important parameter is the CO_2 polar cap albedo, which may vary with intensity of insolation. The major effects anticipated at low obliquity are growth of the polar caps, substantial decrease in surface pressure, cessation of dust storms, release of CO_2 from the regolith, and poleward migration of H_2O ground ice. At high obliquity, the mass of the perennial polar caps decreases and permanent CO_2 frost disappears, CO_2 desorbs from the regolith at high latitudes, the surface pressure may increase to several times its current value, and the atmospheric dust load increases. Because the amount of exchangeable CO_2 now in the condensed state is not well known, predictions to higher atmospheric pressure are less certain. The polar layers are thought to result from the corresponding variation of the amount of dust incorporated in the polar caps, although no quantitative relation *has been established. The time scales for the exchange of CO_2, H_2O and dust between hemispheres are different from one another. Under the current climatic regime, atmospheric CO_2 can be redistributed in about 100 yr, H_2O in polar caps in about 1 Myr, and dust deposits 300 m thick also in about 1 Myr. The rates of exchange between the atmosphere and the regolith are known only within a few orders of magnitude. The different elevations of the two polar regions may contribute to a persistent difference in the nature of the two perennial caps. The topographically lower north polar region is the favored location for long-lived volatile reservoirs.*

I. INTRODUCTION

A. Evidence for quasi-periodic climate changes on Mars

Mariner 9 imaging revealed a complex sequence of layered deposits in the Martian polar regions. These deposits, termed "laminated" terrain by Murray et al. (1973), are thought to have formed by deposition of dust and perhaps H_2O from the atmosphere (Cutts 1973*b*), with cross-cutting relations developed by intervening periods of aeolian erosion (Cutts 1973*a*). Individual layers are visible down to about 30 m in thickness at the limit of imaging resolution, and thinner layers are thought to exist (see figs. 6 to 9 in chapter 23). Detailed study of the layering has led to identification of several stratigraphic relations and the suggestion that both wind scour and poleward slope retreat are important in forming unconformities (Blasius et al. 1982; Howard et al. 1982; see also chapter 23). Shortly after their discovery, the layering was associated with periodicities of the Martian orbital elements, specifically that of the eccentricity (Cutts 1973*a*). Both Cutts and Murray et al. suggested that the layering in polar deposits is tied to the periodic astronomic variations through variable dust storm activity. At nearly the same time, Ward (1973) recognized that Mars must undergo substantial variations in obliquity which would be critical to conditions at the poles (Fig. 1; see Sec. I.C below).

The layered deposits remain enigmatic. They are commonly assumed to be a mixture of ice and "dust" (some combination of aeolian dust, sand, or other nonvolatile material); variations in ice abundance and in dust composition between layers are thought to be related to cyclic changes in atmospheric pressure and dust content and ease of saltation. No quantitative relation to astronomic variations has yet been established, however.

The paucity of craters in the north polar layered terrain led to deposition-rate estimates of 1 km Myr^{-1} (Cutts et al. 1976). Recent identification of several small craters in the south polar layered terrain has led to a substantial increase in its estimated age, with the interpretations that the layering represents periods of a few Myr or longer, or that deposition ceased several 100 Myr ago (Plaut et al. 1988).

The distributions of materials related to atmospheric deposition are markedly different. Dunes and mantle material are extensive in the north and

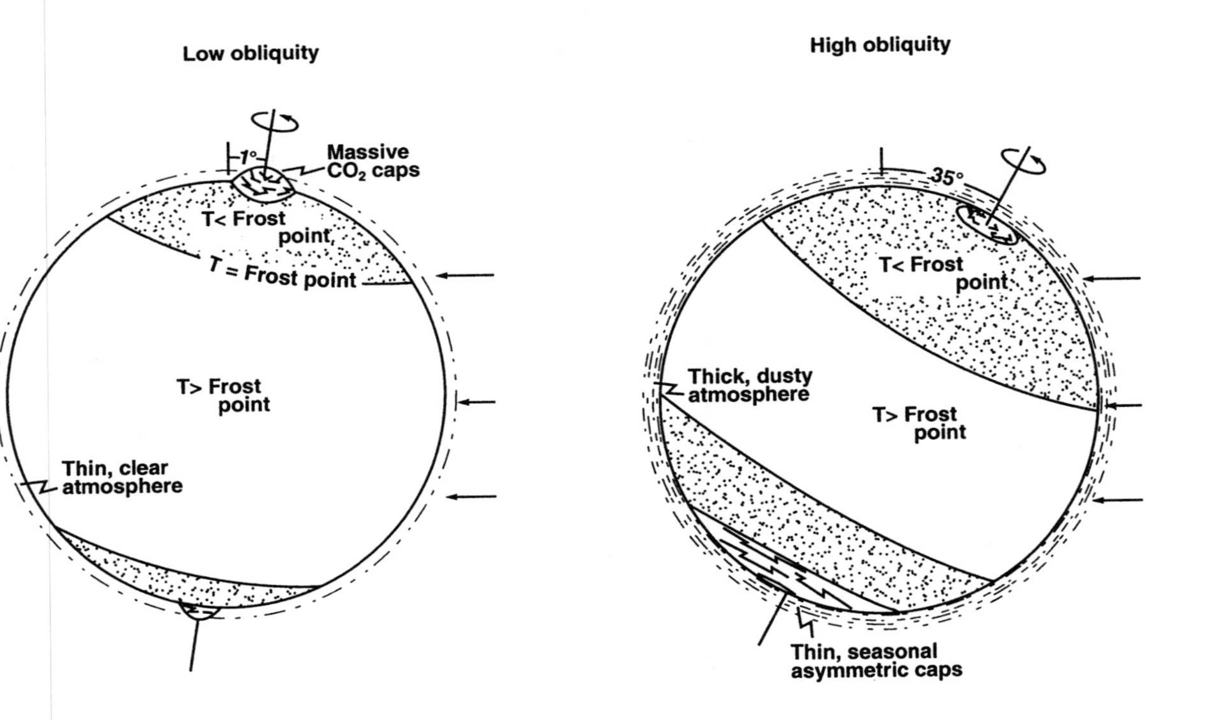

Fig. 1. Cartoon of conditions on Mars at the extremes of obliquity. At low obliquity, the variation of average annual surface temperature T with latitude is relatively great; perennial CO_2 caps have formed at the expense of atmospheric CO_2; atmospheric water vapor is reduced by the permanent cold traps; and permafrost is stable only at high latitudes. The atmospheric pressure may drop to the point that dust is not raised. At high obliquity, both minimum and maximum planetary temperatures are moderated; CO_2 caps are no longer perennial; and the surface pressure may be several tens of millibars, supporting a dusty atmosphere.

much less common in the south (Tanaka and Scott 1987, see Map I-1802-C; Thomas and Weitz 1989); the amounts of material involved suggest that this difference may persist through climate variations. On the other hand, the extents of layered deposits are similar in the two polar regions, although the extents of the perennial caps are different ($\sim 3°$ radius in the south and $\sim 9°$ in the north). Also, the current perennial polar cap and layered deposits in the south are asymmetric from the pole in nearly opposite directions.

A possible indication that long-term climatic variation continues is the apparent net annual southward transport of water vapor noted in Viking observations (Jakosky and Farmer 1982; Jakosky 1985; see also chapter 28).

It should be emphasized here that although periodic climate change on Mars is a plausible working hypothesis, originally devised to explain the formation of the layered terrain and based on the astronomic theory of insolation variation, the magnitude of the planet's response to this variation is quite uncertain and likely to remain so until subsequent investigations examine the evidence on Mars.

B. Climate Change on Earth and Mars

Before considering Mars in detail, it is instructive to compare Mars and Earth (see Christie-Blick 1982). In terms of climate, the major differences between these two planets are the presence of oceans on Earth and the condensation of the major component of the Martian atmosphere on Mars' surface.

The terrestrial oceans influence climate in several major ways. The oceans are large heat reservoirs; if all the solar energy that fell upon Earth were absorbed in the oceans and no reradiation occurred, it would take approximately one year to raise the bulk ocean temperature 1°C; actual response of mean ocean temperature to climate change is more on the order of a fraction of a millidegree per year. A second influence is heat transport; the oceans carry substantial quantities of heat from the equatorial and temperate regions into the polar regions and thus moderate the variation of climate with latitude. A third influence of the oceans, of substantial importance in our comparison, is that they are sinks for dust; the rain-river-ocean system systematically removes from the surface-atmosphere system the fine-grained products of weathering and erosion that are easily moved and suspended. This material exits the ocean system by plate subsidence and lithification, eventually re-emerging at the surface as immobile rock.

On Earth, the perennial polar caps and oceans constitute masses of about 5 and 275, respectively, times the mass of the atmosphere, of which about 0.2% is H_2O. However, the amount of the major component gases is not thought to change appreciably through climate cycles; the major reservoir response is the amount of H_2O stored in the polar caps.

The terrestrial geologic record contains evidence of periodic climate change in the form of glacial deposits indicative of ice ages, variations in the $^{18}O/^{16}O$ isotopic ratio indicative of global (ocean surface) temperature, sea-level variations largely related to changes in size of the massive polar ice caps, and floral and faunal changes indicating generally warm or cool conditions.

The geologic record of the last few million years has been analyzed in both the time and frequency domains in terms of the astronomic variation of the orbit of Earth and the orientation of its spin axis, a theory largely associated with the name Milankovitch (Imbrie and Imbrie 1980; Imbrie 1982). The terrestrial variations are less pronounced than those for Mars; the obliquity varies from 22° to 24°.5 with a period of 41,000 yr, the eccentricity varies from 0.01 to 0.05 with periods of 100,000 and 400,000 yr, the precession of the equinox has a period of 23,000 yr; compare Fig. 1 in chapter 9 (Mars) with Fig. 2 in chapter 9 (Earth). Although these periods can be clearly identified in the geological/geochemical record of temperature indicators, quantitative models of climatic response to the astronomic insolation cycle are far more numerous than they are certain; see, for example, the review by Held (1982).

The condensation of CO_2, the primary component of the Martian atmosphere, in the coldest regions of Mars means that the total atmospheric pressure is in balance with the temperature of the coldest regions. The relative ineffectiveness of a mechanism for permanently removing dust from the atmosphere and converting it to an immobile form has left Mars' surface with an abundant supply of material which, under current conditions, can be periodically raised into the atmosphere to substantially transform the radiative properties of the atmosphere and, upon fallout, the surface albedo.

The probable volatile inventories of some reservoirs are summarized in Table I in Sec. I.D; characteristic times are discussed in Secs. III.B and IV.B. It is the exchange of CO_2 between the rapidly responsive reservoirs (atmosphere and polar caps) and the regolith that constitutes the major aspect of climate change. It is the associated relocation of H_2O and dust that has left a record of climate change.

The difficulty of climate modeling for Earth and the similarity of the Martian problem should not be underemphasized. Both Earth and Mars have long-term geophysical hemispherical asymmetries: Mars in elevation and Earth in land/ocean distribution. Processes or factors held to be important in various terrestrial climate models include the radiative effects of aerosols and condensate clouds and the extent and albedo of polar ice caps; these are equally relevant for Mars. The primary complication that does not need be considered for Mars is the transport of heat by ocean circulation. Pollack (1979) presented a broad review of climate change on all three terrestrial planets, discussing both epochal and periodic changes on Earth and Mars; he has recently reviewed epochal climate change (Pollack 1991). Considering the wealth of information available for Earth on the climatic history of recent geologic time and on the processes that influence climate, it is sobering to consider both how far our understanding of Martian climate has advanced and how much remains to be learned.

C. Astronomic Cycles.

The relative orientation of Mars' orbit and spin axis and the shape (semimajor axis and eccentricity) of the orbit control the distribution of insolation on Mars. The semimajor axis is related to the total energy of the orbital motion and can be considered invariant. The insolation pattern can be characterized by three parameters: obliquity, eccentricity and argument of perihelion. The obliquity θ is the angle between the spin axis and the normal to the orbital plane; it sets the annual range of subsolar latitude and controls the magnitude of the seasons. The eccentricity e, a measure of the deviation of the orbit from circular, sets the range of heliocentric distances each year and controls the magnitude of hemispheric asymmetry of the seasons. The argument of perihelion ω is the angle between the rising node of the orbit on the ecliptic and perihelion, and it and the spin-axis orientation set the phasing of seasonal asymmetry. Currently $\theta = 25.^\circ2$ and $e = 0.093$, so the heliocentric distance ranges from 1.38 to 1.66 AU, and L_s of perihelion is 251°, so southern summer is now hot and brief relative to northern summer.

Ward (1974; chapter 9) has developed in detail the variations of Mars' orbital and spin axis parameters through time. The Laplace-Lagrange solution (Brouwer and van Woerkom 1950) yields obliquity $25.^\circ2 \pm 10.^\circ3$, which oscillates with a period of approximately 125,000 yr and a primary amplitude modulation of about 1.3 Myr. The eccentricity ranges from 0.00 to 0.13, with 0.05 amplitude oscillation of 96,000 yr and 0.1 amplitude oscillation of about 2 Myr. The most rapid term in the argument of perihelion comes from the precession of spin axis in about 175,000 yr.

Using the long-term orbit theory of Bretagnon (1974), Ward (1979) revised the calculation of the variation of Martian obliquity and eccentricity (see Fig. 5 in chapter 9), which yields an obliquity of $24.^\circ4 \pm 13.^\circ6$; the eccentricity range is virtually unchanged, and the detailed pattern of obliquity and eccentricity both changed slightly from those of Ward (1974). Bills (1990) found that uncertainty in the moment of inertia allows a wider range of obliquities in the recent past; models that consider resonances in inclination yield a wide range of obliquity histories. He found that obliquities may have been as low as $0.^\circ2$ or as high as $51.^\circ4$ in the past 10 Myr. The variation in obliquity is currently near a minimum, and hence the recent record of its variation may be less pronounced than what it might have been about 1 Myr ago. The eccentricity is slightly greater than its average value.

D. Volatile Reservoirs on Mars

Five reservoirs of volatiles on Mars, of widely differing magnitudes and exchange rates, are relevant to periodic climate change:

1. The current atmosphere is 95% CO_2 and has an average mass of about 150 kg m^{-2}, about 1.5% of Earth's atmosphere. The average H_2O vapor content of the atmosphere is about 10^{-2} kg m^{-2}.

2. The seasonal polar caps are composed of solid CO_2 in equilibrium with the atmospheric pressure, plus H_2O and dust presumably in approximately their atmospheric proportions, about 100 ppm. Maximum thickness of the caps (apart from local drifts) is known from radiation-balance measurements to be near 1000 kg m^{-2} (Paige and Ingersoll 1985). Their total mass is about one-fourth that of the atmosphere.
3. The perennial ("residual" or "permanent") caps are bright areas that remain after the summer retreat of the seasonal caps. The north and south perennial caps differ markedly in extent and composition. The abundance of volatiles in the perennial polar caps can be crudely constrained based on estimates of their thickness. The south cap is locally mottled, and there are some variations in its minimum extent from year to year and in the pattern of brightest material at the time of minimum extent (see chapters 23 and 27). The surface volatile is solid CO_2, at least when it has been observed by spacecraft, but its thickness is unknown (see chapter 27); a thickness of 1 to 100 m would correspond to 0.6% to 60% of the current global atmospheric mass. Because solid CO_2 is a cold trap for H_2O, a greater amount of H_2O probably exists in or beneath the CO_2 frost. However, most polar H_2O ice is thought to be in the more extensive north perennial cap, which has no CO_2. In the north, local topography suggests that the perennial cap averages about 1 km thick, containing a total amount of H_2O corresponding to 40 times the total atmospheric mass. Thus, the amount of CO_2 in the perennial caps is a modest fraction of that in the atmosphere, while the amount of H_2O in the caps is 5 orders of magnitude greater than that in the atmosphere.
4. The polar layered deposits surround and presumably underlie the perennial caps (see chapter 23). They comprise many layers of alternating light and dark albedo; some layers have unconformable contacts. The layering is thought to be due to differing ice/dust ratios, but the quantitative ratio is not known (the nonvolatile content is discussed in detail by Herkenhoff and Murray 1990*a*). The uppermost few meters are probably free of ice, as suggested by calculations of sublimation rates (Toon et al. 1980).

   In a widely cited analysis of gravity and topography, but one that involved "uncertainties that cannot be evaluated with any confidence," Malin (1986) estimated the mass of the north polar layered terrain (and perennial cap) as 2.0×10^{18} kg and stated that they "consist mostly of ice." This amount of H_2O, equivalent to 1.4×10^4 kg m^{-2} globally or to a layer 14 m thick, is about 10^6 times that currently in the atmosphere.
5. A large inventory of volatiles may currently exist within Mars' regolith. It may contain about 5 to 50 times as much CO_2 as is currently in the atmosphere and on the order of 10^4 to 10^5 kg m^{-2} (global) of H_2O (see Sec IV).

The volatile inventories are best known for those reservoirs that exchange rapidly and are easily observed. The mass of the atmosphere is known to the order of 10%; the mass of the perennial caps to a factor of a few. The amount of volatiles in the layered terrain is constrained by their volume, but the proportion of ice is not well known. The potential volatile storage of the regolith is broadly limited by models of adsorption, effective surface area, and regolith thickness, but the *exchangeable* reservoir, that is, the amount that is able to exchange between the regolith and atmosphere for the actual thermal and gaseous diffusion properties of the regolith, is roughly estimated by matching climate models to current conditions.

For the purposes of estimating reservoir capacities, the thickness of the south perennial CO_2 cap is somewhat arbitrarily taken as 10 m, the thickness of both H_2O perennial caps and all layered deposits as 1 km. The extent of these units was given by Tanaka et al. (1988). The ice content of the layered deposits is assumed to be 50% by volume, or 500 kg m^{-3}. The resulting reservoir capacities are given in Table I.

E. Overview of Mars' climate changes

Periodic climate change is riddled with feedback loops, which contribute to the complexity of the subject (Fig. 1). The fundamental boundary conditions of Martian climate are the luminosity of the Sun, the orbit and orientation of Mars, the global inventory of volatiles accessible to the atmosphere over time scales of years to millions of years, and the inventory of mobile dust and "sand." The driver of climate is *absorbed* solar radiation, and hence the temporal and spatial patterns of insolation and albedo are of primary importance. The distribution of annual average insolation with latitude depends upon the three astronomic orbit and orientation parameters discussed in Sec I.C. The partitioning of insolation by the surface and polar cap albedo, and the scattering of absorption of radiation in the atmosphere by dust and con-

TABLE I

Current Capacity of Volatile Reservoirs Expressed as Amounts Averaged over the Entire Planet.

	Global *kg m^{-2}*		Uncertainty Factor	
	CO_2	H_2O	CO_2	H_2O
Atmosphere	150	.01	.1	0.3[b]
Seasonal Polar Caps	40	.01	.2	2
Perennial Cap, North	0	5800	—	4
Perennial Cap, South	8	600	10	5
Layered Deposits	0?	6200[a]	—	3
Regolith, Adsorbed and Exchangeable H_2O Ice	1000	10	+6–20	10

[a]The exchangeable amount is probably limited to far less by gaseous diffusion through a surficial particulate layer and by thermal diffusion of the change of mean temperature with climate.

[b]Viking MAWD observations. Interannual variations might be a factor of 2.

densate clouds are important controls on the surface and atmospheric temperature. The temperature, volatile inventory and characteristic time scale for exchange between atmosphere, cap and regolith set the total pressure and the condensate cloudiness of the atmosphere. Winds, through determination of the amount and size distribution of suspended fine dust, modify the atmospheric opacity, which strongly influences atmospheric heating, and dust settling upon the polar caps modifies their albedo. Sand and dust can be redistributed on the surface to modify the surface albedo. Atmospheric pressure, distribution of atmospheric and surface heating, topography, and seasonal CO_2 frost condensation/sublimation control the atmospheric dynamics.

The primary process involved in periodic climate change is thought to be a large variation in the total atmospheric pressure (see heavy arrows in Fig. 2), which modifies the magnitude of the greenhouse effect and the dustiness of the atmosphere. As the annual average temperatures change at all latitudes, modest amounts of H_2O may be redistributed over the globe. Small amounts of H_2O condensed in clouds could have an important effect on the planetary albedo.

A fundamental question about periodic climate change on Mars is the extent to which the atmospheric pressure varies with astronomic variation of orbital parameters. This is a two-sided question, as the Martian behavior at obliquities higher than at present strongly depends upon the now largely unseen reservoir of volatiles that can exchange with the atmosphere (discussed in Sec. IV). At lower obliquity, the expected behavior is the growth of the

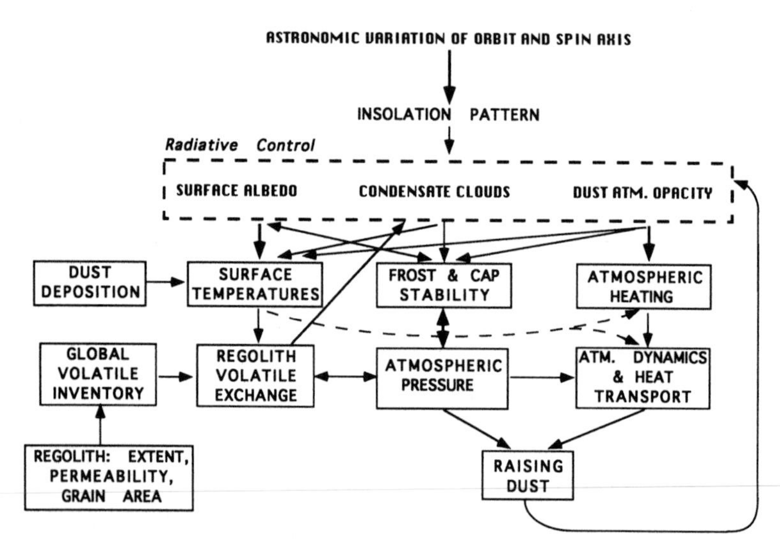

Fig. 2. Primary elements related to climate variation on Mars. Arrows indicate the direction of influence. Heavy arrows indicate the major effects related to the variation of atmospheric pressure with obliquity.

perennial polar caps at the expense of the currently observed atmosphere; this behavior is less dependent upon the regolith reservoir.

As CO_2 exchanges between the atmospheric, regolith, and polar cap reservoirs, substantial changes can occur in the atmospheric pressure as well as in the dustiness of the atmosphere, and the mobility of H_2O (Fig. 3). Substantial hysterisis may occur; polar caps, which grow at decreasing obliquity, may lower the atmospheric pressure to the point that dust storms cease, resulting in clean, high-albedo deposits that are not easily warmed as the obliquity increases.

At present, *seasonal* climate involves the annual progression of seasons (see Fig. 1 in chapter 1), annual cycles of the CO_2 cap (chapter 27), water vapor (chapter 28) and dust (chapter 29). The extent of dust storms varies from year to year (see Figs. 1 and 7 in chapter 29) and appears to involve a major instability in the atmosphere system.

At present, the atmospheric pressure is controlled by the condensation/sublimation cycle of the polar caps, and the polar caps seem to be in equilibrium with the annual radiation balance (Paige and Ingersoll 1985). The criti-

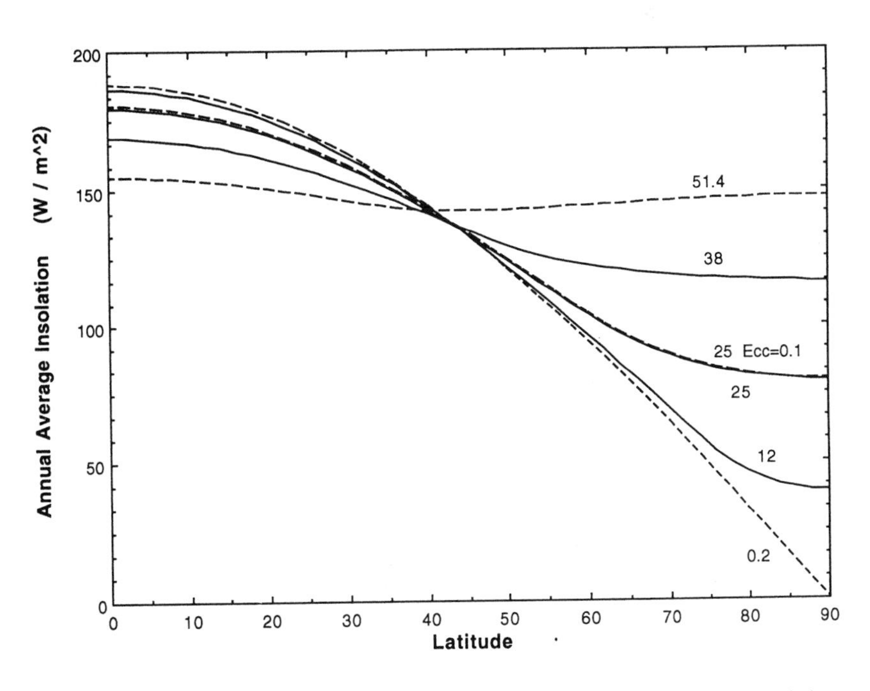

Fig. 3. Effects of obliquity and orbital eccentricity on the distribution of average annual insolation with latitude. The three solid curves are for obliquities of 12°, 25° and 38°, the current value and the limits found by Ward (1979) with eccentricity equal zero except for the slightly higher curve (long dashes) at 25° where $e = 0.1$. The short-dashed curves indicate the extreme obliquities found by Bills (1990).

cal parameter in this radiation balance is the albedo of the polar caps; an important point is that the observed average value for the CO_2 cap albedo of 0.65 is unexpected. This albedo is thought to result from a mix of CO_2 and dust. It is roughly midway between the albedo of pure CO_2 and pure dust, a point that is extremely sensitive to the abundance of included dust. The bolometric albedo of at least the south seasonal polar cap varies in the spring (Kieffer 1979), and over at least the area of the perennial cap the bolometric albedo may depend upon intensity of insolation (Paige 1985). Thus far, all models of Mars' climate change have used a constant value for the albedo of the polar caps. Until we determine the mechanism that controls the polar-cap albedo in the current climate, this approximation may remain an unfortunate, but prudent, necessity.

As models of Martian climate change have incorporated greater complexity, the estimates of surface-pressure variation have decreased. In conjunction with Ward's work on the astronomic variations, Ward et al. (1974) modeled the behavior of CO_2 polar caps, predicting atmospheric pressure variation from ~0.3 mbar to ~30 mbar and the cessation of dust storms at low obliquity.

Regolith temperature distributions resulting from orbital variation and the response of possible regolith volatile inventory were studied by Toon et al. (1980) and Fanale et al. (1982*a*). Toon et al. also developed a quantitative estimate of the exchange of water between polar cap, polar layered terrain, and regolith reservoirs. They incorporated a model of regolith desorption and gas diffusion, and discussed possible greenhouse effects, including clouds; this work also addressed the extent of dust storms at different obliquities and a model for the resulting layered terrain. They concluded that, although gaseous diffusion wavelength in the regolith is decreased about an order of magnitude by adsorption dampening, for pulverized basalt with pore radius $>0.1\mu m$ it is still longer than that for thermal diffusion; hence, the regolith reservoir could respond to the astronomic cycles. Correspondingly, the current atmosphere would be in equilibrium with the regolith. Toon et al. found that exposed polar H_2O ice caps would disappear at high obliquity but that burial by dust would allow H_2O ice caps formed at low obliquity to persist through high-obliquity periods; they suggested that this is the origin of the layered terrain. Fanale et al. (1982*a*) described in detail the subsurface temperature variation and regolith CO_2 exchange through obliquity cycles for a variety of subsurface models. They found that availability of exchangeable CO_2 in the regolith was likely to limit atmospheric pressure at high obliquity to ~15 mbar; if not, a large permanent CO_2 cap should now be present. Pollack and Toon (1982) reviewed prior work on the orbital variations, the response of the CO_2, H_2O and dust annual cycles and the possible record in the layered terrain. They concluded that (1) the current seasonal asymmetry of the dust cycle is a major factor in the differences between the two polar caps; (2) the extent of exchangeable regolith storage of CO_2 is uncertain but that it could dominate the longer-term climate response to orbital variations; and (3) the dependence of dust-storm activity on atmospheric pressure is significant and responsible for the layered nature of the polar deposits.

James and North (1982) developed a two-dimensional model of the annual cap formation and atmospheric pressure variation. They excluded a significant CO_2 reservoir, and they found that permanent CO_2 polar caps form only for obliquity less than about 15°; above this value the average annual surface pressure decreases with increasing obliquity.

Recent work has concentrated on the behavior of the regolith. Fanale et al. (1986) and Zent et al. (1986) modeled in detail the distribution of H_2O (discussed in Sec. IV). Zent et al. (1987) presented new measurements of CO_2 adsorption. They concluded that the constraint of matching current conditions allows the modeling of the effective behavior of the regolith reservoir through an obliquity cycle, even without better information on the individual parameters of available CO_2 inventory or regolith extent and mineralogy.

Incorporation of the diurnal temperature variation, resulting in lower average temperatures away from the poles, and more detailed treatment of the CO_2 greenhouse effect and meridional heat transport by the atmosphere (François et al. 1990) suggest that the current exchangeable CO_2 reservoir is about 1250 kg m^{-2}, or 7 times the current atmosphere. When the exchangeable reservoir was greater than about 5000 kg m^{-2}, atmospheric heat transport prevented the formation of permanent CO_2 polar caps.

II. PROPAGATION OF ENERGY

Climate on Mars is largely an issue of temperature. The temperature of the polar caps can set an upper limit on atmospheric pressure; the temperature distribution in the regolith controls the stability and migration of volatiles; and the temperature distribution in the atmosphere governs the formation of condensate clouds and drives the winds. Because atmospheric transport of heat plays a minor role for atmospheric pressures up to several times the current values, temperatures at the surface are controlled primarily by the geographic (zonal) distribution of solar radiation and surface albedo.

A. Distribution of Insolation

The instantaneous, diurnally averaged, and seasonally averaged insolation can be expressed in terms of the orbital parameters (see also Sec. IV in chapter 9). The instantaneous solar irradiance onto the top of the atmosphere is $S^* \cos i$ ($\cos i > 0$) where i is the incidence angle;

$$\cos i = \sin \delta \sin \delta' + \cos \delta \cos \delta' \cos h \tag{1}$$

where δ and δ' are the latitude and subsolar latitude, and h is the solar hour angle measured from midnight.

$S^* = S_o/r^2$ is the solar flux, where S_o is the solar constant, and r is the heliocentric distance in astronomical units,

$$r = a\frac{1-e^2}{1+e\cos f} \tag{2}$$

where a is the semimajor axis, e is the eccentricity and f is the true anomaly from perihelion.

The average daily insolation at the top of the atmosphere is

$$I_d = S^*/\pi\ (\eta \sin\delta' \sin\delta + \sin\eta \cos\delta' \cos\delta) \tag{3a}$$

(Ward 1974, Eq. 49) where η is the rotational half-angle of daylight.

If $|\delta| < 90 - |\delta'|$ (day and night),

$$\cos\eta = -\tan\delta' \tan\delta; \tag{3b}$$

or if $\delta \geq (90 - \delta')$ or $\delta \leq (-90 - \delta')$ (polar day),

$$\cos\eta = -1 \text{ and } I_d = \sin\delta' \sin\delta; \tag{3c}$$

or if $\delta \leq (-90 + \delta')$ or $\delta \geq (90 + \delta')$ (polar night),

$$\cos\eta = 1 \text{ and } I_d = 0. \tag{3d}$$

It will be convenient to designate the solar flux at the Martian average distance: $S = S_o/a^2$ (590 W m^{-2}).

The average annual insolation at latitude δ is

$$I_a = \frac{S}{2\pi^2}(1-e^2)^{-1/2}\int_o^{2\pi}[1-(\sin\delta\cos\theta\text{-}\cos\delta\sin\theta\sin h)^2]^{1/2}dh \tag{4}$$

(Ward 1974, Eq. 44) where θ is the obliquity. Figure 3 shows the results of numerical integration of this equation for the current and extreme values of obliquity. Over the Martian range of obliquity, the average annual insolation decreases away from the equator and, poleward of about 45°, increases with obliquity; the difference between eccentricity of 0 and 0.15 is only 1%. At the poles, this simplifies to

$$I_a = \frac{S}{\pi}\sin\theta\ (1-e^2)^{-1/2} \tag{5}$$

The maximum insolation at the poles at summer solstice is

$$I_p = S \sin \theta \frac{(1 \pm e \sin \omega)^2}{(1 - e^2)^2} \tag{6}$$

where the plus sign is for the north pole and the minus sign for the south. The north-south difference of this normalized maximum insolation is on the order of $4e$, or up to 50%, and oscillates with a period of about 51,000 yr (Ward 1974). This reversal of the perihelion summer hemisphere is the most rapid of the known major astronomic variations that might leave a regular record in the layered terrain.

B. Atmospheric Radiation and Heat Transport

The absorption of solar energy by Martian atmospheric gas is negligible. An infrared greenhouse due to CO_2 or H_2O is not thought to be significant unless pressures are far higher than those discussed here (Pollack 1979; but also see François et al. 1990). Although heating due to suspended dust can be important and estimates of its effect have been made, it has not been treated explicitly in any climate models.

Qualitatively, atmospheric dust produces two partially offsetting effects on radiative transfer. In the visible, atmospheric dust scatters and absorbs incoming insolation such that the flux of solar photons reaching the surface decreases, but the diffuse component of that flux increases somewhat. In effect, atmospheric dust causes a change in where insolation is absorbed: more absorption takes place in the atmosphere and away from the subsolar point. The planetary albedo increases over dark soil and rock areas and decreases over the bright seasonal polar caps.

In the infrared, the dust particles are too small to be good scatterers; their primary radiative role is adsorption and emission. A dusty atmosphere is a much stronger infrared source than a clear atmosphere. The question is whether the enhanced atmospheric infrared radiation will compensate thermally for the reduced visible flux at the surface. Average temperature observations from the Viking Landers before and immediately after the beginning of the global dust storms indicate that the balance is near zero (Ryan and Henry 1979), but that there may have been some atmospheric cooling due to the increase in atmospheric dust. Lindner (1990) found that dust opacity has little effect on the total (solar plus thermal) absorbed energy for a cap albedo of 0.5 in the daylight, but that atmospheric dust increases the heat load on the cap in the polar night. He found that water-ice clouds increase the heat load on the cap except near its daylight edge when the dust opacity is low.

The effect of condensate cloud formation in different climates has not been considered except for the explicitly biased treatment by Toon et al. (1980) to assess how large a greenhouse enhancement could be accomplished with water-cloud formation. Water clouds increase the planetary albedo and can contribute to the greenhouse by blocking thermal radiation; the net effect can either raise or lower surface temperatures. Currently, Mars commonly displays persistent, thin high-altitude clouds over the winter hemisphere. Considering the intense debate in the mid-1970s over what effect increased high-altitude water clouds resulting from supersonic transport activity might have on terrestrial temperatures, and the continuing debate on the role of clouds in greenhouse and climate-change models, it is not surprising that the net effect of Martian clouds on the radiation balance during periods of higher atmospheric pressure remains an open question.

Net heat transport into the polar regions has been assumed or calculated by all workers to increase with atmospheric pressure. An energy-balance study of Viking observations suggests that the current annual average is on the order of 2 ± 5 W m^{-2} (Paige and Ingersoll 1985). Recent numerical modeling of Mars' general circulation, including effects of dust and clouds, yields an average heating rate at 82° N and S due to transport of about 3 W m^{-2} (see Figs. 4 and 5 in Pollack et al. 1990*a*).

Gierasch and Toon (1973) modeled atmospheric heat advection as a

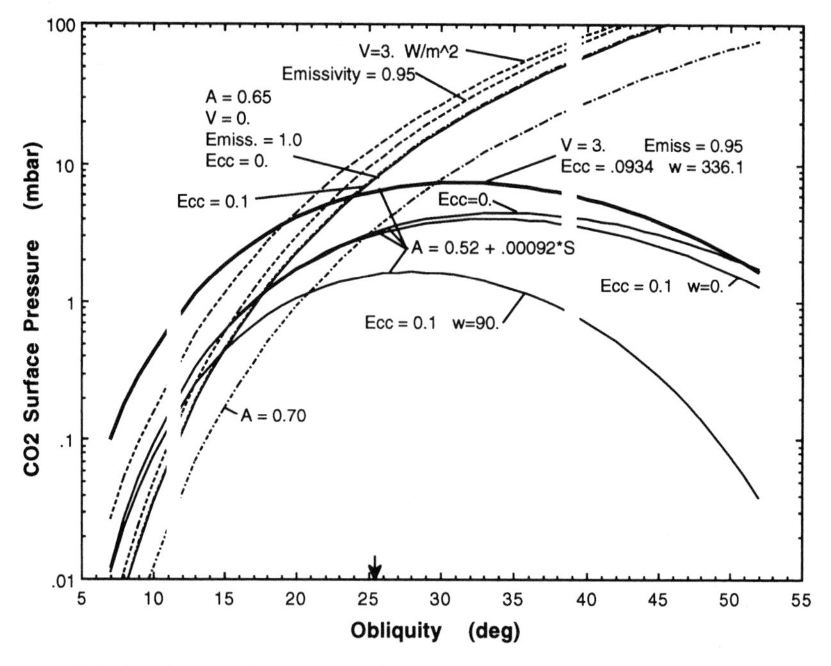

Fig. 4. Variation of CO_2 surface pressure with obliquity for a simple polar energy balance model. The nominal model is for eccentricity of 0, atmospheric advection V of 0 W m^{-2}, frost albedo A of 0.65, and frost emissivity of 1.0. The effects of small independent changes in emissivity, heating, albedo and eccentricity are shown. Three curves show the dramatic effect of an albedo that depends upon instantaneous insolation, in which case eccentricity and argument of perihelion become important. With estimates for the present parameters, $\varepsilon = 0.95$, $H = 3.$, $e = 0.0934$, $\omega = 336.1$, this simple model predicts a CO_2 surface pressure of 5.0 mbar at the current obliquity of 25°.2.

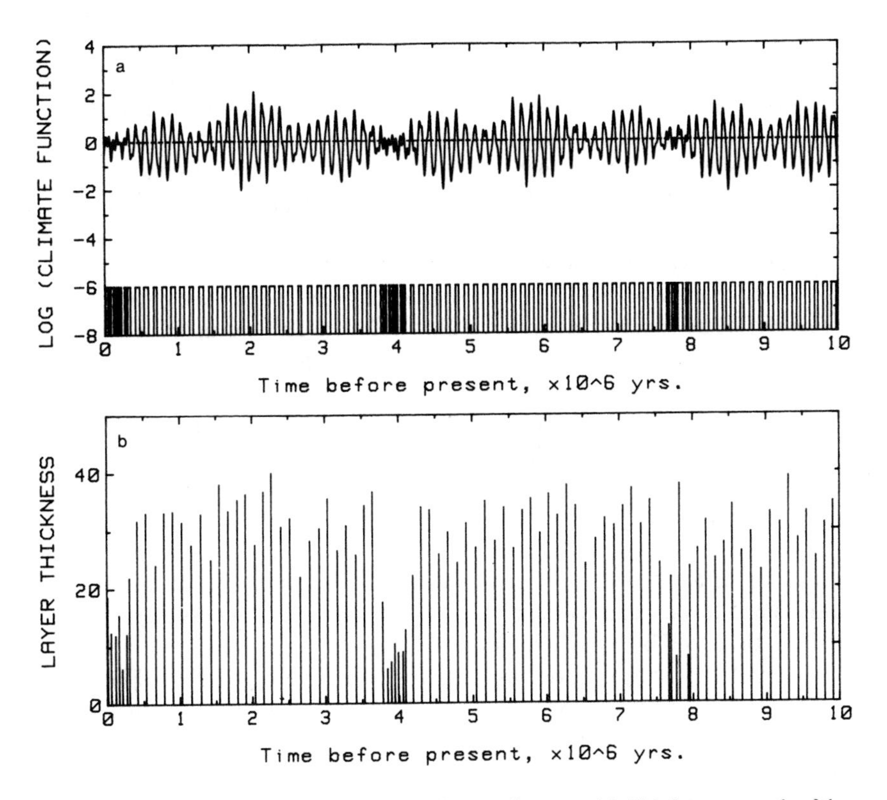

Fig. 5. Simulation of formation of polar layers from a climate model. This is one example of the possible layer-thickness sequence resulting from a climatic response to the variation of orbital parameters (figure from Cutts and Lewis 1982).

function of surface pressure and found that heat transport became significant (10% effect) at about 10 mbar atmospheric pressure, and equivalent to the average absorbed insolation at about 50 mbar. They also found that, if an adequate CO_2 reservoir is available, the surface pressure could be unstable at intermediate values on the order of 0.1 bar. There are two solutions. One has an extensive cap, low pressure, and low heat transport; the other has no cap, high pressure, high heat transport. Gierasch and Toon, however ignored clouds or a greenhouse effect, and they assumed that a combination of a variety of heat-transfer parameters (their β) was independent of pressure. François et al. (1990) adopted a meridional heat-transfer coefficient that was linearly proportional to surface pressure. As mentioned above, they found that at total CO_2 inventories >5000 kg m^{-2}, no cap was allowed. They did not find the large, bright-cap, low-pressure solution.

C. Mobility of Dust

A prominent aspect of Mars' current climate is dust in the atmosphere, which is a pervasive contributor to atmospheric opacity in the form of local dust storms and as occasional major storms that can envelope the globe (see chapter 29). The importance of dust to climate models is primarily in its influence on atmospheric heating and on the albedo of the polar caps, although quantification of these effects is not yet possible.

A critical issue is whether winds in a specific climatic epoch are capable of raising dust. If not, the polar-cap albedos may approach that of pure CO_2, with a major increase in polar-cap stability.

The behavior of dust storms at various surface pressures has been discussed by Ward et al. (1974), Cutts and Lewis (1982), and Toon et al. (1980), among others. Most workers have assumed that dust-storm frequency and intensity increase with surface pressure and that dust storms will be rare or nonexistent at low obliquity.

In a highly theoretical development, Ward et al. (1974) found that the wind speed required to lift dust at minimum obliquity exceeds the speed of sound. The threshold conditions for raising dust on Mars were studied in an important pair of papers (Iversen et al. 1976; Pollack et al. 1976*a*). Particles most easily moved are about 100 μm in diameter. The threshold wind speed at the top of the boundary layer was found to be 100 m s^{-1} for current conditions and to vary approximately as $P^{-1/2}$; abundant roughness elements can reduce this speed by about a factor of 3.

Small dust storms have been observed near the edge of the springtime polar cap (Peterfreund and Kieffer 1979). A combination of thermal winds and the net sublimation wind from the polar cap might raise dust even at low atmospheric pressures.

Climate models have assumed that soil and rock albedos do not change with time, and they are considered uniform over the planet in all but the recent GCM work on the current climate of Pollack et al. (1990*a*). A worst-case estimate of the magnitude of this approximation can be made by assuming that the darker materials on Mars, having an average albedo of 0.15 and covering approximately half the planet, could in extremely dusty conditions be covered by material similar to the current brighter materials, with an albedo of 0.26. This would raise the global albedo by approximately 0.05.

III. POLAR CAP BEHAVIOR

A. A Simple Energy Balance Model

Because of the strong nonlinearity with respect to temperature of thermal radiation and of the vapor pressure of pure or adsorbed volatiles, equilibrium cap/atmosphere models with no temperature variation are much simpler to evaluate than are models that track thermal variations.

To illustrate the influence of several parameters on the atmospheric pressure, a simple model of the energy balance at both polar caps was generated. In this model, the CO_2 surface pressure is assumed to be set by the annual equilibrium temperature at the colder pole. Atmospheric opacity is ignored, and the supply of CO_2 is treated as unlimited. This is essentially the model of Ward et al. (1974), and the basis for all low-obliquity models. Although the total annual insolation is identical at the two poles, any nonlinearity such as a variable albedo will result in one pole absorbing less radiation than the other. The resulting equilibrium vapor pressures are shown in Fig. 4 as a function of obliquity.

The parameters for the idealized case are albedo $A = 0.65$, net advective heating $V = 0$, frost emissivity $\epsilon = 1.0$, and eccentricity $e = 0$. With these parameters, the surface pressure is 7.1 mbar for the current obliquity. Decreasing the frost emissivity to 0.95, corresponding approximately to radiation blockage by the 15-μm atmospheric band, raises the pressure by 31%; increasing the albedo to 0.7 decreases the pressure by more than a factor of 2. A net atmospheric heat transport of 3 W m^{-2} raises the equilibrium pressure by 70%. An eccentricity of 0.1 increases the pressure about 3%. Incorporating the dependence of albedo upon insolation reported by Paige (1985; see Sec. III.C below) has a dramatic effect, decreasing the sensitivity of pressure to obliquity by an order of magnitude but generating a major variation with argument of perihelion ω. With the idealized parameter values (apart from albedo), an eccentricity of 0.1 causes a 5% pressure decrease if ω is 0° (symmetric summers) but cuts the pressure by a factor of 2 for $\omega = 90°$ (asymmetric summers). With the current orbital parameters, $V = 3$, $\varepsilon = 0.95$ and variable albedo, the equilibrium pressure is 5.0 mbar. However, the reported dependence of albedo on insolation is uncertain and its possible cause not adequately understood to apply this relation with any confidence to other obliquities.

The substantial regional relief on Mars modifies the atmospheric and saturation temperatures at the polar surfaces. Because the average elevation of the perennial south polar cap is approximately 5 km (or about ½ the average scale height) higher than the north polar cap (see map I-2160 sheet 3), the average surface pressure is about 65% of that at the north polar cap, and the average annual temperature would need to be about 7 K lower for it to control the global surface pressure. There appears to be nowhere in the literature a calculation of the current average annual surface pressure for the residual CO_2 cap. The elevation of the Viking Lander 1 site (VL-1) is -1.5 km relative to the reference geoid, and an elevation of $+5$ km is representative of the perennial south polar cap (map I-2160-3). Adopting an average annual surface pressure at VL-1 of 7.8 mbar and an average scale height of 10 km (chapter 26), gives a surface pressure at the south polar cap of 4.1 mbar.

B. Characteristic Time Scales

The characteristic time scale of a process or a deposit is the time required for it to recover substantially from a perturbation away from steady-state conditions. Whether the process is linear, asymptotic, or more nonlinear, the time scale can be approximated by the magnitude of a small perturbation divided by the initial rate of recovery. It is a useful concept in determining whether a process or a deposit will "track" a periodic change of conditions (characteristic time $<<$ period) or be largely immune to such a change (characteristic time $>>$ period).

The response of Mars to the astronomic variation of insolation can be characterized to first order by the inventory of exchangeable volatiles and thermal and gas-diffusion coefficients of the near surface and regolith.

Three fundamental time scales are important: diurnal (24.66 hr), annual (687 Earth days or 669.6 Martian "sols"), and "astronomic" (ranging from 24,000 to 2 Myr).

A perennial polar-cap reservoir of CO_2 provides a rapid and strong buffer to atmospheric pressure changes. The governing relations are

$$\ln P = \frac{-L\,m}{R\,T} + b \qquad \text{saturation pressure relation} \qquad (7)$$

$$F = \epsilon\,\sigma\,T^4 \qquad \text{net thermal emission} \qquad (8)$$

$$H = (1-A)\,\langle S\rangle \qquad \text{surface heating from insolation} \qquad (9)$$

$$\frac{dM}{dt} = (F-H)/L \qquad \text{conservation of energy} \qquad (10)$$

$$\frac{dP}{dt} = -C\,g\,\frac{dM}{dt} \qquad \text{conservation of mass} \qquad (11)$$

The symbols and their values are as given below, with Martian parameters chosen to be representative of current conditions; units are in square brackets.

A = polar cap albedo, including effect of atmospheric opacity; 0.65
b = constant term in vapor pressure relation; 27.954
C = fraction of planet covered by perennial polar caps; adopted as 0.01
C_p = specific heat capacity of solid CO_2; 1021 [J kg^{-1} K^{-1}] (at 140 K)
F = net thermal upward flux [W m^{-2}]
g = Mars surface gravity; 3.73 [m s^{-2}]
H = average annual heating of the surface from insolation (and atmospheric heat transport, assumed negligible) [W m^{-2}]
L = latent heat of condensation of CO_2; 6.0×10^5 [J kg^{-1}]
M = mass per unit area of CO_2 polar cap [kg m^{-2}]
m = molecular weight of CO_2; 44.01 [kg kmole^{-1}]
P = CO_2 surface partial pressure (here assume the cap is at Mars average elevation); 650 [Pa = kg m^{-1}s^{-2}]
R = gas constant; 8314.3 [J K^{-1} kmole^{-1}]
$\langle S\rangle$ = average annual insolation at cap surface; 78 [W m^{-2}]
T = polar cap kinetic temperature; 145 [K]

ϵ = effective polar-cap emissivity (through the atmosphere) ; 0.95
σ = Stefan Boltzman constant; 5.67×10^{-8} [W m^{-2} K^{-4}]

The equilibrium condition, no net condensation, is set by a balance of insolation and net thermal radiation (Eq. 10), which determines T_o (Eq. 8) and sets P_o (Eq. 7). If a perturbation in pressure occurs, the increased saturation temperature will result in higher thermal emission; the corresponding net energy loss will cause additional CO_2 condensation, dropping the pressure (Eq. 11) until radiative equilibrium is again attained.

If a perturbation is inserted in the surface energy balance by changing the insolation in the above equations, the characteristic time is found to be

$$\tau = \frac{L^2 m P_o}{4 \epsilon \sigma R C g T_o^5} \tag{12}$$

which is 76 yr for the assumed values: In similar analyses, Leighton and Murray (1966) derived a time constant of 200 yrs, and Gierasch and Toon (1973) derived 75 year for the "time constant for climate relaxation."

If the CO_2 polar cap is permeable, the temperature throughout must change to stay in equilibrium with the atmospheric pressure. The ratio of the heat capacity of the cap to the condensation energy is

$$\frac{E_C}{E_L} = C M \frac{R C_p g T^2}{L^2 m P}. \tag{13}$$

This ratio is one if the columnar mass of the CO_2 polar cap is 1.5×10^5 kg m^{-2}, (approximately 100 m of solid CO_2 ice); in this circumstance the changing cap temperature would approximately double the energy required to bring the surface temperature into equilibrium, and it would double the response time scale. Thus, the characteristic time for polar cap buffering of the atmospheric pressure is short compared with that of the astronomic variations. Note that if the CO_2 cap remains permeable, thermal diffusion is not relevant as long as CO_2 remains the major atmospheric gas.

A characteristic time scale for the perennial H_2O cap is less certain. If the ice of the north polar cap is assumed to be 1000 m thick, and the amount that can be moved in or out of one polar region in one Martian year is taken as the average water vapor content of the entire atmosphere, 1.2×10^{12} kg (8.3×10^{-3} kg m^{-2}), the exchange time is about 1 Myr. Toon et al. (1980) estimated net growth or sublimation rates of the H_2O cap as on the order of 0.1 to 10 mm yr^{-1}, which yield characteristic times an order of magnitude on either side of this estimate. Haberle and Jakosky (1990) used Viking observations to estimate the summertime loss of H_2O from the perennial north cap as 0.1 to 0.8 mm yr^{-1}; they pointed out, however, that even the sign of the annual H_2O budget is unknown.

The rate at which dust might be redistributed can be estimated crudely from probable limits on mass loading of the atmosphere. Pollack et al. (1979) estimated a columnar dust load at the peak of a dust storm to be 4×10^{-2} kg m^{-2}, and, assuming that dust deposition was concentrated in the polar regions and involved an equal mass of H_2O ice, they derived a representative polar deposition rate of 0.4 mm yr^{-1} in the current epoch. Because high dust opacities stabilize the lower atmosphere and appear to inhibit the initiation of dust storms, they considered much higher dust loads unlikely, even for epochs of considerably higher atmospheric pressure. Their estimate yields a minimum characteristic deposition time for an individual polar lamina of about 1 Myr. Table II gives characteristic times.

C. Polar Cap Albedo

Our ability to estimate reliable atmosphere pressures of other epochs is critically limited by our understanding of polar-cap albedo, and the observed values for both the annual and permanent polar caps are puzzling in the sense that they are well away from pure CO_2 frost or H_2O ice (Kieffer 1970*b*) and are very sensitive to dust abundance (Toon et al. 1980; Warren and Wiscombe 1980; Kieffer 1990; Warren et al. 1990).

Although the influence of different polar-cap albedos has been discussed (Toon et al. 1980), all numerical climate models to date have used a polar-cap albedo constant through the year and with epoch (see, e.g., Ward et al. 1974; Toon et al. 1980; Cutts and Lewis 1982; Fanale et al. 1982*a*, 1986; James and North 1982; Jakosky and Carr 1985; Zent et al. 1986; François et al. 1990). It is illustrative to consider the effect of variable albedo.

There is some evidence that the albedo of the seasonal cap varies through the year and is a function of insolation. In an indirect analysis based on energy balance, Paige (1985) found the surface radiometric albedo of the

TABLE II
Characteristic Deposition Time

Process	Years
Response Times	
CO_2 cap stabilization of atmospheric pressure	100
Sublimation of an H_2O cap	0.1–10 Myr
Dust deposition of polar lamina (30 m)	100,000
Regolith (100 m)[a]	10,000
Regolith (1 km)	1 Myr
Driving Functions	
Precession of the equinox	175,000
Oscillation of eccentricity	100,000 and 2 Myr
Variation of obliquity	125,000 and 1.3 Myr

[a]See Sec. IV.B.

south cap to be nearly linear with insolation: $A_c = 0.52 + 9.2 \times 10^{-4} I$, where I is the insolation [W m^{-2}]. This relation corresponds to the maximum albedo of 0.81 observed by Viking and would yield albedo = 0.925 for a summer solstice perihelion at an obliquity of 38°. This relation cannot hold to the full range of obliquities shown in Fig. 5, as albedos greater than 1 would be involved. Although there are substantial uncertainties in correcting for atmospheric opacity and the surface scattering phase function, such a relation is consistent with higher net sublimation rates acting to progressively clean fine dust from the cap surface by lofting it on the vertical sublimation wind (Kieffer and Paige 1987; Kieffer 1990; see also chapter 27). The dependence of albedo upon insolation may be significant only when the atmosphere can raise dust.

D. Formation of layered terrain and size of the polar caps

Processes associated with seasonal polar-cap formation are thought to transport dust and H_2O into the polar regions. Estimates of accumulation rate (sec. I.A) and the observation that laminae are typically 10 to 30 m thick have led to formation rate estimates of 0.1 to 1 Myr per layer, and the cratering age of the north polar layered terrain is on the order of 10 Myr. Dust and H_2O ice may be deposited in comparable amounts during the winter (Kieffer 1990) but may become separated during the sublimation season, so that no direct way has been found to estimate their proportion in the long-term accumulation.

Toon et al. (1980) have addressed the water-loss rate from a perennial H_2O ice cap; their computed rates depend upon albedo, thermal conductivity, and wind speed and are typically 0.05 g cm^{-2} yr^{-1} for the colder pole (aphelion summer) and 2 order of magnitude greater fro the "hot" pole at higher obliquity. Toon et al. noted, however, that a meter-thick layer of dust would insulate an ice deposit from summertime high temperatures and greatly stabilize the ice. Calculated sublimation rates from a free surface of ice in the polar regions at various obliquities (Toon et al. 1980) indicate that ice is probably absent from the upper ~1m of the layered deposits; its absence is further suggested by the dryness of the summer atmosphere over the south polar region (chapter 28). A corollary is that the apparent layering is not due solely to variation in the silicate/ice ratio, but it must represent some variation in the composition or relative abundance of nonvolatile materials. The color of the southern layered deposits indicates that some material darker than aeolian dust is also involved (Herkenhoff and Murray 1990*a*). Herkenhoff and Murray (1990*b*) determined that local slopes are steep enough that the surface layer must have some competence; it is perhaps composed of a rind of "filamentary sublimation residue" (see Saunders et al. 1986; Storrs et al. 1988).

Several studies of potential glacial flow of Martian polar ice sheets (Budd et al. 1986; Clifford 1987*b;* Hofstadter and Murray 1990) have concluded that substantial flow would occur if the layered deposits were primarily water ice. In the last of these studies, the absence of discernable flow features in the layered deposits was interpreted to indicate an ice proportion of less than 40% or more than 90% by volume, but several assumptions were required, including the presence of ice grains no larger than about 10 μm.

The layer sequence in the polar layered terrain has yet to be decoded. Cutts and Lewis (1982) derived scenarios in which the compositions or total deposition rate were modulated by climate, including situations where deposition was triggered at specific levels of a "climate function." Although the resulting patterns of layer thickness (Fig. 5) resemble those of the layered terrain, no specific relation between the model and the actual patterns was developed. Various interpretations of the observed concentration of dust and water vapor in the atmosphere, of the partition of deposition between the polar caps and the rest of the surface, and of the density of the deposits have led to varied estimates of accumulation rates: 10 m Myr^{-1} (Cutts 1973*a*), 400 m Myr^{-1} (Pollack et al. 1979*b*), and 10 to 1000 m Myr^{-1} (Pollack and Toon 1982).

We currently lack an understanding of what determines the geographic extent of the perennial caps under the current climatic regime, let alone under other regimes. At low obliquity, the pole-to-equator temperature gradient would be steeper than at present, and the perennial cap would tend to be concentrated near the geographic pole. Complex dynamical models will likely be required to predict the extent of the perennial caps to conditions other than the present. The latitude extent of the layered terrain is an observational constraint that is not yet understood; Toon et al. (1980) considered the boundary of the laminae to mark the equatorward limit of permanent H_2O caps and the boundary of the polar debris mantle to mark the limit of annual CO_2 caps.

IV. THE REGOLITH RESERVOIR

A. Volatile Inventories

The magnitude of the climate's response to insolation variations depends in part on the inventory of exchangeable volatiles. As an extreme example, if all volatiles were locked in nonexchangeable reservoirs there would be no climatic response to astronomic insolation variations.

The inventory of exchangeable volatiles on all planets can be thought of as the interplay of (1) the condensation history of the planet, which determines its bulk volatile abundance; (2) the planet's energy history, which controls degassing of those volatiles to the surface (chapters 4 and 5); and (3) the sinks available for the volatiles, including exchangeable sinks, such as adsorbate, and nonexchangeable sinks, such as escape to space (chapter 32). Estimates of the volatile inventory that remains accessible to quasi-periodic forcing can be derived from several lines of evidence (chapter 32).

Geochemical Evidence. The first post-Viking estimates of the Martian volatile inventory were based on Viking measurements of ^{36}Ar in the Martian atmosphere, depleted by a factor of 200 relative to Earth (Owen and Biemann 1976). It was originally thought that Mars had a low volatile abundance and had degassed inefficiently, yielding only 10 m of outgassed H_2O (Anders and Owen 1977). With later data from Pioneer Venus, Pollack and Black (1979) showed that noble gases cannot be ratioed in a linear manner to more reactive volatiles. Using N_2 as an index gas, they increased the estimate to 80 to 160 m of H_2O and perhaps 1 to 3 bars of CO_2.

More recent estimates of the volatile inventory are based on identification of Mars as the Shergottite Parent Body (see e.g., Wasson and Wetherill 1979; chapter 4). Using an accretion model of Ringwood (1977, 1979), Dreibus and Wanke (1987) predicted 3.3% H_2O by weight for Mars at the time of accretion, and only about 130 m of H_2O after degassing and loss to space (see also chapters 4 and 6).

Geomorphic Evidence. The Martian surface provides ample evidence that ground ice has been present throughout the regolith (chapter 16). Indeed, the abundance of surface features attributed to flow of H_2O (chapter 15) is difficult to reconcile with the relatively low abundances predicted by some geochemical models (see chapter 6). Carr (1986) estimated that some 500 m of H_2O may have been retained and outgassed by Mars after accretion.

B. Response of the Regolith

Regolith Temperature Variation. Because of the strong nonlinearity of thermal radiation with respect to temperature, diurnal and annual variation in insolation lower the average surface temperature. Hence, thermal models are necessary to assess quantitatively the response of regolith volatile inventories.

The response of a regolith volatile depends upon both the penetration of the climatic thermal wave and the resulting gas pressure wave. The same situation that would allow a large volatile inventory in the regolith—large surface area of material with high adsorption potential—would lower the gaseous diffusion coefficient due to small pore size, tortuosity, and damping caused by repeated cycles of adsorption and desorption.

Thermal diffusivity of the regolith is determined by the thermal conductivity, the specific heat capacity, and the bulk density, the same parameters that determine the thermal inertia of the surface. The thermal skin depth for a period P_j is $Z_t = (\kappa P_j/\pi)^{1/2}$ where $\kappa = k/\rho C_p$ is the thermal diffusivity; ρ is the material density and C_p its specific heat capacity. Because the thermal conductivity k of geologic materials ranges widely, it is useful to express the thermal skin depth in terms of the thermal inertia $I = (k \rho C_p)^{1/2}$. Then

$$Z_t = (P_j/\pi)^{1/2} (I/\rho C_p). \tag{14}$$

Because the product of the heat capacity and the density of most geologic materials varies only over a factor of about 3, the thermal conductivity of the shallow regolith can be well constrained from thermal inertia determinations. The average density of Mars' surface soils is near 1500 kg m^{-3} (see chapter 21), and the specific heat of many geologic materials in the range of Mars' surface temperatures is 586 + 2.3 (T-220) J kg^{-1} K^{-1} (Kieffer 1976). The thermal inertia of soil at the Viking landing sites is 260 J m^{-2} $s^{-1/2}$ K^{-1} (6.2×10^{-3} cal cm^{-2} $s^{-1/2}$ K^{-1}) (Kieffer 1976).

The implied thermal conductivity of 0.077 J m^{-1} s^{-1} K^{-1} is some 30 times lower than the average thermal conductivity of terrestrial frozen ground (cf. Clifford 1987); this probably represents desiccation of the diurnal thermal skin depth, with gas conduction in the very thin atmosphere dominating solid conduction (Jakosky 1986). These physical-property values yield $Z_t = 3. \times 10^{-4}$ m $s^{-1/2}$ $(P_j/\pi)^{1/2}$ (at 220 K), or a diurnal thermal skin depth of 0.05 m and a Martian annual skin depth of 1.3 m. This value of 0.94 m^2 yr^{-1} is a reasonable lower bound on thermal diffusivity of the Martian regolith. At depth, compaction is expected to increase conductivity. Fanale et al. (1982*a*) suggested 15 m^2 yr^{-1} as an upper bound. Thus, propagation of a thermal wave to the bottom of a 100-m-thick regolith requires on the order of 10^4 yr; for a 1 km regolith, about 1 Myr.

Detailed surface and subsurface temperatures for a homogeneous model with nominal physical properties representative of Mars were computed throughout a Martian year in support of the Viking mission (Kieffer et al. 1977, Appendix A); these temperatures have been widely used to estimate current conditions in the Martian regolith. Considering the implication of this model on the current stability of H_2O (Fig. 10 in chapter 28) is illustrative of the behavior of deeper regolith volatiles over orbital-variation time scales. There is an equatorial zone where the temperature exceeds that of the average atmospheric frost point throughout the year and is thus expected to be desiccated. A polar zone, which extends to lower latitudes at depth, is perpetually colder than the atmospheric frost point. In a shallow temperate zone, the temperature oscillates across the atmospheric frost point. Over obliquity-oscillation time scales, the temperature variation to depths of several hundred meters has the potential to pump adsorbed volatiles into and out of the megaregolith (Fig. 6) (Fanale et al. 1986).

Toon et al. (1980) calculated diurnal average temperatures as a function of season, and determined annual average temperature, for several latitudes for obliquities from 15° to 45° and eccentricity of 0.0 and 0.14. Their work emphasized the response of possible regolith-trapped CO_2 to the thermal wave associated with orbital variation. Their work included treatment of the variation of both the thermal and gaseous diffusion coefficient in the regolith with total atmospheric pressure, and showed that adsorption significantly slowed gas diffusion through the regolith.

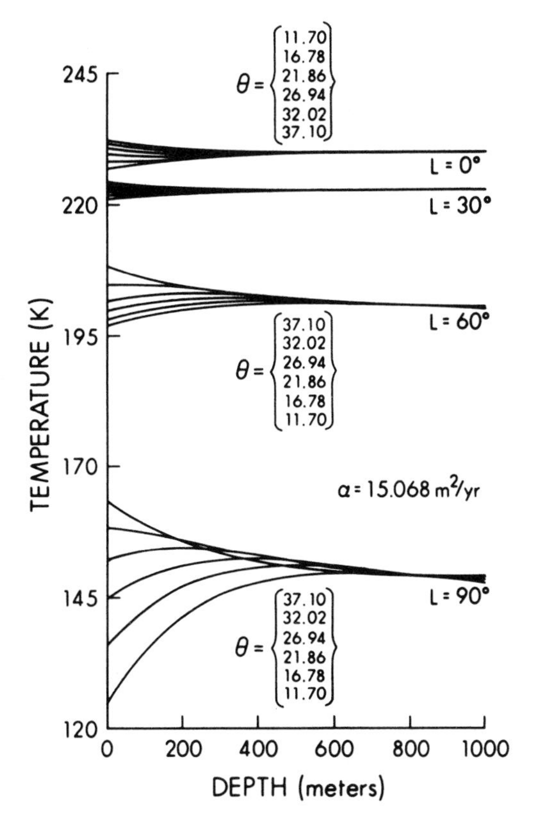

Fig. 6. Temperature in the regolith at four latitudes L during a period of increasing obliquity. Thermal diffusivity is assumed to be 15 m^2/yr^{-1}, and internal heat flow is neglected (figure from Fanale et al. 1982a).

Equilibrium Adsorption Whenever a gas equilibrates with either a solid or liquid interface, electron cloud interactions between the surface and gas molecules cause them to bind, a process known as *adsorption*. Van der Waals forces establish reversible bonds that allow gas molecules to move freely from the adsorbed to the gaseous phase. Davis (1969) and Fanale and Cannon (1971) pointed out that adsorption onto soil grains could result in a substantial volatile reservoir on Mars.

The size and role of the adsorbate reservoir in the Martian soil depends on the many variables that affect the process of adsorption. All have significant uncertainty. The population of adsorbed molecules on a given sample increases with pressure and decreases with temperature. Adsorption data applicable to Mars were measured by Mooney et al. (1952*b*), Fanale and Cannon (1971,1974), Anderson et al. (1978), and Zent et al. (1987); these data are discussed below.

In a given volume of soil, the total adsorbate population increases with the specific surface area ($m^2\ g^{-1}$) of the soil. Specific surface area depends in part on mineralogy, grain size and measurement method. Mineralogy affects adsorptive coverage because different minerals present different surface functional groups to the gas. Surface functional groups are regularly repeated atomic structures that characterize crystal faces and control electronic structure at the molecular scale (Sposito 1984). Certain clay minerals are superb adsorbents; with specific surface areas of a few thousand $m^2\ g^{-1}$, they may hold a few tens of percent H_2O by weight under Martian conditions. They have been suggested repeatedly as Martian soil analogs (see, e.g., Hunt et al. 1973; Toulmin et al. 1977; Banin et al. 1988*a*). Conversely, morphologic evidence of both widespread volcanics (Carr et al. 1977*a*) and permafrost (Carr and Schaber 1977) led Toulmin et al. (1977) to suggest that palagonite, an iron-rich, altered volcanic glass, may be a significant component of the regolith. Palagonite typically has a specific surface area of a few tens of m^2 g^{-1} and will adsorb only a few tenths of a percent H_2O by mass under Martian conditions.

The electronic structure of adsorbate molecules also plays a role in determining adsorptive behavior. It is represented by the molecular dipole moment. Species with permanent dipole moments, such as H_2O, typically adsorb at higher coverages than species without permanent dipole moments, such as CO_2.

The mass and distribution of the Martian adsorbate inventory also depend on the area and distribution of the gas/solid interfaces in the soil. Total surface area is usually represented at the product of two factors: the specific surface area and the depth of the soil. An interpretation of the Viking Lander gas-exchange experiment by Ballou et al. (1978) concluded that the quantities of gases released were consistent with a soil of 17 $m^2\ g^{-1}$, not unreasonable for a poorly sorted soil. Although model dependent, that number is the only estimate of the surface area of any Martian regolith material.

Estimates of the regolith thickness in diffusive contact with the atmosphere depend upon observations of visible debris on the Martian surface and upon comparison with the lunar regolith. Crater studies indicate that 1 to 2 km of ejecta may have covered Mars (Fanale 1976), although by analogy with the Moon, much of this material may be impact breccia. Only about 10 m of fine-grained regolith was found at the Apollo 14 and 16 sites (Cooper et al. 1974); the underlying material is inferred to be impact breccia. The self-compaction depth, or lower limit of the megaregolith, is probably around 10 km on Mars, based on analogy with the Moon and Earth (Carr 1979). In analyzing models of the adsorptive behavior of the regolith, it should be kept in mind that the models assume idealized regoliths with uniform coverage by well-sorted, fine-grained powders. The actual Martian regolith contains blocks,volcanic extrusives and intrusives, impact breccia (MacKinnon and

Tanaka 1989), and possible widespread soil crust (Jakosky and Christensen 1986*a*), all of which decrease the surface area in the regolith column.

There is very little data on the distribution of the regolith as it pertains to climate history. The most critical factor is its latitudinal distribution, because temperature and hence adsorptive coverage are strongly latitude dependent. Some workers have thought that large circumpolar debris mantles are present (Cutts 1973*a;* Soderblom et al. 1973*b*), but nearly ubiquitous polar haze throughout the Viking observations has made quantification difficult (Kahn et al. 1986). Large regional dust deposits do exist, primarily in the northern tropics; however, their low latitude and relative thinness (≤ 5 m) (Christensen 1986*a*) make them unlikely reservoirs for exceptional volatile inventories.

The adsorptive coverage of the soil under Mars-like conditions has been estimated from laboratory data. Typically, a series of adsorption isotherms are measured, and an analytic expression is fit to the laboratory data. The expression can be used to calculate the adsorbed population as a function of temperature and partial pressure throughout the regolith. Most adsorptive behavior can be described reasonably well by an equation of the form

$$\alpha = A\delta P^{\gamma} T^{\beta} \qquad (15)$$

where A is the specific surface area of the adsorbent, measured via BET isotherm; the adsorbate molecule is used as the probe. The constants δ, γ and β are derived by fitting the expression to the data (β is always negative for physical adsorption).

Studies of H_2O adsorption have been made on basalt (Fanale and Cannon 1971,1974) and on montmorillonite (Mooney et al. 1952*b;* Anderson et al. 1978). Fanale and Cannon performed adsorption measurements at temperatures from 250 to 300 K and predicted that basalt at the Martian surface would carry 0.5 to 10 mg g^{-1} H_2O soil. Measured specific surface areas are in the range of 5 to 10 m^2 g^{-1} (see Fig. 7). In adsorption studies on montmorillonite at 270 to 290 K, it was found that 100 to 200 mg g^{-1} H_2O might adsorb under Mars-like conditions. The specific surface area of the montmorillonite averaged 420 m^2 g^{-1}. The difference between the two adsorbents (a factor of 10 to 20) is due to the ability of smectite clays to store H_2O in interlayer sites. Anderson et al. (1967), in a classic analysis, pointed out that ice formation places an upper limit on the amount of H_2O adsorbate at low temperatures. No more than roughly one monolayer of adsorbed H_2O can exist on the Martian regolith; additional H_2O finds ice to be the most stable phase.

An important caveat to the H_2O adsorption data is that all such data have been acquired at temperatures of 250 K and above and at partial pressure a few orders of magnitude greater than on Mars. Inferences regarding the adsorptive behavior of H_2O on the Martian regolith are therefore based on ex-

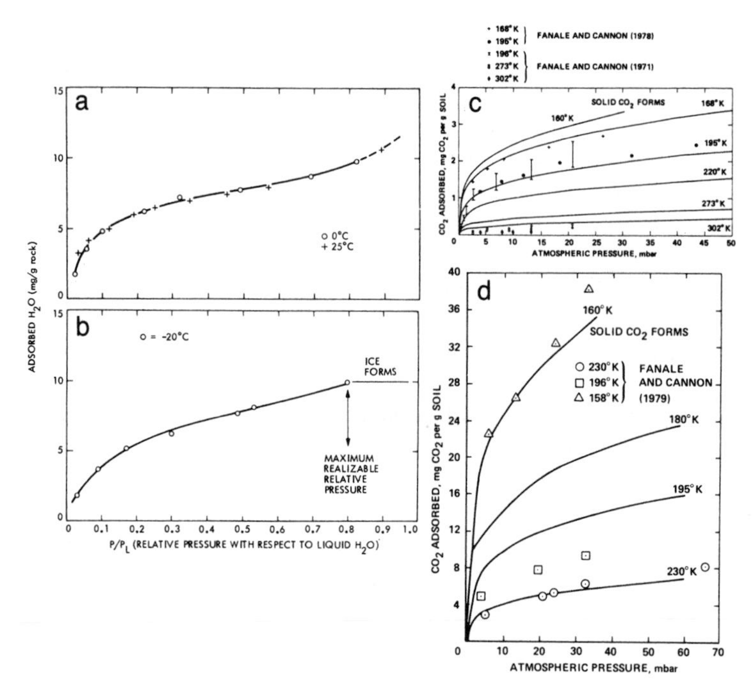

Fig. 7. (a,b) The absorption of H_2O and CO_2 as a function of temperature and pressure for possible Martian soils; H_2O adsorption isotherms for ground basalt. P/P_L is the vapor partial pressure of water relative to the (undercooled) liquid (figure from Fanale and Cannon 1974). (c,d) Measurements of CO_2 adsorption data for basaltic soil (c) and a clay soil (d) by Fanale and Cannon; the solid curves are an empirical fit developed by Toon et al. 1980.

trapolation of analytic expressions as opposed to more reliable interpolation. Small extrapolations have been tested (Anderson et al. 1978), but laboratory confirmation under Mars-like conditions is necessary.

Carbon-dioxide adsorption measurements have been carried out on basalt (Fanale and Cannon 1971,1978), on nontronite (Fanale and Cannon 1978), and on palagonite (Zent et al. 1987). Carbon-dioxide isotherms have been acquired on all adsorbents over a more representative temperature range (160 to 230 K). In general, under Mars-like conditions, basalts adsorb roughly 1 to 2 mg g^{-1}, nontronite 6 to 10 mg g^{-1}, and palagonite 15 to 20 mg g^{-1}. The resulting specific surface areas are 5, 47 and 61 m^2 g^{-1}, respectively. When corrected for differences in specific surface area due to mineralogy and grain size, all of these adsorbents hold around 2×10^{-4} g m^{-2} under Mars-like conditions.

One unexplored aspect of CO_2 adsorption concerns the effects that simultaneous adsorption of two species will have on adsorbed inventory and its exchange with the atmosphere and polar cap. Models of quasi-periodic climate change are based on adsorption isotherm measurements made in single adsorbate systems. In co-adsorption studies, Carter and Husain (1974) found that H_2O was strongly enriched in the adsorbed phase relative to its mole fraction in the vapor phase and that only 10% relative humidity at 273 K decreased the adsorptive coverage of CO_2 by more than an order of magnitude. The dependence of the H_2O enrichment on temperature is unknown.

With analytic expressions in hand for adsorption isotherms, it is still necessary to estimate the regolith thickness and distribution to determine approximately the adsorbed inventory and its subsequent exchange. The problem for CO_2 is posed in the following way: there is a certain inventory of CO_2 available (Σ CO_2), and it is everywhere partitioned between the atmosphere and the regolith via adsorption. The pressure of CO_2 in the atmosphere is everywhere the same, but the temperature of the regolith is not, so coverage must be calculated for individual chunks of regolith. If the polar temperature is lower than the frost point of the atmosphere, then the atmosphere is buffered by the presence of polar caps, and the local adsorptive coverage is determined by the temperature and pressure (Fanale et al. 1982*a*). When these calculations are carried out for the current orbital configuration, some combinations of Σ CO_2 and regolith depth produce conditions that are grossly incompatible with observations; for example, they may predict massive CO_2 polar caps. Indeed, for a given analytic expression for adsorption isotherms, and for a given assumption about the thermal state of the regolith, there are restricted combinations of regolith thickness and Σ CO_2 that are compatible with an atmospheric pressure of ~ 8 mbar and the absence of massive CO_2 polar caps. Models of the atmosphere/regolith/cap system by Fanale et al. (1982*a*), Zent et al. (1987) and François et al. (1990) describe a cloud of points (see Fig. 8) that are compatible with the current atmospheric pressure and the absence of a massive polar cap. We can refine somewhat the conclusions about the current regolith CO_2 adsorbate inventory by recalling that the

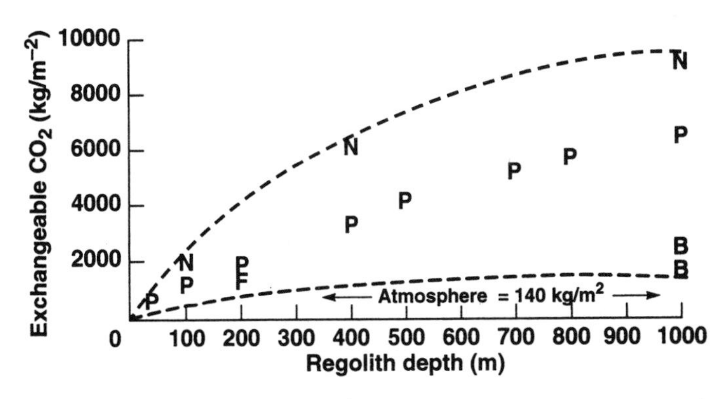

Fig. 8. Combinations of exchangeable CO_2 (Σ CO_2) and regolith thickness that correctly predict the current climate according to several published models; *P*, palagonite (Zent et al. 1987); *B*, basalt; *N*, nontronite (Fanale et al. 1982*a*); and F, palagonite (François et al. 1990).

regolith considered in these models is an idealized, uniform and fine-grained particulate blanket, and that the actual Martian regolith has less surface area per unit mass than the fine powders used for laboratory adsorption studies. If we assume as a basis of comparison that the actual Martian regolith has as much surface area as 40 m of this idealized model regolith, then 2000 kg m^{-2} would be a reasonable upper limit on $\Sigma\ CO_2$.

A primary qualitative conclusion from these measurements is that the regolith may hold as much as, but probably not more than, a few tens of times more CO_2 than the atmosphere/cap system. This conclusion removes the philosophically uncomfortable conclusion that the Martian atmosphere is currently buffered by a south polar cap only a fraction as massive as the atmosphere itself. Instead, we may envision the arrangement as one in which the atmosphere plus cap system is buffered by an ocean of adsorbed CO_2.

Estimating the total adsorbed H_2O inventory is even more problematic, for several reasons. First, as already mentioned, no adsorption measurements discussed in the literature were made under actual Mars-like conditions, so empirical equations must be applied to conditions very different from those at which they were derived. Further, although CO_2 has not been observed to interfere with H_2O adsorption, co-adsorption data are unavailable for Mars-like conditions. Finally, the specific surface area of adsorbents is more varied for water than for CO_2. For example, smectite clays have interlayer sites that at room temperature are accessible to H_2O, but not to CO_2 (Alymore et al. 1970). If H_2O is present in interlayer sites, however, it appears that CO_2 can occupy vacant sites (Thomas and Bohor 1968). The resulting effective increase in specific surface area can be a few orders of magnitude; however, kinetic barriers at 210 K may limit exchange with interlayer sites. As an exercise, let us assume the H_2O adsorption isotherm for basalt, presented in Zent et al. (1986), at 210 K and 10 pr μm. For a regolith equivalent to 40 m of idealized regolith, the adsorbed inventory is 240 kg m^{-2}. The considerable uncertainty in these numbers for both H_2O and CO_2 will be reduced only with laboratory data obtained under conditions appropriate to Mars and with an understanding of regolith composition, grain size and distribution.

Magnitude of Exchange. All volatile exchange between the regolith and atmosphere is ultimately driven by redistribution of insolation across Mars as its orbital elements vary over time scales of 10^5 to 10^6 yr. The climatic consequences of orbital variations have been explored in papers by Toon et al. (1980), Fanale et al. (1982*a*, 1986), Pollack and Toon (1982), Zent et al. 1987) and François et al. (1990), from which most of the results below are taken.

The exchange of volatiles that is thought to cause periodic climate change is controlled by the diffusivity of H_2O, CO_2, and heat through the regolith and by the time history of boundary conditions.

In the regolith, adsorption slows diffusion relative to its rate in the free-gas phase. Because a given molecule spends orders of magnitude more time adsorbed onto the walls of the pore spaces than actually diffusing, the effective diffusion coefficient decreases by a few orders of magnitude. When pore diameters become smaller than the mean free path of the pore gases, collisions with pore walls begin to dominate over collisions with other molecules, a phenomenon known as Knudsen diffusion; diffusion is slowed further in the Knudsen regime (see, e.g., Clifford and Hillel 1986) (Fig. 9).

Laboratory simulations of regolith diffusion carried out by Fanale et al. (1982*b*) supported a theoretical treatment of CO_2 diffusion by Toon et al. (1980). The effective diffusion coefficient of CO_2 through the soil column is on the order to 1 to 5 $\times$ 10^{-2} cm^{-2} s^{-1}, and it is very sensitive to temperature. In the absence of adsorption, the diffusion coefficient is larger by a factor

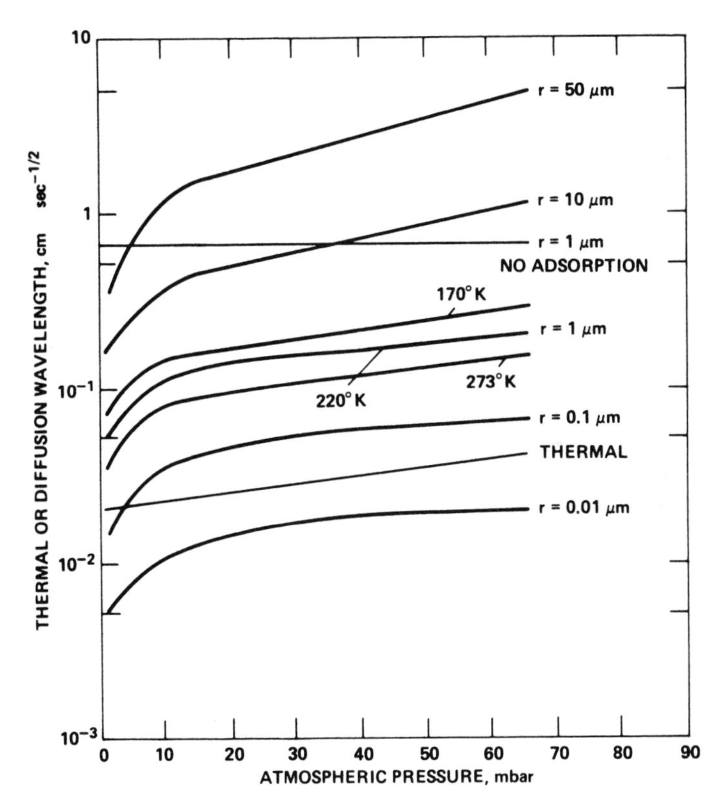

Fig. 9. Thermal- and gas-diffusion "wavelengths" (cm $s^{-1/2}$) for several possible soils as a function of atmospheric pressure; *r* is the mean pore size of the soil. Thermal diffusion has a weak dependence on pressure due to pore conduction. Gas diffusion is independent of pressure if there is no absorption. The presence of adsorption severely dampens the pressure-diffusion wavelength (figure from Toon et al. 1980).

of up to 10^3 and almost independent of temperature. The Martian regolith is likely to be a poor isothermal buffer of atmospheric pressures on a seasonal basis because adsorption substantially slows the atmospheric pressure wave. Nonetheless, CO_2 diffusivity substantially exceeds thermal diffusivity, and CO_2 will diffuse faster than heat through basalt unless pore sizes are as small as 0.1 μm or unless partial pressures go below a millibar or two. Thermal diffusivity may possibly exceed CO_2 vapor diffusivity at lowest obliquities, when polar cold trapping can reduce atmospheric pressures to fractions of a millibar (see below). The consequences of this possible reversal have not been explored.

H_2O diffusion is likewise curtailed by factors of 100 to 1000 due to adsorption. At the upper end of the scale, H_2O diffusion is ultimately limited by its ability to diffuse through CO_2. At the slower end, diffusion is limited by adsorption and Knudsen diffusion. Effective diffusion coefficients can be as low as 10^{-4} for H_2O in high-surface area adsorbents (Fanale and Jakosky 1982; Jakosky 1985).

Both theoretical and laboratory experiments of adsorption and diffusion have relied on analysis of fine-grained, uniform soils. One matter of concern that has not heretofore been explored is the effect of impermeable layers such as salts, intrusive volcanics, or ground ice on volatile exchange on Mars.

The thermal response of the regolith to obliquity variations has already been discussed (Sec. IV.B).

As the obliquity of Mars increases, the average annual insolation increases at high latitudes and decreases slightly at lower latitudes (see Fig. 3 and chapter 9). Both the high-latitude regolith and the surface of the polar caps warm, and the atmospheric pressure increases accordingly (Fig. 10). Eventually polar caps are completely sublimated, leaving only a seasonal cap at the winter pole. The high-latitude regolith holds considerably more CO_2 than the low-latitude regolith, and increased insolation at high latitude causes additional desorption and increase in atmospheric pressure. Atmospheric pressure at high obliquity is buffered at about 10 to 15 mbar unless feedback effects, such as hysteresis of the cap albedo, operate that may push pressures somewhat higher. In models where the regolith poleward of 60° is mathematically removed from the exchange, there is no increase in atmospheric pressure at high obliquity, because the low-latitude regolith is capable of adsorbing only a small amount of CO_2. High-latitude ground ice may seal off the high-latitude regolith, and so prevent CO_2 there from exchanging with the atmosphere (Pollack and Toon 1982). Further uncertainty with respect to the available CO_2 inventory (see above) makes the conclusion of increased atmospheric pressure at high obliquity tentative. François et al. (1990) modeled climate evolution with subannual resolution and predicted seasonal caps so large that annual average atmospheric pressures have been no higher than at present. If the atmospheric pressure does increase at high obliquities, hydrodynamic flow of the atmosphere from one seasonal cap to the other may

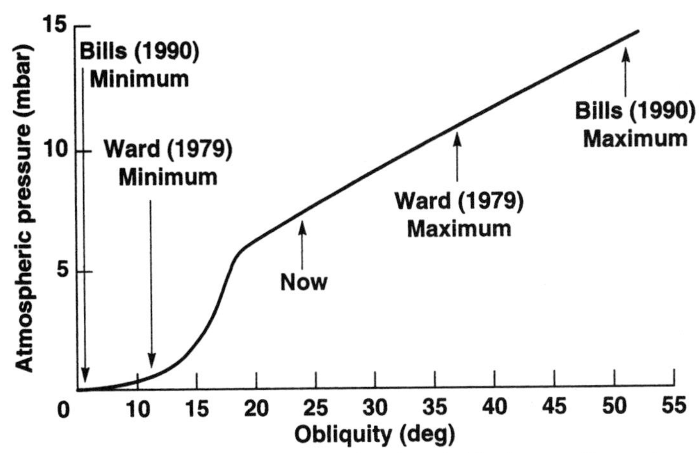

Fig. 10. The atmospheric pressure of Mars as a function of obliquity according to the model of Fanale et al. (1982*a*). At high obliquity, the atmospheric pressure is controlled by adsorptive equilibrium between the atmosphere and regolith. At obliquities just lower than the present configuration, quasi-permanent CO_2 polar caps form and buffer atmospheric pressure through the frost-point relationship. Before the growth of the Tharsis bulge, obliquity oscillations had larger extremes with correspondingly greater oscillations in pressure.

produce very large semiannual pressure excursions. High pressures and condensation/sublimation cycles may result in dust storms that are more continuous (Pollack et al. 1979*b*) introducing the possibility of significant atmospheric optical opacity at high obliquity. The history of surface insolation under such circumstances is uncertain.

Additionally, dust may be deposited onto polar ice, changing the polar radiation balance and making the subsequent evolution of the system even more difficult to predict. There is the possibility of hysteresis involving particularly the albedo of the polar caps. If dust-storm activity increases at high obliquity, dust deposition on the polar caps can decrease the cap albedo and strongly increase vapor pressures buffered by the cap as long as ice remains in diffusive communication with the atmosphere.

The absence of perennial CO_2 polar caps exposes polar H_2O ice to summertime insolation maxima, and consequently the perennial H_2O cap will lose mass at high obliquities, a subject we will return to below.

As the obliquity of the planet decreases, insolation is redistributed toward low latitudes (Fig. 3). Polar temperatures fall, and the temperature of one or both poles eventually falls below the CO_2 frost point of the atmosphere. The result of continued polar cooling will be quasi-permanent polar caps of CO_2 formed by atmospheric condensation. The atmospheric pressure can decrease to as low at 0.1 mbar, depending upon the albedo of the condensing ice (Fig. 4). As the atmospheric pressure is buffered at lower and lower levels, the regolith responds by desorbing CO_2, particularly at low latitudes. The desorbed CO_2 cannot raise the atmospheric pressure, and so it goes into the mass of the perennial cap. The total amount of CO_2 that exchanges between the regolith and cap is difficult to estimate. A few tenths of a bar may exchange, using the atmosphere as a conduit, even though the atmosphere itself never holds more than 10 to 20 mbar.

Dust storms probably cease; because hydrodynamic lift forces are proportional to fluid density, the moving of surface particles becomes much more difficult. At low obliquities, the complete absence of dust storms could lead to very bright caps whose albedo approaches unity. Atmospheric pressures buffered by such bright caps would be the residual 0.03 mb of noncondensing gases and they would remain low until dust-storm generation became possible at higher obliquities.

Exchange of H_2O between the atmosphere and the regolith over obliquity time scales is smaller in magnitude, for several reasons. Because H_2O is a more stable adsorbate, its diffusion is slowed more drastically by adsorption than is CO_2 diffusion. Further, H_2O is never more than a minor constituent in the Martian atmosphere, so that to be cold trapped it must diffuse or be advected to the poles. The partial pressure of H_2O in the atmosphere is likely to be controlled by the CO_2 polar caps, which will act as a "cold finger," incorporating the regolith-derived H_2O into the growing cap (see also chapter 28, especially Sec. V). Models of the kinetics of H_2O exchange through the regolith suggest that no more than about 100 kg m^{-2} H_2O can move into or out of the regolith over an obliquity cycle (Zent et al. 1986).

Redistribution of H_2O Ice. Over Martian volatile history, the overriding process has been ground-ice withdrawal from the low-latitude regolith and ground-ice deposition in the high-latitude regolith and polar caps (see chapter 32). This transfer is forced by cold trapping, a simple consequence of the significantly lower partial pressure of H_2O over the cold polar caps than over ground ice in the relatively warm equatorial regolith. Mass transfer down the concentration gradient will eventually deplete the equatorial regolith of ground ice altogether (see, e.g., Leighton and Murray 1966), although kinetic limitations due to migration through the regolith and atmosphere may have so far prevented the system from achieving equilibrium (Clifford and Hillel 1983).

As boundary conditions are altered by redistribution of insolation during cyclic variations of orbital parameters, H_2O exchanges between the caps and high-latitude regolith. At higher obliquities, insolation onto the polar H_2O caps raises atmospheric P_{H_2O}, and the regolith attempts to accommodate the increased pressure by adsorbing more H_2O. The ground-ice interface may approach the surface as the regolith attempts to accommodate the increased partial pressure of water. Diffusive slowing through the regolith by adsorption may result in atmospheric saturation and in precipitation at low latitudes (Jakosky and Carr 1985). At low obliquity, when extensive and quasi-permanent CO_2 polar cold traps exist, H_2O is desorbed from the regolith and transferred back to the polar caps. The depth of the ground-ice interface increases at lowest obliquities as well, because the polar cold trap becomes more efficient relative to lower latitudes as well as absolutely more efficient when insolation is concentrated at the equator.

Controls on this system are analogous to those that control CO_2 exchange: periodic boundary-condition variations, thermal and water-vapor diffusivity through the regolith, and the adsorptive properties of H_2O in the regolith. The sense of exchange at any location is determined by the partial pressure of H_2O at the base of the atmosphere and in the pores at the ground-ice interface.

In addition to insolation at the poles, which determines local atmospheric P_{H_2O}, atmospheric boundary-layer dynamics play a role in controlling the vapor pressure at the base of the atmosphere. Deposition or withdrawal may be enhanced by sequestering atmospheric H_2O near, or well above, the surface, respectively. Uncertainties involved in predicting the albedo of the polar caps and atmospheric optical depth come into play as well, as polar atmospheric P_{H_2O} is very sensitive to the flux of energy at the surface of the caps. The atmospheric pressure must be considered because H_2O must diffuse through CO_2 in the pore spaces if the mean free path of the CO_2 is less than the characteristic pore size of the regolith.

Ground ice is currently stable with respect to Mars' atmosphere at latitudes poleward of about 40° (Leighton and Murray 1966). Equatorward of 40°, escape to the polar cold traps has probably persisted throughout geologic history. Smoluchowski (1968) argued that only a few meters of fine-grained regolith material could prevent the removal of low-latitude ice, but it now appears unlikely that even 100 m of dry regolith could prevent the escape of significant amounts of low-latitude ground ice over the course of a few billion years (Clifford and Hillel 1983). Fanale et al. (1986) calculated the global distribution and migration of subsurface ice on Mars starting from a uniform initial distribution. Their model accounts explicitly for the obliquity, eccentricity and longitude of perihelion variations, as well as for the details of condensible gas flow through porous media. They assumed no significant optical depth at any orbital configuration and invariant polar cap albedo equal to 0.4 for H_2O ice and 0.65 for CO_2. Despite the extreme variations of the relevant boundary conditions, they found that only about 1 g cm^{-2} of ice should exchange between the regolith and polar caps over each obliquity cycle. Like Clifford and Hillel (1983), they found that the upper 100 to 200 m of the regolith below 30° to 40° latitude should be depleted of ice by polar cold trapping. At greater depths any ice that may have been present early in Mars' history should survive to the present. The persistence of low-latitude, deep ground ice may be supported by the observations of the young-looking channels in Valles Marineris (see chapter 14).

At latitudes higher than about 40°, prediction of H_2O exchange over obliquity cycles is even more complex, because the ice interface is shallow enough to feel the annual thermal wave. Accordingly, annual thermal waves and the P_{H_2O} boundary condition must be tracked with subseasonal temporal resolution through several obliquity cycles to predict the exchange of H_2O and the depth of high-latitude ground ice. The profile of ground ice in equilibrium with the current climate lies within a meter of the surface at latitudes greater than about 50° (Farmer and Doms 1979), and it moves 5° equatorward at high obliquity and probably 3° poleward at low obliquity (Toon et al. 1980). These results were supported by a kinetic model by Zent et al. (1986), which explicitly modeled the penetration of the seasonal thermal wave through several obliquity cycles. Their shallow ground-ice profile was virtually identical to the Viking-based profile of Farmer and Doms (1979) in the northern hemisphere, and it differed only slightly in the southern hemisphere. They found that ground ice is stable to within centimeters of the surface at latitude 80°, but it dropped away to about 30 cm at 60° latitude and 80 cm at 50° latitude. Such calculations must be understood to be preliminary; the mass-transport rates of H_2O are so low that erosion and deposition at the solid surface may play a greater role in determining the depth of ground ice than thermodynamics. About the mean profile, the ground ice migrates perhaps 20 cm per obliquity cycle; it is near the surface at high obliquity and deeper at low obliquity.

V. SUMMARY OF CONDITIONS THROUGH AN OBLIQUITY CYCLE

As can be seen in Figs. 3 and 4, obliquity is the astronomical parameter of primary importance (unless albedo is insolation dependent, as discussed below). Beginning with obliquity at its average value, approximately the current condition, as obliquity decreases the polar-cap temperature will drop, lowering the atmospheric pressure. There will be a small time lag from the relation shown in Figs. 4 or 10 due to outgassing of the regolith. As the atmospheric pressure drops, the ability of the atmosphere to carry dust and the frequency of dust storms both decrease, and major dust storms may cease altogether. The incorporation of dust into the seasonal cap will decrease and the cap albedo will increase, further accelerating the atmospheric change. At minimum obliquity, the pressure will be strongly dependent upon the albedo of the polar cap, probably dropping to the point where CO_2 is a minor component. At a surface pressure of 0.6 mbar, other gases make up half the atmosphere, and the transport and exchange of CO_2 becomes diffusion limited. This condition is reached when polar temperatures drop below 129.5 K, which would occur even at average obliquity if the cap albedo exceeds 0.81. During this period of low obliquity, low pressure, clean atmosphere and clean polar cap, water would be moving from the low-latitude regolith to high latitudes. Because the diffusion coefficient for a minor component such as H_2O is inversely proportional to pressure, H_2O mobility within the regolith is enhanced at low total pressure, and hence the poleward migration of water will proceed more easily at low obliquity than the reverse flow at high obliquity.

As the obliquity increases from its minimum, little change in Martian conditions will occur until dust is raised into the atmosphere and reaches the surface of the polar cap, lowering its albedo. The positive feedback expected between atmospheric pressure and dustiness could yield an instability wherein both the perennial and seasonal polar-cap albedos drop significantly and the pressure increases an order of magnitude in a few hundred years. Climate change through the remainder of the increasing high-obliquity phases is then relatively continuous; the atmosphere is dusty much or all of the year, and the polar regolith desorbs both H_2O and CO_2. During the period of seasonal dust storms, the longitude of perihelion would control the hemispheric asymmetry; extrapolation from present conditions would suggest any perennial CO_2 polar cap would be in the perihelion summer hemisphere.

If the seasonal polar cap albedo is dependent on insolation throughout the obliquity cycle, the hemispheric asymmetry is enhanced with increasing eccentricity, the summer perihelion hemisphere being strongly favored for the perennial CO_2 cap.

Although the perennial CO_2 cap would oscillate between hemispheres with a 51,000-yr period, the favored location for the H_2O cap may not oscillate, but rather it would remain at both poles with the larger deposit generally in the north because elevations are lower there. When the dust opacity is not near unity, the surface solar and thermal radiation fields are similar on an annual basis at both poles. Seasonal CO_2 can condense more easily at the higher surface pressure in the north. Therefore, it yields a higher average annual albedo, providing lower annual average temperatures and a stronger long-term cold trap for water.

VI. PUZZLES AND PROSPECTS

Examination of the current form and distribution of polar volatile and sedimentary deposits (Plates 5 and 6 and map I-1802C), and a glance at Fig. 2 will reveal several open issues. We currently have no good explanation as to why the south polar perennial cap is offset from the geographic pole, nor what causes the arcuate form of the dark lanes in both polar caps.

Correspondences between the extent of layered terrain and the range of perennial caps, and between the extent of the polar debris mantle and the range of seasonal caps, have been suggested but remain uncertain. Apart from a general consensus that the layered terrain is composed of H_2O ice and fine dust, little information is at hand on the proportion of these two components or on what is responsible for the change in appearance between layers. High-resolution imaging by Mars Observer may provide clues.

Other questions related to current conditions include the depth to permafrost as a function of latitude and perhaps local conditions, and the permeability of the soil below the diurnal and annual thermal skin depths. The extent of permafrost in the upper meter or so can be addressed by analysis of detailed thermal mapping over a Martian year, but deeper probing will require appropriate electromagnetic sounding. We do not know if the south polar CO_2 deposit is really permanent under the current climatic regime. If it is observed to disappear, then we will know; otherwise, an estimate of its thickness may result from meter-scale imaging or gamma-ray spectrometer mapping in the late summer.

Better constraints on the extent of dust storms and the magnitude of advective heating through a climate cycle may result from more detailed modeling. Over these long time scales, the influence of condensate clouds remains unmodeled and unknown. Likewise, whether the Martian atmosphere ever becomes so free of dust that the seasonal (or even the perennial) polar-frost albedo approaches the high values of the pure volatiles is unknown. There are no immediate prospects of answers to either of these questions. Confidence in our understanding of the Martian climate over the full range of the astronomic cycle will probably require sampling and detailed chemical analysis of volatiles that were condensed throughout this range of conditions.

PART VI
Biology

34. THE SEARCH FOR EXTANT LIFE ON MARS

HAROLD P. KLEIN
Santa Clara University

NORMAN H. HOROWITZ
California Institute of Technology

and

KLAUS BIEMANN
Massachusetts Institute of Technology

Mars is important to biologists because it may well be the only testable extraterrestrial site where chemical evolution resulted in the production of replicating molecules, as is presumed to be the case for the Earth. Direct tests for the presence of living organisms on Mars became possible as part of the Viking mission, the combined results of which indicated that extant organisms were not present at the two landing sites. Whether "oases" exist anywhere on Mars, capable of supplying liquid water and appropriate energy sources to an extant biota, is dubious but cannot be ruled out, given our present lack of adequate information about Martian micro-environments.

I. INTRODUCTION

The search for life on Mars is, for biologists, part of a deeper question, one that is of fundamental importance to the science of biology. This concerns the central problem of the origin of life. Although general outlines have been advanced to explain the processes by which life may have arisen on the Earth, there are still many significant gaps in our understanding of these events, and this subject remains under intensive investigation by scores of laboratories around the world. It is also recognized that, quite apart from the purely scientific interest in this issue, the question of where and how life originated has been a subject of general human interest throughout recorded history.

In this regard, most biologists currently accept, as a working hypothesis, the theory of chemical evolution (Oparin 1938; Haldane 1929) as the basic framework for an understanding of our origins (cf. Miller and Orgel 1974). Life presumably arose through a long process of evolution going back to the early formation of the "biogenic" elements (i.e., those necessary for terrestrial life) during stellar evolution, through their processing into organic compounds in the interstellar medium, and their further processing during incorporation into planetary nebulae, planetesimals and planets. On Earth, this evolution continued, driven by a variety of readily available energy sources on the primitive planet, leading ultimately to the formation of replicating molecules. This latter event presumably occurred well before 3.5 Gyr ago, since by that time well-developed microbial communities were already widely distributed on the Earth (Schopf 1983). Once a replicating system became established, the process continued through biological evolution, in which the resident organisms were able to adapt to changing environmental conditions over the succeeding billions of years.

Many lines of evidence lend support to this general concept, e.g., astronomical measurements indicating the presence of organic compounds in interstellar space (Greenberg 1982; Irvine 1986; Allamendola et al. 1988; Irvine and Knacke 1989); analysis of meteoritic materials and other solar system objects indicating the presence in them of amino acids and other organic compounds of biological relevance (Hayatsu and Anders 1981; Cronin et al. 1988); and laboratory investigations in which a variety of biologically important organic compounds have been made under conditions meant to simulate the prebiotic terrestrial environment (see, e.g., Oro and Lazcano-Araujo 1981; Miller 1982; Ponnamperuma 1985; Orgel 1987).

The discovery of life on another planet could answer many fundamental questions about the nature and origin of life. Among such questions are: Has life originated more than once? Must life be based on carbon chemistry and, if so, must genes be composed of nucleic acids and enzymes of proteins? Must these polymers be built up out of the familiar nucleotides and amino acids? For questions such as these, the study of extraterrestrial life would prove the best (in some cases the only) means to an answer (Horowitz 1986).

II. MARS AS A TARGET

In any consideration of the question of extant life on Mars, it is important to begin with a little history, because on this question Mars has deluded more than one generation of scientists, and we want our generation to avoid this fate. When the Space Age dawned in 1957, the generally accepted picture of Mars was that of an arid but fundamentally Earth-like planet. At that time,

Mars was considered to have an atmospheric pressure of 85 mbar and ice caps of water ice that waxed and waned with the seasons. Water was transferred seasonally from one pole to the other, and its movement across the planet in the springtime was accompanied by a darkening (often described as a "greening") of large areas of the surface. Spectroscopic evidence suggested the existence of organic matter on the Martian surface.

It was this picture that guided NASA and its advisors in their early planning for the exploration of Mars, and it was still visible years later in the Viking payload. Where did this picture originate? It originated with Percival Lowell who, when he died in 1916, was considered to be the world's leading authority on Mars, thanks to years of observing at the Lowell Observatory, which he had established at Flagstaff, Arizona, for the express purpose of studying Mars. Lowell was convinced, and he convinced the general public, that Mars was the home of a race of intelligent creatures who had built a vast network of irrigation canals which he could see through his telescope. This notion proved to be too much for contemporary astronomers to swallow, and few if any accepted it. As a result, it died with Lowell in 1916, except in science fiction. The rest of Lowellian Mars, including the water, the 85-mbar atmosphere and the vegetation, survived and prospered. Amazingly, these parts of Lowell's dream were repeatedly confirmed, and even embellished, by observers of Mars after 1916.

By the 1950s, Lowell's name was no longer attached to the picture of Mars he had created. The credit for Lowellian Mars, if that is the right phrase, went to those who followed him. Nevertheless, when NASA opened its doors in 1958, the established view of Mars was the fiction that Lowell had invented. This fiction had great influence on the shape of the U.S. planetary program.

The first crack in Lowellian Mars appeared in 1963 with a single infrared spectrogram made with the 100-inch telescope on Mt. Wilson (Kaplan et al. 1964). The spectrogram showed that the Martian atmospheric pressure could not be anything like 85 mbar, but must be much less. The point to be made here is that the 100-inch telescope was installed in 1917, the year after Lowell's death. Allowing twenty years or so for spectroscopy to catch up, it would seem that this observation could have been made around 1940, if anyone had been interested. Thus, it was not inadequate instrumentation that kept Lowellian Mars alive, although this may help explain its birth.

By the time the two Vikings were launched in 1975, Mars had been almost completely "de-Lowellized." Mariners 4, 6, 7 and 9 had revealed a planet quite different from the one that NASA and its advisors had planned on in the beginning. The new Mars had an atmosphere composed mostly of CO_2, with a surface pressure of only 6 mbar. This meant that solar ultraviolet radiation penetrated to the ground almost unfiltered and that liquid water could not exist on the surface. Even saturated salt solutions (even a saturated solution of calcium chloride at its freezing point of $-51°C$) would evaporate on Mars. The seasonal polar caps, it turned out, were frozen carbon dioxide, not water; there was no seasonal transport of water along the planet's surface. The dark areas that are so clear when viewed from Earth, and that were reputed to change color in the spring, were revealed as variable wind streaks on close approach to the planet. Everything was reddish desert. Mars appeared as barren as the Moon.

There was just one slim hope. Mariner 9 had photographed the entire Martian surface, and in doing so discovered that the planet had been geologically active in the past. There were a number of large, extinct volcanoes and, even more surprising, evidence suggesting that water had run on the surface. If liquid water once existed on Mars, then life may have originated there. If so, there was a remote chance that it still survived. It was agreed to continue to emphasize the question of life in the upcoming Viking mission.

As we took stock of what was known about Mars in pre-Viking days, it seemed at least remotely plausible that chemical evolution had also resulted in the production of a replicating system on the early Mars—during the period when this was occurring on the Earth (Pittendrigh et al. 1966; see chapter 35). The uncertainties inherent in this possibility were compounded when considering whether the descendants of such ancient putative organisms might still be present on Mars. That is, could a Martian biota, once having arisen on Mars, have effectively adapted and survived over geologic time? Are there extant organisms on Mars? Despite the uncertainties, especially those that were concerned with possible micro-environments on Mars (Klein 1976), the question of extant biology on Mars was deemed of sufficient interest and importance that the search for extant life became a dominant theme in the Viking mission to that planet.

III. THE VIKING PAYLOAD

Once it was decided to attempt to look for extant organisms on Mars in the Viking mission, we had to face the question of what kinds of measurements were to be made for this purpose. Several strategies for the detection of living organisms had been discussed by numerous authors prior to the Viking mission (see, e.g., Lederberg 1965; Young et al. 1965; Bruch 1966), and the general conclusion from all of these sources was that metabolic experiments were most likely to be of principal importance in the initial searches on that planet. In addition, it was strongly argued that it would be dangerous to rely on any single experimental assay system in the face of the many uncertainties about Mars.

Accordingly, a biological payload for Viking was selected that initially contained four separate elements, each based upon different assumptions about the possible nature of Martian biota (Klein et al. 1972). (Subsequently, during the development of the flight hardware, one of the proposed experimental techniques was dropped in order to reduce complexity and costs. However, one of the remaining experimental concepts was modified so that the remaining hardware could test the hypothesis that was the basis for the dropped experiment.) On the chance that organisms substantially larger and more complex than microbes might be present on Mars (cf. Sagan and Lederberg 1976), the Viking Lander imaging system was to provide panoramic pictures, in color and near infrared, of the local scene over the course of the Martian seasons, at a resolution, near the Landers, of a few mm (Mutch et al. 1972; Patterson et al. 1977). Finally, a sensitive gas chromatograph-mass spectrometer instrument was included for elucidating the nature of carbon-containing compounds that might be present in the Martian surface, and which could yield information relevant to the question of extant life on Mars (Biemann 1974).

IV. THE VIKING RESULTS

The Viking mission yielded an enormous amount of information about Mars, including, of course, data from the various experiments noted above. For the latter, only brief summary discussions are given below, as these have been well documented in the literature.

The Viking Lander Images

Close to 5000 Lander images from the two Viking sites were analyzed for the purpose of obtaining scientific insights into various geologic, climatological and other phenomena on Mars. As part of this process, the Lander imaging team studied pictures for evidence that might suggest the presence of living organisms using a number of criteria. They reported that no structures were seen that could be interpreted as being of biological origins; nor were color changes observed in the panoramic scanning that was conducted over the Martian seasons; nor was movement into or out of the fields of view noted (Levinthal et al. 1977). In short, their conclusions were completely negative with regard to the possible presence of organisms large enough to have been seen by the Lander imaging system. This view has been challenged by Levin and Straat (1978), who have re-examined some of the imaging data, claiming to have noted colored areas on some of the rocks on Mars, which they believe may represent patches of living organisms.

The Viking Gas Chromatograph-Mass Spectrometer Experiments

As indicated above, the Viking Lander carried the Molecular Analysis experiment utilizing a miniaturized combination of a gas chromatograph with a mass spectrometer (GCMS) for the purpose of detecting and identifying organic compounds that may be present at or near the surface of the planet Mars. This technique was chosen because of the high sensitivity of the mass spectrometer. The gas chromatograph provides the capability of separating organic compounds with high efficiency and the elution characteristics of any given compound adds a further identifying parameter.

Knowledge of the type of compounds present in the surface of the planet and their relative and absolute abundances would shed light on the processes that could have led to their formation on the planet or their origin if formed elsewhere. While this experiment was not aimed at the detection of "life," as were the three biology experiments, it had implications for the question of whether there are or ever were living systems present on Mars.

Four scenarios can be invoked in this respect: (1) if there are presently living systems in relatively high abundance, they themselves could be detected on the basis of their constituent organic compounds; (2) if they are present at a much lower level but continuously are formed, grow and die, their organic decomposition products would accumulate at a level that could be detected; (3) if no living systems are present on Mars today but have flourished there in the past (at least at the two Viking landing sites), the Molecular Analysis experiment would be able to detect the organic debris accumulated over thousands or millions of years by constantly growing and dying biota; and (4) organic compounds are now or have been produced in the past by abiotic processes.

On the other hand, if no organic compounds could be detected at all, these four scenarios are ruled out, unless there exists a process that destroys organic compounds at such a rate that they have vanished by now. This process would have to be faster and more efficient, the more recent the compounds were formed. In order to have a reasonable "time resolution," the detection limits of the experiment had to be as low as possible.

The Instrument and Experimental Strategy. In view of the constraints of the Viking mission with respect to weight, volume, power, data storage and transmission, and the desired specificity of the data for the identification of compounds encountered, thermal volatilization and/or pyrolysis of the organic compounds possibly present in the Martian surface material was chosen. A hydrogen stream was used to carry the products into the gas chromatograph, the effluent of which was monitored by a repetitively (every 10 s) scanning mass spectrometer. The gas chromatograph was temperature programmed from 50 to 200 °C, and the mass range of the spectrometer was from M/Z 10–220 with a resolution of about 1:200. With these capabilities, any organic compound that could be vaporized (or pyrolyzed) out of a soil sample at Martian atmospheric pressure could be passed through the chromatograph and detected by the mass spectrometer.

Practical constraints limited the experiment to the analysis of three approximately 0.1 g soil samples, which could be heated successively to 50, 200, 350, and 500 °C (pre-programmed or selected by commands from Earth) to expel or pyrolyze the organic material. Because, in each of the GCMS instruments on both Viking Landers, one of the three sample ovens (2 mm ID, 19 mm long) was believed to be not operative, only two surface samples were analyzed at each landing site and for only one of them were all four temperature levels applied.

Results and Discussion. As has been reported in great detail during and after the primary Viking mission (Biemann et al. 1976*b*; Biemann 1979) no organic compounds were found at either of the two landing sites (Chryse Planitia and Utopia Planitia) at detection limits (for organic compounds with 3 or more carbon atoms) ranging from 0.0015 parts per billion (ppb, 10^{-9}) for naphthalene to 80 ppb of benzene, with most detection limits in the range of <1 ppb. These detection limits were governed by the noise in the mass chromatograms for the *M/Z* (molecular/atomic mass) value of the most abundant ion of the particular compound in the region of its elution time from the gas chromatograph.

For some compounds (such as acetone, benzene and toluene) this limit was unusually high because of the high background signal, due to some contamination of the instrument with traces of remaining solvents that had been used in cleaning the instrument and sampling system before installation in the Viking Lander. While this was at first thought to be unfortunate, it turned out to be an advantage because the evolution of these compounds upon heating one of the sample ovens during interplanetary flight on each of the two spacecraft, and again with each sample on the surface of Mars, demonstrated two extremely important facts: (1) the instruments worked perfectly during each experiment conducted on the planet; and (2) the compounds reproducibly detected were terrestrial contaminants.

The absence of organic compounds at these two very distant (from each other) sites demonstrated that there is presently neither biological nor abiological synthesis of organic compounds occurring, although in the latter case destruction could be as fast as formation. It also rules out carbonaceous remnants of now extinct biological or abiological processes in the surface material of these two sites.

These findings had some obvious impact on the interpretation of the results of the three biology experiments (Klein 1977), all of which produced some signals that initially might be interpreted as positive. Those of the Gas Exchange (GEx) experiment (Oyama and Berdahl 1977) and of the Pyrolytic Release (PR) experiment (Horowitz et al. 1977) were subsequently attributed to nonbiological events. Only the data from the Labeled Release (LR) experiment (Levin and Straat 1977) are still claimed to be indicative of viable organisms in the samples analyzed.

Without arguing the merits and the nature of the data generated by the Labeled Release (LR) experiment (cf. Klein [1978] for a discussion of this), its relationship to the GCMS data needs to be examined in the light of recent re-interpretation of the LR data (Levin and Straat 1981). It is now argued that the fact that the GCMS experiments did not reveal detectable amounts of organic matter does not rule out the co-existence of viable organisms in the surface samples analyzed by both experiments. The key argument here is that a terrestrial soil sample (#726), collected in Antarctica (Cameron 1971) and analyzed by the laboratory test instruments of both the GCMS and LR experiments, gave a "weak, but clearly positive response" in the latter (Levin and Straat 1981) but did not reveal any organic compounds in the former (other than traces of chlorocarbon believed to be derived from the original sampling bags that were lined with polyvinyl chloride) (Biemann and Lavoie 1979). It should be noted that this sample, after it was originally collected, was found to be sterile (Cameron 1971), i.e., "the microbial count was below the detectable limit of approximately one organism per gram in all media tested" (Horowitz et al. 1969). This seems to imply that the weak response in the LR experiment must be due to factors other than microbial activity. Another possibility that cannot be excluded is that sample #726 may have become contaminated in the years intervening between the initial testing and those carried out by Levin and Straat.

The second argument, that the GCMS did not detect organic compounds in this soil sample (even though it was found by Cameron to contain 0.03% organic carbon), and that this instrument therefore was not sensitive enough to have any implications for the biology experiments, is equally fallacious. Cameron (1971) and Horowitz et al. (1969) used the "Allison method" (Allison 1960) to measure "inorganic, organic and total carbon" in this sample. In this regard, it should be noted that the term "organic carbon" was used by Allison very loosely, actually referring to carbon that is not carbonate (inorganic) but that can be oxidized with strong oxidants (heating with a mixture of potassium dichromate, conc. sulfuric and phosphoric acid). Horowitz et al. (1969) noted that all other tests on the sterile antarctic samples indicated that it is *highly improbable that the "organic" (quotation marks added) carbon could be present in any form except as elemental carbon.* For soil #726, it is specifically stated in a footnote to the table listing the result (Cameron 1971) that the value listed in the column "organic carbon" is due to "anthracite coal." Thus, the results from the experiments carried out on antarctic soil #726 do not invalidate the sensitivity of the Viking GCMS and are not relevant to the interpretation of the LR experiments carried out on Mars during the Viking mission.

Implications for Future Experiments. The Viking 1976 Landers were limited to the collection of loose surface material within the limited range (3 m) of the sample acquisition arm and to a depth of ~ 10 cm. During the primary mission a degree of confidence had been acquired in the remote control of the transmission from Earth of complex command sequences for the sampler arm. Thus it was possible to push away one of the nearby rocks ("Badger" rock), which had been protected from incident ultra-violet irradiation for the last Myr and sample the loose surface material underneath (see chapter 21 and nomenclature appendix). Still no organic compounds were detected in this sample.

There may, of course, be niches on the planet where organic compounds may be more protected from destruction, could have been concentrated over millions of years under certain conditions, or may indeed be the products of contemporary or past living systems. Drilling down to the permafrost region would be desirable to sample a region where organic materials of whatever origin might have accumulated over time by a variety of processes. Finally, the near-polar regions where there is an annual cycle of condensation and re-evaporation of water ice and solid carbon dioxide, should be sampled. Such processes could enrich the organic compounds in a narrow region at high latitude by their annual co-precipitation from the atmosphere with water and carbon dioxide, to be left behind on the surface of dust particles when the water and carbon dioxide re-evaporate in the Martian summer (see chapters 27–29). This strategy had been considered during the planning stages of the Viking mission in the early 1970s but was eliminated as being too risky because of insufficient information available at the time about the terrain and the more limited communication capabilities between Earth and these regions of Mars. The complete and redundant Viking Orbiter data which provided extensive high-resolution mapping of the polar regions now eliminates or substantially reduces these constraints.

A roving vehicle with an operating radius of a few hundred km would be a basic requirement of a future unmanned Mars Lander. It remains to be discussed whether remote-controlled instrumentation à la Viking 1976 (but improved by the recent advances in robotics, computer control and telemetry) or sample return to Earth should be chosen. In any event, many generations of unmanned investigations of the Red Planet should precede any of the presently discussed manned missions to Mars, which would have to expend most of the resources for the purpose of maintaining the health and safety of the crew rather than for the acquisition of scientific data.

The Viking Biological Experiments

Many detailed reports concerning the Viking biological experiments are available in the literature (cf., for example, *J. Geophys. Research,* Vol. 82, number 28, 1977; *J. Mol. Evolution,* Vol. 14, numbers 1, 2, 3, 1979), as are summaries and overall interpretations of the data that were obtained (cf., Horowitz 1986; Klein 1979). These are not reviewed here.

What became clear even during the Viking mission was that if the GCMS results were correct (and there was reason to believe that this was the case), the three biological experiments had essentially lost their original purpose. With no detectable trace of organic matter in the surface material, there was no possibility of finding extant life at the two landing sites. The puzzle was that the biological experiments were showing chemical activity of some kind in the Martian soil. The most interesting activity was coming from the two experiments (GEx and LR) that were Lowellian in concept, i.e., experiments that were designed in the era when Mars was thought to have water on its surface.

Both experiments employed aqueous media to moisten Martian soil samples with solutions of organic compounds. As open bodies of liquid water cannot exist at the temperature and pressure of the Martian surface, these solutions had to be heated and pressurized to values well above Martian to prevent their freezing or boiling. If there had been life on Mars (a form of life adapted to the actual Martian environment), it is questionable whether it could have survived exposure to such highly non-Martian conditions as those it would have been subjected to in these experiments. As it turned out, however, the experiments were highly informative, because they confirmed the presence in the surface of very reactive, oxidizing species resembling those predicted earlier by Hunten (1974). This was manifested by the production of oxygen on contact of the Martian soil with water vapor (GEx), and by the production of carbon dioxide when acqueous solutions of organic compounds were added to Martian samples (LR).

The third biological experiment (PR) was performed under actual Martian conditions, or as close to them as it was possible to achieve in the Viking spacecraft. Water was not needed for this experiment, but could be provided on command. The question in this experiment was whether any synthesis of organic matter could be detected in Martian surface material under Martian conditions, starting from CO and CO_2, two gases present in the Martian atmosphere. To our surprise, trace amounts of carbon-containing substances were formed. Although the quantities were very small, even these were unexpected in view of the precautions we had taken to avoid nonbiological responses. Tests showed, however, that the response could not be biological because it was not destroyed by heat. This conclusion was subsequently supported by laboratory experiments showing similar results when iron-rich minerals such as maghemite were exposed to the same two gases. The Martian surface is, as Viking demonstrated, rich in iron-containing minerals.

The Viking findings established that there is no life at the two landing sites, Chryse and Utopia. Although the two sites are 25° apart in latitude and on opposite sides of the planet, they were found to be very similar in their surface chemistry. This similarity reflects the influence of global forces such as extreme dryness, low atmospheric pressure, short-wavelength ultraviolet flux, and planet-wide dust storms in shaping the Martian environment. These same forces virtually guarantee that the Martian surface is lifeless everywhere. We sampled only two sites, and some have argued that one cannot generalize from these to the entire planet. This conclusion is based on more than these samples, however. It has to be remembered that from 1963, the year the modern study of Mars began, until the landings in 1976, Mars was studied intensively from flyby and orbiting spacecraft and from Earth-based observatories. The conclusion is based on *all* these observations. To some of us, the evidence against life on Mars was already strong, even before the Viking landings (Horowitz 1986).

V. DISCUSSION

The general conclusions that we have drawn from these data have been the subject of some debate; opinions on the issue of extant organisms on Mars range from those who feel that the information already accumulated by Viking essentially rules out the possibility of living organisms anywhere on Mars (see, e.g., Mazur et al. 1978; Horowitz 1986), through others who claim that the evidence is not yet conclusive on this issue (see, e.g., Imshenetskii and Mursakov 1977; Aksyonov 1979), to those who feel that there may well be active biology on Mars (see, e.g., Chandler 1979; Feinberg and Shapiro 1980; Adelman 1986; Levin and Straat 1988; Ivanov 1988*b*).

For example, the question has been raised whether some unusual microbial habitats recently discovered on our own planet may have counterparts on Mars. One of these habitats is found in the dry valleys of Antarctica, probably the most Mars-like region of Earth, where the dryness and low temperature limit the population to relatively small numbers of micro-organisms living in the soil. In the dryest parts of the valleys, some 10 to 15% of soil samples are actually sterile. Some years ago, Friedman and colleagues reported finding dense growths of lichens and bacteria inside certain kinds of rocks in the valleys (Friedmann and Ocampo 1976; for review, see Friedmann 1982). The rocks in question are translucent and porous, and the growth occurs in a narrow zone a few mm beneath the rock surface. Friedmann found that the rocks are warmed sufficiently by the summer Sun to melt snow on their north-facing surfaces. The melt water is absorbed by the rock, and a sheltered, moist environment is thus created.

Some hopeful biologists have suggested that similar environments may exist on Mars. This is exceedingly improbable. The difficulty is the same one that stands in the way of all efforts to find a biological habitat on Mars: surface snow cannot melt, and liquid water cannot exist on the Martian surface owing to the low water-vapor content of the atmosphere, combined with the low total atmospheric pressure and the high CO_2 content. In this and in many other ways, there is no real resemblance between the Martian and Antarctic deserts. To name one other difference, the Antarctic desert is within but a few miles of the ocean. The Antarctic, in fact, is just marginally unfit for life. It is a place where a small climatic variation like the one that gives rise to the endolithic habitats can make the difference between a livable environment and a nonlivable one.

It is worth pointing out that although the Antarctic desert is not really Mars-like, we learned an important lesson in planetary biology there. We learned that the adaptability of living organisms is very limited where the need for water is concerned. The discovery of sterile and nearly sterile soils in this desert, which has been exposed for tens of thousands of years to a windborne influx of contaminants and genetic variants from a vast pool of microbial life outside the area, was a revelation. It demonstrated in an unmistakable way the importance of water for life, and it caused some of us to wonder for the first time whether it was reasonable to expect to find life on a planet as dry as Mars.

Another habitat for consideration has recently been invoked by Ivanov (1988*b*). This habitat is the one associated with the vents of hot springs on the floor of the Pacific Ocean. As is now well known, large populations of animals and bacteria live around these vents. Most of these species are not relevant to the problem we are discussing because they are aerobic (that is, they require oxygen) and the source of that oxygen is photosynthesis carried out by green plants and algae growing at the Earth's surface. We are interested in the possibility of organisms that do not depend on life at the surface, but that can live on resources available to them from within the planet.

Just such an organism has been identified among the bacteria growing around one of the deep-sea hydrothermal vents in the Pacific Ocean (Jones et al. 1983). The organism in question is a methanogen, one that produces methane from hydrogen gas and CO_2. This reaction yields sufficient energy to support growth of the bacteria, and this occurs in the dark and without oxygen. (Similar nonphotosynthetic, anaerobic processes are known in other bacteria that can utilize hydrogen to reduce sulfates or oxidized iron in energy-yielding reactions. Examples of such chemo-autotrophic organisms are discussed by Clark [1979]). If Mars has a subsurface ocean and internal sources of hydrogen and carbon dioxide, and if other conditions are satisfactory, it would be possible for bacteria like these to survive there. Of the necessary conditions, it would seem that the most difficult in the case of Mars would be a source of hydrogen since this implies a level of geologic activity that seems inconsistent with the planet's quiescent appearance. In this regard, it should be noted that methane has not been detected on Mars, and that its upper limit has been estimated to be 0.02 ppm (Maguire 1977), this despite the fact that its average lifetime in the Martian atmosphere is 300 yr (Hunten 1979*b*).

Clearly, from a biological point of view, a metabolic scenario involving anaerobic, nonphotosynthetic, fixation of carbon dioxide in which the organisms derive their energy from the reduction of CO_2, iron, or sufates by hydrogen, is feasible. But, whether there are any sites on Mars where microenvironments of this type exist, or are even possible, is open to question.

Such a site would require at least the intermittent presence of liquid water, as well as the availability of CO_2 and (internal?) sources of reductants. From what is known about Mars, there is no evidence that readily supports this idea of a "deep ocean" habitat on Mars, although such a possibility has not been ruled out.

35. THE POSSIBILITY OF LIFE ON MARS DURING A WATER-RICH PAST

C. P. McKAY, R. L. MANCINELLI, C. R. STOKER
NASA Ames Research Center

and

R. A. WHARTON, JR.
University of Nevada

Geomorphological evidence for past liquid water on Mars implies an early, warmer, epoch. In this review we compare this early warm environment to the first Gyr of Earth's history, the time within which we know life originated. We consider the key question about early Mars from the biological standpoint. How long was liquid water present? The range of answers encompasses the time interval for the origin of life on Earth. We use studies of early life on Earth as a guide, albeit a limited one, to the possible forms of evidence for past life on Mars. Presumptive evidence for microbial life on early Earth are stromatolites, layered deposits produced by microorganisms binding and trapping sediment. A search for fossils might be fruitful at sites on Mars that contained standing bodies of water over long periods of time. The ice-covered lakes of the dry valleys of Antarctica may provide analogs to the ultimate lakes on Mars as the surface pressure fell with a concomitant decrease in surface temperatures.

I. INTRODUCTION: THE NATURE OF LIFE

While the results of the spacecraft exploration of Mars have diminished hopes of finding living organisms on its surface, they have also shown that, in the distant past, Mars was a very different place than it is now. The early environment of Mars may have been similar to that on the early Earth and both planets may have enjoyed conditions that were conducive to the origin of life—life that may have long since become extinct on Mars.

On Earth, evidence from stromatolites and microfossils suggests that life was present by 3.5 Gyr ago (Schopf 1983; Schopf and Packer 1987). Pushing the origin of life further back into the geological past is difficult because of the scarcity of unaltered rocks older than 3.5 Gyr. The oldest known rocks of sedimentary origin, which date back to 3.8 Gyr old, are found in Isua (Greenland). Although these rocks are partially metamorphized, they contain organic material that might have resulted from biological activity, suggesting an even earlier origin for life (Schidlowski 1988). The surface of the Earth, i.e., the crystallization of any magma ocean and the formation of a crust, may date from about 4.2 Gyr ago (Compston and Pidgeon 1986). This suggests that the time interval for the origin of life on Earth was between approximately 4.2 and 3.5 Gyr ago or about 700 Myr. Recently, Maher and Stevenson (1988) have pointed out that large objects impacting the Earth during the late phase of planetary accretion would have sterilized the Earth's surface and forced chemical evolution to restart. Such planet-sterilizing impacts may have frustrated the origin of life on Earth as late as 3.7 Gyr ago, implying a further reduction in the times available for the origin of life (cf. Sleep et al. 1989). We conclude that 700 Myr (4.2–3.5 Gyr) may be an upper limit to the time required for the origin of life on an Earth-like planet and a much shorter time period may be adequate.

The standard theory for the origin of life on Earth was outlined by Oparin (1938) and Haldane (1928) (for a modern review see Miller and Orgel 1974). This theory posits the abiological production of organic matter followed by the self-assembly, or chemical evolution, of this material into the first living organism with subsequent biological evolution. The synthesis of living organisms from abiotic organic matter has not been accomplished in the laboratory; nor is there any direct evidence in the fossil record of chemical evolution on the early Earth, or the steps leading to the origin of life. At least two methods have been suggested for the abiotic production of organic material on the early Earth: *in situ* production of organics in an initially reducing atmosphere (as in the classic experiments of Miller 1953); or by the importation of organics via comets from the outer solar system (Oró 1961).

A corollary of the standard theory for the origin of life on Earth is that, on a planet with a similar environment, similar events would ensue, also leading to life. In a general sense the first organisms might be biochemically similar to the first organisms on Earth. This similarity probably would extend to the use of carbon and liquid water, but may not extend to exactly the same basic 20 amino acids, 5 nucleotide bases, and 8 nucleotides that form life on Earth. More specific analysis of the general case is not possible because, on the Earth, we really have only one example of life. It is therefore difficult to separate the fundamental traits that would appear in any life that originated on an Earth-like planet from the peculiarities of historical precedent written into terrestrial biochemistry. Virtually all studies and proposals for possible Martian biology, either extant or extinct, have worked from a basis of terrestrial biochemistry (see, e.g., Pittendrigh et al. 1966; Klein 1979; Clark 1979), and we do also. In this chapter, we make four basic assumptions about early Martian life, to wit: (1) liquid water was required; (2) CO_2 was used as a source of carbon; (3) sunlight was the principal energy source; and (4) fixed nitrogen was required.

II. EARLY MARS: THE QUESTION OF LIQUID WATER

From a biological perspective, the most important information returned from spacecraft exploration of Mars may be the geologic evidence that liquid water was abundant on the Martian surface at some time in the past. Water is the quintessence of life—all life on Earth requires the presence of liquid water. Liquid water is the only compound for which this statement is true; e.g., O_2, CO_2 and NH_3 are not universally required. The lower limit of water activity for growth of most bacteria is 0.95, extreme halophiles can tolerate 0.75 and only certain eukaryotes can withstand down to 0.6 (Mazur 1980; Kushner 1981).

The geologic determination that there was liquid water on the surface of Mars for an extended period during its early history is a pivotal point upon which the discussion of a possible past Martian biota rests. Given liquid water, there is no requirement for warm temperatures or dense atmospheres, *per se,* in order to support life on early Mars.

A. Water on Early Mars

The outflow channels and the valley networks provide two complementary pieces of information about water on Mars (see chapter 15). The size of the outflow channels (such as shown in Fig. 1 in chapter 15) suggests that large-scale fluvial processes were involved, implying that large amounts of liquid water flowed over the surface during some interval or intervals of time. In addition, from the morphology and size of the valley networks it is clear that liquid water must have been fairly stable at the surface (Wallace and Sagan 1979; Carr 1983). The morphology of some of the valley networks (see Fig. 4 in chapter 15) is indicative of sapping of groundwater or melting of ice (see, e.g., Carr and Clow 1981; Carr 1981; Baker 1982). However, some networks (see Fig. 5 in chapter 15) have clearly defined dendritic systems that may have been caused by precipitation and runoff (Masursky et al. 1977). Estimates for the total amount of water on Mars vary from 6 to 500 m spread over the planet (for a discussion, see Carr 1986; McKay and Stoker 1989; chapters 6 and 25), but the general conclusion is that there were aquatic environments on early Mars.

The presence of liquid water has been used to infer atmospheric conditions based on the assumption that surface temperatures must have been near or above freezing when the runoff channels formed (Pollack 1979; Pollack and Yung 1980; Cess et al. 1980; Hoffert et al. 1981; Postawko and Kuhn 1986; Pollack et al. 1987; chapter 32). The major component of the early Martian atmosphere is considered to have been CO_2, as it is in the present Martian atmosphere (see, e.g., Pollack et al. 1987). Similarly, it is believed to have been a major constituent in the early atmosphere of the Earth (see, e.g., Walker 1977,1985; Holland 1984). Increased CO_2 in the atmosphere of the Earth, compared to the present level, could have provided the greenhouse effect (Moroz and Mukhin 1977; Owen et al. 1979; Kasting and Ackerman 1986) required to keep the temperature above freezing even though the early Sun was 30% less luminous than at present (Sagan and Mullen 1972). Pollack et al. (1987) suggest that from 1 to 5 bar CO_2 were required to raise the mean global temperature of early Mars above freezing and that 0.75 bar would be required to raise only the subsolar point at perihelion above freezing

Mars could have lost an initial dense atmosphere of CO_2 to carbonate formation (Kahn 1985; Pollack et al. 1987; chapter 32). The time scale for eliminating atmospheric CO_2 by carbonate formation on early Mars is estimated to be a few times 10 Myr. Thus, in the absence of recycling, the lifetime of a thick early atmosphere would have been very short indeed. It is also possible that Mars lost a considerable fraction of its early atmosphere as a result of erosion by high-velocity impacts (Melosh and Vickery 1989).

To argue for the origin of life on Mars by analogy with the origin of life on Earth, the critical determinant is the length of time liquid water persisted on early Mars compared to the time required for life to have originated on Earth. Pollack et al. (1987) suggest that intense early volcanism on Mars could have recycled carbonate rocks into atmospheric CO_2, thereby maintaining warm moist conditions for as long as 1 Gyr, for optimistic values of the early heat flow and total CO_2 inventory. Carr (1989) has suggested that meteoritic impacts also would have recycled carbonate on early Mars, which ties the formation of dendritic channels with the late phase of planetary accretion. Carr (1989) has suggested that the early thick Martian atmosphere was short lived based upon the low levels of erosion and the absence of infilling of old post-bombardment surfaces and craters. This is an important geomorphological argument *against* a thick early Martian atmosphere lasting for billions of years. However, there is some evidence of episodic channel formation that may have resulted from later episodes of impact recycling of volatiles (see chapters 15 and 32).

B. Nitrogen On Early Mars

Because CO_2 and N_2 are believed to have been carried to the inner planets by the same mechanism (see, e.g., Pollack and Black 1979), then, if there was a dense CO_2 atmosphere on early Mars, this should have been associated with a correspondingly large amount of N_2 with values from $\sim$ 10 mbar to 300 mbar (for an overview see McKay and Stoker 1989). Nitrogen is a key biological element used in the formation of many essential biochemical compounds (e.g., DNA, RNA and amino acids). Although all organisms require nitrogen, N_2 (the predominant form of nitrogen) is not usable by the vast majority of organisms; it must first be transformed or fixed into a more usable form (e.g., NH_3, NH_4^+, NO_x and organically bound N that is hydrolyzable to NH_3, or NH_4^+). In many ecosystems on Earth, fixed nitrogen is the primary nutrient limiting growth. Because of its unique chemistry, and the universality of the importance of nitrogen to life, we have assumed that a possible Martian biota would also require fixed nitrogen.

As previously mentioned, a key step in the nitrogen cycle is fixation, i.e., the transformation of atmospheric N_2 into a useful form. This can be accomplished by abiotic processes such as the thermal shock of a lightning bolt, or by biological fixation (for a review see Mancinelli and McKay 1988). On Earth, abiotic processes of N_2 fixation are insignificant when compared to the amount of N_2 fixed biologically (see, e.g., Walker 1977). Biological fixation is the transformation of N_2 into NH_2 by the enzyme *nitrogenase,* and has only been conclusively demonstrated to occur in prokaryotes. Nitrogenase occurs as a Mo-Fe or a Va-Fe protein (Robson et al. 1986), and appears to be similar in all organisms that possess it, indicating either a recent origin or preservation over evolutionary time (see, e.g., Postgate 1982; Hennecke et al. 1985).

The evolution of the nitrogen cycle on Earth is not fully understood (see, e.g., Mancinelli and McKay 1988). Abiotic factors influenced certain metabolic pathways for the transformation of certain nitrogen compounds above others, which may have differed on early Earth and Mars. As a consequence, direct extrapolation of terrestrial physiology to Mars may not be applicable. If, however, life on Mars evolved to a point that an appreciable biomass was supported, then it is possible that biological nitrogen fixation played a dominant role compared to abiotic fixation. The fact that all terrestrial nitrogen-fixing microorganisms use basically the same enzyme to fix nitrogen, and incur high energy costs to do so, may suggest that there is no practical alternative, and that some type of biological fixation would be inevitable for a possible Martian biota. Thus, the low abundance of N_2 on early Mars may have played a critical role in the possible evolution of life. Recently, it has been shown that certain nitrogen fixing microorganisms (*Azotobacter vinelandii* and *Azomonas agilis*) can fix dinitrogen down to a partial pressure of 5 mbar (Klingler et al. 1989), a pressure lower than that thought to have existed on early Mars, but significantly higher than the level of N_2 in the present atmosphere. Thus, the declining concentration of N_2 on Mars over time may have biological implications if, as a result of its decline, it became unavailable to nitrogen fixing organisms. If there were biological systems on early Mars that utilized nitrogen, it is likely that they played a major role in determining the distribution of nitrogen. Conversely abiotic factors may have promoted the development of certain biological pathways and not others (see, e.g., Mancinelli and McKay 1988). This requires that models of the nitrogen cycle on Mars consider both biotic and abiotic effects.

The variation in the ratio of the stable isotopes of nitrogen ($^{15}N/^{14}N$) on Mars is interesting because it is believed that the isotopic ratio of atmospheric nitrogen has varied monotonically with time due to atmospheric escape, enriching the heavier isotope (McElroy et al. 1977; Fox and Dalgarno 1983). Hence, the N isotopic ratio of organic material or nitrates in the Martian sediments may reflect their time of incorporation.

C. Ice-Covered Lakes on Early Mars

McKay and Nedell (1988) have pointed out that if atmospheric CO_2 was deposited as carbonates, lowering the pressure and temperature, any liquid water present would have become ice-covered. On Earth, ice-covered lakes are found in the Antarctic dry valleys, and they maintain liquid water under a perennial ice cover despite a mean annual temperature of $-20°C$. These Antarctic lakes may be analogs for lakes on early Mars (McKay et al. 1985). One biologically important feature of an ice-covered lake is its capacity to provide a thermally buffered habitat in the underice water column despite the cold external temperature (McKay et al. 1985; Clow et al. 1988*a*). Consequently, microorganisms are capable of living and growing in these thermally stable lakes in regions that are otherwise bereft of life (Parker et al. 1982). The other terrestrial ecosystem in the Antarctic dry valleys are the cryptoendolithic microorganisms studied by Friedmann and coworkers (Friedmann 1982; Friedmann et al. 1987). By analogy to the Antarctic lakes, it is possible that ice-covered lakes on early Mars provided a relatively warm, liquid water environment for early Martian biota long after surface temperatures fell below freezing (McKay et al. 1985; McKay and Nedell 1988). Another feature of perenially ice-covered lakes is their ability to concentrate atmospheric gases in the water column. For example, Lake Hoare, Antarctica has about 3 times the oxygen and 1.5 times the nitrogen that would be in equilibrium with the atmosphere (Wharton et al. 1986, 1987). Thus, as climatic conditions deteriorated on early Mars, ice-covered lakes could have been one of life's final retreats, providing both thermal stability against a cooling external environment and increased concentrations of CO_2 and N_2 against a thinning atmosphere.

McKay and Davis (1991) have estimated the duration of such ice-covered liquid water habitats on early Mars. They conclude that if there was a source of ice to provide melt water, liquid water habitats could have been maintained under relatively thin ice covers for up to 700 Myr *after* mean global temperatures fell below the freezing point. Their analysis suggests that during this time Mars may have had a thick ($>$ 1 bar), but very cold and dry atmosphere. Such a cold, dry atmosphere may be consistent with the inferred weathering rates. Furthermore, episodic flows during this time could also be an explanation for the evidence of fairly late valley network formation (Gulick and Baker 1989).

III. THE FOSSIL RECORD ON MARS

The remnents of any biology that may have existed on Mars can be explored using many of the techniques and approaches that have been developed in the study of the Earth's earliest biosphere (Schopf 1983). Finding evidence of past life on Mars will involve searching for many things (McKay 1986) including: direct traces of life such as microfossils, organically preserved cellular material, altered organic material, certain morphological microstructures and chemical discontinuities associated with life, isotopic signatures due to biochemical reactions, and inorganic mineral deposits attributable to biomineralization.

On the Earth, one way in which biological material is preserved as fossils is by incorporation in aquatic sedimentary deposits and precipitates. As suggested by McKay and Nedell (1988), lakes on Mars could have been the site of considerable carbonate deposition. Such locations may prove to be good places to look for microfossils and possibly organically preserved material (McKay 1986; Nedell et al. 1987; McKay and Stoker 1989). In this context, it is important to note that in many aquatic environments the formation of carbonates is the direct result of, and an indicator of, biological activity.

A. Stromatolites on Mars

One of the most important forms of fossil evidence for microbial life on early Earth are stromatolites (Schopf 1983; Awramik 1989; Cloud 1989). Stromatolites are formed when layers of microbial organisms at the shallow bottom of a lake or tide pool are covered with sediment or precipitating salts (e.g., carbonate). Often this sedimentation occurs on a cyclic basis, tidally or seasonally. The layer of sediment obscures the sunlight and so the photosynthetic organisms that form the basis of the community must migrate toward the light to survive. This phototactic response is not necessarily directed upward and often imparts a characteristic domed shape to biogenic stromatolites. As the photosynthetic organisms migrate, the rest of the community moves toward the light as well. The cellular and detrital material left behind combines with the inorganic sediments to form a microbial mat. In active microbial bottom communities, undisturbed by the presence of larger organisms, these mats can take on a tissue-like consistency. Over time, the compaction of these sediments form stromatolites. On the Earth, stromatolites have been found that are 3.5 Gyr old (see Fig. 1) and thus they are the earliest evidence of life on the Earth.

Fig. 1. A stromatolite from the 3.5 Gyr old Warrawoona Group, Western Australia. Stromatolites are commonly organo-sedimentary structures formed by the lithification of bottom sediments containing microorganisms. They are presumptive indications of past microbial activity. Their formation is typically associated with the phototactic properties of photosynthetic bacteria and algae and might be expected to exist on Mars as well, if life had evolved there. (Photo courtesy S. M. Awramik.)

In present ecosystems, metazoan organisms graze on the microbial communities that form stromatolites and thus contemporary stromatolites form typically in unusual environments such as hypersaline ponds (Javor and Castenholtz 1981) and ice-covered Antarctic lakes (Parker et al. 1981; Wharton et al. 1982) where grazers cannot survive. On the early Earth, disturbance by metazoans would not have been a factor, and well-developed microbial mats, and hence stromatolites, developed frequently. If a similar microbial system evolved on Mars, their remains might be abundant on Mars today.

Stromatolites are an important form of fossil evidence of life because they form macroscopic structures that could be found by geologic exploration of Mars (see, e.g., Awramik 1989; Bauld 1989). It is therefore possible that a search for stromatolites near the shores of an ancient Martian lake or bay could be conducted as part of robotic missions to Mars.

Expecting microbial communities to have formed stromatolites on Mars is not entirely misplaced geocentricism. The properties of a microbial mat community that result in stromatolite formation need only be those associated with photosynthetic uptake of CO_2. These are broad ecological properties that we expect to hold on Mars even if the details of the biochemistry and community structure of Martian microbial mats were quite alien compared to their terrestrial counterparts. Within stromatolites, trace microfossils can sometimes be found (Schopf 1983; Cloud 1989), and indeed must be found to indicate unambiguously a biological origin.

B. Carbon Isotopes of Martian Organic Matter

While the detection of trace microfossils and/or stromatolites would be persuasive evidence for past life, it is possible that a future mission to Mars would find only organic material of uncertain origin. No organic material was found in the soils of Mars at either of the Viking Lander sites (Biemann et al. 1977; Biemann 1979; chapter 34). This observation and the other results of the Viking Biology experiments led to the suggestion that there are oxidants in the Martian soil that have destroyed the organic material to some undetermined depth (Klein 1978, 1979). However, organics have been reported in a meteorite believed to have originated from Mars (Wright et al. 1989). Hence, remanent organic material from an early biological period may exist below the oxidized layer of soil. If such material were recovered, would it be possible to assign a biological origin to it? There are ways in which organic material can be produced nonbiologically on Mars, such as importation by meteor infall and by photochemistry (Biemann et al. 1977; Biemann 1979). On the Earth, a powerful technique has been developed that can be applied to this question. This is based upon the fact that biologically produced organic matter on the Earth usually has a slightly different ratio of the stable carbon isotopes than the inorganic reservoir of carbon. Biological systems preferentially select the lighter isotope ^{12}C over ^{13}C by about 2% (see, e.g., O'Leary 1981). As shown in Fig. 2, the isotopic composition of the organic material in the Earth's sediments has remained roughly constant over geologic time, despite significant changes in the atmospheric composition, other aspects of the environment, and the biota itself.

In studies of the Earth's earliest biosphere, this isotopic shift provides useful evidence for determining if organic matter is of biological origin. An interesting test of the applicability of this method when no other data are available is in the suggestion by Shidlowski (1988), who argues that the pres-

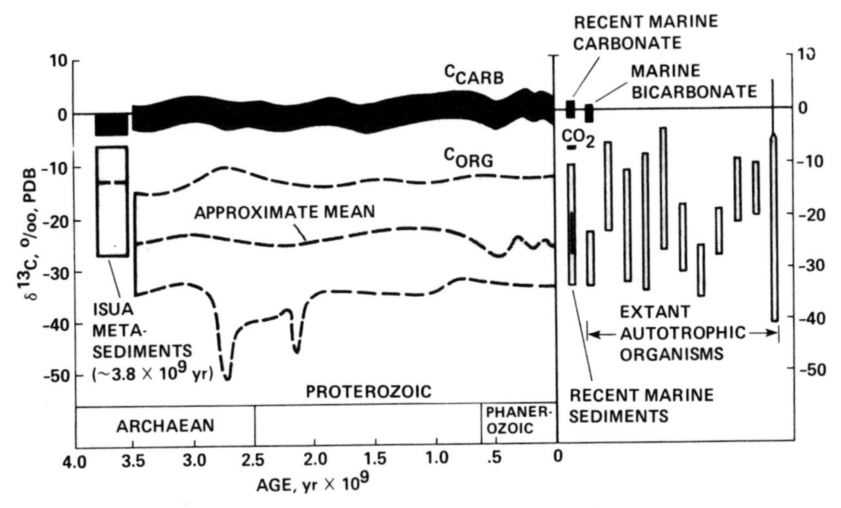

Fig. 2. Carbon isotopic values for carbonate and organic sediments on Earth for the last 3.8 Gyr. (adapted from Schidlowski 1988). Carbon isotope values are reported in units of ‰, which corresponds to the enrichment of the heavy isotope ^{13}C in parts per thousand, when compared to a standard Peedee belemnite. Photosynthetic organisms preferentially take up ^{12}C over ^{13}C resulting in an isotopic shift of about −25‰. This isotopic shift has characterized organic sediments throughout the Earth's history despite significant changes in the Earth's environment and biota. On Mars a narrowly defined shift between organic material and carbonates, possibly at a level different from that of the terrestrial value, could be indicative of biological origin.

ence of the characteristic isotopic shift due to biological activity seen in the Isua metasediments is presumptive proof of biological origin (Fig. 2).

It is uncertain that Martian enzymes which fix CO_2 would have the same characteristic isotopic shift observed in their terrestrial counterparts. However, CO_2 was probably abundant throughout Mars' history, and it is the nature of many enzymes to discriminate between isotopes in the presence of a ready supply of substrate. This has led to the suggestion that a narrowly defined isotopic shift between organic sediments and inorganic sediments, such as carbonates, would be indicative of biological origin, whereas a broad range of isotopic ratios would not allow one to discriminate between abiotic and biotic carbon fixation (Rothschild and DesMarais 1989).

IV. DISCUSSION

If life did evolve on Mars, what are the factors that led to its extinction and is it a foregone conclusion that current Mars is lifeless? The critical parameter limiting the presence of life on Mars today is the apparent absence of liquid water. Surface temperatures on Mars exceed 273 K only during summer at low latitudes (Martin 1981)—latitudes which are apparently devoid of water. Moreover, the present atmospheric pressure on Mars is close to the triple point vapor pressure of water, which is the minimum overburden pressure required to allow water to form in the liquid state. Thus, on present Mars, water cannot form in the liquid state except perhaps briefly under special circumstances (Ingersoll 1970; Kahn 1985). As the atmospheric pressure declined from an initial state, due presumably to the formation of carbonates, and Mars cooled, the sites where liquid water occurred must have become progressively more scarce. Thus, as the habitats disappeared, so would have life.

Another factor that may have caused further stress to life on Mars, even during an early clement period, is the periodic variations of climate due to variations in the orbital parameters (Ward 1974; Ward et al. 1979; chapters 9 and 33). The obliquity of Mars oscillates periodically by $\pm 13°$ on a 10^5-yr time scale with a modulation period of 10^6 yr. Currently, during low obliquity periods, the poles are expected to become cold enough for CO_2 to remain stable throughout the year and the atmospheric pressure drops to 1 mbar (Toon et al. 1980). Once early Mars cooled enough for CO_2 to condense at the polar caps; long periods of reduced atmospheric pressure would have resulted. This would have further limited the liquid-water habitats where life could persist.

How long life on Mars, if it had originated, could have survived the deterioration of climatic conditions depends on how diverse and adaptable it was. However, from the Viking results, it is probable that life on Mars did not adapt to living without liquid water. Presumably, had such an adaptation occurred it should have allowed a readily detectable biosphere on Mars today. It may be that permanent liquid-water habitats such as lakes were the only niches to which life on Mars was adapted, and as these dried up, life was extinguished. On the other hand, it is still possible that life persists on Mars but is restricted to regions where liquid water occurs, at least on a transient basis. Habitats associated with volcanic activity are particularly interesting as it is known that Mars had, and possibly still has, extensive volcanic activity (see, e.g., Carr 1981; chapter 11). Furthermore, volcanic sites on Mars seem to have persisted for billions of years, providing much more stable habitats compared to terrestrial hydrothermal or geothermal zones (see chapter 11). As discussed by McKay and Stoker (1989), volcanic sites on Mars, if associated with a source of water, could be ideal locations for life to maintain itself after the diminution of habitable conditions on the planet as a whole. The main problem with postulating the persistence of such volcanic oases to the present time is the maintenance of a supply of water to the thermal zone.

From the above discussion, it is likely that, if life evolved on Mars, it may have gradually declined along with the atmospheric pressure. However, it is also possible that life on Mars was suddenly and catastrophically rendered extinct. There is now a consensus that impacts of large planetesimals severely altered the environment of the early planets as late as 3.5 Gyr ago (Maher and Stevenson 1988). For large impacts, the size of the oceans (which must be completely evaporated) is the primary determinate of the energy of the object required to completely eliminate life by heating (Sleep et al. 1989). On Mars, because the total water inventory was probably smaller than on the Earth (see chapter 6), a smaller impacter is required to vaporize all the water and hence, a large impact may have heat-sterilized the planet in the late stages of the heavy bombardment. Thus, any life on Mars may have ended in either fire or ice.

The discovery of a past biology on Mars would have a profound implication for the field of biology as well as in the broader context of science. In the discussion below, we have picked a particular example, based upon the current hypothesis for the earliest common ancestor of terrestrial life, that illustrates how biologically important information can be obtained from the geologic and fossil record of an early Martian biota.

Based upon molecular sequencing data, phylogenetic trees have been developed that show the unity and relatedness of all life on Earth (Woese 1987). Although interpreting these trees in an evolutionary context is difficult, a consensus is emerging that the universal genetic ancestor common to all life extant on Earth today was a sulphur metabolizing thermophilic organism (Woese 1987; Lake 1988). With only one example of life to study, it is not possible to determine if there is some biological imperative associated with sulphur hot springs or if this setting was merely an historical accident. If geological analysis of fossils from Mars were to reveal that the earliest examples of life on that planet also arose in a sulphur-rich, hot-spring environment then this would be significant collaborative evidence. This example is but one of many biology-wide questions that probably cannot be answered until we find evidence of life, even extinct life, other than on the Earth.

In the broader context, the discovery of fossil evidence of past life on Mars would have significant implications for our understanding of life as a widespread planetary phenomenon; so too would the determination that life did not originate on Mars. Our understanding of the processes that led to the origin of life on Earth are uncertain at best and we do not know whether life is a singular event or is widespread in the universe.

Acknowledgments. The authors wish to acknowledge numerous helpful discussions concerning geology, climate and life on early Mars with A. Albee, S. Awramik, M. Carr, I. Friedmann, R. Kahn and S. Squyres. We thank L. Rothschild and D. DesMarais for helpful reviews of the manuscript. This work was supported by the NASA Exobiology Program.

PART VII
Satellites of Mars

36. GEODESY AND CARTOGRAPHY OF THE MARTIAN SATELLITES

R. M. BATSON, KATHLEEN EDWARDS
U.S. Geological Survey

and

T. C. DUXBURY
Jet Propulsion laboratory

Conventional maps of Phobos and Deimos are difficult to use because (1) mathematical projections of their highly irregular surfaces fail to convey an accurate visual impression of landforms, complicating the interpretation and testing of theories on the origin of such features as the grooved terrain; and (2) map scales differ according to the shapes of the satellites themselves, complicating the measurement and interpretation of geologic features and their areal distributions. These obstacles are largely overcome by producing maps in digital forms, i.e., by projecting Viking Orbiter images onto a global topographic model made from collections of radii derived by photogrammetry. The resulting digital mosaics are then formatted as arrays of body-centered latitudes, longitudes, radii and brightness values of Viking Orbiter images. The Phobos mapping described here was done with Viking Orbiter data. Significant new coverage was obtained by the Soviet Phobos mission; this has not yet been incorporated into the mapping. The mapping of Deimos is in progress, using the techniques developed for Phobos.

I. INTRODUCTION

Maps of spherical or spheroidal planets are familiar and well understood. They are projections of spherical or ellipsoidal surfaces on sheets of paper, and they have mathematical characteristics that facilitate the measurement of areas or distances. Like many of the small satellites and asteroids of our solar system, however, the satellites of Mars are very irregular in shape. Conformal or equal-area map projections of these bodies fail to convey the true character of their surfaces. Traditional methods of map compilation involve transformation of images or other spatial data to a map projection selected prior to the mapping process. The projected images are treated as map fragments and are assembled as mosaics to produce a final product. Maps of Phobos, on the other hand, are compiled digitally as arrays of Cartesian coordinates consisting of radii registered with images transmitted by spacecraft.

The format of a digital map differs in concept from that of a mathematical projection designed for visual use. As long as a computer can readily access a specified element of a digital map, it does not matter if the array has equal-area or conformal properties. Digital planetary maps are normally stored in the Sinusoidal Equal-Area projection to allow efficient computer access (Batson 1987).

Below we illustrate some of the most useful formats for the Phobos data that allow more reliable mapping of surface features than was possible with previous, manually generated projections used by Thomas (1979). Because mapping of Deimos is not yet complete, it will not be discussed in detail.

II. GEODETIC CONSIDERATIONS

A body-fixed $\overline{\mathbf{xyz}}$ reference system (Fig. 1) was used to define the coordinates of the control points. The axes were chosen to coincide with the three principal moments of inertia of Phobos and Deimos (Duxbury and Callahan 1989). In body-fixed $\overline{\mathbf{xyz}}$, $\bar{\mathbf{z}}$ lies along the spin axis, $\bar{\mathbf{x}}$ is normal to $\bar{\mathbf{z}}$ in the direction of the longest axis and defines the prime meridian, and $\bar{\mathbf{y}}$ completes the orthogonal, right-handed system. In stable dynamical configurations, the three radii of a triaxial ellipsoid that approximates the mean surface have the longest radius (length a) along $\bar{\mathbf{x}}$, the intermediate radius (length b) along $\mathbf{y}$, and the shortest radius (length c) along $\bar{\mathbf{z}}$.

There is no useful constraint on the position of the center of mass relative to the center of figure. The mass of Phobos did not significantly deflect the Mariner 9 or Viking spacecraft orbits; hence, the center of mass is poorly constrained relative to Phobos' surface. In addition, the location of the center of figure relative to the surface varies by up to 1 km depending on how complex a reference surface is utilized.

The pole and prime meridian of Phobos are defined relative to the entire control network (unlike the procedure used for the larger, more regularly shaped bodies, where a prime meridian is defined by the longitude of one surface feature). Phobos' north pole lies on the spin axis vector above the

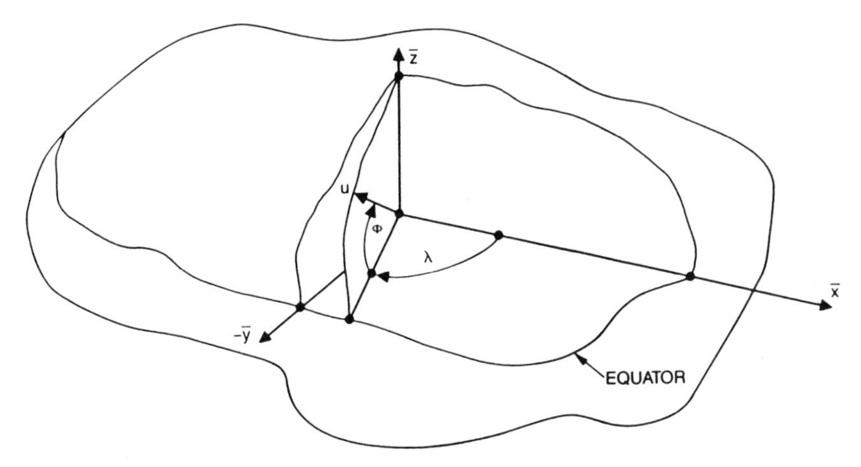

Fig. 1. Cartographic coordinates (φ, λ) of a control point (*u*) on the surface of Phobos in a body-fixed $\overline{\mathbf{xyz}}$ system that aligns the $\bar{\mathbf{x}}$ principal axis of Phobos in the direction to Mars and the $\bar{\mathbf{z}}$ axis with the spin axis.

invariable plane of the solar system, in accordance with IAU convention (Davies et al. 1986). Planetocentric cartographic coordinates were used to define the body-fixed coordinates of the control points as well as the mean surface of the satellite.

III. MAP CONTROL NETWORKS

For Phobos, whose entire surface was viewed by the Viking Orbiters, only 70% of the coverage has sufficient image overlap to yield mapping accuracies of 300 m or better. Lesser accuracies, ranging from 500 to 1000 m, can be achieved in the north-polar and northern leading-edge regions. The Phobos mission of the USSR produced significant new data, particularly in the area west of Stickney. These have not yet been incorporated into our mapping.

The new Phobos control network discussed here was derived by measuring the image coordinates of about 450 landmarks in more than 100 images. These landmarks are craters, grooves and other distinct features such as a "bump" on a crater rim.

The data analysis of the landmarks' image locations involved estimating Phobos' rotational properties, the body-fixed coordinates of the landmarks, and camera pointing for each image. The photogrammetric geometry of narrow-angle imaging systems is too weak to derive camera-station positions; the spacecraft/Phobos positions previously derived by spacecraft tracking were therefore accepted without modification.

Phobos was observed to be in near-synchronous rotation with its spin axis normal to its orbit plane and its prime meridian pointing within a few degrees of the Phobos-Mars line. The deviation from perfect synchronous rotation results from both optical and forced libration, because Phobos has an eccentric orbit and is irregular in shape. Both librations have the same period as the orbital period of Phobos.

The optical libration is twice the orbital eccentricity or $\sim 3°$. The forced libration amplitude was observed to be $0.°8 \pm 0.°4$, which can be computed as a function of the orbit eccentricity and the angular moments of inertia of Phobos.

The control-network coordinates were used to compute a figure of Phobos and the moments of inertia from the figure model. The predicted forced libration amplitude of $1.°0$ is in good agreement with the observed value and supports a homogeneous mass distribution for Phobos. Mean ellipsoidal radii were determined to be $13.3 \times 11.1 \times 9.3 \pm 0.3$ km ($1\ \sigma$) for Phobos and $7.5 \times 6.2 \times 5.4 \pm 0.5$ km ($1\ \sigma$) for Deimos. These radii are in good agreement with those reported by Thomas (1989). Elevation variations from the ellipsoidal surfaces are 1.2 km ($1\ \sigma$) for Phobos and 1.0 km ($1\ \sigma$) for Deimos. These new radii give: $\text{Volume}_{\text{Phobos}} = 5751 \pm 460$ ($1\ \sigma$) km^3; $\text{Volume}_{\text{Deimos}} = 1052 \pm 250$ ($1\ \sigma$) km^3. A detailed surface model that accounts for craters yields a volume for Phobos of 5680 ± 250 km^3 (Duxbury 1991).

The inherent accuracy of map coordinates produced from Viking images is about 50 m in the best observed regions. However, this level of accuracy can be achieved elsewhere only if values for the spacecraft/satellite positions can be determined more precisely than the present, assumed values. The potential for making such improvements with Mariner 9 and Viking Orbiter data has not been verified.

IV. DIGITAL RADIUS MODEL

Radii at control points provide the primary control for modeling (see Fig. 2). Stereophotogrammetry can sometimes be used to make continuous measurements in the form of contour maps or profiles. This method is capable of exceptionally high relative accuracies. Absolute accuracy is a function of overall control accuracy, however, and the absolute accuracy of the Phobos map is still an order of magnitude less than accuracies theoretically achievable by stereophotogrammetry.

The preliminary radius model of Phobos is a grid of radius values interpolated linearly from a manually drawn contour map of the control net. This digital model provides the coarse framework upon which spacecraft images can be projected, allowing first-order correction of topographic distortions in the images so that a mosaic can be made.

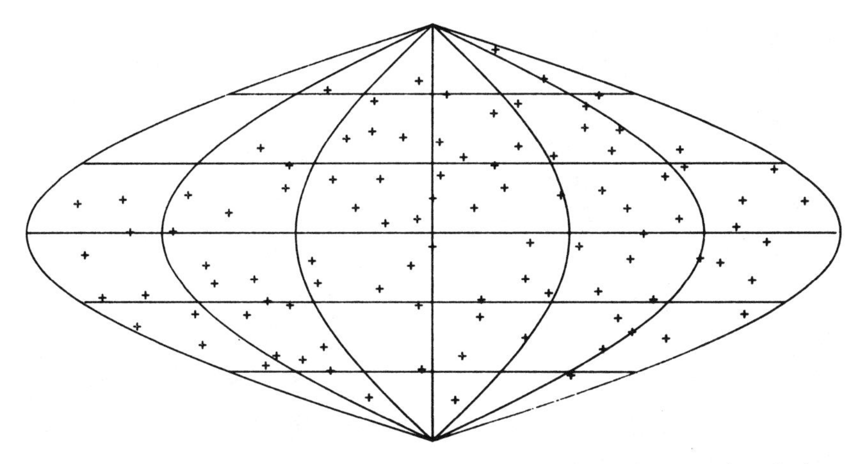

Fig. 2. Primary control network for Phobos in a sinusoidal equal area planetocentric projection. These 109 control points were derived by simultaneous least-squares adjustment of camera pointing and the parameters of a dynamical model of the spacecraft orbit. These points were then used to control the stereographic modeling parameters of all photographs used in establishing the full 450 control points.

V. DIGITAL MOSAICKING

Mosaics are made by projecting images onto a digital radius model. When the figure of a planetary object can be defined mathematically as a sphere or spheroid, topography is usually considered to be negligible and the projection can be done mathematically (Edwards 1987). On irregularly shaped bodies, however, topography is dominant, and reasonably accurate models must exist before useful projections can be made. The first step in mapping these bodies, therefore, is to project a radius model to the plane of a spacecraft image, so that the two digital arrays can be registered. Only after the image is registered can the latitude and longitude of each pixel in the image be determined, allowing the image to be projected into the Sinusoidal Equal-Area mapping array.

This process is repeated for each frame used in the mapping, and each transformed frame is then added to the array. Each latitude/longitude bin in the completed Sinusoidal array contains a brightness value from the best available spacecraft image and a radius from the geometric model.

Figure 3 is a photomosaic of 30 Viking Orbiter images of Phobos, compiled in registry with the geometric model. A three-dimensional reprojection of the entire mosaic to any desired perspective can now be performed by the model-to-image projection process mentioned above, using spacecraft vectors specified in terms of desired viewing geometry (Fig. 4) rather than those constrained by an actual spacecraft trajectory.

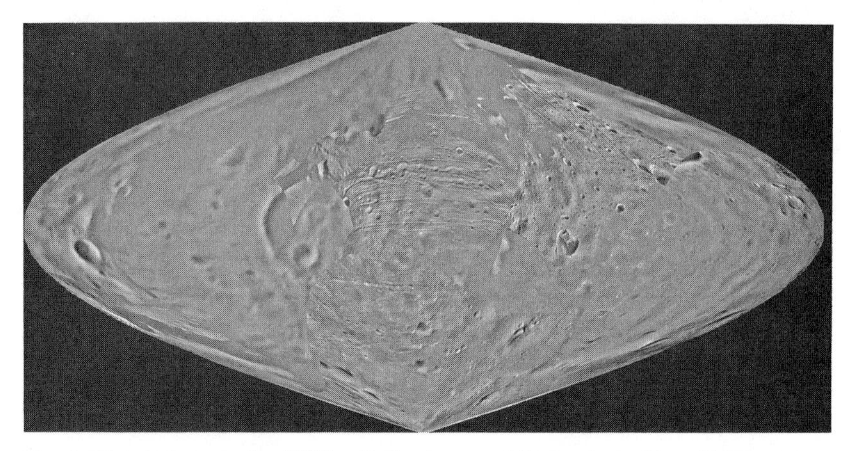

Fig. 3. Digital photomosaic of Viking Orbiter images of Phobos in the Sinusoidal Equal Area projection.

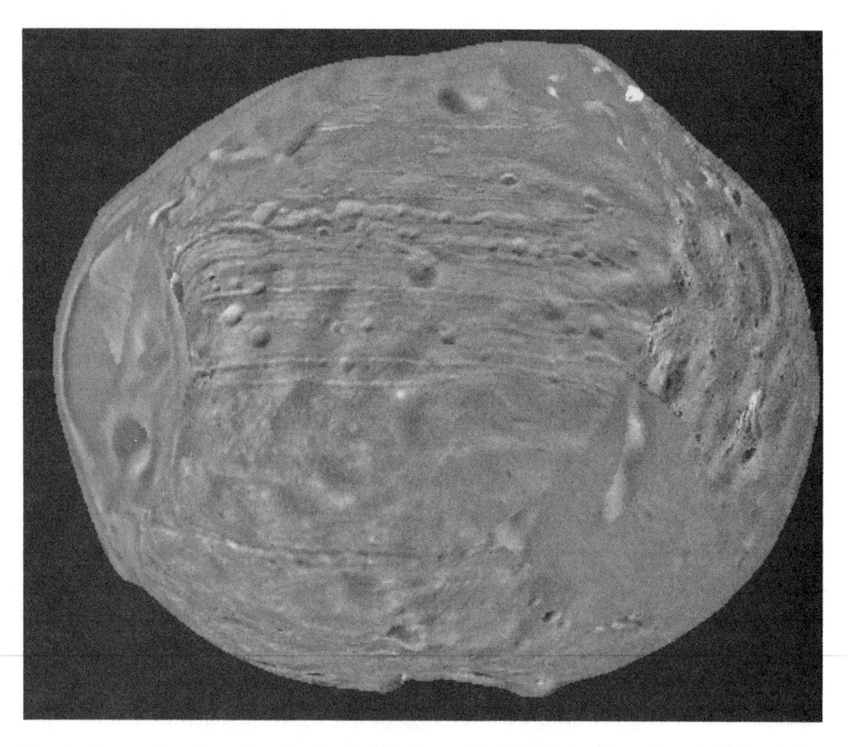

Fig. 4. Perspective view of preliminary digital mosaic of Phobos, centered on the Mars-facing side.

VI. THE AIRBRUSH MAP

Varied resolution and surface illumination present in the Viking Orbiter data set result in inconsistent portrayal of surface features in the mosaic. A shaded relief map was therefore made manually by methods described by Inge and Bridges (1975) and Batson (1978). The photomosaic was used for geometric control, and virtually the entire Viking Orbiter image data set for Phobos was examined in making the portrayal. The map was made in Sinusoidal Equal-Area segments, digitized as a raster image, and the segments merged into a single Sinusoidal array for consistency with the rest of the Phobos database. Figure 5 shows such an airbrush map after digital reprojection.

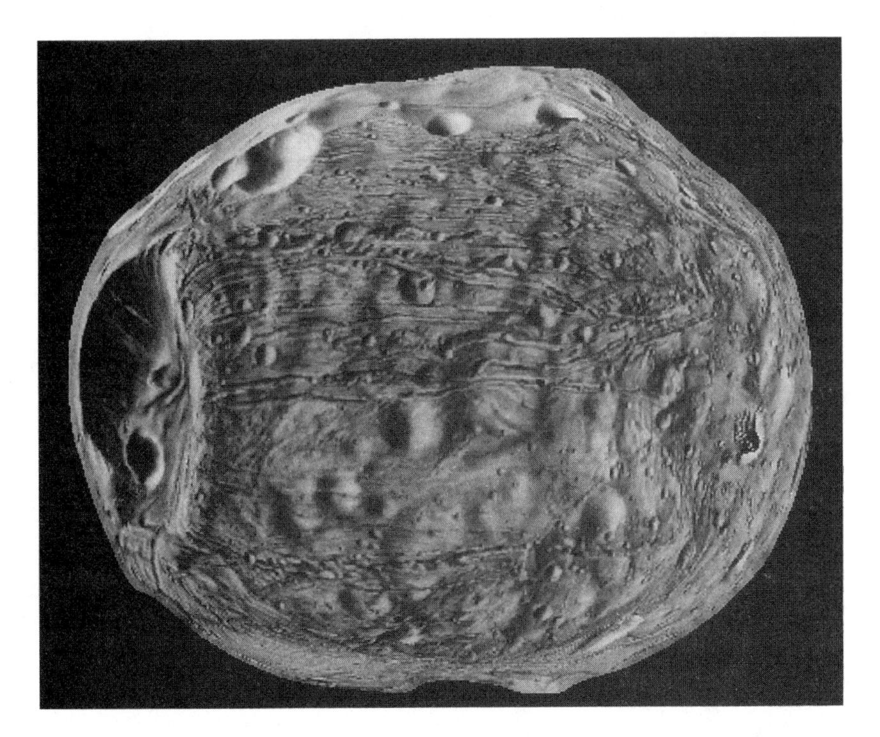

Fig. 5. Perspective view of Phobos' Mars-facing side, made up of 6 digitized, airbrushed map segments. The vertical (north-south) dimension is 18.6 km, and the east-west dimension is 22.2 km. The large crater Stickney is visible on the western (left) limb of the view. Conspicuous east-west grooved terrain is also evident. (Airbrush cartography by P. M. Bridges.)

VII. CONCLUSIONS

Geologic phenomena are strongly correlated with their spatial distributions. The derivation and testing of theories regarding these phenomena are therefore compromised by the absence of maps. The projection of spherical or spheroidal surfaces to conventional map formats is reasonably straightforward, but maps of the satellites of Mars (and other small, irregular bodies) are best compiled digitally, as images registered with radius maps. Image-processing methods can be used to view these digital maps in a variety of projections in hard-copy form and to extract data on size and distributions of surface features, photometric properties, and the like.

Acknowledgments. The authors wish to thank E. M. Lee, who did all of the program testing, digital processing and mosaicking required to develop the mapping techniques described here. Without her perseverance and ingenuity, this operational capability would not yet exist. P. M. Bridges drew the shaded relief map on the basis of digital transformations performed by H. F. Morgan. This work is supported by the Planetary Geology and Geophysics Program, Planetary Division, Office of Space Science, National Aeronautics and Space Administration.

37. SATELLITES OF MARS: GEOLOGIC HISTORY

P. THOMAS, J. VEVERKA
Cornell University

J. BELL
University of Hawaii

J. LUNINE
University of Arizona

and

D. CRUIKSHANK
NASA Ames Research Center

The small, irregularly shaped satellites of Mars, Phobos and Deimos, provide the most detailed view of the geomorphic forms and processes important on small solar system bodies. The satellites appear to be very similar in composition, strongly resembling carbonaceous asteroids; however, recent groundbased spectra suggest that their surfaces have little bound or interlayer water. Despite their similar compositions, sizes and environments, Phobos and Deimos have radically different surface features. Phobos is densely covered by craters that are nearly lunar in appearance; Deimos' craters are subdued and largely filled in by debris. Phobos shows only local downslope movement of regolith; Deimos has it on a global scale. Phobos is criss-crossed by linear depressions; Deimos has none. Crater ejecta appear to be retained near their sources on Phobos while the ejecta are widespread on Deimos. The reasons for the differences between the satellites are not known; imaging of asteroids should tell us which, if either, satellite is typical of the many small bodies that populate the asteroid belt.

The tiny, irregularly-shaped satellites of Mars currently provide our best glimpses into the processes that affected the evolution of small bodies in the solar system. For no other small satellites do we have a comparable wealth of detailed information, and, whatever their actual origins, even after the 1991 October flyby of 951 Gaspra, Phobos and Deimos provide some of the major reference points for study of asteroids.

One of the more exciting and demanding aspects of studying the geology of these satellites is that the surfaces probably record not only the usual processes expected on small bodies such as asteroids, but may also show the influences of an evolving tidal environment, both on the solid-body characteristics as well as on surface processes such as ejecta distribution. The challenge in studying the Martian satellites is to separate the results of processes particular to the Martian environment from those that are ubiquitous on small bodies in general. Only then can one hope to discern any remaining hints of the primordial characteristics of these puzzling objects.

I. COMPOSITION

For more than a decade it has been suspected that the composition of the satellites of Mars is related to that of low-albedo C-type objects in the asteroid belt (see, e.g., Veverka and Burns 1980), but documenting the relationship has been a slow process. The proximity of the satellites to Mars makes conventional spectral observations very difficult from Earth during all but the very best Mars oppositions. Spacecraft data have also been limited; until the Phobos 2 mission, no spacecraft was designed specifically to observe the Martian satellites.

Broadband color measurements over restricted spectral ranges were made by the Mariner 9 and Viking Orbiter cameras (Pollack et al. 1972; Duxbury and Veverka 1978). Additionally, Barth et al. (1972*a*) obtained a spectrum from 0.26 to 0.35 μm with the Mariner 9 ultraviolet spectrometer, and Pollack et al. (1978) measured the reflectances in three broadband filters with the Viking Lander cameras. These data (summarized in Fig. 1) suggested that the spectral reflectance curves and albedos of both satellites are similar to those of asteroid Ceres (Pang et al. 1978,1980). These colors and spectra and the low densities (Phobos $\sim 2.2 \pm 0.5$ g cm^{-3} and Deimos $\sim 1.7 \pm 0.5$ g cm^{-3} [Duxbury and Callahan 1982]; Phobos 2 results suggest a density of about 2.0 g cm^{-3} [Avanesov et al. 1989]), have led to suggestions that the satellites are essentially C-type asteroids somehow displaced from their natural home in the middle asteroid belt. A popular view is that the two bodies were captured into orbit around Mars by gas drag in the very early days of the solar system, after which strong tidal evolution brought them to their present orbital positions (Pollack et al. 1979*a*; Hunten 1979*a*; Zharkov et al. 1984; Hartmann 1987; Sasaki 1990; chapter 38).

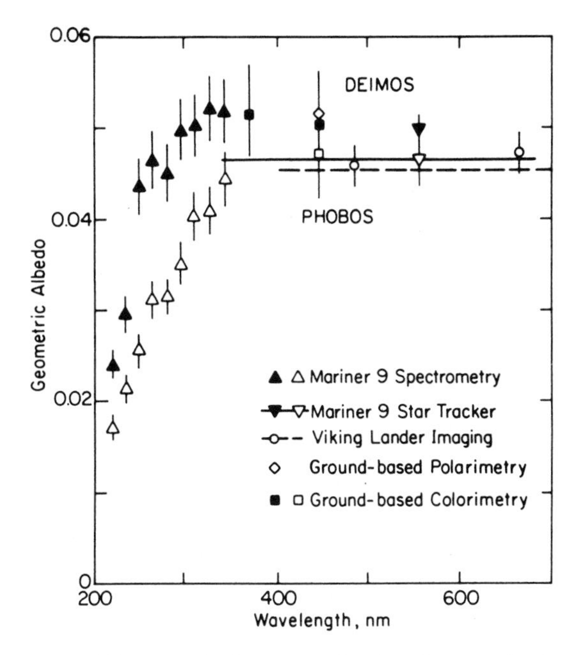

Fig. 1. Visible spectra of Phobos and Deimos, a combination of spacecraft and groundbased data from Pang et al. (1980). Improved scaling factors indicate that the absolute geometric albedos of Phobos and Deimos are somewhat higher: between 0.5 μm and 0.6 μm they are about 0.05 ± 0.01 and 0.06 ± 0.01, respectively (cf. Veverka and Burns 1980; Avanesov et al. 1989).

The generally flat spectral reflectance longward of ≈ 0.4 μm and the downturn toward shorter wavelengths are characteristics of several classes of low-albedo asteroids, specifically the C, B, F and G types (Tholen 1984, Fig. 9). In Tholen's taxonomy, the B, F and G types are essentially subtypes of the broader C class; the G asteroids, for example, have a stronger ultraviolet absorption than the C's, and exhibit an absorption near 3 μm resulting from hydrous minerals in their regoliths (Feierberg et al. 1985).

While the asteroid taxonomy is derived from data in the photovisual spectral region, reflectance data in the near-infrared (0.8 to 2.5 μm) are also useful for identifying the characteristics of the low-albedo objects (Bell et al. 1988; Hartmann et al. 1987), in particular the slopes and inflections in their reflectance curves. Data in this spectral region were obtained for a limited and spatially resolved portion of Phobos from the Infrared Spectrometer (ISM) aboard the Phobos 2 spacecraft (Bibring et al. 1988). Two linear tracks across Phobos were obtained, with full spectra at each point along the track (Bibring et al. 1989), and also a spectral image with 24 × 25 pixels in which the pixels projected on Phobos are 0.7 × 0.7 km (Langevin et al. 1990). Although they are in a preliminary stage of reduction and interpretation, the near-infrared reflectance data show the following (Langevin et al. 1990):

a. A weak and spatially variable "red" slope upward between 0.8 and 3.0 μm;
b. No strong hydration signature in the 3-μm region;
c. Weak, but measurable spectral contrast. The latter has been interpreted by the investigators as possibly indicating compositional variations on a km spatial scale.

To a first approximation, then, the photovisual and near-infrared reflectance data are consistent with "dry" C asteroids that show no 2.9-μm band due to hydrous minerals. Thus, the G asteroids are eliminated from the comparison. The weak red slope in the near infrared and the drop in reflectance in the ultraviolet are both characteristic to varying degrees of the P- and D-type asteroids. These are low-albedo objects which differ from the C's in that their reflectances slope upward toward longer wavelengths; the P's slope weakly and the D's more strongly. Tholen (1984) shows examples in the near-infrared. The work of Jones (1988) and Lebofsky and Jones (1989) suggests that the P and D asteroids typically do not have the 2.9-μm hydration band.

Because there appears to be a continuous gradation in continuum reflectance slope from zero (neutral) to positive ("red") from the C to P to D asteroid classes, probably with a concomitant decrease in ultraviolet absorption, it is consistent with the data to say that Phobos closely resembles the C and "early" P asteroids in reflectance. This comparison with C's has been made for quite some time (see, e.g., Pollack et al. 1978; Pang et al. 1978).

It is common to regard the low-albedo asteroids as carbonaceous on the basis of their reflectance similarities to the carbonaceous meteorites, and because of the paucity of other tenable hypotheses. Some low-albedo asteroids contain hydrated minerals, as do some of the carbonaceous meteorites (CM and CI types). In addition, Cruikshank and Brown (1987) have presented preliminary spectroscopic evidence for the 3.4-μm C-H absorption band on one G-type asteroid (130 Elektra), which, if confirmed, would further draw the ties between the carbonaceous meteorites and the low-albedo asteroids. However, the connection between such meteorites and asteroids still remains to be firmly established.

Both Pang et al. (1980) and Lucey et al. (1989) contend that the similarities in the reflectances of Phobos and Deimos are consistent with their origin as two fragements of the same parent body. Asteroid parent bodies must be expected to exhibit considerable internal heterogeneity due to thermally driven internal processes and the associated variations in the abundance and mobility of water in metamorphic heating processes. Clear examples of such diversity are seen in the Themis asteroid family, which contains examples of classes F, B, C and G plus many objects with intermediate properties (Tholen 1984; Bell 1989). In terms of the actual spectra, these variations correspond to differences in the depth of the 0.3-μm absorption. The difference between the spectral curves of Phobos and Deimos in Fig. 1 is well within the range seen in the Themis family, and considerably less than the vast spectral variation ranges found in the asteroid belt (Chapman and Gaffey 1979; Zellner et al. 1985).

The only infrared spectral data for Deimos are those obtained in 1988 by Bell et al. (1989*b*) (Fig. 2). Comparing the Deimos spectrum with existing asteroid data bases, Bell et al. (1989*b*) concluded that the combination of pronounced ultraviolet absorption, flat visible continuum, and red continuum in the near infrared occur together only in a few asteroids in the transition region between C-class and P-class objects. The closest match appears to be 65 Cybele, which is located at 3.4 AU from the Sun in the outer reaches of the asteroid belt. This asteroid is classed as C by Tholen (1984), but is listed as a P in the classification of Tedesco et al. (1989) due to slight differences in defining the boundary between these classes. Asteroids like Cybele are largely found only within the zone between the main asteroid belt and the Trojan asteroids (Gradie and Tedesco 1982; Bell et al. 1989*a*). The reddish spectral slope of P-class asteroids is conventionally attributed to an increased abundance of complex organic polymers over that preserved in C-class asteroids and the carbonaceous meteorites; no meteorite analogs of P asteroids are

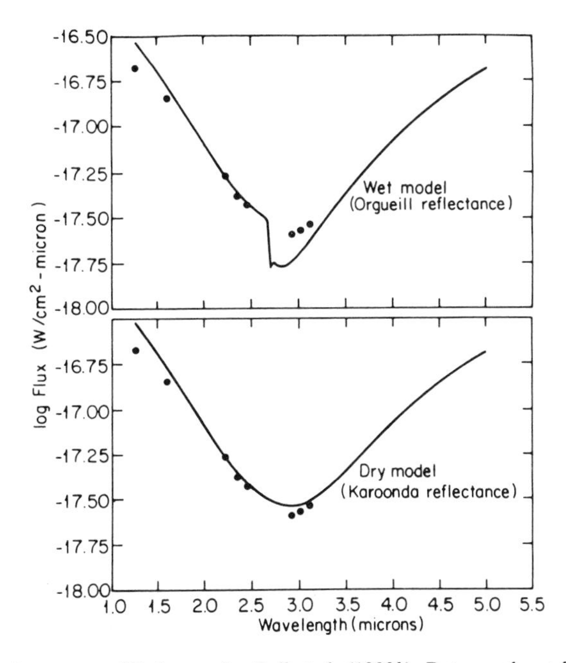

Fig. 2. Infrared spectrum of Deimos, after Bell et al. (1989*b*). Data are shown by dots and are compared to two model curves. The upper curve assumes the reflection spectrum of the CI chondrite, Orgeuil, one of the most hydrated meteorites. The lower curve assumes the spectrum of Karoonda, an anhydrous carbonaceous chondrite.

known. But if the speculation is correct, the satellites of Mars may be richer in organic compounds than any meteorite.

Beyond the issue of composition, the low mean densities of Phobos and Deimos led to early suggestions (see, e.g., Veverka 1978) that they ought to be identified with hydrated carbonaceous chondrites. Bell et al. (1989*b*) searched directly for hydrated silicates on Deimos by means of the clay mineral water band at 3 μm, which has been detected in many dark asteroids (Lebofsky 1980; Lebofsky et al. 1981; Jones et al. 1987). Because the Mars satellites are much warmer than typical main-belt asteroids, it is more difficult to separate thermal emissions from reflected light at these wavelengths. Figure 2 shows the observed Deimos spectrum compared with two model curves generated using the standard asteroid thermal model. The comparisons suggest at most a weak 3-μm band on Deimos of about 5 to 10% relative depth, much less than in any hydrated meteorite and comparable to the driest C-type asteroids. Given the uncertainties in the thermal models, it is quite possible that Deimos has a completely anhydrous surface. Certainly the surface of Deimos is hydrated much less than the average main-belt C-type asteroid. This finding is consistent with the overall resemblance of the spectrum to that of Cybele, which has been shown to lack the 3-μm band. In fact, most P-class asteroids (as well as the more distant D class, most common among the Trojan asteroids in Jupiter's orbit) appear to lack the 3-μm water of hydration band (Jones 1988).

The apparent lack of hydrated clay minerals in the surfaces of the satellites of Mars is connected to a broader, related question: Why do the surfaces of dark, "primitive" asteroids apparently become drier with increasing distance from the Sun? Three possibilities have been suggested (Jones et al. 1987):

1. The clay absorption band is suppressed by an increasing abundance of dark "red" organic polymers. However, subsequent laboratory studies suggest that the clay band is too strong to be completely suppressed in this way (Jones 1988).
2. The abundance of water in the original solar nebula declined with distance from the Sun, at least in the region of the outer asteroid belt. There is no corroborative evidence to support such an *ad hoc* hypothesis.
3. The water in all asteroids was originally ice, but only in the middle asteroid belt were bodies heated enough after accretion to melt the ice and create hydrated silicates through the action of "groundwater." According to this view, in the P-class and D-class asteroids, ice would still be present. In this model, the zone of hydration in the middle belt would be a zone of metamorphism between the igneous rocks of the inner belt and the primitive material of the outer belt.

The balance of the evidence favors the third alternative. If this scenario is true, Phobos and Deimos could represent our first opportunity to study unaltered ultracarbonaceous asteroidal material. It also implies that a considerable amount of ice could have been contained in the Phobos/Deimos parent body. Fanale and Salvail (1989) investigated the behavior of such hypothetical ice since the satellites arrived in their current warm environment and concluded that ice could survive to the present in the deep interior even though the regoliths are almost certainly completely devolatilized (see also Hartmann 1988). In principle, a remnant icy core could account for the low measured densities of the satellites. However, analyses of the observed libration of Phobos suggest that the core of Phobos is not substantially more or less dense than are the outer layers (Duxbury and Callahan 1989). Recent radar results suggest a surface density near 2 g cm^{-3} for Phobos (Ostro et al. 1989), a value very close to the satellite's bulk density. Consequently, the implication is that the density of Phobos is close to being uniform with depth.

Dubinin et al. (1990) have claimed that the Phobos 2 particles and fields data contain indirect evidence of water outgassing from Phobos and Deimos. The rate of water production derived from these observations by Ip and Banaszkiewicz (1990) closely matches that derived by Fanale and Salvail (1990) from theoretical considerations. Unfortunately the subsequent failure of the spacecraft before it could examine Phobos closely prevented confirmation of this exciting result.

In summary, the current spectral data support a captured asteroid origin for the satellites, and suggest that these objects came from the vicinity of 3.5 AU from the Sun.

II. REGOLITH

The surface properties of most airless objects imaged by spacecraft are dominated by loose, fragmental debris formed by impact cratering. (Io, and possibly Europa, are exceptions.) The regolith is both a repository of material mixed upward from the object's "bedrock" as well as a blanket covering many or even all inhomogeneities. The latter characteristic is a prime concern when dealing with small bodies.

Early clues that Phobos and Deimos are covered by extensive regoliths have been reviewed by Veverka and Burns (1980), and involve photometry (Klaasen et al. 1979), polarimetry (Zellner 1972; Noland et al. 1973), thermal measurements (Gatley et al. 1974), as well as the visual appearance of the surfaces (Veverka 1978; Thomas 1979). The most direct demonstration was provided by the eclipse measurements of Gatley et al. which showed that following an eclipse by Mars, the surface of Phobos heats up extremely rapidly and must therefore be covered by a layer of very low thermal inertia.

The most recent theoretical calculations of regolith evolution on the Martian satellites are those by K. Housen (see Veverka et al. 1986). For Phobos and Deimos, the calculated median regolith depths are 35 m and 5 m, respectively, on the assumption that escaping ejecta are not re-

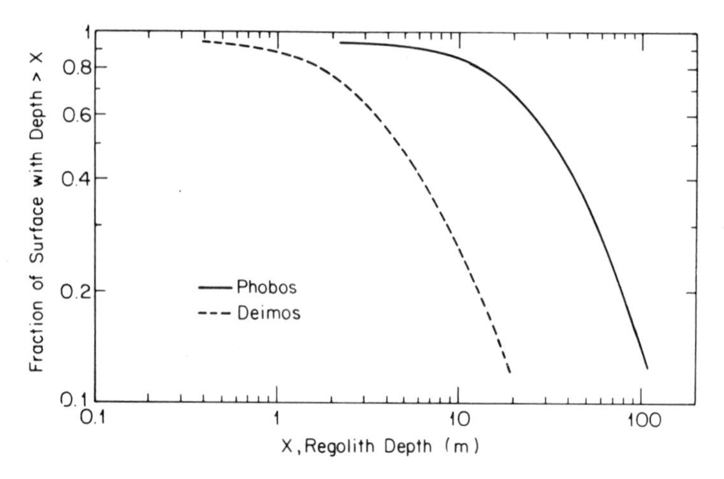

Fig. 3. Ejecta depth distributions on the Martian satellites modeled by K. Housen (Veverka et al. 1986).

accumulated (Fig. 3). Assuming that all ejecta are retained (i.e., 100% effective re-accumulation) does not change the median regolith depths significantly. The calculations predict a thinner regolith on Deimos, because of its smaller crater density.

A. Regolith Properties Inferred from Thermal Infrared Data

Available infrared data on the regoliths of Phobos and Deimos come from the Mariner 9 infrared radiometer (Gatley et al. 1974), the Viking Infrared Thermal Mapper (IRTM) (Lunine et al. 1982), the Phobos 2 mission (Ksanfomality et al. 1989) as well as from groundbased observations (Veeder et al. 1987). The data sets may be divided into whole-disk observations as a function of phase angle, whole-disk eclipse observations, and spatially resolved observations of selected areas and times of day; all at multiple wavelengths. In conjunction with thermal models (Kieffer et al. 1977) and knowledge of the albedo, properties of the regolith over a skin depth (several cm), including thermal inertia, emissivity, particle sizes, and areal coverage by blocky (high-inertia) materials, can be determined from the measurements.

Lunine et al. (1982) employed Viking observations of Phobos and Deimos at 11 and 20 μm as a function of phase angle, combined with eclipse data, to derive thermal inertia I values of 0.6 to 2 (in units of 10^{-3} cal cm^{-2} s$^{-1/2}$ K^{-1}). (Here $I = \sqrt{k\rho c}$, where k is the thermal conductivity, ρ the bulk density of the surface layer and c the specific-heat capacity.) In addition, spatially resolved observations of pre-dawn temperatures on Phobos gave I values of 0.9 to 1.6. From the Phobos 2 data, Ksanfomality et al. (1989) found $I = 4$ in the region where the local time was 11:00–11:40, suggesting an especially thin layer of fine material or local exposures of more solid rocks. Gatley et al. (1974) derived a global thermal inertia of about 0.5 at 10 and 20 μm. Both the Viking and Mariner 9 data can be fitted well by single-layer models, a fact that suggests the low-inertia layer is at least several thermal skin depths in thickness. The thermal inertia values of Phobos and Deimos are comparable to the lowest values found for Mars and are equivalent to or smaller than lunar values. Comparison with laboratory data of fine particulate powders by Lunine et al. (1982) suggests particle sizes of 50 to 100 μm.

The emissivity of the regolith on Deimos was constrained by Lunine et al. (1982) by requiring that the ratio of 9 and 11 μm IRTM fluxes measured at a single solar phase angle give the same thermal inertia as the single channel radiometric phase curves. This procedure gave emissivities between 0.9 and 1. No comparable 9-μm IRTM observations of Phobos were available; however, comparison of the values for I determined from radiometric phase curves with those derived from pre-dawn brightness temperatures on Phobos suggests an emissivity close to unity for the inner satellite.

Sufficient IRTM data at 11 and 20 μm, and Mariner 9 data at 10 and 20 μm exist for Phobos that information on regional slopes and fractional surface coverage by high-inertia material can be derived. Figure 4 shows the

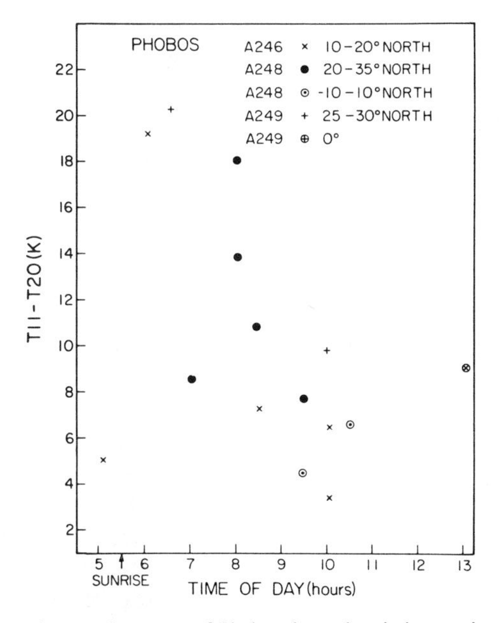

Fig. 4. Dependence of spectral contrast of Phobos thermal emission on time of day interpreted to be largely caused by topographic slopes. Difference between 11- and 20-μm brightness temperatures plotted vs time of day; average sunrise time for these observations is indicated. Symbols refer to different sets of IRTM data (labeled by Viking orbit number). (Figure from Lunine et al. 1982.)

color temperature difference T (11 μm) minus T (20 μm) vs the time of day on Phobos. Based on thermal models of Christensen (1982), the significant temperature contrast before dawn suggests the presence of high-inertia material, perhaps cm-sized particles, on the surface (Lunine et al. 1982). The data, which show that the contrast is a maximum after dawn and decreases toward noon, indicate that steep slopes (30°) relative to an ellipsoidal model of Phobos are responsible. More limited data near the evening terminator of Deimos also suggest spectral contrasts caused by large slopes. Support for the notion that topography plays a dominant role in the infrared emission from the Martian moons comes from the groundbased detection of Deimos at 4.8, 10 and 20 μm by Veeder et al. (1987). This group found that a standard, nonrotating thermal model with a single "beaming factor" for infrared emission could not fit Deimos' flux in all three wavelengths, and concluded that nonsphericity, significant topography and low thermal inertia must be invoked.

A thermal model incorporating these effects was constructed by Kührt and Giese (1989). Unfortunately, they too were unable to account for the high thermal fluxes at Deimos measured by Veeder et al. (1987). The nonspherical model does predict interesting differences in diurnal cooling curves as a function of latitude, but has not yet been compared to Viking IRTM data.

The temperature contrasts of the 11- and 20-μm data can be used to constrain the areal fraction of high-inertia material. Figure 5 shows the predawn temperature contrast as a function of effective surface thermal inertia for different amounts of areal coverage of blocks with $I = 30$, based on the models of Christensen (1982). The data are consistent with some 5% of Phobos' surface being covered by high-inertia, blocky material. The results are changed little by allowing for nonunity emissivities.

In summary, the thermal infrared data show that the surface of Phobos to depths of 1 to 10 cm consists of low-inertia material with some 5% of the surface covered by high-inertia blocks. Topographic slopes, perhaps up to 30°, strongly influence the surface response to insolation. Deimos also possesses a low-inertia surface with strong topographic slope effects; however, the extent of high-inertia material on its surface cannot be assessed.

B. Albedo Features

According to Viking Orbiter data, albedo markings are subdued on Phobos, and color variations are essentially absent on both satellites. Deimos has large brighter patches and streamers with reflectances up to 30% greater than surrounding areas (Fig. 6; Noland and Veverka 1977; French et al. 1988). These areas have normal reflectances of 0.08 compared to 0.06 for the Deimos average. French et al. (1988) have compared the albedo differences and the lack of color variations of the Deimos markings and their surroundings with the characteristics of laboratory samples of pulverized meteorites, and concluded that the brighter areas could simply be composed of finer-grained

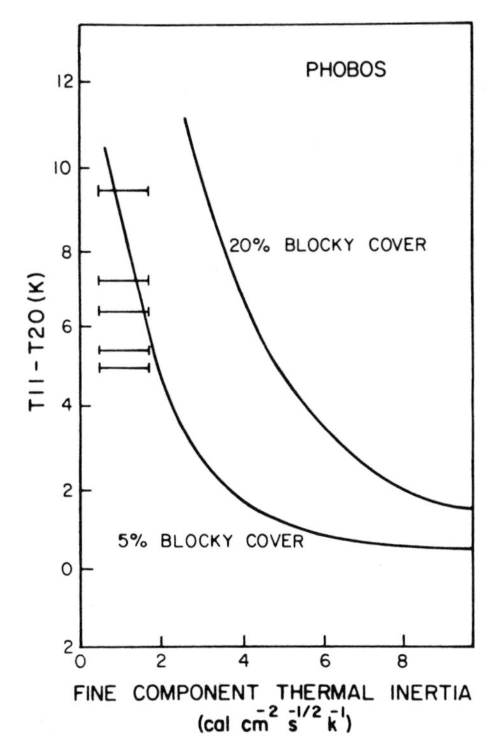

Fig. 5. Pre-dawn brightness temperature difference vs effective thermal inertia on Phobos. Five separate observations are plotted; error bars span the thermal inertia range derived in the text. Model curves for 5% and 20% areal coverage by high-inertia material from Christensen (1982). (Figure from Lunine et al. 1982.)

(<40 μm vs 75–150 μm) carbonaceous materials. These particle sizes are consistent with those inferred from the infrared data. It is not clear, however, why crests and crater rims, the sources of the bright materials, should be finer grained than the general regolith.

Significant color variations on Phobos have been reported from initial analysis of Phobos 2 data (Ksanfomality et al. 1989; Murchie et al. 1990). Assessment of the implications of these possible variations must await a complete photometric calibration of the Phobos 2 observations and the reconciliation of the Phobos 2 and Viking data.

On both Phobos and Deimos some small craters expose material darker than the surroundings (Veverka 1978; Thomas 1979). The phase behavior of these contrasts has not been established, making it impossible to decide whether they are caused primarily by differences in texture or in composition. Other apparently dark markings within craters on Phobos have been shown by Goguen et al. (1978) to be largely due to rougher surface textures and not different normal albedos. Some small craters on Phobos have bright markings

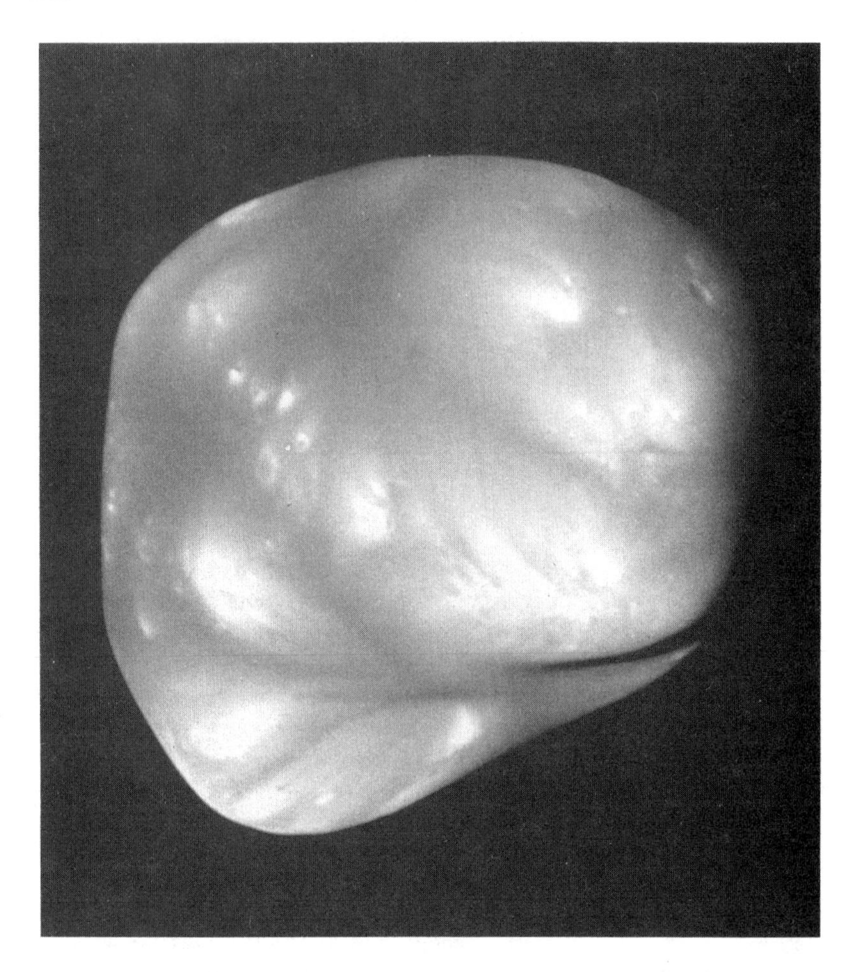

Fig. 6. Deimos at 5°.7 phase angle. Image 507A01, clear filter; subspacecraft point is -1, 217°. Deimos is approximately 12 km across in this view; south is at top. Bright streamers (albedo about 8%) extend up to 30° from ridge crests and crater rims. Streamers near the top show filling of a 2-km crater as well as diversion by what are probably positive relief features a few tens of m high.

associated with rims that resemble crater ejecta patterns or rays both in terms of morphology and in that they become more prominent at lower phase angles. A similar phase behavior is observed for some groove segments and topographic ridges (Veverka and Thomas 1982).

C. Visual Evidence and Morphologic Information on the Regolith

The most obvious visual evidence of regolith occurs on Deimos: high-resolution images clearly show that impact craters are filled in, and there is abundant evidence of downslope movement of surface materials (Fig. 6). On both satellites one can readily detect abundant blocks of ejecta, direct evidence of the efficacy of regolith formation by impact cratering. Both of these important aspects are discussed in more detail in Secs. III and IV. Here we summarize a few observations which give limits to the depth of regolith.

On Phobos, layering is observed in the walls of at least two craters at depths of about 80 and 150 m (Thomas 1979). These layers are expressed as albedo features and not as topographic steps or benches. A few craters on the satellite show flat or hummocky floors. On the basis of the work of Quaide and Oberbeck (1968) on lunar craters, such forms have been interpreted as indicating discontinuities in regolith properties at depths from 35 to 200 m (Thomas 1979). The floor of the largest crater on Phobos is also hummocky, and could indicate a much deeper discontinuity. However, there are other processes that could modify the basic bowl-shaped form of this crater.

Thomas et al. (1979*b*) argued that the morphology of grooves on Phobos suggests that relatively loose regolith up to 100 m in depth occurs in some areas. Grooves that have shallower, smoothly concave profiles also suggest the lack of hard underlying material within 100 m or so of the surface. More recently, pit spacings in the grooves, based on analogy with drainage pit spacings in loose material, were used by Horstmann and Melosh (1989) to infer regolith depths of 300 m in two widely separated parts of Phobos.

Variable regolith depths may be indicated by the presence of bright halos or rays around small craters on ridge crests on Phobos and the lack of such features in the interiors of degraded craters. One interpretation is that a greater depth of loose fill in the low areas prevents excavation of the brighter material; regolith depths would be of order 30 m in these instances (Thomas 1979).

The amount of regolith expected can be approximated by the volume of material excavated by impact craters. The crater density on Phobos (Thomas and Veverka 1980) is such that a simple calculation indicates that about 150 to 200 m of debris is expected. As there remain traces of many degraded features that may be ancient large craters, the expected amount of regolith could be substantially more. In this context, it is important to recall the detailed calculations of Housen referred to previously. These show (Fig. 3) that considerable variations in depths of regolith are expected on the satellites. According to these results, some 10% of the surface area of Phobos is expected to have accumulated $\sim$ 100 m of regolith, with an average of 56 m. If all ejecta that initially escaped were recaptured, depths would be nearly twice as great. Regolith depths on Deimos, according to this calculation, are expected to be substantially smaller than those on Phobos. If the interpretation of Thomas (1989) of an 11-km impact scar encompassing the south polar area of Deimos is correct, regolith depths could be much greater.

D. Radar Data on the Regolith of Phobos

Ostro et al. (1989) have recently detected Phobos with the Goldstone radar. Their results indicate the upper few decimeters of the satellite have a density of about 2.0 g cm^{-3}. These data and the object's bulk density suggest maximum porosities of the upper layers are not radically different from those of the Moon. The data also suggest that there are very few cm- to decimeter-sized particles in the upper regolith. Radar polarization and albedo characteristics of Phobos more closely resemble large main-belt asteroids than they do the Moon or Earth-crossing asteroids, and perhaps thereby strengthen the asteroidal connection of the satellites.

III. IMPACT CRATERS AND EJECTA DISTRIBUTION

Impact craters show a variety of forms on the Martian satellites: much of the variation appears to arise from evolution under subsequent impacts and different styles of filling by crater ejecta. Bowl-shaped craters on Phobos approximate the shape of fresh lunar ones (Thomas 1979). Comparable measurements on Deimos are difficult since there are few fresh, unfilled craters on this satellite (see Sec. IV).

Topographic expressions of ejecta blankets are rare on the Martian moons; only near Stickney is there reasonable evidence of an ejecta blanket (Fig. 7). Some small craters are surrounded by bright albedo patterns at low phase that may be caused by ejecta. A few craters on both Phobos and Deimos excavate darker material (Veverka and Thomas 1982). Attempts to measure rim heights on fresh Phobos craters yield values that suggest the rims could have ejecta as well as structural components (Thomas 1979).

On the Martian satellites, escape velocities are only a few m s^{-1} and depend strongly on location and direction, especially on Phobos, because of substantial tidal and rotational modifications of the satellite's effective gravity field (Dobrovolskis and Burns 1980; Davis et al. 1981). On both satellites, ejecta thrown more than a few m s^{-1} will travel at least several km. Unless crater formation on Phobos is in the gravity-scaling regime (Cintala et al. 1978) where shape of the crater and ejecta blanket are independent of crater size, these satellites (g between $\sim$ 0.0015 and 0.005 that on the Moon) may not show ejecta blankets thick enough to map.

Because of the scarcity of discernible continuous ejecta blankets, individual blocks provide much of our information on ejecta distribution. Blocks are observed near 8 craters on Phobos and 6 craters on Deimos. The largest is a 150 m block on Deimos; others are 10 to 100 m across (Fig. 8). Although these scales are impressive, these large blocks cover far too small a fraction of the surface to contribute to the high-inertia material detected on Phobos by thermal infrared observations (Lunine et al. 1982).

Fig. 7. Phobos: crater Stickney rim, ejecta and grooves. The rim of Stickney is at lower right; the mottled, hummocky region is interpreted as ejecta from Stickney. Image 343A13.

Lee et al. (1986) measured the sizes and radial positions of blocks on the two satellites and found that the relation of largest block to source crater size is similar to that on the Moon. The radial distribution of blocks near 200-m craters is similar on the Moon and Phobos, but is very different on Deimos (Fig. 9): the blocks on Deimos are much more widely dispersed, and are often difficult to relate to source craters.

Blocks are just a particular size fraction of the crater ejecta, but in general regolith properties on the two satellites suggest that ejecta are more closely confined to craters on Phobos than on Deimos. As discussed in the following section, the appearance of Deimos demands not only that there be widespread ballistic filling of craters, but also efficient redistribution of the material by downslope movement.

Why ejecta are more widely dispersed on Deimos than on Phobos is not understood. In part, it may be due to the lower gravity on Deimos, but different material properties might cause cratering on Deimos to be more strength dominated than on Phobos, where a gravity scaling might obtain. Strength-dominated cratering would tend to result in higher ejecta velocities and more dispersed ejecta (Cintala et al. 1978).

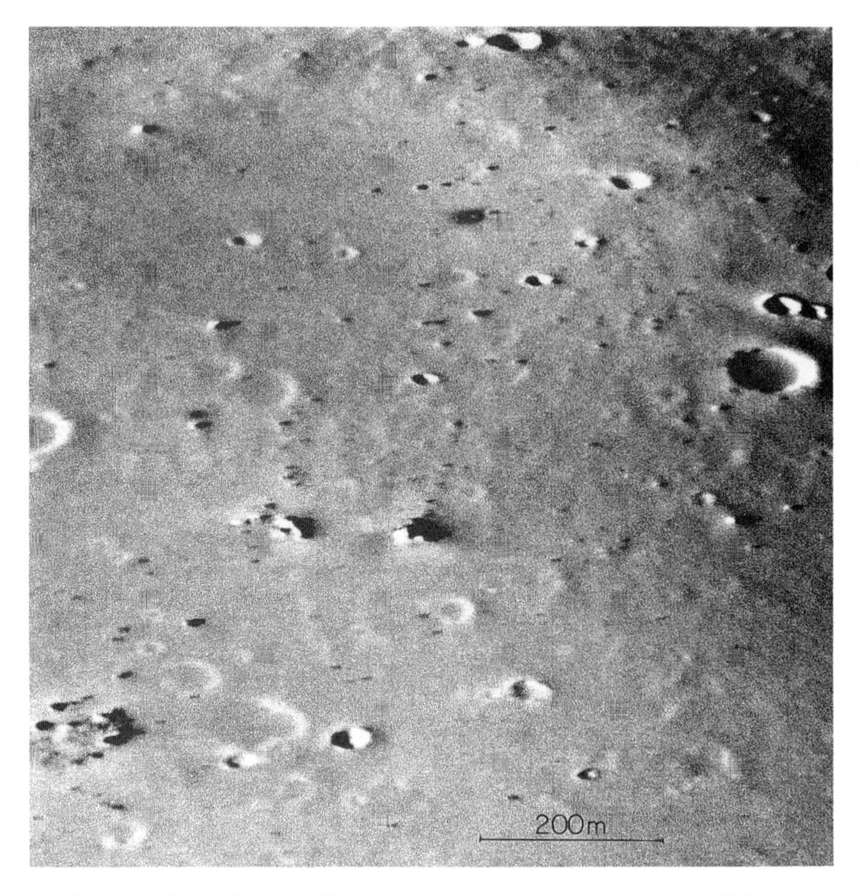

Fig. 8. Deimos: Ejecta blocks and filled craters. Resolution is about 3 m per pixel. This area near the north pole of Deimos is near the crest of a gentle ridge. Clumps of blocks, individually up to 20 m across, litter the surface. A 20-m crater near the top middle exposed darker material. Image 423A63.

IV. DOWNSLOPE MOVEMENT OF REGOLITH ON PHOBOS AND DEIMOS

The appearance of Deimos is dramatically different from that of its companion, with striking global-scale redistribution of material evident in most images (Fig. 6). High-resolution views reveal streamers of brighter material extending from crater rims and ejecta blocks and demonstrate clearly that surface debris is ponded inside craters (Figs. 8 and 10). The sediment in craters has an asymmetric distribution, the asymmetry being consistent with the downslope direction (Thomas and Veverka 1980). Some craters are as full as they can get with the downslope rim covered by debris, but the upslope rim extends above the sediment. Given that there are streamers of material

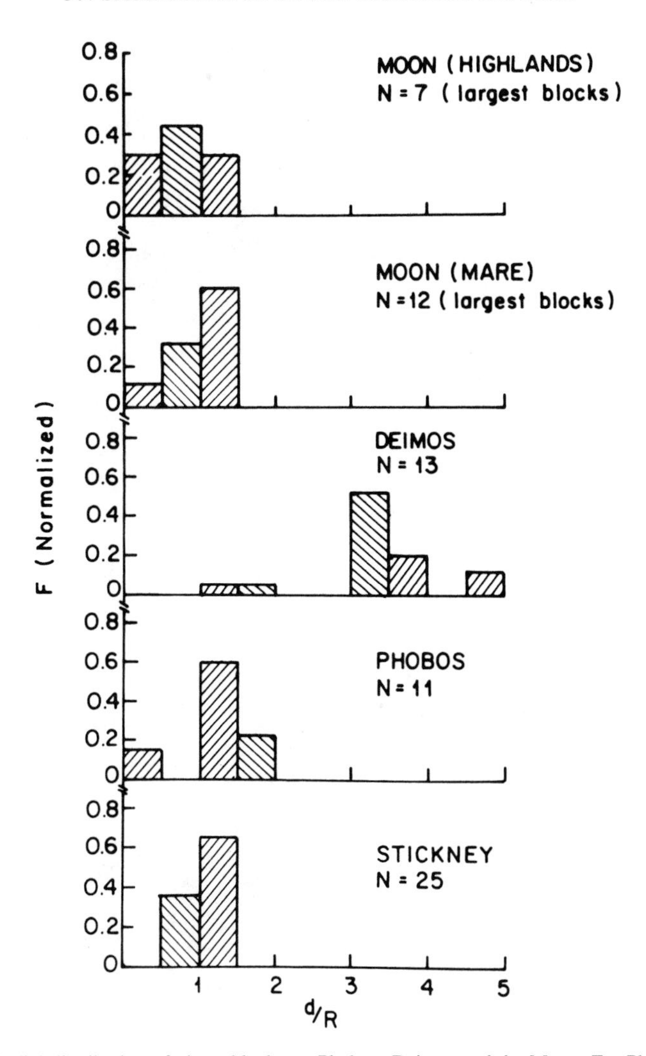

Fig. 9. Radial distribution of ejecta blocks on Phobos, Deimos and the Moon. For Phobos and Deimos the data are all blocks > 20 m in maximum dimension; for the Moon, the largest blocks (10 to 30 m) are included. *F* is fraction of block mass, *R* is the crater radius, *d* is radial distance of ejecta from crater center. Deimos has a significantly different radial distribution of ejecta blocks that parallels the widespread distribution of fine material (figure after Lee et al. 1986).

within the crater fill, the fill in craters does not parallel an equipotential surface. Two important observations indicate that most of the material moves along the surface, rather than on ballistic trajectories: first, craters with raised rims display the smallest amount of fill; second, the flow of material is clearly diverted by relief such as blocks or crater walls (Figs. 6 and 10).

The amounts of material involved in downslope movement on Deimos

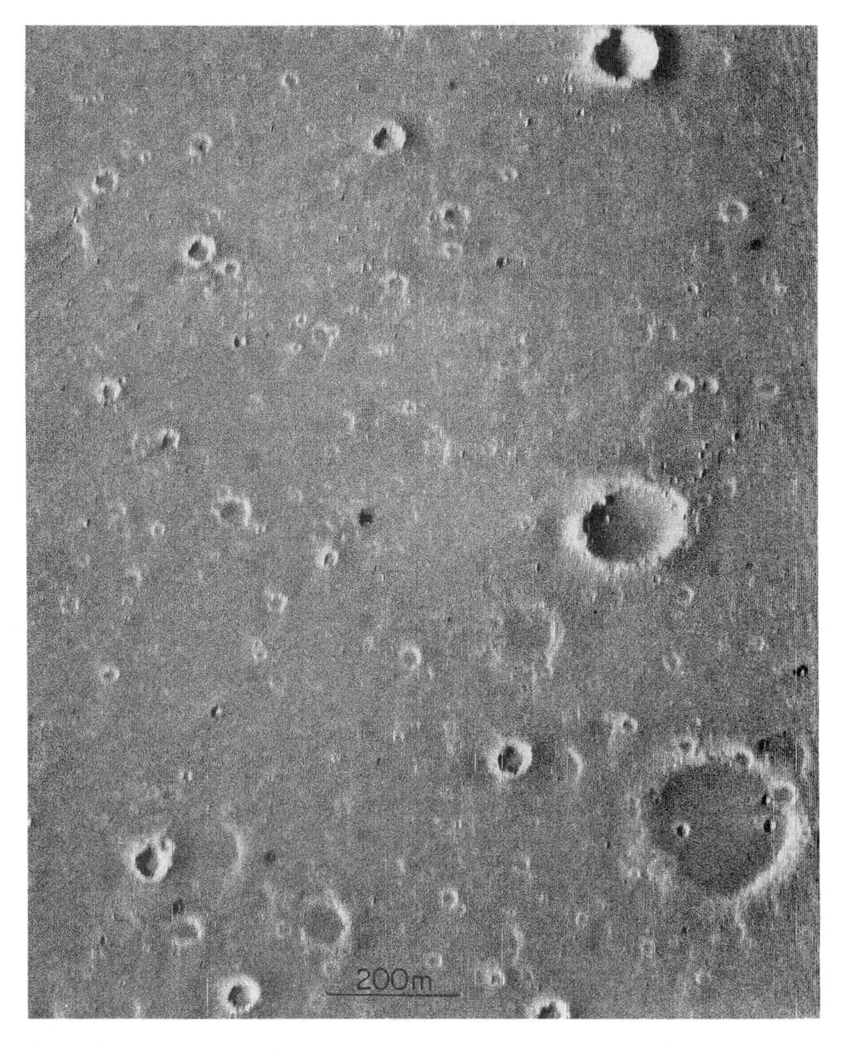

Fig. 10. Crater fill on Deimos. The fill is ponded in the craters asymmetrically. There is a ridge crest at the right of the image; subtle streamers follow slopes to the left. Craters with sharp, raised rims have little fill; older, degraded craters have substantial fill. Image 423A61.

have been calculated by estimating the depth of crater fill seen in the high-resolution images; up to 20 m of debris has been concentrated in some craters (Thomas and Veverka 1980). Lower-resolution images such as Fig. 6 suggest that a 2-km crater on a substantial regional slope has been smoothed by material moving downslope (and by some ballistically emplaced debris?).

Phobos shows no evidence of global downslope movement. No features analogous to the Deimos streamers are observed, nor do craters on the inner moon show signs of ponded sediment. These differences are difficult to explain, but may be due to differences in surface roughness. Whatever the origin of the overall smooth surfaces that define the shape of Deimos, they allow material to move downslope. Phobos's rougher surface would permit material to move downslope only locally. Evidence of localized downslope movement on Phobos, although restricted, has been found (see, e.g., Thomas 1979; Veverka 1978).

It is important to note that despite the downslope movement, neither satellite has reduced its shape to an equipotential surface (equilibrium ellipsoid). Numerical modeling of their shapes combined with the tidal and rotational components of surface gravity have been used to develop effective topography along several tracks across Phobos and Deimos. The results, from Thomas et al. (1986; see Fig. 11), demonstrate that both satellites have substantial regional slopes that show major departures from equilibrium forms.

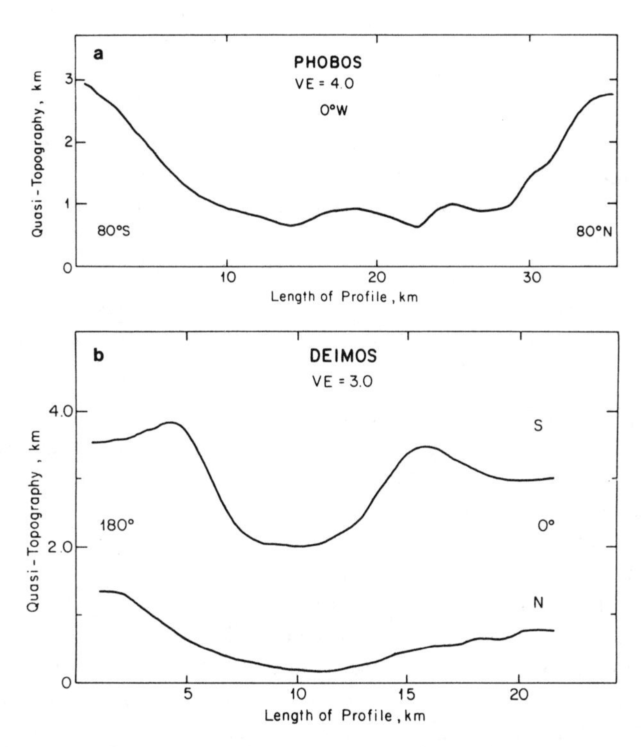

Fig. 11. Calculated effective topography on Phobos and Deimos, after Thomas et al. (1986). The gravitational field has been numerically simulated by estimates of the irregular shape of Phobos and Deimos. The tidal and rotational components are added, and local slopes are calculated by differences of surface normals and resultant surface accelerations. Topography is calculated by integration of slopes along track of a limb profile. Note the very large topography; neither satellite approximates a relaxed, ellipsoidal object. (a) Phobos topography along approximately the 0° longitude. (b) Deimos topography along the 0, 180° longitude. Note that all the surfaces, including the northern one (at bottom) have concave profiles.

The existence of downslope movement on satellites with surface gravity of 0.2 to 0.6 cm s^{-2} is often regarded as amazing. Possible mechanisms were examined by Thomas and Veverka (1980), including thermal creep, micrometeorite bombardment, and impact-related seismic shaking. The effectiveness of thermal cycling depends on gravity affecting the contraction phase of daily thermal cycles. Calculations for Deimos, based on lunar work of Duennebier (1976) and Duennebier and Sutton (1974) showed potential for explaining the volume of material moved on Deimos (Thomas and Veverka 1980). However, the poor knowledge of regolith properties, and the nearly total ignorance of the time scale represented by the features, make it impossible to carry out definitive calculations. Fundamentally the same restrictions apply to modeling the effects of micrometeoritic impacts, secondary impacts and seismic shaking. As in the case of the Moon, the effectiveness of electrostatic transport is probably restricted by the fact that the regoliths almost certainly contain wide distributions of particle sizes (Lindsay 1975).

V. THE GROOVES ON PHOBOS

The grooves on Phobos are among the most puzzling features observed on any satellite. To date, comparable features have not been seen on other objects, although there have been speculations that such features should occasionally occur on other small bodies (Thomas and Veverka 1979*a*). Perhaps the only vague analogs seen elsewhere are the poorly understood linear troughs associated with the large crater Herschel on Saturn's Mimas.

The dramatic global pattern of the grooves on Phobos (Fig. 12) has several important characteristics:

1. Longitudinal symmetry: the geometric pattern of the grooves is largely dependent upon longitude and is symmetric about the 0° longitude. Contrary to some statements in the literature, the grooves do not simply radiate from the crater Stickney.
2. Grooves occur in several sets of parallel members that can cross other sets of grooves.
3. Grooves are deepest near crater Stickney.
4. They are widest near the 0° and 180° longitudes; however, the ones at 180° longitude are very shallow compared to those on the sub-Mars region.
5. An area in the trailing side (± 10°, 250–290° W; Fig. 12) totally lacks grooves. There is no high-resolution coverage of the antipodal area, just west of Stickney, so it cannot be said yet if this area mimicks the trailing side in lacking grooves.

Morphologically, the grooves have straight-to-beaded and pitted appearance, are usually less than 30 m deep and typically 100 to 200 m wide (Thomas et al. 1979*b*); they reach lengths of nearly 20 km (Figs. 7 and 12).

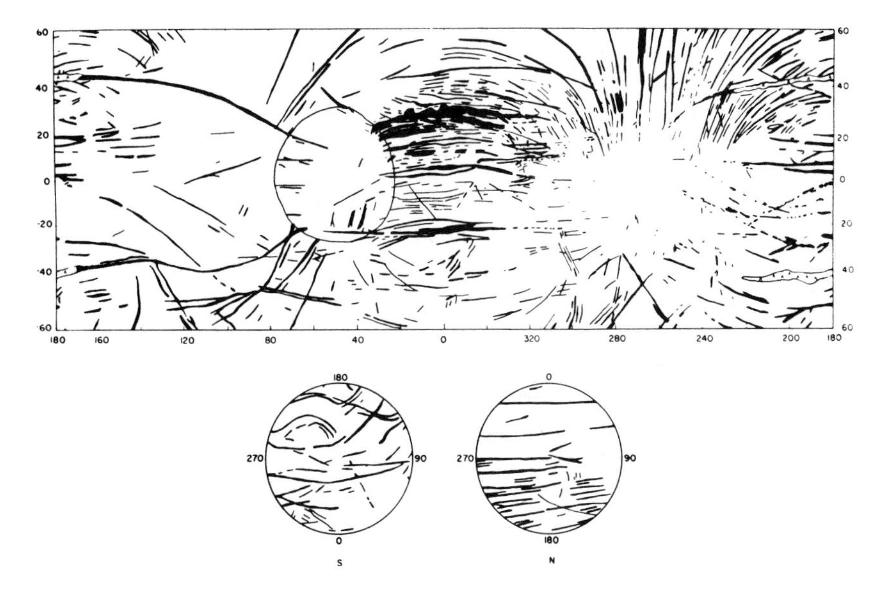

Fig. 12. Map of grooves on Phobos, after Thomas (1979). Note the longitudinal symmetry of most grooves, the morphologic association with Stickney, the crossing sets of grooves, and the large grooves near 180° longitude.

Some of the large grooves near Stickney have complex, hummocky floors and steep, straight walls that suggest material at the angle of repose. There are subtle raised rims on a few grooves, but because so many grooves are nested against other grooves, trustworthy data on groove rims are very scarce (Thomas et al. 1983,1986).

Interpretations of grooves fall into two main groups: surface regolith expressions of deeper fractures (Soter and Harris 1977; Thomas et al. 1979*b*; Weidenschilling 1979; Fujiwara and Asada 1983; Horstman and Melosh 1989) and secondary cratering (Head and Cintala 1979). Combinations of various effects of impacts, layering and degassing have been proposed by Illes and Horvath (1981) and Dobryshevski (1988). The fracture hypotheses derive from the observations of sets of parallel planes defined by the groove surface patterns, scarcity of craterforms in the grooves, the groove-within-groove patterns, and steep walls on the largest grooves. The secondary crater hypotheses rest on some morphologic similarity to secondary crater chains and the difficulty of explaining the putative fracture pattern shown by the grooves. However, the secondary crater hypotheses falter on their inability to explain the geometries of the crater chains in terms of realistic ejecta trajectories (see below).

Fractures with orientations defined by the grooves have not been adequately explained. The orientations of the groove (fracture) planes are clearly related to the principal axes of the satellite: normals to the planes fall very nearly in the plane defined by the long and short axes of the satellites. Unfortunately, available measures of the planarity of the groove traces remain crude. What needs to be explained is how principal stresses that would cause such fractures would maintain their orientations over the irregular surface of the object. Dobrovolskis (1982) examined likely tidal stress patterns and did not find good agreement with the groove patterns. The association of grooves with Stickney prompted experimental impact fracturing of ellipsoidal objects that result in only a very crude approximation of the groove patterns (Fujiwara and Asada 1983).

The linearity and parallelism of the grooves pose severe challenges for models that associate grooves with secondary craters. Many grooves would have to be formed by material that fell at the edge of Stickney's rim as well as nearly going into orbit, all without significant width change along the path of impacts making the groove. Granted the environment for ejecta distribution can be somewhat nonintuitive in Phobos's tidal and rotational environment, but the long sets of intersecting parallel groups seems more than difficult to explain by impacts. Perhaps the best candidates for secondary craters are a few grooves near 180° longitude (Fig. 12). Even these are organized well enough along their lengths that attribution to impacts requires very special conditions.

Because of the morphologic association of grooves with Stickney, fracture hypotheses concentrate on how an impact would form or enhance fractures that in turn disturb the regolith to make grooves. The grooves presumably represent drainage of regolith into fractures, expulsion of regolith from above fractures, or compaction of regolith along fracture traces. The difficulty of establishing the existence of significant constructional rims on the grooves has tempered enthusiasm for expulsion methods. The basic expulsion hypothesis (Thomas et al. 1979*b*) rests on heating of water in hydrated silicates by the Stickney impact with subsequent expulsion of steam along the fractures to fluidize and eject regolith along the fracture traces. This scenario depends on sufficient heating by the impact, adequate water to fluidize regolith, and a plumbing system that allows vigorous degassing around the satellite. While this idea is hard to disprove, it certainly requires several novel processes. One should note that contrary to some recently expressed opinions, the idea is not new. Pollack et al. (1973) identified in Mariner 9 images some of the more conspicuous rows of "pits" associated with Stickney grooves and speculated about "volcanic" processes on this satellite.

The drainage hypothesis has been recently re-examined by Horstmann and Melosh (1989), who regard the pitted appearance of many grooves as crucial indicators of origin by regolith drainage. They point out that this probably means cracks many meters wide must exist in order to accommodate the volume of debris, which in turn would imply a very fragile Phobos. One interesting problem is that many of the groove planes are far from parallel to the local gravity; this might decrease the efficiency of drainage, though it could be argued that unless this angle was lower than a typical angle of repose (35°), the material could still drain away.

In summary, it is difficult to escape the conclusion that grooves on Phobos owe part of their origin to the Stickney impact. An unresolved issue is whether tidal effects played a significant role in developing the groove pattern as suggested by Weidenschilling (1979). In this regard, high-resolution imaging of asteroids will prove helpful: if the impact hypothesis (minus tidal effects) is valid, then groove-like features should be seen on some asteroid surfaces (Thomas and Veverka 1979*a*).

VI. PHOBOS AND DEIMOS: ASTEROIDAL ANALOGS?

Because of their possibly asteroidal compositions, their small sizes and irregular shapes, the Martian satellites have frequently been touted as asteroidal analogs (see, e.g., Veverka and Thomas 1982). The new spectral evidence summarized in Sec. I strengthens this possible analogy. On the other hand, asteroids exist in quite a different environment. Both Phobos and Deimos find themselves deep within the gravitational well of Mars. It has been argued on dynamical grounds that virtually all crater ejecta should be reaccreted somehow, either by not traveling very far initially or by being reaccreted from Mars orbit. One can speculate that the Phobos situation may represent the former case, and the Deimos one a significant component of the latter.

Fortunately, within the next few years we can begin to address these questions. On its route to Jupiter, Galileo will fly by two small asteroids that are in the size range of Phobos and Deimos. In October 1991, the spacecraft is scheduled to fly by asteroid 951 Gaspra (d = 16 km), and two years later will encounter asteroid 243 Ida (d = 32 km). Although both flybys involve S, rather than C objects, the results should help us resolve the Phobos/Deimos dilemma. Additional important data will come from the planned flyby of a C asteroid by CRAF in the late 1990s.

After these flybys, we will be in a much better position to address fundamental questions such as: what is the suite of typical surface processes on a small body? How are these processes affected by material properties (e.g., composition)? How are they modified by the environmental effects of a planetary potential well? Are grooves common features of asteroid surfaces, etc.? But we must come back to a fundamental point: we know that the surfaces of Phobos and Deimos, in spite of similar composition, look dramatically different; and we do not understand why. Seeing the surfaces of a few small asteroids would help set us on the right track. In terms of small asteroid characteristics, is Phobos the anomalous one? Or is it Deimos? Or perhaps, are they both?

VII. THE HISTORIES OF THE MARTIAN SATELLITES

Without getting bogged down in arguments about the origin(s) of Phobos and Deimos, can we establish chronologies for the satellites? The rather uniform covering by craters and regolith on both satellites suggests that records of spallations, catastrophic breakup, capture into orbit, and similar dramatic events do not remain exposed for study by remote sensing techniques.

A first-order fact is that the surfaces of both satellites are heavily cratered and therefore old. While age estimates are very uncertain, there is no reason to believe that the surfaces that we see today differ considerably from those at the epoch of the end of heavy bombardment in the solar system. The grooves appear to have an intermediate age among surface features on Phobos (Thomas et al. 1979*b*). Accurate timing of groove formation is evidently a crucial issue. The impact hypotheses require dates closely contemporaneous with the formation of Stickney. Tidal scenarios would favor later formation as tidal stresses on Phobos have been increasing rapidly (cf. Weidenschilling 1979), although the exact orbital history is not well constrained (see chapter 38). The grooves span at least some interval of time as shown by cross-cutting relationships, but this interval could be very brief, geologically.

Phobos's orbital evolution might help date any features on the surface that are affected by tidal forces, such as asymmetric distribution of ejecta (Dobrovolskis and Burns 1980). At present no such evidence has been identified.

A major problem in the history of Deimos (aside from its origin) is when and how it achieved its overall smooth shape. A smooth ellipsoidal shape would be much easier to attribute to re-accretion of debris following a nearly catastrophic impact than is the arrangement of high-standing ridges. It has been suggested that the saddle in Deimos' southern hemisphere is an impact feature (11 km across; Thomas 1989). Ejecta from such a feature could blanket all of Deimos with over 500 m of material; redistribution could then smooth off features with relief less than 1 to 2 km. However, such a scenario leaves unexplained the ridge geometry. The ridges are themselves smooth and continuous, and at least the northern ridge would predate the large impact and remain unexplained. We do not know spatial variations of crater densities on Deimos, so it is impossible to compare the ages of the major surfaces on the satellite. Knowledge of the real depth of regolith over Deimos would help directly in investigating the satellite's structural history. Regardless of the timing of the largest cratering event on Deimos, the fill in many smaller craters suggests that ejecta continue to be spread over the satellite in the geologic present.

VIII. CONCLUSIONS

The fundamental puzzle about the two satellites of Mars is why their surfaces look so different, given that in terms of compositional and bulk properties they are so similar (Table I). We do not understand to what extent the pronounced surficial differences result from as yet undetected differences in the subsurface makeup of these moons, or to what extent they can be ascribed to the unusual, but different, tidal environments in which these bodies have existed for presumably many billion years.

The Earth-based and spacecraft data continue to suggest a relationship of the satellites to carbonaceous asteroids. The degree of hydration of the satellites as a whole remains uncertain, although the 3-μm data summarized in Sec. I strongly suggest that water is scarce within the topmost layers of the regolith accessible to remote sensing. If so, the characteristics of the interiors must be quite different to explain the relatively low mean densities of around 2 g cm^{-3}. Given the deduction from the libration data that Phobos at least is relatively homogeneous in density (i.e., that it has neither an overdense or an underdense core), the possibility remains that by bulk the two satellites contain considerable water of hydration and that only the surface layers are dry. The most water-rich samples of CI chondrites have densities around 2.2 g cm^{-3} (compared with 2.7 for CM's and 3.4 for CO/CV's). As the density of neither satellite is currently known to better than $\pm$ 0.3 g cm^{-3}, it is impos-

TABLE I
Phobos and Deimos: Similarities and Differences

	Phobos	Deimos
Similarities		
Albedos[a]	0.05 ± 0.01	0.06 ± 0.01
Spectral Reflectance[b]	C type	C type
Mean Density[c] (g cm^{-3})	2.2 ± 0.5	1.7 ± 0.5
Surface Thermal Inertia[d]	0.6–2.0	0.9–1.6
($Jm^{-2} s^{-1/2} K^{-1}$)	25–84	38–67
Differences		
Crater Density[e]	higher	lower
Cratering Behavior	gravity dominated?	strength dominated?
Ejecta Dispersl	localized	widespread
Downslope Movement of Regolith	localized	global
Tectonics/Structure	global groove system	no grooves

[a]French (1980).
[b]Lucey et al. (1989).
[c]Duxbury and Callahan (1982).
[d]Lunine et al. (1982). Units are 10^{-3} cal $cm^{-2} s^{-1/2} K^{-1}$
[e]Thomas and Veverka (1979*a*).

sible to place firm constraints on interior properties. Certainly, if it were demonstrated that the densities really are as low as 2 g cm^{-3}, one would have to consider seriously the possibility of actual water ice within the satellites, or argue for a very uncompacted internal structure.

A better understanding of the material properties of the surface layers and realistic modeling of cratering and ejecta processes are needed to understand the evolution of the surfaces. In this context, we also need further work on the dynamics of ejecta in the vicinity of Mars. Past work has indicated that most ejecta will not escape the Mars system, and will therefore eventually be re-accumulated or will run into the planet. It is fair to say that we lack as yet sufficient understanding of this re-accumulation process to assess its influence on the development of the satellite surfaces. Numerical simulations have shown that ejecta trajectories, especially in the case of Phobos, are extremely complicated. Because most ejecta are produced by major impacts, which according to recent work by Wisdom (1987*b*) will also throw Phobos or Deimos into a temporary chaotic spin state, can we really claim that we understand ejecta redistribution within the Mars satellite system?

In terms of the possible connection of Phobos/Deimos and asteroids, some limited progress has been made. The compositional data certainly allow an "asteroid origin" for the two moons of Mars. Whether the surface processes that we see on either satellite are representative of those on asteroids of comparable size and composition will only be resolved after a few spacecraft flybys of minor planets such as those currently planned for Galileo, CRAF, etc. For example, we should eventually be able to tell whether grooves occur on some asteroids, or whether their development does require strong tidal effects as has been claimed by some. Until a few asteroid flybys have taken place, we will not know which of Phobos or Deimos, if either, looks more like a true asteroid, but our experience with the satellites of Mars has already taught us an important lesson: whole-disk (spatially unresolved) remote sensing data are very poor indicators of diversity of surface characteristics and processes. In terms of whole-disk remote sensing, photometry, spectral reflectance and thermal radiometry, Phobos and Deimos are very similar. Yet one good image of each is enough to reveal that quite different things have happened on their surfaces.

38. CONTRADICTORY CLUES AS TO THE ORIGIN OF THE MARTIAN MOONS

JOSEPH A. BURNS
Cornell University

The meager available information that is pertinent to the origin and evolution of the Martian satellites is contradictory. The known physical properties of the Martian moons (density, albedo, color and spectral reflectivity) are similar to those of many C-type asteroids, the dark "carbonaceous" objects abundant in the outer belt but scarce near Mars; thus this line of physical evidence suggests that Phobos and Deimos are captured bodies. In contrast, calculated histories of orbital evolution due to tides in the planet and in the satellites indicate that these small craggy moons originated on nearly circular, uninclined orbits not far from their current positions; hence dynamicists prefer an origin in circum-Martian orbit. Ways are described in which these apparently contradictory viewpoints may be reconciled, although a definitive answer to the origin of the Martian satellites will almost surely have to await in situ *measurements.*

During the 1989 encounter of the Phobos spacecraft with its namesake satellite, the popular press was fond of writing that the Martian moon was of interest primarily because it is a captured asteroid. However, as will be described here, the scientific jury actually remains divided on the question of this satellite's origin. Dynamicists argue that the present orbits (see Table I) could not be produced following capture, and so the objects must have originated near Mars, where the satellites are found today, even if they do not appear as though they had been born at 1.5 AU. Virtually all other scientists maintain that every observable physical property indicates that the Martian satellites once resided in the outer asteroid belt (at $\sim$ 3 AU), and thus the objects must be captured, although the precise mechanism whereby that occurred is often conveniently unspecified.

TABLE I
Dynamical Parameters for the Martian Satellites[a]

	Phobos	Deimos
a(km)	9378.5 (2.76 $R_{♂}$)	23459 (6.92 $R_{♂}$)
Period	7hr 39 min	30 hr 18 min
n (deg/day)	1128.8446	285.16191
e	0.01515($\pm$0.00004)	0.000196($\pm$0.000034)
i (deg)[b]	1.068($\pm$0.001)	1.789($\pm$0.003)
I (deg)[c]	0.00934	0.8965
Mass (kg)	1.08($\pm$0.01)$\times 10^{16}$	1.80($\pm$0.15)$\times 10^{15}$
Axes (km)[d]	13.4($\pm$.4)$\times$11.2($\pm$.2)$\times$9.2($\pm$.1)	7.5($\pm$.3)$\times$6.1($\pm$.3)$\times$5.2($\pm$.2)

[a]Orbital parameters are taken from Chapront-Touzé (1990) but are similar to those given by Sinclair (1989) and Shor (1988). Phobos' mass comes from Avensov et al. (1989), and the semiaxes of assumed triaxial ellipsoidal shapes from Thomas (1989; cf. Duxbury and Callahan 1989); Duxbury (1989) instead fits a 6th degree and order model to Phobos' surface with a typical error of a few hundred meters. See also Burns (1986*b*).
[b]Referred to the Laplace pole.
[c]Inclination of the local Laplace pole relative to Mars' rotation pole.
[d]Slightly revised values are given in chapter 36.

This chapter will review the lines of evidence for capture as well as for *in situ* origin and will discuss the inferences made therefrom. Throughout, ways will be suggested whereby the nominal models of the two camps may be modified so that, on the one hand, primordial Martian moons could exhibit properties similar to objects from the outer asteroid belt or, on the other hand, orbital evolution scenarios could be misleading. This chapter begins by reviewing ideas concerning the origin of Phobos and Deimos.

I. PROPOSED ORIGINS

Little thought has been devoted to developing detailed models for the origin of Phobos and Deimos, despite the uniqueness of the Martian moons as the only extant small satellites of the terrestrial planets (see, e.g., Stevenson et al. 1986). Nevertheless, as mentioned above, two camps have formed, one that argues for accretion in orbit and the other that maintains that the moons of Mars originated elsewhere only to be later captured by Mars.

According to the *accretion* school (Safronov et al. 1986), the Martian satellites, like other regular satellites (see Burns 1986*b*), are an unavoidable consequence of planetary formation. Circumplanetary swarms of orbiting debris result when heliocentric particles collide inelastically within a planet's sphere of influence. Once a swarm surrounds a planet, some fraction of it is gradually lost as this orbiting material systematically decays down to the planet's surface, while the swarm itself is continually replenished through ongoing collisions with other heliocentric particles. The processes that govern the agglomeration of the Martian satellites might be similar to those that produced the regular satellites of the giant planets, although in the latter case gas-dominated accretion was more probable. The small sizes of Phobos and Deimos compared to the satellites of the giant planets may be attributed to the violent bombardment of Mars' locale by invading planetesimals from the zone of Jupiter (Safronov et al. 1986).

As the final remnant of the nebula from which Mars itself grew, the satellites, according to the accretion school, should be composed of material like Mars. They will, however, contain an especially high proportion of components that were brought in with the last heliocentric objects to be acquired by Mars; of course, similar constituents should also form a disproportionate fraction of the surface layers of Mars. As this late-arriving material might have originated from a different region of the solar nebula than bulk Mars' material, the moons could have a composition separate from that of the planet as a whole. Extending this idea to the very last accreted material, one might think that the satellites themselves could be coated by a thin veneer having an especially distinctive make-up. However, even if this were the case initially, subsequent bombardment by stray comets and asteroids should have disrupted any objects that originally accreted about Mars (Shoemaker 1989) so that upon their re-accretion in orbit, the contemporary Martian moons may have been homogenized. However, this remixing process has not been studied and the only available evidence (from collected meteorites) shows that, at least on the regoliths of the small bodies where meteorites originated, energetic events still allow significant heterogeneity in small samples. Perhaps on the larger scale or with the re-accumulation of lost ejecta by the Martian moons due to their orbits about the planet, the satellites should be homogeneous but we simply do not know.

Permanent *capture,* the alternate proposal for the satellite's origin, requires energy loss; otherwise the time reversibility of Newtonian dynamics would allow escape ultimately (Burns 1986b). Sufficient energy loss transforms a hyperbolic orbit into a bound elliptic one as long as the loss occurs somewhere within Mars's sphere of influence, a region of some several hundred Martian radii surrounding the planet within which the planet has gravitational dominance over the Sun (cf. Hamilton and Burns 1991). *Tidal* capture, which not too many years ago was the only capture mode thought to be relevant to the birth of Earth's Moon (see Burns 1977; Boss and Peale 1986), is especially unlikely for the Martian moons since their small sizes mean that relatively little energy can be drained from a hyperbolic orbit during a single pass of the planet. Hence, by elimination, *aerodynamic* drag has been taken as the preferred process for capture of the Martian satellites; it has also occasionally been considered as a mechanism to capture the Moon (Nakazawa et al. 1983). Usually the drag has been considered to take place in a nebula that surrounds the planet shortly after it formed. Evolution through such a nebula will essentially always cause the orbital semimajor axis to collapse, the orbit to circularize and the inclination to decay (cf. Pollack et al. 1979*a*). To have a reasonable probability for capture at many planetary radii, a fairly substantial nebula is required but, in such a nebula, evolution will be rapid so that captured objects are swiftly transported to the planet's surface. That is, there is a fundamental conundrum associated with capture by nebular drag: if capture occurs close-in, where the nebula is dense and retention more likely, then the captured body is likely to be gone after a few orbits; whereas, when the energy loss transpires at great distances from the planet, where the orbit will not evolve rapidly, the process becomes highly unlikely because the slow evolution means that there is little, if any, nebula to decelerate the particle in the first place. Unfortunately, it is not possible to go beyond these generalities to clarify the limits of plausible gas density histories that would lead to probable captures.

Burns (1978) and Pollack et al. (1979*a*) have described qualitatively the possible capture of a single object by aerodynamic drag in a circumplanetary envelope about Mars. Burns suggested that the current Martian configuration, where individual satellites lie on either side of synchronous orbit, might have arisen by a sequence of two events. First, a heliocentric planetoid is captured by Mars through the action of gas drag and, under the same influence, evolves to near the synchronous orbit position, where, for a sufficiently thin nebula, the evolution will effectively cease because there is little relative velocity between the captured object and the nebula. Then, that captured object is struck by another heliocentric projectile and shattered into two main fragments, the larger of which (by happenstance) lands inside synchronous orbit while the smaller falls beyond the synchronous position; these two objects then evolve tidally through the eons, with the bigger one moving faster. Giving even fewer details, Shoemaker (1989) mentioned a similar scenario in which Phobos and Deimos were split apart near synchronous orbit during the heavy bombardment epoch.

Capture by an extended protoatmosphere about Mars was first considered in detail by Hunten (1979*a*) and more recently was pursued by Sasaki (1990). In the first of these capture scenarios, the atmosphere was initially modeled as a slowly rotating condensation of the solar nebula with a surface pressure of a few tenths of a bar, but under such circumstances capture was found to be improbable. The likelihood improved when Hunten took the atmosphere to be rapidly rotating or its density to be 1 or 2 orders of magnitude higher. Hunten considered the high-density case in some detail and argued that the evolution was halted once the upper atmosphere rapidly dissipated following the removal of the solar nebula. In this scenario, as in that of Pollack et al., many objects were captured by proto-Mars and then orbitally evolved inward and were lost to its surface: Phobos and Deimos are distinguished solely by being the final planetesimals to approach Mars while it still retained its relatively dense surrounding nebula.

Sasaki's study addresses numerically the history of captured objects as their orbits evolve through an extended atmosphere in the context of a full three-body problem. He argues that, at great distances from the planet where the initial capture takes place, solar tides produce important perturbations to the orbit. He considers evolution through an extended H_2-He atmosphere that has condensed from the solar nebula as such an atmosphere will have a larger scale height and gas density than would that of a degassed secondary atmosphere. In these models, the periapsis distance initially increases following capture, and that extends the evolution time so that, for a careful choice of initial conditions, it may reach 10^3 to 10^4 yr. Sasaki expects that this time could be lengthened more by lowering the nebular density even further to the point where the capture time could equal the lifetime of the solar nebula. It will be interesting to learn, through further numerical exploration, how narrow are the windows that allow this fate.

Hartmann (1987,1990) points out that the irregular outer satellites of Jupiter and Saturn, which *are* generally believed to be captured objects (see Burns 1986*b*), are apparently all members of spectral class C, which dominates the central part of the asteroid belt. He maintains, therefore, that these captured satellites originated in the asteroid belt and were scattered by Jupiter throughout the solar system. Since Phobos and Deimos also seem to be similar to C objects, Hartmann claims that they too were part of this scattered and captured population.

II. CLUES TO ORIGIN: PHYSICAL PROPERTIES

Essentially all known physical characteristics of the Martian satellites are consistent with the proposition that Phobos and Deimos are immigrants to the Mars region from the outer portion of the main asteroid belt. As chapter 37 describes in more detail, the limited available data on the bulk densities and on the surface properties of the Martian moons show that these characteristics of the satellites lie in the general range expected for minor planets from the outer asteroid belt, say the region from 2.8 to 3.3 AU. While this viewpoint has been generally accepted for many years, the case was strengthened considerably by several findings in the late 1980s.

The first piece of recent evidence concerns the most fundamental physical property. Estimates of the *bulk density* of the inner Martian satellite were narrowed by the Phobos mission's imaging team (Avensov et al. 1989,1991) to 1.90 ($\pm$ 0.10) g cm^{-3}, which falls at the lower end of the error bound of the most recent Viking determination, 2.2 ($\pm$ 0.3) g cm^{-3} (Duxbury and Callahan 1989). Since the volume of Deimos is poorly defined, this satellite's density is not as constrained, 1.7 ($\pm$ 0.4) g cm^{-3} (Duxbury and Callahan 1989), but once again is relatively small compared to most likely constituents.

These low satellite densities approach those of the most volatile-rich carbonaceous chondrites CIs and CMs, which have the lowest densities of any meteorites. CIs have densities as low as ~ 2.2 g cm^{-3} whereas other classes of carbonaceous chondrites have densities in the range 2.5 to 3.5 g cm^{-3} (Mason 1962; cf. Wasson 1974). These densities are significantly less than those of the minerals out of which Mars is thought to have accumulated; accordingly they have been interpreted by some to suggest that the satellites *are* carbonaceous. However, the densities of the materials that most closely match Phobos's spectral reflectivity (see below), namely, anhydrous and hydrous carbonaceous chondrites or optically blackened mafic material, are not nearly as low as these numbers (Britt and Pieters 1988, Bibring et al. 1989).

If one does not wish to accept that these measured low-bulk densities require unaltered carbonaceous material, another possible interpretation of them is that, in addition to dense Martian material, the satellite interiors contain a substantial component of water ice; this volatile compound can be shielded at all depths greater than tens-to-hundreds of meters over the age of the solar system by an insulating satellite regolith having nominal properties (Fanale and Salvail 1989). The presence of some water ice in the vicinity of Mars at the time of its origin is not unreasonable (Lewis 1974*a*) as temperatures today in that locale are marginally low enough for its condensation and, at the time of the satellites' birth, the Sun is thought to have been cooler; moreover, the immediate neighborhood of Mars as the planet formed may have been screened from direct solar radiation by the opacity of any nebula surrounding the planet.

Another explanation for the reduced densities of the Martian moons may be that the satellites are totally unconsolidated (even at depth) and contain significant open pore space; after all, the highest internal compressive stress experienced in Phobos is < 100 mbar (10^4 Pa) and, in Deimos, it is even smaller (Dobrovolskis 1982). These pressures are scarcely enough to overcome the cohesion of the lunar soil and are certainly not sufficient to eliminate the pore space in unconsolidated geologic materials nor to crush such materials. A final possibility is that the surface layers of the moons are deep and especially porous, much like those suggested for Mimas (Dermott and Thomas 1988). Several comments are in order in connection with these last two ideas. First, lunar rocks that start with densities of approximately 3.0 to 3.3 g cm^{-3} develop densities of about 2.0 g cm^{-3} during the process of brecciation through impact in the lunar regolith (Britt and Pieters 1988). Second, such low densities might result as a consequence of the satellite's primordial accretion (or re-accretion out of orbiting debris following a catastrophic breakup) in the low gravity environment found in circum-Martian orbit. To reconcile the discrepancy between Phobos's measured density and those of suggested meteorite analogs requires porosities of 10 to 30% in the case of lower-density carbonaceous chondrites and 40 to 50% for the higher-density black chondrites (Hartmann 1990; Murchie et al. 1991*a*). The lower porosities are consistent with values of 10 to 24% found in some meteorite breccias (Wasson 1974) and measured for lunar regolith breccias of 30% or more. However, it seems improbable that the high values required by a black chondrite composition can be achieved. Finally, the porous regolith model appears to be inconsistent with the measured amplitude of $0\overset{\circ}{.}8$ ($\pm$ $0\overset{\circ}{.}4$) for Phobos' libration, which implies that this satellite's density is uniform with depth (Duxbury 1989). Borderies and Yoder (1990) point out that Phobos' gravity field induces a secular change in the position of the satellite's pericenter; from a recent model of the observed orbital motion, they argue that the satellite must have a nonuniform density although an even more recent orbital model (Chapront-Touzé 1990) is found to be consistent with a constant density moon. The porous regolith model also does not agree with the interpretation of radar returns from Phobos that indicates the bulk density of its surface is 2.0 ($\pm$ 0.4) g cm^{-3} (Ostro et al. 1989).

So, in summary, the measured densities of the Martian satellites are surprisingly low for any meteorite class. This problem can be eliminated by mixing in some low-density volume, either pore space or ice. However, once one admits the possibility that some low-density material is present, then the densities of the satellites are no longer particularly diagnostic of composition because any intrinsically high density can be accommodated by just stirring in more pore space.

The second piece of evidence pertinent to satellite origin concerns the inferred composition of the surfaces of the Martian satellites. When interpreting this evidence, it is worthwhile to recall first that only the topmost layers are probed by remote-sensing techniques and that these layers may not indicate the true nature of the bulk interior. *Spectral reflectance* data for Phobos and Deimos obtained from the ground and from spacecraft are plotted in Figs. 1 and 2 (cf. Figs. 1 and 2 of chapter 37). Figure 1 compares the reflectances of Phobos and Deimos with those of several selected meteorites and the asteroid Ceres (cf. Fig. 1 of Britt and Pieters 1988, Fig. 1 of chapter 37 and Fig. 2.1.3 of Wetherill and Chapman 1988). Figure 2 shows data obtained by the Phobos spacecraft and illustrates possible matches to some meteorites (Murchie et al. 1991*a*).

The satellites have quite low geometric albedos ($\sim$ 0.05) and generally flat spectra in the visible, although they drop off at the shortest wavelengths. Black, neutral objects such as these are common in the outer part of the main asteroid belt; in contrast, asteroids closer to the Sun are brighter and inferred to be siliceous while materials from farther out in the solar system (e.g., the Trojan and Hilda asteroids) tend to be dark and red (Hartmann 1987). Bibring et al. (1989) argue that the very low albedo of Phobos and the satellite's weak hydration band (which is significantly shallower than that of Mars) imply that the satellite's surface composition is most similar to metamorphic grade 3 meteorites (which show no visible alteration by groundwater) rather than the water-rich C1s or C2s (cf. Clark et al. 1986); these are the most dense of the chondrites and thus do not give a match to the satellite densities.

Murchie et al. (1991*a*) discern at least four recognizable spectral units

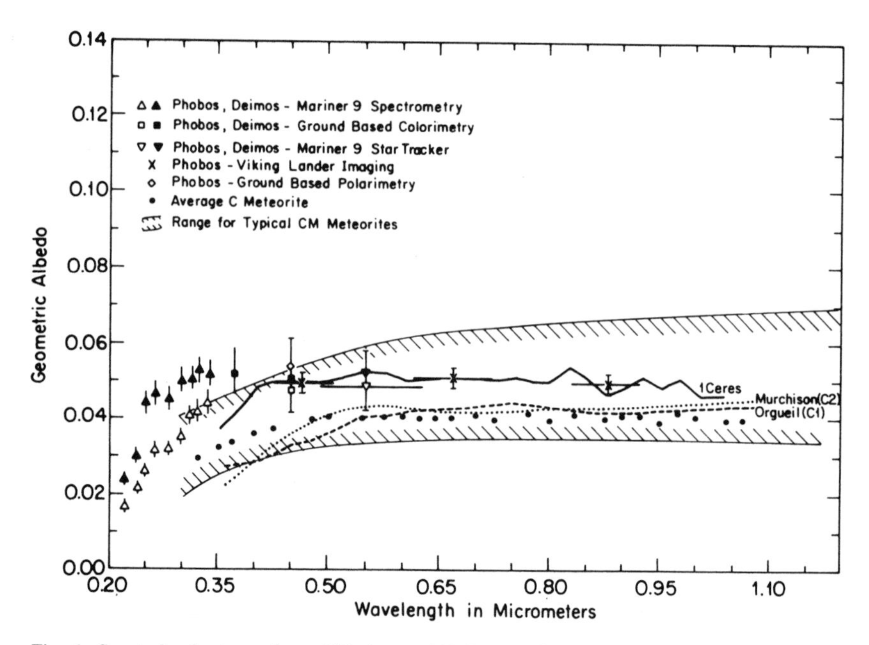

Fig. 1. Spectral reflectance data of Phobos and Deimos, selected carbonaceous meteorites and the asteroid Ceres. Figure adopted from summaries presented by Britt and Pieters (1988, Fig. 1), chapter 37 and Wetherill and Chapman (1988, Fig. 2.13).

on the satellite (see Fig. 2). They claim that the large color ratio exhibited by one of these spectral units, the "blue" material found on Phobos, is inconsistent with a carbonaceous chondritic composition, but is comparable to that of an assemblage of mafic minerals like those forming black chondrites, which are anhydrous, optically darkened, ordinary chondrites. From its color ratio and its ultraviolet-visible spectral properties, the "bluish gray" material also has black chondrites as its closest spectral analog. Both carbonaceous chondrites and black chondrites match the ultraviolet-visible spectra of the "reddish-gray" and the "red" units identified by Phobos investigators. Black chondrites are also anhydrous but, as previously stated, have high densities in the range of 3.4 to 3.8 g cm^{-3} (Wasson 1974).

The near-infrared reflectance data obtained by the Phobos spacecraft (Bibring et al. 1989) indicate that the surface material exhibits a weak and spatially variable red slope upward between 0.8 and 3.0 μm, and demonstrates no strong water of hydration signature in the 3-μm region (Langevin et al. 1990).

Most recently, Murchie et al. (1991*b*) attempt to integrate results from the three multispectral detectors, aboard the Soviet spacecraft (the VSK "visible" and "near infrared" channel [Avenesov et al. 1990], the KRFM 0.3–0.6 μm instrument [Ksanfomality et al. 1991; see Fig. 2] and the ISM infrared imaging spectrometer [Langevin et al. 1990]). They argue that two fundamental spectral units are seen within the region where the detectors have overlapping coverage. The "bluer" of these two materials is considered to be ejecta from the large crater Stickney having a black chondrite-like composition. The "redder" unit is believed to be a mix of a red component (argued to be made up of a low-density carbonaceous chondrite material such as CM chondrites) having a weak or negligible 1-μm absorption and a gray component (averred to be like a black chondrite). Murchie et al. (1991*b*) maintain that this mixture reconciles the measured low density with their preferred spectral analog of black chondrites. They then go on to claim that such a mixture implies formation from more than a single source, whether planetary or asteroidal.

The spectral data on Deimos has been recently surveyed by Bell et al. (1989*b*) who conclude that the satellite's reddish slope in the near infrared, flat visible continuum and strong ultraviolet absorption are seen only in a few asteroids lying between C and P classes. In their own infrared data, Bell et al. (1989*b*) find just a weak 3-μm band on Deimos suggesting that at least this moon's surface is anhydrous. This observation, of course, says nothing about deeper layers which may contain hydrated clay minerals or even water ice.

The close correspondence of the spectral reflectances of the Martian satellites to various meteorites has been used to aver that the bodies **are** the same (see, e.g., Bell et al. 1989*b*), but one should be chary of such claims because existing remote-sensing data can do no more than provide nonunique identifications. While the dark neutral character of these spectra must imply some opaque component, that material could be carbon, Fe-Ni metal, troilite or something else. Furthermore, relatively small amounts of any opaque component could dominate the spectra and confound their interpretation (Clark et al. 1986). The final note to make is that, at present, it is not possible to make a one-to-one match between the spectra of the Martian satellites and those of any specific class of chondrite; all those that match the visible spectrum have a strong clay band at 3 μm, something not seen for the satellites.

Processes in the regoliths of asteroids almost certainly have altered the optical signatures of these surfaces. This has been pointed out particularly by Britt and Pieters (1988) but is contested by J. F. Bell (personal communication, 1990) who argues that meteorite breccias show that the typical surface material on the meteorite source bodies is not highly shock-blackened. In some respects, processing of the surface materials on the Martian moons will be similar to that on minor planets, but in one possibly profound way it will differ. Since Phobos and Deimos reside in Mars's gravitational potential well, almost all the impact ejecta that is thrown off their surfaces will ultimately re-impact them (Soter 1971). This means that the constituents comprising the outer skin of the satellites have suffered innumerable collisions although most of these will occur at low impact velocities; exactly how, or even whether,

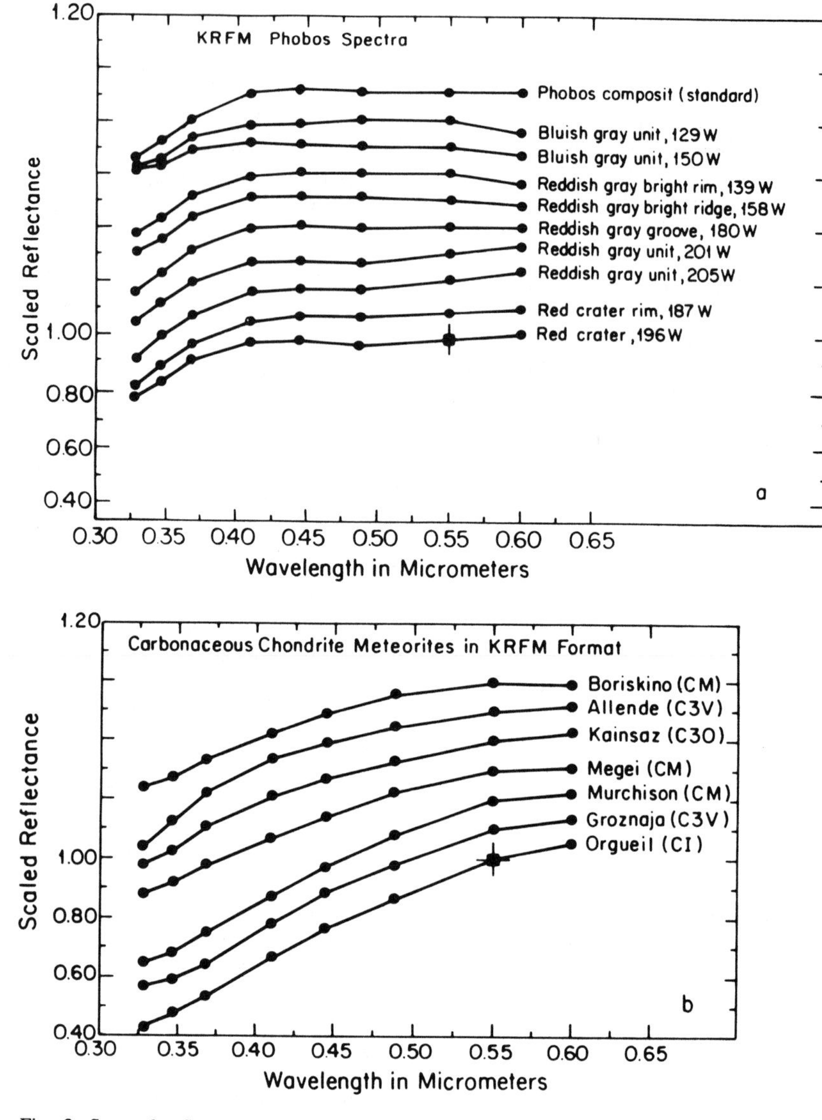

Fig. 2. Spectral reflectance data of Phobos taken by the KRFM instrument on its namesake spacecraft and compared to meteorite data. (a) 0.3–0.6 μm reflectance spectra of particular color-ratio units located in different geologic settings on Phobos, as identified at the right. All curves are normalized at 0.55 μm (square symbols) and are offset from one another by 10% for clarity. The left-hand scale only gives relative values, and the values for 0.4 to 1.0 on the scale pertain to the bottom curve. The curves were calibrated using a representative standard and a composite whole-disk spectrum of Phobos compiled from Mariner 9 UVS, Viking Lander camera and telescopic observations, and then resampled into the frequency response functions of the KRFM channels. (b) Normalized reflectance spectra of carbonaceous chondrite meteorites that are possible spectral analogs to Phobos, resampled into the KRFM bandpasses. (c) Laboratory spectra of blackened and less altered normal ordinary chondrite meteorites. (d) Laboratory spectra of differentiated meteorites, including SNCs and the euchrite assemblage; SNCs are closely related achondrites that are believed to come from Mars, while euchrites, also achondrites, are magmatic rocks. Spectra are ordered approximately by decreasing redness (figure from Murchie et al. 1991*a*).

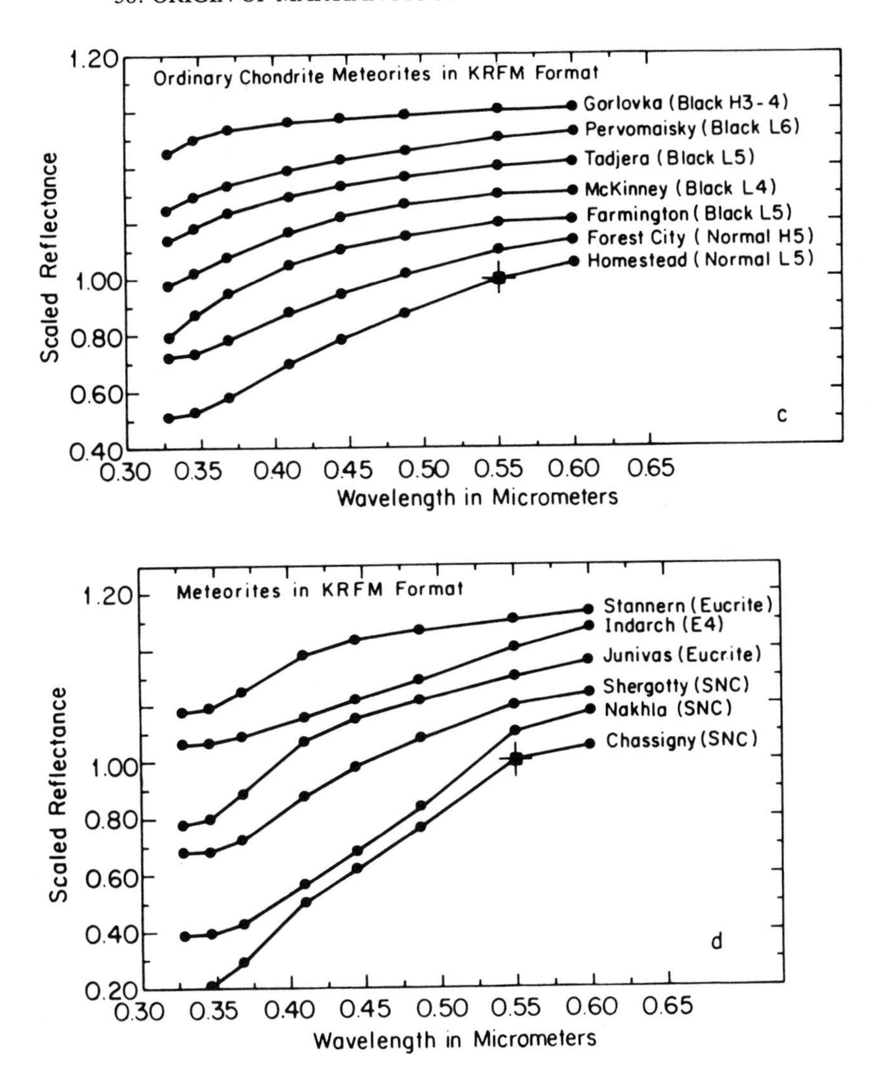

these many low-speed collisions will shock-alter the surface materials in some unique way is unknown.

Besides having experienced this unusual collisional environment, materials on the surfaces of the Martian satellites will have also spent considerable time isolated and exposed in space. In contrast, only a small fraction of the material resident on the surfaces of comparable-sized asteroids will have undergone a similar history; indeed, it is suspected theoretically that most of a small asteroid's surface will be bare rock rather than accumulated layers of recycled regolith (Housen and Wilkening 1982). Thus Britt and Pieters

(1988) caution that measured optical characteristics of the Martian satellites may not reflect those of the underlying material (cf. Murchie et al. 1991*a*).

Among the most interesting findings of the 1989 Soviet space mission insofar as elucidating the satellite's origin is that Phobos has a *heterogeneous surface*. It exhibits 10% variations on a kilometer scale in albedo and in the strength of the hydration band (Bibring et al. 1989) as well as in thermal reflectivity (Ksanfomality et al. 1989). Murchie et al. (1991*a,b*; Murchie 1990; cf. Bell et al. 1990*a*) identify lateral variations up to 45% in the surface color ratio and ultraviolet-visible (KRFM) spectra; they associate these variations with large craters and propose that some other distinct material has been brought up from depth in these regions. Viking measurements had earlier shown some variations in albedo and in thermal inertia (see Veverka and Burns 1980; chapter 37). If the spectral and albedo variation seen on Phobos indicated chemical or mineralogical differences across the surface, how such variations might occur on a satellite that is believed to have re-accreted many times over (Shoemaker 1989; see the discussion in Sec. I) is a mystery. Perhaps a nonuniform Phobos would be easier to understand if the satellite accreted inhomogeneously in the asteroid belt and was subsequently (but gently) captured intact. On the other hand, the observed variations might be caused simply by the effects of particle size which likely would differ in the neighborhood of craters (J. F. Bell, personal communication, 1990).

Essentially all of the discoveries about the Martian moons made during the 1980s have only strengthened the convictions of many planetary scientists that Phobos and Deimos are captured asteroids. However more thought should be devoted to seeing whether we are being misled by nonunique remote sensing data and incomplete information.

III. CLUES TO ORIGINS: ORBITAL HISTORIES

Sharpless (1945), using five sets of position measurements taken in the period 1877–1941, proposed that Phobos's orbit was secularly approaching Mars. Although this interpretation may have been premature at the time it was given, subsequent observations have shown it to be true, albeit with an actual value for the secular acceleration of only about one-half that computed by Sharpless. Modern determinations of the acceleration in the longitudes of the satellites, which combine both groundbased and spacecraft observations of the Martian satellite positions relative to Mars and one another, are listed in Table II. Improvements in these values will come only gradually with refined positions from groundbased observatories and especially from spacecraft missions to Mars (cf. Kolyuka et al. 1990), as well as perhaps radar returns (Ostro et al. 1989). Phobos' measured acceleration corresponds to an added 15° of longitudinal displacement since the satellite's discovery in 1877.

The orbital acceleration of Phobos, compared to its Keplerian rate, is now generally acknowledged to be due to solid-body tides raised by Phobos

TABLE II
Acceleration in Longitude[a]

Phobos	Deimos	Reference	
.002625 ± .000056	.000072 ± .000028	Shor	(1988)
.002474 ± .000034	−.0000056 ± .0000158	Sinclair	(1989)
.002480 ± .000034	−.000004 ± .000016	Jones et al.	(1989)
.002539 ± .000017	−.000101 ± .000013	Chapront-Touzé	(1990)

[a] $\dot{n}$ given in deg-yr^{-2}.

in Mars (see Burns 1977,1978,1986*a*; Pollack 1977). Mechanisms, including some rather bizarre ones, that earlier had been proposed to account for Phobos' acceleration are recalled in Shklovskii and Sagan (1966) and Burns (1972). If the tidal interpretation is correct, then one can predict the future orbit and also infer its past history. We start by considering the simplest evolution model which considers only the lowest-order tidal response, assumes a circular, uninclined orbit and no changes in the dissipative properties of Mars, and ignores all resonant interactions. With this simple model, the time evolution of the orbital semimajor axis a may be written (Burns 1986*a*, Eq. 13)

$$\dot{a} = 3\, m\, k_T (G/aM_{♂})^{1/2} (R_{♂}/a)^5 \sin 2\varepsilon \tag{1}$$

where m is Phobos' mass and $M_{♂}$ that of Mars, G is the Newtonian gravitational constant, $R_{♂}$ is the Martian radius, and ε is the angular displacement (in radians) between Phobos' longitude and the maximum of the tidal bulge on the planet (see Fig. 3). The tidal Love number k_T which depends on Mars' internal rigidity and density distribution, has been estimated to be ~ 0.08 (Ward et al. 1979). For slow motions of a linearly dissipative system, the lag angle can be expressed as $\sin 2\varepsilon = 1/Q$, where $1/Q$ is the specific dissipation function, which is proportional to the energy loss per cycle (see Burns 1977,1986*a*). By differentiating Kepler's third law, $\dot{a} = -2/3(\dot{n}/n)a$, where n is the mean motion of Phobos about Mars. Phobos' measured secular acceleration $\dot{n}$ and Eq. (1) can be employed to obtain Q for Mars, which is found to be $\sim$ 100 to within a factor of about 2; this value is comparable to that of the solid Earth (see references in Burns 1986*a*). We will return to discuss the future and past of Phobos but suffice it to say now that Eq. (1) predicts a collapse time in which Phobos will strike Mars of < 40 Myr if the satellite were able to remain intact all the way to the planet's surface.

The same tidal processes that inexorably drag Phobos towards Mars force Deimos away from the planet because the latter's orbital mean motion is less than Mars's spin rate Ω (i.e., Deimos is beyond synchronous orbit, that orbital distance where $n = \Omega$, but Phobos lies within this distance). This

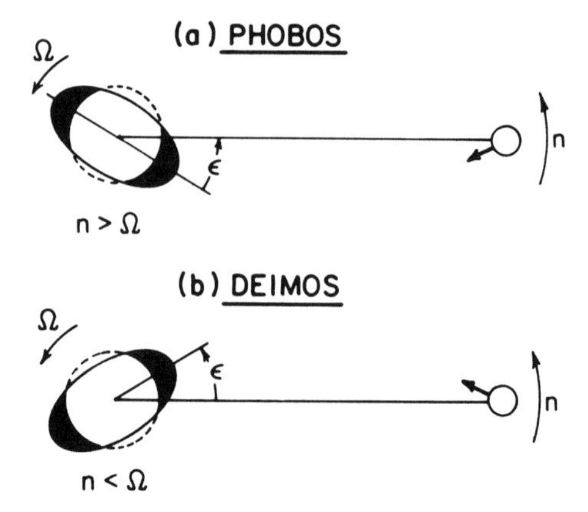

Fig. 3. Tidal distortion of Mars (object on the left) as produced by (*a*) Phobos; and (*b*) Deimos. Due to the planet's inelasticity, the maximum tidal bulge follows (in time) the maximum tidal stress. Thus, for Phobos (whose orbit has $n > \Omega$), the bulge lags the position of the swiftly moving satellite; this configuration allows energy and angular momentum to be taken from Phobos' orbit, causing its collapse, and transferred to the planet's rotation. For Deimos (where $n < \Omega$), the bulge rotates ahead of the perturbation that produced it; energy and angular momentum are given from Mars' rotation to Deimos' orbit, inducing it to expand.

condition means that the time delay between stress and strain that inevitably occurs during any inelastic flexing allows the maximum tidal bulge on Mars due to Deimos to rotate out from under the satellite (i.e., to precede the satellite's orbital longitude; see Fig. 3). The lack of a detectable deceleration for Deimos, as apparent in the orbital position measurements, is consistent with this tidal model since, owing to the outer satellite's smaller size and greater distance from Mars, Eq. (1) would predict that Deimos' deceleration should be $< 1\%$ of Phobos' acceleration. Note that Shor (1988) does not comment on the sign of his solution for Deimos, leading one to suspect that his value, which is listed as positive and which therefore would be contrary to the theoretical prediction, contains a typographical error.

Equation (1) can be integrated to find how distant the two satellites were from Mars 4.6 Gyr ago, presuming that their orbits have always been circular. According to this straightforward but simple-minded approach, Phobos originated at 5.7 $R_♂$ (i.e., its orbit has undergone major change through tidal action) while Deimos, now at 6.92 $R_♂$, would have moved at most a few tenths of a $R_♂$ from Mars (i.e., tides are totally ineffective in altering its path).

Orbital eccentricity e may modify this conclusion significantly, at least for Phobos. Two kinds of tides are important in e's evolution: tides induced in Mars due to Phobos' tug on the planet as well as tides produced in Phobos both due to its rocking (libration) about its synchronous rotation state (Yoder 1982; Burns 1986*a*, Fig. 1d) and due to its changing radial distance from Mars as the satellite moves along its elliptical path. Both types of tides currently circularize this satellite's orbit and thus the small moon's path may have been very elongate many years ago. Starting from Phobos' present eccentricity and naively integrating into the past, Phobos' e would grow to ~ 0.1 to 0.2 about 4.6 Gyr ago owing to planetary tides alone (Singer 1968); tidal dissipation in the satellite would account for more evolution, indicating that the original orbit had yet higher eccentricity. Furthermore, an orbit with greater eccentricity can evolve even more rapidly in a than a circular one because, if e grows swiftly enough as we integrate backwards in time, the orbit's pericenter may approach Mars despite the orbit's overall expansion. With a closer approach to Mars, events at pericenter dominate a's evolution, due to the steep radial dependence of tidal effects (see Eq. 1), such that $\dot{a}$ can actually accelerate even as Phobos withdraws from the planet.

The precise historical track followed by the satellite depends, of course, on the amount of dissipation in the planet as compared to that in the satellite (since tides in the two bodies affect $\dot{a}$ and $\dot{e}$ differently) and on the particular model chosen to represent each of these dissipations (e.g., linearity or nonlinearity with strain, and independence or dependence on forcing frequency). Typical orbital evolution histories resulting from tides in both Mars and the satellite under study can be found in Lambeck (1979*b*), Cazenave et al. (1980), Mignard (1981) and Szeto (1983); a few such paths are shown in Fig. 4. All these authors reach a similar conclusion: under the action of tides alone, Phobos' orbit could be transformed from a primordial large elongate path to the much smaller, nearly circular one seen today. In view of this past orbital history and the previously discussed physical evidence, it then seems as though Phobos' origin is quite simple to discern: the satellite is a captured asteroid. Unfortunately, while this inference may be correct, it cannot be reached so directly for reasons described below.

Deimos' orbit, as already mentioned, expands hardly at all due to tidal action. For similar reasons, primarily the satellite's remoteness from Mars, Deimos' eccentricity, currently small, also has changed little. Thus, by the orbital criteria that were just invoked to argue that Phobos was surely a captured object, clearly Deimos is *not,* because it has always had an orbit much like its current one. Accordingly, the issue of the origin of the Martian moons becomes less clear-cut and, for other reasons to be mentioned presently, it is even more muddied. Before getting into these complications, let us consider the relevant topic of the past histories of the inclinations.

Orbital inclinations are most conveniently measured with respect to the pole of the local Laplace plane, that plane about which a satellite orbit precesses; for distant moons (those principally perturbed by the Sun's gravity), the Laplace plane is very nearly the planet's heliocentric orbital plane but, for close satellites (those mainly affected by the planetary oblateness), it is very nearly the planet's equatorial plane (Burns 1977). The orbits of satellites maintain approximately constant inclinations even as the planet precesses under solar and planetary torques or as the orbit evolves, as long as these

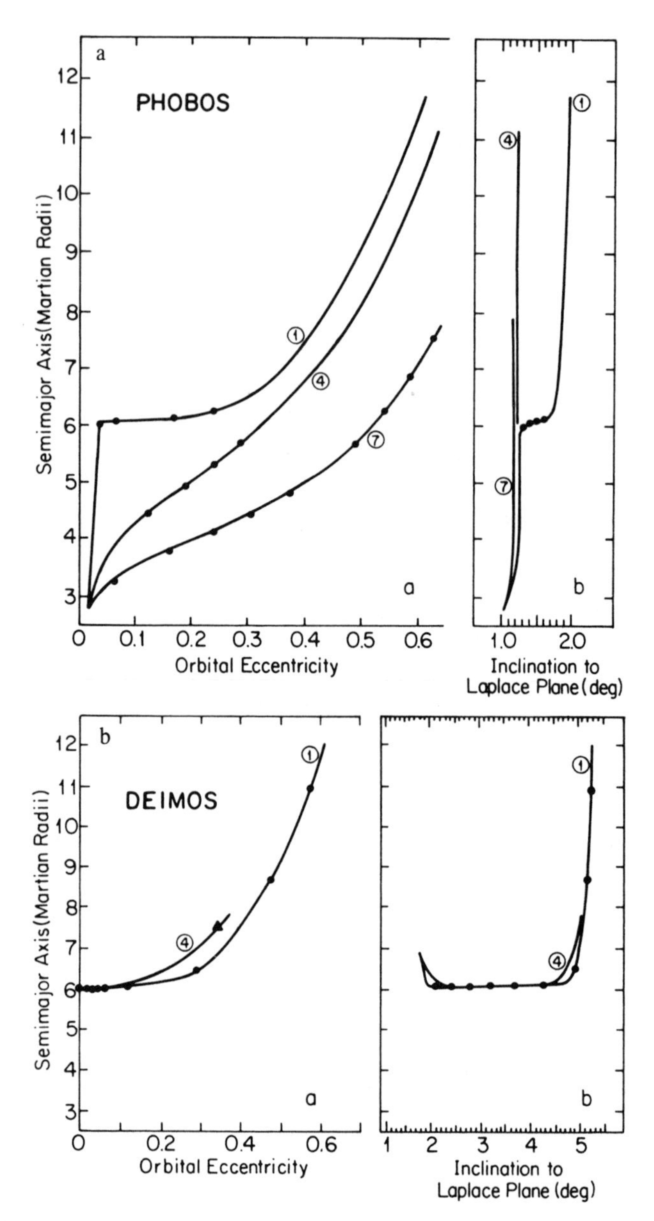

Fig. 4. The Martian satellites' orbital histories as caused by tides in Mars and the satellites. The different curves correspond to different laws for Q's dependence on tidal frequency, in the planet and satellite. Szeto (1983) considered a total of 15 separate cases, of which we show only 3. Curves 1 and 4 have no dissipation in the satellite; for curve 1, Mars' Q is proportional to tidal frequency, while for curve 4, it decreases with increasing tidal frequency. Curve 7 has Q in both planet and satellite increasing with frequency. (a) Phobos. Dots on curves 1 and 4 are time markers that are spaced at 1 Gyr intervals, while those on curve 7 are separated by 50 Myr. The evolution time scale is set by the absolute value of Q chosen, which is fixed by the observed secular acceleration. For all cases, Phobos' current orbit is seen to have evolved from a larger, more eccentric orbit but with little movement in inclination off the Laplace plane. (b) Deimos. Dots on curve 1 are separated by 10 Gyr, the triangle on curve 4 by 1000 Gyr; the points on the inclination history are spaced by 10 Gyr. Deimos' orbit is basically unchanged.

changes occur slowly and as long as these inclinations are measured relative to the Laplace plane (Goldreich 1965). In the specific case of Phobos, which as noted above can develop large eccentricity under tides, Cazenave et al. (1980; see also Szeto 1983) and Mignard (1981) constructed formulations that allowed the inclination of a highly eccentric orbit to be followed relative to its Laplace plane. As expected from Goldreich's analysis, these authors all determined that the two satellites' orbits—each about a degree from the local Laplace plane today—stay close to it throughout the evolution, meaning that the satellite's currently near-equatorial tracks, when evolved back in time, switch fairly abruptly at about 13 $R_♂$ onto a trajectory near Mars' orbital plane.

Thus, if one were arguing that the Martian satellites were captured objects, the fact that their ancient orbital planes may have been close to the ecliptic, near which many solar system objects presently move, might at first seem like strong support for the capture model. However, this notion has two serious failings. First, if capture took place at 13 $R_♂$ or beyond, the amounts of energy that initially bind the protosatellites to Mars are very small. In other words, the capture process would have to be very finely tuned such that the orbits of the soon-to-be-acquired objects would have to closely duplicate Mars' path. But, if the protosatellite-cum-asteroid had a Mars-like orbit, then the original motivation for wanting to capture Phobos and Deimos (namely to explain the apparent differences between their physical properties and those of Mars) is lost. Indeed, if an object were sent along an elliptical orbit directly from the middle of the asteroid belt to the vicinity of Mars, its velocity at Mars would differ from the planet's by nearly 10 km s^{-1}, an amount that could not be removed by a thin nebula. The second failing of the intuitive idea that low inclinations imply capture is that the actual paths along which the objects would move following capture are defined by the relative velocities at which they approach Mars and not by the plane in which they travel prior to their interaction with the planet. For capture at a great distance to be a viable model, it must occur at low approach speed for reasons just mentioned above, but, for an observer on the planet, such small relative velocity vectors are rarely confined to near the planet's orbital plane but instead are almost randomly oriented. Of course all these problems may be alleviated if an unmodeled process, e.g., nebular drag, significantly modifies the orbit following capture. And it is likely that nebula drag will reduce orbital inclinations.

Earlier it was maintained that Deimos' nearly fixed orbit is persuasive evidence against the capture of the outer Martian satellite. It also weighs against Phobos once having a sizeable elongate orbit, which is the principal dynamical point in favor of the inner satellite's capture. This latter restriction occurs because, if Phobos had an orbit with both a and e large, its trajectory would have crossed Deimos's for an extended period of time prior to tidally evolving inward past it. During the interval when the orbits of the Martian satellite pair were interlaced, mutual collisions would have been inevitable as the characteristic collision time ($\sim 10^3$ to 10^6 yr) is much briefer than any plausible orbital evolution time (Lambeck 1979*b*; Cazenave et al. 1980; Szeto 1983). In this way, Deimos' orbit represents a dynamical barrier that firmly limits the total evolution of Phobos. This constraint might be avoided in several ways: (i) the orbits of both moons evolved very swiftly at the outset but then slowed down, perhaps with this evolution history being due to gas drag from a nebula that subsequently, and rapidly, dissipated; (ii) the satellites were captured sequentially with Phobos evolving out of the way before Deimos was acquired; or (iii) the Martian moons originated together as the largest fragments of a catastrophic collision (see below). None of these suggestions can be disproved but, if any were true, then there would be no possibility of following the histories and at least a few planetary scientists would be out of (part-time) work.

The conundrum that Phobos' contemporary orbit suggests a capture that is not permitted according to Deimos' path may be avoided altogether. It is possible that Phobos' present orbital eccentricity (that which, although slight, drives the relatively rapid evolution) is not a tidally damped remnant of an earlier larger eccentricity but rather, at least in part, has been caused by other, more recent events. In this regard, Yoder (1982) has shown that, depending on its evolution, Phobos may have passed through three gravitational resonances at 2.9, 3.2 and 3.9 $R_{♂}$, due respectively to the 3:1 resonance between the satellite's orbital period and Mars' spin period (meaning that the planet's lumpy gravity is experienced at the same inertial position on every third orbit), the 2:1 resonance between the precession period of the satellite's orbit and the planet's orbital period, and the 2:1 resonance between the satellite's orbital period and Mars' spin period. On passage through each of these resonances, Phobos' orbital eccentricity and inclination undergo jumps, whose precise values depend upon Mars' eccentricity, obliquity and phase; related but more complex resonant interactions have been invoked by Tittemore and Wisdom (1988) and by Dermott et al. (1988) to account for Miranda's anomalously high inclination today. If such jumps have occurred, the values of Phobos' e and i measured currently are the tidal remnants of their values after the last resonance was passed rather than values damped from the formation epoch; as such, they should not be used as starting points for integrations into the past. Indeed, since Yoder (1982) found that the magnitudes of the jumps were sensitive functions of the phase at which the resonance was passed, and because the rate of a's evolution at any time depends so critically on how eccentric Phobos' orbit is at that time, the ancient histories of the Martian satellite orbits become untraceable.

Resonances have a place in Deimos' history as well. Yoder (1982) has shown that the outer Martian moon should have an eccentricity of 0.002 if it passed through a 1:2 mean motion resonance with Phobos during the latter's evolution. As the measured eccentricity is considerably less (0.0002), and the time scale for tidal damping of Deimos' eccentricity is much longer than the age of the solar system, this implies that the outer moon must not have gone through the 1:2 resonance. However, if such a resonance passage did not occur, then passages by Phobos through its resonances could not be invoked to account for its eccentricity; the only ways out seem to be to claim that the Martian satellites each has distinctly different physical properties or to hypothesize that Deimos had undergone a propitious impact that just so happened to circularize its orbit. Wisdom (1987*a*) has suggested an alternative solution, namely that Deimos' small orbital eccentricity is a consequence of an episode of chaotic tumbling. He demonstrates that such an interval of chaotic rotation is likely for all irregularly shaped satellites and that, as a result, tidal damping will be much more effective in reducing the eccentricities of such satellites. The same process may have hastened the circularization of Phobos' path.

Much progress has been made in the last few years in understanding the orbital evolution histories of the Martian satellites. This added knowledge has reinforced the opinion of most dynamicists that the orbits of Phobos and Deimos can be explained most directly as being those of objects that originated *in situ*. However, recent studies of the evolution of particles through more realistic circumplanetary nebulae have the promise of allowing outer solar system objects to be captured by Mars and yet have their orbits transformed into the regular ones seen today.

IV. CONCLUSIONS

Where does this leave us? Many years ago Burns (1978) wrote that the pendulum of scientific opinion at that time seemed to be moving toward the idea that the Martian satellites were captured asteroids, but cautioned that pendula have a way of swinging back. The situation has not changed significantly since then: we have more data and better crystallized ideas but they lead in different directions. Had the Phobos spacecraft been able to complete its mission and given a firm indication of the satellite's composition, perhaps we would now know which line of evidence (crude physical measurements or incomplete dynamical theory) led to the correct past. But such was not to be and so we continue to make conjectures.

Acknowledgments. I appreciate the warm hospitality of the Lunar and Planetary Laboratory of the University of Arizona, where this chapter was written while I was on sabbatical leave from Cornell. I thank A. R. Dobrovolskis, W. K. Hartmann, H. H. Kieffer, S. J. Peale, S. Sasaki, C. W. Snyder and P. Thomas for illuminating comments on the manuscript. I especially appreciate the help that I received from J. Bell, D. Britt and S. Murchie in interpreting the spectral and mineralogical data. This research was supported by NASA's Division for Solar System Exploration.

Appendix

APPENDIX: ORIGIN AND USE OF MARTIAN NOMENCLATURE*

JOEL F. RUSSELL
U.S. Geological Survey

CONWAY W. SNYDER
Jet Propulsion Laboratory (retired)

HUGH H. KIEFFER
U.S. Geological Survey

Human beings seem to have an innate urge to give names to things. Names have been assigned to albedo, geographic and topographic features on Mars since they were first observed. Nomenclature is a necessary tool for planetary investigators, and these names are widely used in studies of the surface and in accounts of localized atmospheric phenomena. Initially, the prerogative of mappers, the naming of features has been formalized and now proceeds by international agreement. The 24 specific descriptor terms currently in use for Mars are here defined. The human fondness for informality is evident in the names attached to individual rocks at the Viking Lander sites.

THE HISTORY OF MARTIAN PLACE NAMES

The history of Martian nomenclature begins with the German observers Wilhelm Beer and Johann Mädler, whose 1840 map depicts a dozen unconnected dark regions; they designated some of them rather illogically by the

*This appendix is dedicated to Harold Masursky (1922–1990) and Mary E. Strobell (retired from USGS, 1990), who for many years developed, managed, and championed planetary nomenclature.

letters *a, d, g, mp,* and *efh* (Flammarion 1892, p. 107). From his observations in 1862, the papal astronomer Father Pierre Angelo Secchi named several of these regions for explorers; *a* became the canal of Franklin, *mp* he renamed the sea of Marco Polo, *efh* he called the sea of Cook, and Cabot had a continent named for him.

In 1867 Richard-Anthony Proctor published a map of Mars depicting two hemispheres joined at Dawes Forked Bay (latitude 0°, longitude 0°), on which he placed about 45 place names. He identified bright areas as lands, continents and ice caps, dark areas as seas, oceans, bays, straits, inlets, islands and even one marsh. Most of the features were named for astronomers who had studied Mars, but Newton, Tycho and Copernicus were also honored. Proctor's map was based on drawings made a few years earlier by the Rev. W. R. Dawes, whose name was given to a continent, an ocean, a sea, a strait, an isle and the forked bay, of which only the last has survived. Other astronomers objected to this multiplicity as well as to the repetition of other names, whereas names of more eminent astronomers were omitted. One of the more vocal critics was Flammarion, who in 1876 published his own map. The shapes of his dark regions were similar to Proctor's, and about one-fifth of Proctor's names for them were retained, but the duplicate names were deleted and Dawes' name disappeared entirely. The two largest "seas," Dawes Ocean and De La Rue Ocean, he rechristened Ocean Newton and Ocean Kepler, and he translated all the names into French. However, Flammarion's map received no greater acceptance than Proctor's map.

The next impetus to Martian nomenclature was given by Giovanni Schiaparelli, who studied the planet intensively for 12 years, beginning with the favorable opposition of 1877. He was able to distinguish features more clearly than his predecessors, and he made a series of maps. His 1889 version was in a two-hemisphere format similar to Proctor's except that the circles were joined at longitude 90° and Dawes Forked Bay (renamed Sinus Sabaeus) was at the center of one circle. Good reproductions of this map and Proctor's as well as a modern map in the same format can be found in *The Geology of Mars* (Mutch et al. 1976). Schiaparelli's maps bore little resemblance to earlier ones, and they were dominated in the northern hemisphere by the network of fine straight lines that he called *canali,* whose number steadily increased in later map versions. These features and the new set of place names that Schiaparelli introduced guaranteed that his name should be better known to subsequent generations than that of any of the earlier students of Mars.

Schiaparelli retained the convention that dark areas are oceans and bright areas are continents, but his names were entirely new, all Latin or Greek and based on biblical and mythological geography and history, in which he was an expert. Thus every ocean became a "mare," every bay a "sinus," every lake a "lacus" or a "fons," and every marsh a "palus." Nearly all features had a two-part name (that is, a formal name and a descriptor type) except his "canali," many of which he named after terrestrial rivers. His placement of names was patterned after the geography of the Mediterranean world, with Hellas (Greece) and Ausonia (Italy) to the south and Libya (Africa) and Aeria (Egypt) to the north (his telescopic image was, of course, inverted). Names of mythical people (e.g., Prometheus and Pandora) were also used.

Both Proctor and Schiaparelli insisted that they had named Martian features only for their personal convenience, and they made no claim that others should adopt their systems. Nevertheless, for a decade or so there was a spirited controversy among astronomers, which is described in some detail by Blunck (1977). Flammarion was in favor of Proctor's scheme, although he revised it greatly. Antoniadi adopted Schiaparelli's system for all his maps from 1892 to 1930 and greatly expanded it. Lowell vigorously championed Schiaparelli's nomenclature, and it soon received general acceptance. Despite the decline in popularity of ancient languages, Schiaparelli's basic system of naming is still in vogue.

The origin of Martian place names in mythology and classic literature was described by MacDonald (1971). Discussions of the person, place or event behind many of the Martian names were given in the two editions (1977,1982) of Blunck's book. Most Martian nomenclature appears on the 1:15 million-scale topographic maps (the three sheets of I-2160) of the U.S. Geological Survey (1991), and most regional names are shown in Fig. 1.

Once Mars' topographic features had been revealed by Mariner photography, it became clear that they had little correspondence with the classical regions delineated by their albedo. Major revisions in the nomenclature would be necessary, while maintaining the basic scheme of Schiaparelli.

RECOMMENDATION AND ADOPTION OF NOMENCLATURE

Prior to the founding of the International Astronomical Union (IAU) in 1919, there was no systematic, generally accepted scheme for the naming of extraterrestrial bodies and features, resulting in no small amount of confusion and disagreement. Since then, the IAU has assumed the role of arbiter of planetary nomenclature. One of the first actions of the fledgling IAU was to regularize the lunar nomenclature under the aegis of what would become known as the Working Group on Lunar Nomenclature. In 1958 an *ad hoc* committee of the IAU, chaired by A. Dollfus, recommended adopting the names of 128 features on Mars. In 1970, Commission 16 of the IAU created a Working Group for Martian Nomenclature to meet the growing demand for a systematic Martian naming scheme. In 1973, these two Working Groups were renamed Task Groups and placed under the authority of the newly formed Working Group for Planetary System Nomenclature (WGPSN), along with task groups for Mercury, Venus, and the Outer Solar System. The WGPSN meets annually to formulate policy, define categories, consider sug-

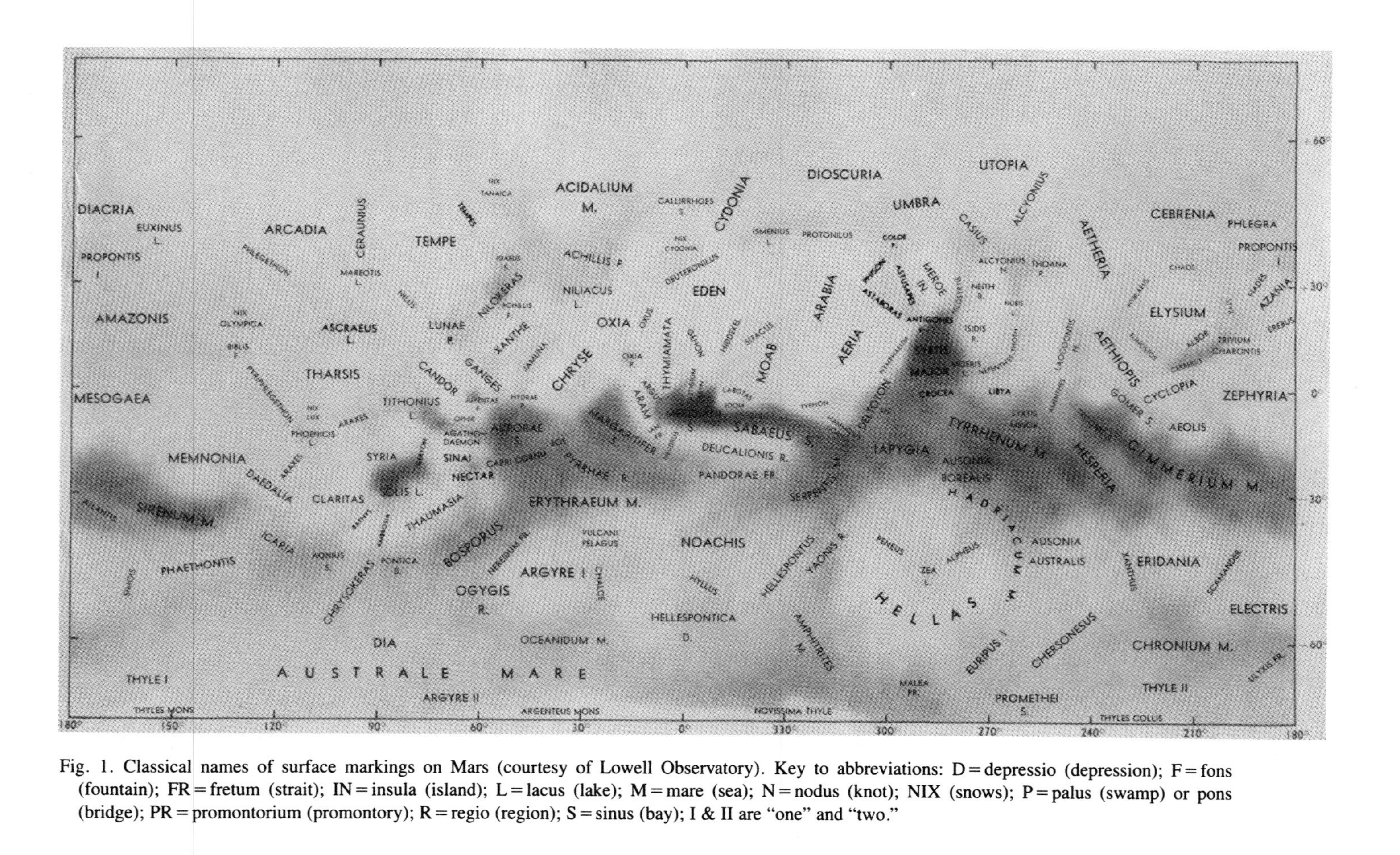

Fig. 1. Classical names of surface markings on Mars (courtesy of Lowell Observatory). Key to abbreviations: D = depressio (depression); F = fons (fountain); FR = fretum (strait); IN = insula (island); L = lacus (lake); M = mare (sea); N = nodus (knot); NIX (snows); P = palus (swamp) or pons (bridge); PR = promontorium (promontory); R = regio (region); S = sinus (bay); I & II are "one" and "two."

gestions from the task groups, and provisionally approve the nomenclature for features on all planets and satellites.

The naming process is begun when anyone, whether scientist or layman, suggests a name for a specific feature (names can be proposed by writing to Cartographic Section, Branch of Astrogeology, U.S. Geological Survey, Flagstaff, Arizona 86001). The Working Group will then review the suggestion and, if it meets IAU guidelines, give the name provisional approval. It can then be formally adopted at the triennial convention of the IAU General Assembly and thus approved for use in planetary cartography.

Certain guidelines must be met before a name can be approved by the IAU. Names must fit the categories defined for each planet, satellite and feature type (see chapter 4, *Planetary Nomenclature,* in Greeley and Batson 1990). Persons to be honored by a feature name must have been deceased for at least 3 years. Names should be clear, simple, unambiguous, easy to pronounce, and as a group, international in scope. For example, small craters on Mars are named for towns and villages around the world. A group of such craters in a given area should represent as large a cross section as possible of the world's countries and continents. Names of military figures, names prominent in any of the six main living religions, and names of specific national significance are not acceptable. Finally, the name should, where possible, fit the traditions of planetary nomenclature, such as the custom of naming certain features on Mars for nearby albedo features or for persons, places, or myths connected with the albedo name. The nomenclature is made international by employing names in many languages.

The policy of the Mars Task Group has been to name features by combining an albedo name for the general area with the Latin or Greek term that described the type of topographic feature. Exceptions to this rule are the names of impact craters and channels. At the suggestion of Gerard Kuiper, large craters have been named after famous people having a relationship to Mars, by analogy with the lunar convention. Enormous channels are given the name "Mars" in various languages, and small ones bear the names of terrestrial rivers. Craters on Phobos are named for scientists who have studied Martian satellites, and those on Deimos are named for authors who have written about the satellites. The total number of Martian names recognized by the IAU now stands at 1334, with no name currently provisional, and the list will doubtless continue to expand.

For further reading on nomenclature rules and history, see Blunck (1982), Masursky et al. (1986), and chapter 4, Planetary Nomenclature, by Strobell and Masursky in Greeley and Batson (1990).

Nomenclature is a tool for planetary investigators, and names applied to Martian as well as other planetary features will likely be used for generations to come. In this context, it is vital that the names chosen today be effective and meaningful, and they should reflect the spirit of fairness and cooperation necessary to further knowledge of the solar system that we inhabit.

CONSTRUCTION OF PLACE NAMES

Martian nomenclature follows the format of other planetary nomenclature, in that names of all features, with the exception of impact craters, are composed of two parts. The first part is the formally adopted name, commonly taken from nearby classical albedo features, mainly those of Schiaparelli (1879). The second part is a one-word descriptor of the feature. This term describes the topography or morphology of the feature in a deliberately general manner and makes no implication about its geology. Describing the feature in topographic rather than geologic terms serves two purposes: (1) a feature can be named before investigations have definitively determined its geology or geomorphology; and (2) similar features can be grouped together under one descriptor term, thus cutting down on the volume of feature types required for a meaningful nomenclature.

Of the 41 feature descriptors now adopted by the IAU, 24 (including the implicit "crater") appear in the Martian nomenclature. Three of these types were originally adopted for use on the Moon; others were added as new features were identified on spacecraft images of Mars. These descriptors follow tradition in that they are derived from Latin or Greek roots, as is much of the planetary nomenclature in general. Listed below are the feature descriptors used on Mars, with a brief description of their derivation, meaning and usage. All examples cited can be found on the 1:15 million-scale topographic maps of Mars (Map I-2160, 3 sheets, U.S. Geological Survey 1991).

FEATURE DESCRIPTORS USED FOR MARS

CATENA (pl. catenae): Latin, meaning a chain or series. Used to describe a string of closely spaced depressions; a chain of craters. As used on Mars, the term does not imply chains of secondary impact craters, but rather features that resemble collapsed lava tubes. Example, Acheron Catena (38°N, 101°W), near Alba Patera.

CAVUS (pl. cavi): Latin, meaning hollow, concave. Used to describe irregular, steep-sided depressions that are apparently not of impact origin. Such depressions are commonly in clusters but may occur singly. Example, Cavi Augusti (3.°7S, 187.°5W).

CHAOS (pl. chaos): Latin, originally from Greek, meaning boundless empty space (personified in Greek cosmology as the Father of Night); in Latin and in Martian nomenclature it carries the meaning familiar to English speakers: an area of distinctive broken or jumbled terrain. Example, Aureum Chaos (4°S, 27°W), at the east end of Valles Marineris.

CHASMA (pl. chasmata): Latin, originally from Greek, has the same roots as our English word "chasm" and has virtually the same meaning for Mars: a steep-walled trough or a large canyon. Example, Candor Chasma (6°S, 71°W) in north-central Valles Marineris.

COLLIS (pl. colles): Latin, meaning a hill or high ground. Used to describe a small hill or knob. Example, Scandia Colles (65°N, 153°W), north and northeast of the crater Milankovič.

DORSUM (pl. dorsa): Latin, meaning the back of an animal, but also used by the Romans to signify a mountain ridge. Used to describe a ridge or other elongate, raised structure. Example, Cerberus Dorsa (16°5S, 255°W).

FLUCTUS (pl. fluctūs): Latin, meaning a wave or billow (literally a streaming, flowing). First used for the Jovian satellite Io to describe terrain that resembles flows. The sole named example of such terrain on Mars is Galaxias Fluctūs (32°N, 217°W), in the Elysium Mons region.

FOSSA (pl. fossae): Latin, meaning a ditch, trench, or channel. Used to describe a long, narrow, shallow depression or a narrow, linear trench. Many fossae on Mars occur in large groups. Example, Tatalus Fossae (44°N, 102°W, near Alba Patera); many occur singly, such as Aganippe Fossa (8°S, 126°W), just west of Arsia Mons.

LABES (pl. labēs): Latin, meaning a falling in or sinking in (especially of the ground). Used to describe a landslide. The Valles Marineris complex contains many examples, such as Coprates Labes and Melas Labes.

LABYRINTHUS: Latin, meaning a labyrinth; originally the labyrinth in Crete constructed by Daedalus. Used to describe a complex of intersecting valleys or canyons. The only named examples of this feature type are found exclusively on Mars: Hyperboreus Labyrinthus (80°5N, 60°W), Adamas Labyrinthus (36°N, 255°W) and Noctis Labyrinthus (7°S, 101°W).

MENSA (pl. mensae): Latin, meaning a table. This feature type has the same root as the Spanish word for table, mesa, and is used in the same manner as mesa is used in the Southwest United States: a flat-topped, steep-sided elevated feature. Examples, Aeolis Mensae (3°S, 218°W) and Nepenthes Mensae (9°N, 251°W), both on the south margin of Elysium Planitia.

MONS (pl. montes): Latin, meaning a mountain. Used to describe such various features as the highlands surrounding a large impact basin, such as Nereidum Montes (41°S, 43°W) on the north side of Argyre Planitia; Geryon Montes (8°S, 80°W), a ridge in Valles Marineris; or a large, solitary edifice such as the well-known Olympus Mons (18°N, 133°W).

PATERA (pl. paterae): Latin, meaning a shallow dish or saucer from which a libation was poured. Used to describe a shallow crater with scalloped, complex walls or edges, particularly craters that do not resemble impact craters, such as Orcus Patera (14°N, 181°W), southeast of Elysium Mons, or those of probable volcanic origin, such as Tyrrhena Patera (21°S, 253°W) in Hesperia Planum.

PLANITIA (pl. planitiae): Latin, meaning a level surface or plain. Used to describe a low plain or a large, level expanse of lowlands. Planitiae are extensive, relatively featureless, low-lying plains, as distinguished from high plains. Examples, Elysium Planitia (20°N, 230°W), Hellas Planitia (43°S, 295°W) and Arcadia Planitia (45°N, 165°W).

PLANUM (pl. plana): Latin, meaning even, level, flat. Used to describe a plateau or high plain, thus distinguishing this type of feature from planitiae. Examples include Sinai Planum (15°S, 87°W), south of Valles Marineris, and Hesperia Planum (20°, 250°W).

RUPES (pl. rupēs): Latin, meaning a rock or cliff. Used to describe a linear scarp or cliff face. Rupes is used generally to identify a relatively straight cliff or scarp, as opposed to a more scalloped cliff. Examples, Bosporos Rupes (42°S, 58°W), northwest of Argyre Planitia, or Chalcoporos Rupēs (54°S, 341°W), southwest of Hellas Planitia.

SCOPULUS (pl. scopuli): Latin, meaning a rock, crag or cliff, especially a rock in the sea. Used to describe a lobate or irregular scarp or cliff, as distinguished from the relatively straighter rupes. Example, Coronae Scopulus (33°S, 295°W), in Hellas Planitia.

SULCUS (pl. sulci): Latin, meaning a furrow, ditch, or wrinkle. Used to describe approximately parallel furrows and ridges that commonly make up terrain that is wrinkled, like a rumpled carpet, usually in a semiparallel arrangement of ridges and troughs. Example, Memnonia Sulci (5°S, 176°W).

TERRA (pl. terrae): Latin, meaning earth, ground, soil, a particular land. Defined by the IAU as an extensive land mass. Used to identify the older, heavily cratered highlands of Mars, which are found mostly in the southern hemisphere. Many terrae coincide with regions that were identified in the nomenclature of Schiaparelli (1879). Examples, Noachis Terra (Land of Noah; 35°S, 335°W) and Terra Cimmeria (40°S, 205°W).

THOLUS (pl. tholi): Latin, meaning a dome or cupola. Used to describe an isolated domical mountain or hill. Although most tholi on Mars appear to be volcanic in origin, the term does not imply a specific geologic origin. Examples, Tharsis Tholus (14°N, 91°W), east of Ascraeus Mons, and Albor Tholus (19°N, 210°W), south of Elysium Mons.

UNDA (pl. undae): Latin, meaning water, especially water in motion, a wave. This recently adopted feature term is used to describe undulatory features. There are only two examples on Mars, both of which are dunelike: Abalos Undae (81°N, 83°W) and Hyperboreae Undae (77°N, 46°W), both on the margins of Planum Boreum.

VALLIS (pl. valles): Latin, meaning valley. Obviously the root of our English word, vallis has the same meaning in Martian nomenclature. Although the Gazetteer of Planetary Nomenclature (Masursky et al. 1986) defines vallis as a sinuous valley, these features on Mars actually assume a variety of shapes, from the gigantic Valles Marineris to the broad, shallow, and topographically ill-defined Simud Vallis in Xanthe Terra. Large valleys are named for the word for Mars in various languages of the world; examples are Ares (Latin), Kasei (Japanese), and Ma'adim (Hebrew). Smaller valleys are named for modern and classical rivers on Earth, such as Loire Vallis and Rubicon Vallis.

VASTITAS: Latin, meaning a waste, emptiness, or desolation; the root word is "vasto" (empty). Defined as widespread lowlands. Vastitas is currently used to describe only one feature in the solar system, the vast, relatively featureless circumpolar plains known as Vastitas Borealis.

INFORMAL NAMES OF INDIVIDUAL ROCKS

The evolution of names for features at Viking landing sites is similar to the early history of Martian names in that the process has been informal and several overlapping sets of names have been used. Moore et al. (1987, p. 221–222) described this process in detail and explained the basis, where known, for the names that survived. Rocks were named for such things as their similarity in appearance to animal and cartoon characters, famous terrestrial rocks (with complete disregard of scale), for some event or happenstance of Lander operation, or for a set of characters in literature. The resulting collection of names includes that of Snow White and all but one of the seven dwarfs and a group of small animals that are characters in Kenneth Grahame's "Wind in the Willows" (Fig. 2). Small, rock-free areas were also given names such as Sandy Flats, Bonneville Salt Flats, Physica Planitia, etc. The locations of the features before and after Viking Lander activity are shown in Plates I and II of Moore et al. (1987). Because a term like "Toad" is much easier to remember than "the 25-cm diameter rock at $Z_L = +3.3$ m, $Y_L = -1.3$ m," these names are common in descriptions of the surface around the Viking Landing sites. Needless to say, these names have not been approved by the IAU.

Fig. 2. Family portrait of some rocks at the Viking 2 Lander site before sampler activity (Viking 2 Lander image 22A05). According to Moore et al. (1987), Alligator is based on a resemblance to Albert the alligator in the Pogo cartoon strip; ICL, an acronym for initial computer load, is located where it would have frustrated the initial sample collection if the Lander had proceeded with its preloaded automatic mission. Toad is perhaps the most life-like rock on Mars; it was initially named "Frawg." Badger was the first named of the four "animals" in the center of this image, and names of the other three were eventually changed to become "Wind in the Willows" characters. Bonneville is so named because the "Bonneville Salt Flats" are immediately in front of it.

Glossary

GLOSSARY*

Compiled by Mary Guerrieri

Definitions are as normally applied to Mars or used in planetary science. Additional meanings are common for many of these terms. Landform terms used in Martian nomenclature are defined in the Appendix.

Aa Lava	a blocky form of lava flow, with a surface made of rubble or clinkers that typically has a few tens of centimeters of relief. Terrestrial aa flows may be 3—20 m thick.
abduction	a scientific reasoning process whereby, based on the respective governing laws, a logical explanation is inferred for a resulting phenomenon. Abduction differs from *deduction* in that it takes into account the possibility that there may be other explanations.
adiabat	the trajectory in pressure-temperature space followed by a parcel of matter which undergoes changes in volume without exchanging heat with its surroundings.
adsorption	physisorption; physical-scale bonding of a gas atom to a surface with a duration of at least one vibrational period.

*We have used some definitions from *Glossary of Astronomy and Astrophysics* by J. Hopkins (by permission of the University of Chicago Press, copyright 1980 by the University of Chicago), from *Astrophysical Quantities* by C. W. Allen (London: Athlone Press, 1973), from *Glossary of Geology*, edited by M. Gary, R. McAfee, and C. L. Wolff (Washington, D.C.:American Geological Institute, 1972), from *Glossary of Geology,* 3rd ed., edited by R. L. Bates and J. A. Jackson (Alexandria, Va.: American Geological Institute, 1987) and from *The Planetary System* by David Morrison and Tobias Owen (Reading, Mass.: Addison-Wesley Publishing Co., 1988). We also acknowledge definitions and helpful comments from various chapter authors.

	Bonding energies are less than 0.5 e V. Chemisorption: chemical-scale bonding of a gas atom to a surface. Bonding energies are greater than 0.5 e V.
advection	transport of heat or material through the atmosphere by mass motions of the atmosphere itself.
Airy isostasy	the gravitational support of surface mass anomalies by horizontal variations in crustal thickness.
albedo, Bond	fraction of the total incident light reflected by a spherical body. Bolometric Bond albedo refers to reflectivity over all wavelengths (*see* radiometric albedo).
aliasing	in a discrete Fourier transform, the overlapping of replicas of the basic transform, usually produced by undersampling.
altitude	vertical distance above the datum of an object or point in space.
amagat	a measure of the amount of gas at 0° C and 1 atm pressure.
Amazonian period	a time-stratigraphic unit (Scott and Carr 1978) which includes events surrounding the formation of lightly cratered smooth plains in Amazonis Planitia to the formation of the layered deposits and dune fields around the poles to the present time. This period encompasses the most recent Martian geologic activity.
anastomosing	pertaining to a network of channels which branch and rejoin, forming a braided pattern.
anastomosis	a branching and reuniting channel morphologic pattern. Term has no genetic connotation in that it can refer to channel morphology whether in alluvial streams or bedrock streams.
aqueous	occuring in water-based solutions.
areo-	a prefix pertaining to the planet Mars.

areocentric coordinates — *see* geocentric coordinates.

areographic coordinates — *see* geographic coordinates.

areoid — a surface of constant gravitational potential on Mars, to which altitudes are often referenced. For topographic mapping, the reference *geoid* has been defined by spherical harmonics of fourth order and degree. *See* geoid.

ascending node — in the orbit of a solar system body, the point where the body crosses the ecliptic from south to north; for a star, out of the plane of the sky toward the observer.

asthenosphere — the region of a planet below the brittle lithosphere in which solids are hot enough and under sufficient pressure to deform plastically rather than by brittle failure.

asymmetry factor (g) — first moment of the phase function, indicating degree of forward or backward scattering; $g = 0$ for isotropic scattering, $g = -1$ for scattering all in the backward hemisphere; $g = 1$ for scattering all in the forward hemisphere; "forward" and "backward" are defined by the direction of the incident radiation.

AU — astronomical unit. Originally the semi-major axis of the Earth's orbit, the precise definition in the J2000 reference system no longer retains this convenient simplicity, and differs by three parts in 10^9. 1.496×10^{11} m.

authigenic — formed or generated in place; specif. said of rock constituents and minerals that have not been transported or that crystallized locally at the spot where they are now found, and of minerals that came into existence at the same time as, or subsequently to, the formation of the rock of which they constitute a part. The term, as used, often refers to a mineral (such as quartz or feldspar) formed after deposition of the original sediment. (*AGI Glossary*, p. 45)

autocorrelation function — the product of a function with a displaced copy of itself, the argument of the autocorrelation function being the

	amount of displacement (often normalized to unity at zero displacement).
backscatter	scattered energy propagating in a direction opposite to that of the incident wave—namely, back toward the transmitter.
barycenter	the center of mass of a system of bodies, e.g., the center of mass of the solar system.
basalt	a fine-grained volcanic rock composed of nearly equal proportions of clinopyroxene and calcic plagioclase with accessory olivine and magnetite. In less strict usage in planetary geology, "basalt" is applied to most low-viscosity, dark silicate lavas.
B class	a subclass of the C asteroids, distinguished by higher albedos than the average C type.
bedrock	solid rock residing where emplaced by geologic processes, and serving as the basement underlying regolith.
bench	a terrace or step-like ledge breaking the continuity of slope.
BET	BET stands for the names of those who first published the seminal paper on using adsorption to measure the specific surface area of a particulate. The appropriate reference is: Brunauer, S., Emmett, P. H., and Teller, E. 1938. Adsorption of gases in multimolecular layers. *J. Amer. Chem. Soc.* 60:309–319.
β-plane	in dynamical meteorology, the approximation of a spherical, rotating planet as a plane upon which the rate of planetary rotation varies linearly with latitude; hence, the Coriolis parameter $f \approx f_o + \beta_y$, where y is the north-south distance and f_o, β are representative constants of a limited latitudinal domain (e.g., northern mid-latitudes).
biogenic elements	elements which constitute the bulk of living things: carbon, hydrogen, oxygen, nitrogen, sulfur, and phosphorus.

bistatic radar	a radar in which transmitting and receiving antennas are physically separate.
blocky material	one of the materials exposed on the surface at Mutch Memorial Station. It has been referred to as duricrust. Its grain size is unknown, although its properties argue for significant amounts of silt-size or smaller grains. It has the largest cohesion of the three soil-like materials observed at the Lander sites, and the large amounts of sulfur and chlorine suggest cementation. Because of its high cohesion, blocky material is not easily eroded by Martian winds.
Bode's "law"	a mnemonic device discovered by Titius in 1766 and advanced by Bode in 1772, used for remembering the distances of the planets (and the asteroid belt) from the Sun. Take the series 0, 3, 6, 12, . . . ; add 4 to each member of the series, and divide by 10. The resulting sequence 0.4, 0.7, 1.0, 1.6 . . . gives the approximate distance from the Sun (in AU) of Mercury, Venus, Earth, Mars, . . . out to Uranus. The "law" fails at Nepture and beyond.
boundary layer	layer of fluid flowing near a surface, which, because of the effects of viscosity, flows at reduced speed compared to the relatively inviscid flow outside the layer.
boustrophedonic	from the Greek, "turning like oxen in plowing." The term describes a raster scan motion with alternate lines in opposite directions.
bow shock	a thin current layer across which the solar wind velocity drops, and the plasma is heated, compressed and deflected around the planetary obstacle. Ahead of the bow shock, the component of velocity along the shock normal is supersonic and behind it, subsonic.
breccia	a coarse-grained clastic rock composed of angular rock fragments and held together by a fine-grained matrix.
Brunt-Väisälä frequency	natural frequency of oscillation for an air parcel displaced vertically in a stably stratified atmosphere; also called the buoyancy frequency.

butte	an isolated, flat-topped hill or small mountain with steep sides, capped by resistant rock. It is generally thought to be an erosional remnant of a pre-existing plateau.
Byerlee's law	a simple bilinear relation between the normal stress on a fracture plane and the shear stress at which frictional resistance is overcome that governs the deformation of fractured, brittle rock.
caldera	a large ($>$ 1 km diameter), basin-shaped volcanic depression, that is more or less circular in form. It can result from collapse, explosion, or a combination of these processes.
canals (pl. canali)	linear features which telescopic observers claimed to have seen on the surface of Mars.
canyon	a steep-sided, deep valley. On Mars, also used for the large equatorial depression Valles Marineris.
cartography	the art and science of expressing graphically, by maps and charts, the known physical features of a planet or satellite.
catena	chain; a string of closely spaced depressions.
C class	a very common asteroid type in the outer part of the main belt, C class asteriods typically have flat spectra longward of 0.4 μm and are presumably similar in surface composition to some carbonaceous chondrites. The relative strength of a UV absorption feature may be correlated with the presence of water of hydration. B, F and G are subclasses of the C class.
center of figure	the point located by the area-weighted average vector to all surface locations.
center of mass	the point located by the mass-weighted average vector to all interior volume elements.
chalcophile	describing the tendency of certain elements to prefer partitioning into sulfide minerals upon melting or differentiation.

Chandler wobble	an aspect of the Earth's rigid body motion departing from simple or pure spin because its angular-momentum vector is not precisely colinear with a principal axis of inertia. (*AGI Glossary,* p. 111).
chaos	*see* chaotic terrain.
chaotic terrain	topographically low regions within the heavily cratered uplands that appear to consist of short, irregular, blocky and fracture terrain. Chaotic terrain is commonly located at the outflow channel source regions and is thought to have formed by the catastrophic removal of large amounts of subsurface fluid.
charge exchange	process in which an ion and neutral atom interact in such a way that the charge on the ion is transferred to the neutral atom.
chemical equilibrium	in a chemically reactive system under fixed temperature and total pressure, the condition beyond which no further net chemical change should occur (because the lowest total Gibbs free energy has been achieved and forward reaction rates are balanced by reverse reaction rates).
chemo-autotroph	organism able to utilize CO_2 as its source of carbon.
chronostratigraphic unit	a body of rock established to serve as the material reference for all rocks formed during the same span of time. Each of its boundaries is synchronous. The body also serves as the basis for defining the specific interval of time, or geochronologic unit, represented by the referent. (*AGI Glossary,* p. 118)
cirque	a deep, semi-circular, bowl-shaped erosional feature which is formed by a glacier and located at the head of a glacial valley.
clay mineral	any crystalline substance occurring in the clay fraction of a soil or sediment. (*AGI Glossary,* p. 123)
compensation depth	depth at which the overlying mass is independent of location; applies to Airy isostasy.

compressional strain	compressional movement of materials in response to compressional stress.
compressional stress	forces which tend to push two locations toward each other.
condensation flow	net atmospheric flow toward one pole due to wintertime condensation of CO_2 onto the surface.
conformal map projection	a projection in which the shapes of small features are preserved, wherever they appear in a map (e.g., a Mercator or polar sterographic projection).
convection	the process of transferring heat by fluid-like motions of material driven by density variations of the material.
convection electric field	electric field associated with a moving, magnetized plasma, equal to minus the cross product of the velocity and magnetic vectors.
core	the central part of the planet below a substantial density discontinuity. Presumed to form by gravitational separation of planetary constitutents.
Coriolis force	a pseudo force that appears in the dynamical equations when set up for a rotating coordinate system. In the Earth's northern hemisphere, it tends to turn all velocity vectors to the right.
creep	slow, forward motion of grains that are too large to be lifted by the wind, whose motion is produced by impact from saltating grains.
crust	the outermost solid layer of a planet or satellite, mostly consisting of crystalline rock and extending no more than a few tens of kilometers from the surface.
crustal dichotomy	strong hemispheric contrast in the physical nature of a planet's crust. On Mars, the division of the planet into heavily cratered highlands and younger lowlands units.
crusty to cloddy material	material observed at the Viking Lander 2 site. It has an inferred grain size near 0.1 to 10 micrometers and relatively high cohesion, possibly due to cementation by sul-

fur and chlorine compounds. On erosion or disturbance, it produces concentrations of mm- to cm-sized clods. Its overall properties suggest that crusty to cloddy materials are not easily eroded by Martian winds.

cryolithosphere — the top part of the crust of Mars in which the temperature is below the freezing point of water (273 K).

cryosphere — the cold upper atmosphere on the night side.

cryptocrystalline — said of the texture of a rock consisting of crystals that are too small to be recognized and separately distinguished even under the ordinary microscope (although crystallinity may be shown by use of the electron microscope); indistinctly crystalline, as evidenced by a confused aggregate effect under polarized light. Also, said of a rock with such a texture. (*AGI Glossary,* p. 158)

darcy — a standard unit of permeability, equivalent to the passage of one cubic centimeter of fluid of one centipoise viscosity flowing in one second under a pressure differential of one atmosphere through a porous medium having an area of cross-section of one square centimeter and a length of one centimeter. (*AGI Glossary,* p. 160)

D class — an asteroid type that is rare in the main belt, but which becomes increasingly dominant beyond the 2:1 Jovian resonance. Their spectra are neutral to slightly reddish shortward of 0.5μm, very red longward of 0.55μm, and for some objects the spectrum tends to flatten longward of 0.95μm. Coloring may be due to kerogen-like materials.

Debye length — a theoretical length that describes the maximum distance at which a given electron can be from a given positive ion and still be influenced by the electric field of that ion in a plasma. Although according to Coulomb's law oppositely charged particles continue to attract each other at infinite distances, Debye showed that there is a cutoff of this force owing to shielding by other charged particles between. This critical separation decreases for increased density: $1_D = (kT_e/4\pi n_e e^2)^{1/2}$.

deflation — removal of loose, granular particles by the wind, caused by less material coming upwind than leaving downwind.

depolarized (echo) component	that part of a radar echo polarized orthogonally to the expected, or "polarized," sense. *See also* polarized (echo) component *and* unpolarized (echo) component.
diabatic	used to describe heating or cooling of an air parcel by radiation.
diapir	a mass of material, less dense than its surroundings, which has risen to its present position due to its buoyancy.
dielectric constant	a numerical value indicating ability of a material to store electric energy relative to the storage ability of a vacuum.
diffuse scatter	that part of radiation (e.g., a radar echo) which has been scattered widely, without a preferred direction.
digital image model	a single array of pixels, organized in latitude and longitude in a manner that can be displayed planimetrically, in which each pixel contains a numerical value related to the brightness of the image of the surface.
dike	an intrusive volcanic formed from magma that exploits a subsurface line or weakness within a volcanic construct. Blade-like in shape, dikes on Earth are typically 0.5–3.0 m wide, can extend for a few tens of kilometers, and are believed to be a few kilometers deep.
dip	the angle that a fault plane makes with respect to the horizontal, measured vertically in a direction perpendicular to the strike of the fault (the line of intersection between the fault plane and the surface).
disconformity	a break in the rock sequence marked by an uneven erosion surface of appreciable relief.
diurnal tide	changes in atmospheric pressure, temperature, or wind velocity resulting from the daily heating and cooling of the atmosphere.
Doppler shift	displacement of spectral lines in the radiation received from a source due to its relative motion in the line of

sight. Sources approaching (−) the observer are shifted toward the blue; those receding (+), toward the red.

drift material — fine grained (0.1 to 10 micrometers), cohesive, porous, unfractured material observed at the Mutch Memorial Station site. Its properties combine to make drift material resistant to aeolian erosion.

drumlins — a low, smooth lemniscate-shaped hill or ridge (in plan view), composed of compact glacial till or, less commonly, drift. They are formed under the margins of large ice sheets and are shaped by their flow. Drumlins occur in large fields with stubby ends pointing in the direction from which the ice approached.

duricrust — case-hardened crust of soil in semi-arid climates, formed by precipitation of aluminous, siliceous, and/or calcareous salts at the surface as rising groundwater evaporates (Stone 1967); often used to describe bonded surfaces at the two Viking lander sites.

dust — material of silt (1/16–1/256 mm) and clay (<1/256 mm) size blown by the wind.

dust devil — swirling, vertical updraft of air developed by local heating of the air above the flat desert floor.

dust storms — atmospheric dust hazes or clouds showing evidence of motion of expansion. Dust storm types include: planet-encircling—those dust storms which are believed to have encircled the planet at some latitude; regional—dust storms, clouds or hazes with a spatial dimension > 2000 km; local—dust storms, clouds or hazes with a spatial dimension < 2000 km; major or great—regional or planet-encircling dust storms.

dynamic pressure — pressure associated with fluid motion, equal to mass density times the velocity squared.

dynamic uplift — the support of a surface mass excess by the pressure exerted by upwelling material.

dynamo — motions of an electrically conducting material which are capable of generating a magnetic field.

eccentricity (of an elliptical orbit)	the amount by which the orbit deviates from circularity: $e = c/a$, where c is the distance from the center to a focus and a is the semimajor axis; or, $e = 1 - q/a$, where q is the periapsis distance.
effusion rate	the rate (volume per unit time) at which lava is erupted from a volcanic vent.
Eh-pH diagram	a two-dimensional plot of oxidation potential (Eh) vs negative logarithm (base 10) of hydronium ion activity (pH) that is used to graphically represent data on water chemistry and stability fields of water-precipitated minerals.
Ekman layer	idealized model of the planetary boundary layer in which a constant interior geostrophic wind is modified near the ground by a turbulent shear stress per unit mass given by the eddy viscosity, $K\, \partial u/\partial z$, with constant eddy coefficient K. The resulting variation of the wind with height is described by the Ekman spiral, in which the wind speed increases with height while the wind direction, angled towards low pressure near the surface, becomes aligned with the geostrophic wind aloft.
Ekman pumping	vertical motion at the top of the planetary boundary layer produced by horizontal convergence due to viscous modification of near-surface horizontal winds. This vertical motion can modify directly the inviscid circulation above it. In the Ekman layer model, the vertical velocity is proportional to the vorticity of the geostrophic flow aloft.
elastic lithosphere	a planet's outer shell of rock, which deforms elastically (rather than plastically) when subject to tectonic forces, i.e., the amount of deformation or strain of the rocks is proportional to the forces exerted in the rocks.
elevation	vertical distance from the datum to a point or object on the planet's surface.
ellipticity (ε)	the ratio of the difference between equatorial and polar radii to their mean value. Usually an indication of how fast a body is rotating.

emission phase function	a relation describing the way in which radiation emerging from a surface varies with changing angle of observation relative to the normal to that surface.
endogenic	said of geologic processes or features resulting from processes originating within the body, e.g., volcanism, volcanoes, geothermal/hydrothermal systems. *Compare* exogenic.
en echelon	textures in an overlapping or staggered arrangement.
equal-area projection	a map projection in which units of area are uniform throughout. For example, even if map images of features appear distorted, one square centimeter on the map always covers the same number of square kilometers on the ground.
erg	a sand desert composed of deep sand, usually containing complex mobile sand dunes.
erosion	physical decomposition of rocks or sediments, by running water or blowing wind, and removal of the resultant debris (usually taken to mean at the scale of outcrops or map units).
eukaryote	a major taxonomic class of life characterized by cells that contain a nucleus carrying genetic information. All metazoa are eukaryotic, as are multicellular plants.
Eulerian (mean meridional) circulation	longitudinally averaged zonal, meridional and vertical wind fields. Typically, only part of the zonally averaged atmospheric transport in the meridional plane is due to advection by these Eulerian-mean winds; zonally averaged flux-divergences associated with atmospheric waves are often comparable, and opposite in sign.
EUV	extreme ultraviolet radiation.
exogenic	geologic processes or features for which the driving force or source of energy is from without the body, e.g., impact cratering.

exosphere — the outermost region of an atmosphere, where atoms and molecules are on ballistic or escaping trajectories and where collisions can usually be neglected.

extension — the increase in distance between two elements of a material in response to forces tending to pull the material apart.

extinction efficiency — ratio of the effective cross-section for extinction (absorption or scattering) of radiation to the geometric cross-section for a particle or distribution of particles.

fault — a plane along which there have been displacements.

F class — a subclass of the C asteroids, distinguished by a weak to nonexistent UV absorption feature.

Ferrel cell, circulation — a thermally indirect zonally symmetric circulation driven by heat and momentum fluxes due to atmospheric waves. On Earth this is a poorly defined, mid-latitude cell in the troposphere, with meridional winds directed polewards near the surface and equatorward aloft. The cell is thermally indirect as its rising motions are poleward of the sinking motions and associated with relatively cold air, given poleward decrease of temperatures in the Earth's troposphere.

fetch — an area over which near-surface wind speed and direction are relatively constant; also, the upstream distance for which wind measurements at a surface station are representative.

fire fountain — plume of incandescent pyroclastic material that erupts from a volcanic vent or fissure. Typical heights of basaltic fire fountains on Earth are 100–600 m.

flexural rigidity — the property of the elastic lithosphere that determines how it bends or deflects in response to vertical forces applied to the lithosphere. Flexural rigidity depends on the thickness of the elastic lithosphere and fundamental properties of its elasticity.

flexure — bending of the elastic lithosphere in response to surface loads.

fluvial	relating to rivers of water or produced by the actions of rivers of water.
flow unit	a successive but essentially contemporaneous layer or portion constituting a single larger flow. Each flow unit represents a separate sheet of liquid lava pouring over another during the course of a single eruption.
fold	a bend in a stratigraphic unit.
forced liberation	rotational oscillation of an irregularly-shaped moon from a synchronous rotation caused by planetary tidal forces acting on the moon in an eccentric orbit.
Fresnel coefficient	ratio of reflected or transmitted to incident field strength at the interface between two media of different dielectric constants; sometimes, also, the squared magnitude of this quantity.
fretted terrain	an erosional landscape on Mars consisting of high-standing remnants of cratered terrain and lower sparsely cratered plains (Sharp 1973) which are separated from the former by abrupt escarpments.
friction velocity, u^*	a measure of the surface stress per unit mass, τ_s/ρ:$u^{*2} = \tau_s/\rho$.
frost streaks	seasonal bright albedo features in the north and south polar caps of Mars that have tapered forms extending from craters. They are interpreted as extra accumulations of frost in the lee of crater rims.
gas dynamics	dynamics of a gas treated as a fluid; often used to describe plasma flow when magnetic field effects on the flow can be neglected.
G class	a subclass of the C asteroids, distinguished by a strong UV absorption feature.
GCM	general circulation model; also, general circulation climate model. Computer models whose objective is the simulation of large-scale, three-dimensional aspects of the general circulation. Typically, a GCM solves the fully nonlinear, time-dependent primitive equations

describing the conservation of momentum, mass and energy, while parameterizing most physical processes thought to be important (e.g., radiation or interaction with the surface).

general circulation — large-scale atmospheric circulation, together with the associated thermal structure and pressure distribution; also, statistical properties of that circulation, particularly when averaged longitudinally and/or for periods long enough to include many daily cycles and individual weather systems, but still short compared to seasonal changes. These properties include zonal-mean temperature, pressure and wind distributions; large-scale eddy wind and temperature variances, and eddy fluxes of momentum, energy and key atmospheric constitutents (e.g., water).

generalized Cassini state — one of the equilibrium states of the pole orientation of a non-spherical planet; it depends on the mass distribution and rotation state of the planet.

geocentric coordinates — geocentric latitude is the angle from the center of a body to a point on the surface relative to the plane of the equator. For oblate bodies, it is equal to geographic latitude only at the equator and at the poles.

geodetic coordinates — geodetic latitude is the angle which the normal at a point on the reference spheroid makes with the plane of the geodetic equator. Geodetic longitude is the angle between the plane of the geodetic meridian and the plane of the prime meridian.

geographic coordinates — geographic latitude is the angle which the normal at a point on a reference ellipsoid makes with the plane of the equator. This is the latitude normally used for maps. Geographic longitude is the angle between the plane of the geographic meridian and the plane of the prime meridian. Planetographic longitudes are numbered such that they increase with time for a fixed remote observer.

geoid — an equipotential surface of a body's gravitational field, for example, "mean sea level" on Earth; commonly approximated by a series of low-order spherical harmonic functions.

geometric albedo	ratio of the flux received from a planet to that expected from a perfectly reflecting Lambert disk of the same size at the same distance at zero phase angle (*compare* Bond albedo).
geostrophic balance	a balance between inertial and Coriolis forces, approximately valid for the atmosphere of a rapidly rotating planet.
goethite	a mineral whose formula is FeO(OH).
graben	extensional tectonic feature consisting of a down-dropped block between two normal faults of sub-parallel strike.
gravitational potential	the amount of work required to move a test mass from infinity to a given location. For a point at a distance R from a mass m, the gravitational potential is given by Gm/R, where G is the universal gravitational constant; for the complex distribution of mass within a planet, the potential is also more complex.
gravity wave	propagating oscillations in atmospheric pressure, density, wind and temperature generated by the vertical displacement of air in a stably stratified atmosphere, from which gravity provides a restoring force. During upward (downward) displacement, expansion (compression) of the air produces adiabatic cooling (warming) which increases (decreases) the density of the air parcel relative to its surroundings.
greenhouse effect	the warming of a planet's surface caused by the trapping of outgoing infrared energy by the planet's atmosphere.
Griffith failure theory	a theory of failure in solids based on growth of microfractures.
gyroradius	radius of gyration of an ion or electron in a magnetic field, given by the moment of the particle divided by the product of the charge times the magnetic field intensity.
Hadley circulation	movement of the atmosphere represented by rising motion near the latitude of greatest solar heating and sink-

ing motion at more poleward latitudes, driven by variations in temperature of the atmosphere.

Hawaiian eruption a type of basaltic volcanic eruption, characterized by highly fluid lavas of low gas content, that gives effusive lava flows and less voluminous pyroclastic deposits consisting of bombs and spatter.

heat flow flow of energy from a planet's interior to its surface.

Hesperian period a time stratigraphic unit (Scott and Carr 1978) which includes events surrounding the formation of moderately cratered, widespread ridged plains (e.g., in Hesperia Planum) to the formation of the plains of Vastitas Borealis. This period encompasses Martian geologic activity of intermediate age.

Hodograph the curve traced out over time, or in the vertical, of the vector wind measured at, or above, a site.

hogback a ridge produced by highly tilted surface layers or strata.

homopause level in an atmosphere where gases cease being uniformly mixed and above which molecules can separate diffusively; often called turbopause.

horst an elongate, relatively uplifted crustal unit or block bounded by normal faults on its long sides.

ignimbrite a deposit of pyroclastic material, typically ash or pumice, that has traveled across the ground as a flow. May be welded (hot on deposition) or non-welded (cold on deposition), depending upon emplacement temperature.

impact ionization process in which an electron is ejected from an atom or ion due to a Coulomb collision with a passing charged particle having kinetic energy in excess of the binding energy of the electron.

inclination the angle between the plane of a planet's or satellite's orbit and the plane of the ecliptic (compare obliquity), or between the plane of a satellite's orbit and the planet's equatorial plane.

index of refraction (n)	the ratio of the speed of light in a vacuum to that in a given medium.
inertial frame	a non-rotating reference frame such as the mean equinox and equator of J2000.0.
interfluve	the ridge or upland area separating adjacent valleys. There is an implication of fluvial origin of the valleys.
invariant plane	the plane perpendicular to the total (spin plus orbital) angular momentum vector.
ionopause	the boundary between the solar wind plasma and the ionospheric plasma.
IRIS	Infrared Interferometer Spectrometer; flew on Mariner 9.
IRR	Infrared Radiometer; flew on Mariner 9.
IRS	Infrared Spectrometer; flew on Mariner 6 and 7.
IRTF	Infrared Telescope Facility (on Mauna Kea).
IRTM	the Infrared Thermal Mapper experiment was a five-channel, multi-spectral radiometer carried on the Viking 1 and 2 orbiters.
ISM	French imaging spectrometer on the PHOBOS mission.
isostasy	the ability of the surface layer to adjust its level according to the load (such as ice caps) that it has to carry.
isostatic	subject to equal pressure from every side; in hydrostatic equilibrium.
J2000	the reference system, including coordinate orientation and definition of constants, adopted by the International Astronomical Union in 1976. Described in detail in "Section S" of the *Astronomical Almanac* for 1984. The standard epoch of this system is 2000 January 1.5 (Julian date 2451545.0) and is denoted J2000.0.

jet, jet stream — relatively narrow, fast-moving wind current flanked by more slowly moving air. On Earth, the subtropical jet marks the poleward limit of the Hadley circulation.

jökuhlaup — an Icelandic term for *glacial outburst flood*. Pron.: yo-kool-loup (the last syllable as in "out"). (*AGI Glossary*, p. 334)

JPL — Jet Propulsion Laboratory; a NASA center in Pasadena, California that is largely responsible for planetary missions.

Julian date or Julian day number (JD) — the number of days that have elapsed since Greenwich noon on January 1,4713 B.C. The Julian period was worked out by Joseph Scaliger and applied in 1582, the year the Gregorian calendar was adopted. January 1.5, 2000, is Julian date 2451545.0

Juvenile (volatiles) — volatiles which have not been previously outgassed to the surface.

KAO — Kuiper Airborne Observatory. A large jet plane and associated telescopes used to make observations from above most of the water vapor in the Earth's atmosphere.

Kelvin mode, waves — a class of eastward propagating, large-scale gravity waves whose associated distributions of pressure and zonal velocity are symmetric about (and, therefore, largest in magnitude at) the equator, but whose meridional velocities are relatively small at low latitudes.

kolks — macroturbulent river phenomena produced by transient eddies which "pluck out" deep depressions in soft rock by highly concentrated, intense suction forces and velocity gradients.

Knudsen diffusion — diffusion of gas through a porous medium in the regime where the mean free path for collisions of gas molecules is larger than the pore space between individual grains.

KREEP — lunar basaltic material rich in radioactive elements (K for potassium, REE for rare Earth elements, P for phosphorus).

KRFM	a Soviet instrument on the Phobos mission, KRFM measures radiation in the visible and infrared portions of the spectrum.
KTT	Thellier-Thellier technique; method to determine the external magnetic field strength in which a sample acquired a magnetic field as it cooled. The technique involves heating and cooling the sample in the absence and presence of a magnetic field and measuring the magnetic remanence lost and gained at successively higher temperatures.
labyrinth	an expanse of intersecting, canyon-like troughs.
laccolith	an intrusive igneous body that has caused the overlying rocks to dome up.
Lagrangian (mean meridional) circulation	velocities averaged with respect to air parcels following the wave motions. Zonal averages can be taken following the motion of the center of mass of a planet-encircling ribbon of fluid. Atmospheric transport is then due solely to advection by the Lagrangian mean motions, and there is no explicit eddy flux-divergence, as in Eulerian-mean transports. Although conceptually straightforward, the Lagrangian-mean motions are difficult to compute in practice.
Lambert albedo	equivalent albedo of a surface for which the brightness depends only on the cosine of the solar incidence angle.
Lambertian surface	a surface which scatters incident light in proportion to the cosine of the emission angle.
layered deposits	*see* polar layered deposits.
lemniscate	a streamlined or "tear-dropped" shaped land feature with rounded prows pointing in the upstream direction and tapering tails pointing in the downstream direction. This form produces the least resistance to flow along a fluid boundary.
limonite	a term describing the occurrence of the hydrous ferric oxide minerals goethite, hematite, or lepidroite.

line-of-sight acceleration (LOS) — the component of the acceleration of a spacecraft, along the line connecting it with the Earth, induced by anomalies in the gravitational field of the planet around which it is orbiting. Measured using radio techniques from the Earth, the acceleration can be used to determine the gravitational field of the planet.

lithosphere — the part of a planet's outer shell that is capable of supporting stresses over geologic time scales.

lithostratigraphic unit — a defined body of sedimentary, extrusive igneous, metasedimentary, or metavolcanic strata that is distinguished and delimited on the basis of lithic characteristics and stratigraphic position. It generally conforms to the Law of Superposition and commonly is stratified and tabular in form. The fundamental unit is the formation. A lithostratigraphic unit has a binomial designation, consisting of a geographic name, derived from the type locality, combined with a descriptive lithologic term; both are capitalized.

longitude of perihelion — the location in a particular coordinate system of the point in a planet's orbit around the Sun at which the planet is closest to the Sun. Normally measured from the referent equinox along the ecliptic to the rising node of the orbit, and then along the orbit to perihelion.

loss tangent — ratio of imaginary to real parts of a dielectric constant; a measure of the attenuating properties of a medium.

lossy — subject to absorption of radiation.

L_s — aerocentric longitude of the Sun as measured in a Mars-centered fixed coordinate system, often used as an angular measure of the Mars year. $L_s = 0° = 360°$ marks vernal equinox. $L_s = 90°$, $180°$ and $270°$ marks the beginning of northern summer, fall and winter (southern winter, spring and summer, respectively). Because of the elliptical orbit of Mars, L_s is not quite linear in time.

mafic — term used to describe a silicate mineral whose cations are predominantly Mg and/or Fe. It is also used for rocks made up principally of such materials.

magma chamber	the subsurface reservoir within a volcanic construct from which molten magma is extruded to form lava flow or dikes. Typically, terrestrial magma chambers may be a few kilometers in diameter and are found 2–5 km beneath the summit of the volcano.
magmatism	the phenomenon of melting of rock beneath the Earth's surface.
magnetic Bode's law	the approximate, linear empirical relation between the rotational angular momentum of a planet and its dipole magnetic moment.
magnetic dipole moment	a parameter describing the strength of the dipole component of a planetary magnetic field. The dipole part of the magnetic field decreases with the cube of the distance from the center of the planet.
magnetic pressure	pressure associated with a magnetic field, proportional to the field magnitude squared.
magnetohydrodynamics	dynamics of a medium in which magnetic forces play a role.
magnetopause	the surface defining an interface between the field of a magnetized body and the interplanetary medium. It is the surface where the ambient magnetic pressure of the field is in balance with the dynamic pressure of the solar wind, decelerated by supersonic collisions at the bow shock.
magnetosheath	the region between a planetary bow shock and magnetopause in which the shocked solar wind plasma flows around the magnetosphere.
magnetotail	region of magnetic field in the wake of a planetary obstacle to a plasma flow.
mantle	the interior zone of a planet or satellite below the crust and above the core.
mascons	mass concentrations.

massif	mountainous mass or group of connected masses, generally delineated by adjacent valleys or plains.
mass-loading	process of adding mass density, via neutral gas ionization, to a plasma.
mass wasting	general term used to describe the dislodgement and downslope transport of rock and soil by gravitational forces (e.g., creep, solufluction, rock falls, rock slides, debris flows, landslides).
MAWD	Mars Atmospheric Water Detectors; flew on the two Viking Orbiters.
megaregolith	rock comminuted by impact that dates from the early history of heavy bombardment. It is formed of overlapping layers of ejecta blankets grading downward into fractured crust.
membrane stress	horizontal stress or force per unit area in the elastic lithosphere due to departures of the surface from a perfect sphere or changes in the radius of curvature of the surface. Such departures from sphericity can arise from the polar flattening and equatorial bulging due to planetary rotation.
meridional	in the direction perpendicular to lines of constant latitude.
mesa (or mensa)	tableland; a flat-topped elevation with steep walls.
mesopause	on Earth, the upper boundary of the mesosphere. It is the region where heat flux transported downward from the thermosphere by conduction is balanced by radiative cooling low in the mesosphere.
metazoa	multicellular animals.
Mie scattering	scattering of light by well-separated spherical particles that are larger than the wavelength of radiation.
mineraloid	a naturally occurring, usually inorganic substace that is not considered to be a mineral because it is amorphous

	and thus lacks characteristic crystal form; e.g., opal. (*AGI Glossary*, p. 402)
moment of inertia	the product of a mass of an element and the square of its radius of gyration, integrated over the mass of the body.
Monin-Obukhov length	characteristic stability parameter of stratified boundary layers. Its magnitude represents the approximate height above the ground where production of turbulence by buoyancy forces becomes comparable to its production by shear stresses. It is inversely proportional to the vertical heat flux and proportional to the cube of the friction velocity u^*.
monostatic radar	a radar which uses the same antenna for both transmitting and receiving.
montmorillonite	a mineral whose formula is $Al_2Si_4O_{10}(OH)_2$, where Mg, Ca, or Na can substitute for the Al.
MORB	mid-ocean-ridge basalt.
Mutch Memorial Station (MMS)	The Viking 1 lander was renamed the Mutch Memorial Station in honor of Dr. Timothy Mutch, Team Leader of the Lander Camera Experiment, following his accidental death in 1980.
M/Z	mass of ionized fragment in mass spectrometer divided by its electric charge.
NCAR	National Center for Atmospheric Research, Boulder, Colorado.
Newtonian fluids	a fluid, such as air or water, that flows (i.e., deforms) when an infinitesimally low shear stress is applied.
Newtonian cooling	a simple linear parameterization of the radiative damping of thermal perturbations in an atmosphere or planetary surface of the form $Q = (T_e - T)/\tau_r$, where T_e is the equilibrium temperature and τr is the radiative time constant.
Noachian period	a time-stratigraphic unit (Scott and Carr 1978) which includes events surrounding the formation of the oldest ex-

posed materials on Mars (e.g., Noachis Terra) to the emplacement of the earlier intercrater plains units. This period includes the end of heavy bombardment.

nomenclature — suite of names given to particular objects, or describing the names given to particular objects (such as features on a planetary surface).

non-thermal escape — loss of volatiles from the atmosphere by means other than thermal evaporation.

nontronite — the mineral montmorillonite when the Al has been largely replaced by Fe.

nucleotides — precursor molecules for genetic materials (RNA and DNA).

Nusselt number (Nu) — dimensionless ratio of heat flow with convection to that carried by conduction alone.

oblateness — *see* ellipticity.

obliquity — the angle between a planet's axis of rotation and the pole of its orbit (*compare* inclination). It is obliquity that is responsible for the seasons of a planet.

obliquity oscillation — time variations in the obliquity. On Mars, there is a quasi-periodic oscillation of the obliquity, with dominant timescales of $\sim 10^5$ and 10^6 yr.

opacity — a measure of the ability of an atmosphere to absorb or scatter radiation. On Mars, atmospheric dust provides much of the opacity in the visible and infrared portions of the spectrum.

opposition — the position of a planet or asteroid when its ecliptic longitude is 180° from that of the Sun. At opposition, the Earth lies approximately between the body and the Sun.

optical depth — the natural logarithm of the factor by which a beam of light incident perpendicular to an atmosphere is attenuated due to both scattering and absorption. Strictly a function of wavelength, it is often applied to the Martian atmosphere at mid-visible wavelengths. The term *opac-*

ity is commonly also used, with a less quantitative meaning.

optical libration — apparent oscillation of a satellite as observed from a planet due to the difference between the satellite's constant (synchronous) rotation rate and its varying orbital rate in an eccentric orbit (*contrast* physical libration).

outflow channel — large channels on the surface of Mars which contain bedform features (e.g., longitudinal grooves, stream-lined islands) and are thought to have formed by the catastrophic release of large-scale fluid flows from the subsurface environment. These channels usually originate in chaotic terrain, extend several thousands of kilometers over the surface, and terminate in the northern plains.

overspread target — a radar target with range depth and limb-to-limb bandwidth which imposes mutually exclusive sampling requirements for range-Doppler mapping; a target for which aliasing in range (time) of Doppler shift (frequency) is inevitable when data acquisition over both is attempted.

oxidation — the chemical process of losing electrons that creates (or increases the net charge on) an ion. In the geologic context, oxidation of iron (from Fe^{2+} to Fe^{3+}) during weathering produces major color changes toward red, brown and black. In some cases, absorption of ultraviolet radiation might eject electrons thereby enhancing oxidation.

Pahoehoe lava — a type of basaltic lava having a glassy, smooth, and billowy or undulating surface. Typically forms in a series of flows each less than 1 m thick; very common in Hawaii.

palagonite — a poorly crystalline, hydrated product of the aqueous alteration of basalt glass. In the strict sense, palagonite is the orange-brown alteration rind on basalt glass and not the total complement of associated weathering products, which may include clay minerals, zeolites, and carbonates. In loose usage, "palagonite" is sometimes applied as a name for altered pyroclastic rocks containing palagonite.

parna — an eolian clay occurring in sheets.

P class — a fairly common asteroid type in the outer main belt with a heliocentric distribution that peaks near the 3:2 Jovain resonance. Their spectra are featureless from 0.3 to 1.1 μm (identical to E and M classes), but the class is distinguishable by low albedos.

periglacial — a term used to describe all nonglacial processes or features that operate or form in cold climates regardless of the presence or absence of glaciers (e.g., permafrost, pingos, frost wedging, patterned ground, thermokarst).

perennial polar cap — high-albedo frost covering within a few degrees of each pole that survives through the Mars summer. The northern one is H_2O ice and the southern one was CO_2 frost during the year observed by the Viking spacecraft; also loosely referred to as the "residual" of "permanent" polar cap.

perihelic opposition — an opposition of Mars in which Mars is at or near perihelion in its orbit.

permafrost (region) — any location where the temperature does not rise above 273 K during the year; often used to describe locations on Mars where ground ice exists.

phase angle — the angle between vectors to the Sun and to an observer.

photoelectric polarimeter — an instrument utilizing a photoelectric tube to determine the degree of polarization.

photoelectric observation — an observation utilizing an instrument with a photoelectric sensor to measure incoming light.

photometric — relating to measuring the intensity of light.

photometric modeling — the use of a physical model of light scattering from a particulate surface to determine the physical properties of the surface, based on observations of the reflected sunlight, or to correct diverse observations to a uniform illumination-observation geometry.

phraetomagmatic eruption	volcanic eruption into a water-rich environment.
phraetomagmatic (geologic) province	a geologic province in which volcanic magma interacts with ground water during the eruption process.
phyllosilicates	a class or structural type of silicate characterized by the sharing of three of the four oxygens in each tetrahedron with a neighboring tetrahedra, to form flat sheets; the Si:O ratio is 2:5. An example is the micas.
physical libration	small variations in the near-synchronous rotation rate of a satellite relative to an inertial frame (*contrast* optical libration).
pickup ions	ions produced in a flowing magnetized plasma which have been accelerated by the convection electric field.
pit chain	aligned conical to irregular depressions.
planetary albedo	the reflectance of the surface-plus-atmosphere system; may be qualified to specific wavelength or angle coverage.
planetary boundary layer	region of the atmosphere near the ground dominated by turbulence generated by thermal convection above the heated surface by mechanical instability of wind shear near the surface.
planetary wave	atmospheric oscillations with characteristic horizontal scales comparable to the planetary radius; waves of this scale are strongly affected by Coriolis forces (i.e., planetary rotation).
planetocentric coordinates	*see* geocentric coordinates.
planetographic coordinates	*see* geographic coordinates.
planetopause	the boundary at which the ion composition in the magnetosheath changes from predominantly solar wind protons to predominantly planetary (oxygen) ions.

planimetric map a map that presents only the horizontal position for the features represented.

plateau an elevated area of land bounded on at least one side by a steep downward slope.

plinian eruption a very explosive volcanic eruption that produces widely dispersed fallout deposits of ash and pumice due to high eruption-rate voluminous gas-rich eruptions, commonly lasting (on Earth) for several hours to 4 days.

polar hood a broad expanse of water-ice or CO_2-ice clouds occurring over the polar regions during the fall or winter seasons.

polar layered deposits deposits approximately centered on each pole extending up to 20° from the poles and displaying nearly horizontal layers, each tens of meters thick.

polarization the division of electromagnetic radiation into two components. Light is usually treated as having two orthogonal polarizations, perpendicular and parallel to the plane of scattering. The degree of polarization is the difference in magnitude of the two components, divided by their sum. Radiation, especially radar, may also be circularly polarized, in which case the plane of polarization rotates rapidly.

polarized (echo) component that part of a radar echo having the polarization expected after reflection from a planar interface between two homogeneous media. *See also* depolarized (echo) *and* unpolarized (echo) component.

polar wander movement of the rotation axis with respect to the body of a planet.

porosity the fraction of volume of a rock or regolith which consists of void space; as distinct from permeability, which determines the ability of a gas molecule to diffuse through the rock or regolith, for example, and which depends also on the connectedness of the pore spaces.

potential temperature temperature that would be reached if air were compressed or expanded adiabatically to a standard pressure.

	Thus, the potential temperature, but generally not temperature, of air parcels displaced vertically and adiabatically remains constant.
Pratt isostasy	the gravitational support of surface mass anomalies by horizontal variations in composition (density) of a crust of constant thickness.
primitive equations	in meteorology, the mathematical equations describing the change with time of the vector wind, pressure, density (mass) and temperature fields. Often, the vertical velocity equation is approximated by hydrostatic balance, in which gravity balances the force per unit mass due to the vertical gradient of pressure. "Primitive" indicates that no further approximation (e.g., steady state, incompressible flow, geostrophic balance) has been made.
prokaryote	a major taxonomic class of life, essentially the bacteria, characterized by cells that do not contain a nucleus.
PVO	Pioneer Venus Orbiter.
pyroclastic	volcanic material that has formed from disrupted magma.
pyrolysis	destruction of organic compounds by combustion.
quasi-specular reflection or scatter	approximately mirror-like scatter; scatter from a surface that is locally mirror-like, but with "rolling" topography that produces many glints rather than a single strong reflection; a radar echo component arising from interaction with such a surface; the reflection of radar energy from a surface which is locally smooth and flat at the scale of the wavelength but which may be rough at much larger scales.
radar cross section	the cross sectional area of perfectly isotropic scatterer which, if positioned at the target location, would give the observed echo strength at the receiver.

radiogenic elements — elements in rocks that produce heat by radioactive decay (e.q., U, Th, K).

radiometric albedo — the fraction of total energy, integrated across all wavelengths, that is reflected by a surface or object. Unless otherwise specified, the incident energy is understood to have a solar spectrum. The incidence angle must be specified unless the object is assumed to be spherical and uniform, or Lambertian.

rampart craters — Martian impact craters whose ejecta deposits are terminated by a distal rampart, as opposed to grading smoothly into the surrounding terrain.

range depth — the apparent (radial) dimension of a radar target, sometimes given as the time required for a radar pulse to traverse that distance; for spherical targets, twice the radius, or that distance divided by the speed of light.

Rayleigh friction — a simple linear parameterization of the eddy viscous damping of momentum in a fluid or atmosphere, of the form $F = (u_e - u)/\tau_f$, where u_e is the equilibrium velocity and τ_f is the eddy "frictional" damping time.

Raleigh mixing formula — an expression, attributed to Lord Rayleigh, which relates densities and dielectric constants of a rock and its powder.

Rayleigh number — a nondimensional parameter involving the temperature gradient and the coefficients of thermal conductivity and kinematic viscosity, which determines when a fluid, under specified geometrical conditions, will become convectively unstable.

REE — rare Earth elements; the lathanide series in the periodic table.

regolith — the unconsolidated layers of soils, sediments, and rock fragments that overlie bedrock.

Reidel shear — staggered or en-echelon (step-like) synthetic (oriented parallel to the main fault) and antithetic (oriented perpendicular to the main fault) shear faults found in association with some strike-slip faults.

remanent magnetization	the inherent magnetism of a rock acquired during cooling, sedimentation, or chemical reactions in the presence of an external magnetic field.
reseau	a mark placed on the detector surface of television imaging systems in order to calibrate the geometry of images.
Reynolds number	a dimensionless number which is proportional to the ratio of inertial force to viscous force in a fluid flow and which is, therefore, a measure of the relative importance of viscosity.
rift zone	(a) a system of crustal fractures; a rift [struc. geol.]; (b) in Hawaii, a zone of volcanic features associated with underlying dike complexes.
rille	a trough structure, commonly found in the volcanic terrain, probably produced by thermal erosion or turbulence caused by volcanic eruptions having effusion rates, or by catastrophic release of water-laden volcanic debris. Martian rilles commonly exceed a kilometer in width and tens of kilometers in length.
riving	a process of physical weathering in which growth of ice crystals or salt crystals in small spaces, such as grain boundaries or microfractures, splits rocks into smaller fragments.
rms slope or tilt	root-mean-square deflection of local surface normals with respect to the normal to the mean surface.
Rossby number	R_o a measure of the importance of planetary rotation on atmospheric motion (small for large R_o).
Rossby waves	cyclonic convection waves in a rotating fluid. Such waves occur in the atmospheres, in the oceans, and in the fluid core of the Earth.
roughness (as used by radar)	a generic term meaning deviations of a planet surface from locally smooth and horizontal; scale of roughness can range from centimeters to hundreds of meters.

saltation	transport of sand by the wind, where individual particles are too heavy to be carried in suspension and continually impact, bouncing up or ejecting other particles.
sand	particles 1/16–2 mm in diameter.
sapping	the process of undermining and removal of material by ground water outflow.
scabland	an elevated region consisting of flat-lying basalt flows with minimal development of soil or vegetation. The surface is erosional in nature and is characterized by large-scale channel forms and dry waterfalls (cataracts), produced by massive, catastrophic flood discharges. The type location for scabland is the Channeled Scablands in eastern Washington.
scabland topography	topography that results from the plucking and scouring of bedrock by catastrophic flood flows (e.g., Channeled Scablands of eastern Washington).
Scorer parameter	ξ^2, where $1/\xi$ is the wave vertical scale (*see* wave vertical scale) derived for mesoscale waves (e.g., gravity waves). When ξ^2 changes sign with height, the wave may be trapped (i.e., confined locally in height) and resonantly amplified. The Scorer parameter given in Ch. 26, Eq. (8) was derived for the idealized case of two-dimensional, incompressible flow on a non-rotating plane.
secular spin-orbit resonance	a correspondence (ratio of small integers) between the spin and orbital periods of a planet or satellite.
sediment	loose material, formed by decomposition of rocks, that is transported from its place of origin and deposited elsewhere.
sedimentation	process of forming sediment.
self-compaction depth	the depth at which crustal porosity becomes $< 1\%$, a value at which pore spaces typically become discontinuous.

semidiurnal tide	the component of changes in atmospheric temperature, pressure, or wind velocity resulting from daily heating and cooling of the atmosphere which appears to occur on a semi-diurnal basis.
shield volcano	a volcanic construct, typically comprised of many fluid lava flows that gradually form a large, broad topographic high that has slopes of a few degrees.
shock foot	region of enhanced magnetic field seen upstream of certain plasma shocks caused by ion reflection at the shock.
sideromelane	an approximate syn. of *tachylyte*. (*AGI Glossary,* p. 612)
siderophile	describing the tendency of certain elements to prefer partitioning into the metallic (iron) phase upon melting or differentiation.
single scattering albedo	ratio of the radiation scattered to the total of radiation scattered and absorbed.
slope	angle of tilt of the local surface with respect to a geopotential surface.
slope wind	winds induced by the thermal or mechanical effects of sloping terrain.
smectite	a family of hydrated, expandable, layer-structured aluminosilicate clay minerals having crystal structures based on octahedral and tetrahedral layers that repeat in a 2:1 ratio. Smectites are formed by precipitation from aqueous solutions produced by weathering.
SNC meteorites	Shergottites, Nakhlites and Chassigny; a group of eight differentiated meteorites for which Mars is suggested as the parent body.
soil	loose or moderately indurated material that forms in place by weathering of rock material, and that may contain zonal structure recording surface-atmosphere interactions; on Earth, the top weathered layer of the terres-

	trial lithosphere which is exposed to atmospheric and biotic effects.
sol	one Martian solar day, equivalent to 24.66 terrestrial hours.
solar wind	the upper ionized hydrogen and helium atmosphere of the Sun that expands radially outward at speeds of ~ 400 km s^{-1}.
specular reflection	scattering which is mirror-like; reflection from a perfectly smooth surface.
splotches	dark markings within Martian craters; applied by Sagan et al. 1972. While some splotches are identified as dunes, and others have thermal characteristics of relatively coarse sediments, the term strictly is one that describes albedo features.
sputtering	ejection of neutral atoms from a solid or gas in response to bombardment by energetic ions.
stereomodel	the three-dimensional model formed by the intersecting homologous rays of an overlapping pair of photographs.
stratigraphy	in planetary usage, the study of the relative age of rock units as determined by superposition and embayment relations and impact crater densities; forms the basis for both local and global correlation of geologic materials, features and events, as well as determination of geologic history.
striae	a series of grooves etched in a rock surface, usually by glaciers, streams or faulting, and oriented parallel to the direction of movement.
strike	the orientation of the line comprising the intersection of the plane of tilt of a surface or stratigraphic unit and the horizontal plane.
strike-slip faults	faults where the crust on either side of the fault plane has moved laterally.

stromatolites	laminated sedimentary buildups, usually of carbonate sediments and commonly of organo-sedimentary origin. Stromatolites may indicate the lithified remains of microbial mat communities. The laminated nature of biogenic stromatolites is due to the phototactic responses of the micro-organisms in the mat moving through periodically deposited sedimentary material or precipitate.
strombolian eruption	a type of basaltic volcanic eruption involving highly fluid lavas of low gas content, in which discrete explosions occur, separated by periods of less than a second to several hours. These eruptions occur in magmas near the surface; named after the volcano Stromboli, Italy.
subcrustal erosion	removal of material from the base of the crust or lithosphere by motions of the underlying mantle.
subduction	the descent of a tectonic plate into the mantle or the downwelling of the lithosphere as a part of the convective mantle circulation.
subradar point	the center of a planet's disk as seen by the observer (radar) at any given time. The locus of such points is called the "subradar track."
subtropical jet	*see* jet.
suspension	mechanism of transport of fine material by the wind, wherein the fall velocity of the individual grains is less than the random component of wind velocity, and grains will be caried large distances before coming to rest on the surface.
swale	a shallow depression in a level or slightly undulating (hilly) landscape.
tableland	*see* mesa.
tachylyte	a volcanic glass that may be black, green, or brown because of abundant crystallites. It is formed from basaltic magma and is commonly found as chilled margins of dikes, sills, or flows. (*AGI Glossary,* p. 670)

talik	a Russian term for a layer of unfrozen ground above, within, or beneath permafrost; occurs in regions of discontinuous permafrost. It may be permanent or temporary. (*AGI Glossary*)
talus	collection of debris which has fallen down a sloped surface and formed a surface with lesser slope.
tectonism	the deformation of the crust and lithosphere by large-scale forces associated with the dynamics of the mantle.
terrigenous	deposited on or in the Earth's crust.
thermal boundary layer	a layer of material in a convecting system adjacent to the boundary of the system across which heat is transported by conduction.
thermal inertia	a material parameter governing the rate at which a body's temperature responds to changing heat input. It is proportional to the square root of the product of thermal conductivity, density and specific heat capacity.
thermal pressure	pressure associated with thermal motion of plasma ions and electrons.
thermal skin depth	the depth within the surface at which the amplitude of a time-varying (sinusoidal) temperature change has been reduced by a factor of $1/e$.
thermal stress	force per unit area in an elastic medium due to strains created by spatially varying temperature.
thermal tide	atmospheric variations with periods of 1 solar day and harmonics, induced by solar heating.
thermokarst	topographic depressions with karst-like features (steep-sided pits) produced in permafrost regions and resulting from the melting of ground ice.
threshold wind speed	value of wind speed at a given reference height, at initiation of grain motion.
tilt (as used in radar)	*see* slope.

topographic datum	the reference surface from which elevations on a topographic map are measured. On the Earth this datum is usually mean sea level. On Mars, this datum is the location of the nominal 6.1-mbar pressure level, as defined during the Mariner 9 mission for a geoid of fourth order and degree.
tributary canyon	deep, locally dendritic valleys having V-shaped cross profiles and merging with the Valles Marineris or other troughs.
triple point	the pressure and temperature at which solid, liquid, and gas phases of a material can coexist in equilibrium.
tropopause	upper boundary of the troposphere, where the temperature gradient goes to zero.
troposphere	region of the atmosphere bounded by the surface and by the tropopause.
trough	a deep, steep-sided depression; used for the Valles Marineris system of grabens.
unpolarized (echo) component	that part of a radar echo which has no deterministic polarization; a randomly polarized echo. *See also* polarized (echo) component *and* depolarized (echo) component.
UV-stimulated oxidation	chemical reaction of minerals with atmospheric gases, stimulated by the energy from the absorption of ultraviolet radiation.
Valles Marineris	a system of depressions located just south of the Martian equator and trending approximately east-west. The system is thousands of kilometers long. Depressions are around 100 km wide and may reach 10 km in depth; named in honor of the Mariner 9 Mission.
valley network	system of interconnected drainages on the surface of Mars which may have carried fluids but lacks the bedform features that are direct indicators of fluid flow.
ventifact	an object shaped by wind erosion, especially rock outcrops or clasts.

VL-1; VL-2 — Viking Lander 1; Viking Lander 2.

VLA — Very Large Array; an array of radio telescopes sited in New Mexico.

VO-1; VO-2 — Viking Orbiter 1; Viking Orbiter 2.

volcanic construct — a hill or mountain made of lava and/or pyroclastic materials that is constructed around a volcanic vent.

vulcanian eruption — a type of volcanic eruption that is highly explosive and short-lived which frequently involves heated external water. Typically on Earth, such eruptions produce black ash- and steam-laden eruption columns. Named after Vulcano, a small volcano in the Eolain Islands, Italy.

Walker-type circulation — large-scale east-west overturnings in which rising and sinking motions are joined by motions in the longitude-height plane. On Earth, the Walker circulation describes air which rises over the western equatorial Pacific Ocean, moves eastward, and then descends over the eastern Pacific.

water activity — the partial pressure of water vapor divided by the partial pressure over pure water. The presence of salt in solution lowers the water activity.

"wave of darkening" — an apparent darkening of the exposed surface after retreat of the seasonal polar cap on Mars, as seen by telescopic observers.

wave vertical scale — many atmospheric waves can usefully be represented mathematically as small perturbations on a background atmosphere state. In such approximations the vertical structure of the wave can be represented as a sum of components proportional to the real part of $e^{(+i\zeta z)}$, with $i \equiv \sqrt{-1}$, height z and vertical scale l/ξ. When $\xi^2 > 0$ and ξ is real, the wave changes phase with height with a vertical wavelength $2\pi/\xi$, while the amplitude (in an idealized incompressible atmosphere) remains constant. When $\xi^2 < 0$, ξ is imaginary and the wave amplitude changes exponentially with height, with an e-folding distance *ie*. The sign $(+)$ is determined by the boundary

	conditions and is generally chosen to yield upward propagation or decay with height as $z \to \infty$.
weathering	physical and chemical interactions between surface materials (including interstitial volatiles) that lead to decomposition or alteration of rocks, minerals, or mineraloids and the possible formation of new phases (usually taken to mean at the scale of individual mineral grains).
Weidermann-Franz law	a relationship between the electronic contribution to the thermal conductivity and the electrical conductivity on iron; used in determining the properties of an iron-rich planetary core.
wind streak	light or dark patches emanating from topographic obstacles on the Martian surface and pointing in the direction which was downwind at the time of their formation. They can be either erosional or depositional features, depending on the conditions which existed at the time of their formation.
wrinkle ridge	a ridge or long quasilinear topographic high on the surface of a planet due to compressional forces acting on the lithosphere.
yardang	streamlined, aerodynamically-shaped elongate hill oriented parallel to the wind; term is derived from the Turkistani word "yar" meaning ridge or steep bank from which material is being removed (McCauley et al. 1977*b*).
yellow cloud	atmospheric dust haze as seen by telescopic observers on Earth; distinguished from "white clouds," which are formed from condensates.
zonal	along the same direction as lines of constant latitude.
Zwan-Wolf effect	tendency of plasma to be squeezed out of a magnetic flux tube draped around an obstacle.

Bibliography

BIBLIOGRAPHY

Compiled by Mary Guerrieri

Abramowitz, M., and Stegun, I. A. 1968. *Handbook of Mathematical Functions,* 7th ed. (Washington, D.C.: Natl. Bureau of Standards).

Adams, J. B. 1968. Lunar and Martian surfaces: Petrologic significance of absorption bands in the near-infrared. *Science* 159:1453–1455.

Adams, J. B. 1974. Visible and near-infrared diffuse reflectance spectra of pyroxenes as applied to remote sensing of solid objects in the solar system. *J. Geophys. Res.* 79:4829–4836.

Adams, J. B., and McCord, T. B. 1969. Mars: Interpretation of spectral reflectivity of light and dark regions. *J. Geophys. Res.* 74:4851–4856.

Adams, J. B., Smith, M. O., and Johnson, P. E. 1986. Spectral mixture modelling: A new analysis of rock and soil types at the Viking Lander site. *J. Geophys. Res.* 91:8098–8112.

Adams, W. S. 1941. Some results with the Coude spectrograph of the Mount Wilson Observatory. Water vapor lines in the spectrum of Mars. *Astrophys. J.* 93:16–17.

Adams, W. S., and Dunham, Th. 1934. The B band of oxygen in the spectrum of Mars. *Astrophys. J.* 79:308–316.

Adams, W. S., and Dunham, Th. 1937. Water-vapor lines in the spectrum of Mars. *Publ. Astron. Soc. Pacific* 49:209–211.

Adams, W. S., and St. John, C. 1926. An attempt to detect water vapor and oxygen lines in the spectrum of Mars with the registering microphotometer. *Astrophys. J.* 63:133–137.

Adelman, B. 1986. The question of life on Mars. *J. Brit. Interplanet. Soc.* 39:256–262.

Afonin, V., McKenna-Lawlor, S., Gringauz, K., Kecskemety, K., Keppler, E., Kirsch, E., Richter, A., O'Sullivan, D., Somogyi, A., Thompson, A., Varga, A., and Witte, M. 1989. Energetic ions in the close environment of Mars and particle shadowing by the planet. *Nature* 341:616–618.

Ahrens, T. J., and O'Keefe, J. D. 1987. Impact on the Earth, ocean and atmosphere. *J. Impact Eng.* 5:13–32.

Ajello, J. M., Hord, C. W., Barth, C. A., Stewart, A. I., and Lane, A. L. 1973. Mariner 9 ultraviolet spectrometer experiment: Afternoon terminator observation of Mars. *J. Geophys. Res.* 78:4279–4290.

Akabane, T., Matusi, K., Ishiura, K., Hattori, A., Iwasaki, K., Saito, Y., Asada, T., Saito, S., and Nakai, Y. 1980. Photographic observations of Mars during the 1977–1978 opposition. *Contrib. Kwasan Hida Obs.* No. 242.

Akabane, T., Matsui, M., Ishiura, K., Iwasaki, K., Saito, Y., and Kitahara, T. 1982. Photographic observations of Mars during the 1979–1980 opposition. *Contrib. Kwasan Hida Obs.* No. 250.

Akabane, T., Iwasaki, K., Saito, Y., and Narumi, Y. 1987. The optical thickness of the blue-white cloud near Nix Olympica of Mars in 1982. *Publ. Astron. Soc. Japan* 39:343–359.

Akaogi, M., Ross, N. L., McMillan, P. F., and Navrotsky, A. 1984. The Mg_2SiO_4 polymorphs (olivine, modified spinel and spinel): Thermodynamic properties from oxide melt solution calorimetry, phase relations, and models of lattice vibrations. *Amer. Mineral.* 69:499–512.

Akimoto, S., Matsui, Y., and Syono, Y. 1976. High pressure crystal chemistry in orthosilicates and formation of the mantle transition zone. In *The Physics and Chemistry of Minerals and Rocks,* ed. R. G. J. Strens (London: Wiley), pp. 327–364.

Aksyonov, S. I. 1979. Some comments on interpretations of Viking biological experiments. *Origins of Life* 9:251–256.

Alaerts, L., Lewis, R. S., and Anders, E. 1979. Isotopic anomalies of noble gases in meteorites and their origins. IV. C3 (Ornans) carbonaceous chondrites. *Geochim. Cosmochim. Acta* 43:1421–1432.

Alexander, C. J., and Russell, C. T. 1985. Solar cycle dependence of the location of the Venus bow shock. *Geophys. Res. Lett.* 12:369–371.

Allamandola, L. J., Sandford, S. A., and Valero, G. J. 1988. The photochemical and thermal evolution of interstellar/precometary ice analogues. *Icarus* 76:225–252.

Allen, C. C. 1975. Central peaks in lunar craters. *Moon* 12:463–474.

Allen, C. C. 1978. Areal distribution of Martian rampart craters. *Icarus* 39:111–123.

Allen, C. C. 1979. Volcano-ice interactions on Mars. *J. Geophys. Res.* 84:8048–8059.

Allen, C. C. 1980. Icelandic subglacial volcanism: Thermal and physical studies. *J. Geol.* 88:108–117.

Allen, C. C., Gooding, J. L., Jercinovic, M., and Keil, K. 1981. Altered basaltic glass: A terrestrial analog to the soil of Mars. *Icarus* 45:347–369.

Allen, C. C., Gooding, J. L., and Keil, K. 1982. Hydrothermally altered impact melt rock and breccia: Contributions to the soil of Mars. *J. Geophys. Res.* 87:10083–10101.

Allison, J. E. 1960. Wet-combustion apparatus and procedure for organic and inorganic carbon in soil. *Soil Sci. Soc. Amer. Proc.* 24:36–41.

Alterman, Z., Jarosch, H., and Pekeris, C. L. 1959. Oscillations of the Earth. *Proc. Roy. Soc. London* 252A:80–95.

Alvarez, W., and Mueller, R. A. 1984. Evidence from crater ages for periodic impacts on Earth. *Nature* 308:718–720.

Alymore, L. A. G., Sills, I. D., and Quirk, J. P. 1970. Surface area of homoionic illito and montmorillonite clay minerals as measured by the adsorption of nitrogen and carbon dioxide. *Clays Clay Mineral.* 18:91–96.

Anders, E., and Ebihara, M. 1982. Solar-system abundances of the elements. *Geochim. Cosmochim. Acta* 46:2363–2380.

Anders, E., and Grevesse, N. 1989. Abundances of the elements: Meteoritic and solar. *Geochim. Cosmochim. Acta* 53:197–214.

Anders, E., and Owen, T. 1977. Mars and Earth: Origin and abundance of volatiles. *Science* 198:453–465.

Andersland, O. B., and Akili, W. 1967. Stress effect on creep rates of a frozen clay soil. *Geotech.* 17:27–39.

Anderson, D. L. 1972. Internal constitution of Mars. *J. Geophys. Res.* 77:789–795.

Anderson, D. L. 1989. Composition of the Earth. *Science* 243:367–370.

Anderson, D. L., Miller, W. F., Latham, G. V., Nakamura, Y., Toksoz, M. N., Dainty, A. M., Duennebier, F. K., Lazarewicz, A. R., Kovach, R. L., and Knight, T. C. D. 1977. Seismology on Mars. *J. Geophys. Res.* 82:4524–4546.

Anderson, D. M. 1968. Undercooling, freezing point depression, and ice nucleation of soil water. *Israel J. Chem.* 6:349–355.

Anderson, D. M., and Tice, A. R. 1973. The unfrozen interfacial phase in frozen soil water systems. In *Ecological Studies Analysis and Synthesis,* vol. 4, eds. A. Hadas, D. Swartzendruber, P. E. Rijtema, M. Fuchs and B. Yaron (New York: Springer-Verlag), pp. 107–124.

Anderson, D. M., and Tice, A. R. 1979. The analysis of water in the Martian regolith. *J. Molec. Evol.* 14:33–38.

Anderson, D. M., Gaffney, E. S., and Low, P. F. 1967. Frost phenomena on Mars. *Science* 155:319–322.

Anderson, D. M., Schwarz, M. J., and Tice, A. R. 1978. Water vapor adsorption by sodium montmorillonite at $-5°C$. *Icarus* 34:638–644.

Anderson, E. M. 1951. *The Dynamics of Faulting and Dyke Formation with Applications to Britain,* 2nd ed. (Edinburgh: Oliver and Boyd).
Anderson, E. M., and Leovy, C. B. 1978. Mariner 9 television limb observations of dust and ice hazes on Mars. *J. Atmos. Sci.* 35:723–734.
André, J. C., De Moor, G., Lacarrère, P., Therry, G., and du Vacht, R. 1978. Modeling the 24 hour evolution of the mean and turbulent structures of the planetary boundary layer. *J. Atmos. Sci.* 35:1861–1883.
Andrews, D. G., and McIntyre, M. E. 1978. Generalized Eliassen-Palm and Charney-Drazin theorems for waves on axisymmetric mean flows in compressible atmospheres. *J. Atmos. Sci.* 35:175–185.
Andrews, D. G., Holton, J. R., and Leovy, C. B. 1987. *Middle Atmosphere Dynamics* (New York: Academic Press).
Andrews, D. J. 1973. Equation of state of the alpha and epsilon phases of iron. *J. Phys. Chem. Solids* 34:825–840.
Antoniadi, E. M. 1930. *La Planète Mars 1659–1929* (Paris: Herman et Cie). Trans. P. Moore, 1975. *The Planet Mars* (Shaldon, Devon: K. Reid).
Arkani-Hamed, J. 1973. Stress differences in the Moon as an evidence for a cold Moon. *Moon* 6:135–163.
Artyushkov, E. V. 1973. Stresses in the lithosphere caused by crustal thickness inhomogeneities. *J. Geophys. Res.* 78:7675–7708.
Artyushkov, E. V. 1974. Can the Earth's crust be in a state of isostasy? *J. Geophys. Res.* 79:741–752.
Arvidson, R. E. 1974. Windblown streaks, splotches, and assorted craters on Mars: Statistical analysis of Mariner 9 photographs. *Icarus* 21:12–27.
Arvidson, R. E., Coradini, M., Caruse, A., Coradini, A., Fulchignoni, M., Federico, C., Funiciello, R., and Salomone, M. 1976. Latitude variations of wind erosion of crater ejecta deposits on Mars. *Icarus* 27:503–516.
Arvidson, R. E., Guinness, E. A., and Lee, S. 1979. Differential aeolian redistribution rates on Mars. *Nature* 278:533–535.
Arvidson, R. E., Goettel, K. A., and Hohenberg, C. M. 1980. A post-Viking view of Martian geologic evolution. *Rev. Geophys. Space Phys.* 18:565–603.
Arvidson, R. E., Guinness, E. A., and Lee, S. 1981*a*. Efficiency of aeolian processes on Mars—present and past. In *Third Intl. Coll. on Mars,* LPI Contrib. 441, pp. 6–8 (abstract).
Arvidson, R. E., Hohenberg, C. M., and Shirck, J. R. 1981*b*. Long-term characterization of the Martian atmosphere and soil from cosmic ray effects in return samples. *Icarus* 45:250–262.
Arvidson, R. E., Guinness, E. A., and Zent, A. P. 1982. Classification of surface units in the equatorial region of Mars based on Viking Orbiter color, albedo, and thermal data. *J. Geophys. Res.* 87:10149–10157.
Arvidson, R. E., Guinness, E. A., Moore, H. J., Tillman, J., and Wall, S. D. 1983. Three Mars years: Viking Lander 1 imaging observations. *Science* 22:463–468.
Arvidson, R. E., Guinness, E. A., Leff, C., Presley, M., Saunders, R., and Roth, L. 1984. Ancient Martian cratered terrain materials exposed by deflation northwest of Baldet and Antoniadi basins. *Lunar Planet. Sci.* XV:19–20 (abstract).
Arvidson, R. E., Gooding, J. L., and Moore, H. J. 1989*a*. The Martian surface as imaged, sampled, and analyzed by the Viking landers. *Rev. Geophys. Space Phys.* 27:39–60.
Arvidson, R. E., Guinness, E. A., Dale-Bannister, M., Adams, J., Smith, M., Christensen, P. R., and Singer, R. 1989*b*. Nature and distribution of surficial deposits in Chryse Planitia and vicinity, Mars. *J. Geophys. Res.* 94:1573–1587.
Arya, S. P. 1988. *Introduction to Micrometeorology* (San Diego: Academic Press).
Atreya, S. K., and Blamont, J. E. 1990. Stability of the Martian atmosphere: Possible role of heterogeneous chemistry. *Geophys. Res. Lett.* 17:287–290.
Atreya, S. K., Donahue, T. M., and Kuhn, W. R. 1978. Evolution of a nitrogen atmosphere on Titan. *Science* 201:611–613.
Aubele, J. C. 1988. Morphologic patterns in lunar mare wrinkle ridges and kinematic implications. *Lunar Planet. Sci.* XIX:19–20 (abstract).
Avanesov, G. A., Bonev, B. I., Kempe, F., Bazilevsky, A. T., Boycheva, V., Chikov, K. N., Danz, M., Dimitrov, D., Duxbury, T., Gromatikov, P., Halmann, D., Head, J., Heifetz,

V. N., Kolev, V., Kostenko, V. I., Kottsov, V. A., Krasevtev, V. M., Krasikov, V. A., Krumov, A., Kuzmin, A. A., Losev, K. D., Lumme, K., Mishev, D. N., Möhlmann, D., Muinonen, K., Murav'ev, V. M., Murchie, S., Murray, B., Neumann, W., Paul, L., Petkov, D., Petuchova, I., Pössel, W., Rebel, B., Shkuratov, Yu. A., Simeonov, S., Smith, B., Totev, A., Uzunov, Yu., Fedotov, V. P., Weide, G. -G., Zapfe, H., Zhukov, B. S., and Ziman, Ya. L. 1989. Television observations of Phobos. *Nature* 341:585–587.

Avanesov, G. A., Zhukov, B., Ziman, Ya., Kostenko, V., Kuzmin, A., Murav'ev, V., Fedotov, V., Bonev, B., Mishev, D., Petkov, D., Krumov, A., Simeonov, S., Boycheva, V., Uzunov, Yu., Weide, G.-G., Halmann, D., Pössel, W., Head, J., Murchie, S., Schkuratov, Yu. G., Berghänel, R., Danz, M., Mangoldt, T., Pihan, U., Weidlich, U., Lumme, K., Muinonen, K., Peltoniemi, J., Duxbury, T., Murray, B., Herkenhoff, K., Fanale, F., Irvine, W., and Smith, B. 1991. Results of TV imaging of Phobos (experiment VSK-FREGAT). *Planet. Space Sci.* 39:281–295.

Avduevsky, V. S., Akim, E. L., Aleshin, V. I., Borodin, N. F., Kerzhanovich, V. V., Malkov, Ya. V., Marov, M. Ya., Morozov, S. F., Rozhdestvenskiy, M. K., Ryabov, O. L., Subbotin, M. I., Suslov, V. M., Cheremukhina, Z. P., and Shkirina, V. I. 1975. Martian atmosphere in the landing site of the descent module of MARS-6 (Preliminary results). *Kosmich. Issled.* 13:21–32. Trans.: NASA TT-F-16336. Microfiche N75–24617.

Awramik, S. M. 1989. Earth's early fossil record: Why not look for similar fossils on Mars? In *Exobiology and Future Mars Missions,* eds. C. P. McKay and W. Davis, NASA CP-10027, pp. 4–5.

Awwiller, D., and Croft, S. K. 1986. Pit craters and their implications concerning volatiles on Mars. In *Advances in Planetary Geology* (NASA TM Series), submitted.

Axford, W. I. 1968. The polar wind and the terrestrial helium budget. *J. Geophys. Res.* 73:6855–6859.

Bagnold, R. A. 1941. *The Physics of Blown Sand and Desert Dunes* (London: Methuen).

Baird, A. K., and Clark, B. C. 1981. On the original igneous source of Martian fines. *Icarus* 45:113–123.

Baird, A. K., Toulmin, P., III, Clark, B. C., Rose, H. J., Jr., Keil, K., Christian, R. P., and Gooding, J. L. 1976. Mineralogic and petrologic implications of Viking geochemical results from Mars: Interim results. *Science* 194:1288–1293.

Baker, B. H., Mohr, P. A., and Williams, L. A. 1972. *Geology of the Eastern Rift System of Africa,* Geological Soc. of America SP-136 (Boulder: Geological Soc. of America).

Baker, R., and Garber, M. 1978. Theoretical analysis of the stability of slopes. *Geotech.* 28:395–411.

Baker, V. R. 1973. *Paleohydrology and Sedimentology of Lake Missoula Flooding of Eastern Washington,* Geological Soc. of America SP-144 (Boulder: Geological Soc. of America).

Baker, V. R. 1978*a*. A preliminary assessment of the fluid erosional processes that shaped the Martian outflow channels. *Proc. Lunar Planet. Sci. Conf.* 9:3205–3223.

Baker, V. R. 1978*b*. The Spokane flood controversy and the Martian outflow channels. *Science* 202:1249–1256.

Baker, V. R. 1979. Erosional processes in channelized water flows on Mars. *J. Geophys. Res.* 84:7985–7993.

Baker, V. R. 1982. *The Channels of Mars* (Austin: Univ. of Texas Press).

Baker, V. R. 1984. Planetary geomorphology. *J. Geol. Educ.* 32:236–246.

Baker, V. R. 1985. Models of fluvial activity on Mars. In *Models in Geomorphology,* ed. M. J. Woldenberg (Boston: Allen and Unwin), pp. 287–312.

Baker, V. R. 1990. Spring sapping and valley network development, with case studies by Kochel, R. C., Baker, V. R., Laity, J. E., and Howard, A. D. In *Groundwater Geomorphology,* eds. C. G. Higgins and D. R. Coates (Boulder: Geological Soc. of America), pp. 235–266.

Baker, V. R., and Costa, J. E. 1987. Flood power. In *Catastrophic Flooding,* eds. L. Mayer and D. Nash (Boston: Allen and Unwin), pp. 1–21.

Baker, V. R., and Kochel, R. C. 1978*a*. Morphological mapping of Martian outflow channels. *Proc. Lunar Planet. Sci. Conf.* 9:3181–3193.

Baker, V. R., and Kochel, R. C. 1978*b*. Morphometry of streamlined forms in terrestrial and Martian channels. *Proc. Lunar Planet. Sci. Conf.* 9:3193–3203.

Baker, V. R., and Kochel, R. C. 1979. Martian channel morphology: Maja and Kasei Valles. *J. Geophys. Res.* 84:7961–7983.

Baker, V. R., and Komar, P. D. 1987. Cataclysmic flood processes and landforms. In *Geomorphic Systems of North America,* ed. W. L. Graf (Boulder: Geological Soc. of America), pp. 423–443.
Baker, V. R., and Milton, D. J. 1974. Erosion by catastrophic floods on Mars and Earth. *Icarus* 23:27–41.
Baker, V. R., and Partridge, J. 1986. Small Martian valleys: Pristine and degraded morphology. *J. Geophys. Res.* 91:3561–3572.
Baker, V. R., Strom, R. G., Croft, S. K., Gulick, V. C., Kargel, J. S., and Komatsu, G. 1991. Ancient oceans, ice sheets and the hydrological cycle on Mars. *Nature* 352:589–594.
Bakhuyzen, H. G. van de Sande. 1897. Untersuchungen uber die rotations-zeit des planeten Mars. *Ann. Leidener Sternwarte* 7:1–13.
Ballou, E. V., Wood, P. C., Wydeven, T., Lehwalt, M. E., and Mack, R. E. 1978. Chemical interpretation of Viking Lander 1 life detection experiment. *Nature* 271:644–645.
Balmino, G., Moynot, G., and Valès, N. 1982. Gravity field model of Mars in spherical harmonics up to degree and order eighteen *J. Geophys. Res.* 87:9735–9746.
Baloga, S. M., and Pieri, D. C. 1985. Estimates of lava eruption rates at Alba Patera, Mars. In *Reports of Planetary Geology and Geophysics Program–1984,* NASA TM-87563, pp. 245–247 (abstract).
Baloga, S., and Pieri, D. 1986. Time-dependent profiles of lava flows. *J. Geophys. Res.* 91:9543–9552.
Banderdt, W. B. 1986. Support of long wavelength loads on Venus and implications for internal structure. *J. Geophys. Res.* 91:403–419.
Banerdt, W. B. 1988. Global dynamic stress modeling on Venus. *Lunar Planet. Sci.* XIX:25–26 (abstract).
Banerdt, W. B., and Golombek, M. P. 1988. Deformational models of rifting and folding on Venus. *J. Geophys. Res.* 93:4759–4772.
Banerdt, W. B., and Golombek, M. P. 1989. Long wavelength stress models for Mars: New and improved. *Lunar Planet. Sci.* XX:40–41 (abstract).
Banderdt, W. B., Phillips, R. J., Sleep, N. H., and Saunders, R. S. 1982. Thick-shell tectonics on one-plate planets: Applications to Mars. *J. Geophys. Res.* 87:9723–9733.
Banin, A. 1989*a*. Mars soil—A sterile regolith or a medium for plant growth? In *Case for Mars III,* ed. C. R. Stoker (San Diego: Univelt), pp. 559–571.
Banin, A. 1989*b*. The mineral components of dust on Mars. In *MECA Workshop on Dust on Mars III,* ed. S. Lee, LPI Tech. Rept. 89-01, pp. 1–3 (abstract).
Banin, A., and Anderson, D. M. 1974. Effects of salt concentration changes during freezing on the unfrozen water content of porous materials. *Water Resources Res.* 10:124–128.
Banin, A., and Margulies, L. 1983. Simulation of Viking biology experiments suggests smectites not palagonite, as Martian soil analogs. *Nature* 305:523–526.
Banin, A., and Rishpon, J. 1978. Experimental simulation of the Viking labeled release (LR) results with iron-adsorbed smectite clay minerals. *Life Sci. Space Res.* 17:59–64.
Banin, A., and Rishpon, J. 1979. Smectite clays in Mars soil: Evidence for their presence and role in Viking biology experimental results. *J. Molec. Evol.* 14:133–152.
Banin, A., Rishpon, J., and Margulies, L. 1981. Composition and properties of the Martian soil as inferred from Viking biology data and simulation experiments with smectite clays. In *Third Intl. Coll. on Mars,* LPI Contrib. 441, pp. 16–18.
Banin, A., Margulies, L., and Chen, Y. 1985. Iron-montmorillonite: A spectral analog of Martian soil. *Proc. Lunar Planet. Sci. Conf.* 15, *J. Geophys. Res. Suppl.* 90:C771-C774.
Banin, A., Carle, G. C., Cheng, S., Coyne, L. M., Orenberg, J. B., and Scattergood, T. W. 1988*a*. Laboratory investigations of Mars: Chemical and spectroscopic characteristics of a suite of clays as Mars soil analogs. *Origins of Life* 18:239–265.
Banin, A., Margulies, L., Ben-Shlomo, T., Carle, G. C., Coyne, L. M., Orenberg, J. B., and Scattergood, T. W. 1988*b*. Constraining Mars soils mineralogical composition: Palagonite vs. iron enriched smectite clays. *Lunar Planet. Sci.* XIX:27–28 (abstract).
Banks, R. J., Parker, R. L., and Huestis, S. P. 1977. Isostatic compensation on a continental scale. *Geophys. J.* 51:431–452.
Barker, E. S. 1971. Variations of the Martian CO_2 abundance with Martian season. In *Planetary Atmospheres,* eds. C. Sagan, T. C. Owen and H. J. Smith (Dordrecht: D. Reidel), pp. 196–202.

Barker, E. S. 1972. Detection of molecular oxygen in the Martian atmosphere. *Nature* 238:447–448.

Barker, E. S. 1974. Ground-based observations of Mars and Venus water vapor during 1972 and 1973. In *Exploration of the Planetary System,* eds. A. Woszczyk and C. Iwaniszewska (Boston: D. Reidel), pp. 203–222.

Barker, E. S. 1976. Martian atmospheric water vapor observations: 1972–74 apparitions. *Icarus* 28:247–268.

Barker, E. S., Schorn, R. A. Woszczyk, A., Tull, R. G., and Little, S. J. 1970. Mars: Detection of atmospheric water vapor during the southern hemisphere spring and summer season. *Science* 170:1308–1310.

Barlow, N. G. 1988*a*. The Martian cratering record and its implications for the early history of the planet. *Lunar Planet. Sci.* XIX:29–30 (abstract).

Barlow, N. G. 1988*b*. Parameters affecting formation of Martian impact crater ejecta morphology. *Lunar Plant. Sci.* XIX:31–32 (abstract).

Barlow, N. G. 1988*c*. Crater size-frequency distribution and a revised Martian chronology. *Icarus* 75:285–305.

Barnes, J. R. 1980. Time spectral analysis of midlatitude disturbances in the Martian atmosphere. *J. Atmos. Sci.* 37:2002–2015.

Barnes, J. R. 1981. Midlatitude disturbances in the Martian atmosphere: A second Mars year. *J. Atmos. Sci.* 38:225–234.

Barnes, J. R. 1984. Linear baroclinic instability in the Martian atmosphere. *J. Atmos. Sci.* 41:1536–1550.

Barnes, J. R. 1986. Finite-amplitude behavior of a single baroclinic wave with multiple vertical modes: Effects of thermal damping. *J. Atmos. Sci.* 43:58–71.

Barnes, J. R. 1990*a*. Possible effects of breaking gravity waves on the circulation of the middle atmosphere of Mars. *J. Geophys. Res.* 95:1401–1422.

Barnes, J. R. 1990*b*. Transport of dust to high northern latitudes in a Martian polar warming. *J. Geophys. Res.* 95:1381–1400.

Barnes, J. R., and Hollingsworth, J. L. 1987. Dynamical modeling of a planetary wave mechanism for a Martian polar warming. *Icarus* 71:313–334.

Barrick, D. E. 1970. Unacceptable height correlation coefficients and the quasi-specular component in rough surface scattering. *Radio Sci.* 5:647–654.

Barth, C. A. 1974. The atmosphere of Mars. *Ann. Rev. Earth Planet. Sci.* 2:333–367.

Barth, C. A. 1985. Photochemistry of the atmosphere of Mars. In *The Photochemistry of the Atmospheres,* ed. J. S. Levine (New York: Academic Press), pp. 337–392.

Barth, C. A., and Hord, C. W. 1971. Mariner ultraviolet spectrometer: Topography and polar cap. *Science* 173:197–201.

Barth, C. A., Hord, C. W., Pearce, J. B., Kelly, K. K., Anderson, G. P., and Stewart, A. I. 1971. Mariner 6 and 7 ultraviolet spectrometer experiment: Upper atmosphere data. *J. Geophys. Res.* 76:2213–2227.

Barth, C. A., Hord, C. W., Stewart, A. I., and Lane, A. L. 1972*a*. Mariner 9 ultraviolet spectrometer experiment: Initial results. *Science* 175:309–312.

Barth, C. A., Stewart, A. I., Hord, C. W., and Lane, A. L. 1972*b*. Mariner 9 ultraviolet spectrometer experiment: Mars airglow spectroscopy and variations in Lyman-alpha. *Icarus* 17:457–468.

Barth, C. A., Hord, C. W., Stewart, A. I., Lane, A. L., Dick, M. L., and Anderson, G. P. 1973. Mariner 9 ultraviolet spectrometer experiment: Seasonal variation of ozone on Mars. *Science* 179:795–796.

Barth, C. A., Tobiska, W. K., Siskind, D. E., and Cleary, D. D. 1988. Solar-terrestrial coupling: Low-latitude thermospheric nitric oxide. *Geophys. Res. Lett.* 15:92–94.

Basaltic Volcanism Study Project. (BVSP) 1981. *Basaltic Volcanism on the Terrestrial Planets* (New York: Pergamon).

Basford, J. R., Dragon, J. C., Pepin, R. O., Coscio, M. R., Jr., and Murthy, V. R. 1973. Krypton and xenon in lunar fines. *Proc. Lunar Sci. Conf.* 4:1915–1955.

Basharinov, A. E., Vetukhnovskaia, Iu. N., Galaktionov, V. N., Egerov, S. T., Kolosov, M. A., Krupenio, N. N., Kuzmin, A. D., Ladygin, V. A., Malafeev, L. I., and Omelchenko, E. I. 1976. Radioastronomical measurements aboard the AMS Mars-5. *Kosmich. Issled.* 14:73–79. Trans.: *Cosmic Res.* 14:65–71.

Batson, R. M. 1973. Cartographic products from the Mariner 9 mission. *J. Geophys. Res.* 78:4424–4435.

Batson, R. M. 1978. Planetary mapping with the airbrush. *Sky & Teles.* 55:109–112.

Batson, R. M. 1982. Planimetric mapping of the planets with spacecraft television pictures. In *Proc. Intl. Soc. for Photogrammetry and Remote Sensing, Commission IV Symp.*, Aug. 25–28, Crystal City, Virginia, pp. 25–34.

Batson, R. M. 1987. Digital cartography of the planets: New methods, its status, and its future. *Photogrammetric Eng. Remote Sensing* 53:1211–1218.

Batson, R. M., Bridges, P. M., and Inge, J. L. 1979. *Atlas of Mars: The 1:5,000,000 Map Series,* NASA SP-438.

Battistini, R. 1985. Hydrolithosphere and problems of subsurface ice in the equatorial zone of Mars. In *Ices in the Solar System,* eds. J. Klinger et al. (Boston: D. Reidel), pp. 607–617.

Bauer, S. J. 1973. *Physics of Planetary Ionospheres* (Berlin: Springer-Verlag).

Bauer, S. J. 1983. Solar control of the Venus ionosphere. *Österreichische Akad. Wissenschaften Sitzungsberichet* 192, 8–10, pp. 309–317.

Bauer, S. J., and Hantsch, M. H. 1989. Solar cycle variation of the upper atmosphere temperature of Mars. *Geophys. Res. Lett.* 5:373–377.

Bauer, S. J., and Hartle, R. E. 1973. On the extent of the Martian ionosphere. *J. Geophys. Res.* 78:3169–3171.

Bauld, J. 1989. Microbial mats, playa lakes and other saline habitats: Early Mars analog? In *Exobiology and Future Mars Missions,* eds. C. P. McKay and W. Davis, NASA CP-10027, pp. 7–8.

Baum, W. A. 1972. Where will the Martian dust be when Viking arrives? *Bull. Amer. Astron. Soc.* 4:374–375 (abstract).

Baum, W. A. 1973. The International Planetary Patrol program: An assessment of the first three years. *Planet. Space Sci.* 21:1511–1519.

Baum, W. A. 1974. Earth-based observations of Martian albedo changes. *Icarus* 22:263–370.

Baum, W. A., Millis, R. L., Jones, S. E., and Martin, L. J. 1970. The International Planetary Patrol program. *Icarus* 12:435–439.

Becker, R. H., and Pepin, R. O. 1984. The case for a Martian origin of the shergottites: Nitrogen and noble gases in EETA 79001. *Earth Planet. Sci. Lett.* 69:225–242.

Becker, R. H., and Pepin, R. O. 1986. Nitrogen and light noble gases in Shergotty. *Geochim. Cosmochim. Acta* 50:993–1000.

Beer, R., Norton, R. H., and Martonchik, J. V. 1971. Astronomical infrared spectroscopy with a Connes-type interferometer. II. Mars, 2500–3500 cm^{-1}. *Icarus* 15:1–10.

Beish, J. D., Capen, C. F., and Parker, D. C. 1987. The meteorology of Mars—Part II. *J. Assoc. Lunar Planet. Obs.* 32:12–21.

Beish, J. D., Parker, D. C., and Hernandez, C. E. 1989. The red planet shows off. *Sky & Teles.* 77:30–35.

Bell, J. F. 1989. Mineralogical clues to the origins of asteroid dynamical families. *Icarus* 78:426–440.

Bell, J. F., Owensby, P. D., Hawke, B. R., and Gaffey, M. J. 1988. The 52-color asteroid survey: Final results and interpretations. *Lunar Planet. Sci.* XIX:57–58 (abstract).

Bell, J. F., Davis, D. R., Hartmann, W. K., and Gaffey, M. J. 1989*a*. Asteroids: The big picture. In *Asteroids II,* eds. R. P. Binzel, T. Gehrels and M. S. Matthews (Tucson: Univ. of Arizona Press), pp. 921–945.

Bell, J. F., Piscitelli, J. R., and Lebofsky, L. A. 1989*b*. Deimos: Hydration state from infrared spectroscopy. In *Lunar and Planet. Sci.* XX:58–59 (abstract).

Bell, J. F., McCord, T. B., and Lucey, P. G. 1989*c*. Mars during the 1988 opposition. *Eos: Trans. AGU* 70:50.

Bell, J. F., Lucey, P. G., Gradie, J. C., Granahan, J. C., Tholen, D. J., Piscitelli, J. R., and Lebofsky, L. A. 1989*d*. Reflection spectroscopy of Phobos and Deimos. *Bull. Amer. Astron. Soc.* 21:991 (abstract)

Bell, J. F., McCord, T. B., and Owensby, P. D. 1989*e*. High spectral resolution 0.3–1.0μm sprectroscopy and imaging of Mars during the 1988 opposition: Characterization of Fe mineralogy. Fourth Intl. Conf. on Mars, Jan. 10–13, Tucson, Arizona, Abstract Book, pp. 67–68.

Bell, J. F., Robinson, M., McCord, T., and Fanale, F. 1990*a*. Comparisons of new ground based and Phobos-2 VSK color ratio data for Mars. *Lunar Planet. Sci.* XXI:63–64 (abstract).

Bell, J. F., McCord, T. B., and Owensby, P. D. 1990*b*. Observational evidence of crystalline iron oxides on Mars. *J. Geophys. Res.* 95:14447–14461.

Bell, J. F., McCord, T. B., and Lucey, P. G. 1990*c*. Imaging spectroscopy of Mars (0.4–1.1μm) during the 1988 opposition. *Lunar Planet. Sci.* XX:479–486 (abstract).

Bell, P. M., Yagi, T., and Mao, H. K. 1979. Iron-magnesium distribution coefficients between spinel [$(Mg,Fe)_2SiO_4$], magnesiowüstite [Mg,Fe)O] and perovskite [$(Mg,Fe)SiO_3$]. *Carnegie Inst. Washington Yearbook* 78:616.

Belotserkovskii, O. M., Breus, T. K., Krymskii, A. M., Mitniskii, V. Ya., Nagy, A. F., and Gombosi, T. I. 1987. The effect of the hot oxygen corona on the interaction of the solar wind with Venus. *Geophys. Res. Lett.* 14:503–506.

Bercovici, D., Schubert, G., Glatzmaier, G. A., and Zebib, A. 1989*a*. Three-dimensional thermal convection in a spherical shell. *J. Fluid Mech.* 206:75–104.

Bercovici, D., Schubert, G., and Glatzmaier, G. A. 1989*b*. Three-dimensional models of convection in the Earth's mantle. *Science* 244:950–955.

Berger, A. L. 1976. Obliquity of precession for the last 5,000,000 years. *Astron. Astrophys.* 51:127–135.

Berkley, J. H., Keil, K., and Prinz, M. 1980. Comparative petrology and origin of Governardor Valadares and other nakhlites. *Proc. Lunar Planet. Sci. Conf.* 11:1089–1102.

Bell, P. M., Yagi, T., and Mao, H. K. 1979. Iron-magnesium distribution coefficients between spinel [$(Mg,Fe)_2SiO_4$], magnesiowüstite [Mg,Fe)O] and perovskite [$(Mg,Fe)SiO_3$]. *Carnegie Inst. Washington Yearbook* 78:616.

Belton, M. J. S., Broadfoot, A. L., and Hunten, D. M. 1968. Abundance and temperature of CO_2 on Mars during the 1967 opposition. *J. Geophys. Res.* 73:4795–4806.

Bertaux, Zh.-L., Blamon, Zh., Babichenko, C. I., Dement'yeva, N. N., D'yachkov, A. V., Kurt, V. G., Sklyankin, V. A., Smirnov, A. S., and Chuvakhin, S. P. 1975. Measurement of intensity and spectral characteristics of radiation in the Lyman-alpha line in the upper Martian atmosphere. *Kosmich. Issled.* 13:42–47. Trans.: NASA TT-F-16339. Microfiche N75–24612.

Bertka, C. M., and Holloway, J. R. 1988. Martian mantle primary melts: An experimental study of iron-rich garnet lherzolite minimum melt composition. *Proc. Lunar Planet. Sci. Conf.* 18:723–739.

Bertka, C. M., and Holloway, J. R. 1989. Martian mantle primary melts: An experimental study of melt density and viscosity at 23 kb. *Lunar Planet. Sci.* XX:69–80 (abstract).

Bezrukikh, V. V., Verigin, M. I., and Shyutte, N. M. 1978. Detection of heavy ions in the interaction region between the solar wind and the planet Mars. *Cosmic Res.* 16:476–479.

Bibring, J.-P., Combes, M., Ksanfomality, L. V., Langevin, Y., Moroz, V. I., and Soufflot, A. 1988. Thermal radiometer (5–50 μm), infrared spectrometer (0.8–3.0 μm), and spectrophotometer (0.3–0.9 μm): The KRFM-ISM experiment, Phobos project. In *Phobos: Scientific and Methodological Aspects of the Phobos Study: Proc. Intl. Workshop,* Nov. 24–28, 1986, Moscow (Moscow: U.S.S.R. Academy of Sci.).

Bibring, J.-P., Combes, M., Langevin, Y., Soufflot, A., Cara, C., Drossart, P., Encrenaz, Th., Erard, S., Forni, O., Gondet, G., Ksanfomality, L., Lellouch, E., Masson, Ph., Moroz, V., Rocard, F., Rosenqvist, J., and Sotin, C. 1989. Results from the ISM experiment. *Nature* 341:591–593.

Bibring, J.-P., Combes, M., Langevin, Y., Cara, C., Drossart, P., Encrenaz, T., Erard, S., Forni, O., Gondet, B., Ksanfomaliti, L., Lellouch, E., Masson, P., Moroz, V., Rocard, F., Rosenqvist, J., Sotin, C., and Soufflot, A. 1990*a*. ISM observations of Mars and Phobos: First results. *Lunar Planet. Sci. Conf.* 20:461–471.

Bibring, J.-P., Langevin, Y., Erard, S., Forni, O., Masson, P., and Sotin, C. 1990*b*. The observation of the surface of Mars by the ISM instrument on board the Phobos 2 spacecraft. *Lunar Planet. Sci.* XXI:79–80 (abstract).

Biemann, K. 1974. Test results on the Viking gas-chromatograph-mass spectrometer experiment. *Origins of Life* 5:417–430.

Biemann, K. 1979. The implications and limitations of the findings of the Viking organic analysis experiment. *J. Molec. Evol.* 14:65–70.

Biemann, K., and Lavoie, J. M. 1979. Some final conclusions and supporting experiments related to the search for organic compounds in the surface of Mars. *J. Geophys. Res.* 84:8385–8390.

Biemann, K., Owen, T., Rushneck, D. R., La Fleur, A. L., and Howarth, D. W. 1976*a*. The atmosphere of Mars near the surface: Isotope ratios and upper limits on noble gases. *Science* 194:76–78.

Biemann, K., Oró, J., Toulmin, J., III, Orgel, L. E., Nier, A. O., Anderson, D. M., Simmonds, P. G., Flory, D., Diaz, A. V., Rushneck, D. R., and Biller, J. E. 1976*b*. Search for organic and volatile inorganic compounds in two surface samples from the Chryse Planitia region of Mars. *Science* 194:72–76.

Biemann, K., Oró, J., Toulmin, P., III, Orgel, L. E., Nier, A. O., Anderson, D. M., Simmonds, P. G., Flory, D., Diaz, A. V., Rushneck, D. R., Biller, J. E., and LaFleur, A. L. 1977. The search for organic substances and inorganic volatile compounds in the surface of Mars. *J. Geophys. Res.* 82:4641–4658.

Bills, B. G. 1978. A Harmonic and Statistical Analysis of the Topography of the Earth, Moon and Mars. Ph.D. Thesis, California Inst. of Tech.

Bills, B. G. 1989*a*. The moments of inertia of Mars. *Geophys. Res. Lett.* 16:385–388.

Bills, B. G. 1989*b*. Comment on "More about the moment of inertia of Mars." *Geophys. Res. Lett.* 16:1337–1338.

Bills, B. G. 1990. The rigid body obliquity history of Mars. *J. Geophys. Res.* 95:14137–14153.

Bills, B. G., and Ferrari, A. J. 1977. A lunar density model consistent with topographic, gravitational, librational and seismic data. *J. Geophys. Res.* 82:1306–1314.

Bills, B. G., and Ferrari, A. J. 1978. Mars topography, harmonics, and geophysical implications. *J. Geophys. Res.* 83:3497–3508.

Bills, B. G., and Ferrari, A. J. 1980. A harmonic analysis of lunar gravity. *J. Geophys. Res.* 85:1013–1025.

Bills, B. G., and Kobrick, M. 1985. Venus topography: A harmonic analysis. *J. Geophys. Res.* 90:827–836.

Bills, B. G., Richards, M. A., and Kiefer, W. S. 1986. Mars: Gravity, topography and dynamic compensation. *Lunar Planet. Sci.* XVII:48–49 (abstract).

Bills, B. G., Kiefer, W. S., and Jones, R. L. 1987. Venus gravity: A harmonic analysis. *J. Geophys. Res.* 92:10335–10351.

Binder, A. B. 1969. Internal structure on Mars. *J. Geophys. Res.* 74:3110–3118.

Binder, A. B., and Cruikshank, D. P. 1963. Comparison of the infrared spectrum of Mars with the spectra of selected terrestrial rocks and minerals. *LPI Commun.* 2:37, 193–196.

Binder, A. B., and Davis, D. R. 1973. Internal structure of Mars. *Phys. Earth Planet. Int.* 7:477–485.

Binder, A. B., and Lange, M. A. 1980. On the thermal history of a moon of fission origin. *J. Geophys. Res.* 85:3194–3208.

Binder, A. B., Arvidson, R. E., Guinness, E. A., Jones, K. L., Morris, E. C., Mutch, T. A., Pieri, D. C., and Sagan, C. 1977. The geology of the Viking Lander 1 site. *J. Geophys. Res.* 82:4439–4451.

Bittner, H., and Fricke, K. H. 1987. Dayside temperatures of the Martian upper atmosphere. *J. Geophys. Res.* 92:12045–12055.

Bjoraker, G. L., Mumma, M. J., and Larson, H. P. 1989. Isotopic abundance ratios for hydrogen and oxygen in the Martian atmosphere. *Bull. Amer. Astron. Soc.* 21:990 (abstract).

Blackburn, T. R., Holland, H. D., and Ceasar, G. P. 1979. Viking gas exchange reaction: Simulation on UV-irradiated manganese dioxide substrate. *J. Geophys. Res.* 84:8391–8394.

Blamont, J. E., Chassefiere, E., Goutail, J. P., Mege, B., Nunes-Pinharanda, M., Souchon, G., Krasnopolsky, V. A., Krysko, A. A., and Moroz, V. I. 1989. Vertical profiles of dust and ozone in the Martian atmosphere deduced from solar occultation measurements. *Nature* 341:600–603.

Blaney, D. L., and McCord, T. B. 1990*a*. Earth-based telescopic observations of Mars in the 4.4 μm to 5.1 μm region. *Lunar Planet. Sci.* XXI:99–100 (abstract).

Blaney, D. L., and McCord, T. B. 1990*b*. An observational search for carbonates on Mars. *J. Geophys. Res.* 94:10159–10166.

Blasius, K. R., and Cutts, J. A. 1976. Shield volcanism and lithospheric structure beneath the Tharsis plateau, Mars. *Proc. Lunar Sci. Conf.* 7:3561–3573.

Blasius, K. R., Cutts, J. A., Guest, J. E., and Masursky, H. 1977. Geology of the Valles Marineris: First analysis of imaging from the Viking 1 Orbiter primary mission. *J. Geophys. Res.* 82:4067–4091.

Blasius, K. R., Cutts, J. A., and Howard, A. D. 1982. Topography and stratigraphy of Martian polar layered deposits. *Icarus* 50:140–160.

Blumsack, S. L. 1971*a*. On the effects of topography on planetary atmospheric circulation. *J. Atmos. Sci.* 28:1134–1143.

Blumsack, S. L. 1971*b*. On the effects of large-scale temperature advection in the Martian atmosphere. *Icarus* 15:429–442.

Blumsack, S. L., and Gierasch, P. J. 1972. Mars: The effects of topography on baroclinic instability. *J. Atmos. Sci.* 29:1081–1089.

Blumsack, S. L., Gierasch, P. J., and Wessel, W. R. 1973. An analytical and numerical study of the Martian planetary boundary layer over slopes. *J. Atmos. Sci.* 30:66–82.

Blunk, J. 1977. *Mars and Its Satellites* (Exposition Press).

Blunk, J. 1982. *Mars and Its Satellites,* 2nd ed. (Exposition Press).

Boctor, N. W., Meyer, H. O., and Kullerud, G. 1976. Lafayette meteorite: Petrology and opaque mineralogy. *Earth Planet. Sci. Lett.* 32:69–76.

Bogard, D. D. 1982. Trapped noble gases in the EETA 79001 shergottite. *Meteoritics* 17:185–186.

Bogard, D. D. 1988. On the origin of Venus' atmosphere: Possible contributions from simple component mixtures and fractionated solar wind. *Icarus* 74:3–20.

Bogard, D. D., and Johnson, P. 1983. Martian gases in an Antarctic meteorite? *Science* 221:651–654.

Bogard, D. D., Nyquist, L. E., and Johnson, P. 1984. Noble gas contents of shergottites and implications for the Martian origin of SNC meteorites. *Geochim. Cosmochim. Acta* 48:1723–1740.

Bogard, D. D., Hörz, F., and Johnson, P. H. 1986. Shock-implanted noble gases: An experimental study with implications for the origin of Martian gases in shergottite meteorites. In *Proc. Lunar Planet. Sci. Conf.* 17, *J. Geophys. Res. Suppl.* 91:E99-E114.

Bogdanov, A. V. 1981. Mars satellite Deimos interaction with the solar wind and its influence on flow around Mars. *J. Geophys. Res.* 86:6926–6932.

Bogdanov, A. V., and Vaisberg, O. L. 1975. Structure and variations of solar wind-Mars interaction region. *J. Geophys. Res.* 80:487–494.

Bonatti, E. 1965. Palagonite, hyaloclastitese and alteration of volcanic glass in the Ocean. *Bull. Volcanol.* 28:3–15.

Booth, M. C., and Kieffer, H. H. 1978. Carbonate formation in Marslike environments. *J. Geophys. Res.* 83:1089–1815.

Borderies, N. 1980. Theory of Mars rotation in Euler angles. *Astron. Astrophys.* 82:129–141.

Borderies, N., and Yoder, C. F. 1990. Phobos' gravity field and its influence on its orbit and physical librations. *Astron. Astrophys.* 233:235–251.

Borderies, N., Balmino, G., Castel, L., and Moynot, B. 1980. Study of Mars dynamics from Lander tracking data analysis. *Moon and Planets* 22:191–200.

Borgia, A., Jeremiah, B., Montero, W., Morales, L. D., and Alvarado, G. E. 1990. Fault propagation folds induced by gravitational failure and slumping of the volcanic edifices. *J. Geophys. Res.* 95:14357–14382.

Born, G. H. 1974. Mars physical parameters as determined from Mariner 9 observations of the natural satellites and Doppler tracking. *J. Geophys. Res.* 79:4837–4844.

Boslough, M. B. 1987. Selective weathering of shocked minerals and chondritic enrichment of the Martian fines. In *Nature and Composition of Surface Units on Mars,* LPI-NASA/MEVTV Workshop, Napa, Calif., pp. 16–18 (abstract).

Boslough, M. B., and Cygan, R. T. 1988. Shock-enhanced dissolution of silicate minerals and chemical weathering on planetary surfaces. *Proc. Lunar Planet. Sci. Conf.* 18:443–453.

Boslough, M. B., Venturini, E. L., Morosin, B., Graham, R. A., and Williamson, D. L. 1986. Physical properties of shocked and thermally altered nontronite: Implications for the Martian surface. *Proc. Lunar Planet. Sci. Conf.* 17, *J. Geophys. Res. Suppl.* 91:E207-E214.

Boss, A. P., and Peale, S. J. 1986. Dynamical constraints on the origin of the Moon. In *The Origin of the Moon,* eds. W. K. Hartmann, R. J. Phillips and G. J. Taylor (Houston: Lunar and Planetary Inst.), pp. 59–101.

Bougher, S. W., and Dickinson, R. E. 1988. Mars mesosphere and thermosphere. I. Global mean heat budget and thermal structure. *J. Geophys. Res.* 94:7325–7337.

Bougher, S. W., Dickinson, R. E., Ridley, E. C., and Roble, R. G. 1988*a*. Venus mesosphere and thermosphere. III. Three dimensional general circulation with coupled dynamics and composition. *Icarus* 73:545–573.

Bougher, S. W., Dickinson, R. E., Roble, R. G., and Ridley, E. C. 1988*b*. Mars thermospheric general circulation model: Calculations for the arrival of Phobos at Mars. *Geophys. Res. Lett.* 15:1511–1514.

Bougher, S. W., Roble, R. G., Ridley, E. C., and Dickinson, R. E. 1989. The Mars thermosphere. II. General circulation with coupled dynamics and composition. *J. Geophys. Res.* 94:14811–14827.

Bougher, S. W., Gérard, J.-C., Stewart, A. I. F., and Fesen, C. G. 1990. The Venus nitric oxide night airglow: Model calculations based on the Venus thermospheric general circulation model. *J. Geophys. Res.* 95:6271–6284.

Boyce, J. M. 1980. Distribution of thermal gradient values in the equatorial region of Mars based on impact crater morphology. In *Reports of Planetary Geology Program-1980*, NASA TM-82385, pp. 140–143.

Boyd, J. P. 1976. The noninteraction of waves with the zonally averaged flow on a spherical Earth and the interrelationships of eddy fluxes of energy, heat, and momentum. *J. Atmos. Sci.* 33:2285–2291.

Bozhinskiy, A. N., Krass, M. S., and Popovnin, V. V. 1986. Role of debris cover in the thermal physics of glaciers. *J. Glaciol.* 32:255–266.

Brace, L. H., Kasparzak, W. T., Taylor, H. A., Jr., Theis, R. F., Russell, C. T., Barnes, A., Mihalov, J. D., and Hunten, D. M. 1987. The ionotail of Venus: Its configuration and evidence for ion escape. *J. Geophys. Res.* 92:15–26.

Brace, W. F. 1980. Permeability of crystalline and argillaceous rocks. *Intl. J. Rock Mech. Mining Sci. Geochem. Abst.* 17:241–251.

Brace, W. F. 1984. Permeability of crystalline rocks: New in situ measurements. *J. Geophys. Res.* 89:4327–4330.

Brace, W. F., and Kohlstedt, D. S. 1980. Limits on lithospheric stress imposed by laboratory experiments. *J. Geophys. Res.* 85:6248–6352.

Bradley, T. L., and Barlow, N. G. 1988. Martian crater interiors: Relationships with ejecta, diameter, latitude, and terrain. *Lunar Planet. Sci.* XIX:128–129 (abstract).

Bragg, S. L. 1977. Characteristics of Martian Soil at Chryse Planitia as Inferred by Reflectance Properties, Magnetic Properties and Dust Accumulation on Viking Lander 1. M.S. Thesis, Washington Univ.

Brakenridge, G. R., Newsom, H. E., and Baker, V. R. 1985. Ancient hot springs on Mars: Origins and paleoenviromental significance of small Martian valleys. *Geology* 13:859–862.

Brass, G. W. 1980. Stability of brines on Mars. *Icarus* 42:20–28.

Bratt, S. R., Solomon, S. C., and Head, J. W. 1985. The evolution of impact basins: Cooling, subsidence, and thermal stress. *J. Geophys. Res.* 90:12415–12433.

Breed, C. S. 1977. Terrestrial analogs of the Hellespontus dunes. *Icarus* 30:326–340.

Breed, C. S., Grolier, M. J., and McCauley, J. F. 1979. Morphology and distribution of common "sand" dunes on Mars: Comparison with the Earth. *J. Geophys. Res.* 84:8183–8204.

Breed, C. S., McCauley, J. F., and Davis, P. A. 1987*a*. Sand sheets of the Eastern Sahara and ripple blankets on Mars. In *Desert Sediments: Ancient and Modern*, eds. L. E. Frostick and I. Reid, Geological Soc. of London SP-35, pp. 337–360.

Breed, C. S., Davis, P. A., and McCauley, J. F. 1987*b*. Accretion mantles on Mars: A new model for Viking lander site characteristics and implications. *Lunar Planet. Sci.* 18:125–126 (abstract).

Breneman, H. H., and Stone, E. C. 1985. Solar coronal and photospheric abundances from solar energetic particle measurements. *Astrophys. J. Lett.* 299:57–61.

Bretagnon, P. 1974. Termes à longues périodes dans le système solaire. *Astron. Astrophys.* 30:141–154.

Bretagnon, P. 1982. Théorie du mouvement de l'ensemble des planètes; solution VSOP82. *Astron. Astrophys.* 114:278–288.

Bretz, J. H., Smith, H. T. V., and Neff, G. E. 1956. Channeled scabland of Washington: New data and interpretations. *Geol. Soc. Amer. Bull.* 67:957–1049.

Breus, T. K. 1986. Mass-loading at Venus: Theoretical expectations. *Adv. Space Res.* 6:167–177.

Breus, T. K., and Gringauz, K. I. 1980. Nature of the obstacles which slow down the solar wind near Venus and Mars, and the properties of the interaction between the solar wind and the atmospheres of these planets. *Cosmic Res.* 18:426–436.

Briggs, G. A. 1974. The nature of the residual Martian polar caps. *Icarus* 23:167–191.

Briggs, G. A., and Leovy, C. B. 1974. Mariner 9 observations of the Mars north polar hood. *Bull. Amer. Meteorol. Soc.* 55:278–296.

Briggs, G. A., Klaasen, K., Thorpe, T., Wellman, J., and Baum, W. 1977. Martian dynamical phenomena during June–November 1976: Viking Orbiter imaging results. *J. Geophys. Res.* 82:4121–4149.

Briggs, G. A., Baum, W. A., and Barnes, J. 1979. Viking Orbiter imaging observations of dust in the Martian atmosphere. *J. Geophys. Res.* 84:2795–2820.

Brinkman, R. T. 1971. Mars: Has nitrogen escaped? *Science* 174:944–945.

Britt, D. T., and Pieters, C. M. 1988. The origin of Phobos: Implications of compositional properties. *Astron. Vestnik* 22:229–239.

Britt, D. T., Murchie, S. L., Pieters, C. M., Head, J. W., Fisher, P. C., Pratt, S. F., Ksanfomality, L. V., Zharkov, A., Zhukov, B. S., and Kuzmin, A. 1990. Phobos KRFM spectral results: Surface heterogeneity and meteorite analogs. *Lunar Planet. Sci.* XXI:129–130 (abstract).

Brotchie, J. F. 1971. Flexure of a liquid-filled spherical shell in a radial gravity field. *Mod. Geol.* 3:15–23.

Brouwer, D., and van Woerkom, A. J. J. 1950. The secular variations of the orbital elements of the planets. *Astron. Papers Amer. Ephemeris Nautical Alm.* 12(2):18

Brown, G. N., and Ziegler, W. T. 1980. Vapor pressure and heats of vaporization and sublimation of liquids and solids of interest in cyrogenics below 1-Atm pressure. *Adv. Cryogen. Eng.* 25:662–670.

Brown, H. 1952. Rare gases and the formation of the Earth's atmosphere. In *The Atmospheres of the Earth and Planets,* 2nd ed., ed. G. P. Kuiper (Chicago: Univ. of Chicago Press), pp. 258–266.

Brown, J. M., and McQueen, R. G. 1986. Phase transitions, Gruneisen parameter and elasticity for shocked iron between 77 and 400 GPa. *J. Geophys. Res.* 91:7485–7494.

Brown, J. M., Ahrens, T. J., and Shampine, D. L. 1984. Hugoniot data for pyrrhotite and the Earth's core. *J. Geophys. Res.* 89:6041–6048.

Bruch, C. W. 1966. Instrumentation for the detection of extraterrestrial life. In *Biology and the Exploration of Mars,* eds. C. S. Pittendrigh, W. Vishniac and J. P. T. Pearman, Natl. Acad. Sci. Publ. 1296 (NRC), pp. 487–502.

Brunauer, S., Emmett, P. H., and Teller, E. 1938. Adsorption of gases in multimolecular layers. *J. Amer. Chem. Soc.* 60:309–319.

Brunauer, S., Copeland, L. E., and Kantro, D. L. 1967. The Langmuir and BET theories. In *The Solid-Gas Interface,* ed. E. A. Flood (New York: M. Dekker), pp. 77–103.

Brutsaert, W. 1982. *Evaporation into the Atmosphere* (Boston: D. Reidel).

Bryan, W. B. 1973. Wrinkle ridges as deformed surface crusts on ponded mare lava. *Proc. Lunar Sci. Conf.* 4:93–106.

Budd, W. F., Jensen, D., Leach, J. H. I., Smith, I. N., and Radok, U. 1986. The north polar ice cap of Mars as a steady-state system. *Polarforschung* 56:43–63.

Bunch, T. E., and Reid, A. M. 1975. The nakhlites. Part I. Petrography and mineral chemistry. *Meteoritics* 10:303–315.

Burghele, A., Dreibus, G., Palme, H., Rammensee, W., Spettel, B., Weckwerth, G., and Wänke, H. 1983. Chemistry of shergottites and the Shergotty parent body (SPB): Further evidence for the two component model of planet formation. *Lunar Planet. Sci.* XIV:80–81 (abstract).

Burk, S. D. 1976. Diurnal winds near the Martian polar caps. *J. Atmos. Sci.* 33:923–939.

Burke, B. F., and Franklin, K. L. 1955. Observations of variable radio source associated with the planet Jupiter. *J. Geophys. Res.* 60:213–217.

Burns, J. A. 1972. The dynamical characteristics of Phobos and Deimos. *Rev. Geophys. Space Phys.* 10:462–483.

Burns, J. A. 1977. Orbital evolution. In *Planetary Satellites,* ed. J. A. Burns (Tucson: Univ. of Arizona Press), pp. 113–156.

Burns, J. A. 1978. On the orbital evolution and origin of the Martian moons. *Vistas in Astron.* 22:193–210.

Burns, J. A. 1986*a*. The evolution of satellite orbits. In *Satellites,* eds. J. A. Burns and M. S. Matthews (Tucson: Univ. of Arizona Press), pp. 117–158.

Burns, J. A. 1986*b*. Some background about satellites. In *Satellites,* eds. J. A. Burns and M. S. Matthews (Tucson: Univ. of Arizona Press), pp. 1–37.

Burns, R. G. 1965. Electronic Spectra of Silicate Minerals: Application of Crystal-Field Theory to Aspects of Geochemistry. Ph.D. Thesis, Univ. of California, Berkeley.

Burns, R. G. 1986. Terrestrial analogs of the surface rocks on Mars? *Nature* 320:55–56.

Burns, R. G. 1987. Ferric sulfates on Mars. *Proc. Lunar Planet. Sci. Conf.* 17, *J. Geophys. Res. Suppl.* 92:E570-E574.

Burns, R. G. 1988. Gossans on Mars. *Proc. Lunar Planet. Sci. Conf.* 18:713–721.

Burns, R. G., and Fisher, D. S. 1990. Evolution of sulfide mineralization on Mars. *J. Geophys. Res.* 95:14169–14173.

Burragato, F., Cavarretta, G., and Funiciello, R. 1975. The new Brazilian achondrite of Governador Valadares (Minas Gerais). *Meteoritics* 10:374–375 (abstract).

Burt, D. M. 1989. Iron-rich clay minerals on Mars: Potential sources or sinks for hydrogen and indicators of hydrogen loss over time. *Proc. Lunar Planet. Sci. Conf.* 19:423–432.

Busse, F. H. 1976. Generation of planetary magnetism by convection. *Phys. Earth Planet. Int.* 12:350–358.

Byerlee, J. D. 1978. Friction of rocks. *Pure Appl. Geophys.* 116:615–626.

Byerlee, J. D., and Brace, W. F. 1968. Stick-slip, stable sliding, and earthquakes—Effect of rock type, pressure, strain rate, and stiffness. *J. Geophys. Res.* 73:6031–6037.

Calvin, W. M., and King, T. V. V. 1991. Spectral evidence for carbonates on Mars. *Lunar and Planet. Sci.* XXII:169–170 (abstract).

Calvin, W. M., Jakosky, B. M., and Christensen, P. R. 1988. A model of diffuse radar scattering from Martian surface rocks. *Icarus* 76:513–524.

Cameron, A. G. W. 1978. Physics of the primitive solar nebula and of giant gaseous protoplanets. In *Protostars and Planets,* ed. T. Gehrels (Tucson: Univ. of Arizona Press), pp. 453–487.

Cameron, A. G. W. 1982. Elemental and nuclidic abundances in the solar system. In *Essays in Nuclear Astrophysics,* eds. C. Barnes, D. Clayton and D. Schramm (Cambridge: Cambridge Univ. Press), pp. 23–43.

Cameron, A. G. W. 1983. Origin of the atmospheres of the terrestrial planets. *Icarus* 56:195–201.

Cameron, A. G. W., Fegley, B., Benz, W., and Slattery, W. L. 1988. The strange density of Mercury: Theoretical considerations. In *Mercury,* eds. F. Vilas, C. R. Chapman and M. S. Matthews (Tucson: Univ. of Arizona Press), pp. 692–709.

Cameron, R. E. 1971. Antarctic soil microbial and ecological investigations. In *Research in the Antarctic,* eds. L. O. Quam and H. D. Porter, Amer. Assoc. Adv. Sci. Publ. 93, pp. 137–189.

Camichel, H. 1954. Détermination photographique de la pôle de Mars, de son diamètre et de ses coordonnées aréographiques. *Bull. Astron.* 18:83–190.

Campbell, D. B., Chandler, J. R., Ostro, S. J., Pettengill, G. H., and Shapiro, I. I. 1978. Galilean satellites: 1976 radar results. *Icarus* 34:254–267.

Campbell, M. J., and Ulrichs, J. 1969. Electrical properties of rocks and their significance for lunar radar observations. *J. Geophys. Res.* 74:5867–5881.

Cann, M. W. P., Davies, W. O., Greenspan, J. A., and Owen, T. C. 1965. *A Review of Recent Determinations of the Composition and Surface Pressure of the Atmosphere of Mars,* NASA CR-298.

Capen, C. F. 1966. *The Mars 1964–1965 Apparition,* JPL Tech. Rept. 32–990.

Capen, C. F. 1971. Martian yellow clouds—past and future. *Sky & Teles.* 41:117–120.

Capen, C. F. 1974. A Martian yellow cloud—July 1971. *Icarus* 22:345–362.

Capen, C. F. 1976. Martian albedo feature variations with season: Data of 1971 and 1973. *Icarus* 28:213–230.

Capen, C. F., and Capen, V. W. 1970. Martian north polar cap, 1962–1968. *Icarus* 13:100–108.
Capen, C. F., and Martin, L. J. 1971. The developing stages of the Martian yellow storm of 1971. *Lowell Obs. Bull.* 7(157):211–216.
Capen, C. F., and Martin, L. J. 1972. Mars' great storm of 1971. *Sky & Teles.* 43:276–279.
Carey, K. L. 1973. Icings Developed from Surface Water and Groundwater. U.S. Army CRREL Monograph III-D3.
Carleton, N. P., and Traub, W. A. 1972. Detection of molecular oxygen on Mars. *Science* 177:988–992.
Carpenter, R. 1967. 1967 radar observations of Mars. *JPL Space Programs Summary 37–48* III:157–160.
Carr, M. H. 1973. Volcanism on Mars. *J. Geophys. Res.* 78:4049–4062.
Carr, M. H. 1974*a*. The role of lava erosion in the formation of lunar rilles and Martian channels. *Icarus* 22:1–23.
Carr, M. H. 1974*b*. Tectonism and volcanism of the Tharsis region of Mars. *J. Geophys. Res.* 79:3943–3949.
Carr, M. H. 1976. Change in height of Martian volcanoes with time. *Geol. Rom.* 15:521–422.
Carr, M. H. 1979. Formation of Martian flood features by release of water from confined aquifers. *J. Geophys. Res.* 84:2995–3007.
Carr, M. H. 1981. *The Surface of Mars* (New Haven, Conn.: Yale Univ. Press).
Carr, M. H. 1983. Stability of streams and lakes on Mars. *Icarus* 56:476–495.
Carr, M. H. 1984. Mars. In *The Geology of the Terrestrial Planets,* ed. M. H. Carr, NASA SP-469, pp. 206–263.
Carr, M. H. 1986. Mars: A water-rich planet? *Icarus* 56:187–216.
Carr, M. H. 1987. Water on Mars. *Nature* 326:30–35.
Carr, M. H. 1989. Recharge of the early atmosphere of Mars by impact-induced release of CO_2. *Icarus* 79:311–327.
Carr, M. H. 1990. D/H on Mars: Effects of floods, volcanism, impacts and polar processes. *Icarus* 41:159–165.
Carr, M. H., and Clow, G. D. 1981. Martian channels and valleys: Their characteristics, distribution, and age. *Icarus* 48:91–117.
Carr, M. H., and Evans, N. 1980. *Images of Mars: The Viking Extended Mission,* NASA SP-444.
Carr, M. H., and Schaber, G. G. 1977. Martian permafrost features. *J. Geophys. Res.* 82:4039–4054.
Carr, M. H., Masursky, H., and Saunders, R. S. 1973. A generalized geologic map of Mars. *J. Geophys. Res.* 78:4031–4036.
Carr, M. H., Masursky, H., Baum, W. A., Blasius, K. R., Briggs, G. A., Cutts, J. A., Duxbury, T., Greeley, R., Guest, J. E., Smith, B. A., Soderblom, L. A., Veverka, J., and Wellman, J. B. 1976. Preliminary results from the Viking orbiter imaging experiment. *Science* 193:766–776.
Carr, M. H., Greeley, R., Blasius, K. R., Guest, J. E., and Murray, J. B. 1977*a*. Some Martian volcanic features as viewed from the Viking orbiters. *J. Geophys. Res.* 82:3985–4015.
Carr, M. H., Crumpler, L. S., Cutts, J. A., Greeley, R., Guest, J. E., and Masursky, H. 1977*b*. Martian impact craters and emplacement of ejecta by surface flow. *J. Geophys. Res.* 82:4055–4065.
Carr, R. H., Grady, M. M., Wright, I. P., and Pillinger, C. T. 1985. Martian atmospheric carbon dioxide and weathering products in SNC meteorites. *Nature* 314:248–250.
Carrol, J. J., and Ryan, J. A. 1970. Atmospheric vorticity and dust devil rotation. *J. Geophys. Res.* 75:5179–5184.
Carter, J. W., and Husain, H. 1974. The simultaneous adsorption of carbon dioxide and water vapor by fixed beds of molecular sieves. *Chem. Eng. Sci.* 29:267–273.
Cary, J. W., and Maryland, H. F. 1972. Salt and water movement in unsaturated frozen soil. *Soil Sci. Soc. America Proc.* 36:549–555.
Cas, R. A. F., and Wright, J. V. 1987. *Volcanic Successions Modern and Ancient* (London: Allen and Unwin).
Cassini. 1666. *J. Savants* 2:316.
Cattermole, P. 1986. Linear volcanic features at Alba Patera, Mars—probable spatter ridges. *Proc. Lunar Planet. Sci. Conf.* 17, *J. Geophys. Res. Suppl.* 91:E159-E165.

Cattermole, P. 1987. Sequence, rheological properties, and effusion rates of volcanic flows at Alba Patera, Mars. *Proc. Lunar Planet. Sci. Conf.* 17, *J. Geophys. Res. Suppl.* 92:E553-E560.

Cazenave, A., Dobrovolskis, A., and Lago, B. 1980. Orbital history of the Martian satellites with inferences on their origin. *Icarus* 44:730–744.

Cess, R. D., Ramanathan, V., and Owen, T. 1980. The Martian paleoclimate and enhanced atmospheric carbon dioxide. *Icarus* 41:159–165.

Chamberlain, J. W., and Hunten, D. M. 1965. Pressure and CO_2 content of the Martian atmosphere: A critical discussion. *Rev. Geophys.* 3:299–317.

Chamberlain, J. W., and Hunten, D. M. 1987. *Theory of Planetary Atmospheres,* 2nd ed. (Orlando: Academic Press).

Chamberlain, T. E., Cole, H. L., Dutton, R. G., Greene, G. C., and Tillman, J. E. 1976. Atmospheric measurements on Mars: The Viking meteorology experiment. *Bull. Amer. Meteorol. Soc.* 57:1094–1104.

Chan, H. L., and Fung, A. K. 1978. A numerical study of the Kirchhoff approximation in horizontally polarized backscattering from a random surface. *Radio Sci.* 13:811–818.

Chandler, D. L. 1979. *Life on Mars* (New York: E. P. Dutton).

Chandrasekhar, S. 1969. *Ellipsoidal Figures of Equilibrium* (New Haven, Conn.: Yale Univ. Press).

Chao, B. F., and Rubincam, D. P. 1990. Variation of Mars gravitational field and rotation due to seasonal CO_2 exchange. *J. Geophys. Res.* 95:14755–14760.

Chapman, C. R. 1974. Cratering on Mars I. Cratering and obliteration history. *Icarus* 22:272–291.

Chapman, C. R., and Gaffey, M. J. 1979. Reflectance spectra for 277 asteroids. In *Asteroids,* ed. T. Gehrels (Tucson: Univ. of Arizona Press), pp. 655–687.

Chapman, C. R., and Jones, K. L. 1977. Cratering and obliteration history of Mars. *Ann. Rev. Earth Planet. Sci.* 5:515–540.

Chapman, C. R., and McKinnon, W. B. 1986. Cratering of planetary satellites. In *Satellites,* eds. J. A. Burns and M. S. Matthews (Tucson: Univ. of Arizona Press), pp. 492–580.

Chapman, C. R., Pollack, J. B., and Sagan, C. 1969. An analysis of Mariner 4 cratering statistics. *Astron. J.* 74:1039–1051.

Chapman, S., and Lindzen, R. S. 1970. *Atmospheric Tides* (Dordrecht: D. Reidel).

Chapront-Touzé, M. 1990. Orbits of the Martian satellites from ESAPHO and ESADE theories. *Astron. Astrphys.* 240:159–172.

Charney, J. G., and Stern, M. E. 1962. On the stability of internal baroclinic jets in a rotating atmosphere. *J. Atmos. Sci.* 19:159–172.

Chase, S. C., Jr., Engel, J. L., Eyerly, H. W., Kieffer, H. H., Palluconi, F. D., and Schofield, D. 1978. Viking infrared thermal mapper. *Appl. Opt.* 17:1243–1251.

Chen, J. H., and Wasserburg, G. J. 1986. Formation ages and evolution of Shergotty and its parent planet from U-Th-Pb systematics. *Geochim. Cosmochim. Acta* 50:955–968.

Chen, R. H., Cravens, T. E., and Nagy, A. F. 1978. The Martian ionosphere in light of the Viking observations. *J. Geophys. Res.* 83:3871–3876.

Chepil, W. S. 1950. Properties of soil which influence wind erosion. I. The governing principle of surface roughness. *Soil Sci.* 69:149–162.

Chepil, W. S., and Woodruff, N. P. 1963. The physics of wind erosion and its control. *Adv. Agron.* 15:211–302.

Chicarro, A. F., Schultz, P. H., and Masson, P. 1985. Global and regional ridge patterns on Mars. *Icarus* 63:153–174.

Childers, J. M., Sloan, C. E., Meckel, J. P., and Nauman, J. W. 1977. Hydrologic Reconnaissance of the Eastern North Slope, Alaska U.S.G.S. Open File Rept. 77–492.

Christensen, E. J. 1975. Martian topography derived from occultation, radar, spectral, and optical measurements. *J. Geophys. Res.* 80:2909–2913.

Christensen, E. J., and Balmino, G. 1979. Development and analysis of a twelfth degree and order gravity model of Mars. *J. Geophys. Res.* 84:7943–7953.

Christensen, P. R. 1982. Martian dust mantling and surface composition: Interpretation of thermophysical properties. *J. Geophys. Res.* 87:9985–9998.

Christensen, P. R. 1983. Eolian intracrater deposits on Mars: Physical properties and global distribution. *Icarus* 56:496–518.

Christensen, P. R. 1986*a*. Regional dust deposits on Mars: Physical properties, age, and history. *J. Geophys. Res.* 91:3533–3545.
Christensen, P. R. 1986*b*. The spatial distribution of rocks on Mars. *Icarus* 68:217–238.
Christensen, P. R. 1988. Global albedo variations of Mars: Implications for active aeolian transport, deposition, and erosion. *J. Geophys. Res.* 93:7611–7624.
Christensen, P. R., and Kieffer, H. H. 1979. Moderate resolution thermal mapping of Mars: The channel terrain around the Chryse basin. *J. Geophys. Res.* 84:8233–8238.
Christensen, P. R., and Malin, M. C. 1988. High resolution thermal imaging of Mars. *Lunar Planet. Sci.* XIX:180–181 (abstract).
Christensen, P. R., and Zurek, R. W. 1984. Martian north polar hazes and surface ice: Results from the Viking Survey/Completion mission. *J. Geophys. Res.* 89:4587–4596.
Christiansen, E. H. 1985. Geology of Hebrus Valles and Hephaestus Fossae, Mars: Evidence for basement control of fluvial patterns. *Geol. Soc. Amer. Abs. w/Prog.* 17:545.
Christiansen, E. H. 1989. Lahars in the Elysium region of Mars. *Geology* 17:203–206.
Christie-Blick, N. 1982. Pre-Pleistocene glaciation on Earth: Implications for climatic history of Mars. *Icarus* 50:423–443.
Chyba, C. F. 1987. The cometary contribution to the oceans of primitive Earth. *Nature* 330:632–635.
Chyba, C. F. 1990. Impact delivery and erosion of planetary oceans in the early inner solar system. *Nature* 343:129–133.
Chyba, C. F., Thomas, P. J., Brookshaw, L., and Sagan, C. 1990. Cometary delivery of organic molecules to the early Earth. *Science* 249:366–373.
Chylek, P., and Grams, G. W. 1978. Scattering by nonspherical particles and optical properties of Martian dust. *Icarus* 36:198–203.
Cintala, M. J., and Mouginis-Mark, P. J. 1980. Martian fresh crater depths: More evidence for substrate volatiles? *Lunar Planet. Sci.* XI:143–145 (abstract).
Cintala, M. J., Head, J. W., and Veverka, J. 1978. Characteristics of the cratering process on small satellites and asteroids. *Proc. Lunar Planet. Sci. Conf.* 9:3803–3830.
Cisowski, S. M. 1986. Magnetic studies on Shergotty and other SNC meteorites. *Geochim. Cosmochim. Acta* 50:1043–1048.
Clancy, R. T., and Lee, S. W. 1990. Derivation of Mars atmospheric dust properties from radiative transfer analysis of Viking IRTM emission phase function sequences. *Lunar Planet. Sci.* XXI:194–195 (abstract).
Clancy, R. T., and Lee, S. W. 1991. A new look at dust and clouds in the Mars atmosphere: Analysis of emission-phase-function sequences from global Viking IRTM observations. *Icarus* 93:135–158.
Clancy, R. T., Muhleman, D. O., and Jakosky, B. M. 1983. Variability of carbon monoxide in the Mars atmosphere. *Icarus* 55:282–301.
Clancy, R. T., Muhleman, D. O., and Berge, G. L. 1989. Global changes in the 0–70 km thermal structure of the Mars atmosphere derived from 1975–1989 microwave CO spectra. *Bull. Amer. Astron. Soc.* 21:977 (abstract).
Clancy, R. T., Muhleman, D. O., and Berge, G. L. 1990. Global changes in the 0–70 km thermal structure of the Mars atmosphere derived from 1975 to 1989 microwave CO spectra. *J. Geophys. Res.* 95:14543–14554.
Clark, B. C. 1978. Implications of abundant hygroscopic minerals in the Martian regolith. *Icarus* 34:645–665.
Clark, B. C. 1979. Solar-driven chemicals energy sources for a Martian biota. *Origins of Life* 9:241–249.
Clark, B. C., and Baird, A. K. 1973. Martian regolith X-ray analyzer: Test results of geochemical performance. *Geology* 1:15–18.
Clark, B. C., and Baird, A. K. 1974. Martian regolith X-ray analyzer: Reply. *Geology* 2:23–24.
Clark, B. C., and Baird, A. K. 1979*a*. Volatiles in the Martian regolith. *Geophys. Res. Lett.* 6:811–814.
Clark, B. C., and Baird, A. K. 1979*b*. Is the Martian lithosphere sulfur rich? *J. Geophys. Res.* 84:8395–8403.
Clark, B. C., and Van Hart, D. C. 1981. The salts of Mars. *Icarus* 45:370–378.

Clark, B. C., Baird, A. K., Rose, H. J., Toulmin, P., III, Keil, K., Castro, A. J., Kelliher, W. C., Rowe, C. D., and Evans, P. H. 1976. Inorganic analysis of Martian surface samples at the Viking landing sites. *Science* 194:1283–1288.

Clark, B. C., Baird, A. K., Rose, H. J., Jr., Toulmin, P., III, Christian, R. P., Kelliher, W. C., Castro, A. J., Rowe, C. D., Kiel, K., and Huss, G. 1977. The Viking X-ray fluorescence experiment: Analytical methods and early results. *J. Geophys. Res.* 82:4577–4594.

Clark, B. C., Kenley, S. L., O'Brien, D. L., Huss, G. R., Mack, R., and Baird, A. K. 1979. Heterogeneous phase reactions of Martian volatiles with putative regolith minerals. *J. Molec. Evol.* 14:91–102.

Clark, B. C., Baird, A. K., Weldon, R. J., Tsusaki, D. M., Schnabel, L., and Candelaria, M. P. 1982. Chemical composition of Martian fines. *J. Geophys. Res.* 87:10059–10067.

Clark, B. D. 1978. Implications of abundant hygroscopic minerals in the Martian regolith. *Icarus* 34:645–665.

Clark, B. R., and Mullin, R. P. 1976. Martian glaciation and the flow of solid CO_2. *Icarus* 27:215–228.

Clark, R. N., and McCord, T. B. 1982. Mars residual north polar cap: Earth-based spectroscopic confirmation of water ice as a major constituent and evidence for hydrated minerals. *J. Geophys. Res.* 87:367–370.

Clark, R. N., Fanale, F. P., and Gaffey, M. J. 1986. Surface composition of natural satellites. In *Satellites,* eds. J. A. Burns and M. S. Matthews (Tucson: Univ. of Arizona Press), pp. 437–491.

Clark, R. N., Singer, R. B., Swayze, G. A., and Tokunaga, A. 1989. Discovery of globally-distributed scapolite on Mars. Fourth Intl. Conf. on Mars, Jan. 10–13, Tucson, Arizona, Abstract Book, pp. 82–83.

Clark, R. N., Swayze, G. A., Singer, R. B., and Pollack, J. 1990. High-resolution reflectance spectra of Mars in the 2.3 μm region: Evidence for the mineral scapolite. *J. Geophys. Res.* 95:14463-14480.

Clayton, R. N., and Mayeda, T. K. 1983. Oxygen isotopes in eucrites, shergottites, nakhlites and chassignites. *Earth Planet. Sci. Lett.* 62:1–6.

Clayton, R. N., and Mayeda, T. K. 1988. Isotopic composition of carbonate in EETA 79001 and its relation to parent body volatiles. *Geochim. Cosmochim. Acta* 52:925–927.

Clayton, R. N., Onuma, N., and Mayeda, T. K. 1976. A classification of meteorites based on oxygen isotopes. *Earth Planet. Sci. Lett.* 30:10–18.

Clifford, S. M. 1981. A pore volume estimate of the Martian megaregolith based on a lunar analog. In *Third Intl. Coll. on Mars,* LPI Contrib. 441, pp. 46–48 (abstract).

Clifford, S. M. 1984. A Model for the Climatic Behavior of Water of Mars. Ph.D. Thesis, Univ. of Massachusetts.

Clifford, S. M. 1987*a*. Mars: Crustal pore volume, cryospheric depth, and the global occurrence of groundwater. In *MECA Symp. on Mars: Evolution of its Climate and Atmosphere,* eds. V. Baker, M. Carr, F. Fanale, R. Greeley, R. Haberle, C. Leovy and T. Maxwell, LPI Tech. Rept. 87-01, p. 32 (abstract).

Clifford, S. M. 1987*b*. Polar basal melting on Mars. *J. Geophys. Res.* 92:9135–9152.

Clifford, S. M. 1988. Basal melting and the Martian polar cap balance. In *Polar Processes on Mars,* MECA Workshop on Polar Processes, May 1988.

Clifford, S. M., and Fanale, F. P. 1985. The thermal conductivity of the Martian dust. *Proc. Lunar Planet. Sci. Conf.* 16, *J. Geophys. Res. Suppl.* 90:D144-D145.

Clifford, S. M., and Hillel, D. 1983. The stability of ground ice in the equatorial regions of Mars. *J. Geophys. Res.* 88:2456–2474.

Clifford, S. M., and Hillel, D. 1986. Knudsen diffusion: The effect of small pore size and low gas pressure on gaseous transport in soils. *Soil Sci.* 141:289–299.

Clifford, S. M., Greeley, R., and Haberle, R. M. 1988. NASA Mars Project: Evolution of climate and atmosphere. *Eos: Trans. AGU* 69:1585, 1595–1596.

Cloud, P. 1989. *Oasis in Space: Earth History from the Beginning* (New York: Norton).

Cloutier, P. A., and Daniell, R. E., Jr. 1973. Ionospheric currents induced by solar wind interaction with planetary atmospheres. *Planet. Space Sci.* 21:463–474.

Cloutier, P. A., and Daniell, R. E., Jr. 1979. An electrodynamic model of the solar wind interaction with the ionospheres of Mars and Venus. *Planet. Space Sci.* 21:1111–1121.

Cloutier, P. A., McElroy, M. B., and Michel, F. C. 1969. Modification of the Martian ionosphere by the solar wind. *J. Geophys. Res.* 74:6251–6228.
Cloutier, P. A., Daniell, R. E., Jr., and Butler, D. M. 1974. Atmospheric ion wakes of Venus and Mars in the solar wind. *Planet. Space Sci.* 22:967–990.
Clow, G. D. 1987. Generation of liquid water on Mars through the melting of a dusty snowpack. *Icarus* 72:95–127.
Clow, G. D., and Moore, H. J. 1988. Stability of chasma walls in the Valles Marineris, Mars. *Lunar Planet. Sci.* XIX:201–202 (abstract).
Clow, G. D., McKay, C. P., Simmons, G. M., Jr., and Wharton, R. A., Jr. 1988. Climatological observations and predicted sublimation rates at Lake Hoare, Antarctica. *J. Climate* 1:715–728.
Clow, G. D., Moore, H. J., Davis, P. A., and Strichartz, L. R. 1988. Stability of chasma walls in the Valles Marineris, Mars. *Lunar Planet. Sci.* XIX:201–202 (abstract).
Coblentz, W. W. 1906. *Investigations of Infrared Spectra. Part III. Infrared Transmission Spectra.* Carnegie Inst. Washington Publ. no. 65, pp. 17–53.
Coblentz, W. W., and Lampland, C. O. 1927. Further radiometric measurements and temperature estimates of the planet Mars. *Sci. Paper Natl. Bur. Stds.* 22:237–276.
Coffin, M. D. 1967. *A Numerical Investigation of the Energy Balance of the Planet Mars.* Lawrence Radiation Lab, Publ. No. UCRL-50309(Livermore, Calif.).
Colburn, D., Pollack, J., and Haberle, R. 1989. Diurnal variations in optical depth at Mars. *Icarus* 79:159–189.
Collins, S. A. 1971. *The Mariner 6 and 7 Pictures of Mars,* NASA SP-263.
Collinson, D. W. 1986. Magnetic properties of Antarctic shergottite meteorites EETA 79001 and ALPHA 77005: Possible relevance to a Martian magnetic field. *Earth Planet. Sci.* 77:159–164.
Colombo, G. 1966. Cassini's second and third laws. *Astron. J.* 71:891–896.
Colton, G. W., Howard, K. A., and Moore, H. J. 1972. Mare ridges and arches in southern Oceanus Procellarum. In *Apollo 16 Prelim. Sci. Rept.,* NASA SP-315, pp. 29/90–29/93.
Comer, R. P., Solomon, S. C., and Head, J. W. 1985. Mars: Thickness of the lithosphere from the tectonic response to volcanic loads. *Rev. Geophys.* 23:61–92.
Compston, W., and Pidgeon, R. T. 1986. Jack Hills, evidence of more very old detrital zircons in Western Australia. *Nature* 321:766–769.
Condit, C. D. 1978. Distribution and relations of 4- to 10-km diameter craters to global geologic units of Mars. *Icarus* 34:465–478.
Condit, C. D., and Soderblom, L. A. 1978. *Geologic Map of the Mare Australe Area of Mars.* U.S.G.S. Misc. Inv. Series Map 1076.
Conel, J. E. 1969. Structural features relating to the origin of lunar wrinkle ridges. In *JPL Space Prog. Summary,* pp. 58–63.
Connes, J., and Connes, P. 1966. Near-infrared planetary spectra by Fourier spectroscopy. I. Instruments and results. *J. Opt. Soc. Amer.* 56:896–910.
Connes, P., Connes, J., and Maillard, J. P. 1969. *Atlas des Spèctres dans le Proche Infrarouge de Vénus, Mars, Jupiter, et Saturne* (Paris: Editions CNRS).
Conrath, B. J. 1975. Thermal structure of the Martian atmosphere during the dissipation of the dust storm of 1971. *Icarus* 24:36–46.
Conrath, B. J. 1976. Influence of planetary-scale topography on the diurnal thermal tide during the 1971 Martian dust storm. *J. Atmos. Sci.* 33:2430–2439.
Conrath, B. J. 1981. Planetary-scale wave structure in the Martian atmosphere. *Icarus* 48:246–255.
Conrath, B. J., Curran, R., Hanel, R., Kunde, V., Maguire, W., Pearl, J. Pirraglia, J., Welker, J., and Burke, T. 1973. Atmospheric surface properties of Mars obtained by infrared spectroscopy on Mariner 9. *J. Geophys. Res.* 78:4267–4278.
Cook, A. H. 1977. The moment of inertia of Mars and the existence of a core. *Geophys. J. Roy. Astron. Soc.* 51:349–356.
Cook, F. A., and Turcotte, D. L. 1981. Parameterized convection and the thermal history of the Earth. *Tectonophys.* 75:1–17.
Cooper, M. R., Kovach, R. L., and Watkins, J. S. 1974. Lunar near-surface structure. *Rev. Geophys. Space Phys.* 12:291–308.

Coradini, A., Federico, C., and Lanciano, P. 1983. Earth and Mars: Early thermal profiles. *Phys. Earth Planet. Int.* 31:145–160.
Cordell, B. M. 1978. Radial thickness variations of Orientale basin ejecta. *Modern Geol.* 6:229–240.
Cordell, B. M. 1980. Martian climatic change: A magnetic trigger? *Geophys. Res. Lett.* 7:1065–1068.
Cordell, B. M., Lingenfelter, R. E., and Schubert, G. 1974. Martian cratering and central peak statistics: Mariner 9 results. *Icarus* 21:448–456.
Costard, F. M. 1987. Quelques modèles liés à des lentilles de glace fossilés sur Mars. *Zeit. Geomorph.* 31:243–251.
Costard, F. M. 1988. Thickness of sedimentary deposits of the mouth of outflow channels. *Lunar Planet.* Sci. XIX:211–212 (abstract).
Costard, F. M. 1989. Asymmetric distribution of volatiles on Mars. *Lunar Planet. Sci.* XX:187–188 (abstract).
Costard, F., and Dollfus, A. 1987. Ice lenses on Mars. In *MECA Special Session at LPSC XVII: Martian Geomorphology and Its Relation to Subsurface Volatiles,* eds. S. M. Clifford, L. A. Rossbacher and J. R. Zimbelman, LPI Tech. Rept. 87-02, pp. 16–17 (abstract).
Courtillot, V. C., Allègre, C. J., and Matteur, M. 1975. On the existence of lateral relative motions on Mars. *Earth Planet. Sci. Lett.* 25:279–285.
Craddock, R. A., and Maxwell, T. A. 1990. Resurfacing of the Martian highlands in the Amenthes and Tyrrhena region. *J. Geophys. Res.* 95:14265–14278.
Craddock, R. A., Greeley, R., and Christensen, P. R. 1990. Evidence from an ancient impact basin in Daedalia Planum, Mars. *J. Geophys. Res.* 95:10729–10741.
Cragin, B. L., Hanson, W. B., and Santani, S. 1982. The solar wind interaction with Mars as seen by the Viking retarding potential analyzers. *J. Geophys. Res.* 87:4395–4404.
Crater Analysis Techniques Working Group. 1979. Standard techniques for presentation and analysis of crater size-frequency data. *Icarus* 37:467–474.
Cravens, T. E., Nagy, A. F., Brace, L. H., Chen, R. H., and Knudsen, W. C. 1979. The energetics of the ionosphere of Venus: A preliminary model based on Pioneer Venus observations. *Geophys. Res. Lett.* 6:341–344.
Cravens, T. E., Gombosi, T. I., Kozyra, J., Nagy, A. F., Brace, L. H., and Knudsen, W. C. 1980. Model calculations of the dayside ionosphere of Venus: Energetics. *J. Geophys. Res.* 85:7778–7786.
Cravens, T. E., Kliore, A. J., Kozyra, J. U., and Nagy, A. F. 1981. The ionospheric peak on the Venus dayside. *J. Geophys. Res.* 86:11323–11329.
Cravens, T. E., Brace, L. H., Taylor, H. A., Jr., Russell, C. T., Knudsen, W. L., Miller, K. L., Barnes, A., Mihalov, J. D., Scarf, F. L., Quenon, S. J., and Nagy, A. F. 1982. Disappearing ionospheres on the nightside of Venus. *Icarus* 51:271–282.
Crescenti, G. H. 1984. Analysis of permafrost depths on Mars. In, NASA TM 86247, pp. 115–132.
Crisp, J. 1984. Rates of magma emplacement and volcanic output. *J. Volcanol. Geothermal Res.* 20:177-211.
Crisp, D. 1990. Infrared radiative transfer in the dust-free martian atmosphere. *J. Geophys. Res.* 94:14577–14588.
Croft, S. K. 1979. Impact Craters from Centimeters to Megameters. Ph.D. Thesis, Univ. of California, Los Angeles.
Croft, S. K. 1980. Cratering flow field: Implications for the excavation and transient expansion stages of crater formation. *Proc. Lunar Planet. Sci. Conf.* 11:2347–2378.
Croft, S. K. 1981*a*. On the origin of pit craters. *Lunar Planet. Sci.* XII:196–198 (abstract).
Croft, S. K. 1981*b*. The modification stage of basin formation: Conditions of ring formation. In *Multi-Ring Basins, Proc. Lunar Planet. Sci. 12A,* eds. P. H. Schultz and R. B. Merrill, pp. 227–257.
Croft, S. K. 1983. A proposed origin for palimpsests and anomalous pit craters on Ganymede and Callisto. *Proc. Lunar Planet. Sci. Conf.* 14, *J. Geophys. Res. Suppl.* 88:B71-B89.
Croft, S. K. 1984. Scaling of complex craters. *Lunar Planet. Sci.* XV:188–189 (abstract).
Croft, S. K. 1985*a*. Rim structures in lunar craters and basins: New constraints on the origin of asymmetric basin rings. *Lunar Planet. Sci.* XVI:154–155 (abstract).

Croft, S. K. 1985*b*. The scaling of complex craters. *Proc. Lunar Planet. Sci. Conf.* 15, *J. Geophys. Res. Suppl.* 90:C828-C842.

Croft, S. K. 1989*c*. Spelunking on Mars: The carbonate-tectonic hypothesis for the origin of Valles Marineris, Mars. In *MEVTV Workshop on Tectonic Features on Mars*, eds. T. R. Watters and M. P. Golombek, LPI Tech. Rept. 89-06, pp. 21–23 (abstract).

Croft, S. K. 1989*a*. Mars' canyon systems: Problems of origin. Fourth Intl. Conf. on Mars, Jan. 10–13, Tucson, Arizona, Abstract Book, pp. 88–89.

Croft, S. K. 1989*b*. Canyon structure in the Hebes-Juventae-Ganges area, Mars. *Lunar Planet. Sci.* XX:203–204 (abstract).

Croft, S. K., Kieffer, S. W., and Ahrens, T. J. 1979. Low-velocity impact craters in ice-saturated sand with implications for Martian crater count ages. *Proc. Lunar Planet. Sci. Conf.* 14, *J. Geophys. Res. Suppl.* 89:B8023-B8032.

Croll, J. 1875. *Climate and Time* (New York: Appleton).

Cronin, J. R., Pizzarello, S., and Cruikshank, D. P. 1988. Organic matter in carbonaceous chondrites, planetary satellites, asteroids and comets. In *Meteorites and the Early Solar System*, eds. J. F. Kerridge and M. S. Matthews (Tucson: Univ. of Arizona Press), pp. 819–857.

Crooker, N. U., Luhmann, J. G., Russell, C. T., Smith, E. J., Spreiter, J. R., and Stahara, S. S. 1985. Magnetic field draping against the dayside magnetopause. *J. Geophys. Res.* 90:3505–3510.

Cross, C. A. 1971*a*. The heat balance of the Martian polar caps. *Icarus* 15:110–114.

Cross, C. A. 1971*b*. A Mariners' 1969 closeup map of Mars. *Sky & Teles.* 7:16–17.

Crovisier, J. L., Honnorez, J., and Eberhart, J. P. 1987. Dissolution of basaltic glass in seawater: Mechanism and rate. *Geochim. Cosmochim. Acta* 51:2977–2990.

Crovisier, J. L., Thomassin, J. H., Juteau, T., Eberhart, J. P., Touray, J. C., and Baillif, P. 1983. Experimental seawater basaltic glass interaction at 50°C. Study of early developed phases by electron microscopy and X-ray photoelectron spectrometry. *Geochim. Cosmochim. Acta* 47:377–387.

Cruikshank, D. P., and Brown, R. H. 1987. Organic matter on asteroid 130 Elektra. *Science* 238:183–185.

Crumpler, L. S., and Aubele, J. C. 1978. Structural evolution of Arsia Mons, Pavonis Mons and Ascraeus Mons: Tharsis region of Mars. *Icarus* 34:496–511.

Cunningham, N. W., and Schurmeier, H. M. 1969. Introduction in Mariner Mars 1969: A preliminary report. NASA SP-225, pp. 1–36.

Curran, R. J., Conrath, B. J., Hanel, R. A., Kunde, V. G., and Pearl, J. C. 1973. Mars: Mariner 9 spectroscopic evidence for H_2O ice clouds. *Science* 182:381–383.

Curtis, S. A., and Ness, N. F. 1988. Remnant magnetics at Mars. *Geophys. Res. Lett.* 15:737–739.

Cutts, J. A. 1973*a*. Nature and origin of layered deposits of the Martian polar regions. *J. Geophys. Res.* 78:4231–4249.

Cutts, J. A. 1973*b*. Wind erosion in the martian polar regions. *J. Geophys. Res.* 78:4211–4221.

Cutts, J. A., and Blasius, K. R. 1981. Origin of Martian outflow channels: The eolian hypothesis. *J. Geophys. Res.* 86:5075–5102.

Cutts, J. A., and Lewis, B. H. 1982. Models of climatic cycles record in Martian polar layered deposits. *Icarus* 50:216–244.

Cutts, J. A., and Smith, R. S. U. 1973. Aeolian deposits and dunes on Mars. *J. Geophys. Res.* 78:4139–4154.

Cutts, J. A., Veverka, J., and Goguen, J. D. 1972. The location of the Mountains of Mitchel and evidence for their nature in Mariner 7 pictures. *Icarus* 16:528–534.

Cutts, J. A., Blasius, K. R., Briggs, G. A., Carr, M. H., Greeley, R., and Masursky, H. 1976. North polar region of Mars: Imaging results from Viking 2. *Science* 194:1329–1337.

Cutts, J. A., Blasius, K. R., and Robert, W. J. 1979. Evolution of Martian polar landscapes: Interplay of long term variations in perennial ice cover and dust storm intensity. *J. Geophys. Res.* 84:2975–2994.

Cuzzi, J. N., and Muhleman, D. O. 1972. The microwave spectrum and the nature of the subsurface of Mars. *Icarus* 17:548–560.

Cuzzi, J. N., and Pollack, J. B. 1978. Saturn's rings: Particle composition and size distribution as constrained by microwave observations. I. Radar observations. *Icarus* 33:233–262.

Czudek, T., and Demek, J. 1970. Thermokarst in Siberia and its influence on the development of lowland relief. *Quat. Res.* 1:103–120.

Dale-Bannister, M., Arvidson, R. E., and Moore, H. J. 1988. On the presence of unweathered lithic fragments in Viking Lander 1 soil. *Lunar Planet. Sci.* XIX:239–240 (abstract).

Darwin, G. H. 1899. The theory of the Earth carried to the second order of small quantities. *Mon. Not. Roy. Astron. Soc.* 60:82–124.

Davies, D. W. 1979*a*. Effects of dust on the heating of Mars' surface and atmosphere. *J. Geophys. Res.* 84:8289–8294.

Davies, D. W. 1979*b*. The relative humidity of Mars' atmosphere. *J. Geophys. Res.* 84:8335–8340.

Davies, D. W. 1979*c*. The vertical distribution of Mars water vapor. *J. Geophys. Res.* 84:2875–2879.

Davies, D. W. 1981. The Mars water cycle. *Icarus* 45:398–414.

Davies, D. W., and Wanio, L. A. 1981. Measurements of water vapor in Mars' antarctic. *Icarus* 45:216–230.

Davies, D. W., Farmer, C. B., and LaPorte, D. D. 1977. Behavior of volatiles in Mars' polar areas: A model incorporating new experimental data. *J. Geophys. Res.* 82:3815–3822.

Davies, G. F., and Arvidson, R. E. 1981. Martian thermal history, core segregation, and tectonics. *Icarus* 45:339–346.

Davies, M. E. 1972. Coordinates of features on the Mariner 6 and 7 pictures of Mars. *Icarus* 17:116–167.

Davies, M. E. 1973. Mariner 9: Primary control net. *Photogrammetric Eng. Remote Sensing* 39:1297–1302.

Davies, M. E. 1974. *Mariner 9 Control Net of Mars: May 1974*. RAND Rept. R-1525-NASA.

Davies, M. E. 1978*a*. The control net of Mars. In *Reports of Planetary Geology Program–1977–1978,* NASA TM-79729, pp. 328–329.

Davies, M. E. 1978*b*. The control net of Mars: May 1977. *J. Geophys. Res.* 83:2311–2312.

Davies, M. E., and Arthur, D. W. G. 1973. Martian surface coordinates. *J. Geophys. Res.* 78:4355–4394.

Davies, M. E., and Berg, R. A. 1971. A preliminary control net of Mars. *J. Geophys. Res.* 76:373–393.

Davies, M. E., and Katayama, F. Y. 1983. The 1982 control network of Mars. *J. Geophys. Res.* 88:7503–7504.

Davies, M. E., Katayama, F. Y., and Roth, J. A. 1978. *Control Net of Mars, February 1978*. RAND Rept. R-2309-NASA.

Davies, M. E., Abalakin, V. K., Cross, C. A., Duncombe, R. L., Masursky, H., Morando, B., Owen, T. C., Seidelmann, P. K., Sinclair, A. T., Wilkins, G. A., and Tjuflin, Y. S. 1980. Report of the IAU Working Group on cartographic coordinates and rotational elements of the planets and satellites. *Celest. Mech.* 22:205–230.

Davies, M. E., Abalakin, V. K., Lieske, J. H., Seidelmann, P. K., Sinclair, A. T., Sinzi, A. M., Smith, B. A., and Tjuflin, Y. S. 1983. Report of the IAU Working Group on cartographic coordinates and rotational elements of the planets and satellites: 1982. *Celest. Mech.* 29:309–321.

Davies, M. E., Abalakin, V. K., Bursa, M., Lederle, T., Lieske, J. H., Rapp, R. H., Seidelman, P. K., Sinclair, A. T., Teifel, V. G., and Tjuflin, Y. S. 1986. Report of the IAU/IAG/COSPAR Working Group on cartographic coordinates and rotational elements of the planets and satellites: 1985. *Celest. Mech.* 39:103–113.

Davies, M. E., Abalakin, V. K., Bursa, M., Hunt, G. E., Lieske, J. H., Morando, B., Rapp, R. H., Seidelmann, P. K., Sinclair, A. T., and Tjuflin, Y. S. 1989. Report of the IAU/IAG/COSPAR Working Group on cartographic coordinates and rotational elements of the planets and satellites: 1988. *Celest. Mech. Dyn. Astron.* 46:187–204.

Davis, B. W. 1969. Some speculations on absorption and desorption of CO_2 in martian bright regions. *Icarus* 11:155–158.

Davis, D. R., Housen, K. R., and Greenberg, R. 1981. The unusual dynamical environment of Phobos and Deimos. *Icarus* 47:220–233.

Davis, D. R., Chapman, C. R., Weidenschilling, S. J., and Greenberg, R. 1985. Collisional history of asteroids: Evidence from Vesta and the Hirayama families. *Icarus* 62:30–53.
Davis, M., Hut, P., and Mueller, R. A. 1984. Extinctions of species by periodic comet showers. *Nature* 308:715–717.
Davis, P. A., and Golombek, M. P. 1989. Discontinuities in the shallow Martian crust. *Lunar Planet. Sci. Conf.* XX:224–225 (abstract).
Davis, P. A., and Golombek, M. P. 1990. Discontinuities in the shallow Martian crust at Lunae, Syria, and Sinai Plana. *J. Geophys. Res.* 95:14231–14248.
Davis, S. N. 1969. Porosity and permeability of natural materials. In *Flow Through Porous Media,* ed. R. J. M. de Wiest (New York: Academic Press), pp. 53–89.
Deardorff, J. W. 1972. Parameterization of the planetary boundary layer for use in general circulation models. *Mon. Weather Rev.* 100:93–106.
de Boer, J. H. 1968. *The Dynamical Character of Adsorption* (Oxford: Clarendon).
DeHon, R. A. 1981. Thickness distribution of tharsis volcanic materials. In *Reports of Planetary Geology Program*—1981, NASA TM-84211, pp. 144–146 (abstract).
DeHon, R. A. 1982. Martian volcanic materials: Preliminary thickness estimates in the Eastern Tharsis region. *J. Geophys. Res.* 87:9821–9828.
DeHon, R. A. 1985. Ridged plains of the Lunae Planum: Thickness distribution revised. *Lunar Planet. Sci.* XVI:171–172 (abstract).
DeHon, R. A. 1988. Progress in determining the thickness distribution of volcanic materials on Mars. In *MEVTV Workshop: Nature and Composition of Surface Units on Mars,* eds. J. R. Zimbelman, S. C. Solomon and V. L. Sharpton, LPI Tech. Rept. 88-05, pp. 54–56 (abstract).
de Marsily, G., Ledoux, E., Barbraeu, A., and Margat, J. 1977. Nuclear waste disposal: Can the geologists guarantee isolation? *Science* 197:519–527.
Dementyeva, N. N., Kurt, V. G., Smirnov, A. S., Titarchuk, L. G., and Chuvahin, S. D. 1972. Preliminary results of measurement of uv-emission scattered in the Martian upper atmosphere. *Icarus* 17:475–483.
Deming, D., Mumma, M. J., Espenak, F., Kostiuk, T., and Zipoy, D. 1986. Polar warming in the middle atmosphere of Mars. *Icarus* 66:366–368.
de Mottoni y Palacios, G. 1969. Synthèse graphique complémentaire des photographies de Mars en 1956. *Publ. Osserv. Astron. Milano* No. 19.
de Mottoni y Palacios, G. 1975. The appearance of Mars from 1907–1971: Graphic synthesis of photographs from the IAU Center at Meudon. *Icarus* 25:296–332.
de Mottoni y Palacios, G., and Dollfus, A. 1982. Surface marking variations of selected areas on Mars. *Astron. Astrophys.* 116:323–331.
Dence, M. R., Grieve, R. A. F., and Robertson, P. B. 1977. Terrestrial impact structures: Principal characteristics and energy considerations. In *Impact and Explosion Cratering,* eds. D. J. Roddy, R. O. Pepin and R. B. Merrill (New York: Pergamon), pp. 247–276.
Denisov, V. P., and Alimov, V. I. 1971. *The Moon and the Planets* (Moscow: Znaniye Press). English trans.: NASA TT F-767.
Dermott, S. F., and Thomas, P. C. 1988. The shape and internal structure of Mimas. *Icarus* 73:25–65.
Dermott, S. F., Malhotra, R., and Murray, C. D. 1988. Dynamics of the Uranian and Saturnian systems: A chaotic route to melting Miranda? *Icarus* 76:295–334.
de Sitter, W. 1924. On the flattening and constitution of the Earth. *Bull. Astron. Inst. Netherlands* 2:97–108.
de Vaucouleurs, G. 1950. *The Planet Mars,* trans. Patrick Moore (London: Faber and Faber).
de Vaucouleurs, G. 1954. *Physics of the Planet Mars* (London: Faber and Faber).
de Vaucouleurs, G. 1958. World-wide observations of Mars in 1956. *Publ. Astron. Soc. Pacific Leaflet* No. 351.
de Vaucouleurs, G. 1962. Precision mapping of Mars in "La Physique des Planètes." *Bull. Soc. Roy. Sci. Liège* 24:369–385.
de Vaucouleurs, G. 1964. The physical ephemeris of Mars. *Icarus* 3:236–247.
de Vaucouleurs, G. 1969. Research directed toward development of a homogeneous Martian coordinate system. Final report. (Bedford, Mass.: U.S. Air Force Aerospace Research).
de Vaucouleurs, G. 1980. Rotation period of Mars from transits of albedo stations at central meridian, 1659–1971. *Astron. J.* 85:945–960.

de Vaucouleurs, G. 1989. The best telescopic pictures of Mars. *Sky & Teles.* 77:15–17.
de Vaucouleurs, G., Davids, M. E., and Sturms, F. M., Jr. 1973. Mariner 9 areographic coordinate system. *J. Geophys. Res.* 78:4395–4404.
de Vaucouleurs, G., Blunck, J., Davies, M., Dollfus, A., Koval, I. K., Kuiper, G. P., Masursky, H., Miyamoto, S., Moroz, V. I., Sagan, C., and Smith, B. 1975. The new Martian nomenclature of the International Astronomical Union. *Icarus* 26:85–98.
Dial, A. L., Jr. 1984. *Geologic Map of the Mare Boreum Area of Mars.* U.S.G.S. Misc. Inv. Series Map I-1640 (MC-1).
Dickinson, R. E., and Bougher, S. W. 1986. Venus mesosphere and thermosphere. I. Heat budget and thermal structure. *J. Geophys. Res.* 91:70–80.
Dickinson, R. E., and Ridley, E. C. 1977. Venus mesosphere and thermosphere temperature structure. II. Day night variations. *Icarus* 30:163–178.
Dickinson, R. E., Ridley, E. C., and Roble, R. G. 1981. A three-dimensional general circulation model of the thermosphere. *J. Geophys. Res.* 86:1499–1512.
Dickinson, R. E., Ridley, E. C., and Roble, R. G. 1984. Thermospheric general circulation with coupled dynamics and composition. *J. Atmos. Sci.* 43:205–219.
Ditteon, R. 1982. Daily temperature variations on Mars. *J. Geophys. Res.* 87:10197–10214.
Ditteon, R., and Kieffer, H. H. 1979. Optical properties of solid CO_2: Application to Mars. *J. Geophys. Res.* 84:8294–8300.
Dobbins, T. A., Parker, D. C., and Capen, C. F. 1988. *Introduction to Observing and Photographing the Solar System* (Richmond, Mass.: Willmann-Bell).
Dobrovolskis, A. R. 1982. Internal stresses in Phobos and other triaxial bodies. *Icarus* 5:136–148.
Dobrovolskis, A. R., and Burns, J. A. 1980. Life near the Roche limit: Behavior of ejecta from satellites close to planets. *Icarus* 42:422–441.
Dobrovolskis, A. R., and Ingersoll, A. P. 1975. Carbon dioxide-water clathrate as a reservoir of CO_2 on Mars. *Icarus* 26:353–357.
Doherty, L. H., Andrew, B. H., and Briggs, F. H. 1979. Confirmation of the longitudinal dependence of the radio brightness of Mars. *Astrophys. J.* 233:L165-L168.
Dolginov, Sh. Sh. 1976. The magnetosphere of Mars. In *Physics of Solar Planetary Environments Vol. 2,* ed. D. J. Williams (Washington, D.C.: American Geophysical Union), pp. 872–888.
Dolginov, Sh. Sh. 1978*a*. On the magnetic field of Mars: Mars 2 and 3 evidence. *Geophys. Res. Lett.* 5:89–92.
Dolginov, Sh. Sh. 1978*b*. On the magnetic field of Mars: Mars 5 evidence. *Geophys. Res. Lett.* 5:93–95.
Dolginov, Sh. Sh. 1986. Comments on "The magnetic field of Mars: Implications from gas dynamic modeling" by C. T. Russell, J. G. Luhmann, J. R. Spreiter and S. S. Stahara. *J. Geophys. Res.* 91:12143–12148.
Dolginov, Sh. Sh. 1987. What have we learned about the Martian magnetic field? *Earth, Moon and Planets* 37:17–52.
Dolginov, Sh. Sh., Yeroskenko, Ye. G., and Zhuzgov, L. N. 1973. The magnetic field in the very close neighborhood of Mars according to data from the Mars 2 and 3 spacecraft. *J. Geophys. Res.* 78:4779–4786.
Dolginov, Sh. Sh., Yeroskenko, Ye. G., and Zhuzgov, L. N. 1976. The magnetic field of Mars according to data from the Mars 3 and Mars 5. *J. Geophys. Res.* 81:3353–3362.
Dollfus, A. 1948. Observation de la structure des "canaux" de la planète Mars. *Compt. Rend.* 226:996–997.
Dollfus, A. 1951*a*. Détermination de la pression atmosphérique sur la planète Mars. *Compt. Rend.* 232:1066–1068.
Dollfus, A. 1951*b*. La polarization de la lumière renvoyée par les différentes régions de la surface de la planète Mars et son interprétation. *Compt. Rend.* 233:467–469.
Dollfus, A. 1953. Etude visuelle de la surface da la planète Mars. *Astronomie* 67:85–106.
Dollfus, A. 1957. Etude des planètes pour la polarisation de leur lumière. *Ann. Astrophys. Suppl. 4*. English trans.: NASA-TT-F-188.
Dollfus, A. 1958. III. The nature of the surface of Mars. *Publ. Astron. Soc. Pacific* 70:56–64.
Dollfus, A. 1961*a*. Polarization studies of the planets. In *The Solar System. III. Planets and*

Satellites, eds. G. P. Kuiper and B. Middlehurst (Chicago: Univ. of Chicago Press), pp. 378–387.

Dollfus, A. 1961*b*. Visual and photographic studies of planets at the Pic du Midi. In *The Solar System. III. Planets and Satellites,* eds. G. P., Kuiper and B. Middlehurst (Chicago: Univ. of Chicago Press), pp. 534–571.

Dollfus, A. 1963. Mésure de la quantité de vapeur d'eau contenue dans l'atmosphère de la planète Mars. *Comptes Rend. Acad. Sci.* 256:3009–3011.

Dollfus, A. 1964. Observations of water vapor on Mars and Venus. In *Origin and Evolution of Atmosphere and Oceans,* eds. P. J. Brancazio and A. G. W. Cameron (New York: Wiley), pp. 257–268.

Dollfus, A. 1965*a*. Etude de la planète Mars de 1954 à 1958. *Ann. Astrophys.* 28:722–747.

Dollfus, A. 1965*b*. Analyse des mésures de la quantité de vapeur d'eau dans l'atmosphère de la planète Mars. *Compt. Rend.* 261:1603–1606.

Dollfus, A. 1972. New optical measurements of planetary diameters. IV. Planet Mars. *Icarus* 17:525–539.

Dollfus, A. 1973. New optical measurements of planetary diameters. V. Size of the north polar cap of Mars. *Icarus* 18:142–155.

Dollfus, A., and Focas, J. 1969. La planète Mars: La nature de sa surface et les propriétés de son atmosphère, d'après la polarisation de la lumière. I. Observations. *Astron. Astrophys.* 2:63–74.

Dollfus, A., Focas, J., and Bowell, E. L. 1969. La planète Mars: La nature de sa surface et les propriétés de son atmosphère, d'après la polarisation de la lumière. II. La nature du sol. *Astron. Astrophys.* 2:105–121.

Dollfus, A., Deschamps, M., and Ksanfomality, L. V. 1983. The surface texture of the Martian soil from the Soviet spacecraft Mars-5 polarimeter. *Astron. Astrophys.* 123:225–237.

Dollfus, A., Ebisawa A., and Bowell, E. 1984*a*. Polarimetric analysis of the Martian dust storms and clouds in 1971. *Astron. Astrophys.* 131:123–136.

Dollfus, A., Bowell, E., and Ebisawa, S. 1984*b*. The Martian dust storms of 1973: A polarimetric analysis. *Astron. Astrophys.* 134:343–353.

Donahue, T. M. 1986. Fractionation of noble gases by thermal escape from accreting planetesimals. *Icarus* 66:195–210.

Donahue, T. M., and Hunten, D. M. 1976. Hydrogen loss from the terrestrial planets. *Ann. Rev. Earth Planet. Sci.* 4:265–292.

Donahue, T. M., and Pollack, J. B. 1983. Origin and evolution of the atmosphere of Venus. In *Venus,* eds. D. Hunten, L. Colin, T. Donahue and V. Moroz (Tucson: Univ. of Arizona Press), pp. 1003–1036.

Dorman, L. M., and Lewis, B. T. R. 1970. Experimental isostasy. *J. Geophys. Res.* 75:3357–3365.

Downs, G. S., Goldstein, R. M., Green, R. R., and Morris, G. A. 1971. Mars radar observations, a preliminary report. *Science* 174:1324–1327.

Downs, G. S., Goldstein, R. M., Green, R. R., Morris, G. A., and Reichley, P. E. 1973. Martian topography and surface properties as seen by radar: The 1971 opposition. *Icarus* 18:8–21.

Downs, G. S., Reichley, P. E., and Green, R. R. 1975. Radar measurements of Martian topography and surface properties: The 1971 and 1973 oppositions. *Icarus* 26:273–312.

Downs, G. S., Green, R. R., and Reichley, P. E. 1978. Radar studies of the Martian surface at centimeter wavelengths: The 1975 opposition. *Icarus* 33:441–453.

Downs, G. S., Mouginis-Mark, P. J., Zisk, S. H., and Thompson, T. W. 1982. New radar-derived topography for the northern hemisphere of Mars. *J. Geophys. Res.* 82:9747–9754.

Drake, D. M., Feldman, W. C., and Jakosky, B. M. 1988. Martian neutron leakage spectra. *J. Geophys. Res.* 934:6353–6368.

Drake, J. J. 1981. The effects of surface dust on snowmelt rates. *Arctic and Alpine Res.* 13:219–223.

Dreibus, G., and Wänke, H. 1984. Accretion of the Earth and the inner planets. In *Proc. 27th Intl. Geol. Conf.*, vol 11 (Utrecht: VNU Science Press), pp. 1–20.

Dreibus, G., and Wänke, H. 1985. Mars, a volatile-rich planet. *Meteoritics* 20:367–381.

Dreibus, G., and Wänke, H. 1987. Volatiles on Earth and Mars: A comparison. *Icarus* 71:225–240.

Dreibus, G., and Wänke, H. 1989. Supply and loss of volatile constituents during the accretion of terrestrial planets. In *Origin and Evolution of Planetary and Satellite Atmospheres* (Tucson: Univ. of Arizona Press), pp. 268–288.

Dreibus, G., Kruse, H., Spettel, B., and Wänke, H. 1977. The bulk composition of the Moon and the eucrite parent body. *Proc. Lunar Planet. Sci. Conf.* 8:211–228.

Dreibus, G., Palme, H., Rammensee, W., Spettel, B., Weckwerth, G., and Wänke, H. 1982. Composition of Shergotty parent body: Further evidence of a two component model of planet formation. *Lunar Planet. Sci.* XIII:186–187 (abstract).

Drobyshevski, E. M. 1988. The origin and specific features of the Martian satellites in the context of the eruption hypothesis. *Earth, Moon and Planets* 40:1–19.

Dryer, M., and Heckman, G. R. 1967. Application of the hypersonic analog to the standing shock of Mars. *Solar Phys.* 2:112–124.

Dubinin, E. M., Lundin, R., Pissarenko, N. F., Barabash, S. V., Zakaharov, A. V., Koskinen, H., Schwingenshuh, K., and Yeroshenko, Ye. G. 1990. Indirect evidence for a dust/gas torus along the Phobos orbit. *Geophys. Res. Lett.* 17:861–864.

Duffield, W. A., Stieltjes, L., and Varet, J. 1982. Huge landslide blocks in the growth of Piton de la Fournaise, La Réunion, and Kilauea Volcano, Hawaii. *J. Volcanol. Geotherm. Res.* 12:147–160.

Duke, M. B. 1968. The Shergotty meteorite: Magmatic and shock metamorphic features. In *Shock Metamorphism of Nature Materials*, eds. B. M. French and N. M. Short (Baltimore: Mono), pp. 613–621.

Dunham, T., Jr. 1952. Spectroscopic observations of the planets at Mt. Wilson. In *The Atmospheres of the Earth and Planets*, 2nd ed., ed. G. P. Kuiper (Chicago: Univ. of Chicago Press), pp. 288–305.

Dunne, T. 1980. Formation and controls of channel networks. *Prog. Phys. Geography* 4:211–239.

Dunnebier, F. K. 1976. Thermal movement of the regolith. *Proc. Lunar Planet. Sci.* 7:1073–1086.

Dunnebier, F. K., and Sutton, G. H. 1974. Thermal moonquakes. *J. Geophys. Res.* 79:4351–4363.

d'Uston, C., Atteia, J.-L., Barat, C., Chernenko, A., Dolidze, V., Dyatchkov, A., Jourdain, E., Khariukova, V., Khavenson, N., Kozlenkov, A., Kucherova, R., Mitrofanov, I., Moskaleva, L., Niel, M., Pozanenko, A., Scheglov, P., Surkov, Yu., and Vilchinskaya, A. 1989. Observation of the gamma-ray emission from the Martian surface by the APEX experiment. *Nature* 341:598–600.

Duxbury, T. C. 1989. The figure of Phobos. *Icarus* 78:169–180.

Duxbury, T. C. 1991. An analytic model for the Phobos surface. *Planet. Space Sci.* 39:355–376.

Duxbury, T. C., and Callahan, J. D. 1982. Phobos and Deimos cartography. *Lunar Planet. Sci.* XIII:190 (abstract).

Duxbury, T. C., and Callahan, J. D. 1989. Phobos and Deimos control networks. *Icarus* 77:275–286.

Duxbury, T. C., and Ocampo, A. C. 1988. Mars: Satellite and ring search from Viking. *Icarus* 76:160–162.

Duxbury, T. C., and Veverka, J. 1978. Deimos encounter by Viking: Preliminary imaging results. *Science* 201:812–814.

Dyce, R. B. 1965. Recent Arecibo observations of Mars and Jupiter. *J. Natl. Bur. Stnds. (Radiosci.)* 69D:1628–1629.

Dyce, R. B., Pettengill, G. H., and Sanchez, A. D. 1967. Radar observations of Mars and Jupiter at 70 cm. *Astron. J.* 72:771–777.

Dzurisin, D., and Blasius, K. R. 1975. Topography of the polar layered deposits of Mars. *J. Geophys. Res.* 89:3286–3306.

Dzurisin, D., and Ingersoll, A. P. 1975. Seasonal buffering of atmospheric pressure on Mars. *Icarus* 26:437–440.

Eberhardt, P., Eugster, O., and Geiss, J. 1965. A redetermination of the isotopic composition of atmospheric neon. *Z. Naturforsch.* 21a:623–624.

Ebisawa, S. 1960. Planisphere of Mars with the list of the names of its surface markings. *Contrib. Inst. Astrophys. Kwasan Obs.* No. 89.

Ebisawa, S. 1963. Some problems in Mars mapping. *Contrib. Inst. Astrophys. Kwasan Obs.* No. 129.

Ebisawa, S. 1973. Analytical study on the secular and seasonal modifications of Martian soil. *Contrib. Kwasan Hida Obs.* No. 210.

Ebisawa, S. 1975. Mars observations and secular change of albedo markings during the 1973 opposition. *Contrib. Kwasan Hida Obs.* No. 223.

Ebisawa, S. 1984. Polarimetric observations of Mars during the 1982 opposition and the dust storm in the northern hemisphere. *Planetary Research Obs. Contrib.* No. 12, Japan.

Ebisawa, S. 1989. On the Martian south pre-polar and polar hoods preceded the formation of polar surface cap. Fourth Intl. Conf. on Mars, Jan. 10–13, Tucson, Arizona, Abstract Book, p. 98.

Ebisawa, S., and Dollfus, A. 1986. Martian dust storms at the early stage of their evolution. *Icarus* 66:75–82.

Ebisawa, S., and Dollfus, A. 1987. Observation de Mars en 1984 et 1986. Apparition des prémices de la couverture polaire et formation de la couverture hévernale. *Astronomie* 181:404–410.

Edmon, H. J., Jr., Hoskins, B. J., and McIntyre, M. E. 1980. Eliassen-Palm cross sections for the troposphere. *J. Atmos. Sci.* 37:2600–2616; corrigendum, *J. Atmos. Sci.* 38:1115 (1981).

Edwards, K. 1987. Geometric processing of digital images of the planets. *Photogrammetric Eng. Remote Sensing* 53:1219–1222.

Egan, W. G., Hilgeman, T., and Pang, K. 1975. Ultraviolet complex refractive index of Martian dust: Laboratory measurements of terrestrial analogs. *Icarus* 25:344–355.

Elachi, C. 1987. *Introduction to the Physics and Techniques of Remote Sensing* (New York: Wiley).

Elachi, C., Blom, R., Daily, M., Farr, T., and Saunders, R. S. 1980. Radar imaging of volcanic fields and sand dune fields: Implications for VOIR. In *Radar Geology: An Assessment,* Rept. of the Radar Geology Workshop, July 16–20, 1979, Snowmass, Colo., JPL Publ. 80-61, pp. 114–150.

Elassal, A. A., and Malhotra, R. C. 1987. *General Integrated Analytical Triangulation (GIANT) User's Guide.* NOAA Tech. Rept. NOS 126, CGS 11.

El-Baz, F., Breed, C. S., Grolier, M., and McCauley, J. F. 1979. Eolian features in the westen desert of Egypt and some applications to Mars. *J. Geophys. Res.* 84:8205–8221.

Elphic, R. C., Russel, C. T., Slavin, J. A., and Brace, L. H. 1980. Observations of the dayside ionopause and ionsphere of Venus. *J. Geophys. Res.* 85:7679–7696.

Emerson, R. W. 1841. Essays. Reproduced in *Bartlett's Familiar Quotation,* 14th ed. (Boston: Little Brown, 1968), p. 606.

Encrenaz, T., and Lellouch, E. 1990. On the atmospheric origin of weak absorption features in the infrared spectrum of Mars. *J. Geophys. Res.* 95:14489–14594.

Engelhardt, W., and Zimmerman, J. 1988. *Theory of Earth Science* (Cambridge: Cambridge Univ. Press).

Eppler, D. B., and Malin, M. C. 1982. Modification of Martian fretted terrain. In NASA TM-85127, pp. 237–238.

Epstein, E. E., Andrew, B. H., Briggs, F. H., Jakosky, B. M., and Palluconi, F. E. 1983. Mars: Subsurface properties from observed longitudinal variation of the 3.5-mm brightness temperature. *Icarus* 56:465–475.

Erard, S., Bibring, J-P., and Langevin, Y. 1990. Determination of spectral units in the Syrtis Major-Isidis Planitia region from Phobos/ISM observations. *Lunar Planet Sci.* XX:327 (abstract)

Ersoy, T., and Togro, E. 1978. Temperature and strain rate effects on the strength of compacted frozen silt-clay. In *Third Intl. Conf. on Permafrost,* pp. 643–647.

Eshleman, V. R. 1970. Atmospheres of Mars and Venus: A review of Mariner 4 and 5 and Venera 4 experiments. *Radio Sci.* 5:325–332.

Eshleman, V.R. 1976. The spacecraft radio occultation technique for the study of planetary atmospheres. *J. Spacecraft and Rockets* 13:768.

Esipov, V. F., and Moroz, V. I. 1963. Spectroscopy of Mars. *Astron. Tsirk.* no. 262, p. 1.

Eugster, O., Eberhardt, P., and Geiss, J. 1967. The isotopic composition of krypton in unequilibrated and gas rich chondrites. *Earth Planet. Sci. Lett.* 2:385–393.

Evans, D. L., and Adams, J. B. 1979. Comparison of Viking Lander multispectral images and laboratory reflectance spectra of terrestrial samples. *Proc. Lunar Planet. Sci. Conf.* 10:1829–1834.

Evans, D. L., and Adams, J. B. 1980. Amorphous gels as possible analogs to Martian weathering products. *Proc. Lunar Planet. Sci. Conf.* 11:757–763.

Evans, D. L., and Evans, J. B. 1979. Comparison of Viking Lander multispectral images and laboratory reflectance spectra of terrestrial samples. *Proc. Lunar Planet. Sci. Conf.* 10:1829–1834.

Evans, D. L., and Freeland, H. J. 1977. Variations in the Earth's orbit: Pacemaker of the ice ages? *Science* 198:528–529.

Evans, J. V. 1962. Radio echo studies of the Moon. In *Physics and Astronomy of the Moon,* ed. Z. Kopal (New York: Academic Press), pp. 429–479.

Evans, J. V. 1969. Radar studies of planetary surfaces. *Ann. Rev. Astron. Astrophys.* 7:201–248.

Evans, J. V., and Hagfors, T. 1968. *Radar Astronomy* (New York: McGraw-Hill).

Evans, J. V., and Pettengill, G. H. 1963. The scattering behavior of the Moon at wavelengths of 3.6, 68, and 784 centimeters. *J. Geophys. Res.* 68:423–477.

Evans, N., and Rossbacher, L. A. 1980. The last picture show: Small-scale patterned ground in Lunae Planum. In *Reports of Planetary Geology Program–1980*, NASA TM-82385, pp. 376–378.

Ezell, E. C., and Ezell, L. N. 1984. *On Mars: Exploration of the Red Planet, 1958–1978.* NASA SP-4212.

Fagin, S. W., Worrall, D. M., and Muehlberger, W. R. 1978. Lunar mare ridge orientations: Implications for lunar tectonic models. *Proc. Lunar Sci. Conf.* 9:3473–3479.

Fanale, F. P. 1976. Martian volatiles: Their degassing history and geochemical fate. *Icarus* 28:179–202.

Fanale, F. P., and Cannon, W. A. 1971. Adsorption on the Martian regolith. *Nature* 230:502–504.

Fanale, F. P., and Cannon, W. A. 1974. Exchange of adsorbed H_2O and CO_2 between the regolith and atmosphere of Mars caused by changes in surface insolation. *J. Geophys. Res.* 79:3397–3402.

Fanale, F. P., and Cannon, W. A. 1978. Mars: The role of the regolith in determining atmospheric pressure and the atmosphere's response to insolation changes. *J. Geophys. Res.* 83:2321–2325.

Fanale, F. P., and Cannon, W. A. 1979. Mars: CO_2 adsorption and capillary condensation on clays—Significance for volatile storage and atmospheric history. *J. Geophys. Res.* 84:8404–8414.

Fanale, F. P., and Jakosky, B. M. 1982. Regolith-atmosphere exchange of water and carbon dioxide on Mars: Effects on atmospheric history and climate change. *Planet. Space Sci.* 30:819–831.

Fanale, F. P., and Salvail, J. R. 1989. Loss of water from Phobos. *Geophys. Res. Lett.* 16:287–290.

Fanale, F. P., and Salvail, J. R. 1990. Evolution of the water regime of Phobos. *Icarus* 88:380–395.

Fanale, F. P., Salvail, J. R., Banerdt, W. B., and Saunders, R. S. 1982*a*. Mars: The regolith-atmosphere-cap system and climate change. *Icarus* 50:381–407.

Fanale, F. P., Banerdt, W. B., Saunders, R. S., Johansen, L. A., and Salvail, J. R. 1982*b*. Seasonal carbon dioxide exchange between the regolith and atmosphere of Mars: Experimental and theoretical studies. *J. Geophys. Res.* 87:10215–10225.

Fanale, F. P., Salvail, J. R., Zent, A. P., and Postawko, S. E. 1986. Global distribution and migration of subsurface ice on Mars. *Icarus* 67:1–18.

Farmer, C. B. 1976. Liquid water on Mars. *Icarus* 28:279–289.

Farmer, C. B., and Doms, P. E. 1979. Global and seasonal variation of water vapor on Mars and the implications for permafrost. *J. Geophys. Res.* 84:2881–2888.

Farmer, C. B., and LaPorte, D. D. 1972. The detection and mapping of water vapor in the Martian atmosphere. *Icarus* 16:34–46.

Farmer, C. B., Davies, D. W., and LaPorte, D. D. 1976. Mars: Northern summer ice cap—water vapor observations from Viking 2. *Science* 194:1339–1341.

Farmer, C. B., Davies, D. W., Holland, A. L., LaPorte, D. D., and Doms, P. E. 1977. Mars: Water vapor observations from the Viking orbiters. *J. Geophys. Res.* 82:4225–4248.

Farmer, V. C. 1974. *The Infrared Spectra of Minerals*. Monograph 4 (London: Mineralogical Society).

Feierberg, M. A., Lebofsky, L. A., and Tholen, D. J. 1985. The nature of C-class asteroids from 3-μm spectrophotometry. *Icarus* 63:183–191.

Feigelson, E. D., Giapapa, M. S., and Vrba, F. J. 1991. Magnetic activity in pre-main sequence stars. In *The Sun in Time,* eds. C. P. Sonett, M. S. Giampapa and M. S. Matthews (Tucson: Univ. of Arizona Press), pp. 658–681.

Feinberg, G., and Shapiro, R. 1980. Mars today. Is there life after Viking? *The Sciences,* May/June, pp. 13–27.

Feissel, M. 1980. Determination of the Earth's rotation parameters by the Bureau International de l'Heure. *Bull. Geod.* 54:81–102.

Ferrari, A. J., Sinclair, W. S., Sjogren, W. L., Williams, J. G., and Yoder, C. F. 1980. Geophysical parameters of the Earth-Moon system. *J. Geophys. Res.* 85:3939–3951.

Fetter, C. W. 1980. *Applied Hydrogeology* (Columbus, Ohio: C. E. Merrill).

Finnerty, A. A., Phillips, R. J., and Banerdt, W. B. 1988. Igneous processes and the closed system evolution of the Tharsis region of Mars. *J. Geophys. Res.* 93:10225–10235.

Firsoff, V. A. 1969. *The World of Mars* (Edinburgh: Oliver and Boyd).

Firsoff, V. A. 1980. *The New Face of Mars* (Hornchurch: I. Henry)

Fischbacher, G. E., Martin, L. J., and Baum, W. A. 1969. Mars Polar Cap Boundaries. Final Rept. under JPL Contract 951547 (Flagstaff: Planetary Research Center, Lowell Obs.).

Fiske, R. S., and Jackson, E. D. 1972. Orientation and growth of Hawaiian rifts: The effect of regional structure and gravitational stresses. *Proc. Roy. Soc. London* A329:299–326.

Fjeldbo, G., and Eshleman, V. R. 1968. The atmosphere of Mars analyzed by integral inversion of the Mariner IV occultation data. *Planet. Space Sci.* 16:1035–1059.

Fjeldbo, G., Kliore, A., and Seidel, B. 1970. The Mariner 1969 occultation measurements of the upper atmosphere of Mars. *Radio Sci.* 5:381–386.

Fjeldbo, G., Kliore, A. J., and Seidel, B. 1972. Bistatic radar measurements of the surface of Mars with Mariner 1969. *Icarus* 16:502–508.

Flammarion, C. 1892. *La Planète Mars et ses Conditions d'Habitabilité,* vol. 1 (Paris: Gauthier-Villars et Fils).

Flammarion, C. 1909. *La Planète Mars et ses Conditions d'Habitabilité,* vol. 2 (Paris: Gauthier-Villars et Fils).

Flasar, F. M., and Goody, R. M. 1976. Diurnal behavior of water on Mars. *Planet. Space Sci.* 24:161–181.

Floran, R. J., Prinz, M., Hlava, P. F., Keil, K., Nehru, C. E., and Hinthorne, J. R. 1978. The Chassigny meteorite: A cumulate dunite with hydrous amphibole bearing melt inclusions. *Geochim. Cosmochim. Acta* 42:1213–1229.

Florensky, K. P., Bazilvskiy, A. T., Kuzmin, R. O., and Chernaya, I. M. 1975. Results of geologo-morphological analysis of some photographs of the Martian surface obtained by the Mars-4 and Mars-5 automatic stations. *Kosmich. Issled.* 13:67–76. Trans.: NASA TT-F-16343. Microfiche N75–24622.

Focas, J. H. 1961. Etude photométrique et polarimétrique des phénomènes saisonniers de la planète Mars. *Ann. Astrophys.* 24:309–325.

Focas, J. H. 1962. Seasonal evolution of the fine structure of the dark areas of Mars. *J. Planet. Space Sci.* 9:371–381.

Forsythe, R. D., and Zimbelman, J. R. 1988. Is the Gordii Dorsum escarpment on Mars an exhumed-transcurrent fault? *Nature* 336:143–146.

Fountain, J. A., and West, E. A. 1970. Thermal conductivity of particulate basalts as a function of density in simulated lunar and Martian environments. *J. Geophys. Res.* 75:4063–4069.

Fox, J. L. 1989. Dissociative recombination in aeronomy. In *Dissociative Recombination: Theory, Experiment, and Applications,* eds. J. B. A. Mitchell and F. L. Guberman (Singapore: World Scientific), pp. 264–285.

Fox, J. L., and Dalgarno, A. 1979. Ionization, luminosity, and heating of the upper atmosphere of Mars. *J. Geophys. Res.* 84:7315–7331.

Fox, J. L., and Dalgarno, A. 1980. The production of nitrogen atoms on Mars and their escape. *Planet. Space Sci.* 28:41–46.

Fox, J. L., and Dalgarno, A. 1983. Nitrogen escape from Mars. *J. Geophys. Res.* 88:9027–9032.

Francis, P. W., and Wadge, G. 1983. The Olympus Mons aureole: Formation by gravitational spreading. *J. Geophys. Res.* 88:8333–8344.

Francis, P. W., and Wood, C. A. 1982. Absence of silicic volcanism of Mars: Implications for crustal composition and volatile abundance. *J. Geophys. Res.* 87:9881–9889.

François, L. M., Walker, J. C. G., and Kuhn, W. R. 1990. A numerical simulation of climate changes during the obliquity cycle on Mars. *J. Geophys. Res.* 95:14761–14778.

French, L. M. 1980. Photometric Properties of Carbonaceous Chondrites and Related Materials. Ph.D. Thesis, Cornell Univ.

French, L. M., Veverka, J., and Thomas, P. 1988. Brighter material on Deimos: A particle size effect? *Icarus* 75:127–132.

French, R. G., and Gierasch, P. J. 1979. The Martian polar vortex: Theory of seasonal variation and observations of eolian features. *J. Geophys. Res.* 84:4634–4642.

French, R. G., Gierasch, P. J., Popp, B. D., and Yerdon, R. J. 1981. Global patterns in cloud forms on Mars. *Icarus* 45:468–493.

Frey, H. V. 1979*a*. Martian canyons and African rifts: Structural comparisons and implications. *Icarus* 37:142–155.

Frey, H. V. 1979*b*. Thaumasia: A fossilized early forming Tharsis uplift. *J. Geophys. Res.* 84:1009–1023.

Frey, H. V., and Grant, T. 1990. Resurfacing history of Tempe Terra and surroundings. *J. Geophys. Res.* 95:14249-14264.

Frey, H. V., and Jarosewich, M. 1982. Subkilometer Martian volcanoes: Properties and possible terrestrial analogs. *J. Geophys. Res.* 87:9867–9879.

Frey, H. V., and Schultz, R. A. 1988. Large impact basins and the mega-impact origin for the crustal dichotomy on Mars. *Geophys. Res. Lett.* 15:229–232.

Frey, H. V., Lowry, B. L., and Chase, S. A. 1979. Pseudocraters on Mars. *J. Geophys. Res.* 84:8075–8086.

Frey, H. V., Semeniuk, A. M., Semeniuk, J. A., and Tokarcik, S. 1988. A widespread common age resurfacing event in the highland-lowland transition zone in eastern Mars. *Proc. Lunar Planet. Sci. Conf.* 18:679–699.

Frey, H. V., Semeniuk, J., and Grant, T. 1989. Lunae Planum age resurfacing of Mars. Fourth Intl. Conf. on Mars, Jan. 10–13, Tucson, Arizona, Abstract Book, pp. 104–105.

Friedmann, E. I. 1982. Endolithic microorganisms in the Antarctic cold desert. *Science* 215:1045–1053.

Freidmann, E. I., and Ocampo, R. 1976. Endolithic blue-green algae in the dry valleys: Primary producers in the Antarctic desert ecosystem. *Science* 193:1247–1249.

Friedmann, E. I., McKay, C. P., and Nienow, J. A. 1987. The cryptoendolithic microbial environment in the Ross Desert of Antarctica: Nanoclimae data, 1984 to 1986. *Polar Biol.* 7:273–287.

Fujiwara, A., and Asada, N. 1983. Impact fracture patterns on Phobos ellipsoids. *Icarus* 56:590–602.

Fuller, A. O., and Hargraves, R. B. 1978. Some consequences of a liquid water saturated regolith in early Martian history. *Icarus* 34:614–621.

Furnes, H. 1980. Chemical changes during palagonization of an alkaline olivine basaltic hyaloclastite, Santa Maria, Azores. *Neues Jahrb. Mineral. Abh.* 138:14–30.

Furnes, H. 1984. Chemical changes during progressive subaerial palonization of a subglacial olivine tholeiite hyaloclastite: A microprobe study. *Chem. Geol.* 43:271–285.

Furnes, H., and El-Anbaaway, M. I. H. 1980. Chemical changes and authigenic mineral formation during palagonization of a basanite hyaloclastite, Gran Canaria, Canary Islands. *Neues Jahrb. Mineral. Abh.* 139:279–302.

Gaddis, L., Mouginis-Mark, P., Singer, R., and Kaupp, V. 1989. Geologic analyses of shuttle Imaging Radar (SIR-B) data of Kilauea Volcano, Hawaii. *Geol. Soc. Amer. Bull.* 101:317–332.

Gadian, A. M. 1978. The dynamics of and the heat transfer by baroclinic eddies and large-scale stationary topographically forced long waves in the Martian atmosphere. *Icarus* 33:454–465.

Gadsden, J. A. 1975. *Infrared Spectra of Minerals and Related Inorganic Compounds* (London: Butterworths).

Gapcynski, J. P., Tolson, R. H., and Michael, W. H., Jr. 1977. Mars gravity field: Combined Viking and Mariner 9 results. *J. Geophys. Res.* 82:4325–4327.

Garcia, R. R., and Hartmann, D. L. 1980. The role of planetary waves in the maintenance of the zonally averaged ozone distribution of the upper stratosphere. *J. Atmos. Sci.* 37:2248–2264.

Garvin, J. B., Mouginis-Mark, P. J., and Head, J. W. 1981. Characterization of rock populations on planetary surfaces: Techniques and a preliminary analysis of Mars and Venus. *Moon and Planets* 24:355–387.

Garvin, J. B., Head, J. W., Pettengill, G. H., and Zisk, S. H. 1985. Venus global radar reflectivity and correlations with elevations. *J. Geophys. Res.* 90:6859–6871.

Gatley, I., Kieffer, H., Miner, E., and Neugebauer, G. 1974. Infrared observations of Phobos from Mariner 9. *Astrophys. J.* 190:497–503.

Gault, D. E., and Greeley, R. 1978. Exploratory experiments of impact craters formed in viscous-liquid targets: Analogs for Martian rampart craters? *Icarus* 33:483–513.

Gault, D. E., and Wedekind, J. A. 1978. Experimental studies of oblique impact. *Proc. Lunar Sci. Conf.* 9:3843–3875.

Geissler, P. E., Singer, R. B., and Lucchitta, B. K. 1989. Valles Marineris: Compositional constraints from Viking multispectral images. Fourth Intl. Conf. on Mars, Jan. 10–13, Tucson, Arizona, Abstract Book, pp. 111–112.

Geissler, P. E., Singer, R. B., and Lucchitta, B. K. 1990. Dark materials in Valles Marineris: Indications of the style of volcanism and magmatism on Mars. *J. Geophys. Res.* 95:14399–14413.

Gerstl, Z., and Banin, A. 1980. $Fe^{2+} - Fe^{3+}$ transformations in clay and resin ion-exchange systems. *Clays Clay Mineral.* 28:335–345.

Ghan, S. J., and Covey, C. 1989. Unstable radiative-dynamic interactions. Part III. Application to Martian dust storms. *J. Atmos. Sci.*, submitted.

Giardino, J. R., Shroder, J. F., and Vitek, J. D., eds. 1987. *Rock Glaciers* (Boston: Allen and Unwin).

Gierasch, P. J. 1974. Martian dust storms. *Rev. Geophys. Space Phys.* 12:730–734.

Gierasch, P. J., and Goody, R. M. 1967. An approximate calculation of radiative heating and radiative equilibrium in the Martian atmosphere. *Planet. Space Sci.* 15:1465–1477.

Gierasch, P. J., and Goody, R. 1968. A study of the thermal and dynamical structure of the Martian lower atmosphere. *Planet. Space Sci.* 16:615–646.

Gierasch, P. J., and Goody, R. M. 1972. The effect of dust on the temperature of the Mars atmosphere. *J. Atmos. Sci.* 29:400–402.

Gierasch, P. J., and Goody, R. M. 1973. A model of a Martian great dust storm. *J. Atmos. Sci.* 30:169–179.

Gierasch, P. J., and Toon, O. B. 1973. Atmospheric pressure variation and the climate of Mars. *J. Atmos. Sci.* 30:1502–1508.

Gierasch, P., Thomas, P., French, R., and Veverka, J. 1979. Spiral clouds on Mars: A new atmospheric phenomenon. *Geophys. Res. Lett.* 6:405–408.

Gifford, A. W. 1981. Ridge systems on Mars. In *Advances in Planetary Geology*, NASA TM-84412, pp. 219–263.

Gifford, F. A. 1964. A study of Martian yellow clouds that display movement. *Mon. Weather Rev.* 92:435–440.

Gilbert, W. 1600. *De Magnete* (London: P. Short). English trans. (London: Chiswick Press, 1900).

Gilbert, G. K. 1886. The inculcation of scientific method by example. *Amer. J. Sci.* 31:248–299.

Gille, J. C., Lyjak, L. V., and Smith, A. 1987. The global residual mean circulation in the middle atmosphere for the northern winter period. *J. Atmos. Sci.* 44:1437–1452.

Gillette, D. A., Adams, J., Muhs, D., and Kihl, R. 1982. Threshold friction velocities and rupture moduli for crusted desert soils for the input of soil particles into the air. *J. Geophys. Res.* 87:9003–9015.

Gislason, S. R., and Eugster, H. P. 1987*a*. Meteoric water-basalt interactions. I. A laboratory study. *Geochim. Cosmochim. Acta* 51:2827–2840.

Gislason, S. R., and Eugster, H. P. 1987*b*. Meteoritic water-basalt interactions. II. A field study in N.E. Iceland. *Geochim. Cosmochim. Acta* 51:2841–2855.
Glasstone, S. 1968. *The Book of Mars* (Washington, D.C.: NASA).
Glatzmaier, G. A. 1988. Numerical simulations of mantle convection: Time-dependent, three-dimensional, compressible, spherical shell. *Geophys. Astrophys. Fluid Dyn.* 43:223–264.
Goettel, K. A. 1981. Density of the mantle of Mars. *Geophys. Res. Lett.* 8:497–500.
Goettel, K. A. 1983. Present constraints on the mantle of Mars. *Carnegie Inst. Washington Yearbook* 82:363–366.
Goetze, C., and Evans, B. 1979. Stress and temperatures in a bending lithosphere as constrained by experimental rock mechanics. *Geophys. J. Roy. Astron. Soc.* 59:463–478.
Goguen, J., Veverka, J., Thomas, P., and Duxbury, T. 1978. Phobos: Photometry and origin of dark markings on crater floors. *Geophys. Res. Lett.* 5:981–984.
Goins, N. R., and Lazerewicz, A. R. 1979. Martian seismicity. *Geophys. Res. Lett.* 6:368–370.
Gold, T. 1955. Instability of the Earth's axis of rotation. *Nature* 175:526–529.
Gold, T., Campbel, M. J., and O'Leary, B. T. 1970. Optical and high-frequency properties of the lunar sample. *Science* 167:707–709.
Goldreich, P. 1965. Inclination of satellite orbits about an oblate precessing planet. *Astron. J.* 70:5–9.
Goldreich, P., and Toomre, A. 1969. Some remarks on polar wandering. *J. Geophys. Res.* 74:2555–2567.
Goldstein, R. M. 1965. Mars: Radar observations. *Science* 150:1715–1717.
Goldstein, R. M., and Gillmore, W. F. 1963. Radar observations of Mars. *Science* 141:1171–1172.
Goldstein, R. M., and Green, R. R. 1980. Ganymede: Radar surface characteristics. *Science* 207:179–180.
Goldstein, R. M., Melbourne, W. G., Morris, G. A., Downs, G. S., and O'Handley, D. A. 1970. Preliminary radar results of Mars. *Radio Sci.* 5:475–478.
Golitsyn, G. S. 1973. On the Martian dust storms. *Icarus* 18:113–119.
Golombek, M. P. 1979. Structural analysis of lunar grabens and the shallow crustal structure of the Moon. *J. Geophys. Res.* 84:4657–4666.
Golombek, M. P. 1985. Fault type predictions from stress distributions on planetary surfaces: Importance of fault initiation depth. *J. Geophys. Res.* 90:3065–3074.
Golombek, M. P. 1989. Geometry of stresses around Tharsis on Mars. *Lunar Planet. Sci.* XX:345–346 (abstract).
Golombek, M. P., and Banerdt, W. B. 1990. Constraints on the subsurface structure of Europa. *Icarus* 83:441–452.
Golombek, M. P., and McGill, G. E. 1983. Grabens, basin tectonics, and the maximum total expansion of the Moon. *J. Geophys. Res.* 88:3563–3578.
Golombek, M. P., and Phillips, R. J. 1983. Tharsis fault sequence as a test of a deformation mechanism. *Lunar Planet. Sci.* XIV:253–254 (abstract).
Golombek, M. P., Suppe, J., Narr, W., Plescia, J., and Banerdt, W. B. 1990. Does wrinkle ridge formation on Mars involve most of the lithosphere? *Lunar Planet. Sci.* XXI:421–422 (abstract).
Golombek, M. P., Plescia, J. B., and Franklin, B. J. 1991. Faulting and folding in the formation of planetary wrinkle ridges. *Proc. Lunar Planet. Sci Conf.* 21:679–693.
Good, J. C., and Schloerb, F. P. 1981. Martian CO abundance from the $J = 0 \rightarrow$ rotational transition: Evidence for temporal variations. *Icarus* 47:166–172.
Gooding, J. L. 1978. Chemical weathering on Mars: Thermodynamic stabilities of primary minerals (and their alteration products) from mafic igneous rocks. *Icarus* 33:483–513.
Gooding, J. L. 1986*a*. Planetary surface weathering. In *The Solar System: Observations and Interpretations,* ed. M. G. Kivelson (Englewood Cliffs, N.J.: Prentice-Hall), pp. 208–229.
Gooding, J. L. 1986*b*. Martian dust particles as condensation nuclei: A preliminary assessment of mineralogical factors. *Icarus* 66:56–74.
Gooding, J. L. 1986*c*. Weathering of stony meteorites in Antarctica. In *Intl. Workshop on Antarctic Meteorites,* eds. J. O. Annexstad, L. Schultz and H. Wänke, LPI Tech. Rept. 86-01, pp. 48–54.
Gooding, J.L. 1989. Chemical weathering under sub-freezing conditions. Special effects of ice

nucleation. In *Reports of Planetary Geology and Geophysics Program–1988*, NASA TM-4130, pp. 222–223 (abstract).
Gooding, J. L., and Keil, K. 1978. Alteration of glass as a possible source of clay minerals on Mars. *Geophys. Res. Lett.* 5:727–730.
Gooding, J. L., and Muenow, D. W. 1986. Martian volatiles in shergottite EETA 79001: New evidence from oxidized sulfur and sulfur-rich aluminosilicates. *Geochim. Cosmochim. Acta* 50:1049–1060.
Gooding, J. L., Wentworth, S. J., and Zolensky, M. E. 1988. Calcium carbonate and sulfate of possible extraterrestrial origin in the EETA 79001 meteorite. *Geochim. Cosmochim. Acta* 52:909–915.
Gooding, J. L., Carr, M. H., and McKay, C. P. 1989. The case for planetary sample return missions. 2. History of Mars. *Eos: Trans. AGU* 70:745, 754-55.
Goody, R. M., and Belton, M. J. S. 1967. Radiative relaxation times for Mars: A discussion of Martian atmospheric dynamics. *Planet. Space Sci.* 15:247–256.
Gough, D. O. 1981. Solar interior structure and luminosity variations. *Solar Phys.* 74:21–34.
Gradie, J. C., and Tedesco, E. F. 1982. The compositional structure of the asteroid belt. *Science* 216:1405–1407.
Grandjean, J., and Goody, R. M. 1955. The concentration of carbon dioxide in the atmosphere of Mars. *Astrophys. J.* 121:548-552.
Grant, J. A. 1987. *The Geomorphic Evolution of Eastern Margaritifer Sinus, Mars.* NASA TM 89871.
Grant, J. A., and Schultz, P. H. 1987. Possible tornado-like tracks on Mars. *Science* 237:883–885.
Grant, J. A., and Schultz, P. H. 1988. The degradation history of etched/channeled terrains west and northwest of Isidis. *Lunar Planet. Sci. Conf.* XIX:411–412 (abstract).
Grard, R., Pedersen, A., Klimov, S., Savin, S., Skalsky, A., Trotignon, J. G., and Kennel, C. 1989. First measurements of plasma waves near Mars. *Nature* 341:607–609.
Gray-Young, L. D. 1971. Interpretation of high-resolution spectra of Mars. II. Calculations of CO_2 abundance, rotational temperature, and surface pressure. *J. Quant. Spectros. Radiat. Transfer* 11:1075–1086.
Greeley, R. 1973. Mariner 9 photographs of small volcanic structures on Mars. *Geology* 1L:173–180.
Greeley, R. 1979. Silt-clay aggregates on Mars. *J. Geophys. Res.* 84:6248–6254.
Greeley, R. 1985. Dust storms on Mars: Mechanisms for dust-raising. In *MECA Workshop on Dust on Mars,* ed. S. Lee, LPI Tech. Rept. 85-02, pp. 3–5 (abstract).
Greeley, R. 1986. Toward an understanding of the Martian dust cycle. In *MECA Workshop on Dust on Mars II,* ed. S. Lee, LPI Tech. Rept. 86-09, pp. 29–31 (abstract).
Greeley, R. 1987. Release of juvenile water on Mars: Estimated amounts and timing associated with volcanism. *Science* 236:1653–1654.
Greeley, R., and Batson, R. M. 1990. *Planetary Mapping* (Cambridge: Cambridge Univ. Press).
Greeley, R., and Crown, D. A. 1990. Volcanic geology of Tyrrhena Patera, Mars. *J. Geophys. Res.* 95:7133–7149.
Greeley, R., and Guest, J. E. 1987. *Geologic Map of the Eastern Equatorial Region of Mars.* U.S. Geol. Surv. Misc. Inv. Series Map I-1802-B.
Greeley, R., and Iversen, J. D. 1985. *Wind as a Geological Process on Earth, Mars, Venus, and Titan* (Cambridge: Cambridge Univ. Press).
Greeley, R., and Leach, R. 1978. A preliminary assessment of the effects of electrostatics on aeolian processes. In *Reports of Planetary Geology Program–1977–78,* NASA TM-79729, pp. 236–237 (abstract).
Greeley, R., and Leach, R. 1979. "Steam" injection of dust on Mars: Laboratory simulation. In *Reports of Planetary Geology Program-1978–79,* NASA TM-80339, pp. 304–307 (abstract).
Greeley, R., and Spudis, P. D. 1978. Volcanism in the cratered terrain hemisphere of Mars. *Geophys. Res. Lett.* 5:453–455.
Greeley, R., and Spudis, P. 1981. Volcanism on Mars. *Rev. Geophys. Space Phys.* 19:13–41.
Greeley, R., Iversen, J. D., Pollack, J. B., Udovich, N., and White, B. 1974. Wind tunnel simulations of light and dark streaks on Mars. *Science* 183:847–849.
Greeley, R., Leach, R., White, B. R., Iversen, J., and Pollack, J. 1980. Threshold windspeeds for sand on Mars: Wind tunnel simulations. *Geophys. Res. Lett.* 7:121–124.

Greeley, R., White, B. R., Pollack, J. B., Iversen, J. D., and Leach, R. N. 1981. Dust storms on Mars: Considerations and simulations. In Geological Soc. of America SP-186 (Boulder: Geological Soc. of America), pp. 101–121.
Greeley, R., Leach, R. N., Williams, S. H., White, B. R., Pollack, J. B., Krinsley, D. H., and Marshall, J. R. 1982. Rate of wind abrasion on Mars. *J. Geophys. Res.* 87:10009–10024.
Greeley, R., Williams, S. H., White, B. R., Pollack, J. B., and Marshall, J. R. 1985. Wind abrasion on Earth and Mars. In *Models in Geomorphology,* ed. M. J. Woldenburg (Boston: Allen and Unwin), pp. 373–422.
Greeley, R., Skypeck, A., and Pollack, J. B. 1989. Comparison of Martian aeolian features and results from the Global Circulation Model. In *MECA Workshop on Dust on Mars III,* ed. S. Lee, LPI Tech. Rept. 89–01, pp. 30–32 (abstract).
Greenberg, M. 1982. Dust in dense clouds; one stage in a cycle. In *Submillimeter Wave Astronomy,* eds. J. E. Beckman and J. P. Phillips (London: Cambridge Univ. Press), pp. 261–301.
Grieve, R. A. F., Robertson, P. B., and Dence, M. R. 1981. Constraints on the formation of ring impact structures, based on terrestrial data. In *Multi-Ring Basins: Proc. Lunar Planet. Sci. Conf. 12A,* eds. P. H. Schultz and R. B. Merrill, pp. 37–57.
Grieve, R. A. F., Sharpton, V. L., Rupert, J. D., and Goodacre, A. K. 1987. Detecting a periodic signal in the terrestrial cratering record. *Proc. Lunar Planet. Sci. Conf.* 18:375–382.
Grimm, R. E., and Solomon, S. C. 1986. Tectonic tests of proposed polar wander paths for Mars and the Moon. *Icarus* 65:110–121.
Gringauz, K. I. 1976. Interaction of solar wind with Mars as seen by charged particle traps on Mars 2, 3 and 5 satellites. *J. Geophys. Res.* 81:391–402.
Gringauz, K. I., Bezrukikh, V. V., Volkov, G. I., Breus, T. K., Musatov, I. S., Havkin, L. P., and Sloutchonkov, G. P. 1973. Preliminary results on plasma electrons from Mars-2 and Mars-3. *Icarus* 18:54–58.
Gringauz, K. I., Bezrukikh, V. V., Breus, T. K., Verigin, M. I., Volkov, G. I., and Dyachkov, A. V. 1974. Study of solar plasma near Mars and on Earth-Mars route using charged particle traps on Soviet spacecraft in 1971–1973 II. Characteristics of electrons along orbits of artificial Mars satellites Mars 2 and Mars 3. *Cosmic Res.* 12:535–546.
Gringauz, K. I., Bezrukikh, V. V., Verigin, M. I., and Remizov, A. P. 1976. On electron and ion components of plasma in the antisolar part of near-Martian space. *J. Geophys. Res.* 81:3349–3352.
Grinspoon, D. H., and Lewis, J. S. 1988. Cometary water on Venus: Implications of stochastic impacts. *Icarus* 74:21–35.
Grizzaffi, P., and Schultz, P. H. 1989. Isidis basin: Site of ancient volatile-rich debris layer. *Icarus* 77:358–381.
Gubbins, D. 1974. Theories of the geomagnetic and solar dynamos. *Rev. Geophys. Space Phys.* 12:137–154.
Gubbins, D., Masters, T. G., and Jacobs, J. A. 1979. Thermal evolution of the Earth's core. *Geophys. J. Roy. Astron. Soc.* 59:57–99.
Guinness, E. A. 1981. Spectral properties of soils exposed at the Viking 1 landing sites. *J. Geophys. Res.* 86:7983–7992.
Guiness, E. A., and Arvidson, R. E. 1977. On the constancy of the lunar cratering flux over the past 3.3×10^9 years. *Proc. Lunar Sci. Conf.* 8:3475–3494.
Guinness, E. A., Leff, C. E., and Arvidson, R. E. 1982. Two Mars years of changes seen at the Viking landing sites. *J. Geophys. Res.* 87:10051–10058.
Guinness, E. A., Arvidson, R. E., Dale-Bannister, M., Singer, R. B., and Bruckenthal, E. A. 1987. On the spectral reflectance properties of materials exposed at the Viking landing sites. *Proc. Lunar Planet. Sci. Conf.* 17, *J. Geophys. Res. Suppl.* 92:E575-E587.
Gulick, V. C. 1986. Structural control of valley networks on Mars. In NASA TM 88784, 2:361–363.
Gulick, V. C. 1987. Origin and Evolution of Valleys on the Martian Volcanoes: The Hawaiian Analog. Master's Thesis, Univ. of Arizona.
Gulick, V. C. 1992. Magmatic intrusions and hydrothermal systems on Mars. *Geophys. Res. Lett.*, in press.
Gulick, V. C., and Baker, V. R. 1986. Evolution of valley networks on Mars: The Hawaiian analog. *Geol. Soc. of America Abs. w/Prog.* 56:623.

Gulick, V. C., and Baker, V. R. 1987*a*. Origin and evolution of valleys on Martian volcanoes: The Hawaiian analog. *Lunar Planet. Sci.* XVIII:376–377 (abstract).

Gulick, V. C., and Baker, V. R. 1987*b*. Valley evolution on the Hawaiian Islands. *Geol. Soc. of America Cordilleran Sec. Abst. w/Prog.* 19:384.

Gulick, V. C., and Baker, V. R. 1989. Fluvial valleys and Martian paleoclimates. *Nature* 341:514–516.

Gulick, V. C., and Baker, V. R. 1990. Origin and evolution of valleys on Martian volcanoes. *J. Geophys. Res.* 95:14325–14344.

Gurnis, M. 1981. Martian cratering revisited: Implications for early geologic evolution. *Icarus* 48:62–75.

Haberle, R. M. 1986*a*. The climate of Mars. *Sci. Amer.* 254:54–62.

Haberle, R. M. 1986*b*. Interannual variability of global dust storms on Mars. *Science* 234:459–461.

Haberle, R. M., and Hertenstein, R. 1987. The influence of dust on the structure of the diurnally varying boundary layer on Mars. *Bull. Amer. Astron. Soc.* 19:815–816 (abstract).

Haberle, R. M., and Jakosky, B. M. 1990. Sublimation and transport of water from the north residual polar cap on Mars. *J. Geophys. Res.* 95:1423–1437.

Haberle, R. M., and Jakosky, B. M. 1991. Atmospheric effects on the remote determination of thermal inertia on Mars. *Icarus* 90:187–204.

Haberle, R. M., Leovy, C. B., and Pollack, J. B. 1979. A numerical model of the Martian polar cap winds. *Icarus* 39:151–183.

Haberle, R. M., Leovy, C. B., and Pollack, J. B. 1982. Some effects of global dust storms on the atmospheric circulation of Mars. *Icarus* 50:322–367.

Hagfors, T. 1964. Backscattering from an undulating surface with applications to radar returns from the Moon. *J. Geophys. Res.* 69:3779–3784.

Hagfors, T. 1967. A study of the depolarization of lunar radar echoes. *Radio Sci.* 2:445–465.

Hagfors, T. 1968. Relations between rough surfaces and their scattering properties as applied to radar astronomy. In *Radar Astronomy,* eds. J. V. Evans and T. Hagfors (New York: McGraw-Hill), pp. 187–218.

Haldane, J. B. S. 1929. The origin of life. *Rationalist Ann.* 148:3–10.

Hale, W. S. 1982. Central pits in Martian craters: Occurrence by substrate, ejecta type, and rim diameter. *Lunar Planet. Sci.* XIII:295–296 (abstract).

Hale, W. S., and Head, J. W. 1981. Central peaks in Martian craters: Comparisons to the Moon and Mercury. *Lunar Planet. Sci.* XII:386–388 (abstract).

Hall, A. 1878. *Observations and Orbits of the Satellites of Mars.* U.S. Government Printing Office.

Hall, J. L., Solomon, S. C., and Head, J. W. 1986. Elysium region, Mars: Tests of lithospheric loading models for the formation of tectonic features. *J. Geophys. Res.* 91:11377–11392.

Hamilton, D. P., and Burns, J. A. 1991. Orbital stability zones about asteroids. *Icarus* 92:118–131.

Hamilton, K. 1982. The effect of solar tides on the general circulation of the Martian atmosphere. *J. Atmos. Sci.* 39:481–485.

Hamilton, K., and Garcia, R. R. 1986. Theory and observations of the short-period-normal mode oscillations of the atmosphere. *J. Geophys. Res.* 91:11867–11875.

Hanel, R. A., Conrath, B. J., Hovis, W. A., Kunde, V. G., Lowman, P. D., Pearl, J. C., Prabhakara, C., and Schlachman, B. 1972*a*. Infrared spectroscopy experiment on the Mariner 9 mission: Preliminary results. *Science* 175:4019, 305–308.

Hanel, R. A., Conrath, B., Hovis, W., Kunde, V., Lowman, P., Maguire, W., Pearl, J., Pirraglia, J., Prabhakara, C., Schlachman, B., Levin, G., Straat, P., and Burke, T. 1972*b*. Investigation of the Martian environment by infrared spectroscopy on Mariner 9. *Icarus* 17:423–442.

Hanson, W. B., and Mantas, G. P. 1988. Viking electron temperature measurements: Evidence for a magnetic field in the Martian ionosphere. *J. Geophys. Res.* 93:7538–7544.

Hanson, W. B., Sanatani, S., and Zuccaro, D. R. 1977. The Martian ionosphere as observed by the Viking retarding potential analyzers. *J. Geophys. Res.* 82:4352–4363.

Hapke, B. 1981. Bidirectional reflectance spectroscopy. I. Theory. *J. Geophys. Res.* 86:3039–3054.

Hargraves, R. B., Collinson, D. W., Arvidson, R. E., and Spitzer, C. R. 1977. The Viking magnetic properties experiment: Primary mission results. *J. Geophys. Res.* 82:4547–4558.

Harmon, J. K., and Ostro, S. J. 1985. Mars: Dual-polarization radar observations with extended coverage. *Icarus* 62:110–128.

Harmon, J. K., Campbell, D. B., and Ostro, S. J. 1982. Dual-polarization radar observations of Mars: Tharsis and environs. *Icarus* 52:171–187.

Harmon, J. K., Campbell, D. B., Bindschadler, D. L., Head, J. W., and Shapiro, I. I. 1986. Radar altimetry of Mercury: A preliminary analysis. *J. Geophys. Res.* 91:385–401.

Harris, A. W., and Ward, W. R. 1982. Dynamical constraints on the formation and evolution of planetary bodies. *Ann. Rev. Earth Planet. Sci.* 10:61–108.

Harris, S. A. 1977. The aureole of Olympus Mons, Mars. *J. Geophys. Res.* 82:3099–3107.

Hart, H. M. 1989. Seasonal Changes in the Abundance and Vertical Distribution of Water Vapor in the Atmosphere of Mars. Ph.D. Thesis, Univ. of Colorado.

Hart, H. M., and Jakosky, B. M. 1986. Composition and stability of the condensate observed at the Viking Lander 2 site on Mars. *Icarus* 66:134–142.

Hart, H. M., and Jakosky, B. M. 1989. Vertical distribution of Martian atmospheric water vapor. *Eos: Trans. AGU* 70:388 (abstract).

Hartmann, D. L. 1976. The dynamic climatology of the stratosphere in the Southern Hemisphere during late winter 1973. *J. Atmos. Sci.* 33:1789–1802.

Hartmann, W. K. 1973*a*. Ancient lunar mega-regolith and subsurface structure. *Icarus* 18:634–636.

Hartmann, W. K. 1973*b*. Martian cratering. 4. Mariner 9 initial analysis of cratering chronology. *J. Geophys. Res.* 78:4096–4116.

Hartmann, W. K. 1973*c*. Martian surface and crust: Review and synthesis. *Icarus* 19:550–575.

Hartmann, W. K. 1974. Geological observations of Martian arroyos. *J. Geophys. Res.* 79:3951–3957.

Hartmann, W. K. 1977. Relative crater production rates on planets. *Icarus* 31:260–276.

Hartmann, W. K. 1978. Martian cratering V: Toward an empirical Martian chronology, and its implications. *Geophys. Res. Lett.* 5:450–452.

Hartmann, W. K. 1980. Dropping stones in magma oceans: Effects of early lunar cratering. In *Proc. Conf. on the Lunar Highlands Crust,* eds. J. J. Papike and R. B. Merrill (New York: Pergamon), pp. 155–171.

Hartmann, W. K. 1984. Does crater "saturation equilibrium" occur in the solar system? *Icarus* 60:56–74.

Hartmann, W. K. 1987. A satellite-asteroid mystery and a possible early flux of scattered C-class asteroids. *Icarus* 71:57–68.

Hartmann, W. K. 1988. Surface evolution of two-component stone/ice bodies in the Jupiter region. *Icarus* 44:441–453.

Hartmann, W. K. 1989. What's new on Mars? *Sky & Teles.* 77:471.

Hartmann, W. K. 1990. Additional evidence about an early intense flux of asteroids and the origin of Phobos. *Icarus* 87:236–240.

Hartmann, W. K., and Raper, O. 1974. *The New Mars: The Discoveries of Mariner 9,* NASA SP-337.

Hartmann, W. K., and Vail, S. M. 1986. Giant impactors: Plausible sizes and populations. In *Origin of the Moon,* eds. W. K. Hartmann, R. J. Phillips and G. J. Taylor (Houston: Lunar and Planetary Inst.), pp. 551–556.

Hartmann, W. K., Strom, R. G., Weidenschilling, S. J., Blasius, K. R., Woronow, A., Dence, M. R., Grieve, R. A. F., Diaz, J., Chapman, C. R., Shoemaker, E. N., and Jones, K. L. 1981. Chronology of planetary volcanism by comparative studies of planetary cratering. In *Basaltic Volcanism on the Terrestrial Planets* (New York: Pergamon), pp. 1049–1127.

Hartmann, W. K., Tholen, D. J., and Cruikshank, D. P. 1987. The relationship of active comets, "extinct" comets, and dark asteroids. *Icarus* 69:33–50.

Hattori, A., and Akabane, T. 1974. Photographic observations of Mars during the 1973 opposition. *Contrib. Kwasan Hida Obs.* No. 221.

Hattori, A., Akabane, T., Matsui, M., and Ishiura, K. 1976. Photographic observations of Mars during the 1975 opposition. *Contrib. Kwasan Hida Obs.* No. 233.

Haxby, W. F., and Turcotte, D. L. 1978. On isostatic geoid anomalies. *J. Geophys. Res.* 83:5473–5478.

Hay, R. L., and Iijima, A. 1968*a*. Nature and origin of palagonite tuffs of the Honolulu Group on Oahu, Hawaii. *Geol. Soc. Amer. Mem.* 116:338–376.

Hay, R. L., and Iijima, A. 1968*b*. Petrology of palagonite tuffs of Koko Craters, Hawaii. *Contrib. Mineral. Petrol.* 17:141–154.

Hayashi, C., Nakazawa, K., and Nakagawa, Y. 1985. Formation of the solar system. In *Protostars and Planets II*, eds. D. C. Black and M. S. Matthews (Tucson: Univ. of Arizona Press), pp. 1100–1153.

Hayatsu, R., and Anders, E. 1981. Organic compounds in meteorites and their origins. *Topics of Current Chem.* 99:1–37.

Haynes, D. F. 1978. Strength and deformation of frozen silt. In *Third Intl. Conf. on Permafrost*, pp. 656–661.

Hays, J. D., Imbrie, J., and Shackleton, N. J. 1976. Variations in the Earth's orbit: Pacemaker of the ice ages. *Science* 194:1121–1132.

Head, J. W. 1976. Lunar volcanism in space and time. *Rev. Geophys. Space Phys.* 14:265–296.

Head, J. W., and Carras, N. G. 1985. Olympus Mons aureole: Evidence for extensional and compressional deformation associated with emplacements. *Lunar Planet. Sci.* XVI:333–334 (abstract).

Head, J. W., and Cintala, M. J. 1979. Grooves on Phobos: Evidence for possible secondary cratering origin. In *Reports of Planetary Geology Program-1978–1979*, NASA TM-80339, pp. 19–21.

Head, J. W., and Roth, R. 1976. Mars pedestal crater escarpments: Evidence for ejecta related emplacement. In *Papers Presented to Symp. on Planetary Cratering Mechanics* (Houston: Lunar and Planetary Inst.), pp. 50–52.

Head, J. W., and Wilson, L. 1981. Lunar sinuous rille formation by thermal erosion: Eruption conditions, rates and durations. *Lunar Planet. Sci.* XII:427–429 (abstract).

Head, J. W., and Wilson, L. 1989. Basaltic pyroclastic eruptions: Influence of gas-release patterns and volume fluxes on fountain structure, and the formation of cinder cones, spatter cones, rootless flows, lava ponds and lava flows. *J. Vol. Geotherm. Res.* 37:261–271.

Head, J. W., Settle, M., and Wood, C. A. 1976. Origin of Olympus Mons escarpment by erosion of pre-volcano substrate. *Nature* 263:667–668.

Held, I. M. 1982. Climate models and the astronomical theory of the ice ages. *Icarus* 50:449–461.

Held, I. M., and Hou, A. Y. 1980. Nonlinear axially symmetric circulations in a nearly inviscid atmosphere. *J. Atmos. Sci.* 37:515–533.

Hellings, R. W., Adams, P. J., Anderson, J. D., Keesey, M. S., Lau, E. L., Standish, E. M., Canuto, V. M., and Goldman, I. 1983. Experimental test of the variability of G using Viking Lander ranging data. *Phys. Rev. Lett.* 51:1609–1612.

Henderson, B. G., and Jakosky, B. M. 1990. The Martian south polar cap: Stability and water transport at low obliquities. *Lunar Planet. Sci.* XXI:493–494 (abstract).

Henneck, H., Kaluza, K., Thöny, B., Fuhrmann, M., Ludwig, W., and Stackerbradt, E. 1985. Concurrent evolution of nitrogenase genes and 16S rRNA in Rhizobium species and other nitrogen fixing bacteria. *Arch. Microbiol.* 142:342–348.

Henrard, J., and Murigande, C. 1987. Colombo's top. *Celest. Mech.* 40:345–366.

Herkenhoff, K. E. 1990. Weathering and erosion of the polar layered deposits on Mars. *Lunar Planet. Sci.* XXI:495–496 (abstract).

Herkenhoff, K. E., and Murray, B. C. 1990*a*. Color and albedo of the south polar layered deposits on Mars. *J. Geophys. Res.* 95:1343–1358.

Herkenhoff, K. E., and Murray, B. C. 1990*b*. High resolution topography and albedo of the south polar layered deposits on Mars. *J. Geophys. Res.* 95:14511–14529.

Herr, K. C., and Pimental, G. C. 1969. Infrared observations near three microns recorded over the polar cap of Mars. *Science* 166:496–499.

Herr, K. C., and Pimentel, G. C. 1970. Evidence for solid carbon dioxide in the upper atmosphere of Mars. *Science* 167:47–49.

Herschel, W. 1784. On the remarkable appearances of the polar regions of the planet Mars, the inclination of its axis, the position of its poles, and its spheroidical figure; with a few hints relating to its real diameter and atmosphere. *Phil. Trans.* 24:233–273.

Hess, S. L. 1950. Some aspects of the meteorology of Mars. *J. Meteorol.* 7:1–13.
Hess, S. L. 1951. Water vapor on Mars. In *The Project for the Study of Planetary Atmospheres, Report No. 9* (Flagstaff, Ariz.: Lowell Obs.), pp. 96–98.
Hess, S. L. 1968. The hydrodynamics of Mars and Venus. In *The Atmospheres of Venus and Mars,* eds. J. C. Brandt and M. B. McElroy (New York: Gordon and Breach), pp. 109–131.
Hess, S. L. 1976. The vertical distribution of water vapor in the atmosphere of Mars. *Icarus* 28:269–278.
Hess, S. L. 1979. Static stability and thermal wind in an atmosphere of variable composition: Applications to Mars. *J. Geophys. Res.* 84:2969–2973.
Hess, S. L., Henry, R. M., Leovy, C. B., Mitchell, J. L., Ryan, J. A., and Tillman, J. E. 1976*a*. Early meteorological results from the Viking 2 lander. *Science* 194:1352–1353.
Hess, S. L., Henry, R. M., Leovy, C. B., Ryan, J. A., Tillman, J. E., Chamberlain, T. E., Cole, H. L., Dutton, R. G., Greene, G. C., Simon, W. E., and Mitchell, J. L. 1976*b*. Mars climatology from Viking after 20 sols. *Science* 194:78–81.
Hess, S. L., Henry, R. M., Leovy, C. B., Ryan, J. A., Tillman, J. E., Chamberlain, T. E., Cole, H. L., Dutton, R. G., Greene, G. C., Simon, W. C., and Mitchell, J. L. 1976*c*. Preliminary meteorological results on Mars from the Viking 1 Lander. *Science* 193:788–791.
Hess, S. L., Henry, R. M., Leovy, C. B., Ryan, J. A., and Tillman, J. E. 1977. Meteorological results from the surface of Mars: Viking 1 and 2. *J. Geophys. Res.* 82:4559–4574.
Hess, S. L., Henry, R. M., and Tillman, J. E. 1979. The seasonal variation of atmospheric pressure on Mars as affected by the south polar cap. *J. Geophys. Res.* 84:2923–2927.
Hess, S. L., Ryan, J. A., Tillman, J. E., Henry, R. M., and Leovy, C. B. 1980. The annual cycle of pressure of Mars measured by Viking Landers 1 and 2. *Geophys. Res. Lett.* 7:197–200.
Hide, R. 1969. Dynamics of the atmospheres of the major planets with an appendix on the viscous boundary layer at the rigid boundary surface of an electrically conducting rotating fluid in the presence of a magnetic field. *J. Atmos. Sci.* 26:841–853.
Higgins, C. G. 1982. Drainage systems developed by sapping on Earth and Mars. *Geology* 10:147–152.
Hitchman, M. H., and Leovy, C. B. 1986. Evolution of the zonal mean state in the equatorial middle atmosphere during October 1978–May 1979. *J. Atmos. Sci.* 43:3159–3176.
Hobbs, P. V. 1974. *Ice Physics* (London: Oxford Univ. Press).
Hodges, C. A. 1978. Central pit craters, peak rings, and the Argyre basin. In *Reports of Planetary Geology Program-1977–1978,* NASA TM 79729, pp. 169–171 (abstract).
Hodges, C. A., and Moore, H. J. 1979. The subglacial birth of Olympus Mons and its aureoles. *J. Geophys. Res.* 84:8061–8074.
Hodges, C. A., Shew, N. B., and Clow, G. 1980. Distribution of central pit craters on Mars. *Lunar Planet. Sci.* XI:450–452 (abstract).
Hoek, E., and Bray, J. W. 1977. *Rock Slope Engineering* (London: Inst. of Mining and Metallurgy).
Hoekstra, P., and Delaney, A. 1974. Dielectric properties of soils at UHF and microwave frequencies. *J. Geophys. Res.* 79:1699–1708.
Hoffert, M. I., Callegar, A. J., Hsieh, C. T., and Ziegler, W. 1981. Liquid water on Mars: An energy balance climate model for CO_2/H_2O atmospheres. *Icarus* 47:112–129.
Hoffman, J. H., Oyama, V. I., and von Zahn, U. 1980. Measurements of the Venus lower atmosphere composition: A comparison of results. *J. Geophys. Res.* 85:7871–7881.
Hofstadter, M. D., and Murray, B. C. 1990. Ice sublimation and rheology: Implications for the martian polar layered deposits. *Icarus* 84:352–361.
Hogan, J. S., Stewart, R. W., and Rasool, S. I. 1972. Radio occultation measurements of the Mars atmosphere with Mariners 6 and 7. *Radio Sci.* 7:525–537.
Holland, H. D. 1978. *The Chemistry of the Atmosphere and Oceans* (New York: Wiley-Interscience).
Holland, H. D. 1984. *The Chemical Evolution of the Atmospheres and Oceans* (Princeton, N.J.: Princeton Univ. Press).
Hollingsworth, J. L., and Barnes, J. R. 1989. Stationary planetary waves in the Mars atmosphere: Response to orographic and thermal forcings in winter. Fourth Intl. Conf. on Mars, Jan. 10–13, Tucson, Arizona, Abstract Book. pp. 126–127.

Holsapple, K. A. 1987. The scaling of impact phenomena. *Intl. J. Impact Eng.* 5:343–355.

Holsapple, K. A., and Schmidt, R. M. 1982. On the scaling of crater dimensions. 2. Impact processes. *J. Geophys. Res.* 87:1849–1870.

Holsapple, K. A., and Schmidt, R. M. 1987. Point source solutions and coupling parameters in cratering mechanics. *J. Geophys. Res.* 92:6350–6376.

Holton, J. R. 1979. *An Introduction to Dynamic Meteorology,* 2nd ed. (New York: Academic Press).

Holton, J. R. 1980. The dynamics of sudden stratospheric warmings. *Ann. Rev. Earth Planet. Sci.* 8:169–190.

Holton, J. R. 1982. The role of gravity wave induced drag and diffusion in the momentum budget of the mesosphere. *J. Atmos. Sci.* 39:791–799.

Honnorez, J. 1978. Generation of phillipsites by palagonitization of basaltic glass in sea water and the origin of K-rich deep-sea sediments. In *Natural Zeolites: Occurrence, Properties, Use,* eds. L. B. Sand and F. A. Mumpton (Oxford: Pergamon), pp. 245–258.

Honnorez, J. 1980. The aging of the oceanic crust at low temperature. In *The Sea,* ed. C. Emiliani (New York: Wiley), pp. 525–587.

Horai, K., and Winkler, J. L. 1980. Thermal diffusivity of two Apollo 11 samples 10020, 44 and 10065, 23: Effect of petrofabrics on the thermal conductivity of porous rocks under vacuum. *Proc. Lunar Sci. Conf.* 11:1777–1788.

Horanyi, M., Burns, J. A., Tatrallyay, M., and Luhmann, J. G. 1990. Toward understanding the fate of dust lost from the Martian satellites. *Geophys. Res. Lett.* 17:853–856.

Hord, C. W., Barth, C. A., and Stewart, A. T. 1972. Mariner 9 Ultraviolet Spectrometer Experiment: Photometry and topography of Mars. *Icarus* 7:443–456.

Hord, C. W., Simmons, K. D., and McLaughlin, L. K. 1974. Mariner 9 ultraviolet spectrometer experiment: Pressure-altitude measurement on Mars. *Icarus* 17:292–302.

Horn, D., McAfee, J. M., Winer, A. M., Herr, K. C., and Pimentel, G. C. 1972. The composition of the Martian atmosphere: Minor constituents. *Icarus* 16:543–556.

Horner, V. M., and Barlow, N. G. 1988. Martian craters: Changes in the diameter range for ejecta fluidization with latitude. *Lunar Planet. Sci.* XIX:505–506 (abstract).

Horner, V. M., and Greeley, R. 1982. Pedestal craters on Ganymede. *Icarus* 51:549–562.

Horner, V. M., and Greeley, R. 1987. Effects of elevation and ridged plains thickness on Martian crater ejecta morphology. *Proc. Lunar Planet. Sci. Conf.* 17, *J. Geophys. Res. Suppl.* 92: E561-E569.

Horowitz, N. H. 1986. *To Utopia and Back: The Search for Life in the Solar System* (New York: Freeman).

Horowitz, N. H., Bauman, A. J., Cameron, R. E., Geiger, P. J., Hubbard, J. S., Shulman, G. P., Simmonds, P. G., and Westberg, K. 1969. Sterile soil from Antarctica: Organic analysis. *Science* 164:1054–1056.

Horowitz, N. H., Hobby, G. L., and Hubbard, J. S. 1977. Viking on Mars: The carbon assimilation experiments. *J. Geophys. Res.* 82:4659–4662.

Horstman, K. C., and Melosh, H. J. 1989. Drainage pits in cohesionless materials: Implications for the surface of Phobos. *Proc. Lunar Planet. Sci. Conf.* 18:439–440.

Houben, H. 1981. A global Martian dust storm model. In *Third Intl. Coll. on Mars,* LPI Contrib. 441 (Houston: Lunar and Planetary Inst.), p. 117 (abstract).

Houben, H., Haberle, R. M., and Young, R. E. 1988. Martian water vapor transport by baroclinic waves. *Bull. Amer. Astron. Soc.* 20:860 (abstract).

Houck, J. R., Pollack, J. B., Sagan, C., Schaack, D., and Decker, J. A., Jr. 1973. High altitude infrared spectroscopic evidence for bound water on Mars. *Icarus* 18:470–480.

Housen, K. R., and Wilkening, L. L. 1982. Regoliths on small bodies in the solar system. *Ann. Rev. Earth Planet. Sci.* 10:355–376.

Housen, K. R., Schmidt, R. M., and Holsapple, K. A. 1983. Crater ejecta scaling laws: Fundamental forms based on dimensional analysis. *J. Geophys. Res.* 88:2385–2499.

Howard, A. D. 1980. Effect of wind on scarp evolution on the Martian poles. In *Reports of Planetary Geology Program–1980,* NASA TM-82385, pp. 333–335 (abstract).

Howard, A. D. 1988. Groundwater sapping experiments and modeling. In *Sapping Features of the Colorado Plateau,* eds. A. Howard, R. Kochel and H. Holt (Washington, D.C.: NASA), pp. 71–83.

Howard, A. D., and McLane, C. F., III. 1988. Erosion of cohesionless sediment by groundwater seepage. *Water Resources Res.* 24:1659–1674.

Howard, A. D., Cutts, J. A., and Blasius, K. R. 1982*a*. Stratigraphic relationships within Martian polar cap deposits. *Icarus* 50:161–215.

Howard, A. D., Cutts, J. A., and Blasius, K. R. 1982*b*. Photoclinometric determination of the topography of the Martian north polar cap. *Icarus* 50:245–258.

Howard, K. A., and Muehlberger, W. R. 1973. Lunar thrust faults in the Taurus-Littrow region. In *Apollo 17 Prelim. Sci. Rept.*, NASA SP-330, pp. 31/22–31/25.

Hoyt, W. G. 1976. *Lowell and Mars* (Tucson: Univ. of Arizona Press).

Hsü, K. J. 1975. Catastrophic debris streams (Sturzstrom): Generated by rock falls. *Geol. Soc. of Amer. Bull.* 86:129–140.

Hubbard, J. S. 1979. Laboratory simulation of the Pyrolytic Release experiments. An interim report. *J. Molec. Evol.* 14:211–222.

Hubert, M. K., and Rubey, W. W. 1959. Role of fluid pressure in mechanics of overthrust faulting. *Bull. Geol. Soc. Amer.* 70:115.

Huguenin, R. L. 1973*a*. Photostimulated oxidation of magnetite. 1. Kinetics and alteration phase identification. *J. Geophys. Res.* 78:8481–8493.

Huguenin, R. L. 1973*b*. Photostimulated oxidation of magnetite. 2. Mechanism. *J. Geophys. Res.* 78:8495–8506.

Huguenin, R. L. 1974. The formation of goethite and hydrated clay minerals on Mars. *J. Geophys. Res.* 79:3895–3905.

Huguenin, R. L. 1976. Mars: Chemical weathering as a massive volatile sink. *Icarus* 28:203–212.

Huguenin, R. L. 1982. Chemical weathering and the Viking biology experiments on Mars. *J. Geophys. Res.* 87:10069–10082.

Huguenin, R. L., and Harris, S. L. 1985. Accreted H_2O inventory on Mars. In *Abstracts of Talks Presented to the Workshop on the Evolution of the Martian Atmosphere,* Honolulu, HI, August (Houston: Lunar and Planetary Inst.), pp. 4–5.

Huguenin, R. L., and Harris, S. L. 1986. Accreted H_2O inventory on Mars. In *Workshop on the Evolution of the Martian Atmosphere,* eds. M. Carr, P. James, C. Leovy and R. Pepin, LPI Tech. Rept. 86-07, pp. 31–32 (abstract).

Huguenin, R. L., Adams, J. B., and McCord, T. B. 1977. Mars: Surface mineralogy from reflectance spectra. *Lunar Sci.* VII:478–480 (abstract).

Hugenin, R. L., Head, J.W., and McGetchin, T. R. 1978. Mars: Petrologic units in the Margaritifer Sinus and Coprates Quadrangle. In *Reports of Planetary Geology Program–1977–1978* NASA TM–79729, 118–120.

Huguenin, R. L., Clifford, S. M., Sullivan, C. A., and Miller, K. J. 1979. Remote sensing evidence for oases on Mars. In *Reports of Planetary Geology Program–1978–1979,* NASA TM-80339, pp. 208–214 (abstract).

Huguenin, R. L., Harris, S. L., and Carter, R. 1986. Injection of dust into the Martian atmosphere: Evidence from the Viking gas exchange experiment. *Icarus* 68:99–119.

Hulme, G. 1973. Turbulent lava flow and the formation of lunar sinuous rilles. *Modern Geol.* 4:107–117.

Hulme, G. 1976. The determination of the rheological properties and effusion rate of an Olympus Mons lava. *Icarus* 27:207–213.

Hulme, G., and Fielder, G. 1977. Effusion rates and rheology of lunar lavas. *Phil. Trans. Roy. Soc. London* A285:227–234.

Hunt, G. E. 1979. Thermal infrared properties of the Martian atmosphere. 4. Predictions of the presence of dust and ice clouds from Viking IRTM spectral measurements. *J. Geophys. Res.* 84:2865–2874.

Hunt, G. E. 1980. On the infrared radiative properties of CO_2 ice clouds: Application to Mars. *Geophys. Res. Lett.* 7:481–484.

Hunt, G. E., and James, P. B. 1979. Martian extratropical cyclones. *Nature* 278:531–532.

Hunt, G. E., Pickersgill, A. O., James, P. B., and Johnson, G. 1980. Some diurnal properties of clouds over the Martian volcanoes. *Nature* 286:362–364.

Hunt, G.R., Logan, L. M., and Salisbury, J. W. 1973. Mars: Component of infrared spectra and composition of the dust cloud. *Icarus* 18:459–469.

Hunten, D. M. 1973. The escape of light gases from planetary atmospheres. *J. Atmos. Sci.* 30:1481–1494.

Hunten, D. M. 1974. Aeronomy of the lower atmosphere of Mars. *Rev. Geophys. Space Phys.* 12:529–535.

Hunten, D. M. 1979*a*. Capture of Phobos and Deimos by protoatmospheric drag. *Icarus* 37:113–123.

Hunten, D. M. 1979*b*. Possible oxidant sources in the atmosphere and surface of Mars. *J. Molec. Evol.* 14:71–78.

Hunten, D. M. 1982. Thermal and nonthermal escape mechanisms for terrestrial bodies. *Planet. Space Sci.* 30:773–783.

Hunten, D. M., and Donahue, T. M. 1976. Hydrogen loss from the terrestrial planets. *Ann. Rev. Earth Planet. Sci.* 4:265–292.

Hunten, D. M., and McElroy, M. B. 1970. Production and escape of hydrogen on Mars. *J. Geophys. Res.* 75:5989–6001.

Hunten, D. M., Pepin, R. O., and Walker, J. C. G. 1987. Mass fractionation in hydrodynamic escape. *Icarus* 69:532–549.

Hunten, D. M., Pepin, R. O., and Owen, T. C. 1988. Planetary atmospheres. In *Meteorites and the Early Solar System,* eds.J. F. Kerridge and M. S. Matthews (Tucson: Univ. of Arizona Press), pp. 565–591.

Hunten, D. M., Donahue, T. M., Kasting, J. F., and Walker, J. C. G. 1989. Escape of atmospheres and loss of water. In *Origin and Evolution of Planetary and Satellite Atmospheres,* eds. S. K. Atreya, J. B. Pollack and M. S. Matthews (Tucson: Univ. of Arizona Press), pp. 386–422.

Hutton, J. 1788. Theory of the Earth; or an investigation of the laws observable in the composition, dissolution, and restoration of land upon the globe. *Trans. Roy. Soc. Edinburgh* 1:209–304.

Hutton, J. 1795. *Theory of the Earth, with Proofs and Illustrations* (Edinburgh).

Hutton, R. E., Moore, H. J., Scott, R. F., Shorthill, R. W., and Spitzer, C. R. 1980. Surface erosion caused on Mars from the Viking descent engine plume. *Moon and Planets* 23:293–305.

Illes, E., and Horvath, A. 1981. On the origin of the grooves on Phobos. *Adv. Space Res.* 1:49–52.

Imbrie, J. 1982. Astronomical theory of the Pleistocene ice ages: A brief historical review. *Icarus* 50:408–422.

Imbrie, J., and Imbrie, J. Z. 1980. Modeling the climatic response to orbital variations. *Science* 207:943–953.

Imbrie, J., and Imbrie, K. P. 1986. *Ice Ages: Solving the Mystery* (Cambridge: Harvard Univ. Press).

Imshenetskii, A. A., and Mursakov, B. G. 1977. The search for life on Mars. *Mikrobiologiya* 46:1103–1113.

Inge, J. L., and Baum, W. A. 1973. A comparison of Martian albedo features with topography. *Icarus* 19:323–328.

Inge, J. L., and Bridges, P. M. 1976. Applied photo interpretation for airbrush cartography. *Photogrammetry and Remote Sensing* 42:749–760.

Inge, J. L., Capen, C. F., Martin, L. J., Faure, B. Q., and Baum, W. A. 1971. A new map of Mars from Planetary Patrol photographs. *Sky & Teles.* 41:336–339.

Ingersoll, A. P. 1970. Mars: Occurrence of liquid water. *Icarus* 168:972–973.

Ingersoll, A. P. 1974. Mars: The case against permanent CO_2 frost caps. *J. Geophys. Res.* 79:3403–3410.

Intriligator, D. S. 1989. Results of the first statistical study of Pioneer Venus Orbiter plasma observations in the distant Venus tail: Evidence for a hemispheric asymmetry in the pickup of ionospheric ions. *Geophys. Res. Lett.* 16:167–170.

Intriligator, D. S., and Smith, E. J. 1979. Mars in the solar wind. *J. Geophys. Res.* 84:8427–8435.

Ip, W.-H. 1988. On a hot oxygen corona of Mars. *Icarus* 76:135–145.

Ip, W.-H., and Banaszkiewicz, M. 1990. On the gas/gust tori of Phobos and Deimos. *Geophys. Res. Lett.* 17:857–860.

Ip, W.-H., and Fernandez, J. A. 1988. Exchange of condensed matter among the outer and terrestrial protoplanets and the effect on surface impact and atmospheric accretion. *Icarus* 74:47–61.

Irvine, W. M. 1986. The chemistry of cold, dark interstellar clouds. In *Astrochemistry,* eds. M. S. Vardya and S. P. Tarafdar (Dordrecht: D. Reidel), pp. 245–252.

Irvine, W. M., and Knacke, R. F. 1989. The chemistry of interstellar gas and grains. In *Origin and Evolution of Planetary and Satellite Atmospheres,* eds. S. K. Atreya, J. B. Pollack and M. S. Matthews (Tucson: Univ. of Arizona Press), pp. 3–34.

Istomin, V. G., and Grechnev, K. V. 1976. Argon in the Martian atmosphere: Evidence from the Mars 6 descent module. *Icarus* 28:155–158.

Istomin, V. G., Grechnev, K. V., Ozerov, L. N., Slutskiy, M. Ye., Pavlenko, V. A., and Tsvetkov, V. N. 1975. Experiment measuring composition of Martian atmosphere of board the descent module of the Mars-6 space station. *Kosmich. Issled.* 13:16–20. Trans.: NASA-TT-F-116335. Microfiche N75–24616.

Ito, E., and Takahashi, E. 1987. Melting of peridotite under the lower mantle condition. *Nature* 328:514–517.

Ivanov, B. A. 1986. *Cratering Mechanics,* NASA TM-88477 (N87–15662).

Ivanov, B. A. 1988*a*. Effect of modification of impact craters on the size-frequency distribution and scaling law. *Lunar Planet. Sci.* XIX:531–532 (abstract).

Ivanov, M. 1988*b*. Potential for searching for chemolithautotrophic microorganisms on Mars. Presented at US-USSR Joint Working Group Mtg., Washington, D.C., Sept. 17.

Ivanov-Kholodny, G. S., Mikhailov, A. V., and Savich, N. A. 1973. Formation of the Martian ionosphere. *Uspekhi Fizich. Nauk* 111:373–375.

Iversen, J. D., Pollack, J. B., Greeley, R., and White, B. R. 1976. Saltation threshold on Mars: The effect of interparticle force, surface roughness, and low atmospheric density. *Icarus* 29:381–393.

Ives, R. L. 1940. Rock glaciers in the Colorado Front Range. *Bull. Geol. Soc. Amer.* 51:1271–1294.

Iwasaki, K., Saito, Y., and Akabane, T. 1979. Behavior of the Martian north polar cap, 1975–1978. *J. Geophys. Res.* 84:8311–8316.

Iwasaki, K., Saito, Y., and Akabane, T. 1982. Martian north polar cap 1979–1980. *J. Geophys. Res.* 87:10265–10269.

Iwasaki, K., Saito, Y., and Akabane, T. 1984. Martian north polar cap and haze 1981–1982. *Publ. Astron. Soc. Japan* 36:347–356.

Iwasaki, K., Saito, Y., and Akabane, T. 1986. Martian south polar cap 1973. *Publ. Astron. Soc. Japan* 38:267–275.

Iwasaki, K., Saito, Y., Akabane, T., Nakai, Y., Panjaitan, E., Radiman, I., and Wiramihardja, S. D. 1988. Martian south polar cap 1986. *Vistas in Astron.* 31:141–146.

Iwasaki, K., Saito, Y., Nakai, Y., Akabane, T., Panjaitan, E., Radiman, I., and Wiramihardja, S. D. 1989. Behaviour of the Martian south polar cap 1986. *Publ. Astron. Soc. Japan* 41:1083–1094.

Iwasaki, K., Saito, Y., Nakai, Y., Akabane, T., Panjaitan, E., Radiman, I., and Wiramihardja, S. D. 1990. Martian south polar cap 1988. *J. Geophys. Res.* 95:14751–14754.

Jacobs, W. C. 1942. On the energy exchange between the sea and atmosphere. *J. Marine Res.* 5:37–66.

Jaeger, J. C., and Cook, N. G. W. 1979. *Fundamentals of Rock Mechanics* (New York: Halsted Press).

Jagoutz, E. 1989. Sr and Nd isotopic systematics in ALHA 77005: Age of shock metamorphism in shergottites and magmatic differentiation on Mars. *Geochim. Cosmochim. Acta,* in press.

Jagoutz, E., and Wänke, H. 1986. Sr and Nd isotopic systematics of Shergotty meteorite. *Geochim. Cosmochim. Acta* 50:939–953.

Jagoutz, E., Palme, H., Badderhausen, H., Blum, R., Cendales, M., Dreibus, G., Spettel, B., Lorenz, V., and Wänke, H. 1979. The abundances of major, minor and trace elements in the Earth's mantle as derived from primitive ultramafic nodules. *Proc. Lunar Planet. Sci. Conf.* 10:2031–2050.

Jakobsson, S. P. 1972. On the consolidation and palagonization of the tephra of the Surtsey volcanic island, Iceland. *Surtsey Prog. Rept.* VI:7–8.

Jakobsson, S. P. 1978. Environmental factors controlling the palagonitization of the Surtsey Tephra, Iceland. *Bull. Soc. Geol. Denmark* 27:91–105.

Jakobsson, S. P., and Moore, J. G. 1986. Hydrothermal minerals and alteration rates at Surtsey volcano, Iceland. *Geol. Soc. Amer. Bull.* 97:648–659.

Jakosky, B. M. 1979. The effects of nonideal surfaces on the derived thermal properties of Mars. *J. Geophys. Res.* 84:8252–8262.

Jakosky, B. M. 1983*a*. The role of seasonal reservoirs in the Mars water cycle. I. Seasonal exchange of water with the regolith. *Icarus* 55:1–18.

Jakosky, B. M. 1983*b*. The role of seasonal reservoirs in the Mars water cycle. II. Coupled models of the regolith, the polar caps, and atmospheric transport. *Icarus* 55:19–39.

Jakosky, B. M. 1983*c*. Comment on "Mars residual north polar cap: Earth-based spectroscopic confirmation of water ice as a major constituent and evidence for hydrated minerals" by Roger N. Clark and Thomas B. McCord. *J. Geophys. Res.* 88:4329–4330.

Jakosky, B. M. 1985. The seasonal cycle of water on Mars. *Space Sci. Rev.* 41:131–200.

Jakosky, B. M. 1986. On the thermal properties of Martian fines. *Icarus* 66:117–124.

Jakosky, B. M. 1989. The Mars atmosphere and climate system. Fourth Intl. Conf. on Mars, Jan. 10–13, Tucson, Arizona, Abstract Book, p. 31.

Jakosky, B. M. 1990*a*. Mars atmospheric D/H: Consistent with polar volatile theory? *J. Geophys. Res.* 95:1475–1480.

Jakosky, B. M. 1990*b*. Mars climate evolution from stable isotopic abundances. *Bull. Amer. Astron. Soc.* 22:1073 (abstract).

Jakosky, B. M. 1991. Mars volatile evolution: Evidence from stable isotopes. *Icarus* 94: 14–31.

Jakosky, B. M., and Barker, E. S. 1984. Comparison of groundbased and Viking Orbiter measurements of Martian water vapor: Variability of the seasonal cycle. *Icarus* 57:322–334.

Jakosky, B. M., and Carr, M. H. 1985. Possible precipitation of ice at low latitudes of Mars during periods of high obliquity. *Nature* 315:559–561.

Jakosky, B. M., and Christensen, P. R. 1986*a*. Are the Viking Lander sites representative of the surface of Mars? *Icarus* 66:125–133.

Jakosky, B. M., and Christensen, P. R. 1986*b*. Global duricrust on Mars: Analysis of remote sensing data. *J. Geophys. Res.* 91:3547–3559.

Jakosky, B. M., and Farmer, C. B. 1982. The seasonal and global behavior of water vapor in the Mars atmosphere: Complete global results of the Viking atmospheric water detector experiment. *J. Geophys. Res.* 87:2999–3019.

Jakosky, B. M., and Haberle, R. M. 1990*a*. The thermal inertia of Mars: R-interpretation using a better atmospheric model. *Lunar Planet. Sci.* XXI:566 (abstract).

Jakosky, B. M., and Haberle, R. M. 1990*b*. Year-to-year instability of the Mars south polar cap. *J. Geophys. Res.* 95:1359–1365.

Jakosky, B. M., and Martin, T. Z. 1987. Mars: North-polar atmospheric temperatures during dust storms. *Icarus* 72:528–534.

Jakosky, B. M., and Muhleman, D. O. 1980. The longitudinal variation of the thermal inertia and of the 2.8μm brightness temperature of Mars. *Astrophys. J.* 239:403–409.

Jakosky, B. M., and Muhlemann, D. O. 1981. A comparison of the thermal and radar characteristics of Mars. *Icarus* 45:25–38.

Jakosky, B. M., Zurek, R. W., and LaPointe, M. R. 1988. The observed day-to-day variability of Mars atmospheric water vapor. *Icarus* 73:513–524.

James, P. B. 1979. Recession of Martian north polar cap: 1977–1978 Viking observations. *J. Geophys. Res.* 84:8332–8334.

James, P. B. 1982. Recession of Martian north polar cap: 1979–1980 Viking observations. *Icarus* 52:565–569.

James, P. B. 1983. Condensation phase of the Martian south polar cap. *Bull. Amer. Astron. Soc.* 15:846–847 (abstract).

James, P. B. 1985. The Martian hydrologic cycle: Effects of CO_2 mass flux on global water distribution. *Icarus* 64:249–264.

James, P. B. 1988. Water trapping by the Martian polar caps. In *Polar Processes on Mars*, MECA Workshop, May 1988.

James, P. B. 1990. The role of water ice clouds in the Martian hydrologic cycle. *J. Geophys. Res.* 95:1439–1445.

James, P. B., and Evans, N. 1981. A local dust storm in the Chryse region of Mars. *Geophys. Res. Lett.* 8:903–906.

James, P. B., and Lumme, K. 1982. Martian south polar cap boundary: 1971 and 1973 data. *Icarus* 50:368–380.

James, P. B., and North, G. R. 1982. The seasonal CO_2 cycle on Mars: An application of an energy balance climate model. *J. Geophys. Res.* 87:10271–10283.

James, P. B., Briggs, G., Barnes, J., and Spruck, A. 1979. Seasonal recession of Mars' south polar cap as seen by Viking. *J. Geophys. Res.* 84:2889–2922.

James, P. B., Pierce, M., and Martin, L. J. 1987*a*. Martian north polar cap and circumpolar clouds: 1975–1980 telescopic observations. *Icarus* 71:306–312.

James, P. B., Malolepszy, K. M., and Martin, L. J. 1987*b*. Interannual variability of Mars' south polar cap. *Icarus* 71:298–305.

James, P. B., Martin, L. J., Henson, J. R., and Birch, P. V. 1990. Seasonal recession of Mars' south polar cap in 1986. *J. Geophys. Res.* 95:1337–1341.

Janes, D. M., and Melosh, H. J. 1990. Tectonics of planetary loading: A general model and results. *J. Geophys. Res.* 95:21345–21356.

Janle, P. 1983. Bouguer gravity profiles across the highland-lowland escarpment on Mars. *Moon and Planets* 28:55–67.

Janle, P., and Jannsen, D. 1986. Isostatic gravity and elastic bending models of Olympus Mons, Mars. *Ann. Geophys.* 4B:537–546.

Janle, P., and Ropers, J. 1983. Investigation of the isostatic state of the Elysium dome on Mars by gravity models. *Phys. Earth Planet. Int.* 32:132–145.

Jaquin, F., Gierasch, P., and Kahn, R. 1986. The vertical structure of limb hazes in the Martian atmosphere. *Icarus* 68:442–461.

Javor, B. J., and Catenholtz, R. W. 1981. Laminated microbial mats, Lagun Guerrero Negro, Mexico. *Geomicrobio. J.* 2:237–273.

Jeanloz, R. 1979. Properties of iron at high pressure. *J. Geophys. Res.* 84:6059–6069.

Jeanloz, R., and Knittle, E. 1986. Reduction of mantle and core properties to a standard state by adiabatic decompression. In *Chemistry and Physics of the Terrestrial Planets*, ed. S. K. Saxena (New York: Springer-Verlag), pp. 275–309.

Jeanloz, R., and Wenk, H.-R. 1988. Convection and anisotropy of the inner core. *Geophys. Res. Lett.* 15:72–75.

Jephcoat, A. P., Mao, H. K., and Bell, P. M. 1986. Static compression of iron to 78 GPa with rare gas solids as pressure-transmitting media. *J. Geophys. Res.* 91:4677–4684.

Johansen, L. A. 1979. The latitude dependence of Martian splosh cratering and its relationship to water. In *Reports of Planetary Geology Program-1978–1979*, NASA TM-80339, pp. 123–125 (abstract).

Johnson, A. M. 1984. Debris flow. In *Slope Instability*, eds. D. Brunsden and D. B. Prior (New York: Wiley), pp. 257–361.

Johnson, D. W., Harteck, P., and Reeves, R. R. 1975. Dust injection into the Martian atmosphere. *Icarus* 26:441–443.

Johnson, M. C., Rutherford, M. J., and Hess, P. C. 1991. Chassigny petrogenesis: Melt compositions, intensive parameters, and water contents of Martian (?) magmas. *Geochim. Cosmochim. Acta.* 55:349–366.

Johnson, N. L. 1979. *Handbook of Soviet Lunar and Planetary Exploration*, American Astronautical Soc. Sci. and Tech. Series, vol. 47 (San Diego: Univelt).

Johnson, R. E. 1978. Comment on ion and electron temperatures in the Martian upper atmosphere. *Geophys. Res. Lett.* 5:989–992.

Johnson, T. V., Soderblom, L. A., Danielson, G. E., Cook, A. F., and Kupferman, P. 1983. Global multispectral mosaics of the icy Galilean satellites. *J. Geophys. Res.* 88:5789–5805.

Johnston, D. H., and Toksöz, M. N. 1977. Internal structure and properties of Mars. *Icarus* 32:73–84.

Johnston, D. H., McGetchin, T. R., and Toksöz, M. N. 1974. The thermal state and internal structure of Mars. *J. Geophys. Res.* 79:3959–3971.

Jones, D. H. P., Sinclair, A. T., and Williams, I. P. 1989. Secular acceleration of Phobos confirmed from positions obtained on La Palma. *Mon. Not. Roy. Astron. Soc.* 237:15p–19p.

Jones, J. G. 1970. Intraglacial volcanoes of the Laugarvatn region, southwest Iceland. *J. Geol.* 78:127–140.

Jones, J. H. 1986. A discussion of isotopic systematics and mineral zoning in the shergottites: Evidence for a 180 m.y. igneous crystallization age. *Geochim. Cosmochim. Acta* 50:969–977.

Jones, K. L. 1974. Evidence for an episode of crater obliteration intermediate in Martian history. *J. Geophys. Res.* 79:3917–3931.

Jones, K. L., Arvidson, R. E., Guinness, E. A., Bragg, S. L., Wall, S. D., Carlston, C. E., and Pidek, D. G. 1979. One Mars year: Viking lander imaging observations. *Science* 204:799–806.

Jones, T. D. 1988. An Infrared Reflectance Study of Water in Outer Belt Asteroids: Clues to Composition and Origin. Ph.D. Thesis, Univ. of Arizona.

Jones, T. D., Lebofsky, L. A., and Lewis, J. S. 1987. Mid-IR reflectance spectra of C-class asteroids. *Bull. Amer. Astron. Soc.* 19:841 (abstract).

Jones, W. J., Leigh, J. A., Mayer, F., Woese, C. R., and Wolfe, R. S. 1983. *Methanococcus jannaschii* sp. nov., an extremely thermophilic methanogen from a submarine thermal vent. *Arch. Microbio.* 136:254–261.

Jordan, J. F., and Lorell, J. 1975. Mariner 9: An instrument of dynamical science. *Icarus* 25:146–165.

Junk, G., and Svec, H. J. 1958. The absolute abundance of the nitrogen isotopes in the atmosphere and compressed gas from various sources. *Geochim. Cosmochim. Acta* 14:234–243.

Jurgens, R. F., Goldstein, R. M., Rumsey, H. R., and Green, R. R. 1980. Images of Venus by three-station interferometry-1977 results. *J. Geophys. Res.* 85:8282–8294.

Kahn, R. 1980. Some Properties of the Martian Atmosphere Obtained from the Viking Experiments. Ph.D. Thesis, Harvard Univ.

Kahn, R. 1983. Some observational constraints on the global-scale wind systems of Mars. *J. Geophys. Res.* 88:10189–10209.

Kahn, R. 1984. The spatial and seasonal distribution of Martian clouds, and some meteorological implications. *J. Geophys. Res.* 89:6671–6688.

Kahn, R. 1985. The evolution of CO_2 on Mars. *Icarus* 62:175–190.

Kahn, R. 1989. Polar cap edge heating in the dusty northern winter on Mars. In *MECA Workshop on Dust on Mars III,* ed. S. Lee, LPI Tech. Rept. 89-01, p. 37 (abstract).

Kahn, R. 1990. Ice haze, snow, and the Mars water cycle. *J. Geophys. Res.* 95:14677–14693.

Kahn, R., and Gierasch, P. 1982. Long cloud observations on Mars and implications for boundary layer characteristics over slopes. *J. Geophys. Res.* 87:867–880.

Kahn, R., Goody, R., and Pollack, J. 1981. The Martian twilight. *J. Geophys. Res.* 86:5833–5838.

Kahn, R., Guinness, E. A., and Arvidson, R. E. 1986. Loss of fine-scale surface texture in the Viking Orbiter images and implications for the inferred distribution of debris mantles. *Icarus* 66:22–38.

Kakar, R. K., Water, J. W., and Wilson, W. J. 1977. Mars: Microwave detection of carbon monoxide. *Science* 196:1090–1091.

Kamra, A. K. 1972. Measurements of the electrical properties of dust storms. *J. Geophys. Res.* 77:5856–5869.

Kaplan, D. 1988. *Environment of Mars,* NASA TM-100470.

Kaplan, L. D., Münch, G., and Spinrad, H. 1964. An analysis of the spectrum of Mars. *Astrophys. J.* 139:1–15.

Kaplan, L. D., Connes, J., and Connes, P. 1969. Carbon monoxide in the Martian atmosphere. *Astrophys. J.* 157:L187-L192.

Kargel, J. S. 1986. Morphologic variations of Martian rampart crater ejecta and their dependencies and implications. *Lunar Planet. Sci.* XVII:410–411 (abstract).

Kargel, J. S., and Strom, R. G. 1992. Ancient glaciation on Mars. *J. Geology* 20:3–7.

Kasting, J. F., and Ackerman, T. P. 1986. Climatic consequences of very high CO_2 levels in Earth's early atmosphere. *Science* 234:1383–1385.

Kaula, W. M. 1963. Elastic models of the mantle corresponding to variations in the external gravity field. *J. Geophys. Res.* 68:4967–4978.

Kaula, W. M. 1968. *An Introduction to Planetary Physics: The Terrestrial Planets* (New York: Wiley).

Kaula, W. M. 1979*a*. The moment of inertia of Mars. *Geophys. Res. Lett.* 6:194–196.

Kaula, W. M. 1979*b*. Thermal evolution of Earth and Moon growing by planetesimal impacts. *J. Geophys. Res.* 84:999–1008.

Kaula, W. M., Sleep, N. H., and Phillips, R. J. 1989. More about the moment of inertia of Mars. *Geophys. Res. Lett.* 16:1333–1336.

Keeler, R. N., and Mitchell, A. C. 1969. *Solid State Commun.* 7:271.

Keihm, S. J. 1982. Effects of subsurface volume scattering on the lunar microwave brightness temperature spectrum. *Icarus* 52:570–584.

Kerridge, J. F. 1985. Carbon, hydrogen, and nitrogen in carbonaceous chondrites: Abundances and isotopic compositions in bulk samples. *Geochim. Cosmochim. Acta* 49:1707–1714.

Kerridge, J. F. 1988. Deuterium in Shergotty and Lafayette (and on Mars?). *Lunar Planet. Sci.* XIX:599–600 (abstract).

Kerzhanovich, V. V. 1977. Improved analysis of the descent module measurements. *Icarus* 30:1–25.

Khan, M. A. 1977. Depth of sources of gravity anomalies. *Geophys. J. Roy. Astron. Soc.* 21:197–209.

Kiefer, W. S., and Hager, B. H. 1989. The role of mantle convection in the origin of the Tharsis and Elysium provinces of Mars. In *MEVTV-LPI Workshop: Early Tectonic and Volcanic Evolution of Mars,* ed. H. Frey, LPI Tech. Rept. 89-04, pp. 48–50 (abstract).

Kieffer, H. H. 1968. Near Infrared Spectral Reflectance of Simulated Martian Frosts. Ph.D. Thesis, California Inst. of Tech.

Kieffer, H. H. 1970*a*. Interpretation of the Martian polar cap spectra. *J. Geophys. Res.* 75:510–514.

Kieffer, H. H. 1970*b*. Spectral reflectance of CO_2-H_2O frosts. *J. Geophys. Res.* 75:501–509.

Kieffer, H. H. 1976. Soil and surface temperatures at the Viking Lander sites. *Science* 194:1344–1346.

Kieffer, H. H. 1979. Mars south polar spring and summer temperatures: A residual CO_2 frost. *J. Geophys. Res.* 84:8263–8288.

Kieffer, H. H. 1990. H_2O grain size and the amount of dust in Mars' residual north polar cap. *J. Geophys. Res.* 95:1481–1493.

Kieffer, H. H., and Paige, D. A. 1987. Influence of polar cap albedo on past and current martian climate. In *MECA Symp. on Mars: Evolution of Its Climate and Atmosphere,* eds. V. Baker, M. Carr, F. Fanale, R. Greeley, R. Haberle, C. Leovy and T. Maxwell, LPI Tech. Rept. 87-01, pp. 69–70 (abstract).

Kieffer, H. H., Chase, S. C., Miner, E. D., Munch, G., and Neugebauer, G. 1973. Preliminary report on infrared radiometric measurements from the Mariner 9 spacecraft. *J. Geophys. Res.* 78:4291–4312.

Kieffer, H. H., Chase, S. C., Miner, E. D., Palluconi, F. D., Munch, G., Neugebauer, G., and Martin, T. Z. 1976*a*. Infrared thermal mapping of the Martian surface and atmosphere: First results. *Science* 193:780–786.

Kieffer, H. H., Chase, S. C., Jr., Martin, T. Z., Miner, E. D., and Palluconi, F. D. 1976*b*. Martian north pole summer temperatures: Dirty water ice. *Science* 194:1341–1344.

Kieffer, H. H., Martin, T. Z., Peterfreund, A. R., Jakosky, B. M., Miner, E. D., and Palluconi, F. D. 1977. Thermal and albedo mapping of Mars during the Viking primary mission. *J. Geophys. Res.* 82:4249–4291.

Kieffer, H. H., Davis, P. A., and Soderblom, L. A. 1981. Mars' global properties: Maps and applications. *Proc. Lunar Planet. Sci. Conf.* 12:1395–1417, and frontispiece, plates 1–15.

Kieffer, S. W., and Ahrens, T. J. 1980. The role of volatiles and lithology in the impact cratering process. *Rev. Geophys. Space Phys.* 18:143–181.

Kiess, C. C., Karrer, S., and Kiess, H. K. 1960. A new interpretation of Martian phenomena. *Publ. Astron. Soc. Pacific* 72:256–267.

Killworth, P. D., and McIntyre, M. E. 1985. Do Rossby-wave critical layers absorb, reflect or over-reflect? *J. Fluid Mech.* 161:449–492.

Kim, J., Nagy, A. F., Cravens, T. E., and Kliore, A. J. 1989. Solar cycle variations in the lower ionosphere of Venus. *J. Geophys. Res.* 94:11997–12001.

King, E. A. 1978. *Geologic Map of the Mare Tyrrhenum Quadrangle of Mars, scale 1:5,000,000*. U.S. Geol. Surv. Misc. Inv. Series Map I-1073.

King, J. S., and Riehle, J. R. 1974. A proposed origin of the Olympus Mons escarpment. *Icarus* 23:300–317.

Kirby, T. B., and Robinson, J. C. 1971. Dust storm observations from New Mexico. *Sky & Teles.* 41:264–265.

Klaasen, K. P., Duxbury, T. C., and Veverka, J. 1979. Photometry of Phobos and Deimos from Viking Orbiter Images. *J. Geophys. Res.* 84:8478–8486.

Klein, H. P. 1976. Microbiology on Mars? *Amer. Soc. Microbiol. News* 42:207–214.

Klein, H. P. 1977. The Viking biological investigation: General aspects. *J. Geophys. Res.* 82:4677–4680.

Klein, H. P. 1978. The Viking biological experiments on Mars. *Icarus* 34:666–674.

Klein, H. P. 1979. The Viking mission and the search for life on Mars. *Rev. Geophys. Space Phys.* 17:1655–1662.

Klein, H. P., Lederberg, J., and Rich, A. 1972. Biological experiments: The Viking lander. *Icarus* 16:139–146.

Klingler, J. M., Mancinelli, R. L., and White, M. R. 1989. Biological nitrogen fixation under primordial Martian partial pressures of dinitrogen. *Adv. Space Res.* 9:173–176.

Kliore, A. J., and Mullen, L. 1989. The long-term behavior of the main peak of the dayside ionosphere of Venus during solar cycle 21 and its implications on the effect of the solar cycle upon the electron temperatures in the main peak region. *J. Geophys. Res.* 94:13339–13351.

Kliore, A. J., Cain, D. L., Levy, G. S., Eshleman, V. R., Fjeldbo, G., and Drake, F. D. 1965. Occultation experiment: Results of the first direct measurement of Mars' atmosphere and ionosphere. *Science* 149:1243–1248.

Kliore, A. J., Cain, D. L., Seidel, B. L., and Fjeldbo, G. 1970. S-band occultation experiment for Mariner Mars 1971. *Icarus* 12:90–92.

Kliore, A. J., Cain, D. L., Fjeldbo, G., Seidel, B. L., Sykes, M. J., and Rasool, S. I., 1972. The atmosphere of Mars from Mariner 9 radio occultation measurements. *Icarus* 17:484–516.

Kliore, A. J., Fjeldbo, G., Seidel, B. L., Sykes, M. J., and Woiceshyn, P. M. 1973. S-band occultation measurements of the atmosphere and topography of Mars with Mariner 9: Extended mission coverage of polar and intermediate latitudes. *J. Geophys. Res.* 78:4331–4351.

Knight, M. D., and Walker, G. P. L. 1988. Magma flow directions in dikes of the Koolau complex, Oahu, determined from magnetic fabric studies. *J. Geophys. Res.* 93:4301–4319.

Knighton, D. 1984. *Fluvial Forms and Processes* (London: E. Arnold).

Knittle, E., and Jeanloz, R. 1986. High-pressure metallization of FeO and implications for the Earth's core. *Geophys. Res. Lett.* 13:1541–1544.

Knudsen, W. C. 1988. Solar cycle changes in the morphology of the Venus ionosphere. *J. Geophys. Res.* 93:8756–8762.

Knudsen, W. C., Spenner, K., Miller, K. L., and Novak, V. 1980. Transport of ionospheric O^+ ions across the Venus terminator and implications. *J. Geophys. Res.* 85:7803–7810.

Knudsen, W. C., Kliore, A. J., and Whitten, R. C. 1987. Solar cycle changes in the ionization sources of the nightside Venus ionosphere. *J. Geophys. Res.* 92:13391–13398.

Kochel, R. C., and Capar, A. P. 1982. Structural control of sapping valley networks along Valles Marineris, Mars. In *Reports of Planetary Geology Program-1982*, NASA TM-85127, pp. 295–297.

Kochel, R. C., and Peake, R. T. 1984. Quantification of waste morphology in Martian fretted terrain. In *Proc. Lunar Planet. Sci.* 15, *J. Geophys. Res. Suppl.* 89:C336-C350.

Kochel, R. C., and Piper, J. 1986. Morphology of large valleys on Hawaii: Evidence for ground water sapping and comparisons with Martian valleys. *Proc. Lunar Planet. Sci. Conf.* 17, *J. Geophys. Res. Suppl.* 91:E175-E192.

Kochel, R. C., Howard, A. D., and McLane, C. 1985. Channel networks developed by groundwater sapping in fine-grained sediments: Analogs to some Martian valleys. In *Models in Geomorphology,* ed. M. J. Woldenberg (Boston: Allen and Unwin), pp. 313–341.

Kochel, R. C., Simmons, D. W., and Piper, J. F. 1988. Ground water sapping experiments in weakly consolidated layered sediments: A qualitative summary. In *Sapping Features of the Colorado Plateau,* eds. A. Howard, R. Kochel and H. Holt (Washington, D.C.: NASA), pp. 84–93.

Kolosov, M. A., Yakovlev, O. I., Yakovleva, G. D., Yefimov, A. I., Trusov, B. P., Timofeyeva, T. S., Kroglov, Yu. M., Vinogradov, T. A., and Oreshkin, V. P. 1975. Results of investigating the Martian atmosphere by radio-occultation using the Mars-2, Mars-4, and Mars-6 craft. *Kosmich. Issled.* 13:54–59. Trans.: NASA TT-F-16341. Microfiche N75–24624.

Kolyuka, Yu., Tikhonov, V., Ivanov, N., Polyakov, V., Avenesov, G., Heifetz, V., Zhukov, B., Akim, V., Stepanyants, V., Papkov, O., and Duxbury, T. 1990. Phobos and Deimos astrometric observations from the Phobos mission. *Astron. Astrophys.*, submitted.

Kolyuka, Yu., Kudryavstev, S., Tarsov, V., Tikhonov, V., Ivanov, N., Polyakov, O., Sukhanov, K., Akim, E., Stepanians, V., and Nasirov, R. 1991. International project PHOBOS experiment "Celestial Mechanics." *Planet. Space Sci.* 39:349–354.

Komar, P. D. 1979. Comparisons of the hydraulics of water flows in Martian outflow channels with flows of similar scale on Earth. *Icarus* 37:156–181.

Komar, P. D. 1980. Modes of sediment transport in channelized water flows with ramifications to the erosion of Martian outflow channels. *Icarus* 42:317–329.

Komar, P. D. 1983. Shapes of streamlined islands on Earth and Mars: Experiments and analyses of the minimum-drag form. *Geology* 11:651–655.

Komar, P. D. 1984. The lemniscate loop—comparisons with the shapes of streamlined landforms. *J. Geol.* 92:133–145.

Komar, P. D. 1985. Experiments and analyses of the formation of erosional scour marks with implications to the origin of martian outflow channels. In *Reports of Planetary Geology and Geophysics Program—1984,* NASA TM-87563, pp. 322–324.

Kong, T. Y., and McElroy, M. B. 1977*a*. The global distribution of O_3 on Mars. *Planet. Space Sci.* 25:839–857.

Kong, T. Y., and McElroy, M. B. 1977*b*. Photochemistry of the Martian atmosphere. *Icarus* 32:168–189.

Konig, B., Neukum, G., and Fechtig, H. 1977. Recent lunar cratering: Absolute ages of Kepler, Aristarchus, and Tycho. *Lunar Sci.* VII:555–557 (abstract).

Koppes, C. R. 1982. *JPL and the American Space Program: A History of the Jet Propulsion Laboratory* (New Haven, Conn.: Yale Univ. Press).

Kotel'nikov, V. A., Alesandrov, Yu. N., Andreev, R. A., Vyshlov, A. S., Dubrovin, V. M., Zaitsev, A. L., Ignatov, S. P., Kaevitser, V. I., Kozlov, A. N., Krymov, A. A., Molotov, E. P., Petrov, G. M., Rzhiga, O. N., Tagaevskii, A. T., Khasyanov, A. F., Shakhovskoi, A. M., and Shschetinnikov, S. A. 1983. 39-cm radar observations of Mars in 1980. *Soviet Astron.* 27:246–250.

Kopal, Z., and Carder, R. W. 1974. *Mapping of the Moon* (Dordrecht: D. Reidel).

Kovach, R. L., and Anderson, D. L. 1965. The interiors of the terrestrial planets. *J. Geophys. Res.* 70:2873–2882.

Krasnopolsky, V. A. 1979. Vertical distribution of water vapor and Mars model lower and middle atmosphere. *Icarus* 37:182–189.

Krasnopolsky, V. I. 1986. In *Photochemistry of the Atmospheres of Mars and Venus* (Berlin: Springer-Verlag), pp. 69–74.

Krasnopolsky, V. A., and Krysko, A. A. 1976. On the night airglow of the Martian atmosphere. *Space Res.* XVI:1005–1008.

Krasnopolsky, V. A., Krysko, A. A., and Rogachev, V. N. 1975. Ozone in the planetary atmosphere from measurements on board the Mars-7 automatic interplanetary station. *Kosmich. Issled.* 13:37–41. Trans.: NASA TT-F-16,338. Microfiche N75-24618.

Krasnopolsky, V. A., Moroz, V. I., Krysko, A. A., Korablev, O. I., Zhegulev, V. S., Grigoriev, A. V., Tkachuk, A. Yu., Parshev, V. A., Blamont, J. E., and Goutail, J. P. 1989. Solar occultation spectroscopic measurements of the martian atmosphere at 1.9 and 3.7 μm. *Nature* 341:603–604.

Kraus, H. 1967. *Thin Elastic Shells: An Introduction to the Theoretical Foundation and the Analysis of Their Static and Dynamic Behavior* (New York: Wiley).

Krupenio, N. N., Ladygin, V. A., and Shapirovskaia, N. Ia. 1974. Accuracy of the determination of the dielectric constant and temperature of a subsurface layer from polarization measurements. *Kosmich. Issled.* 12:740–747. Trans.: *Cosmic Res.* 12:673–679.

Ksanfomality, L. V., and Krasofsky, G. N. 1975. Photometry in the range from 3200 to 9000 Å onboard the Mars-5 spacecraft. *Kosmich. Issled.* 13:87–91. In Russian.

Ksanfomality, L. V., and Moroz, V. I. 1975. Infrared radiometry on board Mars-5. *Cosmic Res.* 13:65–67.

Ksanfomality, L. V., Moroz, V. I., and Kunashev, B. S. 1975. Pressures and altitudes according to CO_2 band altimetry on Mars-3. *Kosmich. Issled.* 13:563–580. Trans.: NASA TT-F-16,346. Microfiche N75-23671.

Ksanfomality, L. V., Moroz, V. I., Bibring, J. P., Combes, M., Soufflot, A., Ganpantzerova, O. F., Goroshkova, N. V., Zharkov, A. V., Nikitin, G. E., and Petrova, E. V. 1989. Spatial variations in the thermal and albedo properties of the surface of Phobos. *Nature* 341:588–591.

Ksanfomality, L. V., Murchie, S., Britt, D., Duxbury, T., Fisher, P., Goroshkova, N., Head, J., Kuhrt, E., Moroz, V., Murray, B., Nitken, G., Petroux, E., Pieters, C., Soufflot, A., Zharkov, A., and Zhukov, B. 1990. Phobos: Spectrophotometry between 0.3 and 0.6 μm and IR radiometry. *Planet Space Sci.*, in press.

Kuhn, W. R., Atreya, S. K., and Postawko, S. E. 1979. The influence of ozone on Martian atmospheric temperatures. *J. Geophys. Res.* 84:8341–8342.

Kührt, E., and Giese, B. 1989. A thermal model of the Martian satellites. *Icarus* 81:102–112.

Kuiper, G. P. 1949. Survey of planetary atmospheres. In *The Atmospheres of the Earth and Planets* (Chicago: Univ. of Chicago Press), pp. 304–345.

Kuiper, G. P. 1952. *The Atmospheres of the Earth and Planets*, rev. ed., (Chicago: Univ. of Chicago Press), esp. pp. 358–361.

Kuiper, G. P. 1957. Visual observations of Mars, 1956. *Astrophys. J.* 125:307–317.

Kuiper, G. P. 1964. Infrared spectra of stars and planets. IV. The spectrum of Mars, 1–2.5 microns, and the structure of its atmosphere. *Commun. Lunar and Planetary Lab* 2:79–112.

Kulhawy, F. H. 1975. Stress deformation properties of rock and rock discontinuities. *Eng. Geol.* 9:327–350.

Kulikov, Y. N., and Rykhletskii, M. V. 1984. Modeling of the vertical distribution of water in the atmosphere of Mars. *Solar System Res.* 17:112–118.

Kushner, D. 1981. Extreme environments: Are there any limits to life? In *Comets and the Origin of Life*, ed. C. Ponnaperuma (Dordrecht: D. Reidel), pp. 241–248.

Kuzmin, R. O. 1977. Questions of the structure of the Martian cryolithosphere. In *Problems of Cryolithology*, vol. 6 (Moscow: Univ. Press), pp. 7–25.

Kuzmin, R. O. 1978. The peculiarities of the Martian cryolithosphere and its display in relief. In *Problems of Cryolithology*, vol. 7 (Moscow: Univ. Press), pp. 7–27.

Kuzmin, R. O. 1980*a*. Determination of the bedding depth of ice rocks on Mars from the morphology of fresh craters. *Dokl. ANSSSP* 252:1445–1448.

Kuzmin, R. O. 1980*b*. Morphology of fresh Martian craters as an indicator of the depth of the upper boundary of the ice-bearing permafrost: A photogeologic study. *Lunar Planet. Sci.* XI:585–586 (abstract).

Kuzmin, R. O. 1983. *Cryolithosphere of Mars* (Izdatel'stvo Nauka.).

Kuzmin, R. O., and Losovskii, B. Y. 1984. Radiometric inhomogeneity of Mars at millimeter radio wavelengths. *Solar System Res.* 17:119–124.

Kuzmin, R. O., Bobina, N. N., Zabalueva, E. V., and Shashkina, V. P. 1988*a*. Inhomogeneities in the upper levels of the Martian cryolithosphere. *Lunar Planet. Sci.* XIX:655–656 (abstract).

Kuzmin, R. O., Bobina, N. N., Zabalueva, E. V., and Shashkina, V. P. 1988*b*. Structure inhomogeneities of the Martian cryolithosphere. *Solar System Res.* 22:195–212.

Kuzmin, R. O., Bobina, N. N., Zabalueva, E. V., and Shashkina, V. P. 1989. Martian cryolithosphere: Structure and relative ice content. *Intl. Geol. Congress*, in press (abstract).

Laity, J. E., and Malin, M. C. 1985. Sapping processes and the development of theater-headed valley networks on the Colorado Plateau. *Geol. Soc. Amer. Bull.* 96:203–217.

Lake, J. A. 1988. Origin of the eukaryotic nucleus determined by rate-invariant analysis of rRNA sequences. *Nature* 331:184–186.

Lambeck, K. 1972. Gravity anomalies over ocean ridges. *Geophys. J.* 30:37–53.

Lambeck, K. 1979*a*. Comments on the gravity and topography of Mars. *J. Geophys. Res.* 84:6241–6247.

Lambeck, K. 1979*b*. On the orbital evolution of the Martian satellites. *J. Geophys. Res.* 84:5651–5658.

Lambeck, K. 1980. *The Earth's Variable Rotation* (Cambridge: Cambridge Univ. Press).

Lancaster, N., and Greeley, R. 1987. Mars: Morphology of southern hemisphere intracrater dune fields. In *Reports of Planetary Geology and Geophysics Program–1986*, NASA TM-89810, pp. 264–265 (abstract).

Lancaster, N., and Greeley, R. 1989. The north polar sand seas: Preliminary estimates of sediment volume. In *MECA Workshop on Dust on Mars III*, ed. S. Lee, LPI Tech. Rept. 89-01, pp. 38–40 (abstract).

Lancaster, N., and Greeley, R. 1990. Sediment volume in the north polar sand seas of Mars. *J. Geophys. Res.* 95:10921–10927.

Lane, A. L., Barth, C. A., Hord, C. W., and Stewart, A. I. 1973. Mariner 9 ultraviolet spectrometer experiment: Observations of ozone on Mars. *Icarus* 18:102–108.

Lange, M. A., and Ahrens, T. J. 1981. The evolution of an impact-generated atmosphere. *Icarus* 51:96–120.

Langevin, Y., Bibring, J. P., Gondet, B., Combes, M., Grigoriev, A. V., Joukov, B., and Nikolsky, Y. V. 1990. Observations of Phobos from 0.8 to 3.15 μm with the ISM experiment on board the Soviet "Phobos II" spacecraft. *Lunar Planet. Sci.* XXI:682–683 (abstract).
Larson, H. P., and Fink, U. 1972. Identification of carbon dioxide frost on the Martian polar caps. *Astrophys. J.* 171:L91-L95.
Larson, S. M., and Minton, R. B. 1971. Some high-resolution photographs of Mars. *Sky & Teles.* 41:260–261.
Lasker, J. 1988. Secular evolution of the solar system over 10 million years. *Astron. Astrophys.* 198:341–362.
Lasker, J. 1989. A numerical experiment on the chaotic behaviour of the solar system. *Nature* 338:237–238.
Laul, J. C. 1986. The Shergotty Consortium and SNC meteorites: An overview. *Geochim. Cosmochim. Acta* 50:875–887.
Laul, J. C., Keays, R. R., Ganapathy, R., Anders, E., and Morgan, J. W. 1972. Chemical fractionations in meteorites. V. Volatile and siderophile elements in achondrites and ocean ridge basalts. *Geochim. Cosmochim. Acta* 37:667–684.
Laul, J. C., Smith, M. R., Wänke, H., Jagoutz, E., Dreibus, G., Palme, H., Spettel, B., Burghele, A., Lipschutz, M. E., and Verkouteren, R. M. 1986. Chemical systematics of the Shergotty meteorite and the composition of its parent body (Mars). *Geochim. Cosmochim. Acta* 50:909–926.
Lebofsky, L. A. 1980. Infrared reflection spectra of asteroids: A search for water of hydration. *Astron. J.* 85:573–585.
Lebofsky, L. A., and Jones, T. D. 1989. The nature of low-albedo asteroids from 3-μm spectrophotometry. *Lunar Planet. Sci.* XX:562–563 (abstract).
Lebofsky, L. A., Feierberg, M. A., Tokunaga, A. T., Larson, H. P., and Johnson, J. R. 1981. The 1.7- to 4.2-micron spectrum of asteroid 1 Ceres: Evidence for structural water in clay minerals. *Icarus* 48:453–459.
Lederberg, J. 1965. Signs of life. The criterion system of exobiology. *Nature* 207:9–13.
Lee, S. W. 1984. Mars: Wind streak production as related to obstacle type and size. *Icarus* 58:339–357.
Lee, S. W. 1985. Influence of atmospheric dust loading and water vapor content on settling velocities of Martian dust/ice grains: Preliminary results. In *MECA Workshop on Dust on Mars,* eds. S. Lee, LPI Tech. Rept. 85–02, pp. 51–52 (abstract).
Lee, S. W. 1986*a*. IRTM thermal and albedo observations of Syrtis Major. *Lunar Planet. Sci.* XVI:470–471 (abstract).
Lee, S. W. 1986*b*. Viking observations of regional sources and sinks of dust on Mars. In *MECA Workshop on Dust on Mars II,* ed. S. Lee, LPI Tech. Rept. 86-09, p. 44 (abstract).
Lee, S. W. 1987. Regional sources and sinks of dust on Mars: Viking observations of Cerberus, Solis Planum, and Syrtis Major. In *MECA Symp. on Mars: Evolution of Its Climate and Atmosphere,* eds. V. Baker, M. Carr, F. Fanale, R. Greeley, R. Haberle, C. Leovy and T. Maxwell, LPI Tech. Rept. 87-01, pp. 57–58 (abstract).
Lee, S. W., Thomas, P. C., and Veverka, J. 1982. Wind streaks in Tharsis and Elysium: Implications for sediment transport by slope winds. *J. Geophys. Res.* 87:10025–10042.
Lee, S. W., Thomas, P., and Veverka, J. 1986. Phobos, Deimos, and the Moon: Size and distribution of crater ejecta blocks. *Icarus* 68:77–86.
Legrand, H. 1979. Evaluation techniques of fractured-rock hydrology. *J. Hydrology,* pp. 333–346.
Leighton, R. B., and Murray, B. C. 1966. Behavior of carbon dioxide and other volatiles on Mars. *Science* 153:136–144.
Leighton, R. B., Horowitz, N. H., Murray, B. C., Sharp, R. P., Herriman, A. G., Young, A. T., Smith, B. A., Davies, M. E., and Leovy, C. G. 1969. Mariner 6 and 7 television pictures: Preliminary analysis. *Science* 166:49–67.
Leovy, C. B. 1966*a*. Mars ice caps. *Science* 154:1178–1179.
Leovy, C. B. 1966*b*. Note on thermal properties of Mars. *Icarus* 5:1–6.
Leovy, C. B. 1969*a*. Bulk transfer coefficient for heat transfer. *J. Geophys. Res.* 74:3313–3321.
Leovy, C. B. 1969*b*. Mars: Theoretical aspects of meteorology. *Appl. Opt.* 8:1279–1286.
Leovy, C. B. 1973. Exchange of water vapor between the atmosphere and surface of Mars. *Icarus* 18:120–125.

Leovy, C. B. 1979. Martian meteorology. *Ann. Rev. Astron. Astrophys.* 17:387–413.
Leovy, C. B. 1981. Observations of Martian tides over two annual cycles. *J. Atmos. Sci.* 38:30–39.
Leovy, C. B. 1982. Martian meteorological variability. *Adv. Space Res.* 2:19–44.
Leovy, C. B. 1985. The general circulation of Mars: Models and observations. *Adv. Geophys.* 28a:327–346.
Leovy, C. B., and Mintz, Y. 1969. Numerical simulation of the atmospheric circulation and climate of Mars. *J. Atmos. Sci.* 26:1167–1190.
Leovy, C. B., and Zurek, R. W. 1979. Thermal tides and Martian dust storms: Direct evidence for coupling. *J. Geophys. Res.* 84:2956–2968.
Leovy, C. B., Smith, B. A., Young, A. T., and Leighton, R. B. 1971. Mariner Mars 1969: Atmospheric results. *J. Geophys. Res.* 76:297–312.
Leovy, C., Briggs, G., Young, A., Smith, B., Pollack, J., Shipley, E., and Widley, R. 1972. The Martian atmosphere: Mariner 9 television experiment progress report. *Icarus* 17:373–393.
Leovy, C. B., Briggs, G. A., and Smith, B. A. 1973*a*. Mars atmosphere during the Mariner 9 extended mission: Television results. *J. Geophys. Res.* 78:4252–4266.
Leovy, C. B., Zurek, R. W., and Pollack, J. B. 1973*b*. Mechanisms for Mars dust storms. *J. Atmos. Sci.* 30:749–762.
Leovy, C. B., Tillman, J. E., Guest, W. R., and Barnes, J. R. 1985. Interannual variability of Martian weather. In *Recent Advances in Planetary Meteorology,* ed. G. E. Hunt (Cambridge: Cambridge Univ. Press), pp. 69–84.
Lerch, F. J., Klosko, S. M., Laubscher, R. E., and Wagner, C. A. 1979. Gravity model improvement using GEOS 3 (GEM 9 and 10). *J. Geophys. Res.* 84:3897–3915.
Lerch, F. J., Klosko, S. M., Wagner, C. A., and Patel, G. B. 1985. On the accuracy of recent Goddard gravity models. *J. Geophys. Res.* 90:9312–9334.
Leshin, L. A., Holloway, J. A., and Bertka, C. 1988. Atmospheric pressure experimental studies of a low-silica Martian mantle composition. *Lunar Planet. Sci.* XIX:677–678 (abstract).
Levin, G. V., and Straat, P. A. 1977. Recent results from the Viking labeled release experiment on Mars. *J. Geophys. Res.* 82:4663–4668.
Levin, G. V., and Straat, P. A. 1981. Antarctic soil No. 726 and implications for the Viking labeled release experiment. *J. Theor. Biol.* 91:41–45.
Levin, G. V., and Straat, P. A. 1988. A reappraisal of life on Mars. In *The NASA Mars Conference,* ed. D. B. Reiber (San Diego: Sci. and Tech. Ser.), pp. 187–208.
Levin, G. V., Straat, P. A., and Benton, W. D. 1978. Color and feature changes at Mars Viking landing site. *J. Theor. Biol.* 76:381–390.
Levine, J. S., ed. 1985. Appendix II. Chemical reaction rates. In *The Photochemistry of Atmospheres* (Orlando: Academic Press), pp. 497–508.
Levinthal, E. C., Jones, K. L., Fox, P., and Sagan, C. 1977. Lander imaging as a detector of life on Mars. *J. Geophys. Res.* 82:4468–4478.
Lewis, J. S. 1973. Chemistry of the planets. *Ann. Rev. Phys. Chem.* 24:339–351.
Lewis, J. S. 1974*a*. The temperature gradient in the solar nebula. *Science* 186:440–443.
Lewis, J. S. 1974*b*. Volatile element influx on Venus from cometary impacts. *Earth Planet. Sci. Lett.* 2:239–244.
Ley, W., and von Braun, W. 1956. *The Exploration of Mars* (London: Sidgwick and Jackson).
Lincoln Laboratories, 1967. Radar studies of the Moon. *MIT Lincoln Laboratories Report,* 31 August 1967.
Lindal, G. F. 1978. Occultation data update, Viking Mars physical properties working group report. NASA/Langley/JPL/MMC, Viking Project, pp. 106–116.
Lindal, G. F., Hotz, H. B., Sweetnam, D. N., Shippony, Z., Brenkle, J. P., Hartsell, G. V., Spear, R. T., and Michael, W. H., Jr. 1979. Viking radio occultation measurements of the atmosphere and topography of Mars: Data acquired during 1 Martian year of tracking. *J. Geophys. Res.* 84:8443–8456.
Lindner, B. L. 1990. The Martian polar cap: Radiative effects of ozone, clouds and airborne dust. *J. Geophys. Res.* 95:1367–1379.
Lindner, B. L., and Jakosky, B. M. 1985. Martian atmospheric photochemistry and composition during periods of low obliquity. *J. Geophys. Res.* 90:3435–3440.
Lindsay, J. F. 1975. *Lunar Stratigraphy and Sedimentology* (New York: Elsevier).

Lindzen, R. S. 1966. On the relation of wave behavior to source strength and distribution in a propagating medium. *J. Atmos. Sci.* 23:630–632.
Lindzen, R. S. 1970. The application and applicability of terrestrial atmospheric tidal theory to Venus and Mars. *J. Atmos. Sci.* 27:536–549.
Lindzen, R. S. 1981. Turbulence and stress owing to gravity wave and tidal breakdown. *J. Geophys. Res.* 86:9707–9714.
Lindzen, R. S., and Hou, A. Y. 1988. Hadley circulations for zonally averaged heating centered off the equator. *J. Atmos. Sci.* 45:2416–2427.
Lingenfelter, R. E., and Schubert, G. 1973. Evidence for convection in planetary interiors from first-order topography. *Moon* 7:172–180.
Lingenfelter, R. E., Peale, S. J., and Schubert, G. 1968. Lunar rivers. *Science* 161:266–269.
Liou, K.-N. 1980. *An Introduction to Atmospheric Radiation* (San Diego: Academic Press).
Liu, L. G. 1979. Phase transformations and the constitution of the deep mantle. In *The Earth: Its Origin, Structure and Evolution,* ed. M. W. McElhinny (New York: Academic Press), pp. 117–202.
Liu, S. C., and Donahue, T. M. 1976. The regulation of hydrogen and oxygen escape from Mars. *Icarus* 28:231–246.
Longhi, J., and Pan, V. 1988. The parent magmas of the SNC meteorites. *Proc. Lunar Planet. Sci. Conf.* 19:451–464.
Longman, I. M. 1963. A Green's function for determining the deformation of the Earth under surface mass loads. 2. Computations and numerical results. *J. Geophys. Res.* 68:485–496.
Lopes, R. M. C., Guest, J. E., and Wilson, C. J. 1980. Origin of the Olympus Mons aureole and perimeter scarp. *Moon and Planets* 22:221–234.
Lopes, R. M. C., Guest, J. E., Hiller, K., and Neukum, G. 1982. Further evidence for a mass movement origin of the Olympus Mons aureole. *J. Geophys. Res.* 87:9917–9928.
Lorell, J. 1972. Estimation of gravity field harmonics in the presence of spin axis direction error using radio tracking data. *J. Astronaut. Sci.* 20:44–54.
Lorell, J., Born, G. H., Christensen, E. J., Jordan, J. F., Laing, P. A., Martin, W. L., Sjogren, W. L., Shapiro, I. I., Reasenberg, R. D., and Slater, G. L. 1972. Mariner 9 celestial mechanics experiment: Gravity field and pole direction of Mars. *Science* 175:317–320.
Lorell, J., Born, G. H., Christensen, E. J., Esposito, P. B., Jordon, J. F., Laing, P. A., Sjorgren, W. L., Wong, S. K., Reasenberg, R. D., Shapiro, I. I., and Slater, G. L. 1973. Gravity field of Mars from Mariner 9 tracking data. *Icarus* 18:304–316.
Lorenz, E. N. 1967. *The Nature and Theory of the General Circulation of the Atmosphere* (Geneva: World Meteor. Org.).
Lorenz, E. N. 1978. Introduction. In *Dynamics of Earth and Planetary Atmospheres,* JPL Publ. 78–046, pp. 1-1–1-4.
Love, A. E. H. 1911. *Some Problems of Geodynamics* (Cambridge: Cambridge Univ. Press).
Lowell, P. 1895. *Mars* (London: Longmans and Green).
Lowell, P. 1896. *Mars* (Boston: Houghton Mifflin).
Lowell, P. 1906. *Mars and Its Canals* (New York: MacMillan).
Lowell, P. 1908. *Mars as the Abode of Life* (New York: MacMillan).
Lucchitta, B. K. 1976. Mare ridges and related highland scarps—Results of vertical tectonism? *Proc. Lunar Sci. Conf.* 7:2761–2782.
Lucchitta, B. K. 1977. Topography, structure, and mare ridges in southern Mare Imbrium and northern Oceanus Procellarum. *Proc. Lunar Sci. Conf.* 8:2691–2703.
Lucchitta, B. K. 1978*a*. Morphology of chasma walls, Mars. *U.S.G.S. J. Res.* 6:651–662.
Lucchitta, B. K. 1978*b*. A large landslide on Mars. *Geol. Soc. Amer. Bull.* 89:1601–1609.
Lucchitta, B. K. 1978*c*. *Geologic Map of the Ismenius Lacus Region of Mars.* U.S. Geol. Survey Misc. Map Series I-1065.
Lucchitta, B. K. 1979. Landslides in Valles Marineris, Mars. *J. Geophys. Res.* 84:8097–8113.
Lucchitta, B. K. 1981. Mars and Earth: Comparison of cold-climate features. *Icarus* 45:264–303.
Lucchitta, B. K. 1982*a*. Ice sculpture in the Martian outflow channels. *J. Geophys. Res.* 87:9951–9973.
Lucchitta, B. K. 1982*b*. Lakes or playas in Valles Marineris. In *Reports of Planetary Geology Program-1982,* NASA TM-85127, pp. 233–234 (abstract).
Lucchitta, B. K. 1983*a*. Ice-lubricated flow in Martian fretted channels and implications for

outflow channel processes. *Proc. Lunar Planet. Sci. Conf.* 14, *J. Geophys. Res. Suppl.* 89:B446-B447.

Lucchitta, B. K. 1983*b*. Permafrost on Mars: Polygonally fractured ground. In *Permafrost: Proc. Fourth Intl. Conf. Proc.* July 17–22, Fairbanks, Al. (Washington, D.C.: National Academy Press), pp. 744–749.

Lucchitta, B. K. 1984*a*. Ice and debris in the fretted terrain, Mars. *Proc. Lunar Planet. Sci. Conf.* 14, *J. Geophys. Res. Suppl.* 89:B409-B418.

Lucchitta, B. K. 1984*b*. A late climatic change on Mars. *Lunar Planet. Sci.* XV:493–494 (abstract).

Lucchitta, B. K. 1985*a*. Geomorphologic evidence for ground ice on Mars. In *Ices in the Solar System,* eds. J. Klinger, D. Benest, A. Dollfus and R. Smoluchowski (New York: D. Reidel), pp. 583–604.

Lucchitta, B. K. 1985*b*. Valles Marineris basin beds: A complex story. In *Reports of Planetary Geology Program 1984,* NASA TM-87563, pp. 506–508 (abstract).

Lucchitta, B. K. 1987*a*. History of Valles Marineris. *Lunar Planet. Sci.* XVIII:572–573 (abstract).

Lucchitta, B. K. 1987*b*. Recent mafic volcanism on Mars: *Science* 235:565–567.

Lucchitta, B. K. 1987*c*. Valles Marineris, Mars: Wet debris flows and ground ice. *Icarus* 72:411–429.

Lucchitta, B. K. 1989. Young volcanic deposits in the Valles Marineris, Mars. *Icarus* 86:476–509.

Lucchitta, B. K., and Anderson, D. M. 1980. Martian outflow channels sculptured by glaciers. In *Reports of Planetary Geology Program—1980,* NASA TM-81776, pp. 271–273 (abstract).

Lucchitta, B. K., and Bertolini, L. M. 1989. Interior structures of Valles Marineris, Mars. *Lunar Planet. Sci.* XX:590–591 (abstract).

Lucchitta, B. K., and Ferguson, H. M. 1983. Chryse Basin channels: Low gradients and ponded flows. *Proc. Lunar Planet. Sci. Conf.* 13, *J. Geophys. Res. Suppl.* 88:A553-A568.

Lucchitta, B. K., and Klockenbrink, J. L. 1981. Ridges and scarps in the equatorial belt of Mars. *Moon and Planets* 24:415–429.

Lucchitta, B. K., and Persky, P. H. 1982. Ground ice and debris flows in the fretted terrain, Mars. In *Reports of Planetary Geology Program-1982,* NASA TM-85127, pp. 268–270 (abstract).

Lucchitta, B. K., Anderson, D. M., and Shoji, H. 1981. Did ice streams carve Martian outflow channels? *Nature* 290:759–763.

Lucchitta, B. K., Ferguson, H. M., and Summers, C. 1986. Sedimentary deposits in the northern lowland plains, Mars. *Proc. Lunar Planet. Sci. Conf.* 17, *J. Geophys. Res. Suppl.* 91:E166-E174.

Lucchitta, B. K., Balser, R. A., and Bertolini, L. M. 1990. Valles Marineris, Mars: Are pit chains formed by erosion and troughs by tectonism? *Lunar Planet. Sci.* XXI:722–723 (abstract).

Lucey, P. G., Bell, J. F., and Piscitelli, J. R. 1989. High spectral resolution spectroscopy of the Martian moons. *Lunar Planet. Sci.* XX:598–599 (abstract).

Luhmann, J. G. 1986. The solar wind interaction with Venus. *Space Sci. Rev.* 44:241–306.

Luhmann, J. G. 1990. A model of the ion wake of Mars. *Geophys. Res. Lett.* 17:869–872.

Luhmann, J. G., and Kozyra, J. U. 1991. Dayside pickup oxygen ion precipitation at Venus and Mars: Spatial distributions, energy deposition and consequences. *J. Geophys. Res.* 96:5457–5468.

Luhmann, J. G., and Schwingenschuh, K. A. 1990. A model of the energetic ion environment of Mars. *J. Geophys. Res.* 95:939–946.

Luhmann, J. G., Tatrallyay, M., Russell, C. T., and Winterhalter, D. 1983. Magnetic field fluctuations in the Venus magnetosheath. *Geophys. Res. Lett.* 10:655–658.

Luhmann, J. G., Warniers, R. J., Russell, C. T., Spreiter, J. R., and Stahara, S. S. 1986*a*. A gas dynamic magnetosheath field model for unsteady interplanetary fields: Application to the solar wind interaction with Venus. *J. Geophys. Res.* 91:3001–3010.

Luhmann, J. G., Russell, C. T., and Elphic, R. C. 1986*b*. Spatial distribution of magnetic field fluctuations in the dayside magnetosheath. *J. Geophys. Res.* 91:1711–1715.

Luhmann, J. G., Russell, C. T., Scarf, F. L., Brace, L. H., and Knudsen, W. C. 1987. Characteristics of the Mars like limit of the Venus-solar wind interaction. *J. Geophys. Res.* 92:8455–8557.

Luhmann, J. G., Kliore, A., Barnes, A., and Brace, L. 1990. Remote sensing of Mars' ionosphere and solar wind interaction: Lessons from Venus. *Adv. Space Res.* 10:43–48.

Lumme, K., and James, P. B. 1984. Some photometric properties of the Martian south polar cap region during the 1971 apparition. *Icarus* 58:363–376.

Lundin, R., Zakharov, A., Pellinen, R., Borg, H., Hultqvist, B., Pissarenko, N., Dubinin, E. M., Barabash, S. W., Liede, I., and Koskinen, H. 1989. First measurements of the ionosphere plasma escape from Mars. *Nature* 341:609–612.

Lundin, R., Zakharov, A., Pellinen, R., Barabasj, S. W., Borg, H., Dubinin, E. M., Hultqvist, B., Koskinen, H., Liede, I., and Pissarenko, N. 1990*a*. Aspera/Phobos measurements of the ion outflow from the Martian ionosphere. *Geophys. Res. Lett.* 17:873–876.

Lundin, R., Zakharov, A., Pellinen, R., Borg, H., Hultqvist, B., Pissarenko, N., Dubinin, E. M., Barabasj, S. W., Liede, I., and Koskinen, H. 1990*b*. Plasma composition measurements of the Martian magnetosphere morphology. *Geophys. Res. Lett.* 17:877–880.

Lunine, J. I., Neugebauer, G., and Jakosky, B. M. 1982. Infrared observations of Phobos and Deimos from Viking. *J. Geophys. Res.* 87:10297–10305.

Lupishko, D. F., and Lupishko, T. A. 1977. *Soviet Astron. Lett.* 3(6):1056–1061.

Luther, G. G., and Towler, W. P. 1982. A redetermination of the Newtonian gravitational constant. *Phys. Rev. Lett.* 48:121–123.

Lyot, B. 1929. Recherches sur la polarisation de la lumière des planètes et de quelques substances terrestres. *Ann. Obs. Paris/Meudon,* vol. 8, no. 1. Trans.: NASA TT-F-187 (1964).

Lyot, B. 1953. L'Aspect des planètes au Pic du Midi: Dans une lunette de 60 cm d'ouverture. *Astronomie* 67:3–21.

Lyot, B., Camichel, H., and Gentili, G. 1943. Observations planétaires au Pic du Midi, en 1941. *Astronomie* 77:67–72.

Ma, M. S., Laul, J. C., and Schmitt, R. A. 1981. Complementary rare Earth element patterns in unique achondrites, such as ALHA 77005 and shergottites, and in the Earth. *Proc. Lunar Planet. Sci. Conf.* 12:1349–1358.

Macdonald, G. A., Abbott, A. T., and Peterson, F. L. 1983. *Volcanoes in the Sea* (Honolulu: Univ. of Hawaii Press).

MacDonald, T. L. 1971. The origins of Martian nomenclature. *Icarus* 15:232–240.

MacKinnon, D. J., and Tanaka, K. L. 1989. The impacted Martian crust: Structure, hydrology, and some geologic implications. *J. Geophys. Res.* 94:17359–17370.

MacRobert, T. M. 1967. *Spherical Harmonics* (New York: Pergamon).

Madsen, F. T., and Muller-Vonmoos, M. 1985. Swelling pressure calculated from mineralogical properties of a Jurassic opaline shale, Switzerland. *Clays Clay Minerals* 33:501–509.

Magalhäes, J. A. 1987. The Martian Hadley circulation: Comparison of "viscous" model predictions to observations. *Icarus* 70:442–468.

Magalhäes, M. J., and Gierasch, P. 1982. A model of Martian slope winds: Implications for eolian transport. *J. Geophys. Res.* 87:9975–9984.

Maggini, M. 1939. *Il Planeta Marte* (Milan: Ulrico Hoepli).

Maguire, W. C. 1977. Martian isotopic ratios and upper limits for possible minor constituents as derived from Mariner 9 infrared spectrometer data. *Icarus* 32:85–97.

Mahajan, K. K., and Kar, J. 1988. Planetary ionospheres. *Space Sci. Rev.* 47:303–397.

Maher, K. A., and Stevenson, D. J. 1988. Impact frustration of the origin of life. *Nature* 331:612–614.

Malin, N. C. 1974. Salt weathering on Mars. *J. Geophys. Res.* 79:3888–3894.

Malin, M. C. 1976*a*. Comparison of large craters and multiringed basin populations on Mars, Mercury, and the Moon. *Proc. Lunar Sci. Conf.* 7:3589–3602.

Malin, M. C. 1976*b*. Nature and Origin of the Intercrater Plains on Mars. Ph.D. Thesis, California Inst. of Tech.

Malin, M. C. 1977. Comparison of volcanic features of Elysium (Mars) and Tibesti (Earth). *Geol. Soc. Amer. Bull.* 88:908–919.

Malin, M. C. 1980. Lengths of Hawaiian lava flows. *Geology* 8:306–308.

Malin, M. C. 1986. Density of Martian north polar layered deposits: Implications for composition. *Geophys. Res. Lett.* 13:444–447.

Mancinelli, R. L., and McKay, C. P. 1988. The evolution of nitrogen cycling. *Origins of Life* 18:311–325.

Manker, J. P., and Johnson, A. P. 1982. Simulation of Martian chaotic terrain and outflow channels. *Icarus* 51:121–132.

Mao, H. K. 1981. High pressure experiments on FeS with bearing on the Earth's core. *Carnegie Inst. Washington Yearbook* 80:267–272.

Marachi, N. D., Chan, C. K., and Seed, H. B. 1972. Evaluation of properties of rockfill materials. *J. Soil Mech.* 98:95–114.

Marangunic, C., and Bull, C. 1968. The landslide on the Sherman Glacier. In *The Great Alaska Earthquake of 1964: Hydrology,* Natl. Acad. of Sci. Publ. 1603, pp. 383–394.

Mariner Mars 1971, Project Final Rept. JPL Tech. Rept. 32-1550, vol. 5, Aug. 20, 1973.

Markov, Y. 1989. Kurz Na Mars. *Mashinostroyeniye.* Quoted in a letter by Bart Hendricks in *Spaceflight* (1991) 33:102–103.

Mars Channel Working Group. 1983. Channels and valleys on Mars. *Geol. Soc. Amer. Bull.* 94:1035–1054.

Marshall, J. K. 1971. Drag measurements in roughness arrays of varying density and distribution. *Agric. Meteorol.* 8:269–292.

Martin, L. J. 1974*a*. The major Martian yellow storm of 1971. *Icarus* 22:175–188.

Martin, L. J. 1974*b*. The major Martian dust storms of 1971 and 1973. *Icarus* 23:108–115.

Martin, L. J. 1975. North polar hood observations during Martian dust storms. *Icarus* 6:341–352.

Martin, L. J. 1976. 1973 dust storm on Mars: Maps from hourly photographs. *Icarus* 29:363–380.

Martin, L. J. 1983. Mars from Earth and space. *Sky & Teles.* 65:559.

Martin, L. J. 1984. Clearing the Martian air: The troubled history of dust storms. *Icarus* 57:317–321.

Martin, L. J., and Baum, W. A. 1969. A Study of Cloud Motions of Mars. Final Report B under JPL Contract 951547 (Flagstaff: Planetary Research Center, Lowell Obs.).

Martin, L. J., and Baum, W. A. 1978. Two look-alike Martian dust storms: Inference from 1977 (Viking) and 1973 (Earth-based) observations. *Bull. Amer. Astron. Soc.* 10:551 (abstract).

Martin, L. J., and James, P. B. 1985. Mars' "orographic" clouds: Where's the "W"? *Bull. Amer. Astron. Soc.* 17:733 (abstract).

Martin, L. J., and McKinney, W. M. 1974. North polar hood of Mars in 1969 (May 18–July 25). I. Blue light. *Icarus* 23:380–387.

Martin, L. J., and Zurek, R. W. 1992. An analysis of the history of dust activity on Mars. *J. Geophys. Res.*, in press.

Martin, L. J., James, P. B., Parker, D., and Beish, J. 1989. Telescopic observations of Mars in 1988. Fourth Intl. Conf. on Mars, Jan. 10–13, Tucson, Arizona, Abstract Book, p. 137.

Martin, T. Z. 1981. Mean thermal and albedo behavior of the Mars surface and atmosphere over a Martian year. *Icarus* 45:427–446.

Martin, T. Z. 1986. Thermal infrared opacity of the Mars atmosphere. *Icarus* 66:2–21.

Martin, T. Z. 1989. Earthbased monitoring of the Martian atmosphere dust opacity. In *MECA Workshop on Dust on Mars III,* ed. S. Lee, LPI Tech. Rept. 89-01, pp. 45–47 (abstract).

Martin, T. Z. 1992. New dust opacity mapping from Viking IR Thermal Mapper data. *J. Geophys. Res.,* in press.

Martin, T. Z., and Kieffer, H. 1979. Thermal infrared properties of the Martian atmosphere. 2. The 15 μm band measurements. *J. Geophys. Res.* 84:2843–2852.

Mason, B. 1962. *Meteorites* (New York: Wiley).

Mass, C., and Sagan, C. 1976. A numerical circulation model with topography for the Martian summer hemisphere. *J. Atmos. Sci.* 33:1418–1430.

Masson, P. 1977. Structure pattern analysis of the Noctis Labyrinthus-Valles Marineris regions of Mars. *Icarus* 30:49–62.

Masson, P. 1980. Contribution to the structural interpretation of the Valles Marineris-Noctis Labyrinthus-Claritas Fossae regions of Mars. *Moon and Planets* 22:211–219.

Masson, P. 1985. Origin and evolution of the Valles Marineris region of Mars. *Adv. Space Res.* 5:83–92.

Masursky, H. 1973. An overview of geologic results from Mariner 9. *J. Geophys. Res.* 78:4009–4030.

Masursky, H., and Crabill, N. L. 1976*a*. The Viking landing sites: Selection and certification. *Science* 193:809–812.
Masursky, H., and Crabill, N. L. 1976*b*. Search for the Viking 2 landing site. *Science* 194:62–68.
Masursky, H., Batson, R. M., Borgeson, W., Carr, M., McCauley, J., Milton, D., Wildey, R., and Wilhelms, D. 1970. Television experiment for Mariner Mars 1971. *Icarus* 12:10–45.
Masursky, H., Boyce, J. V., Dial, A. L., Jr., Schaber, G. G., and Strobell, M. E. 1977. Classification and time of formation of Martian channels based on Viking data. *J. Geophys. Res.* 82:4016–4037.
Masursky, H., Dial, A. L., Jr., and Strobell, M. E. 1978. *Geologic Map of the Phoenicis Lacus Quadrangle of Mars*. U.S. Geol. Surv. Misc. Inv. Map I-896.
Masursky, H., Chapman, M. G., Dial, A. L., Jr., and Strobell, M. E. 1986*a*. Ages of rocks and channels in prospective Martian landing sites of Mangala Valles region. *Lunar Planet. Sci.* XVII:520–521 (abstract).
Masursky, H., Chapman, M. G., Dial, A. L., Jr., and Strobell, M. E. 1986*b*. Episodic channeling punctuated by volcanic flows in Mangala Valles region, Mars. In NASA TM-88383, pp. 459–461 (abstract).
Masursky, H., Aksnes, K., Hunt, G. E., Marov, M. Ya., Millman, P. M., Morrison, D., Owen, T. C., Shevchenko, V. V., Smith, B. A., and Tejfel, V. G. 1986*c*. *Annual Gazetteer of Planetary Nomenclature*. U.S.G.S. Open-File Rept. 84-692.
Masursky, H., Chapman, M. G., Davis, P. A., Dial, A. L., and Strobell, M. E. 1987. Mars Kabder/Rover/Returned Sample sites. *Lunar Planet. Sci.* XVIII:600–601 (abstract).
Matassov, G. 1977. *Lawrence Livermore Natl. Lab. Publ. U.C.R.L.-52322.*
Matson, D. L., and Brown, R. H. 1989. Solid-state greenhouses and their implications for icy satellites. *Icarus* 77:67–81.
Maxwell, J. C. 1964. Influence of depth, temperature, and geologic age on the porosity of quartzose sandstone. *Bull. Amer. Assoc. Petrol. Geol.* 48:697–709.
Maxwell, T. A. 1982. Orientation and origin of ridges in the Lunae Palus-Coprates region of Mars. *Proc. Lunar Planet. Sci. Conf.* 13, *J. Geophys. Res. Suppl.* 87:A97-A108.
Maxwell, T. A., and McGill, G. E. 1988. Ages of fracturing and resurfacing in the Amenthes region, Mars. *Proc. Lunar Planet. Sci. Conf.* 18:701–711.
Maxwell, T. A., El-Baz, F., and Ward, S. H. 1975. Distribution, morphology and origin of ridges and arches in Mare Serenitatis. *Geol. Soc. Amer. Bull.* 86:1273–1278.
Mayo, A. P., Blackshear, W. T., Tolson, R. H., Michael, W. H., Jr., Kelly, G. M., Brenkle, J. P., and Komarek, T. A. 1977. Lander locations, Mars physical ephemeris, and solar system parameters: Determination from Viking Lander tracking data. *J. Geophys. Res.* 82:4297–4303.
Mazur, P. 1980. Limits to life at low temperatures and at reduced water contents and water activities. *Origins of Life* 10:137–159.
Mazur, P., Barghoorn, E. S., Halvorson, H. O., Jukes, T. H., Kaplan, I. R., and Margulis, L. 1978. Biological implications of the Viking mission to Mars. *Space Sci. Rev.* 22:3–34.
McAdoo, D. C., and Burns, J. A. 1975. The Coprates trough assemblage: More evidence for Martian polar wander. *Earth Planet. Sci. Lett.* 25:347–354.
McAdoo, D. C., and Sandwell, D. T. 1985. Folding of oceanic lithosphere. *J. Geophys. Res.* 90:8563–8569.
McCauley, J. F. 1973. Mariner 9 evidence for wind erosion in the Equatorial and mid-latitude regions of Mars. *J. Geophys. Res.* 78:4123–4137.
McCauley, J. F. 1978. *Geologic Map of the Coprates Quadrangle of Mars, scale 1:5,000,000*. U.S. Geol. Surv. Misc. Inv. Series Map I-897.
McCauley, J. F., Carr, M. H., Cutts, J. A., Hartmann, W. K., Masursky, H., Milton, D. J., Sharp, R. P., and Wilhelms, D. E. 1972. Preliminary Mariner 9 report on the geology of Mars. *Icarus* 17:289–327.
McCauley, J. F., Breed, C. S., El-Baz, F., Whitney, M. I., Grolier, M. J., and Ward, A. W. 1979. Pitted and fluted rocks in the Western Desert of Egypt: Viking comparisons. *J. Geophys. Res.* 84:8222–8232.
McCord, T. B., and Adams, J. B. 1969. Spectral reflectivity of Mars. *Science* 163:1058–1060.
McCord, T. B., and Westphal, J. A. 1971. Mars: Narrowband photometry, from 0.3 to 2.5 microns, of surface regions during the 1969 opposition. *Astrophys. J.* 168:141–153.

McCord, T. B., Elias, J. H., and Westphal, J. A. 1971. Mars: The spectral albedo (0.3–2.5μm) of small bright and dark regions. *Icarus* 41:245–251.

McCord, T. B., Huguenin, R. L., Mink, D., and Pieters, C. 1977. Spectral reflectance of Martian areas during the 1973 opposition. Photoelectric filter photometry 0.33–1.10 μm. *Icarus* 31:25–39.

McCord, T. B., Clark, R., and Huguenin, R. L. 1978. Mars: Near-infrared reflectance and spectra of surface regions and compositional implications. *J. Geophys. Res.* 87:3021–3032.

McCord, T. B., Clark, R. N., and Singer, R. B. 1982*a*. Mars: Near-infrared reflectance spectra of surface regions and compositional implications. *J. Geophys. Res.* 78:3021–3032.

McCord, T. B., Singer, R. B., Hawke, B. R., Adams, J. B., Evans, D. L., Head, J. W., Mouginis-Mark, P. J., Pieters, C. M., Huguenin, R. L., and Zisk, S. H. 1982*b*. Mars: Definition and characterization of global surface units with emphasis on composition. *J. Geophys. Res.* 87:10129–10148.

McElroy, M. B. 1972. Mars: An evolving atmosphere. *Science* 175:443–445.

McElroy, M. B., and Donahue, T. M. 1972. Stability of the Martian atmosphere. *Science* 177:986–988.

McElroy, M. B., and McConnell, J. C. 1971. Dissociation of CO_2 in the Martian atmosphere. *J. Atmos. Sci.* 28:879–884.

McElroy, M. B., and Prather, M. J. 1981. Noble gases in the terrestrial planets. *Nature* 293:535–539.

McElroy, M. B., and Yung, Y. L. 1976. Oxygen isotopes in the Martian atmosphere: Implications for the evolution of volatiles. *Planet. Space Sci.* 14:1107–1113.

McElroy, M. B., Kong, T. Y., Yung, Y. L., and Nier, A. O. 1976*a*. Composition and structure of the Martian upper atmosphere: Analysis of results from Viking. *Science* 194:1295–1298.

McElroy, M. B., Yung, Y. L., and Nier, A. O. 1976*b*. Isotopic composition of nitrogen: Implications for the past history of Mars' atmosphere. *Science* 194:70–72.

McElroy, M. B., Kong, T. Y., and Yung, Y. L. 1977. Photochemistry and evolution of Mars' atmosphere: A Viking perspective. *J. Geophys. Res.* 82:4379–4388.

McEwen, A. S. 1985. Topography and albedo of Ius Chasma, Mars. *Lunar Planet. Sci.* XVI:528–529 (abstract).

McEwen, A. S. 1989. Mobility of large rock avalanches: Evidence from Valles Marineris, Mars. *Geology* 17:1111–1114.

McEwen, A. S., and Soderblom, L. A. 1989. Mars color/albedo and bedrock geology. Fourth Intl. Conf. on Mars, Jan. 10–13, Tucson, Arizona, Abstract Book, pp. 138–139.

McGaw, R. W., and Tice, A. R. 1976. A simple procedure to calculate the volume of water remaining unfrozen in a freezing soil. In *Proc. Second Conf. on Soil Water Problems in Cold Regions,* Edmonton, Alb., 1–2 Sept., pp.114–122.

McGetchin, T. R., and Smyth, J. R. 1978. The mantle of Mars: Some possible geological implications of its high density. *Icarus* 34:512–536.

McGill, G. E. 1971. Attitude of fractures bounding straight and arcuate lunar rilles. *Icarus* 14:53–58.

McGill, G. E. 1985. Age and origin of large Martian polygons. *Lunar Planet. Sci.* XVI:534–535 (abstract).

McGill, G. E. 1986. The giant polygons of Utopia, northern Martian plains. *Geophys. Res. Lett.* 13:705–708.

McGill, G. E. 1988*a*. Evidence for a very large basin beneath Utopia Planitia, Mars. *Lunar Planet. Sci.* XIX:752–753 (abstract).

McGill, G. E. 1988*b*. The Martian crustal dichotomy. In *MEVTV-LPI Workshop: Early Tectonic and Volcanic Evolution of Mars,* pp. 42–44 (abstract).

McGill, G. E. 1989*a*. Buried topography of Utopia, Mars: Persistence of a giant impact depression. *J. Geophys. Res.* 94:2753–2759.

McGill, G. E. 1989*b*. Geologic evidence supporting an endogenic origin for the Martian crustal dichotomy. In *Lunar Planet. Sci.* XX:667–668 (abstract).

McGill, G. E., and Dimitriou, A. M. 1990. Origin of the Martian global dichotomy by crustal thinning in the late Noachian or early Hesperian. *J. Geophys. Res.* 95:12595–12605.

McGill, G. E., and Squyres, S. W. 1991. Origin of the martian crustal dichotomy: Evaluating hypotheses. *Icarus* 93:386–393.

McGill, G. E., and Stromquist, A. W. 1979. The grabens of Canyonlands National Park, Utah: Geometry, mechanics and kinematics. *J. Geophys. Res.* 84:4547–4563.
McGovern, P. J., and Schubert, G. 1989. Thermal evolution of the Earth: Effects of volatile exchange between atmosphere and interior. *Earth Planet. Sci. Lett.* 96:27–37.
McGovern, P. J., and Solomon, S. C. 1990. State of stress and eruption characteristics of Martian volcanoes. *Lunar Planet. Sci.* XXI:765–766 (abstract).
McKay, C. P. 1986. Exobiology and future Mars missions: The search for Mars' earliest biosphere. *Adv. Space Res.* 6(12)269–285.
McKay, C. P., and Davis, W. L. 1991. Duration of liquid water habitats on early Mars. *Icarus* 90:214–221.
McKay, C. P., and Nedell, S. S. 1988. Are there carbonate deposits in Valles Marineris, Mars? *Icarus* 73:142–148.
McKay, C. P., and Stoker, C. R. 1989. The early environment and its evolution on Mars: Implications for life. *Rev. Geophys. Space Phys.* 27:189–214.
McKay, C. P., Clow, G. D., Wharton, R. A., Jr., and Squyres, S. W. 1985. Thickness of ice on perenially frozen lakes. *Nature* 313:561–562.
McKim, R. 1986. Observing Mars in the 1980s: Reference maps for the visual observer. *J. Brit. Astron. Soc.* 96:166–169.
McLaughlin, D. 1954. Volcanism and aeolian deposition on Mars. *Bull. Geol. Soc. Amer.* 65:715–717.
McNutt, M. K. 1984. Lithospheric flexure and thermal anomalies. *J. Geophys. Res.* 89:11180–11194.
McNutt, M. K. 1987. Temperature beneath midplate swells: The inverse problem. In *Seamounts, Islands, and Atolls*, eds. B. H. Keating, P. Fryer, R. Batiza and G. W. Boehlert, Geophysical Mono. 43 (Washington, D.C.: Amer. Geophysical Union), pp. 123–132.
McSween, H. Y., Jr. 1984. SNC meteorites: Are they Martian rocks? *Geology* 12:3–6.
McSween, H. Y., Jr. 1985. SNC meteorites: Clues to Martian petrologic evolution? *Rev. Geophys.* 23:391–416.
McSween, H. Y., Jr., and Jarosewich, E. 1983. Petrogenesis of the Elephant Moraine A79001 meteorite: Multiple magma pulses on the shregottite parent body. *Geochim. Cosmochim. Acta* 47:1501–1513.
McSween, H. Y., Jr., and Stolper, E. M. 1980. Basaltic meteorites. *Sci. Amer.* 242:54–63.
McSween, H. Y., Jr., Taylor, L. A., and Stolper, E. 1979. Allan Hills 77005: A new meteorite type found in Antarctica. *Science* 204:1201–1203.
Mehegan, J. M., Robinson, J. M., and Delaney, J. R. 1982. Secondary mineralization and hydrothermal alteration in the Reydarfjordur Drill Core, East Iceland. *J. Geophys. Res.* 87:6511–6524.
Mellor, G. L., and Yamada, T. 1982. Development of a turbulence closure model for geophysical fluid problems. *Rev. Geophys. Space Phys.* 200:851–875.
Melosh, H. J. 1977. Global tectonics of a despun planet. *Icarus* 31:221–242.
Melosh, H. J. 1980*a*. Cratering mechanics—Observational, experimental, and theoretical. *Ann. Rev. Earth Planet. Sci.* 8:65–93.
Melosh, H. J. 1980*b*. Tectonic patterns on a reoriented planet: Mars. *Icarus* 44:745–751.
Melosh, H. J. 1989. *Impact Cratering. A Geologic Process* (New York: Oxford Univ. Press).
Melosh, H. J., and Dzurisin, D. 1978. Mercurian global tectonics: A consequence of tidal despinning? *Icarus* 35:227–236.
Melosh, H. J., and Vickery, A. M. 1989. Impact erosion of the primordial Martian atmosphere. *Nature* 338:487–489.
Meyers, C. H., and Van Dusen, M. S. 1933. *J. Res. Natl. Bur. Stndrds.* 84:2843–2852.
Michael, F. C. 1971. Solar-wind-induced mass loss from magnetic field-free planets. *Planet. Space Sci.* 19:1580–1583.
Michael, W. H., Jr. 1979. Viking Lander tracking contributions to Mars mapping. *Moon and Planets* 20:149–152.
Michaux, C. M. 1967. *Handbook of the Physical Properties of the Planet Mars* (Washington, D.C.: U.S. Government Printing Office).
Michaux, C. M., and Newburn, R. L., Jr. 1972. Mars Scientific Model. JPL Doc. No. 606-1.
Michelangeli, D. V., Zurek, R. W., and Elson, L. S. 1987. Barotropic instability of midlatitude zonal jets on Mars, Earth and Venus. *J. Atmos. Sci.* 44:2031–2041.

Michelangeli, D. V., Toon, O. B., and Haberle, R. M. 1989. A model of water-ice cloud formations on Mars. *Bull. Amer. Astron. Soc.* 21:979–980.
Michels, J. W., Tsong, I. S. T., and Nelson, C. M. 1983. Obsidian dating and East African archeology. *Science* 219:361–366.
Mignard, F. 1981. Evolution of the Martian satellites. *Mon. Not. Roy. Soc.* 194:365–379.
Mihalov, J. D., and Barnes, A. 1981. Evidence for the acceleration of ionospheric O^+ in the magnetosheath of Venus. *Geophys. Res. Lett.* 8:1277–1280.
Mihalov, J. D., and Barnes, A. 1982. The distant interplanetary wake of Venus: Plasma observations from Pioneer Venus. *J. Geophys. Res.* 87:9045–9053.
Mihalov, J. D., Spreiter, J. R., and Stahara, S. S. 1982. Comparison of gas dynamic model with steady state solar wind flow around Venus. *J. Geophys. Res.* 87:10363–10372.
Milanovsky, E. E., and Nikishin, A. M. 1984. Differences in the formation mechanism of rifts of the Earth and of rifts in the Valles Marineris system of Mars. *Lunar Planet. Sci.* XV:548–549 (abstract).
Miller, S. L. 1953. A production of amino acids under possible primitive Earth conditions. *Science* 117:528–529.
Miller, S. L. 1982. Prebiotic synthesis of organic compounds. In *Mineral Deposits and the Evolution of the Biosphere,* eds. H. D. Holland and M. Schidlowski (New York: Springer-Verlag), pp. 155–176.
Miller, S. L., and Orgel, L. E. 1974. *The Origins of Life on the Earth* (Englewood Cliffs, N.J.: Prentice Hall).
Miller, S. L., and Smythe, W. D. 1970. Carbon dioxide clathrate in the Martian ice cap. *Science* 170:531–535.
Milton, D. J. 1973. Water and processes of degradation in the Martian landscape. *J. Geophys. Res.* 78:4037–4047.
Milton, D. J. 1974*a*. Carbon dioxide hydrate and floods on Mars. *Science* 183:654–656.
Milton, D. J. 1974*b*. *Geologic Map of the Lunae Quadrangle of Mars, scale 1:5,000,000.* U.S.G.S. Misc. Inv. Series Map I-894.
Mintz, Y. 1961. The general circulation of planetary atmospheres. In *The Atmospheres of Mars and Venus,* eds. W. W. Kellogg and C. Sagan, NAS-NRC Publ. 944, pp. 107–146.
Mitzuni, H., and Osako, M. 1974. Elastic wave velocities and thermal diffusivities of Apollo 17 rocks and their geophysical implications. *Proc. Lunar Sci. Conf.* 5:2891–2901.
Miyamoto, S. 1957. The great yellow cloud and the atmosphere of Mars. Report of visual observations during the 1956 opposition. *Contrib. Inst. Astrophys. Kwasan Obs.* No. 71.
Miyamoto, S. 1963. Meteorological observations of Mars during the 1962–63 opposition. *Contrib. Inst. Astrophys. Kwasan Obs.* No. 124.
Miyamoto, S. 1965. Meteorological observations of Mars during the 1965 opposition. *Contrib. Inst. Astrophys. Kwasan Obs.* No. 141.
Miyamoto, S. 1966. Martian atmosphere and crust. *Icarus* 5:360–374.
Miyamoto, S. 1968. Meteorological observations of Mars during the 1967 opposition. *Contrib. Inst. Astrophys. Kwasan Obs.* No. 169.
Miyamoto, S. 1970. Meteorological observations of Mars during the 1969 opposition. *Contrib. Inst. Astrophys. Kwasan Obs.* No. 184.
Miyamoto, S. 1972. Meteorological observations of Mars during the 1971 opposition. *Contrib. Kwasan Hida Obs.* No. 206.
Miyamoto, S. 1974. Meteorological observations of Mars during the 1973 opposition. *Contrib. Kwasan Hida Obs.* No. 217.
Miyamoto, S., and Matsui, M. 1960. Report of Mars observations during the 1958 opposition. *Contrib. Inst. Asltrophys. Kwasan Obs.* No. 87.
Miyamoto, S., and Nakai, Y. 1961. Meteorological observations of Mars during the 1960–61 opposition. *Contrib. Inst. Astrophys. Kwasan Obs.* No. 105.
Mooney, R. W., Keenan, A. G., and Wood, L. A. 1952*a*. Adsorption of water by montmorillonite. I. Heat of desorption and application of BET theory. *J. Amer. Chem. Soc.* 74:1367–1371.
Mooney, R. W., Keenan, A. G., and Wood, L. A. 1952*b*. Adsorption of water by montmorillonite. II. Effect of exchangeable ions and lattice swelling as measured by X-ray diffraction. *J. Amer. Chem. Soc.* 74:1371–1374.

Moore, H. J. 1980. *Geologic Map of the Sinus Sabaeus Quadrangle of Mars, scale 1:5,000,000*. U.S.G.S. Misc. Inv. Series Map I-1196 (MC-20)
Moore, H. J. 1985. The Martian dust storm of sol 1742. *Proc. Lunar Planet. Sci. Conf.* 16, *J. Geophys. Res. Suppl.* 90:D163-D174.
Moore, H. J. 1986. Miniature slope failures, Mutch Memorial Station. In *MECA Workshop on Dust on Mars II*, ed. S. Lee, LPI Tech. Rept. 86-09, pp. 53–55 (abstract).
Moore, H. J., and Jakosky, B. M. 1989. Viking landing sites, remote-sensing observations, and physical properties of Martian surface materials. *Icarus* 81:164–184.
Moore, H. J., Hutton, R. E., Scott, R. F., Spitzer, C. R., and Shorthill, R. W. 1977. Surface materials of the Viking landing sites. *J. Geophys. Res.* 82:4497–4523.
Moore, H. J., Liebes, S., Jr., Crouch, D. S., and Clark, L. V. 1978*a*. *Rock Pushing and Sampling Under Rocks on Mars*. U.S.G.S. Prof. Paper 1081.
Moore, H. J., Arthur, D. W. G., and Schaber, G. G. 1978*b*. Yield strengths of flows on Earth, Mars and Moon. *Proc. Lunar Planet. Sci. Conf.* 9:3351–3378.
Moore, H. J., Spitzer, C. R., Bradford, K. Z., Cates, P. M., Hutton, R. E., and Shorthill, R. W. 1979. Sample fields of the Viking landers, physical properties, and aeolian processes. *J. Geophys. Res.* 84:8365–8377.
Moore, H. J., Clow, G. D., and Hutton, R. E. 1982. A summary of Viking sample trench analyses for angles of internal friction and cohesions. *J. Geophys. Res.* 87:10043–10050.
Moore, H. J., Hutton, R. M., Clow, G. D., and Spitzer, C. R. 1987. *Physical Properties of the Surface Materials at the Viking Landing Sites on Mars*. U.S.G.S. Prof. Paper 1389.
Moore, J. G. 1966. Rate of palagonitization of submarine basalt adjacent to Hawaii. In U.S.G.S. Prof. Paper 550-D, pp. 163–171.
Moore, P. 1977. *Guide to Mars* (New York: W. W. Norton).
Morgan, J. W., and Anders, E. 1979. Chemical composition of Mars. *Geochim. Cosmochim. Acta* 43:1601–1610.
Moriyama, S. 1974. Effects of dust on radiation transfer in the Martian atmosphere. I. On infrared cooling. *J. Meteor. Soc. Japan* 52:457–462.
Moriyama, S. 1975. Effects of dust on radiation transfer in the Martian atmosphere. II. Heating due to absorption of the visible solar radiation and importance of radiative effects of dust on the Martian meteorological phenomena. *J. Meteor. Soc. Japan* 53:214–220.
Moriyama, S. 1976. Effects of dust on radiation transfer in the Martian atmosphere. III. Numerical experiments of radiative-convective equilibrium of the Martian atmosphere including the radiative effects due to dust. *J. Meteor. Soc. Japan* 54:52–57.
Moriyama, S., and Iwashima, T. 1980. A spectral model of the atmospheric general circulation of Mars: A numerical experiment including the effects of suspended dust and the topography. *J. Geophys. Res.* 85:2847–2860.
Moroz, V. I. 1964. The infrared spectrum of Mars (1.1–4.1 μm). *Soviet Astron.* 8:273–281.
Moroz, V. I. 1975. Preliminary results of research conducted aboard the Soviet Mars-4, Mars-5, Mars-6 and Mars-7 planetary probes. *Kosmich. Issled.* 13:3–8. Trans.: NASA TT-F-16,333. Microfiche N75–24615.
Moroz, V. I. 1976. Argon in the Martian atmosphere: Do the results of Mars 6 agree with the optical and radio occultation measurement? *Icarus* 28:159–164.
Moroz, V. I., and Ksanfomality, L. V. 1972. Preliminary results of astrophysical observations of Mars from Mars-3. *Icarus* 17:408–422.
Moroz, V. I., and Muhkin, L. M. 1977. Early evolutionary stages in the atmosphere and climate of the terrestrial planets. *Cosmic Res.* 15:769–791.
Moroz, V. I., and Nadzhip, A. E. 1975*a*. Mars-3: Water vapor in the atmosphere of the planet. *Kosmich. Issled.* 13:738–752. Trans.: NASA TT-F-16,337. Microfiche N75–24613.
Moroz, V. I., and Nadzhip, A. E. 1975*b*. Measurements of water vapor densities on Mars 5 orbiter: Preliminary results. *Cosmic Res.* 13:28–30.
Moroz, V. I., and Nadzhip, A. E. 1975*c*. Preliminary results of measuring water vapor contention planetary atmosphere from measurements on board the Mars-5 automatic interplanetary station. *Kosmich. Issled.* 13:33–36. Trans.: NASA TT-F-16,337.
Moroz, V. I., Ksanfomality, L. V., and Nadzhip, A. E. 1973. Mars-3: Astrophysical study of the lower atmosphere and planet surface. *Uspekhi Fizich. Nauk* 111:360–370. Trans.: NASA TT-F-15,543. Microfiche N74–22467.

Moroz, V. I., Ksanfomality, L. V., and Krasovsky, G. N. 1975. Infrared temperature and thermal properties of Martian surface according to Mars-3 measurements. *Kosmich. Issled.* 13:389–404. In Russian.

Moroz, V. I., Ksanfomality, L. V., Krasovskii, G. N., Davydov, V. D., Parfent'ev, N. A., Zhenulev, V. S., and Filippov, G.S. 1976. Infrared temperature and thermal properties of the Martian surface measured by the Mars-3 orbiter. *Cosmic Res.* 13:346–358.

Morozhenko, A. V. 1966. *Publ. Acad. Sci. Ukraine* 45.

Morozhenko, A. V. 1973. Polarimetric observations of the giant planets. *Soviet Astron. J.* 17:105–107.

Morozhenko, A. V. 1975. *Astron. Astrofiz.* L 22, 26:97–107.

Morris, E. C. 1981*a*. The basal scarp of Olympus Mons. In *Reports of Planetary Geology Program,* NASA TM-84211, pp. 389–390 (abstract).

Morris, E. C. 1981*b*. Structure of Olympus Mons and its basal escarpment. In *Proc. Third Intl. Coll. on Mars,* LPI Contrib. 441, pp. 161–162 (abstract).

Morris, E. C. 1982. Aureole deposits of the Martian volcano Olympus Mons. *J. Geophys. Res.* 87:1164–1178.

Morris, E. C., and Dwornik, S. E. 1978. *Geologic Map of the Amazonis Quadrangle of Mars, scale 1:5,000,000.* U.S.G.S. Misc. Inv. Series Map I-1049.

Morris, E. C., and Howard, K. A. 1981. *A Geologic Map of the Diacria Quadrangle of Mars, scale 1:5,000,000.* U.S.G.S. Misc. Inv. Series Map I-1286.

Morris, E. C., and Jones, K. L. 1980. Viking 1 lander on the surface of Mars: Revised location. *Icarus* 44:217–222.

Morris, R. V. 1988. What ferric oxide/oxyhydroxide phases are present on Mars? In *Mars Sample Return Science Workshop,* LPI Tech. Rept. 88-07, pp. 126–127 (abstract).

Morris, R. V., and Lauer, H. V., Jr. 1980. The case against UV photostimulated oxidation of magnetite. *Geophys. Res. Lett.* 7:605–608.

Morris, R. V., and Lauer, H. V., Jr. 1981. Stability of geothite (α-FeOOH) and lepidocrocite (γ-FeOOH) to dehydration by UV radiation: Implications for their occurrence on the Martian surface. *J. Geophys. Res.* 85:10893–10899.

Morris, R. V., and Lauer, H. V., Jr. 1988. Effect of aluminum substitution on visible and near-IR optical properties of hematite with implications for Martian spectral data. *Lunar Planet. Sci.* XIX:811–812 (abstract).

Morris, R. V., and Lauer, H. V., Jr. 1990. Matrix effects for reflectivity spectra of dispersed nanophase (superparamagnetic) hematite with application to Martian spectral data. *J. Geophys. Res.* 95:5101–5109.

Morris, R. V., Lauer, H. V., Jr., Gooding, J. L., and Mendell, W. W. 1983. Spectral evidence and implications for the occurrence of aluminous iron oxides on Mars. *Lunar Planet. Sci.* XIV:526–527 (abstract).

Morris, R. V., Lauer, H. V., Jr., Lawson, C. A., Gibson, E. K., Jr., Nace, G. A., and Stewart, C. 1985. Spectral and other physiochemical properties of submicron powders of hematite $(\gamma - Fe_2O_3)$, maghemite $(\gamma - Fe_2O_3)$, magnetite (Fe_3O_4), goethite $(\gamma - FeOOH)$, and lepidocrocite $(- FeOOH)$. *J. Geophys. Res.* 90:3126–3144.

Morris, R. V., Agresti, D. G., Lauer, H. V., Jr., Newcomb, J. A., Shelfer, T. D., and Murali, A. V. 1989. Evidence for pigmentary hematite on Mars based on optical, magnetic, and Mossbauer studies of superparamagnetic (nanocrystalline) hematite. *J. Geophys. Res.* 94:2760–2778.

Morris, R. V., Gooding, J. L., Lauer, H. V., Jr., and Singer, R. B. 1990. Origins of the Mars-like spectral and magnetic properties of a Hawaiian palagonitic soil. *J. Geophys. Res.* 95:14427–14435.

Moses, S. L., Coroniti, F. V., and Scarf, F. L. 1988. Expectations for the microphysics of the Mars-solar wind interaction. *Geophys. Res. Lett.* 15:429–432.

Moskowitz, M. B., and Hargraves, R. B. 1982. Magnetic changes accompanying the thermal decomposition of nontronite (in air) and its relevance to Martian mineralogy. *J. Geophys. Res.* 87:10115–10128.

Mouginis-Mark, P. J. 1979. Martian fluidized crater morphology: Variations with crater size, latitude, altitude, and target material. *J. Geophys. Res.* 84:8011–8022.

Mouginis-Mark, P. J. 1981. Late-stage summit activity of Martian shield volcanoes. *Proc. Lunar Planet. Sci. Conf.* 12:1431–1447.

Mouginis-Mark, P. J. 1985. Volcano-ground ice interactions in Elysium Planitia. *Icarus* 64:265–284.
Mouginis-Mark, P. J. 1990. Recent melt water release in the Tharsis region of Mars. *Icarus* 84:362–373.
Mouginis-Mark, P. J., and Cloutis, E. A. 1983. Ejecta areas of impact craters on the Martian ridged plains. *Lunar Planet. Sci.* XIV:532–533 (abstract).
Mouginis-Mark, P. J., and Mathews, A. 1987. Geology of the Olympus Mons caldera, Mars. *Eos: Trans. AGU* 68:1342.
Mouginis-Mark, P. J., Sharpton, V. L., and Hawke, B. R. 1981. Schiaparelli Basin, Mars: Morphology, tectonics and infilling history. In *Multi-Ring Basins,* eds. P. H. Schultz and R. B. Merrill (New York: Pergamon), pp. 155–172.
Mouginis-Mark, P. J., Zisk, S. H., and Downs, G. S. 1982*a*. Ancient and modern slopes in the Tharsis region of Mars. *Nature* 297:546–550.
Mouginis-Mark, P. J., Wilson, L., and Head, J. W. 1982*b*. Explosive volcanism of Hecates Tholus, Mars: Investigation of eruption conditions. *J. Geophys. Res.* 87:9890–9904.
Mouginis-Mark, P. J., Wilson, L., Head, J. W., Brown, S. H., Hall, J. L., and Sullivan, K. D. 1984. Elysium Planitia, Mars: Regional geology, volcanology, and evidence for volcano ground ice interactions. *Earth, Moon and Planets* 30:149–173.
Mouginis-Mark, P. J., Wilson, L., and Zimbelman, J. R. 1988. Polygenic eruptions on Alba Patera, Mars: Evidence of channel erosion on pyroclastic flows. *Bull. Vol.* 50:361–379.
Mouginis-Mark, P. J., Robinson, M. S., and Zuber, M. T. 1990. Evolution of the Olympus Mons caldera, Mars. *Lunar Planet. Sci.* XXI:815–816 (abstract).
Muehlberger, W. R. 1961. Conjugate joint sets of small dihedral angle. *J. Geol.* 69:211–219.
Muehlberger, W. R. 1974. Structural history of southeastern Mare Serenitatis and adjacent highlands. *Proc. Lunar Sci. Conf.* 5:101–110.
Muhleman, D. O. 1972. Microwave emission from the Moon. In *Thermal Characteristics of the Moon,* ed. T. Lucas (Cambridge: MIT Press), pp. 51–81.
Muhleman, D. O., Butler, B., Grossman, A. W., Slade, M., and Jurgens, R. 1989. Very Large Array/Goldstone radar response from the Mars south polar residual cap. Fourth Intl. Conf. on Mars, Jan. 10–13, Tucson, Arizona, Abstract Book, pp. 150–151.
Muhleman, D. O., Grossman, A. W., Butler, B. J., and Slade, M. A. 1991. Radar images of Mars. *Science* 253:1508–1513.
Muller, P. M., and Sjogren, W. L. 1968. Mascons: Lunar mass concentration. *Science* 161:680–684.
Mullins, J. C., Kirk, B. S., and Ziegler, W. T. 1963. Calculation of the Vapor Pressure and Heats of Vaporization and Sublimation of Liquids and Solids, Especially Below One Atmosphere. V. Carbon Monoxide and Carbon Dioxide. Tech. Rept. No. 2, Project A-663, Engineering Experiment Station, Georgia Inst. of Tech., August 15, 1963.
Munk, W. H., and MacDonald, G. J. F. 1960. *The Rotation of the Earth* (Cambridge: Cambridge Univ. Press).
Munro, D. C., and Mouginis-Mark, P. J. 1990. Eruptive patterns and structure of Isla Fernandina, Galapagos Islands, from SPOT-1 HRN and large format camera images. *Intl. J. Remote Sensing* 8:1501–1509.
Murchie, S. L. 1990. Summary: The geology of Phobos as revealed by Phobos 2. In *Television Investigations of Phobos,* in press.
Murchie, S. L., Britt, D., Head, J., Pratt, S., Fisher, P., Zhukov, B., Kuzmin, A., Ksanfomality, L., Nikitin, G., Zharkov, A., Fanale, F., Blaney, D., and Robinson, M. 1990. Color variations on the surface of Phobos and their relationship to geologic features. *Lunar Planet. Sci.* XXI:825–826 (abstract).
Murchie, S. L., Britt, D. T., Head, J. W., Pratt, S. F., Fisher, P. C., Zhukov, B. S., Kuzmin, A. A., Ksanfomality, L. V., Zharkov, A. V., Fanale, F. P., Blaney, D. L., Robinson, M. S., and Bell, J. F. 1991*a*. Color heterogeneity of the surface of Phobos: Relationship to geologic features and comparisons to meteortite analogs. *J. Geophys. Res.* 96:5925–5945.
Murchie, S. L., Erard, S., Langevin, Y., Britt, D. T., Bibring, J.-P., Mustard, J. F., Head, J. W., and Pieters, C. M. 1991*b*. Disk-resolved spectral reflectance properties of Phobos from 0.3–3.2 μm: Preliminary integrated results from Phobos 2. *Lunar Planet. Sci.* XXII:943–944 (abstract).
Murphy, J. R., Toon, O. B., Pollack, J. B., and Haberle, R. B. 1989. Numerical simulations of

global dust-storm decay. In *MECA Workshop on Dust on Mars III,* ed. S. Lee, LPI Tech. Rept. 89-01, pp. 28–30 (abstract).

Murphy, J. R., Toon, O. B., Haberle, R. M., and Pollack, J. B. 1990*a*. Numerical simulations of the decay of Martian global dust storms. *J. Geophys. Res.* 95:14629–14648.

Murphy, J. R., Leovy, C. B., and Tillman, J. E. 1990*b*. Observations of Martian surface winds at the Viking Lander 1 site. *J. Geophys. Res.* 95:14555–14576.

Murray, B. C., and Malin, M. C. 1973*a*. Polar volatiles on Mars—Theory versus observation. *Science* 182:437–443.

Murray, B. C., and Malin, M. C. 1973*b*. Polar wandering on Mars. *Science* 179:997–1000.

Murray, B. C., Soderblom, L. A., Sharp, R. P., and Cutts, J. A. 1971. The surface of Mars: 1. The cratered terrains. *J. Geophys. Res.* 76:313–330.

Murray, B. C., Soderblom, L. A., Cutts, J. A., Sharp, R. P., Milton, D. J., and Leighton, R. B. 1972. Geological framework of the south polar region of Mars. *Icarus* 17:328–345.

Murray, B. C., Ward, W. R., and Yeung, S. C. 1973. Periodic insolation variations on Mars. *Science* 180:638–640.

Murtry, T. S., El-Sabh, M. I., Whitney, M. I., and Komar, P. D. 1984. Shapes of streamlined islands on Earth and Mars: Experiments and analyses of the minimum-drag form: Discussion and reply. *Geology* 12:569–572.

Mustard, J. F., Bibring, J.-P., Erard, S., Fischer, E. M., Head, J. W., Hurtrez, S., Langevin, Y., Pieters, C. M., and Sotin, C. J. 1990. Interpretation of spectral units of Isidis-Syrtis Major from ISM-Photos-2 observations. *Lunar Planet. Sci.* XXI:835–836 (abstract).

Mutch, P., and Woronow, A. 1980. Martian rampart and pedestal craters' ejecta emplacement: Coprates Quadrangle. *Icarus* 41:259–268.

Mutch, T. A. 1979. Planetary surfaces. *Rev. Geophys. Space Phys.* 17:1694–1722.

Mutch, T. A., and Morris, E. C. 1979. *Geologic Map of the Memnonia Quadrangle of Mars.* U.S.G.S. Map I-1137 (MC-16).

Mutch, T. A., and Saunders, R. S. 1976. The geologic development of Mars: A review. *Space Sci. Rev.* 19:3–57.

Mutch, T. A., Binder, A. B., Huck, F. O., Levinthal, E. C., Morris, E. C., Sagan, C., and Young, A. T. 1972. The Viking lander imaging investigation. *Icarus* 16:92–110.

Mutch, T. A., Binder, A. B., Huck, F. L., Levinthal, E. C., Liebes, S., Morris, E. C., Patterson, W. R., Pollack, J. B., Sagan, C., and Taylor, G. R. 1976*a*. The surface of Mars: The view from the Viking 1 Lander. *Science* 193:791–801.

Mutch, T. A., Arvidson, R. E., Binder, A. B., Huck, F. O., Levinthal, E. C., Liebes, S., Jr., Morris, E. C., Nummedal, D., Pollack, J. B., and Sagan, C. 1976*b*. Fine particles on Mars: Observations with the Viking 1 Lander cameras. *Science* 194:87–91.

Mutch, T. A., Arvidson, R. E., Head, J. W., III, Jones, K. L., and Saunders, R. S. 1976*c*. *The Geology of Mars* (Princeton, N.J.: Princeton Univ. Press).

Mutch, T. A., Arvidson, R. E., Binder, A. B., Guinness, E. A., and Morris, E. C. 1977. The geology of the Viking 2 Lander site. *J. Geophys. Res.* 82:4452–4467.

Nagy, A. F., and Cravens, T. E. 1988. Hot oxygen atoms in the upper atmospheres of Venus and Mars. *Geophys. Res. Lett.* 15:433–435.

Nagy, A. F., Cravens, T. E., and Gombosi, T. I. 1983. Basic theory and model calculations of the Venus ionosphere. In *Venus,* eds. D. M. Hunten, L. Colin, T. M. Donahue and V. I. Moroz (Tucson: Univ. of Arizona Press), pp. 841–872.

Nagy, A. F., Gombosi, T. I., Szego, K., Sagdeev, R. Z., Shapiro, V. D., and Schevchenko, V. I. 1990. Venus mantle-Mars planetosphere: What are the similarities and differences? *Geophys. Res. Lett.* 17:865–868.

Nakamura, K. 1977. Volcanoes as possible indicators of tectonic stress orientation—Principle and proposal. *J. Vol. Geotherm. Res.* 2:1–16.

Nakamura, N., Unruh, D. M., Tatsumoto, M., and Hutchinson, R. 1982. Origin and evolution of the Nakhla meteorite inferred from the Sm-Nd and U-Pb systematics and REE, Ba, Sr, Rb abundances. *Geochim. Cosmochim. Acta* 46:1555–1583.

Nakawo, M., and Young, G. J. 1981. Field experiments to determine the effect of a debris layer on ablation of glacier ice. *Ann. Glaciol.* 2:85–91.

Nakawo, M., and Young, G. J. 1982. Estimate of glacier ablation under a debris layer from surface temperature and meteorological variable. *J. Glaciol.* 28:85–91.

Nakazawa, K., Komuro, T., and Hayashi, C. 1983. Origin of the Moon—Capture by gas drag of the Earth's primordial atmosphere. *Moon and Planets* 28:311–327.

Nakiboglu, S. M. 1982. Hydrostatic theory of the Earth and its mechanical implications. *Phys. Earth Planet. Int.* 28:302–311.

Nautical Almanac Office. 1989. *Astronomical Almanac for the Year 1990* (Washington, D.C.: U.S. Government Printing Office).

Nedell, S. S., Squyres, S. W., and Andersen, D. W. 1987. Origin and evolution of the layered deposits in the Valles Marineris, Mars. *Icarus* 70:409–441.

Nehru, C. E., Prinz, M., Delaney, J. S., Dreibus, G., Palme, H., Spettel, B., and Wänke, H. 1983. Brachina: A new type of meteorite, not a chassignite. *Proc. Lunar Planet. Sci. Conf.* 14, *J. Geophys. Res. Suppl.* 88:B237-B244.

Ness, N. F., and Bauer, S. J. 1974. *USSR Mars Observations: The Case for an Intrinsic Magnetic Field.* GSFC Rept. X-690–74–69.

Neubauer, F. M. 1966. Thermal convection in the Martian atmosphere. *J. Geophys. Res.* 71:2419–2426.

Neugebauer, G., Munch, G., Chase, S. C., Hatzenbeler, H., Miner, E., and Schofield, D. 1969. Mariner 1969: Preliminary results of the infrared radiometer experiment. *Science* 166:98–99.

Neugebauer, G., Munch, G., Kieffer, H. H., Chase, S. C., Jr., and Miner, E. D. 1971. Mariner 1969 infrared radiometer results: Temperatures and thermal properties of the Martian surface. *Astron. J.* 76:719–749.

Neukum, G., and Hiller, K. 1981. Martian ages. *J. Geophys. Res.* 86:3097–3121.

Neukum, G., and Konig, B. 1976. Dating of individual lunar craters. *Proc. Lunar Sci. Conf.* 7:2867–2881.

Neukum, G., and Wise, D. U. 1976. Mars: A standard crater curve and possible new time scale. *Science* 194:1381–1387.

Neukum, G., Konig, B., Fechtig, H., and Storzer, D. 1975. Cratering in the Earth-Moon system: Consequences for age determination by crater counting. *Proc. Lunar Sci. Conf.* 6:2597–2620.

Neukum, G., Hiller, K., Henkel, J., and Bodechtel, J. 1978. Mars chronology. In *Reports of Planetary Geology and Geophysics Program 1977–1978,* NASA TM-79729, pp. 172–174 (abstract).

Newman, M. J., and Rood, R. T. 1977. Implications of solar evolution for the Earth's early atmosphere. *Science* 198:1035–1037.

Newsom, H. E. 1980. Hydrothermal alteration of impact melt sheets with implications for Mars. *Icarus* 44:207–216.

Newsom, H. E., Graup, G., Sewards, T., and Keil, K. 1986. Fluidization and hydrothermal alteration of the suevite deposit at the Ries crater, West Germany, and implications for Mars. *Proc. Lunar Planet. Sci. Conf.* 17, *J. Geophys. Res. Suppl.* 91:E239-E251.

Nier, A. O. 1950. A redetermination of the relative abundances of the isotopes of carbon, nitrogen, oxygen, argon and potassium. *Phys. Rev.* 77:789–793.

Nier, A. O., and McElroy, M. B. 1977. Composition and structure of Mars' upper atmosphere: Results from the neutral mass spectrometers on Viking 1 and 2. *J. Geophys. Res.* 82:4341–4349.

Nier, A. O., Hanson, W. B., Seiff, A., McElroy, M. B., Spencer, N. W., Duckett, R. J., Knight, T. C. D., and Cook, W. S. 1976*a*. Composition and structure of the Martian atmosphere: Preliminary results from Viking 1. *Science* 193:786–788.

Nier, A. O., McElroy, M. B., and Yung, Y. L. 1976*b*. Isotopic composition of the Martian atmosphere. *Science* 194:68–70.

Nishiizumi, K., Klein, J., Middleton, R., Elmore, D., Kubik, P. W., and Arnold, J. R. 1986. Exposure history of shergottites. *Geochim. Cosmochim. Acta* 50:1017–1022.

Niver, D. S., and Hess, S. L. 1982. Band-pass filtering of one year of daily mean pressures on Mars. *J. Geophys. Res.* 87:10191–10196.

Noland, M., and Veverka, J. 1977. The photometric functions of Phobos and Deimos. II. Surface photometry of Deimos. *Icarus* 30:200–211.

Noland, M., Veverka, J., and Pollack, J. B. 1973. Mariner 9 polarimetry of Phobos and Deimos. *Icarus* 20:490–502.

North American Commission on Stratigraphic Nomenclature. 1983. North American stratigraphic code. *Amer. Assoc. Petrol. Geol. Bull.* 67:841–875.

Noxon, J. F., Traub, W. A., Carleton, N. P., and Connes, P. 1976. Detection of O_2 dayglow emission from Mars and the Martian ozone abundance. *Astrophys. J.* 207:1025–1035.

Null, G. W. 1969. A solution for the mass and dynamical oblateness of Mars using Mariner IV Doppler data. *Bull. Amer. Astron. Soc.* 1:356 (abstract).

Nummedal, D. 1978. The role of liquefaction in channel development on Mars. In *Reports of Planetary Geology and Geophysics Program–1977–1978*, NASA TM-79729, pp. 257–259 (abstract).

Nummedal, D., and Prior, D. B. 1981. Generation of Martian chaos and channels by debris flows. *Icarus* 45:77–86.

Nummedal, D., Masursky, H., Mainguet, M., Cutts, J. A., and Blasius, K. R. 1983. Origin of Martian outflow channels: The eolian hypothesis: Discussion and reply. *J. Geophys. Res.* 88:1243–1247.

Nyquist, L. E. 1983. Do oblique impacts produce Martian meteorites? *Proc. Lunar Planet. Sci. Conf.* 13, *J. Geophys. Res. Suppl.* 88:A785-A798.

Nyquist, L. E., Bogard, D. D., Wooden, J. L., Weismann, H., Shih, C. Y., Bansal, B. M., and McKay, G. A. 1979. Early differentiation, late magmatism and recent bombardment of the Shergottite parent planet. *Meteoritics* 14:502 (abstract).

O'Connor, J. T. 1968*a*. Mineral stability at the Martian surface. *J. Geophys. Res.* 73:5301–5311.

O'Connor, J. T. 1968*b*. "Fossil" Martian weathering. *Icarus* 8:513–517.

Odezynskyj, M. I., and Holloway, J. R. 1989. Carbonate composition and stability in the Martian mantle: Preliminary results. In *Lunar Planet. Sci.* XX:806–807 (abstract).

Ohtani, E., Kato, T., and Sawamoto, H. 1986. Melting of a model chondritic mantle to 20 GPa. *Nature* 322:352–353.

Okal, E. A., and Anderson, D. L. 1978. Theoretical models for Mars and their seismic properties. *Icarus* 33:514–528.

O'Keefe, J. D., and Ahrens, T. J. 1981. Impact cratering: The effect of crustal strength and planetary gravity. *Rev. Geophys. Space Phys.* 19:1–12.

O'Leary, M. H. 1981. Carbon isotope fractionation in plants. *Phytochem.* 20:553–567.

Olson, P., Schubert, G., and Anderson, C. 1987. Plume formation in the D″-layer and the roughness of the core-mantle boundary. *Nature* 327:409–413.

O'Meara, S. J. 1988. Behold, Mars. *Sky & Teles.* 76:614.

Ong, M., Luhmann, J. G., Russell, C. T., Schwingenschuh, K., Reidler, W., and Yeroshenko, Ye. 1989. Phobos observations of the Martian magnetotail. *Eos: Trans. AGU* 70:1174 (abstract).

Oort, A. H. 1983. *Global Atmospheric Circulation Statistics, 1958–1973*. NOAA Prof. Paper 14 (Boulder: Natl. Oceanic and Atmospheric Administration).

Oort, A. H., and Rasmusson, E. M. 1971. *Atmospheric Circulation Statistics*. NOAA Prof. Paper 5 (Boulder: Natl. Oceanic and Atmospheric Administration).

Oparin, A. I. 1938. *The Origin of Life* (New York: MacMillan).

Orgel, L. 1987. Evolution of the genetic apparatus. A review. In *Cold Spring Harbor Symp. Quant. Biology* 52:9–16.

Oró, J. 1961. Comets and the formation of biochemical compounds on the primitive Earth. *Nature* 190:389–390.

Oró, J., and Holzer, G. 1979. The photolytic degradation and oxidation of organic compounds under simulated Martian conditions. *J. Molec. Evol.* 14:153–160.

Oró, J., and Lazcano-Arujo, A. 1981. The role of HCN and its derivatives in pre-biotic evolution. In *Cyanide in Biology,* eds. B. Vennesland, E. E. Conn, C. J. Knowles, J. Westly and F. Wissling (London: Academic Press), pp. 517–541.

Osberg, W. E., and Hornig, D. F. 1952. The vibrational spectra of molecules and complex ions in crystals, 6: Carbon dioxide. *J. Chem. Phys.* 20:1345–1347.

Ostro, S. J. 1987. Planetary radar astronomy. *Enc. Phys. Sci. Tech.* 10:611–634.

Ostro, S. J., Jurgens, R. F., Yeomans, D. K., Standish, E. M., and Grenier, W. 1989. Radar detection of Phobos. *Science* 243:1584–1586.

Ott, U. 1988. Noble gases in SNC meteorites: Shergotty, Nakhla, Chassigny. *Geochim. Cosmochim. Acta* 52:1937–1948.

Ott, U., and Begemann, F. 1985. Are all the "martian" meteorites from Mars? *Nature* 317:509–512.

Owen, T. 1966. The composition and surface pressure of the Martian atmosphere: Results from the 1965 opposition. *Astrophys. J.* 146:257–270.

Owen, T. 1974. Martian climate: An empirical test of possible gross variations. *Science* 183:763–764.

Owen, T. 1986. Update of the Anders-Owen model for Martian volatiles. In *Workshop on the Evolution of the Martian Atmosphere,* eds. M. Carr, P. James, C. Leovy and R. Pepin, LPI Tech. Rept. 86-08, pp. 31–32 (abstract).

Owen, T. 1987. Could icy impacts reconcile Venus with Earth and Mars? In *MECA Symp. on Mars: Evolution of its Climate and Atmosphere,* eds. V. Baker, M. Carr, F. Fanale, R. Greeley, R. Haberle, C. Leovy and T. Maxwell, LPI Tech. Rept. 87-01, pp. 77 (abstract).

Owen, T., and Biemann, K. 1976. Composition of the atmosphere at the surface of Mars: Detection of argon-36 and preliminary analysis. *Science* 193:801–803.

Owen, T., and Kuiper, G. P. 1964. A determination of the composition and surface pressure of the Martian atmosphere. *Commun. Lunar and Planetary Lab* 2:113–132.

Owen, T., and Mason, H. P. 1969. Mars: Water vapor in its atmosphere. *Science* 165:893–895.

Owen, T., and Sagan, C. 1972. Minor constituents in planetary atmospheres: Ultraviolet spectroscopy from the Orbiting Astronomical Observatory. *Icarus* 16:557–568.

Owen, T., Scattergood, T., and Woodman, J. H. 1975. On the abundance of NO_2 in the Martian atmosphere. *Icarus* 24:193–196.

Owen, T., Biemann, K., Rushneck, D. R., Biller, J. E., Howarth, D. W., and LaFleur, A. L. 1976. The atmosphere of Mars: Detection of krypton and xenon. *Science* 194:1293–1295.

Owen, T., Biemann, K., Rushneck, D. R., Biller, J. E., Howarth, D. W., and Lafleur, A. L. 1977. The composition of the atmosphere at the surface of Mars. *J. Geophys. Res.* 82:4635–4639.

Owen, T., Cess, R. D., and Ramanathan, V. 1979. Enhanced CO_2 greenhouse to compensate for reduced solar luminosity on early Earth. *Nature* 277:640–642.

Owen, T., Maillard, J. P., de Bergh, C., and Lutz, B. L. 1988. Deuterium on Mars: The abundance of HDO and the value of D/H. *Science* 240:1767–1770.

Owen, T., Bar-Nun, A., Kleinfeld, I. 1991. Noble gases in terrestrial planets: Evidence for cometary impacts? In *Comets in the Post-Halley Era,* eds. R. Newburn, Jr., M. Neugebauer and J. Rahe (Dordrecht: Kluwer), pp. 429–437.

Oyama, V. I., and Berdahl, B. J. 1977. The Viking gas exchange experiment results from Chryse and Utopia surface samples. *J. Geophys. Res.* 82:4669–4676.

Ozima, M., and Nakazawa, K. 1980. Origin of rare gases in the Earth. *Nature* 284:313–316.

Paige, D. A. 1985. The Annual Heat Balance of the Martian Polar Caps from Viking Observations. Ph.D. Thesis, California Inst. of Technology

Paige, D. A. 1989. Mariner 9 IRIS and Viking observations of Martian CO_2 clouds. Fourth Intl. Conf. on Mars, Jan. 10–13, Tucson, Arizona, Abstract Book, pp. 158–159.

Paige, D. A., and Ingersoll, A. P. 1985. Annual heat balance of Martian polar caps: Viking observations. *Science* 228:1160–1168.

Paige, D. A., and Kieffer, H. H. 1987. The thermal properties of Martian surface materials at high latitudes: Possible evidence for permafrost. In *MECA Symp. on Mars: Evolution of its Climate and Atmosphere,* eds. V. Baker, M. Carr, F. Fanale, R. Greeley, R. Haberle, C. Leovy and T. Maxwell, LPI Tech. Rept. 87–01, pp. 79–81 (abstract).

Paige, D. A., Crisp, D., and Santee, M. L. 1990*a*. It snows on Mars. *Bull. Amer. Astron. Soc.* 22:1075 (abstract).

Paige, D. A., Herkenhoff, K. E., and Murray, B. C. 1990*b*. Mariner 9 observations of the south polar cap of Mars: Evidence for residual CO_2 frost. *J. Geophys. Res.* 95:1319–1335.

Palluconi, F. D. 1977. North polar mapping of Mars with the Viking Thermal Mappers. *Bull. Amer. Astron. Soc.* 9:540 (abstract).

Palluconi, F. D., and Kieffer, H. H. 1981. Thermal inertia mapping of Mars for 60°S to 60°N. *Icarus* 45:415–426.

Pang, K. D., and Ajello, J. M. 1977. Complex refractive index of Martian dust: Wavelength dependence and composition. *Icarus* 30:63–74.

Pang, K. D., Ajello, J. M., Hord, C. W., and Egan, W. S. 1976. Complex refractive index of Martian dust: Mariner 9 ultraviolet observations. *Icarus* 27:55–67.

Pang, K. D., Pollack, J. B., Veverka, J., Lane, A. L., and Ajello, J. M. 1978. The composition of Phobos: Evidence for carbonaceous chondrite surface from spectral analysis. *Science* 199:64–66.

Pang, K. D., Rhoads, J. W., Lane, A. L., and Ajello, J. M. 1980. Spectral evidence for a carbonaceous chondrite surface composition on Deimos. *Nature* 283:277–278.

Parker, B. C., Simmons, G. M., Jr., Love, F. G., Wharton, R. A., Jr., and Seaburg, K. G. 1981. Modern stromatolites in Antarctic dry valley lakes. *BioSci.* 31:656–661.

Parker, B. C., Simmons, G. M., Jr., Seaburg, K. G., Cathey, D. D., and Allnutt, F. T. C. 1982. Comparative ecology of plankton communities in seven Antarctic oasis lakes. *J. Plank. Res.* 4:271–286.

Parker, D. C., Capen, C. F., and Beish, J. D. 1983. Exploring the Martian arctic. *Sky & Teles.* 65:218–220.

Parker, T. S., Saunders, R. S., and Schneeberger, D. M. 1989. Transitional morphology in the west Deuteronilus Mensae region of Mars: Implications for modification of the lowland/upland boundary. *Icarus* 82:111–145.

Parkinson, T. D., and Hunten, D. M. 1972. Spectroscopy and aeronomy of O_2 on Mars. *J. Atmos. Sci.* 29:1380–1390.

Patera, E. S., and Holloway, J. R. 1982. Experimental determination of the spinel-garnet boundary in a Martian mantle composition. *Proc. Lunar Planet. Sci.* 13, *J. Geophys. Res. Suppl.* 87:A31-A36.

Patterson, W. R., Huck, F. O., Wall, S. D., and Wolf, M. R. 1977. Calibration and performance of the Viking lander cameras. *J. Geophys. Res.* 82:4391–4400.

Patton, P. C. 1981. Evolution of the spur and gully topography on the Valles Marineris wall scarps. In *Reports of the Planetary Geology Program—1982,* NASA TM-84211, pp. 324–325 (abstract).

Patton, P. C. 1984. Scarp development in the Valles Marineris. In *Reports of Planetary Geology Program–1983*, NASA TM 86246, pp. 234–236 (abstract).

Patton, P. C. 1985. Lithologic and structural control on slope morphology in the Valles Marineris. In *Reports of Planetary Geology and Geophysics Program–1984,* NASA TM-87563, pp. 503–505 (abstract).

Peacock, M. A., and Fuller, R. E. 1928. Chlorophaeite, sideromelane and palagonite from the Columbia River Plateau. *Amer. Mineral.* 13:360–382.

Peale, S. J. 1969. Generalized Cassini's laws. *Astron. J.* 74:483–489.

Peale, S. J., Schubert, G., and Lingenfelter, R. E. 1975. Origin of Martian channels: Clathrates and water. *Science* 187:273–274.

Pechmann, J. C. 1980. The origin of polygonal troughs on the northern plains of Mars. *Icarus* 42:185–210.

Penner, E. 1970. Thermal conductivity of frozen soils. *Canadian J. Earth Sci.* 7:982–987.

Pepin, R. O. 1985. Evidence of Martian origins. *Nature* 317:473–475.

Pepin, R. O. 1987*a*. Volatile inventories of the terrestrial planets. *Rev. Geophys.* 25:293–296.

Pepin, R. O. 1987*b*. Volatile inventory of Mars. II. Primordial sources and fractionating processes. In *Mars: Evolution of its Climate and Atmosphere,* eds. V. Baker, M. Carr, F. Fanale, R. Greeley, R. Haberle, C. Leovy and T. Maxwell, LPI Tech. Rept. 87-01, pp. 99–101 (abstract).

Pepin, R. O. 1989. Atmospheric compositions: Key similarities and differences. In *Origin and Evolution of Planetary and Satellite Atmospheres,* eds. S. K. Atreya, J. B. Pollack and M. S. Matthews (Tucson: Univ. of Arizona Press), pp. 291–305.

Pepin, R. O. 1991. On the origin and early evolution of terrestrial planet atmospheres and meteoritic volatiles. *Icarus* 92:2–79.

Pepin, R. O., and Phinney, D. 1978. Components of xenon in the solar system. Unpublished preprint.

Perez-de-Tejada, H. 1987. Plasma flow in the Mars magnetosphere. *J. Geophys. Res.* 92:4713–4718.

Peterfreund, A. R. 1981. Visual and infrared observations of wind streaks on Mars. *Icarus* 45:447–467.

Peterfreund, A. R. 1985. Contemporary Aeolian Processes on Mars: Local Dust Storms. Ph.D. Thesis, Arizona State Univ.

Peterfreund, A. R., and Kieffer, H. H. 1979. Thermal infrared properties of the Martian atmosphere. 3. Local dust clouds. *J. Geophys. Res.* 84:2853–2863.

Peterson, C. 1981. A secondary origin for the central plateau of Hebes Chasma. *Proc. Lunar Planet. Sci.* 12:1459–1471.
Peterson, J. E. 1977. *Geologic Map of the Noachis Quadrangle of Mars, scale 1:5000,000.* U.S.G.S. Misc. Inv. Series Map I-910.
Peterson, J. E. 1978. Volcanism in the Noachis-Hellas region of Mars, 2. *Proc. Lunar Planet. Sci. Conf.* 9:3411–3432.
Pettengill, G. H. 1978. Physical properties of the planets and satellites from radar observations. *Ann. Rev. Astron. Astrophys.* 16:265–292.
Pettengill, G. H., Counselan, C. C., Rainville, L. P., and Shapiro, I. I. 1969. Radar measurements of Martian topography. *Astron. J.* 74:461–482.
Pettengill, G. H., Rogers, A. E. E., and Shapiro, I. I. 1971. Martian craters and a scarp as seen by radar. *Science* 174:1321–1324.
Pettengill, G. H., Shapiro, I. I., and Rogers, A. E. E. 1973. Topography and radar scattering properties of Mars. *Icarus* 18:22–28.
Pettengill, G. H., Campbell, D. B., and Masursky, H. 1980. The surface of Venus. *Sci. Amer.* 243:54–65.
Pettengill, G. H., Ford, P. G., and Chapman, B. D. 1988. Venus: Surface electromagnetic properties. *J. Geophys. Res.* 93:14881–14892.
Pettijohn, F. J. 1975. *Sedimentary Rocks,* 3rd ed. (New York: Harper and Row).
Pettit, E., and Richardson, R. S. 1955. Observations of Mars made at Mount Wilson in 1954. *Publ. Astron. Soc. Pacific* 67:62–73.
Peulvast, J. P., and Costard, F. M. 1989. 1:500,00 geomorphological mapping of Mars: Melas Chasma, Valles Marineris. *Lunar Planet. Sci.* XX:840–841 (abstract).
Philip, J. R. 1979. Angular momentum of seasonally condensing atmospheres, with special reference to Mars. *Geophys. Res. Lett.* 6:727–730.
Phillips, J. L., Luhmann, J. G., and Russell, C. T. 1984. Growth and maintenance of large-scale magnetic fields in the dayside Venus ionosphere. *J. Geophys. Res.* 89:10676–10684.
Phillips, J. L., Luhmann, J. G., Russell, C. T., and Moore, K. R. 1987. Finite Larmor radius effect on ion pickup at Venus. *J. Geophys. Res.* 92:9920–9930.
Phillips, R. J. 1988. The geophysical signature of the Martian global dichotomy. *Eos: Trans. AGU* 69:389.
Phillips, R. J., and Lambeck, K. 1980. Gravity fields of the terrestrial planets: Long wavelength anomalies and tectonics. *Rev. Geophys. Space Phys.* 18:27–76.
Phillips, R. J., and Maxwell, T. A. 1978. Lunar sounder revisited: Stratigraphic correlations and structural inferences. *Lunar Planet. Sci.* IX:890–892 (abstract).
Phillips, R. J., and Saunders, R. S. 1975. The isostatic state of Martian topography. *J. Geophys. Res.* 80:2893–2898.
Phillips, R. J., Saunders, R. S., and Conel, J. E. 1973. Mars: Crustal structure inferred from Bouguer gravity anomalies. *J. Geophys. Res.* 78:4815–4820.
Phillips, R. J., Sjorgen, W. L., Abbott, E. A., and Zisk, S. H. 1978. Simulation gravity modeling to spacecraft-tracking data: Analysis and application. *J. Geophys. Res.* 83:5455–5464.
Phillips, R. J., Sleep, N. H., and Banerdt, W. B. 1990. Permanent uplift in magmatic systems with application to the Tharsis region of Mars. *J. Geophys. Res.* 95:5089–5100.
Phukan, A. 1983. Long-term creep deformation of roadway embankment on ice-rich permafrost. In *Proc. Fourth Intl. Conf. on Permafrost,* July 17–22, Fairbanks, Alaska (Washington, D.C.: National Academy Press), pp. 994–999.
Pickering, W. H. 1921. *Mars* (Boston: Gorham Press).
Pickersgill, A. O., and Hunt, G. E. 1981. An examination of the formation of linear lee waves generated by giant Martian volcanoes. *J. Atmos. Sci.* 38:40–51.
Pieri, D. C. 1976. Martian channels: Distribution of small channels on the Martian surface. *Icarus* 27:25–50.
Pieri, D. C. 1979. Geomorphology of Martian Valleys. Ph.D. Thesis, Cornell Univ.
Pieri, D. C. 1980*a*. *Geomorphology of Martian Valleys.* NASA TM-81979.
Pieri, D. C. 1980*b*. Martian valleys: Morphology, distribution, age, and origin. *Science* 210:895–897.
Pieri, D. C., and Baloga, S. M. 1986. Eruption rate, area and length relationships for some Hawaiian flows. *J. Volcan. Geotherm. Res.* 30:29–45.

Pieri, D. C., Schneeberger, D., Baloga, S., and Saunders, S. R. 1986. Dimensions of lava flows at Alba Patera, Mars. In *Reports of Planetary Geology and Program Geophysics–1985*, TM-88383, pp. 318–320 (abstract).

Pike, R. J. 1978. Volcanoes on the inner planets: Some preliminary comparisons of gross topography. *Proc. Lunar Planet. Sci Conf.* 9:3239–3273.

Pike, R. J. 1980*a*. Control of crater morphology by gravity and target type: Mars, Earth, and Moon. *Proc. Lunar Planet. Sci. Conf.* 11:2159–2189.

Pike, R. J. 1980*b*. *Geometric Interpretation of Lunar Craters*. U.S.G.S. Prof. Paper 1046-C.

Pike, R. J. 1988. Geomorphology of impact craters on Mercury. In *Mercury*, eds. F. Vilas, C. R. Chapman and M. S. Matthews (Tucson: Univ. of Arizona Press), pp. 165–273.

Pike, R. J., and Arthur, D. W. G. 1979. Simple to complex impact craters: The transition on Mars. In *Reports of Planetary Geology Program–1978–1979*, NASA TM-80339, pp. 132–134.

Pike, R. J., and Spudis, P. D. 1987. Basin-ring spacing on the Moon, Mercury, and Mars. *Earth, Moon and Planets* 39:129–194.

Pimentel, G. C., Forney, P. B., and Herr, K. C. 1974. Evidence about hydrate and solid water in the Martian surface from the 1969 Mariner infrared spectrometer. *J. Geophys. Res.* 79:1623–1634.

Pinet, P., and Chevrel, S. 1990. Spectral identification of geological units on the surface of Mars related to the presence of silicates from Earth-based near-infrared telescopic charge-coupled device imaging. *J. Geophys. Res.* 95: 14435–14446.

Pinkerton, H., and Wilson, L. 1988. The lengths of lava flows. *Lunar Planet. Sci.* XIX:937–938 (abstract).

Pirraglia, J. 1975. Polar symmetric flow of a viscous compressible atmosphere: An application to Mars. *J. Atmos. Sci.* 32:60–72.

Pirraglia, J. 1976. Martian atmospheric lee waves. *Icarus* 7:517–530.

Pirraglia, J., and Conrath, B. 1974. Martian tidal pressure and wind fields obtained from the Mariner 9 infrared spectroscopy experiment. *J. Atmos. Sci.* 31:318–329.

Pittendrigh, C. S., Vishniac, W., and Pearman, J. P. T. 1966. *Biology and the Exploration of Mars*, Natl. Acad. Sci. Publ. 1296 (Washington, D.C.: Natl. Academy of Sciences).

Plafker, G., and Erikson, G. E. 1978. Mechanism of catastrophic avalanches from Nevados Huascaran, Peru. In *Rock Slides and Avalanches*, ed. B. Voight (Amsterdam: Elsevier), pp. 277–314.

Planetary Reports, 1989 9(9), p. 22, July/Aug. 1989.

Plaut, J. J., Kahn, R., Guinness, E. A., and Arvidson, R. E. 1988. Accumulation of sedimentary debris in the south polar region of Mars and implication for climate history. *Icarus* 75:357–377.

Plescia, J. B. 1981. The Tempe volcanic province of Mars and comparisons with the Snake River Plains of Idaho. *Icarus* 45:586–601.

Plescia, J. B. 1990. Young flood lavas in the Elysium region, Mars. *Lunar Planet. Sci.* XXI:969–970 (abstract).

Plescia, J. B., and Golombek, M. P. 1986. Origin of planetary wrinkle ridges based on the study of terrestrial analogs. *Geol. Soc. Amer. Bull.* 97:1289–1299.

Plescia, J. B., and Saunders, R. S. 1979. The chronology of the Martian volcanoes. *Proc. Lunar Planet. Sci. Conf.* 10:2841–2859.

Plescia, J. B., and Saunders, R. S. 1980. Estimation of the thickness of the Tharsis lava flows and implications for the nature of the topography of the Tharsis plateau. *Proc. Lunar Planet. Sci. Conf.* 11:2423–2426.

Plescia, J. B., and Saunders, R. S. 1982. Tectonic history of the Tharsis region, Mars. *J. Geophys. Res.* 87:9775–9791.

Pleskot, L. K., and Miner, E. D. 1981. Time variability of Martian bolometric albedo. *Icarus* 45:179–201.

Poldervaart, A. 1955. *Chemistry of the Earth's Crust*. Geological Soc. of America. SP-62.

Pollack, J. B. 1977. Phobos and Deimos. In *Planetary Satellites*, ed. J. A. Burns (Tucson: Univ. of Arizona Press), pp. 319–345.

Pollack, J. B. 1979. Climate change on the terrestrial planets. *Icarus* 37:479–553.

Pollack, J. B. 1991. Kuiper prize lecture: Present and past climates of the terrestrial planets. *Icarus* 91:173–198.

Pollack, J. B., and Black, D. C. 1979. Implications of the gas compositional measurements of Pioneer Venus for the origin of planetary atmospheres. *Science* 205:56–59.

Pollack, J. B., and Black, D. C. 1982. Noble gases in planetary atmospheres: Implications for the origin and evolution of atmospheres. *Icarus* 51:169–198.

Pollack, J. B., and Cuzzi, J. N. 1979. Scattering by nonspherical particles of size comparable to the wavelength: A new semi-empirical theory and its application to troposphere aerosols. *J. Atmos. Sci.* 37:868–881.

Pollack, J. B., and Toon, O. B. 1982. Quasi-periodic climate changes on Mars: A review. *Icarus* 50:259–287.

Pollack, J. B., and Whitehill, L. 1972. A multiple scattering model of the diffuse component of lunar radar echoes. *J. Geophys. Res.* 77:4289–4303.

Pollack, J. B., and Yung, Y. L. 1980. Origin and evolution of planetary atmospheres. *Ann. Rev. Earth Planet. Sci.* 8:425–487.

Pollack, J. B., Greenberg, E. H., and Sagan, C. 1967. A statistical analysis of the Martian wave of darkening and related phenomena. *Planet. Space Sci.* 15:817–824.

Pollack, J. B., Veverka, J., Noland, M., Sagan, C., Hartmann, W. K., Duxbury, T. C., Born, G. H., Milton, D. J., and Smith, B. A. 1972. Mariner 9 television observations of Phobos and Deimos. *Icarus* 17:394–407.

Pollack, J. B., Veverka, J., Noland, M., Sagan, C., Duxbury, T. C., Acton, C., Born, G. H., Hartmann, W. K., and Smith, B. A. 1973. Mariner 9 television observations of Phobos and Deimos. 2. *J. Geophys. Res.* 87:4313–4326.

Pollack, J. B., Haberle, R., Greeley, R., and Iversen, J. 1976*a*. Estimates of the wind speeds required for particle motion on Mars. *Icarus* 29:395–417.

Pollack, J. B., Leovy, C. B., Mintz, Y., and Van Camp, W. 1976*b*. Winds on Mars during the Viking season: Predictions based on a general circulation model with topography. *Geophys. Res. Lett.* 3:479–483.

Pollack, J. B., Colburn, D., Kahn, R., Hunter, J., Van Camp, W., Carlston, C. E., and Wolf, M. R. 1977. Properties of aerosols in the Martian atmosphere, as inferred from Viking Lander imaging data. *J. Geophys. Res.* 82:4479–4496.

Pollack, J. B., Veverka, J., Pang, K., Colburn, D., Lane, A. L., and Ajello, J. M. 1978. Multicolor observations of Phobos with the Viking lander cameras: Evidence for a carbonaceous chondrite composition. *Science* 199:66–69.

Pollack, J. B., Burns, J. A., and Tauber, M. E. 1979*a*. Gas drag in primordial circumplanetary envelopes: A mechanism for satellite capture. *Icarus* 37:587–611.

Pollack, J. B., Colburn, D. S., Flasar, F. M., Kahn, R., Carlston, C. E., and Pidek, D. C. 1979*b*. Properties and effects of dust particles suspended in the Martian atmosphere. *J. Geophys. Res.* 84:2929–2945.

Pollack, J. B., Leovy, C. B., Greiman, P. W., and Mintz, Y. 1981. A Martian general circulation experiment with large topography. *J. Atmos. Sci.* 38:3–29.

Pollack, J. B., Kasting, J. F., Richardson, S. M., and Poliakoff, K. 1987. The case for a wet, warm climate on early Mars. *Icarus* 71:203–224.

Pollack, J. B., Haberle, R. M., Schaeffer, J., and Lee, H. 1990*a*. Simulations of the general circulation of the Martian atmosphere. I. Polar processes. *J. Geophys. Res.* 95:1447–1473.

Pollack, J. B., Roush, T., Whitteborn, F., Bregman, J., Wooden, D., Stoker, C., Toon, O. B., Rank, D., Dalton, B., and Freedman, R. 1990*b*. Thermal emission spectra of Mars (5.4–10.5 μm): Evidence for sulfates, carbonates, and hydrates. *J. Geophys. Res.* 95:14595–14628.

Ponnamperuma, C. 1985. Synthesis and analysis in chemical evolution. In *The Search for Extraterrestrial Life: Recent Developments,* ed. M. D. Pappagianis (Dordrecht: D. Reidel), pp. 185–197.

Porter, S. G. 1972. Distribution, morphology, and size-frequency of cinder cones on Mauna Kea volcano, Hawaii. *Geol. Soc. Amer. Bull.* 83:3607–3612.

Posin, S. B. 1989. Yield strength of Martian complex craters. Fourth Intl. Conf. on Mars, Jan. 10–13, Tucson, Arizona, Abstract Book, pp. 162–163.

Postawko, S. E., and Kuhn, W. R. 1986. Effect of the greenhouse gases (CO_2, H_2O, SO_2) on Martian paleoclimate. *Proc. Lunar Planet. Sci. Conf.* 16, *J. Geophys. Res. Suppl.* 91:D431–D438.

Postgate, J. R. 1982. *The Fundamentals of Nitrogen Fixation* (Cambridge: Cambridge Univ. Press).

Potter, D. B. 1976. *Geologic Map of the Hellas Quadrangle of Mars, scale 1:5,000,000.* U.S.G.S. Misc. Inv. Series Map I-941.

Prabhakara, C. P., and Hogan, J. S., Jr. 1965. Ozone and carbon dioxide heating in the Martian atmosphere. *J. Atmos. Sci.* 22:97–106

Prandtl, L. 1952. *Fluid Dynamics* (New York: Hafner).

Presley, M. A., and Arvidson, R. E. 1988. Nature and origin of materials exposed in the Oxia Palus-Western Arabia-Sinus Meridiani region, Mars. *Icarus* 75:499–517.

Pruppacher, H. R., and Klett, J. D. 1978. *Microphysics of Clouds and Precipitation* (Dordrecht: D. Reidel).

Priroda 1963. En route to the planet Mars. Unsigned, Moscow, Priroda No. 1, pp. 16–17.

Quaide, W. L. 1965. Rilles, ridges and domes—Clues to maria history. *Icarus* 4:374–389.

Quaide, W. L., and Oberbeck, V. R. 1968. Thickness determination of the lunar surface layer from lunar impact craters. *J. Geophys. Res.* 73:5247–5270.

Rampino, M. R., and Stothers, R. B. 1984. Terrestrial mass extinctions, cometary impacts, and the Sun's motion perpendicular to the galactic planet. *Nature* 308:709–712.

Rapp, A. 1960. Talus slopes and mountain walls at Tempelfjorden, Spitsbergen. *Norsk Polarinstitutt Skrifter* 119:93.

Rapp, R. H. 1974. Current estimates of mean Earth ellipsoid parameters. *Geophys. Res. Lett.* 1:35–38.

Rassbach, M. E., Wolf, R. A., and Daniell, R. E., Jr. 1974. Convection in a Martian magnetosphere. *J. Geophys. Res.* 79:1125–1127.

Rattcliffe, E. H. 1962. The thermal conductivity of ice: New data on the temperature coefficient. *Phil. Magnet.* 1:1197–1203.

Raup, D. M., and Sepkoski, J. J. 1984. Periodicity of extinctions in the geologic past. *Proc. Natl. Acad. Sci.* 81:801–805.

Rea, D. G. 1964. Evidence for life on Mars. *Nature* 200:114–116.

Rea, D. G., O'Leary, B. T., and Sinton, W. M. 1965. Mars: The origin of the 3.58 and 3.69-micron minima in the infrared spectra. *Science* 147:1286–1288.

Reasenberg, R. D. 1977. The moment of inertia and isostasy of Mars. *J. Geophys. Res.* 82:369–375.

Reasenberg, R. D., and King, R. W. 1979. The rotation of Mars. *J. Geophys. Res.* 84:6231–6240.

Reasenberg, R. D., Shapiro, I. I., and White, R. D. 1975. The gravity field of Mars. *Geophys. Res. Lett.* 2:89–92.

Reber, C. A. 1990. The upper atmosphere research satellite. *Eos: Trans. AGU* 71:1867–68, 1873–74, 1878.

Reid, A. M., and Bunch, T. E. 1975. The nakhlites. II. Where, when, and how? *Meteoritics* 10:317–324.

Reimers, C. E., and Komar, P. D. 1979. Evidence for explosive volcanic density currents on certain Martian volcanoes. *Icarus* 39:88–110.

Ricard, Y., Fleitout, L. M., and Froideveaux, C. 1984. Geoid heights and lithospheric stresses for a dynamic Earth. *Ann. Geophys.* 2:267–286.

Richards, M. A., and Hager, B. H. 1984. Geoid anomalies in a dynamic Earth. *J. Geophys. Res.* 89:5987–6002.

Richardson, R. S. 1954. *Exploring Mars* (New York: McGraw-Hill).

Riedler, W., Möhlmann, D., Oraevsky, V. N., Schwingenschuh, K., Yeroshenko, Ye., Rustenbach, J., Aydoyar, Oe., Berghofer, G., Lichtenegger, H., Delva, M., Schelch, G., Pirsh, K., Fremuth, G., Steller, M., Arnold, H., Raditsch, T., Auster, U., Fornacon, K.-H., Schenk, H. J., Michaelis, H., Motschmann, U., Roatsch, T., Sauer, K., Schröter, R., Kurths, J., Lenners, D., Linthe, J., Kobsev, V., Styashin, V., Achache, J., Slavin, J., Luhmann, J. G., and Russell, C. T. 1989*a*. Magnetic field near Mars: First results. *Nature* 341:604–607.

Riedler, W., Schwingenschuh, K., Yeroshenko, Ye., Luhmann, J. G., Ong, M., and Russell, C. T. 1989*b*. The magnetotail of Mars. *Bull. Amer. Astron. Soc.* 21:1989 (abstract).

Rind, D., Suozzo, R., and Balachandran, N. K. 1988. The GISS global climate-middle atmo-

sphere model. Part II. Model variability due to interactions between planetary waves, the mean circulation and gravity wave drag. *J. Atmos. Sci.* 45:371–386.
Ringwood, A. E. 1975. *Composition and Petrology of the Earth's Mantle* (New York: McGraw-Hill).
Ringwood, A. E. 1977. Composition of the core and implications for the origin of the Earth. *Geochem. J.* 11:111–135.
Ringwood, A. E. 1979. *On the Origin of the Earth and Moon* (New York: Springer-Verlag).
Ringwood, A. E. 1981. In *Basaltic Volcanism on the Terrestrial Planets*, eds. R. B. Merrill and R. Ridings (New York: Pergamon), p. 641.
Ringwood, A. E., and Clark, S. P. 1971. Internal constitution of Mars. *Nature* 234:89–92.
Rittenhouse, G. 1971. Pore-space reduction by solution and cementation. *Amer. Assoc. Petrol. Geol. Bull.* 55:80–91.
Rizk, B., Wells, W. K., Hunten, D. M., Stoker, C. R., Freedman, R. S., Roush, T., Pollack, J. B., and Haberle, R. M. 1991. Meridional Martian water abundance profiles during the 1988–1989 season. *Icarus* 90:205–213.
Robinson, M. S., and Tanaka, K. L. 1988. Stratigraphy on the Kasei Valles region, Mars. In *MEVTV Workshop: Nature and Composition of Surface Units on Mars*, eds. J. R. Zimbelman, S. C. Solomon and V. L. Sharpton, LPI Tech Rept. 88-05, pp. 106–108 (abstract).
Robson, R. L., Eady, R. R., Richardson, T. H., Miller, R. W., Hawkins, M., and Postgate, J. R. 1986. The alternative nitrogenase of *azotobacter chroococcum* is a vanadium enzyme. *Nature* 322:388–390.
Rochester, M. G., and Smylie, D. E. 1974. On the changes in the trace of the Earth's inertia tensor. *J. Geophys. Res.* 79:4948–4951.
Roddy, D. J., Pepin, R. O., and Merrill, R. B. eds. 1977. *Impact and Explosion Cratering* (New York: Pergamon).
Rodrigo, R., Garcia-Alvarez, E., Lopez-Gonzalez, M. J., and Lopez-Moreno, J. J. 1990. A nonsteady one-dimensional theoretical model of Mars' neutral atmospheric composition between 30 and 200 km. *J. Geophys. Res.* 95:14795–14810.
Rogers, A. E. E., Ash, M. E., Counselman, C. C., Shapiro, I. I., and Pettengill, G. H. 1970. Radar measurements of the surface topography and roughness of Mars. *Radio Sci.* 5:465–473.
Rohrbaugh, R. P., Nisbet, J. S., Bleuler, E., and Herman, J. R. 1979. The effect of energetically produced O_2^+ on the ion temperatures of the Martian thermosphere. *J. Geophys. Res.* 84:3327–3338.
Romine, G. L., Reisert, T. D., and Gliozzi, J. 1973. *Site Alternation Effects from Rocket Exhaust Impingement During a Simulated Viking Mars Landing, Part 1 of 2, Nozzle Development and Physical Alteration*, NASA CR-2252.
Ronca, L. B. 1965. A geologic record of mare Humorum. *Icarus* 4:390–395.
Rosenbauer, H., Schutte, N., Apáthy, I., Galeev, A., Gringauz, K., Grünwaldt, H., Hemmerich, P., Jockers, K., Király, P., Kotova, G., Livi, S., Marsch, E., Richter, A., Riedler, W., Remizov, T., Schwenn, R., Schwingenschuh, K., Steller, M., Szegö, K., Verigin, M., and Witte, M. 1989. Ions of martian origin and plasma sheet in the Martian magnetosphere: Initial results of the TAUS experiment. *Nature* 341:612–614.
Rosenqvist, J., Bibring, J. P., Bombes, M., Rossart, P., Encrenaz, T., Erard, S., Forni, O., Gondet, B., Langevin, Y., Lellouch, E., Masson, P., and Soufflot, A. 1990. The vertical distribution of carbon monoxide on Mars from the ISM-Phobos experiment. *Astron. Astrophys.* 231:L29–L32.
Rossbacher, L. A., and Judson, S. 1981. Ground ice on Mars: Inventory, distribution, and resulting landforms. *Icarus* 45:39–59.
Rossow, W. B. 1978. Cloud microphysics: Analysis of the clouds of Earth, Venus, Mars, and Jupiter. *Icarus* 36:1–50.
Roth, L. E., Downs, G. S., Saunders, R. S., and Schubert, G. 1980. Radar altimetry of south Tharsis, Mars. *Icarus* 42:287–316.
Roth, L. E., Kobrick, M., Downs, G. S., Saunders, R. S., and Schubert, G. 1981. Martian center of mass-center of figure offset. In *Reports of Planetary Geology Program–1981*, NASA TM-84211, pp. 372–374.

Roth, L. E., Saunders, R. S., Downs, G. S., and Schubert, G. 1989. Radar altimetry of large Martian craters. *Icarus* 79:289–310.

Rothman, L. S., Gamache, R. R., Goldman, A., Brown, L. R., Toth, R. A., Pickett, H. M., Poynter, R. L., Flaud, J.-M., Camy-Peyret, C., Barbe, A., Husson, N., Rinsland, C. P., and Smith, M. A. H. 1987. The HITRAN database: 1986 edition. *Appl. Opt.* 26:4058–4097.

Rothschild, L. J., and DesMarais, D. 1989. Stable carbon isotope fractionation in the search for life on early Mars. *Adv. Space Res.* 9(6):159–165.

Rotto, S. L., and Tanaka, K. L. 1989. Faulting history of the Alba Patera-Ceraunius Fossae region of Mars. *Lunar Planet. Sci.* XX:926–927 (abstract).

Roush, T. L. 1989. Infrared transmission measurements of Martian soil analogs. In *MECA Workshop on Dust on Mars III,* ed. S. Lee, LPI Tech. Rept. 89-01, pp. 52–54 (abstract).

Roush, T. L., Pollack, J. B., and Banin, A. 1989*a*. Mid-infrared transmission measurements of additional Martian soil analogs. Fourth Intl. Conf. on Mars, Jan. 10–13, Tucson, Arizona, Abstract Book, pp. 173–174.

Roush, T. L., Pollack, J. B., Stoker, C., Witteborn, F., Bregman, J., Wooden, D., and Rank, D. 1989*b*. CO_3^2 and SO_4^2-bearing anionic complexes detected in Martian atmospheric dust. *Lunar Planet Sci.* XX:928–929 (abstract).

Rowland, S. K., and Walker, G. P. L. 1990. Pahoehoe and aa in Hawaii: Volumetric flow rate controls the lava structure. *Bull. Volcanol.* 52:615–628.

Ruck, G. T., Barrick, D. E., Stuart, W. D., and Krichbaum, C. K. 1970. *Radar Cross Section Handbook* (New York: Plenum Press).

Rudy, D. J., Muhleman, D. O., Berge, G. L., Jakosky, B. M., and Christensen, P. R. 1987. Mars: VLA observations of the Northern hemisphere and North Polar region at wavelengths of 2 and 6 cm. *Icarus* 71:159–177.

Runyon, C. J., and Golombek, M. P. 1983. Martian grabens and permafrost thickness on Mars. *Lunar Planet. Sci.* XIV:660–661 (abstract).

Russell, C. T. 1978*a*. The magnetic field of Mars: Mars 3 evidence reexamined. *Geophys. Res. Lett.* 5:81–84.

Russell, C. T. 1978*b*. The magnetic field of Mars: Mars 5 evidence reexamined. *Geophys. Res. Lett.* 5:85–88.

Russell, C. T. 1979. The interaction of the solar wind with Mars, Venus and Mercury. In *Solar System Plasma Physics Vol. II,* eds. C. F. Kennel, L. J. Lanzerotti and E. N. Parker (Amsterdam: North Holland Press), pp. 209–247.

Russell, C. T. 1981. The magnetic fields of Mercury, Venus and Mars. *Adv. Space Res.* 1:3–20.

Russell, C. T. 1985. Planetary bow shocks. In *Collisionless Shocks in the Heliosphere: Reviews of Current Research,* AGU Geophys. Mono. 35 (Washington, D.C.: American Geophysical Union), pp. 109–130.

Russell, C. T. 1986. Reply. *J. Geophys. Res.* 91:12149–12150.

Russell, C. T. 1987. Planetary magnetism. In *Geomagnetism Vol. 2,* ed. J. A. Jacobs (Orlando: Academic Press), pp. 458–523.

Russell, C. T., Elphic, R. C., and Slavin, J. A. 1979. Initial Pioneer Venus magnetic field results: Dayside observations. *Science* 203:745–747.

Russell, C. T., Gombosi, T. I., Horanyi, M., Cravens, T. E., and Nagy, A. F. 1983. Charge exchange in the magnetosheaths of Venus and Mars: A comparison. *Geophys. Res. Lett.* 10:163–164.

Russell, C. T., Luhmann, J. G., Spreiter, J. R., and Stahara, S. S. 1984. The magnetic field of Mars: Implications from gas dynamic modeling. *J. Geophys. Res.* 89: 2997–3003.

Ryan, J. A. 1964. Notes on the Martian yellow clouds. *J. Geophys. Res.* 69:3759–3770.

Ryan, J. A. 1972. Relation of dust devil frequency and diameter to atmospheric temperature. *J. Geophys. Res.* 77:7133–7137.

Ryan, J. A., and Carroll, J. J. 1970. Dust devil wind velocities: Mature state. *J. Geophys. Res.* 75:531–541.

Ryan, J. A., and Henry, R. M. 1979. Mars atmospheric phenomena during major dust storms as measured at the surface. *J. Geophys. Res.* 84:2821–2829.

Ryan, J. A., and Lucich, R. D. 1983. Possible dust devils, vortices on Mars. *J. Geophys. Res.* 88:11005–11011.

Ryan, J. A., and Sharman, R. D. 1981*a*. H_2O frost point detection of Mars? *J. Geophys. Res.* 86:503–511.

Ryan, J. A., and Sharman, R. D. 1981*b*. Two major dust storms, one Mars year apart: Comparison from Viking data. *J. Geophys. Res.* 86:3247–3254.

Ryan, J. A., Henry, R. M., Hess, S. L., Leovy, C. B., Tillman, J. E., and Walcek, C. 1978. Mars meteorology: Three seasons at the surface. *Geophys. Res. Lett.* 5:715–718; corrigendum 5:815.

Ryan, J. A., Sharman, R. D., and Lucich, R. D. 1981. Local Mars dust storm generation mechanism. *Geophys. Res. Lett.* 8:899–902.

Ryan, J. A., Sharman, R. D., and Lucich, R. D. 1982. Mars water vapor, near-surface. *J. Geophys. Res.* 87:7279–7284.

Ryan, M. P. 1987. The elasticity and contractancy of Hawaiian olivine tholeiite and its role in the stability and structural evolution of subcaldera magma reservoirs and rift systems. *U.S. Geol. Survey Prof. Paper 1350,* col. 2, pp. 1395–1447.

Ryan, M. P. 1988. The mechanics and three-dimensional internal structure of active magmatic systems: Kilauea Volcano, Hawaii. *J. Geophys. Res.* 93:4213–4248.

Sadovsky, A. V., and Bondarenko, G. I. 1983. Creep of frozen soils on rock slopes. In *Proc. Fourth Intl. Conf. on Permafrost,* pp. 1101–1104.

Safronov, V. S., Pechernikova, G. V., Ruskol, E. L., and Vitjazev, A. V. 1986. Protosatellite swarms. In *Satellites,* eds. J. A. Burns and M. S. Matthews (Tucson: Univ. of Arizona Press), pp. 89–116.

Sagan, C. 1977. Reducing greenhouses and the temperature history of Earth and Mars. *Nature* 269:224–226.

Sagan, C., and Lederberg, J. 1976. The prospects for life on Mars: A pre-Viking assessment. *Icarus* 28:291–300.

Sagan, C. A., and Mullen, G. 1972. Earth and Mars: Evolution of atmospheres and surface temperatures. *Science* 177:52–56.

Sagan, C. A., and Pollack, J. B. 1969. Windblown dust on Mars. *Nature* 223:791–794.

Sagan, C., Hanst, P. L., and Young, A. T. 1965. Nitrogen oxides on Mars. *Planet. Space Sci.* 13:73–88.

Sagan, C. A., Pollack, J. B., and Goldstein, R. M. 1967. Radar Doppler spectroscopy of Mars. I. Elevation differences between bright and dark areas. *Astron. J.* 72:20–34.

Sagan, C., Veverka, J., and Gierasch, P. 1971*a*. Observational consequences of Martian wind regimes. *Icarus* 15:253–278.

Sagan, C., Owen, T. C., and Smith, H. J., eds. 1971*b*. *Planetary Atmospheres* (Dordrecht: D. Reidel).

Sagan, C. A., Veverka, J., Fox, P., Dubisch, R., Lederberg, J., Levinthal, E., Quam, L., Tucker, R., Pollack, J. B., and Smith, B. A. 1972. Variable features on Mars: Preliminary Mariner 9 television results. *Icarus* 17:346–372.

Sagan, C., Toon, O. B., and Gierasch, P. J. 1973*a*. Climate change on Mars. *Science* 181:1045–1049.

Sagan, C., Veverka, J., Fox, P., Dubisch, R., French, R., Gierasch, P., Quam, L., Lederberg, J., Levinthal, E., Tucker, R., Eross, B., and Pollack, J. B. 1973*b*. Variable features on Mars, 2, Mariner 9. Global results. *J. Geophys. Res.* 78:4163–4196.

Sagan, C., Pieri, D., Fox, P., Arvidson, R., and Guinness, E. 1977. Particle motion on Mars inferred from the Viking lander cameras. *J. Geophys. Res.* 82:4330–4438.

Sagdeev, R. Z., and Zakharov, A. V. 1989. Brief history of the Phobos mission. *Nature* 341:581–585.

Salby, M. L. 1984. Survey of planetary-scale traveling waves: The state of theory and observations. *Rev. Geophys.* 2:209–236.

Salisbury, J. W., Walter, L. S., and Vergo, N. 1987. Mid-Infrared (2.1–25μm) Spectra of Minerals: First Edition. U.S.G.S. Open-File Rept. 87–263.

Sasaki, S. 1990. Origin of Phobos—Aerodynamic drag capture by the primary atmosphere of Mars. *Lunar Planet. Sci.* XXI:1069–1070 (abstract).

Sasaki, S., and Nakazawa, K. 1986. Terrestrial Xe fractionation due to escape of primordial H_2-He atmosphere. In *Abstracts for the Japan-U.S. Seminar on Terrestrial Rare Gases,* ed. J. Reynolds (Berkeley: Univ. of California Dept. of Physics), pp. 68–71 (abstract).

Sasaki, S., and Nakazawa, K. 1988. Origin of isotopic fractionation of terrestrial Xe: Hydrodynamic fractionation during escape of the primordial H_2-He atmosphere. *Earth Planet. Sci. Lett.* 89:323–334.

Saunders, M. A., and Russell, C. T. 1986. Average dimension and magnetic structure of the distant Venus magnetotail. *J. Geophys. Res.* 91:5589–5604.

Saunders, R. S., and Blewett, D. T. 1987. Mars north polar dunes: Possible formation from low density sediment aggregates. *Astron. Vestn.* 21:181–188.

Saunders, R. S., and Gregory, T. E. 1980. Tectonic implications of the Martian ridged plains. In *Reports of Planetary Geology Program–1980,* NASA TM-82385, pp. 93–94 (abstract).

Saunders, R. S., Bills, T. G., and Johansen, L. 1981. The ridged plains of Mars. *Lunar Planet. Sci.* XII:924–925 (abstract).

Saunders, R. S., Parker, T. J., Stephens, J. B., Laue, E. G., and Fanale, F. P. 1985. Transformation of polar ice sublimate residue into Martian circumpolar sand. In *Reports of Planetary Geology Program–1984,* NASA TM-87563, pp. 300–302 (abstract).

Saunders, R. S., Fanale, F. P., Parker, T. J., Stephens, J. B., and Sutton, S. 1986. Properties of filamentary sublimation residues from dispersions of clay in ice. *Icarus* 66:94–104.

Savich, N. A., Samovol, V. A., Vasilyev, M. B., Vyshlov, A. S., Samoznaev, L. N., Sidorenko, A. I., and Shtern, D. Ya. 1976. The nighttime ionosphere of Mars from Mars 4 and Mars 5 radio occultation dual-frequency measurements. In *Solar Wind Interaction with the Planets Mercury, Venus and Mars,* ed. N. F. Ness, NASA SP-397, pp. 41–46.

Scambos, T. A., and Jakosky, B. M. 1990. An outgassing release factor for non-radiogenic volatiles on Mars. *J. Geophys. Res.* 95:14779–14787.

Scarfe, C. M., and Takahashi, E. 1978. Melting of garnet peridotite to 13 GPa: Implications for the early history of the upper mantle. *Nature* 322:354–356.

Schaber, G. G. 1977. *Geologic Map of the Iapygia Quadrangle of Mars.* U.S.G.S. Map I-1020 (MC-21).

Schaber, G. G. 1980. Radar, visual and thermal characteristics of Mars: Rough planar surfaces. *Icarus* 42:159–184.

Schaber, G. G. 1982. Syrtis Major: A low-relief volcanic shield. *J. Geophys. Res.* 87:9852–9866.

Schaber, G. G., Horstman, K. C., and Dial, A. L., Jr. 1978. Lava flow materials in the Tharsis regions of Mars. *Proc. Lunar Planet. Sci. Conf.* 9:3433–3458.

Scheidegger, A. E. 1973. On the prediction of the reach and velocity of catastrophic landslides. *Rock Mech.* 5:231–236.

Schiaparelli, G. V. 1879. Osservazioni astronomiche e fisiche sull'asse di rotazione e sulla topografia del planete Marte. In *Atti della Royale Accademia dei Lincei, Memoria 1, ser.3* 2:308–439.

Schidlowski, M. 1988. A 3,800-million-year isotopic record of life from carbon in sedimentary rocks. *Nature* 333:313–318.

Schmidt, R. M., and Housen, K. R. 1987. Some recent advances in the scaling of impact and explosion cratering. *Intl. J. Impact Eng.* 5:543–560.

Schneeberger, D. M., and Pieri, D. C. 1991. Geomorphology and stratigraphy of Alba Patera, Mars. *J. Geophys. Res.* 96:1907–1930.

Schneider, E. K. 1983. Martian great dust storms: Interpretive axially symmetric models. *Icarus* 55:302–331.

Schneider, E. K. 1984. Response of the annual and zonal mean winds and temperatures to variations in the heat and momentum sources. *J. Atmos. Sci.* 41:1093–1115.

Schonfeld, E. 1976. On the origin of the Martian channels. *Eos: Trans. AGU* 57:948 (abstract).

Schonfeld, E. 1979. Origin of Valles Marineris. *Proc. Lunar Planet. Sci. Conf.* 10:3031–3038.

Schopf, J. W., ed. 1983. *Earth's Earliest Biosphere: Its Origin and Evolution* (Princeton, N.J.: Princeton Univ. Press).

Schopf, J. W., and Packer, B. M. 1987. Early Archean (3.3-billion to 3.5-billion-year-old) microfossils from Warrawoona Group, Australia. *Science* 237:70–73.

Schorn, R. A. 1971. The spectroscopic search for water vapor on Mars: A history. In *Planetary Atmospheres,* eds. C. Sagan, T. C. Owen and H. J. Smith (Dordrecht: D. Reidel), pp. 223–236.

Schorn, R. A., Spinard, H., Moore, R. C., Smith, H. J., and Giver, L. P. 1967. High dispersion spectroscopic observations of Mars. II. The water-vapor variations. *Astrophys. J.* 147:743–752.

Schubert, G., and Lingenfelter, R. E. 1973. Martian center of mass-center of figure offset. *Nature* 242:251–252.

Schubert, G., and Spohn, T. 1990. Thermal history of Mars and the sulfur content of its core. *J. Geophys. Res.* 95:14095–14104.
Schubert, G., Cassen, P., and Young, R. E. 1979. Subsolidus convective cooling histories of terrestrial planets. *Icarus* 38:192–211.
Schubert, G., Bercovici, D., and Glatzmaier, G. A. 1990. Mantle dynamics in Mars and Venus: Influence of an immobile lithosphere on three-dimensional mantle convection. *J. Geophys. Res.* 95:14105–14130.
Schultz, P. H. 1976. *Moon Morphology* (Austin: Univ. of Texas Press).
Schultz, P. H. 1984. Impact basin control of volcanic and tectonic provinces on Mars. *Lunar Planet. Sci.* XV:728–729 (abstract).
Schultz, P. H. 1985. Polar wandering on Mars. *Sci. Amer.* 253:94–102.
Schultz, P. H. 1986. Crater ejecta morphology and the presence of water on Mars. In *Symp. on Mars: Evolution of its Climate and Atmosphere,* LPI Contrib. 599 (Houston: Lunar and Planetary Inst.), pp. 95–97.
Schultz, P. H. 1987. Impact velocity and changes in crater shape, morphology, and statistics. *Lunar Planet. Sci.* XVIII:886–887 (abstract)
Schultz, P. H. 1988. Cratering on Mercury: A relook. In *Mercury,* eds. F. Vilas, C. R. Chapman and M. S. Matthews (Tucson: Univ. of Arizona Press), pp. 274–335.
Schultz, P. H. 1989. Factors controlling impact ejecta emplacement on Mars. Fourth Intl. Conf. on Mars, Jan. 10–13, Tucson, Arizona, Abstract Book, pp. 181–182.
Schultz, P. H., and Britt, D. 1986. Martian gradation history. *Lunar Planet. Sci.* XVII:775–776 (abstract).
Schultz, P. H., and Gault, D. E. 1979. Atmospheric effects on Martian ejecta emplacement. *J. Geophys. Res.* 84:7669–7687.
Schultz, P. H., and Gault, D. E. 1984. On the formation of contiguous ramparts around Martian impact craters. *Lunar Planet. Sci.* XV:732–733 (abstract).
Schultz, P. H., and Glicken, H. 1979. Impact crater and basin control of igneous processes on Mars. *J. Geophys. Res.* 84:8033–8047.
Schultz, P. H., and Ingerson, F. E. 1973. Martian lineaments from Mariner 6 and 7 images. *J. Geophys. Res.* 78:8415–8427.
Schultz, P. H., and Lutz, A. B. 1988. Polar wandering of Mars. *Icarus* 73:91–141.
Schultz, P. H., and Lutz-Garihan, A. B. 1981. Ancient polar locations on Mars: Evidence and implications. In *Proc. Third Intl. Coll. on Mars,* LPI Contrib. 441, pp. 229–232 (abstract).
Schultz, P. H., and Lutz-Garihan, A. B. 1982. Grazing impacts on Mars: A record of lost satellites. *Proc. Lunar Planet. Sci. Conf.* 13, *J. Geophys. Res. Suppl.* 87:A84-A96.
Schultz, P. H., Glicken, H., and McGetchin, T. R. 1979. Genesis of Martian outflow channels by crater-controlled intrusions. NASA CP-2072, p. 72 (abstract).
Schultz, P. H., Schultz, R. A., and Rogers, J. R. 1982. The structure and evolution of ancient impact basins on Mars. *J. Geophys. Res.* 78:9803–9820.
Schultz, R. A. 1985. Assessment of global and regional tectonic models for faulting in the ancient terrains of Mars. *J. Geophys. Res.* 90:7849–7860.
Schultz, R. A. 1989*a*. Do pit-crater chains grow up to be Valles Marineris canyons? In *MEVTV Workshop on Tectonic Features on Mars,* eds. T. R. Watters and M. P. Golombek, LPI Tech. Rept. 89–06, pp. 47–48 (abstract).
Schultz, R. A. 1989*b*. Strike-slip faulting of ridged plains of Mars. *Nature* 341:424–428.
Schultz, R. A. 1991. Structural development of Coprates chasma and western Ophir Planum, Valles Maineris rift, Mars. *J. Geophys. Res.* 96:2277–2279.
Schultz, R. A., and Frey, H. V. 1988. Peripheral grabens associated with Valles Marineris canyons, Mars: Clues to canyon growth and early Tharsis tectonism. *Eos: Trans. AGU* 69:389–390.
Schultz, R. A., and Frey, H. V. 1989. Topography and structure of Valles Marineris, Mars. Fourth Intl. Conf. on Mars, Jan. 10–13, Tucson, Arizona, Abstract Book, pp. 183–184.
Schultz, R. A., and Frey, H. V. 1990. A new survey of large multiring impact basins on Mars. *J. Geophys. Res.* 95:14175–14189.
Schumm, S. A. 1974. Structural origin of large Martian channels. *Icarus* 22:371–384.
Schunk, R. W., and Nagy, A. F. 1980. Ionospheres of the terrestrial planets. *Rev. Geophys. Space Phys.* 18:813–852.

Schurmeier, H. M. 1975. Planetary exploration. Earth's new horizon. *J. Spacecraft and Rockets* 12:385–405.

Schutte, N. M., Király, P., Cravens, T. E., Dyachkov, A. V., Gombosi, T. I., Gringauz, K. I., Nagy, A. F., Sharp, W. E., Sheronova, S. M., Szegö, K., Szemery, T., Szucs, I. T., Tátrallyay, M., Tóth, A., and Verigin, M. 1989. Observation of electron and ion fluxes in the vicinity of Mars with the HARP spectrometer. *Nature* 341:614–616.

Schwarcz, H. P., Hoefs, J., and Welte, D. 1969. Carbon. In *Handbook of Geochemistry, Vol. II/1* (Berlin: Springer-Verlag), pp. 6-B-1–6–O-3.

Schwertmann, U. 1988. Geothites and hematite formation in the presence of clay minerals and gibbsite at 25°C. *Soil Sci. Soc. Amer. J.* 52:288–291.

Schwingenschuh, K., Riedler, W., Lichtenegger, H., Yeroshenko, Ye., Sauer, K., Luhmann, J. G., Ong, M., and Russell, C. T. 1990. The Martian bow shock: Phobos observations. *Geophys. Res. Lett.* 17:889–892.

Scott, D. H. 1973. Small structures of the Taurus-Littrow region. In *Apollo 17 Prelim. Sci. Rept.*, NASA SP-330, pp. 31/25–31/28.

Scott, D. H. 1978. Mars, highlands-lowlands: Viking contributions to Mariner relative age studies. *Icarus* 34:479–485.

Scott, D. H. 1979. Geologic problems in the northern plains of Mars. *Proc. Lunar Planet. Sci. Conf.* 10:3039–3054.

Scott, D. H. 1982. Volcanoes and volcanic provinces: Martian western hemisphere. *J. Geophys. Res.* 87:9839–9851.

Scott, D. H. 1983. Meander relics: Direct evidence of extensive flooding on Mars. In *Conf. on Planetary Volatiles,* eds. R. Pepin and R. O'Connell, LPI Tech. Rept. 83-01, pp. 157–159 (abstract).

Scott, D. H. 1989. New evidence-old problem: Wringle ridge origin. In *MEVTV Workshop on Tectonic Features on Mars,* eds. T. R. Watters and M. P. Golombek, LPI Tech. Rept.89-06, pp. 26–28 (abstract).

Scott, D. H., and Carr, M. H. 1978. *Geologic Map of Mars, scale 1:25,000,000.* U.S.G.S. Misc. Inv. Series Map I-1083.

Scott, D. H., and Dohm, J. M. 1989. Chronology and global distribution of fault and ridge systems on Mars. *Lunar Planet. Sci.* XX:976–977 (abstract).

Scott, D. H., and Dohm, J. M. 1990*a*. Chronology and global distribution of fault and ridge systems on Mars. *Proc. Lunar Planet. Sci. Conf.* 20:487–501.

Scott, D. H., and Dohm, J. M. 1990*b*. Faults and ridges: Historical development in Tempe Terra and Ulysses Patera regions of Mars. *Proc. Lunar Planet. Sci. Conf.* 20:503–513.

Scott, D. H., and King, J. S. 1984. Ancient surfaces of Mars: The basement complex. In *Lunar Planet. Sci.* XV:736–737 (abstract).

Scott, D. H., and Tanaka, K. L. 1980. Mars Tharsis region: Volcanotectonic events in the stratigraphic record. *Proc. Lunar Planet. Sci. Conf.* 11:2403–2421.

Scott, D. H., and Tanaka, K. L. 1981. Mars: Paleostratigraphic restoration of buried surfaces in Tharsis Montes. *Icarus* 45:304–319.

Scott, D. H., and Tanaka, K. L. 1982. Ignimbrites of Amazonis Planitia region of Mars. *J. Geophys. Res.* 87:1179–1190.

Scott, D. H., and Tanaka, K. L. 1986. *Geologic Map of the Western Equatorial Region of Mars, scale 1:15,000,000.* U.S.G.S. Misc. Inv. Series Map I-1802-A.

Scott, D. H., Schaber, G. G., Tanaka, K. L., Horstman, K. C., and Dial, A. L., Jr. 1981*a*. *Map Series Showing Lava-Flow Fronts in the Tharsis Region of Mars.* U.S.G.S. Misc. Inv. Series Maps I-1266 to I-1280.

Scott, D. H., Tanaka, K. L., and Schaber, G. G. 1981*b*. *Map Showing Lava Flows in the Southeast Part of the Diacria Quadrangle of Mars. U.S.G.S. Atlas of Mars, 1:2,000,000* Geol. series I-1276 (MC-2SE).

Seeburger, D. A., and Zoback, M. D. 1982. The distribution of natural fractures and joints at depth in crystalline rock. *J. Geophys. Res.* 87:5517–5534.

Seiff, A., and Kirk, D. B. 1977. Structure of the atmosphere of Mars in summer at mid-latitudes. *J. Geophys. Res.* 82:4364–4378.

Selivanov, A. S., Naraeva, M. K., Panfilov, A. S., Gektin, Yu. M., Kharlamov, V. D., Romanov, A. V., Fomin, D. A., and Miroshnichenko, Ya. Ya. 1989. Thermal imaging of the surface of Mars. *Nature* 341:593–595.

Settle, M. 1979. Formation and deposition of volcanic sulfate aerosols on Mars. *J. Geophys. Res.* 84:8343–8354.

Shaller, P. J., Murray, B. C., and Albee, A. L. 1989. Subaqueous landslides on Mars? *Lunar Planet. Sci.* XX:990–991 (abstract).

Sharman, R. D., and Ryan, J. A. 1980. Mars atmospheric pressure periodicities from Viking observations. *J. Atmos. Sci.* 37:1994–2001.

Sharonov, V. V. 1961. A lithological interpretation of the photometric and colorimetric studies of Mars. *Soviet Astron.-AJ* 5:199–202.

Sharp, R. P. 1973*a*. Mars: Fretted and chaotic terrain. *J. Geophys. Res.* 78:4073–4083.

Sharp, R. P. 1973*b*. Mars: South Polar pits and etched terrain. *J. Geophys. Res.* 78:4222–4230.

Sharp, R. P. 1973*c*. Mars: Troughed terrain. *J. Geophys. Res.* 78:4063–4072.

Sharp, R. P., and Malin, M. C. 1975. Channels on Mars. *Geol. Soc. Amer. Bull.* 86:593–609.

Sharp, R. P., and Malin, M. C. 1984. Surface geology from Viking landers on Mars: A second look. *Geol. Soc. Amer. Bull.* 96:1398–1412.

Sharpe, H. N., and Peltier, W. R. 1979. A thermal history for the Earth with parameterized convection. *Geophys. J. Roy. Astron. Soc.* 59:171–203.

Sharpless, B. P. 1945. Secular accelerations in the longitudes of the satellites of Mars. *Astron. J.* 51:185–195.

Sharpton, V. L., and Head, J. W., III. 1982. Stratigraphy and structural evolution of southern Mare Serenitatis: A reinterpretation based on Apollo lunar sounder experiment data. *J. Geophys. Res.* 87:10983–10998.

Sharpton, V. L., and Head, J. W., III. 1988. Lunar mare ridges: Analysis of ridge-crater intersections and implications for the tectonic origin of mare ridges. *Proc. Lunar Planet. Sci. Conf.* 18:307–317.

Sherman, D. M., Burns, R. G., and Burns, V. M. 1982. Spectral characteristics of the iron oxides with application to the Martian bright region mineralogy. *J. Geophys. Res.* 87:10169–10180.

Sherman, M. W., ed. 1982. *TRW Space Log,* vol. 19 (Redondo Beach, Calif.: TRW Electronics and Defense).

Shih, C.-Y., Nyquist, L. E., Bogard, D. D., McKay, G. A., Wooden, J. L., Bansal, B. M., and Wiesmann, H. 1982. Chronology and petrogenesis of young achondrites, Shergotty, Zagami and ALHA 77005: Late magmatism on the geologically active planet. *Geochim. Cosmochim. Acta* 46:2323–2344.

Shinagawa, H., and Cravens, T. E. 1989. A one-dimensional multi-species magnetohydrodynamic model of the dayside ionosphere of Venus. *J. Geophys. Res.* 94:6506–6516.

Shklovskii, I. S., and Sagan, C. 1966. *Intelligent Life in the Universe* (San Francisco: Holden-Day).

Shoemaker, E. M. 1963. Impact mechanics at Meteor Crater, Arizona. In *The Moon, Meteorites, and Comets,* eds. B. M. Middlehurst and G. P. Kuiper (Chicago: Univ. of Chicago Press), pp. 301–336.

Shoemaker, E. M. 1989. *A.I.A.A. Meeting: "Exploration of the Solar System,"* Pasadena, Calif.

Shoemaker, E. M., and Helin, E. F. 1977. Populations of planet-crossing asteroids and the relationship of Apollo objects to main belt asteroids and comets. In *Comets, Asteroids, Meteorites,* ed. A. H. Delsemme (Toledo: Univ. of Toledo Press), pp. 297–300.

Shoemaker, E. M., and Wolfe, R. F. 1982. Cratering time scales for the Galilean satellites. In *Satellites of Jupiter,* ed. D. Morrison (Tucson: Univ. of Arizona Press), pp. 339–377.

Shoji, H., and Higashi, A. 1978. A deformation mechanism map of ice. *J. Glaciol.* 21:419–427.

Shor, V. A. 1988. Refinement of the orbits of Phobos and Deimos using ground and spacecraft observations. *Soviet Astron. Lett.* 14:477–481.

Short, N. M. 1970. Anatomy of a meteorite impact crater: West Hawk Lake, Manitoba, Canada. *Geol. Soc. Amer. Bull.* 81:609–648.

Shreve, R. L. 1966. Sherman landslide, Alaska. *Science* 154:1639–1643.

Sidorenko, A. V., ed. 1980. *The Surface of Mars* (Moscow: Nauka Press). In Russian.

Sidorov, Yu. I., and Zolotov, M. Yu. 1986. Weathering of Martian surface rocks. In *Chemistry and Physics of Terrestrial Planets,* ed. S. K. Saxena (New York: Springer-Verlag), pp. 191–223.

Sill, G. T., and Wilkening, L. L. 1978. Ice clathrate as a possible source of the atmospheres of the terrestrial planets. *Icarus* 33:13–22.

Simpson, R. A., and Tyler, G. L. 1980. Radar measurement of heterogeneous small-scale texture on Mars: Chryse. *J. Geophys. Res.* 85:6610–6614.
Simpson, R. A., and Tyler, G. L. 1981. Viking bistatic radar experiment: Summary of first-order results emphasizing north polar data. *Icarus* 46:361–389.
Simpson, R. A., and Tyler, G. L. 1982. Radar scattering laws for the lunar surface. *IEEE Trans. Antennas and Propagation* AP-30, pp. 438–449.
Simpson, R. A., Tyler, G. L., and Lipa, B. J. 1977. Mars surface properties observed by earth-based radar at 70-, 12.5-, and 3.8-cm wavelengths. *Icarus* 32:147–167.
Simpson, R. A., Tyler, G. L., and Campbell, D. B. 1978*a*. Arecibo radar observations of Martian surface characteristics near the equator. *Icarus* 33:102–115.
Simpson, R. A., Tyler, G. L., and Campbell, D. B. 1978*b*. Arecibo radar observations of Mars surface characteristics in the northern hemisphere. *Icarus* 36:153–173.
Simpson, R. A., Tyler, G. L., Brenkle, J. P., and Sue, M. 1979. Viking bistatic radar observations of the Hellas Basin on Mars: Preliminary results. *Science* 203:45–46.
Simpson, R. A., Tyler, G. L., Harmon, J. K., and Peterfreund, A. R. 1982. Radar measurement of small-scale surface texture: Syrtis Major. *Icarus* 49:258–283.
Simpson, R. A., Tyler, G. L., and Schaber, G. G. 1984. Viking bistatic radar experiment: Summary of results in near-equatorial regions. *J. Geophys. Res.* 89:10385–10404.
Sinclair, A. T. 1989. The orbits of the satellites of Mars determined from Earth-based and spacecraft observations. *Astron. Astrophys.* 220:321–328.
Sinclair, P. C. 1966. A Quantitative Analysis of the Dust Devils. Ph.D. Thesis, Univ. of Arizona.
Sinclair, P. C. 1969. General characteristics of dust devils. *J. Appl. Meteor.* 8:32–45.
Singer, R. B. 1980. The dark materials on Mars. II. New mineralogic interpretations from reflectance spectroscopy and petrologic implications. *Lunar Planet. Sci.* XI:1048–1050 (abstract).
Singer, R. B. 1982. Spectral evidence for the mineralogy of high albedo soils and dust on Mars. *J. Geophys. Res.* 87:10159–10168.
Singer, R. B. 1985. Spectroscopic observations of Mars. *Adv. Space Res.* 5:59–68.
Singer, R. B., and Geissler, P. E. 1988. An independent assessment of derivative analysis of reflectance spectra. *Lunar Planet. Sci.* XIX:1087–1088 (abstract).
Singer, R. B., and McCord, T. B. 1979. Mars: Large-scale mixing of bright and dark surface materials and implications for analysis of spectral reflectance. *Proc. Lunar Sci. Conf.* 10:1835–1848.
Singer, R. B., and Roush, T. L. 1985. Analysis of Martian crustal petrology. *Amer. Astron. Soc. Bull.* 17:737 (abstract).
Singer, R. B., McCord, T. B., Clark, R. N., Adams, J. B., and Huguenin, R. L. 1979. Mars surface composition from reflectance spectroscopy: A summary. *J. Geophys. Res.* 84:8415–8426.
Singer, R. B., Owensby, P. D., and Clark, R. N. 1984. First direct detection of clay minerals on Mars. *Bull. Amer. Astron. Soc.* 16:679–680 (abstract).
Singer, R. B., Bus, E. S., Wells, K. W., and Miller, J. S. 1989. Spectral imaging of Mars during the 1988 apparition. Fourth Intl. Conf. on Mars, Jan. 10–13, Tucson, Arizona, Abstract Book, p. 185 (abstract).
Singer, R. B., Miller, J. S., and Wells, W. K. 1990*a*. Observed variations in Martian crustal composition. *Bull. Amer. Astron. Soc.* 22:1061 (abstract).
Singer, R. B., Miller, J. S., Wells, K. W., and Bus, E. S. 1990*b*. Visible and near-IR spectral imaging of Mars during the 1988 opposition. *Lunar Planet. Sci.* XXI:1154–1155 (abstract).
Singer, S. F. 1968. The origin of the Moon and its geophysical consequences. *Geophys. J.* 15:205–226.
Sinton, W. M. 1957. Spectroscopic evidence for vegetation on Mars. *Astrophys. J.* 126:231–239.
Sinton, W. M. 1967. On the composition of Martian surface materials. *Icarus* 6:222–228.
Sinton, W. M., and Strong, J. 1960. Radiometric observations of Mars. *Astrophys. J.* 131:459–469.
Sjogren, W. L. 1979. Mars gravity high resolution results from Viking Orbiter 2. *Science* 203:1006–1010.
Sjogren, W. L., and Ritke, S. J. 1982. Mars: Gravity data analysis of the crater Antoniadi. *Geophys. Res. Lett.* 9:739–742.

Sjogren, W. L., and Wimberly, R. N. 1981. Mars: Hellas Planitia gravity analysis. *Icarus* 45:331–338.
Sjogren, W. L., Lorell, J., Wong, L., and Downs, W. 1975. Mars gravity field based on a short arc technique. *J. Geophys. Res.* 80:2899–2908.
Skinner, S., and Zimbelman, J. R. 1986. Thermal inertia data of Lunae Plaus and Coprates quadrangles, Mars. In *Second Annual Summer-Intern Conf.* (Houston: Lunar and Planetary Inst.), pp. 39–41 (abstract).
Slavin, J. A., and Holzer, R. E. 1981. Solar wind flow about the terrestrial planets 1. Modeling bow shock position and shape. *J. Geophys. Res.* 86:11401–11418.
Slavin, J. A., and Holzer, R. E. 1982. The solar wind interaction with Mars revisited. *J. Geophys. Res.* 87:10285–10296.
Slavin, J. A., Holzer, R. E., Spreiter, J. R., Stahara, S. S., and Chausse, D. S. 1983. Solar wind flow about the terrestrial planets. 2. Comparison with gas dynamic theory and implications for solar-planetary interaction. *J. Geophys. Res.* 88:19–35.
Sleep, N. H., and Phillips, R. J. 1979. An isostatic model for the Tharsis province, Mars. *Geophys. Res. Lett.* 6:803–806.
Sleep, N. H., and Phillips, R. J. 1985. Gravity and lithospheric stress on the terrestrial planets with reference to the Tharsis region of Mars. *J. Geophys. Res.* 90:4469–4489.
Sleep, N. H., Zahnle, K., Kasting, J. F., and Morowitz, H. J. 1989. Annihilation of ecosystems by large asteroid impacts on the early Earth. *Nature* 342:139–142.
Slipher, E. C. 1962. *The Photographic Story of Mars* (Flagstaff: Northland Press).
Slipher, E. C. 1964. *Report of the International Mars Committee 1956* (Flagstaff: Lowell Observatory).
Slipher, E. C., and Wilson, A. G. 1955. *Report of International Mars Committee 1954* (Flagstaff: Lowell Observatory).
Sloan, C. E., Zenone, C., and Mayo, I. R. 1976. *Icings Along the Trans-Alaska Pipeline Route.* U.S.G.S. Prof. Paper 979.
Smalley, I. J., and Krinsley, D. H. 1979. Eolian sedimentation on Earth and Mars: Some comparisons. *Icarus* 40:276–288.
Smirnov, V. N., Omelchenko, A. N., and Vaisberg, O. L. 1978. Possible discovery of cusps near Mars. *Cosmic Res.* 16:688–692.
Smith, A. K., and Lyjak, L. V. 1985. An observational estimate of gravity wave drag from the momentum balance in the middle atmosphere. *J. Geophys. Res.* 90:2233–2241.
Smith, B. A., Young, A. T., and Leovy, C. B. 1970. Blue hazes and Mariner 6 pictures of Mars. *Science* 167:908.
Smith, E. J., Davis, L., Jr., Coleman, P. J., Jr., and Jones, D. E. 1965. Magnetic field measurements near Mars. *Science* 149:1241–1242.
Smith, J. V., and Hervig, R. L. 1979. Shergotty meteorite: Mineralogy, petrography and minor elements. *Meteoritics* 14:121–142.
Smith, M. R., Laul, J. C., Ma, M.-S., Huston, T., Verkouteren, R. M., Lipschutz, M. E., and Schmitt, R. A. 1984. Petrogenesis of the SNC (Shergottites, Nakhlites, Chassigny) meteorites: Implications for their origin from a large dynamic planet, possibly Mars. *Proc. Lunar Planet. Sci. Conf.* 14, *J. Geophys. Res. Suppl.* 86:B612-B630.
Smith, S. A., and Smith, B. A. 1972. Diurnal and seasonal behavior of discrete white clouds on Mars. *Icarus* 16:509–521.
Smoluchowski, R. 1968. Mars: Retention of ice. *Science* 159:1348–1350.
Snyder, C. W. 1979*a*. The extended mission of Viking. *J. Geophys. Res.* 84:7917–7933.
Snyder, C. W. 1979*b*. The planet Mars as seen at the end of the Viking mission. *J. Geophys. Res.* 84:8487–8519.
Snyder, J. P. 1982. Map projections used by the U.S. Geological Survey. *Geol. Survey Bull.* 1532.
Snyder, J. P. 1987. Map projections—A working manual. *U.S.G.S. Bulletin* 1532.
Soderblom, L. A. 1977. Historical variations in the density and distribution of impacting debris in the inner solar system: Evidence from planetary imaging. In *Impact and Explosion Cratering,* eds. D. J. Roddy, R. O. Pepin and R. B. Merrill (New York: Pergamon), pp. 629–633.
Soderblom, L. A., and Wenner, D. B. 1978. Possible fossil water liquid-ice interfaces in the Martian crust. *Icarus* 34:622–637.

Soderblom, L. A., Kriedler, T. J., and Masursky, H. 1973*a*. Latitudinal distribution of a debris mantle on the Martian surface. *J. Geophys. Res.* 78:4117–4122.

Soderblom, L. A., Malin, M. C., Cutts, J. A., and Murray, B. C. 1973*b*. Mariner 9 observations of the surface of Mars in the north polar region. *J. Geophys. Res.* 78:4197–4210.

Soderblom, L. A., Condit, C. D., West, R. A., Herman, B. M., and Kriedler, T. J. 1974. Martian planetwide crater distributions: Implications for geologic history and surface processes. *Icarus* 22:239–263.

Soderblom, L. A., Edwards, K., Eliason, E. M., Sanchez, E. M., and Charette, M. P. 1978. Global color variations of the Martian surface. *Icarus* 34:446–464.

Soffen, G. A. 1977. The Viking project. *J. Geophys. Res.* 82:3959–3970.

Soler, T. 1984. A new matrix development of the potential and attraction at exterior points as a function of the inertia tensors. *Celest. Mech.* 32:257–296.

Solomon, S. C. 1978. On volcanism and thermal tectonics on one-plate planets. *Geophys. Res. Lett.* 5:461–464.

Solomon, S. C. 1979. Formation, history and energetics of cores in the terrestrial planets. *Phys. Earth Planet. Int.* 19:168–182.

Solomon, S. C. 1986. On the early thermal state of the Moon. In *Origin of the Moon,* eds. W. K. Hartmann, R. J. Phillips and G. J. Taylor (Houston: Lunar and Planetary Inst.), pp. 435–452.

Solomon, S. C., and Chaiken, J. 1976. Thermal expansion and thermal stress in the Moon and terrestrial planets: Clues to early thermal history. *Proc. Lunar Sci. Conf.* 7:3229–3243.

Solomon, S. C., and Head, J. W. 1982. Evolution of the Tharsis province of Mars: The importance of heterogeneous lithospheric thickness and volcanic construction. *J. Geophys. Res.* 87:9755–9774.

Solomon, S. C., and Head, J. W. 1990. Heterogeneities in the thickness of the elastic lithosphere of Mars: Constraints on heat flow and internal dynamics. *J. Geophys. Res.* 95:11073–11083.

Soloviev, P. A. 1973. Thermokarst phenomena and landforms due to frost heaving in central Yakutia. *Biuletyn Peryglacjalny* 23:135–155.

Sorem, R. K. 1982. Volcanic ash clusters: Tephra rafts and scavengers. *J. Volcan. Geotherm. Res.* 13:63–71.

Soter, S. L. 1971. The Dust Belts of Mars. CRSR Rept. No. 462, Cornell Univ.

Soter, S. L., and Harris, A. 1977. Are striations on Phobos evidence for tidal stress? *Nature* 268:421–422.

Spencer, J. R. 1984. A tectonic geomorphological classification of the walls of Valles Marineris. In *Reports of Planetary Geology Program–1983,* NASA TM-86246, pp. 243–245 (abstract).

Spencer, J. R., and Croft, S. K. 1986. Valles Marineris as karst: Feasibility and implications for Martian atmospheric evolution. In *Reports of Planetary Geology and Geophysics Program–1985,* NASA TM-88383, pp. 193–195 (abstract).

Spencer, J. R., and Fanale, F. P. 1990. New models for the origin of Valles Marineris closed depressions. *J. Geophys. Res.* 95:14301–14313.

Spenner, K., Knudsen, W. C., Miller, K. L., Novak, V., Russell, C. T., and Elphic, R. C. 1980. Observations of the Venus mantle, the boundary region between solar wind and ionosphere. *J. Geophys. Res.* 85:7655–7662.

Spinrad, H., Münch, G., and Kaplan, L. D. 1963. The detection of water vapor on Mars. *Astrophys. J.* 137:1319–1321.

Spinrad, H., Schorn, R. A., Moore, R., Giver, L. P., and Smith, H. J. 1966. High dispersion spectroscopic observations of Mars. I. The CO_2 content and surface pressure. *Astrophys. J.* 146:331–338.

Spohn, T. 1990. Mantle differentiation and thermal evolution of Mars, Mercury and Venus. *Icarus* 90:222–236.

Sposito, G. 1984. *The Surface Chemistry of Soils* (New York: Oxford Press).

Spreiter, J. R., and Rizzi, A. W. 1972. The Martian bow wave—theory and observation. *Planet. Space Sci.* 20:205–208.

Spreiter, J. R., and Stahara, S. S. 1980. Solar wind flow past Venus: Theory and comparisons. *J. Geophys. Res.* 85:7715–7738.

Spreiter, J. R., Marsh, M. C., and Summers, A. L. 1969. Hydromagnetic aspects of solar wind flow past the Moon. *Cosmic Electrodyn.* 1:5–50.

Spreiter, J. R., Summers, A. L., and Rizzi, A. W. 1970. Solar wind flow past nonmagnetic planets—Venus and Mars. *Planet. Space Sci.* 18:1281–1299.
Squyres, S. W. 1978. Martian fretted terrain: Flow of erosional debris. 34:600–613.
Squyres, S. W. 1979*a*. The distribution of lobate debris aprons and similar flows on Mars. *J. Geophys. Res.* 84:8087–8096.
Squyres, S. W. 1979*b*. The evolution of dust deposits in the Martian north polar region. *Icarus* 40:244–261.
Squyres, S. W. 1984. The history of water on Mars. *Ann. Rev. Earth Planet. Sci.* 12:83–106.
Squyres, S. W. 1989*a*. Early Mars: wet and warm, or just wet? Paper presented at the Fourth Intl. Conf. on Mars, Jan. 10–13, Tucson, Arizona.
Squyres, S. W. 1989*b*. Early Mars: Wet and warm, or just wet? *Lunar Planet. Sci.* XX:1044–1045 (abstract).
Squyres, S. W. 1989*c*. Urey prize lecture: Water on Mars. *Icarus* 79:229–288.
Squyres, S. W., and Carr, M. H. 1986. Geomorphic evidence for the distribution of ground ice on Mars. *Science* 231:249–252.
Squyres, S. W., Wilhelms, D. E., and Moosman, A. C. 1987. Large-scale volcano-ground ice interactions on Mars. *Icarus* 70:385–408.
Stacey, F. D. 1977. *Physics of the Earth,* 2nd ed. (New York: Wiley).
Staudigel, H., and Hart, S. 1983. Alteration of basaltic glass: Mechanisms and significance for the oceanic crust-seawater budget. *Geochim. Cosmochim. Acta* 47:337–350.
Stevenson, D. J. 1974. Planetary magnetism. *Icarus* 22:403.
Stevenson, D. J. 1980. Lunar symmetry and paleomagnetism. *Nature* 287:520–521.
Stevenson, D. J. 1981. Models of the Earth's core. *Science* 241:611–619.
Stevenson, D. J., and Turner, J. S. 1979. Fluid models of mantle convection. In *The Earth, Its Origin, Evolution and Structure,* ed. M. W. McElhinny (New York: Academic Press), pp. 227–263.
Stevenson, D. J., Spohn, T., and Schubert, G. 1983. Magnetism and thermal evolution of the terrestrial planets. *Icarus* 54:466–489.
Stevenson, D. J., Harris, A. W., and Lunine, J. I. 1986. Origins of satellites. In *Satellites,* eds. J. A. Burns and M. S. Matthews (Tucson: Univ. of Arizona Press), pp. 39–88.
Stewart, A. I. 1972. Mariner 6 and 7 ultraviolet spectrometer experiment: Implications of CO_2^+, CO, and O airglow. *J. Geophys. Res.* 77:54–68.
Stewart, A. I., and Barth, C. A. 1979. Ultraviolet night airglow of Venus. *Science* 205:59–62.
Stewart, A. I., and Hanson, W. B. 1982. Mars' upper atmosphere: Mean and variations. *Adv. Space Res.* 2:87–101.
Stewart, A. I., Barth, C. A., Hord, C. W., and Lane, A. L. 1972. Mariner 9 ultraviolet spectrometer experiment: Structure of Mars upper atmosphere. *Icarus* 17:469–474.
Stöffler, D., Gault, F. E., Wedekind, J., and Polkowski, G. 1975. Experimental hypervelocity impact into quartz sand: Distribution and shock metamorphism of ejecta. *J. Geophys. Res.* 80:4062–4077.
Stolper, E., and McSween, H. Y., Jr. 1979. Petrology and origin of the shergottite meteorites. *Geochim. Cosmochim. Acta* 43:1475–1498.
Storrs, A. D., Fanale, F. P., Saunders, R. S., and Stephens, J. B. 1988. The formation of filamentary sublimate residues (FSR) from mineral grains. *Icarus* 76:493–512.
Streltsova, T. D. 1976. Hydrodynamics of groundwater flow in a fractured formation. *Water Resources Res.* 12:405–414.
Strickland, D. J., Thomas, G. E., and Sparks, P. R. 1972. Mariner 6 and 7 ultraviolet spectrometer experiment: Analysis of the OI 1304 Å and 1356 Å emissions. *J. Geophys. Res.* 77:4052–4068.
Strom, R. G. 1964. Analysis of lunar lineaments. 1. Tectonic maps of the Moon. *Commun. Lunar and Planetary Lab* 2:205–216.
Strom, R. G. 1972. Lunar mare ridges, rings, and volcanic ring complexes. In *The Moon: Proc. IAU Symp. 47,* eds. S. K. Runcorn and H. C. Urey (Dordrecht: D. Reidel), pp. 187–215.
Strom, R. G. 1977. Origin and relative age of lunar and Mercurian intercrater plains. *Phys. Earth Planet. Int.* 15:156.
Strom, R. G. 1987. The solar system cratering record: Voyager 2 results at Uranus and implications for the origin of impacting objects. *Icarus* 70:517–535.
Strom, R. G., and Neukum, G. 1988. The cratering record on Mercury and the origin of im-

pacting objects. In *Mercury,* eds. F. Vilas, C. R. Chapman and M. S. Matthews (Tucson: Univ. of Arizona Press), pp. 336–373.

Strom, R. G., and Whitaker, E. A. 1976. Populations of impacting bodies in the inner solar system. In *Reports Accompanying Planetary Program,* NASA TM-X-3364, pp. 194–196.

Struve, H. 1895. Bestimmung der Abplattung und des Aequators von Mars. *Astron. Nachr.* 138:217–228.

Sugiura, N., and Strangway, D. W. 1988. Magnetic studies of meteorites. In *Meteorites and the Early Solar System,* eds. J. Kerridge and M. S. Matthews (Tucson: Univ. of Arizona Press), pp. 595–615.

Surkov, Yu. A. 1977. *Gamma Spectrometry in Cosmic Investigations* (Moscow: Atomizdat).

Surkov, Yu. A., Barsukov, V. L., Moskaleva, L. P., Kharyukova, V. P., Zitseva, S. Ye., Smirnov, G. G., and Manvelyan, O. S. 1989. Determination of the elemental composition of Martian rocks from Phobos 2. *Nature* 341:595–598.

Sussman, G. J., and Wisdom, J. 1988. Numerical evidence that the motion of Pluto is chaotic. *Science* 241:433–437.

Sutton, J. L., Leovy, C. B., and Tillman, J. E. 1978. Diurnal variations of the Martian surface layer meteorological parameters during the first 45 sols at two Viking lander sites. *J. Atmos. Sci.* 35:2346–2355.

Svitek, T., and Murray, B. 1990. Winter frost at Viking lander 2 site. *J. Geophys. Res.* 95:1495–1510.

Swift, J. 1726. *Gulliver's Travels* (London), part III, ch. 3. Uncensored, complete version reprinted by Collier Books (1962).

Swindle, T. D., Caffee, M. W., and Hohenberg, C. M. 1986. Xenon and other noble gases in shergottites. *Geochim. Cosmochim. Acta* 50:1001–1005.

Szeto, A. M. K. 1983. Orbital evolution and origin of the Martian satellites. *Icarus* 55:133–168.

Takashi, E., and Ito, E. 1987. Mineralogy of mantle peridotite along a model geotherm up to 700 km depth. In *High-Pressure Research in Mineral Physics,* eds. M. H. Manghnani and S. Syono, AGU Mono. 39 (Washington, D.C.: American Geophysical Union), pp. 427–438.

Tanaka, K. L. 1981. Structure of Olympus Mons aureoles and perimeter escarpment. In *Proc. Third Intl. Coll. on Mars,* LPI Contrib. 441, pp. 261–263 (abstract).

Tanaka, K. L. 1985. Ice-lubricated gravity spreading of the Olympus Mons aureole deposits. *Icarus* 62:191–206.

Tanaka, K. L. 1986. The stratigraphy of Mars. *Proc. Lunar Planet. Sci. Conf.* 17, *J. Geophys. Res. Suppl.* 91:E139-E158.

Tanaka, K. L. 1990. Tectonic history of the Alba Patera-Ceraunius Fossae region of Mars. *Lunar Planet. Sci. Conf.* 20:515–523.

Tanaka, K. L., and Davis, P. A. 1988. Tectonic history of the Syria Planum province of Mars. *J. Geophys. Res.* 93:14893–14917.

Tanaka, K. L., and Golombek, M. P. 1989. Martian tension fractures and the formation of grabens and collapse features at Valles Marineris. *Proc. Lunar Planet. Sci. Conf.* 19:383–396.

Tanaka, K. L., and Scott, D. H. 1987. *Geologic Map of the Polar Regions of Mars, scale 1:15,000,000.* U.S.G.S. Misc. Inv. Series Map I-1802-C.

Tanaka, K. L., Witbeck, N. E., and Scott, D. H. 1986. Gravity-profile analysis over central Valles Marineris, Mars. In *Reports of the Planetary Geology and Geophysics Program–1985,* NASA TM-88383, pp. 603–604 (abstract).

Tanaka, K. L., Isbell, N. K., Scott, D. H., Greeley, R., and Guest, J. E. 1988. The resurfacing history of Mars: A synthesis of digitized, Viking-based geology. *Proc. Lunar Planet. Sci. Conf.* 18:665–678.

Tass. 1972. Soviet-bloc research in geophysics, astronomy, and space. No. 268, 14 Jan., pp. 47–58.

Taylor, S. R. 1986. The origin of the Moon: Geochemical considerations. In *Origin of the Moon,* eds. W. K. Hartmann, R. J. Phillips and G. J. Taylor (Houston: Lunar and Planetary Inst.), pp. 125–143.

Taylor, W. W. L., Scarf, F. L., Russell, C. T., and Brace, L. H. 1979. Absorption of whistler mode waves in the ionosphere of Venus. *Science* 205:112–114.

Tedesco, E. F., Williams, J. G., Matson, D. L., Veeder, G. J., Gradie, J. C., and Lebofsky, L. A. 1989. A three-parameter asteroid taxonomy. *Astron. J.* 97:580–606.

Theilig, E., and Greeley, R. 1979. Plains and channels in the Lunae Planum-Chryse Planitia region of Mars. *J. Geophys. Res.* 4:7994–8010.
Theilig, E., and Greeley, R. 1986. Lava flows on Mars: Analysis of small surface features and comparisons with terrestrial analogs. *Proc. Lunar Planet. Sci. Conf.*, 17, *J. Geophys. Res. Suppl.* 91:E193-E206.
THERMOSCAN Images. In *Planetary Repts.* 9(4):22.
Tholen, D. J. 1984. Asteroid Taxonomy from Cluster Analysis of Photometry. Ph.D. Thesis, Univ. of Arizona.
Thomas, J., Jr., and Bohor, B. F. 1968. Surface area on montmorillonite from the dynamic sorption of nitrogen and carbon dioxide. *Clays Clay Mineral.* 16:83–91.
Thomas, P. C. 1979. Surface features of Phobos and Deimos. *Icarus* 40:223–243.
Thomas, P. C. 1981. North-south asymmetry of eolian features in Martian polar regions: Analysis based on crater-related wind markers. *Icarus* 48:76–90.
Thomas, P. C. 1982. Present wind activity on Mars: Relation to large latitudinally zoned sediment deposits. *J. Geophys. Res.* 87:9999–10008.
Thomas, P. C. 1984. Martian intracrater splotches: Occurrence, morphology, and colors. *Icarus* 57:205–227.
Thomas, P. C. 1989. The shapes of small satellites. *Icarus* 77:248–274.
Thomas, P. C., and Gierasch, P. J. 1985. Dust devils on Mars. *Science* 230:175–177.
Thomas, P. C., and Veverka, J. 1979*a*. Grooves on asteroids. A prediction. *Icarus* 40:395–406.
Thomas, P. C., and Veverka, J. 1979*b*. Seasonal and secular variation of wind streaks on Mars: An analysis of Mariner 9 and Viking data. *J. Geophys. Res.* 84:8131–8146.
Thomas, P. C., and Veverka, J. 1980. Downslope movement of material on Deimos. *Icarus* 42:234–250.
Thomas, P. C., and Veverka, J. 1986. Red/violet contrast reversal on Mars: Significance for eolian sediments. *Icarus* 66:39–55.
Thomas, P. C., and Weitz, C. 1989. Sand dune materials and polar layered deposits on Mars. *Icarus* 81:185–215.
Thomas, P. C., Veverka, J., and Campos-Marquetti, R. 1979*a*. Frost streaks in the south polar cap of Mars. *J. Geophys. Res.* 84:4621–4633.
Thomas, P. C., Veverka, J., Bloom, A., and Duxbury, T. 1979*b*. Grooves on Phobos: Their distribution, morphology, and possible origin. *J. Geophys. Res.* 84:8457–8477.
Thomas, P. C., Veverka, J., Lee, S., and Bloom, A. 1981. Classification of wind streaks on Mars. *Icarus* 45:124–153.
Thomas, P. C., Adams, E., and Veverka, J. 1983. Constructional topography on Phobos: Do grooves give raised rims? *Bull. Amer. Astron. Soc.* 15:237 (abstract).
Thomas, P. C., Veverka, J., Gineris, D., and Wong, L. 1984. "Dust" streaks on Mars. *Icarus* 49:398–415.
Thomas, P. C., Veverka, J., and Dermott, S. 1986. Small satellites. In *Satellites*, eds. J. A. Burns and M. S. Matthews (Tucson: Univ. of Arizona Press), pp. 802–835.
Thomas, P. J., Squyres, S. W., and Carr, M. H. 1990. Flank tectonics of Martian volcanoes. *J. Geophys. Res.* 95:14345–14355.
Thomas, R. J., Barth, C. A., Rusch, D. W., and Sanders, R. W. 1984. Solar Mesosphere Explorer near infrared spectrometer: Measurements of 1.27 μm radiances and the inference of mosspheric ozone. *J. Geophys. Res.* 89:9569–9580.
Thompson, D. E. 1979. Origin of longitudinal grooving in Tiu Vallis, Mars: Isolation of responsible fluid types. *Geophys. Res. Lett.* 6:735–738.
Thompson, D. T. 1973*a*. A new look at the Martian "violet haze" problem. II. "Blue clearing" in 1969. *Icarus* 18:164–170 [paper III].
Thompson, D. T. 1973*b*. Time variation of Martian regional contrasts. *Icarus* 20:42–47.
Thompson, E. G., and Sayles, F. H. 1972. In situ creep analysis of room in frozen soil. *J. Soil Mech.* 98:899–915.
Thompson, T. W., and Moore, H. J. 1989*a*. A model for depolarized radar echoes from Mars. *Proc. Lunar Planet. Sci. Conf.* 19:402–422.
Thompson, T. W., and Moore, H. J. 1989*b*. Martian quasi-specular echoes: Preliminary 1986 results. *Lunar Planet. Sci.* XX:713–714 (abstract).

Thompson, T. W., Pollack, J. B., Campbell, M. J., and O'Leary, B. T. 1970. Radar maps of the Moon at 70-cm wavelength and their interpretation. *Radio Sci.* 5:253–262.

Thompson, W. F. 1962. Preliminary notes on the nature and distribution of rock glaciers relative to true glaciers and other effects of the climate on the ground in North America. In *Symp. at Obergurgl*, Intl. Assoc. Sci. Hydrol. Publ. No. 58 (Obergurgl: Intl. Assoc. for Sci. Hydrology), pp. 212–219.

Thorarinsson, S. 1953. The crater group in Iceland. *Bull. Vol. Ser. 2* 14:3–44.

Thorpe, T. E. 1977. Viking orbiter observations of atmospheric opacity during July–November 1976. *J. Geophys. Res.* 82:4151–4159.

Thorpe, T. E. 1978. Viking Orbiter observations of the Mars opposition effect. *Icarus* 36:204–215.

Thorpe, T. E. 1979. A history of Mars atmospheric opacity in the southern hemisphere during the Viking extended mission. *J. Geophys. Res.* 84:6663–6683.

Thorpe, T. E. 1981. Mars atmospheric opacity effects observed in the northern hemisphere by Viking Orbiter imaging. *J. Geophys. Res.* 86:11419–11429.

Thorpe, T. E. 1982. Martian surface properties indicated by the opposition effect. *Icarus* 49:398–415.

Thurber, C. H., and Toksöz, M. N. 1978. Martian lithospheric thickness from elastic flexure theory. *Geophys. Res. Lett.* 5:977–980.

Tillman, J. E. 1972. The indirect determination of stability, heat and momentum fluxes in the atmospheric boundary layer from simple scalar variables during dry unstable conditions. *J. Appl. Meteor.* 11:783–792.

Tillman, J. E. 1977. Dynamics of the boundary layer of Mars. In *Proc. Symp. on Planetary Atmospheres*, eds. A. Vallance Jones (Ottawa: Royal Soc. of Canada), pp. 145–149.

Tillman, J. E. 1985. Martian meteorology and dust storms from Viking observations. In *Case for Mars II*, ed. C. P. McKay. (San Diego: Univelt), pp. 333–342.

Tillman, J. E. 1988. Mars global atmospheric oscillations: Annually synchronized, transient normal mode oscillations and the triggering of global dust storms. *J. Geophys. Res.* 93:9433–9451.

Tillman, J. E., Henry, R. M., and Hess, S. L. 1979. Frontal systems during passage of the Martian north polar hood over the Viking Lander 2 site prior to the first 1977 dust storm. *J. Geophys. Res.* 84:2947–2955.

Tittemore, W. C., and Wisdom, J. 1988. Tidal evolution of the Uranian satellites. II. An explanation of the anomalously high orbital inclination of Miranda. *Icarus* 78:63–89.

Toksöz, M. N. 1979. Planetary seismology and interiors. *Rev. Geophys. Space Phys.* 17:1641–1655.

Toksöz, M. N., and Hsui, A. T. 1978. Thermal history and evolution of Mars. *Icarus* 34:537–547.

Toksöz, M. N., and Johnston, D. H. 1977. The evolution of the Moon and the terrestrial planets. In *The Soviet-American Conf. on Cosmochemistry of the Moon and Planets*, eds. J. H. Pomeroy and N. J. Hubbard, NASA SP-370, pp. 295–327.

Toksöz, M. N., Hsui, A. T., and Johnson, D. H. 1978. Thermal evolutions of the terrestrial planets. *Moon and Planets* 18:281–320.

Tombaugh, C. W. 1968. A survey of long-term observational behavior of various Martian features that affect some recently proposed interpretations. *Icarus* 8:227–258.

Toon, O. B., and Farlow, N. H. 1981. Particles above the tropopause. *Ann. Rev. Earth Planet. Sci.* 9:19–58.

Toon, O. B., Pollack, J. B., and Sagan, C. 1977. Physical properties of the particles composing the Martian dust storm of 1971–1972. *Icarus* 30:663–696.

Toon, O. B., Pollack, J. B., Ward, W., Burns, J. A., and Bilski, K. 1980. The astronomical theory of climatic change on Mars. *Icarus* 44:552–607.

Torr, M. R., and Torr, D. G. 1985. Ionization frequencies for solar cycle 21: Revised. *J. Geophys. Res.* 90:6675–6678.

Torr, M. R., Torr, D. G., Ong, R. A., and Hinteregger, H. E. 1979. Ionization frequencies for major thermospheric constituents as a function of solar cycle 21. *Geophys. Res. Lett.* 6:771–774.

Toulmin, P., Baird, A. K., Clark, B. C., Keil, K., Rose, H. J., Jr., Christian, R. P., Evans,

P. H., and Kelliher, W. C. 1977. Geochemical and mineralogical interpretation of the Viking inorganic chemical results. *J. Geophys. Res.* 84:4625–4634.
Tozer, D. C. 1985. Heat transfer and planetary evolution. *Geophys. Surv.* 7:213–246.
Traub, W. A., Carleton, N. P., Connes, P., and Noxon, J. F. 1979. The latitude variation of O_2 dayglow and O_3 abundance on Mars. *Astrophys. J.* 229:846–850.
Trauger, J. T., and Lunine, J. I. 1983. Spectroscopy of molecular oxygen in the atmospheres of Venus and Mars. *Icarus* 55:272–281.
Treiman, A. H. 1985. Ambipole and hercynite spinel Shergotty and Zagami: Magmatic water, depth of crystallization, and metasomatism. *Meteoritics* 20:229–243.
Treiman, A. H. 1986. The parental magma of the Nakha achondrite: Ultrabasic volcanism on the shergottite parent body. *Geochim. Cosmochim. Acta* 50:1061–1070.
Treiman, A. H., Drake, M. J., Janssens, M.-J., Wolf, R., and Ebihara, M. 1986. Core formation in the Earth and shergottite parent body (SPB): Chemical evidence from basalts. *Geochim. Cosmochim. Acta* 50:1071–1091.
Treiman, A. H., Jones, J. H., and Drake, M. J. 1987. Core formation in the shergottite parent body and comparison with the Earth. *Proc. Lunar Planet. Sci. Conf.* 17, *J. Geophys. Res. Suppl.* 92:E627-E632.
Trud. 1963. Mars-1 in flight. Moscow, 18 May, pp. 1 and 4.
Tsoar, H., Greeley, R., and Peterfreund, A. R. 1979. Mars: The north polar sand sea and related wind patterns. *J. Geophys. Res.* 84:8167–8180.
Tsytovich, N. A. 1975. *Mechanics of Frozen Ground* (New York: McGraw-Hill).
Tull, R. G. 1966. The reflectivity spectrum of Mars in the near-infrared. *Icarus* 5:505–514.
Turcotte, D. L. 1980. On the thermal evolution of the Earth. *Earth Planet. Sci. Lett.* 48:53–58.
Turcotte, D. L. 1983. Thermal stress in planetary elastic lithospheres. *Proc. Lunar Planet. Sci. Conf.* 13, *J. Geophys. Res. Suppl.* 88:A585–587.
Turcotte, D. L. 1988. Early evolution of the Martian interior. *Eos: Trans. AGU* 69:389.
Turcotte, D. L. 1987. A fractal interpretation of topography and geoid spectra on the Earth, Moon, Venus and Mars. *J. Geophys. Res.* 92:E597-E601.
Turcotte, D. L., and Huang, J. 1990. Implications of crustal fractionation for planetary evolution. *Icarus,* submitted.
Turcotte, D. L., and Schubert, G. 1982. *Geodynamics* (New York: Wiley).
Turcotte, D. L., and Schubert, G. 1988. Tectonic implications of radiogenic gases in planetary atmospheres. *Icarus* 74:36–46.
Turcotte, D. L., Cook, F. A., and Willemann, R. J. 1979. Parameterized convection within the Moon and the terrestrial planets. *Proc. Lunar Sci. Conf.* 10:2375–2392.
Turcotte, D. L., Willemann, R. J., Haxby, W. F., and Norberry, J. 1981. Role of membrane stresses in the support of planetary topography. *J. Geophys. Res.* 86:3951–3959.
Turekian, K. K., and Clark, S. P. 1975. Nonhomogeneous accumulation model for terrestrial planet formation and the consequences for the atmosphere of Venus. *J. Atmos. Sci.* 32:1257–1261.
Tyler, G. L., Campbell, D. B., Downs, G. S., Green, R. R., and Moore, H. J. 1976. Radar characteristics of Viking 1 landing sites. *Science* 193:812–815.
Ugolini, F. C., and Anderson, D. M. 1973. Ionic migration and weathering in frozen Antarctic soils. *Soil Sci.* 115:461–470.
U.S. Geological Survey. 1972. Prototype and Preliminary Geologic Mapping of Mars from Mariner 6 and 7 data. U.S.G.S. Interagency Rept.: Astrogeology 45.
U.S. Geological Survey. 1980. *Topographic Orthophoto Mosaic of the Tithonium Chasma Region of Mars, scale 1:500,000.* U.S.G.S. Misc. Inv. Series Map I-1294.
U.S. Geological Survey. 1986. *Topographic Map of the Coprates Northwest Quadrangle of Mars, scale 1:2,000,000.* U.S.G.S. Misc. Inv. Series Map I-1712.
U.S. Geological Survey. 1989. *Topographic Maps of the Western, Eastern Equatorial and Polar Regions of Mars, scale 1:15,000,000.* U.S.G.S. Misc. Inv. Series Map I-2030.
U.S. Geological Survey. 1991. *Topographic Map of Mars.* U.S.G.S. Misc. Inv. Series Map I-2160.
U.S. Naval Observatory. 1988. *The Astronomical Almanac for the Year 1989* (Washington, D.C.: U.S. Govt. Printing Office).
U.S. Standard Atmosphere. 1976. (Washington, D.C.: U.S. Government Printing Office).

Usselman, T. M. 1975. Experimental approach to the state of the core: Part II. Composition and thermal regime. *Amer. J. Sci.* 275:291–303.

Vaisberg, O. L. 1976. Mars-plasma environment. In *Physics of Solar Planetary Environments,* vol. 2, ed. D. J. Williams (Washington, D.C.: American Geophysical Union), pp. 854–871.

Vaisberg, O. L., and Bogdanov, A. V. 1974. Flow of the solar wind around Mars and Venus—General principals. *Cosmic Res.* 12:253–257.

Vaisberg, O. L., and Smirnov, V. 1986. The Martian magnetotail. *Adv. Space Res.* 6:301–314.

Vaisberg, O. L., and Zeleny, L. M. 1984. Formation of plasma mantle in the Venusian magnetosphere. *Icarus* 58:412–430.

Vaisberg, O. L., Bogodanov, A. V., Smirnov, V. N., and Romanov, S. A. 1975. Preliminary results of ion flux by means of REIP-2801 M instrument on Mars-4, Mars-5. *Kosmich. Issled.* 13:129–130. Trans.: NASA TT-F-16352.

Vaisberg, O. L., Smirnov, V. N., Bogodanov, A. V., Kalinin, A. P., Karpinsky, I. P., Polenov, B. V., and Romanov, S. A. 1976*a*. Ion flux parameters in the region of solar wind interaction with Mars according to measurements of Mars 4 and Mars 5. *Space Res.* XVI:1033–1038.

Vaisberg, O. L., Bogodanov, A. V., Smirnov, V. N., and Romanov, S. A. 1976*b*. On the nature of solar-wind-Mars interaction. In *Solar Wind Interaction with the Planets Mercury, Venus and Mars,* ed. N. F. Ness, NASA SP-397, pp. 21–40.

Vaisberg, O. L., Luhmann, J. G., and Russell, C. T. 1990. Plasma observations of the solar wind interaction with Mars. *J. Geophys. Res.* 95:14841–14852.

Vallis, G. K. 1986. El Niño: A chaotic dynamical system? *Science* 232:243–245.

Vasil'yev, M. B., Vyshlov, A. S., Kolosov, M. A., Savich, N. A., Samovol, V. A., Samoznayev, L. N., Sidorenko, A. I., Aleksandrov, Yu. N., Danilenko, A. I., Dubrovnin, V. M., Zaytsev, A. L., Petrov, G. M., Rzhiga, O. N., Shtern, D. Ya., and Romanova, L. I. 1975. Preliminary results of two-frequency radio-occultation of the Mars ionosphere by means of Mars planetary probes in 1974. *Kosmich. Issled.* 13:48–53. Trans.: NASA TT-F-16,340. Microfiche N75-24625.

Vdovin, V. V., Zhegulev, V. S., Ksanfomality, L. V., Moroz, V. I., and Petrova, E. E. 1980. Some properties of soil in equatorial region of Mars according to thermal radiometry of Mars-5. *Kosmich. Issled.* 18:609–622. In Russian.

Veeder, G. J., Matson, D. L., Tedesco, E. F., Lebofsky, L. A., and Gradie, J. C. 1987. Radiometry of Deimos. *Astron. J.* 94:1361–1363.

Vening-Meinesz, F. A. 1947. Shear patterns of the Earth's crust. *Eos: Trans. AGU* 28:1–61.

Vening-Meinesz, F. A. 1951. A remarkable feature of the Earth's topography. *Proc. K. Ned. Akad. Wet. Ser. B Phys. Sci.* 54:212–228.

Veverka, J. 1978. The surfaces of Phobos and Deimos. *Vistas in Astron.* 22:163–192.

Veverka, J., and Burns, J. A. 1980. The moons of Mars. *Ann. Rev. Earth Planet. Sci.* 8:527–558.

Veverka, J., and Goguen, J. D. 1973. The nonuniform recession of the south polar cap of Mars. *J. Roy. Aston. Soc. Canada* 67:273–290.

Veverka, J., and Thomas, P. 1982. Phobos and Deimos: A preview of what asteroids are like? In *Asteroids,* ed. T. Gehrels (Tucson: Univ. of Arizona Press), pp. 628–651.

Veverka, J., Goguen, J., and Liller, W. 1973. Multicolor Polarimetric Observations of the Great 1971 Martian Dust Storm. Cornell Univ. Internal Report CRSR 550.

Veverka, J., Thomas, P., and Greeley, R. 1977. A study of variable features on Mars during the Viking primary mission. *J. Geophys. Res.* 82:4167–4187.

Veverka, J., Gierasch, P., and Thomas, P. 1981. Wind streaks on Mars: Meteorological control of occurrence and mode of formation. *Icarus* 45:154–166.

Veverka, J., Thomas, P., Johnson, T. V., Matson, D., and Housen, K. 1986. The physical characteristics of satellite surfaces. In *Satellites,* eds. J. A. Burns and M. S. Matthews (Tucson: Univ. of Arizona Press), pp. 342–402.

Vickery, A. M., and Melosh, H. J. 1987. The large crater origin of SNC meteorites. *Science* 237:738–743.

Viking Lander Imaging Team. 1978. *The Martian Landscape,* NASA SP-425.

Viking Orbiter Imaging Team. 1980. *Viking Orbiter Views of Mars,* NASA SP-441.

Vinogradov, A. P., Surkov, Yu. A., Moskaleva, L. P., and Kirnozov, F. F. 1975. Measurements of intensity and spectrum of gamma radiation of Mars on AIS Mars-5. *Doklady Akad. Nauk* 223:1336–1339. Trans.: NASA TT-F-16,850. Microfiche N76-18015.

Vlasov, V. Z. 1964. *General Theory of Shells and Its Applications in Engineering,* NASA TT-F-99.

Vogt, P. R. 1974. Volcano height and plate thickness. *Earth Planet. Sci. Lett.* 23:337–348.

Voight, B., Glicken, H., Janda, R. J., and Douglass, P. M. 1981. Catastrophic rock slide avalanche of May 18. In *The 1980 Eruption of Mount St. Helens, Washington,* U.S.G.S. Prof. Paper 1250, pp. 347–377.

Von Waltershausen, S. 1845. Uber die submarinen vulkanischen ausbruke in der tertiar formation des Val di Nito im vergleich mit verwandten erscheinungen am Aetna. *Gottingen Studien* 1:71–431.

von Zahn, U., Kumar, S., Niemann, H., and Prinn, R. 1983. Composition of the Venus atmosphere. In *Venus,* eds. D. Hunten, L. Colin, T. Donahue and V. Moroz (Tucson: Univ. of Arizona Press), pp. 299–430.

Wahrhaftig, C., and Cox, A. 1959. Rock glaciers in the Alaska Range. *Bull. Geol. Soc. Amer.* 70:383–436.

Walker, G. P. L. 1972. Compound and simple lava flows and flood basalts. *Bull. Volcanol.* 35:579–590.

Walker, G. P. L. 1973*a*. Explosive volcanic eruptions—a new classification scheme. *Geol. Rudsch.* 62:431–446.

Walker, G. P. L. 1973*b*. Lengths of lava flows. *Phil. Trans. Roy. Soc. London* A274:107–118.

Walker, G. P. L. 1987. In *The Dike Complex of Koolaua Volcano, Oahu: Internal Structure of a Hawaiian Rift Zone.* U.S.G.S. Prof. Paper 1350, pp. 961–993.

Walker, G. P. L. 1988. Three Hawaiian calderas: An origin through loading by shallow intrusions? *J. Geophys. Res.* 93:14773–14784.

Walker, G. P. L., Self, S., and Wilson, L. 1984. Tarawera 1886, New Zealand—A basaltic plinian fissure eruption. *J. Volcan. Geotherm. Res.* 21:61–78.

Walker, J. C. G. 1977. *Evolution of the Atmosphere* (New York: MacMillan).

Walker, J. C. G. 1978. Atmospheric evolution on the inner planets. In *Comparative Planetology,* ed. C. Ponnamperuma (New York: Academic Press), pp. 141–163.

Walker, J. C. G. 1985. Carbon dioxide on the early Earth. *Origins of Life* 16:117–127.

Walker, J. C. G. 1986. Impact erosion of planetary atmospheres. *Icarus* 68:87–98.

Wall, S. D. 1981. Analysis of condensates formed at the Viking 2 Lander site: The first winter. *Icarus* 47:173–183.

Wallace, A. R. 1907. *Is Mars Habitable?* (London: Macmillan).

Wallace, D., and Sagan, C. 1979. Evaporation of ice in planetary atmospheres: Ice-covered rivers on Mars. *Icarus* 39:385–400.

Wallis, M. K. 1975. Does Mars have a magnetosphere? *Geophys. J. Roy. Astron. Soc.* 41:349–354.

Wallis, M. K. 1978. Exospheric density and escape fluxes of atomic isotopes on Venus and Mars. *Planet. Space Sci.* 26:949–953.

Wallis, M. K. 1982. Comet-like interaction of Venus with the solar wind III—The atomic oxygen corona. *Geophys. Res. Lett.* 9:427–430.

Wallis, M. K. 1989. C, N, O isotope fractionation of Mars: Implications for crustal H_2O and SNC meteorites. *Earth Planet. Sci. Lett.* 93:321–324.

Wallis, M. K., and Ip, W.-H. 1982. Atmospheric interactions of planetary bodies with the solar wind. *Nature* 298:229–234.

Walter, F. M., Brown, A., Mathieu, R. D., Myers, P. C., and Vrba, F. J. 1988. X-ray sources in regions of star formation. III. Naked T Tauri stars associated with the Taurus-Auriga complex. *Astron. J.* 96:297–325.

Walterschied, R. L. 1981. Inertio-gravity wave induced accelerations of mean flow having an imposed periodic component: Implications for tidal observations in the meteor region. *J. Geophys. Res.* 86:9698–9706.

Walterscheid, R. L., and Schubert, G. 1990. Nonlinear evolution of an upward propagating gravity wave: Overturning, convection, transience and turbulence. *J. Atmos.* Sci. 47:101–125.

Wänke, H., and Dreibus, G. 1988. Chemical composition and accretion history of terrestrial planets. *Phil. Trans. Roy. Soc. London* A235:545–557.
Ward, A. W. 1979. Yardangs on Mars: Evidence of recent wind erosion. *J. Geophys. Res.* 84:8147–8166.
Ward, A. W., and Doyle, K. B. 1983. Speculation on Martian north polar wind circulation and resultant orientations of polar sand dunes. *Icarus* 55:420–431.
Ward, A. W., Doyle, K. B., Helm, P. J., Weisman, M. K., and Witback, N. E. 1985. Global map of aeolian features of Mars. *J. Geol. Res.* 90:2038–2056.
Ward, W. R. 1973. Large-scale variations in the obliquity of Mars. *Science* 181:260–262.
Ward, W. R. 1974. Climatic variations on Mars. I. Astronomical theory of insolation. *J. Geophys. Res.* 79:3375–3386.
Ward, W. R. 1975. Tidal friction and generalized Cassini's laws in the solar system. *Astron. J.* 80:64–70.
Ward, W. R. 1979. Present obliquity oscillations of Mars: Fourth-order accuracy in orbital e and I. *J. Geophys. Res.* 84:237–241.
Ward, W. R. 1981. Solar nebula dispersal and the stability of the planetary system. I. Scanning secular resonance theory. *Icarus* 47:234–264.
Ward, W. R. 1982. Comments of the long-term stability of the Earth's obliquity. *Icarus* 50:444–448.
Ward, W. R., and Rudy, D. 1991. Resonant obliquity of Mars? *Icarus* 94:160–164.
Ward, W. R., Murray, B. C., and Malin, M. C. 1974. Climatic variations on Mars. 2. Evolution of carbon dioxide atmosphere and polar caps. *J. Geophys. Res.* 79:3387–3395.
Ward, W. R., Colombo, G., and Franklin, F. A. 1976. Secular resonance, solar spin down and the orbit of Mercury. *Icarus* 28:441–452.
Ward, W. R., Burns, J. A., and Toon, O. B. 1979. Past obliquity oscillations of Mars: The role of the Tharsis uplift. *J. Geophys. Res.* 84:243–259.
Warner, B. 1962. Some problems of lunar orogeny. *J. Brit. Astron. Assoc.* 72:280–285.
Warren, P. H., and Rasmussen, K. L. 1987. Megaregolith insulation, internal temperatures, and the bulk uranium content of the Moon. *J. Geophys. Res.* 92:3453–3465.
Warren, S. G. 1984. Impurities in snow: Effects on albedo and snowmelt. *Ann. Glaciol.* 5:177–179.
Warren, S. G. 1986. Optical constants of carbon dioxide ice. *Appl. Opt.* 25:2650–2674.
Warren, S. G., and Wiscombe, W. J. 1980. A model for the spectral albedo of snow. II. Snow containing atmospheric aerosols. *J. Atmos. Sci.* 37:2734–2745.
Warren, S. G., Wiscombe, W. J., and Firestone, J. F. 1990. Spectral albedo and emissivity of CO_2 in Martian polar caps: Model results. *J. Geophys. Res.* 95:14717–14741.
Washburn, A. L. 1973. *Periglacial Processes and Environments* (New York: St. Martin's).
Washburn, E. W., ed. 1948. *International Critical Tables of Numerical Data, Physics, Chemistry, and Technology, Volume 3* (New York: McGraw Hill).
Wasson, J. T. 1974. *Meteorites: Classification and Properties* (New York: Springer-Verlag).
Wasson, J. T., and Wetherill, G. W. 1979. Dynamical, chemical and isotopic evidence regarding the formation locations of asteroids and meteorites. In *Asteroids*, ed. T. Gehrels (Tucson: Univ. of Arizona Press), pp. 926–974.
Watkins, G. H., and Lewis, J. S. 1986. Evolution of the atmosphere of Mars as a result of asteroidal and cometary impacts. In *Workshop on the Evolution of the Martian Atmosphere*, eds. M. Carr, P. James, C. Leovy and R. Pepin, LPI Tech. Rept. 86–07, pp. 46–47 (abstract).
Watson, A. J., Donahue, T. M., and Walker, J. C. G. 1981. The dynamics of a rapidly escaping atmosphere: Applications to the evolution of Earth and Venus. *Icarus* 48:150–166.
Watson, C. C., Haff, P. K., and Tombrello, T. A. 1980. Solar wind sputtering effects in the atmospheres of Mars and Venus. *Proc. Lunar Planet. Sci. Conf.* 11:1479–2502.
Watters, T. R. 1988. Wrinkle ridge assemblages on the terrestrial planets. *J. Geophys. Res.* 93:10236–10254.
Watters, T. R. 1991. Origin of periodically spaced wrinkle ridges on the Tharsis region of Mars. *J. Geophys. Res.* 96:15599–15616.
Watters, T. R., and Chadwick, D. G. 1989. Crosscutting periodically spaced first-order ridges in the ridge plains of Hesperia Planum. In *MEVTV Workshop on Tectonic Features on Mars*, eds. T. R. Watters and M. P. Golombek, LPI Tech. Rept. 89-06, pp. 42–44 (abstract).

Watters, T. R., and Chadwick, D. J. 1990. Distribution of strain in the floor of the Olympus Mons caldera. *Lunar Planet. Sci.* XXI:1310–1311 (abstract).
Watters, T. R., and Maxwell, T. A. 1983. Crosscutting relations and relative ages of ridges and faults in the Tharsis region of Mars. *Icarus* 56:278–298.
Watters, T. R., and Maxwell, T. A. 1986. Orientation, relative age, and extent of the Tharsis plateau ridge system. *J. Geophys. Res.* 91:8113–8125.
Watters, T. R., and Tuttle, M. J. 1989. Strike-slip faulting associated with the folded Columbia River basalts: Implications for the deformed ridged plains of Mars. In *MEVTV-LPI Workshop on Tectonic Features on Mars,* eds. T. R. Watters and M. P. Golombeck, LPI Tech. Rept. 89-06, pp. 45–47 (abstract).
Watters, T. R., Tuttle, M. J., and Kiger, F. J. 1990. Symmetry of infrared stress fields in the Tharsis region of Mars. *Lunar Planet. Sci.* XXI:1312–1313 (abstract).
Weast, R. C., ed. 1969. *Handbook of Chemistry and Physics* (Cleveland: Chemical Rubber Co.).
Webster, P. J. 1977. The low-latitude circulation of Mars. *Icarus* 30:626–649.
Wechsler, A. E., and Glaser, P. E. 1965. Pressure effects on postulated lunar materials. *Icarus* 4:335–352.
Wedepohl, K. H. 1981. Der primare Erdmantel (Mp) und die durch die Krustenbildung verarmte Mantelzusammensetzung (Md). *Fortschr. Mineral.* 59:203–205.
Weerdenberg, P. C., and Morgenstern, N. R. 1983. Underground cavities in ice-rich frozen ground. In *Permafrost: Proc. Fourth Intl. Conf.*, July 17–22, Fairbanks, Alaska. (Washington, D.C.: National Academy Press), pp. 1384–1389.
Weidenschilling, S. J. 1976. Accretion of the terrestrial planets. II. *Icarus* 27:161–170.
Weidenschilling, S. J. 1979. A possible origin for the grooves of Phobos. *Nature* 282:697–698.
Weiss, D., and Fagan, J. J. 1982. Possible evidence of hydrocompaction within the fretted terrain of Mars. In *Reports of Planetary Geology Program–1982,* NASA TM-85127, pp. 239–241 (abstract).
Wells, E. N., Veverka, J., and Thomas, P. 1984. Mars: Experimental study of albedo changes caused by dust fallout. *Icarus* 58:331–338.
Wells, R. A. 1979. *Geophysics of Mars* (Amsterdam: Elsevier).
Wentworth, S. J., and Gooding, J. L. 1988*a*. Chloride and sulfate minerals in the Nakhla meteorite. *Lunar Planet. Sci.* XIX:1261–1262 (abstract).
Wentworth, S. J., and Gooding, J. L. 1988*b*. Calcium carbonate in Nakhla: Further evidence for pre-terrestrial secondary minerals in SNC-meteorites. *Meteoritics* 23:310 (abstract).
Wentworth, S. J., and Gooding, J. L. 1989. Calcium carbonate and silicate "rust" in the Nakhla meteorite. *Lunar Planet. Sci.* XX:1193–1194 (abstract).
West, M. 1974. Martian volcanism: Additional observations and evidence for pyroclastic activity. *Icarus* 21:1–11.
Wetherill, G. W. 1974. Problems associated with estimating the relative impact rates on Mars and the Moon. *Moon* 9:227–231.
Wetherill, G. W. 1975. Late heavy bombardent of the Moon and terrestrial planets. *Proc. Lunar Sci. Conf.* 6:1539–1561.
Wetherill, G. W. 1976. The role of large bodies in the formation of the Earth and Moon. *Proc. Lunar Sci. Conf.* 7:3245–3257.
Wetherill, G. W. 1977. Evolution of the Earth's planetesimal swarm subsequent tot the formation of the Earth and Moon. *Proc. Lunar Sci. Conf.* 8:1–16.
Wetherill, G. W. 1981. Solar wind origin of ^{36}Ar on Venus. *Icarus* 46:70–80.
Wetherill, G. W. 1984. Orbital evolution of impact ejecta from Mars. *Meteoritics* 19:1–13.
Wetherill, G. W. 1985. Occurrence of giant impacts during the growth of the terrestrial planets. *Science* 228:877–879.
Wetherill, G. W. 1986. Accumulation of the terrestrial planets and implications concerning lunar origin. In *Origin of the Moon,* eds. W. K. Hartmann, R. J. Phillips and G. J. Taylor (Houston: Lunar and Planetary Inst.), pp. 519–550.
Wetherill, G. W. 1988. Accumulation of Mercury from planetesimals. In *Mercury,* eds. F. Vilas, C. R. Chapman and M. S. Matthews (Tucson: Univ. of Arizona Press), pp. 670–691.
Wetherill, G. W., and Chapman, C. R. 1988. Asteroids and meteorites. In *Meteorites and the Early Solar System,* eds. J. F. Kerridge and M. S. Matthews (Tucson: Univ. of Arizona Press), pp. 35–67.

Wharton, R. A., Jr., Parker, B. C., Simmons, G. M., Jr., Seaburg, K. G., and Love, F. G. 1982. Biogenic calcite structures forming in Lake Fryxell, Antarctica. *Nature* 295:403–405.

Wharton, R. A., Jr., McKay, C. P., Simmons, G. M., Jr., and Parker, B. C. 1986. Oxygen budget of a perennially ice-covered Antarctic dry valley lake. *Limnol. Oceanog.* 31:437–443.

Wharton, R. A., Jr., McKay, C. P., Mancinelli, R. L., and Simmons, G. M., Jr. 1987. Perennial N_2 supersaturation in an Antarctic lake. *Nature* 325:343–345.

White, B. R. 1979. Soil transport by winds on Mars. *J. Geophys. Res.* 84:4643–4651.

White, B. R. 1981. Low-Reynolds-number turbulent boundary layers. *J. Fluids Eng.* 103:624–630.

White, B. R., and Greeley, R. 1989. Martian dust threshold measurements—Simulation under heated surface conditions. In *MECA Workshop on Dust on Mars III*, ed. S. Lee, LPI Tech. Rept. 89-01, pp. 60–61 (abstract).

White, S. E. 1976. Rock glaciers and block fields, review and new data. *Quat. Res.* 6:77–97.

Whitford-Stark, J. L. 1982. Tharsis volcanoes: Separation distances, relative ages, sizes, morphologies, and depths of burial. *J. Geophys. Res.* 87:9829–9838.

Whitmire, D. P., and Jackson, A. A. 1984. Are periodic mass extinctions driven by a distant solar companion? *Nature* 380:713–715.

Whitney. 1983. Eolian features shaped by aerodynamic and vorticity processes. In *Eolian Sediments and Processes*, eds. M. E. Brookfield and T. S. Ahlbrandt, *Developments in Sedimentology*, vol. 38 (New York: Elsevier), pp. 223–245.

Wichman, R. W., and Schultz, P. H. 1986. Timing of ancient extensional tectonic features on Mars. *Lunar Planet. Sci.* XVII:942–943 (abstract).

Wichman, R. W., and Schultz, P. H. 1989. An ancient Valles Marineris? In *MEVTV Workshop on Early Tectonic and Volcanic Evolution of Mars*, eds. H. Frey, LPI Tech. Rept. 89-04, pp. 88–90 (abstract).

Wieler, R., Baur, H., Signer, P., Lewis, R. S., and Anders, E. 1989. Planetary noble gases in "Phase Q" of Allende: Direct determination by closed system etching. *Lunar Planet. Sci.* XX:1201–1202 (abstract).

Wiens, R. C. 1988. On the siting of gases shock-emplaced from internal cavities in basalt. *Geochim. Cosmochim. Acta* 52:2775–2783.

Wiens, R. C., and Pepin, R. O. 1988. Laboratory shock emplacement of noble gases, nitrogen, and carbon dioxide into basalt, and implications for trapped gases in shergottite EETA 79001. *Geochim. Cosmochim. Acta* 52:295–307.

Wiens, R. C., Becker, R. H., and Pepin, R. O. 1986. The case for Martian origin of the shergottites. II. Trapped and indigenous gas components in EETA 79001 glass. *Earth Planet. Sci. Lett.* 77:149–158.

Wildt, R. 1934. *Ozon und Sauerstoff in den Planeten-Atmospharen.* Veroffentl. Univ.-Sternwarte Gottingen no. 38.

Wilhelms, D. E. 1973. Comparison of Martian and lunar multiringed circular basins. *J. Geophys. Res.* 78:4084–4095.

Wilhelms, D. E. 1987. *The Geologic History of the Moon.* U.S.G.S. Prof. Paper 1348.

Wilhelms, D. E. 1990. Geologic mapping. In *Planetary Mapping*, eds. R. Greeley and R. M. Batson (New York: Cambridge Univ. Press), pp. 208–260.

Wilhelms, D. E., and Baldwin, R. J. 1989*a*. The relevance of knobby terrain to the Martian dichotomy. In *MEVTV-LPI Workshop: Early Tectonic and Volcanic Evolution of Mars*, ed. H. Frey, LPI Tech. Rept. 89-04, pp. 91–93 (abstract).

Wilhelms, D. E., and Baldwin, R. J. 1989*b*. The role of igneous sills in shaping the Martian uplands. *Proc. Lunar Planet. Sci. Conf.* 19:355–365.

Wilhelms, D. E., and Squyres, S. W. 1984. The Martian hemispheric dichotomy may be due to a giant impact. *Nature* 309:138–140.

Willemann, R. J. 1984. Reorientation of planets with elastic lithospheres. *Icarus* 60:701–709.

Willemann, R. J., and Turcotte, D. L. 1981. Support of topographic and other loads on the Moon and on the terrestrial planets. *Proc. Lunar Planet. Sci. Conf.* 12B:837–851.

Willemann, R. J., and Turcotte, D. L. 1982. The role of membrane stress in the support of the Tharsis rise. *J. Geophys. Res.* 87:9793–9801.

Williams, B., Duxbury, T., and Hildebrand, C. 1988. Improved determination of Phobos and Deimos masses from Viking flybys. *Lunar Planet. Sci.* XIX:1274 (abstract).

Williams, J. G. 1984. Determining asteroid masses from perturbations on Mars. *Icarus* 57:1–13.

Williams, J. G., Sinclair, W. S., and Yoder, C. F. 1978. Tidal acceleration of the Moon. *Geophys. Res. Lett.* 5:943–946.

Williams, Q., and Jeanloz, R. 1990. Melting relations in the iron-sulfur system at ultra-high pressures: Implications for the thermal state of the Earth. *J. Geophys. Res.* 95:19299–19310.

Williams, Q., Jeanloz, R., Bass, J., Svendsen, B., and Ahrens, T. J. 1987. The melting curve of iron to 250 gigapascals: A constraint on the temperature at the Earth's center. *Science* 236:181–183.

Williams, Q., Knittle, E., and Jeanloz, R. 1989. Geophysical and crystal chemical significance of (Mg,Fe)SiO_3 perovskite. In *Perovskite: A Structure of Great Interest in the Earth and Material Sciences,* eds. A. Navrotsky and D. J. Weidner, AGU Mineral Physics Mono. 3 (Washington, D.C.: American Geophysical Union), pp. 1–12.

Willson, R. C. 1984. Measurements of solar total irradiance and its variability. *Space Sci. Rev.* 38:203–242.

Wilson, I. G. 1971. Desert sandflow basins and a model for the development of ergs. *Geographical J.* 137:180–199.

Wilson, L. 1980. Relationships between pressure, volatile content and ejecta velocity in three types of volcanic explosion. *J. Volcan. Geotherm. Res.* 8:297–313.

Wilson, L. 1984. The influences of planetary environments on the eruption styles of volcanoes. *Vistas in Astron.* 27:333–360.

Wilson, L., and Head, J. W. 1981*a*. Ascent and eruption of basaltic magma on the Earth and Moon. *J. Geophys. Res.* 86:2971–3001.

Wilson, L., and Head, J. W. 1981*b*. Volcanic eruption mechanisms on Mars: Some theoretical constraints. *Lunar Planet. Sci.* XII:1194–1196 (abstract).

Wilson, L., and Head, J. W. 1983. A comparison of volcanic eruption processes on Earth, Moon, Mars, Io and Venus. *Nature* 302:663–669.

Wilson, L., and Head, J. W. 1988*a*. Nature of local magma storage zones and geometry of conduit systems below basaltic eruption sites: Pu'u O'o. Kilauea East Rift, Hawaii, example. *J. Geophys. Res.* 93:14785–14792.

Wilson, L., and Head, J. W. 1988*b*. The influence of gravity on planetary volcanic eruption rates. *Lunar Planet. Sci.* XIX:1283–1284 (abstract).

Wilson, L., and Head, J. W. 1990. Factors controlling the structures of magma chambers in basaltic volcanoes. *Lunar Planet. Sci.* XXI:1343–1344 (abstract).

Wilson, L., and Mouginis-Mark, P. J. 1984. Martian sinuous rilles. *Lunar Planet. Sci.* XV:926–927 (abstract).

Wilson, L., and Mouginis-Mark, P. J. 1987. Volcanic input to the atmosphere from Alba Patera on Mars. *Nature* 330:354–357.

Wilson, L., and Parfitt, E. A. 1989. The influence of gravity on planetary volcanic eruption rates: A reappraisal. *Lunar Planet. Sci.* XX:1213–1214 (abstract).

Wilson, L., Head, J. W., and Mouginis-Mark, P. J. 1982. Theoretical analysis of Martian volcanic eruption mechanisms. In ESA SP-185, pp. 107–113.

Winkler, E. M., and Singer, P. C. 1972. Crystallization pressure of salts in stone and concrete. *Geol. Soc. Amer. Bull.* 83:3509–3514.

Wisdom, J. 1987*a*. Rotational dynamics of irregularly shaped natural satellites. *Astron. J.* 94:1350–1360.

Wisdom, J. 1987*b*. Urey Prize lecture: Chaotic dynamics in the solar system. *Icarus* 72:241–275.

Wise, D. U. 1979. *Geologic Map of the Arcadia Quadrangle of Mars, scale 1:5,000,000.* U.S.G.S. Misc. Inv. Series Map I-1154.

Wise, D. U. 1984. Kasei Vallis of Mars: Dating the interplay of tectonics and geomorphology. *Geol. Soc. Amer. Abs. w/Prog.* 16:699.

Wise, D. U., Golombek, M. P., and McGill, G. E. 1979*a*. Tectonic evolution of Mars. *J. Geophys. Res.* 84:7934–7939.

Wise, D. U., Golombek, M. P., and McGill, G. E. 1979*b*. Tharsis province of Mars: Geologic sequence, geometry, and a deformation mechanism. *Icarus* 38:456–472.

Wislicenus. 1886. *Beitrag zur Bestimmung der Rotationszeit des Planiten Mars.* (Leipsig).

Witbeck, N. E., Tanaka, K. L., and Scott, D. H. 1991. *Geologic Map of the Valles Marineris Region of Mars, scale 1:2,000,000.* U.S.G.S. Misc. Inv. Series Map I-2010.

Woese, C. R. 1987. Bacterial evolution. *Microbiol. Rev.* 51:221–271.
Wohletz, K. H., and Sheridan, M. F. 1983. Martian rampart crater ejecta: Experiments and analysis of melt-water interaction. *Icarus* 56:15–37.
Wood, B. J., and Holloway, J. R. 1982. Theoretical prediction of phase relationships in planetary mantles. *Proc. Lunar Planet. Sci. Conf.* 13, *J. Geophys. Res. Suppl.* 87:A19-A30.
Wood, C. A. 1968. Statistics of central peaks in lunar craters. *Commun. Lunar and Planetary Lab* 7:157–160.
Wood, C. A. 1980. Martian double ring basins: New observations. *Proc. Lunar Planet. Sci. Conf.* 11:2221–2241.
Wood, C. A. 1984*a*. Why Martian lava flows are so long. *Lunar Planet. Sci.* XV:929–930 (abstract).
Wood, C. A. 1984*b*. Calderas: A planetary perspective. *J. Geophys. Res.* 89:8391–8406.
Wood, C. A., and Ashwal, L. D. 1981. SNC meterites: Igneous rocks from Mars? *Proc. Lunar Planet. Sci. Conf.* 12, *Geochim. Cosmochim. Acta Suppl.* 16:1359–1375.
Wood, C. A., and Head, J. W. 1976. Comparison of impact basin on Mercury, Mars, and the Moon. *Proc. Lunar Sci. Conf.* 7:3629–3651.
Wood, C. A., Head, J. W., and Cintala, M. J. 1978. Interior morphology of fresh Martian craters: The effects of target characteristics. *Proc. Lunar Planet. Sci. Conf.* 9:3691–3709.
Woolley, R. 1953. Monochromatic magnitudes of Mars in 1952. *Mon. Not. Roy. Astron. Soc.* 113:521–525.
Woronow, A. 1977. Crater saturation and equilibrium: A Monte Carlo simulation. *J. Geophys. Res.* 82:2447–2465.
Woronow, A. 1978. A general cratering history model and its implications for the lunar highlands. *Icarus* 34:76–88.
Woronow, A. 1988. Variation in the thickness of ejecta cover on Mars with increasing crater density. In *The MEVTV Workshop on Nature and Composition of Surface Units on Mars,* LPI Tech. Rept. 88–05, pp. 135–137.
Woronow, A., and Strom, R. G. 1981. Limits on large-scale crater production and obliteration of Callisto. *Geophys. Res. Lett.* 8:891–894.
Woszczyk, A., and Iwaniszewska, C., eds. 1974. *Exploration of the Planetary System* (Dordrecht: D. Reidel).
Wright, I. P., Carr., R. H., and Pillinger, C. T. 1986. Carbon abundance and isotopic studies of Shergotty and other shergottite meteorites. *Geochim. Cosmochim. Acta* 50:983–991.
Wright, I. P., Grady, M. M., and Pillinger, C. T. 1988. Carbon, oxygen and nitrogen isotopic compositions of possible Martian weathering products in EETA 79001. *Geochim. Cosmochim. Acta* 52:917–924.
Wright, I. P., Grady, M. M., and Pillinger, C. T. 1989. Organic materials in a Martian meteorite. *Nature* 340:220–222.
Wu, S. S. C. 1975. Topographic Mapping of Mars, 1975. U.S. Geological Survey Interagency Rept. *Astrogeology* 63:193.
Wu, S. S. C. 1978. Mars synthetic topographic mapping. *Icarus* 33:417–440.
Wu, S. S. C. 1981. A method of defining topographic datums of planetary bodies. *Ann. Geophys.* 1:147–160.
Wu, S. S. C., and Doyle, F. J. 1990. Topography. In *Mapping the Planets and Satellites,* eds. R. Greeley and R. M Batson (New York: Cambridge Univ. Press), pp. 169–207.
Wu, S. S. C., and Howington, A.-E. 1986. A Mars digital terrain model and sample correlation with a Mars digital image model. In *Reports of Planetary Geology Program–1985,* NASA TM-88383, pp. 608-611 (abstract).
Wu, S. S. C., and Schafer, F. J. 1982. Photogrammetry of the Viking lander imagery. *Photogrammetric Eng. Remote Sensing* 48(5):803–816.
Wu, S. S. C., and Schafer, F. J. 1984. Mars control network: American Society of Photogrammetry. *Proc. 50th Annual Meeting* 2:456–464.
Wu, S. S. C., Garcia, P. A., Jordan, R., Schafer, F. J., and Skiff, B. A. 1984. Topography of the shield volcano, Olympus Mons on Mars. *Nature* 309:432–435.
Wychoff, R. W. G. 1963. *Crystal Structures* (New York: Interscience).
Yang, J., and Epstein, S. 1985. A study of stable isotopes in Shergotty meteorite. *Lunar Planet. Sci.* XVI:25–26 (abstract).

Ye, Z. J., Sega, M., and Pielke, R. A. 1990. A comparative study of daytime thermally induced upslope flow on Mars and Earth. *J. Atmos. Sci.* 47:612–648.

Yeroshenko, Ye., Riedler, W., Schwingenschuh, K., Luhmann, J. G., Ong, M., and Russell, C. T. 1990. The magnetotail of Mars: Phobos observations. *Geophys. Res. Lett.*, in press.

Yoder, C. F. 1982. Tidal rigidity of Phobos. *Icarus* 49:327–346.

Yoder, C. F., and Ward, W. R. 1979. Does Venus wobble? *Astrophys. J.* 233:L33-L37.

Young, L. D. G. 1971. Interpretation of high-resolution spectra of Mars. II. Calculations of CO_2 abundance, rotational temperature, and surface pressure. *J. Quant. Spectros. Radiat. Transfer* 11:1075–1086.

Young, L. D. G., and Young, A. T. 1977. Interpretation of high-resolution spectra of Mars IV. New calculations of the CO abundance. *Icarus* 30:75–79.

Young, R. A. 1977. The lunar impact flux, radiometric age correlation, and dating of specific lunar features. *Proc. Lunar Sci. Conf.* 8:3457–3473.

Young, R. A., Brennan, W. J., Wolfe, R. W., and Nichols, D. J. 1973. Volcanism in the lunar mare. In *Apollo 17 Preliminary Science Report,* NASA SP-330, pp. 31/1–31/11.

Young, R. E., and Schubert, G. 1974. Temperatures inside Mars: Is the core liquid or solid? *Geophys. Res. Lett.* 81:157–160.

Young, R. S., Painter, R. B., and Johnson, R. D. 1965. *An Analysis of the Extraterrestrial Life Detection Problem,* NASA SP-75.

Younkin, R. L. 1966. A search for limonite near-infrared spectral features on Mars. *Astrophys. J.* 144:809–818.

Yung, Y. L., and Pinto, J. P. 1978. Primitive atmosphere and implications for the formation of channels on Mars. *Nature* 288:735–738.

Yung, Y. L., Strobel, D. F., Kong, T. Y., and McElroy, M. B. 1977. Photochemistry of nitrogen in the Martian atmosphere. *Icarus* 30:26–41.

Yung, Y. L., Wen, J. S., Pinto, J. P., Allen, M., Pierce, K., and Panlom, S. 1988. HDO in the Martian atmosphere: Implications for the abundance of crustal water. *Icarus* 76:146–159.

Zahnle, K. J., and Kasting, J. F. 1986. Mass fractionation during transonic hydrodynamic escape and implications for loss of water from Venus and Mars. *Icarus* 68:462–480.

Zahnle, K. J., and Walker, J. C. G. 1982. The evolution of solar ultraviolet luminosity. *Rev. Geophys. Space Phys.* 20:280–292.

Zahnle, K. J., Kasting, J. F., and Pollack, J. B. 1990*a*. Mass fractionation of noble gases in diffusion-limited hydrodynamic hydrogen escape. *Icarus* 84:502–527.

Zahnle, K. J., Pollack, J. B., and Kasting, J. F. 1990*b*. Xenon fractionation in porous planetesimals. *Geochim. Cosmochim. Acta* 54:2577–2586.

Zassova, L. V., Zubkova, V. M., Zhegulev, V. S., Ksanfomality, L. V., Moroz, V. I., and Petrova, E. V. 1982. Altitudes in Memnonia Fossae-Margaritifer Sinus region according to CO_2 altimetry on Mars-5. *Kosmich. Issled.* 20:921–927. In Russian.

Zebib, A., Schubert, G., Dein, J. L., and Paliwal, R. C. 1983. Character and stability of axisymmetric thermal convection in spheres and spherical shells. *Geophys. Astrophys. Fluid Dyn.* 23:1–42.

Zellner, B. H. 1972. Minor planets and related objects. VIII. Deimos. *Astron. J.* 77:183–185.

Zellner, B. H., Tholen, D. J., and Tedesco, E. F. 1985. The eight-color asteroid survey: Results for 589 minor planets. *Icarus* 61:355–416.

Zent, A. P., and Fanale, F. P. 1986. Possible Mars brines: Equilibrium and kinetic considerations. *Proc. Lunar Planet. Sci. Conf.* 16, *J. Geophys. Res. Suppl.* 91:D439-D445.

Zent, A. P., Fanale, F. P., Salvail, J. R., and Postawko, S. E. 1986. Distribution and state of H_2O in the high-latitude shallow subsurface of Mars. *Icarus* 67:19–36.

Zent, A. P., Fanale, F. P., and Postawko, S. E. 1987. Carbon dioxide: Absorption on palagonite and partitioning in the Martian regolith. 71:241–249.

Zent, A. P., Fanale, F. P., and Roth, L. E. 1988. The Goldstone Mars data: is there really evidence for melting? *Lunar Planet. Sci.* XIX:1315–1316 (abstract).

Zhang, M. H. G., Luhmann, J. G., and Kliore, A. J. 1990*a*. An observational study of the nightside ionospheres of Mars and Venus with radio occultation methods. *J. Geophys. Res.* 95:17095–17102.

Zhang, M. H. G., Luhmann, J. G., Kliore, A. J., and Kim, J. 1990*b*. A post-Pioneer, Venus

reassessment of the Martian dayside ionosphere as observed by radio occultation methods. *J. Geophys. Res.* 95:14829–14839.
Zharkov, V. N., Kosenko, A. V., and Maeva, S. V. 1984. Structure and origin of the satellites of Mars. *Astron. Vestnik* 18:88–89.
Zimbelman, J. R. 1984. Geologic Interpretation of Remote Sensing Data for the Martian Volcano Ascraeus Mons. Ph.D. Thesis, Arizona State Univ.
Zimbelman, J. R. 1985. Estimates of rheologic properties for flows on the Martian volcano Ascraeus Mons. *Lunar Planet. Sci. Conf.* 16, *J. Geophys. Res. Suppl.* 90:D157-D162.
Zimbelman, J. R. 1986. Surface properties of the Pettit wind streak on Mars: Implications for sediment transport. *Icarus* 66:83–93.
Zimbelman, J. R., and Greeley, R. 1982. Surface properties of ancient cratered terrain in the northern hemisphere of Mars. *J. Geophys. Res.* 87:10181–10189.
Zimbelman, J. R., and Kieffer, H. H. 1979. Thermal mapping of the northern equatorial and temperate latitudes of Mars. *J. Geophys. Res.* 84:8239–8251.
Zimbelman, J. R., and Leshin, L. A. 1987. A geologic evaluation of thermal properties for the Elysium and Aeolis quadrangles of Mars. *Proc. Lunar Planet. Sci. Conf.* 17, *J. Geophys. Res. Suppl.* 92:E586-E588.
Zimbelman, J. R., and Wells, G. L. 1987. Geomorphic evidence for climatic change at equatorial latitudes on Mars. *Geol Soc. America Abst. w/Prog.* 19(7):905.
Zimbelman, J. R., Leshin, L. A., Edgett, K. S., and Skinner, S. 1987. High-resolution thermal inertias at equatorial latitudes on Mars. *Lunar Planet. Sci.* XVIII:1128–1129 (abstract).
Zimbelman, J. R., Clifford, A. M., and Williams, S. H. 1989. Concentric crater fill on Mars: An aeolian alternative to ice-rich mass wasting. *Proc. Lunar Planet. Sci. Conf.* 19:397–407.
Zisk, S. H. 1972. Lunar topography: First radar-interferometer measurements of the Alphonsus-Ptolemaeus-Arzachel region. *Science* 178:977–988.
Zisk, S. H., and Mouginis-Mark, P. J. 1980. Anomalous region on Mars: Implications for near-surface liquid water. *Nature* 288:735–738.
Zolensky, M. E., Bourcier, W. L., and Gooding, J. L. 1988. Computer modeling of the mineralogy of the Martian surface, as modified by aqueous alteration. In *Workshop on Mars Sample Return Science,* eds. M. J. Drake, R. Greeley, G. A. McKay, D. P. Blanchard, M. H. Carr, J. Gooding, C. P. McKay, P. D. Spudis and S. W. Squyres, LPI Tech Rept. 88-07, pp. 188–189 (abstract).
Zolotov, M. Yu., and Sidorov, Yu. I. 1986. Nitrates in Martian soil? *Lunar Planet. Sci.* XVII:975–976 (abstract).
Zolotov, M. Yu., Sidorov, Yu. I., Volkov, V. P., Borisov, M. V., and Khodakovsky, I. L. 1983. Mineral composition of Martian regolith: Thermodynamic assessment. *Lunar Planet. Sci.* XIV:883–884 (abstract).
Zuber, M. T., and Aist, L. L. 1990. The shallow structure of the Martian lithosphere in the vicinity of the ridged plains. *J. Geophys. Res.* 95:14215–14230.
Zuber, M. T., and Mouginis-Mark, P. J. 1990. Constraints on the depth and geometry of the magma chamber of the Olympus Mons volcano, Mars. *Lunar Planet. Sci.* XXI:1387–1388 (abstract).
Zurek, R. W. 1976. Diurnal tide in the Martian atmosphere. *J. Atmos. Sci.* 33:321–337.
Zurek, R. W. 1978. Solar heating of the Martian dusty atmosphere. *Icarus* 35:196–208.
Zurek, R. W. 1980. Surface pressure response to elevated tidal heating sources: Comparison of Earth and Mars. *J. Atmos. Sci.* 37:1132–1136.
Zurek, R. W. 1981. Inference of dust opacities for the 1977 Martian great dust storms from Viking Lander 1 pressure data. *Icarus* 45:202–215.
Zurek, R. W. 1982. Martian great dust storms: An update. *Icarus* 50:288–310.
Zurek, R. W. 1986. Atmospheric tidal forcing of the zonal-mean circulation: The Martian dusty atmosphere. *J. Atmos. Sci.* 43:652–670.
Zurek, R. W. 1988*a*. Free and forced modes in the Martian atmosphere. *J. Geophys. Res.* 93:9452–9462.
Zurek, R. W. 1988*b*. The interannual variability of atmospheric water vapor on Mars. In *MECA Workshop on Atmospheric H_2O Observations of Earth and Mars,* eds. S. M. Clifford and R. M. Haberle, LPI Tech. Rept. 88-10, p. 91 (abstract).
Zurek, R. W., and Christensen, P. R. 1990. Dust redistribution by east-west circulations on Mars. *Bull. Amer. Astron. Soc.* 22:1075 (abstract).

Zurek, R. W., and Haberle, R. M. 1988. Zonally symmetric response to atmospheric tidal forcing in the dusty Martian atmosphere. *J. Atmos. Sci.* 45:2469–2485.

Zurek, R. W., and Haberle, R. M. 1989. Martian great dust storms: Aperiodic phenomena? In *MECA Workshop on Dust on Mars III,* ed. S. Lee, LPI Tech. Rept. 89-01, pp. 40–41 (abstract).

Zurek, R. W., and Leovy, C. B. 1981. Thermal tides in the dusty Martian atmosphere: A verification of theory. *Science* 213:437–439.

Zurek, R. W., and Martin, L. J. 1989. Interannual variability of the major dust storms on Mars: Evidence from the historical and spacecraft records. *Bull. Amer. Astron. Soc.* 21:980 (abstract).

Zurek, R. W., and Martin, L. J. 1992. Interannual variability of planet-encircling dust storms on Mars. *J. Geophys. Res.*, in press.

Zwan, B. J., and Wolf, R. A. 1976. Depletion of solar wind plasma near a planetary boundary. *J. Geophys. Res.* 81:1636–1648.

COLOR SECTION

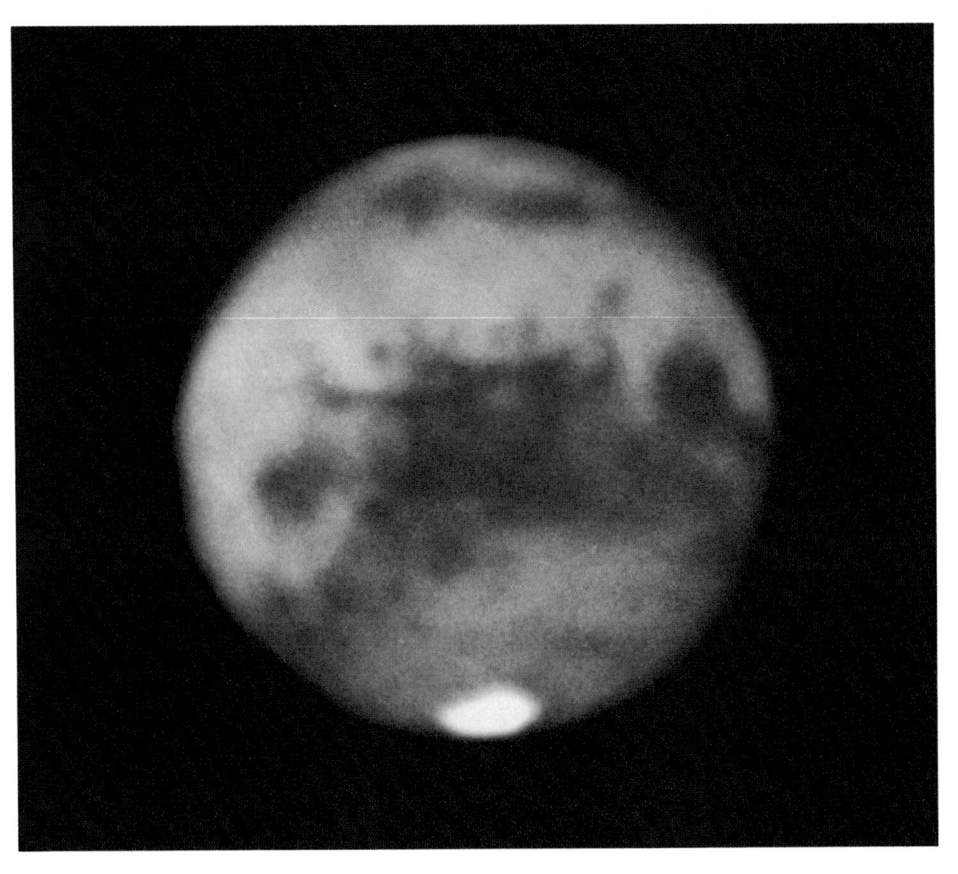

Plate 1. Lowell Observatory photograph taken 1988 September 28 (L_s = 281) with the 61-cm Planetary Patrol Telescope at Mauna Kea Observatory by L. Martin. Central meridian is 43. The narrow dark feature extending left just above the middle of the image is Valles Marineris; the rounder dark area just below is the classic feature Solis Lacus. Acidalia Planitia is the dark feature at the top (north) of the image, with blue atmospheric haze at the north limb. Compare with Plate 10 and Appendix Fig. 1. The Telescopic Monitoring of Mars: 1988 Opposition was conducted by P. James, University of Toledo, and L. Martin, Lowell Observatory with support from the National Geographic Society. Original negatives from red-, green-, and blue-filter images were digitized at Lowell Observatory. Color compositing and processing was done by the U.S. Geological Survey, Branch of Astrogeology. See text p. 58.

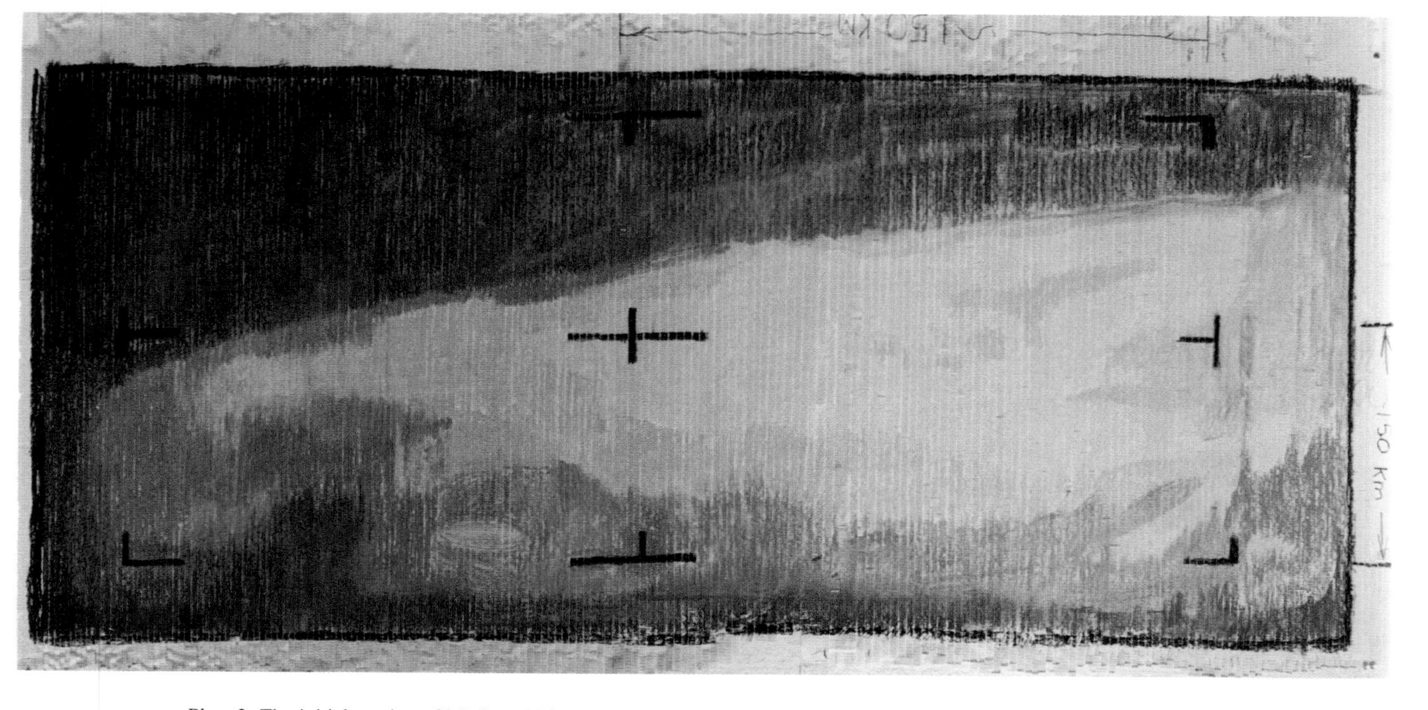

Plate 2. The initial version of Mariner 4 Picture Number 1 — the first close-up photograph of another planet ever taken. With a telemetry transmission rate of 8.33 bits per second, the 22 recorded photographs required four days to be sent back to Earth. As a communications dropout had occurred close to the moment of encounter, it was not known whether any pictures had actually been recorded, so the Mariner Product Manager directed that this picture be assembled in real time as the data came in. The numbers representing the brightness of each pixel in the picture were printed out in a vertical column on a strip of adding-machine tape by a computer at JPL. As the data from each scan line were completed, technicians cut the paper strip and pasted it onto a plywood sheet, overlapping the strips so as to produce one large 200 × 200 matrix of numbers. Using colored chalk, they produced a kind of false-color contour map of the brightness of this historic picture. For many years it hung in the administration building at the laboratory, but recently it was moved into the Space Flight Operations Facility. See text p. 78.

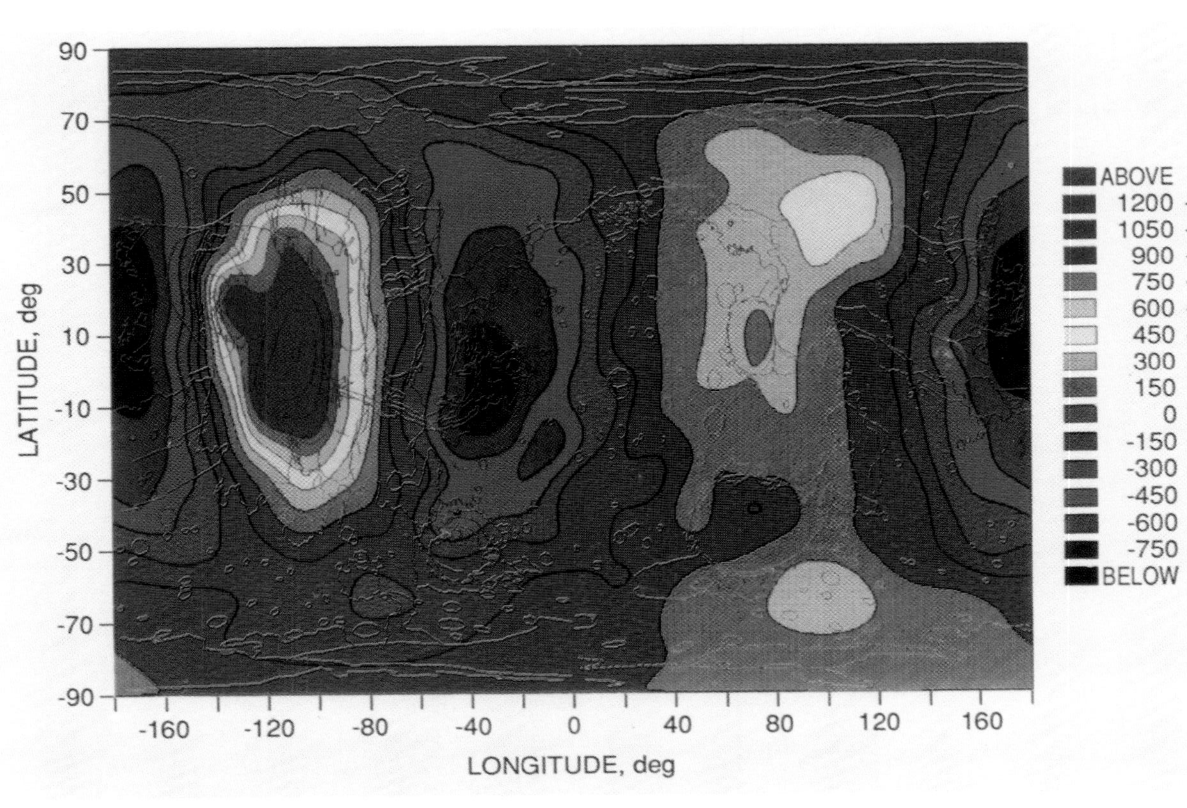

Plate 3. Mars geoid heights (in m) referenced to an ellipsoid with a semimajor axis of 3394 km and a flattening of 1/191.13720. The red lines indicate the outline of topographic features relative to this geoid (figure from Balmino et al. 1982). See text p. 213.

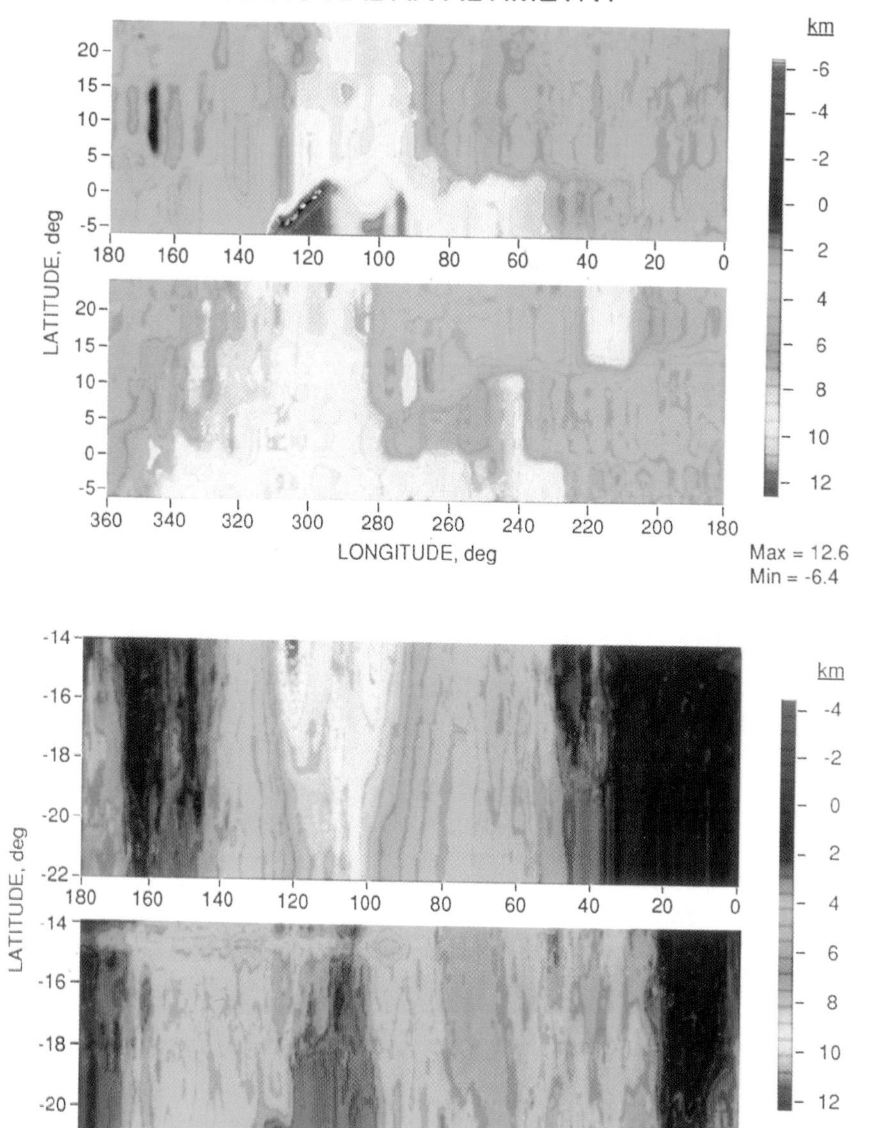

Plate 4. Near-equatorial Mars topography deduced from radar altimetry. The top pair of plots cover all longitudes for latitudes between −7° and 24°; these data were acquired during 1975–82. The bottom pair present the data acquired during the 1971–73 oppositions which cover all longitudes for latitudes between −14° to −22°. See text p. 221.

Plate 5. Geologic map of western hemisphere of Mars based on interpretation of Viking Orbiter images (Scott and Tanaka 1986; Tanaka and Scott 1987). Map base is digitized, shaded-relief Lambert Equal-Area projection. Interval of grid is 10° (except latitudes $\pm > 60°$, where interval of meridians is 30°); extra grid lines at latitudes $\pm 55°$ are joins of polar maps and equatorial map. See Plate 7 for correlation of map units (modified from Tanaka et al. 1988; see Map I-1802-A and -C). See text pp. 283, 348 and maps in accompanying volume.

Plate 6. Geologic map of eastern hemisphere of Mars based on interpretation of Viking Orbiter images (Greeley and Guest 1986; Tanaka and Scott 1987). Map base is digitized, shaded-relief Lambert Equal-Area projection. Interval of grid is 10° (except above latitudes ± >60°, where interval of meridians is 30°); extra grid lines at latitudes ± 55° are joins of polar maps and equatorial map. See Plate 7 for correlation of map units (modified from Tanaka et al. 1988; see Map I-1802-B and -C). See text pp. 283, 348 and maps in accompanying volume.

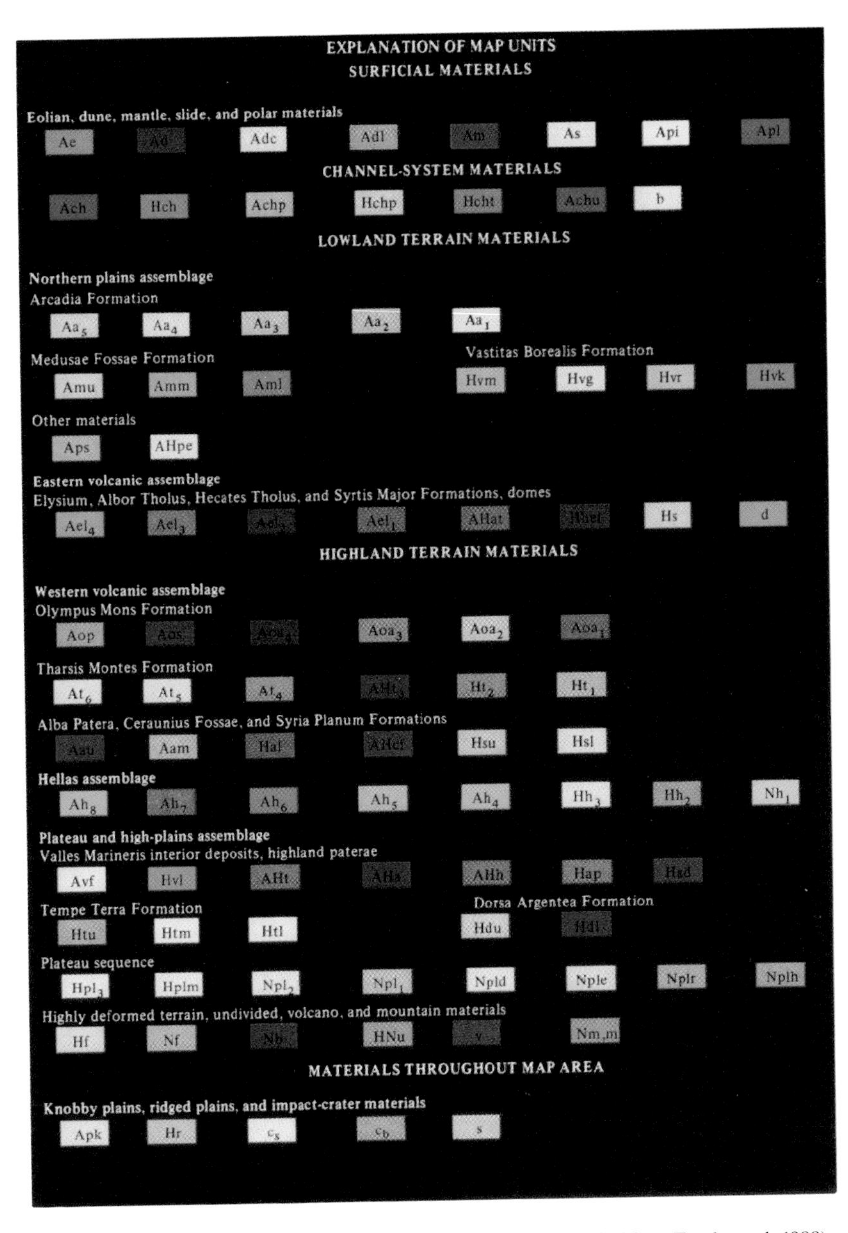

Plate 7. Explanation of geologic units shown on Plates 4 and 5 (modified from Tanaka et al. 1988). For names, relative ages and geologic characteristics of units, see geologic map I-1802 in accompanying volume. See text p. 349.

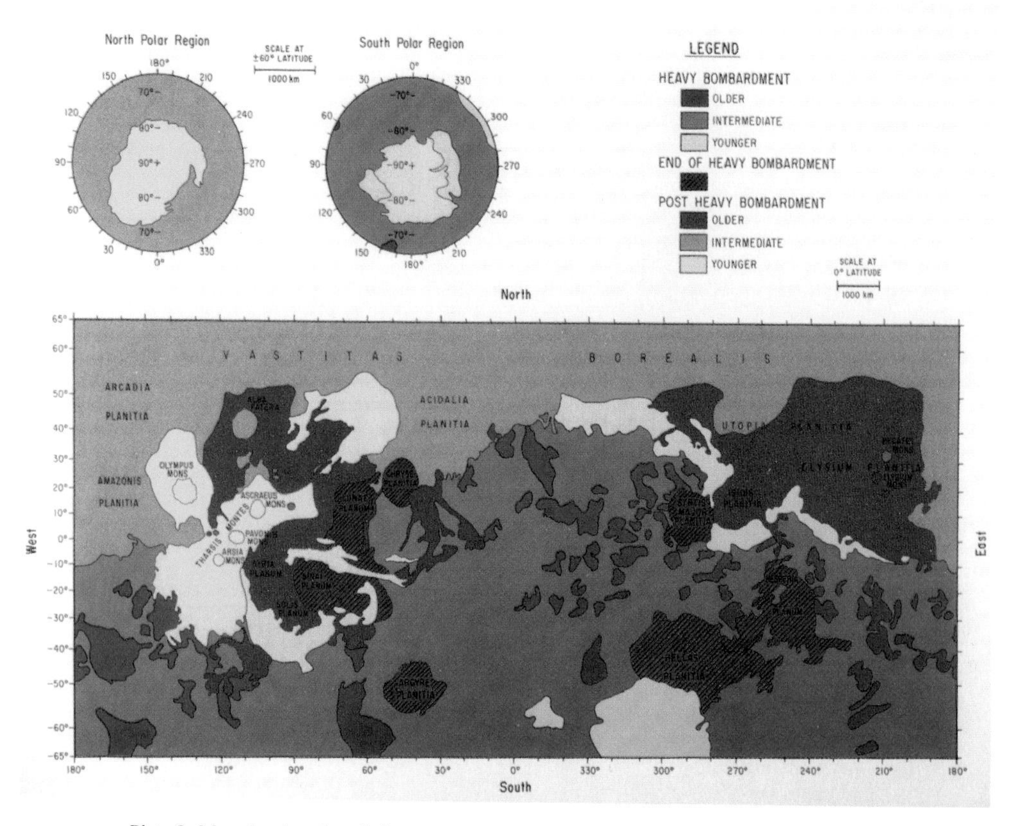

Plate 8. Map showing the relative ages of generalized Martian surface units referenced to the end of late heavy bombardment where the crater production size/frequency distribution changes from a differential −2 slope index to one with a −3 slope index. The bluish areas date from the period of late heavy bombardment whereas the red, orange and yellow areas formed since that time. See text p. 413.

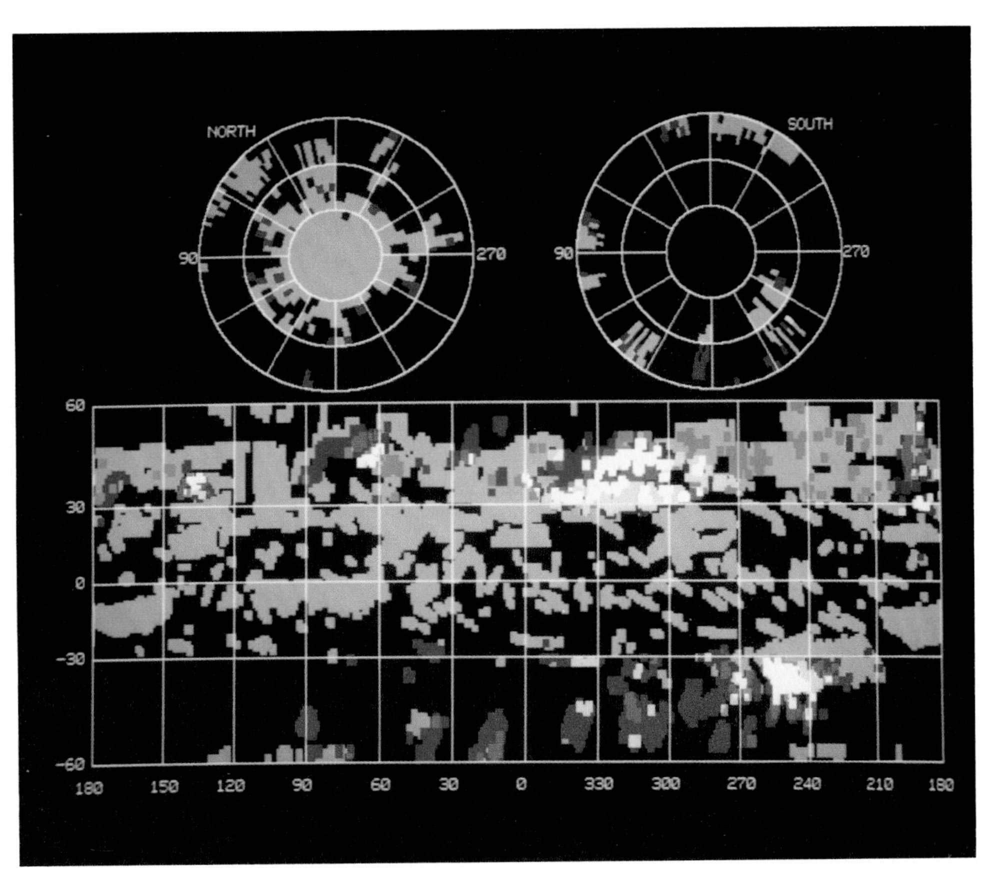

Plate 9. Map showing the distribution of viscous creep features on Mars. Blue: lobate debris aprons and lineated valley fill; green: concentric crater fill; red: terrain softening. Where two or more classes of features occur together, colors are added. Yellow: terrain softening plus concentric crater fill; magenta: terrain softening plus lobate debris aprons/lineated valley fill; cyan: lobate debris aprons/lineated valley fill plus concentric crater fill; white: all three. Gray indicates areas where no creep features are observed, and black indicates areas where no adequate data are available (figure from Squyres and Carr 1986). See text p. 543.

Plate 10. Valles Marineris Hemisphere. Mosaic composed of 100 Viking Orbiter images of Mars, covering nearly a full hemisphere of the planet (approximate latitude −55° to 60°, longitude 30° to 130°). The mosaic is in an orthographic projection with a scale of 1 km/pixel. The color variations have been substantially exaggerated and the large-scale brightness variations (mostly due to illumination-angle variations) have been retained. The center of the scene shows the entire Valles Marineris canyon system, over 3000 km long and up to 8 km deep, extending from Noctis Labyrinthus, the arcuate system of graben to the west, to the chaotic terrain to the east. Huge ancient river channels begin in the chaotic terrain and from north-central canyons and run north. Many of the channels flowed into the Chryse and Acidalia Planitiae basin which includes part of the dark area in the extreme northeast of this picture. The Viking 1 landing site (Mutch Memorial Station) is located in Chryse Planitia, the light area south of Acidalia Planitia. The three Tharsis volcanoes (dark red spots), as much as 25 km high, are visible to the west. The large crater with two prominent rings located at the bottom of this image is Lowell. The images were acquired by Viking Orbiter 1 in 1980 during early northern summer on Mars ($L_s = 70°$); the atmosphere was relatively dust free. A variety of clouds appear as bright blue streaks and hazes, and probably consist of water ice. Long, linear clouds north of central Valles Marineris appear to originate at impact craters. Image processing was done at the U.S. Geological Survey, Flagstaff, Arizona. See text p. 562.

Plate 11. Schiaparelli Hemisphere. This mosaic is composed of 97 red- and violet-filter Viking Orbiter images; projection, scale and processing are the same as for Plate 10. The images were acquired in 1980 during mid northern summer on Mars ($L_s = 89°$). The center of this image is near the impact crater Schiaparelli (latitude −3°, longitude 343°). The large circular area with a bright yellow color (in this rendition) is known as Arabia. The boundary between the ancient, heavily cratered southern highlands and the younger northern plains occurs far to the north (latitude 40°) on this side of the planet, just north of Arabia. The dark streaks with bright margins emanating from craters in the Oxia Palus region (to the left of Arabia) are caused by erosion and/or deposition by the wind. The dark blue area on the far right, called Syrtis Major Planum, is composed of a low-relief volcanic shield and surrounding lava plains of probable basaltic composition. Bright white areas to the south, including the Hellas impact basin at lower right, are covered by carbon dioxide frost. This digital mosaic was produced at the U.S. Geologic Survey in Flagstaff, Arizona. See text p. 562.

Plate 12. Syrtis Major Hemisphere. This mosaic is composed of 99 red- and violet-filter Viking Orbiter images; projection, scale and processing are the same as for Plate 10. The images were acquired in 1980 during early northern summer on Mars ($L_s = 75°$). The center of this image is near latitude 0° N, longitude 310° W, and the limits of this mosaic are approximately latitude −60° to 60° and longitude 260° to 350°. As in Plate 11, the large circular area left of center with a bright yellow color (in this rendition) is Arabia. The bright yellow area to the right of Syrtis Major is Isidis Planitia, a basin formed by an ancient impact and later filled in. Bright white areas to the south, including Hellas impact basin at lower right, are covered by carbon dioxide frost. The CO_2 frost coverage in relatively small craters west of Hellas is less in this image than in Plate 11, which was acquired about 25 days later into southern winter. This digital mosaic was produced at the U.S. Geological Survey in Flagstaff, Arizona. See text p. 562.

Plate 13. Cerberus Hemisphere. This mosaic is composed of 104 Viking Orbiter images acquired on 1980 February 11. At that time, it was early northern summer on Mars (L_s = 65°). The center of the image is at latitude 3°, longitude 185°, and the latitude limits are approximately 60° N and S; processing, projection and scale as described for Plate 10. A major geologic boundary extends across this mosaic, with the lower third of the image showing ancient cratered highlands (mostly dark terrain); north of this boundary are the lowland northern plains. A series of poorly indurated sedimentary deposits occur just north of the highland-lowland boundary; some workers believe that these are explosive volcanic deposits (ignimbrites) whereas others have postulated that they are paleo-polar deposits. Other prominent features in this image include the large dark area left of the image center (named Cerberus), and the Elysium volcanic region (bright yellowish area north of Cerberus). The crater "Mie" is located near the top left, and has a dark spot near its center. The Viking 2 Lander is located about 400 km west of the center of this crater, or about 1½ crater diameters left of the left edge of the crater. Thin white clouds are dispersed over the northern hemisphere, and the opaque cloud in the upper right overlies the Olympus Mons aureole. The arcuate markings west of the aureole are thought to be extended drifts of windblown material. The bright blue area at the bottom of the picture shows the extent of the seasonal carbon dioxide polar cap (these frosts are actually white). This digital mosaic was produced at the U.S. Geological Survey in Flagstaff, Arizona. See text p. 562.

Plate 14. South Polar Residual Ice Cap. This mosaic is composed of 18 Viking Orbiter images (6 each in red, green and violet filters), acquired on 1977 September 28 during revolution 407 of Viking Orbiter 2. [The mosaic is in polar stereographic projection with a scale of 0.5 km/pixel.] The south pole is located just off the lower left edge of the polar cap, and the 0° longitude meridian extends toward the top of the mosaic. The large crater near the right edge (named "South") is about 100 km in diameter. These images were acquired during southern summer on Mars (L_s = 341°); the sub-solar declination was 8° S, and the south polar cap was nearing its final stage of retreat just prior to vernal equinox. The south residual cap is approximately 400 km across, and the exposed surface is thought to consist dominantly of carbon-dioxide frost. This is in contrast to the water-ice surface of the north polar residual cap. It is likely that water ice is present in layers that underlie the south polar cap and that comprise the surrounding layered terrains. Near the top of this image, irregular pits with sharp-rimmed cliffs appear "etched," presumably by wind. A series of rugged mountains (extending toward the upper right corner of the image) are of unknown origin. This digital mosaic was produced at the U.S. Geological Survey in Flagstaff, Arizona. See text p. 562.

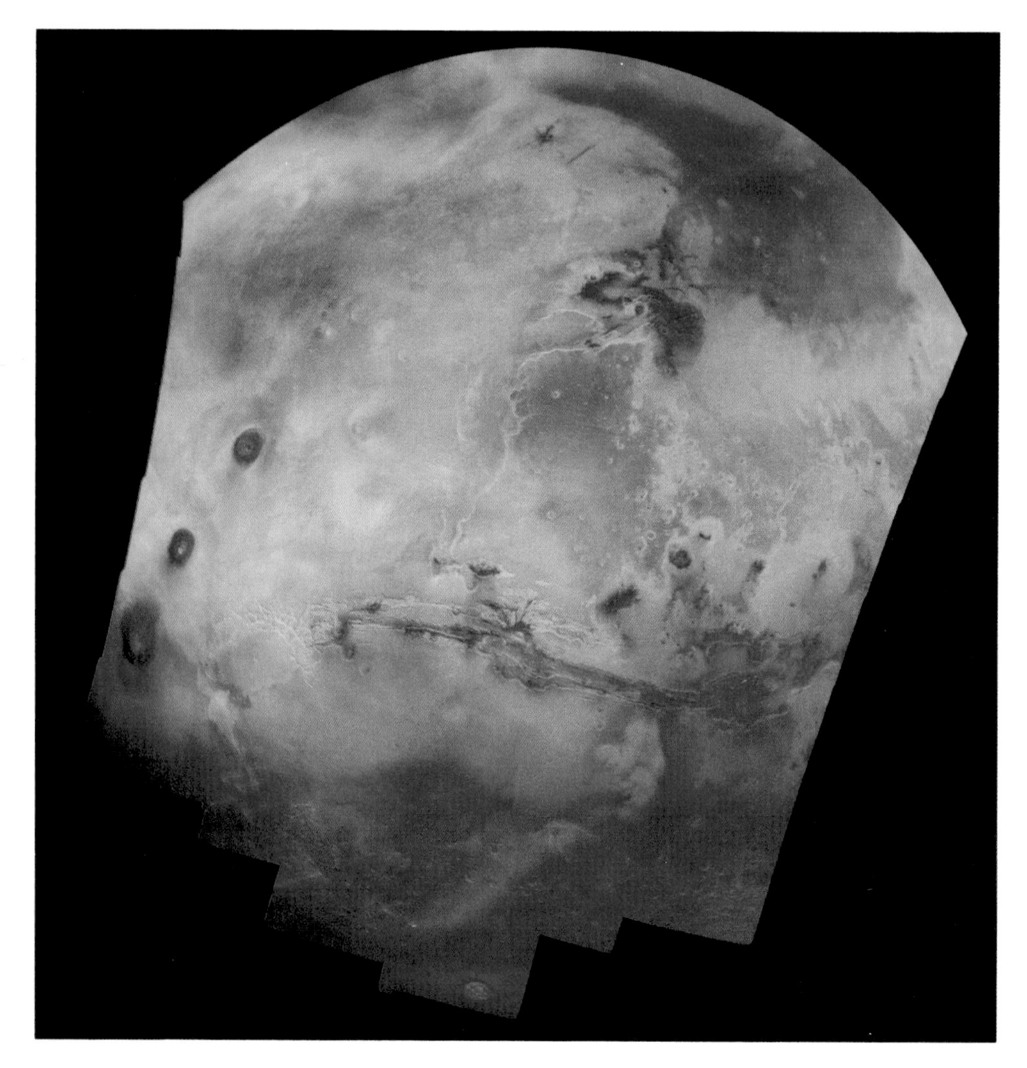

Plate 10.

Plate 11.

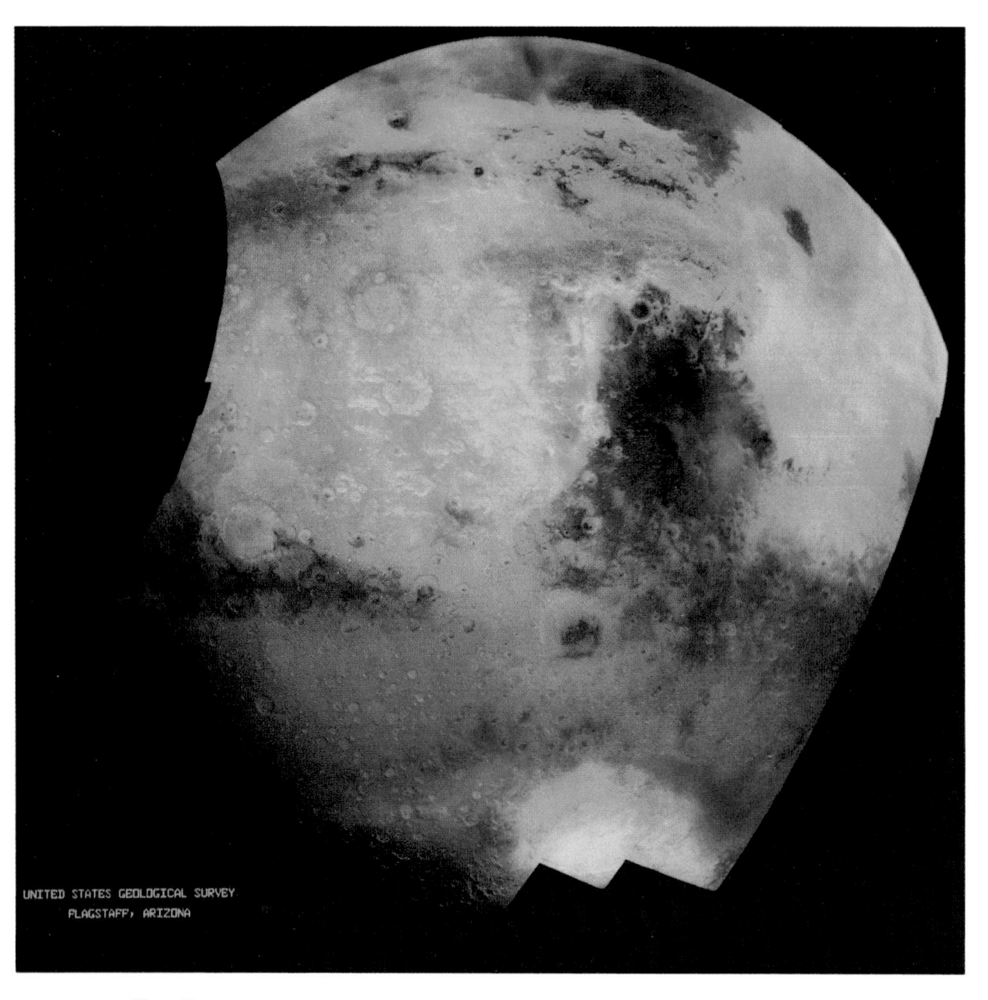

Plate 12.

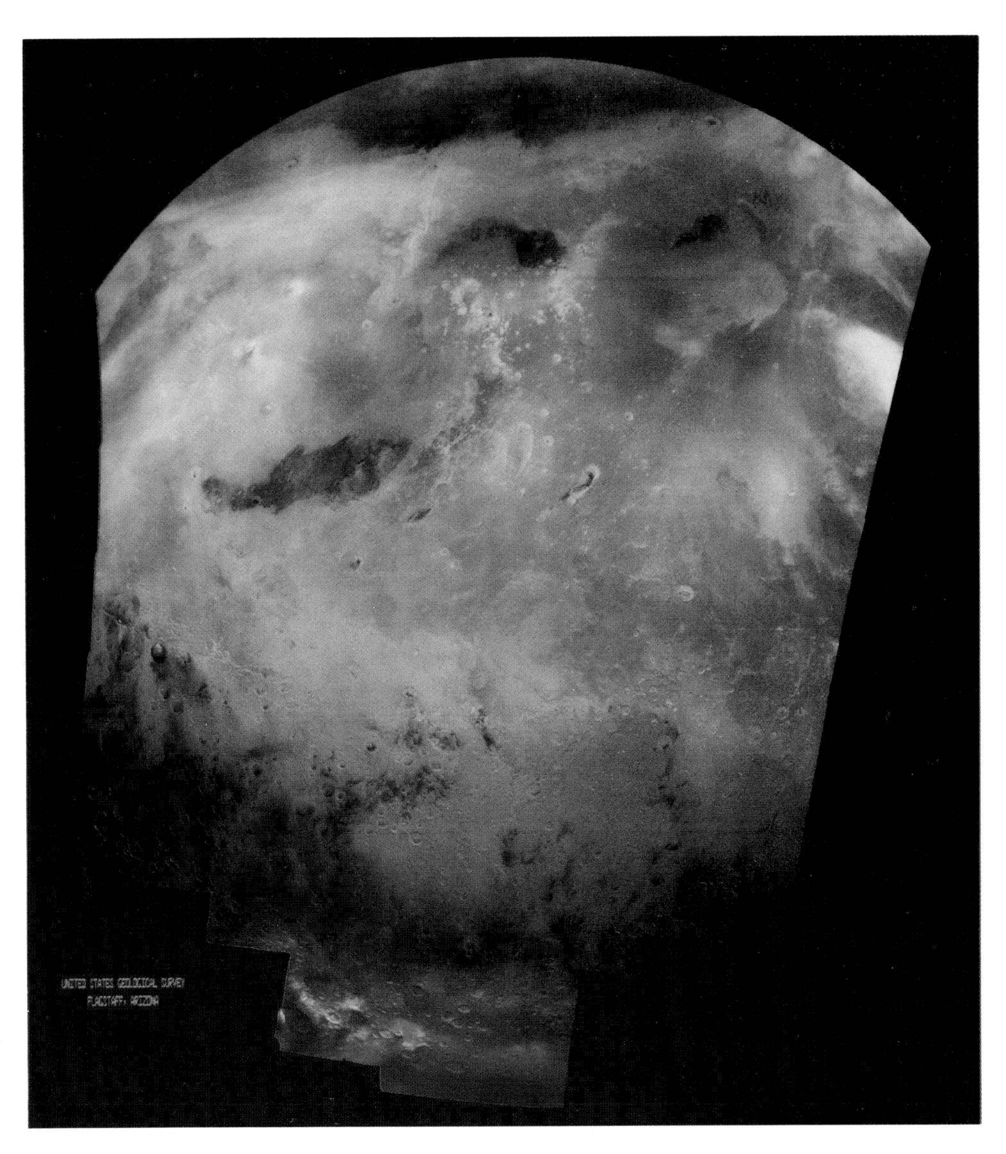

Plate 13.

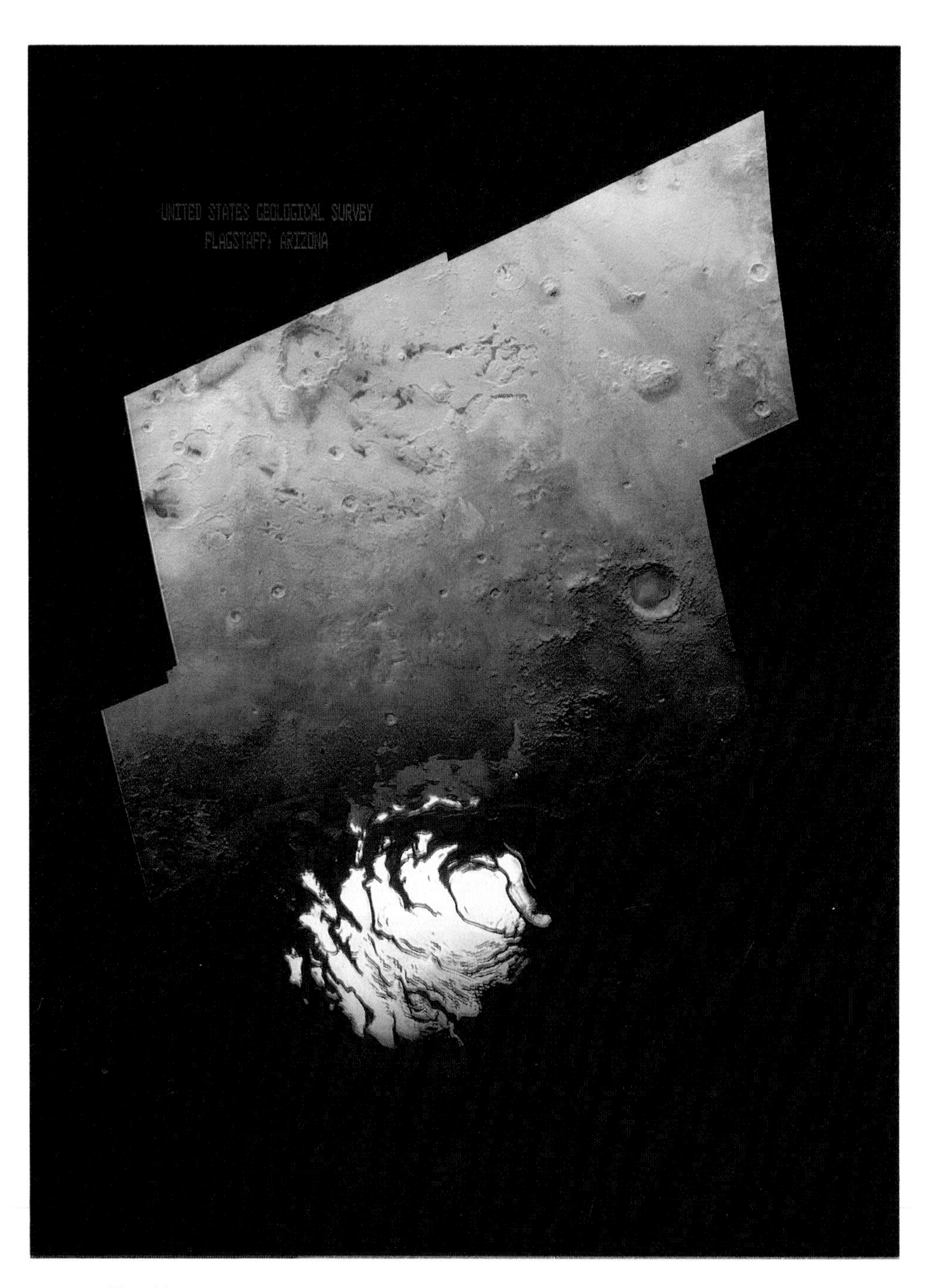

Plate 14.

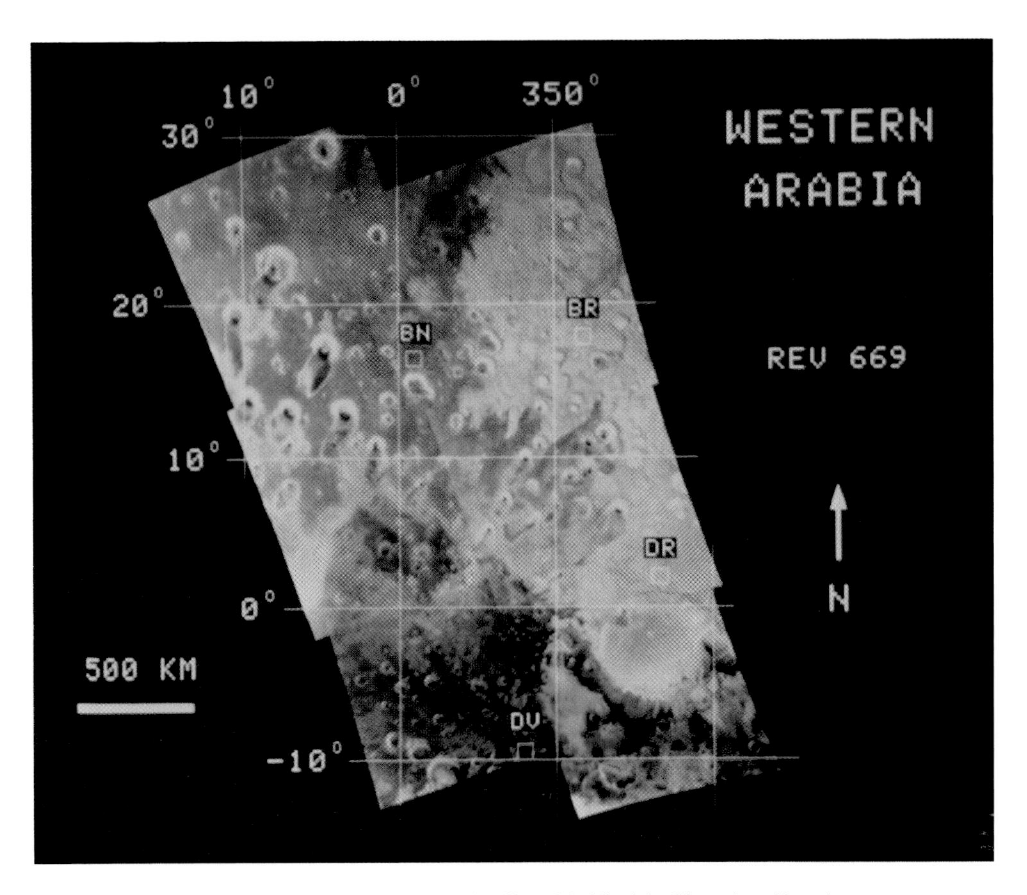

Plate 15. Multispectral mosaic of the Oxia Palus-Sinus Meridiani-Arabia region. Note the presence of three major units: (1) bright red (BR); (2) dark violet (DV); and (3) brown (BN). The bright-red unit is interpreted to be wind blown dust; the dark violet unit is likely coarse, mobile material and the brown unit is interpreted to be indurated. Compare with regional setting shown in Plate 10 (figure from Presley and Arvidson 1988). See text p. 714.

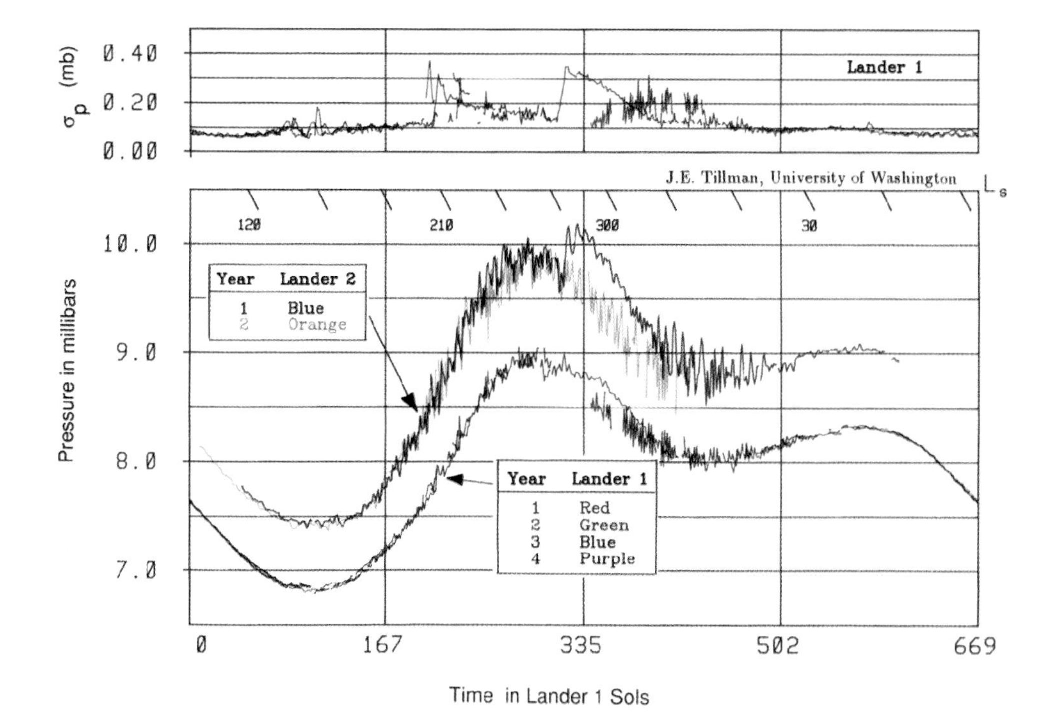

Plate 16. This figure shows superimposed sol-averaged surface pressure [vs time] observations made by the Viking Landers during nearly four Martian years. This illustrates the significant annual variation due to the seasonal asymmetry between hemispheres as well as the close reproducibility from year to year. The major difference, in the VL2 curve near $L_s = 280°$ during the first year, is associated with the larger of the two 1977 global dust storms. Daily (sol) standard deviations of surface pressure at Lander 1 are shown in the top panel (Tillman 1988). This information is also in Fig. 1 in chapter 26. See text pp. 919 and 945.

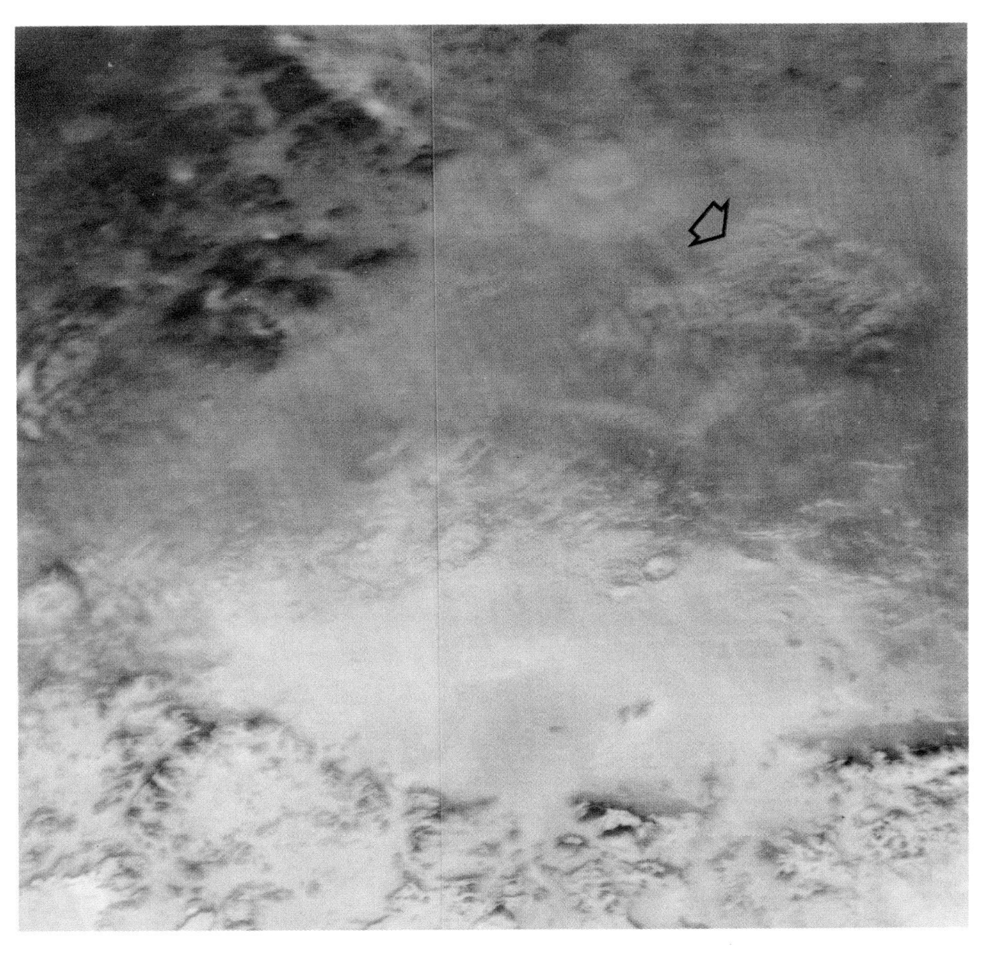

Plate 17. Viking Orbiter 2 image of local dust storm (arrow) in Argyre basin. The storm is more than 300 km across, and was imaged on $L_s = 177°$. The white area is the extension of the south polar seasonal cap. (Plate published in Briggs et al. 1979.) See text pp. 925 and 1041.

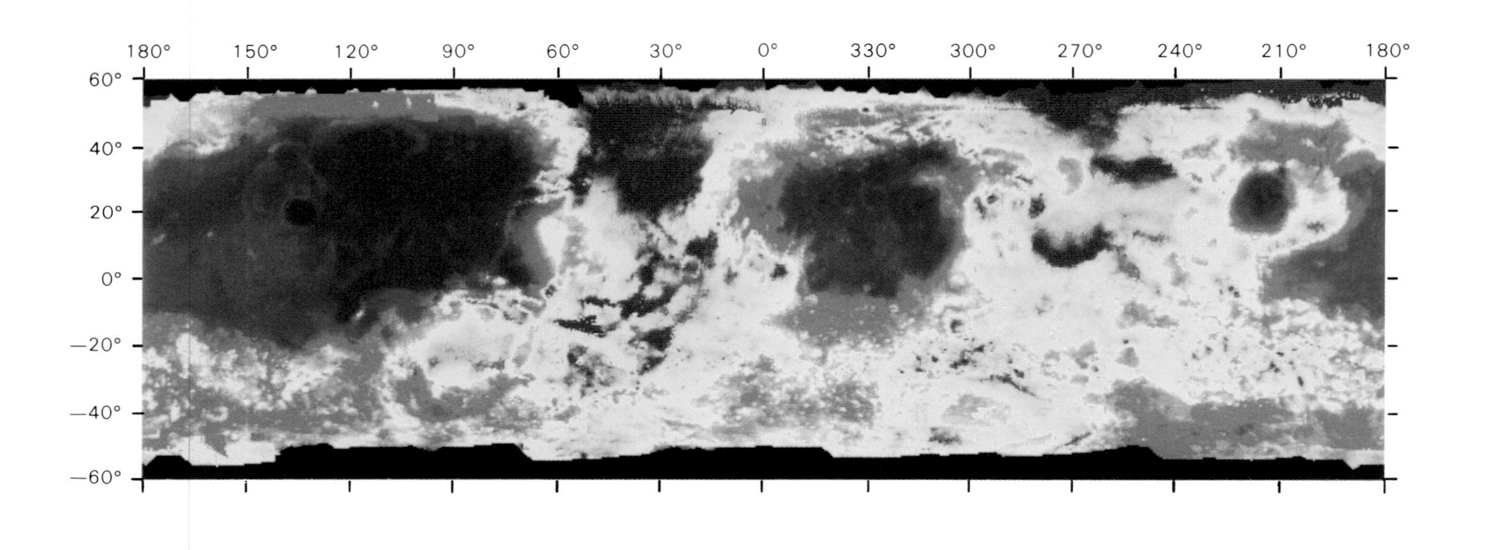

Plate 18. Map of Martian thermal inertia. Data were compiled from Viking IRTM observational sequences when the atmosphere was relatively dust-free and when good global coverage was available. Data were collected into bins of 0°.5 latitude by 0°.5 longitude. Thermal inertia is expressed in units of 10^{-3} cal cm^{-2} s$^{-1/2}$ K^{-1}. Thermal inertia values from 12 up to 24 are rare and are shown in the same color. Diurnal temperature change in the continental-scale low thermal inertia regions (dark blue) is typically ~60 K greater than the high thermal inertia regions (red). (Data are from P. Christensen.) See text pp. 702, 812 and 1025.

Acknowledgments

ACKNOWLEDGMENTS

The editors acknowledge National Aeronautics and Space Administration Grant NAGW-1463, Jet Propulsion Laboratory Contract No. 958584, and The University of Arizona for support in the preparation of this book. They wish to thank J. E. Frecker, who volunteered as one of the proofreaders of this book. The following authors wish to acknowledge specific funds involved in supporting the preparation of their chapters.

Baker, V. R.: NASA Grant NAGW-285
Barnes, J. R.: NASA Grants NAGW-1127 *and* NAGW-2445
Barth, C. A.: NASA Grant NSG-5103
Bougher, S. W.: NASA Planetary Atmospheres Grant NAGW-2493
Burns, J. A.: NASA Grant NAGW-310
Davies, M. E.: NASA Grant NAGW-1718
Drake, M. J.: NASA Grant NAG-9-39
Fanale, F. P.: NASA Planetary Geology Grant NAGW-583
Greeley, R.: NASA Grants NCC 2-346 *and* NAGW-2064
Gulick, V. C.: NASA Grant NGT-50662
Herkenhoff, K.: NASA Grants NAGW-549 *and* NAGW-1373
Holloway, J. R.: NASA Grant NAGW-182, suppl. 6
James, P. B.: NASA Planetary Atmospheres Grant NAGW-1116
Kahn, R.: The work of R. Kahn was performed at the Jet Propulsion Laboratory/California Institute of Technology, under contract with the Planetary Exploration Division, National Aeronautics and Space Administration.
Leovy, C. B.: NASA Grant NS6-7085
Longhi, J.: NASA Grant NAG-9-93
Lucchitta, B. K.: NASA Contract W15,814
Luhmann, J. G.: NASA Grant NAGW-1347
Martin, L. J.: NASA Grant NAGW-2257
Mouginis-Mark, P. J.: NASA Geology and Geophysics Program Grants NAGW-437 *and* NAGW-1084
Murray, B.: NASA Grants NAGW-549 *and* NAGW-1373
Owen, T.: NASA Grants NGR 33-015-141, NAGW-725 *and* NSF INT-8715243
Pepin, R. O.: NASA Grants NAG 9-60 *and* NAGW-1371
Schubert, G.: NASA Grants NSG-7315 *and* NAS7-100 (subcontract)

Sleep, N. H.: NASA Grant NSG-7297 *and* NSF Grant EAR-8719278
Solomon, S. C.: NASA Grants NAG-5-814 *and* NSG-7297
Squyres, S. W.: NASA Grant NAGW-1023
Stewart, A. I. F.: NASA Grant NAGW-389
Strom, R. G.: NASA Grant NSG-7146
Thomas, P. C.: NASA Grant NAGW-111
Tillman, J. E.: NASA Grants NAGW-1341, NAGW-2041 *and* NAGW 2477

Index

INDEX